Hans H. Maurer, Karl Pfleger, and Armin A. Weber

Mass Spectral and GC Data
of Drugs, Poisons, Pesticides, Pollutants
and Their Metabolites

The Maurer/Pfleger/Weber collection is also available as an electronic database

Maurer, H. H., Pfleger, K., Weber, A. A.

Mass Spectral Library
of Drugs, Poisons, Pesticides, Pollutants
and Their Metabolites

2016 DVD with booklet
ISBN: 978-3-527-33951-8

Hans H. Maurer, Karl Pfleger, and Armin A. Weber

Mass Spectral and GC Data
of Drugs, Poisons, Pesticides, Pollutants and Their Metabolites

Volume 1: Methods and Tables, Spectra (m/z 30 – 299)

5th, revised and enlarged edition

WILEY-VCH
Verlag GmbH & Co. KGaA

The Authors

Prof. Dr. Dr. h.c. Hans H. Maurer
Prof. Dr. Karl Pfleger (†)
Armin A. Weber
Department of Experimental and Clinical Toxicology
Saarland University
66421 Homburg (Saar)
Germany

All books published by **Wiley-VCH** are carefully produced. Nevertheless, authors, editors, and publisher do not warrant the information contained in these books, including this book, to be free of errors. Readers are advised to keep in mind that statements, data, illustrations, procedural details or other items may inadvertently be inaccurate.

Library of Congress Card No.:
applied for

British Library Cataloguing-in-Publication Data
A catalogue record for this book is available from the British Library.

Bibliographic information published by the Deutsche Nationalbibliothek
Die Deutsche Nationalbibliothek lists this publication in the Deutsche Nationalbibliografie; detailed bibliographic data are available in the Internet at http://dnb.d-nb.de.

© 2017 WILEY-VCH Verlag GmbH & Co. KGaA, Boschstr. 12, 69469 Weinheim, Germany

All rights reserved (including those of translation into other languages). No part of this book may be reproduced in any form – by photoprinting, microfilm, or any other means – nor transmitted or translated into a machine language without written permission from the publishers. Registered names, trademarks, etc. used in this book, even when not specifically marked as such, are not to be considered unprotected by law.

Printing and Binding Strauss GmbH, Mörlenbach

Printed in the Federal Republic of Germany
Printed on acid-free paper

Print ISBN 978-3-527-34287-7

Dedicated in memory of

Professor Karl Pfleger (1924-2013)

"Unicum signum certi dati veneni

est analysis chimica..."

Josef Jakob von Plencks:
Toxicologica seu doctrina de venenis est antidoti
Vienna, 1785

Preface to the Fith Edition

First of all, we regret to have to inform our readers that our senior author Professor Karl Pfleger passed away in 2013. We have finished this edition in memory of him.

After publication of the first edition in 1985 with 1,500 data sets, the second in 1992 with 4,700, and with the Supplement Volume in 2000 with 6,700 data sets, the third in 2007 with a completely revised book concept and enlarged to 7,840 data sets, the fourth edition in 2011 with 8,650 data sets, the new edition contains 10,430 data sets.

Since 2011, we have recorded and collected in our laboratory mass spectral and gas chromatographic data of new drugs (of abuse), of emerging new psychoactive substances (NPS), poisons and their metabolites bringing the total number of mass spectra to **10,430** among which about 2,000 data sets have been revised or added. Over 7,800 of the data sets are from metabolites. All formulas were drawn in the molefile format allowing their use in the corresponding electronic databases.

The sections on sample preparation and GC-MS method have been updated, but only our classic and new standard procedures have been described in detail, as series of papers on sample preparation have been published during the past years.

The Table of compounds (Table 8.1) covers all the data with directions on which page in the data may be found, as well as the entry numbers of the electronic versions.

The GC and MS data are also available in electronic form. The combination of the handbook with the full data and the electronic versions has proven to be very efficient in analytical toxicology. Using macros and particularly using the AMDIS software, the data evaluation can be automated, but the final decision on the identity of a compound will always rest in the hands of the experienced toxicologist or analyst who undertakes a visual comparison between the full mass spectrum of the measured compound and the reference spectra, by considering further chemical, analytical and toxicological aspects for plausibility.

Again, while many colleagues throughout the world have acknowledged the value of our collection, some have also drawn our attention to errors and have provided suggestions, and for this we would like to express our sincere gratitude.

It is our hope that the new edition will be accepted as well as the previous ones by our colleagues in the fields of clinical and forensic toxicology, doping control, drug metabolism and in all other areas of analytical toxicology.

As LC-MS is established in the meanwhile, a similar collection of MS^2 and MS^3 spectra of drugs, poisons, and their metabolites has been published in 2014 [1] and onother collection with LC-high-resolution-MS/MS spectra will follow in 2016 [2].

Homburg (Saar), July 2016   **Hans H. Maurer**

Acknowledgements

The authors are indebted to many pharmaceutical companies and to many colleagues for supplying reference substances or biosamples containing rare compounds and/or metabolites.

Professor Maurer thanks the following former or present co-workers for the recording and interpretation of new mass spectra during their scientific studies and further support:
J. Arlt, M. Bach, J. Beyer, J. Bickeboeller-Friedrich, A. Bierl, K. Bock, T. Braun, C. Brengel, A.T. Caspar, J. Dinger, P. Du, A.H. Ewald, C. Fritz, P. Goepfert, C. Haas, A.G. Helfer, H.C. Hoerth, A. Holderbaum, J. Jung, C. Gerber, I. Kleff, N. Kneller, T. Kraemer, C. Kratzsch, O. Ledvinka, C. Lindauer, N. Makkinejad, S. Manier, S. Mauer, M.R. Meyer, G.M.J. Meyer, J.A. Michely, L.D. Paul, F.T. Peters, A.A. Philipp, D. Prosser, D. Remane, L.H.J. Richter, A. Robert, S. Roditis, C. Sauer, N. Schaefer, S. Schaefer, C.J. Schmitt, S. Schmitt, C. Schroeder, A. Schütz, A.E. Schwaninger (Steuer), F. Sick, D. Springer, R.F. Staack, E. Strack, A. Turcant, F.X. Tauvel, G. Theis, D.S. Theobald, A. Thomann, S.W. Toennes, G. Ulrich, I. Vernaleken, C. Vollmar, L. Wagmann, J. Welter-Lüdecke, C.S.D. Wink, and D.K. Wissenbach.

Furthermore, the authors wish to thank their families for forbearance and sacrifice of personal interests. Finally, they wish to thank the Wiley-VCH editors Dr. Graeme Whitley, Dr. Bernd Berger, and Dr. Frank O. Weinreich for their understanding and the agreeable collaboration.

Contents of Volume 1 (Methods, Tables, Mass Spectra)

1	**Introduction**	**1**

2 Experimental Section 2
2.1 **Origin and choice of samples** 2
2.2 **Sample preparation** 2
2.2.1 Standard extraction procedures 2
2.2.1.1 Standard liquid-liquid extraction (LLE) for plasma, urine or gastric contents (P, U, G) 2
2.2.1.2 STA procedure (hydrolysis, extraction and microwave-assisted acetylation) for urine (U+UHYAC) 3
2.2.1.3 Extraction of urine after cleavage of conjugates by glucuronidase and arylsulfatase (UGLUC) 3
2.2.1.4 Extractive methylation procedure for urine or plasma (UME, PME) 3
2.2.1.5 Solid-phase extraction for plasma or urine (PSPE, USPE) 4
2.2.1.6 LLE of plasma for determination of drugs for brain death diagnosis 4
2.2.1.7 LLE of plasma for determination of therapeutics relevant in emergency toxicology 4
2.2.1.8 Extraction and silylation of ethylene glycol, 1,2-propylene glycol, 1,3-propylene glycol, diethylene glycol, triethylene glycol, tetraethylene glycol, glycolic acid, lactic acid, and gamma-hydroxybutyric acid (GHB) in plasma or urine 4
2.2.2 Derivatization procedures 5
2.2.2.1 Acetylation (AC) 5
2.2.2.2 Methylation (ME) 5
2.2.2.3 Ethylation (ET) 5
2.2.2.4 tert.-Butyldimethylsilylation (TBDMS) 5
2.2.2.5 Trimethylsilylation (TMS) 5
2.2.2.6 Trimethylsilylation followed by trifluoroacetylation (TMSTFA) 5
2.2.2.7 Trifluoroacetylation (TFA) 6
2.2.2.8 Pentafluoropropionylation (PFP) 6
2.2.2.9 Pentafluoropropylation (PFPOL) 6
2.2.2.10 Heptafluorobutyrylation (HFB) 6
2.2.2.11 Pivalylation (PIV) 6
2.2.2.12 Heptafluorobutyrylprolylation (HFBP) 6
2.3 **GC-MS Apparatus** 7
2.3.1 Apparatus and operation conditions 7
2.3.2 Quality assurance of the apparatus performance 7
2.4 **Determination of retention indices** 9
2.5 **Systematic toxicological analysis (STA) of several classes of drugs and their metabolites by GC-MS** 9
2.5.1 Screening for 200 drugs in blood plasma after LLE 9
2.5.2 Screening for most of the basic and neutral drugs in urine after acid hydrolysis, LLE and acetylation 10
2.5.3 Systematic toxicological analysis procedures for the detection of acidic drugs and/or their metabolites 13
2.5.4 General screening procedure for zwitterionic compounds after SPE and silylation 13
2.6 **Application of the electronic versions of this handbook using macros or AMDIS** 14
2.7 **Quantitative determination** 14

3 Correlation between Structure and Fragmentation 15
3.1 **Principle of electron-ionization mass spectrometry (EI-MS)** 15
3.2 **Correlation between fundamental structures or side chains and fragment ions** 15

4	**Formation of Artifacts 16**	
4.1	**Artifacts formed by oxidation during extraction with diethyl ether 16**	
4.1.1	N-Oxidation of tertiary amines 16	
4.1.2	S-Oxidation of phenothiazines 16	
4.2	**Artifacts formed by thermolysis during GC (GC artifact) 16**	
4.2.1	Decarboxylation of carboxylic acids 16	
4.2.2	Cope elimination of N-oxides (-$(CH_3)_2$NOH, -$(C_2H_5)_2$NOH, -$C_6H_{14}N_2O_2$) 16	
4.2.3	Rearrangement of bis-deethyl flurazepam (-H_2O) 16	
4.2.4	Elimination of various residues 16	
4.2.5	Methylation of carboxylic acids in methanol ((ME), ME in methanol) 17	
4.2.6	Formation of formaldehyde adducts using methanol as solvent (GC artifact in methanol) 17	
4.3	**Artifacts formed by thermolysis during GC and during acid hydrolysis (GC artifact, HY artifact) 17**	
4.3.1	Dehydration of alcohols (-H_2O) 17	
4.3.2	Decarbamoylation of carbamates 17	
4.3.3	Cleavage of morazone to phenmetrazine 17	
4.4	**Artifacts formed during acid hydrolysis 17**	
4.4.1	Cleavage of the ether bridge in beta-blockers and alkanolamine antihistamines (HY) 17	
4.4.2	Cleavage of 1,4-benzodiazepines to aminobenzoyl derivatives (HY) 17	
4.4.3	Cleavage and rearrangement of N-demethyl metabolites of clobazam to benzimidazole derivatives (HY) 18	
4.4.4	Cleavage and rearrangement of bis-deethyl flurazepam (HY -H_2O) 18	
4.4.5	Cleavage and rearrangement of tetrazepam and its metabolites 18	
4.4.6	Dealkylation of ethylenediamine antihistamines (HY) 18	
4.4.7	Hydration of a double bond (+H_2O) 18	
5	**Table of Atomic Masses 19**	
6	**Table of Abbreviations 20**	
7	**References 23**	
8	**Tables of Compounds 33**	
	8.1 Compounds in order of names 35	
9	**Mass Spectra**	
	9.1 Arrangement of spectra 244	
	9.2 Lay-out of spectra 244	
	9.3 Mass spectra (*m/z* 30 - 299) 247 - 903	

Contents of Volume 2 (Mass Spectra)

9.3 (continued) Mass spectra (*m/z* 300 - 1920) 904 - 1736

1 Introduction

After publication of the first edition in 1985 with 1,500 data sets, the second in 1992 with 4,700, and with the Supplement Volume in 2000 with 6,700 data sets, the third in 2007 with a completely revised book concept and enlarged to 7,840 data sets, the fourth edition in 2011 with 8,650 data sets, the new edition contains **10,430 data sets**.

Since 2011, we have recorded and collected in our laboratory mass spectral and gas chromatographic data of new drugs (of abuse), of emerging new psychoactive substances (NPS), poisons and their metabolites bringing the total number of mass spectra to **10,430** among which about 2,000 data sets have been revised or added. Over 7,800 data sets are from metabolites, over 4,400 from acetylated, over 200 from ethylated, over 1,700 from methylated, over 1,600 from trimethylsilylated, over 1,100 from trifluoroacetylated, over 1,600 from pentafluoropropionylated, and over 1,600 heptafluorobutyrylated compounds.

Section 2 on sample preparation and GC-MS methods has been updated, but only our standard operation procedures have been described in detail, as they were used for recording the data.

The structures of the compounds are also reproduced to facilitate the identification of unknown metabolites by correlating the fragmentation patterns with the probable structure, as described in Section 3. Further data are also included, as outlined in the Explanatory notes.

Several artifacts have been detected which are formed during sample preparation and/or the GC procedure. Their formation is described in detail in Section 4.

Section 5 contains the table of atomic masses used for molecular mass calculations and Section 6 the actualized Table of abbreviations (Table 6.1) used.

All the compounds are listed alphabetically in Table 8.1 in order to assist the search for data concerning specific compounds with directions on which page in the data can be found, as well as the entry numbers of the electronic versions.

In electronic versions not using the NIST search algorithm, the name of only one parent compound can be given, but the symbol '@' indicates, that further compounds can form this particular metabolite, derivative or artifact. This information can be found in Table 8.2 under the corresponding library entry number.

The extremely large amount of data presented makes it likely that some errors will be present in this handbook. The authors cannot be held responsible for such errors, nor for any consequences arising from the employment of the published data. Users are requested to report any errors that are found and to suggest other data and other groups of compounds, which are of interest (e-mail: hans.maurer@uks.eu). If possible, these will be included in future editions, along with corrections of any errors.

2 Experimental Section

All mass spectral and gas chromatographic data have been recorded in our laboratory, according to our standard operation procedures outlined below.

2.1 Origin and choice of samples

Most of the metabolite data have been recorded from samples (plasma or urine) of patients suspected of poisoning or intoxication and admitted to various anti-poison centers, from in-patients of several clinics, particularly the University Hospital at Homburg, or from volunteers treated with therapeutic dosages of drugs. If suitable samples from humans were not available, samples from incubations of the drug with human liver preparations [3,4] or from rats were used. Rats were administered the corresponding drugs in aqueous suspension, by gastric intubation [5].

The choice of biosample for toxicological analysis depends on the toxicological problem. Concentrations of drugs are relatively high in urine, so that urine is the sample of choice for a comprehensive screening and identification of unknown drugs or poisons [6-10]. However, the metabolites of these drugs must be identified in addition, or even exclusively. Blood (plasma, serum) is the sample of choice for quantification. However, if the blood concentration is high enough, screening can also be performed therein. This may be advantageous, as sometimes only blood samples are available and some procedures allow simultaneous screening and quantification [7,8,11-17]. GC-MS analysis of drugs in alternative matrices such as hair, sweat and oral fluid, meconium, or nails was also described, but the toxicological interpretation of the analytical results may be difficult.

2.2 Sample preparation

Independent of the biosample, a suitable sample preparation is necessary before GC-MS analysis. This may involve cleavage of conjugates, isolation, clean-up steps and/or derivatization of the drugs and their metabolites. In the following sections, sample preparation and derivatization procedures are described for systematic toxicological analysis, as well as for sensitive procedures for particular drugs of toxicological relevance.

Compounds were isolated by liquid-liquid extraction (LLE) or by solid phase extraction (SPE) preceded or followed by clean-up steps. Universal LLE procedures have still been used for emergency analyses and for systematic toxicological analysis (STA), because substances with very different physico-chemical properties had to be isolated from heterogeneous matrices. SPE is preferred for screening and quantification of lower-dosed drugs in plasma, as this results in considerably cleaner extracts. The information of which compound can be detected in which sample after which type of sample preparation (column *Detected* in Table 8.1 as well as in the legends of the mass spectra) is based on the sample preparation procedures described in Section 2.2. Special procedures are not described here, as series of different (new) extraction procedures are in use around the world.

Many drugs and poisons are excreted in urine in a completely metabolized and conjugated form. Cleavage of conjugates by enzymatic or acid hydrolysis is necessary before extraction, as the analysis of conjugates is difficult or even impossible by GC-MS. The gentle but time-consuming method of enzymatic hydrolysis is preferable in metabolism studies or in forensic and doping analysis, if the drugs are destroyed during acid hydrolysis and the time of analysis is not limited (Section 2.2.1.3). However, fast enzymatic cleavage has also been described as alternative to acid hydrolysis [18]. For emergency cases, it is preferable to cleave the conjugates by rapid acid hydrolysis (Section 2.2.1.2). However, the formation of artifacts during this procedure must be considered (Section 4).

Typical derivatization procedures are given in Section 2.2.2.

2.2.1 Standard extraction procedures

2.2.1.1 Standard liquid-liquid extraction (LLE) for plasma, urine or gastric contents (P, U, G)

This standard extraction procedure can be used for (limited) plasma screening and quantification. For quantification of 40 relevant drugs, a fast and simple multi-analyte procedure using GC-MS and one-point calibration was developed [15] (see Section 2.7).

Plasma (1 mL) mixed with 0.1 mL internal standard (IS, 0.01 mg/mL trimipramine-D3 in methanol) or 2.5 mL of urine or gastric contents were extracted with 5 mL of a mixture of diethyl ether:ethyl acetate (1:1; v/v) after addition of 2 mL saturated sodium sulfate solution. After phase separation by centrifugation, the organic extract was transferred into a flask and evaporated to dryness. The aqueous residue was then basified with 0.5 mL of 1 M sodium hydroxide and extracted a second time with 5 mL of the solvent mixture. This organic extract was transferred to the same flask and evaporated at 60 °C under reduced pressure. The combined residues were dissolved in 100 µL methanol [19].

2.2.1.2 STA procedure (hydrolysis, extraction and microwave-assisted acetylation) for urine (U+UHYAC)

Some analytes are completely destroyed and therefore not detectable after acid hydrolysis. Therefore, the following modification was developed [20]. The same volume of untreated urine was added to the hydrolyzed sample before extraction and derivatization. In addition, the derivatization time could be reduced from 30 to 5 min using microwave irradiation. The reduction of the extract concentrations of compounds excreted in conjugated form could be compensated by the higher sensitivity of newer GC-MS apparatus [21-36].

A 5-mL portion of urine was divided in two equal parts. One 2.5-mL part was refluxed with 1 mL of 37% hydrochloric acid for 15 min. Following hydrolysis, the sample was basified with 2 mL of 2.3 M aqueous ammonium sulfate and 1.5 mL of 10 M aqueous sodium hydroxide to obtain a pH between 8 and 9. Before extraction, the second 2.5 mL part of untreated urine was added. This solution was extracted with 5 mL of a dichloromethane:isopropanol:ethyl acetate mixture (1:1:3; v/v/v). After phase separation by centrifugation, the organic layer was transferred to a flask and evaporated to dryness at 60 °C under reduced pressure. The residue was derivatized by acetylation with 100 µL of an acetic anhydride:pyridine mixture (3:2; v/v) for 5 min under microwave irradiation at about 400 W. After evaporation of the derivatization mixture at 60 °C under reduced pressure, the residue was dissolved in 100 µL of methanol and 1–2 µL were injected into the gas chromatograph [36].

2.2.1.3 Extraction of urine after cleavage of conjugates by glucuronidase and arylsulfatase (UGLUC)

Several enzymatic hydrolysis procedures have been described using solutions of beta-glucuronidase and arylsulfatase from Helix pomatia or Patella vulgata and incubation times from 2 to 24 h [24,37-40]. Microvave-assisted enzymatic hyrdrolysis could be finished in five minutes [18]. In our experience the following procedure is an acceptable compromise between time consumption and cleavage efficiency.

A 5-mL portion of urine was adjusted to pH 5.2 with acetic acid (1 M) and incubated at 50 °C for 1.5 h with 100 µL of a mixture (100,000 Fishman units per mL) of glucuronidase (EC no. 3.2.1.31) and arylsulfatase (EC no. 3.1.6.1) from *Helix Pomatia*. The mixture was then adjusted to pH 8-9 with a mixture of 1 mL of 37% hydrochloric acid, 2 mL of 2.3 M aqueous ammonium sulfate and 1.5 mL of 10 M aqueous sodium hydroxide and extracted with 5 mL of a dichloromethane:isopropanol:ethyl acetate mixture (1:1:3; v/v/v). After phase separation by centrifugation, the organic layer was carefully evaporated to dryness at 60 °C under a stream of nitrogen. The residue was derivatized if necessary. The residue was dissolved in 50 µL of the corresponding solvent and 2 µL were injected into the gas chromatograph [22].

2.2.1.4 Extractive methylation procedure for urine or plasma (UME, PME)

A 2-mL portion of urine or plasma was mixed in a centrifuge tube with 2 mL of the phase-transfer reagent consisting of 0.02 M tetrahexylammonium hydrogen sulfate in 1 M phosphate buffer pH 12. After addition of 6 mL of freshly prepared 1 M methyl iodide in toluene, the closed tube was shaken at 50 °C for 30 min. After phase separation by centrifugation at 3,000 × g for 3 min, the organic phase, containing the analytes and tetrahexylammonium salts, was transferred to a diol SPE column, which was conditioned as follows: 5 mL of methanol at a slow flow rate, drying the column under vacuum for 15 seconds, 5 mL of toluene at a slow flow rate. The organic phase was rinsed through the sorbent at a flow rate of 3 mL/min to adsorb the tetrahexylammonium salts. The part of the analytes also adsorbed on the sorbent was selectively eluted with 5 mL of diethyl ether:ethyl acetate (95:5, v/v) at a flow rate of 3 mL/min. The combined eluates were carefully evaporated to dryness at 60 °C under reduced pressure. The residue was dissolved in 50 µL of ethyl acetate and a 1–2 µL aliquot of this extract was injected into the GC-MS [41-43].

2.2.1.5 Solid-phase extraction for plasma or urine (PSPE, USPE)

SPE is preferred for analysis of lower-dosed drug.

Plasma or urine (0.5-1 mL) was diluted with 2 mL of purified water. After addition of 50-100 µL of a methanolic solution of suitable IS, the samples were mixed for 15 s on a rotary shaker, centrifuged for 3 min at 1,000 × g and loaded on mixed-mode (Bond Elute Certify, 130 mg, 3 mL: C8 and strong cation exchange sorbents) SPE cartridges previously conditioned with 1 mL of methanol and 1 mL of purified water. After extraction, the cartridges were washed with 1 mL of purified water, 1 mL of 0.01 M aqueous hydrochloric acid and 2 mL of methanol. Reduced pressure was applied until the cartridges were dry, and the analytes were eluted with 1 mL of a freshly prepared mixture of methanol-aqueous ammonia (98:2, v/v) into 1.5 mL polypropylene reaction vials. The eluates were evaporated to dryness under a stream of nitrogen at 60 °C and the residue was dissolved in 50-100 µL of methanol [11,12,44-46].

2.2.1.6 LLE of plasma for determination of drugs for brain death diagnosis

After addition of 50 µL of IS solution (4.0 mg/L of pentobarbital-D5, 2.0 mg/L of methohexital-D5, 40.0 mg/L of phenobarbital-D5, 0.8 mg/L of each diazepam-D5 and nordazepam-D5) and 50 µL of butyl acetate, plasma samples (200 µL) were extracted for 2 min on a rotary shaker. After phase separation by centrifugation (1 min, 10,000g, ambient temperature), the organic layer (upper) was transferred to autosampler vials and 2 µL were injected into the GC-MS [47].

2.2.1.7 LLE of plasma for determination of therapeutics relevant in emergency toxicology

Single test immunoassays for quantification of valproic acid, salicylic acid, paracetamol, phenobarbital, phenytoin, and primidone are no longer available. Therefore, a fast, simple and cost-effective GC-MS method was developed and validated for these drugs [17].

Prior to extraction, 50 µL of internal standard (IS) solution (25.0 mg/L of phenobarbital-d_5,) and 450 µL of butyl acetate (ButAc) were added to 200 µL of the plasma sample. After addition of 10 µL of acetic acid conc., plasma samples were extracted for 2 min on a rotary shaker. After centrifugation (2 min, 14,000g) the organic layer was transferred to autosampler vials and 1 µL was injected into the GC-MS system [17].

2.2.1.8 Extraction and silylation of ethylene glycol, 1,2-propylene glycol, 1,3-propylene glycol, diethylene glycol, triethylene glycol, tetraethylene glycol, glycolic acid, lactic acid, and gamma-hydroxybutyric acid (GHB) in plasma or urine.

Among the different methods for the determination of ethylene glycol in plasma or blood, gas chromatographic procedures require isolation and derivatization. Besides our previous GC-MS procedure using pivalylation [48], a new method using a special silylation reagent [49] was modified, validated, and established in our laboratory [16].

To 50 µL of urine or plasma, 10 µL of internal standard (1,3-propyleneglycol, 0.5 g/L) and 50 µL acetonitrile were added in an Eppendorff tube. The mixture was thoroughly shaken and centrifuged (14,000g) for 2 min, respectively. Afterwards, 20 µL of the supernatant were transferred to an autosampler screw-cap vial and mixed-up with 20 µl of dimethylformamide (DMF) and 300 µL of BSTFA. The vial was closed, immediately vortexed and put into the microwave at 450 W for 5 min. After cooling, the vial was placed into the autosampler and 1 µL was injected into the GC-MS with a 1/30 split ratio on an Agilent J&W HP-5MS fused silica capillary column; 30 m, 0.25 mm I.D., 0.25 mm film thickness (WICOM, Heppenheim, Germany) with helium as carrier gas. The used DSQ II (2.0.1) GC system, Trace GC Ultra (2.0), and Tri Plus autosampler (2.0) were from Thermo Fisher Scientific (Dreieich, Germany). The inlet temperature was set at 250°C. The oven temperature profile was as follows: initial temperature of 50°C raised by 10°C/min to 80°C (hold time 1 min), raised by 30°C/min to 230°C and finally by 50°C/min to 310°C (hold time 2 min). The total run time was 12 min. The MS conditions were as follows for identification in urine and plasma: full-scan mode, *m/z* 50-500 u; EI mode, ionization energy, 70 eV; ion source temperature, 240°C, and identity of the peaks was confirmed by computerized comparison of the underlying full mass spectrum with reference spectra depicted in Volume 2. For quantification in the plasma samples, the MS conditions were as follows: EI mode, positive selected ion monitoring (SIM); width, 1.0; dwell time, 20 ms; ionization energy, 70 eV; ion source temperature, 240°C. The monitored masses were as follows (quantifier in italic): ethylene glycol: *m/z* *103*, 147, 191; 1,2 propylene glycol: *117*, 147, 205; 1,3 propylene glycol (IS): *130*, 115, 205; lactic acid: *130*, 117, 219; glycolic acid: *177*, 147, 161; GHB: *233*, 98, 147; diethylene glycol: *117*, 103, 147;

triethylene glycol: *117*, 103, 161; tetraethylene glycol: *117*, 147, 207.

Data were qualitatively processed using Xcalibur 2.0.7 (ThermoFisher Scientific, Dreieich, Germany) and quantitative results were achieved using LCquan 2.5.6 (ThermoFisher Scientific, Dreieich, Germany) software. The settings were as follows: peak detection algorithm ICIS; baseline, 40; area noise factor, 5; peak noise factor, 10. GraphPad Prism 5.00 (GraphPad Software, San Diego, CA, USA) was used for all calculations. Quantification was based on the peak area ratios of the quantifier ions of the different compounds and the internal standard 1,3-propylene glycol. The mass spectra are depicted in the mass spectra part (cf. Table 8.1). The method was linear from 0.05 g/L to 1.0 g/L for glycols and 0.01 g/L to 0.2 g/L for GHB [16].

2.2.2 Derivatization procedures

Derivatization steps are necessary to improve the gas chromatographic characteristics of polar compounds. Furthermore, mass spectra of several compounds are altered on derivatization so that they contain more typical ions, e.g. the molecular ion (cf. the mass spectra of amfetamine and its acetyl derivative). The sensitivity of the method can be improved by introduction of halogen atoms into the molecule if negative-ion chemical ionization (NICI) is used. Finally, enantiomers can be separated on achiral columns after derivatization to the diastereomers with chiral reagents, e.g. S(-)-trifluoroacetylprolylchloride [50] or S(-)-heptafluorobutyrylprolyl chloride [44,51,52].

The use of **microwave irradiation** drastically reduces the incubation time (e.g., of acetylation from 30 min to 5 min [20,53]). However, microwave irradiation should not be used for chiral derivatization, because the enantiomers may racemize. Therefore, derivatization should no longer be renounced due to time consumption. An overview on derivatization procedures for GC-MS can be found in corresponding reviews [53,54]. The following procedures have been used for recording the mass spectra presented in Volume 2.

2.2.2.1 Acetylation (AC)

The evaporated sample was acetylated with 50 µL of an acetic anhydride:pyridine mixture (3:2; v/v) for 5 min under microwave irradiation at about 400 W. After evaporation of the derivatization mixture, the residue was dissolved in 50 µL of methanol and 1–2 µL were injected into the GC-MS.

2.2.2.2 Methylation (ME)

A 50-µL aliquot of extract in methanol was methylated for 30 min (carboxylic acids) or for about 12 h (phenols) at room temperature with 100 µL of an ethanol-free solution of diazomethane in diethyl ether synthesized according to [53,55]. This solution was stable for several months when stored in sealed 10 mL flasks in the freezer at -20 °C. After evaporation of the methylation mixture the residue was dissolved in 50 µL methanol and 1-2 µL were injected into the GC-MS.

(Extractive methylation was described in Section 2.2.1.4.)

2.2.2.3 Ethylation (ET)

The procedure was as for methylation, but using diazoethane synthesized according to the procedure of McKay et al. using 1-methyl-3-nitro-1-nitroso-3-nitroguanidine, KOH and diethyl ether [55].

2.2.2.4 tert.-Butyldimethylsilylation (TBDMS)

A 50-µL aliquot of extract was evaporated and then derivatized with 50 µL of N-methyl-N-(tert.-butyldimethylsilyl)-trifluoroacetamide (MTBSTFA) for 10 min at 60 °C. Aliquots of this mixture (1–2 µL) were injected into the GC-MS with an alcohol- and water-free syringe [56].

(Silylation of glycols using dimethylformamide (DMF) and BSTFA was described in Section 2.2.1.8.)

2.2.2.5 Trimethylsilylation (TMS)

A 50-µL aliquot of extract was evaporated and then silylated with 50 µL N-methyl-N-trimethylsilyl-trifluoroacetamide (MSTFA) for 30 min at 60 °C. Aliquots (1-2 µL) of this mixture were injected into the GC-MS with an alcohol- and water-free syringe.

2.2.2.6 Trimethylsilylation followed by trifluoroacetylation (TMSTFA)

A 50-µL aliquot of extract was evaporated and then silylated with 100 µL MSTFA for 30 min at 60 °C, and then derivatized with 20 µL N-methyl-bis-trifluoroacetamide (MBTFA) for 10 min at 60 °C. After evaporation of the derivatization mixture the residue was dissolved in 50 µL alcohol- and water-free ethyl acetate and 1–2 µL of this mixture were injected into the GC-MS with an alcohol- and water-free syringe [39].

2.2.2.7 Trifluoroacetylation (TFA)

A 50-µL aliquot of extract was evaporated and then derivatized with 50 µL trifluoroacetic anhydride and 50 µL ethyl acetate for 3 min under microwave irradiation at about 400 W. After evaporation of the derivatization mixture the residue was dissolved in 50 µL alcohol- and water-free ethyl acetate and 1–2 µL were injected into the GC-MS.

2.2.2.8 Pentafluoropropionylation (PFP)

A 50-µL aliquot of extract was evaporated and then derivatized with 50 µL pentafluoropropionic anhydride for 5 min under microwave irradiation at about 400 W. After evaporation of the reagent the residue was dissolved in 50 µL alcohol- and water-free ethyl acetate and 1–2 µL were injected into the GC-MS.

2.2.2.9 Pentafluoropropylation (PFPOL)

A 50-µL aliquot of extract was evaporated and then derivatized with 50 µL pentafluoropropanol (PFPOL) for 5 min under microwave irradiation at about 400 W. After evaporation of the reagent the residue was dissolved in 50 µL alcohol- and water-free ethyl acetate and 1–2 µL were injected into the GC-MS.

2.2.2.10 Heptafluorobutyrylation (HFB)

A 50-µL aliquot of extract was evaporated and then derivatized with 50 µL heptafluorobutyric anhydride for 5 min under microwave irradiation at about 400 W. After evaporation of the reagent the residue was dissolved in 50 µL alcohol- and water-free ethyl acetate and 1–2 µL were injected into the GC-MS [11,12,44,57].

2.2.2.11 Pivalylation (PIV)

The sample was dissolved in 50 µL of a freshly prepared (!) mixture of pivalic acid anhydride:triethylamine:methanol (20:1:1; v/v/v) and incubated for 10 min under microwave irradiation at about 400 W. This solution was diluted by 200 µL methanol and 1–2 µL were injected into the GC-MS [48].

2.2.2.12 Heptafluorobutyrylprolylation (HFBP)

The evaporated sample was derivatized by addition of 0.2 mL of aqueous carbonate buffer (sodium bicarbonate/sodium carbonate, 7:3; 5%, w/v; pH 9) and 6 µL of derivatization reagent (0.1 M S-HFBPCl in dichloromethane). The reaction vials were sealed and left on a rotary shaker at room temperature for 30 min. Thereafter, 0.1 mL of cyclohexane were added, the reaction vials were sealed again and left on a rotary shaker for 1 min. The phases were separated by centrifugation (14,000g, 1 min) and the cyclohexane phase (upper) was transferred to autosampler vials. Aliquots (3 µL) were injected into the GC-MS [44,51,52,58].

2.3 GC-MS Apparatus

2.3.1 Apparatus and operation conditions

The following GC-MS apparatus were used for recording the GC and MS data in this handbook:
Agilent Technologies (Waldbronn, Germany) 5890 Series II gas chromatograph combined with an HP 5970 MSD, an HP 5989B MS Engine mass spectrometer, or an Agilent Technologies HP 6890 gas chromatograph combined with HP 5971, 5972, or 5973 MSDs.

In addition, a Thermo Fisher Scientific (Dreieich, Germany) DSQ II or ISQ GC-MS system, Trace GC Ultra (2.0), and Tri Plus autosampler (2.0) were used.

The operating conditions for routine screening were as follows:

Column:	cross-linked methylsilicone capillary Optima-5 MS 12 m × 0.2 I.D., film thickness 0.35 µm (Macherey-Nagel, Düren, Germany)
Column temperature:	solvent delay: 3 min at 100 °C programmed from 100 to 310 °C in 30 °C/min, 5 min at maximum temperature
Injector port temperature:	280 °C
Carrier gas:	helium, flow rate 1 mL/min
Ionization mode:	electron ionization (EI)
Ionization energy:	70 eV
Ion source temperature:	200 °C
Scan rate:	1 scan/sec

2.3.2 Quality assurance of the apparatus performance

For daily checking of apparatus performance, a methanolic solution of typical drugs covering a wide range of physicochemical properties and relevant retention time was injected into the GC-MS. Fig. 2.1 shows a typical total ion chromatogram of this standard solution containing 50 ng/µL of valproic acid, metamfepramone, acetylated amfetamine, pentobarbital, diphenhydramine, phenobarbital, methaqualone, codeine, morphine, nalorphine, quinine, haloperidol, strychnine and the hydrocarbon C40 (injection volume 1 µL).

The quality criteria are as follows: all compounds should clearly be separated and the peaks should be sharp and of sufficient abundance. The relation of the peak areas of underivatized (!) morphine to codeine should be at least 1:10. If these demands are not fulfilled, the liner should be removed, and the GC column should be shortened (10–20 cm) or even replaced. Last, not least, the ion source should be cleaned.

Of course, the current performance should be tested in an analysis series using calibrators and negative control samples. The use of negative control samples is indispensable, as analyte carry-over is a major problem in trace analysis. Fig. 2.2 (upper part) shows a total ion chromatogram of a urine sample (U+UHYAC, cf. Section 2.2.1.2) containing, besides other drugs, dihydrocodeine and its metabolites. After 20 (!) rinsing steps of the autosampler needle, dihydrocodeine could still be detected (lower part). False-positive GC-MS results reported by proficiency test organizers were typically caused by analyte carry-over.

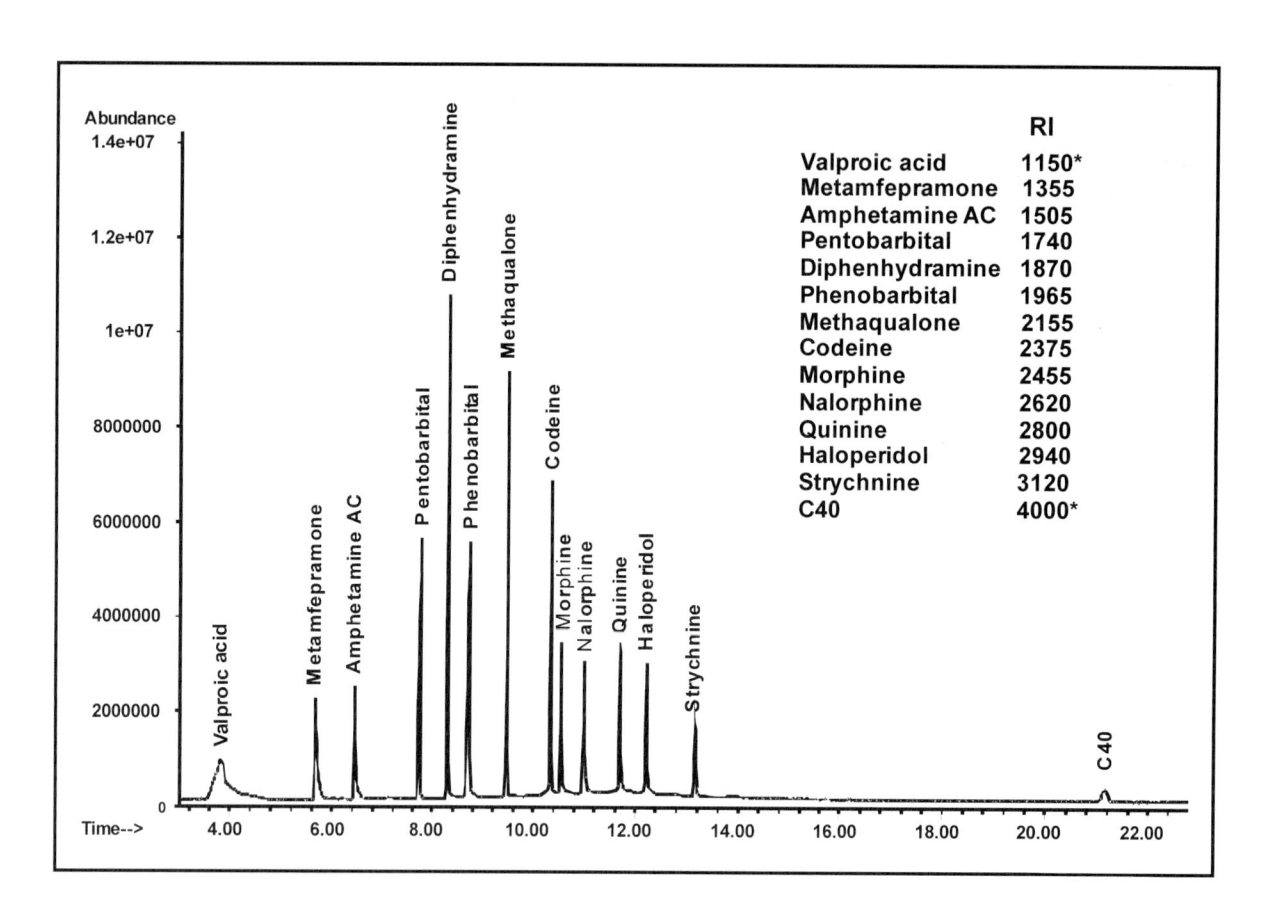

Fig. 2.1: Typical total ion chromatogram of the standard test solution containing 50 ng/μL of the given compounds (injection volume 1 μL).

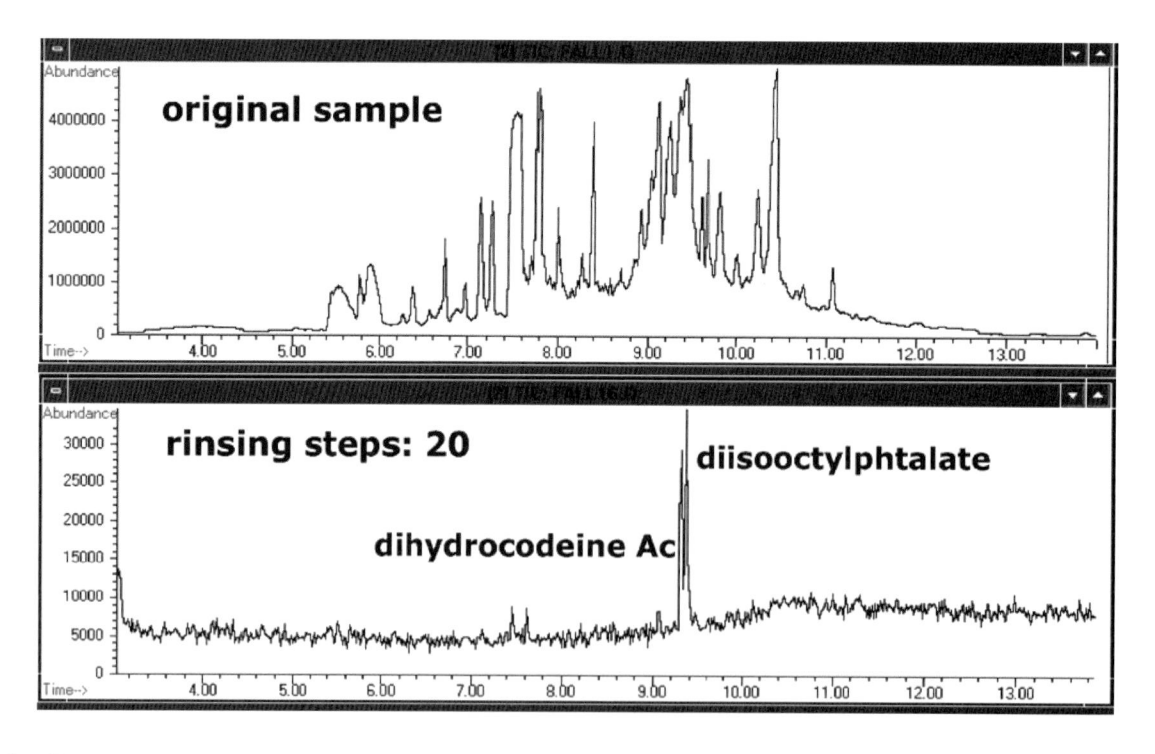

Compound	RI
Valproic acid	1150*
Metamfepramone	1355
Amphetamine AC	1505
Pentobarbital	1740
Diphenhydramine	1870
Phenobarbital	1965
Methaqualone	2155
Codeine	2375
Morphine	2455
Nalorphine	2620
Quinine	2800
Haloperidol	2940
Strychnine	3120
C40	4000*

Fig. 2.2: (Upper part) Total ion chromatogram of a urine sample (U+UHYAC) containing, besides other drugs, dihydrocodeine and its metabolites. Total ion chromatogram of a methanol injection after 20 (!) rinsing steps of the autosampler needle after the injection of the urine sample (lower part).

2.4 Determination of retention indices

The retention indices (RI) were measured by GC-MS on methyl silicone capillary columns using a temperature program. They were calculated in correlation with the Kovats' indices [59] of the components of a standard solution of typical drugs (Section 2.3.2, Fig. 2.1) which is measured daily for testing the GC-MS performance [60]. The RIs of compounds with an asterisk (*) are not detectable by nitrogen-selective flame-ionization detection (N-FID).

2.5 Systematic toxicological analysis (STA) of several classes of drugs and their metabolites by GC-MS

The screening strategy of the systematic toxicological analysis (STA) must be very extensive, because several thousands of drugs or pesticides are to be considered. GC-MS, especially in the full scan electron ionization (EI) mode, is still the best method for such STA procedures [21-36]. Most of the STA procedures cover basic (and neutral) drugs, which are toxicants that are more important. For example, most of the psychotropic drugs have, like neurotransmitters, basic properties. Nevertheless, some classes of acidic drugs or drugs producing acidic metabolites, e.g., angiotensin converting enzyme (ACE) inhibitors, angiotensin receptor II (AT-II) blockers, dihydropyridine calcium channel blockers (metabolites), diuretics, laxatives, coumarin anticoagulants, antidiabetic sulfonylureas, barbiturates, or non-steroidal anti-inflammatory drugs (NSAIDs), are relevant to clinical and forensic toxicology or doping control.

Concentrations of drugs and/or their metabolites are relatively high in urine, so that urine is the sample of choice for a comprehensive screening and identification of unknown drugs or poisons. However, the metabolites of these drugs must be identified in addition, or even exclusively. In horse doping control, urine is also the common sample for screening. If the blood concentration is high enough, screening can also be performed therein. This may be advantageous, as sometimes only blood samples are available and some procedures allow simultaneous screening and quantification.

Our four screening procedures will be presented here, one for screening for 200 drugs in blood plasma after LLE (Section 2.5.1), one for most of the basic and neutral drugs in urine after acid hydrolysis, LLE and acetylation (Section 2.5.2), one for most of the acidic drugs and poisons in urine after extractive methylation (Section 2.5.3), and one for zwitterionic compounds in urine after SPE and silylation.

2.5.1 Screening for 200 drugs in blood plasma after LLE

GC-MS full-scan screening procedures in blood principally allow to detect a wide range of analytes although their concentrations are generally lower than those in urine [61]. Maurer has described a rather comprehensive plasma screening procedure based on LLE with diethyl ether:ethyl acetate (1:1; v/v) at pH 7 and 12 after addition of the universal IS trimipramine-D3 (Section 2.2.1.1) [19,62,63]. This universal extract can be used for GC-MS as well as for LC-MS screening, identification and quantification [19,63,64]. The full scan GC-MS screening is based on reconstructed mass chromatography using macros for selection of suspected drugs followed by identification of the unknown spectra by library search. The selected ions for screening in plasma (and gastric content) have recently been updated using experiences from the daily routine work with this procedure and are summarized in Table 2.1 [61]. Generation of the mass chromatograms can be started by clicking the corresponding pull down menu which executes the user defined macros [65]. The about 200 compounds detected so far by this procedure can be found with a 'P' in the column *Detected* in Table 8.1 as well as in the legends to the mass spectra.

A rather universal SPE procedure (Section 2.2.1.5) has proved to be a good alternative for the LLE procedure leading to cleaner extracts [11,12,44,52,66,67]. With exception of neutral and acidic drugs, this SPE can also be used for the described plasma screening.

Of course, further compounds can be detected, if they are extractable under the conditions applied, volatile in GC and if their mass spectra are in this mass spectral database and reference library. In order to widen the screening window, comprehensive urine screening by full-scan GC-MS allowing detection of several thousand compounds is strongly recommended.

Table 2.1: Classes of basic and neutral drugs simultaneously screened for and identified by the GC-MS STA with LLE of plasma (Section 2.2.1.1) and the ions selected for reconstructed mass chromatography after full-scan GC-MS analysis [61].

Drug classes	Ions (*m/z*) selected for monitoring
Psychotropics I	58, 61 (trimipramine-D3), 70, 72, 86, 98, 112, 192
Psychotropics II	100, 123, 210, 242, 243, 276, 297, 303, 329
Benzodiazepines	239, 253, 269, 283, 300, 305, 308, 310, 312, 315
Other sedative-hypnotics	83, 105, 156, 163, 167, 172, 180, 235, 248, 261
Anticonvulsants	102, 113, 180, 185, 190, 193, 204, 280, 288, 394
Analgesics	120, 139, 151, 161, 188, 214, 217, 230, 231, 299

2.5.2 Screening for most of the basic and neutral drugs in urine after acid hydrolysis, LLE and acetylation

A screening method for the detection of most of basic, neutral and some acidic drugs in urine after acid hydrolysis, LLE and acetylation was developed and has been improved during the past years. Acid hydrolysis has proved to be very rapid and efficient for the cleavage of conjugates, particularly in emergency toxicology. However, some compounds were destroyed or altered during acid hydrolysis (Section 4.4). Therefore, the standard procedure had to be modified [20]. Before extraction, half of the untreated urine volume is added to the previously hydrolyzed part. The extraction solvent used has proved to be very efficient in extracting compounds with very different chemical properties from biomatrices, so that it has been used for a STA procedure for basic and neutral analytes [21-27,30,61,68]. Acetylation has proved to be very suitable for robust derivatization in order to improve the GC properties and thereby the detection limits of thousands of drugs and their metabolites. The use of microwave irradiation reduced the incubation time from 30 min to 5 min [20,69], so that derivatization should no longer be renounced due to expense of time.

As listed in Table 2.2, this comprehensive full scan GC-MS screening procedure allows within one run the simultaneous screening and library-assisted identification of the following categories of drugs: tricyclic antidepressants [70,71], selective serotonin reuptake inhibitors (SSRIs) [20], butyrophenone neuroleptics [72], phenothiazine neuroleptics [73], benzodiazepines [61,74], barbiturates [75] and other sedative-hypnotics [61,75], anticonvulsants [76], anti-parkinsonian drugs [77], phenothiazine antihistamines [78], alkanolamine antihistamines [79], ethylenediamine antihistamines [80], alkylamine antihistamines [81], opiates and opioids [82], non-opioid analgesics [83], stimulants and hallucinogens [50,61,84-89], designer drugs of the amphetamine-type [21-23,84,90], the piperazine-type [90-96,96,97,97], the phenethylamine-type [24-29], the phencyclidine-type [30], the cathinone-type [35,36], *Eschscholtzia californica* ingredients [98], nutmeg ingredients [99], beta-blockers [100], antiarrhythmics [100], and diphenol laxatives [101]. In addition, series of further compounds can be detected [102], if they are present in the extract and their mass spectra are contained in the used reference libraries [103-106]. As shown in Table 2.2, several ions per category were individually selected from the mass spectra of the corresponding drugs and their metabolites identified in authentic urine samples. Generation of the mass chromatograms can be started by clicking the corresponding pull down menu which executes the user defined macros [65].

In the meantime, further series of metabolism and urine screening procedeures have been published and the underlying GC and MS data have been added to this handbook and library [3,5,35,36,107-125].

The more than 3,000 compounds detected so far by this procedure can be found with a 'U+UHYAC' in the column *Detected* in Table 8.1 as well as in the legends to the mass spectra.

The procedure is illustrated in Figs. 2.3 and 2.4. Typical reconstructed mass chromatograms with the given ions of an acetylated extract of a rat urine sample collected over 24 hours after application of 0.8 mg/kg BM of 2C-D which corresponds to a common users' dose of about 50 mg. The identity of peaks in the mass chromatograms was confirmed by computerized comparison of the underlying mass spectrum with reference spectra published in this handbook. The selected ions were used for indication of the 2C-D and its metabolites. Fig. 2.4 illustrates the mass spectrum underlying the marked peak in Fig. 2.3, reference spectrum, structure, and the hit list found by computer library search [103].

Table 2.2: Classes of basic and neutral drugs simultaneously screened for and identified by the GC-MS STA with acid hydrolysis, LLE and acetylation of urine, the monitoring ions, and the corresponding references.

Drug classes	Ions (*m/z*) selected for monitoring	References
Antidepressants, SSRIs	58, 72, 86, 173, 176, 234, 238, 290	[20]
Antidepressants, tricyclic	58, 84, 86, 100, 191, 193, 194, 205, and	[70,71]
	120, 182, 195, 235, 261, 276, 284, 293	
Neuroleptics, butyrophenones	112, 123, 134, 148, 169, 257, 321 and	[72]
	189, 191, 223, 233, 235, 245, 287, 297	
Neuroleptics, phenothiazine	58, 72, 86, 98, 100, 113, 114, 141 and	[73]
	132, 148, 154, 191, 198, 199, 243, 267	
Benzodiazepines	111, 205, 211, 230, 241, 245, 249, 257, 308, 312, 333, 340, 357	[61,74]
Barbiturates	83, 117, 141, 157, 167, 207, 221, 235	[75]
Other sedative-hypnotics	83, 105, 156, 163, 167, 172, 216, 235, 248, 261	[75]
Anticonvulsants	102, 113, 146, 185, 193, 204, 208, 241	[76]
Antiparkinsonian drugs	86.98.136.150.165.196.197.208	[77]
Phenothiazine antihistamines	58.72.100.114.124.128.141.199	[78]
Alkanolamine antihistamines	58.139.165.167.179.182.218.260	[79]
Ethylenediamine antihistamines	58.72.85.125.165.183.198.201	[80]
Alkylamine antihistamines	58.169.203.205.230.233.262.337	[81]
Opiates and opioids	111, 138, 187, 245, 259, 327, 341, 343, 359, 420	[82]
Non-opioid analgesics	120, 139, 151, 161, 188, 217, 230, 231, 258, 308	[83]
Stimulants/Hallucinogens	58, 72, 86, 82, 94, 124, 140, 192, 250	[50,61,84-89]
Designer drugs, amphetamine-type	58, 72, 86, 150, 162, 164, 176, 178	[21-23,84,90]
Designer drugs, cathinone-type	58, 72, 86, 100, 114, 149	[35]
Designer drugs, phenethylamine-type	2C-B: 228, 287, 288	[29]
	2C-D: 178, 238, 164, 223, 236, 295	[28]
	2C-E: 192, 251, 178, 237	[27]
	2C-I: 290, 349, 276, 335	[26]
	2C-T-2: 224, 283, 256, 315	[25]
	2C-T-7: 238, 297, 296, 355	[24]
Designer drugs, phencyclidine-type	232, 274, 273, 290	[30]
Designer drugs, piperazine-type	BZP: 91, 107, 137, 146, 191, 204	[91]
	MDBP: 135, 137, 170, 262, 306	[95]
	mCPP: 143, 145, 166, 182, 238, 254	[93]
	TFMPP: 157, 161, 174, 200, 216, 330	[92]
	MeOPP: 109, 148, 151, 162, 234, 262	[94]
	Drugs with metabolites common with such designer drugs:	
	Dropropizine 132, 148, 175, 233, 320, 378	[97]
	Fipexide: 135, 137, 141, 170, 262, 306	[96]
Designer drugs, pyrrolidinophenone-type	MDPV: 84, 98, 126, 137, 140, 151	[36]
	PVP: 126, 140, 144	[34]
	Others detectable after UGLUC, SPE, TMS (see 2.5.4)	[126-131]
Eschscholtzia californica alkaloids	136, 148, 165, 174, 188, 190	[98]
Nutmeg ingredients	150, 164, 165, 180, 194, 252, 266	[99]
Beta-blockers	72, 86, 98, 140, 151, 159, 200, 335	[100]
Antiarrhythmics	72, 86, 98, 140, 151, 159, 200, 335	[100]
Laxatives	349, 360, 361, 379, 390, 391, 402, 432	[101]

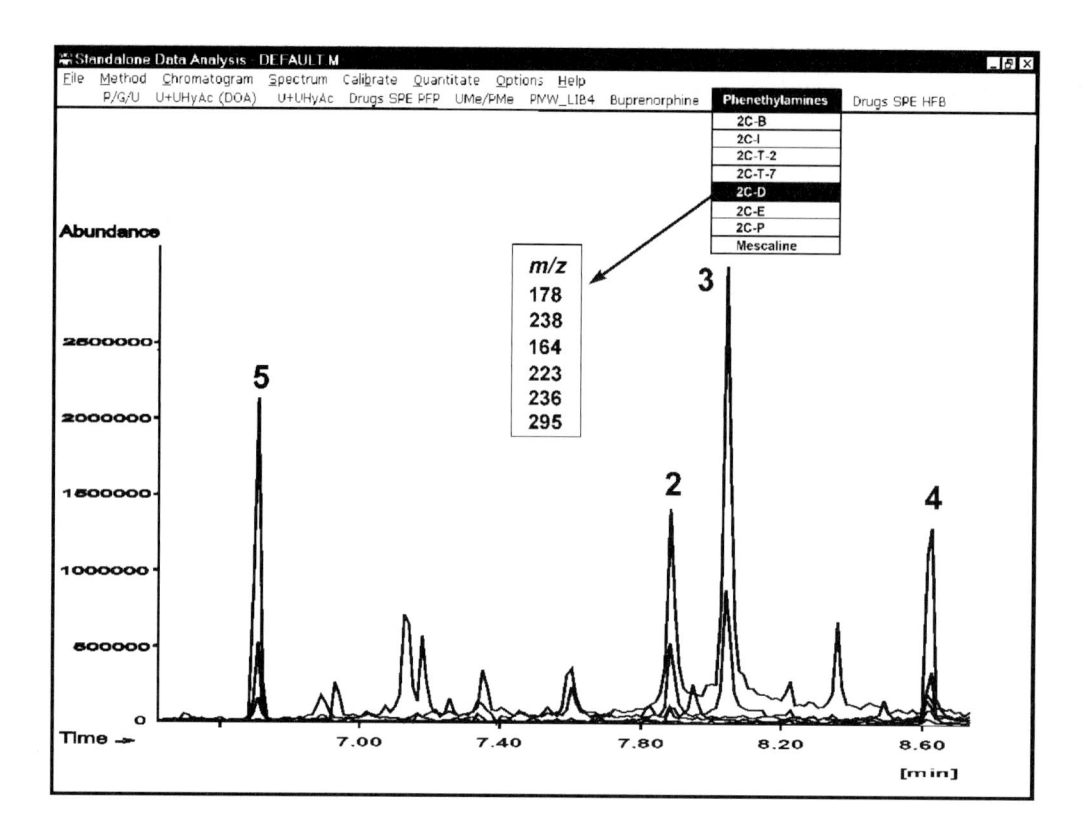

Fig. 2.3: Selective mass chromatograms indicating the new designer drugs 2C-D in a urine extract after acid hydrolysis and acetylation (U+UHYAC). Peaks 2-5 indicate 2C-D metabolites (taken from reference [28]). The merged ion chromatograms can be differentiated by their colors on a color screen.

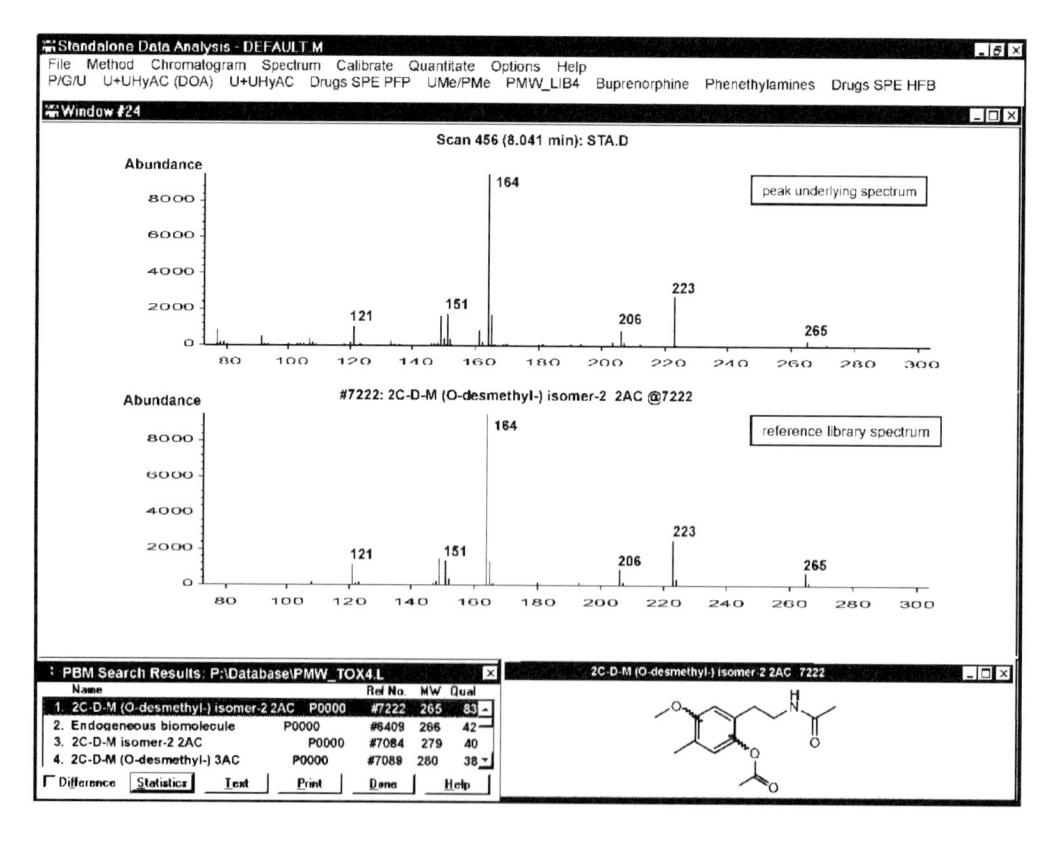

Fig. 2.4: Mass spectrum underlying peak 3 in Fig. 2.3, the reference spectrum, the structure and the hit list found by library search [132] (taken from reference [28]).

2.5.3 Systematic toxicological analysis procedures for the detection of acidic drugs and/or their metabolites

Extractive alkylation has proved to be a powerful procedure for simultaneous extraction and derivatization of acidic compounds. The extracted and derivatized analytes were separated by GC and identified by GC-MS in the full-scan mode. As already described in Section 2.5.2, the possible presence of acidic drugs and/or their metabolites could be indicated using mass chromatography with selective ions followed by peak identification using library search [132].

As listed in Table 2.3, this full scan GC-MS screening procedure allows within one run the simultaneous screening and library-assisted identification of the following categories of drugs: ACE inhibitors and AT_1 blockers [133], coumarin anticoagulants of the first generation [134], dihydropyridine calcium channel blockers [135], NSAIDs [41], barbiturates [136], diuretics [43], antidiabetics of the sulfonylurea type (sulfonamide part), and finally after enzymatic cleavage of the acetalic glucuronides, anthraquinone and diphenol laxatives [42] or buprenorphine [137]. In addition, various other acidic compounds could also be detected (Table 8.1).

Table 2.3: Classes of acidic drugs simultaneously screened for and identified by GC-MS after extractive methylation of urine, the monitoring ions, and the corresponding references.

Drug classes	Ions (*m/z*) selected for monitoring	References
ACE inhibitors and AT-II blocker	157, 160, 172, 192, 204, 220, 234, 248, 249, 262	[133]
Anticoagulants	291, 294, 295, 309, 313, 322, 324, 336, 343, 354	[134]
Calcium channel blockers (dihydropyridines)	139, 284, 297, 298, 310, 312, 313, 318, 324, 332	[135]
NSAIDs	119, 135, 139, 152, 165, 229, 244, 266, 272, 326	[41]
Barbiturates	117, 169, 183, 185, 195, 221, 223, 232, 235, 249	[136]
Diuretics (thiazides)	267, 352, 353, 355, 386, 392	[43]
Diuretics (loop diuretics)	77, 81, 181, 261, 270, 295, 406, 438	[43]
Diuretics (others)	84, 85, 111, 112, 135, 161, 249, 253, 289, 363	[43]
Laxatives (after enzymatic cleavage of the acetalic glucuronides)	305, 290, 335, 320, 365, 350, 311, 326, 271, 346	[42]
Buprenorphine (after enzymatic cleavage of the acetalic glucuronides)	352, 384, 392, 424, 441, 481	[137]

2.5.4 General screening procedure for zwitterionic compounds after SPE and silylation

Pyrrolidinophenone derivatives such as *R,S*-α-pyrrolidinopropiophenone (PPP), *R,S*-4'-methyl-α-pyrrolidinopropiophenone (MPPP), 4'-methyl-α-pyrrolidinobutyrophenone (MPBP), *R,S*-4'-methyl-α-pyrrolidinohexanophenone (MPHP), *R,S*-3',4'-methylenedioxy-α-pyrrolidinopropiophenone (MDPPP), *R,S*-4'-methoxy-α-pyrrolidinopropiophenone (MOPPP), and α-Pyrrolidinovalerophenone (PVP) are new designer drugs which have appeared on the illicit drug market [34,126-130,138]. Unfortunately, these drugs cannot be detected by common screening procedures due to the zwitterionic structure of their main metabolites. Mixed-mode SPE (Section 2.2.1.5) has proven to be suitable even for the extraction of their zwitterionic metabolites.

Trimethylsilylation (Section 2.2.2.5) led to good GC properties. This comprehensive full scan GC-MS screening procedure allows within one run the simultaneous screening and library-assisted identification of the following drugs and/or their metabolites: PPP [129], MPPP [129], MPBP [130], MPHP [126], MDPPP [128], MOPPP [127], and PVP [34]. This procedure has proven to suitable also for detection of many other compounds such as *Kratom* alkaloid [139].

2.6 Application of the electronic versions of this handbook using macros or AMDIS

The mass spectral and GC data are also available in electronic form with reduced data in the library formats of nearly all GC-MS manufacturers. The combination of the handbook presenting the full data with the electronic versions allows their efficient use in analytical toxicology. Using macros, the data evaluation can be automated [61,65,140,141].

An efficient alternative is the use of AMDIS (Automated Mass Spectral Deconvolution and Identification System), a new, easy to use, sophisticated software for GC-MS data interpretation from the U.S. National Institute of Standards and Technology (NIST). It helps to analyze GC-MS data of complex mixtures, even with strong background and coeluting peaks. For efficient use of AMDIS, the optimal parameters for urine drug screening were developed and the applicability tested using over 100 different authentic urine samples [10].

Briefly, the full-scan data files acquired by the GC-MS system were analyzed by AMDIS (http://chemdata.nist.gov/mass-spc/amdis/) in simple mode. The deconvolution and identification parameters were determined to be the best as follows: width, 32; adjacent peak subtraction, two; sensitivity, very high; resolution, high; shape requirement, low. With respect to the minimum match factor, a setting of 50 was found to be the best compromise between "true" hits (targets proposed by the software present in urine) and "false" hits (targets proposed by the software not present in urine).

In the meantime, the urine screening using AMDIS was transferred also to blood screening again using the Maurer/Pfleger/Weber library [142].

2.7 Quantitative determination

GC-MS, especially in the selected ion monitoring (SIM) mode, can be employed for precise and sensitive quantification in biosamples. In contrast to therapeutic drug, monitoring or testing for drugs of abuse, in clinical toxicology several hundred drugs or poisons must be quantified, ideally using one or a few standard procedures (e.g., after standard extraction; Section 2.2.1.1). Even if such procedures cannot be fully validated for all compounds, they can be used in emergency toxicology with sufficient precision. However, when ever possible, quantification procedures should be validated before use in analytical toxicology. In emergency cases, one-point (linear through zero) calibration is often used as a compromise between necessary calibration, workload, and time. After comparing bias and precision data obtained with full and one-point calibration, Peters and Maurer could show that one-point calibration with a calibrator close to the center of the full calibration range can be a feasible alternative to full calibration, especially at medium or high concentrations [143]. Several guidelines for quality assurance and/or method validation have been developed and reviewed [144-148].

Meyer et al. [15] developed and validated a fast and simple multi-analyte procedure for quantification of 40 drugs relevant to emergency toxicology using GC-MS and one-point calibration. Daily one-point calibration with calibrators stored for up to four weeks reduced workload and turn-around time to less than one hour.

3 Correlation between Structure and Fragmentation

This section deals with the correlation between chemical structure and mass fragmentation pattern. Only descriptive explanations are included for illustrating the identification of metabolites using electron-ionization mass spectrometry. Detailed mechanistic explanations for the interpretation of mass spectra are included in specialist texts [149,150].

3.1 Principle of electron-ionization mass spectrometry (EI-MS)

The substance, dispersed in the helium carrier gas, is introduced into the ion source of the mass spectrometer through a GC-MS interface (on-line). Compounds of low volatility can be introduced through a direct insert system (DIS) (off-line). A small fraction of the evaporated substance is ionized by electron bombardment in the high vacuum of the ion source. The resulting radical cations ($M^{+\cdot}$) decompose to defined fragment ions. All the positive ions are accelerated by the positive pusher of the ion source, focused and then separated according to their mass to charge ratio (m/z) in an electrodynamic quadrupole field (quadrupole mass spectrometer). The current of each separated ion is measured. The mass spectrum represents the relation between the mass to charge ratio of the several fragment ions and the relative intensity of their ion currents. The mass spectrum is usually represented as a bar graph, in which the abscissa represents the mass to charge ratio (m/z) in atomic mass units (u) and the ordinate represents the relative intensities of the ion currents of the several fragment ions in %. The fragmentation pattern is reproducible and characteristic for each organic compound. For this reason, mass spectrometry is the most specific method for the identification of organic compounds.

3.2 Correlation between fundamental structures or side chains and fragment ions

Fundamental structures or side chains of organic compounds can be correlated to fragment ions by observing the fragmentation pattern of analogous compounds or drugs. The elemental composition of fragment ions can be calculated from accurate mass measurements (Section 5). This allows confirmation of suspected correlations. These empirical observations are useful for the identification of metabolites of known drugs or poisons. Metabolites usually have the same fundamental structure and/or the same side chains as their parent compounds and therefore their mass spectra contain similar or identical fragment ions. Chemical changes to fundamental structures and side chains by metabolism and/or derivatization procedures lead to typical shifts in the mass spectrum. The most important shifts are summarized in Table 3.1. Consideration of these correlations and of the fundamental principles of the metabolism of xenobiotics allow the identification of metabolites [3,5,107-118,120-124,151]. The precise chemical identification can then be confirmed by chemical or biotechnological synthesis followed by further spectroscopic studies [152].

Table 3.1: Shifts of the fragment ions of derivatized metabolites.

Metabolite	Underivatized	AC	ME	TMS	TFA	PFP	HFB
		+42	+14	+72	+96	+146	+196
HO-	+16	+58	+30	+88	+112	+162	+212
Di-HO-	+32	+116	+60	+176	+224	+324	+424
HO-Methoxy-	+46	+88	+60	+118	+142	+192	+242
Demethyl-	-14	+28	±0	+58	+82	+132	+182
Deethyl-	-28	+14	-14	+44	+68	+118	+168
Bis-demethyl-	-28	+14	±0	+44	+68	+118	+168
Bis-deethyl-	-56	-14	-28	+16	+30	+90	+130
Demethyl-HO-	+2	+86	+30	+146	+194	+294	+394
Demethyl-HO-Methoxy-	+32	+116	+60	+176	+224	+324	+424
Bis-demethyl-HO-Methoxy-	+18	+102	+60	+162	+210	+310	+410
Deethyl-HO-Methoxy-	+18	+102	+46	+162	+210	+310	+410
Bis-deethyl-HO-Methoxy-	-10	+74	+32	+134	+172	+282	+382

4 Formation of Artifacts

Several of the compounds studied were modified during the analytical procedure employed. Sometimes this occurred thermally in the GC and sometimes during acid hydrolysis. Since the artifacts were formed reproducibly and they have been identified by mass spectrometry, they can be used for detection of the parent compounds. The artifacts are indicated in Table 8.1) and in the legends of the mass spectra in Volume 2 (-CO2, -H2O, HY etc.; cf. abbreviation in Section 6). It cannot be excluded that further compounds were modified.

It should be noted that compounds can be acetylated by acetylsalicylic acid in the gastric contents if acetylsalicylic acid was simultaneously ingested.

4.1 Artifacts formed by oxidation during extraction with diethyl ether

N-Oxidation and S-oxidation are metabolic pathways catalyzed by cytochrome P450. Nevertheless, N- and S-oxidation were also observed during extraction with diethyl ether, which contained traces of peroxides.

4.1.1 N-Oxidation of tertiary amines

$$R^3-N\overset{R^1}{\underset{R^2}{}} \xrightarrow{\frac{1}{2}O_2} R^3-N\overset{R^1}{\underset{R^2}{\downarrow O}}$$

(e.g. amitriptyline)

N-Oxides undergo Cope elimination during GC (Section 4.2.2).

4.1.2 S-Oxidation of phenothiazines

(e.g. promazine)

4.2 Artifacts formed by thermolysis during GC (GC artifact)

4.2.1 Decarboxylation of carboxylic acids

$-CO_2$ decarboxylation

$$R-COOH \xrightarrow[-CO_2]{\Delta T} R-H$$

$-C_2H_2O_2$ decarboxylation after hydrolysis of methyl carboxylate

$-C_3H_4O_2$ decarboxylation after hydrolysis of ethyl carboxylate

$$R^2-\overset{O}{\underset{}{C}}-O-R^1 \xrightarrow[-R^1OH,\ -CO_2]{H_2O/\Delta T} R^2-H$$

(e.g. Carbendazim)

4.2.2 Cope elimination of N-oxides ($-(CH_3)_2NOH$, $-(C_2H_5)_2NOH$, $-C_6H_{14}N_2O_2$)

$$R^4-\underset{R^3}{\overset{}{C}}H-CH_2-\underset{O}{\overset{R^1}{N}}R^2 \xrightarrow[-(R^1R^2)NOH]{\Delta T} R^4-\underset{R^3}{\overset{}{C}}H-CH=CH_2$$

4.2.3 Rearrangement of bis-deethyl flurazepam ($-H_2O$)

Reference: [153]

4.2.4 Elimination of various residues

Several artifacts were observed by elimination of various residues.

$-CH_3Br$ elimination of methyl bromide
$-CH_2O$ elimination of formaldehyde
$-HCl$ elimination of hydrochloric acid
$-HCN$ elimination of prussic acid
$-NH_3$ elimination of ammonia
$-SO_2NH$ elimination of a sulfonamide group

4.2.5 Methylation of carboxylic acids in methanol ((ME), ME in methanol)

Some carboxylic acids are methylated when their methanolic solutions are injected into the GC-MS system.

$$R\text{-COOH} \xrightarrow[-H_2O]{+CH_3OH/\Delta T} R\text{-COOCH}_3$$

4.2.6 Formation of formaldehyde adducts using methanol as solvent (GC artifact in methanol)

Primary amines (e.g. amfetamine), beta-blockers and some local anesthetics (e.g. flecainide, prilocaine) are altered during GC by reaction with formaldehyde, which is probably formed by thermal dehydrogenation of methanol in the injection port of the GC. This was confirmed using deuteromethanol [24,25,154].

$$R\text{-NH}_2 + O=CH_2 \xrightarrow[-H_2O]{\Delta T} R\text{-N}=CH_2$$

(e.g. amfetamine)

(e.g. acebutolol)

(e.g. flecainide)

(e.g. prilocaine)

References: [24,25,100,154]

4.3 Artifacts formed by thermolysis during GC and during acid hydrolysis (GC artifact, HY artifact)

4.3.1 Dehydration of alcohols (-H$_2$O)

4.3.2 Decarbamoylation of carbamates

Carbamates (e.g. insecticides) are easily cleaved during gas chromatography and/or acid hydrolysis.

-CHNO   decarbamoylation
-C$_2$H$_3$NO   N-methyl decarbamoylation
-C$_3$H$_5$NO   N,N-dimethyl decarbamoylation
-C$_5$H$_9$NO   N-isobutyl decarbamoylation

4.3.3 Cleavage of morazone to phenmetrazine

References: [155] (GC)
              [156] (HY)

4.4 Artifacts formed during acid hydrolysis

4.4.1 Cleavage of the ether bridge in beta-blockers and alkanolamine antihistamines (HY)

$$R^1\text{-O-}R^2 \xrightarrow{+H_2O\ [H+]} R^1\text{-OH} + HO\text{-}R^2$$

(e.g. acebutolol or diphenhydramine)

References: [79,100]

4.4.2 Cleavage of 1,4-benzodiazepines to aminobenzoyl derivatives (HY)

(e.g. diazepam)

Reference: [74]

4.4.3 Cleavage and rearrangement of N-demethyl metabolites of clobazam to benzimidazole derivatives (HY)

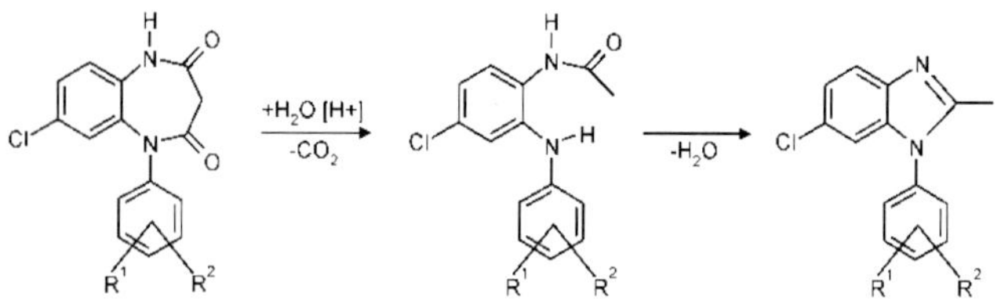

Reference: [157]

4.4.4 Cleavage and rearrangement of bis-deethyl flurazepam (HY -H$_2$O)

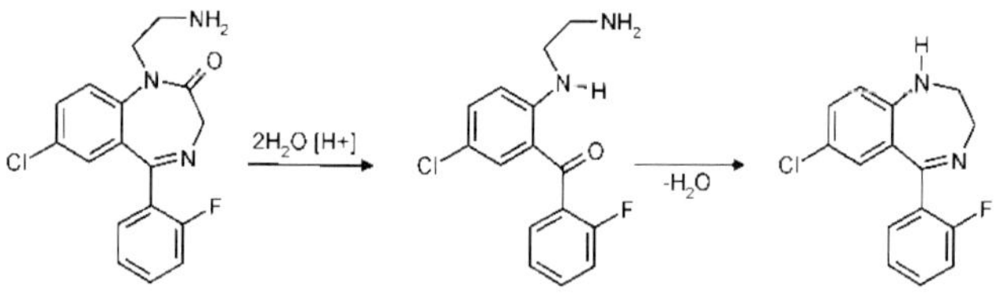

Reference: [157]

4.4.5 Cleavage and rearrangement of tetrazepam and its metabolites

Tetrazepam and its metabolites are transformed into two *cis/trans* isomeric hexahydroacridone derivatives each.

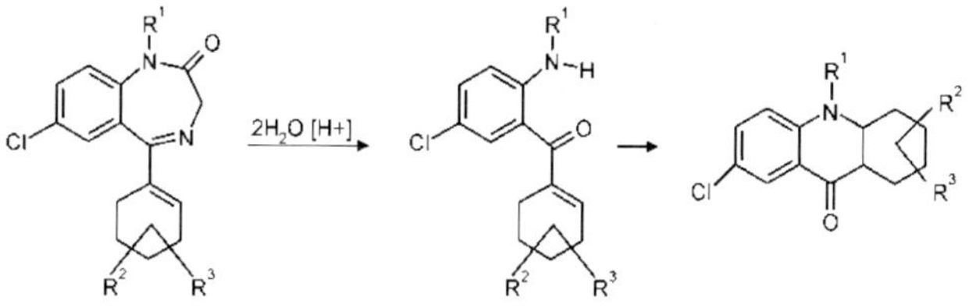

References: [74,158]

4.4.6 Dealkylation of ethylenediamine antihistamines (HY)

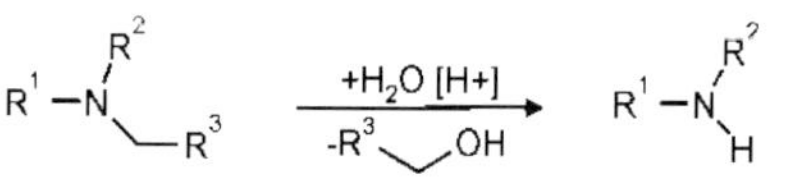

(e.g. antazoline)

Reference: [80]

4.4.7 Hydration of a double bond (+H$_2$O)

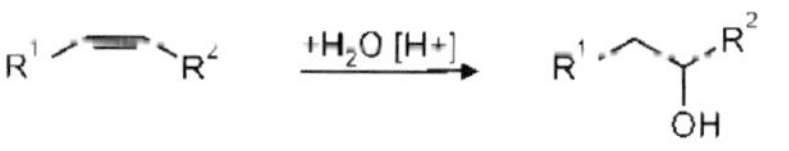

(e.g. pentazocine, alprenolol)

Reference: [100]

5 Table of Atomic Masses

The accurate atomic masses of the most abundant isotopes of the elements, which were employed for the calculations in this handbook, are listed in Table 5.1.

Table 5.1: Atomic masses of elements.

Symbol	Element	Atomic mass
As	Arsenic	74.921595
B	Boron	11.009305
Br	Bromine	78.918336
C	Carbon	12.000000
Cl	Chlorine	34.968853
D	Deuterium	2.014102
F	Fluorine	18.998403
Fe	Iron	55.934939
H	Hydrogen	1.007825
Hg	Mercury	201.970632
I	Iodine	126.904477
N	Nitrogen	14.003074
O	Oxygen	15.994915
P	Phosphorus	30.973763
Pb	Lead	207.976641
S	Sulfur	31.972072
Si	Silicon	27.976928

Reference: [159]

6 Table of Abbreviations

The abbreviations used in this handbook and library are listed in Table 6.1.

Table 6.1: Abbreviations.

Abbreviation	Meaning	see Section (Volume 1)
A	Artifact	4
AC	Acetylated	2.2.2.1
(AC)	Possibly acetylated	
ACE	Angiotensin converting enzyme	
ALHY	Extract after alkaline hydrolysis	
Altered during HY	Altered compound detectable in UHY	4
AMDIS	Automated Mass Spectral Deconvolution and Identification System	2.6
Artifact ( )	( ) artifact	4
BPH	Benzophenone	
BZP	N-Benzylpiperazine	2.5.2
CI	Chemical ionization	
-CH_3Br	Artifact formed by elimination of methyl bromide	4.2.4
-CHNO	Artifact formed by decarbamoylation	4.3.2
-C_2H_3NO	Artifact formed by N-methyl decarbamoylation	4.3.2
-C_3H_5NO	Artifact formed by N,N-dimethyl decarbamoylation	4.3.2
-C_5H_9NO	Artifact formed by N-isobutyl decarbamoylation	4.3.2
-$(CH_3)_2NOH$	Artifact formed by Cope elimination of the N-oxide	4.2.2
-$(C_2H_5)_2NOH$	Artifact formed by Cope elimination of the N-oxide	4.2.2
-$C_6H_{14}N_2O_2$	Artifact formed by Cope elimination of the N-oxide	4.2.2
-CH_2O	Artifact formed by elimination of formaldehyde	4.2.4
-$C_2H_2O_2$	Artifact formed by decarboxylation after hydrolysis of methyl carboxylate	4.2.1
-$C_3H_4O_2$	Artifact formed by decarboxylation after hydrolysis of ethyl carboxylate	4.2.1
-CO_2	Artifact formed by decarboxylation	4.2.1
mCPP	1-(3-Chlorophenyl)piperazine	2.5.2
DIS	Direct insert system used for recording the spectrum	
EG	Ethylene glycol	
EI	Electron ionization	3.1
ET	Ethylated	2.2.2.3
FID	Flame-ionization detector	
G	Standard extract of gastric contents	2.2.1.1
GC	Gas chromatographic, -graph, -graphy	2.3
GC artifact	Artifact formed during GC	4.2-3
GC artifact in methanol	Artifact (of beta-adrenergic blocking agents) formed by reaction with methanol during GC	4.2.6
-HCl	Artifact formed by elimination of hydrogen chloride	4.2.4
-HCN	Artifact formed by elimination of hydrogen cyanide	4.2.4
HFB	Heptafluorobutyrylated	2.2.2.10
HFBP	Heptafluorobutyrylprolylated	2.2.2.12
HO-	Hydroxy	

Abbreviation	Meaning	see Section
+H$_2$O	Artifact formed by hydration (of an alkene)	4.4.7
-H$_2$O	Artifact formed by dehydration (of an alcohol or	4.3.1
	with rearrangement of an amino oxo compound)	4.4.4
HOOC-	Carboxy	
HY	Acid-hydrolyzed or acid hydrolysis	2.2.1.2
HY artifact	Artifact formed during acid hydrolysis	4.4
-I	Intoxication; this compound is only detectable after a toxic dosage	
I.D.	Internal diameter	
INN	International non-proprietary name (WHO)	
IS	Internal standard	
iso	Isomer	
LLE	Liquid-liquid extraction	2.2
LM	Low-resolution mass spectrum	
LM/Q	Low-resolution mass spectrum recorded on a quadrupole MS	
LOD	Limit of detection	
LS	Background subtracted low-resolution mass spectrum	
LS/Q	Background subtracted low-resolution mass spectrum recorded on a quadrupole MS	
M	1 mol/L	
M$^+$	Molecular ion	
-M	Metabolite	
-M ( )	( ) metabolite	
-M (HO-)	Hydroxy metabolite	
-M (HOOC-)	Carboxylated metabolite	
-M (nor-)	N-Demethyl metabolite	
-M (ring)	Ring compound as metabolite (e.g., of phenothiazines)	
-M artifact	Artifact of a metabolite	
-M/artifact	Metabolite or artifact	
m/z	Mass to charge ratio	3.1
MBTFA	N-methyl-bis-trifluoroacetamide	2.2.2.6-7
MDBP	1-(3,4-Methylenedioxybenzyl)piperazine	2.5.2
MDPPP	*R,S*-3',4'-Methylenedioxy-α-pyrrolidinopropiophenone	2.5.4
MDPV	Methylenedioxy-pyrovalerone	2.5.2
ME	Methylated	2.2.2.2
(ME)	Methylated by methanol during GC	4.2.5
ME in methanol	Methylated by methanol during GC	4.2.5
MeOPP	1-(4-Methoxyphenyl)piperazine	2.5.2
MOPPP	*R,S*-4'-Methoxy-α-pyrrolidinopropiophenone	2.5.4
MPBP	4'-Methyl-α-pyrrolidinobutyrophenone	2.5.4
MPHP	*R,S*-4'-Methyl-α-pyrrolidinohexanophenone	2.5.4
MPPP	*R,S*-4'-Methyl-α-pyrrolidinopropiophenone	2.5.4
MS	Mass spectrometric, -meter, -metry, mass spectrum	
MSTFA	N-Methyl-N-trimethylsilyl-trifluoroacetamide	2.2.2.5
MTBSTFA	N-Methyl-N-(tert.-butyldimethylsilyl)-trifluoroacetamide	2.2.2.4
N-FID	Nitrogen-sensitive flame-ionization detector	
-NH$_3$	Artifact formed by elimination of ammonia	4.2.4
NICI	Negative-ion chemical ionization	
NIST	National Institute of Standards and Technology	2.6
Not detectable after HY	Compound destroyed during acid hydrolysis	2.2.1.2

Abbreviation	Meaning	see Section
P	Standard extract of plasma	2.2.1.1
PEGPIV	Pivalylated extract of plasma for determination of glycols	2.2.1.7
PFP	Pentafluoropropionylated	2.2.2.8
PFPA	Pentafluoropropionic anhydride	2.2.2.8
PFPOH	Pentafluoropropanol	2.2.2.9
PIV	Pivalylated	2.2.2.11
PPP	R,S-α-Pyrrolidinopropiophenone	2.5.4
PS	Pure substance	
PSPE	Solid-phase extract of plasma	2.2.1.5
PTHCME	Extract of plasma for detection of tetrahydrocannabinol metabolites [160]	
PVP	α-Pyrrolidinovalerophenone	2.5.4
R	Any unknown substituent	
Rat	Compound found in the urine of rats	2.1
RI	Retention index	2.4
-SO$_2$NH	Artifact formed by elimination of a sulfonamide group	4.2.4
SIM	Selected ion mode	
SPE	Solid phase extraction	2.2.1.5
STA	Systematic toxicological analysis	2.5
TBDMS	Tertiary butyl dimethyl silylated	2.2.2.4
TFA	Trifluoroacetylated	2.2.2.7
TFMPP	1-(3-Trifluoromethylphenyl)piperazine	2.5.2
THC	Tetrahydrocannabinol	
THC-COOH	11-Nor-delta-9-tetrahydrocannabinol-9-carboxylic acid	
TM	Trade mark	
TMS	Trimethylsilylated	2.2.2.5
TMSTFA	Trimethylsilylated followed by trifluoroacetylation	2.2.2.6
u	(Atomic mass) Unit, 1/12 of the mass of the nuclide ^{12}C (*SI* unit)	5
U	Standard extract of urine	2.2.1.1
UA	Extract of urine for detection of amphetamines [50]	
UCO	Extract of urine for detection of cocaine [161]	
UGLUC	Extract of urine after cleavage of conjugates by glucuronidase and arylsulfatase	2.2.1.3
UHY	Extract of urine after acid hydrolysis	2.2.1.2
ULSD	Extract of urine for detection of lysergide (LSD) [162]	
UMAM	Extract of urine for detection of 6-monoacetyl morphine [163]	
USPE	Solid-phase extract of urine	2.2.1.5
UTHCME	Extract of urine for detection of tetrahydrocannabinol metabolites after methylation [160]	
U+UHYAC	Extract of urine with and without acid hydrolysis and acetylation	2.2.1.2
*	Compound contains no nitrogen and cannot be detected by N-FID	
----	RI not determined	
9999	RI > 4000, compound not detectable by GC (-MS).	

7 References

1. H.H.Maurer, D.K.Wissenbach and A.A.Weber. *Maurer/Wissenbach/Weber MWW LC-MSn Library of Drugs, Poisons, and their Metabolites,* Wiley-VCH, Weinheim (Germany), (2014).

2. H.H.Maurer, A.G.Helfer, M.R.Meyer and A.A.Weber. *Maurer/Helfer/Meyer/Weber MHMW LC-HR-MS/MS Library of Drugs, Poisons, and their Metabolites,* Wiley-VCH, Weinheim (Germany), (2016).

3. C.S.D. Wink, G.M.J. Meyer, D.K. Wissenbach, A. Jacobsen-Bauer, M.R. Meyer and H.H. Maurer. Lefetamine-derived designer drugs *N*-ethyl-1,2-diphenylethylamine (NEDPA) and *N-iso*-propyl-1,2-diphenylethylamine (NPDPA): Metabolism and detectability in rat urine using GC-MS, LC-MS^n and LC-high resolution (HR)-MS/MS. *Drug Test. Anal.* 6, 1038-1048 (2014).

4. F.T. Peters and M.R. Meyer. In vitro approaches to studying the metabolism of new psychoactive compounds [review]. *Drug Test. Anal.* 3, 483-495 (2011).

5. J. Welter, P. Kavanagh and H.H. Maurer. GC-MS and LC-(high-resolution)-MS^n studies on the metabolic fate and detectability of camfetamine in rat urine. *Anal. Bioanal. Chem.* 406, 3815-3829 (2014).

6. H.H. Maurer. Analytical toxicology [review]. *Experientia* 100, 317-337 (2010).

7. H.H. Maurer. Perspectives of liquid chromatography coupled to low and high resolution mass spectrometry for screening, identification and quantification of drugs in clinical and forensic toxicology [review]. *Ther. Drug Monit.* 32, 324-327 (2010).

8. H.H. Maurer. How can analytical diagnostics in clinical toxicology be successfully performed today? *Ther. Drug Monit.* 34, 561-564 (2012).

9. H.H. Maurer. What is the future of (ultra)high performance liquid chromatography coupled to low and high resolution mass spectrometry for toxicological drug screening? [review]. *J. Chromatogr. A* 1292, 19-24 (2013).

10. M.R. Meyer, F.T. Peters and H.H. Maurer. Automated mass spectral deconvolution and identification system for GC-MS screening for drugs, poisons, and metabolites in urine. *Clin. Chem.* 56, 575-584 (2010).

11. F.T. Peters, S. Schaefer, R.F. Staack, T. Kraemer and H.H. Maurer. Screening for and validated quantification of amphetamines and of amphetamine- and piperazine-derived designer drugs in human blood plasma by gas chromatography/mass spectrometry. *J. Mass Spectrom.* 38, 659-676 (2003).

12. V. Habrdova, F.T. Peters, D.S. Theobald and H.H. Maurer. Screening for and validated quantification of phenethylamine-type designer drugs and mescaline in human blood plasma by gas chromatography/mass spectrometry. *J. Mass Spectrom.* 40, 785-795 (2005).

13. T. Kraemer and L.D. Paul. Bioanalytical procedures for determination of drugs of abuse in blood [review]. *Anal. Bioanal. Chem.* 388, 1415-1435 (2007).

14. H.H. Maurer. Mass spectrometric approaches in impaired driving toxicology [review]. *Anal. Bioanal. Chem.* 393, 97-107 (2009).

15. G.M.J. Meyer, A.A. Weber and H.H. Maurer. Development and validation of a fast and simple multi-analyte procedure for quantification of 40 drugs relevant to emergency toxicology using GC-MS and one-point calibration. *Drug Test. Anal.* 6, 472-481 (2014).

16. M.R. Meyer, A.A. Weber and H.H. Maurer. A validated GC-MS procedure for fast, simple, and cost-effective quantification of glycols and GHB in human plasma and their identification in urine and plasma developed for emergency toxicology. *Anal. Bioanal. Chem.* 400, 411-414 (2011).

17. M.R. Meyer, J. Welter, A.A. Weber and H.H. Maurer. Development, validation, and application of a fast and simple GC-MS method for determination of some therapeutic drugs relevant in emergency toxicology. *Therap. Drug Monitor.* 33, 649-653 (2011).

18. F. Versace, F. Sporkert, P. Mangin and C. Staub. Rapid sample pre-treatment prior to GC-MS and GC-MS/MS urinary toxicological screening. *Talanta.* 101, 299-306 (2012).

19. H.H. Maurer, C. Kratzsch, T. Kraemer, F.T. Peters and A.A. Weber. Screening, library-assisted identification and validated quantification of oral antidiabetics of the sulfonylurea-type in plasma by atmospheric pressure chemical ionization liquid chromatography-mass spectrometry (APCI-LC-MS). *J. Chromatogr. B* 773, 63-73 (2002).

20. H.H. Maurer and J. Bickeboeller-Friedrich. Screening procedure for detection of antidepressants of the selective serotonin reuptake inhibitor type and their metabolites in urine as part of a modified systematic toxicological analysis procedure using gas chromatography-mass spectrometry. *J. Anal. Toxicol.* 24, 340-347 (2000).

21. A.H. Ewald, F.T. Peters, M. Weise and H.H. Maurer. Studies on the metabolism and toxicological detection of the designer drug 4-methylthioamphetamine (4-MTA) in human urine using gas chromatography-mass spectrometry. *J. Chromatogr. B* 824, 123-131 (2005).

22. A.H. Ewald, G. Fritschi, W.R. Bork and H.H. Maurer. Designer drugs 2,5-dimethoxy-4-bromoamphetamine (DOB) and 2,5-dimethoxy-4-bromomethamphetamine (MDOB): Studies on their metabolism and toxicological detection in rat urine using gas chromatographic/mass spectrometric techniques. *J. Mass Spectrom.* 41, 487-498 (2006).

23. A.H. Ewald, G. Fritschi and H.H. Maurer. Designer drug 2,4,5-trimethoxyamphetamine (TMA-2): Studies on its metabolism and toxicological detection in rat urine using gas chromatographic/mass spectrometric techniques. *J. Mass Spectrom.* 41, 1140-1148 (2006).

24. D.S. Theobald, S. Fehn and H.H. Maurer. New designer drug 2,5-dimethoxy-4-propylthiophenethylamine (2C-T-7): studies on its metabolism and toxicological detection in rat urine using gas chromatography/mass spectrometry. *J. Mass Spectrom.* 40, 105-116 (2005).

25. D.S. Theobald, R.F. Staack, M. Puetz and H.H. Maurer. New designer drug 2,5-dimethoxy-4-ethylthio-beta-phenethylamine (2C-T-2): studies on its metabolism and toxicological detection in rat urine using gas chromatography/mass spectrometry. *J. Mass Spectrom.* 40, 1157-1172 (2005).

26. D.S. Theobald, M. Putz, E. Schneider and H.H. Maurer. New designer drug 4-iodo-2,5-dimethoxy-beta-phenethylamine (2C-I): studies on its metabolism and toxicological detection in rat urine using gas chromatographic/mass spectrometric and capillary electrophoretic/mass spectrometric techniques. *J. Mass Spectrom.* 41, 872-886 (2006).

27. D.S. Theobald and H.H. Maurer. Studies on the metabolism and toxicological detection of the designer drug 4-ethyl-2,5-dimethoxy-beta-phenethylamine (2C-E) in rat urine using gas chromatographic-mass spectrometric techniques. *J. Chromatogr. B* 842, 76-90 (2006).

28. D.S. Theobald and H.H. Maurer. Studies on the metabolism and toxicological detection of the designer drug 2,5-dimethoxy-4-methyl-beta-phenethylamine (2C-D) in rat urine using gas chromatographic-mass spectrometric techniques. *J. Mass Spectrom.* 41, 1509-1519 (2006).

29. D.S. Theobald, G. Fritschi and H.H. Maurer. Studies on the toxicological detection of the designer drug 4-bromo-2,5-dimethoxy-beta-phenethylamine (2C-B) in rat urine using gas chromatography-mass spectrometry. *J. Chromatogr. B* 846, 374-377 (2007).

30. C. Sauer, F.T. Peters, R.F. Staack, G. Fritschi and H.H. Maurer. New designer drug (1-(1-phenylcyclohexyl)-3-ethoxypropylamine (PCEPA): Studies on its metabolism and toxicological detection in rat urine using gas chromatography/mass spectrometry. *J. Mass Spectrom.* 41, 1014-1029 (2006).

31. C. Sauer, F.T. Peters, R.F. Staack, G. Fritschi and H.H. Maurer. Metabolism and toxicological detection of a new designer drug, *N*-(1-phenylcyclohexyl)propanamine, in rat urine using gas chromatography-mass spectrometry. *J. Chromatogr. A* 1186, 380-390 (2008).

32. C. Sauer, F.T. Peters, R.F. Staack, G. Fritschi and H.H. Maurer. New designer drugs N-(1-phenylcyclohexyl)-2-ethoxyethanamine (PCEEA) and N-(1-phenylcyclohexyl)-2-methoxyethanamine (PCMEA): Studies on their metabolism and toxicological detection in rat urine using gas chromatographic/mass spectrometric techniques. *J. Mass Spectrom.* 43, 305-316 (2008).

33. C. Sauer, F.T. Peters, R.F. Staack, G. Fritschi and H.H. Maurer. Metabolism and toxicological detection of the designer drug *N*-(1-phenylcyclohexyl)-3-methoxypropanamine (PCMPA) in rat urine using gas chromatography-mass spectrometry. *Forensic Sci. Int.* 181, 47-51 (2008).

34. C. Sauer, F.T. Peters, C. Haas, M.R. Meyer, G. Fritschi and H.H. Maurer. New designer drug alpha-pyrrolidinovalerophenone (PVP): Studies on its metabolism and toxicological detection in rat urine using gas chromatographic/mass spectrometric techniques. *J. Mass Spectrom.* 44, 952-964 (2009).

35. M.R. Meyer, J. Wilhelm, F.T. Peters and H.H. Maurer. Beta-keto amphetamines: studies on the metabolism of the designer drug mephedrone and toxicological detection of mephedrone, butylone, and methylone in urine using gas chromatography-mass spectrometry. *Anal. Bioanal. Chem.* 397, 1225-1233 (2010).

36. M.R. Meyer, P. Du, F. Schuster and H.H. Maurer. Studies on the metabolism of the alpha-pyrrolidinophenone designer drug methylenedioxy-pyrovalerone (MDPV) in rat urine and human liver microsomes using GC-MS and LC-high-resolution-MS and its detectability in urine by GC-MS. *J. Mass Spectrom.* 45, 1426-1442 (2010).

37. S.W. Toennes and H.H. Maurer. Efficient cleavage of urinary conjugates of drugs or poisons in analytical toxicology using purified and immobilized β-glucuronidase and arylsulfatase packed in columns. *Clin. Chem.* 45, 2173-2182 (1999).

38. R. Meatherall. Optimal enzymatic hydrolysis of urinary benzodiazepine conjugates. *J. Anal. Toxicol.* 18, 382-384 (1994).

39. A. Solans, M. Carnicero, R. de-la-Torre and J. Segura. Comprehensive screening procedure for detection of stimulants, narcotics, adrenergic drugs, and their metabolites in human urine. *J. Anal. Toxicol.* 19, 104-114 (1995).

40. R.W. Romberg and L. Lee. Comparison of the hydrolysis rates of morphine-3-glucuronide and morphine-6-glucuronide with acid and beta-glucuronidase. *J. Anal. Toxicol.* 19, 157-162 (1995).

41. H.H. Maurer, F.X. Tauvel and T. Kraemer. Screening procedure for detection of non-steroidal antiinflammatory drugs (NSAIDs) and their metabolites in urine as part of a systematic toxicological analysis (STA) procedure for acidic drugs and poisons by gas chromatography-mass spectrometry (GC-MS) after extractive methylation. *J. Anal. Toxicol.* 25, 237-244 (2001).

42. J. Beyer, F.T. Peters and H.H. Maurer. Screening procedure for detection of stimulant laxatives and/or their metabolites in human urine using gas chromatography-mass spectrometry after enzymatic cleavage of conjugates and extractive methylation. *Ther. Drug Monit.* 27, 151-157 (2005).

43. J. Beyer, A. Bierl, F.T. Peters and H.H. Maurer. Screening procedure for detection of diuretics and uricosurics and/or their metabolites in human urine using gas chromatography-mass spectrometry after extractive methylation. *Ther. Drug Monit.* 27, 509-520 (2005).

44. F.T. Peters, T. Kraemer and H.H. Maurer. Drug testing in blood: validated negative-ion chemical ionization gas chromatographic-mass spectrometric assay for determination of amphetamine and methamphetamine enantiomers and its application to toxicology cases. *Clin. Chem.* 48, 1472-1485 (2002).

45. J. Beyer, F.T. Peters, T. Kraemer and H.H. Maurer. Detection and validated quantification of herbal phenalkylamines and methcathinone in human blood plasma by LC/MS/MS. *J. Mass Spectrom.* 42, 150-160 (2007).

46. J. Beyer, F.T. Peters, T. Kraemer and H.H. Maurer. Detection and validated quantification of toxic alkaloids in human blood plasma - comparison of LC-APCI-MS with LC-ESI-MS/MS. *J. Mass Spectrom.* 42, 621-633 (2007).

47. F.T. Peters, J. Jung, T. Kraemer and H.H. Maurer. Fast, simple, and validated gas chromatographic-mass spectrometric assay for quantification of drugs relevant to diagnosis of brain death in human blood plasma samples. *Ther. Drug Monit.* 27, 334-344 (2005).

48. H.H. Maurer, F.T. Peters, L.D. Paul and T. Kraemer. Validated GC-MS assay for determination of the antifreezes ethylene glycol and diethylene glycol in human plasma after microwave-assisted pivalylation. *J. Chromatogr. B* 754, 401-409 (2001).

49. P. Van Hee, H. Neels, M. De Doncker, N. Vrydags, K. Schatteman, W. Uyttenbroeck, N. Hamers, D. Himpe and W. Lambert. Analysis of gamma-hydroxybutyric acid, DL-lactic acid, glycolic acid, ethylene glycol and other

glycols in body fluids by a direct injection gas chromatography-mass spectrometry assay for wide use. *Clin. Chem. Lab Med.* 42, 1341-1345 (2004).

50. H.H. Maurer and T. Kraemer. Toxicological detection of selegiline and its metabolites in urine using fluorescence polarization immunoassay (FPIA) and gas chromatography-mass spectrometry (GC-MS) and differentiation by enantioselective GC-MS of the intake of selegiline from abuse of methamphetamine or amphetamine. *Arch. Toxicol.* 66, 675-678 (1992).

51. F.T. Peters, N. Samyn, M. Wahl, T. Kraemer, G. de Boeck and H.H. Maurer. Concentrations and ratios of amphetamine, methamphetamine, MDA, MDMA, and MDEA enantiomers determined in plasma samples from clinical toxicology and driving under the influence of drugs cases by GC-NICI-MS. *J. Anal. Toxicol.* 27, 552-559 (2003).

52. F.T. Peters, N. Samyn, C. Lamers, W. Riedel, T. Kraemer, G. de Boeck and H.H. Maurer. Drug testing in blood: validated negative-ion chemical ionization gas chromatographic-mass spectrometric assay for enantioselective measurement of the designer drugs MDEA, MDMA, and MDA and its application to samples from a controlled study with MDMA. *Clin. Chem.* 51, 1811-1822 (2005).

53. S.L. Soderholm, M. Damm and C.O. Kappe. Microwave-assisted derivatization procedures for gas chromatography/mass spectrometry analysis. *Mol. Divers.* (2010).

54. J. Segura, R. Ventura and C. Jurado. Derivatization procedures for gas chromatographic-mass spectrometric determination of xenobiotics in biological samples, with special attention to drugs of abuse and doping agents [review]. *J. Chromatogr. B* 713, 61-90 (1998).

55. A.F. McKay, W.L. Ott, G.W. Taylor, M.N. Buchanan and J.F. Crooker. Diazohydrocarbons. *Can. J. Res.* 28, 683-688 (1950).

56. S.W. Toennes, A.S. Fandino and G. Kauert. Gas chromatographic-mass spectrometric detection of anhydroecgonine methyl ester (methylecgonidine) in human serum as evidence of recent smoking of crack. *J. Chromatogr. B* 127-132 (1999).

57. H.H. Maurer, J. Bickeboeller-Friedrich, T. Kraemer and F.T. Peters. Toxicokinetics and analytical toxicology of amphetamine-derived designer drugs ("Ecstasy"). *Toxicol. Lett.* 112, 133-142 (2000).

58. F.T. Peters, N. Samyn, T. Kraemer, W. Riedel and H.H. Maurer. Negative-ion chemical ionization gas chromatographic-mass spectrometric assay for enantioselective determination of amphetamines in oral fluid: Application to a controlled study with MDMA and driving under the influence of drugs cases. *Clin. Chem.* 53, 702-710 (2007).

59. E. Kovats. Gaschromatographische Charakterisierung organischer Verbindungen. Teil 1. Retentionsindices aliphatischer Halogenide, Alkohole, Aldehyde und Ketone. *Helv. Chim. Acta* 41, 1915-1932 (1958).

60. R.A.de-Zeeuw, J.P.Franke, H.H.Maurer and K.Pfleger. *Gas Chromatographic Retention Indices of Toxicologically Relevant Substances and their Metabolites (Report of the DFG commission for clinical toxicological analysis, special issue of the TIAFT bulletin),* 3rd ed. VCH publishers, Weinheim, (1992).

61. H.H. Maurer. Position of chromatographic techniques in screening for detection of drugs or poisons in clinical and forensic toxicology and/or doping control [review]. *Clin. Chem. Lab. Med.* 42, 1310-1324 (2004).

62. H.H. Maurer, C. Kratzsch, A.A. Weber, F.T. Peters and T. Kraemer. Validated assay for quantification of oxcarbazepine and its active dihydro metabolite 10-hydroxy carbazepine in plasma by atmospheric pressure chemical ionization liquid chromatography/mass spectrometry. *J. Mass Spectrom.* 37, 687-692 (2002).

63. C. Kratzsch, O. Tenberken, F.T. Peters, A.A. Weber, T. Kraemer and H.H. Maurer. Screening, library-assisted identification and validated quantification of 23 benzodiazepines, flumazenil, zaleplone, zolpidem and zopiclone in plasma by liquid chromatography/mass spectrometry with atmospheric pressure chemical ionization. *J. Mass Spectrom.* 39, 856-872 (2004).

64. H.H. Maurer, T. Kraemer, C. Kratzsch, F.T. Peters and A.A. Weber. Negative ion chemical ionization gas chromatography-mass spectrometry (NICI-GC-MS) and atmospheric pressure chemical ionization liquid

chromatography-mass spectrometry (APCI-LC-MS) of low-dosed and/or polar drugs in plasma. *Ther. Drug Monit.* 24, 117-124 (2002).

65. H.H. Maurer. Toxicological analysis of drugs and poisons by GC-MS [review]. *Spectroscopy Europe* 6, 21-23 (1994).

66. H.H. Maurer, O. Tenberken, C. Kratzsch, A.A. Weber and F.T. Peters. Screening for, library-assisted identification and fully validated quantification of twenty-two beta-blockers in blood plasma by liquid chromatography-mass spectrometry with atmospheric pressure chemical ionization. *J. Chromatogr. A* 1058, 169-181 (2004).

67. C. Kratzsch, A.A. Weber, F.T. Peters, T. Kraemer and H.H. Maurer. Screening, library-assisted identification and validated quantification of fifteen neuroleptics and three of their metabolites in plasma by liquid chromatography/mass spectrometry with atmospheric pressure chemical ionization. *J. Mass Spectrom.* 38, 283-295 (2003).

68. H.H. Maurer and F.T. Peters. Towards High-throughput Drug Screening Using Mass Spectrometry. *Ther. Drug Monit.* 27, 686-688 (2005).

69. T.Kraemer, A.A.Weber and H.H.Maurer. Improvement of sample preparation for the STA - Acceleration of acid hydrolysis and derivatization procedures by microwave irradiation. In *Proceedings of the Xth GTFCh Symposium in Mosbach*, F.Pragst, Ed. Helm-Verlag, Heppenheim, 1997, pp. 200-204.

70. J. Bickeboeller-Friedrich and H.H. Maurer. Screening for detection of new antidepressants, neuroleptics, hypnotics, and their metabolites in urine by GC-MS developed using rat liver microsomes. *Ther. Drug Monit.* 23, 61-70 (2001).

71. H. Maurer and K. Pfleger. Screening procedure for detection of antidepressants and their metabolites in urine using a computerized gas chromatographic-mass spectrometric technique. *J. Chromatogr.* 305, 309-323 (1984).

72. H. Maurer and K. Pfleger. Screening procedure for detecting butyrophenone and bisfluorophenyl neuroleptics in urine using a computerized gas chromatographic-mass spectrometric technique. *J. Chromatogr.* 272, 75-85 (1983).

73. H. Maurer and K. Pfleger. Screening procedure for detection of phenothiazine and analogous neuroleptics and their metabolites in urine using a computerized gas chromatographic-mass spectrometric technique. *J. Chromatogr.* 306, 125-145 (1984).

74. H. Maurer and K. Pfleger. Identification and differentiation of benzodiazepines and their metabolites in urine by computerized gas chromatography-mass spectrometry. *J. Chromatogr.* 422, 85-101 (1987).

75. H.H. Maurer. Identification and differentiation of barbiturates, other sedative-hypnotics and their metabolites in urine integrated in a general screening procedure using computerized gas chromatography- mass spectrometry. *J. Chromatogr.* 530, 307-326 (1990).

76. H.H. Maurer. Detection of anticonvulsants and their metabolites in urine within a "general unknown" analysis procedure using computerized gas chromatography-mass spectrometry. *Arch. Toxicol.* 64, 554-561 (1990).

77. H. Maurer and K. Pfleger. Screening procedure for the detection of antiparkinsonian drugs and their metabolites in urine using a computerized gas chromatographic-mass spectrometric technique. *Fresenius' Z. Anal. Chem.* 321, 363-370 (1985).

78. H. Maurer and K. Pfleger. Identification of phenothiazine antihistamines and their metabolites in urine. *Arch. Toxicol.* 62, 185-191 (1988).

79. H. Maurer and K. Pfleger. Screening procedure for the detection of alkanolamine antihistamines and their metabolites in urine using computerized gas chromatography-mass spectrometry. *J. Chromatogr.* 428, 43-60 (1988).

80. H. Maurer and K. Pfleger. Toxicological detection of ethylenediamine and piperazine antihistamines and their metabolites in urine by computerized gas chromatography-mass spectrometry. *Fresenius' Z. Anal. Chem.* 331, 744-756 (1988).

81. H. Maurer and K. Pfleger. Identification and differentiation of alkylamine antihistamines and their metabolites in urine by computerized gas chromatography-mass spectrometry. *J. Chromatogr.* 430, 31-41 (1988).

82. H. Maurer and K. Pfleger. Screening procedure for the detection of opioids, other potent analgesics and their metabolites in urine using a computerized gas chromatographic-mass spectrometric technique. *Fresenius' Z. Anal. Chem.* 317, 42-52 (1984).

83. H. Maurer and K. Pfleger. Screening procedure for detecting anti-inflammatory analgesics and their metabolites in urine. *Fresenius' Z. Anal. Chem.* 314, 586-594 (1983).

84. H.H. Maurer. On the metabolism and the toxicological analysis of methylenedioxyphenylalkylamine designer drugs by gas chromatography-mass spectrometry. *Ther. Drug Monit.* 18, 465-470 (1996).

85. T. Kraemer, I. Vernaleken and H.H. Maurer. Studies on the metabolism and toxicological detection of the amphetamine-like anorectic mefenorex in human urine by gas chromatography-mass spectrometry and fluorescence polarization immunoassay. *J. Chromatogr. B* 702, 93-102 (1997).

86. T. Kraemer, G.A. Theis, A.A. Weber and H.H. Maurer. Studies on the metabolism and toxicological detection of the amphetamine-like anorectic fenproporex in human urine by gas chromatography-mass spectrometry and fluorescence polarization immunoassay (FPIA). *J. Chromatogr. B* 738, 107-118 (2000).

87. T. Kraemer, R. Wennig and H.H. Maurer. The antispasmodic mebeverine leads to positive amphetamine results with the fluorescence polarization immuno assay (FPIA) - Studies on the toxicological detection in urine by GC-MS and FPIA. *J. Anal. Toxicol.* 25, 1-6 (2001).

88. T. Kraemer, S.K. Roditis, F.T. Peters and H.H. Maurer. Amphetamine concentrations in human urine following single-dose administration of the calcium antagonist prenylamine – Studies using FPIA and GC-MS. *J. Anal. Toxicol.* 27, 68-73 (2003).

89. H.H. Maurer, T. Kraemer, O. Ledvinka, C.J. Schmitt and A.A. Weber. Gas chromatography-mass spectrometry (GC-MS) and liquid chromatography-mass spectrometry (LC-MS) in toxicological analysis. Studies on the detection of clobenzorex and its metabolites within a systematic toxicological analysis procedure by GC-MS and by immunoassay and studies on the detection of alpha- and beta-amanitin in urine by atmospheric pressure ionization electrospray LC-MS. *J. Chromatogr. B* 689, 81-89 (1997).

90. R.F. Staack, J. Fehn and H.H. Maurer. New designer drug para-methoxymethamphetamine: Studies on its metabolism and toxicological detection in urine using gas chromatography-mass spectrometry. *J. Chromatogr. B* 789, 27-41 (2003).

91. R.F. Staack, G. Fritschi and H.H. Maurer. Studies on the metabolism and the toxicological analysis of the new piperazine-like designer drug N-benzylpiperazine in urine using gas chromatography-mass spectrometry. *J. Chromatogr. B* 773, 35-46 (2002).

92. R.F. Staack, G. Fritschi and H.H. Maurer. New designer drug 1-(3-trifluoromethylphenyl)piperazine (TFMPP): gas chromatography/mass spectrometry and liquid chromatography/mass spectrometry studies on its phase I and II metabolism and on its toxicological detection in rat urine. *J. Mass Spectrom.* 38, 971-981 (2003).

93. R.F. Staack and H.H. Maurer. Piperazine-derived designer drug 1-(3-chlorophenyl)piperazine (mCPP): GC-MS studies on its metabolism and its toxicological detection in urine including analytical differentiation from its precursor drugs trazodone and nefazodone. *J. Anal. Toxicol.* 27, 560-568 (2003).

94. R.F. Staack and H.H. Maurer. Toxicological detection of the new designer drug 1-(4-methoxyphenyl)piperazine and its metabolites in urine and differentiation from an intake of structurally related medicaments using gas chromatography-mass spectrometry. *J. Chromatogr. B* 798, 333-342 (2003).

95. R.F. Staack and H.H. Maurer. New designer drug 1-(3,4-methylenedioxybenzyl) piperazine (MDBP): studies on its metabolism and toxicological detection in rat urine using gas chromatography/mass spectrometry. *J. Mass Spectrom.* 39, 255-261 (2004).

96. R.F. Staack and H.H. Maurer. Studies on the metabolism and the toxicological analysis of the nootropic drug fipexide in rat urine using gas chromatography-mass spectrometry. *J. Chromatogr. B* 804, 337-343 (2004).

97. R.F. Staack, D.S. Theobald and H.H. Maurer. Studies on the Human Metabolism and the Toxicologic Detection of the Cough Suppressant Dropropizine in Urine Using Gas Chromatography-Mass Spectrometry. *Ther. Drug Monit.* 26, 441-449 (2004).

98. L.D. Paul and H.H. Maurer. Studies on the metabolism and toxicological detection of the *Eschscholtzia californica* alkaloids californine and protopine in urine using gas chromatography-mass spectrometry. *J. Chromatogr. B* 789, 43-57 (2003).

99. J. Beyer, D. Ehlers and H.H. Maurer. Abuse of Nutmeg (*Myristica fragrans* Houtt.): Studies on the Metabolism and the Toxicological Detection of its Ingredients Elemicin, Myristicin and Safrole in Rat and Human Urine Using Gas Chromatography/Mass Spectrometry. *Ther. Drug Monit.* 28, 568-575 (2006).

100. H. Maurer and K. Pfleger. Identification and differentiation of beta-blockers and their metabolites in urine by computerized gas chromatography-mass spectrometry. *J. Chromatogr.* 382, 147-165 (1986).

101. H.H. Maurer. Toxicological detection of the laxatives bisacodyl, picosulfate, phenolphthalein and their metabolites in urine integrated in a "general-unknown" analysis procedure using gas chromatography-mass spectrometry. *Fresenius' J. Anal. Chem.* 337, 144(1990).

102. H.H.Maurer, T.Kraemer, C.Kratzsch, L.D.Paul, F.T.Peters, D.Springer, R.F.Staack and A.A.Weber. What is the appropriate analytical strategy for effective management of intoxicated patients? In *Proceedings of the 39th International TIAFT Meeting in Prague, 2001*, M.Balikova and E.Navakova, Eds. Charles University, Prague, 2002, pp. 61-75.

103. H.H.Maurer, K.Pfleger and A.A.Weber. *Mass Spectral Library of Drugs, Poisons, Pesticides, Pollutants and their Metabolites,* 5th Rev. ed. Wiley-VCH, Weinheim, (2016).

104. F.W.McLafferty. *Wiley Registry of Mass Spectral Data 10th Edition with NIST 2013 Spectral Data,* John Wiley & Sons, New York NY, (2013).

105. NIST/EPA/NIH. *NIST/EPA/NIH Mass Spectral Library with Search Program Data Version: NIST 14,* John Wiley & Sons, New York NY, (2014).

106. P.Roesner. *Mass Spectra of Designer Drugs 2015,* Wiley-VCH, Weinheim, (2015).

107. M.R. Meyer, C. Vollmar, A.E. Schwaninger and H.H. Maurer. New cathinone-derived designer drugs 3-bromomethcathinone and 3-fluoromethcathinone: studies on their metabolism in rat urine and human liver microsomes using GC-MS and LC-high-resolution MS and their detectability in urine. *J. Mass Spectrom.* 47, 253-262 (2012).

108. G.M.J. Meyer, M.R. Meyer, D.K. Wissenbach and H.H. Maurer. Studies on the metabolism and toxicological detection of glaucine, an isoquinoline alkaloid from *Glaucium flavum* (Papaveraceae), in rat urine using GC-MS, LC-MS^n and LC-high-resolution MS^n. *J. Mass Spectrom.* 48, 24-41 (2013).

109. G.M.J. Meyer and H.H. Maurer. Qualitative metabolism assessment and toxicological detection of xylazine, a veterinary tranquilizer and drug of abuse, in rat and human urine using GC-MS, LC-MSn, and LC-HR-MSn. *Anal. Bioanal. Chem.* 405, 9779-9789 (2013).

110. M.R. Meyer, M. Bach, J. Welter, M. Bovens, A. Turcant and H.H. Maurer. Ketamine-derived designer drug methoxetamine: metabolism including isoenzyme kinetics and toxicological detectability using GC-MS and LC-(HR-)MSn. *Anal. Bioanal. Chem.* 405, 6307-6321 (2013).

111. M.R. Meyer, S. Schmitt and H.H. Maurer. Studies on the metabolism and detectability of the emerging drug of abuse diphenyl-2-pyrrolidinemethanol (D2PM) in rat urine using GC-MS and LC-HR-MS/MS. *J. Mass Spectrom.* 48, 243-249 (2013).

112. M.R. Meyer, D. Posser and H.H. Maurer. Studies on the metabolism and detectability of the designer drug â-naphyrone in rat urine using GC-MS and LC-HR-MS/MS. *Drug Test. Anal.* 5, 259-265 (2013).

113. J. Welter, M.R. Meyer, E. Wolf, W. Weinmann, P. Kavanagh and H.H. Maurer. 2-Methiopropamine, a thiophene analogue of methamphetamine: studies on its metabolism and detectability in the rat and human using GC-MS and LC-(HR)-MS techniques. *Anal. Bioanal. Chem.* 405, 3125-3135 (2013).

114. M.R. Meyer, S. Mauer, G.M.J. Meyer, J. Dinger, B. Klein, F. Westphal and H.H. Maurer. The in vivo and in vitro metabolism and the detectability in urine of 3',4'-methylenedioxy-alpha-pyrrolidinobutyrophenone (MDPBP), a new pyrrolidinophenone-type designer drug, studied by GC-MS and LC-MSn. *Drug Test. Anal.* 6, 746-756 (2014).

115. M.R. Meyer, C. Lindauer, J. Welter and H.H. Maurer. Dimethocaine, a synthetic cocaine derivative: Studies on its in vivo metabolism and its detectability in urine by LC-HR-MS^n and GC-MS using a rat model. *Anal. Bioanal. Chem.* 406, 1845-1854 (2014).

116. J. Welter, M.R. Meyer, P. Kavanagh and H.H. Maurer. Studies on the metabolism and the detectability of 4-methyl-amphetamine and its isomers 2-methyl-amphetamine and 3-methyl-amphetamine in rat urine using GC-MS and LC-(high-resolution)-MSn. *Anal. Bioanal. Chem.* 406, 1957-1974 (2014).

117. A.T. Caspar, A.G. Helfer, J.A. Michely, V. Auwaerter, S.D. Brandt, M.R. Meyer and H.H. Maurer. Studies on the metabolism and toxicological detection of the new psychoactive designer drug 2-(4-iodo-2,5-dimethoxyphenyl)-N-[(2-methoxyphenyl)methyl]ethanamine (25I-NBOMe) in human and rat urine using GC-MS, LC-MSn, and LC-HR-MS/MS. *Anal. Bioanal. Chem.* 407, 6697-6719 (2015).

118. A.G. Helfer, A. Turcant, D. Boels, S. Ferec, B. Lelievre, J. Welter, M.R. Meyer and H.H. Maurer. Elucidation of the metabolites of the novel psychoactive substance 4-methyl-*N*-ethyl-cathinone (4-MEC) in human urine and pooled liver microsomes by GC-MS and LC-HR-MS/MS techniques and of its detectability by GC-MS or LC-MS^n standard screening approaches. *Drug Test. Anal.* 7, 368-375 (2015).

119. G.M.J. Meyer, C.S.D. Wink, J. Zapp and H.H. Maurer. GC-MS, LC-MS^n, LC-high resolution-MS^n, and NMR studies on the metabolism and toxicological detection of mesembrine and mesembrenone, the main alkaloids of the legal high "Kanna" isolated from *Sceletium tortuosum*. *Anal. Bioanal. Chem.* 407, 761-778 (2015).

120. J.A. Michely, A.G. Helfer, S.D. Brandt, M.R. Meyer and H.H. Maurer. Metabolism of the new psychoactive substances N,N-diallyltryptamine (DALT) and 5-methoxy-DALT and their detectability in urine by GC-MS, LC-MSn, and LC-HR-MS/MS. *Anal Bioanal. Chem.* 407, 7831-7842 (2015).

121. J. Welter, P. Kavanagh, M.R. Meyer and H.H. Maurer. Benzofuran analogues of amphetamine and methamphetamine: studies on the metabolism and toxicological analysis of 5-APB and 5-MAPB in urine and plasma using GC-MS and LC-(HR)-MS^n techniques. *Anal. Bioanal. Chem.* 407, 1371-1388 (2015).

122. J. Welter, S.D. Brandt, P. Kavanagh, M.R. Meyer and H.H. Maurer. Metabolic fate, mass spectral fragmentation, detectability, and differentiation in urine of the benzofuran designer drugs 6-APB and 6-MAPB in comparison to their 5-isomers using GC-MS and LC-(HR)-MS techniques. *Anal Bioanal. Chem.* 407, 3457-3470 (2015).

123. C.S.D. Wink, M.R. Meyer, T. Braun, A. Turcant and H.H. Maurer. Biotransformation and detectability of the designer drug 2,5-dimethoxy-4-propylphenethylamine (2C-P) studied in urine by GC-MS, LC-MS^n and LC-high resolution-MS^n. *Anal. Bioanal. Chem.* 407, 831-843 (2015).

124. C.S.D. Wink, G.M.J. Meyer, J. Zapp and H.H. Maurer. Lefetamine, a controlled drug and pharmaceutical lead of new designer drugs: Synthesis, metabolism, and detectability in urine and human liver preparations using GC-MS, LC-MS^n, and LC-high resolution-MS/MS. *Anal. Bioanal. Chem.* 407, 1545-1557 (2015).

125. C.S.D. Wink, J.A. Michely, A. Jacobsen-Bauer, J. Zapp and H.H. Maurer. Diphenidine, a new psychoactive substance: Metabolic fate elucidated with rat urine and human liver preparations and detectability in urine using GC-MS, LC-MS^n, and LC-HR-MS^n. *Drug Test. Anal.* DOI 10.1002/dta.1946(2016).

126. D. Springer, F.T. Peters, G. Fritschi and H.H. Maurer. New designer drug 4'-methyl-alpha-pyrrolidinohexanophenone: Studies on its metabolism and toxicological detection in urine using gas chromatography-mass spectrometry. *J. Chromatogr. B* 789, 79-91 (2003).

127. D. Springer, G. Fritschi and H.H. Maurer. Metabolism and toxicological detection of the new designer drug 4'-methoxy-α-pyrrolidinopropiophenone studied in rat urine using gas chromatography-mass spectrometry. *J. Chromatogr. B* 793, 331-342 (2003).

128. D. Springer, G. Fritschi and H.H. Maurer. Metabolism and toxicological detection of the new designer drug 3',4'-methylenedioxy-alpha-pyrrolidinopropiophenone studied in urine using gas chromatography-mass spectrometry. *J. Chromatogr. B* 793, 377-388 (2003).

129. D. Springer, G. Fritschi and H.H. Maurer. Metabolism of the new designer drug alpha-pyrrolidinopropiophenone (PPP) and the toxicological detection of PPP and 4'methyl-alpha-pyrrolidinopropiophenone (MPPP) studied in urine using gas chromatography-mass spectrometry. *J. Chromatogr. B* 796, 253-266 (2003).

130. F.T. Peters, M.R. Meyer, G. Fritschi and H.H. Maurer. Studies on the metabolism and toxicological detection of the new designer drug 4'-methyl-alpha-pyrrolidinobutyrophenone (MPBP) in urine using gas chromatography-mass spectrometry. *J. Chromatogr. B* 824, 81-91 (2005).

131. C. Sauer, K. Hoffmann, U. Schimmel and F.T. Peters. Acute poisoning involving the pyrrolidinophenone-type designer drug 4'-methyl-alpha-pyrrolidinohexanophenone (MPHP). *Forensic Sci Int* 208, e20-e25(2011).

132. H.H.Maurer, K.Pfleger and A.A.Weber. *Mass spectral library of drugs, poisons, pesticides, pollutants and their metabolites,* 4th Rev. ed. Wiley-VCH, Weinheim, (2007).

133. H.H. Maurer, T. Kraemer and J.W. Arlt. Screening for the detection of angiotensin-converting enzyme inhibitors, their metabolites, and AT II receptor antagonists. *Ther. Drug Monit.* 20, 706-713 (1998).

134. H.H. Maurer and J.W. Arlt. Detection of 4-hydroxycoumarin anticoagulants and their metabolites in urine as part of a systematic toxicological analysis procedure for acidic drugs and poisons by gas chromatography-mass spectrometry after extractive methylation. *J. Chromatogr. B* 714, 181-195 (1998).

135. H.H. Maurer and J.W. Arlt. Screening procedure for detection of dihydropyridine calcium channel blocker metabolites in urine as part of a systematic toxicological analysis procedure for acidics by gas chromatography-mass spectrometry (GC-MS) after extractive methylation. *J. Anal. Toxicol.* 23, 73-80 (1999).

136. H.H.Maurer, F.X.Tauvel and T.Kraemer. Detection of non-steroidal anti-inflammatory drugs (NSAIDs), barbiturates and their metabolites in urine as part of a systematic toxicological analysis (STA) procedure for acidic drugs and poisons by GC-MS. In , I.Rasanen, Ed. TIAFT, Helsinki, 2001, pp. 316-323.

137. A.M. Lisi, R. Kazlauskas and G.J. Trout. Gas chromatographic-mass spectrometric quantitation of urinary buprenorphine and norbuprenorphine after derivatization by direct extractive alkylation. *J. Chromatogr. B* 692, 67-77 (1997).

138. P. Roesner, T. Junge, G. Fritschi, B. Klein, K. Thielert and M. Kozlowski. Neue synthetische Drogen: Piperazin-, Procyclidin- und alpha-Aminopropiophenonderivate. *Toxichem. Krimtech.* 66, 81-90 (1999).

139. A.A. Philipp, M.R. Meyer, D.K. Wissenbach, A.A. Weber, S.W. Zoerntlein, P.G.M. Zweipfenning and H.H. Maurer. Monitoring of *Kratom* or *Krypton* intake in urine using GC-MS relevant in clinical and forensic toxicology. *Anal. Bioanal. Chem.* 400, 127-135 (2011).

140. H.H.Maurer and F.T.Peters. Analyte Identification Using Library Searching in GC-MS and LC-MS. In *Encyclopedia of Mass Spectrometry*, M.Gross and R.M.Caprioli, Eds. Elsevier Science, Oxford, (2006), pp. 115-121.

141. H.H. Maurer. Hyphenated mass spectrometric techniques - indispensable tools in clinical and forensic toxicology and in doping control [review]. *J. Mass Spectrom.* 41, 1399-1413 (2006).

142. M. Grapp, H.H. Maurer and H. Desel. Systematic forensic toxicological analysis by GC-MS in serum using automated mass spectral deconvolution and identification system. *Drug Test. Anal* (2015).

143. F.T. Peters and H.H. Maurer. Systematic comparison of bias and precision data obtained with multiple-point and one-point calibration in six validated assays for quantification of drugs in human plasma. *Anal. Chem.* 79, 4967-4976 (2007).

144. F.T. Peters and H.H. Maurer. Bioanalytical method validation and its implications for forensic and clinical toxicology - A review [review]. *Accred. Qual. Assur.* 7, 441-449 (2002).

145. F.T. Peters, O.H. Drummer and F. Musshoff. Validation of new methods [review]. *Forensic Sci. Int.* 165, 216-224 (2007).

146. F.T. Peters, L.D. Paul, F. Musshoff, B. Aebi, V. Auwaerter, T. Kraemer and G. Skopp. Appendix B - Requirements for the validation of analytical methods. *Toxichem Krimtech* 76, 185-208 (http://www.gtfch.org/cms/images/stories/files/Appendix%20B%20GTFCh%2020090601.pdf) (2009).

147. F.T. Peters, L.D. Paul, F. Musshoff, B. Aebi, V. Auwaerter, T. Kraemer and G. Skopp. Guidelines for quality assurance in forensic-toxicological analyses. *Toxichem Krimtech* 76, 185-208 (http://www.gtfch.org/cms/index.php/guidelines) (2009).

148. S.M. Wille, F.T. Peters, V. Di-Fazio and N. Samyn. Practical aspects concerning validation and quality control for forensic and clinical bioanalytical quantitative methods [review]. *Accred. Qual. Assur.* 16, 279-292 (2011).

149. F.W.McLafferty and F.Turecek. *Interpretation of Mass Spectra,* 4th ed. University Science Books, Mill Valley CA, (1993).

150. R.M.Smith. *Understanding Mass Spectra: A Basic Approach, 2nd Edition,* Wiley, New York NY, (2004).

151. G.M.J. Meyer. Herbal Drugs of Abuse - *Glaucium flavum and Sceletium tortuosum*: Metabolism and toxicological detectability of their alkaloids glaucine, mesembrine and mesembrenone studied in rat urine and human liver preparations using GC-MS, LC-MS, LC-HR-MS^n, and NMR. *Dissertation, Saarland University, Saarbruecken* (2014).

152. F.T. Peters, C.A. Dragan, D.R. Wilde, M.R. Meyer, M. Bureik and H.H. Maurer. Biotechnological synthesis of drug metabolites using human cytochrome P450 2D6 heterologously expressed in fission yeast exemplified for the designer drug metabolite 4'-hydroxymethyl-alpha-pyrrolidinobutyrophenone. *Biochem. Pharmacol.* 74, 511-520 (2007).

153. A.J. Clatworthy, L.V. Jones and M.J. Whitehouse. The gas chromatography mass spectrometry of the major metabolites of flurazepam. *Biomed. Mass Spectrom.* 4, 248-254 (1977).

154. C. Koppel, J. Tenczer and K.M. Peixoto Menezes. Formation of formaldehyde adducts from various drugs by use of methanol in a toxicological screening procedure with gas chromatography-mass spectrometry. *J. Chromatogr.* 563, 73-81 (1991).

155. G.P. Cartoni, A. Cavalli, A. Giarusso and F. Rosati. A gas chromatographic study of the metabolism of tarugan. *J. Chromatogr.* 84, 419-422 (1973).

156. G. Bohn, G. Rucker and H. Kroger. [Investigations of the decomposition and detection of morazone by thin-layer- and gas-liquid-chromatography]. *Arch. Toxicol.* 35, 213-220 (1976).

157. H. Maurer and K. Pfleger. Determination of 1,4- and 1,5-benzodiazepines in urine using a computerized gas chromatographic-mass spectrometric technique. *J. Chromatogr.* 222, 409-419 (1981).

158. H. Schutz, S. Ebel and H. Fitz. [Screening and detection of tetrazepam and its major metabolites]. *Arzneimittelforschung.* 35, 1015-1024 (1985).

159. A.H. Wapstra and K. Bos. *Atomic Data and Nuclear Data Tables* 19, 177-214 (1977).

160. S. Steinmeyer, D. Bregel, S. Warth, T. Kraemer and M.R. Moeller. Improved and validated method for the determination of tetrahydrocannabinol (THC), 11-hydroxy-THC and 11-nor-9-carboxy-THC in serum, and in human liver microsomal preparations using gas chromatography-mass spectrometry. *J. Chromatogr. B* 722, 239-248 (2002).

161. M.R. Moeller, P. Fey and R. Wennig. Simultaneous determination of drugs of abuse (opiates, cocaine and amphetamine) in human hair by GC/MS and its application to a methadone treatment program. *Forensic Sci Int* 63, 185-206 (1993).

162. E.D. Clarkson, D. Lesser and B.D. Paul. Effective GC-MS procedure for detecting iso-LSD in urine after base-catalyzed conversion to LSD. *Clin. Chem.* 44, 287-292 (1998).

163. M.R. Moeller and C. Mueller. The detection of 6-monoacetylmorphine in urine, serum and hair by GC/MS and RIA. *Forensic Sci Int* 70, 125-133 (1995).

8 Tables of Compounds

Table 8-1 is arranged in order to facilitate the search for the data of particular compounds. Derivatives, metabolites and derivatized metabolites are listed under their parent compounds. Metabolites or derivatives common to several substances are listed under all their parent compounds.

The first column contains the compound names (INN for drugs, common names for pesticides, chemical names for chemicals, abbreviations in Table 6.1). If necessary, an internet search will help to find synonyms. The second column contains the information in which biosample and after which sample preparation (Section 2.2, abbreviations in Table 6.1) compounds could be detected. These data have been evaluated from about 100,000 clinical and forensic cases. It should be recognized that the plasma samples were analyzed by our most sensitive GC-MS (HP 5973, details cf. Section 2.3.1).

The third column lists the retention indices (RI, Section 2.4) and the fourth column fragment ions typical for the particular compound and their relative intensities.

The fifth column indicates the page on which the mass spectrum is reproduced under the molecular or pseudomolecular mass. The sixth column indicates the library entry number of the electronic versions.

In electronic versions not using the NIST search algorithm, the name of only one parent compound can be given, but the symbol '@' indicates, that further compounds can form this particular metabolite, derivative or artifact. This information can be found under the mass spectrum on the given page.

Table 8.1: Compounds in order of names

A-796,260

Name	Detected	RI	Typical ions and intensities					Page	Entry
A-796,260		2885	354 $_6$	143 $_3$	129 $_5$	114 $_4$	100 $_{100}$	1234	9669
A-796,260 TMS		2735	426 $_1$	383 $_3$	326 $_5$	283 $_5$	100 $_{100}$	1523	9670
A-834,735		2830	339 $_{32}$	324 $_{65}$	257 $_{100}$	242 $_8$	55 $_{57}$	1153	9671
A-834,735 TMS		2685	411 $_3$	396 $_{11}$	368 $_{17}$	242 $_{17}$	73 $_{100}$	1484	9672
Abacavir	P	2745	286 $_{31}$	271 $_{12}$	189 $_{60}$	175 $_{100}$	162 $_{16}$	815	5867
Abacavir 2AC	U+UHYAC	3210	370 $_{100}$	231 $_{70}$	189 $_{91}$	173 $_{56}$	79 $_{82}$	1312	6558
Abacavir 2HFB		2565	678 $_{23}$	385 $_{100}$	371 $_{74}$	200 $_{21}$	79 $_{22}$	1727	6148
Abacavir 2PFP		2605	578 $_{14}$	335 $_{100}$	321 $_{82}$	200 $_{32}$	79 $_{66}$	1706	6133
Abacavir 2TMS		3090	430 $_{64}$	415 $_{35}$	261 $_{46}$	247 $_{100}$	73 $_{31}$	1535	5869
Abacavir 3HFB		2460	706 $_{15}$	413 $_{45}$	385 $_{100}$	331 $_{11}$	169 $_{17}$	1733	6149
Abacavir AC	U+UHYAC	2780	328 $_{46}$	313 $_{18}$	189 $_{81}$	175 $_{100}$	162 $_{15}$	1082	5868
Abacavir-M (N-dealkyl-) AC	U+UHYAC	3020	232 $_{18}$	217 $_{16}$	189 $_{24}$	175 $_{100}$	148 $_{48}$	516	7960
AB-CHMINACA	P-I	2880	356 $_3$	312 $_{79}$	241 $_{100}$	145 $_{34}$	131 $_5$	1244	9690
AB-CHMINACA -CONH3		2735	311 $_{14}$	283 $_{11}$	241 $_{74}$	145 $_{100}$	55 $_{78}$	978	9673
AB-CHMINACA -CONH3 TMS		2640	383 $_{100}$	368 $_{49}$	314 $_{48}$	210 $_{84}$	73 $_{88}$	1373	9680
AB-CHMINACA TMS		2800	500 $_1$	384 $_{23}$	241 $_{36}$	145 $_{49}$	73 $_{100}$	1660	9674
AB-FUBINACA		2890	368 $_1$	324 $_{57}$	253 $_{83}$	145 $_6$	109 $_{100}$	1302	9691
AB-FUBINACA -CONH3		2730	323 $_3$	253 $_{14}$	214 $_4$	145 $_{14}$	109 $_{100}$	1051	9681
AB-FUBINACA -CONH3 TMS		2610	395 $_{48}$	380 $_{20}$	286 $_{82}$	196 $_{20}$	109 $_{100}$	1426	9679
AB-PINACA		2640	330 $_1$	286 $_{69}$	215 $_{100}$	145 $_{33}$	131 $_6$	1095	9692
AB-PINACA -CONH3		2525	285 $_{12}$	257 $_6$	215 $_{100}$	145 $_{88}$	131 $_{22}$	811	9684
AB-PINACA -CONH3 TMS		2420	357 $_{100}$	342 $_{58}$	288 $_{48}$	210 $_{72}$	73 $_{79}$	1250	9682
AB-PINACA TMS		2615	474 $_5$	358 $_{51}$	268 $_{22}$	215 $_{27}$	73 $_{100}$	1626	9683
Acebutolol	G U	2955	336 $_1$	321 $_4$	221 $_{60}$	151 $_{22}$	72 $_{100}$	1133	1562
Acebutolol 2TMSTFA		2780	504 $_1$	284 $_{60}$	218 $_{19}$	129 $_{61}$	73 $_{61}$	1665	6159
Acebutolol 3TMS		2800	552 $_5$	537 $_{10}$	365 $_{41}$	350 $_{69}$	72 $_{100}$	1696	5465
Acebutolol 4TMS		2870	624 $_1$	609 $_2$	437 $_{13}$	144 $_{100}$	73 $_{79}$	1720	5466
Acebutolol formyl artifact	U	3055	348 $_{18}$	333 $_{73}$	221 $_{93}$	151 $_{100}$	86 $_{70}$	1201	1563
Acebutolol -H2O	G U P	2850	318 $_{19}$	303 $_{56}$	151 $_{26}$	140 $_{67}$	98 $_{100}$	1020	4
Acebutolol -H2O AC		3100	360 $_{21}$	259 $_{80}$	230 $_{78}$	151 $_{100}$	98 $_{71}$	1266	1345
Acebutolol -H2O HY	UHY	2010	248 $_{23}$	233 $_{24}$	140 $_{24}$	98 $_{100}$	56 $_{72}$	591	1565
Acebutolol -H2O HY2AC	U+UHYAC	3055	332 $_8$	289 $_{30}$	231 $_{100}$	202 $_{60}$	98 $_{30}$	1106	1570
Acebutolol HY	UHY	2240	266 $_6$	151 $_{100}$	72 $_{92}$			697	1567
Acebutolol-M (diacetolol) -H2O AC	U+UHYAC	3055	332 $_8$	289 $_{30}$	231 $_{100}$	202 $_{60}$	98 $_{30}$	1106	1570
Acebutolol-M/artifact (phenol)	G U P	2450	221 $_{38}$	151 $_{100}$	136 $_{24}$			473	1
Acebutolol-M/artifact (phenol) HY	UHY	1530	151 $_{100}$	136 $_{62}$	108 $_{36}$	80 $_{34}$		291	1564
Acebutolol-M/artifact (phenol) HYAC	U+UHYAC	1850	193 $_{98}$	151 $_{100}$	136 $_{89}$	133 $_{44}$		377	1568
Acecarbromal	P G U	1720	250 $_{18}$	208 $_6$	165 $_{18}$	129 $_{62}$	69 $_{100}$	766	2
Acecarbromal artifact-1	P	1115	157 $_2$	129 $_{97}$	114 $_{14}$	87 $_{19}$	57 $_{100}$	301	1026
Acecarbromal artifact-2		1210	180 $_6$	129 $_{62}$	69 $_{100}$			344	1328
Acecarbromal artifact-3	G P U	1480	223 $_{29}$	191 $_5$	149 $_{11}$	102 $_{15}$	69 $_{100}$	480	1880
Acecarbromal artifact-4		1510	165 $_{11}$	113 $_{100}$	98 $_{81}$	69 $_{74}$		312	1329
Acecarbromal-M (carbromal)	P G U	1515	208 $_{41}$	191 $_4$	165 $_{16}$	114 $_{14}$	69 $_{100}$	537	652
Acecarbromal-M (debromo-carbromal)		1380	143 $_2$	130 $_{60}$	113 $_{96}$	87 $_{100}$	71 $_{63}$	301	655
Acecarbromal-M/artifact (carbromide)	P G U	1215	165 $_{67}$	150 $_{18}$	114 $_{52}$	69 $_{100}$	55 $_{38}$	377	653
Aceclidine		1460	169 $_{18}$	141 $_6$	126 $_{10}$	110 $_{16}$	98 $_{37}$	322	2785
Aceclofenac ME	P(ME) G(ME)	2540	367 $_{26}$	277 $_8$	242 $_{10}$	214 $_{100}$	179 $_{12}$	1295	6489
Aceclofenac-M (diclofenac)	G P	2205	295 $_{17}$	242 $_{54}$	214 $_{100}$	179 $_{19}$	108 $_{30}$	874	4469
Aceclofenac-M (diclofenac) 2ME		2220	323 $_{53}$	264 $_{10}$	228 $_{100}$	214 $_{21}$		1048	2323
Aceclofenac-M (diclofenac) -H2O	P G U+UHYAC	2135	277 $_{73}$	242 $_{63}$	214 $_{100}$	179 $_{30}$	89 $_{26}$	757	716
Aceclofenac-M (diclofenac) -H2O ET		2130	305 $_{100}$	290 $_{69}$	270 $_{97}$	242 $_{65}$	227 $_{42}$	934	6390
Aceclofenac-M (diclofenac) -H2O ME	G P	2300	291 $_{19}$	263 $_{31}$	228 $_{100}$	200 $_{33}$	109 $_4$	845	2324
Aceclofenac-M (diclofenac) -H2O TMS		2180	349 $_{100}$	314 $_{27}$	241 $_{11}$	190 $_{48}$	73 $_{81}$	1202	4538
Aceclofenac-M (diclofenac) ME	P(ME) G(ME)	2195	309 $_{12}$	277 $_3$	242 $_{30}$	214 $_{100}$	179 $_{17}$	961	717
Aceclofenac-M (diclofenac) TMS		2170	367 $_{19}$	352 $_{11}$	242 $_{38}$	214 $_{100}$	73 $_{86}$	1295	5467
Aceclofenac-M (HO-diclofenac) -H2O	P U+UHYAC	2400	293 $_{39}$	258 $_{100}$	230 $_{46}$	195 $_{34}$	166 $_{40}$	859	6467
Aceclofenac-M (HO-diclofenac) -H2O isomer-1 AC	U+UHYAC	2520	335 $_{88}$	293 $_{100}$	258 $_{93}$	230 $_{93}$	166 $_{23}$	1122	2321
Aceclofenac-M (HO-diclofenac) -H2O isomer-2 AC	U+UHYAC	2540	335 $_{22}$	293 $_{100}$	258 $_{29}$	230 $_{48}$	195 $_{17}$	1122	1212
Aceclofenac-M (HO-diclofenac) -H2O ME	G P	2365	307 $_{65}$	272 $_{97}$	244 $_{100}$	209 $_{15}$	201 $_{47}$	946	6490
Aceclofenac-M/artifact	U+UHYAC	2980	355 $_{88}$	320 $_{100}$	292 $_{10}$	228 $_{16}$	75 $_6$	1234	2322
Acemetacin artifact-1 2ME	PME UME	2090	247 $_{38}$	188 $_{100}$	173 $_{10}$	145 $_8$		585	6294
Acemetacin artifact-1 ME	PME UME	2130	233 $_{38}$	174 $_{100}$				521	1230
Acemetacin artifact-2 ME		2390	291 $_{17}$	233 $_{16}$	174 $_{100}$	159 $_{13}$	131 $_{14}$	847	1384
Acemetacin ET		3220	443 $_{39}$	312 $_{12}$	158 $_5$	139 $_{100}$	111 $_{16}$	1568	3167
Acemetacin ME	PME UME	3150	429 $_{44}$	312 $_{14}$	158 $_6$	139 $_{100}$	111 $_{13}$	1529	1374
Acemetacin-M (chlorobenzoic acid)	G UHY UHYAC	1400*	156 $_{61}$	139 $_{100}$	111 $_{54}$	85 $_4$	75 $_{39}$	299	2726
Acemetacin-M/artifact (HO-indometacin) 2ME	UME	2880	401 $_{27}$	262 $_4$	139 $_{100}$	111 $_{23}$		1448	6293
Acemetacin-M/artifact (indometacin)	G P-I	2550	313 $_{30}$	139 $_{100}$	111 $_{24}$			1246	1038
Acemetacin-M/artifact (indometacin) ET		2820	385 $_{40}$	312 $_{19}$	158 $_6$	139 $_{100}$	111 $_{20}$	1380	3168
Acemetacin-M/artifact (indometacin) ME	P(ME) G(ME) U(ME)	2770	371 $_9$	312 $_6$	139 $_{100}$	111 $_{17}$		1316	1039
Acemetacin-M/artifact (indometacin) TMS		2650	429 $_{23}$	370 $_4$	312 $_{17}$	139 $_{100}$	73 $_{22}$	1530	5462
Acenaphthene		1440*	154 $_{100}$	153 $_{93}$	126 $_3$	87 $_5$	76 $_{31}$	296	3700
Acenaphthylene		1380*	152 $_{100}$	126 $_3$	98 $_1$	76 $_9$	63 $_5$	293	2558
Acenocoumarol AC	U+UHYAC	3105	395 $_4$	353 $_{26}$	335 $_{12}$	310 $_{100}$	121 $_{13}$	1423	4788

Acenocoumarol ET   Table 8.1: Compounds in order of names

Name	Detected	RI	Typical ions and intensities					Page	Entry
Acenocoumarol ET	UET	3040	381 $_{18}$	338 $_{100}$	310 $_{93}$	189 $_9$	121 $_{37}$	1360	4781
Acenocoumarol ME	UME UGLUCME	3035	367 $_{15}$	324 $_{100}$	278 $_7$	189 $_8$	121 $_{10}$	1296	1372
Acenocoumarol TMS		3110	425 $_{23}$	382 $_{87}$	261 $_{16}$	219 $_{24}$	73 $_{100}$	1519	4885
Acenocoumarol-M (acetamido-) 2ET	UET	3200	421 $_{28}$	378 $_{100}$	350 $_{84}$	292 $_8$	121 $_{19}$	1511	4787
Acenocoumarol-M (acetamido-) 2ME	UME UGLUCME	3265	393 $_{21}$	350 $_{100}$	336 $_9$	278 $_{17}$	56 $_{30}$	1415	4434
Acenocoumarol-M (acetamido-) ME	UME UGLUCME	3520	379 $_{33}$	336 $_{100}$	322 $_{13}$	280 $_{11}$	201 $_{20}$	1350	4433
Acenocoumarol-M (amino-) 2ET	UET	3040	379 $_{74}$	322 $_{100}$	308 $_{30}$	148 $_{29}$	121 $_{23}$	1352	4784
Acenocoumarol-M (amino-) 2ME	UME UHYME	2980	351 $_{34}$	294 $_{100}$	278 $_{24}$	120 $_{23}$		1213	4430
Acenocoumarol-M (amino-) 3ET	UET	3070	407 $_{57}$	392 $_{26}$	350 $_{100}$	306 $_{25}$	121 $_{10}$	1468	4785
Acenocoumarol-M (amino-) 3ME	UME UHYME	2985	365 $_{39}$	308 $_{100}$	292 $_{26}$	249 $_5$	121 $_7$	1286	4431
Acenocoumarol-M (amino-dihydro-) 3ET	UET	3065	409 $_{64}$	394 $_{40}$	362 $_{28}$	350 $_{100}$	176 $_6$	1475	4786
Acenocoumarol-M (amino-dihydro-) 3ME	UME	3060	367 $_{51}$	334 $_{25}$	308 $_{100}$	292 $_{35}$		1298	4432
Acenocoumarol-M (HO-) isomer-1 2ET	UET	3435	425 $_{16}$	382 $_{100}$	354 $_{65}$	233 $_6$	165 $_7$	1519	4782
Acenocoumarol-M (HO-) isomer-1 2ME	UME UGLUCME	3350	397 $_{11}$	354 $_{100}$	308 $_4$	219 $_2$	151 $_8$	1433	4428
Acenocoumarol-M (HO-) isomer-2 2ET	UET	3630	425 $_{15}$	382 $_{100}$	354 $_{61}$	233 $_6$	165 $_9$	1520	4783
Acenocoumarol-M (HO-) isomer-2 2ME	UME UGLUCME	3500	397 $_{14}$	354 $_{100}$	308 $_3$	219 $_2$	151 $_5$	1433	4429
Acephate		1470	183 $_5$	142 $_{13}$	136 $_{100}$	94 $_{57}$	79 $_{17}$	353	3504
Acephate -C2H2O TFA		1110	237 $_{15}$	168 $_{22}$	125 $_{55}$	96 $_{100}$	69 $_{58}$	541	4031
Acepromazine	G U UHY UHYAC	2755	326 $_4$	241 $_3$	198 $_2$	86 $_8$	58 $_{100}$	1069	3
Acepromazine-M (dihydro-) AC	UHYAC	2765	370 $_3$	310 $_3$	225 $_7$	86 $_{28}$	58 $_{100}$	1312	1307
Acepromazine-M (dihydro-) -H2O	UHYAC	2720	310 $_6$	225 $_6$	86 $_{10}$	58 $_{100}$		971	1306
Acepromazine-M (HO-) AC	U+UHYAC	3040	384 $_7$	256 $_2$	86 $_{23}$	58 $_{100}$		1376	1309
Acepromazine-M (HO-dihydro-) 2AC	UHYAC	3000	428 $_3$	343 $_{16}$	154 $_{14}$	86 $_{25}$	58 $_{100}$	1528	1308
Acepromazine-M (nor-) AC	U+UHYAC	3145	354 $_{20}$	241 $_{17}$	114 $_{100}$	100 $_{34}$		1232	1235
Acepromazine-M (nor-dihydro-) -H2O AC	UHYAC	3150	338 $_5$	114 $_{100}$	100 $_{13}$			1143	1310
Acepromazine-M (ring)	UHY UHYAC	2525	241 $_{100}$	226 $_{18}$	198 $_{45}$	166 $_4$	154 $_5$	559	6804
Aceprometazine		2625	326 $_5$	255 $_{10}$	222 $_7$	197 $_7$	72 $_{100}$	1069	5
Aceprometazine-M (dihydro-) AC	UHYAC	2690	370 $_3$	299 $_6$	224 $_8$	72 $_{100}$		1312	1236
Aceprometazine-M (HO-) AC	U+UHYAC	3025	384 $_3$	313 $_3$	256 $_3$	72 $_{100}$		1376	1238
Aceprometazine-M (methoxy-dihydro-) AC	UHYAC	3165	400 $_1$	329	270 $_1$	225	72 $_{100}$	1446	1239
Aceprometazine-M (methoxy-dihydro-) -H2O	UHYAC	2920	340 $_3$	238 $_6$	72 $_{100}$			1156	1237
Aceprometazine-M (nor-) AC	U+UHYAC	2940	354 $_6$	254 $_{23}$	114 $_{24}$	72 $_{23}$	58 $_{100}$	1232	1311
Aceprometazine-M (nor-HO-) 2AC	UHYAC	3205	412 $_3$	254 $_{16}$	114 $_{45}$	100 $_{32}$	58 $_{100}$	1405	1312
Aceprometazine-M (ring)	UHY UHYAC	2525	241 $_{100}$	226 $_{18}$	198 $_{45}$	166 $_4$	154 $_5$	559	6804
Acetaldehyde		<1000*	44 $_{83}$	29 $_{100}$				248	4193
Acetaminophen	G P U	1780	151 $_{34}$	109 $_{100}$	81 $_{16}$	80 $_{22}$		291	825
Acetaminophen 2AC	U+UHYAC	2085	235 $_{10}$	193 $_{11}$	151 $_{30}$	109 $_{100}$		532	827
Acetaminophen 2TMS		1780	295 $_{50}$	280 $_{68}$	206 $_{83}$	116 $_{15}$	73 $_{100}$	878	4578
Acetaminophen AC	PAC U+UHYAC	1765	193 $_{10}$	151 $_{53}$	109 $_{100}$	80 $_{24}$		377	188
Acetaminophen Cl-artifact AC	UHYAC	2030	227 $_6$	185 $_{74}$	143 $_{100}$	114 $_4$	79 $_{12}$	498	2993
Acetaminophen HFB	UHYHFB PHFB	1735	347 $_{24}$	305 $_{39}$	169 $_{13}$	108 $_{100}$	69 $_{31}$	1192	5099
Acetaminophen HY	UHY	1240	109 $_{100}$	80 $_{41}$	53 $_{82}$	53 $_8$	52 $_{90}$	260	826
Acetaminophen HYME	UHYME	1100	123 $_{27}$	109 $_{100}$	94 $_7$	80 $_{96}$	53 $_{47}$	267	3766
Acetaminophen ME	PME UME	1630	165 $_{59}$	123 $_{74}$	108 $_{100}$	95 $_{10}$	80 $_{20}$	314	5046
Acetaminophen PFP		1675	297 $_{19}$	255 $_{31}$	119 $_{38}$	108 $_{100}$	80 $_{28}$	888	5095
Acetaminophen TFA		1630	247 $_{11}$	205 $_{30}$	108 $_{100}$	80 $_{19}$	69 $_{34}$	583	5092
Acetaminophen-D4		1770	155 $_{36}$	113 $_{100}$	85 $_{14}$	57 $_8$		298	8068
Acetaminophen-M (HO-) 3AC	U+UHYAC	2150	251 $_6$	209 $_{23}$	167 $_{87}$	125 $_{100}$		608	2384
Acetaminophen-M (HO-methoxy-) AC	U+UHYAC	2170	239 $_{12}$	197 $_{86}$	155 $_{100}$	140 $_{42}$	110 $_9$	551	2383
Acetaminophen-M (methoxy-) AC	U+UHYAC	1940	223 $_{12}$	181 $_{79}$	139 $_{100}$			482	201
Acetaminophen-M (methoxy-) Cl-artifact AC	UHYAC	2060	257 $_6$	215 $_{77}$	173 $_{100}$	158 $_{21}$	130 $_5$	641	2994
Acetaminophen-M 2AC	U+UHYAC	2270	262 $_{20}$	220 $_{35}$	188 $_{17}$	160 $_{74}$	146 $_{100}$	667	2387
Acetaminophen-M 3AC	U+UHYAC	2340	304 $_{15}$	261 $_{31}$	219 $_{46}$	160 $_{100}$	146 $_{72}$	928	2388
Acetaminophen-M conjugate 2AC	U+UHYAC	3050	396 $_{20}$	354 $_7$	246 $_{100}$	204 $_{73}$	162 $_{71}$	1427	2389
Acetaminophen-M conjugate 3AC	U+UHYAC	3030	438 $_{35}$	353 $_{40}$	246 $_{72}$	204 $_{97}$	162 $_{100}$	1553	1387
Acetaminophen-M isomer-1 3AC	U+UHYAC	2200	305 $_{26}$	263 $_{57}$	221 $_{14}$	160 $_{69}$	146 $_{100}$	934	2385
Acetaminophen-M isomer-2 3AC	U+UHYAC	2220	305 $_{34}$	263 $_{100}$	221 $_{82}$	162 $_{54}$	146 $_{99}$	934	2386
Acetanilide	G	1380	135 $_{35}$	93 $_{100}$				274	222
Acetazolamide 3ME	UEXME	2040	264 $_{22}$	249 $_{100}$	108 $_{21}$	92 $_2$	83 $_{18}$	680	6844
Acetazolamide ME	UEXME	1995	236 $_{11}$	129 $_{21}$	108 $_9$	88 $_{16}$	70 $_{100}$	536	6843
Acetic acid		<1000*	60 $_{47}$	45 $_{78}$	43 $_{100}$			249	1548
Acetic acid anhydride		<1000*	102 $_1$	60 $_3$	43 $_{100}$			259	2756
Acetic acid ET		<1000*	88 $_5$	70 $_{14}$	61 $_{16}$	43 $_{100}$	29 $_{33}$	255	60
Acetic acid ME		<1000*	74 $_{23}$	59 $_{10}$	43 $_{100}$	29 $_{23}$		252	3777
Acetochlor		1845	269 $_8$	223 $_{39}$	174 $_{39}$	146 $_{64}$	59 $_{100}$	712	3507
Acetone		<1000*	58 $_{44}$	43 $_{100}$				249	1547
Acetonitrile		<1000	41 $_{100}$					247	2752
4-Acetoxy-N,N-diallyl-tryptamine		2350	298 $_2$	269 $_2$	160 $_5$	146 $_{13}$	110 $_{100}$	897	9396
4-Acetoxy-N,N-diisopropyl-tryptamine		2390	302 $_1$	160 $_4$	146 $_{12}$	114 $_{100}$	72 $_{19}$	920	8875
4-Acetoxy-N,N-diisopropyl-tryptamine HFB		2140	160 $_4$	146 $_{10}$	114 $_{100}$	72 $_{18}$		1657	9532
4-Acetoxy-N,N-diisopropyl-tryptamine PFP		2145	222 $_1$	160 $_4$	146 $_9$	114 $_{100}$	72 $_{16}$	1577	9562
4-Acetoxy-N,N-diisopropyl-tryptamine TFA		2150	242 $_2$	160 $_6$	146 $_{17}$	114 $_{100}$	72 $_{17}$	1438	9561
4-Acetoxy-N,N-diisopropyl-tryptamine TMS		2350	374 $_3$	274 $_4$	218 $_{15}$	114 $_{100}$	72 $_{25}$	1332	9532
Acetylmethadol	G P U+UHYAC	2230	353 $_1$	338 $_1$	225 $_4$	91 $_6$	72 $_{100}$	1230	5616

Table 8.1: Compounds in order of names N-Acetyl-proline ME

Name	Detected	RI	Typical ions and intensities					Page	Entry
N-Acetyl-proline ME		1465	171_8	128_1	112_{32}	70_{100}	68_7	325	2708
Acetylsalicylic acid	G P-I U+UHYAC	1545*	180_7	138_{80}	120_{100}	92_{38}		344	1443
Acetylsalicylic acid ME	P(ME)	1400*	194_{60}	179_{40}	135_{100}	91_{10}		383	2637
Acetylsalicylic acid-M	U	1825	195_{32}	177_{17}	121_{98}	120_{100}	92_{43}	385	956
Acetylsalicylic acid-M (deacetyl-)	G P UHY	1295*	138_{40}	120_{90}	92_{100}	64_{52}		277	953
Acetylsalicylic acid-M (deacetyl-) 2ME	PME UME	1210*	166_{28}	135_{100}	133_{47}	92_{30}	77_{52}	316	6391
Acetylsalicylic acid-M (deacetyl-) 2TMS		1195*	267_{61}	193_7	135_{29}	91_{19}	73_{100}	793	4523
Acetylsalicylic acid-M (deacetyl-) artifact (trimer)	G U+UHYAC	3190*	360_{39}	240_{58}	152_{36}	120_{100}	92_{75}	1263	4496
Acetylsalicylic acid-M (deacetyl-) ET		1350*	166_{34}	120_{100}	92_{41}	65_{19}		315	955
Acetylsalicylic acid-M (deacetyl-) ME	P U+UHYAC	1200*	152_{39}	120_{94}	92_{100}	65_{53}		292	954
Acetylsalicylic acid-M (deacetyl-3-HO-) 3ME	UME	1385*	196_{73}	165_{81}	163_{100}	122_{26}	107_{20}	390	6393
Acetylsalicylic acid-M (deacetyl-5-HO-) 3ME	UME	1530*	196_{100}	181_{34}	165_{72}	163_{66}	107_{29}	389	6394
Acetylsalicylic acid-M (deacetyl-HO-) 2ME	PME UME	1210*	182_{36}	150_{100}	122_{12}	107_{30}	79_{18}	350	6392
Acetylsalicylic acid-M 2ME	UME	1845	223_{12}	135_{100}	90_{58}	77_{44}		482	958
Acetylsalicylic acid-M ME	U	1810	209_{20}	149_{12}	121_{100}	92_{17}	65_{22}	427	957
Acetylsalicylic acid-M MEAC	U+UHYAC	1885	251_3	209_{69}	177_{23}	149_{29}	121_{100}	608	2976
Acetyltriethylcitrate		1880*	318_1	273_8	213_{20}	203_{55}	157_{100}	1018	4478
Aclidinium-M/artifact		1860*	194_{43}	166_{10}	121_4	111_{100}	83_{18}	381	9584
Aclidinium-M/artifact (HOOC-) 2ME		2160*	268_1	237_3	209_{100}	195_{22}	111_{99}	705	7371
Aclidinium-M/artifact (HOOC-) 2TMS		2030	384_1	311_5	267_{100}	177_{13}	111_{20}	1373	9466
Aclidinium-M/artifact (HOOC-) ME		2140*	254_2	195_{86}	177_2	111_{100}	83_{13}	625	7369
Aclidinium-M/artifact (HOOC-) MEAC		2240*	296_9	237_{36}	195_{100}	177_8	111_{61}	882	7370
4-AcO-DALT		2350	298_2	269_2	160_5	146_{13}	110_{100}	897	9396
4-AcO-DALT-M/artifact (deacetyl-)		2245	256_{10}	160_2	146_7	117_3	110_{100}	639	9395
4-AcO-DALT-M/artifact (deacetyl-) 2TMS		2355	400_2	359_5	304_5	290_{100}	110_{100}	1446	9398
4-AcO-DALT-M/artifact (deacetyl-) ME		2300	270_3	229_5	160_{16}	130_{23}	110_{100}	720	9397
4-AcO-DALT-M/artifact (deacetyl-) TFA		2125	323_3	256_3	242_8	145_4	110_{100}	1220	10140
4-AcO-DiPT		2390	302_1	160_6	146_{12}	114_{100}	72_{19}	920	8875
4-AcO-DiPT HFB		2140	160_4	146_{10}	114_{100}	72_{16}		1657	9563
4-AcO-DiPT PFP		2145	222_1	160_4	146_9	114_{100}	72_{16}	1577	9562
4-AcO-DiPT TFA		2150	242_2	160_6	146_{17}	114_{100}	72_{17}	1438	9561
4-AcO-DiPT TMS		2350	374_3	274_4	218_{15}	114_{100}	72_{25}	1332	9532
4-AcO-DMT	U+UHYAC	2270	246_5	160_3	146_7	130_3	58_{100}	580	2471
4-AcO-DMT AC	U+UHYAC	2340	288_7	246_1	202_1	122_3	58_{100}	828	2472
ADB-CHMICA		3150	369_4	296_6	240_{100}	144_{20}	129_7	1308	9498
ADB-CHMINACA		2900	370_1	326_{40}	297_{14}	241_{100}	145_{65}	1314	9585
ADB-FUBINACA		2900	338_{19}	309_3	253_{82}	145_{10}	109_{100}	1365	9704
ADB-FUBINACA 2TMS		2800	526_1	410_8	302_{23}	109_{100}	73_{89}	1681	9700
ADB-FUBINACA -CONH3		2610	337_1	253_{63}	109_{100}	88_{14}	57_{25}	1138	9699
ADB-FUBINACA -CONH3 TMS		2950	409_3	297_3	253_{82}	109_{100}	57_{86}	1475	9701
ADBICA		2850	343_6	299_5	270_9	214_{100}	144_{22}	1176	9707
ADBICA -CONH3		2530	298_9	214_{100}	144_{31}	129_{10}	116_{10}	898	9708
ADB-PINACA		2790	344_1	300_{76}	271_{23}	215_{100}	145_{28}	1181	9652
ADB-PINACA 2TMS		2680	488_1	474_1	372_{76}	215_{50}	73_{100}	1646	9645
ADB-PINACA artifact -CONH3		2480	299_2	215_{100}	145_{51}	131_{17}	103_{21}	903	9646
ADB-PINACA-M/artifact (HOOC-) (ME)		2610	359_1	303_{21}	271_{24}	215_{100}	145_{35}	1260	9653
Adefovir		2700	489_8	474_{37}	250_{100}	235_{87}	220_{70}	1648	8237
Adefovir dipivoxyl-M/artifact (Adefovir)		2700	489_8	474_{37}	250_{100}	235_{87}	220_{70}	1648	8237
Adeptolon	U UHY UHYAC	2375	347_1	263_3	169_{11}	86_{11}	72_{100}	1193	7
Adeptolon-M (HO-)	UHY	2760	363_2	325_7	169_{29}	90_{20}	72_{100}	1277	2164
Adeptolon-M (HO-) AC	UHYAC	2780	405_1	333_1	169_{13}	135_3	72_{100}	1461	2160
Adeptolon-M (N-dealkyl-)	UHY UHYAC	1920	262_{97}	184_{100}	169_{32}	90_{67}	78_{70}	667	2156
Adeptolon-M (N-dealkyl-) AC	UHYAC	2200	304_{12}	261_{100}	245_{24}	90_{23}	78_{27}	929	2157
Adeptolon-M (N-dealkyl-HO-)	UHY	2510	278_{99}	184_{83}	169_{100}	90_{94}		765	2163
Adeptolon-M (N-dealkyl-HO-) AC	UHYAC	2500	320_{26}	278_{100}	184_{53}	169_{79}	90_{55}	1031	2158
Adeptolon-M (N-deethyl-) AC	UHYAC	2470	283_{57}	198_{56}	169_{100}	100_{53}	90_{60}	1268	2165
Adeptolon-M (N-deethyl-HO-) 2AC	UHYAC	3010	419_{10}	333_{40}	177_{64}	169_{100}	100_{15}	1505	2162
Adeptolon-M (nor-) AC	UHYAC	2530	297_{18}	253_9	198_{20}	169_{39}	58_{100}	1333	2159
Adeptolon-M (nor-HO-) 2AC	UHYAC	3030	433_7	333_{35}	177_{61}	169_{100}	58_{36}	1541	2161
Adinazolam		2955	351_1	308_{100}	280_5	205_5	58_8	1212	3068
Adiphenine		2215	311_1	239_2	167_{15}	99_{20}	86_{100}	978	6
Adiphenine-M/artifact (-COOH) (ME)	P-I	1715*	226_{24}	167_{100}	152_{13}			495	120
Agomelatine	P-I	2210	243_{25}	184_{100}	171_{14}	153_{26}	128_{42}	568	8369
Agomelatine AC	U+UHYAC	2300	285_{17}	184_{100}	171_{62}	153_{18}	128_{28}	811	8370
Agomelatine-M (di-HO-aryl-) isomer-1 2AC	U+UHYAC	2675	359_{11}	317_{14}	275_{45}	216_{100}	203_{55}	1258	8499
Agomelatine-M (di-HO-aryl-) isomer-2 2AC	U+UHYAC	2750	359_{17}	317_{22}	258_{24}	216_{100}	203_{16}	1258	8500
Agomelatine-M (HO-aryl-) 2AC	U+UHYAC	2575	343_{12}	301_{12}	242_{20}	200_{100}	187_{39}	1174	8497
Agomelatine-M (HO-aryl-) AC	U+UHYAC	2600	301_{16}	259_{23}	200_{100}	187_{38}	128_{11}	913	8494
Agomelatine-M (HO-aryl-alkyl-) 2AC	U+UHYAC	2715	359_{10}	317_{24}	242_8	200_{100}	187_{33}	1258	8498
Agomelatine-M (O-demethyl-) 2AC	U+UHYAC	2460	313_6	237_7	212_{17}	170_{100}	157_{37}	988	8502
Agomelatine-M (O-demethyl-) AC	U+UHYAC	2340	271_4	229_{22}	170_{100}	157_{50}	128_{21}	724	8493
Agomelatine-M (O-demethyl-HO-aryl-) 2AC	U+UHYAC	2630	329_5	287_{25}	245_{30}	186_{100}	173_{35}	1087	8495
Agomelatine-M (O-demethyl-HO-aryl-) AC	U+UHYAC	2595	287_{21}	245_{25}	186_{100}	173_{29}		820	8498
Agomelatine-M (O-demethyl-HO-aryl-HO-alkyl-) 3AC	U+UHYAC	2795	387_3	345_{17}	303_{29}	186_{100}	173_{22}	1390	8501

Air

Table 8.1: Compounds in order of names

Name	Detected	RI	Typical ions and intensities					Page	Entry
Air		<1000	44 $_2$	40 $_2$	32 $_{26}$	28 $_{100}$		247	3773
Air with Helium and Water		<1000	44 $_1$	40 $_2$	32 $_{26}$	28 $_{100}$	18 $_{36}$	247	4251
Ajmaline		2880	326 $_{65}$	297 $_8$	220 $_8$	182 $_9$	144 $_{100}$	1070	2719
Ajmaline 2AC	UHYAC	2890	410 $_{100}$	368 $_{16}$	353 $_{30}$	307 $_{15}$	182 $_{24}$	1479	2720
Ajmaline 2TMS		2565	470 $_{100}$	455 $_{83}$	246 $_{32}$	182 $_{43}$	73 $_{77}$	1620	6273
Ajmaline-M (dihydro-) 3AC	UHYAC	3065	454 $_{100}$	412 $_{87}$	397 $_{37}$	184 $_{31}$	146 $_{15}$	1590	2858
Ajmaline-M (HO-) isomer-1 3AC	U+UHYAC	3100	468 $_{100}$	426 $_{93}$	384 $_{14}$	198 $_{16}$	160 $_{15}$	1614	2859
Ajmaline-M (HO-) isomer-2 3AC	U+UHYAC	3130	468 $_{100}$	426 $_{64}$	197 $_{29}$	160 $_{65}$		1614	6786
Ajmaline-M (HO-methoxy-) 3AC	U+UHYAC	3160	498 $_{100}$	456 $_{67}$	441 $_{67}$	399 $_{12}$	228 $_{20}$	1657	6785
Ajmaline-M (nor-) 3AC	UHYAC	2980	438 $_{40}$	396 $_{100}$	354 $_{63}$	222 $_{53}$	196 $_{44}$	1554	2857
Alachlor		1850	269 $_{10}$	237 $_{23}$	188 $_{94}$	160 $_{100}$	77 $_{32}$	712	3505
Albendazole artifact (decarbamoyl-)		2510	207 $_{100}$	178 $_{16}$	165 $_{91}$	134 $_{21}$	122 $_{19}$	418	6073
Albendazole artifact (decarbamoyl-) AC		2410	249 $_{71}$	207 $_{86}$	165 $_{100}$	164 $_{99}$	134 $_{26}$	594	6072
Albendazole ME		2485	279 $_{100}$	236 $_{41}$	204 $_{21}$	178 $_{44}$	150 $_{16}$	772	6071
Aldicarb		1320	144 $_{55}$	100 $_{47}$	86 $_{100}$	76 $_{42}$	58 $_{56}$	369	3316
Aldrin		1945*	362 $_1$	329 $_2$	293 $_{11}$	263 $_{39}$	66 $_{100}$	1271	1330
Alfentanil	P-I	2990	416 $_1$	359 $_5$	289 $_{100}$	268 $_{36}$	140 $_{20}$	1497	1773
Alimemazine	P G U+UHYAC	2315	298 $_7$	198 $_{12}$	100 $_7$	84 $_5$	58 $_{100}$	896	8
Alimemazine-M (bis-nor-) AC	U+UHYAC	2765	312 $_{60}$	212 $_{100}$	114 $_{57}$			981	1240
Alimemazine-M (HO-)	UHY	2650	314 $_{18}$	228 $_4$	214 $_5$	100 $_{10}$	58 $_{100}$	995	11
Alimemazine-M (HO-) AC	U+UHYAC	2600	356 $_8$	228 $_2$	214 $_7$	100 $_6$	58 $_{100}$	1242	13
Alimemazine-M (nor-)	UHY	2335	284 $_{91}$	252 $_{21}$	212 $_{55}$	199 $_{100}$	180 $_{39}$	803	2243
Alimemazine-M (nor-) AC	UHYAC	2710	326 $_{73}$	212 $_{63}$	198 $_{38}$	180 $_{31}$	128 $_{100}$	1069	14
Alimemazine-M (nor-HO-) 2AC	U+UHYAC	2930	384 $_{15}$	270 $_{12}$	214 $_{16}$	128 $_{100}$	86 $_{30}$	1376	15
Alimemazine-M (ring)	P G U+UHYAC	2010	199 $_{100}$	167 $_{44}$				396	10
Alimemazine-M 2AC	U+UHYAC	2865	315 $_{54}$	273 $_{34}$	231 $_{100}$	202 $_{11}$		999	2618
Alimemazine-M AC	U+UHYAC	2550	257 $_{23}$	215 $_{100}$	183 $_7$			641	12
Alimemazine-M/artifact (sulfoxide)	G P U	2665	314 $_2$	298 $_5$	212 $_{32}$	199 $_{10}$	58 $_{100}$	995	9
Aliskiren artifact		3100	477 $_1$	436 $_3$	308 $_{100}$	291 $_{76}$	209 $_{39}$	1629	8592
Aliskiren artifact AC		3160	477 $_{45}$	291 $_{100}$	209 $_{82}$	137 $_{56}$		1630	8590
Aliskiren artifact HFB		2805	631 $_{16}$	209 $_{100}$	163 $_{19}$	137 $_{48}$	73 $_{34}$	1721	8595
Aliskiren artifact PFP		2800	581 $_{33}$	209 $_{100}$	137 $_{50}$	73 $_{38}$		1708	8594
Aliskiren artifact TFA		2855	531 $_{10}$	460 $_7$	386 $_{10}$	209 $_{100}$	137 $_{80}$	1685	8593
Aliskiren artifact TMS		3620	591 $_1$	380 $_{100}$	308 $_{15}$	291 $_{10}$		1712	8591
Alizapride		2855	190 $_7$	162 $_5$	147 $_{11}$	132 $_9$	110 $_{100}$	1001	7816
Alizapride AC		2855	176 $_1$	148 $_4$	133 $_{15}$	110 $_{100}$	70 $_9$	1248	7817
Alizapride ME		2700	329 $_2$	190 $_{22}$	147 $_7$	110 $_{100}$	70 $_{14}$	1089	7818
Alizapride TMS		2785	387 $_1$	372 $_1$	248 $_5$	162 $_4$	110 $_{100}$	1391	7819
AL-LAD		3100	349 $_{100}$	308 $_{12}$	247 $_{42}$	207 $_{75}$	72 $_{23}$	1206	10188
AL-LAD TMS		3210	421 $_{30}$	380 $_{11}$	319 $_{32}$	279 $_{77}$	253 $_{100}$	1512	9873
Allethrin		2105*	302 $_2$	168 $_6$	136 $_{23}$	123 $_{100}$	79 $_{34}$	920	2786
Allidochlor		1140	173 $_1$	132 $_{28}$	70 $_{37}$	56 $_{100}$		328	4041
Allobarbital	G P U+UHYAC	1595	208 $_2$	193 $_{23}$	167 $_{100}$	124 $_{94}$	80 $_{68}$	424	16
Allobarbital 2ME	UME	1505	236 $_6$	195 $_{100}$	138 $_{100}$	80 $_{49}$		540	643
Allopurinol	U+UHYAC	2700	136 $_{100}$	120 $_3$	109 $_6$	80 $_4$	67 $_7$	275	5241
Allylestrenol		2370*	300 $_{34}$	259 $_{53}$	241 $_{100}$	201 $_{38}$	91 $_{81}$	909	1376
Almotriptan		2890	335 $_1$	156 $_2$	143 $_5$	115 $_2$	58 $_{100}$	1125	8503
Almotriptan HFB		2685	156 $_1$	142 $_3$	58 $_{100}$			1685	8509
Almotriptan PFP		2675	156 $_1$	142 $_4$	58 $_{100}$			1636	8508
Almotriptan TFA		2700	156 $_1$	142 $_5$	69 $_2$	58 $_{100}$		1537	8507
Almotriptan TMS		2865	407 $_1$	215 $_2$	143 $_1$	73 $_{18}$	58 $_{100}$	1468	8504
Aloe-emodin		2660*	270 $_{100}$	241 $_{92}$	213 $_{12}$	139 $_{17}$	121 $_{19}$	715	3552
Aloe-emodin 2AC		3000*	354 $_{45}$	312 $_{79}$	270 $_{100}$	241 $_{13}$	139 $_4$	1231	3560
Aloe-emodin -2H		2530*	268 $_{100}$	239 $_{31}$	183 $_8$	155 $_{13}$	127 $_9$	705	3553
Aloe-emodin 2ME		2705*	298 $_{100}$	267 $_{60}$	239 $_{28}$	209 $_6$	155 $_9$	894	3562
Aloe-emodin 2TMS		2785*	399 $_{100}$	310 $_6$	184 $_9$	95 $_{21}$	73 $_{21}$	1490	3577
Aloe-emodin 3TMS		2900*	471 $_{100}$	399 $_9$	367 $_3$	220 $_2$	73 $_{58}$	1643	3576
Aloe-emodin AC		2735*	312 $_{25}$	270 $_{100}$	241 $_{38}$	139 $_7$	121 $_4$	979	3559
Aloe-emodin ME		2900*	284 $_{100}$	266 $_{32}$	238 $_{28}$	209 $_{23}$	139 $_{19}$	799	3561
Aloe-emodin TMS		2695*	342 $_{16}$	311 $_{100}$	225 $_5$	139 $_6$	75 $_{18}$	1168	3575
alpha-Hexachlorocyclohexane (HCH)		1690*	252 $_1$	217 $_{43}$	181 $_{95}$	109 $_{100}$	51 $_{97}$	824	3853
Alphamethrin		2790	415 $_1$	209 $_{23}$	181 $_{84}$	163 $_{100}$	91 $_{43}$	1493	3509
Alpha-Methyltryptamine 2HFB		1785	566 $_4$	353 $_{69}$	326 $_{83}$	240 $_{100}$	169 $_{57}$	1702	9538
Alpha-Methyltryptamine 2PFP		1750	466 $_1$	303 $_{55}$	276 $_{98}$	190 $_{86}$	119 $_{100}$	1609	9537
Alpha-Methyltryptamine 2TFA		1795	366 $_7$	253 $_{78}$	226 $_{100}$	156 $_{25}$	140 $_{60}$	1289	9536
Alpha-Methyltryptamine AC		2150	216 $_{10}$	157 $_{59}$	130 $_{100}$	103 $_{13}$	86 $_5$	452	9534
Alpha-Methyltryptamine HFB		1920	370 $_{14}$	240 $_{16}$	169 $_{14}$	157 $_{14}$	130 $_{100}$	1310	9539
Alpha-Methyltryptamine TFA		1795	270 $_{12}$	157 $_{10}$	140 $_9$	130 $_{100}$	103 $_9$	717	9535
alpha-Pyrrolidinohexiophenone		1930	188 $_1$	140 $_{100}$	105 $_8$	96 $_5$	84 $_6$	577	10416
alpha-Pyrrolidinohexiophenone-M (dihydro-oxo-) AC		2150	154 $_{100}$	107 $_1$	98 $_{49}$	86 $_{26}$		926	10418
alpha-Pyrrolidinohexiophenone-M (oxo-)		2125	259 $_1$	154 $_{100}$	105 $_{17}$	98 $_{56}$	86 $_{29}$	653	10417
alpha-Tocopherol	G P UHY	3030*	430 $_{35}$	205 $_{11}$	165 $_{100}$	57 $_{11}$		1536	2403
alpha-Tocopherol AC	G UHYAC	3070*	472 $_{10}$	430 $_{100}$	247 $_3$	165 $_{58}$	57 $_{16}$	1623	2402
Alprazolam	G P-I U+UHYAC-I	3100	308 $_{61}$	279 $_{100}$	273 $_{45}$	239 $_{22}$	204 $_{96}$	955	1730

Table 8.1: Compounds in order of names **Alprazolam-M (HO-)**

Name	Detected	RI	Typical ions and intensities					Page	Entry
Alprazolam-M (HO-)		3245	324 $_{19}$	322 $_{58}$	293 $_{12}$	287 $_{100}$		1056	1704
Alprazolam-M (HO-) AC	U+UHYAC-I	3180	366 $_{33}$	323 $_{100}$	295 $_{12}$	271 $_{18}$	77 $_{12}$	1290	1765
Alprazolam-M (HO-) artifact HYAC	UHYAC-I	2580	399 $_{2}$	356 $_{83}$	312 $_{50}$	284 $_{100}$	77 $_{58}$	1440	2046
Alprazolam-M (HO-) -CH2O		3070	294 $_{82}$	259 $_{100}$	239 $_{43}$	205 $_{61}$	101 $_{54}$	869	2392
Alprazolam-M/artifact HY	U+UHYAC-I	2500	341 $_{3}$	298 $_{100}$	105 $_{4}$	77 $_{9}$		1159	2045
Alprenolol	G U	1825	249 $_{6}$	234 $_{2}$	205 $_{3}$	100 $_{7}$	72 $_{100}$	601	17
Alprenolol 2AC	UHYAC	2275	333 $_{1}$	273 $_{2}$	200 $_{100}$	98 $_{20}$	72 $_{35}$	1114	1575
Alprenolol 2TMS		2205	393 $_{1}$	378 $_{1}$	144 $_{100}$	101 $_{7}$	73 $_{60}$	1418	5450
Alprenolol AC		2185	291 $_{1}$	273 $_{1}$	158 $_{100}$	116 $_{14}$	72 $_{96}$	851	1348
Alprenolol TFATMS		2080	402 $_{2}$	284 $_{100}$	228 $_{14}$	126 $_{44}$	73 $_{84}$	1500	6153
Alprenolol TMS		1940	321 $_{4}$	306 $_{6}$	205 $_{8}$	101 $_{18}$	72 $_{100}$	1042	5449
Alprenolol-M (deamino-di-HO-) +H2O 4AC	UHYAC	2450*	410 $_{4}$	350 $_{9}$	159 $_{100}$	99 $_{8}$		1478	1576
Alprenolol-M (deamino-di-HO-) 3AC	U+UHYAC	2220*	350 $_{4}$	308 $_{1}$	159 $_{100}$	99 $_{16}$		1208	1574
Alprenolol-M (deamino-HO-) +H2O 3AC	UHYAC	2100*	352 $_{1}$	292 $_{15}$	159 $_{100}$	99 $_{12}$	91 $_{6}$	1221	1573
Alprenolol-M (deamino-HO-) 2AC	UHYAC	1850*	292 $_{6}$	159 $_{100}$	131 $_{12}$	99 $_{14}$		856	1572
Alprenolol-M (HO-) 2AC	UHYAC	2510	349 $_{2}$	331	200 $_{8}$	158 $_{100}$	98 $_{13}$	1205	1577
Alprenolol-M (HO-) 3AC	UHYAC	2575	391 $_{1}$	332 $_{7}$	200 $_{100}$	98 $_{19}$	72 $_{16}$	1408	1578
Alprenolol-M/artifact (phenol) AC	U+UHYAC	1520*	176 $_{26}$	134 $_{100}$	119 $_{26}$	107 $_{34}$	77 $_{17}$	332	1571
AM-2201		3165	359 $_{30}$	342 $_{16}$	284 $_{28}$	232 $_{39}$	127 $_{100}$	1258	8532
AM-694		3015	435 $_{28}$	360 $_{12}$	232 $_{100}$	220 $_{55}$	144 $_{50}$	1547	8531
AM-694 (chloropentyl analog)		3190	451 $_{63}$	360 $_{38}$	248 $_{100}$	220 $_{80}$	203 $_{48}$	1581	9620
AM-694 (chlorophenyl analog)		2770	343 $_{78}$	308 $_{6}$	268 $_{62}$	232 $_{100}$	139 $_{42}$	1173	9621
AM-694 4-iodo isomer		2940	435 $_{100}$	360 $_{28}$	232 $_{65}$	220 $_{38}$	203 $_{17}$	1547	9619
Amantadine	G P U UHY	1240	151 $_{12}$	134 $_{4}$	94 $_{100}$	57 $_{24}$		292	18
Amantadine AC	PAC U+UHYAC	1640	193 $_{35}$	136 $_{100}$	94 $_{57}$			380	22
Amantadine formyl artifact	G P U UHY	1190	163 $_{5}$	135 $_{100}$	107 $_{13}$	93 $_{35}$	79 $_{42}$	309	8940
Amantadine TMS		1525	223 $_{20}$	208 $_{19}$	166 $_{100}$	150 $_{19}$	73 $_{42}$	485	4524
AMB		2445	345 $_{6}$	286 $_{38}$	231 $_{7}$	215 $_{100}$	145 $_{21}$	1186	9658
AMB TMS		2495	417 $_{12}$	374 $_{14}$	268 $_{26}$	215 $_{100}$	145 $_{49}$	1501	9660
AMB-M/artifact (HOOC-) (ET)		2485	359 $_{6}$	286 $_{55}$	231 $_{6}$	215 $_{100}$	145 $_{25}$	1260	9659
Ambroxol	P G U UHY	2665	376 $_{8}$	279 $_{77}$	264 $_{100}$	262 $_{53}$	114 $_{69}$	1338	19
Ambroxol 2AC	U+UHYAC	3015	460 $_{22}$	419 $_{100}$	417 $_{51}$	279 $_{74}$	264 $_{56}$	1600	20
Ambroxol 2TMS		2800	520 $_{1}$	391 $_{5}$	351 $_{100}$	186 $_{50}$	73 $_{66}$	1677	4528
Ambroxol 3AC	U+UHYAC	3100	502 $_{4}$	461 $_{100}$	459 $_{47}$	401 $_{22}$	279 $_{41}$	1661	2228
Ambroxol AC		2850	418 $_{7}$	279 $_{100}$	262 $_{41}$	156 $_{48}$	97 $_{30}$	1502	2226
Ambroxol formyl artifact	P G U UHY	2780	387 $_{28}$	331 $_{100}$	329 $_{53}$	289 $_{42}$	195 $_{82}$	1392	6315
Ambroxol -H2O	P G U UHY	2395	358 $_{13}$	289 $_{42}$	264 $_{94}$	262 $_{55}$	68 $_{100}$	1251	6314
Ambroxol -H2O 2AC	U+UHYAC	3030	444 $_{100}$	442 $_{50}$	303 $_{70}$	301 $_{34}$	81 $_{63}$	1563	2227
Ambroxol TMS		2665	448 $_{4}$	319 $_{23}$	279 $_{100}$	264 $_{92}$	186 $_{64}$	1576	4527
Ambroxol-M (HO-) 4AC	U+UHYAC	3375	560 $_{3}$	519 $_{88}$	303 $_{91}$	279 $_{73}$	264 $_{100}$	1700	4446
Ambroxol-M (HOOC-) ME	P U	1770	309 $_{68}$	307 $_{36}$	277 $_{100}$	275 $_{53}$	249 $_{27}$	945	5131
Ambroxol-M/artifact AC	U+UHYAC	1890	319 $_{100}$	317 $_{50}$	304 $_{11}$	277 $_{57}$		1008	21
Ambucetamide		2330	248 $_{100}$	192 $_{6}$	164 $_{6}$	136 $_{11}$		858	2287
Ametryne		1890	227 $_{88}$	212 $_{61}$	170 $_{41}$	68 $_{91}$	58 $_{100}$	499	3308
Amfebutamone	P-I	1695	239 $_{1}$	224 $_{11}$	139 $_{16}$	111 $_{25}$	100 $_{100}$	552	4699
Amfebutamone AC		2210	264 $_{40}$	225 $_{19}$	208 $_{52}$	183 $_{100}$	57 $_{43}$	787	5700
Amfebutamone formyl artifact		1755	237 $_{5}$	222 $_{4}$	139 $_{19}$	98 $_{38}$	57 $_{100}$	609	4700
Amfebutamone-M (3-chlorobenzoic acid)		1430*	156 $_{75}$	139 $_{100}$	111 $_{56}$	75 $_{44}$	65 $_{6}$	299	6024
Amfebutamone-M (3-chlorobenzoic acid) (ME)	U+UHYAC	1100*	170 $_{26}$	139 $_{100}$	111 $_{57}$	75 $_{50}$		323	6953
Amfebutamone-M (3-chlorobenzyl alcohol)		1560*	142 $_{53}$	125 $_{7}$	113 $_{28}$	107 $_{33}$	77 $_{100}$	281	6025
Amfebutamone-M (dihydro-) AC	U+UHYAC	1780	268 $_{2}$	208 $_{13}$	115 $_{6}$	100 $_{100}$		796	8520
Amfebutamone-M (dihydro-HO-) isomer-1 2AC	U+UHYAC P-I	2100	326 $_{1}$	266 $_{7}$	224 $_{4}$	100 $_{100}$	57 $_{25}$	1161	8537
Amfebutamone-M (dihydro-HO-) isomer-2 2AC	U+UHYAC	2120	326 $_{1}$	266 $_{7}$	224 $_{13}$	163 $_{12}$	100 $_{100}$	1162	8538
Amfebutamone-M (dihydro-HO-methoxy-) 2AC	U+UHYAC	2190	356 $_{1}$	296 $_{3}$	254 $_{1}$	187 $_{2}$	100 $_{100}$	1318	8539
Amfebutamone-M (HO-)	P-I	2040	240 $_{3}$	224 $_{30}$	166 $_{7}$	139 $_{29}$	116 $_{100}$	632	7660
Amfebutamone-M (HO-) AC		2130	224 $_{12}$	166 $_{4}$	158 $_{100}$	115 $_{44}$	98 $_{8}$	889	7661
Amfebutamone-M (HO-) TMS		2075	240 $_{17}$	224 $_{79}$	188 $_{97}$	145 $_{53}$	73 $_{100}$	1074	7662
Amfebutamone-M (N-dealkyl-)		1430*	183 $_{100}$	166 $_{17}$	141 $_{75}$	139 $_{41}$	111 $_{33}$	353	10297
Amfebutamone-M/artifact		2075*	309 $_{34}$	224 $_{34}$	174 $_{100}$	139 $_{21}$	119 $_{38}$	960	10298
Amfebutamone-M/artifact		2430*	309 $_{100}$	280 $_{17}$	224 $_{27}$	111 $_{26}$	57 $_{66}$	960	10299
Amfebutamone-M/artifact		1350*	223 $_{4}$	208 $_{5}$	166 $_{100}$	139 $_{38}$	103 $_{26}$	481	10300
Amfepramone	G U+UHYAC	1505	205 $_{1}$	160	100 $_{100}$	77 $_{9}$	72 $_{12}$	414	25
Amfepramone-M (deethyl-)	SPE	1355	105 $_{11}$	77 $_{37}$	72 $_{100}$			335	6685
Amfepramone-M (deethyl-) AC	SPEAC	1705	219 $_{2}$	114 $_{50}$	105 $_{19}$	77 $_{62}$	72 $_{100}$	466	6691
Amfepramone-M (deethyl-) HFB	SPEHFB	1565	373 $_{1}$	268 $_{100}$	240 $_{34}$	105 $_{64}$	77 $_{57}$	1326	6689
Amfepramone-M (deethyl-dihydro-) 2AC	SPEAC	1845	263 $_{1}$	114 $_{62}$	105 $_{16}$	91 $_{10}$	72 $_{100}$	676	6690
Amfepramone-M (deethyl-dihydro-) 2HFB	SPEHFB	1540	358 $_{5}$	268 $_{100}$	240 $_{20}$	169 $_{15}$	105 $_{9}$	1704	6688
Amfepramone-M (deethyl-dihydro-) TMS	SPETMS	1435	236 $_{2}$	179 $_{2}$	163 $_{8}$	149 $_{8}$	72 $_{100}$	614	6684
Amfepramone-M (deethyl-hydroxy-) 2AC	SPEAC	2095	277 $_{1}$	192 $_{2}$	121 $_{21}$	114 $_{74}$	72 $_{100}$	759	6681
Amfepramone-M (deethyl-hydroxy-) 2HFB	SPEHFB	1725	516 $_{2}$	317 $_{17}$	268 $_{100}$	240 $_{31}$	169 $_{27}$	1709	6680
Amfepramone-M (deethyl-hydroxy-) HFB	SPEHFB	1910	268 $_{55}$	240 $_{24}$	169 $_{7}$	121 $_{100}$		1397	6679
Amfepramone-M (deethyl-hydroxy-methoxy-) 2AC	SPEAC	2190	307 $_{1}$	151 $_{9}$	123 $_{5}$	114 $_{65}$	72 $_{100}$	948	6682
Amfepramone-M (deethyl-hydroxy-methoxy-) 2HFB	SPEHFB	1830	476 $_{2}$	347 $_{20}$	268 $_{100}$	240 $_{17}$	169 $_{11}$	1718	6678
Amfepramone-M (deethyl-hydroxy-methoxy-) HFB	SPEHFB	1890	419 $_{2}$	268 $_{71}$	240 $_{37}$	151 $_{48}$	121 $_{100}$	1505	6677

Amfepramone-M (dihydro-)

Table 8.1: Compounds in order of names

Name	Detected	RI	Typical ions and intensities					Page	Entry
Amfepramone-M (dihydro-)	SPE	1565	206 $_1$	105 $_4$	100 $_{100}$	77 $_{14}$	72 $_{13}$	422	6683
Amfepramone-M (dihydro-) AC	SPEAC	1605	248 $_1$	117 $_6$	105 $_9$	100 $_{100}$	77 $_9$	601	6692
Amfepramone-M (dihydro-) HFB	SPEHFB	1525	403 $_1$	303 $_1$	190 $_{10}$	169 $_{24}$	100 $_{100}$	1456	6687
Amfepramone-M (dihydro-) TMS	SPETMS	1550	264 $_1$	179 $_1$	163 $_4$	149 $_5$	100 $_{100}$	779	6686
Amfetamine		1160	134 $_4$	120 $_{15}$	91 $_{100}$	77 $_{18}$	65 $_{73}$	275	54
Amfetamine	U	1160	134 $_1$	120 $_1$	91 $_6$	65 $_4$	44 $_{100}$	275	5514
Amfetamine AC	U+UHYAC	1505	177 $_4$	118 $_{60}$	91 $_{35}$	86 $_{100}$	65 $_{16}$	336	55
Amfetamine AC	U+UHYAC	1505	177 $_1$	118 $_{19}$	91 $_{11}$	86 $_{31}$	44 $_{100}$	335	5515
Amfetamine formyl artifact		1100	147 $_2$	146 $_6$	125 $_5$	91 $_{12}$	56 $_{100}$	286	3261
Amfetamine HFB		1355	240 $_{79}$	169 $_{21}$	118 $_{100}$	91 $_{53}$		1099	5047
Amfetamine intermediate		1560	163 $_{16}$	146 $_{12}$	115 $_{100}$	105 $_{59}$	91 $_{81}$	307	2839
Amfetamine PFP		1330	281 $_1$	190 $_{73}$	118 $_{100}$	91 $_{36}$	65 $_{12}$	786	4379
Amfetamine precursor (phenylacetone)		<1000*	134 $_{23}$	91 $_{100}$	65 $_{40}$			274	3240
Amfetamine precursor (phenylacetone)		<1000*	134 $_{13}$	91 $_{54}$	65 $_{22}$	43 $_{100}$		274	5516
Amfetamine R-(-)-enantiomer HFBP		1160	337 $_{15}$	294 $_{16}$	266 $_{100}$	118 $_{11}$	91 $_{12}$	1527	6514
Amfetamine S-(+)-enantiomer HFBP		1190	337 $_{13}$	294 $_{21}$	266 $_{100}$	118 $_{16}$	91 $_{19}$	1527	6515
Amfetamine TFA		1095	231 $_1$	140 $_{100}$	118 $_{92}$	91 $_{45}$	69 $_{19}$	513	4000
Amfetamine TMS		1190	192 $_6$	116 $_{100}$	100 $_{10}$	91 $_{11}$	73 $_{87}$	421	5581
Amfetamine-D11 PFP		1610	194 $_{100}$	128 $_{82}$	98 $_{43}$	70 $_{14}$		856	7284
Amfetamine-D11 R-(-)-enantiomer HFBP		1995	341 $_{66}$	294 $_{49}$	266 $_{100}$	128 $_{28}$	98 $_{31}$	1557	6518
Amfetamine-D11 S-(+)-enantiomer HFBP		2000	341 $_{51}$	294 $_{36}$	266 $_{100}$	128 $_{39}$	98 $_{26}$	1556	6519
Amfetamine-D11 TFA		1615	242 $_1$	144 $_{100}$	128 $_{82}$	98 $_{43}$	70 $_{14}$	566	7283
Amfetamine-D5 AC		1480	182 $_3$	122 $_{46}$	92 $_{30}$	90 $_{100}$	66 $_{16}$	352	5690
Amfetamine-D5 HFB		1330	244 $_{100}$	169 $_{14}$	122 $_{46}$	92 $_{41}$	69 $_{40}$	1130	6316
Amfetamine-D5 PFP		1320	194 $_{100}$	123 $_{42}$	119 $_{32}$	92 $_{46}$	69 $_{11}$	815	5566
Amfetamine-D5 TFA		1085	144 $_{100}$	123 $_{53}$	122 $_{56}$	92 $_{51}$	69 $_{28}$	540	5570
Amfetamine-D5 TMS		1180	212 $_1$	197 $_8$	120 $_{100}$	92 $_{11}$	73 $_{57}$	441	5582
Amfetamine-M (3-HO-) 2AC	UHYAC	1930	235 $_2$	176 $_{48}$	134 $_{52}$	107 $_{21}$	86 $_{100}$	532	4387
Amfetamine-M (3-HO-) 2HFB		1620	330 $_{30}$	303 $_{11}$	240 $_{100}$	169 $_{15}$	69 $_{29}$	1691	5737
Amfetamine-M (3-HO-) 2PFP		1520	280 $_{36}$	253 $_9$	190 $_{100}$	119 $_{19}$	69 $_8$	1567	5738
Amfetamine-M (3-HO-) 2TFA		<1000	230 $_{33}$	203 $_8$	140 $_{100}$	115 $_6$		1172	6224
Amfetamine-M (3-HO-) 2TMS	UHYTMS	1850	280 $_{11}$	179 $_3$	116 $_{100}$	100 $_{12}$	73 $_{72}$	881	5693
Amfetamine-M (3-HO-) formyl artifact ME		1290	177 $_2$	162 $_4$	121 $_4$	77 $_5$	56 $_{100}$	335	5129
Amfetamine-M (3-HO-) TMSTFA		1630	319 $_8$	206 $_{86}$	191 $_{32}$	140 $_{100}$	73 $_{58}$	1024	6141
Amfetamine-M (4-HO-)		1480	151 $_{10}$	107 $_{69}$	91 $_{10}$	77 $_{42}$	56 $_{100}$	292	1802
Amfetamine-M (4-HO-) 2AC	U+UHYAC	1900	235 $_1$	176 $_{72}$	134 $_{100}$	107 $_{46}$	86 $_{70}$	533	1804
Amfetamine-M (4-HO-) 2HFB		<1000	330 $_{48}$	303 $_{15}$	240 $_{100}$	169 $_{44}$	69 $_{42}$	1691	6326
Amfetamine-M (4-HO-) 2PFP		<1000	280 $_{77}$	253 $_{16}$	190 $_{100}$	119 $_{56}$	69 $_{16}$	1567	6325
Amfetamine-M (4-HO-) 2TFA		<1000	230 $_{72}$	203 $_{11}$	140 $_{100}$	92 $_{12}$	69 $_{59}$	1172	6324
Amfetamine-M (4-HO-) 2TMS		<1000	280 $_7$	179 $_9$	149 $_8$	116 $_{100}$	73 $_{78}$	880	6327
Amfetamine-M (4-HO-) AC	U+UHYAC	1890	193 $_1$	134 $_{100}$	107 $_{26}$	86 $_{24}$	77 $_{16}$	379	1803
Amfetamine-M (4-HO-) formyl art.		1220	163 $_3$	148 $_4$	107 $_{30}$	77 $_{12}$	56 $_{100}$	308	6323
Amfetamine-M (4-HO-) formyl artifact ME		1255	177 $_6$	162 $_4$	121 $_{60}$	77 $_{12}$	56 $_{100}$	336	3250
Amfetamine-M (4-HO-) ME		1225	165 $_3$	122 $_{100}$	107 $_{11}$	91 $_{22}$	77 $_{45}$	314	3249
Amfetamine-M (4-HO-) ME		1225	165 $_1$	122 $_{16}$	91 $_3$	77 $_7$	44 $_{100}$	314	5517
Amfetamine-M (4-HO-) TFA		1670	247 $_4$	140 $_{15}$	134 $_{54}$	107 $_{100}$	77 $_{15}$	584	6335
Amfetamine-M (deamino-oxo-di-HO-) 2AC	U+UHYAC	1735*	250 $_3$	208 $_{15}$	166 $_{100}$	123 $_{100}$		603	4210
Amfetamine-M (deamino-oxo-HO-methoxy-)	UHY	1510*	180 $_{19}$	137 $_{100}$	122 $_{19}$	107 $_2$	94 $_{16}$	346	4247
Amfetamine-M (deamino-oxo-HO-methoxy-) ME	UHYME	1540*	194 $_{25}$	151 $_{100}$	135 $_4$	107 $_{18}$	65 $_4$	384	4353
Amfetamine-M (di-HO-) 3AC	U+UHYAC	2150	293 $_1$	234 $_{26}$	192 $_{42}$	150 $_{89}$	86 $_{100}$	862	3725
Amfetamine-M (HO-methoxy-)	UHY	1465	181 $_9$	138 $_{100}$	122 $_{18}$	94 $_{24}$	77 $_{16}$	350	4351
Amfetamine-M (HO-methoxy-deamino-HO-) 2AC	U+UHYAC	1820*	266 $_3$	206 $_9$	164 $_{100}$	150 $_{10}$	137 $_{30}$	695	6409
Amfetamine-M (norephedrine) 2AC	U+UHYAC	1805	235 $_1$	176 $_5$	134 $_7$	107 $_{13}$	86 $_{100}$	532	2476
Amfetamine-M (norephedrine) 2HFB	UHYHFB	1455	543 $_1$	330 $_{14}$	240 $_{100}$	169 $_{44}$	69 $_{57}$	1691	5098
Amfetamine-M (norephedrine) 2PFP	UHYPFP	1380	443 $_1$	280 $_9$	190 $_{100}$	119 $_{59}$	105 $_{26}$	1567	5094
Amfetamine-M (norephedrine) 2TFA	UTFA	1355	343 $_1$	230 $_6$	203 $_5$	140 $_{100}$	69 $_{29}$	1172	5091
Amfetamine-M (norephedrine) TMSTFA		1890	240 $_8$	198 $_3$	179 $_{100}$	117 $_5$	73 $_{88}$	1024	6146
Amfetamine-M 2AC	U+UHYAC	2065	265 $_3$	206 $_{27}$	164 $_{100}$	137 $_{23}$	86 $_{33}$	688	3498
Amfetamine-M 2HFB	UHFB	1690	360 $_{82}$	333 $_{15}$	240 $_{100}$	169 $_{42}$	69 $_{39}$	1704	6512
Amfetamine-M AC	U+UHYAC	1600*	222 $_2$	180 $_{22}$	137 $_{100}$			479	4211
Amfetamine-M ME	UHYME	1550	195 $_1$	152 $_{100}$	137 $_{17}$	107 $_{16}$	77 $_{14}$	387	4352
Amfetamine-N-formyl		1490	163 $_1$	118 $_{72}$	91 $_{30}$	72 $_{100}$	65 $_{23}$	308	6428
Amfetaminil		1755	132 $_{100}$	105 $_{51}$	91 $_{38}$	77 $_{17}$	65 $_{18}$	606	56
Amfetaminil-M/artifact (AM)		1160	134 $_4$	120 $_{15}$	91 $_{100}$	77 $_{18}$	65 $_{73}$	275	54
Amfetaminil-M/artifact (AM)	U	1160	134 $_1$	120 $_1$	91 $_6$	65 $_4$	44 $_{100}$	275	5514
Amfetaminil-M/artifact (AM) AC	U+UHYAC	1505	177 $_4$	118 $_{60}$	91 $_{35}$	86 $_{100}$	65 $_{16}$	336	55
Amfetaminil-M/artifact (AM) AC	U+UHYAC	1505	177 $_1$	118 $_{19}$	91 $_{11}$	86 $_{31}$	44 $_{100}$	335	5515
Amfetaminil-M/artifact (AM) formyl artifact		1100	147 $_2$	146 $_6$	125 $_5$	91 $_{12}$	56 $_{100}$	286	3261
Amfetaminil-M/artifact (AM) HFB		1355	240 $_{79}$	169 $_{21}$	118 $_{100}$	91 $_{53}$		1099	5047
Amfetaminil-M/artifact (AM) PFP		1330	281 $_1$	190 $_{73}$	118 $_{100}$	91 $_{36}$	65 $_{12}$	786	4379
Amfetaminil-M/artifact (AM) TFA		1095	231 $_1$	140 $_{100}$	118 $_{92}$	91 $_{45}$	69 $_{19}$	513	4000
Amfetaminil-M/artifact (AM) TMS		1190	192 $_6$	116 $_{100}$	100 $_{10}$	91 $_{11}$	73 $_{87}$	421	5581
Amfetaminil-M/artifact-D11 TFA		1615	242 $_1$	144 $_{100}$	128 $_{82}$	98 $_{43}$	70 $_{14}$	566	7283
Amfetaminil-M/artifact-D11 PFP		1610	194 $_{100}$	128 $_{82}$	98 $_{43}$	70 $_{14}$		856	7284

Table 8.1: Compounds in order of names

Name	Detected	RI	Typical ions and intensities					Page	Entry
Amfetaminil-M/artifact-D5 AC		1480	182_3	122_{46}	92_{30}	90_{100}	66_{16}	352	5690
Amfetaminil-M/artifact-D5 HFB		1330	244_{100}	169_{14}	122_{46}	92_{41}	69_{40}	1130	6316
Amfetaminil-M/artifact-D5 PFP		1320	194_{100}	123_{42}	119_{32}	92_{46}	69_{11}	815	5566
Amfetaminil-M/artifact-D5 TFA		1085	144_{100}	123_{53}	122_{56}	92_{51}	69_{28}	540	5570
Amfetaminil-M/artifact-D5 TMS		1180	212_1	197_8	120_{100}	92_{11}	73_{57}	441	5582
Amidithion		1930	273_6	131_{74}	125_{100}	93_{94}	59_{77}	732	3317
Amidotrizoic acid 2ME	UME	3000	642_8	569_3	515_{100}	483_{21}	314_2	1722	3708
Amidotrizoic acid 3ME		2920	656_{10}	625_5	529_{100}	471_4	386_3	1724	3709
Amidotrizoic acid -CO2 2ME		2680	598_5	471_{100}	403_{17}	328_3	287_1	1714	3711
Amidotrizoic acid -CO2 ME		2725	584_6	516_{28}	457_{100}	389_{39}	288_3	1708	3710
Amiloride-M/artifact (HOOC-) 2ME		1860	216_2	187_{100}	170_{33}	142_{20}	116_{17}	451	2629
Amiloride-M/artifact (HOOC-) 3ME		1930	230_{100}	201_{55}	169_{26}	129_{21}	114_7	509	6878
Amiloride-M/artifact (HOOC-) ME		1840	202_{100}	171_{51}	144_{68}	116_{32}	101_{22}	403	2628
Amineptine (ME)AC		2885	393_{59}	250_{26}	208_{37}	192_{100}	178_{43}	1418	6050
Amineptine 2ME		2570	365_2	192_{100}	178_{27}	174_{21}	165_{15}	1288	6042
Amineptine artifact (ring)		1775*	194_{100}	179_{74}	165_{23}	152_{12}	115_{32}	384	6036
Amineptine HY(ME)		1930	223_2	192_{100}	178_{35}	165_{36}	115_{14}	484	6046
Amineptine ME		2610	351_1	192_{100}	178_{21}	165_{12}	115_{11}	1217	6041
Amineptine TMS		2750	409_1	309_7	218_{12}	192_{100}	178_{19}	1476	6051
Amineptine TMSTFA		2770	505_1	304_{19}	300_{19}	193_{100}	178_{39}	1665	6052
Amineptine-M (dealkyl-) ME		1930	223_2	192_{100}	178_{35}	165_{36}	115_{14}	484	6046
Amineptine-M (N-pentanoic acid) 2ME		2490	337_1	192_{100}	178_{34}	165_{20}	115_{22}	1141	6049
Amineptine-M (N-pentanoic acid) -H2O		2585	291_{10}	206_6	192_{100}	178_{28}	165_{20}	849	6045
Amineptine-M (N-pentanoic acid) ME		2550	323_1	192_{100}	178_{27}	165_{17}	115_{13}	1053	6043
Amineptine-M (N-propionic acid) (ME)AC		2585	337_{12}	294_{100}	208_{34}	192_{100}	178_{23}	1139	6044
Amineptine-M (N-propionic acid) 2ME		2350	309_2	192_{100}	178_{36}	165_{26}	115_{19}	965	6048
Amineptine-M (N-propionic acid) ME		2400	295_1	192_{100}	178_{24}	165_{17}	115_{12}	879	6047
1-Amino-1,2-diphenylethane		1670	178_2	118_2	106_{100}	91_{11}	79_{25}	392	8423
1-Amino-1,2-diphenylethane AC		2020	180_3	148_{34}	106_{100}	91_{16}	79_{17}	553	8425
1-Amino-1,2-diphenylethane formyl artifact		1660	209_1	181_5	165_7	118_{100}	91_{66}	389	8424
1-Amino-1,2-diphenylethane HFB		1760	302_{100}	180_{28}	107_{19}	91_{34}	79_{44}	1413	8429
1-Amino-1,2-diphenylethane PFP		1730	252_{100}	180_{35}	165_7	91_{24}	79_{24}	1173	8428
1-Amino-1,2-diphenylethane TFA		1740	202_{100}	180_{27}	107_{25}	91_{24}	79_{33}	860	8427
1-Amino-1,2-diphenylethane TMS		1760	254_4	178_{100}	162_{15}	91_{19}	73_{29}	713	8426
2-Aminobenzoic acid ME		1290	151_{60}	119_{100}	92_{68}	65_{21}		291	4939
4-Aminobenzoic acid 2TMS		1645	281_{91}	236_{14}	148_{24}	73_{100}		788	5487
4-Aminobenzoic acid AC	U+UHYAC	2145	179_{31}	137_{100}	120_{92}	92_{16}	65_{24}	340	3298
4-Aminobenzoic acid ET	G	1820	165_{37}	137_{11}	120_{100}	92_{19}	65_{22}	313	1457
4-Aminobenzoic acid ETAC		1990	207_{62}	165_{66}	137_{26}	120_{100}	92_{18}	419	1440
4-Aminobenzoic acid ME		1550	151_{55}	120_{100}	92_{28}	65_{26}		291	23
4-Aminobenzoic acid MEAC		1985	193_{32}	151_{60}	120_{100}	92_{18}	65_{18}	378	24
Aminocarb		1720	208_{12}	151_{100}	136_{51}	120_{22}	77_{17}	425	3753
Aminocarb -C2H3NO		1215	151_{100}	150_{83}	136_{56}	120_{15}	77_{20}	291	3911
Aminocarb TFA		1700	304_{35}	247_{34}	232_{13}	150_{100}	69_{80}	931	4032
Aminoethanol		<1000	61_5	42_9	30_{100}			250	4189
4-(1-Aminoethyl-)phenol		<1000	137_{100}	121_{29}	103_4	91_8	77_1	277	7597
4-(1-Aminoethyl-)phenol 2AC		1740	221_{14}	179_{46}	164_{59}	136_1	122_{100}	473	7600
4-(1-Aminoethyl-)phenol 2HFB		1370	529_{52}	514_{100}	319_{37}	316_{42}	169_{38}	1684	7605
4-(1-Aminoethyl-)phenol 2PFP		1225	429_{57}	414_{100}	269_{35}	266_{33}	119_{49}	1529	7604
4-(1-Aminoethyl-)phenol 2TFA		1200	329_{79}	314_{100}	219_{25}	216_{22}	103_{10}	1085	7602
4-(1-Aminoethyl-)phenol 2TMS		1125	281_1	266_{100}	223_7	194_3	73_{30}	789	7599
4-(1-Aminoethyl-)phenol TFA		1430	233_{45}	218_{100}	148_{11}	120_{32}	95_{16}	520	7603
4-(1-Aminoethyl-)phenol TMS		1125	209_{100}	193_{19}	177_8	151_5	73_{20}	430	7598
Aminoglutethimide	P-I	2340	232_{49}	203_{100}	175_{56}	132_{56}	117_{20}	517	2741
Aminoglutethimide AC	UHYAC	2900	274_{78}	245_{56}	203_{100}	175_{26}	132_{21}	740	2249
Aminoglutethimide ME		2310	246_{55}	217_{72}	189_{100}	132_{44}	117_{20}	581	2742
Aminoglutethimide MEAC		2880	288_{100}	259_{49}	231_{84}	217_{46}	189_{54}	829	2250
1-Amino-naphtalene		1530	143_{100}	115_{42}	89_8	72_{10}	63_7	283	9194
Aminophenazone	P G U-I	1895	231_{36}	123_7	111_{17}	97_{56}	56_{100}	514	189
Aminophenazone-M (bis-nor-)	P U UHY	1955	203_{23}	93_{14}	84_{59}	56_{100}		406	219
Aminophenazone-M (bis-nor-) 2AC	UHYAC	2280	287_8	245_{31}	203_{15}	84_{56}	56_{100}	820	3333
Aminophenazone-M (bis-nor-) AC	P U U+UHYAC	2270	245_{30}	203_{13}	84_{50}	56_{100}		575	183
Aminophenazone-M (bis-nor-) artifact	U UHY	1945	180_{13}	119_{100}	91_{45}			344	424
Aminophenazone-M (deamino-HO-)	U UHY	1855	204_{35}	120_{18}	85_{100}	56_{50}		409	218
Aminophenazone-M (deamino-HO-) AC	U+UHYAC	2095	246_2	204_{19}	119_1	91_3	56_{100}	579	190
Aminophenazone-M (nor-)	P U UHY	1980	217_{16}	123_{14}	98_7	83_{16}	56_{100}	455	220
Aminophenazone-M (nor-) AC	P U+UHYAC	2395	259_{20}	217_8	123_9	56_{100}		652	184
3-Aminophenol	U UHY	1290	109_{86}	80_{100}				261	216
3-Aminophenol AC		1860	151_{33}	109_{100}	81_{52}	80_{54}		290	223
4-Aminophenol	UHY	1240	109_{100}	80_{41}	53_{82}	53_{82}	52_{90}	260	826
4-Aminophenol 2AC	PAC U+UHYAC	1765	193_{10}	151_{53}	109_{100}	80_{24}		377	188
4-Aminophenol 3AC	U+UHYAC	2085	235_{10}	193_{11}	151_{30}	109_{100}		532	827
4-Aminophenol ME	UHYME	1100	123_{27}	109_{100}	94_7	80_{96}	53_{47}	267	3766
5-(2-Aminopropyl)benzofuran		1450	175_1	131_{13}	102_3	77_{10}	44_{100}	330	9083

5-(2-Aminopropyl)benzofuran formyl artifact

Table 8.1: Compounds in order of names

Name	Detected	RI	Typical ions and intensities					Page	Entry
5-(2-Aminopropyl)benzofuran formyl artifact		1505	187_6	172_8	131_{41}	77_{19}	56_{100}	363	9084
5-(2-Aminopropyl)benzofuran HFB		1670	371_1	240_{15}	158_{57}	131_{100}	77_{28}	1316	9088
5-(2-Aminopropyl)benzofuran PFP		1640	321_2	190_{16}	158_{46}	131_{100}	77_{31}	1038	9087
5-(2-Aminopropyl)benzofuran TFA		1655	271_2	158_{40}	131_{100}	77_{33}	69_{40}	722	9086
5-(2-Aminopropyl)benzofuran TMS		1650	232_6	131_{23}	116_{100}	100_{11}	73_{57}	586	9085
6-(2-Aminopropyl)benzofuran		1585	175_3	131_{25}	102_5	77_{14}	44_{100}	330	8704
6-(2-Aminopropyl)benzofuran		1585	175_{14}	131_{100}	102_{21}	77_{56}	63_{14}	330	8705
6-(2-Aminopropyl)benzofuran AC		1890	217_8	158_{100}	131_{69}	86_{42}	77_{42}	455	9089
6-(2-Aminopropyl)benzofuran formyl artifact		1585	187_{17}	172_8	131_{56}	77_{15}	56_{100}	363	8706
6-(2-Aminopropyl)benzofuran HFB		1600	371_1	240_{15}	158_{62}	131_{100}	77_{26}	1316	9093
6-(2-Aminopropyl)benzofuran PFP		1610	321_2	190_{15}	158_{54}	131_{100}	77_{31}	1038	9092
6-(2-Aminopropyl)benzofuran TFA		1670	271_3	158_{42}	131_{100}	77_{34}	69_{42}	723	9091
6-(2-Aminopropyl)benzofuran TMS		1655	232_5	131_{22}	116_{100}	100_9	73_{60}	586	9090
5-Aminopropylbenzofuran		1450	175_9	160_5	131_{100}	102_{25}	77_{77}	330	8951
5-Aminopropylbenzofuran AC		1870	217_3	158_{100}	131_{67}	86_{59}	77_{51}	454	8950
5-Aminopropylindole		1765	174_5	131_{100}	103_{10}	77_{21}	63_6	329	9095
5-Aminopropylindole		1765	174_5	131_{85}	103_9	77_{16}	44_{100}	329	9096
5-Aminopropylindole 2AC		2345	199_{100}	157_{31}	130_{54}	103_{20}	86_{29}	647	9099
5-Aminopropylindole 2HFB		1880	353_{18}	326_{32}	240_{87}	129_{100}		1702	9107
5-Aminopropylindole 2ME		1850	202_1	130_7	115_2	72_{100}		404	9102
5-Aminopropylindole 2PFP		2230	466_4	303_{59}	276_{100}	190_{51}	184_{41}	1609	9106
5-Aminopropylindole 2TFA		2255	366_1	297_5	253_{75}	226_{100}	129_{73}	1289	9105
5-Aminopropylindole 2TMS		1950	303_1	202_8	116_{100}	100_9	73_{67}	1021	9103
5-Aminopropylindole AC		2145	216_4	157_{100}	130_{88}	103_{14}	77_{25}	453	9098
5-Aminopropylindole formyl artifact		1745	186_{16}	130_{100}	103_7	77_{13}	56_{41}	361	9097
5-Aminopropylindole formyl artifact ME		1780	200_{14}	144_{100}	129_4	115_4	56_{11}	400	9100
5-Aminopropylindole HFB		1970	370_1	169_4	157_{18}	130_{100}	104_1	1310	9108
5-Aminopropylindole ME		1810	151_8	130_{13}	98_8	70_{15}	58_{100}	366	9101
5-Aminopropylindole TMS		1645	246_2	203_{100}	188_{19}	130_9	73_{55}	582	9104
5-Aminopropylindole-M (HO-) 2AC		2360	274_2	215_{80}	173_{100}	146_{42}	86_{49}	740	9213
5-Aminopropylindole-M (HO-) AC		2340	232_4	173_{100}	146_{26}	118_{16}	86_{16}	517	9212
6-Aminopropylindole		1765	174_5	131_{100}	103_{10}	77_{19}	63_5	329	9109
6-Aminopropylindole		1765	174_3	131_{75}	103_8	77_{16}	44_{100}	329	9110
6-Aminopropylindole 2HFB		2245	566_1	353_{10}	326_{19}	184_{71}	129_{100}	1702	9116
6-Aminopropylindole 2PFP		2220	466_1	341_7	303_{92}	276_{100}	184_{28}	1609	9115
6-Aminopropylindole 2TFA		2285	366_8	297_6	253_{56}	226_{100}	102_{35}	1290	9114
6-Aminopropylindole AC		2140	216_7	157_{100}	130_{86}	103_{13}	86_8	453	9112
6-Aminopropylindole formyl artifact		1740	186_{19}	130_{100}	103_8	77_{17}	56_{52}	361	9111
6-Aminopropylindole HFB		1870	370_2	240_3	157_{20}	130_{100}	103_5	1310	9117
6-Aminopropylindole TFA		1885	270_{13}	157_{14}	130_{100}	103_9	77_{19}	717	9113
6-Aminopropylindole-M (di-HO-) 3AC		2440	332_6	273_{100}	231_{87}	189_{62}	86_{72}	1106	9216
6-Aminopropylindole-M (HO-) 2AC		2310	274_2	215_{38}	173_{100}	146_{29}	86_{40}	740	9215
6-Aminopropylindole-M (HO-) AC		2320	232_{10}	173_{100}	146_{25}	117_6	86_{28}	518	9214
Aminorex		2065	162_{14}	145_6	118_{22}	91_{11}	56_{100}	306	3197
Aminorex isomer-1 2AC		1990	246_4	203_{100}	189_{14}	161_{94}	72_{33}	579	3203
Aminorex isomer-2 2AC		2115	246_6	231_{25}	189_{43}	146_{100}	56_{86}	579	3204
4-Aminosalicylic acid 2ME		1735	181_{100}	149_{80}	121_{51}			348	215
4-Aminosalicylic acid acetyl conjugate ME		1995	209_{39}	167_{32}	135_{100}			427	213
4-Aminosalicylic acid ME		1600	167_{60}	135_{100}	107_{81}	79_{100}		319	214
4-Aminosalicylic acid-M (3-aminophenol)	U UHY	1290	109_{86}	80_{100}				261	216
4-Aminosalicylic acid-M acetyl conjugate		1860	151_{33}	109_{100}	81_{52}	80_{54}		290	223
4-Aminothiophenol		1025	125_{100}	98_{12}	93_{23}	80_{28}	65_7	268	6351
Amiodarone artifact		2800*	420_{11}	294_{91}	265_{83}	142_{100}	121_{93}	1693	1386
Amiodarone artifact AC		2965*	588_{83}	546_{100}	517_{57}	461_{50}	373_{36}	1710	7587
Amiodarone artifact HFB		3670*	517_1	268_{100}	240_8	201_4		1731	7591
Amiodarone artifact PFP		3650*	517_1	391_1	218_{100}	190_9	119_4	1728	7590
Amiodarone artifact TFA		3740*	642_1	529_2	251_3	168_{100}	140_{18}	1722	7589
Amiodarone artifact TMS		3055*	618_{89}	335_{34}	320_{57}	201_{66}	73_{100}	1718	7588
Amiodarone-M (N-deethyl-) artifact AC		2965*	588_{83}	546_{100}	517_{57}	461_{50}	373_{36}	1710	7587
Amiodarone-M (N-deethyl-) artifact HFB		3670*	517_1	268_{100}	240_8	201_4		1731	7591
Amiodarone-M (N-deethyl-) artifact PFP		3650*	517_1	391_1	218_{100}	190_9	119_4	1728	7590
Amiodarone-M (N-deethyl-) artifact TFA		3740*	642_1	529_2	251_3	168_{100}	140_{18}	1722	7589
Amiodarone-M (N-deethyl-) artifact TMS		3055*	618_{89}	335_{34}	320_{57}	201_{66}	73_{100}	1718	7588
Amiphenazole		2170	191_{100}	149_{24}	121_{58}	104_{20}	77_{38}	371	34
Amiphenazole 2AC		2575	275_{27}	233_{59}	191_{100}	121_{42}		744	35
Amiphenazole 2ME		1925	219_2	191_{100}	147_{50}	121_{71}	77_{62}	463	36
Amisulpride	U+UHYAC P-I	3260	369_1	242_6	196_3	149_3	98_{100}	1307	5409
Amisulpride 2TMS	U+UHYTMS	3065	513_1	498_2	314_{10}	196_4	98_{100}	1672	5839
Amisulpride PFP	U+UHYPFP	2880	515_1	388_2	266_2	98_{100}	70_{12}	1673	5838
Amisulpride TFA	U+UHYTFA	2905	338_2	216_2	187_1	98_{100}	70_6	1608	5837
Amisulpride TMS		3400	441_1	426	314_5	115_1	98_{100}	1562	5840
Amisulpride-M (O-demethyl-)	U+UHYAC	2960	355_1	228_4	182_3	135_4	98_{100}	1237	5410
Amitraz artifact-1		1570	162_{29}	149_{63}	120_{100}	106_{63}	77_{31}	306	4043
Amitraz artifact-2		2570	252_{14}	132_6	121_{100}	106_{18}	77_{13}	614	4042

Table 8.1: Compounds in order of names   **Amitriptyline**

Name	Detected	RI	Typical ions and intensities					Page	Entry
Amitriptyline	G P U+UHYAC	2205	277_1	215_5	202_{10}	189_5	58_{100}	764	37
Amitriptyline-M (bis-nor-HO-) -H2O AC	U+UHYAC	2710	289_{15}	230_{100}	215_{70}	202_{31}	189_5	835	1873
Amitriptyline-M (di-HO-N-oxide) -H2O -(CH3)2NOH	UHY	2280*	246_{100}	228_{35}	215_{50}	202_{36}	178_{33}	580	2698
Amitriptyline-M (di HO-N-oxide) -H2O -(CH3)2NOH AC	U+UHYAC	2530*	288_{35}	246_{100}	229_{58}	215_{89}	202_{37}	827	2541
Amitriptyline-M (HO-)	P-I U UGLUC	2380	293_1	215_3	202_2	91_1	58_{100}	866	27
Amitriptyline-M (HO-) AC	UGLUCAC	2500	335_1	273_1	215_5	202_5	58_{100}	1126	44
Amitriptyline-M (HO-) -H2O	P UHY U+UHYAC	2235	275_1	215_{14}	202_8	189_5	58_{100}	748	40
Amitriptyline-M (HO-N-oxide) -(CH3)2NOH AC	U+UHYAC	2490*	290_{31}	248_{100}	230_{50}	215_{36}	202_{30}	842	1874
Amitriptyline-M (HO-N-oxide) -H2O -(CH3)2NOH	U UHY U+UHYAC	2000*	230_{100}	215_{40}				510	46
Amitriptyline-M (nor-)	P-I G U UHY	2255	263_{27}	220_{67}	202_{100}	189_{39}	91_{30}	677	38
Amitriptyline-M (nor-) AC	PAC UHYAC	2660	305_9	232_{100}	217_{31}	202_{25}	86_{53}	939	41
Amitriptyline-M (nor-) HFB		2420	459_5	240_{100}	232_{49}	217_{36}	202_{36}	1599	7685
Amitriptyline-M (nor-) PFP		2405	409_3	232_{100}	217_{71}	203_{69}	190_{69}	1473	7684
Amitriptyline-M (nor-) TFA		2410	359_3	232_{76}	217_{54}	202_{70}	140_{100}	1258	7683
Amitriptyline-M (nor-) TMS		2340	335_1	320_1	203_5	116_{100}	73_{52}	1127	5440
Amitriptyline-M (nor-)-D3		2250	266_6	220_{41}	215_{51}	202_{100}	189_{45}	698	7794
Amitriptyline-M (nor-)-D3 AC		2655	308_{11}	232_{100}	217_{46}	202_{47}	89_{23}	960	7795
Amitriptyline-M (nor-)-D3 HFB		2415	462_2	243_{58}	232_{100}	217_{40}	203_{33}	1604	7798
Amitriptyline-M (nor-)-D3 PFP		2400	412_2	232_{100}	217_{53}	203_{47}	193_{46}	1485	7797
Amitriptyline-M (nor-)-D3 TFA		2405	362_2	232_{100}	217_{53}	202_{48}	143_{47}	1274	7796
Amitriptyline-M (nor-)-D3 TMS		2335	338_1	323_{10}	202_{33}	119_{100}	73_{73}	1146	7799
Amitriptyline-M (nor-di-oxo-) AC	U+UHYAC	2790	333_5	291_{41}	246_{35}	217_{100}	86_{65}	1111	9705
Amitriptyline-M (nor-HO-)	U-I UGLUC	2390	279_8	261_6	218_{100}	203_{39}	91_{10}	777	39
Amitriptyline-M (nor-HO-) -H2O	UHY	2600	261_{14}	218_{99}	215_{100}	202_{66}	189_{23}	665	2270
Amitriptyline-M (nor-HO-) -H2O AC	U+UHYAC	2670	303_{20}	230_{100}	215_{74}	202_{34}	86_{18}	925	42
Amitriptyline-M (N-oxide) -(CH3)2NOH	P G U+UHYAC	1975*	232_{100}	215_{72}	202_{72}	189_{39}	165_{26}	518	45
Amitriptylinoxide -(CH3)2NOH	P G U+UHYAC	1975*	232_{100}	215_{72}	202_{72}	189_{39}	165_{26}	518	45
Amitriptylinoxide-M (deoxo-bis-nor-HO-) -H2O AC	U+UHYAC	2710	289_{15}	230_{100}	215_{70}	202_{31}	189_5	835	1873
Amitriptylinoxide-M (deoxo-HO-)	P-I U UGLUC	2380	293_1	215_3	202_2	91_1	58_{100}	866	27
Amitriptylinoxide-M (deoxo-HO-) AC	UGLUCAC	2500	335_1	273_1	215_5	202_5	58_{100}	1126	44
Amitriptylinoxide-M (deoxo-HO-) -H2O	P UHY U+UHYAC	2235	275_1	215_{14}	202_8	189_5	58_{100}	748	40
Amitriptylinoxide-M (deoxo-nor-HO-) -H2O	UHY	2600	261_{14}	218_{99}	215_{100}	202_{66}	189_{23}	665	2270
Amitriptylinoxide-M (deoxo-nor-HO-) -H2O AC	U+UHYAC	2670	303_{20}	230_{100}	215_{74}	202_{34}	86_{18}	925	42
Amitriptylinoxide-M (di-HO-) -H2O -(CH3)2NOH	UHY	2280*	246_{100}	228_{35}	215_{50}	202_{36}	178_{33}	580	2698
Amitriptylinoxide-M (di-HO-) -H2O -(CH3)2NOH AC	U+UHYAC	2530*	288_{35}	246_{100}	229_{58}	215_{89}	202_{37}	827	2541
Amitriptylinoxide-M (HO-) -(CH3)2NOH AC	U+UHYAC	2490*	290_{31}	248_{100}	230_{50}	215_{36}	202_{30}	842	1874
Amitriptylinoxide-M (HO-) -H2O -(CH3)2NOH	U UHY U+UHYAC	2000*	230_{100}	215_{40}				510	46
Amitrole		<1000	84_{100}	75_3	57_{14}			254	4509
Amitrole 2ME		1050	112_{25}	111_{28}	98_{100}	84_9	56_{75}	262	3121
Amitrole AC	UHYAC	1010	126_{18}	108_3	84_{100}	57_{35}		269	4233
Amlodipine 2ME		2815	436_2	325_{52}	208_7	72_{64}	58_{62}	1550	4842
Amlodipine 2TMS		3130	552_{16}	441_{38}	174_{100}	116_{79}	73_{74}	1696	5014
Amlodipine AC	U+UHYAC	3170	450_1	347_{20}	339_{87}	208_{39}	86_{100}	1580	4844
Amlodipine ME		2820	422_3	311_{100}	254_{53}	208_{12}	88_{20}	1513	4843
Amlodipine TMS		2935	480_{20}	369_{100}	326_{53}	208_{17}	73_{17}	1635	5013
Amlodipine-M (deamino-HOOC-) ME	PME UME	2830	437_3	326_{100}	312_{30}	280_{48}	208_{41}	1551	4846
Amlodipine-M (deethyl-deamino-HOOC-) 2ME	UME	2800	423_3	392_4	312_{100}	280_{15}	222_{52}	1514	4847
Amlodipine-M (dehydro-2-HOOC-) ME	UME	2430	391_1	356_{100}	296_{85}	268_{15}	224_{20}	1405	4850
Amlodipine-M (dehydro-deamino-HOOC-) ME	UME	2635	435_1	400_{16}	347_{98}	318_{61}	260_{100}	1548	4848
Amlodipine-M (dehydro-deethyl-O-dealkyl-) -H2O	UME	2300	317_1	282_{100}	267_8	250_6	139_5	1009	4849
Amlodipine-M/artifact (dehydro-) 2ME	PME	2825	434_5	323_{100}	277_{12}	88_{20}	86_{21}	1544	4845
Amlodipine-M/artifact (dehydro-) 2TMS		2925	550_{11}	535_{58}	477_{100}	447_9	359_{10}	1695	5015
Amlodipine-M/artifact (dehydro-) AC	UHYAC	2910	448_3	363_{34}	347_{100}	260_{83}	86_{41}	1576	4851
Amobarbital	p G U UHY U+UHYAC	1710	211_2	198_6	197_9	156_{100}	141_{73}	496	47
Amobarbital 2ME	UME	1595	239_2	225_7	184_{100}	169_{90}		629	51
Amobarbital 2TMS		1530	370_1	355_{100}	300_{40}	100_{54}	73_{95}	1314	5498
Amobarbital-M (HO-)	U	1915	227_{26}	195_6	157_{72}	156_{100}	141_{60}	565	49
Amobarbital-M (HO-) 2ME	UME	1750	270_4	255_{22}	184_{95}	169_{64}	137_{100}	719	52
Amobarbital-M (HO-) -H2O	UHY U+UHYAC	1830	224_9	195_{13}	156_{73}	141_{37}	69_{100}	488	48
Amobarbital-M (HOOC-)	U	1960	212_{24}	183_{24}	156_{100}	141_{93}	55_{59}	636	50
Amobarbital-M (HOOC-) 3ME	UME	1850	240_8	184_{99}	169_{100}	137_{72}		896	53
Amodiaquine 2AC		3000	356_{13}	314_{23}	282_{100}	253_{27}	218_{23}	1556	7838
Amodiaquine 2TMS		2780	499_8	428_4	354_5	86_{41}	73_{100}	1658	7787
Amodiaquine AC		2875	326_{17}	286_{29}	284_{100}	205_7	99_{11}	1435	6889
Amodiaquine artifact		2850	284_{100}	268_{11}	248_{26}	234_{14}		798	7459
Amodiaquine artifact AC		2875	326_{13}	284_{100}	248_9	205_4	99_5	1066	7839
Amodiaquine artifact ME		2905	296_{100}	260_{14}	232_7	99_{17}		882	7191
Amodiaquine ME		3030	369_{20}	354_{17}	297_{100}	269_{29}	252_{19}	1307	7840
Amodiaquine TMS		3090	427_9	412_9	355_{76}	86_{25}	73_{100}	1526	7836
Amoxapine		2665	313_8	257_{46}	245_{100}	228_{20}	193_{34}	987	8231
Amoxapine AC		2940	355_{17}	269_{50}	257_{65}	193_{100}	56_{76}	1236	8232
Amoxapine HFB		2745	509_{37}	257_{75}	229_{55}	193_{100}	69_{81}	1668	8236
Amoxapine ME	G U+UHYAC	2555	327_4	257_{36}	193_{24}	83_{57}	70_{100}	1074	549
Amoxapine PFP		2730	459_{46}	257_{69}	229_{53}	193_{100}	69_{41}	1598	8235

Amoxapine TFA **Table 8.1:** Compounds in order of names

Name	Detected	RI	Typical ions and intensities					Page	Entry
Amoxapine TFA		2745	409$_{52}$	257$_{65}$	229$_{52}$	193$_{100}$	69$_{71}$	1473	8234
Amoxapine TMS		2770	385$_{7}$	370$_{7}$	317$_{10}$	128$_{100}$	73$_{73}$	1381	8233
Amoxaprine-M (HO-) 2AC	UHYAC	3450	413$_{31}$	207$_{100}$	112$_{19}$			1487	1275
Amoxicilline-M/artifact 4TMS		1215	455$_{8}$	440$_{7}$	216$_{23}$	172$_{65}$	73$_{100}$	1592	7655
Amoxicilline-M/artifact ME2AC		1930	314$_{35}$	230$_{38}$	198$_{100}$	156$_{43}$	97$_{77}$	993	7652
Amoxicilline-M/artifact ME2AC		1900	265$_{5}$	233$_{37}$	164$_{43}$	122$_{100}$	120$_{59}$	686	7653
Amoxicilline-M/artifact ME2TFA		1755	422$_{28}$	326$_{100}$	267$_{36}$	196$_{54}$	165$_{32}$	1512	7656
Amoxicilline-M/artifact ME3AC		2025	307$_{5}$	265$_{39}$	233$_{67}$	180$_{100}$	120$_{69}$	947	7654
Amoxicilline-M/artifact MEAC		1980	272$_{40}$	230$_{91}$	215$_{38}$	100$_{53}$	97$_{100}$	729	7651
Amoxicilline-M/artifact MEPFP		1750	376$_{100}$	317$_{65}$	246$_{70}$	243$_{62}$	215$_{53}$	1338	7657
Amperozide artifact (methylpiperazine)		2415	344$_{14}$	201$_{9}$	183$_{12}$	113$_{100}$	70$_{35}$	1180	6097
Amperozide-M (deamino-carboxy-)	P-I UHY UHYAC	2230*	276$_{7}$	216$_{17}$	203$_{100}$	183$_{22}$		753	169
Amperozide-M (deamino-carboxy-) ME	P-I UHYME U+UHYAC	2125*	290$_{12}$	258$_{13}$	216$_{30}$	203$_{100}$	183$_{22}$	841	3372
Amperozide-M (deamino-HO-) AC	UHYAC	2150*	304$_{7}$	244$_{22}$	216$_{41}$	203$_{100}$	183$_{16}$	931	307
Amperozide-M (N-dealkyl-) AC	UHYAC	2970	372$_{29}$	300$_{13}$	201$_{11}$	141$_{100}$	109$_{9}$	1324	3370
Amprenavir artifact (HOOC-) -H2O		3950	417$_{1}$	374$_{4}$	241$_{31}$	156$_{100}$	92$_{58}$	1500	7959
Amprenavir artifact (HOOC-) -H2O 2TMS		3240	561$_{1}$	313$_{48}$	228$_{100}$	180$_{29}$	164$_{30}$	1701	8238
Amprenavir artifact 2AC		3660	458$_{3}$	416$_{3}$	350$_{9}$	283$_{40}$	198$_{100}$	1597	8242
Amprenavir artifact HFB		3480	570$_{2}$	437$_{33}$	288$_{34}$	91$_{100}$	86$_{62}$	1703	8241
Amprenavir artifact PFP		3500	520$_{2}$	387$_{53}$	238$_{48}$	91$_{100}$	86$_{61}$	1677	8240
Amprenavir artifact TFA		3570	470$_{2}$	337$_{54}$	252$_{48}$	91$_{100}$	86$_{65}$	1617	8239
AMT 2HFB		1785	566$_{4}$	353$_{69}$	326$_{70}$	240$_{100}$	169$_{57}$	1702	9538
AMT 2PFP		1750	466$_{1}$	303$_{55}$	276$_{98}$	190$_{86}$	119$_{100}$	1609	9537
AMT 2TFA		1795	366$_{7}$	253$_{78}$	226$_{100}$	156$_{25}$	140$_{60}$	1289	9536
AMT AC		2150	216$_{10}$	157$_{59}$	130$_{100}$	103$_{13}$	86$_{5}$	452	9534
AMT HFB		1920	370$_{14}$	240$_{16}$	157$_{14}$	130$_{100}$		1310	9539
AMT TFA		1795	270$_{12}$	157$_{10}$	140$_{9}$	130$_{100}$	103$_{9}$	717	9535
Amylnitrite		<1000	85$_{5}$	70$_{18}$	57$_{54}$	41$_{100}$		264	58
Anastrozole		2270	293$_{22}$	266$_{17}$	225$_{28}$	209$_{100}$	70$_{52}$	866	9571
Ancymidol		2220	256$_{4}$	228$_{76}$	215$_{17}$	121$_{61}$	107$_{100}$	638	4144
Androst-4-ene-3,17-dione		2600*	286$_{100}$	244$_{19}$	148$_{26}$	124$_{59}$	79$_{36}$	816	3762
Androst-4-ene-3,17-dione enol 2TMS		2650*	430$_{98}$	415$_{11}$	234$_{16}$	209$_{15}$	73$_{100}$	1536	3803
Androstane-3,17-dione		2555*	288$_{100}$	255$_{22}$	244$_{32}$	217$_{27}$	124$_{21}$	830	3761
Androstane-3,17-dione enol 2TMS		2600*	432$_{60}$	417$_{41}$	290$_{47}$	275$_{71}$	73$_{100}$	1541	3001
Androsterone	UHY	2475*	290$_{100}$	246$_{24}$	147$_{16}$	107$_{39}$	67$_{36}$	844	59
Androsterone AC	UHYAC	2580*	332$_{28}$	272$_{100}$	257$_{36}$	201$_{24}$	79$_{54}$	1109	61
Androsterone enol 2TMS		2500*	434$_{46}$	419$_{52}$	329$_{35}$	169$_{37}$	73$_{100}$	1546	3002
Androsterone -H2O	UHY UHYAC	2240*	272$_{100}$	218$_{87}$	190$_{34}$	161$_{50}$	79$_{49}$	731	2481
Anilazine		2050	274$_{8}$	239$_{100}$	178$_{31}$	143$_{26}$	75$_{11}$	736	3426
Aniline		<1000	93$_{100}$	66$_{28}$				256	1550
Aniline AC	G	1380	135$_{35}$	93$_{100}$				274	222
4-Anisic acid		1320*	152$_{95}$	135$_{100}$	107$_{11}$	92$_{16}$	77$_{23}$	293	10270
4-Anisic acid ET		1415*	180$_{21}$	152$_{19}$	135$_{100}$	107$_{12}$	77$_{21}$	345	6447
4-Anisic acid ME		1270*	166$_{36}$	135$_{100}$	107$_{11}$	92$_{15}$	77$_{18}$	316	6446
p-Anisidine		<1000	123$_{50}$	108$_{100}$	95$_{3}$	80$_{44}$	65$_{8}$	267	7638
p-Anisidine AC	PME UME	1630	165$_{59}$	123$_{74}$	108$_{100}$	95$_{10}$	80$_{20}$	314	5046
p-Anisidine formyl artifact		1080	135$_{84}$	120$_{100}$	92$_{33}$	77$_{8}$	65$_{11}$	275	7639
p-Anisidine HFB	U+UHYHFB	1400	319$_{100}$	304$_{6}$	300$_{6}$	150$_{6}$	122$_{78}$	1023	6620
p-Anisidine TFA	U+UHYTFA	1335	219$_{100}$	204$_{16}$	149$_{11}$	122$_{76}$	109$_{19}$	463	6615
p-Anisidine TMS		<1000	195$_{57}$	180$_{100}$	164$_{9}$	147$_{8}$	73$_{46}$	387	7640
Antazoline		2350	265$_{1}$	182$_{6}$	91$_{22}$	84$_{100}$		689	62
Antazoline +H2O AC	UHYAC	2650	325$_{4}$	196$_{30}$	182$_{15}$	91$_{100}$	77$_{8}$	1064	2068
Antazoline AC		2610	307$_{4}$	274$_{12}$	182$_{44}$	91$_{100}$	84$_{82}$	951	2053
Antazoline artifact AC	UHYAC	2260	255$_{3}$	196$_{14}$	104$_{7}$	91$_{100}$	77$_{14}$	632	2067
Antazoline HY	UHY	1930	183$_{25}$	106$_{16}$	91$_{100}$	77$_{24}$	65$_{25}$	355	2065
Antazoline HYAC	UHYAC	2080	225$_{22}$	183$_{23}$	106$_{10}$	91$_{100}$	77$_{24}$	491	2066
Antazoline TMS		2450	322$_{1}$	180$_{5}$	155$_{100}$	91$_{43}$	73$_{53}$	1140	5459
Antazoline-M (HO-) AC	UHYAC	2620	323$_{1}$	254$_{6}$	212$_{6}$	91$_{68}$	84$_{100}$	1052	2069
Antazoline-M (HO-) HY	UHY	1920	199$_{53}$	163$_{21}$	91$_{100}$	76$_{24}$	65$_{34}$	398	2143
Antazoline-M (HO-) HY2AC	UHYAC	2340	283$_{18}$	241$_{16}$	199$_{34}$	91$_{100}$	65$_{22}$	795	2072
Antazoline-M (HO-) HYAC	UHYAC	2300	241$_{12}$	199$_{68}$	108$_{30}$	91$_{100}$	65$_{29}$	560	2071
Antazoline-M (HO-methoxy-) HY2AC	UHYAC	2460	313$_{10}$	271$_{8}$	254$_{10}$	212$_{18}$	91$_{100}$	988	2074
Antazoline-M (HO-methoxy-) HYAC	UHYAC	2370	271$_{12}$	212$_{17}$	120$_{8}$	91$_{100}$	65$_{18}$	724	2073
Antazoline-M (methoxy-) HYAC	UHYAC	2290	255$_{22}$	213$_{64}$	136$_{22}$	122$_{52}$	91$_{100}$	633	2070
Anthracene		1760*	178$_{100}$	176$_{20}$	152$_{7}$	89$_{6}$	76$_{6}$	337	2562
Anthranilic acid ME		1290	151$_{60}$	119$_{100}$	92$_{68}$	65$_{21}$		291	4939
Anthraquinone		2090*	208$_{93}$	180$_{100}$	152$_{79}$	126$_{10}$	76$_{47}$	422	4048
ANTU 3ME		2090	244$_{26}$	197$_{100}$	182$_{38}$	154$_{50}$	127$_{40}$	571	3972
4-APB-NBOMe		2375	164$_{78}$	131$_{46}$	121$_{100}$	91$_{57}$	77$_{19}$	879	10368
4-APB-NBOMe AC		2660	337$_{1}$	164$_{17}$	131$_{50}$	121$_{100}$	91$_{55}$	1139	10370
4-APB-NBOMe HFB		2415	360$_{2}$	332$_{5}$	158$_{67}$	131$_{82}$	121$_{100}$	1649	10374
4-APB-NBOMe ME		2405	178$_{69}$	131$_{30}$	121$_{100}$	91$_{53}$	77$_{15}$	965	10369
4-APB-NBOMe PFP		2410	310$_{2}$	158$_{40}$	131$_{42}$	121$_{100}$	91$_{36}$	1561	10373
4-APB-NBOMe TFA		2450	260$_{9}$	158$_{68}$	131$_{84}$	121$_{100}$	91$_{71}$	1406	10372

Table 8.1: Compounds in order of names   4-APB-NBOMe TMS

Name	Detected	RI	Typical ions and intensities					Page	Entry
4-APB-NBOMe TMS		2495	236_{70}	164_{7}	131_{38}	121_{100}	91_{55}	1299	10371
5-APB		1450	175_{9}	160_{5}	131_{100}	102_{25}	77_{77}	330	8951
5-APB		1450	175_{1}	131_{13}	102_{3}	77_{10}	44_{100}	330	9083
5-APB 2ME		1570	202_{1}	188_{1}	131_{6}	77_{4}	72_{100}	407	8945
5-APB AC		1870	217_{3}	158_{100}	131_{67}	86_{59}	77_{51}	454	8950
5-APB formyl artifact		1505	187_{6}	172_{8}	131_{41}	77_{19}	56_{100}	363	9084
5-APB HFB		1670	371_{1}	240_{15}	158_{57}	131_{100}	77_{28}	1316	9088
5-APB ME		1550	188_{1}	174_{1}	131_{14}	77_{14}	58_{100}	368	8943
5-APB PFP		1640	321_{2}	190_{16}	158_{46}	131_{100}	77_{31}	1038	9087
5-APB TFA		1655	271_{2}	158_{40}	131_{100}	77_{33}	69_{40}	722	9086
5-APB TMS		1650	232_{6}	131_{23}	116_{100}	100_{11}	73_{57}	586	9085
5-APB-M (di-HO-) 2AC		2240	291_{3}	232_{2}	190_{6}	163_{11}	75_{100}	847	9229
5-APB-M (di-HO-) AC		2050	249_{2}	190_{100}	163_{42}	131_{20}	86_{13}	595	9226
5-APB-M (HO-) isomer-1 2AC		2070	275_{4}	216_{68}	174_{71}	147_{100}	86_{95}	746	9227
5-APB-M (HO-) isomer-2 2AC		2120	275_{5}	252_{31}	233_{22}	174_{100}	86_{95}	745	9225
5-APB-M (HO-deamino-dihydro-) isomer-1 2AC		1810*	276_{1}	216_{14}	189_{19}	174_{16}	147_{100}	754	9235
5-APB-M (HO-deamino-dihydro-) isomer-2 2AC		1830*	276_{1}	216_{11}	189_{22}	174_{13}	147_{100}	754	9236
5-APB-M (ring cleavage-carboxy-) ME2AC		2230	291_{6}	248_{44}	206_{87}	174_{50}	86_{100}	949	9233
5-APB-M (ring cleavage-di-HO-) 4AC		2400	379_{1}	320_{22}	278_{54}	218_{92}	86_{100}	1351	9228
5-APB-M (ring cleavage-HO-) 3AC		2300	321_{3}	262_{60}	220_{90}	160_{100}	86_{85}	1039	9230
6-APB		1585	175_{3}	131_{25}	102_{5}	77_{14}	44_{100}	330	8704
6-APB		1585	175_{14}	131_{100}	102_{21}	77_{56}	63_{14}	330	8705
6-APB 2ME		1630	201_{1}	131_{16}	102_{2}	77_{6}	72_{100}	407	9094
6-APB AC		1890	217_{8}	158_{100}	131_{69}	86_{42}	77_{42}	455	9089
6-APB formyl artifact		1585	187_{17}	172_{8}	131_{56}	77_{15}	56_{100}	363	8706
6-APB HFB		1600	371_{1}	240_{15}	158_{62}	131_{100}	77_{26}	1316	9093
6-APB ME		1540	174_{1}	131_{23}	102_{7}	77_{17}	58_{100}	368	9206
6-APB PFP		1610	321_{2}	190_{15}	158_{54}	131_{100}	77_{31}	1038	9092
6-APB TFA		1670	271_{3}	158_{42}	131_{100}	77_{34}	69_{42}	723	9091
6-APB TMS		1655	232_{5}	131_{22}	116_{100}	100_{9}	73_{60}	586	9090
6-APB-M (HO-) 2AC		2065	276_{1}	216_{100}	174_{35}	147_{95}	86_{99}	754	9220
6-APB-M (HO-deamino-dihydro-) 2AC		1830*	276_{1}	216_{19}	189_{15}	174_{26}	147_{100}	753	9217
6-APB-M (ring cleavage-carboxy-) ME2AC		2200	307_{2}	248_{66}	206_{53}	174_{35}	86_{100}	949	9219
6-APB-M (ring cleavage-di-HO-) -H2O 3AC		2400	319_{25}	260_{11}	218_{28}	176_{18}	86_{100}	1026	9224
6-APB-M (ring cleavage-HO-) 3AC		2270	321_{1}	262_{12}	202_{46}	160_{85}	86_{100}	1039	9218
5-API		1765	174_{5}	131_{100}	103_{10}	77_{21}	63_{6}	329	9095
5-API		1765	174_{5}	131_{85}	103_{9}	77_{16}	44_{100}	329	9096
5-API 2AC		2345	199_{100}	157_{31}	130_{54}	103_{20}	86_{29}	647	9099
5-API 2HFB		1880	353_{18}	326_{32}	240_{87}	129_{100}		1702	9107
5-API 2ME		1850	202_{1}	130_{7}	115_{2}	72_{100}		404	9102
5-API 2PFP		2230	466_{4}	303_{59}	276_{100}	190_{51}	184_{41}	1609	9106
5-API 2TFA		2255	366_{1}	297_{5}	253_{75}	226_{100}	129_{73}	1289	9105
5-API 2TMS		1950	303_{1}	202_{8}	116_{100}	100_{9}	73_{67}	1021	9103
5-API AC		2145	216_{4}	157_{100}	130_{88}	103_{14}	77_{25}	453	9098
5-API formyl artifact		1745	186_{16}	130_{100}	103_{7}	77_{13}	56_{41}	361	9097
5-API formyl artifact ME		1780	200_{14}	144_{100}	129_{4}	115_{4}	56_{11}	400	9100
5-API HFB		1970	370_{1}	169_{4}	157_{18}	130_{100}	104_{1}	1310	9108
5-API ME		1810	151_{8}	130_{13}	98_{8}	70_{15}	58_{100}	366	9101
5-API TMS		1645	246_{2}	203_{100}	188_{19}	130_{9}	73_{55}	582	9104
5-API-M (HO-) 2AC		2360	274_{2}	215_{80}	173_{100}	146_{42}	86_{49}	740	9213
5-API-M (HO-) AC		2340	232_{4}	173_{100}	146_{26}	118_{16}	86_{16}	517	9212
6-API		1765	174_{5}	131_{100}	103_{10}	77_{21}	63_{5}	329	9109
6-API		1765	174_{3}	131_{75}	103_{9}	77_{16}	44_{100}	329	9110
6-API 2HFB		2245	566_{1}	353_{10}	326_{19}	184_{71}	129_{100}	1702	9116
6-API 2PFP		2220	466_{1}	347_{7}	303_{92}	276_{100}	184_{28}	1609	9115
6-API 2TFA		2285	366_{8}	297_{6}	253_{56}	226_{100}	102_{35}	1290	9114
6-API AC		2140	216_{7}	157_{100}	130_{86}	103_{13}	86_{8}	453	9112
6-API formyl artifact		1740	186_{19}	130_{100}	103_{8}	77_{17}	56_{52}	361	9111
6-API HFB		1870	370_{2}	240_{3}	157_{20}	130_{100}	103_{5}	1310	9117
6-API TFA		1885	270_{13}	157_{14}	130_{100}	103_{9}	77_{19}	717	9113
6-API-M (di-HO-) 3AC		2440	332_{6}	273_{100}	231_{87}	189_{62}	86_{72}	1106	9216
6-API-M (HO-) 2AC		2310	274_{2}	215_{38}	173_{100}	146_{29}	86_{40}	740	9215
6-API-M (HO-) AC		2320	232_{10}	173_{100}	146_{25}	117_{6}	86_{28}	518	9214
APINACA		3025	365_{32}	337_{34}	294_{45}	215_{100}	145_{84}	1288	9720
A-PINACA		3025	365_{32}	337_{34}	294_{45}	215_{100}	145_{84}	1288	9720
Apomorphine		2715	267_{64}	266_{100}	224_{21}	220_{16}	152_{11}	700	3988
Apomorphine 2AC	U+UHYAC	2830	351_{100}	308_{67}	266_{78}	224_{12}	165_{9}	1213	2286
Apomorphine 2TMS		2715	411_{67}	410_{100}	368_{15}	322_{45}	73_{47}	1484	4525
Aprindine	G U UHY UHYAC	2460	322_{3}	249_{8}	206_{13}	113_{74}	86_{100}	1047	1378
Aprindine-M (4-aminophenol)	UHY	1240	109_{100}	80_{41}	53_{82}	53_{82}	52_{90}	260	826
Aprindine-M (4-aminophenol) 2AC	PAC U+UHYAC	1765	193_{10}	151_{53}	109_{100}	80_{24}		377	188
Aprindine-M (aniline) AC	G	1380	135_{35}	93_{100}				274	222
Aprindine-M (deethyl-HO-) 2AC	UHYAC	3220	394_{25}	280_{60}	190_{53}	117_{100}	58_{66}	1422	2889
Aprindine-M (deindane) AC	UHYAC	1880	248_{2}	219_{7}	176_{8}	86_{100}	72_{19}	593	2881

Aprindine-M (deindane-HO-) 2AC

Table 8.1: Compounds in order of names

Name	Detected	RI	Typical ions and intensities	Page	Entry
Aprindine-M (deindane-HO-) 2AC	UHYAC	2205	306_1 277_5 219_9 86_{100} 58_{29}	945	2883
Aprindine-M (dephenyl-) AC	UHYAC	2300	288_4 216_5 117_9 86_{100} 72_{19}	831	2884
Aprindine-M (dephenyl-HO-) 2AC	UHYAC	2680	346_2 187_{35} 128_{15} 116_{43} 86_{100}	1190	2886
Aprindine-M (HO-) AC	UHYAC	2850	380_4 307_6 264_5 113_{50} 86_{100}	1358	2887
Aprindine-M (HO-methoxy-) AC	UHYAC	2995	410_1 206_{26} 162_{45} 113_{47} 86_{100}	1479	2888
Aprindine-M (N-dealkyl-)	UHY UHYAC	1920	209_{48} 166 104_{100} 94_{18} 77_{33}	430	2882
Aprindine-M (N-dealkyl-HO-) 2AC	UHYAC	2410	267_{36} 225_{100} 120_{74} 115_{29} 91_{21}	700	2885
Aprobarbital	P G U+UHYAC	1610	210_6 195_{18} 167_{100} 124_{43}	433	63
Aprobarbital 2ME		1520	238_2 220_7 195_{100} 138_{50} 111_{26}	549	1145
Aprobarbital 2TMS		1620	354_3 339_{81} 297_{40} 100_{47} 73_{100}	1233	5458
Aprobarbital-M (HO-)	U	1800	226_4 183_{100} 154_{62} 97_{49} 69_{76}	494	2960
Arabinose 4AC		1760*	259_8 170_{61} 128_{100} 115_{73} 103_{50}	1017	1963
Arabinose 4HFB		1235*	478_{12} 465_8 293_{64} 265_{11} 169_{100}	1734	5799
Arabinose 4PFP		1310*	411_4 378_{13} 243_{33} 219_{43} 119_{100}	1731	5798
Arabinose 4TFA		1290*	311_{10} 278_5 265_{11} 169_{63} 69_{100}	1687	5797
Arachidonic acid-M (15-HETE) -H2O ME		2360*	316_4 189_{37} 119_{54} 105_{100} 91_{92}	1008	4355
Arachidonic acid-M (15-HETE) METFA		2390*	430_1 316_6 131_{43} 117_{66} 91_{100}	1535	4354
Aramite		2400*	334_{23} 319_{37} 185_{100} 135_{30} 63_{83}	1118	4049
Aramite -C2H3ClSO2		1650*	208_{14} 193_{45} 135_{100} 107_{19} 91_{10}	426	4050
Arecaidine		1325	141_{39} 96_{100} 81_{12} 68_{11} 53_{17}	280	5938
Arecaidine ME		<1000	155_{55} 140_{100} 124_{33} 96_{90} 81_{64}	298	5870
Arecaidine TMS		1460	213_{32} 198_{48} 155_{78} 96_{100}	443	5939
Arecoline		<1000	155_{55} 140_{100} 124_{33} 96_{90} 81_{64}	298	5870
Arecoline-M/artifact (HOOC-)		1325	141_{39} 96_{100} 81_{12} 68_{11} 53_{17}	280	5938
Arecoline-M/artifact (HOOC-) TMS		1460	213_{32} 198_{48} 155_{78} 96_{100}	443	5939
Aripiprazole		3400	447_1 285_6 243_{100} 84_{26}	1575	7261
Aripiprazole-M (N-dealkyl-) AC	U+UHYAC	2255	272_{12} 229_9 200_{100} 188_{35} 56_{46}	728	7123
Aripiprazole-M (N-dealkyl-HO-) 2AC	U+UHYAC	2555	330_{23} 288_{43} 216_{100} 203_{89} 56_{44}	1092	7884
Aripiprazole-M/artifact	U+UHYAC	2720	330_7 287_{45} 244_{100} 216_{41} 203_{34}	1091	10294
Aripiprazole-M/artifact	U+UHYAC	2400	273_{21} 247_8 230_{100} 195_6 151_{18}	731	10296
Artemether		2030*	298_1 267_7 209_6 165_{12} 138_{100}	898	9444
Articaine		2170	284_2 171_{12} 139_4 86_{100} 56_4	803	2342
Articaine AC	U+UHYAC	2455	295_1 171_2 156_{40} 128_{33} 86_{100}	1068	4442
Articaine artifact	U	2230	296_{28} 201_{100} 06_{66} 84_{58} 56_{79}	885	4443
Articaine -CO2 AC	UHYAC	2250	268_2 222_5 156_{50} 128_{42} 86_{100}	707	4444
Articaine-M (HO-) 2AC	UHYAC	2470	369_1 229_2 156_{27} 128_{26} 86_{100}	1376	4445
Artifact of roasted food (cyclo (Phe-Pro)) AC		2360	286_{29} 153_{62} 125_{100} 91_{57} 70_{73}	815	5217
Artifact of roasted food (cyclo (Phe-Pro)) isomer-1	U+UHYAC P-I	2335	244_{36} 153_{50} 125_{100} 91_{38} 70_{50}	571	4495
Artifact of roasted food (cyclo (Phe-Pro)) isomer-2	U+UHYAC P-I	2365	244_{54} 153_{21} 125_{100} 91_{41} 70_{37}	571	8443
Ascorbic acid	U	2120*	176_7 116_{100} 85_{25}		331	64
Ascorbic acid 2AC		2065*	260_2 242_{18} 200_{100} 158_{42} 85_{44}	655	3307
Ascorbic acid 2ME		1700*	144_{100} 129_{29} 117_9 101_{14}	408	2634
Ascorbic acid isomer-1 3ME		1600*	218_6 200_4 144_{100} 129_{28} 101_{12}	458	2635
Ascorbic acid isomer-2 3ME		1720*	218_{11} 158_{100} 130_{77} 115_{21} 101_{16}	457	2636
Astemizole	G	3900	458_5 337_{88} 294_{13} 109_{18} 96_{100}	1598	1774
Astemizole-M (N-dealkyl-) 2AC	UHYAC	3170	408_1 366_{21} 268_{22} 242_{52} 109_{100}	1472	4505
Astemizole-M (N-dealkyl-) AC	UHYAC	3150	366_{14} 268_{17} 242_{50} 109_{100} 82_{32}	1292	4506
Astemizole-M/artifact (N-dealkyl-)		2470	241_{60} 132_{14} 109_{100} 83_{11}		560	1775
Astemizole-M/artifact (N-dealkyl-) AC	U+UHYAC	2490	283_{30} 241_{13} 240_{17} 109_{100} 83_{11}	795	1776
Asulam -C2H2O2	G P UHY	2185	172_{56} 156_{55} 108_{50} 92_{74} 65_{100}	326	973
Asulam -C2H2O2 2TMS		2210	316_{19} 301_{35} 222_5 163_8 73_{100}	1005	10331
Asulam -C2H2O2 3TMS		2125	388_5 373_{28} 210_{16} 180_{12} 147_{100}	1394	10330
Asulam -COOCH3 4ME		2095	228_{44} 184_{30} 136_{100} 120_{70} 77_{29}	501	4098
Atazanavir artifact-1		1370	144_{63} 132_{11} 115_{46} 100_{100}		283	7932
Atazanavir artifact-2 (formylphenylpyridine)		1700	183_{100} 154_{69} 127_{33} 101_7 77_{18}	354	7930
Atazanavir artifact-3		1870	229_7 198_{100} 182_{23} 154_{27} 127_{12}	504	7931
Atenolol	G P-I U	2380	251_2 222_5 107_6 72_{100}		697	1721
Atenolol 2TMS		2250	410_1 395_{11} 294_{11} 188_{10} 72_{100}	1479	5471
Atenolol 3TMS (amide/amide/HO-)		2220	467_3 295_{46} 188_{81} 73_{100} 72_{58}	1637	5474
Atenolol 3TMS (amide/amine/HO-)		2460	467_5 295_{27} 144_{100} 101_7 73_{90}	1638	5473
Atenolol 4TMS		2430	539_8 277_4 188_{19} 144_{100} 73_{99}	1697	5472
Atenolol artifact (formyl-HOOC-) ME	U	2175	293_{36} 278_{100} 127_{75} 112_{74} 56_{66}	864	2682
Atenolol artifact (HOOC-) ME		2140	281_1 267 237_5 107_{11} 72_{100}	789	2681
Atenolol formyl artifact	G P U	2400	278_9 263_{38} 127_{67} 86_{82} 56_{100}	770	65
Atenolol -H2O	U	2150	248_{17} 218_2 190_3 100_{100} 56_{76}	591	2680
Atenolol -H2O AC	U+UHYAC	2975	290_{26} 205_{86} 188_{38} 140_{100} 98_{52}	843	1349
Atenolol TMSTFA		2600	434_1 377_4 332_{15} 284_{100} 73_{65}	1545	6037
Atomoxetine		2000	255_{12} 148_{33} 104_{48} 91_{39} 77_{100}	634	7192
Atomoxetine		2000	255_1 151_2 148_5 77_{13} 44_{100}	634	7247
Atomoxetine AC		2310	297_1 190_{65} 117_{34} 86_{100} 77_{16}	891	7236
Atomoxetine -H2O HYAC	UHYAC-I	1680	189_6 146_{30} 115_{56} 98_{100} 70_8	368	4339
Atomoxetine -H2O HYHFB		1470	343_9 252_{100} 174_{63} 146_5 115_{25}	1172	7240
Atomoxetine -H2O HYPFP		1450	293_{17} 202_{87} 174_{100} 117_{51} 115_{90}	860	7242
Atomoxetine -H2O HYTMS		1580	219_{27} 204_{44} 161_{30} 103_{49} 75_{100}	466	7246

Table 8.1: Compounds in order of names   **Atomoxetine HFB**

Name	Detected	RI	Typical ions and intensities					Page	Entry
Atomoxetine HFB		2190	451_1	344_{49}	240_{100}	169_{33}	117_{84}	1582	7239
Atomoxetine HY2AC	U+UHYAC	1890	249_2	206_{100}	146_{36}	98_{78}	86_{34}	597	4340
Atomoxetine HY2HFB		1490	557_1	434_4	360_7	343_{30}	241_{100}	1698	7241
Atomoxetine HY2PFP		1430	457_2	334_4	310_{16}	239_{28}	190_{100}	1595	7243
Atomoxetine HY2TFA		1435	357_4	243_{20}	174_{32}	140_{100}	117_{42}	1246	7244
Atomoxetine ME		1950	269_2	163_2	115_3	77_{13}	58_{100}	713	7193
Atomoxetine PFP		2250	294_{48}	190_{100}	117_{44}			1449	7238
Atomoxetine TFA		2000	351_1	244_{44}	140_{100}	117_{78}	77_{19}	1212	7237
Atomoxetine TMS		2055	327_7	208_5	116_{100}	104_5	73_{44}	1077	7245
Atomoxetine-D6 -H2O HYPFP		1420	463_3	334_9	298_{41}	190_{100}	119_{87}	1605	7791
Atomoxetine-M (nor-) -H2O HYAC		1700	175_{35}	132_{100}	116_{59}	105_{15}	84_{64}	331	7880
Atomoxetine-M (nor-) -H2O HYTFA		1290	229_{26}	160_8	138_9	132_{13}	116_{100}	505	7878
Atomoxetine-M (nor-) -H2O HYTMS		1290	205_{100}	190_{33}	161_{24}	100_{46}	73_{85}	413	7879
Atomoxetine-M (nor-) HY2AC		1870	192_{100}	133_{67}	84_{41}	72_{42}		533	5342
Atomoxetine-M (nor-) HY2PFP		1400	443_8	296_{40}	280_7	239_{100}	177_{14}	1567	7711
Atomoxetine-M (nor-HO-) -H2O HYAC		2080	233_8	191_{23}	148_{47}	132_{100}	84_{59}	522	8069
Atracurium-M (N-demethyl-O-tri-demethyl-)/artifact 3AC	U+UHYAC	3370	427_{50}	385_9	354_{55}	312_{100}	137_{76}	1526	6789
Atracurium-M (O-tri-demethyl-)/artifact 3AC	U+UHYAC	3020	262_{40}	234_{100}	220_{59}	192_{78}	178_{42}	1562	6788
Atracurium-M (O-tri-demethyl-)/artifact AC	U+UHYAC	2595	357_1	315_4	234_{88}	192_{100}	177_{16}	1248	6787
Atracurium-M (tri-demethyl-)/artifact isomer-1 2AC	U+UHYAC	2950	399_{58}	326_9	295_{40}	282_{28}	151_{100}	1441	7857
Atracurium-M (tri-demethyl-)/artifact isomer-2 2AC	U+UHYAC	3210	399_{55}	326_{45}	313_{19}	295_{24}	151_{100}	1441	6790
Atracurium-M/artifact	P U+UHYAC	2575	357_1	206_{100}	190_{23}	162_8	151_7	1249	6106
Atrazine	P G	1720	215_{44}	200_{75}	173_{27}	68_{68}	58_{100}	449	66
Atrazine-M (deethyl-)	U	1680	187_{22}	172_{72}	70_{77}	58_{100}		363	68
Atrazine-M (deethyl-dechloro-methoxy-)	U	1670	183_{49}	168_{95}	141_{54}	70_{79}	58_{100}	355	67
Atrazine-M (deisopropyl-)	U	1730	173_{100}	158_{97}	145_{77}	130_{18}	68_{78}	328	4236
Atropine	P G U	2215	289_9	272_1	140_5	124_{100}	94_6	837	69
Atropine AC	U+UHYAC	2275	331_4	140_8	124_{100}	94_{22}	82_{34}	1102	71
Atropine -CH2O	U+UHYAC	1980	259_{25}	221_4	140_6	124_{100}	91_{31}	652	2343
Atropine -H2O	P G UHY U+UHYAC	2085	271_{33}	140_8	124_{100}	96_{21}	82_{13}	726	70
Atropine HFB		2060	485_1	271_{23}	140_{10}	124_{100}	96_{37}	1641	8125
Atropine PFP		2050	435_1	271_{26}	140_{14}	124_{100}	96_{39}	1548	8124
Atropine TFA		2070	385_1	271_{27}	140_{15}	124_{100}	96_{38}	1381	8123
Atropine TMS		2295	361_5	140_5	124_{100}	82_{20}	73_{50}	1270	4526
Atropine-M/artifact (HOOC-) -H2O ME		1510*	162_{100}	150_{38}	118_{48}	103_{38}	77_{18}	306	3196
Atropine-M/artifact (tropine) AC		1240	183_{25}	140_{11}	124_{100}	94_{42}	82_{63}	355	5125
Atropine-M/artifact (tropine) TFA		1020	237_{19}	124_{100}	94_{34}	82_{44}	67_{25}	543	7914
Azamethiphos artifact		1655	184_{100}	143_{12}	129_{12}	101_{26}	64_{21}	356	4038
Azaperone		2650	327_6	309_{10}	233_{23}	165_{22}	107_{100}	1076	6098
Azaperone enol TMS		2655	399_1	176_{100}	147_{22}	121_{36}	107_{22}	1443	6277
Azaperone-M (dihydro-)		2730	329_7	235_{10}	165_{16}	121_{29}	107_{100}	1089	6115
Azaperone-M (dihydro-) AC		2775	371_3	311_3	222_{17}	121_{25}	107_{100}	1320	6116
Azaperone-M (dihydro-) -H2O		2625	311_3	176_{100}	147_{62}	121_{69}	107_{38}	978	6117
Azapropazone		2610	300_{75}	189_{44}	160_{100}	145_{36}		908	1955
Azatadine	U UHY UHYAC	2375	290_{85}	246_{100}	232_{69}	96_{15}	70_{25}	843	1379
Azatadine-M (di-HO-aryl-) 2AC	UHYAC	2620	406_4	346_{13}	304_{15}	287_{100}	230_{13}	1465	2105
Azatadine-M (HO-alkyl-) AC	UHYAC	2520	348_{55}	305_{55}	288_{100}	244_{62}	230_{53}	1200	2103
Azatadine-M (HO-alkyl-) -H2O	UHYAC	2410	288_{66}	244_{100}	230_{46}	216_{79}	70_{24}	829	2102
Azatadine-M (HO-alkyl-HO-aryl-) 2AC	UHYAC	2640	406_{100}	363_{75}	347_{27}	304_{70}	287_{40}	1465	2106
Azatadine-M (HO-aryl-) AC	UHYAC	2540	348_{100}	305_{76}	262_{72}	244_{74}	230_{54}	1200	2104
Azatadine-M (nor-) AC	UHYAC	2720	318_{100}	258_{49}	246_{52}	232_{62}	217_{19}	1020	2107
Azatadine-M (nor-HO-alkyl-) 2AC	UHYAC	2810	376_{55}	316_{90}	256_{52}	244_{100}	230_{67}	1339	2109
Azatadine-M (nor-HO-alkyl-) -H2O AC	UHYAC	2750	316_{100}	256_{37}	244_{45}	230_{47}	217_{16}	1006	2108
Azelastine		3180	381_6	271_{24}	256_{22}	130_{31}	110_{100}	1361	4626
Azidocilline-M/artifact ME2AC		1930	314_{35}	230_{38}	198_{100}	156_{43}	97_{77}	993	7652
Azidocilline-M/artifact ME2TFA		1755	422_{28}	326_{100}	267_{36}	196_{54}	165_{32}	1512	7656
Azidocilline-M/artifact MEAC		1980	272_{40}	230_{91}	215_{38}	100_{53}	97_{100}	729	7651
Azidocilline-M/artifact MEPFP		1750	376_{100}	317_{65}	246_{70}	243_{62}	215_{53}	1338	7657
Azinphos-ethyl		2570	345_1	186_{10}	160_{77}	132_{100}	77_{56}	1181	1380
Azinphos-methyl	G P-I U+UHYAC	2460	160_{58}	132_{79}	93_{56}	77_{100}		1009	1412
Aziprotryne		1765	225_{100}	182_{67}	139_{85}	115_{42}	68_{75}	490	3506
Azosemide-M (N-dealkyl-) -SO2NH ME		1960	209_{64}	180_{44}	152_{100}	138_{83}	102_{55}	427	4279
Azosemide-M (thiophenecarboxylic acid)		<1000*	128_5	127_{55}	111_{100}	83_7		270	4282
Azosemide-M (thiophenecarboxylic acid) glycine conjugate ME		1720	199_{10}	167_5	117_{100}	83_7		396	4281
Azosemide-M (thiophenylmethanol)		<1000*	114_{100}	97_{62}	85_{78}	81_{24}		262	4280
Baclofen -H2O	U P-I	1990	195_{48}	138_{100}	103_{26}	77_{14}	63_7	385	4456
Baclofen -H2O AC	UMEAC	1975	237_5	195_5	138_{100}	103_{23}	77_{23}	542	4458
Baclofen ME	PME UME P-I	1715	196_{30}	138_{100}	103_{19}	77_{10}	63_8	498	4457
Bambuterol		2930	367_2	352_1	282_2	86_{89}	72_{100}	1300	7546
Bambuterol AC		2900	409_4	394_1	334_5	86_{83}	72_{100}	1475	7548
Bambuterol -C2H8 AC		2760	335_8	293_1	221	72_{100}	55_{16}	1124	7549
Bambuterol formyl artifact		2930	379_1	364_{14}	334_{12}	99_{65}	72_{100}	1353	7547
Bambuterol HY2AC		2500	380_2	305_7	211_2	86_{100}	72_{63}	1357	7550
Bambuterol HY3AC		2200	351_1	336_1	276_{15}	234_4	86_{100}	1214	7551

Bambuterol TFA Table 8.1: Compounds in order of names

Name	Detected	RI	Typical ions and intensities					Page	Entry
Bambuterol TFA		2395	389 $_4$	267 $_{38}$	212 $_2$	153 $_4$	72 $_{100}$	1605	7553
Bambuterol TMS		2600	439 $_2$	354 $_7$	282 $_4$	86 $_{73}$	72 $_{100}$	1557	7554
Bamethan 3AC		2330	335 $_1$	275 $_{66}$	233 $_{100}$	191 $_{82}$	148 $_{60}$	1125	1402
Bamethan 3TMS		1865	410 $_3$	267 $_4$	158 $_{100}$	116 $_{17}$	73 $_{90}$	1521	5483
Bamethan formyl artifact		2020	148 $_7$	120 $_7$	107 $_{11}$	98 $_{84}$	57 $_{100}$	475	4654
Bamethan -H2O 2AC	U+UHYAC	2310	275 $_{66}$	233 $_{100}$	191 $_{82}$	148 $_{60}$	98 $_{52}$	747	1385
Bamipine	G P U	2250	280 $_4$	182 $_{13}$	97 $_{100}$	91 $_{61}$	70 $_{39}$	784	28
Bamipine-M (HO-)	UHY	2580	296 $_{46}$	198 $_{40}$	97 $_{100}$	91 $_{75}$	70 $_{70}$	887	2139
Bamipine-M (HO-) AC	UHYAC	2620	338 $_{20}$	240 $_{16}$	97 $_{100}$	91 $_{44}$	70 $_{55}$	1145	2138
Bamipine-M (N-dealkyl-)	UHY	1930	183 $_{25}$	106 $_{16}$	91 $_{100}$	77 $_{24}$	65 $_{25}$	355	2065
Bamipine-M (N-dealkyl-) AC	UHYAC	2080	225 $_{22}$	183 $_{23}$	106 $_{10}$	91 $_{100}$	77 $_{24}$	491	2066
Bamipine-M (N-dealkyl-HO-)	UHY	1920	199 $_{53}$	163 $_{21}$	91 $_{100}$	76 $_{24}$	65 $_{34}$	398	2143
Bamipine-M (N-dealkyl-HO-) 2AC	UHYAC	2340	283 $_{18}$	241 $_{16}$	199 $_{34}$	91 $_{100}$	65 $_{22}$	795	2072
Bamipine-M (nor-) AC	UHYAC	2675	308 $_{20}$	182 $_{43}$	91 $_{100}$	77 $_{33}$		960	2141
Bamipine-M (nor-HO-) 2AC	U+UHYAC	3020	366 $_{34}$	324 $_7$	240 $_{26}$	199 $_{34}$	91 $_{100}$	1293	2142
Barban ME		2335	271 $_{25}$	256 $_{100}$	152 $_{11}$	111 $_{26}$	75 $_{18}$	721	4091
Barban-M/artifact (chloroaniline) 2ME		1180	155 $_{62}$	154 $_{100}$	140 $_{11}$	118 $_{16}$	75 $_{17}$	297	4090
Barban-M/artifact (chloroaniline) AC	U+UHYAC	1580	169 $_{31}$	127 $_{100}$	111 $_2$	99 $_8$		322	6593
Barban-M/artifact (chloroaniline) HFB		1310	323 $_{60}$	304 $_8$	154 $_{100}$	126 $_{68}$	111 $_{59}$	1047	6607
Barban-M/artifact (chloroaniline) ME		1100	141 $_{74}$	140 $_{100}$	111 $_9$	105 $_{11}$	77 $_{31}$	280	4089
Barban-M/artifact (chloroaniline) TFA		1125	223 $_{99}$	154 $_{100}$	126 $_{51}$	111 $_{55}$	69 $_{42}$	481	4124
Barban-M/artifact (HOOC-) ME		1500	185 $_{100}$	153 $_{64}$	140 $_{87}$	99 $_{46}$	59 $_{69}$	358	4123
Barbital	P G U+UHYAC	1500	156 $_{100}$	141 $_{97}$	112 $_{20}$	98 $_{22}$	83 $_{12}$	357	72
Barbital 2ME		1420	184 $_{96}$	169 $_{100}$	126 $_{38}$	112 $_{25}$		440	74
Barbital ME	P G UHY UHYAC	1455	170 $_{100}$	155 $_{97}$	126 $_{12}$	112 $_{34}$		395	73
Barbituric acid 3ME		1645	170 $_{100}$	113 $_{12}$	98 $_{22}$	82 $_{46}$	55 $_{75}$	323	75
Barnidipine	G	4140	491 $_1$	315 $_2$	210 $_{17}$	159 $_{100}$	91 $_{50}$	1650	4507
BB-22		4000	384 $_1$	259 $_7$	240 $_{100}$	144 $_{55}$	116 $_{78}$	1377	9667
BB-22-M/artifact (8-hydroxyquinoline)		1550	145 $_{100}$	117 $_{83}$	90 $_{24}$	63 $_{16}$		284	9574
BB-22-M/artifact (8-hydroxyquinoline) TMS		1570	217 $_2$	202 $_{100}$	172 $_{39}$	128 $_6$	77 $_6$	454	9650
BB-22-M/artifact (HOOC-) (ET)		2340	285 $_{42}$	240 $_{24}$	202 $_{100}$	174 $_{53}$	130 $_{60}$	811	9678
BB-22-M/artifact (HOOC-) (ME)		2320	271 $_{24}$	240 $_7$	188 $_{100}$	144 $_{21}$	130 $_{35}$	725	9677
BB-22-M/artifact (HOOC-) TMS		4000	329 $_9$	270 $_{19}$	246 $_{33}$	129 $_{42}$	55 $_{100}$	1089	9668
2,3-BDB		1550	193 $_1$	164 $_2$	135 $_2$	77 $_6$	58 $_{100}$	380	5414
2,3-BDB AC		1895	235 $_7$	176 $_{43}$	135 $_{10}$	100 $_{21}$	58 $_{100}$	533	5504
2,3-BDB formyl artifact		1575	205 $_9$	176 $_3$	135 $_9$	77 $_9$	70 $_{100}$	413	5415
2,3-BDB HFB		1660	389 $_{23}$	345 $_8$	254 $_{59}$	176 $_{100}$	135 $_{33}$	1397	5505
2,3-BDB PFP		1615	339 $_1$	204 $_7$	176 $_{43}$	135 $_{100}$	119 $_{14}$	1147	5544
2,3-BDB TFA		1705	289 $_{30}$	176 $_{100}$	154 $_{74}$	135 $_{62}$	77 $_{24}$	833	5506
2,3-BDB TMS		1670	250 $_{11}$	236 $_{14}$	135 $_{23}$	130 $_{100}$	73 $_{61}$	689	5603
BDB		1570	193 $_1$	164 $_2$	136 $_{12}$	77 $_6$	58 $_{100}$	378	3253
BDB AC	UHYAC	1950	235 $_3$	176 $_{23}$	162 $_{14}$	100 $_{13}$	58 $_{100}$	533	3262
BDB formyl artifact		1585	205 $_8$	176 $_3$	135 $_{25}$	77 $_{16}$	70 $_{100}$	413	3246
BDB HFB		1690	389 $_4$	254 $_6$	176 $_{42}$	135 $_{100}$	77 $_{11}$	1398	5288
BDB intermediate-1 (1-(1,3-benzodioxol-5-yl)-butan-1-ol)		1560*	194 $_{17}$	151 $_{100}$	123 $_{21}$	93 $_{72}$	65 $_{37}$	383	3290
BDB intermediate-1 AC		1670*	236 $_{21}$	193 $_{10}$	151 $_{100}$	135 $_{30}$	93 $_{16}$	538	3294
BDB intermediate-2		1385*	176 $_{59}$	131 $_{100}$	103 $_{80}$	77 $_{41}$	63 $_{26}$	332	3291
BDB intermediate-3 (1-(1,3-benzodioxol-5-yl)-butan-2-one)		1525*	192 $_{20}$	135 $_{100}$	105 $_6$	77 $_{22}$	57 $_{21}$	375	3292
BDB PFP		1700	339 $_4$	204 $_7$	176 $_{43}$	135 $_{100}$	119 $_{14}$	1147	5287
BDB precursor (piperonal)		1160*	150 $_{80}$	149 $_{100}$	121 $_{34}$	91 $_{12}$	63 $_{49}$	289	3275
BDB TFA		1705	289 $_4$	176 $_{33}$	154 $_{11}$	135 $_{100}$	77 $_{12}$	833	5286
BDB TMS		1650	250 $_4$	236 $_3$	135 $_{36}$	130 $_{100}$	73 $_{64}$	689	8375
BDB-M (demethylenyl-) 2AC	UHYAC	2205	265 $_1$	206 $_{11}$	164 $_{29}$	100 $_{23}$	58 $_{100}$	688	8253
BDB-M (demethylenyl-) 3AC	UHYAC	2235	307 $_1$	248 $_{16}$	164 $_{38}$	100 $_{53}$	58 $_{100}$	950	5551
BDB-M (demethylenyl-methyl-) 2AC	UHYAC	2140	279 $_4$	220 $_{32}$	178 $_{74}$	100 $_{24}$	58 $_{100}$	775	5550
BDB-M (demethylenyl-methyl-) 2HFB		1695	587 $_1$	374 $_{40}$	333 $_{15}$	254 $_{89}$	69 $_{100}$	1710	8475
BDB-M (demethylenyl-methyl-) 2PFP		1600	487 $_1$	324 $_{39}$	283 $_{13}$	204 $_{100}$	176 $_{20}$	1644	8473
BDB-M (demethylenyl-methyl-) 2TFA		1590	387 $_2$	274 $_{40}$	233 $_{10}$	154 $_{100}$	126 $_{23}$	1389	8471
BDB-M (demethylenyl-methyl-) 2TMS		1800	324 $_2$	209 $_3$	179 $_7$	130 $_{100}$	73 $_{89}$	1153	8469
BDB-M (demethylenyl-methyl-) HFB		1675	391 $_1$	178 $_{28}$	163 $_4$	137 $_{100}$	122 $_{10}$	1406	8476
BDB-M (demethylenyl-methyl-) PFP		1630	341 $_2$	204 $_3$	178 $_{24}$	137 $_{100}$	122 $_{12}$	1160	8474
BDB-M (demethylenyl-methyl-) TFA		1640	291 $_4$	178 $_{23}$	137 $_{100}$	122 $_{12}$	94 $_{13}$	847	8472
BDB-M (demethylenyl-methyl-) TMS		1655	252 $_1$	210 $_{10}$	180 $_4$	75 $_8$	58 $_{100}$	703	8470
BDMPEA		1785	259 $_{15}$	230 $_{100}$	215 $_{29}$	199 $_{10}$	77 $_{37}$	569	3254
BDMPEA 2AC		2230	343 $_7$	242 $_{100}$	229 $_{31}$	201 $_{12}$	148 $_{29}$	1171	6924
BDMPEA 2TMS		2195	403 $_1$	388 $_{14}$	272 $_7$	207 $_{11}$	174 $_{100}$	1455	6926
BDMPEA AC		2180	301 $_{15}$	242 $_{100}$	229 $_{31}$	199 $_{12}$	148 $_{39}$	910	3267
BDMPEA formyl artifact		1840	271 $_{18}$	240 $_{100}$	229 $_{41}$	199 $_{17}$	77 $_{37}$	722	3245
BDMPEA formyl artifact		1840	271 $_{12}$	240 $_{65}$	229 $_{27}$	199 $_{11}$	42 $_{100}$	722	5522
BDMPEA HFB		2030	455 $_{32}$	242 $_{100}$	229 $_{41}$	199 $_{29}$	148 $_{33}$	1591	6941
BDMPEA intermediate-1 (2,5-dimethoxyphenyl-2-nitroethene)		1900	209 $_{100}$	162 $_{51}$	147 $_{52}$	133 $_{61}$	77 $_{62}$	427	3286
BDMPEA intermediate-2 (2,5-dimethoxyphenethylamine)		1630	181 $_{15}$	162 $_{33}$	152 $_{100}$	137 $_{47}$	121 $_{27}$	350	3287
BDMPEA intermediate-2 (2,5-dimethoxyphenethylamine)		1630	181 $_{15}$	162 $_{33}$	152 $_{100}$	137 $_{47}$	44 $_{95}$	350	5523
BDMPEA intermediate-2 (2,5-dimethoxyphenethylamine) 2AC		1935	265 $_{11}$	164 $_{100}$	149 $_{24}$	121 $_{25}$	91 $_{13}$	688	9162

Table 8.1: Compounds in order of names

Name	Detected	RI	Typical ions and intensities					Page	Entry
BDMPEA intermediate-2 (2,5-dimethoxyphenethylamine) AC		1935	223_{15}	164_{100}	149_{29}	121_{31}	91_{15}	484	3288
BDMPEA intermediate-2 (2,5-dimethoxyphenethylamine) formyl artifact		1540	193_{12}	162_{100}	151_{30}	121_{43}	91_{29}	378	3293
BDMPEA intermediate-2 (2,5-dimethoxyphenethylamine) formyl artifact		1540	193_{12}	162_{100}	151_{30}	121_{43}	42_{64}	379	5524
BDMPEA PFP		1995	405_{36}	242_{100}	229_{84}	199_{33}	148_{35}	1461	6936
BDMPEA precursor (2,5-dimethoxybenzaldehyde)		1345*	166_{100}	151_{39}	120_{36}	95_{61}	63_{62}	316	3278
BDMPEA TFA		2000	355_{42}	242_{100}	229_{84}	199_{37}	148_{36}	1235	6931
BDMPEA TMS		1935	331_{1}	272_{3}	229_{2}	102_{100}	73_{50}	1098	6925
BDMPEA-M (deamino-di-HO-) 2AC	U+UHYAC	2230*	360_{13}	300_{27}	258_{43}	245_{100}	138_{17}	1262	7214
BDMPEA-M (deamino-di-HO-) 2TFA		1790*	468_{100}	354_{14}	341_{77}	311_{14}	276_{76}	1613	7208
BDMPEA-M (deamino-HO-) AC	U+UHYAC	2300*	302_{20}	242_{100}	227_{25}	183_{33}	148_{42}	916	7198
BDMPEA-M (deamino-HO-) TFA		1880*	356_{49}	341_{15}	242_{100}	229_{72}	148_{71}	1240	7209
BDMPEA-M (deamino-HOOC-) ME		2030*	288_{100}	273_{12}	241_{8}	229_{67}	199_{19}	824	7212
BDMPEA-M (deamino-oxo-)		2020*	258_{35}	229_{100}	215_{9}	199_{35}	186_{8}	644	7215
BDMPEA-M (O-demethyl- N-acetyl-) iso-1 AC	U+UHYAC	2410	329_{4}	287_{33}	228_{100}	215_{15}	165_{5}	1085	7196
BDMPEA-M (O-demethyl- N-acetyl-) iso-2 AC	U+UHYAC	2440	329_{8}	287_{21}	228_{100}	215_{17}	72_{13}	1085	7197
BDMPEA-M (O-demethyl- N-acetyl-) isomer-1 TFA		2090	383_{8}	324_{100}	255_{7}	148_{21}	72_{23}	1368	7204
BDMPEA-M (O-demethyl- N-acetyl-) isomer-2 TFA		2130	383_{7}	324_{100}	311_{11}	227_{11}	72_{17}	1368	7205
BDMPEA-M (O-demethyl-) isomer-1 2AC	U+UHYAC	2410	329_{4}	287_{33}	228_{100}	215_{15}	165_{5}	1085	7196
BDMPEA-M (O-demethyl-) isomer-1 2TFA		1900	437_{12}	324_{100}	311_{10}	255_{7}	148_{32}	1550	7206
BDMPEA-M (O-demethyl-) isomer-2 2AC	U+UHYAC	2440	329_{8}	287_{21}	228_{100}	215_{17}	72_{13}	1085	7197
BDMPEA-M (O-demethyl-) isomer-2 2TFA		1950	437_{6}	324_{100}	311_{32}	253_{3}	227_{15}	1550	7207
BDMPEA-M (O-demethyl-deamino-di-HO-) 3AC	U+UHYAC	2280*	388_{6}	346_{43}	286_{55}	244_{67}	231_{100}	1393	7201
BDMPEA-M (O-demethyl-deamino-HO-) 2TFA		1800*	438_{31}	341_{16}	324_{100}	311_{35}	227_{55}	1553	7210
BDMPEA-M (O-demethyl-deamino-HO-) iso-1 2AC	U+UHYAC	2160*	330_{5}	288_{17}	270_{6}	228_{100}	213_{12}	1092	7199
BDMPEA-M (O-demethyl-deamino-HO-) iso-2 2AC	U+UHYAC	2180*	330_{4}	288_{15}	246_{10}	228_{100}	213_{14}	1092	7200
BDMPEA-M (O-demethyl-deamino-HOOC-) -H2O	U+UHYAC	1980*	242_{80}	214_{100}	186_{37}			562	7203
BDMPEA-M (O-demethyl-deamino-HOOC-) MEAC		2120*	316_{14}	274_{100}	242_{92}	214_{87}	186_{18}	1003	7213
BDMPEA-M (O-demethyl-deamino-HOOC-) METFA		1890*	370_{100}	311_{46}	257_{55}	241_{60}	148_{21}	1309	7211
BDMPEA-M (O-demethyl-deamino-HO-oxo-) 2AC	U+UHYAC	2160*	344_{11}	302_{73}	260_{46}	242_{100}	214_{25}	1177	7202
Beclamide	U	1720	197_{20}	162_{15}	148_{10}	106_{80}	91_{100}	391	76
Beclamide artifact (-HCl)		1680	161_{90}	117_{38}	106_{46}	91_{53}	55_{100}	303	104
Beclamide artifact (-HCl) TMS		1160	233_{28}	232_{30}	218_{22}	91_{100}	73_{64}	522	5469
Beclamide TMS		1690	269_{3}	254_{7}	234_{37}	91_{100}	73_{47}	711	5468
Beclobrate		2430*	346_{22}	273_{18}	218_{83}	183_{100}	125_{35}	1189	2247
Bedaquiline artifact-1		1825*	182_{78}	155_{100}				352	9450
Bedaquiline artifact-2		2580	329_{100}	327_{95}	312_{51}	236_{62}	216_{89}	1072	9448
Bedaquiline artifact-3		2670	343_{100}	341_{97}	312_{46}	250_{86}	105_{81}	1159	9449
Befunolol		2610	291_{1}	276_{2}	247_{5}	161_{3}	72_{100}	848	2400
Befunolol formyl artifact		2630	303_{3}	288_{4}	247_{8}	161_{2}	72_{100}	924	2401
Befunolol -H2O AC		2730	315_{13}	230_{22}	140_{41}	98_{100}	56_{67}	1000	2427
Befunolol TMSTFA		2430	444_{1}	402_{3}	284_{100}	129_{47}	73_{84}	1600	6181
Behenic acid ME		2460*	354_{13}	311_{6}	143_{21}	87_{63}	74_{100}	1234	2669
Bemegride		1350	155_{18}	127_{24}	113_{26}	82_{37}	55_{100}	298	77
Bemetizide 2ME		3100	429_{70}	333_{36}	324_{100}	101_{2}		1529	2854
Bemetizide 3ME		3070	445_{36}	443_{100}	348_{17}	338_{75}	240_{36}	1567	2855
Bemetizide 4ME	UEXME	3700	457_{1}	352_{100}	244_{15}	145_{6}	105_{7}	1595	6845
Bemetizide -SO2NH ME		2800	336_{8}	240_{8}	231_{100}	105_{9}	77_{8}	1129	2853
Benactyzine		2270	327_{2}	239_{9}	182_{19}	105_{50}	86_{100}	1077	1391
Benactyzine TMS		2230	399_{1}	384_{2}	255_{58}	100_{64}	86_{100}	1443	6272
Benactyzine-M (HOOC-) ME		1840*	242_{2}	183_{100}	105_{72}	77_{64}		565	78
Benazepril 2ET	UET	3040	480_{2}	407_{100}	289_{10}	218_{42}	91_{30}	1635	4723
Benazepril 2ME	UME	3015	452_{1}	379_{100}	204_{60}	144_{37}	91_{70}	1585	4715
Benazepril ET	UET	3080	452_{4}	406_{9}	379_{100}	218_{23}	91_{16}	1585	4722
Benazepril isopropylester		3165	466_{3}	420_{6}	393_{100}	232_{32}	91_{44}	1611	4724
Benazepril ME	G PME UME	3030	438_{3}	392_{8}	365_{100}	204_{45}	91_{53}	1554	4714
Benazepril TMS		3070	496_{5}	423_{100}	262_{35}	91_{42}	73_{38}	1656	4973
Benazeprilate 2ET	UET	3080	452_{4}	406_{9}	379_{100}	218_{23}	91_{16}	1585	4722
Benazeprilate 2ME	UME	2975	424_{4}	392_{12}	365_{100}	204_{68}	91_{60}	1517	4716
Benazeprilate 2TMS	UTMS	3130	540_{1}	525_{3}	423_{100}	262_{22}	73_{42}	1689	4974
Benazeprilate 3ET	UET	3040	480_{2}	407_{100}	289_{10}	218_{42}	91_{30}	1635	4723
Benazeprilate 3ME	UME	2985	438_{2}	379_{100}	204_{39}	144_{12}	91_{16}	1554	4717
Benazeprilate-M (HO-) isomer-1 2ET	UET	3330	496_{4}	423_{100}	361_{18}	218_{22}	135_{33}	1656	4725
Benazeprilate-M (HO-) isomer-1 3ME	UME	3160	454_{3}	395_{100}	333_{38}	204_{60}	121_{65}	1589	4718
Benazeprilate-M (HO-) isomer-2 2ET	UET	3330	496_{8}	423_{89}	361_{25}	218_{39}	135_{100}	1656	4726
Benazeprilate-M (HO-) isomer-2 3ME	UME	3235	454_{7}	395_{58}	333_{33}	204_{68}	121_{100}	1589	4719
Benazeprilate-M (HO-) isomer-2 4ME	UME	3240	468_{5}	409_{100}	347_{13}	261_{20}	204_{68}	1614	4721
Benazepril-M (HO-) isomer-1 2ET	UET	3330	496_{4}	423_{100}	361_{18}	218_{22}	135_{33}	1656	4725
Benazepril-M (HO-) isomer-1 4ME	UME	3165	468_{4}	409_{100}	347_{8}	204_{65}	121_{25}	1614	4720
Benazepril-M (HO-) isomer-2 2ET	UET	3330	496_{8}	423_{89}	361_{25}	218_{39}	135_{100}	1656	4726
Benazepril-M/artifact (deethyl-) 2ET	UET	3080	452_{4}	406_{9}	379_{100}	218_{23}	91_{16}	1585	4722
Benazepril-M/artifact (deethyl-) 2ME	UME	2975	424_{4}	392_{12}	365_{100}	204_{68}	91_{60}	1517	4716
Benazepril-M/artifact (deethyl-) 2TMS	UTMS	3130	540_{1}	525_{3}	423_{100}	262_{22}	73_{42}	1689	4974
Benazepril-M/artifact (deethyl-) 3ET	UET	3040	480_{2}	407_{100}	289_{10}	218_{42}	91_{30}	1635	4723
Benazepril-M/artifact (deethyl-) 3ME	UME	2985	438_{2}	379_{100}	204_{39}	144_{12}	91_{16}	1554	4717

Benazepril-M/artifact (deethyl-HO-) isomer-1 3ET

Table 8.1: Compounds in order of names

Name	Detected	RI	Typical ions and intensities					Page	Entry
Benazepril-M/artifact (deethyl-HO-) isomer-1 3ET	U E T	3330	496_{4}	423_{100}	361_{18}	218_{22}	135_{33}	1656	4725
Benazepril-M/artifact (deethyl-HO-) isomer-1 3ME	U M E	3160	454_{4}	395_{100}	333_{38}	204_{60}	121_{65}	1589	4718
Benazepril-M/artifact (deethyl-HO-) isomer-1 4ME	U M E	3165	468_{4}	409_{100}	347_{8}	204_{65}	121_{25}	1614	4720
Benazepril-M/artifact (deethyl-HO-) isomer-2 3ET	U E T	3330	496_{8}	423_{89}	361_{25}	218_{39}	135_{100}	1656	4726
Benazepril-M/artifact (deethyl-HO-) isomer-2 3ME	U M E	3235	454_{7}	395_{58}	333_{33}	204_{68}	121_{100}	1589	4719
Benazepril-M/artifact (deethyl-HO-) isomer-2 4ME	U M E	3240	468_{5}	409_{100}	347_{13}	261_{20}	204_{68}	1614	4721
Benazolin		2055	243_{61}	198_{59}	170_{100}	134_{48}	108_{16}	566	3623
Benazolin ET		2045	271_{40}	198_{54}	170_{100}	134_{47}	108_{20}	721	3625
Benazolin ME		2000	257_{61}	198_{59}	170_{100}	134_{48}	108_{16}	640	3624
Benazolin-ethyl		2045	271_{40}	198_{54}	170_{100}	134_{47}	108_{20}	721	3625
Bencyclane	G U	2120	198_{6}	102_{34}	86_{44}	58_{100}		838	79
Bencyclane-M (bis-nor-) AC	U+UHYAC	2545	100_{100}	72_{24}				928	2306
Bencyclane-M (bis-nor-HO-) isomer-1 2AC	U+UHYAC	2670	114_{18}	100_{100}	91_{15}	72_{37}		1271	2307
Bencyclane-M (bis-nor-HO-) isomer-2 2AC	U+UHYAC	2700	114_{5}	100_{100}	91_{12}	72_{26}		1271	2308
Bencyclane-M (deamino-di-HO-) isomer-1 2AC	U+UHYAC	2640*	129_{2}	101_{100}	91_{9}	73_{7}		1275	2310
Bencyclane-M (deamino-di-HO-) isomer-2 2AC	U+UHYAC	2660*	129_{1}	101_{100}	91_{8}	73_{6}		1275	2311
Bencyclane-M (deamino-HO-) AC	U+UHYAC	2345*	128_{1}	101_{100}	91_{13}	73_{6}		933	2309
Bencyclane-M (deamino-HO-oxo-) isomer-1 2AC	U+UHYAC	2440*	115_{2}	101_{100}	91_{19}			1020	2312
Bencyclane-M (deamino-HO-oxo-) isomer-2 2AC	U+UHYAC	2560*	129_{1}	101_{100}	91_{10}			1020	2313
Bencyclane-M (HO-) isomer-1	U	2350	214_{16}	102_{69}	86_{56}	58_{100}		940	2297
Bencyclane-M (HO-) isomer-1 AC	U+UHYAC	2420	256_{12}	129_{1}	102_{62}	86_{57}	58_{100}	1197	2301
Bencyclane-M (HO-) isomer-2	P U	2370	214_{12}	185_{5}	102_{52}	86_{39}	58_{100}	940	80
Bencyclane-M (HO-) isomer-2 AC	U+UHYAC	2430	256_{4}	117_{2}	102_{51}	86_{48}	58_{100}	1197	2302
Bencyclane-M (HO-oxo-) -H2O HYAC	UHYAC	1920*	258_{13}	227_{58}	190_{81}	129_{54}	91_{100}	647	2318
Bencyclane-M (HO-oxo-) HY	UHY	2280*	234_{71}	190_{29}	147_{23}	107_{100}	77_{24}	528	2320
Bencyclane-M (HO-oxo-) HY2AC	UHYAC	2240*	318_{17}	276_{100}	229_{14}	187_{27}	107_{15}	1019	2317
Bencyclane-M (HO-oxo-) HYAC	UHYAC	2080*	276_{13}	234_{50}	206_{53}	127_{30}	107_{100}	755	2319
Bencyclane-M (nor-)	U	2130	198_{3}	184_{22}	91_{32}	88_{37}	72_{100}	751	2300
Bencyclane-M (nor-) AC	U+UHYAC	2570	130_{2}	114_{100}	91_{13}	86_{14}		1015	2303
Bencyclane-M (nor-HO-) isomer-1 2AC	U+UHYAC	2690	130_{2}	114_{100}	91_{14}	86_{27}		1337	2304
Bencyclane-M (nor-HO-) isomer-2 2AC	U+UHYAC	2730	130_{2}	114_{100}	91_{7}	86_{11}		1337	2305
Bencyclane-M (oxo-) isomer-1	U	2340	212_{6}	102_{40}	86_{21}	58_{100}		928	2298
Bencyclane-M (oxo-) isomer-1 HY	UHY	1380*	218_{34}	190_{6}	127_{42}	99_{30}	91_{100}	460	81
Bencyclane-M (oxo-) isomer-1 HYAC	UHYAC	1750*	260_{12}	200_{33}	171_{100}	109_{56}	91_{71}	658	83
Bencyclane-M (oxo-) isomer-2	U	2380	303_{1}	212_{20}	102_{63}	86_{38}	58_{100}	928	2299
Bencyclane-M (oxo-) isomer-2 HY	UHY	1415*	218_{57}	189_{10}	107_{100}	77_{14}		459	82
Bencyclane-M (oxo-) isomer-2 HYAC	U+UHYAC	1780*	260_{26}	218_{100}	189_{10}	107_{39}		657	2316
Bendiocarb		1640	223_{6}	166_{33}	151_{100}	126_{47}	58_{22}	482	3912
Bendiocarb -C2H3NO		1110*	166_{42}	151_{92}	126_{100}	108_{18}	80_{15}	316	3913
Bendiocarb -C2H3NO TFA		<1000*	262_{47}	247_{100}	205_{34}	125_{51}	79_{25}	668	4131
Bendiocarb TFA		1560	319_{52}	247_{100}	222_{19}	125_{40}	69_{88}	1023	3607
Bendroflumethiazide 3ME		3360	463_{16}	372_{100}	264_{9}	260_{6}	91_{10}	1605	3106
Bendroflumethiazide 4ME		3360	477_{1}	386_{100}	278_{20}	145_{3}	91_{7}	1630	6890
Benfluorex		2175	350_{1}	216_{6}	192_{100}	159_{24}	105_{65}	1213	4707
Benfluorex AC		2530	374_{1}	234_{57}	192_{100}	159_{22}	105_{51}	1414	4709
Benfluorex ME		2220	364_{1}	230_{8}	206_{100}	159_{33}	149_{82}	1286	4708
Benfluorex-M (-COOH) MEAC	UHYAC	1870	298_{1}	258_{2}	158_{38}	116_{100}	56_{37}	1011	4711
Benfluorex-M (deamino-oxo-HO-) enol 2AC		2150*	303_{1}	216_{10}	159_{40}	143_{29}	101_{100}	918	4712
Benfluorex-M (hippuric acid)	U	1745	179_{1}	161_{2}	135_{22}	105_{100}	77_{9}	340	96
Benfluorex-M (hippuric acid) 2TMS	UTMS	2070	323_{10}	308_{16}	280_{11}	206_{50}	105_{100}	1051	5812
Benfluorex-M (hippuric acid) ME	UME	1660	193_{5}	161_{7}	134_{19}	105_{100}	77_{45}	377	97
Benfluorex-M (hippuric acid) TMS	UTMS	1925	251_{1}	236_{8}	206_{71}	105_{100}	73_{85}	609	5813
Benfluorex-M (N-dealkyl-) AC	UHYAC	1510	245_{2}	226_{5}	186_{6}	159_{12}	86_{100}	574	782
Benfluorex-M/artifact (alcohol) 2AC	UHYAC	1890	312_{1}	250_{5}	172_{50}	130_{100}	87_{35}	1100	4710
Benfluorex-M/artifact (benzoic acid)	P U UHY	1235*	122_{77}	105_{100}	77_{72}			266	95
Benfluorex-M/artifact (benzoic acid) ME	P(ME)	1180*	136_{30}	105_{100}	77_{73}			275	1211
Benfluorex-M/artifact (benzoic acid) TBDMS		1295*	221_{1}	179_{100}	135_{30}	105_{55}	77_{51}	540	6247
Benomyl artifact (debutylcarbamoyl-) 2ME		1875	219_{37}	160_{100}	132_{23}	119_{16}	77_{13}	464	4078
Benomyl-M/artifact (aminobenzimidazole) 3ME		1715	175_{78}	160_{100}	146_{80}	131_{71}	119_{22}	331	4101
Benoxaprofen	P	2550	301_{52}	256_{100}	119_{31}	91_{47}	65_{25}	910	1458
Benoxaprofen ME		2485	315_{39}	256_{100}	119_{25}	91_{47}		999	1392
Benoxaprofen-M (HO-) ME	UME	2580	331_{8}	272_{93}	230_{100}	195_{6}	91_{12}	1098	6286
Benperidol	G U +UHYAC	3440	381_{2}	363_{50}	230_{100}	109_{88}	82_{65}	1362	84
Benperidol-M	U+UHYAC	1490*	180_{21}	125_{49}	123_{35}	95_{17}	56_{100}	345	85
Benperidol-M (N-dealkyl-)	UHY	2415	217_{10}	134_{100}	106_{42}	79_{87}		455	87
Benperidol-M (N-dealkyl-) 2AC	U+UHYAC	2750	301_{28}	259_{43}	134_{28}	125_{28}	82_{100}	913	88
Benperidol-M (N-dealkyl-) AC	UHYAC	2770	259_{60}	216_{15}	134_{64}	125_{42}	82_{100}	651	89
Benperidol-M (N-dealkyl-) ME	UHY	2290	231_{12}	134_{100}	106_{38}	79_{82}		514	86
Benproperine		2425	309_{2}	181_{2}	165_{3}	112_{100}	91_{5}	966	1749
Bentazone		2040	240_{6}	198_{61}	161_{31}	119_{100}	92_{43}	554	3626
Bentazone artifact		1675	178_{38}	120_{100}	92_{43}	65_{38}	58_{54}	338	3627
Bentazone ME		1910	254_{23}	212_{100}	175_{22}	133_{48}	105_{83}	626	3628
Benzaldehyde		<1000*	106_{76}	105_{75}	77_{100}	51_{48}		259	4215
Benzalkonium chloride compound-1 -C7H8Cl	G P U	1380	213_{2}	156	114	84_{1}	58_{100}	443	1057

Table 8.1: Compounds in order of names

Name	Detected	RI	Typical ions and intensities					Page	Entry
Benzalkonium chloride compound-1 -CH3Cl	G P U	1965	289_3	160_1	134_{100}	91_{50}	58_5	838	1059
Benzalkonium chloride compound-2 -C7H8Cl	G P U	1595	241_1	170	128	84	58_{100}	562	1058
Benzalkonium chloride compound-2 -CH3Cl	G P U	2150	317_1	253_3	206_3	134_{100}	91_{47}	1015	1060
Benzamide		1400	121_{83}	105_{100}	77_{98}			266	90
Benzarone	UHY	2405*	266_{100}	251_{57}	223_{15}	121_{66}	93_{27}	693	1978
Benzarone AC	UHYAC	2405*	308_{100}	266_{81}	249_{80}	224_{27}	121_{58}	956	1986
Benzarone-M (di-HO-) 3AC	UHYAC	2550*	424_{28}	294_{26}	267_{100}	223_8	101_8	1516	2644
Benzarone-M (di-HO-) -H2O 2AC	UHYAC	2840*	364_{100}	322_{100}	280_{91}	173_{54}	121_{31}	1281	2647
Benzarone-M (HO-) isomer-1 2AC	UHYAC	2650*	366_5	324_{100}	282_{82}	187_{95}	121_{25}	1290	2649
Benzarone-M (HO-) isomer-2 2AC	UHYAC	2680*	366_{33}	324_{17}	282_{100}	265_{37}	137_{22}	1290	2650
Benzarone-M (HO-) isomer-3 2AC	UHYAC	2730*	366_{76}	324_{100}	282_{98}	265_{67}	121_{52}	1290	2651
Benzarone-M (HO-) isomer-4 2AC	UHYAC	2790*	366_{41}	324_{100}	282_{69}	265_{34}	121_{45}	1291	2652
Benzarone-M (HO-ethyl-) -H2O AC	UHYAC	2440*	306_{49}	264_{100}	235_{30}	171_{14}	115_{21}	942	2643
Benzarone-M (HO-methoxy-) isomer-1 2AC	UHYAC	2710*	396_9	354_{93}	312_{66}	187_{100}	145_{17}	1429	2654
Benzarone-M (HO-methoxy-) isomer-2 2AC	UHYAC	2740*	396_{57}	354_{57}	312_{100}	197_4	120_{20}	1428	2655
Benzarone-M (HO-methoxy-) isomer-3 2AC	UHYAC	2910*	396_{47}	354_{64}	312_{100}	269_{25}	151_{19}	1428	2656
Benzarone-M (HO-methoxy-) isomer-4 2AC	UHYAC	2950*	396_{46}	354_{67}	312_{100}	187_{21}	151_{25}	1428	2657
Benzarone-M (methoxy-) AC	UHYAC	2570*	338_{84}	296_{100}	279_{55}	253_{54}	151_{57}	1142	2645
Benzarone-M (oxo-) AC	UHYAC	2620*	322_{20}	280_{100}	237_{54}	187_{29}	121_{35}	1043	2646
Benzatropine	G U	2315	307_4	167_{26}	140_{59}	124_{27}	83_{100}	952	91
Benzatropine HY	UHY	1645*	184_{45}	165_{14}	152_7	105_{100}	77_{63}	357	1333
Benzatropine HYAC	U+UHYAC	1700*	226_{20}	184_{20}	165_{100}	105_{14}	77_{35}	495	1241
Benzatropine HYHFB		1475*	380_3	183_4	166_{100}	152_{22}	83_{25}	1355	8146
Benzatropine HYME	UHY	1655*	198_{70}	167_{94}	121_{100}	105_{56}	77_{71}	395	6779
Benzatropine HYPFP		1410*	330_4	183_4	166_{100}	152_{24}	83_{30}	1093	8145
Benzatropine HYTFA		1420*	280_{19}	183_7	166_{100}	152_{24}	83_{21}	780	8144
Benzatropine HYTMS		1540*	256_{28}	241_{14}	179_{38}	167_{100}	152_{17}	638	8159
Benzbromarone	G U UHY	2750*	424_{100}	422_{51}	344_{32}	279_{32}	264_{58}	1512	1393
Benzbromarone AC	UHYAC	2820*	464_{17}	424_{100}	422_{51}	264_{28}	173_{15}	1607	2255
Benzbromarone ET		2760*	452_{100}	450_{54}	423_{19}	264_{71}	173_{46}	1580	2262
Benzbromarone ME		2730*	438_{100}	436_{52}	342_{14}	278_{57}	173_{21}	1550	2258
Benzbromarone-M (HO-aryl-) isomer-1 2AC	UHYAC	2950*	522_1	482_{100}	480_{55}	440_{57}	187_{57}	1678	2659
Benzbromarone-M (HO-aryl-) isomer-2 2AC	U+UHYAC	3080*	522_8	482_{66}	440_{100}	438_{53}	279_{32}	1678	2660
Benzbromarone-M (HO-ethyl-) -H2O AC	U+UHYAC	2850*	462_{17}	422_{100}	420_{51}	297_{65}	255_{45}	1604	2257
Benzbromarone-M (HO-methoxy-) 2AC	UHYAC-I	3120*	552_8	512_{62}	510_{31}	470_{100}	468_{54}	1695	2256
Benzbromarone-M (methoxy-) AC	UHYAC	3070*	494_5	454_{74}	452_{41}	372_{77}	284_{100}	1652	2661
Benzbromarone-M (oxo-) AC	U+UHYAC	2900*	478_3	438_{100}	436_{51}	395_{40}	187_{79}	1631	2261
Benzene		<1000*	78_{100}	63_3	51_{20}	50_{17}	39_{12}	253	1542
1,4-Benzenediamine	G	1280	108_{100}	91_3	80_{35}	53_{13}		260	5330
1,4-Benzenediamine 2AC	UHYAC	2690	192_{39}	150_{15}	108_{100}	80_{59}	52_{45}	375	5331
1,4-Benzenediamine 2HFB		1775	500_{57}	481_{16}	331_{12}	303_{100}	108_{51}	1659	5332
1,4-Benzenediamine 2ME		1060	136_{16}	122_{10}	108_{100}	93_6	80_{36}	276	5334
1,4-Benzenediamine 2PFP		1600	400_{65}	281_{17}	253_{82}	119_{28}	108_{100}	1444	5858
1,4-Benzenediamine 2TFA		1800	300_{100}	203_{59}	133_{16}	108_{54}	69_{30}	904	5397
1,4-Benzenediamine ME		1000	122_{38}	108_{100}	93_6	80_{40}		267	5333
Benzene-M (hydroquinone)	UHY	<1000*	110_{100}	81_{27}				261	814
Benzene-M (hydroquinone) 2AC	UHYAC	1395*	194_8	152_{26}	110_{100}			382	815
Benzene-M (hydroquinone) 2ME		<1000*	138_{56}	123_{100}	95_{54}	63_{22}		278	3282
Benzene-M (hydroxyhydroquinone)	UHY	1460*	126_{100}	109_{18}	79_{26}	53_9		269	3163
Benzene-M (hydroxyhydroquinone) 3AC	UHYAC	1710*	252_1	210_7	168_{46}	126_{100}	97_7	616	4336
Benzene-M (methoxyhydroquinone) 2AC	UHYAC	1450*	224_3	182_{23}	140_{100}	125_{71}	97_9	486	4337
Benzene-M (phenol)	UHY	<1000*	94_{100}	66_{41}				256	4219
Benzhydrol	UHY	1645*	184_{45}	165_{14}	152_7	105_{100}	77_{63}	357	1333
Benzhydrol AC	U+UHYAC	1700*	226_{20}	184_{20}	165_{100}	105_{14}	77_{35}	495	1241
Benzhydrol HFB		1475*	380_3	183_4	166_{100}	152_{22}	83_{25}	1355	8146
Benzhydrol ME	UHY	1655*	198_{70}	167_{94}	121_{100}	105_{56}	77_{71}	395	6779
Benzhydrol PFP		1410*	330_4	183_4	166_{100}	152_{24}	83_{30}	1093	8145
Benzhydrol TFA		1420*	280_{19}	183_7	166_{100}	152_{24}	83_{21}	780	8144
Benzhydrol TMS		1540*	256_{28}	241_{14}	179_{38}	167_{100}	152_{17}	638	8159
Benzil	U UHY UHYAC	1825*	210_3	105_{100}	77_{45}	51_{26}		432	1233
Benzilic acid ME		1840*	242_2	183_{100}	105_{72}	77_{64}		565	78
Benzil-M (HO-) AC	UHYAC	2160*	268_1	226_1	163_{31}	121_{100}	105_{30}	706	2546
Benzil-M (HO-) ME	UHYME	2290*	240_3	135_{100}	105_{18}	77_{10}		555	2545
Benzo[a]anthracene		2410*	228_{100}	164_{10}	131_{10}	114_{20}		501	3701
Benzo[a]pyrene		2775*	252_{100}	224_3	126_{17}	113_7		617	3703
Benzo[b]fluoranthene		2815*	252_{100}	224_3	126_{19}	113_{11}		617	3704
Benzo[g,h,i]perylene		3125*	276_{100}	138_{23}	124_4	100		753	3707
Benzo[k]fluoranthene		2750*	252_{100}	224_3	126_{19}	113_9		617	3702
Benzocaine	G	1820	165_{37}	137_{11}	120_{100}	92_{19}	65_{22}	313	1457
Benzocaine AC		1990	207_{62}	165_{66}	137_{26}	120_{100}	92_{18}	419	1440
Benzocaine TMS		1500	237_{72}	222_{100}	192_{50}	149_{61}	73_{83}	544	5486
Benzocaine-M (PABA) 2TMS		1645	281_{91}	236_{14}	148_{24}	73_{100}		788	5487
Benzocaine-M (PABA) AC	U+UHYAC	2145	179_{31}	137_{100}	120_{92}	92_{16}	65_{24}	340	3298
Benzocaine-M (PABA) ME		1550	151_{55}	120_{100}	92_{28}	65_{26}		291	23

Benzocaine-M (PABA) MEAC   **Table 8.1:** Compounds in order of names

Name	Detected	RI	Typical ions and intensities					Page	Entry
Benzocaine-M (PABA) MEAC		1985	193 $_{32}$	151 $_{60}$	120 $_{100}$	92 $_{18}$	65 $_{18}$	378	24
Benzoctamine	UHY	2070	249 $_2$	218 $_{100}$	203 $_{29}$	191 $_{68}$	178 $_{55}$	599	94
Benzoctamine AC	UHYAC	2540	291 $_{39}$	263 $_{100}$	218 $_{63}$	191 $_{77}$		850	1245
Benzoctamine TMS		2240	306 $_1$	218 $_7$	191 $_{14}$	116 $_{100}$	73 $_{70}$	1041	5460
Benzoctamine-M (deamino-di-HO-) 2AC	UHYAC	2470*	336 $_7$	266 $_{30}$	249 $_{100}$	191 $_{36}$		1130	1244
Benzoctamine-M (deamino-di-HO-methoxy-) 2AC	UHYAC	2685*	366 $_{27}$	324 $_{51}$	296 $_{100}$	249 $_{56}$	237 $_{52}$	1291	1246
Benzoctamine-M (deamino-HO-) AC	UHYAC	2145*	278 $_{29}$	250 $_{100}$	191 $_{86}$			768	1242
Benzoctamine-M (HO-) 2AC	UHYAC	2890	349 $_{23}$	321 $_{70}$	279 $_{100}$	207 $_{68}$		1205	1250
Benzoctamine-M (nor-) AC	UHYAC	2420	277 $_{12}$	249 $_{100}$	207 $_{17}$	191 $_{13}$	178 $_{13}$	761	1243
Benzoctamine-M (nor-HO-) isomer-1 2AC	UHYAC	2725	335 $_5$	293 $_{33}$	265 $_{100}$			1125	1247
Benzoctamine-M (nor-HO-) isomer-2 2AC	UHYAC	2790	335 $_{11}$	307 $_{29}$	265 $_{100}$	207 $_{16}$		1125	1248
Benzoctamine-M (nor-HO-methoxy-) 2AC	UHYAC	2875	365 $_{15}$	323 $_{24}$	295 $_{100}$			1286	1249
1-(1,3-Benzodioxol-6-yl)butane-2-yl-azane		1550	193 $_1$	164 $_2$	135 $_2$	77 $_5$	58 $_{100}$	380	5414
1-(1,3-Benzodioxol-6-yl)butane-2-yl-azane AC		1895	235 $_7$	176 $_{43}$	135 $_{10}$	100 $_{21}$	58 $_{100}$	533	5504
1-(1,3-Benzodioxol-6-yl)butane-2-yl-azane formyl artifact		1575	205 $_9$	176 $_3$	135 $_9$	77 $_9$	70 $_{100}$	413	5415
1-(1,3-Benzodioxol-6-yl)butane-2-yl-azane HFB		1660	389 $_{23}$	345 $_8$	254 $_{59}$	176 $_{100}$	135 $_{57}$	1397	5505
1-(1,3-Benzodioxol-6-yl)butane-2-yl-azane PFP		1615	339 $_4$	204 $_7$	176 $_{43}$	135 $_{100}$	119 $_{14}$	1147	5544
1-(1,3-Benzodioxol-6-yl)butane-2-yl-azane TFA		1705	289 $_{30}$	176 $_{100}$	154 $_{74}$	135 $_{62}$	77 $_{24}$	833	5506
1-(1,3-Benzodioxol-6-yl)butane-2-yl-dimethylazane		1660	192 $_3$	135 $_2$	96 $_4$	86 $_{100}$	71 $_6$	475	5418
1-(1,3-Benzodioxol-6-yl)butane-2-yl-ethylazane		1670	192 $_5$	135 $_5$	86 $_{100}$	77 $_5$	58 $_{11}$	476	5417
1-(1,3-Benzodioxol-6-yl)butane-2-yl-ethylazane AC		2000	263 $_3$	192 $_4$	176 $_{18}$	128 $_{42}$	86 $_{100}$	677	5511
1-(1,3-Benzodioxol-6-yl)butane-2-yl-ethylazane HFB		1790	417 $_1$	282 $_{100}$	176 $_{25}$	135 $_{24}$	77 $_{17}$	1499	5594
1-(1,3-Benzodioxol-6-yl)butane-2-yl-ethylazane PFP		1755	367 $_3$	232 $_{100}$	176 $_{16}$	119 $_{30}$	69 $_{10}$	1296	5595
1-(1,3-Benzodioxol-6-yl)butane-2-yl-ethylazane TFA		1780	317 $_5$	182 $_{100}$	176 $_{37}$	154 $_{16}$	135 $_{14}$	1011	5512
1-(1,3-Benzodioxol-6-yl)butane-2-yl-ethylazane TMS		1825	278 $_2$	264 $_5$	158 $_{100}$	135 $_{16}$	73 $_{48}$	866	5596
1-(1,3-Benzodioxol-6-yl)butane-2-yl-methylazane		1610	178 $_3$	135 $_3$	89 $_4$	72 $_{100}$	57 $_7$	420	5416
1-(1,3-Benzodioxol-6-yl)butane-2-yl-methylazane AC		1965	249 $_4$	176 $_{17}$	135 $_7$	114 $_{35}$	72 $_{100}$	596	5507
1-(1,3-Benzodioxol-6-yl)butane-2-yl-methylazane HFB		1735	403 $_3$	268 $_{100}$	210 $_{17}$	176 $_8$	135 $_5$	1455	5591
1-(1,3-Benzodioxol-6-yl)butane-2-yl-methylazane PFP		1710	353 $_8$	218 $_{100}$	176 $_{23}$	160 $_{14}$	135 $_{12}$	1225	5592
1-(1,3-Benzodioxol-6-yl)butane-2-yl-methylazane TFA		1725	303 $_4$	176 $_{43}$	168 $_{100}$	135 $_{14}$	110 $_{23}$	923	5508
1-(1,3-Benzodioxol-6-yl)butane-2-yl-methylazane TMS		1730	264 $_2$	250 $_5$	144 $_{100}$	135 $_{13}$	73 $_{54}$	778	5593
Benzoflavone		2810	272 $_{64}$	244 $_8$	170 $_{100}$	122 $_{12}$	114 $_{51}$	729	6460
Benzofluorene		2220*	216 $_{100}$	215 $_{70}$	213 $_{19}$	108 $_8$	95 $_6$	452	2568
Benzoic acid	P UHY	1235*	122 $_{77}$	105 $_{100}$	77 $_{72}$			266	95
Benzoic acid anhydride		1880*	226 $_7$	198 $_{32}$	105 $_{100}$	77 $_{92}$		493	1742
Benzoic acid butylester		1275*	178 $_2$	123 $_{70}$	105 $_{100}$	77 $_{37}$	56 $_{19}$	337	98
Benzoic acid ethylester		1225*	150 $_{20}$	122 $_{26}$	105 $_{100}$	77 $_{52}$		289	99
Benzoic acid glycine conjugate	U	1745	179 $_1$	161 $_2$	135 $_{22}$	105 $_{100}$	77 $_9$	340	96
Benzoic acid glycine conjugate 2TMS	UTMS	2070	323 $_{10}$	308 $_{16}$	280 $_{11}$	206 $_{50}$	105 $_{100}$	1051	5812
Benzoic acid glycine conjugate TMS	UTMS	1925	251 $_1$	236 $_8$	206 $_{71}$	105 $_{100}$	73 $_{85}$	609	5813
Benzoic acid methylester	P(ME)	1180*	136 $_{30}$	105 $_{100}$	77 $_{73}$			275	1211
Benzoic acid TBDMS		1295*	221 $_1$	179 $_{100}$	135 $_{30}$	105 $_{55}$	77 $_{51}$	540	6247
Benzoic acid-M (glycine conjugate ME)	UME	1660	193 $_5$	161 $_7$	134 $_{19}$	105 $_{100}$	77 $_{45}$	377	97
Benzophenone	U+UHYAC	1610*	182 $_{31}$	152 $_3$	105 $_{100}$	77 $_{70}$	51 $_{39}$	351	1624
Benzoresorcinol	UHY	2280*	214 $_{61}$	213 $_{83}$	137 $_{100}$	105 $_{21}$	77 $_{33}$	445	3660
Benzoresorcinol 2AC	UHYAC	2315*	298 $_3$	256 $_{45}$	213 $_{100}$	137 $_{21}$	77 $_{18}$	893	3661
Benzquinamide	U UHYAC	2980	404 $_{16}$	345 $_{20}$	244 $_{44}$	205 $_{100}$	100 $_{10}$	1460	1777
Benzquinamide artifact		2880	339 $_{100}$	325 $_{10}$	268 $_3$	224 $_2$		1147	1778
Benzquinamide HY	UHY	3000	362 $_{12}$	317 $_6$	218 $_{27}$	205 $_{100}$	100 $_{10}$	1275	2135
Benzquinamide-M (N-deethyl-)	UHY UHYAC	2960	376 $_{17}$	317 $_{70}$	244 $_{100}$	205 $_{75}$	176 $_{16}$	1340	2136
Benzquinamide-M (O-demethyl-)	UHYAC	2990	390 $_6$	303 $_{48}$	272 $_{73}$	230 $_{100}$	191 $_{67}$	1403	2137
Benzthiazuron 2ME		1985	235 $_8$	136 $_4$	109 $_4$	72 $_{100}$		531	3941
Benzydamine	U UHY UHYAC	2400	309 $_2$	225 $_9$	91 $_{24}$	85 $_{54}$	58 $_{100}$	965	1394
Benzydamine-M (deamino-HO-) AC	UHYAC	2450	324 $_1$	273 $_1$	162 $_1$	101 $_{100}$	91 $_{47}$	1057	4375
Benzydamine-M (HO-) AC	UHYAC	2670	367 $_3$	283 $_4$	265 $_7$	85 $_{82}$	58 $_{100}$	1299	4376
Benzydamine-M (nor-) AC	U+UHYAC	2780	337 $_1$	114 $_{100}$	91 $_{28}$	86 $_{12}$		1139	1875
Benzydamine-M (nor-HO-) 2AC	UHYAC	3220	395 $_1$	269 $_2$	158 $_2$	114 $_{100}$	91 $_{24}$	1426	4377
Benzydamine-M (O-dealkyl-) AC	UHYAC	2150	266 $_4$	224 $_{47}$	146 $_7$	117 $_6$	91 $_{100}$	693	4378
Benzylacetamide	U+UHYAC	1410	149 $_{62}$	106 $_{100}$	91 $_{33}$	79 $_{19}$	77 $_{17}$	287	5160
Benzylacetamide AC		1450	191 $_5$	148 $_{36}$	106 $_{100}$	91 $_{24}$	79 $_{17}$	372	5161
Benzylalcohol		<1000*	108 $_{74}$	107 $_{55}$	91 $_{18}$	79 $_{100}$	77 $_{64}$	260	4447
Benzylamine		<1000	107 $_{100}$	91 $_{50}$	79 $_{72}$	77 $_{39}$	65 $_{14}$	260	100
Benzylamine 2AC		1450	191 $_5$	148 $_{36}$	106 $_{100}$	91 $_{24}$	79 $_{17}$	372	5161
Benzylamine AC	U+UHYAC	1410	149 $_{62}$	106 $_{100}$	91 $_{33}$	79 $_{19}$	77 $_{17}$	287	5160
Benzylamine artifact		1730	195 $_{22}$	194 $_{26}$	117 $_9$	91 $_{100}$	65 $_{20}$	386	5159
Benzylamine HFB		1220	303 $_{56}$	184 $_6$	169 $_6$	134 $_{11}$	91 $_{100}$	921	6577
Benzylamine TFA		1155	203 $_{69}$	134 $_{35}$	104 $_{11}$	91 $_{100}$	69 $_{46}$	406	6572
Benzylbenzoate		1740*	212 $_{15}$	194 $_6$	105 $_{100}$	91 $_{50}$	77 $_{42}$	438	4450
Benzylbutanoate		1065*	178 $_{17}$	108 $_{63}$	91 $_{100}$	79 $_{14}$	71 $_{36}$	338	4448
Benzylbutylphthalate		2270*	312 $_1$	206 $_{24}$	149 $_{100}$	91 $_{85}$	65 $_{31}$	982	3540
Benzylether		1600*	107 $_{13}$	92 $_{100}$	91 $_{79}$	79 $_{18}$	65 $_{25}$	395	4449
N-Benzylethylenediamine 3TMS		2215	366 $_1$	351 $_4$	259 $_9$	192 $_{100}$	174 $_{29}$	1294	7635
N-Benzylidenebenzylamine		1730	195 $_{22}$	194 $_{26}$	117 $_9$	91 $_{100}$	65 $_{20}$	386	5159
Benzylnicotinate		1800	213 $_{48}$	168 $_3$	106 $_{93}$	91 $_{100}$		442	1400

Table 8.1: Compounds in order of names   5-Benzyloxy-N,N-diallyl-tryptamine

Name	Detected	RI	Typical ions and intensities					Page	Entry
5-Benzyloxy-N,N-diallyl-tryptamine		2920	346_1	145_3	130_3	110_{100}	91_{19}	1190	8849
5-Benzyloxy-N,N-diallyl-tryptamine HFB		2615	515_2	451_4	432_2	341_4	110_{100}	1690	10027
5-Benzyloxy-N,N-diallyl-tryptamine PFP		2640	382_3	291_2	144_5	110_{100}	91_{39}	1651	10195
5-Benzyloxy-N,N-diallyl-tryptamine TFA		2645	442_1	332_2	144_6	110_{100}	91_{51}	1566	10026
5-Benzyloxy-N,N-diallyl-tryptamine TMS		2880	418_1	377_5	308_{11}	202_{14}	110_{100}	1503	10025
5-Benzyloxy-N,N-diisopropyl-tryptamine		2920	350_1	250_2	145_9	114_{100}	91_{42}	1210	8871
5-Benzyloxy-N,N-diisopropyl-tryptamine AC		2540	202_4	160_9	146_{27}	114_{100}	72_{30}	1412	9526
5-Benzyloxy-N,N-diisopropyl-tryptamine HFB		2750	531_1	432_2	341_2	114_{100}	72_{29}	1693	9525
5-Benzyloxy-N,N-diisopropyl-tryptamine PFP		3120	408_6	317_{10}	114_{100}	91_{60}	72_{30}	1656	9524
5-Benzyloxy-N,N-diisopropyl-tryptamine TFA		2785	445_2	431_8	346_9	114_{100}	72_{56}	1573	9523
5-Benzyloxy-N,N-diisopropyl-tryptamine TMS		2800	422_2	322_6	308_{11}	202_{14}	114_{100}	1514	9531
5-Benzyloxy-N,N-dimethyl-tryptamine		2680	294_3	145_3	130_3	91_{19}	58_{100}	871	8870
5-Benzyloxy-N,N-dipropyl-tryptamine		2920	350_2	250_3	145_8	114_{100}	91_{43}	1210	8869
Benzylpenicilline artifact-1		1260	175_{21}	147_9	119_8	91_{100}	65_{15}	330	8355
Benzylpenicilline artifact-1 TMS		1125	231_2	216_{17}	172_{43}	75_{32}	73_{100}	513	8358
Benzylpenicilline artifact-2 (ME)		1590	207_{18}	118_{30}	92_{76}	91_{100}	88_{70}	420	8356
Benzylpenicilline artifact-2 TMS		1485	247_{36}	218_{41}	202_{17}	91_{83}	73_{100}	585	8357
2-Benzylphenol		1680*	184_{100}	165_{34}	106_{54}	78_{40}		357	1395
4-Benzylphenol		1720*	184_{100}	165_{19}	91_{18}	77		357	1396
Benzylpiperazine		1530	176_9	146_5	134_{55}	91_{100}	56_{29}	334	5880
Benzylpiperazine AC		1915	218_4	146_{27}	132_{17}	91_{100}	85_{20}	460	5881
Benzylpiperazine HFB		1730	372_{13}	295_6	281_{15}	175_{13}	91_{100}	1322	5884
Benzylpiperazine PFP		1690	322_{18}	245_9	231_{21}	175_{15}	91_{100}	1044	5883
Benzylpiperazine TFA		1665	272_3	195_4	181_{10}	146_9	91_{100}	730	5882
Benzylpiperazine TMS		1860	248_{31}	157_{27}	102_{100}	91_{69}	73_{97}	592	5885
Benzylpiperazine-M (benzylamine)		<1000	107_{100}	91_{50}	79_{72}	77_{39}	65_{14}	260	100
Benzylpiperazine-M (benzylamine) 2AC		1450	191_5	148_{36}	106_{100}	91_{24}	79_{17}	372	5161
Benzylpiperazine-M (benzylamine) AC	U+UHYAC	1410	149_{62}	106_{100}	91_{33}	79_{19}	77_{17}	287	5160
Benzylpiperazine-M (benzylamine) HFB		1220	303_{56}	184_6	169_6	134_{11}	91_{100}	921	6577
Benzylpiperazine-M (benzylamine) TFA		1155	203_{69}	134_{35}	104_{11}	91_{100}	69_{46}	406	6572
Benzylpiperazine-M (deethylene-) 2AC	U+UHYAC	2080	234_1	191_{49}	175_{17}	120_{83}	91_{100}	529	6507
Benzylpiperazine-M (deethylene-) 2HFB	U+UHYHFB	1705	542_1	524_1	345_{79}	226_6	91_{100}	1690	6576
Benzylpiperazine-M (deethylene-) 2PFP	U+UHYTFA	1875	311_9	295_{20}	190_3	119_9	91_{100}	1564	7636
Benzylpiperazine-M (deethylene-) 2TFA	U+UHYTFA	1670	342_2	324_{20}	245_{42}	126_{11}	91_{100}	1167	6571
Benzylpiperazine-M (deethylene-) 3AC	U+UHYAC	2125	276_1	233_{44}	175_9	120_{100}	91_{76}	755	6513
Benzylpiperazine-M (deethylene-) HFB		1870	302_5	295_{61}	190_8	119_{12}	91_{100}	1187	7637
Benzylpiperazine-M (HO-) isomer-1 2AC	U+UHYAC	2245	276_9	204_{43}	149_{73}	107_{100}	85_{36}	755	6506
Benzylpiperazine-M (HO-) isomer-1 2HFB	U+UHYHFB	1930	584_{10}	387_{31}	358_{20}	303_{100}	169_{95}	1709	6574
Benzylpiperazine-M (HO-) isomer-1 2TFA	U+UHYTFA	1830	384_{30}	287_{23}	203_{100}	181_{37}	69_{95}	1374	6569
Benzylpiperazine-M (HO-) isomer-2 2AC	U+UHYAC	2290	276_3	204_{16}	149_{20}	107_{100}	85_{57}	755	6505
Benzylpiperazine-M (HO-) isomer-2 2HFB	U+UHYHFB	1970	584_9	387_{42}	358_{25}	303_{100}	281_{36}	1709	6573
Benzylpiperazine-M (HO-) isomer-2 2TFA	U+UHYTFA	1870	384_{100}	287_{47}	258_{29}	203_{92}	181_{28}	1374	6568
Benzylpiperazine-M (HO-methoxy-) 2AC	U+UHYAC	2380	306_2	234_9	179_{13}	137_{100}	85_{64}	944	6508
Benzylpiperazine-M (HO-methoxy-) AC	U+UHYAC	2410	264_9	192_8	137_{100}	122_{18}	85_{42}	682	6509
Benzylpiperazine-M (HO-methoxy-) HFB	U+UHYHFB	2135	418_{22}	295_{15}	281_{27}	138_{87}	137_{100}	1502	6575
Benzylpiperazine-M (HO-methoxy-) TFA	U+UHYTFA	2120	318_8	181_{10}	137_{100}	122_{21}	69_{12}	1018	6570
Benzylpiperazine-M (piperazine) 2AC		1750	170_9	85_{33}	69_{25}	56_{100}		324	879
Benzylpiperazine-M (piperazine) 2HFB		1290	478_3	459_9	309_{100}	281_{22}	252_{41}	1631	6634
Benzylpiperazine-M (piperazine) 2TFA		1005	278_{10}	209_{59}	152_{25}	69_{56}	56_{100}	766	4129
Betahistine AC		1575	178_{13}	135_{40}	106_{100}	93_{45}	86_{36}	338	5173
Betahistine impurity/artifact-1 AC		1700	192_{13}	149_{58}	120_{100}	107_{48}	86_{25}	374	5174
Betahistine impurity/artifact-2 AC		1755	206_{11}	163_{56}	134_{100}	121_{45}	86_{28}	415	5175
beta-keto-2,5-Dimethoxy-4-bromophenethylamine		1720	258_{47}	243_{100}	228_7	200_9	185_9	732	10202
beta-keto-2,5-Dimethoxy-4-bromophenethylamine AC		2320	315_{18}	243_{100}	228_5	200_9	185_8	998	10203
beta-keto-2,5-Dimethoxy-4-bromophenethylamine enol 2AC		2320	357_{17}	315_{12}	243_{100}	228_5	200_8	1245	10204
beta-keto-2,5-Dimethoxy-4-bromophenethylamine enol 2TMS		2330	417_1	402_8	287_{14}	174_{100}	86_{36}	1498	10206
beta-keto-2,5-Dimethoxy-4-bromophenethylamine HFB		2095	469_7	243_{100}	226_{12}	200_{12}	157_8	1615	10209
beta-keto-2,5-Dimethoxy-4-bromophenethylamine PFP		2075	419_{13}	243_{100}	228_8	200_{13}	185_{12}	1504	10208
beta-keto-2,5-Dimethoxy-4-bromophenethylamine TFA		2090	369_{20}	243_{100}	228_9	200_{13}	185_{11}	1304	10207
beta-keto-2,5-Dimethoxy-4-bromophenethylamine TMS		2035	345_{84}	315_{25}	287_{12}	240_{100}	73_{89}	1182	10205
beta-keto-2,5-Dimethoxy-4-iodophenethylamine		2105	321_{13}	291_{100}	248_{15}	233_{12}	194_{22}	1036	10189
beta-keto-2,5-Dimethoxy-4-iodophenethylamine 2TMS		2440	450_4	335_4	291_5	174_{100}	147_5	1608	9874
beta-keto-2,5-Dimethoxy-4-iodophenethylamine AC		2400	363_{21}	317_2	291_{100}	248_5	233_{11}	1276	9814
beta-keto-2,5-Dimethoxy-4-iodophenethylamine HFB		2190	517_{13}	291_{100}	276_7	248_{10}	233_7	1674	9817
beta-keto-2,5-Dimethoxy-4-iodophenethylamine PFP		2180	467_{25}	291_{100}	276_8	248_{17}	233_{13}	1611	9816
beta-keto-2,5-Dimethoxy-4-iodophenethylamine TFA		2180	417_{25}	332_2	291_{100}	248_{17}	233_{16}	1497	9815
beta-keto-2,5-Dimethoxy-4-iodophenethylamine TMS		2295	393_5	349_5	291_{19}	127_{20}	102_{100}	1413	9875
Beta-keto-EBDB		1860	206_3	149_{20}	121_{15}	91_6	86_{100}	533	9149
Beta-keto-EBDB AC		2200	277_1	192_4	149_{12}	128_{58}	86_{100}	760	9150
Beta-keto-EBDB HFB		1920	431_1	282_{50}	254_8	149_{100}	121_{21}	1537	9153
Beta-keto-EBDB PFP		1890	381_1	232_{44}	204_6	149_{100}	121_{11}	1359	9152
Beta-keto-EBDB TFA		1950	331_2	182_{57}	154_{10}	149_{100}	121_{19}	1099	9151
Beta-keto-MBDB		1740	192_2	162_4	149_{21}	121_{19}	72_{100}	473	8319
Beta-keto-MBDB AC	U+UHYAC	2215	263_{10}	149_{27}	121_{15}	114_{84}	72_{100}	674	7872

Beta-keto-MBDB HFB **Table 8.1:** Compounds in order of names

Name	Detected	RI	Typical ions and intensities					Page	Entry
Beta-keto-MBDB HFB		1800	417 $_2$	268 $_{29}$	210 $_{16}$	149 $_{100}$	121 $_{21}$	1498	8323
Beta-keto-MBDB ME		1800	206 $_1$	149 $_{12}$	121 $_{13}$	91 $_6$	86 $_{100}$	532	8320
Beta-keto-MBDB PFP		1790	367 $_7$	268 $_2$	218 $_{28}$	149 $_{100}$	121 $_{17}$	1295	8324
Beta-keto-MBDB TFA		1815	317 $_{11}$	168 $_{40}$	149 $_{100}$	121 $_{25}$	110 $_{16}$	1010	8325
Beta-keto-MBDB TMS		1890	293 $_1$	278 $_{10}$	249 $_{14}$	179 $_4$	144 $_{100}$	863	8321
Beta-keto-MBDB-M (demethylenyl-methyl-) 2AC	U+UHYAC	2250	307 $_2$	151 $_{13}$	123 $_3$	114 $_{80}$	72 $_{100}$	948	7966
Beta-keto-MBDB-M (dihydro-) 2AC	U+UHYAC	2105	307 $_2$	193 $_3$	151 $_7$	114 $_{73}$	72 $_{100}$	948	7976
Beta-keto-MBDB-M (nor-) AC	U+UHYAC	2200	249 $_7$	181 $_5$	149 $_{34}$	100 $_{42}$	58 $_{100}$	595	7873
Beta-keto-MBDB-M (nor-demethylenyl-methyl-) 2AC	U+UHYAC	2020	293 $_2$	234 $_{48}$	151 $_{50}$	100 $_{50}$	58 $_{100}$	862	7975
Beta-keto-MDMA		1775	205 $_2$	149 $_{21}$	121 $_{21}$	91 $_{11}$	58 $_{100}$	420	8331
Beta-keto-MDMA 2TMS		1880	351 $_4$	264 $_8$	249 $_6$	149 $_{15}$	130 $_{100}$	1215	8333
Beta-keto-MDMA AC	U+UHYAC	1950	249 $_6$	149 $_{26}$	121 $_{18}$	100 $_{44}$	58 $_{100}$	595	7971
Beta-keto-MDMA HFB		1845	403 $_4$	254 $_{38}$	210 $_{23}$	149 $_{100}$	121 $_{20}$	1454	8336
Beta-keto-MDMA ME		1765	221 $_1$	149 $_6$	121 $_7$	91 $_6$	72 $_{100}$	474	8332
Beta-keto-MDMA PFP		1815	353 $_2$	204 $_{20}$	160 $_{18}$	149 $_{100}$	121 $_{16}$	1225	8337
Beta-keto-MDMA TFA		1835	303 $_{17}$	154 $_{42}$	149 $_{100}$	121 $_{32}$	110 $_{25}$	922	8338
Beta-keto-MDMA TMS		1920	279 $_1$	264 $_{18}$	249 $_{14}$	149 $_{17}$	130 $_{100}$	773	8334
Beta-kcto-MDMA-M (demethylenyl-methyl-) 2AC	U+UHYAC	2045	293 $_7$	237 $_7$	151 $_{38}$	100 $_{67}$	58 $_{100}$	862	7974
Beta-keto-MDMA-M (nor-) AC	U+UHYAC	1930	235 $_{18}$	192 $_6$	149 $_{100}$	121 $_{19}$	86 $_{42}$	531	7972
Beta-keto-MDMA-M (nor-demethylenyl-methyl-) 2AC	U+UHYAC	1990	279 $_5$	220 $_{78}$	178 $_{21}$	151 $_{100}$	86 $_{54}$	773	7973
Betamethasone		2795*	312 $_{23}$	268 $_7$	160 $_{10}$	122 $_{100}$	91 $_{29}$	1412	5220
Betamethasone -2H2O		2910*	356 $_{21}$	253 $_6$	147 $_8$	122 $_{100}$	91 $_{19}$	1243	5221
Betaxolol	G	2355	307 $_2$	292 $_5$	263 $_{11}$	100 $_4$	72 $_{100}$	954	1579
Betaxolol 2AC	UHYAC	2770	331 $_{14}$	200 $_{100}$	140 $_{52}$	98 $_{71}$	72 $_{48}$	1409	1582
Betaxolol 2TMS		2400	436 $_2$	264 $_9$	144 $_{100}$	101 $_{11}$	73 $_{77}$	1583	5494
Betaxolol formyl artifact	P-I G	2410	319 $_{34}$	304 $_{67}$	127 $_{99}$	112 $_{58}$	55 $_{100}$	1031	1580
Betaxolol -H2O		2400	289 $_2$	158 $_2$	98 $_{30}$	72 $_{100}$	55 $_{16}$	838	1583
Betaxolol -H2O AC		2720	331 $_3$	288 $_4$	140 $_{84}$	98 $_{46}$	55 $_{100}$	1103	1581
Betaxolol TMS		2220	364 $_4$	263 $_6$	188 $_4$	101 $_{15}$	72 $_{100}$	1354	5493
Betaxolol TMSTFA		2485	460 $_1$	284 $_{100}$	129 $_{46}$	73 $_{69}$	55 $_{88}$	1628	6179
Betaxolol-M (O-dealkyl-) 3AC	U+UHYAC	2620	319 $_{12}$	200 $_{100}$	140 $_{55}$	98 $_{50}$	72 $_{60}$	1352	1585
Betaxolol-M (O-dealkyl-) -H2O 2AC	UHYAC	2570	319 $_{46}$	234 $_{40}$	217 $_{56}$	140 $_{61}$	98 $_{100}$	1027	1584
Bezafibrate	G U UHY UHYAC	3100	316 $_3$	269 $_7$	205 $_{53}$	139 $_{44}$	120 $_{100}$	1268	2494
Bezafibrate -CO2	G P U+UHYAC	2800	317 $_1$	275 $_2$	139 $_{35}$	120 $_{100}$	107 $_{16}$	1011	1745
Bezafibrate ME	PME UME	2910	375 $_2$	316 $_{16}$	220 $_{58}$	139 $_{56}$	120 $_{100}$	1334	1746
Bezafibrate-M (chlorobenzoic acid)	G UHY UHYAC	1400*	156 $_{61}$	139 $_{100}$	111 $_{54}$	05 $_4$	75 $_{39}$	299	2726
BHB 2TMS		1095*	233 $_{16}$	191 $_{32}$	147 $_{100}$	117 $_{51}$	73 $_{62}$	590	8923
Bifenox		2500	341 $_{75}$	310 $_{25}$	189 $_{39}$	173 $_{53}$	75 $_{100}$	1158	5685
Bifonazole		3070	310 $_3$	243 $_{100}$	228 $_8$	165 $_{14}$	91 $_6$	971	2347
BIM-2201		2880	360 $_{19}$	331 $_8$	271 $_{72}$	155 $_{30}$	127 $_{100}$	1265	9511
Binapacryl		2270	292 $_4$	210 $_{46}$	133 $_2$	83 $_{100}$	55 $_{19}$	1044	3510
Bioallethrin		2105*	302 $_2$	168 $_6$	136 $_{23}$	123 $_{100}$	79 $_{34}$	920	2786
Bioresmethrin		2300*	338 $_7$	171 $_{50}$	143 $_{34}$	128 $_{44}$	123 $_{100}$	1145	4035
Biperiden	P-I G U+UHYAC	2280	311 $_1$	218 $_{15}$	98 $_{100}$			979	101
Biperiden TMS		2420	383 $_1$	294 $_2$	205 $_3$	98 $_{100}$	73 $_{12}$	1373	4529
Biperiden-M (HO-)	U UHY	2645	327 $_3$	218 $_5$	114 $_6$	98 $_{100}$		1077	102
Biperiden-M (HO-) AC	U+UHYAC	2620	369 $_4$	257 $_8$	112 $_5$	98 $_{100}$	84 $_5$	1308	103
Biphenyl		1320*	154 $_{100}$	128 $_6$	102 $_6$	76 $_{25}$	63 $_{16}$	297	3318
Biphenylol	G P U+UHYAC	1550*	170 $_{100}$	141 $_{31}$	115 $_{26}$	77 $_{16}$		324	217
Biphenylol AC	U+UHYAC	1690*	212 $_7$	170 $_{100}$	141 $_{15}$	115 $_{20}$		438	2280
Biphenylol ME		1540*	184 $_1$	170 $_{100}$	141 $_{35}$	115 $_{30}$		357	2281
Biphenylol-M (HO-) 2AC	UHYAC	1900*	270 $_4$	228 $_{21}$	186 $_{100}$	105 $_{36}$		716	2349
2,2'-Bipyridine		1460	156 $_{100}$	128 $_{34}$	102 $_6$	78 $_{31}$	51 $_{54}$	300	105
1,1-Bis-(2-hydroxy-3,5-dimethylphenyl-)-2-methylpropane		2050*	298 $_{21}$	255 $_{100}$	237 $_7$	209 $_4$	179 $_4$	898	5658
Bis-(2-hydroxy-3-tert-butyl-5-ethylphenyl)methane		2450*	368 $_{49}$	312 $_{17}$	191 $_{100}$	175 $_{61}$	163 $_{46}$	1303	2870
Bis-(4-chlorophenyl-)sulfone		2240*	286 $_{26}$	159 $_{100}$	131 $_9$	111 $_{23}$	75 $_{34}$	812	5739
2,4-Bis-(tert.-butyl-)-phenol		1440*	206 $_{14}$	191 $_{100}$	163 $_7$	91 $_{14}$	57 $_{71}$	418	8528
2,4-Bis(tert-butyl)-phenol		1430	206 $_{15}$	191 $_{100}$	163 $_6$	91 $_8$	57 $_{47}$	417	644
Bis-(trimethylsilyl-)trifluoroacetamide		1100	257 $_1$	192 $_{18}$	188 $_{24}$	100 $_{29}$	73 $_{100}$	642	5431
Bisacodyl	G PAC-I U+UHYAC	2835	361 $_{100}$	319 $_{63}$	277 $_{75}$	199 $_{46}$		1268	106
Bisacodyl HY	UHY	2655	277 $_{100}$	199 $_{52}$				759	107
Bisacodyl HY2ME	UGLUCEXME	2595	305 $_{100}$	290 $_{27}$	227 $_{49}$	182 $_6$	169 $_6$	937	6811
Bisacodyl-M (bis-deacetyl-)	UHY	2655	277 $_{100}$	199 $_{52}$				759	107
Bisacodyl-M (bis-deacetyl-) 2ME	UGLUCEXME	2595	305 $_{100}$	290 $_{27}$	227 $_{49}$	182 $_6$	169 $_6$	937	6811
Bisacodyl-M (bis-methoxy-bis-deacetyl-)	UHY	2820	337 $_{100}$	322 $_{69}$	307 $_8$	259 $_{14}$		1136	2458
Bisacodyl-M (bis-methoxy-bis-deacetyl-) 2AC	U+UHYAC	2950	421 $_{83}$	379 $_{100}$	364 $_{54}$	337 $_{25}$	322 $_{46}$	1510	2456
Bisacodyl-M (bis-methoxy-bis-deacetyl-) 2ME	UGLUCEXME	2760	365 $_{100}$	350 $_{61}$	287 $_{41}$	249 $_{13}$	220 $_{11}$	1286	6813
Bisacodyl-M (bis-methoxy-deacetyl-)	U+UHYAC	2890	379 $_{100}$	364 $_{34}$	336 $_{25}$	322 $_{41}$	259 $_8$	1350	2457
Bisacodyl-M (deacetyl-)	UHYAC	2750	319 $_{100}$	277 $_{65}$	276 $_{17}$	199 $_{31}$	153 $_7$	1024	2459
Bisacodyl-M (methoxy-bis-deacetyl-)	UHY	2680	307 $_{100}$	306 $_{49}$	292 $_{19}$	229 $_{35}$	69 $_{22}$	948	109
Bisacodyl-M (methoxy-bis-deacetyl-) 2AC	U+UHYAC	2870	391 $_{46}$	349 $_{100}$	307 $_{48}$	292 $_{12}$	229 $_{23}$	1406	1750
Bisacodyl-M (methoxy-bis-deacetyl-) 2ME	UGLUCEXME	2695	335 $_{100}$	320 $_{40}$	257 $_{57}$	220 $_7$	139 $_{13}$	1125	6812
Bisacodyl-M (methoxy-deacetyl-)	UHYAC	2810	349 $_{100}$	307 $_{43}$	306 $_{54}$	292 $_{17}$	229 $_{30}$	1204	210
Bisacodyl-M (trimethoxy-bis-deacetyl-) 2AC	UHYAC	3060	451 $_{72}$	409 $_{100}$	367 $_{81}$	329 $_{77}$	203 $_{62}$	1582	3425

Table 8.1: Compounds in order of names

Name	Detected	RI	Typical ions and intensities					Page	Entry
Bisoctylphenylamine		2910	393 $_6$	378 $_5$	322 $_{100}$	250 $_{27}$		1419	4950
Bisoprolol	G P U	2570	325 $_1$	310	281 $_2$	116 $_{15}$	72 $_{100}$	1065	2787
Bisoprolol 2AC	U+UHYAC	2770	349 $_1$	245 $_{15}$	200 $_{100}$	98 $_{23}$	72 $_{55}$	1476	2791
Bisoprolol AC	U+UHYAC	2880	352 $_1$	158 $_8$	98 $_{29}$	72 $_{100}$		1300	2790
Bisoprolol formyl artifact	G P U	2595	337 $_{19}$	322 $_{34}$	234 $_{23}$	127 $_{100}$	112 $_{77}$	1141	2788
Bisoprolol -H2O	U	2400	307 $_9$	220 $_1$	204 $_2$	98 $_{100}$	56 $_{88}$	954	2933
Bisoprolol -H2O AC	U+UHYAC	2900	349 $_1$	306 $_4$	262 $_{14}$	140 $_{24}$	98 $_{100}$	1207	2789
Bisoprolol N-AC		2730	349 $_1$	245 $_6$	158 $_{100}$	139 $_9$	72 $_{57}$	1300	6408
Bisoprolol TMSTFA		2570	493 $_1$	332 $_{10}$	284 $_{100}$	221 $_2$	73 $_{60}$	1652	6134
Bisoprolol-M (phenol)	U	1690*	210 $_4$	167 $_9$	123 $_{42}$	107 $_{100}$	77 $_{19}$	434	2932
Bisphenol A	G U UHY	2155*	228 $_{32}$	213 $_{100}$				502	108
Bisphenol A 2AC	UHYAC	2380*	312 $_{11}$	270 $_{28}$	228 $_{31}$	213 $_{100}$	119 $_{11}$	982	3360
Bis-tert.-butylmethylenecyclohexanone		1480*	218 $_{36}$	203 $_{58}$	189 $_{22}$	175 $_{43}$	161 $_{100}$	461	5132
Bis-tert.-butylquinone		1465*	220 $_{53}$	177 $_{100}$	149 $_{36}$	135 $_{44}$	67 $_{39}$	470	4949
Bis-tert-butyl-methoxymethylphenol		1710*	250 $_{26}$	235 $_{100}$	219 $_{20}$	193 $_{13}$	91 $_{11}$	607	6367
Bitertanol		2650	337 $_2$	170 $_{100}$	141 $_{11}$	112 $_{17}$	57 $_{27}$	1140	4146
Biuret		<1000	103 $_{59}$	102 $_{58}$	75 $_{16}$	70 $_{31}$	59 $_{100}$	259	8462
bk-2C-B		1720	258 $_{47}$	243 $_{100}$	228 $_7$	200 $_9$	185 $_9$	732	10202
bk-2C-B AC		2320	315 $_{18}$	243 $_{100}$	228 $_5$	200 $_9$	185 $_8$	998	10203
bk-2C-B enol 2AC		2320	357 $_{17}$	315 $_{12}$	243 $_{100}$	228 $_5$	200 $_8$	1245	10204
bk-2C-B enol 2TMS		2330	417 $_1$	402 $_8$	287 $_{14}$	174 $_{100}$	86 $_{36}$	1498	10206
bk-2C-B HFB		2095	469 $_7$	243 $_{100}$	226 $_{12}$	200 $_{12}$	157 $_8$	1615	10209
bk-2C-B PFP		2075	419 $_{13}$	243 $_{100}$	228 $_9$	200 $_{13}$	185 $_{12}$	1504	10208
bk-2C-B TFA		2090	369 $_{20}$	243 $_{100}$	228 $_9$	200 $_{13}$	185 $_{11}$	1304	10207
bk-2C-B TMS		2035	345 $_{84}$	315 $_{25}$	287 $_{12}$	240 $_{100}$	73 $_{89}$	1182	10205
bk-2C-I		2105	321 $_{13}$	291 $_{100}$	248 $_{15}$	233 $_{12}$	194 $_{20}$	1036	10189
bk-2C-I 2TMS		2440	450 $_4$	335 $_4$	291 $_5$	174 $_{100}$	147 $_5$	1608	9874
bk-2C-I AC		2400	363 $_{21}$	317 $_2$	291 $_{100}$	248 $_5$	233 $_{11}$	1276	9814
bk-2C-I HFB		2190	517 $_{13}$	291 $_{100}$	276 $_7$	248 $_{10}$	233 $_7$	1674	9817
bk-2C-I PFP		2180	467 $_{25}$	291 $_{100}$	276 $_8$	248 $_{17}$	233 $_{13}$	1611	9816
bk-2C-I TFA		2180	417 $_{25}$	332 $_2$	291 $_{100}$	248 $_{17}$	233 $_{16}$	1497	9815
bk-2C-I TMS		2295	393 $_5$	349 $_5$	291 $_{100}$	127 $_{20}$	102 $_{10}$	1413	9875
bk-EBDB		1860	206 $_3$	149 $_{20}$	121 $_{15}$	91 $_6$	86 $_{100}$	533	9149
bk-EBDB AC		2200	277 $_1$	192 $_4$	149 $_{12}$	128 $_{58}$	86 $_{100}$	760	9150
bk-EBDB HFB		1920	431 $_1$	282 $_{50}$	254 $_8$	149 $_{100}$	121 $_{21}$	1537	9153
bk-EBDB PFP		1890	381 $_1$	232 $_{44}$	204 $_6$	149 $_{100}$	121 $_{11}$	1359	9152
bk-EBDB TFA		1950	331 $_2$	182 $_{57}$	154 $_{10}$	149 $_{100}$	121 $_{19}$	1099	9151
bk-MBDB		1740	192 $_2$	162 $_4$	149 $_{21}$	121 $_{19}$	72 $_{100}$	473	8319
bk-MBDB AC	U+UHYAC	2215	263 $_{10}$	149 $_{27}$	121 $_{15}$	114 $_{84}$	72 $_{100}$	674	7872
bk-MBDB HFB		1800	417 $_2$	268 $_{29}$	210 $_{16}$	149 $_{100}$	121 $_{21}$	1498	8323
bk-MBDB ME		1800	206 $_1$	149 $_{12}$	121 $_{13}$	91 $_6$	86 $_{100}$	532	8320
bk-MBDB PFP		1790	367 $_7$	268 $_2$	218 $_{28}$	149 $_{100}$	121 $_{17}$	1295	8324
bk-MBDB TFA		1815	317 $_{11}$	168 $_{40}$	149 $_{100}$	121 $_{25}$	110 $_{16}$	1010	8325
bk-MBDB TMS		1890	293 $_1$	278 $_{10}$	249 $_{14}$	179 $_4$	144 $_{100}$	863	8321
bk-MBDB-M (demethylenyl-methyl-) 2AC	U+UHYAC	2250	307 $_2$	151 $_{13}$	123 $_3$	114 $_{80}$	72 $_{100}$	948	7966
bk-MBDB-M (dihydro-) 2AC	U+UHYAC	2105	307 $_2$	193 $_3$	151 $_7$	114 $_{73}$	72 $_{100}$	948	7976
bk-MBDB-M (nor-) AC	U+UHYAC	2200	249 $_7$	181 $_5$	149 $_{34}$	100 $_{42}$	58 $_{100}$	595	7873
bk-MBDB-M (nor-demethylenyl-methyl-) 2AC	U+UHYAC	2020	293 $_2$	234 $_4$	151 $_{50}$	100 $_{50}$	58 $_{100}$	862	7975
bk-MDMA		1775	205 $_4$	149 $_{21}$	121 $_{21}$	91 $_{11}$	58 $_{100}$	420	8331
bk-MDMA 2TMS		1880	351 $_4$	264 $_8$	249 $_6$	149 $_{15}$	130 $_{100}$	1215	8333
bk-MDMA AC	U+UHYAC	1950	249 $_6$	149 $_{26}$	121 $_{18}$	100 $_{44}$	58 $_{100}$	595	7971
bk-MDMA HFB		1845	403 $_4$	254 $_{38}$	210 $_{23}$	149 $_{100}$	121 $_{20}$	1454	8336
bk-MDMA ME		1765	221 $_1$	149 $_6$	121 $_7$	91 $_6$	72 $_{100}$	474	8332
bk-MDMA PFP		1815	353 $_2$	204 $_{20}$	160 $_{18}$	149 $_{100}$	121 $_{16}$	1225	8337
bk-MDMA TFA		1835	303 $_{17}$	154 $_{42}$	149 $_{100}$	121 $_{32}$	110 $_{25}$	922	8338
bk-MDMA TMS		1920	279 $_1$	264 $_{18}$	249 $_{14}$	149 $_{17}$	130 $_{100}$	773	8334
bk-MDMA-M (demethylenyl-methyl-) 2AC	U+UHYAC	2045	293 $_7$	237 $_7$	151 $_{38}$	100 $_{67}$	58 $_{100}$	862	7974
bk-MDMA-M (nor-) AC	U+UHYAC	1930	235 $_{18}$	192 $_6$	149 $_{100}$	121 $_{19}$	86 $_{42}$	531	7972
bk-MDMA-M (nor-demethylenyl-methyl-) 2AC	U+UHYAC	1990	279 $_5$	220 $_7$	178 $_{21}$	151 $_{100}$	86 $_{54}$	773	7973
25B-NBOMe		2650	379 $_1$	346 $_{46}$	229 $_{11}$	199 $_{10}$	121 $_{100}$	1349	9317
25B-NBOMe AC		2920	421 $_2$	242 $_{33}$	229 $_{12}$	150 $_9$	121 $_{100}$	1509	9319
25B-NBOMe HFB		2640	575 $_2$	242 $_{52}$	229 $_{21}$	199 $_8$	121 $_{100}$	1705	9323
25B-NBOMe ME		2610	393 $_1$	229 $_{16}$	199 $_{17}$	164 $_{67}$	121 $_{100}$	1413	9318
25B-NBOMe PFP		2650	525 $_1$	242 $_{55}$	229 $_{22}$	199 $_8$	121 $_{100}$	1680	9322
25B-NBOMe TFA		2680	475 $_2$	242 $_{56}$	229 $_{18}$	199 $_7$	121 $_{100}$	1627	9321
25B-NBOMe TMS		2720	451 $_1$	229 $_7$	222 $_{74}$	150 $_{48}$	121 $_{100}$	1582	9320
25B-NBOMe-M (2C-B) AC		2180	301 $_{15}$	242 $_{100}$	229 $_{31}$	199 $_{12}$	148 $_{39}$	910	3267
25B-NBOMe-M (deamino-HO-2C-B) AC	U+UHYAC	2300*	302 $_{20}$	242 $_{100}$	227 $_{25}$	183 $_{33}$	148 $_{42}$	916	7198
25B-NBOMe-M (deamino-HOOC-2C-B) ME		2030*	288 $_{100}$	273 $_{12}$	241 $_8$	229 $_{67}$	199 $_{19}$	824	7212
25B-NBOMe-M (O,O-bis-demethyl-) 3AC		3020	477 $_1$	435 $_3$	270 $_{12}$	228 $_{42}$	178 $_{100}$	1630	9389
25B-NBOMe-M (O-demethyl-) isomer-1 2AC		2960	449 $_1$	242 $_{53}$	178 $_{69}$	122 $_{70}$	107 $_{100}$	1578	9388
25B-NBOMe-M (O-demethyl-) isomer-2 2AC		3000	449 $_2$	270 $_3$	228 $_{19}$	192 $_{22}$	121 $_{100}$	1578	9387
25B-NBOMe-M (O-demethyl-2C-B) isomer-1 2AC	U+UHYAC	2410	329 $_4$	287 $_{33}$	228 $_{100}$	215 $_{15}$	165 $_5$	1085	7196
25B-NBOMe-M (O-demethyl-2C-B) isomer-2 2AC	U+UHYAC	2440	329 $_8$	287 $_{21}$	228 $_{100}$	215 $_{17}$	72 $_{13}$	1085	7197

25B-NBOMe-M (O-demethyl-deamino-HO-2C-B) iso-1 2AC   Table 8.1: Compounds in order of names

Name	Detected	RI	Typical ions and intensities					Page	Entry
25B-NBOMe-M (O-demethyl-deamino-HO-2C-B) iso-1 2AC	U+UHYAC	2160*	330 $_5$	288 $_{17}$	270 $_6$	228 $_{100}$	213 $_{12}$	1092	7199
25B-NBOMe-M (O-demethyl-deamino-HO-2C-B) iso-2 2AC	U+UHYAC	2180*	330 $_4$	288 $_{15}$	246 $_{10}$	228 $_{100}$	213 $_{14}$	1092	7200
25B-NBOMe-M (O-demethyl-HO-) 3AC		3210	507 $_1$	237 $_{10}$	208 $_{67}$	179 $_{100}$	137 $_{79}$	1667	9390
5-BnO-DALT		2920	346 $_1$	145 $_3$	130 $_3$	110 $_{100}$	91 $_{19}$	1190	8849
5-BnO-DALT HFB		2615	515 $_2$	451 $_4$	432 $_9$	341 $_4$	110 $_{100}$	1690	10027
5-BnO-DALT PFP		2640	382 $_3$	291 $_2$	144 $_5$	110 $_{100}$	91 $_{39}$	1651	10195
5-BnO-DALT TFA		2645	442 $_1$	332 $_2$	144 $_6$	110 $_{100}$	91 $_{51}$	1566	10026
5-BnO-DALT TMS		2880	418 $_1$	377 $_5$	308 $_{11}$	202 $_{14}$	110 $_{100}$	1503	10025
5-BnO-DiPT		2920	350 $_1$	250 $_2$	145 $_9$	114 $_{100}$	91 $_{42}$	1210	8871
5-BnO-DiPT AC		2540	202 $_4$	160 $_9$	146 $_{27}$	114 $_{100}$	72 $_{30}$	1412	9526
5-BnO-DiPT HFB		2750	531 $_1$	432 $_2$	341 $_2$	114 $_{100}$	72 $_{29}$	1693	9525
5-BnO-DiPT HYAC		2390	302 $_1$	160 $_6$	146 $_{12}$	114 $_{100}$	72 $_{19}$	920	8875
5-BnO-DiPT PFP		3120	408 $_6$	317 $_{10}$	114 $_{100}$	91 $_{60}$	72 $_{30}$	1656	9524
5-BnO-DiPT TFA		2785	445 $_2$	431 $_8$	346 $_9$	114 $_{100}$	72 $_{56}$	1573	9523
5-BnO-DiPT TMS		2800	422 $_2$	322 $_6$	308 $_{11}$	202 $_{14}$	114 $_{100}$	1514	9531
5-BnO-DMT		2680	294 $_3$	145 $_3$	130 $_3$	91 $_{19}$	58 $_{100}$	871	8870
5-BnO-DPT		2920	350 $_2$	250 $_3$	145 $_8$	114 $_{100}$	91 $_{43}$	1210	8869
Boldine	U+UHYAC	2870	327 $_{69}$	326 $_{100}$	312 $_{41}$	296 $_{19}$	269 $_{16}$	1075	8104
Boldine 2AC		3230	411 $_{77}$	396 $_{21}$	368 $_{100}$	354 $_{36}$	326 $_{75}$	1483	8543
Boldine 2HFB		2730	719 $_{86}$	718 $_{100}$	704 $_{44}$	645 $_{32}$	522 $_{27}$	1730	8548
Boldine 2ME	U+UHYAC	2680	355 $_{94}$	354 $_{100}$	340 $_{57}$	324 $_{31}$	281 $_{79}$	1238	5775
Boldine 2PFP		2670	619 $_{91}$	618 $_{100}$	604 $_{53}$	545 $_{32}$	472 $_{27}$	1718	8547
Boldine 2TFA		2655	519 $_{96}$	518 $_{100}$	504 $_{45}$	422 $_{61}$	363 $_{55}$	1676	8546
Boldine isomer-1 AC		3080	369 $_{100}$	354 $_{64}$	326 $_{75}$	312 $_{47}$	224 $_{34}$	1306	8544
Boldine isomer-2 AC		3100	369 $_{82}$	354 $_{19}$	326 $_{100}$	312 $_{59}$	284 $_{47}$	1306	8545
Bornaprine	G U+UHYAC	2260	329 $_{10}$	314 $_9$	257 $_2$	171 $_4$	86 $_{100}$	1091	110
Bornaprine-M (deethyl-HO-) isomer-1 2AC	UHYAC	2790	401 $_{15}$	358 $_9$	142 $_{41}$	112 $_{56}$	58 $_{100}$	1450	1252
Bornaprine-M (deethyl-HO-) isomer-2 2AC	U+UHYAC	2875	401 $_{11}$	358 $_9$	169 $_{39}$	128 $_{49}$	58 $_{100}$	1450	1253
Bornaprine-M (deethyl-HO-) isomer-3 2AC	U+UHYAC	2890	401 $_{11}$	358 $_9$	169 $_{39}$	128 $_{49}$	58 $_{100}$	1450	918
Bornaprine-M (HO-) isomer-1 AC	UHYAC	2385	387 $_3$	372 $_8$	169 $_3$	143 $_5$	86 $_{100}$	1392	1251
Bornaprine-M (HO-) isomer-2 AC	UHYAC	2465	387 $_3$	372 $_9$	169 $_2$	91 $_3$	86 $_{100}$	1392	632
Bornaprine-M (HO-) isomer-3 AC	UHYAC	2565	387 $_7$	372 $_9$	233 $_4$	169 $_5$	86 $_{100}$	1392	683
Bornyl salicylate		1870*	274 $_4$	137 $_{54}$	121 $_{58}$	81 $_{100}$		740	1403
Bornyl salicylate ME		2110*	288 $_{10}$	135 $_{100}$	81 $_{12}$			829	1405
Bortezomib		2650	384 $_1$	338 $_5$	226 $_{77}$	85 $_{100}$	79 $_{73}$	1377	8281
Bortezomib artifact-1 (-COOH) (ME)		2245	285 $_2$	226 $_{14}$	194 $_{24}$	162 $_{100}$	107 $_{57}$	809	8283
Bortezomib artifact-2 (-HB(OH)2)		2585	338 $_4$	226 $_{71}$	85 $_{100}$	79 $_{81}$	70 $_{67}$	1144	8282
Bortezomib artifact-3		2620	340 $_2$	281 $_{15}$	136 $_{69}$	120 $_{63}$	79 $_{100}$	1157	8284
Bortezomib artifact-4		2675	338 $_3$	281 $_{47}$	231 $_{75}$	136 $_{43}$	79 $_{100}$	1145	8285
Bortezomib artifact-4 TMS		2545	410 $_2$	395 $_7$	353 $_{54}$	319 $_{86}$	91 $_{100}$	1478	8288
Bortezomib artifact-5		2425	270 $_4$	226 $_{62}$	147 $_{100}$	107 $_{65}$	79 $_{93}$	718	8286
Brallobarbital	P G UHY U+UHYAC	1850	245 $_1$	207 $_{100}$	165 $_{18}$	124 $_{15}$	91 $_{14}$	812	111
Brallobarbital (ME)	P	1780	259 $_1$	221 $_{100}$	176 $_1$	136 $_{15}$	91 $_{29}$	904	3996
Brallobarbital 2ET		1830	263 $_{100}$	221 $_{23}$	121 $_4$	91 $_7$	77 $_6$	1167	2598
Brallobarbital 2ME		1725	235 $_{100}$	193 $_{30}$				992	645
Brallobarbital-M (debromo-HO-)	U UHY UHYAC	1795	224 $_6$	181 $_{13}$	167 $_{100}$	141 $_{13}$	124 $_{19}$	486	114
Brallobarbital-M (dihydro-)	U UHY UHYAC	1970	209 $_{100}$	167 $_{50}$	141 $_{45}$	120 $_{38}$	67 $_{39}$	825	119
Brallobarbital-M (HO-)	U UHY UHYAC	2040	223 $_{100}$					916	118
Brassidic acid ME		2610*	352 $_3$	320 $_{20}$	97 $_{27}$	69 $_{55}$	55 $_{100}$	1224	3795
5-Br-DALT		2350	318 $_2$	222 $_4$	208 $_{26}$	129 $_{20}$	110 $_{100}$	1017	10131
5-Br-DALT artifact PFP		2140	368 $_1$	354 $_5$	207 $_7$	128 $_4$	110 $_{100}$	1607	10200
5-Br-DALT HFB		2150	418 $_3$	404 $_8$	207 $_{12}$	128 $_9$	110 $_{100}$	1672	10201
5-Br-DALT PFP		2340	289 $_4$	208 $_{53}$	110 $_{100}$			1607	10134
5-Br-DALT TFA		2120	387 $_2$	318 $_3$	304 $_7$	207 $_5$	110 $_{100}$	1489	10133
5-Br-DALT TMS		2455	390 $_1$	349 $_5$	294 $_4$	280 $_9$	110 $_{100}$	1401	10132
Brivudine 2AC		2840	416 $_1$	416 $_1$	217 $_5$	137 $_{36}$	81 $_{100}$	1494	8219
Brivudine 2HFB		2505	724 $_1$	645 $_3$	216 $_7$	137 $_{100}$	81 $_{79}$	1730	8218
Brivudine 2PFP		2505	624 $_1$	545 $_3$	216 $_2$	137 $_{100}$	81 $_{55}$	1720	8217
Brivudine 2TFA		2530	524 $_1$	445 $_3$	216 $_6$	137 $_{100}$	81 $_{76}$	1680	8216
Brivudine 2TMS		2775	476 $_1$	261 $_4$	171 $_{30}$	103 $_{100}$	73 $_{84}$	1629	8214
Brivudine artifact 2TMS		1880	360 $_7$	345 $_{34}$	281 $_{100}$	193 $_{49}$	73 $_{70}$	1263	8215
Brofaromine AC	UHYAC	2780	351 $_{59}$	308 $_{38}$	266 $_{41}$	125 $_{30}$	56 $_{100}$	1211	2405
Brofaromine-M (HO-) 2AC	UHYAC	2980	409 $_{33}$	369 $_{97}$	367 $_{100}$	324 $_{20}$	284 $_{11}$	1473	2710
Brofaromine-M (O-demethyl-) 2AC	UHYAC	2830	379 $_{49}$	337 $_{73}$	294 $_{52}$	125 $_{28}$	56 $_{100}$	1349	2404
Brofaromine-M/artifact (pyridyl-) AC	UHYAC	2650	331 $_{12}$	291 $_{58}$	289 $_{100}$	182 $_7$	153 $_{10}$	1097	2406
Brolamfetamine		1800	273 $_4$	232 $_{82}$	230 $_{87}$	199 $_{12}$	77 $_{100}$	732	2548
Brolamfetamine		1800	273 $_1$	230 $_6$	105 $_3$	77 $_7$	44 $_{100}$	732	5527
Brolamfetamine AC		2150	315 $_{15}$	256 $_{100}$	229 $_7$	162 $_{22}$	86 $_{71}$	999	2549
Brolamfetamine AC		2150	315 $_3$	256 $_{20}$	162 $_4$	86 $_{22}$	44 $_{100}$	998	5528
Brolamfetamine formyl artifact		1790	285 $_3$	254 $_{15}$	229 $_5$	199 $_3$	56 $_{100}$	807	3242
Brolamfetamine HFB		1945	469 $_{24}$	256 $_{90}$	240 $_{55}$	229 $_{100}$	199 $_{29}$	1616	6008
Brolamfetamine PFP		1905	419 $_{22}$	256 $_{73}$	229 $_{100}$	190 $_{59}$	119 $_{93}$	1504	6007
Brolamfetamine precursor		1345*	166 $_{100}$	151 $_{39}$	120 $_{36}$	95 $_{61}$	63 $_{62}$	316	3278
Brolamfetamine TFA		1935	369 $_{28}$	256 $_{81}$	229 $_{100}$	199 $_{40}$	69 $_{88}$	1304	6006

Table 8.1: Compounds in order of names   Brolamfetamine TMS

Name	Detected	RI	Typical ions and intensities					Page	Entry
Brolamfetamine TMS		1920	345_1	272_2	229_2	116_{100}	73_{79}	1182	6009
Brolamfetamine-M (bis-O-demethyl-) 3AC	U+UHYAC	2325	371_2	329_{35}	287_{54}	228_{100}	86_{44}	1315	7075
Brolamfetamine-M (bis-O-demethyl-) artifact 2AC	U+UHYAC	2225	311_{23}	269_{100}	227_{60}	212_{40}	133_{18}	974	7184
Brolamfetamine-M (deamino-HO-) AC	U+UHYAC	1950*	316_7	274_{23}	214_{100}	186_{18}		1003	7061
Brolamfetamine-M (deamino-oxo-)	U+UHYAC	1835*	272_{59}	229_{100}				727	7062
Brolamfetamine-M (HO-) 2AC	U+UHYAC	2270	373_3	313_{14}	271_{37}	86_{100}		1325	7081
Brolamfetamine-M (HO-) -H2O	U+UHYAC	1960*	273_{72}	271_{68}	258_{100}	256_{98}		721	7073
Brolamfetamine-M (HO-) -H2O AC	U+UHYAC	2130*	313_{24}	271_{79}	256_{100}			985	7074
Brolamfetamine-M (O-demethyl-) isomer-1 2AC	U+UHYAC	2235	343_{12}	301_{23}	284_{57}	242_{100}	86_{81}	1172	7065
Brolamfetamine-M (O-demethyl-) isomer-1 AC	U+UHYAC	2120	301_{18}	242_{100}	215_{11}	185_{13}	86_{20}	910	7070
Brolamfetamine-M (O-demethyl-) isomer-2 2AC	U+UHYAC	2275	343_2	284_{56}	242_{100}	215_{13}	86_{23}	1171	7066
Brolamfetamine-M (O-demethyl-) isomer-2 AC	U+UHYAC	2180	301_{29}	242_{100}	215_{14}	86_{36}		910	7071
Brolamfetamine-M (O-demethyl-deamino-oxo-) AC	U+UHYAC	1930*	300_8	258_{94}	215_{100}			904	7063
Brolamfetamine-M (O-demethyl-deamino-oxo-) isomer-1	U+UHYAC	1870*	260_{66}	258_{72}	217_{99}	215_{100}		644	7068
Brolamfetamine-M (O-demethyl-deamino-oxo-) isomer-2	U+UHYAC	1885*	260_{99}	258_{100}	217_{97}	215_{93}		644	7069
Brolamfetamine-M (O-demethyl-HO-) 3AC	U+UHYAC	2385	401_5	359_{49}	317_{74}	258_{100}	86_{71}	1447	7067
Brolamfetamine-M (O-demethyl-HO-) -H2O 2AC	U+UHYAC	2280	341_{32}	299_{62}	257_{100}	242_{72}		1159	7072
Brolamfetamine-M (O-demethyl-HO-deamino-HO-) 3AC	U+UHYAC	2145*	402_8	360_{39}	315_{52}	300_{34}	231_{100}	1451	7064
Bromacil	G U	1900	260_4	231_9	205_{100}	188_{19}	162_{16}	655	124
Bromadiolone artifact		1985*	260_{99}	258_{100}	178_{71}	152_{21}	76_{39}	645	3629
Bromantane		2420	305_{73}	184_{30}	171_{38}	135_{100}	130_{58}	935	6130
Bromantane AC		2515	347_{23}	288_{39}	213_{20}	135_{100}	67_{69}	1193	6202
Bromantane HFB		2305	501_6	367_3	169_{12}	135_{100}	67_{59}	1660	6145
Bromantane ME		2310	319_{44}	198_{35}	135_{100}	93_{79}	67_{95}	1023	6201
Bromantane PFP		2295	451_{23}	317_7	155_{16}	135_{100}	93_{74}	1581	6131
Bromantane TFA		2250	401_{10}	267_5	155_{11}	135_{100}	67_{57}	1447	6203
Bromazepam	P G U UGLUC	2670	315_{91}	286_{57}	236_{100}	208_{46}	179_{43}	998	125
Bromazepam artifact-3	U+UHYAC	2500	303_{99}	301_{100}	222_{82}			909	2117
Bromazepam HY	UHY	2250	276_{27}	247_{100}	198_{17}	168_{28}		751	127
Bromazepam HYAC	U+UHYAC	2490	318_8	289_8	247_{66}	121_{100}	78_{50}	1016	129
Bromazepam isomer-1 ME		2385	329_{84}	250_{100}	208_{33}	179_{49}		1085	130
Bromazepam isomer-2 ME		2540	329_{100}	300_{53}	250_{59}	78_{50}		1084	131
Bromazepam TMS		2450	387_{68}	372_{12}	272_{10}	179_{19}	73_{100}	1388	4530
Bromazepam-M (3-HO-)	UGLUC-I	2470	313_{100}	284_{22}	206_{44}	179_{13}		1097	126
Bromazepam-M (3-HO-) 2TMS		2475	475_1	460_2	386_{28}	360_{18}	73_{100}	1627	5441
Bromazepam-M (3-HO-) artifact-1	P-I UHY-I UHYAC-I	2255	285_{100}	206_{97}	179_{18}			807	128
Bromazepam-M (3-HO-) artifact-2	U+UHYAC	2265	299_{100}	220_{90}	179_{19}	152_7		899	2116
Bromazepam-M (3-HO-) HY	UHY	2250	276_{27}	247_{100}	198_{17}	168_{28}		751	127
Bromazepam-M (3-HO-) HYAC	U+UHYAC	2490	318_8	289_8	247_{66}	121_{100}	78_{50}	1016	129
Bromazepam-M (HO-) HYAC	U+UHYAC	2580	334_{54}	292_{66}	264_{17}	247_{100}	78_{36}	1117	1876
Bromazepam-M (HO-) HYME	UEXME	2250	306_{21}	277_6	247_{100}	184_{23}	78_{34}	941	7703
Bromazepam-M/artifact	UHYAC	2310	316_{55}	288_{32}	248_{100}	238_{85}	210_{40}	1003	2700
Bromazepam-M/artifact	UHYAC	2670	352_{100}	325_{31}	296_{16}	273_{34}	216_{17}	1217	3059
Bromazepam-M/artifact AC	U+UHYAC	2260	319_{24}	277_{100}	249_{46}	198_{23}	170_{58}	1022	1877
Bromhexine		2375	374_{12}	293_{50}	262_{38}	112_{66}	70_{100}	1329	132
Bromhexine-M (HO-)	UHY	2660	390_9	293_{60}	262_{34}	128_{100}	86_{38}	1401	133
Bromhexine-M (HO-) 2AC	UHYAC	2930	474_5	417_{27}	335_{100}	304_{24}	264_{29}	1625	134
Bromhexine-M (HOOC-) ME	P U	1770	309_{68}	307_{36}	277_{100}	275_{53}	249_{27}	945	5131
Bromhexine-M (nor-HO-)	P G U UHY	2665	376_8	279_{77}	264_{100}	262_{53}	114_{69}	1338	19
Bromhexine-M (nor-HO-) 2TMS		2800	520_1	391_5	351_{100}	186_{50}	73_{66}	1677	4528
Bromhexine-M (nor-HO-) formyl artifact	P G U UHY	2780	387_{28}	331_{100}	329_{53}	289_{42}	195_{82}	1392	6315
Bromhexine-M (nor-HO-) -H2O	P G U UHY	2395	358_{13}	289_{42}	264_{100}	262_{55}	68_{100}	1251	6314
Bromhexine-M (nor-HO-) isomer-1 2AC	UHYAC	2935	460_{14}	417_{37}	279_{47}	262_{25}	81_{100}	1601	135
Bromhexine-M (nor-HO-) isomer-2 2AC	U+UHYAC	3015	460_{22}	419_{100}	417_{51}	279_{74}	264_{56}	1600	20
Bromhexine-M (nor-HO-) isomer-3 2AC	UHYAC	3165	460_{16}	417_{30}	279_{48}	262_{24}	81_{100}	1601	136
Bromhexine-M (nor-HO-) TMS		2665	448_4	319_{23}	279_{100}	264_{92}	186_{64}	1576	4527
Bromisoval	P-I G U	1540	222_{40}	163_{73}	70_{80}	55_{100}		478	137
Bromisoval artifact	P G U	1510	137_{100}	120_{13}	100_{70}			276	138
Bromisoval-M (Br-isovalerianic acid)		1190*	165_1	140_{100}	138_{10}	120_9	101_{20}	343	2395
Bromisoval-M (HO-isovalerianic acid)		1140*	118_1	89_1	76_{100}	73_{66}	55_{35}	264	2394
Bromisoval-M (isovalerianic acid carbamide)		1850	129_4	112_5	102_{100}	85_{46}	59_{71}	284	139
Bromisoval-M/artifact (bromoisovalerianic acid)	G	1570*	180_{29}	163_{85}	137_{54}	70_{93}	55_{100}	343	2393
4-Bromo-2,5-dimethoxyphenylethylamine		1785	259_{15}	230_{100}	215_{29}	199_{10}	77_{37}	650	3254
4-Bromo-2,5-dimethoxyphenylethylamine 2AC		2230	343_7	242_{100}	229_{32}	201_{12}	148_{29}	1171	6924
4-Bromo-2,5-dimethoxyphenylethylamine 2TMS		2195	403_1	388_{14}	272_7	207_{11}	174_{100}	1455	6926
4-Bromo-2,5-dimethoxyphenylethylamine AC		2180	301_{15}	242_{100}	229_{31}	199_{12}	148_{39}	910	3267
4-Bromo-2,5-dimethoxyphenylethylamine formyl artifact		1840	271_{18}	240_{100}	229_{41}	199_{17}	77_{37}	722	3245
4-Bromo-2,5-dimethoxyphenylethylamine formyl artifact		1840	271_{12}	240_{65}	229_{27}	199_{11}	42_{100}	722	5522
4-Bromo-2,5-dimethoxyphenylethylamine HFB		2030	455_{32}	242_{100}	229_{81}	199_{29}	148_{33}	1591	6941
4-Bromo-2,5-dimethoxyphenylethylamine intermediate-1		1900	209_{100}	162_{51}	147_{52}	133_{61}	77_{62}	427	3286
4-Bromo-2,5-dimethoxyphenylethylamine intermediate-2		1630	181_{15}	162_{33}	152_{100}	137_{47}	121_{27}	350	3287
4-Bromo-2,5-dimethoxyphenylethylamine intermediate-2		1630	181_{15}	162_{33}	152_{100}	137_{47}	44_{95}	350	5523
4-Bromo-2,5-dimethoxyphenylethylamine intermediate-2 2AC		1935	265_{11}	164_{100}	149_{24}	121_{25}	91_{13}	688	9162
4-Bromo-2,5-dimethoxyphenylethylamine intermediate-2 AC		1935	223_{15}	164_{100}	149_{29}	121_{31}	91_{15}	484	3288

4-Bromo-2,5-dimethoxyphenylethylamine intermediate-2

Table 8.1: Compounds in order of names

Name	Detected	RI	Typical ions and intensities					Page	Entry
4-Bromo-2,5-dimethoxyphenylethylamine intermediate-2 formyl artifact		1540	193 $_{12}$	162 $_{100}$	151 $_{30}$	121 $_{43}$	91 $_{29}$	378	3293
4-Bromo-2,5-dimethoxyphenylethylamine intermediate-2 formyl artifact		1540	193 $_{12}$	162 $_{100}$	151 $_{30}$	121 $_{43}$	42 $_{64}$	379	5524
4-Bromo-2,5-dimethoxyphenylethylamine PFP		1995	405 $_{36}$	242 $_{100}$	229 $_{84}$	199 $_{33}$	148 $_{35}$	1461	6936
4-Bromo-2,5-dimethoxyphenylethylamine TFA		2000	355 $_{42}$	242 $_{100}$	229 $_{84}$	199 $_{37}$	148 $_{36}$	1235	6931
4-Bromo-2,5-dimethoxyphenylethylamine TMS		1935	331 $_{1}$	272 $_{3}$	229 $_{2}$	102 $_{100}$	73 $_{50}$	1098	6925
4-Bromo-2,5-dimethoxyphenylethylamine-M (deamino-di-HO-) 2AC	U+UHYAC	2230*	360 $_{13}$	300 $_{27}$	258 $_{43}$	245 $_{100}$	138 $_{17}$	1262	7214
4-Bromo-2,5-dimethoxyphenylethylamine-M (deamino-di-HO-) 2TFA		1790*	468 $_{100}$	354 $_{14}$	341 $_{77}$	311 $_{14}$	276 $_{76}$	1613	7208
4-Bromo-2,5-dimethoxyphenylethylamine-M (deamino-HO-) AC	U+UHYAC	2300*	302 $_{20}$	242 $_{100}$	227 $_{25}$	183 $_{33}$	148 $_{42}$	916	7198
4-Bromo-2,5-dimethoxyphenylethylamine-M (deamino-HO-) TFA		1880*	356 $_{49}$	341 $_{15}$	242 $_{100}$	229 $_{2}$	148 $_{71}$	1240	7209
4-Bromo-2,5-dimethoxyphenylethylamine-M (deamino-HOOC-) ME		2030*	288 $_{100}$	273 $_{12}$	241 $_{8}$	229 $_{67}$	199 $_{19}$	824	7212
4-Bromo-2,5-dimethoxyphenylethylamine-M (deamino-oxo-)		2020*	258 $_{35}$	229 $_{100}$	215 $_{9}$	199 $_{35}$	186 $_{8}$	644	7215
4-Bromo-2,5-dimethoxyphenylethylamine-M (O-demethyl- N-acetyl-) isomer-1 TFA		2090	383 $_{8}$	324 $_{100}$	255 $_{7}$	148 $_{21}$	72 $_{23}$	1368	7204
4-Bromo-2,5-dimethoxyphenylethylamine-M (O-demethyl- N-acetyl-) isomer-2 TFA		2130	383 $_{7}$	324 $_{100}$	311 $_{11}$	227 $_{11}$	72 $_{17}$	1368	7205
4-Bromo-2,5-dimethoxyphenylethylamine-M (O-demethyl-) iso-1 2AC	U+UHYAC	2410	329 $_{4}$	287 $_{33}$	228 $_{100}$	215 $_{15}$	165 $_{5}$	1085	7196
4-Bromo-2,5-dimethoxyphenylethylamine-M (O-demethyl-) iso-2 2AC	U+UHYAC	2440	329 $_{8}$	287 $_{20}$	228 $_{100}$	215 $_{17}$	72 $_{13}$	1085	7197
4-Bromo-2,5-dimethoxyphenylethylamine-M (O-demethyl-) isomer-1 2TFA		1900	437 $_{12}$	324 $_{100}$	311 $_{10}$	255 $_{7}$	148 $_{32}$	1550	7206
4-Bromo-2,5-dimethoxyphenylethylamine-M (O-demethyl-) isomer-2 2TFA		1950	437 $_{6}$	324 $_{100}$	311 $_{32}$	253 $_{3}$	227 $_{15}$	1550	7207
4-Bromo-2,5-dimethoxyphenylethylamine-M (O-demethyl-deamino-di-HO-) 3AC	U+UHYAC	2280*	388 $_{6}$	346 $_{43}$	286 $_{55}$	244 $_{67}$	231 $_{100}$	1393	7201
4-Bromo-2,5-dimethoxyphenylethylamine-M (O-demethyl-deamino-HO-) 2TFA		1800*	438 $_{31}$	341 $_{16}$	324 $_{100}$	311 $_{35}$	227 $_{55}$	1553	7210
4-Bromo-2,5-dimethoxyphenylethylamine-M (O-demethyl-deamino-HO-) iso-1 2AC	U+UHYAC	2160	330 $_{5}$	288 $_{17}$	270 $_{6}$	228 $_{100}$	213 $_{12}$	1092	7199
4-Bromo-2,5-dimethoxyphenylethylamine-M (O-demethyl-deamino-HO-) iso-2 2AC	U+UHYAC	2180*	330 $_{4}$	288 $_{15}$	246 $_{10}$	228 $_{100}$	213 $_{14}$	1092	7200
4-Bromo-2,5-dimethoxyphenylethylamine-M (O-demethyl-deamino-HOOC-) -H2O	U+UHYAC	1980*	242 $_{80}$	214 $_{100}$	186 $_{17}$			562	7203
4-Bromo-2,5-dimethoxyphenylethylamine-M (O-demethyl-deamino-HOOC-) MEAC		2120*	316 $_{14}$	274 $_{100}$	242 $_{92}$	214 $_{87}$	186 $_{18}$	1003	7213
4-Bromo-2,5-dimethoxyphenylethylamine-M (O-demethyl-deamino-HOOC-) METFA		1890*	370 $_{100}$	311 $_{46}$	257 $_{55}$	241 $_{60}$	148 $_{21}$	1309	7211
4-Bromo-2,5-dimethoxyphenylethylamine-M (O-demethyl-deamino-HO-oxo-) 2AC	U+UHYAC	2160*	344 $_{11}$	302 $_{73}$	260 $_{46}$	242 $_{100}$	214 $_{25}$	1177	7202
2-Bromo-4-cyclohexylphenol		1915*	254 $_{100}$	198 $_{34}$	185 $_{56}$	132 $_{78}$	107 $_{28}$	625	5165
2-Bromo-4-cyclohexylphenol AC		1925*	296 $_{6}$	254 $_{100}$	198 $_{17}$	185 $_{18}$	132 $_{24}$	883	5169
2-Bromo-4-cyclohexylphenol ME		1800*	268 $_{100}$	199 $_{54}$	146 $_{92}$	118 $_{63}$	90 $_{40}$	705	5172
5-Bromo-AMT		2060	252 $_{3}$	209 $_{100}$	154 $_{4}$	129 $_{57}$	102 $_{45}$	615	10251
5-Bromo-AMT		2060	252 $_{1}$	209 $_{38}$	129 $_{19}$	102 $_{13}$	44 $_{100}$	615	10253
5-Bromo-AMT 2HFB		2005	644 $_{3}$	431 $_{45}$	404 $_{27}$	240 $_{100}$	209 $_{18}$	1722	9840
5-Bromo-AMT 2PFP		1975	544 $_{4}$	381 $_{43}$	354 $_{25}$	207 $_{17}$	190 $_{100}$	1692	9838
5-Bromo-AMT 2TFA		2000	444 $_{5}$	331 $_{33}$	304 $_{24}$	207 $_{12}$	140 $_{100}$	1569	9836
5-Bromo-AMT 2TMS		2290	396 $_{1}$	381 $_{1}$	281 $_{9}$	116 $_{100}$	100 $_{12}$	1428	9881
5-Bromo-AMT AC		2475	294 $_{10}$	235 $_{100}$	208 $_{58}$	156 $_{11}$	129 $_{43}$	068	9834
5-Bromo-AMT formyl artifact		2270	264 $_{12}$	247 $_{5}$	221 $_{100}$	142 $_{15}$	115 $_{12}$	680	10252
5-Bromo-AMT HFB		2230	448 $_{26}$	240 $_{19}$	235 $_{18}$	208 $_{100}$	129 $_{29}$	1576	9839
5-Bromo-AMT TFA		2240	348 $_{18}$	254 $_{6}$	235 $_{16}$	208 $_{100}$	140 $_{14}$	1197	9837
5-Bromo-AMT TMS		2670	324 $_{1}$	309 $_{1}$	281 $_{6}$	116 $_{100}$	100 $_{16}$	1055	9882
Bromobenzene		<1000*	158 $_{59}$	156 $_{58}$	77 $_{100}$	51 $_{36}$		299	3611
3-Bromo-d-camphor		1450*	230 $_{12}$	151 $_{34}$	123 $_{100}$	83 $_{82}$	55 $_{55}$	508	2985
Bromofenoxim artifact-1		1520	184 $_{100}$	154 $_{28}$	107 $_{41}$	91 $_{45}$	63 $_{88}$	356	728
Bromofenoxim artifact-2		1690	277 $_{100}$	275 $_{56}$	168 $_{18}$	117 $_{14}$	88 $_{59}$	743	3630
Bromofenoxim artifact-2 ME		1650	289 $_{15}$	248 $_{12}$	202 $_{26}$	86 $_{100}$	72 $_{59}$	831	3631
3-Bromomethcathinone		1680	183 $_{1}$	155 $_{5}$	75 $_{9}$	58 $_{100}$	56 $_{23}$	559	8092
3-Bromomethcathinone AC		1965	183 $_{2}$	155 $_{6}$	100 $_{61}$	75 $_{10}$	58 $_{100}$	794	8095
3-Bromomethcathinone artifact (dehydro-)		1595	183 $_{1}$	155 $_{4}$	75 $_{11}$	56 $_{100}$		550	8093
3-Bromomethcathinone HFB		1775	437 $_{1}$	254 $_{100}$	210 $_{38}$	183 $_{17}$	155 $_{14}$	1551	8098
3-Bromomethcathinone PFP		1725	387 $_{1}$	204 $_{100}$	183 $_{17}$	160 $_{36}$	119 $_{24}$	1387	8097
3-Bromomethcathinone precursor		1595*	226 $_{4}$	183 $_{100}$	155 $_{55}$	76 $_{62}$		493	8094
3-Bromomethcathinone TFA		1740	337 $_{1}$	183 $_{28}$	154 $_{100}$	110 $_{50}$	76 $_{25}$	1134	8096
3-Bromomethcathinone-M (nor-) AC	U+UHYAC	1925	269 $_{1}$	183 $_{6}$	177 $_{11}$	155 $_{6}$	86 $_{100}$	710	8099
3-Bromomethcathinone-M (nor-dihydro-) isomer-1 2AC	U+UHYAC	2030	313 $_{1}$	212 $_{4}$	183 $_{6}$	155 $_{3}$	86 $_{100}$	985	8100
3-Bromomethcathinone-M (nor-dihydro-) isomer-2 2AC	U+UHYAC	2045	313 $_{1}$	212 $_{3}$	183 $_{5}$	155 $_{3}$	86 $_{100}$	985	8101
3-Bromomethcathinone-M (nor-HO-dihydro-) 3AC	U+UHYAC	2305	371 $_{1}$	312 $_{1}$	228 $_{2}$	201 $_{6}$	86 $_{100}$	1315	8103
3-Bromomethcathinone-M (nor-HO-dihydro-) -H2O 2AC	U+UHYAC	2140	311 $_{8}$	269 $_{59}$	227 $_{100}$	210 $_{5}$	147 $_{18}$	974	8102
5-Bromo-N,N-diallyl-tryptamine		2350	318 $_{2}$	222 $_{4}$	208 $_{26}$	129 $_{20}$	110 $_{100}$	1017	10131
5-Bromo-N,N-diallyl-tryptamine artifact PFP		2140	368 $_{1}$	354 $_{5}$	207 $_{7}$	128 $_{4}$	110 $_{100}$	1607	10200
5-Bromo-N,N-diallyl-tryptamine HFB		2150	418 $_{3}$	404 $_{6}$	207 $_{12}$	128 $_{9}$	110 $_{100}$	1672	10201
5-Bromo-N,N-diallyl-tryptamine PFP		2340	289 $_{4}$	208 $_{53}$	110 $_{100}$			1607	10134
5-Bromo-N,N-diallyl-tryptamine TFA		2120	387 $_{2}$	318 $_{3}$	304 $_{7}$	207 $_{5}$	110 $_{100}$	1489	10133
5-Bromo-N,N-diallyl-tryptamine TMS		2455	390 $_{1}$	349 $_{5}$	294 $_{4}$	280 $_{9}$	110 $_{100}$	1401	10132
5-Bromonicotinic acid		1020	201 $_{84}$	183 $_{49}$	156 $_{30}$	76 $_{68}$	51 $_{100}$	401	5252
5-Bromonicotinic acid ME	UME	1095	215 $_{42}$	184 $_{79}$	156 $_{70}$	136 $_{30}$	76 $_{100}$	448	5250
4-Bromophenol		1310*	174 $_{94}$	172 $_{100}$	93 $_{21}$	65 $_{56}$		326	1995
Bromophos	P-I	1995*	331 $_{100}$	329 $_{74}$	213 $_{5}$	125 $_{68}$		1280	1406
Bromophos-ethyl		2060*	359 $_{79}$	357 $_{60}$	301 $_{69}$	240 $_{20}$	97 $_{100}$	1409	3508
Bromopride		2850	343 $_{1}$	245 $_{3}$	228 $_{11}$	99 $_{23}$	86 $_{100}$	1173	1407
Bromopride AC		3080	385 $_{1}$	313 $_{1}$	270 $_{2}$	228 $_{4}$	86 $_{100}$	1380	2607
Bromopropylate		2425*	426 $_{1}$	341 $_{100}$	339 $_{53}$	183 $_{73}$	76 $_{40}$	1522	4142
3-Bromoquinoline		1490	209 $_{98}$	207 $_{100}$	128 $_{83}$	101 $_{40}$	75 $_{19}$	418	2638
5-Bromosalicylic acid		1530*	216 $_{32}$	198 $_{100}$	170 $_{35}$	142 $_{8}$	63 $_{40}$	450	1996
5-Bromosalicylic acid 2ET		1600*	272 $_{100}$	257 $_{33}$	213 $_{97}$	198 $_{32}$	170 $_{16}$	727	1998
5-Bromosalicylic acid 2ME		1500*	244 $_{32}$	213 $_{100}$	183 $_{10}$	170 $_{16}$	155 $_{14}$	569	2031

Table 8.1: Compounds in order of names — 5-Bromosalicylic acid -CO2

Name	Detected	RI	Typical ions and intensities	Page	Entry
5-Bromosalicylic acid -CO2		1310*	174_{94} 172_{100} 93_{21} 65_{56}	326	1995
5-Bromosalicylic acid ME		1465*	230_{36} 198_{100} 170_{30} 143_{8} 63_{34}	507	1997
5-Bromosalicylic acid MEAC		1600*	272_{2} 230_{95} 198_{100} 170_{22} 142_{7}	727	2032
Bromothiophene		<1000*	164_{99} 162_{100} 117_{4} 83_{80} 57_{31}	305	3609
Bromoxynil		1690	277_{100} 275_{56} 168_{18} 117_{14} 88_{59}	743	3630
Bromoxynil ME		1650	289_{15} 248_{12} 202_{26} 86_{100} 72_{59}	831	3631
Bromperidol	G U+UHYAC	3050	419_{2} 281_{100} 268_{93} 250_{21} 123_{77}	1505	2110
Bromperidol 2TMS		2840	548_{7} 340_{94} 250_{35} 103_{37} 73_{100}	1701	5480
Bromperidol -H2O	U+UHYAC	3020	401_{7} 263_{24} 250_{31} 236_{100} 123_{57}	1448	2115
Bromperidol TMS		2730	478_{4} 340_{80} 250_{70} 123_{100} 73_{100}	1649	5479
Bromperidol-M	U+UHYAC	1490*	180_{25} 125_{49} 123_{35} 95_{17} 56_{100}	345	85
Bromperidol-M	UHY	1890	267_{15} 233_{96} 127_{38} 94_{44} 56_{100}	698	141
Bromperidol-M (N-dealkyl-) AC	UHYAC	2335	297_{35} 254_{36} 183_{22} 99_{28} 57_{100}	888	166
Bromperidol-M (N-dealkyl-oxo-) -2H2O	U UHY U+UHYAC	1850	233_{100} 154_{21} 127_{37}	520	140
Bromperidol-M 4 AC	UHYAC	2260	293_{68} 279_{85} 251_{100} 222_{56}	859	142
Brompheniramine	U UHY UHYAC	2105	247_{28} 167_{12} 72_{38} 58_{100}	1017	144
Brompheniramine-M (bis-nor-) AC	UHYAC	2170	332_{4} 260_{100} 247_{38} 180_{22} 167_{35}	1104	2812
Brompheniramine-M (nor-) AC	U+UHYAC	2195	346_{12} 260_{48} 247_{100} 180_{20} 167_{50}	1187	145
Brotizolam	G U+UHYAC-I	3090	394_{100} 392_{74} 363_{12} 316_{28} 245_{48}	1410	1408
Brotizolam-M (HO-) AC		3140	450_{34} 409_{100} 407_{69} 289_{11} 245_{14}	1580	2052
Brotizolam-M (HO-) -CH2O		3050	380_{100} 378_{79} 299_{24} 245_{16}	1345	2051
Brucine	U	3275	394_{100} 379_{16} 355_{5}	1421	146
BSTFA		1100	257_{1} 192_{18} 188_{24} 100_{29} 73_{100}	642	5431
Bucetin		2020	223_{4} 179_{46} 137_{62} 109_{86} 108_{100}	483	147
Bucetin AC	UGLUCAC	2095	265_{25} 205_{30} 137_{100} 109_{47} 108_{52}	687	185
Bucetin HYAC	G U+UHYAC	1680	179_{66} 137_{51} 109_{97} 108_{100} 80_{18}	341	186
Bucetin-M	UHY	1240	109_{100} 80_{41} 53_{82} 53_{82} 52_{90}	260	826
Bucetin-M (deethyl-) HYME	UHYME	1100	123_{27} 109_{100} 94_{7} 80_{96} 53_{47}	267	3766
Bucetin-M (HO-) HY2AC	UHYAC	1755	237_{14} 195_{31} 153_{100} 124_{55}	544	187
Bucetin-M (O-deethyl-) 2AC	UGLUCAC	2110	279_{4} 237_{12} 177_{9} 151_{10} 109_{100}	772	30
Bucetin-M (p-phenetidine)	UHY	1280	137_{68} 108_{100} 80_{39} 65_{10}	277	844
Bucetin-M HY2AC	PAC U+UHYAC	1765	193_{10} 151_{53} 109_{100} 80_{24}	377	188
Buclizine	G U UHY UHYAC	3360	432_{5} 285_{37} 231_{100} 165_{35} 147_{53}	1540	2414
Buclizine artifact-1	G U+UHYAC	1600*	202_{30} 167_{100} 165_{52} 152_{17} 125_{7}	403	2442
Buclizine HY		1830	232_{62} 190_{61} 147_{100} 117_{41}	520	2416
Buclizine HYAC		2020	274_{18} 202_{24} 188_{25} 147_{100} 85_{30}	742	2415
Buclizine-M	UHYAC	2210	280_{100} 201_{35} 165_{57}	779	770
Buclizine-M (carbinol)	UHY	1750*	218_{17} 183_{7} 139_{39} 105_{100} 77_{87}	457	2239
Buclizine-M (carbinol) AC	U+UHYAC	1890*	260_{8} 200_{40} 165_{100} 139_{10} 77_{29}	656	1270
Buclizine-M (Cl-benzophenone)	U+UHYAC	1850*	216_{43} 139_{58} 105_{100} 77_{44}	451	1343
Buclizine-M (HO-Cl-benzophenone)	UHY	2300*	232_{36} 197_{7} 139_{23} 121_{100} 111_{23}	515	2240
Buclizine-M (HO-Cl-BPH) isomer-1 AC	UHYAC	2200*	274_{18} 232_{75} 139_{100} 121_{44} 111_{51}	737	2229
Buclizine-M (HO-Cl-BPH) isomer-2 AC	U+UHYAC	2230*	274_{7} 232_{43} 139_{25} 121_{100} 111_{27}	737	2230
Buclizine-M (N-dealkyl-)	UHY	2520	286_{13} 241_{48} 201_{50} 165_{65} 56_{100}	815	2241
Buclizine-M (N-dealkyl-) AC	U+UHYAC	2620	328_{7} 242_{19} 201_{48} 165_{66} 85_{100}	1081	1271
Buclizine-M (N-dealkyl-HO-) 2AC	UHYAC	2640	332_{12} 260_{37} 205_{100} 117_{34} 85_{59}	1107	2433
Buclizine-M (N-dealkyl-HO-) AC-conj.	U	2580	290_{20} 204_{36} 163_{100} 117_{50} 85_{63}	844	2432
Buclizine-M/artifact HYAC	U+UHYAC	2935	280_{4} 201_{100} 165_{26}	780	1272
Budesonide 2TMS		3635	574_{8} 486_{7} 329_{7} 147_{28} 73_{100}	1705	9564
Budesonide AC		3400*	472_{3} 371_{52} 281_{20} 121_{100} 91_{55}	1622	9583
Budipine		2300	293_{2} 278_{100} 178_{5} 165_{5} 70_{28}	867	6114
Bufexamac 2ME		2005	251_{23} 190_{79} 163_{51} 107_{100} 77_{10}	613	6398
Bufexamac artifact (deoxo-)		1970	207_{12} 163_{25} 107_{100} 89_{6} 77_{16}	421	6083
Bufexamac artifact (deoxo-formyl-)		1780	219_{3} 163_{36} 107_{100} 89_{4} 77_{11}	466	6084
Bufexamac ME		1995	237_{34} 222_{42} 166_{100} 122_{51} 107_{60}	545	6086
Bufexamac-M/artifact (HOOC-) ME		1720*	222_{37} 166_{49} 163_{25} 107_{100} 77_{37}	480	6085
Buflomedil	G P U UHY UHYAC	2390	307_{2} 210_{6} 195_{31} 97_{73} 84_{100}	952	2907
Buflomedil TMS		2275	379_{3} 295_{16} 181_{17} 84_{100} 73_{42}	1353	6274
Buflomedil-M (O-demethyl-)	UHY	2375	293_{13} 181_{19} 97_{83} 84_{100} 55_{14}	865	3980
Buflomedil-M (O-demethyl-) AC	UHYAC	2530	335_{2} 181_{16} 97_{83} 84_{100} 55_{15}	1126	3981
Bufotenin		2150	204_{3} 160_{1} 146_{5} 117_{2} 58_{100}	410	8858
Bulbocapnine	UHY	2960	325_{100} 310_{98} 282_{27} 178_{30} 162_{44}	1062	4249
Bulbocapnine AC	UHYAC	2990	367_{100} 324_{32} 310_{83} 280_{28} 162_{23}	1297	4250
Bumadizone		2270	282_{10} 184_{100} 183_{69} 93_{31} 77_{80}	1070	5184
Bumadizone artifact (azobenzene)		1620	182_{14} 152_{5} 105_{15} 77_{100} 63_{2}	352	5186
Bumadizone artifact (hexanilide)		1755	191_{4} 135_{6} 93_{100} 77_{7} 65_{7}	372	5187
Bumadizone ME		2280	296_{13} 184_{72} 183_{100} 77_{87} 57_{49}	1157	5185
Bumatizone artifact AC		2435	366_{67} 184_{69} 183_{100} 105_{33} 77_{94}	1289	5188
Bumetanide 2ME		3180	392_{100} 349_{76} 318_{44} 254_{42} 77_{25}	1411	2780
Bumetanide 2MEAC		3120	434_{100} 379_{11} 349_{23} 254_{21} 56_{57}	1544	2781
Bumetanide 3ME		2970	406_{100} 363_{68} 318_{28} 298_{10} 254_{28}	1465	2779
Bumetanide 3MEAC		3190	448_{24} 383_{100} 328_{48} 110	1576	2783
Bumetanide -SO2NH ME		2340	299_{38} 256_{100} 178_{11} 91_{22} 77_{23}	902	2778
Bumetanide -SO2NH MEAC		3150	341_{100} 285_{7} 254_{8} 195_{2} 91_{4}	1164	2782

Bunazosin   Table 8.1: Compounds in order of names

Name	Detected	RI	Typical ions and intensities					Page	Entry
Bunazosin		3330	373_{73}	260_{58}	247_{100}	233_{65}	221_{38}	1329	4690
Bunitrolol		1960	233_{35}	204_{6}	86_{100}	57_{32}		592	2608
Bunitrolol AC		2070	275_{46}	119_{22}	98_{45}	86_{100}	56_{80}	843	1351
Bunitrolol formyl artifact		1980	260_{2}	245_{100}	86_{13}	70_{28}	57_{33}	659	1350
Bunitrolol TMS		2025	305_{6}	204_{3}	176_{5}	86_{100}	73_{15}	1035	6165
Bunitrolol-M (deisobutyl-) 2AC	UHYAC	2040	276_{45}	233_{64}	158_{32}	96_{35}	86_{100}	754	1586
Bunitrolol-M (HO-) 2AC	UHYAC	2300	333_{50}	291_{16}	174_{11}	98_{28}	86_{100}	1199	1587
Bunitrolol-M (HO-) artifact AC	UHYAC	2370	318_{1}	303_{100}	261_{4}	174_{2}	70_{20}	1020	1588
Bunitrolol-M (HO-methoxy-) 2AC	UHYAC	2480	363_{50}	321_{5}	204_{5}	98_{20}	86_{100}	1347	1589
Buphanamine		----	301_{100}	256_{35}	231_{45}	218_{39}	204_{40}	913	4689
Buphedrone		1290	148_{3}	105_{7}	77_{20}	72_{100}	57_{12}	336	9721
Buphedrone AC		1695	219_{3}	148_{2}	114_{52}	77_{33}	72_{100}	465	9722
Buphedrone HFB		1450	373_{10}	269_{41}	240_{11}	210_{62}	105_{100}	1326	9724
Buphedrone PFP		1400	323_{1}	218_{100}	160_{29}	105_{76}	77_{45}	1049	9723
Buphedrone TFA		1365	273_{4}	168_{100}	140_{7}	110_{21}	105_{58}	733	10262
Buphedrone TMS		1530	249_{1}	234_{3}	205_{5}	144_{100}	73_{59}	600	9760
Buphenine		2420	176_{100}	121_{16}	91_{83}	71_{38}		903	1409
Bupirimate		2165	316_{34}	273_{100}	208_{92}	166_{64}	108_{79}	1006	3319
Bupivacaine	P U	2260	288_{1}	245	140_{100}	98_{3}	84_{8}	831	148
Bupranolol		1900	271_{1}	256_{17}	227_{6}	86_{100}	57_{27}	725	2609
Bupranolol AC		2370	298_{30}	142_{8}	112_{26}	86_{100}		988	1346
Bupranolol formyl artifact	P-I	1915	283_{10}	268_{100}	142_{16}	86_{68}	70_{73}	796	1347
Bupranolol TMS		2000	343_{1}	328_{6}	227_{6}	86_{100}	73_{36}	1175	6147
Bupranolol-M (HO-) 2AC	U+UHYAC	2260	356_{60}	314_{7}	112_{18}	98_{18}	86_{100}	1318	1569
Bupranolol-M (HO-) AC	UHYAC	2150	314_{36}	272_{4}	197_{4}	112_{11}	86_{100}	1087	1590
Bupranolol-M (HO-) formyl artifact AC	UHYAC	2380	341_{4}	326_{100}	197_{6}	86_{30}	70_{30}	1162	1591
Bupranolol-M (HO-methoxy-) 2AC	UHYAC	2500	386_{26}	112_{20}	86_{100}	70_{17}		1449	1592
Buprenorphine	G	3360	467_{12}	435_{27}	410_{47}	378_{100}	55_{82}	1612	212
Buprenorphine 2HFB		2820	645_{18}	630_{20}	604_{15}	562_{68}	55_{100}	1733	6345
Buprenorphine 2PFP		2775	595_{12}	580_{27}	554_{24}	512_{71}	55_{100}	1732	6343
Buprenorphine 2TFA		2800	545_{25}	530_{28}	504_{22}	462_{81}	55_{100}	1724	6341
Buprenorphine AC	U+UHYAC-I	3410	509_{6}	452_{60}	420_{100}	408_{15}	55_{50}	1669	211
Buprenorphine -CH3OH AC		3380	477_{49}	435_{38}	393_{50}	273_{64}	55_{100}	1631	9355
Buprenorphine -CH3OH HFB		2770	631_{38}	590_{22}	547_{45}	273_{39}	55_{100}	1721	6339
Buprenorphine -CH3OH TFA		2785	531_{31}	490_{15}	447_{36}	273_{40}	55_{100}	1685	6338
Buprenorphine -CH6O AC		3530	475_{74}	433_{29}	391_{76}	272_{35}	55_{100}	1629	9356
Buprenorphine -H2O		3240	449_{100}	434_{95}	408_{42}	392_{18}	55_{64}	1580	3421
Buprenorphine -H2O AC		3320	491_{100}	476_{95}	450_{67}	434_{29}	55_{71}	1650	3418
Buprenorphine -H2O HFB		2800	645_{37}	630_{55}	604_{27}	84_{19}	55_{100}	1722	6344
Buprenorphine -H2O PFP		2730	595_{9}	580_{14}	555_{3}	498_{2}	55_{100}	1713	6342
Buprenorphine -H2O TFA		2770	545_{41}	530_{56}	504_{33}	434_{7}	55_{100}	1693	6340
Buprenorphine HFB		2960	663_{2}	606_{31}	574_{100}	548_{16}	55_{79}	1725	6346
Buprenorphine ME	UME	3330	481_{4}	448_{7}	424_{30}	392_{100}	55_{50}	1637	6318
Buprenorphine PFP		3040	613_{2}	556_{38}	524_{100}	512_{21}	55_{51}	1717	6123
Buprenorphine TFA		2920	563_{1}	548_{6}	506_{42}	474_{91}	55_{100}	1702	6337
Buprenorphine TMS		3890	539_{4}	506_{17}	482_{30}	450_{100}	55_{49}	1688	5698
Buprenorphine-D4 ME	UME	3315	485_{7}	428_{37}	396_{100}	370_{14}	59_{9}	1642	6354
Buprenorphine-M (nor-)		3420	413_{3}	395_{19}	356_{54}	338_{100}	324_{91}	1488	7774
Buprenorphine-M (nor-) 2AC		3870	497_{1}	482_{2}	440_{100}	408_{45}	366_{26}	1657	7776
Buprenorphine-M (nor-) 2ME	UGLUCME	3100	441_{4}	409_{11}	384_{70}	352_{100}	326_{15}	1563	6328
Buprenorphine-M (nor-) -CH3OH 2AC		3730	465_{32}	450_{100}	423_{29}	367_{21}	282_{43}	1609	9357
Buprenorphine-M (nor-) ME		3330	427_{6}	409_{19}	395_{33}	370_{46}	338_{100}	1527	7775
Buprenorphine-M (nor-)-D3		3080	416_{3}	398_{18}	359_{44}	341_{100}	324_{91}	1497	7301
Buprenorphine-M (nor-)-D3 2AC		3690	500_{1}	485_{2}	443_{100}	408_{41}	366_{29}	1659	7304
Buprenorphine-M (nor-)-D3 2ME		3050	444_{3}	426_{7}	408_{14}	387_{84}	352_{100}	1571	7303
Buprenorphine-M (nor-)-D3 2TMS		3110	560_{2}	527_{9}	503_{29}	468_{95}	73_{100}	1700	7307
Buprenorphine-M (nor-)-D3 AC		3670	458_{2}	443_{2}	401_{83}	383_{84}	366_{100}	1598	7305
Buprenorphine-M (nor-)-D3 -H2O 2TFA		2740	590_{10}	555_{17}	533_{100}	478_{19}	81_{55}	1712	7306
Buprenorphine-M (nor-)-D3 ME		3070	430_{6}	412_{25}	373_{46}	355_{39}	338_{100}	1536	7302
Buprenorphine-M (nor-)-D3 TMS		3080	488_{1}	470_{3}	455_{10}	396_{56}	73_{100}	1646	7308
Bupropion	P-I	1695	239_{1}	224_{11}	139_{16}	111_{25}	100_{100}	552	4699
Bupropion AC		2210	264_{40}	225_{19}	208_{52}	183_{100}	57_{43}	787	5700
Bupropion formyl artifact		1755	237_{5}	222_{4}	139_{19}	98_{38}	57_{100}	609	4700
Bupropion-M (3-chlorobenzoic acid)		1430*	156_{75}	139_{100}	115_{46}	75_{44}	65_{6}	299	6024
Bupropion-M (3-chlorobenzoic acid) (ME)	U+UHYAC	1100*	170_{26}	139_{100}	111_{57}	75_{50}		323	6953
Bupropion-M (3-chlorobenzyl alcohol)		1560*	142_{53}	125_{7}	113_{28}	107_{33}	77_{100}	281	6025
Bupropion-M (dihydro-) AC	U+UHYAC	1780	268_{2}	208_{13}	115_{6}	100_{100}		796	8520
Bupropion-M (dihydro-HO-) isomer-1 2AC	U+UHYAC P-I	2100	326_{1}	266_{7}	224_{4}	100_{100}	57_{25}	1161	8537
Bupropion-M (dihydro-HO-) isomer-2 2AC	U+UHYAC	2120	326_{1}	266_{7}	224_{3}	163_{13}	100_{100}	1162	8538
Bupropion-M (dihydro-HO-methoxy-) 2AC	U+UHYAC	2190	356_{1}	296_{3}	254_{7}	187_{2}	100_{100}	1318	8539
Bupropion-M (HO-)	P-I	2040	240_{3}	224_{30}	166_{7}	139_{29}	116_{100}	632	7660
Bupropion-M (HO-) AC		2130	224_{12}	166_{4}	158_{100}	115_{44}	98_{8}	889	7661
Bupropion-M (HO-) TMS		2075	240_{17}	224_{79}	188_{97}	145_{53}	73_{100}	1074	7662
Bupropion-M (N-dealkyl-)		1430*	183_{100}	166_{17}	141_{75}	139_{45}	111_{33}	353	10297

Table 8.1: Compounds in order of names **Bupropion-M/artifact**

Name	Detected	RI	Typical ions and intensities					Page	Entry
Bupropion-M/artifact		2075*	309 $_{34}$	224 $_{34}$	174 $_{100}$	139 $_{21}$	119 $_{38}$	960	10298
Bupropion-M/artifact		2430*	309 $_{100}$	280 $_{17}$	224 $_{27}$	111 $_{26}$	57 $_{66}$	960	10299
Bupropion-M/artifact		1350*	223 $_{4}$	208 $_{5}$	166 $_{100}$	139 $_{38}$	103 $_{26}$	481	10300
Buspirone	G U+UHYAC	3300	385 $_{42}$	290 $_{21}$	277 $_{100}$	265 $_{95}$	177 $_{78}$	1382	1779
Butabarbital	P G U+UHYAC	1655	183 $_{6}$	156 $_{100}$	141 $_{84}$			439	149
Butabarbital 2ME		1565	211 $_{10}$	184 $_{75}$	169 $_{100}$			557	646
Butabarbital-M (HO-)	U	1925	213 $_{4}$	199 $_{8}$	181 $_{6}$	156 $_{100}$	141 $_{64}$	502	150
Butabarbital-M (HO-) -H2O	UHY U+UHYAC	1905	210 $_{6}$	181 $_{51}$	156 $_{40}$	141 $_{48}$	55 $_{100}$	433	2952
Butalamine		2590	273 $_{1}$	188 $_{6}$	155 $_{7}$	142 $_{100}$	100 $_{21}$	1007	2285
Butalbital	P G UHY U+UHYAC	1690	209 $_{3}$	181 $_{30}$	168 $_{100}$	167 $_{88}$	141 $_{24}$	489	151
Butalbital (ME)	P U	1630	223 $_{3}$	195 $_{18}$	182 $_{100}$	181 $_{62}$	155 $_{10}$	548	153
Butalbital 2ME	PME	1655	237 $_{6}$	209 $_{23}$	196 $_{100}$	195 $_{80}$	169 $_{18}$	619	154
Butalbital 2TMS		1790	368 $_{1}$	353 $_{100}$	312 $_{54}$	100 $_{37}$	73 $_{83}$	1303	4531
Butalbital-M (HO-)	U UHY U+UHYAC	1940	240 $_{2}$	168 $_{100}$	167 $_{41}$	141 $_{18}$		556	152
Butallylonal	G P U	1990	223 $_{38}$	167 $_{100}$	124 $_{20}$			917	1916
Butane		400*	58 $_{13}$	43 $_{100}$	41 $_{40}$	29 $_{42}$	27 $_{42}$	249	3808
1,2-Butane diol dibenzoate		2300*	298 $_{1}$	227 $_{10}$	193 $_{15}$	105 $_{100}$	77 $_{84}$	895	1762
1,2-Butane diol phenylboronate		1350*	176 $_{40}$	147 $_{100}$	105 $_{28}$	91 $_{62}$	77 $_{16}$	332	1900
1,2-Butane diol dipivalate		1425*	157 $_{2}$	143 $_{5}$	103 $_{10}$	85 $_{29}$	57 $_{100}$	648	6425
1,3-Butane diol dibenzoate		2300*	269 $_{1}$	241 $_{4}$	227 $_{6}$	176 $_{24}$	105 $_{100}$	894	1763
1,3-Butane diol phenylboronate		1390*	176 $_{44}$	161 $_{50}$	104 $_{30}$	91 $_{30}$	77 $_{16}$	333	1901
1,3-Butane diol dipivalate		1420*	173 $_{1}$	157 $_{10}$	103 $_{21}$	85 $_{17}$	57 $_{100}$	648	6424
1,4-Butane diol dibenzoate		2400*	298 $_{1}$	193 $_{8}$	176 $_{8}$	105 $_{100}$	77 $_{51}$	895	1764
1,4-Butane diol phenylboronate		1420*	176 $_{58}$	146 $_{22}$	105 $_{100}$	91 $_{10}$		333	1902
1,4-Butane diol dipivalate		1520*	156 $_{8}$	103 $_{14}$	101 $_{13}$	85 $_{20}$	57 $_{100}$	648	1906
Butanilicaine		2030	254 $_{1}$	219 $_{7}$	141 $_{22}$	86 $_{100}$	72 $_{35}$	628	1410
1-Butanol		<1000*	74 $_{1}$	73 $_{1}$	56 $_{76}$	41 $_{71}$	31 $_{100}$	252	2448
2-Butanol		<1000*	74 $_{2}$	59 $_{20}$	45 $_{100}$			252	2447
Butaperazine	G U UHY UHYAC	3190	409 $_{6}$	269 $_{4}$	141 $_{20}$	113 $_{50}$	70 $_{100}$	1475	155
Butaperazine-M (nor-) AC	U+UHYAC	3800	437 $_{70}$	269 $_{100}$	141 $_{98}$	99 $_{66}$		1552	1254
1-Butene		<1000*	56 $_{34}$	41 $_{100}$	39 $_{47}$	27 $_{58}$		248	3807
2-Butene		<1000*	56 $_{44}$	41 $_{100}$	39 $_{52}$	27 $_{47}$		248	3806
Butethamate		1760	263 $_{15}$	248 $_{15}$	191 $_{42}$	99 $_{26}$	86 $_{100}$	678	156
Butethamate-M/artifact (HOOC-)	U UHY UHYAC	1300*	164 $_{11}$	119 $_{29}$	91 $_{100}$	77 $_{10}$		311	2912
Butethamate-M/artifact (HOOC-) ME	UME	1200*	178 $_{11}$	150 $_{3}$	119 $_{51}$	91 $_{100}$	77 $_{4}$	338	2911
Butinoline	P G U	2285	291 $_{53}$	290 $_{51}$	115 $_{70}$	105 $_{96}$	70 $_{100}$	850	3237
Butinoline artifact-1	U	1990	221 $_{5}$	175 $_{19}$	147 $_{100}$	115 $_{44}$	77 $_{27}$	471	3239
Butinoline artifact-2	U	2045*	236 $_{15}$	207 $_{8}$	165 $_{100}$	105 $_{47}$	77 $_{34}$	537	3238
Butinoline-M (benzophenone)	U+UHYAC	1610*	182 $_{31}$	152 $_{3}$	105 $_{100}$	77 $_{70}$	51 $_{39}$	351	1624
Butinoline-M/artifact	U UHY UHYAC	1850*	209 $_{4}$	167 $_{100}$	152 $_{22}$	121 $_{8}$	115 $_{7}$	426	2081
Butinoline-M/artifact	U	2675	304 $_{48}$	220 $_{69}$	115 $_{48}$	105 $_{100}$	98 $_{51}$	928	3236
Butizide 2ME		3785	381 $_{6}$	366 $_{13}$	324 $_{100}$	246 $_{2}$	230 $_{2}$	1359	3094
Butizide 3ME		3455	395 $_{7}$	380 $_{12}$	338 $_{100}$	313 $_{3}$	230 $_{4}$	1423	3095
Butizide 4ME	UEXME	3100	409 $_{48}$	352 $_{100}$	309 $_{3}$	244 $_{10}$	145 $_{4}$	1473	3096
Butobarbital	P G U UHY UHYAC	1665	197 $_{2}$	184 $_{10}$	156 $_{96}$	141 $_{100}$	98 $_{19}$	440	157
Butobarbital 2ME		1585	212 $_{6}$	184 $_{60}$	169 $_{100}$	112 $_{24}$		557	647
Butobarbital 2TMS		1720	356 $_{2}$	341 $_{89}$	300 $_{29}$	100 $_{40}$	73 $_{100}$	1244	5464
Butobarbital-M (HO-)	U UHY	1920	213 $_{2}$	199 $_{5}$	156 $_{100}$	141 $_{73}$	98 $_{12}$	502	159
Butobarbital-M (HO-) AC	UHYAC	1940	227 $_{25}$	198 $_{33}$	181 $_{43}$	156 $_{100}$	87 $_{25}$	718	2953
Butobarbital-M (oxo-)	U UHY UHYAC	1880	211 $_{4}$	198 $_{39}$	156 $_{100}$	141 $_{65}$	128 $_{13}$	494	158
Butocarboxim		1595	144 $_{39}$	133 $_{4}$	87 $_{100}$	75 $_{47}$	55 $_{56}$	369	1327
Butoxycarboxim		1940	165 $_{16}$	108 $_{5}$	86 $_{97}$	85 $_{100}$	55 $_{78}$	478	4382
Butoxycarboxim artifact		1405	149 $_{1}$	108 $_{7}$	86 $_{100}$			287	2271
Buturon		2135	236 $_{28}$	152 $_{9}$	111 $_{14}$	75 $_{19}$	56 $_{100}$	537	4138
Butyl-2-ethylhexylphthalate		1950*	223 $_{7}$	205 $_{4}$	149 $_{100}$	104 $_{4}$	57 $_{7}$	1121	713
Butyl-2-methylpropylphthalate		1970*	278 $_{1}$	223 $_{4}$	205 $_{4}$	149 $_{100}$	76 $_{4}$	769	2995
N-Butyl-3-(2-methoxybenzoyl-)indole		2620	307 $_{100}$	290 $_{70}$	264 $_{62}$	200 $_{56}$	144 $_{58}$	951	9622
1-Butylamine		<1000	73 $_{14}$	39 $_{9}$	30 $_{100}$			252	4183
2-Butylamine		<1000	73 $_{7}$	39 $_{5}$	30 $_{100}$			251	4190
tert.-Butylamine		<1000	73 $_{1}$	58 $_{100}$	41 $_{23}$	30 $_{13}$		251	4184
Butylhexadecanoate		2340*	312 $_{2}$	257 $_{8}$	239 $_{4}$	129 $_{7}$	56 $_{100}$	985	160
Butyloctadecanoate		2380*	340 $_{12}$	285 $_{35}$	267 $_{19}$	129 $_{20}$	56 $_{100}$	1158	161
Butyloctylphthalate		1950*	223 $_{6}$	205 $_{4}$	149 $_{100}$	122 $_{4}$	104 $_{11}$	1121	2361
Butylone		1740	192 $_{9}$	162 $_{4}$	149 $_{21}$	121 $_{19}$	72 $_{100}$	473	8319
Butylone AC	U+UHYAC	2215	263 $_{10}$	149 $_{27}$	121 $_{15}$	114 $_{84}$	72 $_{100}$	674	7872
Butylone HFB		1800	417 $_{2}$	268 $_{29}$	210 $_{16}$	149 $_{100}$	121 $_{21}$	1498	8323
Butylone ME		1800	206 $_{1}$	149 $_{12}$	121 $_{13}$	91 $_{6}$	86 $_{100}$	532	8320
Butylone PFP		1790	367 $_{7}$	268 $_{2}$	218 $_{28}$	149 $_{100}$	121 $_{17}$	1295	8324
Butylone TFA		1815	317 $_{11}$	168 $_{40}$	149 $_{100}$	121 $_{25}$	110 $_{16}$	1010	8325
Butylone TMS		1890	293 $_{1}$	278 $_{10}$	179 $_{4}$	179 $_{4}$	144 $_{100}$	863	8321
Butylone-M (demethylenyl-methyl-) 2AC	U+UHYAC	2250	307 $_{2}$	151 $_{13}$	123 $_{3}$	114 $_{80}$	72 $_{100}$	948	7966
Butylone-M (dihydro-) 2AC	U+UHYAC	2105	307 $_{2}$	193 $_{3}$	151 $_{7}$	114 $_{73}$	72 $_{100}$	948	7976
Butylone-M (nor-) AC	U+UHYAC	2200	249 $_{7}$	181 $_{5}$	149 $_{34}$	100 $_{42}$	58 $_{100}$	595	7873
Butylone-M (nor-demethylenyl-methyl-) 2AC	U+UHYAC	2020	293 $_{2}$	234 $_{48}$	151 $_{50}$	100 $_{50}$	58 $_{100}$	862	7975

Butylparaben
Table 8.1: Compounds in order of names

Name	Detected	RI	Typical ions and intensities					Page	Entry
Butylparaben		1700*	194_{16}	138_{100}	121_{98}			383	162
Butylscopolaminium bromide-M/artifact (scopolamine) AC	U+UHYAC	2450	345_{13}	154_{22}	138_{59}	108_{41}	94_{100}	1185	1526
Butylscopolaminium bromide-M/artifact (scopolamine) -H2O	U+UHYAC	2230	285_{18}	154_{22}	138_{38}	108_{43}	94_{100}	810	960
Byproduct 1 of APAAN hydrolysis		1980*	232_{100}	215_{45}	202_{27}	189_{7}	152_{6}	518	10164
Byproduct 2 of APAAN hydrolysis		2075*	232_{100}	217_{78}	215_{51}	202_{33}	189_{7}	518	10210
Byproduct 3 of APAAN hydrolysis		2540	275_{70}	274_{100}	246_{11}	230_{6}	202_{6}	746	10211
Byproduct 3 of APAAN hydrolysis AC		2320	275_{69}	274_{100}	246_{8}	230_{6}	202_{5}	1012	10212
BZP		1530	176_{9}	146_{5}	134_{55}	91_{100}	56_{29}	334	5880
BZP AC		1915	218_{4}	146_{27}	132_{17}	91_{100}	85_{20}	460	5881
BZP HFB		1730	372_{13}	295_{6}	281_{15}	175_{13}	91_{100}	1322	5884
BZP PFP		1690	322_{18}	245_{9}	231_{21}	175_{15}	91_{100}	1044	5883
BZP TFA		1665	272_{3}	195_{4}	181_{10}	146_{6}	91_{100}	730	5882
BZP TMS		1860	248_{31}	157_{27}	102_{100}	91_{69}	73_{97}	592	5885
BZP-M (piperazine) 2AC		1750	170_{15}	85_{33}	69_{25}	56_{100}		324	879
BZP-M (piperazine) 2HFB		1290	478_{3}	459_{9}	309_{100}	281_{22}	252_{41}	1631	6634
BZP-M (piperazine) 2TFA		1005	278_{10}	209_{59}	152_{25}	69_{56}	56_{100}	766	4129
Cabergoline artifact		1550	173_{5}	158_{4}	129_{5}	72_{21}	58_{100}	329	8191
Cabergoline artifact (-COOH) 2ME		2685	324_{36}	324_{36}	209_{40}	167_{54}	154_{100}	1058	8194
Cabergoline artifact (-COOH) ME		2730	310_{100}	279_{12}	269_{18}	209_{40}	154_{61}	972	8193
Cabergoline artifact (-COOH) MEHFB		2700	506_{100}	405_{24}	340_{34}	232_{44}	154_{43}	1666	8198
Cabergoline artifact (-COOH) MEPFP		2730	456_{100}	355_{29}	290_{43}	232_{43}	154_{84}	1593	8197
Cabergoline artifact (-COOH) METFA		2695	406_{80}	305_{20}	240_{31}	154_{100}	127_{45}	1464	8196
Cabergoline artifact (-COOH) METMS		2750	382_{92}	341_{14}	281_{17}	226_{28}	73_{100}	1366	8195
Cabergoline artifact TMS		1290	245_{19}	147_{18}	127_{12}	73_{34}	58_{100}	577	8192
3-CAF		3300	382_{9}	239_{100}	210_{14}	184_{6}	115_{14}	1364	10185
3-CAF-M/artifact (2-naphthol)		1470*	144_{100}	115_{40}	89_{6}	72_{6}		283	9500
3-CAF-M/artifact (HOOC-) (ET)		2170	284_{51}	239_{55}	212_{100}	191_{11}		800	9663
3-CAF-M/artifact (HOOC-) (ME)		2150	270_{100}	239_{78}	212_{58}	184_{11}	75_{11}	716	9499
Cafedrine -H2O	G P UHY	2960	339_{2}	277_{3}	250_{100}	207_{40}	70_{10}	1151	1313
Cafedrine -H2O AC	U+UHYAC	3285	381_{1}	339_{2}	292_{52}	250_{100}	207_{38}	1362	1739
Cafedrine -H2O PFP		2790	485_{1}	339_{6}	206_{36}	146_{59}	132_{100}	1641	6118
Cafedrine TMS		2815	415_{1}	250_{100}	207_{32}	73_{40}	70_{20}	1532	6216
Cafedrine-M (cathinone) AC		1610	191_{2}	134_{2}	105_{35}	86_{100}	77_{48}	371	5901
Cafedrine-M (cathinone) HFB		1395	345_{1}	240_{6}	169_{4}	105_{100}	77_{36}	1182	5904
Cafedrine-M (cathinone) PFP		1335	190_{6}	119_{7}	105_{100}	77_{40}	69_{5}	874	5903
Cafedrine-M (cathinone) TFA		1350	245_{1}	140_{7}	105_{100}	77_{48}	69_{10}	573	5902
Cafedrine-M (cathinone) TMS		1590	206_{14}	191_{15}	116_{100}	77_{27}	73_{74}	475	5905
Cafedrine-M (etofylline)	UHY	2125	224_{47}	194_{16}	180_{100}	109_{34}	95_{74}	487	771
Cafedrine-M (etofylline) AC	U+UHYAC	2200	266_{79}	206_{59}	180_{34}	122_{31}	87_{100}	693	772
Cafedrine-M (etofylline) TMS	UHYTMS	2160	296_{10}	281_{26}	252_{6}	180_{100}	73_{47}	886	5696
Cafedrine-M (N-dealkyl-) AC	U+UHYAC	2480	265_{35}	206_{100}	180_{34}	122_{18}	86_{26}	687	1886
Cafedrine-M (norpseudoephedrine)	U UHY	1360	132_{4}	117_{9}	105_{22}	79_{54}	77_{100}	291	1154
Cafedrine-M (norpseudoephedrine) 2AC	U+UHYAC	1740	235_{2}	176_{4}	129_{8}	107_{9}	86_{100}	534	1155
Cafedrine-M (norpseudoephedrine) 2HFB		1335	330_{16}	303_{6}	240_{100}	169_{19}	119_{12}	1691	7418
Cafedrine-M (norpseudoephedrine) formyl artifact		1280	117_{2}	105_{2}	91_{4}	77_{6}	57_{100}	307	4649
Cafedrine-M (norpseudoephedrine) TMSTFA		1630	213_{7}	191_{7}	179_{100}	149_{5}	73_{80}	1025	6260
Caffeic acid 2AC		2240*	264_{2}	222_{12}	180_{100}	163_{8}	134_{15}	680	5968
Caffeic acid 2ME		1930*	208_{100}	177_{36}	145_{53}	133_{19}	117_{24}	424	5966
Caffeic acid 3ME	UME	1850*	222_{100}	207_{22}	191_{75}	164_{16}	147_{15}	479	4945
Caffeic acid 3TMS		2115*	396_{86}	381_{22}	219_{78}	191_{17}	73_{100}	1430	6014
Caffeic acid artifact (dihydro-)		2400*	182_{17}	136_{13}	123_{100}	77_{55}	51_{67}	351	5763
Caffeic acid artifact (dihydro-) 3TMS		2250*	398_{67}	280_{16}	267_{39}	179_{98}	73_{100}	1438	5996
Caffeic acid artifact (dihydro-) -CO2		1295*	138_{23}	123_{100}	91_{28}	77_{47}	51_{92}	278	5756
Caffeic acid artifact (dihydro-) ME		1870*	196_{24}	136_{45}	123_{100}	91_{11}	77_{16}	389	5764
Caffeic acid artifact (dihydro-) ME2HFB		1720*	588_{11}	528_{100}	349_{32}	169_{37}	69_{58}	1711	5994
Caffeic acid artifact (dihydro-) ME2PFP		1590*	488_{19}	428_{100}	299_{26}	281_{26}	119_{73}	1645	5993
Caffeic acid artifact (dihydro-) ME2TMS		2220*	340_{44}	267_{36}	193_{10}	179_{100}	73_{97}	1156	5995
Caffeic acid artifact (dihydro-) MEAC		1980*	280_{2}	238_{15}	196_{100}	136_{65}	123_{68}	780	5992
Caffeic acid artifact (dihydro-) METFA		1540*	292_{88}	233_{100}	195_{27}	107_{46}	69_{62}	854	5969
Caffeic acid -CO2		1375*	136_{53}	89_{71}	77_{48}	63_{7}	51_{100}	276	5757
Caffeic acid ME2AC		2170*	278_{3}	236_{15}	194_{100}	163_{40}	134_{21}	766	5967
Caffeic acid ME2HFB		1985*	586_{56}	555_{74}	389_{18}	169_{93}	69_{100}	1709	5971
Caffeic acid ME2PFP		1985*	486_{61}	455_{89}	323_{14}	119_{100}	77_{31}	1643	5970
Caffeic acid ME2TMS		1930*	338_{72}	297_{9}	219_{100}	191_{21}	73_{97}	1143	6013
Caffeine	P G UHY U+UHYAC	1820	194_{100}	109_{81}	82_{33}	67_{44}	55_{67}	383	191
Caffeine-M (1-nor-)	P G U+UHYAC	1980	180_{100}	137_{12}	109_{41}	82_{38}		345	989
Caffeine-M (1-nor-) TMS		2020	252_{20}	237_{100}	109_{15}	100_{12}	73_{25}	618	5452
Caffeine-M (7-nor-)	P G U+UHYAC	2025	180_{100}	95_{85}	68_{69}			345	990
Caffeine-M (7-nor-) TMS		1920	252_{61}	237_{100}	223_{14}	135_{7}	73_{37}	618	4600
Caffeine-M (HO-) ME	UME	1930	224_{100}	209_{56}	139_{7}	124_{29}	83_{66}	487	5044
Californine		2615	323_{12}	322_{13}	188_{100}	165_{4}	130_{7}	1049	6735
Californine-M (bis-(demethylene-methyl-)) isomer-1		2860	327_{35}	311_{5}	190_{100}	175_{7}		1076	6734
Californine-M (bis-(demethylene-methyl-)) isomer-1 2AC		2920	411_{7}	368_{6}	348_{8}	232_{100}	190_{59}	1483	6729
Californine-M (bis-(demethylene-methyl-)) isomer-2 2AC		3040	411_{11}	368_{6}	326_{5}	232_{100}	190_{63}	1483	6730

Table 8.1: Compounds in order of names   **Californine-M (bis-(demethylene-methyl-)) isomer-3 2AC**

Name	Detected	RI	Typical ions and intensities					Page	Entry
Californine-M (bis-(demethylene-methyl-)) isomer-3 2AC		3055	411_{19}	368_{15}	326_{11}	232_{98}	190_{100}	1482	6731
Californine-M (demethylene-) 2AC		3025	395_{22}	310_{16}	218_{16}	188_{100}	176_{13}	1425	6724
Californine-M (demethylene-) AC		2960	353_{37}	310_{23}	218_{21}	188_{100}	176_{23}	1226	6723
Californine-M (demethylene-methyl-) isomer-1		2810	325_{42}	310_{4}	294_{4}	190_{39}	188_{100}	1062	6725
Californine-M (demethylene-methyl-) isomer-1 AC		2910	367_{33}	324_{13}	250_{5}	232_{30}	188_{100}	1297	6727
Californine-M (demethylene-methyl-) isomer-2		2820	325_{67}	309_{13}	294_{10}	190_{91}	188_{100}	1062	6726
Californine-M (demethylene-methyl-) isomer-2 AC		2920	367_{33}	324_{21}	310_{6}	232_{18}	188_{100}	1297	6728
Californine-M (nor-)		2625	309_{39}	174_{100}	147_{11}	95_{6}		961	6732
Californine-M (nor-) AC		3090	351_{45}	308_{28}	292_{12}	216_{25}	174_{100}	1211	6733
Californine-M (nor-demethylene-) 3AC		3350	423_{15}	339_{16}	280_{22}	216_{37}	174_{100}	1514	6735
Californine-M (nor-demethylene-methyl-) 2AC		3220	395_{37}	353_{37}	310_{41}	216_{36}	174_{100}	1425	6736
Californine-M/artifact (reframidine)		2735	323_{33}	322_{42}	280_{40}	188_{100}		1049	6737
Camazepam	G	2960	371_{9}	299_{6}	271_{33}	255_{16}	72_{100}	1316	416
Camazepam HY	UHY U+UHYAC	2100	245_{95}	228_{38}	193_{29}	105_{38}	77_{100}	573	272
Camazepam HYAC	U+UHYAC	2260	287_{11}	244_{100}	228_{39}	182_{49}	77_{70}	818	2542
Camazepam-M	P G UGLUC	2320	268_{98}	239_{56}	233_{52}	205_{66}	77_{100}	813	579
Camazepam-M (temazepam)	P UGLUC	2625	300_{33}	271_{100}	256_{23}	228_{16}	77_{30}	905	417
Camazepam-M (temazepam) AC	UGLUCAC	2730	342_{6}	300_{40}	271_{100}	255_{16}	77_{17}	1167	2099
Camazepam-M (temazepam) artifact-1	G	2475	256_{19}	241_{7}	179_{100}	163_{7}	77_{8}	636	5780
Camazepam-M (temazepam) artifact-2	G	2815	270_{64}	269_{100}	254_{12}	228_{26}	191_{5}	715	5779
Camazepam-M (temazepam) ME		2600	314_{60}	271_{100}	255_{46}			993	418
Camazepam-M (temazepam) TMS		2665	372_{23}	343_{100}	283_{26}	257_{38}	73_{54}	1322	4598
Camazepam-M 2TMS		2200	430_{51}	429_{89}	340_{15}	313_{19}	73_{100}	1533	5499
Camazepam-M HY	UHY	2050	231_{80}	230_{95}	154_{23}	105_{38}	77_{100}	512	419
Camazepam-M HYAC	PHYAC U+UHYAC	2245	273_{30}	230_{100}	154_{13}	105_{23}	77_{50}	732	273
Camazepam-M TMS		2635	356_{9}	341_{100}	312_{56}	239_{12}	135_{21}	1252	4577
Camfetamine		1620	201_{83}	172_{34}	91_{45}	84_{100}	70_{67}	402	8952
Camfetamine AC		2100	243_{1}	170_{82}	142_{100}	84_{35}	70_{36}	569	8954
Camfetamine HFB		1860	328_{3}	266_{11}	210_{49}	170_{90}	142_{100}	1434	8958
Camfetamine ME		1640	215_{63}	172_{13}	98_{100}	91_{38}	84_{36}	450	8953
Camfetamine PFP		1840	216_{12}	201_{62}	170_{100}	142_{96}	84_{47}	1194	8957
Camfetamine TFA		1860	296_{1}	228_{6}	170_{100}	142_{97}	110_{38}	890	8956
Camfetamine TMS		1830	273_{68}	258_{46}	245_{18}	156_{91}	73_{100}	736	8955
Camfetamine-M (di-HO-) 3AC		2620	359_{1}	317_{2}	258_{36}	216_{80}	188_{100}	1259	8969
Camfetamine-M (HO-alkyl-) isomer-1 2AC		2440	301_{1}	228_{11}	198_{9}	168_{29}	142_{100}	915	8963
Camfetamine-M (HO-alkyl-) isomer-2 2AC		2500	301_{1}	228_{37}	142_{100}	117_{16}	91_{23}	915	8965
Camfetamine-M (HO-aryl-)		2010	217_{61}	188_{21}	98_{40}	84_{100}	70_{57}	455	8961
Camfetamine-M (HO-aryl-) isomer-1 2AC		2600	301_{4}	258_{6}	228_{86}	200_{83}	186_{100}	915	8966
Camfetamine-M (HO-aryl-) isomer-2 2AC		2650	301_{1}	258_{5}	228_{93}	186_{76}	158_{100}	915	8967
Camfetamine-M (HO-methoxy-aryl-)		2060	247_{58}	216_{14}	137_{37}	84_{100}	70_{79}	587	8962
Camfetamine-M (HO-methoxy-aryl-) 2AC		2680	331_{1}	258_{72}	216_{100}	188_{73}	137_{11}	1102	8970
Camfetamine-M (nor-)		1600	187_{100}	170_{22}	158_{49}	70_{69}	56_{83}	364	8959
Camfetamine-M (nor-) AC	U+UHYAC	2005	229_{1}	186_{3}	170_{98}	142_{100}	91_{49}	506	776
Camfetamine-M (nor-HO-alkyl-)		1850	203_{39}	159_{40}	115_{40}	91_{58}	56_{100}	407	8960
Camfetamine-M (nor-HO-alkyl-) isomer-1 2AC		2470	287_{2}	228_{24}	168_{66}	142_{100}	94_{31}	822	8964
Camfetamine-M (nor-HO-alkyl-) isomer-2 2AC		2520	287_{1}	228_{29}	168_{25}	142_{100}	94_{11}	822	777
Camfetamine-M (nor-HO-aryl-) 2AC		2540	287_{2}	228_{64}	200_{59}	186_{90}	158_{100}	822	8968
Camfetamine-M (nor-HO-methoxy-aryl-) 2AC		2660	317_{2}	258_{34}	216_{77}	188_{100}	137_{17}	1013	8971
Camylofine	G U UHY UHYAC	2085	320_{3}	205_{2}	118_{6}	86_{100}	58_{15}	1035	1411
Cannabidiol	G U-I	2400*	314_{5}	246_{16}	231_{100}	174_{21}	121_{27}	997	648
Cannabidiol 2AC		2450*	398_{14}	355_{28}	273_{30}	231_{100}	121_{38}	1439	649
Cannabidiol 2TMS		2330*	458_{9}	390_{100}	337_{41}	301_{22}	73_{87}	1598	4679
Cannabidiol AC		2420*	356_{12}	273_{26}	231_{100}	174_{22}	121_{32}	1244	6461
Cannabidiol ME		2670*	328_{16}	313_{41}	285_{46}	272_{100}	229_{21}	1083	10136
Cannabidivarol		2165*	286_{7}	218_{14}	203_{100}	174_{9}	121_{8}	816	4071
Cannabidivarol 2AC		2630*	370_{19}	327_{100}	285_{70}	257_{12}		1314	4072
Cannabielsoic acid -CO2		2405*	330_{67}	247_{76}	205_{100}	148_{43}	108_{20}	1096	4073
Cannabielsoic acid -CO2 2AC		2540*	414_{3}	330_{51}	312_{100}	247_{50}	205_{70}	1491	4074
Cannabigerol		2500*	316_{17}	247_{12}	231_{31}	193_{100}	123_{29}	1008	4075
Cannabigerol 2AC		2595*	400_{7}	247_{73}	193_{67}	123_{90}	69_{100}	1447	4076
Cannabinol	G UHY	2555*	310_{13}	295_{100}	238_{23}	223_{9}		972	650
Cannabinol AC	UHYAC	2540*	352_{28}	337_{100}	295_{71}	238_{20}		1223	651
Cannabinol ME		2335*	324_{23}	309_{100}	252_{13}	238_{17}	209_{14}	1059	10137
Cannabinol TMS		2485*	382_{10}	367_{100}	310_{6}	238_{5}	73_{17}	1367	4532
Cannabispirol AC		2350*	290_{18}	248_{19}	189_{92}	176_{100}	115_{16}	842	6462
Cannabispirone AC		2350*	288_{7}	189_{10}	176_{100}	115_{9}		828	6463
Canrenoic acid		3100*	358_{100}	329_{35}	274_{13}	201_{14}	85_{68}	1255	2743
Canrenoic acid -H2O	P UHY UHYAC	3250*	340_{100}	325_{18}	267_{80}	227_{15}		1157	2344
Canrenoic acid -H2O ME		3130*	354_{100}	339_{19}	173_{9}	149_{42}	115_{8}	1234	2744
Canrenone	P UHY UHYAC	3250*	340_{100}	325_{18}	267_{80}	227_{15}		1157	2344
Capric acid		1340*	172_{3}	129_{34}	87_{15}	73_{99}	60_{100}	327	5629
Capric acid ET		1370*	200_{3}	157_{23}	155_{24}	101_{41}	88_{100}	400	5399
Capric acid ME		1360*	186_{3}	155_{8}	143_{15}	87_{45}	74_{100}	362	2665
Caprylic acid cetylester		2500*	368_{3}	224_{4}	145_{100}	88_{26}	57_{65}	1304	6565

Caprylic acid ET   **Table 8.1:** Compounds in order of names

Name	Detected	RI	Typical ions and intensities					Page	Entry
Caprylic acid ET		1185*	172 $_2$	143 $_6$	127 $_{22}$	101 $_{40}$	88 $_{100}$	327	5398
Caprylic acid ME		1170*	158 $_1$	127 $_{12}$	115 $_{10}$	87 $_{39}$	74 $_{100}$	302	2664
Capsaicine		2415	305 $_{16}$	195 $_8$	152 $_{21}$	137 $_{100}$	122 $_{11}$	940	6780
Capsaicine AC		2490	347 $_5$	305 $_{29}$	195 $_{33}$	152 $_{45}$	137 $_{100}$	1196	6782
Captafol		2355	347 $_3$	311 $_6$	183 $_5$	107 $_9$	79 $_{100}$	1192	3320
Captafol artifact-1 (cyclohexenedicarboxylic acid) 2ME		1190*	198 $_1$	167 $_{18}$	138 $_{47}$	107 $_{14}$	79 $_{100}$	394	4206
Captafol artifact-2 (cyclohexenedicarboximide)		1450	151 $_{41}$	123 $_9$	80 $_{63}$	79 $_{100}$		290	3321
Captan		2030	299 $_1$	264 $_5$	149 $_{19}$	117 $_{14}$	79 $_{100}$	899	2614
Captan artifact-1 (cyclohexenedicarboxylic acid) 2ME		1190*	198 $_1$	167 $_{18}$	138 $_{47}$	107 $_{14}$	79 $_{100}$	394	4206
Captan artifact-2 (cyclohexenedicarboximide)		1450	151 $_{41}$	123 $_9$	80 $_{63}$	79 $_{100}$		290	3321
Captopril		1925	217 $_{12}$	199 $_{11}$	140 $_{17}$	126 $_{16}$	70 $_{100}$	454	6417
Captopril 2ME		1810	245 $_{25}$	198 $_{17}$	128 $_{38}$	89 $_{39}$	70 $_{100}$	575	6418
Captopril artifact (disulfide) 2ME		3200	460 $_7$	230 $_{100}$	198 $_{76}$	128 $_{68}$	70 $_{99}$	1602	6419
Captopril ME		1730	231 $_{13}$	199 $_{12}$	172 $_{16}$	128 $_{18}$	70 $_{100}$	513	3005
Carazolol	U-I	2810	298 $_{16}$	183 $_{100}$	154 $_{16}$	72 $_{58}$		897	1593
Carazolol 2TMSTFA		2880	538 $_{29}$	284 $_{100}$	255 $_{22}$	129 $_{28}$	73 $_{44}$	1688	6177
Carazolol formyl artifact	U-I	2830	310 $_{29}$	183 $_{100}$	154 $_{22}$	127 $_{51}$	86 $_{26}$	972	1352
Carazolol H2O AC		3130	322 $_{54}$	220 $_{100}$	140 $_{100}$	98 $_{64}$		1046	1353
Carazolol ME		2815	312 $_6$	183 $_{20}$	154 $_{12}$	86 $_{100}$	72 $_{12}$	983	1595
Carazolol TMSTFA		2755	466 $_{24}$	284 $_{100}$	183 $_{34}$	129 $_{80}$	73 $_{58}$	1610	6178
Carazolol-M (deamino-di-HO-) 2AC	UGLUCAC-I	3050	341 $_4$	199 $_7$	159 $_{100}$	99 $_{15}$		1161	1594
Carazolol-M (deamino-tri-HO-) 3AC	UGLUCAC	3290	399 $_{43}$	372 $_{31}$	199 $_{17}$	159 $_{100}$		1441	4253
Carazolol-M/artifact (4-hydroxycarbazole)		2160	183 $_{100}$	154 $_{50}$	127 $_{11}$	77 $_8$		354	7886
Carazolol-M/artifact (4-hydroxycarbazole) AC		2210	225 $_{35}$	207 $_2$	183 $_{100}$	154 $_{38}$	127 $_{11}$	490	7885
Carbamazepine	P G U+UHYAC	2285	236 $_{83}$	193 $_{100}$	165 $_{31}$			538	420
Carbamazepine TMS		2285	308 $_1$	293 $_4$	193 $_{100}$	165 $_{13}$	73 $_{15}$	958	4533
Carbamazepine-M (acridine)	U UHY U+UHYAC	1800	179 $_{100}$	151 $_{14}$				340	421
Carbamazepine-M (formyl-acridine)	U UHY UHYAC	2025	207 $_{98}$	179 $_{100}$	151 $_{36}$			418	422
Carbamazepine-M (HO-methoxy-ring)	U UHY	2340	239 $_{100}$	224 $_{47}$	209 $_{42}$	180 $_{74}$		551	423
Carbamazepine-M (HO-methoxy-ring) AC	U+UHYAC	2420	281 $_{42}$	239 $_{100}$	224 $_{28}$	196 $_{29}$	162 $_{16}$	786	2506
Carbamazepine-M (HO-ring)	UHY	2240	209 $_{100}$	180 $_{16}$	152 $_7$			428	2511
Carbamazepine-M (HO-ring) 2AC	U+UHYAC	2490	293 $_{21}$	251 $_{25}$	209 $_{79}$	208 $_{100}$	178 $_{17}$	861	2672
Carbamazepine-M (HO-ring) AC	UHYAC	2450	251 $_{33}$	209 $_{100}$	180 $_{74}$	152 $_{11}$		609	425
Carbamazepine-M AC	UHYAC	3195	340 $_{81}$	298 $_{95}$	297 $_{100}$	241 $_6$	179 $_{16}$	1154	426
Carbamazepine-M cysteine-conjugate (ME)	U	2715	326 $_{34}$	283 $_{44}$	180 $_{100}$	152 $_{32}$		1067	428
Carbamazepine-M/artifact (ring)	P U UHY U+UHYAC	1985	193 $_{100}$	165 $_{19}$	139 $_5$	113 $_3$	96 $_9$	378	309
Carbamazepine-M/artifact AC	U+UHYAC	2040	235 $_{27}$	193 $_{100}$	192 $_{68}$	165 $_{17}$		532	2671
Carbaryl		1865	201 $_2$	144 $_{100}$	115 $_{59}$	89 $_{10}$	63 $_{11}$	402	3751
Carbaryl TFA		1785	297 $_{11}$	240 $_{54}$	143 $_{55}$	115 $_{82}$	69 $_{100}$	889	4134
Carbaryl-M/artifact (1-naphthol)		1500*	144 $_{100}$	115 $_{83}$	89 $_{14}$	74 $_6$	63 $_{17}$	283	928
Carbaryl-M/artifact (1-naphthol) AC	U+UHYAC	1555*	186 $_{13}$	144 $_{100}$	115 $_{47}$	89 $_8$	63 $_7$	361	932
Carbaryl-M/artifact (1-naphthol) HFB		1310*	340 $_{46}$	169 $_{25}$	143 $_{28}$	115 $_{100}$	89 $_{11}$	1155	7476
Carbaryl-M/artifact (1-naphthol) PFP		1510*	290 $_{45}$	171 $_{100}$	143 $_{20}$	115 $_{49}$	89 $_8$	839	7468
Carbaryl-M/artifact (1-naphthol) TMS		1525*	216 $_{100}$	201 $_{95}$	185 $_{51}$	115 $_{39}$	73 $_{21}$	452	7460
Carbendazim -C2H2O2		1930	133 $_{100}$	105 $_{26}$	79 $_{13}$	63 $_6$		273	4033
Carbetamide		1975	236 $_{12}$	165 $_3$	119 $_{100}$	93 $_{35}$	72 $_{28}$	539	3172
Carbetamide 2ME		1965	264 $_3$	158 $_{30}$	134 $_{66}$	86 $_{73}$	58 $_{100}$	682	4095
Carbetamide TFA		1870	332 $_{17}$	196 $_{73}$	124 $_{80}$	119 $_{100}$	77 $_{34}$	1105	4127
Carbidopa 2ME	U+UHYAC	1660	224 $_{28}$	164 $_3$	137 $_{100}$	122 $_6$		629	1805
Carbidopa 2MEAC		1990	266 $_5$	138 $_{36}$	129 $_{100}$	97 $_{25}$	69 $_{26}$	886	1807
Carbidopa 3ME		1680	238 $_{19}$	222 $_{20}$	162 $_{27}$	151 $_{100}$	137 $_{17}$	708	1806
Carbidopa 3MEAC		2100	294 $_6$	280 $_{41}$	221 $_{22}$	157 $_{42}$	143 $_{100}$	1058	1810
Carbidopa isomer-1 3MEAC		2060	280 $_6$	221 $_6$	152 $_{44}$	129 $_{100}$	97 $_{28}$	971	1808
Carbidopa isomer-2 3MEAC		2080	280 $_{19}$	221 $_9$	143 $_{100}$	115 $_8$	56 $_{52}$	972	1809
Carbidopa-M (di-HO-phenylacetone) 2AC	U+UHYAC	1735*	250 $_3$	208 $_{15}$	166 $_{45}$	123 $_{100}$		603	4210
Carbidopa-M (HO-methoxy-phenylacetone) AC	U+UHYAC	1600*	222 $_2$	180 $_{22}$	137 $_{100}$			479	4211
Carbidopa-M (HO-methoxy-phenylacetone) ME	UHYME	1540*	194 $_{25}$	151 $_{100}$	135 $_4$	107 $_{18}$	65 $_4$	384	4353
Carbimazole	G U+UHYAC	1705	186 $_{74}$	114 $_{88}$	109 $_{23}$	81 $_{35}$	72 $_{100}$	360	4705
Carbimazole-M/artifact (thiamazole)	G P-I	1615	114 $_{100}$	99 $_5$	81 $_{16}$	72 $_{30}$	69 $_{18}$	263	4703
Carbimazole-M/artifact (thiamazole) AC	GAC PAC-I U+UHYAC	1440	156 $_{40}$	114 $_{100}$	86 $_{10}$	81 $_{18}$	72 $_{20}$	300	4704
Carbimazole-M/artifact (thiamazole) ME	GME PME-I	1205	128 $_{40}$	113 $_8$	95 $_{22}$	72 $_{18}$	59 $_7$	270	4687
Carbimazole-M/artifact (thiamazole) TMS	GTMS PTMS-I	1400	186 $_{51}$	171 $_{100}$	116 $_7$	113 $_8$	73 $_{23}$	360	4688
Carbinoxamine	G U+UHYAC	2120	218 $_1$	203 $_2$	167 $_8$	71 $_{62}$	58 $_{100}$	841	1780
Carbinoxamine-M (bis-nor-) AC	UHYAC	2430	304 $_1$	218 $_{100}$	203 $_{65}$	167 $_{97}$	86 $_{50}$	931	2171
Carbinoxamine-M (carbinol)	UHY	1670	219 $_{54}$	139 $_{15}$	108 $_{37}$	79 $_{100}$		463	2173
Carbinoxamine-M (carbinol) AC	UHYAC	1700	261 $_4$	218 $_{100}$	201 $_{32}$	167 $_{30}$	78 $_{31}$	661	2167
Carbinoxamine-M (Cl-benzoylpyridine)	UHY UHYAC	1645	217 $_{35}$	189 $_{98}$	139 $_{100}$	111 $_{88}$	75 $_{80}$	454	2166
Carbinoxamine-M (deamino-HO-) AC	UHYAC	2240	305 $_3$	218 $_{89}$	203 $_{86}$	167 $_{92}$	87 $_{100}$	935	2169
Carbinoxamine-M (nor-)	UHY	2150	276 $_1$	220 $_{40}$	203 $_{100}$	167 $_{61}$	139 $_{13}$	754	2174
Carbinoxamine-M (nor-) AC	U+UHYAC	2400	318 $_1$	218 $_{41}$	203 $_{55}$	167 $_{89}$	100 $_{100}$	1018	2170
Carbinoxamine-M/artifact	UHYAC	2170	239 $_4$	218 $_{100}$	202 $_6$	167 $_{64}$	78 $_{25}$	550	2168
Carbinoxamine-M/artifact	UHY	1600	202 $_{100}$	167 $_{90}$	139 $_{11}$			403	2172
Carbochromene	G U UHY UHYAC	2850	360 $_1$	316	289 $_3$	86 $_{100}$	58 $_{25}$	1269	2586
Carbofuran		1660	221 $_5$	164 $_{100}$	149 $_{56}$	123 $_{17}$	58 $_{16}$	474	3899

Table 8.1: Compounds in order of names   **Carbofuran -C2H3NO**

Name	Detected	RI	Typical ions and intensities					Page	Entry
Carbofuran -C2H3NO		1060*	164_{100}	149_{85}	131_{35}	122_{33}	103_{22}	311	3900
Carbon disulfide		<1000*	76_{100}	44_{17}				253	2754
Carbophenothion		2320*	342_{17}	199_{17}	157_{100}	121_{46}	97_{48}	1166	3322
Carboxin		2410	235_{59}	143_{100}	115_{9}	87_{33}	77_{5}	531	3884
Carbromal	P G U	1515	208_{41}	191_{4}	165_{16}	114_{14}	69_{100}	537	652
Carbromal artifact-1	P	1115	157_{2}	129_{97}	114_{14}	87_{19}	57_{100}	301	1026
Carbromal artifact-2		----	171_{3}	143_{50}	57_{100}			325	739
Carbromal artifact-3		1450	179_{24}	105_{22}	69_{100}			339	1878
Carbromal artifact-4		1470	191_{8}	149_{10}	140_{18}	112_{9}	69_{100}	370	1879
Carbromal artifact-5	G P U	1480	223_{29}	191_{5}	149_{11}	102_{15}	69_{100}	480	1880
Carbromal-M		----	113_{92}	98_{100}	69_{34}	55_{75}		262	658
Carbromal-M (cyamuric acid)		----	129_{3}	114_{36}	98_{27}	85_{31}	57_{100}	271	657
Carbromal-M (debromo-)		1380	143_{2}	130_{60}	113_{96}	87_{100}	71_{63}	301	655
Carbromal-M (debromo-HO-) -H2O		----	156_{9}	139_{37}	113_{54}	98_{51}	69_{100}	300	656
Carbromal-M (ethyl-HO-butyric acid) ME		<1000*	117_{10}	87_{100}	69_{20}	57_{72}		285	659
Carbromal-M (HO-carbromide)	U	1340	194_{3}	181_{4}	165_{79}	150_{100}	69_{28}	426	654
Carbromal-M/artifact (carbromide)	P G U	1215	165_{67}	150_{18}	114_{52}	69_{100}	55_{38}	377	653
Carbutamide artifact	G P UHY	2185	172_{56}	156_{55}	108_{50}	92_{74}	65_{100}	326	973
Carbutamide artifact 2TMS		2210	316_{19}	301_{35}	222_{5}	163_{8}	73_{100}	1005	10331
Carbutamide artifact 3TMS		2125	388_{5}	373_{28}	210_{16}	180_{12}	147_{100}	1394	10330
Carbutaminde-A ME	UME	2135	186_{68}	156_{61}	108_{52}	92_{100}	65_{78}	360	3136
Carglumic acid -2CO2 2AC		1775	186_{77}	154_{100}	142_{35}	111_{60}	100_{36}	361	8170
Carisoprodol	P U+UHYAC	2150	260_{1}	245_{5}	158_{48}	97_{62}	55_{100}	660	2792
Carisoprodol artifact		1585	202_{5}	104_{100}	84_{21}	69_{20}	55_{66}	403	5682
Carisoprodol-M (dealkyl-)	P G U+UHYAC	1785	144_{20}	114_{28}	96_{34}	83_{89}	55_{100}	459	1088
Carisoprodol-M (dealkyl-) artifact-1	P G U	1535*	84_{100}	56_{81}				254	1089
Carisoprodol-M (dealkyl-) artifact-2	P U UHY UHYAC	1720*	173_{2}	101_{9}	84_{100}	56_{90}		328	580
Carphedone		2170	218_{17}	174_{100}	160_{76}	145_{47}	104_{60}	458	5912
Carphedone 2TMS		2460	362_{1}	347_{4}	247_{33}	188_{31}	73_{100}	1274	6031
Carphedone TMS		2400	290_{1}	275_{5}	175_{100}	104_{73}	73_{28}	842	6030
Carprofen		2280	227_{100}	201_{5}	191_{26}	164_{6}		733	1999
Carprofen 2ME		2630	301_{69}	242_{100}	227_{8}	207_{42}	191_{8}	911	5134
Carprofen -CO2		2250	229_{71}	214_{100}	193_{8}	178_{12}	152_{8}	505	2000
Carprofen ME		2750	287_{48}	228_{100}	193_{46}	165_{4}	114_{4}	819	2001
Carprofen-M (HO-) isomer-1 2ME	UME	2740	317_{58}	258_{73}	216_{100}	181_{16}	129_{8}	1010	6285
Carprofen-M (HO-) isomer-1 3ME	UME	2805	331_{100}	316_{21}	272_{81}	256_{25}	222_{16}	1099	6288
Carprofen-M (HO-) isomer-2 2ME	UME	2810	317_{49}	258_{100}	223_{26}	208_{13}	180_{10}	1010	6287
Carprofen-M (HO-) isomer-2 3ME	UME	2865	331_{68}	272_{100}	257_{6}	237_{19}	194_{9}	1099	6289
Carteolol		2670	292_{6}	277_{13}	202_{3}	86_{100}	57_{15}	857	2610
Carteolol AC		2700	334_{4}	319_{5}	163_{6}	86_{25}	57_{100}	1120	1355
Carteolol formyl artifact		2690	304_{22}	289_{100}	202_{38}	141_{73}	70_{57}	932	1354
Carteolol-M (deisobutyl-) -H2O AC	UHYAC	2430	260_{54}	188_{18}	161_{100}	99_{32}	57_{78}	657	1596
Carteolol-M (HO-) 2AC	UHYAC	2800	392_{20}	377_{50}	335_{13}	218_{25}	86_{100}	1412	1597
Carvedilol		2210	406_{13}	269_{12}	183_{100}	180_{90}	154_{13}	1465	7887
Carvedilol artifact (N-dealkyl-) -H2O AC		2595	280_{66}	197_{7}	183_{100}	166_{7}	154_{20}	781	7888
Carvedilol TMSTFA		2970	574_{16}	451_{32}	392_{100}	183_{50}	73_{65}	1705	6140
Carvedilol-M/artifact (4-hydroxycarbazole)		2160	183_{100}	154_{50}	127_{11}	77_{8}		354	7886
Carvedilol-M/artifact (4-hydroxycarbazole) AC		2210	225_{3}	207_{2}	183_{100}	154_{38}	127_{11}	490	7885
Carveol		1050*	152_{6}	137_{11}	119_{22}	109_{100}	84_{80}	294	9178
Carzenide 2ME		1920	229_{50}	198_{48}	135_{100}	103_{40}	76_{55}	504	2479
Carzenide 3ME	UME	1850	243_{54}	199_{18}	135_{100}	104_{46}	76_{51}	567	2480
Catechol 2TMS		1245*	254_{17}	239_{5}	151_{8}	136_{7}	73_{100}	628	6021
Cathine	U UHY	1360	132_{4}	117_{9}	105_{22}	79_{54}	77_{100}	291	1154
Cathine 2AC	U+UHYAC	1740	235_{2}	176_{4}	129_{8}	107_{9}	86_{100}	534	1155
Cathine 2HFB		1335	330_{16}	303_{6}	240_{100}	169_{19}	119_{12}	1691	7418
Cathine formyl artifact		1280	117_{2}	105_{2}	91_{4}	77_{6}	57_{100}	307	4649
Cathine TMSTFA		1630	213_{7}	191_{7}	179_{100}	149_{5}	73_{80}	1025	6260
Cathinone AC		1610	191_{2}	134_{2}	105_{35}	86_{100}	77_{48}	371	5901
Cathinone HFB		1395	345_{1}	240_{6}	169_{4}	105_{100}	77_{36}	1182	5904
Cathinone PFP		1335	190_{6}	119_{7}	105_{100}	77_{40}	69_{5}	874	5903
Cathinone precursor 1b		1585	191_{82}	142_{5}	123_{8}	105_{100}	77_{35}	535	9819
Cathinone precursor 1b TMS		1780	307_{1}	206_{6}	191_{78}	132_{17}	116_{100}	953	9878
Cathinone precursor 4		2275	279_{4}	174_{100}	130_{21}	105_{35}	77_{24}	772	9820
Cathinone precursor 4n		2500	280_{3}	191_{100}	174_{36}	130_{25}	105_{79}	1286	9821
Cathinone TFA		1350	245_{1}	140_{7}	105_{100}	77_{48}	69_{10}	573	5902
Cathinone TMS		1590	206_{14}	191_{15}	116_{100}	77_{27}	73_{74}	475	5905
Caulophyllin		1995	204_{16}	160_{5}	146_{10}	117_{7}	58_{100}	411	5597
2C-B		1785	259_{15}	230_{100}	215_{29}	199_{10}	77_{37}	650	3254
2C-B 2AC		2230	343_{7}	242_{100}	229_{32}	201_{12}	148_{29}	1171	6924
2C-B 2TMS		2195	403_{5}	388_{14}	272_{7}	207_{11}	174_{100}	1455	6926
2C-B AC		2180	301_{15}	242_{100}	229_{31}	199_{12}	148_{39}	910	3267
2C-B formyl artifact		1840	271_{18}	240_{100}	229_{41}	199_{17}	77_{37}	722	3245
2C-B formyl artifact		1840	271_{12}	240_{65}	229_{27}	199_{11}	42_{100}	722	5522
2C-B HFB		2030	455_{32}	242_{100}	229_{81}	199_{29}	148_{33}	1591	6941

2C-B intermediate-1 (2,5-dimethoxyphenyl-2-nitroethene)

Table 8.1: Compounds in order of names

Name	Detected	RI	Typical ions and intensities					Page	Entry
2C-B intermediate-1 (2,5-dimethoxyphenyl-2-nitroethene)		1900	209_{100}	162_{51}	147_{52}	133_{61}	77_{62}	427	3286
2C-B intermediate-2 (2,5-dimethoxyphenethylamine)		1630	181_{15}	162_{33}	152_{100}	137_{47}	121_{27}	350	3287
2C-B intermediate-2 (2,5-dimethoxyphenethylamine)		1630	181_{15}	162_{33}	152_{100}	137_{47}	44_{95}	350	5523
2C-B intermediate-2 (2,5-dimethoxyphenethylamine) 2AC		1935	265_{11}	164_{100}	149_{24}	121_{25}	91_{13}	688	9162
2C-B intermediate-2 (2,5-dimethoxyphenethylamine) AC		1935	223_{15}	164_{100}	149_{29}	121_{31}	91_{15}	484	3288
2C-B intermediate-2 formyl artifact		1540	193_{12}	162_{100}	151_{30}	121_{43}	91_{29}	378	3293
2C-B intermediate-2 formyl artifact		1540	193_{12}	162_{100}	151_{30}	121_{43}	42_{64}	379	5524
2C-B PFP		1995	405_{36}	242_{100}	229_{84}	199_{33}	148_{35}	1461	6936
2C-B precursor (2,5-dimethoxybenzaldehyde)		1345*	166_{100}	151_{39}	120_{36}	95_{61}	63_{62}	316	3278
2C-B TFA		2000	355_{42}	242_{100}	229_{84}	199_{37}	148_{36}	1235	6931
2C-B TMS		1935	331_{1}	272_{3}	229_{2}	102_{100}	73_{50}	1098	6925
2C-B-M (deamino-di-HO-) 2AC	U+UHYAC	2230*	360_{13}	300_{27}	258_{43}	245_{100}	138_{17}	1262	7214
2C-B-M (deamino-di-HO-) 2TFA		1790*	468_{100}	354_{14}	341_{77}	311_{14}	276_{76}	1613	7208
2C-B-M (deamino-HO-) AC	U+UHYAC	2300*	302_{20}	242_{100}	227_{25}	183_{33}	148_{42}	916	7198
2C-B-M (deamino-HO-) TFA		1880*	356_{49}	341_{15}	242_{100}	229_{72}	148_{71}	1240	7209
2C-B-M (deamino-HOOC-) ME		2030*	288_{100}	273_{12}	241_{8}	229_{67}	199_{19}	824	7212
2C-B-M (deamino-oxo-)		2020*	258_{35}	229_{100}	215_{9}	199_{35}	186_{8}	644	7215
2C-B-M (O-demethyl- N-acetyl-) isomer-1 AC	U+UHYAC	2410	329_{4}	287_{33}	228_{100}	215_{15}	165_{5}	1085	7196
2C-B-M (O-demethyl- N-acetyl-) isomer-1 TFA		2090	383_{8}	324_{100}	255_{7}	148_{21}	72_{23}	1368	7204
2C-B-M (O-demethyl- N-acetyl-) isomer-2 AC	U+UHYAC	2440	329_{8}	287_{21}	228_{100}	215_{17}	72_{13}	1085	7197
2C-B-M (O-demethyl- N-acetyl-) isomer-2 TFA		2130	383_{7}	324_{100}	311_{11}	227_{11}	72_{17}	1368	7205
2C-B-M (O-demethyl-) isomer-1 2AC	U+UHYAC	2410	329_{4}	287_{33}	228_{100}	215_{15}	165_{5}	1085	7196
2C-B-M (O-demethyl-) isomer-1 2TFA		1900	437_{12}	324_{100}	311_{10}	255_{7}	148_{32}	1550	7206
2C-B-M (O-demethyl-) isomer-2 2AC	U+UHYAC	2440	329_{8}	287_{21}	228_{100}	215_{17}	72_{13}	1085	7197
2C-B-M (O-demethyl-) isomer-2 2TFA		1950	437_{6}	324_{100}	311_{32}	253_{3}	227_{15}	1550	7207
2C-B-M (O-demethyl-deamino-di-HO-) 3AC		2280*	388_{6}	346_{43}	286_{55}	244_{67}	231_{100}	1393	7201
2C-B-M (O-demethyl-deamino-HO-) 2TFA		1800*	438_{31}	341_{16}	324_{100}	311_{35}	227_{55}	1553	7210
2C-B-M (O-demethyl-deamino-HO-) iso-1 2AC	U+UHYAC	2160*	330_{5}	288_{17}	270_{6}	228_{100}	213_{12}	1092	7199
2C-B-M (O-demethyl-deamino-HO-) iso-2 2AC	U+UHYAC	2180*	330_{4}	288_{15}	246_{10}	228_{100}	213_{14}	1092	7200
2C-B-M (O-demethyl-deamino-HOOC-) -H2O	U+UHYAC	1980*	242_{80}	214_{100}	186_{37}			562	7203
2C-B-M (O-demethyl-deamino-HOOC-) MEAC		2120*	316_{14}	274_{100}	242_{92}	214_{87}	186_{18}	1003	7213
2C-B-M (O-demethyl-deamino-HOOC-) METFA		1890*	370_{100}	311_{46}	257_{55}	241_{60}	148_{21}	1309	7211
2C-B-M (O-demethyl-deamino-HO-oxo-) 2AC	U+UHYAC	2160*	344_{11}	302_{73}	260_{46}	242_{100}	214_{25}	1177	7202
2C-D		1605	195_{20}	166_{100}	151_{60}	135_{27}	91_{19}	387	6904
2C-D 2AC		2010	279_{11}	170_{100}	163_{34}	135_{21}	72_{9}	774	6913
2C-D 2TMS		2020	339_{2}	324_{13}	174_{100}	100_{23}	86_{36}	1153	6915
2C-D AC		1940	237_{14}	178_{100}	165_{35}	163_{40}	135_{51}	546	6912
2C-D formyl artifact		1530	207_{25}	176_{100}	165_{69}	135_{39}	91_{16}	421	6909
2C-D HFB		1710	391_{49}	226_{6}	178_{85}	165_{100}	135_{46}	1405	6937
2C-D PFP		1680	341_{53}	178_{79}	165_{100}	135_{54}	91_{21}	1160	6932
2C-D TFA		1685	291_{43}	178_{73}	165_{100}	135_{57}	91_{22}	846	6927
2C-D TMS		1735	267_{7}	237_{8}	166_{20}	102_{100}	73_{91}	703	6914
2C-D-M (deamino-COOH) ME		1755*	224_{100}	209_{19}	177_{12}	165_{35}	135_{8}	488	7229
2C-D-M (deamino-HO-) AC	U+UHYAC	1740*	238_{27}	178_{100}	163_{57}	135_{33}	79_{27}	548	7216
2C-D-M (deamino-oxo-)		1730*	194_{54}	165_{100}	151_{25}	135_{85}	91_{51}	384	7232
2C-D-M (HO-) 2AC		2390	295_{33}	236_{100}	223_{6}	193_{35}	163_{12}	877	7219
2C-D-M (HO-) 2TFA		1950	403_{42}	290_{100}	277_{32}	177_{57}	163_{25}	1455	7228
2C-D-M (HO-) 3AC	U+UHYAC	2400	337_{27}	244_{23}	236_{100}	193_{46}	125_{30}	1138	7220
2C-D-M (O-demethyl- N-acetyl-) 2AC	U+UHYAC	2250	307_{5}	265_{7}	206_{25}	164_{100}	149_{13}	950	7223
2C-D-M (O-demethyl- N-acetyl-) isomer-1 AC	U+UHYAC	2130	265_{6}	223_{36}	164_{100}	151_{14}	91_{4}	687	7221
2C-D-M (O-demethyl- N-acetyl-) isomer-1 TFA		1990	319_{12}	260_{100}	247_{4}	191_{18}	163_{26}	1024	7224
2C-D-M (O-demethyl- N-acetyl-) isomer-2 AC	U+UHYAC	2200	265_{7}	223_{25}	164_{100}	151_{14}	91_{6}	688	7222
2C-D-M (O-demethyl- N-acetyl-) isomer-2 TFA		2050	319_{10}	260_{100}	245_{22}	217_{4}	163_{39}	1024	7225
2C-D-M (O-demethyl-) 3AC		2250	307_{5}	265_{7}	206_{25}	164_{100}	149_{13}	950	7223
2C-D-M (O-demethyl-) isomer-1 2AC	U+UHYAC	2130	265_{6}	223_{36}	164_{100}	151_{14}	91_{4}	687	7221
2C-D-M (O-demethyl-) isomer-1 2TFA		1780	373_{28}	260_{100}	247_{24}	191_{30}	163_{49}	1325	7226
2C-D-M (O-demethyl-) isomer-2 2AC	U+UHYAC	2200	265_{7}	223_{25}	164_{100}	151_{14}	91_{6}	688	7222
2C-D-M (O-demethyl-) isomer-2 2TFA		1850	373_{18}	260_{100}	247_{48}	217_{15}	163_{39}	1325	7227
2C-D-M (O-demethyl-deamino-COOH) isomer-1 MEAC		1860*	252_{28}	210_{100}	178_{40}	150_{100}	122_{12}	618	7230
2C-D-M (O-demethyl-deamino-COOH) isomer-2 MEAC		1900*	252_{21}	210_{100}	193_{10}	163_{7}	151_{55}	618	7231
2C-D-M (O-demethyl-deamino-HO-) isomer-1 2AC	U+UHYAC	1875*	266_{5}	224_{13}	164_{100}	154_{46}	114_{15}	695	7217
2C-D-M (O-demethyl-deamino-HO-) isomer-2 2AC	U+UHYAC	1890*	266_{8}	224_{12}	206_{6}	164_{100}	121_{10}	695	7218
2C-E		1660	209_{20}	180_{100}	165_{52}	149_{9}	91_{17}	431	6905
2C-E 2AC		2075	293_{9}	192_{100}	177_{34}	149_{11}	91_{15}	865	6917
2C-E 2TMS		2065	353_{2}	338_{15}	174_{100}	100_{18}	86_{24}	1229	6919
2C-E AC		2000	251_{12}	192_{100}	177_{25}	149_{13}	91_{13}	613	6916
2C-E formyl artifact		1630	221_{24}	190_{100}	179_{72}	149_{12}	91_{18}	476	6910
2C-E HFB		1790	405_{54}	226_{7}	192_{90}	179_{100}	149_{21}	1462	6938
2C-E PFP		1760	355_{55}	192_{88}	179_{100}	149_{22}	119_{20}	1236	6933
2C-E TFA		1765	305_{42}	192_{71}	179_{100}	149_{22}	91_{22}	936	6928
2C-E TMS		1790	281_{2}	251_{3}	180_{25}	102_{100}	73_{47}	790	6918
2C-E-M (-COOH N-acetyl-) ME		2605	295_{30}	236_{100}	223_{28}	193_{11}	163_{11}	878	7093
2C-E-M (-COOH) MEAC		2605	295_{30}	236_{100}	223_{28}	193_{11}	163_{11}	878	7093
2C-E-M (deamino-COOH) ME		1820*	238_{100}	223_{22}	192_{11}	179_{39}	163_{6}	548	7091

Table 8.1: Compounds in order of names

Name	Detected	RI	Typical ions and intensities					Page	Entry
2C-E-M (deamino-HO-) AC		1850*	252 $_{18}$	192 $_{100}$	177 $_{55}$	149 $_{19}$	91 $_{23}$	619	7082
2C-E-M (deamino-HO-) TFA		1680*	306 $_{65}$	192 $_{76}$	177 $_{100}$	149 $_{43}$	91 $_{59}$	943	7092
2C-E-M (deamino-oxo-)		1745*	208 $_{57}$	179 $_{100}$	149 $_{24}$	91 $_{89}$	77 $_{66}$	425	7704
2C-E-M (HO- N-acetyl-) 2TFA		2080	459 $_{27}$	345 $_{19}$	304 $_{100}$	276 $_{26}$	69 $_{23}$	1599	7105
2C-E-M (HO- N-acetyl-) -H2O		2175	249 $_{33}$	190 $_{100}$	175 $_{23}$	147 $_{31}$	91 $_{16}$	598	7120
2C-E-M (HO- N-acetyl-) isomer-1 propionylated		2370	323 $_{14}$	264 $_{83}$	249 $_{28}$	208 $_{36}$	191 $_{100}$	1052	7127
2C-E-M (HO- N-acetyl-) isomer-1 TMS		2230	339 $_{26}$	324 $_{10}$	280 $_{49}$	265 $_{100}$	191 $_{18}$	1152	7125
2C-E-M (HO- N-acetyl-) isomer-2 propionylated		2570	323 $_{13}$	252 $_{34}$	208 $_{31}$	196 $_{71}$	57 $_{100}$	1052	7128
2C-E-M (HO- N-acetyl-) isomer-2 TMS		2380	339 $_{100}$	294 $_{5}$	251 $_{10}$	249 $_{7}$	73 $_{42}$	1152	7126
2C-E-M (HO-) 2TFA		2035	417 $_{43}$	304 $_{87}$	291 $_{18}$	190 $_{53}$	177 $_{100}$	1498	7121
2C-E-M (HO-) -H2O AC		2175	249 $_{33}$	190 $_{100}$	175 $_{23}$	147 $_{31}$	91 $_{16}$	598	7120
2C-E-M (HO-) -H2O TFA		1945	303 $_{64}$	190 $_{100}$	177 $_{85}$	175 $_{21}$	147 $_{65}$	923	7119
2C-E-M (HO-) isomer-1 AC		2340	309 $_{22}$	250 $_{100}$	237 $_{6}$	207 $_{50}$	191 $_{77}$	964	7096
2C-E-M (HO-) isomer-2 AC		2420	309 $_{16}$	250 $_{16}$	190 $_{100}$	161 $_{18}$	135 $_{9}$	964	7097
2C-E-M (HO-) isomer-3 2AC		2595	351 $_{2}$	309 $_{22}$	280 $_{32}$	238 $_{56}$	196 $_{100}$	1214	7099
2C-E-M (HO-) isomer-3 AC		2500	309 $_{22}$	250 $_{32}$	238 $_{32}$	208 $_{28}$	196 $_{100}$	965	7098
2C-E-M (HO-deamino-COOH) isomer-1 AC		2070*	296 $_{88}$	253 $_{38}$	237 $_{100}$	222 $_{21}$	177 $_{33}$	885	7103
2C-E-M (HO-deamino-COOH) isomer-2 AC		2150*	296 $_{12}$	236 $_{100}$	177 $_{59}$	161 $_{29}$	147 $_{13}$	885	7104
2C-E-M (O-demethyl- N-acetyl-) isomer-1 2TFA		1860	429 $_{15}$	316 $_{6}$	274 $_{100}$	259 $_{24}$	205 $_{39}$	1530	7110
2C-E-M (O-demethyl- N-acetyl-) isomer-1 TFA		1950	333 $_{16}$	274 $_{100}$	259 $_{12}$	205 $_{14}$	177 $_{11}$	1111	7108
2C-E-M (O-demethyl- N-acetyl-) isomer-2 2TFA		1870	429 $_{4}$	274 $_{100}$	261 $_{32}$	259 $_{23}$	231 $_{9}$	1530	7111
2C-E-M (O-demethyl- N-acetyl-) isomer-2 TFA		2020	333 $_{12}$	274 $_{100}$	259 $_{18}$	177 $_{9}$	91 $_{18}$	1110	7109
2C-E-M (O-demethyl-) isomer-1 2AC		2205	279 $_{6}$	237 $_{31}$	178 $_{100}$	165 $_{17}$	122 $_{5}$	775	7083
2C-E-M (O-demethyl-) isomer-1 2TFA		1740	387 $_{13}$	274 $_{100}$	259 $_{20}$	205 $_{20}$	177 $_{31}$	1389	7106
2C-E-M (O-demethyl-) isomer-2 2TFA		1805	387 $_{11}$	274 $_{100}$	261 $_{41}$	231 $_{7}$	177 $_{27}$	1389	7107
2C-E-M (O-demethyl-) isomer-2 AC		2240	279 $_{46}$	237 $_{28}$	178 $_{100}$	163 $_{15}$	135 $_{6}$	774	7084
2C-E-M (O-demethyl-deamino-COOH) -H2O		1690*	192 $_{78}$	164 $_{100}$	136 $_{32}$	121 $_{27}$	91 $_{17}$	375	7122
2C-E-M (O-demethyl-deamino-COOH) isomer-1 MEAC		1940*	266 $_{8}$	224 $_{50}$	192 $_{73}$	164 $_{100}$	136 $_{17}$	695	7100
2C-E-M (O-demethyl-deamino-COOH) isomer-1 METFA		1710*	320 $_{93}$	305 $_{14}$	261 $_{53}$	207 $_{68}$	191 $_{100}$	1032	7094
2C-E-M (O-demethyl-deamino-COOH) isomer-2 MEAC		1980*	266 $_{9}$	224 $_{100}$	207 $_{6}$	165 $_{60}$	135 $_{15}$	695	7101
2C-E-M (O-demethyl-deamino-COOH) isomer-2 METFA		1730*	320 $_{51}$	261 $_{100}$	223 $_{17}$	163 $_{13}$	91 $_{50}$	1032	7095
2C-E-M (O-demethyl-deamino-HO-) isomer-1 2AC		1990*	280 $_{7}$	238 $_{18}$	178 $_{100}$	163 $_{40}$	145 $_{23}$	782	7089
2C-E-M (O-demethyl-deamino-HO-) isomer-1 2TFA		1540	388 $_{39}$	274 $_{100}$	259 $_{46}$	205 $_{28}$	177 $_{42}$	1393	7116
2C-E-M (O-demethyl-deamino-HO-) isomer-2 2AC		2000*	280 $_{6}$	238 $_{9}$	220 $_{8}$	178 $_{100}$	163 $_{26}$	782	7090
2C-E-M (O-demethyl-deamino-HO-) isomer-2 2TFA		1580	388 $_{18}$	274 $_{100}$	259 $_{24}$	177 $_{65}$	69 $_{29}$	1393	7117
2C-E-M (O-demethyl-HO- N-acetyl-) 2AC		2425	337 $_{2}$	309 $_{2}$	277 $_{22}$	235 $_{41}$	176 $_{100}$	1138	7085
2C-E-M (O-demethyl-HO- N-acetyl-) isomer-1 -H2O AC		2255	277 $_{10}$	235 $_{37}$	176 $_{100}$	161 $_{12}$	133 $_{8}$	760	7086
2C-E-M (O-demethyl-HO- N-acetyl-) isomer-1 -H2O TFA		2015	331 $_{16}$	272 $_{100}$	259 $_{22}$	205 $_{21}$	177 $_{41}$	1099	7112
2C-E-M (O-demethyl-HO- N-acetyl-) isomer-2 -H2O AC		2280	277 $_{5}$	235 $_{32}$	176 $_{100}$	161 $_{12}$	133 $_{9}$	759	7087
2C-E-M (O-demethyl-HO- N-acetyl-) isomer-2 -H2O TFA		2050	331 $_{9}$	272 $_{100}$	259 $_{16}$	203 $_{10}$	192 $_{21}$	1099	7113
2C-E-M (O-demethyl-HO-) 3AC		2425	337 $_{2}$	309 $_{2}$	277 $_{22}$	235 $_{41}$	176 $_{100}$	1138	7085
2C-E-M (O-demethyl-HO-) 3TFA		1750	499 $_{9}$	386 $_{39}$	373 $_{100}$	343 $_{11}$		1657	7124
2C-E-M (O-demethyl-HO-) -H2O 2TFA		1810	385 $_{29}$	272 $_{100}$	259 $_{16}$	203 $_{15}$	175 $_{21}$	1380	7114
2C-E-M (O-demethyl-HO-) isomer-1 -H2O 2AC		2255	277 $_{10}$	235 $_{37}$	176 $_{100}$	161 $_{12}$	133 $_{8}$	760	7086
2C-E-M (O-demethyl-HO-) isomer-2 -H2O 2AC		2280	277 $_{5}$	235 $_{32}$	176 $_{100}$	161 $_{12}$	133 $_{9}$	759	7087
2C-E-M (O-demethyl-oxo- N-acetyl-)		2320	251 $_{26}$	192 $_{100}$	177 $_{38}$	151 $_{11}$	137 $_{18}$	610	7088
2C-E-M (O-demethyl-oxo- N-acetyl-) AC		2430	293 $_{3}$	251 $_{40}$	192 $_{100}$	176 $_{53}$	137 $_{9}$	863	7118
2C-E-M (O-demethyl-oxo- N-acetyl-) TFA		2115	347 $_{46}$	233 $_{24}$	192 $_{100}$	177 $_{68}$	69 $_{68}$	1193	7115
2C-E-M (O-demethyl-oxo-) AC		2320	251 $_{26}$	192 $_{100}$	177 $_{38}$	151 $_{11}$	137 $_{18}$	610	7088
2C-E-M (oxo-deamino-COOH) ME		2025*	252 $_{54}$	237 $_{100}$	193 $_{25}$	177 $_{23}$	163 $_{29}$	617	7102
Cefadroxil-M/artifact 4TMS		1215	455 $_{8}$	440 $_{7}$	216 $_{23}$	172 $_{65}$	73 $_{100}$	1592	7655
Cefadroxil-M/artifact ME2AC		1900	265 $_{5}$	233 $_{37}$	164 $_{43}$	122 $_{100}$	120 $_{59}$	686	7653
Cefadroxil-M/artifact ME3AC		2025	307 $_{5}$	265 $_{39}$	233 $_{67}$	180 $_{100}$	120 $_{69}$	947	7654
Cefalexine artifact MEAC	U+UHYAC	1590	207 $_{1}$	175 $_{10}$	164 $_{11}$	148 $_{27}$	106 $_{100}$	419	5143
Cefazoline artifact		1430	132 $_{100}$	91 $_{5}$	76 $_{15}$	64 $_{26}$	56 $_{61}$	272	7314
Cefazoline artifact ME		1075	146 $_{100}$	105 $_{44}$	91 $_{26}$	76 $_{16}$	59 $_{77}$	284	7315
Celecoxib	P-I G	2770	381 $_{100}$	300 $_{32}$	281 $_{11}$	204 $_{4}$	115 $_{7}$	1359	6537
Celiprolol		2610	280 $_{1}$	265 $_{12}$	151 $_{88}$	86 $_{100}$	57 $_{41}$	1354	2846
Celiprolol AC		2370	307 $_{64}$	219 $_{5}$	151 $_{83}$	112 $_{36}$	86 $_{100}$	1512	2849
Celiprolol artifact-1		2350	333 $_{100}$	216 $_{9}$	151 $_{7}$	112 $_{19}$	86 $_{96}$	1109	2847
Celiprolol artifact-2		2650	291 $_{21}$	277 $_{14}$	151 $_{53}$	114 $_{16}$	86 $_{100}$	845	2850
Celiprolol artifact-3		2740	323 $_{14}$	294 $_{6}$	209 $_{24}$	114 $_{12}$	86 $_{100}$	1047	2848
Celiprolol artifact-3 AC		2800	365 $_{60}$	248 $_{6}$	209 $_{18}$	112 $_{28}$	86 $_{100}$	1284	2851
Cetirizine artifact-1	G U+UHYAC	1600*	202 $_{30}$	167 $_{100}$	165 $_{52}$	152 $_{17}$	125 $_{7}$	403	2442
Cetirizine artifact-2		1900*	232 $_{60}$	201 $_{62}$	165 $_{64}$	105 $_{100}$	77 $_{54}$	516	1344
Cetirizine ME	G PME UME U+UHYAC	2910	402 $_{10}$	229 $_{6}$	201 $_{100}$	165 $_{46}$	146 $_{23}$	1453	4323
Cetirizine-M	UHYAC	2210	280 $_{100}$	201 $_{35}$	165 $_{57}$			779	770
Cetirizine-M (amino-) AC	U+UHYAC	2310	259 $_{36}$	217 $_{100}$	182 $_{14}$	152 $_{12}$	75 $_{5}$	650	4324
Cetirizine-M (amino-HO-) 2AC	UGLUCAC	2550	317 $_{85}$	275 $_{100}$	216 $_{65}$	181 $_{99}$	121 $_{78}$	1010	4325
Cetirizine-M (carbinol)	UHY	1750*	218 $_{17}$	183 $_{7}$	139 $_{39}$	105 $_{100}$	77 $_{87}$	457	2239
Cetirizine-M (carbinol) AC	U+UHYAC	1890*	260 $_{8}$	200 $_{40}$	165 $_{100}$	139 $_{10}$	77 $_{29}$	656	1270
Cetirizine-M (Cl-benzophenone)	U+UHYAC	1850*	216 $_{43}$	139 $_{58}$	105 $_{100}$	77 $_{44}$		451	1343
Cetirizine-M (HO-Cl-benzophenone)	UHY	2300*	232 $_{36}$	197 $_{7}$	139 $_{23}$	121 $_{100}$	111 $_{23}$	515	2240
Cetirizine-M (HO-Cl-BPH) isomer-1 AC	UHYAC	2200*	274 $_{18}$	232 $_{75}$	139 $_{100}$	121 $_{44}$	111 $_{51}$	737	2229

Cetirizine-M (HO-Cl-BPH) isomer-2 AC

Table 8.1: Compounds in order of names

Name	Detected	RI	Typical ions and intensities					Page	Entry
Cetirizine-M (HO-Cl-BPH) isomer-2 AC	U+UHYAC	2230*	274_7	232_{43}	139_{25}	121_{100}	111_{27}	737	2230
Cetirizine-M (N-dealkyl-)	UHY	2520	286_{13}	241_{48}	201_{50}	165_{65}	56_{100}	815	2241
Cetirizine-M (N-dealkyl-) AC	U+UHYAC	2620	328_7	242_{19}	201_{48}	165_{66}	85_{100}	1081	1271
Cetirizine-M (piperazine) 2AC		1750	170_{15}	85_{33}	69_{25}	56_{100}		324	879
Cetirizine-M (piperazine) 2HFB		1290	478_3	459_9	309_{100}	281_{22}	252_{41}	1631	6634
Cetirizine-M (piperazine) 2TFA		1005	278_{10}	209_{59}	152_{25}	69_{56}	56_{100}	766	4129
Cetirizine-M/artifact	P-I U+UHYAC UME	2220	300_{17}	228_{38}	165_{52}	99_{100}	56_{62}	908	670
Cetirizine-M/artifact HYAC	U+UHYAC	2935	280_4	201_{100}	165_{26}			780	1272
Cetobemidone	UHY	2045	247_6	218_1	190_3	119_6	70_{100}	587	429
Cetobemidone AC	U+UHYAC	2095	289_7	247_6	190_7	70_{100}		836	1181
Cetobemidone HFB		1915	443_1	386_3	128_8	96_7	70_{100}	1568	6144
Cetobemidone ME		1950	261_{12}	204_{16}	70_{100}			666	430
Cetobemidone PFP	UHYPFP	1865	393_4	336_7	265_2	128_6	70_{100}	1414	4303
Cetobemidone TFA		1925	343_4	286_{10}	215_3	128_8	70_{100}	1174	6210
Cetobemidone TMS	UHYTMS	2070	319_{13}	304_6	262_{20}	71_{81}	70_{100}	1030	4302
Cetobemidone-M (methoxy-) AC	U+UHYAC	2265	319_5	220_6	70_{100}			1027	1182
Cetobemidone-M (nor-) 2AC	U+UHYAC	2545	317_{11}	261_{100}	218_{32}	70_{54}	58_{99}	1012	1183
2C-H		1630	181_{15}	162_{33}	152_{100}	137_{47}	44_{95}	350	5523
2C-H 2AC		1935	265_{11}	164_{100}	149_{24}	121_{25}	91_{13}	688	9162
2C-H 2TMS		1970	310_4	194_7	174_{100}	100_9	86_{21}	1064	9161
2C-H AC		1935	223_{15}	164_{100}	149_{29}	121_{31}	91_{15}	484	3288
2C-H formyl artifact		1540	193_{12}	162_{100}	151_{30}	121_{43}	91_{29}	378	3293
2C-H formyl artifact		1540	193_{12}	162_{100}	151_{30}	121_{43}	42_{64}	379	5524
2C-H HFB		1700	377_{27}	164_{100}	151_{50}	121_{61}	91_{23}	1342	9167
2C-H PFP		1660	327_{40}	164_{100}	151_{50}	121_{72}	91_{29}	1073	9166
2C-H TFA		1670	277_{48}	210_5	164_{100}	151_{55}	121_{82}	758	9165
2C-H TMS		1665	253_3	223_5	194_8	152_7	102_{100}	624	9163
Chavicine	G P	2900	285_{58}	201_{71}	173_{36}	115_{100}	84_{31}	810	660
Chelerythrine artifact (dihydro-)		2965	349_{100}	348_{88}	332_{15}	318_{12}	290_{12}	1204	5772
Chelerythrine artifact (N-demethyl-)		3160	333_{100}	318_{19}	290_{42}	275_{22}	188_{10}	1110	5771
Chenodesoxycholic acid -2H2O ME	UHYAC	2680*	370_7	355_{10}	255_{100}	147_{10}	105_{14}	1315	4474
Chenodesoxycholic acid ME2AC	UHYAC	3435*	430_1	370_{75}	355_{24}	315_{17}	255_{100}	1649	4473
Chloral hydrate	G	<1000*	146_8	111_{32}	82_{100}			309	1470
Chloral hydrate-M (trichloroethanol)	P UHY	<1000*	148_3	119_{20}	113_{60}	82_{46}	77_{100}	286	1413
Chloralose 3AC	UIIYAC I	2260*	399_2	361_{17}	317_{63}	272_{88}	115_{100}	1543	2128
Chloralose artifact	G U-I UHY-I	2155*	349_1	333_2	279_1	247_{11}	71_{100}	1202	2129
Chloralose-M/artifact (detrichloroethylidenyl-) 5HFB		2030*	583_9	269_{25}	169_9	72_{86}	69_{100}	1735	5895
Chloralose-M/artifact (detrichloroethylidenyl-) 5PFP		1925*	483_{13}	395_9	273_{12}	119_{100}	72_{78}	1734	5894
Chloralose-M/artifact (detrichloroethylidenyl-) 5TFA		1795*	479_{13}	319_{20}	223_{35}	109_{76}	69_{100}	1725	5893
Chloramben isomer-1 2ME		1795	233_{100}	202_{57}	174_{26}	139_{15}	100_{11}	520	4140
Chloramben isomer-2 2ME		1815	233_{75}	205_{38}	188_{100}	161_{49}	124_{41}	520	4141
Chloramben ME		1730	219_{66}	188_{100}	160_{50}	124_{55}	97_{25}	461	4139
Chlorambucil		2420	303_7	254_{100}	230_5	132_{18}	118_{63}	922	1414
Chlorambucil ME		2340	317_9	268_{100}	230_6	131_6	118_{30}	1010	1781
Chloramphenicol 2AC	U+UHYAC	2630	273_4	212_{50}	170_{36}	153_{100}	118_{30}	1464	1383
Chlorazanil		2650	221_{100}	220_{96}	193_{10}	152_{47}	99_9	471	3081
Chlorazepate artifact	P G U	2430	284_{81}	283_{91}	256_{100}	221_{31}	77_7	799	481
Chlorbenside		2035*	268_{18}	143_4	125_{100}	108_8	89_{16}	704	3512
Chlorbenzoxamine		3350	434_1	218_8	203_{100}	165_{31}	105_{64}	1545	2417
Chlorbenzoxamine artifact-1		2060	276_5	216_{66}	203_{68}	160_{22}	105_{100}	756	2419
Chlorbenzoxamine artifact-2		2580	291_{37}	171_{14}	134_{36}	105_{100}		845	2420
Chlorbenzoxamine artifact-2 HY		1900	234_5	216_{12}	203_{64}	105_{100}	77_{10}	530	2422
Chlorbenzoxamine HY	UHY	1790*	218_{91}	165_{26}	139_{83}	105_{57}	77_{100}	456	2421
Chlorbenzoxamine HYAC	UHYAC	1890*	260_6	218_{11}	200_{23}	165_{100}	77_{13}	656	2418
Chlorbenzoxamine-M (HO-phenyl-) HY	UHY	1900*	234_{91}	197_{62}	155_{67}	105_{100}	77_{63}	525	2437
Chlorbenzoxamine-M (HO-phenyl-) HY2AC	UHYAC	2170*	318_2	276_9	216_{41}	181_{100}	152_{31}	1017	2435
Chlorbenzoxamine-M (N-dealkyl-)	UHY	2150	190_{26}	163_{46}	134_{49}	105_{100}	91_{44}	370	2438
Chlorbenzoxamine-M (N-dealkyl-) AC	U UHYAC	2110	232_7	160_{26}	146_{19}	105_{100}	85_{33}	519	2434
Chlorbenzoxamine-M (N-dealkyl-HO-methyl-) 2AC	UHYAC	2390	290_{10}	218_{69}	163_{95}	121_{100}	85_{73}	843	2436
Chlorbenzoxamine-M (N-dealkyl-HO-methyl-) AC-conj.	U	2130	248_{20}	160_{39}	146_{23}	105_{100}	85_{24}	591	2439
Chlorbromuron 2ME		1880	306_9	248_{100}	246_{86}	220_{97}	218_{86}	941	3935
Chlorbufam		1720	223_{29}	171_{12}	164_{18}	127_{35}	53_{100}	481	3515
Chlorbufam TFA		1510	319_2	274_4	223_9	154_9	53_{100}	1023	4122
Chlorcarvacrol		1505*	184_{61}	169_{100}	134_{22}	133_{20}	105_{30}	356	1979
Chlorcarvacrol AC		1520*	226_{15}	184_{68}	169_{100}	133_{10}	105_{15}	493	1987
Chlorcyclizine	P-I U+UHYAC UME	2220	300_{17}	228_{38}	165_{52}	99_{100}	56_{62}	908	670
Chlorcyclizine-M (nor-)	UHY	2520	286_{13}	241_{48}	201_{50}	165_{65}	56_{100}	815	2241
Chlorcyclizine-M (nor-) AC	U+UHYAC	2620	328_7	242_{19}	201_{48}	165_{66}	85_{100}	1081	1271
Chlordecone		2320*	486_1	455_{22}	355_{20}	272_{100}	237_{42}	1642	3324
Chlordiazepoxide	P G	2820	299_{10}	282_{100}	241_{19}	124_8	77_{30}	900	431
Chlordiazepoxide artifact (deoxo-)	P G	2535	283_{83}	282_{100}	247_{14}	220_{13}	124_9	794	432
Chlordiazepoxide HY	UHY	2050	231_{80}	230_{95}	154_{23}	105_{38}	77_{100}	512	419
Chlordiazepoxide HYAC	PHYAC U+UHYAC	2245	273_{30}	230_{100}	154_{13}	105_{23}	77_{50}	732	273
Chlordimeform		1635	196_{89}	181_{74}	152_{54}	117_{100}	89_{68}	391	5196

Table 8.1: Compounds in order of names **Chlordimeform artifact-1 (chloromethylbenzamine)**

Name	Detected	RI	Typical ions and intensities					Page	Entry
Chlordimeform artifact-1 (chloromethylbenzamine)		1030	141 83	106 100	89 10	77 37	52 21	280	5194
Chlordimeform artifact-1 (chloromethylbenzamine) AC		1620	183 39	141 100	106 66	77 26	51 15	353	5197
Chlordimeform artifact-2		1550	169 53	152 6	140 57	106 100	77 48	322	5195
Chlorfenson		2150*	302 13	175 72	111 100	99 43	75 71	916	3325
Chlorfenvinphos		2080*	358 1	323 22	267 47	109 49	81 100	1251	3169
Chlorfenvinphos-M/artifact		1495*	222 1	173 100	145 24	109 19	74 26	477	3170
Chlorflurenol impurity (dechloro-) ME		1950*	240 7	181 100	152 31	126 2	76 6	555	3633
Chlorflurenol ME		2095*	274 9	215 100	180 4	152 65	76 11	737	3632
Chloridazone TFA		1170	317 52	282 4	105 19	77 100	69 24	1009	3749
Chlormadinone AC		3360*	404 3	319 14	301 100	267 15		1460	2477
Chlormadinone -H2O		3340*	344 44	234 45	175 100	147 14		1179	2478
Chlormephos	G	1385*	234 28	154 46	121 100	97 96	65 52	524	3299
Chlormezanone	G P U	2210	209 8	152 100	98 70			732	671
Chlormezanone artifact	G P U	1235	153 57	152 100	75 35			295	672
Chlormezanone-M (chlorobenzoic acid)	G UHY UHYAC	1400*	156 61	139 100	111 54	85 4	75 39	299	2726
Chlormezanone-M/artifact (N-methyl-4-chlorobenzamide)	U+UHYAC	1555	169 36	139 100	111 42	75 34		321	673
4-Chloro-2,5-dimethoxy-amfetamine		1770	229 5	186 100	171 20	91 18	77 39	505	7847
4-Chloro-2,5-dimethoxy-amfetamine AC	U+UHYAC	2055	271 21	212 100	197 11	185 12	86 68	723	7849
4-Chloro-2,5-dimethoxy-amfetamine formyl artifact		1750	241 5	210 26	185 13	155 9	56 100	559	7848
4-Chloro-2,5-dimethoxy-amfetamine HFB		1875	425 27	240 28	212 91	185 100	155 35	1519	7853
4-Chloro-2,5-dimethoxy-amfetamine PFP		1850	375 14	212 81	185 100	155 36	119 20	1332	7852
4-Chloro-2,5-dimethoxy-amfetamine TFA		1875	325 10	212 69	185 100	155 40	140 27	1059	7851
4-Chloro-2,5-dimethoxy-amfetamine TMS		1885	301 1	286 8	228 12	116 100	73 81	912	7850
4-Chloro-2,5-dimethoxy-amfetamine-M (O-demethyl-) AC		2315	257 30	215 100	180 30	150 7	91 7	641	7854
4-Chloro-2,5-dimethoxy-amfetamine-M (O-demethyl-) isomer-1 2AC	U+UHYAC	2300	299 6	257 14	240 50	198 100	86 55	901	7855
4-Chloro-2,5-dimethoxy-amfetamine-M (O-demethyl-) isomer-2 2AC	U+UHYAC	2305	299 1	269 5	240 27	198 100	86 85	900	7856
2-Chloro-4-cyclohexylphenol		1820*	210 87	167 45	154 70	141 100	107 34	432	5164
2-Chloro-4-cyclohexylphenol AC		1830*	252 5	210 100	167 23	154 35	141 48	616	5168
2-Chloro-4-cyclohexylphenol ME		1750*	224 100	181 66	168 46	155 82	125 80	487	5171
5-Chloro-AB-PINACA 3TMS		2755	580 2	565 13	392 16	188 87	73 100	1708	9687
5-Chloro-AB-PINACA -CONH3		2750	319 25	291 17	249 100	213 31	145 74	1026	9686
5-Chloro-AB-PINACA -CONH3 TMS		2640	391 84	376 44	322 21	210 53	73 100	1408	9685
5-Chloro-AMT		2025	208 2	165 57	128 12	101 10	44 100	424	10216
5-Chloro-AMT		2025	208 3	165 100	128 20	101 17	70 14	424	10217
5-Chloro-AMT 2HFB		2030	600 3	387 71	360 52	240 100		1714	9787
5-Chloro-AMT 2PFP		1995	500 2	337 65	310 56	190 100	163 30	1659	9786
5-Chloro-AMT 2TFA		2035	400 4	287 48	260 42	232 11	140 100	1444	9784
5-Chloro-AMT 2TMS		2220	337 4	237 16	116 100	100 14	73 60	1221	9782
5-Chloro-AMT AC		2525	250 12	191 100	164 75	128 18	86 17	603	9781
5-Chloro-AMT ethylimine artifact		2015	234 4	164 23	128 5	70 100	44 51	526	10222
5-Chloro-AMT ethylimine artifact		2015	234 1	164 17	129 4	102 4	70 100	527	10223
5-Chloro-AMT formyl artifact ME		2365	234 7	220 4	203 6	177 100		527	9780
5-Chloro-AMT HFB		2250	404 34	240 21	191 22	164 100	128 7	1459	9788
5-Chloro-AMT PFP		2530	354 28	191 15	164 100	128 8	101 5	1230	9785
5-Chloro-AMT TFA		2525	304 34	191 17	164 100	140 7	128 10	930	9783
3-Chloroaniline 2ME		1180	155 62	154 100	140 11	118 16	75 17	297	4090
3-Chloroaniline AC	U+UHYAC	1580	169 31	127 100	111 2			322	6593
3-Chloroaniline HFB		1310	323 60	304 8	154 100	126 68	111 59	1047	6607
3-Chloroaniline ME		1100	141 74	140 100	111 9	105 11	77 31	280	4089
3-Chloroaniline TFA		1125	223 99	154 100	126 51	111 55	69 42	481	4124
4-Chlorobenzaldehyde		1105*	140 67	139 100	111 67	75 55		279	3171
Chlorobenzilate		2210*	324 1	251 100	152 5	139 85	111 32	1055	3511
Chlorobenzilate-M/artifact (HOOC-) ME		2230*	310 1	251 56	139 100	111 47	75 20	968	3634
3-Chlorobenzoic acid		1430*	156 75	139 100	111 56	75 44	65 6	299	6024
4-Chlorobenzoic acid	G UHY UHYAC	1400*	156 61	139 100	111 54	85 4	75 39	299	2726
3-Chlorobenzoic acid (ME)	U+UHYAC	1100*	170 26	139 100	111 57	75 50		323	6953
3-Chlorobenzyl alcohol		1560*	142 53	125 7	113 28	107 100	77 100	281	6025
4-Chlorobenzyl alcohol		1200*	142 40	125 9	107 57	79 69	77 100	281	2727
4-Chlorobenzylchloride	UHYAC	1150*	160 24	125 100	99 8	89 25	63 19	302	5601
4-Chlorobiphenyl	U+UHYAC	1645*	188 100	152 47	126 4	94 4	76 16	364	4702
2-Chlorobiphenyl (PCB 1)		1540*	188 100	152 62	126 8	94 5	76 32	364	9196
Chlorocresol	G U UHY	1400*	142 58	107 100	77 86			281	674
Chlorocresol AC	U+UHYAC	1345*	184 12	142 100	124 2	107 94	77 53	356	2345
Chlorocresol-M (HO-) 2AC	U+UHYAC	1560*	242 2	200 12	158 100	123 53	65 29	562	2346
Chloroform		<1000*	118 2	83 100	47 40	35 36		264	675
5-Chloro-MDMB-PINACA		2770	393 4	337 48	305 47	249 100	145 45	1416	10421
5-Chloro-N,N-diallyl-tryptamine		2245	274 1	178 6	164 26	143 8	110 100	739	10149
5-Chloro-N,N-diallyl-tryptamine HFB		2005	443 2	374 7	360 8	163 8	110 100	1617	10153
5-Chloro-N,N-diallyl-tryptamine PFP		1995	324 2	310 4	177 2	163 8	110 100	1507	10152
5-Chloro-N,N-diallyl-tryptamine TFA		2015	370 1	274 5	260 9	163 15	110 100	1311	10151
5-Chloro-N,N-diallyl-tryptamine TMS		2380	346 1	305 10	250 8	236 16	110 100	1189	10150
5-Chloro-NNEI		4000	390 8	325 4	248 100	212 5	144 19	1402	10186
5-Chloro-NNEI TMS		3200	462 60	447 31	321 79	248 56	73 100	1604	9603
5-Chloro-NNEI-M/artifact (1-naphthylamine) HFB		1455	339 100	170 17	142 46	127 30	115 80	1147	9628

5-Chloro-NNEI-M/artifact (1-naphthylamine) PFP

Table 8.1: Compounds in order of names

Name	Detected	RI	Typical ions and intensities					Page	Entry
5-Chloro-NNEI-M/artifact (1-naphthylamine) PFP		1400	289 $_{100}$	170 $_{14}$	142 $_{40}$	127 $_{22}$	115 $_{75}$	832	9631
5-Chloro-NNEI-M/artifact (1-naphthylamine) TFA		1440	239 $_{100}$	169 $_{13}$	142 $_{25}$	127 $_{12}$	115 $_{58}$	551	9633
Chlorophacinone		3280*	374 $_{21}$	201 $_6$	173 $_{100}$	165 $_{22}$	89 $_{13}$	1330	2382
2-Chlorophenol		1035*	128 $_{100}$	100 $_{13}$	92 $_{27}$	64 $_{96}$		270	3173
3-Chlorophenol		1750*	128 $_{100}$	100 $_{19}$	92 $_5$	73 $_{10}$	65 $_{41}$	270	2728
4-Chlorophenol	U UHY	1390*	128 $_{100}$	100 $_{20}$	65 $_{54}$			270	676
4-Chlorophenoxyacetic acid		1770*	186 $_{100}$	141 $_{92}$	128 $_{74}$	111 $_{87}$	99 $_{50}$	359	1881
4-Chlorophenoxyacetic acid ME		1510*	200 $_{98}$	141 $_{100}$	111 $_{58}$	99 $_{17}$	75 $_{38}$	399	1077
m-Chlorophenylpiperazine		1910	196 $_{24}$	154 $_{100}$	138 $_{12}$	111 $_9$	75 $_{12}$	390	6885
m-Chlorophenylpiperazine AC	U+UHYAC	2265	238 $_{32}$	195 $_{15}$	166 $_{100}$	154 $_{31}$	111 $_{27}$	547	405
m-Chlorophenylpiperazine HFB	U+UHYHFB	1960	392 $_{100}$	195 $_{38}$	166 $_{41}$	139 $_{36}$	111 $_{25}$	1410	6604
m-Chlorophenylpiperazine ME		1820	210 $_{100}$	166 $_{19}$	139 $_{81}$	111 $_{33}$	70 $_{99}$	433	6886
m-Chlorophenylpiperazine TFA	U+UHYTFA	1920	292 $_{100}$	250 $_{12}$	195 $_{77}$	166 $_{79}$	139 $_{66}$	854	6597
m-Chlorophenylpiperazine TMS		2035	268 $_{85}$	253 $_{41}$	128 $_{88}$	101 $_{96}$	86 $_{100}$	707	6888
m-Chlorophenylpiperazine-M (chloroaniline) 2ME		1180	155 $_{62}$	154 $_{100}$	140 $_{11}$	118 $_{15}$	75 $_{17}$	297	4090
m-Chlorophenylpiperazine-M (chloroaniline) AC	U+UHYAC	1580	169 $_{31}$	127 $_{100}$	111 $_2$	99 $_8$		322	6593
m-Chlorophenylpiperazine-M (chloroaniline) HFB		1310	323 $_{60}$	304 $_8$	154 $_{100}$	126 $_{68}$	111 $_{59}$	1047	6607
m-Chlorophenylpiperazine-M (chloroaniline) ME		1100	141 $_{74}$	140 $_{100}$	111 $_9$	105 $_{11}$	77 $_{31}$	280	4089
m-Chlorophenylpiperazine-M (chloroaniline) TFA		1125	223 $_{99}$	154 $_{100}$	126 $_{51}$	111 $_{55}$	69 $_{42}$	481	4124
m-Chlorophenylpiperazine-M (deethylene-) 2AC	U+UHYAC	2080	254 $_1$	195 $_{50}$	153 $_{31}$	140 $_{100}$	111 $_{10}$	627	6592
m-Chlorophenylpiperazine-M (deethylene-) 2HFB	U+UHYHFB	1705	562 $_1$	349 $_{42}$	336 $_{73}$	240 $_{26}$	139 $_{100}$	1701	6606
m-Chlorophenylpiperazine-M (deethylene-) 2TFA	U+UHYTFA	1670	362 $_3$	249 $_{77}$	236 $_{100}$	139 $_{66}$	111 $_{28}$	1272	6601
m-Chlorophenylpiperazine-M (HO-) isomer-1 2AC	U+UHYAC	2515	296 $_{30}$	254 $_{56}$	211 $_{18}$	182 $_{100}$	154 $_{36}$	884	406
m-Chlorophenylpiperazine-M (HO-) isomer-1 2TFA	U+UHYTFA	2040	404 $_{38}$	307 $_{100}$	278 $_{47}$	265 $_{17}$	154 $_{55}$	1458	6600
m-Chlorophenylpiperazine-M (HO-) isomer-1 AC	U+UHYAC	2335	254 $_{65}$	211 $_{23}$	182 $_{100}$	166 $_{71}$	154 $_{33}$	627	5308
m-Chlorophenylpiperazine-M (HO-) isomer-2 2AC	U+UHYAC	2525	296 $_{20}$	254 $_{74}$	182 $_{100}$	169 $_{72}$	154 $_{24}$	884	32
m-Chlorophenylpiperazine-M (HO-) isomer-2 2HFB	U+UHYHFB	2145	604 $_{43}$	585 $_9$	407 $_{100}$	378 $_{22}$	154 $_{22}$	1715	6605
m-Chlorophenylpiperazine-M (HO-) isomer-2 2TFA	U+UHYTFA	2045	404 $_{39}$	307 $_{100}$	278 $_{31}$	265 $_{11}$	154 $_{67}$	1458	6598
m-Chlorophenylpiperazine-M (HO-) isomer-2 AC	U+UHYAC	2345	254 $_{79}$	211 $_{15}$	182 $_{100}$	169 $_{40}$	154 $_{40}$	627	5307
m-Chlorophenylpiperazine-M (HO-) TFA	U+UHYTFA	2035	308 $_{100}$	272 $_{19}$	211 $_{46}$	182 $_{36}$	155 $_{46}$	955	6599
m-Chlorophenylpiperazine-M (HO-chloroaniline N-acetyl-) HFB	U+UHYHFB	1820	381 $_{10}$	339 $_{22}$	169 $_{15}$	142 $_{100}$	69 $_{23}$	1359	6796
m-Chlorophenylpiperazine-M (HO-chloroaniline N-acetyl-) TFA	U+UHYTFA	1765	281 $_{29}$	239 $_{58}$	142 $_{100}$	114 $_{10}$	69 $_{40}$	785	6797
m-Chlorophenylpiperazine-M (HO-chloroaniline) 2HFB	U+UHYHFB	1540	535 $_{36}$	516 $_8$	338 $_{72}$	169 $_{100}$	143 $_{82}$	1688	6608
m-Chlorophenylpiperazine-M (HO-chloroaniline) isomer-1 2AC	U+UHYAC	1980	227 $_{14}$	185 $_{18}$	167 $_{33}$	143 $_{100}$	114 $_7$	497	6594
m-Chlorophenylpiperazine-M (HO-chloroaniline) isomer-1 2TFA	U+UHYTFA	1440	335 $_{100}$	238 $_{59}$	168 $_{16}$	69 $_{99}$		1122	6603
m-Chlorophenylpiperazine-M (HO-chloroaniline) isomer-1 3AC	U+UHYAC	1940	269 $_2$	227 $_{13}$	185 $_{17}$	167 $_{33}$	143 $_{100}$	710	6596
m-Chlorophenylpiperazine-M (HO-chloroaniline) isomer-2 2AC	U+UHYAC	2020	227 $_4$	185 $_{57}$	143 $_{100}$	114 $_5$	79 $_{18}$	497	404
m-Chlorophenylpiperazine-M (HO-chloroaniline) isomer-2 2TFA	U+UHYTFA	1440	335 $_{85}$	266 $_7$	238 $_{100}$	210 $_{27}$	143 $_{81}$	1121	6602
m-Chlorophenylpiperazine-M (HO-chloroaniline) isomer-2 3AC	U+UHYAC	1900	269 $_4$	227 $_{37}$	185 $_{75}$	143 $_{100}$	79 $_7$	710	6595
o-Chlorophenylpiperazine		1800	196 $_{19}$	161 $_{14}$	154 $_{100}$	138 $_{18}$	111 $_{12}$	390	8561
o-Chlorophenylpiperazine AC		2260	238 $_{36}$	195 $_{15}$	166 $_{100}$	154 $_{45}$	138 $_{38}$	547	8562
o-Chlorophenylpiperazine HFB		2045	392 $_{24}$	195 $_{38}$	166 $_{31}$	138 $_{59}$	56 $_{100}$	1410	8566
o-Chlorophenylpiperazine PFP		1985	342 $_{27}$	195 $_{24}$	166 $_{43}$	139 $_{44}$	56 $_{100}$	1166	8565
o-Chlorophenylpiperazine TFA		2010	292 $_{33}$	195 $_{20}$	166 $_{47}$	138 $_{46}$	56 $_{100}$	854	8564
Chloropicrin		<1000	119 $_{97}$	117 $_{100}$	82 $_{35}$	61 $_{12}$		307	3730
Chloropropham		1620	213 $_{34}$	171 $_{28}$	154 $_{25}$	127 $_{100}$	99 $_{12}$	441	3327
Chloropropylate		2230*	338 $_1$	251 $_{100}$	152 $_5$	139 $_{78}$	111 $_{30}$	1142	3513
Chloropropylate-M/artifact (HOOC-) ME		2230*	310 $_1$	251 $_{56}$	139 $_{100}$	111 $_{47}$	75 $_{20}$	968	3634
Chloropyramine	U UHY UHYAC	2190	289 $_3$	231 $_5$	125 $_{37}$	72 $_{13}$	58 $_{100}$	835	1416
Chloropyramine-M (bis-nor-) AC	U UHYAC	2420	303 $_{10}$	231 $_{43}$	217 $_{11}$	125 $_{100}$	89 $_{16}$	923	2180
Chloropyramine-M (HO-) AC	UHYAC	2440	347 $_2$	289 $_6$	234 $_5$	125 $_{31}$	58 $_{100}$	1194	2177
Chloropyramine-M (N-dealkyl-)	UHY UHYAC	1900	218 $_{19}$	181 $_6$	140 $_{100}$	125 $_{49}$	79 $_{53}$	457	2175
Chloropyramine-M (N-dealkyl-) AC	UHYAC	2160	260 $_1$	246 $_{19}$	217 $_{100}$	125 $_{21}$	78 $_{37}$	656	2176
Chloropyramine-M (nor-)	U UHY	2210	275 $_4$	232 $_{25}$	219 $_{32}$	125 $_{100}$	107 $_{69}$	746	2179
Chloropyramine-M (nor-) AC	UHYAC	2470	317 $_{13}$	231 $_{36}$	217 $_9$	125 $_{100}$	119 $_{38}$	1011	2178
Chloroquine	P G U	2595	319 $_{11}$	290 $_4$	245 $_4$	112 $_5$	86 $_{100}$	1028	677
Chloroquine-M (deethyl-) AC	U+UHYAC	3010	333 $_{62}$	219 $_{100}$	205 $_{78}$	58 $_9$		1113	1759
Chlorothalonil		1775	266 $_{100}$	264 $_{77}$	229 $_{11}$	168 $_{10}$	109 $_{33}$	679	3326
8-Chlorotheophylline	P G U	2500	214 $_{100}$	157 $_{16}$	129 $_{56}$	68 $_{52}$		444	681
8-Chlorotheophylline ET		1910	242 $_{100}$	214 $_{61}$	185 $_8$	157 $_{13}$	129 $_{50}$	563	2399
8-Chlorotheophylline ME	UME	1900	228 $_{100}$	199 $_6$	171 $_{10}$	143 $_{33}$	67 $_{55}$	500	2195
8-Chlorotheophylline TMS		2105	286 $_{92}$	271 $_{100}$	251 $_{42}$	214 $_4$	73 $_{40}$	813	4612
Chlorothiazide artifact 3ME	UEXME	2750	339 $_{12}$	275 $_{72}$	248 $_{100}$	220 $_{26}$	167 $_{27}$	1147	6847
Chlorothiazide artifact 5ME	UEXME	2710	355 $_{100}$	263 $_{17}$	248 $_{26}$	220 $_{59}$	139 $_{39}$	1235	6846
4-Chlorotoluene		1165*	126 $_{28}$	91 $_{100}$	65 $_{10}$	63 $_{12}$		269	3192
Chlorotrimethoxyhippuric acid ME		2405	317 $_{39}$	286 $_2$	229 $_{100}$	186 $_5$	100 $_5$	1009	5181
Chloro-UR-144		2725	345 $_{15}$	330 $_{34}$	263 $_{74}$	248 $_{100}$	144 $_{59}$	1185	9592
Chloroxuron		2245	290 $_{45}$	232 $_{18}$	136 $_5$	105 $_{21}$	72 $_{100}$	840	4137
Chloroxuron ME		2430	304 $_{25}$	232 $_7$	168 $_2$	85 $_{19}$	72 $_{100}$	930	4136
Chloroxylenol		1420*	156 $_{95}$	121 $_{100}$	91 $_{32}$	77 $_{22}$		299	678
Chloroxylenol AC	U+UHYAC	1450*	198 $_{15}$	156 $_{100}$	121 $_{48}$	91 $_{19}$		393	121
Chlorphenamine	G P U+UHYAC	2020	274 $_1$	203 $_{100}$	167 $_{19}$	72 $_{18}$	58 $_{64}$	739	679
Chlorphenamine-M (bis-nor-) AC	U+UHYAC	2535	288 $_4$	216 $_{100}$	203 $_{47}$	181 $_{18}$	167 $_{30}$	827	2183
Chlorphenamine-M (deamino-HO-) AC	U+UHYAC	2130	289 $_{27}$	230 $_{27}$	216 $_{100}$	203 $_{92}$	167 $_{68}$	832	2181

Table 8.1: Compounds in order of names **Chlorphenamine-M (HO-) AC**

Name	Detected	RI	Typical ions and intensities					Page	Entry
Chlorphenamine-M (HO-) AC	UHYAC	2405	332 $_1$	261 $_{61}$	219 $_{100}$	72 $_{11}$	58 $_{55}$	1105	2182
Chlorphenamine-M (nor-) AC	U+UHYAC	2530	302 $_8$	216 $_{49}$	203 $_{100}$	167 $_{29}$	78 $_7$	919	2040
Chlorphenesin		1690*	202 $_{10}$	153 $_5$	128 $_{100}$	111 $_9$	99 $_9$	403	2768
Chlorphenesin 2AC		2070*	286 $_1$	159 $_{100}$	128 $_{37}$	99 $_{35}$	75 $_{14}$	813	2770
Chlorphenesin AC		2030*	244 $_5$	141 $_5$	128 $_{66}$	117 $_{100}$	111 $_{14}$	570	2769
Chlorphenethazine		2420	304 $_8$	246 $_3$	214 $_5$	152 $_2$	58 $_{100}$	930	4262
Chlorphenoxamine	U	2095	303 $_1$	230 $_8$	178 $_6$	165 $_{11}$	58 $_{100}$	924	1417
Chlorphenoxamine artifact	G U+UHYAC	1700*	214 $_{48}$	179 $_{100}$	152 $_8$	139 $_3$	89 $_{14}$	445	1217
Chlorphenoxamine HY	UHY	1750*	232 $_{12}$	217 $_{80}$	139 $_{81}$	105 $_{75}$	77 $_{100}$	516	1079
Chlorphenoxamine HYAC	UHYAC	2180*	274 $_{18}$	232 $_{75}$	197 $_{14}$	139 $_{35}$	121 $_{100}$	738	2185
Chlorphenoxamine-M (HO-)	U	2470	319 $_1$	231 $_3$	195 $_2$	165 $_4$	58 $_{100}$	1025	2188
Chlorphenoxamine-M (HO-) -H2O HY	U UHY	2050*	230 $_{100}$	215 $_{27}$	195 $_{60}$	177 $_{33}$	165 $_{64}$	509	2187
Chlorphenoxamine-M (HO-) isomer-1 -H2O HYAC	UHYAC	2030*	272 $_{27}$	230 $_{100}$	195 $_{34}$	165 $_{56}$	152 $_{10}$	728	2184
Chlorphenoxamine-M (HO-) isomer-2 -H2O HYAC	U+UHYAC	2090*	272 $_{16}$	230 $_{100}$	215 $_{15}$	195 $_{34}$	165 $_{47}$	728	2189
Chlorphenoxamine-M (HO-methoxy-) -H2O HYAC	UHYAC	2210*	302 $_{10}$	260 $_{100}$	182 $_{10}$	152 $_{16}$	75 $_4$	918	2186
Chlorphenoxamine-M (HO-methoxy-carbinol) -H2O	U UHY	2220*	262 $_{36}$	260 $_{100}$				656	2194
Chlorphenoxamine-M (nor-) AC	U	2580	215 $_{37}$	179 $_{26}$	116 $_{100}$	86 $_{38}$	74 $_{33}$	1100	2191
Chlorphenphos-methyl		1540*	232 $_3$	196 $_{72}$	165 $_{80}$	137 $_{27}$	125 $_{100}$	515	4039
Chlorphenphos-methyl -HCl		1455*	196 $_{53}$	165 $_{10}$	137 $_{100}$	101 $_{39}$	75 $_{26}$	389	4040
Chlorphentermine		1355	168 $_3$	125 $_4$	107 $_8$	58 $_{100}$		354	680
Chlorphentermine AC		1730	225 $_1$	166 $_{13}$	100 $_{53}$	86 $_{30}$	58 $_{100}$	491	1418
Chlorphentermine HFB		1560	364 $_1$	254 $_{100}$	214 $_{11}$	166 $_{15}$	125 $_{23}$	1349	5048
Chlorphentermine PFP		1515	329 $_1$	204 $_{100}$	166 $_8$	154 $_7$	125 $_{11}$	1086	5049
Chlorphentermine TFA		1520	279 $_1$	166 $_{14}$	154 $_{100}$	125 $_{16}$	114 $_{14}$	771	5050
Chlorphentermine TMS		1520	255 $_1$	240 $_2$	130 $_{100}$	114 $_{30}$	73 $_{88}$	632	5447
Chlorpromazine	P-I G U+UHYAC	2500	318 $_{10}$	272 $_4$	232 $_4$	86 $_{15}$	58 $_{100}$	1018	310
Chlorpromazine chloro artifact isomer-1		2645	352 $_{19}$	306 $_{15}$	268 $_8$	86 $_{24}$	58 $_{100}$	1218	7647
Chlorpromazine chloro artifact isomer-2		2660	352 $_{17}$	306 $_{15}$	268 $_5$	86 $_{23}$	58 $_{100}$	1218	7648
Chlorpromazine-M (bis-nor-) AC	U+UHYAC	2990	332 $_8$	233 $_{19}$	100 $_{100}$			1104	1255
Chlorpromazine-M (HO-) ME	UME	2590	348 $_{20}$	302 $_4$	262 $_8$	86 $_{44}$	58 $_{100}$	1198	434
Chlorpromazine-M (nor-) AC	U+UHYAC	3070	346 $_8$	232 $_7$	114 $_{100}$	86 $_9$		1187	1256
Chlorpromazine-M (ring)	U-I UHY-I UHYAC-I	2100	233 $_{100}$	198 $_{54}$				520	311
Chlorpromazine-M/artifact (sulfoxide)	G P U	2900	334 $_3$	318 $_3$	246 $_{12}$	86 $_6$	58 $_{100}$	1118	433
Chlorpropamide 2ME		2275	304 $_1$	197 $_{13}$	175 $_{34}$	129 $_{51}$	72 $_{100}$	930	4899
Chlorpropamide artifact-1		1685	217 $_{29}$	175 $_{80}$	111 $_{100}$	75 $_{60}$		453	4900
Chlorpropamide artifact-2		1730	191 $_{63}$	175 $_{41}$	128 $_{43}$	111 $_{100}$	75 $_{81}$	370	4901
Chlorpropamide artifact-2 2ME	UME	1690	219 $_{43}$	175 $_{27}$	111 $_{100}$	75 $_{37}$		462	3124
Chlorpropamide artifact-2 ME	UME	1825	205 $_{30}$	175 $_{29}$	141 $_{12}$	111 $_{100}$	75 $_{42}$	412	3123
Chlorpropamide artifact-3 ME		1860	199 $_{37}$	175 $_{14}$	111 $_{66}$	75 $_{35}$	72 $_{100}$	396	3125
Chlorpropamide artifact-4 2ME	UME	2150	262 $_{10}$	197 $_7$	125 $_{22}$	111 $_{56}$	87 $_{100}$	667	4903
Chlorpropamide artifact-4 ME	UME	2135	248 $_{23}$	141 $_{22}$	125 $_{88}$	111 $_{100}$	75 $_{84}$	589	4902
Chlorpropamide ME	UME	2250	290 $_{19}$	175 $_{54}$	115 $_{100}$	111 $_{72}$	58 $_{64}$	840	3122
Chlorpropamide TMS		2205	348 $_1$	333 $_4$	173 $_{19}$	73 $_{47}$	58 $_{100}$	1198	5024
Chlorprothixene	P-I G U+UHYAC	2510	315 $_2$	255 $_4$	221 $_{17}$	58 $_{100}$		999	312
Chlorprothixene artifact (dihydro-)	G UHY U+UHYAC	2490	317 $_{27}$	231 $_{28}$	152 $_6$	73 $_7$	58 $_{100}$	1010	3732
Chlorprothixene-M (bis-nor-) AC	U+UHYAC	2910	329 $_{16}$	270 $_{87}$	255 $_{22}$	235 $_{77}$	221 $_{100}$	1086	3736
Chlorprothixene-M (bis-nor-dihydro-) AC	U+UHYAC	2870	331 $_{12}$	231 $_{100}$	195 $_7$	152 $_6$	100 $_{15}$	1098	3734
Chlorprothixene-M (bis-nor-HO-) isomer-1 2AC	U+UHYAC	3150	387 $_6$	328 $_{100}$	286 $_{56}$	269 $_{45}$	238 $_{78}$	1389	4167
Chlorprothixene-M (bis-nor-HO-) isomer-2 2AC	U+UHYAC	3190	387 $_{20}$	328 $_{91}$	269 $_{48}$	238 $_{100}$	72 $_{13}$	1388	4169
Chlorprothixene-M (bis-nor-HO-dihydro-) isomer-1 2AC	UHYAC	3170	389 $_{34}$	289 $_{65}$	247 $_{100}$	100 $_{14}$	72 $_{10}$	1397	3737
Chlorprothixene-M (bis-nor-HO-dihydro-) isomer-2 2AC	UHYAC	3210	389 $_{26}$	289 $_{50}$	247 $_{100}$	184 $_2$	100 $_{11}$	1397	3738
Chlorprothixene-M (bis-nor-HO-methoxy-) 2AC	U+UHYAC	3360	417 $_{20}$	358 $_{91}$	303 $_{49}$	238 $_{100}$		1498	4171
Chlorprothixene-M (bis-nor-HO-methoxy-dihydro-) 2AC	UHYAC	3380	419 $_{28}$	319 $_{37}$	277 $_{100}$	234 $_8$	100 $_{10}$	1505	3740
Chlorprothixene-M (HO-) isomer-1 AC	U+UHYAC	2750	373 $_1$	273 $_2$	237 $_7$	58 $_{100}$		1325	4163
Chlorprothixene-M (HO-) isomer-2 AC	U+UHYAC	2760	373 $_1$	342 $_5$	237 $_6$	58 $_{100}$		1326	4164
Chlorprothixene-M (HO-dihydro-) isomer-1	UHY	2750	333 $_{23}$	247 $_8$	58 $_{100}$			1110	437
Chlorprothixene-M (HO-dihydro-) isomer-1 AC	UHYAC	2770	375 $_{11}$	247 $_{15}$	184 $_4$	58 $_{100}$		1334	313
Chlorprothixene-M (HO-dihydro-) isomer-2	UHY	2790	333 $_{18}$	247 $_{12}$	58 $_{100}$			1110	3742
Chlorprothixene-M (HO-dihydro-) isomer-2 AC	U+UHYAC	2800	375 $_{11}$	247 $_{20}$	101 $_1$	58 $_{100}$		1333	3733
Chlorprothixene-M (HO-methoxy-) AC	U+UHYAC	2870	403 $_2$	358 $_2$	267 $_5$	261 $_6$	58 $_{100}$	1455	4165
Chlorprothixene-M (HO-methoxy-dihydro-)	UHY	2810	363 $_{12}$	277 $_{10}$	101 $_1$	58 $_{100}$		1278	3743
Chlorprothixene-M (HO-methoxy-dihydro-) AC	UHYAC	2890	405 $_{11}$	277 $_{10}$	73 $_8$	58 $_{100}$		1462	3735
Chlorprothixene-M (HO-N-oxide) isomer-1 -(CH3)2NOH AC	U+UHYAC	2590*	328 $_{75}$	293 $_{54}$	269 $_{85}$	251 $_{100}$	250 $_{91}$	1078	4160
Chlorprothixene-M (HO-N-oxide) isomer-2 -(CH3)2NOH AC	U+UHYAC	2620*	328 $_{49}$	285 $_{47}$	268 $_{100}$	251 $_{50}$	221 $_{29}$	1078	4161
Chlorprothixene-M (nor-) AC	U+UHYAC	2945	343 $_{12}$	270 $_{100}$	257 $_{51}$	235 $_{71}$	221 $_{86}$	1172	1259
Chlorprothixene-M (nor-dihydro-) AC	U+UHYAC	2930	345 $_{16}$	231 $_{100}$	195 $_8$	152 $_6$	114 $_{11}$	1182	1258
Chlorprothixene-M (nor-HO-) isomer-1 2AC	U+UHYAC	3175	401 $_{16}$	328 $_{100}$	269 $_{43}$	238 $_{66}$	86 $_{17}$	1448	4168
Chlorprothixene-M (nor-HO-) isomer-2 2AC	U+UHYAC	3220	401 $_{20}$	328 $_{100}$	273 $_{91}$	237 $_{60}$	86 $_{11}$	1448	4170
Chlorprothixene-M (nor-HO-dihydro-) isomer-1 2AC	UHYAC	3195	403 $_{34}$	289 $_{65}$	247 $_{100}$	114 $_{22}$	86 $_8$	1455	314
Chlorprothixene-M (nor-HO-dihydro-) isomer-2 2AC	U+UHYAC	3240	403 $_{46}$	289 $_{46}$	247 $_{100}$	114 $_9$	86 $_4$	1455	3739
Chlorprothixene-M (nor-HO-methoxy-) 2AC	U+UHYAC	3390	431 $_{26}$	358 $_{100}$	303 $_{97}$	243 $_{54}$		1536	4172
Chlorprothixene-M (nor-HO-methoxy-dihydro-) 2AC	UHYAC	3410	433 $_{28}$	319 $_{37}$	277 $_{100}$	234 $_8$	114 $_4$	1541	3741
Chlorprothixene-M (nor-sulfoxide) AC	U+UHYAC	2960	359 $_{30}$	270 $_{94}$	257 $_{46}$	235 $_{76}$	221 $_{100}$	1256	4166
Chlorprothixene-M (N-oxide) -(CH3)2NOH	P-I U+UHYAC	2410*	270 $_{40}$	255 $_{21}$	234 $_{100}$	202 $_{23}$	117 $_{27}$	714	438

Chlorprothixene-M (N-oxide-sulfoxide) -(CH3)2NOH   Table 8.1: Compounds in order of names

Name	Detected	RI	Typical ions and intensities					Page	Entry
Chlorprothixene-M (N-oxide-sulfoxide) -(CH3)2NOH	P-I U UGLUC UGLUCAC	2560*	286_{21}	251_{20}	234_{57}	203_{100}	101_{10}	812	436
Chlorprothixene-M / artifact (Cl-thioxanthenone)	U	2260*	246_{100}	218_{46}	183_{9}	139_{25}	91_{10}	578	2641
Chlorprothixene-M/artifact (sulfoxide)	G P U+UHYAC	2720	331_{1}	314_{1}	221_{6}	189_{4}	58_{100}	1098	4162
Chlorpyrifos	G P-I	1980	349_{2}	314_{46}	258_{37}	197_{96}	97_{100}	1202	1397
Chlorpyrifos HY		1440	197_{100}	169_{67}	161_{15}	134_{21}	107_{39}	391	7439
Chlorpyrifos HYAC		1420	239_{3}	197_{100}	169_{13}	140_{8}	98_{17}	550	7440
Chlorpyrifos-methyl		1840	321_{2}	286_{82}	125_{100}	79_{69}	63_{48}	1036	3328
Chlortalidone 3ME		3015	380_{1}	349_{100}	255_{4}	176_{8}		1355	3103
Chlortalidone 4ME		2830	394_{39}	379_{100}	363_{95}	285_{53}	176_{62}	1419	3104
Chlortalidone artifact 3ME	UME	2950	363_{100}	287_{60}	255_{17}	220_{3}	176_{23}	1277	3105
Chlorthal-methyl		1965*	330_{24}	301_{100}	299_{75}	221_{16}	142_{20}	1091	3329
Chlorthiamid		1870	205_{43}	170_{100}	134_{12}	100_{14}	75_{27}	412	3752
Chlorthiamid artifact	U UHY UHYAC	1300	171_{100}	136_{21}	100_{35}			324	736
Chlorthiophos isomer-1		2210*	360_{16}	289_{21}	257_{37}	222_{61}	97_{100}	1261	3300
Chlorthiophos isomer-2		2230*	360_{3}	325_{15}	269_{35}	97_{100}	65_{36}	1261	3301
Chlorthiophos isomer-3		2250*	360_{26}	325_{42}	269_{77}	208_{20}	97_{100}	1261	3302
Chlortoluron ME		1695	226_{4}	154_{3}	89_{4}	72_{100}		494	3973
Chlorzoxazone	U	1800	169_{100}	113_{26}	78_{25}	63_{4}		321	4372
Chlorzoxazone AC	U+UHYAC	1595	211_{10}	169_{100}	125_{7}	113_{13}	76_{11}	435	6362
Chlorzoxazone artifact Me		1820	201_{43}	142_{100}	78_{36}			401	4373
Chlorzoxazone HY2AC	U+UHYAC	1850	227_{17}	185_{15}	167_{34}	143_{100}	114_{17}	498	6364
Chlorzoxazone HY3AC	U+UHYAC	2160	269_{6}	227_{20}	185_{100}	129_{10}	86_{15}	710	6363
Chlorzoxazone ME		1750	183_{100}	154_{45}	92_{65}	76_{20}	63_{16}	353	2440
Cholesta-3,5-dien-7-one		2860*	382_{53}	269_{12}	187_{24}	174_{100}	161_{29}	1367	4347
Cholestenone	U UME	3150*	384_{47}	342_{16}	261_{27}	229_{32}	124_{100}	1379	6353
Cholesterol	P U UHY	3085*	386_{100}	368_{30}	353_{18}	301_{35}	275_{32}	1387	682
Cholesterol -H2O	P UHY U+UHYAC	3050*	368_{100}	353_{19}	260_{18}	147_{40}		1304	143
Cholesterol TMS		3110*	458_{82}	368_{100}	329_{100}	129_{93}	73_{73}	1598	3209
Chrysene		2420*	228_{100}	226_{26}	202_{3}	113_{8}	101_{4}	502	2570
Chrysophanol		2410*	254_{100}	226_{18}	197_{13}	152_{13}	115_{10}	626	3554
Chrysophanol 2ME		2600*	282_{35}	267_{100}	165_{35}	152_{26}	76_{22}	792	3564
Chrysophanol ME		2540*	268_{100}	250_{45}	222_{5}	165_{22}	152_{20}	705	3563
2C-I		2330	307_{16}	278_{100}	263_{20}	247_{9}	232_{3}	946	6954
2C-I 2AC	U+UHYAC	2340	391_{16}	290_{100}	275_{19}	247_{8}	148_{10}	1405	6958
2C-I 2ME		2320	335_{18}	290_{100}	275_{16}	247_{21}	148_{10}	1123	6962
2C-I AC	U+UHYAC	2260	349_{25}	290_{100}	275_{21}	247_{14}	148_{21}	1202	6957
2C-I deuteroformyl artifact		1850	321_{23}	290_{100}	277_{42}	247_{18}	232_{4}	1036	6956
2C-I formyl artifact		1860	319_{20}	288_{100}	277_{50}	247_{23}	232_{6}	1022	6955
2C-I HFB		2110	503_{32}	290_{100}	277_{51}	247_{30}	148_{21}	1663	6948
2C-I intermediate-2 (2,5-dimethoxyphenethylamine)		1630	181_{15}	162_{33}	152_{100}	137_{47}	121_{27}	350	3287
2C-I intermediate-2 (2,5-dimethoxyphenethylamine) 2AC		1935	265_{11}	164_{100}	149_{24}	121_{25}	91_{13}	688	9162
2C-I intermediate-2 (2,5-dimethoxyphenethylamine) AC		1935	223_{15}	164_{100}	149_{29}	121_{31}	91_{15}	484	3288
2C-I PFP		2080	453_{43}	290_{100}	277_{93}	247_{44}	148_{33}	1586	6960
2C-I TFA	UGLUCTFA	2100	403_{20}	290_{100}	277_{69}	247_{49}	148_{31}	1454	6959
2C-I TMS		2070	379_{5}	320_{9}	278_{7}	102_{100}	73_{33}	1349	6961
2C-I-M (deamino-HO-)	UGLUC	2020	308_{100}	277_{74}	263_{7}	247_{16}	150_{9}	954	6966
2C-I-M (deamino-HO-) AC	UGLUCAC	2150	350_{16}	290_{100}	275_{28}	247_{10}	148_{25}	1207	6969
2C-I-M (deamino-HO-) TFA	UGLUCTFA	1980	404_{100}	290_{83}	275_{24}	247_{21}	148_{36}	1458	6978
2C-I-M (deamino-HOOC-) ME	U+UHYAC	2115	336_{100}	321_{12}	289_{8}	277_{57}	247_{18}	1129	6982
2C-I-M (deamino-HOOC-O-demethyl-) -H2O	UGLUC	2080	290_{100}	262_{69}	234_{42}	191_{11}	127_{16}	839	6965
2C-I-M (deamino-HOOC-O-demethyl-) ME	UGLUCMETFA	2160	322_{62}	290_{87}	262_{100}	234_{21}	191_{6}	1042	6984
2C-I-M (deamino-HOOC-O-demethyl-) MEAC	UGLUCMEAC	2170	364_{26}	322_{100}	290_{53}	262_{44}	234_{10}	1281	6981
2C-I-M (deamino-HOOC-O-demethyl-) METFA	UGLUCMETFA	1980	418_{100}	404_{25}	359_{32}	305_{49}	289_{53}	1501	6983
2C-I-M (deamino-HO-O-demethyl-) isomer-1 2AC	U+UHYAC	2240	378_{2}	336_{16}	276_{100}	261_{10}	134_{16}	1346	6970
2C-I-M (deamino-HO-O-demethyl-) isomer-1 2TFA	UGLUCTFA	1865	486_{100}	372_{98}	303_{28}	261_{9}		1642	6979
2C-I-M (deamino-HO-O-demethyl-) isomer-2 2AC	UGLUCAC	2275	378_{8}	336_{33}	276_{100}	261_{34}	150_{24}	1345	6971
2C-I-M (deamino-HO-O-demethyl-) isomer-2 2TFA	UGLUCTFA	1890	486_{35}	372_{100}	275_{47}	261_{25}	245_{22}	1642	6980
2C-I-M (deamino-oxo-)		1965*	306_{100}	277_{82}	263_{18}	247_{29}	232_{10}	941	7233
2C-I-M (O-demethyl- N-acetyl-) isomer-1	UGLUC	2370	335_{15}	276_{100}	263_{11}	233_{10}	134_{14}	1122	6963
2C-I-M (O-demethyl- N-acetyl-) isomer-1 AC	U+UHYAC	2480	377_{5}	335_{46}	276_{100}	259_{16}	233_{20}	1341	6967
2C-I-M (O-demethyl- N-acetyl-) isomer-2	UGLUC	2520	335_{26}	276_{100}	261_{18}	220_{3}	121_{7}	1122	6964
2C-I-M (O-demethyl- N-acetyl-) isomer-2 AC	U+UHYAC	2500	377_{8}	335_{43}	276_{100}	263_{17}	236_{7}	1341	6968
2C-I-M (O-demethyl- N-acetyl-) TFA	UGLUCTFA	2270	431_{10}	389_{100}	276_{94}	263_{43}	148_{7}	1536	6974
2C-I-M (O-demethyl-) isomer-1 2AC	U+UHYAC	2480	377_{5}	335_{46}	276_{100}	259_{16}	233_{20}	1341	6967
2C-I-M (O-demethyl-) isomer-1 2TFA	UGLUCTFA	1970	485_{15}	372_{100}	359_{14}	303_{11}	234_{9}	1641	6972
2C-I-M (O-demethyl-) isomer-1 TFA	UGLUCTFA	2100	389_{39}	276_{100}	263_{35}	233_{19}	134_{15}	1396	6976
2C-I-M (O-demethyl-) isomer-2 2AC	U+UHYAC	2500	377_{8}	335_{43}	276_{100}	263_{17}	236_{7}	1341	6968
2C-I-M (O-demethyl-) isomer-2 2TFA	UGLUCTFA	2010	485_{6}	372_{100}	359_{23}	275_{11}	126_{10}	1640	6973
2C-I-M (O-demethyl-) isomer-2 TFA	UGLUCTFA	2275	389_{62}	276_{100}	263_{58}	261_{18}		1396	6977
2C-I-M (O-demethyl-deamino-di-HO-) 3AC	UGLUCAC	2310*	436_{27}	394_{100}	334_{68}	292_{62}	279_{62}	1550	7130
2C-I-M (O-demethyl-deamino-di-HO-) 3AC	UGLUCAC	2310*	436_{27}	394_{100}	334_{68}	292_{62}	279_{62}	1550	7130
2C-I-M (O-demethyl-deamino-HO-oxo-) 2AC	UGLUCAC	2200*	392_{8}	350_{61}	308_{37}	290_{100}	262_{87}	1410	7129
2C-I-M (O-demethyl-deamino-HO-oxo-) 2AC	UGLUCAC	2200*	392_{8}	350_{61}	308_{37}	290_{100}	262_{87}	1410	7129
Cianidanol 5TMS		2805*	650_{2}	368_{100}	355_{38}	179_{11}	73_{57}	1723	5817

Table 8.1: Compounds in order of names   Cianidanol -H2O 4AC

Name	Detected	RI	Typical ions and intensities					Page	Entry
Cianidanol -H2O 4AC		3025*	440 $_{34}$	398 $_{87}$	356 $_{73}$	314 $_{100}$	272 $_{80}$	1558	5818
Cicloprofen		2305*	238 $_{37}$	193 $_{100}$	178 $_{46}$	165 $_{44}$	96 $_{10}$	547	4275
Cicloprofen ME		2220*	252 $_{27}$	193 $_{100}$	178 $_{36}$	165 $_{24}$	95 $_{9}$	618	4376
Cilazapril 2ET	UET	2980	473 $_{2}$	445 $_{7}$	274 $_{90}$	239 $_{90}$	171 $_{100}$	1624	4732
Cilazapril 2ME	UME	2945	445 $_{1}$	417 $_{3}$	311 $_{26}$	225 $_{79}$	157 $_{100}$	1572	4728
Cilazapril ET	UET	3055	445 $_{2}$	417 $_{6}$	311 $_{75}$	239 $_{43}$	171 $_{100}$	1572	4731
Cilazapril ME	PME UME	3010	431 $_{1}$	358 $_{13}$	297 $_{41}$	225 $_{49}$	157 $_{100}$	1539	4727
Cilazapril METMS		3125	503 $_{1}$	488 $_{6}$	369 $_{51}$	215 $_{100}$	73 $_{52}$	1664	4976
Cilazapril TMS		3030	489 $_{1}$	474 $_{10}$	355 $_{77}$	215 $_{100}$	73 $_{71}$	1648	4975
Cilazaprilate 2ET	UET	3055	445 $_{2}$	417 $_{6}$	311 $_{75}$	239 $_{43}$	171 $_{100}$	1572	4731
Cilazaprilate 2ME	UME	2945	417 $_{1}$	389 $_{5}$	283 $_{57}$	225 $_{42}$	157 $_{100}$	1501	4729
Cilazaprilate 2TMS	UTMS	3055	533 $_{1}$	518 $_{4}$	283 $_{69}$	215 $_{100}$	73 $_{77}$	1686	4977
Cilazaprilate 3ET	UET	2980	473 $_{2}$	445 $_{7}$	274 $_{90}$	239 $_{90}$	171 $_{100}$	1624	4732
Cilazaprilate 3ME	UME	2960	431 $_{1}$	372 $_{22}$	297 $_{45}$	225 $_{93}$	157 $_{100}$	1538	4730
Cilazapril-M/artifact (deethyl-) 2ET	UET	3055	445 $_{2}$	417 $_{6}$	311 $_{75}$	239 $_{43}$	171 $_{100}$	1572	4731
Cilazapril-M/artifact (deethyl-) 2ME	UME	2945	417 $_{1}$	389 $_{5}$	283 $_{57}$	225 $_{42}$	157 $_{100}$	1501	4729
Cilazapril-M/artifact (deethyl-) 2TMS	UTMS	3055	533 $_{1}$	518 $_{4}$	283 $_{69}$	215 $_{100}$	73 $_{77}$	1686	4977
Cilazapril-M/artifact (deethyl-) 3ET	UET	2980	473 $_{2}$	445 $_{7}$	274 $_{90}$	239 $_{90}$	171 $_{100}$	1624	4732
Cilazapril-M/artifact (deethyl-) 3ME	UME	2960	431 $_{1}$	372 $_{22}$	297 $_{45}$	225 $_{93}$	157 $_{100}$	1538	4730
Cinchocaine		2890	343 $_{1}$	326	271 $_{2}$	116 $_{8}$	86 $_{100}$	1176	2126
Cinchonidine		2575	294 $_{1}$	159 $_{6}$	136 $_{100}$	95 $_{4}$	81 $_{7}$	872	1980
Cinchonidine AC		2740	336 $_{2}$	294	277 $_{6}$	159 $_{19}$	136 $_{100}$	1132	1988
Cinchonine	P-I G U	2590	294 $_{56}$	253 $_{12}$	159 $_{32}$	136 $_{100}$	81 $_{26}$	871	684
Cinchonine AC		2750	336 $_{34}$	277 $_{34}$	253 $_{8}$	159 $_{25}$	136 $_{100}$	1132	2002
Cinnamolaurine	U	2855	297 $_{1}$	190 $_{100}$	175 $_{2}$	131 $_{3}$	91 $_{3}$	891	5659
Cinnamolaurine-M (nor-)	U	2955	283 $_{1}$	176 $_{100}$	149 $_{3}$	118 $_{4}$	91 $_{6}$	796	5660
Cinnamolaurine-M (nor-) 2AC	UAC	2930	367 $_{4}$	324 $_{9}$	218 $_{100}$	176 $_{94}$	118 $_{4}$	1296	5662
Cinnamoylcocaine isomer-1		2345	329 $_{8}$	238 $_{10}$	182 $_{60}$	96 $_{62}$	82 $_{100}$	1088	4402
Cinnamoylcocaine isomer-2		2450	329 $_{10}$	238 $_{13}$	182 $_{54}$	96 $_{58}$	82 $_{100}$	1089	4403
Cinnarizine	G	3040	368 $_{2}$	251 $_{16}$	201 $_{100}$	167 $_{22}$	117 $_{39}$	1303	1934
Cinnarizine-M (benzophenone)	U+UHYAC	1610*	182 $_{31}$	152 $_{3}$	105 $_{100}$	77 $_{70}$	51 $_{39}$	351	1624
Cinnarizine-M (carbinol)	UHY	1645*	184 $_{45}$	165 $_{14}$	152 $_{7}$	105 $_{100}$	77 $_{63}$	357	1333
Cinnarizine-M (carbinol) AC	U+UHYAC	1700*	226 $_{20}$	184 $_{20}$	165 $_{100}$	105 $_{14}$	77 $_{35}$	495	1241
Cinnarizine-M (carbinol) HFB		1475*	380 $_{3}$	183 $_{4}$	166 $_{100}$	152 $_{22}$	83 $_{25}$	1355	8146
Cinnarizine-M (carbinol) ME	UHY	1655*	198 $_{70}$	167 $_{94}$	121 $_{100}$	105 $_{56}$	77 $_{71}$	395	6779
Cinnarizine-M (carbinol) PFP		1410*	330 $_{4}$	183 $_{4}$	166 $_{100}$	152 $_{24}$	83 $_{30}$	1093	8145
Cinnarizine-M (carbinol) TFA		1420*	280 $_{19}$	183 $_{7}$	166 $_{100}$	152 $_{24}$	83 $_{21}$	780	8144
Cinnarizine-M (carbinol) TMS		1540*	256 $_{28}$	241 $_{14}$	179 $_{38}$	167 $_{100}$	152 $_{17}$	638	8159
Cinnarizine-M (HO-BPH) isomer-1	UHY	2065*	198 $_{93}$	121 $_{72}$	105 $_{100}$	93 $_{22}$	77 $_{66}$	393	1627
Cinnarizine-M (HO-BPH) isomer-1 AC	U+UHYAC	2010*	240 $_{27}$	198 $_{100}$	121 $_{47}$	105 $_{85}$	77 $_{80}$	555	2196
Cinnarizine-M (HO-BPH) isomer-2	P-I U UHY	2080*	198 $_{50}$	121 $_{100}$	105 $_{17}$	93 $_{14}$	77 $_{28}$	394	732
Cinnarizine-M (HO-BPH) isomer-2 AC	U+UHYAC	2050*	240 $_{19}$	198 $_{100}$	121 $_{94}$	105 $_{41}$	77 $_{51}$	555	2197
Cinnarizine-M (HO-methoxy-BPH)	UHY	2050*	228 $_{46}$	197 $_{6}$	151 $_{100}$	105 $_{22}$	77 $_{41}$	501	1625
Cinnarizine-M (HO-methoxy-BPH) AC	U+UHYAC	2100*	270 $_{3}$	228 $_{92}$	151 $_{100}$	105 $_{25}$	77 $_{40}$	716	1622
Cinnarizine-M (HO-methoxy-BPH) AC	U+UHYAC	2090*	284 $_{5}$	242 $_{15}$	224 $_{17}$	182 $_{100}$	153 $_{19}$	800	2425
Cinnarizine-M (N-dealkyl-) AC	U+UHYAC	2350	244 $_{29}$	201 $_{12}$	172 $_{48}$	117 $_{100}$	85 $_{52}$	572	2198
Cinnarizine-M (N-dealkyl-HO-) 2AC	U+UHYAC	2580	302 $_{3}$	243 $_{20}$	141 $_{100}$	99 $_{21}$	56 $_{8}$	919	2199
Cinnarizine-M (norcyclizine) AC	U+UHYAC	2525	294 $_{16}$	208 $_{56}$	167 $_{100}$	152 $_{30}$	85 $_{78}$	871	1601
Cinnarizine-M (piperazine) 2AC		1750	170 $_{15}$	85 $_{33}$	69 $_{25}$	56 $_{100}$		324	879
Cinnarizine-M (piperazine) 2HFB		1290	478 $_{3}$	459 $_{9}$	309 $_{100}$	281 $_{22}$	252 $_{41}$	1631	6634
Cinnarizine-M (piperazine) 2TFA		1005	278 $_{10}$	209 $_{59}$	152 $_{25}$	69 $_{56}$	56 $_{100}$	766	4129
Cinnarizine-M/artifact	UHY	2070*	228 $_{8}$	186 $_{100}$	157 $_{10}$	128 $_{7}$	77 $_{4}$	499	1626
Cinnarizine-M/artifact AC	U+UHYAC	2200*	270 $_{7}$	228 $_{17}$	186 $_{100}$	157 $_{10}$	128 $_{7}$	714	1625
Cisapride		3895	433 $_{16}$	280 $_{12}$	232 $_{100}$	201 $_{19}$	184 $_{88}$	1608	5607
Cisapride AC		3970	475 $_{8}$	322 $_{11}$	232 $_{100}$	226 $_{26}$	184 $_{43}$	1667	5608
Cisapride-M (N-dealkyl-) -CH3OH 2AC	U+UHYAC	3195	365 $_{91}$	322 $_{12}$	243 $_{14}$	226 $_{82}$	184 $_{100}$	1285	5609
Cisapride-M -CH3OH 2AC	U+UHYAC	3195	365 $_{91}$	322 $_{12}$	243 $_{14}$	226 $_{82}$	184 $_{100}$	1285	5609
Citalopram	G P U+UHYAC	2525	324 $_{9}$	238 $_{5}$	208 $_{4}$	190 $_{3}$	58 $_{100}$	1057	4452
Citalopram-M (bis-nor-) AC	U+UHYAC	2780	338 $_{1}$	320 $_{3}$	261 $_{7}$	238 $_{100}$	100 $_{8}$	1143	4454
Citalopram-M (nor-)	UHY	2500	310 $_{37}$	238 $_{100}$	208 $_{32}$	190 $_{26}$	138 $_{39}$	971	4453
Citalopram-M (nor-) AC	U+UHYAC	2820	352 $_{1}$	261 $_{8}$	238 $_{100}$	114 $_{16}$	86 $_{11}$	1221	4455
Citric Acid 3ETAC		1880*	318 $_{1}$	273 $_{8}$	213 $_{20}$	203 $_{55}$	157 $_{100}$	1018	4478
Citric Acid 3ME	UME	1410*	175 $_{14}$	143 $_{100}$	101 $_{86}$	69 $_{18}$	59 $_{44}$	526	4451
Citric Acid 4ME		1445*	189 $_{11}$	157 $_{100}$	133 $_{4}$	59 $_{16}$		590	5705
Citric Acid 4TMS	UTMS	1410*	465 $_{3}$	375 $_{6}$	273 $_{28}$	147 $_{26}$	73 $_{100}$	1635	6566
5-Cl-DALT		2245	274 $_{1}$	178 $_{6}$	164 $_{26}$	143 $_{8}$	110 $_{100}$	739	10149
5-Cl-DALT HFB		2005	443 $_{2}$	374 $_{7}$	360 $_{8}$	163 $_{8}$	110 $_{100}$	1617	10153
5-Cl-DALT PFP		1995	324 $_{2}$	310 $_{4}$	177 $_{2}$	163 $_{8}$	110 $_{100}$	1507	10152
5-Cl-DALT TFA		2015	370 $_{1}$	274 $_{5}$	260 $_{9}$	163 $_{15}$	110 $_{100}$	1311	10151
5-Cl-DALT TMS		2380	346 $_{1}$	305 $_{10}$	250 $_{8}$	236 $_{16}$	110 $_{100}$	1189	10150
Clemastine	G U	2445	342 $_{1}$	215 $_{3}$	178 $_{4}$	128 $_{34}$	84 $_{100}$	1175	1222
Clemastine artifact	G U+UHYAC	1700*	214 $_{48}$	179 $_{100}$	152 $_{8}$	139 $_{3}$	89 $_{14}$	445	1217
Clemastine HY	UHY	1750*	232 $_{12}$	217 $_{80}$	139 $_{81}$	105 $_{75}$	77 $_{100}$	516	1079
Clemastine HYAC	UHYAC	2180*	274 $_{18}$	232 $_{75}$	197 $_{14}$	139 $_{35}$	121 $_{100}$	738	2185

Clemastine-M (di-HO-) -H2O HY2AC Table 8.1: Compounds in order of names

Name	Detected	RI	Typical ions and intensities					Page	Entry
Clemastine-M (di-HO-) -H2O HY2AC	UHYAC	2440*	330 $_9$	288 $_{23}$	246 $_{100}$	211 $_6$	152 $_{11}$	1093	2190
Clemastine-M (HO-) -H2O HY	U UHY	2050*	230 $_{100}$	215 $_{27}$	195 $_{60}$	177 $_{33}$	165 $_{64}$	509	2187
Clemastine-M (HO-) isomer-1 -H2O HYAC	UHYAC	2030*	272 $_{27}$	230 $_{100}$	195 $_{34}$	165 $_{56}$	152 $_{10}$	728	2184
Clemastine-M (HO-) isomer-2 -H2O HYAC	U+UHYAC	2090*	272 $_{16}$	230 $_{100}$	215 $_{15}$	195 $_{34}$	165 $_{47}$	728	2189
Clemastine-M (HO-methoxy-) -H2O HYAC	UHYAC	2210*	302 $_{10}$	260 $_{100}$	182 $_{10}$	152 $_{16}$	75 $_4$	918	2186
Clemastine-M (HO-methoxy-carbinol) -H2O	U UHY	2220*	262 $_{36}$	260 $_{100}$				656	2194
Clemizole	G U+UHYAC	2620	325 $_1$	256 $_{100}$	255 $_{90}$	131 $_{65}$	125 $_{35}$	1062	1613
Clemizole artifact	U+UHYAC	2300	242 $_{49}$	127 $_{30}$	125 $_{100}$	89 $_{21}$		563	1611
Clemizole-M (di-HO-) 2AC	UHYAC	3200	371 $_{35}$	286 $_{17}$	204 $_{60}$	162 $_{100}$	125 $_{60}$	1562	5648
Clemizole-M (di-HO-) artifact 2AC	UHYAC	2805	358 $_{22}$	316 $_{30}$	274 $_{100}$	146 $_{68}$	125 $_{65}$	1252	5652
Clemizole-M (di-HO-methoxy-) 2AC	UHYAC	3300	401 $_{18}$	359 $_{38}$	316 $_{21}$	192 $_{100}$	125 $_{63}$	1620	5653
Clemizole-M (di-HO-methoxy-) -H2O AC	UHYAC	3750	411 $_{21}$	369 $_{43}$	244 $_{100}$	162 $_{12}$	125 $_{48}$	1481	5655
Clemizole-M (HO-) artifact-1 AC	UHYAC	2585	300 $_{19}$	258 $_{100}$	125 $_{91}$	91 $_{30}$		905	5651
Clemizole-M (HO-) artifact-2 AC	UHYAC	3080	314 $_{100}$	272 $_{28}$	236 $_7$	147 $_{96}$	125 $_{31}$	992	5649
Clemizole-M (HO-deamino-HO-) 2AC	UHYAC	2995	372 $_{33}$	330 $_{83}$	287 $_{100}$	245 $_{29}$	125 $_{93}$	1322	5654
Clemizole-M (HO-methoxy-deamino-HO-) 2AC	UHYAC	2970	402 $_{13}$	360 $_{73}$	299 $_{14}$	175 $_{22}$	125 $_{100}$	1451	5650
Clemizole-M (HO-methoxy-oxo-) AC	UHYAC	3190	427 $_{31}$	385 $_{29}$	302 $_{34}$	260 $_{100}$	125 $_{50}$	1525	2861
Clemizole-M (HO-oxo-) AC	UHYAC	3120	397 $_{70}$	314 $_{47}$	272 $_{73}$	230 $_{100}$	125 $_{70}$	1433	2860
Clemizole-M (oxo-)	U+UHYAC	2965	339 $_{75}$	255 $_{100}$	214 $_{80}$	131 $_{70}$	125 $_{58}$	1148	1612
Clemizole-M/artifact	UHYAC	3050	353 $_{80}$	228 $_{100}$	200 $_{20}$	146 $_{36}$	125 $_{52}$	1224	5647
Clenbuterol		2100	276 $_1$	243 $_3$	127 $_{16}$	86 $_{100}$	57 $_{32}$	752	3990
Clenbuterol AC		2090	318 $_1$	243 $_4$	190 $_4$	86 $_{100}$	57 $_{22}$	1017	3992
Clenbuterol formyl artifact		2160	288 $_1$	243 $_{16}$	188 $_9$	99 $_{100}$	57 $_{45}$	826	3989
Clenbuterol -H2O		1895	258 $_9$	202 $_9$	174 $_{32}$	102 $_5$	57 $_{100}$	646	3991
Clenbuterol -H2O AC		2285	300 $_{12}$	244 $_{21}$	202 $_{100}$	166 $_{29}$	57 $_{48}$	906	3993
Climbazole		2205	292 $_5$	207 $_{100}$	109 $_{50}$	69 $_{65}$	57 $_{89}$	855	6087
Clindamycin	G P U	2750	388 $_1$	341	126 $_{100}$	82 $_5$		1517	4481
Clindamycin 3AC	U+UHYAC	2850	549 $_1$	514 $_{35}$	471 $_{20}$	417 $_{92}$	126 $_{100}$	1695	4479
Clindamycin-M (nor-) 4AC	U+UHYAC	2940	531 $_{11}$	452 $_{30}$	428 $_{28}$	154 $_{84}$	112 $_{100}$	1707	4480
Clionasterol		3265*	414 $_{69}$	329 $_{43}$	303 $_{49}$	105 $_{97}$	55 $_{100}$	1492	5622
Clionasterol -H2O		3300*	396 $_{100}$	381 $_{25}$	147 $_{80}$	105 $_{64}$	81 $_{65}$	1432	5626
Clobazam	P G U	2610	300 $_{100}$	283 $_{33}$	255 $_{39}$	231 $_{23}$	77 $_{55}$	905	439
Clobazam HY	UHY U+UHYAC	2225	274 $_{50}$	257 $_{100}$	231 $_{24}$	215 $_{24}$	77 $_{66}$	738	275
Clobazam-M (HO-)	UGLUC	3000	316 $_{100}$	299 $_{24}$	271 $_{35}$			1004	441
Clobazam-M (HO-) AC	UGLUCAC	2900	358 $_{33}$	316 $_{100}$	299 $_{22}$	271 $_{34}$		1252	443
Clobazam-M (HO-methoxy-)	UGLUC UGLUCAC	3255	346 $_{100}$	316 $_{83}$	301 $_{38}$	271 $_{26}$	245 $_{17}$	1187	442
Clobazam-M (HO-methoxy-) HY	UHY UHYAC	2905	320 $_{100}$	240 $_{23}$	206 $_{25}$			1033	277
Clobazam-M (nor-)	P U	2740	286 $_{100}$	244 $_{32}$	215 $_{45}$	77 $_{60}$		813	440
Clobazam-M (nor-) HY	UHY U+UHYAC	2210	242 $_{100}$	206 $_8$	166 $_8$	77 $_{48}$		563	276
Clobazam-M (nor-HO-) HY	UHY	2650	258 $_{100}$					645	445
Clobazam-M (nor-HO-) HYAC	U+UHYAC	3000	300 $_{40}$	258 $_{100}$				906	279
Clobazam-M (nor-HO-methoxy-) HY	UHY UHYAC	2405	288 $_{100}$					826	444
Clobazam-M (nor-HO-methoxy-) HYAC	UHYAC	2615	330 $_{13}$	288 $_{100}$				1093	278
Clobenzorex	G	1940	259 $_1$	244 $_1$	168 $_{100}$	125 $_{88}$	91 $_{24}$	650	4409
Clobenzorex AC		2290	301 $_1$	266 $_3$	210 $_{60}$	168 $_{100}$	125 $_{70}$	912	4410
Clobenzorex HFB		2075	364 $_{13}$	125 $_{100}$	118 $_{20}$	91 $_{33}$		1591	5051
Clobenzorex PFP		2040	314 $_{21}$	125 $_{100}$	118 $_{21}$	91 $_{18}$		1461	5052
Clobenzorex TFA		2075	355 $_1$	264 $_{24}$	125 $_{100}$	118 $_{38}$	91 $_{42}$	1235	5053
Clobenzorex-M	UHY	1465	181 $_9$	138 $_{100}$	122 $_{18}$	94 $_{24}$	77 $_{16}$	350	4351
Clobenzorex-M (4-HO-amfetamine)		1480	151 $_{10}$	107 $_{69}$	91 $_{10}$	77 $_{42}$	56 $_{100}$	292	1802
Clobenzorex-M (4-HO-amfetamine) 2AC	U+UHYAC	1900	235 $_1$	176 $_{72}$	134 $_{100}$	107 $_{46}$	86 $_{70}$	533	1804
Clobenzorex-M (4-HO-amfetamine) 2HFB		<1000	330 $_{48}$	303 $_{15}$	240 $_{100}$	169 $_{44}$	69 $_{42}$	1691	6326
Clobenzorex-M (4-HO-amfetamine) 2PFP		<1000	280 $_{77}$	253 $_{16}$	190 $_{100}$	119 $_{56}$	69 $_{61}$	1567	6325
Clobenzorex-M (4-HO-amfetamine) 2TFA		<1000	230 $_{72}$	203 $_{11}$	140 $_{100}$	92 $_{12}$	69 $_{59}$	1172	6324
Clobenzorex-M (4-HO-amfetamine) 2TMS		<1000	280 $_7$	179 $_9$	149 $_8$	116 $_{100}$	73 $_{78}$	880	6327
Clobenzorex-M (4-HO-amfetamine) AC	U+UHYAC	1890	193 $_1$	134 $_{100}$	107 $_{26}$	86 $_{24}$	77 $_{16}$	379	1803
Clobenzorex-M (4-HO-amfetamine) formyl art.		1220	163 $_3$	148 $_4$	107 $_{30}$	77 $_{12}$	56 $_{100}$	308	6323
Clobenzorex-M (4-HO-amfetamine) TFA		1670	247 $_4$	140 $_{15}$	134 $_{54}$	107 $_{100}$	77 $_{15}$	584	6335
Clobenzorex-M (AM)		1160	134 $_4$	120 $_{15}$	91 $_{100}$	77 $_{18}$	65 $_{73}$	275	54
Clobenzorex-M (AM)	U	1160	134 $_1$	120 $_1$	91 $_6$	65 $_4$	44 $_{100}$	275	5514
Clobenzorex-M (AM) AC	U+UHYAC	1505	177 $_4$	118 $_{60}$	91 $_{35}$	86 $_{100}$	65 $_{16}$	336	55
Clobenzorex-M (AM) AC	U+UHYAC	1505	177 $_1$	118 $_{19}$	91 $_{11}$	86 $_{31}$	44 $_{100}$	335	5515
Clobenzorex-M (AM) formyl artifact		1100	147 $_2$	146 $_6$	125 $_5$	91 $_{12}$	56 $_{100}$	286	3261
Clobenzorex-M (AM) HFB		1355	240 $_{79}$	169 $_{21}$	118 $_{100}$	91 $_{53}$		1099	5047
Clobenzorex-M (AM) PFP		1330	281 $_1$	190 $_{73}$	118 $_{100}$	91 $_{36}$	65 $_{12}$	786	4379
Clobenzorex-M (AM) TFA		1095	231 $_1$	140 $_{100}$	118 $_{92}$	91 $_{45}$	69 $_{19}$	513	4000
Clobenzorex-M (AM) TMS		1190	192 $_6$	116 $_{100}$	100 $_{10}$	91 $_{11}$	73 $_{87}$	421	5581
Clobenzorex-M (AM)-D11 TFA		1615	242 $_1$	144 $_{100}$	128 $_{82}$	98 $_{43}$	70 $_{14}$	566	7283
Clobenzorex-M (AM)-D11PFP		1610	194 $_{100}$	128 $_{82}$	98 $_{43}$	70 $_{14}$		856	7284
Clobenzorex-M (AM)-D5 AC		1480	182 $_3$	122 $_{46}$	92 $_{30}$	90 $_{100}$	66 $_{16}$	352	5690
Clobenzorex-M (AM)-D5 HFB		1330	244 $_{100}$	169 $_{14}$	122 $_{46}$	92 $_{41}$	69 $_{40}$	1130	6316
Clobenzorex-M (AM)-D5 PFP		1320	194 $_{100}$	123 $_{42}$	119 $_{32}$	92 $_{46}$	69 $_{11}$	815	5566
Clobenzorex-M (AM)-D5 TFA		1085	144 $_{100}$	123 $_{53}$	122 $_{56}$	92 $_{51}$	69 $_{28}$	540	5570
Clobenzorex-M (AM)-D5 TMS		1180	212 $_1$	197 $_8$	120 $_{100}$	92 $_{11}$	73 $_{57}$	441	5582

Table 8.1: Compounds in order of names

Name	Detected	RI	Typical ions and intensities					Page	Entry
Clobenzorex-M (di-HO-) 3AC	UHYAC	2765	234_8	210_{36}	168_{100}	125_{51}		1500	4415
Clobenzorex-M (HO-) isomer-1 2AC	UHYAC	2585	359_1	324_6	210_{56}	168_{100}	125_{64}	1258	4412
Clobenzorex-M (HO-) isomer-2 2AC	UHYAC	2630	359_1	324_5	210_{79}	168_{100}	125_{89}	1257	4413
Clobenzorex-M (HO-chlorobenzyl-) 2AC	UHYAC	2565	324_{10}	268_{23}	226_{100}	183_{30}	141_{29}	1257	4411
Clobenzorex-M (HO-HO-alkyl-) 3AC	UHYAC	2725	417_1	210_{27}	168_{100}	168_{100}	125_{70}	1499	5106
Clobenzorex-M (HO-HO-chlorobenzyl-) isomer-1 3AC	UHYAC	2705	417_1	268_{15}	226_{100}	183_{31}	141_{50}	1499	5105
Clobenzorex-M (HO-HO-chlorobenzyl-) isomer-2 3AC	UHYAC	2725	268_{38}	226_{77}	183_{23}	141_{100}		1499	5104
Clobenzorex-M (HO-HO-chlorobenzyl-) isomer-3 3AC	UHYAC	2775	417_1	268_{15}	226_{100}	183_{31}	141_{39}	1499	5103
Clobenzorex-M (HO-HO-chlorobenzyl-) isomer-4 3AC	UHYAC	2795	417_1	268_{18}	226_{67}	183_{40}	141_{100}	1500	4416
Clobenzorex-M (HO-methoxy-) 2AC	UHYAC	2690	389_1	210_{28}	206_{38}	168_{100}	125_{56}	1398	4414
Clobenzorex-M (norephedrine) 2AC	U+UHYAC	1805	235_1	176_5	134_7	107_{13}	86_{100}	532	2476
Clobenzorex-M (norephedrine) 2HFB	UHYHFB	1455	543_1	330_{14}	240_{100}	169_{44}	69_{57}	1691	5098
Clobenzorex-M (norephedrine) 2PFP	UHYPFP	1380	443_1	280_9	190_{100}	119_{59}	105_{26}	1567	5094
Clobenzorex-M (norephedrine) 2TFA	UTFA	1355	343_1	230_6	203_5	140_{100}	69_{29}	1172	5091
Clobenzorex-M (norephedrine) TMSTFA		1890	240_8	198_3	179_{100}	117_5	73_{88}	1024	6146
Clobenzorex-M 2AC	U+UHYAC	2065	265_3	206_{27}	164_{100}	137_{23}	86_{33}	688	3498
Clobenzorex-M 2AC	U+UHYAC	1820*	266_3	206_9	164_{100}	150_{10}	137_{30}	695	6409
Clobenzorex-M 2HFB	UHFB	1690	360_{82}	333_{15}	240_{100}	169_{42}	69_{39}	1704	6512
Clobutinol	G P U	1895	255_1	240	130_{28}	125_{31}	58_{100}	633	2793
Clobutinol AC	U+UHYAC	1980	238_1	222_1	125_{18}	89_9	58_{100}	891	3060
Clofedanol	U UHY	2105	274_1	254_7	111_2	77_6	58_{100}	834	1935
Clofedanol AC	UHYAC	2120	296_6	236_2	165_3	58_{100}		1100	1936
Clofedanol -H2O	UHY UHYAC	2085	271_{12}	270_{21}	236_{100}	160_{41}	58_{68}	723	1639
Clofedanol-M (2-Cl-benzophenone)	U UHY UHYAC	1720*	216_{86}	139_{57}	111_{20}	105_{100}	77_{60}	451	1636
Clofedanol-M (aldehyde)	U UHY UHYAC	1900*	207_{100}	179_{27}				563	1632
Clofedanol-M (HO-) artifact	U UHY	2040*	230_{100}	195_{70}	177_{56}	165_{65}	152_{20}	509	1637
Clofedanol-M (HO-) -H2O	UHY	2130	287_8	286_{16}	252_{100}	222_{33}	58_{85}	819	1640
Clofedanol-M (HO-) -H2O AC	UHYAC	2370	329_7	294_{90}	226_{50}	178_{57}	58_{100}	1087	1635
Clofedanol-M (nor-) -H2O	U UHY	2090	257_{13}	256_{17}	222_{100}	163_{85}	134_{72}	642	1641
Clofedanol-M (nor-) -H2O AC	UHYAC	2400	299_{34}	256_{16}	226_{65}	191_{54}	98_{100}	901	1633
Clofedanol-M (nor-HO-) -H2O 2AC	UHYAC	2800	357_{15}	242_{54}	178_{16}	152_8	98_{100}	1247	1634
Clofedanol-M/artifact	U UHY UHYAC	1700*	214_{50}	200_5	179_{100}	178_{98}	151_7	445	1631
Clofedanol-M/artifact	UHY	2060*	244_{100}	209_{42}	194_{26}	165_{22}	115_{32}	569	1638
Clofibrate	U	1540*	242_8	169_{16}	128_{100}			564	685
Clofibrate-M (clofibric acid)	U	1640*	214_2	168_9	128_{100}	86_9	65_{17}	445	686
Clofibrate-M (clofibric acid) artifact	U+UHYAC	1580*	168_{35}	128_{100}				320	1373
Clofibrate-M (clofibric acid) ME	U	1500*	228_3	169_{16}	128_{100}	99_5	75_8	500	687
Clofibrate-M/artifact (4-chlorophenol)	U UHY	1390*	128_{100}	100_{20}	65_{54}			270	676
Clofibric acid	U	1640*	214_2	168_9	128_{100}	86_9	65_{17}	445	686
Clofibric acid artifact	U+UHYAC	1580*	168_{35}	128_{100}				320	1373
Clofibric acid ME	U	1500*	228_3	169_{16}	128_{100}	99_5	75_8	500	687
Clofibric acid-M/artifact (4-chlorophenol)	U UHY	1390*	128_{100}	100_{20}	65_{54}			270	676
Clomethiazole	P G U+UHYAC	1230	161_{29}	112_{100}	85_{28}			303	446
Clomethiazole-M (1-HO-ethyl-)	UHY	1560	177_{27}	159_{27}	142_{43}	124_{100}	100_{68}	334	3311
Clomethiazole-M (1-HO-ethyl-) AC	U+UHYAC	1430	219_2	183_{15}	160_{11}	141_{55}	128_{100}	462	452
Clomethiazole-M (1-HO-ethyl-) TMS		1560	249_1	234_8	200_{100}	93_{18}	73_{49}	593	4622
Clomethiazole-M (2-HO-)	P UHY	1440	177_5	128_{100}	100_7	73_{29}		334	450
Clomethiazole-M (2-HO-) AC	U+UHYAC	1590	219_3	183_{16}	176_{67}	141_{38}	128_{100}	462	3310
Clomethiazole-M (dechloro-2-HO-)	P U UHY	1160	143_{23}	128_{100}	100_{52}	73_{47}		282	449
Clomethiazole-M (dechloro-2-HO-ethyl-)	UHY	1380	143_{31}	113_{42}	112_{100}	85_{30}	71_9	282	448
Clomethiazole-M (dechloro-2-HO-ethyl-) AC	U+UHYAC	1050	185_7	143_5	128_{32}	125_{100}	98_{32}	359	451
Clomethiazole-M (dechloro-di-HO-)	UHY	1685	159_7	128_{100}	100_8	73_{38}		302	3312
Clomethiazole-M (dechloro-di-HO-) -H2O AC	U+UHYAC	1420	183_{19}	170_9	141_{40}	128_{100}		353	1461
Clomethiazole-M (dechloro-HOOC-)	U	1235	157_{50}	128_{17}	112_{100}	85_{24}		300	447
Clomethiazole-M (dechloro-HOOC-2-HO-)	U+UHYAC	1690	155_{50}	125_{100}	97_{61}	70_{32}		297	6560
Clomiphene		2885	405_1	252_2	239_1	100_9	86_{100}	1463	7533
Clomipramine	P G U+UHYAC	2455	314_{17}	269_{50}	227_{11}	85_{42}	58_{100}	995	315
Clomipramine-D3		2440	317_4	268_{21}	130_7	88_{49}	61_{100}	1013	5425
Clomipramine-M (bis-nor-) AC	U+UHYAC	2960	328_{42}	242_{100}	227_{38}	100_{45}	72_{10}	1081	1177
Clomipramine-M (bis-nor-HO-) 2AC	UHYAC	3120	386_{76}	300_{81}	258_{100}	243_{25}	100_{77}	1384	3414
Clomipramine-M (HO-) isomer-1	U UHY	2540	330_8	285_{16}	245_7	85_{20}	58_{100}	1094	453
Clomipramine-M (HO-) isomer-1 AC	U+UHYAC	2805	372_5	327_2	285_{17}	85_{17}	58_{100}	1323	317
Clomipramine-M (HO-) isomer-2	U UHY	2800	330_8	285_{15}	245_7	85_{19}	58_{100}	1094	33
Clomipramine-M (HO-) isomer-2 AC	U+UHYAC	2905	372_5	327_3	285_{22}	85_{23}	58_{100}	1323	122
Clomipramine-M (HO-ring) AC	UHYAC	2645	287_{30}	245_{100}	230_{11}	210_{19}	180_{10}	818	4159
Clomipramine-M (nor-)		2620	300_{31}	268_{100}	229_{78}	192_{38}	71_{38}	908	7663
Clomipramine-M (nor-) AC	U+UHYAC	2980	342_{16}	256_{19}	242_{100}	227_{69}	114_{81}	1169	1176
Clomipramine-M (nor-) HFB		2650	496_{16}	268_{33}	242_{100}	228_{38}	169_{71}	1655	7666
Clomipramine-M (nor-) PFP		2690	446_4	268_4	242_{100}	227_{41}	190_{19}	1573	7665
Clomipramine-M (nor-) TFA		2650	396_{14}	242_{100}	227_{53}	191_{22}	69_{74}	1429	7664
Clomipramine-M (nor-) TMS		2575	372_{21}	269_{100}	242_{32}	227_{36}	116_{46}	1324	7707
Clomipramine-M (nor-) TMS		2505	372_{15}	269_{100}	242_{15}	227_{16}	73_{83}	1324	7785
Clomipramine-M (nor-HO-) 2AC	U+UHYAC	3205	400_{47}	300_{44}	258_{33}	114_{100}	86_{13}	1446	318
Clomipramine-M (N-oxide) -(CH3)2NOH	G P UHY UHYAC	2160	269_{42}	228_{100}	193_{60}	165_{14}	89_{15}	711	4346

Clomipramine-M (ring)

Table 8.1: Compounds in order of names

Name	Detected	RI	Typical ions and intensities					Page	Entry
Clomipramine-M (ring)	U+UHYAC	2230	229$_{100}$	214$_{30}$	194$_{41}$	165$_8$	152$_3$	504	316
Clonazepam	P-I G U-I	2840	315$_{89}$	314$_{97}$	286$_{55}$	280$_{100}$	234$_{53}$	998	454
Clonazepam HY	UHY-I U+UHYAC-I	2470	276$_{45}$	241$_{100}$	195$_{26}$	139$_{51}$	111$_{31}$	752	280
Clonazepam isomer-1 ME		2555	329$_{100}$	294$_{54}$	248$_{36}$			1085	460
Clonazepam isomer-2 ME		2760	329$_{100}$	302$_{49}$	294$_{60}$	248$_{38}$		1086	461
Clonazepam TMS		2795	387$_{78}$	372$_{33}$	352$_{58}$	306$_{36}$	73$_{100}$	1389	5463
Clonazepam-M (amino-)	UGLUC-I	2880	285$_{100}$	256$_{62}$	250$_{24}$	222$_{18}$	111$_{44}$	807	455
Clonazepam-M (amino-) AC	UGLUCAC-I	3190	327$_{100}$	299$_{40}$	292$_{36}$	256$_{30}$	220$_{16}$	1073	457
Clonazepam-M (amino-) HY	UHY-I	2285	246$_{100}$	211$_{71}$	139$_{18}$	111$_{23}$	107$_{45}$	579	458
Clonazepam-M (amino-) HY2AC	U+UHYAC-I	2845	330$_{100}$	288$_{90}$	246$_{94}$	211$_{44}$	139$_{28}$	1093	281
Clonazepam-M (amino-HO-)		2935	283$_{100}$	255$_{41}$	220$_{48}$			911	456
Clonazepam-M (amino-HO-) artifact	UHY-I UHYAC-I	2325	255$_{100}$	220$_{95}$				631	459
Clonazolam		3190	353$_{66}$	324$_{100}$	278$_{22}$	249$_{78}$	203$_{76}$	1224	9698
Clonidine	G	2090	229$_{100}$	200$_{14}$	194$_{34}$	172$_{33}$	109$_7$	504	1785
Clonidine 2AC	UHYAC	2315	313$_{16}$	278$_{57}$	236$_{100}$	194$_{66}$	85$_{17}$	986	688
Clonidine 2TMS		2000	373$_{18}$	358$_{16}$	338$_{100}$	214$_{44}$	73$_{62}$	1326	6303
Clonidine AC	U+UHYAC	2060	271$_{34}$	236$_{100}$	229$_{12}$	194$_{50}$	172$_{12}$	722	1786
Clonidine artifact (dehydro-) AC	U+UHYAC	1820	269$_{18}$	227$_{66}$	192$_{100}$	157$_{21}$	109$_8$	710	1790
Clonidine artifact (dichloroaniline) AC	U+UHYAC	1550	203$_1$	168$_{58}$	161$_{100}$	133$_{14}$	125$_{14}$	405	1789
Clonidine artifact (dichlorophenylisocyanate)	G P	1350	187$_{100}$	159$_{18}$	124$_{40}$			362	1787
Clonidine artifact (dichlorophenylmethylcarbamate)		1500	219$_{10}$	184$_{100}$	174$_{74}$	160$_{14}$	133$_{18}$	461	1788
Clonidine artifact-5		2110	283$_{10}$	248$_{100}$	243$_{18}$	229$_{17}$	194$_{56}$	794	1791
Clonidine TMS		1925	301$_{19}$	286$_{19}$	266$_{100}$	142$_{40}$	73$_{37}$	911	6302
Clopamide		2880	345$_1$	330$_{12}$	218$_{30}$	127$_{71}$	111$_{100}$	1182	6879
Clopamide 2ME		2805	373$_1$	358$_2$	246$_2$	127$_{64}$	111$_{100}$	1327	3097
Clopamide 3ME	UEXME	2800	372$_1$	246$_6$	141$_{90}$	112$_{100}$	83$_{20}$	1390	3098
Clopamide ME		2850	359$_1$	344$_7$	232$_{11}$	127$_{60}$	111$_{100}$	1257	6880
Clopamide -SO2NH		2195	251$_8$	139$_{26}$	127$_{51}$	111$_{100}$		696	3099
Clopenthixol (cis)	G U	3360	400$_1$	221$_{12}$	143$_{100}$	100$_{18}$	70$_{24}$	1445	462
Clopenthixol (cis) AC	U+UHYAC	3460	442$_1$	221$_9$	185$_{100}$	98$_{24}$	70$_{11}$	1564	319
Clopenthixol (cis) TMS		3490	472$_1$	457$_6$	221$_{19}$	215$_{100}$	98$_{23}$	1622	4534
Clopenthixol (trans)		3400	400$_1$	221$_{15}$	143$_{100}$	100$_{21}$	70$_{30}$	1445	4619
Clopenthixol (trans) AC	U+UHYAC	3570	442$_1$	221$_{12}$	185$_{100}$	98$_{22}$	70$_{10}$	1564	4680
Clopenthixol (trans) TMS		3555	472$_1$	457$_6$	221$_{20}$	215$_{100}$	98$_9$	1622	4535
Clopenthixol-M (dealkyl-) AC	U+UHYAC	3490	398$_2$	268$_7$	141$_{100}$	99$_{30}$		1437	1261
Clopenthixol-M (dealkyl-dihydro-) AC	U+UHYAC	3450	400$_{46}$	231$_{44}$	141$_{100}$	128$_{16}$	99$_{25}$	1445	1260
Clopenthixol-M (N-oxide) -C6H14N2O2	P-I U+UHYAC	2410*	270$_{40}$	255$_{21}$	234$_{100}$	202$_{23}$	117$_{27}$	714	438
Clopenthixol-M (N-oxide-sulfoxide) -C6H14N2O2	P-I U UGLUC UGLUCAC	2560*	286$_{21}$	251$_{20}$	234$_{57}$	203$_{100}$	101$_{10}$	812	436
Clopenthixol-M / artifact (Cl-thioxanthenone)	U	2260*	246$_{100}$	218$_{46}$	183$_9$	139$_{25}$	91$_{10}$	578	2641
Clopidogrel	P	2320	320$_1$	262$_{100}$	152$_{41}$	138$_{46}$	125$_{45}$	1037	5704
Clopidogrel artifact (-COOCH3)	P U+UHYAC UHYME	2110	263$_{12}$	125$_{18}$	110$_{100}$			673	996
Clopyralide ME		1320	205$_{15}$	174$_{27}$	147$_{100}$	110$_{50}$	75$_{31}$	412	4119
Clorazepate -H2O -CO2	P G U	2520	270$_{86}$	269$_{97}$	242$_{100}$	241$_{82}$	77$_{17}$	715	463
Clorazepate -H2O -CO2 enol AC		2545	312$_{55}$	270$_{34}$	241$_{100}$	227$_8$	205$_9$	980	6102
Clorazepate -H2O -CO2 enol ME		2225	284$_{79}$	283$_{100}$	110$_3$	91$_{62}$		799	464
Clorazepate -H2O -CO2 TMS		2300	342$_{62}$	341$_{100}$	327$_{19}$	269$_4$	73$_{30}$	1168	4573
Clorazepate HY	UHY	2050	231$_{80}$	230$_{95}$	154$_{23}$	105$_{38}$	77$_{100}$	512	419
Clorazepate HYAC	PHYAC U+UHYAC	2245	273$_{50}$	230$_{100}$	154$_{13}$	105$_{23}$	77$_{54}$	732	273
Clorazepate-M	P G UGLUC	2320	268$_{98}$	239$_{56}$	233$_{52}$	205$_{66}$	77$_{100}$	813	579
Clorazepate-M (HO-) artifact AC	UHYAC	2515	312$_{30}$	270$_{100}$	253$_{46}$	235$_{76}$	206$_9$	980	1747
Clorazepate-M (HO-) -H2O -CO2	UGLUC	2750	286$_{82}$	258$_{100}$	230$_{11}$	166$_7$	139$_8$	813	2113
Clorazepate-M (HO-) -H2O -CO2 AC	U+UHYAC	3000	328$_{22}$	286$_{90}$	258$_{100}$	166$_8$	139$_7$	1079	2111
Clorazepate-M (HO-) HY	UHY	2400	247$_{72}$	246$_{100}$	230$_{11}$	121$_{26}$	65$_{22}$	583	2112
Clorazepate-M (HO-) HYAC	U+UHYAC	2270	289$_{18}$	247$_{86}$	246$_{100}$	105$_7$	77$_{35}$	832	3143
Clorazepate-M (HO-) isomer-1 HY2AC	U+UHYAC	2560	331$_{48}$	289$_{64}$	247$_{100}$	230$_{41}$	154$_{13}$	1098	2125
Clorazepate-M (HO-) isomer-2 HY2AC	U+UHYAC	2610	331$_{46}$	289$_{54}$	246$_{100}$	154$_{11}$	121$_{11}$	1098	1751
Clorazepate-M (HO-methoxy-) HY2AC	U+UHYAC	2700	361$_{17}$	319$_{72}$	276$_{100}$	260$_{14}$	246$_{10}$	1267	1752
Clorazepate-M 2TMS		2200	430$_{51}$	429$_{89}$	340$_{15}$	313$_{19}$	73$_{100}$	1533	5499
Clorazepate-M TMS		2635	356$_9$	341$_{100}$	239$_{12}$	135$_{21}$	73$_{100}$	1252	4577
Clorazepate-M/artifact AC	UHYAC	3000	356$_{16}$	314$_{100}$	297$_{22}$	256$_{48}$	219$_{17}$	1240	1748
Clorofene	G U UHY	1950*	218$_{100}$	183$_{73}$	140$_{82}$			456	689
Clorofene AC	U+UHYAC	1885*	260$_{24}$	218$_{100}$	183$_{50}$	152$_{24}$	140$_{47}$	656	690
Clostebol AC		2965*	364$_{25}$	328$_{100}$	287$_{26}$	269$_{23}$	147$_{32}$	1283	3945
Clostebol acetate		2965*	364$_{25}$	328$_{100}$	287$_{26}$	269$_{23}$	147$_{32}$	1283	3945
Clostebol acetate TMS		2870*	436$_{62}$	401$_{12}$	230$_6$	133$_{12}$	73$_{100}$	1550	3952
Clostebol enol 2TMS		2830*	466$_{12}$	358$_{26}$	268$_{24}$	129$_{69}$	73$_{100}$	1611	3953
Clostebol -HCl AC		2700*	328$_{20}$	286$_{27}$	253$_{18}$	133$_{100}$	91$_{34}$	1082	3951
Clostebol -HCl enol 2TMS		2640*	430$_{38}$	415$_8$	231$_6$	207$_{10}$	73$_{100}$	1535	3955
Clostebol -HCl TMS		2675*	358$_{32}$	268$_{34}$	253$_{23}$	145$_{10}$	73$_{100}$	1255	3954
Clotiapine	U UHY UHYAC	2590	343$_{41}$	285$_{43}$	273$_{100}$	244$_{87}$	209$_{45}$	1173	2373
Clotiapine artifact (desulfo-)	U UHY UHYAC	2600	311$_7$	241$_{100}$	177$_{24}$	83$_{45}$	70$_{40}$	976	2377
Clotiapine-M (HO-) AC	UHYAC	3000	401$_{31}$	331$_{100}$	302$_{56}$	260$_{82}$	70$_{80}$	1448	2375
Clotiapine-M (nor-) AC	U UHYAC	3030	371$_{87}$	285$_{85}$	273$_{73}$	244$_{85}$	209$_{100}$	1316	2374
Clotiapine-M (nor-) artifact AC	U UHY UHYAC	3070	339$_{22}$	253$_{58}$	241$_{100}$	228$_{26}$	177$_{57}$	1149	2379

Table 8.1: Compounds in order of names

Name	Detected	RI	Typical ions and intensities					Page	Entry
Clotiapine-M (nor-HO-) 2AC	UHYAC	3400	429_{86}	344_{46}	302_{91}	260_{100}	112_{23}	1529	2376
Clotiapine-M (oxo-)	U	3030	357_{100}	285_{33}	244_{49}	209_{74}		1246	2380
Clotiapine-M (oxo-) artifact	U UHY	3040	325_{23}	253_{30}	241_{100}	213_{23}	177_{27}	1061	2378
Clotiazepam	P-I g UGLUC	2540	318_{72}	289_{100}	275_{20}			1016	267
Clotiazepam artifact		2280	274_{100}	259_{48}	245_{32}	223_{10}	139_9	736	2350
Clotiazepam-M (di-HO-) 2AC	UGLUCAC	2995	434_{36}	374_{81}	332_{100}	319_{61}	291_{52}	1543	271
Clotiazepam-M (HO-)	UGLUC	2705	316_{95}	287_{100}				1118	269
Clotiazepam-M (HO-) AC	UGLUCAC	2870	376_{14}	316_{84}	271_{100}	256_{76}		1338	270
Clotiazepam-M (oxo-)		2660	332_{75}	303_{100}	297_{28}			1104	268
Clotrimazole		2800	277_{100}	239_{15}	199_4	165_{30}		1178	1753
Clotrimazole artifact-1	U+UHYAC	2240*	278_{38}	243_{32}	201_8	165_{100}		767	1756
Clotrimazole artifact-2	U+UHYAC	2530*	294_{100}	217_{60}	183_{54}	139_{50}	105_{55}	869	1757
Clotrimazole artifact-3	U+UHYAC	2550*	308_{100}	277_{76}	231_{86}	165_{62}	139_{60}	955	1758
Cloxazolam		2775	318_4	305_{100}	261_{48}	226_{29}	191_{21}	1198	2264
Cloxazolam HY	UHY	2180	265_{62}	230_{100}	139_{43}	111_{50}		684	543
Cloxazolam HYAC	U+UHYAC	2300	307_{42}	265_{58}	230_{100}	139_{16}	111_{14}	946	290
Cloxiquine		1565	179_{100}	151_{84}	116_{62}	89_{42}		339	2003
Cloxiquine AC		1790	221_2	179_{100}	151_{82}	116_{42}	89_{28}	471	2004
Clozapine	P G U UHY U+UHYAC	2895	326_{27}	256_{75}	243_{100}	192_{36}	70_{31}	1068	320
Clozapine AC	U+UHYAC	2870	368_{24}	298_{70}	256_{40}	83_{100}	70_{97}	1302	2604
Clozapine TMS		2895	398_{41}	328_{88}	315_{100}	299_{50}	73_{52}	1437	4536
Clozapine-M (HO-) 2AC	U+UHYAC	2980	426_{24}	356_{71}	314_{47}	83_{100}	70_{93}	1523	2606
Clozapine-M (HO-) AC	U+UHYAC	3050	384_{38}	314_{70}	301_{83}	259_{100}	70_{35}	1376	2605
Clozapine-M (nor-)	UHY	3105	312_{44}	269_{22}	256_{42}	243_{100}	192_{37}	981	321
Clozapine-M (nor-) 2AC	U+UHYAC	3490	396_{100}	310_{60}	298_{61}	227_{42}	192_{56}	1429	323
Clozapine-M (nor-) AC	UHYAC	3650	354_{100}	243_{51}	228_{23}	192_{51}	112_{21}	1231	322
Clozapine-M/artifact	U+UHYAC	3875	378_{100}	280_{48}	225_{71}	209_{54}	112_{61}	1346	6766
Clozapine-M/artifact AC	U+UHYAC	3855	420_{83}	335_{100}	322_{90}	251_{37}	209_{45}	1507	7802
2C-N		2030	226_7	197_{100}	180_{15}	167_{28}	149_7	494	9177
2C-N 2TMS		2330	355_4	174_{100}	100_9	86_{23}	73_{52}	1312	9157
2C-N AC		2300	268_{17}	209_{100}	197_{15}	180_{10}	167_{18}	707	9156
2C-N artifact (-CH3N)		1710	209_{100}	162_9	148_{27}	133_{26}	118_{32}	427	9155
2C-N formyl artifact		2030	238_8	221_4	207_{100}	160_5	91_{13}	547	9154
CN gas (chloroacetophenone)		1020*	154_3	105_{100}	91_4	77_{61}	51_{26}	296	3731
2C-N HFB		2100	422_{10}	226_9	209_{100}	196_6	148_{13}	1512	9160
2C-N PFP		2050	372_{17}	209_{100}	196_6	176_{12}	148_{14}	1322	9159
2C-N TFA		2070	322_{10}	209_{100}	196_6	148_{13}	136_{15}	1043	9158
2C-N TMS		2085	283_2	269_{14}	197_2	102_{100}	73_{46}	896	9164
25C-NBOMe		2560	335_1	302_9	185_7	150_{33}	121_{100}	1124	10310
25C-NBOMe AC		2785	377_{11}	198_{64}	185_{15}	150_{19}	121_{100}	1343	10312
25C-NBOMe HFB		2540	531_9	198_{100}	185_{55}	155_{21}	121_{86}	1685	10266
25C-NBOMe HY artifact (dimer)		2780	413_2	380_{19}	226_{51}	199_{100}	185_{21}	1487	10314
25C-NBOMe HY artifact (dimer) AC		3030	455_2	270_3	228_{18}	198_{100}	169_{24}	1591	10313
25C-NBOMe HY artifact (dimer) HFB		2690	609_2	198_{100}	185_{44}	169_{25}	155_{12}	1717	10320
25C-NBOMe HY artifact (dimer) PFP		2700	559_7	198_{100}	185_{27}	169_{13}	90_8	1700	10319
25C-NBOMe HY artifact (dimer) TFA		2750	509_3	198_{100}	185_{46}	169_{24}	90_{23}	1668	10318
25C-NBOMe ME		2520	349_1	302_9	185_9	164_{83}	121_{100}	1204	10311
25C-NBOMe PFP		2550	481_9	198_{100}	185_{28}	121_{82}	91_{53}	1636	10317
25C-NBOMe TFA		2580	431_{18}	198_{100}	185_{34}	155_{13}	121_{45}	1537	10316
25C-NBOMe TMS		2600	392_3	222_{69}	185_{17}	150_{35}	121_{100}	1468	10315
25C-NBOMe-M (O,O-bis-demethyl-) isomer-1 3AC		3010	433_4	391_9	184_{54}	178_{100}	107_{65}	1542	10420
25C-NBOMe-M (O,O-bis-demethyl-) isomer-2 3AC		3060	433_2	270_9	192_{37}	150_{19}	121_{100}	1542	10429
25C-NBOMe-M (O-demethyl-) isomer-1 2AC		2930	405_3	198_7	192_{23}	121_{100}	91_{36}	1463	10430
25C-NBOMe-M (O-demethyl-) isomer-2 2AC		2980	405_{15}	363_{10}	192_{21}	121_{100}	91_{36}	1462	10419
25C-NBOMe-M (O-demethyl-HO-) -H2O AC		2770	361_9	319_{41}	198_{59}	167_{51}	107_{100}	1268	10428
Cocaethylene	U+UHYAC	2250	317_{58}	272_{24}	196_{100}	82_{74}		1013	466
Cocaethylene-D3		2240	320_{18}	275_{12}	215_8	199_{81}	85_{100}	1034	9331
Cocaethylene-M (ethylecgonine) AC		1675	255_{12}	196_{60}	168_{14}	94_{44}	82_{100}	633	6231
Cocaethylene-M (ethylecgonine) TBDMS		1685	327_1	270_7	204_{10}	196_9	82_{100}	1078	6249
Cocaethylene-M (ethylecgonine) TFA		1520	309_5	264_6	196_{56}	94_{53}	82_{100}	963	6241
Cocaethylene-M (ethylecgonine) TMS		1485	285_8	240_{12}	196_8	96_{68}	82_{100}	811	6257
Cocaethylene-M (nor-)		2115	303_6	182_{100}	136_{53}	105_{46}	68_{88}	925	6253
Cocaethylene-M (nor-) AC		2535	345_{10}	182_{99}	136_{54}	109_{98}	105_{100}	1184	6233
Cocaethylene-M (nor-) TFA		2245	399_2	277_{43}	208_{20}	164_{26}	105_{100}	1441	6245
Cocaine	P-I G U+UHYAC	2200	303_6	272_3	198_8	182_{66}	82_{100}	924	465
Cocaine-D3		2180	306_{30}	275_{11}	201_{11}	185_{100}	85_{86}	944	5565
Cocaine-M (benzoylecgonine)	U+UHYAC	2570	289_{19}	168_{34}	124_{100}	82_{42}	77_{50}	834	2120
Cocaine-M (benzoylecgonine) ET	U+UHYAC	2250	317_{58}	272_{24}	196_{100}	82_{74}		1013	466
Cocaine-M (benzoylecgonine) HFPOL		2125	471_{25}	366_{13}	350_{100}	272_{11}	82_{67}	1620	10322
Cocaine-M (benzoylecgonine) ME	P-I G U+UHYAC	2200	303_6	272_3	198_8	182_{66}	82_{100}	924	465
Cocaine-M (benzoylecgonine) PFP		2275	421_9	316_5	300_{38}	94_{52}	82_{100}	1510	4381
Cocaine-M (benzoylecgonine) TBDMS		2465	403_{20}	346_{25}	282_{35}	105_{38}	82_{100}	1457	6236
Cocaine-M (benzoylecgonine) TMS		2285	361_{12}	256_8	240_{47}	105_{39}	82_{100}	1269	5579
Cocaine-M (benzoylecgonine)-D3 ET		2240	320_{18}	275_{12}	215_8	199_{81}	85_{100}	1034	9331

Cocaine-M (benzoylecgonine)-D3 HFPOL

Table 8.1: Compounds in order of names

Name	Detected	RI	Typical ions and intensities					Page	Entry
Cocaine-M (benzoylecgonine)-D3 HFPOL		2115	474_{24}	369_{11}	353_{100}	275_{13}	85_{73}	1625	10321
Cocaine-M (benzoylecgonine)-D3 ME		2180	306_{30}	275_{11}	201_{11}	185_{100}	85_{86}	944	5565
Cocaine-M (benzoylecgonine)-D3 TMS		2275	364_{31}	349_{9}	243_{66}	105_{34}	85_{100}	1283	5580
Cocaine-M (cocaethylene)	U+UHYAC	2250	317_{58}	272_{24}	196_{100}	82_{74}		1013	466
Cocaine-M (cocaethylene)-D3		2240	320_{18}	275_{12}	215_{8}	199_{81}	85_{100}	1034	9331
Cocaine-M (ecgonine) 2TBDMS		1970	398_{1}	356_{12}	275_{4}	96_{28}	82_{100}	1488	6251
Cocaine-M (ecgonine) 2TMS		1680	329_{2}	314_{4}	96_{45}	82_{100}	73_{49}	1089	5445
Cocaine-M (ecgonine) ACTMS		1680	299_{11}	240_{41}	122_{9}	94_{29}	82_{100}	903	6238
Cocaine-M (ecgonine) TBDMS		1700	299_{5}	242_{21}	205_{5}	96_{34}	82_{100}	903	6250
Cocaine-M (ecgonine) TMSTFA		1395	353_{15}	267_{35}	240_{82}	94_{38}	82_{100}	1226	6255
Cocaine-M (ecgonine)-D3 2TMS		1670	332_{13}	317_{15}	99_{64}	85_{100}	73_{56}	1107	5576
Cocaine-M (ethylecgonine) AC		1675	255_{12}	196_{60}	168_{14}	94_{44}	82_{100}	633	6231
Cocaine-M (ethylecgonine) TBDMS		1685	327_{1}	270_{5}	204_{10}	196_{9}	82_{100}	1078	6249
Cocaine-M (ethylecgonine) TFA		1520	309_{5}	264_{6}	196_{56}	94_{53}	82_{100}	963	6241
Cocaine-M (ethylecgonine) TMS		1485	285_{8}	240_{12}	196_{8}	96_{68}	82_{100}	811	6257
Cocaine-M (HO-)		2460	319_{19}	182_{52}	121_{21}	82_{100}		1025	468
Cocaine-M (HO-) ME		2450	333_{29}	182_{85}	135_{42}	94_{41}	82_{100}	1112	470
Cocaine-M (HO-benzoylecgonine) 2TBDMS		2940	533_{37}	476_{25}	282_{50}	235_{14}	82_{100}	1686	6237
Cocaine-M (HO-benzoylecgonine) 2TMS		2505	449_{21}	240_{55}	193_{16}	82_{100}	73_{46}	1579	6258
Cocaine-M (HO-benzoylecgonine) ACTBDMS		2765	461_{13}	404_{7}	282_{18}	121_{14}	82_{100}	1603	6235
Cocaine-M (HO-benzoylecgonine) ACTMS		2565	419_{17}	240_{46}	163_{9}	94_{25}	82_{100}	1506	6239
Cocaine-M (HO-di-methoxy-) AC	UGLUCAC	2750	421_{13}	198_{13}	182_{71}	151_{7}	82_{100}	1510	5945
Cocaine-M (HO-di-methoxy-) HFB	UGLUCHFB	2585	575_{4}	377_{9}	182_{85}	94_{37}	82_{100}	1706	5947
Cocaine-M (HO-di-methoxy-) ME	UGLUCME	2550	393_{26}	212_{21}	182_{99}	94_{23}	82_{100}	1415	5678
Cocaine-M (HO-di-methoxy-) PFP	UGLUCPFP	2555	525_{5}	327_{11}	182_{71}	94_{31}	82_{100}	1681	5948
Cocaine-M (HO-di-methoxy-) TFA	UGLUCTFA	2530	473_{12}	277_{19}	182_{87}	94_{37}	82_{100}	1628	5953
Cocaine-M (HO-di-methoxy-) TMS	UGLUCTMS	2970	451_{15}	198_{7}	182_{61}	94_{29}	82_{100}	1583	5951
Cocaine-M (HO-methoxy-)		2670	349_{58}	198_{16}	182_{100}	151_{19}	82_{61}	1204	469
Cocaine-M (HO-methoxy-) AC	UGLUCAC	2695	391_{13}	198_{10}	182_{67}	151_{19}	82_{100}	1407	5944
Cocaine-M (HO-methoxy-) HFB	UGLUCHFB	2500	545_{3}	347_{8}	182_{64}	94_{34}	82_{100}	1693	5946
Cocaine-M (HO-methoxy-) ME		2650	363_{18}	198_{8}	182_{86}	94_{37}	82_{100}	1279	471
Cocaine-M (HO-methoxy-) PFP	UGLUCPFP	2470	495_{2}	297_{13}	182_{76}	94_{40}	82_{100}	1654	5949
Cocaine-M (HO-methoxy-) TFA	UGLUCTFA	2470	445_{11}	247_{13}	182_{73}	94_{35}	82_{100}	1572	5952
Cocaine-M (HO-methoxy-) TMS	UGLUCTMS	2850	421_{16}	198_{8}	182_{71}	94_{35}	82_{100}	1511	5950
Cocaine-M (HO-methoxy-benzoylecgonine) ACTMS		2505	449_{23}	240_{56}	193_{17}	82_{100}	73_{46}	1579	6240
Cocaine-M (nor-)		2080	289_{5}	168_{100}	136_{45}	77_{87}	68_{89}	835	6252
Cocaine-M (nor-) AC		2495	331_{18}	209_{39}	194_{36}	168_{100}	136_{49}	1101	6232
Cocaine-M (nor-) TFA		2185	385_{4}	263_{16}	194_{17}	105_{100}	77_{64}	1380	6244
Cocaine-M (nor-benzoylecgonine) ET		2115	303_{6}	182_{100}	136_{53}	105_{46}	68_{88}	925	6253
Cocaine-M (nor-benzoylecgonine) ME		2080	289_{5}	168_{100}	136_{45}	77_{87}	68_{89}	835	6252
Cocaine-M (nor-benzoylecgonine) MEAC		2495	331_{18}	209_{39}	194_{36}	168_{100}	136_{49}	1101	6232
Cocaine-M (nor-benzoylecgonine) METFA		2185	385_{4}	263_{16}	194_{17}	105_{100}	77_{64}	1380	6244
Cocaine-M (nor-benzoylecgonine) TBDMS		2375	389_{12}	268_{66}	210_{7}	136_{84}	68_{100}	1400	6254
Cocaine-M (nor-benzoylecgonine) TFATBDMS		2460	428_{24}	306_{5}	179_{20}	105_{100}	77_{39}	1641	6246
Cocaine-M (nor-cocaethylene)		2115	303_{6}	182_{100}	136_{53}	105_{46}	68_{88}	925	6253
Cocaine-M (nor-cocaethylene) AC		2535	345_{10}	182_{99}	136_{54}	109_{98}	105_{100}	1184	6233
Cocaine-M (nor-cocaethylene) TFA		2245	399_{2}	277_{23}	208_{20}	164_{26}	105_{100}	1441	6245
Cocaine-M/artifact (anhydroecgonine) TBDMS		1520	281_{16}	252_{29}	224_{100}	150_{44}	122_{30}	790	6242
Cocaine-M/artifact (anhydroecgonine) TMS		1345	239_{31}	224_{25}	210_{100}	183_{10}	122_{28}	553	6256
Cocaine-M/artifact (anhydromethylecgonine)	UHY-I UHYAC-I	1280	181_{31}	152_{100}	138_{9}	122_{15}	82_{18}	349	3574
Cocaine-M/artifact (benzoic acid)	P U UHY	1235*	122_{77}	105_{100}	77_{72}			266	95
Cocaine-M/artifact (benzoic acid) ME	P(ME)	1180*	136_{30}	105_{100}	77_{73}			275	1211
Cocaine-M/artifact (benzoic acid) TBDMS		1295*	221_{1}	179_{100}	135_{30}	105_{55}	77_{51}	540	6247
Cocaine-M/artifact (ecgonine) ACTBDMS		2010	341_{16}	284_{34}	282_{41}	142_{17}	82_{100}	1165	6234
Cocaine-M/artifact (ecgonine) ETAC		1675	255_{12}	196_{60}	168_{14}	94_{44}	82_{100}	633	6231
Cocaine-M/artifact (ecgonine) ETPFP		1620	359_{34}	314_{28}	196_{100}	96_{21}	82_{90}	1257	5563
Cocaine-M/artifact (ecgonine) -H2O TBDMS		1520	281_{16}	252_{29}	224_{100}	150_{44}	122_{30}	790	6242
Cocaine-M/artifact (ecgonine) -H2O TMS		1345	239_{31}	224_{25}	210_{100}	183_{10}	122_{28}	553	6256
Cocaine-M/artifact (ecgonine) MEAC	U+UHYAC	1595	241_{6}	182_{56}	96_{37}	94_{51}	82_{100}	561	472
Cocaine-M/artifact (ecgonine) MEHFB		1620	395_{14}	364_{9}	182_{100}	94_{53}	82_{95}	1423	5676
Cocaine-M/artifact (ecgonine) MEPFP		1530	345_{15}	314_{12}	182_{79}	96_{26}	82_{100}	1183	5562
Cocaine-M/artifact (ecgonine) METBDMS		1625	313_{2}	256_{9}	182_{15}	96_{38}	82_{100}	991	6248
Cocaine-M/artifact (ecgonine) METFA		1490	295_{27}	264_{13}	182_{80}	96_{26}	82_{100}	875	5564
Cocaine-M/artifact (ecgonine) METMS		1580	271_{9}	212_{18}	182_{9}	96_{40}	82_{100}	726	5583
Cocaine-M/artifact (ecgonine) TFATBDMS		1585	395_{4}	338_{18}	282_{18}	94_{23}	82_{100}	1425	6243
Cocaine-M/artifact (methylecgonine)		1465	199_{18}	168_{12}	96_{58}	82_{100}		398	467
Cocaine-M/artifact (methylecgonine) AC	U+UHYAC	1595	241_{6}	182_{56}	96_{37}	94_{51}	82_{100}	561	472
Cocaine-M/artifact (methylecgonine) -H2O	UHY-I UHYAC-I	1280	181_{31}	152_{100}	138_{9}	122_{15}	82_{18}	349	3574
Cocaine-M/artifact (methylecgonine) HFB		1620	395_{14}	364_{9}	182_{100}	94_{53}	82_{95}	1423	5676
Cocaine-M/artifact (methylecgonine) PFP		1530	345_{15}	314_{12}	182_{79}	96_{26}	82_{100}	1183	5562
Cocaine-M/artifact (methylecgonine) TBDMS		1625	313_{2}	256_{9}	182_{15}	96_{38}	82_{100}	991	6248
Cocaine-M/artifact (methylecgonine) TFA		1490	295_{27}	264_{13}	182_{80}	96_{26}	82_{100}	875	5564
Cocaine-M/artifact (methylecgonine) TMS		1580	271_{9}	212_{18}	182_{9}	96_{40}	82_{100}	726	5583
Codeine	P G U UHY	2375	299_{100}	229_{26}	162_{46}	124_{23}		902	473

Table 8.1: Compounds in order of names

Name	Detected	RI	Typical ions and intensities					Page	Entry
Codeine AC	PAC U+UHYAC	2500	341 $_{100}$	282 $_{40}$	229 $_{20}$	204 $_{17}$	162 $_{5}$	1163	224
Codeine Cl-artifact AC	UHYAC	2630	375 $_{86}$	316 $_{100}$	263 $_{12}$	204 $_{42}$	162 $_{14}$	1334	2991
Codeine HFB		2320	495 $_{8}$	438 $_{3}$	282 $_{100}$	266 $_{9}$	225 $_{11}$	1654	6142
Codeine PFP		2430	445 $_{100}$	388 $_{6}$	282 $_{73}$	266 $_{6}$	119 $_{10}$	1571	2252
Codeine TFA		2280	395 $_{64}$	338 $_{5}$	282 $_{100}$	115 $_{20}$	69 $_{34}$	1424	4011
Codeine TMS		2520	371 $_{50}$	196 $_{34}$	178 $_{51}$	146 $_{36}$	73 $_{100}$	1320	2464
Codeine-D3		2370	302 $_{100}$	232 $_{26}$	165 $_{47}$	127 $_{24}$		920	7295
Codeine-D3 AC		2495	344 $_{100}$	301 $_{9}$	285 $_{75}$	232 $_{38}$	165 $_{20}$	1180	7300
Codeine-D3 HFB		2310	498 $_{33}$	438 $_{3}$	285 $_{100}$	269 $_{8}$	225 $_{9}$	1657	9333
Codeine-D3 PFP		2420	448 $_{45}$	388 $_{4}$	285 $_{100}$	269 $_{9}$	225 $_{10}$	1576	9332
Codeine-M (hydrocodone)	G UHY UHYAC	2440	299 $_{100}$	242 $_{51}$	185 $_{23}$	96 $_{24}$	59 $_{23}$	902	238
Codeine-M (hydrocodone) Cl-artifact		2630	375 $_{100}$	340 $_{47}$	318 $_{28}$	146 $_{13}$	115 $_{10}$	1334	4401
Codeine-M (hydrocodone) enol AC		2500	341 $_{100}$	298 $_{65}$	242 $_{32}$	162 $_{26}$		1164	258
Codeine-M (hydrocodone) enol TMS		2475	371 $_{31}$	356 $_{14}$	313 $_{9}$	234 $_{30}$	73 $_{100}$	1320	6215
Codeine-M (nor-) 2AC	U+UHYAC	2945	369 $_{14}$	327 $_{3}$	223 $_{37}$	87 $_{100}$	72 $_{36}$	1305	226
Codeine-M (nor-) 2HFB		2580	677 $_{30}$	451 $_{18}$	405 $_{78}$	223 $_{100}$	169 $_{72}$	1727	9342
Codeine-M (nor-) 2PFP		2540	577 $_{59}$	401 $_{20}$	355 $_{86}$	223 $_{100}$	119 $_{97}$	1706	9341
Codeine-M (O-demethyl-)	G UHY	2455	285 $_{100}$	268 $_{15}$	162 $_{59}$	124 $_{21}$		810	474
Codeine-M (O-demethyl-) 2AC	G PHYAC U+UHYAC	2620	369 $_{59}$	327 $_{100}$	310 $_{36}$	268 $_{47}$	162 $_{11}$	1306	225
Codeine-M (O-demethyl-) 2HFB		2375	677 $_{10}$	480 $_{10}$	464 $_{100}$	407 $_{9}$	169 $_{8}$	1726	6120
Codeine-M (O-demethyl-) 2PFP		2360	577 $_{51}$	558 $_{7}$	430 $_{8}$	414 $_{100}$	119 $_{22}$	1706	2251
Codeine-M (O-demethyl-) 2TFA		2250	477 $_{71}$	364 $_{100}$	307 $_{6}$	115 $_{8}$	69 $_{31}$	1630	4008
Codeine-M (O-demethyl-) 2TMS	UHYTMS	2560	429 $_{19}$	236 $_{21}$	196 $_{15}$	146 $_{21}$	73 $_{100}$	1531	2463
Codeine-M (O-demethyl-) Cl-artifact 2AC	UHYAC	2680	403 $_{59}$	361 $_{100}$	344 $_{63}$	302 $_{90}$	204 $_{55}$	1456	2992
Codeine-M (O-demethyl-) TFA		2285	381 $_{55}$	268 $_{100}$	146 $_{11}$	115 $_{13}$	69 $_{23}$	1360	5569
Codeine-M (O-demethyl-)-D3 2HFB		2375	680 $_{4}$	483 $_{9}$	467 $_{100}$	414 $_{7}$	169 $_{23}$	1727	6126
Codeine-M (O-demethyl-)-D3 2PFP		2350	580 $_{16}$	433 $_{7}$	417 $_{100}$	269 $_{5}$	119 $_{8}$	1708	5567
Codeine-M (O-demethyl-)-D3 2TFA		2240	480 $_{32}$	383 $_{6}$	367 $_{100}$	314 $_{6}$	307 $_{6}$	1634	5571
Codeine-M (O-demethyl-)-D3 2TMS		2550	432 $_{91}$	290 $_{27}$	239 $_{60}$	199 $_{40}$	73 $_{100}$	1540	5578
Codeine-M (O-demethyl-)-D3 TFA		2275	384 $_{39}$	271 $_{100}$	211 $_{8}$	165 $_{6}$	152 $_{7}$	1376	5572
Codeine-M 2PFP		2440	563 $_{100}$	400 $_{10}$	355 $_{38}$	327 $_{7}$	209 $_{15}$	1701	3534
Codeine-M 3AC	U+UHYAC	2955	397 $_{8}$	355 $_{9}$	209 $_{41}$	87 $_{100}$	72 $_{33}$	1434	1194
Codeine-M 3PFP	UHYPFP	2405	709 $_{80}$	533 $_{28}$	388 $_{29}$	367 $_{51}$	355 $_{100}$	1729	3533
Codeine-M 3TMS	UHYTMS	2605	487 $_{17}$	416 $_{19}$	222 $_{36}$	131 $_{19}$	73 $_{100}$	1645	3525
Colchicine		3200	399 $_{51}$	371 $_{20}$	312 $_{100}$	297 $_{36}$	281 $_{31}$	1442	2852
Colecalciferol		3150*	384 $_{43}$	351 $_{100}$	325 $_{38}$	143 $_{37}$	57 $_{35}$	1379	2794
Colecalciferol AC		3300*	426 $_{16}$	398 $_{84}$	382 $_{100}$	351 $_{44}$	145 $_{78}$	1524	2796
Colecalciferol -H2O	P	3130*	366 $_{100}$	351 $_{16}$	271 $_{9}$	158 $_{11}$	143 $_{19}$	1294	2795
Coniine		1610	127 $_{2}$	98 $_{2}$	84 $_{100}$	70 $_{4}$	56 $_{13}$	270	4459
Coniine AC	U+UHYAC	1405	169 $_{5}$	154 $_{3}$	126 $_{43}$	98 $_{3}$	84 $_{100}$	323	4460
Cotinine	P U+UHYAC	1715	176 $_{36}$	118 $_{12}$	98 $_{100}$			332	692
Coumachlor artifact	UME	1575*	180 $_{32}$	165 $_{100}$	145 $_{41}$	137 $_{49}$	102 $_{51}$	344	4427
Coumachlor enol 2TMS		2990*	486 $_{9}$	443 $_{77}$	247 $_{14}$	193 $_{11}$	73 $_{100}$	1643	4963
Coumachlor ET		2780*	370 $_{35}$	327 $_{100}$	299 $_{78}$	187 $_{38}$	139 $_{45}$	1311	4812
Coumachlor isomer-1 AC		2810*	384 $_{14}$	342 $_{34}$	299 $_{100}$	187 $_{20}$	121 $_{44}$	1374	4816
Coumachlor isomer-2 AC		2810*	384 $_{40}$	342 $_{25}$	299 $_{100}$	187 $_{8}$	121 $_{27}$	1374	4817
Coumachlor ME	UME UGLUCME	2770*	356 $_{21}$	313 $_{100}$	201 $_{9}$	189 $_{8}$	125 $_{19}$	1240	4143
Coumachlor TMS		2870*	414 $_{38}$	371 $_{100}$	261 $_{21}$	75 $_{41}$	73 $_{56}$	1489	4962
Coumachlor-M (di-HO-) 3ME	UME	3195*	416 $_{27}$	373 $_{100}$	359 $_{11}$	180 $_{3}$	125 $_{7}$	1495	4425
Coumachlor-M (HO-) 2TMS	UTMS	3150*	502 $_{26}$	459 $_{100}$	446 $_{34}$	281 $_{31}$	73 $_{85}$	1661	4964
Coumachlor-M (HO-) enol 3TMS	UTMS	3240*	574 $_{19}$	531 $_{100}$	335 $_{36}$	281 $_{19}$	73 $_{78}$	1704	4965
Coumachlor-M (HO-) isomer-1 2ET	UET	3020*	414 $_{33}$	371 $_{100}$	343 $_{69}$	231 $_{16}$	139 $_{22}$	1489	4813
Coumachlor-M (HO-) isomer-1 2ME	UME UHYME	2990*	386 $_{31}$	343 $_{100}$	231 $_{6}$	151 $_{20}$	125 $_{13}$	1384	4422
Coumachlor-M (HO-) isomer-2 2ET	UET	3095*	414 $_{29}$	371 $_{100}$	343 $_{56}$	231 $_{12}$	139 $_{21}$	1489	4814
Coumachlor-M (HO-) isomer-2 2ME	UME UHYME	3035*	386 $_{42}$	343 $_{100}$	231 $_{8}$	151 $_{14}$	125 $_{19}$	1384	4423
Coumachlor-M (HO-dihydro-) 2ME	UME	3095*	388 $_{100}$	343 $_{94}$	329 $_{69}$	245 $_{48}$	125 $_{57}$	1394	4426
Coumachlor-M (HO-dihydro-) 3TMS	UME	3170*	576 $_{8}$	459 $_{77}$	446 $_{100}$	281 $_{60}$	73 $_{73}$	1706	4966
Coumachlor-M (HO-methoxy-) 2ET	UET	3320*	444 $_{29}$	401 $_{100}$	373 $_{37}$	263 $_{11}$	139 $_{16}$	1569	4815
Coumachlor-M (HO-methoxy-) 2ME	UME	3195*	416 $_{27}$	373 $_{100}$	359 $_{11}$	180 $_{3}$	125 $_{7}$	1495	4425
Coumaphos		2575*	362 $_{73}$	226 $_{58}$	210 $_{36}$	109 $_{100}$	97 $_{100}$	1272	3330
m-Coumaric acid		1940*	164 $_{100}$	147 $_{20}$	118 $_{33}$	91 $_{49}$	65 $_{39}$	310	5765
m-Coumaric acid 2TMS		1910*	308 $_{78}$	293 $_{90}$	249 $_{49}$	203 $_{77}$	73 $_{100}$	958	6004
m-Coumaric acid AC		1970*	206 $_{14}$	164 $_{100}$	147 $_{29}$	118 $_{20}$	91 $_{21}$	416	5998
m-Coumaric acid HFB		1820*	360 $_{100}$	169 $_{36}$	147 $_{43}$	91 $_{35}$	69 $_{93}$	1262	6003
m-Coumaric acid ME		1720*	178 $_{61}$	147 $_{100}$	119 $_{28}$	91 $_{49}$	65 $_{33}$	337	5997
m-Coumaric acid MEAC		1760*	220 $_{13}$	178 $_{96}$	147 $_{100}$	119 $_{19}$	91 $_{25}$	468	5999
m-Coumaric acid MEHFB		1665*	374 $_{45}$	343 $_{100}$	169 $_{32}$	101 $_{36}$	69 $_{44}$	1329	6002
m-Coumaric acid MEPFP		1580*	324 $_{49}$	293 $_{100}$	119 $_{33}$	101 $_{36}$	69 $_{26}$	1055	6001
m-Coumaric acid METMS		1750*	250 $_{76}$	235 $_{76}$	203 $_{100}$	89 $_{45}$	73 $_{64}$	604	6005
m-Coumaric acid PFP		1670*	310 $_{100}$	293 $_{19}$	146 $_{25}$	119 $_{58}$	69 $_{33}$	968	6000
p-Coumaric acid		2225*	164 $_{34}$	118 $_{32}$	91 $_{82}$	65 $_{100}$	63 $_{96}$	310	5760
p-Coumaric acid 2TMS		2040*	308 $_{35}$	293 $_{48}$	249 $_{29}$	219 $_{63}$	73 $_{100}$	958	6019
p-Coumaric acid AC		1910*	206 $_{6}$	164 $_{100}$	147 $_{29}$	118 $_{17}$	89 $_{12}$	415	5981
p-Coumaric acid -CO2		1045*	120 $_{51}$	91 $_{100}$	65 $_{62}$	63 $_{56}$	51 $_{52}$	264	5761

p-Coumaric acid HFB

Table 8.1: Compounds in order of names

Name	Detected	RI	Typical ions and intensities					Page	Entry
p-Coumaric acid HFB		1855*	360_{100}	343_{15}	169_{29}	147_{20}	69_{81}	1262	5986
p-Coumaric acid ME		1800*	178_{56}	147_{100}	119_{39}	91_{37}	65_{31}	337	5979
p-Coumaric acid MEAC		1785*	220_{5}	178_{89}	147_{100}	119_{29}	89_{18}	468	5980
p-Coumaric acid MEHFB		1695*	374_{57}	343_{100}	315_{13}	129_{26}	69_{51}	1329	5985
p-Coumaric acid METFA		1540*	274_{65}	243_{100}	215_{17}	99_{47}	69_{61}	738	5982
p-Coumaric acid METMS		2750*	250_{100}	235_{18}	203_{18}	179_{13}	73_{98}	604	6020
p-Coumaric acid PFP		1720*	310_{100}	293_{13}	163_{20}	119_{55}	69_{61}	968	5984
p-Coumaric acid TFA		1665*	260_{95}	243_{18}	101_{29}	89_{33}	69_{100}	655	5983
Coumarin	G	1550*	146_{66}	118_{100}	90_{31}	63_{23}		285	4365
Coumarin-M (HO-)	UHY	1780*	162_{94}	134_{100}	105_{23}	78_{26}	63_{9}	305	4366
Coumarin-M (HO-) AC	UHYAC	1840*	204_{16}	162_{100}	134_{85}	105_{9}	77_{11}	408	4367
Coumarin-M (HO-) HFB		1685*	358_{100}	330_{51}	169_{25}	133_{80}	105_{23}	1251	7614
Coumarin-M (HO-) ME		1750*	176_{100}	148_{76}	133_{86}	105_{12}	77_{16}	332	7611
Coumarin-M (HO-) PFP		1550*	308_{100}	280_{50}	261_{5}	161_{3}	133_{49}	954	7613
Coumarin-M (HO-) TFA		1540*	258_{100}	230_{50}	133_{49}	119_{14}	105_{8}	645	7615
Coumarin-M (HO-) TMS		1925*	234_{87}	219_{100}	191_{20}	163_{58}	73_{25}	526	7612
Coumatetralyl	G	2660*	292_{100}	188_{68}	130_{41}	121_{69}	91_{24}	856	1431
Coumatetralyl AC	UHYAC	2725*	334_{11}	292_{100}	188_{47}	175_{23}	121_{44}	1120	4789
Coumatetralyl HY		2250*	266_{3}	248_{13}	220_{15}	130_{100}	121_{39}	696	4809
Coumatetralyl HYAC		2350*	308_{1}	265_{7}	248_{12}	130_{64}	121_{100}	959	4811
Coumatetralyl HYME		2300*	280_{4}	135_{70}	130_{100}	115_{9}	77_{28}	783	4810
Coumatetralyl isomer-1 ET		2680*	320_{54}	291_{62}	175_{100}	129_{21}	121_{32}	1034	4800
Coumatetralyl isomer-1 ME	UME	2655*	306_{100}	291_{47}	175_{72}	121_{34}	115_{41}	943	4790
Coumatetralyl isomer-2 ET		2705*	320_{58}	291_{69}	175_{100}	129_{26}	121_{33}	1034	4801
Coumatetralyl isomer-2 ME	UME	2690*	306_{100}	291_{28}	202_{23}	175_{27}	115_{25}	943	2084
Coumatetralyl TMS		2765*	364_{76}	349_{16}	260_{43}	193_{16}	73_{100}	1282	5026
Coumatetralyl-M (di-HO-) 3ET	UET	3290*	408_{44}	379_{100}	219_{14}			1472	4807
Coumatetralyl-M (di-HO-) isomer-1 2ME	UME	3005*	352_{1}	333_{100}	319_{9}	205_{11}	151_{6}	1220	4798
Coumatetralyl-M (di-HO-) isomer-1 3TMS	UTMS	2955*	540_{6}	348_{100}	333_{33}	73_{36}		1689	5030
Coumatetralyl-M (di-HO-) isomer-2 2ME	UME	3085*	352_{2}	333_{100}	205_{15}	177_{14}	151_{20}	1219	4797
Coumatetralyl-M (di-HO-) isomer-2 3TMS	UTMS	3230*	540_{1}	525_{3}	449_{39}	348_{100}	73_{82}	1689	5031
Coumatetralyl-M (di-HO-) isomer-3 3ME	UME	3105*	366_{100}	351_{92}	232_{33}	193_{24}	159_{45}	1291	4794
Coumatetralyl-M (HO-) isomer-1 2TMS	UTMS	2835*	452_{100}	437_{19}	348_{20}	193_{17}	131_{45}	1584	5028
Coumatetralyl-M (HO-) isomer-1 ET	UET	2905*	336_{13}	318_{52}	289_{81}	217_{100}	121_{62}	1131	4802
Coumatetralyl-M (HO-) isomer-1 ME	UME	2910*	322_{11}	303_{91}	203_{100}	187_{28}	121_{18}	1044	4795
Coumatetralyl-M (HO-) isomer-2 2ET	UET	2910*	364_{51}	335_{100}	219_{80}	161_{55}	91_{60}	1282	4803
Coumatetralyl-M (HO-) isomer-2 2ME	UME	2925*	336_{100}	321_{12}	232_{34}	217_{23}	205_{62}	1131	4791
Coumatetralyl-M (HO-) isomer-2 2TMS	UTMS	2880*	452_{8}	362_{100}	361_{99}	233_{33}	73_{98}	1585	5029
Coumatetralyl-M (HO-) isomer-3 2ET	UET	2920*	364_{54}	335_{100}	187_{16}	175_{50}	121_{28}	1283	4804
Coumatetralyl-M (HO-) isomer-3 2ME	UME	2935*	336_{51}	321_{100}	305_{29}	175_{50}	121_{40}	1131	4793
Coumatetralyl-M (HO-) isomer-3 2TMS	UTMS	3015*	452_{68}	437_{15}	348_{29}	333_{25}	73_{100}	1585	5027
Coumatetralyl-M (HO-) isomer-4 2ET	UET	3000*	364_{95}	335_{100}	245_{37}	219_{69}	165_{17}	1282	4805
Coumatetralyl-M (HO-) isomer-4 2ME	UME	2990*	336_{100}	321_{26}	232_{50}	217_{29}	205_{61}	1131	4792
Coumatetralyl-M (HO-methoxy-) 2ET	UET	3070*	394_{100}	378_{37}	365_{70}	349_{51}	275_{16}	1421	4806
Coumatetralyl-M (HO-methoxy-) 2ME	UME	3070*	366_{100}	351_{37}	262_{42}	235_{37}	181_{14}	1291	4796
Coumatetralyl-M (tri-HO-) -H2O 2ET	UET	3320*	378_{82}	349_{73}	206_{100}	165_{84}	137_{71}	1347	4808
Coumatetralyl-M (tri-HO-) -H2O 2ME	UME	3175*	350_{100}	335_{11}	205_{28}	177_{30}	151_{21}	1208	4799
2C-P		1720	223_{21}	194_{100}	179_{13}	165_{51}	135_{8}	485	6909
2C-P 2AC		2160	307_{20}	206_{100}	193_{24}	177_{42}	135_{8}	952	6921
2C-P 2TMS		2130	367_{1}	352_{5}	174_{100}	100_{11}	86_{28}	1300	6923
2C-P AC		2090	265_{14}	206_{100}	193_{28}	177_{60}	135_{13}	690	6920
2C-P formyl artifact		1755	235_{23}	204_{100}	193_{62}	163_{9}	135_{11}	536	6908
2C-P HFB		1895	419_{49}	206_{76}	193_{100}	177_{42}	163_{15}	1506	6940
2C-P PFP		1865	369_{54}	206_{77}	193_{100}	177_{42}	119_{15}	1305	6935
2C-P TFA		1870	319_{36}	206_{69}	193_{100}	177_{39}	149_{38}	1025	6930
2C-P TMS		1860	295_{5}	265_{5}	194_{20}	102_{100}	73_{64}	881	6922
2C-P-M (bis-O-demethyl-) 3AC	UGLUCSPEAC	2280	321_{1}	279_{13}	237_{33}	178_{100}	165_{15}	1040	8800
2C-P-M (deamino-HO-) AC		2090*	266_{18}	206_{100}	193_{8}	177_{70}	91_{16}	696	9205
2C-P-M (di-HO- N-acetyl-) 2TFA	UGLUCSPETFA	2340	489_{7}	430_{42}	345_{12}	206_{100}	198_{23}	1647	8806
2C-P-M (di-HO-) 3AC	U+UHYAC	2505	381_{19}	322_{14}	252_{26}	207_{59}	193_{100}	1362	8794
2C-P-M (HO-) 2TFA	UGLUCSPETFA	1925	431_{31}	318_{38}	204_{53}	191_{100}	177_{71}	1537	8807
2C-P-M (HO-) -H2O AC	U+UHYAC	2195	263_{19}	204_{100}	191_{20}			676	8795
2C-P-M (HO-) -H2O TFA	UGLUCSPETFA	1950	317_{39}	256_{6}	204_{55}	191_{100}	161_{16}	1011	8809
2C-P-M (HO-) isomer-1 2AC	UGLUCSPEAC	2225	323_{4}	294_{19}	281_{32}	235_{28}	193_{100}	1052	8791
2C-P-M (HO-) isomer-2 2AC	UGLUCSPEAC	2245	323_{1}	279_{13}	220_{47}	177_{100}	165_{21}	1053	8792
2C-P-M (HO-) isomer-3 2AC	U+UHYAC	2330	323_{18}	263_{8}	204_{100}	191_{20}	177_{15}	1052	8790
2C-P-M (HO-deamino-COOH-) (ME)AC	UGLUCSPEAC	2020*	310_{11}	250_{100}	223_{10}	191_{46}	175_{19}	971	8799
2C-P-M (HO-deamino-COOH-) (ME)TFA		1835*	364_{68}	305_{17}	250_{78}	223_{44}	191_{100}	1281	8808
2C-P-M (HO-deamino-COOH-) (ME-D4)AC	UGLUCSPEAC	2010*	314_{13}	254_{100}	227_{9}	192_{38}	175_{15}	996	8811
2C-P-M (HO-deamino-HO-) 2AC	U+UHYAC	1990*	324_{1}	264_{2}	204_{100}	177_{23}	91_{6}	1057	8937
2C-P-M (HO-N-acetyl-) TFA	UGLUCSPETFA	2145	377_{26}	318_{100}	231_{28}	216_{45}	204_{76}	1343	8801
2C-P-M (HOOC-) (ME)AC	UGLUCSPEAC	2365	309_{18}	278_{4}	250_{100}	177_{37}	161_{10}	965	8796
2C-P-M (HOOC-) (ME-D4)AC		2355	313_{14}	253_{100}	190_{8}	177_{43}	161_{13}	991	8812
2C-P-M (O-demethyl- N-acetyl-) isomer-1 TFA	UGLUCSPETFA	1930	347_{8}	318_{4}	288_{100}	259_{59}	191_{16}	1194	8802

Table 8.1: Compounds in order of names

Name	Detected	RI	Typical ions and intensities					Page	Entry
2C-P-M (O-demethyl- N-acetyl-) isomer-2 TFA	UGLUCSPETFA	1990	347_3	288_{100}	259_{18}	191_{10}	163_{10}	1194	8803
2C-P-M (O-demethyl-) isomer-1 2AC	U+UHYAC	2170	293_3	251_{28}	192_{100}	179_{17}	163_{24}	866	8788
2C-P-M (O-demethyl-) isomer-2 2AC	U+UHYAC	2210	293_3	251_{20}	192_{100}	179_{10}	163_{20}	865	8789
2C-P-M (O-demethyl-deamino-COOH-) (ME-D4)AC		1910*	284_8	242_{100}	213_{32}	180_{29}		805	8810
2C-P-M (O-demethyl-deamino-COOH-) isomer-1 (ME)AC	UGLUCSPEAC	1875*	280_{14}	250_{11}	238_{75}	206_{86}	178_{100}	782	8797
2C-P-M (O-demethyl-deamino-COOH-) isomer-2 (ME)AC	UGLUCSPEAC	1915*	280_7	238_{100}	209_{32}	179_{32}	91_{11}	782	8798
2C-P-M (O-demethyl-deamino-HO-) 2AC	UGLUCSPEAC	1930*	294_8	252_{21}	210_{12}	192_{100}	163_{46}	870	8936
2C-P-M (O-demethyl-deamino-HO-) isomer-1 2AC	U+UHYAC	1870*	294_1	252_{21}	210_{14}	192_{100}	163_{40}	870	8938
2C-P-M (O-demethyl-deamino-HO-) isomer-2 2AC	U+UHYAC	1880*	294_3	252_5	210_1	192_{100}	163_{42}	871	8939
2C-P-M (O-demethyl-HO- N-acetyl-) isomer-1 2TFA	UGLUCSPETFA	2000	459_7	400_{28}	346_8	286_{100}	259_{24}	1599	8804
2C-P-M (O-demethyl-HO- N-acetyl-) isomer-2 2TFA	UGLUCSPETFA	2030	459_7	417_{27}	400_{73}	286_{100}	190_{92}	1599	8805
2C-P-M (O-demethyl-HO-) 3AC	U+UHYAC	2395	351_1	309_{10}	249_{11}	190_{100}	177_{13}	1214	8793
CRA-13		3230*	368_{27}	297_{21}	281_{18}	171_{70}	127_{100}	1303	8534
m-Cresol TMS		1040*	180_{38}	165_{100}	149_7	135_7	91_{31}	346	5674
p-Cresol	UHY	1060*	108_{76}	107_{100}	77_{32}	53_{17}		260	4220
p-Cresol AC	UHYAC	1110*	150_{10}	110	108_{100}	77_{18}		289	4225
Crimidine	G	1560	171_{76}	156_{73}	142_{100}	120_{31}	93_{36}	325	693
Crinosterol		3135*	398_{100}	300_{62}	271_{71}	255_{85}	69_{96}	1440	5619
Crinosterol -H2O		3210*	380_{100}	255_{39}	81_{53}	69_{53}	55_{67}	1358	5623
Croconazole		2390	310_{60}	243_{13}	185_{59}	125_{100}	89_{75}	969	5686
Cropropamide		1725	195_4	168_{14}	100_{100}	69_{54}		558	694
Crotamiton (cis)	P G U	1560	203_{12}	188_{16}	135_{14}	120_{40}	69_{100}	407	5347
Crotamiton (trans)	P G U	1600	203_9	188_{14}	135_{13}	120_{27}	69_{100}	406	695
Crotamiton-M (4-HO-crotyl-) (cis)	UGLUC	1790	219_{14}	135_{21}	120_{100}	91_{39}	85_{29}	465	5357
Crotamiton-M (4-HO-crotyl-) (trans)	UGLUC	1865	219_{31}	201_{38}	133_{64}	120_{100}	85_{78}	465	5356
Crotamiton-M (4-HO-crotyl-) (trans) AC	UGLUC	1940	261_{71}	219_{81}	133_{80}	118_{79}	85_{100}	664	5358
Crotamiton-M (4-HO-crotyl-) (trans) TMS	UGLUCTMS	1800	291_{78}	276_{49}	162_{26}	91_{38}	73_{100}	850	5359
Crotamiton-M (di-HO-) 2AC	UGLUCAC	2215	319_{32}	260_{36}	246_{86}	118_{100}	85_{86}	1026	5362
Crotamiton-M (di-HO-) 2TMS	UGLUCTMS	2050	379_2	364_{16}	276_{52}	132_{73}	73_{100}	1352	5363
Crotamiton-M (di-HO-dihydro-)	UGLUC	1900	237_6	219_{19}	206_{47}	162_{82}	120_{100}	545	5360
Crotamiton-M (di-HO-dihydro-) 2AC	UGLUCAC	2105	321_{36}	219_{56}	162_{61}	135_{98}	120_{100}	1039	5361
Crotamiton-M (HO-ethyl-) (cis)	UGLUC	1805	219_2	188_{25}	150_{15}	118_{85}	69_{100}	465	5354
Crotamiton-M (HO-ethyl-) (trans)	P U UGLUC	1830	219_2	188_{25}	150_{17}	118_{80}	69_{100}	465	5353
Crotamiton-M (HO-ethyl-) (trans) AC	UGLUC	1905	261_{18}	188_{80}	150_{33}	118_{100}	69_{92}	664	5355
Crotamiton-M (HO-ethyl-HOOC-) MEAC	UGLUCAC	2135	305_6	245_{15}	232_{100}	132_{90}	118_{79}	936	5374
Crotamiton-M (HO-methyl-disulfide)	UGLUC	2235	299_{13}	253_{100}	174_{77}	162_{62}	134_{81}	901	5370
Crotamiton-M (HO-methyl-disulfide) AC	UGLUCAC	2315	341_{28}	295_{85}	202_{100}	162_{79}	134_{75}	1160	5371
Crotamiton-M (HO-methylthio-)	UGLUC	2025	267_2	221_{90}	190_{62}	162_{92}	134_{100}	701	5351
Crotamiton-M (HO-methylthio-) AC	UGLUCAC	2115	309_{19}	263_{100}	190_{89}	162_{95}	134_{99}	963	5352
Crotamiton-M (HOOC-)	U	1940	233_5	188_{39}	134_{100}	120_{67}	99_{49}	522	697
Crotamiton-M (HOOC-) (cis) TMS	UTMS	1855	305_9	290_{42}	187_{60}	134_{84}	73_{100}	937	5350
Crotamiton-M (HOOC-) (trans) TMS	UTMS	1875	305_{13}	290_{53}	187_{58}	134_{71}	73_{100}	937	5349
Crotamiton-M (HOOC-) ME	UME	1865	247_{27}	216_{30}	188_{83}	134_{100}	113_{72}	586	5348
Crotamiton-M (HOOC-dihydro-)	UGLUC	1900	235_{40}	202_{26}	190_{82}	162_{40}	134_{100}	534	5364
Crotamiton-M (HOOC-dihydro-) ME	UGLUCME	1845	249_{42}	218_{63}	162_{61}	120_{96}	115_{100}	598	5365
Crotamiton-M (HOOC-methyl-thio-) ME	UGLUCME	2010	295_1	249_{75}	234_{27}	162_{100}	134_{95}	877	5376
Crotamiton-M (HOOC-thio-)	UGLUC	2150	267_{80}	208_{21}	174_{61}	162_{100}	135_{24}	699	5375
Crotamiton-M (HO-thio-)	UGLUC	1970	253_7	209_{66}	134_{59}	120_{45}	91_{45}	622	5367
Crotamiton-M (HO-thio-) 2AC	UGLUCAC	2210	337_{27}	295_{89}	262_{37}	209_{59}	162_{100}	1137	5369
Crotamiton-M (HO-thio-) AC	UGLUCAC	2070	295_{69}	262_{42}	209_{71}	176_{83}	162_{100}	876	5368
Crotamiton-M (N-deethyl-)	UHY	1415	175_{27}	107_{100}	96_{24}	83_{32}	69_{77}	331	5373
Crotamiton-M (N-deethyl-HO-methyl-)	UGLUC	1995	191_{60}	123_{100}	94_{10}	69_{72}		371	696
Crotamiton-M (N-deethyl-HO-methyl-) AC	UGLUCAC	2055	233_{53}	191_{91}	123_{100}	69_{83}		521	5372
Crotamiton-M/artifact (methyl-thio-chloro-)	UGLUC	1985	285_{12}	239_{69}	190_{40}	162_{95}	134_{100}	808	5366
Crotethamide		1675	181_5	154_{20}	86_{100}	69_{48}		496	698
Crotylbarbital	P G U UHY U+UHYAC	1620	210_7	181_{85}	156_{93}	141_{39}	55_{100}	433	699
Crotylbarbital-M (HO-) -H2O	U UHY UHYAC	1600	208_3	179_{32}	157_{100}	141_{65}		424	700
CS gas (o-chlorobenzylidenemalonitrile)		1500	188_{43}	161_{15}	110_{14}	99_{11}	75_{29}	364	3539
2C-T-2		1980	241_{22}	212_{100}	197_{20}	183_{35}	153_{22}	560	5035
2C-T-2 2AC	U+UHYAC	2395	325_{17}	224_{100}	211_{50}	181_{16}	153_{14}	1063	5038
2C-T-2 2TMS		2405	385_1	370_3	254_4	211_4	174_{100}	1382	6815
2C-T-2 AC	U+UHYAC	2310	283_{28}	224_{100}	211_{34}	181_{12}	153_{12}	796	5037
2C-T-2 deuteroformyl artifact		1935	255_{37}	224_{31}	211_{100}	181_{14}	153_{16}	633	5036
2C-T-2 HFB		2040	437_{24}	224_{21}	211_{100}	181_{19}	169_{17}	1551	6816
2C-T-2 PFP		2090	387_{34}	224_{29}	211_{100}	181_{18}	153_{15}	1389	6817
2C-T-2 TFA	UGLUCTFA	2210	337_{26}	224_{17}	211_{100}	181_{13}	153_{10}	1134	6818
2C-T-2 TMS		2405	313_1	299_3	174_{100}	147_4	86_7	989	6814
2C-T-2-M (aryl-HOOC-)	UGLUC	1970	242_{100}	227_{20}	183_{68}	153_{14}		563	6893
2C-T-2-M (aryl-HOOC-) ME	USPEME	1960*	256_{100}	241_{10}	197_{49}	181_{12}	167_{15}	636	6842
2C-T-2-M (deamino-HO-)	UGLUC	1905*	242_{71}	211_{19}	181_{15}	153_{11}		565	6839
2C-T-2-M (deamino-HO-) AC	U+UHYAC	2050	284_{40}	224_{100}	209_{30}	167_{20}	150_{10}	801	6892
2C-T-2-M (deamino-HOOC-)		2130*	256_{100}	242_{31}	211_{60}	195_{30}	181_{32}	636	6840
2C-T-2-M (deamino-HOOC-) ME	UHYME	1910*	270_{100}	255_{24}	211_{82}	195_{44}	181_{46}	717	6838
2C-T-2-M (deamino-HOOC-) TMS	USPETMS	2075*	328_{100}	313_{29}	298_{25}	255_{61}	211_{57}	1080	6841

2C-T-2-M (deamino-oxo-) Table 8.1: Compounds in order of names

Name	Detected	RI	Typical ions and intensities					Page	Entry
2C-T-2-M (deamino-oxo-)		2130*	240 $_{33}$	211 $_{100}$	181 $_{11}$	153 $_{19}$	122 $_{11}$	555	7234
2C-T-2-M (HO- N-acetyl-) TFA	UGLUCTFA	2270	427 $_{24}$	367 $_3$	259 $_{100}$	167 $_6$		1524	6834
2C-T-2-M (HO- sulfone) 2AC	U+UHYAC	2780	373 $_{35}$	314 $_{84}$	302 $_{75}$	272 $_{54}$	259 $_{100}$	1327	6833
2C-T-2-M (HO- sulfone) AC	U+UHYAC	2730	331 $_7$	272 $_{100}$	259 $_{74}$	238 $_{26}$	165 $_{47}$	1100	6828
2C-T-2-M (O-demethyl- N-acetyl-) 2TFA	U+UHYTFA	2180	461 $_{16}$	306 $_{100}$	293 $_{43}$	209 $_{25}$		1602	6894
2C-T-2-M (O-demethyl- N-acetyl-) TFA	U+UHYTFA	2250	365 $_{11}$	323 $_{19}$	306 $_{100}$	293 $_{18}$	197 $_9$	1285	6942
2C-T-2-M (O-demethyl- sulfone) 2AC	U+UHYAC	2510	343 $_3$	301 $_{90}$	242 $_{100}$	230 $_{48}$	153 $_9$	1173	6835
2C-T-2-M (O-demethyl-) 2AC	U+UHYAC	2120	311 $_{48}$	297 $_{31}$	269 $_{100}$	252 $_{46}$	210 $_{78}$	976	6837
2C-T-2-M (O-demethyl-) 2TFA	UGLUCTFA	1980	419 $_{28}$	306 $_{100}$	293 $_{92}$	209 $_{40}$	69 $_{32}$	1504	6821
2C-T-2-M (O-demethyl-) 3AC	U+UHYAC	2290	353 $_{22}$	311 $_{32}$	252 $_{33}$	210 $_{100}$	197 $_{20}$	1226	6836
2C-T-2-M (O-demethyl-sulfone N-acetyl-) TFA	UGLUCTFA	2450	397 $_4$	355 $_{68}$	242 $_{100}$	153 $_{14}$		1433	6820
2C-T-2-M (S-deethyl-) 3AC	U+UHYAC	2420	339 $_{22}$	297 $_5$	238 $_{21}$	196 $_{100}$	183 $_{28}$	1149	6827
2C-T-2-M (S-deethyl-) AC	U+UHYAC	2170	255 $_{18}$	196 $_{100}$	183 $_{41}$	181 $_{34}$	153 $_{21}$	631	6831
2C-T-2-M (S-deethyl-) isomer-1 2AC		2240	297 $_{29}$	210 $_{14}$	196 $_{100}$	183 $_{35}$	181 $_{29}$	889	6823
2C-T-2-M (S-deethyl-) isomer-2 2AC	U+UHYAC	2360	297 $_{16}$	255 $_{11}$	238 $_{20}$	196 $_{100}$	183 $_{37}$	889	6826
2C-T-2-M (S-deethyl-methyl- N-acetyl-)	U+UHYAC	2230	269 $_{19}$	210 $_{100}$	197 $_{35}$	195 $_{21}$	167 $_{27}$	712	6832
2C-T-2-M (S-deethyl-methyl- sulfone) AC	U+UHYAC	2580	301 $_7$	242 $_{100}$	230 $_4$	196 $_7$	124 $_7$	911	6829
2C-T-2-M (S-deethyl-methyl- sulfoxide) AC	U+UHYAC	2460	285 $_{16}$	268 $_{23}$	226 $_{33}$	211 $_{100}$	197 $_{31}$	808	6830
2C-T-2-M (sulfone N-acetyl-) TFA	UGLUCTFA	2400	411 $_{23}$	256 $_{100}$	242 $_4$	181 $_{10}$	167 $_{11}$	1480	6822
2C-T-2-M (sulfone) 2AC	U+UHYAC	2640	357 $_{10}$	256 $_{100}$	244 $_6$	167 $_7$	91 $_7$	1247	6824
2C-T-2-M (sulfone) AC	U+UHYAC	2600	315 $_{15}$	256 $_{100}$	244 $_9$	167 $_{12}$	91 $_8$	1000	6825
2C-T-2-M (sulfone) TFA	UGLUCTFA	2310	369 $_{44}$	256 $_{100}$	243 $_7$	211 $_4$	167 $_{23}$	1305	6819
2C-T-7		2470	255 $_{26}$	226 $_{100}$	183 $_{63}$	169 $_{31}$	153 $_{34}$	633	6855
2C-T-7 2AC		2470	339 $_{14}$	238 $_{100}$	225 $_{30}$	181 $_{22}$	153 $_{17}$	1150	6859
2C-T-7 2TMS		2395	399 $_1$	384 $_4$	369 $_4$	225 $_7$	174 $_{100}$	1443	6860
2C-T-7 AC		2410	297 $_{66}$	238 $_{100}$	225 $_{37}$	196 $_{56}$	183 $_{55}$	891	6858
2C-T-7 deuteroformyl artifact		2060	269 $_{27}$	238 $_{25}$	225 $_{100}$	183 $_{13}$	153 $_{24}$	713	6857
2C-T-7 formyl artifact		2050	267 $_{35}$	236 $_{27}$	225 $_{100}$	183 $_{26}$	153 $_{46}$	701	6856
2C-T-7 HFB		2175	451 $_{14}$	238 $_{24}$	225 $_{100}$	181 $_{23}$	153 $_{21}$	1582	6861
2C-T-7 PFP		2160	401 $_{23}$	238 $_{17}$	225 $_{100}$	181 $_{17}$	153 $_{19}$	1449	6862
2C-T-7 TFA		2170	351 $_{26}$	238 $_{17}$	225 $_{100}$	181 $_{24}$	153 $_{23}$	1212	6863
2C-T-7-M (deamino-HO-)	UGLUC	2000*	256 $_{69}$	225 $_{100}$	183 $_{23}$	150 $_{56}$	135 $_{23}$	637	6864
2C-T-7-M (deamino-HO-) AC	U+UHYAC	2080*	298 $_{56}$	238 $_{100}$	225 $_{10}$	196 $_{80}$	181 $_{69}$	895	6869
2C-T-7-M (deamino-HOOC-)		2110*	270 $_{100}$	225 $_{55}$	213 $_{46}$	181 $_{34}$	153 $_{21}$	717	6872
2C-T-7-M (deamino-HOOC-) ME		1950*	284 $_{100}$	227 $_{50}$	225 $_{74}$	183 $_{24}$	153 $_{25}$	801	6873
2C-T-7-M (deamino-oxo-)		2190*	254 $_{42}$	225 $_{100}$	183 $_{14}$	153 $_{24}$	137 $_8$	628	7235
2C-T-7-M (HO- N-acetyl-)	UGLUC	2525	313 $_{40}$	254 $_{100}$	242 $_{44}$	210 $_{38}$	183 $_{21}$	988	6866
2C-T-7-M (HO- N-acetyl-) TFA		2345	409 $_{27}$	350 $_{100}$	337 $_7$	236 $_5$	181 $_{13}$	1473	6871
2C-T-7-M (HO- sulfone N-acetyl-)	UGLUC	2740	345 $_{31}$	286 $_{73}$	164 $_{100}$	151 $_{27}$	120 $_{18}$	1184	6865
2C-T-7-M (HO- sulfone) 2AC	U+UHYAC	2760	387 $_{31}$	340 $_{42}$	328 $_{100}$	268 $_{36}$	108 $_{33}$	1390	6868
2C-T-7-M (HO-) 2AC	U+UHYAC	2585	355 $_{51}$	296 $_{72}$	283 $_{10}$	236 $_{92}$	101 $_{100}$	1237	6867
2C-T-7-M (HO-) 2TFA		2110	463 $_{80}$	434 $_{43}$	350 $_{60}$	337 $_{100}$	231 $_{67}$	1605	6870
2C-T-7-M (HO-) 3AC	U+UHYAC	2630	397 $_{69}$	296 $_{99}$	283 $_{12}$	236 $_{100}$	101 $_{64}$	1435	6875
2C-T-7-M (S-depropyl-) AC	U+UHYAC	2170	255 $_{18}$	196 $_{100}$	183 $_{41}$	181 $_{34}$	153 $_{21}$	631	6831
2C-T-7-M (S-depropyl-) isomer-1 2AC		2240	297 $_{29}$	210 $_{14}$	196 $_{100}$	183 $_{35}$	181 $_{29}$	889	6823
2C-T-7-M (S-depropyl-) isomer-2 2AC	U+UHYAC	2360	297 $_{16}$	255 $_{11}$	238 $_{20}$	196 $_{100}$	183 $_{37}$	889	6826
2C-T-7-M (S-depropyl-methyl- N-acetyl-)	U+UHYAC	2230	269 $_{19}$	210 $_{100}$	197 $_{35}$	195 $_{21}$	167 $_{27}$	712	6832
2C-T-7-M (S-depropyl-methyl- sulfone) AC	U+UHYAC	2580	301 $_7$	242 $_{100}$	230 $_4$	196 $_7$	124 $_7$	911	6829
2C-T-7-M (S-depropyl-methyl- sulfoxide) AC	U+UHYAC	2460	285 $_{16}$	268 $_{23}$	226 $_{33}$	211 $_{100}$	197 $_{31}$	808	6830
CUMYL-BICA		2800	334 $_{12}$	216 $_{45}$	200 $_{100}$	173 $_{17}$	144 $_{31}$	1121	9489
CUMYL-PICA		2880	348 $_{15}$	230 $_{49}$	214 $_{100}$	173 $_{16}$	144 $_{29}$	1201	9490
CUMYL-PINACA		2750	349 $_8$	334 $_{26}$	231 $_6$	215 $_{100}$	145 $_{47}$	1206	9492
CUMYL-PINACA TMS		2730	421 $_1$	406 $_{11}$	215 $_{71}$	119 $_{82}$	73 $_{100}$	1511	9493
CUMYL-PINACA-5F		2675	367 $_4$	352 $_{25}$	249 $_{10}$	233 $_{100}$	145 $_{59}$	1299	9501
CUMYL-PINACA-5F TMS		2715	439 $_1$	424 $_8$	233 $_{39}$	119 $_{71}$	73 $_{100}$	1557	9502
CUMYL-THPINACA		2950	377 $_{10}$	362 $_{39}$	258 $_{12}$	243 $_{100}$	145 $_{34}$	1344	9494
CUMYL-THPINACA TMS		2980	449 $_1$	434 $_4$	364 $_{44}$	119 $_{74}$	73 $_{100}$	1579	9495
Cyamemazine		2565	323 $_2$	277 $_1$	223 $_1$	100 $_2$	58 $_{100}$	1051	4248
Cyamemazine-M (bis-nor-) AC	UHYAC	3035	337 $_{65}$	237 $_{93}$	205 $_{49}$	114 $_{100}$	72 $_{25}$	1136	4393
Cyamemazine-M (bis-nor-HO-) 2AC	UHYAC	3300	395 $_{86}$	295 $_{25}$	253 $_{59}$	114 $_{100}$	72 $_{48}$	1424	4396
Cyamemazine-M (HO-) AC	UHYAC	3000	381 $_{20}$	294 $_5$	239 $_6$	100 $_4$	58 $_{100}$	1361	4391
Cyamemazine-M (HO-methoxy-) AC	UHYAC	3110	411 $_{35}$	324 $_5$	269 $_{27}$	100 $_{10}$	58 $_{100}$	1481	4392
Cyamemazine-M (nor-) AC	UHYAC	3080	351 $_{45}$	237 $_{47}$	205 $_{31}$	128 $_{100}$	86 $_{45}$	1212	4394
Cyamemazine-M (nor-HO-) 2AC	UHYAC	3320	409 $_{48}$	295 $_{13}$	253 $_{27}$	128 $_{100}$	86 $_{35}$	1473	4395
Cyamemazine-M (nor-HO-methoxy-) 2AC	UHYAC	3500	439 $_{27}$	269 $_{19}$	128 $_{100}$	86 $_{44}$		1556	4397
Cyamemazine-M (nor-sulfoxide) AC	UHYAC	3285	367 $_4$	350 $_{32}$	277 $_{100}$	237 $_{87}$	128 $_{27}$	1296	4398
Cyamemazine-M (sulfoxide)		2960	339 $_1$	322 $_{22}$	237 $_{68}$	224 $_{10}$	58 $_{100}$	1150	4399
Cyamemazine-M/artifact (ring)	U UHY UHYAC	2555	224 $_{100}$	192 $_{32}$				485	1281
Cyamemazine-M/artifact (ring) TMS		2310	296 $_{34}$	281 $_3$	223 $_6$	73 $_{100}$		884	5437
Cyamemazine-M/artifact (ring-COOH) METMS		2430	329 $_{21}$	314 $_5$	77 $_{39}$	73 $_{100}$		1086	5438
Cyanazine		1960	240 $_{24}$	225 $_{45}$	198 $_{30}$	172 $_{35}$	68 $_{100}$	555	3175
Cyanophenphos		2310	303 $_{11}$	185 $_{22}$	169 $_{48}$	157 $_{100}$	63 $_{68}$	921	3331
Cyanophos		1720	243 $_{35}$	125 $_{53}$	109 $_{100}$	79 $_{38}$	63 $_{35}$	567	3332
Cyanuric acid	U UHY UHYAC	2880	129 $_{100}$	86 $_{20}$	70 $_{13}$			271	4424

Table 8.1: Compounds in order of names

Name	Detected	RI	Typical ions and intensities					Page	Entry
Cyclamate 2TMS		1680	323_7	280_{66}	210_9	147_{51}	73_{100}	1051	4537
Cyclamate-M AC	U+UHYAC	1290	141_{26}	98_{19}	67_{15}	60_{100}	56_{94}	280	1229
Cyclandelate		1975	276_1	125_{20}	107_{80}	83_{31}	69_{100}	756	7524
Cyclandelate AC		2080	149_8	125_{15}	107_{29}	83_{26}	69_{100}	1020	7525
Cyclandelate-M/artifact (mandelic acid)		1485*	166_7	107_{100}	79_{59}	77_{40}		315	1071
Cyclandelate-M/artifact (mandelic acid)		1890*	152_{10}	107_{100}	79_{75}	77_{55}	51_{63}	292	5759
Cyclizine	G U UHY UHYAC	2045	266_{46}	207_{50}	194_{54}	165_{38}	99_{100}	698	1782
Cyclizine-M (benzophenone)	U+UHYAC	1610*	182_{31}	152_3	105_{100}	77_{70}	51_{39}	351	1624
Cyclizine-M (carbinol)	UHY	1645*	184_{45}	165_{14}	152_7	105_{100}	77_{63}	357	1333
Cyclizine-M (carbinol) AC	U+UHYAC	1700*	226_{20}	184_{20}	165_{100}	105_{14}	77_{35}	495	1241
Cyclizine-M (carbinol) HFB		1475*	380_3	183_4	166_{100}	152_{22}	83_{25}	1355	8146
Cyclizine-M (carbinol) ME	UHY	1655*	198_{70}	167_{94}	121_{100}	105_{56}	77_{71}	395	6779
Cyclizine-M (carbinol) PFP		1410*	330_4	183_4	166_{100}	152_{24}	83_{30}	1093	8145
Cyclizine-M (carbinol) TFA		1420*	280_{19}	183_7	166_{100}	152_{24}	83_{21}	780	8144
Cyclizine-M (carbinol) TMS		1540*	256_{28}	241_{14}	179_{38}	167_{100}	152_{17}	638	8159
Cyclizine-M (HO-BPH) isomer-1	UHY	2065*	198_{93}	121_{72}	105_{100}	93_{22}	77_{66}	393	1627
Cyclizine-M (HO-BPH) isomer-1 AC	U+UHYAC	2010*	240_{27}	198_{100}	121_{47}	105_{85}	77_{80}	555	2196
Cyclizine-M (HO-BPH) isomer-2	P-I U UHY	2080*	198_{50}	121_{100}	105_{17}	93_{14}	77_{28}	394	732
Cyclizine-M (HO-BPH) isomer-2 AC	U+UHYAC	2050*	240_{20}	198_{100}	121_{94}	105_{41}	77_{51}	555	2197
Cyclizine-M (HO-methoxy-BPH)	UHY	2050*	228_{43}	197_6	151_{100}	105_{22}	77_{41}	501	1625
Cyclizine-M (HO-methoxy-BPH) AC	U+UHYAC	2100*	270_3	228_{92}	151_{100}	105_{25}	77_{40}	716	1622
Cyclizine-M (nor-)	U UHY	2120	252_{12}	207_{58}	167_{100}	152_{33}	85_{49}	620	1602
Cyclizine-M (nor-) AC	U+UHYAC	2525	294_{16}	208_{56}	167_{100}	152_{30}	85_{78}	871	1601
Cyclizine-M/artifact	UHY	2070*	228_8	186_{100}	157_{10}	128_7	77_4	499	1626
Cyclizine-M/artifact AC	U+UHYAC	2200*	270_7	228_{17}	186_{100}	157_{10}	128_7	714	1623
Cycloate		1610	215_5	154_{54}	83_{100}	72_{26}	55_{93}	449	3174
Cyclobarbital	P G U+UHYAC	1970	236_1	207_{100}	157_4	141_{22}	79_{16}	540	701
Cyclobarbital (ME)	P	1940	221_{100}	155_{44}	143_9	87_{18}		605	2288
Cyclobarbital 2ME	PME	1845	264_1	235_{100}	178_7	169_{45}	79_{12}	682	705
Cyclobarbital 2TMS		1890	380_{17}	365_{27}	351_{89}	150_{70}	73_{100}	1358	5496
Cyclobarbital-M (di-HO-) -2H2O	P G U+UHYAC	1965	232_4	204_{100}	161_{18}	146_{12}	117_{37}	516	854
Cyclobarbital-M (di-HO-) -2H2O 2ME	PME UME	1860	260_2	232_{100}	175_{20}	146_{24}	117_{34}	657	1121
Cyclobarbital-M (di-HO-) -2H2O 2TMS		2015	376_2	361_{34}	261_{15}	146_{100}	73_{46}	1339	4582
Cyclobarbital-M (di-HO-) -2H2O ME	P G U+UHYAC	1895	246_{10}	218_{100}	146_{23}	117_{39}		580	1120
Cyclobarbital-M (di-oxo-)	U UHY UHYAC	1980	264_4	235_{100}	207_{26}	193_{12}	79_{11}	680	4461
Cyclobarbital-M (di-oxo-) 2ME	UME UHYME	2100	292_{35}	263_{100}	235_{50}	207_{18}	178_{25}	855	4462
Cyclobarbital-M (HO-) 3TMS	UTMS	2600	468_1	453_{32}	439_{100}	349_{57}	73_{100}	1615	4463
Cyclobarbital-M (HO-) -H2O	U UHY U+UHYAC	2170	234_{57}	205_{100}	156_{45}	141_{76}	79_{21}	527	702
Cyclobarbital-M (oxo-)	U+UHYAC	2190	250_{11}	221_{100}	193_{35}	179_{25}	150_{15}	604	703
Cyclobarbital-M (oxo-) 2ME	U UHY	2050	278_1	249_{100}	221_{23}	164_{11}	79_8	768	706
Cyclobarbital-M (oxo-) 2TMS	UTMS	2570	394_{11}	379_{100}	264_{49}	164_{15}	73_{95}	1421	4464
Cyclobenzaprine	P UHY U+UHYAC	2235	275_1	215_{14}	202_8	189_5	58_{100}	748	40
Cyclobenzaprine-M (bis-nor-) AC	U+UHYAC	2710	289_{15}	230_{100}	215_{70}	202_{31}	189_5	835	1873
Cyclobenzaprine-M (HO-N-oxide) -(CH3)2NOH	UHY	2280*	246_{100}	228_{35}	215_{50}	202_{36}	178_{33}	580	2698
Cyclobenzaprine-M (HO-N-oxide) -(CH3)2NOH AC	U+UHYAC	2530*	288_{35}	246_{100}	229_{58}	215_{89}	202_{37}	827	2541
Cyclobenzaprine-M (nor-)	UHY	2600	261_{14}	218_{99}	215_{100}	202_{66}	189_{23}	665	2270
Cyclobenzaprine-M (nor-) AC	U+UHYAC	2670	303_{20}	230_{100}	215_{74}	202_{34}	86_{18}	925	42
Cyclobenzaprine-M (N-oxide) -(CH3)2NOH	U UHY U+UHYAC	2000*	230_{100}	215_{40}				510	46
Cyclocumarol		2670*	322_{100}	265_{70}	249_{39}	148_{17}	72_{87}	1044	4047
Cyclofenil		2710*	364_{71}	322_{76}	280_{100}	263_{25}	199_{11}	1282	2282
Cyclofenil artifact (deacetyl-)		2680*	322_{76}	280_{100}	263_{25}	199_{11}	107_{15}	1045	3210
Cyclofenil HY		2700*	280_{100}	237_{17}	199_{32}			783	2278
Cyclohexadecane		1950*	224_1	196_3	97_{59}	83_{73}	55_{100}	489	2355
Cyclohexane		<1000*	84_{10}	69_{42}	56_{100}	41_{80}	27_{41}	254	3774
Cyclohexanol		<1000*	100_4	82_{57}	71_{17}	67_{38}	57_{100}	258	707
Cyclohexanol -H2O		<1000*	82_{66}	67_{58}	43_{100}			253	1629
Cyclohexanone		<1000*	98_{30}	83_7	69_{27}	55_{100}	42_{86}	257	3610
Cyclohexene		<1000*	82_{66}	67_{58}	43_{100}			253	1629
2-Cyclohexylphenol		1580*	176_{61}	133_{87}	120_{70}	107_{100}	91_{21}	333	5162
2-Cyclohexylphenol AC		1615*	218_9	176_{100}	133_{47}	120_{40}	107_{48}	459	5166
2-Cyclohexylphenol ME		1565*	190_{39}	147_{100}	134_{23}	121_{60}	91_{45}	370	5170
4-Cyclohexylphenol		1595*	176_{45}	133_{100}	120_{38}	107_{52}	91_{12}	333	5163
4-Cyclohexylphenol AC		1720*	218_6	176_{100}	133_{88}	120_{30}	107_{36}	459	5167
Cyclopentamine		1230	141_1	126_2	67_2	58_{100}		281	2771
Cyclopentamine AC		1680	183_1	168_{17}	100_{60}	58_{100}		356	2284
Cyclopentaphenanthrene		2000*	190_{100}	189_{82}	163_4	161_1	95_{21}	369	2565
Cyclopenthiazide 4ME	UEXME	3660	435_1	352_{100}	309_3	244_{10}	145_2	1547	6849
Cyclopentobarbital	P G U UHY UHYAC	1865	193_{51}	169_{46}	67_{100}			527	708
Cyclopentobarbital 2ME		1775	221_{75}	196_{34}	67_{100}			669	709
Cyclopentolate		2025	175_1	163_1	91_{15}	71_{13}	58_{100}	852	2760
Cyclopentolate -H2O		2000	273_1	129_2	91_6	71_{34}	58_{100}	735	2772
Cyclophosphamide		2065	260_1	211_{15}	175_{100}	147_{24}	69_{18}	655	1496
Cyclophosphamide -HCl	P	1975	224_{13}	175_{100}	147_{28}	69_{18}		486	1489
Cyclotetradecane		1860*	196_1	111_{25}	97_{51}	83_{66}	55_{100}	391	2354

Cyclothiazide 4ME

Table 8.1: Compounds in order of names

Name	Detected	RI	Typical ions and intensities					Page	Entry
Cyclothiazide 4ME	UEXME	3730	445 $_3$	352 $_{100}$	244 $_{15}$	145 $_8$		1571	6850
Cycloxydim		2580	279 $_7$	251 $_4$	178 $_{100}$	149 $_8$	108 $_{21}$	1064	3635
Cycloxydim ME		2380	293 $_9$	192 $_{100}$	164 $_3$	123 $_{15}$	95 $_{27}$	1152	3636
Cycluron		1760	198 $_3$	127 $_{10}$	99 $_7$	89 $_{17}$	72 $_{100}$	396	3936
Cycluron ME		1720	212 $_2$	141 $_{13}$	113 $_{22}$	102 $_8$	72 $_{100}$	441	3937
Cyfluthrin		2755	433 $_1$	226 $_{35}$	206 $_{58}$	163 $_{100}$	127 $_{29}$	1541	3514
Cypermethrin		2815	415 $_3$	209 $_{15}$	181 $_{78}$	163 $_{100}$	91 $_{72}$	1493	3176
Cypermethrin-M/artifact (deacyl-) -HCN		1700*	198 $_{100}$	169 $_{64}$	141 $_{74}$	115 $_{50}$	77 $_{68}$	394	2797
Cypermethrin-M/artifact (deacyl-) ME		2590	239 $_2$	197 $_{100}$	141 $_{30}$	115 $_{30}$	77 $_{19}$	551	2819
Cypermethrin-M/artifact (HOOC-) ME		1170*	222 $_4$	187 $_{64}$	163 $_{70}$	127 $_{65}$	91 $_{100}$	478	4207
Cyphenothrin		2960	375 $_6$	181 $_{27}$	167 $_{13}$	123 $_{100}$	81 $_{30}$	1335	3881
Cyprazepam artifact		2505	265 $_{70}$	264 $_{100}$	230 $_{79}$	177 $_{24}$	75 $_{33}$	684	4010
Cyprazepam artifact (deoxo-)		2730	323 $_{38}$	294 $_{100}$	241 $_{21}$	91 $_{68}$	55 $_{91}$	1049	4012
Cyprazepam HY	UHY	2050	231 $_{80}$	230 $_{95}$	154 $_{23}$	105 $_{38}$	77 $_{100}$	512	419
Cyprazepam HYAC	PHYAC U+UHYAC	2245	273 $_{30}$	230 $_{100}$	154 $_{13}$	105 $_{23}$	77 $_{50}$	732	273
Cyroheptadine	G U+UHYAC	2340	287 $_{100}$	215 $_{52}$	96 $_{80}$	70 $_{27}$		823	710
Cyroheptadine-M (HO-)	UHY-I	3060	303 $_{100}$	243 $_6$	217 $_{26}$	202 $_{41}$	178 $_9$	925	1620
Cyroheptadine-M (nor-)	U-I UHY-I	2400	273 $_{100}$	229 $_{29}$	215 $_{84}$	165 $_3$	82 $_{16}$	735	1619
Cyroheptadine-M (nor-) AC	UHYAC	2920	315 $_{100}$	300 $_{41}$	243 $_{39}$	229 $_{48}$	215 $_{57}$	1001	1614
Cyroheptadine-M (nor-HO-) 2AC	UHYAC-I	3000	373 $_2$	331 $_{29}$	303 $_{100}$	202 $_{61}$	82 $_{24}$	1327	1615
Cyroheptadine-M (nor-HO-) AC	UHYAC-I	2980	331 $_{100}$	241 $_{19}$	229 $_{35}$	215 $_{54}$	202 $_{58}$	1101	1616
Cyroheptadine-M (nor-HO-) -H2O	U-I UHY-I	2450	271 $_{100}$	241 $_{11}$	213 $_7$	193 $_{23}$	165 $_{25}$	725	1618
Cyroheptadine-M (nor-HO-) -H2O AC	UHYAC-I	2940	313 $_{59}$	243 $_{47}$	229 $_{80}$	215 $_{100}$	202 $_{29}$	989	1617
Cyroheptadine-M (nor-HO-aryl-) 2AC	UHYAC-I	3060	373 $_{100}$	358 $_{36}$	316 $_{38}$	259 $_{66}$	72 $_7$	1328	2691
Cyroheptadine-M (oxo-)	U-I UHY-I UHYAC-I	2960	301 $_{56}$	258 $_{25}$	229 $_{100}$	215 $_{82}$	202 $_{53}$	914	1621
Cyproterone AC		3340*	416 $_8$	356 $_{51}$	313 $_{100}$	246 $_{34}$	175 $_{70}$	1496	1415
Cyproterone -H2O	UHYAC	3310*	356 $_{43}$	246 $_{44}$	175 $_{100}$			1242	1208
Cyproterone-M/artifact-1 AC	UHYAC	3320*	374 $_8$	356 $_{33}$	339 $_{100}$	175 $_{44}$		1329	1209
Cyproterone-M/artifact-2 AC	UHYAC	3330*	372 $_{15}$	354 $_{39}$	339 $_{100}$			1321	1210
Cytisine		2100	190 $_{73}$	160 $_{24}$	146 $_{38}$	134 $_{25}$		369	1630
Cytisine AC		2480	232 $_{36}$	189 $_{10}$	160 $_{13}$	146 $_{100}$	134 $_{14}$	517	7442
Cytisine HFB		2255	386 $_{26}$	240 $_{10}$	217 $_6$	189 $_{17}$	146 $_{100}$	1384	7445
Cytisine PFP		2245	336 $_{29}$	292 $_3$	217 $_4$	189 $_{12}$	146 $_{100}$	1129	7444
Cytisine TFA		2230	286 $_{31}$	242 $_4$	189 $_{12}$	146 $_{100}$	69 $_{39}$	814	7443
Cytisine TMS		2110	262 $_{39}$	218 $_{34}$	146 $_{22}$	110 $_{96}$	73 $_{100}$	670	7446
Cytosine 2TMS		1480	255 $_{38}$	254 $_{91}$	240 $_{100}$	170 $_{23}$	73 $_{62}$	632	7555
Cytosine 3TMS		1795	327 $_{66}$	326 $_{94}$	312 $_{100}$	197 $_{20}$	73 $_{81}$	1076	8138
DALT		2070	240 $_1$	144 $_6$	130 $_{24}$	110 $_{100}$	77 $_{11}$	558	8853
DALT HFB		1895	409 $_1$	340 $_9$	326 $_{16}$	143 $_9$	110 $_{100}$	1550	9128
DALT PFP		1980	359 $_1$	290 $_6$	276 $_{11}$	143 $_6$	110 $_{100}$	1385	9127
DALT TFA		1935	309 $_1$	240 $_3$	226 $_7$	143 $_5$	110 $_{100}$	1131	9126
DALT TMS		2025	312 $_4$	271 $_{14}$	216 $_4$	202 $_{14}$	110 $_{100}$	984	9125
DALT-D4		2070	244 $_1$	148 $_7$	132 $_{27}$	112 $_{100}$	79 $_{10}$	572	8856
DALT-M (HO-) isomer-1		2370	256 $_2$	230 $_2$	160 $_3$	146 $_{11}$	110 $_{100}$	638	9255
DALT-M (HO-) isomer-1 AC		2310	298 $_1$	257 $_1$	160 $_3$	146 $_9$	110 $_{100}$	897	9251
DALT-M (HO-) isomer-2		2540	256 $_1$	202 $_3$	160 $_6$	146 $_{15}$	110 $_{100}$	639	9257
DALT-M (HO-) isomer-2 AC		2460	298 $_1$	188 $_3$	160 $_4$	146 $_{17}$	110 $_{100}$	897	9252
DALT-M (N-dealkyl-HO-) isomer-1 2AC		2530	300 $_6$	201 $_{61}$	159 $_{100}$	146 $_{74}$	70 $_{77}$	908	9253
DALT-M (N-dealkyl-HO-) isomer-2 2AC		2690	300 $_{10}$	201 $_{53}$	159 $_{100}$	146 $_{56}$	70 $_{59}$	908	9254
DALT-M (tri-HO-) 3AC		2400	414 $_1$	373 $_{15}$	160 $_3$	146 $_{10}$	110 $_{100}$	1491	9256
Danazole		2880	337 $_{25}$	270 $_{100}$	121 $_{81}$	105 $_{71}$	79 $_{59}$	1141	6112
Danazole AC		2820	379 $_2$	337 $_{45}$	173 $_{53}$	146 $_{57}$	91 $_{100}$	1353	6113
Danthron		2330*	240 $_{100}$	212 $_{16}$	184 $_{18}$	138 $_{10}$	92 $_{14}$	554	3555
Danthron 2AC		2595*	324 $_1$	282 $_{27}$	240 $_{100}$	212 $_7$	155 $_8$	1055	3679
Danthron 2ET		2560*	296 $_{100}$	253 $_{60}$	237 $_{12}$	165 $_{10}$	139 $_9$	884	3696
Danthron 2ME		2475*	268 $_{100}$	239 $_{39}$	180 $_{10}$	139 $_{12}$	126 $_{23}$	706	3694
Danthron 2TMS		2530*	369 $_{100}$	297 $_8$	268 $_2$	210 $_2$	73 $_{50}$	1375	3698
Danthron AC		2460*	282 $_{11}$	240 $_{100}$	212 $_9$	184 $_9$	127 $_7$	792	3678
Danthron ET		2500*	268 $_{32}$	253 $_{100}$	236 $_9$	152 $_{14}$	139 $_{13}$	705	3695
Danthron ME		2435*	254 $_{100}$	236 $_{29}$	208 $_{45}$	168 $_{15}$	139 $_{20}$	626	3693
Danthron TMS		2465*	312 $_2$	297 $_{100}$	253 $_5$	240 $_4$	127 $_6$	980	3697
Dantrolene		1900	214 $_{100}$	184 $_{36}$	156 $_{28}$	140 $_{47}$	113 $_{36}$	992	2033
Dantrolene artifact		1880	184 $_{92}$	155 $_{100}$	130 $_{20}$	102 $_8$	92 $_{11}$	356	2034
Dapaglifozin		*3275	408 $_{19}$	274 $_{39}$	165 $_{36}$	153 $_{100}$	107 $_{80}$	1471	10325
Dapaglifozin 4AC		*3320	576 $_9$	533 $_4$	341 $_{100}$	139 $_{24}$	107 $_{23}$	1706	10323
Dapaglifozin 4TMS		*2985	696 $_1$	517 $_2$	461 $_{24}$	388 $_{56}$	217 $_{100}$	1729	10324
Dapoxetine		2730	305 $_4$	134 $_{100}$	115 $_{30}$	91 $_{21}$	58 $_{24}$	938	8421
Dapoxetine artifact (-N(CH3)2)		2530*	262 $_{15}$	144 $_{56}$	115 $_{28}$	91 $_{100}$	65 $_{15}$	670	8422
Dapoxetine-M/artifact (1-naphthol)		1500*	144 $_{100}$	115 $_{83}$	89 $_{14}$	74 $_6$	63 $_{17}$	283	928
Dapoxetine-M/artifact (1-naphthol) AC	U+UHYAC	1555*	186 $_{13}$	144 $_{100}$	115 $_{47}$	89 $_8$	63 $_7$	361	932
Dapsone	P-I	2865	248 $_{100}$	184 $_{13}$	140 $_{58}$	108 $_{86}$	92 $_{40}$	589	6534
Dapsone 2AC	U+UHYAC-I	3960	332 $_{100}$	290 $_{61}$	248 $_{51}$	140 $_{19}$	108 $_{31}$	1105	6535
Dapsone 2HFB		2695	640 $_{44}$	336 $_{100}$	304 $_{53}$	141 $_{55}$	118 $_{46}$	1721	6563
Dapsone 2PFP		2670	540 $_{71}$	286 $_{100}$	254 $_{51}$	141 $_{49}$	119 $_{64}$	1689	6562

Table 8.1: Compounds in order of names   Dapsone 2TFA

Name	Detected	RI	Typical ions and intensities					Page	Entry
Dapsone 2TFA		2700	440 96	236 100	204 47	188 29	109 49	1557	6564
Darunavir artifact (HOOC-) -H2O		3950	417 1	374 4	241 31	156 100	92 58	1500	7959
Darunavir artifact (HOOC-) -H2O 2TMS		3240	561 1	313 48	228 100	180 29	164 30	1701	8238
Darunavir artifact 2AC		3660	458 3	416 3	350 9	283 40	198 100	1597	8242
Darunavir artifact HFB		3480	570 2	437 33	288 34	91 100	86 62	1703	8241
Darunavir artifact PFP		3500	520 2	387 53	238 48	91 100	86 61	1677	8240
Darunavir artifact TFA		3570	470 2	337 54	252 48	91 100	86 65	1617	8239
Darunavir artifact-1		1040	111 1	100 100	82 22	71 59	55 24	261	7958
Dazomet		1660	162 54	129 3	89 100	72 16	57 35	305	3915
o,p'-DDD	G P U	2230*	318 6	235 100	199 12	165 25		1015	1783
o,p'-DDD -HCl	P U	1800*	282 49	247 18	212 100	176 34		791	1888
o,p'-DDD-M (dichlorophenylmethane)	P U	1900*	236 54	201 100	165 82	82 30		537	1743
o,p'-DDD-M (HO-) -2HCl	P U	1790*	264 8	235 100	199 19	165 46		679	1884
o,p'-DDD-M (HO-HOOC-)	P U	2040*	296 1	251 100	139 88	111 28		882	1893
o,p'-DDD-M (HOOC-) ME	P U	2530*	294 16	259 14	235 100	199 26	165 66	868	1889
o,p'-DDD-M/artifact (dehydro-)	g p u	2100*	316 46	281 6	246 100	210 6	176 11	1003	1784
o,p'-DDE	g p u	2100*	316 46	281 6	246 100	210 6	176 11	1003	1784
o,p'-DDT		2275*	352 1	235 100	199 23	165 65	75 31	1217	3178
p,p'-DDD		2240*	318 2	235 100	199 16	165 74	75 32	1015	1954
p,p'-DDD -HCl		2390*	282 61	247 17	212 100	176 49	75 31	791	3177
p,p'-DDE	U	2150*	316 54	246 100	210 17	176 52	75 38	1002	1931
p,p'-DDT	U	2320*	352 1	235 100	199 17	165 66	75 31	1217	1932
Decamethrin		2900	503 1	253 57	181 100	93 40	77 55	1663	2818
Decamethrin-M/artifact (deacyl-) -HCN		1700*	198 100	169 64	141 74	115 50	77 68	394	2797
Decamethrin-M/artifact (deacyl-) ME		2590	239 2	197 100	141 30	115 30	77 19	551	2819
Decamethrin-M/artifact (HOOC-) ME		1540*	310 3	253 46	231 44	172 37	91 100	967	2798
Decamethylcyclopentasiloxane		<1000*	355 81	267 66	251 10	193 10	73 100	1310	9741
Decamethyltetrasiloxane		1300*	295 19	207 100	191 17	147 20	73 37	970	5429
Decane		1000*	142 4	120 14	105 31	71 33	57 100	282	3776
Decyldodecylphthalate		2990*	474 1	335 6	307 8	149 100	57 11	1627	3542
Decylhexylphthalate		2665*	390 1	307 5	251 10	233 2	149 100	1404	6402
Decyloctylphthalate		2675*	418 1	307 9	279 12	149 100	57 28	1503	3544
Decyltetradecylphthalate		3250*	363 4	307 7	149 100	57 14		1663	3543
DEET		1550	190 40	162 4	119 100	91 48	65 24	374	4501
4,5-Dehydro-4'-methyl-sceletone		2170	259 57	188 100	174 51	96 56	70 61	652	8989
Dehydroabietic acid	P	2590*	300 20	285 73	239 100	197 39	141 24	909	4493
Dehydroepiandrosterone		2530*	288 100	270 43	255 56	203 31	91 55	830	3760
Dehydroepiandrosterone enol 2TMS		2580*	432 44	417 36	327 27	169 25	73 100	1541	3800
Dehydroepiandrosterone -H2O	U UHY UHYAC	2595*	270 100	255 23	121 74	91 46	79 39	720	3770
4,5-Dehydro-sceletone		2265	245 46	188 23	174 100	96 30	70 56	576	8990
1-Dehydrotestosterone		2610*	286 4	147 13	122 100	91 28	55 17	816	3892
1-Dehydrotestosterone AC		2690*	328 4	147 23	122 100	91 26	55 15	1083	3922
1-Dehydrotestosterone enol 2TMS		2600*	430 38	415 12	325 14	206 58	73 100	1535	3965
1-Dehydrotestosterone TMS		2640*	358 5	268 7	147 39	122 100	73 93	1255	3926
Deiquate artifact		1460	156 100	128 34	102 6	78 31	51 54	300	105
Delavirdine artifact (indole part)		2400	210 22	161 4	131 100	104 26	87 21	432	9414
Delavirdine artifact (indole part) AC		2530	252 7	210 33	158 4	131 100	103 14	615	9415
Delavirdine artifact (piperazine part) 2AC		2610	304 100	289 59	261 71	176 95	164 94	933	9416
Delavirdine artifact (piperazine part) AC		2360	262 100	247 33	190 25	164 95	162 72	671	9417
Delorazepam HY	UHY	2180	265 62	230 100	139 43	111 50		684	543
Delorazepam HYAC	U+UHYAC	2300	307 42	265 58	230 100	139 16	111 14	946	290
Delorazepam-M (HO-)	P-I G UGLUC	2440	302 45	274 62	239 91	75 100		1031	539
Delorazepam-M (HO-) 2AC		2730	345 17	307 41	265 53	230 100		1458	540
delta-Hexachlorocyclohexane (HCH)		1710*	252 2	217 32	181 64	109 100	51 81	824	3854
Deltamethrin		2900	503 1	253 57	181 100	93 40	77 55	1663	2818
Deltamethrin-M/artifact (deacyl-) -HCN		1700*	198 100	169 64	141 74	115 50	77 68	394	2797
Deltamethrin-M/artifact (deacyl-) ME		2590	239 2	197 100	141 30	115 30	77 19	551	2819
Deltamethrin-M/artifact (HOOC-) ME		1540*	310 3	253 46	231 44	172 37	91 100	967	2798
Demethylmesembranol isomer-1	UGLUCSPE	2290	277 89	276 100	260 36	205 91	70 48	762	8998
Demethylmesembranol isomer-1 2AC	UGLUCSPE	2350	361 22	318 32	302 100	260 24	205 28	1269	9004
Demethylmesembranol isomer-2	UGLUCSPE	2310	277 93	276 100	260 33	218 21	205 90	762	8999
Demethylmesembranol isomer-2 2AC	UGLUCSPE	2535	361 100	302 55	260 52	163 70	121 34	1270	9002
Demethylmesembranol isomer-3	UGLUCSPE	2375	277 61	276 100	260 22	218 13	205 90	763	9000
Demethylmesembranol isomer-3 2AC	UGLUCSPE	2590	361 100	302 23	260 53	163 83	121 34	1270	9003
Demethylmesembrine 2AC	UGLUCSPE	2650	359 8	317 50	231 35	217 100	100 78	1259	9005
Demeton-S-methyl	G P-I U-I	1635*	230 3	142 12	109 23	88 100	60 64	508	1112
Demeton-S-methylsulfone	G	1865*	262 1	169 100	125 45	109 87	79 21	667	3428
Demeton-S-methylsulfoxide	G P-I	1860*	218 1	169 60	125 47	109 100	79 26	578	1500
Denaverine		2225	283 2	267 3	183 26	71 67	58 100	1373	8364
Denaverine artifact		2230*	267 19	183 100	165 27	105 49	77 35	698	8365
Denaverine HYAC		2010	183 4	105 23	77 27	71 29	58 100	353	8366
Deschloroetizolam		2930	308 82	279 100	252 35	239 42	211 19	956	9695
Desipramine	UHY	2225	266 28	235 61	208 61	195 100	71 59	698	324
Desipramine AC	PAC U+UHYAC	2670	308 40	208 100	193 55	114 62		959	325

Desipramine HFB Table 8.1: Compounds in order of names

Name	Detected	RI	Typical ions and intensities					Page	Entry
Desipramine HFB		2450	462_{23}	268_{13}	240_{20}	208_{100}	193_{54}	1604	7706
Desipramine PFP		2450	412_{38}	234_{16}	218_{19}	208_{100}	193_{47}	1485	7667
Desipramine TFA		2430	362_{14}	208_{100}	193_{51}	140_{17}	69_{21}	1273	7786
Desipramine TMS		2470	338_{7}	235_{89}	143_{41}	116_{72}	73_{100}	1146	5461
Desipramine-M (di-HO-) 3AC	UHYAC	3380	424_{44}	324_{35}	282_{34}	240_{27}	114_{100}	1517	3315
Desipramine-M (di-HO-ring)	UHY	2600	227_{100}	196_{7}				499	2296
Desipramine-M (di-HO-ring) 2AC	U+UHYAC	2750	311_{28}	269_{23}	227_{100}	196_{7}		975	2292
Desipramine-M (HO-) 2AC	U+UHYAC	3065	366_{27}	266_{39}	114_{100}			1293	1175
Desipramine-M (HO-methoxy-ring)	UHY	2390	241_{100}	226_{17}	210_{12}	180_{14}		560	2315
Desipramine-M (HO-methoxy-ring) AC	U+UHYAC	2370	283_{10}	241_{100}	226_{17}	210_{12}	180_{14}	796	2867
Desipramine-M (HO-ring)	UHY	2240	211_{100}	196_{15}	180_{10}	152_{4}		436	2295
Desipramine-M (HO-ring) AC	U+UHYAC	2535	253_{26}	211_{100}	196_{19}	180_{11}	152_{4}	621	1218
Desipramine-M (nor-) AC	UHYAC	2640	294_{23}	208_{100}	193_{43}	152_{6}	100_{17}	871	3313
Desipramine-M (nor-HO-) 2AC	UHYAC	2980	352_{60}	266_{88}	224_{100}	180_{15}	100_{48}	1223	3314
Desipramine-M (ring)	U U+UHYAC	1930	195_{100}	180_{40}	167_{9}	96_{33}	83_{22}	386	308
Desipramine-M (ring) ME		1915	209_{70}	194_{100}	178_{13}	165_{11}		429	6352
Desloratadine AC	UHYAC	3120	352_{100}	294_{32}	280_{33}	266_{60}	245_{29}	1220	5610
Desmedipham TFA		2460	396_{52}	277_{100}	218_{59}	205_{99}	119_{73}	1428	4125
Desmedipham-M/artifact (phenol)		1740	181_{60}	122_{62}	109_{100}	81_{43}	53_{28}	348	3750
Desmedipham-M/artifact (phenol) 2ME		1640	209_{100}	150_{52}	136_{72}	108_{57}	77_{35}	429	4099
Desmedipham-M/artifact (phenol) 3ME		1560	195_{100}	164_{8}	136_{47}	108_{34}	72_{57}	386	4093
Desmedipham-M/artifact (phenol) TFA		1540	277_{100}	218_{70}	205_{92}	91_{25}	69_{34}	758	4126
Desmedipham-M/artifact (phenylcarbamic acid) 2ME		1190	165_{100}	134_{13}	120_{40}	106_{56}	77_{72}	313	4100
Desmedipham-M/artifact (phenylcarbamic acid) ME	G	1320	151_{100}	119_{56}	106_{76}	92_{36}	65_{66}	290	3909
Desmetryn		1800	213_{89}	198_{64}	171_{44}	82_{80}	58_{100}	443	3829
Desomorphine		2300	271_{100}	256_{15}	228_{18}	214_{30}	148_{18}	725	9381
Desomorphine 2TMS		2520	431_{33}	373_{7}	236_{23}	146_{15}	73_{100}	1538	2469
Desomorphine AC	P UHYAC	2370	313_{100}	271_{63}	228_{14}	214_{27}	148_{13}	990	9382
Desomorphine HFB		2155	467_{100}	452_{19}	424_{22}	410_{20}	148_{16}	1612	9386
Desomorphine PFP		2130	417_{100}	402_{16}	374_{13}	360_{13}	148_{6}	1500	9385
Desomorphine TFA		2145	367_{100}	352_{17}	324_{23}	310_{13}	148_{19}	1296	9384
Desomorphine TMS		2250	343_{78}	328_{71}	286_{33}	271_{43}	73_{100}	1176	9383
Desomorphine-M (HO-)	UHY	2400	287_{100}	230_{14}	164_{35}	115_{28}	70_{54}	821	484
Desomorphine M (HO-) 2AC	U+UHYAC	2545	371_{83}	329_{100}	286_{34}	212_{21}	70_{33}	1319	234
Desomorphine-M (HO-) 2HFB		2260	679_{41}	482_{53}	466_{100}	360_{13}	169_{21}	1727	6197
Desomorphine-M (HO-) 2PFP		2330	579_{60}	432_{21}	416_{49}	310_{4}	119_{100}	1707	2460
Desomorphine-M (HO-) 2TFA		2190	479_{91}	382_{25}	366_{61}	260_{7}	69_{100}	1632	6198
Desomorphine-M (HO-) AC	UHYAC	2490	329_{100}	287_{56}	230_{10}	164_{20}	70_{21}	1089	3055
Desomorphine-M (HO-) TFA		2250	383_{100}	286_{19}	270_{44}	213_{19}	69_{50}	1370	6199
Desomorphine-M (nor-) 2AC	P UHYAC	2700	341_{34}	299_{48}	255_{25}	213_{100}	87_{65}	1164	9582
Desomorphine-M (nor-HO-) 3AC	UHYAC	2790	399_{20}	357_{50}	229_{19}	87_{100}	72_{22}	1442	3050
Desoxycholic acid -H2O ME		3630*	388_{5}	370_{27}	273_{76}	255_{100}	55_{28}	1396	3126
Desoxycortone		2785*	330_{52}	288_{60}	245_{38}	147_{60}	124_{100}	1096	6069
Desoxycortone AC		3175*	372_{6}	299_{100}	271_{54}	253_{54}	147_{38}	1324	6068
Desoxycortone acetate		3175*	372_{6}	299_{100}	271_{54}	253_{54}	147_{38}	1324	6068
Desoxypipradrol		1960	250_{1}	165_{7}	152_{3}	84_{100}	56_{11}	613	9278
Desoxypipradrol AC		2300	250_{1}	165_{18}	152_{8}	126_{33}	84_{100}	866	9279
Desoxypipradrol HFB		2070	447_{1}	280_{100}	252_{6}	226_{11}	165_{24}	1575	9379
Desoxypipradrol ME		2000	264_{4}	165_{16}	152_{6}	98_{100}	70_{19}	691	9380
Desoxypipradrol PFP		2060	397_{1}	230_{100}	202_{5}	176_{10}	165_{21}	1434	9378
Desoxypipradrol TFA		2080	347_{2}	278_{5}	180_{100}	165_{36}	152_{22}	1194	9377
Desoxypipradrol TMS		1960	322_{2}	308_{14}	165_{31}	156_{100}	152_{12}	1054	9376
Desoxypipradrol-M (HO-aryl-) isomer-1 2AC		2710	266_{1}	183_{5}	165_{6}	126_{42}	84_{100}	1216	9283
Desoxypipradrol-M (HO-aryl-) isomer-2 2AC		2740	308_{1}	183_{9}	165_{8}	126_{37}	84_{100}	1216	9284
Desoxypipradrol-M (HO-piperidyl-) isomer-1 2AC		2600	184_{100}	165_{30}	142_{75}	124_{33}	82_{87}	1216	9285
Desoxypipradrol-M (HO-piperidyl-) isomer-2 2AC		2920	184_{100}	165_{16}	142_{72}	124_{27}	82_{64}	1216	9286
Desoxypipradrol-M (HO-piperidyl-) isomer-3 2AC		2940	184_{100}	165_{18}	142_{76}	124_{30}	82_{61}	1215	9287
Desoxypipradrol-M (HO-piperidyl-) isomer-4 2AC		2960	184_{100}	165_{10}	142_{69}	124_{25}	82_{58}	1215	9288
Desoxypipradrol-M (HO-piperidyl-) isomer-5 2AC		3000	213_{15}	184_{100}	142_{70}	124_{27}	82_{60}	1216	9289
Desoxypipradrol-M (oxo-)		2500	178_{1}	165_{22}	152_{8}	98_{100}	55_{37}	688	9281
Desoxypipradrol-M (oxo-) AC		2430	307_{3}	165_{18}	140_{18}	98_{100}	55_{19}	951	9280
Desoxypipradrol-M/artifact AC		2550	266_{21}	172_{100}	165_{47}	130_{90}	98_{60}	1359	9282
Detajmium bitartrate artifact -H2O		3700	437_{1}	365_{24}	196_{8}	112_{30}	86_{100}	1552	4263
Detajmium bitartrate artifact -H2O AC		3680	479_{1}	407_{31}	144_{20}	112_{79}	86_{100}	1633	4272
DET-D4		1910	220_{1}	148_{5}	132_{22}	105_{4}	88_{100}	471	8863
DET-D4 HFB		1795	401_{1}	344_{10}	328_{11}	131_{12}	88_{100}	1496	10126
DET-D4 PFP		1785	351_{1}	294_{5}	278_{6}	131_{13}	88_{100}	1292	10125
DET-D4 TFA		1780	301_{1}	244_{7}	228_{13}	131_{18}	88_{100}	1007	10124
DET-D4 TMS		2055	292_{4}	220_{3}	204_{6}	147_{6}	88_{100}	858	10123
Dextromethorphan	G P-I U+UHYAC	2145	271_{31}	214_{16}	171_{14}	150_{29}	59_{100}	726	227
Dextromethorphan-M (bis-demethyl-) 2AC	UHYAC	2710	327_{100}	240_{6}	199_{22}	87_{100}	72_{62}	1077	228
Dextromethorphan-M (nor-) AC	UHYAC	2590	299_{13}	213_{42}	171_{22}	87_{100}	72_{52}	903	4477
Dextromethorphan-M (O-demethyl-)	UHY	2255	257_{38}	200_{17}	150_{28}	59_{100}		643	475
Dextromethorphan-M (O-demethyl-) AC	U+UHYAC	2280	299_{100}	231_{42}	200_{20}	150_{48}	59_{15}	903	230

Table 8.1: Compounds in order of names   **Dextromethorphan-M (O-demethyl-) HFB**

Name	Detected	RI	Typical ions and intensities					Page	Entry
Dextromethorphan-M (O-demethyl-) HFB		2100	453 $_{100}$	396 $_{18}$	385 $_{91}$	169 $_{19}$	150 $_{27}$	1586	6151
Dextromethorphan-M (O-demethyl-) PFP	UHYPFP	2060	403 $_{93}$	335 $_{78}$	303 $_{14}$	150 $_{100}$	119 $_{58}$	1457	4305
Dextromethorphan-M (O-demethyl-) TFA		2015	353 $_{69}$	285 $_{80}$	150 $_{100}$	115 $_{26}$	69 $_{72}$	1227	4006
Dextromethorphan-M (O-demethyl-) TMS	UHYTMS	2230	329 $_{31}$	272 $_{20}$	150 $_{39}$	73 $_{26}$	59 $_{100}$	1090	4304
Dextromethorphan-M (O-demethyl-HO-) 2AC	U+UHYAC	2580	357 $_{68}$	247 $_{22}$	215 $_{17}$	150 $_{100}$	59 $_{30}$	1249	1187
Dextromethorphan-M (O-demethyl-methoxy-) AC	U+UHYAC	2520	329 $_{48}$	261 $_{23}$	229 $_{23}$	150 $_{100}$	59 $_{28}$	1090	4476
Dextromethorphan-M (O-demethyl-oxo-) AC	U+UHYAC	2695	313 $_{16}$	240 $_{11}$	199 $_{98}$	157 $_{12}$	73 $_{100}$	990	4475
Dextromoramide	G P-I U UHY UHYAC	2920	306 $_{1}$	265 $_{35}$	165 $_{6}$	128 $_{42}$	100 $_{100}$	1412	229
Dextromoramide-M (HO-)	UHY	3095	322 $_{2}$	281 $_{54}$	165 $_{6}$	128 $_{43}$	100 $_{100}$	1472	1185
Dextromoramide-M (HO-) AC	U+UHYAC	3210	364 $_{1}$	323 $_{33}$	194 $_{13}$	128 $_{41}$	100 $_{100}$	1581	1184
Dextropropoxyphene	G P	2205	250 $_{2}$	193 $_{3}$	178 $_{2}$	91 $_{15}$	58 $_{100}$	1153	476
Dextropropoxyphene artifact		1755*	208 $_{56}$	193 $_{41}$	130 $_{38}$	115 $_{100}$	91 $_{42}$	426	477
Dextropropoxyphene-M (HY)	UHY	2395	281 $_{9}$	190 $_{76}$	119 $_{96}$	105 $_{100}$	56 $_{97}$	785	480
Dextropropoxyphene-M (nor-) -H2O	UHY	2240	251 $_{30}$	217 $_{96}$	119 $_{100}$	91 $_{72}$	77 $_{40}$	613	479
Dextropropoxyphene-M (nor-) -H2O AC	UHYAC	2365	293 $_{18}$	220 $_{100}$	205 $_{39}$			866	232
Dextropropoxyphene-M (nor-) -H2O N-prop.	U UHY UHYAC	2555	307 $_{8}$	234 $_{76}$	105 $_{100}$	100 $_{74}$	91 $_{67}$	953	231
Dextropropoxyphene-M (nor-) N-prop.	P U	2400	307 $_{16}$	220 $_{68}$	100 $_{100}$	57 $_{83}$		1065	478
Dextrorphan	UHY	2255	257 $_{38}$	200 $_{17}$	150 $_{28}$	59 $_{100}$		643	475
Dextrorphan AC	U+UHYAC	2280	299 $_{100}$	231 $_{42}$	200 $_{20}$	150 $_{48}$	59 $_{15}$	903	230
Dextrorphan HFB		2100	453 $_{100}$	396 $_{18}$	385 $_{91}$	169 $_{19}$	150 $_{27}$	1586	6151
Dextrorphan PFP	UHYPFP	2060	403 $_{93}$	335 $_{78}$	303 $_{14}$	150 $_{100}$	119 $_{58}$	1457	4305
Dextrorphan TFA		2015	353 $_{69}$	285 $_{80}$	150 $_{100}$	115 $_{26}$	69 $_{72}$	1227	4006
Dextrorphan TMS	UHYTMS	2230	329 $_{31}$	272 $_{20}$	150 $_{39}$	73 $_{26}$	59 $_{100}$	1090	4304
Dextrorphan-M (methoxy-) AC	U+UHYAC	2520	329 $_{48}$	261 $_{23}$	229 $_{23}$	150 $_{100}$	59 $_{28}$	1090	4476
Dextrorphan-M (nor-) 2AC	UHYAC	2710	327 $_{11}$	240 $_{8}$	199 $_{12}$	87 $_{100}$	72 $_{62}$	1077	228
Dextrorphan-M (oxo-) AC	U+UHYAC	2695	313 $_{16}$	240 $_{11}$	199 $_{98}$	157 $_{12}$	73 $_{100}$	990	4475
DFBDB		1335	228 $_{1}$	200 $_{3}$	171 $_{5}$	105 $_{4}$	58 $_{100}$	505	8251
DFBDB AC		1755	271 $_{1}$	212 $_{28}$	171 $_{14}$	100 $_{46}$	58 $_{100}$	723	8252
DFBDB formyl artifact		1585	241 $_{1}$	171 $_{12}$	105 $_{6}$	77 $_{15}$	70 $_{100}$	559	8250
DFBDB HFB		1550	425 $_{1}$	406 $_{3}$	254 $_{100}$	212 $_{94}$	171 $_{65}$	1519	8257
DFBDB HY2AC	UHYAC	2205	265 $_{1}$	206 $_{11}$	164 $_{29}$	100 $_{23}$	58 $_{100}$	688	8253
DFBDB HY3AC	UHYAC	2235	307 $_{1}$	248 $_{16}$	164 $_{38}$	100 $_{53}$	58 $_{100}$	950	5551
DFBDB PFP		1510	375 $_{1}$	356 $_{1}$	212 $_{86}$	204 $_{100}$	171 $_{50}$	1333	8256
DFBDB TFA		1500	325 $_{1}$	306 $_{2}$	212 $_{76}$	171 $_{43}$	154 $_{100}$	1060	8255
DFBDB TMS		1455	286 $_{8}$	272 $_{8}$	171 $_{26}$	130 $_{100}$	73 $_{81}$	912	8254
DFMBDB		1390	242 $_{1}$	214 $_{20}$	171 $_{20}$	105 $_{15}$	72 $_{100}$	568	8258
DFMBDB AC		1800	266 $_{1}$	212 $_{10}$	171 $_{12}$	114 $_{62}$	72 $_{100}$	809	8259
DFMBDB HFB		1570	268 $_{88}$	212 $_{36}$	210 $_{37}$	77 $_{65}$	69 $_{100}$	1555	8389
DFMBDB PFP		1605	370 $_{1}$	218 $_{100}$	212 $_{42}$	171 $_{28}$	160 $_{42}$	1398	8263
DFMBDB TFA		1615	339 $_{1}$	212 $_{41}$	168 $_{100}$	110 $_{42}$	77 $_{31}$	1148	8262
DFMBDB TMS		1520	300 $_{2}$	171 $_{17}$	144 $_{100}$	105 $_{8}$	73 $_{77}$	1000	8261
DFMDA		1120	214 $_{1}$	200 $_{3}$	171 $_{12}$	77 $_{61}$	51 $_{100}$	448	8377
DFMDA AC		1705	257 $_{4}$	238 $_{4}$	198 $_{100}$	171 $_{34}$	86 $_{26}$	641	8265
DFMDA formyl artifact		1170	227 $_{4}$	171 $_{30}$	105 $_{22}$	77 $_{41}$	56 $_{100}$	498	8264
DFMDA HFB		1450	240 $_{44}$	198 $_{43}$	171 $_{25}$	77 $_{67}$	51 $_{100}$	1480	8388
DFMDA HY3AC	U+UHYAC	2150	293 $_{1}$	234 $_{26}$	192 $_{42}$	150 $_{89}$	86 $_{100}$	862	3725
DFMDA PFP		1450	361 $_{2}$	342 $_{3}$	198 $_{84}$	190 $_{100}$	171 $_{45}$	1267	8268
DFMDA TFA		1435	311 $_{1}$	292 $_{2}$	198 $_{82}$	171 $_{44}$	140 $_{100}$	974	8267
DFMDA TMS		1375	287 $_{1}$	272 $_{9}$	171 $_{21}$	116 $_{100}$	73 $_{81}$	819	8266
DFMDE		1560	242 $_{1}$	228 $_{2}$	171 $_{14}$	105 $_{10}$	72 $_{100}$	568	8269
DFMDE AC		1770	266 $_{1}$	198 $_{28}$	171 $_{16}$	114 $_{68}$	72 $_{100}$	809	8270
DFMDE HFB		1610	420 $_{2}$	268 $_{100}$	240 $_{66}$	198 $_{72}$	171 $_{38}$	1555	8274
DFMDE HYAC	U+UHYAC	2200	321 $_{1}$	234 $_{8}$	150 $_{27}$	114 $_{80}$	72 $_{100}$	1040	4208
DFMDE PFP		1570	370 $_{1}$	218 $_{100}$	198 $_{53}$	190 $_{55}$	171 $_{11}$	1398	8273
DFMDE TFA		1580	320 $_{1}$	198 $_{54}$	168 $_{100}$	140 $_{59}$	77 $_{32}$	1148	8272
DFMDE TMS		1540	300 $_{3}$	171 $_{12}$	144 $_{100}$	105 $_{8}$	73 $_{81}$	1000	8271
DFMDMA		1535	228 $_{1}$	214 $_{3}$	171 $_{10}$	77 $_{23}$	58 $_{100}$	505	8275
DFMDMA AC		1745	271 $_{1}$	198 $_{24}$	171 $_{16}$	100 $_{79}$	58 $_{100}$	723	8276
DFMDMA HFB		1580	406 $_{2}$	254 $_{100}$	210 $_{66}$	198 $_{67}$	171 $_{29}$	1519	8280
DFMDMA HY3AC	U+UHYAC	2190	307 $_{1}$	234 $_{7}$	150 $_{12}$	100 $_{59}$	58 $_{100}$	950	4244
DFMDMA PFP		1540	356 $_{2}$	204 $_{100}$	198 $_{58}$	171 $_{24}$	160 $_{61}$	1333	8279
DFMDMA TFA		1535	306 $_{2}$	198 $_{66}$	171 $_{22}$	154 $_{100}$	110 $_{74}$	1059	8278
DFMDMA TMS		1500	300 $_{1}$	286 $_{15}$	171 $_{17}$	130 $_{100}$	73 $_{80}$	912	8277
DFMDP		1540	201 $_{20}$	172 $_{30}$	152 $_{17}$	77 $_{57}$	51 $_{100}$	401	8339
DFMDP 2AC		1780	285 $_{1}$	266	184 $_{100}$	171 $_{11}$	72 $_{28}$	807	8342
DFMDP AC		1670	243 $_{6}$	184 $_{100}$	171 $_{14}$	72 $_{23}$	51 $_{34}$	567	8341
DFMDP formyl artifact		1205	213 $_{31}$	184 $_{10}$	171 $_{59}$	77 $_{83}$	51 $_{100}$	442	8340
DFMDP HFB		1485	397 $_{2}$	378 $_{2}$	226 $_{13}$	184 $_{100}$	171 $_{55}$	1432	8346
DFMDP PFP		1445	347 $_{4}$	328 $_{2}$	184 $_{100}$	171 $_{46}$	105 $_{26}$	1192	8345
DFMDP TFA		1440	297 $_{3}$	184 $_{100}$	171 $_{24}$	126 $_{22}$	105 $_{35}$	888	8344
DFMDP TMS		1400	272 $_{1}$	258 $_{36}$	171 $_{20}$	153 $_{13}$	102 $_{100}$	734	8343
Diacetolol -H2O AC	U+UHYAC	3055	332 $_{8}$	289 $_{30}$	231 $_{100}$	202 $_{60}$	98 $_{30}$	1106	1570
Diacetolol -H2O HY	UHY	2010	248 $_{23}$	233 $_{24}$	140 $_{38}$	98 $_{100}$	56 $_{72}$	591	1565
Diacetolol HY	UHY	2240	266 $_{6}$	151 $_{100}$	72 $_{92}$			697	1567

Dialifos Table 8.1: Compounds in order of names

Name	Detected	RI	Typical ions and intensities					Page	Entry
Dialifos		2545	357$_4$	208$_{100}$	129$_{14}$	97$_{14}$	76$_{21}$	1413	3833
Diallate		1670	254$_1$	234$_{35}$	152$_4$	128$_{18}$	86$_{100}$	710	3429
N,N-Diallyl-tryptamine		2070	240$_1$	144$_6$	130$_{24}$	110$_{100}$	77$_{11}$	558	8853
N,N-Diallyl-tryptamine HFB		1895	409$_1$	340$_9$	326$_{16}$	143$_6$	110$_{100}$	1550	9128
N,N-Diallyl-tryptamine PFP		1980	359$_1$	290$_6$	276$_{11}$	143$_6$	110$_{100}$	1385	9127
N,N-Diallyl-tryptamine TFA		1935	309$_1$	240$_3$	226$_7$	143$_5$	110$_{100}$	1131	9126
N,N-Diallyl-tryptamine TMS		2025	312$_4$	271$_{14}$	216$_4$	202$_{14}$	110$_{100}$	984	9125
N,N-Diallyl-tryptamine-D4		2070	244$_1$	148$_7$	132$_{27}$	112$_{100}$	79$_{10}$	572	8856
N,N-Diallyl-tryptamine-M (HO-) isomer-1		2370	256$_2$	230$_2$	160$_3$	146$_{11}$	110$_{100}$	638	9255
N,N-Diallyl-tryptamine-M (HO-) isomer-1 AC		2310	298$_1$	257$_1$	160$_3$	146$_9$	110$_{100}$	897	9251
N,N-Diallyl-tryptamine-M (HO-) isomer-2		2540	256$_1$	202$_3$	160$_6$	146$_{15}$	110$_{100}$	639	9257
N,N-Diallyl-tryptamine-M (HO-) isomer-2 AC		2460	298$_1$	188$_3$	160$_4$	146$_{17}$	110$_{100}$	897	9252
N,N-Diallyl-tryptamine-M (N-dealkyl-HO-) isomer-1 2AC		2530	300$_6$	201$_{61}$	159$_{100}$	146$_{74}$	70$_{77}$	908	9253
N,N-Diallyl-tryptamine-M (N-dealkyl-HO-) isomer-2 2AC		2690	300$_{10}$	201$_{53}$	159$_{100}$	146$_{56}$	70$_{59}$	908	9254
N,N-Diallyl-tryptamine-M (tri-HO-) 3AC		2400	414$_1$	373$_{15}$	160$_3$	146$_{10}$	110$_{100}$	1491	9256
Diazedine		3480	335$_{100}$	292$_6$	221$_{80}$	207$_{40}$	70$_9$	1127	10187
Diazepam	P G U	2430	284$_{81}$	283$_{91}$	256$_{100}$	221$_{31}$	77$_7$	799	481
Diazepam HY	UHY U+UHYAC	2100	245$_{95}$	228$_{38}$	193$_{29}$	105$_{38}$	77$_{100}$	573	272
Diazepam HYAC	U+UHYAC	2260	287$_{11}$	244$_{100}$	228$_{39}$	182$_{49}$	77$_{70}$	818	2542
Diazepam-D5		2425	289$_{81}$	287$_{89}$	261$_{100}$	226$_{18}$		833	6848
Diazepam-M (3-HO-)	P UGLUC	2625	300$_{33}$	271$_{100}$	256$_{23}$	228$_{16}$	77$_{30}$	905	417
Diazepam-M (3-HO-) AC	UGLUCAC	2730	342$_6$	300$_{40}$	271$_{100}$	255$_{16}$	77$_{17}$	1167	2099
Diazepam-M (3-HO-) artifact-1	G	2475	256$_{19}$	241$_7$	179$_{100}$	163$_7$	77$_8$	636	5780
Diazepam-M (3-HO-) artifact-2	G	2815	270$_{64}$	269$_{100}$	254$_{12}$	228$_{26}$	191$_5$	715	5779
Diazepam-M (3-HO-) ME		2600	314$_{60}$	271$_{100}$	255$_{46}$			993	418
Diazepam-M (3-HO-) TMS		2665	372$_{23}$	343$_{100}$	283$_{26}$	257$_{38}$	73$_{54}$	1322	4598
Diazepam-M (HO-)	UGLUC	2670	300$_{67}$	272$_{100}$	237$_{10}$			905	619
Diazepam-M (HO-) AC	UGLUCAC	2790	342$_{16}$	300$_{61}$	272$_{100}$	237$_9$		1167	621
Diazepam-M (HO-) HY	UHY	2580	261$_{91}$	260$_{100}$	244$_{42}$	209$_{21}$	121$_{17}$	662	2048
Diazepam-M (HO-) HYAC	U+UHYAC	2600	303$_{77}$	260$_{100}$	244$_{47}$	121$_{11}$		921	2060
Diazepam-M (nor-)	P G U	2520	270$_{86}$	269$_{97}$	242$_{100}$	241$_{82}$	77$_{17}$	715	463
Diazepam-M (nor-) TMS		2300	342$_{62}$	341$_{100}$	327$_{19}$	269$_4$	73$_{34}$	1168	4573
Diazepam-M (nor-HO-)	UGLUC	2750	286$_{82}$	258$_{100}$	230$_{11}$	166$_7$	139$_8$	813	2113
Diazepam-M (nor-HO-) AC	U+UHYAC	3000	328$_{22}$	286$_{90}$	258$_{100}$	166$_8$	139$_7$	1079	2111
Diazepam-M (nor-HO-) HY	UHY	2400	247$_{72}$	246$_{100}$	230$_{11}$	121$_{26}$	65$_{22}$	583	2112
Diazepam-M (nor-HO-) HYAC	U+UHYAC	2270	289$_{18}$	247$_{86}$	246$_{100}$	105$_7$	77$_{35}$	832	3143
Diazepam-M (nor-HO-) isomer-1 HY2AC	U+UHYAC	2560	331$_{48}$	289$_{64}$	247$_{100}$	230$_{41}$	154$_{13}$	1098	2125
Diazepam-M (nor-HO-) isomer-2 HY2AC	U+UHYAC	2610	331$_{46}$	289$_{54}$	246$_{100}$	154$_{10}$	121$_{11}$	1098	1751
Diazepam-M (nor-HO-methoxy-) HY2AC	U+UHYAC	2700	361$_{17}$	319$_{72}$	276$_{100}$	260$_{14}$	246$_{10}$	1267	1752
Diazepam-M (oxazepam)	P G UGLUC	2320	268$_{98}$	239$_{56}$	233$_{52}$	205$_{66}$	77$_{100}$	813	579
Diazepam-M (oxazepam) TMS		2635	356$_9$	341$_{100}$	312$_{56}$	239$_{12}$	135$_{21}$	1252	4577
Diazepam-M 2TMS		2200	430$_{51}$	429$_{89}$	340$_{15}$	313$_{19}$	73$_{100}$	1533	5499
Diazepam-M artifact-3	P-I UHY U+UHYAC	2060	240$_{59}$	239$_{100}$	205$_{81}$	177$_{16}$	151$_9$	554	300
Diazepam-M artifact-4	UHY UHYAC	2070	254$_{77}$	253$_{100}$	219$_{98}$	111$_5$		626	301
Diazepam-M HY	UHY	2050	231$_{80}$	230$_{95}$	154$_{23}$	105$_{38}$	77$_{100}$	512	419
Diazepam-M HYAC	PHYAC U+UHYAC	2245	273$_{30}$	230$_{100}$	154$_{13}$	105$_{23}$	77$_{50}$	732	273
Diazinon	P G	1760	304$_{27}$	199$_{50}$	179$_{81}$	152$_{64}$	137$_{100}$	931	2784
Diazinon artifact-1		1140	166$_{41}$	151$_{100}$	138$_{50}$	109$_{42}$	93$_{39}$	318	1399
Diazinon artifact-2	P-I U	1400*	198$_{26}$	170$_{38}$	138$_{57}$	111$_{100}$	81$_{80}$	393	1442
Diazinon artifact-3		1685	152$_{49}$	137$_{100}$	124$_{18}$	109$_{19}$	84$_{38}$	294	1375
Dibenzepin	P-I G U UHY UHYAC	2465	295$_1$	224$_{21}$	180$_2$	72$_6$	58$_{100}$	880	326
Dibenzepin-M (bis-nor-)	UHY	2700	267$_1$	235$_{100}$	207$_{20}$	179$_{10}$	103$_8$	701	2221
Dibenzepin-M (bis-nor-) AC	UHYAC	2870	309$_{13}$	236$_{65}$	223$_{100}$	195$_{62}$	100$_{16}$	964	327
Dibenzepin-M (HO-) isomer-1 AC	PAC UHYAC	2600	353$_1$	282$_{11}$	240$_5$	71$_8$	58$_{100}$	1228	3335
Dibenzepin-M (HO-) isomer-2 AC	PAC UHYAC	2770	353$_3$	282$_{18}$	240$_{12}$	209$_{18}$	58$_{100}$	1228	3337
Dibenzepin-M (N5-demethyl-)	U+UHYAC	2460	281$_1$	237$_1$	210$_{51}$	72$_5$	58$_{100}$	788	482
Dibenzepin-M (N5-demethyl-HO-) isomer-1 AC	UHYAC	2680	339$_1$	268$_{15}$	226$_8$	71$_{13}$	58$_{100}$	1150	3336
Dibenzepin-M (N5-demethyl-HO-) isomer-2 AC	UHYAC	2825	339$_1$	268$_{14}$	226$_{12}$	71$_5$	58$_{100}$	1150	3338
Dibenzepin-M (nor-) AC	PAC U+UHYAC	2800	323$_{30}$	250$_{83}$	237$_{47}$	209$_{100}$	100$_{15}$	1052	1165
Dibenzepin-M (nor-HO-) isomer-1 2AC	UHYAC	3110	381$_{31}$	308$_{76}$	266$_{87}$	253$_{100}$	100$_{45}$	1361	3309
Dibenzepin-M (nor-HO-) isomer-2 2AC	UHYAC	3290	381$_{41}$	308$_{61}$	266$_{100}$	225$_{69}$	100$_{32}$	1361	3339
Dibenzepin-M (ter-nor-)	UHY	2680	235$_{100}$	207$_{23}$	179$_{12}$	117$_7$	103$_7$	622	2222
Dibenzepin-M (ter-nor-) AC	UHYAC	2825	295$_{29}$	236$_{45}$	223$_{100}$	195$_{65}$	167$_{26}$	877	328
Dibenzo[a,h]anthracene		3055*	278$_{100}$	250$_3$	139$_{35}$	125$_{11}$	113$_6$	767	3705
Dibenzofuran		1520*	168$_{100}$	139$_{27}$	113$_7$	84$_6$	70$_4$	320	2559
Dibutyladipate		2385*	258$_2$	185$_{92}$	129$_{100}$	111$_{64}$		648	722
Dibutylpentylpyridine		1930	261$_1$	232$_{20}$	190$_{35}$	163$_{100}$	120$_{13}$	667	5133
Dicamba		1795*	220$_{62}$	191$_{35}$	173$_{100}$	113$_{26}$	73$_{41}$	467	3637
Dicamba -CO2		1200*	176$_{100}$	161$_{25}$	133$_{72}$	75$_{24}$	63$_{37}$	331	3638
Dicamba ME	G P-I	1525*	234$_2$	203$_{100}$	188$_{23}$	97$_{13}$	75$_{10}$	525	3639
Dicamba TMS	UTMS	1735*	292$_{22}$	277$_{42}$	203$_{100}$	188$_{37}$	73$_{61}$	853	6464
4,4'-Dicarbonitrile-1,1'-biphenyl	U+UHYAC	1960	204$_{100}$	177$_8$	150$_4$	102$_6$		409	2408
Dichlobenil	U UHY UHYAC	1300	171$_{100}$	136$_{21}$	100$_{35}$			324	736
Dichlobenil-M (HO-)	U UHY	1540	187$_{100}$	159$_{62}$	88$_{53}$	86$_{57}$		362	2986

Table 8.1: Compounds in order of names

Name	Detected	RI	Typical ions and intensities					Page	Entry
Dichlobenil-M (HO-) AC	UHYAC	1660	229 $_{24}$	187 $_{100}$	159 $_{26}$	120 $_{44}$	88 $_{40}$	503	2987
Dichlofenthion		1870*	314 $_1$	279 $_{72}$	223 $_{81}$	162 $_{43}$	97 $_{100}$	991	3431
Dichlofluanid		1950	332 $_7$	224 $_{24}$	167 $_{34}$	123 $_{100}$	77 $_{27}$	1103	2999
Dichloran		1730	206 $_{100}$	176 $_{75}$	160 $_{44}$	124 $_{89}$	62 $_{33}$	415	3432
2,3-Dichloroaniline	G P U UHY	1400	161 $_{100}$	126 $_{15}$	99 $_{18}$	90 $_{24}$	63 $_{25}$	303	3427
3,4-Dichloroaniline	P-I U UHY UHYAC	1420	161 $_{100}$	126 $_{14}$	99 $_{20}$	90 $_{18}$	63 $_{18}$	303	4234
3,4-Dichloroaniline AC	UHYAC	1990	203 $_{28}$	161 $_{100}$	133 $_{13}$	90 $_9$	63 $_{26}$	405	4235
1,2-Dichlorobenzene		1040*	146 $_{100}$	111 $_{51}$	84 $_7$	75 $_{51}$		284	3179
1,3-Dichlorobenzene		1040*	146 $_{100}$	128 $_{53}$	111 $_{43}$	75 $_{38}$	64 $_{33}$	284	3180
p,p'-Dichlorobenzophenone (DCBP)		2340*	250 $_{25}$	215 $_9$	139 $_{100}$	111 $_{38}$	75 $_{26}$	601	1953
2,2'-Dichlorobiphenyl		1630*	222 $_{73}$	187 $_{46}$	152 $_{100}$	126 $_8$	75 $_{27}$	478	9197
Dichlorodifluoromethane		<1000*	120 $_1$	101 $_{20}$	85 $_{100}$	66 $_{10}$	50 $_{36}$	264	3793
Dichloromethane		<1000*	84 $_{74}$	49 $_{100}$				254	1543
2,5-Dichloromethoxybenzene		1200*	176 $_{100}$	161 $_{25}$	133 $_{72}$	75 $_{24}$	63 $_{37}$	331	3638
Dichlorophen 2AC		2250*	352 $_{11}$	310 $_{32}$	268 $_{100}$	233 $_{12}$	128 $_{32}$	1218	2035
Dichlorophen 2ET		2225*	324 $_{100}$	309 $_{44}$	289 $_{30}$	273 $_6$	215 $_8$	1055	2005
Dichlorophen 2ME		2245*	296 $_{51}$	261 $_{19}$	155 $_{47}$	141 $_{40}$	121 $_{100}$	883	2721
2,4-Dichlorophenol	U	1320*	164 $_{58}$	162 $_{100}$	126 $_{15}$	98 $_{37}$	63 $_{64}$	305	712
2,4-Dichlorophenoxyacetic acid (D)	P U	1800*	220 $_{55}$	175 $_{22}$	162 $_{100}$	133 $_{29}$	111 $_{22}$	467	711
2,4-Dichlorophenoxyacetic acid (D) ME	P U PME UME	1580*	234 $_{57}$	199 $_{100}$	175 $_{51}$	145 $_{26}$	111 $_{32}$	525	2370
2,4-Dichlorophenoxyacetic acid (D)-M (dichlorophenol)	U	1320*	164 $_{58}$	162 $_{100}$	126 $_{15}$	98 $_{37}$	63 $_{64}$	305	712
2,4-Dichlorophenoxybutyric acid ME		1835*	262 $_2$	231 $_9$	162 $_{32}$	101 $_{100}$	59 $_{43}$	667	4118
p,p'-Dichlorophenylacetate ME		2160*	294 $_{19}$	235 $_{100}$	199 $_{20}$	165 $_{70}$	82 $_{16}$	868	3184
p,p'-Dichlorophenylethanol		2185*	266 $_4$	235 $_{55}$	199 $_{19}$	165 $_{100}$	75 $_{19}$	692	3181
o,p'-Dichlorophenylmethane	P U	1900*	236 $_{54}$	201 $_{100}$	165 $_{82}$	82 $_{30}$		537	1743
p,p'-Dichlorophenylmethane		1855*	236 $_{35}$	201 $_{98}$	165 $_{100}$	125 $_{20}$	82 $_{50}$	536	3182
p,p'-Dichlorophenylmethanol		2080*	252 $_7$	217 $_6$	139 $_{100}$	111 $_{15}$	77 $_{40}$	615	3183
Dichlorophenylpiperazine isomer-1 AC		2380	272 $_{17}$	229 $_8$	200 $_{100}$	188 $_{48}$	172 $_{36}$	728	8563
Dichlorophenylpiperazine isomer-2 AC		2440	272 $_{20}$	229 $_{11}$	200 $_{100}$	187 $_{40}$	172 $_{38}$	728	8567
Dichloroquinolinol	P G UHY UHYAC	1850	213 $_{100}$	185 $_9$	150 $_{11}$			441	714
Dichlorprop	G P-I U-I	1840*	234 $_{19}$	220 $_5$	162 $_{100}$	133 $_{11}$	109 $_{11}$	524	2371
Dichlorprop ME		1630*	248 $_{34}$	189 $_{43}$	162 $_{100}$	133 $_{11}$	109 $_{12}$	588	2372
Dichlorprop-M (2,4-dichlorophenol)	U	1320*	164 $_{58}$	162 $_{100}$	126 $_{15}$	98 $_{37}$	63 $_{64}$	305	712
Dichlorvos		1275*	220 $_4$	185 $_{19}$	145 $_7$	109 $_{100}$	79 $_{26}$	467	1423
Diclazepam		2530	318 $_{92}$	290 $_{70}$	283 $_{100}$	255 $_{49}$	177 $_{25}$	1016	9572
Diclazepam HY	UHY UHYAC	2220	279 $_{100}$	244 $_{76}$	229 $_{75}$	111 $_{53}$	75 $_{38}$	771	291
Diclazepam-M (HO-)	P-I G UGLUC	2735	334 $_8$	307 $_{66}$	305 $_{100}$	111 $_6$	75 $_{10}$	1117	547
Diclazepam-M (HO-) AC		2740	376 $_6$	334 $_{22}$	305 $_{100}$	291 $_{20}$	255 $_8$	1338	5604
Diclazepam-M (nor-) HY	UHY	2180	265 $_{62}$	230 $_{100}$	139 $_{43}$	111 $_{50}$		684	543
Diclazepam-M (nor-) HYAC	U+UHYAC	2300	307 $_{47}$	265 $_{58}$	230 $_{100}$	139 $_{16}$	111 $_{14}$	946	290
Diclazepam-M (nor-HO-)	P-I G UGLUC	2440	302 $_{45}$	274 $_{62}$	239 $_{91}$	75 $_{100}$		1031	539
Diclazepam-M (nor-HO-) 2AC		2730	345 $_{17}$	307 $_{41}$	265 $_{53}$	230 $_{100}$		1458	540
Diclofenac	G P	2205	295 $_{17}$	242 $_{54}$	214 $_{100}$	179 $_{19}$	108 $_{30}$	874	4469
Diclofenac 2ME		2220	323 $_{53}$	264 $_{10}$	228 $_{100}$	214 $_{21}$		1048	2323
Diclofenac ET		2240	323 $_{33}$	277 $_{12}$	242 $_{52}$	214 $_{100}$	179 $_{11}$	1048	6488
Diclofenac -H2O	P G U+UHYAC	2135	277 $_{73}$	242 $_{63}$	214 $_{100}$	179 $_{30}$	89 $_{26}$	757	716
Diclofenac -H2O ET		2130	305 $_{100}$	290 $_{69}$	270 $_{97}$	242 $_{65}$	227 $_{42}$	934	6390
Diclofenac -H2O ME	G P	2300	291 $_{19}$	263 $_{31}$	228 $_{100}$	200 $_{33}$	109 $_8$	845	2324
Diclofenac -H2O TMS		2180	349 $_{100}$	314 $_{27}$	241 $_{11}$	190 $_{48}$	73 $_{81}$	1202	4538
Diclofenac ME	P(ME) G(ME)	2195	309 $_{12}$	277 $_3$	242 $_{30}$	214 $_{100}$	179 $_{17}$	961	717
Diclofenac TMS		2170	367 $_{19}$	352 $_{11}$	242 $_{38}$	214 $_{100}$	73 $_{86}$	1295	5467
Diclofenac-M (di-HO-) 3ME	UME	2490	369 $_{100}$	337 $_{18}$	322 $_{56}$	274 $_{59}$	231 $_{16}$	1305	6388
Diclofenac-M (di-HO-) -H2O 2AC	U+UHYAC	2880	393 $_{16}$	351 $_{100}$	309 $_{69}$	274 $_{21}$	246 $_{30}$	1413	4467
Diclofenac-M (glycine conjugate) ME	P-I	2550	366 $_{15}$	331 $_{28}$	242 $_6$	214 $_{100}$	179 $_9$	1289	6411
Diclofenac-M (HO-) 2ME	UHYME	2460	339 $_{78}$	272 $_{62}$	244 $_{100}$	201 $_{30}$	166 $_{16}$	1147	2325
Diclofenac-M (HO-) -H2O	P U+UHYAC	2400	293 $_{39}$	258 $_{100}$	230 $_{46}$	195 $_{34}$	166 $_{40}$	859	6467
Diclofenac-M (HO-) -H2O isomer-1 AC	U+UHYAC	2520	335 $_{68}$	293 $_{100}$	258 $_{93}$	230 $_{93}$	166 $_{23}$	1122	2321
Diclofenac-M (HO-) -H2O isomer-2 AC	U+UHYAC	2540	335 $_{22}$	293 $_{100}$	258 $_{29}$	230 $_{48}$	195 $_{17}$	1122	1212
Diclofenac-M (HO-) -H2O ME	G P	2365	307 $_{65}$	272 $_{97}$	244 $_{100}$	209 $_{15}$	201 $_{47}$	946	6490
Diclofenac-M (HO-) ME	P	2540	325 $_{23}$	258 $_{29}$	230 $_{100}$	201 $_5$	166 $_{12}$	1059	5958
Diclofenac-M (HO-methoxy-) 2ME	UME	2490	369 $_{100}$	337 $_{18}$	322 $_{56}$	274 $_{59}$	231 $_{16}$	1305	6388
Diclofenac-M (HO-methoxy-) 2ME	UME	2550	369 $_{96}$	337 $_{30}$	302 $_{42}$	274 $_{100}$	260 $_{86}$	1389	6389
Diclofenac-M (HO-methoxy-) -H2O	U+UHYAC	2505	323 $_{43}$	288 $_{100}$	260 $_{33}$	245 $_{21}$	89 $_{29}$	1047	6466
Diclofenac-M (HO-methoxy-) isomer-1 -H2O AC	U+UHYAC	2595	365 $_{15}$	323 $_{32}$	288 $_{100}$	260 $_{25}$	180 $_{31}$	1284	4468
Diclofenac-M (HO-methoxy-) isomer-2 -H2O AC	U+UHYAC	2640	365 $_{17}$	323 $_{46}$	288 $_{100}$	260 $_{37}$	180 $_{27}$	1284	6465
Diclofenac-M/artifact	U+UHYAC	2980	355 $_{68}$	320 $_{100}$	292 $_{10}$	228 $_{16}$	75 $_6$	1234	2322
Diclofenac-M/artifact AC	UHYAC	3225	413 $_{30}$	371 $_{100}$	336 $_{50}$	214 $_8$		1486	26
Diclofenac-M/artifact AC	U+UHYAC	2680	347 $_{84}$	305 $_{100}$	270 $_{53}$	258 $_{52}$	89 $_{61}$	1192	6468
Diclofenamide 4ME		2540	360 $_{42}$	316 $_6$	253 $_{100}$	144 $_{13}$	108 $_{37}$	1262	3127
Diclofensine		2675	321 $_{38}$	278 $_{24}$	243 $_{100}$	208 $_{85}$	165 $_{56}$	1037	8549
Diclofop-methyl		2360*	340 $_{48}$	281 $_{40}$	253 $_{100}$	120 $_{59}$	59 $_{67}$	1154	3832
Dicloxacillin artifact-1	U UHY UHYAC	1300	171 $_{100}$	136 $_{21}$	100 $_{35}$			324	736
Dicloxacillin artifact-10 HYAC	UHYAC	2030	294 $_8$	259 $_{100}$	252 $_{44}$	197 $_{28}$	102 $_{48}$	868	3015
Dicloxacillin artifact-11 HYAC	UHYAC	2220	310 $_9$	275 $_{100}$	97 $_{43}$	70 $_{46}$	58 $_{53}$	968	3016

Dicloxacillin artifact-12 HYAC **Table 8.1:** Compounds in order of names

Name	Detected	RI	Typical ions and intensities					Page	Entry
Dicloxacillin artifact-12 HYAC	UHYAC	2640	335_{100}	293_{30}	247_{7}	100		1122	3017
Dicloxacillin artifact-13 HYAC	UHYAC	2460	354_{6}	312_{89}	277_{42}	254_{29}	212_{100}	1230	3018
Dicloxacillin artifact-14 HYAC	UHYAC	2560	368_{8}	333_{100}	326_{20}	266_{10}	70_{10}	1301	3019
Dicloxacillin artifact-15 HYAC	UHYAC	2785	386_{57}	351_{37}	254_{19}	214_{66}	212_{100}	1383	3020
Dicloxacillin artifact-16 HYAC	UHYAC	3370	398_{2}	216_{28}	174_{100}	114_{11}		1436	3021
Dicloxacillin artifact-17 HYAC	UHYAC	3340	467_{35}	393_{35}	212_{100}	139_{56}	97_{66}	1611	3022
Dicloxacillin artifact-2	G U UHY UHYAC	1800	229_{4}	214_{18}	194_{100}	171_{23}	123_{8}	504	2978
Dicloxacillin artifact-3	G U UHY UHYAC	1845	250_{100}	212_{12}	183_{7}			602	3006
Dicloxacillin artifact-4	U UHY UHYAC	2060	266_{47}	254_{26}	214_{64}	212_{100}	75_{11}	692	3007
Dicloxacillin artifact-5	G P U UHY UHYAC	2095	235_{100}	212_{13}	100_{2}	75_{6}		530	3008
Dicloxacillin artifact-5 AC	UHYAC	2105	277_{50}	235_{100}	212_{17}	98_{7}		757	3013
Dicloxacillin artifact-6	G U UHY UHYAC	2295	307_{100}	254_{8}	247_{18}	212_{30}		946	3009
Dicloxacillin artifact-7		2340	364_{29}	321_{90}	247_{24}	212_{100}	100_{19}	1281	3010
Dicloxacillin artifact-8 HY	UHY	2710	382_{38}	254_{13}	212_{40}	127_{98}	100_{100}	1364	3011
Dicloxacillin artifact-8 HYAC	UHYAC	3500	424_{12}	249_{16}	212_{19}	155_{34}	142_{100}	1516	3014
Dicloxacillin artifact-9 HY	UHY	2905	407_{100}	372_{15}	254_{38}	212_{99}	153_{60}	1466	3012
Dicloxacillin-M (HO-) artifact-1 AC	UHYAC	2090	308_{100}	270_{7}	211_{11}	172_{5}	148_{5}	954	3023
Dicloxacillin-M (HO-) artifact-2 AC	UHYAC	2210	324_{26}	291_{45}	289_{25}	254_{24}	212_{100}	1055	3024
Dicloxacillin-M/artifact-1 HY	UHY UHYAC	1795	220_{100}	185_{7}	102_{9}	100_{4}		467	3025
Dicloxacillin-M/artifact-10 HYAC	U+UHYAC	2830	423_{100}	310_{8}	254_{11}	212_{65}	169_{52}	1514	3034
Dicloxacillin-M/artifact-2 HY	UHY UHYAC	1970	252_{100}	220_{53}	172_{32}	152_{24}		615	3026
Dicloxacillin-M/artifact-3 HY	UHY UHYAC	2155	274_{100}	241_{24}	192_{18}	148_{40}	94_{37}	736	3027
Dicloxacillin-M/artifact-4 HYAC	UHYAC	2015	219_{100}	172_{9}	141_{5}	100_{3}		462	3028
Dicloxacillin-M/artifact-5 HYAC	UHYAC	2110	287_{3}	252_{14}	214_{100}	171_{24}	123_{8}	817	3029
Dicloxacillin-M/artifact-6 HYAC	UHYAC	2295	350_{38}	293_{100}	251_{41}	212_{60}	184_{11}	1207	3030
Dicloxacillin-M/artifact-7 HYAC	UHYAC	2300	354_{9}	319_{57}	254_{10}	212_{100}	183_{9}	1230	3031
Dicloxacillin-M/artifact-8 HYAC	UHYAC	2520	397_{10}	369_{98}	254_{16}	212_{100}	59_{70}	1432	3032
Dicloxacillin-M/artifact-9 HYAC	UHYAC	2790	449_{4}	391_{100}	356_{16}	254_{19}	212_{50}	1577	3033
Dicofol		2485*	368_{1}	251_{59}	199_{4}	139_{100}	111_{36}	1301	4147
Dicofol artifact (DCBP)		2340*	250_{25}	215_{9}	139_{100}	111_{38}	75_{26}	601	1953
Dicrotophos		1645	237_{7}	193_{10}	127_{100}	109_{10}	67_{46}	542	3433
Dicycloverine		2120	309_{3}	294_{3}	165_{5}	99_{14}	86_{100}	967	710
Didanosine 2TMS		2240	380_{3}	365_{5}	209_{100}	193_{53}	157_{16}	1357	8353
Didanosine artifact 2TMS		1675	280_{53}	265_{100}	206_{14}	193_{20}	73_{51}	781	8354
Diethylallylacetamide	P G U	1285	155_{1}	140_{24}	126_{100}	69_{94}	55_{86}	299	719
Diethylallylacetamide-M	U	1510*	144_{40}	113_{41}	95_{52}	69_{84}	55_{100}	327	720
Diethylallylacetamide-M AC	UHYAC-I	1725*	186_{18}	141_{56}	126_{54}	95_{91}	69_{100}	447	4245
Diethylamine		<1000	73_{17}	58_{94}	44_{25}	30_{100}		251	4188
Diethyldithiocarbamic acid ME	P	1340	163_{82}	116_{60}	91_{49}	88_{100}	60_{76}	307	6458
Diethylene glycol 2TMS		1170*	191_{1}	147_{10}	117_{33}	103_{5}	73_{100}	606	8584
Diethylene glycol dibenzoate		2445*	227_{1}	149_{71}	105_{100}	77_{46}		993	1755
Diethylene glycol dipivalate	PPIV	1520*	159_{1}	129_{73}	113_{4}	85_{16}	57_{100}	741	1904
Diethylene glycol monoethylether pivalate	PPIV	1345*	129_{26}	85_{13}	72_{66}	57_{100}		460	6422
Diethylether		<1000	74_{58}	59_{100}				252	2755
Diethylphthalate		1495*	222_{3}	177_{18}	149_{100}			479	721
Diethylstilbestrol		2295*	268_{100}	239_{46}	159_{15}	145_{42}	107_{60}	708	1419
Diethylstilbestrol 2AC		2450*	352_{27}	310_{70}	268_{100}	239_{39}	107_{48}	1222	1420
Diethylstilbestrol 2ME		2190*	296_{100}	267_{26}	159_{32}	121_{43}		886	1421
N,N-Diethyl-tryptamine-D4		1910	220_{2}	148_{5}	132_{2}	105_{4}	88_{100}	471	8863
N,N-Diethyl-tryptamine-D4 HFB		1795	401_{2}	344_{10}	328_{11}	131_{12}	88_{100}	1496	10126
N,N-Diethyl-tryptamine-D4 PFP		1785	351_{1}	294_{5}	278_{6}	131_{13}	88_{100}	1292	10125
N,N-Diethyl-tryptamine-D4 TFA		1780	301_{1}	244_{7}	228_{13}	131_{18}	88_{100}	1007	10124
N,N-Diethyl-tryptamine-D4 TMS		2055	292_{4}	220_{3}	204_{6}	147_{6}	88_{100}	858	10123
Difenzoquate -C2H6SO4		1665	234_{100}	189_{11}	165_{6}	118_{16}	77_{48}	528	3958
Diflubenzuron 2ME		2290	338_{8}	154_{8}	141_{100}	113_{19}	63_{10}	1142	3974
Diflufenicam		2670	394_{9}	266_{100}	246_{12}	169_{9}	101_{9}	1419	3891
Diflunisal		2095*	250_{57}	232_{100}	204_{30}	175_{34}	151_{19}	602	1478
Diflunisal 2ME		1990*	278_{100}	247_{82}	204_{26}	188_{27}	175_{33}	766	1432
Diflunisal -CO2		1950*	206_{100}	177_{29}	151_{18}	115_{18}		415	2225
Diflunisal ME		2050*	264_{61}	232_{100}	204_{30}	175_{35}	151_{8}	680	2223
Diflunisal MEAC		2060*	306_{100}	247_{74}	199_{4}	175_{8}	143_{9}	942	2224
Difluoro-BDB		1335	228_{1}	200_{3}	171_{5}	105_{4}	58_{100}	505	8251
Difluoro-BDB AC		1755	271_{1}	212_{28}	171_{14}	100_{46}	58_{100}	723	8252
Difluoro-BDB formyl artifact		1585	241_{1}	171_{12}	105_{6}	77_{15}	70_{100}	559	8250
Difluoro-BDB HFB		1550	425_{1}	406_{3}	254_{100}	212_{41}	171_{65}	1519	8257
Difluoro-BDB PFP		1510	375_{1}	356_{1}	212_{86}	204_{100}	171_{50}	1333	8256
Difluoro-BDB TFA		1500	325_{1}	306_{2}	212_{76}	171_{43}	154_{100}	1060	8255
Difluoro-BDB TMS		1455	286_{8}	272_{8}	171_{26}	130_{100}	73_{81}	912	8254
Difluoro-MBDB		1390	242_{1}	214_{20}	171_{20}	105_{15}	72_{100}	568	8258
Difluoro-MBDB AC		1800	266_{1}	212_{10}	171_{12}	114_{62}	72_{100}	809	8259
Difluoro-MBDB HFB		1570	268_{88}	212_{36}	210_{37}	77_{65}	69_{100}	1555	8389
Difluoro-MBDB PFP		1605	370_{1}	218_{100}	212_{42}	171_{28}	160_{42}	1398	8263
Difluoro-MBDB TFA		1615	339_{1}	212_{41}	168_{100}	110_{42}	77_{31}	1148	8262
Difluoro-MBDB TMS		1520	300_{2}	171_{17}	144_{100}	105_{8}	73_{77}	1000	8261

Table 8.1: Compounds in order of names

Name	Detected	RI	Typical ions and intensities						Page	Entry
Difluoro-MDA		1120	214_1	200_3	171_{12}	77_{61}	51_{100}		448	8377
Difluoro-MDA AC		1705	257_4	238_4	198_{100}	171_{34}	86_{26}		641	8265
Difluoro-MDA formyl artifact		1170	227_4	171_{30}	105_{22}	77_{41}	56_{100}		498	8264
Difluoro-MDA HFB		1450	240_{44}	198_{43}	171_{25}	77_{67}	51_{100}		1480	8388
Difluoro-MDA PFP		1450	361_2	342_3	198_{84}	190_{100}	171_{45}		1267	8268
Difluoro-MDA TFA		1435	311_1	292_2	198_{82}	171_{44}	140_{100}		974	8267
Difluoro-MDA TMS		1375	287_1	272_9	171_{21}	116_{100}	73_{81}		819	8266
Difluoro-MDE		1560	242_1	228_2	171_{14}	105_{10}	72_{100}		568	8269
Difluoro-MDE AC		1770	266_1	198_{28}	171_{16}	114_{68}	72_{100}		809	8270
Difluoro-MDE HFB		1610	420_2	268_{100}	240_{66}	198_{72}	171_{38}		1555	8274
Difluoro-MDE PFP		1570	370_1	218_{100}	198_{53}	190_{55}	171_{11}		1398	8273
Difluoro-MDE TFA		1580	320_1	198_{54}	168_{100}	140_{59}	77_{32}		1148	8272
Difluoro-MDE TMS		1540	300_3	171_{12}	144_{100}	105_8	73_{81}		1000	8271
Difluoro-MDMA		1535	228_1	214_3	171_{10}	77_{23}	58_{100}		505	8275
Difluoro-MDMA AC		1745	271_1	198_{24}	171_{16}	100_{79}	58_{100}		723	8276
Difluoro-MDMA HFB		1580	406_2	254_{100}	210_{66}	198_{67}	171_{29}		1519	8280
Difluoro-MDMA PFP		1540	356_2	204_{100}	198_{58}	171_{24}	160_{61}		1333	8279
Difluoro-MDMA TFA		1535	306_2	198_{66}	171_{22}	154_{100}	110_{74}		1059	8278
Difluoro-MDMA TMS		1500	300_1	286_{15}	171_{17}	130_{100}	73_{80}		912	8277
3,4-Difluoromethylenedioxyphenethylamine		1540	201_{20}	172_{30}	152_{17}	77_{57}	51_{100}		401	8339
3,4-Difluoromethylenedioxyphenethylamine 2AC		1780	285_1	266	184_{100}	171_{11}	72_{28}		807	8342
3,4-Difluoromethylenedioxyphenethylamine AC		1670	243_6	184_{100}	171_{14}	72_{23}	51_{34}		567	8341
3,4-Difluoromethylenedioxyphenethylamine formyl artifact		1205	213_{31}	184_{10}	171_{59}	77_{83}	51_{100}		442	8340
3,4-Difluoromethylenedioxyphenethylamine HFB		1485	397_2	378_2	226_{13}	184_{100}	171_{55}		1432	8346
3,4-Difluoromethylenedioxyphenethylamine PFP		1445	347_1	328_2	184_{100}	171_{46}	105_{26}		1192	8345
3,4-Difluoromethylenedioxyphenethylamine TFA		1440	297_3	184_{100}	171_{64}	126_{22}	105_{35}		888	8344
3,4-Difluoromethylenedioxyphenethylamine TMS		1400	272_1	258_{36}	171_{20}	153_{13}	102_{100}		734	8343
Digitoxigenin -2H2O		2840*	338_{100}	323_{43}	282_{37}	228_{74}	91_{92}		1146	5243
Digitoxigenin -H2O AC		3180*	398_5	338_{52}	323_{100}	145_{81}	91_{81}		1439	5242
Digitoxin -2H2O HY		2840*	338_{100}	323_{43}	282_{37}	228_{74}	91_{92}		1146	5243
Digitoxin -H2O HYAC		3180*	398_5	338_{52}	323_{100}	145_{81}	91_{81}		1439	5242
Dihexylamine		1380	185_4	114_{100}	100_3	79_5	57_8		359	4947
Dihydrobrassicasterol		3190*	400_{60}	315_{33}	289_{41}	105_{84}	55_{100}		1447	5620
Dihydrobrassicasterol -H2O		3270*	382_{100}	213_{22}	147_{78}	105_{62}	55_{56}		1368	5624
Dihydrocapsaicine		2430	307_{10}	195_9	151_{13}	137_{100}	122_9		953	5927
Dihydrocapsaicine 2TMS		2700	451_5	436_4	339_{23}	209_{87}	73_{100}		1583	6035
Dihydrocapsaicine AC		2540	349_3	308_7	195_{17}	151_{15}	137_{100}		1207	5928
Dihydrocapsaicine HFB		2490	503_{31}	404_{22}	391_{40}	347_{100}	333_{93}		1664	5931
Dihydrocapsaicine ME		2470	321_{20}	195_{23}	151_{25}	137_{100}	122_9		1042	6781
Dihydrocapsaicine MEAC		2510	363_2	321_{27}	195_{27}	151_{19}	137_{100}		1280	6783
Dihydrocapsaicine PFP		2410	453_{34}	354_{22}	341_{45}	297_{100}	283_{92}		1587	5930
Dihydrocapsaicine TFA		2410	403_6	304_{22}	291_{41}	247_{100}	233_{93}		1457	5929
Dihydrocapsaicine TMS		2700	379_{17}	364_{10}	209_{100}	179_{36}	73_{63}		1354	6034
Dihydrocodeine	P G UHY	2410	301_{100}	244_{10}	164_{28}	115_{25}	70_{37}		914	483
Dihydrocodeine AC	U+UHYAC	2435	343_{100}	300_{33}	284_{30}	226_{14}	70_{10}		1176	233
Dihydrocodeine Br-artifact	UHYAC	2485	379_{100}	362_{21}	322_{30}	265_{22}	164_{73}		1349	2988
Dihydrocodeine Cl-artifact AC		2500	377_{100}	334_{31}	318_{41}	260_{17}	164_{10}		1342	2989
Dihydrocodeine HFB		2315	497_{100}	440_8	300_{24}	284_{79}	227_{18}		1656	6143
Dihydrocodeine PFP		2360	447_{100}	390_{16}	300_{17}	284_{55}	119_{45}		1575	2248
Dihydrocodeine TFA		2265	397_{100}	340_8	284_{29}	70_{28}	59_{47}		1434	4001
Dihydrocodeine TMS		2480	373_{59}	236_{15}	178_{14}	146_{30}	73_{100}		1328	2468
Dihydrocodeine-M (dehydro-)	G UHY UHYAC	2440	299_{100}	242_{51}	185_{23}	96_{24}	59_{23}		902	238
Dihydrocodeine-M (dehydro-) enol AC		2500	341_{100}	298_{65}	242_{32}	162_{26}			1164	258
Dihydrocodeine-M (dehydro-) enol Cl-artifact AC		2630	375_{100}	340_{47}	318_{28}	146_{13}	115_{10}		1334	4401
Dihydrocodeine-M (dehydro-) enol TMS		2475	371_{31}	356_{14}	313_9	234_{30}	73_{100}		1320	6215
Dihydrocodeine-M (N,O-bis-demethyl-) 3AC	UHYAC	2790	399_{20}	357_{50}	229_{19}	87_{100}	72_{22}		1442	3050
Dihydrocodeine-M (nor-)	UHY	2440	287_{100}	244_{24}	242_{22}	150_{32}	115_{24}		820	4368
Dihydrocodeine-M (nor-) 2AC	UHYAC	2750	371_{20}	285_7	243_{26}	87_{100}	72_{33}		1319	235
Dihydrocodeine-M (nor-) AC	UHYAC	2700	329_{40}	243_{42}	183_{26}	87_{100}	72_{44}		1088	3054
Dihydrocodeine-M (nor-) Cl-artifact 2AC	UHYAC	2820	405_8	320_2	259_4	87_{100}	72_{26}		1463	2990
Dihydrocodeine-M (O-demethyl-)	UHY	2400	287_{100}	230_{14}	164_{35}	115_{28}	70_{54}		821	484
Dihydrocodeine-M (O-demethyl-) 2AC	U+UHYAC	2545	371_{83}	329_{100}	286_{34}	212_{21}	70_{33}		1319	234
Dihydrocodeine-M (O-demethyl-) 2HFB		2260	679_{41}	482_{53}	466_{100}	360_{13}	169_{21}		1727	6197
Dihydrocodeine-M (O-demethyl-) 2PFP		2330	579_{60}	432_{24}	416_{49}	310_4	119_{100}		1707	2460
Dihydrocodeine-M (O-demethyl-) 2TFA		2190	479_{91}	382_{25}	366_{61}	260_7	69_{100}		1632	6198
Dihydrocodeine-M (O-demethyl-) 2TMS		2520	431_{33}	373_7	236_{23}	146_{15}	73_{100}		1538	2469
Dihydrocodeine-M (O-demethyl-) AC	UHYAC	2490	329_{100}	287_{56}	230_{10}	164_{20}	70_{21}		1089	3055
Dihydrocodeine-M (O-demethyl-) TFA		2250	383_{100}	286_{19}	270_{44}	213_{19}	69_{50}		1370	6199
Dihydrocodeine-M (O-demethyl-dehydro-)	UHY	2445	285_{100}	228_{18}	214_{10}	171_{10}	96_{82}		810	527
Dihydrocodeine-M (O-demethyl-dehydro-) AC	UHYAC	2595	327_{34}	285_{100}	229_{36}	200_{14}	171_{13}		1075	240
Dihydroergotamine artifact-1 (cyclo (Phe-Pro)) isomer-1	U+UHYAC P-I	2335	244_{36}	153_{50}	125_{100}	91_{38}	70_{50}		571	4495
Dihydroergotamine artifact-1 (cyclo (Phe-Pro)) isomer-2	U+UHYAC P-I	2365	244_{54}	153_{21}	125_{100}	91_{41}	70_{37}		571	8443
Dihydroergotamine artifact-1 AC		2360	286_{29}	153_{62}	125_{100}	91_{57}	70_{73}		815	5217
Dihydroergotamine artifact-2		2440	314_{45}	244_{35}	153_{84}	125_{75}	70_{100}		994	4494

Dihydromorphine

Table 8.1: Compounds in order of names

Name	Detected	RI	Typical ions and intensities					Page	Entry
Dihydromorphine	UHY	2400	287 100	230 14	164 35	115 28	70 54	821	484
Dihydromorphine 2AC	U+UHYAC	2545	371 83	329 100	286 34	212 21	70 33	1319	234
Dihydromorphine 2HFB		2260	679 41	482 53	466 100	360 13	169 21	1727	6197
Dihydromorphine 2PFP		2330	579 60	432 21	416 49	310 4	119 100	1707	2460
Dihydromorphine 2TFA		2190	479 91	382 25	366 61	260 7	69 100	1632	6198
Dihydromorphine 2TMS		2520	431 33	373 7	236 23	146 15	73 100	1538	2469
Dihydromorphine AC	UHYAC	2490	329 100	287 56	230 10	164 20	70 21	1089	3055
Dihydromorphine TFA		2250	383 100	286 19	270 44	213 19	69 50	1370	6199
Dihydromorphine-M (nor-) 3AC	UHYAC	2790	399 20	357 50	229 19	87 100	72 22	1442	3050
Dihydrotestosterone		2510*	290 95	247 49	220 86	161 44	55 100	845	3896
Dihydrotestosterone AC		2620*	332 73	272 100	257 44	201 46	79 67	1109	3918
Dihydrotestosterone enol 2TMS		2450*	434 100	405 12	202 7	143 41	73 90	1547	3964
Dihydrotestosterone TMS		2485*	362 15	347 10	246 24	129 70	73 100	1276	3963
Dihydroxybenzoic acid 3ME		1600*	196 100	165 97	137 12	125 18	79 26	390	4942
3,4-Dihydroxybenzoic acid ME2AC		1750*	252 1	210 19	168 100	137 60	109 8	615	5254
3,4-Dihydroxybenzylamine 3AC		2100	265 5	223 30	181 100	138 30	122 15	686	5692
3,4-Dihydroxycinnamic acid 2AC		2240*	264 2	222 12	180 100	163 8	134 15	680	5968
3,4-Dihydroxycinnamic acid 3TMS		2115*	396 86	381 22	219 78	191 17	73 100	1430	6014
3,4-Dihydroxycinnamic acid -CO2		1375*	136 53	89 71	77 48	63 71	51 100	276	5757
3,4-Dihydroxycinnamic acid ME2AC		2170*	278 3	236 15	194 100	163 40	134 21	766	5967
3,4-Dihydroxycinnamic acid ME2HFB		1985*	586 56	555 74	389 18	169 93	69 100	1709	5971
3,4-Dihydroxycinnamic acid ME2PFP		1985*	486 61	455 89	323 14	119 100	77 31	1643	5970
3,4-Dihydroxycinnamic acid ME2TMS		1930*	338 72	297 9	219 100	191 21	73 97	1143	6013
Dihydroxynorcholanoic acid -H2O MEAC	UHYAC	2980*	416 31	356 21	343 53	255 100	145 36	1497	2455
3,4-Dihydroxyphenethylamine 3AC	U+UHYAC	2150	279 3	237 22	220 21	178 30	136 100	773	5284
3,4-Dihydroxyphenethylamine 4AC		2245	321 1	220 33	178 36	136 100	123 32	1038	5285
3,4-Dihydroxyphenylacetic acid		2440*	168 14	123 100	105 6	77 48	51 47	320	5754
3,4-Dihydroxyphenylacetic acid 3TMS		1880*	384 17	297 100	237 13	209 16	73 53	1377	6012
3,4-Dihydroxyphenylacetic acid ME		1870*	182 21	123 100	105 4	77 20	51 20	351	5755
3,4-Dihydroxyphenylacetic acid ME2AC		2105*	266 2	224 14	182 100	123 96	94 10	692	5960
3,4-Dihydroxyphenylacetic acid ME2HFB		1680*	574 25	515 33	302 39	69 87	59 100	1704	5964
3,4-Dihydroxyphenylacetic acid ME2PFP		1590*	474 42	415 52	252 47	119 86	59 100	1625	5962
3,4-Dihydroxyphenylacetic acid ME2TFA		1560*	374 45	315 47	202 56	69 97	59 100	1329	5961
3,4-Dihydroxyphenylacetic acid ME2TMS		1695*	326 52	267 31	179 100	149 12	73 90	1069	6011
3,4-Dihydroxyphenylacetic acid MEHFB		1905*	378 61	319 100	169 32	94 39	69 65	1346	5965
3,4-Dihydroxyphenylacetic acid MEPFP		1680*	328 49	269 100	137 17	119 46	59 47	1078	5963
Diisodecylphthalate		2800*	446 1	307 24	167 2	149 100	57 7	1574	3541
Diisohexylphthalate		2380*	334 1	251 10	233 2	149 100	104 3	1121	6397
Diisononylphthalate		2700*	418 1	293 19	167 4	149 100	71 13	1503	1232
Diisooctylphthalate		2520*	390 1	279 15	167 41	149 100	57 29	1404	723
Diisopropylidene-fructopyranose	P	1680*	245 100	229 22	171 46	127 54	69 81	657	5707
Diisopropylidene-fructopyranose TMS		1900*	317 31	257 13	229 65	199 20	171 100	1106	5709
N,N-Diisopropyl-tryptamine-D4		2050	248 1	148 13	132 40	116 100	74 24	593	8861
N,N-Diisopropyl-tryptamine-D4 HFB		1865	429 4	344 15	328 10	147 8	116 100	1570	10118
N,N-Diisopropyl-tryptamine-D4 PFP		1855	379 4	294 8	278 6	147 4	116 100	1421	10117
N,N-Diisopropyl-tryptamine-D4 TFA		1855	329 3	244 16	228 8	147 8	116 100	1180	10116
N,N-Diisopropyl-tryptamine-D4 TMS		2170	320 1	220 7	204 13	116 100	74 21	1036	10115
Dilaurylthiodipropionate		3970*	514 8	329 19	178 36	143 42	55 100	1673	3532
Dilazep-M/artifact (trimethoxybenzoic acid)		1780*	212 100	197 57	169 13	141 27		438	1949
Dilazep-M/artifact (trimethoxybenzoic acid) ET		1770*	240 100	225 44	212 17	195 45	141 24	556	5219
Dilazep-M/artifact (trimethoxybenzoic acid) ME		1740*	226 100	211 48	195 22	155 21		494	1950
Diltiazem	G P U+UHYAC	2960	414 1	150 2	121 4	71 27	58 100	1490	2504
Diltiazem-M (deacetyl-)	P UHY	2990	372 1	178 1	150 2	71 24	58 100	1323	2505
Diltiazem-M (deacetyl-) TMS		2835	444 1	429 1	374 9	222 27	58 100	1570	4539
Diltiazem-M (deamino-HO-) AC	U+UHYAC	3060	429 2	341 25	240 15	150 100	121 48	1530	2705
Diltiazem-M (deamino-HO-) -H2O	UHYAC	3310	369 1	309 100	150 71	121 44	100 19	1305	2703
Diltiazem-M (deamino-HO-) HY	UHY	3020	345 11	316 88	208 35	150 31	121 100	1183	2706
Diltiazem-M (O-demethyl-) AC	UHYAC	3080	442 1	178 1	136 7	71 34	58 100	1565	2701
Diltiazem-M (O-demethyl-) HY	UHY	3050	358 1	136 7	107 9	71 25	58 100	1253	2707
Diltiazem-M (O-demethyl-deamino-HO-) 2AC	UHYAC	3170	457 2	369 15	178 34	136 100	87 20	1595	2702
Diltiazem-M (O-demethyl-deamino-HO-) -H2O AC	UHYAC	3540	397 2	337 93	178 34	136 100	100 37	1433	2704
Dimefuron +H2O 3ME		2600	398 10	314 3	255 8	72 100	57 55	1438	3939
Dimefuron ME		2520	352 5	269 3	225 4	127 2	72 100	1219	3938
Dimetacrine	G U	2315	294 18	279 28	86 100	58 80		873	329
Dimetacrine-M (N-oxide) -(CH3)2NOH	U	2020	249 18	234 100	194 73			599	1170
Dimetacrine-M (ring)	U	1905	209 11	194 100				430	1169
Dimetamfetamine		1250	163 1	148 1	117 1	91 10	72 100	309	1427
Dimetamfetamine-M (nor-)	U	1195	148 1	134 2	115 1	91 9	58 100	288	1093
Dimetamfetamine-M (nor-) AC	U+UHYAC	1575	191 1	117 2	100 42	91 6	58 100	373	1094
Dimetamfetamine-M (nor-) HFB		1460	254 100	210 44	169 15	118 41	91 38	1183	5069
Dimetamfetamine-M (nor-) PFP		1415	204 100	160 46	118 35	91 25	69 4	874	5070
Dimetamfetamine-M (nor-) TFA		1300	245 1	154 100	118 48	110 55	91 23	574	3998
Dimetamfetamine-M (nor-) TMS		1325	206 4	130 100	91 17	73 83	59 13	477	6214
Dimethachlor		1565	255 1	210 10	197 32	134 100	77 18	632	3830

Table 8.1: Compounds in order of names Dimethoate

Name	Detected	RI	Typical ions and intensities					Page	Entry
Dimethoate	P G U	1725	229 $_7$	125 $_{45}$	93 $_{62}$	87 $_{100}$		504	724
Dimethoate-M (HO-)	U	1430	245 $_{19}$	218 $_{20}$	125 $_{45}$	93 $_{100}$		573	2119
Dimethoate-M (HOOC-) ME	U	1400*	230 $_{49}$	198 $_{66}$	125 $_{67}$	93 $_{100}$	79 $_{43}$	507	2118
Dimethoate-M (oxo-)	G P-I	1585	213 $_7$	156 $_{91}$	110 $_{100}$	79 $_{52}$	58 $_{70}$	441	1501
Dimethocaine	U UGLUC USPE	2415	278 $_1$	137 $_4$	120 $_{27}$	86 $_{100}$	58 $_{11}$	770	8550
Dimethocaine AC	UGLUCAC USPEAC	2730	320 $_1$	162 $_5$	120 $_9$	86 $_{100}$	58 $_6$	1035	8551
Dimethocaine HFB		2405	316 $_{13}$	169 $_2$	164 $_1$	86 $_{100}$	58 $_6$	1625	8556
Dimethocaine ME		2430	292 $_1$	134 $_{11}$	120 $_3$	86 $_{100}$	58 $_6$	858	8553
Dimethocaine PFP		2380	266 $_{24}$	168 $_2$	146 $_1$	86 $_{100}$	58 $_7$	1517	8555
Dimethocaine TFA		2390	216 $_{17}$	168 $_2$	146 $_2$	86 $_{100}$	58 $_8$	1331	8554
Dimethocaine TMS		2370	350 $_1$	335 $_1$	194 $_4$	86 $_{100}$	58 $_{16}$	1210	8572
Dimethocaine-M (bis-nor-) 2AC	USPEAC	2405	306 $_1$	162 $_8$	120 $_{11}$	72 $_{100}$		943	8820
Dimethocaine-M (bis-nor-HO-) 2AC	USPEAC	2560	322 $_1$	178 $_3$	153 $_4$	136 $_7$	72 $_{100}$	1045	8821
Dimethocaine-M (bis-nor-HO-N-acetyl-) formyl artifact	UGLUC	2530	219 $_{13}$	191 $_3$	162 $_{100}$	120 $_{29}$	92 $_6$	856	8823
Dimethocaine-M (HO-)	UGLUC USPE	2395	294 $_1$	153 $_4$	136 $_{18}$	86 $_{100}$		872	8816
Dimethocaine-M (HO-) 2AC	UGLUCAC	2590	378 $_1$	178 $_4$	167 $_5$	86 $_{100}$		1348	8822
Dimethocaine-M (nor-)	U UGLUC USPE	2180	250 $_2$	194 $_2$	137 $_{100}$	120 $_{29}$	58 $_{49}$	607	8813
Dimethocaine-M (nor-) 2AC	U+UHYAC	2710	334 $_1$	319 $_2$	291 $_{10}$	100 $_{79}$	58 $_{100}$	1120	8819
Dimethocaine-M (nor-) AC	USPE USPEAC	2370	292 $_1$	179 $_{26}$	162 $_{15}$	137 $_{52}$	58 $_{100}$	857	8818
Dimethocaine-M (nor-HO-) isomer-1	U UGLUC USPE	2390	266 $_9$	179 $_2$	153 $_{84}$	136 $_{35}$	58 $_{100}$	697	8814
Dimethocaine-M (nor-HO-) isomer-2	U UGLUC USPE	2410	266 $_{13}$	179	153 $_{90}$	136 $_{30}$	58 $_{100}$	697	8815
Dimethocaine-M (nor-HO-N-acetyl-)	USPE	2530	308 $_1$	251 $_4$	195 $_{17}$	153 $_{54}$	58 $_{100}$	959	8817
Dimethocaine-M (nor-N-acetyl-)	USPE USPEAC	2370	292 $_1$	179 $_{26}$	162 $_{15}$	137 $_{52}$	58 $_{100}$	857	8818
Dimethocaine-M/artifact (alcohol)	U+UHY	<1000	159 $_2$	128 $_3$	86 $_{100}$	72 $_4$	58 $_{19}$	302	8824
Dimethocaine-M/artifact (alcohol) AC	U+UHYAC	1250	201 $_1$	128 $_2$	86 $_{100}$	72 $_4$	58 $_{12}$	402	8552
2,5-Dimethoxy-4-iodophenethylamine		2330	307 $_{16}$	278 $_{100}$	263 $_{20}$	247 $_9$	232 $_3$	946	6954
2,5-Dimethoxy-4-iodophenethylamine	UGLUCAC	2200*	392 $_8$	350 $_{61}$	308 $_{37}$	290 $_{100}$	262 $_{87}$	1410	7129
2,5-Dimethoxy-4-iodophenethylamine	UGLUCAC	2310*	436 $_{27}$	394 $_{100}$	334 $_{68}$	292 $_{62}$	279 $_{62}$	1550	7130
2,5-Dimethoxy-4-iodophenethylamine (O-demethyl- N-acetyl-) isomer-1	UGLUC	2370	335 $_{15}$	276 $_{100}$	263 $_{11}$	233 $_{10}$	134 $_{14}$	1122	6963
2,5-Dimethoxy-4-iodophenethylamine (O-demethyl- N-acetyl-) isomer-1 AC	U+UHYAC	2480	377 $_5$	335 $_{46}$	276 $_{100}$	259 $_{16}$	233 $_{20}$	1341	6967
2,5-Dimethoxy-4-iodophenethylamine (O-demethyl- N-acetyl-) isomer-2 AC	U+UHYAC	2500	377 $_8$	335 $_{43}$	276 $_{100}$	263 $_{17}$	236 $_7$	1341	6968
2,5-Dimethoxy-4-iodophenethylamine (O-demethyl-) isomer-1 2AC	U+UHYAC	2480	377 $_5$	335 $_{46}$	276 $_{100}$	259 $_{16}$	233 $_{20}$	1341	6967
2,5-Dimethoxy-4-iodophenethylamine (O-demethyl-) isomer-2 2AC	U+UHYAC	2500	377 $_8$	335 $_{43}$	276 $_{100}$	263 $_{17}$	236 $_7$	1341	6968
2,5-Dimethoxy-4-iodophenethylamine 2AC	U+UHYAC	2340	391 $_{16}$	290 $_{100}$	275 $_{19}$	247 $_8$	148 $_{10}$	1405	6958
2,5-Dimethoxy-4-iodophenethylamine 2ME		2320	335 $_{18}$	290 $_{100}$	275 $_{16}$	247 $_{21}$	148 $_{10}$	1123	6962
2,5-Dimethoxy-4-iodophenethylamine AC	U+UHYAC	2260	349 $_{25}$	290 $_{100}$	275 $_{21}$	247 $_{14}$	148 $_{12}$	1202	6957
2,5-Dimethoxy-4-iodophenethylamine deuteroformyl artifact		1850	321 $_{23}$	290 $_{100}$	277 $_{42}$	247 $_{18}$	232 $_4$	1036	6956
2,5-Dimethoxy-4-iodophenethylamine formyl artifact		1860	319 $_{20}$	288 $_{100}$	277 $_{50}$	247 $_{23}$	232 $_6$	1022	6955
2,5-Dimethoxy-4-Iodophenethylamine HFB		2110	503 $_{32}$	290 $_{100}$	277 $_{51}$	247 $_{30}$	148 $_{21}$	1663	6948
2,5-Dimethoxy-4-iodophenethylamine PFP		2080	453 $_{43}$	290 $_{100}$	277 $_{93}$	247 $_{44}$	148 $_{33}$	1586	6960
2,5-Dimethoxy-4-iodophenethylamine TFA	UGLUCTFA	2100	403 $_{20}$	290 $_{100}$	277 $_{69}$	247 $_{49}$	148 $_{31}$	1454	6959
2,5-Dimethoxy-4-iodophenethylamine TMS		2070	379 $_5$	320 $_9$	278 $_7$	102 $_{100}$	73 $_{33}$	1349	6961
2,5-Dimethoxy-4-iodophenethylamine-M (deamino-HO-)	UGLUC	2020	308 $_{100}$	277 $_{74}$	263 $_7$	247 $_{16}$	150 $_9$	954	6966
2,5-Dimethoxy-4-iodophenethylamine-M (deamino-HO-) AC	UGLUCAC	2150	350 $_{16}$	290 $_{100}$	275 $_{28}$	247 $_{10}$	148 $_{25}$	1207	6969
2,5-Dimethoxy-4-iodophenethylamine-M (deamino-HO-) TFA	UGLUCTFA	1980	404 $_{100}$	290 $_{83}$	275 $_{24}$	247 $_{21}$	148 $_{36}$	1458	6978
2,5-Dimethoxy-4-iodophenethylamine-M (deamino-HOOC-) ME	U+UHYAC	2115	336 $_{100}$	321 $_{12}$	289 $_8$	277 $_{57}$	247 $_{18}$	1129	6982
2,5-Dimethoxy-4-iodophenethylamine-M (deamino-HOOC-O-demethyl) -H2O	UGLUC	2080	290 $_{100}$	262 $_{42}$	234 $_{42}$	191 $_{11}$	127 $_{16}$	839	6965
2,5-Dimethoxy-4-iodophenethylamine-M (deamino-HOOC-O-demethyl-) ME	UGLUCMETFA	2160	322 $_{62}$	290 $_{87}$	262 $_{100}$	234 $_{21}$	191 $_6$	1042	6984
2,5-Dimethoxy-4-iodophenethylamine-M (deamino-HOOC-O-demethyl-) MEAC	UGLUCMEAC	2170	364 $_{26}$	322 $_{100}$	290 $_{53}$	262 $_{44}$	234 $_{10}$	1281	6981
2,5-Dimethoxy-4-iodophenethylamine-M (deamino-HOOC-O-demethyl-) METFA	UGLUCMETFA	1980	418 $_{100}$	404 $_{25}$	359 $_{32}$	305 $_{49}$	289 $_{53}$	1501	6983
2,5-Dimethoxy-4-iodophenethylamine-M (deamino-HO-O-demethyl-) isomer-1 2AC	U+UHYAC	2240	378 $_2$	336 $_{16}$	276 $_{100}$	261 $_{10}$	134 $_{16}$	1346	6970
2,5-Dimethoxy-4-iodophenethylamine-M (deamino-HO-O-demethyl-) isomer-1 2TFA	UGLUCTFA	1865	486 $_{100}$	372 $_{98}$	303 $_{28}$	261 $_9$		1642	6979
2,5-Dimethoxy-4-iodophenethylamine-M (deamino-HO-O-demethyl-) isomer-2 2AC	UGLUCAC	2275	378 $_8$	336 $_{33}$	276 $_{100}$	261 $_{34}$	150 $_{24}$	1345	6971
2,5-Dimethoxy-4-iodophenethylamine-M (deamino-HO-O-demethyl-) isomer-2 2TFA	UGLUCTFA	1890	486 $_{35}$	372 $_{100}$	275 $_{47}$	261 $_{25}$	245 $_{22}$	1642	6980
2,5-Dimethoxy-4-iodophenethylamine-M (deamino-oxo-)		1965*	306 $_{100}$	277 $_{82}$	263 $_{18}$	247 $_{29}$	232 $_{10}$	941	7233
2,5-Dimethoxy-4-iodophenethylamine-M (O-demethyl- N-acetyl-) isomer-2	UGLUC	2520	335 $_{26}$	276 $_{100}$	261 $_{18}$	220 $_3$	121 $_7$	1122	6964
2,5-Dimethoxy-4-iodophenethylamine-M (O-demethyl- N-acetyl-) TFA	UGLUCTFA	2270	431 $_{10}$	389 $_{100}$	276 $_{94}$	263 $_{43}$	148 $_7$	1536	6974
2,5-Dimethoxy-4-iodophenethylamine-M (O-demethyl-) isomer-1 2TFA	UGLUCTFA	1970	485 $_{17}$	372 $_{100}$	359 $_{14}$	303 $_{11}$	234 $_9$	1641	6972
2,5-Dimethoxy-4-iodophenethylamine-M (O-demethyl-) isomer-1 TFA	UGLUCTFA	2100	389 $_{39}$	276 $_{100}$	263 $_{35}$	233 $_{19}$	134 $_{15}$	1396	6976
2,5-Dimethoxy-4-iodophenethylamine-M (O-demethyl-) isomer-2 2TFA	UGLUCTFA	2010	485 $_6$	372 $_{100}$	359 $_{23}$	275 $_{11}$	126 $_{10}$	1640	6973
2,5-Dimethoxy-4-iodophenethylamine-M (O-demethyl-) isomer-2 TFA	UGLUCTFA	2275	389 $_{62}$	276 $_{100}$	263 $_{58}$	261 $_{18}$		1396	6977
2,5-Dimethoxy-4-nitro-phenethylamine		2030	226 $_7$	197 $_{100}$	180 $_{15}$	167 $_{28}$	149 $_7$	494	9177
2,5-Dimethoxy-4-nitro-phenethylamine 2TMS		2330	355 $_4$	174 $_{100}$	100 $_9$	86 $_{23}$	73 $_{52}$	1312	9157
2,5-Dimethoxy-4-nitro-phenethylamine AC		2300	268 $_{17}$	209 $_{100}$	197 $_{15}$	180 $_{10}$	167 $_{18}$	707	9156
2,5-Dimethoxy-4-nitro-phenethylamine artifact (-CH3N)		1710	209 $_{100}$	162 $_9$	148 $_{27}$	133 $_{26}$	118 $_{32}$	427	9155
2,5-Dimethoxy-4-nitro-phenethylamine formyl artifact		2030	238 $_8$	221 $_4$	207 $_{100}$	160 $_5$	91 $_{13}$	547	9154
2,5-Dimethoxy-4-nitro-phenethylamine HFB		2100	422 $_{10}$	226 $_9$	209 $_{100}$	196 $_6$	148 $_{13}$	1512	9160
2,5-Dimethoxy-4-nitro-phenethylamine PFP		2050	372 $_{17}$	209 $_{100}$	196 $_6$	176 $_{12}$	148 $_{14}$	1322	9159
2,5-Dimethoxy-4-nitro-phenethylamine TFA		2070	322 $_{25}$	209 $_{100}$	196 $_6$	148 $_{13}$	136 $_{15}$	1043	9158
2,5-Dimethoxy-4-nitro-phenethylamine TMS		2085	283 $_2$	269 $_{14}$	197 $_2$	102 $_{100}$	73 $_{46}$	896	9164
2,5-Dimethoxybenzaldehyde		1615*	166 $_{100}$	151 $_{30}$	123 $_{22}$	95 $_{36}$	63 $_{41}$	317	7705
Dimethoxyethane		<1000	89 $_1$	75 $_{38}$	59 $_{100}$	31 $_{73}$	29 $_{100}$	256	3778
3,4-Dimethoxyhydrocinnamic acid ME	UME	1705*	224 $_{38}$	164 $_{15}$	151 $_{100}$	121 $_9$	107 $_{10}$	487	4943
2,5-Dimethoxyphenethylamine 2TMS		1970	310 $_4$	194 $_7$	174 $_{100}$	100 $_9$	86 $_{21}$	1064	9161

2,5-Dimethoxyphenethylamine HFB

Table 8.1: Compounds in order of names

Name	Detected	RI	Typical ions and intensities					Page	Entry
2,5-Dimethoxyphenethylamine HFB		1700	377_{27}	164_{100}	151_{50}	121_{61}	91_{23}	1342	9167
2,5-Dimethoxyphenethylamine PFP		1660	327_{40}	164_{100}	151_{50}	121_{72}	91_{29}	1073	9166
2,5-Dimethoxyphenethylamine TFA		1670	277_{48}	210_{5}	164_{100}	151_{55}	121_{82}	758	9165
2,5-Dimethoxyphenethylamine TMS		1665	253_{3}	223_{5}	194_{8}	152_{7}	102_{100}	624	9163
2,5-Dimethoxyphenethylamine-M (O-demethyl- N-acetyl-)	UGLUCTFA	2270*	209_{20}	150_{100}	135_{31}	107_{25}	77_{12}	428	6975
3,4-Dimethoxyphenethylamine		1530	181_{9}	152_{100}	137_{21}	107_{17}	91_{8}	349	7350
3,4-Dimethoxyphenethylamine 2AC		1995	265_{5}	164_{100}	151_{62}	107_{13}	91_{7}	688	7353
3,4-Dimethoxyphenethylamine 2TMS		1945	325_{1}	310_{4}	174_{100}	86_{34}	73_{62}	1064	7358
3,4-Dimethoxyphenethylamine AC		1900	223_{6}	164_{100}	151_{57}	107_{10}	91_{9}	484	7352
3,4-Dimethoxyphenethylamine formyl artifact		1510	193_{17}	164_{3}	151_{100}	107_{11}	91_{8}	379	7351
3,4-Dimethoxyphenethylamine HFB		1665	377_{13}	164_{75}	151_{100}	107_{15}	91_{11}	1342	7356
3,4-Dimethoxyphenethylamine PFP		1630	327_{22}	164_{44}	151_{100}	107_{7}	91_{5}	1073	7355
3,4-Dimethoxyphenethylamine TFA		1645	277_{34}	164_{64}	151_{100}	107_{21}	91_{9}	759	7354
3,4-Dimethoxyphenethylamine TMS		1650	253_{1}	238_{7}	151_{11}	102_{100}	73_{57}	624	7357
Dimethoxyphthalic acid 2ME		1870*	254_{37}	223_{100}	207_{16}	191_{62}	77_{30}	627	5152
1,2-Dimethyl-3-phenyl-aziridine		1145	147_{1}	121_{2}	105_{2}	77_{5}	58_{100}	286	7526
N,N-Dimethyl-4-aminophenol	UHY	1220	137_{68}	136_{100}	121_{23}	94_{9}	65_{14}	277	3415
N,N-Dimethyl-4-aminophenol AC	UHYAC	1370	179_{20}	137_{96}	136_{100}	121_{7}	65_{13}	341	3416
N,N-Dimethyl-4-aminophenol-M	UHY	1240	109_{100}	80_{41}	53_{82}	53_{82}	52_{90}	260	826
N,N-Dimethyl-4-aminophenol-M (nor-) 2AC	UHYAC	1615	207_{9}	193_{3}	165_{40}	123_{100}	94_{10}	419	3417
N,N-Dimethyl-4-aminophenol-M 2AC	PAC U+UHYAC	1765	193_{10}	151_{53}	109_{100}	80_{24}		377	188
N,N-Dimethyl-5-methoxy-tryptamine	G U UHY UHYAC	2040	218_{16}	160_{10}	145_{7}	117_{10}	58_{100}	460	4059
N,N-Dimethyl-5-methoxy-tryptamine-M (HO-)	U UHY	2335	234_{16}	175_{4}	163_{6}	72_{12}	58_{100}	529	4060
N,N-Dimethyl-5-methoxy-tryptamine-M (O-demethyl-HO-) 2AC	UHYAC	2400	304_{3}	234_{5}	175_{10}	149_{12}	58_{100}	932	4061
Dimethylamine		<1000	45_{50}	44_{100}	28_{70}			248	3618
4,4'-Dimethylaminorex (cis)		1680	190_{10}	175_{13}	146_{20}	91_{24}	70_{100}	370	9244
4,4'-Dimethylaminorex (cis) 2TMS		1850	334_{3}	205_{40}	171_{88}	130_{69}	73_{100}	1121	9247
4,4'-Dimethylaminorex (cis) AC		1980	232_{3}	217_{24}	174_{28}	112_{76}	70_{100}	517	9246
4,4'-Dimethylaminorex (cis) HFB		2010	266_{4}	217_{84}	174_{100}	131_{36}	70_{62}	1383	9250
4,4'-Dimethylaminorex (cis) ME		1560	204_{4}	190_{33}	146_{18}	85_{19}	70_{100}	411	9245
4,4'-Dimethylaminorex (cis) PFP		1990	336_{1}	217_{100}	174_{99}	121_{46}	70_{61}	1129	9249
4,4'-Dimethylaminorex (cis) TFA		2010	286_{1}	217_{100}	174_{83}	91_{62}	70_{67}	814	9248
4,4'-Dimethylaminorex (trans)		1660	190_{11}	175_{15}	146_{23}	91_{26}	70_{100}	370	9237
4,4'-Dimethylaminorex (trans) 2TMS		1770	334_{8}	193_{45}	171_{60}	141_{47}	73_{100}	1120	9240
4,4'-Dimethylaminorex (trans) AC		1970	232_{4}	217_{22}	174_{27}	112_{53}	70_{100}	517	9239
4,4'-Dimethylaminorex (trans) HFB		1990	217_{81}	174_{100}	156_{24}	131_{37}	70_{60}	1383	9243
4,4'-Dimethylaminorex (trans) ME		1540	204_{5}	190_{43}	146_{24}	85_{22}	70_{100}	411	9238
4,4'-Dimethylaminorex (trans) PFP		1970	336_{1}	217_{92}	174_{100}	121_{51}	70_{69}	1129	9242
4,4'-Dimethylaminorex (trans) TFA		1980	286_{1}	217_{100}	174_{89}	91_{70}	70_{80}	814	9241
2,6-Dimethylaniline		1180	121_{100}	106_{77}				266	725
2,6-Dimethylaniline AC	U+UHYAC	1470	163_{33}	121_{100}	106_{66}	91_{17}	77_{30}	308	57
Dimethylbromophenol		1470*	200_{82}	121_{100}	91_{61}	77_{57}		398	1424
2,2-Dimethylbutane		<1000*	71_{45}	57_{61}	43_{100}	41_{71}		255	3815
1,3-Dimethylcyclopentane		<1000*	98_{9}	83_{17}	70_{82}	56_{100}	41_{95}	257	3821
1,2-Dimethylcyclopropane		<1000*	70_{29}	55_{100}	42_{73}	39_{73}	29_{53}	250	3813
Dimethylformamide		<1000	73_{81}	58_{7}	44_{100}	42_{60}	28_{53}	251	3781
1,5-Dimethylnaphthalene		1340*	156_{100}	153_{34}	141_{63}	128_{10}	115_{10}	300	2557
1,3-Dimethylpentylamine		<1000	115_{2}	100_{9}	69_{8}	56_{13}	44_{100}	263	8622
1,3-Dimethylpentylamine		<1000	115_{12}	100_{67}	83_{14}	69_{51}	57_{100}	263	8623
2,6-Dimethylphenol		1155*	122_{100}	107_{90}				266	726
2,6-Dimethylphenol AC		1130*	164_{13}	122_{100}	107_{48}	91_{12}	77_{17}	311	857
Dimethylphenylthiazolanimin		1760	206_{42}	191_{8}	132_{15}	118_{27}	58_{100}	416	1426
Dimethylphthalate		1450*	194_{7}	163_{100}	133_{9}	104_{7}	77_{15}	382	4948
Dimethylsulfoxide		<1000*	78_{72}	63_{100}	45_{24}			253	1469
N,N-Dimethyl-tryptamine		1870	188_{3}	143_{2}	130_{8}	115_{3}	58_{100}	365	8873
N,N-Dimethyl-tryptamine HFB		1685	340_{2}	326_{10}	169_{11}	115_{12}	58_{100}	1375	9548
N,N-Dimethyl-tryptamine TFA		1675	240_{3}	226_{6}	129_{14}	115_{10}	58_{100}	802	9547
N,N-Dimethyl-tryptamine TMS		1810	260_{1}	202_{6}	200_{5}	186_{4}	58_{100}	660	9549
Dimetindene		2290	292_{3}	218_{3}	58_{100}			858	727
Dimetindene-M (nor-) AC	U+UHYAC	2775	320_{8}	218_{6}	100_{100}	86_{19}	58_{37}	1035	1331
Dimetindene-M (nor-HO-) 2AC	U+UHYAC	3090	378_{12}	276_{3}	234_{6}	100_{100}		1348	1332
Dimetotiazine	G U UHY UHYAC	3060	391_{1}	320_{2}	276	179	72_{100}	1406	1937
Dimetotiazine-M (bis-nor-) AC	UHYAC	3380	405_{48}	346_{39}	319_{100}	210_{29}	100_{18}	1462	1644
Dimetotiazine-M (HO-) AC	UHYAC	3200	449_{3}	398_{5}	245_{18}	198_{8}	72_{100}	1578	1645
Dimetotiazine-M (nor-)	U UHY	3150	377_{3}	320_{25}	306_{23}	198_{15}	72_{100}	1342	1642
Dimetotiazine-M (nor-) AC	UHYAC	3360	419_{41}	346_{55}	319_{75}	114_{92}	58_{100}	1506	1643
Dimpylate	P G	1760	304_{27}	199_{50}	179_{81}	152_{64}	137_{100}	931	2784
Dimpylate artifact-1		1140	166_{41}	151_{100}	138_{50}	109_{42}	93_{39}	318	1399
Dimpylate artifact-2	P-I U	1400*	198_{26}	170_{38}	138_{95}	111_{100}	81_{80}	393	1442
Dimpylate artifact-3		1685	152_{49}	137_{100}	124_{18}	109_{19}	84_{38}	294	1375
2,4-Dinitrophenol		1520	184_{100}	154_{28}	107_{41}	91_{45}	63_{88}	356	728
Dinobuton		2060	267_{4}	240_{12}	211_{100}	163_{26}	147_{19}	1068	3516
Dinocap		2460	364_{1}	197	130	103_{1}	69_{100}	1282	3828
Dinoseb		1780	240_{16}	211_{100}	163_{48}	147_{35}	117_{33}	554	3640

Table 8.1: Compounds in order of names   Dinoterb

Name	Detected	RI	Typical ions and intensities					Page	Entry
Dinoterb		1760	240_{12}	225_{100}	177_{40}	131_{22}	77_{22}	555	3641
Dioctylphthalate		2655*	390_{1}	279_{12}	261_{2}	167_{2}	149_{100}	1404	6401
Dioctylsebacate	U UHY UHYAC	2705*	426_{1}	315_{2}	297_{4}	185_{100}	112_{19}	1524	5408
Diosgenin		3150*	414_{9}	342_{21}	300_{27}	282_{58}	139_{100}	1492	8751
Dioxacarb		1825	193_{1}	166_{63}	149_{22}	121_{100}	73_{35}	482	3914
Dioxacarb -C2H3NO	U	1325*	166_{30}	149_{3}	121_{100}	104_{15}	73_{28}	315	729
Dioxane		<1000*	88_{100}	58_{89}	43_{33}	31_{44}		256	730
Dioxathion		1705	270_{11}	197_{6}	125_{39}	97_{100}	73_{37}	1593	3831
Dioxethedrine 4AC		2090	319_{8}	150_{8}	114_{52}	72_{100}	70_{26}	1351	1795
Dioxethedrine -H2O 2AC		1950	277_{26}	235_{8}	193_{20}	114_{12}	70_{100}	760	1792
Dioxethedrine -H2O 3AC		2075	319_{37}	277_{52}	235_{49}	193_{32}	70_{100}	1025	1794
Dioxethedrine ME3AC		2060	351_{1}	222_{2}	153_{8}	114_{70}	72_{100}	1214	1793
Diphenhydramine	P G U	1870	227_{1}	165_{16}	152_{6}	73_{29}	58_{100}	634	731
Diphenhydramine HY	UHY	1645*	184_{45}	165_{14}	152_{7}	105_{100}	77_{63}	357	1333
Diphenhydramine HYAC	U+UHYAC	1700*	226_{20}	184_{20}	165_{100}	105_{14}	77_{35}	495	1241
Diphenhydramine HYHFB		1475*	380_{3}	183_{4}	166_{100}	152_{22}	83_{25}	1355	8146
Diphenhydramine HYME	UHY	1655*	198_{70}	167_{94}	121_{100}	105_{56}	77_{71}	395	6779
Diphenhydramine HYPFP		1410*	330_{4}	183_{4}	166_{100}	152_{24}	83_{30}	1093	8145
Diphenhydramine HYTFA		1420*	280_{19}	183_{7}	166_{100}	152_{24}	83_{21}	780	8144
Diphenhydramine HYTMS		1540*	256_{28}	241_{14}	179_{38}	167_{100}	152_{17}	638	8159
Diphenhydramine-M (benzophenone)	U+UHYAC	1610*	182_{31}	152_{3}	105_{100}	77_{70}	51_{39}	351	1624
Diphenhydramine-M (bis-nor-) AC	U+UHYAC	2240	183_{30}	167_{100}	87_{45}	72_{26}		712	2080
Diphenhydramine-M (deamino-HO-)	P U	1760*	228_{9}	183_{46}	167_{100}	152_{26}	105_{27}	503	2049
Diphenhydramine-M (deamino-HO-) AC	U+UHYAC	1820*	270_{1}	183_{98}	167_{100}	152_{31}	87_{67}	719	2079
Diphenhydramine-M (di-HO-)	U	1895*	244_{61}	213_{100}	167_{78}			571	733
Diphenhydramine-M (HO-)	P U	1890	213_{11}	183_{25}	167_{19}	58_{100}		726	734
Diphenhydramine-M (HO-) HY2AC	U+UHYAC	2090*	284_{5}	242_{15}	224_{17}	182_{100}	153_{19}	800	2425
Diphenhydramine-M (HO-BPH) isomer-1	UHY	2065*	198_{93}	121_{72}	105_{100}	93_{22}	77_{66}	393	1627
Diphenhydramine-M (HO-BPH) isomer-1 AC	U+UHYAC	2010*	240_{27}	198_{100}	121_{47}	105_{85}	77_{80}	555	2196
Diphenhydramine-M (HO-BPH) isomer-2	P-I U UHY	2080*	198_{50}	121_{100}	105_{17}	93_{14}	77_{28}	394	732
Diphenhydramine-M (HO-methoxy-BPH)	UHY	2050*	228_{46}	197_{6}	151_{100}	105_{22}	77_{41}	501	1625
Diphenhydramine-M (HO-methoxy-BPH) AC	U+UHYAC	2100*	270_{3}	228_{92}	151_{100}	105_{25}	77_{40}	716	1622
Diphenhydramine-M (methoxy-)	U	2010	285_{1}	183_{4}	165_{3}	73_{12}	58_{100}	811	2078
Diphenhydramine-M (methoxy-) HY	UHY	1875*	214_{41}	183_{100}	137_{80}	121_{39}	105_{30}	446	4483
Diphenhydramine-M (methoxy-) HYAC	U+UHYAC	1780*	256_{11}	214_{58}	183_{100}	105_{42}	77_{47}	637	2077
Diphenhydramine-M (nor-)	P U	1520	167_{100}	165_{41}	152_{21}			561	2047
Diphenhydramine-M (nor-) AC	P U+UHYAC	2265	241_{1}	167_{100}	152_{30}	101_{85}	86_{27}	797	735
Diphenhydramine-M i-2 AC	U+UHYAC	2050*	240_{20}	198_{100}	121_{94}	105_{41}	77_{51}	555	2197
Diphenhydramine-M/artifact	UHY	2070*	228_{8}	186_{100}	157_{10}	128_{7}	77_{4}	499	1626
Diphenhydramine-M/artifact	U UHY UHYAC	1850*	209_{4}	167_{100}	152_{22}	121_{8}	115_{7}	426	2081
Diphenhydramine-M/artifact AC	U+UHYAC	2200*	270_{7}	228_{17}	186_{100}	157_{10}	128_{7}	714	1623
Diphenidine		2030	264_{1}	174_{100}	165_{4}	103_{4}	91_{25}	691	9118
Diphenidine artifact/impurity (dehydro-)		2050	263_{3}	172_{100}	144_{5}	104_{17}	91_{16}	677	9119
Diphenidine-M (bis-HO-piperidine) 2AC		2970	290_{100}	230_{8}	170_{6}	107_{5}	91_{5}	1363	9299
Diphenidine-M (bis-nor-) AC		2020	180_{3}	148_{34}	106_{100}	91_{16}	79_{17}	553	8425
Diphenidine-M (bis-nor-) HFB		1760	302_{100}	180_{28}	107_{29}	91_{34}	79_{44}	1413	8429
Diphenidine-M (bis-nor-) PFP		1730	252_{100}	180_{35}	165_{7}	91_{24}	79_{24}	1173	8428
Diphenidine-M (bis-nor-) TFA		1740	202_{100}	180_{27}	107_{25}	91_{24}	79_{33}	860	8427
Diphenidine-M (bis-nor-) TMS		1760	254_{4}	178_{100}	162_{15}	91_{19}	73_{29}	713	8426
Diphenidine-M (bis-nor-HO-benzyl-) 2AC	U+UHYAC	2380	238_{13}	196_{14}	148_{57}	106_{100}	79_{16}	890	8654
Diphenidine-M (HO-benzyl-) AC		2540	322_{1}	174_{100}	152_{3}	107_{12}	91_{22}	1054	9294
Diphenidine-M (HO-methoxy-benzyl-)		2500	174_{100}	153_{2}	137_{9}	122_{3}	91_{13}	978	9295
Diphenidine-M (HO-methoxy-benzyl-HO-piperidine) isomer-1 2AC		2920	232_{100}	179_{2}	172_{15}	137_{9}	91_{11}	1484	9302
Diphenidine-M (HO-methoxy-benzyl-HO-piperidine) isomer-2 2AC		2930	232_{100}	172_{18}	137_{14}	118_{11}	91_{21}	1484	9303
Diphenidine-M (HO-phenyl-) AC		2550	324_{11}	232_{77}	190_{100}	107_{10}	91_{20}	1053	9290
Diphenidine-M (HO-phenyl-HO-piperidine) isomer-1 2AC		2820	290_{100}	248_{66}	188_{82}			1363	9300
Diphenidine-M (HO-phenyl-HO-piperidine) isomer-2 2AC		3120	290_{100}	248_{79}	188_{15}	107_{28}		1363	9301
Diphenidine-M (HO-piperidine) isomer-1 AC		2510	266_{1}	232_{100}	172_{22}	118_{8}	91_{43}	1053	9291
Diphenidine-M (HO-piperidine) isomer-2 AC		2780	232_{100}	172_{14}	118_{6}	91_{8}		1054	9292
Diphenidine-M (HO-piperidine) isomer-3 AC		2830	322_{1}	232_{100}	197_{3}	118_{8}	91_{12}	1054	9293
Diphenidine-M (oxo-)		2490	262_{4}	188_{100}	178_{13}	160_{12}	91_{59}	777	9296
Diphenidine-M (oxo-HO-benzyl-) AC		2810	290_{7}	238_{21}	196_{32}	188_{100}	160_{14}	1138	9298
Diphenidine-M (oxo-HO-phenyl-) AC		2825	337_{3}	246_{88}	204_{100}	151_{17}		1139	9297
Diphenoxylate		3415	452_{8}	377_{21}	246_{100}	193_{15}	165_{18}	1586	236
1,1-Diphenyl-1-butene	UHYAC	1900*	208_{100}	193_{68}	178_{31}	165_{33}	130_{49}	425	5294
Diphenylamine		1595	169_{100}	168_{74}	141_{4}	84_{24}	77_{15}	322	3434
1-(1,2-Diphenylethyl)piperidine		2030	264_{1}	174_{100}	165_{4}	103_{4}	91_{25}	691	9118
1-(1,2-Diphenylethyl)piperidine artifact/impurity (dehydro-)		2050	263_{3}	172_{100}	144_{5}	104_{17}	91_{16}	677	9119
1-(1,2-Diphenylethyl)piperidine-M (bis-HO-piperidine) 2AC		2970	290_{100}	230_{8}	170_{6}	107_{5}	91_{5}	1363	9299
1-(1,2-Diphenylethyl)piperidine-M (HO-benzyl-) AC		2540	322_{1}	174_{100}	152_{3}	107_{12}	91_{22}	1054	9294
1-(1,2-Diphenylethyl)piperidine-M (HO-methoxy-benzyl-)		2500	174_{100}	153_{2}	137_{9}	122_{3}	91_{13}	978	9295
1-(1,2-Diphenylethyl)piperidine-M (HO-methoxy-benzyl-HO-piperidine) isomer-1 2AC		2920	232_{100}	179_{2}	172_{15}	137_{9}	91_{11}	1484	9302
1-(1,2-Diphenylethyl)piperidine-M (HO-methoxy-benzyl-HO-piperidine) isomer-2 2AC		2930	232_{100}	172_{18}	137_{14}	118_{11}	91_{21}	1484	9303
1-(1,2-Diphenylethyl)piperidine-M (HO-phenyl-) AC		2550	324_{11}	232_{77}	190_{100}	107_{10}	91_{20}	1053	9290

1-(1,2-Diphenylethyl)piperidine-M isomer-1 2AC

Table 8.1: Compounds in order of names

Name	Detected	RI	Typical ions and intensities					Page	Entry
1-(1,2-Diphenylethyl)piperidine-M (HO-phenyl-HO-piperidine) isomer-1 2AC		2820	290_{100}	248_{66}	188_{82}			1363	9300
1-(1,2-Diphenylethyl)piperidine-M (HO-phenyl-HO-piperidine) isomer-2 2AC		3120	290_{100}	248_{79}	188_{15}	107_{28}		1363	9301
1-(1,2-Diphenylethyl)piperidine-M (HO-piperidine) isomer-1 AC		2510	266_1	232_{100}	172_{22}	118_8	91_{43}	1053	9291
1-(1,2-Diphenylethyl)piperidine-M (HO-piperidine) isomer-2 AC		2780	232_{100}	172_{14}	118_6	91_8		1054	9292
1-(1,2-Diphenylethyl)piperidine-M (HO-piperidine) isomer-3 AC		2830	322_1	232_{100}	197_3	118_8	91_{12}	1054	9293
1-(1,2-Diphenylethyl)piperidine-M (oxo-)		2490	262_4	188_{100}	178_{13}	160_{12}	91_{59}	777	9296
1-(1,2-Diphenylethyl)piperidine-M (oxo-HO-benzyl-) AC		2810	290_7	238_{21}	196_{32}	188_{100}	160_{14}	1138	9298
1-(1,2-Diphenylethyl)piperidine-M (oxo-HO-phenyl-) AC		2825	337_3	246_{88}	204_{100}	151_{17}		1139	9297
1-(1,2-Diphenylethyl)pyrolidine		1880	250_1	178_4	160_{100}	103_6	91_{35}	613	10181
1,2-Diphenylethylamine-M (deamino-HO-bis-HO-benzyl-) 3AC		2350*	355_2	296_9	208_{27}	166_{100}	123_{93}	1242	8987
1,2-Diphenylethylamine-M (deamino-HO-phenyl-) AC		2230*	238_{28}	196_{31}	165_{30}	150_{27}	107_{100}	895	8932
1,2-Diphenylethylamine-M (deamino-oxo-bis-HO-benzyl-) 2AC		2345*	207_{44}	165_{80}	123_{100}	105_{81}	77_{45}	981	8933
1,2-Diphenylethylamine-M (deamino-oxo-bis-HO-benzyl-) AC		2360*	270_7	228_{19}	153_{21}	123_{15}	105_{100}	716	8986
1,2-Diphenylethylamine-M (deamino-oxo-HO-benzyl-) isomer-1 AC		2080*	254_4	212_4	105_{100}	77_{34}		628	8934
1,2-Diphenylethylamine-M (deamino-oxo-HO-benzyl-) isomer-2 AC		2120*	254_2	212_{12}	105_{100}	77_{31}		628	8935
1,2-Diphenylethylamine-M (deamino-oxo-HO-methoxy-benzyl-) AC		2250*	284_1	242_{22}	226_{27}	137_{100}	105_{84}	800	8985
2,2-Diphenylethylamine 2TMS		1950	341_1	326_3	174_{100}	86_{15}	73_{20}	1165	7625
2,2-Diphenylethylamine formyl artifact		1510	209_1	178_3	167_{100}	152_{18}	105_{12}	430	7623
2,2-Diphenylethylamine HFB		1720	226_3	180_{62}	167_{100}	165_{52}	152_{23}	1413	7626
2,2-Diphenylethylamine PFP		1650	224_1	180_{60}	167_{100}	165_{49}	152_{14}	1173	7627
2,2-Diphenylethylamine TFA		1665	224_1	180_{54}	167_{100}	165_{59}	152_{28}	860	7628
2,2-Diphenylethylamine TMS		1650	269_1	254_{13}	165_{26}	102_{100}	73_{77}	713	7624
Diphenylethylamine		1670	178_2	118_2	106_{100}	91_{11}	79_{25}	392	8423
Diphenylethylamine AC		2020	180_3	148_{34}	106_{100}	91_{16}	79_{17}	553	8425
Diphenylethylamine formyl artifact		1660	209_1	181_5	165_7	118_{100}	91_{66}	430	8424
Diphenylethylamine HFB		1760	302_{100}	180_{29}	107_{29}	91_{34}	79_{44}	1413	8429
Diphenylethylamine PFP		1730	252_{100}	180_{35}	165_7	91_{24}	79_{24}	1173	8428
Diphenylethylamine TFA		1740	202_{100}	180_{27}	107_{25}	91_{24}	79_{33}	860	8427
Diphenylethylamine TMS		1760	254_4	178_{100}	162_{15}	91_{19}	73_{29}	713	8426
1,1-Diphenylethylpiperidine		1980	264_1	178_3	165_{32}	152_{10}	98_{100}	691	10182
1,1-Diphenylethylpyrrolidine		1915	250_1	178_4	165_{37}	152_{11}	84_{100}	614	10183
Diphenyloctylamine		2330	281_{11}	210_{100}	194_7	180_4	92_9	790	5145
Diphenylprolinol		2120	181_7	165_5	105_{22}	77_{31}	70_{100}	624	7804
Diphenylprolinol 2TMS		2160	382_1	255_4	239_{12}	142_{100}	73_{66}	1436	7814
Diphenylprolinol AC		2405	181_4	165_3	113_{44}	77_{27}	70_{100}	879	7805
Diphenylprolinol -H2O		2095	206_2	165_{14}	105_{25}	83_{100}	55_{46}	534	7803
Diphenylprolinol -H2O AC		2265	277_{85}	234_{100}	167_{52}	165_{55}	152_{20}	761	7809
Diphenylprolinol -H2O HFB		2065	431_{100}	354_{10}	262_{13}	234_{57}	206_{82}	1537	7812
Diphenylprolinol -H2O PFP		2050	381_{100}	262_{17}	234_{52}	206_{80}	119_{57}	1360	7810
Diphenylprolinol -H2O TFA		2075	331_{100}	262_{34}	234_{41}	206_{86}	69_{83}	1100	7808
Diphenylprolinol HFB		2185	266_{36}	239_{14}	183_{100}	105_{83}	77_{54}	1578	7813
Diphenylprolinol ME		2070	181_8	165_5	152_6	105_{19}	84_{100}	702	7806
Diphenylprolinol PFP		2160	216_{49}	183_{100}	119_{60}	105_{89}	77_{69}	1441	7811
Diphenylprolinol TFA		2185	183_{77}	166_{31}	139_{17}	105_{100}	77_{81}	1203	7807
Diphenylprolinol-M (di-HO-) 3AC		2560	279_{32}	241_{83}	237_{25}	199_{100}		1482	8683
Diphenylprolinol-M (HO-BPH) isomer-1	UHY	2065*	198_{93}	121_{72}	105_{100}	93_{22}	77_{66}	393	1627
Diphenylprolinol-M (HO-BPH) isomer-1 AC	U+UHYAC	2010*	240_{27}	198_{100}	121_{47}	105_{85}	77_{80}	555	2196
Diphenylprolinol-M (HO-BPH) isomer-2	P-I U UHY	2080*	198_{50}	121_{100}	105_{17}	93_{14}	77_{28}	394	732
Diphenylprolinol-M (HO-phenyl-) -H2O		2350	251_{100}	222_{19}	183_{32}	165_{29}	152_{14}	611	8680
Diphenylprolinol-M (HO-phenyl-) isomer-1 2AC		2710	241_1	199_{10}	121_{16}	113_{100}	70_{92}	1228	8681
Diphenylprolinol-M (HO-phenyl-) isomer-2 2AC		2740	241_1	199_9	121_{13}	113_{100}	70_{83}	1228	8682
Diphenylprolinol-M (HO-phenyl-oxo-)		2140	198_3	121_2	105_7	84_{100}		795	8700
Diphenylprolinol-M (HO-phenyl-oxo-) AC		2760	241_{62}	199_{100}	121_{42}	105_{31}	84_{35}	1062	8684
Diphenylprolinol-M (HO-pyrrolidinyl-) 2AC		2355	293_{10}	250_5	183_{82}	105_{55}	85_{100}	1227	8685
Diphenylprolinol-M (HO-pyrrolidinyl-) -H2O isomer-1		2430	183_{100}	165_5	105_{68}	85_{36}		611	8678
Diphenylprolinol-M (HO-pyrrolidinyl-) -H2O isomer-2		2460	183_{45}	105_{100}	85_{70}	77_{55}		610	8679
Diphenylprolinol-M (oxo-)		2490	267_1	183_{100}	165_5	77_{40}		700	8701
Diphenylprolinol-M i-2 AC	U+UHYAC	2050*	240_{20}	198_{100}	121_{94}	105_{41}	77_{51}	555	2197
Diphenylprolinol-M/artif. (benzophenone)	U+UHYAC	1610*	182_{31}	152_3	105_{100}	77_{70}	51_{39}	351	1624
Diphenylpyraline	G U+UHYAC	2115	281_1	167_{30}	114_{41}	99_{100}	70_{19}	789	737
Diphenylpyraline HY	UHY	1645*	184_{45}	165_{14}	152_7	105_{100}	77_{63}	357	1333
Diphenylpyraline HYAC	U+UHYAC	1700*	226_{20}	184_{20}	165_{100}	105_{14}	77_{35}	495	1241
Diphenylpyraline HYHFB		1475*	380_3	183_4	166_{100}	152_{22}	83_{25}	1355	8146
Diphenylpyraline HYME	UHY	1655*	198_{70}	167_{94}	121_{100}	105_{56}	77_{71}	395	6779
Diphenylpyraline HYPFP		1410*	330_4	183_4	166_{100}	152_{22}	83_{30}	1093	8145
Diphenylpyraline HYTFA		1420*	280_{19}	183_7	166_{100}	152_{24}	83_{21}	780	8144
Diphenylpyraline HYTMS		1540*	256_{28}	241_{14}	179_{38}	167_{100}	152_{17}	638	8159
Diphenylpyraline-M (benzophenone)	U+UHYAC	1610*	182_{31}	152_3	105_{100}	77_{70}	51_{39}	351	1624
Diphenylpyraline-M (HO-BPH) isomer-1	UHY	2065*	198_{93}	121_{72}	105_{100}	93_{22}	77_{66}	393	1627
Diphenylpyraline-M (HO-BPH) isomer-1 AC	U+UHYAC	2010*	240_{27}	198_{100}	121_{47}	105_{85}	77_{80}	555	2196
Diphenylpyraline-M (HO-BPH) isomer-2	P-I U UHY	2080*	198_{50}	121_{100}	105_{17}	93_{14}	77_{28}	394	732
Diphenylpyraline-M (HO-methoxy-BPH)	UHY	2050*	228_{46}	197_6	151_{100}	105_{22}	77_{41}	501	1625
Diphenylpyraline-M (HO-methoxy-BPH) AC	U+UHYAC	2100*	270_3	228_{92}	151_{100}	105_{25}	77_{40}	716	1622
Diphenylpyraline-M i-2 AC	U+UHYAC	2050*	240_{20}	198_{100}	121_{94}	105_{41}	77_{51}	555	2197

Table 8.1: Compounds in order of names — Diphenylpyraline-M/artifact

Name	Detected	RI	Typical ions and intensities	Page	Entry
Diphenylpyraline-M/artifact	UHY	2070*	228 $_8$ 186 $_{100}$ 157 $_{10}$ 128 $_7$ 77 $_4$	499	1626
Diphenylpyraline-M/artifact AC	U+UHYAC	2200*	270 $_7$ 228 $_{17}$ 186 $_{100}$ 157 $_{10}$ 128 $_7$	714	1623
Dipivefrin 2AC		2760	435 $_1$ 362 $_{10}$ 307 $_{17}$ 86 $_{41}$ 57 $_{100}$	1549	2747
Dipivefrin 2TMS		2410	495 $_1$ 480 $_1$ 116 $_{100}$ 73 $_{37}$ 57 $_{24}$	1655	6332
Dipivefrin -H2O		2505	333 $_1$ 249 $_2$ 205 $_1$ 85 $_8$ 57 $_{100}$	1115	2745
Dipivefrin -H2O AC		2720	375 $_1$ 362 $_9$ 307 $_{18}$ 115 $_{20}$ 57 $_{100}$	1336	2746
Dipivefrin TFATMS		2400	519 $_1$ 379 $_{56}$ 295 $_{41}$ 211 $_{22}$ 57 $_{100}$	1676	6333
Diprophylline 2AC	U+UHYAC	2455	338 $_{52}$ 236 $_{14}$ 194 $_{30}$ 180 $_{100}$ 159 $_{26}$	1142	1433
Dipropylbarbital	P G U UHY UHYAC	1650	170 $_{95}$ 141 $_{100}$ 98 $_{16}$	440	1428
Dipropylbarbital 2ME	UME PME	1580	198 $_{44}$ 183 $_4$ 169 $_{100}$ 140 $_5$ 112 $_{12}$	558	6406
Dipropylbarbital-M (HO-) isomer-1	U UHY	1930	213 $_3$ 171 $_{22}$ 141 $_{100}$ 112 $_{12}$ 98 $_{31}$	502	2955
Dipropylbarbital-M (HO-) isomer-1 AC	UHYAC	1950	226 $_9$ 184 $_{38}$ 168 $_{97}$ 141 $_{100}$ 101 $_{74}$	718	2957
Dipropylbarbital-M (HO-) isomer-2	U UHY	1980	210 $_1$ 186 $_{17}$ 168 $_{54}$ 141 $_{100}$ 98 $_{35}$	502	2956
Dipropylbarbital-M (HO-) isomer-2 AC	UHYAC	2000	227 $_5$ 210 $_3$ 168 $_{100}$ 141 $_{26}$ 97 $_{10}$	718	2958
Dipropylbarbital-M (oxo-)	U UHY UHYAC	1870	226 $_2$ 184 $_{24}$ 169 $_{100}$ 141 $_{39}$ 98 $_{20}$	494	2954
N,N-Dipropyl-tryptamine		2090	244 $_1$ 144 $_{18}$ 130 $_{37}$ 114 $_{100}$ 86 $_{17}$	572	8829
N,N-Dipropyl-tryptamine HFB		1915	411 $_3$ 340 $_{10}$ 326 $_8$ 143 $_9$ 114 $_{100}$	1558	10093
N,N-Dipropyl-tryptamine PFP		1905	361 $_1$ 290 $_6$ 276 $_6$ 143 $_{12}$ 114 $_{100}$	1403	10092
N,N-Dipropyl-tryptamine TFA		1925	311 $_4$ 240 $_{10}$ 226 $_{10}$ 143 $_7$ 114 $_{100}$	1157	10091
N,N-Dipropyl-tryptamine TMS		2195	316 $_3$ 216 $_{10}$ 202 $_{14}$ 143 $_{11}$ 114 $_{100}$	1007	10090
DiPT-D4		2050	248 $_1$ 148 $_{13}$ 132 $_{40}$ 116 $_{100}$ 74 $_{24}$	593	8861
DiPT-D4 HFB		1865	429 $_4$ 344 $_{15}$ 328 $_{10}$ 147 $_8$ 116 $_{100}$	1570	10118
DiPT-D4 PFP		1855	379 $_4$ 294 $_8$ 278 $_6$ 147 $_4$ 116 $_{100}$	1421	10117
DiPT-D4 TFA		1855	329 $_3$ 244 $_{16}$ 228 $_8$ 147 $_8$ 116 $_{100}$	1180	10116
DiPT-D4 TMS		2170	320 $_1$ 220 $_7$ 204 $_{13}$ 116 $_{100}$ 74 $_{21}$	1036	10115
Dipyrone	G P U	1995	215 $_{35}$ 123 $_{10}$ 91 $_{18}$ 56 $_{41}$	975	197
Dipyrone-M (bis-dealkyl-)	P U UHY	1955	203 $_{23}$ 93 $_{14}$ 84 $_{59}$ 56 $_{100}$	406	219
Dipyrone-M (bis-dealkyl-) 2AC	UHYAC	2280	287 $_8$ 245 $_{31}$ 203 $_{15}$ 84 $_{56}$ 56 $_{100}$	820	3333
Dipyrone-M (bis-dealkyl-) AC	P U U+UHYAC	2270	245 $_{30}$ 203 $_{13}$ 84 $_{50}$ 56 $_{100}$	575	183
Dipyrone-M (bis-dealkyl-) artifact	U UHY	1945	180 $_{13}$ 119 $_{100}$ 91 $_{45}$	344	424
Dipyrone-M (dealkyl-)	P U UHY	1980	217 $_{16}$ 123 $_{14}$ 98 $_7$ 83 $_{16}$ 56 $_{100}$	455	220
Dipyrone-M (dealkyl-) AC	P U+UHYAC	2395	259 $_{20}$ 217 $_8$ 123 $_9$ 56 $_{100}$	652	184
Dipyrone-M (dealkyl-) ME artifact	P G U-I	1895	231 $_{36}$ 123 $_7$ 111 $_{17}$ 97 $_{56}$ 56 $_{100}$	514	189
Disopyramide	P G U UHY UHYAC	2490	239 $_{14}$ 212 $_{65}$ 195 $_{100}$ 167 $_{42}$ 114 $_{52}$	1153	2872
Disopyramide artifact	P G U UHY UHYAC	1980	193 $_{100}$ 165 $_6$	377	330
Disopyramide -CHNO	UHY UHYAC	2030	296 $_1$ 253 $_6$ 196 $_{32}$ 169 $_{100}$ 128 $_{38}$	887	2873
Disopyramide-M (bis-dealkyl-) -NH3	UHYAC	2245	238 $_{60}$ 194 $_{23}$ 182 $_{70}$ 167 $_{100}$ 152 $_8$	548	2874
Disopyramide-M (N-dealkyl-) 2TMS		2200	426 $_1$ 336 $_2$ 284 $_{100}$ 195 $_{34}$ 73 $_{54}$	1563	7582
Disopyramide-M (N-dealkyl-) AC	UHYAC	2640	339 $_1$ 296 $_4$ 212 $_{70}$ 195 $_{100}$ 167 $_{61}$	1152	2876
Disopyramide-M (N-dealkyl-) -CHNO AC	UHYAC	2330	296 $_1$ 196 $_7$ 182 $_{13}$ 169 $_{100}$ 72 $_{11}$	887	2875
Disopyramide-M (N-dealkyl-) -H2O		2075	279 $_{51}$ 236 $_{19}$ 193 $_{72}$ 182 $_{100}$ 167 $_{64}$	778	1926
Disopyramide-M (N-dealkyl-) -H2O HFB		2075	475 $_1$ 226 $_4$ 221 $_{14}$ 207 $_{37}$ 194 $_{100}$	1628	7586
Disopyramide-M (N-dealkyl-) -H2O PFP		2080	425 $_4$ 382 $_{32}$ 306 $_{46}$ 278 $_{31}$ 193 $_{100}$	1520	7584
Disopyramide-M (N-dealkyl-) -H2O TFA		2120	375 $_1$ 332 $_{26}$ 306 $_{27}$ 278 $_{26}$ 193 $_{100}$	1335	7585
Disopyramide-M (N-dealkyl-) ME		2345	280 $_8$ 224 $_{60}$ 194 $_{62}$ 167 $_{55}$ 98 $_{100}$	979	7581
Disopyramide-M (N-dealkyl-) -NH3		2100	280 $_{76}$ 209 $_{29}$ 194 $_{100}$ 180 $_{43}$ 167 $_{58}$	783	1925
Disopyramide-M (N-dealkyl-) TMS		2155	369 $_1$ 354 $_5$ 284 $_{100}$ 195 $_{53}$ 167 $_{51}$	1308	2155
Disopyramide-M (N-dealkyl-) TMS		2200	354 $_6$ 284 $_{100}$ 195 $_{65}$ 167 $_{49}$ 73 $_{22}$	1308	7583
Disopyramide-M/artifact AC		2300	321 $_1$ 278 $_1$ 221 $_{15}$ 194 $_{100}$ 167 $_3$	1036	1929
Disugram	G P-I	1525*	234 $_{22}$ 203 $_{100}$ 188 $_{23}$ 97 $_{13}$ 75 $_{10}$	525	3639
Disulfiram	G P-I	2470	296 $_4$ 148 $_{19}$ 116 $_{100}$ 88 $_{76}$ 60 $_{60}$	883	1494
Disulfiram artifact	P	1340	163 $_{82}$ 116 $_{60}$ 91 $_{49}$ 88 $_{100}$ 60 $_{76}$	307	6458
Disulfiram-M/artifact (di-oxo-)		2215	264 $_8$ 116 $_{100}$ 88 $_{86}$ 76 $_{33}$ 60 $_{47}$	681	4471
Disulfoton		1780*	274 $_{15}$ 186 $_{10}$ 125 $_{12}$ 97 $_{28}$ 88 $_{100}$	737	6429
Ditalimfos		2095	299 $_{39}$ 243 $_{31}$ 209 $_{26}$ 148 $_{45}$ 130 $_{100}$	899	3435
Ditazol		2900	324 $_{100}$ 293 $_{36}$ 249 $_{76}$ 165 $_{52}$ 77 $_{82}$	1057	1430
Ditazol 2AC	PAC UHYAC	2985	408 $_{100}$ 365 $_4$ 322 $_{17}$ 262 $_{15}$ 87 $_{71}$	1471	738
Ditazol-M (benzil)	U UHY UHYAC	1825*	210 $_3$ 105 $_{100}$ 77 $_{45}$ 51 $_{26}$	432	1233
Ditazol-M (bis-dealkyl-)	UHY	2280	236 $_{30}$ 165 $_{78}$ 105 $_{78}$ 104 $_{77}$ 77 $_{44}$	537	2544
Ditazol-M (bis-dealkyl-) AC	UHYAC	2560	278 $_{36}$ 236 $_{100}$ 165 $_9$ 105 $_{72}$ 77 $_{39}$	767	1234
Ditazol-M (bis-dealkyl-HO-) 2AC	UHYAC	2845	336 $_{26}$ 294 $_{22}$ 252 $_{100}$ 121 $_{35}$ 105 $_{20}$	1129	1202
Ditazol-M (bis-dealkyl-HO-) MEAC	UHYMEAC	2960	308 $_{42}$ 266 $_{100}$ 135 $_{54}$ 134 $_{45}$ 77 $_{24}$	956	1205
Ditazol-M (dealkyl-) 2AC	UHYAC	2620	364 $_{17}$ 322 $_{100}$ 249 $_{44}$ 105 $_{10}$ 87 $_{25}$	1282	2547
Ditazol-M (dealkyl-HO-) 3AC	UHYAC	3020	422 $_{21}$ 380 $_{78}$ 338 $_{75}$ 252 $_{100}$ 87 $_{33}$	1513	1203
Ditazol-M (dealkyl-HO-) ME2AC	UHYMEAC	2970	394 $_{35}$ 352 $_{100}$ 279 $_{13}$ 135 $_{20}$ 87 $_{25}$	1420	1206
Ditazol-M (deamino-HO-)	UHY UHYAC	2580	237 $_{60}$ 165 $_{16}$ 105 $_{44}$ 104 $_{100}$ 77 $_{53}$	542	2543
Ditazol-M (HO-) 3AC	UHYAC	3250	466 $_{100}$ 424 $_{37}$ 338 $_{16}$ 278 $_{13}$ 87 $_{76}$	1610	1204
Ditazol-M (HO-) ME2AC	UHYMEAC	3200	438 $_{100}$ 352 $_{15}$ 279 $_{10}$ 135 $_{20}$ 87 $_{75}$	1554	1207
Ditazol-M (HO-benzil) AC	UHYAC	2160*	268 $_1$ 226 $_1$ 163 $_{31}$ 121 $_{100}$ 105 $_{30}$	706	2546
Ditazol-M (HO-benzil) ME	UHYME	2290*	240 $_3$ 135 $_{100}$ 105 $_{18}$ 77 $_{10}$	555	2545
Diuron ME		1880	246 $_{14}$ 174 $_5$ 145 $_4$ 109 $_5$ 72 $_{100}$	578	4092
Diuron-M (3,4-dichloroaniline)	P-I U UHY UHYAC	1420	161 $_{100}$ 126 $_{14}$ 99 $_{20}$ 90 $_{18}$ 63 $_{18}$	303	4234
Diuron-M (3,4-dichloroaniline) AC	UHYAC	1990	203 $_{28}$ 161 $_{100}$ 133 $_{13}$ 90 $_9$ 63 $_{26}$	405	4235
Diuron-M/artifact (3,4-dichlorocarbanilic acid) ME	G P-I U UHY UHYAC	1850	219 $_{100}$ 187 $_{77}$ 174 $_{86}$ 160 $_{47}$ 133 $_{56}$	461	850

Diuron-M/artifact (dichlorophenylisocyanate)   Table 8.1: Compounds in order of names

Name	Detected	RI	Typical ions and intensities					Page	Entry
Diuron-M/artifact (dichlorophenylisocyanate)	P U	1960	187 $_{100}$	159 $_{30}$	124 $_{74}$	62 $_{21}$		362	4508
Dixyrazine	UHY	3220	427 $_{73}$	352 $_{33}$	212 $_{100}$	187 $_{63}$		1526	485
Dixyrazine AC	UHYAC	3530	469 $_{44}$	366 $_{28}$	229 $_{77}$	212 $_{100}$	180 $_{26}$	1617	331
Dixyrazine-M (amino-) AC	U+UHYAC	2765	312 $_{60}$	212 $_{100}$	114 $_{57}$			981	1240
Dixyrazine-M (N-dealkyl-) AC	UHYAC	3355	381 $_{16}$	339 $_{1}$	199 $_{20}$	141 $_{100}$	99 $_{21}$	1362	1263
Dixyrazine-M (O-dealkyl-) AC	UHYAC	3350	425 $_{55}$	365 $_{8}$	199 $_{66}$	185 $_{100}$	98 $_{15}$	1521	1262
Dixyrazine-M (ring)	P G U+UHYAC	2010	199 $_{100}$	167 $_{44}$				396	10
Dixyrazine-M 2AC	U+UHYAC	2865	315 $_{54}$	273 $_{34}$	231 $_{100}$	202 $_{11}$		999	2618
Dixyrazine-M AC	U+UHYAC	2550	257 $_{23}$	215 $_{100}$	183 $_{7}$			641	12
DMA		1535	195 $_{5}$	152 $_{100}$	137 $_{28}$	121 $_{12}$	77 $_{16}$	387	3255
DMA		1535	195 $_{1}$	152 $_{18}$	137 $_{5}$	121 $_{2}$	44 $_{100}$	388	5525
DMA AC		1870	237 $_{14}$	178 $_{100}$	152 $_{12}$	121 $_{15}$	86 $_{30}$	546	3268
DMA AC		1870	237 $_{6}$	178 $_{39}$	121 $_{6}$	86 $_{12}$	44 $_{100}$	545	5526
DMA formyl artifact		1550	207 $_{5}$	176 $_{45}$	151 $_{11}$	121 $_{12}$	56 $_{100}$	420	3243
DMA intermediate (2,5-dimethoxyphenyl-2-nitropropene)		1860	223 $_{98}$	176 $_{54}$	161 $_{100}$	147 $_{78}$	91 $_{97}$	483	3284
DMA precursor (2,5-dimethoxybenzaldehyde)		1345*	166 $_{100}$	151 $_{39}$	120 $_{36}$	95 $_{61}$	63 $_{62}$	316	3278
3,4-DMA-NBOMe		2425	313 $_{2}$	179 $_{6}$	164 $_{58}$	151 $_{61}$	121 $_{100}$	1002	10353
3,4-DMA-NBOMe AC		2685	357 $_{2}$	206 $_{44}$	178 $_{79}$	151 $_{68}$	121 $_{100}$	1249	10354
3,4-DMA-NBOMe HFB		2450	511 $_{7}$	332 $_{16}$	178 $_{71}$	151 $_{83}$	121 $_{100}$	1670	10358
3,4-DMA-NBOMe ME		2450	178 $_{86}$	151 $_{34}$	121 $_{100}$	107 $_{20}$	91 $_{72}$	1090	10362
3,4-DMA-NBOMe PFP		2455	461 $_{1}$	282 $_{4}$	178 $_{65}$	151 $_{59}$	121 $_{100}$	1603	10357
3,4-DMA-NBOMe TFA		2495	411 $_{12}$	232 $_{9}$	178 $_{39}$	151 $_{42}$	121 $_{100}$	1482	10356
3,4-DMA-NBOMe TMS		2540	372 $_{1}$	236 $_{100}$	151 $_{52}$	121 $_{79}$	91 $_{47}$	1392	10355
4,4'-DMAR (cis)		1680	190 $_{10}$	175 $_{13}$	146 $_{20}$	91 $_{24}$	70 $_{100}$	370	9244
4,4'-DMAR (cis) 2TMS		1850	334 $_{3}$	205 $_{40}$	171 $_{88}$	130 $_{69}$	73 $_{100}$	1121	9247
4,4'-DMAR (cis) AC		1980	232 $_{3}$	217 $_{24}$	174 $_{28}$	112 $_{76}$	70 $_{100}$	517	9246
4,4'-DMAR (cis) HFB		2010	266 $_{4}$	217 $_{84}$	174 $_{100}$	131 $_{36}$	70 $_{62}$	1383	9250
4,4'-DMAR (cis) ME		1560	204 $_{4}$	190 $_{33}$	146 $_{18}$	85 $_{19}$	70 $_{100}$	411	9245
4,4'-DMAR (cis) PFP		1990	336 $_{1}$	217 $_{100}$	174 $_{99}$	121 $_{46}$	70 $_{61}$	1129	9249
4,4'-DMAR (cis) TFA		2010	286 $_{1}$	217 $_{100}$	174 $_{83}$	91 $_{62}$	70 $_{67}$	814	9248
4,4'-DMAR (trans)		1660	190 $_{11}$	175 $_{15}$	146 $_{23}$	91 $_{26}$	70 $_{100}$	370	9237
4,4'-DMAR (trans) 2TMS		1770	334 $_{8}$	193 $_{45}$	171 $_{60}$	141 $_{47}$	73 $_{100}$	1120	9240
4,4'-DMAR (trans) AC		1970	232 $_{4}$	217 $_{22}$	174 $_{27}$	112 $_{53}$	70 $_{100}$	517	9239
4,4'-DMAR (trans) HFB		1990	217 $_{81}$	174 $_{100}$	156 $_{24}$	131 $_{37}$	70 $_{60}$	1383	9243
4,4'-DMAR (trans) ME		1540	204 $_{5}$	190 $_{43}$	146 $_{24}$	85 $_{22}$	70 $_{100}$	411	9238
4,4'-DMAR (trans) PFP		1970	336 $_{1}$	217 $_{92}$	174 $_{100}$	121 $_{51}$	70 $_{69}$	1129	9242
4,4'-DMAR (trans) TFA		1980	286 $_{1}$	217 $_{100}$	174 $_{89}$	91 $_{70}$	70 $_{80}$	814	9241
DMCC		<1000	152 $_{7}$	151 $_{13}$	137 $_{18}$	109 $_{100}$	84 $_{12}$	294	3580
DMT		1870	188 $_{3}$	143 $_{2}$	130 $_{8}$	115 $_{3}$	58 $_{100}$	365	8873
DMT HFB		1685	340 $_{2}$	326 $_{10}$	169 $_{11}$	115 $_{12}$	58 $_{100}$	1375	9548
DMT TFA		1675	240 $_{3}$	226 $_{6}$	129 $_{14}$	115 $_{10}$	58 $_{100}$	802	9547
DMT TMS		1810	260 $_{1}$	202 $_{6}$	200 $_{5}$	186 $_{4}$	58 $_{100}$	660	9549
DNOC		1660	198 $_{100}$	168 $_{39}$	121 $_{60}$	105 $_{63}$	53 $_{60}$	393	2508
DOB		1800	273 $_{4}$	232 $_{82}$	230 $_{87}$	199 $_{12}$	77 $_{100}$	732	2548
DOB		1800	273 $_{1}$	230 $_{6}$	105 $_{3}$	77 $_{7}$	44 $_{100}$	732	5527
DOB AC		2150	315 $_{15}$	256 $_{100}$	229 $_{7}$	162 $_{22}$	86 $_{71}$	999	2549
DOB AC		2150	315 $_{3}$	256 $_{20}$	162 $_{4}$	86 $_{22}$	44 $_{100}$	998	5528
DOB formyl artifact		1790	285 $_{2}$	254 $_{15}$	229 $_{5}$	199 $_{3}$	56 $_{100}$	807	3242
DOB HFB		1945	469 $_{24}$	256 $_{90}$	240 $_{55}$	229 $_{100}$	199 $_{29}$	1616	6008
DOB PFP		1905	419 $_{22}$	256 $_{73}$	229 $_{100}$	190 $_{59}$	119 $_{93}$	1504	6007
DOB precursor (2,5-dimethoxybenzaldehyde)		1345*	166 $_{100}$	151 $_{39}$	120 $_{36}$	95 $_{61}$	63 $_{62}$	316	3278
DOB TFA		1935	369 $_{28}$	256 $_{81}$	229 $_{100}$	199 $_{40}$	69 $_{88}$	1304	6006
DOB TMS		1920	345 $_{1}$	272 $_{2}$	229 $_{2}$	116 $_{100}$	73 $_{79}$	1182	6009
DOB-M (bis-O-demethyl-) 3AC	U+UHYAC	2325	371 $_{2}$	329 $_{35}$	287 $_{54}$	228 $_{100}$	86 $_{44}$	1315	7075
DOB-M (bis-O-demethyl-) artifact 2AC	U+UHYAC	2225	311 $_{23}$	269 $_{100}$	227 $_{60}$	212 $_{40}$	133 $_{18}$	974	7184
DOB-M (deamino-HO-) AC	U+UHYAC	1950*	316 $_{7}$	274 $_{23}$	214 $_{100}$	186 $_{18}$		1003	7061
DOB-M (deamino-oxo-)	U+UHYAC	1835*	272 $_{59}$	229 $_{100}$				727	7062
DOB-M (HO-) 2AC	U+UHYAC	2270	373 $_{3}$	313 $_{14}$	271 $_{37}$	86 $_{100}$		1325	7081
DOB-M (HO-) -H2O	U+UHYAC	1960*	273 $_{72}$	271 $_{68}$	258 $_{100}$	256 $_{98}$		721	7073
DOB-M (HO-) -H2O AC	U+UHYAC	2130*	313 $_{24}$	271 $_{79}$	256 $_{100}$			985	7074
DOB-M (O-demethyl-) isomer-1 2AC	U+UHYAC	2235	343 $_{12}$	301 $_{23}$	284 $_{57}$	242 $_{100}$	86 $_{81}$	1172	7065
DOB-M (O-demethyl-) isomer-1 AC	U+UHYAC	2120	301 $_{18}$	242 $_{100}$	215 $_{11}$	185 $_{13}$	86 $_{20}$	910	7070
DOB-M (O-demethyl-) isomer-2 2AC	U+UHYAC	2275	343 $_{2}$	284 $_{56}$	242 $_{100}$	215 $_{13}$	86 $_{23}$	1171	7066
DOB-M (O-demethyl-) isomer-2 AC	U+UHYAC	2180	301 $_{29}$	242 $_{100}$	215 $_{14}$	86 $_{36}$		910	7071
DOB-M (O-demethyl-deamino-oxo-) AC	U+UHYAC	1930*	300 $_{8}$	258 $_{94}$	215 $_{100}$			904	7063
DOB-M (O-demethyl-deamino-oxo-) isomer-1	U+UHYAC	1870*	260 $_{66}$	258 $_{72}$	217 $_{99}$	215 $_{100}$		644	7068
DOB-M (O-demethyl-deamino-oxo-) isomer-2	U+UHYAC	1885*	260 $_{99}$	258 $_{100}$	217 $_{97}$	215 $_{93}$		644	7069
DOB-M (O-demethyl-HO-) 3AC	U+UHYAC	2385	401 $_{5}$	359 $_{49}$	317 $_{74}$	258 $_{100}$	86 $_{71}$	1447	7067
DOB-M (O-demethyl-HO-) -H2O 2AC	U+UHYAC	2280	341 $_{32}$	299 $_{62}$	257 $_{100}$	242 $_{72}$		1159	7072
DOB-M (O-demethyl-HO-deamino-HO-) 3AC	U+UHYAC	2145*	402 $_{8}$	360 $_{39}$	315 $_{52}$	300 $_{34}$	231 $_{100}$	1451	7064
Dobutamine 3TMS		2875	517 $_{1}$	502 $_{6}$	250 $_{26}$	73 $_{60}$	58 $_{100}$	1675	4540
Dobutamine 3TMSTFA		2780	613 $_{7}$	280 $_{100}$	267 $_{56}$	179 $_{64}$	73 $_{71}$	1717	6182
Dobutamine 4AC	U+UHYAC	3495	469 $_{2}$	262 $_{100}$	220 $_{57}$	107 $_{51}$	58 $_{68}$	1617	3531
Dobutamine 4TMS		3025	589 $_{1}$	574 $_{2}$	322 $_{52}$	130 $_{100}$	73 $_{75}$	1711	4541

Table 8.1: Compounds in order of names

Name	Detected	RI	Typical ions and intensities					Page	Entry
Dobutamine-M (N-dealkyl-O-methyl-) 2AC	U+UHYAC	2070	251_2	209_7	150_{100}	137_{15}		609	1273
Dobutamine-M (N-dealkyl-O-methyl-) AC	UHYAC	2330	209_{100}	180_{16}	150_{43}	138_{17}	58_{10}	429	2980
Dobutamine-M (O-methyl-)	UHY	3200	315_1	178_{24}	151_7	107_{30}	58_{100}	1002	2979
Dobutamine-M (O-methyl-) 2AC	UHYAC	3100	399_2	250_{87}	220_{29}	150_{100}	58_{58}	1443	2981
Dobutamine-M (O-methyl-) 3AC	UHYAC	3350	441_1	262_{51}	220_{34}	150_{82}	58_{100}	1563	2484
DOC		1770	229_5	186_{100}	171_{20}	91_{18}	77_{39}	505	7847
DOC AC	U+UHYAC	2055	271_{21}	212_{100}	197_{11}	185_{12}	86_{68}	723	7849
DOC formyl artifact		1750	241_5	210_{26}	185_{13}	155_9	56_{100}	559	7848
DOC HFB		1875	425_{27}	240_{28}	212_{91}	185_{100}	155_{35}	1519	7853
DOC PFP		1850	375_{14}	212_{81}	185_{100}	155_{36}	119_{20}	1332	7852
DOC TFA		1875	325_{10}	212_{69}	185_{100}	155_{40}	140_{27}	1059	7851
DOC TMS		1885	301_1	286_8	228_{12}	116_{100}	73_{81}	912	7850
DOC-M (O-demethyl-) AC		2315	257_{30}	215_{100}	180_{30}	150_7	91_7	641	7854
DOC-M (O-demethyl-) isomer-1 2AC	U+UHYAC	2300	299_6	257_{14}	240_{50}	198_{100}	86_{65}	901	7855
DOC-M (O-demethyl-) isomer-2 2AC	U+UHYAC	2305	299_1	269_5	240_{27}	198_{100}	86_{85}	900	7856
Docosane		2200*	310_1	99_{11}	85_{45}	71_{64}	57_{100}	973	4946
Dodecamethylcyclohexasiloxane		1085*	429_{23}	341_{51}	325_{20}	147_{13}	73_{100}	1569	9739
Dodecane		1200*	170_6	99_7	85_{39}	71_{64}	57_{100}	324	4701
Dodemorph		2020	281_{10}	238_9	210_4	154_{100}	55_{30}	791	4034
DOET		1610	223_3	180_{100}	165_{31}	91_{21}	77_{13}	485	3260
DOET		1610	223_1	180_{33}	165_{10}	91_7	44_{100}	485	5529
DOET AC		1990	265_{15}	206_{100}	179_{20}	165_{11}	86_{22}	690	3269
DOET AC		1990	265_9	206_{60}	179_{12}	86_{13}	44_{100}	690	5530
DOET formyl artifact		1600	235_{12}	204_{73}	179_{51}	91_{23}	56_{100}	535	3247
DOET precursor (2,5-dimethoxyacetophenone)		1280*	180_{42}	165_{100}	150_{12}	107_{32}	77_{36}	346	3283
DOET precursor (2,5-dimethoxyacetophenone)		1280*	180_{42}	165_{100}	107_{32}	77_{36}	43_{61}	346	5531
DOI		2025	321_2	278_{100}	263_{13}	247_9	77_{32}	1036	7172
DOI 2AC		2360	405_{12}	304_{100}	277_{15}	247_5	86_{39}	1461	7175
DOI 2ME		2305	349_{23}	304_{100}	277_{18}	162_{11}	72_{65}	1203	7569
DOI 2TFA		1940	513_7	304_{25}	277_{100}	247_{18}	69_{53}	1672	7177
DOI AC	U+UHYAC	2295	363_8	304_{100}	277_6	247_6	86_{43}	1277	7174
DOI formyl artifact		1960	333_1	302_{10}	277_5	247_2	56_{100}	1110	7173
DOI HFB		2070	517_{10}	304_{58}	277_{100}	247_{36}	69_{71}	1675	7179
DOI PFP		2055	467_{31}	304_{53}	277_{100}	247_{23}	190_{20}	1612	7178
DOI TFA		2075	417_{24}	304_{32}	277_{100}	247_{12}	140_{10}	1498	7176
DOI-M (bis-O-demethyl-) 3AC		2480	419_4	377_{32}	360_{36}	276_{78}	86_{100}	1504	7837
DOI-M (bis-O-demethyl-) artifact 2AC		2425	359_{24}	317_{100}	275_{68}	260_{43}	133_{28}	1256	7182
DOI-M (O-demethyl-) isomer-1 2AC	U+UHYAC	2395	391_2	349_9	332_{36}	290_{100}	86_{55}	1405	7180
DOI-M (O-demethyl-) isomer-2 2AC	U+UHYAC	2410	391_3	349_{40}	332_{10}	290_{100}	86_{30}	1405	7181
DOM		1660	209_6	166_{100}	151_{58}	135_{11}	91_{19}	431	2573
DOM		1660	209_3	166_{42}	151_{24}	135_5	44_{100}	431	5532
DOM 2AC		2090	293_6	192_{100}	165_{12}	135_{12}	86_{36}	865	2575
DOM AC	U+UHYAC	2020	251_8	192_{100}	165_{19}	135_{17}	86_{29}	613	2574
DOM AC	U+UHYAC	2020	251_4	192_{50}	165_9	86_{15}	44_{100}	612	5533
DOM formyl artifact		1565	221_{10}	190_{63}	165_{51}	135_{33}	56_{100}	475	3248
DOM intermediate (2,5-dimethoxytoluene)		1020*	152_{61}	137_{100}	109_{17}	77_{17}	65_{11}	294	3289
DOM PFP	UPFP	1730	355_{19}	192_{60}	165_{100}	135_{34}	119_{29}	1236	2591
DOM precursor-1 (hydroquinone dimethylether)		<1000*	138_{56}	123_{100}	95_{54}	63_{22}		278	3282
DOM precursor-2 (2-methylhydroquinone)		1210*	124_{100}	107_{12}	95_{33}	77_{23}	67_{30}	268	3280
DOM precursor-2 (2-methylhydroquinone) 2AC		1440*	208_3	166_{13}	124_{100}	95_3	77_4	423	3281
DOM precursor-2 (2-methylhydroquinone) 2AC		1440*	208_3	166_{13}	124_{100}	95_3	43_{58}	423	5534
DOM-M (deamino-oxo-HO-) 2AC	U+UHYAC	2560*	308_{17}	249_{65}	223_{29}	206_{30}	164_{100}	957	2587
DOM-M (deamino-oxo-HO-) 2PFP	UPFP	2045*	516_{37}	353_{100}	233_{38}	206_{35}		1674	2590
DOM-M (HO-) 2AC	U+UHYAC	2260	309_{18}	250_{100}	191_{11}	164_{14}	86_{15}	964	2588
DOM-M (HO-) 2PFP	UPFP	1830	517_{20}	354_{100}	327_{57}	190_{20}	119_{15}	1675	2589
DOM-M (O-demethyl-) 2PFP	UGLUCPFP	1780	487_4	324_{100}	297_{55}	190_{24}	119_{15}	1644	2592
DOM-M (O-demethyl-) isomer-1 2AC		2140	279_9	237_{40}	178_{100}	152_{10}	86_{14}	776	7844
DOM-M (O-demethyl-) isomer-2 2AC		2190	279_{12}	237_{18}	220_{76}	178_{100}	86_{18}	776	7845
DOM-M (O-demethyl-HO-) 3AC		2140	337_5	295_{48}	250_{49}	236_{100}	86_{23}	1138	7846
Donepezil	U+UHYAC	3150	379_{46}	288_{44}	188_{25}	175_{58}	91_{100}	1353	6548
Donepezil-M (O-demethyl-)	U+UHYAC	3180	365_{31}	274_{44}	175_{38}	146_{30}	91_{100}	1288	6549
Dopamine 3AC		2150	279_3	237_{22}	220_{21}	178_{30}	136_{100}	773	5284
Dopamine 4AC		2245	321_1	220_{33}	178_{36}	136_{100}	123_{32}	1038	5285
Dopamine-M (O-methyl-) 2AC	U+UHYAC	2070	251_2	209_7	150_{100}	137_{15}		609	1273
Dopamine-M (O-methyl-) AC	UHYAC	2330	209_{100}	180_{16}	150_{43}	138_{17}	58_{10}	429	2980
Doripenem artifact-1		1465	169_{46}	138_{94}	137_{100}	122_{41}	82_{41}	322	9463
Doripenem artifact-2		1875	213_{62}	181_{100}	166_{33}	152_{39}	148_{84}	442	9462
Doripenem artifact-2 2TMS		1740	357_{18}	342_8	310_9	240_{100}	208_{53}	1248	9465
Doripenem artifact-2 AC		1875	255_4	195_{33}	169_{21}	137_{100}	96_{24}	631	9464
Doripenem -H2O 2TMS		2730	531_2	392_{26}	174_7	140_{100}	102_{20}	1693	9425
Dorzolamide		2715	307_1	282_9	218_{20}	203_{19}	138_{100}	1055	7427
Dorzolamide 2TMS		2695	453_3	381_{12}	290_{37}	275_{28}	138_{100}	1613	7428
Dorzolamide isomer-1 2ME		2640	335_1	310_7	246_{21}	231_9	138_{100}	1218	7426
Dorzolamide isomer-2 2ME		2660	352_1	310_8	246_{10}	199_5	152_{100}	1218	7424

Dorzolamide ME

Table 8.1: Compounds in order of names

Name	Detected	RI	Typical ions and intensities					Page	Entry
Dorzolamide ME		2670	321 $_2$	296 $_{10}$	232 $_{28}$	217 $_{18}$	138 $_{100}$	1142	7425
Dosulepin	P G U UHY UHYAC	2385	295 $_3$	234 $_2$	221 $_5$	202 $_7$	58 $_{100}$	877	435
Dosulepin-M (bis-nor-) AC	U+UHYAC	2800	309 $_{53}$	250 $_{38}$	235 $_{43}$	217 $_{100}$	202 $_{43}$	963	2943
Dosulepin-M (HO-)	U UHY	2500	311 $_2$	217 $_{27}$	202 $_{23}$	165 $_2$	58 $_{100}$	976	2939
Dosulepin-M (HO-) isomer-1 AC	U+UHYAC	2660	353 $_1$	272	219 $_1$	202 $_1$	58 $_{100}$	1227	2942
Dosulepin-M (HO-) isomer-2 AC	U+UHYAC	2690	353 $_1$	266 $_1$	219	150 $_4$	58 $_{100}$	1227	2944
Dosulepin-M (HO-N-oxide) -(CH3)2NOH	U UHY	2130*	266 $_{100}$	251 $_{28}$	233 $_{28}$	206 $_{28}$	165 $_{77}$	692	2937
Dosulepin-M (HO-N-oxide) -(CH3)2NOH AC	U+UHYAC	2480*	308 $_{70}$	266 $_{100}$	233 $_{28}$	206 $_{28}$	165 $_{40}$	955	2941
Dosulepin-M (nor-)	U UHY	2370	281 $_{21}$	238 $_{10}$	204 $_{100}$	178 $_8$	165 $_9$	787	2940
Dosulepin-M (nor-) AC	U+UHYAC	2820	323 $_{24}$	250 $_{72}$	217 $_{100}$	202 $_{36}$	86 $_{47}$	1050	2934
Dosulepin-M (nor-HO-) isomer-1 2AC	U+UHYAC	3110	381 $_6$	308 $_{14}$	266 $_{17}$	203 $_{16}$	86 $_{100}$	1360	2935
Dosulepin-M (nor-HO-) isomer-2 2AC	U+UHYAC	3150	381 $_7$	308 $_{64}$	266 $_{35}$	235 $_{100}$	86 $_{57}$	1360	2945
Dosulepin-M (N-oxide) -(CH3)2NOH	U+UHYAC	2100*	250 $_{76}$	235 $_{67}$	217 $_{100}$	202 $_{53}$	165 $_{10}$	603	2938
Doxepin	P-I G U+UHYAC	2240	279 $_1$	219 $_3$	178 $_4$	165 $_5$	58 $_{100}$	777	332
Doxepin artifact	G U+UHYAC	1905*	210 $_{100}$	181 $_{74}$	165 $_6$	152 $_{26}$	89 $_8$	432	4470
Doxepin-M (HO-) isomer-1	U UHY	2535	295 $_5$	178 $_4$	165 $_8$	58 $_{100}$		879	488
Doxepin-M (HO-) isomer-1 AC	U+UHYAC	2540	337 $_1$	178 $_1$	165 $_1$	58 $_{100}$		1139	336
Doxepin-M (HO-) isomer-2	U UHY	2560	295 $_5$	178 $_4$	165 $_8$	58 $_{100}$		879	920
Doxepin-M (HO-) isomer-2 AC	U+UHYAC	2585	337 $_1$	165 $_1$	152 $_1$	58 $_{100}$		1139	883
Doxepin-M (HO-dihydro-)	UHY	2530	297 $_4$	71 $_6$	58 $_{100}$			891	487
Doxepin-M (HO-dihydro-) AC	U+UHYAC	2340	339 $_4$	211 $_7$	165 $_2$	58 $_{100}$		1152	334
Doxepin-M (HO-methoxy-) isomer-1 AC	U+UHYAC	2735	367 $_1$	178 $_1$	165 $_1$	58 $_{100}$		1298	6777
Doxepin-M (HO-methoxy-) isomer-2 AC	U+UHYAC	2780	367 $_1$	178 $_1$	165 $_2$	58 $_{100}$		1299	6778
Doxepin-M (HO-N-oxide) -(CH3)2NOH	UHY	2120*	250 $_{100}$	231 $_{44}$	203 $_{37}$			604	557
Doxepin-M (HO-N-oxide) -(CH3)2NOH AC	U+UHYAC	2360*	292 $_{79}$	250 $_{84}$	233 $_{100}$	165 $_{66}$		855	335
Doxepin-M (nor-)	UHY	2270	265 $_{11}$	222 $_{73}$	204 $_{100}$	178 $_6$	115 $_{43}$	689	486
Doxepin-M (nor-) AC	U+UHYAC	2700	307 $_{22}$	234 $_{100}$	219 $_{44}$	86 $_{48}$		950	337
Doxepin-M (nor-) HFB		2395	461 $_6$	240 $_{43}$	234 $_{100}$	219 $_{76}$	178 $_{56}$	1603	7709
Doxepin-M (nor-) PFP		2580	411 $_{15}$	234 $_{100}$	219 $_{55}$	190 $_{40}$	178 $_{56}$	1480	7669
Doxepin-M (nor-) TFA		2495	361 $_{28}$	234 $_{100}$	219 $_{86}$	202 $_{52}$	178 $_{69}$	1268	7668
Doxepin-M (nor-) TMS		2340	337 $_1$	322 $_2$	219 $_{14}$	178 $_{21}$	116 $_{100}$	1140	7708
Doxepin-M (nor-HO-)	U UHY	2540	281 $_{25}$	238 $_{59}$	220 $_{100}$	165 $_{62}$	152 $_{46}$	788	489
Doxepin-M (nor-HO-) isomer-1 2AC	U+UHYAC	2995	365 $_{17}$	292 $_{27}$	250 $_{100}$	237 $_{39}$	86 $_{69}$	1286	338
Doxepin-M (nor-HO-) isomer-2 2AC	UHYAC	3035	365 $_{20}$	292 $_{100}$	250 $_{85}$	233 $_{91}$	86 $_{79}$	1287	31
Doxepin-M (N-oxide) -(CH3)2NOH	P U+UHYAC	1970*	234 $_{100}$	219 $_{71}$	165 $_{42}$			527	333
Doxylamine	P-I G U+UHYAC	1920	270 $_1$	182 $_4$	167 $_8$	71 $_{62}$	58 $_{100}$	720	740
Doxylamine HY	U+UHYAC	1630	199 $_{100}$	184 $_{48}$				398	743
Doxylamine-M	UHY UHYAC	1520	183 $_{57}$	182 $_{100}$	167 $_{43}$			355	741
Doxylamine-M (bis-nor-) AC	U+UHYAC	2280	284 $_1$	198 $_{32}$	182 $_{100}$	167 $_{82}$	86 $_{84}$	804	746
Doxylamine-M (bis-nor-HO-) 2AC	UHYAC	2720	284 $_6$	241 $_{51}$	198 $_{100}$	183 $_{41}$	86 $_{61}$	1170	2696
Doxylamine-M (carbinol) -H2O	U+UHYAC	1560	181 $_{35}$	180 $_{100}$	152 $_{11}$	77 $_{16}$		348	742
Doxylamine-M (deamino-HO-) AC	UHYAC	1960	285 $_6$	198 $_{100}$	182 $_{50}$	167 $_{41}$	87 $_{81}$	810	2692
Doxylamine-M (HO-) AC	U+UHYAC	2300	258 $_1$	198 $_{11}$	183 $_{18}$	71 $_{61}$	58 $_{100}$	1082	744
Doxylamine-M (HO-carbinol) AC	UHYAC	2980	257 $_{98}$	242 $_{55}$	200 $_{79}$	137 $_{100}$	78 $_{93}$	642	2693
Doxylamine-M (HO-carbinol) -H2O	UHY	1800	197 $_{44}$	196 $_{100}$	167 $_{16}$	139 $_4$	89 $_3$	392	2688
Doxylamine-M (HO-carbinol) -H2O AC	UHYAC	1940	239 $_{14}$	197 $_{44}$	196 $_{100}$	167 $_{16}$	139 $_4$	551	2689
Doxylamine-M (HO-methoxy-) AC	U+UHYAC	2320	258 $_1$	198 $_4$	183 $_5$	71 $_{59}$	58 $_{100}$	1254	745
Doxylamine-M (HO-methoxy-carbinol) AC	UHYAC	2030	287 $_{47}$	245 $_{86}$	230 $_{100}$	167 $_{48}$	106 $_{39}$	820	2695
Doxylamine-M (HO-methoxy-carbinol) -H2O AC	UHYAC	2010	269 $_7$	227 $_{58}$	226 $_{100}$	211 $_{25}$	154 $_{21}$	711	2694
Doxylamine-M (nor-) AC	U UHYAC	2340	298 $_1$	212 $_2$	182 $_{100}$	167 $_{75}$	100 $_{68}$	897	2690
Doxylamine-M (nor-HO-) 2AC	UHYAC	2760	257 $_4$	241 $_{31}$	198 $_{53}$	183 $_{35}$	100 $_{100}$	1243	2697
DPT		2090	244 $_1$	144 $_{18}$	130 $_{37}$	114 $_{100}$	86 $_{17}$	572	8829
DPT HFB		1915	411 $_3$	340 $_{10}$	326 $_8$	143 $_9$	114 $_{100}$	1558	10093
DPT PFP		1905	361 $_1$	290 $_6$	276 $_5$	143 $_{12}$	114 $_{100}$	1403	10092
DPT TFA		1925	311 $_4$	240 $_{10}$	226 $_{10}$	143 $_7$	114 $_{100}$	1157	10091
DPT TMS		2195	316 $_3$	216 $_{10}$	202 $_{14}$	143 $_{11}$	114 $_{100}$	1007	10090
Drofenine		2180	317 $_1$	173 $_1$	99 $_{16}$	86 $_{100}$		1015	747
Dronabinol	G P-I	2470*	314 $_{85}$	299 $_{100}$	271 $_{41}$	243 $_{29}$	231 $_{45}$	997	981
Dronabinol AC	U+UHYAC-I	2450*	356 $_{13}$	313 $_{35}$	297 $_{100}$	243 $_{12}$	231 $_{24}$	1244	982
Dronabinol ET		2390*	342 $_{90}$	327 $_{100}$	313 $_{39}$	271 $_{27}$	259 $_{30}$	1171	2531
Dronabinol isomer-1 PFP		2150*	460 $_{100}$	445 $_{14}$	417 $_{70}$	392 $_{30}$	377 $_{100}$	1602	5669
Dronabinol isomer-2 PFP		2170*	460 $_{100}$	445 $_{65}$	417 $_{75}$	389 $_{86}$	297 $_{80}$	1602	5668
Dronabinol ME		2360*	328 $_{82}$	313 $_{100}$	285 $_{28}$	257 $_{24}$	245 $_{27}$	1083	2530
Dronabinol TMS		2405*	386 $_{100}$	371 $_{86}$	315 $_{37}$	303 $_{31}$	73 $_{80}$	1387	4599
Dronabinol-D3		2450*	317 $_{96}$	302 $_{100}$	274 $_{43}$	258 $_{26}$	234 $_{55}$	1015	5663
Dronabinol-D3 AC		2750*	359 $_9$	316 $_{18}$	300 $_{100}$	274 $_9$	234 $_{17}$	1261	7309
Dronabinol-D3 isomer-1 PFP		2130*	463 $_{60}$	420 $_{54}$	395 $_{25}$	380 $_{100}$	342 $_{13}$	1606	5665
Dronabinol-D3 isomer-1 TFA		2160*	413 $_{51}$	370 $_{31}$	345 $_{14}$	330 $_{100}$	232 $_{10}$	1488	5667
Dronabinol-D3 isomer-2 PFP		2150*	463 $_{100}$	448 $_{65}$	420 $_{70}$	389 $_{85}$	300 $_{81}$	1606	5664
Dronabinol-D3 isomer-2 TFA		2180*	413 $_{100}$	398 $_{74}$	370 $_{71}$	300 $_{73}$		1488	5666
Dronabinol-D3 ME		2355*	331 $_{81}$	316 $_{100}$	288 $_{34}$	257 $_{29}$	248 $_{39}$	1103	6040
Dronabinol-D3 TMS		2385*	389 $_{100}$	374 $_{96}$	346 $_{26}$	315 $_{59}$	306 $_{41}$	1400	5670
Dronabinol-M (11-HO-)	P-I	2775*	330 $_{12}$	299 $_{100}$	231 $_{10}$	217 $_9$	193 $_7$	1096	4661
Dronabinol-M (11-HO-) 2ME		2580*	358 $_{13}$	313 $_{100}$	257 $_3$	231 $_3$		1255	4659

Table 8.1: Compounds in order of names Dronabinol-M (11-HO-) 2PFP

Name	Detected	RI	Typical ions and intensities					Page	Entry
Dronabinol-M (11-HO-) 2PFP		2350*	622_{15}	607_5	551_9	458_{100}	415_{24}	1719	4658
Dronabinol-M (11-HO-) 2TFA		2450*	522_{13}	451_8	408_{100}	395_{13}	365_{24}	1678	4657
Dronabinol-M (11-HO-) 2TMS		2630*	474_5	459_4	403_2	371_{100}	73_{14}	1626	4656
Dronabinol-M (11-HO-) -H2O AC	U+UHYAC-I	2740*	354_{48}	312_{100}	297_{19}	269_{31}	91_{21}	1233	4660
Dronabinol-M (HO-nor-delta-9-HOOC-) 2ME	UTHCME-I	2840*	388_{42}	373_{65}	329_{100}	201_{24}	189_{28}	1396	3466
Dronabinol-M (nor-delta-9-HOOC-) 2ME	UTHCME UGlucExMe	2620*	372_{52}	357_{79}	341_9	313_{100}	245_6	1324	1439
Dronabinol-M (nor-delta-9-HOOC-) 2PFP		2440*	622_{35}	607_{49}	459_{100}	445_{76}	69_{82}	1719	4380
Dronabinol-M (nor-delta-9-HOOC-) 2TMS		2470*	488_{40}	473_5	398_{12}	371_{100}	73_{66}	1646	5671
Dronabinol-M (nor-delta-9-HOOC-)-D3 2ME		2590*	375_{39}	360_{68}	356_{19}	316_{100}	301_9	1337	6187
Dronabinol-M (nor-delta-9-HOOC-)-D3 2PFP		2425*	625_{40}	610_{57}	462_{100}	448_{62}	432_{43}	1720	6039
Dronabinol-M (nor-delta-9-HOOC-)-D3 2TMS		2660*	491_{44}	476_{55}	374_{100}	300_{15}	73_{28}	1650	5672
Dronabinol-M (oxo-nor-delta-9-HOOC-) 2ME	UTHCME-I	2860*	386_{55}	371_{73}	327_{100}	314_{22}	189_{11}	1386	3467
Dronedarone		3000	556_1	513_{18}	142_{83}	100_{100}	70_{53}	1698	8450
Dronedarone-M (debutyl-)		3900	500_2	471_5	100_{64}	86_{76}	58_{100}	1659	8451
Dronedarone-M (debutyl-) ME		3900	514_1	485_4	100_{94}	86_{87}	58_{100}	1673	8452
Droperidol		9999	379_6	246_{100}	165_{40}	134_{57}	123_{96}	1351	1495
Droperidol 2TMS		3485	523_6	300_{73}	271_{45}	255_{29}	73_{100}	1679	4542
Droperidol ME		3370	393_6	246_{83}	165_{81}	123_{100}		1416	490
Droperidol-M	U+UHYAC	1490*	180_{25}	125_{49}	123_{35}	95_{17}	56_{100}	345	85
Droperidol-M (benzimidazolone)	UHY-I	1950	134_{100}	106_{36}	79_{62}	67_{20}		273	491
Droperidol-M (benzimidazolone) 2AC	UHYAC-I	1730	218_{11}	176_{19}	134_{100}	106_{11}		457	171
Dropropizine		2205	236_{14}	175_{100}	132_{41}	104_{46}	70_{83}	541	2775
Dropropizine 2AC		2330	320_6	260_6	175_{100}	132_{44}	70_{51}	1034	2777
Dropropizine AC		2390	278_{14}	175_{100}	132_{35}	104_{23}	70_{53}	769	2776
Dropropizine-M (HO-) 3AC		2675	378_{17}	259_{17}	233_{100}	216_{20}	191_{35}	1348	6805
Dropropizine-M (HO-phenylpiperazine) 2AC	U+UHYAC	2350	262_{39}	220_{60}	177_{21}	148_{100}	135_{59}	670	6610
Dropropizine-M (phenylpiperazine) AC	UHYAC	1920	204_{48}	161_{21}	132_{100}	56_{77}		410	1276
Drostanolone		2555*	304_{49}	245_{58}	177_{21}	95_{57}	55_{100}	934	2773
Drostanolone AC		2700*	346_{26}	286_{48}	271_{48}	149_{64}	55_{100}	1192	2774
Drostanolone enol 2TMS		2625*	448_{100}	405_{14}	157_{13}	141_{25}	73_{74}	1577	3957
Drostanolone propionate		2985*	360_3	286_{13}	271_{12}	149_{16}	57_{100}	1266	2761
Drostanolone TMS		2575*	376_8	361_{18}	286_{26}	129_{81}	73_{100}	1340	3956
Duloxetine		2500	297_{91}	265_{51}	181_{56}	115_{100}	97_{50}	889	7461
Duloxetine 2AC		3160	381_{62}	266_{100}	239_{74}	221_{45}	87_{75}	1360	7464
Duloxetine 2HFB		2435	689_3	476_{13}	435_{50}	249_{100}	237_{94}	1728	7477
Duloxetine 2PFP		2425	589_{25}	385_{72}	249_{89}	237_{91}	119_{100}	1711	7470
Duloxetine isomer-1 2TMS		2545	441_{19}	369_{14}	338_9	116_{79}	73_{100}	1562	7482
Duloxetine isomer-1 AC		3050	339_{68}	266_{52}	237_{100}	182_{27}	87_{62}	1149	7463
Duloxetine isomer-1 HFB		2650	493_{46}	266_{14}	239_{100}	182_{44}	169_{49}	1651	7478
Duloxetine isomer-1 PFP		2300	443_{44}	239_{100}	190_{49}	182_{43}	119_{55}	1567	7471
Duloxetine isomer-1 TFA		2690	393_{71}	266_{17}	239_{100}	140_{50}	69_{51}	1414	7473
Duloxetine isomer-1 TMS		2510	369_{46}	338_{11}	311_{24}	249_{33}	73_{100}	1307	7480
Duloxetine isomer-2 2TMS		2620	441_{19}	369_4	337_{10}	116_{63}	73_{100}	1562	7483
Duloxetine isomer-2 AC		3150	339_{57}	266_{80}	239_{100}	221_{82}	87_{93}	1149	7474
Duloxetine isomer-2 HFB		2725	493_{39}	266_6	239_{100}	221_{34}	169_{33}	1652	7479
Duloxetine isomer-2 PFP		2700	443_{44}	239_{100}	190_{49}	182_{43}	119_{55}	1568	7472
Duloxetine isomer-2 TFA		2700	393_{45}	265_7	239_{100}	221_{41}	69_{30}	1414	7467
Duloxetine isomer-2 TMS		2550	369_{56}	337_{26}	311_{28}	249_{19}	73_{100}	1307	7481
Duloxetine ME		2490	311_{14}	237_{13}	209_{11}	141_{14}	58_{100}	976	7462
Duloxetine-M (1-naphthol)		1500*	144_{100}	115_{83}	89_{14}	74_6	63_{17}	283	928
Duloxetine-M (1-naphthol) AC	U+UHYAC	1555*	186_{13}	144_{100}	115_{47}	89_8	63_7	361	932
Duloxetine-M (1-naphthol) HFB		1310*	340_{46}	169_{25}	143_{28}	115_{100}	89_{11}	1155	7476
Duloxetine-M (1-naphthol) PFP		1510*	290_{45}	171_{100}	143_{20}	115_{49}	89_8	839	7468
Duloxetine-M (1-naphthol) TMS		1525*	216_{100}	201_{95}	185_{51}	115_{39}	73_{21}	452	7460
Duloxetine-M (4-HO-1-naphthol) 2AC	U+UHYAC	1900*	244_{14}	202_{19}	160_{100}	131_{21}	103_8	570	933
Duloxetine-M/artifact -H2O AC		1760	195_{20}	152_{27}	138_{52}	123_{27}	98_{100}	386	7465
Duloxetine-M/artifact -H2O HFB		1560	349_{23}	252_{62}	180_{56}	123_{100}	69_{73}	1202	7475
Duloxetine-M/artifact -H2O PFP		1535	299_{47}	202_{65}	180_{72}	123_{100}	119_{79}	899	7469
Duloxetine-M/artifact -H2O TFA		1545	249_{43}	180_{25}	152_{60}	123_{63}	69_{100}	593	7466
Dutasteride		3620	528_5	300_{31}	272_{100}	236_{11}	110_{46}	1683	8212
Dutasteride 2TMS		3010	672_{37}	356_{54}	182_{62}	142_{94}	73_{100}	1726	8213
EAM-2201		3435	387_{100}	370_{70}	312_{72}	232_{99}	144_{68}	1391	9615
4-EA-NBOMe		2235	164_{12}	121_{100}	119_{23}	104_{21}	91_{65}	798	10359
4-EA-NBOMe AC		2535	325_2	206_{77}	164_{28}	121_{100}	119_{45}	1065	10360
4-EA-NBOMe HFB		2285	360_1	146_{53}	121_{100}	119_{37}	91_{75}	1633	10366
4-EA-NBOMe ME		2270	178_{82}	121_{100}	119_{13}	104_{14}	91_{58}	892	10361
4-EA-NBOMe PFP		2280	310_2	282_2	146_{33}	121_{100}	91_{50}	1531	10365
4-EA-NBOMe TFA		2320	379_1	260_{10}	146_{57}	121_{100}	91_{67}	1352	10364
4-EA-NBOMe TMS		2370	340_5	236_{100}	121_{96}	119_{37}	91_{64}	1240	10363
5-EAPB		1370	188_1	131_{26}	102_6	77_{15}	72_{100}	407	9363
5-EAPB AC		1560	245_4	158_{28}	131_{34}	114_{35}	72_{100}	575	9364
5-EAPB HFB		1350	399_1	268_{100}	240_{57}	158_{53}	131_{47}	1440	9368
5-EAPB PFP		1355	349_1	218_{100}	190_{52}	158_{51}	131_{58}	1203	9367
5-EAPB TFA		1385	299_1	168_{100}	158_{51}	140_{46}	131_{48}	901	9366

5-EAPB TMS

Table 8.1: Compounds in order of names

Name	Detected	RI	Typical ions and intensities					Page	Entry
5-EAPB TMS		1380	260 $_7$	144 $_{100}$	131 $_{31}$	102 $_5$	73 $_{47}$	749	9365
Ebastine		3940	469 $_1$	293 $_7$	280 $_7$	167 $_{100}$	111 $_{71}$	1617	8158
Ebastine artifact ME	G U+UHYAC	2115	281 $_1$	167 $_{30}$	114 $_{41}$	99 $_{100}$	70 $_{19}$	789	737
Ebastine HY	UHY	1645*	184 $_{45}$	165 $_{14}$	152 $_7$	105 $_{100}$	77 $_{63}$	357	1333
Ebastine HYAC	U+UHYAC	1700*	226 $_{20}$	184 $_{20}$	165 $_{100}$	105 $_{14}$	77 $_{35}$	495	1241
Ebastine HYHFB		1475*	380 $_3$	183 $_4$	166 $_{100}$	152 $_{22}$	83 $_{25}$	1355	8146
Ebastine HYME	UHY	1655*	198 $_{70}$	167 $_{94}$	121 $_{100}$	105 $_{56}$	77 $_{71}$	395	6779
Ebastine HYPFP		1410*	330 $_4$	183 $_4$	166 $_{100}$	152 $_{24}$	83 $_{30}$	1093	8145
Ebastine HYTFA		1420*	280 $_{19}$	183 $_7$	166 $_{100}$	152 $_{24}$	83 $_{21}$	780	8144
Ebastine HYTMS		1540*	256 $_{28}$	241 $_{14}$	179 $_{38}$	167 $_{100}$	152 $_{17}$	638	8159
2,3-EBDB		1670	192 $_5$	135 $_5$	86 $_{100}$	77 $_6$	58 $_{11}$	476	5417
2,3-EBDB AC		2000	263 $_3$	192 $_4$	176 $_{18}$	128 $_{42}$	86 $_{100}$	677	5511
2,3-EBDB HFB		1790	417 $_1$	282 $_{100}$	176 $_{25}$	135 $_{24}$	77 $_{17}$	1499	5594
2,3-EBDB PFP		1755	367 $_3$	232 $_{100}$	176 $_{16}$	119 $_{30}$	69 $_{10}$	1296	5595
2,3-EBDB TFA		1780	317 $_5$	182 $_{100}$	176 $_{37}$	154 $_{16}$	135 $_{14}$	1011	5512
2,3-EBDB TMS		1825	278 $_2$	264 $_5$	178 $_{30}$	135 $_{16}$	73 $_{48}$	866	5596
ECC -HCN		<1000	125 $_4$	110 $_{13}$	96 $_{45}$	82 $_{34}$	56 $_{100}$	269	3598
ECC -HCN		<1000	125 $_4$	110 $_{12}$	96 $_{44}$	56 $_{97}$	41 $_{100}$	268	5535
Ecgonidine ME	UHY-I UHYAC-I	1280	181 $_{31}$	152 $_{100}$	138 $_9$	122 $_{15}$	82 $_{18}$	349	3574
Ecgonidine TBDMS		1520	281 $_{16}$	252 $_{29}$	224 $_{100}$	150 $_{44}$	122 $_{30}$	790	6242
Ecgonidine TMS		1345	239 $_{31}$	224 $_{25}$	210 $_{100}$	183 $_{10}$	122 $_{28}$	553	6256
Ecgonine 2TMS		1680	329 $_2$	314 $_4$	96 $_{45}$	82 $_{100}$	73 $_{49}$	1089	5445
Ecgonine ACTMS		1680	299 $_{11}$	240 $_{41}$	122 $_9$	94 $_{29}$	82 $_{100}$	903	6238
Ecgonine TMSTFA		1395	353 $_{15}$	267 $_{35}$	240 $_{82}$	94 $_{38}$	82 $_{100}$	1226	6255
Ecgonine-D3 2TMS		1670	332 $_{13}$	317 $_{15}$	99 $_{64}$	85 $_{100}$	73 $_{56}$	1107	5576
Econazole	U	3550	380 $_3$	299 $_9$	206 $_4$	125 $_{100}$	81 $_{27}$	1355	2550
EDDP	U UHY U+UHYAC	2040	277 $_{100}$	276 $_{95}$	262 $_{41}$	220 $_{22}$	165 $_{10}$	764	242
EDDP-M (HO-) AC	UHYAC	2350	335 $_{100}$	304 $_{27}$	292 $_{26}$	276 $_{16}$	234 $_{18}$	1126	5297
EDDP-M (nor-) AC	U+UHYAC	2220	305 $_{100}$	290 $_{74}$	262 $_{26}$	236 $_{29}$	220 $_{25}$	938	5292
EDDP-M (nor-HO-) 2AC	UHYAC	2645	363 $_{100}$	348 $_{43}$	320 $_{63}$	278 $_{22}$	149 $_{15}$	1279	5299
5,6-EDO-DALT		2400	298 $_9$	257 $_5$	202 $_5$	188 $_{23}$	110 $_{100}$	897	10154
5,6-EDO-DALT HFB		2280	467 $_1$	398 $_7$	384 $_{15}$	187 $_{23}$	110 $_{100}$	1653	10158
5,6-EDO-DALT PFP		2275	444 $_1$	348 $_4$	334 $_{10}$	187 $_{14}$	110 $_{100}$	1569	10157
5,6-EDO-DALT TFA		2305	394 $_2$	367 $_3$	298 $_9$	284 $_{22}$	110 $_{100}$	1420	10156
5,6-EDO-DALT TMS		2570	370 $_3$	329 $_7$	274 $_5$	260 $_{29}$	110 $_{100}$	1314	10155
EDTA 3ME1ET		2125*	362 $_{14}$	303 $_{28}$	289 $_{18}$	188 $_{100}$	174 $_{64}$	1274	6452
EDTA 4ME		2105*	348 $_8$	289 $_{25}$	188 $_{23}$	174 $_{100}$	146 $_{40}$	1199	6451
Efavirenz		2100	315 $_{43}$	246 $_{100}$	243 $_{79}$	180 $_{69}$	167 $_{28}$	998	7841
Efavirenz AC		2045	357 $_2$	315 $_{100}$	256 $_{25}$	243 $_{92}$	202 $_{42}$	1245	7945
Efavirenz artifact		1560	271 $_{58}$	270 $_{100}$	245 $_{10}$	201 $_{26}$	167 $_{24}$	722	7933
Efavirenz artifact AC		1650	313 $_{18}$	270 $_4$	256 $_7$	244 $_{100}$	167 $_{14}$	986	7934
Efavirenz ME		2010	329 $_{45}$	270 $_{14}$	260 $_{100}$	257 $_{90}$	188 $_{30}$	1085	7842
Efavirenz TMS		1880	387 $_9$	343 $_{11}$	318 $_{16}$	250 $_{33}$	73 $_{100}$	1388	7843
Efavirenz-M (HO-) artifact isomer-1	U+UHYAC	1875	287 $_{68}$	286 $_{100}$	261 $_{14}$	218 $_{20}$	183 $_{16}$	817	7944
Efavirenz-M (HO-) artifact isomer-1 AC	U+UHYAC	1990	329 $_3$	287 $_{100}$	286 $_{86}$	218 $_{25}$	183 $_{13}$	1083	7961
Efavirenz-M (HO-) artifact isomer-2	U+UHYAC	1925	287 $_{70}$	286 $_{100}$	251 $_{18}$	217 $_{39}$	183 $_{22}$	817	7962
Efavirenz-M (HO-) artifact isomer-2 AC	U+UHYAC	2035	329 $_{18}$	286 $_{100}$	261 $_{12}$	218 $_{28}$	183 $_{13}$	1083	7963
EG-018		>4000	391 $_{64}$	334 $_{100}$	264 $_{18}$	179 $_{34}$	127 $_{48}$	1408	9581
Eicosane		2000*	282 $_6$	99 $_{15}$	85 $_{38}$	71 $_{65}$	57 $_{100}$	793	2352
Eicosanoic acid ME		2275*	326 $_{14}$	283 $_6$	143 $_{14}$	87 $_{61}$	74 $_{100}$	1072	3035
Elemicin		1435*	208 $_{100}$	193 $_{60}$	133 $_{25}$	118 $_{22}$	77 $_{32}$	425	7136
Elemicin-M (1-HO-) AC	U+UHYAC	2035*	266 $_{100}$	223 $_{28}$	207 $_{27}$	195 $_{21}$	176 $_{40}$	694	7142
Elemicin-M (1-HO-) ME	UME	2085*	238 $_{100}$	223 $_{50}$	207 $_{15}$	195 $_{13}$	163 $_{29}$	548	7151
Elemicin-M (bis-demethyl-) 2AC	U+UHYAC	1880*	264 $_1$	222 $_{30}$	180 $_{100}$	147 $_{15}$	91 $_{34}$	681	7148
Elemicin-M (demethyl-)	U+UHYAC	1600*	194 $_{10}$	151 $_{100}$	135 $_5$	107 $_{17}$	91 $_{10}$	384	10267
Elemicin-M (demethyl-) isomer-1 AC	U+UHYAC	1755*	236 $_{35}$	194 $_{100}$	179 $_{43}$	119 $_{15}$	91 $_{12}$	538	7140
Elemicin-M (demethyl-) isomer-2 AC	U+UHYAC	1790*	236 $_9$	194 $_{100}$	179 $_9$	133 $_{14}$	119 $_7$	539	7141
Elemicin-M (demethyl-dihydroxy-) isomer-1 3AC	U+UHYAC	2275*	354 $_8$	294 $_{27}$	280 $_{25}$	252 $_{100}$	210 $_{30}$	1231	7138
Elemicin-M (demethyl-dihydroxy-) isomer-2 3AC	U+UHYAC	2300*	354 $_5$	312 $_{20}$	252 $_{100}$	210 $_8$	167 $_{18}$	1231	7139
Elemicin-M (dihydroxy-) 2AC	U+UHYAC	2195*	326 $_{28}$	266 $_{100}$	223 $_{28}$	207 $_{23}$	181 $_{59}$	1069	7137
Eletriptan		3650	380 $_1$	156 $_5$	129 $_1$	84 $_{100}$	82 $_4$	1365	7491
Eletriptan HFB		3370	576 $_1$	352 $_1$	156 $_2$	129 $_1$	84 $_{100}$	1707	7494
Eletriptan TFA		3650	476 $_1$	252 $_2$	156 $_3$	129 $_1$	84 $_{100}$	1631	7493
Eletriptan TMS		3580	371 $_1$	228 $_1$	156 $_3$	84 $_{100}$		1590	7492
Embutramide		2240	293 $_{12}$	190 $_{100}$	135 $_{63}$	121 $_{83}$	98 $_{37}$	866	8310
Embutramide 2TMS		2200	422 $_2$	365 $_6$	350 $_{13}$	190 $_{82}$	159 $_{100}$	1552	8315
Embutramide AC	U+UHYAC	2435	335 $_7$	190 $_{100}$	177 $_{26}$	135 $_{62}$	121 $_{78}$	1127	7917
Embutramide artifact (amine formyl)		1545	219 $_7$	190 $_4$	177 $_{41}$	135 $_{84}$	121 $_{100}$	466	8312
Embutramide -H2O		2025	275 $_1$	191 $_{26}$	162 $_7$	121 $_{17}$	98 $_{100}$	750	8311
Embutramide -H2O TFA		2185	371 $_6$	190 $_{47}$	165 $_{66}$	135 $_{67}$	121 $_{100}$	1319	8316
Embutramide ME		2205	307 $_2$	190 $_{45}$	121 $_{58}$	101 $_{100}$	98 $_{27}$	954	8313
Embutramide TMS		2290	365 $_2$	350 $_5$	190 $_{73}$	159 $_{100}$	121 $_{59}$	1288	8314
Embutramide-M/artifact (amine) 2AC	U+UHYAC	2080	291 $_6$	218 $_{40}$	163 $_{60}$	135 $_{59}$	121 $_{100}$	852	7916
Embutramide-M/artifact (amine) AC	U+UHYAC	2045	249 $_{13}$	190 $_{57}$	177 $_{23}$	135 $_{78}$	121 $_{100}$	601	7915

Table 8.1: Compounds in order of names
Embutramide-M/artifact (amine) HFB

Name	Detected	RI	Typical ions and intensities					Page	Entry
Embutramide-M/artifact (amine) HFB		1650	403 10	226 14	177 87	135 83	121 100	1456	8318
Embutramide-M/artifact (amine) PFP		1625	353 5	177 50	135 60	121 100	91 14	1226	8317
Emedastine		2395	302 12	245 45	174 75	147 84	97 100	920	8185
Emetine	G	4055	480 12	288 21	272 33	206 32	192 100	1635	5611
Emetine ET		3320	508 2	302 3	206 100	190 7	150 6	1668	5614
Emetine ME		4010	494 2	288 4	272 3	206 100	190 9	1654	5612
Emtricitabine		2555	229 3	190 24	130 55	100 47	87 100	583	7485
Emtricitabine 2AC		2580	188 5	154 12	130 18	100 100	87 26	1098	7486
Emtricitabine 2TFA		2350	250 3	200 7	182 26	154 69	100 100	1555	7487
Emtricitabine 2TMS		2455	376 4	190 40	100 65	87 95	73 100	1406	7484
Enalapril 2ET		2745	432 1	359 13	262 100	188 9	91 13	1540	4739
Enalapril 2ME	UME	2690	404 1	331 11	248 100	174 14	91 36	1460	3201
Enalapril 2TMS		2790	505 20	447 14	306 100	73 8		1677	4979
Enalapril ET		2715	404 1	331 12	234 100	160 17	91 31	1460	4738
Enalapril -H2O	P-I G UHY UHYAC UME	2770	358 22	254 90	208 100	160 25	91 59	1254	3199
Enalapril ME	P-I PME UME	2675	390 1	317 12	234 100	91 50	70 41	1403	3200
Enalapril METMS		2800	462 1	447 2	375 10	248 100	91 18	1604	4984
Enalapril TMS		2740	448 1	433 3	375 20	234 100	73 14	1577	4608
Enalaprilate 2ET		2715	404 1	331 12	234 100	160 17	91 31	1460	4738
Enalaprilate 2ME	UME	2620	376 1	317 8	220 100	116 21	91 48	1339	3198
Enalaprilate 2TMS	UME	2780	492 1	477 6	375 39	278 100	234 23	1651	4978
Enalaprilate 3ET		2745	432 1	359 13	262 100	188 9	91 13	1540	4739
Enalaprilate 3ME	UME	2680	390 1	331 11	234 100	174 13	130 24	1404	4733
Enalaprilate -H2O ME	UME	2735	344 24	240 87	208 100	91 84	70 70	1179	3202
Enalaprilate METMS		2730	434 1	419 5	375 10	220 100	91 23	1546	4609
Enalaprilate-M/artifact (HOOC-) 2ET	UET	2025	307 2	234 100	160 12	117 17	91 32	952	4740
Enalaprilate-M/artifact (HOOC-) 2ME	UME	1870	279 2	220 100	160 10	117 28	91 57	777	4734
Enalaprilate-M/artifact (HOOC-) 3ET	UET	2095	335 1	262 100	234 7	188 7	91 27	1127	4741
Enalaprilate-M/artifact (HOOC-) 3ME	UME	1935	293 2	234 100	174 8	130 16	91 50	865	4735
Enalapril-M/artifact (deethyl-) 2ET		2715	404 1	331 12	234 100	160 17	91 31	1460	4738
Enalapril-M/artifact (deethyl-) 2ME	UME	2620	376 1	317 8	220 100	116 21	91 48	1339	3198
Enalapril-M/artifact (deethyl-) 2TMS	UME	2780	492 1	477 6	375 39	278 100	234 23	1651	4978
Enalapril-M/artifact (deethyl-) 3ET		2745	432 1	359 13	262 100	188 9	91 13	1540	4739
Enalapril-M/artifact (deethyl-) 3ME	UME	2680	390 1	331 11	234 100	174 13	130 24	1404	4733
Enalapril-M/artifact (deethyl-) -H2O ME	UME	2735	344 24	240 87	208 100	91 84	70 70	1179	3202
Enalapril-M/artifact (deethyl-) METMS		2730	434 1	419 5	375 10	220 100	91 23	1546	4609
Enalapril-M/artifact (deethyl-HOOC-) 2ME	UME	1870	279 2	220 100	160 10	117 28	91 57	777	4734
Enalapril-M/artifact (deethyl-HOOC-) 3ET	UET	2095	335 1	262 100	234 7	188 7	91 27	1127	4741
Enalapril-M/artifact (deethyl-HOOC-) 3ME	UME	1935	293 2	234 100	174 8	130 16	91 50	865	4735
Enalapril-M/artifact (HOOC-) 2ET	UET	2095	335 1	262 100	234 7	188 7	91 27	1127	4741
Enalapril-M/artifact (HOOC-) 2ME	UME	1985	307 2	248 34	234 100	174 5	91 6	952	4737
Enalapril-M/artifact (HOOC-) ET	UET	2025	307 2	234 100	160 12	117 17	91 32	952	4740
Enalapril-M/artifact (HOOC-) ME	P-I UME	1930	293 3	234 36	220 100	160 11	91 23	864	4736
25E-NBOMe		2460	329 1	298 2	180 22	150 25	121 100	1090	9324
25E-NBOMe AC		2790	371 13	192 100	177 17	150 11	121 79	1320	9326
25E-NBOMe HFB		2420	525 8	192 100	179 48	149 13	121 46	1681	9330
25E-NBOMe ME		2390	343 1	179 16	164 61	134 7	121 100	1176	9325
25E-NBOMe PFP		2430	475 13	192 100	179 45	150 15	121 48	1628	9329
25E-NBOMe TFA		2560	425 14	192 100	179 47	149 13	121 39	1520	9328
25E-NBOMe TMS		2570	401 1	386 1	222 67	179 15	121 100	1451	9327
Endogenous biomolecule	UHYAC	1550	200 3	97 16	86 100	71 28	55 50	398	492
Endogenous biomolecule	UHY	2520*	288 65	255 100	197 42	134 57	91 74	825	715
Endogenous biomolecule	UHY UHYAC	1790*	192 6	153 13	137 100	121 78	82 95	374	1947
Endogenous biomolecule	UHY UHYAC	2545*	310 51	267 8	197 100	153 9		967	2368
Endogenous biomolecule	UME	1640*	179 11	171 88	130 20	115 100	99 36	339	4951
Endogenous biomolecule	UME	1750*	268 15	208 91	195 40	179 100	165 10	704	4952
Endogenous biomolecule	UME	2050*	278 74	246 17	203 16	151 100	150 91	765	4954
Endogenous biomolecule	UME	2510*	320 16	265 20	239 100	210 17	83 29	1031	4957
Endogenous biomolecule	UME	2140*	294 27	238 56	235 47	206 100	195 18	868	4958
Endogenous biomolecule	U+UHYAC	2025	329 19	287 63	244 100	228 13	137 5	1084	10280
Endogenous biomolecule	U+UHYAC	2195	257 11	215 100	186 8	138 14	115 19	640	10292
Endogenous biomolecule (ME)	UME	2100	252 32	192 100	179 39	147 14	84 17	614	5040
Endogenous biomolecule 2AC	UHYAC	2910*	400 14	340 100	265 33	172 20	157 29	1444	985
Endogenous biomolecule 2AC	UHYAC	2280*	292 3	250 19	208 100	123 39	85 13	853	1002
Endogenous biomolecule 2AC	UHYAC	1920	247 8	205 27	163 100	135 25		583	1135
Endogenous biomolecule 2AC	UHYAC	2000	263 29	221 100	177 57	162 45	133 11	672	1508
Endogenous biomolecule 2AC	UHYAC	1875	235 12	193 66	151 100	136 52	108 19	530	1566
Endogenous biomolecule 2AC	UHYAC	2650*	394 3	352 9	310 100	197 43		1419	2369
Endogenous biomolecule 2AC	UHYAC	1800*	296 5	236 9	193 100	149 46	135 77	882	2482
Endogenous biomolecule 2AC	UHYAC	1695	193 1	151 8	133 19	109 100	80 14	377	3212
Endogenous biomolecule 2AC	UHYAC	2800	361 22	319 12	277 100	127 91	84 33	1267	3744
Endogenous biomolecule 2AC	U+UHYAC	1820*	266 3	206 9	164 100	150 10	137 30	695	6409
Endogenous biomolecule 3AC	UHYAC	2060*	276 7	234 15	192 32	150 100	122 5	751	493
Endogenous biomolecule 3AC	UHYAC	1950	305 23	263 65	221 100	179 41	161 66	934	1481

Endogenous biomolecule 3AC

Table 8.1: Compounds in order of names

Name	Detected	RI	Typical ions and intensities					Page	Entry
Endogenous biomolecule 3AC	UHYAC	1760	239 $_2$	197 $_7$	155 $_{58}$	140 $_{22}$	113 $_{100}$	550	2453
Endogenous biomolecule 3AC	UHYAC	1710	235 $_1$	193 $_6$	151 $_{34}$	109 $_{100}$	80 $_{11}$	530	3213
Endogenous biomolecule AC	UHYAC	2620*	370 $_{30}$	310 $_{100}$	254 $_{12}$	239 $_{14}$	161 $_{29}$	1309	43
Endogenous biomolecule AC	UHYAC	2400	315 $_{100}$	255 $_{77}$	214 $_4$	161 $_7$	147 $_{13}$	998	622
Endogenous biomolecule AC	UHYAC	2575*	330 $_{62}$	270 $_{14}$	255 $_{100}$	197 $_{42}$	117 $_{45}$	1091	984
Endogenous biomolecule AC	UHYAC	1350*	155 $_{49}$	140 $_{100}$	112 $_{36}$	69 $_{11}$		297	2367
Endogenous biomolecule AC	UHYAC	2240*	264 $_1$	222 $_{30}$	137 $_{100}$	122 $_6$	85 $_{23}$	679	2452
Endogenous biomolecule AC	UHYAC	3040*	474 $_{16}$	414 $_{10}$	294 $_{55}$	269 $_{100}$	173 $_{55}$	1624	2454
Endogenous biomolecule AC	UHYAC	1640*	208 $_6$	166 $_{49}$	151 $_{100}$	123 $_{12}$		423	2483
Endogenous biomolecule -H2O AC	UHYAC	2830*	340 $_{100}$	265 $_{37}$	172 $_{22}$	157 $_{33}$	145 $_{17}$	1154	802
Endogenous biomolecule isomer-1 2AC	UHYAC	2700*	326 $_{31}$	284 $_{90}$	242 $_{64}$	123 $_{100}$	120 $_{70}$	1065	2428
Endogenous biomolecule isomer-1 AC	UHYAC	2750	340 $_1$	265 $_6$	144 $_{31}$	102 $_{23}$	84 $_{100}$	1154	3664
Endogenous biomolecule isomer-2 2AC	UHYAC	2750*	326 $_{29}$	284 $_{65}$	242 $_{36}$	123 $_{55}$	120 $_{100}$	1066	2429
Endogenous biomolecule isomer-2 AC	UHYAC	2825	352 $_2$	144 $_{52}$	102 $_{35}$	84 $_{100}$	60 $_{31}$	1217	3665
Endogenous biomolecule ME	UME	1945*	296 $_{12}$	236 $_{85}$	223 $_{32}$	179 $_{100}$	147 $_{21}$	882	4953
Endogenous biomolecule ME	UME	2160	292 $_{31}$	203 $_8$	165 $_{100}$	121 $_{10}$	91 $_{24}$	853	4955
Endogenous biomolecule ME	UME	2235*	306 $_{52}$	259 $_{10}$	217 $_9$	179 $_{100}$	91 $_{21}$	941	4956
Endosulfan		2080*	339 $_{17}$	265 $_{51}$	237 $_{84}$	195 $_{100}$	159 $_{77}$	1457	3834
Endosulfan sulfate		2260*	420 $_3$	387 $_{33}$	272 $_{100}$	227 $_{76}$	85 $_{48}$	1507	3835
Endothal		1370*	140 $_9$	100 $_{29}$	81 $_{13}$	68 $_{100}$	53 $_{18}$	360	4154
Endrin		2175*	345 $_6$	281 $_{24}$	263 $_{53}$	113 $_{27}$	81 $_{100}$	1345	3836
Enilconazole		2140	296 $_{10}$	240 $_{12}$	215 $_{100}$	173 $_{74}$	81 $_{32}$	883	2054
Enoximone	U+UHYAC	2770	248 $_{100}$	247 $_{92}$	201 $_{43}$	151 $_{72}$	124 $_{22}$	589	5212
Enoximone 2AC		2560	332 $_8$	290 $_{25}$	248 $_{100}$	201 $_{70}$	151 $_{60}$	1105	5210
Enoximone AC	U+UHYAC	2600	290 $_{13}$	248 $_{100}$	201 $_{43}$	151 $_{44}$	108 $_{18}$	840	5211
Ephedrine	G UHY	1375	146 $_1$	131	105 $_2$	77 $_{12}$	58 $_{100}$	314	748
Ephedrine 2AC	PAC U+UHYAC	1795	249 $_1$	148 $_2$	117 $_2$	100 $_{57}$	58 $_{100}$	597	749
Ephedrine 2HFB	UHYHFB	1500	344 $_{10}$	254 $_{100}$	210 $_{27}$	169 $_{18}$	69 $_{36}$	1698	5097
Ephedrine 2PFP		1370	338 $_1$	294 $_3$	204 $_{100}$	160 $_{25}$	119 $_{14}$	1595	2577
Ephedrine 2TFA		1345	338 $_1$	244 $_4$	154 $_{100}$	110 $_{72}$	69 $_{47}$	1246	3997
Ephedrine 2TMS		1620	294 $_3$	163 $_4$	147 $_8$	130 $_{100}$	73 $_{84}$	966	4543
Ephedrine formyl artifact	G U	1430	177 $_1$	121 $_{10}$	107 $_6$	71 $_{100}$	56 $_{76}$	336	4500
Ephedrine -H2O AC	UHYAC	1560	189 $_1$	148 $_3$	121 $_8$	100 $_{49}$	58 $_{100}$	368	5646
Ephedrine TMSTFA		1620	318 $_1$	227 $_9$	179 $_{100}$	110 $_8$	73 $_{79}$	1111	6038
Ephedrine-M (HO-)		1875	148 $_1$	95 $_1$	77 $_4$	71 $_6$	58 $_{100}$	349	1971
Ephedrine-M (HO-) 3AC	U+UHYAC	2145	307 $_1$	247	100 $_{72}$	58 $_{100}$		950	750
Ephedrine-M (HO-) formyl artifact		1790	133 $_2$	121 $_{10}$	107 $_6$	71 $_{100}$	56 $_{76}$	380	4499
Ephedrine-M (HO-) -H2O 2AC	U+UHYAC	1990	247 $_9$	205 $_8$	163 $_{24}$	107 $_8$	56 $_{100}$	585	1972
Ephedrine-M (HO-) ME2AC		2000	279 $_1$	247	206 $_3$	100 $_{60}$	58 $_{100}$	776	2348
Ephedrine-M (nor-)	P U	1370	132 $_2$	118 $_8$	107 $_{11}$	91 $_{10}$	77 $_{100}$	292	2475
Ephedrine-M (nor-) 2AC	U+UHYAC	1805	235 $_1$	176 $_5$	134 $_7$	107 $_{13}$	86 $_{100}$	532	2476
Ephedrine-M (nor-) 2HFB	UHYHFB	1455	543 $_1$	330 $_{14}$	240 $_{100}$	169 $_{44}$	69 $_{57}$	1691	5098
Ephedrine-M (nor-) 2PFP	UHYPFP	1380	443 $_1$	280 $_9$	190 $_{100}$	119 $_{59}$	105 $_{26}$	1567	5094
Ephedrine-M (nor-) 2TFA	UTFA	1355	343 $_1$	230 $_6$	203 $_5$	140 $_{100}$	69 $_{29}$	1172	5091
Ephedrine-M (nor-) 2TMS		1555	280 $_5$	163 $_4$	147 $_{10}$	116 $_{100}$	73 $_{83}$	880	4574
Ephedrine-M (nor-) formyl artifact		1240	117 $_2$	105 $_4$	91 $_4$	77 $_{10}$	57 $_{100}$	308	4650
Ephedrine-M (nor-) TMSTFA		1890	240 $_8$	198 $_3$	179 $_{100}$	117 $_5$	73 $_{88}$	1024	6146
Ephedrine-M (nor-HO-) 3AC	U+UHYAC	2135	234 $_6$	165 $_8$	123 $_{11}$	86 $_{100}$	58 $_{45}$	862	4961
Ephenidine		1710	224 $_1$	181 $_2$	178 $_3$	134 $_{100}$	79 $_{21}$	492	8434
Ephenidine AC		2140	267 $_1$	196 $_3$	176 $_{46}$	165 $_7$	134 $_{100}$	702	8435
Ephenidine HFB		1920	330 $_{59}$	224 $_{100}$	180 $_{27}$	165 $_{12}$	91 $_{35}$	1510	8440
Ephenidine ME		1750	186 $_1$	165 $_3$	148 $_{100}$	127 $_4$	91 $_8$	553	8437
Ephenidine PFP		1890	280 $_{50}$	180 $_{16}$	174 $_{100}$	165 $_8$	91 $_{31}$	1318	8439
Ephenidine TFA		1910	230 $_{56}$	180 $_{19}$	124 $_{100}$	91 $_{18}$	79 $_{16}$	1038	8438
Ephenidine TMS		1950	282 $_1$	206 $_{100}$	190 $_8$	165 $_8$	134 $_5$	892	8436
Ephenidine-M (di-HO-benzyl-) 3AC	U+UHYAC	2790	296 $_8$	254 $_{11}$	212 $_{15}$	176 $_{42}$	134 $_{100}$	1371	8653
Ephenidine-M (HO-benzyl-) isomer-1 2AC	U+UHYAC	2580	238 $_6$	196 $_{44}$	176 $_{30}$	134 $_{100}$		1063	8983
Ephenidine-M (HO-benzyl-) isomer-2 2AC	U+UHYAC	2620	324 $_1$	238 $_{21}$	196 $_{31}$	176 $_{50}$	134 $_{100}$	1063	8651
Ephenidine-M (HO-methoxy-benzyl-) 2AC	U+UHYAC	2710	296 $_2$	268 $_{21}$	226 $_{44}$	176 $_{28}$	134 $_{100}$	1238	8652
Ephenidine-M (nor-HO-alkyl-) 2AC	U+UHYAC	1740	206 $_{42}$	164 $_{90}$	122 $_{30}$	104 $_{100}$	91 $_{33}$	890	8659
Ephenidine-M 2AC	U+UHYAC	2500	296 $_1$	206 $_{98}$	164 $_{79}$	122 $_{100}$	91 $_{25}$	890	8655
Ephenidine-M 2AC	U+UHYAC	2540	268 $_{15}$	236 $_{28}$	194 $_{39}$	152 $_{45}$	106 $_{100}$	1075	8656
Ephenidine-M 2AC	U+UHYAC	2560	236 $_{29}$	194 $_{45}$	152 $_{73}$	137 $_{10}$	106 $_{100}$	1076	8825
Ephenidine-M 2AC	U+UHYAC	2680	284 $_3$	242 $_9$	153 $_9$	148 $_{49}$	106 $_{100}$	1175	8827
Ephenidine-M 2AC		2345*	207 $_{44}$	165 $_{80}$	123 $_{100}$	105 $_{81}$	77 $_{45}$	981	8933
Ephenidine-M 2AC	U+UHYAC	2560	254 $_2$	212 $_3$	178 $_7$	148 $_{62}$	106 $_{100}$	988	9120
Ephenidine-M 3AC		2350*	355 $_2$	296 $_9$	208 $_{27}$	166 $_{100}$	123 $_{93}$	1242	8987
Ephenidine-M AC		2230*	238 $_{28}$	196 $_{31}$	165 $_{20}$	150 $_{27}$	107 $_{100}$	895	8932
Ephenidine-M AC		2250*	284 $_1$	242 $_{22}$	226 $_{27}$	137 $_{100}$	105 $_{84}$	800	8985
Ephenidine-M AC		2360*	270 $_7$	228 $_{19}$	153 $_{15}$	123 $_{15}$	105 $_{100}$	716	8986
Ephenidine-M isomer-1 2AC	U+UHYAC	2350	238 $_7$	196 $_{12}$	148 $_{47}$	106 $_{100}$	79 $_{11}$	890	8984
Ephenidine-M isomer-1 3AC	U+UHYAC	2610	296 $_{35}$	254 $_{11}$	212 $_{10}$	148 $_{46}$	106 $_{100}$	1237	9121
Ephenidine-M isomer-1 AC		2080*	254 $_1$	212 $_4$	105 $_{100}$	77 $_{34}$		628	8934
Ephenidine-M isomer-2 2AC	U+UHYAC	2380	238 $_{13}$	196 $_{14}$	148 $_{57}$	106 $_{100}$	79 $_{16}$	890	8654

Table 8.1: Compounds in order of names

Name	Detected	RI	Typical ions and intensities					Page	Entry
Ephenidine-M isomer-2 3AC	U+UHYAC	2630	296_{7}	254_{3}	212_{8}	148_{55}	106_{100}	1237	8826
Ephenidine-M isomer-2 AC		2120*	254_{2}	212_{12}	105_{100}	77_{31}		628	8935
Epiandrosterone		2520*	290_{100}	246_{28}	147_{25}	107_{71}	67_{57}	844	3898
Epiandrosterone AC		2630*	332_{19}	272_{14}	218_{41}	201_{57}	107_{80}	1108	3919
Epiandrosterone enol 2TMS		2570*	434_{78}	419_{100}	329_{9}	239_{3}	73_{50}	1547	3960
Epiandrosterone TMS		2500*	362_{11}	347_{60}	272_{11}	155_{11}	75_{100}	1276	3959
Epinastine		2430	249_{44}	194_{100}	178_{34}	165_{30}	116_{22}	595	7262
Epinastine 2TMS		2470	393_{100}	378_{94}	279_{74}	171_{88}	73_{97}	1417	7270
Epinastine AC		2600	291_{36}	276_{48}	248_{43}	194_{100}	178_{42}	847	7264
Epinastine HFB		2530	445_{7}	276_{100}	248_{7}	178_{27}	165_{28}	1571	7267
Epinastine ME		2380	263_{74}	262_{100}	194_{46}	178_{44}	165_{51}	675	7263
Epinastine PFP		2520	395_{31}	276_{100}	249_{27}	194_{33}	165_{34}	1423	7266
Epinastine TFA		2580	345_{32}	276_{100}	178_{28}	165_{29}	69_{72}	1184	7265
Epinastine TMS		2450	321_{13}	249_{28}	194_{100}	178_{29}	165_{23}	1040	7268
Epinephrine artifact (3,4-dihydroxybenzoic acid) ME2AC		1750*	252_{1}	210_{19}	168_{100}	137_{60}	109_{8}	615	5254
Epitestosterone enol 2TMS		2620*	432_{91}	417_{8}	327_{8}	209_{16}	73_{100}	1541	3802
Eplerenone	U+UHYAC	3455*	414_{11}	399_{5}	355_{51}	194_{23}	55_{100}	1491	7270
Eplerenone HFB		3015*	610_{1}	551_{5}	533_{2}	169_{60}	69_{100}	1717	7273
Eplerenone PFP		2985*	560_{1}	501_{6}	483_{2}	119_{100}	55_{52}	1700	7272
Eplerenone TFA		2995*	510_{1}	451_{15}	433_{6}	111_{44}	69_{100}	1669	7271
Eplerenone TMS		3430*	486_{4}	427_{21}	291_{3}	111_{17}	73_{100}	1644	7274
Eprazinone		2820	380_{1}	245_{100}	139_{9}	111_{30}	105_{60}	1358	1938
Eprosartan 2ME		3335	452_{100}	410_{48}	351_{43}	149_{59}	121_{77}	1584	7592
Eprosartan 2TMS		3480	568_{69}	409_{54}	361_{57}	271_{54}	73_{100}	1703	7593
EPTC		1350	189_{14}	160_{5}	132_{25}	128_{78}	86_{100}	368	3188
Ergine		2820	267_{100}	221_{51}	207_{53}	180_{49}	154_{48}	701	8441
Ergocristine artifact-1 (cyclo (Phe-Pro)) isomer-1	U+UHYAC P-I	2335	244_{36}	153_{50}	125_{100}	91_{38}	70_{50}	571	4495
Ergocristine artifact-1 (cyclo (Phe-Pro)) isomer-2	U+UHYAC P-I	2365	244_{54}	153_{21}	125_{100}	91_{41}	70_{37}	571	8443
Ergocristine artifact-2		2440	314_{45}	244_{35}	153_{84}	125_{75}	70_{100}	994	4494
Ergocristine artifact-3 (- LSA)		2450	342_{32}	273_{11}	243_{38}	153_{24}	70_{100}	1170	8442
Ergocristine artifact-4 (LSA)		2820	267_{100}	221_{51}	207_{53}	180_{49}	154_{48}	701	8441
Ergometrine		3120	325_{100}	267_{13}	221_{85}	207_{85}	181_{70}	1064	751
Ergometrine AC		3235	367_{60}	281_{10}	221_{100}	207_{69}	181_{50}	1299	8514
Ergometrine artifact (-COOH) (ME)		2770	282_{2}	252_{100}	196_{34}	180_{53}	154_{69}	793	8513
Ergometrine -H2O		2860	307_{35}	264_{18}	221_{100}	196_{57}	112_{51}	951	8515
Ergometrine TMS		3020	397_{67}	307_{13}	221_{100}	196_{64}	112_{56}	1436	8515
Ergost-3,5-ene		3270*	382_{100}	213_{22}	147_{78}	105_{62}	55_{56}	1368	5624
Ergost-5-en-3-ol		3190*	400_{60}	315_{33}	289_{41}	105_{84}	55_{100}	1447	5620
Ergost-5-en-3-ol -H2O		3270*	382_{100}	213_{22}	147_{78}	105_{62}	55_{56}	1368	5624
Ergosta-3,5,22-triene		3210*	380_{100}	255_{39}	81_{53}	69_{53}	55_{67}	1358	5623
Ergosta-5,22-dien-3-ol		3135*	398_{100}	300_{62}	271_{11}	255_{85}	69_{96}	1440	5619
Ergosta-5,22-dien-3-ol -H2O		3210*	380_{100}	255_{39}	81_{53}	69_{53}	55_{67}	1358	5623
Ergosterol	G	3130*	396_{64}	363_{100}	337_{34}	253_{30}	143_{36}	1432	5137
Ergotamine artifact-1 (cyclo (Phe-Pro)) isomer-1	U+UHYAC P-I	2335	244_{36}	153_{50}	125_{100}	91_{38}	70_{50}	571	4495
Ergotamine artifact-1 (cyclo (Phe-Pro)) isomer-2	U+UHYAC P-I	2365	244_{54}	153_{21}	125_{100}	91_{41}	70_{37}	571	8443
Ergotamine artifact-1 AC		2360	286_{29}	153_{62}	125_{100}	91_{57}	70_{73}	815	5217
Ergotamine artifact-2		2440	314_{45}	244_{35}	153_{84}	125_{75}	70_{100}	994	4494
Erlotinib		3370	393_{30}	392_{35}	334_{20}	276_{31}	59_{100}	1415	8167
Erlotinib TFA		2780	489_{10}	420_{41}	362_{10}	304_{12}	59_{100}	1647	8169
Erlotinib TMS		2970	465_{9}	450_{28}	350_{25}	73_{100}	59_{72}	1608	8168
Erucic acid ME		2490*	352_{5}	320_{45}	97_{44}	69_{58}	55_{100}	1224	2670
Erythritol 4AC		1595*	217_{11}	145_{94}	128_{35}	115_{100}	103_{88}	841	5605
Esmolol		2225	294_{1}	251_{2}	116_{4}	107_{5}	72_{100}	880	6266
Esmolol 2AC		2400	291_{2}	200_{100}	140_{26}	98_{40}	72_{82}	1352	5136
Esmolol 2HFB		2005	687_{1}	508_{56}	466_{100}	252_{39}	226_{26}	1728	6269
Esmolol 2PFP		2115	587_{1}	408_{28}	366_{100}	202_{40}	176_{35}	1710	6268
Esmolol 2TFA		1990	487_{1}	308_{72}	266_{100}	152_{19}	126_{9}	1645	6271
Esmolol formyl artifact		2290	307_{16}	292_{53}	127_{89}	112_{68}	56_{100}	952	5135
Esmolol -H2O AC	G	2575	319_{3}	200_{100}	140_{19}	98_{18}	72_{48}	1028	6267
Esmolol TMSTFA		2130	463_{1}	448_{2}	284_{100}	270_{10}	131_{19}	1606	6270
Estazolam		3070	294_{82}	259_{100}	239_{43}	205_{61}	101_{54}	869	2392
Estradiol		2550*	272_{100}	213_{30}	172_{24}	160_{30}	146_{25}	731	1434
Estradiol 2AC	U+UHYAC	2780*	356_{10}	314_{100}	172_{18}	146_{18}		1244	1435
Estradiol undecylate		3900*	440_{60}	255_{19}	159_{77}	133_{52}	57_{100}	1560	5244
Estriol		2940*	288_{100}	213_{25}	172_{21}	160_{32}	146_{25}	829	1436
Estriol 3AC	U+UHYAC	3010*	414_{14}	372_{100}	330_{4}	270_{8}	160_{30}	1491	1476
Estriol-M (HO-) 4AC	UHYAC	3280*	472_{8}	430_{100}	268_{69}	250_{20}	107_{11}	1622	4290
Estrone		2580*	270_{100}	213_{19}	185_{40}	172_{36}	146_{55}	720	5178
Estrone AC	U+UHYAC	2630*	312_{5}	270_{100}	213_{19}	185_{40}	146_{39}	983	5207
Estrone ME		2530*	284_{100}	227_{18}	199_{82}	160_{72}	115_{40}	805	5206
Etacrinic acid ET		2230*	330_{27}	315_{37}	263_{67}	261_{100}	203_{6}	1092	2631
Etacrinic acid ME		2195*	316_{11}	281_{5}	263_{66}	261_{100}	243_{40}	1003	2630
Etafenone		2680	325_{1}	310	99_{20}	86_{100}	58_{15}	1064	2503
Etafenone-M (di-HO-) 2AC	UHYAC	3070	441_{1}	426_{5}	99_{9}	86_{100}	58_{6}	1563	3358

Etafenone-M (HO-) isomer-1   Table 8.1: Compounds in order of names

Name	Detected	RI	Typical ions and intensities					Page	Entry
Etafenone-M (HO-) isomer-1	UHY	2800	341_1	326_1	99_{10}	86_{100}	58_7	1165	3347
Etafenone-M (HO-) isomer-1 AC	UHYAC	2775	383_1	368_2	99_{16}	86_{100}	58_{10}	1372	3355
Etafenone-M (HO-) isomer-2	UHY	2820	341_2	326_1	99_{15}	86_{100}	58_7	1165	3348
Etafenone-M (HO-) isomer-2 AC	UHYAC	2810	383_1	368_2	99_{15}	86_{100}	58_{11}	1372	3356
Etafenone-M (HO-methoxy-)	U UHY	2830	371_9	137_5	99_8	86_{100}	58_6	1320	3349
Etafenone-M (HO-methoxy-) AC	UHYAC	2955	413_1	398_4	137_7	99_{15}	86_{100}	1488	3357
Etafenone-M (O-dealkyl-)	G P-I U+UHYAC	1830*	226_{35}	207_{14}	121_{100}	91_{28}	65_{32}	495	896
Etafenone-M (O-dealkyl-) AC	UHYAC	2130*	268_6	225_{38}	208_{32}	121_{100}	91_{15}	707	3726
Etafenone-M (O-dealkyl-di-HO-) 2AC	UHYAC	2620*	342_6	300_{40}	258_{87}	136_{37}	121_{100}	1168	3354
Etafenone-M (O-dealkyl-HO-) 2AC	UHYAC	2515*	326_{15}	284_{26}	242_{20}	224_{27}	121_{100}	1068	3352
Etafenone-M (O-dealkyl-HO-) isomer-1	UHY	2345*	242_{35}	223_{22}	121_{100}	107_{67}	65_{26}	565	3344
Etafenone-M (O-dealkyl-HO-) isomer-1 AC	U+UHYAC	2215*	284_{30}	242_{96}	137_{100}	91_{79}		801	899
Etafenone-M (O-dealkyl-HO-) isomer-2	UHY	2355*	242_{43}	223_{27}	121_{100}	107_{78}	65_{30}	564	3345
Etafenone-M (O-dealkyl-HO-) isomer-2 AC	UHYAC	2370*	284_{17}	242_{26}	224_{38}	121_{100}	65_{17}	800	3350
Etafenone-M (O-dealkyl-HO-) isomer-3 AC	UHYAC	2410*	284_8	242_{31}	224_{26}	121_{100}	107_{81}	801	3351
Etafenone-M (O-dealkyl-HO-methoxy-)	UHY	2400*	272_{36}	151_9	137_{100}	121_{52}	65_{32}	730	3346
Etafenone-M (O-dealkyl-HO-methoxy-) isomer-1 AC	UHYAC	2525*	314_6	272_{58}	137_{100}	121_{61}	65_{32}	994	3353
Etafenone-M (O-dealkyl-HO-methoxy-) isomer-2 AC	U+UHYAC	2580*	314_{19}	272_{90}	167_{100}	137_{17}	91_{53}	994	900
Etamiphylline	G U UHY UHYAC	2210	279_2	99_6	86_{100}	58_9		778	1201
Etamiphylline-M (deethyl-) AC	UHYAC	2560	293_{10}	250_{10}	206_{50}	114_{26}	58_{100}	864	1723
Etamivan	G UHY	1900	223_{35}	151_{100}	72_{25}			483	752
Etamivan AC	UHYAC	1970	265_{11}	222_{38}	194_{12}	151_{100}	72_{26}	687	753
7-Et-DALT		2200	268_1	172_3	158_{12}	143_{13}	110_{100}	709	8852
7-Et-DALT HFB		2020	368_1	354_4	199_2	156_{16}	110_{100}	1608	10106
7-Et-DALT PFP		2025	414_3	318_{31}	304_{100}	198_{14}	157_{23}	1491	10105
7-Et-DALT TFA		2065	364_1	337_2	268_5	254_{11}	110_{100}	1283	10104
7-Et-DALT TMS		2345	340_1	299_2	230_3	156_2	110_{100}	1158	10103
Ethacridine	G	3000	253_{84}	224_{100}	196_{33}	179_3	169_7	622	6376
Ethadione		1120	157_{51}	70_{37}	58_{100}			301	221
Ethambutol 4AC	U+UHYAC	2455	329_1	299_2	199_{37}	144_{100}	84_{33}	1265	0440
Ethanol		<1000*	46_{24}	45_{38}	31_{100}	28_{50}		248	1545
Ethaverine	P G U+UHYAC	2940	395_{57}	366_{100}	352_{11}	252_5	236_7	1426	754
Ethaverine-M (bis-deethyl-) isomer-1 2AC	UHYAC	3050	423_{88}	381_{96}	352_{79}	310_{100}	133_{13}	1515	3668
Ethaverine-M (bis-deethyl-) isomer-2 2AC	UHYAC	3085	423_{45}	380_{86}	352_{17}	310_{100}	133_{13}	1515	3669
Ethaverine-M (HO-) AC	UHYAC	3160	453_{31}	424_{22}	410_{33}	394_{100}	382_{69}	1587	3670
Ethaverine-M (HO-) ME	UHYME	2905	425_{51}	396_{100}	368_5	228_3	213_8	1521	3713
Ethaverine-M (O-deethyl-) isomer-1	UHY	2900	367_{63}	338_{100}	310_{14}	236_8	196_5	1299	3666
Ethaverine-M (O-deethyl-) isomer-1 AC	UHYAC	2980	409_{94}	380_{42}	366_{41}	338_{100}	310_{13}	1474	3074
Ethaverine-M (O-deethyl-) isomer-1 ME	UHYME	2850	381_{69}	352_{100}	324_7	236_6	196_4	1362	3715
Ethaverine-M (O-deethyl-) isomer-2	UHY	2930	367_{60}	338_{100}	310_{12}	236_6	208_4	1298	3667
Ethaverine-M (O-deethyl-) isomer-2 AC	UHYAC	3020	409_{46}	380_{37}	366_{51}	338_{100}	310_6	1474	3075
Ethaverine-M (O-deethyl-) isomer-2 ME	UHYME	2880	381_{94}	352_{100}	324_8	236_5	196_3	1363	3716
Ethaverine-M (O-deethyl-HO-) 2AC	UHYAC	3210	467_{13}	424_{36}	408_{100}	382_{28}	366_{46}	1612	3671
Ethaverine-M (O-deethyl-HO-) 2ME	UHYME	2980	411_{100}	396_{75}	382_{51}	358_{21}	233_4	1484	3714
Ethchlorvynol		<1000*	115_{100}	109_{20}	89_{20}	53_{34}		283	2407
Ethenzamide	G P	1575	165_9	150_{38}	120_{100}	105_{53}	92_{67}	313	192
Ethenzamide-M (deethyl-)	P G UHY	1460	137_{90}	120_{100}	92_{80}	65_{56}		276	755
Ethenzamide-M (deethyl-) 2ME		1480	165_{11}	135_{100}	105_6	92_{15}	77_{30}	312	6395
Ethenzamide-M (deethyl-) 2TMS		1725	281_3	266_{100}	250_{80}	176_{40}	73_{88}	788	4596
Ethenzamide-M (deethyl-) AC	U+UHYAC	1660	179_{39}	137_{65}	120_{100}	92_{39}	63_{20}	340	193
Ethinamate	P G U	1395	124_{65}	95_{100}	91_{84}	81_{77}		319	756
Ethinylestradiol		2525*	296_{40}	228_{13}	213_{100}	160_{40}	133_{24}	887	5177
Ethinylestradiol AC		2610*	338_{15}	296_{89}	228_{25}	213_{100}	160_{40}	1145	5180
Ethinylestradiol -HCCH		2580*	270_{100}	213_{19}	185_{40}	172_{36}	146_{25}	720	5178
Ethinylestradiol -HCCH AC	U+UHYAC	2630*	312_5	270_{100}	213_{19}	185_{40}	146_{39}	983	5207
Ethinylestradiol -HCCH ME		2530*	284_{100}	227_{18}	199_{82}	160_{72}	115_{40}	805	5206
Ethiofencarb		1835	225_4	168_{62}	139_7	107_{100}	77_{44}	490	3444
Ethiofencarb-M/artifact (decarbamoyl-)		1390*	168_{54}	137_2	107_{100}	77_{55}		321	3445
Ethion	G	2235*	384_5	231_{67}	153_{55}	125_{51}	97_{100}	1373	3837
Ethirimol		2080	209_{48}	194_3	166_{100}	96_{37}	55_{17}	431	3642
Ethofumesate		1985*	286_{23}	207_{100}	161_{87}	137_{52}	79_{49}	813	4080
Ethoprofos		1700*	242_{12}	200_{27}	158_{83}	139_{45}	97_{100}	563	4081
Ethosuximide	P G U+UHYAC	1225	141_8	113_{95}	70_{86}	55_{100}		280	757
Ethosuximide ME		1130	155_1	127_{89}	112_{12}	70_{69}	55_{100}	298	2922
Ethosuximide-M (3-HO-)	U UHY	1325	157_{12}	129_{76}	86_{78}	71_{100}		301	758
Ethosuximide-M (3-HO-) AC	U+UHYAC	1350	199_3	171_{100}	129_{54}	86_{95}	84_{73}	397	760
Ethosuximide-M (HO-ethyl-)	U UHY	1370	142_4	113_{100}	85_{46}	69_{40}		301	759
Ethosuximide-M (HO-ethyl-) AC	U+UHYAC	1390	171_{20}	155_{35}	139_{33}	113_{100}		397	761
Ethosuximide-M (oxo-)	U+UHYAC	1270	155_{82}	113_{46}	98_{31}	70_{100}	55_{77}	298	2913
7-Ethoxy-4-trifluoromethylumbelliferone		1540*	258_{44}	230_{28}	202_{100}	173_{13}	145_{18}	645	8151
7-Ethoxycoumarin		----*	190_{100}	162_{50}	134_{94}			369	762
5-Ethoxy-N,N-diallyl-tryptamine		2330	284_2	174_8	146_{12}	130_5	110_{100}	805	8846
5-Ethoxy-N,N-diallyl-tryptamine AC		2570	326_{16}	283_3	174_7	146_4	110_{100}	1071	10017
5-Ethoxy-N,N-diallyl-tryptamine HFB		2210	480_1	453_2	384_8	370_{21}	110_{100}	1634	10016

Table 8.1: Compounds in order of names

5-Ethoxy-N,N-diallyl-tryptamine PFP

Name	Detected	RI	Typical ions and intensities					Page	Entry
5-Ethoxy-N,N-diallyl-tryptamine PFP		2160	403 $_2$	334 $_6$	320 $_{12}$	110 $_{100}$		1534	10015
5-Ethoxy-N,N-diallyl-tryptamine TFA		2185	353 $_3$	284 $_9$	270 $_{15}$	145 $_{27}$	110 $_{100}$	1357	10014
5-Ethoxy-N,N-diallyl-tryptamine TMS		2400	356 $_4$	315 $_5$	246 $_{17}$	202 $_8$	110 $_{100}$	1244	10013
5-Ethoxy-N,N-diallyl-tryptamine-D4		2330	288 $_2$	176 $_5$	148 $_7$	119 $_5$	112 $_{100}$	831	8847
5-Ethoxy-N,N-diallyl-tryptamine-D4 HFB		2165	457 $_2$	388 $_7$	372 $_{14}$	147 $_{29}$	112 $_{100}$	1639	10020
5-Ethoxy-N,N-diallyl-tryptamine-D4 PFP		2150	434 $_1$	407 $_2$	322 $_8$	147 $_{26}$	112 $_{100}$	1545	10019
5-Ethoxy-N,N-diallyl-tryptamine-D4 TFA		2255	384 $_1$	357 $_3$	272 $_{14}$	147 $_9$	112 $_{100}$	1377	10193
5-Ethoxy-N,N-diallyl-tryptamine-D4 TMS		2435	360 $_6$	319 $_7$	248 $_{24}$	204 $_{11}$	112 $_{100}$	1266	10018
5-Ethoxy-N,N-diethyl-tryptamine-D4		2190	264 $_8$	176 $_9$	148 $_{15}$	119 $_{11}$	88 $_{100}$	684	8872
5-Ethoxy-N,N-diethyl-tryptamine-D4 HFB		2000	445 $_1$	388 $_4$	372 $_{10}$	147 $_{22}$	88 $_{100}$	1602	9530
5-Ethoxy-N,N-diethyl-tryptamine-D4 PFP		1990	322 $_1$	175 $_2$	147 $_7$	119 $_6$	88 $_{100}$	1478	9529
5-Ethoxy-N,N-diethyl-tryptamine-D4 TFA		2010	288 $_3$	272 $_8$	244 $_4$	147 $_{21}$	88 $_{100}$	1265	9528
5-Ethoxy-N,N-diethyl-tryptamine-D4 TMS		2300	336 $_{27}$	264 $_8$	248 $_{25}$	204 $_{16}$	88 $_{100}$	1134	9527
5-Ethoxy-N-allyl-N-cyclohexyl-tryptamine-D4		2690	330 $_2$	176 $_9$	154 $_{100}$	148 $_{13}$	72 $_{49}$	1096	8848
5-Ethoxy-N-allyl-N-cyclohexyl-tryptamine-D4 AC		2900	372 $_{16}$	329 $_{10}$	190 $_4$	176 $_5$	154 $_{100}$	1325	10024
5-Ethoxy-N-allyl-N-cyclohexyl-tryptamine-D4 HFB		2485	388 $_3$	372 $_4$	344 $_3$	154 $_{100}$		1681	10023
5-Ethoxy-N-allyl-N-cyclohexyl-tryptamine-D4 PFP		2475	338 $_7$	322 $_{12}$	294 $_6$	154 $_{100}$	72 $_{35}$	1629	10194
5-Ethoxy-N-allyl-N-cyclohexyl-tryptamine-D4 TFA		2485	425 $_1$	288 $_8$	272 $_{14}$	244 $_6$	154 $_{100}$	1523	10022
5-Ethoxy-N-allyl-N-cyclohexyl-tryptamine-D4 TMS		2715	402 $_5$	264 $_5$	248 $_{13}$	204 $_8$	154 $_{100}$	1454	10021
Ethoxyphenyldiethylphenyl butyramine		2350	325 $_1$	252 $_4$	206 $_{100}$	178 $_8$	105 $_{34}$	1065	763
Ethoxyquin		1720	217 $_{11}$	202 $_{100}$	174 $_{56}$	145 $_{25}$	115 $_7$	455	3851
N-Ethyl-1,2-diphenylethylamine		1710	224 $_1$	181 $_2$	178 $_3$	134 $_{100}$	79 $_{21}$	492	8434
N-Ethyl-1,2-diphenylethylamine AC		2140	267 $_1$	196 $_3$	176 $_{46}$	165 $_7$	134 $_{100}$	702	8435
N-Ethyl-1,2-diphenylethylamine HFB		1920	330 $_{59}$	224 $_{100}$	180 $_{27}$	165 $_{12}$	91 $_{35}$	1510	8440
N-Ethyl-1,2-diphenylethylamine ME		1750	186 $_1$	165 $_3$	148 $_{100}$	127 $_4$	91 $_8$	553	8437
N-Ethyl-1,2-diphenylethylamine PFP		1890	280 $_{50}$	180 $_{16}$	174 $_{100}$	165 $_8$	91 $_{31}$	1318	8439
N-Ethyl-1,2-diphenylethylamine TFA		1910	230 $_{56}$	180 $_{19}$	124 $_{100}$	91 $_{18}$	79 $_{16}$	1038	8438
N-Ethyl-1,2-diphenylethylamine TMS		1950	282 $_1$	206 $_{100}$	190 $_8$	165 $_8$	134 $_5$	892	8436
N-Ethyl-1,2-diphenylethylamine-M (deamino-HO-phenyl-) AC		2230*	238 $_{28}$	196 $_{31}$	165 $_{20}$	150 $_{27}$	107 $_{100}$	895	8932
N-Ethyl-1,2-diphenylethylamine-M (di-HO-benzyl-) 3AC	U+UHYAC	2790	296 $_8$	254 $_{11}$	212 $_{15}$	176 $_{42}$	134 $_{100}$	1371	8653
N-Ethyl-1,2-diphenylethylamine-M (HO-benzyl-) isomer-1 2AC	U+UHYAC	2580	238 $_6$	196 $_{44}$	176 $_{30}$	134 $_{100}$		1063	8983
N-Ethyl-1,2-diphenylethylamine-M (HO-benzyl-) isomer-2 2AC	U+UHYAC	2620	324 $_1$	238 $_{21}$	196 $_{31}$	176 $_{50}$	134 $_{100}$	1063	8651
N-Ethyl-1,2-diphenylethylamine-M (HO-methoxy-benzyl-) 2AC	U+UHYAC	2710	296 $_2$	268 $_{21}$	226 $_{44}$	176 $_{28}$	134 $_{100}$	1238	8652
N-Ethyl-1,2-diphenylethylamine-M (nor-HO-alkyl-) 2AC	U+UHYAC	1740	206 $_{42}$	164 $_{90}$	122 $_{30}$	104 $_{100}$	91 $_{33}$	890	8659
N-Ethyl-1,2-diphenylethylamine-M (nor-HO-benzyl-) isomer-1 2AC	U+UHYAC	2350	238 $_7$	196 $_{14}$	148 $_{47}$	106 $_{100}$	79 $_{11}$	890	8984
N-Ethyl-1,2-diphenylethylamine-M 2AC	U+UHYAC	2500	296 $_1$	206 $_{98}$	164 $_{79}$	122 $_{100}$	91 $_{25}$	890	8655
N-Ethyl-1,2-diphenylethylamine-M 2AC		2540	268 $_{15}$	236 $_{28}$	194 $_{39}$	152 $_{45}$	106 $_{100}$	1075	8656
N-Ethyl-1,2-diphenylethylamine-M 2AC	U+UHYAC	2560	236 $_{29}$	194 $_{45}$	152 $_{73}$	137 $_{10}$	106 $_{100}$	1076	8825
N-Ethyl-1,2-diphenylethylamine-M 2AC	U+UHYAC	2680	284 $_3$	242 $_9$	153 $_9$	148 $_{49}$	106 $_{100}$	1175	8827
N-Ethyl-1,2-diphenylethylamine-M 2AC		2345*	207 $_{44}$	165 $_{42}$	123 $_{100}$	105 $_{81}$	77 $_{45}$	981	8933
N-Ethyl-1,2-diphenylethylamine-M 2AC	U+UHYAC	2560	254 $_2$	212 $_3$	178 $_7$	148 $_{62}$	106 $_{100}$	988	9120
N-Ethyl-1,2-diphenylethylamine-M 3AC		2350*	355 $_2$	296 $_9$	208 $_{27}$	166 $_{11}$	123 $_{93}$	1242	8987
N-Ethyl-1,2-diphenylethylamine-M AC		2250*	284 $_1$	242 $_{22}$	226 $_{27}$	137 $_{100}$	105 $_{84}$	800	8985
N-Ethyl-1,2-diphenylethylamine-M AC		2360*	270 $_7$	228 $_{19}$	153 $_{21}$	123 $_{15}$	105 $_{100}$	716	8986
N-Ethyl-1,2-diphenylethylamine-M isomer-1 3AC	U+UHYAC	2610	296 $_1$	254 $_{11}$	212 $_{10}$	148 $_{46}$	106 $_{100}$	1237	9121
N-Ethyl-1,2-diphenylethylamine-M isomer-1 AC		2080*	254 $_1$	212 $_4$	105 $_{100}$	77 $_{34}$		628	8934
N-Ethyl-1,2-diphenylethylamine-M isomer-2 2AC	U+UHYAC	2380	238 $_{13}$	196 $_{14}$	148 $_{57}$	106 $_{100}$	79 $_{16}$	890	8654
N-Ethyl-1,2-diphenylethylamine-M isomer-2 3AC	U+UHYAC	2630	296 $_7$	254 $_3$	212 $_8$	148 $_{55}$	106 $_{100}$	1237	8826
N-Ethyl-1,2-diphenylethylamine-M isomer-2 AC		2120*	254 $_2$	212 $_{12}$	105 $_{100}$	77 $_{31}$		628	8935
4-Ethyl-2,5-dimethoxyphenethylamine		1660	209 $_{20}$	180 $_{100}$	165 $_{52}$	149 $_9$	91 $_{17}$	431	6905
4-Ethyl-2,5-dimethoxyphenethylamine 2AC		2075	293 $_9$	192 $_{100}$	177 $_{34}$	149 $_{11}$	91 $_{15}$	865	6917
4-Ethyl-2,5-dimethoxyphenethylamine 2TMS		2065	353 $_2$	338 $_{15}$	174 $_{100}$	100 $_{18}$	86 $_{24}$	1229	6919
4-Ethyl-2,5-dimethoxyphenethylamine AC		2000	251 $_{12}$	192 $_{100}$	177 $_{25}$	149 $_{13}$	91 $_{13}$	613	6916
4-Ethyl-2,5-dimethoxyphenethylamine formyl artifact		1630	221 $_{24}$	190 $_{100}$	179 $_{72}$	149 $_{12}$	91 $_{18}$	476	6910
4-Ethyl-2,5-dimethoxyphenethylamine HFB		1790	405 $_{54}$	226 $_7$	192 $_{90}$	179 $_{100}$	149 $_{21}$	1462	6938
4-Ethyl-2,5-dimethoxyphenethylamine PFP		1760	355 $_{55}$	192 $_{88}$	179 $_{100}$	149 $_{22}$	119 $_{20}$	1236	6933
4-Ethyl-2,5-dimethoxyphenethylamine TFA		1765	305 $_{42}$	192 $_{71}$	179 $_{100}$	149 $_{19}$	91 $_{22}$	936	6928
4-Ethyl-2,5-dimethoxyphenethylamine TMS		1790	281 $_2$	251 $_3$	180 $_{25}$	102 $_{100}$	73 $_{47}$	790	6918
4-Ethyl-2,5-dimethoxyphenethylamine-M (-COOH) MEAC		2605	295 $_{30}$	236 $_{100}$	223 $_{28}$	193 $_{11}$	163 $_{11}$	878	7093
4-Ethyl-2,5-dimethoxyphenethylamine-M (deamino-COOH) ME		1820*	238 $_{100}$	223 $_{22}$	192 $_{11}$	179 $_{39}$	163 $_6$	548	7091
4-Ethyl-2,5-dimethoxyphenethylamine-M (deamino-HO-) AC		1850*	252 $_{18}$	192 $_{100}$	177 $_{55}$	149 $_{19}$	91 $_{23}$	619	7082
4-Ethyl-2,5-dimethoxyphenethylamine-M (deamino-HO-) TFA		1680*	306 $_{65}$	192 $_{76}$	177 $_{100}$	149 $_{43}$	91 $_{59}$	943	7092
4-Ethyl-2,5-dimethoxyphenethylamine-M (deamino-oxo-)		1745*	208 $_{57}$	179 $_{100}$	149 $_{24}$	91 $_{89}$	77 $_{66}$	425	7704
4-Ethyl-2,5-dimethoxyphenethylamine-M (HO- N-acetyl-) 2TFA		2080	459 $_{27}$	345 $_{19}$	304 $_{100}$	276 $_{26}$	69 $_{23}$	1599	7105
4-Ethyl-2,5-dimethoxyphenethylamine-M (HO- N-acetyl-) isomer-1 propionylated		2370	323 $_{14}$	264 $_{83}$	249 $_{28}$	208 $_{36}$	191 $_{100}$	1052	7127
4-Ethyl-2,5-dimethoxyphenethylamine-M (HO- N-acetyl-) isomer-1 TMS		2230	339 $_{26}$	324 $_{10}$	280 $_{49}$	265 $_{100}$	191 $_{18}$	1152	7125
4-Ethyl-2,5-dimethoxyphenethylamine-M (HO- N-acetyl-) isomer-2 propionylated		2570	323 $_{13}$	252 $_{34}$	208 $_{31}$	196 $_{71}$	57 $_{100}$	1052	7128
4-Ethyl-2,5-dimethoxyphenethylamine-M (HO- N-acetyl-) isomer-2 TMS		2380	339 $_{100}$	294 $_5$	251 $_{10}$	249 $_7$	73 $_{42}$	1152	7126
4-Ethyl-2,5-dimethoxyphenethylamine-M (HO-) 2TFA		2035	417 $_{43}$	304 $_{87}$	291 $_8$	190 $_{53}$	177 $_{100}$	1498	7121
4-Ethyl-2,5-dimethoxyphenethylamine-M (HO-) -H2O AC		2175	249 $_{33}$	190 $_{100}$	175 $_{23}$	147 $_{31}$	91 $_{16}$	598	7120
4-Ethyl-2,5-dimethoxyphenethylamine-M (HO-) -H2O TFA		1945	303 $_{64}$	190 $_{100}$	177 $_{85}$	175 $_{21}$	147 $_{65}$	923	7119
4-Ethyl-2,5-dimethoxyphenethylamine-M (HO-) isomer-1 AC		2340	309 $_{22}$	250 $_{100}$	237 $_6$	207 $_{50}$	191 $_{77}$	964	7096
4-Ethyl-2,5-dimethoxyphenethylamine-M (HO-) isomer-2 AC		2420	309 $_{16}$	250 $_{16}$	190 $_{100}$	161 $_{18}$	135 $_9$	964	7097
4-Ethyl-2,5-dimethoxyphenethylamine-M (HO-) isomer-3 2AC		2595	351 $_2$	309 $_{22}$	280 $_{32}$	238 $_{56}$	196 $_{100}$	1214	7099

4-Ethyl-2,5-dimethoxyphenethylamine-M (HO-) isomer-3 AC

Table 8.1: Compounds in order of names

Name	Detected	RI	Typical ions and intensities					Page	Entry
4-Ethyl-2,5-dimethoxyphenethylamine-M (HO-) isomer-3 AC		2500	309_{22}	250_{32}	238_{32}	208_{28}	196_{100}	965	7098
4-Ethyl-2,5-dimethoxyphenethylamine-M (HO-deamino-COOH) isomer-1 AC		2070*	296_{88}	253_{38}	237_{100}	222_{21}	177_{33}	885	7103
4-Ethyl-2,5-dimethoxyphenethylamine-M (HO-deamino-COOH) isomer-2 AC		2150*	296_{12}	236_{100}	177_{59}	161_{20}	147_{13}	885	7104
4-Ethyl-2,5-dimethoxyphenethylamine-M (O-demethyl- N-acetyl-) isomer-1 2TFA		1860	429_{15}	316_6	274_{100}	259_{24}	205_{39}	1530	7110
4-Ethyl-2,5-dimethoxyphenethylamine-M (O-demethyl- N-acetyl-) isomer-1 TFA		1950	333_{16}	274_{100}	259_{12}	205_{14}	177_{11}	1111	7108
4-Ethyl-2,5-dimethoxyphenethylamine-M (O-demethyl- N-acetyl-) isomer-2 2TFA		1870	429_4	274_{100}	261_{32}	259_{19}	231_9	1530	7111
4-Ethyl-2,5-dimethoxyphenethylamine-M (O-demethyl- N-acetyl-) isomer-2 TFA		2020	333_{12}	274_{100}	259_{18}	177_9	91_{18}	1110	7109
4-Ethyl-2,5-dimethoxyphenethylamine-M (O-demethyl-) isomer-1 2AC		2205	279_6	237_{31}	178_{100}	165_{17}	122_5	775	7083
4-Ethyl-2,5-dimethoxyphenethylamine-M (O-demethyl-) isomer-1 2TFA		1740	387_{13}	274_{100}	259_{20}	205_{20}	177_{31}	1389	7106
4-Ethyl-2,5-dimethoxyphenethylamine-M (O-demethyl-) isomer-2 2TFA		1805	387_{11}	274_{100}	261_{41}	231_7	177_{27}	1389	7107
4-Ethyl-2,5-dimethoxyphenethylamine-M (O-demethyl-) isomer-2 AC		2240	279_6	237_{28}	178_{100}	163_{15}	135_6	774	7084
4-Ethyl-2,5-dimethoxyphenethylamine-M (O-demethyl-deamino-COOH) -H2O		1690*	192_{78}	164_{100}	136_{32}	121_{27}	91_{17}	375	7122
4-Ethyl-2,5-dimethoxyphenethylamine-M (O-demethyl-deamino-COOH) iso.-1 MEAC		1940	266_8	224_{50}	192_{73}	164_{100}	136_{17}	695	7100
4-Ethyl-2,5-dimethoxyphenethylamine-M (O-demethyl-deamino-COOH) iso.-1 METFA		1710	320_{93}	305_{14}	261_{53}	207_{68}	191_{100}	1032	7094
4-Ethyl-2,5-dimethoxyphenethylamine-M (O-demethyl-deamino-COOH) iso.-2 MEAC		1980	266_9	224_{100}	207_6	165_{60}	135_{15}	695	7101
4-Ethyl-2,5-dimethoxyphenethylamine-M (O-demethyl-deamino-COOH) iso.-2 METFA		1730	320_{51}	261_{100}	223_{17}	163_{13}	91_{50}	1032	7095
4-Ethyl-2,5-dimethoxyphenethylamine-M (O-demethyl-deamino-HO-) isomer-1 2AC		1990	280_7	238_{18}	178_{100}	163_{40}	145_{23}	782	7089
4-Ethyl-2,5-dimethoxyphenethylamine-M (O demethyl-deamino-HO-) isomer-1 2TFA		1540	388_{39}	274_{100}	259_{46}	205_{28}	177_{42}	1393	7116
4-Ethyl-2,5-dimethoxyphenethylamine-M (O-demethyl-deamino-HO-) isomer-2 2AC		2000	280_6	238_9	220_8	178_{100}	163_{26}	782	7090
4-Ethyl-2,5-dimethoxyphenethylamine-M (O-demethyl-deamino-HO-) isomer-2 2TFA		1580	388_{18}	274_{100}	259_{24}	177_{65}	69_{29}	1393	7117
4-Ethyl-2,5-dimethoxyphenethylamine-M (O-demethyl-HO- N-acetyl-) iso-1 -H2O TFA		2015	331_{16}	272_{100}	259_{22}	205_{20}	177_{41}	1099	7112
4-Ethyl-2,5-dimethoxyphenethylamine-M (O-demethyl-HO- N-acetyl-) iso-2 -H2O TFA		2050	331_9	272_{100}	259_{16}	203_{10}	192_{21}	1099	7113
4-Ethyl-2,5-dimethoxyphenethylamine-M (O-demethyl-HO-) 3AC		2425	337_2	309_2	277_{22}	235_{41}	176_{100}	1138	7085
4-Ethyl-2,5-dimethoxyphenethylamine-M (O-demethyl-HO-) 3TFA		1750	499_9	386_{39}	373_{100}	343_{11}		1657	7124
4-Ethyl-2,5-dimethoxyphenethylamine-M (O-demethyl-HO-) -H2O 2TFA		1810	385_{29}	272_{100}	259_{16}	203_{13}	175_{21}	1380	7114
4-Ethyl-2,5-dimethoxyphenethylamine-M (O-demethyl-HO-) isomer-1 -H2O 2AC		2255	277_{10}	235_{37}	176_{100}	161_{12}	133_8	760	7086
4-Ethyl-2,5-dimethoxyphenethylamine-M (O-demethyl-HO-) isomer-2 -H2O 2AC		2280	277_5	235_{32}	176_{100}	161_{12}	133_9	759	7087
4-Ethyl-2,5-dimethoxyphenethylamine-M (O-demethyl-oxo- N-acetyl-) AC		2430	293_3	251_{40}	192_{100}	176_{53}	137_9	863	7118
4-Ethyl-2,5-dimethoxyphenethylamine-M (O-demethyl-oxo- N-acetyl-) TFA		2115	347_{46}	233_{24}	192_{100}	177_{68}	69_{68}	1193	7115
4-Ethyl-2,5-dimethoxyphenethylamine-M (O-demethyl-oxo-) AC		2320	251_{26}	192_{100}	177_{38}	151_{11}	137_{18}	610	7088
4-Ethyl-2,5-dimethoxyphenethylamine-M (oxo-deamino-COOH) ME		2025*	252_{54}	237_{100}	193_{25}	177_{77}	163_{23}	617	7102
1-Ethyl-2-methylbenzene		<1000*	120_{27}	105_{100}	91_{15}	77_{15}	63_7	265	3787
2-Ethyl-3-methyl-1-butene		<1000*	98_{32}	83_{64}	69_{100}	55_{80}	41_{92}	257	3824
1-Ethyl-4-methylbenzene		<1000*	120_{40}	105_{100}	91_{12}	77_{17}	65_7	265	3827
N-Ethyl-4-methyl-norpentedrone		1710	219_1	142_{15}	119_{10}	100_{100}	91_{21}	466	9578
N-Ethyl-4-methyl-norpentedrone AC		1970	261_3	142_{61}	119_{14}	100_{100}	91_{27}	665	9577
N-Ethyl-5-aminopropylbenzofuran		1370	188_1	131_{26}	102_6	77_{15}	72_{100}	407	9363
N-Ethyl-5-aminopropylbenzofuran AC		1560	245_4	158_{28}	131_{34}	114_{35}	72_{100}	575	9364
N-Ethyl-5-aminopropylbenzofuran HFB		1350	399_1	268_{100}	240_{57}	158_{53}	131_{47}	1440	9368
N-Ethyl-5-aminopropylbenzofuran PFP		1355	349_1	218_{100}	190_{52}	158_{51}	131_{58}	1203	9367
N-Ethyl-5-aminopropylbenzofuran TFA		1385	299_1	168_{100}	158_{51}	140_{46}	131_{48}	901	9366
N-Ethyl-5-aminopropylbenzofuran TMS		1380	260_1	144_{100}	131_{31}	102_5	73_{47}	749	9365
2-Ethyl-5-methyl-3,3-diphenyl-1-pyrroline (EMDP)	UHYAC	1940	263_1	208_{100}	193_{78}	179_{38}	130_{50}	677	5295
Ethylacetate		<1000*	88_5	70_{14}	61_{16}	43_{100}	29_{33}	255	60
Ethylamine		<1000	45_{19}	30_{100}				248	3617
N-Ethyl-Buphedrone		1365	162_2	105_{10}	86_{100}	77_{25}	58_{15}	372	9725
N-Ethyl-Buphedrone AC		1760	233_2	190_3	128_{69}	105_{18}	86_{100}	523	9726
N-Ethyl-Buphedrone HFB		1520	387_1	282_{100}	254_9	105_{34}	77_{29}	1389	9729
N-Ethyl-Buphedrone PFP		1490	337_1	232_{100}	204_{12}	105_{43}	77_{35}	1135	9728
N-Ethyl-Buphedrone TFA		1500	287_2	182_{100}	154_6	105_{39}	77_{33}	819	9727
N-Ethyl-Buphedrone TMS		1600	248_3	219_2	158_{40}	77_{100}	58_{60}	678	9761
N-Ethylcarboxamido-adenosine 2AC		2735	392_5	333_{13}	262_{66}	136_{75}	85_{100}	1411	3092
N-Ethylcarboxamido-adenosine -2H2O		2930	272_{27}	228_{100}	172_{30}	136_{28}	66_{30}	729	3093
N-Ethylcarboxamido-adenosine 3AC		3265	434_5	375_{15}	363_{16}	304_{50}	85_{100}	1544	3091
Ethyldimethylbenzene		1065*	134_{32}	119_{100}	105_{26}	91_{44}	77_{20}	274	3790
Ethylene glycol		<1000*	62_{33}	43_{64}	33_{100}	31_{100}		250	765
Ethylene glycol 2AC		<1000*	116_{25}	103_{24}	86_{100}	73_{41}		285	766
Ethylene glycol 2TMS		<1000*	191_{14}	147_{100}	133_8	103_8	73_{73}	416	8589
Ethylene glycol dibenzoate		2120*	270_1	227_{14}	162_{10}	105_{100}	77_{87}	716	1741
Ethylene glycol dipivalate		1320*	185_1	143_2	129_9	85_{28}	57_{100}	512	1903
Ethylene glycol monomethylether		<1000*	76_5	58_4	45_{100}	31_{30}	29_{66}	253	3779
Ethylene glycol phenylboronate		1210*	148_{84}	118_{34}	91_{100}	77_{14}		286	1896
Ethylene glycol-M (glycolic acid) 2TMS		<1000*	205_8	177_{12}	161	147_{86}	73_{100}	469	8585
Ethylene oxide		<1000*	44_{57}	29_{100}				248	4195
Ethylene thiourea		2080	102_{100}	100_1	73_{13}	60_4		258	3910
Ethylenediaminetetraacetic acid 3ME1ET		2125*	362_{14}	303_{28}	289_{18}	188_{100}	174_{64}	1274	6452
Ethylenediaminetetraacetic acid 4ME		2105*	348_8	289_{25}	188_{23}	174_{100}	146_{40}	1199	6451
5,6-Ethylenedioxy-N,N-diallyl-tryptamine		2400	298_9	257_5	202_5	188_{23}	110_{100}	897	10154
5,6-Ethylenedioxy-N,N-diallyl-tryptamine HFB		2280	467_1	398_3	384_{15}	187_{23}	110_{100}	1653	10158
5,6-Ethylenedioxy-N,N-diallyl-tryptamine PFP		2275	444_1	348_4	334_{10}	187_{14}	110_{100}	1569	10157
5,6-Ethylenedioxy-N,N-diallyl-tryptamine TFA		2305	394_2	367_3	298_9	284_{22}	110_{100}	1420	10156
5,6-Ethylenedioxy-N,N-diallyl-tryptamine TMS		2570	370_3	329_7	274_5	260_{29}	110_{100}	1314	10155
2-Ethylhexyldiphenylphosphate	P G U UHY UHYAC	2450*	362_{10}	251_{100}	170_5	94_{14}	77_{14}	1273	3053
Ethylhexylmethylphthalate		2010*	181_{24}	163_{100}	149_{48}	83_{11}	70_{34}	857	5319

Table 8.1: Compounds in order of names

Name	Detected	RI	Typical ions and intensities					Page	Entry
Ethylloflazepate artifact	U+UHYAC	2050	272 70	271 100	237 78	151 11	110 8	728	2409
Ethylloflazepate -C3H4O2	G P-I UGLUC	2470	288 100	287 64	260 93	259 75		825	508
Ethylloflazepate -C3H4O2 TMS		2470	360 63	359 69	341 43	197 17	73 100	1263	4621
Ethylloflazepate HY	UHY	2030	249 100	154 34	123 46	95 42		593	512
Ethylloflazepate HYAC	U+UHYAC	2195	291 52	249 100	123 57	95 61		846	286
Ethylloflazepate-M (HO-) artifact-1	UHYAC	2380	330 26	288 100	287 90	271 39	253 82	1093	2410
Ethylloflazepate-M (HO-) artifact-2	UHYAC	2420	316 24	258 100	221 15	95 19	75 16	1003	2412
Ethylloflazepate-M (HO-) HY2AC	UHYAC	2500	349 47	307 82	265 100	264 81	139 46	1203	2411
Ethylmethylphthalate		1520*	208 2	176 11	163 100	149 58	77 25	423	4940
Ethylmorphine	U UHY	2420	313 100	284 19	162 54	124 29		990	494
Ethylmorphine AC	U+UHYAC	2530	355 100	327 24	204 18	162 10	124 14	1238	237
Ethylmorphine PFP		2430	459 63	430 8	402 7	296 100	119 21	1599	2461
Ethylmorphine TFA		2320	409 100	380 15	296 97	115 21	59 39	1474	4014
Ethylmorphine TMS		2540	385 45	234 17	192 31	146 31	73 100	1382	2455
Ethylmorphine-M (nor-) 2AC	U+UHYAC	2930	383 23	237 29	209 47	87 100	72 34	1371	1193
Ethylmorphine-M (O-deethyl-)	G UHY	2455	285 100	268 15	162 59	124 21		810	474
Ethylmorphine-M (O-deethyl-) 2AC	G PHYAC U+UHYAC	2620	369 59	327 100	310 36	268 47	162 11	1306	225
Ethylmorphine-M (O-deethyl-) 2HFB		2375	677 10	480 10	464 100	407 9	169 8	1726	6120
Ethylmorphine-M (O-deethyl-) 2PFP		2360	577 51	558 7	430 8	414 100	119 22	1706	2251
Ethylmorphine-M (O-deethyl-) 2TFA		2250	477 71	364 100	307 6	115 8	69 31	1630	4008
Ethylmorphine-M (O-deethyl-) 2TMS	UHYTMS	2560	429 19	236 21	196 15	146 21	73 100	1531	2463
Ethylmorphine-M (O-deethyl-) Cl-artifact 2AC	UHYAC	2680	403 59	361 100	344 63	302 90	204 55	1456	2992
Ethylmorphine-M (O-deethyl-) TFA		2285	381 55	268 100	146 14	115 13	69 23	1360	5569
Ethylmorphine-M (O-deethyl-)-D3 2HFB		2375	680 4	483 9	467 100	414 7	169 23	1727	6126
Ethylmorphine-M (O-deethyl-)-D3 2PFP		2350	580 16	433 7	417 100	269 5	119 8	1708	5567
Ethylmorphine-M (O-deethyl-)-D3 2TFA		2240	480 32	383 6	367 100	314 6	307 6	1634	5571
Ethylmorphine-M (O-deethyl-)-D3 2TMS		2550	432 91	290 27	239 60	199 40	73 100	1540	5578
Ethylmorphine-M (O-deethyl-)-D3 TFA		2275	384 39	271 100	211 8	165 6	152 7	1376	5572
Ethylmorphine-M 2PFP		2440	563 100	400 10	355 38	327 7	209 15	1701	3534
Ethylmorphine-M 3AC	U+UHYAC	2955	397 8	355 9	209 41	87 100	72 33	1434	1194
Ethylmorphine-M 3PFP	UHYPFP	2405	709 80	533 28	388 29	367 51	355 100	1729	3533
Ethylmorphine-M 3TMS	UHYTMS	2605	487 17	416 19	222 36	131 19	73 100	1645	3525
7-Ethyl-N,N-diallyl-tryptamine		2200	268 1	172 3	158 12	143 13	110 100	709	8852
7-Ethyl-N,N-diallyl-tryptamine HFB		2020	368 1	354 4	199 2	156 16	110 100	1608	10106
7-Ethyl-N,N-diallyl-tryptamine PFP		2025	414 3	318 31	304 100	198 14	157 23	1491	10105
7-Ethyl-N,N-diallyl-tryptamine TFA		2065	364 1	337 2	268 5	254 11	110 100	1283	10104
7-Ethyl-N,N-diallyl-tryptamine TMS		2345	340 1	299 2	230 3	156 2	110 100	1158	10103
4-Ethyl-naphthalen-1-yl-(1-pentylindol-3-yl)methanone		3320	369 100	352 53	312 51	214 60	144 46	1308	8524
4-Ethyl-naphthalen-1-yl-(1-pentylindol-3-yl)methanone-M (5-HOOC-) ME		4015	413 27	298 100	152 70	139 73	128 56	1488	10402
4-Ethyl-naphthalen-1-yl-(1-pentylindol-3-yl)methanone-M (5-HOOC-) TMS		4260	471 16	298 41	152 27	128 43	75 100	1621	10403
4-Ethyl-naphthalen-1-yl-(1-pentylindol-3-yl)methanone-M (5-HO-pentyl-)		4050	385 51	368 28	152 19	139 99	128 68	1382	10388
4-Ethyl-naphthalen-1-yl-(1-pentylindol-3-yl)methanone-M (5-HO-pentyl-) AC		4200	427 58	310 47	254 62	152 90	61 100	1526	10389
4-Ethyl-naphthalen-1-yl-(1-pentylindol-3-yl)methanone-M (5-HO-pentyl-) HFB		3585	581 73	426 45	312 85	183 100	153 92	1708	10393
4-Ethyl-naphthalen-1-yl-(1-pentylindol-3-yl)methanone-M (5-HO-pentyl-) PFP		3570	531 79	312 62	254 56	183 64	119 100	1685	10392
4-Ethyl-naphthalen-1-yl-(1-pentylindol-3-yl)methanone-M (5-HO-pentyl-) TFA		3600	481 74	254 60	183 56	153 77	69 100	1636	10391
4-Ethyl-naphthalen-1-yl-(1-pentylindol-3-yl)methanone-M (5-HO-pentyl-) TMS		3960	457 53	298 57	254 29	183 58	75 100	1596	10390
Ethylparaben		1580*	166 32	138 29	121 100			316	767
Ethylparaben-M (4-hydroxyhippuric acid) ME	U	1820	209 100	177 32	149 34	121 87		427	817
Ethylphenidate		1760	172 2	130 2	115 6	84 100	56 8	587	9145
Ethylphenidate AC		2090	244 2	174 2	126 48	91 19	84 100	837	9359
Ethylphenidate artifact (Phenylacetic acid ethylester)		1050*	164 27	119 3	105 4	91 100	65 31	311	9358
Ethylphenidate ET		1785	274 1	130 4	112 100	91 11	84 12	750	9369
Ethylphenidate HFB		1900	280 50	252 5	226 9	164 9	91 100	1568	9362
Ethylphenidate PFP		1885	230 100	202 2	176 5	164 8	119 10	1414	9361
Ethylphenidate TFA		1750	298 2	180 100	164 14	126 7	91 11	1174	9360
Ethylphenidate-M (ritalinic acid) HFB HFBOL		1815	280 100	226 4	169 3			1714	9347
Ethylphenidate-M (ritalinic acid) HFB PFPOL		1805	280 100	226 4	169 5			1694	9346
Ethylphenidate-M (ritalinic acid)-D9 isomer-1 HFB HFBOL		1800	289 100	229 4	169 6	119 6		1716	9350
Ethylphenidate-M (ritalinic acid)-D9 isomer-1 HFB PFPOL		1790	289 100	229 4	169 6	119 7		1698	9348
Ethylphenidate-M (ritalinic acid)-D9 isomer-2 HFB HFBOL		1810	289 100	229 4	169 5	119 6		1716	9351
Ethylphenidate-M (ritalinic acid)-D9 isomer-2 HFB PFPOL		1795	289 100	229 4	169 6	119 8		1697	9349
1-Ethylpiperidine		<1000	113 12	98 100	70 10	58 39	42 76	262	3613
4-Ethylthio-2,5-dimethoxyphenethylamine		1980	241 22	212 100	197 20	183 35	153 22	560	5035
4-Ethylthio-2,5-dimethoxyphenethylamine 2AC	U+UHYAC	2395	325 17	224 100	211 50	181 16	153 14	1063	5038
4-Ethylthio-2,5-dimethoxyphenethylamine 2TMS		2405	385 1	370 3	254 4	211 4	174 100	1382	6815
4-Ethylthio-2,5-dimethoxyphenethylamine AC	U+UHYAC	2310	283 28	224 100	211 34	181 20	153 12	796	5037
4-Ethylthio-2,5-dimethoxyphenethylamine deuteroformyl artifact		1935	255 37	224 31	211 100	181 14	153 16	633	5036
4-Ethylthio-2,5-dimethoxyphenethylamine HFB		2040	437 24	224 21	211 100	181 19	169 17	1551	6816
4-Ethylthio-2,5-dimethoxyphenethylamine PFP		2090	387 4	224 29	211 100	181 18	153 15	1389	6817
4-Ethylthio-2,5-dimethoxyphenethylamine TFA	UGLUCTFA	2210	337 26	224 17	211 100	181 13	153 10	1134	6818
4-Ethylthio-2,5-dimethoxyphenethylamine TMS		2405	313 1	299 3	174 100	147 4	86 7	989	6814
4-Ethylthio-2,5-dimethoxyphenethylamine-M (aryl-HOOC-)	UGLUC	1970	242 100	227 20	183 68	153 14		563	6893
4-Ethylthio-2,5-dimethoxyphenethylamine-M (aryl-HOOC-) ME	USPEME	1960*	256 100	241 10	197 49	181 12	167 15	636	6842
4-Ethylthio-2,5-dimethoxyphenethylamine-M (deamino-HO-)	UGLUC	1905*	242 71	211 100	181 15	153 11		565	6839

4-Ethylthio-2,5-dimethoxyphenethylamine-M (deamino-HO-) AC

Table 8.1: Compounds in order of names

Name	Detected	RI	Typical ions and intensities					Page	Entry
4-Ethylthio-2,5-dimethoxyphenethylamine-M (deamino-HO-) AC	U+UHYAC	2050*	284 $_{40}$	224 $_{100}$	209 $_{30}$	167 $_{20}$	150 $_{10}$	801	6892
4-Ethylthio-2,5-dimethoxyphenethylamine-M (deamino-HOOC-)		2130*	256 $_{100}$	242 $_{31}$	211 $_{60}$	195 $_{30}$	181 $_{32}$	636	6840
4-Ethylthio-2,5-dimethoxyphenethylamine-M (deamino-HOOC-) ME	UHYME	1910*	270 $_{100}$	255 $_{24}$	211 $_{82}$	195 $_{44}$	181 $_{46}$	717	6838
4-Ethylthio-2,5-dimethoxyphenethylamine-M (deamino-HOOC-) TMS	USPETMS	2075*	328 $_{100}$	313 $_{29}$	298 $_{25}$	255 $_{61}$	211 $_{57}$	1080	6841
4-Ethylthio-2,5-dimethoxyphenethylamine-M (deamino-oxo-)		2130*	240 $_{33}$	211 $_{100}$	181 $_{11}$	153 $_{19}$	122 $_{11}$	555	7234
4-Ethylthio-2,5-dimethoxyphenethylamine-M (HO- N-acetyl-) TFA	UGLUCTFA	2270	427 $_{24}$	367 $_{3}$	259 $_{100}$	167 $_{6}$		1524	6834
4-Ethylthio-2,5-dimethoxyphenethylamine-M (HO- sulfone) 2AC	U+UHYAC	2780	373 $_{35}$	314 $_{84}$	302 $_{75}$	272 $_{54}$	259 $_{100}$	1327	6833
4-Ethylthio-2,5-dimethoxyphenethylamine-M (HO- sulfone) AC	U+UHYAC	2730	331 $_{7}$	272 $_{100}$	259 $_{74}$	238 $_{26}$	165 $_{47}$	1100	6828
4-Ethylthio-2,5-dimethoxyphenethylamine-M (O-demethyl- N-acetyl-) 2TFA	U+UHYTFA	2180	461 $_{16}$	306 $_{100}$	293 $_{43}$	209 $_{25}$		1602	6894
4-Ethylthio-2,5-dimethoxyphenethylamine-M (O-demethyl- N-acetyl-) TFA	U+UHYTFA	2250	365 $_{11}$	323 $_{19}$	306 $_{100}$	293 $_{18}$	197 $_{9}$	1285	6942
4-Ethylthio-2,5-dimethoxyphenethylamine-M (O-demethyl- sulfone) 2AC	U+UHYAC	2510	343 $_{3}$	301 $_{90}$	242 $_{100}$	230 $_{48}$	153 $_{9}$	1173	6835
4-Ethylthio-2,5-dimethoxyphenethylamine-M (O-demethyl-) 2AC	U+UHYAC	2120	311 $_{48}$	297 $_{31}$	269 $_{100}$	252 $_{46}$	210 $_{78}$	976	6837
4-Ethylthio-2,5-dimethoxyphenethylamine-M (O-demethyl-) 2TFA	UGLUCTFA	1980	419 $_{28}$	306 $_{100}$	293 $_{92}$	209 $_{20}$	69 $_{32}$	1504	6821
4-Ethylthio-2,5-dimethoxyphenethylamine-M (O-demethyl-) 3AC	U+UHYAC	2290	353 $_{22}$	311 $_{32}$	252 $_{33}$	210 $_{100}$	197 $_{20}$	1226	6836
4-Ethylthio-2,5-dimethoxyphenethylamine-M (O-demethyl-sulfone N-acetyl-) TFA	UGLUCTFA	2450	397 $_{4}$	355 $_{68}$	242 $_{100}$	153 $_{14}$		1433	6820
4-Ethylthio-2,5-dimethoxyphenethylamine-M (S-deethyl-) 3AC	U+UHYAC	2420	339 $_{22}$	297 $_{5}$	238 $_{21}$	196 $_{100}$	183 $_{28}$	1149	6827
4-Ethylthio-2,5-dimethoxyphenethylamine-M (S-deethyl-) AC	U+UHYAC	2170	255 $_{18}$	196 $_{100}$	183 $_{41}$	181 $_{34}$	153 $_{21}$	631	6831
4-Ethylthio-2,5-dimethoxyphenethylamine-M (S-deethyl-) isomer-1 2AC		2240	297 $_{29}$	210 $_{14}$	196 $_{100}$	183 $_{35}$	181 $_{29}$	889	6823
4-Ethylthio-2,5-dimethoxyphenethylamine-M (S-deethyl-) isomer-2 2AC	U+UHYAC	2360	297 $_{16}$	255 $_{11}$	238 $_{20}$	196 $_{100}$	183 $_{37}$	889	6826
4-Ethylthio-2,5-dimethoxyphenethylamine-M (S-deethyl-methyl- N-acetyl-)	U+UHYAC	2230	269 $_{19}$	210 $_{100}$	197 $_{35}$	195 $_{21}$	167 $_{27}$	712	6832
4-Ethylthio-2,5-dimethoxyphenethylamine-M (S-deethyl-methyl- sulfone) AC	U+UHYAC	2580	301 $_{7}$	242 $_{100}$	230 $_{4}$	196 $_{7}$	124 $_{7}$	911	6829
4-Ethylthio-2,5-dimethoxyphenethylamine-M (S-deethyl-methyl- sulfoxide) AC	U+UHYAC	2460	285 $_{16}$	268 $_{23}$	226 $_{33}$	211 $_{100}$	197 $_{31}$	808	6830
4-Ethylthio-2,5-dimethoxyphenethylamine-M (sulfone N-acetyl-) TFA	UGLUCTFA	2400	411 $_{23}$	256 $_{100}$	242 $_{4}$	181 $_{10}$	167 $_{11}$	1480	6822
4-Ethylthio-2,5-dimethoxyphenethylamine-M (sulfone) 2AC	U+UHYAC	2640	357 $_{10}$	256 $_{100}$	244 $_{6}$	167 $_{7}$	91 $_{7}$	1247	6824
4-Ethylthio-2,5-dimethoxyphenethylamine-M (sulfone) AC	U+UHYAC	2600	315 $_{15}$	256 $_{100}$	244 $_{9}$	167 $_{12}$	91 $_{8}$	1000	6825
4-Ethylthio-2,5-dimethoxyphenethylamine-M (sulfone) TFA	UGLUCTFA	2310	369 $_{44}$	256 $_{100}$	243 $_{7}$	211 $_{4}$	167 $_{23}$	1305	6819
Ethyltolylbarbital 2ET		2010	302 $_{15}$	274 $_{100}$	246 $_{11}$	160 $_{15}$	117 $_{18}$	919	2597
Eticyclidine		1545	203 $_{18}$	160 $_{100}$	146 $_{11}$	117 $_{16}$	91 $_{22}$	408	3602
Eticyclidine intermediate (ECC) -HCN		<1000	125 $_{4}$	110 $_{13}$	96 $_{45}$	82 $_{34}$	56 $_{100}$	269	3598
Eticyclidine intermediate (ECC) -HCN		<1000	125 $_{4}$	110 $_{12}$	96 $_{44}$	56 $_{97}$	41 $_{100}$	268	5535
Eticyclidine precursor (ethylamine)		<1000	45 $_{19}$	30 $_{100}$				248	3617
Etidocaine		2040	276 $_{1}$	259 $_{3}$	245 $_{8}$	128 $_{100}$	86 $_{12}$	757	1437
Etifelmin		1880	237 $_{6}$	208 $_{100}$	191 $_{40}$	165 $_{16}$	91 $_{29}$	546	1796
Etifelmin AC		2220	279 $_{37}$	220 $_{53}$	205 $_{73}$	191 $_{84}$	112 $_{100}$	777	1441
Etilamfetamine	U	1230	162 $_{1}$	148 $_{2}$	117 $_{2}$	91 $_{26}$	72 $_{100}$	309	764
Etilamfetamine AC	U+UHYAC	1675	205 $_{1}$	114 $_{53}$	91 $_{12}$	72 $_{100}$		414	1438
Etilamfetamine HFB		1485	268 $_{100}$	240 $_{46}$	118 $_{18}$	91 $_{25}$		1257	5085
Etilamfetamine PFP		1450	309 $_{1}$	218 $_{100}$	190 $_{48}$	118 $_{34}$	91 $_{35}$	962	5082
Etilamfetamine TFA		1450	213 $_{1}$	168 $_{100}$	140 $_{38}$	118 $_{36}$	69 $_{56}$	651	4004
Etilamfetamine-M	UHY	1465	181 $_{9}$	138 $_{100}$	122 $_{18}$	94 $_{24}$	77 $_{16}$	350	4351
Etilamfetamine-M (AM)		1160	134 $_{4}$	120 $_{15}$	91 $_{100}$	77 $_{18}$	65 $_{73}$	275	54
Etilamfetamine-M (AM)	U	1160	134 $_{1}$	120 $_{1}$	91 $_{6}$	65 $_{4}$	44 $_{100}$	275	5514
Etilamfetamine-M (AM) AC	U+UHYAC	1505	177 $_{4}$	118 $_{60}$	91 $_{35}$	86 $_{100}$	65 $_{19}$	336	55
Etilamfetamine-M (AM) AC	U+UHYAC	1505	177 $_{1}$	118 $_{19}$	91 $_{11}$	86 $_{31}$	44 $_{100}$	335	5515
Etilamfetamine-M (AM) formyl artifact		1100	147 $_{2}$	146 $_{6}$	125 $_{5}$	91 $_{12}$	56 $_{100}$	286	3261
Etilamfetamine-M (AM) HFB		1355	240 $_{79}$	169 $_{21}$	118 $_{100}$	91 $_{53}$		1099	5047
Etilamfetamine-M (AM) PFP		1330	281 $_{1}$	190 $_{73}$	118 $_{100}$	91 $_{36}$	65 $_{12}$	786	4379
Etilamfetamine-M (AM) TFA		1095	231 $_{1}$	140 $_{100}$	118 $_{92}$	91 $_{45}$	69 $_{19}$	513	4000
Etilamfetamine-M (AM) TMS		1190	192 $_{6}$	116 $_{100}$	100 $_{9}$	91 $_{11}$	73 $_{87}$	421	5581
Etilamfetamine-M (AM)-D11 PFP		1610	194 $_{100}$	128 $_{82}$	98 $_{43}$	70 $_{100}$		856	7284
Etilamfetamine-M (AM)-D11 TFA		1615	242 $_{1}$	144 $_{100}$	128 $_{82}$	98 $_{43}$	70 $_{14}$	566	7283
Etilamfetamine-M (AM)-D5 AC		1480	182 $_{3}$	122 $_{46}$	92 $_{30}$	90 $_{100}$	66 $_{16}$	352	5690
Etilamfetamine-M (AM)-D5 HFB		1330	244 $_{100}$	169 $_{14}$	122 $_{46}$	92 $_{41}$	69 $_{40}$	1130	6316
Etilamfetamine-M (AM)-D5 PFP		1320	194 $_{100}$	123 $_{42}$	119 $_{32}$	92 $_{46}$	69 $_{11}$	815	5566
Etilamfetamine-M (AM)-D5 TFA		1085	144 $_{100}$	123 $_{53}$	122 $_{56}$	92 $_{51}$	69 $_{28}$	540	5570
Etilamfetamine-M (AM)-D5 TMS		1180	212 $_{1}$	197 $_{8}$	120 $_{100}$	92 $_{11}$	73 $_{57}$	441	5582
Etilamfetamine-M (AM-4-HO-)		1480	151 $_{10}$	107 $_{69}$	91 $_{10}$	77 $_{42}$	56 $_{100}$	292	1802
Etilamfetamine-M (AM-4-HO-) 2AC	U+UHYAC	1900	235 $_{1}$	176 $_{72}$	134 $_{100}$	107 $_{46}$	86 $_{70}$	533	1804
Etilamfetamine-M (AM-4-HO-) 2HFB		<1000	330 $_{48}$	303 $_{15}$	240 $_{100}$	169 $_{46}$	69 $_{42}$	1691	6326
Etilamfetamine-M (AM-4-HO-) 2PFP		<1000	280 $_{77}$	253 $_{16}$	190 $_{100}$	119 $_{56}$	69 $_{16}$	1567	6325
Etilamfetamine-M (AM-4-HO-) 2TFA		<1000	230 $_{72}$	203 $_{11}$	140 $_{100}$	92 $_{12}$	69 $_{59}$	1172	6324
Etilamfetamine-M (AM-4-HO-) 2TMS		<1000	280 $_{7}$	179 $_{9}$	149 $_{8}$	116 $_{100}$	73 $_{78}$	880	6327
Etilamfetamine-M (AM-4-HO-) AC	U+UHYAC	1890	193 $_{1}$	134 $_{100}$	107 $_{26}$	86 $_{24}$	77 $_{16}$	379	1803
Etilamfetamine-M (AM-4-HO-) formyl art.		1220	163 $_{1}$	148 $_{4}$	107 $_{30}$	77 $_{12}$	56 $_{100}$	308	6323
Etilamfetamine-M (AM-4-HO-) TFA		1670	247 $_{4}$	140 $_{15}$	134 $_{54}$	107 $_{100}$	77 $_{15}$	584	6335
Etilamfetamine-M (deamino-oxo-HO-methoxy-)	UHY	1510*	180 $_{19}$	137 $_{100}$	122 $_{19}$	107 $_{2}$	94 $_{16}$	346	4247
Etilamfetamine-M (di-HO-) 3AC	U+UHYAC	2200	321 $_{1}$	234 $_{8}$	150 $_{27}$	114 $_{80}$	72 $_{100}$	1040	4208
Etilamfetamine-M (HO-) 2AC	UHYAC	1995	263 $_{1}$	176 $_{8}$	134 $_{15}$	114 $_{44}$	72 $_{100}$	676	5323
Etilamfetamine-M (HO-) 2ME		1780	206 $_{1}$	121 $_{5}$	86 $_{100}$	72 $_{3}$	58 $_{20}$	422	5835
Etilamfetamine-M (HO-) ME		1660	192 $_{1}$	149 $_{1}$	121 $_{5}$	91 $_{2}$	72 $_{100}$	380	5831
Etilamfetamine-M (HO-) MEAC	UHYAC	1855	235 $_{1}$	148 $_{32}$	121 $_{7}$	114 $_{26}$	72 $_{100}$	535	5322
Etilamfetamine-M (HO-) MEHFB	UHFB	1785	389 $_{2}$	268 $_{100}$	240 $_{46}$	148 $_{62}$	121 $_{43}$	1398	5834
Etilamfetamine-M (HO-) MEPFP	UPFP	1765	339 $_{1}$	218 $_{100}$	190 $_{44}$	148 $_{59}$	121 $_{49}$	1149	5833
Etilamfetamine-M (HO-) METFA	UTFA	1775	289 $_{1}$	168 $_{100}$	148 $_{63}$	140 $_{41}$	121 $_{62}$	834	5832

110

Table 8.1: Compounds in order of names   **Etilamfetamine-M (HO-) METMS**

Name	Detected	RI	Typical ions and intensities					Page	Entry
Etilamfetamine-M (HO-) METMS		2065	264 1	250 14	144 100	121 17	73 84	691	5836
Etilamfetamine-M (HO-methoxy-)	UHY	1640	209 1	137 12	122 7	94 9	72 100	431	4364
Etilamfetamine-M (HO-methoxy-) 2AC	U+UHYAC	2080	293 1	206 20	164 38	114 77	72 100	865	4209
Etilamfetamine-M (HO-methoxy-) 2HFB		1695	360 15	284 8	218 100	190 31	163 8	1715	8483
Etilamfetamine-M (HO-methoxy-) 2PFP		1645	310 19	218 100	190 36	119 37	72 25	1660	8482
Etilamfetamine-M (HO-methoxy-) 2TFA		1650	401 1	260 24	233 4	168 100	140 42	1449	8479
Etilamfetamine-M (HO-methoxy-) 2TMS		1885	338 2	209 2	179 7	144 91	73 100	1229	8477
Etilamfetamine-M (HO-methoxy-) AC	UHYAC	2000	251 1	164 46	137 7	114 33	72 100	612	4274
Etilamfetamine-M (HO-methoxy-) HFB		1735	360 23	333 6	268 100	240 55	107 23	1462	8484
Etilamfetamine-M (HO-methoxy-) ME	UHYME	1930	223 17	194 7	151 36	94 12	72 100	485	4350
Etilamfetamine-M (HO-methoxy-) PFP		1710	355 5	218 100	190 49	164 70	137 59	1236	8481
Etilamfetamine-M (HO-methoxy-) TFA		1730	305 5	168 100	164 70	140 50	137 58	936	8480
Etilamfetamine-M (HO-methoxy-) TMS		1645	266 1	210 2	179 3	149 2	72 100	790	8478
Etilamfetamine-M (HO-methoxy-AM) ME	UHYME	1550	195 1	152 100	137 17	107 16	77 14	387	4352
Etilamfetamine-M 2AC	U+UHYAC	2065	265 3	206 27	164 100	137 23	86 33	688	3498
Etilamfetamine-M 2AC	U+UHYAC	1735*	250 3	208 15	166 45	123 100		603	4210
Etilamfetamine-M 2AC	U+UHYAC	1820*	266 3	206 9	164 100	150 10	137 30	695	6409
Etilamfetamine-M 2HFB	UHFB	1690	360 82	333 15	240 100	169 42	69 39	1704	6512
Etilamfetamine-M AC	U+UHYAC	1600*	222 2	180 22	137 100			479	4211
Etilamfetamine-M ME	UHYME	1540*	194 25	151 100	135 4	107 18	65 4	384	4353
Etilefrine		1690	181 1	121 1	95 3	77 5	58 100	349	4667
Etilefrine 3AC	U+UHYAC	2150	307 2	264 8	247 4	100 76	58 100	949	768
Etilefrine 3TMS		1885	397 1	382 5	147 10	130 100	73 89	1436	4544
Etilefrine formyl artifact		1860	193 19	178 9	135 36	107 16	58 100	378	1969
Etilefrine ME2AC		2000	279 1	247 4	192 25	100 32	58 100	776	1970
3-alpha-Etiocholanolone		2515*	290 100	257 16	246 18	107 27	79 28	844	3759
3-alpha-Etiocholanolone 2TMS		2520*	434 54	419 52	329 41	169 33	73 100	1546	3799
3-alpha-Etiocholanolone AC		2585*	332 10	272 100	257 47	108 60	67 56	1109	3769
3-beta-Etiocholanolone		2465*	290 63	244 50	201 27	93 66	67 100	845	3897
3-beta-Etiocholanolone 2TMS		2485*	434 46	419 43	329 29	169 19	73 100	1547	3962
3-beta-Etiocholanolone AC		2540*	332 18	272 100	257 57	79 89	67 50	1109	3921
3-beta-Etiocholanolone TMS		2430*	362 9	347 15	272 100	244 61	75 99	1276	3961
Etiroxate artifact ME		3700	490 1	448 1	387 3	130 100	102 11	1444	2750
Etiroxate artifact-1		2285	416 10	288 1	132 3	116 100	88 17	1494	2749
Etiroxate artifact-1 2AC		2690	559 3	500 4	458 42	158 56	116 100	1700	2763
Etiroxate artifact-2 AC		3300	651 12	609 21	550 100	158 38	116 49	1724	2764
Etiroxate artifact-3		3360	506 1	451	337 2	116 100	88 14	1666	2748
Etiroxate artifact-4 AC		3800	777 10	735 57	676 100	158 95	116 97	1732	2765
Etizolam		2980	342 100	313 37	266 32	224 24	137 14	1167	4022
5-EtO-ALCHT-D4		2690	330 2	176 9	154 100	148 13	72 49	1096	8848
5-EtO-ALCHT-D4 AC		2900	372 16	329 10	190 4	176 5	154 100	1325	10024
5-EtO-ALCHT-D4 HFB		2485	388 3	372 4	344 3	154 100		1681	10023
5-EtO-ALCHT-D4 PFP		2475	338 7	322 12	294 6	154 100	72 35	1629	10194
5-EtO-ALCHT-D4 TFA		2485	425 1	288 8	272 11	244 6	154 100	1523	10022
5-EtO-ALCHT-D4 TMS		2715	402 2	264 5	248 13	204 8	154 100	1454	10021
5-EtO-DALT		2330	284 2	174 8	146 12	130 5	110 100	805	8846
5-EtO-DALT AC		2570	326 16	283 3	174 7	146 4	110 100	1071	10017
5-EtO-DALT HFB		2210	480 1	453 2	384 8	370 21	110 100	1634	10016
5-EtO-DALT PFP		2160	403 2	334 6	320 12	110 100		1534	10015
5-EtO-DALT TFA		2185	353 3	284 9	270 15	145 27	110 100	1357	10014
5-EtO-DALT TMS		2400	356 4	315 5	246 17	202 8	110 100	1244	10013
5-EtO-DALT-D4		2330	288 2	176 5	148 7	119 5	112 100	831	8847
5-EtO-DALT-D4 HFB		2165	457 2	388 7	372 14	147 29	112 100	1639	10020
5-EtO-DALT-D4 PFP		2150	434 1	407 2	322 8	147 26	112 100	1545	10019
5-EtO-DALT-D4 TFA		2255	384 1	357 3	272 14	147 7	112 100	1377	10193
5-EtO-DALT-D4 TMS		2435	360 6	319 7	248 24	204 11	112 100	1266	10018
5-EtO-DET-D4		2190	264 8	176 9	148 15	119 11	88 100	684	8872
5-EtO-DET-D4 HFB		2000	445 2	388 4	372 10	147 22	88 100	1602	9530
5-EtO-DET-D4 PFP		1990	322 1	175 2	147 7	119 6	88 100	1478	9529
5-EtO-DET-D4 TFA		2010	288 3	272 8	244 4	147 21	88 100	1265	9528
5-EtO-DET-D4 TMS		2300	336 27	264 8	248 25	204 16	88 100	1134	9527
Etodolac ME		2225	301 21	272 62	228 100	198 32	115 9	915	6128
Etodolac TMS		2350	359 26	330 62	309 27	228 100	73 49	1259	6129
Etodroxizine	G UHY	3155	418 8	299 14	201 100	165 17		1503	769
Etodroxizine AC	U+UHYAC	3180	460 12	299 36	201 100	165 18	87 12	1602	1797
Etodroxizine artifact-1	G U+UHYAC	1600*	202 30	167 100	165 52	152 17	125 7	403	2442
Etodroxizine artifact-2		1900*	232 60	201 62	165 64	105 100	77 54	516	1344
Etodroxizine-M	UHYAC	2210	280 100	201 35	165 57			779	770
Etodroxizine-M (carbinol)	UHY	1750*	218 17	183 7	139 39	105 100	77 87	457	2239
Etodroxizine-M (carbinol) AC	U+UHYAC	1890*	260 8	200 40	165 100	139 10	77 29	656	1270
Etodroxizine-M (Cl-benzophenone)	U+UHYAC	1850*	216 43	139 58	105 100	77 44		451	1343
Etodroxizine-M (HO-Cl-benzophenone)	UHY	2300*	232 36	197 7	139 23	121 100	111 23	515	2240
Etodroxizine-M (HO-Cl-BPH) isomer-1 AC	UHYAC	2200*	274 18	232 75	139 100	121 44	111 51	737	2229
Etodroxizine-M (HO-Cl-BPH) isomer-2 AC	U+UHYAC	2230*	274 7	232 43	139 25	121 100	111 27	737	2230

Etodroxizine-M (N-dealkyl-)

Table 8.1: Compounds in order of names

Name	Detected	RI	Typical ions and intensities					Page	Entry
Etodroxizine-M (N-dealkyl-)	UHY	2520	286_{13}	241_{48}	201_{50}	165_{65}	56_{100}	815	2241
Etodroxizine-M (N-dealkyl-) AC	U+UHYAC	2620	328_{7}	242_{19}	201_{48}	165_{66}	85_{100}	1081	1271
Etodroxizine-M/artifact	P-I U+UHYAC UME	2220	300_{17}	228_{38}	165_{52}	99_{100}	56_{62}	908	670
Etodroxizine-M/artifact 2AC	U+UHYAC	2300	302_{1}	199_{1}	154_{4}	141_{100}	99_{24}	920	2445
Etodroxizine-M/artifact HYAC	U+UHYAC	2935	280_{4}	201_{100}	165_{26}			780	1272
Etofenamate		2510	369_{32}	263_{100}	243_{5}	235_{13}	167_{11}	1305	6093
Etofenamate AC		2590	411_{15}	263_{100}	235_{17}	167_{13}	87_{20}	1480	6094
Etofenamate-M/artifact (flufenamic acid) ME	PME	1880	295_{51}	263_{100}	235_{15}	166_{10}	92_{7}	874	5147
Etofenamate-M/artifact (HO-flufenamic acid) 2ME	PME	2115	325_{100}	293_{91}	278_{88}	250_{67}	202_{19}	1060	6377
Etofenamate-M/artifact (oxoethyl-)		2125	323_{36}	263_{100}	243_{6}	235_{12}	167_{8}	1048	6092
Etofibrate		2520	363_{4}	236_{64}	128_{100}	106_{78}	78_{54}	1277	2762
Etofibrate-M (clofibric acid)	U	1640*	214_{2}	168_{9}	128_{100}	86_{9}	65_{17}	445	686
Etofibrate-M (clofibric acid) ME	U	1500*	228_{8}	169_{16}	128_{100}	99_{5}	75_{8}	500	687
Etofibrate-M artifact	U+UHYAC	1580*	168_{35}	128_{100}				320	1373
Etofibrate-M/artifact (denicotinyl-)		2030*	258_{5}	169_{13}	128_{100}	111_{8}	69_{100}	646	2751
Etofylline	UHY	2125	224_{47}	194_{16}	180_{100}	109_{34}	95_{74}	487	771
Etofylline AC	U+UHYAC	2200	266_{79}	206_{59}	180_{34}	122_{31}	87_{100}	693	772
Etofylline clofibrate	G	3125	420_{9}	293_{43}	206_{24}	113_{35}	69_{100}	1507	1939
Etofylline clofibrate-M (clofibric acid)	U	1640*	214_{2}	168_{9}	128_{100}	86_{9}	65_{17}	445	686
Etofylline clofibrate-M (clofibric acid) ME	U	1500*	228_{8}	169_{16}	128_{100}	99_{5}	75_{8}	500	687
Etofylline clofibrate-M (etofylline)	UHY	2125	224_{47}	194_{16}	180_{100}	109_{34}	95_{74}	487	771
Etofylline clofibrate-M (etofylline) AC	U+UHYAC	2200	266_{79}	206_{59}	180_{34}	122_{31}	87_{100}	693	772
Etofylline clofibrate-M (etofylline) TMS	UHYTMS	2160	296_{10}	281_{26}	252_{6}	180_{100}	73_{47}	886	5696
Etofylline clofibrate-M artifact	U+UHYAC	1580*	168_{35}	128_{100}				320	1373
Etofylline TMS	UHYTMS	2160	296_{10}	281_{26}	252_{6}	180_{100}	73_{47}	886	5696
Etoloxamine		2120	283_{2}	268	181_{2}	165_{4}	86_{100}	797	4264
Etomidate	G P U	1870	244_{16}	199_{2}	105_{100}	77_{22}		571	1924
Etomidate-M (HOOC-) ME	UME	1840	230_{11}	199_{1}	105_{100}	77_{18}		510	3371
Etonitazene		3375	396_{1}	135_{4}	107_{4}	86_{100}	58_{30}	1431	3655
Etonitazene intermediate-1		2515	267_{1}	196	117_{1}	86_{100}	58_{15}	793	2843
Etonitazene intermediate-2		2540	252_{3}	164_{4}	118_{4}	86_{100}	58_{38}	620	2844
Etonitazene intermediate-2 2AC		2745	336_{1}	321	118_{1}	86_{100}	58_{8}	1132	2845
Etonogestrel		2770*	324_{24}	295_{26}	257_{47}	133_{64}	91_{100}	1059	8177
Etonogestrel 2TFA		2590*	516_{1}	488	389_{5}	277_{70}	91_{100}	1674	8183
Etonogestrel 2TMS		2820*	468_{27}	439_{6}	349_{5}	180_{37}	73_{100}	1615	8181
Etonogestrel AC		2895*	366_{7}	337_{22}	324_{37}	133_{55}	91_{100}	1294	8179
Etonogestrel -C2H2		2705*	298_{38}	256_{19}	213_{38}	133_{72}	91_{100}	898	8178
Etonogestrel PFP		2675*	470_{1}	441_{27}	319_{13}	173_{64}	119_{100}	1618	8184
Etonogestrel TFA		2800*	420_{3}	391_{5}	293_{40}	172_{60}	91_{100}	1509	8182
Etonogestrel TMS		2885*	396_{10}	367_{58}	329_{34}	153_{45}	73_{100}	1431	8180
Etoricoxib	G P U+UHYAC	2750	357_{100}	278_{29}	263_{3}	243_{4}	202_{3}	1251	7447
Etoricoxib-M (HO-) AC	U+UHYAC	3500	415_{73}	373_{100}	356_{14}	294_{27}	263_{14}	1495	7883
Etoricoxib-M (HOOC-) -CO2	U+UHYAC	2710	343_{100}	264_{41}	229_{14}	202_{5}	175_{6}	1177	7882
Etozoline		2390	284_{14}	251_{11}	211_{44}	154_{20}	84_{100}	803	3107
Etravirine		3590	434_{28}	419_{6}	355_{100}	212_{16}	144_{51}	1543	7943
Etridiazole		1480	246_{11}	211_{100}	183_{83}	140_{48}	108_{34}	578	4051
Etridiazole artifact (dechloro-)		1320	212_{25}	184_{79}	149_{100}	141_{66}	106_{44}	437	4052
Etrimfos		1850	292_{40}	181_{52}	153_{59}	125_{61}	56_{100}	854	2509
Etryptamine		1860	188_{7}	131_{91}	130_{71}	103_{12}	58_{100}	366	5552
Etryptamine 2HFB		1830	580_{10}	367_{95}	326_{100}	254_{62}	129_{17}	1707	6195
Etryptamine 2PFP		1840	480_{15}	317_{92}	276_{100}	204_{61}	129_{45}	1634	5554
Etryptamine 2TFA		1860	380_{27}	267_{98}	226_{100}	154_{64}	129_{32}	1355	5557
Etryptamine 2TMS		1880	332_{1}	317_{3}	203_{17}	130_{100}	73_{60}	1108	5559
Etryptamine AC		2380	230_{10}	171_{82}	156_{24}	130_{100}	58_{74}	511	4694
Etryptamine ACPFP		2150	376_{18}	213_{24}	184_{42}	172_{100}	130_{32}	1339	5556
Etryptamine formyl artifact		1890	200_{20}	169_{7}	143_{100}	115_{9}	85_{6}	400	5553
Etryptamine HFB		1945	384_{16}	171_{9}	130_{100}	103_{4}	77_{4}	1374	6196
Etryptamine PFP		1880	334_{17}	171_{10}	130_{100}	103_{6}	77_{6}	1119	5555
Etryptamine TFA		1950	284_{19}	171_{9}	156_{4}	130_{100}	103_{7}	802	5558
Eutylone		1860	206_{3}	149_{20}	121_{15}	91_{6}	86_{100}	533	9149
Eutylone AC		2200	277_{1}	192_{4}	149_{12}	128_{58}	86_{100}	760	9150
Eutylone HFB		1920	431_{1}	282_{50}	254_{8}	149_{100}	121_{21}	1537	9153
Eutylone PFP		1890	381_{1}	232_{44}	204_{6}	149_{100}	121_{11}	1359	9152
Eutylone TFA		1950	331_{2}	182_{57}	154_{10}	149_{100}	121_{19}	1099	9151
Exemestane		2580*	296_{4}	268_{11}	211_{9}	148_{100}	133_{18}	886	7621
Exemestane AC		2880*	338_{75}	296_{100}	281_{28}	211_{30}	198_{39}	1145	9146
Exemestane TMS		2590*	368_{60}	353_{35}	233_{11}	221_{29}	73_{100}	1303	7622
5-F-2-Me-DALT		2090	272_{3}	231_{6}	176_{7}	162_{25}	110_{100}	730	10141
5-F-2-Me-DALT artifact HFB		1700	467_{34}	425_{3}	372_{9}	122_{17}	110_{100}	1611	10148
5-F-2-Me-DALT artifact PFP		1665	417_{31}	322_{5}	174_{4}	160_{7}	110_{100}	1498	10146
5-F-2-Me-DALT artifact TFA		1920	367_{32}	272_{5}	174_{3}	160_{7}	110_{100}	1295	10144
5-F-2-Me-DALT HFB		1905	441_{1}	372_{5}	358_{12}	161_{29}	110_{100}	1613	10147
5-F-2-Me-DALT PFP		1885	391_{1}	391_{1}	308_{15}	161_{33}	110_{100}	1502	10145
5-F-2-Me-DALT TFA		1920	272_{2}	258_{5}	175_{3}	161_{15}	110_{100}	1302	10143

Table 8.1: Compounds in order of names

Name	Detected	RI	Typical ions and intensities					Page	Entry
5-F-2-Me-DALT TMS		2320	344_1	303_4	234_6	218_2	110_{100}	1180	10142
FAB-144		2320	330_{11}	287_{33}	247_{50}	173_{60}	145_{100}	1095	9503
Famciclovir		2430	321_{27}	278_{21}	262_{100}	202_{54}	136_{62}	1038	7739
Famciclovir AC		2645	363_{16}	304_{90}	262_{100}	202_{73}	135_{77}	1278	7741
Famciclovir artifact (deacetyl) 2TMS		2430	423_{64}	364_{67}	348_{71}	220_{49}	73_{100}	1516	7750
Famciclovir artifact (deacetyl) HFB		2299	475_{68}	432_{41}	416_{100}	262_{53}	202_{31}	1627	7747
Famciclovir artifact (deacetyl) ME		2280	293_{78}	278_{32}	251_{41}	163_{74}	135_{100}	863	7740
Famciclovir artifact (deacetyl) PFP		2340	425_{47}	366_{100}	262_{58}	202_{48}	135_{54}	1519	7744
Famciclovir artifact (deacetyl) TFA		2350	375_{68}	332_{52}	316_{100}	262_{47}	202_{40}	1334	7745
Famciclovir artifact (deacetyl) TMS		2375	351_{69}	292_{100}	202_{74}	163_{74}	135_{69}	1215	7749
Famciclovir HFB		2405	517_{11}	458_{100}	412_{45}	398_{42}	332_{38}	1675	7746
Famciclovir PFP		2380	467_2	407_{10}	348_{100}	334_{82}	308_{39}	1612	7743
Famciclovir TFA		2400	417_5	348_{76}	298_{76}	284_{100}	162_{68}	1499	7742
Famciclovir TMS		2485	393_{47}	378_{26}	334_{62}	318_{100}	276_{63}	1416	7748
Famciclovir-M (bis-deacetyl-)		2675	237_{56}	220_{19}	206_{45}	148_{80}	136_{100}	545	9435
Famciclovir-M (bis-deacetyl-) 3TMS		2560	453_{41}	438_{39}	350_{51}	220_{46}	73_{100}	1587	9437
Famciclovir-M (deacetyl-)		2640	279_{27}	220_{72}	148_{49}	136_{73}	135_{100}	773	9436
Famciclovir-M (deacetyl-) 2HFB		2280	629_{13}	432_{21}	416_{100}	202_{32}	148_{39}	1720	9439
Famciclovir-M (deacetyl-) 2PFP		2250	529_{29}	382_{23}	366_{100}	202_{40}	148_{49}	1684	9438
Famotidine artifact (sulfurylamine)		1625	96_{63}	82_7	80_{100}	64_{29}		256	6055
Famotidine artifact (sulfurylamine) 2ME		1140	124_{64}	94_{100}	78_{13}	60_{33}		267	6056
Famotidine artifact (sulfurylamine) ME		1345	110_{82}	109_{69}	94_{100}	80_{42}	64_{54}	261	6057
Famprofazone		2965	377_2	286_{100}	229_{22}	136_6	91_{12}	1345	1968
Famprofazone-M (AM)		1160	134_4	120_{15}	91_{100}	77_{18}	65_{73}	275	54
Famprofazone-M (AM)	U	1160	134_1	120_1	91_6	65_4	44_{100}	275	5514
Famprofazone-M (AM) AC	U+UHYAC	1505	177_4	118_{60}	91_{35}	86_{100}	65_{16}	336	55
Famprofazone-M (AM) AC	U+UHYAC	1505	177_1	118_{19}	91_{11}	86_{31}	44_{100}	335	5515
Famprofazone-M (AM) HFB		1355	240_{79}	169_{21}	118_{100}	91_{53}		1099	5047
Famprofazone-M (AM) PFP		1330	281_1	190_{73}	118_{100}	91_{36}	65_{12}	786	4379
Famprofazone-M (AM) TFA		1095	231_1	140_{100}	118_{92}	91_{45}	69_{19}	513	4000
Famprofazone-M (AM) TMS		1190	192_6	116_{100}	100_{10}	91_{11}	73_{87}	421	5581
Famprofazone-M (HO-metamfetamine)		1885	150_1	135	107_4	77_4	58_{100}	314	1766
Famprofazone-M (HO-metamfetamine) 2AC	U+UHYAC	1995	249_1	176_6	134_7	100_{43}	58_{100}	599	1767
Famprofazone-M (HO-metamfetamine) 2HFB		1670	538_1	330_{17}	254_{100}	210_{32}	169_{22}	1698	5076
Famprofazone-M (HO-metamfetamine) 2PFP		1605	280_{18}	204_{100}	160_{47}	119_{39}		1595	5077
Famprofazone-M (HO-metamfetamine) 2TFA		1585	357_1	230_{22}	154_{100}	110_{42}	69_{29}	1246	5078
Famprofazone-M (HO-metamfetamine) 2TMS		1620	309_{10}	206_{70}	179_{100}	154_{53}	73_{40}	966	6190
Famprofazone-M (HO-metamfetamine) TFA		1770	261_1	154_{100}	134_{68}	110_{42}	107_{41}	662	6180
Famprofazone-M (HO-metamfetamine) TMSTFA		1690	333_3	206_{72}	179_{100}	154_{53}	73_{50}	1111	6228
Famprofazone-M (HO-propyphenazone)	UHY	2410	246_{51}	231_{100}	215_9	77_{16}		580	912
Famprofazone-M (HO-propyphenazone) AC	UHYAC	2240	288_{63}	273_{83}	245_{100}	232_{95}	190_{39}	828	206
Famprofazone-M (metamfetamine)	U	1195	148_1	134_2	115_1	91_9	58_{100}	288	1093
Famprofazone-M (metamfetamine) AC	U+UHYAC	1575	191_1	117_2	100_{42}	91_6	58_{100}	373	1094
Famprofazone-M (metamfetamine) HFB		1460	254_{100}	210_{44}	169_{15}	118_{41}	91_{38}	1183	5069
Famprofazone-M (metamfetamine) PFP		1415	204_{100}	160_{46}	118_{35}	91_{25}	69_4	874	5070
Famprofazone-M (metamfetamine) TFA		1300	245_1	154_{100}	118_{48}	110_{55}	91_{23}	574	3998
Famprofazone-M (metamfetamine) TMS		1325	206_{41}	130_{100}	91_{17}	73_{83}	59_{13}	477	6214
5-F-CUMYL-PICA		3000	366_{11}	248_{41}	232_{100}	173_{22}	144_{19}	1293	9491
5-F-DALT		2095	258_1	162_9	148_{36}	133_6	110_{100}	647	9124
5-F-DALT PFP		1955	404_4	308_2	294_7	285_3	110_{100}	1459	9131
5-F-DALT TFA		1900	327_2	258_7	244_9	147_{15}	110_{100}	1231	9130
5-F-DALT TMS		2040	330_2	289_{10}	220_9	110_{100}	73_{32}	1095	9129
5-F-DALT-M (HO-) isomer-1 AC		2450	316_1	210_5	178_4	164_{13}	110_{100}	1006	9258
5-F-DALT-M (HO-) isomer-2 AC		2300	316_1	178_3	164_7	110_{100}		1006	9264
5-F-DALT-M (HO-) isomer-3 AC		2430	316_4	178_5	164_{10}	110_{100}		1006	9265
5-F-DALT-M (N-dealkyl-HO-) isomer-1 2AC		2560	318_8	219_{50}	177_{100}	164_{66}	70_{31}	1018	9259
5-F-DALT-M (N-dealkyl-HO-) isomer-2 2AC		2610	318_{17}	219_{53}	177_{100}	164_{41}	70_{59}	1019	9260
5-F-DALT-M (N-dealkyl-HO-) isomer-3 2AC		2690	318_{12}	219_{51}	177_{100}	164_{58}	70_{52}	1019	9261
5-F-DALT-M (N-dealkyl-HO-) isomer-4 2AC		2710	318_5	261_{34}	219_{100}	177_{84}	70_{77}	1019	9262
6-F-DALT		2045	258_2	162_{11}	148_{46}	133_8	110_{100}	647	10159
6-F-DALT HFB		1840	427_1	358_8	344_{12}	147_{27}	110_{100}	1588	9132
6-F-DALT PFP		1830	308_2	294_4	147_{10}	110_{100}		1459	10162
6-F-DALT TFA		1845	327_1	258_2	244_5	147_{11}	110_{100}	1232	10161
6-F-DALT TMS		2150	330_2	289_3	220_8	110_{100}	73_{40}	1095	10160
5-F-DALT-M (tri-HO-) 3AC		2290	391_{15}	278_{11}	218_{15}	176_{14}	110_{100}	1539	9263
FDU-PB-22		3780	395_1	252_{91}	143_8	115_{12}	109_{100}	1424	9643
FDU-PB-22-M/artifact (HOOC-) (ET)		2420	297_{47}	252_{13}	222_3	143_2	109_{100}	889	9665
FDU-PB-22-M/artifact (HOOC-) (ME)		2400	283_{44}	252_9	222_4	146_6	109_{100}	794	9635
FDU-PB-22-M/artifact (HOOC-) TMS		2535	341_1	282_6	217_5	173_7	109_{100}	1161	9649
Felbamate -C2H3NO2		1890	162_1	134_9	104_{100}	91_{23}	77_{21}	313	4696
Felbamate -CH3NO2		2210	177_9	134_{100}	104_{63}	91_{26}	77_{22}	334	4695
Felbamate-M/artifact (bis-decarbamoyl-) -H2O		1450*	134_1	121_{11}	104_{100}	91_{26}	77_{25}	273	4698
Felbamate-M/artifact (bis-decarbamoyl-) -H2O AC		2010*	176_{11}	134_{100}	104_{75}	91_{19}	77_{20}	332	4697
Felbinac ET		1980*	240_{46}	167_{100}	165_{29}	152_{14}	83_4	557	6075

Felbinac ME   **Table 8.1:** Compounds in order of names

Name	Detected	RI	Typical ions and intensities					Page	Entry
Felbinac ME		1960*	226_{33}	167_{100}	165_{23}	152_9	83_5	495	6074
Felodipine	UME	2670	383_7	354_{11}	238_{100}	210_{36}	150_7	1368	4627
Felodipine HY		2240*	254_6	210_4	101_3	82_{100}	54_{15}	625	6064
Felodipine ME		2725	397_7	338_{16}	324_{51}	252_{100}	224_{59}	1433	4853
Felodipine-M (dehydro-COOH) ET	UET	2665	439_1	404_{100}	376_{40}	344_{25}	309_{18}	1555	4862
Felodipine-M (dehydro-COOH) ME	UME	2570	425_1	390_{100}	362_{53}	309_8	245_9	1518	4854
Felodipine-M (dehydro-COOH) TMS	UTMS	2840	448_{33}	434_{46}	343_{79}	287_{100}	117_{26}	1638	5007
Felodipine-M (dehydro-deethyl-COOH) 2ME	UME	2520	376_{100}	352_8	295_3	172_8		1480	4857
Felodipine-M (dehydro-deethyl-HO-) -H2O	UME U+UHYAC-I	2235	351_1	316_{100}	301_8	284_4		1211	4858
Felodipine-M (dehydro-demethyl-COOH) 2ET	UET	2600	453_1	418_{100}	390_{40}	344_{78}	244_{21}	1586	4864
Felodipine-M (dehydro-demethyl-HO-) -H2O	UME U+UHYAC-I	2560	330_{81}	302_{100}	267_9	164_7		1284	4859
Felodipine-M (dehydro-HO-)	UET	2430	397_5	362_{96}	334_{100}	295_{22}	260_{22}	1432	4863
Felodipine-M/artifact (dehydro-)	UME P-I	2280	381_1	346_{100}	318_{76}	286_{18}	173_{32}	1359	4855
Felodipine-M/artifact (dehydro-deethyl-) -CO2	UME P-I	2235	309_1	274_{100}	259_{11}	215_4	139_5	961	4860
Felodipine-M/artifact (dehydro-deethyl-) ME	UME	2235	367_1	332_{100}	300_7	258_6	173_7	1295	4856
Felodipine-M/artifact (dehydro-deethyl-) TMS	UTMS	2610	390_{100}	380_8	362_{41}	164_5	139_2	1518	5005
Felodipine-M/artifact (dehydro-demethyl-) ET	UET	2375	360_{96}	332_{100}	286_{21}	173_{10}		1423	4861
Felodipine-M/artifact (dehydro-demethyl-deethyl-) -CO2 TMS	UTMS	2250	332_{100}	300_4	257_2	173_3	137_2	1295	5006
Fenamiphos		2020	303_{100}	260_{26}	217_{27}	154_{43}	80_{27}	922	3436
Fenarimol		2605	330_{38}	251_{44}	219_{49}	139_{100}	107_{80}	1092	3437
Fenazepam		2440	350_{92}	348_{72}	321_{100}	313_{71}	177_{31}	1197	5850
Fenazepam artifact-1	U UHY UHYAC	2230	320_{100}	318_{79}	283_{47}	239_{66}	75_{34}	1015	2152
Fenazepam artifact-2	U UHY UHYAC	2250	334_{98}	332_{79}	297_{50}	253_{100}	75_{86}	1103	2153
Fenazepam HY	UHY	2270	311_{100}	309_{71}	276_{88}	274_{90}	195_{41}	960	2151
Fenazepam HYAC	UHYAC	2500	353_{44}	351_{32}	311_{85}	276_{100}	274_{100}	1211	2149
Fenazepam isomer-1 ME		2395	364_{100}	362_{82}	327_{81}	212_{21}	125_{24}	1271	5851
Fenazepam isomer-2 ME		2530	364_{100}	362_{85}	336_{86}	327_{95}	299_{45}	1271	5852
Fenazepam TMS		2790	422_{100}	420_{72}	405_{48}	385_{80}	73_{68}	1507	5853
Fenazepam-M HY	UHY	2270	311_{100}	309_{71}	276_{88}	274_{90}	195_{41}	960	2151
Fenazepam-M HYAC	UHYAC	2500	353_{44}	351_{32}	311_{85}	276_{100}	274_{100}	1211	2149
Fenbendazole 2ME		2935	327_{100}	268_{75}	254_9	230_{14}	59_{26}	1073	7409
Fenbendazole artifact (decarbamoyl-) 2ME		2700	269_{100}	254_{13}	241_7	227_4	184_9	711	7410
Fenbendazole artifact (decarbamoyl-) AC		2930	283_{79}	241_{100}	209_{11}	199_{11}	171_{17}	794	7411
Fenbendazole artifact (decarbamoyl-) AC		2910	297_{59}	255_{100}	225_6	208_{10}	195_5	889	7412
Fenbendazole artifact (decarbamoyl-) ME		2985	255_{100}	239_{12}	225_{11}	199_6	171_9	631	7408
Fenbendazole ME		2965	313_{100}	281_{12}	254_{31}	225_7		987	7407
Fenbuconazole		2665	336_1	211_6	198_{48}	129_{100}	125_{39}	1130	6089
Fenbufen		2010*	254_{12}	181_{100}	152_{38}	127_4	76_{10}	627	5245
Fenbufen ME		1975*	268_{14}	237_9	181_{100}	152_{43}	76_{10}	707	5246
Fenbufen-M (acetic acid HO-) 2ME	UME	2200*	256_{50}	197_{100}	182_{13}	154_{30}	128_{20}	637	6292
Fenbufen-M (dihydro-) ME	UME	1995*	270_{50}	211_{100}	178_{21}	165_{23}	152_{16}	719	6291
Fenbutrazate	U	2680	367_{11}	261_{34}	190_{28}	91_{80}	69_{100}	1300	773
Fencamfamine	G U UA UHY	1685	215_{43}	186_{18}	98_{100}	84_{49}	58_{68}	450	774
Fencamfamine AC	U+UHYAC	2085	257_4	170_{100}	142_{97}	58_{74}		643	775
Fencamfamine HFB		1795	342_2	280_8	170_{100}	142_{95}	91_{47}	1481	6305
Fencamfamine PFP		1755	292_2	230_{11}	170_{100}	142_{88}	91_{34}	1269	6304
Fencamfamine TFA		1970	242_2	180_9	170_{95}	142_{100}	91_{39}	977	3699
Fencamfamine TMS		1780	287_{71}	272_{45}	258_{94}	170_{97}	73_{100}	823	6306
Fencamfamine-M (deethyl-) AC	U+UHYAC	2005	229_1	186_3	170_{98}	142_{100}	91_{49}	506	776
Fencamfamine-M (deethyl-HO-) 2AC		2520	287_1	228_{29}	168_{25}	142_{100}	94_{11}	822	777
Fencamfamine-M (deethyl-HO-aryl) 2AC		2540	287_2	228_{64}	200_{59}	186_{90}	158_{100}	822	8968
Fencamfamine-M (deethyl-HO-methoxy-aryl-) 2AC		2660	317_2	258_{34}	216_{77}	188_{100}	137_{17}	1013	8971
Fencarbamide		2470	326_1	196_6	169_{37}	99_{23}	86_{100}	1082	1444
Fenchlorphos		1905*	320_{12}	285_{94}	167_6	125_{100}	79_{19}	1031	3438
Fendiline	U UHY	2450	315_{27}	181_{14}	167_{19}	132_{33}	105_{100}	1002	1445
Fendiline AC	UHYAC	2825	357_{47}	162_{25}	120_{26}	105_{100}	72_{80}	1250	1446
Fendiline-M (deamino-HO-) -H2O	UHY UHYAC	1940*	194_{49}	167_{100}	165_{67}	152_{34}	116_{17}	384	3388
Fendiline-M (HO-)	UHY	2785	331_{53}	316_{14}	197_{35}	120_{28}	105_{100}	1102	3389
Fendiline-M (HO-) 2AC	UHYAC	3275	415_{35}	251_{13}	177_{52}	105_{100}	72_{95}	1494	3394
Fendiline-M (HO-methoxy-)	UHY	2820	361_{32}	227_{45}	120_{35}	105_{100}	91_{21}	1270	3389
Fendiline-M (HO-methoxy-) 2AC	UHYAC	3410	445_{54}	239_{59}	177_{76}	105_{100}	72_{84}	1572	3395
Fendiline-M (N-dealkyl-) AC	UHYAC	2320	253_{15}	193_{12}	165_{19}	152_{10}	73_{100}	623	3391
Fendiline-M (N-dealkyl-HO-) 2AC	UHYAC	2635	311_{54}	269_{14}	239_{21}	183_{63}	73_{100}	977	3392
Fendiline-M (N-dealkyl-HO-methoxy-) 2AC	UHYAC	2700	341_{10}	299_{54}	213_{74}	152_{14}	73_{100}	1163	3393
Fenetylline	G P-I U UHY	2830	326_1	250_{100}	207_{34}	91_{23}	70_{28}	1165	778
Fenetylline AC	U+UHYAC	3110	383_9	292_{25}	250_{100}	207_{51}		1372	779
Fenetylline HFB		2815	537_7	446_{54}	266_{35}	207_{22}	91_{100}	1688	5054
Fenetylline PFP		2790	487_9	396_{90}	369_{59}	207_{43}	91_{100}	1645	5055
Fenetylline TFA		2840	437_9	346_{52}	319_{40}	166_{65}	91_{100}	1551	5056
Fenetylline-M (AM)		1160	134_4	120_{15}	91_{100}	77_{18}	65_{73}	275	54
Fenetylline-M (AM)	U	1160	134_1	120_1	91_6	65_4	44_{100}	275	5514
Fenetylline-M (AM) AC	U+UHYAC	1505	177_4	118_{60}	91_{35}	86_{100}	65_{16}	336	55
Fenetylline-M (AM) AC	U+UHYAC	1505	177_1	118_{19}	91_{11}	86_{31}	44_{100}	335	5515
Fenetylline-M (AM) HFB		1355	240_{79}	169_{21}	118_{100}	91_{53}		1099	5047

Table 8.1: Compounds in order of names — Fenetylline-M (AM) PFP

Name	Detected	RI	Typical ions and intensities					Page	Entry
Fenetylline-M (AM) PFP		1330	281 $_1$	190 $_{73}$	118 $_{100}$	91 $_{36}$	65 $_{12}$	786	4379
Fenetylline-M (AM) TFA		1095	231 $_1$	140 $_{100}$	118 $_{92}$	91 $_{45}$	69 $_{19}$	513	4000
Fenetylline-M (AM) TMS		1190	192 $_6$	116 $_{100}$	100 $_{10}$	91 $_{11}$	73 $_{87}$	421	5581
Fenetylline-M (AM)-D11 PFP		1610	194 $_{100}$	128 $_{82}$	98 $_{43}$	70 $_{14}$		856	7284
Fenetylline-M (AM)-D11 TFA		1615	242 $_1$	144 $_{100}$	128 $_{82}$	98 $_{43}$	70 $_{14}$	566	7283
Fenetylline-M (AM)-D5 AC		1480	182 $_3$	122 $_{46}$	92 $_{30}$	90 $_{100}$	66 $_{16}$	352	5690
Fenetylline-M (AM)-D5 HFB		1330	244 $_{100}$	169 $_{14}$	122 $_{46}$	92 $_{41}$	69 $_{40}$	1130	6316
Fenetylline-M (AM)-D5 PFP		1320	194 $_{100}$	123 $_{42}$	119 $_{32}$	92 $_{46}$	69 $_{11}$	815	5566
Fenetylline-M (AM)-D5 TFA		1085	144 $_{100}$	123 $_{53}$	122 $_{56}$	92 $_{51}$	69 $_{11}$	540	5570
Fenetylline-M (AM)-D5 TMS		1180	212 $_1$	197 $_8$	120 $_{100}$	92 $_{11}$	73 $_{57}$	441	5582
Fenetylline-M (etofylline)	UHY	2125	224 $_{47}$	194 $_{16}$	180 $_{100}$	109 $_{34}$	95 $_{74}$	487	771
Fenetylline-M (etofylline) AC	U+UHYAC	2200	266 $_{79}$	206 $_{59}$	180 $_{34}$	122 $_{31}$	87 $_{100}$	693	772
Fenetylline-M (etofylline) TMS	UHYTMS	2160	296 $_{10}$	281 $_{26}$	252 $_6$	180 $_{100}$	73 $_{47}$	886	5696
Fenetylline-M (N-dealkyl-) AC	U+UHYAC	2480	265 $_{35}$	206 $_{100}$	180 $_{34}$	122 $_{18}$	86 $_{26}$	687	1886
Fenfluramine	G P U	1250	230 $_1$	216 $_2$	159 $_7$	72 $_{100}$		513	780
Fenfluramine AC	U+UHYAC	1580	254 $_1$	216 $_1$	159 $_6$	114 $_{33}$	72 $_{100}$	734	781
Fenfluramine HFB		1495	408 $_6$	268 $_{100}$	240 $_{54}$	159 $_{45}$		1525	5057
Fenfluramine PFP		1455	377 $_1$	358 $_3$	218 $_{100}$	190 $_{59}$	159 $_{35}$	1342	5058
Fenfluramine TFA		1455	327 $_1$	308 $_3$	186 $_7$	168 $_{100}$	140 $_{48}$	1074	5059
Fenfluramine-M (deethyl-) AC	UHYAC	1510	245 $_2$	226 $_5$	186 $_5$	159 $_{12}$	86 $_{100}$	574	782
Fenfluramine-M (deethyl-HO-) 2AC	UHYAC	1980	176 $_4$	159 $_{12}$	133 $_{10}$	107 $_{15}$	86 $_{100}$	923	5657
Fenfluramine-M (di-HO-) 3AC	UHYAC	2585	191 $_1$	150 $_2$	114 $_{49}$	72 $_{100}$		1398	5656
Fenfluramine-M (HO-) 2AC	UHYAC	1895	175 $_1$	162 $_1$	134 $_2$	114 $_{47}$	72 $_{100}$	1100	4472
Fenfuram		1900	201 $_{26}$	184 $_1$	144	109 $_{100}$	65 $_7$	402	2532
Fenitrothion		1925	277 $_{39}$	260 $_{24}$	125 $_{100}$	109 $_{90}$	79 $_{53}$	757	2510
Fenitrothion-M/artifact (3-methyl-4-nitrophenol)		1560	153 $_{34}$	136 $_{68}$	108 $_{11}$	77 $_{100}$	53 $_{77}$	295	7537
Fenitrothion-M/artifact (3-methyl-4-nitrophenol) AC		1455	195 $_{10}$	153 $_{42}$	136 $_{100}$	108 $_9$	77 $_{30}$	385	7538
Fenofibrate		2515*	360 $_{29}$	273 $_{62}$	232 $_{60}$	139 $_{61}$	121 $_{100}$	1264	1940
Fenofibrate-M (HOOC-) ME	P UME U UHY UHYAC	2430*	332 $_{19}$	273 $_{13}$	232 $_{39}$	139 $_{29}$	121 $_{100}$	1104	3039
Fenofibrate-M (O-dealkyl-)	UHY	2300*	232 $_{36}$	197 $_7$	139 $_{23}$	121 $_{100}$	111 $_{23}$	515	2240
Fenofibrate-M (O-dealkyl-) AC	U+UHYAC	2230*	274 $_7$	232 $_{43}$	139 $_{25}$	121 $_{100}$	111 $_{27}$	737	2230
Fenoprofen		2035*	242 $_{85}$	197 $_{100}$	104 $_{25}$	91 $_{44}$	77 $_{46}$	564	5112
Fenoprofen artifact		1765*	212 $_{64}$	197 $_{100}$	169 $_{19}$	141 $_{40}$	115 $_{23}$	437	5113
Fenoprofen ME		1970*	256 $_{81}$	197 $_{100}$	181 $_{25}$	103 $_{29}$	91 $_{36}$	637	5111
Fenoprofen-M (HO-) 2ME	UME	2130*	286 $_{100}$	227 $_{51}$	152 $_7$	123 $_{17}$	91 $_8$	815	6290
Fenoprop	P-I G U	1760*	268 $_{17}$	223 $_8$	196 $_{100}$	167 $_{13}$	97 $_{33}$	704	783
Fenoprop ME		1720*	282 $_{31}$	223 $_{35}$	196 $_{100}$	87 $_{31}$	59 $_{92}$	791	2397
Fenoprop-M (2,4,5-trichlorophenol)	U	1440*	198 $_{93}$	196 $_{100}$	132 $_{34}$	97 $_{64}$	73 $_{14}$	388	784
Fenoterol -H2O 4AC	U+UHYAC	3440	453 $_7$	304 $_{19}$	262 $_{100}$	220 $_{19}$	107 $_{11}$	1587	3146
Fenoxaprop-ethyl	U+UHYAC	2615	361 $_{77}$	288 $_{100}$	261 $_{28}$	119 $_{35}$	76 $_{58}$	1267	4120
Fenoxaprop-ethyl-M/artifact (phenol)		1630*	210 $_{30}$	137 $_{64}$	110 $_{100}$	81 $_{35}$	65 $_{27}$	433	4121
Fenpipramide		2690	322 $_1$	238 $_2$	211 $_6$	112 $_{40}$	98 $_{100}$	1047	785
Fenpipramide TMS		2690	394 $_1$	283 $_{38}$	112 $_{56}$	98 $_{100}$	73 $_{66}$	1422	4614
Fenpropathrin		2450	349 $_3$	265 $_{15}$	208 $_{13}$	181 $_{48}$	97 $_{100}$	1205	3843
Fenpropemorph		2010	303 $_9$	147 $_2$	128 $_{100}$	91 $_5$	70 $_9$	928	3439
Fenproporex	U	1585	173 $_1$	132 $_2$	97 $_{100}$	91 $_{18}$	56 $_{31}$	365	786
Fenproporex AC	U+UHYAC	1915	139 $_{18}$	118 $_5$	97 $_{100}$	91 $_{13}$	56 $_{23}$	511	787
Fenproporex HFB		1730	293 $_{100}$	240 $_{33}$	118 $_{61}$	91 $_{48}$	56 $_{54}$	1375	5060
Fenproporex PFP		1685	243 $_{100}$	190 $_{50}$	118 $_{77}$	91 $_{44}$	56 $_{52}$	1119	5061
Fenproporex TFA		1705	193 $_{100}$	140 $_{47}$	118 $_{62}$	91 $_{42}$	56 $_{37}$	802	5062
Fenproporex-M	UHY	1465	181 $_9$	138 $_{100}$	122 $_{18}$	94 $_{24}$	77 $_{16}$	350	4351
Fenproporex-M (AM)		1160	134 $_4$	120 $_{15}$	91 $_{100}$	77 $_{18}$	65 $_{73}$	275	54
Fenproporex-M (AM)	U	1160	134 $_1$	120 $_1$	91 $_6$	65 $_4$	44 $_{100}$	275	5514
Fenproporex-M (AM) AC	U+UHYAC	1505	177 $_4$	118 $_{60}$	91 $_{35}$	86 $_{100}$	65 $_{16}$	336	55
Fenproporex-M (AM) AC	U+UHYAC	1505	177 $_1$	118 $_{19}$	91 $_{11}$	86 $_{31}$	44 $_{100}$	335	5515
Fenproporex-M (AM) formyl artifact		1100	147 $_2$	146 $_6$	125 $_5$	91 $_{12}$	56 $_{100}$	286	3261
Fenproporex-M (AM) HFB		1355	240 $_{79}$	169 $_{21}$	118 $_{100}$	91 $_{53}$		1099	5047
Fenproporex-M (AM) PFP		1330	281 $_1$	190 $_{73}$	118 $_{100}$	91 $_{36}$	65 $_{12}$	786	4379
Fenproporex-M (AM) TFA		1095	231 $_1$	140 $_{100}$	118 $_{92}$	91 $_{45}$	69 $_{19}$	513	4000
Fenproporex-M (AM) TMS		1190	192 $_6$	116 $_{100}$	100 $_{10}$	91 $_{11}$	73 $_{87}$	421	5581
Fenproporex-M (di-HO-) 3AC	UHYAC	2575	346 $_1$	234 $_7$	192 $_7$	150 $_{11}$	97 $_{100}$	1189	4386
Fenproporex-M (HO-) isomer-1 2AC	UHYAC	2260	288 $_1$	176 $_7$	139 $_{13}$	134 $_9$	97 $_{100}$	828	4383
Fenproporex-M (HO-) isomer-2 2AC	UHYAC	2350	176 $_{13}$	139 $_{13}$	134 $_{18}$	97 $_{100}$		828	4384
Fenproporex-M (HO-methoxy-) 2AC	UHYAC	2495	318 $_1$	206 $_{12}$	164 $_{31}$	137 $_{11}$	97 $_{100}$	1019	4385
Fenproporex-M (N-dealkyl-3-HO-) 2AC	UHYAC	1930	235 $_2$	176 $_{48}$	134 $_{52}$	107 $_{21}$	86 $_{100}$	532	4387
Fenproporex-M (N-dealkyl-3-HO-) 2HFB		1620	330 $_{30}$	303 $_{11}$	240 $_{100}$	169 $_{15}$	69 $_{29}$	1691	5737
Fenproporex-M (N-dealkyl-3-HO-) 2PFP		1520	280 $_{36}$	253 $_9$	190 $_{100}$	119 $_{19}$	69 $_8$	1567	5738
Fenproporex-M (N-dealkyl-3-HO-) 2TFA		<1000	230 $_{33}$	203 $_8$	140 $_{100}$	115 $_6$		1172	6224
Fenproporex-M (N-dealkyl-3-HO-) 2TMS	UHYTMS	1850	280 $_{11}$	179 $_3$	116 $_{100}$	100 $_{12}$	73 $_{72}$	881	5693
Fenproporex-M (N-dealkyl-3-HO-) TMSTFA		1630	319 $_8$	206 $_{86}$	191 $_{32}$	140 $_{100}$	73 $_{58}$	1024	6141
Fenproporex-M (N-dealkyl-4-HO-)		1480	151 $_{10}$	107 $_{69}$	91 $_{10}$	77 $_{42}$	56 $_{100}$	292	1802
Fenproporex-M (N-dealkyl-4-HO-) 2AC	U+UHYAC	1900	235 $_1$	176 $_{72}$	134 $_{100}$	107 $_{46}$	86 $_{70}$	533	1804
Fenproporex-M (N-dealkyl-4-HO-) 2HFB		<1000	330 $_{48}$	303 $_{15}$	240 $_{100}$	169 $_{44}$	69 $_{42}$	1691	6326
Fenproporex-M (N-dealkyl-4-HO-) 2PFP		<1000	280 $_{77}$	253 $_{16}$	190 $_{100}$	119 $_{56}$	69 $_{16}$	1567	6325

Fenproporex-M (N-dealkyl-4-HO-) 2TFA Table 8.1: Compounds in order of names

Name	Detected	RI	Typical ions and intensities					Page	Entry
Fenproporex-M (N-dealkyl-4-HO-) 2TFA		<1000	230_{72}	203_{11}	140_{100}	92_{12}	69_{59}	1172	6324
Fenproporex-M (N-dealkyl-4-HO-) 2TMS		<1000	280_{7}	179_{9}	149_{8}	116_{100}	73_{78}	880	6327
Fenproporex-M (N-dealkyl-4-HO-) AC	U+UHYAC	1890	193_{1}	134_{100}	107_{26}	86_{24}	77_{16}	379	1803
Fenproporex-M (N-dealkyl-4-HO-) TFA		1670	247_{4}	140_{15}	134_{54}	107_{100}	77_{15}	584	6335
Fenproporex-M (N-dealkyl-di-HO-) 3AC	U+UHYAC	2150	293_{1}	234_{26}	192_{42}	150_{89}	86_{100}	862	3725
Fenproporex-M (N-dealkyl-HO-methoxy-) 2HFB	UHFB	1690	360_{82}	333_{15}	240_{100}	169_{42}	69_{39}	1704	6512
Fenproporex-M (norephedrine) 2AC	U+UHYAC	1805	235_{1}	176_{5}	134_{7}	107_{13}	86_{100}	532	2476
Fenproporex-M (norephedrine) 2HFB	UHYHFB	1455	543_{1}	330_{14}	240_{100}	169_{44}	69_{57}	1691	5098
Fenproporex-M (norephedrine) 2PFP	UHYPFP	1380	443_{1}	280_{7}	190_{100}	119_{59}	105_{26}	1567	5094
Fenproporex-M (norephedrine) 2TFA	UTFA	1355	343_{1}	230_{6}	203_{5}	140_{100}	69_{29}	1172	5091
Fenproporex-M (norephedrine) TMSTFA		1890	240_{8}	198_{3}	179_{100}	117_{5}	73_{88}	1024	6146
Fenproporex-M 2AC	U+UHYAC	2065	265_{3}	206_{27}	164_{100}	137_{23}	86_{33}	688	3498
Fenproporex-M 2AC	U+UHYAC	1820*	266_{3}	206_{9}	164_{100}	150_{10}	137_{30}	695	6409
Fenproporex-M formyl art.		1220	163_{3}	148_{4}	107_{30}	77_{12}	56_{100}	308	6323
Fenproporex-M-D11 PFP		1610	194_{100}	128_{82}	98_{43}	70_{14}		856	7284
Fenproporex-M-D11 TFA		1615	242_{1}	144_{100}	128_{82}	98_{43}	70_{14}	566	7283
Fenproporex-M-D5 AC		1480	182_{3}	122_{46}	92_{30}	90_{100}	66_{16}	352	5690
Fenproporex-M-D5 HFB		1330	244_{100}	169_{14}	122_{46}	92_{41}	69_{40}	1130	6316
Fenproporex-M-D5 PFP		1320	194_{100}	123_{42}	119_{32}	92_{46}	69_{11}	815	5566
Fenproporex-M-D5 TFA		1085	144_{100}	123_{53}	122_{56}	92_{51}	69_{28}	540	5570
Fenproporex-M-D5 TMS		1180	212_{1}	197_{8}	120_{100}	92_{11}	73_{57}	441	5582
Fenson		1980*	268_{28}	141_{76}	99_{14}	77_{100}	51_{35}	704	3440
Fensulfothion		2250*	308_{58}	293_{100}	141_{48}	125_{68}	97_{80}	954	1447
Fensulfothion impurity		1910*	292_{100}	156_{52}	140_{46}	125_{31}	97_{46}	854	1452
Fentanyl	P-I U+UHYAC	2720	245_{100}	189_{62}	146_{94}			1133	788
Fentanyl-D5		2710	341_{1}	250_{100}	207_{13}	194_{29}	151_{53}	1166	7368
Fenthion	G	1930*	278_{100}	169_{35}	125_{77}	109_{63}	79_{33}	766	3838
Fenticonazole		3410	454_{5}	209_{8}	199_{100}	185_{7}	81_{10}	1588	6088
Fenuron ME		1405	178_{13}	133_{10}	106_{28}	72_{100}		338	3967
Fenvalerate isomer-1		2890	419_{5}	225_{23}	181_{44}	167_{72}	125_{100}	1505	3839
Fenvalerate isomer-2		3839	419_{5}	225_{25}	181_{43}	167_{73}	125_{100}	1506	3840
Ferulic acid 2ME	UME	1850*	222_{100}	207_{22}	191_{75}	164_{14}	147_{19}	479	4945
Ferulic acid 2TMS		2160*	338_{92}	323_{56}	308_{48}	249_{45}	73_{100}	1143	5815
Ferulic acid -CO2		1195*	150_{34}	135_{33}	107_{68}	77_{100}	51_{68}	289	5752
Ferulic acid glycine conjugate 2ME		2450	279_{41}	191_{100}	163_{12}	148_{6}	133_{5}	772	5825
Ferulic acid glycine conjugate 3TMS		2540	467_{61}	453_{8}	336_{22}	249_{67}	73_{100}	1612	5826
Ferulic acid glycine conjugate ME		2380	265_{75}	204_{7}	177_{100}	145_{47}	117_{13}	686	5766
Ferulic acid ME		1930*	208_{100}	177_{80}	145_{53}	133_{19}	117_{24}	424	5966
Ferulic acid MEAC		1950*	250_{3}	208_{100}	177_{47}	145_{26}	117_{9}	603	5814
Fesoterodine		2735	411_{5}	396_{33}	223_{5}	114_{100}	72_{19}	1484	9405
Fesoterodine AC		2785	453_{3}	438_{30}	223_{5}	114_{100}	72_{11}	1588	9413
Fesoterodine HFB		2555	607_{2}	592_{10}	409_{8}	223_{36}	114_{100}	1716	9411
Fesoterodine ME		2700	425_{11}	410_{40}	223_{2}	114_{100}	72_{7}	1522	9406
Fesoterodine PFP		2550	557_{2}	542_{23}	394_{8}	223_{45}	114_{100}	1698	9410
Fesoterodine TFA		2555	507_{2}	492_{32}	394_{5}	223_{24}	114_{100}	1667	9409
Fesoterodine TMS		2700	483_{7}	468_{38}	223_{14}	114_{100}	72_{13}	1639	9407
Fesoterodine-M/A (phenol) 2AC		2690	425_{1}	410_{13}	223_{8}	114_{100}	72_{23}	1521	9412
Fesoterodine-M/A (phenol) 2TMS		2555	485_{2}	470_{9}	308_{4}	223_{5}	114_{100}	1642	9408
Fexofenadine 2TMS		3950	645_{18}	280_{67}	183_{18}	105_{25}	73_{100}	1723	7731
Fexofenadine -H2O 2TMS		3690	627_{11}	262_{82}	248_{25}	129_{8}	73_{100}	1720	7732
Fexofenadine -H2O -CO2	U+UHYAC	3650	439_{6}	280_{100}	262_{12}	131_{10}	105_{17}	1557	5223
Fexofenadine-M (benzophenone)	U+UHYAC	1610*	182_{31}	152_{3}	105_{100}	77_{70}	51_{39}	351	1624
Fexofenadine-M (N-dealkyl-oxo-) -2H2O	U+UHYAC	2190	245_{62}	167_{100}	152_{17}	139_{21}	115_{16}	575	2218
Fipexide		3090	388_{7}	261_{8}	253_{6}	176_{17}	135_{100}	1394	6718
Fipexide Cl-artifact		1295*	170_{25}	135_{100}	105_{8}	77_{41}		323	6635
Fipexide-M (deethylene-MDBP) 2AC	U+UHYAC	2320	278_{4}	235_{64}	177_{11}	150_{25}	135_{100}	767	6626
Fipexide-M (deethylene-MDBP) 2HFB	U+UHYHFB	2080	586_{3}	389_{24}	346_{7}	240_{7}	135_{100}	1709	6632
Fipexide-M (deethylene-MDBP) 2TFA	U+UHYTFA	2230	386_{8}	289_{19}	246_{10}	150_{4}	135_{100}	1383	6629
Fipexide-M (HO-methoxy-BZP) 2AC	U+UHYAC	2380	306_{2}	234_{9}	179_{13}	137_{100}	85_{64}	944	6508
Fipexide-M (HO-methoxy-BZP) AC	U+UHYAC	2410	264_{9}	192_{8}	137_{100}	122_{18}	85_{42}	682	6509
Fipexide-M (HO-methoxy-BZP) HFB	U+UHYHFB	2135	418_{22}	295_{15}	281_{27}	138_{87}	137_{100}	1502	6575
Fipexide-M (HO-methoxy-BZP) TFA	U+UHYTFA	2120	318_{8}	181_{10}	137_{100}	122_{21}	69_{12}	1018	6570
Fipexide-M (N-dealkyl-) AC		2460	296_{20}	169_{23}	155_{62}	113_{100}	85_{77}	884	6809
Fipexide-M (N-dealkyl-deethylene-) AC		2280	270_{14}	211_{45}	185_{48}	111_{51}	87_{100}	716	6810
Fipexide-M (piperazine) 2AC		1750	170_{15}	85_{33}	69_{25}	56_{100}		324	879
Fipexide-M (piperazine) 2HFB		1290	478_{3}	459_{9}	309_{100}	281_{22}	252_{41}	1631	6634
Fipexide-M (piperazine) 2TFA		1005	278_{10}	209_{59}	152_{25}	69_{56}	56_{100}	766	4129
Fipexide-M/artifact (HOOC-)		1770*	186_{100}	141_{92}	128_{74}	111_{87}	99_{50}	359	1881
Fipexide-M/artifact (HOOC-) ME		1510*	200_{98}	141_{100}	111_{58}	99_{17}	75_{38}	399	1077
Fipexide-M/artifcat (MDBP)		1890	220_{21}	178_{14}	164_{13}	135_{100}	85_{36}	469	6624
Fipexide-M/artifcat (MDBP) AC	U+UHYAC	2350	262_{16}	190_{11}	176_{21}	135_{100}	85_{45}	669	6625
Fipexide-M/artifcat (MDBP) TMS		2080	292_{87}	157_{53}	135_{100}	102_{57}	73_{85}	857	6887
Flamprop-isopropyl		2225	363_{1}	276_{4}	156_{3}	105_{100}	77_{27}	1278	3844
Flamprop-methyl		2155	276_{1}	156_{3}	105_{100}	77_{48}		1123	3845

116

Table 8.1: Compounds in order of names   Flecainide

Name	Detected	RI	Typical ions and intensities					Page	Entry
Flecainide	P-I G U UHY	2520	395 $_1$	301 $_6$	209 $_3$	97 $_{10}$	84 $_{100}$	1490	2822
Flecainide 2TMS		2520	558 $_1$	543 $_3$	301 $_8$	156 $_{100}$	73 $_{58}$	1699	4545
Flecainide AC	U+UHYAC	2515	456 $_3$	301 $_8$	218 $_2$	126 $_{74}$	84 $_{100}$	1593	1449
Flecainide formyl artifact	P G U UHY	2500	426 $_{71}$	301 $_{79}$	218 $_{19}$	125 $_{100}$	97 $_{37}$	1522	1448
Flecainide-M (HO-) 2AC	UHYAC	2680	514 $_2$	301 $_{23}$	184 $_{39}$	142 $_{100}$	100 $_{29}$	1673	2868
Flecainide-M (O-dealkyl-) 2AC	U+UHYAC	2780	416 $_2$	301 $_7$	219 $_{16}$	126 $_{84}$	84 $_{100}$	1496	2390
Flephedrone AC	U+UHYAC	1660	223 $_2$	123 $_{21}$	100 $_{76}$	95 $_{30}$	58 $_{100}$	483	8660
Flephedrone-M (dihydro-) isomer-1 2AC	U+UHYAC	1940	166 $_5$	151 $_4$	123 $_{10}$	100 $_{64}$	58 $_{100}$	700	8089
Flephedrone-M (dihydro-) isomer-2 2AC	U+UHYAC	1960	166 $_3$	123 $_8$	100 $_{58}$	58 $_{100}$		700	8090
Flipiperazine		1785	180 $_{27}$	138 $_{100}$	122 $_{19}$	95 $_{16}$	58 $_{31}$	347	9168
Flipiperazine AC		2135	222 $_{45}$	179 $_{17}$	150 $_{100}$	137 $_{32}$	122 $_{50}$	480	9170
Flipiperazine HFB		1885	376 $_{100}$	179 $_{82}$	150 $_{64}$	123 $_{62}$	95 $_{33}$	1338	9174
Flipiperazine ME		1600	194 $_{100}$	163 $_{24}$	150 $_{17}$	123 $_{72}$	95 $_{36}$	384	9169
Flipiperazine PFP		1855	326 $_{100}$	179 $_{62}$	150 $_{45}$	123 $_{53}$	95 $_{35}$	1067	9173
Flipiperazine TFA		1850	276 $_{100}$	179 $_{62}$	150 $_{52}$	123 $_{83}$	95 $_{53}$	753	9172
Flipiperazine TMS		1835	252 $_{100}$	237 $_{47}$	210 $_{19}$	101 $_{45}$	86 $_{45}$	619	9171
Fluanisone	U UHY UHYAC	2795	356 $_{79}$	218 $_{100}$	205 $_{100}$	162 $_{50}$	123 $_{69}$	1243	172
Fluanisone-M	U+UHYAC	1490*	180 $_{25}$	125 $_{49}$	123 $_{35}$	95 $_{17}$	56 $_{100}$	345	85
Fluanisone-M (N,O-bis-dealkyl-) 2AC	UHYAC	2140	262 $_{10}$	220 $_{18}$	148 $_{100}$	120 $_{51}$	86 $_{12}$	669	170
Fluanisone-M (O-demethyl-)	UHY	2715	342 $_{52}$	194 $_{100}$	165 $_{84}$	123 $_{83}$		1171	495
Fluanisone-M (O-demethyl-) AC	UHYAC	2830	384 $_{33}$	246 $_{88}$	233 $_{100}$	123 $_{100}$		1377	173
Fluanison-M (N-dealkyl-) AC	U+UHYAC	2070	234 $_{36}$	162 $_{100}$	149 $_{29}$	134 $_{38}$	120 $_{33}$	529	6808
Fluanison-M (N-dealkyl-HO-) 2AC	U+UHYAC	2490	292 $_{45}$	250 $_{25}$	207 $_{15}$	178 $_{100}$	165 $_{39}$	856	496
Fluanison-M (N-dealkyl-HO-) AC	U+UHYAC	2420	250 $_{69}$	207 $_{12}$	178 $_{100}$	165 $_{27}$	150 $_{40}$	605	8763
Fluazifop-butyl		2200	383 $_{11}$	282 $_{78}$	146 $_{97}$	91 $_{100}$	57 $_{98}$	1370	3846
Flubenzimine		2430	416 $_{100}$	212 $_{27}$	186 $_{64}$	135 $_{66}$	77 $_{51}$	1495	3847
Flubromazepam		2625	332 $_{61}$	305 $_{100}$	276 $_{12}$	223 $_{34}$	197 $_{30}$	1103	9714
Flubromazepam ET		2415	360 $_{49}$	331 $_{77}$	315 $_{48}$	304 $_{41}$	210 $_{100}$	1262	9715
Flubromazepam HY	UHY U+UHYAC	2230	293 $_{60}$	275 $_{21}$	198 $_{45}$	123 $_{100}$	95 $_{94}$	859	9717
Flubromazepam HYAC	U+UHYAC	2370	335 $_{47}$	292 $_{100}$	273 $_{31}$	213 $_{67}$	123 $_{94}$	1122	9716
Flubromazepam TMS		2410	404 $_{100}$	385 $_{39}$	331 $_4$	73 $_9$		1458	9718
Flubromazolam		2945	370 $_{34}$	341 $_{65}$	301 $_{12}$	222 $_{100}$	181 $_{19}$	1309	9697
Fluchloralin		1800	355 $_1$	326 $_{30}$	306 $_{34}$	264 $_{25}$	63 $_{100}$	1235	3841
Flucloxacilline		2800	452 $_5$	238 $_{22}$	196 $_{100}$	139 $_{56}$	97 $_{86}$	1586	8161
Flucloxacilline artifact		2155	291 $_{82}$	238 $_{25}$	231 $_{30}$	196 $_{100}$	168 $_{17}$	845	8162
Flucloxacilline artifact HYAC	UHYAC	3370	398 $_2$	216 $_{28}$	174 $_{100}$	114 $_{11}$		1436	3021
Flucloxacilline HYAC		2240	296 $_{19}$	261 $_6$	238 $_{22}$	196 $_{100}$	168 $_{10}$	883	8160
Flucloxacilline MEAC		3000	509 $_2$	334 $_4$	238 $_{34}$	196 $_{100}$	142 $_{81}$	1668	8163
Flucloxacilline MEHFB		2650	424 $_2$	238 $_{25}$	196 $_{81}$	115 $_{100}$		1725	8166
Flucloxacilline MEPFP		2700	613 $_1$	325 $_1$	238 $_{33}$	196 $_{100}$	114 $_{20}$	1717	8165
Flucloxacilline METFA		2800	563 $_1$	325 $_{18}$	270 $_{54}$	238 $_{16}$	196 $_{100}$	1701	8164
Fluconazole	P U+UHYAC	2210	224 $_1$	155 $_6$	141 $_{22}$	127 $_{64}$	82 $_{61}$	943	4349
Fludiazepam		2530	302 $_{100}$	301 $_{90}$	274 $_{99}$	239 $_{18}$	183 $_{16}$	917	3069
Fludiazepam HY		2180	263 $_{100}$	246 $_{51}$	211 $_{33}$	95 $_{30}$	75 $_{26}$	672	3070
Fludiazepam-M (nor-)	G P-I UGLUC	2470	288 $_{100}$	287 $_{64}$	260 $_{93}$	259 $_{75}$		825	508
Fludiazepam-M (nor-) artifact	U+UHYAC	2050	272 $_{70}$	271 $_{100}$	237 $_{78}$	151 $_{11}$	110 $_8$	728	2409
Fludiazepam-M (nor-) HY	UHY	2030	249 $_{100}$	154 $_{34}$	123 $_{46}$	95 $_{42}$		593	512
Fludiazepam-M (nor-) HYAC	U+UHYAC	2195	291 $_{52}$	249 $_{100}$	123 $_{57}$	95 $_{61}$		846	286
Fludiazepam-M (nor-) TMS		2470	360 $_{63}$	359 $_{69}$	341 $_{43}$	197 $_{17}$	73 $_{100}$	1263	4621
Flufenamic acid		1935	281 $_{69}$	263 $_{100}$	235 $_{29}$	166 $_{26}$	92 $_{29}$	785	5149
Flufenamic acid 2ME		1785	309 $_{100}$	276 $_{70}$	248 $_{58}$	180 $_{32}$	77 $_{35}$	961	5148
Flufenamic acid ME	PME	1880	295 $_{51}$	263 $_{100}$	235 $_{15}$	166 $_{10}$	92 $_7$	874	5147
Flufenamic acid MEAC		1950	337 $_6$	306 $_{11}$	295 $_{100}$	263 $_{98}$	235 $_{18}$	1134	5150
Flufenamic acid TMS		2095	353 $_{28}$	263 $_{100}$	235 $_9$	167 $_{11}$	75 $_{11}$	1225	6331
Flufenamic acid-M (HO-) 2ME	PME	2115	325 $_{100}$	293 $_{91}$	278 $_{88}$	250 $_{67}$	202 $_{19}$	1060	6377
Flumazenil		2660	303 $_{23}$	257 $_{43}$	229 $_{100}$	201 $_{22}$	94 $_{13}$	922	3674
Flumazenil-M (HOOC-) -CO2		2245	231 $_{100}$	203 $_{70}$	189 $_{27}$	147 $_{26}$	94 $_{19}$	512	3676
Flumazenil-M (HOOC-) ME		2555	289 $_{42}$	257 $_{75}$	229 $_{100}$	201 $_{38}$	94 $_8$	832	3675
Flunarizine	G P U+UHYAC	3135	404 $_2$	287 $_{18}$	201 $_{100}$	183 $_{19}$	117 $_{67}$	1460	789
Flunarizine-M (bis-4-fluorophenylcarbinol)		1690*	220 $_{17}$	201 $_{13}$	183 $_8$	123 $_{100}$	95 $_{31}$	468	3378
Flunarizine-M (bis-4-fluorophenylcarbinol) PFP		1555*	203 $_{100}$	201 $_{68}$	183 $_{44}$	170 $_9$		1289	9479
Flunarizine-M (bis-4-fluorophenylcarbinol) PFP		1555*	416 $_2$	203 $_{100}$	201 $_{79}$	183 $_{30}$	170 $_7$	1494	9480
Flunarizine-M (bis-4-fluorophenylcarbinol) TFA		1550	316 $_7$	203 $_{100}$	201 $_{87}$	183 $_{46}$	170 $_{12}$	1004	9477
Flunarizine-M (carbinol) AC	UHYAC	1740*	262 $_5$	202 $_{100}$	201 $_{98}$	158 $_{61}$	116 $_{44}$	668	3374
Flunarizine-M (difluoro-benzophenone)	U UHY UHYAC	1595*	218 $_{38}$	123 $_{100}$	109	95 $_{62}$	75 $_{28}$	457	3373
Flunarizine-M (HO-difluoro-benzophenone)	UHY	1965*	234 $_{64}$	139 $_{52}$	123 $_{100}$	95 $_{53}$	75 $_{17}$	525	3379
Flunarizine-M (HO-difluoro-benzophenone) AC	UHYAC	1995*	276 $_{18}$	234 $_{82}$	139 $_{45}$	123 $_{100}$	95 $_{47}$	752	3375
Flunarizine-M (HO-methoxy-difluoro-benzophenone) AC	UHYAC	2565*	306 $_{29}$	264 $_{60}$	185 $_{100}$	143 $_{38}$		942	3377
Flunarizine-M (N-dealkyl-) AC	U+UHYAC	2350	244 $_{29}$	201 $_{12}$	172 $_{48}$	117 $_{100}$	85 $_{52}$	572	2198
Flunarizine-M (N-dealkyl-HO-) 2AC	U+UHYAC	2580	302 $_3$	243 $_{20}$	141 $_{100}$	99 $_{27}$	56 $_{10}$	919	2199
Flunarizine-M (N-deciannamyl-) AC	UHYAC	2545	330 $_4$	244 $_{29}$	203 $_{59}$	146 $_{47}$	85 $_{100}$	1094	3376
Flunitrazepam	G P-I	2610	313 $_{89}$	312 $_{98}$	285 $_{100}$	266 $_{46}$	238 $_{37}$	986	497
Flunitrazepam HY	UHY-I U+UHYAC-I	2370	274 $_{100}$	257 $_{22}$	211 $_{22}$	123 $_{15}$	95 $_9$	738	282
Flunitrazepam-D7		2600	320 $_{48}$	318 $_{69}$	301 $_{37}$	292 $_{100}$	272 $_{40}$	1033	7777

Flunitrazepam-D7 HY   **Table 8.1:** Compounds in order of names

Name	Detected	RI	Typical ions and intensities					Page	Entry
Flunitrazepam-D7 HY		2360	281_{100}	263_{21}	217_{37}	127_{40}	99_{29}	787	7778
Flunitrazepam-M (amino-)	P-I U-I UGLUC-I	2615	283_{100}	255_{64}	254_{52}	240_{12}		795	498
Flunitrazepam-M (amino-) AC	UGLUCAC	2950	325_{100}	306_{22}	297_{66}	255_{24}		1061	501
Flunitrazepam-M (amino-) formyl artifact		2580	295_{97}	276_{24}	267_{100}	239_{11}	183_{11}	876	6322
Flunitrazepam-M (amino-) HY	UHY	2795	244_{100}	227_{44}				570	504
Flunitrazepam-M (amino-) HY2AC	UHYAC-I	2870	328_{64}	286_{80}	244_{43}	205_{100}		1080	285
Flunitrazepam-M (amino-) TMS		2585	355_{100}	336_{12}	327_{48}	312_{10}	73_{70}	1237	7502
Flunitrazepam-M (nor-)		2705	299_{65}	298_{100}	272_{76}	252_{40}	224_{60}	900	500
Flunitrazepam-M (nor-) HY	UHY	2335	260_{100}	241_{13}	213_{24}	123_{78}	95_{52}	655	283
Flunitrazepam-M (nor-) HYAC	UHYAC	2380	302_{21}	260_{87}	259_{100}	241_{18}	123_{31}	918	6321
Flunitrazepam-M (nor-) TMS		2450	371_{100}	356_{34}	352_{48}	324_{23}	73_{96}	1317	7501
Flunitrazepam-M (nor-amino-)		2690	269_{100}	241_{52}	240_{64}			711	499
Flunitrazepam-M (nor-amino-) AC		3035	311_{100}	283_{59}	241_{33}			975	502
Flunitrazepam-M (nor-amino-) HY		2165	230_{100}	211_{34}				509	503
Flunitrazepam-M (nor-amino-) HY2AC	U+UHYAC-I	2715	314_{100}	272_{81}	230_{98}	123_{23}		993	284
Flunitrazepam-M (nor-HO-)		2510	297_{47}	280_{61}	269_{84}	250_{81}	223_{100}	999	9709
Flunixin		2070	296_{42}	281_{79}	263_{100}	251_{47}	235_{28}	883	8645
Flunixin 2ME		2040	324_{31}	309_{74}	277_{25}	263_{100}	251_{55}	1056	8647
Flunixin 2TMS		1980	440_{13}	425_{57}	335_{67}	267_{69}	73_{100}	1558	8649
Flunixin ME		1985	310_{33}	295_{80}	263_{100}	251_{50}	181_{46}	969	8646
Flunixin MEAC		2005	352_{8}	310_{44}	295_{100}	263_{64}	251_{49}	1219	8650
Flunixin TMS		2110	368_{17}	353_{64}	263_{100}	251_{53}	77_{80}	1301	8648
Fluocortolone		3225*	345_{100}	299_{38}	279_{6}	171_{12}	139_{14}	1340	1798
Fluocortolone 2AC		3400*	460_{12}	418_{24}	387_{100}	299_{62}		1602	1799
Fluocortolone AC		3420*	418_{10}	398_{64}	345_{100}	299_{63}	279_{58}	1503	1800
Fluoperazine		1785	180_{27}	138_{100}	122_{19}	95_{16}	58_{31}	347	9168
Fluoperazine AC		2135	222_{45}	179_{17}	150_{100}	137_{32}	122_{50}	480	9170
Fluoperazine HFB		1885	376_{100}	179_{82}	150_{64}	123_{62}	95_{33}	1338	9174
Fluoperazine ME		1600	194_{100}	163_{24}	150_{17}	123_{72}	95_{36}	384	9169
Fluoperazine PFP		1855	326_{100}	179_{62}	150_{45}	123_{53}	95_{35}	1067	9173
Fluoperazine TFA		1850	276_{100}	179_{62}	150_{52}	123_{83}	95_{52}	753	9172
Fluoperazine TMS		1835	252_{100}	237_{47}	210_{10}	101_{45}	06_{45}	619	9171
Fluoranthene		1970*	202_{100}	200_{20}	174_{2}	150_{1}	101_{7}	403	2566
Fluorene		1570*	166_{100}	165_{94}	139_{5}	115_{3}	82_{12}	317	2560
5-Fluoro-2-Me-AMT		1760	206_{2}	163_{100}	148_{20}	146_{7}	133_{8}	417	10248
5-Fluoro-2-Me-AMT		1760	206_{2}	163_{100}	148_{20}	133_{10}	44_{92}	417	10254
5-Fluoro-2-Me-AMT 2TMS		2165	350_{1}	335_{8}	235_{33}	116_{100}	100_{10}	1209	9879
5-Fluoro-2-Me-AMT AC		2145	248_{18}	189_{58}	162_{100}	146_{7}	133_{8}	591	9829
5-Fluoro-2-Me-AMT ethylimine artifact		1750	232_{6}	189_{3}	162_{100}	146_{5}	133_{4}	518	10249
5-Fluoro-2-Me-AMT formyl artifact		1740	218_{11}	174_{4}	162_{100}	146_{5}	133_{5}	459	9828
5-Fluoro-2-Me-AMT HFB		1910	402_{10}	240_{3}	189_{3}	162_{100}	131_{4}	1451	9833
5-Fluoro-2-Me-AMT PFP		1890	352_{14}	190_{5}	162_{100}	146_{4}	119_{7}	1218	9832
5-Fluoro-2-Me-AMT TFA		1870	302_{13}	189_{2}	162_{100}	146_{3}	140_{3}	918	10250
5-Fluoro-2-Me-AMT TMS		2030	278_{1}	263_{2}	162_{17}	116_{100}	100_{9}	769	9880
5-Fluoro-2-methyl-N,N-diallyl-tryptamine		2090	272_{3}	231_{6}	176_{7}	162_{25}	110_{100}	730	10141
5-Fluoro-2-methyl-N,N-diallyl-tryptamine artifact HFB		1700	467_{34}	425_{3}	372_{9}	122_{17}	110_{100}	1611	10148
5-Fluoro-2-methyl-N,N-diallyl-tryptamine artifact PFP		1665	417_{31}	322_{5}	174_{4}	160_{7}	110_{100}	1498	10146
5-Fluoro-2-methyl-N,N-diallyl-tryptamine artifact TFA		1920	367_{32}	272_{5}	174_{3}	160_{7}	110_{100}	1295	10144
5-Fluoro-2-methyl-N,N-diallyl-tryptamine HFB		1905	441_{1}	372_{5}	358_{12}	161_{29}	110_{100}	1613	10147
5-Fluoro-2-methyl-N,N-diallyl-tryptamine PFP		1885	391_{1}	391_{1}	308_{15}	161_{33}	110_{100}	1502	10145
5-Fluoro-2-methyl-N,N-diallyl-tryptamine TFA		1920	272_{2}	258_{5}	175_{3}	161_{15}	110_{100}	1302	10143
5-Fluoro-2-methyl-N,N-diallyl-tryptamine TMS		2320	344_{1}	303_{4}	234_{6}	218_{2}	110_{100}	1180	10142
5-Fluoro-ABICA		2925	347_{4}	303_{2}	248_{9}	232_{100}	144_{18}	1196	9595
5-Fluoro-ABICA -CONH3		2675	302_{10}	232_{100}	144_{15}	130_{2}	116_{6}	920	9597
5-Fluoro-ABICA -CONH3 TMS		2925	374_{30}	359_{10}	305_{54}	232_{100}	144_{44}	1331	9596
5-Fluoro-ABICA-M/artifact (HOOC-) (ME)		2725	362_{4}	248_{40}	232_{100}	173_{7}	144_{27}	1274	9598
5-Fluoro-AB-PINACA		2720	348_{1}	304_{59}	233_{100}	145_{23}	131_{7}	1200	9703
5-Fluoro-AB-PINACA -CONH3		2690	303_{36}	275_{12}	233_{100}	213_{17}	145_{61}	926	9661
5-Fluoro-AB-PINACA -CONH3 TMS		2500	375_{100}	360_{44}	285_{55}	210_{56}	73_{77}	1336	9693
5-Fluoro-AB-PINACA TMS		2690	492_{4}	376_{72}	286_{33}	233_{23}	73_{100}	1651	9694
5-Fluoro-ADBICA		3150	361_{5}	317_{3}	288_{5}	232_{100}	144_{19}	1271	9644
5-Fluoro-ADBICA -CONH3		2620	316_{8}	232_{100}	144_{23}	129_{6}	116_{8}	1007	9651
5-Fluoro-ADB-PINACA		2725	362_{1}	318_{73}	289_{20}	233_{100}	145_{38}	1275	9702
5-Fluoro-ADB-PINACA 2TMS		2675	506_{4}	491_{9}	401_{19}	390_{100}	73_{90}	1667	9654
5-Fluoro-ADB-PINACA-M/artifact (HOOC-) (ET)		2170	278_{44}	249_{100}	175_{35}	145_{48}	131_{81}	768	9666
5-Fluoro-ADB-PINACA-M/artifact (HOOC-) (ME)		2150	264_{21}	249_{28}	189_{100}	176_{21}	129_{18}	682	9623
5-Fluoro-ADB-PINACA-M/artifact (HOOC-) TMS		2475	322_{1}	233_{100}	213_{12}	145_{73}	103_{35}	1045	9655
5-Fluoro-AKB-48		3045	383_{45}	355_{47}	294_{49}	233_{100}	145_{76}	1373	9656
5-Fluoro-AKB-48 TMS		2985	455_{27}	366_{91}	308_{74}	277_{56}	73_{100}	1593	9657
5-Fluoro-AKB-48-M/artifact (HOOC-) (ET)		2170	278_{44}	249_{100}	175_{35}	145_{48}	131_{81}	768	9666
5-Fluoro-AKB-48-M/artifact (HOOC-) (ME)		2150	264_{21}	249_{28}	189_{100}	176_{21}	129_{18}	682	9623
5-Fluoro-AKB-48-M/artifact (HOOC-) TMS		2475	322_{1}	233_{100}	213_{12}	145_{73}	103_{35}	1045	9655
5-Fluoro-AMB		2525	363_{5}	304_{34}	249_{11}	233_{100}	145_{53}	1279	9647
5-Fluoro-AMB TMS		2575	435_{20}	376_{34}	286_{34}	233_{100}	145_{47}	1549	9662

Table 8.1: Compounds in order of names

Name	Detected	RI	Typical ions and intensities					Page	Entry
4-Fluoroamphetamine		1400	152_5	138_{19}	109_{100}	83_{67}	63_{23}	295	8624
4-Fluoroamphetamine		1400	152_1	138_4	109_{20}	83_{13}	44_{100}	296	8625
4-Fluoroamphetamine AC		1495	195_2	136_{59}	109_{52}	86_{100}	83_{30}	386	8627
4-Fluoroamphetamine formyl artifact		1120	165_1	109_{25}	83_{11}	63_4	56_{100}	314	8626
4-Fluoroamphetamine HFB		1155	349_1	240_{100}	192_{10}	136_{88}	109_{81}	1203	8632
4-Fluoroamphetamine PFP		1080	299_1	190_{100}	136_{93}	109_{71}	83_{21}	900	8631
4-Fluoroamphetamine TFA		1220	249_1	140_{100}	136_{71}	109_{57}	83_{19}	594	8630
4-Fluoroamphetamine TMS		1065	224_1	210_{21}	116_{100}	109_{32}	73_{79}	491	8628
4-Fluoroamphetamine-M (hydroxy-) isomer-1 2AC	U+UHYAC	1850	252_2	194_4	152_{16}	123_{28}	86_{100}	622	8087
4-Fluoroamphetamine-M (hydroxy-) isomer-2 2AC	U+UHYAC	1870	253_1	194_2	152_{12}	123_{32}	86_{100}	621	8088
5-Fluoro-AMT		1980	192_4	175_3	149_{84}	101_{15}	44_{100}	376	10233
5-Fluoro-AMT		1980	192_3	176_2	149_{100}	120_8	101_{17}	375	10234
6-Fluoro-AMT		1780	192_4	176_2	149_{100}	120_8	101_{17}	376	10255
6-Fluoro-AMT		1780	192_1	149_{54}	120_4	101_7	44_{100}	376	10256
6-Fluoro-AMT 2HFB		1740	584_3	371_{77}	344_{58}	240_{100}	147_{10}	1708	9848
6-Fluoro-AMT 2PFP		1705	484_{17}	321_{83}	294_{66}	190_{100}	174_5	1639	9846
6-Fluoro-AMT 2TFA		1740	384_8	271_{49}	244_{53}	147_{28}	140_{100}	1374	9844
5-Fluoro-AMT 2TMS		2030	336_1	321_7	221_{20}	116_{100}	100_{12}	1132	9812
6-Fluoro-AMT 2TMS		2015	336_1	321_4	221_{22}	116_{100}	100_{16}	1132	9843
5-Fluoro-AMT AC		2305	234_7	175_{100}	148_{77}	101_{15}	86_{10}	528	9811
6-Fluoro-AMT AC		2175	234_9	175_{100}	148_{91}	101_{19}	86_{13}	528	9842
5-Fluoro-AMT ethylimine artifact		2055	218_{17}	201_5	148_{100}	120_7	101_{10}	459	9810
6-Fluoro-AMT formyl artifact		1970	204_{13}	187_6	161_{100}	133_9	107_2	409	9841
6-Fluoro-AMT HFB		1930	388_{22}	240_8	175_{15}	148_{100}	128_4	1393	9849
6-Fluoro-AMT PFP		1900	338_{16}	190_5	175_{10}	148_{100}	101_6	1142	9847
6-Fluoro-AMT TFA		1935	288_5	174_5	148_{100}	140_6	101_6	826	9845
5-Fluoro-APP-PICA		3740	395_5	248_{21}	232_{100}	173_4	144_{19}	1426	9675
5-Fluoro-APP-PICA-M/artifact (HOOC-) (ME)		3220	410_6	248_{38}	232_{100}	173_8	144_{19}	1478	9676
3-Fluoromethcathinone		1185	166_1	123_4	95_{11}	75_9	58_{100}	348	8072
3-Fluoromethcathinone AC	U+UHYAC	1645	223_1	123_6	100_{43}	95_{16}	58_{100}	483	8074
3-Fluoromethcathinone artifact (dehydro-)		1220	179_1	123_4	95_{14}	75_{13}	56_{100}	340	8073
3-Fluoromethcathinone HFB		1440	377_1	254_{100}	210_{51}	123_{47}	95_{43}	1341	8077
3-Fluoromethcathinone PFP		1360	254_{100}	210_{61}	123_{57}	95_{51}		1073	8076
3-Fluoromethcathinone TFA		1385	277_1	234_7	154_{100}	123_{43}	110_{59}	758	8075
3-Fluoromethcathinone-M (dihydro-)	U+UHYAC	1960	208_1	166_4	123_8	100_{46}	58_{100}	700	8661
3-Fluoromethcathinone-M (dihydro-HO-) 3AC	U+UHYAC	2175	298_1	144_2	100_4	74_{11}	58_{100}	1062	8081
3-Fluoromethcathinone-M (dihydro-HO-) isomer-1 artifact 2AC		1930*	252_1	207_6	180_{32}	164_8	138_{100}	616	8085
3-Fluoromethcathinone-M (dihydro-HO-) isomer-1 artifact AC		1830*	209_{13}	167_{28}	150_{10}	124_{100}	97_{10}	432	8083
3-Fluoromethcathinone-M (dihydro-HO-) isomer-2 artifact 2AC	U+UHYAC	1945*	252_1	207_6	180_{32}	164_8	138_{100}	616	8086
3-Fluoromethcathinone-M (dihydro-HO-) isomer-2 artifact AC		1860*	209_{12}	167_{30}	150_9	124_{100}	97_5	432	8084
3-Fluoromethcathinone-M (HO-) 2AC	U+UHYAC	2035	281_1	196_2	139_{16}	100_{73}	58_{100}	786	8078
3-Fluoromethcathinone-M (HO-) AC	U+UHYAC	1950	239_1	162_4	120_{18}	100_{31}	58_{100}	552	8079
3-Fluoromethcathinone-M (HO-) artifact AC		1740*	181_{32}	138_{100}	109_9	83_5		422	8082
3-Fluoromethcathinone-M (nor-) AC	U+UHYAC	1620	209_1	123_{19}	95_{35}	86_{100}	75_{21}	428	8080
3-Fluoromethcathinone-M (nor-dihydro-) isomer-1 2AC	U+UHYAC	1850	252_2	194_4	152_{16}	123_{28}	86_{100}	622	8087
3-Fluoromethcathinone-M (nor-dihydro-) isomer-2 2AC	U+UHYAC	1870	253_1	194_2	152_{12}	123_{32}	86_{100}	621	8088
3-Fluoromethcathinone-M (nor-dihydro-) -NH3 HFB		1370*	348_{10}	320_{100}	169_{57}	135_{25}	69_{72}	1198	8091
4-Fluoromethcathinone AC	U+UHYAC	1660	223_2	123_{21}	100_{76}	95_{30}	58_{100}	483	8660
4-Fluoromethcathinone-M (dihydro-) isomer-1 2AC		1940	166_5	151_4	123_{10}	100_{64}	58_{100}	700	8089
4-Fluoromethcathinone-M (dihydro-) isomer-2 2AC	U+UHYAC	1960	166_3	123_8	100_{58}	58_{100}		700	8090
5-Fluoro-N,N-diallyl-tryptamine		2095	258_1	162_9	148_{36}	133_6	110_{100}	647	9124
6-Fluoro-N,N-diallyl-tryptamine		2045	258_2	162_{11}	148_{46}	133_8	110_{100}	647	10159
6-Fluoro-N,N-diallyl-tryptamine HFB		1840	427_1	358_8	344_{19}	147_{27}	110_{100}	1588	9132
5-Fluoro-N,N-diallyl-tryptamine PFP		1955	404_4	308_9	294_7	285_3	110_{100}	1459	9131
6-Fluoro-N,N-diallyl-tryptamine PFP		1830	308_2	294_4	147_{10}	110_{100}		1459	10162
5-Fluoro-N,N-diallyl-tryptamine TFA		1900	327_2	258_7	244_9	147_{15}	110_{100}	1231	9130
6-Fluoro-N,N-diallyl-tryptamine TFA		1845	327_1	258_2	244_5	147_{11}	110_{100}	1232	10161
5-Fluoro-N,N-diallyl-tryptamine TMS		2040	330_2	289_{10}	220_9	110_{100}	73_{32}	1095	9129
6-Fluoro-N,N-diallyl-tryptamine TMS		2150	330_2	289_3	220_8	110_{100}	73_{40}	1095	10160
5-Fluoro-N,N-diallyl-tryptamine-M (HO-) isomer-1 AC		2450	316_1	210_5	178_4	164_{13}	110_{100}	1006	9258
5-Fluoro-N,N-diallyl-tryptamine-M (HO-) isomer-2 AC		2300	316_1	178_3	164_7	110_{100}		1006	9264
5-Fluoro-N,N-diallyl-tryptamine-M (HO-) isomer-3 AC		2430	316_4	178_5	164_{10}	110_{100}		1006	9265
5-Fluoro-N,N-diallyl-tryptamine-M (N-dealkyl-HO-) isomer-1 2AC		2560	318_8	219_{50}	177_{100}	164_{66}	70_{31}	1018	9259
5-Fluoro-N,N-diallyl-tryptamine-M (N-dealkyl-HO-) isomer-2 2AC		2610	318_{17}	219_{53}	177_{100}	164_{41}	70_{59}	1019	9260
5-Fluoro-N,N-diallyl-tryptamine-M (N-dealkyl-HO-) isomer-3 2AC		2690	318_{12}	219_{51}	177_{100}	164_{58}	70_{52}	1019	9261
5-Fluoro-N,N-diallyl-tryptamine-M (N-dealkyl-HO-) isomer-4 2AC		2710	318_5	261_{34}	219_{100}	177_{84}	70_{77}	1019	9262
5-Fluoro-N,N-diallyl-tryptamine-M (tri-HO-) 3AC		2290	391_{15}	278_{11}	218_{15}	176_{14}	110_{100}	1539	9263
5-Fluoro-NNEI		3770	374_{13}	232_{100}	144_{22}	129_4	115_{13}	1331	9640
5-Fluoro-NNEI TMS		3150	374_{13}	232_{100}	144_{22}	129_4	115_{13}	1574	9641
5-Fluoro-NNEI-M/artifact (1-naphthylamine) HFB		1455	339_{100}	170_{17}	142_{46}	127_{30}	115_{80}	1147	9628
5-Fluoro-NNEI-M/artifact (1-naphthylamine) PFP		1400	289_{100}	170_{14}	142_{40}	127_{22}	115_{75}	832	9631
5-Fluoro-NNEI-M/artifact (1-naphthylamine) TFA		1440	239_{100}	169_{13}	142_{25}	127_{12}	115_{58}	551	9633
5-Fluoro-NPB-22-M/artifact (HOOC-) (ET)		2170	278_{44}	249_{100}	175_{35}	145_{48}	131_{81}	768	9666
5-Fluoro-NPB-22-M/artifact (HOOC-) (ME)		2150	264_{21}	249_{28}	189_{100}	176_{21}	129_{18}	682	9623

5-Fluoro-NPB-22-M/artifact (HOOC-) TMS

Table 8.1: Compounds in order of names

Name	Detected	RI	Typical ions and intensities					Page	Entry
5-Fluoro-NPB-22-M/artifact (HOOC-) TMS		2475	322_{1}	233_{100}	213_{12}	145_{73}	103_{35}	1045	9655
1-(5-Fluoropentyl)-3-(2-iodobenzoyl)indole		3015	435_{28}	360_{12}	232_{100}	220_{55}	144_{50}	1547	8531
1-(5-Fluoropentyl)-3-(4-iodobenzoyl)indole		2940	435_{100}	360_{28}	232_{65}	220_{38}	203_{17}	1547	9619
1-(5-Fluoropentyl)-3-(naphthalen-1-oyl)indole		3165	359_{30}	342_{16}	284_{28}	232_{39}	127_{100}	1258	8532
3-Fluoro-phenmetrazine		1500	195_{19}	135_{11}	123_{29}	71_{100}	56_{54}	386	9457
3-Fluoro-phenmetrazine AC		1880	237_{30}	194_{38}	113_{78}	71_{100}	56_{79}	544	9458
3-Fluoro-phenmetrazine PFP		1640	341_{1}	217_{81}	190_{16}	123_{38}	70_{100}	1159	9460
3-Fluoro-phenmetrazine PFP		1640	391_{1}	267_{62}	240_{14}	123_{35}	70_{100}	1159	9461
3-Fluoro-phenmetrazine TFA		1630	291_{4}	290_{15}	167_{82}	123_{40}	70_{100}	846	9459
3-Fluoro-phenmetrazine-M (HO-) isomer-1 2AC		2160	295_{44}	252_{65}	113_{53}	85_{34}	71_{100}	876	10240
3-Fluoro-phenmetrazine-M (HO-) isomer-2 2AC		2225	295_{24}	252_{42}	113_{90}	85_{47}	71_{100}	876	10241
3-Fluoro-phenmetrazine-M (HO-methoxy-) 2AC		2320	325_{2}	283_{30}	269_{18}	113_{95}	71_{100}	1062	10242
3-Fluoro-phenmetrazine-M (O,N-bisdealkyl-) 2AC		1770	253_{1}	193_{4}	152_{19}	123_{32}	86_{100}	622	10243
3-Fluoro-phenmetrazine-M (O,N-bisdealkyl-HO-) isomer-1 3AC		2080	311_{1}	252_{1}	168_{5}	139_{9}	86_{100}	975	10244
3-Fluoro-phenmetrazine-M (O,N-bisdealkyl-HO-) isomer-2 3AC		2130	311_{1}	252_{6}	139_{11}	86_{100}		976	10245
3-Fluoro-phenmetrazine-M (oxo-HO-) isomer-1 2AC		1850	267_{5}	152_{6}	139_{59}	85_{51}	57_{100}	699	10247
3-Fluoro-phenmetrazine-M (oxo-HO-) isomer-2 2AC		1880	267_{7}	152_{4}	139_{5}	85_{82}	57_{100}	699	10246
4-Fluorophenylacetic acid		<1000*	154_{22}	109_{100}	83_{17}	63_{4}	57_{7}	296	5156
4-Fluorophenylacetic acid ME		1005*	168_{19}	109_{100}	89_{5}	83_{8}	63_{3}	321	5157
4-Fluorophenyl-piperazine		1785	180_{27}	138_{100}	122_{19}	95_{16}	58_{31}	347	9168
4-Fluorophenyl-piperazine AC		2135	222_{45}	179_{17}	150_{100}	137_{32}	122_{50}	480	9170
4-Fluorophenyl-piperazine HFB		1885	376_{100}	179_{82}	150_{64}	123_{62}	95_{33}	1338	9174
4-Fluorophenyl-piperazine ME		1600	194_{100}	163_{24}	150_{17}	123_{72}	95_{36}	384	9169
4-Fluorophenyl-piperazine PFP		1855	326_{100}	179_{62}	150_{45}	123_{53}	95_{35}	1067	9173
4-Fluorophenyl-piperazine TFA		1850	276_{100}	179_{62}	150_{52}	123_{83}	95_{53}	753	9172
4-Fluorophenyl-piperazine TMS		1835	252_{100}	237_{47}	210_{19}	101_{45}	86_{45}	619	9171
5-Fluoro-SDB-005		3100	376_{10}	233_{100}	213_{11}	145_{45}	115_{32}	1339	9607
5-Fluoro-SDB-006		2970	338_{29}	263_{9}	232_{100}	205_{12}	144_{20}	1145	9486
5-Fluoro-SDB-00M/A		1500*	144_{100}	115_{83}	89_{14}	74_{6}	63_{17}	283	928
5-Fluoro-SDB-00M/A TMS		1525*	216_{100}	201_{95}	185_{51}	115_{39}	73_{21}	452	7460
5-Fluoro-SDB-00M/artifact (HOOC-) (ET)		2170	278_{44}	249_{100}	175_{35}	145_{48}	131_{81}	768	9666
5-Fluoro-SDB-00M/artifact (HOOC-) (ME)		2150	264_{71}	249_{79}	189_{100}	176_{21}	129_{18}	882	9623
5-Fluoro-SDB-00M/artifact (HOOC-) TMS		2475	322_{1}	233_{100}	213_{12}	145_{73}	103_{35}	1045	9655
5-Fluoro-THJ		3445	376_{67}	348_{6}	233_{100}	171_{30}	145_{73}	1339	9624
Fluorotropacocaine		1940	263_{12}	140_{13}	124_{100}	94_{44}	82_{43}	675	9573
Fluoro-UR-144		2580	329_{36}	314_{64}	247_{88}	232_{100}	144_{42}	1090	9591
Fluorouracil		2090	130_{100}	87_{64}	60_{64}			272	4174
Fluoxetine	G	1920	309_{22}	183_{17}	162_{16}	104_{100}	91_{67}	963	4277
Fluoxetine	P-I	1950	309_{2}	183_{2}	162_{4}	104_{10}	44_{100}	963	7249
Fluoxetine AC	P-I U+UHYAC	2250	351_{1}	190_{45}	145_{2}	117_{35}	86_{100}	1213	4278
Fluoxetine -H2O HYAC	UHYAC-I	1680	189_{6}	146_{30}	115_{56}	98_{100}	70_{8}	368	4339
Fluoxetine -H2O HYHFB		1470	343_{9}	252_{100}	174_{63}	146_{5}	115_{25}	1172	7240
Fluoxetine -H2O HYPFP		1450	293_{17}	202_{87}	174_{100}	117_{51}	115_{90}	860	7242
Fluoxetine -H2O HYTMS		1580	219_{27}	204_{44}	161_{30}	103_{49}	75_{100}	466	7246
Fluoxetine HFB		1980	344_{66}	252_{9}	240_{100}	169_{56}	117_{85}	1665	7672
Fluoxetine HY2AC	U+UHYAC	1890	249_{2}	206_{100}	146_{36}	98_{78}	86_{34}	597	4340
Fluoxetine HY2HFB		1490	557_{1}	434_{4}	360_{7}	343_{30}	241_{100}	1698	7241
Fluoxetine HY2PFP		1430	457_{2}	334_{4}	310_{16}	239_{28}	190_{100}	1595	7243
Fluoxetine HY2TFA		1435	357_{4}	243_{20}	174_{32}	140_{100}	117_{42}	1246	7244
Fluoxetine ME		1920	323_{1}	183_{2}	162_{3}	104_{8}	58_{100}	1051	7248
Fluoxetine PFP		2080	294_{21}	202_{4}	190_{85}	174_{5}	117_{100}	1591	7671
Fluoxetine TFA		1950	244_{64}	183_{9}	162_{24}	140_{100}	117_{77}	1462	7670
Fluoxetine TMS		2060	381_{5}	262_{7}	219_{13}	116_{100}	73_{58}	1361	4546
Fluoxetine-D6		1890	315_{52}	257_{8}	162_{72}	110_{100}	83_{44}	1001	7788
Fluoxetine-D6 AC		1900	357_{1}	196_{100}	123_{3}	110_{32}	86_{44}	1249	7789
Fluoxetine-D6 -H2O HYPFP		1420	463_{3}	334_{9}	298_{41}	190_{100}	119_{87}	1605	7791
Fluoxetine-D6 HFB		1750	350_{52}	252_{6}	240_{100}	169_{38}	123_{70}	1670	7790
Fluoxetine-D6 TFA		1730	250_{58}	162_{8}	140_{100}	123_{63}	110_{18}	1481	7792
Fluoxetine-D6 TMS		1670	387_{9}	262_{15}	219_{21}	116_{100}	73_{45}	1391	7793
Fluoxetine-M (nor-) 2TMS		2010	439_{6}	320_{19}	219_{14}	174_{100}	104_{16}	1556	7713
Fluoxetine-M (nor-) AC	U+UHYAC	2190	176_{51}	117_{100}	104_{20}	72_{91}		1136	4338
Fluoxetine-M (nor-) formyl artifact		1750	307_{5}	183_{23}	162_{25}	146_{100}	119_{33}	948	7710
Fluoxetine-M (nor-) -H2O HYAC		1700	175_{35}	132_{100}	116_{59}	105_{15}	84_{64}	331	7880
Fluoxetine-M (nor-) -H2O HYTFA		1290	229_{26}	160_{8}	138_{9}	132_{13}	116_{100}	505	7878
Fluoxetine-M (nor-) -H2O HYTMS		1290	205_{100}	190_{33}	161_{24}	100_{46}	73_{85}	413	7879
Fluoxetine-M (nor-) HFB		1895	330_{62}	226_{48}	169_{61}	162_{50}	117_{100}	1649	7674
Fluoxetine-M (nor-) HY2AC		1870	192_{100}	133_{67}	84_{41}	72_{42}		533	5342
Fluoxetine-M (nor-) HY2PFP		1400	443_{8}	296_{40}	280_{7}	239_{100}	177_{14}	1567	7711
Fluoxetine-M (nor-) TFA		1900	230_{61}	183_{10}	162_{45}	126_{65}	117_{100}	1405	7673
Fluoxetine-M (nor-) TMS		1830	367_{7}	248_{17}	219_{37}	102_{100}	73_{32}	1298	7712
Fluoxetine-M (nor-HO-) -H2O HYAC		2080	233_{8}	191_{23}	148_{47}	132_{100}	84_{59}	522	8069
Fluoxymesterone		2835*	336_{51}	279_{33}	175_{27}	123_{49}	71_{100}	1133	3893
Fluoxymesterone AC		2850*	378_{1}	336_{53}	279_{33}	175_{24}	71_{100}	1348	3923
Fluoxymesterone enol 3TMS		2840*	552_{80}	462_{14}	407_{7}	319_{4}	73_{100}	1696	3966

Table 8.1: Compounds in order of names

Fluoxymesterone TMS

Name	Detected	RI	Typical ions and intensities					Page	Entry
Fluoxymesterone TMS		2785*	408_{100}	335_4	207_7	111_9	73_{94}	1472	3928
Flupentixol		3055	434_2	403_4	289_8	143_{100}	100_{14}	1545	1314
Flupentixol AC	U+UHYAC	3045	476_2	457_2	291_7	221_{10}	185_{100}	1629	1315
Flupentixol TMS		3360	506_1	491_4	403_4	215_{100}	98_{49}	1666	5697
Flupentixol-M (dealkyl-dihydro-) AC	UHYAC	3055	434_{18}	265_{32}	185_{100}	141_{67}	99_{27}	1545	1265
Flupentixol-M (dihydro-) AC	UHYAC	3005	478_{47}	265_{100}	185_{70}	125_{59}	98_{32}	1632	1264
Flupentixol-M/artifact (N-oxide) -C6H14N2O2		2120*	304_{100}	303_{93}	289_4	234_{72}	202_{16}	929	1891
Fluphedrone		1185	166_1	123_4	95_{11}	75_9	58_{100}	348	8072
Fluphedrone AC	U+UHYAC	1645	223_1	123_6	100_{43}	95_{16}	58_{100}	483	8074
Fluphedrone artifact (dehydro-)		1220	179_1	123_4	95_{14}	75_{13}	56_{100}	340	8073
Fluphedrone HFB		1440	377_1	254_{100}	210_{51}	123_{47}	95_{43}	1341	8077
Fluphedrone PFP		1360	254_{100}	210_{61}	123_{57}	95_{51}		1073	8076
Fluphedrone TFA		1385	277_1	234_7	154_{100}	123_{43}	110_{59}	758	8075
Fluphedrone-M (dihydro-)	U+UHYAC	1960	208_1	166_4	123_8	100_{46}	58_{100}	700	8661
Fluphedrone-M (dihydro-HO-) 3AC	U+UHYAC	2175	298_1	144_2	100_{44}	74_{11}	58_{100}	1062	8081
Fluphedrone-M (dihydro-HO-) isomer-1 artifact 2AC		1930*	252_1	207_6	180_{32}	164_8	138_{100}	616	8085
Fluphedrone-M (dihydro-HO-) isomer-1 artifact AC		1830*	209_{13}	167_{28}	150_{10}	124_{100}	97_{10}	432	8083
Fluphedrone-M (dihydro-HO-) isomer-2 artifact 2AC	U+UHYAC	1945*	252_1	207_6	180_{32}	164_8	138_{100}	616	8086
Fluphedrone-M (dihydro-HO-) isomer-2 artifact AC		1860*	209_{12}	167_{30}	150_9	124_{100}	97_5	432	8084
Fluphedrone-M (HO-) 2AC	U+UHYAC	2035	281_1	196_2	139_{16}	100_{73}	58_{100}	786	8078
Fluphedrone-M (HO-) AC	U+UHYAC	1950	239_1	162_4	120_{18}	100_{31}	58_{100}	552	8079
Fluphedrone-M (HO-) artifact AC		1740*	181_{32}	138_{100}	109_9	83_5		422	8082
Fluphedrone-M (nor-) AC	U+UHYAC	1620	209_1	123_{19}	95_{35}	86_{100}	75_{21}	428	8080
Fluphedrone-M (nor-dihydro-) isomer-1 2AC	U+UHYAC	1850	252_2	194_8	152_{16}	123_{28}	86_{100}	622	8087
Fluphedrone-M (nor-dihydro-) isomer-2 2AC	U+UHYAC	1870	253_1	194_2	152_{12}	123_2	86_{100}	621	8088
Fluphedrone-M (nor-dihydro-) -NH3 HFB		1370*	348_{10}	320_{100}	169_{57}	135_{25}	69_{72}	1198	8091
Fluphenazine	G UHY	3050	437_{18}	280_{100}	143_{47}	113_{42}	70_{100}	1551	505
Fluphenazine AC	G U UHY UHYAC	3170	479_{48}	419_{41}	280_{100}	185_{52}	125_{56}	1633	339
Fluphenazine TMS		3155	509_5	494_4	406_{25}	280_{100}	73_{18}	1669	4547
Fluphenazine-M (amino-) AC	U+UHYAC	2765	366_{16}	280_6	266_{13}	248_6	100_{100}	1290	1267
Fluphenazine-M (dealkyl-) AC	U+UHYAC	3145	435_{49}	267_{90}	141_{100}	99_{59}		1548	1268
Fluphenazine-M (ring)	U+UHYAC	2190	267_{100}	235_{19}				698	1266
Flupirtine	P-I	2880	304_{89}	258_{17}	231_{33}	124_{28}	109_{100}	932	1811
Flupirtine 2AC	U+UHYAC	2900	388_3	346_{41}	303_{21}	258_{91}	109_{100}	1395	1815
Flupirtine -C2H5OH	G	2930	258_{95}	163_6	135_{14}	109_{100}		646	1812
Flupirtine -C2H5OH 2AC	U+UHYAC	2860	342_{20}	300_{35}	257_{69}	135_{13}	109_{100}	1168	1814
Flupirtine -C2H5OH 2TMS		2640	402_{100}	387_{13}	293_{20}	109_{47}	73_{69}	1453	4548
Flupirtine -C2H5OH 3TMS		2600	474_{41}	459_{99}	401_{55}	109_{29}	73_{100}	1626	4673
Flupirtine -C2H5OH AC	UHYAC	2840	300_{10}	258_{76}	163_7	124_{15}	109_{100}	906	1813
Flupirtine-M (decarbamoyl-) 3AC	U+UHYAC	2700	358_2	315_{65}	273_{75}	231_{59}	109_{100}	1253	4342
Flupirtine-M (decarbamoyl-) formyl artifact 3AC	U+UHYAC	2570	370_{29}	328_{29}	285_{82}	271_{33}	109_{100}	1312	4341
Flupirtine-M (decarbamoyl-) -H2O 2AC	UHYAC	2780	340_7	298_5	256_{47}	133_{31}	109_{100}	1155	4343
Flupirtine-M/artifact (fluorobenzylamine) 2AC		1500	209_{11}	166_{55}	124_{100}	109_{31}	97_8	428	7948
Flupirtine-M/artifact (fluorobenzylamine) AC		1420	167_{100}	124_{95}	109_{48}	97_{14}	83_{10}	319	7947
Flurazepam	G P-I	2780	387_4	315_1	245_1	99_7	86_{100}	1390	506
Flurazepam HY		2555	348_7	86_{100}				1199	511
Flurazepam-M (bis-deethyl-) AC	U+UHYAC	3025	373_{74}	314_{66}	286_{54}	273_{100}	246_{66}	1326	1451
Flurazepam-M (bis-deethyl-) -H2O	P-I U+UHYAC-I	2650	313_{100}					986	1450
Flurazepam-M (bis-deethyl-) -H2O HY	UHY	2295	274_{61}	246_{80}	211_{100}			738	513
Flurazepam-M (bis-deethyl-) -H2O HYAC	UHYAC	2460	316_{44}	315_{50}	274_{53}	246_{100}	211_{57}	1004	287
Flurazepam-M (dealkyl-)	G P-I UGLUC	2470	288_{100}	287_{64}	260_{93}	259_{75}		825	508
Flurazepam-M (dealkyl-) artifact	U+UHYAC	2050	272_{70}	271_{100}	237_{78}	151_{11}	110_8	728	2409
Flurazepam-M (dealkyl-) HY	UHY	2030	249_{100}	154_{34}	123_{46}	95_{42}		593	512
Flurazepam-M (dealkyl-) HYAC	U+UHYAC	2195	291_{52}	249_{100}	123_{57}	95_{61}		846	286
Flurazepam-M (dealkyl-) ME		2530	302_{100}	301_{90}	274_{99}	239_{18}	183_{16}	917	3069
Flurazepam-M (dealkyl-) TMS		2470	360_{63}	359_{69}	341_{43}	197_{17}	73_{100}	1263	4621
Flurazepam-M (dealkyl-HO-)		2265	286_{100}	258_{92}	223_{97}	75_{92}		929	507
Flurazepam-M (deethyl-) AC	U+UHYAC	2990	401_{23}	358_5	314_{20}	100_{15}	58_{100}	1449	1845
Flurazepam-M (HO-ethyl-)	UGLUC	2660	332_{39}	288_{100}	273_{61}	211_{22}		1104	509
Flurazepam-M (HO-ethyl-) AC	UGLUCAC	2725	374_{54}	346_{32}	314_{100}	287_{60}	87_{12}	1330	510
Flurazepam-M (HO-ethyl-) HY	UHY	2385	293_{34}	262_{100}	166_{64}	109_{87}		860	514
Flurazepam-M (HO-ethyl-) HYAC	UHYAC	2470	335_{32}	275_{54}	262_{100}	166_{75}	109_{89}	1123	288
Flurazepam-M/artifact	G P-I	2510	314_{79}	285_{100}	258_{23}	223_{10}	75_{13}	991	3545
Flurazepam-M/artifact AC	U+UHYAC	2430	287_{46}	245_{100}	217_3	210_4	181_9	818	5735
Flurbiprofen	G	1900*	244_{52}	199_{100}	183_{17}	170_{15}		570	1453
Flurbiprofen ME	UME	1880*	258_{22}	199_{100}	183_{16}	178_{23}	170_{10}	646	1456
Flurbiprofen-M (di-HO-) 3ME		2310*	318_{100}	259_{68}	215_8			1018	1455
Flurbiprofen-M (HO-) 2ME	UME	2180*	288_{78}	229_{100}				827	1454
Flurbiprofen-M (HO-methoxy-) 2ME		2310*	318_{100}	259_{68}	215_8			1018	1455
Flurenol artifact		1790*	180_{100}	152_{60}	126_7	76_{31}	63_{16}	345	3186
Flurenol ME		1950*	240_7	181_{100}	152_{31}	126_2	76_6	555	3633
Flurochloridone		2005	311_{58}	187_{87}	145_{83}	103_{60}	75_{100}	974	3187
Flurodifen		2120	328_6	309_6	190_{100}	126_{40}	75_{28}	1078	3842
Fluroxypyr 2ME		1890	282_{68}	237_{10}	209_{100}	181_{52}	152_{12}	792	4150

Fluroxypyr ME **Table 8.1:** Compounds in order of names

Name	Detected	RI	Typical ions and intensities					Page	Entry
Fluroxypyr ME		1830	268 $_{40}$	237 $_7$	209 $_{100}$	181 $_{51}$	152 $_{17}$	704	4149
Flusilazole		2150	315 $_{20}$	300 $_9$	246 $_7$	233 $_{100}$	206 $_{46}$	999	7523
Fluspirilene		9999	475 $_{64}$	418 $_{13}$	244 $_{100}$	187 $_{34}$	72 $_{50}$	1628	1499
Fluspirilene AC		3340	517 $_{31}$	475 $_{10}$	286 $_{72}$	109 $_{39}$	72 $_{100}$	1675	519
Fluspirilene-M (deamino-carboxy-)	P-I UHY UHYAC	2230*	276 $_7$	216 $_{17}$	203 $_{100}$	183 $_{22}$		753	169
Fluspirilene-M (deamino-carboxy-) ME	P-I UHYME U+UHYAC	2125*	290 $_{12}$	258 $_{13}$	216 $_{30}$	203 $_{100}$	183 $_{22}$	841	3372
Fluspirilene-M (deamino-HO-)	UHY-I	2120*	262 $_{16}$	203 $_{100}$	201 $_{16}$	183 $_{12}$		668	515
Fluspirilene-M (deamino-HO-) AC	UHYAC	2150*	304 $_1$	244 $_{22}$	216 $_{41}$	203 $_{100}$	183 $_{16}$	931	307
Fluspirilene-M (N-dealkyl-) ME	UHY-I	2500	245 $_{16}$	71 $_{100}$	57 $_{60}$			576	518
Fluspirilene-M (N-dealkyl-oxo-)	UHY-I	2405	245 $_{49}$	57 $_{100}$				575	517
Fluspirilene-M (N-dealkyl-oxo-) AC	UHYAC-I	2730	287 $_8$	245 $_{16}$	57 $_{100}$			820	180
Fluspirilene-M (N-dealkyl-oxo-) ME	UHYME-I	2350	259 $_{12}$	71 $_{100}$				651	516
Flutazolam		2460	289 $_{100}$	259 $_{38}$	245 $_{72}$	210 $_{55}$	183 $_{34}$	1338	4026
Flutazolam AC		2475	331 $_{11}$	289 $_7$	245 $_9$	210 $_6$	87 $_{100}$	1502	4027
Flutazolam artifact		2185	259 $_{100}$	209 $_{24}$	183 $_{16}$	130 $_{14}$	111 $_{30}$	649	4028
Flutazolam HY	UHY	2385	293 $_{34}$	262 $_{100}$	166 $_{64}$	109 $_{87}$		860	514
Flutazolam HYAC	UHYAC	2470	335 $_{32}$	275 $_{54}$	262 $_{100}$	166 $_{75}$	109 $_{89}$	1123	288
Fluvoxamine	G	1890	299 $_{10}$	276 $_{57}$	187 $_{34}$	172 $_{23}$	71 $_{100}$	1019	1819
Fluvoxamine AC	U+UHYAC	2240	360 $_8$	341 $_9$	258 $_{27}$	102 $_{60}$	86 $_{100}$	1265	1820
Fluvoxamine artifact	G U+UHYAC	1895*	329 $_{12}$	311 $_{18}$	258 $_{72}$	226 $_{48}$	71 $_{100}$	1084	1818
Fluvoxamine artifact (imine)		1560	259 $_6$	244 $_{12}$	200 $_{60}$	187 $_{100}$	172 $_{88}$	650	1817
Fluvoxamine artifact (ketone)	P-I G U+UHYAC	1525*	260 $_1$	242 $_3$	228 $_{38}$	173 $_{100}$	145 $_{32}$	657	1816
Fluvoxamine HFB		1990	514 $_1$	495 $_8$	258 $_{77}$	240 $_{100}$	226 $_{94}$	1672	7677
Fluvoxamine PFP		1930	464 $_1$	445 $_6$	258 $_{90}$	226 $_{87}$	190 $_{100}$	1607	7676
Fluvoxamine TFA		1950	414 $_1$	395 $_8$	258 $_{81}$	140 $_{100}$	71 $_{100}$	1490	7675
Fluvoxamine TMS		1925	390 $_1$	185 $_{10}$	145 $_{21}$	102 $_{100}$	73 $_{62}$	1403	7678
Fluvoxamine-M (HO-HOOC-) (ME)2AC	UHYAC-I	2655	330 $_2$	198 $_{20}$	102 $_{75}$	86 $_{100}$	60 $_{71}$	1539	5341
Fluvoxamine-M (HOOC-) (ME)AC	U+UHYAC	2355	355 $_1$	272 $_3$	102 $_{77}$	86 $_{100}$	60 $_{83}$	1330	5338
Fluvoxamine-M (HOOC-) artifact (ketone)	U+UHYAC	1545*	260 $_3$	241 $_{25}$	214 $_{10}$	173 $_{100}$	145 $_{42}$	656	5339
Fluvoxamine-M (HOOC-) artifact (ketone) (ME)	U+UHYAC	1550*	274 $_1$	255 $_3$	242 $_6$	173 $_{100}$	145 $_{51}$	738	5336
Fluvoxamine-M (O-demethyl-) 2AC	UHYAC-I	2355	244 $_9$	187 $_{20}$	102 $_{90}$	86 $_{66}$	60 $_{100}$	1395	6300
Fluvoxamine-M (O-demethyl-) artifact (ketone) AC	UHYAC-I	2010*	243 $_{23}$	214 $_{23}$	173 $_{100}$	145 $_{34}$	101 $_{23}$	826	5340
Fluvoxate	G	3210	391 $_1$	234 $_5$	147 $_{24}$	111 $_{39}$	98 $_{100}$	1407	4520
Fluvoxate artifact (dehydro-)		3230	389 $_{20}$	234 $_5$	207 $_7$	109 $_{15}$	96 $_{100}$	1399	4647
Fluvoxate artifact (dihydro-)	G	2940	393 $_2$	111 $_{14}$	98 $_{100}$	70 $_4$	55 $_7$	1416	4645
Fluvoxate-M/artifact (alcohol)	UHY	1270	129 $_3$	98 $_{100}$	84 $_3$	70 $_8$	55 $_{12}$	272	4516
Fluvoxate-M/artifact (alcohol) AC	UHYAC	1530	171 $_1$	138 $_3$	111 $_8$	98 $_{100}$	70 $_8$	325	4517
Fluvoxate-M/artifact (HOOC-)	G UHY UHYAC	2770*	280 $_{46}$	279 $_{100}$	205 $_4$	147 $_7$	115 $_{32}$	780	4519
Fluvoxate-M/artifact (HOOC-) ET		2615*	308 $_{39}$	307 $_{100}$	279 $_{44}$	263 $_{16}$	147 $_{16}$	956	4646
Fluvoxate-M/artifact (HOOC-) isopropylester	G UHY UHYAC	2625*	322 $_{56}$	321 $_{77}$	279 $_{100}$	147 $_{26}$	115 $_{22}$	1045	4521
Fluvoxate-M/artifact (HOOC-) ME	G P UHY UHYAC	2580*	294 $_{49}$	293 $_{100}$	263 $_9$	147 $_{15}$	115 $_{17}$	869	4518
5F-MDMB-PINACA	P	2585	377 $_{10}$	321 $_{66}$	289 $_{67}$	233 $_{100}$	145 $_{64}$	1345	10422
5F-MDMB-PINACA (ethyl homolog) TMS		2350	463 $_{42}$	406 $_{51}$	361 $_{28}$	233 $_{100}$		1607	10427
5F-MDMB-PINACA artifact (-HF)		2460	357 $_8$	301 $_{22}$	269 $_{12}$	213 $_{100}$	145 $_{29}$	1250	10425
5F-MDMB-PINACA TMS		2290	449 $_{18}$	434 $_{15}$	392 $_{67}$	233 $_{100}$	145 $_{61}$	1580	10426
5F-MDMB-PINACA-M (HOOC-) (ME)	P	2585	377 $_{10}$	321 $_{66}$	289 $_{67}$	233 $_{100}$	145 $_{64}$	1345	10422
5F-MDMB-PINACA-M (HOOC-) ET		2620	391 $_6$	335 $_{57}$	318 $_{71}$	289 $_{61}$	233 $_{100}$	1409	10424
5F-MMB-PICA		2660	362 $_5$	248 $_{45}$	232 $_{100}$	173 $_8$	144 $_{30}$	1274	9602
Folpet		2000	295 $_{19}$	260 $_{91}$	130 $_{62}$	104 $_{94}$	76 $_{100}$	873	3441
Fonazepam		2705	299 $_{65}$	298 $_{100}$	272 $_{76}$	252 $_{40}$	224 $_{60}$	900	500
Fonazepam HY	UHY	2335	260 $_{100}$	241 $_{13}$	213 $_{24}$	123 $_{78}$	95 $_{52}$	655	283
Fonazepam HYAC	UHYAC	2380	302 $_{21}$	260 $_{87}$	259 $_{100}$	241 $_{18}$	123 $_{31}$	918	6321
Fonazepam TMS		2450	371 $_{100}$	356 $_{34}$	352 $_{48}$	324 $_{23}$	73 $_{96}$	1317	7501
Fonazepam-M (amino-)		2690	269 $_{100}$	241 $_{52}$	240 $_{64}$			711	499
Fonazepam-M (amino-) AC		3035	311 $_{100}$	283 $_{59}$	241 $_{33}$			975	502
Fonazepam-M (amino-) HY		2165	230 $_{100}$	211 $_{34}$				509	503
Fonazepam-M (amino-) HY2AC	U+UHYAC-I	2715	314 $_{100}$	272 $_{81}$	230 $_{98}$	123 $_{23}$		993	284
Fonazepam-M (HO-)		2510	297 $_{47}$	280 $_{61}$	269 $_{84}$	250 $_{81}$	223 $_{100}$	999	9709
Fonofos		1750*	246 $_{34}$	137 $_{48}$	109 $_{100}$	81 $_{17}$	63 $_{19}$	578	3442
Formaldehyde		<1000*	30 $_{90}$	29 $_{100}$				247	4192
Formetanate		2100	221 $_{100}$	163 $_{84}$	149 $_{79}$	122 $_{66}$	92 $_{66}$	474	3901
Formetanate -C2HNO		1660	166 $_9$	121 $_4$	109 $_{100}$	80 $_{28}$	65 $_5$	318	3902
Formoterol HY		1225	165 $_3$	122 $_{100}$	107 $_{11}$	91 $_{22}$	77 $_{45}$	314	3249
Formoterol HY		1225	165 $_{11}$	122 $_{16}$	91 $_3$	77 $_7$	44 $_{100}$	314	5517
Formoterol HY -2H		1320	163 $_4$	148 $_4$	107 $_{27}$	77 $_{11}$	56 $_{100}$	307	4079
Formoterol HY AC	U+UHYAC	1720	207 $_1$	148 $_{100}$	121 $_{40}$	91 $_{11}$	86 $_{24}$	421	3265
Formoterol HY AC	U+UHYAC	1720	207 $_1$	148 $_{41}$	121 $_{16}$	86 $_{10}$	44 $_{100}$	421	5537
Formoterol HY formyl artifact		1255	177 $_6$	162 $_4$	121 $_{60}$	77 $_{12}$	56 $_{100}$	336	3250
Formoterol HYHFB		1560	361 $_2$	240 $_5$	169 $_5$	148 $_{13}$	121 $_{100}$	1268	6769
Formoterol HYPFP		1460	311 $_2$	190 $_4$	148 $_{43}$	121 $_{100}$	91 $_7$	975	6775
Formoterol HYTFA		1460	261 $_6$	148 $_{33}$	140 $_4$	121 $_{100}$	91 $_8$	662	6774
Formothion		1820	257 $_9$	224 $_{11}$	170 $_{20}$	125 $_{89}$	93 $_{100}$	640	3443
Formothion -CO	P G U	1725	229 $_7$	125 $_{45}$	93 $_{62}$	87 $_{100}$		504	724
4-Formyl-phenazone		2285	216 $_{22}$	188 $_{60}$	121 $_{100}$	77 $_{37}$	56 $_{54}$	452	4214

Table 8.1: Compounds in order of names — Fosazepam

Name	Detected	RI	Typical ions and intensities					Page	Entry
Fosazepam		3070	360$_{25}$	283$_{17}$	255$_{100}$	227$_{63}$	91$_{84}$	1263	4025
Foxy		2260	274$_{1}$	174$_{7}$	160$_{31}$	114$_{100}$	72$_{31}$	742	8867
Foxy HFB		2050	455$_{2}$	370$_{8}$	356$_{10}$	159$_{16}$	114$_{100}$	1618	9516
Foxy PFP		2040	405$_{1}$	306$_{3}$	159$_{13}$	144$_{8}$	114$_{100}$	1508	9515
Foxy TFA		2065	355$_{1}$	270$_{6}$	256$_{10}$	159$_{21}$	114$_{100}$	1313	9514
Foxy TMS		2365	346$_{8}$	246$_{18}$	232$_{34}$	114$_{100}$	72$_{51}$	1191	9513
5-F-PB-22		3150	376$_{2}$	232$_{100}$	144$_{30}$	116$_{27}$	89$_{12}$	1339	9575
5-F-PB-22-M/artifact (8-hydroxyquinoline)		1550	145$_{100}$	117$_{83}$	90$_{24}$	63$_{16}$		284	9574
5-F-PB-22-M/artifact (8-hydroxyquinoline) TMS		1570	217$_{2}$	202$_{100}$	172$_{39}$	128$_{6}$	77$_{6}$	454	9650
5-F-PB-22-M/artifact (HOOC-) (ET)		2300	277$_{74}$	232$_{79}$	202$_{100}$	174$_{71}$	144$_{46}$	762	9576
5-F-PB-22-M/artifact (HOOC-) (ME)		2100	263$_{55}$	232$_{33}$	188$_{100}$	144$_{24}$	130$_{24}$	674	9484
5-F-PCN		3500	375$_{47}$	233$_{100}$	213$_{10}$	145$_{62}$	115$_{27}$	1335	9487
5-F-PCN TMS		2930	447$_{80}$	374$_{4}$	287$_{2}$	240$_{11}$	73$_{100}$	1575	9488
3-FPM		1500	195$_{19}$	135$_{11}$	123$_{29}$	71$_{100}$	56$_{54}$	386	9457
3-FPM AC		1880	237$_{30}$	194$_{38}$	113$_{78}$	71$_{100}$	56$_{79}$	544	9458
3-FPM PFP		1640	341$_{1}$	217$_{81}$	190$_{16}$	123$_{38}$	70$_{100}$	1159	9460
3-FPM PFP		1640	391$_{1}$	267$_{62}$	240$_{14}$	123$_{35}$	70$_{100}$	1159	9461
3-FPM TFA		1630	291$_{4}$	290$_{15}$	167$_{82}$	123$_{40}$	70$_{100}$	846	9459
3-FPM-M (HO-) isomer-1 2AC		2160	295$_{44}$	252$_{65}$	113$_{53}$	85$_{34}$	71$_{100}$	876	10240
3-FPM-M (HO-) isomer-2 2AC		2225	295$_{24}$	252$_{42}$	113$_{90}$	85$_{47}$	71$_{100}$	876	10241
3-FPM-M (HO-methoxy-) 2AC		2320	325$_{2}$	283$_{30}$	269$_{18}$	113$_{95}$	71$_{100}$	1062	10242
3-FPM-M (O,N-bisdealkyl-) 2AC		1770	253$_{1}$	193$_{4}$	152$_{19}$	123$_{32}$	86$_{100}$	622	10243
3-FPM-M (O,N-bisdealkyl-HO-) isomer-1 3AC		2080	311$_{1}$	252$_{1}$	168$_{5}$	139$_{9}$	86$_{100}$	975	10244
3-FPM-M (O,N-bisdealkyl-HO-) isomer-2 3AC		2130	311$_{1}$	252$_{6}$	139$_{11}$	86$_{100}$		976	10245
3-FPM-M (oxo-HO-) isomer-1 2AC		1850	267$_{5}$	152$_{6}$	139$_{59}$	85$_{51}$	57$_{100}$	699	10247
3-FPM-M (oxo-HO-) isomer-2 2AC		1880	267$_{7}$	152$_{4}$	139$_{5}$	85$_{82}$	57$_{100}$	699	10246
4-F-PVP		1750	206$_{2}$	126$_{100}$	123$_{25}$	95$_{37}$	84$_{12}$	600	10415
4-F-Pyrrolidinovalerophenone		1750	206$_{2}$	126$_{100}$	123$_{25}$	95$_{37}$	84$_{12}$	600	10415
Frangula-emodin		2620*	270$_{100}$	242$_{8}$	213$_{5}$	185$_{6}$	139$_{6}$	715	3565
Frangula-emodin 2ME		2775*	298$_{100}$	280$_{48}$	252$_{55}$	135$_{42}$	115$_{12}$	893	3567
Frangula-emodin 3ME		2845*	312$_{65}$	297$_{100}$	295$_{27}$	267$_{7}$	142$_{13}$	981	3568
Frangula-emodin AC		2740*	312$_{19}$	270$_{100}$	242$_{8}$	213$_{6}$	139$_{5}$	980	3566
Frigen 11		<1000*	117$_{3}$	101$_{100}$	82$_{6}$	66$_{17}$		275	3794
Frigen 12		<1000*	120$_{1}$	101$_{20}$	85$_{100}$	66$_{10}$	50$_{36}$	264	3793
Frovatriptan		2960	243$_{43}$	212$_{9}$	186$_{100}$	170$_{36}$	142$_{15}$	568	7751
Frovatriptan isomer-1 2TMS		2985	387$_{26}$	372$_{11}$	330$_{100}$	214$_{35}$	73$_{77}$	1392	7643
Frovatriptan isomer-1 3TMS		2745	459$_{58}$	401$_{57}$	330$_{72}$	287$_{49}$	147$_{100}$	1600	7645
Frovatriptan isomer-2 2TMS		3000	387$_{20}$	258$_{100}$	243$_{49}$	129$_{61}$	73$_{62}$	1392	7646
Frovatriptan isomer-2 3TMS		3075	459$_{12}$	444$_{9}$	330$_{100}$	214$_{22}$	73$_{39}$	1600	7642
Frovatriptan ME		2785	257$_{26}$	212$_{4}$	186$_{100}$	170$_{24}$	71$_{29}$	643	7647
Frovatriptan TMS		2800	315$_{34}$	258$_{100}$	243$_{49}$	200$_{26}$	75$_{24}$	1001	7644
Fructose 5AC		1995*	331$_{4}$	317$_{11}$	275$_{68}$	187$_{24}$	101$_{100}$	1402	1958
Fructose 5HFB		1620*	750$_{1}$	322$_{52}$	309$_{75}$	169$_{100}$	69$_{80}$	1735	5793
Fructose 5PFP		1250*	419$_{13}$	405$_{7}$	203$_{16}$	147$_{23}$	119$_{100}$	1733	5792
Fructose 5TFA		1470*	450$_{2}$	222$_{40}$	209$_{75}$	125$_{24}$	69$_{100}$	1724	5791
FUB-144		2680	349$_{14}$	334$_{26}$	267$_{21}$	252$_{77}$	109$_{100}$	1205	9599
FUB-144 TMS		2570	421$_{9}$	406$_{18}$	378$_{41}$	324$_{7}$	109$_{100}$	1511	9600
FUB-AKB48		3410	403$_{6}$	375$_{9}$	294$_{24}$	253$_{37}$	109$_{100}$	1457	9504
FUB-AKB48 TMS		3160	475$_{1}$	366$_{4}$	217$_{3}$	135$_{27}$	109$_{100}$	1629	9505
FUB-AMB TMS		2680	455$_{10}$	412$_{19}$	346$_{30}$	253$_{26}$	109$_{100}$	1592	9507
Fuberidazole		1940	184$_{100}$	155$_{24}$	129$_{13}$	92$_{10}$	63$_{10}$	356	3643
FUB-PB-22		3945	396$_{1}$	287$_{2}$	252$_{43}$	116$_{45}$	109$_{100}$	1429	9648
FUB-PB-22-M/artifact (8-hydroxyquinoline)		1550	145$_{100}$	117$_{83}$	90$_{24}$	63$_{16}$		284	9574
FUB-PB-22-M/artifact (8-hydroxyquinoline) TMS		1570	217$_{2}$	202$_{100}$	172$_{39}$	128$_{6}$	77$_{6}$	454	9650
FUB-PB-22-M/artifact (HOOC-) (ET)		2420	297$_{47}$	252$_{13}$	222$_{3}$	143$_{2}$	109$_{100}$	889	9665
FUB-PB-22-M/artifact (HOOC-) (ME)		2400	283$_{44}$	252$_{9}$	222$_{4}$	146$_{6}$	109$_{100}$	794	9635
FUB-PB-22-M/artifact (HOOC-) TMS		2535	341$_{1}$	282$_{6}$	217$_{5}$	173$_{7}$	109$_{100}$	1161	9649
Furalaxyl		1960	301$_{11}$	242$_{30}$	152$_{14}$	95$_{100}$	77$_{9}$	913	4044
Furmecyclox		1850	251$_{13}$	138$_{2}$	123$_{100}$	81$_{29}$	53$_{33}$	613	2998
Furosemide 2ME	PME ume	2850	358$_{26}$	325$_{8}$	297$_{4}$	96$_{8}$	81$_{100}$	1251	2330
Furosemide 2TMS		2895	474$_{19}$	459$_{7}$	355$_{12}$	81$_{100}$	73$_{30}$	1625	4549
Furosemide 3ME	PME ume	2800	372$_{35}$	339$_{7}$	311$_{4}$	96$_{8}$	81$_{100}$	1321	2331
Furosemide 3TMS		2805	546$_{3}$	531$_{13}$	147$_{23}$	81$_{100}$	73$_{36}$	1693	4550
Furosemide ME	P pme UME	2890	344$_{26}$	311$_{9}$	283$_{4}$	96$_{9}$	81$_{100}$	1177	2329
Furosemide -SO2NH	P U	2040	251$_{23}$	233$_{3}$	96$_{12}$	81$_{100}$	53$_{19}$	607	3367
Furosemide -SO2NH 2ME	UME	2050	279$_{32}$	250$_{24}$	232$_{36}$	204$_{12}$	81$_{100}$	771	2333
Furosemide -SO2NH ME	PME-I UME UHYME	2020	265$_{33}$	232$_{10}$	204$_{7}$	96$_{14}$	81$_{100}$	684	2332
Furosemide-M (N-dealkyl-) 2ME	UME	2450	278$_{100}$	248$_{32}$	200$_{60}$	185$_{41}$	169$_{32}$	765	2335
Furosemide-M (N-dealkyl-) 2MEAC		2375	320$_{18}$	278$_{42}$	200$_{35}$	169$_{22}$	56$_{100}$	1032	2337
Furosemide-M (N-dealkyl-) ME	UME	2750	264$_{100}$	232$_{86}$	200$_{10}$	169$_{11}$	141$_{8}$	679	2334
Furosemide-M (N-dealkyl-) MEAC		2440	306$_{1}$	263$_{1}$	200$_{1}$	169$_{2}$	56$_{100}$	941	2336
Furosemide-M (N-dealkyl-) -SO2NH 2ME		1500	199$_{19}$	185$_{64}$	153$_{100}$	126$_{38}$	90$_{22}$	396	2339
Furosemide-M (N-dealkyl-) -SO2NH ME		1470	185$_{52}$	153$_{100}$	126$_{43}$	99$_{13}$	63$_{20}$	358	2338
Furosemide-M (N-dealkyl-) -SO2NH MEAC		1650	227$_{19}$	185$_{60}$	153$_{100}$	126$_{33}$	63$_{50}$	497	2340

Gabapentin -H2O

Table 8.1: Compounds in order of names

Name	Detected	RI	Typical ions and intensities					Page	Entry
Gabapentin -H2O	G U+UHYAC	1750	153_{82}	110_{39}	96_{35}	81_{100}	67_{74}	296	3112
Gabapentin -H2O AC	U+UHYAC	1730	195_{100}	167_{49}	153_{43}	81_{96}	67_{47}	388	6555
Gabapentin -H2O ME	UME U+UHYAC	1560	167_{79}	124_{42}	110_{21}	81_{100}	67_{81}	320	3113
Galactose 5AC	U+UHYAC	1995*	331_{14}	245_{42}	168_{77}	143_{100}	103_{83}	1401	1959
Galactose 5HFB		1505*	519_{25}	465_{8}	277_{12}	249_{8}	169_{100}	1735	5796
Galactose 5PFP		1200*	419_{13}	365_{5}	227_{8}	147_{22}	119_{100}	1734	5795
Galactose 5TFA		1190*	547_{1}	407_{6}	319_{31}	265_{26}	69_{100}	1724	5794
Galantamine	U+UHYAC	2340	287_{83}	286_{100}	270_{13}	244_{21}	216_{23}	822	6710
Galantamine AC		2450	329_{31}	328_{34}	270_{100}	216_{27}	165_{14}	1088	6712
Galantamine -H2O		2180	269_{100}	268_{46}	226_{25}	211_{5}	165_{59}	713	6711
Galantamine HFB		2330	483_{36}	482_{32}	270_{100}	216_{22}	174_{24}	1638	6717
Galantamine HYAC	U+UHYAC	2280	311_{36}	268_{10}	226_{100}	211_{19}	165_{32}	977	6713
Galantamine PFP		2295	433_{43}	432_{43}	270_{100}	216_{19}	174_{21}	1542	6716
Galantamine TFA		2300	383_{89}	382_{67}	270_{100}	216_{25}	174_{26}	1370	6715
Galantamine TMS		2420	359_{100}	358_{99}	344_{16}	316_{23}	216_{18}	1259	6714
Galantamine-M (nor-) HYAC	U+UHYAC	2410	339_{42}	296_{17}	280_{18}	212_{100}	165_{25}	1150	7385
Gallopamil	G P-I U UHY UHYAC	3190	484_{1}	483_{1}	333_{100}	151_{15}	58_{50}	1640	2520
Gallopamil-M (N-bis-dealkyl-) AC	UHYAC	2500	348_{54}	305_{70}	263_{100}	194_{47}	100_{17}	1201	2908
Gallopamil-M (N-dealkyl-)		2180	320_{7}	289_{12}	194_{85}	70_{70}	57_{100}	1035	2522
Gallopamil-M (N-dealkyl-) AC	U+UHYAC	2520	362_{58}	319_{83}	277_{100}	114_{68}	86_{56}	1275	2524
Gallopamil-M (N-dealkyl-bis-O-demethyl-) 2AC	UHYAC	2650	376_{19}	334_{100}	291_{44}	114_{92}	86_{40}	1340	2910
Gallopamil-M (N-dealkyl-O-demethyl-) 2AC	UHYAC	2600	390_{5}	348_{80}	305_{100}	263_{58}	114_{87}	1403	2909
Gallopamil-M (nor-)		3260	470_{1}	319_{100}	290_{10}	151_{32}	57_{9}	1620	2521
Gallopamil-M (nor-) AC	U+UHYAC	3520	512_{3}	348_{11}	319_{23}	164_{100}	151_{16}	1671	2523
Gallopamil-M (O-demethyl-) AC	U+UHYAC	3300	511_{1}	361_{100}	319_{85}	276_{11}	58_{79}	1671	1927
gamma-Butyrolactone		<1000*	86_{11}	56_{16}	42_{100}			255	7275
gamma-Hexachlorocyclohexane (HCH)	P-I	1740*	288_{2}	252_{8}	217_{67}	181_{100}	109_{48}	824	1067
gamma-Hydroxybutyric acid 2TMS		1160*	248_{1}	233_{26}	147_{100}	117_{30}	73_{32}	591	5430
gamma-Tocopherol	P	2990*	416_{53}	203_{6}	191_{20}	151_{100}	57_{6}	1497	5816
GC septum bleed		----	503_{25}	355_{83}	281_{54}	147_{83}	73_{100}	1663	2220
GC stationary phase (methylsilicone)		----	429_{10}	355_{31}	281_{67}	207_{100}	73_{27}	1529	2627
GC stationary phase (OV-101)		----	356_{10}	201_{4b}	207_{100}	73_{35}		1234	1016
GC stationary phase (OV-17)		----	452_{20}	394_{34}	315_{100}	198_{51}	135_{46}	1584	1017
GC stationary phase (UCC-W-982)		----	429_{5}	355_{9}	281_{36}	207_{100}	73_{45}	1529	1018
Gefitinib		3450	446_{6}	318_{2}	128_{30}	100_{100}	70_{16}	1573	8447
Gefitinib TMS		2870	518_{4}	503_{3}	376_{6}	128_{62}	100_{100}	1676	8448
Gemfibrozil ME		1855*	264_{6}	143_{100}	122_{55}	83_{92}	59_{57}	683	2799
Genkwanin-4-methylether		2830*	298_{100}	269_{26}	255_{19}	166_{8}	135_{16}	893	10423
Gepefrine 2AC	UHYAC	1930	235_{2}	176_{48}	134_{52}	107_{21}	86_{100}	532	4387
Gepefrine 2HFB		1620	330_{30}	303_{11}	240_{100}	169_{15}	69_{29}	1691	5737
Gepefrine 2PFP		1520	280_{36}	253_{9}	190_{100}	119_{19}	69_{8}	1567	5738
Gepefrine 2TFA		<1000	230_{33}	203_{8}	140_{100}	115_{6}		1172	6224
Gepefrine 2TMS	UHYTMS	1850	280_{11}	179_{3}	116_{100}	100_{12}	73_{72}	881	5693
Gepefrine formyl artifact ME		1290	177_{2}	162_{4}	121_{4}	77_{5}	56_{100}	335	5129
Gepefrine TMSTFA		1630	319_{8}	206_{86}	191_{32}	140_{100}	73_{58}	1024	6141
Geranamine		<1000	115_{2}	100_{9}	69_{8}	56_{13}	44_{100}	263	8622
Geranamine		<1000	115_{12}	100_{67}	83_{14}	69_{51}	57_{100}	263	8623
Gestonorone caproate		3440*	414_{2}	371_{17}	316_{9}	273_{100}	99_{59}	1492	2279
Gestonorone -H2O		3410*	298_{64}	283_{41}	255_{100}	161_{17}	91_{30}	898	2075
GHB 2TMS		1160*	248_{1}	233_{26}	147_{100}	117_{30}	73_{32}	591	5430
GHB -H2O		<1000*	86_{11}	56_{16}	42_{100}			255	7275
Glaucine	U+UHYAC	2680	355_{94}	354_{100}	340_{57}	324_{31}	281_{29}	1238	5775
Glaucine artifact (dehydro-)		3235	353_{100}	338_{77}	307_{12}	280_{26}	176_{13}	1227	6744
Glaucine-M (bis-O-demethyl-)	U+UHYAC	2870	327_{69}	326_{100}	312_{41}	296_{19}	269_{16}	1075	8104
Glaucine-M (bis-O-demethyl-) 2AC		3230	411_{77}	396_{21}	368_{100}	354_{36}	326_{75}	1483	8543
Glaucine-M (bis-O-demethyl-) 2HFB		2730	719_{86}	718_{100}	704_{44}	645_{32}	522_{27}	1730	8548
Glaucine-M (bis-O-demethyl-) 2PFP		2670	619_{91}	618_{100}	604_{53}	545_{32}	472_{27}	1718	8547
Glaucine-M (bis-O-demethyl-) 2TFA		2655	519_{96}	518_{100}	504_{45}	422_{61}	363_{55}	1676	8546
Glaucine-M (bis-O-demethyl-) isomer-1 2AC	U+UHYAC	3050	411_{95}	368_{100}	354_{41}	326_{70}	224_{32}	1482	8109
Glaucine-M (bis-O-demethyl-) isomer-2 2AC	U+UHYAC	3070	411_{87}	396_{87}	368_{100}	354_{37}	326_{77}	1482	8110
Glaucine-M (HO-) AC	U+UHYAC	3135	411_{19}	369_{7}	339_{56}	310_{60}	297_{100}	1488	8752
Glaucine-M (O-demethyl-) isomer-1		2850	341_{76}	340_{100}	326_{44}	310_{19}	267_{21}	1163	8105
Glaucine-M (O-demethyl-) isomer-1 AC	U+UHYAC	2980	383_{97}	382_{100}	368_{47}	340_{72}	326_{35}	1371	8107
Glaucine-M (O-demethyl-) isomer-2	U+UHYAC	2900	341_{82}	340_{100}	326_{50}	310_{29}	267_{33}	1164	8106
Glaucine-M (O-demethyl-) isomer-2 AC	U+UHYAC	3000	383_{83}	368_{100}	352_{35}	340_{55}	326_{16}	1371	8108
Glaucine-M (tri-O-demethyl-) 3AC	U+UHYAC	3140	439_{72}	424_{41}	396_{100}	354_{67}	312_{63}	1556	8111
Glibenclamide artifact-1		2035	224_{25}	143_{22}	99_{27}	83_{7}	56_{100}	489	4904
Glibenclamide artifact-2		2480	289_{25}	198_{15}	169_{100}	126_{16}	111_{10}	832	4905
Glibenclamide artifact-3 2ME	UME	3355	353_{8}	289_{42}	198_{5}	169_{100}	111_{8}	1427	4906
Glibenclamide artifact-3 ME	UME	3445	382_{1}	353_{3}	287_{13}	198_{13}	169_{100}	1364	3128
Glibenclamide artifact-4 ME		3460	468_{2}	287_{15}	198_{21}	169_{100}	126_{6}	1613	4907
Glibornuride 2TMS		2855	510_{2}	495_{22}	355_{64}	155_{49}	73_{100}	1670	5020
Glibornuride AC		1923	408_{14}	393_{35}	315_{100}	229_{72}	91_{76}	1471	2013
Glibornuride artifact-1		1390	169_{4}	154_{9}	140_{20}	98_{42}	84_{100}	323	2006

Table 8.1: Compounds in order of names

Name	Detected	RI	Typical ions and intensities					Page	Entry
Glibornuride artifact-1 2AC	U+UHYAC	1800	253_2	238_4	193_{41}	168_{40}	95_{100}	624	2012
Glibornuride artifact-1 2TMS		1555	313_4	298_{10}	170_{28}	156_{66}	73_{100}	991	5021
Glibornuride artifact-1 -H2O AC		1370	193_6	150_{11}	137_{50}	134_{70}	95_{100}	381	2010
Glibornuride artifact-2		1405	181_3	152_{100}	123_6	95_{18}	70_{25}	347	2007
Glibornuride artifact-3	UME	1620	197_{33}	155_{50}	91_{100}	65_{36}		391	4910
Glibornuride artifact-4	G P-I U+UHYAC UME	1700	171_{25}	155_{19}	107_6	91_{100}	65_{25}	325	2008
Glibornuride artifact-4 2ME		1690	199_{34}	155_{15}	91_{100}	65_{23}		397	3130
Glibornuride artifact-4 AC		1550	237_1	195_{28}	151_{22}	134_{26}	95_{100}	542	2011
Glibornuride artifact-4 ME	UME	1740	185_{25}	155_{19}	121_6	91_{100}	65_{25}	359	3131
Glibornuride artifact-4 TMS	UTMS	1875	243_3	228_{100}	167_6	149_{12}	91_9	567	5022
Glibornuride artifact-5	U+UHYAC UME	1840	195_{17}	164_{12}	134_{14}	109_{44}	95_{100}	387	2009
Glibornuride artifact-5 ME	UME	1715	209_{53}	139_{46}	109_{49}	100_{62}	95_{100}	431	4916
Glibornuride artifact-6 ME		1845	212_1	179_{46}	122_{21}	91_{100}	72_{53}	437	3132
Glibornuride -H2O ME		2670	362_5	207_{66}	150_{65}	134_{20}	91_{100}	1274	3129
Glibornuride-M (HO-) artifact 2ME	UME	2030	215_{60}	171_{29}	107_{100}	89_{80}	77_{72}	448	4914
Glibornuride-M (HO-) artifact 3TMS	UTMS	2000	403_{25}	388_{98}	272_{44}	258_{69}	73_{100}	1456	5018
Glibornuride-M (HO-) artifact AC	U+UHYAC	2180	229_{16}	187_{81}	107_{81}	89_{100}	77_{45}	504	4915
Glibornuride-M (HO-) artifact ME	UME	2265	201_{40}	172_{45}	107_{100}	89_{81}	77_{91}	401	4913
Glibornuride-M (HO-bornyl-) artifact	UME	2305	211_3	181_7	125_{13}	108_{22}	95_{100}	436	4917
Glibornuride-M (HO-bornyl-) artifact 2TMS	UTMS	1955	355_{11}	340_{36}	265_{67}	107_{56}	73_{100}	1239	5023
Glibornuride-M (HOOC-) artifact 2ME		1920	229_{50}	198_{48}	135_{100}	103_{40}	76_{55}	504	2479
Glibornuride-M (HOOC-) artifact 3ME	UME	1850	243_{54}	199_{18}	135_{100}	104_{46}	76_{51}	567	2480
Gliclazide		2440	278_{14}	168_{72}	125_{62}	110_{100}	81_{91}	1050	4908
Gliclazide artifact-1 ME	UME	1545	184_{13}	125_{44}	110_{100}	81_{70}	67_{32}	358	4909
Gliclazide artifact-2	UME	1620	197_{33}	155_{50}	91_{100}	65_{36}		391	4910
Gliclazide artifact-3	U+UHYAC UME	1670	169_1	125_{68}	110_{79}	81_{100}	67_{46}	323	4911
Gliclazide artifact-4	G P-I U+UHYAC UME	1700	171_{25}	155_{19}	107_6	91_{100}	65_{25}	325	2008
Gliclazide artifact-4 2ME		1690	199_{34}	155_{15}	91_{100}	65_{23}		397	3130
Gliclazide artifact-4 ME	UME	1740	185_{25}	155_{19}	121_6	91_{100}	65_{25}	359	3131
Gliclazide artifact-4 TMS	UTMS	1875	243_3	228_{100}	167_6	149_{12}	91_9	567	5022
Gliclazide artifact-5 AC	U+UHYAC	1535	210_7	168_{69}	125_{61}	110_{100}	81_{78}	431	4912
Gliclazide-M (HO-) artifact 2ME	UME	2030	215_{60}	171_{29}	107_{100}	89_{80}	77_{72}	448	4914
Gliclazide-M (HO-) artifact 3TMS	UTMS	2000	403_{25}	388_{98}	272_{44}	258_{69}	73_{100}	1456	5018
Gliclazide-M (HO-) artifact AC	U+UHYAC	2180	229_{16}	187_{81}	107_{81}	89_{100}	77_{45}	504	4915
Gliclazide-M (HO-) artifact ME	UME	2265	201_{40}	172_{45}	107_{100}	89_{81}	77_{91}	401	4913
Gliclazide-M (HOOC-) artifact 2ME		1920	229_{50}	198_{48}	135_{100}	103_{40}	76_{55}	504	2479
Gliclazide-M (HOOC-) artifact 3ME	UME	1850	243_{54}	199_{18}	135_{100}	104_{46}	76_{51}	567	2480
Glimepiride artifact-1 ME		1195	171_{11}	156_3	114_{100}	101_9	76_{21}	326	4918
Glimepiride artifact-2 ME		1265	183_6	151_{41}	124_{64}	96_{87}	81_{100}	354	4925
Glimepiride artifact-3		1275	125_{100}	110_{46}	96_{67}	82_{22}	67_{27}	268	4919
Glimepiride artifact-3 TMS		1360	197_{67}	182_{100}	166_{35}	126_{19}	73_{46}	393	5025
Glimepiride artifact-4		2130	252_{32}	157_{33}	113_{15}	95_7	56_{100}	620	4922
Glimepiride artifact-5 2ME		2690	254_4	240_{45}	184_{34}	125_{38}	120_{100}	626	4921
Glimepiride artifact-5 ME		2325	240_{41}	184_{32}	146_{24}	120_{100}	89_{28}	554	4920
Glimepiride-M (HO-) artifact ME		1630	187_4	155_8	128_{16}	114_{100}	76_{34}	363	4923
Glimepiride-M (HOOC-) artifact 2ME		1670	215_6	184_{10}	156_{60}	114_{100}	101_{32}	449	4904
Glipizide artifact-1		2035	224_{25}	143_{22}	99_{27}	83_7	56_{100}	489	4904
Glipizide artifact-2		3100	320_9	183_{23}	150_{100}	121_{47}	93_{42}	1033	4926
Glipizide artifact-2 2ME	UME	3005	348_3	241_{92}	150_{100}	121_{96}	93_{81}	1198	4927
Glipizide artifact-2 ME	UME	3020	334_{35}	239_{10}	150_{100}	121_{82}	93_{67}	1119	3133
Glipizide artifact-2 TMS	UTMS	3195	392_{19}	377_8	240_{37}	150_{100}	121_{43}	1410	5019
Gliquidone artifact-1		1845	219_{54}	204_{100}	158_{25}	133_{39}	62_{100}	464	4928
Gliquidone artifact-2		2035	224_{25}	143_{22}	99_{27}	83_7	56_{100}	489	4904
Gliquidone artifact-3	U+UHYAC UME	2555	323_{25}	219_{94}	204_{100}	191_{60}	176_{96}	1051	4929
Gliquidone artifact-4		3440	402_{26}	321_{38}	219_{74}	204_{100}	175_{76}	1452	4930
Gliquidone artifact-4 2ME	UME	3415	430_9	323_{72}	220_{100}	204_{63}	175_{47}	1534	3134
Gliquidone artifact-4 ME	UME	3460	416_{30}	321_{28}	219_{87}	204_{100}	175_{70}	1495	4931
Gliquidone artifact-4 TMS	UTMS	3585	474_{11}	459_{21}	219_{61}	204_{100}	176_{63}	1625	5016
Gliquidone artifact-5 ME		3415*	502_{12}	321_{100}	219_{65}	204_{77}	175_{52}	1661	4932
Glisoxepide artifact-1 ME	UME	1315	172_{12}	113_{76}	98_{80}	68_{70}	59_{100}	327	3140
Glisoxepide artifact-3 2ME	UME	2840	337_1	294_{16}	230_{100}	199_{30}	110_{59}	1135	4934
Glisoxepide artifact-3 ME	UME	2855	323_1	228_{42}	197_{43}	139_{87}	110_{100}	1048	4933
Gluconic acid ME5AC		1975*	361_3	289_{14}	173_{58}	145_{72}	115_{100}	1508	5227
Glucose 5AC	UHYAC	2010*	331_2	242_{40}	157_{66}	115_{100}	98_{85}	1401	790
Glucose 5HFB		1460*	519_{21}	321_6	277_{11}	197_{14}	169_{100}	1735	5784
Glucose 5PFP		1180*	747_1	419_{26}	227_{15}	147_{30}	119_{100}	1733	5783
Glucose 5TFA		1200*	547_1	413_4	319_{33}	265_{25}	69_{100}	1724	5782
Glucose 5TMS		2050*	435_1	217_{17}	204_{100}	191_{67}	73_{91}	1690	4333
Glutethimide	P G U UHY UHYAC	1830	217_{20}	189_{100}	160_{36}	132_{64}	117_{67}	454	791
Glutethimide 2TMS		1845	361_{20}	346_{68}	332_{62}	147_{34}	73_{100}	1270	5482
Glutethimide TMS		1800	274_{13}	245_{10}	174_{26}	132_{100}	117_{39}	836	5481
Glutethimide-M (HO-ethyl-)	U UHY	1865	233_{68}	205_{14}	189_{32}	146_{100}	104_{78}	521	792
Glutethimide-M (HO-ethyl-) AC	UHYAC	2060	275_{20}	247_{37}	233_{47}	189_{100}	187_{96}	745	794
Glutethimide-M (HO-phenyl-)	U UHY	1875	233_{91}	204_{100}	176_{81}	148_{44}	133_{96}	522	793

Glutethimide-M (HO-phenyl-) AC

Table 8.1: Compounds in order of names

Name	Detected	RI	Typical ions and intensities					Page	Entry
Glutethimide-M (HO-phenyl-) AC	UHYAC	2250	275_{17}	233_{100}	204_{84}	189_{83}	176_{46}	746	795
Glycerol 3AC	U+UHYAC	1485*	158_{2}	145_{86}	116_{54}	103_{100}	86_{30}	458	2014
Glycerol 3TMS		1125*	293_{2}	218_{20}	205_{57}	147_{100}	73_{93}	959	7451
Glyceryl dimyristate -H2O	G	3830*	494_{6}	285_{25}	98_{49}	71_{65}	57_{100}	1654	5631
Glyceryl monomyristate	G	2260*	285_{6}	271_{10}	211_{66}	98_{83}	55_{100}	921	5587
Glyceryl monooleate 2AC	UHYAC	2790*	380_{34}	264_{71}	159_{100}	81_{49}	69_{85}	1559	5602
Glyceryl monopalmitate	G	2420*	313_{3}	299_{11}	239_{49}	98_{100}	57_{99}	1096	5588
Glyceryl monopalmitate 2AC	UHYAC	2645*	354_{4}	239_{34}	159_{100}	98_{29}	84_{18}	1492	5412
Glyceryl monopalmitate 2TMS	G	2620*	460_{2}	372_{46}	205_{15}	147_{36}	73_{100}	1626	7449
Glyceryl monostearate 2AC	UHYAC	2790*	382_{6}	267_{26}	159_{100}	98_{30}	84_{19}	1567	5413
Glyceryl monostearate 2TMS		2780*	488_{2}	400_{45}	205_{15}	147_{42}	73_{100}	1663	7450
Glyceryl triacetate	U+UHYAC	1485*	158_{2}	145_{86}	116_{54}	103_{100}	86_{30}	458	2014
Glyceryl tridecanoate	G	3280*	383_{16}	355_{49}	155_{57}	127_{42}	57_{100}	1697	4466
Glyceryl trioctanoate	G	2850*	452_{1}	327_{24}	201_{8}	127_{73}	57_{100}	1620	4465
Glycolic acid 2TMS		<1000*	205_{8}	177_{12}	161	147_{86}	73_{100}	469	8585
Glycophen		2470	329_{5}	187_{18}	142_{58}	127_{84}	56_{100}	1085	3848
Glyphosate 3ME		1410	211_{5}	179_{15}	152_{75}	102_{100}	74_{70}	435	4153
Glyphosate 4ME		1390	225_{6}	166_{67}	116_{100}	93_{71}	58_{39}	490	4152
Granisetron	P U UHY UHYAC	2880	312_{24}	159_{80}	136_{40}	110_{70}	96_{100}	983	3185
Grepafloxacin 2ME		3520	387_{72}	317_{100}	242_{20}	85_{17}	70_{40}	1391	7734
Grepafloxacin -CO2		3120	315_{25}	259_{100}	245_{5}	215_{3}	174_{5}	1001	7738
Grepafloxacin -CO2 ME		3130	329_{21}	294_{4}	259_{100}	215_{8}	189_{5}	1090	7737
Grepafloxacin -CO2 TMS		3120	387_{34}	372_{22}	331_{100}	301_{16}	273_{7}	1391	7736
Grepafloxacin ME		3540	373_{37}	317_{100}	242_{16}	85_{11}	70_{30}	1328	7733
Grepafloxacin TMS		3570	431_{69}	416_{34}	375_{100}	331_{22}	214_{8}	1538	7735
Guaifenesin	P G	1610*	198_{15}	124_{100}	109_{74}			394	796
Guaifenesin 2AC	U+UHYAC	1865*	282_{3}	159_{100}	124_{14}	99_{18}		792	799
Guaifenesin 2TMS		1850*	342_{16}	196_{33}	149_{32}	103_{47}	73_{100}	1170	4551
Guaifenesin AC		2000*	240_{10}	124_{43}	117_{100}	109_{42}	77_{14}	556	1992
Guaifenesin-M (HO-) 3AC	U+UHYAC	2235*	340_{2}	298_{1}	159_{100}	140_{10}	99_{8}	1155	797
Guaifenesin-M (HO-methoxy-) 2AC	U+UHYAC	2290*	328_{2}	245_{5}	170_{15}	159_{100}	99_{9}	1080	001
Guaifenesin-M (HO-methoxy-) 3AC	U+UHYAC	2265*	370_{9}	230_{7}	212_{6}	170_{26}	159_{100}	1311	798
Guaifenesin-M (O-demethyl-)	UHY	1700*	184_{26}	135_{4}	110_{100}	81_{8}	64_{5}	357	2683
Guaifenesin-M (O-demethyl-) 3AC	U+UHYAC	1920*	310_{1}	268_{13}	159_{100}	117_{12}	110_{19}	970	800
Guanfacine		1890	225_{2}	159_{9}	123_{7}	101_{100}	89_{7}	573	7561
Guanfacine AC		2020	287_{1}	267_{1}	159_{17}	143_{45}	101_{100}	818	7567
Guanfacine artifact (-COONH2)		1680	203_{1}	168_{100}	159_{29}	125_{70}	89_{29}	405	7562
Guanfacine artifact (-COONH2) 2AC		2150	272_{1}	252_{4}	159_{21}	128_{87}	86_{100}	818	7568
Guanfacine artifact (-COONH2) 2TMS		1635	347_{2}	332_{4}	188_{64}	147_{43}	73_{100}	1193	7563
Guanfacine artifact (-COONH2) TMS		1685	275_{1}	260_{17}	160_{25}	116_{50}	73_{100}	743	7564
Guanfacine artifact (HOOC-) ME		1390*	218_{6}	183_{69}	159_{100}	123_{23}	89_{25}	456	7560
Guanfacine artifact (HOOC-) TMS		1510	261_{27}	241_{16}	232_{8}	159_{13}	73_{100}	752	7565
Guanfacine artifact (-NH3) TMS		1880	285_{6}	265_{30}	237_{27}	141_{77}	73_{100}	904	7566
Guanfacine HFB		1985	406_{1}	282_{19}	272_{72}	159_{76}	86_{100}	1560	7572
Guanfacine PFP		1965	391_{1}	356_{4}	272_{91}	159_{79}	86_{100}	1404	7571
Guanfacine TFA		1995	341_{1}	306_{11}	306_{11}	272_{82}	159_{100}	1158	7570
Halazepam	P U+UHYAC	2335	352_{60}	324_{100}	289_{14}	241_{11}		1218	2083
Halazepam HY	UHY UHYAC	2380	313_{100}	296_{6}	244_{11}	105_{15}	77_{16}	986	2091
Halazepam-M (HO-) isomer-1 HYAC	UHYAC	2350	371_{100}	328_{73}	312_{40}	260_{9}	208_{7}	1316	2121
Halazepam-M (HO-) isomer-2 HYAC	UHYAC	2370	371_{100}	328_{76}	312_{28}	260_{16}		1315	2122
Halazepam-M (HO-methoxy-) HYAC	UHYAC	2500	401_{47}	358_{100}	342_{40}	273_{21}	85_{37}	1448	2123
Halazepam-M (N-dealkyl-HO-)	UGLUC	2750	286_{82}	258_{100}	230_{11}	166_{7}	139_{8}	813	2113
Halazepam-M (N-dealkyl-HO-) AC	U+UHYAC	3000	328_{22}	286_{90}	258_{100}	166_{8}	139_{7}	1079	2111
Halazepam-M (N-dealkyl-HO-) HY	UHY	2400	247_{72}	246_{100}	230_{11}	121_{26}	65_{22}	583	2112
Halazepam-M (N-dealkyl-HO-) HYAC	U+UHYAC	2270	289_{18}	247_{86}	246_{100}	105_{7}	77_{35}	832	3143
Halazepam-M (N-dealkyl-HO-) isomer-1 HY2AC	U+UHYAC	2560	331_{48}	289_{64}	247_{100}	230_{41}	154_{13}	1098	2125
Halazepam-M (N-dealkyl-HO-) isomer-2 HY2AC	U+UHYAC	2610	331_{46}	289_{54}	246_{100}	154_{11}	121_{11}	1098	1751
Halazepam-M (N-dealkyl-HO-methoxy-) HY2AC	U+UHYAC	2700	361_{17}	319_{72}	276_{100}	260_{14}	246_{10}	1267	1752
Halazepam-M artifact	P-I UHY U+UHYAC	2060	240_{59}	239_{100}	205_{81}	177_{16}	151_{6}	554	300
Halazepam-M HYAC	PHYAC U+UHYAC	2245	273_{30}	230_{100}	154_{13}	105_{23}	77_{50}	732	273
Haloperidol	G P-I U UHY	2940	375_{2}	237_{92}	224_{100}	123_{50}	95_{31}	1335	340
Haloperidol 2TMS		3055	519_{1}	504_{3}	296_{100}	206_{55}	73_{75}	1676	4553
Haloperidol -H2O	U+UHYAC	2915	357_{13}	206_{39}	192_{100}	123_{36}	95_{23}	1248	523
Haloperidol TMS		2965	447_{1}	432_{7}	296_{100}	206_{97}	123_{98}	1575	4552
Haloperidol-D4		2930	379_{1}	316_{7}	237_{89}	224_{100}	127_{51}	1351	5427
Haloperidol-D4 2TMS		3050	523_{1}	508_{2}	296_{100}	206_{77}	73_{74}	1679	7285
Haloperidol-D4 -H2O		2900	361_{6}	206_{36}	192_{100}	127_{36}		1269	5428
Haloperidol-D4 TMS		2960	451_{1}	436_{7}	309_{75}	296_{100}	127_{91}	1583	7286
Haloperidol-M	U+UHYAC	1490*	180_{25}	125_{49}	123_{35}	95_{17}	56_{100}	345	85
Haloperidol-M	U UHY	1750	223_{20}	189_{19}	139_{23}	84_{40}	56_{100}	481	520
Haloperidol-M	U	2250	239_{9}	189_{20}	139_{24}	100_{9}	56_{100}	550	522
Haloperidol-M (N-dealkyl-)	UHY	1800	211_{6}	139_{20}	84_{42}	56_{100}		436	521
Haloperidol-M (N-dealkyl-) AC	U UHYAC	2235	253_{31}	210_{53}	193_{14}	139_{42}	57_{100}	621	524
Haloperidol-M (N-dealkyl-) -H2O AC	U+UHYAC	2155	235_{100}	192_{58}	158_{51}	129_{53}	82_{33}	531	182

Table 8.1: Compounds in order of names
Haloperidol-M (N-dealkyl-oxo-) -2H2O

Name	Detected	RI	Typical ions and intensities					Page	Entry
Haloperidol-M (N-dealkyl-oxo-) -2H2O	U UHY U+UHYAC	1650	189 $_{100}$	154 $_{32}$	127 $_{33}$	101 $_8$	75 $_{17}$	367	181
Haloperidol-M/artifact	U+UHYAC	1350*	164 $_{43}$	133 $_{38}$	123 $_{100}$	95 $_{41}$		310	1914
Halothane		<1000*	196 $_{50}$	177 $_{16}$	117 $_{100}$	98 $_{35}$	67 $_{47}$	388	2996
Harmaline	G	2430	214 $_{91}$	213 $_{100}$	198 $_{29}$	170 $_{25}$	115 $_{12}$	447	4062
Harmaline 2AC		2800	298 $_{33}$	255 $_{100}$	241 $_{17}$	212 $_{11}$	141 $_5$	895	4064
Harmaline -2H	G U UHY U+UHYAC	2460	212 $_{100}$	197 $_{29}$	169 $_{67}$	140 $_6$	115 $_8$	439	4066
Harmaline -2H AC		2545	254 $_{42}$	212 $_{100}$	197 $_{18}$	169 $_{26}$	140 $_4$	628	4067
Harmaline AC		2670	256 $_{73}$	213 $_{100}$	186 $_{13}$	170 $_{21}$	115 $_9$	637	4063
Harmaline artifact (dihydro-)		2150	216 $_{39}$	201 $_{100}$	186 $_{14}$	172 $_{18}$	144 $_5$	453	4065
Harmaline artifact (dihydro-) 2TFA		2115	408 $_{22}$	393 $_{100}$	296 $_{22}$	280 $_{17}$	199 $_{75}$	1470	9552
Harmaline artifact (dihydro-) AC		2525	258 $_{58}$	243 $_{75}$	215 $_{13}$	201 $_{100}$	172 $_{28}$	647	9550
Harmaline artifact (dihydro-) HFB		2300	412 $_{47}$	397 $_{100}$	243 $_{21}$	199 $_{34}$	172 $_{22}$	1485	9554
Harmaline artifact (dihydro-) PFP		2280	362 $_{46}$	347 $_{100}$	243 $_{21}$	199 $_{54}$	172 $_{37}$	1272	9553
Harmaline artifact (dihydro-) TFA		2295	312 $_{57}$	297 $_{100}$	282 $_4$	243 $_7$	199 $_{18}$	981	9551
Harmaline HFB		2590	410 $_{28}$	241 $_{100}$	226 $_7$	198 $_6$	170 $_5$	1477	5925
Harmaline PFP		2540	360 $_{40}$	241 $_{100}$	198 $_8$	184 $_7$	121 $_{14}$	1263	5917
Harmaline TFA		2525	310 $_{58}$	241 $_{100}$	198 $_6$	184 $_7$	169 $_7$	969	5919
Harmaline-M (O-demethyl-) -2H	UHY	2550	198 $_{100}$	170 $_{17}$	140 $_2$	99 $_{11}$	75 $_5$	394	4068
Harmaline-M (O-demethyl-) -2H AC	UHYAC	2600	240 $_{22}$	198 $_{100}$	169 $_{10}$	140 $_3$	115 $_5$	556	4069
Harmine	G U UHY U+UHYAC	2460	212 $_{100}$	197 $_{29}$	169 $_{67}$	140 $_6$	115 $_8$	439	4066
Harmine AC		2545	254 $_{42}$	212 $_{100}$	197 $_{18}$	169 $_{26}$	140 $_4$	628	4067
Harmine-M (O-demethyl-)	UHY	2550	198 $_{100}$	170 $_{17}$	140 $_2$	99 $_{11}$	75 $_5$	394	4068
Harmine-M (O-demethyl-) AC	UHYAC	2600	240 $_{22}$	198 $_{100}$	169 $_{10}$	140 $_3$	115 $_5$	556	4069
HDMP-28 artifact (2-Naphthaleneacetic acid) ME		1820*	200 $_{44}$	141 $_{100}$	115 $_{35}$	89 $_5$	70 $_5$	399	9467
HDMP-28 isomer-1 AC		2695	325 $_2$	294 $_1$	224 $_3$	126 $_{83}$	84 $_{100}$	1063	9471
HDMP-28 isomer-1 HFB		2445	479 $_2$	448 $_1$	280 $_{100}$	226 $_{11}$	200 $_{16}$	1632	9475
HDMP-28 isomer-1 PFP		2445	429 $_1$	230 $_{100}$	200 $_{13}$	176 $_5$	139 $_6$	1531	9473
HDMP-28 isomer-1 TFA		2485	379 $_1$	347 $_2$	200 $_{22}$	180 $_{100}$	139 $_{13}$	1350	9470
HDMP-28 isomer-2 AC		2745	325 $_2$	294 $_4$	224 $_5$	126 $_{62}$	84 $_{100}$	1063	9468
HDMP-28 isomer-2 HFB		2540	479 $_2$	420 $_3$	280 $_{100}$	200 $_{40}$	139 $_{20}$	1632	9474
HDMP-28 isomer-2 PFP		2520	429 $_2$	370 $_2$	230 $_{100}$	200 $_{16}$	139 $_4$	1530	9472
HDMP-28 isomer-2 TFA		2545	379 $_3$	347 $_2$	200 $_{30}$	180 $_{100}$	139 $_{19}$	1350	9469
Heptabarbital	P G UHY U+UHYAC	2070	221 $_{100}$	141 $_{32}$				605	803
Heptabarbital (ME)	G P	1800	235 $_{100}$	221 $_{32}$	169 $_8$	155 $_{48}$	141 $_{10}$	682	1885
Heptabarbital 2ME		1915	249 $_{100}$	169 $_{42}$	133 $_{13}$			769	806
Heptabarbital 2TMS		1980	394 $_3$	379 $_{17}$	365 $_{97}$	100 $_{36}$	73 $_{100}$	1422	5492
Heptabarbital-M (HO-)	U	2275	266 $_6$	237 $_{46}$	219 $_{96}$	141 $_{51}$	93 $_{100}$	696	804
Heptabarbital-M (HO-) -H2O	U+UHYAC	2300	248 $_8$	219 $_{52}$	157 $_{43}$	141 $_{39}$	93 $_{100}$	590	805
Heptachlor		1860*	370 $_1$	337 $_1$	272 $_{15}$	135 $_{14}$	100 $_{100}$	1309	3849
Heptachlorepoxide		2015*	351 $_2$	253 $_{19}$	183 $_{49}$	135 $_{48}$	81 $_{100}$	1382	3850
2,2',3,4,4',5,5'-Heptachlorobiphenyl		2460*	394 $_{100}$	392 $_{44}$	324 $_{48}$	252 $_{14}$	162 $_{15}$	1409	885
Heptadecane		1700*	240 $_4$	127 $_6$	85 $_{43}$	71 $_{60}$	57 $_{100}$	559	2977
Heptadecane		1700*	240 $_{15}$	85 $_{43}$	71 $_{52}$	57 $_{100}$	43 $_{86}$	558	7687
Heptadecanoic acid ET		2035*	298 $_5$	255 $_8$	157 $_{18}$	101 $_{54}$	88 $_{100}$	899	5404
Heptadecanoic acid ME		2025*	284 $_{20}$	241 $_{14}$	143 $_{19}$	87 $_{66}$	74 $_{100}$	806	3037
Heptafluorobutanoic acid		<1000*	197 $_3$	169 $_{34}$	150 $_{78}$	119 $_{90}$	69 $_{100}$	444	5545
Heptafluorobutanoic acid		<1000*	197 $_3$	169 $_{31}$	119 $_{81}$	69 $_{91}$	45 $_{100}$	444	5548
Heptaminol		1125	127 $_{17}$	113 $_{43}$	110 $_{26}$	69 $_{54}$	59 $_{100}$	284	1459
Heptaminol 2AC		1530	172 $_{14}$	169 $_{30}$	114 $_{22}$	95 $_{28}$	86 $_{100}$	507	1460
Heptane		700*	100 $_{11}$	71 $_{36}$	57 $_{53}$	43 $_{100}$	29 $_{69}$	258	3823
Heptenophos		1570*	250 $_2$	141 $_6$	124 $_{100}$	109 $_{48}$	89 $_{91}$	602	3852
Heroin	G PHYAC U+UHYAC	2620	369 $_{59}$	327 $_{100}$	310 $_{36}$	268 $_{47}$	162 $_{11}$	1306	225
Heroin Cl-artifact	UHYAC	2680	403 $_{22}$	361 $_{100}$	344 $_{63}$	302 $_{90}$	204 $_{55}$	1456	2992
Heroin-D3		2510	372 $_{87}$	330 $_{100}$	313 $_{57}$	271 $_{87}$	218 $_{56}$	1324	7294
Heroin-M (3-acetyl-morphine)		2500	327 $_{100}$	310 $_{10}$	285 $_{73}$	215 $_{19}$	162 $_{22}$	1075	2341
Heroin-M (3-acetyl-morphine) PFP		2490	473 $_{25}$	431 $_{45}$	310 $_{60}$	268 $_{100}$	119 $_{16}$	1624	2462
Heroin-M (3-acetyl-morphine) TMS		2570	399 $_{51}$	357 $_{45}$	234 $_{31}$	164 $_{34}$	73 $_{100}$	1442	2466
Heroin-M (6-acetyl-morphine)	U-I	2535	327 $_{100}$	268 $_{53}$	162 $_{11}$	124 $_{13}$		1074	525
Heroin-M (6-acetyl-morphine) HFB		2425	523 $_{42}$	464 $_{100}$	411 $_{39}$	204 $_{56}$	69 $_{45}$	1679	6121
Heroin-M (6-acetyl-morphine) PFP	UMAMPFP	2650	473 $_{90}$	430 $_{11}$	414 $_{100}$	361 $_{27}$	204 $_{24}$	1624	2253
Heroin-M (6-acetyl-morphine) TFA		2630	423 $_{65}$	380 $_{15}$	364 $_{100}$	311 $_{36}$	204 $_{30}$	1514	5575
Heroin-M (6-acetyl-morphine) TMS	UMAMTMS	2590	399 $_{92}$	340 $_{50}$	287 $_{32}$	204 $_{25}$	73 $_{100}$	1442	2465
Heroin-M (6-acetyl-morphine)-D3		2515	330 $_{100}$	287 $_{10}$	271 $_{99}$	218 $_{37}$	165 $_{24}$	1094	5574
Heroin-M (6-acetyl-morphine)-D3 HFB		2415	526 $_{41}$	467 $_{100}$	414 $_{36}$	207 $_{45}$	69 $_{45}$	1681	6122
Heroin-M (6-acetyl-morphine)-D3 PFP		2640	476 $_{63}$	417 $_{100}$	364 $_{31}$	207 $_{46}$	165 $_{15}$	1629	5568
Heroin-M (6-acetyl-morphine)-D3 TFA		2630	426 $_{53}$	367 $_{100}$	314 $_{28}$	207 $_{43}$	69 $_{25}$	1523	5573
Heroin-M (6-acetyl-morphine)-D3 TMS		2580	402 $_{100}$	343 $_{75}$	290 $_{43}$	207 $_{22}$	73 $_{90}$	1454	5577
Heroin-M (morphine)	G UHY	2455	285 $_{100}$	268 $_{15}$	162 $_{59}$	124 $_{21}$		810	474
Heroin-M (morphine) 2HFB		2375	677 $_{10}$	480 $_{10}$	464 $_{100}$	407 $_9$	169 $_8$	1726	6120
Heroin-M (morphine) 2PFP		2360	577 $_{51}$	558 $_7$	430 $_8$	414 $_{100}$	119 $_{22}$	1706	2251
Heroin-M (morphine) 2TFA		2250	477 $_{71}$	364 $_{100}$	307 $_6$	115 $_8$	69 $_{31}$	1630	4008
Heroin-M (morphine) 2TMS	UHYTMS	2560	429 $_{19}$	236 $_{21}$	196 $_{15}$	146 $_{21}$	73 $_{100}$	1531	2463
Heroin-M (morphine) TFA		2285	381 $_{55}$	268 $_{100}$	146 $_{14}$	115 $_{13}$	69 $_{23}$	1360	5569
Heroin-M (morphine)-D3 2HFB		2375	680 $_4$	483 $_9$	467 $_{100}$	414 $_7$	169 $_{23}$	1727	6126

Heroin-M (morphine)-D3 2PFP

Table 8.1: Compounds in order of names

Name	Detected	RI	Typical ions and intensities					Page	Entry
Heroin-M (morphine)-D3 2PFP		2350	580_{16}	433_7	417_{100}	269_5	119_8	1708	5567
Heroin-M (morphine)-D3 2TFA		2240	480_{32}	383_6	367_{100}	314_6	307_6	1634	5571
Heroin-M (morphine)-D3 2TMS		2550	432_{91}	290_{27}	239_{60}	199_{40}	73_{100}	1540	5578
Heroin-M (morphine)-D3 TFA		2275	384_{39}	271_{100}	211_8	165_6	152_7	1376	5572
Heroin-M 2PFP		2440	563_{100}	400_{10}	355_{38}	327_7	209_{15}	1701	3534
Heroin-M 3AC	U+UHYAC	2955	397_8	355_9	209_{41}	87_{100}	72_{33}	1434	1194
Heroin-M 3PFP	UHYPFP	2405	709_{80}	533_{28}	388_{29}	367_{51}	355_{100}	1729	3533
Heroin-M 3TMS	UHYTMS	2605	487_{17}	416_{19}	222_{36}	131_{19}	73_{100}	1645	3525
Hexachlorobenzene		1690*	284_{100}	282_{53}	247_{11}	212_7	142_{15}	791	1462
2,2',3,4,4',5'-Hexachlorobiphenyl		2290*	362_{80}	360_{100}	358_{52}	288_{34}	218_{12}	1250	884
2,2',4,4',5,5'-Hexachlorobiphenyl		2330*	362_{79}	360_{100}	358_{45}	288_{27}	218_9	1251	2633
1,2,3,4,7,8-Hexachlorodibenzofuran (HXCDF)		----*	374_{100}	372_{56}	309_{19}	239_{34}	156_{39}	1321	3497
1,2,3,6,7,8-Hexachlorodibenzofuran (HXCDF)		----*	374_{100}	372_{50}	309_{15}	239_{31}	187_{28}	1321	3495
2,3,4,6,7,8-Hexachlorodibenzofuran (HXCDF)		----*	374_{100}	372_{58}	309_{19}	239_{34}	156_{41}	1321	3496
Hexachlorophene		2790*	404_{37}	369_{15}	335_{17}	209_{56}	196_{100}	1458	3644
Hexacosane		2600*	366_3	99_{13}	85_{36}	71_{56}	57_{100}	1294	2365
Hexadecane		1600*	226_4	99_{10}	85_{35}	71_{59}	57_{100}	497	2353
Hexamid	U	2380	331_2	259_4	117_5	86_{100}	58_7	1102	1908
Hexamid-M (bis-deethyl-) AC	U+UHYAC	2570	317_2	258_{12}	117_{21}	85_{100}	72_{34}	1011	1910
Hexamid-M (deethyl-)	U+UHYAC	2200	303_1	117_{15}	71_{51}	58_{100}		925	1912
Hexamid-M (deethyl-) AC	U+UHYAC	2780	345_{10}	302_{61}	113_{44}	100_{45}	58_{100}	1185	1909
Hexamid-M (deethyl-HO-) 2AC	U+UHYAC	3140	403_{16}	360_{66}	318_{27}	100_{36}	58_{100}	1457	1911
Hexamid-M (phenobarbital)	P G U+UHYAC	1965	232_{14}	204_{100}	161_{18}	146_{12}	117_{37}	516	854
Hexamid-M (phenobarbital) 2TMS		2015	376_2	361_{34}	261_{15}	146_{100}	73_{46}	1339	4582
Hexane		600*	86_{15}	57_{100}	43_{85}	41_{74}	29_{76}	255	3775
Hexazinone		2295	252_4	171_{100}	128_{24}	83_{57}	71_{47}	620	4053
Hexethal	P G U	1835	211_6	156_{100}	141_{93}	55_{36}		557	807
Hexethal 2ME		1745	210_6	184_{88}	169_{100}	112_{18}	55_{20}	709	808
Hexobarbital	P G U+UHYAC	1855	236_{10}	221_{100}	157_{43}	155_{26}	81_{70}	539	809
Hexobarbital ME		1805	250_3	235_{100}	169_{27}	81_{59}		604	811
Hexobarbital-M (HO-) -H2O		1970	234_{36}	219_{59}	156_{96}	91_{73}	79_{100}	527	2265
Hexobarbital-M (nor-)		1980	222_7	207_{40}	143_{50}	81_{100}		480	1917
Hexobarbital-M (oxo-)	U+UHYAC	2055	250_{60}	235_{80}	193_{27}	156_{28}	95_{100}	604	810
Hexobarbital-M (oxo-) ME	PME UME	2020	264_{28}	249_{100}	221_{20}	207_{17}	95_{37}	681	2759
Hexobendine-M/artifact (trimethoxybenzoic acid)		1780*	212_{100}	197_{57}	169_{13}	141_{27}		438	1949
Hexobendine-M/artifact (trimethoxybenzoic acid) FT		1770*	240_{100}	225_{44}	212_{17}	195_{45}	141_{24}	556	5219
Hexobendine-M/artifact (trimethoxybenzoic acid) ME		1740*	226_{100}	211_{48}	195_{22}	155_{21}		101	1950
Hexyloctylphthalate		2500*	362_1	279_4	251_5	149_{100}	85_5	1276	6053
Hexylresorcinol	P	1830*	194_{41}	123_{100}	95_9	77_{10}		385	1981
Hexylresorcinol 2AC		1935*	278_6	236_{22}	194_{90}	123_{100}		769	1990
Hexylresorcinol AC		1875*	236_6	194_{30}	123_{100}			541	1989
Hippuric acid	U	1745	179_1	161_2	135_{22}	105_{100}	77_9	340	96
Hippuric acid 2TMS	UTMS	2070	323_{10}	308_{16}	280_{11}	206_{50}	105_{100}	1051	5812
Hippuric acid ME	UME	1660	193_5	161_7	134_{19}	105_{100}	77_{45}	377	97
Hippuric acid TMS	UTMS	1925	251_1	236_8	206_{71}	105_{100}	73_{85}	609	5813
Histapyrrodine	G U UHY UHYAC	2240	280_8	196_{27}	91_{62}	84_{100}	65_9	784	1646
Histapyrrodine-M (HO-)	UHY	2500	296_{23}	212_{81}	120_7	91_{84}	84_{100}	887	1650
Histapyrrodine-M (HO-) AC	UHYAC	2630	338_2	254_{23}	120_5	91_{56}	84_{100}	1146	1652
Histapyrrodine-M (N-dealkyl-)	UHY	1930	183_{25}	106_{16}	91_{100}	77_{24}	65_{25}	355	2065
Histapyrrodine-M (N-dealkyl-) AC	UHYAC	2080	225_{22}	183_{23}	106_{10}	91_{100}	77_{24}	491	2066
Histapyrrodine-M (N-debenzyl-)	UHY	1800	190_{12}	120_3	106_6	84_{100}	77_{10}	370	1654
Histapyrrodine-M (N-debenzyl-oxo-)	UHY	2120	204_{19}	119_{54}	106_{100}	98_{17}	77_{23}	411	1653
Histapyrrodine-M (N-debenzyl-oxo-) AC	UHYAC	2160	246_4	161_{59}	119_{49}	106_{100}	98_{29}	581	1649
Histapyrrodine-M (N-dephenyl-) AC	UHYAC	2120	246_2	176_3	91_{19}	84_{100}	65_5	583	1647
Histapyrrodine-M (N-dephenyl-HO-) -H2O	UHY	2100	216_{29}	159_{29}	97_{10}	91_{100}	69_{23}	452	1655
Histapyrrodine-M (N-dephenyl-oxo-) AC	UHYAC	2260	260_2	217_{45}	175_{12}	120_{79}	91_{100}	658	1648
Histapyrrodine-M (oxo-)	U UHY UHYAC	2570	294_7	209_9	196_{44}	120_5	91_{100}	872	1651
4-HO-DALT		2245	256_{10}	160_2	146_7	117_3	110_{100}	639	9395
4-HO-DALT 2TMS		2355	400_4	359_9	304_5	290_{26}	110_{100}	1446	9398
4-HO-DALT AC		2350	298_2	269_2	160_5	146_{13}	110_{100}	897	9396
4-HO-DALT isomer-1 HFB		2230	452_2	425_4	356_{28}	342_{62}	110_{100}	1584	9403
4-HO-DALT isomer-1 PFP		2395	402_{31}	292_{52}	186_{36}	145_{64}	110_{100}	1452	9401
4-HO-DALT isomer-1 TFA		2410	352_{18}	256_{40}	172_{38}	146_{42}	110_{100}	1220	9399
4-HO-DALT isomer-2 HFB		2450	452_3	425_3	356_{15}	342_{47}	110_{100}	1584	9404
4-HO-DALT isomer-2 PFP		2440	402_8	292_{58}	186_{33}	145_{52}	110_{100}	1452	9402
4-HO-DALT isomer-2 TFA		2460	352_{37}	305_{15}	242_{39}	172_{55}	110_{100}	1220	9400
4-HO-DALT ME		2300	270_3	229_5	160_{16}	130_{23}	110_{100}	720	9397
4-HO-DALT TFA		2125	323_3	256_3	242_8	145_4	110_{100}	1220	10140
4-HO-DiPT		2280	260_7	160_4	146_{12}	114_{100}	72_{18}	660	8830
4-HO-DiPT 2TMS		3025	404_3	304_5	290_{16}	114_{100}	72_{21}	1460	9481
4-HO-DiPT AC		2390	302_1	160_6	146_{10}	114_{100}	72_{19}	920	8875
4-HO-DMT		1995	204_{15}	159_3	146_5	130_3	58_{100}	410	2470
4-HO-DMT 2TMS		2250	348_6	333_1	290_{29}	73_{29}	58_{100}	1201	6348
4-HO-DMT AC	U+UHYAC	2270	246_5	160_3	146_7	130_3	58_{100}	580	2471

Table 8.1: Compounds in order of names

Name	Detected	RI	Typical ions and intensities					Page	Entry
4-HO-DMT AC	U+UHYAC	2340	288_7	246_1	202_1	122_3	58_{100}	828	2472
4-HO-DMT HFB		2110	400_2	342_4	145_5	117_8	58_{100}	1445	6317
4-HO-DMT PFP		2095	350_3	292_2	186_2	145_5	58_{100}	1208	6350
4-HO-DMT TFA		2080	300_2	242_2	117_3	69_4	58_{100}	906	6349
5-HO-DMT		2150	204_3	160_1	146_5	117_2	58_{100}	410	8858
Homatropine		2340	275_6	142_4	124_{100}	94_{21}	82_{23}	747	6259
Homatropine AC		2250	317_3	245_2	124_{100}	94_{18}	82_{22}	1012	6264
Homatropine TMS		2090	347_4	179_{23}	124_{100}	94_{15}	73_{30}	1196	6307
Homatropine-M (mandelic acid)		1890*	152_{10}	107_{100}	79_{75}	77_{55}	51_{63}	292	5759
Homatropine-M (mandelic acid) ME		1485*	166_7	107_{100}	79_{59}	77_{40}		315	1071
Homatropine-M (nor-) 2AC		2565	345_{11}	302_{10}	152_{39}	110_{100}	68_{24}	1185	6265
Homatropine-M/artifact (tropine) AC		1240	183_{25}	140_{11}	124_{100}	94_{42}	82_{63}	355	5125
Homatropine-M/artifact (tropine) TFA		1020	237_{19}	124_{100}	94_{34}	82_{44}	67_{25}	543	7914
4-HO-MET		2100	218_{24}	159_4	146_{12}	117_5	72_{100}	460	9555
4-HO-MET 2TMS		2060	362_{15}	347_6	290_{24}	216_4	72_{100}	1276	9560
4-HO-MET AC		2190	260_4	160_5	146_{10}	117_5	72_{100}	659	9556
4-HO-MET HFB		2225	414_5	356_5	342_{25}	145_{33}	72_{100}	1489	9559
4-HO-MET PFP		2225	364_2	306_4	292_{11}	145_{22}	72_{100}	1281	9558
4-HO-MET TFA		2185	314_{10}	267_4	242_6	186_{12}	72_{100}	994	9557
4-HO-MiPT		2150	232_{18}	160_5	146_{15}	130_6	86_{100}	519	8828
4-HO-MiPT 2TMS		3025	205_1	147_7	103_4	86_{100}	73_{26}	1340	9483
4-HO-MiPT AC		2240	274_1	160_4	146_7	130_3	86_{100}	741	9482
Homofenazine	G	3165	433_8	280_{50}	167_{63}	58_{100}		1583	526
Homofenazine AC		3260	493_3	433_{29}	280_{65}	167_{100}	87_{83}	1652	341
Homofenazine-M (amino-) AC	U+UHYAC	2765	366_{16}	280_6	266_{13}	248_6	100_{100}	1290	1267
Homofenazine-M (dealkyl-) AC	UHYAC	3240	449_{80}	267_{100}	112_{26}			1578	1269
Homofenazine-M (ring)	U+UHYAC	2190	267_{100}	235_{19}				698	1266
Homovanillic acid	U	1610*	182_{35}	137_{100}	122_{12}	94_7	65_9	350	3368
Homovanillic acid 2ME	UME	1720*	210_{31}	151_{100}	123_4	107_6		433	5959
Homovanillic acid 2TMS		1760*	326_{32}	267_{20}	209_{46}	179_{34}	73_{100}	1069	6015
Homovanillic acid HFB		1770*	378_{79}	333_{39}	181_{60}	107_{100}	69_{72}	1346	5975
Homovanillic acid ME	UME	1750*	196_{25}	137_{100}	122_{16}	107_2	94_{12}	389	812
Homovanillic acid MEAC	U+UHYAC	1700*	238_2	196_{47}	137_{100}	122_6	107_6	547	2973
Homovanillic acid MEHFB		1570*	392_{100}	333_{84}	169_{24}	107_{78}	69_{60}	1410	5974
Homovanillic acid MEPFP		1570*	342_{96}	283_{100}	195_{35}	119_{37}	107_{51}	1166	5972
Homovanillic acid METMS		1670*	268_{41}	238_{69}	209_{52}	179_{100}	73_{95}	707	6016
Homovanillic acid PFP		1685*	328_{85}	283_{45}	181_{58}	119_{33}	107_{100}	1079	5973
Hydrocaffeic acid		2400*	182_{17}	136_{13}	123_{100}	77_{55}	51_{67}	351	5763
Hydrocaffeic acid 3TMS		2250*	398_{67}	280_{16}	267_{39}	179_{98}	73_{100}	1438	5996
Hydrocaffeic acid -CO2		1295*	138_{23}	123_{100}	91_{28}	77_{47}	51_{92}	278	5756
Hydrocaffeic acid ME		1870*	196_{24}	136_{45}	123_{100}	91_{11}	77_{16}	389	5764
Hydrocaffeic acid ME2AC		1980*	280_2	238_{15}	196_{100}	136_{65}	123_{68}	780	5992
Hydrocaffeic acid ME2HFB		1720*	588_{11}	528_{100}	349_{32}	169_{37}	69_{58}	1711	5994
Hydrocaffeic acid ME2PFP		1590*	488_{19}	428_{100}	299_{26}	281_{26}	119_{73}	1645	5993
Hydrocaffeic acid ME2TMS		2220*	340_{44}	267_{36}	193_{10}	179_{100}	73_{97}	1156	5995
Hydrocaffeic acid METFA		1540*	292_{88}	233_{100}	195_{27}	107_{46}	69_{62}	854	5969
Hydrochlorothiazide		9999	297_{53}	269_{100}	221_{38}			888	813
Hydrochlorothiazide 4ME	UME	2905	353_{100}	310_{90}	288_{45}	218_{38}	138_{44}	1224	6536
Hydrochlorothiazide artifact ME	UME	1980	220_{31}	191_{10}	142_{25}	127_{100}	99_{21}	467	3003
Hydrochlorothiazide -SO2NH ME	UME	2170	232_{94}	167_{48}	139_{100}	127_{85}	125_{70}	515	3002
Hydrocodone	G UHY UHYAC	2440	299_{100}	242_{51}	185_{23}	96_{24}	59_{23}	902	238
Hydrocodone enol AC		2500	341_{100}	298_{65}	242_{32}	162_{26}		1164	258
Hydrocodone enol Cl-artifact AC		2630	375_{100}	340_{47}	318_{28}	146_{13}	115_{10}	1334	4401
Hydrocodone enol TMS		2475	371_{31}	356_{14}	313_9	234_{30}	73_{100}	1320	6215
Hydrocodone-M (dihydro-) 6-beta isomer TMS		2495	373_{50}	316_{13}	236_3	146_4	73_{100}	1329	6762
Hydrocodone-M (N,O-bis-demethyl-) enol 3TMS		2635	487_{50}	472_{22}	357_{32}	292_{16}	73_{100}	1645	6764
Hydrocodone-M (N,O-bis-demethyl-dihydro-) 3AC	UHYAC	2790	399_{20}	357_{50}	229_{19}	87_{100}	72_{22}	1442	3050
Hydrocodone-M (N,O-bis-demethyl-dihydro-) 6-beta isomer 3TMS		2600	489_{50}	474_{10}	374_{27}	294_{12}	73_{100}	1648	6765
Hydrocodone-M (N-demethyl-) enol 2TMS		2610	429_{25}	414_{17}	314_{25}	292_{20}	73_{100}	1532	6763
Hydrocodone-M (N-demethyl-dihydro-) 6-beta isomer 2TMS		2560	431_{50}	316_{16}	234_6	144_7	73_{100}	1538	6761
Hydrocodone-M (nor-) AC	UHYAC	2760	327_{18}	241_{16}	87_{100}	72_{36}		1074	239
Hydrocodone-M (nor-dihydro-)	UHY	2440	287_{100}	244_{24}	242_{22}	150_{32}	115_{24}	820	4368
Hydrocodone-M (nor-dihydro-) 2AC	UHYAC	2750	371_{20}	285_7	243_{26}	87_{100}	72_{33}	1319	235
Hydrocodone-M (nor-dihydro-) AC	UHYAC	2700	329_{40}	243_{42}	183_{26}	87_{100}	72_{44}	1088	3054
Hydrocodone-M (O-demethyl-) enol 2AC	U+UHYAC	2625	369_{39}	327_{100}	284_{54}	228_{37}	162_{40}	1306	1186
Hydrocodone-M (O-demethyl-) enol 2TFA		2230	477_{100}	380_{32}	258_{18}	223_7	117_{10}	1630	4009
Hydrocodone-M (O-demethyl-) enol 2TMS		2520	429_{22}	414_{22}	357_{10}	234_{23}	73_{100}	1532	6208
Hydrocodone-M (O-demethyl-) TMS		2475	357_{51}	342_{18}	300_{38}	96_{29}	73_{100}	1248	6209
Hydrocodone-M (O-demethyl-dihydro-)	UHY	2400	287_{100}	230_{14}	164_{35}	115_{28}	70_{54}	821	484
Hydrocodone-M (O-demethyl-dihydro-) 2AC	U+UHYAC	2545	371_{83}	329_{100}	286_{34}	212_{21}	70_{33}	1319	234
Hydrocodone-M (O-demethyl-dihydro-) 2HFB		2260	679_{41}	482_{53}	466_{100}	360_{13}	169_{21}	1727	6197
Hydrocodone-M (O-demethyl-dihydro-) 2PFP		2330	579_{60}	432_{21}	416_{49}	310_4	119_{100}	1707	2460
Hydrocodone-M (O-demethyl-dihydro-) 2TFA		2190	479_{91}	382_{25}	366_{61}	260_7	69_{100}	1632	6198
Hydrocodone-M (O-demethyl-dihydro-) 2TMS		2520	431_{33}	373_7	236_{23}	146_{15}	73_{100}	1538	2469

Hydrocodone-M (O-demethyl-dihydro-) 6-beta isomer 2TMS

Table 8.1: Compounds in order of names

Name	Detected	RI	Typical ions and intensities					Page	Entry
Hydrocodone-M (O-demethyl-dihydro-) 6-beta isomer 2TMS		2540	431 44	374 14	236 4	146 4	73 100	1538	6760
Hydrocodone-M (O-demethyl-dihydro-) AC	UHYAC	2490	329 100	287 56	230 10	164 20	70 21	1089	3055
Hydrocodone-M (O-demethyl-dihydro-) TFA		2250	383 100	286 19	270 44	213 19	69 50	1370	6199
Hydrocortisone	UME	2740*	302 53	189 31	163 100	123 62	91 59	1275	3295
Hydrocotarnine	U+UHYAC	1790	221 60	220 100	205 23	178 66	163 21	473	2862
Hydrocotarnine-M (demethyl-) AC	U+UHYAC	1975	248 2	206 45	190 80	164 100	125 57	594	9375
Hydromorphone	UHY	2445	285 100	228 18	214 10	171 10	96 30	810	527
Hydromorphone AC	UHYAC	2595	327 34	285 100	229 36	200 14	171 13	1075	240
Hydromorphone enol 2AC	U+UHYAC	2625	369 39	327 100	284 54	228 37	162 40	1306	1186
Hydromorphone enol 2HFB		2325	677 100	480 31	464 18	358 10	169 7	1726	6138
Hydromorphone enol 2PFP		2320	577 48	430 100	414 62	372 30	308 68	1706	2663
Hydromorphone enol 2TFA		2230	477 100	380 32	258 18	223 7	117 10	1630	4009
Hydromorphone enol 2TMS		2520	429 22	414 22	357 10	234 23	73 100	1532	6208
Hydromorphone HFB		2385	481 100	452 17	425 77	410 22	284 8	1636	6137
Hydromorphone PFP		2250	431 100	375 78	360 33	346 25	119 50	1537	2662
Hydromorphone TMS		2475	357 51	342 18	300 38	96 29	73 100	1248	6209
Hydromorphone-M (dihydro-)	UHY	2400	287 100	230 14	164 35	115 28	70 54	821	484
Hydromorphone-M (dihydro-) 2AC	U+UHYAC	2545	371 83	329 100	286 34	212 21	70 33	1319	234
Hydromorphone-M (dihydro-) 2HFB		2260	679 41	482 53	466 100	360 13	169 21	1727	6197
Hydromorphone-M (dihydro-) 2PFP		2330	579 60	432 21	416 49	310 4	119 100	1707	2460
Hydromorphone-M (dihydro-) 2TFA		2190	479 91	382 25	366 61	260 7	69 100	1632	6198
Hydromorphone-M (dihydro-) 2TMS		2520	431 33	373 7	236 23	146 15	73 100	1538	2469
Hydromorphone-M (dihydro-) 6-beta isomer 2TMS		2540	431 44	374 14	236 4	146 4	73 100	1538	6760
Hydromorphone-M (dihydro-) AC	UHYAC	2490	329 100	287 56	230 10	164 20	70 21	1089	3055
Hydromorphone-M (dihydro-) TFA		2250	383 100	286 19	270 44	213 19	69 50	1370	6199
Hydromorphone-M (N-demethyl-) enol 3TMS		2635	487 50	472 22	357 32	292 16	73 100	1645	6764
Hydromorphone-M (N-demethyl-dihydro-) 6-beta isomer 3TMS		2600	489 50	474 10	374 27	294 12	73 100	1648	6765
Hydroquinone	UHY	<1000*	110 100	81 27				261	814
Hydroquinone 2AC	UHYAC	1395*	194 8	152 26	110 100			382	815
Hydroquinone 2ME		<1000*	138 56	123 100	95 54	63 22		278	3282
Hydroquinone-M (2-HO-)	UHY	1460*	126 100	109 18	79 26	53 9		269	3163
Hydroquinone-M (2-HO-) 3AC	UHYAC	1710*	252 1	210 7	168 46	126 100	97 7	616	4336
Hydroquinone-M (2-methoxy-) 2AC	UHYAC	1450*	224 3	182 23	140 100	125 71	97 9	486	4337
3-Hydroxybutyric acid 2TMS		1095*	233 16	191 32	147 100	117 51	73 62	590	8923
4-Hydroxy-3-methoxy-benzylamino 2AC		1995	237 9	195 100	152 48	137 43	122 33	544	5691
4-Hydroxy-3-methoxy-cinnamic acid 2ME	UME	1850*	222 100	207 72	191 75	164 14	147 19	479	4945
4-Hydroxy-3-methoxy-cinnamic acid 2TMS		2160*	338 92	323 56	308 48	249 45	73 100	1143	5015
4-Hydroxy-3-methoxy-cinnamic acid -CO2		1195*	150 34	135 33	107 68	77 100	51 68	289	5752
4-Hydroxy-3-methoxy-cinnamic acid glycine conjugate		2380	265 75	204 7	177 100	145 47	117 13	686	5766
4-Hydroxy-3-methoxy-cinnamic acid glycine conjugate 2ME		2450	279 41	191 100	163 12	148 6	133 5	772	5825
4-Hydroxy-3-methoxy-cinnamic acid glycine conjugate 3TMS		2540	467 61	453 8	336 22	249 67	73 100	1612	5826
4-Hydroxy-3-methoxy-cinnamic acid ME		1930*	208 100	177 80	145 53	133 19	117 24	424	5966
4-Hydroxy-3-methoxy-cinnamic acid MEAC		1950*	250 3	208 100	177 47	145 26	117 9	603	5814
4-Hydroxy-3-methoxyhydrocinnamic acid 2TMS		1940*	340 71	310 33	209 100	192 56	73 89	1156	5824
4-Hydroxy-3-methoxyhydrocinnamic acid ME		1670*	210 26	150 20	137 100	122 9	107 8	433	5822
4-Hydroxy-3-methoxyhydrocinnamic acid MEAC		1860*	252 3	210 59	179 8	150 29	137 100	617	5823
4-Hydroxy-3-methoxy-phenethylamine		1410	167 10	138 100	123 34	94 19	77 10	319	5615
15-Hydroxy-5,8,11,13-eicosatetraenoic acid -H2O ME		2360*	316 4	189 37	119 54	105 100	91 92	1008	4313
15-Hydroxy-5,8,11,13-eicosatetraenoic acid METFA		2390*	430 1	316 6	131 43	117 66	91 100	1535	4354
N-Hydroxy-Amfetamine		1180	149 2	116 3	91 31	65 12	60 100	291	5906
N-Hydroxy-Amfetamine 2AC		1720	144 27	118 26	102 100	91 56	60 39	534	5908
N-Hydroxy-Amfetamine AC		1300	149 7	107 12	102 74	91 62	60 100	379	5907
N-Hydroxy-Amfetamine TFA		1195	156 88	118 63	91 100	69 41	65 33	584	5909
Hydroxyandrostanedione	UHY	2530*	304 75	286 73	232 100	191 55		933	816
Hydroxyandrostanedione AC	UHYAC	2630*	346 25	286 76	271 100	232 61	191 89	1190	2699
Hydroxyandrostene	U	2300*	274 100	259 75	241 66	148 86	94 85	743	614
Hydroxyandrostene AC	UGLUCAC	2860*	316 100	256 58	241 24	215 24		1008	266
11-Hydroxyandrosterone		2640*	306 100	288 52	270 41	147 45	55 86	945	3763
11-Hydroxyandrosterone AC		2760*	348 45	273 44	270 39	105 68	55 100	1202	3771
11-Hydroxyandrosterone enol 3TMS		2705*	522 24	417 23	256 25	168 48	73 100	1678	3805
3-Hydroxybenzoic acid		1620*	138 65	121 63	93 42	65 100		277	5758
3-Hydroxybenzoic acid 2ME		1490*	166 60	135 100	107 44	92 24	77 41	315	1110
3-Hydroxybenzoic acid 2TMS		1535*	282 38	267 100	223 42	193 40	73 86	793	6017
3-Hydroxybenzoic acid AC		1560*	180 10	138 100	121 50	93 11	63 17	344	5978
3-Hydroxybenzoic acid ME		1330*	152 46	121 100	93 43	65 33	53 7	293	5976
3-Hydroxybenzoic acid MEAC		1375*	194 6	163 8	152 72	121 100	93 21	382	5977
4-Hydroxybenzoic acid		1320*	152 95	135 100	107 11	92 16	77 23	293	10270
4-Hydroxybenzoic acid 2ME		1270*	166 36	135 100	107 11	92 15	77 18	316	6446
3-Hydroxybenzylalcohol		1310*	124 100	107 15	95 60	77 66	65 22	268	4663
3-Hydroxybutyric acid		1095*	233 16	191 32	147 100	117 51	73 62	590	8923
4-Hydroxybutyric acid 2TMS		1160*	248 1	233 26	147 100	117 30	73 32	591	5430
Hydroxyethylsalicylate		1540*	182 13	164 32	120 100	92 45	65 21	351	5224
Hydroxyethylsalicylate 2AC		1800*	266 1	224 7	164 6	120 32	87 100	692	5225
Hydroxyethylurea		9999	104 1	86 6	74 100	61 7		259	1551

Table 8.1: Compounds in order of names — 11-Hydroxyetiocholanolone

Name	Detected	RI	Typical ions and intensities					Page	Entry
11-Hydroxyetiocholanolone		2675*	306 $_{100}$	273 $_{64}$	147 $_{30}$	107 $_{40}$	55 $_{53}$	945	3764
11-Hydroxyetiocholanolone AC		2770*	348 $_{20}$	288 $_{80}$	270 $_{71}$	255 $_{62}$	55 $_{100}$	1202	3772
11-Hydroxyetiocholanolone enol 3TMS		2735*	522 $_{22}$	417 $_{12}$	327 $_{23}$	168 $_{72}$	73 $_{100}$	1678	3798
4-Hydroxyhippuric acid ME	U	1820	209 $_{100}$	177 $_{32}$	149 $_{34}$	121 $_{87}$		427	817
5-Hydroxyindole	UHY	1340	133 $_{100}$	105 $_{25}$	78 $_{16}$	51 $_{12}$		273	3285
5-Hydroxyindole AC	UHYAC	1370	175 $_{8}$	133 $_{100}$	106 $_{13}$	78 $_{17}$	51 $_{11}$	330	4273
5-Hydroxyindoleacetic acid 2ME		1995	219 $_{55}$	160 $_{100}$	145 $_{23}$	117 $_{16}$	89 $_{7}$	464	5042
5-Hydroxyindolepropanoic acid 2ME		1695	233 $_{35}$	174 $_{16}$	160 $_{100}$	149 $_{12}$	130 $_{19}$	521	5041
N-Hydroxy-MDA 2AC		2010	279 $_{1}$	162 $_{100}$	135 $_{76}$	102 $_{58}$	61 $_{24}$	773	5910
N-Hydroxy-MDA TFA		1665	291 $_{12}$	162 $_{37}$	135 $_{100}$	105 $_{5}$	77 $_{22}$	846	5911
Hydroxy-methoxy-acetophenone AC	UHYAC	1640*	208 $_{6}$	166 $_{49}$	151 $_{100}$	123 $_{12}$		423	2483
Hydroxymethoxyflavone		2610*	268 $_{100}$	253 $_{21}$	225 $_{9}$	132 $_{78}$	117 $_{17}$	706	5598
Hydroxymethoxyflavone AC		2610*	310 $_{64}$	268 $_{100}$	253 $_{17}$	132 $_{40}$	117 $_{9}$	969	5599
Hydroxymethoxyflavone ME		2600*	282 $_{100}$	267 $_{15}$	150 $_{15}$	132 $_{64}$	89 $_{14}$	792	5600
4-Hydroxy-N,N-diallyl-tryptamine		2245	256 $_{100}$	160 $_{2}$	146 $_{1}$	117 $_{3}$	110 $_{100}$	639	9395
4-Hydroxy-N,N-diallyl-tryptamine 2TMS		2355	400 $_{2}$	359 $_{6}$	304 $_{5}$	290 $_{26}$	110 $_{100}$	1446	9398
4-Hydroxy-N,N-diallyl-tryptamine isomer-1 HFB		2230	452 $_{2}$	425 $_{4}$	356 $_{28}$	342 $_{62}$	110 $_{100}$	1584	9403
4-Hydroxy-N,N-diallyl-tryptamine isomer-1 PFP		2395	402 $_{31}$	292 $_{52}$	186 $_{36}$	145 $_{64}$	110 $_{100}$	1452	9401
4-Hydroxy-N,N-diallyl-tryptamine isomer-1 TFA		2410	352 $_{18}$	256 $_{40}$	172 $_{38}$	146 $_{42}$	110 $_{100}$	1220	9399
4-Hydroxy-N,N-diallyl-tryptamine isomer-2 HFB		2450	452 $_{3}$	425 $_{3}$	356 $_{15}$	342 $_{47}$	110 $_{100}$	1584	9404
4-Hydroxy-N,N-diallyl-tryptamine isomer-2 PFP		2440	402 $_{8}$	292 $_{58}$	186 $_{33}$	145 $_{52}$	110 $_{100}$	1452	9402
4-Hydroxy-N,N-diallyl-tryptamine isomer-2 TFA		2460	352 $_{37}$	305 $_{15}$	242 $_{39}$	172 $_{55}$	110 $_{100}$	1220	9400
4-Hydroxy-N,N-diallyl-tryptamine ME		2300	270 $_{3}$	229 $_{5}$	160 $_{16}$	130 $_{23}$	110 $_{100}$	720	9397
4-Hydroxy-N,N-diallyl-tryptamine TFA		2125	323 $_{3}$	256 $_{3}$	242 $_{8}$	145 $_{4}$	110 $_{100}$	1220	10140
4-Hydroxy-N,N-diisopropyl-tryptamine		2280	260 $_{7}$	160 $_{4}$	146 $_{12}$	114 $_{100}$	72 $_{18}$	660	8830
4-Hydroxy-N,N-diisopropyl-tryptamine 2TMS		3025	404 $_{3}$	304 $_{5}$	290 $_{16}$	114 $_{100}$	72 $_{21}$	1460	9481
4-Hydroxy-N,N-dimethyltryptamine		1995	204 $_{15}$	159 $_{3}$	146 $_{5}$	130 $_{3}$	58 $_{100}$	410	2470
5-Hydroxy-N,N-dimethyl-tryptamine		2150	204 $_{3}$	160 $_{1}$	146 $_{5}$	117 $_{2}$	58 $_{100}$	410	8858
4-Hydroxy-N-methyl-N-ethyltryptamine		2100	218 $_{24}$	159 $_{4}$	146 $_{12}$	117 $_{5}$	72 $_{100}$	460	9555
4-Hydroxy-N-methyl-N-ethyltryptamine 2TMS		2060	362 $_{15}$	347 $_{6}$	290 $_{24}$	216 $_{4}$	72 $_{100}$	1276	9560
4-Hydroxy-N-methyl-N-ethyltryptamine AC		2190	260 $_{4}$	160 $_{5}$	146 $_{10}$	117 $_{5}$	72 $_{100}$	659	9556
4-Hydroxy-N-methyl-N-ethyltryptamine HFB		2225	414 $_{5}$	356 $_{5}$	342 $_{25}$	145 $_{33}$	72 $_{100}$	1489	9559
4-Hydroxy-N-methyl-N-ethyltryptamine PFP		2225	364 $_{2}$	306 $_{4}$	292 $_{11}$	145 $_{22}$	72 $_{100}$	1281	9558
4-Hydroxy-N-methyl-N-ethyltryptamine TFA		2185	314 $_{10}$	267 $_{4}$	242 $_{6}$	186 $_{12}$	72 $_{100}$	994	9557
4-Hydroxy-N-methyl-N-isopropyl-tryptamine		2150	232 $_{18}$	160 $_{5}$	146 $_{15}$	130 $_{6}$	86 $_{100}$	519	8828
4-Hydroxy-N-methyl-N-isopropyl-tryptamine 2TMS		3025	205 $_{1}$	147 $_{1}$	103 $_{4}$	86 $_{100}$	73 $_{26}$	1340	9483
4-Hydroxy-N-methyl-N-isopropyl-tryptamine AC		2240	274 $_{1}$	160 $_{4}$	146 $_{7}$	130 $_{3}$	86 $_{100}$	741	9482
Hydroxypethidine AC	U+UHYAC	2205	305 $_{9}$	230 $_{6}$	188 $_{10}$	71 $_{100}$		938	1195
3-Hydroxyphenylacetic acid 2TMS		1695*	296 $_{20}$	281 $_{19}$	164 $_{50}$	147 $_{39}$	73 $_{100}$	886	6010
4-Hydroxyphenylacetic acid	U	1565*	152 $_{26}$	107 $_{100}$	77 $_{18}$			293	818
4-Hydroxyphenylacetic acid 2ME	UME	1420*	180 $_{17}$	121 $_{100}$	91 $_{9}$	77 $_{10}$		346	4228
4-Hydroxyphenylacetic acid 2PFP		1340*	430 $_{2}$	253 $_{100}$	225 $_{13}$	119 $_{45}$	69 $_{52}$	1532	5675
4-Hydroxyphenylacetic acid 2TMS		1675*	296 $_{16}$	281 $_{18}$	252 $_{18}$	179 $_{35}$	73 $_{100}$	886	5821
4-Hydroxyphenylacetic acid AC		1565*	194 $_{4}$	152 $_{46}$	107 $_{100}$	77 $_{18}$		381	5819
4-Hydroxyphenylacetic acid HFB		1495*	348 $_{40}$	303 $_{100}$	275 $_{35}$	169 $_{30}$	69 $_{70}$	1197	5956
4-Hydroxyphenylacetic acid ME	UME	1570*	166 $_{18}$	121 $_{3}$	107 $_{100}$	77 $_{19}$	59 $_{3}$	317	4224
4-Hydroxyphenylacetic acid MEAC		1550*	208 $_{3}$	166 $_{37}$	149 $_{3}$	107 $_{100}$	77 $_{12}$	423	5820
4-Hydroxyphenylacetic acid MEHFB		1405*	362 $_{36}$	303 $_{100}$	275 $_{33}$	78 $_{49}$	59 $_{56}$	1272	5957
4-Hydroxyphenylacetic acid MEPFP		1220*	312 $_{34}$	253 $_{100}$	225 $_{28}$	119 $_{39}$	78 $_{43}$	979	5955
4-Hydroxyphenylacetic acid METFA		1120*	262 $_{33}$	203 $_{100}$	175 $_{21}$	69 $_{56}$	59 $_{35}$	668	5750
4-Hydroxyphenylacetic acid METMS		1485*	238 $_{48}$	223 $_{17}$	179 $_{100}$	163 $_{68}$	73 $_{78}$	548	6018
4-Hydroxyphenylacetic acid TFA		1450*	248 $_{36}$	203 $_{100}$	175 $_{29}$	77 $_{37}$	69 $_{57}$	589	5954
Hydroxyprogesterone -H2O		2650*	312 $_{84}$	297 $_{35}$	269 $_{100}$	227 $_{34}$	91 $_{39}$	984	5182
Hydroxyproline ME2AC		1690	229 $_{1}$	198	169 $_{9}$	110 $_{3}$	68 $_{100}$	506	2709
Hydroxyproline MEAC		1635	187 $_{7}$	128 $_{54}$	86 $_{100}$	68 $_{28}$		363	2283
Hydroxyquinoline		1550	145 $_{100}$	117 $_{83}$	90 $_{24}$	63 $_{16}$		284	9574
Hydroxyquinoline TMS		1570	217 $_{2}$	202 $_{100}$	172 $_{39}$	128 $_{6}$	77 $_{6}$	454	9650
2-Hydroxyquinoxaline		2020	146 $_{91}$	118 $_{91}$	91 $_{42}$	76 $_{3}$	64 $_{100}$	285	7413
2-Hydroxyquinoxaline ME		1750	160 $_{100}$	131 $_{71}$	104 $_{24}$	90 $_{11}$	77 $_{17}$	302	7414
3-Hydroxytyramine 3AC	U+UHYAC	2150	279 $_{3}$	237 $_{22}$	220 $_{21}$	178 $_{30}$	136 $_{100}$	773	5284
3-Hydroxytyramine 4AC		2245	321 $_{1}$	220 $_{33}$	178 $_{36}$	136 $_{100}$	123 $_{32}$	1038	5285
Hydroxyzine	G P-I U	2900	374 $_{13}$	299 $_{17}$	201 $_{100}$	165 $_{32}$		1330	820
Hydroxyzine AC	U+UHYAC	3000	416 $_{5}$	299 $_{16}$	201 $_{100}$	165 $_{49}$	87 $_{19}$	1497	1463
Hydroxyzine artifact-1	G U+UHYAC	1600*	202 $_{30}$	167 $_{100}$	165 $_{52}$	152 $_{17}$	125 $_{7}$	403	2442
Hydroxyzine artifact-2		1900*	232 $_{60}$	201 $_{62}$	165 $_{64}$	105 $_{100}$	77 $_{54}$	516	1344
Hydroxyzine-M	UHYAC	2210	280 $_{100}$	201 $_{35}$	165 $_{57}$			779	770
Hydroxyzine-M (carbinol)	UHY	1750*	218 $_{17}$	183 $_{7}$	139 $_{39}$	105 $_{100}$	77 $_{87}$	457	2239
Hydroxyzine-M (carbinol) AC	U+UHYAC	1890*	260 $_{8}$	200 $_{40}$	165 $_{100}$	139 $_{10}$	77 $_{29}$	656	1270
Hydroxyzine-M (Cl-benzophenone)	U+UHYAC	1850*	216 $_{43}$	139 $_{58}$	105 $_{100}$	77 $_{44}$		451	1343
Hydroxyzine-M (HO-Cl-benzophenone)	UHY	2300*	232 $_{36}$	197 $_{7}$	139 $_{23}$	121 $_{100}$	111 $_{23}$	515	2240
Hydroxyzine-M (HO-Cl-BPH) isomer-1 AC	UHYAC	2200*	274 $_{18}$	232 $_{75}$	139 $_{100}$	121 $_{44}$	111 $_{51}$	737	2229
Hydroxyzine-M (HO-Cl-BPH) isomer-2 AC	U+UHYAC	2230*	274 $_{7}$	232 $_{43}$	139 $_{25}$	121 $_{100}$	111 $_{27}$	737	2230
Hydroxyzine-M (HOOC-) ME	G PME UME U+UHYAC	2910	402 $_{10}$	229 $_{6}$	201 $_{100}$	165 $_{46}$	146 $_{23}$	1453	4323
Hydroxyzine-M (N-dealkyl-)	UHY	2520	286 $_{13}$	241 $_{48}$	201 $_{50}$	165 $_{65}$	56 $_{100}$	815	2241

Hydroxyzine-M (N-dealkyl-) AC **Table 8.1:** Compounds in order of names

Name	Detected	RI	Typical ions and intensities					Page	Entry
Hydroxyzine-M (N-dealkyl-) AC	U+UHYAC	2620	328 $_7$	242 $_{19}$	201 $_{48}$	165 $_{66}$	85 $_{100}$	1081	1271
Hydroxyzine-M AC	U+UHYAC	2380	299 $_4$	285 $_{61}$	226 $_{100}$	191 $_{58}$	84 $_{45}$	899	821
Hydroxyzine-M/artifact	P-I U+UHYAC UME	2220	300 $_{17}$	228 $_{38}$	165 $_{52}$	99 $_{100}$	56 $_{62}$	908	670
Hydroxyzine-M/artifact 2AC		2005	258 $_2$	199 $_6$	141 $_{100}$	112 $_9$	99 $_{20}$	647	2443
Hydroxyzine-M/artifact HYAC	U+UHYAC	2935	280 $_4$	201 $_{100}$	165 $_{26}$			780	1272
Hymecromone	G UHY	2015*	176 $_{74}$	148 $_{100}$	147 $_{71}$	120 $_{12}$	91 $_{27}$	332	2571
Hymecromone AC	U+UHYAC	2005*	218 $_{11}$	176 $_{92}$	148 $_{100}$	120 $_8$	91 $_{29}$	457	2572
Hymexazol		1300	100 $_5$	99 $_{100}$	71 $_{18}$	54 $_{14}$		257	3645
Hyoscyamine	P G U	2215	289 $_9$	272 $_1$	140 $_5$	124 $_{100}$	94 $_6$	837	69
Hyoscyamine AC	U+UHYAC	2275	331 $_4$	140 $_8$	124 $_{100}$	94 $_{22}$	82 $_{34}$	1102	71
Hyoscyamine -H2O	P G UHY U+UHYAC	2085	271 $_{33}$	140 $_8$	124 $_{100}$	96 $_{21}$	82 $_{13}$	726	70
Hyoscyamine HFB		2060	485 $_1$	271 $_{23}$	140 $_{10}$	124 $_{100}$	96 $_{37}$	1641	8125
Hyoscyamine PFP		2050	435 $_1$	271 $_{26}$	140 $_{14}$	124 $_{100}$	96 $_{39}$	1548	8124
Hyoscyamine TFA		2070	385 $_1$	271 $_{27}$	140 $_{15}$	124 $_{100}$	96 $_{38}$	1381	8123
Hyoscyamine TMS		2295	361 $_5$	140 $_5$	124 $_{100}$	82 $_{20}$	73 $_{50}$	1270	4526
5-IAI		1810	259 $_{100}$	242 $_{18}$	130 $_{13}$	115 $_{26}$	105 $_{15}$	649	8702
5-IAI formyl artifact		1825	271 $_{68}$	259 $_{26}$	242 $_{56}$	231 $_8$	115 $_{100}$	721	8703
Ibogaine		2870	310 $_{62}$	225 $_{51}$	149 $_{34}$	136 $_{100}$	122 $_{40}$	973	8874
Ibuprofen	G P U+UHYAC	1615*	206 $_{44}$	163 $_{92}$	161 $_{100}$	119 $_{44}$	91 $_{75}$	417	1941
Ibuprofen ammonia artifact	U+UHYAC	1630*	205 $_{35}$	161 $_{100}$	119 $_{39}$	105 $_{27}$	91 $_{33}$	414	7881
Ibuprofen ME	PME UME UHYME	1505*	220 $_{25}$	177 $_{28}$	161 $_{100}$	119 $_{18}$	91 $_{18}$	470	1942
Ibuprofen TMS		1665*	278 $_5$	263 $_{17}$	160 $_{31}$	117 $_{21}$	73 $_{100}$	770	4554
Ibuprofenamide	U+UHYAC	1630*	205 $_{35}$	161 $_{100}$	119 $_{39}$	105 $_{27}$	91 $_{33}$	414	7881
Ibuprofen-M (HO-) -H2O	U+UHYAC	1700*	204 $_{43}$	159 $_{100}$	128 $_{13}$	117 $_{30}$	91 $_{14}$	410	3382
Ibuprofen-M (HO-) -H2O ME	UHYME	1585*	218 $_{23}$	159 $_{100}$	128 $_{11}$	117 $_{25}$	91 $_{11}$	459	3380
Ibuprofen-M (HO-) isomer-1 ME	UME	1680*	236 $_1$	178 $_{52}$	119 $_{100}$	118 $_{99}$	91 $_{72}$	540	3381
Ibuprofen-M (HO-) isomer-2 ME	UME	1770*	236 $_2$	193 $_{100}$	133 $_{32}$	105 $_{66}$	77 $_{16}$	541	6386
Ibuprofen-M (HO-) isomer-3 ME	PME UME	1830*	236 $_{13}$	205 $_6$	177 $_{100}$	159 $_{33}$	117 $_{54}$	540	3383
Ibuprofen-M (HO-) isomer-4 ME	UME	1925*	236 $_{20}$	161 $_{12}$	118 $_{100}$	91 $_9$	59 $_{24}$	541	6387
Ibuprofen-M (HO-) MEAC	UHYAC	1880*	278 $_1$	218 $_{41}$	177 $_{12}$	159 $_{100}$	117 $_{30}$	769	3385
Ibuprofen-M (HO-HOOC-) -H2O 2ME	UME	1900*	262 $_{17}$	203 $_{100}$	157 $_{15}$	143 $_{43}$	128 $_{59}$	668	3386
Ibuprofen-M (HOOC-) 2ME	UME	1810*	264 $_{24}$	205 $_{100}$	177 $_{65}$	145 $_{95}$	117 $_{42}$	682	3384
Idobutal	P G U	1700	181 $_{14}$	167 $_{100}$	124 $_{36}$			488	1036
Idobutal 2ME		1610	223 $_6$	195 $_{100}$	181 $_{21}$	169 $_{20}$	138 $_{41}$	620	1037
Imazalil		2140	296 $_{10}$	240 $_{12}$	215 $_{100}$	173 $_{74}$	81 $_{32}$	883	2054
Imidapril artifact		2000	234 $_{10}$	220 $_{100}$	160 $_{13}$	117 $_{27}$	91 $_{79}$	525	6280
Imidapril ME		2700	419 $_2$	346 $_{28}$	234 $_{100}$	159 $_{21}$	91 $_{41}$	1506	6279
Imidapril TMS		2700	477 $_2$	404 $_{31}$	234 $_{100}$	160 $_{32}$	91 $_{70}$	1631	6281
Imidaprilate 2ME		2695	405 $_2$	346 $_{14}$	220 $_{100}$	159 $_{20}$	91 $_{56}$	1463	6282
Imidaprilate 2TMS		2770	521 $_1$	506 $_3$	404 $_{72}$	278 $_{100}$	160 $_{37}$	1677	6284
Imidaprilate 3ME		2710	419 $_1$	360 $_{15}$	234 $_{100}$	159 $_{15}$	91 $_{33}$	1506	6283
Imidapril-M (deethyl-) 2ME		2695	405 $_2$	346 $_{14}$	220 $_{100}$	159 $_{20}$	91 $_{56}$	1463	6282
Imidapril-M (deethyl-) 2TMS		2770	521 $_1$	506 $_3$	404 $_{72}$	278 $_{100}$	160 $_{37}$	1677	6284
Imidapril-M (deethyl-) 3ME		2710	419 $_1$	360 $_{15}$	234 $_{100}$	159 $_{15}$	91 $_{33}$	1506	6283
2,2'-Iminodibenzyl	U U+UHYAC	1930	195 $_{100}$	180 $_{40}$	167 $_9$	96 $_{33}$	83 $_{22}$	386	308
2,2'-Iminodibenzyl ME		1915	209 $_{70}$	194 $_{100}$	178 $_{13}$	165 $_{11}$		429	6352
Imipramine	P-I G U+UHYAC	2215	280 $_{11}$	234 $_{99}$	193 $_{24}$	85 $_{65}$	58 $_{100}$	784	342
Imipramine-M (bis-nor-) AC	UHYAC	2640	294 $_{23}$	208 $_{100}$	193 $_{43}$	152 $_6$	100 $_{17}$	871	3313
Imipramine-M (bis-nor-HO-) 2AC	UHYAC	2980	352 $_{60}$	266 $_{88}$	224 $_{100}$	180 $_{15}$	100 $_{48}$	1223	3314
Imipramine-M (di-HO-ring)	UHY	2600	227 $_{100}$	196 $_7$				499	2296
Imipramine-M (di-HO-ring) 2AC	U+UHYAC	2750	311 $_{28}$	269 $_{23}$	227 $_{100}$	196 $_7$		975	2292
Imipramine-M (HO-)	UHY	2565	296 $_{34}$	251 $_{44}$	85 $_{40}$	58 $_{100}$		887	528
Imipramine-M (HO-) AC	U+UHYAC	2610	338 $_2$	251 $_{15}$	211 $_{18}$	85 $_{30}$	58 $_{100}$	1146	343
Imipramine-M (HO-) ME		2480	310 $_{16}$	265 $_{13}$	225 $_8$	85 $_8$	58 $_{100}$	973	529
Imipramine-M (HO-methoxy-ring)	UHY	2390	241 $_{100}$	226 $_{17}$	210 $_{12}$	180 $_{14}$		560	2315
Imipramine-M (HO-methoxy-ring) AC	U+UHYAC	2370	283 $_{10}$	241 $_{100}$	226 $_{17}$	210 $_{12}$	180 $_{14}$	796	2867
Imipramine-M (HO-ring)	UHY	2240	211 $_{100}$	196 $_{15}$	180 $_{10}$	152 $_6$		436	2295
Imipramine-M (HO-ring) AC	U+UHYAC	2535	253 $_{26}$	211 $_{100}$	196 $_{19}$	180 $_{11}$	152 $_4$	621	1218
Imipramine-M (nor-)	UHY	2225	266 $_{28}$	235 $_{61}$	208 $_{61}$	195 $_{100}$	71 $_{59}$	698	324
Imipramine-M (nor-) AC	PAC U+UHYAC	2670	308 $_{40}$	208 $_{100}$	193 $_{55}$	114 $_{62}$		959	325
Imipramine-M (nor-) HFB		2450	462 $_{23}$	268 $_{13}$	240 $_{20}$	208 $_{100}$	193 $_{54}$	1604	7706
Imipramine-M (nor-) PFP		2450	412 $_{38}$	234 $_{16}$	218 $_{19}$	208 $_{100}$	193 $_{47}$	1485	7667
Imipramine-M (nor-) TFA		2430	362 $_{14}$	208 $_{100}$	193 $_{51}$	140 $_{17}$	69 $_{21}$	1273	7786
Imipramine-M (nor-) TMS		2470	238 $_7$	235 $_{89}$	143 $_{41}$	116 $_{72}$	73 $_{100}$	1146	5461
Imipramine-M (nor-di-HO-) 3AC	UHYAC	3380	424 $_{44}$	324 $_{35}$	282 $_{34}$	240 $_{27}$	114 $_{100}$	1517	3315
Imipramine-M (nor-HO-) 2AC	U+UHYAC	3065	366 $_{27}$	266 $_{39}$	114 $_{100}$			1293	1175
Imipramine-M (ring)	U U+UHYAC	1930	195 $_{100}$	180 $_{40}$	167 $_9$	96 $_{33}$	83 $_{22}$	386	308
Imipramine-M (ring) ME		1915	209 $_{70}$	194 $_{100}$	178 $_{13}$	165 $_{11}$		429	6352
Impurity		1730*	249 $_1$	180 $_{12}$	123 $_{27}$	97 $_{100}$	57 $_{92}$	593	116
Impurity		3580	441 $_{100}$	385 $_1$	308 $_6$	147 $_{14}$	57 $_{91}$	1560	3573
Impurity		3420	662 $_6$	648 $_6$	316 $_{38}$	191 $_{38}$	57 $_{100}$	1725	9711
Impurity		3160	646 $_1$	441 $_{19}$	191 $_{21}$	147 $_{37}$	57 $_{100}$	1723	9719
Impurity		2430	375 $_{38}$	360 $_{100}$	217 $_3$	203 $_7$	128 $_4$	1332	9818
Impurity AC	U+UHYAC	1430*	195 $_1$	152 $_6$	122 $_{20}$	92 $_6$	80 $_{100}$	385	2358

Table 8.1: Compounds in order of names   **Impurity AC**

Name	Detected	RI	Typical ions and intensities					Page	Entry
Impurity AC		1625*	181 3	122 18	92 6	80 100		347	2359
Impurity AC	UHYAC	1800*	194 1	179 2	117 1	87 100	58 4	381	2495
Impurity AC	UHYAC	2095*	240 1	179	131 1	87 100	73 3	554	2496
Impurity AC	UHYAC	2340*	219 1	175 1	131 4	87 100	73 8	461	2497
Impurity AC	UHYAC	2570*	219 1	175 3	131 6	87 100	73 13	461	2498
Impurity AC	UHYAC	2780*	219 1	175 3	131 7	87 100	73 13	462	2499
Impurity AC	UHYAC	2950*	219 1	175 2	131 5	87 100	73 10	462	2500
Impurity AC	UHYAC	3020*	219 1	175 2	131 3	87 100	73 6	461	2501
Impurity TMS		2555	484 51	469 8	427 39	233 26	73 100	1639	4613
25I-NB2OMe AC		3040	469 6	290 100	192 8	192 8	121 84	1616	9391
25I-NB2OMe-M (HO-) 2AC		3010	527 1	290 42	208 54	179 100	107 87	1682	9394
25I-NB2OMe-M (O,O-bis-demethyl-) 3AC		3090	525 3	483 4	318 20	276 43	178 100	1680	9393
25I-NB2OMe-M (O-demethyl-) 2AC		3040	497 10	318 26	276 100	192 78		1656	9392
25I-NB3B		2725	475 2	442 4	278 85	198 70	169 100	1627	9730
25I-NB3B AC		2900	517 3	290 100	277 9	198 29	169 25	1674	9733
25I-NB3B dehydro artifact		2700	473 4	442 36	277 84	196 100	169 96	1623	9731
25I-NB3B HFB		2620	290 100	277 64	247 25	169 44		1726	9736
25I-NB3B ME		2615	489 1	277 5	212 100	169 58	133 6	1647	9732
25I-NB3B PFP		2615	290 100	277 62	247 24	169 43		1719	9735
25I-NB3B TFA		2660	290 100	277 56	247 21	169 33		1703	9734
25I-NB3B TMS		2780	363 3	278 26	270 100	198 18	169 31	1694	9762
25I-NB3OMe		2700	427 2	394 10	278 26	150 70	121 100	1524	9747
25I-NB3OMe AC		2915	469 7	290 100	192 18	150 43	121 67	1616	9750
25I-NB3OMe dehydro artifact		2685	425 5	394 58	277 12	247 7	148 100	1518	9748
25I-NB3OMe HFB		2655	623 10	290 100	277 55	247 22	121 65	1719	9753
25I-NB3OMe ME		2640	441 1	277 5	247 5	164 100	121 79	1560	9749
25I-NB3OMe PFP		2650	573 11	290 100	277 51	247 21	121 60	1704	9752
25I-NB3OMe TFA		2695	523 6	290 100	277 43	247 18	121 65	1679	9751
25I-NB3OMe TMS		2745	484 2	394 3	278 17	222 100	121 71	1658	9763
25I-NB4B		2760	475 2	444 3	278 86	198 91	169 100	1627	9737
25I-NB4B AC		3000	517 4	290 100	275 11	198 33	169 41	1675	9742
25I-NB4B dehydro artifact		2735	473 3	442 18	277 83	196 73	169 100	1623	9738
25I-NB4B HFB		2720	290 100	277 73	247 34	169 53		1726	9745
25I-NB4B ME		2700	489 1	277 10	212 100	169 72	133 4	1647	9746
25I-NB4B PFP		2725	290 100	277 67	247 25	169 39		1719	9744
25I-NB4B TFA		3040	290 100	277 69	247 32	169 41		1703	9743
25I-NB4B TMS		2800	532 1	442 1	363 4	270 100	169 57	1694	9765
25I-NB4OMe		2725	427 1	394 15	278 9	150 18	121 100	1524	9754
25I-NB4OMe AC		2975	469 15	290 100	277 8	247 7	121 81	1616	9756
25I-NB4OMe HFB		2685	623 8	290 100	277 42	247 13	121 51	1720	9759
25I-NB4OMe ME		2660	277 2	247 2	164 38	121 100		1560	9755
25I-NB4OMe PFP		2685	573 9	290 100	277 30	247 10	121 71	1704	9758
25I-NB4OMe TFA		2645	523 11	290 100	277 23	247 8	121 50	1679	9757
25I-NB4OMe TMS		2745	484 1	394 3	278 13	222 64	121 100	1658	9764
25I-NBOMe AC		3040	469 6	290 100	192 8	192 8	121 84	1616	9391
25I-NBOMe-M (deamino-HOOC-2C-I)	U+UHYAC	2115	336 100	321 12	289 8	277 57	247 14	1129	6982
25I-NBOMe-M (deamino-HO-O-demethyl-2C-I) isomer-1 2AC	U+UHYAC	2240	378 2	336 16	276 100	261 10	134 16	1346	6970
25I-NBOMe-M (deamino-HO-O-demethyl-2C-I) isomer-2 2AC	UGLUCAC	2275	378 8	336 33	276 100	261 34	150 24	1345	6971
25I-NBOMe-M (HO-) 2AC		3010	527 1	290 42	208 54	179 100	107 87	1682	9394
25I-NBOMe-M (O,O-bis-demethyl-) 3AC		3090	525 3	483 4	318 20	276 43	178 100	1680	9393
25I-NBOMe-M (O-demethyl- N-acetyl-2C-I)	UGLUC	2370	335 15	276 100	263 11	233 10	134 14	1122	6963
25I-NBOMe-M (O-demethyl-) 2AC		3040	497 10	318 26	276 100	192 78		1656	9392
Indacaterol artifact (-NH2) formyl artifact		1530	201 58	172 50	157 72	128 72	115 70	402	8459
Indanazoline AC		2415	243 49	200 100	130 16	115 45	86 33	568	2800
Indapamide -2H 3ME		2940	405 88	298 100	246 5	130 41	77 10	1461	3115
Indapamide 3ME		3035	407 27	246 1	161 100	132 35	91 7	1467	3114
Indapamide artifact (ME)		2215	234 44	199 11	127 100	99 46	90 39	525	3116
Indapamide-M/artifact (H2N-)		1100	147 30	132 100	117 44	91 12	77 8	286	3117
Indapamide-M/artifact (HOOC-) 3ME		2130	277 71	169 100	138 61	110 67	75 93	757	3118
Indeloxazine		2085	231 63	132 47	115 38	100 71	56 100	514	6109
Indeloxazine AC		2400	273 48	142 44	132 37	100 100	86 52	734	6111
Indeloxazine ME		2030	245 38	131 9	114 100	84 28	70 46	575	6110
Indeloxazine TFA		2080	327 41	196 15	140 9	132 100	103 29	1074	7755
Indeloxazine TMS		2080	303 23	172 58	157 46	73 100	56 96	926	7754
Indene		1050*	116 100	115 82	89 12	63 10		263	2553
Indeno[1,2,3-c,d]pyrene		3075*	276 100	248 2	138 9	125 7		753	3706
Indinavir artifact -H2O HFB		1450	327 100	169 17	130 36	115 38	103 17	1072	7324
Indinavir artifact -H2O PFP		1450	277 100	130 52	119 40	103 54	77 36	758	7323
Indinavir artifact -H2O TFA		1485	227 100	146 11	130 4	103 54	69 51	498	7322
Indinavir artifact-1		1300	161 23	144 72	132 100	115 69	104 74	303	7317
Indinavir artifact-1 AC		1800	203 5	173 100	131 40	116 29	103 12	406	7935
Indinavir artifact-2 2AC	U+UHYAC	1780	173 40	148 10	131 100	118 11	103 26	521	7321
Indinavir artifact-2 AC		1850	173 75	131 100	104 31	91 14	77 17	371	7937
Indinavir artifact-2 -H2O AC		1385	173 81	144 23	131 40	104 100	77 39	328	7936

Indinavir artifact-2 isomer-2 2AC   Table 8.1: Compounds in order of names

Name	Detected	RI	Typical ions and intensities					Page	Entry
Indinavir artifact-2 isomer-2 2AC	U+UHYAC	2005	215 $_{37}$	173 $_{100}$	148 $_{21}$	131 $_{94}$		522	7938
Indinavir artifact-3	U+UHYAC	3435*	464 $_1$	364 $_{100}$	272 $_{29}$	174 $_{33}$	92 $_{73}$	1607	7316
Indinavir TFA		3170	459 $_{55}$	367 $_{93}$	284 $_{14}$	193 $_{19}$	91 $_{100}$	1730	7320
Indinavir-M artifact-1		1300	161 $_{23}$	144 $_{72}$	132 $_{100}$	115 $_{69}$	104 $_{74}$	303	7317
Indinavir-M artifact-1 AC		1800	203 $_5$	173 $_{100}$	131 $_{40}$	116 $_{29}$	103 $_{12}$	406	7935
Indinavir-M artifact-2 2AC	U+UHYAC	1780	173 $_{40}$	148 $_{10}$	131 $_{100}$	118 $_{11}$	103 $_{26}$	521	7321
Indinavir-M artifact-2 -H2O AC		1385	173 $_{81}$	144 $_{23}$	131 $_{40}$	104 $_{100}$	77 $_{39}$	328	7936
Indinavir-M artifact-2 isomer-2 2AC	U+UHYAC	2005	215 $_{37}$	173 $_{100}$	148 $_{21}$	131 $_{94}$		522	7938
Indole		1350	117 $_{100}$	90 $_{38}$	63 $_{18}$			264	1466
Indole acetic acid ME	UME	1900	189 $_{25}$	130 $_{100}$	103 $_5$	77 $_8$		367	1011
Indole propionic acid ME	P	1910	203 $_{18}$	143 $_{16}$	130 $_{100}$	115 $_{17}$	91 $_{12}$	406	6375
Indometacin	G P-I	2550	313 $_{30}$	139 $_{100}$	111 $_{24}$			1246	1038
Indometacin artifact 2ME	PME UME	2090	247 $_{38}$	188 $_{100}$	173 $_{10}$	145 $_8$		585	6294
Indometacin artifact ME	PME UME	2130	233 $_{38}$	174 $_{100}$				521	1230
Indometacin ET		2820	385 $_{40}$	312 $_{19}$	158 $_6$	139 $_{100}$	111 $_{20}$	1380	3168
Indometacin ME	P(ME) G(ME) U(ME)	2770	371 $_9$	312 $_6$	139 $_{100}$	111 $_{17}$		1316	1039
Indometacin TMS		2650	429 $_{23}$	370 $_4$	312 $_{17}$	139 $_{100}$	73 $_{22}$	1530	5462
Indometacin-M (chlorobenzoic acid)	G UHY UHYAC	1400*	156 $_{61}$	139 $_{100}$	111 $_{54}$	85 $_4$	75 $_{39}$	299	2726
Indometacin-M (HO-) 2ME	UME	2880	401 $_{27}$	262 $_4$	139 $_{100}$	111 $_{23}$		1448	6293
Inositol 6AC		2060*	373 $_2$	270 $_6$	210 $_{68}$	168 $_{100}$	126 $_{58}$	1539	5677
Instillagel (TM) ingredient	G U UHY UHYAC	1830	232 $_{14}$	218 $_{36}$	132 $_{19}$	85 $_{32}$	71 $_{100}$	519	1040
5-Iodo-2,3-dihydro-1H-inden-2-amine		1810	259 $_{100}$	242 $_{18}$	130 $_{13}$	115 $_{26}$	105 $_{15}$	649	8702
5-Iodo-2,3-dihydro-1H-inden-2-amine formyl artifact		1825	271 $_{68}$	259 $_{26}$	242 $_{56}$	231 $_8$	115 $_{100}$	721	8703
4-Iodo-2,5-dimethoxy-amfetamine		2025	321 $_2$	278 $_{100}$	263 $_{13}$	247 $_9$	77 $_{32}$	1036	7172
4-Iodo-2,5-dimethoxy-amfetamine 2AC		2360	405 $_{12}$	304 $_{100}$	277 $_{15}$	247 $_5$	86 $_{39}$	1461	7175
4-Iodo-2,5-dimethoxy-amfetamine 2ME		2305	349 $_{23}$	304 $_{100}$	277 $_{18}$	162 $_{11}$	72 $_{65}$	1203	7569
4-Iodo-2,5-dimethoxy-amfetamine 2TFA		1940	513 $_7$	304 $_{25}$	277 $_{100}$	247 $_{18}$	69 $_{53}$	1672	7177
4-Iodo-2,5-dimethoxy-amfetamine AC	U+UHYAC	2295	363 $_8$	304 $_{100}$	277 $_6$	247 $_6$	86 $_{43}$	1277	7174
4-Iodo-2,5-dimethoxy-amfetamine formyl artifact		1960	333 $_1$	302 $_{10}$	277 $_5$	247 $_2$	56 $_{100}$	1110	7173
4-Iodo-2,5-dimethoxy-amfetamine HFB		2070	517 $_{10}$	304 $_{58}$	277 $_{100}$	247 $_{36}$	69 $_{71}$	1675	7179
4-Iodo-2,5-dimethoxy-amfetamine PFP		2055	467 $_{31}$	304 $_{53}$	277 $_{100}$	247 $_{23}$	190 $_{20}$	1612	7178
4-Iodo-2,5-dimethoxy-amfetamine TFA		2075	417 $_{24}$	304 $_{32}$	277 $_{100}$	247 $_{12}$	140 $_{10}$	1498	7176
4-Iodo-2,5-dimethoxy-amfetamine-M (bis-O-demethyl-) 3AC		2180	410 $_4$	377 $_{38}$	360 $_{30}$	276 $_{70}$	86 $_{100}$	1504	7837
4-Iodo-2,5-dimethoxy-amfetamine-M (bis-O-demethyl-) artifact 2AC		2425	359 $_{20}$	317 $_{100}$	275 $_{68}$	260 $_{43}$	133 $_{28}$	1256	7182
4-Iodo-2,5-dimethoxy-amfetamine-M (O-demethyl-) isomer-1 2AC	U+UHYAC	2395	391 $_2$	349 $_9$	332 $_{36}$	290 $_{100}$	86 $_{55}$	1405	7180
4-Iodo-2,5-dimethoxy-amfetamine-M (O-demethyl-) isomer-2 2AC	U+UHYAC	2410	391 $_3$	349 $_{40}$	332 $_{10}$	290 $_{100}$	86 $_{30}$	1405	7181
Iodofenphos		2150*	412 $_2$	377 $_{100}$	362 $_5$	125 $_{37}$	79 $_{24}$	1485	3448
Ionol		1515*	220 $_{16}$	205 $_{100}$	177 $_{15}$	145 $_{16}$	57 $_{22}$	471	1041
Ionol-4		1710*	250 $_{26}$	235 $_{100}$	219 $_{20}$	193 $_{13}$	91 $_{11}$	607	6367
Ionol-acetamide		2070	277 $_{49}$	262 $_{40}$	220 $_{18}$	203 $_{100}$	178 $_{45}$	765	5751
Ioxynil ME		1885	385 $_{100}$	370 $_{43}$	243 $_{23}$	127 $_7$	88 $_{26}$	1379	4145
IPCC		<1000	166 $_2$	151 $_{61}$	123 $_{61}$	81 $_{46}$	54 $_{51}$	318	3584
IPCC		<1000	166 $_1$	151 $_{62}$	123 $_{38}$	54 $_{31}$	44 $_{100}$	318	5536
IPCC -HCN		<1000	139 $_{18}$	124 $_{59}$	96 $_{32}$	82 $_{26}$	54 $_{100}$	279	3586
IPCC -HCN		<1000	139 $_{18}$	124 $_{59}$	96 $_{32}$	54 $_{100}$	41 $_{83}$	279	5538
Irbesartan ME	U+UHYAC UME, PME	3500	442 $_2$	413 $_{12}$	400 $_{100}$	192 $_{30}$	165 $_{23}$	1566	5039
Irganox	G P U UHY UHYAC	3390*	530 $_{100}$	515 $_{51}$	219 $_{57}$	203 $_{22}$	57 $_{67}$	1684	4648
Irinotecan artifact (bipiperidine) AC		1945	210 $_{24}$	167 $_{32}$	124 $_{100}$	110 $_{52}$	84 $_{38}$	434	9420
Irinotecan artifact (bipiperidine) HFB		1770	364 $_{13}$	294 $_6$	167 $_{73}$	124 $_{100}$	110 $_{29}$	1281	9424
Irinotecan artifact (bipiperidine) PFP		1755	314 $_{42}$	285 $_{11}$	167 $_{100}$	124 $_{89}$	110 $_{36}$	995	9423
Irinotecan artifact (bipiperidine) TFA		1750	264 $_{14}$	167 $_{65}$	124 $_{100}$	110 $_{28}$	85 $_{35}$	682	9421
Irinotecan artifact (bipiperidine-COOH) (ME)		1930	226 $_{28}$	167 $_{12}$	124 $_{100}$	110 $_{90}$	84 $_{29}$	496	9419
Irinotecan impurity (piperidine) AC		1050	127 $_{51}$	99 $_{10}$	84 $_{100}$	70 $_{43}$	56 $_{58}$	269	9422
Isoaminile	U UHY UHYAC	1705	244 $_1$	229 $_3$	158 $_1$	115 $_7$	72 $_{100}$	572	4389
Isoaminile-M (nor-)	U UHY UHYAC	1725	229 $_2$	215 $_{11}$	188 $_{100}$	173 $_{27}$	91 $_7$	512	4390
Isoamyltiglate		1000*	155 $_1$	101 $_{57}$	83 $_{91}$	70 $_{90}$	55 $_{100}$	324	9175
Isobutylbenzene		1050*	134 $_{21}$	119 $_8$	105 $_{100}$	91 $_{13}$	77 $_{17}$	274	3789
Isocarbamide 2ME		1685	213 $_1$	170 $_{32}$	127 $_{100}$	86 $_{69}$	70 $_{54}$	443	4157
Isocitric acid 3ME		1495*	175 $_{33}$	143 $_{76}$	115 $_{100}$	83 $_{24}$	55 $_{65}$	526	6453
Isoconazole	U+UHYAC	3150	414 $_4$	333 $_{19}$	159 $_{100}$	123 $_6$	81 $_{14}$	1489	2055
Isofenphos		2005	345 $_2$	255 $_{18}$	213 $_{45}$	121 $_{34}$	58 $_{100}$	1184	3446
Isofenphos-M/artifact (HOOC-) ME		1980	317 $_2$	259 $_5$	227 $_{57}$	121 $_{30}$	58 $_{100}$	1010	3447
Isofentanyl	USPEAC	2665	336 $_2$	279 $_6$	146 $_{16}$	96 $_{94}$	91 $_{100}$	1133	8024
Isofentanyl artifact (depropionyl-)	U+UHYAC USPEAC	2330	280 $_2$	187 $_{16}$	172 $_{21}$	119 $_{76}$	91 $_{100}$	785	8026
Isofentanyl-M (alkyl-HO-) AC	USPEAC	2960	279 $_4$	188 $_7$	146 $_{13}$	96 $_{87}$	91 $_{100}$	1422	8027
Isofentanyl-M (alkyl-HO-) TFA	USPEAC	2725	448 $_1$	240 $_3$	188 $_7$	96 $_{100}$	91 $_{96}$	1577	8030
Isofentanyl-M (aryl-HO-) AC	USPEAC	3005	245 $_6$	204 $_5$	120 $_5$	107 $_{84}$	96 $_{100}$	1422	8028
Isofentanyl-M (nor-) AC	U+UHYAC USPEAC	2430	288 $_4$	245 $_{36}$	172 $_{60}$	132 $_{100}$	98 $_{63}$	830	8016
Isofentanyl-M (nor-) ME	USPEME	1955	260 $_3$	223 $_{12}$	203 $_{26}$	112 $_{56}$	96 $_{100}$	661	8021
Isofentanyl-M (nor-) TFA	USPETFA	2170	342 $_1$	269 $_6$	194 $_{28}$	150 $_{69}$	57 $_{100}$	1170	8023
Isofentanyl-M (nor-) TMS	USPETMS	2180	318 $_{15}$	294 $_{57}$	261 $_{32}$	154 $_{55}$	73 $_{100}$	1021	8031
Isofentanyl-M (nor-alkyl-HO-) 2TMS	USPEAC	2415	406 $_4$	391 $_7$	261 $_{73}$	169 $_{85}$	73 $_{100}$	1466	8032
Isofentanyl-M (nor-alkyl-HO-) isomer-1 2AC	USPEAC	2635	346 $_2$	172 $_{57}$	132 $_{69}$	98 $_{100}$		1190	8017
Isofentanyl-M (nor-alkyl-HO-) isomer-1 2TFA	USPETFA	2280	454 $_4$	262 $_{63}$	193 $_{88}$	132 $_{70}$	55 $_{100}$	1589	8029

Table 8.1: Compounds in order of names

Name	Detected	RI	Typical ions and intensities					Page	Entry
Isofentanyl-M (nor-alkyl-HO-) isomer-2 2AC	USPEAC	2730	346 $_2$	303 $_{29}$	172 $_{88}$	98 $_{81}$	55 $_{100}$	1190	8025
Iso-LSD TMS		3515	395 $_{90}$	293 $_{43}$	279 $_{24}$	253 $_{39}$	73 $_{100}$	1426	6222
Iso-Lysergide (iso-LSD) TMS		3515	395 $_{90}$	293 $_{43}$	279 $_{24}$	253 $_{39}$	73 $_{100}$	1426	6222
Isoniazid	P-I G U	1650	137 $_{38}$	122 $_5$	106 $_{77}$	78 $_{100}$	51 $_{73}$	276	1043
Isoniazid 2AC	U+UHYAC	1825	221 $_7$	179 $_{26}$	161 $_{44}$	137 $_{55}$	106 $_{100}$	472	1045
Isoniazid AC	U+UHYAC	1950	179 $_{96}$	137 $_{67}$	106 $_{100}$	78 $_{66}$	51 $_{57}$	340	1044
Isoniazid acetone derivate	U+UHYAC	1840	177 $_{11}$	162 $_{73}$	106 $_{75}$	78 $_{100}$		334	1046
Isoniazid artifact (HOOC-) TMS		1295	195 $_4$	180 $_{100}$	136 $_{36}$	106 $_{30}$	78 $_{37}$	386	4555
Isoniazid formyl artifact	P G U+UHYAC	1510	149 $_5$	122 $_{55}$	106 $_{76}$	78 $_{100}$	51 $_{69}$	287	4057
Isoniazid formyl artifact AC		1785	191 $_{17}$	149 $_{100}$	106 $_{51}$	78 $_{39}$	51 $_{30}$	371	4058
Isoniazid-M glycine conjugate	U	----	180 $_{100}$	165 $_{34}$	137 $_{39}$	106 $_{41}$	78 $_{46}$	345	1047
Isonicotinic acid TMS		1295	195 $_4$	180 $_{100}$	136 $_{36}$	106 $_{30}$	78 $_{37}$	386	4555
Isooctane		<1000*	99 $_6$	57 $_{100}$				263	2753
Isopaynantheine		3090	396 $_{36}$	381 $_{43}$	253 $_{23}$	214 $_{100}$	200 $_{64}$	1431	8053
Isopaynantheine TMS		3030	468 $_2$	339 $_9$	286 $_{65}$	272 $_{52}$	73 $_{100}$	1615	8058
Isopaynantheine-M (9-O-demethyl-) 2TMS	USPETMS	3035	526 $_{32}$	344 $_{100}$	329 $_{62}$	73 $_{50}$		1682	8038
Isopaynantheine-M (9-O-demethyl-16-COOH) 3TMS	USPETMS	3040	584 $_{10}$	344 $_{31}$	147 $_{35}$	73 $_{100}$		1709	8037
Isoprenaline 4AC		2460	365 $_6$	319 $_{24}$	277 $_{46}$	193 $_{50}$	84 $_{100}$	1351	1468
Isopropanol		<1000*	60 $_1$	45 $_{100}$				249	1546
N-Isopropyl-1,2-diphenylethane AC		2160	281 $_1$	180 $_{11}$	148 $_{100}$	106 $_{41}$		789	8431
N-Isopropyl-1,2-diphenylethylamine		1720	181 $_{13}$	165 $_8$	148 $_{100}$	132 $_8$	106 $_{57}$	553	8430
N-Isopropyl-1,2-diphenylethylamine HFB		1925	344 $_{38}$	302 $_{100}$	180 $_{43}$	165 $_{20}$	91 $_{61}$	1548	8433
N-Isopropyl-1,2-diphenylethylamine PFP		1910	294 $_{41}$	252 $_{100}$	180 $_{16}$	91 $_{15}$	79 $_{14}$	1381	8444
N-Isopropyl-1,2-diphenylethylamine TFA		1930	244 $_{62}$	202 $_{100}$	180 $_{22}$	165 $_8$	107 $_{11}$	1124	8432
N-Isopropyl-1,2-diphenylethylamine-M (deamino-HO-bis-HO-benzyl-) 3AC		2350*	355 $_2$	296 $_9$	208 $_{27}$	166 $_{100}$	123 $_{93}$	1242	8987
N-Isopropyl-1,2-diphenylethylamine-M (deamino-HO-phenyl-) AC		2230*	238 $_{28}$	196 $_{31}$	165 $_{20}$	150 $_{27}$	107 $_{100}$	895	8932
N-Isopropyl-1,2-diphenylethylamine-M (deamino-oxo-bis-HO-benzyl-) 2AC		2345*	207 $_{44}$	165 $_{80}$	123 $_{100}$	105 $_{81}$	77 $_{45}$	981	8933
N-Isopropyl-1,2-diphenylethylamine-M (deamino-oxo-bis-HO-benzyl-) AC		2360*	270 $_7$	228 $_{19}$	153 $_{21}$	123 $_{15}$	105 $_{100}$	716	8986
N-Isopropyl-1,2-diphenylethylamine-M (deamino-oxo-HO-benzyl-) isomer-1 AC		2080*	254 $_1$	212 $_4$	105 $_{100}$	77 $_{34}$		628	8934
N-Isopropyl-1,2-diphenylethylamine-M (deamino-oxo-HO-benzyl-) isomer-2 AC		2120*	254 $_2$	212 $_{12}$	105 $_{100}$	77 $_{31}$		628	8935
N-Isopropyl-1,2-diphenylethylamine-M (deamino-oxo-HO-methoxy-benzyl-) AC		2250*	284 $_1$	242 $_{22}$	226 $_{27}$	137 $_{100}$	105 $_{84}$	800	8985
N-Isopropyl-1,2-diphenylethylamine-M (di-HO-benzyl-) 3AC	U+UHYAC	2700	296 $_9$	254 $_{10}$	212 $_{12}$	148 $_{100}$	106 $_{30}$	1435	8657
N-Isopropyl-1,2-diphenylethylamine-M (HO-benzyl-) 2AC	U+UHYAC	2500	254 $_1$	238 $_{25}$	196 $_{29}$	190 $_{42}$	148 $_{100}$	1151	8658
N-Isopropyl-1,2-diphenylethylamine-M (nor-di-HO-benzyl-) 2AC	U+UHYAC	2560	254 $_2$	212 $_3$	178 $_7$	148 $_{62}$	106 $_{100}$	988	9120
N-Isopropyl-1,2-diphenylethylamine-M (nor-di-HO-benzyl-) isomer-1 3AC	U+UHYAC	2610	296 $_1$	254 $_{11}$	212 $_{10}$	148 $_{46}$	106 $_{100}$	1237	9121
N-Isopropyl-1,2-diphenylethylamine-M (nor-di-HO-benzyl-) isomer-2 3AC	U+UHYAC	2630	296 $_1$	254 $_3$	212 $_8$	148 $_{55}$	106 $_{100}$	1237	8826
N-Isopropyl-1,2-diphenylethylamine-M (nor-di-HO-methoxy-benzyl-) 2AC	U+UHYAC	2680	284 $_3$	242 $_6$	153 $_9$	148 $_{49}$	106 $_{100}$	1175	8827
N-Isopropyl-1,2-diphenylethylamine-M (nor-HO-alkyl-) 2AC	U+UHYAC	1740	206 $_{42}$	164 $_{90}$	122 $_{30}$	104 $_{100}$	91 $_{33}$	890	8659
N-Isopropyl-1,2-diphenylethylamine-M (nor-HO-benzyl-) isomer-1 2AC	U+UHYAC	2350	238 $_7$	196 $_{12}$	148 $_{47}$	106 $_{100}$	79 $_{11}$	890	8984
N-Isopropyl-1,2-diphenylethylamine-M (nor-HO-benzyl-) isomer-2 2AC	U+UHYAC	2380	238 $_{13}$	196 $_{14}$	148 $_{57}$	106 $_{100}$	79 $_{16}$	890	8654
N-Isopropyl-1,2-diphenylethylamine-M (nor-HO-methoxy-benzyl-) 2AC	U+UHYAC	2540	268 $_{15}$	236 $_{28}$	194 $_{39}$	152 $_{45}$	106 $_{100}$	1075	8656
N-Isopropyl-1,2-diphenylethylamine-M (nor-HO-methoxy-phenyl-) 2AC	U+UHYAC	2560	236 $_{29}$	194 $_{45}$	152 $_{73}$	137 $_{10}$	106 $_{100}$	1076	8825
N-Isopropyl-1,2-diphenylethylamine-M (nor-HO-phenyl-) 2AC	U+UHYAC	2500	296 $_1$	206 $_{98}$	164 $_{79}$	122 $_{89}$	91 $_{25}$	890	8655
N-Isopropyl-BDB		1720	206 $_2$	135 $_{10}$	100 $_{100}$	77 $_7$	58 $_{77}$	535	5419
N-Isopropyl-BDB AC		2095	206 $_1$	176 $_{36}$	142 $_{27}$	100 $_{100}$	58 $_{35}$	762	5509
N-Isopropyl-BDB TFA		1895	331 $_3$	196 $_{30}$	176 $_{38}$	154 $_{100}$	135 $_{22}$	1101	5510
Isopropylbenzene		<1000*	120 $_{30}$	105 $_{100}$	77 $_{16}$	65 $_8$	51 $_{11}$	265	3785
N-isopropyl-tryptamine		2185	202 $_2$	131 $_{49}$	72 $_{100}$			404	8868
N-isopropyl-tryptamine 2HFB		1835	339 $_{100}$	326 $_{46}$	268 $_{15}$	226 $_{73}$	129 $_{15}$	1712	9520
N-isopropyl-tryptamine 2PFP		1815	289 $_{100}$	276 $_{27}$	176 $_{74}$	119 $_{29}$		1652	9522
N-isopropyl-tryptamine 2TFA		1860	394 $_8$	239 $_{100}$	226 $_{40}$	168 $_{13}$	126 $_{41}$	1420	9518
N-isopropyl-tryptamine AC		2270	244 $_3$	143 $_{100}$	130 $_{50}$	115 $_5$	72 $_{55}$	572	9521
N-isopropyl-tryptamine HFB		2015	398 $_2$	226 $_9$	169 $_5$	143 $_{31}$	130 $_{100}$	1437	9519
N-isopropyl-tryptamine TMS		2200	202 $_{28}$	186 $_8$	144 $_{70}$	130 $_7$	73 $_{100}$	741	9517
Isoproturon ME		1685	220 $_8$	205 $_5$	148 $_8$	132 $_5$	72 $_{100}$	470	3968
Isopyrin	G	2045	245 $_{28}$	230 $_{12}$	137 $_{31}$	83 $_{25}$	56 $_{100}$	577	530
Isopyrin AC		2400	287 $_{35}$	244 $_{37}$	137 $_{31}$	56 $_{100}$		822	194
Isopyrin-M (nor-) 2AC	UHYAC	2365	315 $_5$	273 $_{45}$	231 $_{48}$	123 $_{100}$	70 $_{94}$	1000	195
Isopyrin-M (nor-HO-) -H2O 2AC	UHYAC	2160	313 $_3$	271 $_8$	229 $_{28}$	77 $_{16}$	56 $_{100}$	988	531
Isosteviol		2620*	318 $_{25}$	300 $_{21}$	121 $_{66}$	109 $_{71}$	55 $_{100}$	1022	3680
Isosteviol ME		2520*	332 $_{73}$	300 $_{71}$	273 $_{100}$	121 $_{81}$	55 $_{71}$	1109	3681
Isothipendyl	P-I G U+UHYAC	2245	285 $_1$	214 $_1$	200 $_2$	181 $_1$	72 $_{100}$	809	1467
Isothipendyl-M (bis-nor-)	UHY	2230	257 $_{75}$	214 $_{100}$	181 $_{24}$	58 $_{58}$		642	1666
Isothipendyl-M (bis-nor-) AC	UHYAC	2520	299 $_{41}$	213 $_{100}$	200 $_{25}$	181 $_{43}$	100 $_{10}$	901	1662
Isothipendyl-M (HO-)	UHY	2450	301 $_{22}$	218 $_{11}$	197 $_{10}$	72 $_{100}$		912	1665
Isothipendyl-M (HO-) AC	UHYAC	2640	343 $_2$	272 $_3$	229 $_3$	197 $_4$	72 $_{100}$	1174	1663
Isothipendyl-M (HO-ring)	U UHY	2800	216 $_{100}$	187 $_{68}$	168 $_{50}$	140 $_9$		451	2272
Isothipendyl-M (HO-ring) AC	U+UHYAC	2575	258 $_{30}$	216 $_{100}$	187 $_{13}$	183 $_{13}$	155 $_{10}$	645	2275
Isothipendyl-M (nor-)	UHY	2220	271 $_3$	214 $_{43}$	199 $_3$	181 $_6$	58 $_{100}$	724	1664
Isothipendyl-M (nor-) AC	U+UHYAC	2600	313 $_{16}$	213 $_{100}$	181 $_{48}$	114 $_{58}$	58 $_{89}$	987	1661
Isothipendyl-M (nor-HO-) 2AC	UHYAC	2940	371 $_{31}$	271 $_{45}$	229 $_{78}$	114 $_{100}$	58 $_{58}$	1318	2441
Isothipendyl-M (nor-sulfone) AC	UHYAC	2900	345 $_1$	272 $_{26}$	257 $_{15}$	100 $_{55}$	58 $_{100}$	1184	2687
Isothipendyl-M (nor-sulfoxide) AC	UHYAC	2880	329 $_3$	312 $_{26}$	213 $_{99}$	100 $_{28}$	58 $_{100}$	1087	2686
Isothipendyl-M (ring)	U+UHYAC	2045	200 $_{100}$	168 $_{36}$	156 $_{13}$			399	386

Isovanillic acid MEAC

Table 8.1: Compounds in order of names

Name	Detected	RI	Typical ions and intensities					Page	Entry
Isovanillic acid MEAC		1630*	224 $_6$	193 $_2$	182 $_{92}$	151 $_{100}$	123 $_{13}$	486	5329
Isoxaben		2910	332 $_1$	250	165 $_{100}$	150 $_8$	107 $_5$	1107	3885
Isradipine	UME	2680	371 $_9$	284 $_{14}$	252 $_{42}$	210 $_{100}$	150 $_{17}$	1318	4628
Isradipine ME	UME	2670	385 $_{38}$	326 $_{27}$	298 $_{100}$	268 $_{51}$	224 $_{70}$	1381	4852
Isradipine-M (dehydro-demethyl-HO-) -H2O	UME	2635	353 $_8$	311 $_{100}$	294 $_{49}$	267 $_{49}$	237 $_{61}$	1225	4869
Isradipine-M/artifact (dehydro-)	UME	2360	369 $_{21}$	327 $_{85}$	295 $_{100}$	265 $_{46}$	251 $_{61}$	1305	4865
Isradipine-M/artifact (dehydro-deisopropyl-) ME	UME	2270	341 $_{51}$	309 $_{100}$	294 $_{36}$	279 $_{34}$	264 $_{63}$	1160	4868
Isradipine-M/artifact (dehydro-deisopropyl-) TMS	UTMS	2395	399 $_{46}$	384 $_{100}$	309 $_{77}$	264 $_{18}$	164 $_{13}$	1440	5009
Isradipine-M/artifact (dehydro-demethyl-) TMS	UTMS	2535	427 $_{20}$	385 $_{28}$	370 $_{100}$	295 $_{62}$	251 $_{23}$	1525	5010
Isradipine-M/artifact (deisopropyl-) 2ME	UME	2655	357 $_{40}$	326 $_{14}$	298 $_{100}$	268 $_{52}$	238 $_{62}$	1248	4867
Isradipine-M/artifact (deisopropyl-) ME	UME	2610	343 $_{10}$	284 $_{18}$	254 $_{11}$	224 $_{100}$	192 $_9$	1174	4866
5-IT		1765	174 $_5$	131 $_{100}$	103 $_{10}$	77 $_{21}$	63 $_6$	329	9095
5-IT		1765	174 $_5$	131 $_{85}$	103 $_9$	77 $_{16}$	44 $_{100}$	329	9096
5-IT 2AC		2345	199 $_{100}$	157 $_{31}$	130 $_{54}$	103 $_{20}$	86 $_{29}$	647	9099
5-IT 2HFB		1880	353 $_{18}$	326 $_{32}$	240 $_{87}$	129 $_{100}$		1702	9107
5-IT 2ME		1850	202 $_1$	130 $_7$	115 $_2$	72 $_{100}$		404	9102
5-IT 2PFP		2230	466 $_4$	303 $_{59}$	276 $_{100}$	190 $_{51}$	184 $_{41}$	1609	9106
5-IT 2TFA		2255	366 $_1$	297 $_5$	253 $_{75}$	226 $_{100}$	129 $_{13}$	1289	9105
5-IT 2TMS		1950	303 $_1$	202 $_8$	116 $_{100}$	100 $_9$	73 $_{67}$	1021	9103
5-IT AC		2145	216 $_4$	157 $_{100}$	130 $_{88}$	103 $_{14}$	77 $_{25}$	453	9098
5-IT formyl artifact		1745	186 $_{16}$	130 $_{100}$	103 $_7$	77 $_{13}$	56 $_{41}$	361	9097
5-IT formyl artifact ME		1780	200 $_{14}$	144 $_{100}$	129 $_4$	115 $_4$	56 $_{11}$	400	9100
5-IT HFB		1970	370 $_1$	169 $_4$	157 $_{18}$	130 $_{100}$	104 $_1$	1310	9108
5-IT ME		1810	151 $_8$	130 $_{13}$	98 $_8$	70 $_{15}$	58 $_{100}$	366	9101
5-IT TMS		1645	246 $_2$	203 $_{100}$	188 $_{19}$	130 $_9$	73 $_{55}$	582	9104
5-IT-M (HO-) 2AC		2360	274 $_2$	215 $_{80}$	173 $_{100}$	146 $_{42}$	86 $_{49}$	740	9213
5-IT-M (HO-) AC		2340	232 $_4$	173 $_{100}$	146 $_{26}$	118 $_{16}$	86 $_{16}$	517	9212
6-IT		1765	174 $_5$	131 $_{100}$	103 $_{10}$	77 $_{19}$	63 $_5$	329	9109
6-IT		1765	174 $_3$	131 $_{75}$	103 $_8$	77 $_{16}$	44 $_{100}$	329	9110
6-IT 2HFB		2245	566 $_1$	353 $_{10}$	326 $_{19}$	184 $_{71}$	129 $_{100}$	1702	9116
6-IT 2PFP		2220	466 $_2$	347 $_7$	303 $_{92}$	276 $_{100}$	184 $_{28}$	1609	9115
6-IT 2TFA		2285	366 $_8$	297 $_6$	253 $_{56}$	226 $_{100}$	102 $_{25}$	1290	9114
6-IT AC		2140	216 $_7$	157 $_{100}$	130 $_{96}$	103 $_{13}$	86 $_9$	453	9112
6-IT formyl artifact		1740	186 $_{19}$	130 $_{100}$	103 $_8$	77 $_{17}$	56 $_{52}$	361	9111
6-IT HFB		1870	370 $_2$	240 $_3$	157 $_{20}$	130 $_{100}$	103 $_5$	1310	9117
6-IT TFA		1885	270 $_{13}$	157 $_{14}$	130 $_{100}$	103 $_9$	77 $_{19}$	717	9113
6-IT-M (di-HO-) 3AC		2440	332 $_6$	273 $_{100}$	231 $_{87}$	189 $_{62}$	86 $_{72}$	1106	9216
6-IT-M (HO-) 2AC		2310	274 $_2$	215 $_{38}$	173 $_{100}$	146 $_{29}$	86 $_{40}$	740	9215
6-IT-M (HO-) AC		2320	232 $_{10}$	173 $_{100}$	146 $_{25}$	117 $_6$	86 $_{28}$	518	9214
JWH-015		2995	327 $_{100}$	310 $_{30}$	270 $_{27}$	200 $_{21}$	127 $_{36}$	1076	8521
JWH-018		3240	341 $_{100}$	324 $_{43}$	284 $_{69}$	214 $_{76}$	127 $_{74}$	1164	7875
JWH-018-M/artifact (N-dealkyl-) AC		2720	313 $_{73}$	254 $_{62}$	241 $_{30}$	127 $_{100}$	89 $_{39}$	987	7876
JWH-018-M/artifact (N-dealkyl-HO-naphthalen-) 2AC		3065	371 $_{35}$	329 $_{46}$	286 $_{100}$	270 $_{34}$	144 $_{29}$	1317	8510
JWH-018-M/artifact (N-dealkyl-HO-naphthalen-) 2AC		3200	371 $_{20}$	329 $_{100}$	270 $_{84}$	160 $_{57}$	127 $_{40}$	1317	8511
JWH-019		3220	355 $_{100}$	338 $_{44}$	284 $_{67}$	228 $_{56}$	127 $_{68}$	1239	8522
JWH-073		3150	327 $_{100}$	310 $_{49}$	284 $_{65}$	200 $_{59}$	127 $_{56}$	1076	7874
JWH-073-M/artifact (N-dealkyl-) AC		2720	313 $_{73}$	254 $_{62}$	241 $_{30}$	127 $_{100}$	89 $_{39}$	987	7876
JWH-073-M/artifact (N-dealkyl-HO-indol-) 2AC		3200	371 $_{20}$	329 $_{100}$	270 $_{84}$	160 $_{57}$	127 $_{40}$	1317	8511
JWH-073-M/artifact (N-dealkyl-HO-naphthalen-) 2AC		3065	371 $_{35}$	329 $_{46}$	286 $_{100}$	270 $_{34}$	144 $_{29}$	1317	8510
JWH-081		3400	371 $_{100}$	354 $_{47}$	314 $_{54}$	214 $_{66}$	185 $_{64}$	1320	8526
JWH-122		3270	355 $_{100}$	338 $_{50}$	298 $_{59}$	214 $_{48}$	115 $_{47}$	1239	8523
JWH-122 (pentenyl analog)		3160	353 $_{77}$	298 $_{53}$	212 $_{46}$	169 $_{100}$	115 $_{50}$	1229	9618
JWH-20		3220	369 $_{100}$	352 $_{41}$	284 $_{74}$	242 $_{52}$	127 $_{77}$	1308	8536
JWH-200		3500	384 $_2$	155 $_3$	127 $_4$	100 $_{100}$	70 $_5$	1377	8527
JWH-203		2895	339 $_1$	214 $_{100}$	144 $_{35}$	116 $_{18}$	89 $_{21}$	1149	8530
JWH-210		3320	369 $_{100}$	352 $_{53}$	312 $_{51}$	214 $_{60}$	144 $_{46}$	1308	8524
JWH-210-M (5-HOOC-) ME		4015	413 $_{27}$	298 $_{100}$	152 $_{70}$	139 $_{73}$	128 $_{56}$	1488	10402
JWH-210-M (5-HOOC-) TMS		4260	471 $_{16}$	298 $_{41}$	152 $_{27}$	128 $_{43}$	75 $_{100}$	1621	10403
JWH-210-M (5-HO-pentyl-)		4050	385 $_{51}$	368 $_{28}$	152 $_{100}$	139 $_{99}$	128 $_{88}$	1382	10388
JWH-210-M (5-HO-pentyl-) AC		4200	427 $_{58}$	310 $_{47}$	254 $_{62}$	152 $_{90}$	61 $_{100}$	1526	10389
JWH-210-M (5-HO-pentyl-) HFB		3585	581 $_{73}$	426 $_{45}$	312 $_{85}$	183 $_{100}$	153 $_{92}$	1708	10393
JWH-210-M (5-HO-pentyl-) PFP		3570	531 $_{79}$	312 $_{62}$	254 $_{56}$	183 $_{64}$	119 $_{100}$	1685	10392
JWH-210-M (5-HO-pentyl-) TFA		3600	481 $_{74}$	254 $_{60}$	183 $_{56}$	153 $_{77}$	69 $_{100}$	1636	10391
JWH-210-M (5-HO-pentyl-) TMS		3960	457 $_{53}$	298 $_{52}$	254 $_{29}$	183 $_{58}$	75 $_{100}$	1596	10390
JWH-250		2910	335 $_4$	214 $_{100}$	144 $_{24}$	116 $_8$	91 $_{12}$	1126	8445
JWH-250 E/Z isomer-1 TMS		2655	407 $_{82}$	364 $_{14}$	300 $_{62}$	286 $_{100}$	73 $_{67}$	1469	9688
JWH-250 E/Z isomer-2 TMS		2820	407 $_{86}$	364 $_{14}$	300 $_{60}$	286 $_{100}$	73 $_{80}$	1469	9689
JWH-251		2835	319 $_3$	214 $_{100}$	144 $_{35}$	116 $_{17}$	105 $_{16}$	1029	8529
JWH-368		3030	385 $_{38}$	328 $_{38}$	314 $_{64}$	155 $_{66}$	127 $_{100}$	1381	9512
Kadethrin		3190*	396 $_1$	170 $_{100}$	143 $_{56}$	128 $_{66}$	91 $_{41}$	1429	2801
Karbutilate -C3H5NO		1640	208 $_{31}$	164 $_5$	136 $_{13}$	92 $_{23}$	72 $_{100}$	425	4151
Karbutilate -C5H9NO		1890	167 $_{63}$	135 $_{100}$	122 $_{31}$	81 $_{28}$	52 $_{36}$	346	3907
Kavain	G P	2235*	230 $_{32}$	202 $_{40}$	104 $_{31}$	98 $_{98}$	68 $_{100}$	509	1048
Kavain -CO2	P	1705*	186 $_{100}$	155 $_{33}$	128 $_{51}$	95 $_{74}$	77 $_{18}$	361	1049

Table 8.1: Compounds in order of names   **Kavain-M (O-demethyl-) -CO2**

Name	Detected	RI	Typical ions and intensities					Page	Entry
Kavain-M (O-demethyl-) -CO2	U UHY	1680*	172 84	157 58	128 100	95 39	77 60	327	2936
Kebuzone		2525	322 3	264 6	183 20	93 53	77 100	1045	4265
Kebuzone artifact		2150	266 37	183 70	118 8	105 28	77 100	694	4266
Kebuzone enol ME	UME	2510	336 58	266 16	183 51	105 52	77 100	1132	6378
Kebuzone-M (HO-) enol 2ME	UME	2690	366 77	296 2	213 8	107 74	77 100	1292	6379
Kelevan		2895*	488 1	455 15	357 23	272 100	270 49	1720	4045
Kelevan artifact		2320*	486 1	455 22	355 20	272 100	237 42	1642	3324
Ketamine	P U UHY	1835	237 1	209 22	180 100	152 14	102 15	543	1050
Ketamine AC	U+UHYAC	2170	279 2	216 100	208 91	180 93	152 80	772	1056
Ketamine artifact (bicyclo-) AC	U+UHYAC	2360	303 36	260 100	232 14	192 33	152 64	922	8786
Ketamine isomer	G P	1735	237 1	180 8	152 100	138 19	102 11	543	5561
Ketamine TMS		1800	309 1	294 11	157 100	152 63	73 81	963	4556
Ketamine-D4		1825	241 1	213 26	184 100	156 14	142 15	561	7779
Ketamine-D4 AC		2165	283 1	255 11	220 100	212 95	184 63	796	7780
Ketamine-D4 HFB		1895	437 1	402 4	374 29	366 34	210 100	1551	7784
Ketamine-D4 ME		1840	255 1	236 4	227 20	198 100	156 12	633	7781
Ketamine-D4 TFA		1900	337 2	309 12	274 81	240 51	110 100	1135	7783
Ketamine-D4 TMS		1795	313 3	298 11	229 6	157 100	73 48	989	7782
Ketamine-M (nor-)	P U	1810	223 3	195 34	166 100	138 10	131 14	482	1055
Ketamine-M (nor-) AC	U+UHYAC	2035	265 1	230 89	202 100	166 74	138 29	685	7826
Ketamine-M (nor-di-HO-) -2H2O	U	1920	219 90	190 100	184 22	156 43	129 25	463	1054
Ketamine-M (nor-di-HO-) -2H2O AC	UHYAC	1970	261 5	219 85	190 100	184 32	157 46	661	3672
Ketamine-M (nor-HO-) -H2O	U UHY	1960	221 28	193 15	166 100	131 44	102 27	472	1051
Ketamine-M (nor-HO-) -H2O AC	UHYAC	2080	263 3	228 41	160 100	153 23	102 11	673	3673
Ketamine-M (nor-HO-) -NH3	P U UHY	1740*	222 24	187 100	159 11	115 18	77 21	478	1053
Ketamine-M (nor-HO-) -NH3 -H2O	U UHY	1620*	204 49	169 100	139 32	115 18	70 37	408	1052
Ketamine-M (nor-HO-) -NH3 -H2O AC	U+UHYAC	1670*	246 17	204 100	169 63	139 19	107 10	579	1231
Ketamine-M/artifact	U UHY UHYAC	1630	189 50	154 100	127 17	75 8		367	3683
Ketanserin-M/artifact		2470	201 31	123 100	95 40	75 16	51 21	401	4232
Ketazolam artifact	P G U	2430	284 81	283 91	256 100	221 31	77 7	799	481
Ketazolam HY	UHY U+UHYAC	2100	245 95	228 38	193 29	105 38	77 100	573	272
Ketazolam HYAC	U+UHYAC	2260	287 11	244 100	228 39	182 49	77 70	818	2542
Ketazolam-M	P G U	2520	270 86	269 97	242 100	241 82	77 17	715	463
Ketazolam-M	P G U	2430	284 81	283 91	256 100	221 31	77 7	799	481
Ketazolam-M	P G UGLUC	2320	268 98	239 56	233 52	205 66	77 100	813	579
Ketazolam-M 2TMS		2200	430 51	429 89	340 15	313 19	73 100	1533	5499
Ketazolam-M artifact-1	P-I UHY U+UHYAC	2060	240 59	239 100	205 81	177 16	151 9	554	300
Ketazolam-M artifact-2	UHY UHYAC	2070	254 77	253 100	219 98	111 5		626	301
Ketazolam-M HY	UHY	2050	231 80	230 95	154 23	105 38	77 100	512	419
Ketazolam-M HYAC	PHYAC U+UHYAC	2245	273 30	230 100	154 13	105 23	77 50	732	273
Ketazolam-M TMS		2300	342 62	341 100	327 19	269 4	73 30	1168	4573
Ketazolam-M TMS		2635	356 9	341 100	312 56	239 12	135 21	1252	4577
Ketoprofen		2245*	254 40	209 50	177 62	105 100	77 56	627	1425
Ketoprofen ME	PME UME	2090*	268 50	209 100	191 22	105 76	77 63	707	1471
Ketoprofen-M (HO-) isomer-1 2ME	UME	2250*	298 80	239 50	191 34	135 54	77 30	895	5213
Ketoprofen-M (HO-) isomer-2 2ME	UME	2295*	298 70	239 50	191 7	135 100	77 18	894	5214
Ketoprofen-M (HO-) ME	UME	2345*	284 59	225 71	191 6	121 100	65 26	800	5215
Ketorolac ME		2265	269 51	210 84	132 9	105 100	77 57	711	4625
Ketotifen	G U+UHYAC	2600	309 100	237 10	208 12	96 29	70 6	962	1472
Ketotifen-M (dihydro-) -H2O	UHY UHYAC	2480	293 100	249 27	221 44	202 9	96 19	861	4482
Ketotifen-M (nor-)	U-I UHY-I	2700	295 100	253 84	208 43	165 24	152 24	875	2202
Ketotifen-M (nor-) AC	UHYAC-I	3180	337 100	277 24	265 47	221 36	208 39	1135	2203
LAAM	G P U+UHYAC	2230	353 1	338 1	225 4	91 6	72 100	1230	5616
Labetalol 2TMS		2530	472 1	439 4	162 62	73 63	58 100	1623	5489
Labetalol 3AC	UHYAC	3400	376 32	335 100	293 24	133 32	91 50	1590	1357
Labetalol 3TMS		2620	511 2	365 4	234 100	130 48	91 35	1693	5490
Labetalol artifact (1-methyl-3-phenylpropylamine)		1320	149 14	132 60	117 43	91 100	57 35	287	1356
Labetalol artifact (1-methyl-3-phenylpropylamine) AC	UHYAC	1780	191 21	132 15	117 43	87 100	72 32	372	1701
Labetalol-M (HO-) isomer-1 artifact 2AC	UHYAC	1940	249 4	206 100	147 72	104 46	86 47	597	1702
Labetalol-M (HO-) isomer-2 artifact 2AC	U+UHYAC	2000	249 11	207 18	148 42	133 39	87 100	597	1703
Lacidipine		2955	455 1	382 12	326 25	252 71	57 100	1592	5749
Lacosamide	P	1960	250 12	191 20	106 100	85 89	74 99	605	8347
Lacosamide 2TMS		1950	394 1	349 14	232 52	218 52	91 100	1421	8350
Lacosamide artifact (-CH3OH)		1920	218 7	175 7	130 7	106 100	91 53	458	8348
Lacosamide artifact (-CH3OH) 2TFA		1880	410 3	297 1	130 2	91 100	65 16	1476	8351
Lacosamide HYAC	U+UHYAC	1410	149 62	106 100	91 33	79 19	77 17	287	5160
Lacosamide TMS		1985	322 3	277 21	158 24	106 34	91 100	1046	8349
Lactic acid 2TMS		<1000*	219 3	191 12	147 70	117 56	73 100	528	8586
Lactose 8AC	U+UHYAC	3100*	331 1	169 100	127 23	109 46	81 28	1727	1960
Lactose 8HFB		2070*	537 4	519 19	293 16	169 100	81 73	1736	5787
Lactose 8PFP		1950*	437 6	419 55	273 12	119 100		1736	5786
Lactose 8TFA		1980*	337 11	319 87	223 30	193 43	69 100	1735	5785
Lactose 8TMS		2730*	361 8	217 31	204 83	191 86	73 100	1734	4334
Lactylphenetidine	UGLUC	1885	209 59	137 74	109 84	108 100		428	532

Lactylphenetidine AC

Table 8.1: Compounds in order of names

Name	Detected	RI	Typical ions and intensities					Page	Entry
Lactylphenetidine AC	UGLUCAC	1960	251 $_{51}$	137 $_{100}$	109 $_{38}$	108 $_{45}$		610	196
Lactylphenetidine HYAC	G U+UHYAC	1680	179 $_{66}$	137 $_{51}$	109 $_{97}$	108 $_{100}$	80 $_{18}$	341	186
Lactylphenetidine-M	UHY	1240	109 $_{100}$	80 $_{41}$	53 $_{82}$	53 $_{82}$	52 $_{90}$	260	826
Lactylphenetidine-M (deethyl-) HYME	UHYME	1100	123 $_{27}$	109 $_{100}$	94 $_{7}$	80 $_{96}$	53 $_{47}$	267	3766
Lactylphenetidine-M (HO-) HY2AC	UHYAC	1755	237 $_{14}$	195 $_{31}$	153 $_{100}$	124 $_{55}$		544	187
Lactylphenetidine-M (O-deethyl-) 2AC	UGLUCAC	1975	265 $_{5}$	223 $_{20}$	151 $_{9}$	109 $_{100}$		686	533
Lactylphenetidine-M (p-phenetidine)	UHY	1280	137 $_{68}$	108 $_{100}$	80 $_{39}$	65 $_{10}$		277	844
Lactylphenetidine-M HY2AC	PAC U+UHYAC	1765	193 $_{10}$	151 $_{53}$	109 $_{100}$	80 $_{24}$		377	188
Lambda		3480	335 $_{100}$	292 $_{6}$	221 $_{80}$	207 $_{40}$	70 $_{9}$	1127	10187
Lamivudine 2TMS		2780	358 $_{3}$	190 $_{52}$	184 $_{67}$	100 $_{100}$	87 $_{90}$	1327	8137
Lamivudine artifact (cytosine) 2TMS		1480	255 $_{38}$	254 $_{91}$	240 $_{100}$	170 $_{23}$	73 $_{62}$	632	7555
Lamivudine artifact (cytosine) 3TMS		1795	327 $_{66}$	326 $_{94}$	312 $_{100}$	197 $_{30}$	73 $_{81}$	1076	8138
Lamivudine -H2O		2700	211 $_{5}$	118 $_{59}$	112 $_{100}$	100 $_{71}$	87 $_{62}$	435	3929
Lamivudine -H2O AC		2670	253 $_{1}$	211 $_{1}$	154 $_{5}$	112 $_{17}$	100 $_{100}$	621	8136
Lamivudine -H2O HFB		2300	407 $_{3}$	378 $_{5}$	332 $_{23}$	164 $_{100}$	100 $_{78}$	1466	8139
Lamivudine -H2O PFP		2320	357 $_{1}$	282 $_{8}$	164 $_{66}$	135 $_{66}$	100 $_{100}$	1245	8140
Lamivudine -H2O TFA		2295	307 $_{2}$	278 $_{3}$	232 $_{15}$	164 $_{100}$	100 $_{94}$	946	8141
Lamotrigine	P	2635	255 $_{45}$	185 $_{100}$	157 $_{16}$	123 $_{30}$	114 $_{26}$	630	4636
Lamotrigine 2AC	PAC U+UHYAC	2855	339 $_{6}$	297 $_{14}$	255 $_{11}$	185 $_{100}$	114 $_{16}$	1147	4638
Lamotrigine AC	PAC U+UHYAC	2665	297 $_{7}$	268 $_{14}$	185 $_{100}$	157 $_{21}$	114 $_{12}$	888	4637
LAMPA TMS		3740	395 $_{87}$	337 $_{15}$	293 $_{39}$	253 $_{55}$	73 $_{100}$	1427	6263
Laudanosine	P U+UHYAC	2575	357 $_{1}$	206 $_{100}$	190 $_{23}$	162 $_{8}$	151 $_{7}$	1249	6106
Laudanosine-M (N-demethyl-O-tri-demethyl-) 3AC	U+UHYAC	3370	427 $_{50}$	385 $_{9}$	354 $_{55}$	312 $_{100}$	137 $_{76}$	1526	6789
Laudanosine-M (O-tri-demethyl-) 3AC	U+UHYAC	3020	262 $_{40}$	234 $_{100}$	220 $_{59}$	192 $_{78}$	178 $_{42}$	1562	6788
Laudanosine-M (O-tri-demethyl-) AC	U+UHYAC	2595	357 $_{1}$	315 $_{4}$	234 $_{88}$	192 $_{100}$	177 $_{16}$	1248	6787
Laudanosine-M (tri-demethyl-) isomer-1 2AC	U+UHYAC	2950	399 $_{58}$	326 $_{59}$	295 $_{40}$	282 $_{28}$	151 $_{100}$	1441	7857
Laudanosine-M (tri-demethyl-) isomer-2 2AC	U+UHYAC	3210	399 $_{55}$	326 $_{45}$	313 $_{19}$	295 $_{24}$	151 $_{100}$	1441	6790
Lauric acid		1670*	200 $_{7}$	157 $_{16}$	129 $_{25}$	73 $_{100}$	60 $_{87}$	400	5630
Lauric acid ET		1570*	228 $_{5}$	185 $_{13}$	157 $_{16}$	101 $_{51}$	88 $_{100}$	503	5400
Lauric acid ME		1550*	214 $_{3}$	183 $_{6}$	143 $_{12}$	87 $_{52}$	74 $_{100}$	448	2666
Lauric acid TMS		1670*	272 $_{5}$	257 $_{86}$	117 $_{68}$	75 $_{81}$	73 $_{100}$	731	5716
Lauroscholtzine		2665	341 $_{89}$	340 $_{100}$	326 $_{42}$	310 $_{77}$	267 $_{74}$	1162	5773
Lauroscholtzine AC		2750	383 $_{100}$	382 $_{92}$	368 $_{56}$	340 $_{64}$	326 $_{34}$	1371	5774
Lauroscholtzine artifact (dehydro-)		3180	339 $_{100}$	324 $_{77}$	296 $_{22}$	281 $_{17}$	238 $_{14}$	1150	6742
Lauroscholtzine artifact (dehydro-) AC		3285	381 $_{100}$	339 $_{16}$	324 $_{90}$	296 $_{17}$	280 $_{12}$	1361	6743
Lauroscholtzine artifact (dehydro-) ME		3235	353 $_{100}$	338 $_{71}$	307 $_{12}$	280 $_{26}$	176 $_{13}$	1227	6744
Lauroscholtzine ME	U+UHYAC	2680	355 $_{94}$	354 $_{100}$	340 $_{57}$	324 $_{31}$	281 $_{29}$	1238	5775
Lauroscholtzine-M (bis-O-demethyl-) 3AC		3170	439 $_{78}$	424 $_{39}$	396 $_{100}$	354 $_{46}$	312 $_{85}$	1556	6754
Lauroscholtzine-M (O-demethyl-) isomer-1 2AC		3095	411 $_{91}$	396 $_{100}$	368 $_{62}$	326 $_{47}$	295 $_{32}$	1483	6752
Lauroscholtzine-M (O-demethyl-) isomer-2 2AC		3055	411 $_{95}$	368 $_{100}$	354 $_{37}$	326 $_{73}$	224 $_{36}$	1482	6753
Lauroscholtzine-M (seco-O-demethyl-) 3AC		3650	453 $_{38}$	380 $_{63}$	338 $_{100}$	296 $_{85}$	283 $_{69}$	1587	6755
Lauroscholtzine-M/artifact (nor-seco-) 2AC		3315	411 $_{100}$	352 $_{40}$	339 $_{88}$	310 $_{74}$	297 $_{71}$	1483	6751
Lauroscholtzine-M/artifact (nor-seco-) AC		3230	369 $_{60}$	310 $_{45}$	297 $_{100}$	283 $_{14}$	263 $_{9}$	1306	6750
Lauroscholtzine-M/artifact (seco-) 2AC		3470	425 $_{75}$	352 $_{100}$	339 $_{91}$	310 $_{95}$	297 $_{45}$	1521	6746
Lauroscholtzine-M/artifact (seco-) 2ME		3030	369 $_{17}$	311 $_{6}$	265 $_{7}$	165 $_{3}$	58 $_{100}$	1307	6749
Lauroscholtzine-M/artifact (seco-) AC		3405	383 $_{52}$	310 $_{100}$	297 $_{65}$	263 $_{20}$	251 $_{11}$	1372	6745
Lauroscholtzine-M/artifact (seco-) ME		3035	355 $_{5}$	297 $_{3}$	165 $_{1}$	152 $_{2}$	58 $_{100}$	1238	6747
Lauroscholtzine-M/artifact (seco-) MEAC		3120	397 $_{4}$	339 $_{3}$	297 $_{6}$	251 $_{5}$	58 $_{100}$	1435	6748
Laurylmethylthiodipropionate		2550*	360 $_{13}$	192 $_{58}$	175 $_{27}$	146 $_{69}$	55 $_{100}$	1266	4400
Lefetamine		2190	165 $_{2}$	134 $_{100}$	118 $_{6}$	91 $_{7}$	77 $_{4}$	493	8927
Lefetamine-M (bis-HO-benzyl-) AC		2410	254 $_{4}$	212 $_{22}$	178 $_{9}$	165 $_{5}$	134 $_{100}$	902	8925
Lefetamine-M (bis-nor-)		1670	178 $_{2}$	118 $_{2}$	106 $_{100}$	91 $_{11}$	79 $_{25}$	392	8423
Lefetamine-M (bis-nor-) AC		2020	180 $_{3}$	148 $_{34}$	106 $_{100}$	91 $_{16}$	79 $_{17}$	553	8425
Lefetamine-M (bis-nor-) formyl artifact		1660	209 $_{1}$	181 $_{5}$	165 $_{7}$	118 $_{100}$	91 $_{66}$	430	8424
Lefetamine-M (bis-nor-) HFB		1760	302 $_{100}$	180 $_{28}$	107 $_{29}$	91 $_{34}$	79 $_{44}$	1413	8429
Lefetamine-M (bis-nor-) PFP		1730	252 $_{100}$	180 $_{35}$	165 $_{7}$	91 $_{24}$	79 $_{24}$	1173	8428
Lefetamine-M (bis-nor-) TFA		1740	202 $_{100}$	180 $_{27}$	107 $_{25}$	91 $_{24}$	79 $_{33}$	860	8427
Lefetamine-M (bis-nor-) TMS		1760	249 $_{4}$	178 $_{100}$	162 $_{15}$	91 $_{19}$	73 $_{29}$	713	8426
Lefetamine-M (bis-nor-di-HO-benzyl-) 2AC	U+UHYAC	2560	254 $_{2}$	212 $_{3}$	178 $_{2}$	148 $_{62}$	106 $_{100}$	988	9120
Lefetamine-M (bis-nor-di-HO-benzyl-) isomer-1 3AC	U+UHYAC	2610	296 $_{1}$	254 $_{11}$	212 $_{10}$	148 $_{46}$	106 $_{100}$	1237	9121
Lefetamine-M (bis-nor-di-HO-benzyl-) isomer-2 3AC	U+UHYAC	2630	296 $_{7}$	254 $_{3}$	212 $_{8}$	148 $_{55}$	106 $_{100}$	1237	8826
Lefetamine-M (bis-nor-di-HO-methoxy-benzyl-) 2AC	U+UHYAC	2680	284 $_{3}$	242 $_{9}$	153 $_{9}$	148 $_{49}$	106 $_{100}$	1175	8827
Lefetamine-M (bis-nor-HO-alkyl-) 2AC	U+UHYAC	1740	206 $_{42}$	164 $_{90}$	122 $_{30}$	104 $_{100}$	91 $_{33}$	890	8659
Lefetamine-M (bis-nor-HO-benzyl-) isomer-1 2AC	U+UHYAC	2350	238 $_{5}$	196 $_{12}$	148 $_{47}$	106 $_{100}$	79 $_{11}$	890	8984
Lefetamine-M (bis-nor-HO-benzyl-) isomer-2 2AC	U+UHYAC	2380	238 $_{13}$	196 $_{14}$	148 $_{57}$	106 $_{100}$	79 $_{16}$	890	8654
Lefetamine-M (bis-nor-HO-methoxy-benzyl-) 2AC	U+UHYAC	2540	268 $_{15}$	236 $_{28}$	194 $_{39}$	152 $_{45}$	106 $_{100}$	1075	8656
Lefetamine-M (bis-nor-HO-methoxy-phenyl-) 2AC	U+UHYAC	2560	236 $_{29}$	194 $_{45}$	152 $_{73}$	137 $_{10}$	106 $_{100}$	1076	8825
Lefetamine-M (bis-nor-HO-phenyl-) 2AC	U+UHYAC	2500	296 $_{1}$	206 $_{98}$	164 $_{79}$	122 $_{100}$	91 $_{25}$	890	8655
Lefetamine-M (deamino-HO-bis-HO-benzyl-) 3AC		2350*	355 $_{2}$	296 $_{9}$	208 $_{27}$	166 $_{100}$	123 $_{93}$	1242	8987
Lefetamine-M (deamino-HO-phenyl-) AC		2230*	238 $_{17}$	196 $_{31}$	165 $_{20}$	150 $_{27}$	107 $_{100}$	895	8932
Lefetamine-M (deamino-oxo-bis-HO-benzyl-) 2AC		2345*	207 $_{44}$	165 $_{80}$	123 $_{100}$	105 $_{81}$	77 $_{45}$	981	8933
Lefetamine-M (deamino-oxo-bis-HO-benzyl-) AC		2360*	270 $_{7}$	228 $_{19}$	153 $_{21}$	123 $_{15}$	105 $_{100}$	716	8986
Lefetamine-M (deamino-oxo-HO-benzyl-) isomer-1 AC		2080*	254 $_{1}$	212 $_{4}$	105 $_{100}$	77 $_{34}$		628	8934
Lefetamine-M (deamino-oxo-HO-benzyl-) isomer-2 AC		2120*	254 $_{2}$	212 $_{12}$	105 $_{100}$	77 $_{31}$		628	8935

Table 8.1: Compounds in order of names **Lefetamine-M (deamino-oxo-HO-methoxy-benzyl-) AC**

Name	Detected	RI	\multicolumn{5}{c}{Typical ions and intensities}	Page	Entry
Lefetamine-M (deamino-oxo-HO-methoxy-benzyl-) AC		2250*	284_1	242_{22}	226_{27}	137_{100}	105_{84}	800	8985
Lefetamine-M (HO-methoxy-benzyl-) AC	U+UHYAC	2370	268_7	226_{50}	165_{10}	153_4	134_{100}	990	8926
Lefetamine-M (nor-) AC		2200	180_6	162_{60}	120_{100}	91_7	77_5	623	8928
Lefetamine-M (nor-bis-HO-benzyl-) 3AC		2730	296_{12}	254_5	212_{12}	162_{40}	120_{100}	1306	8929
Lefetamine-M (nor-HO-benzyl-) 2AC	U+UHYAC	2530	238_{23}	196_{26}	162_{54}	120_{100}	107_{12}	977	8931
Lefetamine-M (nor-HO-methoxy-benzyl-) 2AC	U+UHYAC	2530	268_{23}	226_{36}	162_{24}	137_9	120_{100}	1163	8930
Leflunomide		1860	270_{25}	251_3	161_{100}	110_{71}	68_{38}	715	8395
Leflunomide artifact (4-trifluoromethylaniline) TFA		1050	257_{30}	188_{37}	160_{40}	145_{53}	69_{100}	641	8397
Leflunomide HYAC	U+UHYAC	1420	203_{22}	184_4	161_{100}	142_{20}	111_{12}	405	7372
Leflunomide TMS		1675	342_2	327_{13}	309_{11}	75_{62}	73_{100}	1168	8396
Lenacil		2275	153_{100}	136_8	110_{13}	67_9	53_{24}	529	3855
Lenacil 2ME		2280	262_3	181_{100}	165_{13}	138_{12}	67_{21}	671	3970
Lenacil ME		2260	248_2	167_{100}	124_{14}	95_{13}	67_{22}	592	3969
Leptaflorine		2150	216_{39}	201_{100}	186_{14}	172_{18}	144_5	453	4065
Leptaflorine 2TFA		2115	408_{22}	393_{100}	296_{22}	280_{17}	199_{75}	1470	9552
Leptaflorine AC		2525	258_{58}	243_{75}	215_{13}	201_{100}	172_{28}	647	9550
Leptaflorine HFB		2300	412_{47}	397_{100}	243_{25}	199_{34}	172_{22}	1485	9554
Leptaflorine PFP		2280	362_{46}	347_{100}	243_{21}	199_{54}	172_{37}	1272	9553
Leptaflorine TFA		2295	312_{57}	297_{100}	282_4	243_7	199_{18}	981	9551
Lercanidipine-M (deamino-HO-) -H2O	UHY UHYAC	1940*	194_{49}	167_{100}	165_{67}	152_{34}	116_{17}	384	3388
Lercanidipine-M (N-dealkyl-) AC	UHYAC	2320	253_{15}	193_{12}	165_{19}	152_{10}	73_{100}	623	3391
Lercanidipine-M (N-dealkyl-HO-) AC	UHYAC	2635	311_{54}	269_{14}	239_{21}	183_{63}	73_{100}	977	3392
Lercanidipine-M (N-dealkyl-HO-methoxy-) 2AC	UHYAC	2700	341_{10}	299_{54}	213_{74}	152_{14}	73_{100}	1163	3393
Lercanidipine-M/artifact (alcohol)		2010	238_{16}	165_9	152_7	91_{14}	58_{100}	892	7596
Lercanidipine-M/artifact (alcohol) AC		2080	339_1	282_1	238_{34}	165_{15}	58_{100}	1153	7595
Lercanidipine-M/artifact (alcohol) -H2O		1845	279_3	165_8	115_5	98_{100}	58_{15}	779	7594
Lercanidipine-M/artifact -CO2	U UHY UHYAC UME	2175	286_{30}	269_{100}	239_{48}	180_8	139_{13}	814	3656
Letrozole		2630	285_{49}	217_{100}	190_{61}	156_{15}	102_{22}	808	7510
Levacetylmethadol	G P U+UHYAC	2230	353_1	338_1	225_4	91_6	72_{100}	1230	5616
Levallorphan	UHY	2355	283_{100}	256_{54}	176_{63}	157_{70}	85_{80}	798	534
Levallorphan AC	UHYAC	2390	325_{72}	298_{36}	257_{20}	176_{30}	85_{100}	1064	1473
Levallorphan HFB		2205	479_{100}	452_{79}	411_{46}	353_{14}	176_{14}	1633	6227
Levallorphan PFP		2120	429_{95}	402_{82}	361_{57}	176_{38}	119_{100}	1531	6226
Levallorphan TFA		2110	379_{100}	352_{79}	311_{52}	176_{35}	69_{100}	1352	6225
Levallorphan TMS		2375	355_{16}	272_{16}	176_{32}	85_{68}	73_{100}	1240	6213
Levamisole	U+UHYAC	2025	204_{79}	148_{100}	127_{24}	121_{26}	101_{39}	409	8601
Levamisole artifact (dehydro-)	U+UHYAC	2320	202_{100}	174_{17}	147_{24}	116_{13}	103_{21}	403	8605
Levamisole artifact-1 (+H2O) AC	U+UHYAC	2200	264_{35}	217_{29}	179_{25}	175_{62}	132_{100}	679	8602
Levamisole artifact-2 (+H2O) AC	U+UHYAC	2390	264_{18}	231_{58}	217_{33}	175_{58}	132_{100}	679	8606
Levamisole artifact-3 (+H2O) AC	U+UHYAC	2405	264_1	205_{15}	161_9	148_{67}	106_{100}	679	8604
Levamisole-M/artifact	U+UHYAC	2250	236_{37}	175_{97}	132_{100}	105_{49}	91_{42}	536	8603
Levetiracetam	P G U+UHYAC	1740	170_3	126_{100}	112_2	98_{12}	69_{33}	324	6876
Levetiracetam 2HFB		1670	562_1	349_4	322_{100}	265_{18}	152_{27}	1701	7363
Levetiracetam 2TFA		1190	319_4	250_3	222_{100}	165_{39}	69_{33}	1272	7360
Levetiracetam 2TMS		1700	314_1	299_7	199_{44}	184_{46}	73_{100}	997	7364
Levetiracetam AC	U+UHYAC	1780	212_1	153_2	126_{100}	98_9	69_{21}	439	6877
Levetiracetam HFB		1590	366_1	169_4	153_4	126_{100}	69_{33}	1290	7362
Levetiracetam PFP		1540	316_1	153_3	126_{100}	98_6	69_{24}	1004	7361
Levetiracetam TFA		1500	266_1	153_4	126_{100}	98_{11}	69_{32}	693	7359
Levetiracetam TMS		1655	242_1	227_8	126_{100}	112_{81}	98_{44}	566	7365
Levobunolol		2430	291_1	276_{17}	115_5	86_{100}	57_{18}	852	2611
Levobunolol AC		2460	333_8	318_{26}	259_{15}	200_9	86_{100}	1114	1540
Levobunolol formyl artifact		2450	303_{10}	288_{100}	201_{11}	141_{13}	70_{42}	926	1539
Levobunolol -H2O AC		2570	315_2	259_{44}	200_{29}	160_{20}	57_{100}	1001	1541
Levodopa 3ME	UME	1870	239_7	194_6	180_6	151_{100}	102_{36}	552	2903
Levodopa-M (homovanillic acid)	U	1610*	182_{35}	137_{100}	122_{12}	94_7	65_9	350	3368
Levodopa-M (homovanillic acid) 2ME	UME	1720*	210_{31}	151_{100}	123_4	107_6		433	5959
Levodopa-M (homovanillic acid) 2TMS		1760*	326_{32}	267_{20}	209_{46}	179_{34}	73_{100}	1069	6015
Levodopa-M (homovanillic acid) HFB		1770*	378_{79}	333_{39}	181_{60}	107_{100}	69_{72}	1346	5975
Levodopa-M (homovanillic acid) ME	UME	1750*	196_{25}	137_{100}	122_{16}	107_2	94_{12}	389	812
Levodopa-M (homovanillic acid) MEAC	U+UHYAC	1700*	238_2	196_{47}	137_{100}	122_6	107_6	547	2973
Levodopa-M (homovanillic acid) MEHFB		1570*	392_{100}	333_{84}	169_{24}	107_{78}	69_{60}	1410	5974
Levodopa-M (homovanillic acid) MEPFP		1570*	342_{96}	283_{100}	195_{35}	119_{37}	107_{51}	1166	5972
Levodopa-M (homovanillic acid) METMS		1670*	268_{41}	238_{69}	209_{52}	179_{40}	73_{95}	707	6016
Levodopa-M (homovanillic acid) PFP		1685*	328_{85}	283_{45}	181_{58}	119_{33}	107_{100}	1079	5973
Levodopa-M (O-methyl-dopamine) 2AC	U+UHYAC	2070	251_2	209_7	150_{100}	137_{15}		609	1273
Levodopa-M (O-methyl-dopamine) AC	UHYAC	2330	209_{100}	180_{16}	150_{43}	138_{17}	58_{10}	429	2980
Levofloxacin 2TMS		3300	505_{15}	490_{16}	435_{37}	407_{12}	93_{100}	1665	8246
Levofloxacin -CO2	U+UHYAC	3285	317_{25}	247_{16}	231_{12}	121_7	71_{100}	1012	4691
Levofloxacin ME	UME	3750	375_{24}	305_{12}	290_{11}	246_{28}	71_{100}	1335	4692
Levomepromazine	P-I G U UHY U+UHYAC	2540	328_{10}	228_6	185_7	100_7	58_{100}	1082	344
Levomepromazine-M (di-HO-) 2AC	U+UHYAC	3100	444_{11}	316_3	260_8	100_9	58_{100}	1570	3052
Levomepromazine-M (HO-)	UHY	2735	344_6	100_6	58_{100}			1179	537
Levomepromazine-M (HO-) isomer-1 AC	U+UHYAC	2745	386_{15}	299_2	244_3	100_6	58_{100}	1385	345

Levomepromazine-M (HO-) isomer-2 AC

Table 8.1: Compounds in order of names

Name	Detected	RI	Typical ions and intensities					Page	Entry
Levomepromazine-M (HO-) isomer-2 AC	U+UHYAC	2850	386 $_{11}$	299 $_2$	244 $_8$	100 $_8$	58 $_{100}$	1385	8516
Levomepromazine-M (N,O-bis-demethyl-HO-) 3AC	U+UHYAC	3750	442 $_{14}$	328 $_7$	230 $_{19}$	128 $_{100}$	86 $_{30}$	1565	8519
Levomepromazine-M (nor-)	UHY	2600	314 $_{24}$	229 $_{69}$	213 $_{100}$	72 $_{88}$		995	536
Levomepromazine-M (nor-) AC	U+UHYAC	2970	356 $_6$	242 $_5$	228 $_4$	128 $_{100}$		1242	346
Levomepromazine-M (nor-HO-)	UHY	2750	330 $_{45}$	258 $_{37}$	245 $_{80}$	86 $_{49}$	72 $_{100}$	1094	538
Levomepromazine-M (nor-HO-) AC	U+UHYAC	3140	372 $_{46}$	258 $_{28}$	244 $_{44}$	128 $_{100}$	86 $_{28}$	1323	6415
Levomepromazine-M (nor-HO-) isomer-1 2AC	U+UHYAC	3220	414 $_{23}$	300 $_5$	244 $_{24}$	128 $_{100}$	86 $_{19}$	1490	347
Levomepromazine-M (nor-HO-) isomer-2 2AC	U+UHYAC	3220	414 $_{17}$	300 $_{15}$	244 $_{22}$	128 $_{100}$	86 $_{33}$	1490	8518
Levomepromazine-M (nor-O-demethyl-) 2AC	U+UHYAC	2930	384 $_{10}$	270 $_{12}$	214 $_{16}$	128 $_{100}$	86 $_{30}$	1376	15
Levomepromazine-M (O-demethyl-)	UHY	2650	314 $_{18}$	228 $_4$	214 $_5$	100 $_{10}$	58 $_{100}$	995	11
Levomepromazine-M (O-demethyl-) AC	U+UHYAC	2600	356 $_8$	228 $_2$	214 $_7$	100 $_6$	58 $_{100}$	1242	13
Levomepromazine-M (O-demethyl-HO-) 2AC	U+UHYAC	2960	414 $_7$	286 $_2$	230 $_6$	100 $_8$	58 $_{100}$	1490	8517
Levomepromazine-M/artifact (sulfoxide)	G P U+UHYAC	2940	344 $_4$	242 $_{41}$	229 $_{14}$	210 $_7$	58 $_{100}$	1179	535
Levorphanol	UHY	2255	257 $_{38}$	200 $_{17}$	150 $_{28}$	59 $_{100}$		643	475
Levorphanol AC	U+UHYAC	2280	299 $_{100}$	231 $_{42}$	200 $_{20}$	150 $_{48}$	59 $_{15}$	903	230
Levorphanol HFB		2100	453 $_{100}$	396 $_{18}$	385 $_{91}$	169 $_{19}$	150 $_{27}$	1586	6151
Levorphanol PFP	UHYPFP	2060	403 $_{93}$	335 $_{78}$	303 $_{14}$	150 $_{100}$	119 $_{50}$	1457	4305
Levorphanol TFA		2015	353 $_{69}$	285 $_{80}$	150 $_{100}$	115 $_{26}$	69 $_{72}$	1227	4006
Levorphanol TMS	UHYTMS	2230	329 $_{31}$	272 $_{20}$	150 $_{39}$	73 $_{26}$	59 $_{100}$	1090	4304
Levorphanol-M (HO-) 2AC	U+UHYAC	2580	357 $_{68}$	247 $_{22}$	215 $_{17}$	150 $_{100}$	59 $_{30}$	1249	1187
Levorphanol-M (methoxy-) AC	U+UHYAC	2520	329 $_{48}$	261 $_{23}$	229 $_{23}$	150 $_{100}$	59 $_{28}$	1090	4476
Levorphanol-M (nor-) 2AC	UHYAC	2710	327 $_{11}$	240 $_8$	199 $_{12}$	87 $_{100}$	72 $_{62}$	1077	228
Levorphanol-M (oxo-) AC	U+UHYAC	2695	313 $_{16}$	240 $_{11}$	199 $_{98}$	157 $_{12}$	73 $_{100}$	990	4475
Levosimendan		2640	280 $_{100}$	265 $_{19}$	223 $_{27}$	195 $_{25}$	115 $_{45}$	781	10337
Lidocaine	P G U+UHYAC	1875	234 $_6$	120 $_6$	86 $_{100}$	72 $_{10}$	58 $_{26}$	530	1061
Lidocaine AC	UHYAC	1860	276 $_5$	204 $_1$	120 $_2$	86 $_{100}$	58 $_{17}$	756	2585
Lidocaine artifact		1855	232 $_5$	217 $_{23}$	132 $_{11}$	85 $_{100}$	70 $_{21}$	519	6784
Lidocaine TMS		1785	306 $_1$	291 $_7$	235 $_{27}$	220 $_{50}$	86 $_{100}$	945	4557
Lidocaine-M (deethyl-)	U UHY	1790	206 $_3$	163 $_{12}$	121 $_{10}$	58 $_{100}$		417	1063
Lidocaine-M (deethyl-) AC	U+UHYAC	2115	248 $_4$	128 $_{26}$	100 $_{15}$	58 $_{100}$		592	1066
Lidocaine-M (dimethylaniline)		1180	121 $_{100}$	106 $_{77}$				266	725
Lidocaine-M (dimethylaniline) AC	U+UHYAC	1470	163 $_{33}$	121 $_{100}$	106 $_{66}$	91 $_{17}$	77 $_{30}$	308	57
Lidocaine-M (dimethylhydroxyanilino)	UHY	1460	137 $_{100}$	122 $_{30}$	107 $_{76}$			277	1062
Lidocaine-M (dimethylhydroxyaniline) 2AC	U+UHYAC	1885	221 $_9$	179 $_{39}$	137 $_{100}$			473	1064
Lidocaine-M (dimethylhydroxyaniline) 3AC	U+UHYAC	1900	263 $_{26}$	221 $_{40}$	179 $_{100}$	137 $_{80}$	108 $_4$	674	1065
Lidocaine-M (HO-)	UHY	2350	250 $_3$	194 $_3$	86 $_{100}$	58 $_{19}$		607	4070
Lidocaine-M (HO-) AC	UHYAC	2300	292 $_4$	204 $_3$	127 $_4$	86 $_{100}$	58 $_8$	857	3361
Lidoflazine		3870	491 $_{18}$	343 $_{100}$	288 $_6$	260 $_4$	109 $_6$	1650	2725
Lidoflazine-M (deamino-carboxy-)	P-I UHY UHYAC	2230*	276 $_7$	216 $_{17}$	203 $_{100}$	183 $_{22}$		753	169
Lidoflazine-M (deamino-carboxy-) ME	P-I UHYME U+UHYAC	2125*	290 $_{12}$	258 $_{13}$	216 $_{30}$	203 $_{100}$	183 $_{22}$	841	3372
Lidoflazine-M (deamino-HO-) AC	UHYAC	2150*	304 $_7$	244 $_{22}$	216 $_{41}$	203 $_{100}$	183 $_{16}$	931	307
Lidoflazine-M (N-dealkyl-) AC	UHYAC	2970	372 $_{29}$	300 $_{13}$	201 $_{11}$	141 $_{100}$	109 $_9$	1324	3370
Lignoceric acid ME		2745*	382 $_{25}$	339 $_6$	143 $_{15}$	87 $_{66}$	74 $_{100}$	1368	3796
Lincomycin (4)AC	U	2660	126 $_{100}$	82 $_4$				1705	5126
Lincomycin (4)AC	U	2695	527 $_1$	126 $_{100}$	82 $_2$			1705	5127
Lincomycin (4)AC	U	2725	416 $_1$	126 $_{100}$	82 $_3$			1705	5128
Lindane	P-I	1740*	288 $_2$	252 $_8$	217 $_{67}$	181 $_{100}$	109 $_{48}$	824	1067
Lindane-M (2,3,4,5-tetrachlorophenol)	U	1500*	232 $_{100}$	230 $_{79}$	194 $_{15}$	166 $_{20}$	131 $_{26}$	507	3366
Lindane-M (2,4,5-trichlorophenol)	U	1440*	198 $_{93}$	196 $_{100}$	132 $_{34}$	97 $_{64}$	73 $_{14}$	388	784
Lindane-M (2,4,6-trichlorophenol)	U	1420*	198 $_{98}$	196 $_{100}$	160 $_{11}$	132 $_{37}$	97 $_{31}$	389	3363
Lindane-M (dichloro-HO-thiophenol)	U	1470*	194 $_{100}$	159 $_{84}$	131 $_{27}$	95 $_{34}$		381	3365
Lindane-M (dichlorothiophenol)	U	1250*	178 $_{100}$	143 $_{88}$	107 $_{26}$	69 $_{19}$		336	3362
Lindane-M (tetrachlorocyclohexene)	U	1470*	218 $_1$	183 $_9$	147 $_{100}$	111 $_{56}$	77 $_{39}$	456	3369
Lindane-M (trichlorothiophenol)	U	1450*	212 $_{100}$	177 $_{89}$	142 $_{23}$	106 $_{16}$		437	3364
Linezolide		2770	337 $_{100}$	293 $_{30}$	234 $_{43}$	209 $_{60}$	164 $_{48}$	1137	7318
Linezolide artifact		2270	293 $_{100}$	234 $_{18}$	209 $_{82}$	151 $_{26}$	85 $_{15}$	861	7319
Linezolide artifact (deacetyl-) HFB		2580	491 $_{100}$	433 $_{33}$	175 $_{25}$	164 $_{28}$	149 $_{36}$	1649	7328
Linezolide artifact (deacetyl-) PFP		2550	441 $_{100}$	383 $_{43}$	176 $_{39}$	149 $_{42}$	119 $_{31}$	1560	7326
Linezolide artifact (deacetyl-) TFA		2575	391 $_{100}$	333 $_{35}$	175 $_{25}$	149 $_{36}$		1406	7327
Linezolide TMS		2380	409 $_{98}$	312 $_8$	281 $_{32}$	150 $_{21}$	73 $_{100}$	1474	7325
Linoleic acid	G	2140*	280 $_4$	95 $_{47}$	81 $_{65}$	67 $_{92}$	55 $_{100}$	785	2551
Linoleic acid ET		2150*	308 $_{10}$	263 $_6$	95 $_{51}$	81 $_{85}$	67 $_{100}$	960	5642
Linoleic acid ME		2110*	294 $_{10}$	263 $_6$	95 $_{43}$	81 $_{67}$	67 $_{100}$	873	1068
Linolenic acid ME		2130*	292 $_{16}$	191 $_6$	121 $_{26}$	95 $_{74}$	79 $_{100}$	859	2668
Linuron ME		1785	262 $_{11}$	231 $_{10}$	202 $_{87}$	174 $_{100}$	109 $_{22}$	667	3940
Lisinopril 3TMS		3165	621 $_1$	606 $_6$	361 $_{95}$	324 $_{100}$	73 $_{84}$	1719	4982
Lisinopril 4TMS		3260	693 $_1$	678 $_{13}$	479 $_{100}$	318 $_{38}$	73 $_{53}$	1728	4983
Lisofylline	G P UHY	2505	280 $_{25}$	236 $_{20}$	193 $_{32}$	180 $_{100}$	109 $_{31}$	783	1213
Lisofylline AC	U+UHYAC	2560	322 $_{32}$	262 $_{15}$	193 $_{35}$	180 $_{100}$		1045	1214
Lisofylline -H2O	U+UHYAC	2300	262 $_{24}$	193 $_{32}$	181 $_{100}$	137 $_{24}$	109 $_{45}$	670	1732
Lobeline		1820	337 $_{100}$	216 $_5$	120 $_{43}$	105 $_{100}$	77 $_{79}$	1141	1474
Lobeline artifact		1880	217 $_{16}$	97 $_{84}$	96 $_{100}$	77 $_{14}$		455	1821
Lobeline artifact AC		1900	259 $_{20}$	200 $_4$	97 $_{42}$	96 $_{100}$		652	1822
Lodoxamide artifact		1790	167 $_{100}$	139 $_{11}$	132 $_7$	105 $_{25}$	77 $_8$	318	7519

Table 8.1: Compounds in order of names   **Lodoxamide artifact 2AC**

Name	Detected	RI	Typical ions and intensities					Page	Entry
Lodoxamide artifact 2AC		2325	251_{7}	216_{100}	174_{70}	167_{80}	139_{8}	607	7522
Lodoxamide artifact 3AC		2235	293_{1}	258_{16}	251_{19}	216_{100}	167_{26}	860	7521
Lodoxamide artifact AC		2080	209_{17}	174_{83}	167_{100}	139_{11}	105_{14}	426	7520
Lofepramine-M (dealkyl-)	UHY	2225	266_{28}	235_{61}	208_{61}	195_{100}	71_{59}	698	324
Lofepramine-M (dealkyl-) AC	PAC U+UHYAC	2670	308_{40}	208_{100}	193_{55}	114_{62}		959	325
Lofepramine-M (dealkyl-) HFB		2450	462_{23}	268_{13}	240_{20}	208_{100}	193_{54}	1604	7706
Lofepramine-M (dealkyl-) PFP		2450	412_{38}	234_{16}	218_{19}	208_{100}	193_{47}	1485	7667
Lofepramine-M (dealkyl-) TFA		2430	362_{14}	208_{100}	193_{51}	140_{17}	69_{21}	1273	7786
Lofepramine-M (dealkyl-) TMS		2470	338_{7}	235_{89}	143_{41}	116_{72}	73_{100}	1146	5461
Lofepramine-M (dealkyl-HO-) 2AC	U+UHYAC	3065	366_{27}	266_{39}	114_{100}			1293	1175
Lofepramine-M (di-HO-ring)	UHY	2600	227_{100}	196_{7}				499	2296
Lofepramine-M (di-HO-ring) 2AC	U+UHYAC	2750	311_{28}	269_{23}	227_{100}	196_{7}		975	2292
Lofepramine-M (HO-methoxy-ring)	UHY	2390	241_{100}	226_{17}	210_{12}	180_{14}		560	2315
Lofepramine-M (HO-methoxy-ring) AC	U+UHYAC	2370	283_{10}	241_{100}	226_{17}	210_{12}	180_{14}	796	2867
Lofepramine-M (HO-ring)	UHY	2240	211_{100}	196_{15}	180_{10}	152_{4}		436	2295
Lofepramine-M (HO-ring) AC	U+UHYAC	2535	253_{26}	211_{100}	196_{19}	180_{11}	152_{4}	621	1218
Lofepramine-M (ring)	U U+UHYAC	1930	195_{100}	180_{40}	167_{9}	96_{33}	83_{22}	386	308
Lofepramine-M (ring) ME		1915	209_{70}	194_{100}	178_{13}	165_{11}		429	6352
Lofexidine		1910	258_{1}	243_{100}	223_{71}	95_{69}	67_{91}	645	5208
Lofexidine AC		2200	300_{1}	265_{12}	257_{20}	139_{58}	86_{100}	905	5209
Lonazolac	G	3000	312_{82}	267_{100}	232_{30}	104_{20}	77_{69}	980	1913
Lonazolac -CO2		2400	268_{100}	232_{4}	164_{6}	130_{38}	77_{32}	706	1975
Lonazolac ET		2950	340_{52}	267_{100}	232_{20}	164_{7}	77_{25}	1155	1994
Lonazolac ME	PME UME U+UHYAC	2685	326_{88}	267_{100}	232_{22}	164_{9}	77_{51}	1066	1377
Lonazolac-M (HO-) 2ME	UME	2880	356_{100}	297_{79}	262_{78}	247_{13}	117_{8}	1241	6296
Loperamide AC		3370	476_{1}	432	266_{10}	238_{100}	224_{44}	1676	1824
Loperamide artifact	G U+UHYAC	3380	401_{90}	250_{100}	238_{76}	222_{38}	115_{37}	1450	1825
Loperamide -H2O	U+UHYAC	3000	458_{13}	266_{100}	239_{24}	192_{81}	72_{17}	1597	1823
Loperamide-M (N-dealkyl-)	UHY	1800	211_{6}	139_{20}	84_{42}	56_{100}		436	521
Loperamide-M (N-dealkyl-) AC	U UHYAC	2235	253_{31}	210_{53}	193_{14}	139_{42}	57_{100}	621	524
Loperamide-M (N-dealkyl-oxo-) -2H2O	U UHY U+UHYAC	1650	189_{100}	154_{32}	127_{13}	101_{8}	75_{17}	367	181
Lopinavir artifact-1		1185	141_{35}	98_{11}	70_{8}	60_{79}	56_{100}	280	7955
Lopinavir artifact-2		1840	214_{6}	171_{14}	155_{100}	102_{8}	84_{9}	444	7956
Lopinavir artifact-3		2085	224_{53}	143_{37}	99_{42}	70_{22}	56_{100}	485	7957
Loprazolam HY	UHY-I U+UHYAC-I	2470	276_{45}	241_{100}	195_{26}	139_{51}	111_{31}	752	280
Loratadine	G U+UHYAC	3050	382_{100}	292_{33}	280_{34}	266_{50}	244_{33}	1365	5283
Loratadine-M/artifact (-COOCH2CH3) AC	UHYAC	3120	352_{100}	294_{32}	280_{33}	266_{60}	245_{29}	1220	5610
Lorazepam	P-I G UGLUC	2440	302_{45}	274_{62}	239_{91}	75_{100}		1031	539
Lorazepam 2AC		2730	345_{17}	307_{41}	265_{53}	230_{100}		1458	540
Lorazepam 2TMS		2380	464_{2}	449_{6}	429_{100}	347_{13}	73_{34}	1607	4607
Lorazepam artifact-1	UHY UHYAC	2140	288_{57}	287_{54}	253_{100}	177_{16}	150_{13}	825	2526
Lorazepam artifact-2	UHY-I UHYAC-I	2170	274_{44}	273_{36}	239_{100}	177_{7}	110_{10}	737	289
Lorazepam artifact-3	U+UHYAC	2325	290_{77}	255_{100}				839	544
Lorazepam HY	UHY	2180	265_{62}	230_{100}	139_{43}	111_{50}		684	543
Lorazepam HYAC	U+UHYAC	2300	307_{42}	265_{58}	230_{100}	139_{16}	111_{14}	946	290
Lorazepam isomer-1 2ME		2485	348_{15}	305_{43}	75_{100}			1198	541
Lorazepam isomer-2 2ME		2525	330_{10}	316_{46}	274_{38}	239_{40}	75_{100}	1091	542
Lorazepam-M (HO-) artifact	UHY-I	2400	304_{98}	303_{98}	277_{62}	275_{100}	203_{31}	929	2529
Lorazepam-M (HO-) artifact AC	UHYAC-I	2550	346_{9}	305_{79}	303_{100}	275_{25}	239_{8}	1186	2527
Lorazepam-M (HO-) HY	UHY-I	2360	281_{62}	246_{100}	228_{26}	154_{44}	126_{42}	785	545
Lorazepam-M (HO-) HY2AC	UHYAC	2600	365_{28}	323_{30}	281_{39}	246_{100}		1284	2528
Lorazepam-M (HO-methoxy-) HY	UHY-I	2780	311_{23}	281_{46}	246_{100}			974	546
Lorcainide	G U UHY UHYAC	2815	370_{5}	355_{33}	110_{74}	91_{69}	82_{100}	1313	1477
Lorcainide-M (deacyl-)	UHY UHYAC	2100	252_{18}	237_{9}	125_{23}	110_{100}	58_{47}	619	2890
Lorcainide-M (deacyl-) AC	UHYAC	2200	294_{5}	279_{55}	110_{64}	82_{94}	56_{100}	871	2891
Lorcainide-M (HO-) AC	UHYAC	2880	428_{3}	413_{30}	251_{23}	110_{64}	82_{100}	1528	2893
Lorcainide-M (HO-di-methoxy-) AC	UHYAC	3010	488_{10}	473_{65}	251_{32}	110_{66}	82_{100}	1645	2895
Lorcainide-M (HO-methoxy-) AC	UHYAC	2940	458_{7}	443_{62}	251_{30}	110_{66}	82_{100}	1597	2894
Lorcainide-M (N-dealkyl-deacyl-) 2AC	UHYAC	2490	294_{7}	251_{35}	192_{24}	125_{37}	83_{100}	870	2892
Lormetazepam	P-I G UGLUC	2735	334_{8}	307_{66}	305_{100}	111_{6}	75_{10}	1117	547
Lormetazepam AC		2740	376_{6}	334_{22}	305_{100}	291_{20}	255_{8}	1338	5604
Lormetazepam artifact-1	UHY-I UHYAC-I	2170	274_{44}	273_{36}	239_{100}	177_{7}	110_{10}	737	289
Lormetazepam artifact-2		2585	291_{72}	289_{100}	253_{6}	179_{44}	109_{6}	832	2381
Lormetazepam artifact-3	G	2850	304_{98}	269_{100}	262_{87}	254_{22}	75_{31}	929	5640
Lormetazepam artifact-4	G	3120	334_{79}	262_{47}	228_{100}	195_{84}	75_{39}	1117	5641
Lormetazepam HY	UHY UHYAC	2220	279_{100}	244_{76}	229_{75}	111_{53}	75_{38}	771	291
Lormetazepam isomer-1 TMS		2735	406_{29}	363_{66}	267_{100}	228_{51}	73_{66}	1464	4606
Lormetazepam isomer-2 TMS		2735	406_{10}	391_{22}	377_{100}	291_{25}	73_{34}	1464	4558
Lormetazepam-M (HO-) HY	UHY	2470	295_{86}	260_{100}	245_{70}			873	548
Lormetazepam-M (nor-)	P-I G UGLUC	2440	302_{45}	274_{62}	239_{91}	75_{100}		1031	539
Lormetazepam-M (nor-) 2AC		2730	345_{17}	307_{41}	265_{53}	230_{100}		1458	540
Lormetazepam-M (nor-) 2TMS		2380	464_{2}	449_{6}	429_{100}	347_{13}	73_{34}	1607	4607
Lormetazepam-M (nor-) HY	UHY	2180	265_{62}	230_{100}	139_{43}	111_{50}		684	543
Lormetazepam-M (nor-) HYAC	U+UHYAC	2300	307_{42}	265_{58}	230_{100}	139_{16}	111_{14}	946	290

Losartan 2ME **Table 8.1:** Compounds in order of names

Name	Detected	RI	Typical ions and intensities					Page	Entry
Losartan 2ME	UME	3555	450_{18}	249_{92}	201_{93}	192_{100}	165_{72}	1581	4841
Lovastatin -H2O -C5H10O2	G P-I	2775*	284_{23}	199_{100}	198_{90}	172_{61}	157_{80}	805	6449
Loxapine	G U+UHYAC	2555	327_{4}	257_{36}	193_{24}	83_{57}	70_{100}	1074	549
Loxapine-M (HO-) AC	UHYAC	2935	385_{5}	315_{10}	83_{65}	70_{100}		1380	1274
Loxapine-M (nor-)		2665	313_{8}	257_{46}	245_{20}	228_{20}	193_{34}	987	8231
Loxapine-M (nor-) AC		2940	355_{17}	269_{50}	257_{65}	193_{100}	56_{76}	1236	8232
Loxapine-M (nor-) HFB		2745	509_{37}	257_{75}	229_{55}	193_{100}	69_{81}	1668	8236
Loxapine-M (nor-) PFP		2730	459_{46}	257_{69}	229_{53}	193_{100}	69_{41}	1598	8235
Loxapine-M (nor-) TFA		2745	409_{52}	257_{65}	229_{52}	193_{100}	69_{71}	1473	8234
Loxapine-M (nor-) TMS		2770	385_{7}	370_{7}	317_{10}	128_{100}	73_{73}	1381	8233
Loxapine-M (nor-HO-) 2AC	UHYAC	3450	413_{31}	207_{100}	112_{77}			1487	1275
LSA		2820	267_{100}	221_{51}	207_{53}	180_{49}	154_{48}	701	8441
LSD		3445	323_{100}	221_{73}	207_{53}	181_{48}	72_{50}	1054	1069
LSD TMS		3595	395_{100}	293_{53}	253_{84}	73_{97}		1426	1070
LSD-M (2-oxo-3-HO-) 2TMS		3430	499_{71}	325_{12}	309_{100}	235_{21}	73_{93}	1658	6223
LSD-M (nor-) 2TMS		3515	453_{92}	351_{33}	279_{19}	253_{51}	73_{100}	1588	6261
LSD-M (nor-) TMS		3705	381_{91}	279_{66}	254_{31}	100_{8}	73_{100}	1363	6262
LSM-775		3480	337_{100}	294_{6}	221_{73}	207_{47}	181_{50}	1140	9870
LSM-775 TMS		3535	409_{75}	351_{19}	293_{90}	279_{62}	253_{100}	1475	9869
LSZ		3480	335_{100}	292_{6}	221_{80}	207_{40}	70_{9}	1127	10187
LSZ TMS		3240	407_{35}	364_{5}	293_{100}	279_{42}	268_{28}	1470	9871
Lumefantrine		4480	384_{1}	286_{4}	250_{3}	142_{100}	100_{21}	1682	8607
Lumefantrine -H2O		4460	384_{2}	286_{4}	250_{3}	142_{100}	100_{26}	1668	8608
Lupanine	P U UHY UHYAC	2230	248_{42}	149_{51}	136_{100}	98_{26}	84_{18}	592	2877
Lynestrenol	G	2260*	284_{26}	201_{50}	159_{32}	91_{100}	79_{96}	806	2242
Lynestrenol AC		2280*	326_{28}	266_{39}	201_{74}	159_{49}	91_{100}	1071	2263
Lysergic acid 2,4-dimethylazetidide		3480	335_{100}	292_{6}	221_{80}	207_{40}	70_{9}	1127	10187
Lysergic acid 2,4-dimethylazetidide TMS		3240	407_{35}	364_{5}	293_{100}	279_{42}	268_{28}	1470	9871
Lysergic acid amide		2820	267_{100}	221_{51}	207_{53}	180_{49}	154_{48}	701	8441
Lysergic acid N,N-methylpropylamine TMS		3740	395_{87}	337_{15}	293_{39}	253_{55}	73_{100}	1427	6263
Lysergide		3445	323_{100}	221_{73}	207_{53}	181_{48}	72_{50}	1054	1069
Lysergide alpha isomer (iso-LSD) TMS		3515	395_{90}	293_{43}	279_{24}	253_{39}	73_{100}	1426	6222
Lysergide TMS		3595	395_{100}	293_{53}	253_{84}	73_{97}		1426	1070
Lysergide-M (2-oxo-3-HO-) 2TMS		3430	499_{71}	325_{12}	309_{100}	235_{21}	73_{93}	1658	6223
Lysergide-M (nor-) 2TMS		3515	453_{92}	351_{33}	279_{19}	253_{51}	73_{100}	1588	6261
Lysergide-M (nor-) TMS		3705	381_{91}	279_{66}	254_{31}	100_{8}	73_{100}	1363	6262
M-144		2415	343_{14}	328_{35}	246_{100}	158_{27}	144_{16}	1176	9625
MAB-CHMINACA 2TMS		2750	514_{2}	499_{5}	398_{100}	384_{23}	241_{32}	1673	9533
MAB-CHMINACA -CONH2		2520	325_{1}	241_{100}	145_{65}	131_{28}	103_{14}	1065	9587
Mafenide		2340	186_{3}	185_{21}	141_{11}	106_{100}	77_{36}	360	5228
Mafenide 2ME		1920	214_{17}	133_{3}	89_{10}	74_{11}	58_{100}	446	5230
Mafenide 3ME		1900	228_{13}	214_{4}	133_{4}	89_{13}	58_{100}	501	5229
Mafenide 4ME		1870	242_{18}	134_{7}	107_{10}	89_{19}	58_{100}	565	5231
Mafenide AC		2425	228_{57}	185_{22}	147_{60}	106_{74}	105_{100}	500	5232
Mafenide MEAC		2300	242_{42}	199_{26}	185_{50}	161_{58}	119_{100}	564	5233
Malaoxon		1890*	314_{1}	268_{16}	195_{17}	127_{100}	99_{51}	992	3449
Malathion		1940*	330_{2}	285_{9}	173_{87}	127_{100}	93_{56}	1092	1401
Malathion-M (malaoxon)		1890*	314_{1}	268_{16}	195_{17}	127_{100}	99_{51}	992	3449
Maleic acid 2TMS		1080*	245_{15}	147_{100}	133_{7}	83_{20}	73_{80}	656	4674
Maleic hydrazide (MH)		1735	112_{100}	97_{5}	82_{80}	68_{4}	55_{75}	261	3916
Malic acid 3TMS		1415*	335_{1}	245_{3}	233_{5}	147_{20}	73_{100}	1208	8505
MAM-2201		3340	373_{100}	356_{60}	298_{71}	232_{82}	115_{58}	1328	9616
MAM-2201 (chloro analog)		3775	389_{100}	372_{53}	298_{92}	248_{83}	115_{82}	1399	9617
MAM-2201 artifact (-HF)		3160	353_{77}	298_{53}	212_{46}	169_{100}	115_{50}	1229	9618
Mandelic acid		1890*	152_{10}	107_{100}	79_{75}	77_{55}	51_{63}	292	5759
Mandelic acid ME		1485*	166_{7}	107_{100}	79_{59}	77_{40}		315	1071
Mannitol		2180*	133_{10}	117_{5}	103_{47}	73_{100}	61_{90}	352	8449
Mannitol 6AC	U+UHYAC	2080*	361_{26}	289_{34}	187_{53}	139_{70}	115_{100}	1544	1965
Mannitol 6HFB		1510*	521_{3}	478_{5}	307_{52}	240_{29}	169_{100}	1736	5802
Mannitol 6PFP		1510*	378_{3}	257_{17}	219_{34}	190_{24}	119_{100}	1735	5801
Mannitol 6TFA		1370*	321_{3}	278_{13}	265_{12}	140_{56}	69_{100}	1732	5800
Mannose 5AC	PAC U+UHYAC	2000*	331_{10}	242_{23}	157_{66}	115_{100}	98_{66}	1402	1964
Mannose 5HFB		1805*	519_{1}	465_{11}	321_{21}	257_{24}	169_{100}	1736	5805
Mannose 5PFP		1285*	419_{13}	365_{7}	227_{11}	147_{29}	119_{100}	1733	5804
Mannose 5TFA		1650*	290_{10}	265_{34}	221_{32}	157_{36}	69_{100}	1724	5803
Mannose isomer-1 5TMS		1885*	435_{1}	217_{26}	204_{100}	191_{57}	73_{96}	1690	4559
Mannose isomer-2 5TMS		1990*	435_{2}	217_{32}	204_{100}	191_{66}	73_{98}	1689	4560
5-MAPB		1550	188_{1}	174_{1}	131_{14}	77_{14}	58_{100}	368	8943
5-MAPB AC		1960	231_{1}	158_{16}	131_{12}	100_{36}	58_{100}	514	8944
5-MAPB HFB		1770	385_{1}	254_{100}	210_{45}	158_{70}	131_{43}	1380	8949
5-MAPB ME		1570	202_{1}	188_{1}	131_{6}	77_{4}	72_{100}	407	8945
5-MAPB PFP		1720	335_{1}	204_{100}	160_{37}	158_{52}	131_{27}	1123	8948
5-MAPB TFA		1750	285_{1}	158_{56}	154_{100}	131_{46}	110_{41}	808	8947
5-MAPB TMS		1710	260_{1}	246_{6}	130_{100}	115_{3}	73_{53}	665	8946

Table 8.1: Compounds in order of names   5-MAPB-M (di-HO-) 2AC

Name	Detected	RI	Typical ions and intensities					Page	Entry
5-MAPB-M (di-HO-) 2AC		2310	305_{14}	190_5	100_{76}	75_{72}	58_{100}	936	9231
5-MAPB-M (di-HO-nor-) 2AC		2240	291_3	232_2	190_6	163_{11}	75_{100}	847	9229
5-MAPB-M (HO-deamino-dihydro-) isomer-1 2AC		1810*	276_1	216_{14}	189_{19}	174_{16}	147_{100}	754	9235
5-MAPB-M (HO-deamino-dihydro-) isomer-2 2AC		1830*	276_1	216_{11}	189_{22}	174_{13}	147_{100}	754	9236
5-MAPB-M (HO-nor-) 2AC		2070	275_4	216_{68}	174_{21}	147_{100}	86_{95}	746	9227
5-MAPB-M (nor-)		1450	175_9	160_5	131_{100}	102_{25}	77_{77}	330	8951
5-MAPB-M (nor-)		1450	175_1	131_{13}	102_3	77_{10}	44_{100}	330	9083
5-MAPB-M (nor-) AC		1870	217_3	158_{100}	131_{67}	86_{59}	77_{51}	454	8950
5-MAPB-M (nor-) formyl artifact		1505	187_6	172_8	131_{41}	77_{19}	56_{100}	363	9084
5-MAPB-M (nor-) HFB		1670	371_1	240_{15}	158_{57}	131_{100}	77_{28}	1316	9088
5-MAPB-M (nor-) PFP		1640	321_2	190_{16}	158_{46}	131_{100}	77_{31}	1038	9087
5-MAPB-M (nor-) TFA		1655	271_2	158_{40}	131_{100}	77_{33}	69_{40}	722	9086
5-MAPB-M (nor-) TMS		1650	232_6	131_{23}	116_{100}	100_{11}	73_{57}	586	9085
5-MAPB-M (ring cleavage-carboxy-) ME2AC		2300	321_1	248_{14}	206_7	100_{67}	58_{100}	1039	9230
5-MAPB-M (ring cleavage-carboxy-nor-) ME2AC		2230	291_6	248_{44}	206_{87}	174_{50}	86_{100}	949	9233
5-MAPB-M (ring cleavage-HO-) 3AC		2360	335_1	262_8	160_8	100_{73}	58_{100}	1126	9232
5-MAPB-M (ring cleavage-HO-nor-) 3AC		2300	321_3	262_{60}	220_{90}	160_{100}	86_{85}	1039	9234
6-MAPB		1540	174_1	131_{23}	102_7	77_{17}	58_{100}	368	9206
6-MAPB AC		1960	231_1	158_{32}	131_{28}	100_{29}	58_{100}	513	9207
6-MAPB HFB		1730	385_1	254_{100}	210_{45}	158_{66}	131_{62}	1380	9211
6-MAPB ME		1630	201_1	131_{16}	102_2	77_6	72_{100}	407	9094
6-MAPB PFP		1690	335_3	204_{100}	160_{52}	158_{55}	131_{53}	1123	9210
6-MAPB TFA		1750	285_1	158_{73}	154_{100}	131_{57}	110_{41}	808	9209
6-MAPB TMS		1650	246_8	131_{53}	130_{100}	115_4	73_{63}	665	9208
6-MAPB-M (HO-) 2AC		2210	289_1	216_{21}	174_{14}	100_{39}	58_{100}	835	9223
6-MAPB-M (HO-deamino-dihydro-) 2AC		1830*	276_1	216_{15}	189_{15}	174_{26}	147_{100}	753	9217
6-MAPB-M (HO-nor-) 2AC		2065	276_1	216_{100}	174_{33}	147_{95}	86_{99}	754	9220
6-MAPB-M (nor-) AC		1890	217_8	158_{100}	131_{69}	86_{42}	77_{42}	455	9089
6-MAPB-M (nor-) HFB		1600	371_2	240_{15}	158_{62}	131_{100}	77_{26}	1316	9093
6-MAPB-M (nor-) PFP		1610	321_2	190_{15}	158_{54}	131_{100}	77_{31}	1038	9092
6-MAPB-M (nor-) TFA		1670	271_1	158_{42}	131_{100}	77_{34}	69_{42}	723	9091
6-MAPB-M (nor-) TMS		1655	232_5	131_{22}	116_{100}	100_9	73_{60}	586	9090
6-MAPB-M (ring cleavage-carboxy-) ME2AC		2270	321_1	248_{12}	206_6	100_{79}	58_{100}	1039	9221
6-MAPB-M (ring cleavage-carboxy-nor-) ME2AC		2200	307_2	248_{66}	206_{53}	174_{35}	86_{100}	949	9219
6-MAPB-M (ring cleavage-di-HO-nor-) -H2O 3AC		2400	319_{25}	260_{11}	218_{28}	176_{18}	86_{100}	1026	9224
6-MAPB-M (ring cleavage-HO-) 3AC		2350	335_1	262_2	160_7	100_{76}	58_{100}	1126	9222
6-MAPB-M (ring cleavage-HO-nor-) 3AC		2270	321_1	262_{12}	202_{46}	160_{85}	86_{100}	1039	9218
Maprotiline	P-I G UHY	2390	277_{24}	204_{100}	189_{79}	70_{56}	59_{63}	764	550
Maprotiline (ME)		2360	291_2	203_{24}	189_{18}	178_{12}	58_{100}	852	2254
Maprotiline AC	U+UHYAC	2800	319_5	291_{100}	218_{75}	203_{74}	191_{66}	1029	349
Maprotiline HFB		2525	473_2	445_{62}	240_{64}	203_{74}	191_{100}	1624	7681
Maprotiline PFP		2530	423_2	395_{77}	203_{61}	191_{100}	119_{71}	1514	7680
Maprotiline TFA		2430	373_3	345_{100}	203_{67}	191_{100}	140_{62}	1327	7679
Maprotiline TMS		2565	349_{11}	203_9	191_{15}	116_{100}	73_{46}	1206	4561
Maprotiline-M (deamino-di-HO-)	UHY	2570*	280_{15}	252_{100}	207_{64}			783	551
Maprotiline-M (deamino-di-HO-) 2AC	UHYAC	2820*	364_3	336_{67}	294_{100}	234_{11}	207_{26}	1282	351
Maprotiline-M (deamino-HO-propyl-) AC	U+UHYAC	2425*	306_2	278_{100}	218_{12}	203_{16}	191_{49}	944	350
Maprotiline-M (deamino-tri-HO-) 3AC	U+UHYAC	3200*	422_2	394_{84}	352_{56}	310_{100}	223_{21}	1513	355
Maprotiline-M (HO-anthryl-) 2AC	U+UHYAC	3095	377_8	349_{100}	307_{45}	234_{35}	207_{16}	1344	353
Maprotiline-M (HO-anthryl-) AC	U+UHYAC	2995	335_{56}	307_{100}	234_{32}	207_{41}	100_9	1127	6478
Maprotiline-M (HO-ethanediyl-) 2AC	U+UHYAC	2995	377_1	291_{100}	218_{44}	191_{25}	100_9	1344	352
Maprotiline-M (nor-) AC	P U+UHYAC	2760	305_4	277_{100}	218_{35}	191_{36}		938	348
Maprotiline-M (nor-di-HO-anthryl-) 3AC	UHYAC	3100	421_2	393_{100}	351_{38}	309_{49}	223_{15}	1510	3359
Maprotiline-M (nor-HO-anthryl-) 2AC	U+UHYAC	3150	363_3	335_{100}	293_{77}	234_{20}	207_{25}	1279	354
Maprotiline-M (nor-HO-anthryl-) AC	U+UHYAC	3020	321_{42}	293_{100}	234_{20}	222_{30}	207_{25}	1040	6479
Maprotiline-M (nor-HO-ethanediyl-) 2AC	U+UHYAC	2970	363_1	277_{100}	218_{32}	191_{38}	179_{10}	1279	6477
Maraviroc		3030	513_{14}	280_{14}	261_{100}	247_{31}	99_{36}	1672	7911
Maraviroc artifact (isopropylmethyltriazole)		1090	125_{12}	124_{16}	110_{100}	69_8	56_{24}	268	7909
Maraviroc artifact (isopropylmethyltriazole) HFB		1560	321_{14}	306_8	152_{100}	136_{11}	124_{17}	1037	7913
Maraviroc artifact (isopropylmethyltriazole) ME		1080	139_{19}	138_{11}	124_{100}	83_{14}	56_{17}	278	7910
Maraviroc artifact (isopropylmethyltriazole) TFA		1470	221_{19}	206_{20}	152_{100}	136_{14}	124_{17}	472	7912
Maropitant		3780	466_1	301_{100}	291_{38}	274_{25}	177_{56}	1615	9176
Mazindol AC		2705	326_{73}	284_{27}	256_{100}	220_{77}		1067	1073
Mazindol -H2O		2345	266_{102}	231_{22}	102_{27}			692	1072
2,3-MBDB		1610	178_3	135_3	89_4	72_{100}	57_7	420	5416
2,3-MBDB AC		1965	249_4	176_{17}	135_7	114_{35}	72_{100}	596	5507
2,3-MBDB HFB		1735	403_3	268_{100}	210_{17}	176_8	135_5	1455	5591
2,3-MBDB PFP		1710	353_8	218_{100}	176_{23}	160_{14}	135_{12}	1225	5592
2,3-MBDB TFA		1725	303_4	176_{43}	168_{100}	135_{14}	110_{23}	923	5508
2,3-MBDB TMS		1730	264_2	250_5	144_{100}	135_{13}	73_{54}	778	5593
2,3-MBDB-M (nor-)		1550	193_1	164_2	135_2	77_5	58_{100}	380	5414
2,3-MBDB-M (nor-) AC		1895	235_7	176_{43}	135_{10}	100_{21}	58_{100}	533	5504
2,3-MBDB-M (nor-) formyl artifact		1575	205_9	176_3	135_9	77_9	70_{100}	413	5415
2,3-MBDB-M (nor-) HFB		1660	389_{23}	345_8	254_{59}	176_{100}	135_{57}	1397	5505

MDBP-M (piperonylamine) 2AC　　　　　　　　　　　　　　　　　　Table 8.1: Compounds in order of names

Name	Detected	RI	Typical ions and intensities					Page	Entry
MDBP-M (piperonylamine) 2AC		2230	235_{23}	192_{35}	150_{100}	135_{17}	93_9	532	7631
MDBP-M (piperonylamine) 2TMS		2130	295_{27}	280_{100}	206_{15}	179_{50}	73_{51}	878	7630
MDBP-M (piperonylamine) AC	U+UHYAC	2015	193_{100}	150_{80}	135_{54}	121_{37}	93_{27}	377	6627
MDBP-M (piperonylamine) formyl artifact		1560	163_{22}	135_{100}	121_3	105_{10}	77_{37}	307	7629
MDBP-M (piperonylamine) HFB	U+UHYHFB	1640	347_{100}	317_6	289_{10}	178_6	135_{75}	1192	6633
MDBP-M (piperonylamine) PFP		1755	297_{100}	267_8	239_{13}	178_7	135_{51}	888	7632
MDBP-M (piperonylamine) TFA	U+UHYTFA	1775	247_{100}	217_9	189_{16}	148_{12}	135_{80}	583	6630
MDBP-M/artifact (piperazine) 2AC		1750	170_{15}	85_{33}	69_{25}	56_{100}		324	879
MDBP-M/artifact (piperazine) 2HFB		1290	478_3	459_9	309_{100}	281_{22}	252_{41}	1631	6634
MDBP-M/artifact (piperazine) 2TFA		1005	278_{10}	209_{59}	152_{25}	69_{56}	56_{100}	766	4129
5,6-MD-DALT		2450	284_2	188_2	174_{13}	116_7	110_{100}	804	8855
5,6-MD-DALT HFB		2350	480_1	384_4	370_8	173_{16}	110_{100}	1634	9144
5,6-MD-DALT PFP		2345	430_1	334_4	320_8	173_{14}	110_{100}	1533	9143
5,6-MD-DALT TFA		2385	380_1	284_6	270_9	200_7	110_{100}	1356	9142
5,6-MD-DALT TMS		2380	356_{13}	315_3	259_4	246_9	110_{100}	1243	9141
5,6-MD-DALT-M (demethylenyl-methyl-) AC		2550	328_4	287_5	244_{11}	176_8	110_{100}	1082	9277
5,6-MD-DALT-M AC		2520	403_4	343_5	301_{21}	110_{100}		1454	9274
MDDM		1760	206_1	192_2	135_{15}	105_6	72_{100}	421	2869
MDEA	G	1560	207_1	163_1	135_4	77_4	72_{100}	420	3257
MDEA AC	U+UHYAC	1985	249_1	162_{41}	135_6	114_{23}	72_{100}	597	3271
MDEA HFB		1790	403_5	268_{100}	240_{53}	162_{62}	135_{47}	1456	5087
MDEA PFP		1755	353_{16}	218_{100}	190_{73}	162_{71}	135_{38}	1225	5083
MDEA precursor-1 (piperonal)		1160*	150_{80}	149_{100}	121_{34}	91_{12}	63_{49}	289	3275
MDEA precursor-2 (isosafrole)		1215*	162_{100}	131_{51}	104_{71}	77_{52}	63_{28}	306	3276
MDEA precursor-3 (piperonylacetone)		1365*	178_{20}	135_{100}	105_7	77_{41}	51_{38}	337	3274
MDEA R-(-)-enantiomer HFBP		2460	500_1	365_6	266_{100}	162_{46}	135_8	1659	6644
MDEA S-(+)-enantiomer HFBP		2470	500_1	365_6	266_{100}	162_{46}	135_6	1659	6645
MDEA TFA		1770	303_{14}	168_{100}	162_{76}	140_{63}	135_{49}	923	5080
MDEA TMS		1825	264_{14}	144_{100}	135_{23}	100_4	73_{79}	778	4604
MDEA-D5		1555	212_1	135_8	105_2	77_{100}		440	7287
MDEA-D5 HFB		1770	408_3	273_{100}	241_{35}	162_{47}	135_{20}	1471	6772
MDEA-D5 PFP		1750	358_4	223_{93}	191_{53}	162_{100}	135_{44}	1253	7283
MDEA-D5 R-(-)-enantiomer HFBP		2455	505_1	370_{17}	266_{100}	162_{31}		1665	6802
MDEA-D5 S-(+)-enantiomer HFBP		2465	505_1	370_8	266_{100}	182_{38}		1666	6803
MDEA-D5 TFA		1765	308_9	173_{100}	162_{71}	141_{35}	135_{29}	958	7288
MDEA-D5 TMS		1820	284_1	269_3	149_{100}	135_6	73_{32}	806	7290
MDEA-M	UHY	1465	181_9	138_{100}	122_{18}	94_{24}	77_{16}	350	4351
MDEA-M (deamino-HO-) AC	U+UHYAC	1620*	222_9	162_{100}	135_{56}	104_{10}	77_{17}	479	6410
MDEA-M (deamino-oxo-)		1365*	178_{20}	135_{100}	105_7	77_{41}	51_{38}	337	3274
MDEA-M (deethyl-)	U UHY	1495	179_7	136_{100}	105_7	77_{29}		342	3241
MDEA-M (deethyl-)	U UHY	1495	179_1	136_{18}	77_5	51_5	44_{100}	340	5518
MDEA-M (deethyl-) AC	U+UHYAC	1860	221_7	162_{100}	135_{26}	86_{18}	77_{27}	474	3263
MDEA-M (deethyl-) AC	U+UHYAC	1860	221_3	162_{46}	135_{12}	86_8	44_{100}	474	5519
MDEA-M (deethyl-) HFB	UHFB	1650	375_4	240_6	169_{10}	162_{50}	135_{100}	1333	5291
MDEA-M (deethyl-) PFP	UPFP	1605	325_4	190_{10}	162_{46}	135_{100}	119_{18}	1060	5290
MDEA-M (deethyl-) R-(-)-enantiomer HFBP		2280	472_2	294_{15}	266_{100}	162_{66}	135_8	1622	6640
MDEA-M (deethyl-) S-(+)-enantiomer HFBP		2290	472_2	294_{18}	266_{100}	162_{76}	135_{10}	1621	6641
MDEA-M (deethyl-) TFA	UTFA	1615	275_4	162_{39}	135_{100}	105_6	77_{12}	744	5289
MDEA-M (deethyl-) TMS		1735	251_1	236_4	135_8	116_{100}	73_{47}	611	6334
MDEA-M (deethyl-)-D5 2AC		1910	268_3	167_{100}	166_{81}	136_{47}	90_{59}	708	5689
MDEA-M (deethyl-)-D5 AC		1840	226_{13}	167_{100}	166_{93}	136_{48}	78_{28}	496	5688
MDEA-M (deethyl-demethylenyl-) 3AC	U+UHYAC	2150	293_1	234_{26}	192_{42}	150_{89}	86_{100}	862	3725
MDEA-M (deethyl-demethylenyl-methyl-) 2AC	U+UHYAC	2065	265_3	206_{27}	164_{100}	137_{23}	86_{33}	688	3498
MDEA-M (demethylenyl-) 3AC	U+UHYAC	2200	321_2	234_8	150_{27}	114_{80}	72_{100}	1040	4208
MDEA-M (demethylenyl-methyl-)	UHY	1640	209_1	137_{12}	122_7	94_9	72_{100}	431	4364
MDEA-M (demethylenyl-methyl-) 2AC	U+UHYAC	2080	293_1	206_{20}	164_{38}	114_{77}	72_{100}	865	4209
MDEA-M (demethylenyl-methyl-) 2HFB		1695	360_{15}	284_8	218_{100}	190_{31}	163_8	1715	8483
MDEA-M (demethylenyl-methyl-) 2PFP		1645	310_{19}	218_{100}	190_{36}	119_{37}	72_{25}	1660	8482
MDEA-M (demethylenyl-methyl-) 2TFA		1650	401_1	260_{24}	233_4	168_{40}	140_{42}	1449	8479
MDEA-M (demethylenyl-methyl-) 2TMS		1885	338_2	209_4	179_7	144_{91}	73_{100}	1229	8477
MDEA-M (demethylenyl-methyl-) AC	UHYAC	2000	251_1	164_{46}	137_7	114_{33}	72_{100}	612	4274
MDEA-M (demethylenyl-methyl-) HFB		1735	360_{23}	333_6	268_{100}	240_{35}	107_{23}	1462	8484
MDEA-M (demethylenyl-methyl-) ME	UHYME	1930	223_{17}	194_7	151_{36}	94_{12}	72_{100}	485	4350
MDEA-M (demethylenyl-methyl-) PFP		1710	355_5	218_{100}	190_{49}	164_{70}	137_{59}	1236	8481
MDEA-M (demethylenyl-methyl-) TFA		1730	305_5	168_{100}	164_{70}	140_{50}	137_{58}	936	8480
MDEA-M (demethylenyl-methyl-) TMS		1645	266_1	210_2	179_3	149_2	72_{100}	790	8478
MDEA-M (HO-methoxy-hippuric acid) ME	UHYME	2165	239_9	151_{100}				551	4213
MDEA-M (methylenedioxy-hippuric acid) ME	UME UHYME	2065	237_{16}	178_4	149_{100}	121_{15}	65_{13}	542	4212
MDEA-M 2AC	U+UHYAC	1735*	250_3	208_{15}	166_{45}	123_{100}		603	4210
MDEA-M 2AC	U+UHYAC	1820*	266_3	206_9	164_{100}	150_{10}	137_{30}	695	6409
MDEA-M 2HFB	UHFB	1690	360_{82}	333_{100}	240_{100}	169_{42}	69_{39}	1704	6512
MDEA-M AC	U+UHYAC	1600*	222_2	180_{22}	137_{100}			479	4211
MDEA-M ME	UHYME	1550	195_1	152_{100}	137_{17}	107_{16}	77_{14}	387	4352
MDEA-M ME	UHYME	1540*	194_{25}	151_{100}	135_4	107_{18}	65_4	384	4353

Table 8.1: Compounds in order of names — 2,3-MDEA-M (deethyl-)

Name	Detected	RI	Typical ions and intensities					Page	Entry
2,3-MDEA-M (deethyl-)		1470	179_{34}	164_{13}	135_{96}	77_{100}	51_{91}	341	5420
2,3-MDEA-M (deethyl-)		1470	179_{3}	135_{7}	77_{11}	51_{8}	44_{100}	342	5513
2,3-MDEA-M (deethyl-) AC		1770	221_{30}	162_{100}	135_{35}	86_{8}	77_{40}	472	5589
2,3-MDEA-M (deethyl-) AC		1770	221_{18}	162_{71}	135_{35}	77_{64}	44_{100}	474	6310
2,3-MDEA-M (deethyl-) formyl artifact		1490	191_{11}	135_{6}	105_{4}	77_{11}	56_{100}	371	5421
2,3-MDEA-M (deethyl-) HFB		1595	375_{12}	240_{42}	169_{11}	162_{100}	135_{61}	1333	5502
2,3-MDEA-M (deethyl-) PFP		1545	325_{8}	190_{42}	162_{100}	135_{71}	119_{46}	1060	5542
2,3-MDEA-M (deethyl-) TFA		1585	275_{32}	162_{100}	140_{56}	135_{77}	77_{26}	744	5503
2,3-MDEA-M (deethyl-) TMS		1655	251_{1}	236_{7}	135_{19}	116_{100}	73_{71}	611	5590
MDMA	G P-I	1790	193_{4}	177_{3}	135_{19}	77_{10}	58_{100}	379	2599
MDMA AC	U+UHYAC	2140	235_{1}	162_{20}	100_{14}	77_{7}	58_{100}	534	2600
MDMA HFB	PHFB	1740	389_{7}	254_{100}	210_{43}	162_{82}	135_{56}	1397	5086
MDMA intermediate		2025	207_{35}	160_{50}	131_{11}	103_{100}	77_{69}	418	2842
MDMA PFP		1750	339_{23}	204_{100}	162_{64}	135_{38}	119_{17}	1148	2601
MDMA precursor-1 (piperonal)		1160*	150_{80}	149_{100}	121_{34}	91_{12}	63_{49}	289	3275
MDMA precursor-2 (isosafrole)		1215*	162_{100}	131_{51}	104_{71}	77_{52}	63_{29}	306	3276
MDMA precursor-3 (piperonylacetone)		1365*	178_{20}	135_{100}	105_{7}	77_{41}	51_{38}	337	3274
MDMA R-(-)-enantiomer HFBP		2450	486_{1}	351_{5}	266_{100}	162_{31}	135_{4}	1643	6642
MDMA S-(+)-enantiomer HFBP		2460	486_{1}	351_{5}	266_{100}	162_{60}	135_{7}	1643	6643
MDMA TFA		1720	289_{5}	162_{75}	154_{100}	135_{67}	110_{44}	833	5079
MDMA TMS		1710	250_{4}	135_{16}	130_{100}	77_{15}	73_{67}	689	4562
MDMA-D5		1770	198_{1}	136_{5}	78_{4}	62_{100}		395	6356
MDMA-D5 AC		2130	240_{2}	164_{35}	136_{9}	104_{26}	62_{100}	558	6355
MDMA-D5 HFB		1750	394_{13}	258_{100}	213_{32}	164_{45}	136_{27}	1420	6359
MDMA-D5 PFP		1740	344_{15}	208_{100}	163_{62}	136_{33}	119_{10}	1178	6358
MDMA-D5 R-(-)-enantiomer HFBP		2445	491_{1}	355_{6}	266_{100}	164_{38}		1650	6800
MDMA-D5 S-(+)-enantiomer HFBP		2455	491_{3}	355_{11}	266_{100}	164_{43}		1650	6801
MDMA-D5 TFA		1700	294_{13}	164_{48}	158_{100}	136_{55}	113_{28}	870	6357
MDMA-D5 TMS		1700	255_{6}	134_{100}	73_{30}			720	6360
MDMA-M	UHY	1465	181_{9}	138_{100}	122_{18}	94_{24}	77_{16}	350	4351
MDMA-M (deamino-HO-) AC	U+UHYAC	1620*	222_{1}	162_{100}	135_{56}	104_{10}	77_{17}	479	6410
MDMA-M (deamino-oxo-)		1365*	178_{20}	135_{100}	105_{7}	77_{41}	51_{38}	337	3274
MDMA-M (demethylenyl-) 3AC	U+UHYAC	2190	307_{1}	234_{7}	150_{12}	100_{59}	58_{100}	950	4244
MDMA-M (demethylenyl-methyl-)	UHY	1810	195_{1}	137_{4}	122_{2}	94_{3}	58_{100}	387	4246
MDMA-M (demethylenyl-methyl-) 2HFB	UHFB	1760	360_{28}	333_{6}	254_{100}	210_{23}	169_{18}	1710	6492
MDMA-M (demethylenyl-methyl-) isomer-1 2AC	U+UHYAC	2095	279_{1}	206_{16}	164_{26}	100_{43}	58_{100}	776	6757
MDMA-M (demethylenyl-methyl-) isomer-2 2AC	U+UHYAC	2115	279_{1}	206_{19}	164_{26}	100_{48}	58_{100}	775	4243
MDMA-M (HO-methoxy-hippuric acid) ME	UHYME	2165	239_{9}	151_{100}				551	4213
MDMA-M (methylenedioxy-hippuric acid) ME	UME UHYME	2065	237_{16}	178_{4}	149_{100}	121_{15}	65_{13}	542	4212
MDMA-M (nor-)	U UHY	1495	179_{7}	136_{100}	105_{7}	77_{29}		342	3241
MDMA-M (nor-)	U UHY	1495	179_{1}	136_{18}	77_{5}	51_{5}	44_{100}	340	5518
MDMA-M (nor-) AC	U+UHYAC	1860	221_{7}	162_{100}	135_{26}	86_{10}	77_{27}	474	3263
MDMA-M (nor-) AC	U+UHYAC	1860	221_{3}	162_{46}	135_{12}	86_{8}	44_{100}	474	5519
MDMA-M (nor-) HFB	UHFB	1650	375_{4}	240_{6}	169_{10}	162_{50}	135_{100}	1333	5291
MDMA-M (nor-) PFP	UPFP	1605	325_{4}	190_{10}	162_{46}	135_{100}	119_{18}	1060	5290
MDMA-M (nor-) R-(-)-enantiomer HFBP		2280	472_{2}	294_{15}	266_{100}	162_{66}	135_{8}	1622	6640
MDMA-M (nor-) S-(+)-enantiomer HFBP		2290	472_{2}	294_{16}	266_{100}	162_{76}	135_{10}	1621	6641
MDMA-M (nor-) TFA	UTFA	1615	275_{4}	162_{39}	135_{100}	105_{5}	77_{12}	744	5289
MDMA-M (nor-) TMS		1735	251_{1}	236_{4}	135_{8}	116_{100}	73_{47}	611	6334
MDMA-M (nor-)-D5 2AC		1910	268_{3}	167_{100}	166_{81}	136_{47}	90_{59}	708	5689
MDMA-M (nor-)-D5 AC		1840	226_{13}	167_{100}	166_{93}	136_{48}	78_{28}	496	5688
MDMA-M (nor-demethylenyl-) 3AC	U+UHYAC	2150	293_{1}	234_{26}	192_{42}	150_{89}	86_{100}	862	3725
MDMA-M (nor-demethylenyl-methyl-) 2AC	U+UHYAC	2065	265_{3}	206_{27}	164_{100}	137_{23}	86_{33}	688	3498
MDMA-M (nor-demethylenyl-methyl-) 2HFB	UHFB	1690	360_{82}	333_{16}	240_{100}	169_{42}	69_{39}	1704	6512
MDMA-M 2AC	U+UHYAC	1735*	250_{3}	208_{15}	166_{45}	123_{100}		603	4210
MDMA-M 2AC	U+UHYAC	1820*	266_{3}	206_{9}	164_{100}	150_{10}	137_{30}	695	6409
MDMA-M AC	U+UHYAC	1600*	222_{2}	180_{22}	137_{100}			479	4211
MDMA-M ME	UHYME	1550	195_{1}	152_{100}	137_{17}	107_{16}	77_{14}	387	4352
MDMA-M ME	UHYME	1540*	194_{25}	151_{100}	135_{4}	107_{15}	65_{9}	384	4353
MDMB-CHMICA	P-I	2930	384_{14}	328_{16}	268_{8}	240_{100}	144_{22}	1379	9589
MDMB-CHMICA TMS		2670	456_{5}	399_{72}	328_{5}	240_{100}	144_{41}	1594	9590
MDMB-CHMINACA		2725	385_{5}	329_{35}	326_{20}	297_{20}	241_{100}	1382	9586
MDMB-CHMINACA -COOCH3		2520	325_{1}	241_{100}	145_{65}	131_{28}	103_{14}	1065	9587
MDMB-CHMINACA TMS		2620	457_{9}	442_{8}	400_{34}	241_{100}	145_{28}	1596	9588
2,3-MDMA-M (nor-)		1470	179_{34}	164_{13}	135_{96}	77_{100}	51_{91}	341	5420
2,3-MDMA-M (nor-)		1470	179_{3}	135_{7}	77_{11}	51_{8}	44_{100}	342	5513
2,3-MDMA-M (nor-) AC		1770	221_{30}	162_{100}	135_{35}	86_{8}	77_{40}	472	5589
2,3-MDMA-M (nor-) AC		1770	221_{18}	162_{71}	135_{35}	77_{64}	44_{100}	474	6310
2,3-MDMA-M (nor-) formyl artifact		1490	191_{11}	135_{6}	105_{4}	77_{11}	56_{100}	371	5421
2,3-MDMA-M (nor-) HFB		1595	375_{12}	240_{42}	169_{11}	162_{100}	135_{61}	1333	5502
2,3-MDMA-M (nor-) PFP		1545	325_{8}	190_{42}	162_{100}	135_{71}	119_{46}	1060	5542
2,3-MDMA-M (nor-) TFA		1585	275_{32}	162_{100}	140_{56}	135_{77}	77_{26}	744	5503
2,3-MDMA-M (nor-) TMS		1655	251_{1}	236_{7}	135_{19}	116_{100}	73_{71}	611	5590
MDPBP		2080	261_{1}	149_{10}	121_{10}	112_{100}	96_{4}	664	8709

MDPBP artifact (-2H)   **Table 8.1:** Compounds in order of names

Name	Detected	RI	Typical ions and intensities					Page	Entry
MDPBP artifact (-2H)		2100	259_{19}	190_{20}	149_{11}	121_{10}	110_{100}	651	8717
MDPBP artifact (-4H)		2050	257_{14}	149_{63}	121_{10}	108_{100}		642	8708
MDPBP-M (carboxy-oxo-) AC	UGLUCAC	2570	349_1	276_{11}	200_{68}	158_{100}	149_{21}	1203	8741
MDPBP-M (deamino-oxo-)		1670*	206_4	149_{100}	121_{18}	91_4	65_8	415	8750
MDPBP-M (demethylenyl-) 2AC		2315	333_1	137_9	112_{100}	96_3	70_6	1112	8735
MDPBP-M (demethylenyl-deamino-oxo-)	UGLUC	1640*	194_{19}	137_{100}	57_{17}			382	8718
MDPBP-M (demethylenyl-deamino-oxo-) 2AC	UGLUCAC	1920*	236_2	221_{28}	179_{74}	137_{100}	109_5	766	8721
MDPBP-M (demethylenyl-deamino-oxo-dihydro-) 3AC	UGLUCAC	2090*	322_1	265_{25}	223_{53}	181_{100}	139_{40}	1043	8724
MDPBP-M (demethylenyl-methyl-) artifact (-2H)	UGLUC	2170	261_{20}	192_{30}	151_7	110_{100}	70_{47}	664	8726
MDPBP-M (demethylenyl-methyl-) artifact (-4H)	UGLUC	2090	259_{19}	230_3	151_{100}	123_{11}	108_{84}	651	8743
MDPBP-M (demethylenyl-methyl-) artifact (-4H) AC	UGLUC	2190	301_3	151_{46}	123_5	108_{100}		913	8727
MDPBP-M (demethylenyl-methyl-) isomer-1	UGLUC	2110	263_1	191_8	151_6	123_4	112_{100}	675	8725
MDPBP-M (demethylenyl-methyl-) isomer-1 AC	UGLUC	2210	305_1	151_9	123_6	112_{100}	70_9	938	8730
MDPBP-M (demethylenyl-methyl-) isomer-1 artifact (-2H) AC	UGLUC	2265	303_{24}	192_{34}	151_6	110_{100}	70_{37}	925	8732
MDPBP-M (demethylenyl-methyl-) isomer-2	UGLUC	2185	263_1	191	151_7	123_4	112_{100}	675	8744
MDPBP-M (demethylenyl-methyl-) isomer-2 AC	UGLUC	2245	305_1	151_6	123_6	112_{100}	70_5	937	8731
MDPBP-M (demethylenyl-methyl-) isomer-2 artifact (-2H) AC	UGLUC	2290	303_{12}	192_{31}	151_{10}	110_{100}	70_{46}	925	8733
MDPBP-M (demethylenyl-methyl-carboxy-oxo-) 2AC	UGLUCAC	2635	320_8	200_{72}	158_{100}	151_{18}	101_{31}	1414	8742
MDPBP-M (demethylenyl-methyl-deamino-oxo-)	UGLUC	1670*	208_1	151_{100}	123_{15}	108_5		422	8719
MDPBP-M (demethylenyl-methyl-deamino-oxo-) AC	UGLUCAC	1790*	250_1	193_{18}	151_{100}	123_6	108_3	603	8720
MDPBP-M (demethylenyl-methyl-deamino-oxo-dihydro-) isomer-1 2AC	UGLUCAC	1970*	294_2	252_6	237_{12}	195_{36}	153_{100}	869	8722
MDPBP-M (demethylenyl-methyl-deamino-oxo-dihydro-) isomer-2 2AC	UGLUCAC	1990*	237_{20}	219_3	195_{100}	153_{52}		869	8723
MDPBP-M (demethylenyl-methyl-HO-alkyl-)	UGLUC	2300	151_6	128_{100}	123_4	110_9		774	8747
MDPBP-M (demethylenyl-methyl-HO-alkyl-) isomer-1 2AC		2390	170_{100}	151_{11}	123_4	110_{40}		1278	8737
MDPBP-M (demethylenyl-methyl-HO-alkyl-) isomer-2 2AC		2470	192_1	170_{100}	151_7	123_2	110_{35}	1278	8738
MDPBP-M (demethylenyl-methyl-HO-phenyl-)	UGLUC	2335	167_3	139_3	112_{100}	96_2	70_3	775	8749
MDPBP-M (demethylenyl-methyl-HO-phenyl-) 2AC		2410	167_3	139_1	112_{100}	96_2	70_3	1278	8739
MDPBP-M (demethylenyl-methyl-HO-phenyl-deamino-oxo-dihydro-) 3AC	UGLUC	2200*	352_1	295_{16}	253_{100}	211_{51}	169_{41}	1219	8728
MDPBP-M (demethylenyl-methyl-N,N-bis-dealkyl-) 2AC	UGLUCAC	2205	293_3	251_{17}	151_{100}	100_{66}	58_{80}	863	8729
MDPBP-M (demethylenyl-methyl-oxo-)	UGLUC	2250	194_{10}	151_{10}	126_{100}	98_2	69_8	760	8746
MDPBP-M (demethylenyl-methyl-oxo-) artifact (-4H)	UGLUC	2145	273_{19}	151_{26}	122_{100}	108_{15}		734	8745
MDPBP-M (demethylenyl-methyl-oxo-pyrrolidinyl-) AC		2355	319_1	236_5	151_6	126_{100}	98_7	1026	8736
MDPBP-M (HO-alkyl-) AC		2630	310_2	170_{22}	149_{100}	121_0	110_{11}	1026	8740
MDPBP-M (oxo-carboxy-) (ME)	UGLUC	2325	172_{16}	158_{100}	149_{21}	126_{34}	101_{16}	1038	8748
MDPBP-M (oxo-pyrrolidinyl-)		2275	275_1	192_{15}	149_{15}	126_{100}	121_{10}	745	8734
2,3-MDPEA		1410	165_{64}	136_{100}	106_{17}	89_5	77_{45}	313	8415
2,3-MDPEA AC		1785	207_{32}	148_{100}	135_{33}	105_8	77_{28}	419	8417
2,3-MDPEA formyl artifact		1475	177_{83}	148_{14}	135_{100}	91_{29}	77_{61}	334	8416
2,3-MDPEA HFB		1620	361_{32}	226_4	148_{100}	135_{67}	77_{35}	1267	8420
2,3-MDPEA PFP		1580	311_{30}	148_{100}	135_{74}	105_{15}	77_{44}	974	8419
2,3-MDPEA TFA		1585	261_{32}	194_3	148_{100}	135_{54}	77_{31}	662	8418
2,3-MDPEA TMS		1565	222_1	194_3	135_6	102_{98}	73_{100}	544	8465
MDPPP		1995	178_1	149_2	121_2	98_{100}	56_{19}	585	5422
MDPPP-M (4-HO-3-methoxy-benzoic acid) 2ET		1675*	224_{44}	196_{17}	179_{68}	151_{100}		488	6531
MDPPP-M (deamino-oxo-)		1525*	192_{12}	149_{100}	121_{24}	65_7		374	6529
MDPPP-M (demethylene-) 2ET		2165	290_1	193_1	165_2	137_3	98_{100}	851	6525
MDPPP-M (demethylene-deamino-oxo-) 2ET		1720*	236_4	193_{100}	165_{48}	137_{45}	109_{12}	538	6530
MDPPP-M (demethylene-methyl-) ET		2135	277_1	208_1	179_2	151_7	98_{100}	762	6524
MDPPP-M (demethylene-methyl-) HFB		1960	445_1	347_4	98_{100}	69_{19}		1571	6532
MDPPP-M (demethylene-methyl-) ME		2070	165_1	98_{100}	79_3	69_4	56_{11}	676	6538
MDPPP-M (demethylene-methyl-) TMS		1960	321_1	306_3	223_3	165_1	98_{100}	1041	6533
MDPPP-M (demethylene-methyl-deamino-oxo-) ET		1680*	222_4	179_{100}	151_{81}	123_{19}		479	6523
MDPPP-M (demethylene-methyl-oxo-) ET		2290	290_1	208_{12}	179_9	151_{11}	112_{100}	848	6527
MDPPP-M (demethylene-oxo-) 2ET		2325	305_2	222_3	193_4	151_{11}	112_{100}	937	6526
MDPPP-M (dihydro-)		2040	248_1	149_3	121_2	98_{100}	56_{19}	598	6698
MDPPP-M (dihydro-) AC		2065	232_3	149_6	121_2	98_{100}	56_{15}	848	6703
MDPPP-M (dihydro-) TMS		1965	306_2	232_1	149_3	98_{100}	73_{24}	1041	6708
MDPPP-M (oxo-)		2290	261_2	178_{26}	149_{19}	121_{10}	112_{100}	663	6528
MDPV		2110	275_1	232_1	149_7	126_{100}	84_5	747	7980
MDPV artifact (bis-dehydro-)		2095	271_{38}	149_{56}	122_{100}	80_{51}		724	7982
MDPV artifact (dehydro-)		2130	273_{43}	230_{11}	204_{21}	124_{100}	70_{45}	734	7981
MDPV-M (deamino-oxo-)	U+UHYAC	1675*	220_8	176_3	149_{100}	121_{26}	99_9	468	8877
MDPV-M (demethylenyl-)	UGLSPE	2205	248_1	193_3	165_{10}	126_{100}	96_6	676	7993
MDPV-M (demethylenyl-) 2AC	UGLSPEAC	2340	347_1	262_1	137_7	126_{100}	84_8	1195	7986
MDPV-M (demethylenyl-) 2TMS	UGLSPETMS	2300	392_5	281_4	223_3	126_{100}	96_7	1469	8005
MDPV-M (demethylenyl-deamino-oxo-dihydro-) 3AC	U+UHYAC	2175*	336_1	265_{26}	223_{60}	181_{100}	139_{44}	1130	8880
MDPV-M (demethylenyl-HO-) 3TMS	UGLSPETMS	2455	480_2	214_{100}	137_1	124_2		1655	8008
MDPV-M (demethylenyl-methyl-) AC	UGLSPEAC	2155	234_2	151_{14}	126_{100}	84_{14}		1028	7983
MDPV-M (demethylenyl-methyl-) isomer-1	UGLSPE	2150	208_1	193_2	151_{10}	126_{100}	96_9	762	7994
MDPV-M (demethylenyl-methyl-) isomer-2	UGLSPE	2240	234_2	193_1	151_{16}	126_{100}	96_{11}	763	7995
MDPV-M (demethylenyl-methyl-) TMS	UGLSPE	2260	334_3	223_{15}	165_3	126_{100}	84_6	1206	8006
MDPV-M (demethylenyl-methyl-deamino-oxo-) AC	U+UHYAC	1835*	264_1	193_{30}	151_{100}	137_{13}	123_{10}	681	8876
MDPV-M (demethylenyl-methyl-deamino-oxo-dihydro-) isomer-1 2AC	U+UHYAC	1990*	308_4	266_6	237_{20}	195_{56}	153_{100}	957	8878
MDPV-M (demethylenyl-methyl-deamino-oxo-dihydro-) isomer-2 2AC	U+UHYAC	2000*	308_2	237_{25}	205_{10}	195_{100}	153_{41}	957	8879

Table 8.1: Compounds in order of names   **MDPV-M (demethylenyl-methyl-HO-)**

Name	Detected	RI	Typical ions and intensities					Page	Entry
MDPV-M (demethylenyl-methyl-HO-)	UGLSPE	2385	165_1	151_2	142_{100}	124_5		865	7996
MDPV-M (demethylenyl-methyl-HO-) 2TMS	UGLSPETMS	2430	422_2	223_2	214_{100}	124_5	73_9	1552	8007
MDPV-M (demethylenyl-methyl-HO-) isomer-1 2AC	UGLSPEAC	2530	377_1	184_{100}	151_6	124_{55}	95_7	1343	7988
MDPV-M (demethylenyl-methyl-HO-) isomer-1 TFA	UGLSPETFA	2175	370_1	238_{100}	151_8	124_{38}	95_7	1399	8001
MDPV-M (demethylenyl-methyl-HO-) isomer-2 2AC	UGLSPEAC	2570	184_{100}	151_{10}	124_{45}	95_7		1344	7992
MDPV-M (demethylenyl-methyl-HO-) isomer-2 TFA	UGLSPETFA	2240	238_{100}	151_{11}	124_{44}	95_3		1398	8002
MDPV-M (demethylenyl-methyl-HO-phenyl-) 2TMS	UGLSPE	2320	422_2	311_1	209_1	126_{100}	96_3	1552	8013
MDPV-M (demethylenyl-methyl-N,N-bis-dealkyl-) 2AC	UGLSPEAC	2245	307_4	265_{16}	193_8	114_{78}	72_{100}	949	7984
MDPV-M (demethylenyl-methyl-N,N-bis-dealkyl-) 2TMS	UGLSPETMS	2070	352_6	223_4	208_2	144_{100}	73_{36}	1299	8010
MDPV-M (demethylenyl-methyl-oxo-) AC	UGLSPEAC	2400	250_{10}	151_4	140_{100}	98_{36}	86_{12}	1112	7987
MDPV-M (demethylenyl-methyl-oxo-) isomer-1	UGLSPE	2300	291_1	208_7	151_{11}	140_{100}	98_{25}	848	7997
MDPV-M (demethylenyl-methyl-oxo-) isomer-2	UGLSPE	2440	291_1	208_9	151_{12}	140_{100}	98_{25}	848	7998
MDPV-M (demethylenyl-methyl-oxo-) TMS	UGLSPETMS	2410	363_3	348_{10}	280_{19}	223_{52}	140_{100}	1279	8012
MDPV-M (demethylenyl-methyl-oxo-carboxy-) isomer-1 2AC	UGLSPEAC	2590	279_{20}	214_{63}	172_{100}	151_7	140_{23}	1467	7989
MDPV-M (demethylenyl-methyl-oxo-carboxy-) isomer-2 2AC	UGLSPEAC	2610	279_1	214_{73}	172_{100}	151_7	140_{17}	1467	7991
MDPV-M (demethylenyl-methyl-oxo-carboxy-) ME	UGLSPEME	2380	186_2	172_{100}	165_{11}	140_5	101_{16}	1138	8000
MDPV-M (demethylenyl-methyl-oxo-carboxy-) TFA	UGLSPETFA	2320	419_5	268_{13}	236_{30}	194_7	151_{100}	1505	8004
MDPV-M (demethylenyl-N,N-bis-dealkyl-) 3TMS	UGLSPETMS	2115	425_1	410_2	365_2	144_{100}	72_{24}	1521	8009
MDPV-M (demethylenyl-oxo-) 2TMS	UGLSPETMS	2450	406_5	338_{16}	281_{56}	140_{100}	98_{15}	1511	8011
MDPV-M (oxo-)	UGLSPEAC	2330	289_7	206_{59}	140_{100}	98_{84}	86_{41}	835	7985
MDPV-M (oxo-carboxy-) AC	UGLSPEAC	2645	290_{12}	214_{85}	172_{100}	140_{48}	101_{55}	1278	7990
MDPV-M (oxo-carboxy-) ME	UGLSPEME	2380	305_2	222_{16}	172_{100}	140_{64}	101_{48}	1124	7999
MDPV-M (oxo-carboxy-) TFA	UGLSPETFA	2320	268_{14}	236_{41}	194_9	149_{100}	101_{15}	1499	8003
1-Me-2-Ph-AMT		2335	264_3	221_{100}	204_{27}	178_7	144_6	683	9769
1-Me-2-Ph-AMT AC		2640	306_{19}	247_{57}	220_{100}	204_{43}	178_{14}	944	9771
1-Me-2-Ph-AMT ethylimine artifact		2235	290_9	220_{100}	204_{18}	178_5	70_8	843	10213
1-Me-2-Ph-AMT formyl artifact		2335	276_{15}	220_{100}	204_{23}	178_6	56_{14}	756	9770
1-Me-2-Ph-AMT HFB		2360	460_2	240_{10}	220_{100}	204_{27}	178_6	1601	9775
1-Me-2-Ph-AMT PFP		2350	410_9	247_4	220_{100}	204_{33}	178_3	1478	9773
1-Me-2-Ph-AMT TFA		2385	360_{13}	247_3	220_{100}	204_{27}	178_9	1265	9772
1-Me-2-Ph-AMT TMS		2700	321_6	221_{61}	204_{25}	116_{100}	73_{44}	1133	9774
1-Me-AMT		1695	188_4	144_{100}	128_7	102_7	44_{27}	366	10218
1-Me-AMT		1695	188_3	144_{100}	128_7	102_7	77_7	365	10220
1-Me-AMT AC		2155	230_{24}	171_{64}	144_{100}	128_{10}	115_9	511	9789
1-Me-AMT ethylimine artifact		1705	214_{12}	144_{100}	128_4	102_3	70_{11}	447	10219
1-Me-AMT ethylimine artifact		1705	214_{12}	144_{100}	128_4	102_4	70_{12}	447	10221
1-Me-AMT HFB		1920	384_{31}	240_{10}	171_7	144_{100}	128_5	1374	9792
1-Me-AMT PFP		1905	334_{30}	190_{10}	171_{11}	144_{100}	128_{11}	1119	9791
1-Me-AMT TFA		1925	284_{26}	171_7	144_{100}	128_9	115_9	802	9790
1-Me-AMT TMS		1925	260_2	245_5	144_{34}	116_{100}	100_{11}	660	9793
2-Me-AMT		1810	188_4	145_{100}	130_{21}	115_7	44_{43}	365	10235
2-Me-AMT		1810	188_2	145_{100}	130_{17}	115_7	77_7	365	10237
2-Me-AMT 2PFP		1880	480_3	317_{10}	290_{100}	190_6	143_{18}	1634	9825
2-Me-AMT 2TMS		2145	317_2	217_{30}	200_4	144_4	116_{100}	1107	9823
2-Me-AMT AC		2150	230_{21}	171_{49}	144_{100}	128_6	115_8	510	9822
2-Me-AMT ethylimine artifact		1800	214_6	171_3	144_{100}	128_4	70_{28}	447	10236
2-Me-AMT ethylimine artifact		1800	214_{11}	171_3	144_{100}	128_3	70_{22}	448	10238
2-Me-AMT HFB		2050	384_7	240_{10}	172_4	144_{100}	130_6	1375	9827
2-Me-AMT PFP		2030	334_8	190_4	144_{100}	128_5	119_{14}	1119	9826
2-Me-AMT TFA		1950	284_{14}	217_5	170_6	144_{100}	128_{10}	802	9824
2-Me-AMT TMS		2010	245_2	144_{24}	116_{100}	100_8	73_{51}	660	9824
5-Me-AMT		1770	188_3	162_4	145_{32}	115_4	44_{100}	366	10258
5-Me-AMT 2HFB		1830	580_{12}	367_{69}	340_{100}	240_{49}	143_{40}	1707	9858
5-Me-AMT 2PFP		1800	480_{17}	317_{70}	290_{100}	190_{37}	143_{41}	1633	9855
5-Me-AMT 2TFA		1835	380_{20}	267_{35}	240_{100}	143_{17}	140_{18}	1355	9853
5-Me-AMT 2TMS		2075	332_1	317_6	217_{54}	116_{100}	100_{11}	1108	9852
5-Me-AMT AC		2155	230_{20}	171_{77}	144_{100}	128_4	115_{18}	510	9851
5-Me-AMT ethylimine artifact		1800	214_{13}	171_{12}	144_{100}	115_5	70_{32}	448	10257
5-Me-AMT formyl artifact		2050	200_{14}	183_6	157_{100}	128_5	115_6	400	9850
5-Me-AMT HFB		1960	384_{15}	240_5	171_6	144_{100}	115_8	1375	9857
5-Me-AMT PFP		1940	334_{20}	190_6	171_6	144_{100}	115_7	1119	9856
5-Me-AMT TFA		1970	284_4	171_5	144_{100}	140_8	115_9	802	9854
7-Me-AMT		1765	188_3	145_{100}	130_5	115_{10}	91_7	366	10259
7-Me-AMT		1765	188_3	145_{100}	130_4	115_{13}	44_{78}	366	10261
7-Me-AMT 2HFB		1805	580_8	367_{38}	340_{100}	240_{28}	143_{23}	1707	9868
7-Me-AMT 2PFP		1765	480_7	317_{37}	290_{100}	190_{63}	143_{65}	1634	9866
7-Me-AMT 2TFA		1820	380_{33}	267_{36}	240_{100}	143_{25}	140_{24}	1355	9863
7-Me-AMT 2TMS		2085	332_1	317_{11}	217_{60}	116_{100}	100_8	1108	9862
7-Me-AMT AC		2165	230_{19}	171_{79}	144_{100}	128_4	115_{19}	511	9860
7-Me-AMT ethylimine artifact		1780	214_5	171_3	144_{100}	115_{17}	70_{35}	447	10260
7-Me-AMT formyl artifact		1985	200_{11}	183_7	157_{100}	128_3	115_7	400	9859
7-Me-AMT HFB		1935	384_7	240_6	171_7	144_{100}	115_8	1375	9867
7-Me-AMT PFP		1915	334_{23}	190_6	171_7	144_{100}	115_9	1119	9865
7-Me-AMT TFA		1950	284_{19}	171_6	144_{100}	140_6	115_9	802	9864

7-Me-AMT TMS

Table 8.1: Compounds in order of names

Name	Detected	RI	Typical ions and intensities					Page	Entry
7-Me-AMT TMS		1980	260_1	245_{10}	144_{45}	116_{100}	100_{23}	660	9861
Mebendazole artifact (amine) 3ME		2930	279_{100}	264_{70}	249_{22}	173_{15}	77_{23}	774	7543
Mebendazole artifact (amine) isomer-1 2ME		2930	265_{100}	250_{10}	188_{58}	160_{11}	77_{10}	687	7542
Mebendazole artifact (amine) isomer-2 2ME		2950	265_{100}	250_{12}	188_{60}	160_{24}	77_{30}	687	7545
Mebendazole isomer-1 2ME		2785	323_{62}	264_{100}	246_{16}	159_{13}	77_{35}	1049	7544
Mebendazole isomer-2 2ME		2930	323_{62}	264_{100}	246_{16}	159_{13}	77_{35}	1050	7541
Mebendazole ME		2950	309_{100}	250_{50}	232_{96}	200_{42}	77_{67}	962	7540
Mebeverine		3045	428_1	308_{100}	165_{18}	121_7	84_2	1532	4404
Mebeverine-M (3,4-dihydroxybenzoic acid) ME2AC		1750*	252_1	210_{19}	168_{100}	137_{60}	109_8	615	5254
Mebeverine-M (HO-phenyl-alcohol) 2AC		2415	186_{100}	165_4	137_6	98_3	72_4	1288	5326
Mebeverine-M (HO-phenyl-O-demethyl-alcohol) 3AC		2525	193_2	186_{100}	151_4	123_4	72_4	1417	5327
Mebeverine-M (isovanillic acid) MEAC		1630*	224_6	193_2	182_{92}	151_{100}	123_{13}	486	5329
Mebeverine-M (N-dealkyl-)		1660	192_1	149_1	121_5	91_2	72_{100}	380	5831
Mebeverine-M (N-dealkyl-) AC	UHYAC	1855	235_1	148_{32}	121_7	114_{26}	72_{100}	535	5322
Mebeverine-M (N-dealkyl-) HFB	UHFB	1785	389_2	268_{100}	240_{46}	148_{62}	121_{43}	1398	5834
Mebeverine-M (N-dealkyl-) ME		1780	206_1	121_6	86_{100}	72_3	58_{20}	422	5835
Mebeverine-M (N-dealkyl-) PFP	UPFP	1765	339_1	218_{100}	190_{44}	148_{59}	121_{49}	1149	5033
Mebeverine-M (N-dealkyl-) TFA	UTFA	1775	289_1	168_{100}	148_{63}	140_{41}	121_{62}	834	5832
Mebeverine-M (N-dealkyl-) TMS		2065	264_1	250_{14}	144_{100}	121_{17}	73_{84}	691	5836
Mebeverine-M (N-dealkyl-N-deethyl-) AC	U+UHYAC	1720	207_1	148_{100}	121_{40}	91_{11}	86_{24}	421	3265
Mebeverine-M (N-dealkyl-N-deethyl-) AC	U+UHYAC	1720	207_1	148_{41}	121_{16}	86_{10}	44_{100}	421	5537
Mebeverine-M (N-dealkyl-O-demethyl-) 2AC	UHYAC	1995	263_1	176_8	134_{15}	114_{44}	72_{100}	676	5323
Mebeverine-M (N-deethyl-alcohol) 2AC	UHYAC	2390	321_1	200_{25}	158_{100}	148_{68}	98_{65}	1041	5321
Mebeverine-M (N-deethyl-O-demethyl-alcohol) 3AC	UHYAC	2535	200_{37}	158_{100}	134_{21}	107_{20}	98_{44}	1206	5320
Mebeverine-M (O-demethyl-alcohol) 2AC		2305	234_3	186_{100}	135_4	107_9	72_7	1128	5324
Mebeverine-M (O-demethyl-alcohol) AC		2245	186_{100}	135_3	107_{17}	72_8		867	5325
Mebeverine-M (vanillic acid) ME		1455*	182_{50}	151_{100}	123_{14}	108_{16}	77_{16}	350	5216
Mebeverine-M (vanillic acid) MEAC		1640*	224_6	193_2	182_{96}	151_{100}	123_{11}	486	5328
Mebeverine-M/artifact (alcohol)		2110	264_1	144_{100}	121_{15}	72_{52}	55_{19}	691	4405
Mebeverine-M/artifact (alcohol) AC	U+UHYAC	2210	292_1	186_{100}	121_{12}	72_{16}	55_{20}	953	4406
Mebeverine-M/artifact (veratric acid)	P UHY	1730*	182_{100}	167_{29}	121_{20}	111_{29}	77_{38}	351	4407
Mebeverine-M/artifact (veratric acid) ME	UHYME	1585*	196_{97}	181_{11}	165_{100}	137_{15}	137_{15}	390	4408
Mebhydroline	U UHY UHYAC	2445	276_{43}	233_{89}	115_{22}	91_{100}	65_{29}	756	1667
Mebhydroline-M (nor-) AC	UHYAC	2820	304_{75}	232_{35}	213_{28}	91_{100}	65_{20}	932	1668
Mebhydroline-M (nor-HO-) 2AC	UHYAC	3130	362_{56}	320_{35}	249_{49}	187_{12}	91_{100}	1273	1669
4-MEC AC	U+UHYAC	1820	233_1	148_3	114_{69}	91_{12}	72_{100}	524	8770
4-MEC HFB		1505	359_1	268_{34}	240_1	119_{100}	91_{23}	1256	8772
MECC		<1000	138_3	137_{10}	123_{13}	95_{100}	82_{15}	278	3593
MECC -HCN		<1000	111_{25}	91_{13}	82_{27}	68_{100}	55_{67}	261	3597
4-MEC-M (carboxy-) (ME)AC	U+UHYAC	2115	277_1	246_5	163_4	114_{84}	72_{100}	760	8771
4-MEC-M (deethyl-) AC	U+UHYAC	1875	205_{14}	177_4	119_{100}	91_{46}	86_{97}	413	7871
4-MEC-M (deethyl-dihydro-) isomer-1 2AC	U+UHYAC	1885	249_3	190_{56}	129_{27}	121_{52}	86_{100}	598	7968
4-MEC-M (deethyl-dihydro-) isomer-2 2AC	U+UHYAC	1900	249_8	190_{59}	129_{33}	121_{62}	86_{100}	598	7967
4-MEC-M (deethyl-HO-tolyl-) 2AC	U+UHYAC	2310	263_1	177_{44}	118_{11}	105_2	86_{100}	674	7969
4-MEC-M (dihydro-) 2AC	U+UHYAC	1950	277_1	218_4	176_7	114_{79}	72_{100}	762	8924
4-MEC-M (HO-tolyl-) 2AC	U+UHYAC	2180	291_1	177_3	114_{79}	89_7	72_{100}	848	8769
Meclofenamic acid		2350	295_{32}	242_{100}	214_{13}	179_{11}	151_8	873	5768
Meclofenamic acid 2ME		2275	323_{24}	277_{10}	242_{100}	214_{10}	180_7	1048	5702
Meclofenamic acid AC		2320	337_{33}	295_{13}	242_{100}	214_{14}	180_{10}	1134	5769
Meclofenamic acid -CO2		2035	251_{34}	216_{13}	181_{100}	152_9	77_{13}	607	5767
Meclofenamic acid ME		2240	309_{36}	277_{15}	242_{100}	214_{14}	179_{11}	961	5701
Meclofenamic acid TMS		2750	367_{22}	352_6	277_{14}	242_{100}	214_9	1295	5703
Meclofenoxate	G	1790	257_4	141_6	111_{22}	71_{37}	58_{100}	641	1076
Meclofenoxate-M (HOOC-)		1770*	186_{100}	141_{92}	128_{74}	111_{87}	99_{50}	359	1881
Meclofenoxate-M (HOOC-) ME		1510*	200_{98}	141_{100}	111_{58}	99_{17}	75_{38}	399	1077
Meclonazepam		2815	329_{12}	294_{88}	288_{100}	286_{79}	240_{69}	1086	9710
Meclonazepam		2815	286_{100}	240_{43}	204_{67}	178_{30}	102_{34}	812	9712
Meclonazepam HY	UHY-I U+UHYAC-I	2470	276_{45}	241_{100}	195_{26}	139_{51}	111_{31}	752	280
Meclonazepam TMS		2610	401_{41}	386_{16}	366_{100}	320_{43}	281_9	1448	9713
Mecloxamine	G	2180	317_1	215_1	179_3	165_4	72_{100}	1012	1078
Mecloxamine artifact	G U+UHYAC	1700*	214_{48}	179_{100}	152_8	139_3	89_{14}	445	1217
Mecloxamine HY	UHY	1750*	232_{12}	217_{80}	139_{81}	105_{75}	77_{100}	516	1079
Mecloxamine HYAC	UHYAC	2180*	274_{18}	232_{75}	197_{14}	139_{35}	121_{100}	738	2185
Mecloxamine-M (HO-) -H2O HY	U UHY	2050	230_{100}	215_{27}	195_{60}	177_{33}	165_{64}	509	2187
Mecloxamine-M (HO-) isomer-1 -H2O HYAC	UHYAC	2030*	272_{27}	230_{100}	195_{34}	165_{56}	152_{10}	728	2184
Mecloxamine-M (HO-) isomer-2 -H2O HYAC	U+UHYAC	2090*	272_{16}	230_{100}	215_{15}	195_{34}	165_{47}	728	2189
Mecloxamine-M (HO-methoxy-) -H2O HYAC	UHYAC	2210*	302_{10}	260_{100}	182_{10}	152_{16}	75_4	918	2186
Mecloxamine-M (HO-methoxy-carbinol) -H2O	U UHY	2220*	262_{36}	260_{100}				656	2194
Mecloxamine-M (nor-)	U	2440	303_1	233_7	179_4	165_7	58_{100}	924	2192
Mecloxamine-M (nor-) AC	U	2580	310_{10}	215_{19}	130_{29}	100_{67}	58_{100}	1184	2193
Meclozine	G	3040	390_{13}	285_{12}	189_{89}	105_{100}		1403	1080
Meclozine artifact-1	G U+UHYAC	1600*	202_{30}	167_{100}	165_{52}	152_{17}	125_7	403	2442
Meclozine artifact-2		1900*	232_{60}	201_{62}	165_{64}	105_{100}	77_{54}	516	1344
Meclozine-M (carbinol)	UHY	1750*	218_{17}	183_7	139_{39}	105_{100}	77_{87}	457	2239

Table 8.1: Compounds in order of names   **Meclozine-M (carbinol) AC**

Name	Detected	RI	Typical ions and intensities					Page	Entry
Meclozine-M (carbinol) AC	U+UHYAC	1890*	260_8	200_{40}	165_{100}	139_{10}	77_{29}	656	1270
Meclozine-M (Cl-benzophenone)	U+UHYAC	1850*	216_{43}	139_{58}	105_{100}	77_{44}		451	1343
Meclozine-M (HO-Cl-benzophenone)	UHY	2300*	232_{36}	197_7	139_{23}	121_{100}	111_{23}	515	2240
Meclozine-M (HO-Cl-BPH) isomer-1 AC	UHYAC	2200*	274_{18}	232_{75}	139_{100}	121_{44}	111_{51}	737	2229
Meclozine-M (HO-Cl-BPH) isomer-2 AC	U+UHYAC	2230*	274_7	232_{43}	139_{25}	121_{100}	111_{27}	737	2230
Meclozine-M (N-dealkyl-)	UHY	2520	286_{13}	241_{48}	201_{50}	165_{65}	56_{100}	815	2241
Meclozine-M (N-dealkyl-) AC	U+UHYAC	2620	328_7	242_{19}	201_{48}	165_{66}	85_{100}	1081	1271
Meclozine-M AC	U+UHYAC	2380	299_4	285_{61}	226_{100}	191_{58}	84_{45}	899	821
Meclozine-M/artifact	P-I U+UHYAC UME	2220	300_{17}	228_{38}	165_{52}	99_{100}	56_{62}	908	670
Meclozine-M/artifact AC		2010	232_7	160_{26}	146_{17}	105_{100}	85_{32}	518	2444
Meconin	U+UHYAC	1780*	194_{92}	176_{52}	165_{100}	147_{69}	77_{29}	383	2326
Meconin-M (O-demethyl-) isomer-1 AC	U+UHYAC	1780*	222_2	180_{100}	162_{33}	151_{41}	134_{56}	478	9370
Meconin-M (O-demethyl-) isomer-2 AC	U+UHYAC	1825*	222_4	180_{100}	162_{26}	151_{30}	134_{32}	478	9418
Mecoprop	U	1540*	214_{38}	169_{22}	142_{100}	107_{36}	77_{31}	445	1081
Mecoprop ME		1500*	228_{100}	169_{81}	142_{64}	107_{58}	77_{32}	500	2268
5-Me-DALT		2150	254_1	158_6	144_{23}	114_{18}	110_{100}	630	8851
5-Me-DALT HFB		1970	450_1	354_5	340_{10}	143_{19}	110_{100}	1580	10101
5-Me-DALT PFP		1955	400_1	304_6	290_{13}	143_{26}	110_{100}	1446	10100
5-Me-DALT TFA		1980	323_3	254_{11}	240_{23}	143_{29}	110_{100}	1209	10099
5-Me-DALT TMS		2255	326_4	285_{11}	230_8	216_{23}	110_{100}	1071	10098
7-Me-DALT		2140	254_1	158_6	144_{23}	115_{13}	110_{100}	630	8854
7-Me-DALT HFB		1990	423_2	354_{10}	340_{24}	143_{16}	110_{100}	1580	9136
7-Me-DALT PFP		1975	373_1	304_4	290_{12}	143_{23}	110_{100}	1446	9135
7-Me-DALT TFA		2025	350_1	323_2	254_4	240_{11}	110_{100}	1209	9134
7-Me-DALT TMS		2200	326_4	285_{12}	216_8	200_7	110_{100}	1071	9133
7-Me-DALT-M (HO-alkyl-) AC		2440	312_2	253_6	160_9	130_{12}	110_{100}	983	9267
7-Me-DALT-M (HO-aryl-) AC		2450	312_2	283_3	160_{10}	110_{100}		983	9268
7-Me-DALT-M (N-dealkyl-) AC		2360	256_6	157_{100}	144_{71}	115_{13}	70_{32}	639	9266
7-Me-DALT-M (N-dealkyl-HO-alkyl-) 2AC		2750	314_2	254_7	188_3	157_{100}	144_{43}	996	9271
7-Me-DALT-M (N-dealkyl-HO-aryl-) isomer-1 2AC		2630	314_{12}	215_{62}	202_{25}	173_{100}	160_{55}	996	9269
7-Me-DALT-M (N-dealkyl-HO-aryl-) isomer-2 2AC		2660	314_{16}	215_{46}	202_{20}	173_{100}	160_{51}	996	9270
Medazepam	G P-I U+UHYAC-I	2235	270_{38}	242_{98}	207_{100}	165_{24}		717	292
Medazepam-M	P G U	2520	270_{86}	269_{97}	242_{100}	241_{82}	77_{17}	715	463
Medazepam-M (nor-)	P-I U UHY	2280	256_{73}	228_{100}	193_{85}	165_{40}		635	293
Medazepam-M (nor-) AC	U+UHYAC	2470	298_{41}	297_{80}	256_{51}	228_{100}	193_{52}	894	1324
Medazepam-M (nor-HO-) 2AC	UHYAC	2590	356_{59}	355_{100}	314_{44}	271_{38}	244_{77}	1241	3046
Medazepam-M (oxo-)	P G U	2430	284_{81}	283_{91}	256_{100}	221_{31}	77_7	799	481
Medazepam-M (oxo-) HY	UHY U+UHYAC	2100	245_{95}	228_{38}	193_{29}	105_{38}	77_{100}	573	272
Medazepam-M (oxo-) HYAC	U+UHYAC	2260	287_{11}	244_{100}	228_{39}	182_{49}	77_{70}	818	2542
Medazepam-M HY	UHY	2050	231_{80}	230_{95}	154_{23}	105_{38}	77_{100}	512	419
Medazepam-M HYAC	PHYAC U+UHYAC	2245	273_{30}	230_{100}	154_{13}	105_{23}	77_{50}	732	273
Medazepam-M TMS		2300	342_{62}	341_{100}	327_{19}	269_4	73_{30}	1168	4573
Medroxyprogesterone AC		3050*	386_3	344_{16}	301_{29}	283_{100}	243_{13}	1387	2803
Medroxyprogesterone -H2O		3010*	326_{100}	311_{35}	283_{70}	138_{54}	91_{65}	1071	2802
Medrylamine	G U	2230	285_1	257_2	213_{13}	73_{41}	58_{100}	811	2423
Medrylamine HY	UHY	1930*	214_{34}	135_{46}	109_{100}	105_{56}	77_{71}	446	2426
Medrylamine HYAC	UHYAC	1980	256_{13}	196_{100}	181_{46}	153_{46}	77_{26}	637	2424
Medrylamine-M (HO-benzophenone) AC	U+UHYAC	2050*	240_{20}	198_{100}	121_{94}	105_{41}	77_{51}	555	2197
Medrylamine-M (methoxy-benzophenone)	UHY UHYAC	1930*	212_{30}	135_{100}	105_{16}	92_{16}	77_{47}	438	2431
Medrylamine-M (nor-) AC	U	2450	313_1	213_{40}	197_{60}	101_{100}	86_{51}	990	2430
Medrylamine-M (O-demethyl-) HY2AC	U+UHYAC	2090*	284_5	242_{15}	224_{17}	182_{100}	153_{19}	800	2425
Mefenamic acid		2195	241_{96}	223_{100}	208_{69}	194_{33}	180_{36}	560	5189
Mefenamic acid 2ME		2065	269_{100}	238_{25}	222_{45}	208_{24}	194_{44}	712	5191
Mefenamic acid ET		2160	269_{85}	223_{100}	208_{46}	194_{24}	180_{29}	712	5192
Mefenamic acid ME		2115	255_{69}	223_{100}	208_{70}	194_{33}	180_{42}	632	5190
Mefenamic acid MEAC		2260	297_{25}	255_{79}	223_{100}	208_{40}	194_{36}	890	5193
Mefenamic acid TMS		1980	313_{48}	223_{100}	208_{57}	180_{37}	73_{37}	989	5495
Mefenamic acid-M (HO-) 2ME	UME	2345	285_{100}	252_{54}	238_{79}	224_{72}	210_{44}	809	6301
Mefenamic acid-M (HO-) ME	UME	2400	271_{63}	224_{20}	209_{100}	194_{15}	180_{25}	724	6300
Mefenorex	U UHY	1575	196_1	120_{100}	91_{20}	84_3	58_6	436	1719
Mefenorex AC		1935	162_{40}	120_{100}	91_{22}	84_{11}		622	1083
Mefenorex -HCl	U UHY	1190	174_1	160	91_{20}	84_{100}	56_{70}	331	1082
Mefenorex HFB		1735	316_{100}	240_{61}	118_{55}	91_{84}		1466	5063
Mefenorex PFP		1710	266_{100}	204_8	190_{57}	118_{45}	91_{51}	1247	5064
Mefenorex TFA		1715	216_{100}	154_5	140_{51}	118_{41}	91_{39}	947	5065
Mefenorex-M (AM)		1160	134_4	120_{15}	91_{100}	77_{18}	65_{73}	275	54
Mefenorex-M (AM)	U	1160	134_1	120_1	91_6	65_4	44_{100}	275	5514
Mefenorex-M (AM) AC	U+UHYAC	1505	177_4	118_{60}	91_{35}	86_{100}	65_{16}	336	55
Mefenorex-M (AM) AC	U+UHYAC	1505	177_1	118_{19}	91_{11}	86_{31}	44_{100}	335	5515
Mefenorex-M (AM) formyl artifact		1100	147_2	146_6	125_5	91_{12}	56_{100}	286	3261
Mefenorex-M (AM) HFB		1355	240_{79}	169_{21}	118_{100}	91_{53}		1099	5047
Mefenorex-M (AM) PFP		1330	281_1	190_{73}	118_{100}	91_{36}	65_{12}	786	4379
Mefenorex-M (AM) TFA		1095	231_1	140_{100}	118_{92}	91_{45}	69_{19}	513	4000
Mefenorex-M (AM) TMS		1190	192_6	116_{100}	100_{10}	91_{11}	73_{87}	421	5581

Mefenorex-M (HO-) -HCl   Table 8.1: Compounds in order of names

Name	Detected	RI	Typical ions and intensities					Page	Entry
Mefenorex-M (HO-) -HCl	UHY	1590	190_1	133_1	107_8	84_{100}	56_{30}	373	1725
Mefenorex-M (HO-) -HCl AC	UHYAC	1630	232_1	218_2	176_5	107_9	84_{100}	523	1729
Mefenorex-M (HO-) isomer-1 2AC	U+UHYAC	2115	311_1	162_{39}	120_{100}	107_{24}	58_{47}	976	1731
Mefenorex-M (HO-methoxy-)	UHY UHYAC	2145	256_1	220	137_{10}	120_{100}	84_{59}	642	1728
Mefenorex-M (HO-methoxy-) AC	U+UHYAC	2360	298_1	257_1	162_{33}	137_8	120_{100}	902	1727
Mefenorex-M (HO-methoxy-) -HCl	UHY	1775	220_1	137_5	120_6	84_{100}	56_{20}	476	1726
Mefenorex-M-D11 PFP		1610	194_{100}	128_{82}	98_{43}	70_{14}		856	7284
Mefenorex-M-D11 TFA		1615	242_1	144_{100}	128_{82}	98_{43}	70_{14}	566	7283
Mefenorex-M-D5 AC		1480	182_3	122_{46}	92_{30}	90_{100}	66_{16}	352	5690
Mefenorex-M-D5 HFB		1330	244_{100}	169_{14}	122_{46}	92_{41}	69_{40}	1130	6316
Mefenorex-M-D5 PFP		1320	194_{100}	123_{42}	119_{32}	92_{45}	69_{11}	815	5566
Mefenorex-M-D5 TFA		1085	144_{100}	123_{53}	122_{56}	92_{51}	69_{28}	540	5570
Mefenorex-M-D5 TMS		1180	212_1	197_8	120_{100}	92_{11}	73_{57}	441	5582
Mefexamide		2185	280_2	263_2	155_3	99_4	86_{100}	784	1480
Mefloquine		2280	359_1	246_2	196_1	84_{100}	56_{19}	1347	3205
Mefloquine -H2O		2220	224_1	196_1	97_{100}	69_{28}		1264	3206
Mefloquine -H2O AC		2420	402_{28}	360_7	318_9	126_{50}	84_{100}	1452	3207
Mefruside 2ME	UME	2860	367_1	325_2	218_1	110_3	85_{100}	1477	3056
Mefruside ME		2880	353_1	311_1	204_1	110_3	85_{100}	1427	3057
Mefruside -SO2NH	UME	2150	260_1	218_2	175_3	111_8	85_{100}	922	3058
Melatonin		2450	232_{17}	173_{86}	160_{100}	145_{25}	117_{25}	518	5913
Melatonin 2TFA		2070	424_{13}	269_{100}	256_{44}	159_{53}	144_{29}	1516	5915
Melatonin 2TMS		2640	376_{11}	361_5	245_{30}	232_{100}	73_{82}	1340	6032
Melatonin artifact (deacetyl-) 2HFB		2295	582_{17}	369_{100}	356_{72}	159_{74}	69_{45}	1708	5922
Melatonin artifact (deacetyl-) 2PFP		2030	482_{40}	360_{12}	319_{100}	306_{76}	159_{35}	1637	5923
Melatonin artifact (deacetyl-) 2TFA		2020	382_{34}	269_{100}	256_{86}	159_{87}	144_{38}	1364	5924
Melatonin artifact-1 HFB		2065	410_{100}	213_{31}	186_{22}	169_{16}	69_{30}	1477	5926
Melatonin artifact-1 PFP		2010	360_{100}	213_{38}	186_{34}	119_{37}	69_{29}	1263	5920
Melatonin artifact-1 TFA		1990	310_{100}	213_{38}	186_{31}	170_{25}	69_{47}	969	5918
Melatonin artifact-2 HFB		2590	410_{28}	241_{100}	226_7	198_6	170_5	1477	5925
Melatonin artifact-2 PFP		2540	360_{40}	241_{100}	198_6	104_7	121_{14}	1263	5917
Melatonin artifact-2 TFA		2525	310_{58}	241_{100}	198_6	184_7	169_7	969	5919
Melatonin HFB		2295	428_9	369_{100}	356_{12}	159_{15}	144_8	1527	5921
Melatonin PFP		2240	378_8	319_{100}	172_{13}	159_{35}	144_{23}	1347	5916
Melatonin TFA		2260	328_{11}	269_{100}	256_{13}	159_{29}	144_{19}	1080	5914
Melatonin TMS		2610	304_{14}	245_{48}	232_{64}	202_4	73_{100}	932	6033
Melitracene	G U UHY UHYAC	2285	291_1	217_1	202_1	58_{100}		852	356
Melitracene-M (nor-) AC	U+UHYAC	2760	319_7	246_{100}	231_{37}	86_{25}		1029	1179
Melitracene-M (nor-HO-dihydro-) 2AC	U+UHYAC	3030	379_{39}	265_{44}	223_{100}	114_{26}		1353	1180
Melitracene-M (ring)	U+UHYAC	1900*	208_{11}	193_{100}	178_{61}			426	1178
Meloxicam artifact		1745	211_3	169_{23}	152_{57}	104_{92}	76_{100}	435	6077
Meloxicam artifact AC		2065	253_{17}	211_{100}	152_{43}	118_{46}	77_{53}	621	6076
Melperone	G P-I U+UHYAC	1890	263_2	125_{58}	112_{100}			677	174
Melperone-M	U+UHYAC	1490*	180_{25}	125_{49}	123_{35}	95_{17}	56_{100}	345	85
Melperone-M (dihydro-) AC	UHYAC	2050	307_2	264_9	246_4	123_9	112_{100}	953	176
Melperone-M (dihydro-) -H2O	UHY UHYAC	1835	247_1	228_1	133_3	112_{100}		588	175
Melperone-M (dihydro-oxo-) -H2O	UHY U+UHYAC	2220	261_{15}	148_{100}	137_{23}	126_{85}	98_{81}	665	6511
Melperone-M (HO-) -H2O	UHY UHYAC	1900	261_6	125_{25}	112_{100}			665	552
Melperone-M/artifact	U+UHYAC	1350*	164_{43}	133_{28}	123_{100}	95_{41}		310	1914
Memantine	G U UHY	1250	179_{12}	164_7	122_{86}	108_{100}		343	1557
Memantine AC	U+UHYAC	1600	221_{100}	164_{72}	150_{68}	122_{12}	107_{19}	477	1482
Memantine-M (4-HO-)	U UHY	1550	195_{18}	138_{36}	108_{100}			388	1560
Memantine-M (7-HO-)	UHY	1540	195_{12}	180_2	122_{35}	108_{100}		388	1559
Memantine-M (deamino-HO-)	U UHY	1525*	180_{20}	165_{18}	123_{85}	109_{90}	71_{100}	347	1558
Memantine-M (HO-) 2AC	U+UHYAC	1995	279_{81}	237_{30}	219_{60}	164_{80}	150_{100}	779	1555
Memantine-M (HO-) AC	U+UHYAC	1860	237_{100}	204_{28}	164_{68}	150_{90}	107_{58}	546	1554
Memantine-M (HO-methyl-)	U UHY	1570	195_{10}	164_{55}	138_{30}	120_{20}	108_{100}	388	1561
Memantine-M (HO-methyl-) 2AC	U+UHYAC	2090	279_{100}	206_{86}	164_{27}	150_{68}		779	1556
Menthol	G	1225*	138_{21}	123_{30}	95_{74}	81_{86}	71_{100}	300	1826
5-MeO-2-Me-2-MALET		2330	286_8	188_4	174_{19}	112_{100}	55_{24}	816	8832
5-MeO-2-Me-2-MALET artifact HFB		1925	481_{36}	425_2	384_9	172_4	112_{100}	1635	10072
5-MeO-2-Me-2-MALET artifact PFP		2150	431_{30}	375_4	334_{10}	172_8	112_{100}	1536	10071
5-MeO-2-Me-2-MALET artifact TFA		1990	381_{30}	325_4	284_{24}	172_{22}	112_{100}	1358	10191
5-MeO-2-Me-2-MALET HFB		2165	441_2	384_6	370_{15}	173_{12}	112_{100}	1637	10031
5-MeO-2-Me-2-MALET PFP		2150	334_2	320_6	173_{23}	158_{21}	112_{100}	1540	10030
5-MeO-2-Me-2-MALET TFA		2205	284_{12}	270_{21}	173_{30}	158_{25}	112_{100}	1365	10190
5-MeO-2-Me-2-MALET TMS		2480	358_3	246_{10}	112_{100}	73_{24}	55_{15}	1255	10029
5-MeO-2-Me-ALCHT		2690	326_2	188_6	174_{21}	152_{100}	70_{50}	1071	8839
5-MeO-2-Me-ALCHT artifact HFB		2215	521_{13}	384_5	370_3	172_{25}	152_{100}	1677	10059
5-MeO-2-Me-ALCHT artifact PFP		2460	472_2	334_3	320_{10}	152_{100}	70_{24}	1621	10056
5-MeO-2-Me-ALCHT artifact TFA		2275	421_{33}	337_{10}	284_9	172_{20}	152_{100}	1509	10055
5-MeO-2-Me-ALCHT HFB		2465	522_3	384_4	370_{11}	152_{100}	70_{18}	1678	10057
5-MeO-2-Me-ALCHT PFP		2700	472_1	360_2	346_7	152_{100}	70_{17}	1622	9835
5-MeO-2-Me-ALCHT TFA		2500	422_1	284_3	270_9	173_{13}	152_{100}	1513	10058

Table 8.1: Compounds in order of names

Name	Detected	RI	Typical ions and intensities					Page	Entry
5-MeO-2-Me-ALCHT TMS		2780	398$_{6}$	259$_{3}$	246$_{9}$	152$_{100}$	70$_{15}$	1440	10054
5-MeO-2-Me-ALCHT-M (deallyl-)		2615	286$_{14}$	187$_{59}$	174$_{100}$	159$_{41}$	70$_{43}$	817	10412
5-MeO-2-Me-ALCHT-M (deallyl-HO-aryl-) 2AC		3250	386$_{4}$	232$_{51}$	203$_{100}$	190$_{82}$	175$_{78}$	1386	10414
5-MeO-2-Me-ALCHT-M (deallyl-HO-cyclohexyl-) 2AC		3165	386$_{2}$	187$_{42}$	174$_{100}$	159$_{32}$	131$_{26}$	1386	10410
5-MeO-2-Me-ALCHT-M (decyclohexyl-HO-aryl-) 2AC		2850	344$_{17}$	232$_{45}$	203$_{88}$	190$_{100}$	175$_{70}$	1180	10405
5-MeO-2-Me-ALCHT-M (HO-aryl-) AC		2950	384$_{1}$	232$_{11}$	190$_{46}$	175$_{44}$	152$_{100}$	1378	10406
5-MeO-2-Me-ALCHT-M (HO-aryl-HO-cyclohexyl-) 2AC		3230	442$_{1}$	232$_{10}$	210$_{100}$	190$_{51}$	175$_{51}$	1566	10409
5-MeO-2-Me-ALCHT-M (HO-cyclohexyl-) AC		2900	384$_{3}$	210$_{100}$	174$_{78}$	131$_{26}$	70$_{33}$	1378	10408
5-MeO-2-Me-ALCHT-M (O-demethyl-) AC		2800	354$_{2}$	160$_{58}$	152$_{100}$	119$_{38}$	70$_{58}$	1234	10411
5-MeO-2-Me-ALCHT-M (O-demethyl-deallyl-HO-cyclohexyl-) 3AC		3300	414$_{1}$	215$_{47}$	202$_{49}$	160$_{100}$	131$_{26}$	1491	10413
5-MeO-2-Me-ALCHT-M (O-demethyl-decyclohexyl-) 2AC		2710	314$_{23}$	215$_{89}$	202$_{87}$	173$_{93}$	160$_{100}$	996	10404
5-MeO-2-Me-ALCHT-M (O-demethyl-HO-cyclohexyl-) 2AC		3030	412$_{2}$	210$_{100}$	202$_{31}$	173$_{28}$	160$_{76}$	1486	10407
5-MeO-2-Me-DALT		2330	284$_{1}$	174$_{17}$	159$_{9}$	131$_{11}$	110$_{100}$	805	8838
5-MeO-2-Me-DALT artifact HFB		1945	479$_{32}$	384$_{17}$	369$_{5}$	172$_{13}$	110$_{100}$	1632	10089
5-MeO-2-Me-DALT artifact PFP		1985	429$_{27}$	334$_{10}$	269$_{4}$	172$_{12}$	110$_{100}$	1529	10088
5-MeO-2-Me-DALT artifact TFA		1985	379$_{26}$	284$_{10}$	269$_{4}$	172$_{11}$	110$_{100}$	1348	10087
5-MeO-2-Me-DALT HFB		2175	480$_{1}$	453$_{2}$	384$_{7}$	370$_{17}$	110$_{100}$	1634	10053
5-MeO-2-Me-DALT PFP		2165	430$_{1}$	334$_{2}$	320$_{7}$	173$_{9}$	110$_{100}$	1534	10052
5-MeO-2-Me-DALT TFA		2200	284$_{3}$	270$_{7}$	173$_{11}$	158$_{9}$	110$_{100}$	1357	10051
5-MeO-2-Me-DALT TMS		2475	356$_{4}$	315$_{3}$	246$_{31}$	173$_{2}$	110$_{100}$	1244	10050
5-MeO-2-Me-DALT-M (deallyl-) AC		2585	286$_{11}$	187$_{72}$	174$_{100}$	159$_{33}$	70$_{28}$	816	10376
5-MeO-2-Me-DALT-M (deallyl-HO-aryl-) 2AC		2845	344$_{22}$	245$_{31}$	232$_{48}$	203$_{100}$	190$_{91}$	1179	10380
5-MeO-2-Me-DALT-M (HO-aryl-) AC		2660	342$_{1}$	301$_{3}$	190$_{19}$	175$_{22}$	110$_{100}$	1171	10377
5-MeO-2-Me-DALT-M (O-demethyl-) AC		2515	312$_{1}$	271$_{10}$	202$_{7}$	160$_{35}$	110$_{100}$	983	10375
5-MeO-2-Me-DALT-M (O-demethyl-deallyl-) 2AC		2700	314$_{11}$	215$_{67}$	202$_{62}$	173$_{80}$	160$_{100}$	996	10378
5-MeO-2-Me-DALT-M (O-demethyl-HO-aryl-) 2AC		2760	370$_{2}$	218$_{8}$	190$_{11}$	176$_{27}$	110$_{100}$	1314	10379
5-MeO-2-Me-DiPT		2330	288$_{1}$	188$_{7}$	174$_{21}$	114$_{100}$	72$_{36}$	831	8840
5-MeO-2-Me-DiPT artifact HFB		2165	469$_{2}$	384$_{4}$	370$_{7}$	173$_{7}$	114$_{100}$	1616	10065
5-MeO-2-Me-DiPT artifact PFP		2150	419$_{1}$	334$_{4}$	320$_{7}$	173$_{8}$	114$_{100}$	1504	10064
5-MeO-2-Me-DiPT artifact TFA		1940	383$_{19}$	353$_{7}$	284$_{24}$	127$_{37}$	114$_{100}$	1368	10062
5-MeO-2-Me-DiPT HFB		2165	483$_{26}$	425$_{9}$	384$_{11}$	127$_{36}$	114$_{100}$	1639	10066
5-MeO-2-Me-DiPT PFP		2450	419$_{1}$	334$_{4}$	320$_{7}$	173$_{8}$	114$_{100}$	1545	9830
5-MeO-2-Me-DiPT TFA		2185	369$_{1}$	284$_{3}$	270$_{4}$	173$_{12}$	114$_{100}$	1378	10063
5-MeO-2-Me-DiPT TMS		2480	360$_{1}$	188	174$_{1}$	114$_{100}$	72$_{9}$	1266	10061
5-MeO-2-Me-DiPT-M (deisopropyl-) AC		2600	288$_{13}$	187$_{71}$	174$_{100}$	159$_{29}$	72$_{26}$	829	10383
5-MeO-2-Me-DiPT-M (deisopropyl-HO-aryl-) 2AC		2845	346$_{17}$	300$_{17}$	232$_{57}$	203$_{100}$	190$_{96}$	1189	10387
5-MeO-2-Me-DiPT-M (HO-aryl-) isomer-1 AC		2570	232$_{4}$	190$_{27}$	175$_{8}$	114$_{100}$	72$_{21}$	1191	10382
5-MeO-2-Me-DiPT-M (HO-aryl-) isomer-2 AC		2650	346$_{2}$	246$_{4}$	190$_{32}$	175$_{31}$	114$_{100}$	1190	10384
5-MeO-2-Me-DiPT-M (O-demethyl-) AC		2500	316$_{4}$	216$_{11}$	160$_{43}$	114$_{100}$	72$_{60}$	1007	10381
5-MeO-2-Me-DiPT-M (O-demethyl-deisopropyl-) 2AC		2700	316$_{10}$	215$_{68}$	202$_{71}$	160$_{100}$	72$_{50}$	1007	10385
5-MeO-2-Me-DiPT-M (O-demethyl-HO-aryl-) 2AC		2745	374$_{2}$	260$_{4}$	176$_{28}$	114$_{100}$	72$_{29}$	1331	10386
5-MeO-2-Me-DMT		2060	232$_{12}$	174$_{16}$	159$_{6}$	131$_{6}$	58$_{100}$	519	8837
5-MeO-2-Me-DMT artifact HFB		1765	427$_{24}$	384$_{7}$	186$_{3}$	172$_{7}$	58$_{100}$	1524	10086
5-MeO-2-Me-DMT artifact PFP		1745	377$_{23}$	334$_{4}$	319$_{2}$	172$_{12}$	58$_{100}$	1341	10085
5-MeO-2-Me-DMT artifact TFA		1770	327$_{18}$	284$_{3}$	186$_{2}$	172$_{4}$	58$_{100}$	1072	10084
5-MeO-2-Me-DMT HFB		1965	384$_{1}$	370$_{5}$	173$_{17}$	158$_{20}$	58$_{100}$	1528	10049
5-MeO-2-Me-DMT PFP		1960	334$_{1}$	320$_{6}$	173$_{14}$	158$_{18}$	58$_{100}$	1347	10048
5-MeO-2-Me-DMT TFA		1975	328$_{1}$	284$_{2}$	270$_{4}$	158$_{15}$	58$_{100}$	1081	10047
5-MeO-2-Me-DMT TMS		2235	304$_{7}$	246$_{35}$	174$_{1}$	73$_{22}$	58$_{100}$	933	10046
5-MeO-2-Me-DPT		2330	288$_{3}$	188$_{8}$	174$_{20}$	159$_{9}$	114$_{100}$	831	8836
5-MeO-2-Me-DPT artifact HFB		1940	483$_{45}$	425$_{10}$	384$_{8}$	186$_{10}$	114$_{100}$	1638	10045
5-MeO-2-Me-DPT artifact PFP		2160	433$_{2}$	405$_{4}$	334$_{9}$	320$_{14}$	114$_{100}$	1543	10044
5-MeO-2-Me-DPT artifact TFA		1975	383$_{26}$	284$_{8}$	186$_{13}$	172$_{11}$	114$_{100}$	1368	10083
5-MeO-2-Me-DPT HFB		2170	484$_{1}$	384$_{4}$	370$_{7}$	173$_{8}$	114$_{100}$	1640	10060
5-MeO-2-Me-DPT PFP		2480	433$_{2}$	405$_{4}$	334$_{9}$	320$_{14}$	114$_{100}$	1545	9831
5-MeO-2-Me-DPT TFA		2195	284$_{3}$	270$_{6}$	173$_{10}$	158$_{11}$	114$_{100}$	1378	10043
5-MeO-2-Me-DPT TMS		2515	360$_{4}$	260$_{3}$	246$_{13}$	216$_{2}$	114$_{100}$	1266	10042
5-MeO-2-Me-EiPT		2240	346$_{2}$	260$_{2}$	246$_{11}$	100$_{100}$	58$_{38}$	742	8857
5-MeO-2-Me-EiPT HFB		2075	384$_{8}$	370$_{12}$	173$_{13}$	100$_{100}$	58$_{24}$	1618	10097
5-MeO-2-Me-EiPT PFP		2065	405$_{1}$	334$_{9}$	320$_{13}$	100$_{100}$	58$_{41}$	1508	10096
5-MeO-2-Me-EiPT TFA		2100	355$_{1}$	284$_{6}$	270$_{11}$	100$_{100}$	58$_{38}$	1313	10095
5-MeO-2-Me-EiPT TMS		2400	346$_{2}$	260$_{2}$	246$_{11}$	100$_{100}$	58$_{38}$	1191	10094
5-MeO-2-Me-EPT		2250	274$_{3}$	188$_{6}$	174$_{16}$	100$_{100}$	58$_{25}$	742	8835
5-MeO-2-Me-EPT artifact HFB		1890	469$_{20}$	384$_{9}$	271$_{6}$	186$_{20}$	100$_{100}$	1616	10082
5-MeO-2-Me-EPT artifact PFP		1870	419$_{43}$	334$_{11}$	186$_{12}$	172$_{17}$	100$_{100}$	1504	10081
5-MeO-2-Me-EPT artifact TFA		1840	369$_{9}$	284$_{7}$	186$_{13}$	170$_{11}$	100$_{100}$	1304	10080
5-MeO-2-Me-EPT HFB		2110	384$_{7}$	370$_{10}$	173$_{21}$	158$_{16}$	100$_{100}$	1618	10041
5-MeO-2-Me-EPT PFP		2100	335$_{1}$	320$_{6}$	173$_{17}$	158$_{15}$	100$_{100}$	1508	10040
5-MeO-2-Me-EPT TFA		2135	284$_{1}$	270$_{7}$	173$_{3}$	158$_{11}$	100$_{100}$	1313	10039
5-MeO-2-Me-EPT TMS		2435	346$_{7}$	260$_{3}$	246$_{20}$	216$_{2}$	100$_{100}$	1191	10038
5-MeO-2-Me-MiPT		2190	260$_{13}$	174$_{22}$	159$_{13}$	131$_{14}$	86$_{100}$	661	8841
5-MeO-2-Me-MiPT HFB		2065	455$_{1}$	441$_{1}$	384$_{5}$	370$_{8}$	86$_{100}$	1594	9768
5-MeO-2-Me-MiPT PFP		2055	405$_{1}$	391$_{2}$	334$_{7}$	320$_{8}$	86$_{100}$	1465	9767
5-MeO-2-Me-MiPT TFA		2090	355$_{1}$	341$_{2}$	284$_{10}$	270$_{15}$	86$_{100}$	1243	9766

5-MeO-2-Me-MiPT TMS

Table 8.1: Compounds in order of names

Name	Detected	RI	Typical ions and intensities					Page	Entry
5-MeO-2-Me-MiPT TMS		2190	332_6	246_{23}	174_2	159_1	86_{100}	1108	9876
5-MeO-2-Me-PIP-T		2400	272_{13}	174_{15}	159_{10}	131_9	98_{100}	731	8833
5-MeO-2-Me-PIP-T artifact HFB		2030	467_{40}	384_{16}	186_7	172_{22}	98_{100}	1611	10076
5-MeO-2-Me-PIP-T artifact PFP		2010	417_{28}	334_3	205_3	187_4	98_{100}	1498	10075
5-MeO-2-Me-PIP-T artifact TFA		2070	367_{18}	284_7	172_{15}	111_{20}	98_{100}	1294	10073
5-MeO-2-Me-PIP-T HFB		2230	384_2	370_6	173_{25}	158_{24}	98_{100}	1613	10034
5-MeO-2-Me-PIP-T PFP		2240	418_1	320_5	173_{11}	158_{11}	98_{100}	1503	10033
5-MeO-2-Me-PIP-T TFA		2280	284_2	270_5	173_{12}	158_{12}	98_{100}	1302	10074
5-MeO-2-Me-PIP-T TMS		2575	344_8	246_{19}	216_4	174_2	98_{100}	1181	10032
5-MeO-2-Me-PYR-T		2330	258_{13}	174_{18}	159_8	131_9	84_{100}	648	8834
5-MeO-2-Me-PYR-T artifact HFB		1960	453_{25}	384_8	186_{12}	172_{32}	84_{100}	1586	10079
5-MeO-2-Me-PYR-T artifact PFP		1945	403_{32}	172_6	119_{12}	97_{18}	84_{100}	1454	10078
5-MeO-2-Me-PYR-T artifact TFA		2010	353_{40}	284_8	172_8	97_{14}	84_{100}	1224	10077
5-MeO-2-Me-PYR-T HFB		2170	384_3	370_{10}	173_{14}	158_{13}	84_{100}	1589	10037
5-MeO-2-Me-PYR-T PFP		2175	334_2	320_6	173_{12}	158_{12}	84_{100}	1459	10036
5-MeO-2-Me-PYR-T TFA		2230	351_1	204_2	270_8	158_{21}	84_{100}	1232	10192
5-MeO-2-Me-PYR-T TMS		2510	330_{14}	246_{35}	216_3	174_2	84_{100}	1095	10035
5-MeO-2-TMT		2060	232_{12}	174_{16}	159_6	131_6	58_{100}	519	8837
5-MeO-2-TMT artifact HFB		1765	427_{24}	384_7	186_3	172_7	58_{100}	1524	10086
5-MeO-2-TMT artifact PFP		1745	377_{23}	334_4	319_2	172_{12}	58_{100}	1341	10085
5-MeO-2-TMT artifact TFA		1770	327_{18}	284_3	186_2	172_4	58_{100}	1072	10084
5-MeO-2-TMT HFB		1965	384_1	370_5	173_{17}	158_{20}	58_{100}	1528	10049
5-MeO-2-TMT PFP		1960	334_1	320_6	173_{14}	158_{18}	58_{100}	1347	10048
5-MeO-2-TMT TFA		1975	328_1	284_2	270_4	158_{15}	58_{100}	1081	10047
5-MeO-2-TMT TMS		2235	304_7	246_{35}	174_1	73_{22}	58_{100}	933	10046
4-MeO-AMT		1955	204_5	161_{100}	146_{33}	130_{21}	44_{25}	410	10214
4-MeO-AMT		1955	204_4	161_{100}	146_{34}	130_{20}	70_6	410	10215
4-MeO-AMT 2PFP		2150	496_2	333_{12}	306_{100}	276_9	190_7	1655	9779
4-MeO-AMT 2TMS		2155	348_1	333_3	275_6	233_{67}	116_{100}	1201	9778
4-MeO-AMT AC		2400	246_{18}	187_{93}	160_{100}	130_{52}	117_{17}	581	9777
4-MeO-AMT ethylimine artifact		1965	230_{18}	160_{100}	130_{35}	70_{35}	44_{22}	511	10224
4-MeO-AMT formyl artifact ME		2175	230_{24}	160_{100}	130_{37}	117_{10}	70_{11}	512	9776
5-MoO-AMT		2010	204_3	188_5	161_{100}	146_{20}	44_{57}	411	10229
5-MeO-AMT		2010	204_3	188_5	161_{100}	146_{20}		411	10232
5-MeO-AMT 2HFB		2060	596_2	383_{38}	356_{100}	240_{49}	144_{25}	1713	9808
5-MeO-AMT 2PFP		2030	496_6	333_{55}	306_{100}	190_{22}	144_{11}	1655	9806
5-MeO-AMT 2TFA		2070	396_{13}	283_{52}	256_{100}	159_{41}	140_{26}	1427	9804
5-MeO-AMT 2TMS		2220	348_1	333_5	233_{50}	202_8	116_{100}	1201	9803
5-MeO-AMT AC		2470	246_{26}	187_{77}	160_{100}	145_{22}	117_{18}	580	9802
5-MeO-AMT ethylimine artifact		2000	230_9	187_8	160_{100}	70_{36}	44_{34}	511	10230
5-MeO-AMT ethylimine artifact		2000	230_{22}	187_6	160_{100}	145_{22}	70_{23}	511	10231
5-MeO-AMT HFB		2200	400_{12}	240_{12}	187_5	160_{100}	145_{13}	1445	9809
5-MeO-AMT PFP		2220	350_{14}	229_5	187_5	160_{100}	145_{15}	1208	9807
5-MeO-AMT TFA		2220	300_{19}	187_3	160_{100}	145_{11}	117_{10}	906	9805
5-MeO-DALT		2270	270_1	174_2	160_{10}	145_7	110_{100}	720	8831
5-MeO-DALT HFB		2180	439_1	370_4	356_8	159_{22}	110_{100}	1609	9140
5-MeO-DALT PFP		2130	389_1	320_6	306_{12}	159_{22}	110_{100}	1496	9139
5-MeO-DALT TFA		2145	339_1	270_6	256_{10}	159_{20}	110_{100}	1291	9138
5-MeO-DALT TMS		2300	270_1	174_1	160_1	145_1	110_{100}	1171	9137
5-MeO-DALT-D4		2270	274_2	178_4	162_{16}	147_{11}	112_{100}	742	8850
5-MeO-DALT-D4 HFB		2120	470_1	443_3	374_{11}	358_{20}	112_{100}	1618	10070
5-MeO-DALT-D4 PFP		2115	420_1	393_6	324_9	308_{17}	112_{100}	1508	10069
5-MeO-DALT-D4 TFA		2135	370_2	343_7	274_4	258_{25}	112_{100}	1313	10068
5-MeO-DALT-D4 TMS		2375	346_7	305_6	250_4	234_{16}	112_{100}	1191	10067
5-MeO-DALT-M (N-dealkyl-) AC		2480	272_9	173_{100}	160_{68}	145_{16}	70_{24}	730	9273
5-MeO-DALT-M (O-demethyl-) AC		2310	298_1	257_1	160_3	146_9	110_{100}	897	9251
5-MeO-DALT-M (O-demethyl-di-HO-) 2AC		2440	298_1	146_9	110_{100}			1323	9272
5-MeO-DALT-M (O-demethyl-di-HO-) 3AC		2400	414_1	373_{15}	160_3	146_{10}	110_{100}	1491	9256
5-MeO-DALT-M (O-demethyl-N-dealkyl-) 2AC		2610	300_8	201_{63}	159_{100}	146_{58}	70_{41}	908	9276
5-MeO-DALT-M (oxo-)		2540	284_2	242_2	176_7	161_9	110_{100}	804	9275
5-MeO-DALT-M AC		2520	403_4	343_5	301_{21}	110_{100}		1454	9274
5-MeO-DET		2110	246_3	174_3	160_{19}	145_{14}	86_{100}	582	10177
5-MeO-DET HFB		1960	427_1	370_3	356_5	144_{11}	86_{100}	1565	10180
5-MeO-DET PPF		1950	391_2	377_1	320_5	306_9	86_{100}	1411	10179
5-MeO-DET TFA		1970	341_1	327_1	270_4	256_7	86_{100}	1170	10178
5-MeO-DET TMS		2235	318_7	246_2	232_{10}	202_4	86_{100}	1021	10199
2-MeO-diphenidine		2200	294_1	204_{100}	188_{33}	121_{17}	91_{37}	881	9565
3-MeO-diphenidine		2210	294_1	204_{100}	178_5	121_7	91_{20}	881	9567
4-MeO-diphenidine		2230	294_1	204_{100}	178_7	121_{20}	91_{37}	881	9569
2-MeO-diphenidine artifact		2005*	210_{100}	194_7	179_{13}	165_{46}	152_{23}	434	9566
3-MeO-diphenidine artifact		1995*	210_{100}	194_{28}	179_{38}	165_{53}	152_{24}	434	9568
4-MeO-diphenidine artifact		2010*	210_{100}	195_{17}	179_{11}	165_{67}	152_{43}	434	9570
5-MeO-DiPT		2260	274_1	174_7	160_{31}	114_{100}	72_{31}	742	8867
5-MeO-DiPT HFB		2050	455_2	370_8	356_{10}	159_{16}	114_{100}	1618	9516

Table 8.1: Compounds in order of names

Name	Detected	RI	Typical ions and intensities					Page	Entry
5-MeO-DiPT PFP		2040	405_{1}	306_{3}	159_{13}	144_{8}	114_{100}	1508	9515
5-MeO-DiPT TFA		2065	355_{1}	270_{6}	256_{10}	159_{21}	114_{100}	1313	9514
5-MeO-DiPT TMS		2365	346_{8}	246_{18}	232_{34}	114_{100}	72_{51}	1191	9513
5-MeO-DiPT-D4		2260	278_{1}	178_{9}	162_{33}	116_{100}	74_{29}	771	8862
5-MeO-DiPT-D4 HFB		2105	459_{2}	374_{9}	358_{8}	161_{20}	116_{100}	1626	10122
5-MeO-DiPT-D4 PFP		2100	409_{5}	324_{8}	308_{8}	161_{8}	116_{100}	1518	10121
5-MeO-DiPT-D4 TFA		2115	359_{2}	274_{6}	258_{7}	161_{15}	116_{100}	1331	10120
5-MeO-DiPT-D4 TMS		2350	350_{1}	234_{13}	204_{4}	116_{100}	74_{44}	1210	10119
5-MeO-DPT		2300	274_{3}	174_{10}	160_{35}	145_{14}	114_{100}	742	8859
5-MeO-DPT HFB		2120	441_{4}	370_{13}	356_{15}	159_{12}	114_{100}	1618	10110
5-MeO-DPT PFP		2110	391_{2}	320_{5}	306_{7}	159_{13}	114_{100}	1508	10109
5-MeO-DPT TFA		2145	369_{3}	341_{15}	270_{21}	256_{24}	114_{100}	1313	10108
5-MeO-DPT TMS		2380	346_{5}	246_{8}	232_{15}	202_{5}	114_{100}	1191	10107
5-MeO-DPT-D4		2280	278_{3}	178_{8}	162_{28}	147_{13}	116_{100}	771	8864
5-MeO-DPT-D4 HFB		2120	445_{5}	374_{11}	358_{11}	161_{10}	116_{100}	1626	10130
5-MeO-DPT-D4 PFP		2115	395_{5}	324_{5}	308_{6}	161_{9}	116_{100}	1518	10129
5-MeO-DPT-D4 TFA		2130	345_{4}	275_{1}	258_{9}	161_{12}	116_{100}	1331	10128
5-MeO-DPT-D4 TMS		2345	350_{4}	250_{4}	234_{7}	177_{2}	116_{100}	1210	10127
5-MeO-EPT-D4		2210	264_{3}	178_{5}	162_{16}	147_{9}	102_{100}	684	8866
4-MeO-MiPT		2090	246_{5}	174_{4}	160_{13}	130_{27}	86_{100}	582	10169
4-MeO-MiPT HFB		1930	427_{2}	370_{4}	356_{4}	129_{6}	86_{100}	1565	10172
4-MeO-MiPT PFP		2100	390_{2}	320_{7}	306_{5}	129_{6}	86_{100}	1411	10171
4-MeO-MiPT TFA		1940	340_{17}	327_{57}	270_{100}	256_{66}	226_{45}	1169	10170
4-MeO-MiPT TMS		2215	318_{6}	246_{2}	232_{11}	86_{100}	73_{26}	1021	10197
5-MeO-MiPT		2120	246_{4}	174_{5}	160_{24}	145_{17}	86_{100}	582	10173
5-MeO-MiPT HFB		1970	427_{1}	370_{5}	356_{6}	144_{14}	86_{100}	1565	10176
5-MeO-MiPT PFP		1965	391_{2}	377_{4}	320_{7}	306_{8}	86_{100}	1411	10175
5-MeO-MiPT TFA		1980	341_{1}	327_{3}	270_{7}	256_{8}	86_{100}	1169	10174
5-MeO-MiPT TMS		2255	318_{6}	262_{5}	232_{11}	202_{3}	86_{100}	1021	10198
MEOP		2065	234_{20}	190_{9}	135_{100}	99_{47}	70_{69}	529	9304
MEOP artifact (dehydro-)		2150	232_{54}	135_{100}	97_{71}	92_{19}	70_{23}	516	9305
3-MeO-PCP		2120	273_{32}	272_{35}	230_{100}	121_{36}	84_{19}	736	9451
3-MeO-PCP artifact	G P U+UHYAC	1630*	188_{100}	173_{23}	159_{41}	129_{23}	115_{19}	365	4436
3-MeO-PCP-M (O-demethyl-) AC	U+UHYAC	2210	301_{25}	258_{100}	244_{12}	166_{22}	84_{13}	916	10271
3-MeO-PCP-M (O-demethyl-bis-HO-) isomer-1 3AC	U+UHYAC	2670	417_{15}	374_{100}	282_{20}	175_{16}	107_{18}	1500	10283
3-MeO-PCP-M (O-demethyl-bis-HO-) isomer-2 3AC	U+UHYAC	2690	417_{7}	389_{7}	358_{100}	288_{80}	82_{16}	1501	10278
3-MeO-PCP-M (O-demethyl-bis-HO-) isomer-3 3AC	U+UHYAC	2720	417_{12}	374_{13}	358_{100}	316_{39}	82_{12}	1501	10277
3-MeO-PCP-M (O-demethyl-bis-HO-) isomer-4 3AC	U+UHYAC	2780	417_{4}	357_{60}	316_{100}	107_{8}	82_{11}	1501	10276
3-MeO-PCP-M (O-demethyl-bis-HO-) isomer-5 3AC	U+UHYAC	2800	417_{5}	357_{60}	316_{100}	107_{14}	82_{18}	1501	10284
3-MeO-PCP-M (O-demethyl-HO-) isomer-1 2AC	U+UHYAC	2420	359_{14}	316_{11}	300_{100}	258_{18}	164_{10}	1260	10279
3-MeO-PCP-M (O-demethyl-HO-) isomer-2 2AC	U+UHYAC	2440	359_{13}	316_{13}	300_{100}	258_{34}	84_{14}	1260	10282
3-MeO-PCP-M (O-demethyl-HO-) isomer-3 2AC	U+UHYAC	2460	359_{16}	316_{100}	256_{12}	224_{12}	107_{18}	1260	10275
3-MeO-PCP-M (O-demethyl-HO-) isomer-4 2AC	U+UHYAC	2490	359_{9}	316_{11}	299_{41}	258_{100}	84_{12}	1260	10273
3-MeO-PCP-M (O-demethyl-HO-) isomer-5 2AC	U+UHYAC	2510	359_{25}	316_{100}	256_{25}	107_{28}	82_{23}	1260	10272
3-MeO-PCP-M (oxo-)	U+UHYAC	2150	287_{20}	244_{100}	230_{6}	222_{5}	98_{8}	823	10281
3-MeO-PCP-M/artifact 2AC	U+UHYAC	2485	361_{11}	318_{100}	246_{8}	175_{12}	107_{11}	1267	10274
4-MeO-PCP		2150	273_{29}	272_{30}	230_{100}	121_{35}	84_{22}	736	3594
4-MeO-PCP artifact		1800*	188_{100}	173_{22}	159_{42}	145_{20}	129_{23}	365	9452
3-MeO-PCPy		2100	259_{30}	216_{100}	202_{24}	152_{32}	121_{22}	653	9455
3-MeO-PCPy artifact	G P U+UHYAC	1630*	188_{100}	173_{23}	159_{41}	129_{23}	115_{19}	365	4436
3-MeO-PCPy-M (O-demethyl-) AC	U+UHYAC	2155	287_{23}	244_{100}	152_{40}	107_{23}	70_{33}	823	10285
3-MeO-PCPy-M (O-demethyl-amino-) 2AC	U+UHYAC	2185	275_{6}	233_{9}	216_{80}	190_{31}	174_{100}	748	10291
3-MeO-PCPy-M (O-demethyl-di-HO-) 3AC	U+UHYAC	2670	403_{1}	343_{52}	302_{100}	175_{41}	107_{33}	1457	10290
3-MeO-PCPy-M (O-demethyl-HO-) isomer-1 2AC	U+UHYAC	2385	345_{10}	302_{11}	286_{100}	244_{25}	70_{6}	1185	10286
3-MeO-PCPy-M (O-demethyl-HO-) isomer-2 2AC	U+UHYAC	2400	345_{13}	302_{11}	286_{100}	244_{33}	70_{15}	1185	10287
3-MeO-PCPy-M (O-demethyl-HO-) isomer-3 2AC	U+UHYAC	2420	345_{11}	302_{100}	285_{12}	242_{17}	107_{13}	1186	10288
3-MeO-PCPy-M (O-demethyl-HO-) isomer-4 2AC	U+UHYAC	2440	345_{6}	285_{36}	244_{100}	107_{5}	70_{8}	1185	10289
3-MeO-PCPy-M (O-demethyl-HO-amino-) 2AC	U+UHYAC	2435	273_{5}	231_{16}	214_{94}	188_{11}	172_{100}	734	10293
4-MeO-PCPy		2120	259_{35}	216_{100}	202_{19}	189_{36}	121_{70}	653	9456
4-MeO-PCPy artifact		1800*	188_{100}	173_{22}	159_{42}	145_{20}	129_{23}	365	9452
MEOP-M (4-methoxybenzoic acid)		1320*	152_{95}	135_{100}	107_{11}	92_{16}	77_{23}	293	10270
MEOP-M (4-methoxybenzoic acid) ME		1270*	166_{36}	135_{100}	107_{11}	92_{15}	77_{18}	316	6446
MEOP-M (N,N-bisdealkyl-) AC		2335	250_{1}	177_{12}	164_{8}	135_{100}	99_{35}	605	9314
MEOP-M (N,N-bisdealkyl-nor-) AC		2290	236_{1}	177_{6}	151_{25}	135_{100}	92_{10}	539	9315
MEOP-M (N,N-bisdealkyl-nor-dihydro-) 2AC		2320	280_{1}	221_{7}	135_{62}	129_{72}	116_{100}	783	9316
MEOP-M (N,N-bisdealkyl-oxo-) AC		2280	264_{1}	179_{22}	135_{100}	92_{17}	72_{19}	681	9311
MEOP-M (N,O-bisdemethyl-) 2AC		2525	290_{5}	222_{4}	163_{16}	121_{100}	85_{46}	842	9309
MEOP-M (N-methylpiperazine) AC		1230	142_{13}	99_{26}	83_{16}	70_{100}	56_{69}	282	9345
MEOP-M (nor-)		2040	220_{18}	191_{2}	135_{100}	92_{14}	85_{8}	469	9312
MEOP-M (nor-) AC		2440	262_{9}	203_{4}	194_{7}	135_{100}	85_{32}	670	9308
MEOP-M (nor-) acetyl conjugate		2440	262_{9}	203_{4}	194_{7}	135_{100}	85_{32}	670	9308
MEOP-M (nor-oxo-) dehydro artifact AC		2505	274_{13}	139_{50}	135_{100}	92_{19}	77_{23}	739	9310
MEOP-M (O-demethyl-)		2135	220_{6}	121_{50}	99_{39}	83_{46}	70_{100}	469	9313
MEOP-M (O-demethyl-) AC		2170	262_{7}	176_{7}	121_{65}	99_{55}	70_{100}	669	9306

MEOP-M (oxo-)

Table 8.1: Compounds in order of names

Name	Detected	RI	Typical ions and intensities					Page	Entry
MEOP-M (oxo-)		2180	248 $_{11}$	204 $_4$	193 $_{10}$	135 $_{100}$	113 $_{78}$	590	9307
MeOPP		1880	192 $_{34}$	150 $_{100}$	135 $_{20}$	120 $_{28}$	92 $_9$	376	6622
MeOPP AC	U+UHYAC	2185	234 $_{81}$	162 $_{100}$	149 $_{35}$	134 $_{36}$	120 $_{35}$	529	6609
MeOPP HFB	U+UHYHFB	1965	388 $_{100}$	373 $_{34}$	191 $_{75}$	135 $_{74}$	120 $_{66}$	1394	6617
MeOPP ME		1840	206 $_{100}$	191 $_{17}$	162 $_{14}$	135 $_{77}$	120 $_{62}$	417	6623
MeOPP TFA	U+UHYTFA	1940	288 $_{100}$	273 $_{36}$	191 $_{20}$	135 $_{22}$	120 $_{28}$	827	6612
MeOPP TMS		2070	264 $_{100}$	249 $_{29}$	101 $_{45}$	86 $_{46}$	73 $_{81}$	683	6884
MeOPP-M (4-aminophenol N-acetyl-) HFB	UHYHFB PHFB	1735	347 $_{24}$	305 $_{39}$	169 $_{13}$	108 $_{100}$	69 $_{31}$	1192	5099
MeOPP-M (4-aminophenol N-acetyl-) TFA		1630	247 $_{11}$	205 $_{30}$	108 $_{100}$	80 $_{19}$	69 $_{34}$	583	5092
MeOPP-M (4-aminophenol)	UHY	1240	109 $_{100}$	80 $_{41}$	53 $_{82}$	53 $_{82}$	52 $_{90}$	260	826
MeOPP-M (4-aminophenol) 2AC	PAC U+UHYAC	1765	193 $_{10}$	151 $_{53}$	109 $_{100}$	80 $_{24}$		377	188
MeOPP-M (4-methoxyaniline) HFB	U+UHYHFB	1400	319 $_{100}$	304 $_8$	300 $_6$	150 $_6$	122 $_{78}$	1023	6620
MeOPP-M (4-methoxyaniline) TFA	U+UHYTFA	1335	219 $_{100}$	204 $_{16}$	149 $_{11}$	122 $_{76}$	109 $_{19}$	463	6615
MeOPP-M (aminophenol) 2HFB	U+UHYHFB	1405	501 $_{25}$	482 $_6$	304 $_{81}$	169 $_{100}$	109 $_{81}$	1660	6621
MeOPP-M (aminophenol) 2TFA	U+UHYTFA	1280	301 $_{49}$	204 $_{47}$	176 $_{19}$	109 $_{49}$	69 $_{100}$	910	6616
MeOPP-M (deethylene-) 2AC	U+UHYAC	2120	250 $_7$	191 $_{51}$	165 $_4$	149 $_{33}$	136 $_{100}$	605	6611
MeOPP-M (deethylene-) 2HFB	U+UHYHFB	1765	558 $_7$	345 $_{13}$	332 $_{28}$	240 $_{19}$	135 $_{100}$	1699	6619
MeOPP-M (deethylene-) 2TFA	U+UHYTFA	1765	358 $_{20}$	245 $_{19}$	232 $_{48}$	135 $_{100}$	120 $_{33}$	1252	6614
MeOPP-M (methoxyaniline) AC	PME UME	1630	165 $_{59}$	123 $_{74}$	108 $_{100}$	95 $_{10}$	80 $_{20}$	314	5046
MeOPP-M (O-demethyl-) 2AC	U+UHYAC	2350	262 $_{39}$	220 $_{60}$	177 $_{21}$	148 $_{100}$	135 $_{59}$	670	6610
MeOPP-M (O-demethyl-) 2HFB	U+UHYHFB	1990	570 $_{83}$	551 $_{16}$	373 $_{100}$	344 $_{46}$	317 $_{100}$	1703	6618
MeOPP-M (O-demethyl-) 2TFA	U+UHYTFA	1915	370 $_{90}$	273 $_{100}$	244 $_{60}$	217 $_{34}$	120 $_{75}$	1310	6613
3-Me-PCP		2110	257 $_{40}$	256 $_{49}$	214 $_{100}$	200 $_{35}$		644	9453
3-Me-PCP artifact		1500*	172 $_{100}$	157 $_{66}$	143 $_{35}$	129 $_{81}$	115 $_{32}$	327	9454
3-Me-PCPy		2115	243 $_{40}$	200 $_{100}$	186 $_{36}$	152 $_{39}$	70 $_{19}$	569	10184
3-Me-PCPy artifact		1500*	172 $_{100}$	157 $_{66}$	143 $_{35}$	129 $_{81}$	115 $_{32}$	327	9454
Mephedrone		1560	177 $_1$	119 $_3$	91 $_{10}$	65 $_8$	58 $_{100}$	335	8326
Mephedrone 2TMS		1605	321 $_6$	234 $_{11}$	219 $_8$	130 $_{100}$	91 $_{28}$	1042	8335
Mephedrone AC	U+UHYAC	1915	219 $_2$	119 $_8$	100 $_{63}$	91 $_{19}$	58 $_{100}$	466	7870
Mephedrone HFB		1605	373 $_1$	254 $_{61}$	210 $_{29}$	119 $_{100}$	91 $_{44}$	1326	8329
Mephedrone PFP		1580	323 $_1$	204 $_{35}$	160 $_{28}$	119 $_{100}$	91 $_{38}$	1049	8328
Mephedrone TFA		1580	273 $_1$	154 $_{60}$	119 $_{100}$	110 $_{29}$	91 $_{44}$	733	8330
Mephedrone TMS		1605	249 $_4$	234 $_6$	219 $_5$	130 $_{100}$	91 $_{30}$	600	8327
Mephedrone-M (carboxytolyl-dihydro-) 2AC	U+UHYAC	2235	293 $_2$	251 $_{13}$	151 $_{67}$	100 $_{53}$	58 $_{100}$	862	7965
Mephedrone-M (HO-tolyl-) 2AC	U+UHYAC	2345	277 $_3$	177 $_5$	100 $_{86}$	89 $_{11}$	58 $_{100}$	760	7970
Mephedrone-M (nor-) AC	U+UHYAC	1875	205 $_{14}$	177 $_4$	119 $_{100}$	91 $_{46}$	86 $_{97}$	413	7871
Mephedrone-M (nor-dihydro-) isomer-1 2AC	U+UHYAC	1885	249 $_3$	190 $_{56}$	129 $_{27}$	121 $_{52}$	86 $_{100}$	598	7968
Mephedrone-M (nor-dihydro-) isomer-2 2AC	U+UHYAC	1900	249 $_3$	190 $_{59}$	129 $_{33}$	121 $_{62}$	86 $_{100}$	598	7967
Mephedrone-M (nor-HO-tolyl-) 2AC	U+UHYAC	2310	263 $_1$	177 $_{44}$	118 $_{11}$	105 $_2$	86 $_{100}$	674	7969
Mephenesin	P-I G	1660*	182 $_{13}$	133 $_6$	108 $_{100}$	91 $_{23}$	77 $_{19}$	352	2804
Mephenesin 2AC	UHYAC	1805*	266 $_1$	159 $_{100}$	108 $_{46}$	91 $_{38}$	57 $_{29}$	695	2805
Mephenesin 2TMS		1755*	326 $_{12}$	205 $_{16}$	147 $_{42}$	133 $_{48}$	73 $_{100}$	1070	4563
Mephentermine		1235	148 $_2$	133 $_1$	91 $_7$	72 $_{100}$	56 $_5$	308	3721
Mephentermine AC		1505	148 $_1$	132 $_4$	114 $_{40}$	91 $_{10}$	72 $_{100}$	414	3722
Mephentermine TFA		1335	244 $_1$	168 $_{100}$	110 $_{86}$	91 $_{21}$	56 $_{29}$	650	3727
Mephenytoin	P G U+UHYAC	1780	218 $_6$	189 $_{93}$	104 $_{100}$			458	1084
Mephenytoin-M (HO-)	U UHY	2400	234 $_2$	205 $_{100}$	152 $_{15}$	120 $_{92}$	109 $_{19}$	527	2926
Mephenytoin-M (HO-) isomer-1 AC	UHYAC	2390	276 $_1$	247 $_{15}$	205 $_{100}$	120 $_{46}$	91 $_7$	754	2924
Mephenytoin-M (HO-) isomer-2 AC	UHYAC	2540	247 $_{84}$	205 $_{100}$	134 $_{26}$	107 $_{15}$		754	4191
Mephenytoin-M (HO-methoxy-)	U UHY	2380	264 $_{10}$	235 $_{100}$	150 $_{83}$	135 $_{29}$		681	2927
Mephenytoin-M (HO-methoxy-) 2AC	UHYAC	2630	348 $_1$	306 $_2$	264 $_6$	191 $_{100}$	120 $_{53}$	1199	2925
Mephenytoin-M (nor-)	U UHY UHYAC	1950	204 $_5$	175 $_{74}$	132 $_6$	104 $_{100}$	77 $_{31}$	409	2928
Mephenytoin-M (nor-) AC	UHYAC	1900	246 $_{25}$	175 $_{100}$	144 $_{53}$	104 $_{53}$	77 $_{20}$	579	2929
Mephenytoin-M (nor-HO-) 2AC	UHYAC	2495	304 $_4$	262 $_{52}$	191 $_{100}$	160 $_7$	120 $_{19}$	931	4173
Mepindolol		2390	262 $_{18}$	147 $_{100}$	114 $_{18}$	100 $_{18}$	72 $_{40}$	671	1358
Mepindolol 2AC		2750	346 $_2$	286 $_{50}$	184 $_{100}$	140 $_{85}$	98 $_{93}$	1190	1359
Mepindolol 2TMSTFA		2565	502 $_2$	284 $_{64}$	218 $_{14}$	129 $_{73}$	73 $_{100}$	1662	6170
Mepindolol formyl artifact	U	2410	274 $_{77}$	186 $_{17}$	147 $_{100}$	86 $_{36}$	72 $_{16}$	741	1722
Mepindolol -H2O AC		2680	286 $_{38}$	184 $_{100}$	140 $_{95}$	98 $_{80}$		816	1705
Mepindolol TMS		2320	334 $_2$	188 $_3$	147 $_{100}$	118 $_{11}$	72 $_{36}$	1121	6171
Mepindolol TMSTFA		2455	430 $_{14}$	284 $_{93}$	146 $_{63}$	129 $_{100}$	73 $_{94}$	1534	6169
Mepirapim		2595	313 $_{17}$	256 $_{10}$	230 $_{35}$	214 $_{100}$	144 $_{28}$	991	9510
Mepivacaine	P G U+UHYAC	2075	246 $_1$	176	120 $_1$	98 $_{100}$	70 $_{10}$	582	1085
Mepivacaine TMS		1980	318 $_2$	261 $_{10}$	248 $_4$	98 $_{100}$	73 $_{21}$	1021	4564
Mepivacaine-M (HO-)	UHY	2410	262 $_2$	98 $_{100}$	96 $_8$	70 $_{16}$		671	1086
Mepivacaine-M (HO-) AC	U+UHYAC	2450	304 $_1$	98 $_{100}$	96 $_9$	70 $_{20}$		932	1087
Mepivacaine-M (HO-piperidyl-) AC	UHYAC	2590	304 $_8$	156 $_5$	129 $_{15}$	114 $_{100}$	86 $_{11}$	932	2970
Mepivacaine-M (nor-) AC	UHYAC	2170	274 $_2$	154 $_{58}$	126 $_{69}$	84 $_{100}$		741	2968
Mepivacaine-M (oxo-)	U UHY UHYAC	2400	260 $_1$	218 $_3$	112 $_{100}$			659	2969
Mepivacaine-M (oxo-HO-piperidyl-) AC	UHYAC	2630	318 $_{21}$	258 $_1$	170 $_{64}$	128 $_{100}$	111 $_{12}$	1019	3049
Meprobamate	P G U+UHYAC	1785	144 $_{20}$	114 $_{28}$	96 $_{34}$	83 $_{89}$	55 $_{100}$	459	1088
Meprobamate artifact-1	P G	1535*	84 $_{100}$	56 $_{81}$				254	1089
Meprobamate artifact-2	P U UHY UHYAC	1720*	173 $_2$	101 $_9$	84 $_{100}$	56 $_{90}$		328	580
Meptazinol		1920	233 $_7$	107 $_7$	98 $_{19}$	84 $_{82}$	58 $_{100}$	524	3546

Table 8.1: Compounds in order of names

Name	Detected	RI	Typical ions and intensities					Page	Entry
Meptazinol AC		1945	275_3	107_5	98_{33}	84_{89}	58_{100}	749	3549
Meptazinol HFB		1810	429_2	303_1	98_{34}	84_{85}	58_{100}	1531	6136
Meptazinol PFP		1655	379_{10}	253_4	98_{35}	84_{77}	58_{100}	1351	6127
Meptazinol TFA		1795	329_{17}	203_6	98_{41}	84_{100}	58_{86}	1088	6206
Meptazinol TMS		2005	305_{11}	98_{32}	84_{100}	73_{32}	58_{100}	940	6207
Meptazinol-M (nor-)		1995	219_{13}	159_{11}	107_{15}	84_{23}	70_{100}	467	3547
Meptazinol-M (nor-) 2AC		2395	303_{41}	159_{27}	126_{38}	87_{100}	70_{71}	927	3551
Meptazinol-M (oxo-)		2410	247_{89}	204_{27}	148_{100}	87_{61}	55_{62}	586	3548
Meptazinol-M (oxo-) AC		2350	289_{49}	204_{100}	176_{46}	148_{66}	87_{58}	836	3550
Mepyramine	G U	2220	285_4	215_8	121_{100}	78_{20}	58_{50}	811	1656
Mepyramine HY	UHY UHYAC	1690	208_1	163_1	137_4	71_{50}	58_{100}	426	1660
Mepyramine-M (N-dealkyl-)	U	2120	214_{61}	136_{11}	121_{100}	78_{32}		447	1657
Mepyramine-M (N-dealkyl-) AC	UHYAC	2150	256_{42}	214_{62}	163_{38}	107_{100}	78_{40}	638	1659
Mepyramine-M (N-demethoxybenzyl-)	U	1580	165_3	119_8	107_{12}	78_{26}	58_{100}	315	1658
Mequitazine	G U UHY UHYAC	2765	322_{32}	212_6	198_9	180_6	124_{100}	1045	1483
Mequitazine-M (HO-sulfoxide) AC	UHYAC	3230	396_6	354_5	180_3	124_{100}	70_{11}	1430	1672
Mequitazine-M (ring)	P G U+UHYAC	2010	199_{100}	167_{44}				396	10
Mequitazine-M (sulfone)	U UHY UHYAC	3250	354_7	244_2	180_3	124_{100}	70_9	1232	1671
Mequitazine-M (sulfoxide)	U UHY UHYAC	3120	338_8	321_5	198_{10}	124_{100}	70_{11}	1143	1670
Mequitazine-M 2AC	U+UHYAC	2865	315_{54}	273_{34}	231_{100}	202_{11}		999	2618
Mequitazine-M AC	U+UHYAC	2550	257_{23}	215_{100}	183_7			641	12
Mercaptodimethur		1915	225_{12}	184_7	168_{100}	153_{54}	109_{20}	490	3450
Mercaptodimethur-M/artifact (decarbamoyl-)		1535*	168_{100}	153_{88}	109_{87}	91_{32}	77_{16}	321	3451
Mesalazine ME2AC	UHYAC	1890	251_{13}	209_{100}	177_{91}	135_{91}	107_{20}	608	4485
Mesalazine MEAC	UHYAC	1945	209_{45}	177_{46}	135_{100}	107_{24}		427	4486
Mescaline		1690	211_{24}	182_{100}	167_{55}	151_{15}	148_{13}	437	1090
Mescaline 2AC		2125	295_{15}	194_{100}	181_{55}	179_{54}	151_{10}	878	6943
Mescaline 2TMS		2080	355_1	340_{13}	174_{100}	100_6	73_{28}	1239	5683
Mescaline AC		2070	253_{16}	194_{100}	179_{53}	151_9	77_5	623	1484
Mescaline formyl artifact		1700	223_{25}	181_{100}	167_6	148_{10}	77_4	484	3244
Mescaline HFB		1865	407_{35}	226_4	194_{36}	181_{100}	69_{11}	1466	5066
Mescaline PFP		1835	357_{33}	194_{37}	181_{100}	151_{10}	119_{18}	1247	5067
Mescaline precursor (trimethoxyphenylacetonitrile)		1610	207_{85}	192_{100}	164_{39}	124_{24}	78_{45}	419	3273
Mescaline TFA		1830	307_{23}	194_{46}	181_{100}	148_{18}	126_{10}	947	5068
Mescaline TMS		1895	283_1	268_9	181_9	102_{100}	73_{64}	797	4959
Mescaline-D9		1685	220_{26}	191_{100}	173_{50}	152_{11}		471	6907
Mescaline-D9 2AC		2120	304_9	203_{100}	190_{44}	185_{46}	157_8	933	6945
Mescaline-D9 2TMS		2070	364_1	349_4	190_{12}	174_{100}	73_{30}	1284	6947
Mescaline-D9 AC		2065	262_{18}	203_{100}	190_{51}	185_{51}	157_{10}	671	6944
Mescaline-D9 formyl artifact		1690	232_{34}	203_2	190_{100}	152_7	140_4	519	6911
Mescaline-D9 HFB		1855	416_{28}	203_{40}	190_{100}	185_{22}	157_7	1496	6939
Mescaline-D9 PFP		1820	366_{30}	203_{40}	190_{100}	185_{26}	119_{11}	1291	6934
Mescaline-D9 TFA		1825	316_{29}	203_{30}	190_{100}	185_{20}	157_6	1006	6929
Mescaline-D9 TMS		1885	292_1	277_5	190_9	102_{100}	73_{39}	858	6946
Mescaline-M (deamino-COOH) ME	UGLUCSPEMEAC	1840*	240_{80}	225_{42}	181_{100}	148_{17}	137_{13}	556	7135
Mesembranol	UGLUCSPE	2295	291_{56}	290_{100}	274_{27}	218_{43}	70_{42}	851	8995
Mesembranol AC	UGLUCSPE	2360	333_{57}	290_{49}	274_{100}	219_{60}	70_{28}	1115	9001
Mesembrenone		2335	287_{28}	258_3	219_{10}	115_8	70_{100}	821	8014
Mesembrenone (7-delta)		2585	285_{100}	270_{20}	257_{19}	242_{47}	115_6	810	8993
Mesembrenone AC		2730	329_{15}	256_{30}	229_{91}	100_{100}	58_{66}	1088	8997
Mesembrenone-M	UGLUCSPE	2215	288_{84}	274_{100}	258_{32}	219_{25}	205_{21}	824	9034
Mesembrenone-M 1	UGLUCSPE	1890	219_{59}	205_{100}	190_{70}	174_{16}	162_{86}	462	9037
Mesembrenone-M 2	UGLUCSPE	1935	203_{100}	188_{44}	160_{87}	130_{14}	117_{15}	405	9038
Mesembrenone-M 3	UGLUCSPE	1955	217_{51}	203_{93}	188_{83}	160_{100}	130_{32}	453	9039
Mesembrenone-M 4	UGLUCSPE	1980	297_{15}	262_{18}	233_{23}	205_{100}	70_{19}	888	9040
Mesembrenone-M 5	UGLUCSPE	2045	259_{18}	205_{13}	152_{63}	109_{100}	70_{57}	649	9041
Mesembrenone-M 6	UGLUCSPE	2055	257_{100}	242_{10}	214_{14}	152_{39}	134_{48}	640	9042
Mesembrenone-M 7	UGLUCSPE	2075	230_{100}	215_{58}	187_{46}	172_{19}	115_{32}	508	9043
Mesembrenone-M 8	UGLUCSPE	2095	230_{100}	215_{92}	187_{79}	144_{26}	115_{30}	508	9044
Mesembrenone-M 9	UGLUCSPE	2100	244_{100}	229_{37}	201_{22}	186_{16}	115_{26}	569	9045
Mesembrenone-M 10	UGLUCSPE	2160	258_6	216_{100}	201_{25}	173_{70}	115_{34}	644	9046
Mesembrenone-M 11	UGLUCSPE	2245	287_{64}	272_{100}	256_{18}	230_{22}	115_{23}	817	9047
Mesembrenone-M 12	UGLUCSPE	2270	275_{31}	218_{15}	205_{100}	115_{12}	70_{83}	743	9048
Mesembrenone-M 13	UGLUCSPE	2275	289_8	232_7	219_{58}	70_{100}		831	9049
Mesembrenone-M 14	UGLUCSPE	2290	276_{100}	260_{43}	218_{26}	205_{82}	70_{28}	751	9050
Mesembrenone-M 15	UGLUCSPE	2300	287_{20}	272_{88}	255_{52}	115_{12}	58_{100}	817	9051
Mesembrenone-M 16	UGLUCSPE	2320	273_{72}	244_{10}	205_{22}	115_{21}	70_{100}	731	9052
Mesembrenone-M 17	UGLUCSPE	2325	287_{39}	273_7	205_{11}	115_{11}	70_{100}	817	9053
Mesembrenone-M 18	UGLUCSPE	2335	257_{73}	242_{100}	227_{24}	205_{10}	184_{14}	640	9054
Mesembrenone-M 19	UGLUCSPE	2350	307_4	292_3	276_3	256_5	58_{100}	945	9055
Mesembrenone-M 20	UGLUCSPE	2365	273_{22}	242_{12}	205_7	115_7	70_{100}	732	9056
Mesembrenone-M 21 isomer-1	UGLUCSPE	2395	317_4	302_{12}	253_{100}	210_{24}	115_5	1008	9057
Mesembrenone-M 21 isomer-2	UGLUCSPE	2460	317_4	302_{11}	253_{100}	210_{24}	115_4	1008	9058
Mesembrenone-M 22	UGLUCSPE	2025	247_{58}	205_{100}	190_8	162_{13}	132_{15}	583	9059

Mesembrenone-M 23

Table 8.1: Compounds in order of names

Name	Detected	RI	Typical ions and intensities					Page	Entry
Mesembrenone-M 23	UGLUCSPE	2200	272_{33}	230_{100}	215_{59}	187_{29}	115_{25}	727	9060
Mesembrenone-M 24	UGLUCSPE	2210	272_{59}	230_{100}	215_{71}	187_{36}	115_{14}	727	9061
Mesembrenone-M 25	UGLUCSPE	2240	301_{27}	286_{100}	270_{9}	164_{12}	115_{8}	909	9062
Mesembrenone-M 26	UGLUCSPE	2260	258_{11}	216_{100}	201_{14}	173_{27}	115_{13}	644	9063
Mesembrenone-M 27	UGLUCSPE	2260	292_{6}	258_{20}	231_{36}	216_{100}	189_{51}	853	9064
Mesembrenone-M 28	UGLUCSPE	2305	300_{10}	258_{28}	216_{100}	201_{11}	115_{9}	904	9065
Mesembrenone-M 29	UGLUCSPE	2330	331_{30}	288_{43}	247_{41}	205_{71}	70_{100}	1097	9066
Mesembrenone-M 30	UGLUCSPE	2345	329_{70}	314_{100}	272_{48}	230_{29}	115_{17}	1084	9067
Mesembrenone-M 31	UGLUCSPE	2370	415_{2}	314_{6}	247_{28}	205_{34}	70_{100}	1492	9068
Mesembrenone-M 32	UGLUCSPE	2410	363_{3}	348_{7}	315_{57}	115_{13}	70_{100}	1277	9069
Mesembrenone-M 33	UGLUCSPE	2430	359_{53}	316_{85}	300_{18}	247_{31}	70_{100}	1256	9070
Mesembrenone-M 34	UGLUCSPE	2440	359_{11}	316_{19}	247_{6}	70_{100}		1256	9071
Mesembrenone-M 35	UGLUCSPE	2450	361_{7}	318_{12}	302_{100}	205_{16}	70_{9}	1267	9072
Mesembrenone-M 36	UGLUCSPE	2490	391_{14}	376_{56}	248_{100}	205_{69}	70_{83}	1404	9073
Mesembrenone-M 37	UGLUCSPE	2500	387_{14}	344_{33}	328_{8}	275_{5}	70_{100}	1388	9074
Mesembrenone-M 38	UGLUCSPE	2520	387_{4}	344_{14}	328_{7}	275_{5}	70_{100}	1388	9075
Mesembrenone-M 39	UGLUCSPE	2560	329_{59}	301_{100}	259_{69}	244_{43}	192_{61}	1084	9076
Mesembrenone-M 40	UGLUCSPE	2595	329_{35}	270_{39}	239_{100}	226_{53}	211_{26}	1084	9077
Mesembrenone-M 41	UGLUCSPE	2655	359_{100}	316_{30}	271_{34}	214_{42}	56_{42}	1255	9078
Mesembrenone-M 42	UGLUCSPE	2730	371_{25}	329_{23}	256_{100}	100_{66}	58_{57}	1315	9079
Mesembrenone-M 43	UGLUCSPE	2755	387_{35}	345_{100}	302_{35}	257_{56}	56_{65}	1387	9080
Mesembrenone-M 44	UGLUCSPE	2850	357_{5}	315_{18}	242_{61}	100_{100}	58_{67}	1245	9081
Mesembrenone-M 45	UGLUCSPE	2855	357_{4}	315_{17}	242_{39}	100_{100}	58_{33}	1440	9082
Mesembrine		2310	289_{51}	218_{100}	204_{31}	96_{78}	70_{72}	837	8033
Mesembrine (7-delta)		2585	287_{100}	259_{60}	244_{90}	214_{9}	115_{7}	821	8992
Mesembrine AC		2675	331_{54}	258_{18}	245_{35}	231_{100}	100_{51}	1101	8996
Mesembrine-M (bis-demethyl-dihydro-) isomer-1 3AC	UGLUCSPE	2630	389_{18}	347_{98}	287_{100}	245_{43}	163_{70}	1400	9006
Mesembrine-M (bis-demethyl-dihydro-) isomer-2 3AC	UGLUCSPE	2695	389_{23}	347_{100}	287_{68}	244_{36}	163_{67}	1399	9007
Mesembrine-M (bis-demethyl-dihydro-) isomer-3 3AC	UGLUCSPE	2700	389_{9}	347_{100}	287_{82}	244_{55}	163_{44}	1400	9008
Mesembrine-M (demethyl-) 2AC	UGLUCSPE	2650	359_{8}	317_{50}	231_{35}	217_{100}	100_{78}	1259	9005
Mesembrine-M (demethyl-) AC	UGLUCSPE	2550	317_{32}	244_{21}	231_{35}	217_{100}	100_{68}	1013	9025
Mesembrine-M (demethyl-dihydro-) isomer-1	UGLUCSPE	2290	277_{09}	276_{100}	260_{36}	205_{91}	70_{48}	762	8998
Mesembrine-M (demethyl-dihydro-) isomer-1 2AC	UGLUCSPE	2350	361_{22}	318_{32}	302_{100}	260_{24}	205_{28}	1269	9004
Mesembrine-M (demethyl-dihydro-) isomer-2	UGLUCSPE	2310	277_{93}	276_{100}	260_{33}	218_{21}	205_{90}	762	8999
Mesembrine-M (demethyl-dihydro-) isomer-2 2AC	UGLUCSPE	2535	361_{100}	302_{55}	260_{52}	163_{70}	121_{34}	1270	9002
Mesembrine-M (demethyl-dihydro-) isomer-3	UGLUCSPE	2375	277_{61}	276_{100}	260_{22}	218_{13}	205_{31}	763	9000
Mesembrine-M (demethyl-dihydro-) isomer-3 2AC	UGLUCSPE	2590	361_{100}	302_{23}	260_{53}	163_{83}	121_{34}	1270	9003
Mesembrine-M (dihydro-)	UGLUCSPE	2295	291_{56}	290_{100}	274_{27}	218_{43}	70_{42}	851	8995
Mesembrine-M (dihydro-) AC	UGLUCSPE	2360	333_{57}	290_{49}	274_{100}	219_{60}	70_{28}	1115	9001
Mesembrine-M (HO-) isomer-1 2AC	UGLUCSPE	2390	389_{6}	346_{17}	330_{100}	288_{12}	190_{19}	1400	9010
Mesembrine-M (HO-) isomer-2 2AC	UGLUCSPE	2410	389_{12}	346_{24}	330_{100}	288_{18}	190_{24}	1400	9009
Mesembrine-M 1	UGLUCSPE	2050	301_{100}	273_{23}	230_{12}	134_{33}	109_{22}	910	9011
Mesembrine-M 2	UGLUCSPE	2050	301_{100}	273_{15}	230_{7}	134_{27}	109_{39}	909	9012
Mesembrine-M 3	UGLUCSPE	2095	313_{60}	288_{69}	274_{100}	258_{17}	205_{22}	985	9013
Mesembrine-M 4	UGLUCSPE	2108	302_{27}	288_{42}	249_{33}	205_{70}	86_{100}	916	9014
Mesembrine-M 5	UGLUCSPE	2115	300_{55}	286_{100}	270_{8}	176_{30}	86_{41}	904	9015
Mesembrine-M 6	UGLUCSPE	2125	301_{10}	217_{74}	175_{100}			910	9016
Mesembrine-M 7	UGLUCSPE	2210	330_{81}	316_{100}	300_{29}	274_{14}	164_{18}	1091	9017
Mesembrine-M 8	UGLUCSPE	2220	331_{13}	316_{7}	247_{94}	205_{100}		1097	9018
Mesembrine-M 9	UGLUCSPE	2240	333_{37}	319_{28}	274_{100}	260_{81}	219_{52}	1109	9019
Mesembrine-M 10	UGLUCSPE	2280	319_{16}	276_{12}	260_{100}	205_{23}	190_{18}	1022	9020
Mesembrine-M 11	UGLUCSPE	2305	318_{100}	303_{18}	260_{34}	205_{41}	70_{43}	1016	9021
Mesembrine-M 12	UGLUCSPE	2385	347_{6}	304_{11}	288_{100}	213_{31}	190_{20}	1192	9022
Mesembrine-M 13	UGLUCSPE	2390	327_{100}	285_{59}	243_{60}	228_{12}		1072	9023
Mesembrine-M 14	UGLUCSPE	2470	331_{100}	303_{49}	274_{56}	193_{31}	150_{17}	1097	9024
Mesembrine-M 15	UGLUCSPE	2550	207_{56}	164_{7}	150_{10}	135_{12}	57_{100}	1009	9026
Mesembrine-M 16	UGLUCSPE	1865	205_{100}	190_{23}	172_{12}	162_{42}	144_{10}	412	9027
Mesembrine-M 17	UGLUCSPE	1930	260_{4}	203_{100}	188_{38}	160_{84}	57_{42}	654	9028
Mesembrine-M 18	UGLUCSPE	1980	260_{10}	207_{15}	57_{100}			654	9029
Mesembrine-M 19	UGLUCSPE	2035	259_{100}	231_{51}	216_{11}	135_{43}	109_{33}	649	9030
Mesembrine-M 20	UGLUCSPE	2055	273_{33}	260_{100}	246_{25}	204_{54}	137_{27}	731	9031
Mesembrine-M 21	UGLUCSPE	2070	274_{16}	257_{100}	242_{22}	163_{31}	138_{73}	736	9032
Mesembrine-M 22	UGLUCSPE	2100	259_{100}	231_{37}	216_{21}	198_{19}	170_{21}	649	9033
Mesembrine-M 23	UGLUCSPE	2215	288_{84}	274_{100}	258_{32}	219_{25}	205_{21}	824	9034
Mesembrine-M 24	UGLUCSPE	2225	289_{12}	274_{7}	205_{100}	190_{10}	162_{12}	832	9035
Mesembrine-M 25	UGLUCSPE	2485	310_{24}	252_{12}	205_{100}	190_{11}		968	9036
Mesoridazine	G P U+UHYAC	3350	386_{11}	126_{6}	98_{100}	70_{11}	42_{43}	1385	2200
Mesoridazine	P-I G U+UHYAC	3350	386_{12}	370_{4}	126_{7}	98_{100}	70_{11}	1385	4484
Mesoridazine-M (side chain sulfone)	G P-I U+UHYAC	3415	402_{13}	290_{5}	197_{6}	98_{100}	70_{10}	1453	394
Mesterolone		2545*	304_{100}	218_{90}	200_{42}			933	1091
Mesterolone enol 2TMS		2530*	448_{3}	433_{8}	157_{42}	141_{100}	73_{71}	1577	3982
Mestranol		2630*	310_{35}	242_{13}	227_{100}	174_{46}	115_{48}	973	2806
Mestranol AC		2690*	352_{62}	242_{9}	227_{100}	173_{29}	147_{22}	1223	2807
Mesulphen	U UHY UHYAC	2250*	244_{100}	227_{6}	211_{75}	184_{11}	121_{10}	570	5377

Table 8.1: Compounds in order of names

Name	Detected	RI	Typical ions and intensities					Page	Entry
Mesulphen-M (di-HO-) 2AC	UGLUCAC	2830*	360_{100}	277_{25}	258_{35}	227_{33}	184_{23}	1263	5387
Mesulphen-M (di-HOOC-) 2ME	UME	2805*	332_{74}	304_{100}	273_{20}	214_{14}	184_{14}	1104	5392
Mesulphen-M (di-HOOC-) -CO2 ME	UME	2380*	274_{100}	243_{53}	215_{61}	171_{43}	121_{27}	737	5388
Mesulphen-M (HO-)	U UHY	2430*	260_{100}	242_{12}	227_{11}	197_{17}	184_{17}	655	5378
Mesulphen-M (HO-) AC	UGLUCAC	2535*	302_{100}	259_{13}	242_{26}	227_{35}	198_{24}	917	5379
Mesulphen-M (HO-aryl-sulfoxide)	U UHY	2585*	276_{100}	243_{60}	227_{7}	211_{20}	165_{32}	752	5385
Mesulphen-M (HO-aryl-sulfoxide) ME	UGLUCME	2625*	290_{31}	242_{97}	211_{100}	183_{48}	139_{33}	840	5386
Mesulphen-M (HO-di-sulfoxide)	UHY	2785*	292_{37}	275_{100}	258_{41}	197_{48}	184_{67}	854	5383
Mesulphen-M (HO-di-sulfoxide) AC	UGLUCAC	2895*	334_{69}	291_{100}	275_{25}	209_{24}	184_{64}	1118	5384
Mesulphen-M (HO-HOOC-) MEAC	UMEAC	2825*	346_{100}	303_{16}	271_{21}	227_{54}	184_{23}	1187	5393
Mesulphen-M (HO-HOOC-di-sulfoxide) MEAC	UMEAC	3025*	378_{12}	314_{35}	303_{100}	272_{61}	255_{45}	1346	5395
Mesulphen-M (HO-HOOC-sulfoxide) MEAC	UMEAC	2995*	362_{23}	314_{91}	272_{100}	255_{64}	196_{30}	1272	5394
Mesulphen-M (HOOC-) -CO2	U UHY UHYAC	2235*	230_{100}	197_{82}	171_{13}	152_{15}		508	5396
Mesulphen-M (HOOC-) ME	UME	2545*	288_{100}	257_{40}	229_{32}	214_{13}	185_{17}	825	5389
Mesulphen-M (HOOC-di-sulfoxide) ME	UME	2895*	320_{100}	304_{17}	289_{42}	272_{65}	241_{6}	1032	5391
Mesulphen-M (HOOC-sulfoxide) ME	UME	2665*	304_{13}	256_{100}	225_{58}	197_{35}	152_{10}	929	5390
Mesulphen-M (HO-sulfoxide)	UGLUC	2705*	276_{39}	260_{13}	228_{100}	199_{50}	184_{30}	752	5381
Mesulphen-M (HO-sulfoxide) AC	UGLUCAC	2725*	318_{30}	270_{84}	228_{80}	211_{100}	184_{31}	1016	5382
Mesulphen-M (sulfoxide)	U	2400*	260_{35}	244_{27}	212_{100}			655	5380
Mesuximide	P G U+UHYAC	1705	203_{44}	118_{100}	103_{23}	91_{12}	77_{20}	406	1827
Mesuximide-M (di-HO-) 2AC	UHYAC	2260	319_{4}	277_{23}	235_{100}	185_{21}	150_{78}	1024	2920
Mesuximide-M (HO-)	U UHY	2220	219_{38}	134_{100}	119_{26}	91_{11}	65_{12}	464	2915
Mesuximide-M (HO-) isomer-1 AC	UHYAC	1960	261_{3}	219_{38}	134_{100}	105_{5}	77_{8}	663	2916
Mesuximide-M (HO-) isomer-2 AC	UHYAC	1995	261_{3}	219_{47}	134_{100}	119_{18}	77_{12}	663	2917
Mesuximide-M (nor-)	P U+UHYAC	1750	189_{18}	118_{100}	103_{23}	77_{22}	58_{3}	367	2914
Mesuximide-M (nor-) TMS		1730	261_{1}	246_{13}	146_{6}	118_{100}	103_{23}	663	7423
Mesuximide-M (nor-HO-)	U UHY	2300	205_{37}	134_{100}	119_{35}	91_{5}	65_{11}	412	2921
Mesuximide-M (nor-HO-) isomer-1 AC	UHYAC	2120	247_{3}	205_{45}	134_{100}	105_{5}	94_{10}	585	2918
Mesuximide-M (nor-HO-) isomer-2 AC	UHYAC	2200	247_{3}	205_{43}	134_{100}	119_{14}	77_{8}	585	2919
MET		1815	202_{5}	144_{6}	130_{27}	115_{7}	72_{100}	404	10165
MET HFB		1675	397_{1}	340_{3}	326_{4}	129_{13}	72_{100}	1437	10168
MET PFP		1660	347_{1}	290_{3}	276_{3}	129_{8}	72_{100}	1199	10167
MET TFA		1770	297_{1}	240_{3}	226_{5}	129_{9}	72_{100}	895	10166
MET TMS		1910	274_{9}	216_{5}	202_{13}	186_{5}	72_{100}	741	10196
Metaclazepam	U+UHYAC	2640	392_{6}	349_{100}	347_{76}	319_{16}	163_{4}	1410	2144
Metaclazepam-M (amino-Br-Cl-benzophenone)	UHY	2270	311_{100}	309_{71}	276_{88}	274_{90}	195_{41}	960	2151
Metaclazepam-M (amino-Br-Cl-benzophenone) AC	UHYAC	2500	353_{44}	351_{32}	311_{85}	276_{100}	274_{100}	1211	2149
Metaclazepam-M (amino-Br-Cl-HO-benzophenone) 2AC	UHYAC	2685	409_{28}	367_{50}	325_{45}	292_{96}	290_{100}	1473	2150
Metaclazepam-M (amino-Br-Cl-HO-benzophenone) AC	UHYAC	2570	369_{32}	367_{31}	327_{98}	325_{100}	290_{34}	1294	7415
Metaclazepam-M (nor-)	U UHY UHYAC	2690	378_{22}	349_{39}	335_{100}	333_{78}	305_{14}	1346	2145
Metaclazepam-M (O-demethyl-)		2730	378_{26}	347_{18}	321_{100}	319_{65}	227_{13}	1346	2146
Metaclazepam-M (O-demethyl-) AC	U+UHYAC	2820	420_{11}	337_{33}	335_{100}	333_{79}	163_{13}	1507	2147
Metaclazepam-M/artifact-1	U UHY UHYAC	2230	320_{100}	318_{79}	283_{47}	239_{66}	75_{34}	1015	2152
Metaclazepam-M/artifact-2	U UHY UHYAC	2250	334_{98}	332_{79}	297_{50}	253_{100}	75_{86}	1103	2153
Metaclazepam-M/artifact-3	U UHY UHYAC	2590	334_{35}	299_{63}	227_{100}	163_{18}	75_{22}	1117	2154
Metalaxyl		1890	279_{72}	249_{61}	206_{100}	160_{68}	130_{45}	776	3452
Metaldehyde		1020*	131_{3}	117_{8}	89_{100}	87_{19}		333	1092
Metamfepramone	G U+UHYAC	1355	177_{1}	105_{3}	77_{12}	72_{100}	56_{6}	335	1398
Metamfepramone isomer-1 TMS		1470	249_{41}	219_{11}	176_{10}	158_{22}	73_{100}	600	4565
Metamfepramone isomer-2 TMS		1490	249_{43}	219_{16}	176_{11}	158_{24}	73_{100}	600	4566
Metamfepramone-M (dihydro-)	G P U UHY	1430	161_{1}	115_{1}	105_{2}	77_{6}	72_{100}	343	1113
Metamfepramone-M (dihydro-) AC	U+UHYAC	1495	162_{1}	117_{2}	105_{2}	91_{2}	72_{100}	475	1114
Metamfepramone-M (dihydro-) TFA		1185	260_{1}	162_{2}	134_{4}	91_{5}	72_{100}	745	4003
Metamfepramone-M (dihydro-) TMS		1485	251_{1}	236_{1}	163_{5}	149_{5}	72_{100}	614	4568
Metamfepramone-M (HO-norephedrine) 3AC	U+UHYAC	2135	234_{6}	165_{8}	123_{11}	86_{100}	58_{45}	862	4961
Metamfepramone-M (nor-)		1130	163_{1}	148_{1}	105_{4}	77_{15}	58_{100}	308	5935
Metamfepramone-M (nor-) AC	UHYAC	1650	205_{2}	105_{6}	100_{94}	77_{16}	58_{100}	413	5932
Metamfepramone-M (nor-) HFB		1440	359_{1}	254_{65}	210_{24}	105_{100}	77_{45}	1256	5936
Metamfepramone-M (nor-) PFP		1390	204_{3}	105_{7}	77_{13}	58_{100}		961	5934
Metamfepramone-M (nor-) TFA		1370	259_{1}	154_{68}	110_{31}	105_{100}	77_{49}	650	5933
Metamfepramone-M (nor-) TMS		1570	220_{5}	205_{5}	130_{100}	105_{6}	73_{90}	534	5937
Metamfepramone-M (nor-dihydro-)	G UHY	1375	146_{1}	131	105_{2}	77_{12}	58_{100}	314	748
Metamfepramone-M (nor-dihydro-) 2AC	PAC U+UHYAC	1795	249_{1}	148_{2}	117_{2}	100_{57}	58_{100}	597	749
Metamfepramone-M (nor-dihydro-) 2HFB	UHYHFB	1500	344_{10}	254_{100}	210_{27}	169_{44}	69_{36}	1698	5097
Metamfepramone-M (nor-dihydro-) 2PFP		1370	338_{1}	294_{3}	204_{100}	160_{25}	119_{14}	1595	2577
Metamfepramone-M (nor-dihydro-) 2TFA		1345	338_{1}	244_{4}	154_{100}	110_{72}	69_{47}	1246	3997
Metamfepramone-M (nor-dihydro-) 2TMS		1620	294_{3}	163_{4}	147_{8}	130_{100}	73_{84}	966	4543
Metamfepramone-M (nor-dihydro-) formyl artifact	G U	1430	177_{1}	121_{10}	107_{6}	71_{100}	56_{76}	336	4500
Metamfepramone-M (nor-dihydro-) -H2O AC	UHYAC	1560	189_{1}	148_{3}	121_{8}	100_{49}	58_{100}	368	5646
Metamfepramone-M (nor-dihydro-) TMSTFA		1620	318_{1}	227_{9}	179_{100}	110_{8}	73_{79}	1111	6038
Metamfepramone-M (norephedrine) 2AC	U+UHYAC	1805	235_{1}	176_{5}	134_{7}	107_{13}	86_{100}	532	2476
Metamfepramone-M (norephedrine) 2HFB	UHYHFB	1455	543_{1}	330_{14}	240_{100}	169_{44}	69_{57}	1691	5098
Metamfepramone-M (norephedrine) 2PFP	UHYPFP	1380	443_{1}	280_{9}	190_{100}	119_{59}	105_{26}	1567	5094
Metamfepramone-M (norephedrine) 2TFA	UTFA	1355	343_{1}	230_{6}	203_{5}	140_{100}	69_{29}	1172	5091

Metamfepramone-M (norephedrine) TMSTFA

Table 8.1: Compounds in order of names

Name	Detected	RI	Typical ions and intensities					Page	Entry
Metamfepramone-M (norephedrine) TMSTFA		1890	240_8	198_3	179_{100}	117_5	73_{88}	1024	6146
Metamfepramone-M (nor-HO-) 2AC	UHYAC	1885	263_4	250_2	150_6	100_{48}	58_{100}	673	4960
Metamfetamine	U	1195	148_1	134_2	115_1	91_9	58_{100}	288	1093
Metamfetamine AC	U+UHYAC	1575	191_1	117_2	100_{42}	91_6	58_{100}	373	1094
Metamfetamine HFB		1460	254_{100}	210_{44}	169_{15}	118_{41}	91_{38}	1183	5069
Metamfetamine PFP		1415	204_{100}	160_{46}	118_{35}	91_{25}	69_4	874	5070
Metamfetamine R-(-)-enantiomer HFBP		2000	351_{50}	294_4	266_{100}	169_{11}	121_7	1565	6516
Metamfetamine S-(+)-enantiomer HFBP		2120	351_{31}	294_4	266_{100}	118_6	91_9	1564	6517
Metamfetamine TFA		1300	245_1	154_{100}	118_{48}	110_{55}	91_{23}	574	3998
Metamfetamine TMS		1325	206_4	130_{100}	91_{17}	73_{83}	59_{13}	477	6214
Metamfetamine-D5		1190	154_1	139_2	119_1	92_9	62_{100}	297	7291
Metamfetamine-D5 HFB		1440	258_{100}	213_{28}	169_{10}	120_{14}	92_{13}	1208	6771
Metamfetamine-D5 R-(-)-enantiomer HFBP		2105	355_{56}	294_4	266_{100}	169_8	92_8	1575	6520
Metamfetamine-D5 S-(+)-enantiomer HFBP		2105	355_{48}	294_5	266_{100}	191_7		1575	6521
Metamfetamine-D5 TFA		1295	250_1	158_{100}	120_{22}	113_{35}	92_{16}	606	7292
Metamfetamine-D5 TMS		1320	211_6	134_{100}	118_1	92_6	73_{33}	497	7293
Metamfetamine-M (4-HO-) ME		1475	178_1	121_4	91_2	77_3	58_{100}	343	6719
Metamfetamine-M (4-HO-) MEAC		1820	221_1	148_{24}	121_7	100_{22}	58_{100}	476	6720
Metamfetamine-M (4-HO-) MEHFB		1665	375_1	254_{55}	210_{32}	148_{67}	121_{100}	1334	6722
Metamfetamine-M (4-HO-) MEPFP		1510	325_3	204_{60}	160_{29}	148_{48}	121_{100}	1061	7601
Metamfetamine-M (4-HO-) METFA		1645	275_2	154_{59}	148_{51}	121_{100}	110_{31}	745	6721
Metamfetamine-M (deamino-oxo-di-HO-) 2AC	U+UHYAC	1735*	250_3	208_{15}	166_{45}	123_{100}		603	4210
Metamfetamine-M (deamino-oxo-HO-methoxy-)	UHY	1510*	180_{19}	137_{100}	122_{19}	107_2	94_{16}	346	4247
Metamfetamine-M (deamino-oxo-HO-methoxy-) AC	U+UHYAC	1600*	222_2	180_{22}	137_{100}			479	4211
Metamfetamine-M (deamino-oxo-HO-methoxy-) ME	UHYME	1540*	194_{25}	151_{100}	135_4	107_{18}	65_4	384	4353
Metamfetamine-M (di-HO-) 3AC	U+UHYAC	2190	307_1	234_7	150_{12}	100_{59}	58_{100}	950	4244
Metamfetamine-M (HO-)		1885	150_1	135	107_4	77_4	58_{100}	314	1766
Metamfetamine-M (HO-) 2AC	U+UHYAC	1995	249_1	176_6	134_7	100_{43}	58_{100}	599	1767
Metamfetamine-M (HO-) 2HFB		1670	538_1	330_{17}	254_{100}	210_{32}	169_{22}	1698	5076
Metamfetamine-M (HO-) 2PFP		1605	280_1	204_{100}	160_{47}	119_{39}		1595	5077
Metamfetamine-M (HO-) 2TFA		1585	357_1	230_{22}	154_{100}	110_{42}	69_{29}	1246	5078
Metamfetamine-M (HO-) 2TMS		1620	309_{10}	206_{70}	179_{100}	154_{32}	73_{40}	988	6190
Metamfetamine-M (HO-) TFA		1770	261_1	154_{100}	134_{68}	110_{42}	107_{41}	662	6180
Metamfetamine-M (HO-) TMSTFA		1690	333_3	206_{72}	179_{100}	154_{53}	73_{50}	1111	6228
Metamfetamine-M (HO-methoxy-)	UHY	1810	195_1	137_4	122_2	94_3	58_{100}	387	4246
Metamfetamine-M (HO-methoxy-) 2HFB	UHFB	1760	360_{28}	333_6	254_{100}	210_{23}	169_{18}	1710	6492
Metamfetamine-M (HO-methoxy-) isomer-1 2AC	U+UHYAC	2095	279_1	206_{16}	164_{26}	100_{43}	58_{100}	776	6757
Metamfetamine-M (HO-methoxy-) isomer-2 2AC	U+UHYAC	2115	279_1	206_{19}	164_{26}	100_{48}	58_{100}	775	4243
Metamfetamine-M (nor-)		1160	134_4	120_{15}	91_{100}	77_{18}	65_{73}	275	54
Metamfetamine-M (nor-)	U	1160	134_1	120_1	91_6	65_4	44_{100}	275	5514
Metamfetamine-M (nor-) AC	U+UHYAC	1505	177_4	118_{60}	91_{35}	86_{100}	65_{16}	336	55
Metamfetamine-M (nor-) AC	U+UHYAC	1505	177_1	118_{19}	86_{31}	44_{100}		335	5515
Metamfetamine-M (nor-) formyl artifact		1100	147_2	146_6	125_5	91_{12}	56_{100}	286	3261
Metamfetamine-M (nor-) HFB		1355	240_{79}	169_{21}	118_{100}	91_{53}		1099	5047
Metamfetamine-M (nor-) PFP		1330	281_1	190_{73}	118_{100}	91_{36}	65_{12}	786	4379
Metamfetamine-M (nor-) TFA		1095	231_1	140_{100}	118_{92}	91_{45}	69_{19}	513	4000
Metamfetamine-M (nor-) TMS		1190	192_6	116_{100}	100_{10}	91_{11}	73_{87}	421	5581
Metamfetamine-M (nor-)-D11 PFP		1610	194_{100}	128_{82}	98_{43}	70_{14}		856	7284
Metamfetamine-M (nor-)-D11 TFA		1615	242_1	144_{100}	128_{82}	98_{43}	70_{14}	566	7283
Metamfetamine-M (nor-)-D5 AC		1480	182_3	122_{46}	92_{30}	90_{100}	66_{16}	352	5690
Metamfetamine-M (nor-)-D5 HFB		1330	244_{100}	169_{14}	122_{46}	92_{41}	69_{40}	1130	6316
Metamfetamine-M (nor-)-D5 PFP		1320	194_{100}	123_{42}	119_{32}	92_{46}	69_{11}	815	5566
Metamfetamine-M (nor-)-D5 TFA		1085	144_{100}	123_{53}	122_{56}	92_{51}	69_{28}	540	5570
Metamfetamine-M (nor-)-D5 TMS		1180	212_1	197_8	120_{100}	92_{11}	73_{57}	441	5582
Metamfetamine-M (nor-3-HO-) 2AC	UHYAC	1930	235_2	176_{48}	134_{52}	107_{21}	86_{100}	532	4387
Metamfetamine-M (nor-3-HO-) 2HFB		1620	330_{30}	303_{11}	240_{100}	169_{15}	69_{29}	1691	5737
Metamfetamine-M (nor-3-HO-) 2PFP		1520	280_{36}	253_9	190_{100}	119_{19}	69_8	1567	5738
Metamfetamine-M (nor-3-HO-) 2TFA		<1000	230_{33}	203_8	140_{100}	115_6		1172	6224
Metamfetamine-M (nor-3-HO-) 2TMS	UHYTMS	1850	280_{11}	179_3	116_{100}	100_{12}	73_{72}	881	5693
Metamfetamine-M (nor-3-HO-) formyl artifact ME		1290	177_2	162_4	121_4	77_5	56_{100}	335	5129
Metamfetamine-M (nor-3-HO-) TMSTFA		1630	319_8	206_{86}	191_{32}	140_{100}	73_{58}	1024	6141
Metamfetamine-M (nor-4-HO-)		1480	151_{10}	107_{69}	91_{10}	77_{42}	56_{100}	292	1802
Metamfetamine-M (nor-4-HO-) 2AC	U+UHYAC	1900	235_1	176_{72}	134_{100}	107_{46}	86_{70}	533	1804
Metamfetamine-M (nor-4-HO-) 2HFB		<1000	330_{48}	303_{15}	240_{100}	169_{44}	69_{42}	1691	6326
Metamfetamine-M (nor-4-HO-) 2PFP		<1000	280_{77}	253_{16}	190_{100}	119_{56}	69_{16}	1567	6325
Metamfetamine-M (nor-4-HO-) 2TFA		<1000	230_{72}	203_{11}	140_{100}	92_{12}	69_{59}	1172	6324
Metamfetamine-M (nor-4-HO-) 2TMS		<1000	280_7	179_9	149_8	116_{100}	73_{78}	880	6327
Metamfetamine-M (nor-4-HO-) AC	U+UHYAC	1890	193_1	134_{100}	107_{26}	86_{24}	77_{16}	379	1803
Metamfetamine-M (nor-4-HO-) formyl art.		1220	163_3	148_4	107_{30}	77_{12}	56_{100}	308	6323
Metamfetamine-M (nor-4-HO-) formyl artifact ME		1255	177_6	162_4	121_6	77_{12}	56_{100}	336	3250
Metamfetamine-M (nor-4-HO-) ME		1225	165_3	122_{100}	107_{11}	91_{22}	77_{45}	314	3249
Metamfetamine-M (nor-4-HO-) ME		1225	165_1	122_{16}	91_3	77_7	44_{100}	314	5517
Metamfetamine-M (nor-4-HO-) TFA		1670	247_4	140_{15}	134_{54}	107_{100}	77_{15}	584	6335
Metamfetamine-M (nor-HO-methoxy-)	UHY	1465	181_9	138_{100}	122_{18}	94_{24}	77_{16}	350	4351

Table 8.1: Compounds in order of names   **Metamfetamine-M (nor-HO-methoxy-) ME**

Name	Detected	RI	Typical ions and intensities					Page	Entry
Metamfetamine-M (nor-HO-methoxy-) ME	UHYME	1550	195 $_1$	152 $_{100}$	137 $_{17}$	107 $_{16}$	77 $_{14}$	387	4352
Metamfetamine-M 2AC	U+UHYAC	2065	265 $_3$	206 $_{27}$	164 $_{100}$	137 $_{23}$	86 $_{33}$	688	3498
Metamfetamine-M 2AC	U+UHYAC	1820*	266 $_3$	206 $_9$	164 $_{100}$	150 $_{10}$	137 $_{30}$	695	6409
Metamfetamine-M 2HFB	UHFB	1690	360 $_{82}$	333 $_{15}$	240 $_{100}$	169 $_{42}$	69 $_{39}$	1704	6512
Metamitron		2195	202 $_{85}$	174 $_{40}$	133 $_{10}$	104 $_{100}$	77 $_{37}$	404	3860
Metamizol	G P U	1995	215 $_{35}$	123 $_{100}$	91 $_{18}$	56 $_{41}$		975	197
Metamizol-M (bis-dealkyl-)	P U UHY	1955	203 $_{23}$	93 $_{14}$	84 $_{59}$	56 $_{100}$		406	219
Metamizol-M (bis-dealkyl-) 2AC	UHYAC	2280	287 $_8$	245 $_{31}$	203 $_{15}$	84 $_{56}$	56 $_{100}$	820	3333
Metamizol-M (bis-dealkyl-) AC	P U U+UHYAC	2270	245 $_{30}$	203 $_{13}$	84 $_{50}$	56 $_{100}$		575	183
Metamizol-M (bis-dealkyl-) artifact	U UHY	1945	180 $_{13}$	119 $_{100}$	91 $_{45}$			344	424
Metamizol-M (dealkyl-)	P U UHY	1980	217 $_{16}$	123 $_{14}$	98 $_7$	83 $_{16}$	56 $_{100}$	455	220
Metamizol-M (dealkyl-) AC	P U+UHYAC	2395	259 $_{20}$	217 $_8$	123 $_9$	56 $_{100}$		652	184
Metamizol-M (dealkyl-) ME artifact	P G U-I	1895	231 $_{36}$	123 $_7$	111 $_{17}$	97 $_{56}$	56 $_{100}$	514	189
Metamizol-M (deamino-HO-) AC	U+UHYAC	2095	246 $_2$	204 $_{19}$	119 $_7$	91 $_3$	56 $_{100}$	579	190
Metamizol-M/artifact (ME)	G	1320	151 $_{100}$	119 $_{56}$	106 $_{76}$	92 $_{36}$	65 $_{66}$	290	3909
Metamizol-M/artifact AC	U+UHYAC	1690	230 $_7$	188 $_{100}$	159 $_6$	91 $_{11}$	77 $_{55}$	509	3520
Metandienone	U+UHYAC	2690*	300 $_1$	282 $_2$	242 $_3$	161 $_{16}$	122 $_{100}$	909	2813
Metandienone enol 2TMS		2670*	444 $_{33}$	339 $_9$	206 $_{39}$	143 $_{11}$	73 $_{100}$	1571	3985
Metaraminol		1670	167 $_4$	121 $_{25}$	95 $_{56}$	77 $_{100}$	65 $_{53}$	319	4655
Metaraminol 3AC		2065	293 $_1$	233 $_6$	191 $_5$	86 $_{22}$	69 $_{100}$	861	1486
Metaraminol formyl artifact		1840	179 $_{20}$	160 $_9$	135 $_{100}$	107 $_{56}$	77 $_{52}$	342	4651
Metaraminol -H2O 2AC		1745	233 $_{16}$	93 $_{11}$	69 $_{100}$			521	1479
Metazachlor		2260	277 $_{13}$	228 $_8$	209 $_{100}$	133 $_{93}$	81 $_{84}$	759	3878
Metenolone		2800*	302 $_{22}$	287 $_{16}$	136 $_{72}$	123 $_{100}$	82 $_{25}$	921	2825
Metenolone acetate	U+UHYAC	2825*	344 $_5$	302 $_5$	161 $_{22}$	136 $_{78}$	123 $_{100}$	1181	2815
Metenolone enantate		2835*	344 $_5$	302 $_8$	161 $_{24}$	136 $_{75}$	123 $_{100}$	1492	2814
Metenolone enol 2TMS		2530*	446 $_{73}$	208 $_{17}$	193 $_{12}$	129 $_7$	73 $_{100}$	1574	3986
Metenolone TMS		2580*	374 $_{13}$	359 $_{14}$	331 $_{10}$	136 $_{29}$	73 $_{100}$	1332	3987
Metformine 2PFP		1250	403 $_{100}$	388 $_{61}$	284 $_{18}$	175 $_{27}$	69 $_{43}$	1509	5742
Metformine 2TFA		1220	303 $_{100}$	288 $_{44}$	234 $_{10}$	125 $_{16}$	69 $_{32}$	1037	5723
Metformine artifact-1	U+UHYAC	1380	153 $_{100}$	138 $_{54}$	124 $_{45}$	110 $_{20}$	69 $_{71}$	296	6311
Metformine artifact-1 AC	U+UHYAC	1660	195 $_{80}$	153 $_{20}$	138 $_{100}$	124 $_{32}$	110 $_{34}$	387	6510
Metformine artifact-2		1650	154 $_{100}$	139 $_{39}$	125 $_{30}$	111 $_{37}$	69 $_{43}$	297	6312
Metformine artifact-3		1675	182 $_{100}$	167 $_{78}$	153 $_{41}$	138 $_{29}$	124 $_{33}$	352	6313
Metformine artifact-4		1485	167 $_{100}$	152 $_{51}$	138 $_{38}$	124 $_{25}$	69 $_{70}$	320	6638
Metformine artifact-4 propionylated		1840	223 $_{87}$	167 $_{73}$	152 $_{100}$	138 $_{30}$	96 $_{31}$	484	6639
Metformine HFB		1350	307 $_{100}$	292 $_{67}$	278 $_{38}$	95 $_{18}$	69 $_{57}$	1060	5740
Metformine PFP		1300	257 $_{100}$	242 $_{49}$	228 $_{32}$	175 $_{15}$	69 $_{36}$	744	5741
Metformine TFA		1285	207 $_{100}$	192 $_{51}$	178 $_{29}$	125 $_{13}$	69 $_{44}$	490	5724
Methabenzthiazuron ME		1985	235 $_8$	136 $_4$	109 $_4$	72 $_{100}$		531	3941
Methacrylic acid methylester		<1000*	100 $_{44}$	69 $_{100}$				257	4283
Methadol		2185	296 $_1$	253 $_3$	165 $_3$	115 $_2$	72 $_{100}$	979	5617
Methadol AC	G P U+UHYAC	2230	353 $_1$	338 $_1$	225 $_4$	91 $_6$	72 $_{100}$	1230	5616
Methadone	P G U UHY U+UHYAC	2160	309 $_1$	294 $_3$	223 $_2$	165 $_2$	72 $_{100}$	967	241
Methadone intermediate-1		1750	193 $_{100}$	165 $_{67}$	115 $_{14}$	105 $_{29}$	77 $_{32}$	378	2835
Methadone intermediate-2		2095	278 $_3$	190 $_5$	165 $_{12}$	115 $_8$	58 $_{100}$	770	2838
Methadone intermediate-3		2130	278 $_2$	263 $_3$	192 $_{14}$	165 $_{31}$	72 $_{100}$	770	2836
Methadone intermediate-3 artifact		1920	253 $_2$	167 $_6$	165 $_4$	91 $_2$	72 $_{100}$	625	2837
Methadone TMS	U UHY U+UHYAC	2260	381 $_2$	296 $_{15}$	165 $_4$	73 $_{37}$	72 $_{100}$	1364	4567
Methadone-D9		2150	318 $_1$	303 $_3$	178 $_6$	165 $_7$	78 $_{100}$	1022	7820
Methadone-M (bis-nor-) -H2O	UHYAC	1940	263 $_1$	208 $_{100}$	193 $_{78}$	179 $_{38}$	130 $_{50}$	677	5295
Methadone-M (bis-nor-) -H2O AC	U+UHYAC	2220	305 $_{100}$	290 $_{74}$	262 $_{25}$	236 $_{29}$	220 $_{25}$	938	5292
Methadone-M (bis-nor-HO-) -H2O 2AC	UHYAC	2645	363 $_{100}$	348 $_{43}$	320 $_{63}$	278 $_{22}$	149 $_{15}$	1279	5299
Methadone-M (bis-nor-HO-) -H2O AC	UHYAC	2380	321 $_1$	266 $_{61}$	224 $_{100}$	209 $_{35}$	207 $_{32}$	1040	5298
Methadone-M (EDDP)	U UHY U+UHYAC	2040	277 $_{100}$	276 $_{95}$	262 $_{41}$	220 $_{22}$	165 $_{10}$	764	242
Methadone-M (HO-) AC	UHYAC	2540	367 $_1$	352 $_4$	239 $_{12}$	222 $_{12}$	72 $_{100}$	1300	6026
Methadone-M (HO-EDDP) AC	UHYAC	2350	335 $_{100}$	304 $_{27}$	292 $_{26}$	276 $_{16}$	234 $_{18}$	1126	5297
Methadone-M (nor-) -H2O	U UHY U+UHYAC	2040	277 $_{100}$	276 $_{95}$	262 $_{41}$	220 $_{22}$	165 $_{10}$	764	242
Methadone-M (nor-EDDP) AC	U+UHYAC	2220	305 $_{100}$	290 $_{74}$	262 $_{25}$	236 $_{29}$	220 $_{25}$	938	5292
Methadone-M (nor-HO-) -H2O AC	UHYAC	2350	335 $_{100}$	304 $_{27}$	292 $_{26}$	276 $_{16}$	234 $_{18}$	1126	5297
Methadone-M (nor-HO-EDDP) 2AC	UHYAC	2645	363 $_{100}$	348 $_{43}$	320 $_{63}$	278 $_{22}$	149 $_{15}$	1279	5299
Methadone-M (N-oxide) artifact	UHYAC	1900*	208 $_{100}$	193 $_{68}$	178 $_{31}$	165 $_{33}$	130 $_{49}$	425	5294
Methadone-M/artifact	UHYAC	1960	277 $_8$	235 $_{53}$	208 $_{86}$	193 $_{100}$	130 $_{71}$	757	5296
Methadone-M/artifact		2120	265 $_{100}$	193 $_{67}$	179 $_{36}$	130 $_{46}$	115 $_{50}$	688	5715
Methadone-M/artifact AC	U+UHYAC	2260	319 $_{100}$	304 $_{39}$	276 $_{42}$	234 $_{18}$	115 $_{13}$	1022	5293
Methamidophos		1195	141 $_{30}$	110 $_6$	94 $_{100}$	79 $_{11}$	64 $_{25}$	280	4088
Methanol		<1000*	32 $_{75}$	31 $_{100}$	30 $_{75}$	28 $_{50}$		247	1628
Methapyrilene		2015	261 $_9$	203 $_5$	191 $_{22}$	97 $_{91}$	58 $_{100}$	663	8404
Methapyrilene HY		1665	163 $_2$	137 $_6$	107 $_{12}$	71 $_{37}$	58 $_{100}$	308	8405
Methaqualone	P G U+UHYAC	2155	250 $_{38}$	235 $_{100}$	132 $_9$	91 $_{25}$	65 $_{14}$	604	1095
Methaqualone HFB		2360	446 $_{38}$	427 $_{10}$	399 $_7$	277 $_{100}$	235 $_{88}$	1572	5071
Methaqualone PFP		2345	396 $_{16}$	277 $_{69}$	235 $_{100}$	130 $_{19}$	91 $_{26}$	1427	5072
Methaqualone TFA		2360	346 $_{23}$	277 $_{46}$	235 $_{100}$	160 $_5$	130 $_{10}$	1187	5073
Methaqualone-M (2-carboxy-)	U	2400	280 $_{32}$	235 $_{100}$	146 $_{73}$	132 $_{16}$		780	1099

Methaqualone-M (2-carboxy-) -CO2

Table 8.1: Compounds in order of names

Name	Detected	RI	Typical ions and intensities					Page	Entry
Methaqualone-M (2-carboxy-) -CO2	U	2165	236_{69}	219_{100}	132_{18}	91_{39}	65_{36}	538	1096
Methaqualone-M (2-formyl-)	U UHY UHYAC	2240	264_{15}	235_{100}	132_{21}	91_{12}		681	1097
Methaqualone-M (2'-HO-methyl-)	U	2410	266_{65}	251_{33}	235_{49}	160_{100}	132_{19}	694	1100
Methaqualone-M (2-HO-methyl-)	U UHY	2360	266_{53}	235_{100}	132_{20}	91_{38}		694	1098
Methaqualone-M (2'-HO-methyl-) AC	UHYAC	2505	308_{41}	265_{100}	247_{30}	132_{6}	77_{12}	956	3755
Methaqualone-M (2-HO-methyl-) AC	UHYAC	2475	308_{31}	265_{73}	235_{100}			956	1104
Methaqualone-M (3'-HO-)	UHY	2490	266_{56}	251_{100}	249_{35}	148_{17}		693	1101
Methaqualone-M (3'-HO-) AC	UHYAC	2555	308_{48}	266_{83}	251_{100}	143_{11}	77_{9}	957	3757
Methaqualone-M (4'-HO-)	U UHY	2500	266_{60}	251_{100}	249_{41}	143_{24}		693	1102
Methaqualone-M (4'-HO-) AC	U+UHYAC	2570	308_{40}	266_{46}	251_{100}	143_{39}	77_{36}	957	1105
Methaqualone-M (4'-HO-5'-methoxy-)		2560	296_{100}	281_{91}	279_{33}	249_{24}	143_{43}	885	1106
Methaqualone-M (5'-HO-) AC	UHYAC	2540	308_{39}	266_{54}	251_{100}	143_{7}	77_{12}	957	3756
Methaqualone-M (6-HO-)		2525	266_{76}	251_{100}	249_{47}	132_{24}	91_{47}	694	1103
Methaqualone-M (HO-methoxy-) AC	UHYAC	2640	338_{21}	296_{100}	281_{78}	191_{37}	143_{27}	1143	3758
Metharbital	P G UHY UHYAC	1455	170_{100}	155_{97}	126_{12}	112_{34}		395	73
Metharbital ME		1420	184_{96}	169_{100}	126_{38}	112_{25}		440	74
Metharbital-M (HO-)	U UHY	1800	186_{10}	170_{60}	155_{100}	128_{30}	113_{24}	446	2961
Metharbital-M (HO-) AC	UHYAC	1870	228_{3}	196_{11}	170_{100}	155_{89}	112_{15}	636	2962
Metharbital-M (nor-)	P G U+UHYAC	1500	156_{100}	141_{97}	112_{20}	98_{22}	83_{12}	357	72
Methcathinone		1130	163_{1}	148_{1}	105_{4}	77_{15}	58_{100}	308	5935
Methcathinone AC	UHYAC	1650	205_{2}	105_{6}	100_{49}	77_{16}	58_{100}	413	5932
Methcathinone HFB		1440	359_{1}	254_{65}	210_{24}	105_{100}	77_{45}	1256	5936
Methcathinone PFP		1390	204_{3}	105_{7}	77_{13}	58_{100}		961	5934
Methcathinone TFA		1370	259_{1}	154_{68}	110_{31}	105_{100}	77_{49}	650	5933
Methcathinone TMS		1570	220_{5}	205_{5}	130_{100}	105_{6}	73_{90}	534	5937
Methcathinone-M (HO-) 2AC	UHYAC	1885	263_{4}	250_{2}	150_{6}	100_{48}	58_{100}	673	4960
Methedrone		1620	193_{1}	135_{5}	92_{9}	77_{13}	58_{100}	379	8378
Methedrone AC	U+UHYAC	1880	235_{3}	164_{2}	135_{17}	100_{44}	58_{100}	533	8379
Methedrone HFB		1680	389_{1}	254_{8}	135_{100}	92_{10}	77_{14}	1397	8383
Methedrone PFP		1630	339_{1}	204_{15}	135_{100}	92_{14}	77_{26}	1148	8382
Methedrone TFA		1630	289_{1}	154_{16}	135_{100}	92_{15}	77_{25}	833	8381
Methedrone TMS		1736	260_{1}	220_{2}	130_{100}	92_{16}	58_{80}	689	8380
Methedrone-M (O-demethyl-) 2AC	U+UHYAC	1970	235_{1}	164	135	100_{73}	58_{100}	674	8540
Methedrone-M (O-demethyl-) 2HFB	U+UHYHFB	1655	358_{2}	317_{47}	254_{100}	210_{24}	169_{14}	1703	8542
Methenamine		1210	140_{100}	112_{20}				279	1107
Methidathion		2120	302_{3}	145_{100}	125_{14}	93_{16}	85_{67}	916	3856
Methiomeprazine	U+UHYAC	2725	344_{9}	298_{1}	245_{1}	100_{2}	58_{100}	1179	1828
2-Methiopropamine		1015	154_{1}	111_{1}	97_{6}	58_{100}		298	8633
2-Methiopropamine AC		1660	197_{3}	124_{26}	100_{60}	97_{19}	58_{100}	392	8634
2-Methiopropamine HFB		1480	351_{1}	254_{100}	210_{53}	124_{82}	97_{33}	1211	8638
2-Methiopropamine impurity (oxo-) AC		1760	211_{2}	111_{7}	100_{48}	83_{2}	58_{100}	435	8671
2-Methiopropamine PFP		1425	301_{1}	204_{100}	160_{55}	124_{66}	97_{32}	911	8637
2-Methiopropamine TFA		1410	251_{1}	154_{100}	124_{74}	110_{64}	97_{35}	608	8636
2-Methiopropamine TMS		1420	212_{11}	155_{6}	130_{100}	97_{13}	73_{69}	499	8635
2-Methiopropamine-M (HO-) 2AC		2040	255_{1}	213_{10}	182_{16}	140_{19}	58_{100}	631	8672
2-Methiopropamine-M (nor-) AC		1540	183_{3}	124_{77}	97_{100}	86_{27}	69_{11}	354	8670
3-Methiopropamine		1025	154_{1}	111_{1}	97_{9}	58_{100}		298	8639
3-Methiopropamine AC		1670	197_{3}	124_{26}	100_{60}	97_{19}	58_{100}	392	8640
3-Methiopropamine HFB		1500	351_{1}	254_{100}	210_{68}	124_{70}	97_{49}	1211	8644
3-Methiopropamine PFP		1435	301_{1}	204_{100}	160_{59}	124_{70}	97_{27}	911	8643
3-Methiopropamine TFA		1430	251_{1}	154_{100}	124_{63}	110_{56}	97_{30}	608	8642
3-Methiopropamine TMS		1435	212_{8}	155_{4}	130_{100}	97_{11}	73_{70}	499	8641
3-Methiopropamine-M (di-HO-) 2AC		2125	271_{1}	229_{13}	156_{6}	100_{33}	58_{100}	723	8669
3-Methiopropamine-M (HO-) isomer-1 AC		1780	213_{4}	140_{17}	125_{14}	100_{30}	58_{100}	442	8663
3-Methiopropamine-M (HO-) isomer-2 AC		1820	213_{4}	140_{21}	125_{12}	100_{33}	58_{100}	442	8664
3-Methiopropamine-M (HO-methoxy-) isomer-1 AC		1990	243_{1}	170_{51}	155_{18}	100_{33}	58_{100}	567	8665
3-Methiopropamine-M (HO-methoxy-) isomer-2 AC		2020	243_{3}	170_{13}	155_{4}	100_{30}	58_{100}	567	8666
3-Methiopropamine-M (HO-methoxy-) isomer-3 AC		2040	243_{7}	170_{40}	143_{15}	58_{100}		567	8667
3-Methiopropamine-M (nor-) AC		1600	183_{10}	124_{100}	97_{80}	86_{30}	69_{10}	354	8662
3-Methiopropamine-M (tri-HO-) AC		2085	245_{4}	172_{18}	125_{31}	100_{35}	58_{100}	573	8668
Methitural		2240	288_{17}	214_{28}	171_{64}	155_{27}	74_{100}	826	1487
Methocarbamol	P G	2050	241_{8}	198_{9}	124_{78}	118_{100}	109_{56}	560	1982
Methocarbamol AC	U+UHYAC	2145	283_{2}	240	160_{100}	124_{25}	57_{68}	795	1991
Methocarbamol -CHNO AC		2000*	240_{10}	124_{43}	117_{100}	109_{42}	77_{14}	556	1992
Methocarbamol-M (guaifenesin)	P G	1610*	198_{15}	124_{100}	109_{74}			394	796
Methocarbamol-M (guaifenesin) 2AC	U+UHYAC	1865*	282_{3}	159_{100}	124_{14}	99_{18}		792	799
Methocarbamol-M (guaifenesin) 2TMS		1850*	342_{16}	196_{33}	149_{32}	103_{47}	73_{100}	1170	4551
Methocarbamol-M (HO-) 2AC	U+UHYAC	2560	341_{1}	160_{80}	140_{14}	99_{13}	57_{100}	1160	4504
Methocarbamol-M (HO-guaifensin) 3AC	U+UHYAC	2235*	340_{2}	298_{1}	159_{100}	140_{10}	99_{8}	1155	797
Methocarbamol-M (HO-methoxy-) 2AC	UHYAC	2620	371_{5}	170_{16}	160_{100}	69_{19}	57_{87}	1317	4502
Methocarbamol-M (HO-methoxy-guaifensin) 3AC	U+UHYAC	2265*	370_{9}	230_{7}	212_{6}	170_{26}	159_{100}	1311	798
Methocarbamol-M (O-demethyl-) 2AC		2430	311_{1}	269_{3}	160_{96}	121_{17}	57_{100}	975	4503
Methocarbamol-M (O-demethyl-guaifensin) 3AC	U+UHYAC	1920*	310_{1}	268_{13}	159_{100}	117_{12}	110_{19}	970	800
Methohexital	P U	1780	261_{10}	247_{29}	221_{63}	178_{36}	79_{100}	669	1108

Table 8.1: Compounds in order of names

Name	Detected	RI	Typical ions and intensities					Page	Entry
Methohexital ME		1735	276_9	235_{85}	178_{69}	79_{98}	53_{100}	755	1109
Methohexital-D5		1775	266_{10}	252_{37}	238_{28}	221_{100}	178_{48}	703	6881
Methohexital-M (HO-)	UHY	1880	278_7	245_{19}	219_{80}	79_{100}	53_{86}	767	2959
Methomyl		1515	162_1	115_2	105_{64}	88_{27}	58_{100}	305	3903
Methoprotryne		2235	271_{17}	256_{100}	226_{29}	212_{34}	171_{28}	725	3857
Methorphan	G P-I U+UHYAC	2145	271_{31}	214_{16}	171_{14}	150_{29}	59_{100}	726	227
Methorphan-M (bis-demethyl-) 2AC	UHYAC	2710	327_{11}	240_8	199_{12}	87_{100}	72_{62}	1077	228
Methorphan-M (nor-) AC	UHYAC	2590	299_{13}	213_{42}	171_{22}	87_{100}	72_{52}	903	4477
Methorphan-M (O-demethyl-)	UHY	2255	257_{38}	200_{17}	150_{28}	59_{100}		643	475
Methorphan-M (O-demethyl-) AC	U+UHYAC	2280	299_{100}	231_{42}	200_{20}	150_{48}	59_{15}	903	230
Methorphan-M (O-demethyl-) HFB		2100	453_{100}	396_{18}	385_{91}	169_{19}	150_{27}	1586	6151
Methorphan-M (O-demethyl-) PFP	UHYPFP	2060	403_{93}	335_{78}	303_{14}	150_{100}	119_{58}	1457	4305
Methorphan-M (O-demethyl-) TFA		2015	353_{69}	285_{80}	150_{100}	115_{26}	69_{72}	1227	4006
Methorphan-M (O-demethyl-) TMS	UHYTMS	2230	329_{31}	272_{20}	150_{39}	73_{24}		1090	4304
Methorphan-M (O-demethyl-HO-) 2AC	U+UHYAC	2580	357_{68}	247_{22}	215_{17}	150_{100}	59_{30}	1249	1187
Methorphan-M (O-demethyl-methoxy-) AC	U+UHYAC	2520	329_{48}	261_{23}	229_{23}	150_{100}	59_{28}	1090	4476
Methorphan-M (O-demethyl-oxo-) AC	U+UHYAC	2695	313_{16}	240_{11}	199_{98}	157_{12}	73_{100}	990	4475
Methoxetamine		1915	247_2	219_{32}	190_{100}	176_{16}	134_{22}	588	8506
Methoxetamine AC	U+UHYAC	2240	289_1	218_{38}	190_{30}	174_{100}	166_{49}	836	8783
Methoxetamine artifact (bicyclo-) AC	USPEAC	2370	313_{47}	270_{100}	206_{29}	173_{38}	160_{36}	989	8784
Methoxetamine-M (HO-) 2AC	USPEAC	2410	347_1	260_3	235_{10}	218_{18}	176_{100}	1196	8781
Methoxetamine-M (HO-) AC	U+UHYAC	2180	305_1	277_{56}	248_{100}	234_{21}	206_{71}	938	8776
Methoxetamine-M (HO-alkyl-) -H2O AC	U+UHYAC	2180	287_{10}	244_8	219_{22}	177_{100}	135_{90}	820	8787
Methoxetamine-M (N,O-bisdealkyl-) 2AC	U+UHYAC	2150	289_{18}	246_{10}	202_{45}	190_{31}	160_{100}	835	8775
Methoxetamine-M (N,O-bisdealkyl-HO-) 3AC	U+UHYAC	2670	347_2	304_{34}	219_{91}	218_{100}	176_{74}	1194	8782
Methoxetamine-M (N-deethyl-) AC	U+UHYAC	2030	261_{17}	218_4	190_{12}	174_{100}	162_{40}	664	8785
Methoxetamine-M (N-deethyl-HO-) 2AC	USPEAC	2100	319_{21}	285_{60}	276_{22}	260_{24}	218_{100}	1025	8774
Methoxetamine-M (O-demethyl-) 2AC	U+UHYAC	2370	317_2	289_4	246_{42}	218_{33}	160_{100}	1012	8780
Methoxetamine-M (O-demethyl-) AC	U+UHYAC	2000	275_2	247_{55}	218_{100}	176_{69}	120_{45}	746	8773
Methoxetamine-M (O-demethyl-HO-) 2AC	USPEAC	2300	333_1	305_{45}	276_{100}	234_{91}	192_{64}	1112	8778
Methoxetamine-M (O-demethyl-HO-) -NH3 2AC	USPEAC	2230*	290_6	248_{31}	206_{100}	178_{26}	162_{38}	841	8777
Methoxetamine-M (O-demethyl-oxo-) -NH3 2AC	USPEAC	2320*	288_9	246_{38}	204_{100}	176_{19}	148_{18}	827	8779
Methoxphenidine		2200	294_1	204_{100}	188_{33}	121_{17}	91_{37}	881	9565
Methoxphenidine artifact		2005*	210_{100}	194_7	179_{13}	165_{46}	152_{23}	434	9566
5-Methoxy-2-methyl-2-N-methylallyl-N-ethyl-tryptamine		2330	286_8	188_4	174_{19}	112_{100}	55_{24}	816	8832
5-Methoxy-2-methyl-2-N-methylallyl-N-ethyl-tryptamine artifact HFB		1925	481_{36}	425_2	384_9	172_7	112_{100}	1635	10072
5-Methoxy-2-methyl-2-N-methylallyl-N-ethyl-tryptamine artifact PFP		2150	431_{30}	375_4	334_{10}	172_8	112_{100}	1536	10071
5-Methoxy-2-methyl-2-N-methylallyl-N-ethyl-tryptamine artifact TFA		1990	381_{30}	325_4	284_{24}	172_{22}	112_{100}	1358	10191
5-Methoxy-2-methyl-2-N-methylallyl-N-ethyl-tryptamine HFB		2165	441_2	384_6	370_{15}	173_{12}	112_{100}	1637	10031
5-Methoxy-2-methyl-2-N-methylallyl-N-ethyl-tryptamine PFP		2150	334_2	320_6	173_{23}	158_{21}	112_{100}	1540	10030
5-Methoxy-2-methyl-2-N-methylallyl-N-ethyl-tryptamine TFA		2205	284_{12}	270_{21}	173_{30}	158_{25}	112_{100}	1365	10190
5-Methoxy-2-methyl-2-N-methylallyl-N-ethyl-tryptamine TMS		2480	358_3	246_{10}	112_{100}	73_{24}	55_{15}	1255	10029
5-Methoxy-2-methyl-N,N-diallyl-tryptamine		2330	284_1	174_{17}	159_9	131_{11}	110_{100}	805	8838
5-Methoxy-2-methyl-N,N-diallyl-tryptamine artifact HFB		1945	479_{32}	384_{17}	369_5	172_{13}	110_{100}	1632	10089
5-Methoxy-2-methyl-N,N-diallyl-tryptamine artifact PFP		1985	429_{27}	334_{10}	269_4	172_{12}	110_{100}	1529	10088
5-Methoxy-2-methyl-N,N-diallyl-tryptamine artifact TFA		1985	379_{26}	284_{10}	269_4	172_{11}	110_{100}	1348	10087
5-Methoxy-2-methyl-N,N-diallyl-tryptamine HFB		2175	480_1	453_2	384_7	370_{19}	110_{100}	1634	10053
5-Methoxy-2-methyl-N,N-diallyl-tryptamine PFP		2165	430_1	334_2	320_7	173_9	110_{100}	1534	10052
5-Methoxy-2-methyl-N,N-diallyl-tryptamine TFA		2200	284_2	270_7	173_{11}	158_9	110_{100}	1357	10051
5-Methoxy-2-methyl-N,N-diallyl-tryptamine TMS		2475	356_4	315_3	246_{31}	173_2	110_{100}	1244	10050
5-Methoxy-2-methyl-N,N-diallyl-tryptamine-M (deallyl-) AC		2585	286_{11}	187_{72}	174_{100}	159_{33}	70_{28}	816	10376
5-Methoxy-2-methyl-N,N-diallyl-tryptamine-M (deallyl-HO-aryl-) 2AC		2845	344_{22}	245_{12}	232_{48}	203_{100}	190_{91}	1179	10380
5-Methoxy-2-methyl-N,N-diallyl-tryptamine-M (HO-aryl-) AC		2660	342_1	301_3	190_9	175_{22}	110_{100}	1171	10377
5-Methoxy-2-methyl-N,N-diallyl-tryptamine-M (O-demethyl-) AC		2515	312_1	271_{10}	202_7	160_{35}	110_{100}	983	10375
5-Methoxy-2-methyl-N,N-diallyl-tryptamine-M (O-demethyl-deallyl-) 2AC		2700	314_{11}	215_{67}	202_{62}	173_{80}	160_{100}	996	10378
5-Methoxy-2-methyl-N,N-diallyl-tryptamine-M (O-demethyl-HO-aryl-) 2AC		2760	370_2	218_8	190_{11}	176_{27}	110_{100}	1314	10379
5-Methoxy-2-methyl-N,N-diisopropyl-tryptamine		2330	288_1	188_7	174_{21}	114_{100}	72_{36}	831	8840
5-Methoxy-2-methyl-N,N-diisopropyl-tryptamine artifact HFB		2165	469_2	384_4	370_7	173_7	114_{100}	1616	10065
5-Methoxy-2-methyl-N,N-diisopropyl-tryptamine artifact PFP		2150	419_3	334_4	320_7	173_8	114_{100}	1504	10064
5-Methoxy-2-methyl-N,N-diisopropyl-tryptamine artifact TFA		1940	383_{19}	353_7	284_{24}	127_{37}	114_{100}	1368	10062
5-Methoxy-2-methyl-N,N-diisopropyl-tryptamine HFB		2165	483_{26}	425_9	384_{11}	127_{36}	114_{100}	1639	10066
5-Methoxy-2-methyl-N,N-diisopropyl-tryptamine PFP		2450	419_1	334_4	320_7	173_8	114_{100}	1545	9830
5-Methoxy-2-methyl-N,N-diisopropyl-tryptamine TFA		2185	369_1	284_3	270_4	173_{12}	114_{100}	1378	10063
5-Methoxy-2-methyl-N,N-diisopropyl-tryptamine TMS		2480	360_1	188	174_1	114_{100}	72_9	1266	10061
5-Methoxy-2-methyl-N,N-diisopropyl-tryptamine-M (deisopropyl-) AC		2600	288_{13}	187_{71}	174_{100}	159_{29}	72_{26}	829	10383
5-Methoxy-2-methyl-N,N-diisopropyl-tryptamine-M (deisopropyl-HO-aryl-) 2AC		2845	346_{17}	300_{17}	232_{57}	203_{100}	190_{96}	1189	10387
5-Methoxy-2-methyl-N,N-diisopropyl-tryptamine-M (HO-aryl-) isomer-1 AC		2570	232_4	190_{27}	175_8	114_{100}	72_{21}	1191	10382
5-Methoxy-2-methyl-N,N-diisopropyl-tryptamine-M (HO-aryl-) isomer-2 AC		2650	346_2	246_4	190_{32}	175_{31}	114_{100}	1190	10384
5-Methoxy-2-methyl-N,N-diisopropyl-tryptamine-M (O-demethyl-) AC		2500	316_{11}	216_{11}	160_{43}	114_{100}	72_{60}	1007	10381
5-Methoxy-2-methyl-N,N-diisopropyl-tryptamine-M (O-demethyl-deisopropyl-) 2AC		2700	316_{10}	215_{68}	202_{71}	160_{100}	72_{50}	1007	10385
5-Methoxy-2-methyl-N,N-diisopropyl-tryptamine-M (O-demethyl-HO-aryl-) 2AC		2745	374_2	260_4	176_{28}	114_{100}	72_{29}	1331	10386
5-Methoxy-2-methyl-N,N-dimethyl-tryptamine		2060	232_{12}	174_{16}	159_6	131_6	58_{100}	519	8837
5-Methoxy-2-methyl-N,N-dimethyl-tryptamine artifact HFB		1765	427_{24}	384_7	186_3	172_7	58_{100}	1524	10086
5-Methoxy-2-methyl-N,N-dimethyl-tryptamine artifact PFP		1745	377_{23}	334_4	319_2	172_{12}	58_{100}	1341	10085

5-Methoxy-2-methyl-N,N-dimethyl-tryptamine artifact TFA

Table 8.1: Compounds in order of names

Name	Detected	RI	Typical ions and intensities					Page	Entry
5-Methoxy-2-methyl-N,N-dimethyl-tryptamine artifact TFA		1770	327$_{18}$	284$_3$	186$_2$	172$_4$	58$_{100}$	1072	10084
5-Methoxy-2-methyl-N,N-dimethyl-tryptamine HFB		1965	384$_1$	370$_5$	173$_{17}$	158$_{20}$	58$_{100}$	1528	10049
5-Methoxy-2-methyl-N,N-dimethyl-tryptamine PFP		1960	334$_1$	320$_6$	173$_{18}$	158$_{18}$	58$_{100}$	1347	10048
5-Methoxy-2-methyl-N,N-dimethyl-tryptamine TFA		1975	328$_1$	284$_2$	270$_4$	158$_{15}$	58$_{100}$	1081	10047
5-Methoxy-2-methyl-N,N-dimethyl-tryptamine TMS		2235	304$_7$	246$_{35}$	174$_1$	73$_{22}$	58$_{100}$	933	10046
5-Methoxy-2-methyl-N,N-dipropyl-tryptamine		2330	288$_3$	188$_8$	174$_{20}$	159$_9$	114$_{100}$	831	8836
5-Methoxy-2-methyl-N,N-dipropyl-tryptamine artifact HFB		1940	483$_{45}$	425$_{10}$	384$_8$	186$_{10}$	114$_{100}$	1638	10045
5-Methoxy-2-methyl-N,N-dipropyl-tryptamine artifact PFP		2160	433$_2$	405$_4$	334$_9$	320$_{14}$	114$_{100}$	1543	10044
5-Methoxy-2-methyl-N,N-dipropyl-tryptamine artifact TFA		1975	383$_{26}$	284$_8$	186$_{13}$	172$_{11}$	114$_{100}$	1368	10083
5-Methoxy-2-methyl-N,N-dipropyl-tryptamine HFB		2170	484$_1$	384$_4$	370$_7$	173$_8$	114$_{100}$	1640	10060
5-Methoxy-2-methyl-N,N-dipropyl-tryptamine PFP		2480	433$_2$	405$_4$	334$_9$	320$_{14}$	114$_{100}$	1545	9831
5-Methoxy-2-methyl-N,N-dipropyl-tryptamine TFA		2195	284$_3$	270$_6$	173$_{10}$	158$_{11}$	114$_{100}$	1378	10043
5-Methoxy-2-methyl-N,N-dipropyl-tryptamine TMS		2515	360$_4$	260$_3$	246$_{13}$	216$_2$	114$_{100}$	1266	10042
5-Methoxy-2-methyl-N-allyl-N-cyclohexyl-tryptamine		2690	326$_2$	188$_6$	174$_{21}$	152$_{100}$	70$_{50}$	1071	8839
5-Methoxy-2-methyl-N-allyl-N-cyclohexyl-tryptamine artifact HFB		2215	521$_{13}$	384$_5$	370$_3$	172$_{25}$	152$_{100}$	1677	10059
5-Methoxy-2-methyl-N-allyl-N-cyclohexyl-tryptamine artifact PFP		2460	472$_2$	334$_3$	320$_{10}$	152$_{100}$	70$_{24}$	1621	10056
5-Methoxy-2-methyl-N-allyl-N-cyclohexyl-tryptamine artifact TFA		2275	421$_{33}$	337$_{10}$	284$_9$	172$_{20}$	152$_{100}$	1509	10055
5-Methoxy-2-methyl-N-allyl-N-cyclohexyl-tryptamine HFB		2465	522$_3$	304$_4$	370$_{11}$	152$_{100}$	70$_{18}$	1678	10057
5-Methoxy-2-methyl-N-allyl-N-cyclohexyl-tryptamine PFP		2700	472$_1$	360$_2$	346$_7$	152$_{100}$	70$_{17}$	1622	9835
5-Methoxy-2-methyl-N-allyl-N-cyclohexyl-tryptamine TFA		2500	422$_1$	284$_3$	270$_9$	173$_{13}$	152$_{100}$	1513	10058
5-Methoxy-2-methyl-N-allyl-N-cyclohexyl-tryptamine TMS		2780	398$_6$	259$_3$	246$_9$	152$_{100}$	70$_{15}$	1440	10054
5-Methoxy-2-methyl-N-allyl-N-cyclohexyl-tryptamine-M (deallyl-)		2615	286$_{14}$	187$_{59}$	174$_{100}$	159$_{41}$	70$_{43}$	817	10412
5-Methoxy-2-methyl-N-allyl-N-cyclohexyl-tryptamine-M (deallyl-HO-aryl-) 2AC		3250	386$_4$	232$_{51}$	203$_{100}$	190$_{82}$	175$_{78}$	1386	10414
5-Methoxy-2-methyl-N-allyl-N-cyclohexyl-tryptamine-M (deallyl-HO-cyclohexyl-)		3165	386$_2$	187$_{42}$	174$_{100}$	159$_{32}$	131$_{26}$	1386	10410
5-Methoxy-2-methyl-N-allyl-N-cyclohexyl-tryptamine-M (decyclohexyl-HO-aryl-) 2AC		2850	344$_{17}$	232$_{45}$	203$_{88}$	190$_{100}$	175$_{70}$	1180	10405
5-Methoxy-2-methyl-N-allyl-N-cyclohexyl-tryptamine-M (demethyl-deallyl-HO-) 3AC		3300	414$_1$	215$_{47}$	202$_{49}$	160$_{100}$	131$_{26}$	1491	10413
5-Methoxy-2-methyl-N-allyl-N-cyclohexyl-tryptamine-M (demethyldecyclohexyl-) 2AC		2710	314$_{23}$	215$_{89}$	202$_{87}$	173$_{93}$	160$_{100}$	996	10404
5-Methoxy-2-methyl-N-allyl-N-cyclohexyl-tryptamine-M (di-HO-) 2AC		3230	442$_1$	232$_{10}$	210$_{100}$	190$_{51}$	175$_{51}$	1566	10409
5-Methoxy-2-methyl-N-allyl-N-cyclohexyl-tryptamine-M (HO-aryl-) AC		2950	384$_1$	232$_{11}$	190$_{46}$	175$_{44}$	152$_{100}$	1378	10406
5-Methoxy-2-methyl-N-allyl-N-cyclohexyl-tryptamine-M (HO-cyclohexyl-) AC		2900	384$_3$	210$_{100}$	174$_{78}$	131$_{26}$	70$_{33}$	1378	10408
5-Methoxy-2-methyl-N-allyl-N-cyclohexyl-tryptamine-M (O-demethyl-) AC		2800	354$_2$	160$_{58}$	152$_{100}$	119$_{38}$	70$_{58}$	1234	10411
5-Methoxy-2-methyl-N-allyl-N-cyclohexyl-tryptamine-M 2AC		3030	412$_2$	210$_{100}$	202$_{31}$	173$_{28}$	160$_{76}$	1486	10407
5-Methoxy-2-methyl-N-ethyl-N-isopropyl-tryptamine		2240	346$_2$	260$_2$	246$_{11}$	100$_{100}$	58$_{18}$	742	8857
5-Methoxy-2-methyl-N-ethyl-N-isopropyl-tryptamine HFB		2075	384$_8$	370$_{12}$	173$_{13}$	100$_{100}$	58$_{24}$	1618	10097
5-Methoxy-2-methyl-N-ethyl-N-isopropyl-tryptamine PFP		2065	405$_1$	334$_9$	320$_{13}$	100$_{100}$	58$_{41}$	1508	10096
5-Methoxy-2-methyl-N-ethyl-N-isopropyl-tryptamine TFA		2100	355$_1$	284$_6$	270$_{11}$	100$_{100}$	58$_{38}$	1313	10095
5-Methoxy-2-methyl-N-ethyl-N-isopropyl-tryptamine TMS		2400	346$_2$	260$_2$	246$_{16}$	100$_{100}$	58$_{38}$	1191	10094
5-Methoxy-2-methyl-N-ethyl-N-propyl-tryptamine		2250	274$_3$	188$_6$	174$_{16}$	100$_{100}$	58$_{25}$	742	8835
5-Methoxy-2-methyl-N-ethyl-N-propyl-tryptamine artifact HFB		1890	469$_{20}$	384$_9$	271$_6$	186$_{20}$	100$_{100}$	1616	10082
5-Methoxy-2-methyl-N-ethyl-N-propyl-tryptamine artifact PFP		1870	419$_{43}$	334$_{11}$	186$_{12}$	172$_{17}$	100$_{100}$	1504	10081
5-Methoxy-2-methyl-N-ethyl-N-propyl-tryptamine artifact TFA		1840	369$_9$	284$_7$	186$_{13}$	170$_{11}$	100$_{100}$	1304	10080
5-Methoxy-2-methyl-N-ethyl-N-propyl-tryptamine HFB		2110	384$_7$	370$_{10}$	173$_{21}$	158$_{16}$	100$_{100}$	1618	10041
5-Methoxy-2-methyl-N-ethyl-N-propyl-tryptamine PFP		2100	335$_1$	320$_5$	173$_{17}$	158$_{15}$	100$_{100}$	1508	10040
5-Methoxy-2-methyl-N-ethyl-N-propyl-tryptamine TFA		2135	284$_3$	270$_7$	173$_{13}$	158$_{11}$	100$_{100}$	1313	10039
5-Methoxy-2-methyl-N-ethyl-N-propyl-tryptamine TMS		2435	346$_7$	260$_3$	246$_{20}$	216$_2$	100$_{100}$	1191	10038
5-Methoxy-2-methyl-N-isopropyl-N-methyl-tryptamine		2190	260$_{13}$	174$_{22}$	159$_{13}$	131$_{14}$	86$_{100}$	661	8841
5-Methoxy-2-methyl-N-isopropyl-N-methyl-tryptamine HFB		2065	455$_1$	441$_1$	384$_5$	370$_8$	86$_{100}$	1594	9768
5-Methoxy-2-methyl-N-isopropyl-N-methyl-tryptamine PFP		2055	405$_1$	391$_2$	334$_7$	320$_8$	86$_{100}$	1465	9767
5-Methoxy-2-methyl-N-isopropyl-N-methyl-tryptamine TFA		2090	355$_1$	341$_2$	284$_{10}$	270$_{15}$	86$_{100}$	1243	9766
5-Methoxy-2-methyl-N-isopropyl-N-methyl-tryptamine TMS		2190	332$_6$	246$_{23}$	174$_2$	159$_1$	86$_{100}$	1108	9876
5-Methoxy-2-methyl-piperidine-tryptamine		2400	272$_{13}$	174$_{15}$	159$_{10}$	131$_9$	98$_{100}$	731	8833
5-Methoxy-2-methyl-piperidine-tryptamine artifact HFB		2030	467$_{40}$	384$_{16}$	186$_7$	172$_{22}$	98$_{100}$	1611	10076
5-Methoxy-2-methyl-piperidine-tryptamine artifact PFP		2010	417$_{28}$	334$_3$	205$_3$	187$_4$	98$_{100}$	1498	10075
5-Methoxy-2-methyl-piperidine-tryptamine artifact TFA		2070	367$_{18}$	284$_7$	172$_{15}$	111$_{20}$	98$_{100}$	1294	10073
5-Methoxy-2-methyl-piperidine-tryptamine HFB		2230	384$_2$	370$_6$	173$_{25}$	158$_{24}$	98$_{100}$	1613	10034
5-Methoxy-2-methyl-piperidine-tryptamine PFP		2240	418$_1$	320$_5$	173$_{11}$	158$_{11}$	98$_{100}$	1503	10033
5-Methoxy-2-methyl-piperidine-tryptamine TFA		2280	284$_2$	270$_5$	173$_{12}$	158$_{12}$	98$_{100}$	1302	10074
5-Methoxy-2-methyl-piperidine-tryptamine TMS		2575	344$_8$	246$_{19}$	216$_4$	174$_2$	98$_{100}$	1181	10032
5-Methoxy-2-methyl-pyrrolidine-tryptamine		2330	258$_{13}$	174$_{19}$	159$_8$	131$_9$	84$_{100}$	648	8834
5-Methoxy-2-methyl-pyrrolidine-tryptamine artifact HFB		1960	453$_{25}$	384$_8$	186$_{12}$	172$_{32}$	84$_{100}$	1586	10079
5-Methoxy-2-methyl-pyrrolidine-tryptamine artifact PFP		1945	403$_{32}$	172$_6$	119$_{12}$	97$_{18}$	84$_{100}$	1454	10078
5-Methoxy-2-methyl-pyrrolidine-tryptamine artifact TFA		2010	353$_{40}$	284$_8$	172$_8$	97$_{14}$	84$_{100}$	1224	10077
5-Methoxy-2-methyl-pyrrolidine-tryptamine HFB		2170	384$_3$	370$_{10}$	173$_{14}$	158$_{13}$	84$_{100}$	1589	10037
5-Methoxy-2-methyl-pyrrolidine-tryptamine PFP		2175	334$_2$	320$_8$	173$_{12}$	158$_{12}$	84$_{100}$	1459	10036
5-Methoxy-2-methyl-pyrrolidine-tryptamine TFA		2230	354$_1$	284$_2$	270$_8$	158$_{21}$	84$_{100}$	1232	10192
5-Methoxy-2-methyl-pyrrolidine-tryptamine TMS		2510	330$_{14}$	246$_{35}$	216$_3$	174$_2$	84$_{100}$	1095	10035
p-Methoxyamfetamine		1225	165$_3$	122$_{100}$	107$_{11}$	91$_{22}$	77$_{45}$	314	3249
p-Methoxyamfetamine		1225	165$_1$	122$_{16}$	91$_3$	77$_7$	44$_{100}$	314	5517
p-Methoxyamfetamine HFB		1560	361$_2$	240$_5$	169$_5$	148$_{53}$	121$_{100}$	1268	6769
p-Methoxyamfetamine PFP		1460	311$_7$	190$_4$	148$_{43}$	121$_{100}$	91$_7$	975	6775
p-Methoxyamfetamine TFA		1460	261$_6$	148$_{33}$	140$_4$	121$_{100}$	91$_8$	662	6774
4-Methoxyaniline AC	PME UME	1630	165$_{59}$	123$_{74}$	108$_{100}$	95$_{10}$	80$_{20}$	314	5046
4-Methoxyaniline HFB	U+UHYHFB	1400	319$_{100}$	304$_8$	300$_6$	150$_6$	122$_{78}$	1023	6620
4-Methoxyaniline TFA	U+UHYTFA	1335	219$_{100}$	204$_{16}$	149$_{11}$	122$_{76}$	109$_{19}$	463	6615

Table 8.1: Compounds in order of names

Name	Detected	RI	Typical ions and intensities					Page	Entry
4-Methoxybenzoic acid		1320*	152_{95}	135_{100}	107_{11}	92_{16}	77_{23}	293	10270
4-Methoxybenzoic acid ET		1415*	180_{21}	152_{19}	135_{100}	107_{12}	77_{21}	345	6447
4-Methoxybenzoic acid ME		1270*	166_{36}	135_{100}	107_{11}	92_{15}	77_{18}	316	6446
3-Methoxybenzoic acid methylester		1490*	166_{60}	135_{100}	107_{44}	92_{24}	77_{41}	315	1110
Methoxychlor		2450*	344_{2}	274_{3}	227_{100}	212_{5}	152_{5}	1177	1488
Methoxychlor -HCl		2340*	308_{100}	273_{13}	238_{80}	223_{28}	152_{24}	955	3858
Methoxydine		2150	273_{29}	272_{30}	230_{100}	121_{35}	84_{22}	736	3594
Methoxydine artifact		1800*	188_{100}	173_{42}	159_{42}	145_{20}	129_{23}	365	9452
p-Methoxyetilamfetamine		1660	192_{1}	149_{1}	121_{5}	91_{2}	72_{100}	380	5831
p-Methoxyetilamfetamine AC	UHYAC	1855	235_{1}	148_{32}	121_{7}	114_{26}	72_{100}	535	5322
p-Methoxyetilamfetamine HFB	UHFB	1785	389_{2}	268_{100}	240_{46}	148_{62}	121_{43}	1398	5834
p-Methoxyetilamfetamine ME		1780	206_{1}	121_{5}	86_{100}	72_{3}	58_{20}	422	5835
p-Methoxyetilamfetamine PFP	UPFP	1765	339_{1}	218_{100}	190_{44}	148_{59}	121_{49}	1149	5833
p-Methoxyetilamfetamine TFA	UTFA	1775	289_{1}	168_{100}	148_{63}	140_{41}	121_{62}	834	5832
p-Methoxyetilamfetamine TMS		2065	264_{1}	250_{14}	144_{100}	121_{17}	73_{84}	691	5836
Methoxyhydroxyphenylglycol (MHPG) 3AC		2030*	310_{9}	268_{14}	208_{52}	166_{100}	153_{91}	970	1111
p-Methoxymetamfetamine		1475	178_{1}	121_{4}	91_{2}	77_{3}	58_{100}	343	6719
p-Methoxymetamfetamine AC		1820	221_{1}	148_{24}	121_{7}	100_{22}	58_{100}	476	6720
p-Methoxymetamfetamine ET		1780	206_{1}	121_{5}	86_{100}	72_{3}	58_{20}	422	5835
p-Methoxymetamfetamine HFB		1665	375_{1}	254_{55}	210_{32}	148_{67}	121_{100}	1334	6722
p-Methoxymetamfetamine PFP		1510	325_{3}	204_{60}	160_{29}	148_{48}	121_{100}	1061	7601
p-Methoxymetamfetamine TFA		1645	275_{2}	154_{59}	148_{51}	121_{100}	110_{31}	745	6721
5-Methoxy-N,N-diallyl-tryptamine		2270	270_{1}	174_{2}	160_{10}	145_{7}	110_{100}	720	8831
5-Methoxy-N,N-diallyl-tryptamine HFB		2180	439_{1}	370_{4}	356_{9}	159_{22}	110_{100}	1609	9140
5-Methoxy-N,N-diallyl-tryptamine PFP		2130	389_{1}	320_{6}	306_{10}	159_{22}	110_{100}	1496	9139
5-Methoxy-N,N-diallyl-tryptamine TFA		2145	339_{1}	270_{6}	256_{11}	159_{20}	110_{100}	1291	9138
5-Methoxy-N,N-diallyl-tryptamine TMS		2300	270_{1}	174_{1}	160_{1}	145_{1}	110_{100}	1171	9137
5-Methoxy-N,N-diallyl-tryptamine-D4		2270	274_{2}	178_{4}	162_{16}	147_{11}	112_{100}	742	8850
5-Methoxy-N,N-diallyl-tryptamine-D4 HFB		2120	470_{1}	443_{3}	374_{11}	358_{20}	112_{100}	1618	10070
5-Methoxy-N,N-diallyl-tryptamine-D4 PFP		2115	420_{3}	393_{6}	324_{9}	308_{17}	112_{100}	1508	10069
5-Methoxy-N,N-diallyl-tryptamine-D4 TFA		2135	370_{2}	343_{7}	274_{14}	258_{25}	112_{100}	1313	10068
5-Methoxy-N,N-diallyl-tryptamine-D4 TMS		2375	346_{7}	305_{6}	250_{4}	234_{16}	112_{100}	1191	10067
5-Methoxy-N,N-diallyl-tryptamine-M (N-dealkyl-) AC		2480	272_{9}	173_{100}	160_{68}	145_{16}	70_{24}	730	9273
5-Methoxy-N,N-diallyl-tryptamine-M (O-demethyl-) AC		2310	298_{1}	257_{1}	160_{3}	146_{9}	110_{100}	897	9251
5-Methoxy-N,N-diallyl-tryptamine-M (O-demethyl-di-HO-) 2AC		2440	298_{1}	146_{9}	110_{100}			1323	9272
5-Methoxy-N,N-diallyl-tryptamine-M (O-demethyl-di-HO-) 3AC		2400	414_{1}	373_{15}	160_{3}	146_{10}	110_{100}	1491	9256
5-Methoxy-N,N-diallyl-tryptamine-M (O-demethyl-N-dealkyl-) 2AC		2610	300_{8}	201_{63}	159_{100}	146_{58}	70_{41}	908	9276
5-Methoxy-N,N-diallyl-tryptamine-M (oxo-)		2540	284_{2}	242_{2}	176_{7}	161_{9}	110_{100}	804	9275
5-Methoxy-N,N-diallyl-tryptamine-M AC		2520	403_{4}	343_{5}	301_{21}	110_{100}		1454	9274
5-Methoxy-N,N-diethyltryptamine		2110	246_{3}	174_{3}	160_{19}	145_{14}	86_{100}	582	10177
5-Methoxy-N,N-diethyltryptamine HFB		1960	427_{1}	370_{3}	356_{5}	144_{11}	86_{100}	1565	10180
5-Methoxy-N,N-diethyltryptamine PFP		1950	391_{2}	377_{1}	320_{5}	306_{9}	86_{100}	1411	10179
5-Methoxy-N,N-diethyltryptamine TFA		1970	341_{1}	327_{1}	270_{4}	256_{7}	86_{100}	1170	10178
5-Methoxy-N,N-diethyltryptamine TMS		2235	318_{7}	246_{2}	232_{10}	202_{4}	86_{100}	1021	10199
5-Methoxy-N,N-diisopropyl-tryptamine		2260	274_{1}	174_{7}	160_{31}	114_{100}	72_{31}	742	8867
5-Methoxy-N,N-diisopropyl-tryptamine HFB		2050	455_{2}	370_{8}	356_{10}	159_{16}	114_{100}	1618	9516
5-Methoxy-N,N-diisopropyl-tryptamine PFP		2040	405_{1}	306_{3}	159_{13}	144_{8}	114_{100}	1508	9515
5-Methoxy-N,N-diisopropyl-tryptamine TFA		2065	355_{1}	270_{6}	256_{10}	159_{21}	114_{100}	1313	9514
5-Methoxy-N,N-diisopropyl-tryptamine TMS		2365	346_{8}	246_{18}	232_{34}	114_{100}	72_{51}	1191	9513
5-Methoxy-N,N-diisopropyl-tryptamine-D4		2260	278_{1}	178_{9}	162_{33}	116_{100}	74_{29}	771	8862
5-Methoxy-N,N-diisopropyl-tryptamine-D4 HFB		2105	459_{2}	374_{9}	358_{8}	161_{20}	116_{100}	1626	10122
5-Methoxy-N,N-diisopropyl-tryptamine-D4 PFP		2100	409_{5}	324_{9}	308_{8}	161_{8}	116_{100}	1518	10121
5-Methoxy-N,N-diisopropyl-tryptamine-D4 TFA		2115	359_{2}	274_{6}	258_{7}	161_{6}	116_{100}	1331	10120
5-Methoxy-N,N-diisopropyl-tryptamine-D4 TMS		2350	350_{1}	234_{13}	204_{4}	116_{100}	74_{44}	1210	10119
5-Methoxy-N,N-dipropyl-tryptamine		2300	274_{3}	174_{10}	160_{35}	145_{14}	114_{100}	742	8859
5-Methoxy-N,N-dipropyl-tryptamine HFB		2120	441_{4}	370_{13}	356_{15}	159_{12}	114_{100}	1618	10110
5-Methoxy-N,N-dipropyl-tryptamine PFP		2110	391_{2}	320_{5}	306_{7}	159_{13}	114_{100}	1508	10109
5-Methoxy-N,N-dipropyl-tryptamine TFA		2145	369_{3}	341_{15}	270_{21}	256_{24}	114_{100}	1313	10108
5-Methoxy-N,N-dipropyl-tryptamine TMS		2380	346_{5}	246_{2}	232_{15}	202_{5}	114_{100}	1191	10107
5-Methoxy-N,N-dipropyl-tryptamine-D4		2280	278_{3}	178_{8}	162_{28}	147_{13}	116_{100}	771	8864
5-Methoxy-N,N-dipropyl-tryptamine-D4 HFB		2120	445_{5}	374_{11}	358_{11}	161_{10}	116_{100}	1626	10130
5-Methoxy-N,N-dipropyl-tryptamine-D4 PFP		2115	395_{5}	324_{5}	308_{6}	161_{9}	116_{100}	1518	10129
5-Methoxy-N,N-dipropyl-tryptamine-D4 TFA		2130	345_{4}	275_{1}	258_{5}	161_{12}	116_{100}	1331	10128
5-Methoxy-N,N-dipropyl-tryptamine-D4 TMS		2345	350_{4}	250_{4}	234_{7}	177_{2}	116_{100}	1210	10127
4-Methoxy-naphthalen-1-yl-(1-pentylindol-3-yl)methanone		3400	371_{100}	354_{47}	314_{54}	214_{66}	185_{64}	1320	8526
5-Methoxy-N-ethyl-N-propyl-tryptamine-D4		2210	264_{3}	178_{5}	162_{16}	147_{9}	102_{100}	684	8866
4-Methoxy-N-isopropyl-N-methyl-tryptamine		2090	246_{5}	174_{4}	160_{13}	130_{27}	86_{100}	582	10169
4-Methoxy-N-isopropyl-N-methyl-tryptamine HFB		1930	427_{2}	370_{4}	356_{4}	129_{6}	86_{100}	1565	10172
4-Methoxy-N-isopropyl-N-methyl-tryptamine PFP		2100	390_{2}	306_{5}	129_{6}	86_{100}		1411	10171
4-Methoxy-N-isopropyl-N-methyl-tryptamine TFA		1940	340_{17}	327_{57}	270_{100}	256_{66}	226_{45}	1169	10170
4-Methoxy-N-isopropyl-N-methyl-tryptamine TMS		2215	318_{6}	246_{2}	232_{11}	86_{100}	73_{26}	1021	10197
5-Methoxy-N-isopropyl-N-methyl-tryptamine		2120	246_{4}	174_{5}	160_{24}	145_{17}	86_{100}	582	10173
5-Methoxy-N-isopropyl-N-methyl-tryptamine HFB		1970	427_{1}	370_{5}	356_{6}	144_{14}	86_{100}	1565	10176
5-Methoxy-N-isopropyl-N-methyl-tryptamine PFP		1965	391_{2}	377_{4}	320_{7}	306_{8}	86_{100}	1411	10175

5-Methoxy-N-isopropyl-N-methyl-tryptamine TFA Table 8.1: Compounds in order of names

Name	Detected	RI	Typical ions and intensities					Page	Entry
5-Methoxy-N-isopropyl-N-methyl-tryptamine TFA		1980	341_1	327_3	270_7	256_8	86_{100}	1169	10174
5-Methoxy-N-isopropyl-N-methyl-tryptamine TMS		2255	318_6	262_5	232_{11}	202_3	86_{100}	1021	10198
Methoxyphenamine		1585	178_1	164_2	121_5	91_{20}	58_{100}	343	8112
Methoxyphenamine AC		1800	221_5	148_{49}	100_{61}	91_{43}	58_{100}	476	8118
Methoxyphenamine HFB		1585	375_1	254_{100}	210_{41}	148_{57}	121_{16}	1334	8117
Methoxyphenamine PFP		1560	325_1	204_{100}	160_{33}	148_{39}	121_9	1061	8116
Methoxyphenamine precursor (2-methoxyphenylacetone)		1270*	164_{28}	146_{51}	121_{31}	91_{100}	77_{12}	312	8113
Methoxyphenamine TFA		1575	275_3	154_{100}	148_{55}	110_{38}	91_{38}	745	8115
Methoxyphenamine TMS		1560	236_5	146_8	130_{100}	91_{30}	73_{60}	614	8114
Methoxyphenamine-M (di-HO-) 3AC	U+UHYAC	2175	222_3	180_{12}	153_2	100_{49}	58_{100}	1137	8121
Methoxyphenamine-M (HO-) 2AC	U+UHYAC	2230	279_2	206_{27}	164_{16}	100_{44}	58_{100}	777	8120
Methoxyphenamine-M (O-demethyl-) 2AC	U+UHYAC	1900	249_1	134_4	107_6	100_{52}	58_{100}	598	8119
Methoxyphenamine-M (O-demethyl-HO-) 3AC	U+UHYAC	2280	265_{10}	150_{11}	123_{12}	100_{57}	58_{100}	949	8122
3-Methoxy-phencyclidine		2120	273_{32}	272_{35}	230_{100}	121_{36}	84_{19}	736	9451
4-Methoxy-phencyclidine		2150	273_{29}	272_{30}	230_{100}	121_{35}	84_{22}	736	3594
3-Methoxy-phencyclidine-M (O-demethyl-) AC	U+UHYAC	2210	301_{25}	258_{100}	244_{12}	166_{22}	84_{13}	916	10271
3-Methoxy-phencyclidine-M (O-demethyl-bis-HO-) isomer-1 3AC	U+UHYAC	2670	417_{15}	374_{100}	282_{20}	175_{16}	107_{18}	1500	10283
3-Methoxy-phencyclidine-M (O-demethyl-bis-HO-) isomer-2 3AC	U+UHYAC	2690	417_7	389_7	358_{100}	288_{80}	82_{16}	1501	10278
3-Methoxy-phencyclidine-M (O-demethyl-bis-HO-) isomer-3 3AC	U+UHYAC	2720	417_{12}	374_{13}	358_{100}	316_{39}	82_{12}	1501	10277
3-Methoxy-phencyclidine-M (O-demethyl-bis-HO-) isomer-4 3AC	U+UHYAC	2780	417_4	357_{60}	316_{100}	107_8	82_{11}	1501	10276
3-Methoxy-phencyclidine-M (O-demethyl-bis-HO-) isomer-5 3AC	U+UHYAC	2800	417_5	357_{60}	316_{100}	107_{14}	82_{18}	1501	10284
3-Methoxy-phencyclidine-M (O-demethyl-HO-) isomer-1 2AC	U+UHYAC	2420	359_{14}	316_{11}	300_{100}	258_{18}	164_{10}	1260	10279
3-Methoxy-phencyclidine-M (O-demethyl-HO-) isomer-2 2AC	U+UHYAC	2440	359_{14}	316_{13}	300_{100}	258_{34}	84_{14}	1260	10282
3-Methoxy-phencyclidine-M (O-demethyl-HO-) isomer-3 2AC	U+UHYAC	2460	359_{16}	316_{100}	256_{12}	224_{12}	107_{18}	1260	10275
3-Methoxy-phencyclidine-M (O-demethyl-HO-) isomer-4 2AC	U+UHYAC	2490	359_9	316_1	299_{41}	258_{100}	84_{12}	1260	10273
3-Methoxy-phencyclidine-M (O-demethyl-HO-) isomer-5 2AC	U+UHYAC	2510	359_{25}	316_{100}	256_{25}	107_{28}	82_{23}	1260	10272
3-Methoxy-phencyclidine-M (oxo-)	U+UHYAC	2150	287_{20}	244_{100}	230_6	222_5	98_8	823	10281
3-Methoxy-phencyclidine-M/artifact 2AC	U+UHYAC	2485	361_{15}	318_{100}	246_8	175_{12}	107_{11}	1267	10274
2-Methoxyphenylpiperazine AC	U+UHYAC	2070	234_{36}	162_{100}	149_{29}	134_{38}	120_{33}	529	6808
2-Methoxyphenylpiperazine-M (HO-) 2AC	U+UHYAC	2490	292_{45}	250_{25}	207_{15}	178_{100}	165_{39}	856	496
2-Methoxyphenylpiperazine-M (HO-) AC	U+UHYAC	2420	250_{69}	207_4	178_{100}	165_2	150_{40}	605	8763
2-Methoxyphenylpiperazine-M (O-demethyl-) 2AC	UHYAC	2140	262_{10}	220_{18}	148_{100}	120_{51}	86_{12}	669	170
4-Methoxyphenylpiperazine		1880	192_{34}	150_{100}	135_{20}	120_{28}	92_9	376	6622
4-Methoxyphenylpiperazine 2AC	U+UHYAC	2185	234_{81}	162_{100}	149_{35}	134_{36}	120_{35}	529	6609
4-Methoxyphenylpiperazine HFB	U+UHYHFB	1965	388_{100}	373_{34}	191_{75}	135_{74}	120_{66}	1394	6617
4-Methoxyphenylpiperazine ME		1840	206_{100}	191_{17}	162_{14}	135_{77}	120_{62}	417	6623
4-Methoxyphenylpiperazine TFA	U+UHYTFA	1940	288_{100}	273_{36}	191_{20}	135_{22}	120_{28}	827	6612
4-Methoxyphenylpiperazine TMS		2070	264_{100}	249_{29}	101_{45}	86_{46}	73_{81}	683	6884
4-Methoxyphenylpiperazine-M (aminophenol) 2HFB	U+UHYHFB	1405	501_{25}	482_6	304_{81}	169_{100}	109_{83}	1660	6621
4-Methoxyphenylpiperazine-M (aminophenol) 2TFA	U+UHYTFA	1280	301_{49}	204_{47}	176_{16}	109_{48}	69_{100}	910	6616
4-Methoxyphenylpiperazine-M (deethylene-) 2AC	U+UHYAC	2120	250_7	191_{51}	165_4	149_{33}	136_{100}	605	6611
4-Methoxyphenylpiperazine-M (deethylene-) 2HFB	U+UHYHFB	1765	558_7	345_{13}	332_{28}	240_{19}	135_{100}	1699	6619
4-Methoxyphenylpiperazine-M (deethylene-) 2TFA	U+UHYTFA	1765	358_{20}	245_{19}	232_{48}	135_{100}	120_{33}	1252	6614
4-Methoxyphenylpiperazine-M (methoxyaniline) HFB	U+UHYHFB	1400	319_{100}	304_8	300_6	150_6	122_{78}	1023	6620
4-Methoxyphenylpiperazine-M (methoxyaniline) TFA	U+UHYTFA	1335	219_{100}	204_{16}	149_{11}	122_{76}	109_{19}	463	6615
4-Methoxyphenylpiperazine-M (O-demethyl-) 2AC	U+UHYAC	2350	262_{39}	220_{60}	177_{21}	148_{100}	135_{59}	670	6610
4-Methoxyphenylpiperazine-M (O-demethyl-) 2HFB	U+UHYHFB	1990	570_{83}	551_{16}	373_{100}	344_{46}	317_9	1703	6618
4-Methoxyphenylpiperazine-M (O-demethyl-) 2TFA	U+UHYTFA	1915	370_{90}	273_{100}	244_{60}	217_{34}	120_{75}	1310	6613
Methoxypiperamide		2065	234_{20}	190_9	135_{100}	99_{47}	70_{69}	529	9304
Methoxypiperamide artifact (dehydro-)		2150	232_{54}	135_{100}	97_{71}	92_{19}	70_{23}	516	9305
Methoxypiperamide-M (4-methoxybenzoic acid)		1320*	152_{95}	135_{100}	107_{11}	92_{16}	77_{23}	293	10270
Methoxypiperamide-M (4-methoxybenzoic acid) ME		1270*	166_{36}	135_{100}	107_{11}	92_{15}	77_{18}	316	6446
Methoxypiperamide-M (N,N-bisdealkyl-) AC		2335	250_1	177_{12}	164_8	135_{100}	99_{35}	605	9314
Methoxypiperamide-M (N,N-bisdealkyl-nor-) AC		2290	236_1	177_6	151_{25}	135_{100}	92_{10}	539	9315
Methoxypiperamide-M (N,N-bisdealkyl-nor-dihydro-) 2AC		2320	280_1	221_7	135_{62}	129_{72}	116_{100}	783	9316
Methoxypiperamide-M (N,N-bisdealkyl-oxo-) AC		2280	264_1	179_{22}	135_{100}	92_{17}	72_{19}	681	9311
Methoxypiperamide-M (N,O-bisdemethyl-) 2AC		2525	290_5	222_4	163_{16}	121_{100}	85_{46}	842	9309
Methoxypiperamide-M (N-methylpiperazine) AC		1230	142_{13}	99_{39}	83_{16}	70_{100}	56_{69}	282	9345
Methoxypiperamide-M (nor-)		2040	220_{18}	191_2	135_{100}	92_{14}	85_6	469	9312
Methoxypiperamide-M (nor-) AC		2440	262_9	203_4	194_7	135_{100}	85_{32}	670	9308
Methoxypiperamide-M (nor-) acetyl conjugate		2440	262_9	203_4	194_7	135_{100}	85_{32}	670	9308
Methoxypiperamide-M (nor-oxo-) dehydro artifact AC		2505	274_{13}	139_{50}	135_{100}	92_{19}	77_{23}	739	9310
Methoxypiperamide-M (O-demethyl-)		2135	220_6	121_{50}	99_{39}	83_{46}	70_{100}	469	9313
Methoxypiperamide-M (O-demethyl-) AC		2170	262_7	176_7	121_{65}	99_{55}	70_{100}	669	9306
Methoxypiperamide-M (oxo-)		2180	248_{11}	204_4	193_{10}	135_{100}	113_{78}	590	9307
3-Methoxy-rolicyclidine		2100	259_{30}	216_{100}	202_{24}	152_{32}	121_{22}	653	9455
3-Methoxy-rolicyclidine-M (O-demethyl-) AC	U+UHYAC	2155	287_{23}	244_{100}	152_{40}	107_{23}	70_{33}	823	10285
3-Methoxy-rolicyclidine-M (O-demethyl-amino-) 2AC	U+UHYAC	2185	275_6	233_9	216_{80}	190_{37}	174_{100}	748	10291
3-Methoxy-rolicyclidine-M (O-demethyl-di-HO-) 3AC	U+UHYAC	2670	403_1	343_{52}	302_{100}	175_{41}	107_{33}	1457	10290
3-Methoxy-rolicyclidine-M (O-demethyl-HO-) isomer-1 2AC	U+UHYAC	2385	345_{10}	302_{11}	286_{100}	244_{25}	70_6	1185	10286
3-Methoxy-rolicyclidine-M (O-demethyl-HO-) isomer-2 2AC	U+UHYAC	2400	345_{13}	302_{11}	286_{100}	244_{33}	70_{15}	1185	10287
3-Methoxy-rolicyclidine-M (O-demethyl-HO-) isomer-3 2AC	U+UHYAC	2420	345_{11}	302_{100}	285_{12}	242_{17}	107_{13}	1186	10288
3-Methoxy-rolicyclidine-M (O-demethyl-HO-) isomer-4 2AC	U+UHYAC	2440	345_6	285_{36}	244_{100}	107_5	70_8	1185	10289
3-Methoxy-rolicyclidine-M (O-demethyl-HO-amino-) 2AC	U+UHYAC	2435	273_5	231_{16}	214_{94}	188_{11}	172_{100}	734	10293

Table 8.1: Compounds in order of names   **4-Methoxy-rolicyclidine**

Name	Detected	RI	Typical ions and intensities					Page	Entry
4-Methoxy-rolicyclidine		2120	259 $_{35}$	216 $_{100}$	202 $_{19}$	189 $_{36}$	121 $_{70}$	653	9456
3-Methoxytyramine 2AC	U+UHYAC	2070	251 $_2$	209 $_7$	150 $_{100}$	137 $_{15}$		609	1273
Methsuximide	P G U+UHYAC	1705	203 $_{44}$	118 $_{100}$	103 $_{23}$	91 $_{12}$	77 $_{20}$	406	1827
4-Methycatechol 2TMS		1325*	268 $_{63}$	180 $_{27}$	165 $_{18}$	149 $_{21}$	73 $_{100}$	708	6022
3-Methyl-1-butene		<1000*	70 $_{23}$	55 $_{100}$	42 $_{30}$	39 $_{50}$	27 $_{49}$	250	3810
5-Methyl-1-hexene		<1000*	98 $_9$	83 $_{12}$	70 $_{32}$	56 $_{97}$	41 $_{100}$	257	3822
2-Methyl-1-pentene		<1000*	84 $_{24}$	69 $_{28}$	56 $_{94}$	41 $_{100}$	27 $_{76}$	254	3817
N-Methyl-1-phenylethylamine		1460	134 $_3$	120 $_{100}$	105 $_9$	77 $_{13}$	58 $_{25}$	275	6221
N-Methyl-1-phenylethylamine AC		1430	177 $_{50}$	162 $_9$	120 $_{100}$	105 $_{45}$	77 $_{30}$	335	6229
2-Methyl-1-propanol (isobutanol)		<1000*	74 $_{13}$	55 $_4$	43 $_{100}$			252	1042
N-Methyl-2,3-methylenedioxyphenethylamine		1475	179 $_{24}$	136 $_{39}$	91 $_{20}$	77 $_{59}$	51 $_{100}$	342	8410
N-Methyl-2,3-methylenedioxyphenethylamine AC		1835	221 $_{23}$	148 $_{100}$	135 $_{14}$	86 $_{29}$	77 $_{19}$	473	8411
N-Methyl-2,3-methylenedioxyphenethylamine HFB		1685	375 $_{21}$	240 $_{49}$	169 $_{32}$	148 $_{100}$	135 $_{45}$	1333	8414
N-Methyl-2,3-methylenedioxyphenethylamine PFP		1650	325 $_{24}$	190 $_{44}$	148 $_{100}$	135 $_{44}$	119 $_{35}$	1060	8413
N-Methyl-2,3-methylenedioxyphenethylamine TFA		1655	275 $_{21}$	208 $_2$	148 $_{100}$	140 $_{43}$	135 $_{43}$	744	8412
N-Methyl-2,3-methylenedioxyphenethylamine TMS		1620	236 $_1$	191 $_2$	135 $_5$	116 $_{78}$	73 $_{100}$	611	8464
4-Methyl-2,5-dimethoxyphenethylamine		1605	195 $_{20}$	166 $_{100}$	151 $_{60}$	135 $_{27}$	91 $_{19}$	387	6904
4-Methyl-2,5-dimethoxyphenethylamine 2AC		2010	279 $_{11}$	178 $_{100}$	163 $_{34}$	135 $_{21}$	72 $_9$	774	6913
4-Methyl-2,5-dimethoxyphenethylamine 2TMS		2020	339 $_2$	324 $_{13}$	174 $_{100}$	100 $_{23}$	86 $_{36}$	1153	6915
4-Methyl-2,5-dimethoxyphenethylamine AC		1940	237 $_{14}$	178 $_{100}$	165 $_{35}$	163 $_{40}$	135 $_{44}$	546	6912
4-Methyl-2,5-dimethoxyphenethylamine formyl artifact		1530	207 $_{25}$	176 $_{100}$	165 $_{69}$	135 $_{39}$	91 $_{16}$	421	6909
4-Methyl-2,5-dimethoxyphenethylamine HFB		1710	391 $_{49}$	226 $_6$	178 $_{85}$	165 $_{100}$	135 $_{46}$	1405	6937
4-Methyl-2,5-dimethoxyphenethylamine PFP		1680	341 $_{53}$	178 $_{79}$	165 $_{100}$	135 $_{54}$	91 $_{21}$	1160	6932
4-Methyl-2,5-dimethoxyphenethylamine TFA		1685	291 $_{43}$	178 $_{73}$	165 $_{100}$	135 $_{57}$	91 $_{22}$	846	6927
4-Methyl-2,5-dimethoxyphenethylamine TMS		1735	267 $_1$	237 $_8$	166 $_{20}$	102 $_{100}$	73 $_{91}$	703	6914
4-Methyl-2,5-dimethoxyphenethylamine-M (deamino-COOH) ME		1755*	224 $_{100}$	209 $_{19}$	177 $_{12}$	165 $_{35}$	135 $_8$	488	7229
4-Methyl-2,5-dimethoxyphenethylamine-M (deamino-HO-) AC	U+UHYAC	1740*	238 $_{27}$	178 $_{100}$	163 $_{57}$	135 $_{33}$	79 $_{27}$	548	7216
4-Methyl-2,5-dimethoxyphenethylamine-M (deamino-oxo-)		1730*	194 $_{54}$	165 $_{100}$	151 $_{25}$	135 $_{85}$	91 $_{51}$	384	7232
4-Methyl-2,5-dimethoxyphenethylamine-M (HO-) 2AC	U+UHYAC	2390	295 $_{33}$	236 $_{100}$	223 $_6$	193 $_{35}$	163 $_{12}$	877	7219
4-Methyl-2,5-dimethoxyphenethylamine-M (HO-) 2TFA		1950	403 $_{42}$	290 $_{100}$	277 $_{32}$	177 $_{57}$	163 $_{25}$	1455	7228
4-Methyl-2,5-dimethoxyphenethylamine-M (HO-) 3AC	U+UHYAC	2400	337 $_{27}$	244 $_{23}$	236 $_{100}$	193 $_{46}$	125 $_{30}$	1138	7220
4-Methyl-2,5-dimethoxyphenethylamine-M (O-demethyl- N-acetyl-) 2AC	U+UHYAC	2250	307 $_5$	265 $_7$	206 $_{25}$	164 $_{100}$	149 $_{13}$	950	7223
4-Methyl-2,5-dimethoxyphenethylamine-M (O-demethyl- N-acetyl-) isomer-1 AC	U+UHYAC	2130	265 $_6$	223 $_{36}$	164 $_{100}$	151 $_{14}$	91 $_4$	687	7221
4-Methyl-2,5-dimethoxyphenethylamine-M (O-demethyl- N-acetyl-) isomer-1 TFA		1990	319 $_{12}$	260 $_{100}$	247 $_4$	191 $_{18}$	163 $_{26}$	1024	7224
4-Methyl-2,5-dimethoxyphenethylamine-M (O-demethyl- N-acetyl-) isomer-2 AC	U+UHYAC	2200	265 $_7$	223 $_{25}$	164 $_{100}$	151 $_{14}$	91 $_6$	688	7222
4-Methyl-2,5-dimethoxyphenethylamine-M (O-demethyl- N-acetyl-) isomer-2 TFA		2050	319 $_{10}$	260 $_{100}$	245 $_{22}$	217 $_4$	163 $_{39}$	1024	7225
4-Methyl-2,5-dimethoxyphenethylamine-M (O-demethyl-) 3AC		2250	307 $_5$	265 $_7$	206 $_{25}$	164 $_{100}$	149 $_{13}$	950	7223
4-Methyl-2,5-dimethoxyphenethylamine-M (O-demethyl-) isomer-1 2AC	U+UHYAC	2130	265 $_6$	223 $_{36}$	164 $_{100}$	151 $_{14}$	91 $_4$	687	7221
4-Methyl-2,5-dimethoxyphenethylamine-M (O-demethyl-) isomer-1 2TFA		1780	373 $_{28}$	260 $_{100}$	247 $_{24}$	191 $_{30}$	163 $_{49}$	1325	7226
4-Methyl-2,5-dimethoxyphenethylamine-M (O-demethyl-) isomer-2 2AC	U+UHYAC	2200	265 $_7$	223 $_{25}$	164 $_{100}$	151 $_{14}$	91 $_6$	688	7222
4-Methyl-2,5-dimethoxyphenethylamine-M (O-demethyl-) isomer-2 2TFA		1850	373 $_{18}$	260 $_{100}$	247 $_{48}$	217 $_{15}$	163 $_{93}$	1325	7227
4-Methyl-2,5-dimethoxyphenethylamine-M (O-demethyl-deamino-COOH) iso.-1 MEAC		1860*	252 $_{28}$	210 $_{100}$	178 $_{40}$	150 $_{100}$	122 $_{12}$	618	7230
4-Methyl-2,5-dimethoxyphenethylamine-M (O-demethyl-deamino-COOH) iso.-2 MEAC		1900*	252 $_{21}$	210 $_{100}$	193 $_{10}$	163 $_7$	151 $_{55}$	618	7231
4-Methyl-2,5-dimethoxyphenethylamine-M (O-demethyl-deamino-HO-) isomer-1 2AC	U+UHYAC	1875*	266 $_5$	224 $_{13}$	164 $_{100}$	154 $_{46}$	114 $_{15}$	695	7217
4-Methyl-2,5-dimethoxyphenethylamine-M (O-demethyl-deamino-HO-) isomer-2 2AC	U+UHYAC	1890*	266 $_8$	224 $_{12}$	206 $_6$	164 $_{100}$	121 $_{10}$	695	7218
1-Methyl-2-phenyl-alpha-methyltryptamine		2335	264 $_3$	221 $_{100}$	204 $_{27}$	178 $_7$	144 $_6$	683	9769
1-Methyl-2-phenyl-alpha-methyltryptamine AC		2640	306 $_{19}$	247 $_{57}$	220 $_{100}$	204 $_{43}$	178 $_{14}$	944	9771
1-Methyl-2-phenyl-alpha-methyltryptamine ethylimine artifact		2235	290 $_9$	220 $_{100}$	204 $_{18}$	178 $_5$	70 $_8$	843	10213
1-Methyl-2-phenyl-alpha-methyltryptamine formyl artifact		2335	276 $_{15}$	220 $_{100}$	204 $_{23}$	178 $_6$	56 $_{14}$	756	9770
1-Methyl-2-phenyl-alpha-methyltryptamine HFB		2360	460 $_2$	240 $_{10}$	220 $_{100}$	204 $_{27}$	178 $_9$	1601	9775
1-Methyl-2-phenyl-alpha-methyltryptamine PFP		2350	410 $_9$	247 $_4$	220 $_{100}$	204 $_{33}$	178 $_9$	1478	9773
1-Methyl-2-phenyl-alpha-methyltryptamine TFA		2385	360 $_{13}$	247 $_3$	220 $_{100}$	204 $_{27}$	178 $_9$	1265	9772
1-Methyl-2-phenyl-alpha-methyltryptamine TMS		2700	321 $_6$	221 $_6$	204 $_{25}$	116 $_{20}$	73 $_{44}$	1133	9774
N-Methyl-4,4-difluoro-modafenil		2520	306 $_1$	288 $_2$	214 $_3$	203 $_{100}$	183 $_{50}$	1048	9476
N-Methyl-4,4-difluoro-modafenil artifact (bis-4-fluorophenylcarbinol)		1690*	220 $_{17}$	201 $_{13}$	183 $_8$	123 $_{100}$	95 $_{31}$	468	3378
N-Methyl-4,4-difluoro-modafenil artifact (bis-4-fluorophenylcarbinol) PFP		1555*	203 $_{100}$	201 $_{68}$	183 $_{44}$	170 $_9$		1289	9479
N-Methyl-4,4-difluoro-modafenil artifact (bis-4-fluorophenylcarbinol) PFP		1555*	416 $_2$	203 $_{100}$	201 $_{79}$	183 $_{30}$	170 $_7$	1494	9480
N-Methyl-4,4-difluoro-modafenil artifact (bis-4-fluorophenylcarbinol) TFA		1550	316 $_2$	203 $_{100}$	201 $_{87}$	183 $_{46}$	170 $_{12}$	1004	9477
N-Methyl-4,4-difluoro-modafenil artifact (difluoro-benzophenone)	U UHY UHYAC	1595*	218 $_{38}$	123 $_{100}$	109	95 $_{62}$	75 $_{28}$	457	3373
N-Methyl-4,4-difluoro-modafenil TFA		3650	315 $_7$	281 $_3$	237 $_{14}$	203 $_{100}$	183 $_{20}$	1504	9478
N-Methyl-5-(2-aminopropyl)benzofuran-M (nor-)		1450	175 $_1$	131 $_{13}$	102 $_3$	77 $_{10}$	44 $_{100}$	330	9083
N-Methyl-5-(2-aminopropyl)benzofuran-M (nor-) formyl artifact		1505	187 $_6$	172 $_8$	131 $_{41}$	77 $_{19}$	56 $_{100}$	363	9084
N-Methyl-5-(2-aminopropyl)benzofuran-M (nor-) HFB		1670	371 $_1$	240 $_{15}$	158 $_{57}$	131 $_{100}$	77 $_{28}$	1316	9088
N-Methyl-5-(2-aminopropyl)benzofuran-M (nor-) PFP		1640	321 $_2$	190 $_{16}$	158 $_{46}$	131 $_{100}$	77 $_{31}$	1038	9087
N-Methyl-5-(2-aminopropyl)benzofuran-M (nor-) TFA		1655	271 $_2$	158 $_{40}$	131 $_{100}$	77 $_{33}$	69 $_{40}$	722	9086
N-Methyl-5-(2-aminopropyl)benzofuran-M (nor-) TMS		1650	232 $_6$	131 $_{23}$	116 $_{100}$	100 $_{11}$	73 $_{57}$	586	9085
N-Methyl-5-aminopropylbenzofuran		1550	188 $_1$	174 $_1$	131 $_{14}$	77 $_{14}$	58 $_{100}$	368	8943
N-Methyl-5-aminopropylbenzofuran AC		1960	231 $_1$	158 $_{16}$	131 $_{12}$	100 $_{36}$	58 $_{100}$	514	8944
N-Methyl-5-aminopropylbenzofuran HFB		1770	385 $_1$	254 $_{100}$	210 $_{45}$	158 $_{70}$	131 $_{43}$	1380	8949
N-Methyl-5-aminopropylbenzofuran ME		1570	202 $_1$	188 $_1$	131 $_6$	77 $_4$	72 $_{100}$	407	8945
N-Methyl-5-aminopropylbenzofuran PFP		1720	335 $_1$	204 $_{100}$	160 $_{37}$	158 $_{52}$	131 $_{27}$	1123	8948
N-Methyl-5-aminopropylbenzofuran TFA		1750	285 $_1$	158 $_{56}$	154 $_{100}$	131 $_{46}$	110 $_{41}$	808	8947
N-Methyl-5-aminopropylbenzofuran TMS		1710	260 $_1$	246 $_6$	130 $_{100}$	115 $_3$	73 $_{53}$	665	8946
N-Methyl-5-aminopropylbenzofuran-M (nor-)		1450	175 $_9$	160 $_5$	131 $_{100}$	102 $_{25}$	77 $_{77}$	330	8951

N-Methyl-5-aminopropylbenzofuran-M (nor-) AC

Table 8.1: Compounds in order of names

Name	Detected	RI	Typical ions and intensities					Page	Entry
N-Methyl-5-aminopropylbenzofuran-M (nor-) AC		1870	217_3	158_{100}	131_{67}	86_{59}	77_{51}	454	8950
N-Methyl-6-(2-aminopropyl)benzofuran-M (nor-) AC		1890	217_8	158_{100}	131_{69}	86_{42}	77_{42}	455	9089
N-Methyl-6-(2-aminopropyl)benzofuran-M (nor-) HFB		1600	371_2	240_{15}	158_{62}	131_{100}	77_{26}	1316	9093
N-Methyl-6-(2-aminopropyl)benzofuran-M (nor-) PFP		1610	321_2	190_{15}	158_{54}	131_{100}	77_{31}	1038	9092
N-Methyl-6-(2-aminopropyl)benzofuran-M (nor-) TFA		1670	271_3	158_{42}	131_{100}	77_{34}	69_{42}	723	9091
N-Methyl-6-(2-aminopropyl)benzofuran-M (nor-) TMS		1655	232_5	131_{22}	116_{100}	100_9	73_{60}	586	9090
N-Methyl-6-aminopropylbenzofuran		1540	174_1	131_{23}	102_7	77_{17}	58_{100}	368	9206
N-Methyl-6-aminopropylbenzofuran AC		1960	231_1	158_{32}	131_{28}	100_{29}	58_{100}	513	9207
N-Methyl-6-aminopropylbenzofuran HFB		1730	385_1	254_{100}	210_{45}	158_{66}	131_{62}	1380	9211
N-Methyl-6-aminopropylbenzofuran ME		1630	201_1	131_{16}	102_2	77_6	72_{100}	407	9094
N-Methyl-6-aminopropylbenzofuran PFP		1690	335_3	204_{100}	160_{52}	158_{55}	131_{53}	1123	9210
N-Methyl-6-aminopropylbenzofuran TFA		1750	285_1	158_{73}	154_{100}	131_{57}	110_{41}	808	9209
N-Methyl-6-aminopropylbenzofuran TMS		1650	246_8	131_{53}	130_{100}	115_4	73_{63}	665	9208
Methylacetate		<1000*	74_{23}	59_{10}	43_{100}	29_{23}		252	3777
4'-Methyl-alpha-pyrrolidinohexiophenone		1965	140_{100}	119_4	91_8	84_4	65_5	653	6647
2-Methyl-amfetamine		1450	148_1	105_3	91_4	77_4	44_{100}	288	8881
2-Methyl-amfetamine		1450	148_4	134_{11}	105_{83}	91_{99}	77_{100}	289	8884
2-Methyl-amfetamine AC		1570	191_3	132_{37}	105_{18}	86_{40}	44_{100}	373	8883
2-Methyl-amfetamine AC		1570	191_8	132_{100}	117_{32}	105_{34}	86_{51}	374	8886
2-Methyl-amfetamine formyl artifact		1145	161_2	146_{10}	105_{15}	56_{100}	44_{39}	304	8882
2-Methyl-amfetamine formyl artifact		1145	161_3	146_{10}	118_9	105_{29}	56_{100}	304	8885
2-Methyl-amfetamine HFB		1460	345_1	240_{100}	169_{26}	132_{78}	105_{65}	1183	8890
2-Methyl-amfetamine PFP		1390	295_1	190_{100}	132_{72}	119_{40}	105_{56}	875	8889
2-Methyl-amfetamine TFA		1370	245_1	140_{100}	132_{82}	117_{16}	105_{69}	574	8888
2-Methyl-amfetamine TMS		1410	220_1	206_{11}	116_{100}	105_{16}	73_{55}	477	8887
2-Methyl-amfetamine-M (di-HO-) isomer-1 3AC		2130	265_2	223_{37}	206_{24}	164_{67}	86_{100}	948	8896
2-Methyl-amfetamine-M (di-HO-) isomer-2 3AC		2150	265_6	223_{100}	206_{36}	164_{74}	86_{80}	949	8897
2-Methyl-amfetamine-M (HO-alkyl-) 2AC		1930	249_1	189_{47}	130_{29}	104_{52}	86_{100}	596	8891
2-Methyl-amfetamine-M (HO-aryl-) isomer-1 2AC		1960	249_{12}	190_{46}	148_{50}	121_{30}	86_{100}	596	8892
2-Methyl-amfetamine-M (HO-aryl-) isomer-2 2AC		1995	249_2	190_{88}	148_{100}	121_{80}	86_{81}	597	8893
2-Methyl-amfetamine-M (HO-methoxy-) isomer-1 2AC		2120	279_1	220_{36}	178_{100}	151_{39}	86_{62}	774	8894
2-Methyl-amfetamine-M (HO-methoxy-) isomer-2 2AC		2140	279_4	220_{18}	178_{100}	151_{46}	86_{74}	774	8895
3-Methyl-amfetamine		1450	148_1	105_3	91_4	77_4	44_{100}	288	8898
3-Methyl-amfetamine		1450	148_4	105_{81}	91_{100}	77_{95}		288	8906
3-Methyl-amfetamine AC		1660	191_1	132_{28}	117_8	86_{24}	44_{100}	373	8900
3-Methyl-amfetamine AC		1660	191_4	132_{100}	117_{26}	105_{29}	86_{81}	373	8907
3-Methyl-amfetamine formyl artifact		1150	161_1	146_9	105_{12}	56_{100}	44_4	304	8899
3-Methyl-amfetamine formyl artifact		1150	161_1	146_8	105_{19}	77_6	56_{100}	304	8905
3-Methyl-amfetamine HFB		1440	345_1	240_{100}	169_{21}	132_{84}	105_{39}	1183	8904
3-Methyl-amfetamine PFP		1400	295_1	190_{100}	132_{80}	119_{50}	105_{50}	875	8903
3-Methyl-amfetamine TFA		1410	245_1	140_{100}	132_{92}	105_{57}	91_{27}	574	8902
3-Methyl-amfetamine TMS		1405	220_1	206_{11}	116_{100}	105_{17}	73_{56}	477	8901
3-Methyl-amfetamine-M (di-HO-) 3AC		2200	265_3	206_{43}	164_{13}	131_{24}	86_{100}	948	8941
3-Methyl-amfetamine-M (HO-alkyl-) 2AC		1960	189_3	148_{68}	121_{11}	91_{31}	86_{100}	596	8908
3-Methyl-amfetamine-M (HO-aryl-) isomer-1 2AC		2040	249_1	190_{40}	148_{27}	91_{18}	86_{100}	596	8909
3-Methyl-amfetamine-M (HO-aryl-) isomer-2 2AC		2090	190_{40}	148_{23}	104_{35}	86_{100}	84_{53}	597	8910
3-Methyl-amfetamine-M (HO-methoxy-) 2AC		2085	220_4	178_{17}	91_{20}	86_{100}		774	8911
4-Methyl-amfetamine		1450	148_1	105_7	91_8	77_7	44_{100}	288	8912
4-Methyl-amfetamine		1450	148_4	134_{16}	105_{100}	91_{97}	77_{99}	288	8915
4-Methyl-amfetamine AC		1610	191_1	132_{57}	105_{19}	86_{22}	44_{100}	374	8914
4-Methyl-amfetamine AC		1610	191_2	132_{100}	117_{26}	105_{32}	86_{40}	372	8921
4-Methyl-amfetamine formyl artifact		1165	161_2	146_{14}	105_{25}	56_{100}	44_{28}	304	8913
4-Methyl-amfetamine formyl artifact		1165	161_3	146_{14}	105_{22}	77_{10}	56_{100}	304	8916
4-Methyl-amfetamine HFB		1410	345_1	240_{53}	169_{20}	132_{100}	105_{85}	1183	8920
4-Methyl-amfetamine PFP		1400	295_1	190_{64}	132_{100}	119_{44}	105_{81}	875	8919
4-Methyl-amfetamine TFA		1415	245_1	140_{62}	132_{100}	117_{23}	105_{86}	574	8918
4-Methyl-amfetamine TMS		1400	206_{12}	116_{100}	105_{19}	73_{56}		476	8917
4-Methyl-amfetamine-M (HO-) 2AC		1710	190_{55}	148_{18}	131_{19}	104_{66}	86_{100}	598	8922
4-Methyl-amfetamine-M (HOOC-) (ME)AC		1950	235_1	204_6	176_{30}	145_{29}	86_{100}	533	8942
Methylamine		<1000	31_{40}	30_{82}	28_{100}			247	3619
17-Methylandrostane-17-ol-3-one		2555*	304_{22}	289_{46}	247_{41}	231_{45}	55_{100}	933	3895
17-Methylandrostane-ol-3-one enol 2TMS		2580*	448_{13}	358_4	216_{10}	143_{81}	73_{100}	1577	3978
17-Methylandrostane-ol-3-one enol TMS		2565*	376_{74}	347_{23}	143_{100}	127_{67}	73_{53}	1341	3924
17-Methylandrostane-ol-3-one TMS		2610*	376_1	361_6	306_6	143_{100}	73_{40}	1341	3925
4-Methylbenzoic acid ET		1350*	164_{16}	136_{22}	119_{100}	91_{44}	65_{15}	311	6473
4-Methylbenzoic acid ME		1210*	150_{31}	119_{100}	91_{33}	89_8	65_{22}	289	6472
N-Methyl-Brolamfetamine		1885	230_2	199_2	143_2	77_5	58_{100}	818	6429
N-Methyl-Brolamfetamine AC	U+UHYAC	2225	329_1	256_{15}	199_3	100_{57}	58_{100}	1086	6430
N-Methyl-Brolamfetamine-M (demethyl-)		1800	273_4	232_{82}	230_{87}	199_{12}	77_{100}	732	2548
N-Methyl-Brolamfetamine-M (demethyl-)		1800	273_1	230_6	105_3	77_7	44_{100}	732	5527
N-Methyl-Brolamfetamine-M (demethyl-) AC		2150	315_{15}	256_{100}	229_7	162_{22}	86_{71}	999	2549
N-Methyl-Brolamfetamine-M (demethyl-) AC		2150	315_3	256_{20}	162_4	86_{22}	44_{100}	998	5528
N-Methyl-Brolamfetamine-M (demethyl-) formyl artifact		1790	285_1	254_{15}	229_5	199_3	56_{100}	807	3242
N-Methyl-Brolamfetamine-M (demethyl-) HFB		1945	469_{24}	256_{90}	240_{55}	229_{100}	199_{29}	1616	6008

Table 8.1: Compounds in order of names   N-Methyl-Brolamfetamine-M (demethyl-) PFP

Name	Detected	RI	Typical ions and intensities					Page	Entry
N-Methyl-Brolamfetamine-M (demethyl-) PFP		1905	419 $_{22}$	256 $_{73}$	229 $_{100}$	190 $_{59}$	119 $_{93}$	1504	6007
N-Methyl-Brolamfetamine-M (demethyl-) TFA		1935	369 $_{28}$	256 $_{81}$	229 $_{100}$	199 $_{40}$	69 $_{88}$	1304	6006
N-Methyl-Brolamfetamine-M (demethyl-) TMS		1920	345 $_{1}$	272 $_{2}$	229 $_{2}$	116 $_{100}$	73 $_{79}$	1182	6009
N-Methyl-Brolamfetamine-M (demethyl-deamino-HO-) AC	U+UHYAC	1950*	316 $_{7}$	274 $_{23}$	214 $_{100}$	186 $_{18}$		1003	7061
N-Methyl-Brolamfetamine-M (demethyl-deamino-oxo-)	U+UHYAC	1835*	272 $_{59}$	229 $_{100}$				727	7062
N-Methyl-Brolamfetamine-M (demethyl-HO-) 2AC	U+UHYAC	2270	373 $_{3}$	313 $_{14}$	271 $_{37}$	86 $_{100}$		1325	7081
N-Methyl-Brolamfetamine-M (demethyl-HO-) -H2O	U+UHYAC	1960*	273 $_{72}$	271 $_{68}$	258 $_{100}$	256 $_{98}$		721	7073
N-Methyl-Brolamfetamine-M (demethyl-HO-) -H2O AC	U+UHYAC	2130*	313 $_{24}$	271 $_{79}$	256 $_{100}$			985	7074
N-Methyl-Brolamfetamine-M (HO-) 2AC	U+UHYAC	2350	387 $_{4}$	314 $_{29}$	242 $_{57}$	100 $_{59}$	58 $_{100}$	1388	7059
N-Methyl-Brolamfetamine-M (N,O-bis-demethyl-) isomer-1 2AC	U+UHYAC	2235	343 $_{12}$	301 $_{23}$	284 $_{57}$	242 $_{100}$	86 $_{81}$	1172	7065
N-Methyl-Brolamfetamine-M (N,O-bis-demethyl-) isomer-1 AC	U+UHYAC	2120	301 $_{18}$	242 $_{100}$	215 $_{11}$	185 $_{13}$	86 $_{20}$	910	7070
N-Methyl-Brolamfetamine-M (N,O-bis-demethyl-) isomer-2 2AC	U+UHYAC	2275	343 $_{2}$	284 $_{56}$	242 $_{100}$	215 $_{13}$	86 $_{23}$	1171	7066
N-Methyl-Brolamfetamine-M (N,O-bis-demethyl-) isomer-2 AC	U+UHYAC	2180	301 $_{29}$	242 $_{100}$	215 $_{14}$	86 $_{36}$		910	7071
N-Methyl-Brolamfetamine-M (N,O-bis-demethyl-deamino-oxo-) AC	U+UHYAC	1930*	300 $_{8}$	258 $_{94}$	215 $_{100}$			904	7063
N-Methyl-Brolamfetamine-M (N,O-bis-demethyl-deamino-oxo-) isomer-1	U+UHYAC	1870*	260 $_{66}$	258 $_{72}$	217 $_{99}$	215 $_{100}$		644	7068
N-Methyl-Brolamfetamine-M (N,O-bis-demethyl-deamino-oxo-) isomer-2	U+UHYAC	1885*	260 $_{99}$	258 $_{100}$	217 $_{97}$	215 $_{93}$		644	7069
N-Methyl-Brolamfetamine-M (N,O-bis-demethyl-HO-) 3AC	U+UHYAC	2385	401 $_{5}$	359 $_{49}$	317 $_{74}$	258 $_{100}$	86 $_{71}$	1447	7067
N-Methyl-Brolamfetamine-M (N,O-bis-demethyl-HO-) -H2O 2AC	U+UHYAC	2280	341 $_{32}$	299 $_{62}$	257 $_{100}$	242 $_{72}$		1159	7072
N-Methyl-Brolamfetamine-M (N,O-bis-demethyl-HO-deamino-oxo-) 3AC	U+UHYAC	2145*	402 $_{8}$	360 $_{39}$	315 $_{52}$	300 $_{34}$	231 $_{100}$	1451	7064
N-Methyl-Brolamfetamine-M (O,O-bis-demethyl-) 3AC	U+UHYAC	2330	385 $_{4}$	270 $_{8}$	228 $_{15}$	100 $_{100}$	58 $_{70}$	1379	7058
N-Methyl-Brolamfetamine-M (O-demethyl-) isomer-1 2AC	U+UHYAC	2285	357 $_{6}$	284 $_{24}$	242 $_{17}$	100 $_{83}$	58 $_{100}$	1245	7056
N-Methyl-Brolamfetamine-M (O-demethyl-) isomer-2 2AC	U+UHYAC	2295	357 $_{4}$	284 $_{12}$	242 $_{17}$	100 $_{90}$	58 $_{100}$	1246	7057
N-Methyl-Brolamfetamine-M (O-demethyl-HO-) 3AC	U+UHYAC	2430	415 $_{1}$	373 $_{9}$	258 $_{13}$	100 $_{74}$	58 $_{100}$	1493	7060
N-Methyl-Brolamfetamine-M (tri-demethyl-) 3AC	U+UHYAC	2325	371 $_{2}$	329 $_{35}$	287 $_{54}$	228 $_{100}$	86 $_{44}$	1315	7075
N-Methyl-Brolamfetamine-M (tri-demethyl-) artifact 2AC	U+UHYAC	2225	311 $_{23}$	269 $_{100}$	227 $_{64}$	212 $_{40}$	133 $_{18}$	974	7184
4-Methyl-buphedrone		1670	162 $_{2}$	119 $_{5}$	91 $_{11}$	72 $_{100}$	65 $_{7}$	372	9579
4-Methyl-buphedrone AC		2000	233 $_{2}$	162 $_{2}$	119 $_{20}$	114 $_{70}$	72 $_{100}$	522	9884
4-Methyl-buphedrone HFB		1745	387 $_{1}$	268 $_{52}$	210 $_{19}$	119 $_{100}$	91 $_{33}$	1390	9887
4-Methyl-buphedrone PFP		1720	337 $_{1}$	218 $_{37}$	190 $_{2}$	119 $_{100}$	91 $_{24}$	1135	9886
4-Methyl-buphedrone TFA		1735	287 $_{1}$	168 $_{24}$	119 $_{100}$	110 $_{10}$	91 $_{24}$	819	9885
2-Methylbutane		<1000*	72 $_{7}$	57 $_{57}$	43 $_{100}$	41 $_{95}$	29 $_{44}$	251	3811
2-Methyl-butene		<1000*	70 $_{35}$	55 $_{100}$	42 $_{33}$	39 $_{42}$	29 $_{28}$	250	3814
4-Methylcatechol		1155*	124 $_{100}$	106 $_{15}$	95 $_{11}$	78 $_{84}$	51 $_{31}$	268	5762
4-Methylcatechol 2AC	UHYAC	1450*	208 $_{2}$	166 $_{13}$	124 $_{100}$	106 $_{6}$	78 $_{13}$	423	2451
4-Methylcatechol 2HFB		1165*	516 $_{52}$	319 $_{35}$	263 $_{49}$	169 $_{61}$	69 $_{100}$	1673	5991
4-Methylcatechol 2PFP		<1000*	416 $_{62}$	269 $_{42}$	253 $_{36}$	213 $_{47}$	119 $_{100}$	1494	5989
4-Methylcatechol HFB		1035*	320 $_{50}$	169 $_{28}$	151 $_{25}$	123 $_{100}$	95 $_{50}$	1032	5990
4-Methylcatechol PFP		1035*	270 $_{64}$	151 $_{27}$	123 $_{100}$	95 $_{68}$	77 $_{39}$	714	5988
4-Methylcatechol TFA		<1000*	220 $_{53}$	151 $_{13}$	123 $_{100}$	95 $_{38}$	69 $_{56}$	468	5987
Methylcybin		2100	218 $_{24}$	159 $_{4}$	146 $_{12}$	117 $_{5}$	72 $_{100}$	460	9555
Methylcybin 2TMS		2060	362 $_{15}$	347 $_{6}$	290 $_{4}$	216 $_{4}$	72 $_{100}$	1276	9560
Methylcybin AC		2190	260 $_{4}$	160 $_{5}$	146 $_{10}$	117 $_{5}$	72 $_{100}$	659	9556
Methylcybin HFB		2225	414 $_{5}$	356 $_{5}$	342 $_{25}$	145 $_{33}$	72 $_{100}$	1489	9559
Methylcybin PFP		2225	364 $_{2}$	306 $_{4}$	292 $_{11}$	145 $_{22}$	72 $_{100}$	1281	9558
Methylcybin TFA		2185	314 $_{10}$	267 $_{4}$	242 $_{6}$	186 $_{12}$	72 $_{100}$	994	9557
N-Methylcytisine		1995	204 $_{16}$	160 $_{5}$	146 $_{10}$	117 $_{7}$	58 $_{100}$	411	5597
4-Methyldibenzofuran		1620*	182 $_{100}$	181 $_{91}$	152 $_{14}$	127 $_{3}$	91 $_{6}$	351	2561
N-Methyl-DOB		1885	230 $_{2}$	199 $_{2}$	143 $_{2}$	77 $_{5}$	58 $_{100}$	818	6429
N-Methyl-DOB AC	U+UHYAC	2225	329 $_{1}$	256 $_{15}$	199 $_{3}$	100 $_{57}$	58 $_{100}$	1086	6430
N-Methyl-DOB-M (demethyl-)		1800	273 $_{4}$	232 $_{82}$	230 $_{87}$	199 $_{12}$	77 $_{100}$	732	2548
N-Methyl-DOB-M (demethyl-)		1800	273 $_{1}$	230 $_{6}$	105 $_{3}$	77 $_{7}$	44 $_{100}$	732	5527
N-Methyl-DOB-M (demethyl-) AC		2150	315 $_{15}$	256 $_{100}$	271 $_{?}$	162 $_{2}$	86 $_{71}$	999	2549
N-Methyl-DOB-M (demethyl-) AC		2150	315 $_{3}$	256 $_{20}$	162 $_{4}$	86 $_{22}$	44 $_{100}$	998	5528
N-Methyl-DOB-M (demethyl-) formyl artifact		1790	285 $_{3}$	254 $_{15}$	229 $_{5}$	199 $_{3}$	56 $_{100}$	807	3242
N-Methyl-DOB-M (demethyl-) HFB		1945	469 $_{24}$	256 $_{90}$	240 $_{55}$	229 $_{100}$	199 $_{29}$	1616	6008
N-Methyl-DOB-M (demethyl-) PFP		1905	419 $_{22}$	256 $_{73}$	229 $_{100}$	190 $_{59}$	119 $_{93}$	1504	6007
N-Methyl-DOB-M (demethyl-) TFA		1935	369 $_{28}$	256 $_{81}$	229 $_{100}$	199 $_{40}$	69 $_{88}$	1304	6006
N-Methyl-DOB-M (demethyl-) TMS		1920	345 $_{1}$	272 $_{2}$	229 $_{2}$	116 $_{100}$	73 $_{79}$	1182	6009
N-Methyl-DOB-M (demethyl-deamino-HO-) AC	U+UHYAC	1950*	316 $_{7}$	274 $_{23}$	214 $_{100}$	186 $_{18}$		1003	7061
N-Methyl-DOB-M (demethyl-deamino-oxo-)	U+UHYAC	1835*	272 $_{59}$	229 $_{100}$				727	7062
N-Methyl-DOB-M (demethyl-HO-) 2AC	U+UHYAC	2270	373 $_{3}$	313 $_{14}$	271 $_{37}$	86 $_{100}$		1325	7081
N-Methyl-DOB-M (demethyl-HO-) -H2O	U+UHYAC	1960*	273 $_{72}$	271 $_{68}$	258 $_{100}$	256 $_{98}$		721	7073
N-Methyl-DOB-M (demethyl-HO-) -H2O AC	U+UHYAC	2130*	313 $_{24}$	271 $_{79}$	256 $_{100}$			985	7074
N-Methyl-DOB-M (HO-) 2AC	U+UHYAC	2350	387 $_{4}$	314 $_{29}$	242 $_{57}$	100 $_{59}$	58 $_{100}$	1388	7059
N-Methyl-DOB-M (N,O-bis-demethyl-) isomer-1 2AC	U+UHYAC	2235	343 $_{12}$	301 $_{23}$	284 $_{57}$	242 $_{100}$	86 $_{81}$	1172	7065
N-Methyl-DOB-M (N,O-bis-demethyl-) isomer-1 AC	U+UHYAC	2120	301 $_{18}$	242 $_{100}$	215 $_{11}$	185 $_{13}$	86 $_{20}$	910	7070
N-Methyl-DOB-M (N,O-bis-demethyl-) isomer-2 2AC	U+UHYAC	2275	343 $_{2}$	284 $_{56}$	242 $_{100}$	215 $_{13}$	86 $_{23}$	1171	7066
N-Methyl-DOB-M (N,O-bis-demethyl-) isomer-2 AC	U+UHYAC	2180	301 $_{29}$	242 $_{100}$	215 $_{14}$	86 $_{36}$		910	7071
N-Methyl-DOB-M (N,O-bis-demethyl-deamino-oxo-) AC	U+UHYAC	1930*	300 $_{8}$	258 $_{94}$	215 $_{100}$			904	7063
N-Methyl-DOB-M (N,O-bis-demethyl-deamino-oxo-) isomer-1	U+UHYAC	1870*	260 $_{66}$	258 $_{72}$	217 $_{99}$	215 $_{100}$		644	7068
N-Methyl-DOB-M (N,O-bis-demethyl-deamino-oxo-) isomer-2	U+UHYAC	1885*	260 $_{99}$	258 $_{100}$	217 $_{97}$	215 $_{93}$		644	7069
N-Methyl-DOB-M (N,O-bis-demethyl-HO-) 3AC	U+UHYAC	2385	401 $_{5}$	359 $_{49}$	317 $_{74}$	258 $_{100}$	86 $_{71}$	1447	7067
N-Methyl-DOB-M (N,O-bis-demethyl-HO-) -H2O 2AC	U+UHYAC	2280	341 $_{32}$	299 $_{62}$	257 $_{100}$	242 $_{72}$		1159	7072
N-Methyl-DOB-M (N,O-bis-demethyl-HO-deamino-oxo-) 3AC	U+UHYAC	2145*	402 $_{8}$	360 $_{39}$	315 $_{52}$	300 $_{34}$	231 $_{100}$	1451	7064

Methylpyrrolidinobutyrophenone impurity-1

Table 8.1: Compounds in order of names

Name	Detected	RI	Typical ions and intensities					Page	Entry
Methylpyrrolidinobutyrophenone impurity-1		1760	227 $_{23}$	119 $_{28}$	108 $_{100}$	91 $_{20}$	80 $_{25}$	499	6991
Methylpyrrolidinobutyrophenone impurity-2		1820	229 $_{16}$	145 $_{10}$	110 $_{100}$	91 $_{26}$	70 $_{64}$	506	6992
Methylpyrrolidinobutyrophenone-M (carboxy-) ET		2210	260 $_{1}$	177 $_{2}$	149 $_{6}$	112 $_{100}$	70 $_{11}$	836	6994
Methylpyrrolidinobutyrophenone-M (carboxy-) ME		2080	163 $_{4}$	135 $_{3}$	112 $_{100}$	104 $_{6}$	70 $_{9}$	747	7001
Methylpyrrolidinobutyrophenone-M (carboxy-) TMS		2220	318 $_{1}$	221 $_{1}$	178 $_{3}$	112 $_{100}$	104 $_{4}$	1113	7005
Methylpyrrolidinobutyrophenone-M (carboxy-deamino-oxo-) ET		1720*	234 $_{4}$	189 $_{7}$	177 $_{100}$	149 $_{33}$	104 $_{13}$	526	6995
Methylpyrrolidinobutyrophenone-M (carboxy-deamino-oxo-) ME		1650*	220 $_{6}$	163 $_{100}$	135 $_{19}$	120 $_{6}$	104 $_{12}$	468	7002
Methylpyrrolidinobutyrophenone-M (carboxy-dihydro-) 2TMS		2140	392 $_{1}$	280 $_{1}$	178 $_{6}$	163 $_{5}$	112 $_{100}$	1469	7006
Methylpyrrolidinobutyrophenone-M (carboxy-oxo-) ET		2390	303 $_{1}$	258 $_{2}$	149 $_{6}$	126 $_{100}$	104 $_{10}$	924	6996
Methylpyrrolidinobutyrophenone-M (carboxy-oxo-) ME		2280	289 $_{3}$	258 $_{4}$	163 $_{5}$	126 $_{100}$	104 $_{7}$	834	6998
Methylpyrrolidinobutyrophenone-M (carboxy-oxo-) TMS		2400	347 $_{1}$	332 $_{6}$	221 $_{3}$	178 $_{8}$	126 $_{100}$	1195	7003
Methylpyrrolidinobutyrophenone-M (carboxy-oxo-dihydro-) 2TMS		2430	406 $_{10}$	332 $_{3}$	280 $_{3}$	178 $_{5}$	126 $_{100}$	1511	7004
Methylpyrrolidinobutyrophenone-M (carboxy-oxo-dihydro-) ET		2470	305 $_{1}$	260 $_{2}$	142 $_{11}$	126 $_{100}$	98 $_{6}$	937	6997
Methylpyrrolidinobutyrophenone-M (carboxy-oxo-dihydro-) ETAC		2545	268 $_{2}$	226 $_{4}$	179 $_{2}$	126 $_{100}$	98 $_{5}$	1195	7054
Methylpyrrolidinobutyrophenone-M (carboxy-oxo-dihydro-) ME		2350	260 $_{2}$	165 $_{1}$	126 $_{100}$	98 $_{9}$	69 $_{13}$	847	6999
Methylpyrrolidinobutyrophenone-M (HO-) AC		2170	238 $_{1}$	177 $_{3}$	112 $_{100}$	89 $_{3}$	70 $_{4}$	836	7024
Methylpyrrolidinobutyrophenone-M (HO-) TMS		2145	319 $_{1}$	304 $_{1}$	178 $_{2}$	112 $_{100}$	104 $_{2}$	1029	7055
Methylpyrrolidinobutyrophenone-M (oxo-)		2010	245 $_{2}$	162 $_{8}$	126 $_{100}$	119 $_{7}$	91 $_{11}$	576	6993
3-Methyl-rolicyclidine		2115	243 $_{40}$	200 $_{100}$	186 $_{36}$	152 $_{39}$	70 $_{19}$	569	10184
3-Methyl-rolicyclidine artifact		1500*	172 $_{100}$	157 $_{66}$	143 $_{35}$	129 $_{81}$	115 $_{32}$	327	9454
Methylsalicylate	P U+UHYAC	1200*	152 $_{39}$	120 $_{94}$	92 $_{100}$	65 $_{53}$		292	954
Methylstearate	G P	2130*	298 $_{2}$	255 $_{3}$	143 $_{11}$	87 $_{65}$	74 $_{100}$	899	970
17-Methyltestosterone		2645*	302 $_{100}$	229 $_{37}$	161 $_{36}$	124 $_{80}$	91 $_{40}$	921	3894
17-Methyltestosterone AC		2770*	344 $_{23}$	302 $_{100}$	284 $_{78}$	269 $_{63}$	91 $_{64}$	1181	3920
17-Methyltestosterone enol 2TMS		2665*	446 $_{44}$	356 $_{5}$	301 $_{33}$	143 $_{9}$	73 $_{100}$	1574	3979
17-Methyltestosterone TMS		2590*	374 $_{73}$	302 $_{100}$	229 $_{30}$	124 $_{85}$	79 $_{80}$	1332	3927
Methylthalidomide		2470	272 $_{43}$	229 $_{57}$	130 $_{40}$	104 $_{74}$	76 $_{100}$	729	2114
Methylthalidomide ME		2330	286 $_{91}$	255 $_{38}$	213 $_{45}$	130 $_{100}$	102 $_{76}$	814	2082
4-Methylthio-amfetamine		1300	181 $_{1}$	138 $_{100}$	122 $_{43}$	91 $_{37}$	78 $_{36}$	348	5941
4-Methylthio-amfetamine		1300	181 $_{1}$	138 $_{36}$	122 $_{15}$	91 $_{13}$	44 $_{100}$	349	5942
4-Methylthio-amfetamine 2AC		1760	265 $_{1}$	164 $_{100}$	137 $_{28}$	122 $_{14}$	86 $_{43}$	686	5940
4-Methylthio-amfetamine AC		1700	223 $_{5}$	164 $_{100}$	137 $_{22}$	122 $_{13}$	86 $_{26}$	483	5717
4-Methylthio-amfetamine derivative ME		1940	239 $_{11}$	164 $_{12}$	138 $_{20}$	102 $_{100}$	58 $_{24}$	552	5719
4-Methylthio-amfetamine formyl artifact		1560	193 $_{14}$	137 $_{34}$	122 $_{8}$	78 $_{4}$	56 $_{100}$	378	5718
4-Methylthio-amfetamine HFB		1775	377 $_{19}$	240 $_{14}$	164 $_{69}$	137 $_{100}$	69 $_{24}$	1342	5743
4-Methylthio-amfetamine PFP		1760	327 $_{5}$	190 $_{8}$	164 $_{44}$	137 $_{100}$	122 $_{17}$	1073	5744
4-Methylthio-amfetamine TFA		1750	277 $_{7}$	164 $_{40}$	137 $_{100}$	122 $_{18}$	69 $_{17}$	758	5720
4-Methylthio-amfetamine TMS		1750	238 $_{13}$	137 $_{14}$	116 $_{100}$	100 $_{13}$	73 $_{73}$	623	5721
4-Methylthio-amfetamine-M (deamino-HO-) AC		1460*	224 $_{7}$	164 $_{100}$	137 $_{36}$	122 $_{15}$	117 $_{21}$	487	6898
4-Methylthio-amfetamine-M (deamino-HO-) PFP	U+UHYPFP	1560*	328 $_{45}$	191 $_{4}$	164 $_{100}$	137 $_{94}$	119 $_{26}$	1079	6952
4-Methylthio-amfetamine-M (deamino-oxo-)		1335*	180 $_{28}$	137 $_{100}$	122 $_{21}$			345	6899
4-Methylthio-amfetamine-M (HO-) formyl artifact 2AC	U+UHYAC	2240	251 $_{53}$	195 $_{8}$	152 $_{65}$	137 $_{100}$	122 $_{16}$	609	6902
4-Methylthio-amfetamine-M (HO-) isomer-1 2PFP	U+UHYPFP	1780	475 $_{5}$	326 $_{4}$	285 $_{100}$	256 $_{5}$	190 $_{5}$	1647	6949
4-Methylthio-amfetamine-M (HO-) isomer-2 2AC	U+UHYAC	2260	281 $_{2}$	222 $_{100}$	150 $_{50}$	123 $_{17}$	86 $_{37}$	787	6896
4-Methylthio-amfetamine-M (HO-) isomer-2 2PFP	U+UHYPFP	1790	475 $_{4}$	326 $_{3}$	285 $_{100}$	190 $_{6}$	152 $_{6}$	1647	6950
4-Methylthio-amfetamine-M (methylthiobenzoic acid)		1995*	168 $_{100}$	151 $_{48}$	135 $_{7}$	125 $_{13}$	108 $_{15}$	320	7313
4-Methylthio-amfetamine-M (methylthiobenzoic acid) ME		1610*	182 $_{36}$	151 $_{100}$	123 $_{21}$	108 $_{19}$	79 $_{16}$	350	6900
4-Methylthio-amfetamine-M (methylthiobenzoic acid) TMS		1770*	240 $_{39}$	225 $_{80}$	181 $_{87}$	151 $_{100}$	108 $_{34}$	554	6901
4-Methylthio-amfetamine-M (ring-HO-) 2AC	U+UHYAC	2240	281 $_{7}$	222 $_{100}$	180 $_{17}$	153 $_{92}$	86 $_{58}$	787	6895
4-Methylthio-amfetamine-M (ring-HO-) 2PFP	U+UHYPFP	1860	475 $_{18}$	326 $_{5}$	312 $_{100}$	285 $_{40}$	190 $_{16}$	1647	6951
4-Methylthio-amfetamine-M/artifact (sulfoxide) AC		2360	239 $_{1}$	222 $_{69}$	165 $_{52}$	137 $_{100}$	86 $_{17}$	552	6897
4-Methylthio-amfetamine-M/artifcat (sulfone) AC		2455	255 $_{3}$	196 $_{4}$	180 $_{10}$	107 $_{23}$	86 $_{100}$	631	6903
4-Methylthiobenzoic acid		1995*	168 $_{100}$	151 $_{48}$	135 $_{7}$	125 $_{13}$	108 $_{15}$	320	7313
Methylthionium chloride artifact		2680	285 $_{100}$	270 $_{38}$	254 $_{6}$	225 $_{5}$	142 $_{20}$	809	3387
N-Methyltrifluoroacetaminde		<1000	127 $_{14}$	78 $_{30}$	69 $_{66}$	58 $_{100}$		269	8629
N-Methyl-trimethylsilyl-trifluoroacetamide		<1000	199 $_{2}$	184 $_{14}$	134 $_{39}$	77 $_{100}$	73 $_{88}$	397	5694
N-Methyltryptamine 2HFB		1855	566 $_{3}$	339 $_{100}$	326 $_{64}$	240 $_{84}$	169 $_{76}$	1702	9546
N-Methyltryptamine 2PFP		1830	466 $_{1}$	289 $_{100}$	276 $_{72}$	190 $_{52}$	119 $_{37}$	1609	9543
N-Methyltryptamine 2TFA		1855	366 $_{2}$	239 $_{100}$	226 $_{73}$	156 $_{12}$	140 $_{63}$	1289	9541
N-Methyltryptamine AC		2210	216 $_{14}$	172 $_{2}$	143 $_{100}$	130 $_{92}$	115 $_{10}$	452	9540
N-Methyltryptamine HFB		1830	370 $_{3}$	240 $_{18}$	169 $_{20}$	143 $_{40}$	130 $_{100}$	1310	9545
N-Methyltryptamine PFP		1830	320 $_{15}$	190 $_{24}$	143 $_{33}$	130 $_{100}$	119 $_{24}$	1033	9544
N-Methyltryptamine TFA		1795	270 $_{25}$	143 $_{38}$	130 $_{100}$	115 $_{6}$	102 $_{8}$	717	9542
Methyprylone	P G U	1525	183 $_{2}$	155 $_{92}$	140 $_{98}$	98 $_{62}$	83 $_{100}$	355	1123
Methyprylone enol AC	UHYAC	1610	225 $_{11}$	183 $_{100}$	155 $_{74}$	127 $_{29}$	83 $_{13}$	492	112
Methyprylone-M (HO-) AC	UHYAC	1720	241 $_{1}$	213 $_{6}$	166 $_{22}$	153 $_{68}$	98 $_{100}$	561	115
Methyprylone-M (HO-) -H2O	U UHY	1540	181 $_{4}$	166 $_{47}$	153 $_{57}$	98 $_{29}$	83 $_{100}$	349	1124
Methyprylone-M (HO-) -H2O enol AC	UHYAC	1470	223 $_{2}$	195 $_{15}$	166 $_{87}$	153 $_{100}$	83 $_{80}$	484	123
Methyprylone-M (oxo-)	U UHY UHYAC	1870	197 $_{2}$	182 $_{7}$	168 $_{26}$	98 $_{53}$	83 $_{100}$	392	113
Metipranolol		2220	309 $_{1}$	294	265 $_{2}$	152 $_{18}$	72 $_{100}$	966	4257
Metipranolol 2AC		2670	393 $_{3}$	333 $_{13}$	200 $_{100}$	140 $_{80}$	98 $_{67}$	1417	1361
Metipranolol AC	UHYAC	2260	351 $_{13}$	336 $_{20}$	152 $_{50}$	98 $_{50}$	72 $_{100}$	1217	1600
Metipranolol formyl artifact		2240	321 $_{4}$	306 $_{16}$	127 $_{100}$	112 $_{87}$	86 $_{69}$	1042	1360
Metipranolol -H2O AC		2660	333 $_{24}$	248 $_{22}$	152 $_{32}$	140 $_{100}$	98 $_{59}$	1115	1388

Table 8.1: Compounds in order of names

Name	Detected	RI	Typical ions and intensities					Page	Entry
Metipranolol TMS		2260	366_3	308_2	265_5	152_{18}	72_{100}	1363	6176
Metipranolol TMSTFA		2395	477_1	284_{100}	242_{11}	129_{50}	73_{58}	1631	6175
Metipranolol-M (deamino-HO-) 2AC	U+UHYAC	2240*	352_6	310_2	159_{100}	152_{10}	99_{11}	1221	1599
Metipranolol-M/artifact (deacetyl-)		2190	267_1	223	152_{32}	116_{15}	72_{100}	703	4258
Metipranolol-M/artifact (phenol) AC	UHYAC	1610*	236_{15}	194_{19}	152_{100}			538	1598
Metixene	G U+UHYAC-I	2500	309_{23}	197_{40}	165_9	99_{100}		964	553
Metixene-M (nor-) AC	UHYAC	2960	337_8	197_{100}	165_{10}	152_3	112_2	1137	554
Metobromuron		2040	258_2	197_2	170_4	91_{11}	61_{100}	644	3887
Metobromuron ME		1735	272_{11}	212_{100}	184_{70}	105_{41}	76_{36}	728	3975
Metobromuron-M/artifact (HOOC-) ME		1800	229_{100}	197_{69}	170_{30}	91_{93}	63_{71}	503	3888
Metocin		2100	218_{24}	159_4	146_{12}	117_5	72_{100}	460	9555
Metocin 2TMS		2060	362_{15}	347_6	290_{24}	216_4	72_{100}	1276	9560
Metocin AC		2190	260_4	160_5	146_{10}	117_5	72_{100}	659	9556
Metocin HFB		2225	414_5	356_5	342_{25}	145_{33}	72_{100}	1489	9559
Metocin PFP		2225	364_2	306_4	292_{11}	145_{22}	72_{100}	1281	9558
Metocin TFA		2185	314_{10}	267_4	242_6	186_{12}	72_{100}	994	9557
Metoclopramide	P-I G UHY	2610	299_1	227_2	184_{12}	99_{24}	86_{100}	902	1125
Metoclopramide 2TMS		2400	443_1	428_3	414_5	256_{29}	86_{100}	1568	4569
Metoclopramide AC	PAC U+UHYAC	2735	341_1	269_2	226_4	184_5	86_{100}	1162	1126
Metoclopramide TMS		2655	371_1	273_{10}	256_{15}	99_{20}	86_{100}	1319	4615
Metoclopramide-M (deethyl-)	UHY	2095	71_{43}	58_{100}				723	1127
Metoclopramide-M (deethyl-) 2AC	U+UHYAC	2900	355_1	312_2	226_{28}	184_{20}	58_{100}	1236	1897
Metofenazate-M/artifact (deacyl-)	UHY-I	3360	403_{59}	372_{21}	246_{83}	70_{75}	42_{100}	1456	592
Metofenazate-M/artifact (deacyl-)		3360	403_{27}	246_{72}	171_{31}	143_{84}	70_{100}	1456	4252
Metofenazate-M/artifact (deacyl-) AC	UHYAC-I	3470	445_{33}	246_{41}	185_{51}	125_{60}	70_{100}	1572	373
Metofenazate-M/artifact (deacyl-) TMS		3340	475_{10}	372_{22}	246_{100}	232_{22}	73_{41}	1628	5444
Metofenazate-M/artifact (trimethoxybenzoic acid)		1780*	212_{100}	197_{57}	169_{13}	141_{27}		438	1949
Metofenazate-M/artifact (trimethoxybenzoic acid) ET		1770*	240_{100}	225_{44}	212_{17}	195_{45}	141_{24}	556	5219
Metofenazate-M/artifact (trimethoxybenzoic acid) ME		1740*	226_{100}	211_{48}	195_{22}	155_{21}		494	1950
Metolazone 2ME		3910	393_{15}	378_{100}	287_7	179_8	91_{16}	1413	3108
Metolazone 3ME		3780	407_3	392_{100}	284_{15}	249_4	118_9	1467	6891
Metolazone artifact 2ME		3245	391_{64}	376_{100}	283_{24}	268_{12}	91_{24}	1405	3110
Metolazone artifact ME		3310	377_{45}	362_{100}	282_{18}	267_{21}	91_{25}	1341	3109
Metonitazene		3350	382_3	380_5	352_3	121_{31}	86_{100}	1366	1128
Metoprolol	P-I G U UHY	2080	267_1	252_3	223_{10}	107_{15}	72_{100}	703	1129
Metoprolol 2AC	U+UHYAC	2480	351_1	291_6	200_{100}	98_{27}	72_{52}	1217	1133
Metoprolol 2TMS		2330	396_2	224_6	144_{100}	101_8	73_{52}	1484	4571
Metoprolol formyl artifact	P G U UHY	2120	279_{15}	264_{43}	127_{90}	112_{74}	56_{100}	779	1130
Metoprolol -H2O AC	U+UHYAC	2330	291_{13}	206_{56}	189_{60}	140_{78}	98_{100}	851	1134
Metoprolol TMS		2115	339_1	324_6	223_{11}	101_{11}	72_{100}	1153	4570
Metoprolol TMSTFA		2255	435_1	420_2	284_{100}	235_{19}	73_{34}	1548	6150
Metoprolol-M	U UHY	2200	295_1	280_2	251_4	107_7	72_{100}	873	1132
Metoprolol-M (HO-) 3AC	U+UHYAC	2730	349_3	200_{100}	140_{67}	72_{76}		1475	1136
Metoprolol-M (HO-) artifact	U	2240	295_{36}	280_{50}	250_{52}	128_{100}	56_{61}	880	1131
Metoprolol-M (O-demethyl-) 3AC	U+UHYAC	2620	319_{12}	200_{100}	140_{55}	98_{50}	72_{60}	1352	1585
Metoxuron artifact (HOOC-)		1810	183_{97}	168_{100}	140_{54}	112_{22}	76_{38}	401	2515
Metoxuron artifact (HOOC-) ME		1920	215_{100}	200_{52}	183_{54}	156_{46}	59_{83}	448	2516
Metoxuron ME		1855	242_{11}	170_7	155_4	85_{12}	72_{100}	564	4156
Metribuzin		1870	214_4	198_{11}	144_{13}	103_{13}	57_{21}	446	3859
Metronidazole	G P U	1725	171_{35}	124_{72}	81_{100}	54_{97}		325	1137
Metronidazole AC	U+UHYAC	1695	213_{21}	171_{37}	87_{100}			442	1138
Metronidazole TMS		1665	243_{10}	228_{47}	182_{50}	167_{62}	73_{100}	568	4572
Metronidazole-M (HO-methyl-)		2010	187_3	170_{100}	140_{73}	126_{22}	97_{23}	362	1830
Metronidazole-M (HO-methyl-) 2AC	U+UHYAC	1870	271_8	229_{23}	212_{11}	170_8	87_{100}	722	1832
Metronidazole-M (HO-methyl-) AC	U+UHYAC	1875	229_1	212_{15}	170_{14}	123_7	87_{100}	505	1831
Metronidazole-M (HOOC-) ME		1515	199_{38}	153_{82}	125_{15}	109_{18}	53_{100}	396	1833
Metyrapone		1930	226_{19}	120_{96}	106_{100}	92_{55}	78_{52}	495	5235
Mevinphos		1415*	224_1	192_{25}	164_7	127_{100}	109_{24}	486	4054
Mexazolam		2600	319_{79}	263_{73}	262_{100}	191_{44}	163_{38}	1272	4023
Mexazolam artifact AC		2550	361_{12}	261_{11}	191_6	163_5	101_{100}	1266	4024
Mexazolam HY	UHY	2180	265_{62}	230_{100}	139_{43}	111_{50}		684	543
Mexazolam HYAC	U+UHYAC	2300	307_{42}	265_{58}	230_{100}	139_{16}	111_{14}	946	290
Mexiletine		1425	179_8	122_8	105_8	91_{10}	58_{100}	342	1490
Mexiletine AC	U+UHYAC	1780	221_1	122_6	100_{100}	77_8	58_{62}	475	1491
Mexiletine-M (deamino-di-HO-) isomer-1 2AC	U+UHYAC	1910*	280_1	238	138_9	101_{100}	91_6	782	2899
Mexiletine-M (deamino-di-HO-) isomer-2 2AC	U+UHYAC	1930*	280_1	238	138_9	101_{100}	91_6	782	2900
Mexiletine-M (deamino-di-HO-) isomer-3 2AC	U+UHYAC	1940*	280_2	238	138_{22}	101_{100}	91_6	782	3042
Mexiletine-M (deamino-HO-) AC	U+UHYAC	1530*	222_1	122_{19}	101_{100}	91_9	77_{13}	480	3041
Mexiletine-M (deamino-oxo-)	U HY UHYAC	1350*	178_{75}	135_{40}	121_{28}	105_{100}	91_{40}	338	3040
Mexiletine-M (deamino-oxo-HO-) isomer-1 AC	UHYAC	1700*	236_7	194_{50}	176_{26}	136_{100}	121_{96}	539	2898
Mexiletine-M (deamino-oxo-HO-) isomer-2 AC	UHYAC	1735*	236_{21}	194_{90}	151_{49}	136_{100}	121_{39}	539	3044
Mexiletine-M (deamino-oxo-HO-) isomer-3 AC	UHYAC	1760*	236_{10}	194_{36}	137_{100}	121_{12}	91_3	538	3045
Mexiletine-M (HO-) isomer-1 2AC	U+UHYAC	2100	279_1	160_6	120_9	100_{100}	58_{50}	776	2901
Mexiletine-M (HO-) isomer-2 2AC	UHYAC	2180	279_1	178_2	138_{14}	100_{100}	58_{44}	775	3043

Moexipril-M/artifact (deethyl-) 3ME   Table 8.1: Compounds in order of names

Name	Detected	RI	Typical ions and intensities					Page	Entry
Moexipril-M/artifact (deethyl-) 3ME	UME	3580	512 $_1$	453 $_2$	305 $_{12}$	234 $_{100}$	190 $_{11}$	1671	4745
Moexipril-M/artifact (deethyl-) -H2O ME	UME	3775	466 $_{72}$	449 $_{49}$	330 $_{53}$	190 $_{100}$	91 $_{70}$	1610	4747
Moexipril-M/artifact (deethyl-) -H2O TMS	UTMS	3630	524 $_{49}$	509 $_{35}$	190 $_{85}$	91 $_{100}$	73 $_{48}$	1680	4981
Moexipril-M/artifact (deethyl-HOOC-) 2ME	UME	1870	279 $_2$	220 $_{100}$	160 $_{10}$	117 $_{28}$	91 $_{57}$	777	4734
Moexipril-M/artifact (deethyl-HOOC-) 3ME	UME	1935	293 $_2$	234 $_{100}$	174 $_8$	130 $_{16}$	91 $_{50}$	865	4735
Moexipril-M/artifact (HOOC-) 2ME	UME	1985	307 $_2$	248 $_{34}$	234 $_{100}$	174 $_5$	91 $_6$	952	4737
Moexipril-M/artifact (HOOC-) ET	UET	2025	307 $_2$	234 $_{100}$	160 $_{12}$	117 $_{17}$	91 $_{32}$	952	4740
Moexipril-M/artifact (HOOC-) ME	P-I UME	1930	293 $_3$	234 $_{36}$	220 $_{100}$	160 $_{11}$	91 $_{23}$	864	4736
Mofebutazone		2240	232 $_{96}$	189 $_{48}$	176 $_{49}$	108 $_{100}$	77 $_{98}$	517	2015
Mofebutazone 2AC	UHYAC	2220	316 $_{11}$	274 $_{63}$	232 $_{100}$	189 $_{95}$	108 $_{55}$	1005	2021
Mofebutazone 2ME		1960	260 $_{45}$	204 $_{58}$	121 $_{100}$	83 $_{33}$	77 $_{74}$	659	6403
Mofebutazone AC	UHYAC	2060	274 $_4$	232 $_{100}$	189 $_{24}$	176 $_{36}$	108 $_{48}$	740	2020
Mofebutazone-M (4-HO-) 2AC	UHYAC	2110	332 $_8$	290 $_{100}$	220 $_{83}$	125 $_{84}$	108 $_{80}$	1106	2017
Mofebutazone-M (4-HO-) 2ME		2075	276 $_{62}$	220 $_{22}$	121 $_{66}$	77 $_{100}$	71 $_{64}$	755	6404
Mofebutazone-M (4-HO-) AC	UHYAC	2210	290 $_{92}$	220 $_{92}$	125 $_{46}$	108 $_{100}$	57 $_{60}$	842	2016
Mofebutazone-M (4-HO-) ME		2065	262 $_{26}$	234 $_{10}$	122 $_{98}$	121 $_{100}$	77 $_{46}$	669	2036
Mofebutazone-M (HOOC-)		1930	206 $_{40}$	108 $_{100}$	92 $_{20}$	77 $_{28}$	65 $_{19}$	605	2019
Mofebutazone-M (HOOC-) 2ME		2100	278 $_{26}$	264 $_{16}$	232 $_{12}$	121 $_{100}$	105 $_{50}$	769	2023
Mofebutazone-M (HOOC-) -CO2	U+UHYAC	1600	206 $_4$	120 $_{14}$	99 $_{100}$	77 $_{78}$	71 $_{77}$	417	2018
Mofebutazone-M (HOOC-) ME		2070	264 $_{62}$	232 $_{26}$	204 $_{32}$	134 $_{100}$	108 $_{99}$	683	2022
Mofebutazone-M (HOOC-) MEAC		2250	306 $_1$	264 $_{100}$	232 $_{29}$	134 $_{69}$	108 $_{66}$	943	2024
Monalazone artifact 2ME		1920	229 $_{50}$	198 $_{48}$	135 $_{100}$	103 $_{40}$	76 $_{55}$	504	2479
Monalazone artifact 3ME	UME	1850	243 $_{54}$	199 $_{18}$	135 $_{100}$	104 $_{46}$	76 $_{51}$	567	2480
Monalide		1995	239 $_{15}$	197 $_{38}$	168 $_5$	127 $_{55}$	85 $_{100}$	552	2723
Monocrotophos		1665	223 $_3$	192 $_{13}$	127 $_{100}$	97 $_{25}$	67 $_{49}$	482	4132
Monocrotophos TFA		1540	319 $_1$	236 $_3$	193 $_{27}$	127 $_{100}$	67 $_{73}$	1023	4133
Monoisooctyladipate		2280*	259 $_7$	241 $_{15}$	147 $_{27}$	129 $_{100}$	57 $_{68}$	648	2360
Monolinuron		1910	214 $_9$	153 $_4$	126 $_{11}$	99 $_{10}$	61 $_{100}$	445	3889
Monolinuron ME		1675	228 $_{12}$	168 $_{100}$	140 $_{69}$	111 $_{21}$	77 $_{21}$	501	3976
Monolinuron-M/artifact (HOOC-) ME		1690	185 $_{100}$	153 $_{90}$	140 $_{63}$	126 $_{38}$	99 $_{53}$	358	3890
Monuron ME		1610	212 $_6$	140 $_4$	111 $_4$	72 $_{100}$		438	3942
Moperone	UHY UHYAC	2800	355 $_3$	337 $_3$	217 $_{89}$	204 $_{100}$	123 $_{28}$	1239	177
Moperone -H2O	UHY UHYAC	2710	337 $_{21}$	199 $_{15}$	186 $_{34}$	172 $_{100}$	123 $_{29}$	1140	178
Moperone-M	U+UHYAC	1490*	180 $_{25}$	125 $_{49}$	123 $_{35}$	95 $_{17}$	56 $_{100}$	345	85
Moperone-M	U UHY UHYAC	3110	329 $_{42}$	234 $_{30}$	185 $_{44}$	123 $_{100}$		1084	556
Moperone-M (N-dealkyl-) -H2O AC	UHYAC	2105	215 $_{100}$	173 $_{35}$				449	559
Moperone-M (N-dealkyl-oxo-) -2H2O	U UHY UHYAC	1600	169 $_{100}$	91 $_{22}$				322	163
Moperone-M (N-dealkyl-oxo-HO-) -2H2O	UHY	1875	185 $_{100}$	156 $_{76}$				359	555
Moperone-M (N-dealykl-oxo-HO-) -2H2O AC	UHYAC	2055	227 $_{76}$	185 $_{100}$				498	558
MOPPP		1705	233 $_1$	135 $_7$	98 $_{100}$	92 $_{12}$	77 $_{13}$	523	6547
MOPPP-M (deamino-oxo-)		1440*	178 $_3$	135 $_{100}$	107 $_{12}$	92 $_{12}$	77 $_{22}$	337	6540
MOPPP-M (demethyl-)		2010	219 $_1$	121 $_3$	98 $_{100}$	69 $_3$	56 $_{11}$	465	6545
MOPPP-M (demethyl-) ET		1955	247 $_1$	149 $_2$	121 $_5$	98 $_{100}$	69 $_4$	586	6543
MOPPP-M (demethyl-) HFB		1805	317 $_4$	169 $_7$	98 $_{100}$	69 $_{14}$		1493	6544
MOPPP-M (demethyl-) TMS		2005	276 $_2$	193 $_2$	135 $_3$	98 $_{100}$	73 $_4$	850	6776
MOPPP-M (demethyl-3-HO-) 2ET		2165	290 $_1$	193 $_1$	165 $_2$	137 $_3$	98 $_{100}$	851	6525
MOPPP-M (demethyl-3-methoxy-) ET		2135	277 $_1$	208 $_1$	179 $_2$	151 $_7$	98 $_{100}$	762	6524
MOPPP-M (demethyl-3-methoxy-) HFB		1960	445 $_1$	347 $_4$	98 $_{100}$	69 $_{19}$		1571	6532
MOPPP-M (demethyl-3-methoxy-) ME		2070	165 $_1$	98 $_{100}$	79 $_3$	69 $_4$	56 $_{11}$	676	6538
MOPPP-M (demethyl-3-methoxy-) TMS		1960	321 $_1$	306 $_3$	223 $_3$	165 $_1$	98 $_{100}$	1041	6533
MOPPP-M (demethyl-3-methoxy-deamino-oxo-) ET		1680*	222 $_4$	179 $_{100}$	151 $_{81}$	123 $_{19}$		479	6523
MOPPP-M (demethyl-deamino-oxo-) ET		1530*	192 $_3$	149 $_{100}$	121 $_{95}$	93 $_{25}$	65 $_{21}$	375	6539
MOPPP-M (dihydro-)		1935	234 $_1$	135 $_3$	98 $_{100}$	77 $_9$	56 $_{21}$	535	6697
MOPPP-M (dihydro-) AC		1970	218 $_3$	135 $_9$	98 $_{100}$	77 $_7$	56 $_{18}$	763	6702
MOPPP-M (dihydro-) TMS		1880	292 $_2$	218 $_1$	135 $_2$	98 $_{100}$	73 $_{17}$	953	6707
MOPPP-M (oxo-)		2120	164 $_{15}$	135 $_{15}$	121 $_{14}$	112 $_{100}$		585	6542
MOPPP-M (parahydroxybenzoic acid) 2ET		1520*	194 $_{17}$	166 $_7$	149 $_{100}$	121 $_{76}$	93 $_{17}$	384	6646
MOPPP-M (parahydroxybenzoic acid) ET		1585*	166 $_{53}$	151 $_4$	138 $_{65}$	121 $_{100}$	93 $_{10}$	316	6541
Morazone		----	377 $_3$	201 $_{100}$	176 $_{36}$	56 $_{64}$		1344	1226
Morazone-M (carboxy-phenazone) -CO2	P G U+UHYAC	1845	188 $_{100}$	96 $_{81}$	77 $_{51}$			364	199
Morazone-M/artifact (HO-methoxy-phenmetrazine)	UHY	1900	223 $_6$	151 $_3$	107 $_5$	71 $_{100}$	56 $_5$	484	3518
Morazone-M/artifact (HO-methoxy-phenmetrazine) 2AC	U+UHYAC	2320	307 $_6$	265 $_{22}$	113 $_{86}$	86 $_{24}$	71 $_{100}$	950	1887
Morazone-M/artifact (HO-phenmetrazine) isomer-1	UHY	1830	193 $_{11}$	121 $_7$	107 $_6$	71 $_{100}$	56 $_{62}$	380	562
Morazone-M/artifact (HO-phenmetrazine) isomer-1 2AC	UHYAC	2150	277 $_9$	234 $_{17}$	113 $_{55}$	85 $_{26}$	71 $_{100}$	761	849
Morazone-M/artifact (HO-phenmetrazine) isomer-2	UHY	1865	193 $_6$	163 $_6$	121 $_6$	71 $_{100}$	56 $_{43}$	379	3517
Morazone-M/artifact (HO-phenmetrazine) isomer-2 2AC	UHYAC	2200	277 $_3$	234 $_8$	113 $_{83}$	85 $_{39}$	71 $_{100}$	761	848
Morazone-M/artifact (phenmetrazine)	U UHY	1440	177 $_8$	105 $_5$	77 $_{10}$	71 $_{100}$	56 $_{54}$	336	851
Morazone-M/artifact (phenmetrazine) AC	U+UHYAC	1810	219 $_5$	176 $_{10}$	113 $_{77}$	86 $_{50}$	71 $_{100}$	465	198
Morazone-M/artifact (phenmetrazine) TFA		1530	273 $_4$	167 $_{85}$	105 $_{36}$	98 $_{47}$	70 $_{100}$	733	4002
Morazone-M/artifact (phenmetrazine) TMS		1620	249 $_6$	143 $_{19}$	115 $_{36}$	100 $_{100}$	73 $_{64}$	600	5446
Morazone-M/artifact-1	UHY UHYAC	1670	204 $_{72}$	176 $_{78}$	92 $_{100}$	77 $_{17}$	65 $_{18}$	408	560
Morazone-M/artifact-2	UHY UHYAC	1680	188 $_{100}$	159 $_8$	91 $_{30}$	77 $_{65}$	55 $_{47}$	364	561
Morazone-M/artifact-2 AC	U+UHYAC	1690	230 $_7$	188 $_{100}$	159 $_6$	91 $_{11}$	77 $_{55}$	509	3520
Morazone-M/artifact-3	UHY	1920	202 $_{64}$	110 $_{72}$	82 $_{23}$	77 $_{52}$	56 $_{100}$	404	3519

Table 8.1: Compounds in order of names   Morphine

Name	Detected	RI	Typical ions and intensities					Page	Entry
Morphine	G UHY	2455	285_{100}	268_{15}	162_{59}	124_{21}		810	474
Morphine 2AC	G PHYAC U+UHYAC	2620	369_{59}	327_{100}	310_{36}	268_{47}	162_{11}	1306	225
Morphine 2HFB		2375	677_{10}	480_{10}	464_{100}	407_{9}	169_{8}	1726	6120
Morphine 2PFP		2360	577_{51}	558_{7}	430_{8}	414_{100}	119_{22}	1706	2251
Morphine 2TFA		2250	477_{71}	364_{100}	307_{6}	115_{8}	69_{31}	1630	4008
Morphine 2TMS	UHYTMS	2560	429_{19}	236_{21}	196_{15}	146_{21}	73_{100}	1531	2463
Morphine Cl-artifact 2AC	UHYAC	2680	403_{59}	361_{100}	344_{63}	302_{90}	204_{55}	1456	2992
Morphine ME	P G U UHY	2375	299_{100}	229_{26}	162_{46}	124_{23}		902	473
Morphine TFA		2285	381_{55}	268_{100}	146_{14}	115_{13}	69_{23}	1360	5569
Morphine-D3 2AC		2510	372_{87}	330_{100}	313_{57}	271_{87}	218_{56}	1324	7294
Morphine-D3 2HFB		2375	680_{4}	483_{9}	467_{100}	414_{7}	169_{23}	1727	6126
Morphine-D3 2PFP		2350	580_{16}	433_{7}	417_{100}	269_{5}	119_{8}	1708	5567
Morphine-D3 2TFA		2240	480_{32}	383_{6}	367_{100}	314_{6}	307_{6}	1634	5571
Morphine-D3 2TMS		2550	432_{91}	290_{27}	239_{60}	199_{40}	73_{100}	1540	5578
Morphine-D3 ME		2370	302_{100}	232_{26}	165_{47}	127_{24}		920	7295
Morphine-D3 MEAC		2495	344_{100}	301_{9}	285_{75}	232_{38}	165_{20}	1180	7300
Morphine-D3 MEHFB		2310	498_{33}	438_{3}	285_{100}	269_{8}	225_{9}	1657	9333
Morphine-D3 MEPFP		2420	448_{45}	388_{4}	285_{100}	269_{9}	225_{10}	1576	9332
Morphine-D3 TFA		2275	384_{39}	271_{100}	211_{8}	165_{6}	152_{7}	1376	5572
Morphine-M (nor-) 2PFP		2440	563_{10}	400_{10}	355_{38}	327_{7}	209_{15}	1701	3534
Morphine-M (nor-) 3AC	U+UHYAC	2955	397_{8}	355_{9}	209_{41}	87_{100}	72_{33}	1434	1194
Morphine-M (nor-) 3PFP	UHYPFP	2405	709_{80}	533_{28}	388_{29}	367_{51}	355_{100}	1729	3533
Morphine-M (nor-) 3TMS	UHYTMS	2605	487_{17}	416_{19}	222_{36}	131_{19}	73_{100}	1645	3525
Morpholine		<1000	87_{54}	57_{100}	42_{27}			255	3612
N-Morpholinyllysergamide		3480	337_{100}	294_{6}	221_{73}	207_{47}	181_{50}	1140	9870
N-Morpholinyllysergamide TMS		3535	409_{75}	351_{19}	293_{90}	279_{62}	253_{100}	1475	9869
Moxaverine	P U+UHYAC	2530	307_{61}	292_{100}	248_{11}	91_{10}		950	1493
Moxaverine-M (O-demethyl-) isomer-1	UHY	2560	293_{29}	278_{100}	250_{43}	232_{8}	139_{8}	863	3215
Moxaverine-M (O-demethyl-) isomer-1 AC	UHYAC	2610	335_{26}	320_{16}	292_{30}	278_{100}	250_{39}	1125	3219
Moxaverine-M (O-demethyl-) isomer-2	UHY	2645	293_{40}	292_{100}	276_{11}	248_{5}	204_{2}	863	3216
Moxaverine-M (O-demethyl-) isomer-2 AC	UHYAC	2630	335_{6}	292_{100}	276_{29}	248_{5}	204_{3}	1124	3220
Moxaverine-M (O-demethyl-di-HO-) isomer-1 3AC	UHYAC	2910	451_{33}	393_{55}	336_{100}	306_{62}	290_{47}	1582	3231
Moxaverine-M (O-demethyl-di-HO-) isomer-2 3AC	UHYAC	3075	451_{20}	408_{58}	392_{98}	349_{84}	306_{100}	1582	3233
Moxaverine-M (O-demethyl-di-HO-methoxy-) 3AC	UHYAC	3530	481_{18}	438_{43}	422_{52}	379_{100}	364_{42}	1636	3235
Moxaverine-M (O-demethyl-HO-ethyl-) -H2O isomer-1	UHY	2625	291_{49}	290_{38}	276_{100}	248_{63}	230_{17}	847	3217
Moxaverine-M (O-demethyl-HO-ethyl-) -H2O isomer-1 AC	UHYAC	2660	333_{53}	318_{19}	290_{31}	276_{100}	248_{44}	1111	3221
Moxaverine-M (O-demethyl-HO-ethyl-) -H2O isomer-2	UHY	2710	291_{45}	290_{100}	274_{14}	246_{7}	230_{6}	847	3218
Moxaverine-M (O-demethyl-HO-ethyl-) -H2O isomer-2 AC	UHYAC	2680	333_{10}	290_{100}	274_{24}	246_{8}	230_{8}	1111	3222
Moxaverine-M (O-demethyl-HO-ethyl-) isomer-1 2AC	UHYAC	2815	393_{52}	378_{29}	350_{73}	308_{100}	276_{58}	1415	3227
Moxaverine-M (O-demethyl-HO-ethyl-) isomer-1 AC	UHYAC	2760	351_{32}	336_{54}	308_{100}	276_{52}	248_{38}	1213	3223
Moxaverine-M (O-demethyl-HO-ethyl-) isomer-2 2AC	UHYAC	2830	393_{11}	350_{44}	308_{77}	290_{100}	274_{36}	1415	3228
Moxaverine-M (O-demethyl-HO-ethyl-) isomer-2 AC	UHYAC	2795	351_{54}	336_{100}	308_{94}	276_{75}	91_{6}	1213	3226
Moxaverine-M (O-demethyl-HO-methoxy-phenyl-) isomer-1 2AC	UHYAC	2860	423_{17}	381_{33}	350_{56}	338_{100}	306_{48}	1515	3229
Moxaverine-M (O-demethyl-HO-methyl-) isomer-2 2AC	UHYAC	3120	423_{34}	408_{28}	380_{100}	348_{38}	321_{25}	1515	3234
Moxaverine-M (O-demethyl-HO-phenyl-) isomer-1 2AC	UHYAC	2895	393_{11}	350_{46}	334_{36}	290_{100}	274_{20}	1415	3230
Moxaverine-M (O-demethyl-HO-phenyl-) isomer-2 2AC	UHYAC	2930	393_{12}	350_{100}	334_{21}	308_{4}	292_{7}	1415	3232
Moxaverine-M (O-demethyl-oxo-ethyl-) isomer-1 AC	UHYAC	2775	349_{11}	306_{100}	290_{33}	264_{17}	91_{7}	1204	3224
Moxaverine-M (O-demethyl-oxo-ethyl-) isomer-2 AC	UHYAC	2785	349_{55}	306_{99}	292_{100}	264_{42}	91_{7}	1204	3225
Moxonidine 2AC	U+UHYAC	2455	325_{23}	290_{24}	248_{33}	128_{100}	86_{47}	1061	1277
Moxonidine AC	U+UHYAC	2380	283_{62}	248_{77}	206_{79}	176_{42}	86_{100}	794	6806
MPBP		1790	231_{1}	202_{1}	119_{10}	112_{100}	91_{21}	515	6990
MPBP impurity-1		1760	227_{23}	119_{28}	108_{100}	91_{20}	80_{25}	499	6991
MPBP impurity-2		1820	229_{16}	145_{10}	110_{100}	91_{26}	70_{64}	506	6992
MPBP-M (carboxy-) ET		2210	260_{1}	177_{2}	149_{6}	112_{100}	70_{11}	836	6994
MPBP-M (carboxy-) ME		2080	163_{4}	135_{3}	112_{100}	104_{6}	70_{9}	747	7001
MPBP-M (carboxy-) TMS		2220	318_{1}	221_{1}	178_{3}	112_{100}	104_{4}	1113	7005
MPBP-M (carboxy-deamino-oxo-) ET		1720*	234_{4}	189_{7}	177_{100}	149_{33}	104_{13}	526	6995
MPBP-M (carboxy-deamino-oxo-) ME		1650*	220_{1}	163_{100}	135_{19}	120_{6}	104_{12}	468	7002
MPBP-M (carboxy-dihydro-) 2TMS		2140	392_{1}	280_{1}	178_{4}	163_{5}	112_{100}	1469	7006
MPBP-M (carboxy-oxo-) ET		2390	303_{1}	258_{2}	149_{6}	126_{100}	104_{10}	924	6996
MPBP-M (carboxy-oxo-) ME		2280	289_{3}	258_{4}	163_{5}	126_{100}	104_{7}	834	6998
MPBP-M (carboxy-oxo-) TMS		2400	347_{1}	332_{6}	221_{3}	178_{8}	126_{100}	1195	7003
MPBP-M (carboxy-oxo-dihydro-) 2TMS		2430	406_{2}	332_{3}	280_{3}	178_{5}	126_{100}	1511	7004
MPBP-M (carboxy-oxo-dihydro-) ET		2470	305_{1}	260_{2}	142_{11}	126_{100}	98_{6}	937	6997
MPBP-M (carboxy-oxo-dihydro-) ETAC		2545	268_{2}	226_{4}	179_{2}	126_{100}	98_{5}	1195	7054
MPBP-M (carboxy-oxo-dihydro-) ME		2350	260_{2}	165_{1}	126_{100}	98_{9}	69_{13}	847	6999
MPBP-M (HO-) AC		2170	238_{1}	177_{3}	112_{100}	89_{3}	70_{4}	836	7024
MPBP-M (HO-) TMS		2145	319_{1}	304_{1}	178_{2}	112_{100}	104_{2}	1029	7055
MPBP-M (oxo-)		2010	245_{2}	162_{8}	126_{100}	119_{7}	91_{11}	576	6993
MPHP		1965	140_{100}	119_{4}	91_{8}	84_{4}	65_{5}	653	6647
MPHP-M (carboxy-)		2305	289_{1}	202_{1}	149_{5}	140_{100}	121_{2}	838	6651
MPHP-M (carboxy-) ET		2335	260_{1}	177_{3}	149_{4}	140_{100}	104_{8}	1014	6666
MPHP-M (carboxy-) ME		2260	246_{1}	163_{4}	140_{100}	104_{8}	84_{5}	926	6662
MPHP-M (carboxy-) TMS		2390	304_{1}	221_{2}	178_{7}	140_{100}	104_{6}	1270	6655

MPHP-M (carboxy-HO-alkyl-) ET Table 8.1: Compounds in order of names

Name	Detected	RI	Typical ions and intensities					Page	Entry
MPHP-M (carboxy-HO-alkyl-) ET		2545	177 $_3$	156 $_{100}$	149 $_4$	138 $_4$	104 $_8$	1114	6667
MPHP-M (carboxy-HO-alkyl-) isomer-1 2TMS		2625	434 $_1$	228 $_{100}$	221 $_4$	178 $_5$	138 $_{23}$	1579	6657
MPHP-M (carboxy-HO-alkyl-) isomer-2 2TMS		2635	434 $_2$	228 $_{100}$	221 $_4$	178 $_{10}$	138 $_{62}$	1579	6759
MPHP-M (carboxy-HO-alkyl-) ME		2460	163 $_6$	156 $_{100}$	149 $_4$	138 $_5$	104 $_7$	1028	6663
MPHP-M (carboxy-HO-alkyl-) MEAC		2715	198 $_{100}$	163 $_4$	138 $_{49}$	104 $_4$		1269	6672
MPHP-M (di-HO-) 2AC		2600	198 $_{100}$	177 $_5$	138 $_{47}$			1336	6649
MPHP-M (di-HO-) 2TMS		2525	420 $_4$	228 $_{100}$	204 $_7$	138 $_{60}$	73 $_{25}$	1549	6654
MPHP-M (dihydro-)		1965	260 $_1$	140 $_{100}$	119 $_2$	91 $_{13}$	77 $_9$	666	6699
MPHP-M (dihydro-) AC		1990	244 $_2$	140 $_{100}$	119 $_5$	91 $_7$	77 $_4$	927	6704
MPHP-M (dihydro-) TMS		1900	318 $_2$	244 $_1$	140 $_{100}$	98 $_8$	73 $_{14}$	1117	6709
MPHP-M (HO-alkyl-) isomer-1 AC		2250	316 $_1$	198 $_{100}$	138 $_{60}$	119 $_{12}$	91 $_{17}$	1014	6693
MPHP-M (HO-alkyl-) isomer-2 AC		2445	198 $_{100}$	138 $_{65}$	119 $_{10}$	91 $_{19}$		1013	6694
MPHP-M (HO-tolyl-)		2250	218 $_1$	140 $_{100}$	135 $_4$	77 $_6$		750	6673
MPHP-M (HO-tolyl-) AC		2315	260 $_1$	177 $_4$	140 $_{100}$	89 $_{10}$		1014	6675
MPHP-M (HO-tolyl-) TFA		2085	314 $_2$	231 $_6$	140 $_{100}$	89 $_{11}$		1319	6674
MPHP-M (oxo-)		2165	190 $_3$	154 $_{100}$	119 $_9$	98 $_{34}$	91 $_{18}$	735	6652
MPHP-M (oxo-carboxy-) ET		2525	286 $_2$	177 $_3$	154 $_{100}$	98 $_{31}$	86 $_{11}$	1101	6665
MPHP-M (oxo-carboxy-) ME		2445	286 $_1$	221 $_6$	163 $_3$	154 $_{100}$	98 $_{32}$	1013	6659
MPHP-M (oxo-carboxy-) TMS		2160	360 $_1$	221 $_1$	178 $_5$	154 $_{100}$	104 $_6$	1336	6656
MPHP-M (oxo-carboxy-dihydro-) 2ME		2430	179 $_8$	154 $_{100}$	148 $_4$	120 $_{10}$	98 $_{37}$	1114	6668
MPHP-M (oxo-carboxy-dihydro-) ET		2620	288 $_1$	179 $_1$	154 $_{100}$	112 $_{10}$	98 $_{30}$	1114	6664
MPHP-M (oxo-carboxy-dihydro-) ME		2555	165 $_3$	154 $_{100}$	98 $_{28}$	86 $_{14}$		1028	6660
MPHP-M (oxo-carboxy-dihydro-) MEAC		2725	330 $_1$	154 $_{100}$	98 $_{21}$	86 $_7$		1269	6671
MPHP-M (oxo-carboxy-HO-alkyl-) 2TMS		2695	448 $_{12}$	242 $_{100}$	221 $_{14}$	214 $_{33}$	98 $_{51}$	1606	6658
MPHP-M (oxo-carboxy-HO-alkyl-) ET		2640	177 $_4$	170 $_{100}$	149 $_4$	142 $_{22}$	104 $_{10}$	1195	6653
MPHP-M (oxo-carboxy-HO-alkyl-) ME		2575	170 $_{100}$	163 $_{22}$	142 $_{25}$	104 $_{22}$	98 $_{22}$	1112	6661
MPHP-M (oxo-carboxy-HO-alkyl-) MEAC		2890	344 $_2$	261 $_6$	212 $_{53}$	170 $_{20}$	152 $_{100}$	1335	6670
MPHP-M (oxo-HO-alkyl-) AC		2425	331 $_1$	212 $_{69}$	170 $_{24}$	152 $_{20}$	98 $_{11}$	1101	6648
MPHP-M (oxo-HO-tolyl-) AC		2515	248 $_1$	177 $_4$	154 $_{100}$	98 $_{32}$		1102	6650
MPHP-M (oxo-HO-tolyl-) HFB		2305	331 $_3$	154 $_{100}$	98 $_{31}$			1641	6669
MPPP		1725	216 $_1$	119 $_4$	98 $_{100}$	91 $_{17}$	56 $_{53}$	455	5736
MPPP-M (carboxy-)		2200	247 $_1$	149 $_2$	121 $_1$	98 $_{100}$	56 $_{13}$	586	6500
MPPP-M (carboxy-) ET		2320	275 $_2$	230 $_6$	177 $_{27}$	149 $_{13}$	98 $_{100}$	747	6498
MPPP-M (carboxy-) ME		2030	163 $_1$	135 $_1$	104 $_4$	98 $_{100}$	56 $_{15}$	664	6502
MPPP-M (carboxy-) TMS		2195	304 $_1$	221 $_1$	178 $_4$	104 $_6$	98 $_{100}$	1026	6793
MPPP-M (carboxy-deamino-oxo-) ET		1620*	220 $_1$	177 $_{100}$	149 $_{58}$	121 $_{17}$	104 $_{15}$	469	6494
MPPP-M (carboxy-deamino-oxo-) ME		1635*	206 $_7$	177 $_{100}$	149 $_{31}$	121 $_8$	104 $_9$	416	6496
MPPP-M (carboxy-oxo-) ET		2335	289 $_1$	244 $_3$	149 $_7$	112 $_{10}$	84 $_4$	834	6499
MPPP-M (dihydro-)		1765	218 $_1$	105 $_1$	98 $_{100}$	77 $_{10}$	56 $_{21}$	467	6696
MPPP-M (dihydro-) AC		1815	202 $_2$	119 $_5$	98 $_{100}$	91 $_7$	56 $_{15}$	666	6701
MPPP-M (dihydro-) TMS		1730	276 $_3$	202 $_1$	115 $_3$	98 $_{100}$	73 $_{17}$	853	6706
MPPP-M (HO-)		2020	233 $_1$	135 $_4$	98 $_{100}$	77 $_{19}$	56 $_{35}$	523	6503
MPPP-M (HO-) AC		2115	177 $_3$	98 $_{100}$	89 $_{17}$	56 $_{34}$		746	6504
MPPP-M (HO-) TMS		2095	290 $_8$	135 $_4$	98 $_{100}$			939	6794
MPPP-M (oxo-)		1920	231 $_1$	119 $_7$	112 $_{100}$	84 $_4$	69 $_9$	513	6501
MPPP-M (p-dicarboxy-) 2ET		1645*	222 $_{12}$	194 $_{18}$	177 $_{100}$	166 $_{22}$	149 $_{60}$	479	6495
MPPP-M (p-dicarboxy-) ET		1715*	194 $_{19}$	166 $_{40}$	149 $_{100}$	121 $_{17}$	65 $_{15}$	383	6497
MPPP-M (p-dicarboxy-) ETME		1560*	208 $_1$	193 $_{11}$	180 $_{37}$	163 $_{100}$	149 $_{43}$	423	6493
MSTFA		<1000	199 $_2$	184 $_{14}$	134 $_{39}$	77 $_{100}$	73 $_{88}$	397	5694
4-MTA		1300	181 $_1$	138 $_{100}$	122 $_{43}$	91 $_{37}$	78 $_{36}$	348	5941
4-MTA		1300	181 $_1$	138 $_{36}$	122 $_{15}$	91 $_{13}$	44 $_{100}$	349	5942
4-MTA 2AC		1760	265 $_2$	164 $_{100}$	137 $_{28}$	122 $_{14}$	86 $_{43}$	686	5940
4-MTA AC		1700	223 $_5$	164 $_{100}$	137 $_{22}$	122 $_{13}$	86 $_{26}$	483	5717
4-MTA formyl artifact		1560	193 $_{14}$	137 $_{34}$	122 $_8$	78 $_4$	56 $_{100}$	378	5718
4-MTA HFB		1775	377 $_{19}$	240 $_{14}$	164 $_{69}$	137 $_{100}$	69 $_{24}$	1342	5743
4-MTA PFP		1760	327 $_5$	190 $_8$	164 $_{44}$	137 $_{100}$	122 $_{17}$	1073	5744
4-MTA TFA		1750	277 $_7$	164 $_{40}$	137 $_{100}$	122 $_{18}$	69 $_{17}$	758	5720
4-MTA TMS		1750	238 $_{13}$	137 $_{14}$	116 $_{100}$	100 $_{13}$	73 $_{73}$	623	5721
4-MTA-M (deamino-HO-) AC		1460*	224 $_7$	164 $_{100}$	137 $_{36}$	122 $_{15}$	117 $_{21}$	487	6898
4-MTA-M (deamino-HO-) PFP	U+UHYPFP	1560*	328 $_{45}$	191 $_4$	164 $_{100}$	137 $_{94}$	119 $_{26}$	1079	6952
4-MTA-M (deamino-oxo-)		1335*	180 $_{28}$	137 $_{100}$	122 $_{21}$			345	6899
4-MTA-M (HO-) formyl artifact 2AC	U+UHYAC	2240	251 $_{53}$	195 $_8$	152 $_{65}$	137 $_{100}$	122 $_{16}$	609	6902
4-MTA-M (HO-) isomer-1 2PFP	U+UHYPFP	1780	475 $_5$	326 $_4$	285 $_{100}$	256 $_5$	190 $_5$	1647	6949
4-MTA-M (HO-) isomer-2 2AC	U+UHYAC	2260	281 $_2$	222 $_{100}$	150 $_{50}$	123 $_{17}$	86 $_{37}$	787	6896
4-MTA-M (HO-) isomer-2 2PFP	U+UHYPFP	1790	475 $_4$	326 $_3$	285 $_{100}$	190 $_6$	152 $_6$	1647	6950
4-MTA-M (methylthiobenzoic acid)		1995*	168 $_{100}$	151 $_{48}$	135 $_7$	125 $_{13}$	108 $_{15}$	320	7313
4-MTA-M (methylthiobenzoic acid) ME		1610*	182 $_{36}$	151 $_{100}$	123 $_{21}$	108 $_{19}$	79 $_{16}$	350	6900
4-MTA-M (methylthiobenzoic acid) TMS		1770*	240 $_{39}$	225 $_{80}$	181 $_{87}$	151 $_{100}$	108 $_{34}$	554	6901
4-MTA-M (ring-HO-) 2AC	U+UHYAC	2240	281 $_7$	222 $_{100}$	180 $_{17}$	153 $_{92}$	86 $_{58}$	787	6895
4-MTA-M (ring-HO-) 2PFP	U+UHYPFP	1860	475 $_{18}$	326 $_5$	312 $_{100}$	285 $_{40}$	190 $_{16}$	1647	6951
4-MTA-M/artifact (sulfone) AC		2455	255 $_3$	196 $_6$	180 $_{10}$	107 $_{23}$	86 $_{10}$	631	6903
4-MTA-M/artifact (sulfoxide) AC		2360	239 $_1$	222 $_{69}$	165 $_{52}$	137 $_{100}$	86 $_{17}$	552	6897
5MT-NB2B		3105	358 $_2$	198 $_{25}$	169 $_{44}$	161 $_{100}$	145 $_{17}$	1252	9889
5MT-NB2B 2HFB		2590	750 $_4$	369 $_{100}$	356 $_{55}$	171 $_{58}$	159 $_{44}$	1732	9897

Table 8.1: Compounds in order of names   **5MT-NB2B 2PFP**

Name	Detected	RI	Typical ions and intensities					Page	Entry
5MT-NB2B 2PFP		2570	650_2	319_{100}	306_{34}	169_{43}	159_{27}	1723	9895
5MT-NB2B 2TFA		2645	550_4	269_{100}	256_{58}	169_{51}	159_{40}	1695	9893
5MT-NB2B 2TMS		2925	502_1	270_{100}	246_5	232_{28}	169_{29}	1662	9891
5MT-NB2B AC		3085	400_5	198_{11}	173_{100}	160_{44}	145_{14}	1445	9888
5MT-NB2B HFB		2765	554_8	173_{31}	160_{100}	145_{14}	117_{10}	1697	9896
5MT-NB2B PFP		2770	504_2	173_{29}	160_{100}	145_{14}	117_9	1664	9894
5MT-NB2B TFA		2820	454_{24}	173_{36}	160_{100}	145_{16}	117_{12}	1588	9892
5MT-NB2B TMS		2870	430_2	246_2	233_{100}	198_{16}	169_{34}	1533	9890
5MT-NB2OMe 2HFB		2525	702_8	369_{100}	356_{42}	159_9	121_{22}	1729	9936
5MT-NB2OMe 2PFP		2525	602_5	319_{100}	306_{25}	159_{16}	121_{53}	1715	9934
5MT-NB2OMe 2TFA		2595	502_6	269_{63}	256_{14}	159_{18}	121_{100}	1661	9932
5MT-NB2OMe 2TMS		2870	454_2	233_{25}	222_{100}	202_4	121_{28}	1590	9930
5MT-NB2OMe AC		3015	352_3	173_{100}	160_{28}	145_8	121_{58}	1222	9928
5MT-NB2OMe artifact		2850	322_6	201_3	173_{100}	158_{29}	121_{15}	1046	10264
5MT-NB2OMe HFB		2735	506_{10}	173_{38}	160_{100}	145_{10}	121_{15}	1666	9935
5MT-NB2OMe ME		2715	324_1	164_{55}	160_9	121_{100}	91_{38}	1058	9927
5MT-NB2OMe PFP		2735	456_{13}	173_{44}	160_{100}	145_8	121_{13}	1593	9933
5MT-NB2OMe TFA		2780	406_{21}	173_{35}	160_{100}	145_{16}	121_{17}	1464	9931
5MT-NB2OMe TMS		2840	382_2	232_{100}	202_7	150_9	121_{44}	1367	9929
5MT-NB3B		2860	358_2	198_{14}	169_{33}	161_{100}	145_{15}	1252	9898
5MT-NB3B 2HFB		2590	750_4	369_{100}	356_{55}	171_{10}	159_8	1731	9907
5MT-NB3B 2PFP		2595	650_2	319_{100}	306_{54}	169_{34}	159_{31}	1723	9905
5MT-NB3B 2TFA		2665	550_9	269_{100}	256_{49}	169_{15}	159_{14}	1694	9903
5MT-NB3B 2TMS		2940	502_1	270_{100}	246_4	232_{34}	169_{20}	1662	9901
5MT-NB3B AC		3130	400_4	198_5	173_{100}	160_{57}	145_9	1444	9899
5MT-NB3B HFB		2785	554_3	173_{24}	160_{100}	145_{11}	117_6	1696	9906
5MT-NB3B PFP		2790	504_4	173_{21}	160_{100}	145_{12}	117_8	1664	9904
5MT-NB3B TFA		2835	454_{16}	173_{24}	160_{100}	145_{13}	117_7	1588	9902
5MT-NB3B TMS		2930	430_1	246_2	233_{100}	198_{12}	169_{31}	1533	9900
5MT-NB3CF3		2545	348_6	188_{27}	161_{100}	159_{77}	145_{16}	1199	9909
5MT-NB3CF3 2HFB		2335	740_8	384_2	369_{100}	356_{75}	159_{48}	1731	10000
5MT-NB3CF3 2PFP		2335	640_4	319_{100}	306_{61}	159_{74}	144_{13}	1721	9998
5MT-NB3CF3 2TFA		2395	540_8	269_{100}	256_{67}	159_{60}	144_{12}	1689	9996
5MT-NB3CF3 2TMS		2670	492_2	260_{100}	232_{27}	168_{28}	141_{27}	1651	9994
5MT-NB3CF3 AC		2755	390_3	188_{13}	173_{100}	160_{70}	145_{18}	1403	9992
5MT-NB3CF3 artifact		2695	360_6	199_5	173_{100}	158_{40}	130_6	1262	9990
5MT-NB3CF3 HFB		2515	544_4	173_{17}	160_{100}	159_{23}	144_3	1692	9999
5MT-NB3CF3 ME		2500	362_9	202_{100}	159_{71}	145_5	109_4	1273	9991
5MT-NB3CF3 PFP		2505	494_8	173_{18}	160_{100}	159_{27}	145_{12}	1653	9997
5MT-NB3CF3 TFA		2545	444_8	173_{18}	160_{100}	159_{23}	145_{13}	1569	9995
5MT-NB3CF3 TMS		2650	420_3	260_4	233_{100}	218_{16}	159_{56}	1508	9993
5MT-NB3Cl		2760	314_4	161_{100}	154_{28}	145_{15}	125_{67}	994	9966
5MT-NB3Cl 2HFB		2530	706_6	369_{100}	356_{52}	159_{14}	125_{22}	1729	9977
5MT-NB3Cl 2PFP		2520	606_2	319_{100}	306_{65}	159_{38}	125_{47}	1716	9975
5MT-NB3Cl 2TFA		2590	506_{10}	269_{100}	256_{67}	159_{42}	125_{49}	1666	9973
5MT-NB3Cl 2TMS		2875	458_2	232_{24}	226_{100}	202_6	125_6	1597	9971
5MT-NB3Cl AC		3000	356_5	173_{100}	160_{59}	145_{15}	125_{19}	1242	9969
5MT-NB3Cl artifact		2930	326_6	199_4	173_{100}	158_{32}	125_{13}	1066	9967
5MT-NB3Cl HFB		2715	510_8	173_{22}	160_{100}	145_{11}	125_{18}	1669	9976
5MT-NB3Cl ME		2700	328_6	168_{100}	160_{12}	145_9	125_{63}	1081	9968
5MT-NB3Cl PFP		2710	460_{20}	173_{24}	160_{100}	145_{13}	125_{16}	1601	9974
5MT-NB3Cl TFA		2760	410_{13}	173_{20}	160_{100}	145_{14}	125_{14}	1477	9972
5MT-NB3Cl TMS		2850	386_4	233_{100}	218_{15}	154_{12}	125_{25}	1385	9970
5MT-NB3F		2610	298_3	173_{12}	161_{100}	138_{28}	109_{84}	896	10001
5MT-NB3F 2HFB		2390	690_2	369_{100}	356_{63}	159_{45}	109_{75}	1728	10012
5MT-NB3F 2PFP		2385	590_5	319_{100}	306_{62}	159_{26}	109_{51}	1711	10010
5MT-NB3F 2TFA		2450	490_{10}	269_{100}	256_{56}	159_{30}	109_{43}	1648	10008
5MT-NB3F 2TMS		2735	442_2	232_{19}	210_{100}	202_6	109_{21}	1566	10006
5MT-NB3F AC		2835	340_{16}	173_{100}	160_{62}	145_{17}	109_{25}	1156	10004
5MT-NB3F artifact		2755	310_{19}	199_6	173_{100}	158_{36}	109_{17}	968	10002
5MT-NB3F HFB		2575	494_{10}	173_{20}	160_{100}	145_{11}	109_{22}	1652	10011
5MT-NB3F ME		2560	312_9	160_9	152_{100}	145_8	109_{68}	982	10003
5MT-NB3F PFP		2570	444_{13}	173_{19}	160_{100}	145_{14}	109_{25}	1569	10009
5MT-NB3F TFA		2615	394_{22}	173_{28}	160_{100}	145_{15}	109_{30}	1420	10007
5MT-NB3F TMS		2700	370_6	233_{100}	218_{18}	138_{16}	109_{55}	1313	10005
5MT-NB3I		2950	406_2	246_{17}	217_{48}	161_{100}	145_{13}	1464	9917
5MT-NB3I 2HFB		2670	798_2	446_5	369_{100}	356_{65}	217_{31}	1733	9926
5MT-NB3I 2PFP		2680	698_2	319_{100}	306_{48}	217_{53}	159_{27}	1729	9924
5MT-NB3I 2TFA		2750	598_9	269_{100}	256_{52}	217_{52}	159_{33}	1714	9922
5MT-NB3I 2TMS		3000	550_1	318_{100}	246_{11}	233_{76}	217_{47}	1695	9920
5MT-NB3I AC		3300	448_4	246_{10}	217_{17}	173_{100}	160_{46}	1576	9918
5MT-NB3I HFB		2865	602_4	217_{20}	173_{22}	160_{100}	145_{12}	1715	9925
5MT-NB3I PFP		2875	552_{10}	217_{19}	173_{19}	160_{100}	145_{10}	1695	9923
5MT-NB3I TFA		2930	502_{10}	217_{16}	173_{24}	160_{100}	145_{13}	1661	9921

5MT-NB3I TMS

Table 8.1: Compounds in order of names

Name	Detected	RI	Typical ions and intensities					Page	Entry
5MT-NB3I TMS		3020	478$_{4}$	318$_{21}$	246$_{11}$	233$_{100}$	217$_{29}$	1631	9919
5MT-NB3Me 2HFB		2445	686$_{2}$	369$_{100}$	356$_{43}$	159$_{30}$	105$_{60}$	1728	9965
5MT-NB3Me 2PFP		2445	586$_{8}$	319$_{100}$	306$_{29}$	159$_{24}$	105$_{66}$	1710	9963
5MT-NB3Me 2TFA		2520	486$_{15}$	269$_{100}$	256$_{49}$	159$_{26}$	144$_{14}$	1643	9961
5MT-NB3Me 2TMS		2790	438$_{2}$	246$_{3}$	232$_{18}$	206$_{100}$	105$_{71}$	1555	9959
5MT-NB3Me AC		2915	336$_{9}$	173$_{100}$	160$_{39}$	145$_{12}$	105$_{35}$	1132	9957
5MT-NB3Me artifact		2830	306$_{16}$	199$_{4}$	173$_{100}$	158$_{29}$	105$_{12}$	944	9956
5MT-NB3Me HFB		2640	490$_{6}$	173$_{32}$	160$_{100}$	145$_{12}$	105$_{19}$	1648	9964
5MT-NB3Me PFP		2640	440$_{20}$	173$_{31}$	160$_{100}$	145$_{13}$	105$_{20}$	1558	9962
5MT-NB3Me TFA		2685	390$_{15}$	173$_{32}$	160$_{100}$	145$_{14}$	105$_{20}$	1402	9960
5MT-NB3Me TMS		2765	366$_{2}$	233$_{100}$	218$_{7}$	134$_{12}$	105$_{44}$	1293	9958
5MT-NB3OMe		2750	310$_{13}$	161$_{100}$	150$_{34}$	121$_{95}$	117$_{22}$	972	10265
5MT-NB3OMe 2HFB		2530	702$_{5}$	369$_{100}$	356$_{47}$	159$_{22}$	121$_{43}$	1729	9946
5MT-NB3OMe 2PFP		2540	602$_{5}$	319$_{100}$	306$_{36}$	159$_{10}$	121$_{20}$	1715	9944
5MT-NB3OMe 2TFA		2615	502$_{10}$	269$_{100}$	256$_{59}$	159$_{38}$	121$_{55}$	1661	9942
5MT-NB3OMe 2TMS		2890	454$_{2}$	232$_{18}$	222$_{100}$	202$_{7}$	121$_{72}$	1590	9940
5MT-NB3OMe AC		3045	352$_{3}$	173$_{100}$	160$_{41}$	145$_{10}$	121$_{30}$	1223	9938
5MT-NB3OMe artifact		2960	322$_{12}$	201$_{9}$	173$_{100}$	158$_{40}$	121$_{15}$	1046	9937
5MT-NB3OMe HFB		2740	506$_{14}$	173$_{25}$	160$_{100}$	145$_{9}$	121$_{17}$	1666	9945
5MT-NB3OMe PFP		2740	456$_{3}$	173$_{29}$	160$_{100}$	145$_{14}$	121$_{19}$	1593	9943
5MT-NB3OMe TFA		2780	406$_{21}$	173$_{35}$	160$_{100}$	145$_{15}$	121$_{17}$	1465	9941
5MT-NB3OMe TMS		2875	382$_{4}$	233$_{100}$	218$_{14}$	202$_{8}$	150$_{11}$	1366	9939
5MT-NB3SMe		2960	325$_{5}$	166$_{16}$	161$_{100}$	145$_{20}$	137$_{70}$	1070	9978
5MT-NB3SMe 2HFB		2670	718$_{9}$	369$_{100}$	356$_{57}$	349$_{39}$	137$_{74}$	1730	9988
5MT-NB3SMe 2PFP		2675	618$_{9}$	319$_{100}$	306$_{47}$	299$_{29}$	137$_{61}$	1718	9986
5MT-NB3SMe 2TFA		2750	518$_{19}$	421$_{3}$	269$_{100}$	256$_{61}$	249$_{44}$	1676	9984
5MT-NB3SMe 2TMS		3025	470$_{2}$	238$_{100}$	233$_{49}$	202$_{7}$	137$_{58}$	1619	9982
5MT-NB3SMe AC		3280	368$_{5}$	173$_{100}$	160$_{43}$	137$_{21}$	117$_{9}$	1302	9980
5MT-NB3SMe HFB		2870	522$_{10}$	173$_{28}$	160$_{100}$	137$_{16}$	117$_{7}$	1678	9987
5MT-NB3SMe ME		2885	340$_{6}$	180$_{100}$	160$_{16}$	145$_{8}$	137$_{76}$	1156	9979
5MT-NB3SMe PFP		2880	472$_{14}$	173$_{27}$	160$_{100}$	137$_{14}$	117$_{8}$	1622	9985
5MT-NB3SMe TFA		2935	422$_{21}$	173$_{26}$	160$_{100}$	137$_{12}$	117$_{8}$	1513	9983
5MT-NB3SMe TMS		3080	398$_{2}$	238$_{29}$	233$_{100}$	218$_{6}$	137$_{44}$	1438	9981
5MT-NB4B 2HFB		2600	750$_{4}$	369$_{100}$	356$_{67}$	171$_{41}$	159$_{35}$	1731	9916
5MT-NB4B 2PFP		2615	650$_{1}$	319$_{100}$	306$_{66}$	169$_{40}$	159$_{39}$	1723	9914
5MT-NB4B 2TFA		2685	550$_{5}$	269$_{100}$	256$_{56}$	169$_{26}$	159$_{22}$	1695	9912
5MT-NB4B 2TMS		2960	502$_{1}$	270$_{100}$	246$_{2}$	232$_{22}$	169$_{19}$	1662	9910
5MT-NB4B AC		3165	400$_{3}$	198$_{1}$	173$_{100}$	160$_{49}$	145$_{13}$	1445	9908
5MT-NB4B formyl artifact		2985	370$_{21}$	199$_{6}$	173$_{100}$	158$_{40}$	130$_{7}$	1309	10263
5MT-NB4B HFB		2800	554$_{4}$	173$_{24}$	160$_{100}$	145$_{11}$	117$_{6}$	1697	9915
5MT-NB4B PFP		2800	504$_{9}$	173$_{26}$	160$_{100}$	145$_{15}$	117$_{11}$	1664	9913
5MT-NB4B TFA		2855	454$_{10}$	173$_{23}$	160$_{100}$	145$_{11}$	117$_{8}$	1588	9911
5MT-NB4B TMS		2940	430$_{2}$	246$_{2}$	233$_{100}$	198$_{43}$	169$_{43}$	1533	9909
5MT-NB4OMe 2HFB		2560	702$_{4}$	369$_{100}$	356$_{33}$	159$_{16}$	121$_{64}$	1729	9955
5MT-NB4OMe 2PFP		2560	602$_{3}$	319$_{54}$	306$_{13}$	159$_{17}$	121$_{100}$	1715	9953
5MT-NB4OMe 2TFA		2635	502$_{15}$	405$_{11}$	269$_{100}$	256$_{33}$	121$_{79}$	1661	9951
5MT-NB4OMe 2TMS		2910	454$_{2}$	233$_{23}$	222$_{70}$	202$_{7}$	121$_{100}$	1590	9949
5MT-NB4OMe AC		3090	352$_{6}$	173$_{100}$	160$_{42}$	145$_{8}$	121$_{55}$	1223	9947
5MT-NB4OMe HFB		2260	506$_{8}$	173$_{29}$	160$_{100}$	145$_{10}$	121$_{23}$	1666	9954
5MT-NB4OMe PFP		2780	456$_{10}$	173$_{30}$	160$_{100}$	145$_{12}$	121$_{31}$	1593	9952
5MT-NB4OMe TFA		2710	406$_{24}$	173$_{28}$	160$_{100}$	145$_{11}$	121$_{21}$	1465	9950
5MT-NB4OMe TMS		2895	382$_{3}$	233$_{100}$	218$_{8}$	202$_{6}$	121$_{46}$	1366	9948
Muzolimine		2445	271$_{17}$	256$_{5}$	173$_{16}$	137$_{10}$	99$_{100}$	722	4175
Muzolimine 2AC		2625	355$_{2}$	313$_{20}$	173$_{33}$	141$_{52}$	99$_{100}$	1235	4176
Muzolimine 2ME		2190	299$_{40}$	173$_{11}$	127$_{100}$	98$_{15}$	55$_{19}$	900	4179
Muzolimine 2TFA		2020	463$_{32}$	448$_{8}$	173$_{100}$	102$_{28}$	69$_{40}$	1605	4177
Muzolimine 2TMS		2265	415$_{26}$	400$_{7}$	242$_{18}$	214$_{21}$	73$_{100}$	1493	4182
Muzolimine 3ME		2235	313$_{57}$	298$_{22}$	173$_{8}$	141$_{100}$	84$_{35}$	986	4180
Muzolimine ME		2170	285$_{31}$	173$_{14}$	137$_{9}$	113$_{100}$	84$_{13}$	807	4178
Muzolimine MEAC		2520	327$_{14}$	312$_{7}$	173$_{23}$	155$_{18}$	113$_{100}$	1073	4231
Muzolimine METFA		2290	381$_{8}$	209$_{46}$	173$_{100}$	137$_{43}$	102$_{58}$	1359	4230
Muzolimine TMS		2210	343$_{76}$	328$_{18}$	171$_{100}$	156$_{28}$	73$_{77}$	1172	4181
MXP		2200	294$_{1}$	204$_{100}$	188$_{33}$	121$_{17}$	91$_{37}$	881	9565
MXP artifact		2005*	210$_{100}$	194$_{7}$	179$_{13}$	165$_{46}$	152$_{23}$	434	9566
m-Xylene		<1000*	106$_{42}$	91$_{100}$	77$_{12}$	65$_{6}$	51$_{12}$	259	2966
Mycophenolic acid	U+UHYAC	3000*	320$_{79}$	302$_{34}$	247$_{69}$	207$_{100}$	159$_{31}$	1033	6421
Mycophenolic acid 2ME		2270*	348$_{54}$	316$_{19}$	275$_{26}$	243$_{48}$	221$_{100}$	1199	6795
Mycophenolic acid ME	P	2260*	334$_{45}$	316$_{53}$	247$_{100}$	229$_{39}$	207$_{73}$	1120	6420
Myristic acid	P	1760*	228$_{14}$	185$_{18}$	129$_{34}$	73$_{100}$	60$_{98}$	503	1140
Myristic acid ET		1720*	256$_{6}$	213$_{16}$	157$_{24}$	101$_{53}$	88$_{100}$	639	5401
Myristic acid glycerol ester	G	2260*	285$_{10}$	271$_{10}$	211$_{66}$	98$_{63}$	55$_{100}$	921	5587
Myristic acid isopropyl ester	G	1830*	270$_{2}$	228$_{55}$	211$_{31}$	102$_{80}$	60$_{100}$	720	6469
Myristic acid ME	PME	1710*	242$_{3}$	199$_{8}$	143$_{11}$	87$_{65}$	74$_{100}$	566	1141
Myristic acid TMS		2280*	300$_{5}$	285$_{63}$	149$_{38}$	117$_{71}$	73$_{100}$	909	4644

Table 8.1: Compounds in order of names   **Myristicin**

Name	Detected	RI	Typical ions and intensities					Page	Entry
Myristicin	G	1400*	192_{100}	165_{23}	147_{17}	119_{29}	91_{31}	375	4374
Myristicin-M (1-HO-) AC	U+UHYAC	2020*	250_{100}	208_{30}	154_{47}	149_{49}	133_{79}	603	7150
Myristicin-M (demethyl-) AC	U+UHYAC	1655*	220_{13}	178_{100}	147_{30}	119_{20}	91_{29}	469	7145
Myristicin-M (demethylenyl-) 2AC	U+UHYAC	1880*	264_{1}	222_{30}	180_{100}	147_{15}	91_{34}	681	7148
Myristicin-M (demethylenyl-methyl-) AC	U+UHYAC	1755*	236_{35}	194_{100}	179_{43}	119_{15}	91_{12}	538	7140
Myristicin-M (di-HO-) 2AC	U+UHYAC	2210*	310_{7}	250_{70}	208_{41}	165_{100}	77_{33}	970	7149
N-1-Naphthylphthalimide		2545	273_{100}	228_{42}	202_{10}	140_{13}	76_{49}	733	3646
Nabumetone		1875*	228_{60}	185_{21}	171_{100}	141_{25}	128_{38}	502	7534
Nabumetone-M/artifact (O-demethyl-)		1925*	214_{54}	171_{28}	157_{100}	128_{16}	115_{14}	446	7536
Nabumetone-M/artifact (O-demethyl-) AC		1990*	256_{25}	214_{85}	171_{41}	157_{100}	128_{21}	637	7535
N-Acetyl-2-amino-octanoic acid ME	UME	1560	215_{1}	172_{5}	156_{41}	114_{100}	88_{21}	450	4941
Nadolol		2540	309_{1}	294_{10}	265_{3}	86_{100}	57_{11}	966	2612
Nadolol 3AC	UHYAC	2650	435_{1}	420_{34}	183_{6}	112_{17}	86_{100}	1548	1363
Nadolol 3TMS		2250	525_{1}	510_{3}	147_{8}	86_{100}	73_{30}	1681	5488
Nadolol formyl artifact		2560	321_{8}	306_{100}	201_{14}	141_{24}	70_{34}	1041	1362
Nadolol-M/artifact (deisobutyl-) -2H2O 2AC		2540	301_{56}	259_{10}	241_{36}	98_{100}	57_{51}	913	1706
Naftidrofuryl	G P U+UHYAC	2840	383_{1}	368	141_{10}	99_{29}	86_{100}	1373	2826
Naftidrofuryl-M (deethyl-)	U UHY	2780	355_{3}	296_{16}	198_{47}	141_{100}	58_{85}	1240	2827
Naftidrofuryl-M (di-oxo-HOOC-) ME	UHYME	2810*	326_{42}	198_{54}	153_{67}	141_{100}	71_{20}	1068	2830
Naftidrofuryl-M (HO-HOOC-) MEAC	UHYAC	2740*	356_{1}	283_{35}	153_{63}	141_{100}	73_{75}	1243	2831
Naftidrofuryl-M (HO-oxo-HOOC-) MEAC	U+UHYAC	2920*	297_{36}	198_{47}	153_{62}	141_{100}	115_{30}	1312	2832
Naftidrofuryl-M (oxo-HOOC-) ME	UME UHYAC	2760*	312_{44}	198_{50}	153_{74}	141_{100}	115_{30}	982	2829
Naftidrofuryl-M/artifact (HOOC-) ME	UHYME UHYAC	2390*	298_{25}	153_{52}	141_{100}	84_{63}	71_{94}	897	2828
Nalbuphine	G	2960	357_{15}	302_{100}	284_{4}	110		1249	3061
Nalbuphine 2AC	U+UHYAC	3110	441_{9}	386_{100}	344_{8}	296_{3}		1563	3064
Nalbuphine 2HFB		2560	680_{10}	662_{23}	405_{33}	263_{67}	169_{100}	1731	6135
Nalbuphine 2PFP		2700	649_{11}	594_{100}	486_{11}	400_{10}	119_{26}	1723	6124
Nalbuphine 3AC	U+UHYAC	3080	483_{14}	440_{8}	428_{100}	368_{22}	326_{10}	1639	3065
Nalbuphine 3PFP		2510	795_{4}	740_{87}	576_{97}	412_{100}	357_{89}	1733	6125
Nalbuphine 3TMS		2860	573_{20}	518_{13}	429_{7}	101_{6}	73_{100}	1704	6205
Nalbuphine AC	U+UHYAC	3030	399_{10}	344_{100}	326_{8}	302_{5}		1443	3063
Nalbuphine-M (N-dealkyl-)		2930	289_{100}	272_{36}	242_{12}	202_{20}	115_{13}	834	3062
Nalbuphine-M (N-dealkyl-) 2AC		2970	373_{39}	331_{50}	313_{19}	227_{41}	87_{100}	1327	3066
Nalbuphine-M (N-dealkyl-) 3AC		3020	415_{34}	373_{66}	296_{25}	227_{56}	87_{100}	1493	3067
Naled		1640*	301_{17}	299_{10}	189_{12}	145_{83}	109_{100}	1345	3430
N-Allyl-nor-LSD TMS		3210	421_{30}	380_{11}	319_{32}	279_{77}	253_{100}	1512	9873
N-Allyl-nor-lysergic acid diethylamide		3100	349_{100}	308_{12}	247_{42}	207_{75}	72_{23}	1206	10188
Nalmefen		2640	339_{100}	298_{27}	284_{18}	242_{15}	55_{23}	1152	10345
Nalmefen 2AC		2710	423_{26}	381_{100}	340_{38}	242_{28}	55_{54}	1515	10346
Nalmefen 2TMS		2620	483_{88}	468_{26}	386_{26}	73_{100}	55_{45}	1639	10349
Nalmefen AC		2730	381_{100}	340_{32}	326_{12}	242_{13}	55_{32}	1362	10347
Nalmefen HFB		2485	535_{100}	494_{43}	480_{23}	438_{7}	55_{42}	1688	10352
Nalmefen ME		2610	353_{100}	312_{40}	298_{29}	256_{40}	55_{36}	1229	10348
Nalmefen PFP		2470	485_{100}	444_{66}	430_{45}	388_{24}	55_{71}	1641	10351
Nalmefen TFA		2480	435_{100}	394_{60}	380_{37}	338_{23}	55_{64}	1548	10350
Nalorphine	UHY	2620	311_{100}	294_{11}	282_{6}	241_{12}	188_{14}	977	1736
Nalorphine 2AC	UHYAC	2820	395_{86}	353_{100}	336_{50}	294_{30}	230_{18}	1425	1737
Nalorphine 2TMS		2400	455_{84}	440_{36}	414_{37}	324_{25}	73_{100}	1592	5497
Nalorphine AC		2800	353_{100}	294_{55}	241_{16}	230_{9}		1227	1738
Naloxone	G P-I UHY	2715	327_{100}	286_{14}	242_{24}			1075	563
Naloxone 2AC	U+UHYAC	2750	411_{29}	369_{100}	352_{21}	310_{23}	285_{30}	1482	2982
Naloxone 2ET		2830	383_{99}	270_{100}				1372	564
Naloxone 2ME		2885	355_{100}	256_{28}	82_{86}			1239	566
Naloxone 2PFP		2470	619_{8}	472_{8}	284_{3}	119_{69}	82_{100}	1718	4327
Naloxone 2TMS	UHYTMS	2680	471_{29}	456_{17}	355_{11}	96_{10}	73_{100}	1621	4308
Naloxone AC	UHYAC	2840	369_{54}	327_{100}	286_{14}	242_{18}		1306	361
Naloxone enol 2AC	U+UHYAC	2810	411_{93}	369_{83}	330_{76}	270_{71}	82_{100}	1483	2984
Naloxone enol 2PFP		2360	619_{100}	472_{55}	456_{50}	371_{10}	119_{61}	1718	4326
Naloxone enol 2TMS	UHYTMS	2700	471_{100}	456_{25}	366_{13}	82_{16}	73_{71}	1621	4309
Naloxone enol 3AC	U+UHYAC	2770	453_{5}	411_{14}	369_{33}	327_{100}	242_{11}	1587	2983
Naloxone enol 3PFP		2270	765_{83}	618_{48}	602_{65}	454_{22}	119_{100}	1732	4328
Naloxone enol 3TMS	UHYTMS	2645	543_{50}	528_{59}	438_{57}	355_{29}	73_{100}	1691	4306
Naloxone ME		2825	341_{100}	300_{14}	256_{22}			1163	565
Naloxone MEAC		2890	383_{100}	340_{30}	324_{22}	242_{15}		1371	567
Naloxone PFP		2530	473_{76}	388_{36}	119_{75}	96_{85}	70_{100}	1624	4329
Naloxone TMS		2660	399_{100}	358_{13}	316_{17}	166_{16}	73_{54}	1442	4307
Naloxone-M (dihydro-) 2AC	U+UHYAC	2820	413_{15}	371_{6}	242_{4}	82_{100}		1487	1188
Naloxone-M (dihydro-) 3AC	UHYAC	2855	455_{11}	413_{41}	327_{16}	254_{17}	82_{100}	1592	3720
Naltrexol (beta-) 3TMS		2720	559_{84}	544_{11}	372_{43}	73_{100}	55_{33}	1700	6491
Naltrexone	UHY	2880	341_{88}	300_{24}	256_{20}	243_{18}	55_{100}	1163	4310
Naltrexone 2AC	UHYAC	2870	425_{36}	383_{43}	341_{25}	324_{13}	55_{100}	1520	4311
Naltrexone 2TMS		2760	485_{100}	470_{18}	388_{12}	73_{94}	55_{75}	1642	6276
Naltrexone AC		2980	383_{44}	341_{88}	300_{12}	243_{19}	55_{100}	1372	4313
Naltrexone enol 2AC	UHYAC	3060	425_{58}	383_{43}	342_{12}	110_{12}	55_{100}	1520	4314

Naltrexone enol 3AC   **Table 8.1:** Compounds in order of names

Name	Detected	RI	Typical ions and intensities					Page	Entry
Naltrexone enol 3AC	UHYAC	2960	467 $_{60}$	425 $_{64}$	408 $_{14}$	324 $_{10}$	55 $_{100}$	1612	4312
Naltrexone enol 3TMS		2700	557 $_{100}$	542 $_{68}$	484 $_9$	73 $_{90}$	55 $_{29}$	1698	6275
Naltrexone-M (dihydro-) 3AC	UHYAC	2990	469 $_{26}$	427 $_{36}$	413 $_{39}$	228 $_{13}$	55 $_{100}$	1616	4331
Naltrexone-M (dihydro-) 3TMS		2720	559 $_{84}$	544 $_{11}$	372 $_{43}$	73 $_{100}$	55 $_{33}$	1700	6491
Naltrexone-M (dihydro-methoxy-) 3AC	UHYAC	3200	499 $_{83}$	457 $_{57}$	440 $_{24}$	303 $_{17}$	55 $_{100}$	1658	4332
Naltrexone-M (methoxy-)	UHY	2920	371 $_{92}$	330 $_{19}$	286 $_{23}$	274 $_{22}$	55 $_{100}$	1319	4330
Naltrexone-M (methoxy-) 2AC	UHYAC	3130	455 $_{89}$	412 $_{27}$	396 $_{20}$	273 $_9$	55 $_{100}$	1591	4315
Naltrexone-M (methoxy-) AC	UHYAC	3150	413 $_{83}$	372 $_{22}$	328 $_{15}$	274 $_{31}$	55 $_{100}$	1487	4316
Naltrexone-M (methoxy-) enol 2AC	UHYAC	3300	455 $_{69}$	414 $_{30}$	384 $_8$	110 $_{11}$	55 $_{100}$	1592	4318
Naltrexone-M (methoxy-) enol 3AC	UHYAC	3180	497 $_{78}$	454 $_{27}$	396 $_9$	256 $_{13}$	55 $_{100}$	1656	4317
Nandrolone		2395*	274 $_{93}$	256 $_{17}$	110 $_{100}$	91 $_{70}$	79 $_{62}$	741	3748
Nandrolone TMS		2760*	346 $_5$	255 $_{14}$	237 $_9$	108 $_{95}$	91 $_{100}$	1191	3004
Naphazoline	G	2100	210 $_{46}$	209 $_{100}$	141 $_{24}$			434	1142
Naphthalen-1-yl-(1-butylindol-3-yl)methanone		3150	327 $_{100}$	310 $_{49}$	284 $_{65}$	200 $_{59}$	127 $_{56}$	1076	7874
Naphthalen-1-yl-(1-butylindol-3-yl)methanone-M/artifact (N-dealkyl-) AC		2720	313 $_{73}$	254 $_{62}$	241 $_{30}$	127 $_{100}$	89 $_{39}$	987	7076
Naphthalen-1-yl-(1-butylindol-3-yl)methanone-M/artifact (N-dealkyl-HO-indol) 2AC		3200	371 $_{20}$	329 $_{100}$	270 $_{84}$	160 $_{57}$	127 $_{40}$	1317	8511
Naphthalen-1-yl-(1-butylindol-3-yl)methanone-M/artifact (N-dealkyl HO-napht) 2AC		3065	371 $_{35}$	329 $_{46}$	286 $_{100}$	270 $_{34}$	144 $_{29}$	1317	8510
Naphthalen 1-yl-(1-hexylindol-3-yl)methanone		3220	355 $_{100}$	338 $_{44}$	284 $_{67}$	228 $_{56}$	127 $_{68}$	1239	8522
Naphthalen-1-yl-(1-morpholinoethylindol-3-yl)methanone		3500	384 $_2$	155 $_3$	127 $_8$	100 $_{100}$	70 $_5$	1377	8527
Naphthalen-1-yl-(1-pentylindol-3-yl)methanone		3240	341 $_{100}$	324 $_{43}$	284 $_{69}$	214 $_{76}$	127 $_{74}$	1164	7875
Naphthalen-1-yl-(1-pentylindol-3-yl)methanone-M/artifact (N-dealkyl-) AC		2720	313 $_{73}$	254 $_{62}$	241 $_{30}$	127 $_{100}$	89 $_{39}$	987	7876
Naphthalen-1-yl-(1-pentylindol-3-yl)methanone-M/artifact (N-dealkyl-HO-indo) 2AC		3200	371 $_{20}$	329 $_{100}$	270 $_{84}$	160 $_{57}$	127 $_{40}$	1317	8511
Naphthalen-1-yl-(1-pentylindol-3-yl)methanone-M/artifact (N-dealkyl-HO-naph) 2AC		3065	371 $_{35}$	329 $_{46}$	286 $_{100}$	270 $_{34}$	144 $_{29}$	1317	8510
Naphthalen-1-yl-(2-methyl-1-propylindol-3-yl)methanone		2995	327 $_{100}$	310 $_{30}$	270 $_{27}$	200 $_{21}$	127 $_{36}$	1076	8521
Naphthalene		1190*	128 $_{100}$	102 $_7$	77 $_3$	64 $_6$	51 $_7$	271	2554
1-Naphthaleneacetic acid		1805*	186 $_{30}$	141 $_{100}$	115 $_{27}$	63 $_6$		361	3647
1-Naphthaleneacetic acid ME		1720*	200 $_{26}$	141 $_{100}$	115 $_{19}$	70 $_4$		399	3648
Naphthalene-M (1,4-di-HO-) 2AC	U+UHYAC	1900*	244 $_{14}$	202 $_{19}$	160 $_{100}$	131 $_{21}$	103 $_8$	570	933
Naphthalene-M (1-HO-) AC	U+UHYAC	1555*	186 $_{13}$	144 $_{100}$	115 $_{47}$	89 $_8$	63 $_7$	361	932
Naphthoflavone (alpha-)		2810	272 $_{64}$	244 $_8$	170 $_{100}$	122 $_{12}$	114 $_{51}$	729	6460
1-Naphthol		1500*	144 $_{100}$	115 $_{83}$	89 $_{14}$	74 $_6$	63 $_{17}$	283	928
1-Naphthol AC	U+UHYAC	1555*	186 $_{13}$	144 $_{100}$	115 $_{47}$	89 $_8$	63 $_7$	361	932
1-Naphthol HFB		1310*	340 $_{46}$	169 $_{25}$	143 $_{28}$	115 $_{100}$	89 $_{11}$	1155	7476
1-Naphthol PFP		1510*	290 $_{45}$	171 $_{100}$	143 $_{20}$	115 $_{49}$	89 $_8$	839	7468
1-Naphthol TMS		1525*	216 $_{100}$	201 $_{95}$	185 $_{51}$	115 $_{39}$	73 $_{21}$	452	7460
Naphthoxyacetic acid methylester		1765*	216 $_{100}$	157 $_{29}$	127 $_{68}$	115 $_{71}$	63 $_{11}$	451	4046
Naphyrone		2330	238 $_1$	155 $_2$	126 $_{100}$	96 $_3$		790	8409
Naphyrone-M (HO-alkyl-oxo) AC		2760	270 $_6$	198 $_{52}$	156 $_{56}$	138 $_{100}$	86 $_{84}$	1228	8692
Naphyrone-M (HO-alkyl-oxo) PFP		2540	374 $_3$	302 $_{88}$	155 $_{55}$	127 $_{100}$	69 $_{39}$	1595	8697
Naphyrone-M (HO-naphtyl-) isomer-1 AC		2595	339 $_1$	171 $_2$	143 $_2$	126 $_{100}$	115 $_5$	1151	8695
Naphyrone-M (HO-naphtyl-) isomer-2 AC		2615	171 $_2$	143 $_4$	126 $_{100}$	115 $_3$		1151	8696
Naphyrone-M (HO-naphtyl-HO-alkyl-) isomer-1 2AC		2845	397 $_1$	184 $_{100}$	171 $_9$	124 $_{36}$	91 $_9$	1436	8691
Naphyrone-M (HO-naphtyl-HO-alkyl-) isomer-1 2PFP		2375	317 $_2$	288 $_{100}$	205 $_9$	192 $_{14}$	124 $_{45}$	1716	8698
Naphyrone-M (HO-naphtyl-HO-alkyl-) isomer-2 2AC		2880	184 $_{100}$	157 $_6$	124 $_{37}$	70 $_{10}$		1436	8693
Naphyrone-M (HO-naphtyl-HO-alkyl-) isomer-2 2PFP		2390	317 $_4$	288 $_{100}$	170 $_7$	124 $_{17}$		1716	8699
Naphyrone-M (HO-naphtyl-N,N-bis-dealkyl-) isomer-1 2AC		2580	327 $_3$	285 $_4$	213 $_{11}$	114 $_{65}$	72 $_{100}$	1075	8689
Naphyrone-M (HO-naphtyl-N,N-bis-dealkyl-) isomer-2 2AC		2665	327 $_1$	213 $_{11}$	171 $_{14}$	114 $_{80}$	72 $_{100}$	1075	8694
Naphyrone-M (HO-naphtyl-oxo-) isomer-1 AC		2735	270 $_2$	171 $_5$	140 $_{100}$	115 $_9$	98 $_{20}$	1227	8687
Naphyrone-M (HO-naphtyl-oxo-) isomer-2 AC		2770	270 $_3$	171 $_{100}$	140 $_{80}$	115 $_4$	98 $_{20}$	1228	8688
Naphyrone-M (N,N-bis-dealkyl-) AC		2295	269 $_2$	210 $_3$	155 $_{22}$	114 $_{51}$	72 $_{100}$	712	8690
Naphyrone-M (oxo-)		2470	212 $_3$	155 $_5$	140 $_{100}$	127 $_{21}$	98 $_{21}$	879	8686
Napropamide		2145	271 $_{12}$	171 $_8$	128 $_{50}$	100 $_{58}$	72 $_{100}$	725	3189
Naproxen	G P U+UHYAC	1780*	230 $_{42}$	185 $_{100}$	170 $_{18}$	141 $_{18}$	115 $_{19}$	510	1733
Naproxen -CO2	G P U+UHYAC	1660*	184 $_{100}$	169 $_{24}$	141 $_{68}$	115 $_{32}$		358	1735
Naproxen ET		1830*	258 $_{45}$	185 $_{100}$	153 $_7$	141 $_8$	115 $_7$	646	4356
Naproxen ME	PME UME U+UHYAC	1800*	244 $_{41}$	185 $_{100}$	170 $_{14}$	141 $_{16}$	115 $_{12}$	571	1734
Naproxen TMS		1735*	302 $_{34}$	287 $_{20}$	243 $_{55}$	185 $_{100}$	73 $_{75}$	919	5218
Naproxen-M (HO-) 2ME	UME	2120*	274 $_{100}$	259 $_{43}$	215 $_{89}$	184 $_{31}$	171 $_{36}$	739	6295
Naproxen-M (O-demethyl-) 2ME	PME UME U+UHYAC	1800*	244 $_{41}$	185 $_{100}$	170 $_{14}$	141 $_{16}$	115 $_{12}$	571	1734
Naproxen-M (O-demethyl-) -CH2O2 AC	UHYAC	1810*	212 $_{14}$	170 $_{100}$	153 $_3$	141 $_{11}$	115 $_{13}$	439	4357
Naproxen-M (O-demethyl-) MEAC	UHYAC	2085*	272 $_{12}$	230 $_9$	171 $_{100}$	141 $_5$	115 $_5$	729	4358
Naptalam -H2O		2545	273 $_{100}$	228 $_{42}$	202 $_{10}$	140 $_{13}$	76 $_{49}$	733	3646
Naratriptan		3210	335 $_{28}$	320 $_{15}$	170 $_{15}$	97 $_{53}$	70 $_{100}$	1125	7505
Naratriptan 2TMS		3360	479 $_{20}$	464 $_{24}$	242 $_{40}$	97 $_{100}$	70 $_{79}$	1633	7504
Naratriptan HFB		2970	531 $_{30}$	516 $_4$	438 $_1$	97 $_8$	70 $_{100}$	1685	7507
Naratriptan TFA		2995	431 $_{16}$	416 $_2$	168 $_7$	96 $_7$	70 $_{100}$	1537	7506
Naratriptan TMS		3220	407 $_{24}$	392 $_{25}$	242 $_{40}$	96 $_{100}$	70 $_{74}$	1468	7503
Narceine artifact 2ME		1870*	254 $_{37}$	223 $_{100}$	207 $_{16}$	191 $_{62}$	77 $_{30}$	627	5152
Narceine -H2O	U+UHYAC	3260	427 $_3$	234 $_5$	58 $_{100}$			1525	5153
Narceine ME		2960	459 $_1$	234 $_1$	178 $_1$	58 $_{100}$		1600	5151
Narcobarbital	P U UHY	1805	223 $_{100}$	181 $_{86}$	138 $_{10}$	124 $_{16}$		917	1143
Narconumal	P G U	1560	209 $_5$	181 $_{100}$	167 $_{11}$	124 $_{24}$	97 $_{19}$	489	1144
Narconumal ME		1520	238 $_2$	220 $_7$	195 $_{100}$	138 $_{50}$	111 $_{26}$	549	1145
Nateglinide 2ME		2695	345 $_2$	272 $_5$	176 $_{100}$	125 $_{32}$	120 $_{44}$	1186	9191

Table 8.1: Compounds in order of names | **Nateglinide 2TMS**

Name	Detected	RI	Typical ions and intensities					Page	Entry
Nateglinide 2TMS		2270	461_{15}	418_{16}	370_{85}	344_{77}	73_{100}	1603	9445
Nateglinide ME		2650	331_{4}	272_{3}	162_{100}	125_{23}	120_{25}	1103	9190
Nateglinide TMS		2430	389_{8}	374_{39}	220_{94}	205_{100}	120_{64}	1400	9446
Nealbarbital	P G U	1720	223_{7}	181_{18}	167_{45}	141_{56}	57_{100}	549	1146
Nealbarbital 2ME		1620	250_{15}	209_{52}	195_{78}	169_{100}	57_{62}	697	1147
Nebivolol 2TMSTFA		2900	645_{1}	494_{16}	404_{26}	177_{77}	73_{100}	1722	6204
Nebivolol 3AC		3540	531_{5}	471_{27}	428_{100}	412_{13}	233_{15}	1685	6107
Neburon ME		2070	288_{10}	202_{5}	174_{8}	114_{63}	57_{100}	826	4158
NECA 2AC		2735	392_{5}	333_{13}	262_{66}	136_{75}	85_{100}	1411	3092
NECA -2H2O		2930	272_{27}	228_{100}	172_{30}	136_{28}	66_{30}	729	3093
NECA 3AC		3265	434_{5}	375_{15}	363_{16}	304_{50}	85_{100}	1544	3091
NEDPA		1710	224_{1}	181_{2}	178_{3}	134_{100}	79_{21}	492	8434
NEDPA AC		2140	267_{1}	196_{3}	176_{46}	165_{7}	134_{100}	702	8435
NEDPA HFB		1920	330_{59}	224_{100}	180_{27}	165_{20}	91_{35}	1510	8440
NEDPA ME		1750	186_{1}	165_{3}	148_{100}	127_{4}	91_{8}	553	8437
NEDPA PFP		1890	280_{50}	180_{16}	174_{100}	165_{8}	91_{31}	1318	8439
NEDPA TFA		1910	230_{56}	180_{19}	124_{100}	91_{18}	79_{16}	1038	8438
NEDPA TMS		1950	282_{1}	206_{100}	190_{8}	165_{8}	134_{5}	892	8436
NEDPA-M (di-HO-benzyl-) 3AC	U+UHYAC	2790	296_{8}	254_{11}	212_{15}	176_{42}	134_{100}	1371	8653
NEDPA-M (HO-benzyl-) isomer-1 2AC	U+UHYAC	2580	238_{6}	196_{44}	176_{30}	134_{100}		1063	8983
NEDPA-M (HO-benzyl-) isomer-2 2AC	U+UHYAC	2620	324_{1}	238_{21}	196_{31}	176_{50}	134_{100}	1063	8651
NEDPA-M (HO-methoxy-benzyl-) 2AC	U+UHYAC	2710	296_{2}	268_{21}	226_{44}	176_{28}	134_{100}	1238	8652
NEDPA-M (nor-di-HO-benzyl-) 2AC	U+UHYAC	2560	254_{2}	212_{3}	178_{7}	148_{62}	106_{100}	988	9120
NEDPA-M (nor-di-HO-benzyl-) isomer-1 3AC	U+UHYAC	2610	296_{1}	254_{11}	212_{10}	148_{46}	106_{100}	1237	9121
NEDPA-M (nor-di-HO-benzyl-) isomer-2 3AC	U+UHYAC	2630	296_{7}	254_{3}	212_{8}	148_{55}	106_{100}	1237	8826
NEDPA-M (nor-di-HO-methoxy-benzyl-) 2AC	U+UHYAC	2680	284_{3}	242_{9}	153_{9}	148_{49}	106_{100}	1175	8827
NEDPA-M (nor-HO-alkyl-) 2AC	U+UHYAC	1740	206_{42}	164_{90}	122_{30}	104_{100}	91_{33}	890	8659
NEDPA-M (nor-HO-benzyl-) isomer-1 2AC	U+UHYAC	2350	238_{7}	196_{12}	148_{47}	106_{100}	79_{11}	890	8984
NEDPA-M (nor-HO-methoxy-benzyl-) 2AC	U+UHYAC	2540	268_{15}	236_{28}	194_{39}	152_{45}	106_{100}	1075	8656
NEDPA-M (nor-HO-methoxy-phenyl-) 2AC	U+UHYAC	2560	236_{29}	194_{45}	152_{73}	137_{10}	106_{100}	1076	8825
NEDPA-M (nor-HO-phenyl-) 2AC	U+UHYAC	2500	296_{1}	206_{98}	164_{79}	122_{100}	91_{25}	890	8655
NEDPA-M 2AC		2345*	207_{44}	165_{80}	123_{100}	105_{81}	77_{45}	981	8933
NEDPA-M 3AC		2350*	355_{2}	296_{9}	208_{27}	166_{100}	123_{93}	1242	8987
NEDPA-M AC		2230*	238_{28}	196_{31}	165_{20}	150_{27}	107_{100}	895	8932
NEDPA-M AC		2080*	254_{1}	212_{4}	105_{100}	77_{34}		628	8934
NEDPA-M AC		2120*	254_{2}	212_{12}	105_{100}	77_{31}		628	8935
NEDPA-M AC		2250*	284_{1}	242_{22}	226_{27}	137_{100}	105_{84}	800	8985
NEDPA-M AC		2360*	270_{7}	228_{19}	153_{21}	123_{15}	105_{100}	716	8986
NEDPA-M isomer-2 2AC	U+UHYAC	2380	238_{13}	196_{14}	148_{57}	106_{100}	79_{16}	890	8654
Nefazodone	U+UHYAC	4510	469_{4}	454_{20}	305_{100}	274_{42}	260_{36}	1617	5305
Nefazodone-M (deamino-HO-)		2340	291_{32}	198_{57}	171_{23}	120_{100}	91_{18}	849	5301
Nefazodone-M (deamino-HO-) AC	U+UHYAC	2500	333_{32}	240_{45}	120_{100}	91_{32}	77_{22}	1113	5302
Nefazodone-M (HO-ethyl-deamino-HO-) 2AC	U+UHYAC	2650	391_{11}	298_{23}	238_{9}	120_{100}	91_{13}	1407	5303
Nefazodone-M (HO-phenyl-) AC	U+UHYAC	4890	527_{1}	512_{4}	361_{100}	332_{19}	209_{50}	1682	5306
Nefazodone-M (HO-phenyl-deamino-HO-) 2AC	U+UHYAC	2830	391_{8}	349_{14}	240_{100}	178_{19}	136_{55}	1407	5304
Nefazodone-M (N-dealkyl-)		1910	196_{24}	154_{100}	138_{12}	111_{9}	75_{12}	390	6885
Nefazodone-M (N-dealkyl-) AC	U+UHYAC	2265	238_{32}	195_{15}	166_{100}	154_{31}	111_{27}	547	405
Nefazodone-M (N-dealkyl-) HFB	U+UHYHFB	1960	392_{100}	195_{38}	166_{41}	139_{36}	111_{25}	1410	6604
Nefazodone-M (N-dealkyl-) ME		1820	210_{100}	166_{19}	139_{81}	111_{33}	70_{99}	433	6886
Nefazodone-M (N-dealkyl-) TFA	U+UHYTFA	1920	292_{100}	250_{12}	195_{77}	166_{79}	139_{66}	854	6597
Nefazodone-M (N-dealkyl-) TMS		2035	268_{85}	253_{41}	128_{88}	101_{96}	86_{100}	707	6888
Nefazodone-M (N-dealkyl-HO-) isomer-1 2AC	U+UHYAC	2515	296_{30}	254_{6}	211_{18}	182_{100}	154_{36}	884	406
Nefazodone-M (N-dealkyl-HO-) isomer-1 2TFA	U+UHYTFA	2040	404_{38}	307_{100}	278_{47}	265_{17}	154_{55}	1458	6600
Nefazodone-M (N-dealkyl-HO-) isomer-1 AC	U+UHYAC	2335	254_{65}	211_{23}	182_{100}	166_{71}	154_{33}	627	5308
Nefazodone-M (N-dealkyl-HO-) isomer-2 2AC	U+UHYAC	2525	296_{20}	254_{74}	182_{100}	169_{72}	154_{24}	884	32
Nefazodone-M (N-dealkyl-HO-) isomer-2 2HFB	U+UHYHFB	2145	604_{43}	585_{9}	407_{100}	378_{22}	154_{22}	1715	6605
Nefazodone-M (N-dealkyl-HO-) isomer-2 2TFA	U+UHYTFA	2045	404_{39}	307_{100}	278_{31}	265_{11}	154_{67}	1458	6598
Nefazodone-M (N-dealkyl-HO-) isomer-2 AC	U+UHYAC	2345	254_{79}	211_{15}	182_{100}	169_{40}	154_{40}	627	5307
Nefazodone-M (N-dealkyl-HO-) TFA	U+UHYTFA	2035	308_{100}	272_{19}	211_{46}	182_{36}	155_{46}	955	6599
Nefopam	G P	2035	253_{2}	225_{8}	179_{21}	165_{9}	58_{100}	623	243
Nefopam-M (HO-) isomer-1 AC	U+UHYAC	2250	311_{11}	238_{61}	195_{100}	165_{55}	87_{25}	977	1164
Nefopam-M (HO-) isomer-2 AC	UHYAC	2285	311_{59}	268_{23}	208_{64}	195_{43}	178_{100}	977	245
Nefopam-M (nor-) AC	UHYAC	2080	281_{9}	208_{100}	194_{30}	179_{22}	87_{61}	788	244
Nefopam-M (nor-di-HO-) -H2O isomer-1 2AC	U+UHYAC	2610	337_{26}	295_{14}	266_{67}	87_{100}		1136	1166
Nefopam-M (nor-di-HO-) -H2O isomer-2 2AC	U+UHYAC	2640	337_{54}	295_{27}	195_{35}	87_{100}		1136	1167
Nelfinavir artifact-1		1940	238_{1}	138_{100}	95_{2}	56_{21}		546	7952
Nelfinavir artifact-2		2025	250_{26}	207_{43}	193_{38}	151_{100}	138_{79}	602	7953
Nelfinavir artifact-3		2100	232_{5}	217_{74}	160_{41}	147_{43}	132_{100}	691	7902
Nelfinavir artifact-4		2240	250_{20}	222_{9}	207_{100}	137_{17}	114_{15}	601	7954
Nelfinavir-M artifact-1		1940	238_{1}	138_{100}	95_{2}	56_{21}		546	7952
Nevirapine		2520	266_{57}	265_{100}	251_{87}	237_{38}	133_{28}	695	7436
Nevirapine AC		2465	308_{15}	265_{53}	251_{100}	133_{74}	78_{55}	958	7437
Nevirapine TMS		2435	338_{44}	337_{49}	323_{68}	249_{32}	73_{100}	1144	7438
Nevirapine-M (HO-) AC	U+UHYAC	2760	324_{25}	295_{31}	264_{76}	249_{100}	236_{61}	1056	7951

Nitrendipine-M/artifact (dehydro-demethyl-) TMS Table 8.1: Compounds in order of names

Name	Detected	RI	Typical ions and intensities					Page	Entry
Nitrendipine-M/artifact (dehydro-demethyl-) TMS	UTMS	2530	416 $_{16}$	401 $_{100}$	371 $_{10}$	327 $_{19}$	178 $_{12}$	1495	5002
5-Nitroacenaphthalene		2065	199 $_{79}$	169 $_{33}$	152 $_{100}$	141 $_{33}$	115 $_{25}$	397	9202
5-Nitroanthracene		2065	223 $_{83}$	193 $_{71}$	176 $_{100}$	165 $_{81}$	151 $_{36}$	482	9203
3-Nitrobiphenyl		1860	199 $_{72}$	152 $_{100}$	141 $_{10}$	126 $_{9}$	76 $_{24}$	397	9198
Nitrofen		2205	283 $_{100}$	253 $_{16}$	202 $_{51}$	139 $_{25}$	75 $_{12}$	793	3861
2-Nitrofluorene		2160	211 $_{62}$	194 $_{25}$	165 $_{100}$	152 $_{17}$	139 $_{13}$	435	9193
Nitrofurantoin ME		2250	252 $_{30}$	206 $_{11}$	167 $_{16}$	140 $_{32}$	114 $_{100}$	615	5226
1-Nitro-naphtalene		1620	173 $_{61}$	145 $_{17}$	127 $_{95}$	115 $_{100}$	101 $_{19}$	328	9195
4-Nitrophenol	P-I UHY	1530	139 $_{100}$	109 $_{22}$	93 $_{18}$	65 $_{57}$		278	829
4-Nitrophenol AC	U+UHYAC	1500	181 $_{67}$	139 $_{100}$	109 $_{78}$	65 $_{76}$	63 $_{78}$	348	830
4-Nitrophenol ME		1455	153 $_{100}$	123 $_{37}$	92 $_{55}$	77 $_{72}$		295	831
1-Nitropyrene		2535	247 $_{100}$	217 $_{39}$	200 $_{93}$	189 $_{69}$	100 $_{40}$	584	9201
Nitrothal-isopropyl		2005	295 $_{8}$	254 $_{59}$	236 $_{100}$	212 $_{66}$	194 $_{79}$	875	3455
Nitroxoline	G P U+UHYAC	1750	190 $_{100}$	160 $_{83}$	116 $_{87}$	89 $_{94}$	63 $_{48}$	369	1918
NM2201		3100	375 $_{1}$	232 $_{100}$	212 $_{2}$	144 $_{14}$	115 $_{9}$	1335	9485
NM2201-M/A (1-naphthol) AC	U+UHYAC	1555*	186 $_{13}$	144 $_{100}$	115 $_{47}$	89 $_{8}$	63 $_{7}$	361	932
NM2201-M/artifact (1-naphthol)		1500*	144 $_{100}$	115 $_{83}$	89 $_{14}$	74 $_{6}$	63 $_{17}$	283	928
NM2201-M/artifact (1 naphthol)		1310*	340 $_{46}$	169 $_{25}$	143 $_{28}$	115 $_{100}$	89 $_{11}$	1155	7476
NM2201-M/artifact (1-naphthol) PFP		1510*	290 $_{45}$	171 $_{19}$	143 $_{20}$	115 $_{49}$	89 $_{8}$	839	7468
NM2201-M/artifact (1-naphthol) TMS		1525*	216 $_{100}$	201 $_{95}$	185 $_{51}$	115 $_{39}$	73 $_{21}$	452	7460
NM2201-M/artifact (HOOC-) (ME)		2100	263 $_{55}$	232 $_{33}$	188 $_{100}$	144 $_{24}$	130 $_{24}$	674	9484
NMPEA		1460	134 $_{3}$	120 $_{100}$	105 $_{9}$	77 $_{13}$	58 $_{25}$	275	6221
NMPEA AC		1430	177 $_{50}$	162 $_{9}$	120 $_{100}$	105 $_{45}$	77 $_{30}$	335	6229
NMT 2HFB		1855	566 $_{3}$	339 $_{100}$	326 $_{64}$	240 $_{84}$	169 $_{76}$	1702	9546
NMT 2PFP		1830	466 $_{1}$	289 $_{100}$	276 $_{72}$	190 $_{52}$	119 $_{37}$	1609	9543
NMT 2TFA		1855	366 $_{2}$	239 $_{100}$	226 $_{73}$	156 $_{12}$	140 $_{63}$	1289	9541
NMT AC		2210	216 $_{14}$	172 $_{2}$	143 $_{100}$	130 $_{92}$	115 $_{10}$	452	9540
NMT HFB		1830	370 $_{3}$	240 $_{18}$	169 $_{20}$	143 $_{40}$	130 $_{100}$	1310	9545
NMT PFP		1830	320 $_{15}$	190 $_{24}$	143 $_{33}$	130 $_{100}$	119 $_{24}$	1033	9544
NMT TFA		1795	270 $_{25}$	143 $_{38}$	130 $_{100}$	115 $_{6}$	102 $_{8}$	717	9542
NNEI		3630	356 $_{14}$	214 $_{100}$	144 $_{24}$	115 $_{14}$		1243	9638
NNEI artifact (1-pentyl-indole-3-carboxylate) (ME)		2150	245 $_{45}$	214 $_{26}$	188 $_{100}$	144 $_{37}$	130 $_{54}$	575	9614
NNEI TMS		3070	428 $_{62}$	339 $_{22}$	287 $_{82}$	214 $_{64}$	73 $_{100}$	1528	9639
NNEI-M/artifact (1-naphthylamine) HFB		1455	339 $_{100}$	170 $_{17}$	142 $_{46}$	127 $_{30}$	115 $_{80}$	1147	9628
NNEI-M/artifact (1-naphthylamine) PFP		1400	289 $_{100}$	170 $_{14}$	142 $_{40}$	127 $_{22}$	115 $_{75}$	832	9631
NNEI-M/artifact (1-naphthylamine) TFA		1440	239 $_{100}$	169 $_{13}$	142 $_{25}$	127 $_{14}$	115 $_{58}$	551	9633
Nomifensine	UHY	2150	238 $_{25}$	194 $_{100}$	178 $_{23}$	165 $_{18}$		549	574
Nomifensine AC	UHYAC	2470	280 $_{36}$	222 $_{100}$	194 $_{30}$	178 $_{34}$		783	362
Nomifensine TMS		2065	310 $_{25}$	266 $_{64}$	237 $_{98}$	193 $_{56}$	73 $_{100}$	972	5478
Nomifensine-M (HO-)	UHY	2450	254 $_{31}$	210 $_{62}$	194 $_{29}$	86 $_{100}$		629	575
Nomifensine-M (HO-) isomer-1 2AC	UHYAC	2850	338 $_{54}$	310 $_{48}$	280 $_{100}$	268 $_{66}$	226 $_{56}$	1144	363
Nomifensine-M (HO-) isomer-2 2AC	UHYAC	2880	338 $_{34}$	308 $_{56}$	280 $_{100}$	268 $_{72}$	194 $_{90}$	1144	364
Nomifensine-M (HO-methoxy-) 2AC	UHYAC	2970	368 $_{31}$	310 $_{94}$	268 $_{100}$	224 $_{81}$		1302	365
Nomifensine-M (HO-methoxy-) isomer-1	UHY	2505	284 $_{84}$	241 $_{80}$	210 $_{100}$	86 $_{74}$		804	576
Nomifensine-M (HO-methoxy-) isomer-2	UHY	2590	284 $_{100}$	241 $_{79}$	210 $_{73}$	86 $_{90}$		804	577
Nonadecane		1900*	268 $_{3}$	99 $_{12}$	85 $_{35}$	71 $_{58}$	57 $_{100}$	709	2363
Nonadecanoic acid ME		2200*	312 $_{35}$	269 $_{16}$	143 $_{18}$	87 $_{60}$	74 $_{100}$	984	3038
Nonane		900*	128 $_{3}$	85 $_{19}$	57 $_{69}$	43 $_{100}$	29 $_{17}$	271	3784
Nonivamide		2530	293 $_{23}$	195 $_{13}$	151 $_{16}$	137 $_{100}$	122 $_{9}$	867	5896
Nonivamide 2TMS		2640	437 $_{11}$	422 $_{8}$	339 $_{56}$	209 $_{100}$	73 $_{53}$	1552	6027
Nonivamide AC		2585	335 $_{3}$	293 $_{39}$	195 $_{25}$	151 $_{19}$	137 $_{100}$	1127	5897
Nonivamide HFB		2385	489 $_{25}$	404 $_{17}$	391 $_{35}$	347 $_{100}$	333 $_{86}$	1648	5900
Nonivamide PFP		2320	439 $_{8}$	354 $_{22}$	341 $_{41}$	297 $_{100}$	283 $_{85}$	1556	5899
Nonivamide TFA		2305	389 $_{6}$	304 $_{20}$	291 $_{53}$	247 $_{100}$	233 $_{98}$	1399	5898
Nonivamide TMS		2880	365 $_{47}$	350 $_{9}$	209 $_{100}$	179 $_{39}$	73 $_{97}$	1288	6028
Noradrenaline (3,4-dihydroxybenzoic acid) ME2AC		1750*	252 $_{1}$	210 $_{19}$	168 $_{100}$	137 $_{60}$	109 $_{8}$	615	5254
Noradrenaline 4AC		2175	337 $_{1}$	294 $_{3}$	181 $_{44}$	139 $_{42}$	73 $_{100}$	1136	8360
Noradrenaline 4TMS		1830	457 $_{1}$	442 $_{1}$	355 $_{13}$	102 $_{29}$	73 $_{100}$	1596	8359
Noradrenaline -H2O 3AC		2170	277 $_{6}$	235 $_{18}$	193 $_{35}$	150 $_{70}$	55 $_{100}$	759	2734
Noradrenaline -H2O 3HFB		1680	739 $_{5}$	542 $_{11}$	328 $_{5}$	169 $_{54}$	69 $_{100}$	1731	8362
Noradrenaline -H2O 3PFP		1595	589 $_{12}$	442 $_{11}$	278 $_{5}$	223 $_{5}$	119 $_{100}$	1711	8363
Noradrenaline -H2O 3TFA		1565	439 $_{10}$	342 $_{8}$	228 $_{3}$	201 $_{4}$	69 $_{100}$	1555	8361
Norcinnamolaurine	U	2955	283 $_{1}$	176 $_{100}$	149 $_{3}$	118 $_{4}$	91 $_{6}$	796	5660
Norcinnamolaurine 2AC	UAC	2930	367 $_{4}$	324 $_{9}$	218 $_{100}$	176 $_{94}$	118 $_{4}$	1296	5662
Norcodeine 2AC	U+UHYAC	2945	369 $_{14}$	327 $_{3}$	223 $_{37}$	87 $_{100}$	72 $_{36}$	1305	226
Norcodeine 2HFB		2580	677 $_{30}$	451 $_{18}$	405 $_{78}$	223 $_{100}$	169 $_{72}$	1727	9342
Norcodeine 2PFP		2540	577 $_{59}$	401 $_{20}$	355 $_{86}$	223 $_{100}$	119 $_{97}$	1706	9341
Norcodeine-M (O-demethyl-) 2PFP		2440	563 $_{100}$	400 $_{10}$	355 $_{38}$	327 $_{7}$	209 $_{15}$	1701	3534
Norcodeine-M (O-demethyl-) 3AC	U+UHYAC	2955	397 $_{8}$	355 $_{9}$	209 $_{41}$	87 $_{100}$	72 $_{33}$	1434	1194
Norcodeine-M (O-demethyl-) 3PFP	UHYPFP	2405	709 $_{80}$	533 $_{28}$	388 $_{29}$	367 $_{51}$	355 $_{100}$	1729	3533
Norcodeine-M (O-demethyl-) 3TMS	UHYTMS	2605	487 $_{17}$	416 $_{19}$	222 $_{36}$	131 $_{19}$	73 $_{100}$	1645	3525
Nordazepam	P G U	2520	270 $_{86}$	269 $_{97}$	242 $_{100}$	241 $_{82}$	77 $_{17}$	715	463
Nordazepam enol AC		2545	312 $_{55}$	270 $_{34}$	241 $_{100}$	227 $_{8}$	205 $_{9}$	980	6102
Nordazepam enol ME		2225	284 $_{79}$	283 $_{100}$	110 $_{3}$	91 $_{62}$		799	464

Table 8.1: Compounds in order of names — Nordazepam HY

Name	Detected	RI	Typical ions and intensities					Page	Entry
Nordazepam HY	UHY	2050	231_{80}	230_{95}	154_{23}	105_{38}	77_{100}	512	419
Nordazepam HYAC	PHYAC U+UHYAC	2245	273_{30}	230_{100}	154_{13}	105_{23}	77_{50}	732	273
Nordazepam TMS		2300	342_{62}	341_{100}	327_{19}	269_{4}	73_{30}	1168	4573
Nordazepam-D5		2515	275_{83}	273_{72}	247_{100}	218_{11}	212_{11}	744	6851
Nordazepam-M (HO-)	UGLUC	2750	286_{82}	258_{100}	230_{11}	166_{7}	139_{8}	813	2113
Nordazepam-M (HO-) AC	U+UHYAC	3000	328_{22}	286_{90}	258_{100}	166_{8}	139_{7}	1079	2111
Nordazepam-M (HO-) HY	UHY	2400	247_{72}	246_{100}	230_{11}	121_{26}	65_{22}	583	2112
Nordazepam-M (HO-) HYAC	U+UHYAC	2270	289_{18}	247_{86}	246_{100}	105_{7}	77_{35}	832	3143
Nordazepam-M (HO-) isomer-1 HY2AC	U+UHYAC	2560	331_{48}	289_{64}	247_{100}	230_{41}	154_{13}	1098	2125
Nordazepam-M (HO-) isomer-2 HY2AC	U+UHYAC	2610	331_{46}	289_{54}	246_{100}	154_{11}	121_{11}	1098	1751
Nordazepam-M (HO-methoxy-) HY2AC	U+UHYAC	2700	361_{17}	319_{72}	276_{100}	260_{14}	246_{10}	1267	1752
Norephedrine	P U	1370	132_{2}	118_{8}	107_{11}	91_{10}	77_{100}	292	2475
Norephedrine 2AC	U+UHYAC	1805	235_{1}	176_{5}	134_{7}	107_{13}	86_{100}	532	2476
Norephedrine 2HFB	UHYHFB	1455	543_{1}	330_{14}	240_{100}	169_{44}	69_{57}	1691	5098
Norephedrine 2PFP	UHYPFP	1380	443_{1}	280_{9}	190_{100}	119_{59}	105_{26}	1567	5094
Norephedrine 2TFA	UTFA	1355	343_{1}	230_{6}	203_{5}	140_{100}	69_{29}	1172	5091
Norephedrine 2TMS		1555	280_{5}	163_{4}	147_{10}	116_{100}	73_{83}	880	4574
Norephedrine formyl artifact		1240	117_{2}	105_{4}	91_{4}	77_{10}	57_{100}	308	4650
Norephedrine TMSTFA		1890	240_{8}	198_{3}	179_{100}	117_{5}	73_{68}	1024	6146
Norephedrine-M (HO-) 3AC	U+UHYAC	2135	234_{6}	165_{8}	123_{11}	86_{100}	58_{45}	862	4961
Norepinephrine 4AC		2175	337_{1}	294_{3}	181_{44}	139_{42}	73_{100}	1136	8360
Norepinephrine 4TMS		1830	457_{1}	442_{1}	355_{13}	102_{29}	73_{100}	1596	8359
Norepinephrine artifact ME2AC		1750*	252_{1}	210_{19}	168_{100}	137_{60}	109_{8}	615	5254
Norepinephrine -H2O 3AC		2170	277_{6}	235_{18}	193_{35}	150_{10}	55_{100}	759	2734
Norepinephrine -H2O 3HFB		1680	739_{5}	542_{15}	328_{5}	169_{54}	69_{100}	1731	8362
Norepinephrine -H2O 3PFP		1595	589_{12}	442_{11}	278_{9}	223_{5}	119_{100}	1711	8363
Norepinephrine -H2O 3TFA		1565	439_{10}	342_{8}	228_{3}	201_{4}	69_{100}	1555	8361
Norethisterone AC		2720*	340_{65}	298_{93}	283_{70}	91_{92}	56_{100}	1158	1498
Norethisterone acetate		2720*	340_{65}	298_{93}	283_{70}	91_{92}	56_{100}	1158	1498
Norethisterone -H2O		2480*	280_{24}	265_{77}	149_{75}	91_{100}	77_{94}	784	4260
Norfenefrine	G	1670	153_{7}	124_{100}	95_{60}	77_{66}	65_{22}	295	4662
Norfenefrine 3AC	U+UHYAC	2085	279_{1}	236_{11}	220_{26}	165_{27}	73_{100}	772	1152
Norfenefrine 3TMS		1785	369_{1}	354_{10}	267_{13}	102_{100}	73_{92}	1307	4575
Norfenefrine formyl artifact	G	2040	165_{22}	146_{22}	136_{100}	107_{39}	77_{51}	313	4664
Norfenefrine-M (deamino-HO-) 3AC	U+UHYAC	1790*	280_{3}	220_{19}	178_{52}	136_{100}	123_{80}	780	1153
Norgestrel		2780*	312_{80}	245_{78}	229_{56}	135_{52}	91_{100}	984	4631
Norgestrel AC		2820*	354_{8}	325_{50}	245_{20}	91_{100}	77_{80}	1233	5234
Norgestrel -H2O		2760*	294_{100}	265_{17}	185_{83}	159_{24}	131_{24}	872	4632
Normethadone	UHYAC	2105	295_{1}	224_{2}	165_{2}	72_{5}	58_{100}	881	246
Normethadone-M (HO-) AC	UHYAC	2505	353_{1}	294_{6}	72_{36}	58_{100}		1229	1198
Normethadone-M (nor-) enol 2AC	UHYAC	2665	365_{2}	323_{4}	267_{52}	193_{41}	86_{100}	1288	1199
Normethadone-M (nor-) -H2O	U UHY UHYAC	2030	263_{100}	220_{15}				677	1197
Normethadone-M (nor-dihydro-) -H2O AC	UHYAC	2850	307_{6}	266_{33}	193_{33}	86_{100}		953	1200
d-Norpseudoephedrine	U UHY	1360	132_{4}	117_{9}	105_{22}	79_{54}	77_{100}	291	1154
d-Norpseudoephedrine 2AC	U+UHYAC	1740	235_{2}	176_{4}	129_{8}	107_{9}	86_{100}	534	1155
d-Norpseudoephedrine 2HFB		1335	330_{16}	303_{6}	240_{100}	169_{19}	119_{12}	1691	7418
d-Norpseudoephedrine formyl artifact		1280	117_{2}	105_{2}	91_{4}	77_{6}	57_{100}	307	4649
d-Norpseudoephedrine TMSTFA		1630	213_{7}	191_{7}	179_{100}	149_{5}	73_{80}	1025	6260
Nortriptyline	P-I G U UHY	2255	263_{27}	220_{67}	202_{100}	189_{39}	91_{30}	677	38
Nortriptyline AC	PAC UHYAC	2660	305_{9}	232_{100}	217_{31}	202_{25}	86_{53}	939	41
Nortriptyline HFB		2420	459_{5}	240_{100}	232_{49}	217_{36}	202_{36}	1599	7685
Nortriptyline PFP		2405	409_{3}	232_{100}	217_{71}	203_{69}	190_{69}	1473	7684
Nortriptyline TFA		2410	359_{3}	232_{76}	217_{54}	202_{70}	140_{100}	1258	7683
Nortriptyline TMS		2340	335_{1}	320_{1}	203_{5}	116_{100}	73_{52}	1127	5440
Nortriptyline-D3		2250	266_{6}	220_{41}	215_{51}	202_{100}	189_{45}	698	7794
Nortriptyline-D3 AC		2655	308_{11}	232_{100}	217_{46}	202_{47}	89_{23}	960	7795
Nortriptyline-D3 HFB		2415	462_{2}	243_{58}	232_{100}	217_{40}	203_{33}	1604	7798
Nortriptyline-D3 PFP		2400	412_{2}	232_{100}	217_{53}	203_{47}	193_{46}	1485	7797
Nortriptyline-D3 TFA		2405	362_{2}	232_{100}	217_{53}	202_{48}	143_{47}	1274	7796
Nortriptyline-D3 TMS		2335	338_{1}	323_{10}	202_{33}	119_{100}	73_{73}	1146	7799
Nortriptyline-M (di-oxo-) AC	U+UHYAC	2790	333_{5}	291_{41}	246_{35}	217_{100}	86_{65}	1111	9705
Nortriptyline-M (HO-)	U-I UGLUC	2390	279_{8}	261_{6}	218_{100}	203_{39}	91_{10}	777	39
Nortriptyline-M (HO-) -H2O	UHY	2600	261_{14}	218_{99}	215_{100}	202_{34}	189_{23}	665	2270
Nortriptyline-M (HO-) -H2O AC	U+UHYAC	2670	303_{20}	230_{100}	215_{74}	202_{34}	86_{18}	925	42
Nortriptyline-M (nor-HO-) -H2O AC	U+UHYAC	2710	289_{15}	230_{100}	215_{70}	202_{31}	189_{5}	835	1873
Noscapine	G U+UHYAC	3130	412_{1}	220_{100}	205_{14}	147_{2}	77_{2}	1487	2525
Noscapine artifact (Meconin)	U+UHYAC	1780*	194_{92}	176_{52}	165_{100}	147_{69}	77_{29}	383	2326
Noscapine-artifact (hydrocotarnine)	U+UHYAC	1790	221_{60}	220_{100}	205_{23}	178_{66}	163_{21}	473	2862
Noscapine-M (demethyl-) artifact AC	U+UHYAC	1975	248_{2}	206_{45}	190_{80}	164_{100}	125_{57}	594	9375
Noscapine-M (demethyl-) isomer-1 AC	UGLUCAC	3380	248_{76}	206_{100}	191_{32}	179_{8}		1561	9374
Noscapine-M (demethyl-) isomer-2 AC	UGLUCAC	3540	248_{91}	206_{100}	191_{30}	179_{12}		1561	9371
Noscapine-M (demethyl-) isomer-3 AC	UGLUCAC	3600	248_{83}	206_{100}	191_{27}	179_{14}		1561	9372
Noscapine-M (demethyl-demethylenyl-methyl-) 2AC	UGLUCAC	3225	292_{77}	250_{100}	208_{78}	193_{56}		1641	9373
Noscapine-M (O-demethyl-) isomer-1 artifact AC	U+UHYAC	1780*	222_{2}	180_{100}	162_{33}	151_{41}	134_{56}	478	9370

Noscapine-M (O-demethyl-) isomer-2 artifact AC

Table 8.1: Compounds in order of names

Name	Detected	RI	Typical ions and intensities					Page	Entry
Noscapine-M (O-demethyl-) isomer-2 artifact AC	U+UHYAC	1825*	222_{4}	180_{100}	162_{26}	151_{30}	134_{32}	478	9418
Noxiptyline		2270	224_{2}	208_{11}	178_{3}	71_{17}	58_{100}	872	366
Noxiptyline-M (HO-dibenzocycloheptanone) -H2O	U+UHYAC	2000*	206_{37}	178_{100}	152_{15}			416	1172
Noxiptyline-M (nor-di-HO-) -H2O 2AC	U+UHYAC	3020	336_{4}	220_{52}	178_{100}	100_{91}		1347	1174
Noxiptyline-M (nor-HO-) -H2O AC	U+UHYAC	2750	221_{9}	205_{57}	178_{100}	100_{31}		1034	1173
Noxiptyline-M/artifact (dibenzocycloheptanone)	G U+UHYAC	1850*	208_{100}	193_{10}	180_{86}	165_{45}	152_{24}	424	1171
NPB-22-M/artifact (HOOC-) (ET)		1990	260_{21}	231_{75}	203_{100}	187_{41}	131_{42}	659	9664
NPB-22-M/artifact (HOOC-) (ME)		1970	246_{23}	231_{40}	189_{100}	176_{35}	145_{23}	581	9497
NPDPA		1720	181_{13}	165_{8}	148_{100}	132_{8}	106_{57}	553	8430
NPDPA AC		2160	281_{1}	180_{11}	148_{100}	106_{41}		789	8431
NPDPA HFB		1925	344_{38}	302_{100}	180_{43}	165_{20}	91_{61}	1548	8433
NPDPA PFP		1910	294_{41}	252_{100}	180_{16}	91_{15}	79_{14}	1381	8444
NPDPA TFA		1930	244_{62}	202_{100}	180_{22}	165_{8}	107_{16}	1124	8432
NPDPA-M (di-HO-benzyl-) 3AC	U+UHYAC	2700	296_{9}	254_{10}	212_{12}	148_{100}	106_{30}	1435	8657
NPDPA-M (HO-benzyl-) 2AC	U+UHYAC	2500	254_{1}	238_{25}	196_{29}	190_{42}	148_{100}	1151	8658
NPDPA-M (nor-di-HO-benzyl-) 2AC	U+UHYAC	2560	254_{2}	212_{3}	178_{7}	148_{62}	106_{100}	988	9120
NPDPA-M (nor-di-HO-benzyl-) iso-1 3AC	U+UHYAC	2610	296_{1}	254_{11}	212_{10}	148_{48}	106_{100}	1237	9121
NPDPA-M (nor-di-HO benzyl-) iso-2 3AC	U+UHYAC	2630	296_{7}	254_{3}	212_{8}	148_{55}	106_{100}	1237	8826
NPDPA-M (nor-di-HO-methoxybenzyl-) 2AC	U+UHYAC	2680	284_{3}	242_{9}	153_{9}	148_{49}	106_{100}	1175	8827
NPDPA-M (nor-HO-alkyl-) 2AC	U+UHYAC	1740	206_{42}	164_{90}	122_{30}	104_{100}	91_{33}	890	8659
NPDPA-M (nor-HO-benzyl-) isomer-1 2AC	U+UHYAC	2350	238_{7}	196_{12}	148_{47}	106_{100}	79_{11}	890	8984
NPDPA-M (nor-HO-methoxy-benzyl-) 2AC	U+UHYAC	2540	268_{15}	236_{28}	194_{39}	152_{45}	106_{100}	1075	8656
NPDPA-M (nor-HO-methoxy-phenyl-) 2AC	U+UHYAC	2560	236_{29}	194_{45}	152_{73}	137_{10}	106_{100}	1076	8825
NPDPA-M (nor-HO-phenyl-) 2AC	U+UHYAC	2500	296_{1}	206_{98}	164_{79}	122_{100}	91_{25}	890	8655
NPDPA-M 2AC		2345*	207_{44}	165_{40}	123_{100}	105_{81}	77_{45}	981	8933
NPDPA-M 3AC		2350*	355_{2}	296_{9}	208_{27}	166_{100}	123_{93}	1242	8987
NPDPA-M AC		2230*	238_{28}	196_{31}	165_{20}	150_{27}	107_{100}	895	8932
NPDPA-M AC		2080*	254_{1}	212_{4}	105_{100}	77_{34}		628	8934
NPDPA-M AC		2120*	254_{2}	212_{12}	105_{100}	77_{31}		628	8935
NPDPA-M AC		2250*	284_{1}	242_{22}	226_{27}	137_{100}	105_{84}	800	8985
NPDPA-M AC		2360*	270_{7}	228_{19}	153_{21}	123_{15}	105_{100}	716	8986
NPDPA-M isomer-2 2AC	U+UHYAC	2380	238_{13}	196_{14}	148_{57}	106_{100}	79_{16}	890	8654
Nuarimol		2390	314_{34}	235_{51}	203_{51}	139_{72}	107_{100}	992	3649
oCPP		1800	196_{19}	161_{14}	154_{100}	138_{18}	111_{12}	390	8561
oCPP AC		2260	238_{36}	195_{15}	166_{100}	154_{45}	138_{38}	547	8562
oCPP HPB		2045	392_{24}	195_{28}	166_{31}	138_{59}	56_{100}	1410	8566
oCPP PFP		1985	342_{27}	195_{24}	166_{43}	139_{44}	56_{100}	1166	8565
oCPP TFA		2010	292_{33}	195_{20}	166_{47}	138_{46}	56_{100}	854	8564
Octacosane		2800*	394_{1}	99_{11}	85_{38}	71_{57}	57_{100}	1423	3797
Octadecane		1800*	254_{4}	141_{5}	85_{42}	71_{66}	57_{100}	630	2351
2-Octadecyloxyethanol		2085*	283_{5}	224_{4}	111_{19}	97_{36}	57_{100}	997	2357
Octamethyldiphenylbicyclohexasiloxane	U	2110*	539_{100}	389_{54}	327_{29}	197_{28}	135_{57}	1697	6457
Octamylamine AC		1570	186_{3}	183_{3}	128_{8}	100_{45}	58_{100}	562	5144
Octane		800*	114_{4}	85_{27}	57_{30}	43_{100}		263	3782
Octanoic acid hexadecylester		2500*	368_{3}	224_{4}	145_{100}	88_{26}	57_{65}	1304	6565
Octodrine AC		1140	171_{2}	156_{2}	128_{5}	86_{100}	60_{43}	326	5255
Octopamine		1720	153_{2}	123_{100}	107_{20}	95_{52}	77_{64}	295	4665
Octopamine 3AC		2245	236_{2}	220_{20}	165_{28}	123_{91}	73_{100}	773	2808
O-Demethyl-Tramadol	U	1995	249_{1}	121_{3}	107_{2}	93_{2}	58_{100}	601	634
O-Demethyl-Tramadol-D6		1980	255_{6}	145_{1}	121_{4}	107_{1}	64_{100}	635	9343
O-Demethyl-Tramadol-D6 2AC		2070	174_{4}	145_{1}	121_{2}	64_{100}		1154	9344
Ofloxacin 2TMS		3300	505_{15}	490_{21}	435_{37}	407_{12}	93_{100}	1665	8246
Ofloxacin -CO2	U+UHYAC	3285	317_{33}	247_{16}	231_{12}	121_{7}	71_{100}	1012	4691
Ofloxacin ME	UME	3750	375_{24}	305_{12}	290_{11}	246_{28}	71_{100}	1335	4692
Olanzapine	P-I G U+UHYAC	2765	312_{24}	242_{100}	229_{85}	213_{55}	198_{21}	982	4675
Olanzapine AC	U+UHYAC	2780	354_{18}	284_{100}	242_{87}	83_{79}	70_{50}	1232	4676
Olanzapine-M (N-methylpiperazine) AC		1230	142_{13}	99_{39}	83_{16}	70_{100}	56_{69}	282	9345
Olanzapine-M (nor-) 2AC	U+UHYAC	3200	382_{100}	339_{59}	284_{84}	254_{59}	213_{46}	1365	4677
Oleamide	P U UHY UHYAC	2385	281_{1}	128_{7}	114_{5}	72_{59}	59_{100}	791	5345
Oleic acid ET		2095*	310_{5}	264_{33}	101_{82}	88_{100}	55_{72}	973	5405
Oleic acid glycerol ester 2AC	UHYAC	2790*	380_{34}	264_{71}	159_{100}	81_{49}	69_{85}	1559	5602
Oleic acid ME		2085*	296_{3}	264_{15}	222_{9}	97_{30}	55_{100}	887	2667
Oleic acid TMS		2620*	354_{4}	339_{34}	129_{45}	117_{56}	73_{100}	1234	4522
Olmesartan		2875	341_{23}	312_{15}	192_{100}	165_{26}	121_{29}	1573	8152
Olmesartan AC		3210	398_{49}	383_{21}	249_{94}	208_{31}	149_{100}	1646	8154
Olmesartan -CO2 TMS		2835	474_{5}	459_{6}	368_{4}	192_{100}	165_{31}	1626	8153
Olmesartan HFB		2860	552_{54}	524_{35}	403_{100}	149_{23}	121_{23}	1722	8157
Olmesartan PFP		2830	502_{56}	474_{41}	353_{100}	149_{59}	121_{72}	1712	8156
Olmesartan TFA		2885	452_{59}	424_{43}	303_{100}	149_{55}	121_{63}	1690	8155
Olsalazine 4TMS		2845	590_{3}	575_{31}	266_{10}	191_{13}	73_{100}	1712	8147
Omeprazole artifact-1		1180	151_{100}	136_{62}	121_{25}	106_{13}	77_{20}	290	8618
Omeprazole artifact-2	P-I	1365	165_{100}	150_{14}	137_{92}	122_{75}	107_{77}	312	8617
Omeprazole artifact-3	P-I	1510	167_{100}	152_{31}	138_{29}	123_{38}	92_{31}	318	8619
Omeprazole artifact-4		1680	179_{100}	164_{35}	150_{26}	136_{32}	121_{14}	339	8620

Table 8.1: Compounds in order of names — Omeprazole artifact-5

Name	Detected	RI	Typical ions and intensities					Page	Entry
Omeprazole artifact-5	P-I	1750	162_{100}	147_{85}	119_{28}	92_{27}	65_{25}	305	8615
Omeprazole artifact-6	P-I	1790	148_{100}	133_{95}	120_{5}	105_{57}	78_{19}	286	8621
Omeprazole artifact-7		1960	183_{100}	150_{66}	120_{54}	93_{22}	77_{25}	353	8616
Omeprazole artifact-8		2030	227_{33}	212_{100}	180_{67}	165_{68}	136_{53}	497	8613
Omeprazole artifact-9	P-I	2080	194_{89}	179_{53}	161_{100}	152_{30}	133_{20}	381	8612
Omeprazole artifact-9 ME	P-I	2050	208_{84}	193_{45}	175_{100}	160_{30}	147_{21}	422	8614
Omethoate	G P-I	1585	213_{7}	156_{91}	110_{100}	79_{52}	58_{70}	441	1501
4'-O-Methyl-sceletone		2190	257_{28}	214_{8}	189_{10}	107_{10}	70_{100}	643	8988
Omoconazole		2925	422_{13}	387_{21}	267_{44}	111_{75}	69_{100}	1512	5560
Omoconazole HY		2110	268_{15}	233_{12}	173_{100}	145_{18}	95_{95}	705	6079
Omoconazole HYAC		2185	310_{14}	268_{52}	233_{24}	95_{53}	69_{100}	969	6078
Opipramol	G P UHY	3055	363_{100}	232_{45}	218_{67}	206_{85}	70_{56}	1280	578
Opipramol AC	U+UHYAC	3170	405_{45}	232_{31}	218_{52}	206_{100}	70_{36}	1463	367
Opipramol TMS		3150	435_{47}	232_{14}	206_{49}	113_{32}	73_{43}	1549	4576
Opipramol-M (HO-) 2AC	U+UHYAC	3330	463_{100}	403_{24}	264_{20}	185_{35}	70_{37}	1606	2675
Opipramol-M (HO-methoxy-ring)	U UHY	2340	239_{100}	224_{47}	209_{42}	180_{74}		551	423
Opipramol-M (HO-methoxy-ring) AC	U+UHYAC	2420	281_{42}	239_{100}	224_{28}	196_{29}	162_{16}	786	2506
Opipramol-M (HO-ring)	UHY	2240	209_{100}	180_{16}	152_{7}			428	2511
Opipramol-M (HO-ring) 2AC	U+UHYAC	2490	293_{21}	251_{25}	209_{79}	208_{100}	178_{17}	861	2672
Opipramol-M (HO-ring) AC	UHYAC	2450	251_{33}	209_{100}	180_{74}	152_{11}		609	425
Opipramol-M (N-dealkyl-) AC	U+UHYAC PHYAC-I	3190	361_{100}	232_{45}	193_{84}	141_{44}	99_{43}	1271	427
Opipramol-M (N-dealkyl-) ME		2685	333_{40}	232_{23}	218_{24}	113_{57}	70_{100}	1117	3193
Opipramol-M (N-dealkyl-di-HO-oxo-) 2AC	U+UHYAC	3300	449_{100}	264_{27}	222_{34}	171_{56}	98_{31}	1579	2674
Opipramol-M (N-dealkyl-HO-oxo-) AC	U+UHYAC	3050	391_{100}	232_{37}	206_{49}	171_{41}	98_{35}	1408	2673
Opipramol-M (ring)	P U UHY U+UHYAC	1985	193_{100}	165_{19}	139_{5}	113_{3}	96_{9}	378	309
Opipramol-M (ring) AC	U+UHYAC	2040	235_{27}	193_{100}	192_{68}	165_{17}		532	2671
Opipramol-M/artifact (acridine)	U UHY U+UHYAC	1800	179_{100}	151_{14}				340	421
Orciprenaline 2TMSTFA		2150	451_{1}	436_{1}	283_{100}	126_{10}	73_{93}	1582	6167
Orciprenaline 3TMS		1740	412_{4}	356_{89}	322_{2}	147_{2}	72_{100}	1527	5484
Orciprenaline 3TMSTFA		2100	523_{1}	355_{100}	126_{13}	73_{92}		1679	6166
Orciprenaline 4AC		2370	379_{1}	319_{47}	277_{100}	235_{38}	72_{78}	1351	1342
Orciprenaline 4TMS		1975	484_{4}	144_{100}	102_{5}	73_{83}		1658	5485
Orciprenaline TMSTFA		2180	379_{1}	322_{3}	241_{12}	211_{100}	73_{79}	1350	6168
Orlistat-M/artifact (alcohol) -H2CO3		2820*	292_{40}	160_{24}	142_{25}	114_{100}	69_{60}	859	5862
Orlistat-M/artifact (alcohol) -H2O		2540*	336_{27}	181_{50}	155_{100}	109_{39}	55_{44}	1134	5861
Ornidazole		1825	219_{28}	172_{36}	112_{56}	81_{84}	53_{100}	463	1834
Ornidazole AC	U+UHYAC	1815	261_{36}	219_{69}	173_{34}	135_{83}	53_{100}	661	1836
Ornidazole -HCl		1730	183_{66}	166_{16}	152_{24}	108_{28}	54_{100}	354	1835
Orphenadrine	P-I G U	1935	181_{4}	165_{4}	73_{22}	58_{100}		713	1156
Orphenadrine HY	UHY	1760*	198_{45}	180_{58}	165_{28}	119_{87}	77_{100}	395	1159
Orphenadrine HYAC	U+UHYAC	1750*	240_{2}	180_{100}	165_{12}			557	1161
Orphenadrine-M	UHY	1560*	182_{82}	167_{100}	108_{43}	107_{45}		352	1157
Orphenadrine-M (methyl-benzophenone)	UHY UHYAC	1700*	196_{60}	195_{100}	165_{13}	91_{67}	77_{78}	391	1158
Orphenadrine-M (nor-)	UHY	1900	255_{47}	180_{100}	165_{41}	86_{72}		634	1160
Orphenadrine-M HYAC	U+UHYAC	2005	239_{34}	180_{100}	165_{33}			550	1162
Oryzalin		2680	346_{8}	317_{100}	275_{50}	258_{9}	75_{9}	1188	4055
Oryzalin -SO2NH		2025	267_{12}	238_{100}	222_{11}	196_{64}	138_{8}	699	4056
Oseltamivir 2TMS		2330	441_{5}	325_{44}	312_{28}	254_{34}	73_{100}	1594	7435
Oseltamivir AC		2590	354_{1}	212_{18}	142_{100}	100_{46}	96_{47}	1233	7429
Oseltamivir formyl artifact		2350	324_{12}	253_{54}	142_{60}	112_{86}	96_{100}	1058	7433
Oseltamivir formyl artifact ME		2465	338_{8}	250_{61}	225_{60}	126_{100}	83_{75}	1146	7434
Oseltamivir HFB		2375	421_{6}	333_{12}	212_{26}	142_{72}	96_{100}	1668	7432
Oseltamivir PFP		2385	412_{9}	371_{16}	212_{34}	142_{78}	96_{100}	1597	7431
Oseltamivir TFA		2410	362_{5}	321_{11}	212_{26}	142_{67}	96_{100}	1471	7430
Oxabolone		2640*	290_{68}	147_{39}	126_{88}	91_{77}	55_{100}	843	3947
Oxabolone 2TMS		2695*	434_{1}	419_{100}	329_{2}	303_{2}	73_{45}	1546	3950
Oxabolone AC		2820*	332_{100}	290_{63}	272_{92}	147_{48}	79_{42}	1107	3948
Oxabolone cipionate		3660*	414_{30}	290_{71}	147_{34}	125_{86}	55_{100}	1492	3946
Oxabolone cipionate TMS		3580*	486_{1}	471_{100}	329_{2}	181_{4}	73_{40}	1644	3949
Oxaceprol ME		1635	187_{7}	128_{54}	86_{100}	68_{28}		363	2283
Oxaceprol MEAC		1690	229_{1}	198	169_{9}	110_{37}	68_{100}	506	2709
Oxadiazon		2125	344_{32}	302_{37}	258_{58}	175_{100}	57_{54}	1177	4036
Oxadixyl		2280	278_{7}	233_{12}	163_{75}	132_{54}	105_{100}	768	2517
Oxamyl -C2H3NO		1630	162_{13}	145_{13}	115_{13}	99_{9}	72_{100}	305	3904
Oxapadol		2625	278_{100}	248_{89}	219_{82}	105_{80}	77_{82}	767	1502
Oxatomide		3200	426_{7}	219_{27}	204_{66}	167_{93}	125_{100}	1523	1673
Oxatomide-M (carbinol)	UHY	1645*	184_{45}	165_{14}	152_{7}	105_{100}	77_{63}	357	1333
Oxatomide-M (carbinol) AC	U+UHYAC	1700*	226_{10}	184_{20}	165_{100}	105_{14}	77_{35}	495	1241
Oxatomide-M (carbinol) HFB		1475*	380_{3}	183_{4}	166_{100}	152_{22}	83_{25}	1355	8146
Oxatomide-M (carbinol) ME	UHY	1655*	198_{70}	167_{94}	121_{100}	105_{56}	77_{71}	395	6779
Oxatomide-M (carbinol) PFP		1410*	330_{4}	183_{4}	166_{100}	152_{24}	83_{30}	1093	8145
Oxatomide-M (carbinol) TFA		1420*	280_{19}	183_{7}	166_{100}	152_{24}	83_{21}	780	8144
Oxatomide-M (carbinol) TMS		1540*	256_{28}	241_{14}	179_{38}	167_{100}	152_{17}	638	8159
Oxatomide-M (HO-BPH) isomer-1	UHY	2065*	198_{93}	121_{72}	105_{100}	93_{22}	77_{66}	393	1627

Oxatomide-M (HO-BPH) isomer-1 AC

Table 8.1: Compounds in order of names

Name	Detected	RI	Typical ions and intensities					Page	Entry
Oxatomide-M (HO-BPH) isomer-1 AC	U+UHYAC	2010*	240_{27}	198_{100}	121_{47}	105_{85}	77_{80}	555	2196
Oxatomide-M (HO-BPH) isomer-2	P-I U UHY	2080*	198_{50}	121_{100}	105_{17}	93_{14}	77_{28}	394	732
Oxatomide-M (HO-BPH) isomer-2 AC	U+UHYAC	2050*	240_{20}	198_{100}	121_{94}	105_{41}	77_{51}	555	2197
Oxatomide-M (N-dealkyl-)	U UHY	2120	252_{12}	207_{58}	167_{100}	152_{33}	85_{49}	620	1602
Oxatomide-M (norcyclizine) AC	U+UHYAC	2525	294_{16}	208_{56}	167_{100}	152_{30}	85_{78}	871	1601
Oxazepam	P G UGLUC	2320	268_{98}	239_{56}	233_{52}	205_{66}	77_{100}	813	579
Oxazepam 2ME		2425	314_{25}	271_{100}	239_{45}	205_{26}		993	581
Oxazepam 2TMS		2200	430_{51}	429_{89}	340_{15}	313_{19}	73_{100}	1533	5499
Oxazepam artifact-1	P-I UHY U+UHYAC	2060	240_{59}	239_{100}	205_{81}	177_{16}	151_{9}	554	300
Oxazepam artifact-2	UHY UHYAC	2070	254_{77}	253_{100}	219_{98}	111_{5}		626	301
Oxazepam artifact-3	G P U	2500	298_{78}	240_{100}	203_{65}			892	1257
Oxazepam -H2O 2ME		2575	296_{80}	295_{100}	267_{97}	239_{58}	205_{61}	882	582
Oxazepam HY	UHY	2050	231_{80}	230_{95}	154_{23}	105_{38}	77_{100}	512	419
Oxazepam HYAC	PHYAC U+UHYAC	2245	273_{30}	230_{100}	154_{13}	105_{23}	77_{50}	732	273
Oxazepam TMS		2635	356_{9}	341_{100}	312_{56}	239_{12}	135_{21}	1252	4577
Oxazepam-M	P G U	2520	270_{86}	269_{97}	242_{100}	241_{82}	77_{17}	715	463
Oxazepam-M (HO-) artifact AC	UHYAC	2515	312_{30}	270_{100}	253_{46}	235_{76}	206_{9}	980	1747
Oxazepam-M (HO-) HYAC	U+UHYAC	2270	289_{18}	247_{86}	246_{100}	105_{7}	77_{35}	832	3143
Oxazepam-M TMS		2300	342_{62}	341_{100}	327_{19}	269_{4}	73_{30}	1168	4573
Oxazolam		2540	328_{4}	283_{11}	251_{100}	70_{43}		1079	1168
Oxazolam HYAC	PHYAC U+UHYAC	2245	273_{30}	230_{100}	154_{13}	105_{23}	77_{50}	732	273
Oxazolam-M	P G UGLUC	2320	268_{98}	239_{56}	233_{52}	205_{66}	77_{100}	813	579
Oxazolam-M 2TMS		2200	430_{51}	429_{89}	340_{15}	313_{19}	73_{100}	1533	5499
Oxazolam-M HY	UHY	2050	231_{80}	230_{95}	154_{23}	105_{38}	77_{100}	512	419
Oxazolam-M TMS		2635	356_{9}	341_{100}	312_{56}	239_{12}	135_{21}	1252	4577
Oxcarbazepine		2375	252_{35}	209_{65}	180_{100}	152_{28}	89_{19}	616	6065
Oxcarbazepine artifact (acridinecarboxylic acid) (ME)		2165	237_{100}	206_{58}	178_{81}	151_{27}	75_{9}	543	6066
Oxcarbazepine artifact (carbamazepine)	P G U+UHYAC	2285	236_{83}	193_{100}	165_{31}			538	420
Oxcarbazepine enol AC		2575	294_{9}	252_{84}	209_{100}	180_{87}	152_{25}	869	6067
Oxcarbazepine-M (dihydro-)	P	2385	254_{23}	210_{30}	193_{100}	180_{66}	167_{28}	628	8609
Oxcarbazepine-M (dihydro-) artifact (ring)	P	2110	211_{100}	194_{47}	182_{42}	180_{43}	167_{25}	436	8610
Oxcarbazepine-M (dihydro-) artifact (ring) ME	P	2060	225_{78}	210_{11}	194_{100}	180_{43}	167_{15}	491	8611
Oxcarbazepine-M/artifact	UHY	2240	209_{100}	180_{16}	152_{7}			428	2511
Oxcarbazepine-M/artifact (HO-ring) AC	U+UHYAC	2490	293_{21}	251_{25}	209_{79}	208_{100}	178_{17}	861	2672
Oxcarbazepine-M/artifact (ring)	P U UHY U+UHYAC	1985	193_{100}	165_{19}	139_{5}	113_{3}	96_{9}	378	309
Oxcarbazepine-M/artifact (ring) AC	UHYAC	2450	251_{33}	209_{100}	180_{74}	152_{11}		609	425
Oxcarbazepine-M/artifact AC	UHYAC	3195	340_{81}	298_{95}	297_{100}	241_{6}	179_{16}	1154	426
Oxcarbazepine-M/artifact AC	U+UHYAC	2040	235_{27}	193_{100}	192_{68}	165_{17}		532	2671
Oxedrine 3AC		2175	293_{6}	233_{44}	191_{66}	149_{100}	86_{93}	861	1530
Oxeladin	P U+UHYAC	2180	335_{1}	320_{1}	219_{1}	144_{8}	86_{100}	1128	1163
Oxetacaine AC		2550	318_{3}	287_{2}	188_{69}	91_{20}	87_{100}	1669	6070
Oxiconazole		3290	427_{2}	392_{18}	240_{8}	159_{100}	81_{55}	1524	2824
Oxilofrine (erythro-)		1875	148_{1}	95_{1}	77_{4}	71_{6}	58_{100}	349	1971
Oxilofrine (erythro-) 3AC	U+UHYAC	2145	307_{1}	247	100_{72}	58_{100}		950	750
Oxilofrine (erythro-) formyl artifact		1790	133_{2}	121_{10}	107_{6}	71_{100}	56_{76}	380	4499
Oxilofrine (erythro-) -H2O 2AC	U+UHYAC	1990	247_{9}	205_{8}	163_{24}	107_{8}	56_{100}	585	1972
Oxilofrine (erythro-) ME2AC		2000	279_{1}	247	206_{3}	100_{60}	58_{100}	776	2348
Oxomemazine	G U UHY UHYAC	2830	330_{11}	271_{3}	180_{2}	152_{2}	58_{100}	1094	1768
Oxomemazine-M (bis-nor-)	UHY	2785	272_{100}	244_{22}	231_{8}	180_{8}	152_{6}	918	1770
Oxomemazine-M (bis-nor-) AC	UHYAC	3035	344_{42}	272_{66}	244_{100}	231_{68}	114_{24}	1178	1772
Oxomemazine-M (nor-)	UHY	2720	316_{38}	271_{63}	231_{100}	180_{10}	152_{8}	1005	1769
Oxomemazine-M (nor-) AC	UHYAC	3125	358_{18}	272_{17}	244_{41}	128_{100}	86_{18}	1253	1771
Oxprenolol	P-I G	1970	265_{1}	250_{1}	221_{5}	150_{3}	72_{100}	690	4256
Oxprenolol 2AC	PAC-I	2390	349_{1}	289_{14}	200_{94}	98_{26}	72_{100}	1206	1336
Oxprenolol 2TMS		2070	394_{1}	222_{16}	144_{100}	101_{9}	73_{78}	1476	5476
Oxprenolol formyl artifact	P-I G	1985	277_{17}	262_{77}	248_{44}	148_{48}	56_{100}	763	1339
Oxprenolol -H2O AC	PAC-I UHYAC	2260	289_{12}	188_{16}	140_{16}	98_{16}	72_{100}	837	1335
Oxprenolol TMS		1850	337_{1}	322_{1}	221_{7}	150_{9}	72_{100}	1141	5475
Oxprenolol TMSTFA		2135	433_{1}	418_{1}	284_{100}	129_{77}	73_{85}	1542	6163
Oxprenolol-M (deamino-HO-) 2AC	UHYAC	1900*	308_{2}	249_{8}	159_{100}	99_{16}		957	1334
Oxprenolol-M (deamino-HO-dealkyl-)	UHY	1700*	184_{26}	135_{4}	110_{100}	81_{8}	64_{5}	357	2683
Oxprenolol-M (deamino-HO-dealkyl-) 3AC	U+UHYAC	1920*	310_{1}	268_{13}	159_{100}	117_{12}	110_{19}	970	800
Oxprenolol-M (HO-) -H2O isomer-1 2AC	UHYAC	2520	347_{5}	305_{7}	200_{44}	72_{100}		1195	1337
Oxprenolol-M (HO-) -H2O isomer-2 2AC	UHYAC	2570	347_{13}	305_{5}	204_{19}	72_{100}		1196	1338
Oxprenolol-M (HO-) isomer-1 3AC	UHYAC	3050	407_{5}	347_{50}	305_{17}	204_{70}	72_{100}	1468	1340
Oxprenolol-M (HO-) isomer-2 3AC	UHYAC	3100	407_{10}	347_{27}	200_{72}	140_{40}	72_{100}	1468	1341
Oxybenzone	UHY	2135*	228_{64}	227_{90}	151_{100}	105_{11}	77_{26}	501	3662
Oxybenzone AC	UHYAC	2225*	270_{5}	227_{100}	151_{56}	105_{12}	77_{29}	716	3663
Oxybenzone-M (O-demethyl-)	UHY	2280*	214_{61}	213_{83}	137_{100}	105_{21}	77_{33}	445	3660
Oxybenzone-M (O-demethyl-) 2AC	UHYAC	2315*	298_{3}	256_{45}	213_{100}	137_{21}	77_{18}	893	3661
Oxyberberine		2995	351_{100}	336_{70}	322_{49}	308_{28}	292_{26}	1212	5661
Oxybuprocaine		2425	236_{8}	192_{18}	136_{8}	99_{42}	86_{100}	960	1943
Oxybuprocaine AC		2640	335_{1}	278_{7}	234_{11}	99_{48}	86_{100}	1209	1944
Oxybuprocaine-M (HOOC-) AC		2060	251_{51}	220_{10}	195_{22}	167_{100}	136_{82}	610	1946

Table 8.1: Compounds in order of names Oxybuprocaine-M (HOOC-) MEAC

Name	Detected	RI	Typical ions and intensities					Page	Entry
Oxybuprocaine-M (HOOC-) MEAC		2100	265 40	234 6	223 14	167 100	136 29	687	1945
Oxybutynine		2505	357 5	342 88	189 41	107 55	55 100	1250	3724
Oxycodone	G	2540	315 100	258 18	230 35	140 16	70 37	1000	583
Oxycodone AC	U+UHYAC	2555	357 100	314 42	298 15	240 15		1248	247
Oxycodone enol 2AC		2560	399 100	357 33	314 15	240 21		1442	248
Oxycodone enol 2TMS		2510	459 100	444 22	368 12	312 16	73 74	1600	4321
Oxycodone HFB		2330	511 100	314 70	240 62	115 52	69 77	1670	6152
Oxycodone PFP		2350	461 100	314 56	240 45	212 31	119 31	1603	6119
Oxycodone TFA		2290	411 100	314 33	240 28	115 34	54 67	1480	4013
Oxycodone TMS		2555	387 100	372 23	330 6	229 11	73 34	1391	4322
Oxycodone-D3 AC	U+UHYAC	2545	360 100	317 59	301 18	233 16	215 17	1265	9340
Oxycodone-D6		2535	321 100	236 44	204 33	143 24	115 26	1041	7296
Oxycodone-D6 TMS		2555	393 97	378 19	276 17	236 30	73 100	1417	7297
Oxycodone-M (dihydro-) 2AC	U+UHYAC	2570	401 60	359 100	242 64	70 60		1450	1189
Oxycodone-M (nor-) enol 3AC	U+UHYAC	2680	427 63	385 100	343 23	87 26		1525	1190
Oxycodone-M (nor-dihydro-) 2AC	U+UHYAC	2900	387 14	343 100	258 59	87 24	72 21	1390	1191
Oxycodone-M (nor-dihydro-) 3AC	U+UHYAC	2935	429 3	387 37	242 100	87 19	72 14	1531	1192
Oxycodone-M (O-demethyl-)		2555	301 100	244 9	216 55	203 22	70 50	913	7166
Oxycodone-M (O-demethyl-) 2AC	U+UHYAC	2620	385 28	343 100	300 49	284 29	70 24	1381	7168
Oxycodone-M (O-demethyl-) 2TMS		2570	445 18	288 9	229 10	216 7	73 100	1572	7170
Oxycodone-M (O-demethyl-) 3TMS		2525	517 7	502 8	412 3	355 10	73 100	1675	7171
Oxycodone-M (O-demethyl-) AC	U+UHYAC	2650	343 16	301 100	216 43	203 30	70 42	1174	7167
Oxycodone-M (O-demethyl-) TMS		2560	373 94	288 51	259 41	73 100	70 91	1328	7169
Oxydemeton-S-Methyl	G P-I	1860*	218 1	169 60	125 47	109 100	79 26	578	1500
Oxyfedrine-M (N-dealkyl-)	U UHY	1360	132 4	117 9	105 22	79 54	77 100	291	1154
Oxyfedrine-M (N-dealkyl-) 2AC	U+UHYAC	1740	235 2	176 4	129 8	107 9	86 100	534	1155
Oxyfedrine-M (N-dealkyl-) 2HFB		1335	330 16	303 6	240 100	169 19	119 12	1691	7418
Oxyfedrine-M (N-dealkyl-) formyl artifact		1280	117 2	105 2	91 4	77 6	57 100	307	4649
Oxyfedrine-M (N-dealkyl-) TMSTFA		1630	213 7	191 7	179 100	149 5	73 80	1025	6260
Oxymetazoline	U+UHYAC	2195	260 100	245 93	217 44	81 51		660	1503
Oxymetazoline 2AC		2760	344 60	302 100	287 50	230 40	203 40	1180	1504
Oxymetholone		3005*	332 19	275 21	174 58	91 84	55 100	1108	2823
Oxymetholone enol 3TMS		2870*	548 100	490 23	405 7	281 10	73 75	1694	3983
Oxymorphone		2555	301 100	244 9	216 55	203 22	70 50	913	7166
Oxymorphone 2AC	U+UHYAC	2620	385 28	343 100	300 49	284 29	70 24	1381	7168
Oxymorphone 2TMS		2570	445 18	288 9	229 10	216 7	73 100	1572	7170
Oxymorphone 3TMS		2525	517 7	502 8	412 3	355 10	73 100	1675	7171
Oxymorphone AC	U+UHYAC	2650	343 16	301 100	216 43	203 30	70 42	1174	7167
Oxymorphone TMS		2560	373 94	288 51	259 41	73 100	70 91	1328	7169
Oxypertine		3445	379 11	217 10	175 100	132 16	70 30	1353	368
Oxypertine-M (HO-phenylpiperazine) 2AC	U+UHYAC	2350	262 39	220 60	177 21	148 100	135 59	670	6610
Oxypertine-M (phenylpiperazine) AC	UHYAC	1920	204 48	161 21	132 100	56 77		410	1276
Oxyphenbutazone		9999	324 30	199 90	135 25	93 100	77 70	1057	1513
Oxyphenbutazone AC	U+UHYAC	2700	366 33	324 49	199 100	93 80	77 72	1292	1506
Oxyphenbutazone artifact (phenyldiazophenol)		2070	198 46	121 41	93 100	77 72	65 54	394	1027
Oxyphenbutazone artifact (phenyldiazophenol) ME		2020	212 41	135 35	107 91	77 100	64 26	439	4205
Oxyphenbutazone isomer-1 2ME	UME	2545	352 100	213 66	148 18	107 36	77 65	1222	1505
Oxyphenbutazone isomer-2 2ME		2720	352 100	309 26	190 41	160 46	77 63	1222	1507
Oxyphencyclimine -H2O		2405	326 90	243 100	171 16	127 31	105 41	1070	6308
Oxyphencyclimine-M/artifact (HOOC-) ME		1755	248 1	189 81	166 23	105 100	77 69	591	6309
Palmitamide	P U UHY UHYAC	2130	255 1	212 2	128 5	72 37	59 100	635	5344
Palmitic acid	G P U UHY UHYAC	1965*	256 29	213 20	185 15	129 38	73 100	639	822
Palmitic acid ET		1950*	284 4	241 6	157 18	101 56	88 100	806	5403
Palmitic acid glycerol ester	G	2420*	313 3	299 11	239 49	98 100	57 99	1096	5588
Palmitic acid glycerol ester 2AC	UHYAC	2645*	354 4	239 34	159 100	98 29	84 18	1492	5412
Palmitic acid glycerol ester 2TMS	G	2620*	460 2	372 46	205 15	147 36	73 100	1626	7449
Palmitic acid ME	G P U UHY UHYAC	1940*	270 6	227 4	143 9	87 39	74 100	721	1801
Palmitic acid TMS		2470*	328 6	313 62	132 46	117 93	73 100	1083	4668
Palmitoleic acid TMS		2450*	326 3	311 33	129 52	117 69	73 100	1071	4669
Pangamic acid-M/artifact (gluconic acid) ME5AC		1975*	361 3	289 14	173 58	145 72	115 100	1508	5227
Panthenol		1920	175 5	157 14	133 100	102 24		413	1522
Panthenol 3AC	U+UHYAC	2045	331 1	217 92	175 77	145 51	115 100	1101	1509
Panthenol artifact		1920	189 26	159 58	145 16	71 100		367	823
Papaverine	G P U+UHYAC	2820	339 83	338 100	324 44	308 20	293 10	1150	824
Papaverine-M (bis-demethyl-) isomer-1 2AC	UHYAC	2970	395 31	353 56	310 100	294 15	179 10	1425	3689
Papaverine-M (bis-demethyl-) isomer-2 2AC	UHYAC	2995	395 29	353 50	310 100	294 25	196 8	1424	3690
Papaverine-M (bis-demethyl-) isomer-3 2AC	UHYAC	3050	395 28	353 50	310 100	294 21	179 4	1424	3691
Papaverine-M (bis-demethyl-) isomer-4 2AC	UHYAC	3065	395 29	353 61	310 100	294 22	179 10	1425	3692
Papaverine-M (O-demethyl-)	UHY	2805	325 77	324 100	310 80	266 13	153 9	1061	3684
Papaverine-M (O-demethyl-) isomer-1 AC	UHYAC	2860	367 72	324 100	310 50	278 7	153 13	1297	3685
Papaverine-M (O-demethyl-) isomer-2 AC	UHYAC	2895	367 73	324 100	310 60	296 45	254 53	1296	3686
Papaverine-M (O-demethyl-) isomer-3 AC	UHYAC	2910	367 74	324 100	308 26	254 11	153 10	1297	3687
Papaverine-M (O-demethyl-) isomer-4 AC	U+UHYAC	2940	367 47	324 100	310 77	294 14	137 5	1297	3688
Paracetamol	G P U	1780	151 34	109 100	81 16	80 22		291	825

Paracetamol 2AC

Table 8.1: Compounds in order of names

Name	Detected	RI	Typical ions and intensities					Page	Entry
Paracetamol 2AC	U+UHYAC	2085	235_{10}	193_{11}	151_{30}	109_{100}		532	827
Paracetamol 2TMS		1780	295_{50}	280_{68}	206_{83}	116_{15}	73_{100}	878	4578
Paracetamol AC	PAC U+UHYAC	1765	193_{10}	151_{53}	109_{100}	80_{24}		377	188
Paracetamol Cl-artifact AC	UHYAC	2030	227_{6}	185_{74}	143_{100}	114_{4}	79_{12}	498	2993
Paracetamol HFB	UHYHFB PHFB	1735	347_{24}	305_{39}	169_{13}	108_{100}	69_{31}	1192	5099
Paracetamol HY	UHY	1240	109_{100}	80_{41}	53_{82}	53_{82}	52_{90}	260	826
Paracetamol HYME	UHYME	1100	123_{27}	109_{100}	94_{7}	80_{96}	53_{47}	267	3766
Paracetamol ME	PME UME	1630	165_{59}	123_{74}	108_{100}	95_{10}	80_{20}	314	5046
Paracetamol PFP		1675	297_{19}	255_{31}	119_{38}	108_{100}	80_{28}	888	5095
Paracetamol TFA		1630	247_{11}	205_{30}	108_{100}	80_{19}	69_{34}	583	5092
Paracetamol-D4		1770	155_{36}	113_{100}	85_{14}	57_{8}		298	8068
Paracetamol-D4 2TMS		1775	299_{35}	284_{47}	210_{57}	116_{15}	73_{100}	903	6551
Paracetamol-D4 AC		1760	197_{4}	155_{30}	113_{100}	84_{8}		392	6550
Paracetamol-D4 HFB		1730	351_{21}	309_{26}	169_{8}	112_{100}	69_{25}	1211	6552
Paracetamol-D4 ME		1625	169_{30}	127_{47}	112_{100}	99_{9}	84_{13}	322	6554
Paracetamol-D4 PFP		1675	301_{13}	259_{16}	119_{9}	112_{100}	84_{19}	911	6553
Paracetamol-D4 TFA		1625	251_{30}	209_{44}	112_{100}	84_{13}	69_{33}	608	6559
Paracetamol-M (HO-) 3AC	U+UHYAC	2150	251_{6}	209_{23}	167_{87}	125_{100}		608	2384
Paracetamol-M (HO-methoxy-) AC	U+UHYAC	2170	239_{12}	197_{86}	155_{100}	140_{42}	110_{9}	551	2383
Paracetamol-M (methoxy-) AC	U+UHYAC	1940	223_{12}	181_{79}	139_{100}			482	201
Paracetamol-M (methoxy-) Cl-artifact AC	UHYAC	2060	257_{6}	215_{77}	173_{100}	158_{21}	130_{5}	641	2994
Paracetamol-M 2AC	U+UHYAC	2270	262_{20}	220_{35}	188_{17}	160_{74}	146_{100}	667	2387
Paracetamol-M 3AC	U+UHYAC	2340	304_{15}	261_{31}	219_{46}	160_{100}	146_{72}	928	2388
Paracetamol-M conjugate 2AC	U+UHYAC	3050	396_{20}	354_{7}	246_{40}	204_{73}	162_{71}	1427	2389
Paracetamol-M conjugate 3AC	U+UHYAC	3030	438_{35}	353_{40}	246_{72}	204_{97}	162_{100}	1553	1387
Paracetamol-M isomer-1 3AC	U+UHYAC	2200	305_{26}	263_{57}	221_{14}	160_{69}	146_{100}	934	2385
Paracetamol-M isomer-2 3AC	U+UHYAC	2220	305_{34}	263_{100}	221_{82}	162_{54}	146_{99}	934	2386
Parafluorofentanyl		2560	354_{1}	263_{100}	220_{11}	207_{25}	164_{45}	1233	6029
Parahydroxybenzoic acid 2ET		1520*	194_{17}	166_{7}	149_{100}	121_{76}	93_{17}	384	6646
Parahydroxybenzoic acid ET		1585*	166_{53}	151_{4}	138_{65}	121_{100}	93_{10}	316	6541
Paraldehyde		<1000*	131_{20}	117_{60}	89_{100}	87_{53}		273	1915
Paramethadione		1110	157_{1}	129_{59}	72_{17}	57_{100}		301	274
Paraoxon	P-I	1890	275_{52}	149_{53}	109_{100}	99_{45}	81_{58}	743	1464
Parathion-ethyl	P-I G U	1970	291_{49}	186_{21}	139_{53}	109_{100}	97_{96}	846	828
Parathion-ethyl-M (4-nitrophenol)	P-I UHY	1530	139_{100}	109_{22}	93_{18}	65_{57}		278	829
Parathion-ethyl-M (4-nitrophenol) AC	U+UHYAC	1500	181_{67}	139_{100}	109_{78}	65_{76}	63_{78}	348	830
Parathion-ethyl-M (4-nitrophenol) ME		1455	153_{100}	123_{37}	92_{55}	77_{72}		295	831
Parathion-ethyl-M (amino-)	P U	1900	261_{84}	125_{100}	109_{81}	80_{40}		662	1325
Parathion-ethyl-M (paraoxon)	P-I	1890	275_{52}	149_{53}	109_{100}	99_{45}	81_{58}	743	1464
Parathion-methyl		1855	263_{95}	233_{10}	125_{88}	109_{100}	79_{36}	672	1510
Parathion-methyl-M (4-nitrophenol)	P-I UHY	1530	139_{100}	109_{22}	93_{18}	65_{57}		278	829
Parathion-methyl-M (4-nitrophenol) AC	U+UHYAC	1500	181_{67}	139_{100}	109_{78}	65_{76}	63_{78}	348	830
Parathion-methyl-M (4-nitrophenol) ME		1455	153_{100}	123_{37}	92_{55}	77_{72}		295	831
Parecoxib -C3H4O		2885	314_{45}	299_{15}	272_{47}	191_{57}	77_{100}	992	8199
Parecoxib -C3H4O 2ME		2780	342_{86}	300_{30}	250_{26}	191_{100}	77_{90}	1168	8201
Parecoxib -C3H4O 2TMS		2770	458_{1}	443_{66}	386_{19}	210_{36}	147_{100}	1596	8204
Parecoxib -C3H4O AC		2900	356_{44}	251_{71}	209_{80}	191_{93}	77_{100}	1240	8203
Parecoxib -C3H4O MF		2815	328_{74}	313_{18}	286_{50}	191_{91}	77_{100}	1079	8200
Parecoxib -C3H4O TMS		2850	386_{32}	371_{24}	329_{8}	190_{25}	75_{100}	1384	8202
Paroxetine	G	2850	329_{34}	192_{43}	138_{34}	109_{36}	70_{100}	1087	5264
Paroxetine AC	U+UHYAC	2980	371_{32}	234_{100}	138_{53}	86_{46}	70_{58}	1318	5265
Paroxetine HFB		2685	525_{33}	266_{17}	138_{100}	135_{79}	109_{81}	1680	7686
Paroxetine ME		2600	343_{6}	206_{15}	191_{12}	84_{15}	58_{100}	1175	5275
Paroxetine PFP		2680	475_{7}	338_{5}	175_{19}	138_{100}	109_{77}	1628	6320
Paroxetine TFA		2700	425_{10}	288_{8}	166_{30}	138_{100}	109_{81}	1519	6319
Paroxetine TMS		2710	401_{12}	264_{30}	249_{80}	116_{100}	73_{98}	1450	4579
Paroxetine-M (demethylenyl-3-methyl-) 2AC	U+UHYAC	3030	415_{14}	373_{42}	234_{100}	192_{20}	86_{38}	1493	5263
Paroxetine-M (demethylenyl-4-methyl-) 2AC	U+UHYAC	3020	415_{8}	373_{22}	234_{100}	192_{21}	86_{24}	1494	5343
Paroxetine-M/artifact (dephenyl-) 2AC		2230	293_{19}	233_{69}	220_{100}	123_{53}	87_{45}	863	5309
Paynantheine		3220	396_{50}	381_{34}	253_{33}	214_{100}	200_{89}	1431	8052
Paynantheine HFB		3090	592_{16}	410_{74}	277_{76}	185_{75}	75_{100}	1712	8061
Paynantheine PFP		3090	542_{24}	360_{100}	332_{64}	198_{52}	185_{54}	1690	8060
Paynantheine TFA		3140	492_{17}	349_{19}	310_{100}	296_{75}	282_{73}	1651	8059
Paynantheine TMS		3110	468_{6}	453_{3}	339_{20}	286_{100}	272_{74}	1615	8057
Paynantheine-M (16-COOH) 2TMS	USPETMS	3200	526_{24}	339_{30}	286_{100}	271_{70}	73_{42}	1682	8040
Paynantheine-M (9-O-demethyl-) 2TMS	USPETMS	3220	526_{12}	344_{100}	329_{22}	73_{40}		1682	8041
Paynantheine-M (9-O-demethyl-16-COOH) 3TMS	USPETMS	3140	584_{10}	344_{28}	147_{41}	73_{100}		1709	8039
PB-22		3300	358_{3}	214_{100}	144_{25}	116_{17}	89_{8}	1254	9642
PB-22 artifact (1-pentyl-indole-3-carboxylate) (ME)		2150	245_{45}	214_{26}	188_{100}	144_{37}	130_{54}	575	9614
PCB 4		1630*	222_{73}	187_{46}	152_{100}	126_{8}	75_{27}	478	9197
PCB 54		1945*	292_{100}	290_{78}	255_{8}	220_{78}	184_{15}	839	9204
PCB No 77		2200*	292_{100}	290_{80}	220_{41}	184_{10}	150_{14}	839	9192
PCC		1525	192_{4}	191_{7}	164_{8}	149_{100}	122_{13}	376	3581
PCC -HCN		1190	165_{85}	164_{80}	150_{100}	136_{64}	122_{39}	315	3582

Table 8.1: Compounds in order of names   **PCDI**

Name	Detected	RI	Typical ions and intensities					Page	Entry
PCDI		1570	203 $_{25}$	160 $_{100}$	146 $_{26}$	91 $_{73}$	77 $_{17}$	408	3599
PCDI intermediate (DMCC)		<1000	152 $_7$	151 $_{13}$	137 $_{18}$	109 $_{100}$	84 $_{12}$	294	3580
PCDI precursor (dimethylamine)		<1000	45 $_{50}$	44 $_{100}$	28 $_{70}$			248	3618
PCE AC		1920	245 $_8$	188 $_{42}$	158 $_{78}$	117 $_{17}$	91 $_{100}$	577	3622
PCE artifact (phenylcyclohexene)	UHYAC	1270*	158 $_{100}$	143 $_{56}$	129 $_{93}$	115 $_{60}$	91 $_{30}$	301	3606
PCEEA		1755	247 $_{34}$	204 $_{79}$	188 $_{38}$	159 $_{100}$	91 $_{93}$	588	7076
PCEEA AC		2110	289 $_1$	246 $_7$	232 $_8$	159 $_{100}$	91 $_{65}$	838	7367
PCEEA-M (carboxy-) TMS		1975	305 $_{21}$	262 $_{100}$	188 $_{19}$	159 $_{58}$	91 $_{58}$	939	7376
PCEEA-M (carboxy-3'-HO-) 2TMS		2200	393 $_1$	378 $_3$	350 $_{38}$	246 $_{60}$	157 $_{100}$	1417	7379
PCEEA-M (carboxy-4'-cis-HO-) 2TMS		2250	393 $_3$	262 $_{100}$	246 $_{22}$	157 $_{42}$	91 $_{23}$	1417	7378
PCEEA-M (carboxy-4'-trans-HO-) 2TMS		2285	393 $_1$	276 $_4$	262 $_{100}$	247 $_4$	157 $_{55}$	1417	7377
PCEEA-M (N-dealkyl-) AC		1850	217 $_6$	174 $_{19}$	158 $_{100}$	132 $_{78}$	104 $_{27}$	456	7016
PCEEA-M (N-dealkyl-) TFA		1630	271 $_{10}$	228 $_{31}$	202 $_{22}$	158 $_{100}$	115 $_{61}$	724	7039
PCEEA-M (N-dealkyl-3'-HO-) isomer-1 2AC		2055	275 $_2$	216 $_{25}$	190 $_{16}$	174 $_{32}$	156 $_{100}$	748	7012
PCEEA-M (N-dealkyl-3'-HO-) isomer-1 2TFA		1690	383 $_8$	270 $_{47}$	240 $_{17}$	172 $_{17}$	156 $_{100}$	1369	7041
PCEEA-M (N-dealkyl-3'-HO-) isomer-2 2AC		2065	275 $_2$	233 $_3$	216 $_{14}$	190 $_{15}$	156 $_{100}$	748	7013
PCEEA-M (N-dealkyl-3'-HO-) isomer-2 2TFA		1730	383 $_4$	270 $_{35}$	240 $_{17}$	172 $_{13}$	156 $_{100}$	1369	7040
PCEEA-M (N-dealkyl-4'-HO-) -H2O AC		1680	215 $_{15}$	172 $_{100}$	156 $_9$	119 $_{15}$	103 $_{19}$	449	7021
PCEEA-M (N-dealkyl-4'-HO-) isomer-1 2AC		2090	275 $_1$	215 $_{10}$	172 $_{34}$	156 $_{100}$	132 $_{48}$	747	7014
PCEEA-M (N-dealkyl-4'-HO-) isomer-1 2TFA		1700	383 $_2$	269 $_7$	240 $_{25}$	172 $_{17}$	156 $_{100}$	1369	7042
PCEEA-M (N-dealkyl-4'-HO-) isomer-2 2AC		2100	275 $_3$	215 $_{10}$	172 $_{29}$	156 $_{100}$	132 $_{36}$	748	7015
PCEEA-M (N-dealkyl-4'-HO-) isomer-2 2TFA		1735	383 $_2$	269 $_7$	240 $_{19}$	172 $_{16}$	156 $_{100}$	1369	7043
PCEEA-M (O-deethyl-) AC		1905	261 $_{17}$	218 $_{100}$	159 $_{15}$	91 $_{32}$	87 $_{29}$	666	7077
PCEEA-M (O-deethyl-) TFA		1690	315 $_{64}$	286 $_{10}$	272 $_{100}$	238 $_{16}$	91 $_{54}$	1000	7387
PCEEA-M (O-deethyl-) TMS		1860	291 $_{24}$	248 $_{60}$	188 $_{32}$	159 $_{100}$	91 $_{58}$	852	7380
PCEEA-M (O-deethyl-3'-HO-) 2AC		2225	319 $_{12}$	276 $_{15}$	260 $_{100}$	157 $_{67}$	87 $_{38}$	1027	7078
PCEEA-M (O-deethyl-3'-HO-) 2TFA		1775	427 $_{12}$	314 $_{100}$	272 $_{17}$	157 $_3$	91 $_1$	1525	7389
PCEEA-M (O-deethyl-3'-HO-) 2TMS		2110	336 $_{16}$	276 $_{24}$	247 $_{19}$	157 $_{100}$	129 $_{12}$	1354	7381
PCEEA-M (O-deethyl-3'-HO-HO-phenyl-) 3AC		2470	377 $_1$	318 $_{14}$	276 $_{56}$	234 $_{75}$	155 $_{100}$	1344	7375
PCEEA-M (O-deethyl-4'-cis-HO-) 2TMS		2160	379 $_3$	364 $_2$	248 $_{100}$	246 $_{16}$	157 $_{75}$	1354	7382
PCEEA-M (O-deethyl-4'-HO-) 2TFA		1825	427 $_4$	314 $_{12}$	272 $_{100}$	157 $_{28}$	91 $_{40}$	1525	7388
PCEEA-M (O-deethyl-4'-HO-) -H2O AC		1860	259 $_{38}$	230 $_{45}$	200 $_{94}$	186 $_{100}$		652	7386
PCEEA-M (O-deethyl-4'-HO-) -H2O TFA		1650	313 $_{73}$	284 $_{100}$	200 $_{91}$	170 $_{34}$	141 $_{20}$	987	7390
PCEEA-M (O-deethyl-4'-HO-) isomer-1 2AC		2270	319 $_5$	259 $_{90}$	218 $_{100}$	157 $_{49}$	87 $_{70}$	1027	7079
PCEEA-M (O-deethyl-4'-HO-) isomer-2 2AC		2280	319 $_3$	259 $_{86}$	218 $_{100}$	157 $_{54}$	87 $_{58}$	1027	7080
PCEEA-M (O-deethyl-4'-HO-HO-phenyl-) 3AC		2650	377 $_3$	317 $_{77}$	276 $_{100}$	234 $_{54}$	173 $_{77}$	1343	7374
PCEEA-M (O-deethyl-4'-trans-HO-) 2TMS		2180	379 $_2$	276 $_{13}$	248 $_{90}$	157 $_{100}$	91 $_{25}$	1354	7383
PCEEA-M (O-deethyl-HO-phenyl-) 2AC		2340	319 $_{17}$	276 $_{100}$	234 $_{39}$	175 $_{30}$	107 $_{30}$	1027	7373
PCEEA-M (O-deethyl-HO-phenyl-) 2TMS		2225	379 $_{14}$	336 $_{35}$	247 $_{100}$	207 $_{52}$	179 $_{40}$	1354	7384
PCEPA		1915	261 $_{13}$	232 $_{12}$	218 $_{100}$	117 $_{29}$	91 $_{53}$	666	5877
PCEPA AC		2210	303 $_2$	260 $_{11}$	246 $_{14}$	158 $_{48}$	91 $_{100}$	927	5878
PCEPA TFA		2040	357 $_1$	260 $_4$	159 $_{100}$	91 $_{94}$	81 $_{22}$	1249	5879
PCEPA-M (3'-HO-) AC		2080	319 $_{16}$	276 $_{16}$	260 $_{100}$	234 $_{12}$	218 $_{24}$	1030	7007
PCEPA-M (3'-HO-) TFA		1980	373 $_8$	260 $_{100}$	218 $_{19}$	186 $_{14}$	157 $_{14}$	1328	7052
PCEPA-M (4'-HO-) -H2O		1870	259 $_{35}$	230 $_{100}$	216 $_{20}$	203 $_{25}$	186 $_{32}$	653	7009
PCEPA-M (4'-HO-) isomer-1 AC		2140	319 $_5$	259 $_{50}$	244 $_7$	218 $_{100}$	91 $_{65}$	1030	7010
PCEPA-M (4'-HO-) isomer-2 AC		2145	319 $_4$	259 $_{70}$	244 $_{13}$	218 $_{100}$	87 $_{15}$	1030	7011
PCEPA-M (4'-HO-) TFA		2010	373 $_{10}$	260 $_8$	218 $_{100}$	186 $_4$	157 $_{16}$	1328	7053
PCEPA-M (carboxy-) -H2O		1930	229 $_{100}$	200 $_9$	186 $_{90}$	158 $_{57}$	144 $_{52}$	507	7018
PCEPA-M (carboxy-) TMS		2045	319 $_{25}$	276 $_{100}$	188 $_8$	159 $_{18}$	144 $_{26}$	1029	7027
PCEPA-M (carboxy-2''-HO-) 2TMS		2210	407 $_{12}$	364 $_7$	317 $_9$	290 $_5$	276 $_{100}$	1470	7032
PCEPA-M (carboxy-2''-HO-) -H2O AC		1975	287 $_{26}$	244 $_{100}$	172 $_{17}$	159 $_{15}$	117 $_{17}$	820	7026
PCEPA-M (carboxy-2''-HO-) -H2O TFA		1905	341 $_{12}$	200 $_5$	186 $_{65}$	159 $_{95}$	91 $_{100}$	1161	7048
PCEPA-M (carboxy-3'-HO-) 2TMS		2275	407 $_3$	364 $_{100}$	276 $_{32}$	246 $_{85}$	157 $_{36}$	1470	7028
PCEPA-M (carboxy-3'-HO-) isomer-1 -H2O AC		2080	287 $_{34}$	244 $_{34}$	228 $_{26}$	202 $_{21}$	157 $_{100}$	821	7022
PCEPA-M (carboxy-3'-HO-) isomer-1 -H2O TFA		1960	341 $_7$	271 $_3$	228 $_{29}$	186 $_{31}$	157 $_{100}$	1160	7044
PCEPA-M (carboxy-3'-HO-) isomer-2 -H2O AC		2105	287 $_{32}$	244 $_{43}$	228 $_{41}$	202 $_{26}$	157 $_{100}$	821	7023
PCEPA-M (carboxy-3'-HO-) isomer-2 -H2O TFA		1985	341 $_7$	271 $_5$	228 $_{29}$	186 $_{38}$	157 $_{100}$	1161	7045
PCEPA-M (carboxy-4'-cis-HO-) 2TMS		2310	407 $_4$	276 $_{100}$	246 $_{13}$	157 $_{14}$	144 $_{12}$	1470	7029
PCEPA-M (carboxy-4'-HO-) 2TMS		2370	407 $_{12}$	364 $_{100}$	275 $_{14}$	247 $_{16}$	179 $_{13}$	1470	7031
PCEPA-M (carboxy-4'-HO-) isomer-1 -H2O AC		2160	287 $_6$	227 $_{100}$	198 $_{44}$	184 $_{10}$	156 $_{71}$	821	7020
PCEPA-M (carboxy-4'-HO-) isomer-1 -H2O TFA		1970	341 $_4$	271 $_5$	227 $_{48}$	157 $_{74}$	91 $_{100}$	1161	7046
PCEPA-M (carboxy-4'-HO-) isomer-2 -H2O AC		2175	287 $_2$	227 $_{100}$	198 $_{33}$	157 $_{83}$	144 $_{24}$	821	7019
PCEPA-M (carboxy-4'-HO-) isomer-2 -H2O TFA		2010	341 $_2$	271 $_5$	227 $_{28}$	157 $_{64}$	91 $_{100}$	1161	7047
PCEPA-M (carboxy-4'-trans-HO-) 2TMS		2335	407 $_2$	276 $_{100}$	157 $_{14}$	144 $_{11}$	91 $_{14}$	1469	7030
PCEPA-M (carboxy-HO-phenyl-) 2TMS		2470	495 $_2$	364 $_{100}$	335 $_5$	245 $_{12}$	179 $_5$	1655	7131
PCEPA-M (HO-phenyl-) AC		2150	319 $_{29}$	276 $_{100}$	234 $_{33}$	175 $_{18}$	107 $_{26}$	1030	7000
PCEPA-M (N-dealkyl-) AC		1850	217 $_6$	174 $_{19}$	158 $_{100}$	132 $_{78}$	104 $_{27}$	456	7016
PCEPA-M (N-dealkyl-) TFA		1630	271 $_{10}$	228 $_{31}$	202 $_{22}$	158 $_{100}$	115 $_{61}$	724	7039
PCEPA-M (N-dealkyl-3'-HO-) isomer-1 2AC		2055	275 $_2$	216 $_{25}$	190 $_{16}$	174 $_{32}$	156 $_{100}$	748	7012
PCEPA-M (N-dealkyl-3'-HO-) isomer-1 2TFA		1690	383 $_8$	270 $_{47}$	240 $_{17}$	172 $_{17}$	156 $_{100}$	1369	7041
PCEPA-M (N-dealkyl-3'-HO-) isomer-2 2AC		2065	275 $_2$	233 $_3$	216 $_{14}$	190 $_{15}$	156 $_{100}$	748	7013
PCEPA-M (N-dealkyl-3'-HO-) isomer-2 2TFA		1730	383 $_4$	270 $_{35}$	240 $_{17}$	172 $_{13}$	156 $_{100}$	1369	7040
PCEPA-M (N-dealkyl-4'-HO-) -H2O AC		1680	215 $_{15}$	172 $_{100}$	156 $_9$	119 $_{15}$	103 $_{19}$	449	7021

PCEPA-M (N-dealkyl-4'-HO-) isomer-1 2AC Table 8.1: Compounds in order of names

Name	Detected	RI	Typical ions and intensities					Page	Entry
PCEPA-M (N-dealkyl-4'-HO-) isomer-1 2AC		2090	275_1	215_{10}	172_{34}	156_{100}	132_{48}	747	7014
PCEPA-M (N-dealkyl-4'-HO-) isomer-1 2TFA		1700	383_2	269_7	240_{25}	172_{17}	156_{100}	1369	7042
PCEPA-M (N-dealkyl-4'-HO-) isomer-2 2AC		2100	275_3	215_{10}	172_{29}	156_{100}	132_{36}	748	7015
PCEPA-M (N-dealkyl-4'-HO-) isomer-2 2TFA		1735	383_2	269_7	240_{19}	172_{16}	156_{100}	1369	7043
PCEPA-M (O-deethyl-) 2AC		2590	317_5	274_{14}	260_{28}	158_{69}	91_{100}	1014	7835
PCEPA-M (O-deethyl-) AC		1980	275_{23}	232_{100}	172_{23}	101_{16}	91_{46}	749	6985
PCEPA-M (O-deethyl-) TFA		1830	329_{19}	286_{100}	216_4	172_{14}	159_{11}	1088	7038
PCEPA-M (O-deethyl-) TMS		1955	305_{22}	262_{100}	189_{19}	172_8	159_{13}	940	7033
PCEPA-M (O-deethyl-3'-HO-) 2AC		2165	333_{10}	274_{100}	157_{51}	118_{10}		1116	6988
PCEPA-M (O-deethyl-3'-HO-) 2TFA		1900	441_{16}	328_{100}	286_{26}	172_{10}		1561	7051
PCEPA-M (O-deethyl-3'-HO-) 2TMS		2195	393_6	350_{100}	262_{23}	246_{13}	157_{28}	1418	7034
PCEPA-M (O-deethyl-3'-HO-HO-phenyl-) 3AC		2495	391_{18}	332_{40}	290_{92}	248_{100}	101_{36}	1408	7025
PCEPA-M (O-deethyl-4'-cis-HO-) 2TMS		2240	393_2	262_{100}	246_{11}	157_{11}	132_{12}	1418	7035
PCEPA-M (O-deethyl-4'-HO-) 2TFA		1940	441_7	328_{11}	286_{100}	172_{11}	157_{16}	1561	7050
PCEPA-M (O-deethyl-4'-HO-) -H2O AC		1955	273_{39}	244_{56}	200_{39}	186_{46}	158_{100}	735	7017
PCEPA-M (O-deethyl-4'-HO-) -H2O TFA		1795	327_{30}	298_{41}	214_{32}	186_{35}	158_{100}	1074	7049
PCEPA-M (O-deethyl-4'-HO-) isomer-1 2AC		2200	333_4	273_{73}	232_{100}	172_{41}	91_{53}	1114	6986
PCEPA-M (O-deethyl-4'-HO-) isomer-2 2AC		2210	333_3	273_{76}	232_{100}	172_{43}	91_{46}	1116	6987
PCEPA-M (O-deethyl-4'-HO-HO-phenyl-) 3AC		2730	391_7	331_{86}	290_{100}	248_{48}	173_{37}	1408	7008
PCEPA-M (O-deethyl-4'-trans-HO-) 2TMS		2255	393_2	262_{100}	157_{11}	132_{10}	117_6	1418	7036
PCEPA-M (O-deethyl-HO-phenyl-) 2AC		2230	333_{17}	290_{100}	248_{25}	107_{46}	101_{16}	1116	6989
PCEPA-M (O-deethyl-HO-phenyl-) 2TMS		2300	393_{22}	350_{100}	322_{26}	247_{31}	179_{36}	1419	7037
PCM		1960	245_{29}	202_{100}	168_{14}	117_{23}	91_{86}	577	3592
PCM intermediate (MCC)		1560	194_9	164_9	151_{100}	124_{41}	56_{21}	385	3578
PCM intermediate (MCC) -HCN		1260	167_{81}	152_{52}	108_{100}	94_{36}	81_{77}	320	3579
PCM precursor (morpholine)		<1000	87_{54}	57_{100}	42_{27}			255	3612
PCME		1480	189_{17}	146_{100}	132_{26}	117_{14}	91_{16}	368	3595
PCME AC		1870	231_4	174_{42}	158_{97}	91_{100}	74_{50}	514	3620
PCME artifact (phenylcyclohexene)	UHYAC	1270*	158_{100}	143_{56}	129_{93}	115_{60}	91_{30}	301	3606
PCME intermediate (MECC)		<1000	138_3	137_{10}	123_{13}	95_{100}	82_{15}	278	3593
PCME intermediate (MECC) -HCN		<1000	111_{25}	91_{13}	82_{27}	68_{100}	55_{67}	261	3597
PCME precursor (methylamine)		<1000	31_{40}	30_{82}	28_{100}			247	3619
PCMEA		1790	233_{20}	190_{85}	159_{60}	117_{23}	91_{100}	524	5871
PCMEA AC		2120	275_1	232_3	159_{56}	118_{20}	91_{100}	749	5872
PCMEA TFA		1915	329_1	159_{88}	117_{13}	91_{100}	81_{22}	1088	5873
PCMPA		1895	247_{13}	204_{100}	132_{17}	117_{22}	91_{36}	588	5874
PCMPA AC		2200	289_2	246_9	232_{15}	158_{48}	91_{100}	838	5875
PCMPA TFA		1960	343_1	246_2	159_{82}	91_{100}	81_{22}	1175	5876
PCPIP		2020	258_{20}	215_{45}	99_{100}	70_{57}	56_{68}	649	3605
PCPIP artifact (phenylcyclohexene)	UHYAC	1270*	158_{100}	143_{56}	129_{93}	115_{60}	91_{30}	301	3606
PCPIP intermediate (PICC)		1680	207_9	180_9	123_8	99_{100}	70_{42}	422	3587
PCPIP intermediate (PICC) -HCN		1380	180_{26}	165_8	123_{13}	110_{17}	70_{100}	347	3588
PCPIP precursor (1-methylpiperazine)		<1000	100_{31}	70_4	58_{100}	42_{38}		258	3614
PCPR		1625	217_{15}	174_{100}	104_{23}	91_{48}	58_{20}	456	3604
PCPR AC		1965	259_9	202_{44}	158_{83}	102_{42}	91_{100}	653	3621
PCPR artifact (phenylcyclohexene)	UHYAC	1270*	158_{100}	143_{56}	129_{93}	115_{60}	91_{30}	301	3606
PCPR intermediate (PRCC) -HCN		<1000	139_{23}	110_{74}	96_{45}	69_{22}	54_{100}	279	3600
PCPR intermediate (PRCC) -HCN		<1000	139_{21}	110_{65}	96_{40}	54_{88}	41_{100}	279	5539
PCPR precursor (propylamine)		<1000	59_7	42_{27}	30_{100}			249	3616
PCPR-M (2''-HO-) AC		1965	275_{25}	232_{100}	188_{19}	159_{55}	91_{35}	750	7391
PCPR-M (2''-HO-3'-HO-) 2AC		2250	333_7	290_9	274_{59}	216_{17}	157_{100}	1116	7402
PCPR-M (2''-HO-4'-HO-) isomer-1 2AC		2290	333_1	273_{29}	232_{43}	157_{100}	91_{61}	1115	7403
PCPR-M (2''-HO-4'-HO-) isomer-2 2AC		2300	333_4	273_{82}	232_{85}	174_{66}	157_{100}	1116	7404
PCPR-M (2''-HO-4'-HO-HO-phenyl-) 3AC		2610	391_1	331_{16}	290_{34}	215_{54}	173_{100}	1408	7401
PCPR-M (3'-HO-) isomer-1 AC		1975	275_{15}	232_{13}	216_{100}	174_{25}	157_{24}	750	7392
PCPR-M (3'-HO-) isomer-2 AC		1985	275_{15}	232_{15}	216_{100}	174_{34}	157_{30}	750	7393
PCPR-M (3'-HO-HO-phenyl-) isomer-1 2AC		2345	333_{12}	290_{14}	274_{100}	232_{92}	173_{39}	1115	7397
PCPR-M (3'-HO-HO-phenyl-) isomer-2 2AC		2360	333_{13}	290_{18}	274_{100}	232_{76}	173_{46}	1115	7398
PCPR-M (4'-HO-) isomer-1 AC		2020	275_{79}	215_{79}	174_{100}	157_{31}	91_{31}	750	7394
PCPR-M (4'-HO-) isomer-2 AC		2030	275_6	215_{70}	174_{100}	157_3	91_{32}	749	7395
PCPR-M (4'-HO-HO-phenyl-) isomer-1 2AC		2385	333_6	273_{60}	232_{100}	190_{40}	173_{29}	1115	7399
PCPR-M (4'-HO-HO-phenyl-) isomer-2 2AC		2400	333_4	273_{67}	232_{100}	190_{90}	173_{64}	1116	7400
PCPR-M (HO-phenyl-) AC		2070	275_{17}	232_{100}	190_{56}	175_{20}	107_{44}	749	7396
PCPR-M (N-dealkyl-) AC		1850	217_6	174_{19}	158_{100}	132_{78}	104_{27}	456	7016
PCPR-M (N-dealkyl-) TFA		1630	271_{10}	228_{31}	202_{22}	158_{100}	115_{61}	724	7039
PCPR-M (N-dealkyl-3'-HO-) isomer-1 2AC		2055	275_2	216_{25}	190_{16}	174_{32}	156_{100}	748	7012
PCPR-M (N-dealkyl-3'-HO-) isomer-1 2TFA		1690	383_8	270_{47}	240_{17}	172_{17}	156_{100}	1369	7041
PCPR-M (N-dealkyl-3'-HO-) isomer-2 2AC		2065	275_2	233_3	216_{14}	190_{15}	156_{100}	748	7013
PCPR-M (N-dealkyl-3'-HO-) isomer-2 2TFA		1730	383_4	270_{35}	240_{17}	172_{13}	156_{100}	1369	7040
PCPR-M (N-dealkyl-4'-cis-HO-) 2TMS		1985	335_6	320_4	246_{11}	204_{100}	73_{40}	1128	7405
PCPR-M (N-dealkyl-4'-HO-) -H2O AC		1680	215_{15}	172_{100}	156_9	119_{15}	103_{19}	449	7021
PCPR-M (N-dealkyl-4'-HO-) isomer-1 2AC		2090	275_1	215_{10}	172_{34}	156_{100}	132_{48}	747	7014
PCPR-M (N-dealkyl-4'-HO-) isomer-1 2TFA		1700	383_2	269_7	240_{25}	172_{17}	156_{100}	1369	7042
PCPR-M (N-dealkyl-4'-HO-) isomer-2 2AC		2100	275_3	215_{10}	172_{29}	156_{100}	132_{36}	748	7015

Table 8.1: Compounds in order of names **PCPR-M (N-dealkyl-4'-HO-) isomer-2 2TFA**

Name	Detected	RI	Typical ions and intensities					Page	Entry
PCPR-M (N-dealkyl-4'-HO-) isomer-2 2TFA		1735	383 2	269 7	240 19	172 16	156 100	1369	7043
PCPR-M (N-dealkyl-4'-trans-HO-) 2TMS		2000	335 5	320 1	246	204 100	73 32	1128	7406
Pecazine	G U UHY UHYAC	2545	310 86	199 67	112 68	96 66	58 100	971	369
Pecazine-M (HO-) AC	UHYAC	2750	368 58	326 4	215 40	112 92	58 100	1302	1278
Pecazine-M (nor-) AC	U+UHYAC	2985	338 61	212 100	198 58	98 28		1143	1279
Pecazine-M (nor-HO-) 2AC	UHYAC	3415	396 50	354 21	228 100	214 45	98 16	1430	1280
Pecazine-M (ring)	P G U+UHYAC	2010	199 100	167 44				396	10
Pecazine-M 2AC	U+UHYAC	2865	315 54	273 34	231 100	202 11		999	2618
Pecazine-M AC	U+UHYAC	2550	257 23	215 100	183 7			641	12
Pemoline 2ME		1590	204 46	190 20	118 100	90 44		409	832
Pemoline-M (mandelic acid)		1890*	152 10	107 100	79 75	77 55	51 63	292	5759
Pemoline-M (mandelic acid) ME		1485*	166 7	107 100	79 59	77 40		315	1071
Penbutolol	G	2130	291 5	276 8	161 6	86 100	57 25	853	2596
Penbutolol 2AC		2205	375 2	315 7	158 74	98 60	56 100	1336	1367
Penbutolol formyl artifact		2150	303 10	288 100	159 3	141 9	91 9	927	1366
Penbutolol TMS		2100	363 3	348 16	247 5	101 16	86 100	1280	5491
Penbutolol-M (deisobutyl-HO-) -H2O 2AC	UHYAC	2240	317 8	275 34	216 34	178 100	98 32	1013	1708
Penbutolol-M (di-HO-) 3AC	UHYAC	2890	449 2	434 35	374 13	332 11	86 100	1579	1709
Penbutolol-M (HO-) 2AC		2520	391 2	376 10	158 61	86 52	56 100	1409	1382
Penbutolol-M (HO-) artifact		2425	319 4	304 15	178 7	86 100	57 28	1030	1381
Penciclovir 4TMS		2495	541 19	526 16	438 83	308 60	73 100	1690	9443
Penciclovir 5TMS		2530	613 2	598 49	510 8	147 46	73 100	1717	9442
Pencycuron ME		2575	342 4	273 4	125 100	106 19	77 22	1169	3971
Penfluridol		3350	523 24	292 100	201 26	109 27		1679	584
Penfluridol-M	U	----	291 18	274 7	154 35	84 41	56 100	845	585
Penfluridol-M (deamino-carboxy-)	P-I UHY UHYAC	2230*	276 7	216 14	203 100	183 22		753	169
Penfluridol-M (deamino-carboxy-) ME	P-I UHYME U+UHYAC	2125*	290 12	258 13	216 30	203 100	183 22	841	3372
Penfluridol-M (deamino-HO-)	UHY-I	2120*	262 16	203 100	201 16	183 12		668	515
Penfluridol-M (deamino-HO-) AC	UHYAC	2150*	304 7	244 22	216 41	203 100	183 16	931	307
Penfluridol-M (N-dealkyl-)	UHY	2210	279 2	261 14	82 6	56 100		771	586
Penfluridol-M (N-dealkyl-) AC	U+UHYAC	2240	321 20	303 8	278 23	99 26	57 100	1037	165
Penfluridol-M (N-dealkyl-oxo-) -2H2O	U UHY UHYAC	1920	257 100	222 12	167 11			640	164
Penoxalin		2020	281 15	252 100				788	1221
Pentachloroaniline		1845	265 100	263 58	230 6	192 8	132 8	672	3470
Pentachlorobenzene		1515*	250 100	248 72	213 16	178 8	108 21	588	3471
2,2',4,5,5'-Pentachlorobiphenyl		2155*	326 100	324 60	289 11	254 37	184 11	1054	882
1,2,3,7,8-Pentachlorodibenzo-p-dioxin (PCDD)		----*	356 100	354 60	291 38	228 32	178 20	1230	3494
Pentachlorophenol		1760*	266 100	264 58	228 8	200 8	165 13	678	833
Pentachlorophenol ME	UME	1815*	280 100	278 62	265 100	263 64	235 54	765	834
Pentadecane	P	1500*	212 10	169 6	85 43	71 61	57 100	441	2766
Pentadecanoic acid ET		1840*	270 6	227 9	157 20	101 54	88 100	721	5402
Pentadecanoic acid ME		1830*	256 6	213 6	143 12	87 58	74 100	639	3036
Pentafluoropropionic acid		<1000*	147 9	119 49	100 87	97 31	69 100	310	5543
Pentafluoropropionic acid		<1000*	147 9	119 49	100 87	69 100	45 77	310	5549
Pentamidine		3010	306 14	188 17	132 25	102 26	69 100	1157	1948
Pentane		500*	72 10	57 13	43 100	41 66	29 34	251	3812
Pentanochlor		1935	239 15	197 12	141 100	106 13	71 60	552	4037
Pentazocine	G P-I UHY	2280	285 32	217 84	110 72	70 100		812	587
Pentazocine AC	UHYAC	2330	327 37	312 26	259 100	110 49	70 31	1078	249
Pentazocine artifact (+H2O)	UHY	2375	303 14	288 7	230 100	58 11		927	588
Pentazocine artifact (+H2O) AC	UHYAC	2435	345 7	330 7	272 100	229 3	173 4	1186	252
Pentazocine PFP		2120	431 18	363 100	348 15	110 51	69 59	1538	4320
Pentazocine TFA		2075	381 10	366 16	313 62	110 46	69 100	1362	4007
Pentazocine TMS		2320	357 29	342 33	289 91	244 48	73 100	1250	4319
Pentazocine-M (dealkyl-) 2AC	UHYAC	2380	301 9	87 100	72 29			914	251
Pentazocine-M (HO-)	U	2545	301 21	268 23	217 100	110 46	70 32	915	589
Pentazocine-M/artifact AC	UHYAC	2350*	323 6	109 100	94 26			1047	250
Pentedrone		1420	159 1	148 2	105 7	100 16	86 100	372	10268
Pentedrone AC		1710	233 1	148 2	128 45	105 16	86 100	524	10269
Pentetrazole		1540	138 23	109 16	82 33	55 100		278	835
Pentifylline	G U	2240	264 45	193 24	180 100	137 18	109 33	683	836
Pentifylline-M (di-HO-) -H2O	UHY UHYAC	2285	278 50	261 24	207 24	194 100	123 22	768	1930
Pentifylline-M (di-HO-) isomer-1 2AC	U+UHYAC	2680	380 14	251 19	181 100	180 71		1357	1927
Pentifylline-M (di-HO-) isomer-2 2AC	U+UHYAC	2820	380 20	278 7	265 22	193 27	180 100	1357	1928
Pentifylline-M (HO-)	G P UHY	2505	280 25	236 20	193 32	180 100	109 31	783	1213
Pentifylline-M (HO-) AC	U+UHYAC	2560	322 32	262 15	193 35	180 100		1045	1214
Pentobarbital	P G U+UHYAC	1740	197 4	156 100	141 84	98 10	69 12	496	837
Pentobarbital (ME)	P G	1700	211 4	170 99	155 100	141 11	112 10	557	2584
Pentobarbital 2ME	PME UME	1630	225 6	184 100	169 85			630	839
Pentobarbital 2TMS		1850	370 1	355 40	300 51	285 100	73 39	1314	4580
Pentobarbital-D5		1735	197 7	161 100	143 58	100 7		515	6882
Pentobarbital-D5 2TMS		1845	375 2	360 43	305 63	290 100	100 62	1337	7299
Pentobarbital-M (HO-)	U	1955	227 2	197 26	195 21	156 100	141 70	566	838
Pentobarbital-M (HO-) (ME)	P U	1865	209 19	170 100	155 100	112 10	69 59	638	3341

Pentobarbital-M (HO-) 2ME Table 8.1: Compounds in order of names

Name	Detected	RI	Typical ions and intensities					Page	Entry
Pentobarbital-M (HO-) 2ME	PME UME	1820	223 $_{25}$	184 $_{95}$	169 $_{100}$	112 $_{13}$	69 $_{40}$	719	3340
Pentobarbital-M (HO-) -H2O	U+UHYAC	1890	224 $_{2}$	195 $_{31}$	156 $_{30}$	141 $_{30}$	69 $_{100}$	488	840
Pentobarbital-M (HO-) -H2O (ME)	UHYAC	1870	209 $_{23}$	170 $_{24}$	155 $_{16}$	69 $_{100}$		549	3825
Pentorex	U	1250	148 $_{2}$	105 $_{4}$	91 $_{4}$	58 $_{100}$		309	841
Pentorex AC	U+UHYAC	1580	148 $_{3}$	131 $_{3}$	105 $_{6}$	100 $_{52}$	58 $_{100}$	414	842
Pentoxifylline	P G U	2435	278 $_{56}$	221 $_{100}$	193 $_{62}$	180 $_{68}$	109 $_{37}$	768	843
Pentoxifylline TMS		2505	350 $_{12}$	253 $_{60}$	237 $_{35}$	143 $_{56}$	73 $_{100}$	1209	4581
Pentoxifylline-M (dihydro-)	G P UHY	2505	280 $_{25}$	236 $_{20}$	193 $_{32}$	180 $_{100}$	109 $_{31}$	783	1213
Pentoxifylline-M (dihydro-) AC	U+UHYAC	2560	322 $_{32}$	262 $_{15}$	193 $_{35}$	180 $_{100}$		1045	1214
Pentoxifylline-M (dihydro-) -H2O	U+UHYAC	2300	262 $_{24}$	193 $_{35}$	181 $_{100}$	137 $_{24}$	109 $_{45}$	670	1732
Pentoxifylline-M (dihydro-HO-) 2AC	U+UHYAC	2680	380 $_{14}$	251 $_{19}$	181 $_{100}$	180 $_{71}$		1357	1215
Pentoxyverine	G U+UHYAC	2390	318 $_{1}$	144 $_{8}$	115 $_{5}$	91 $_{25}$	86 $_{100}$	1117	6480
Pentoxyverine-M (deethyl-) AC	G U+UHYAC	2600	347 $_{2}$	217 $_{12}$	145 $_{43}$	100 $_{60}$	58 $_{100}$	1196	6485
Pentoxyverine-M (deethyl-di-HO-) 3AC	G U+UHYAC	3120	463 $_{4}$	289 $_{18}$	141 $_{61}$	100 $_{87}$	58 $_{100}$	1606	6487
Pentoxyverine-M (deethyl-HO-) 2AC	G U+UHYAC	2860	405 $_{2}$	231 $_{5}$	143 $_{48}$	100 $_{50}$	58 $_{100}$	1463	6486
Pentoxyverine-M (HO-) AC	G U+UHYAC	2575	376 $_{3}$	143 $_{13}$	128 $_{7}$	86 $_{100}$		1409	6484
Pentoxyverine-M/artifact (alcohol) AC	G U+UHYAC	1115	203 $_{1}$	188 $_{1}$	144 $_{2}$	100 $_{4}$	86 $_{100}$	408	6481
Pentoxyverine-M/artifact (HOOC-)	G U+UHYAC	1765*	190 $_{12}$	145 $_{100}$	115 $_{22}$	103 $_{17}$	91 $_{73}$	369	6482
Pentoxyverine-M/artifact (HOOC-) ME	G U+UHYAC	1485*	204 $_{11}$	145 $_{100}$	128 $_{8}$	115 $_{9}$	91 $_{44}$	410	6483
1-Pentyl-3-(2-chlorophenylacetyl)indole		2895	339 $_{1}$	214 $_{100}$	144 $_{35}$	116 $_{18}$	89 $_{21}$	1149	8530
1-Pentyl-3-(2-methoxyphenylacetyl)indole		2910	335 $_{4}$	214 $_{100}$	144 $_{24}$	116 $_{8}$	91 $_{12}$	1126	8445
1-Pentyl-3-(2-methoxyphenylacetyl)indole E/Z isomer-1 TMS		2655	407 $_{82}$	364 $_{14}$	300 $_{62}$	286 $_{100}$	73 $_{67}$	1469	9688
1-Pentyl-3-(2-methoxyphenylacetyl)indole E/Z isomer-2 TMS		2820	407 $_{86}$	364 $_{14}$	300 $_{60}$	286 $_{100}$	73 $_{80}$	1469	9689
1-Pentyl-3-(2-methylphenylacetyl)indole		2835	319 $_{3}$	214 $_{100}$	144 $_{35}$	116 $_{17}$	105 $_{16}$	1029	8529
1-Pentyl-3-(4-methoxybenzoyl)indole		2965	321 $_{46}$	264 $_{43}$	214 $_{34}$	135 $_{100}$	77 $_{51}$	1040	8446
1-Pentyl-3-(4-methoxybenzoyl)indole-M (5-HOOC-) ME		3235	365 $_{51}$	264 $_{38}$	222 $_{33}$	135 $_{100}$	92 $_{39}$	1286	10400
1-Pentyl-3-(4-methoxybenzoyl)indole-M (5-HOOC-) TMS		3370	423 $_{40}$	278 $_{21}$	264 $_{34}$	135 $_{83}$	75 $_{100}$	1515	10401
1-Pentyl-3-(4-methoxybenzoyl)indole-M (5-HO-pentyl-)		3265	337 $_{38}$	264 $_{35}$	222 $_{59}$	135 $_{100}$	128 $_{35}$	1138	10394
1-Pentyl-3-(4-methoxybenzoyl)indole-M (5-HO-pentyl-) AC		3365	379 $_{53}$	264 $_{42}$	222 $_{50}$	135 $_{100}$	128 $_{27}$	1352	10395
1-Pentyl-3-(4-methoxybenzoyl)indole-M (5-HO-pentyl-) HFB		3025	533 $_{63}$	426 $_{26}$	264 $_{62}$	222 $_{67}$	135 $_{100}$	1686	10399
1-Pentyl-3-(4-methoxybenzoyl)indole-M (5-HO-pentyl-) PFP		3010	483 $_{68}$	376 $_{28}$	264 $_{58}$	222 $_{66}$	135 $_{100}$	1638	10398
1-Pentyl-3-(4-methoxybenzoyl)indole-M (5-HO-pentyl-) TFA		3030	433 $_{62}$	326 $_{38}$	264 $_{62}$	222 $_{73}$	135 $_{100}$	1542	10397
1-Pentyl-3-(4-methoxybenzoyl)indole-M (5-HO-pentyl-) TMS		3200	409 $_{54}$	394 $_{36}$	264 $_{39}$	135 $_{100}$	75 $_{52}$	1475	10396
1-Pentyl-indole		1650	187 $_{28}$	130 $_{100}$	117 $_{8}$	103 $_{7}$	77 $_{11}$	363	9612
1-Pentyl-indole-3-carboxylate (ME)		2150	245 $_{45}$	214 $_{26}$	188 $_{100}$	144 $_{37}$	130 $_{54}$	575	9614
Perazine	P G U+UHYAC	2790	339 $_{44}$	238 $_{22}$	141 $_{33}$	113 $_{80}$	70 $_{100}$	1151	370
Perazine-M (aminoethyl-aminopropyl-) 2AC	U+UHYAC	3310	383 $_{54}$	198 $_{71}$	185 $_{60}$	100 $_{97}$	86 $_{100}$	1370	2678
Perazine-M (aminopropyl-) AC	U+UHYAC	2720	298 $_{56}$	212 $_{39}$	198 $_{100}$	180 $_{31}$	100 $_{87}$	894	2076
Perazine-M (aminopropyl-HO-) 2AC	U+UHYAC	3100	356 $_{88}$	215 $_{65}$	214 $_{96}$	100 $_{100}$	72 $_{19}$	1242	2677
Perazine-M (di-HO-) 2AC	U+UHYAC	3600	455 $_{83}$	230 $_{52}$	141 $_{49}$	113 $_{71}$	70 $_{100}$	1591	2679
Perazine-M (HO-)	UHY	3175	355 $_{42}$	215 $_{43}$	155 $_{48}$	113 $_{60}$	70 $_{100}$	1238	590
Perazine-M (HO-) AC	U+UHYAC	3190	397 $_{31}$	214 $_{18}$	141 $_{43}$	113 $_{91}$	70 $_{100}$	1435	371
Perazine-M (HO-methoxy-) AC	U+UHYAC	3230	427 $_{84}$	258 $_{28}$	244 $_{45}$	113 $_{100}$	70 $_{80}$	1526	2684
Perazine-M (N-deethyl-) 2AC	U+UHYAC	3400	397 $_{39}$	238 $_{8}$	212 $_{12}$	198 $_{26}$	100 $_{100}$	1435	1323
Perazine-M (nor-) AC	U+UHYAC	3210	367 $_{62}$	238 $_{47}$	199 $_{59}$	141 $_{100}$	99 $_{79}$	1298	1316
Perazine-M (nor-HO-) 2AC	U+UHYAC	3700	425 $_{100}$	214 $_{24}$	141 $_{50}$	99 $_{30}$	56 $_{20}$	1520	2685
Perazine-M (ring)	P G U+UHYAC	2010	199 $_{100}$	167 $_{44}$				396	10
Perazine-M 2AC	U+UHYAC	2865	315 $_{54}$	273 $_{34}$	231 $_{100}$	202 $_{11}$		999	2618
Perazine-M AC	U+UHYAC	2550	257 $_{23}$	215 $_{100}$	183 $_{7}$			641	12
Perfluorotributylamine (PFTBA)	----		614 $_{2}$	502 $_{6}$	219 $_{28}$	131 $_{23}$	69 $_{100}$	1726	2134
Pergolide		2820	314 $_{100}$	285 $_{41}$	267 $_{12}$	194 $_{12}$	154 $_{50}$	997	5627
Pergolide HFB		2835	510 $_{86}$	482 $_{100}$	350 $_{17}$	232 $_{24}$	87 $_{65}$	1669	5856
Pergolide PFP		2830	460 $_{92}$	431 $_{100}$	300 $_{14}$	232 $_{11}$	87 $_{27}$	1601	5855
Pergolide TFA		2835	410 $_{73}$	381 $_{100}$	250 $_{17}$	154 $_{22}$	87 $_{45}$	1478	5854
Pergolide TMS		3205	386 $_{100}$	357 $_{23}$	226 $_{13}$	87 $_{16}$	73 $_{63}$	1387	5857
Perhexiline		2245	277 $_{1}$	194 $_{7}$	98 $_{1}$	84 $_{100}$	55 $_{11}$	765	3303
Perhexiline AC		2540	319 $_{1}$	236 $_{2}$	126 $_{100}$	84 $_{71}$	55 $_{20}$	1031	3304
Perhexiline-M (di-HO-)	U UHY	2660	309 $_{1}$	210 $_{9}$	98 $_{6}$	84 $_{100}$	56 $_{9}$	967	3398
Perhexiline-M (di-HO-) 3AC	UHYAC	3285	435 $_{1}$	294 $_{3}$	126 $_{100}$	84 $_{78}$		1549	3401
Perhexiline-M (di-HO-) -H2O	U UHY	2510	291 $_{6}$	208 $_{6}$	192 $_{11}$	84 $_{100}$	56 $_{10}$	853	3397
Perhexiline-M (di-HO-) -H2O 2AC	UHYAC	2820	375 $_{9}$	315 $_{6}$	234 $_{6}$	126 $_{100}$	84 $_{82}$	1337	3400
Perhexiline-M (HO-)	U UHY	2485	293 $_{1}$	210 $_{6}$	97 $_{5}$	84 $_{100}$	56 $_{10}$	868	3396
Perhexiline-M (HO-) 2AC	UHYAC	2790	377 $_{1}$	294 $_{1}$	236 $_{3}$	126 $_{100}$	84 $_{72}$	1345	3399
Periciazine	G UHY	3265	365 $_{100}$	264 $_{39}$	223 $_{32}$	142 $_{42}$	114 $_{85}$	1285	591
Periciazine AC	UHYAC	3390	407 $_{28}$	263 $_{12}$	184 $_{38}$	156 $_{100}$	114 $_{35}$	1468	372
Periciazine TMS		3250	437 $_{22}$	263 $_{19}$	223 $_{32}$	186 $_{100}$	73 $_{64}$	1552	5436
Periciazine-M/artifact (-COOH) METMS		3285	470 $_{26}$	296 $_{25}$	214 $_{49}$	186 $_{100}$	73 $_{72}$	1618	5439
Periciazine-M/artifact (ring)	U UHY UHYAC	2555	224 $_{100}$	192 $_{32}$				485	1281
Periciazine-M/artifact (ring) TMS		2310	296 $_{34}$	281 $_{3}$	223 $_{6}$	73 $_{100}$		884	5437
Periciazine-M/artifact (ring-COOH) METMS		2430	329 $_{14}$	314 $_{5}$	197 $_{39}$	73 $_{100}$		1086	5438
Perindopril 2ET	UET	2440	424 $_{1}$	351 $_{6}$	200 $_{100}$	172 $_{17}$	126 $_{16}$	1518	4755
Perindopril 2ME	UME	2495	396 $_{1}$	323 $_{11}$	186 $_{100}$	158 $_{4}$	112 $_{11}$	1432	4749
Perindopril 2TMS		2595	512 $_{1}$	497 $_{16}$	439 $_{16}$	244 $_{100}$	240 $_{26}$	1671	4987
Perindopril ET	UET	2415	396 $_{1}$	323 $_{16}$	172 $_{100}$	124 $_{7}$	98 $_{38}$	1432	4754

Table 8.1: Compounds in order of names

Name	Detected	RI	Typical ions and intensities					Page	Entry
Perindopril ME	UME	2450	382_1	309_{12}	172_{100}	124_5	98_{35}	1367	4748
Perindopril METMS		2620	454_1	439_9	367_{28}	186_{100}	98_{58}	1591	4986
Perindopril TMS		2480	440_1	425_7	367_{23}	172_{100}	98_{59}	1559	4985
Perindoprilate 2ET	UET	2415	396_1	323_{16}	172_{100}	124_7	98_{38}	1432	4754
Perindoprilate 2ME	UME	2435	368_1	309_9	158_{100}	124_5	98_{40}	1303	4750
Perindoprilate 2TMS	UTMS	2590	484_1	469_{14}	367_{32}	216_{100}	98_{29}	1640	4988
Perindoprilate 3ET	UET	2440	424_1	351_5	200_{100}	172_{17}	126_{16}	1518	4755
Perindoprilate 3ME	UME	2470	382_1	323_7	172_{100}	112_{11}	86_9	1367	4751
Perindoprilate -H2O isopropylate		2440	364_{78}	277_{79}	249_{100}	222_{68}	98_{67}	1283	4756
Perindoprilate-M/artifact -H2O ME	UME	2560	336_{53}	277_{64}	249_{76}	222_{100}	98_{22}	1133	4753
Perindoprilate-M/artifact -H2O TMS	UTMS	2645	394_{56}	379_{42}	277_{93}	249_{100}	98_{53}	1422	4989
Perindopril-M/artifact (deethyl-) 2ET	UET	2415	396_1	323_{16}	172_{100}	124_7	98_{38}	1432	4754
Perindopril-M/artifact (deethyl-) 2ME	UME	2435	368_1	309_9	158_{100}	124_5	98_{40}	1303	4750
Perindopril-M/artifact (deethyl-) 2TMS	UTMS	2590	484_1	469_{14}	367_{32}	216_{100}	98_{29}	1640	4988
Perindopril-M/artifact (deethyl-) 3ET	UET	2440	424_1	351_5	200_{100}	172_{17}	126_{16}	1518	4755
Perindopril-M/artifact (deethyl-) 3ME	UME	2470	382_1	323_7	172_{100}	112_{11}	86_9	1367	4751
Perindopril-M/artifact (deethyl-) -H2O isopropylate		2440	364_{78}	277_{79}	249_{100}	222_{68}	98_{67}	1283	4756
Perindopril-M/artifact (deethyl-) -H2O ME	UME	2560	336_{53}	277_{64}	249_{76}	222_{100}	98_{22}	1133	4753
Perindopril-M/artifact (deethyl-) -H2O TMS	UTMS	2645	394_{56}	379_{42}	277_{93}	249_{100}	98_{53}	1422	4989
Perindopril-M/artifact -H2O	G UME	2590	350_{66}	277_{92}	249_{100}	222_{90}	98_{56}	1210	4752
Permethrin isomer-1		2640*	390_4	183_{100}	163_{17}	127_6	77_8	1401	3000
Permethrin isomer-2		2670*	390_4	183_{100}	163_{30}	127_6		1401	3001
Perphenazine	UHY-I	3360	403_{59}	372_{21}	246_{83}	70_{75}	42_{100}	1456	592
Perphenazine		3360	403_{27}	246_{72}	171_{31}	143_{84}	70_{100}	1456	4252
Perphenazine AC	UHYAC-I	3470	445_{33}	246_{41}	185_{51}	125_{60}	70_{100}	1572	373
Perphenazine TMS		3340	475_{10}	372_{22}	246_{100}	232_{22}	73_{41}	1628	5444
Perphenazine-M (amino-) AC	U+UHYAC	2990	332_8	233_{19}	100_{100}			1104	1255
Perphenazine-M (dealkyl-) AC	U+UHYAC	3500	401_{59}	233_{61}	141_{100}	99_{69}		1449	1282
Perphenazine-M (ring)	U-I UHY-I UHYAC-I	2100	233_{100}	198_{54}				520	311
Perthane		2225*	306_5	223_{100}	193_{12}	178_{15}	165_{18}	942	3473
Perthane -HCl		2095*	270_8	223_{100}	193_6	179_8	165_{10}	718	3474
Pethidine	P G U UHY U+UHYAC	1760	247_{36}	218_{20}	172_{46}	71_{100}		587	253
Pethidine-M (deethyl-) (ME)	U	1800	233_{14}	218_8	158_{65}	71_{100}		523	593
Pethidine-M (HO-) AC	U+UHYAC	2205	305_9	230_6	188_{10}	71_{100}		938	1195
Pethidine-M (nor-)	U UHY	1885	233_{21}	158_{28}	91_{37}	77_{38}	57_{100}	523	594
Pethidine-M (nor-) AC	UHYAC	2240	275_{34}	232_{36}	202_{30}	187_{100}	158_{37}	747	254
Pethidine-M (nor-) HFB		1690	429_{33}	410_6	356_{42}	341_{100}	143_{16}	1530	7823
Pethidine-M (nor-) PFP		1660	379_{37}	360_7	306_{51}	291_{100}	143_{54}	1350	7822
Pethidine-M (nor-) TFA		1680	329_{42}	256_{47}	241_{100}	143_{68}	103_{36}	1087	7821
Pethidine-M (nor-) TMS		1650	305_{38}	290_{24}	276_{34}	232_{22}	73_{100}	939	7824
Pethidine-M (nor-HO-) 2AC	U+UHYAC	2600	333_{48}	290_{24}	245_{74}	203_{100}	57_{78}	1112	1196
pFPP		1785	180_{27}	138_{100}	122_{19}	95_{16}	58_{31}	347	9168
pFPP AC		2135	222_{45}	179_{17}	150_{100}	137_{32}	122_{50}	480	9170
pFPP HFB		1885	376_{100}	179_{82}	150_{64}	123_{62}	95_{33}	1338	9174
pFPP ME		1600	194_{100}	163_{24}	150_{17}	123_{72}	95_{36}	384	9169
pFPP PFP		1855	326_{100}	179_{62}	150_{45}	123_{53}	95_{35}	1067	9173
pFPP TFA		1850	276_{100}	179_{62}	150_{52}	123_{83}	95_{53}	753	9172
pFPP TMS		1835	252_{100}	237_{47}	210_{19}	101_{45}	86_{45}	619	9171
2-Ph-AMT		2350	250_2	207_{100}	178_{13}	130_5	44_{31}	606	10225
2-Ph-AMT		2350	250_2	207_{100}	204_{28}	178_{11}	130_4	606	10227
2-Ph-AMT 2TMS		2440	394_1	379_3	279_{58}	262_{11}	116_{100}	1422	9796
2-Ph-AMT AC		2400	292_{15}	233_{57}	206_{100}	178_{18}	128_5	857	9798
2-Ph-AMT ethylimine artifact-1		2320	276_9	206_{100}	204_{20}	178_9	70_{18}	755	10226
2-Ph-AMT ethylimine artifact-1		2320	276_8	206_{100}	204_{20}	178_{12}	70_{23}	756	10228
2-Ph-AMT ethylimine artifact-2		2355	276_9	261_{11}	233_{26}	218_{15}	71_{100}	756	9794
2-Ph-AMT formyl artifact		2400	262_{14}	218_6	206_{100}	178_{12}		670	9795
2-Ph-AMT HFB		2455	446_7	240_{10}	206_{100}	178_{15}	128_4	1573	9801
2-Ph-AMT PFP		2440	396_{10}	232_{12}	206_{100}	178_{11}	128_4	1429	9800
2-Ph-AMT TFA		2485	346_{19}	206_{100}	178_{21}	140_{10}	128_5	1189	9799
2-Ph-AMT TMS		2480	322_1	307_3	204_{17}	178_7	116_{100}	1046	9797
2-Ph-DALT		2660	316_1	275_3	204_{17}	178_8	110_{100}	1007	8844
2-Ph-DALT TMS		2615	388_1	347_8	278_{19}	204_6	110_{100}	1396	10028
2-Ph-DET		2535	204_3	178_3	86_{100}	58_9		858	8845
2-Ph-DiPT		2640	320_1	220_3	204_{17}	114_{100}	72_{28}	1035	8842
2-Ph-DiPT TMS		2610	392_1	292_4	278_{12}	262_6	114_{100}	1412	9877
2-Ph-DMT		2320	264_5	218_2	204_{12}	178_4	58_{100}	683	8843
2-Ph-DMT TMS		2380	336_4	278_{18}	204_2	73_{22}	58_{100}	1133	10102
Phenacetin	G U+UHYAC	1680	179_{66}	137_{51}	109_{97}	108_{100}	80_{18}	341	186
Phenacetin TMS		1535	251_{34}	236_{51}	222_{16}	162_{54}	73_{100}	611	5451
Phenacetin-M	UHY	1240	109_{100}	80_{41}	53_{82}	53_{82}	52_{90}	260	826
Phenacetin-M (deethyl-)	G P U	1780	151_{34}	109_{100}	81_{16}	80_{22}		291	825
Phenacetin-M (deethyl-) 2TMS		1780	295_{50}	280_{68}	206_{83}	116_{15}	73_{100}	878	4578
Phenacetin-M (deethyl-) Cl-artifact AC	UHYAC	2030	227_6	185_{74}	143_{100}	114_4	79_{12}	498	2993
Phenacetin-M (deethyl-) HYME	UHYME	1100	123_{27}	109_{100}	94_7	80_{96}	53_{47}	267	3766

Phenacetin-M (deethyl-) ME

Table 8.1: Compounds in order of names

Name	Detected	RI	Typical ions and intensities					Page	Entry
Phenacetin-M (deethyl-) ME	PME UME	1630	165_{59}	123_{74}	108_{100}	95_{10}	80_{20}	314	5046
Phenacetin-M (HO-) AC	UHYAC	1755	237_{14}	195_{31}	153_{100}	124_{55}		544	187
Phenacetin-M (hydroquinone)	UHY	<1000*	110_{100}	81_{27}				261	814
Phenacetin-M (hydroquinone) 2AC	UHYAC	1395*	194_{8}	152_{26}	110_{100}			382	815
Phenacetin-M (p-phenetidine)	UHY	1280	137_{68}	108_{100}	80_{39}	65_{10}		277	844
Phenacetin-M AC	PAC U+UHYAC	1765	193_{10}	151_{53}	109_{100}	80_{24}		377	188
Phenacetin-M HFB	UHYHFB PHFB	1735	347_{24}	305_{39}	169_{13}	108_{100}	69_{31}	1192	5099
Phenacetin-M PFP		1675	297_{19}	255_{31}	119_{38}	108_{100}	80_{28}	888	5095
Phenacetin-M TFA		1630	247_{11}	205_{30}	108_{100}	80_{19}	69_{34}	583	5092
Phenalenone		1790*	180_{100}	152_{60}	126_{7}	76_{31}	63_{16}	345	3186
Phenallymal	P G U UHY UHYAC	2045	244_{22}	215_{100}	141_{5}	104_{57}		570	845
Phenanthrene		1780*	178_{100}	176_{17}	152_{6}	89_{6}	76_{7}	337	2563
Phenazepam		2440	350_{92}	348_{72}	321_{100}	313_{71}	177_{31}	1197	5850
Phenazepam artifact-1	U UHY UHYAC	2230	320_{100}	318_{79}	283_{47}	239_{66}	75_{34}	1015	2152
Phenazepam artifact-2	U UHY UHYAC	2250	334_{98}	332_{79}	297_{50}	253_{100}	75_{86}	1103	2153
Phenazepam HY	UHY	2270	311_{100}	309_{71}	276_{88}	274_{90}	195_{41}	960	2151
Phenazepam HYAC	UHYAC	2500	353_{44}	351_{32}	311_{85}	276_{100}	274_{100}	1211	2149
Phenazepam isomer-1 ME		2395	364_{100}	362_{82}	327_{81}	212_{21}	125_{24}	1271	5851
Phenazepam isomer-2 ME		2530	364_{100}	362_{85}	336_{86}	327_{95}	299_{45}	1271	5852
Phenazepam TMS		2790	422_{100}	420_{72}	405_{48}	385_{80}	73_{68}	1507	5853
Phenazepam-M HY	UHY	2270	311_{100}	309_{71}	276_{88}	274_{90}	195_{41}	960	2151
Phenazepam-M HYAC	UHYAC	2500	353_{44}	351_{32}	311_{85}	276_{100}	274_{100}	1211	2149
Phenazone	P G U+UHYAC	1845	188_{100}	96_{81}	77_{51}			364	199
Phenazone artifact	p	3390	388_{24}	269_{19}	177_{46}	77_{50}	56_{100}	1395	4713
Phenazone-M (HO-)	U UHY	1855	204_{35}	120_{18}	85_{100}	56_{50}		409	218
Phenazone-M (HO-) isomer-1 AC	U+UHYAC	2095	246_{2}	204_{19}	119_{1}	91_{3}	56_{100}	579	190
Phenazone-M (HO-) isomer-2 AC	U+UHYAC	2190	246_{88}	204_{33}	159_{37}	112_{68}	77_{100}	579	3214
Phenazopyridine	G	2480	213_{100}	184_{8}	136_{26}	108_{84}	81_{76}	442	846
Phenazopyridine AC		2700	255_{100}	213_{12}	150_{14}	108_{69}	77_{30}	632	1837
Phencyclidine	P	1910	243_{21}	242_{23}	200_{100}	91_{62}	84_{36}	569	255
Phencyclidine artifact (phenylcyclohexene)	UHYAC	1270*	158_{100}	143_{56}	129_{93}	115_{60}	91_{30}	301	3606
Phencyclidine intermediate (PCC)		1525	192_{4}	191_{7}	164_{8}	149_{100}	122_{13}	376	3581
Phencyclidine intermediate (PCC) -HCN		1190	165_{85}	164_{80}	150_{100}	136_{64}	122_{39}	315	3582
Phencyclidine precursor (piperidine)		<1000	85_{43}	84_{100}	70_{18}	56_{69}		254	3615
Phencyclidine-M (3'-HO-4"-HO-) 2AC	UGlucSPEAC	2550	359_{13}	316_{12}	300_{100}	157_{23}	91_{38}	1259	7132
Phencyclidine-M (4'-HO-4"-HO-) isomer-1 2AC	UGlucSPEAC	2600	359_{9}	299_{72}	258_{100}	157_{16}	91_{49}	1259	7133
Phencyclidine-M (4'-HO-4"-HO-) isomer-2 2AC	UGlucSPEAC	2610	359_{7}	299_{58}	258_{100}	157_{30}	91_{81}	1259	7134
Phendimetrazine	G U UHY UHYAC	1480	191_{10}	85_{55}	57_{100}			374	847
Phendimetrazine-M (nor-)	U UHY	1440	177_{8}	105_{5}	77_{10}	71_{100}	56_{54}	336	851
Phendimetrazine-M (nor-) AC	U+UHYAC	1810	219_{5}	176_{10}	113_{77}	86_{50}	71_{100}	465	198
Phendimetrazine-M (nor-) TFA		1530	273_{4}	167_{85}	105_{36}	98_{47}	70_{100}	733	4002
Phendimetrazine-M (nor-) TMS		1620	249_{6}	143_{19}	115_{36}	100_{100}	73_{64}	600	5446
Phendimetrazine-M (nor-HO-) isomer-1	UHY	1830	193_{11}	121_{7}	107_{6}	71_{100}	56_{62}	380	562
Phendimetrazine-M (nor-HO-) isomer-1 2AC	UHYAC	2150	277_{9}	234_{17}	113_{55}	85_{26}	71_{100}	761	849
Phendimetrazine-M (nor-HO-) isomer-2	UHY	1865	193_{8}	163_{6}	121_{6}	71_{100}	56_{43}	379	3517
Phendimetrazine-M (nor-HO-) isomer-2 2AC	UHYAC	2200	277_{3}	234_{8}	113_{83}	85_{39}	71_{100}	761	848
Phendimetrazine-M (nor-HO-methoxy-)	UHY	1900	223_{6}	151_{3}	107_{5}	71_{100}	56_{5}	484	3518
Phendimetrazine-M (nor-HO-methoxy-) 2AC	U+UHYAC	2320	307_{6}	265_{22}	113_{86}	86_{24}	71_{100}	950	1887
Phendipham-M/artifact (phenol) TFA		1460	263_{100}	231_{22}	218_{49}	69_{53}	59_{88}	672	4128
p-Phenetidine	UHY	1280	137_{68}	108_{100}	80_{39}	65_{10}		277	844
p-Phenetidine AC	G U+UHYAC	1680	179_{66}	137_{51}	109_{97}	108_{100}	80_{18}	341	186
Phenglutarimide	U UHY UHYAC	2235	288_{3}	216_{6}	98_{11}	86_{100}		830	595
Phenglutarimide-M (deethyl-)	UHY	2370	260_{82}	189_{100}				658	1283
Phenglutarimide-M (deethyl-) AC	UHYAC	2530	302_{12}	260_{24}	189_{100}			919	1284
Phenibut -H2O		1810	161_{34}	131_{9}	117_{6}	104_{100}	78_{26}	303	9122
Phenibut -H2O AC	U+UHYAC	1710	203_{16}	161_{2}	118_{10}	104_{100}	78_{10}	406	9123
Phenindamine	U UHY UHYAC	2180	261_{51}	260_{100}	218_{8}	202_{20}	182_{13}	665	1674
Phenindamine-M (HO-)	UHY	2300	277_{67}	276_{100}	233_{11}	200_{15}	189_{10}	761	1678
Phenindamine-M (HO-) AC	UHYAC	2580	319_{86}	318_{100}	276_{61}	234_{24}	57_{36}	1026	1675
Phenindamine-M (nor-)	UHY	2210	247_{80}	246_{100}	217_{43}	202_{68}	168_{46}	586	1679
Phenindamine-M (nor-) AC	UHYAC	2640	289_{100}	259_{44}	246_{48}	218_{42}	202_{64}	835	1676
Phenindamine-M (nor-HO-)	UHY	2590	263_{60}	262_{100}	233_{22}	191_{20}	184_{23}	674	1681
Phenindamine-M (nor-HO-) 2AC	UHYAC	3000	347_{100}	305_{63}	262_{44}	234_{51}	189_{26}	1194	1677
Phenindamine-M (N-oxide)	UHY	2230	277_{38}	260_{100}	215_{13}	202_{14}	189_{16}	761	1680
Pheniramine	P G U+UHYAC	1805	240_{1}	196_{2}	169_{100}	72_{21}	58_{75}	558	852
Pheniramine-M (bis-nor-) AC	U+UHYAC	2210	254_{2}	194_{7}	182_{10}	169_{56}	167_{46}	629	10295
Pheniramine-M (bis-nor-) HFB	UGLUCHFB	1970	408_{3}	239_{7}	196_{10}	182_{100}	169_{64}	1471	10163
Pheniramine-M (bis-nor-) PFP	UGLUCPFP	1945	357_{6}	239_{7}	196_{9}	182_{100}	169_{57}	1253	10138
Pheniramine-M (nor-)	U UHY	2080	226_{7}	182_{10}	169_{100}	168_{79}		496	2148
Pheniramine-M (nor-) AC	U+UHYAC	2250	268_{9}	225_{3}	182_{42}	169_{100}		709	853
Pheniramine-M (nor-) HFB	UGLUCHFB	2030	422_{4}	240_{5}	196_{15}	182_{100}	169_{70}	1513	9883
Pheniramine-M (nor-) PFP	UGLUCPFP	2010	372_{3}	196_{12}	182_{100}	169_{75}		1322	10139
Pheniramine-M (nor-HO-) 2PFP	UGLUCPFP	1915	534_{2}	385_{5}	196_{16}	182_{100}	169_{74}	1687	10135
Phenkapton		2535*	376_{23}	341_{14}	153_{54}	121_{82}	97_{100}	1337	3475

Table 8.1: Compounds in order of names

Name	Detected	RI	Typical ions and intensities					Page	Entry
Phenmedipham-M/artifact (HOOC-) ME		1370	165_{100}	133_{75}	120_{74}	106_{42}	77_{61}	313	3905
Phenmedipham-M/artifact (phenol)		1625	167_{100}	135_{78}	122_{57}	108_{20}	81_{40}	318	3906
Phenmedipham-M/artifact (phenol) 2ME		1560	195_{100}	164_{8}	136_{47}	108_{34}	72_{57}	386	4093
Phenmedipham-M/artifact (tolylcarbamic acid) 2ME		1340	179_{100}	134_{24}	120_{52}	91_{51}	72_{40}	341	4094
Phenmetrazine	U UHY	1440	177_{8}	105_{5}	77_{10}	71_{100}	56_{54}	336	851
Phenmetrazine AC	U+UHYAC	1810	219_{5}	176_{10}	113_{77}	86_{50}	71_{100}	465	198
Phenmetrazine TFA		1530	273_{4}	167_{85}	105_{36}	98_{47}	70_{100}	733	4002
Phenmetrazine TMS		1620	249_{6}	143_{19}	115_{36}	100_{100}	73_{64}	600	5446
Phenmetrazine-M (HO-) isomer-1	UHY	1830	193_{11}	121_{7}	107_{6}	71_{100}	56_{62}	380	562
Phenmetrazine-M (HO-) isomer-1 2AC	UHYAC	2150	277_{9}	234_{17}	113_{55}	85_{26}	71_{100}	761	849
Phenmetrazine-M (HO-) isomer-2	UHY	1865	193_{8}	163_{6}	121_{6}	71_{100}	56_{43}	379	3517
Phenmetrazine-M (HO-) isomer-2 2AC	UHYAC	2200	277_{3}	234_{8}	113_{83}	85_{39}	71_{100}	761	848
Phenmetrazine-M (HO-methoxy-)	UHY	1900	223_{6}	151_{3}	107_{5}	71_{100}	56_{5}	484	3518
Phenmetrazine-M (HO-methoxy-) 2AC	U+UHYAC	2320	307_{6}	265_{22}	113_{86}	86_{24}	71_{100}	950	1887
Phenobarbital	P G U+UHYAC	1965	232_{14}	204_{100}	161_{18}	146_{12}	117_{37}	516	854
Phenobarbital 2ET		1920	288_{2}	260_{100}	232_{17}	146_{35}	117_{27}	829	2450
Phenobarbital 2ME	PME UME	1860	260_{2}	232_{100}	175_{20}	146_{24}	117_{34}	657	1121
Phenobarbital 2TMS		2015	376_{2}	361_{34}	261_{15}	146_{100}	73_{46}	1339	4582
Phenobarbital ME	P G U+UHYAC	1895	246_{10}	218_{100}	146_{23}	117_{39}		580	1120
Phenobarbital-D5		1960	237_{24}	209_{100}	179_{11}	166_{20}	122_{45}	544	6883
Phenobarbital-D5 2TMS		2015	382_{1}	366_{6}	266_{5}	151_{100}	122_{17}	1363	7298
Phenobarbital-M (HO-)	U UHY	2295	248_{70}	220_{61}	219_{100}	148_{54}		589	855
Phenobarbital-M (HO-) 3ME	UME	2200	290_{95}	261_{100}	233_{78}	176_{26}	148_{92}	842	856
Phenobarbital-M (HO-) AC	U+UHYAC	2360	290_{8}	248_{100}	219_{54}	148_{8}	120_{5}	841	2507
Phenobarbital-M (HO-methoxy-) 3ME	UME	2300	320_{100}	291_{91}	263_{49}	206_{10}	178_{49}	1033	6407
Phenol	UHY	<1000*	94_{100}	66_{41}				256	4219
Phenolphthalein 2AC	U+UHYAC	3375*	402_{10}	360_{100}	318_{98}	274_{84}	225_{24}	1451	3077
Phenolphthalein 2ME	UME	3060*	346_{75}	302_{87}	271_{100}	239_{23}	135_{8}	1189	3078
Phenolphthalein-M (methoxy-) 2AC	UHYAC	3395*	432_{5}	390_{100}	348_{30}	304_{22}	273_{35}	1539	3402
Phenopyrazone 2AC		2475	336_{3}	294_{36}	252_{100}	145_{47}	77_{85}	1129	5130
Phenothiazine	P G U+UHYAC	2010	199_{100}	167_{44}				396	10
Phenothiazine-M (di-HO-) 2AC	U+UHYAC	2865	315_{54}	273_{34}	231_{100}	202_{11}		999	2618
Phenothiazine-M AC	U+UHYAC	2550	257_{23}	215_{100}	183_{7}			641	12
Phenothrin		2835*	350_{7}	250_{2}	183_{75}	123_{100}	81_{25}	1209	3882
1-Phenoxy-2-propanol	G	1280*	152_{19}	108_{14}	94_{100}	77_{27}	66_{14}	294	6450
Phenoxyacetic acid methylester	U	1495*	166_{44}	107_{100}	77_{41}			317	858
Phenoxybenzamine		2240	303_{1}	268	254_{1}	196_{83}	91_{100}	923	2037
Phenoxybenzamine artifact-1		2225	268_{1}	254_{15}	192_{54}	182_{11}	91_{100}	704	2038
Phenoxybenzamine artifact-2		2270	268_{1}	254_{3}	220_{51}	91_{100}	77_{11}	704	2039
Phenoxymethylpenicilline-M/artifact	UHY	<1000*	94_{100}	66_{41}				256	4219
Phenoxymethylpenicilline-M/artifact ME	U	1495*	166_{44}	107_{100}	77_{41}			317	858
Phenoxymethylpenicilline-M/artifact ME2AC		1930	314_{35}	230_{38}	198_{100}	156_{43}	97_{77}	993	7652
Phenoxymethylpenicilline-M/artifact ME2AC		1900	265_{5}	233_{37}	164_{43}	122_{100}	120_{59}	686	7653
Phenprocoumon	G P U	2440*	280_{43}	251_{100}	189_{22}	121_{26}	91_{34}	781	859
Phenprocoumon AC	U+UHYAC	2475*	322_{11}	280_{69}	251_{100}	189_{36}	121_{21}	1044	860
Phenprocoumon HY		1980*	254_{22}	225_{75}	136_{33}	121_{100}	91_{54}	629	4822
Phenprocoumon HYAC	U+UHYAC	2095*	296_{1}	278_{24}	225_{66}	121_{100}	91_{49}	886	4824
Phenprocoumon HYME		2025*	268_{13}	239_{50}	150_{18}	135_{100}	91_{27}	708	4823
Phenprocoumon isomer-1 ME	PME UME UHYME	2375*	294_{62}	279_{96}	265_{78}	203_{100}	121_{19}	870	4417
Phenprocoumon isomer-2 ME	PME UME UHYME	2395*	294_{64}	265_{88}	203_{52}	121_{13}	91_{100}	870	861
Phenprocoumon TMS		2585*	352_{35}	323_{100}	261_{15}	193_{53}	73_{73}	1221	4583
Phenprocoumon-M (di-HO-) 3ET	UET	2730*	396_{27}	352_{62}	323_{100}	295_{35}	201_{33}	1430	4821
Phenprocoumon-M (di-HO-) 3ME	UME UGLUCME	2770*	354_{32}	325_{100}	279_{8}	201_{2}	151_{12}	1232	4421
Phenprocoumon-M (di-HO-) 3TMS	UTMS	2730*	528_{23}	499_{100}	484_{21}	412_{10}	73_{24}	1682	5034
Phenprocoumon-M (HO-) isomer-1 2ET	UET	2745*	352_{47}	323_{100}	295_{21}	201_{15}	121_{9}	1222	4818
Phenprocoumon-M (HO-) isomer-1 2ME	UME UHYME	2655*	324_{80}	309_{33}	295_{100}	233_{62}	91_{55}	1056	4418
Phenprocoumon-M (HO-) isomer-1 2TMS	UTMS	2650*	440_{14}	425_{4}	411_{100}	193_{14}	73_{21}	1558	5033
Phenprocoumon-M (HO-) isomer-2 2ET	UET	2760*	352_{40}	337_{79}	323_{100}	295_{32}	165_{22}	1222	4819
Phenprocoumon-M (HO-) isomer-2 2ME	UME UGLUCME	2675*	324_{26}	295_{100}	279_{8}	201_{3}	121_{23}	1056	4420
Phenprocoumon-M (HO-) isomer-2 2TMS	UTMS	2675*	440_{23}	425_{12}	411_{100}	281_{22}	73_{21}	1559	5032
Phenprocoumon-M (HO-) isomer-3 2ET	UET	2770*	352_{63}	323_{100}	295_{75}	165_{14}	137_{21}	1222	4820
Phenprocoumon-M (HO-) isomer-3 2ME	UME UHYME	2705*	324_{38}	295_{100}	233_{9}	151_{4}	91_{17}	1057	4419
Phenprocoumon-M (HO-methoxy-) 2ME	UME UGLUCME	2770*	354_{32}	325_{100}	279_{8}	201_{2}	151_{13}	1232	4421
Phentermine		1170	134_{3}	91_{42}	65_{12}	58_{100}		288	1511
Phentermine AC		1510	134_{6}	117_{8}	100_{59}	58_{100}		373	1512
Phentermine HFB		1365	330_{1}	254_{100}	214_{7}	132_{14}	91_{30}	1183	5074
Phentermine PFP		1335	280_{1}	204_{100}	164_{9}	132_{14}	91_{21}	875	5075
Phentermine TFA		1100	230_{1}	154_{100}	132_{16}	114_{10}	59_{43}	574	3999
Phentermine TMS		1195	221_{1}	206_{74}	130_{100}	114_{17}	73_{46}	477	5102
Phentolamine 2ME		2500	309_{100}	202_{18}	189_{13}	146_{12}	85_{27}	966	5205
Phentolamine artifact AC		2310	313_{43}	254_{100}	212_{68}	167_{12}	91_{46}	991	5201
Phentolamine ME		2475	295_{90}	136_{65}	120_{100}	91_{63}	65_{51}	880	5204
Phentolamine-M/artifact (N-alkyl-)		2080	199_{100}	183_{13}	154_{9}	91_{25}	77_{19}	398	5203
Phentolamine-M/artifact (N-alkyl-) 2AC	U+UHYAC	2280	283_{7}	241_{22}	199_{100}	183_{10}	154_{10}	795	5200

Phosalone-M/artifact HY3AC Table 8.1: Compounds in order of names

Name	Detected	RI	Typical ions and intensities					Page	Entry
Phosalone-M/artifact HY3AC	U+UHYAC	2160	269 $_6$	227 $_{20}$	185 $_{100}$	129 $_{10}$	86 $_{15}$	710	6363
Phosalone-M/artifact ME		1750	183 $_{100}$	154 $_{45}$	92 $_{65}$	76 $_{20}$	63 $_{16}$	353	2440
Phosdrin		1415*	224 $_1$	192 $_{25}$	164 $_8$	127 $_{100}$	109 $_{24}$	486	4054
Phosmet		2380	317 $_{21}$	160 $_{100}$	133 $_{16}$	104 $_{15}$	77 $_{30}$	1008	3477
Phosphamidon isomer-1	G P U	1820	264 $_{63}$	227 $_8$	193 $_{10}$	127 $_{100}$	72 $_{42}$	900	2533
Phosphamidon isomer-2	G	1900	264 $_{67}$	138 $_{38}$	127 $_{100}$	109 $_{55}$	72 $_{76}$	900	2534
Phosphine		<1000*	34 $_{100}$	31 $_{33}$				247	4194
Phosphoric acid 3TMS		1060*	314 $_{17}$	299 $_{100}$	211 $_{15}$	133 $_{15}$	73 $_{61}$	993	4678
Phoxim	G	2005	298 $_{16}$	168 $_{16}$	135 $_{47}$	109 $_{100}$	81 $_{57}$	893	4077
Phoxim artifact-1	P-I U	1400*	198 $_{26}$	170 $_{38}$	138 $_{95}$	111 $_{100}$	81 $_{80}$	393	1442
Phoxim artifact-2	G	1670	306 $_5$	278 $_{24}$	222 $_{64}$	194 $_{100}$	99 $_{70}$	941	4087
Phoxim-M/artifact	P U UHY UHYAC	1350	196 $_9$	171 $_{96}$	143 $_{100}$	111 $_{42}$	97 $_{50}$	389	4369
Phoxim-M/artifact	U UHY UHYAC	1480	146 $_{63}$	116 $_{97}$	89 $_{100}$	51 $_{30}$		285	4370
PHP		1930	188 $_1$	140 $_{100}$	105 $_8$	96 $_6$	84 $_6$	577	10416
PHP-M (dihydro-oxo-) AC		2150	154 $_{100}$	107 $_7$	98 $_{49}$	86 $_{26}$		926	10418
PHP-M (oxo-)		2125	259 $_1$	154 $_{100}$	105 $_{17}$	98 $_{56}$	86 $_{29}$	653	10417
PHPP		1950	258 $_1$	188 $_2$	154 $_{100}$	96 $_5$	84 $_5$	653	9580
Phthalic acid butyl-2-ethylhexyl ester		1950*	223 $_7$	205 $_4$	149 $_{100}$	104 $_4$	57 $_7$	1121	713
Phthalic acid butyl-2-methylpropyl ester		1970*	278 $_1$	223 $_4$	205 $_4$	149 $_{100}$	76 $_4$	769	2995
Phthalic acid butyloctyl ester		1950*	223 $_6$	205 $_4$	149 $_{100}$	122 $_4$	104 $_{11}$	1121	2361
Phthalic acid decyldodecyl ester		2990*	474 $_1$	335 $_6$	307 $_8$	149 $_{100}$	57 $_{11}$	1627	3542
Phthalic acid decylhexyl ester		2665*	390 $_1$	307 $_5$	251 $_{10}$	233 $_2$	149 $_{100}$	1404	6402
Phthalic acid decyloctyl ester		2675*	418 $_1$	307 $_9$	279 $_{12}$	149 $_{100}$	57 $_{28}$	1503	3544
Phthalic acid decyltetradecyl ester		3250*	363 $_4$	307 $_7$	149 $_{100}$	57 $_{14}$		1663	3543
Phthalic acid diisodecyl ester		2800*	446 $_1$	307 $_{24}$	167 $_2$	149 $_{100}$	57 $_7$	1574	3541
Phthalic acid diisohexyl ester		2380*	334 $_1$	251 $_{10}$	233 $_2$	149 $_{100}$	104 $_3$	1121	6397
Phthalic acid diisononyl ester		2700*	418 $_1$	293 $_{19}$	167 $_4$	149 $_{100}$	71 $_{13}$	1503	1232
Phthalic acid diisooctyl ester		2520*	390 $_1$	279 $_{15}$	167 $_{41}$	149 $_{100}$	57 $_{29}$	1404	723
Phthalic acid dimethyl ester		1450*	194 $_7$	163 $_{100}$	133 $_9$	104 $_7$	77 $_{15}$	382	4948
Phthalic acid dioctyl ester		2655*	390 $_1$	279 $_{12}$	261 $_2$	167 $_2$	149 $_{100}$	1404	6401
Phthalic acid ethyl methyl ester		1520*	208 $_2$	176 $_1$	163 $_{100}$	149 $_{58}$	77 $_{25}$	423	4940
Phthalic acid ethylhexyl methyl ester		2010*	181 $_{24}$	163 $_{100}$	149 $_{48}$	83 $_{11}$	70 $_{34}$	857	5319
Phthalic acid hexyloctyl ester		2500*	362 $_1$	279 $_4$	251 $_5$	149 $_{100}$	85 $_5$	1276	6053
Physcion		2660*	284 $_{100}$	255 $_{12}$	241 $_9$	213 $_6$	128 $_{13}$	799	3556
Physcion 2AC		2920*	368 $_1$	326 $_{24}$	284 $_{100}$	255 $_6$	128 $_3$	1301	3569
Physcion 2ME		2845*	312 $_{65}$	297 $_{100}$	295 $_{27}$	267 $_7$	142 $_{13}$	981	3568
Physcion ME		2775*	298 $_{100}$	280 $_{48}$	252 $_{55}$	135 $_{22}$	115 $_{12}$	893	3567
Physostigmine	G U	2240	275 $_{33}$	218 $_{100}$	174 $_{95}$	160 $_{76}$	132 $_{13}$	748	875
Physostigmine-M/artifact	G UHY	1835	218 $_{84}$	188 $_{13}$	174 $_{97}$	160 $_{100}$	146 $_{13}$	460	876
Physostigmine-M/artifact AC	UHYAC	2010	260 $_{89}$	218 $_{100}$	174 $_{83}$	160 $_{76}$	132 $_{19}$	658	2616
Phytanic acid		2035*	312 $_5$	250 $_{20}$	157 $_{44}$	87 $_{100}$	71 $_{87}$	985	6063
Phytanic acid ME		2015*	326 $_8$	171 $_{39}$	143 $_{21}$	101 $_{100}$	74 $_{89}$	1072	6062
PIA 3AC		3730	511 $_1$	420 $_{84}$	259 $_{10}$	162 $_{100}$	139 $_{46}$	1670	3090
PICC		1680	207 $_9$	180 $_9$	123 $_8$	99 $_{100}$	70 $_{42}$	422	3587
PICC -HCN		1380	180 $_{26}$	165 $_8$	123 $_{13}$	110 $_{17}$	70 $_{100}$	347	3588
Picloram -CO2		1440	196 $_{100}$	161 $_{33}$	134 $_{22}$	98 $_{13}$	86 $_{11}$	389	3650
Picloram ME		1875	254 $_{23}$	223 $_{19}$	196 $_{100}$	168 $_{17}$	86 $_{14}$	625	3651
Picosulfate-M (bis-methoxy-bis-phenol)	UHY	2820	337 $_{100}$	322 $_{69}$	307 $_8$	259 $_{14}$		1136	2458
Picosulfate M (bis methoxy bis phenol) 2AC	U+UHYAC	2950	421 $_{83}$	379 $_{100}$	364 $_{54}$	337 $_{25}$	322 $_{46}$	1510	2456
Picosulfate-M (bis-methoxy-bis-phenol) 2ME	UGLUCEXME	2760	365 $_{100}$	350 $_{61}$	287 $_{41}$	249 $_{13}$	220 $_{11}$	1286	6813
Picosulfate-M (bis-phenol)	UHY	2655	277 $_{100}$	199 $_{52}$				759	107
Picosulfate-M (bis-phenol) 2AC	G PAC-I U+UHYAC	2835	361 $_{100}$	319 $_{63}$	277 $_{75}$	199 $_{46}$		1268	106
Picosulfate-M (bis-phenol) 2ME	UGLUCEXME	2595	305 $_{100}$	290 $_{27}$	227 $_{49}$	182 $_5$	169 $_6$	937	6811
Picosulfate-M (methoxy-bis-phenol)	UHY	2680	307 $_{100}$	306 $_{49}$	292 $_{19}$	229 $_{35}$	69 $_{22}$	948	109
Picosulfate-M (methoxy-bis-phenol) 2AC	U+UHYAC	2870	391 $_{46}$	349 $_{100}$	307 $_{48}$	292 $_{12}$	229 $_{23}$	1406	1750
Picosulfate-M (methoxy-bis-phenol) 2ME	UGLUCEXME	2695	335 $_{100}$	320 $_{40}$	257 $_{57}$	220 $_7$	139 $_{13}$	1125	6812
Pilocarpine	G U+UHYAC	2160	208 $_9$	121 $_4$	109 $_9$	95 $_{100}$		425	2233
Pilocarpine-M (1-HO-ethyl-) AC	UHYAC	2390	266 $_7$	206 $_{38}$	177 $_{31}$	95 $_{100}$	82 $_{19}$	696	4360
Pilocarpine-M (2-HO-ethyl-) AC	UHYAC	2200	266 $_{14}$	206 $_{16}$	124 $_{14}$	95 $_{100}$	87 $_{30}$	696	4359
Pimozide		3870	461 $_{41}$	230 $_{100}$	187 $_{37}$	133 $_{35}$	82 $_{38}$	1603	596
Pimozide TMS		4155	533 $_{53}$	302 $_{91}$	259 $_{36}$	203 $_{53}$	73 $_{100}$	1686	4586
Pimozide-M (benzimidazolone)	UHY-I	1950	134 $_{100}$	106 $_{36}$	79 $_{62}$	67 $_{20}$		273	491
Pimozide-M (benzimidazolone) 2AC	UHYAC-I	1730	218 $_{11}$	176 $_{19}$	134 $_{100}$	106 $_{11}$		457	171
Pimozide-M (deamino-carboxy-)	P-I UHY UHYAC	2230*	276 $_7$	216 $_{17}$	203 $_{100}$	183 $_{22}$		753	169
Pimozide-M (deamino-carboxy-) ME	P-I UHYME U+UHYAC	2125*	290 $_{12}$	258 $_{13}$	216 $_{30}$	203 $_{100}$	183 $_{22}$	841	3372
Pimozide-M (deamino-HO-)	UHY-I	2120*	262 $_{16}$	203 $_{100}$	201 $_{16}$	183 $_{12}$		668	515
Pimozide-M (deamino-HO-) AC	UHYAC	2150*	304 $_7$	244 $_{22}$	216 $_{41}$	203 $_{100}$	183 $_{16}$	931	307
Pimozide-M (N-dealkyl-)	UHY	2415	217 $_{10}$	134 $_{100}$	106 $_{42}$	79 $_{87}$		455	87
Pimozide-M (N-dealkyl-) AC	UHYAC	2770	259 $_{60}$	216 $_{15}$	134 $_{64}$	125 $_{42}$	82 $_{100}$	651	89
Pimozide-M (N-dealkyl-) ME	UHY	2290	231 $_{12}$	134 $_{100}$	106 $_{39}$	79 $_{82}$		514	86
Pimozide-M (N-dealkyl-) 2AC	U+UHYAC	2750	301 $_{28}$	259 $_{43}$	134 $_{28}$	125 $_{28}$	82 $_{100}$	913	88
Pinaverium bromide artifact-1		2450	281 $_6$	212 $_3$	114 $_5$	100 $_{100}$	70 $_7$	791	6441
Pinaverium bromide artifact-2		2110	310 $_7$	231 $_{100}$	229 $_{99}$	185 $_{12}$	107 $_{15}$	968	6442
Pinaverium bromide artifact-3		1975	266 $_{24}$	231 $_{100}$	229 $_{98}$	185 $_{11}$	107 $_{13}$	692	6443

Table 8.1: Compounds in order of names   **Pinaverium bromide artifact-4**

Name	Detected	RI	Typical ions and intensities					Page	Entry
Pinaverium bromide artifact-4		1915	260 $_{43}$	231 $_{91}$	229 $_{100}$	181 $_{77}$	107 $_{23}$	654	6444
Pinaverium bromide artifact-5		1695	230 $_{100}$	215 $_{21}$	187 $_{6}$	108 $_{49}$		507	6445
Pinazepam		2585	308 $_{93}$	307 $_{96}$	280 $_{100}$	217 $_{28}$	91 $_{76}$	955	3072
Pinazepam HY		2330	269 $_{19}$	268 $_{22}$	227 $_{82}$	190 $_{23}$	77 $_{100}$	710	3073
Pinazepam HYAC		2400	311 $_{11}$	268 $_{41}$	227 $_{70}$	190 $_{24}$	77 $_{100}$	975	3076
Pinazepam-M	P G U	2520	270 $_{86}$	269 $_{97}$	242 $_{100}$	241 $_{82}$	77 $_{17}$	715	463
Pinazepam-M HY	UHY	2050	231 $_{80}$	230 $_{95}$	154 $_{23}$	105 $_{38}$	77 $_{100}$	512	419
Pinazepam-M TMS		2300	342 $_{62}$	341 $_{100}$	327 $_{19}$	269 $_{4}$	73 $_{30}$	1168	4573
Pindolol	G	2240	248 $_{27}$	204 $_{9}$	133 $_{100}$	116 $_{15}$	72 $_{73}$	592	1227
Pindolol 2AC		2750	332 $_{7}$	200 $_{100}$	186 $_{59}$	140 $_{58}$	98 $_{51}$	1107	878
Pindolol 2TMSTFA		2485	488 $_{7}$	318 $_{6}$	284 $_{87}$	129 $_{71}$	73 $_{100}$	1646	6161
Pindolol formyl artifact		2260	260 $_{34}$	133 $_{86}$	127 $_{100}$	86 $_{56}$		658	877
Pindolol TMSTFA		2415	416 $_{9}$	284 $_{100}$	246 $_{12}$	129 $_{90}$	73 $_{79}$	1496	6160
Pindone		1825*	230 $_{29}$	173 $_{100}$	146 $_{45}$	105 $_{43}$	89 $_{49}$	509	3652
Pioglitazone 2TMS		2870	500 $_{67}$	366 $_{28}$	134 $_{100}$	121 $_{62}$	73 $_{78}$	1659	9447
Pioglitazone artifact (phenol)		2185	223 $_{5}$	151 $_{1}$	107 $_{100}$	91 $_{4}$	91 $_{4}$	481	7730
Pioglitazone artifact (phenol) 3TMS		2235	439 $_{76}$	424 $_{17}$	274 $_{8}$	223 $_{18}$	73 $_{100}$	1555	7727
Pioglitazone artifact (phenol) ME		2160	237 $_{3}$	176 $_{2}$	151 $_{1}$	107 $_{100}$	77 $_{15}$	542	7729
Pipamperone	P-I G U UHY U+UHYAC	3040	331 $_{36}$	194 $_{32}$	165 $_{100}$	138 $_{75}$	123 $_{79}$	1336	179
Pipamperone 2TMS		3100	519 $_{1}$	403 $_{36}$	296 $_{56}$	211 $_{76}$	73 $_{100}$	1677	4587
Pipamperone-M	U+UHYAC	1490*	180 $_{25}$	125 $_{49}$	123 $_{35}$	95 $_{17}$	56 $_{100}$	345	85
Pipamperone-M (dihydro-) -H2O	UHY UHYAC	3000	315 $_{33}$	224 $_{59}$	139 $_{100}$	98 $_{75}$	70 $_{28}$	1261	5586
Pipamperone-M (HO-)	UHY	3250	347 $_{49}$	292 $_{15}$	165 $_{100}$	154 $_{75}$	123 $_{40}$	1409	597
Pipamperone-M (HO-) AC	U+UHYAC	3290	389 $_{63}$	292 $_{27}$	194 $_{37}$	165 $_{100}$	123 $_{70}$	1543	599
Pipamperone-M (N-dealkyl-) AC	PAC-I UHYAC	2500	209 $_{100}$	150 $_{66}$	124 $_{64}$	84 $_{24}$	82 $_{46}$	624	598
Pipamperone-M/artifact	U+UHYAC	1350*	164 $_{43}$	133 $_{38}$	123 $_{100}$	95 $_{41}$		310	1914
Pipazetate-M (alcohol)	U UHY	1830	156 $_{2}$	112 $_{5}$	98 $_{100}$	96 $_{2}$	70 $_{4}$	329	2274
Pipazetate-M (alcohol) AC	UHYAC	1710	215 $_{1}$	156 $_{3}$	142 $_{1}$	98 $_{100}$		450	2276
Pipazetate-M (HO-alcohol) AC	UHYAC	1800	231 $_{1}$	156 $_{2}$	142 $_{1}$	112 $_{5}$	98 $_{100}$	514	2277
Pipazetate-M (HO-ring)	U UHY	2800	216 $_{100}$	187 $_{68}$	168 $_{50}$	140 $_{9}$		451	2272
Pipazetate-M (HO-ring) AC	U+UHYAC	2575	258 $_{30}$	216 $_{100}$	187 $_{13}$	183 $_{13}$	155 $_{10}$	645	2275
Pipazetate-M (ring-sulfone)	U UHY UHYAC	2750	232 $_{100}$	200 $_{42}$	184 $_{20}$	168 $_{22}$		516	2273
Pipazetate-M/artifact (ring)	U+UHYAC	2045	200 $_{100}$	168 $_{36}$	156 $_{13}$			399	386
Piperacilline 2TMS		2780	661 $_{14}$	646 $_{2}$	369 $_{7}$	147 $_{11}$	73 $_{100}$	1725	4616
Piperacilline TMS		2900	589 $_{11}$	574 $_{8}$	486 $_{15}$	446 $_{12}$	73 $_{100}$	1711	4617
Piperacilline-M/artifact 2AC	UHYAC	2530	331 $_{4}$	288 $_{22}$	113 $_{20}$	100 $_{28}$	58 $_{100}$	1100	4289
Piperacilline-M/artifact AC	UHYAC	2660	289 $_{5}$	246 $_{25}$	132 $_{3}$	100 $_{16}$	58 $_{100}$	833	4288
Piperazine 2AC		1750	170 $_{15}$	85 $_{33}$	69 $_{25}$	56 $_{100}$		324	879
Piperazine 2HFB		1290	478 $_{3}$	459 $_{9}$	309 $_{100}$	281 $_{22}$	252 $_{41}$	1631	6634
Piperazine 2TFA		1005	278 $_{10}$	209 $_{59}$	152 $_{25}$	69 $_{56}$	56 $_{100}$	766	4129
Piperidine		<1000	85 $_{43}$	84 $_{100}$	70 $_{18}$	56 $_{69}$		254	3615
Piperonol		1420*	152 $_{100}$	135 $_{47}$	122 $_{37}$	93 $_{70}$	65 $_{56}$	293	7616
Piperonol AC		1530*	194 $_{63}$	152 $_{66}$	135 $_{100}$	122 $_{17}$	104 $_{18}$	382	6637
Piperonol HFB		1400*	348 $_{28}$	271 $_{2}$	135 $_{100}$	105 $_{6}$	77 $_{19}$	1197	7620
Piperonol PFP		1325*	298 $_{30}$	149 $_{8}$	135 $_{100}$	105 $_{7}$	77 $_{28}$	892	7619
Piperonol TFA		1295*	248 $_{40}$	149 $_{4}$	135 $_{100}$	105 $_{8}$	77 $_{24}$	589	7618
Piperonol TMS		1560*	224 $_{45}$	209 $_{27}$	179 $_{17}$	135 $_{100}$	73 $_{24}$	487	7617
Piperonyl butoxide	G P	2375*	338 $_{5}$	193 $_{11}$	176 $_{100}$	149 $_{35}$	57 $_{49}$	1146	3478
Piperonylacetate		1530*	194 $_{63}$	152 $_{66}$	135 $_{100}$	122 $_{17}$	104 $_{18}$	382	6637
Piperonylic acid ET		1560*	194 $_{48}$	166 $_{27}$	149 $_{100}$	121 $_{23}$	65 $_{16}$	382	6471
Piperonylic acid ME		1445*	180 $_{52}$	149 $_{100}$	121 $_{27}$	65 $_{18}$	63 $_{19}$	344	6470
Piperonylpiperazine		1890	220 $_{21}$	178 $_{14}$	164 $_{13}$	135 $_{100}$	85 $_{36}$	469	6624
Piperonylpiperazine AC	U+UHYAC	2350	262 $_{16}$	190 $_{11}$	176 $_{21}$	135 $_{100}$	85 $_{45}$	669	6625
Piperonylpiperazine Cl-artifact		1295*	170 $_{25}$	135 $_{100}$	105 $_{8}$	77 $_{41}$		323	6635
Piperonylpiperazine HFB	U+UHYHFB	2190	416 $_{8}$	281 $_{15}$	148 $_{9}$	135 $_{100}$	105 $_{11}$	1495	6631
Piperonylpiperazine TFA	U+UHYTFA	2350	316 $_{9}$	181 $_{14}$	148 $_{6}$	135 $_{100}$	105 $_{6}$	1005	6628
Piperonylpiperazine TMS		2080	292 $_{67}$	157 $_{53}$	135 $_{100}$	102 $_{57}$	73 $_{85}$	857	6887
Piperonylpiperazine-M (deethylene-) 2AC	U+UHYAC	2320	278 $_{4}$	235 $_{64}$	177 $_{11}$	150 $_{25}$	135 $_{100}$	767	6626
Piperonylpiperazine-M (deethylene-) 2HFB	U+UHYHFB	2080	586 $_{3}$	389 $_{24}$	346 $_{7}$	240 $_{7}$	135 $_{100}$	1709	6632
Piperonylpiperazine-M (deethylene-) 2TFA	U+UHYTFA	2230	386 $_{8}$	289 $_{19}$	246 $_{10}$	150 $_{4}$	135 $_{100}$	1383	6629
Piperonylpiperazine-M (piperonylamine) 2AC		2230	235 $_{23}$	192 $_{35}$	150 $_{100}$	135 $_{17}$	93 $_{9}$	532	7631
Piperonylpiperazine-M (piperonylamine) AC	U+UHYAC	2015	193 $_{100}$	150 $_{80}$	135 $_{54}$	121 $_{37}$	93 $_{27}$	377	6627
Piperonylpiperazine-M (piperonylamine) formyl artifact		1560	163 $_{22}$	135 $_{100}$	121 $_{3}$	105 $_{10}$	77 $_{37}$	307	7629
Piperonylpiperazine-M (piperonylamine) HFB	U+UHYHFB	1640	347 $_{100}$	317 $_{6}$	289 $_{10}$	178 $_{6}$	135 $_{75}$	1192	6633
Piperonylpiperazine-M (piperonylamine) PFP		1755	297 $_{100}$	267 $_{8}$	239 $_{13}$	178 $_{7}$	135 $_{51}$	888	7632
Piperonylpiperazine-M (piperonylamine) TFA	U+UHYTFA	1775	247 $_{100}$	217 $_{9}$	189 $_{16}$	148 $_{12}$	135 $_{80}$	583	6630
Piperonylpiperazine-M (piperonylamine) TMS		2130	295 $_{27}$	280 $_{100}$	206 $_{15}$	179 $_{50}$	73 $_{51}$	878	7630
Pipradol-M (HO-BPH) isomer-1	UHY	2065*	198 $_{93}$	121 $_{72}$	105 $_{100}$	93 $_{22}$	77 $_{66}$	393	1627
Pipradol-M (HO-BPH) isomer-1 AC	U+UHYAC	2010*	240 $_{27}$	198 $_{100}$	121 $_{47}$	105 $_{85}$	77 $_{80}$	555	2196
Pipradol-M (HO-BPH) isomer-2	P-I U UHY	2080*	198 $_{50}$	121 $_{100}$	105 $_{17}$	93 $_{14}$	77 $_{28}$	394	732
Pipradol-M (HO-BPH) isomer-2 AC	U+UHYAC	2050*	240 $_{45}$	198 $_{100}$	121 $_{94}$	105 $_{41}$	77 $_{51}$	555	2197
Pipradol		2400	248 $_{1}$	182 $_{2}$	165 $_{4}$	105 $_{18}$	84 $_{100}$	702	7337
Pipradol 2AC		2630	249 $_{1}$	183 $_{2}$	165 $_{2}$	126 $_{49}$	84 $_{100}$	1216	7339
Pipradol -H2O AC	U+UHYAC	2520	291 $_{26}$	249 $_{100}$	206 $_{11}$	191 $_{16}$	165 $_{24}$	850	7338

PPP-M 2HFB

Table 8.1: Compounds in order of names

Name	Detected	RI	Ion1	Ion2	Ion3	Ion4	Ion5	Page	Entry
PPP-M 2HFB	UHYHFB	1455	543_1	330_{14}	240_{100}	169_{44}	69_{57}	1691	5098
PPP-M 2PFP	UHYPFP	1380	443_1	280_9	190_{100}	119_{59}	105_{26}	1567	5094
PPP-M 2TFA	UTFA	1355	343_1	230_6	203_5	140_{100}	69_{29}	1172	5091
PPP-M TMSTFA		1890	240_8	198_3	179_{100}	117_5	73_{88}	1024	6146
Prajmaline artifact	G P-I U UHY	2925	368_1	340_4	224_{100}	196_{26}	126_{28}	1303	2711
Prajmaline artifact 2AC	UHYAC	3050	452_8	409_4	393_{13}	308_{100}	126_{34}	1586	7575
Prajmaline artifact 2TMS		2680	512_2	368_{66}	296_{48}	198_{100}	73_{84}	1672	7576
Prajmaline artifact AC	UHYAC	2950	410_2	382_{16}	266_{100}	238_{30}	144_{35}	1479	2715
Prajmaline artifact HFB		2545	564_1	420_{100}	393_8	280_{17}	194_{43}	1702	7580
Prajmaline artifact PFP		2370	514_2	486_{13}	370_{100}	342_{49}	279_{73}	1673	7579
Prajmaline artifact TFA		2390	464_3	436_{24}	350_9	320_{100}	279_{38}	1608	7578
Prajmaline artifact TMS		2690	440_2	425_6	296_{100}	268_{18}	73_{35}	1559	7577
Prajmaline-M (HO-) artifact	UHY	3130	384_{12}	313_7	224_{100}	196_{15}	126_{15}	1378	2713
Prajmaline-M (HO-) artifact 2AC	UHYAC	3060	468_2	440_8	266_{100}	238_{76}	126_{17}	1615	2717
Prajmaline-M (HO-methoxy-) artifact	UHY	3200	414_{13}	343_7	224_{100}	196_{18}	126_{12}	1491	2714
Prajmaline-M (HO-methoxy-) artifact 2AC	UHYAC	3300	498_{13}	470_8	303_{28}	266_{100}	126_{11}	1657	2718
Prajmaline-M (methoxy-) artifact	P-I U UHY	2895	398_{11}	370_{15}	297_4	254_{100}	126_{13}	1440	2712
Prajmaline-M (methoxy-) artifact AC	UHYAC	2920	440_7	398_4	340_{16}	296_{100}	126_{22}	1559	2716
Pramipexole		1920	211_{42}	151_{42}	127_{33}	70_{40}	56_{100}	436	7495
Pramipexole 2AC		2550	194_{56}	152_{100}	126_{17}	110_4	99_3	877	7496
Pramipexole 2HFB		2300	348_{100}	300_5	179_9	135_{12}		1715	7499
Pramipexole 2PFP		2270	298_{100}	272_{12}	179_{10}	153_{10}	135_{11}	1663	7498
Pramipexole 2TFA		2220	248_{27}	222_5	179_5	135_{14}	69_{100}	1454	7497
Pramipexole 2TMS		2230	355_3	270_9	198_{25}	115_{100}	73_{97}	1239	7500
Pramiverine		2270	292_3	278_{11}	215_{59}	98_{100}	70_{48}	867	2653
Pramiverine AC	U+UHYAC	2705	335_{24}	292_{12}	234_{18}	180_{100}	98_{51}	1128	2658
Pratol		2610*	268_{100}	253_9	225_1	132_{78}	117_{17}	706	5598
Pratol AC		2610*	310_{64}	268_{100}	253_1	132_{40}	117_9	969	5599
Pratol ME		2600*	282_{100}	267_{15}	150_{15}	132_{64}	89_{14}	792	5600
Pravadoline		3370	378_2	278_1	135_8	100_{100}	70_4	1348	8525
Prazepam	G P-I U+UHYAC	2650	324_{26}	295_{21}	269_{29}	91_{75}	55_{100}	1056	600
Prazepam HY	UHY U+UHYAC	2410	285_{64}	270_{45}	105_{67}	77_{83}	56_{100}	808	302
Prazepam HYAC	PHYAC U+UHYAC	2245	273_{30}	230_{100}	154_{13}	105_{23}	77_{50}	732	273
Prazepam-M	P G U	2520	270_{86}	269_{97}	242_{100}	241_{82}	77_{17}	715	463
Prazepam-M (dealkyl-HO-)	UGLUC	2750	286_{82}	258_{100}	230_{11}	166_7	139_8	813	2113
Prazepam-M (dealkyl-HO-) AC	U+UHYAC	3000	328_{22}	286_{90}	258_{100}	166_8	139_7	1079	2111
Prazepam-M (dealkyl-HO-) HY	UHY	2400	247_{72}	246_{100}	230_{11}	121_{26}	65_{22}	583	2112
Prazepam-M (dealkyl-HO-) HYAC	U+UHYAC	2270	289_{18}	247_{86}	246_{100}	105_7	77_{35}	832	3143
Prazepam-M (dealkyl-HO-) isomer-1 HY2AC	U+UHYAC	2560	331_8	289_{64}	247_{100}	230_{41}	154_{13}	1098	2125
Prazepam-M (dealkyl-HO-) isomer-2 HY2AC	U+UHYAC	2610	331_{46}	289_{54}	246_{100}	154_{11}	121_{11}	1098	1751
Prazepam-M (dealkyl-HO-methoxy-) HY2AC	U+UHYAC	2700	361_{17}	319_{72}	276_{100}	260_{14}	246_{10}	1267	1752
Prazepam-M (HO-) AC	UGLUCAC	2920	382_{18}	340_{40}	311_{99}	257_{100}	55_{42}	1364	2512
Prazepam-M (HO-) HYAC	U+UHYAC	2595	343_8	283_{11}	257_{100}	241_{28}	228_{11}	1173	2513
Prazepam-M HY	UHY	2050	231_{80}	230_{95}	154_{23}	105_{38}	77_{100}	512	419
Prazepam-M TMS		2300	342_{62}	341_{100}	327_{19}	269_4	73_{30}	1168	4573
PRCC -HCN		<1000	139_{23}	110_{74}	96_{45}	69_{22}	54_{100}	279	3600
PRCC -HCN		<1000	139_{21}	110_{65}	96_{40}	54_{88}	41_{100}	279	5539
Prednisolone	G P U	2800*	300_{19}	122_{100}	91_{14}			1265	886
Prednisolone 3AC		3400*	372_5	314_2	147_{12}	122_{100}		1643	704
Prednisolone acetate		3560*	402_{11}	342_8	147_{31}	122_{94}	121_{100}	1453	3296
Prednisone		2610*	358_{22}	256_{12}	160_{30}	121_{100}	91_{90}	1254	5256
Prednisone -C2H4O2		2610*	298_{37}	254_{20}	160_{60}	121_{72}	91_{100}	896	5257
Prednylidene		3330*	342_{38}	309_{11}	147_{25}	122_{100}	121_{71}	1324	2809
Prednylidene artifact		3100*	312_7	159_{10}	122_{100}	91_{34}	77_{17}	979	2810
Pregabaline 2TMS		1995	303_1	288_4	147_{13}	102_{100}	73_{69}	927	7280
Pregabaline -H2O	P	1440	141_{24}	111_{14}	98_{11}	84_{43}	56_{100}	281	7276
Pregabaline -H2O AC		1500	183_3	142_{28}	126_{76}	124_{89}	84_{100}	355	7277
Pregabaline -H2O PFP		1450	287_{15}	246_{18}	202_{22}	176_{55}	55_{100}	819	7279
Pregabaline -H2O TFA		1520	237_2	196_{36}	126_{51}	83_{79}	69_{100}	543	7278
Pregabaline -H2O TMS		1445	213_1	198_{100}	156_{21}	140_7	102_6	443	7281
Pregnandiol -H2O AC	UHYAC	2910*	344_{24}	284_{46}	107_{68}	93_{100}	67_{92}	1181	5585
Prenalterol		1990	225_2	210_2	181_6	110_6	72_{100}	492	1857
Prenalterol 3AC		2430	351_1	291_{12}	200_{100}	140_{20}	72_{46}	1215	1860
Prenalterol formyl artifact		2040	237_{43}	222_{76}	86_{50}	72_{46}	56_{100}	545	1858
Prenalterol -H2O 2AC		2410	291_{70}	207_{64}	150_{34}	140_{46}	98_{100}	848	1859
Prenylamine	U UHY	2560	329_1	238_{64}	165_{14}	91_{40}	58_{100}	1090	1518
Prenylamine AC		2925	371_1	280_{32}	238_{29}	100_{43}	58_{100}	1320	1519
Prenylamine-M (AM)		1160	134_4	120_{15}	91_{100}	77_{18}	65_{73}	275	54
Prenylamine-M (AM)	U	1160	134_1	120_1	91_6	65_4	44_{100}	275	5514
Prenylamine-M (AM) AC	U+UHYAC	1505	177_4	118_{60}	91_{35}	86_{100}	65_{16}	336	55
Prenylamine-M (AM) AC	U+UHYAC	1505	177_1	118_{19}	91_{11}	86_{31}	44_{100}	335	5515
Prenylamine-M (AM) formyl artifact		1100	147_2	146_6	125_5	91_{12}	56_{100}	286	3261
Prenylamine-M (AM) HFB		1355	240_{79}	169_{21}	118_{100}	91_{53}		1099	5047
Prenylamine-M (AM) PFP		1330	281_1	190_{73}	118_{100}	91_{36}	65_{12}	786	4379

Table 8.1: Compounds in order of names

Prenylamine-M (AM) TFA

Name	Detected	RI	Typical ions and intensities					Page	Entry
Prenylamine-M (AM) TFA		1095	231$_1$	140$_{100}$	118$_{92}$	91$_{45}$	69$_{19}$	513	4000
Prenylamine-M (AM) TMS		1190	192$_6$	116$_{100}$	100$_{10}$	91$_{11}$	73$_{87}$	421	5581
Prenylamine-M (AM)-D11 TFA		1615	242$_1$	144$_{100}$	128$_{82}$	98$_{43}$	70$_{14}$	566	7283
Prenylamine-M (AM)-D11 TFA		1610	194$_{100}$	128$_{82}$	98$_{43}$	70$_{14}$		856	7284
Prenylamine-M (AM)-D5 AC		1480	182$_3$	122$_{46}$	92$_{30}$	90$_{100}$	66$_{16}$	352	5690
Prenylamine-M (AM)-D5 HFB		1330	244$_{100}$	169$_{14}$	122$_{46}$	92$_{41}$	69$_{40}$	1130	6316
Prenylamine-M (AM)-D5 PFP		1320	194$_{100}$	123$_{42}$	119$_{32}$	92$_{46}$	69$_{11}$	815	5566
Prenylamine-M (AM)-D5 TFA		1085	144$_{100}$	123$_{53}$	122$_{56}$	92$_{51}$	69$_{28}$	540	5570
Prenylamine-M (AM)-D5 TMS		1180	212$_1$	197$_8$	120$_{100}$	92$_{11}$	73$_{57}$	441	5582
Prenylamine-M (deamino-HO-) -H2O		1940*	194$_{49}$	167$_{100}$	165$_{67}$	152$_{34}$	116$_{17}$	384	3388
Prenylamine-M (HO-) 2AC	UHY UHYAC	3200	429$_2$	338$_{46}$	296$_{77}$	100$_{14}$	58$_{100}$	1532	3403
Prenylamine-M (HO-methoxy-) 2AC	UHYAC	3310	459$_4$	368$_{35}$	326$_{78}$	270$_7$	58$_{100}$	1600	3404
Prenylamine-M (N-dealkyl-) AC	UHYAC	2320	253$_{15}$	193$_{12}$	165$_{19}$	152$_{10}$	73$_{100}$	623	3391
Prenylamine-M (N-dealkyl-HO-) 2AC	UHYAC	2635	311$_{54}$	269$_{14}$	239$_{21}$	183$_{63}$	73$_{100}$	977	3392
Prenylamine-M (N-dealkyl-HO-methoxy-) 2AC	UHYAC	2700	341$_{10}$	299$_{54}$	213$_{74}$	152$_{14}$	73$_{100}$	1163	3393
Pridinol		2290	295$_5$	180$_6$	113$_{13}$	98$_{100}$		881	601
Pridinol -H2O	UHY UHYAC	2220	277$_{24}$	163$_{31}$	110$_{100}$			764	1285
Pridinol-M (amino-) -H2O AC	UHYAC	2250	251$_{14}$	208$_{11}$	192$_{100}$	84$_8$		611	1286
Pridinol-M (amino-HO-) -H2O 2AC	UHYAC	2645	309$_{15}$	208$_{100}$				963	1288
Pridinol-M (di-HO-) -H2O 2AC	UHYAC	2980	393$_{10}$	309$_{10}$	208$_{100}$			1416	1289
Pridinol-M (HO-) -H2O AC	UHYAC	2615	335$_{71}$	292$_{18}$	209$_{52}$	110$_{100}$		1126	1287
Prilocaine	G P UHY	1850	220$_1$	107$_4$	86$_{100}$	65$_2$		470	1216
Prilocaine 2TMS		1910	349$_1$	235$_8$	206$_{11}$	158$_{88}$	73$_{100}$	1283	4618
Prilocaine AC	U+UHYAC	2060	262$_5$	156$_{66}$	128$_{34}$	107$_{14}$	86$_{100}$	671	1520
Prilocaine artifact	G P U+UHYAC	1840	232$_8$	217$_9$	118$_{41}$	84$_{100}$	56$_{63}$	519	4259
Prilocaine TMS		1850	292$_1$	235$_{18}$	207$_{24}$	86$_{100}$	73$_{55}$	858	4589
Prilocaine-M (deacyl-) AC		1300	149$_{53}$	106$_{100}$	91$_{33}$	79$_{18}$	77$_{16}$	287	5198
Prilocaine-M (HO-)	UHY	2155	236$_1$	123$_8$	86$_{100}$			541	3934
Prilocaine-M (HO-) 2AC	U+UHYAC	2435	320$_1$	156$_{66}$	128$_{52}$	86$_{100}$	56$_9$	1034	3932
Prilocaine-M (HO-deacyl-)	UHY	1160	123$_{100}$	106$_{16}$	94$_{29}$	78$_{51}$		267	3933
Prilocaine-M (HO-deacyl-) 2AC	UHYAC	1810	207$_4$	165$_{41}$	123$_{100}$	94$_7$	77$_5$	419	3931
Prilocaine-M (HO-deacyl-) 3AC	UHYAC	1770	249$_2$	207$_{12}$	165$_{44}$	151$_1$	123$_{100}$	595	3930
Primidone	P G U+UHYAC	2260	218$_{10}$	190$_{100}$	161$_{28}$	146$_{93}$	117$_{50}$	458	887
Primidone 2ME	UME PME	2060	246$_7$	218$_{100}$	146$_{91}$	117$_{53}$	103$_{19}$	581	6405
Primidone AC	U+UHYAC	2115	260$_6$	232$_{30}$	189$_{12}$	146$_{100}$	117$_{42}$	657	889
Primidone-M (diamide)	P U+UHYAC	1935	163$_{100}$	148$_{61}$	103$_{27}$	91$_{36}$		416	888
Primidone-M (HO-methoxy-phenobarbital) 3ME	UME	2300	320$_{100}$	291$_{91}$	263$_{49}$	206$_{10}$	178$_{49}$	1033	6407
Primidone-M (HO-phenobarbital)	U UHY	2295	248$_{70}$	220$_{61}$	219$_{100}$	148$_{54}$		589	855
Primidone-M (HO-phenobarbital) 3ME	UME	2200	290$_{95}$	261$_{100}$	233$_{78}$	176$_{26}$	148$_{92}$	842	856
Primidone-M (HO-phenobarbital) AC	U+UHYAC	2360	290$_8$	248$_{100}$	219$_{54}$	148$_8$	120$_5$	841	2507
Primidone-M (phenobarbital)	P G U+UHYAC	1965	232$_{14}$	204$_{100}$	161$_{18}$	146$_{12}$	117$_{37}$	516	854
Primidone-M (phenobarbital) 2ET		1920	288$_2$	260$_{100}$	232$_{17}$	146$_{35}$	117$_{27}$	829	2450
Primidone-M (phenobarbital) 2ME	PME UME	1860	260$_2$	232$_{100}$	175$_{20}$	146$_{24}$	117$_{34}$	657	1121
Primidone-M (phenobarbital) 2TMS		2015	376$_2$	361$_{34}$	261$_{15}$	146$_{100}$	73$_{46}$	1339	4582
Probarbital	P G U	1555	169$_{12}$	156$_{93}$	141$_{100}$	98$_{22}$		395	890
Probarbital 2ME		1485	197$_{18}$	184$_{92}$	169$_{100}$	112$_{23}$		495	891
Probenecide ET		2220	313$_3$	284$_{100}$	213$_{29}$	149$_{31}$	103$_8$	988	3080
Probenecide ME		2205	299$_4$	270$_{100}$	199$_{51}$	135$_{69}$	76$_{17}$	902	3079
Probucol		3195*	279$_{86}$	263$_{11}$	223$_{21}$	73$_{12}$	57$_{100}$	1674	7531
Probucol artifact AC		2680*	442$_1$	410$_9$	238$_{37}$	223$_{67}$	57$_{100}$	1564	7532
Probucol artifact-1		1850*	278$_{89}$	263$_{68}$	219$_8$	207$_{15}$	57$_{100}$	765	7530
Probucol artifact-2		2680*	410$_{23}$	395$_5$	190$_8$	162$_{10}$	57$_{100}$	1476	7529
Probucol artifact-3		2800*	442$_{54}$	427$_5$	237$_{11}$	178$_{18}$	57$_{100}$	1564	7528
Procainamide	P U+UHYAC	2270	235$_2$	120$_{20}$	99$_{48}$	86$_{100}$		536	893
Procainamide AC	UHYAC	2550	277$_1$	275$_1$	120$_4$	86$_{100}$	58$_{15}$	763	2896
Procaine	U+UHYAC	2025	221$_1$	164$_5$	120$_{24}$	99$_{39}$	86$_{100}$	541	892
Procaine AC	U+UHYAC	2350	278$_1$	206$_2$	120$_9$	99$_{23}$	86$_{100}$	770	3297
Procaine-M (PABA) 2TMS		1645	281$_{91}$	236$_{14}$	148$_{24}$	73$_{100}$		788	5487
Procaine-M (PABA) AC	U+UHYAC	2145	179$_{31}$	137$_{100}$	120$_{92}$	92$_{16}$	65$_{24}$	340	3298
Procaine-M (PABA) ME		1550	151$_{55}$	120$_{100}$	92$_{28}$	65$_{26}$		291	23
Procaine-M (PABA) MEAC		1985	193$_{32}$	151$_{60}$	120$_{100}$	92$_{18}$	65$_{18}$	378	24
Procarterol 2TMS		2295	419$_2$	335$_{100}$	100$_{62}$	73$_{30}$	58$_{28}$	1546	6230
Procarterol 3TMS		2390	491$_2$	407$_{76}$	100$_{100}$	73$_{90}$	58$_{83}$	1667	6217
Procarterol -H2O AC		2610	314$_{19}$	272$_{18}$	247$_{20}$	100$_{100}$	58$_{26}$	996	1861
Prochloraz		2405	310$_3$	235$_{50}$	143$_{100}$	130$_9$	87$_{33}$	1332	3886
Prochlorperazine	G U UHY UHYAC	2970	373$_{15}$	141$_{37}$	113$_{69}$	70$_{100}$		1327	376
Prochlorperazine-M (amino-) AC	U+UHYAC	2990	332$_8$	233$_{19}$	100$_{100}$			1104	1255
Prochlorperazine-M (nor-) AC	U+UHYAC	3500	401$_{59}$	233$_{61}$	141$_{100}$	99$_{69}$		1449	1282
Prochlorperazine-M (ring)	U-I UHY-I UHYAC-I	2100	233$_{100}$	198$_{54}$				520	311
Procyclidine	P-I	2320	287$_1$	269$_2$	204$_{12}$	84$_{100}$	55$_8$	824	602
Procyclidine artifact (dehydro-)		2290	285$_3$	202$_{100}$	105$_{28}$	82$_{60}$	55$_{31}$	812	4238
Procyclidine artifact (dehydro-) TMS		2420	357$_{24}$	272$_{40}$	182$_{100}$	115$_{42}$	73$_{69}$	1250	5454
Procyclidine -H2O		2160	269$_{13}$	268$_{40}$	186$_{100}$	96$_{56}$	84$_{50}$	714	4237
Procyclidine TMS		2305	344$_1$	269$_1$	186$_2$	84$_{100}$	73$_{14}$	1261	5453

Procyclidine-M (amino-HO-) isomer-1 -H2O 2AC

Table 8.1: Compounds in order of names

Name	Detected	RI	Typical ions and intensities					Page	Entry
Procyclidine-M (amino-HO-) isomer-1 -H2O 2AC	UHYAC	2560	315_3	255_{20}	196_{100}	168_{35}	155_{22}	1001	1290
Procyclidine-M (amino-HO-) isomer-2 -H2O 2AC	UHYAC	2625	315_8	255_{60}	196_{100}	132_{66}	115_{60}	1002	4242
Procyclidine-M (HO-) -H2O	UHY	2360	285_{23}	284_{31}	186_{100}	96_{42}	84_{46}	812	4239
Procyclidine-M (HO-) isomer-1 -H2O AC	UHYAC	2450	327_{34}	326_{42}	186_{100}	96_{60}	84_{59}	1077	1291
Procyclidine-M (HO-) isomer-2 -H2O AC	UHYAC	2500	327_{27}	326_{35}	186_{100}	96_{46}	84_{63}	1077	4241
Procyclidine-M (oxo-) -H2O	UHY UHYAC	2490	283_8	200_{100}	130_{17}	115_{61}	86_{55}	797	4240
Procymidone		2065	283_{53}	255_5	124_5	96_{100}	67_{43}	794	3481
Procymidone artifact (dechloro-)		1935	249_{54}	220_3	111_5	96_{100}	67_{41}	593	3482
Profenamine	G P-I U+UHYAC	2335	312_1	213_5	199_{11}	100_{100}		982	1317
Profenamine-M (bis-deethyl-) AC	U+UHYAC	2450	298_{43}	212_{100}	180_{42}	100_7	58_5	894	1319
Profenamine-M (bis-deethyl-HO-) 2AC	U+UHYAC	2900	356_{65}	270_{100}	228_{83}	196_{28}	100_{13}	1242	2619
Profenamine-M (deethyl-) AC	UHYAC	2515	326_{23}	212_{100}	128_{80}	72_{93}		1069	1318
Profenamine-M (deethyl-HO-) 2AC	UHYAC	2880	384_{11}	270_{27}	128_{76}	72_{100}		1376	1320
Profenofos		2155*	372_{27}	337_{62}	206_{47}	139_{83}	97_{100}	1321	3483
Profluralin		1830	347_{16}	330_{46}	318_{100}	264_{20}	55_{69}	1193	3880
Progesterone		2780*	314_{68}	272_{44}	124_{100}			997	894
Proglumetacin artifact 2ME	PMF UME	2090	247_{38}	188_{100}	173_{10}	145_8		585	6294
Proglumetacin artifact ME	PME UME	2130	233_{38}	174_{100}				521	1230
Proglumetacin-M/artifact (HO-indometacin) 2ME	UME	2880	401_{27}	262_4	139_{100}	111_{23}		1448	6293
Proglumetacin-M/artifact (HOOC-) ME		2445	348_1	247_2	220_{13}	105_{100}	98_{54}	1201	5258
Proglumetacin-M/artifact (indometacin)	G P-I	2550	313_{30}	139_{100}	111_{24}			1246	1038
Proglumetacin-M/artifact (indometacin) ET		2820	385_{40}	312_{19}	158_6	139_{100}	111_{20}	1380	3168
Proglumetacin-M/artifact (indometacin) ME	P(ME) G(ME) U(ME)	2770	371_9	312_6	139_{100}	111_{17}		1316	1039
Proglumetacin-M/artifact (indometacin) TMS		2650	429_{23}	370_4	312_{17}	139_{100}	73_{22}	1530	5462
Proglumetacin-M/artifact -H2O isomer-1 AC		1765	212_7	139_{83}	98_{40}	70_{100}	56_{98}	440	5260
Proglumetacin-M/artifact -H2O isomer-2 AC		1900	212_{20}	139_{85}	125_{64}	98_{62}	70_{100}	440	5259
Proline MEAC		1465	171_8	128_1	112_{32}	70_{100}	68_7	325	2708
Proline-M (HO-) ME2AC		1690	229_1	198	169_9	110_{37}	68_{100}	506	2709
Proline-M (HO-) MEAC		1635	187_7	128_{54}	86_{100}	68_{28}		363	2283
Prolintane	G U UHY UHYAC	1720	216_1	174_{10}	126_{100}	91_{30}	65_{12}	456	2729
Prolintane-M (di-HO-phenyl-) 2AC	UHYAC	2295	332_1	290_2	248_1	126_{100}	123_2	1114	4112
Prolintane-M (HO-methoxy-phenyl-) AC	UHYAC	2215	304_1	262_2	137_2	126_{100}	55_3	940	4109
Prolintane-M (HO-phenyl-)	UHY	2135	232_1	190_3	126_{100}	107_3	96_3	524	4103
Prolintane-M (HO-phenyl-) AC	UHYAC	2110	274_1	232_4	190_2	126_{100}	107_6	749	4108
Prolintane-M (oxo-)	U UHY UHYAC	1895	231_1	188_2	140_{100}	98_{37}	91_{18}	515	4102
Prolintane-M (oxo-di-HO-) 2AC	UHYAC	2485	347_1	279_7	198_{100}	156_{27}	128_{27}	1196	4115
Prolintane-M (oxo-di-HO-methoxy-) 2AC	UHYAC	2560	377_5	234_{32}	198_{100}	192_{81}	156_{27}	1344	4116
Prolintane-M (oxo-di-HO-phenyl-)		2475	263_2	178_{14}	140_{100}	98_{48}	86_{24}	675	4107
Prolintane-M (oxo-di-HO-phenyl-) 2AC	UHYAC	2460	347_1	220_5	178_6	140_{100}	98_{17}	1195	4114
Prolintane-M (oxo-di-HO-phenyl-) 2ME	UHYME	2260	291_3	206_{35}	140_{100}	98_{32}	86_{12}	851	4106
Prolintane-M (oxo-HO-alkyl-)		2200	188_2	156_{32}	91_{17}	86_{100}	71_{34}	588	4104
Prolintane-M (oxo-HO-alkyl-) AC	U+UHYAC	2255	198_{67}	156_{58}	138_{100}	91_{32}	86_{81}	837	4110
Prolintane-M (oxo-HO-methoxy-phenyl-)	UHY	2240	277_2	192_{28}	140_{100}	98_{35}	86_{13}	763	4105
Prolintane-M (oxo-HO-methoxy-phenyl-) AC	UHYAC	2360	319_2	234_4	192_{18}	140_{100}	98_{19}	1028	4113
Prolintane-M (oxo-HO-phenyl-) AC	UHYAC	2275	289_1	204_3	140_{100}	98_{24}	86_{10}	837	4111
Prolintane-M (oxo-tri-HO-) 3AC	UHYAC	2630	405_2	198_{100}	156_{23}	128_{21}	84_8	1463	4117
Promazine	P G U UHY U+UHYAC	2315	284_{15}	199_{19}	86_{23}	58_{100}		803	377
Promazine-M (bis-nor-) AC	U+UHYAC	2720	298_{56}	212_{39}	198_{100}	180_{31}	100_{87}	894	2076
Promazine-M (bis-nor-HO-) 2AC	U+UHYAC	3100	356_{88}	215_{65}	214_{96}	100_{100}	72_{19}	1242	2677
Promazine-M (HO-)	UHY	2685	300_9	254_3	215_7	86_{16}	58_{100}	907	605
Promazine-M (HO-) AC	U+UHYAC	2710	342_{38}	257_{24}	215_{48}	86_{46}	58_{100}	1169	378
Promazine-M (nor-)	UHY	2405	270_{65}	238_{27}	213_{35}	199_{100}		718	604
Promazine-M (nor-) AC	U+UHYAC	2805	312_{32}	198_{19}	180_{12}	114_{100}		981	379
Promazine-M (nor-HO-) 2AC	UHYAC	3195	370_{21}	328_4	214_{29}	114_{100}	86_{18}	1311	380
Promazine-M (ring)	P G U+UHYAC	2010	199_{100}	167_{44}				396	10
Promazine-M (sulfoxide)	U	2705	300_{10}	284_8	212_{53}	58_{100}		907	603
Promazine-M 2AC	U+UHYAC	2865	315_{54}	273_{34}	231_{100}	202_{11}		999	2618
Promazine-M AC	U+UHYAC	2550	257_{23}	215_{100}	183_7			641	12
Promecarb		1665	207_1	150_{69}	135_{100}	91_{16}	58_{10}	420	3484
Promecarb-M/artifact (decarbamoyl-)		1290*	150_{42}	135_{100}	107_{25}	91_{42}	77_{18}	290	3485
Promethazine	P G U+UHYAC	2270	284_2	213_4	198_5	180_4	72_{100}	803	381
Promethazine-M (bis-nor-) AC	U+UHYAC	2450	298_{43}	212_{100}	180_{42}	100_7	58_5	894	1319
Promethazine-M (bis-nor-HO-) 2AC	U+UHYAC	2900	356_{65}	270_{100}	228_{83}	196_{28}	100_{13}	1242	2619
Promethazine-M (di-HO-) 2AC	U+UHYAC	3075	400_7	329_{10}	244_3	230_3	72_{100}	1446	2621
Promethazine-M (HO-)	UHY	2590	300_5	229_4	214_3	196_3	72_{100}	907	609
Promethazine-M (HO-) AC	U+UHYAC	2690	342_{13}	271_{12}	214_5	196_6	72_{100}	1169	383
Promethazine-M (HO-methoxy-) AC	U+UHYAC	2800	372_6	301_5	253_6	226_2	72_{100}	1323	2617
Promethazine-M (nor-)	P UHY	2250	270_3	213_{100}	198_{21}	180_{16}	58_{52}	718	607
Promethazine-M (nor-) AC	U+UHYAC	2540	312_{52}	212_{100}	180_{73}	114_{71}	58_{48}	981	382
Promethazine-M (nor-di-HO-) 3AC	UHYAC	3360	428_{24}	328_{41}	244_{19}	114_{100}	58_{69}	1528	3334
Promethazine-M (nor-HO-)	UHY	2580	286_7	229_{30}	212_{100}	180_{34}	58_{46}	814	608
Promethazine-M (nor-HO-) 2AC	U+UHYAC	3015	370_{44}	270_{100}	228_{79}	114_{63}		1311	384
Promethazine-M (nor-HO-) AC	U+UHYAC	2960	328_{56}	228_{100}	196_{42}	114_{55}	58_{82}	1081	2620
Promethazine-M (nor-sulfoxide) AC	U+UHYAC	2810	328_9	212_{76}	180_{32}	114_{100}	58_{79}	1081	610

Table 8.1: Compounds in order of names

Name	Detected	RI	Typical ions and intensities					Page	Entry
Promethazine-M (ring)	P G U+UHYAC	2010	199$_{100}$	167$_{44}$				396	10
Promethazine-M 2AC	U+UHYAC	2865	315$_{54}$	273$_{34}$	231$_{100}$	202$_{11}$		999	2618
Promethazine-M AC	U+UHYAC	2550	257$_{23}$	215$_{100}$	183$_{7}$			641	12
Promethazine-M/artifact (sulfoxide)	G P U+UHYAC	2710	300$_{2}$	284$_{4}$	213$_{18}$	72$_{100}$		907	606
Prometryn		1930	241$_{100}$	226$_{58}$	184$_{97}$	106$_{31}$	58$_{70}$	561	3862
Propachlor		1600	211$_{6}$	196$_{7}$	176$_{29}$	120$_{100}$	77$_{44}$	436	3486
Propafenone	P-I G	2740	312$_{2}$	297$_{4}$	121$_{2}$	91$_{9}$	72$_{100}$	1165	2391
Propafenone 2AC	U+UHYAC	2980	425$_{3}$	322$_{63}$	200$_{71}$	140$_{100}$	72$_{92}$	1521	2259
Propafenone 2TMS		2860	485$_{1}$	283$_{2}$	144$_{100}$	91$_{12}$	73$_{51}$	1642	4590
Propafenone 3TMS		2840	557$_{1}$	370$_{9}$	206$_{3}$	144$_{100}$	73$_{66}$	1699	4591
Propafenone artifact	P-I G	2760	353$_{10}$	324$_{16}$	128$_{100}$	98$_{68}$	91$_{77}$	1230	895
Propafenone -H2O	G UHY	2300	323$_{14}$	294$_{22}$	230$_{20}$	98$_{92}$	91$_{100}$	1053	897
Propafenone -H2O AC	U+UHYAC	2930	365$_{7}$	322$_{18}$	140$_{82}$	98$_{100}$	91$_{87}$	1287	902
Propafenone-M (deamino-di-HO-) 3AC	U+UHYAC	2950*	442$_{6}$	224$_{3}$	159$_{100}$	137$_{13}$	91$_{26}$	1565	903
Propafenone-M (deamino-HO-) 2AC	U+UHYAC	2715*	384$_{26}$	159$_{100}$	121$_{78}$	91$_{56}$		1377	901
Propafenone-M (HO-) -H2O	UHY	2720	339$_{21}$	310$_{14}$	98$_{100}$	91$_{72}$		1151	898
Propafenone-M (HO-) -H2O 2AC	U+UHYAC	3050	423$_{10}$	282$_{61}$	140$_{57}$	98$_{100}$	72$_{83}$	1515	904
Propafenone-M (O-dealkyl-)	G P-I U+UHYAC	1830*	226$_{35}$	207$_{14}$	121$_{100}$	91$_{28}$	65$_{32}$	495	896
Propafenone-M (O-dealkyl-) AC	UHYAC	2130*	268$_{6}$	225$_{38}$	208$_{32}$	121$_{100}$	91$_{15}$	707	3726
Propafenone-M (O-dealkyl-HO-) isomer-1	UHY	2345*	242$_{35}$	223$_{22}$	121$_{100}$	107$_{67}$	65$_{26}$	565	3344
Propafenone-M (O-dealkyl-HO-) isomer-1 AC	U+UHYAC	2215*	284$_{30}$	242$_{96}$	137$_{100}$	91$_{79}$		801	899
Propafenone-M (O-dealkyl-HO-) isomer-2	UHY	2355*	242$_{43}$	223$_{27}$	121$_{100}$	107$_{78}$	65$_{30}$	564	3345
Propafenone-M (O-dealkyl-HO-) isomer-2 AC	UHYAC	2370*	284$_{17}$	242$_{26}$	224$_{38}$	121$_{100}$	65$_{17}$	800	3350
Propafenone-M (O-dealkyl-HO-methoxy-)	UHY	2400*	272$_{36}$	151$_{2}$	137$_{100}$	121$_{52}$	65$_{31}$	730	3346
Propafenone-M (O-dealkyl-HO-methoxy-) AC	U+UHYAC	2580*	314$_{19}$	272$_{90}$	167$_{100}$	137$_{17}$	91$_{53}$	994	900
Propallylonal	P G U UHY UHYAC	1875	209$_{84}$	167$_{100}$	124$_{27}$			825	921
Propallylonal 2ME	PME	1745	237$_{99}$	195$_{100}$	138$_{43}$			1004	923
Propallylonal-M (debromo-)	P G U+UHYAC	1610	210$_{6}$	195$_{18}$	167$_{100}$	124$_{43}$		433	63
Propallylonal-M (debromo-) 2TMS		1620	354$_{3}$	339$_{81}$	297$_{40}$	100$_{47}$	73$_{100}$	1233	5458
Propallylonal-M (debromo-dihydro-HO-) 2ME		----	241$_{5}$	214$_{21}$	198$_{53}$	183$_{91}$	169$_{100}$	638	924
Propallylonal-M (debromo-HO-)	U UHY UHYAC	1770	226$_{5}$	184$_{20}$	169$_{100}$	141$_{64}$		494	922
Propallylonal-M (debromo-oxo-) 2ME		1720	239$_{2}$	212$_{29}$	197$_{58}$	169$_{100}$	112$_{39}$	629	925
Propamocarb		1875	188$_{2}$	143$_{2}$	129$_{3}$	72$_{3}$	58$_{100}$	366	2730
Propamocarb TFA		1290	284$_{1}$	225	126	69$_{9}$	58$_{100}$	803	4135
1,2-Propane diol		<1000*	76$_{1}$	61$_{2}$	45$_{29}$	40$_{6}$	32$_{100}$	253	6454
1,2-Propane diol 2TMS		<1000*	205$_{4}$	147$_{50}$	117$_{100}$	101$_{2}$	73$_{90}$	470	8582
1,2-Propane diol dibenzoate		2240*	284$_{1}$	227$_{10}$	162$_{20}$	105$_{100}$	77$_{90}$	801	1760
1,2-Propane diol dipivalate	PPIV	1350*	143$_{8}$	127$_{2}$	103$_{10}$	85$_{27}$	57$_{100}$	572	6423
1,2-Propane diol phenylboronate		1240*	162$_{67}$	147$_{100}$	118$_{22}$	104$_{47}$	91$_{48}$	306	1898
1,3-Propane diol 2TMS		<1000*	205$_{10}$	147$_{100}$	130$_{30}$	114$_{21}$	73$_{78}$	470	8583
1,3-Propane diol dibenzoate		2300*	227$_{4}$	162$_{4}$	105$_{100}$	77$_{35}$		801	1761
1,3-Propane diol dipivalate		1420*	143$_{14}$	103$_{22}$	85$_{39}$	57$_{100}$		572	1905
1,3-Propane diol phenylboronate		1370*	162$_{92}$	132$_{5}$	104$_{100}$	91$_{27}$	77$_{22}$	306	1899
1-Propanol		<1000*	60$_{8}$	59$_{15}$	42$_{15}$	31$_{100}$		250	6456
Propazine		1740	229$_{61}$	214$_{100}$	187$_{33}$	172$_{73}$	58$_{36}$	506	2398
Propetamphos		1780	281$_{2}$	236$_{24}$	194$_{37}$	138$_{100}$	110$_{35}$	786	2518
Propetamphos-M/artifact (HOOC-) ME		1675	253$_{1}$	208$_{24}$	138$_{100}$	122$_{17}$	110$_{38}$	621	7539
Propham		1430	179$_{50}$	137$_{46}$	120$_{34}$	93$_{100}$	65$_{26}$	341	3487
Propicillin artifact-1		1430	179$_{40}$	135$_{100}$	107$_{67}$	94$_{92}$	77$_{83}$	341	8467
Propicillin artifact-2		1830	251$_{10}$	158$_{8}$	135$_{100}$	107$_{59}$	94$_{64}$	610	8466
Propicillin MEAC		1830	434$_{1}$	251$_{4}$	184$_{23}$	142$_{100}$	107$_{15}$	1544	8468
Propiconazole		2330	340$_{1}$	259$_{60}$	191$_{23}$	173$_{83}$	69$_{100}$	1159	3488
Propiconazole artifact (dichlorophenylethanone)	U+UHYAC	1280*	188$_{16}$	173$_{100}$	145$_{19}$	109$_{11}$	75$_{14}$	364	3489
Propionic acid anhydride		<1000*	79$_{2}$	57$_{100}$	44$_{6}$			272	2757
1-Propionyl-LSD		3350	379$_{84}$	277$_{100}$	221$_{79}$	181$_{42}$	72$_{49}$	1353	9813
1-Propionyl-LSD TMS		3210	451$_{17}$	349$_{69}$	309$_{60}$	73$_{100}$	72$_{37}$	1583	9872
Propiophenone		<1000*	134$_{19}$	105$_{100}$	77$_{60}$	74$_{3}$	51$_{24}$	273	7282
Propivan		1840	277$_{1}$	205$_{3}$	99$_{13}$	86$_{100}$	58$_{15}$	764	1523
Propiverine	P U U+UHYAC	2460	367$_{1}$	309$_{31}$	225$_{85}$	183$_{100}$	105$_{65}$	1300	6080
Propiverine-M/artifact	UHY	2070*	228$_{8}$	186$_{100}$	157$_{10}$	128$_{7}$	77$_{4}$	499	1626
Propiverine-M/artifact (benzhydrol) AC	U+UHYAC	1700*	226$_{20}$	184$_{20}$	165$_{100}$	105$_{14}$	77$_{35}$	495	1241
Propiverine-M/artifact (carbinol)		2430	325$_{18}$	183$_{100}$	105$_{89}$	98$_{79}$	77$_{70}$	1063	6081
Propiverine-M/artifact (carbinol) AC		2455	367$_{3}$	183$_{47}$	165$_{8}$	98$_{58}$	96$_{100}$	1298	6082
Propiverine-M/artifact AC	U+UHYAC	2100*	270$_{3}$	228$_{92}$	151$_{100}$	105$_{25}$	77$_{40}$	716	1622
Propiverine-M/artifact AC	U+UHYAC	2200*	270$_{7}$	228$_{17}$	186$_{100}$	157$_{10}$	128$_{7}$	714	1623
Propiverine-M/artifact AC	U+UHYAC	2090*	284$_{5}$	242$_{15}$	224$_{17}$	182$_{100}$	153$_{19}$	800	2425
Propiverine-M/artifact isomer-1 AC	U+UHYAC	2010*	240$_{27}$	198$_{100}$	121$_{47}$	105$_{85}$	77$_{80}$	555	2196
Propiverine-M/artifact isomer-2 AC	U+UHYAC	2050*	240$_{20}$	198$_{100}$	121$_{94}$	105$_{41}$	77$_{51}$	555	2197
Propofol	G P U	1320*	178$_{24}$	163$_{100}$	121$_{15}$	117$_{17}$	91$_{13}$	339	3305
Propofol AC		1510*	220$_{9}$	178$_{49}$	163$_{100}$	135$_{9}$	91$_{17}$	470	3306
Propofol ME		1290*	192$_{38}$	177$_{100}$	149$_{22}$	119$_{36}$	91$_{31}$	376	3521
Propofol TMS		1305*	250$_{33}$	235$_{100}$	219$_{6}$	161$_{6}$	73$_{86}$	607	6874
Propoxur	G P U	1585	209$_{1}$	152$_{17}$	110$_{100}$	81$_{7}$		429	926
Propoxur HYAC	U+UHYAC	1390*	194$_{2}$	152$_{21}$	110$_{100}$	81$_{5}$	52$_{10}$	383	1223

Propoxur HYME **Table 8.1:** Compounds in order of names

Name	Detected	RI	Typical ions and intensities					Page	Entry
Propoxur HYME	UHYME	1380*	166_5	151_{10}	110_{100}	81_{15}	64_{43}	317	2536
Propoxur impurity-M (HO-)	UHY	1440*	186_{10}	146_{32}	144_{100}	79_{30}		360	2537
Propoxur impurity-M (HO-) AC	U+UHYAC	1520*	228_3	186_{22}	146_{47}	144_{100}	79_{10}	500	1225
Propoxur impurity-M (HO-) ME	UHYME	1530*	200_{16}	185_{20}	144_{100}	98_{24}	63_{35}	399	2540
Propoxur impurity-M (O-dealkyl-HO-)	UHY	1490*	146_{34}	144_{100}	115_8	98_{27}	63_{43}	283	2539
Propoxur TFA		1530	305_8	263_{13}	206_{62}	109_{47}	69_{100}	935	4130
Propoxur-M (HO-) HY	P UHY	1470	168_{16}	126_{100}	108_3	97_{11}		321	2538
Propoxur-M (HO-) HY2AC	U+UHYAC	1680*	252_8	210_{17}	168_{42}	126_{100}	97_{11}	618	1224
Propoxur-M (O-dealkyl-) HY	P UHY	<1000*	110_{100}	92_{11}	81_{15}	64_{42}	53_{15}	261	2535
Propoxur-M/artifact (isopropoxyphenol)	P G U	1070*	152_{13}	137	110_{100}	81_7	64_7	294	2632
Propoxyphene	G P	2205	250_2	193_3	178_2	91_{15}	58_{100}	1153	476
Propoxyphene artifact		1755*	208_{56}	193_{41}	130_{38}	115_{100}	91_{42}	426	477
Propoxyphene-M (HY)	UHY	2395	281_9	190_{76}	119_{96}	105_{100}	56_{97}	785	480
Propoxyphene-M (nor-) -H2O	UHY	2240	251_{30}	217_{96}	119_{100}	91_{72}	77_{40}	613	479
Propoxyphene-M (nor-) -H2O AC	UHYAC	2365	293_{18}	220_{100}	205_{39}			866	232
Propoxyphene-M (nor-) -H2O N-prop.	U UHY UHYAC	2555	307_8	234_{76}	105_{100}	100_{74}	91_{67}	953	231
Propoxyphene-M (nor-) N-prop.	P U	2400	307_{16}	220_{68}	100_{100}	57_{83}		1065	478
Propranolol	P-I G U UHY	2160	259_5	215_4	144_{11}	115_{12}	72_{100}	652	927
Propranolol 2AC	U+UHYAC	2605	343_5	283_{14}	200_{100}	140_{80}	98_{80}	1175	931
Propranolol formyl artifact	P G U	2205	271_{65}	256_{41}	183_{27}	127_{100}	112_{44}	725	3413
Propranolol -H2O	UHY	2220	241_5	98_{45}	56_{100}			561	930
Propranolol -H2O AC	U+UHYAC	2330	283_{24}	198_{52}	140_{100}	127_{81}	98_{87}	797	935
Propranolol TMSTFA		2320	427_4	284_{100}	242_{20}	129_{74}	73_{40}	1526	6154
Propranolol-M (1-naphthol)		1500*	144_{100}	115_{83}	89_{14}	74_6	63_{17}	283	928
Propranolol-M (1-naphthol) AC	U+UHYAC	1555*	186_{13}	144_{100}	115_{47}	89_8	63_7	361	932
Propranolol-M (1-naphthol) HFB		1310*	340_{46}	169_{25}	143_{28}	115_{100}	89_{11}	1155	7476
Propranolol-M (1-naphthol) PFP		1510*	290_{45}	171_{100}	143_{20}	115_{49}	89_8	839	7468
Propranolol-M (1-naphthol) TMS		1525*	216_{100}	201_{95}	185_{51}	115_{39}	73_{21}	452	7460
Propranolol-M (4-HO-1-naphthol) 2AC	U+UHYAC	1900*	244_{14}	202_{19}	160_{100}	131_{21}	103_8	570	933
Propranolol-M (deamino-di-HO-) 3AC	U+UHYAC	2565*	360_3	318_1	159_{100}			1264	936
Propranolol-M (deamino-HO-)	UHY	2065*	218_{21}	144_{100}	115_{37}			458	929
Propranolol-M (deamino-HO-) 2AC	U+UHYAC	2195*	302_{10}	159_{100}	144_{48}	115_{44}		918	934
Propranolol-M (HO-) 3AC	U+UHYAC	2940	401_1	341_{12}	186_{66}	140_{100}	98_{75}	1450	939
Propranolol-M (HO-) -H2O isomer-1 2AC	U+UHYAC	2750	341_8	197_{20}	140_{100}	98_{55}		1162	937
Propranolol-M (HO-) -H2O isomer-2 2AC	U+UHYAC	2900	341_{12}	197_{27}	140_{100}	98_{74}		1162	938
4-Propyl-2,5-dimethoxyphenethylamine		1720	223_{21}	194_{100}	179_{13}	165_{51}	135_{15}	485	6906
4-Propyl-2,5-dimethoxyphenethylamine 2AC		2160	307_{20}	206_{100}	193_{24}	177_{42}	135_8	952	6921
4-Propyl-2,5-dimethoxyphenethylamine 2TMS		2130	367_1	352_5	174_{100}	100_{11}	86_{28}	1300	6923
4-Propyl-2,5-dimethoxyphenethylamine AC		2090	265_{14}	206_{100}	193_{28}	177_{60}	135_{13}	690	6920
4-Propyl-2,5-dimethoxyphenethylamine formyl artifact		1755	235_{23}	204_{100}	193_{62}	163_9	135_{11}	536	6908
4-Propyl-2,5-dimethoxyphenethylamine HFB		1895	419_{49}	206_{76}	193_{100}	177_{42}	163_{13}	1506	6940
4-Propyl-2,5-dimethoxyphenethylamine PFP		1865	369_{54}	206_{77}	193_{100}	177_{42}	119_{15}	1305	6935
4-Propyl-2,5-dimethoxyphenethylamine TFA		1870	319_{36}	206_{69}	193_{100}	177_{39}	149_{38}	1025	6930
4-Propyl-2,5-dimethoxyphenethylamine TMS		1860	295_5	265_5	194_{20}	102_{100}	73_{64}	881	6922
4-Propyl-2,5-dimethoxyphenethylamine-M (deamino-HO-) AC		2090*	266_{18}	206_{100}	193_8	177_{70}	91_{16}	696	9205
Propylamine		<1000	59_7	42_{27}	30_{100}			249	3616
Propylbenzene		<1000*	120_{20}	105_4	91_{100}	65_{17}		265	3786
Propylhexedrine	U UHY	1170	155_1	140_3	58_{100}			299	940
Propylhexedrine AC	U+UHYAC	1570	197_{12}	182_{52}	140_{16}	100_{100}	58_{67}	393	942
Propylhexedrine HFB	UHFB UHYHFB	1440	351_1	254_{100}	210_{24}	182_{53}	69_{45}	1212	5100
Propylhexedrine PFP	UPFP UHYPFP	1385	301_1	204_{100}	182_{31}	160_{24}	119_{18}	914	5096
Propylhexedrine TFA	UTFA	1385	251_1	182_{97}	154_{100}	110_{22}	69_{16}	612	5093
Propylhexedrine-M (HO-)	U UHY	1475	171_3	156_4	58_{100}			326	941
Propylhexedrine-M (HO-) 2AC	U+UHYAC	1915	255_2	240_3	195_2	100_{64}	58_{100}	634	943
Propylparaben	U UHY	1630*	180_{11}	138_{67}	121_{100}	93_{12}	65_{14}	346	2971
Propylparaben AC	UHYAC	1610*	222_8	180_{32}	138_{100}	121_{84}	93_9	479	2972
4-Propylthio-2,5-dimethoxyphenethylamine		2470	255_{26}	226_{100}	183_{63}	169_{31}	153_{34}	633	6855
4-Propylthio-2,5-dimethoxyphenethylamine 2AC		2470	339_{14}	238_{100}	225_{30}	181_{22}	153_{17}	1150	6859
4-Propylthio-2,5-dimethoxyphenethylamine 2TMS		2395	399_1	384_4	369_4	225_7	174_{100}	1443	6860
4-Propylthio-2,5-dimethoxyphenethylamine AC		2410	297_{66}	238_{100}	225_{37}	196_{56}	183_{55}	891	6858
4-Propylthio-2,5-dimethoxyphenethylamine deuteroformyl artifact		2060	269_{27}	238_{25}	225_{100}	183_{13}	153_{24}	713	6857
4-Propylthio-2,5-dimethoxyphenethylamine formyl artifact		2050	267_{35}	236_{27}	225_{100}	183_{26}	153_{46}	701	6856
4-Propylthio-2,5-dimethoxyphenethylamine HFB		2175	451_{14}	238_{24}	225_{100}	181_{23}	153_{21}	1582	6861
4-Propylthio-2,5-dimethoxyphenethylamine PFP		2160	401_{23}	238_{17}	225_{100}	181_{17}	153_{19}	1449	6862
4-Propylthio-2,5-dimethoxyphenethylamine TFA		2170	351_{26}	238_{17}	225_{100}	181_{24}	153_{23}	1212	6863
4-Propylthio-2,5-dimethoxyphenethylamine-M (deamino-HO-)	UGLUC	2000*	256_{69}	225_{100}	183_{23}	150_{56}	135_{23}	637	6864
4-Propylthio-2,5-dimethoxyphenethylamine-M (deamino-HO-) AC	U+UHYAC	2080*	298_{56}	238_{100}	225_{10}	196_{80}	181_{69}	895	6869
4-Propylthio-2,5-dimethoxyphenethylamine-M (deamino-HOOC-)		2110*	270_{100}	225_{55}	213_{46}	181_{34}	153_{21}	717	6872
4-Propylthio-2,5-dimethoxyphenethylamine-M (deamino-HOOC-) ME		1950*	284_{100}	227_{50}	225_{74}	183_{24}	153_{25}	801	6873
4-Propylthio-2,5-dimethoxyphenethylamine-M (deamino-oxo-)		2190*	254_{42}	225_{100}	183_{14}	153_{24}	137_8	628	7235
4-Propylthio-2,5-dimethoxyphenethylamine-M (HO- N-acetyl-)	UGLUC	2525	313_{40}	254_{100}	242_{44}	210_{38}	183_{21}	988	6866
4-Propylthio-2,5-dimethoxyphenethylamine-M (HO- N-acetyl-) TFA		2345	409_{27}	350_{100}	337_9	236_5	181_{13}	1473	6871
4-Propylthio-2,5-dimethoxyphenethylamine-M (HO- sulfone N-acetyl-)	UGLUC	2740	345_{31}	286_{73}	164_{100}	151_{27}	120_{18}	1184	6865
4-Propylthio-2,5-dimethoxyphenethylamine-M (HO- sulfone) 2AC	U+UHYAC	2760	387_{31}	340_{42}	328_{100}	268_{36}	108_{33}	1390	6868

Table 8.1: Compounds in order of names

4-Propylthio-2,5-dimethoxyphenethylamine-M (HO-) 2AC

Name	Detected	RI	Typical ions and intensities					Page	Entry
4-Propylthio-2,5-dimethoxyphenethylamine-M (HO-) 2AC	U+UHYAC	2585	355 $_{51}$	296 $_{72}$	283 $_{10}$	236 $_{92}$	101 $_{100}$	1237	6867
4-Propylthio-2,5-dimethoxyphenethylamine-M (HO-) 2TFA		2110	463 $_{80}$	434 $_{43}$	350 $_{60}$	337 $_{100}$	231 $_{67}$	1605	6870
4-Propylthio-2,5-dimethoxyphenethylamine-M (HO-) 3AC	U+UHYAC	2630	397 $_{69}$	296 $_{99}$	283 $_{12}$	236 $_{100}$	101 $_{64}$	1435	6875
4-Propylthio-2,5-dimethoxyphenethylamine-M (S-depropyl-) AC	U+UHYAC	2170	255 $_{18}$	196 $_{100}$	183 $_{41}$	181 $_{34}$	153 $_{21}$	631	6831
4-Propylthio-2,5-dimethoxyphenethylamine-M (S-depropyl-) isomer-1 2AC		2240	297 $_{29}$	210 $_{14}$	196 $_{100}$	183 $_{35}$	181 $_{29}$	889	6823
4-Propylthio-2,5-dimethoxyphenethylamine-M (S-depropyl-) isomer-2 2AC	U+UHYAC	2360	297 $_{16}$	255 $_{11}$	238 $_{20}$	196 $_{100}$	183 $_{37}$	889	6826
4-Propylthio-2,5-dimethoxyphenethylamine-M (S-depropyl-methyl- N-acetyl-)	U+UHYAC	2230	269 $_{19}$	210 $_{100}$	197 $_{35}$	195 $_{21}$	167 $_{27}$	712	6832
4-Propylthio-2,5-dimethoxyphenethylamine-M (S-depropyl-methyl- sulfone) AC	U+UHYAC	2580	301 $_{7}$	242 $_{100}$	230 $_{4}$	196 $_{7}$	124 $_{7}$	911	6829
4-Propylthio-2,5-dimethoxyphenethylamine-M (S-depropyl-methyl- sulfoxide) AC	U+UHYAC	2460	285 $_{16}$	268 $_{23}$	226 $_{33}$	211 $_{100}$	197 $_{31}$	808	6830
Propyphenazone	G P U+UHYAC	1910	230 $_{41}$	215 $_{100}$	56 $_{65}$			510	202
Propyphenazone-M (di-HO-) 2AC	UHYAC	2680	346 $_{26}$	303 $_{28}$	273 $_{100}$	231 $_{44}$	56 $_{62}$	1189	2594
Propyphenazone-M (HO-methyl-)	UHY	2410	246 $_{51}$	231 $_{100}$	215 $_{9}$	77 $_{16}$		580	912
Propyphenazone-M (HO-methyl-) AC	UHYAC	2240	288 $_{63}$	273 $_{83}$	245 $_{100}$	232 $_{95}$	190 $_{39}$	828	206
Propyphenazone-M (HOOC-) ME	UME	2160	274 $_{31}$	215 $_{100}$	56 $_{50}$			739	917
Propyphenazone-M (HO-phenyl-)	UHY	2300	246 $_{56}$	231 $_{100}$	96 $_{38}$	56 $_{100}$		581	911
Propyphenazone-M (HO-phenyl-) AC	U+UHYAC	2530	288 $_{45}$	273 $_{43}$	246 $_{46}$	231 $_{100}$	56 $_{84}$	828	208
Propyphenazone-M (HO-phenyl-) ME	UME	2310	260 $_{36}$	245 $_{64}$	122 $_{26}$	96 $_{22}$	56 $_{100}$	659	915
Propyphenazone-M (HO-propyl-)	P U UHY	2210	246 $_{24}$	231 $_{12}$	215 $_{100}$	124 $_{28}$	56 $_{74}$	581	910
Propyphenazone-M (HO-propyl-) AC	UHYAC	2305	288 $_{23}$	245 $_{19}$	228 $_{24}$	215 $_{100}$	56 $_{51}$	828	207
Propyphenazone-M (isopropanolyl-)	UGLUC	2020	246 $_{13}$	231 $_{100}$	213 $_{12}$			580	913
Propyphenazone-M (isopropenyl-)	P U UHY	1970	228 $_{39}$	136 $_{100}$	95 $_{51}$			503	907
Propyphenazone-M (nor-)	P G U UHY	1765	216 $_{52}$	174 $_{100}$	77 $_{68}$			453	905
Propyphenazone-M (nor-) AC	U+UHYAC	1820	258 $_{6}$	216 $_{35}$	201 $_{100}$	185 $_{6}$	77 $_{17}$	647	203
Propyphenazone-M (nor-) ME	UME	1735	230 $_{30}$	215 $_{100}$	200 $_{20}$	185 $_{17}$	77 $_{55}$	512	914
Propyphenazone-M (nor-) TMS		1860	288 $_{18}$	273 $_{100}$	185 $_{10}$	77 $_{14}$	73 $_{17}$	829	4620
Propyphenazone-M (nor-di-HO-)	UHY	2090	248 $_{100}$	206 $_{19}$	136 $_{19}$	109 $_{48}$		590	909
Propyphenazone-M (nor-di-HO-) 2AC	UHYAC	2400	332 $_{26}$	290 $_{100}$	274 $_{10}$	232 $_{16}$	206 $_{26}$	1106	2593
Propyphenazone-M (nor-di-HO-) 3ME	UHYME	2240	290 $_{54}$	275 $_{100}$	260 $_{40}$	110 $_{2}$		843	3768
Propyphenazone-M (nor-di-HO-) AC	U+UHYAC	2250	290 $_{31}$	248 $_{100}$	206 $_{21}$	136 $_{8}$	109 $_{33}$	841	1882
Propyphenazone-M (nor-HO-)	UHY	1780	232 $_{94}$	190 $_{48}$	121 $_{35}$	93 $_{100}$	77 $_{64}$	517	906
Propyphenazone-M (nor-HO-) AC	UHYAC	1895	274 $_{97}$	232 $_{4}$	214 $_{25}$	190 $_{100}$		740	204
Propyphenazone-M (nor-HO-phenyl-)	UHY	2080	232 $_{81}$	190 $_{100}$	121 $_{51}$	93 $_{68}$	65 $_{60}$	516	908
Propyphenazone-M (nor-HO-phenyl-) 2AC	U+UHYAC	2165	316 $_{11}$	274 $_{48}$	259 $_{12}$	232 $_{82}$	217 $_{100}$	1005	205
Propyphenazone-M (nor-HO-phenyl-) AC	U+UHYAC	2190	274 $_{19}$	232 $_{100}$	190 $_{69}$	121 $_{10}$	93 $_{14}$	740	2595
Propyphenazone-M (nor-HO-phenyl-) isomer-1 2ME	UME	2030	260 $_{45}$	245 $_{100}$	230 $_{50}$	215 $_{28}$		659	916
Propyphenazone-M (nor-HO-phenyl-) isomer-2 2ME		2060	260 $_{44}$	245 $_{100}$	230 $_{48}$	215 $_{13}$	77 $_{35}$	658	3767
Propyphenazone-M (nor-HO-propyl-) 2AC	UHYAC	2120	316 $_{11}$	274 $_{63}$	214 $_{100}$	201 $_{96}$	77 $_{60}$	1005	1933
Propyzamide		1790	255 $_{26}$	173 $_{100}$	145 $_{34}$	109 $_{19}$	84 $_{16}$	630	3490
Propyzamide artifact (dechloro-)		1645	221 $_{24}$	206 $_{7}$	139 $_{100}$	111 $_{29}$	75 $_{16}$	472	3491
Proquazone		2670	278 $_{89}$	235 $_{100}$	221 $_{57}$	77 $_{40}$		768	944
Prothiofos		2190*	344 $_{1}$	309 $_{58}$	267 $_{63}$	113 $_{100}$	63 $_{92}$	1176	3492
Prothipendyl	P G U+UHYAC	2350	285 $_{20}$	227 $_{13}$	200 $_{20}$	86 $_{20}$	58 $_{100}$	809	385
Prothipendyl-M (bis-nor-) AC	U U+UHYAC	2830	299 $_{45}$	227 $_{20}$	213 $_{25}$	200 $_{100}$	100 $_{39}$	901	387
Prothipendyl-M (bis-nor-HO-) 2AC	U+UHYAC	3030	357 $_{58}$	315 $_{7}$	258 $_{22}$	216 $_{100}$	100 $_{40}$	1247	1883
Prothipendyl-M (HO-)	U UHY	2720	301 $_{16}$	230 $_{7}$	216 $_{11}$	86 $_{18}$	58 $_{100}$	912	612
Prothipendyl-M (HO-) AC	U+UHYAC	2780	343 $_{24}$	230 $_{13}$	216 $_{8}$	86 $_{17}$	58 $_{100}$	1174	388
Prothipendyl-M (HO-ring)	U UHY	2800	216 $_{100}$	187 $_{68}$	168 $_{50}$	140 $_{9}$		451	2272
Prothipendyl-M (HO-ring) AC	U+UHYAC	2575	258 $_{30}$	216 $_{100}$	187 $_{13}$	183 $_{13}$	155 $_{10}$	645	2275
Prothipendyl-M (nor-) AC	U+UHYAC	2880	313 $_{34}$	227 $_{28}$	200 $_{29}$	181 $_{19}$	114 $_{100}$	987	389
Prothipendyl-M (nor-HO-) 2AC	UHYAC	3070	371 $_{14}$	258 $_{9}$	216 $_{26}$	114 $_{100}$	86 $_{20}$	1317	390
Prothipendyl-M (ring)	U+UHYAC	2045	200 $_{100}$	168 $_{36}$	156 $_{13}$			399	386
Prothipendyl-M (sulfoxide)	P U	2750	301 $_{2}$	285 $_{3}$	216 $_{9}$	86 $_{13}$	58 $_{100}$	912	611
Protocatechuic acid ME2AC		1750*	252 $_{1}$	210 $_{19}$	168 $_{100}$	137 $_{60}$	109 $_{8}$	615	5254
Protopine		2730	353 $_{1}$	190 $_{7}$	163 $_{20}$	148 $_{100}$	89 $_{27}$	1226	5776
Protopine-M (demethylene-methyl-) isomer-1		2990	355 $_{4}$	190 $_{5}$	165 $_{14}$	148 $_{100}$	136 $_{5}$	1237	6738
Protopine-M (demethylene-methyl-) isomer-1 AC		3050	397 $_{2}$	312 $_{4}$	190 $_{10}$	165 $_{13}$	148 $_{100}$	1434	6740
Protopine-M (demethylene-methyl-) isomer-2		3010	355 $_{8}$	190 $_{8}$	165 $_{21}$	148 $_{100}$	136 $_{36}$	1237	6739
Protopine-M (demethylene-methyl-) isomer-2 AC		3070	397 $_{2}$	312 $_{2}$	190 $_{9}$	165 $_{100}$	148 $_{100}$	1435	6741
Protriptyline	G UHY	2250	263 $_{2}$	191 $_{100}$	84 $_{6}$	70 $_{60}$		677	613
Protriptyline AC	UHYAC	2690	305 $_{11}$	191 $_{100}$	114 $_{15}$			939	391
Protriptyline TMS	G UHY	2350	335 $_{1}$	320 $_{2}$	191 $_{23}$	142 $_{42}$	116 $_{100}$	1127	5455
Protriptyline-M (HO-) 2AC	UHYAC	2895	363 $_{11}$	321 $_{6}$	249 $_{38}$	207 $_{100}$	114 $_{14}$	1279	393
Protriptyline-M (nor-) AC	UHYAC	2780	291 $_{48}$	218 $_{45}$	191 $_{100}$	100 $_{25}$	86 $_{20}$	850	392
Proxyphylline		2080	238 $_{63}$	194 $_{100}$	180 $_{90}$	137 $_{23}$	109 $_{68}$	548	945
Proxyphylline AC	U+UHYAC	2180	280 $_{100}$	237 $_{37}$	220 $_{68}$	193 $_{46}$	180 $_{65}$	781	946
Proxyphylline TMS		2080	310 $_{13}$	295 $_{30}$	180 $_{29}$	117 $_{55}$	73 $_{100}$	971	4592
Proxyphylline-M (HO-) 2AC	U+UHYAC	2455	338 $_{52}$	236 $_{14}$	194 $_{30}$	180 $_{100}$	159 $_{26}$	1142	1433
Pseudoephedrine	G P U	1385	146 $_{1}$	105 $_{4}$	91 $_{6}$	77 $_{9}$	58 $_{100}$	314	2473
Pseudoephedrine 2AC	U+UHYAC	1820	189 $_{1}$	148 $_{1}$	117 $_{2}$	100 $_{15}$	58 $_{100}$	596	2474
Pseudoephedrine 2PFP		1430	438 $_{1}$	338	294 $_{4}$	204 $_{100}$	160 $_{27}$	1595	2578
Pseudoephedrine 2TFA		1440	338 $_{1}$	244 $_{5}$	154 $_{100}$	110 $_{61}$	69 $_{48}$	1246	4016
Pseudoephedrine 2TMS		1605	294 $_{2}$	163 $_{5}$	149 $_{6}$	130 $_{100}$	73 $_{46}$	966	4593
Pseudoephedrine formyl artifact	G	1300	162 $_{1}$	117 $_{3}$	91 $_{7}$	71 $_{100}$	56 $_{30}$	335	4653
Pseudoephedrine TMSTFA		1460	213 $_{1}$	191 $_{7}$	179 $_{100}$	140 $_{4}$	73 $_{80}$	1112	6155

Pseudotropine AC **Table 8.1:** Compounds in order of names

Name	Detected	RI	Typical ions and intensities					Page	Entry
Pseudotropine AC		1230	183 $_{15}$	140 $_5$	124 $_{100}$	94 $_{59}$	82 $_{92}$	355	5435
Pseudotropine benzoate		2040	245 $_{13}$	124 $_{100}$	94 $_{44}$	82 $_{79}$	77 $_{42}$	576	5124
Psilocine		1995	204 $_{15}$	159 $_3$	146 $_5$	130 $_3$	58 $_{100}$	410	2470
Psilocine 2AC	U+UHYAC	2340	288 $_7$	246 $_1$	202 $_1$	122 $_3$	58 $_{100}$	828	2472
Psilocine 2TMS		2250	348 $_6$	333 $_1$	290 $_{29}$	73 $_{29}$	58 $_{100}$	1201	6348
Psilocine AC	U+UHYAC	2270	246 $_5$	160 $_3$	146 $_7$	130 $_3$	58 $_{100}$	580	2471
Psilocine HFB		2110	400 $_2$	342 $_4$	145 $_5$	117 $_8$	58 $_{100}$	1445	6317
Psilocine PFP		2095	350 $_3$	292 $_2$	186 $_2$	145 $_5$	58 $_{100}$	1208	6350
Psilocine TFA		2080	300 $_2$	242 $_2$	117 $_3$	69 $_4$	58 $_{100}$	906	6349
Psilocine-M (4-hydroxyindoleacetic acid) MEAC		2315	247 $_{32}$	205 $_{60}$	173 $_{40}$	145 $_{100}$	117 $_{30}$	584	6346
Psilocine-M (4-hydroxytryptophol) 2AC		2370	261 $_8$	201 $_{19}$	159 $_{100}$	146 $_{45}$	117 $_{30}$	663	6347
Psilocybin artifact		1995	204 $_{15}$	159 $_3$	146 $_5$	130 $_3$	58 $_{100}$	410	2470
Psilocybin artifact 2AC	U+UHYAC	2340	288 $_7$	246 $_1$	202 $_1$	122 $_3$	58 $_{100}$	828	2472
Psilocybin artifact 2TMS		2250	348 $_6$	333 $_1$	290 $_{29}$	73 $_{29}$	58 $_{100}$	1201	6348
Psilocybin artifact AC	U+UHYAC	2270	246 $_5$	160 $_3$	146 $_7$	130 $_3$	58 $_{100}$	580	2471
Psilocybin artifact HFB		2110	400 $_2$	342 $_4$	145 $_5$	117 $_8$	58 $_{100}$	1445	6317
Psilocybin artifact PFP		2095	350 $_3$	292 $_2$	186 $_2$	145 $_5$	58 $_{100}$	1208	6350
Psilocybin artifact TFA		2080	300 $_2$	242 $_2$	117 $_3$	69 $_4$	58 $_{100}$	906	6349
Psilocybin-M (4-hydroxyindoleacetic acid) MEAC		2315	247 $_{32}$	205 $_{60}$	173 $_{40}$	145 $_{100}$	117 $_{30}$	584	6346
Psilocybin-M (4-hydroxytryptophol) 2AC		2370	261 $_8$	201 $_{19}$	159 $_{100}$	146 $_{45}$	117 $_{30}$	663	6347
PV8		1950	258 $_1$	188 $_2$	154 $_{100}$	96 $_5$	84 $_5$	653	9580
PVP		2185	231 $_1$	188 $_1$	126 $_{100}$	97 $_6$	77 $_{14}$	515	7441
PVP-M (carboxy-oxo-) 2TFA	UGLUCSPETFA	2010	373 $_1$	342 $_6$	268 $_{58}$	236 $_{100}$	101 $_{34}$	1326	7832
PVP-M (carboxy-oxo-) AC	UGLUCSPEAC	2215	319 $_1$	246 $_{13}$	214 $_{73}$	172 $_{100}$	101 $_{32}$	1025	7833
PVP-M (carboxy-oxo-) HFB	UGLUCSPEHFB	1980	473 $_1$	442 $_4$	368 $_{43}$	336 $_{100}$	101 $_{66}$	1623	7828
PVP-M (carboxy-oxo-) ME	UGLUCSPEME	1980	260 $_3$	186 $_{100}$	105 $_9$	101 $_{48}$		849	7834
PVP-M (carboxy-oxo-) TMS	UGLUCSPETMS	2025	349 $_1$	334 $_8$	318 $_7$	244 $_{100}$	173 $_6$	1205	7827
PVP-M (di-HO-) 2AC		2440	269 $_5$	251 $_3$	184 $_{100}$	124 $_{63}$	121 $_{11}$	1195	7766
PVP-M (di-HO-) isomer-1 2TMS		2345	392 $_2$	214 $_{100}$	193 $_3$	124 $_6$	73 $_{17}$	1469	7771
PVP-M (di-HO-) isomer-2 2TMS		2350	392 $_2$	214 $_{100}$	193 $_4$	124 $_6$	73 $_5$	1469	7825
PVP-M (HO-alkyl-) AC		2025	227 $_2$	184 $_{100}$	124 $_{51}$	105 $_{12}$	95 $_5$	836	7760
PVP-M (HO-alkyl-) TMS		1950	304 $_3$	214 $_{100}$	124 $_9$	105 $_4$	73 $_8$	1029	7773
PVP-M (HO-alkyl-oxo-) AC		2170	303 $_1$	198 $_{100}$	138 $_{87}$	110 $_{19}$	96 $_{24}$	925	7764
PVP-M (HO-alkyl-oxo-) TMS		2260	318 $_8$	228 $_{100}$	214 $_8$	138 $_{92}$	105 $_{10}$	1113	7772
PVP-M (HO-phenyl-) AC		2110	126 $_{100}$	121 $_4$	96 $_4$	84 $_4$		837	7763
PVP-M (HO-phenyl-) ME		1990	261 $_1$	135 $_8$	126 $_{100}$	110 $_{12}$	96 $_4$	666	7759
PVP-M (HO-phenyl-) TMS		2095	304 $_2$	193 $_2$	150 $_3$	126 $_{100}$	73 $_8$	1029	7770
PVP-M (HO-phenyl-carboxy-oxo-) 2AC	UGLUCSPEAC	2635	377 $_1$	304 $_9$	214 $_{56}$	172 $_{100}$	101 $_{31}$	1343	7831
PVP-M (HO-phenyl-carboxy-oxo-) 2ME	UGLUCSPEME	2360	290 $_2$	186 $_{100}$	135 $_8$	101 $_{31}$	59 $_{17}$	1039	7829
PVP-M (HO-phenyl-carboxy-oxo-) MEAC	UGLUCSPEMEAC	2550	349 $_1$	276 $_8$	214 $_{67}$	172 $_{100}$	135 $_{11}$	1204	7830
PVP-M (HO-phenyl-N,N-bis-dealkyl-) 2AC		2080	277 $_3$	163 $_3$	121 $_{32}$	114 $_{60}$	72 $_{100}$	761	7762
PVP-M (HO-phenyl-N,N-bis-dealkyl-) 2TMS		1860	337 $_1$	322 $_3$	144 $_{100}$	98 $_{61}$	86 $_{29}$	1140	7768
PVP-M (HO-phenyl-N,N-bis-dealkyl-) MEAC		1970	249 $_8$	186 $_{24}$	135 $_{100}$	114 $_{29}$	72 $_{81}$	596	7757
PVP-M (HO-phenyl-oxo-) AC		2320	303 $_1$	220 $_4$	140 $_{100}$	121 $_{11}$	98 $_{32}$	924	7765
PVP-M (HO-phenyl-oxo-) ME		2225	275 $_2$	192 $_{14}$	140 $_{100}$	135 $_{13}$	98 $_{23}$	748	7758
PVP-M (HO-phenyl-oxo-) TMS		2320	318 $_2$	250 $_{11}$	193 $_{19}$	140 $_{100}$	98 $_{19}$	1113	7769
PVP-M (N,N-bis-dealkyl-) AC		1590	219 $_3$	134 $_6$	114 $_{64}$	105 $_{21}$	72 $_{100}$	466	7761
PVP-M (N,N-bis-dealkyl-) TMS		1375	234 $_5$	191 $_4$	156 $_3$	144 $_{100}$	113 $_9$	600	7767
PVP-M (oxo-)		1875	245 $_3$	140 $_{100}$	105 $_8$	98 $_{38}$	86 $_{11}$	576	7756
PX-1		3740	395 $_5$	248 $_{21}$	232 $_{100}$	173 $_4$	144 $_{19}$	1426	9675
PX-1-M/artifact (HOOC-) (ME)		2100	263 $_{55}$	232 $_{33}$	188 $_{100}$	144 $_{24}$	130 $_{24}$	674	9484
PX-1-M/artifact (HOOC-) (ME)		3220	410 $_6$	248 $_{38}$	232 $_{100}$	173 $_8$	144 $_{19}$	1478	9676
PYCC		1255	178 $_9$	150 $_{14}$	135 $_{100}$	121 $_{13}$	70 $_{25}$	339	3583
PYCC -HCN		1180	151 $_{93}$	150 $_{100}$	136 $_{76}$	122 $_{71}$	95 $_{70}$	292	3585
Pyranocoumarin		2670*	322 $_{100}$	265 $_{70}$	249 $_{39}$	148 $_{17}$	72 $_{87}$	1044	4047
Pyranocoumarin-M (demethyl-HO-dihydro-) isomer-1 -H2O 2ME	UME	2780*	336 $_7$	293 $_{100}$	165 $_6$	150 $_{41}$	115 $_{13}$	1130	4827
Pyranocoumarin-M (demethyl-HO-dihydro-) isomer-2 -H2O 2ME	UME	2805*	336 $_2$	293 $_{100}$	173 $_{10}$	145 $_{12}$	121 $_{13}$	1130	4828
Pyranocoumarin-M (demethyl-HO-dihydro-) isomer-3 -H2O 2ME	UME	2830*	336 $_3$	293 $_{100}$	217 $_3$	151 $_{17}$	115 $_{10}$	1130	4829
Pyranocoumarin-M (di-HO-) 2ET	UET	2990*	410 $_{100}$	367 $_{78}$	339 $_{45}$	179 $_{36}$	151 $_{19}$	1478	4838
Pyranocoumarin-M (O-demethyl-) artifact enol 2TMS		2790*	452 $_{12}$	437 $_{10}$	409 $_{51}$	247 $_{13}$	73 $_{100}$	1584	4971
Pyranocoumarin-M (O-demethyl-) artifact ME	UME	2580*	322 $_{18}$	279 $_{100}$	189 $_7$	121 $_{11}$	91 $_{15}$	1044	1030
Pyranocoumarin-M (O-demethyl-) artifact TMS		2675*	380 $_{53}$	337 $_{100}$	261 $_{26}$	193 $_{28}$	73 $_{62}$	1356	4970
Pyranocoumarin-M (O-demethyl-di-HO-) artifact 3ME	UME	3150*	382 $_{33}$	339 $_{100}$	325 $_{47}$	231 $_{44}$	151 $_{14}$	1365	4830
Pyranocoumarin-M (O-demethyl-dihydro-) artifact 2TMS		2785*	454 $_{12}$	439 $_{40}$	364 $_{100}$	335 $_{74}$	73 $_{79}$	1589	4972
Pyranocoumarin-M (O-demethyl-dihydro-) artifact ME	UME	2660*	324 $_{18}$	291 $_{62}$	215 $_{47}$	177 $_{62}$	91 $_{100}$	1057	1032
Pyranocoumarin-M (O-demethyl-dihydro-) -H2O	UHY UHYAC UHYME	2550*	292 $_{100}$	263 $_{72}$	249 $_{43}$	198 $_{17}$	121 $_{32}$	855	1031
Pyranocoumarin-M (O-demethyl-HO-) artifact 2TMS	UTMS	3015*	468 $_{48}$	425 $_{100}$	337 $_{47}$	115 $_{14}$	73 $_{34}$	1614	4967
Pyranocoumarin-M (O-demethyl-HO-) artifact enol 3TMS	UTMS	3105*	540 $_{26}$	497 $_{100}$	395 $_{27}$	335 $_{20}$	73 $_{92}$	1689	4969
Pyranocoumarin-M (O-demethyl-HO-) isomer-1 artifact 2ET	UET	2810*	380 $_{44}$	337 $_{100}$	309 $_{61}$	233 $_{12}$	165 $_{13}$	1356	4832
Pyranocoumarin-M (O-demethyl-HO-) isomer-1 artifact 2ME	UME	2810*	352 $_{21}$	309 $_{100}$	277 $_7$	151 $_5$	91 $_{14}$	1220	1033
Pyranocoumarin-M (O-demethyl-HO-) isomer-1 artifact 2TMS	UTMS	2795*	468 $_{33}$	425 $_{100}$	268 $_{23}$	193 $_{49}$	73 $_{28}$	1614	4968
Pyranocoumarin-M (O-demethyl-HO-) isomer-2 artifact 2ET	UET	2870*	380 $_{73}$	337 $_{100}$	309 $_{66}$	187 $_{41}$	121 $_{59}$	1356	4833
Pyranocoumarin-M (O-demethyl-HO-) isomer-2 artifact 2ME	UME	2830*	352 $_{32}$	309 $_{100}$	295 $_{31}$	201 $_{43}$	121 $_{56}$	1219	4825
Pyranocoumarin-M (O-demethyl-HO-) isomer-3 artifact 2ET	UET	2870*	380 $_{35}$	337 $_{100}$	309 $_{54}$	165 $_{18}$	137 $_{28}$	1356	4834

Table 8.1: Compounds in order of names   **Pyranocoumarin-M (O-demethyl-HO-) isomer-3 artifact 2ME**

Name	Detected	RI	Typical ions and intensities					Page	Entry
Pyranocoumarin-M (O-demethyl-HO-) isomer-3 artifact 2ME	UME	2870*	352 21	309 100	295 12	206 8	91 21	1220	4826
Pyrazinamide	P-I	1460	123 92	80 100	53 81			267	947
Pyrazolam		3010	353 100	286 33	274 82	205 87	78 77	1224	9696
Pyrazophos		2590	373 10	265 10	232 34	221 100	97 11	1325	3863
Pyrene		1990*	202 100	200 21	174 2	150 1	101 10	404	2567
Pyridate		2985	378 2	350 6	283 19	205 100	57 91	1347	3864
Pyridine		<1000	79 100	52 66				253	1549
Pyridostigmine bromide -CH3Br		1320	166 9	95 1	78 1	72 100	56 3	317	4348
Pyridoxic acid lactone		1700	165 100	147 25	136 66	119 21	108 26	312	5645
Pyridoxine	PAC U+UHYAC	1945	295 1	253 23	193 54	151 100	123 49	876	5089
Pyrilamine	G U	2220	285 4	215 8	121 100	78 20	58 50	811	1656
Pyrilamine HY	UHY UHYAC	1690	208 1	163 1	137 4	71 50	58 100	426	1660
Pyrilamine-M (N-dealkyl-)	U	2120	214 61	136 11	121 100	78 32		447	1657
Pyrilamine-M (N-dealkyl-) AC	UHYAC	2150	256 42	214 62	163 38	107 100	78 40	638	1659
Pyrilamine-M (N-demethoxybenzyl-)	U	1580	165 3	119 8	107 12	78 26	58 100	315	1658
Pyrimethamine		2185	248 86	247 100	219 20	212 17		590	2025
Pyrimethamine 2AC	U+UHYAC	2710	332 32	289 69	247 100	219 9	212 10	1105	7977
Pyrimethamine AC	U+UHYAC	2580	290 56	289 80	247 100	219 8	212 8	841	2026
Pyrithyldione	P G U UHY UHYAC	1520	167 3	152 23	139 86	98 77	83 100	319	948
Pyritinol		9999	368 6	199 24	166 100	151 52	106 57	1301	950
Pyritinol 3ME		9999	410 8	165 90	136 100			1477	951
Pyritinol-M	U UHY UHYAC	1800	207 100					418	949
Pyritinol-M		9999	199 6	151 100	122 17	106 22		397	952
Pyrocatechol	P UHY	<1000*	110 100	92 11	81 5	64 42	53 15	261	2535
Pyrrobutamine	U UHY UHYAC	2370	311 2	240 43	205 100	125 25	91 30	976	2204
Pyrrobutamine-M (oxo-)	U UHY UHYAC	2920	325 9	240 65	205 100	115 88	98 71	1061	2205
Pyrrocaine	G U UHY UHYAC	1830	232 14	218 36	132 19	85 32	71 100	519	1040
Pyrrolidine		<1000	71 24	70 33	43 100			250	3608
Pyrrolidine AC		1320	113 61	98 3	85 19	70 100	60 26	262	6459
Pyrrolidinoheptanophenone		1950	258 1	188 2	154 100	96 5	84 5	653	9580
Pyrrolidinopropiophenone		1595	202 1	188	133	98 100	56 34	407	5943
Pyrrolidinopropiophenone-M (oxo-)		1820	217 1	112 100	105 4	84 7	69 24	454	6546
Pyrrolidinovalerophenone		2185	231 1	188 1	126 100	97 6	77 14	515	7441
Pyrrolidinovalerophenone-M (carboxy-oxo-) 2TFA	UGLUCSPETFA	2010	373 1	342 6	268 58	236 100	101 34	1326	7832
Pyrrolidinovalerophenone-M (carboxy-oxo-) AC	UGLUCSPEAC	2215	319 1	246 13	214 71	172 100	101 32	1025	7833
Pyrrolidinovalerophenone-M (carboxy-oxo-) HFB	UGLUCSPEHFB	1980	473 1	442 4	368 48	336 100	101 66	1623	7828
Pyrrolidinovalerophenone-M (carboxy-oxo-) ME	UGLUCSPEME	1980	260 3	186 100	105 9	101 48		849	7834
Pyrrolidinovalerophenone-M (carboxy-oxo-) TMS	UGLUCSPETMS	2025	349 1	334 8	318 7	244 100	173 6	1205	7827
Pyrrolidinovalerophenone-M (di-HO-) 2AC		2440	269 5	251 3	184 100	124 63	121 11	1195	7766
Pyrrolidinovalerophenone-M (di-HO-) isomer-1 2TMS		2345	392 2	214 100	193 3	124 6	73 17	1469	7771
Pyrrolidinovalerophenone-M (di-HO-) isomer-2 2TMS		2350	392 2	214 100	193 4	124 6	73 5	1469	7825
Pyrrolidinovalerophenone-M (HO-alkyl-) AC		2025	227 2	184 100	124 51	105 12	95 5	836	7760
Pyrrolidinovalerophenone-M (HO-alkyl-) TMS		1950	304 3	214 100	124 9	105 4	73 8	1029	7773
Pyrrolidinovalerophenone-M (HO-alkyl-oxo-) AC		2170	303 1	198 100	138 87	110 19	96 24	925	7764
Pyrrolidinovalerophenone-M (HO-alkyl-oxo-) TMS		2260	318 8	228 100	214 9	138 92	105 10	1113	7772
Pyrrolidinovalerophenone-M (HO-phenyl-) AC		2110	126 10	121 4	96 4	84 4		837	7763
Pyrrolidinovalerophenone-M (HO-phenyl-) ME		1990	261 1	135 8	126 100	110 2	96 4	666	7759
Pyrrolidinovalerophenone-M (HO-phenyl-) TMS		2095	304 2	193 2	150 3	126 100	73 8	1029	7770
Pyrrolidinovalerophenone-M (HO-phenyl-carboxy-oxo-) 2AC	UGLUCSPEAC	2635	377 1	304 9	214 56	172 100	101 31	1343	7831
Pyrrolidinovalerophenone-M (HO-phenyl-carboxy-oxo-) 2ME	UGLUCSPEME	2360	290 2	186 100	135 8	101 31	59 17	1039	7829
Pyrrolidinovalerophenone-M (HO-phenyl-carboxy-oxo-) MEAC	UGLUCSPEMEAC	2550	349 1	276 8	214 67	172 100	135 11	1204	7830
Pyrrolidinovalerophenone-M (HO-phenyl-N,N-bis-dealkyl-) 2AC		2080	277 1	163 3	121 32	114 60	72 100	761	7762
Pyrrolidinovalerophenone-M (HO-phenyl-N,N-bis-dealkyl-) 2TMS		1860	337 1	322 3	144 100	98 61	86 29	1140	7768
Pyrrolidinovalerophenone-M (HO-phenyl-N,N-bis-dealkyl-) MEAC		1970	249 8	186 24	135 100	114 29	72 81	596	7757
Pyrrolidinovalerophenone-M (HO-phenyl-oxo-) AC		2320	303 1	220 4	140 100	121 11	98 32	924	7765
Pyrrolidinovalerophenone-M (HO-phenyl-oxo-) ME		2225	275 2	192 14	140 100	135 13	98 23	748	7758
Pyrrolidinovalerophenone-M (HO-phenyl-oxo-) TMS		2320	318 2	250 11	193 19	140 100	98 11	1113	7769
Pyrrolidinovalerophenone-M (N,N-bis-dealkyl-) AC		1590	219 3	134 6	114 64	105 21	72 100	466	7761
Pyrrolidinovalerophenone-M (N,N-bis-dealkyl-) TMS		1375	234 5	191 4	156 8	144 100	113 9	600	7767
Pyrrolidinovalerophenone-M (oxo-)		1875	245 3	140 100	105 8	98 38	86 11	576	7756
Quazepam	U U+UHYAC	2440	386 100	359 64	323 49	303 36	245 40	1383	2130
Quazepam HY	UHY UHYAC	1985	331 100	312 10	262 34	166 16	123 16	1097	2131
Quazepam-M (dealkyl-oxo-)	G P-I UGLUC	2470	288 100	287 64	260 93	259 75		825	508
Quazepam-M (dealkyl-oxo-) HY	UHY	2030	249 100	154 34	123 46	95 42		593	512
Quazepam-M (dealkyl-oxo-) HYAC	U+UHYAC	2195	291 52	249 100	123 57	95 61		846	286
Quazepam-M (dealkyl-oxo-) TMS		2470	360 63	359 69	341 43	197 17	73 100	1263	4621
Quazepam-M (HO-) HYAC	UHYAC	2250	389 81	347 100	278 53	166 62	125 61	1396	2133
Quazepam-M (oxo-)	U UHY UHYAC	2255	370 58	342 100	307 10	259 20	109 7	1309	2132
Quazepam-M (oxo-) HY	UHY UHYAC	1985	331 100	312 10	262 34	166 16	123 16	1097	2131
Quazepam-M/artifact	U	2480	400 44	323 21	244 100	209 22		1444	2140
Quercetin 4AC		3510*	470 1	428 24	386 48	344 57	302 100	1617	4671
Quercetin 4ME		3510*	358 100	343 5	329 5	100		1252	4672
Quercetin 5TMS		3090*	662 1	647 100	575 27	559 15	487 6	1725	2514
Quetiapine	G	3280	383 1	321 24	239 57	210 100	144 48	1371	6448

Sceletone Table 8.1: Compounds in order of names

Name	Detected	RI	Typical ions and intensities					Page	Entry
Sceletone		2275	243_{26}	214_3	175_9	115_7	70_{100}	568	8994
Scopolamine	G U	2315	303_{24}	154_{32}	138_{66}	108_{48}	94_{100}	924	959
Scopolamine AC	U+UHYAC	2450	345_{13}	154_{22}	138_{59}	108_{41}	94_{100}	1185	1526
Scopolamine -H2O	U+UHYAC	2230	285_{18}	154_{22}	138_{38}	108_{43}	94_{100}	810	960
Scopolamine HFB		2140	499_3	285_{14}	154_{27}	138_{83}	94_{100}	1658	8128
Scopolamine PFP		2120	449_2	285_{24}	154_{37}	138_{59}	94_{100}	1578	8127
Scopolamine TFA		2130	399_1	285_{38}	154_{45}	138_{60}	94_{100}	1441	8126
Scopolamine-M/artifact (deacyl-)		1210	155_{48}	126_{27}	96_{100}	94_{61}	81_{72}	299	3194
Scopolamine-M/artifact (deacyl-) AC		1410	197_9	154_5	138_{39}	94_{29}	81_{100}	392	3195
Scopolamine-M/artifact (HOOC-) -H2O ME		1510*	162_{100}	150_{38}	118_{48}	103_{38}	77_{18}	306	3196
SDB-005		3160	358_4	240_{100}	144_{27}	130_7	55_{16}	1254	9496
SDB-005-M/A AC	U+UHYAC	1555*	186_{13}	144_{100}	115_{47}	89_8	63_7	361	932
SDB-005-M/artifact (1-naphthol)		1500*	144_{100}	115_{83}	89_{14}	74_6	63_{17}	283	928
SDB-005-M/artifact (HOOC-) (ET)		1990	260_{21}	231_{75}	203_{100}	187_{41}	131_{42}	659	9664
SDB-005-M/artifact (HOOC-) (ME)		1970	246_{23}	231_{40}	189_{100}	176_{35}	145_{23}	581	9497
SDB-006		2820	320_{49}	263_{17}	214_{50}	187_{18}	144_{37}	1035	9608
SDB-006 TMS		2690	392_{100}	377_{32}	214_{68}	204_{51}	91_{89}	1412	9609
Sebaic acid bisoctyl ester	U UHY UHYAC	2705*	426_1	315_2	297_4	185_{100}	112_{19}	1524	5408
Sebuthylazine		1855	229_{13}	214_{12}	200_{100}	173_8	132_{10}	506	3866
Secobarbital	P G U+UHYAC	1795	209_4	195_{25}	168_{100}	167_{79}	141_{11}	549	961
Secobarbital (ME)	P	1970	252_1	209_{26}	182_{100}	181_{91}	167_{35}	619	2289
Secobarbital 2ME		1690	248_3	196_{100}	181_{25}	138_{25}	111_{22}	697	964
Secobarbital 2TMS		1670	382_1	367_{40}	339_{34}	297_{67}	73_{100}	1367	5470
Secobarbital-M (deallyl-)	U	1665	169_2	154_9	129_{100}			395	962
Secobarbital-M (HO-) -H2O	U+UHYAC	1970	236_7	168_{100}	167_{77}	69_{85}		539	963
Selegiline	G U+UHYAC	1450	172_1	115_1	96_{100}	91_{13}	56_{35}	363	2502
Selegiline-M (4-HO-amfetamine)		1480	151_{10}	107_{69}	91_{10}	77_{42}	56_{100}	292	1802
Selegiline-M (4-HO-amfetamine) 2AC	U+UHYAC	1900	235_1	176_{72}	134_{100}	107_{46}	86_{70}	533	1804
Selegiline-M (4-HO-amfetamine) 2HFB		<1000	330_{48}	303_{15}	240_{100}	169_{44}	69_{42}	1691	6326
Selegiline-M (4-HO-amfetamine) 2PFP		<1000	280_{77}	253_{16}	190_{100}	119_{56}	69_{16}	1567	6325
Selegiline-M (4-HO-amfetamine) 2TFA		<1000	230_{72}	203_{11}	140_{100}	92_{12}	69_{59}	1172	6324
Selegiline-M (4-HO-amfetamine) 2TMS		<1000	280_7	179_4	149_8	116_{100}	73_{78}	880	6327
Selegiline-M (4-HO-amfetamine) AC	U+UHYAC	1890	193_1	134_{100}	107_{26}	86_{24}	77_{16}	379	1803
Selegiline-M (bis-dealkyl-)		1160	134_4	120_{15}	91_{100}	77_{18}	65_{73}	275	54
Selegiline-M (bis-dealkyl-)	U	1160	134_1	120_1	91_6	65_4	44_{100}	275	5514
Selegiline-M (bis-dealkyl-) AC	U+UHYAC	1505	177_4	118_{60}	91_{35}	86_{100}	65_{16}	336	55
Selegiline-M (bis-dealkyl-) AC	U+UHYAC	1505	177_1	118_{19}	91_{11}	86_{31}	44_{100}	335	5515
Selegiline-M (bis-dealkyl-) HFB		1355	240_{79}	169_{21}	118_{100}	91_{53}		1099	5047
Selegiline-M (bis-dealkyl-) PFP		1330	281_1	190_{73}	118_{100}	91_{36}	65_{12}	786	4379
Selegiline-M (bis-dealkyl-) TFA		1095	231_1	140_{100}	118_{92}	91_{45}	69_{19}	513	4000
Selegiline-M (bis-dealkyl-) TMS		1190	192_6	116_{100}	100_{10}	91_{11}	73_{87}	421	5581
Selegiline-M (bis-dealkyl-)-D11 TFA		1615	242_1	144_{100}	128_{82}	98_{43}	70_{14}	566	7283
Selegiline-M (bis-dealkyl-)-D11 TFA		1610	194_{100}	128_{82}	98_{43}	70_{14}		856	7284
Selegiline-M (bis-dealkyl-)-D5 AC		1480	182_3	122_{46}	92_{30}	90_{100}	66_{16}	352	5690
Selegiline-M (bis-dealkyl-)-D5 HFB		1330	244_{100}	169_{14}	122_{46}	92_{41}	69_{40}	1130	6316
Selegiline-M (bis-dealkyl-)-D5 PFP		1320	194_{100}	123_{42}	119_{32}	92_{46}	69_{11}	815	5566
Selegiline-M (bis-dealkyl-)-D5 TFA		1085	144_{100}	123_{53}	122_{56}	92_{51}	69_{28}	540	5570
Selegiline-M (bis-dealkyl-)-D5 TMS		1180	212_1	179_8	120_{100}	92_{11}	73_{57}	441	5582
Selegiline-M (bis-dealkyl-4-HO-) formyl art.		1220	163_3	148_4	107_{30}	77_{12}	56_{100}	308	6323
Selegiline-M (bis-dealkyl-4-HO-) TFA		1670	247_4	140_{15}	134_{54}	107_{100}	77_{15}	584	6335
Selegiline-M (dealkyl-)	U	1195	148_1	134_2	115_1	91_9	58_{100}	288	1093
Selegiline-M (dealkyl-) AC	U+UHYAC	1575	191_1	117_2	100_{42}	91_6	58_{100}	373	1094
Selegiline-M (dealkyl-) HFB		1460	254_{100}	210_{44}	169_{15}	118_{41}	91_{38}	1183	5069
Selegiline-M (dealkyl-) PFP		1415	204_{100}	160_{46}	118_{35}	91_{25}	69_4	874	5070
Selegiline-M (dealkyl-) TFA		1300	245_1	154_{100}	118_{48}	110_{55}	91_{23}	574	3998
Selegiline-M (dealkyl-) TMS		1325	206_4	130_{100}	91_{17}	73_{83}	59_{13}	477	6214
Selegiline-M (dealkyl-HO-)		1885	150_1	135	107_4	77_4	58_{100}	314	1766
Selegiline-M (dealkyl-HO-) 2AC	U+UHYAC	1995	249_1	176_6	134_{100}	100_{43}	58_{100}	599	1767
Selegiline-M (dealkyl-HO-) 2HFB		1670	538_1	330_{17}	254_{100}	210_{32}	169_{22}	1698	5076
Selegiline-M (dealkyl-HO-) 2PFP		1605	280_{18}	204_{100}	160_{47}	119_{39}		1595	5077
Selegiline-M (dealkyl-HO-) 2TFA		1585	357_1	230_{22}	154_{100}	110_{42}	69_{29}	1246	5078
Selegiline-M (dealkyl-HO-) 2TMS		1620	309_{10}	206_{70}	179_{100}	154_{32}	73_{40}	966	6190
Selegiline-M (dealkyl-HO-) TFA		1770	261_1	154_{100}	134_{68}	110_{42}	107_{41}	662	6180
Selegiline-M (dealkyl-HO-) TMSTFA		1690	333_1	206_{72}	179_{100}	154_{53}	73_{50}	1111	6228
Selegiline-M (HO-)	UHY	1580	107_3	96_{100}	56_{21}			407	2948
Selegiline-M (HO-) AC	U+UHYAC	1860	230_1	107_4	96_{100}	56_{17}		576	2950
Selegiline-M (nor-)	UHY	1350	128_1	115_1	91_9	82_{100}	65_6	328	2946
Selegiline-M (nor-) AC	U+UHYAC	1735	214_1	124_{31}	91_{10}	82_{100}	65_6	449	2949
Selegiline-M (nor-HO-)	UHY	1550	135_1	107_5	82_{100}			368	2947
Selegiline-M (nor-HO-) 2AC	U+UHYAC	2030	272_1	176_{10}	134_{19}	124_{33}	82_{100}	734	2951
Serotonin 3ME	G U UHY UHYAC	2040	218_{16}	160_{10}	145_7	117_{10}	58_{100}	460	4059
Sertraline	G P-I U	2260	304_{15}	274_{100}	262_{34}	159_{59}	115_{27}	935	4641
Sertraline AC	U+UHYAC	2760	347_{65}	290_{100}	274_{88}	159_{56}	74_{46}	1193	4640
Sertraline -CH5N	G P-I U+UHYAC	2275*	274_{100}	239_{43}	202_{57}	159_{57}	128_{85}	737	4682

Table 8.1: Compounds in order of names   Sertraline HFB

Name	Detected	RI	Typical ions and intensities					Page	Entry
Sertraline HFB		2525	501(95)	332(26)	274(100)	159(57)	128(49)	1660	7690
Sertraline PFP		2515	451(61)	436(19)	274(100)	202(40)	159(66)	1581	7689
Sertraline TFA		2520	401(99)	400(100)	274(90)	202(44)	159(56)	1447	7688
Sertraline TMS		2530	377(7)	362(14)	348(36)	334(22)	274(100)	1342	7691
Sertraline-M (di-HO-ketone) -H2O enol 2AC	U+UHYAC	2890*	388(8)	346(24)	304(100)	275(8)	176(10)	1393	4685
Sertraline-M (HO-) 2AC	U+UHYAC	3015	405(36)	348(24)	332(67)	290(100)	159(12)	1461	4681
Sertraline-M (HO-ketone) AC	UHYAC	2660*	348(22)	290(75)	288(100)	261(57)	227(35)	1198	5311
Sertraline-M (HO-ketone) -H2O enol AC	U+UHYAC	2600*	330(17)	290(63)	288(100)	218(64)	189(55)	1092	4683
Sertraline-M (ketone)	UHYAC	2480*	290(99)	248(38)	227(100)	199(47)	163(30)	839	5310
Sertraline-M (ketone) enol AC	U+UHYAC	2530*	332(5)	290(100)	247(5)	212(15)	189(6)	1104	4684
Sertraline-M (nor-)	UHY	2400	290(14)	274(26)	159(45)	130(67)	119(100)	846	4643
Sertraline-M (nor-) AC	U+UHYAC	2700	333(6)	274(100)	239(28)	159(32)	115(37)	1110	4642
Sertraline-M (nor-) HFB		2325	487(7)	274(100)	203(23)	159(31)	128(43)	1644	7194
Sertraline-M (nor-) PFP	UHYPFP	2350	437(11)	274(100)	203(28)	159(38)	128(60)	1551	7189
Sertraline-M (nor-) TFA	UHYTFA	2300	387(10)	274(100)	202(25)	159(32)	128(48)	1388	7188
Sertraline-M (nor-) TMS		2350	362(12)	348(14)	274(100)	217(37)	73(67)	1277	7190
Sertraline-M/artifact	U+UHYAC	2320*	272(87)	236(30)	202(100)	118(11)	100(36)	727	4686
Sethoxydim		2390	281(8)	219(31)	178(100)	149(62)	108(24)	1077	3653
Sibutramine		1870	137(2)	128(2)	114(100)	72(33)	58(13)	778	5725
Sibutramine-M (bis-nor-)		1950	194(1)	165(3)	137(3)	130(4)	86(100)	612	5729
Sibutramine-M (bis-nor-) AC	U+UHYAC	2155	293(1)	165(2)	137(5)	128(67)	86(100)	864	5892
Sibutramine-M (bis-nor-) formyl artifcat		1920	263(1)	221(7)	179(4)	165(10)	98(100)	675	5730
Sibutramine-M (bis-nor-) HFB		1940	363(12)	240(23)	165(100)	137(42)	69(47)	1574	5747
Sibutramine-M (bis-nor-) PFP		1900	313(10)	190(16)	165(100)	137(36)	69(28)	1434	5748
Sibutramine-M (bis-nor-) TFA		1875	263(13)	165(100)	137(46)	102(2)	69(40)	1193	5731
Sibutramine-M (bis-nor-) TMS		2450	308(2)	266(2)	158(100)	102(12)	73(62)	1053	5732
Sibutramine-M (nor-)		1840	128(2)	115(2)	100(100)	58(33)		690	5726
Sibutramine-M (nor-) AC	U+UHYAC	2160	307(1)	165(1)	142(75)	100(100)	58(20)	951	5891
Sibutramine-M (nor-) HFB		1990	296(100)	254(31)	240(50)	210(12)	69(51)	1603	5745
Sibutramine-M (nor-) PFP		1975	246(100)	204(35)	190(47)	160(11)	69(36)	1481	5746
Sibutramine-M (nor-) TFA		1950	196(100)	154(31)	140(34)	128(19)	69(40)	1268	5727
Sibutramine-M (nor-) TMS		2460	322(1)	172(100)	116(2)	73(21)		1140	5728
Sigmodal	PGU	2055	237(24)	193(11)	167(100)	122(19)	78(32)	1004	965
Sigmodal 2ME		1910	265(28)	195(100)	138(18)			1178	966
Sildenafil		3400	474(1)	404(16)	99(100)	70(8)	56(31)	1625	5713
Sildenafil ME		3390	488(1)	418(22)	99(100)	70(8)	56(31)	1646	6522
Sildenafil TMS		4030	476(37)	454(1)	99(100)	73(23)	56(33)	1694	5714
Simazine	G P-I U	1690	201(100)	186(67)	173(46)	158(28)	68(83)	402	1326
Simazine-M (deethyl-)	U	1730	173(100)	158(97)	145(77)	130(18)	68(78)	328	4236
Simvastatin -H2O -C6H12O2	G P-I	2775*	284(23)	199(100)	198(90)	172(61)	157(80)	805	6449
Sitagliptin	P U+UHYAC	2465	390(8)	245(2)	171(69)	151(80)	145(100)	1467	8453
Sitagliptin AC	P U+UHYAC	2700	449(5)	304(25)	262(56)	191(42)	70(100)	1578	8454
Sitagliptin PFP		2510	553(5)	408(7)	234(26)	191(100)	70(54)	1696	8456
Sitagliptin TMS		2540	464(1)	334(41)	265(11)	145(23)	73(100)	1632	8455
Skatole	U	1340	131(55)	130(100)	103(10)	77(16)	65(11)	272	4218
Skatole-M (HO-)	U	1370	147(75)	146(100)	117(15)			285	819
Solifenacin		2845	362(15)	236(8)	178(36)	126(100)	109(90)	1274	8247
Solifenacin HY		1935	209(19)	208(26)	179(17)	165(9)	132(100)	429	8249
Solifenacin HYAC		2240	251(29)	208(15)	193(28)	178(25)	132(100)	610	8248
Sorbitol 6AC	U+UHYAC	2090*	361(15)	289(39)	187(55)	145(68)	115(100)	1544	1966
Sorbitol 6HFB		1540*	521(2)	478(5)	307(58)	240(34)	169(100)	1736	5808
Sorbitol 6PFP		1530*	378(4)	257(23)	219(48)	190(29)	119(100)	1734	5807
Sorbitol 6TFA		1435*	435(1)	321(3)	278(14)	265(12)	69(100)	1732	5806
Sorbitol 6TMS		1880*	421(6)	319(100)	217(55)	205(62)	73(92)	1717	8287
Sotalol		9999	272(4)	239(1)	199(3)	122(4)	72(100)	730	1368
Sotalol -H2O AC	U+UHYAC	2675	296(39)	217(29)	175(100)	133(47)	84(42)	885	1369
Sotalol TMSTFA		2410	425(1)	272(100)	193(4)	126(7)	73(66)	1558	6173
Sotalol-M/artifact (amino-) -H2O 2AC	U+UHYAC	2500	260(100)	218(71)	203(54)	133(36)	84(32)	658	1710
Sparfloxacin		3455	392(27)	348(7)	322(100)	278(40)	70(6)	1412	6104
Sparfloxacin -CO2		3190	348(33)	313(7)	278(100)	235(6)	208(4)	1200	6105
Sparteine	G U	1785	234(25)	193(28)	137(95)	98(100)		530	967
Sparteine-M (oxo-)	P U UHY UHYAC	2230	248(42)	149(51)	136(100)	98(26)	84(18)	592	2877
Sparteine-M (oxo-HO-)	U	2290	264(74)	165(13)	150(45)	136(100)	98(36)	684	2878
Sparteine-M (oxo-HO-) enol 2AC	UHYAC	2550	348(5)	306(10)	264(62)	134(100)	121(50)	1200	2880
Sparteine-M (oxo-HO-) -H2O	U	2205	246(13)	148(4)	134(6)	98(100)	84(7)	582	2879
Speciociliatine		3210	398(25)	383(39)	269(27)	214(100)	199(29)	1438	8050
Speciociliatine HFB		3060	594(20)	579(43)	465(18)	410(100)	382(9)	1713	8067
Speciociliatine PFP		3070	544(7)	529(15)	415(17)	360(100)	332(17)	1692	8066
Speciociliatine TFA		3130	494(8)	479(19)	365(16)	310(100)	282(21)	1653	8065
Speciociliatine TMS		3150	470(10)	455(6)	327(6)	286(100)	271(27)	1619	8055
Speciociliatine-M (16-COOH) 2TMS	USPETMS	3320	528(18)	286(100)	271(71)			1683	8046
Speciociliatine-M (9-O-demethyl-) 2TMS	USPETMS	3290	528(13)	513(7)	344(100)	329(11)	73(33)	1684	8047
Speciociliatine-M (9-O-demethyl-16-COOH) 3TMS	USPETMS	3230	586(8)	344(66)	147(23)	73(100)		1710	8045
Speciociliatine-M/artifact (HO-aryl-) 2TMS	USPETMS	3350	558(18)	374(57)	73(100)			1699	8044

Speciogynine   **Table 8.1:** Compounds in order of names

Name	Detected	RI	Typical ions and intensities					Page	Entry
Speciogynine		3240	398_{63}	383_{53}	255_{36}	214_{100}	200_{54}	1439	7869
Speciogynine TMS		3140	470_{18}	455_{8}	327_{13}	286_{100}	271_{33}	1619	8054
Speciogynine-M (16-COOH) 2TMS	USPETMS	3170	528_{24}	286_{100}	271_{37}	147_{56}	73_{92}	1683	8048
Speciogynine-M (9-O-demethyl-) 2TMS	USPETMS	3240	528_{37}	513_{13}	344_{100}	329_{28}	73_{45}	1684	8049
Spirapril ET		3440	421_{10}	289_{8}	234_{100}	160_{21}	91_{22}	1653	7513
Spirapril -H2O		3595	448_{10}	344_{78}	298_{100}	117_{49}	91_{80}	1576	7511
Spirapril ME		3390	407_{5}	275_{6}	234_{100}	160_{16}	91_{47}	1635	7512
Spironolactone -CH3COSH	P UHY UHYAC	3250*	340_{100}	325_{18}	267_{40}	227_{15}		1157	2344
Squalene	G P U UHY UHYAC	2800*	410_{1}	341_{1}	137_{10}	81_{42}	69_{100}	1480	968
Stanozolol		3085	328_{22}	175_{6}	133_{21}	96_{100}	94_{95}	1083	2816
Stanozolol 2TMS		3025	472_{13}	342_{3}	168_{7}	143_{100}	75_{8}	1623	3984
Stanozolol AC		2120	370_{18}	257_{19}	138_{97}	96_{100}	94_{95}	1315	2817
Stavudine		2250	206_{1}	193_{19}	150_{20}	126_{66}	69_{100}	487	7889
Stavudine AC		2265	206_{2}	193_{8}	140_{17}	126_{12}	81_{100}	693	7891
Stavudine artifact (thymine)		1815	126_{100}	97_{3}	83_{14}	70_{6}	55_{79}	269	7890
Stavudine artifact (thymine) 2ME		1450	154_{100}	126_{2}	96_{10}	68_{72}	56_{13}	296	9427
Stavudine artifact (thymine) 2TMS		1380	270_{59}	255_{100}	166_{12}	147_{27}	73_{59}	719	7897
Stavudine artifact (thymine) ME		1550	140_{100}	110_{4}	83_{18}	70_{3}	55_{74}	279	7892
Stavudine HFB		2090	420_{1}	294_{2}	206_{3}	126_{45}	81_{100}	1507	7895
Stavudine ME		2205	207_{15}	150_{11}	140_{100}	83_{27}	69_{88}	547	7893
Stavudine PFP		2150	244_{6}	206_{9}	126_{87}	119_{22}	81_{100}	1309	7896
Stavudine TFA		2190	206_{6}	194_{8}	150_{4}	126_{39}	81_{100}	1032	7894
Stearamide	P U UHY UHYAC	2400	283_{1}	240_{2}	128_{6}	72_{35}	59_{100}	798	5346
Stearic acid	P G U UHY UHYAC	2170*	284_{60}	241_{25}	185_{27}	129_{40}	73_{100}	806	969
Stearic acid ET		2140*	312_{9}	269_{10}	157_{28}	101_{62}	88_{100}	985	5406
Stearic acid glycerol ester 2AC	UHYAC	2790*	382_{6}	267_{26}	159_{100}	98_{30}	84_{19}	1567	5413
Stearic acid glycerol ester 2TMS		2780*	488_{2}	400_{45}	205_{15}	147_{42}	73_{100}	1663	7450
Stearic acid ME	G P	2130*	298_{2}	255_{3}	143_{11}	87_{65}	74_{100}	899	970
Stearic acid TMS		2640*	356_{15}	341_{83}	145_{27}	117_{77}	73_{100}	1245	4017
Stearyl alcohol		2020*	270_{1}	252_{2}	224_{4}	97_{65}	55_{100}	721	2356
Stephanamine		1550	188_{1}	174_{1}	131_{14}	77_{14}	58_{100}	368	8943
Stephanamine AC		1960	231_{1}	158_{16}	131_{12}	100_{36}	58_{100}	514	8944
Stephanamine HFB		1770	385_{1}	254_{100}	210_{45}	158_{70}	131_{43}	1380	8949
Stephanamine ME		1570	202_{1}	188_{1}	131_{6}	77_{4}	72_{100}	407	8945
Stephanamine PFP		1720	335_{1}	204_{100}	160_{37}	158_{52}	131_{27}	1123	8948
Stephanamine TFA		1750	285_{1}	158_{56}	154_{100}	131_{46}	110_{41}	808	8947
Stephanamine TMS		1710	260_{1}	246_{6}	130_{100}	115_{3}	73_{53}	665	8946
Stephanamine-M (nor-)		1450	175_{9}	160_{5}	131_{100}	102_{25}	77_{77}	330	8951
Stephanamine-M (nor-)		1450	175_{1}	131_{13}	102_{3}	77_{10}	44_{100}	330	9083
Stephanamine-M (nor-) AC		1870	217_{3}	158_{100}	131_{67}	86_{59}	77_{51}	454	8950
Stephanamine-M (nor-) formyl artifact		1505	187_{6}	172_{8}	131_{41}	77_{19}	56_{100}	363	9084
Stephanamine-M (nor-) HFB		1670	371_{1}	240_{15}	158_{57}	131_{100}	77_{28}	1316	9088
Stephanamine-M (nor-) PFP		1640	321_{2}	190_{16}	158_{46}	131_{100}	77_{31}	1038	9087
Stephanamine-M (nor-) TFA		1655	271_{2}	158_{40}	131_{100}	77_{33}	69_{40}	722	9086
Stephanamine-M (nor-) TMS		1650	232_{6}	131_{23}	116_{100}	100_{11}	73_{57}	586	9085
Steviol		2600*	318_{41}	300_{17}	260_{17}	121_{100}	55_{45}	1022	3342
Steviol ME		2530*	332_{22}	274_{10}	254_{11}	146_{20}	121_{100}	1108	3343
Steviol MEAC		2580*	374_{25}	332_{32}	314_{40}	146_{18}	121_{100}	1332	4300
Stevioside artifact (isosteviol)		2620*	318_{25}	300_{21}	121_{66}	109_{71}	55_{100}	1022	3680
Stevioside artifact (isosteviol) ME		2520*	332_{73}	300_{71}	273_{100}	121_{81}	55_{71}	1109	3681
Stevioside-M (steviol)		2600*	318_{41}	300_{17}	260_{17}	121_{100}	55_{45}	1022	3342
Stevioside-M (steviol) ME		2530*	332_{22}	274_{10}	254_{11}	146_{20}	121_{100}	1108	3343
Stevioside-M (steviol) MEAC		2580*	374_{25}	332_{32}	314_{40}	146_{18}	121_{100}	1332	4300
Stigma-3,5-dien-7-one		3630*	410_{22}	269_{11}	187_{25}	174_{100}	161_{22}	1479	5584
Stigmast-3,5-ene		3300*	396_{100}	381_{25}	147_{80}	105_{64}	81_{65}	1432	5626
Stigmast-5-en-3-ol		3265*	414_{69}	329_{43}	303_{49}	105_{97}	55_{100}	1492	5622
Stigmast-5-en-3-ol -H2O		3300*	396_{100}	381_{25}	147_{80}	105_{64}	81_{65}	1432	5626
Stigmasterol		3210*	412_{33}	271_{34}	255_{40}	69_{94}	55_{100}	1486	5621
Stigmasterol -H2O		3285*	394_{100}	255_{49}	145_{53}	81_{75}	55_{98}	1423	5625
Stiripentol		1940*	234_{16}	177_{100}	159_{12}	147_{37}	119_{16}	528	8406
Stiripentol ME		1835*	248_{3}	191_{100}	161_{17}	159_{18}	133_{12}	591	8407
Stiripentol TMS		1900*	306_{2}	291_{2}	249_{100}	217_{12}	159_{14}	944	8408
Strychnine		3120	334_{100}	167_{32}	130_{36}	107_{28}	79_{25}	1120	971
Sublimate		9999*	272_{74}	202_{100}				727	972
Sufentanil		2730	289_{100}	158_{7}	140_{25}	110_{21}	77_{13}	1386	6791
Sufentanil HY		2650	330_{1}	233_{100}	158_{23}	140_{16}	96_{18}	1095	6792
Sulazepam		2640	300_{100}	273_{68}	237_{71}	227_{69}	74_{71}	905	4029
Sulazepam HY	UHY U+UHYAC	2100	245_{95}	228_{38}	193_{29}	105_{38}	77_{100}	573	272
Sulazepam HYAC	U+UHYAC	2260	287_{71}	244_{100}	228_{39}	182_{49}	77_{71}	818	2542
Sulfabenzamide 2ME		2770	304_{16}	240_{4}	118_{100}	105_{48}	77_{39}	930	3150
Sulfabenzamide 2MEAC		2650	346_{38}	212_{12}	118_{100}	105_{30}	77_{29}	1188	3166
Sulfabenzamide AC		2720	318_{3}	282_{4}	118_{100}	105_{54}	77_{34}	1017	3164
Sulfabenzamide ME		2700	290_{4}	226_{8}	118_{100}	105_{51}	77_{39}	840	3149
Sulfabenzamide MEAC		2750	332_{38}	184_{14}	118_{100}	105_{54}	77_{52}	1105	3165

Table 8.1: Compounds in order of names

Name	Detected	RI	Typical ions and intensities					Page	Entry
Sulfabenzamide-M	G P UHY	2185	172_{56}	156_{55}	108_{50}	92_{74}	65_{100}	326	973
Sulfabenzamide-M 2TMS		2210	316_{19}	301_{35}	222_{5}	163_{8}	73_{100}	1005	10331
Sulfabenzamide-M 3TMS		2125	388_{5}	373_{28}	210_{16}	180_{12}	147_{100}	1394	10330
Sulfabenzamide-M 4ME		2095	228_{44}	184_{30}	136_{100}	120_{70}	77_{29}	501	4098
Sulfabenzamide-M AC	U+UHYAC	2690	214_{31}	172_{100}	156_{54}	108_{42}	92_{46}	445	974
Sulfabenzamide-M ME	UME	2135	186_{68}	156_{61}	108_{52}	92_{100}	65_{78}	360	3136
Sulfabenzamide-M MEAC		2600	228_{59}	186_{100}	156_{62}	108_{30}	92_{33}	500	3148
Sulfadiazine	P	2640	185_{100}	170_{5}	108_{22}	92_{44}	65_{49}	602	7979
Sulfadiazine AC		2925	292_{2}	227_{100}	185_{58}	108_{15}	92_{26}	854	7978
Sulfadiazine ME		2625	199_{100}	184_{2}	108_{16}	92_{24}	65_{31}	680	3135
Sulfadiazine MEAC		3710	241_{100}	199_{14}	108_{7}	92_{7}		942	3158
Sulfadimethoxine 2TMS		3030	439_{2}	390_{35}	375_{100}	212_{26}	89_{25}	1589	5866
Sulfaethidole		2620	284_{69}	220_{21}	156_{33}	108_{38}	92_{100}	799	1862
Sulfaethidole 2ME		2840	234_{28}	161_{31}	106_{100}	92_{60}	65_{37}	980	3152
Sulfaethidole 2MEAC		3410	354_{27}	276_{57}	203_{71}	148_{100}	106_{86}	1231	3159
Sulfaethidole AC		2490	326_{34}	283_{12}	213_{100}	136_{22}	108_{62}	1066	1863
Sulfaethidole ME		3060	298_{100}	234_{11}	190_{16}	92_{27}	83_{27}	893	3151
Sulfaethidole-M	G P UHY	2185	172_{56}	156_{55}	108_{50}	92_{74}	65_{100}	326	973
Sulfaethidole-M 2TMS		2210	316_{19}	301_{35}	222_{5}	163_{8}	73_{100}	1005	10331
Sulfaethidole-M 3TMS		2125	388_{5}	373_{28}	210_{16}	180_{12}	147_{100}	1394	10330
Sulfaethidole-M 4ME		2095	228_{44}	184_{30}	136_{100}	120_{70}	77_{29}	501	4098
Sulfaethidole-M AC	U+UHYAC	2690	214_{31}	172_{100}	156_{54}	108_{42}	92_{46}	445	974
Sulfaethidole-M ME	UME	2135	186_{68}	156_{61}	108_{52}	92_{100}	65_{78}	360	3136
Sulfaethidole-M MEAC		2600	228_{59}	186_{100}	156_{62}	108_{30}	92_{33}	500	3148
Sulfaguanole ME		2905	323_{49}	249_{14}	203_{100}	178_{8}	57_{80}	1049	3153
Sulfaguanole-M	G P UHY	2185	172_{56}	156_{55}	108_{50}	92_{74}	65_{100}	326	973
Sulfaguanole-M 2TMS		2210	316_{19}	301_{35}	222_{5}	163_{8}	73_{100}	1005	10331
Sulfaguanole-M 3TMS		2125	388_{5}	373_{28}	210_{16}	180_{12}	147_{100}	1394	10330
Sulfaguanole-M 4ME		2095	228_{44}	184_{30}	136_{100}	120_{70}	77_{29}	501	4098
Sulfaguanole-M AC	U+UHYAC	2690	214_{31}	172_{100}	156_{54}	108_{42}	92_{46}	445	974
Sulfaguanole-M ME	UME	2135	186_{68}	156_{61}	108_{52}	92_{100}	65_{78}	360	3136
Sulfaguanole-M MEAC		2600	228_{59}	186_{100}	156_{62}	108_{30}	92_{33}	500	3148
Sulfamerazine		2625	199_{100}	140_{3}	108_{28}	92_{55}	65_{56}	680	4267
Sulfamethizole ME	UME	2660	284_{86}	156_{49}	92_{100}	65_{87}		799	1322
Sulfamethizole-M	G P UHY	2185	172_{56}	156_{55}	108_{50}	92_{74}	65_{100}	326	973
Sulfamethizole-M 2TMS		2210	316_{19}	301_{35}	222_{5}	163_{8}	73_{100}	1005	10331
Sulfamethizole-M 3TMS		2125	388_{5}	373_{28}	210_{16}	180_{12}	147_{100}	1394	10330
Sulfamethizole-M 4ME		2095	228_{44}	184_{30}	136_{100}	120_{70}	77_{29}	501	4098
Sulfamethizole-M AC	U+UHYAC	2690	214_{31}	172_{100}	156_{54}	108_{42}	92_{46}	445	974
Sulfamethizole-M ME	UME	2135	186_{68}	156_{61}	108_{52}	92_{100}	65_{78}	360	3136
Sulfamethizole-M MEAC		2600	228_{59}	186_{100}	156_{62}	108_{30}	92_{33}	500	3148
Sulfamethoxazole 2ME	P	2460	281_{2}	203_{16}	162_{44}	108_{71}	92_{100}	786	3155
Sulfamethoxazole 2TMS		2515	397_{1}	382_{7}	228_{58}	178_{89}	73_{100}	1434	4597
Sulfamethoxazole impurity		1025	125_{100}	98_{12}	93_{23}	80_{28}	65_{7}	268	6351
Sulfamethoxazole ME	P	2500	267_{2}	203_{18}	162_{46}	108_{71}	92_{100}	699	3154
Sulfamethoxazole MEAC		3255	309_{8}	245_{48}	230_{67}	161_{100}	134_{86}	961	3160
Sulfamethoxazole-M	G P UHY	2185	172_{56}	156_{55}	108_{50}	92_{74}	65_{100}	326	973
Sulfamethoxazole-M 2TMS		2210	316_{19}	301_{35}	222_{5}	163_{8}	73_{100}	1005	10331
Sulfamethoxazole-M 3TMS		2125	388_{5}	373_{28}	210_{16}	180_{12}	147_{100}	1394	10330
Sulfamethoxazole-M 4ME		2095	228_{44}	184_{30}	136_{100}	120_{70}	77_{29}	501	4098
Sulfamethoxazole-M AC	U+UHYAC	2690	214_{31}	172_{100}	156_{54}	108_{42}	92_{46}	445	974
Sulfamethoxazole-M ME	UME	2135	186_{68}	156_{61}	108_{52}	92_{100}	65_{78}	360	3136
Sulfamethoxazole-M MEAC		2600	228_{59}	186_{100}	156_{62}	108_{30}	92_{33}	500	3148
Sulfametoxydiazine 3ME	PME	2925	322_{1}	229_{100}	138_{6}	92_{21}	65_{21}	1043	3156
Sulfametoxydiazine MEAC		3620	271_{100}	229_{8}	139_{4}	92_{6}	65_{9}	1207	3161
Sulfametoxydiazine-M	G P UHY	2185	172_{56}	156_{55}	108_{50}	92_{74}	65_{100}	326	973
Sulfametoxydiazine-M 2TMS		2210	316_{19}	301_{35}	222_{5}	163_{8}	73_{100}	1005	10331
Sulfametoxydiazine-M 3TMS		2125	388_{5}	373_{28}	210_{16}	180_{12}	147_{100}	1394	10330
Sulfametoxydiazine-M 4ME		2095	228_{44}	184_{30}	136_{100}	120_{70}	77_{29}	501	4098
Sulfametoxydiazine-M AC	U+UHYAC	2690	214_{31}	172_{100}	156_{54}	108_{42}	92_{46}	445	974
Sulfametoxydiazine-M ME	UME	2135	186_{68}	156_{61}	108_{52}	92_{100}	65_{78}	360	3136
Sulfametoxydiazine-M MEAC		2600	228_{59}	186_{100}	156_{62}	108_{30}	92_{33}	500	3148
Sulfanilamide	G P UHY	2185	172_{56}	156_{55}	108_{50}	92_{74}	65_{100}	326	973
Sulfanilamide 2TMS		2210	316_{19}	301_{35}	222_{5}	163_{8}	73_{100}	1005	10331
Sulfanilamide 3TMS		2125	388_{5}	373_{28}	210_{16}	180_{12}	147_{100}	1394	10330
Sulfanilamide 4ME		2095	228_{44}	184_{30}	136_{100}	120_{70}	77_{29}	501	4098
Sulfanilamide AC	U+UHYAC	2690	214_{31}	172_{100}	156_{54}	108_{42}	92_{46}	445	974
Sulfanilamide ME		2135	186_{68}	156_{61}	108_{52}	92_{100}	65_{78}	360	3136
Sulfanilamide MEAC		2600	228_{59}	186_{100}	156_{62}	108_{30}	92_{33}	500	3148
Sulfaperin 2MEAC		3420	255_{100}	213_{7}	124_{2}	93_{5}	65_{7}	1118	3162
Sulfaperin 3ME		2795	306_{1}	213_{100}	198_{1}	92_{18}	65_{21}	943	3157
Sulfaperin-M	G P UHY	2185	172_{56}	156_{55}	108_{50}	92_{74}	65_{100}	326	973
Sulfaperin-M 2TMS		2210	316_{19}	301_{35}	222_{5}	163_{8}	73_{100}	1005	10331
Sulfaperin-M 3TMS		2125	388_{5}	373_{28}	210_{16}	180_{12}	147_{100}	1394	10330

Tenamfetamine HFB **Table 8.1:** Compounds in order of names

Name	Detected	RI	Typical ions and intensities					Page	Entry
Tenamfetamine HFB	UHFB	1650	375 $_4$	240 $_6$	169 $_{10}$	162 $_{50}$	135 $_{100}$	1333	5291
Tenamfetamine PFP	UPFP	1605	325 $_4$	190 $_{10}$	162 $_{46}$	135 $_{100}$	119 $_{18}$	1060	5290
Tenamfetamine R-(-)-enantiomer HFBP		2280	472 $_2$	294 $_{15}$	266 $_{100}$	162 $_{66}$	135 $_8$	1622	6640
Tenamfetamine S-(+)-enantiomer HFBP		2290	472 $_2$	294 $_{16}$	266 $_{100}$	162 $_{76}$	135 $_{10}$	1621	6641
Tenamfetamine TFA	UTFA	1615	275 $_4$	162 $_{39}$	135 $_{100}$	105 $_6$	77 $_{12}$	744	5289
Tenamfetamine TMS		1735	251 $_1$	236 $_4$	135 $_8$	116 $_{100}$	73 $_{47}$	611	6334
Tenamfetamine-D5 2AC		1910	268 $_3$	167 $_{100}$	166 $_{81}$	136 $_{47}$	90 $_{59}$	708	5689
Tenamfetamine-D5 AC		1840	226 $_{13}$	167 $_{100}$	166 $_{93}$	136 $_{48}$	78 $_{28}$	496	5688
Tenamfetamine-D5 HFB		1630	380 $_{10}$	244 $_{10}$	167 $_{31}$	166 $_{29}$	136 $_{100}$	1356	6773
Tenamfetamine-D5 R-(-)-enantiomer HFBP		2275	477 $_{18}$	341 $_6$	294 $_{27}$	266 $_{100}$	167 $_{59}$	1630	6798
Tenamfetamine-D5 S-(+)-enantiomer HFBP		2285	477 $_9$	341 $_6$	294 $_{26}$	266 $_{100}$	167 $_{51}$	1630	6799
Tenamfetamine-M (deamino-HO-) AC	U+UHYAC	1620*	222 $_9$	162 $_{100}$	135 $_{56}$	104 $_{10}$	77 $_{17}$	479	6410
Tenocyclidine		1910	249 $_{20}$	206 $_{30}$	165 $_{52}$	97 $_{100}$	84 $_{23}$	601	3589
Tenocyclidine artifact/impurity		1310*	164 $_{100}$	149 $_{36}$	135 $_{75}$	97 $_{22}$	91 $_{21}$	310	3590
Tenocyclidine intermediate (PCC)		1525	192 $_4$	191 $_7$	164 $_8$	149 $_{100}$	122 $_{13}$	376	3581
Tenocyclidine intermediate (PCC) -HCN		1190	165 $_{85}$	164 $_{80}$	150 $_{100}$	136 $_{64}$	122 $_{39}$	315	3582
Tenocyclidine precursor (bromothiophene)		<1000*	164 $_{99}$	162 $_{100}$	117 $_4$	83 $_{80}$	57 $_{31}$	305	3609
Tenocyclidine precursor (piperidine)		<1000	85 $_{43}$	84 $_{100}$	70 $_{18}$	56 $_{69}$		254	3615
Tenoxicam 2ME		2690	365 $_7$	350 $_{81}$	176 $_{32}$	135 $_{76}$	78 $_{100}$	1284	4030
TEPP	G	1590*	290 $_3$	263 $_{89}$	235 $_{52}$	179 $_{65}$	161 $_{100}$	840	4086
Terbacil		1850	216 $_4$	201 $_2$	161 $_{100}$	160 $_{73}$	117 $_{37}$	451	3869
Terbinafine		2230	291 $_6$	276 $_{22}$	234 $_{11}$	141 $_{100}$	115 $_{36}$	852	7488
Terbinafine-M (1-naphthol)		1500*	144 $_{100}$	115 $_{83}$	89 $_{14}$	74 $_6$	63 $_{17}$	283	928
Terbinafine-M (1-naphthol) AC	U+UHYAC	1555*	186 $_{13}$	144 $_{100}$	115 $_{47}$	89 $_8$	63 $_7$	361	932
Terbinafine-M (1-naphthol) HFB		1310*	340 $_{46}$	169 $_{25}$	143 $_{28}$	115 $_{100}$	89 $_{11}$	1155	7476
Terbinafine-M (1-naphthol) PFP		1510*	290 $_{45}$	171 $_{100}$	143 $_{20}$	115 $_{49}$	89 $_6$	839	7468
Terbinafine-M (1-naphthol) TMS		1525*	216 $_{100}$	201 $_{95}$	185 $_{51}$	115 $_{39}$	73 $_{21}$	452	7460
Terbufos		1795*	288 $_2$	231 $_{38}$	186 $_9$	97 $_{22}$	57 $_{100}$	825	3872
Terbumeton		1790	225 $_{39}$	210 $_{100}$	169 $_{78}$	154 $_{27}$	141 $_{19}$	493	3874
Terbutaline		2430	225 $_1$	192 $_{10}$	111 $_{11}$	86 $_{100}$	57 $_{34}$	492	2731
Terbutaline 2ME		2120	253 $_1$	220 $_4$	168 $_{11}$	139 $_{11}$	86 $_{100}$	624	2735
Terbutaline 2TMS		2050	354 $_2$	284 $_{32}$	264 $_4$	86 $_{100}$	73 $_{62}$	1308	6184
Terbutaline 3AC	UHYAC	2375	351 $_1$	276 $_9$	192 $_4$	150 $_{10}$	86 $_{100}$	1214	2732
Terbutaline 3TMS		2010	426 $_4$	356 $_{100}$	147 $_5$	86 $_{93}$	73 $_{48}$	1563	6183
Terbutaline artifact 2ME		2250	265 $_5$	250 $_{20}$	220 $_{53}$	164 $_{21}$	99 $_{100}$	691	2736
Terbutaline -H2O 2AC		2040	291 $_{10}$	249 $_4$	192 $_{14}$	150 $_{100}$	57 $_{12}$	849	2733
Terbutaline-M/artifact (N-dealkyl-) 3AC		2170	277 $_6$	235 $_{18}$	193 $_{35}$	150 $_{100}$	55 $_{100}$	759	2734
Terbutaline-M/artifact (N-dealkyl-) 3HFB		1680	739 $_5$	542 $_{11}$	328 $_5$	169 $_{54}$	69 $_{100}$	1731	8362
Terbutaline-M/artifact (N-dealkyl-) 3PFP		1595	589 $_{12}$	442 $_{11}$	278 $_5$	223 $_5$	119 $_{100}$	1711	8363
Terbutaline-M/artifact (N-dealkyl-) 3TFA		1565	439 $_{10}$	342 $_8$	228 $_3$	201 $_4$	69 $_{100}$	1555	8361
Terbutryn		1960	241 $_{72}$	226 $_{100}$	185 $_{74}$	170 $_{49}$	157 $_{11}$	561	3867
Terbutylazine		1805	229 $_{33}$	214 $_{100}$	173 $_{40}$	132 $_{15}$	68 $_{14}$	506	3875
Terephthalic acid diethyl ester		1645*	222 $_{12}$	194 $_{18}$	177 $_{100}$	166 $_{22}$	149 $_{60}$	479	6495
Terephthalic acid ethyl methyl ester		1560*	208 $_{17}$	193 $_{11}$	180 $_{37}$	163 $_{100}$	149 $_{43}$	423	6493
Terephthalic acid monoethyl ester		1715*	194 $_{19}$	166 $_{40}$	149 $_{100}$	121 $_{17}$	65 $_{15}$	383	6497
Terfenadine	G	3700	471 $_{20}$	280 $_{100}$	262 $_{10}$	183 $_{14}$	105 $_{48}$	1621	2237
Terfenadine -2H2O	U+UHYAC	3460	435 $_1$	262 $_{100}$	115 $_4$	91 $_{10}$	57 $_{16}$	1549	2235
Terfenadine AC	U+UHYAC	3600	452 $_1$	280 $_{100}$	262 $_9$	105 $_{22}$	57 $_{21}$	1672	2236
Terfenadine-M (N-dealkyl-) -H2O	UHY	2600	249 $_{53}$	248 $_{100}$	191 $_{24}$	165 $_{24}$	129 $_{64}$	599	2219
Terfenadine-M (N-dealkyl-) -H2O AC	U+UHYAC	2550	291 $_{100}$	205 $_{24}$	191 $_{26}$	91 $_{80}$	72 $_{36}$	849	2217
Terfenadine-M (N-dealkyl-oxo-) -2H2O	U+UHYAC	2190	245 $_{82}$	167 $_{100}$	152 $_{17}$	139 $_{21}$	115 $_{16}$	575	2218
Tertatolol		2310	295 $_4$	280 $_{17}$	251 $_7$	166 $_{53}$	86 $_{100}$	880	4362
Tertatolol AC		2350	337 $_{10}$	322 $_{47}$	166 $_{39}$	112 $_{20}$	86 $_{100}$	1139	4361
Tertatolol formyl artifact	P	2400	307 $_{12}$	292 $_{100}$	141 $_{13}$	96 $_{14}$	57 $_{23}$	951	4363
Tertatolol TMSTFA		2510	463 $_{14}$	392 $_5$	242 $_{35}$	191 $_{10}$	166 $_{100}$	1605	6139
Testosterone		2620*	288 $_{88}$	246 $_{42}$	124 $_{100}$			830	979
Testosterone AC	U+UHYAC	2750*	330 $_{66}$	288 $_{35}$	228 $_{34}$	147 $_{64}$	124 $_{100}$	1096	1864
Testosterone acetate	U+UHYAC	2750*	330 $_{66}$	288 $_{35}$	228 $_{34}$	147 $_{64}$	124 $_{100}$	1096	1864
Testosterone dipropionate		3350*	400 $_{72}$	358 $_{46}$	288 $_{21}$	147 $_{66}$	124 $_{100}$	1447	1865
Testosterone enol 2TMS		2690*	432 $_{91}$	417 $_9$	209 $_{11}$	195 $_3$	73 $_{100}$	1541	3804
Testosterone propionate		2815*	344 $_{100}$	330 $_{23}$	288 $_{26}$	246 $_{14}$	124 $_{36}$	1181	1866
Testosterone propionate enol AC		3020*	386 $_{36}$	344 $_{100}$	329 $_{31}$	302 $_{50}$	284 $_{48}$	1387	1867
Tetrabenazine	G	2490	317 $_{25}$	274 $_{30}$	261 $_{100}$	191 $_{43}$		1014	395
Tetrabenazine-M (O-bis-demethyl-) AC	UHYAC	2510	331 $_{13}$	296 $_{72}$	232 $_{36}$	191 $_{100}$	177 $_{49}$	1102	396
Tetrabenazine-M (O-bis-demethyl-HO-) 2AC	UHYAC	2665	389 $_{77}$	330 $_{89}$	302 $_{100}$	288 $_{89}$	233 $_{68}$	1399	398
Tetrabenazine-M (O-demethyl-HO-)	U UHY	2500	319 $_{62}$	318 $_{65}$	274 $_{57}$	205 $_{100}$	191 $_{90}$	1028	615
Tetrabenazine-M (O-demethyl-HO-) AC	UHYAC	2585	361 $_{60}$	302 $_{100}$	274 $_{70}$	246 $_{48}$	205 $_{66}$	1270	397
Tetrabromo-o-cresol	P	2190*	424 $_{100}$	420 $_{15}$	343 $_{40}$	263 $_{11}$	234 $_8$	1506	2738
Tetrabromo-o-cresol AC	U+UHYAC	2465*	466 $_8$	462 $_1$	424 $_{100}$	420 $_{17}$	343 $_{15}$	1604	2739
Tetrabromo-o-cresol ME	UME	2350*	438 $_{105}$	436 $_{68}$	423 $_{61}$	314 $_{32}$	74 $_{52}$	1543	2740
Tetracaine	G	2350	264 $_1$	221 $_2$	193 $_6$	71 $_{53}$	58 $_{100}$	684	1868
Tetracaine-M/artifact (HOOC-) ME		2015	207 $_{17}$	176 $_7$	164 $_{100}$	120 $_3$	105 $_7$	420	1869
1,2,3,5-Tetrachlorobenzene		1370*	216 $_{100}$	214 $_{76}$	179 $_{19}$	143 $_{11}$	108 $_{20}$	444	3472
2,2',5,5'-Tetrachlorobiphenyl		1945*	292 $_{100}$	290 $_{75}$	255 $_{27}$	220 $_{65}$	184 $_{10}$	839	881

Table 8.1: Compounds in order of names — 2,2',6,6'-Tetrachlorobiphenyl

Name	Detected	RI	Typical ions and intensities					Page	Entry
2,2',6,6'-Tetrachlorobiphenyl		1945*	292 100	290 78	255 8	220 78	184 13	839	9204
3,3',4,4'-Tetrachlorobiphenyl		2200*	292 100	290 80	220 41	184 10	150 14	839	9192
2,3,7,8-Tetrachlorodibenzofuran (TCDF)		----*	306 100	304 85	241 24	171 36	152 21	928	3493
2,3,7,8-Tetrachlorodibenzo-p-dioxin (TCDD)		----*	322 100	320 80	257 33	194 32	161 27	1031	1465
Tetrachloroethylene		<1000*	166 100	164 76	129 74	94 48	47 54	309	3783
Tetrachloromethane		<1000*	117 100	82 51	47 52			292	980
2,3,4,5-Tetrachlorophenol	U	1500*	232 100	230 79	194 15	166 20	131 26	507	3366
Tetrachlorvinphos		2120*	364 1	329 89	240 10	109 100	79 38	1281	3190
Tetrachlorvinphos-M/artifact		1710*	256 5	207 100	179 26	143 10	109 15	635	3191
Tetradecamethylcycloheptasiloxane		1345*	503 11	415 14	327 30	281 50	73 100	1676	9740
Tetradecane	P	1400*	198 10	99 14	85 33	71 56	57 100	396	2767
Tetradifon		2505*	354 19	227 59	159 100	111 68	75 49	1230	3868
Tetraethylene glycol 2TMS		1710*	235 5	191 6	161 15	117 65	73 100	1145	8587
Tetraethylene glycol dipivalate	PPIV	1820*	175 1	129 100	113 5	85 18	57 75	1276	6427
Tetrahexylammoniumhydrogensulfate artifact-1		1380	185 4	114 100	100 3	79 5	57 8	359	4947
Tetrahexylammoniumhydrogensulfate artifact-2		1725	269 2	198 100	128 42	98 14	58 37	714	4491
Tetrahydrocannabinol	G P-I	2470*	314 85	299 100	271 41	243 29	231 45	997	981
Tetrahydrocannabinol AC	U+UHYAC-I	2450*	356 13	313 35	297 100	243 12	231 24	1244	982
Tetrahydrocannabinol ET		2390*	342 90	327 100	313 39	271 27	259 30	1171	2531
Tetrahydrocannabinol isomer-1 PFP		2150*	460 100	445 14	417 70	392 30	377 100	1602	5669
Tetrahydrocannabinol isomer-2 PFP		2170*	460 100	445 65	417 75	389 86	297 80	1602	5668
Tetrahydrocannabinol ME		2360*	328 82	313 100	285 28	257 24	245 27	1083	2530
Tetrahydrocannabinol TMS		2405*	386 100	371 86	315 37	303 31	73 80	1387	4599
Tetrahydrocannabinol-D3		2450*	317 96	302 100	274 43	258 26	234 55	1015	5663
Tetrahydrocannabinol-D3 AC		2750*	359 9	316 18	300 100	274 9	234 17	1261	7309
Tetrahydrocannabinol-D3 isomer-1 PFP		2130*	463 60	420 54	395 25	380 20	342 13	1606	5665
Tetrahydrocannabinol-D3 isomer-1 TFA		2160*	413 51	370 31	345 14	330 100	232 10	1488	5667
Tetrahydrocannabinol-D3 isomer-2 PFP		2150*	463 100	448 65	420 70	389 85	300 81	1606	5666
Tetrahydrocannabinol-D3 isomer-2 TFA		2180*	413 100	398 74	370 71	339 78	300 73	1488	5664
Tetrahydrocannabinol-D3 ME		2355*	331 81	316 100	288 34	257 29	248 39	1103	6040
Tetrahydrocannabinol-D3 TMS		2385*	389 100	374 96	346 26	315 59	306 41	1400	5670
Tetrahydrocannabinolic acid 2TMS		2635*	502 2	487 100	413 3	147 6	73 41	1662	4605
Tetrahydrocannabinol-M (11-HO-)	P-I	2775*	330 12	299 100	231 10	217 9	193 7	1096	4661
Tetrahydrocannabinol-M (11-HO-) 2ME		2580*	358 13	313 100	257 3	231 3		1255	4659
Tetrahydrocannabinol-M (11-HO-) 2PFP		2350*	622 15	607 5	551 8	458 100	415 24	1719	4658
Tetrahydrocannabinol-M (11-HO-) 2TFA		2450*	522 13	451 8	408 100	395 13	365 24	1678	4657
Tetrahydrocannabinol-M (11-HO-) 2TMS		2630*	474 5	459 4	403 2	371 100	73 14	1626	4656
Tetrahydrocannabinol-M (11-HO-) -H2O AC	U+UHYAC-I	2740*	354 48	312 100	297 19	269 31	91 21	1233	4660
Tetrahydrocannabinol-M (HO-nor-delta-9-HOOC-) 2ME	UTHCME-I	2840*	388 42	373 65	329 100	201 24	189 28	1396	3466
Tetrahydrocannabinol-M (nor-delta-9-HOOC-) 2ME	UTHCME UGlucExMe	2620*	372 52	357 79	341 9	313 100	245 6	1324	1439
Tetrahydrocannabinol-M (nor-delta-9-HOOC-) 2PFP		2440*	622 35	607 49	459 100	445 76	69 82	1719	4380
Tetrahydrocannabinol-M (nor-delta-9-HOOC-) 2TMS		2470*	488 40	473 51	398 12	371 100	73 66	1646	5671
Tetrahydrocannabinol-M (nor-delta-9-HOOC-)-D3 2ME		2590*	375 39	360 68	356 19	316 100	301 9	1337	6187
Tetrahydrocannabinol-M (nor-delta-9-HOOC-)-D3 2PFP		2425*	625 40	610 57	462 100	448 62	432 43	1720	6039
Tetrahydrocannabinol-M (nor-delta-9-HOOC-)-D3 2TMS		2660*	491 44	476 55	374 100	300 15	73 28	1650	5672
Tetrahydrocannabinol-M (oxo-nor-delta-9-HOOC-) 2ME	UTHCME-I	2860*	386 55	371 73	327 100	314 22	189 11	1386	3467
Tetrahydrofuran		<1000*	72 40	71 38	42 100	27 31		251	4185
Tetrahydrogestrinone		2660*	312 57	265 52	240 43	227 100	211 34	984	7573
Tetrahydrogestrinone TMS		2490*	384 77	299 100	281 39	270 52	73 92	1379	7574
Tetrahydroharmine		2150	216 39	201 100	186 14	172 18	144 5	453	4065
Tetrahydroharmine 2TFA		2115	408 22	393 100	296 20	280 15	199 75	1470	9550
Tetrahydroharmine AC		2525	258 58	243 75	215 13	201 100	172 28	647	9552
Tetrahydroharmine HFB		2300	412 47	397 100	243 21	199 34	172 22	1485	9554
Tetrahydroharmine PFP		2280	362 46	347 100	243 21	199 54	172 37	1272	9553
Tetrahydroharmine TFA		2295	312 57	297 100	282 4	243 7	199 18	981	9551
Tetramethrin		2735	331 1	164 10	123 31	107 9	81 13	1102	3883
Tetramethylbenzene		1080*	134 25	119 100	105 12	91 23	77 16	274	3791
Tetramethylcitrate		1445*	189 11	157 100	133 4	125 38	59 16	590	5705
Tetrasul		2310*	324 55	322 44	252 100	217 7	108 25	1042	3879
Tetrazepam	G P U	2400	288 52	259 16	253 100	225 13		827	616
Tetrazepam +H2O isomer-1 ALHY		2350	267 76	196 14	179 35	168 100	140 11	699	2094
Tetrazepam +H2O isomer-1 ALHYAC		2420	309 40	249 9	168 100	140 14		962	2095
Tetrazepam +H2O isomer-2 ALHY		2370	267 65	168 100	140 14	77 15		699	2093
Tetrazepam +H2O isomer-2 ALHYAC		2480	309 37	249 7	168 100	140 14	111 10	962	2096
Tetrazepam AC		2590	330 9	288 100	259 28	244 9	180 8	1093	5699
Tetrazepam isomer-1 HY	UHY U+UHYAC	2220	249 46	234 22	220 35	207 100		594	303
Tetrazepam isomer-2 HY	G P U+UHYAC	2280	249 52	220 37	207 100	178 14	165 16	594	2059
Tetrazepam-M (di-HO-) -2H2O HY	UHY U+UHYAC	2100	245 95	228 38	193 29	105 39	77 100	573	272
Tetrazepam-M (di-HO-) -2H2O HYAC	U+UHYAC	2260	287 11	244 100	228 39	182 49	77 70	818	2542
Tetrazepam-M (di-HO-) isomer-1 HY2AC	U+UHYAC	2600	365 37	264 76	246 100	206 53		1285	2063
Tetrazepam-M (di-HO-) isomer-2 HY2AC	U+UHYAC	2640	365 56	264 56	246 100	220 34	206 42	1285	2061
Tetrazepam-M (HO-) -H2O	U+UHYAC	2430	286 100	228 32				813	2089
Tetrazepam-M (HO-) -H2O HY	U+UHYAC	2200	247 100	230 19	192 28	168 19	138 22	584	2062
Tetrazepam-M (HO-) isomer-1	UGLUC	2570	304 43	275 52	261 19	235 100		930	618

Trihexyphenidyl-M (amino-HO-) isomer-1 -H2O 2AC

Table 8.1: Compounds in order of names

Name	Detected	RI	Typical ions and intensities					Page	Entry
Trihexyphenidyl-M (amino-HO-) isomer-1 -H2O 2AC	UHYAC	2560	315 $_3$	255 $_{20}$	196 $_{100}$	168 $_{35}$	155 $_{22}$	1001	1290
Trihexyphenidyl-M (amino-HO-) isomer-2 -H2O 2AC	UHYAC	2625	315 $_8$	255 $_{60}$	196 $_{100}$	132 $_{66}$	115 $_{60}$	1002	4242
Trihexyphenidyl-M (di-HO-) -H2O isomer-1 2AC	UHYAC	2555	399 $_4$	357 $_3$	98 $_{100}$			1443	1303
Trihexyphenidyl-M (di-HO-) -H2O isomer-2 2AC	UHYAC	2665	399 $_5$	338 $_3$	194 $_{22}$	98 $_{100}$		1443	1304
Trihexyphenidyl-M (HO-)	U	2500	317 $_1$	299 $_1$	218 $_6$	98 $_{100}$		1014	93
Trihexyphenidyl-M (HO-) AC	U+UHYAC	2635	359 $_1$	316 $_5$	218 $_{12}$	98 $_{100}$		1261	1553
Trihexyphenidyl-M (HO-) -H2O AC	U+UHYAC	2505	341 $_{16}$	298 $_2$	200 $_{30}$	98 $_{100}$		1166	1552
Trihexyphenidyl-M (tri-HO-) -H2O 3AC	UHYAC	2965	457 $_6$	398 $_3$	336 $_2$	194 $_{25}$	156 $_{100}$	1596	1305
Trihexyphenidyl-M -2H2O -CO2 AC	UHYAC	2095*	242 $_{27}$	200 $_{27}$	182 $_{100}$	167 $_{58}$		566	1302
Trimebutine		2660	195 $_5$	162 $_1$	152 $_1$	109 $_1$	58 $_{100}$	1391	7634
Trimebutine-M (TMBA)		1780*	212 $_{100}$	197 $_{57}$	169 $_{13}$	141 $_{27}$		438	1949
Trimebutine-M (TMBA) ME		1740*	226 $_{100}$	211 $_{48}$	195 $_{22}$	155 $_{21}$		494	1950
Trimebutine-M/artifact (alcohol)		1070	175 $_{12}$	160 $_{13}$	115 $_7$	91 $_{11}$	58 $_{100}$	380	7633
Trimethadion		1080	143 $_{73}$	128 $_{26}$	100 $_4$	70 $_{14}$	58 $_{100}$	282	1003
Trimethadion-M (nor-)	U	1060	129 $_{15}$	107 $_2$	70 $_7$	59 $_{100}$		271	2923
Trimethoprim	P G U UHY	2590	290 $_{100}$	259 $_{36}$	123 $_{27}$			842	1004
Trimethoprim 2AC	U+UHYAC	3000	374 $_{100}$	359 $_{54}$	332 $_{53}$	317 $_{82}$	275 $_{58}$	1330	1006
Trimethoprim 2TMS		2650	434 $_{50}$	419 $_{100}$	331 $_{10}$	210 $_6$	73 $_{42}$	1546	4602
Trimethoprim 3TMS		2805	506 $_{36}$	491 $_{100}$	403 $_{11}$	246 $_{10}$	73 $_{90}$	1667	4603
Trimethoprim isomer-1 AC	PAC U+UHYAC	2700	332 $_{44}$	317 $_{44}$	289 $_{85}$	275 $_{66}$	259 $_{17}$	1106	1005
Trimethoprim isomer-2 AC	U+UHYAC	2880	332 $_{100}$	317 $_{21}$	290 $_{28}$	275 $_{32}$	259 $_{34}$	1106	2576
2,3,5-Trimethoxyamfetamine		2040	225 $_2$	182 $_{100}$	167 $_{30}$	151 $_7$	107 $_5$	491	2622
2,3,5-Trimethoxyamfetamine 2ME		1990	253 $_{26}$	208 $_{95}$	181 $_{100}$	167 $_{15}$	72 $_{15}$	624	2624
2,3,5-Trimethoxyamfetamine AC		2285	267 $_{15}$	208 $_{100}$	193 $_{30}$	181 $_{37}$	86 $_{15}$	702	2625
2,3,5-Trimethoxyamfetamine intermediate (propenyltrimethoxybenzene)		1620*	208 $_{193}$	193 $_{65}$	165 $_9$	150 $_6$	133 $_{12}$	424	2626
2,4,5-Trimethoxyamfetamine 2AC	U+UHYAC	2200	309 $_{10}$	208 $_{100}$	181 $_{65}$	151 $_{21}$	86 $_{11}$	964	7161
2,4,5-Trimethoxyamfetamine AC	U+UHYAC	2140	267 $_{23}$	208 $_{100}$	181 $_{66}$	151 $_{23}$	86 $_6$	702	7152
2,4,5-Trimethoxyamfetamine-M (deamino-HO-) AC	U+UHYAC	1670*	268 $_8$	208 $_{100}$	193 $_{39}$	181 $_{31}$		708	7157
2,4,5-Trimethoxyamfetamine-M (O-bis-demethyl-) artifact 2AC	U+UHYAC	2200	263 $_{28}$	221 $_{100}$	179 $_{65}$	164 $_{44}$	132 $_{10}$	673	7183
2,4,5-Trimethoxyamfetamine-M (O-bis-demethyl-) isomer-1 3AC	U+UHYAC	2300	323 $_2$	281 $_{27}$	180 $_{100}$	153 $_{50}$	86 $_{22}$	1050	7162
2,4,5-Trimethoxyamfetamine-M (O-bis-demethyl-) isomer-2 3AC	U+UHYAC	2305	323 $_2$	281 $_6$	180 $_{100}$	153 $_{52}$	86 $_{20}$	1050	7163
2,4,5-Trimethoxyamfetamine-M (O-bis-demethyl-) isomer-3 3AC	U+UHYAC	2330	323 $_4$	281 $_{12}$	180 $_{100}$	153 $_{52}$	86 $_{25}$	1050	7164
2,4,5-Trimethoxyamfetamine-M (O-deamino-oxo-)	U+UHYAC	1540*	224 $_{25}$	181 $_{100}$	151 $_{37}$	136 $_{28}$		488	7165
2,4,5-Trimethoxyamfetamine-M (O-demethyl-) isomer-1 2AC	U+UHYAC	2215	295 $_9$	253 $_{23}$	194 $_{100}$	167 $_{47}$	86 $_{13}$	878	7154
2,4,5-Trimethoxyamfetamine-M (O-demethyl-) isomer-2 2AC	U+UHYAC	2230	295 $_{17}$	236 $_{32}$	194 $_{100}$	167 $_{54}$	86 $_{14}$	878	7153
2,4,5-Trimethoxyamfetamine-M (O-demethyl-) isomer-2 3AC	U+UHYAC	2280	337 $_5$	236 $_{29}$	194 $_{100}$	167 $_{60}$	86 $_{22}$	1137	7156
2,4,5-Trimethoxyamfetamine-M (O-demethyl-) isomer-3 2AC	U+UHYAC	2250	295 $_8$	236 $_{52}$	194 $_{100}$	167 $_{40}$	86 $_7$	878	7155
2,4,5-Trimethoxyamfetamine-M (O-demethyl-deamino-oxo-) isomer-1 AC	U+UHYAC	1680*	252 $_8$	210 $_{40}$	167 $_{100}$			617	7158
2,4,5-Trimethoxyamfetamine-M (O-demethyl-deamino-oxo-) isomer-2 AC	U+UHYAC	1705*	252 $_{11}$	210 $_{32}$	167 $_{100}$	137 $_9$		617	7159
2,4,5-Trimethoxyamfetamine-M (O-demethyl-deamino-oxo-) isomer-3 AC	U+UHYAC	1760*	252 $_{10}$	210 $_{32}$	167 $_{100}$	137 $_{11}$		618	7160
3,4,5-Trimethoxyamfetamine		1680	225 $_2$	182 $_{100}$	167 $_{34}$	151 $_8$	107 $_6$	492	3259
3,4,5-Trimethoxyamfetamine		1680	225 $_1$	182 $_{24}$	167 $_8$	151 $_2$	44 $_{100}$	492	5540
3,4,5-Trimethoxyamfetamine AC		2020	267 $_{18}$	208 $_{100}$	193 $_{35}$	181 $_{34}$	86 $_{21}$	701	3266
3,4,5-Trimethoxyamfetamine AC		2020	267 $_9$	208 $_{47}$	193 $_{17}$	86 $_{10}$	44 $_{100}$	702	5541
3,4,5-Trimethoxyamfetamine formyl artifact		1680	237 $_{25}$	181 $_{93}$	148 $_9$	77 $_6$	56 $_{100}$	546	3251
3,4,5-Trimethoxyamfetamine intermediate-1		2050	253 $_{100}$	206 $_{30}$	191 $_{44}$	161 $_{27}$	77 $_{48}$	621	2840
3,4,5-Trimethoxyamfetamine intermediate-2		2145	239 $_{100}$	192 $_{34}$	177 $_{44}$	149 $_{32}$	92 $_{26}$	551	2841
3,4,5-Trimethoxybenzaldehyd		1550*	196 $_{100}$	181 $_{51}$	125 $_{52}$	110 $_{51}$	93 $_{37}$	390	3279
Trimethoxybenzoic acid		1780*	212 $_{100}$	197 $_{57}$	169 $_{13}$	141 $_{27}$		438	1949
Trimethoxybenzoic acid ET		1770*	240 $_{100}$	225 $_{44}$	212 $_{17}$	195 $_{45}$	141 $_{24}$	556	5219
Trimethoxybenzoic acid ME		1740*	226 $_{100}$	211 $_{48}$	195 $_{22}$	155 $_{21}$		494	1950
Trimethoxybenzoic acid-M (glycine conjugate) ME		2350	283 $_{62}$	268 $_4$	195 $_{100}$	152 $_8$		795	1952
3,4,5-Trimethoxybenzyl alcohol		1650*	198 $_{100}$	183 $_{22}$	127 $_{51}$	95 $_{39}$	77 $_{22}$	394	6059
3,4,5-Trimethoxybenzyl alcohol AC		1650*	240 $_{80}$	198 $_{52}$	181 $_{100}$	169 $_{29}$	123 $_{27}$	556	6060
Trimethoxycocaine	UGLUCME	2550	393 $_{26}$	212 $_{21}$	182 $_{99}$	94 $_{23}$	82 $_{100}$	1415	5678
Trimethoxyhippuric acid ME		2350	283 $_{62}$	268 $_4$	195 $_{100}$	152 $_8$		795	1952
2,3,5-Trimethoxymetamfetamine AC		2310	281 $_9$	224 $_{12}$	208 $_{65}$	100 $_{47}$	58 $_{100}$	789	2623
3,4,5-Trimethoxyphenyl-2-nitroethene		2145	239 $_{100}$	192 $_{34}$	177 $_{44}$	149 $_{32}$	92 $_{26}$	551	2841
3,4,5-Trimethoxyphenyl-2-nitropropene		2050	253 $_{100}$	206 $_{30}$	191 $_{44}$	161 $_{27}$	77 $_{48}$	621	2840
Trimethylamine		<1000	59 $_{38}$	58 $_{100}$	42 $_{59}$	30 $_{34}$		249	4187
1,2,3-Trimethylbenzene		<1000*	120 $_{42}$	105 $_{100}$	91 $_{11}$	77 $_{16}$	65 $_7$	265	3788
1,2,4-Trimethylbenzene		<1000*	120 $_{45}$	105 $_{100}$	91 $_{11}$	77 $_{15}$	65 $_6$	265	3826
Trimethylcitrate	UME	1410*	175 $_{14}$	143 $_{100}$	101 $_{86}$	69 $_{18}$	59 $_{44}$	526	4451
Trimetozine		2260	281 $_{44}$	195 $_{100}$				787	1529
Trimipramine	P G U+UHYAC	2225	294 $_7$	249 $_{33}$	193 $_{26}$	99 $_{13}$	58 $_{100}$	873	410
Trimipramine artifact	G	2025	235 $_{100}$	220 $_{60}$	206 $_{43}$	192 $_{19}$	178 $_{27}$	534	6561
Trimipramine-D3		2215	297 $_7$	249 $_{29}$	208 $_{19}$	102 $_{16}$	61 $_{100}$	892	5426
Trimipramine-D3 artifact		2045*	249 $_{29}$	234 $_7$	208 $_{23}$	194 $_{100}$	167 $_9$	599	6329
Trimipramine-M (bis-nor-) AC	U+UHYAC	2650	308 $_{13}$	208 $_{100}$	193 $_{32}$	114 $_{40}$	72 $_{20}$	959	2865
Trimipramine-M (bis-nor-di-HO-) 3AC	U+UHYAC	3400	424 $_{14}$	324 $_{100}$	282 $_{30}$	114 $_{30}$	72 $_{20}$	1517	2856
Trimipramine-M (bis-nor-HO-) 2AC	U+UHYAC	3050	366 $_{21}$	266 $_{100}$	224 $_{42}$	209 $_{20}$	114 $_5$	1293	2676
Trimipramine-M (bis-nor-HO-methoxy-) 2AC	UHYAC	3130	396 $_{13}$	296 $_{100}$	254 $_{48}$	114 $_{30}$	72 $_{20}$	1430	2866
Trimipramine-M (di-HO-) 2AC	U+UHYAC	2900	410 $_{10}$	365 $_{12}$	323 $_{60}$	99 $_6$	58 $_{100}$	1479	2293
Trimipramine-M (di-HO-ring)	UHY	2600	227 $_{100}$	196 $_7$				499	2296

Table 8.1: Compounds in order of names

Name	Detected	RI	Typical ions and intensities					Page	Entry
Trimipramine-M (di-HO-ring) 2AC	U+UHYAC	2750	311 $_{28}$	269 $_{23}$	227 $_{100}$	196 $_{7}$		975	2292
Trimipramine-M (HO-)	P-I UHY	2575	310 $_{11}$	265 $_{34}$	250 $_{16}$	224 $_{20}$	58 $_{100}$	973	640
Trimipramine-M (HO-) AC	PAC U+UHYAC	2660	352 $_{12}$	307 $_{26}$	265 $_{38}$	99 $_{7}$	58 $_{100}$	1223	411
Trimipramine-M (HO-methoxy-)	UHY	2590	340 $_{20}$	295 $_{29}$	254 $_{23}$	99 $_{10}$	58 $_{100}$	1158	2314
Trimipramine-M (HO-methoxy-) AC	U+UHYAC	2700	382 $_{4}$	337 $_{7}$	295 $_{23}$	99 $_{8}$	58 $_{100}$	1367	2291
Trimipramine-M (HO-methoxy-ring)	UHY	2390	241 $_{100}$	226 $_{17}$	210 $_{12}$	180 $_{14}$		560	2315
Trimipramine-M (HO-methoxy-ring) AC	U+UHYAC	2370	283 $_{10}$	241 $_{100}$	226 $_{17}$	210 $_{12}$	180 $_{14}$	796	2867
Trimipramine-M (HO-ring)	UHY	2240	211 $_{100}$	196 $_{15}$	180 $_{19}$	152 $_{4}$		436	2295
Trimipramine-M (HO-ring) AC	U+UHYAC	2535	253 $_{26}$	211 $_{100}$	196 $_{19}$	180 $_{11}$	152 $_{4}$	621	1218
Trimipramine-M (nor-)	U UHY	2245	280 $_{13}$	249 $_{31}$	234 $_{14}$	208 $_{100}$	193 $_{56}$	784	6330
Trimipramine-M (nor-) AC	U+UHYAC	2680	322 $_{17}$	208 $_{100}$	193 $_{32}$	128 $_{13}$	86 $_{6}$	1047	2290
Trimipramine-M (nor-di-HO-) 3AC	U+UHYAC	3555	438 $_{15}$	324 $_{100}$	282 $_{33}$	240 $_{16}$	128 $_{32}$	1554	413
Trimipramine-M (nor-HO-) 2AC	UHYAC	3155	380 $_{22}$	266 $_{100}$	224 $_{24}$	128 $_{14}$	86 $_{7}$	1358	412
Trimipramine-M (nor-HO-) -H2O AC	U+UHYAC	2670	320 $_{11}$	206 $_{100}$	128 $_{6}$			1034	991
Trimipramine-M (nor-HO-methoxy-) 2AC	U+UHYAC	3180	410 $_{37}$	296 $_{100}$	254 $_{48}$	128 $_{14}$	86 $_{5}$	1479	2294
Trimipramine-M (N-oxide) -(CH3)2NOH		2045*	249 $_{29}$	234 $_{7}$	208 $_{23}$	194 $_{100}$	167 $_{9}$	599	6329
Trimipramine-M (ring)	U U+UHYAC	1930	195 $_{100}$	180 $_{40}$	167 $_{9}$	96 $_{33}$	83 $_{22}$	386	308
Trimipramine-M (ring) ME		1915	209 $_{70}$	194 $_{100}$	178 $_{13}$	165 $_{11}$		429	6352
Tripelenamine	U UHY UHYAC	1970	255 $_{2}$	197 $_{18}$	185 $_{11}$	91 $_{79}$	58 $_{100}$	634	2030
Tripelenamine-M (benzylpyridylamine)	UHY UHYAC	1650	184 $_{100}$	106 $_{90}$	91 $_{57}$	79 $_{47}$	65 $_{27}$	358	1603
Tripelenamine-M (HO-)	UHY	2400	271 $_{11}$	213 $_{33}$	91 $_{80}$	72 $_{45}$	58 $_{100}$	726	1609
Tripelenamine-M (HO-) AC	UHYAC	2390	313 $_{5}$	255 $_{17}$	91 $_{80}$	72 $_{19}$	58 $_{100}$	990	1606
Tripelenamine-M (nor-)	U UHY	2420	241 $_{8}$	197 $_{28}$	129 $_{81}$	112 $_{18}$	91 $_{100}$	562	1610
Tripelenamine-M (nor-) AC	UHYAC	2420	283 $_{8}$	197 $_{41}$	183 $_{7}$	91 $_{100}$	78 $_{4}$	797	1607
Tripelenamine-M (nor-HO-) 2AC	UHYAC	2860	341 $_{15}$	255 $_{37}$	213 $_{24}$	177 $_{29}$	91 $_{100}$	1164	1608
Tripelenamine-M/artifact-1	UHY UHYAC	1845	212 $_{30}$	183 $_{100}$	107 $_{26}$	91 $_{39}$	78 $_{47}$	437	1604
Tripelenamine-M/artifact-2	UHY UHYAC	2220	239 $_{100}$	210 $_{9}$	148 $_{49}$	134 $_{25}$	91 $_{19}$	550	1605
Triphenylphosphate		2340*	326 $_{100}$	325 $_{80}$	233 $_{15}$	170 $_{14}$	77 $_{35}$	1066	2871
Triphenylphosphine oxide	G	2460*	278 $_{38}$	277 $_{100}$	199 $_{19}$	183 $_{16}$	152 $_{11}$	766	6676
Triprolidine		2315	278 $_{36}$	208 $_{100}$	193 $_{26}$	96 $_{22}$	84 $_{20}$	770	6103
TRIS 4AC	UHYAC	1910	216 $_{3}$	156 $_{36}$	127 $_{26}$	114 $_{66}$	72 $_{100}$	834	4635
Tris-(2-chloroethyl-)phosphate		1870*	249 $_{32}$	205 $_{17}$	143 $_{24}$	63 $_{100}$		798	4255
Trisalicyclide	G U+UHYAC	3190*	360 $_{39}$	240 $_{58}$	152 $_{36}$	120 $_{100}$	92 $_{75}$	1263	4496
Tritoqualine artifact-1		2130	281 $_{100}$	252 $_{94}$	224 $_{58}$	196 $_{60}$	168 $_{57}$	787	5236
Tritoqualine artifact-1 2AC		2350	365 $_{9}$	323 $_{36}$	281 $_{100}$	252 $_{71}$	224 $_{29}$	1285	5240
Tritoqualine artifact-1 AC		2325	323 $_{17}$	281 $_{96}$	252 $_{100}$	224 $_{37}$	196 $_{34}$	1050	5239
Tritoqualine artifact-2		2170	311 $_{100}$	282 $_{51}$	254 $_{29}$	222 $_{27}$	166 $_{20}$	973	5236
Tritoqualine artifact-2 AC		2335	353 $_{17}$	311 $_{100}$	282 $_{34}$	254 $_{21}$	222 $_{17}$	974	5238
Trometamol 4AC	UHYAC	1910	216 $_{3}$	156 $_{36}$	127 $_{26}$	114 $_{66}$	72 $_{100}$	834	4635
Tropacocaine		2040	245 $_{13}$	124 $_{100}$	94 $_{44}$	82 $_{79}$	77 $_{42}$	576	5124
Tropenol		<1000	139 $_{1}$	120 $_{2}$	94 $_{100}$	82 $_{21}$	67 $_{7}$	278	8399
Tropenol AC		<1000	181 $_{3}$	138 $_{13}$	122 $_{18}$	94 $_{100}$	81 $_{37}$	349	8400
Tropenol HFB		1020	335 $_{2}$	138 $_{3}$	122 $_{33}$	94 $_{100}$	81 $_{74}$	1123	8403
Tropenol PFP		1020	285 $_{4}$	138 $_{3}$	122 $_{39}$	94 $_{100}$	81 $_{79}$	807	8402
Tropenol TFA		<1000	235 $_{6}$	138 $_{3}$	122 $_{35}$	94 $_{100}$	81 $_{66}$	531	8401
Tropenol TMS		1050	211 $_{1}$	120 $_{4}$	94 $_{100}$	82 $_{86}$	81 $_{48}$	437	8398
Tropicamide		2340	284 $_{1}$	266 $_{8}$	254 $_{43}$	163 $_{15}$	92 $_{100}$	804	1983
Tropicamide AC	U+UHYAC	2410	326 $_{8}$	266 $_{26}$	163 $_{18}$	104 $_{30}$	92 $_{100}$	1070	2238
Tropicamide -CH2O		2230	254 $_{20}$	163 $_{16}$	121 $_{4}$	92 $_{100}$	65 $_{21}$	629	1985
Tropicamide -H2O		2250	266 $_{86}$	251 $_{16}$	103 $_{100}$	92 $_{28}$	77 $_{26}$	696	1984
Tropine AC		1240	183 $_{25}$	140 $_{11}$	124 $_{100}$	94 $_{42}$	82 $_{63}$	355	5125
Tropine TFA		1020	237 $_{19}$	124 $_{100}$	94 $_{34}$	82 $_{44}$	67 $_{25}$	543	7914
Tropisetrone		2720	284 $_{22}$	144 $_{27}$	124 $_{100}$	94 $_{45}$	82 $_{52}$	804	4633
Tropisetrone AC		2800	326 $_{14}$	144 $_{11}$	124 $_{100}$	94 $_{44}$	82 $_{52}$	1070	4634
Trovafloxacine TMS		3400	473 $_{7}$	431 $_{100}$	207 $_{12}$	165 $_{43}$	57 $_{15}$	1645	5712
Tryptamine		1730	160 $_{19}$	130 $_{100}$	103 $_{14}$	77 $_{19}$		303	1007
Tryptamine 2AC		2440	244 $_{12}$	143 $_{100}$	130 $_{84}$	103 $_{7}$	77 $_{11}$	571	2906
Tryptamine AC		2390	202 $_{16}$	143 $_{100}$	130 $_{95}$	103 $_{10}$	77 $_{13}$	404	2905
Tryptophan 2TMS		2045	348 $_{7}$	333 $_{3}$	231 $_{8}$	202 $_{100}$	130 $_{13}$	1200	9441
Tryptophan 3TMS		1770	405 $_{3}$	377 $_{2}$	291 $_{22}$	218 $_{11}$	202 $_{100}$	1509	9440
Tryptophan ME2AC		2170	302 $_{8}$	243 $_{25}$	201 $_{32}$	130 $_{100}$		919	1009
Tryptophan MEAC		2150	260 $_{13}$	201 $_{33}$	130 $_{100}$			657	1008
Tryptophan-M (HO-skatole)	U	1370	147 $_{75}$	146 $_{100}$	117 $_{15}$			285	819
Tryptophan-M (hydroxy indole acetic acid) ME		----	205 $_{32}$	146 $_{100}$				412	1010
Tryptophan-M (indole acetic acid) ME	UME	1900	189 $_{25}$	130 $_{100}$	103 $_{5}$	77 $_{8}$		367	1011
Tryptophan-M (indole formic acid) 2ME	UME	1760	189 $_{62}$	158 $_{100}$	130 $_{14}$	103 $_{10}$	77 $_{15}$	367	4944
Tryptophan-M (indole formic acid) ME	UME	1940	175 $_{47}$	144 $_{100}$	116 $_{23}$	89 $_{19}$		330	1012
Tryptophan-M (indole lactic acid) ME		----	219 $_{16}$	130 $_{100}$				463	1013
Tryptophan-M (indole pyruvic acid) 2ME		----	231 $_{100}$	216 $_{30}$	188 $_{79}$	129 $_{48}$		513	1014
Tryptophan-M (tryptamine)		1730	160 $_{19}$	130 $_{100}$	103 $_{14}$	77 $_{19}$		303	1007
Tryptophan-M (tryptamine) 2AC		2440	244 $_{12}$	143 $_{100}$	130 $_{84}$	103 $_{7}$	77 $_{11}$	571	2906
Tryptophan-M (tryptamine) AC		2390	202 $_{16}$	143 $_{100}$	130 $_{95}$	103 $_{10}$	77 $_{13}$	404	2905
Tyramine		1745	137 $_{26}$	108 $_{100}$	77 $_{23}$			277	1485
Tyramine 2AC	UHYAC	1950	221 $_{2}$	162 $_{26}$	120 $_{100}$	107 $_{30}$	77 $_{9}$	474	1015

Warfarin-M (HO-) isomer-3 2ET

Table 8.1: Compounds in order of names

Name	Detected	RI	Typical ions and intensities					Page	Entry
Warfarin-M (HO-) isomer-3 2ET	UET	2870*	380_{35}	337_{100}	309_{54}	165_{18}	137_{28}	1356	4834
Warfarin-M (HO-) isomer-3 2ME	UME	2870*	352_{21}	309_{100}	295_{12}	206_{8}	91_{21}	1220	4826
Warfarin-M (HO-dihydro-) isomer-1 -H2O 2ME	UME	2780*	336_{7}	293_{100}	165_{6}	150_{41}	115_{13}	1130	4827
Warfarin-M (HO-dihydro-) isomer-2 -H2O 2ME	UME	2805*	336_{2}	293_{100}	173_{10}	145_{12}	121_{13}	1130	4828
Warfarin-M (HO-dihydro-) isomer-3 -H2O 2ME	UME	2830*	336_{3}	293_{100}	217_{3}	151_{17}	115_{10}	1130	4829
WIN 48.098		3370	378_{2}	278_{1}	135_{8}	100_{100}	70_{4}	1348	8525
WIN 55.212-2		3760*	426_{1}	155_{3}	127_{10}	100_{100}	56_{17}	1523	8535
Xanthinol 2AC		2870	335_{2}	322_{11}	156_{28}	130_{100}	87_{80}	1426	2724
Xipamide 2ME	UME	3350	382_{57}	262_{100}	168_{12}	120_{9}	91_{8}	1364	3082
Xipamide 4ME	UME	2780	410_{34}	289_{100}	276_{67}	168_{19}	134_{23}	1477	3085
Xipamide isomer-1 3ME	UME	2800	396_{100}	365_{48}	276_{58}	121_{27}	77_{24}	1428	3083
Xipamide isomer-2 3ME	UME	3320	396_{67}	276_{100}	233_{12}	168_{24}	77_{10}	1428	3084
Xipamide -SO2NH	P U+UHYAC UME	2385	275_{16}	155_{37}	121_{100}			743	3088
Xipamide -SO2NH 2ME	UME	2115	303_{89}	272_{60}	183_{100}	118_{30}	77_{40}	922	3089
Xipamide -SO2NH ME	PME ume	2480	289_{20}	169_{100}	126_{12}	111_{7}	77_{8}	833	3086
Xipamide-M (HO-) 4ME	UME	3000	426_{100}	395_{36}	275_{26}	151_{22}	134_{18}	1522	3087
Xipamide-M (HO-) -SO2NH 2ME	ume	2550	319_{11}	256_{1}	169_{100}	150_{7}	126_{9}	1023	3419
XLR-11		2580	329_{36}	314_{64}	247_{88}	232_{100}	144_{42}	1090	9591
XLR-11 2-methyl analog		2415	343_{14}	328_{35}	246_{100}	158_{27}	144_{16}	1176	9625
XLR-12		2310	351_{21}	336_{40}	269_{76}	254_{100}	144_{27}	1215	9593
XLR-12 TMS		2215	423_{14}	408_{30}	380_{59}	269_{29}	254_{100}	1516	9594
Xylazine		1970	220_{56}	205_{100}	177_{20}	145_{25}	130_{25}	469	5423
Xylazine AC		2150	262_{4}	247_{2}	220_{30}	205_{100}	77_{20}	668	5424
Xylazine ME		1960	234_{2}	220_{58}	205_{100}	177_{22}	145_{24}	528	8760
Xylazine TMS		1830	292_{43}	277_{100}	203_{22}	173_{24}	73_{56}	856	8761
Xylazine-M	U+UHYAC	1765	221_{14}	179_{100}	164_{6}	146_{23}	91_{12}	472	8753
Xylazine-M (dimethylhydroxyaniline) 2AC	U+UHYAC	1885	221_{9}	179_{39}	137_{100}			473	1064
Xylazine-M (HO-xylyl-) AC	U+UHYAC	2380	278_{55}	263_{57}	236_{70}	221_{100}	130_{15}	767	8756
Xylazine-M (HO-xylyl-oxo-) 2AC	U+UHYAC	2580	334_{12}	292_{18}	250_{100}	204_{14}	162_{99}	1118	8759
Xylazine-M (HO-xylyl-oxo-) isomer-1 AC	U+UHYAC	2480	292_{32}	250_{85}	217_{11}	162_{100}	55_{27}	855	8757
Xylazine-M (HO-xylyl-oxo-) isomer-2 AC	U+UHYAC	2525	292_{19}	250_{73}	217_{1}	162_{100}	55_{30}	855	8758
Xylazine-M (oxo-)	U+UHYAC	2080	234_{43}	219_{3}	146_{100}	131_{11}	77_{10}	526	8754
Xylazine-M (oxo-) AC	U+UHYAC	2170	276_{14}	219_{34}	165_{23}	146_{100}	55_{61}	753	8755
o-Xylene		<1000*	106_{40}	91_{100}	77_{12}	65_{7}	51_{12}	260	2967
pXylene		<1000*	106_{25}	91_{100}	77_{8}	65_{9}	51_{11}	259	2965
Xylitol 5AC		1950*	289_{4}	217_{27}	145_{84}	115_{100}	103_{71}	1273	5606
Xylometazoline		2020	244_{100}	229_{90}	214_{22}	119_{58}	91_{51}	573	1525
Xylometazoline AC		2260	286_{100}	271_{66}	229_{40}	214_{39}	128_{50}	816	1521
Xylose 4AC	U+UHYAC	1745*	259_{1}	170_{61}	157_{31}	128_{100}	115_{60}	1018	1967
Xylose 4HFB		1235*	478_{12}	465_{7}	293_{62}	265_{13}	169_{100}	1734	5811
Xylose 4PFP		1230*	378_{17}	365_{9}	147_{15}	119_{100}	69_{92}	1730	5810
Xylose 4TFA		1315*	311_{3}	278_{14}	265_{10}	197_{13}	69_{100}	1687	5809
Yohimbine		3140	354_{72}	353_{100}	184_{20}	169_{20}	156_{11}	1233	3995
Yohimbine AC		3190	396_{81}	395_{100}	353_{57}	277_{7}	169_{38}	1431	4018
Zaleplone		2960	305_{41}	277_{5}	263_{38}	248_{100}	119_{8}	936	5859
Zaleplone-M/artifact (deacetyl-)		2850	263_{48}	248_{100}	231_{4}	221_{5}	130_{4}	674	5860
Zidovudine 2TMS		2390	411_{8}	216_{23}	203_{90}	147_{87}	73_{100}	1483	6212
Zidovudine AC		2540	309_{1}	184_{11}	126_{38}	96_{58}	81_{100}	962	7946
Zidovudine artifact		2250	206_{1}	193_{19}	150_{20}	126_{66}	69_{100}	487	7889
Zidovudine artifact (thymine)		1815	126_{100}	97_{3}	83_{14}	70_{6}	55_{79}	269	7890
Zidovudine artifact (thymine) 2ME		1450	154_{100}	126_{2}	96_{10}	68_{72}	56_{13}	296	9427
Zidovudine artifact (thymine) 2TMS		1380	270_{59}	255_{100}	166_{12}	147_{27}	73_{59}	719	7897
Zidovudine artifact (thymine) ME		1550	140_{100}	110_{4}	83_{18}	70_{3}	55_{74}	279	7892
Zidovudine artifact AC		2265	206_{2}	193_{6}	140_{17}	126_{12}	81_{100}	693	7891
Zidovudine HFB		2240	463_{1}	338_{4}	310_{6}	126_{100}	81_{73}	1605	8245
Zidovudine PFP		2220	413_{1}	288_{5}	260_{9}	126_{100}	95_{44}	1486	8244
Zidovudine TFA		2230	363_{1}	238_{7}	210_{15}	126_{100}	95_{52}	1277	8243
Zidovudine TMS		2280	339_{42}	255_{18}	129_{37}	117_{70}	73_{100}	1149	6211
Zimelidine		2270	316_{54}	238_{24}	193_{62}	70_{74}	58_{100}	1004	1475
Zinophos		1600	248_{42}	192_{37}	143_{59}	107_{87}	97_{100}	589	3877
Zofenopril artifact (benzoic acid)	P U UHY	1235*	122_{77}	105_{100}	77_{72}			266	95
Zofenopril artifact (benzoic acid) ME	P(ME)	1180*	136_{30}	105_{100}	77_{73}			275	1211
Zofenopril artifact (debenzoyl-) 2ME		2655	353_{23}	306_{16}	244_{64}	178_{100}	89_{90}	1225	8374
Zofenopril artifact (debenzoyl-) ME		2595	339_{4}	280_{15}	230_{100}	178_{83}	68_{77}	1148	8373
Zofenopril artifact (debenzoyl-) MEAC		2745	381_{5}	338_{14}	230_{26}	178_{85}	68_{100}	1360	8372
Zofenopril ME		3400	443_{1}	338_{11}	334_{6}	238_{5}	105_{100}	1568	8371
Zofenoprilate artifact (debenzoyl-) 2ME		2655	353_{23}	306_{16}	244_{64}	178_{100}	89_{90}	1225	8374
Zofenoprilate artifact (debenzoyl-) ME		2595	339_{4}	280_{15}	230_{100}	178_{83}	68_{77}	1148	8373
Zofenoprilate artifact (debenzoyl-) MEAC		2745	381_{5}	338_{14}	230_{26}	178_{85}	68_{100}	1360	8372
Zofenoprilate-M/artifact (-CCOH) ME		3400	443_{1}	338_{11}	334_{6}	238_{5}	105_{100}	1568	8371
Zolazepam		2400	286_{62}	285_{100}	267_{76}	257_{97}	145_{12}	815	7448
Zolmitriptan		2850	287_{2}	156_{2}	143_{4}	115_{2}	58_{100}	822	7508
Zolmitriptan 2HFB		2745	678_{1}	621_{2}	142_{18}	58_{100}		1727	8134
Zolmitriptan 2PFP		2770	156_{2}	143_{5}	115_{2}	58_{100}		1707	8387

Table 8.1: Compounds in order of names

Name	Detected	RI	Typical ions and intensities						Page	Entry
Zolmitriptan 2TMS		2745	431 $_1$	373 $_2$	309 $_6$	73 $_{24}$	58 $_{100}$		1539	8384
Zolmitriptan AC		2755	329 $_3$	156 $_3$	143 $_8$	115 $_2$	58 $_{100}$		1089	7509
Zolmitriptan HFB		2940	482 $_1$	425 $_2$	142 $_{10}$	86 $_8$	58 $_{100}$		1638	8135
Zolmitriptan PFP		2730	156 $_3$	143 $_4$	100 $_2$	86 $_2$	58 $_{100}$		1542	8386
Zolmitriptan TFA		2610	142 $_2$	99 $_4$	85 $_{13}$	71 $_{22}$	58 $_{100}$		1370	8385
Zolpidem	P G U+UHYAC	2715	307 $_{14}$	235 $_{100}$	219 $_9$	92 $_9$	65 $_{10}$		951	5280
Zolpidem-M (4'-HO-) -C2H6N MEAC	U+UHYAC	2670	352 $_{28}$	293 $_{100}$	235 $_{52}$	201 $_{35}$	187 $_{26}$		1221	5281
Zolpidem-M (4'-HOOC-) ME	U+UHYAC	2905	351 $_7$	279 $_{100}$	251 $_{10}$	219 $_8$	72 $_{11}$		1214	5733
Zolpidem-M (6-HO-) -C2H6N MEAC	U+UHYAC	2720	352 $_{36}$	293 $_{100}$	233 $_{52}$	207 $_{25}$	92 $_{14}$		1221	5282
Zolpidem-M (6-HOOC-) ME	U+UHYAC	2950	351 $_{11}$	279 $_{100}$	269 $_{28}$	219 $_{32}$	72 $_{11}$		1213	5734
Zolpidem-M (HO-) isomer-1 AC	U+UHYAC	3095	365 $_{13}$	293 $_{100}$	234 $_8$	219 $_5$	72 $_8$		1287	5107
Zolpidem-M (HO-) isomer-2 AC	U+UHYAC	3150	365 $_{18}$	293 $_{100}$	233 $_{56}$	219 $_{13}$	72 $_{10}$		1287	5108
Zomepirac -CO2	P-I G U UHY	2040	247 $_{75}$	246 $_{100}$	211 $_{21}$	136 $_{16}$			584	1034
Zomepirac ME	PME-I UME	1835	305 $_{32}$	246 $_{100}$	139 $_{40}$	111 $_{38}$			935	1035
Zonisamide	P U+UHYAC	1950	212 $_{36}$	132 $_{79}$	119 $_{20}$	104 $_{29}$	77 $_{100}$		438	7720
Zonisamide 2TMS		1965	356 $_1$	341 $_{90}$	269 $_{42}$	206 $_{29}$	132 $_{100}$		1241	7724
Zonisamide AC	U+UHYAC	2100	254 $_6$	212 $_{100}$	195 $_{22}$	132 $_{48}$	77 $_{80}$		626	7723
Zonisamide ME	P	1930	226 $_{14}$	133 $_{100}$	119 $_3$	104 $_{12}$	77 $_{28}$		493	7721
Zonisamide MEAC		1980	268 $_{21}$	162 $_9$	132 $_{53}$	77 $_{100}$	56 $_{43}$		705	7722
Zopiclone	P-I G U+UHYAC	2950	245 $_{69}$	217 $_{28}$	143 $_{100}$	112 $_{37}$	99 $_{44}$		1394	5314
Zopiclone-M (amino-chloro-pyridine)	U+UHYAC	1200	128 $_{100}$	101 $_{70}$	93 $_{15}$	73 $_{34}$			270	5315
Zopiclone-M (amino-chloro-pyridine) AC	U+UHYAC	1505	170 $_{22}$	128 $_{100}$	101 $_{34}$	93 $_3$	73 $_7$		323	5316
Zopiclone-M (HO-amino-chloro-pyridine) 2AC	U+UHYAC	1720	228 $_{12}$	186 $_{56}$	144 $_{100}$	111 $_9$	81 $_4$		499	6556
Zopiclone-M (HO-amino-chloro-pyridine) AC	U+UHYAC	1680	186 $_{20}$	144 $_{100}$	116 $_{16}$	109 $_{13}$	81 $_{13}$		360	6557
Zopiclone-M (piperazine) 2AC		1750	170 $_{15}$	85 $_{33}$	69 $_{25}$	56 $_{100}$			324	879
Zopiclone-M (piperazine) 2HFB		1290	478 $_3$	459 $_9$	309 $_{100}$	281 $_{22}$	252 $_{41}$		1631	6634
Zopiclone-M (piperazine) 2TFA		1005	278 $_{10}$	209 $_{59}$	152 $_{25}$	69 $_{56}$	56 $_{100}$		766	4129
Zopiclone-M/artifact	P-I U+UHYAC	2060	246 $_{83}$	217 $_{78}$	191 $_{100}$	139 $_{34}$	113 $_{70}$		578	7801
Zopiclone-M/artifact (alcohol) AC	U+UHYAC	2390	304 $_{16}$	261 $_{100}$	217 $_{24}$	155 $_{43}$	112 $_{31}$		929	5317
Zopiclone-M/artifact (alcohol) ME	U+UHYAC	2080	276 $_{12}$	261 $_{65}$	246 $_{100}$	217 $_{54}$	191 $_{23}$		752	5318
Zotepine	P G U	2660	331 $_2$	299 $_3$	199 $_4$	72 $_{30}$	58 $_{100}$		1099	4291
Zotepine artifact (desulfo-) HYAC	U+UHYAC	2395*	270 $_{13}$	228 $_{100}$	199 $_{20}$	165 $_{49}$	115 $_6$		715	6416
Zotepine HY	UHY	2310*	260 $_{100}$	231 $_{29}$	227 $_{56}$	199 $_{30}$	152 $_{22}$		654	4292
Zotepine HYAC	U+UHYAC	2440*	302 $_{19}$	260 $_{100}$	231 $_{12}$	199 $_{44}$	152 $_{18}$		917	4293
Zotepine-M (bis-nor-) HY	UHY	2310*	260 $_{100}$	231 $_{29}$	227 $_{56}$	199 $_{30}$	152 $_{22}$		654	4292
Zotepine-M (bis-nor-) HYAC	U+UHYAC	2440*	302 $_{19}$	260 $_{100}$	231 $_{12}$	199 $_{44}$	152 $_{18}$		917	4293
Zotepine-M (bis-nor-HO-) HY2AC	U+UHYAC	2735*	360 $_{24}$	318 $_{39}$	276 $_{100}$	243 $_{34}$	215 $_{15}$		1262	4294
Zotepine-M (bis-nor-HO-) HYAC	UHYAC	2555*	318 $_{34}$	276 $_{100}$	259 $_{22}$	247 $_{42}$	184 $_{23}$		1016	6278
Zotepine-M (bis-nor-HO-) isomer-1 HY	UHY	2460*	276 $_{21}$	231 $_{41}$	228 $_{100}$	199 $_{14}$	165 $_{93}$		751	4296
Zotepine-M (bis-nor-HO-) isomer-2 HY	UHY	2650*	276 $_{100}$	247 $_{37}$	243 $_{28}$	213 $_{10}$	184 $_{26}$		751	4297
Zotepine-M (bis-nor-HO-methoxy-) HY	UHY	2700*	306 $_{100}$	276 $_{39}$	264 $_{52}$	247 $_{23}$	171 $_{24}$		941	4298
Zotepine-M (bis-nor-HO-methoxy-) HY2AC	U+UHYAC	2915*	390 $_{30}$	348 $_{24}$	306 $_{100}$	273 $_{79}$	245 $_9$		1401	4295
Zotepine-M (HO-) AC	U+UHYAC	2960	341 $_2$	72 $_{56}$	58 $_{100}$				1397	4299
Zotepine-M (HO-) HY2AC	U+UHYAC	2735*	360 $_{24}$	318 $_{39}$	276 $_{100}$	243 $_{34}$	215 $_{15}$		1262	4294
Zotepine-M (HO-) HYAC	UHYAC	2555*	318 $_{34}$	276 $_{100}$	259 $_{22}$	247 $_{42}$	184 $_{23}$		1016	6278
Zotepine-M (HO-) isomer-1 HY	UHY	2460*	276 $_{21}$	231 $_{41}$	228 $_{100}$	199 $_{14}$	165 $_{93}$		751	4296
Zotepine-M (HO-) isomer-2 HY	UHY	2650*	276 $_{100}$	247 $_{37}$	243 $_{28}$	213 $_{10}$	184 $_{26}$		751	4297
Zotepine-M (HO-methoxy-) HY	UHY	2700*	306 $_{100}$	276 $_{39}$	264 $_{52}$	247 $_{23}$	171 $_{24}$		941	4298
Zotepine-M (HO-methoxy-) HY2AC	U+UHYAC	2915*	390 $_{30}$	348 $_{24}$	306 $_{100}$	273 $_{79}$	245 $_9$		1401	4295
Zotepine-M (nor-) HY	UHY	2310*	260 $_{100}$	231 $_{29}$	227 $_{56}$	199 $_{30}$	152 $_{22}$		654	4292
Zotepine-M (nor-) HYAC	U+UHYAC	2440*	302 $_{19}$	260 $_{100}$	231 $_{12}$	199 $_{44}$	152 $_{18}$		917	4293
Zotepine-M (nor-HO-) HY2AC	U+UHYAC	2735*	360 $_{24}$	318 $_{39}$	276 $_{100}$	243 $_{34}$	215 $_{15}$		1262	4294
Zotepine-M (nor-HO-) HYAC	UHYAC	2555*	318 $_{34}$	276 $_{100}$	259 $_{22}$	247 $_{42}$	184 $_{23}$		1016	6278
Zotepine-M (nor-HO-) isomer-1 HY	UHY	2460*	276 $_{21}$	231 $_{41}$	228 $_{100}$	199 $_{14}$	165 $_{93}$		751	4296
Zotepine-M (nor-HO-) isomer-2 HY	UHY	2650*	276 $_{100}$	247 $_{37}$	243 $_{28}$	213 $_{10}$	184 $_{26}$		751	4297
Zotepine-M (nor-HO-methoxy-) HY	UHY	2700*	306 $_{100}$	276 $_{39}$	264 $_{52}$	247 $_{23}$	171 $_{24}$		941	4298
Zotepine-M (nor-HO-methoxy-) HY2AC	U+UHYAC	2915*	390 $_{30}$	348 $_{24}$	306 $_{100}$	273 $_{79}$	245 $_9$		1401	4295
Zuclopenthixol	G U	3360	400 $_1$	221 $_{12}$	143 $_{100}$	100 $_{18}$	70 $_{24}$		1445	462
Zuclopenthixol AC	U+UHYAC	3460	442 $_{11}$	221 $_9$	185 $_{100}$	98 $_{24}$	70 $_{11}$		1564	319
Zuclopenthixol TMS		3490	472 $_1$	457 $_6$	221 $_{19}$	215 $_{100}$	98 $_{23}$		1622	4534
Zuclopenthixol-M (dealkyl-) AC	U+UHYAC	3490	398 $_2$	268 $_7$	141 $_{100}$	99 $_{30}$			1437	1261
Zuclopenthixol-M (dealkyl-dihydro-) AC	U+UHYAC	3450	400 $_{46}$	231 $_{44}$	141 $_{100}$	128 $_{16}$	99 $_{25}$		1445	1260
Zuclopenthixol-M (N-oxide) -C6H14N2O2	P-I U+UHYAC	2410*	270 $_{40}$	255 $_{21}$	234 $_{100}$	202 $_{23}$	117 $_{27}$		714	438
Zuclopenthixol-M (N-oxide-sulfoxide) -C6H14N2O2	P-I U UGLUC UGLUCAC	2560*	286 $_{21}$	251 $_{20}$	234 $_{57}$	203 $_{100}$	101 $_{10}$		812	436
Zuclopenthixol-M / artifact (Cl-thioxanthenone)	U	2260*	246 $_{100}$	218 $_{46}$	183 $_9$	139 $_{25}$	91 $_{10}$		578	2641

9 Mass Spectra

9.1 Arrangement of spectra

The 10,430 different mass spectra are arranged in ascending molecular mass or, if not known, the pseudo molecular mass. For each nominal mass value the spectra are arranged in order of name.

9.2 Lay-out of spectra

For easier visualization of the data, the mass spectra are presented as bar graphs, in which the abscissa represents the mass to charge ratio (m/z) in atomic mass units (u), and the ordinate indicates the relative abundances of the ion currents of the various fragment ions. Predominant ions are labeled with their m/z value.

Some spectra contain molecular ions with a relative intensity of less than 1 %. In these cases, the M^+ is labeled, although it cannot be seen in the spectrum. In our experience, the detection of this low-intensity M^+ can be necessary for the identification of the compound, when the other fragment ions are not typical.

Fig. 9.1 explains the information provided with each spectrum, and the abbreviations used are listed in Table 6.1.

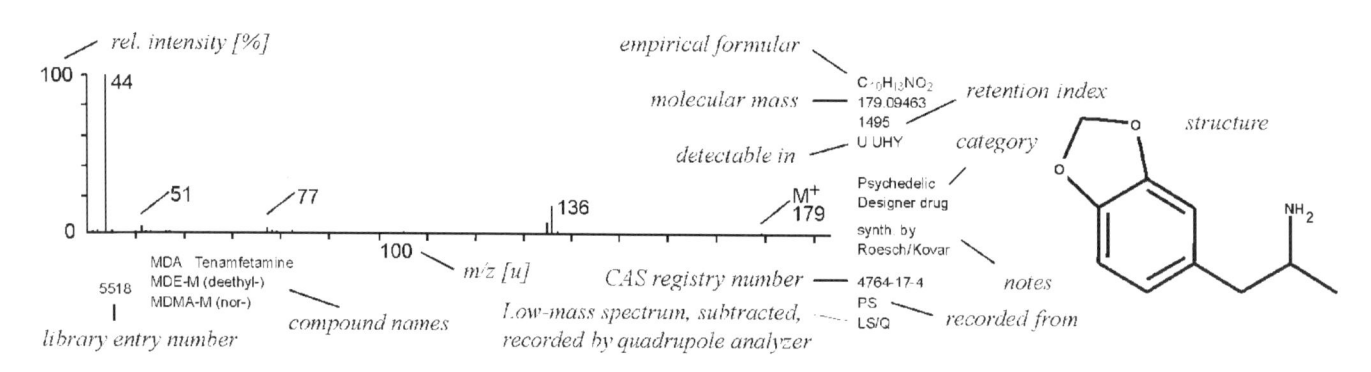

Fig. 9.1: Sample spectrum with explanations

Library entry number:
Entry number of the spectrum in the electronic versions.

Compound names:
The international non-proprietary names for drugs (INN), the common names for pesticides and the chemical names for chemicals are used. If necessary, a synonym index, [156-159] should be consulted. If the compound is a common metabolite or derivative of several parent compounds, then all known parent compounds are given.

Structure:
The formulas are redrawn in the molefile format allowing their use in electronic databases. They are zoomed to fit the available space. Formulas of metabolites or artifacts are those of their probable structures (cf. Section 3). If the position of a substituent is unknown, the substituent is fixed with a tilde. Unknown substituents are named 'R'.

Empirical formula:
The empirical formulas are given to facilitate the identification of new metabolites or derivatives.

Molecular mass:
The molecular masses were calculated from the atomic masses of the most abundant isotopes as given in Table 5.1.

Retention index:
The RIs were measured by GC-MS on methyl silicone capillary columns using a temperature program (Section 2.4). The RIs of compounds with an asterisk (*) are not detectable by nitrogen-selective flame-ionization detection (N-FID).

Detected:
The compound could be detected in the given samples after the given sample preparation (abbreviations in Table 6.1). These data have been evaluated from about 80,000 clinical and forensic cases.

CAS Registry Number:
The *Registry Number* of the Chemical Abstracts Services (CAS) is given here. If only derivatives or metabolites of a compound were included in this handbook, the CAS number of the corresponding parent compound with the prefix '#' is given.

Recorded from:
Type of sample from which the spectrum was recorded (abbreviations in Table 6.1). If the spectrum was recorded from samples of biological origin, it should be taken into consideration that fragment ions from sample impurities may be present in the spectrum. With experience, it is possible to decide whether these ions can be ignored.

LM/Q or LS/Q:
This indicates whether the low-resolution mass spectrum recorded by a quadrupole mass spectrometer (LM/Q) was background-subtracted (LS/Q). Relative ion intensities can be falsified by background-subtraction. This should be taken into account when comparing the spectra. Such variations do not impair the use of the library in our experience. With some experience, it is possible to decide whether the variation is acceptable within two spectra of the same compound. If in doubt, investigators should record a reference spectrum of the suspected compound on their own GC-MS.

Categories:
The major pharmacological category is given.

Notes:
If necessary, notes have been added (abbreviations in Table 6.1).

9.3 Mass spectra (*m/z* 30 – 299)

82

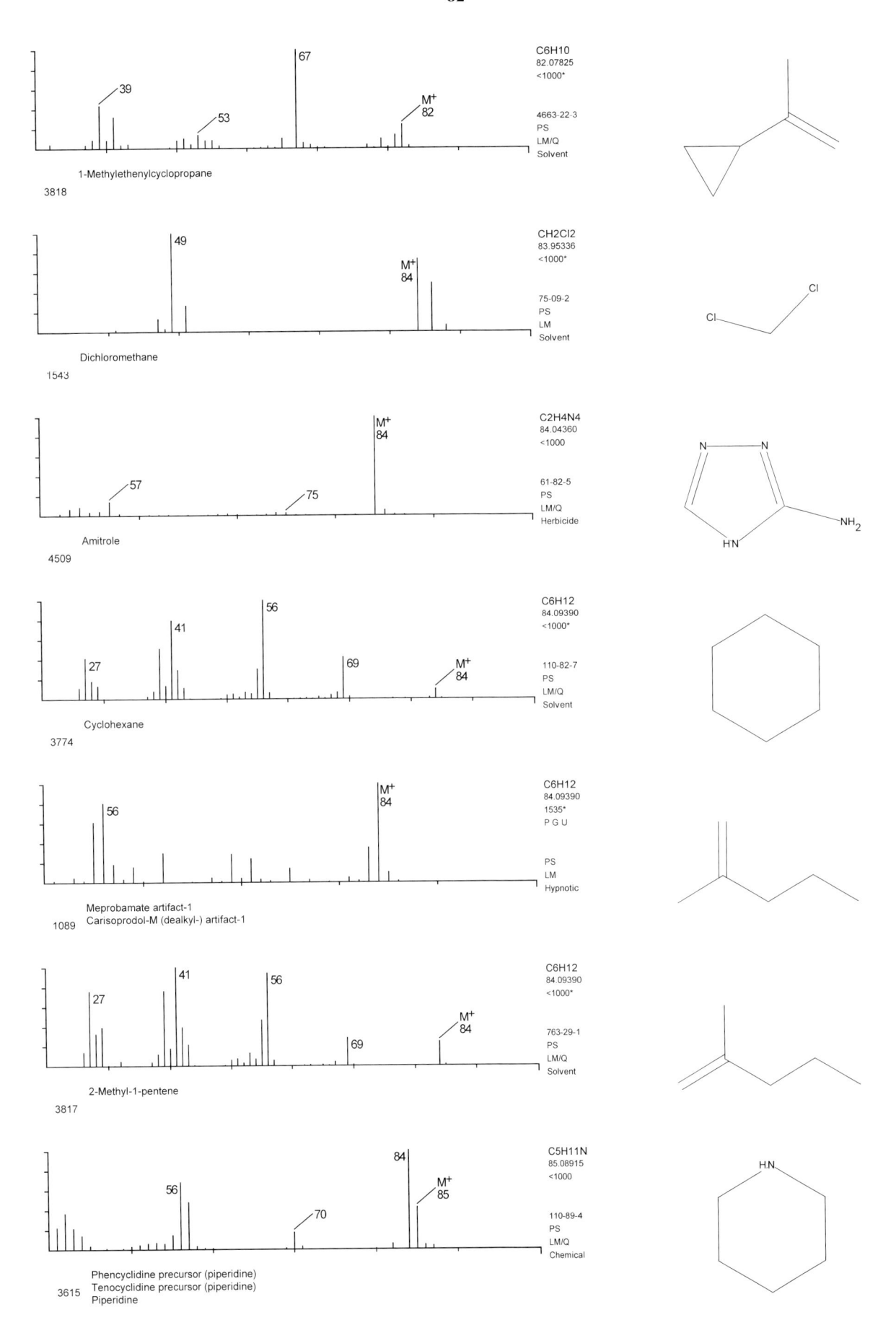

1-Methylethenylcyclopropane
Peaks: 39, 53, 67, M+ 82
C6H10
82.07825
<1000*
4663-22-3
PS
LM/Q
Solvent
3818

Dichloromethane
Peaks: 49, M+ 84
CH2Cl2
83.95336
<1000*
75-09-2
PS
LM
Solvent
1543

Amitrole
Peaks: 57, 75, M+ 84
C2H4N4
84.04360
<1000
61-82-5
PS
LM/Q
Herbicide
4509

Cyclohexane
Peaks: 27, 41, 56, 69, M+ 84
C6H12
84.09390
<1000*
110-82-7
PS
LM/Q
Solvent
3774

Meprobamate artifact-1 / Carisoprodol-M (dealkyl-) artifact-1
Peaks: 56, M+ 84
C6H12
84.09390
1535*
P G U
PS
LM
Hypnotic
1089

2-Methyl-1-pentene
Peaks: 27, 41, 56, 69, M+ 84
C6H12
84.09390
<1000*
763-29-1
PS
LM/Q
Solvent
3817

Phencyclidine precursor (piperidine) / Tenocyclidine precursor (piperidine) / Piperidine
Peaks: 56, 70, 84, M+ 85
C5H11N
85.08915
<1000
110-89-4
PS
LM/Q
Chemical
3615

86

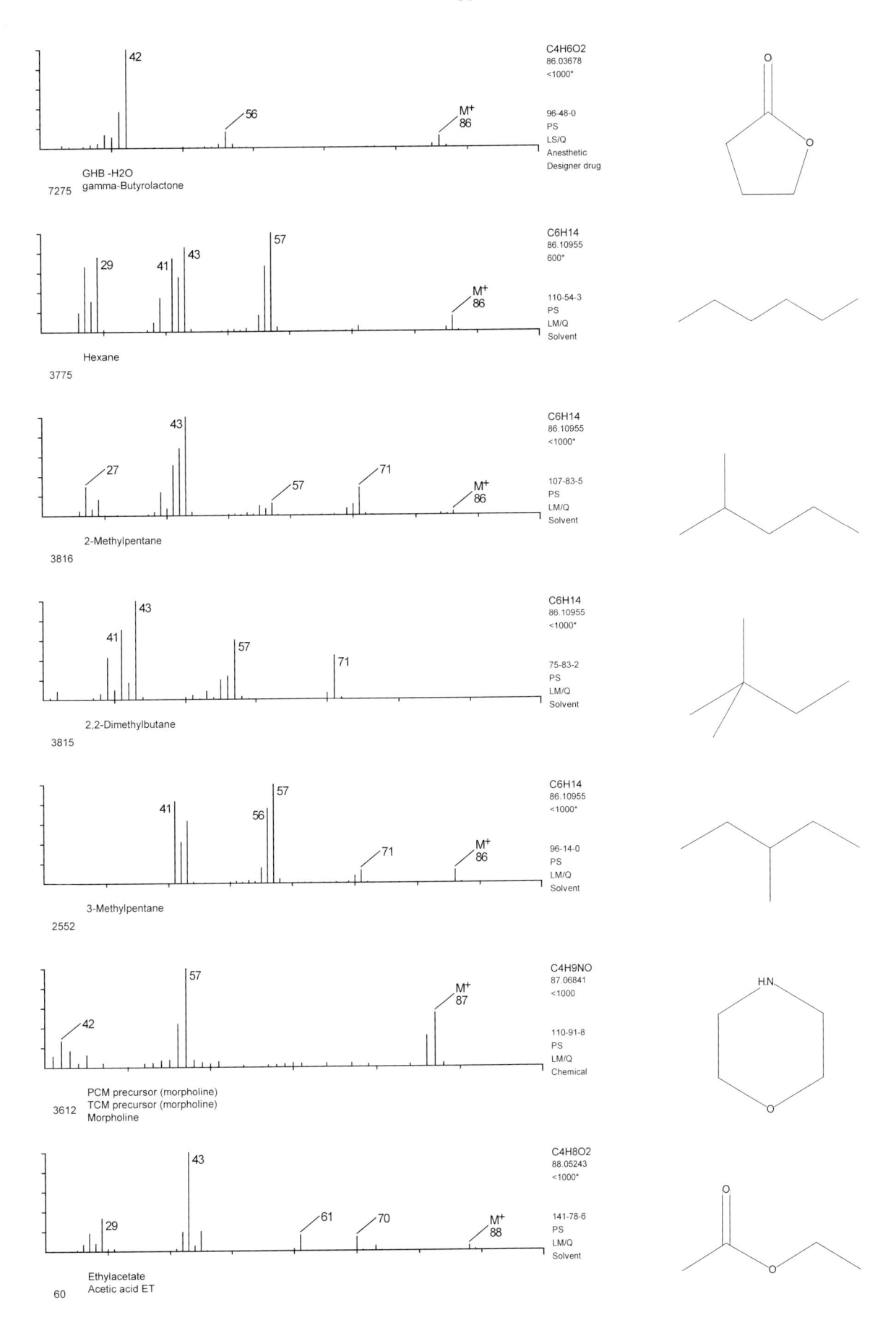

GHB -H2O gamma-Butyrolactone 7275	C4H6O2 86.03678 <1000* 96-48-0 PS LS/Q Anesthetic Designer drug
Hexane 3775	C6H14 86.10955 600* 110-54-3 PS LM/Q Solvent
2-Methylpentane 3816	C6H14 86.10955 <1000* 107-83-5 PS LM/Q Solvent
2,2-Dimethylbutane 3815	C6H14 86.10955 <1000* 75-83-2 PS LM/Q Solvent
3-Methylpentane 2552	C6H14 86.10955 <1000* 96-14-0 PS LM/Q Solvent
PCM precursor (morpholine) TCM precursor (morpholine) Morpholine 3612	C4H9NO 87.06841 <1000 110-91-8 PS LM/Q Chemical
Ethylacetate Acetic acid ET 60	C4H8O2 88.05243 <1000* 141-78-6 PS LM/Q Solvent

255

100

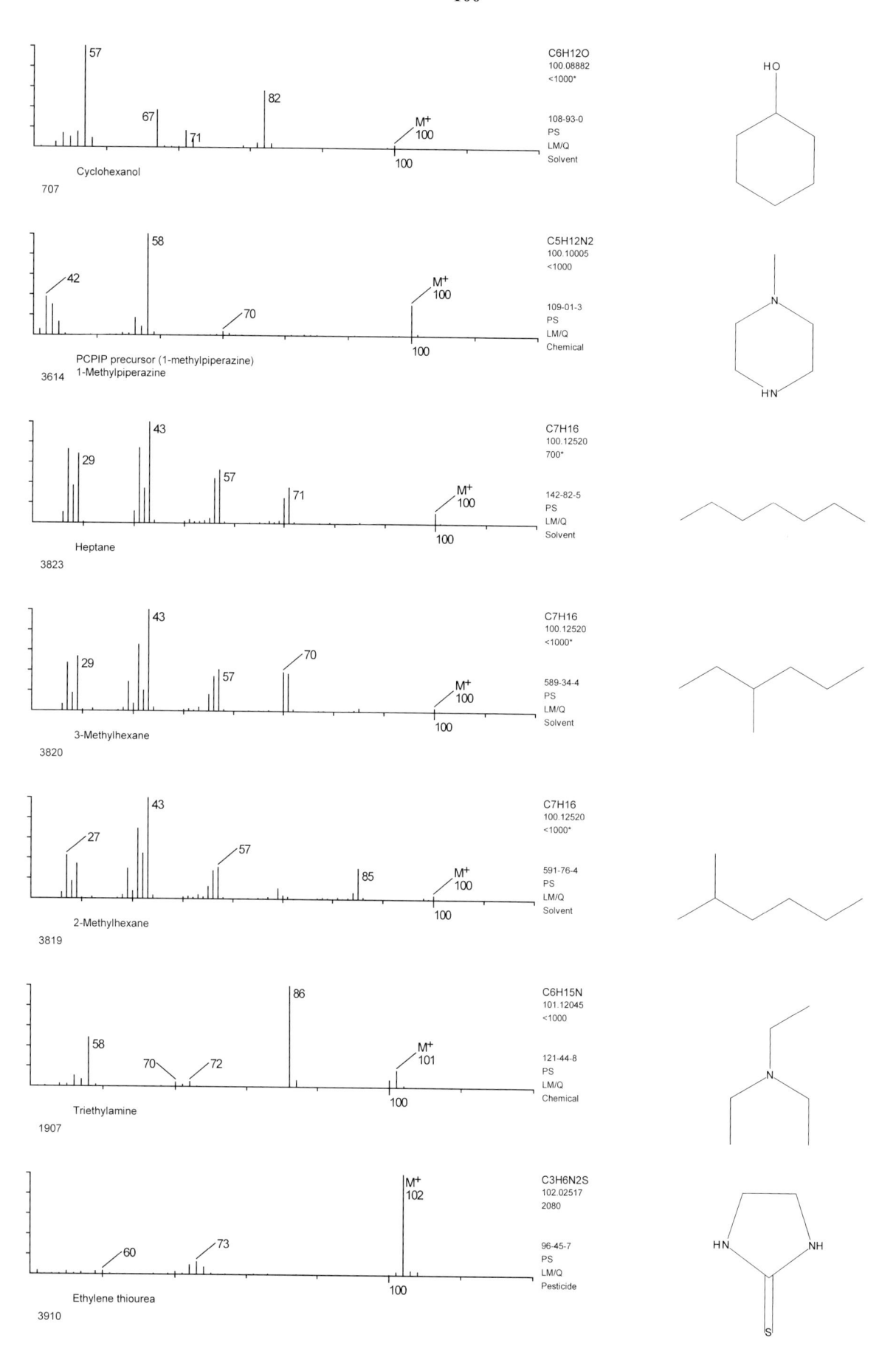

Cyclohexanol 57, 67, 71, 82, M+ 100	C6H12O 100.08882 <1000* 108-93-0 PS LM/Q Solvent	cyclohexanol structure
707		

PCPIP precursor (1-methylpiperazine) 1-Methylpiperazine 42, 58, 70, M+ 100	C5H12N2 100.10005 <1000 109-01-3 PS LM/Q Chemical	1-methylpiperazine structure
3614		

Heptane 29, 43, 57, 71, M+ 100	C7H16 100.12520 700* 142-82-5 PS LM/Q Solvent	heptane structure
3823		

3-Methylhexane 29, 43, 57, 70, M+ 100	C7H16 100.12520 <1000* 589-34-4 PS LM/Q Solvent	3-methylhexane structure
3820		

2-Methylhexane 27, 43, 57, 85, M+ 100	C7H16 100.12520 <1000* 591-76-4 PS LM/Q Solvent	2-methylhexane structure
3819		

Triethylamine 58, 70, 72, 86, M+ 101	C6H15N 101.12045 <1000 121-44-8 PS LM/Q Chemical	triethylamine structure
1907		

Ethylene thiourea 60, 73, M+ 102	C3H6N2S 102.02517 2080 96-45-7 PS LM/Q Pesticide	ethylene thiourea structure
3910		

102

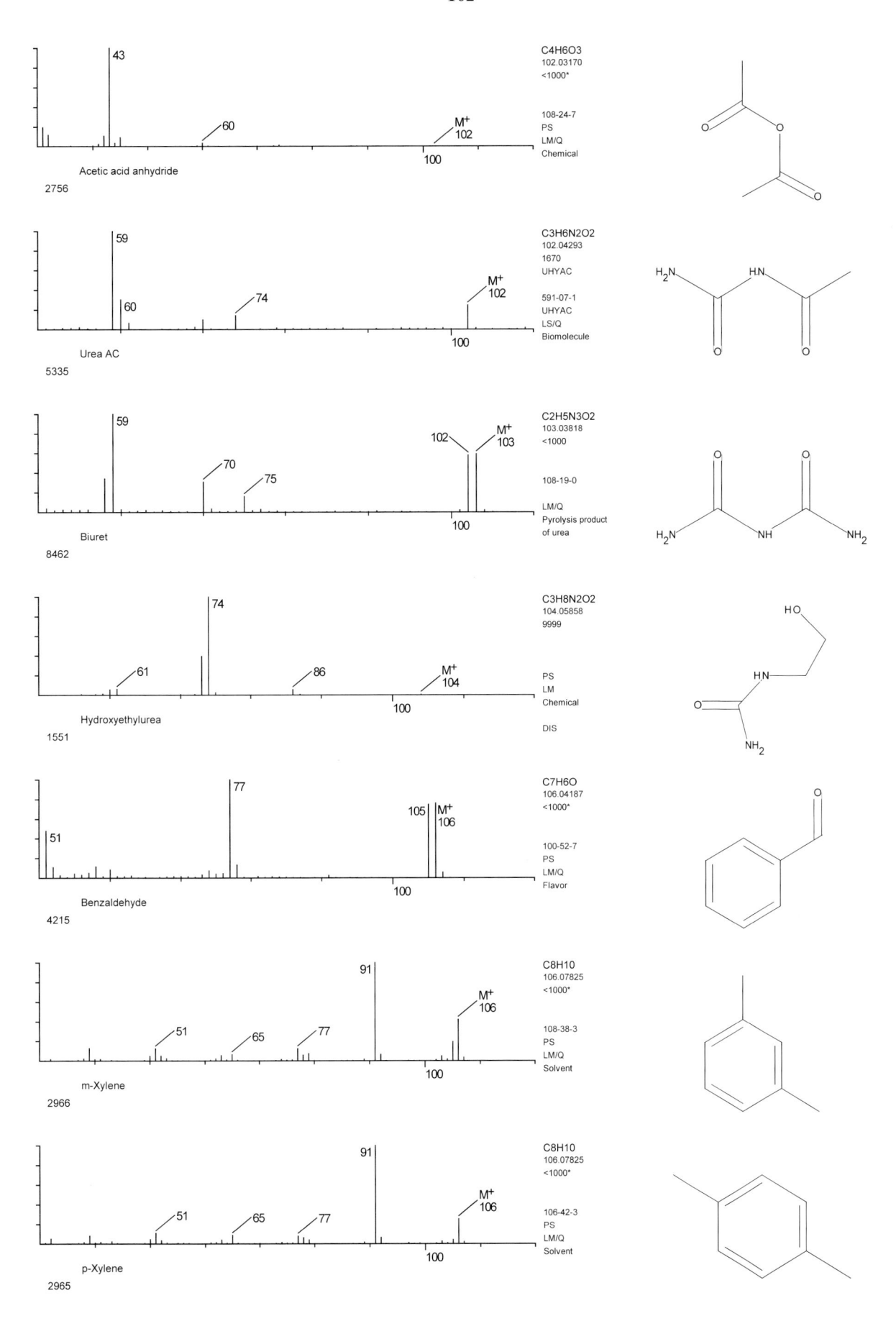

Acetic acid anhydride 2756	Peaks: 43, 60, M+ 102	C4H6O3 102.03170 <1000* 108-24-7 PS LM/Q Chemical
Urea AC 5335	Peaks: 59, 60, 74, M+ 102	C3H6N2O2 102.04293 1670 UHYAC 591-07-1 UHYAC LS/Q Biomolecule
Biuret 8462	Peaks: 59, 70, 75, 102, M+ 103	C2H5N3O2 103.03818 <1000 108-19-0 LM/Q Pyrolysis product of urea
Hydroxyethylurea 1551	Peaks: 61, 74, 86, M+ 104	C3H8N2O2 104.05858 9999 PS LM Chemical DIS
Benzaldehyde 4215	Peaks: 51, 77, 105, M+ 106	C7H6O 106.04187 <1000* 100-52-7 PS LM/Q Flavor
m-Xylene 2966	Peaks: 51, 65, 77, 91, M+ 106	C8H10 106.07825 <1000* 108-38-3 PS LM/Q Solvent
p-Xylene 2965	Peaks: 51, 65, 77, 91, M+ 106	C8H10 106.07825 <1000* 106-42-3 PS LM/Q Solvent

259

106

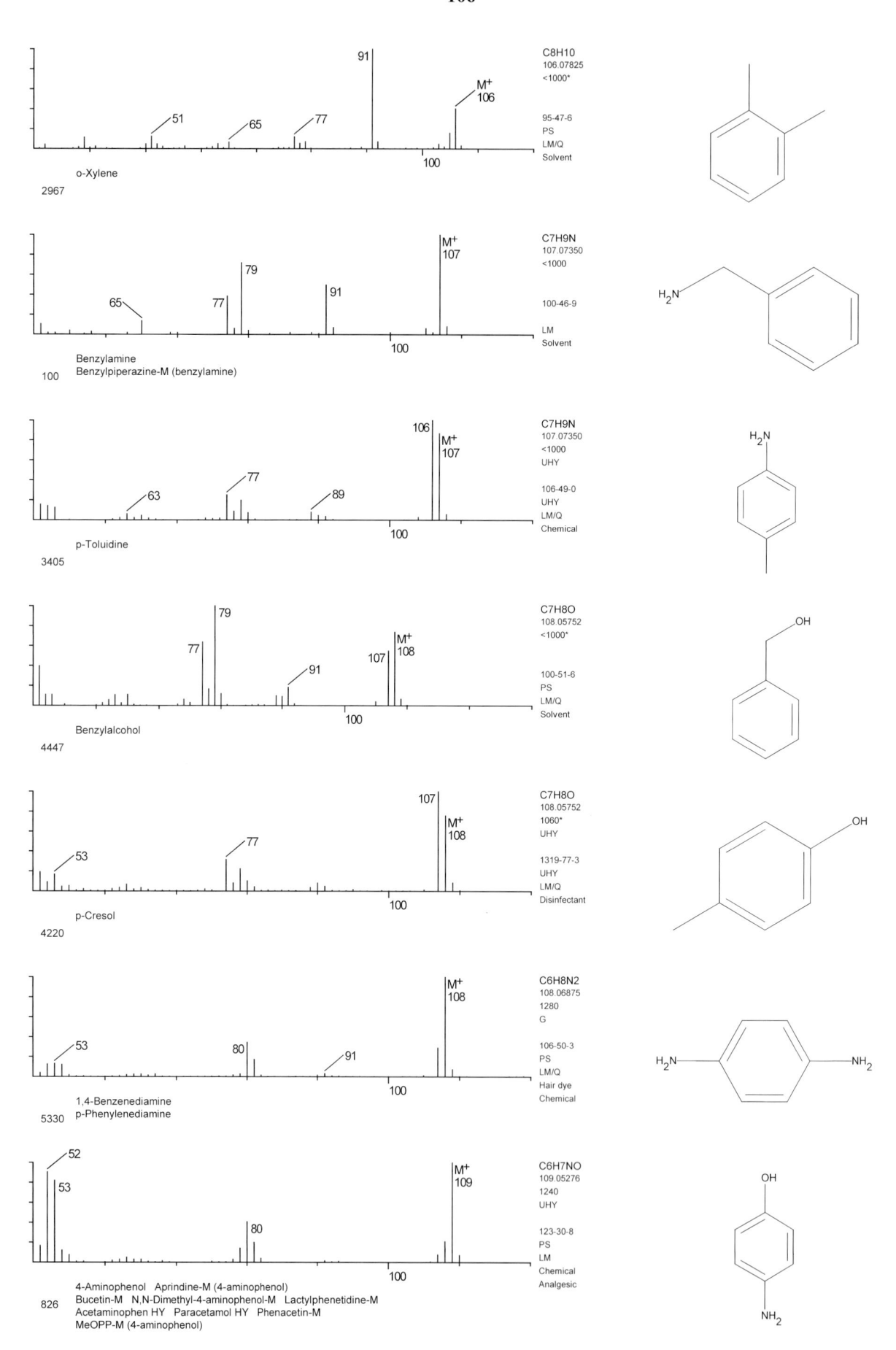

o-Xylene
2967
C8H10
106.07825
<1000*
95-47-6
PS
LM/Q
Solvent

Peaks: 51, 65, 77, 91, M+ 106

Benzylamine
Benzylpiperazine-M (benzylamine)
100
C7H9N
107.07350
<1000
100-46-9
LM
Solvent

Peaks: 65, 77, 79, 91, M+ 107

p-Toluidine
3405
C7H9N
107.07350
<1000
UHY
106-49-0
UHY
LM/Q
Chemical

Peaks: 63, 77, 89, 106, M+ 107

Benzylalcohol
4447
C7H8O
108.05752
<1000*
100-51-6
PS
LM/Q
Solvent

Peaks: 77, 79, 91, 107, M+ 108

p-Cresol
4220
C7H8O
108.05752
1060*
UHY
1319-77-3
UHY
LM/Q
Disinfectant

Peaks: 53, 77, 107, M+ 108

1,4-Benzenediamine
p-Phenylenediamine
5330
C6H8N2
108.06875
1280
G
106-50-3
PS
LM/Q
Hair dye
Chemical

Peaks: 53, 80, 91, M+ 108

4-Aminophenol   Aprindine-M (4-aminophenol)
Bucetin-M   N,N-Dimethyl-4-aminophenol-M   Lactylphenetidine-M
Acetaminophen HY   Paracetamol HY   Phenacetin-M
MeOPP-M (4-aminophenol)
826
C6H7NO
109.05276
1240
UHY
123-30-8
PS
LM
Chemical
Analgesic

Peaks: 52, 53, 80, M+ 109

109

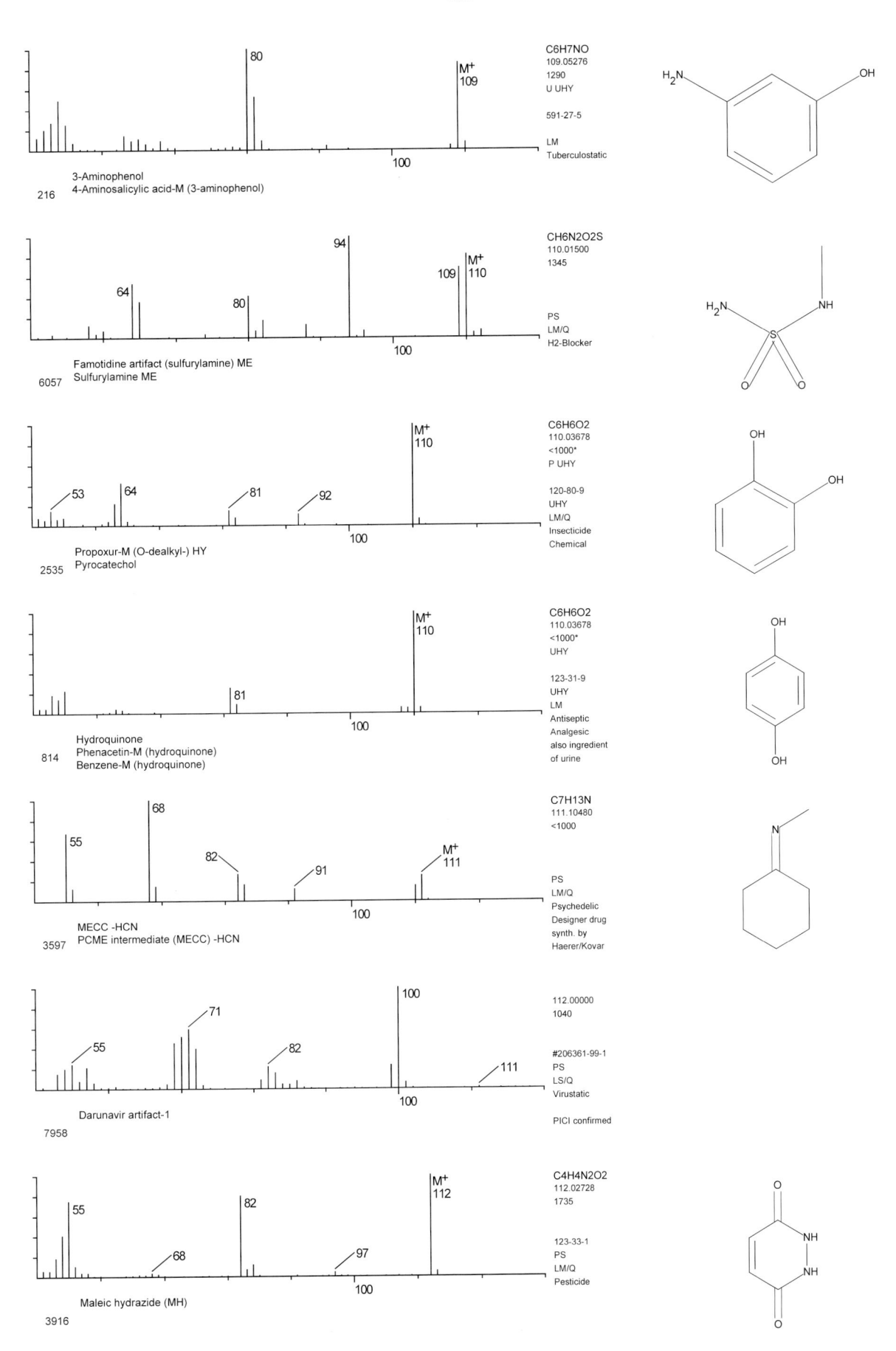

3-Aminophenol
4-Aminosalicylic acid-M (3-aminophenol)
216

C6H7NO
109.05276
1290
U UHY

591-27-5

LM
Tuberculostatic

Famotidine artifact (sulfurylamine) ME
Sulfurylamine ME
6057

CH6N2O2S
110.01500
1345

PS
LM/Q
H2-Blocker

Propoxur-M (O-dealkyl-) HY
Pyrocatechol
2535

C6H6O2
110.03678
<1000*
P UHY

120-80-9
UHY
LM/Q
Insecticide
Chemical

Hydroquinone
Phenacetin-M (hydroquinone)
Benzene-M (hydroquinone)
814

C6H6O2
110.03678
<1000*
UHY

123-31-9
UHY
LM
Antiseptic
Analgesic
also ingredient
of urine

MECC -HCN
PCME intermediate (MECC) -HCN
3597

C7H13N
111.10480
<1000

PS
LM/Q
Psychedelic
Designer drug
synth. by
Haerer/Kovar

Darunavir artifact-1
7958

112.00000
1040

#206361-99-1
PS
LS/Q
Virustatic

PICI confirmed

Maleic hydrazide (MH)
3916

C4H4N2O2
112.02728
1735

123-33-1
PS
LM/Q
Pesticide

261

112

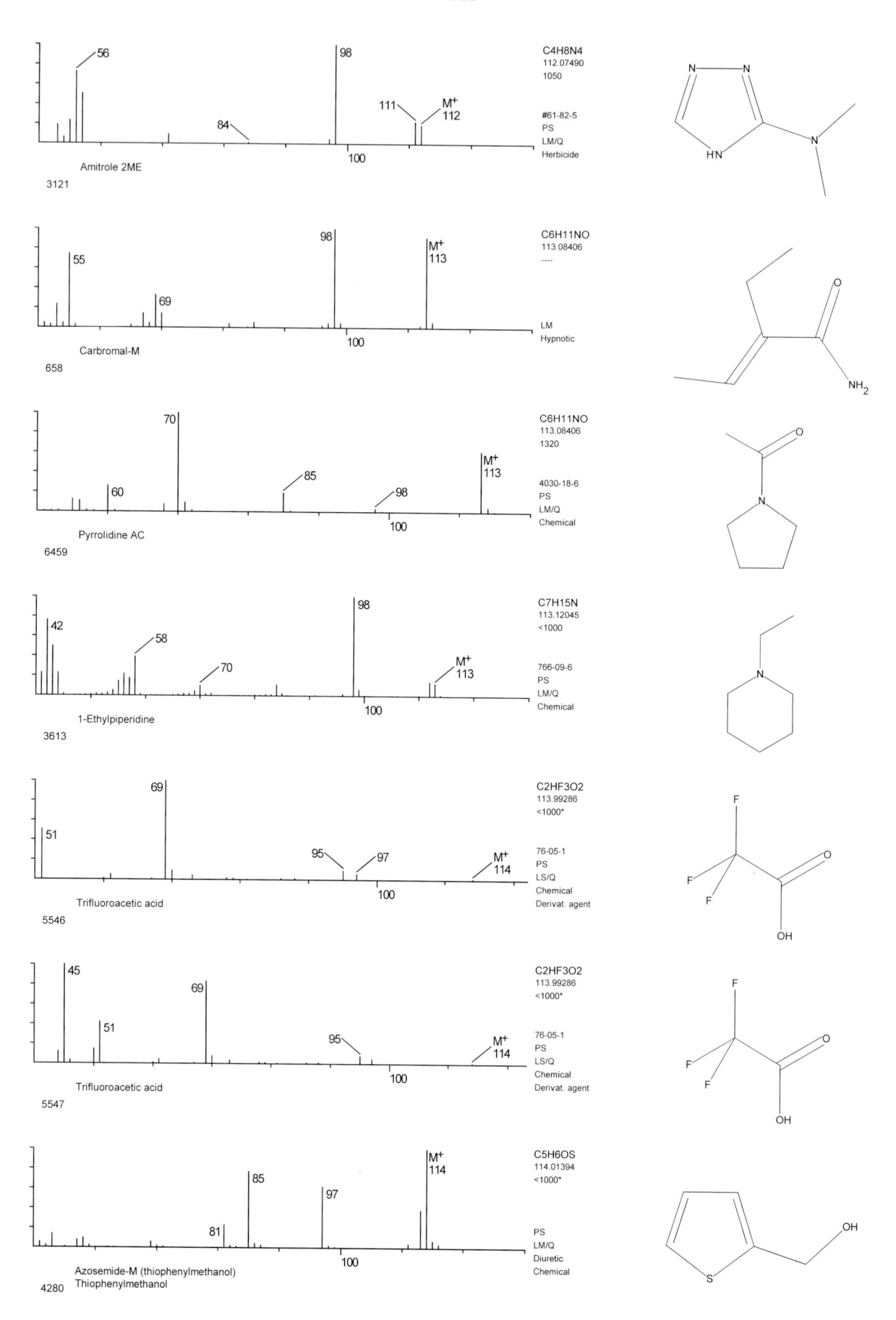

Spectrum	Compound	Formula/Info
56, 84, 98, 111, M⁺ 112	Amitrole 2ME — 3121	C4H8N4, 112.07490, 1050, #61-82-5, PS, LM/Q, Herbicide
55, 69, 98, M⁺ 113	Carbromal-M — 658	C6H11NO, 113.08406, ----, LM, Hypnotic
60, 70, 85, 98, M⁺ 113	Pyrrolidine AC — 6459	C6H11NO, 113.08406, 1320, 4030-18-6, PS, LM/Q, Chemical
42, 58, 70, 98, M⁺ 113	1-Ethylpiperidine — 3613	C7H15N, 113.12045, <1000, 766-09-6, PS, LM/Q, Chemical
51, 69, 95, 97, M⁺ 114	Trifluoroacetic acid — 5546	C2HF3O2, 113.99286, <1000*, 76-05-1, PS, LS/Q, Chemical, Derivat. agent
45, 51, 69, 95, M⁺ 114	Trifluoroacetic acid — 5547	C2HF3O2, 113.99286, <1000*, 76-05-1, PS, LS/Q, Chemical, Derivat. agent
81, 85, 97, M⁺ 114	Azosemide-M (thiophenylmethanol) Thiophenylmethanol — 4280	C5H6OS, 114.01394, <1000*, PS, LM/Q, Diuretic, Chemical

114

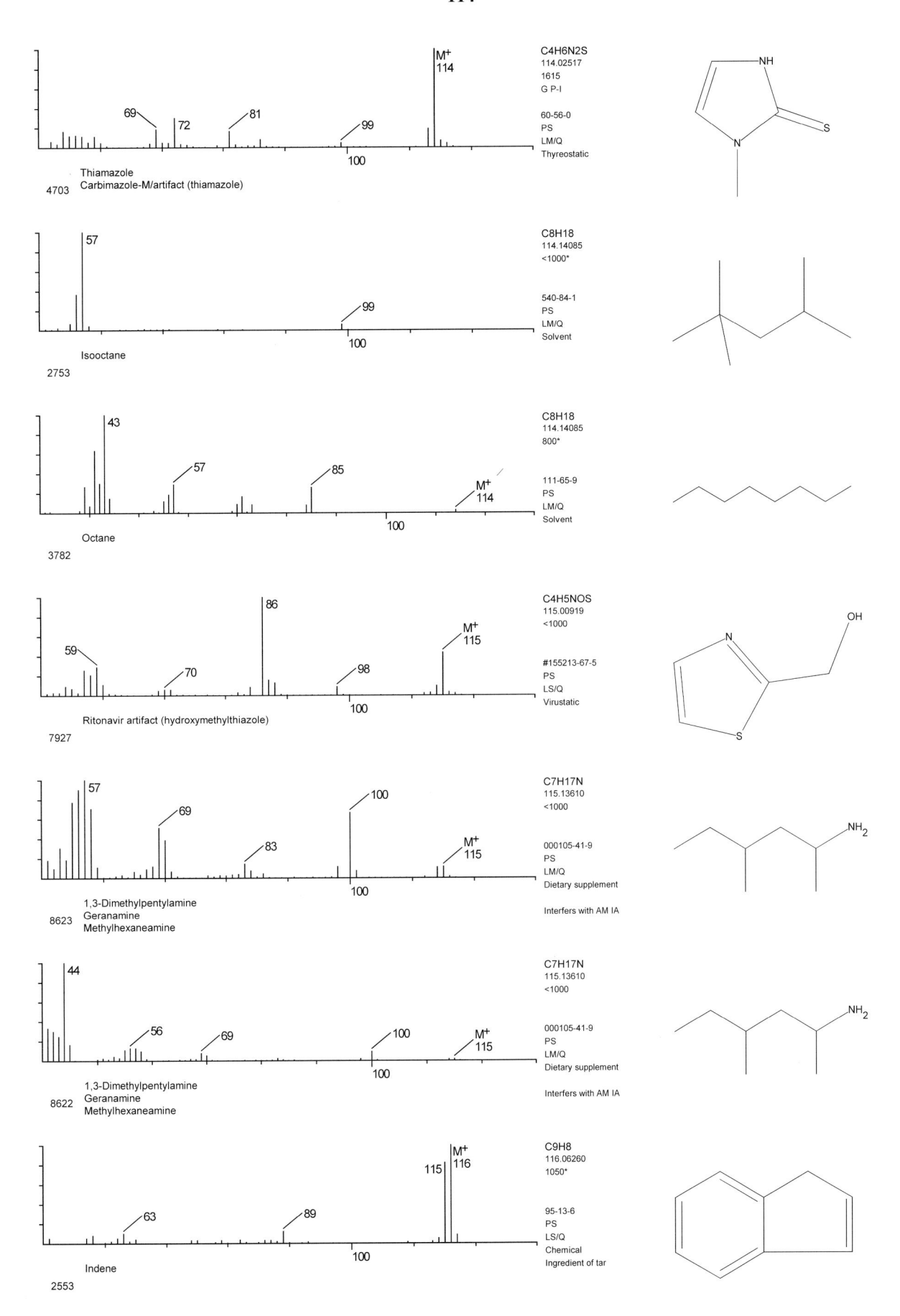

Peak 114 M+, 99, 81, 72, 69	C4H6N2S 114.02517 1615 G P-I 60-56-0 PS LM/Q Thyreostatic	(structure: 1-methyl-imidazole-2-thione)

4703 Thiamazole
Carbimazole-M/artifact (thiamazole)

Peak 57 base, 99	C8H18 114.14085 <1000* 540-84-1 PS LM/Q Solvent	(structure: isooctane)

2753 Isooctane

Peak 43 base, 57, 85, 114 M+	C8H18 114.14085 800* 111-65-9 PS LM/Q Solvent	(structure: n-octane)

3782 Octane

Peak 86 base, 115 M+, 98, 70, 59	C4H5NOS 115.00919 <1000 #155213-67-5 PS LS/Q Virustatic	(structure: hydroxymethylthiazole)

7927 Ritonavir artifact (hydroxymethylthiazole)

Peak 57, 69, 83, 100, 115 M+	C7H17N 115.13610 <1000 000105-41-9 PS LM/Q Dietary supplement Interfers with AM IA	(structure: 1,3-dimethylpentylamine)

8623 1,3-Dimethylpentylamine
Geranamine
Methylhexaneamine

Peak 44 base, 56, 69, 100, 115 M+	C7H17N 115.13610 <1000 000105-41-9 PS LM/Q Dietary supplement Interfers with AM IA	(structure: 1,3-dimethylpentylamine)

8622 1,3-Dimethylpentylamine
Geranamine
Methylhexaneamine

Peak 116 M+, 115, 89, 63	C9H8 116.06260 1050* 95-13-6 PS LS/Q Chemical Ingredient of tar	(structure: indene)

2553 Indene

117

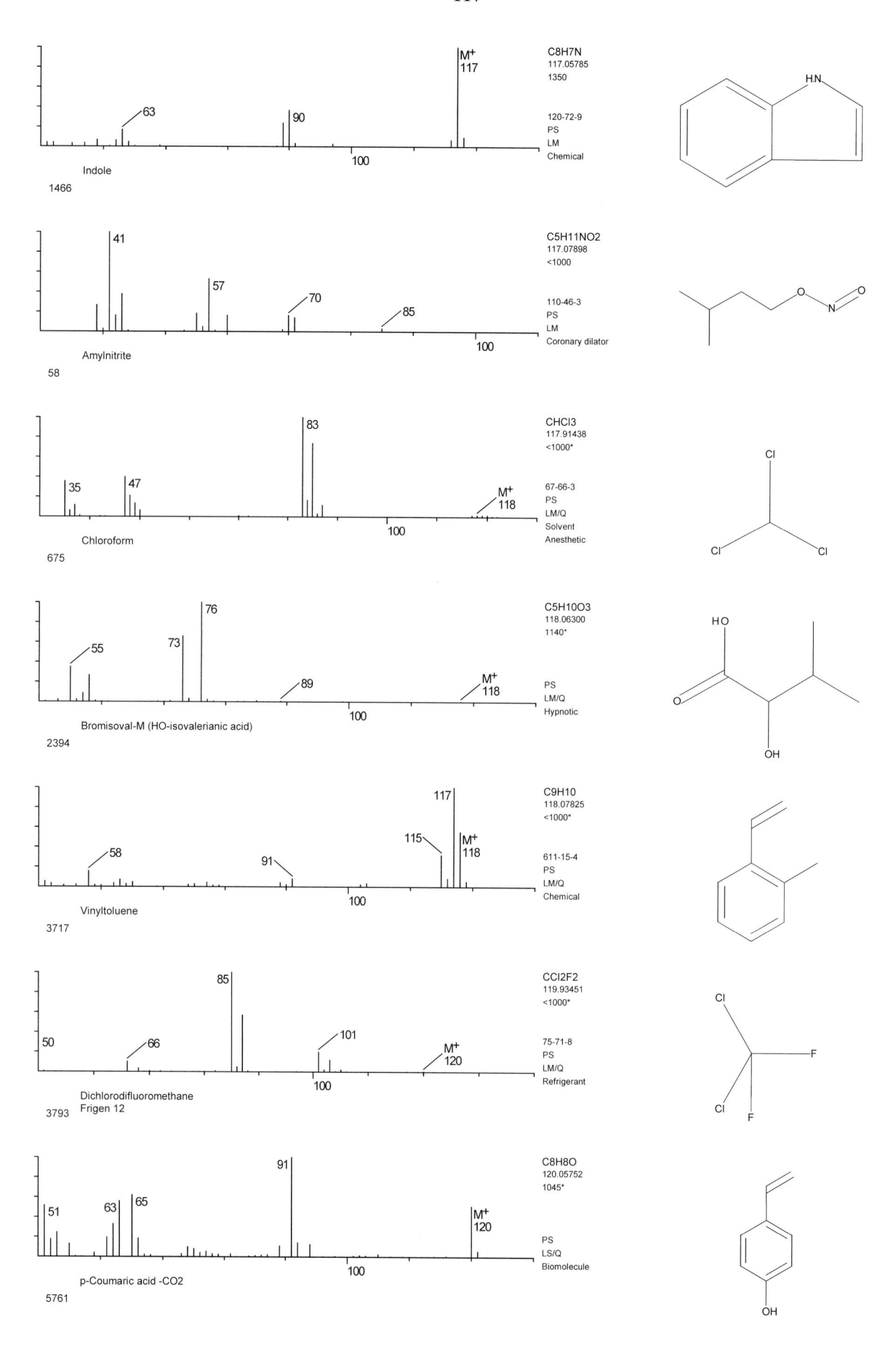

Indole — C8H7N, 117.05785, 1350, 120-72-9, PS, LM, Chemical
Peaks: 63, 90, M+ 117
1466

Amylnitrite — C5H11NO2, 117.07898, <1000, 110-46-3, PS, LM, Coronary dilator
Peaks: 41, 57, 70, 85
58

Chloroform — CHCl3, 117.91438, <1000*, 67-66-3, PS, LM/Q, Solvent Anesthetic
Peaks: 35, 47, 83, M+ 118
675

Bromisoval-M (HO-isovalerianic acid) — C5H10O3, 118.06300, 1140*, PS, LM/Q, Hypnotic
Peaks: 55, 73, 76, 89, M+ 118
2394

Vinyltoluene — C9H10, 118.07825, <1000*, 611-15-4, PS, LM/Q, Chemical
Peaks: 58, 91, 115, 117, M+ 118
3717

Dichlorodifluoromethane Frigen 12 — CCl2F2, 119.93451, <1000*, 75-71-8, PS, LM/Q, Refrigerant
Peaks: 50, 66, 85, 101, M+ 120
3793

p-Coumaric acid -CO2 — C8H8O, 120.05752, 1045*, PS, LS/Q, Biomolecule
Peaks: 51, 63, 65, 91, M+ 120
5761

120

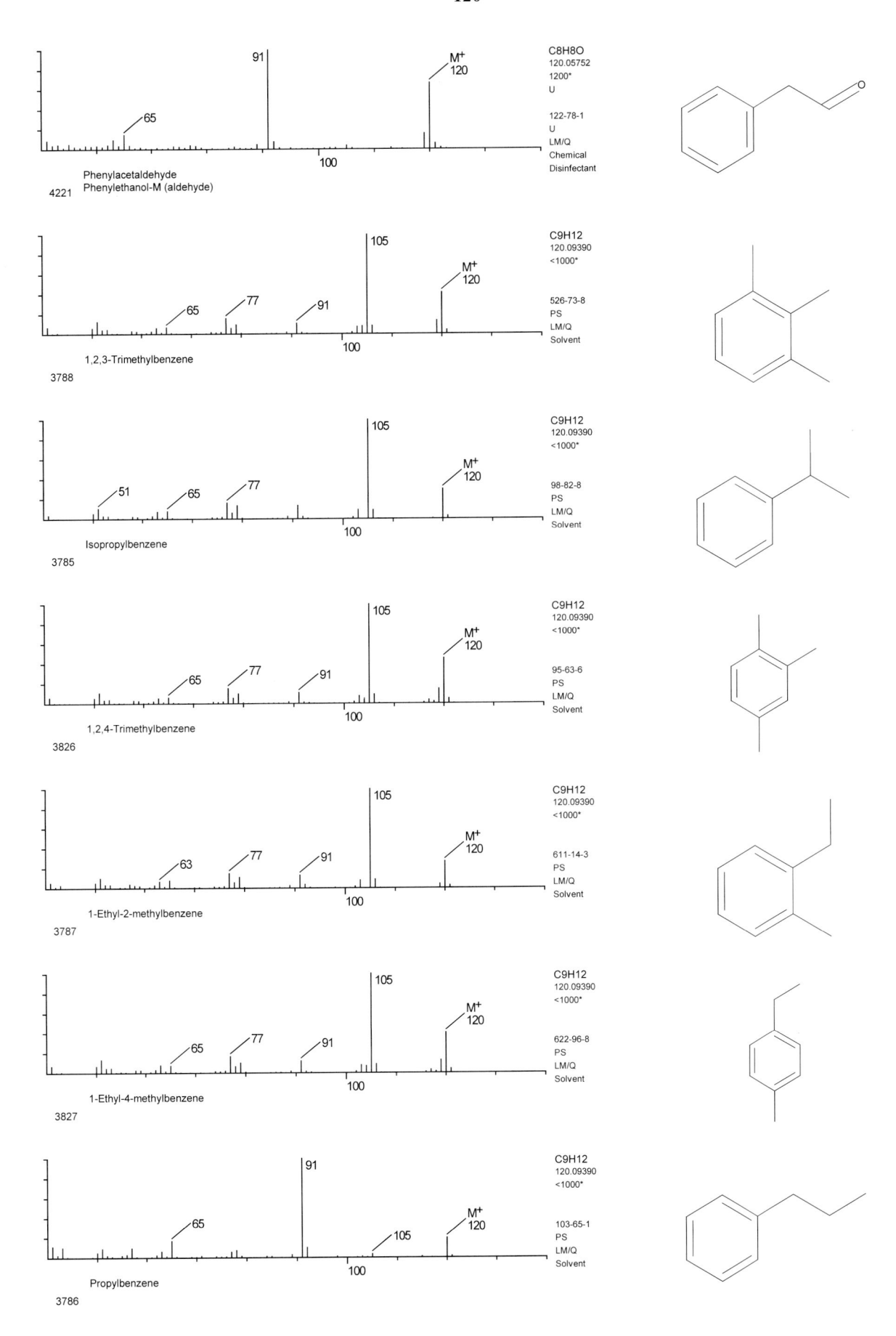

Spectrum	Compound	Formula / Data
4221	Phenylacetaldehyde / Phenylethanol-M (aldehyde)	C8H8O, 120.05752, 1200*, U, 122-78-1, U, LM/Q, Chemical Disinfectant
3788	1,2,3-Trimethylbenzene	C9H12, 120.09390, <1000*, 526-73-8, PS, LM/Q, Solvent
3785	Isopropylbenzene	C9H12, 120.09390, <1000*, 98-82-8, PS, LM/Q, Solvent
3826	1,2,4-Trimethylbenzene	C9H12, 120.09390, <1000*, 95-63-6, PS, LM/Q, Solvent
3787	1-Ethyl-2-methylbenzene	C9H12, 120.09390, <1000*, 611-14-3, PS, LM/Q, Solvent
3827	1-Ethyl-4-methylbenzene	C9H12, 120.09390, <1000*, 622-96-8, PS, LM/Q, Solvent
3786	Propylbenzene	C9H12, 120.09390, <1000*, 103-65-1, PS, LM/Q, Solvent

121

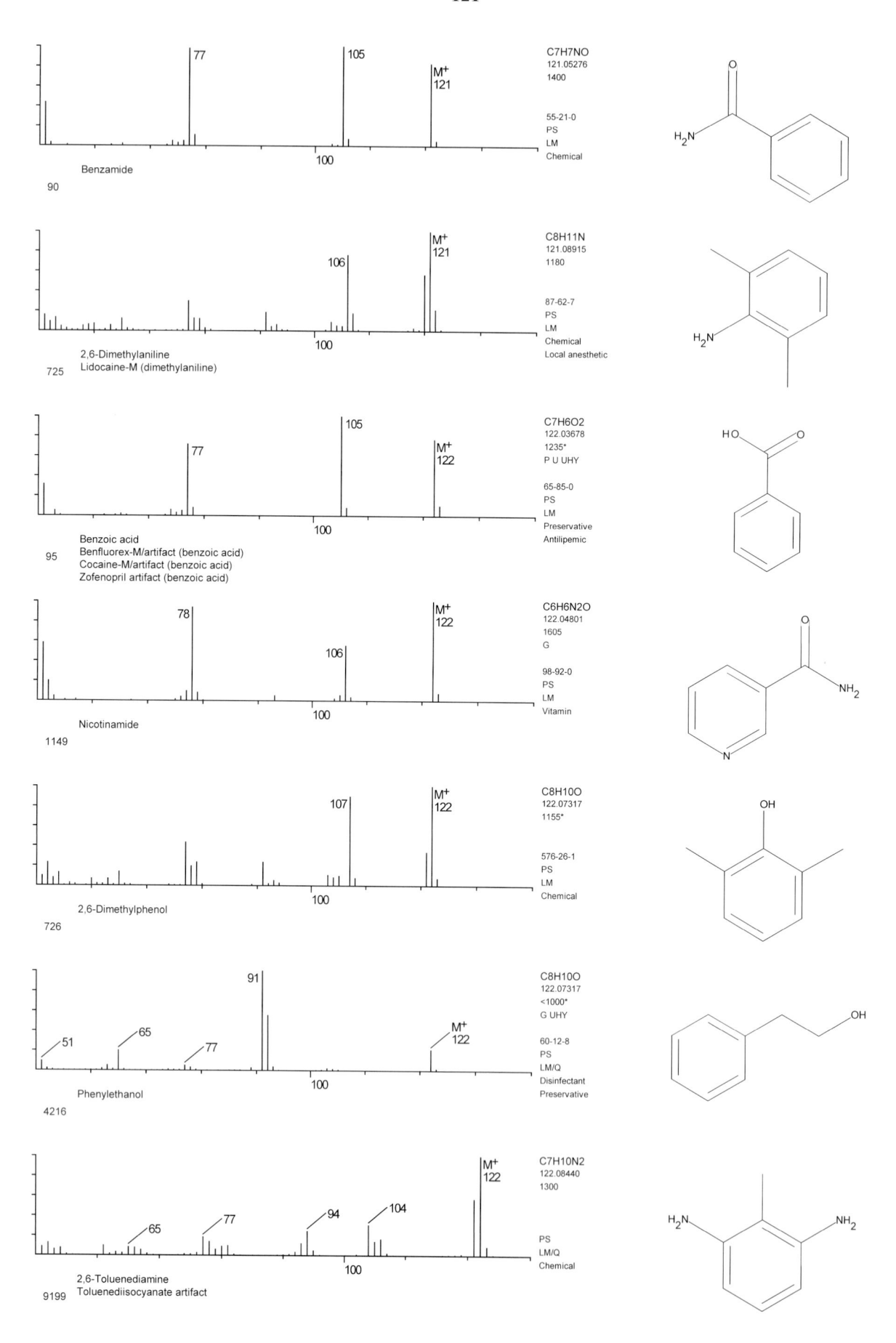

Benzamide
Peaks: 77, 105, M+ 121
C7H7NO
121.05276
1400
55-21-0
PS
LM
Chemical
90

2,6-Dimethylaniline
Lidocaine-M (dimethylaniline)
Peaks: 106, M+ 121
C8H11N
121.08915
1180
87-62-7
PS
LM
Chemical
Local anesthetic
725

Benzoic acid
Benfluorex-M/artifact (benzoic acid)
Cocaine-M/artifact (benzoic acid)
Zofenopril artifact (benzoic acid)
Peaks: 77, 105, M+ 122
C7H6O2
122.03678
1235*
P U UHY
65-85-0
PS
LM
Preservative
Antilipemic
95

Nicotinamide
Peaks: 78, 106, M+ 122
C6H6N2O
122.04801
1605
G
98-92-0
PS
LM
Vitamin
1149

2,6-Dimethylphenol
Peaks: 107, M+ 122
C8H10O
122.07317
1155*
576-26-1
PS
LM
Chemical
726

Phenylethanol
Peaks: 51, 65, 77, 91, M+ 122
C8H10O
122.07317
<1000*
G UHY
60-12-8
PS
LM/Q
Disinfectant
Preservative
4216

2,6-Toluenediamine
Toluenediisocyanate artifact
Peaks: 65, 77, 94, 104, M+ 122
C7H10N2
122.08440
1300
PS
LM/Q
Chemical
9199

266

122

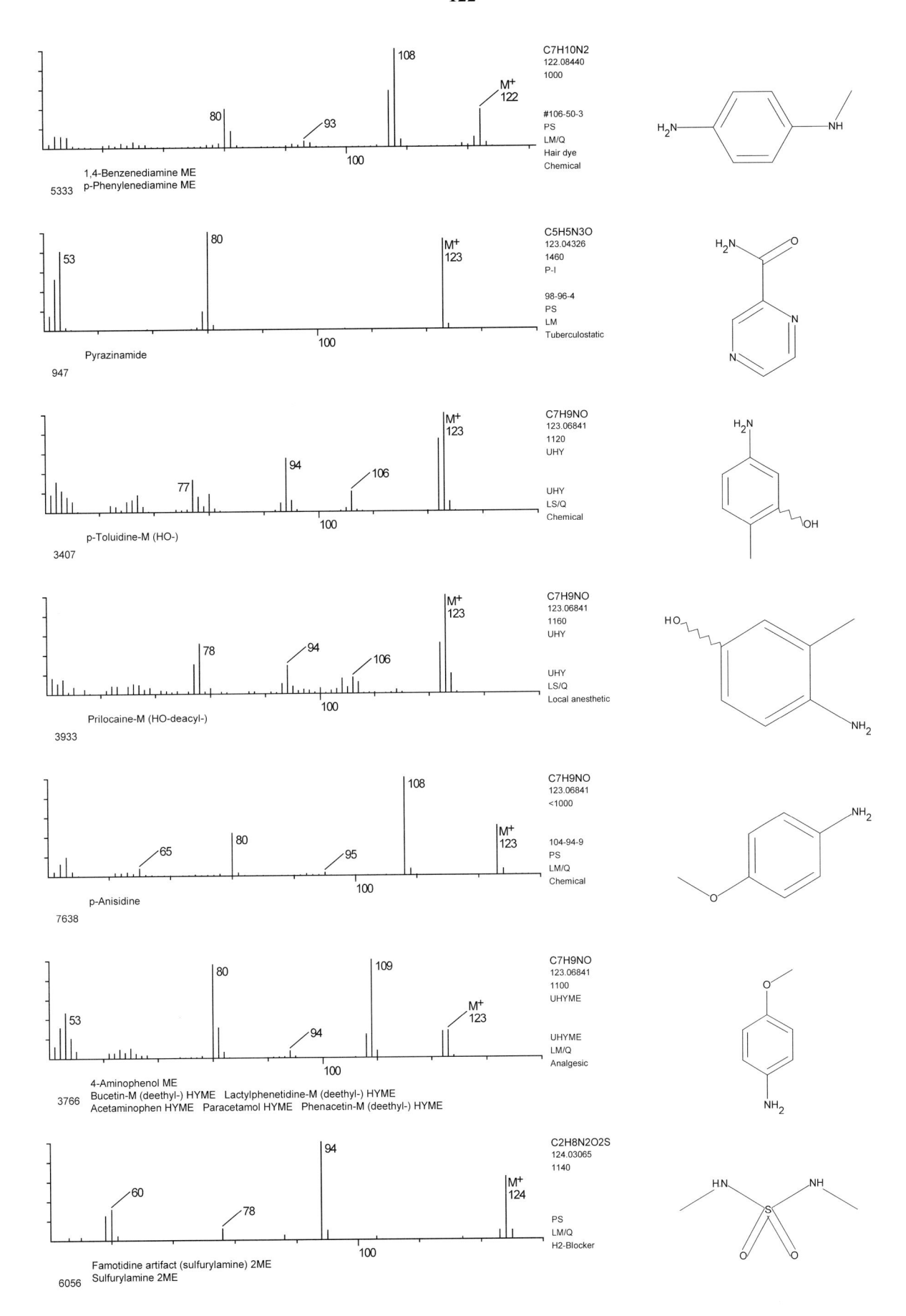

1,4-Benzenediamine ME / p-Phenylenediamine ME — peaks: 80, 93, 108, M+ 122	C7H10N2 / 122.08440 / 1000 / #106-50-3 / PS / LM/Q / Hair dye / Chemical	5333
Pyrazinamide — peaks: 53, 80, M+ 123	C5H5N3O / 123.04326 / 1460 / P-I / 98-96-4 / PS / LM / Tuberculostatic	947
p-Toluidine-M (HO-) — peaks: 77, 94, 106, M+ 123	C7H9NO / 123.06841 / 1120 / UHY / UHY / LS/Q / Chemical	3407
Prilocaine-M (HO-deacyl-) — peaks: 78, 94, 106, M+ 123	C7H9NO / 123.06841 / 1160 / UHY / UHY / LS/Q / Local anesthetic	3933
p-Anisidine — peaks: 65, 80, 95, 108, M+ 123	C7H9NO / 123.06841 / <1000 / 104-94-9 / PS / LM/Q / Chemical	7638
4-Aminophenol ME / Bucetin-M (deethyl-) HYME / Lactylphenetidine-M (deethyl-) HYME / Acetaminophen HYME / Paracetamol HYME / Phenacetin-M (deethyl-) HYME — peaks: 53, 80, 94, 109, M+ 123	C7H9NO / 123.06841 / 1100 / UHYME / UHYME / LM/Q / Analgesic	3766
Famotidine artifact (sulfurylamine) 2ME / Sulfurylamine 2ME — peaks: 60, 78, 94, M+ 124	C2H8N2O2S / 124.03065 / 1140 / PS / LM/Q / H2-Blocker	6056

267

124

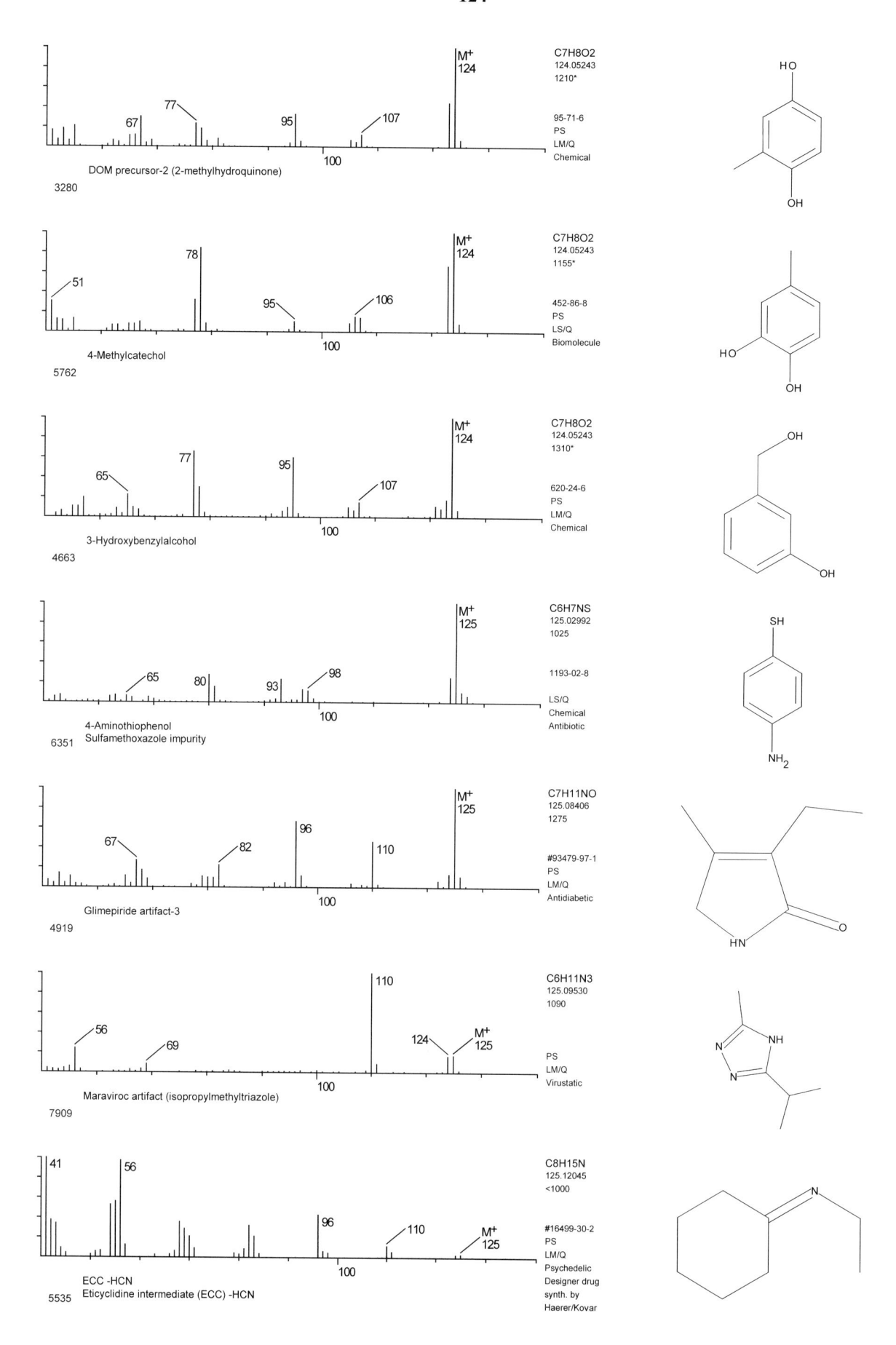

Spectrum	Compound	Formula / Info
3280	DOM precursor-2 (2-methylhydroquinone) — peaks 67, 77, 95, 107, M+ 124	C7H8O2, 124.05243, 1210*, 95-71-6, PS, LM/Q, Chemical
5762	4-Methylcatechol — peaks 51, 78, 95, 106, M+ 124	C7H8O2, 124.05243, 1155*, 452-86-8, PS, LS/Q, Biomolecule
4663	3-Hydroxybenzylalcohol — peaks 65, 77, 95, 107, M+ 124	C7H8O2, 124.05243, 1310*, 620-24-6, PS, LM/Q, Chemical
6351	4-Aminothiophenol / Sulfamethoxazole impurity — peaks 65, 80, 93, 98, M+ 125	C6H7NS, 125.02992, 1025, 1193-02-8, LS/Q, Chemical, Antibiotic
4919	Glimepiride artifact-3 — peaks 67, 82, 96, 110, M+ 125	C7H11NO, 125.08406, 1275, #93479-97-1, PS, LM/Q, Antidiabetic
7909	Maraviroc artifact (isopropylmethyltriazole) — peaks 56, 69, 110, 124, M+ 125	C6H11N3, 125.09530, 1090, PS, LM/Q, Virustatic
5535	ECC -HCN / Eticyclidine intermediate (ECC) -HCN — peaks 41, 56, 96, 110, M+ 125	C8H15N, 125.12045, <1000, #16499-30-2, PS, LM/Q, Psychedelic, Designer drug synth. by Haerer/Kovar

125

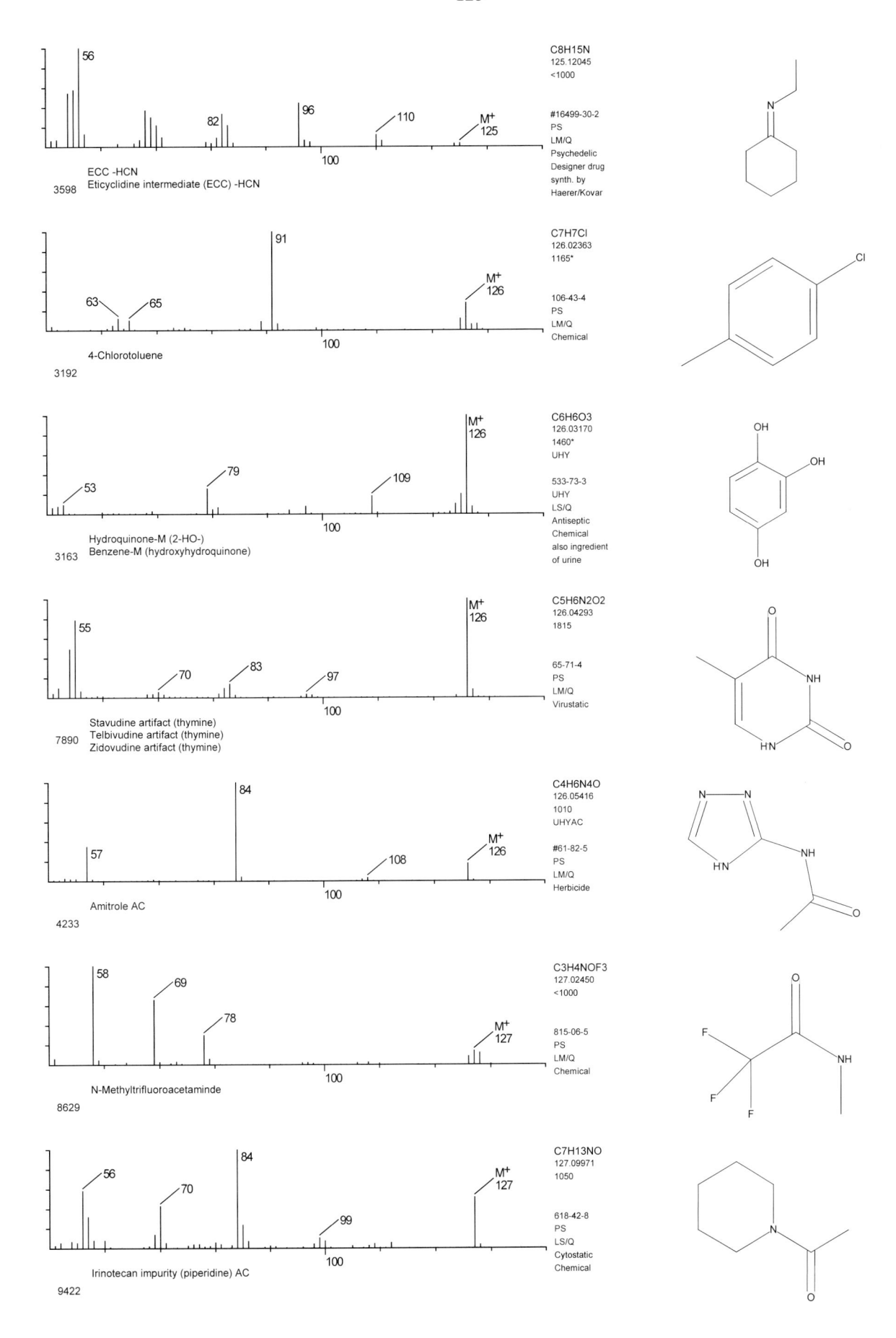

ECC -HCN Eticyclidine intermediate (ECC) -HCN 3598	C8H15N 125.12045 <1000 #16499-30-2 PS LM/Q Psychedelic Designer drug synth. by Haerer/Kovar
4-Chlorotoluene 3192	C7H7Cl 126.02363 1165* 106-43-4 PS LM/Q Chemical
Hydroquinone-M (2-HO-) Benzene-M (hydroxyhydroquinone) 3163	C6H6O3 126.03170 1460* UHY 533-73-3 UHY LS/Q Antiseptic Chemical also ingredient of urine
Stavudine artifact (thymine) Telbivudine artifact (thymine) Zidovudine artifact (thymine) 7890	C5H6N2O2 126.04293 1815 65-71-4 PS LM/Q Virustatic
Amitrole AC 4233	C4H6N4O 126.05416 1010 UHYAC #61-82-5 PS LM/Q Herbicide
N-Methyltrifluoroacetaminde 8629	C3H4NOF3 127.02450 <1000 815-06-5 PS LM/Q Chemical
Irinotecan impurity (piperidine) AC 9422	C7H13NO 127.09971 1050 618-42-8 PS LS/Q Cytostatic Chemical

127

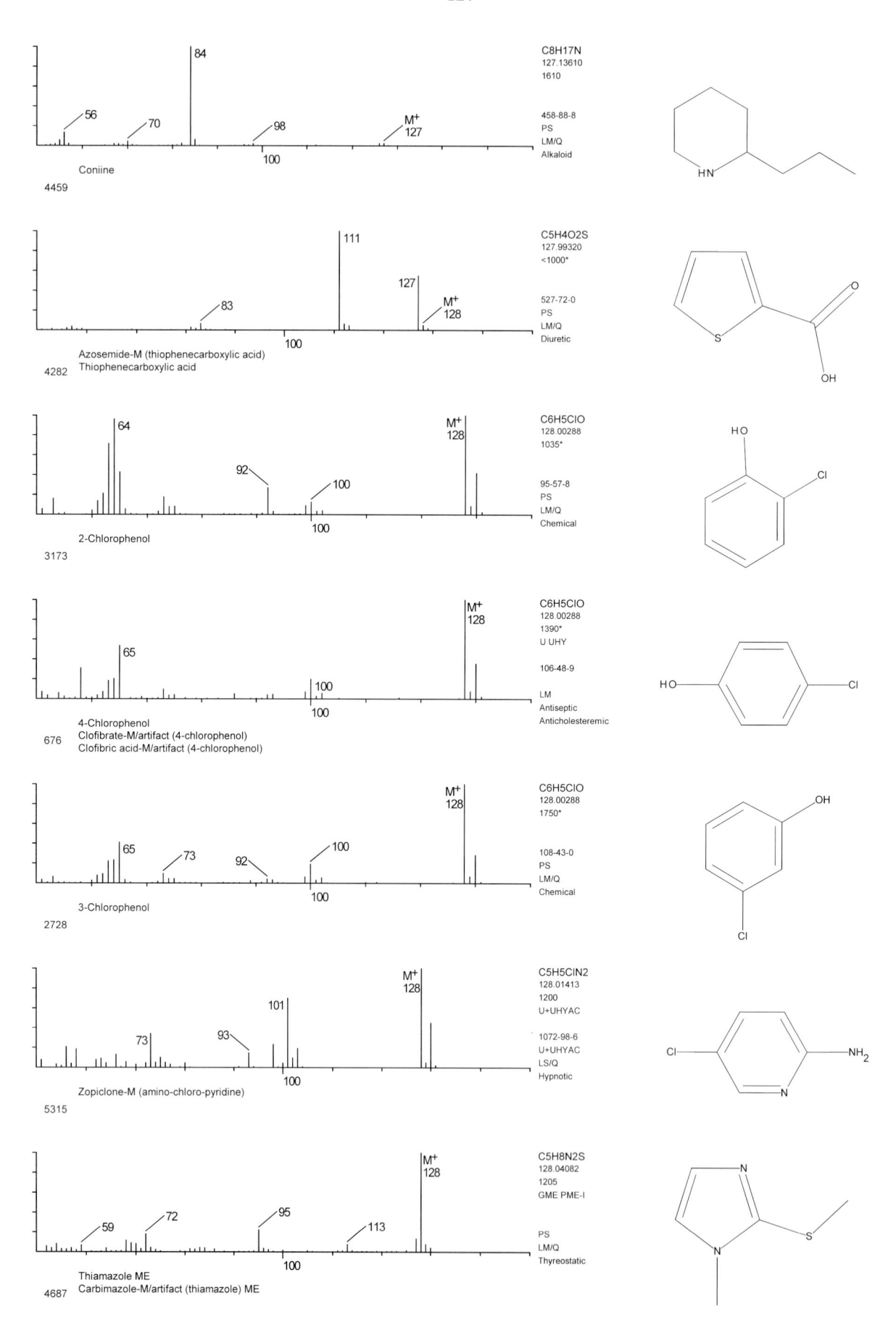

m/z 84 (base), 56, 70, 98, M+ 127	Coniine	C8H17N 127.13610 1610 / 458-88-8 PS LM/Q Alkaloid
4459		

m/z 111 (base), 83, 127, M+ 128	Azosemide-M (thiophenecarboxylic acid) Thiophenecarboxylic acid	C5H4O2S 127.99320 <1000* / 527-72-0 PS LM/Q Diuretic
4282		

m/z 64 (base), 92, 100, M+ 128	2-Chlorophenol	C6H5ClO 128.00288 1035* / 95-57-8 PS LM/Q Chemical
3173		

m/z M+ 128 (base), 65, 100	4-Chlorophenol Clofibrate-M/artifact (4-chlorophenol) Clofibric acid-M/artifact (4-chlorophenol)	C6H5ClO 128.00288 1390* U UHY / 106-48-9 LM Antiseptic Anticholesteremic
676		

m/z M+ 128 (base), 65, 73, 92, 100	3-Chlorophenol	C6H5ClO 128.00288 1750* / 108-43-0 PS LM/Q Chemical
2728		

m/z M+ 128 (base), 73, 93, 101	Zopiclone-M (amino-chloro-pyridine)	C5H5ClN2 128.01413 1200 U+UHYAC / 1072-98-6 U+UHYAC LS/Q Hypnotic
5315		

m/z M+ 128 (base), 59, 72, 95, 113	Thiamazole ME Carbimazole-M/artifact (thiamazole) ME	C5H8N2S 128.04082 1205 GME PME-I PS LM/Q Thyreostatic
4687		

128

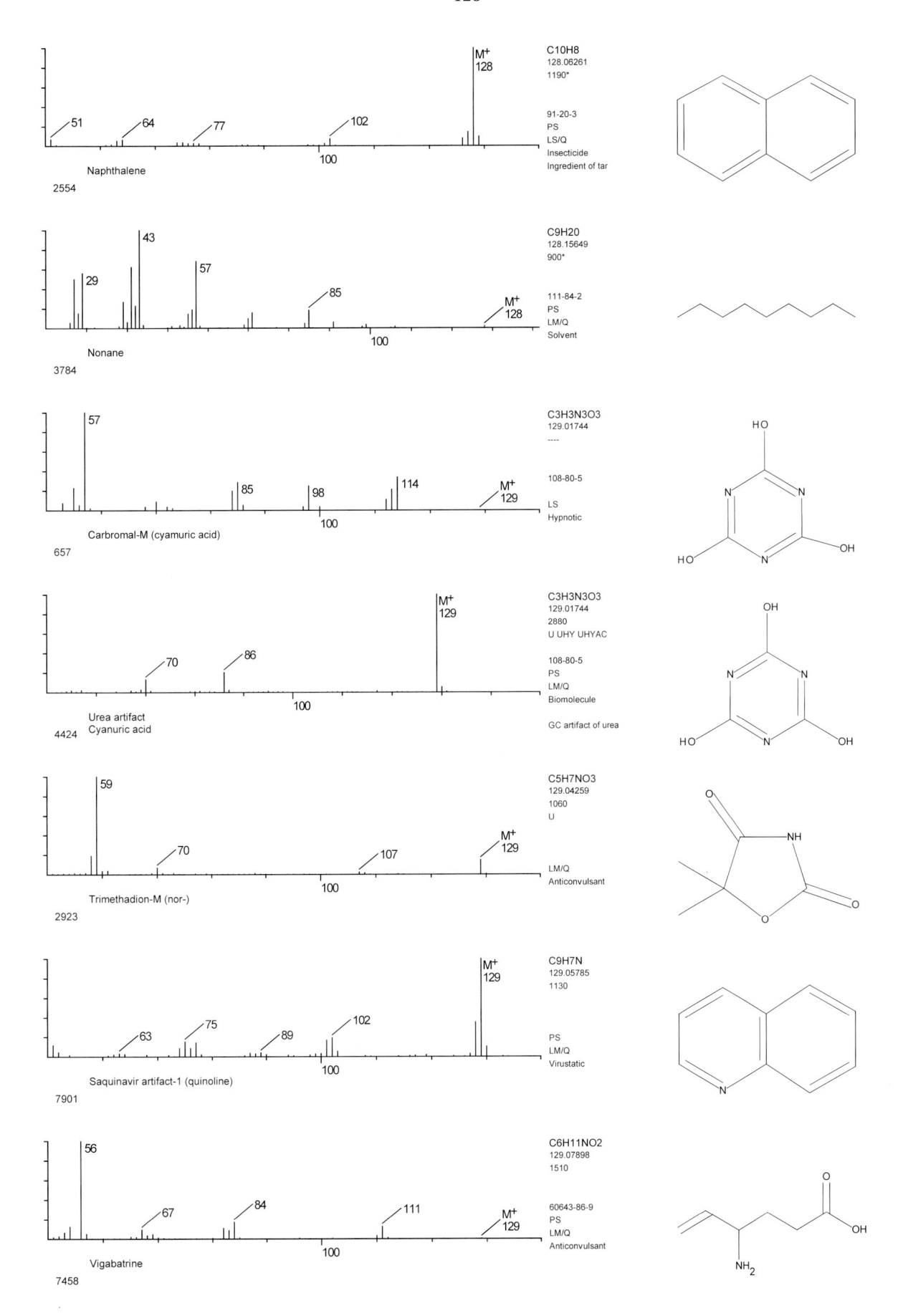

Naphthalene	51, 64, 77, 102, M+ 128	C10H8 128.06261 1190* 91-20-3 PS LS/Q Insecticide Ingredient of tar
2554		

Nonane	29, 43, 57, 85, M+ 128	C9H20 128.15649 900* 111-84-2 PS LM/Q Solvent
3784		

Carbromal-M (cyamuric acid)	57, 85, 98, 114, M+ 129	C3H3N3O3 129.01744 ---- 108-80-5 LS Hypnotic
657		

Urea artifact Cyanuric acid	70, 86, M+ 129	C3H3N3O3 129.01744 2880 U UHY UHYAC 108-80-5 PS LM/Q Biomolecule GC artifact of urea
4424		

Trimethadion-M (nor-)	59, 70, 107, M+ 129	C5H7NO3 129.04259 1060 U LM/Q Anticonvulsant
2923		

Saquinavir artifact-1 (quinoline)	63, 75, 89, 102, M+ 129	C9H7N 129.05785 1130 PS LM/Q Virustatic
7901		

Vigabatrine	56, 67, 84, 111, M+ 129	C6H11NO2 129.07898 1510 60643-86-9 PS LM/Q Anticonvulsant
7458		

271

129

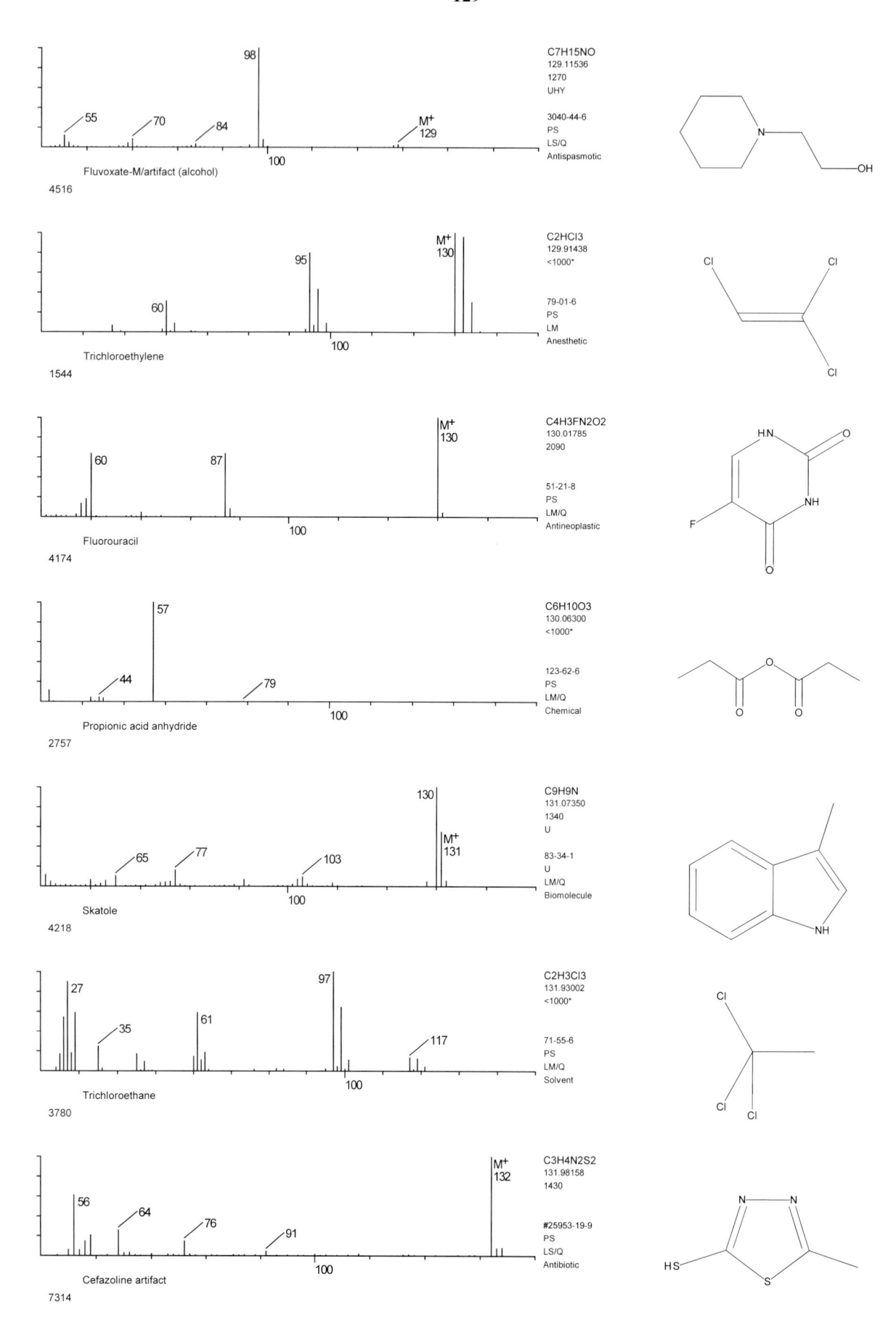

Fluvoxate-M/artifact (alcohol) 4516	C7H15NO 129.11536 1270 UHY 3040-44-6 PS LS/Q Antispasmotic
Trichloroethylene 1544	C2HCl3 129.91438 <1000* 79-01-6 PS LM Anesthetic
Fluorouracil 4174	C4H3FN2O2 130.01785 2090 51-21-8 PS LM/Q Antineoplastic
Propionic acid anhydride 2757	C6H10O3 130.06300 <1000* 123-62-6 PS LM/Q Chemical
Skatole 4218	C9H9N 131.07350 1340 U 83-34-1 U LM/Q Biomolecule
Trichloroethane 3780	C2H3Cl3 131.93002 <1000* 71-55-6 PS LM/Q Solvent
Cefazoline artifact 7314	C3H4N2S2 131.98158 1430 #25953-19-9 PS LS/Q Antibiotic

132

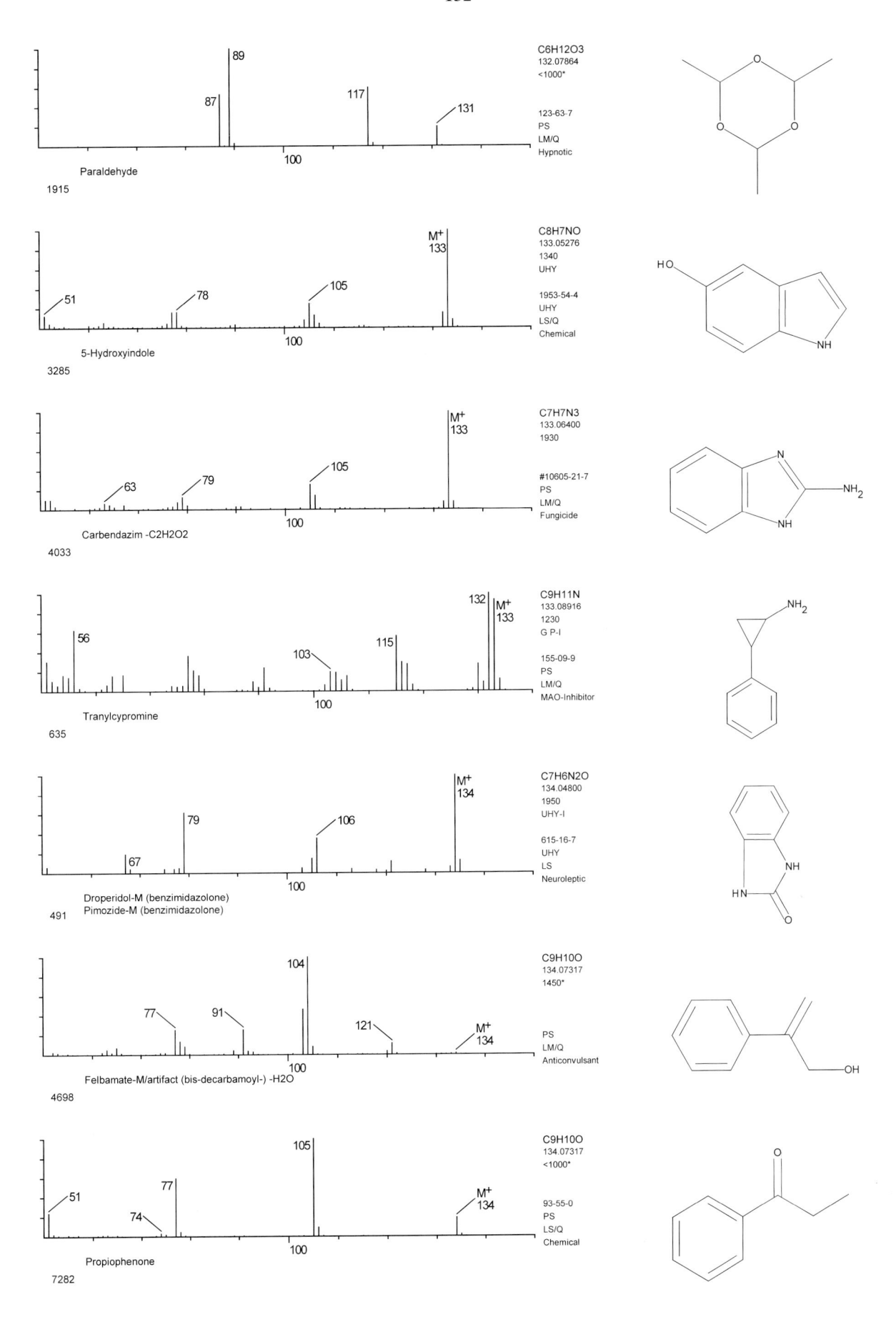

Paraldehyde — peaks: 87, 89, 117, 131	C6H12O3 132.07864 <1000* 123-63-7 PS LM/Q Hypnotic	
1915		

5-Hydroxyindole — peaks: 51, 78, 105, M+ 133	C8H7NO 133.05276 1340 UHY 1953-54-4 UHY LS/Q Chemical	
3285		

Carbendazim -C2H2O2 — peaks: 63, 79, 105, M+ 133	C7H7N3 133.06400 1930 #10605-21-7 PS LM/Q Fungicide	
4033		

Tranylcypromine — peaks: 56, 103, 115, 132, M+ 133	C9H11N 133.08916 1230 G P-I 155-09-9 PS LM/Q MAO-Inhibitor	
635		

Droperidol-M (benzimidazolone) Pimozide-M (benzimidazolone) — peaks: 67, 79, 106, M+ 134	C7H6N2O 134.04800 1950 UHY-I 615-16-7 UHY LS Neuroleptic	
491		

Felbamate-M/artifact (bis-decarbamoyl-) -H2O — peaks: 77, 91, 104, 121, M+ 134	C9H10O 134.07317 1450* PS LM/Q Anticonvulsant	
4698		

Propiophenone — peaks: 51, 74, 77, 105, M+ 134	C9H10O 134.07317 <1000* 93-55-0 PS LS/Q Chemical	
7282		

273

134

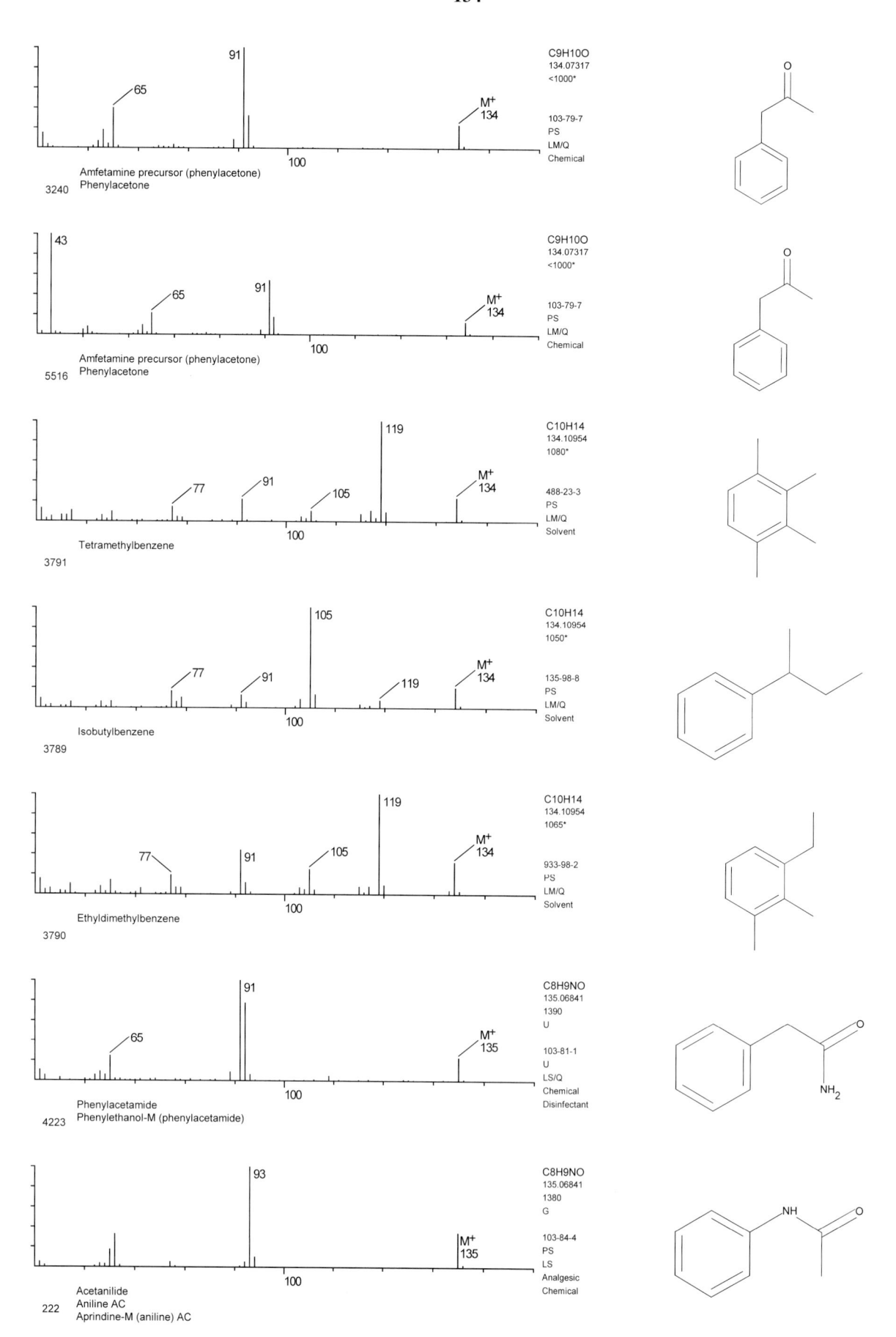

Spectrum peaks: 65, 91, M+ 134	Amfetamine precursor (phenylacetone) Phenylacetone 3240	C9H10O 134.07317 <1000* 103-79-7 PS LM/Q Chemical	
Spectrum peaks: 43, 65, 91, M+ 134	Amfetamine precursor (phenylacetone) Phenylacetone 5516	C9H10O 134.07317 <1000* 103-79-7 PS LM/Q Chemical	
Spectrum peaks: 77, 91, 105, 119, M+ 134	Tetramethylbenzene 3791	C10H14 134.10954 1080* 488-23-3 PS LM/Q Solvent	
Spectrum peaks: 77, 91, 105, 119, M+ 134	Isobutylbenzene 3789	C10H14 134.10954 1050* 135-98-8 PS LM/Q Solvent	
Spectrum peaks: 77, 91, 105, 119, M+ 134	Ethyldimethylbenzene 3790	C10H14 134.10954 1065* 933-98-2 PS LM/Q Solvent	
Spectrum peaks: 65, 91, M+ 135	Phenylacetamide Phenylethanol-M (phenylacetamide) 4223	C8H9NO 135.06841 1390 U 103-81-1 U LS/Q Chemical Disinfectant	
Spectrum peaks: 93, M+ 135	Acetanilide Aniline AC Aprindine-M (aniline) AC 222	C8H9NO 135.06841 1380 G 103-84-4 PS LS Analgesic Chemical	

274

135

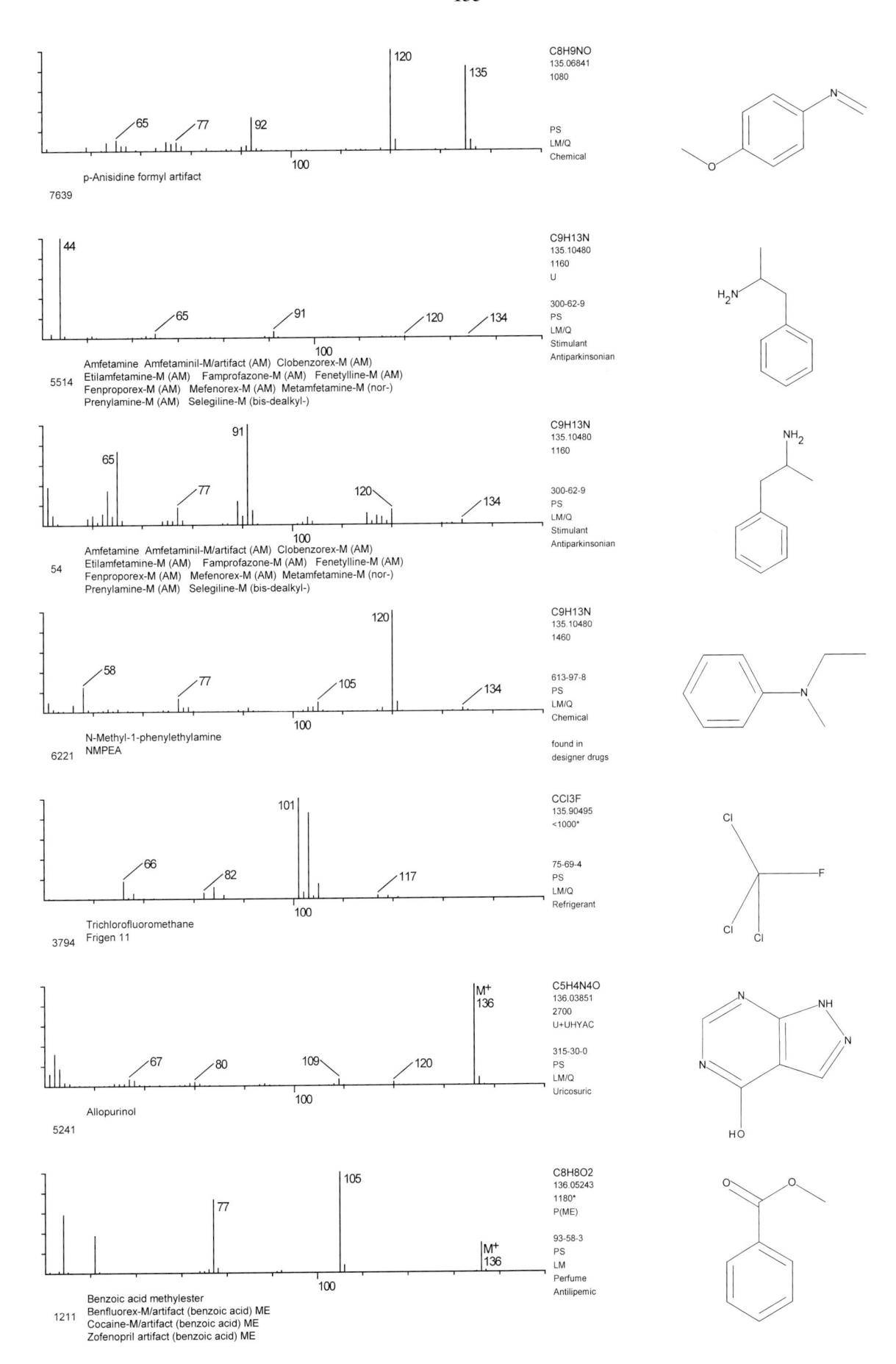

p-Anisidine formyl artifact	C8H9NO 135.06841 1080 PS LM/Q Chemical
7639	

Amfetamine Amfetaminil-M/artifact (AM) Clobenzorex-M (AM) Etilamfetamine-M (AM) Famprofazone-M (AM) Fenetylline-M (AM) Fenproporex-M (AM) Mefenorex-M (AM) Metamfetamine-M (nor-) Prenylamine-M (AM) Selegiline-M (bis-dealkyl-)	C9H13N 135.10480 1160 U 300-62-9 PS LM/Q Stimulant Antiparkinsonian
5514	

Amfetamine Amfetaminil-M/artifact (AM) Clobenzorex-M (AM) Etilamfetamine-M (AM) Famprofazone-M (AM) Fenetylline-M (AM) Fenproporex-M (AM) Mefenorex-M (AM) Metamfetamine-M (nor-) Prenylamine-M (AM) Selegiline-M (bis-dealkyl-)	C9H13N 135.10480 1160 300-62-9 PS LM/Q Stimulant Antiparkinsonian
54	

N-Methyl-1-phenylethylamine NMPEA	C9H13N 135.10480 1460 613-97-8 PS LM/Q Chemical found in designer drugs
6221	

Trichlorofluoromethane Frigen 11	CCl3F 135.90495 <1000* 75-69-4 PS LM/Q Refrigerant
3794	

Allopurinol	C5H4N4O 136.03851 2700 U+UHYAC 315-30-0 PS LM/Q Uricosuric
5241	

Benzoic acid methylester Benfluorex-M/artifact (benzoic acid) ME Cocaine-M/artifact (benzoic acid) ME Zofenopril artifact (benzoic acid) ME	C8H8O2 136.05243 1180* P(ME) 93-58-3 PS LM Perfume Antilipemic
1211	

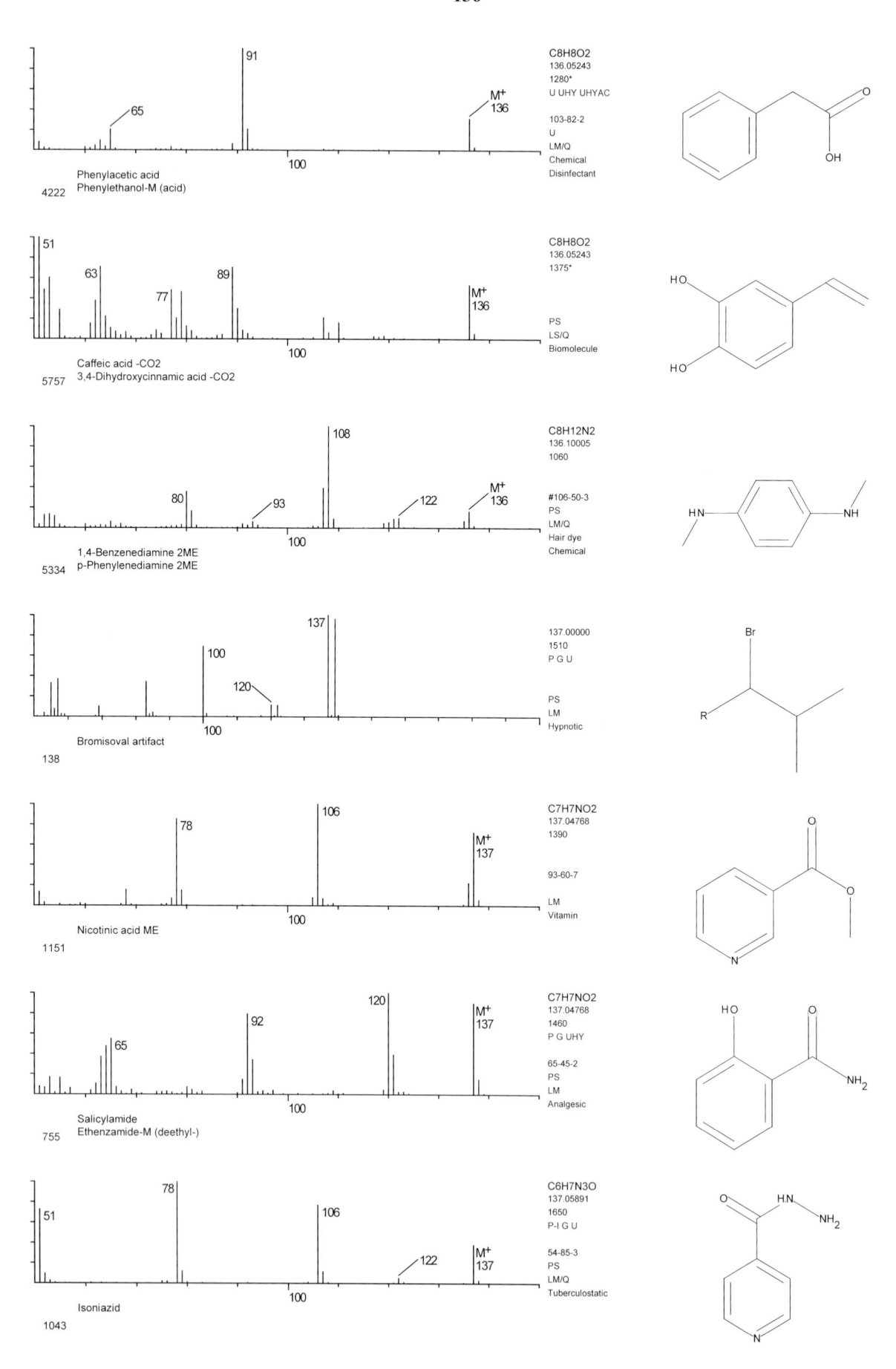

Phenylacetic acid Phenylethanol-M (acid) 4222	C8H8O2 136.05243 1280* U UHY UHYAC 103-82-2 U LM/Q Chemical Disinfectant
Caffeic acid -CO2 3,4-Dihydroxycinnamic acid -CO2 5757	C8H8O2 136.05243 1375* PS LS/Q Biomolecule
1,4-Benzenediamine 2ME p-Phenylenediamine 2ME 5334	C8H12N2 136.10005 1060 #106-50-3 PS LM/Q Hair dye Chemical
Bromisoval artifact 138	137.00000 1510 P G U PS LM Hypnotic
Nicotinic acid ME 1151	C7H7NO2 137.04768 1390 93-60-7 LM Vitamin
Salicylamide Ethenzamide-M (deethyl-) 755	C7H7NO2 137.04768 1460 P G UHY 65-45-2 PS LM Analgesic
Isoniazid 1043	C6H7N3O 137.05891 1650 P-I G U 54-85-3 PS LM/Q Tuberculostatic

137

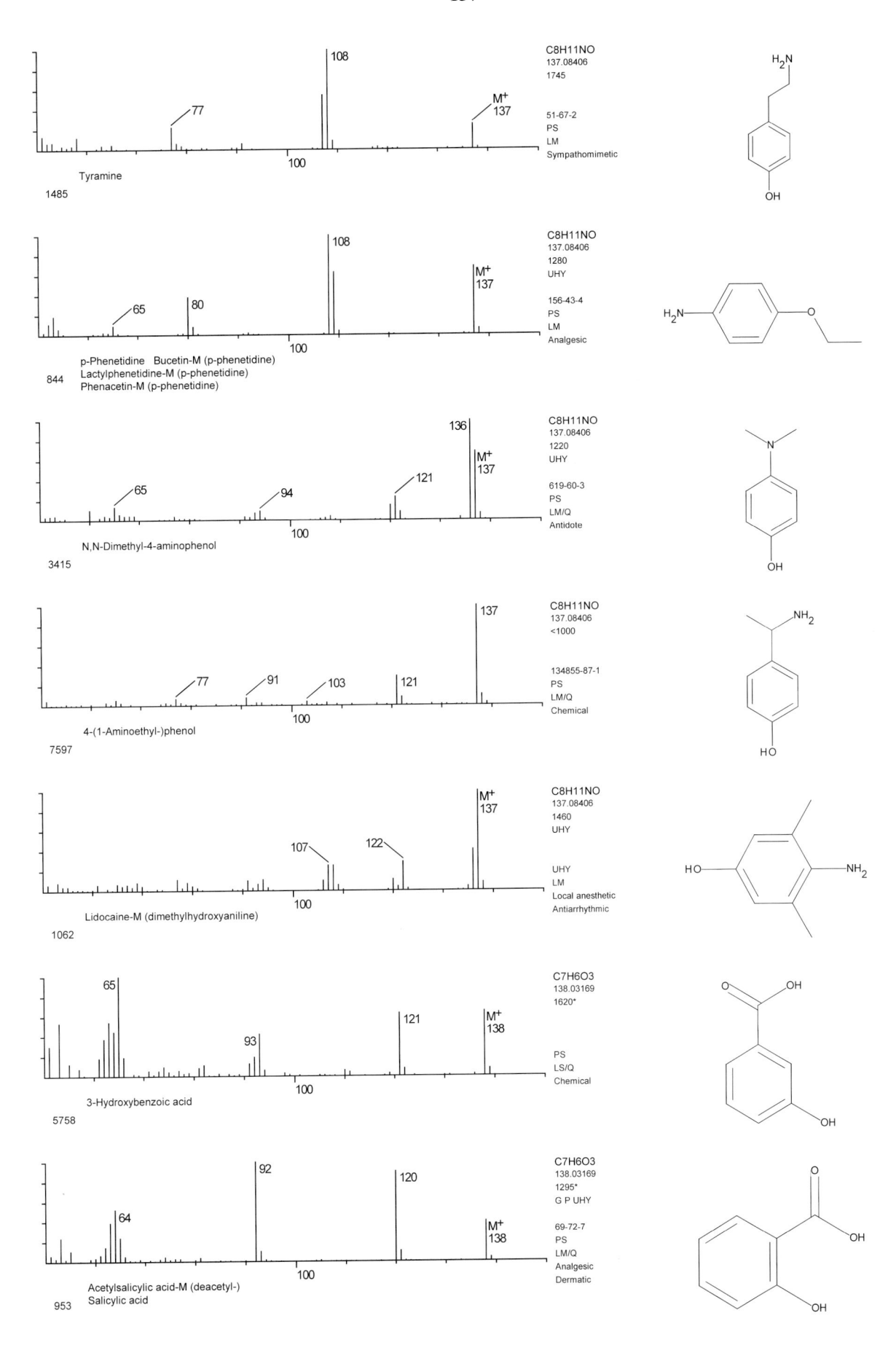

Tyramine
1485

C8H11NO
137.08406
1745

51-67-2
PS
LM
Sympathomimetic

p-Phenetidine Bucetin-M (p-phenetidine)
Lactylphenetidine-M (p-phenetidine)
844 Phenacetin-M (p-phenetidine)

C8H11NO
137.08406
1280
UHY

156-43-4
PS
LM
Analgesic

N,N-Dimethyl-4-aminophenol
3415

C8H11NO
137.08406
1220
UHY

619-60-3
PS
LM/Q
Antidote

4-(1-Aminoethyl-)phenol
7597

C8H11NO
137.08406
<1000

134855-87-1
PS
LM/Q
Chemical

Lidocaine-M (dimethylhydroxyaniline)
1062

C8H11NO
137.08406
1460
UHY

UHY
LM
Local anesthetic
Antiarrhythmic

3-Hydroxybenzoic acid
5758

C7H6O3
138.03169
1620*

PS
LS/Q
Chemical

Acetylsalicylic acid-M (deacetyl-)
953 Salicylic acid

C7H6O3
138.03169
1295*
G P UHY

69-72-7
PS
LM/Q
Analgesic
Dermatic

277

138

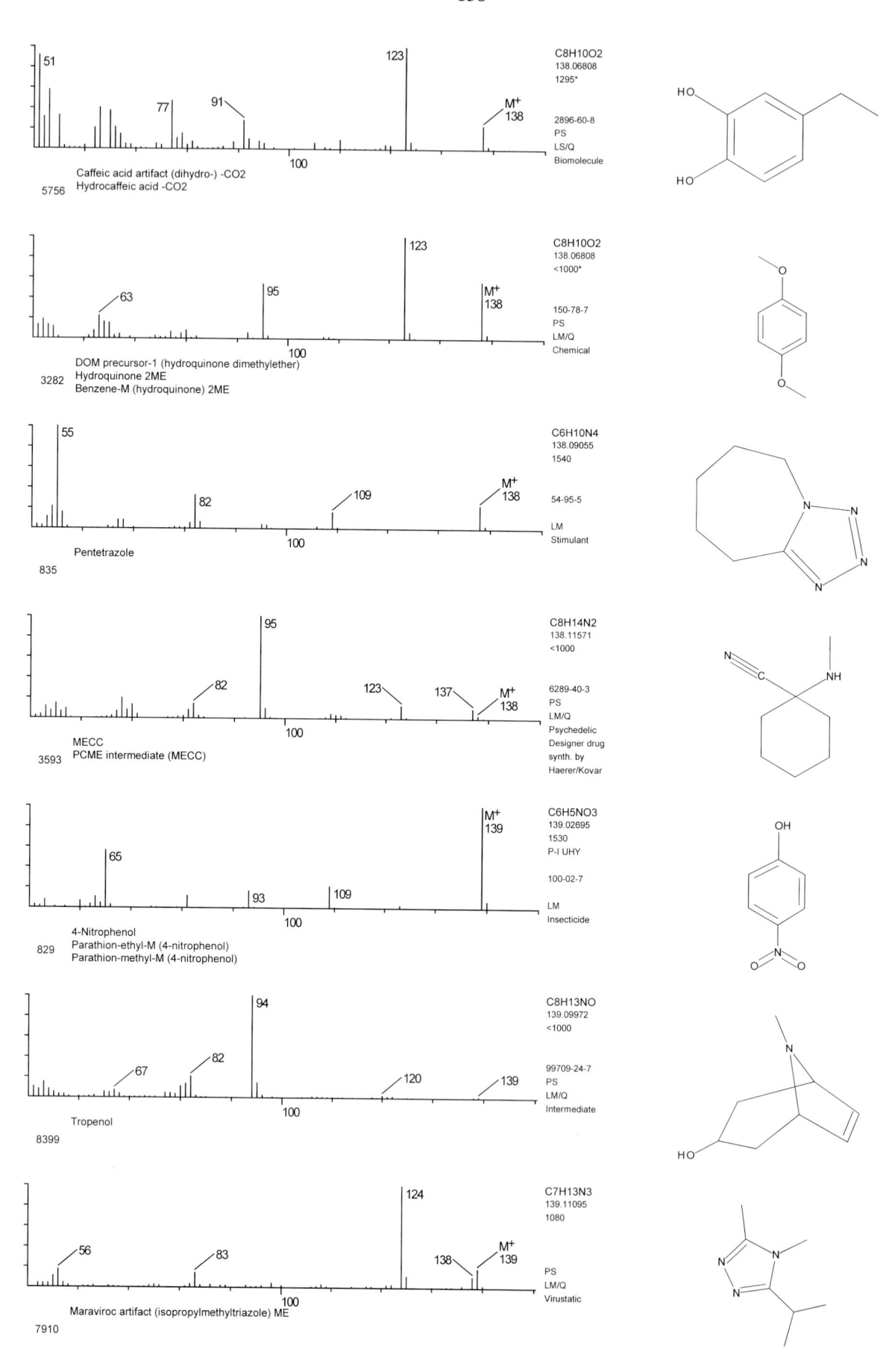

51, 77, 91, 123, M+ 138	C8H10O2 138.06808 1295* 2896-60-8 PS LS/Q Biomolecule

5756 Caffeic acid artifact (dihydro-) -CO2
Hydrocaffeic acid -CO2

63, 95, 123, M+ 138	C8H10O2 138.06808 <1000* 150-78-7 PS LM/Q Chemical

3282 DOM precursor-1 (hydroquinone dimethylether)
Hydroquinone 2ME
Benzene-M (hydroquinone) 2ME

55, 82, 109, M+ 138	C6H10N4 138.09055 1540 54-95-5 LM Stimulant

Pentetrazole
835

82, 95, 123, 137, M+ 138	C8H14N2 138.11571 <1000 6289-40-3 PS LM/Q Psychedelic Designer drug synth. by Haerer/Kovar

3593 MECC
PCME intermediate (MECC)

65, 93, 109, M+ 139	C6H5NO3 139.02695 1530 P-I UHY 100-02-7 LM Insecticide

4-Nitrophenol
829 Parathion-ethyl-M (4-nitrophenol)
Parathion-methyl-M (4-nitrophenol)

67, 82, 94, 120, 139	C8H13NO 139.09972 <1000 99709-24-7 PS LM/Q Intermediate

Tropenol
8399

56, 83, 124, 138, M+ 139	C7H13N3 139.11095 1080 PS LM/Q Virustatic

Maraviroc artifact (isopropylmethyltriazole) ME
7910

139

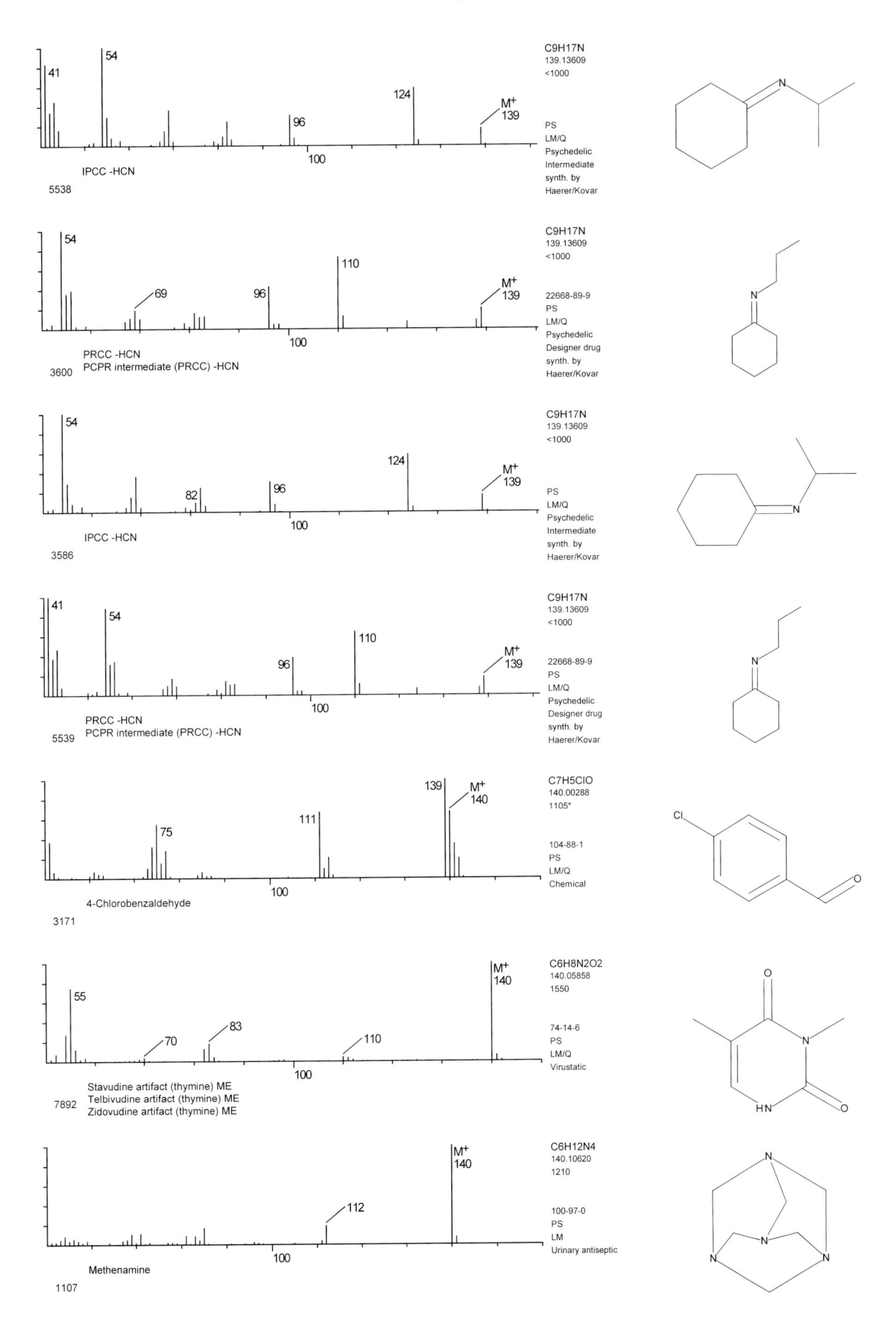

C9H17N
139.13609
<1000

PS
LM/Q
Psychedelic
Intermediate
synth. by
Haerer/Kovar

IPCC -HCN
5538

C9H17N
139.13609
<1000

22668-89-9
PS
LM/Q
Psychedelic
Designer drug
synth. by
Haerer/Kovar

PRCC -HCN
PCPR intermediate (PRCC) -HCN
3600

C9H17N
139.13609
<1000

PS
LM/Q
Psychedelic
Intermediate
synth. by
Haerer/Kovar

IPCC -HCN
3586

C9H17N
139.13609
<1000

22668-89-9
PS
LM/Q
Psychedelic
Designer drug
synth. by
Haerer/Kovar

PRCC -HCN
PCPR intermediate (PRCC) -HCN
5539

C7H5ClO
140.00288
1105*

104-88-1
PS
LM/Q
Chemical

4-Chlorobenzaldehyde
3171

C6H8N2O2
140.05858
1550

74-14-6
PS
LM/Q
Virustatic

Stavudine artifact (thymine) ME
Telbivudine artifact (thymine) ME
Zidovudine artifact (thymine) ME
7892

C6H12N4
140.10620
1210

100-97-0
PS
LM
Urinary antiseptic

Methenamine
1107

141

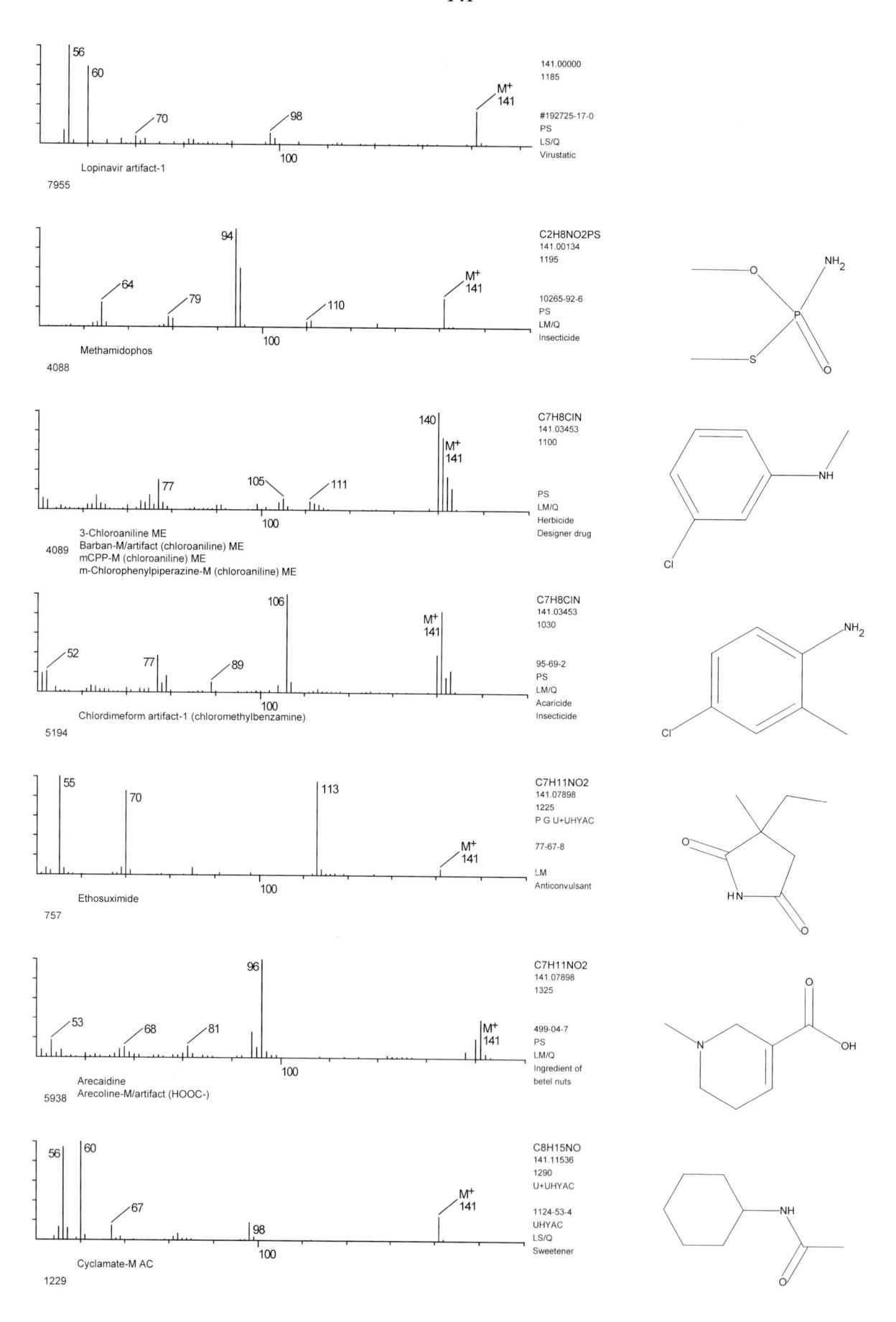

Lopinavir artifact-1
7955
141.00000
1185
#192725-17-0
PS
LS/Q
Virustatic

Methamidophos
4088
C2H8NO2PS
141.00134
1195
10265-92-6
PS
LM/Q
Insecticide

3-Chloroaniline ME
Barban-M/artifact (chloroaniline) ME
mCPP-M (chloroaniline) ME
m-Chlorophenylpiperazine-M (chloroaniline) ME
4089
C7H8ClN
141.03453
1100
PS
LM/Q
Herbicide
Designer drug

Chlordimeform artifact-1 (chloromethylbenzamine)
5194
C7H8ClN
141.03453
1030
95-69-2
PS
LM/Q
Acaricide
Insecticide

Ethosuximide
757
C7H11NO2
141.07898
1225
P G U+UHYAC
77-67-8
LM
Anticonvulsant

Arecaidine
Arecoline-M/artifact (HOOC-)
5938
C7H11NO2
141.07898
1325
499-04-7
PS
LM/Q
Ingredient of betel nuts

Cyclamate-M AC
1229
C8H15NO
141.11536
1290
U+UHYAC
1124-53-4
UHYAC
LS/Q
Sweetener

141

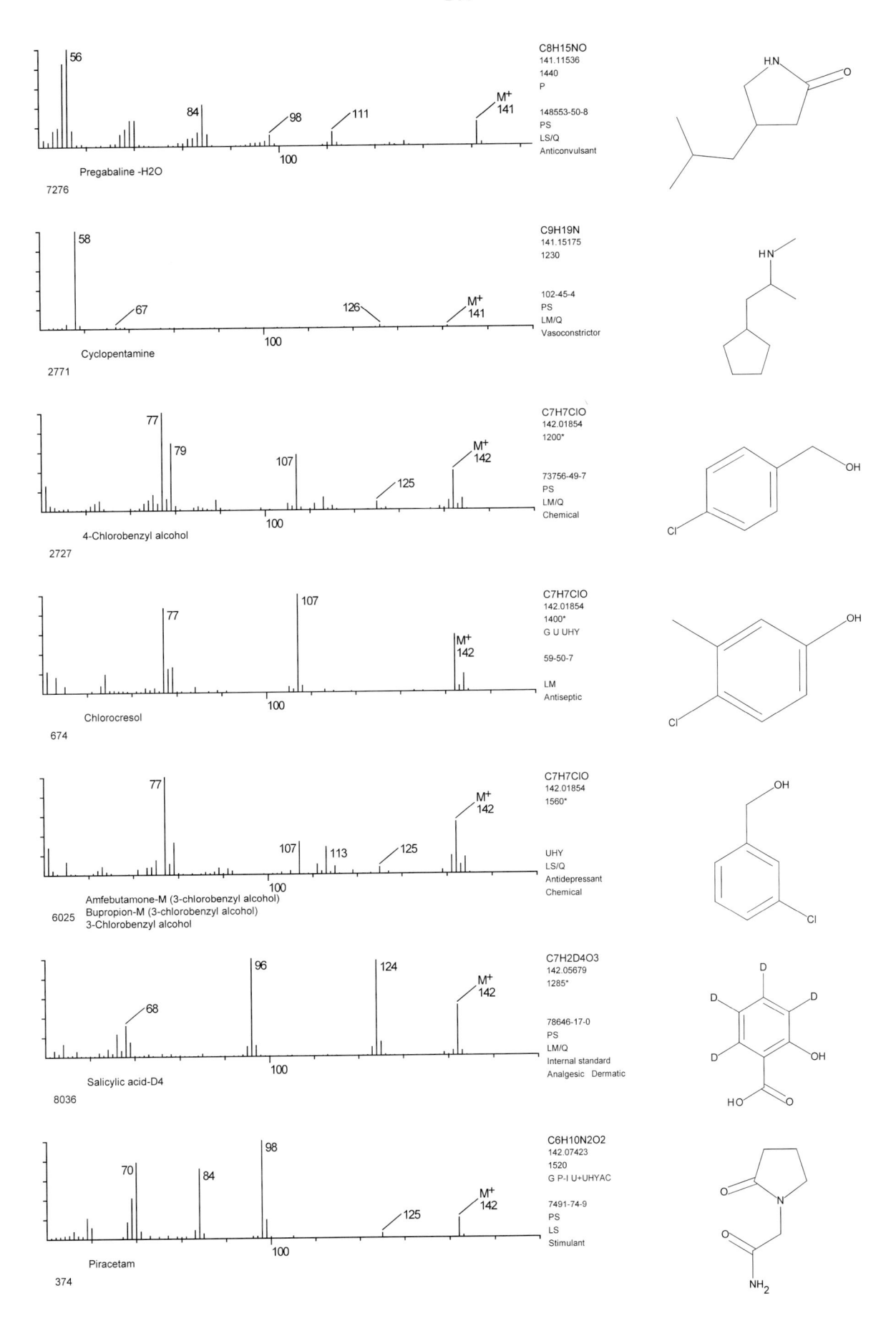

Pregabaline -H2O
7276

C8H15NO
141.11536
1440
P

148553-50-8
PS
LS/Q
Anticonvulsant

Cyclopentamine
2771

C9H19N
141.15175
1230

102-45-4
PS
LM/Q
Vasoconstrictor

4-Chlorobenzyl alcohol
2727

C7H7ClO
142.01854
1200*

73756-49-7
PS
LM/Q
Chemical

Chlorocresol
674

C7H7ClO
142.01854
1400*
G U UHY

59-50-7

LM
Antiseptic

Amfebutamone-M (3-chlorobenzyl alcohol)
Bupropion-M (3-chlorobenzyl alcohol)
3-Chlorobenzyl alcohol
6025

C7H7ClO
142.01854
1560*

UHY
LS/Q
Antidepressant
Chemical

Salicylic acid-D4
8036

C7H2D4O3
142.05679
1285*

78646-17-0
PS
LM/Q
Internal standard
Analgesic Dermatic

Piracetam
374

C6H10N2O2
142.07423
1520
G P-I U+UHYAC

7491-74-9
PS
LS
Stimulant

142

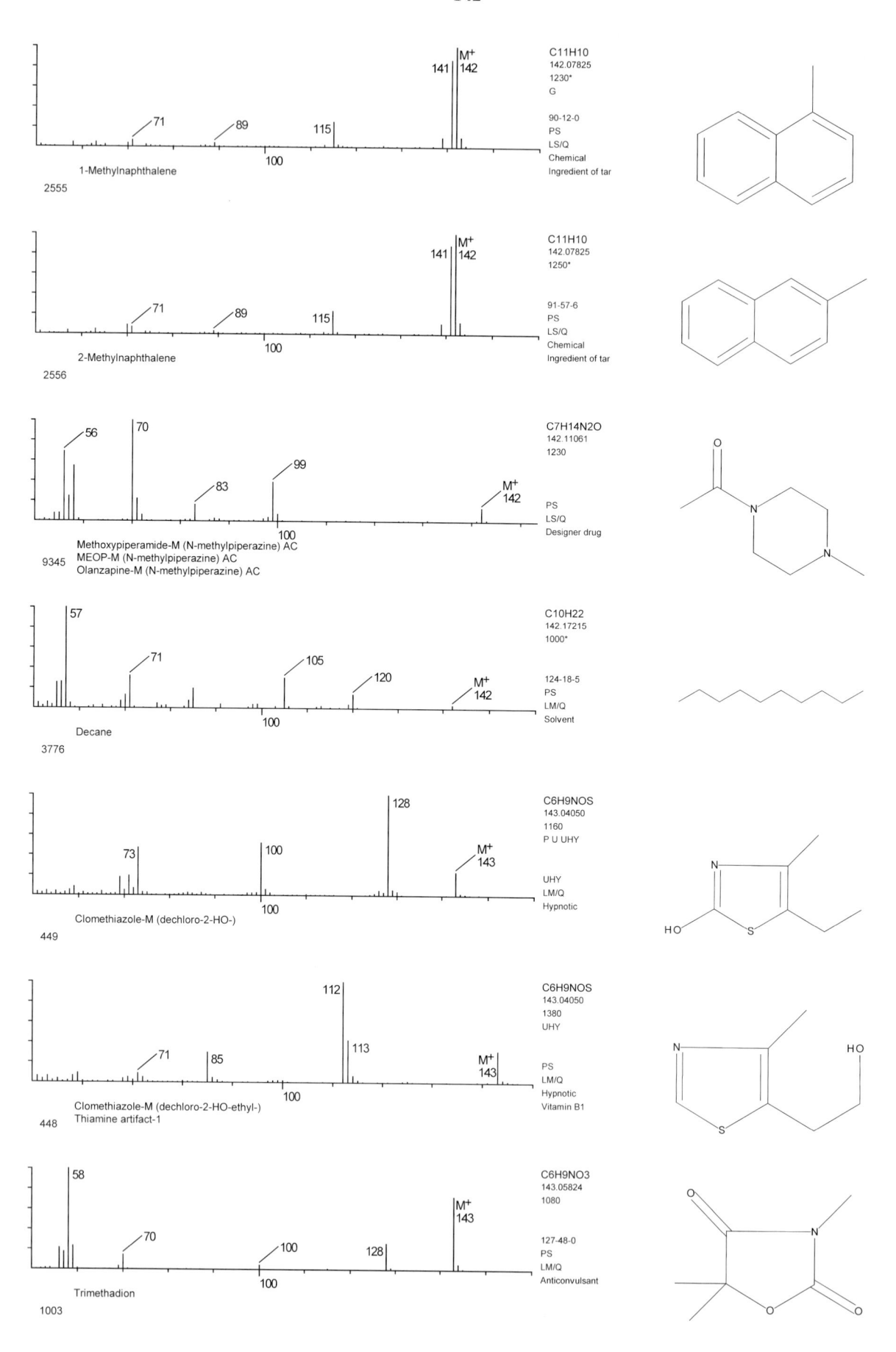

1-Methylnaphthalene
2555
C11H10
142.07825
1230*
G
90-12-0
PS
LS/Q
Chemical
Ingredient of tar

2-Methylnaphthalene
2556
C11H10
142.07825
1250*
91-57-6
PS
LS/Q
Chemical
Ingredient of tar

Methoxypiperamide-M (N-methylpiperazine) AC
MEOP-M (N-methylpiperazine) AC
Olanzapine-M (N-methylpiperazine) AC
9345
C7H14N2O
142.11061
1230
PS
LS/Q
Designer drug

Decane
3776
C10H22
142.17215
1000*
124-18-5
PS
LM/Q
Solvent

Clomethiazole-M (dechloro-2-HO-)
449
C6H9NOS
143.04050
1160
P U UHY
UHY
LM/Q
Hypnotic

Clomethiazole-M (dechloro-2-HO-ethyl-)
Thiamine artifact-1
448
C6H9NOS
143.04050
1380
UHY
PS
LM/Q
Hypnotic
Vitamin B1

Trimethadion
1003
C6H9NO3
143.05824
1080
127-48-0
PS
LM/Q
Anticonvulsant

143

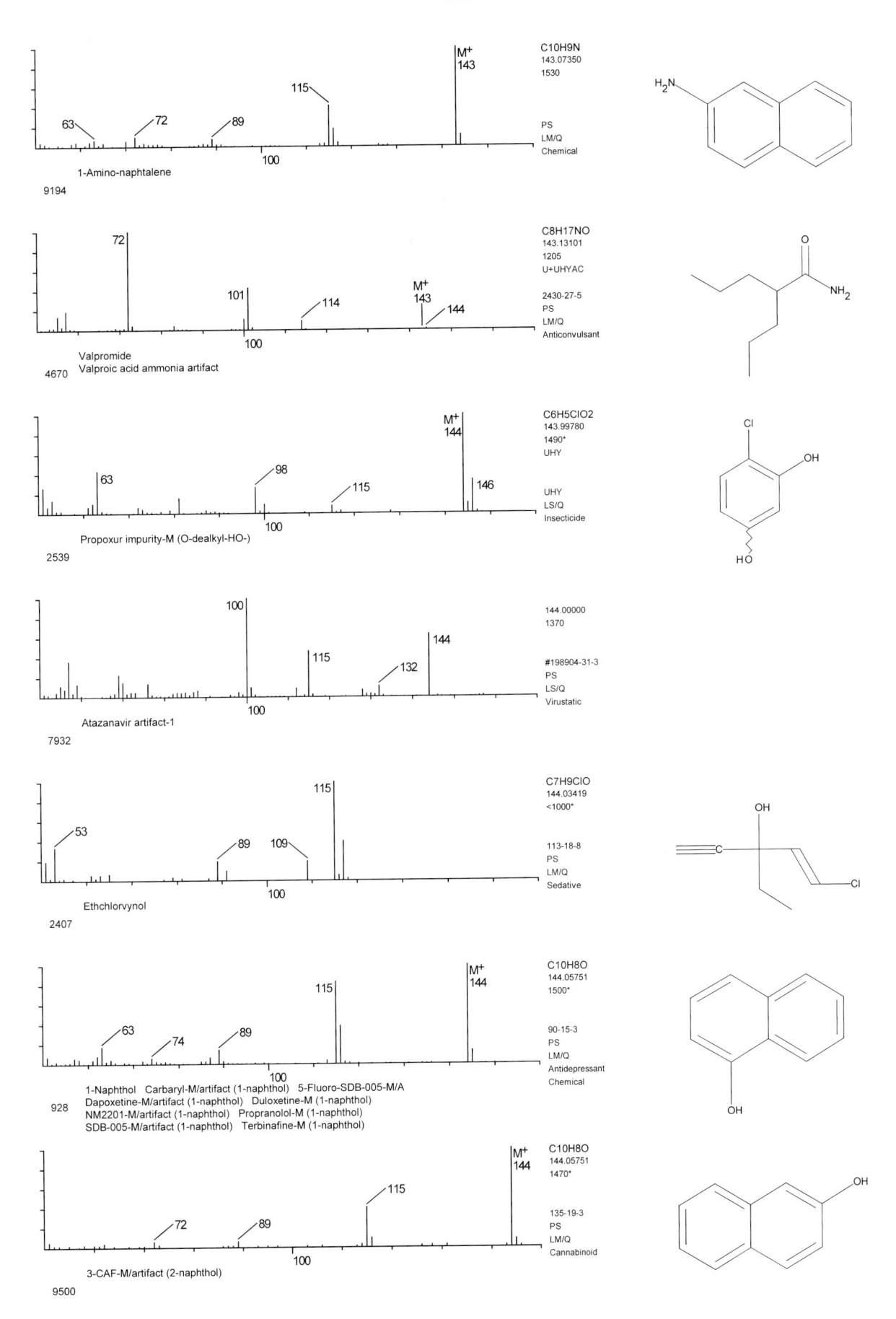

1-Amino-naphtalene 9194	C10H9N 143.07350 1530 PS LM/Q Chemical
Valpromide Valproic acid ammonia artifact 4670	C8H17NO 143.13101 1205 U+UHYAC 2430-27-5 PS LM/Q Anticonvulsant
Propoxur impurity-M (O-dealkyl-HO-) 2539	C6H5ClO2 143.99780 1490* UHY UHY LS/Q Insecticide
Atazanavir artifact-1 7932	144.00000 1370 #198904-31-3 PS LS/Q Virustatic
Ethchlorvynol 2407	C7H9ClO 144.03419 <1000* 113-18-8 PS LM/Q Sedative
1-Naphthol Carbaryl-M/artifact (1-naphthol) 5-Fluoro-SDB-005-M/A Dapoxetine-M/artifact (1-naphthol) Duloxetine-M (1-naphthol) NM2201-M/artifact (1-naphthol) Propranolol-M (1-naphthol) SDB-005-M/artifact (1-naphthol) Terbinafine-M (1-naphthol) 928	C10H8O 144.05751 1500* 90-15-3 PS LM/Q Antidepressant Chemical
3-CAF-M/artifact (2-naphthol) 9500	C10H8O 144.05751 1470* 135-19-3 PS LM/Q Cannabinoid

144

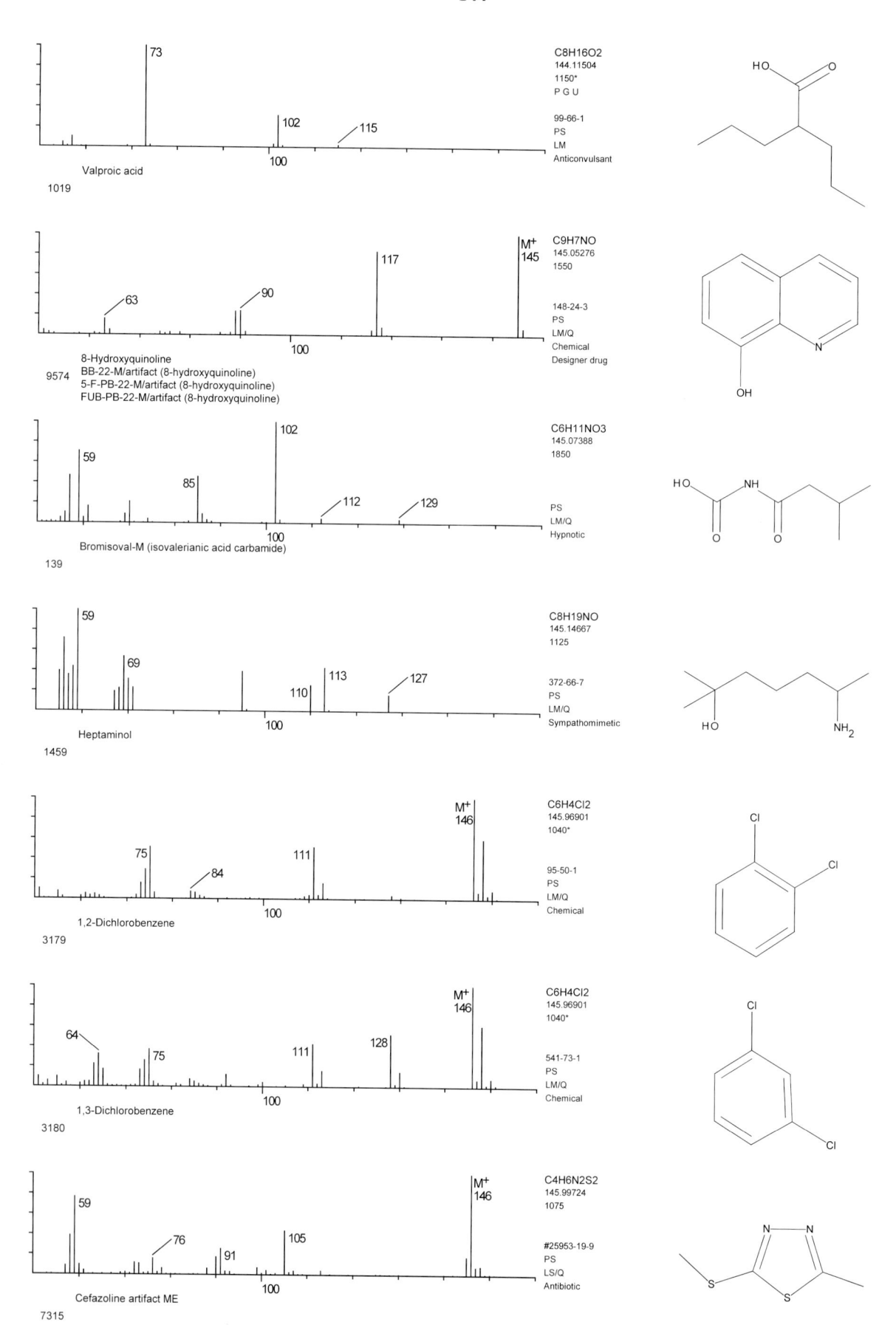

Valproic acid — peaks 73, 102, 115	C8H16O2 145.11504 1150* 99-66-1 PS LM Anticonvulsant	

1019

8-Hydroxyquinoline — peaks 63, 90, 117, M+ 145
BB-22-M/artifact (8-hydroxyquinoline)
5-F-PB-22-M/artifact (8-hydroxyquinoline)
FUB-PB-22-M/artifact (8-hydroxyquinoline)

C9H7NO
145.05276
1550
148-24-3
PS
LM/Q
Chemical
Designer drug

9574

Bromisoval-M (isovalerianic acid carbamide) — peaks 59, 85, 102, 112, 129

C6H11NO3
145.07388
1850
PS
LM/Q
Hypnotic

139

Heptaminol — peaks 59, 69, 110, 113, 127

C8H19NO
145.14667
1125
372-66-7
PS
LM/Q
Sympathomimetic

1459

1,2-Dichlorobenzene — peaks 75, 84, 111, M+ 146

C6H4Cl2
145.96901
1040*
95-50-1
PS
LM/Q
Chemical

3179

1,3-Dichlorobenzene — peaks 64, 75, 111, 128, M+ 146

C6H4Cl2
145.96901
1040*
541-73-1
PS
LM/Q
Chemical

3180

Cefazoline artifact ME — peaks 59, 76, 91, 105, M+ 146

C4H6N2S2
145.99724
1075
#25953-19-9
PS
LS/Q
Antibiotic

7315

284

146

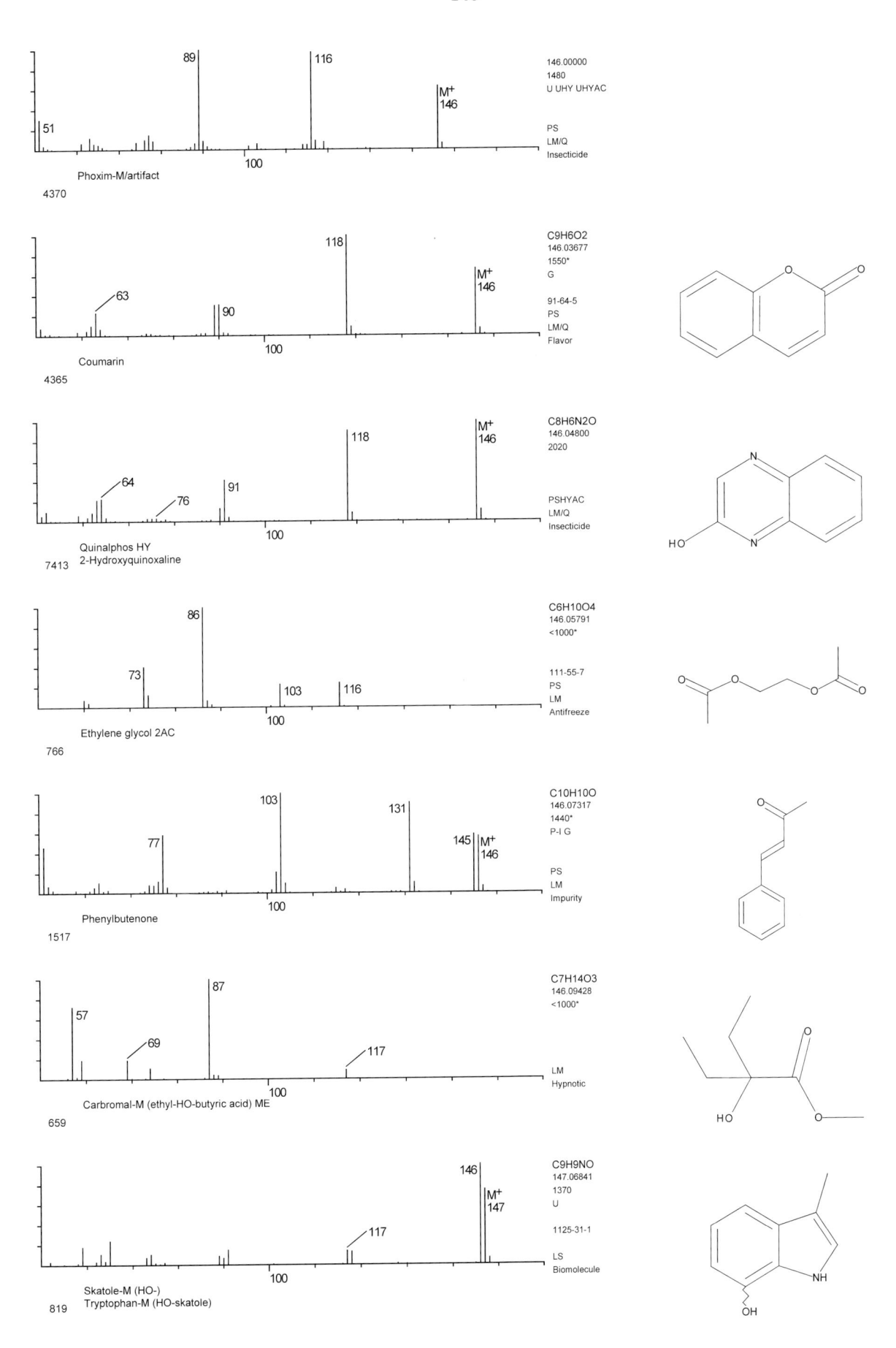

Phoxim-M/artifact
4370
- 146.00000
- 1480
- U UHY UHYAC
- PS
- LM/Q
- Insecticide
- Peaks: 51, 89, 116, M+ 146

Coumarin
4365
- C9H6O2
- 146.03677
- 1550*
- G
- 91-64-5
- PS
- LM/Q
- Flavor
- Peaks: 63, 90, 118, M+ 146

Quinalphos HY
2-Hydroxyquinoxaline
7413
- C8H6N2O
- 146.04800
- 2020
- PSHYAC
- LM/Q
- Insecticide
- Peaks: 64, 76, 91, 118, M+ 146

Ethylene glycol 2AC
766
- C6H10O4
- 146.05791
- <1000*
- 111-55-7
- PS
- LM
- Antifreeze
- Peaks: 73, 86, 103, 116

Phenylbutenone
1517
- C10H10O
- 146.07317
- 1440*
- P-I G
- PS
- LM
- Impurity
- Peaks: 77, 103, 131, 145, M+ 146

Carbromal-M (ethyl-HO-butyric acid) ME
659
- C7H14O3
- 146.09428
- <1000*
- LM
- Hypnotic
- Peaks: 57, 69, 87, 117

Skatole-M (HO-)
Tryptophan-M (HO-skatole)
819
- C9H9NO
- 147.06841
- 1370
- U
- 1125-31-1
- LS
- Biomolecule
- Peaks: 117, 146, M+ 147

147

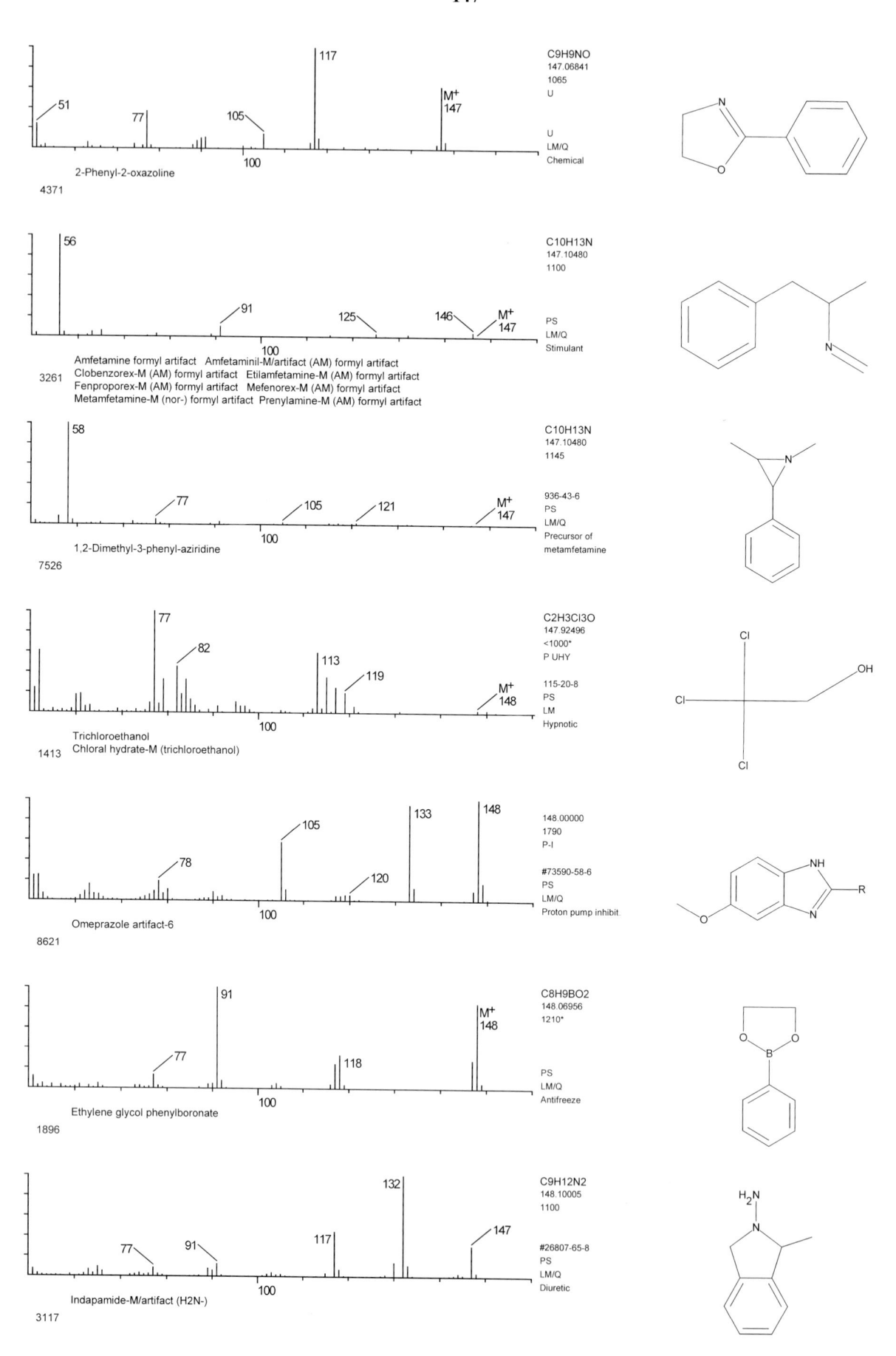

2-Phenyl-2-oxazoline
4371
C9H9NO
147.06841
1065
U

U
LM/Q
Chemical

Amfetamine formyl artifact   Amfetaminil-M/artifact (AM) formyl artifact
Clobenzorex-M (AM) formyl artifact   Etilamfetamine-M (AM) formyl artifact
Fenproporex-M (AM) formyl artifact   Mefenorex-M (AM) formyl artifact
Metamfetamine-M (nor-) formyl artifact   Prenylamine-M (AM) formyl artifact
3261
C10H13N
147.10480
1100

PS
LM/Q
Stimulant

1,2-Dimethyl-3-phenyl-aziridine
7526
C10H13N
147.10480
1145

936-43-6
PS
LM/Q
Precursor of metamfetamine

Trichloroethanol
Chloral hydrate-M (trichloroethanol)
1413
C2H3Cl3O
147.92496
<1000*
P UHY

115-20-8
PS
LM
Hypnotic

Omeprazole artifact-6
8621
148.00000
1790
P-I

#73590-58-6
PS
LM/Q
Proton pump inhibit.

Ethylene glycol phenylboronate
1896
C8H9BO2
148.06956
1210*

PS
LM/Q
Antifreeze

Indapamide-M/artifact (H2N-)
3117
C9H12N2
148.10005
1100

#26807-65-8
PS
LM/Q
Diuretic

149

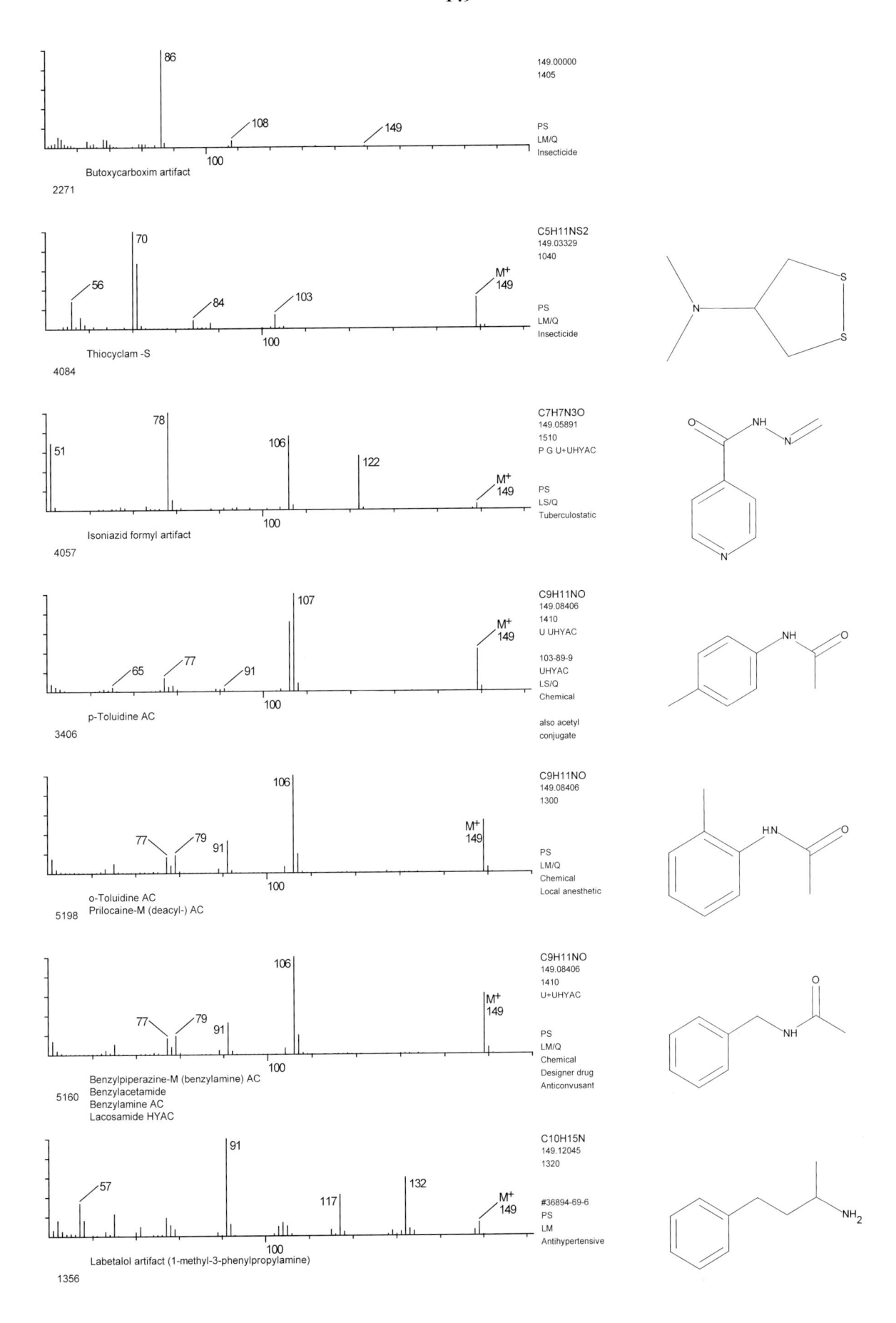

Butoxycarboxim artifact
2271
- 86, 108, 149
- 149.00000
- 1405
- PS
- LM/Q
- Insecticide

Thiocyclam -S
4084
- 56, 70, 84, 103, M+ 149
- C5H11NS2
- 149.03329
- 1040
- PS
- LM/Q
- Insecticide

Isoniazid formyl artifact
4057
- 51, 78, 106, 122, M+ 149
- C7H7N3O
- 149.05891
- 1510
- P G U+UHYAC
- PS
- LS/Q
- Tuberculostatic

p-Toluidine AC
3406
- 65, 77, 91, 107, M+ 149
- C9H11NO
- 149.08406
- 1410
- U UHYAC
- 103-89-9
- UHYAC
- LS/Q
- Chemical
- also acetyl conjugate

o-Toluidine AC
Prilocaine-M (deacyl-) AC
5198
- 77, 79, 91, 106, M+ 149
- C9H11NO
- 149.08406
- 1300
- PS
- LM/Q
- Chemical
- Local anesthetic

Benzylpiperazine-M (benzylamine) AC
Benzylacetamide
Benzylamine AC
Lacosamide HYAC
5160
- 77, 79, 91, 106, M+ 149
- C9H11NO
- 149.08406
- 1410
- U+UHYAC
- PS
- LM/Q
- Chemical
- Designer drug
- Anticonvusant

Labetalol artifact (1-methyl-3-phenylpropylamine)
1356
- 57, 91, 117, 132, M+ 149
- C10H15N
- 149.12045
- 1320
- #36894-69-6
- PS
- LM
- Antihypertensive

287

149

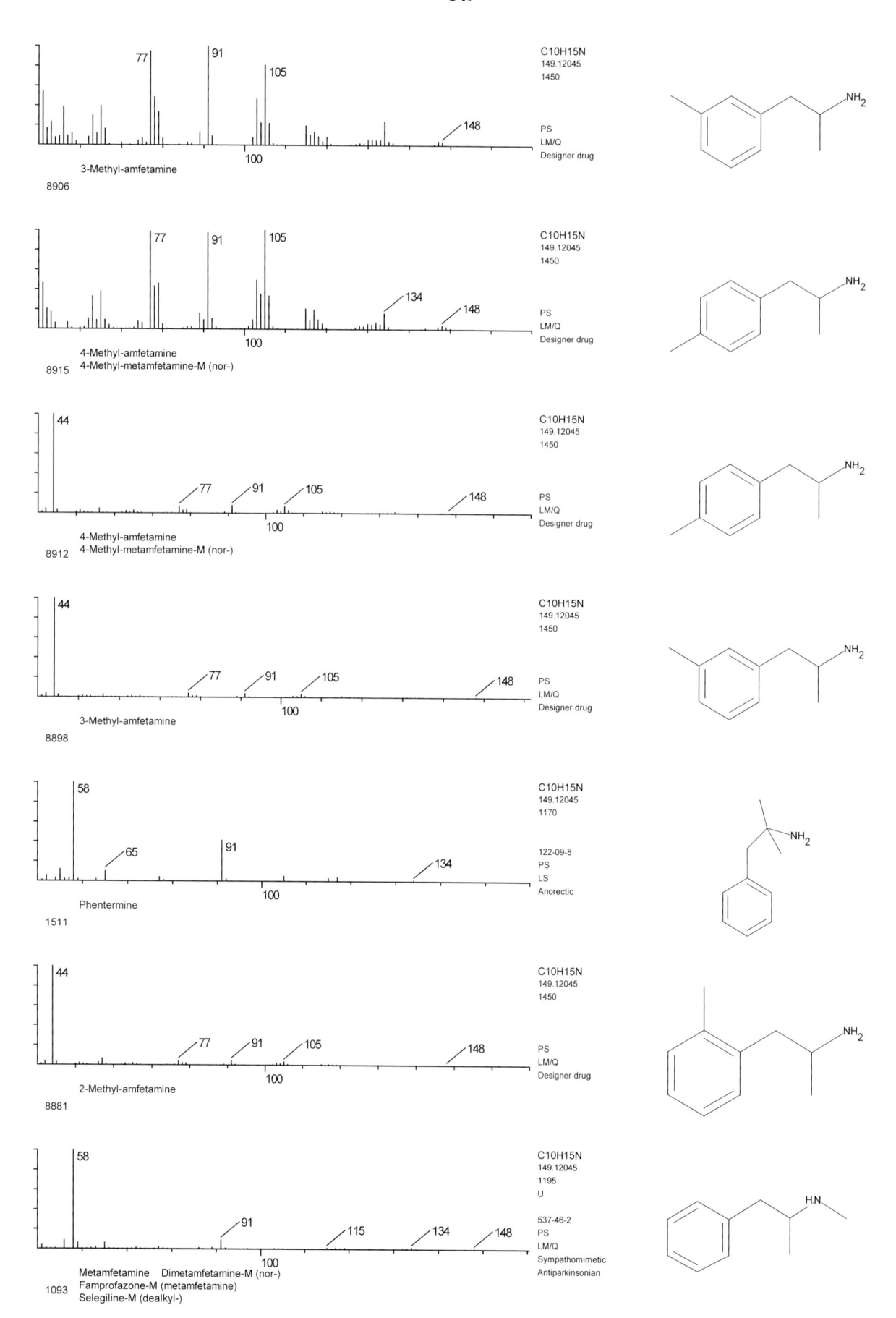

3-Methyl-amfetamine 8906	C10H15N 149.12045 1450 PS LM/Q Designer drug	
4-Methyl-amfetamine 4-Methyl-metamfetamine-M (nor-) 8915	C10H15N 149.12045 1450 PS LM/Q Designer drug	
4-Methyl-amfetamine 4-Methyl-metamfetamine-M (nor-) 8912	C10H15N 149.12045 1450 PS LM/Q Designer drug	
3-Methyl-amfetamine 8898	C10H15N 149.12045 1450 PS LM/Q Designer drug	
Phentermine 1511	C10H15N 149.12045 1170 122-09-8 PS LS Anorectic	
2-Methyl-amfetamine 8881	C10H15N 149.12045 1450 PS LM/Q Designer drug	
Metamfetamine Dimetamfetamine-M (nor-) Famprofazone-M (metamfetamine) Selegiline-M (dealkyl-) 1093	C10H15N 149.12045 1195 U 537-46-2 PS LM/Q Sympathomimetic Antiparkinsonian	

288

149

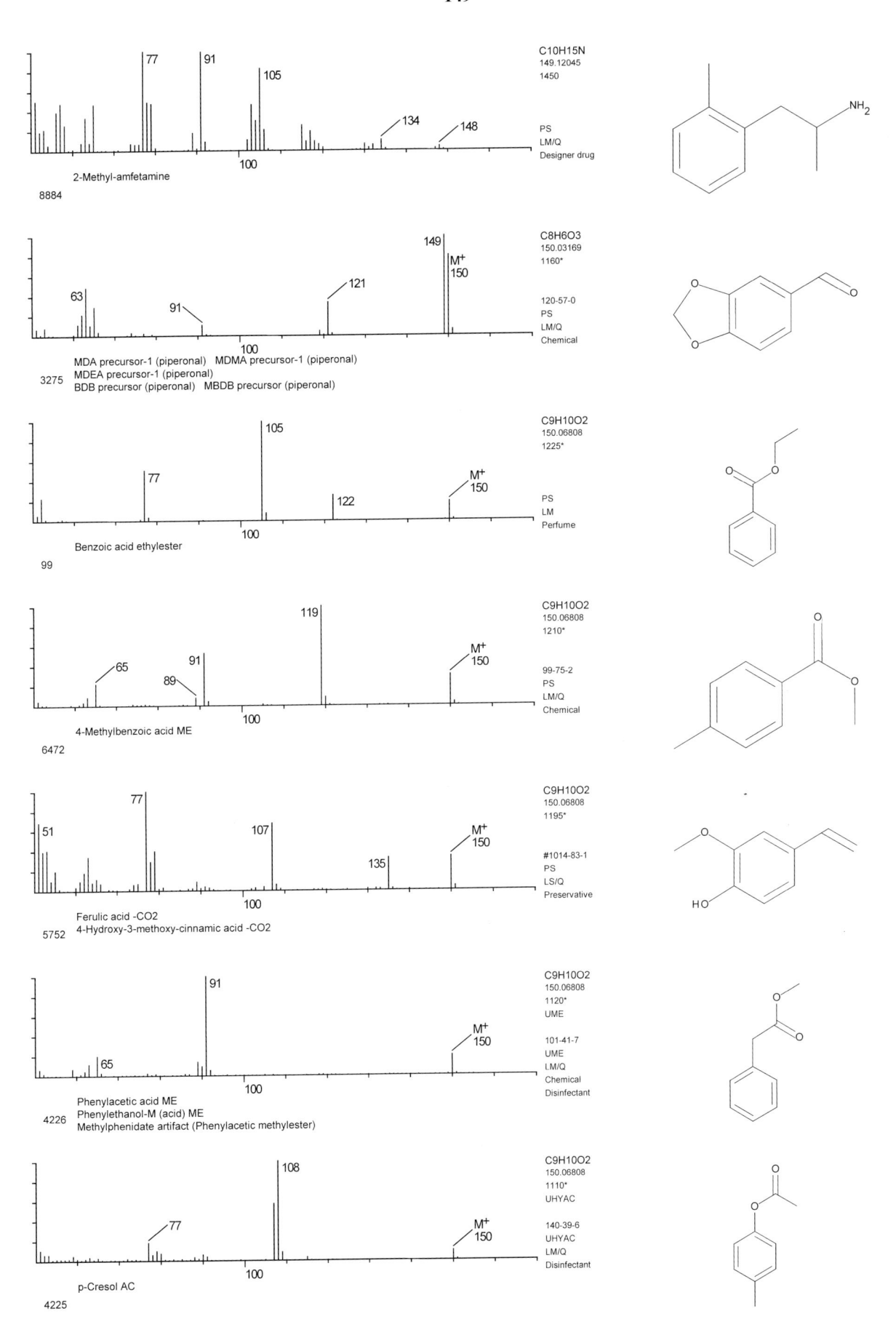

2-Methyl-amfetamine
8884
C10H15N
149.12045
1450
PS
LM/Q
Designer drug
Peaks: 77, 91, 105, 134, 148

MDA precursor-1 (piperonal) MDMA precursor-1 (piperonal)
MDEA precursor-1 (piperonal)
BDB precursor (piperonal) MBDB precursor (piperonal)
3275
C8H6O3
150.03169
1160*
120-57-0
PS
LM/Q
Chemical
Peaks: 63, 91, 121, 149, M+ 150

Benzoic acid ethylester
99
C9H10O2
150.06808
1225*
PS
LM
Perfume
Peaks: 77, 105, 122, M+ 150

4-Methylbenzoic acid ME
6472
C9H10O2
150.06808
1210*
99-75-2
PS
LM/Q
Chemical
Peaks: 65, 89, 91, 119, M+ 150

Ferulic acid -CO2
4-Hydroxy-3-methoxy-cinnamic acid -CO2
5752
C9H10O2
150.06808
1195*
#1014-83-1
PS
LS/Q
Preservative
Peaks: 51, 77, 107, 135, M+ 150

Phenylacetic acid ME
Phenylethanol-M (acid) ME
Methylphenidate artifact (Phenylacetic methylester)
4226
C9H10O2
150.06808
1120*
UME
101-41-7
UME
LM/Q
Chemical
Disinfectant
Peaks: 65, 91, M+ 150

p-Cresol AC
4225
C9H10O2
150.06808
1110*
UHYAC
140-39-6
UHYAC
LM/Q
Disinfectant
Peaks: 77, 108, M+ 150

150

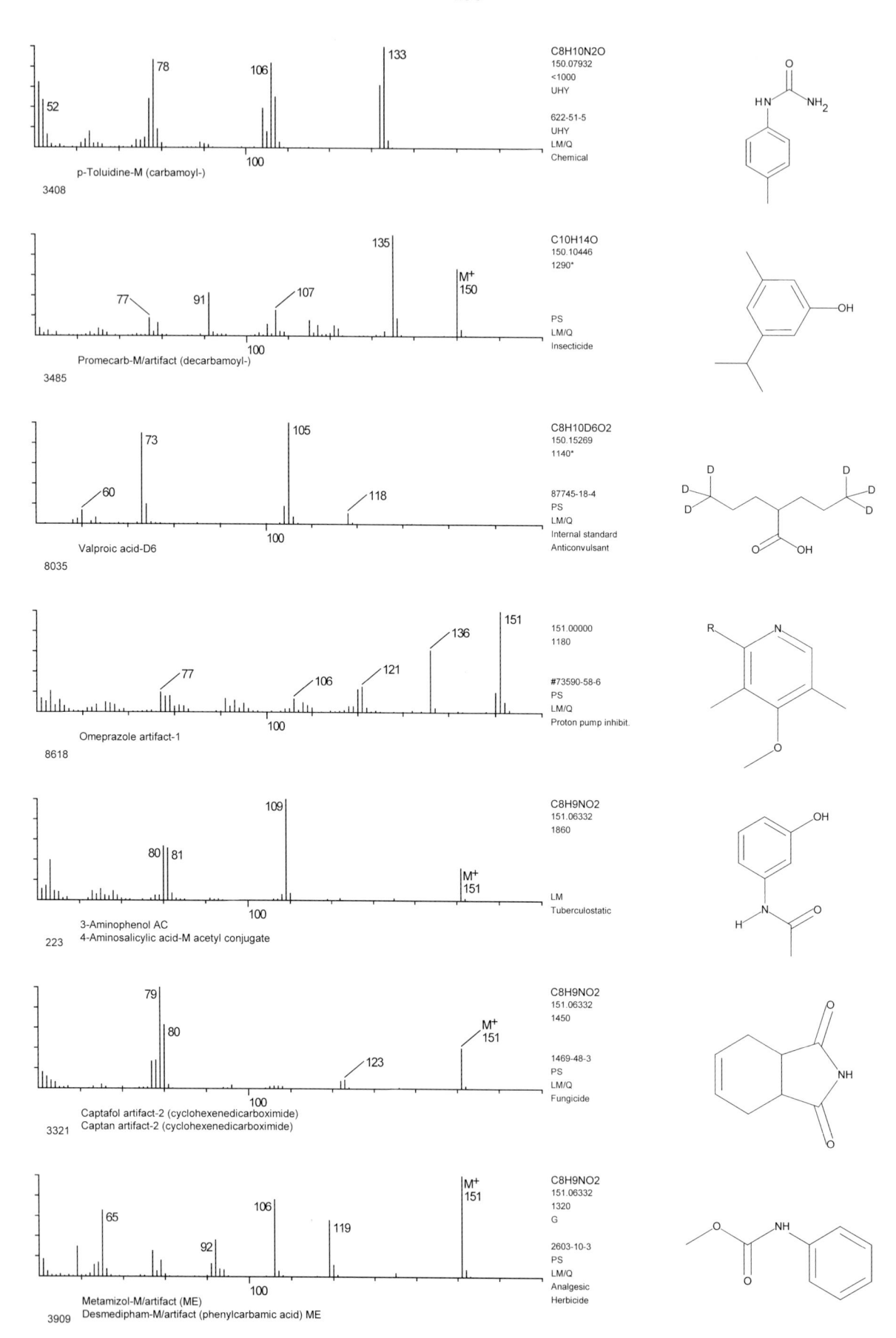

p-Toluidine-M (carbamoyl-)
3408

C8H10N2O
150.07932
<1000
UHY

622-51-5
UHY
LM/Q
Chemical

Promecarb-M/artifact (decarbamoyl-)
3485

C10H14O
150.10446
1290*

PS
LM/Q
Insecticide

Valproic acid-D6
8035

C8H10D6O2
150.15269
1140*

87745-18-4
PS
LM/Q
Internal standard
Anticonvulsant

Omeprazole artifact-1
8618

151.00000
1180

#73590-58-6
PS
LM/Q
Proton pump inhibit.

3-Aminophenol AC
4-Aminosalicylic acid-M acetyl conjugate
223

C8H9NO2
151.06332
1860

LM
Tuberculostatic

Captafol artifact-2 (cyclohexenedicarboximide)
Captan artifact-2 (cyclohexenedicarboximide)
3321

C8H9NO2
151.06332
1450

1469-48-3
PS
LM/Q
Fungicide

Metamizol-M/artifact (ME)
Desmedipham-M/artifact (phenylcarbamic acid) ME
3909

C8H9NO2
151.06332
1320
G

2603-10-3
PS
LM/Q
Analgesic
Herbicide

290

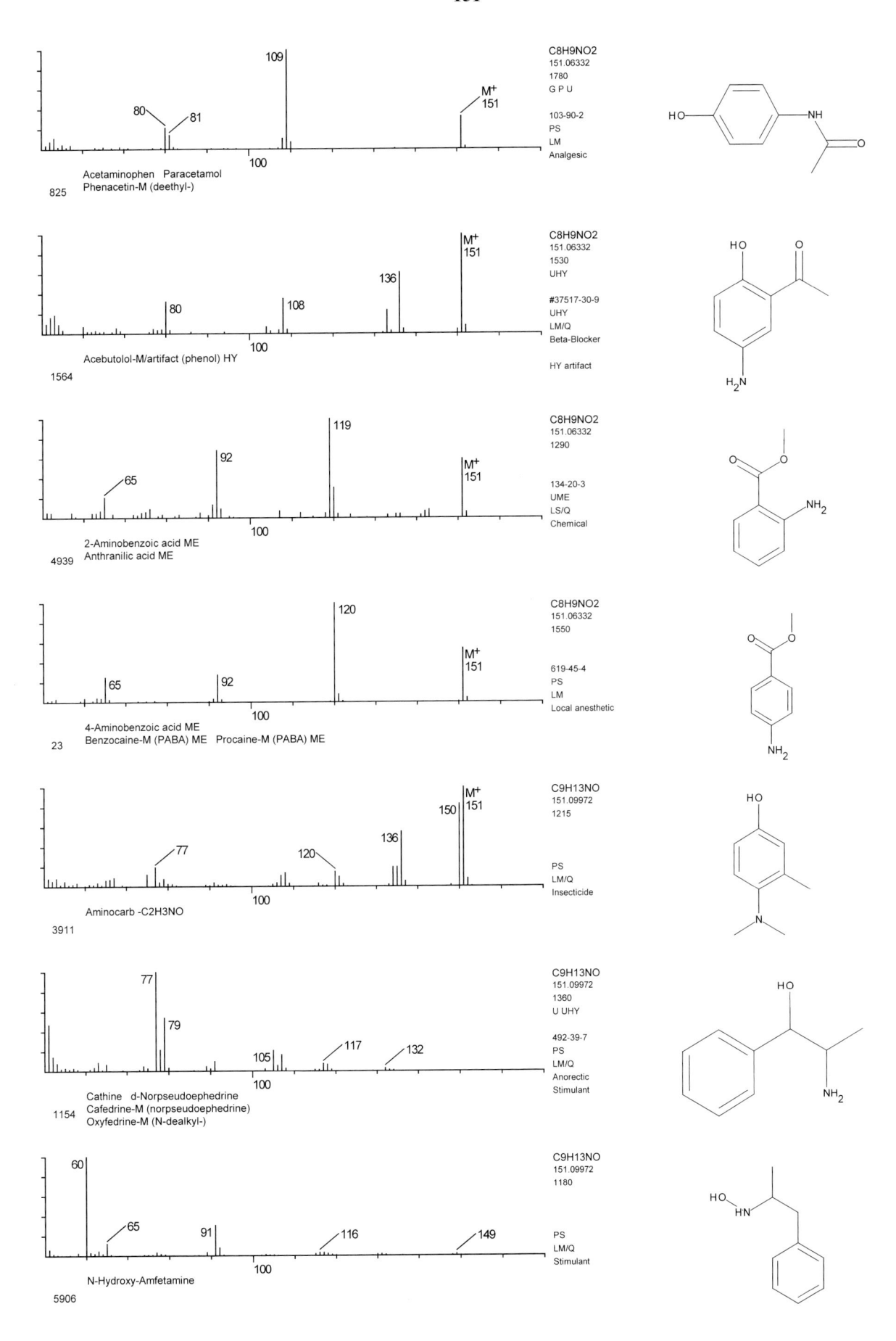

825	Acetaminophen Paracetamol Phenacetin-M (deethyl-) Peaks: 80, 81, 109, 151 M+ C8H9NO2 151.06332 1780 G P U 103-90-2 PS LM Analgesic
1564	Acebutolol-M/artifact (phenol) HY Peaks: 80, 108, 136, 151 M+ C8H9NO2 151.06332 1530 UHY #37517-30-9 UHY LM/Q Beta-Blocker HY artifact
4939	2-Aminobenzoic acid ME Anthranilic acid ME Peaks: 65, 92, 119, 151 M+ C8H9NO2 151.06332 1290 134-20-3 UME LS/Q Chemical
23	4-Aminobenzoic acid ME Benzocaine-M (PABA) ME Procaine-M (PABA) ME Peaks: 65, 92, 120, 151 M+ C8H9NO2 151.06332 1550 619-45-4 PS LM Local anesthetic
3911	Aminocarb -C2H3NO Peaks: 77, 120, 136, 150, 151 M+ C9H13NO 151.09972 1215 PS LM/Q Insecticide
1154	Cathine d-Norpseudoephedrine Cafedrine-M (norpseudoephedrine) Oxyfedrine-M (N-dealkyl-) Peaks: 77, 79, 105, 117, 132 C9H13NO 151.09972 1360 U UHY 492-39-7 PS LM/Q Anorectic Stimulant
5906	N-Hydroxy-Amfetamine Peaks: 60, 65, 91, 116, 149 C9H13NO 151.09972 1180 PS LM/Q Stimulant

151

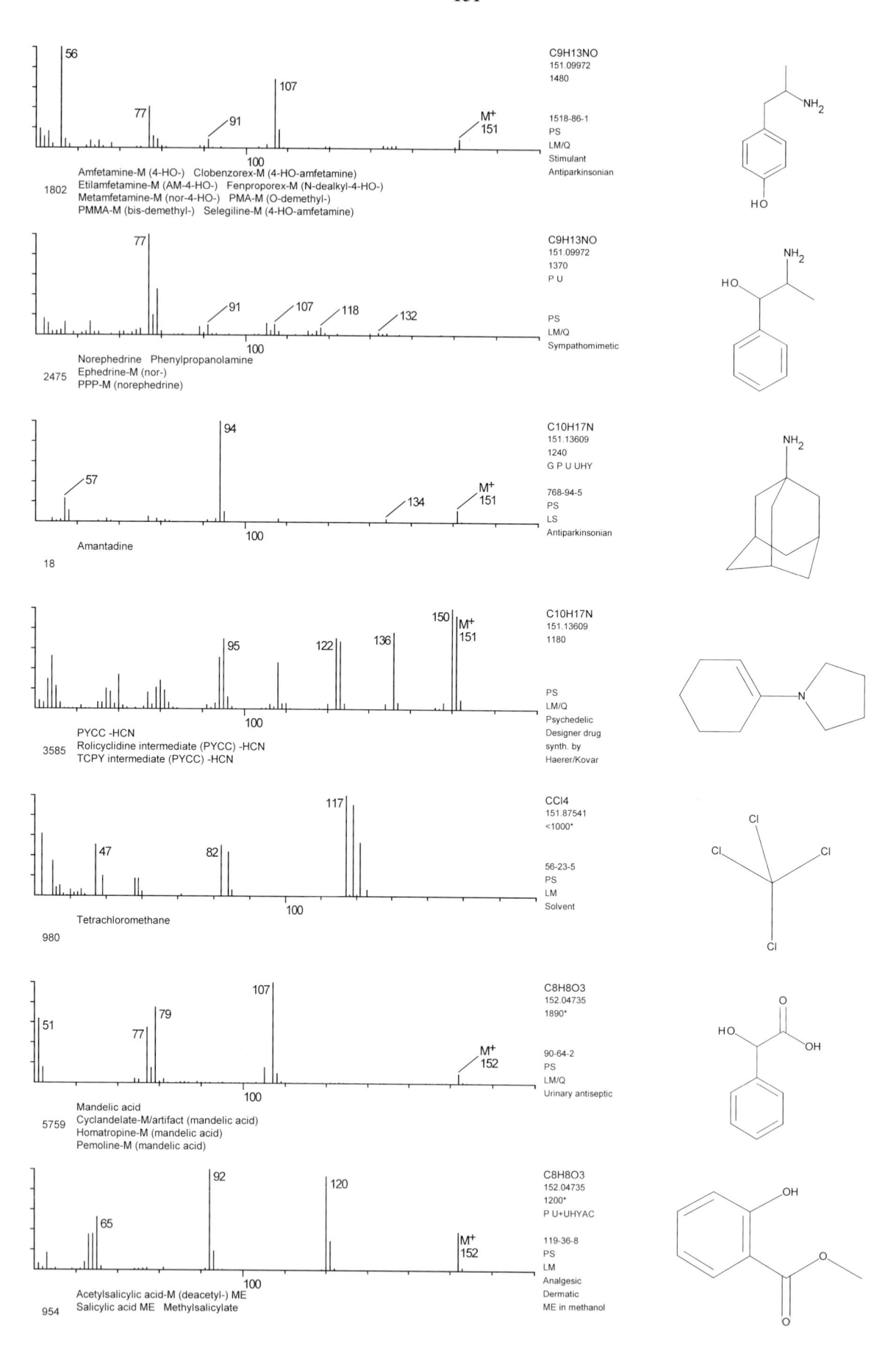

C9H13NO
151.09972
1480

1518-86-1
PS
LM/Q
Stimulant
Antiparkinsonian

Amfetamine-M (4-HO-)   Clobenzorex-M (4-HO-amfetamine)
Etilamfetamine-M (AM-4-HO-)   Fenproporex-M (N-dealkyl-4-HO-)
Metamfetamine-M (nor-4-HO-)   PMA-M (O-demethyl-)
PMMA-M (bis-demethyl-)   Selegiline-M (4-HO-amfetamine)
1802

C9H13NO
151.09972
1370
P U

PS
LM/Q
Sympathomimetic

Norephedrine   Phenylpropanolamine
Ephedrine-M (nor-)
PPP-M (norephedrine)
2475

C10H17N
151.13609
1240
G P U UHY

768-94-5
PS
LS
Antiparkinsonian

Amantadine
18

C10H17N
151.13609
1180

PS
LM/Q
Psychedelic
Designer drug
synth. by
Haerer/Kovar

PYCC -HCN
Rolicyclidine intermediate (PYCC) -HCN
TCPY intermediate (PYCC) -HCN
3585

CCl4
151.87541
<1000*

56-23-5
PS
LM
Solvent

Tetrachloromethane
980

C8H8O3
152.04735
1890*

90-64-2
PS
LM/Q
Urinary antiseptic

Mandelic acid
Cyclandelate-M/artifact (mandelic acid)
Homatropine-M (mandelic acid)
Pemoline-M (mandelic acid)
5759

C8H8O3
152.04735
1200*
P U+UHYAC

119-36-8
PS
LM
Analgesic
Dermatic
ME in methanol

Acetylsalicylic acid-M (deacetyl-) ME
Salicylic acid ME   Methylsalicylate
954

152

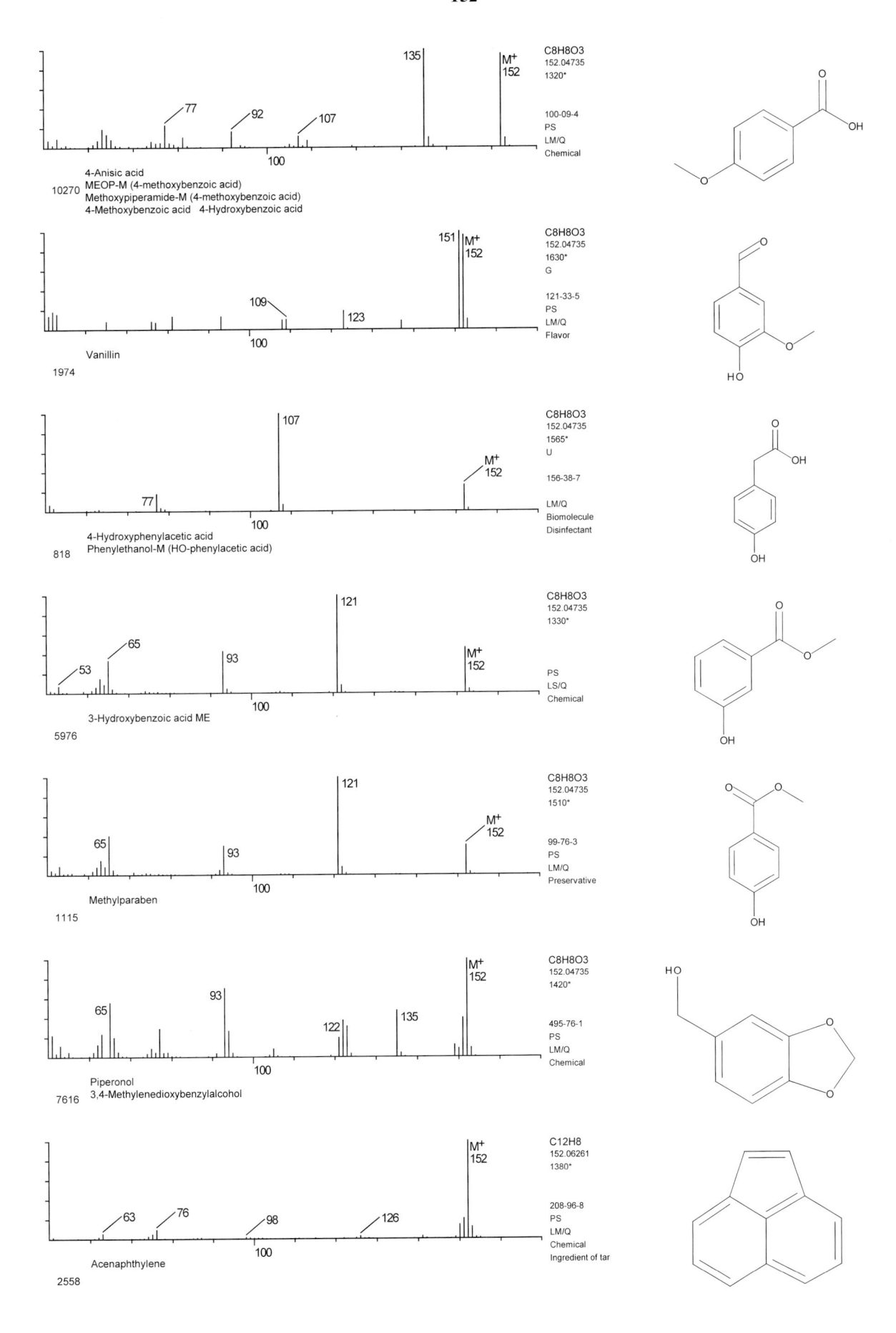

4-Anisic acid	C8H8O3 152.04735 1320*	
MEOP-M (4-methoxybenzoic acid)	100-09-4 PS LM/Q Chemical	
10270 Methoxypiperamide-M (4-methoxybenzoic acid) 4-Methoxybenzoic acid 4-Hydroxybenzoic acid		

Peaks: 77, 92, 107, 135, M+ 152

Vanillin — 1974
C8H8O3 152.04735 1630* G
121-33-5 PS LM/Q Flavor
Peaks: 109, 123, 151, M+ 152

4-Hydroxyphenylacetic acid
818 Phenylethanol-M (HO-phenylacetic acid)
C8H8O3 152.04735 1565* U
156-38-7 LM/Q Biomolecule Disinfectant
Peaks: 77, 107, M+ 152

3-Hydroxybenzoic acid ME — 5976
C8H8O3 152.04735 1330*
PS LS/Q Chemical
Peaks: 53, 65, 93, 121, M+ 152

Methylparaben — 1115
C8H8O3 152.04735 1510*
99-76-3 PS LM/Q Preservative
Peaks: 65, 93, 121, M+ 152

Piperonol
7616 3,4-Methylenedioxybenzylalcohol
C8H8O3 152.04735 1420*
495-76-1 PS LM/Q Chemical
Peaks: 65, 93, 122, 135, M+ 152

Acenaphthylene — 2558
C12H8 152.06261 1380*
208-96-8 PS LM/Q Chemical Ingredient of tar
Peaks: 63, 76, 98, 126, M+ 152

293

152

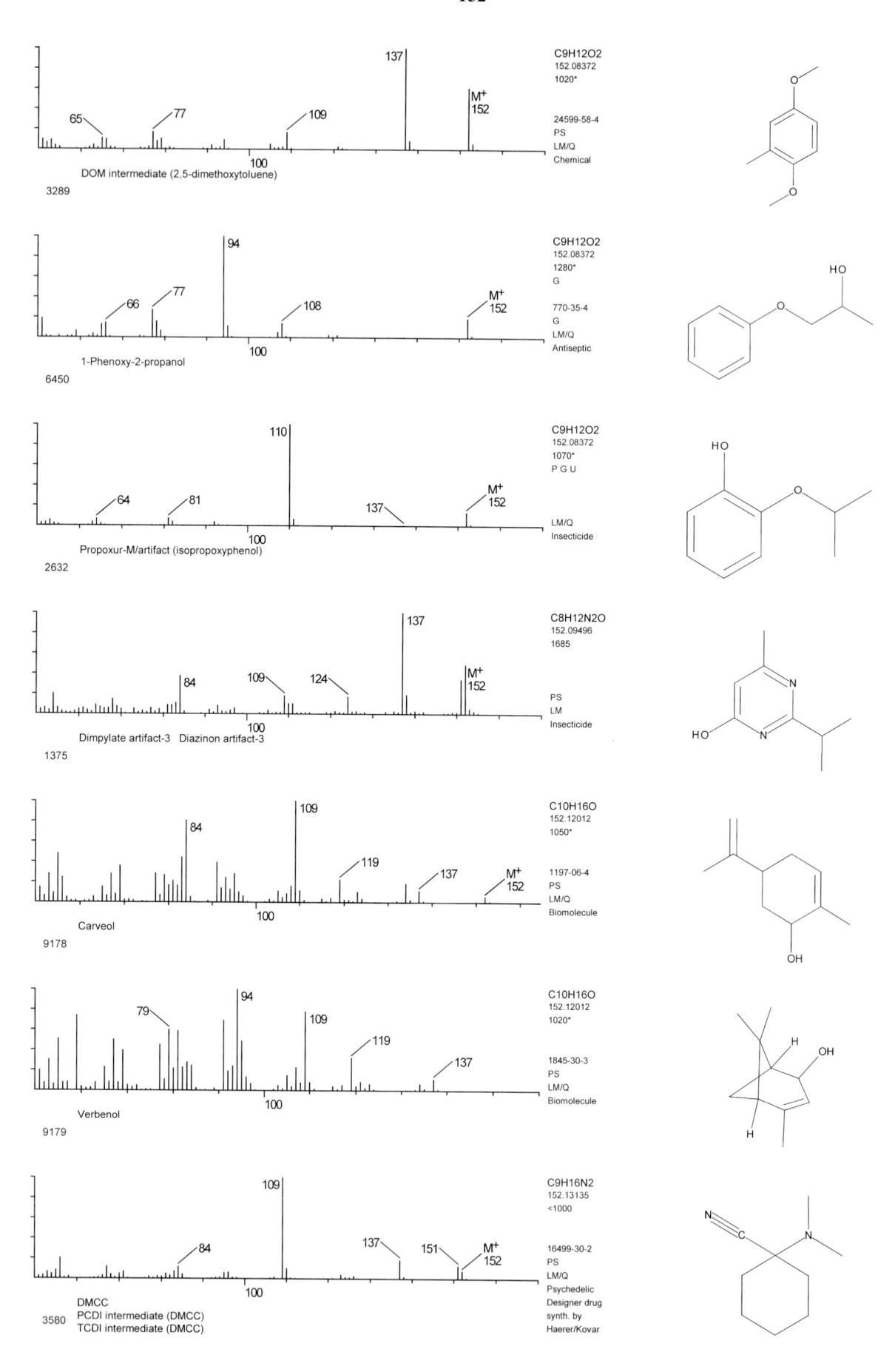

137, 65, 77, 109, M+ 152	C9H12O2 152.08372 1020* 24599-58-4 PS LM/Q Chemical	
DOM intermediate (2,5-dimethoxytoluene) 3289		

94, 66, 77, 108, M+ 152	C9H12O2 152.08372 1280* G 770-35-4 G LM/Q Antiseptic	
1-Phenoxy-2-propanol 6450		

110, 64, 81, 137, M+ 152	C9H12O2 152.08372 1070* P G U LM/Q Insecticide	
Propoxur-M/artifact (isopropoxyphenol) 2632		

137, 84, 109, 124, M+ 152	C8H12N2O 152.09496 1685 PS LM Insecticide	
Dimpylate artifact-3 Diazinon artifact-3 1375		

109, 84, 119, 137, M+ 152	C10H16O 152.12012 1050* 1197-06-4 PS LM/Q Biomolecule	
Carveol 9178		

94, 79, 109, 119, 137	C10H16O 152.12012 1020* 1845-30-3 PS LM/Q Biomolecule	
Verbenol 9179		

109, 84, 137, 151, M+ 152	C9H16N2 152.13135 <1000 16499-30-2 PS LM/Q Psychedelic Designer drug synth. by Haerer/Kovar	
DMCC PCDI intermediate (DMCC) TCDI intermediate (DMCC) 3580		

153

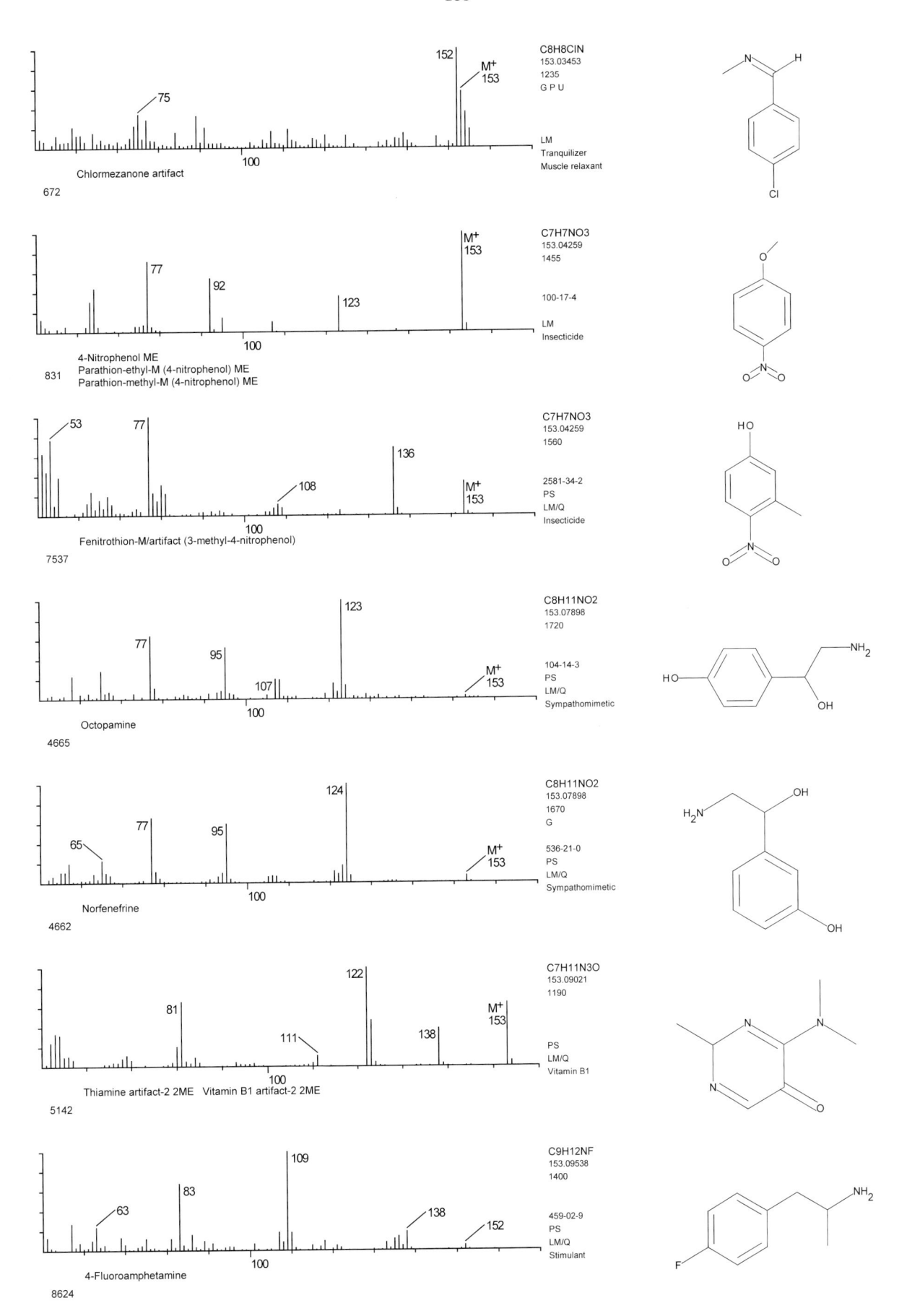

Chlormezanone artifact
672
Peaks: 75, 100, 152, M+ 153
C8H8ClN
153.03453
1235
G P U
LM
Tranquilizer
Muscle relaxant

4-Nitrophenol ME
Parathion-ethyl-M (4-nitrophenol) ME
Parathion-methyl-M (4-nitrophenol) ME
831
Peaks: 77, 92, 100, 123, M+ 153
C7H7NO3
153.04259
1455
100-17-4
LM
Insecticide

Fenitrothion-M/artifact (3-methyl-4-nitrophenol)
7537
Peaks: 53, 77, 100, 108, 136, M+ 153
C7H7NO3
153.04259
1560
2581-34-2
PS
LM/Q
Insecticide

Octopamine
4665
Peaks: 77, 95, 100, 107, 123, M+ 153
C8H11NO2
153.07898
1720
104-14-3
PS
LM/Q
Sympathomimetic

Norfenefrine
4662
Peaks: 65, 77, 95, 100, 124, M+ 153
C8H11NO2
153.07898
1670
G
536-21-0
PS
LM/Q
Sympathomimetic

Thiamine artifact-2 2ME   Vitamin B1 artifact-2 2ME
5142
Peaks: 81, 100, 111, 122, 138, M+ 153
C7H11N3O
153.09021
1190
PS
LM/Q
Vitamin B1

4-Fluoroamphetamine
8624
Peaks: 63, 83, 100, 109, 138, 152, M+ (153)
C9H12NF
153.09538
1400
459-02-9
PS
LM/Q
Stimulant

153

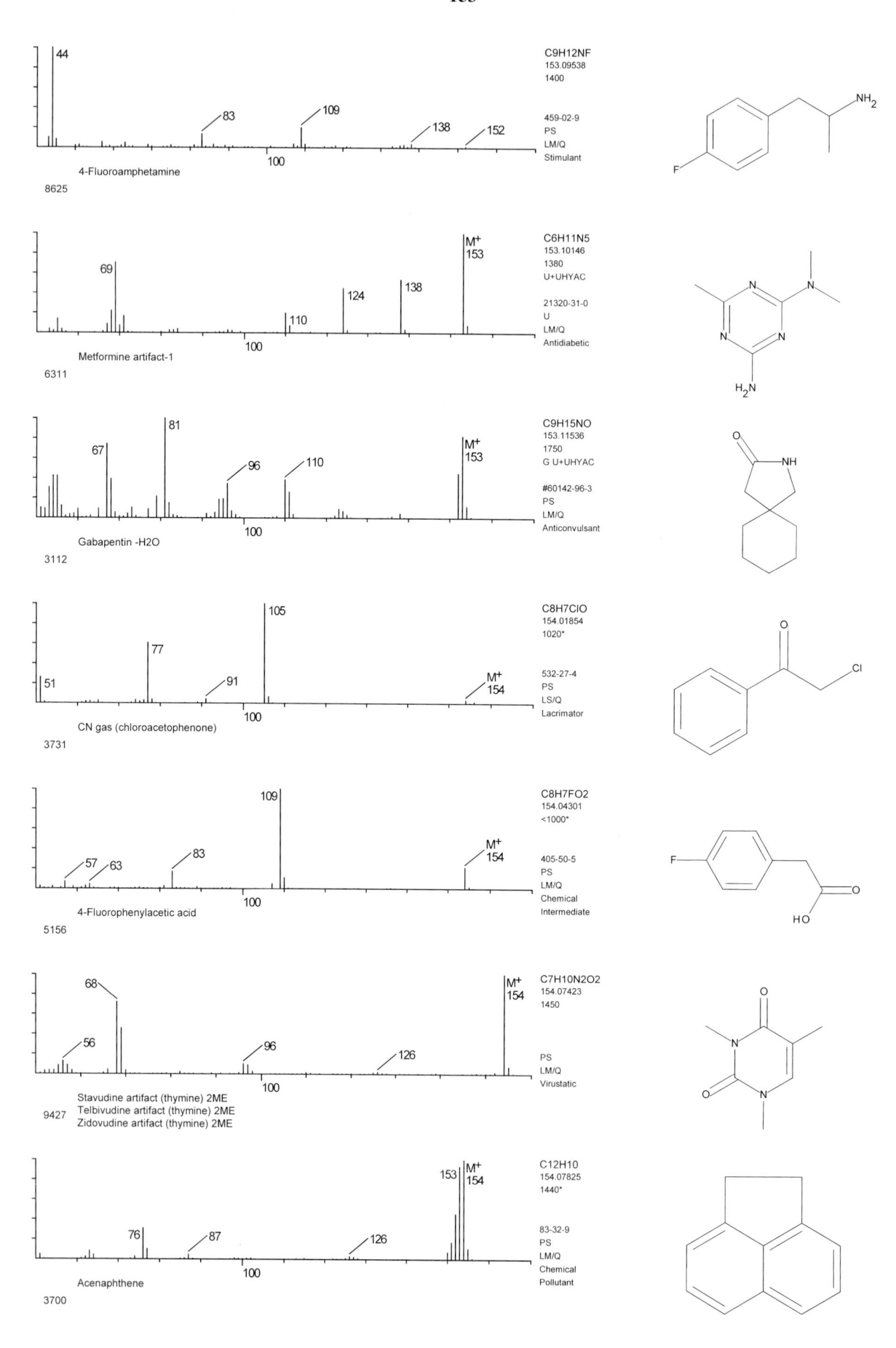

4-Fluoroamphetamine
8625

44, 83, 109, 138, 152

C9H12NF
153.09538
1400

459-02-9
PS
LM/Q
Stimulant

Metformine artifact-1
6311

69, 110, 124, 138, M+ 153

C6H11N5
153.10146
1380
U+UHYAC

21320-31-0
U
LM/Q
Antidiabetic

Gabapentin -H2O
3112

67, 81, 96, 110, M+ 153

C9H15NO
153.11536
1750
G U+UHYAC

#60142-96-3
PS
LM/Q
Anticonvulsant

CN gas (chloroacetophenone)
3731

51, 77, 91, 105, M+ 154

C8H7ClO
154.01854
1020*

532-27-4
PS
LS/Q
Lacrimator

4-Fluorophenylacetic acid
5156

57, 63, 83, 109, M+ 154

C8H7FO2
154.04301
<1000*

405-50-5
PS
LM/Q
Chemical Intermediate

Stavudine artifact (thymine) 2ME
Telbivudine artifact (thymine) 2ME
Zidovudine artifact (thymine) 2ME
9427

56, 68, 96, 126, M+ 154

C7H10N2O2
154.07423
1450

PS
LM/Q
Virustatic

Acenaphthene
3700

76, 87, 126, 153, M+ 154

C12H10
154.07825
1440*

83-32-9
PS
LM/Q
Chemical Pollutant

296

154

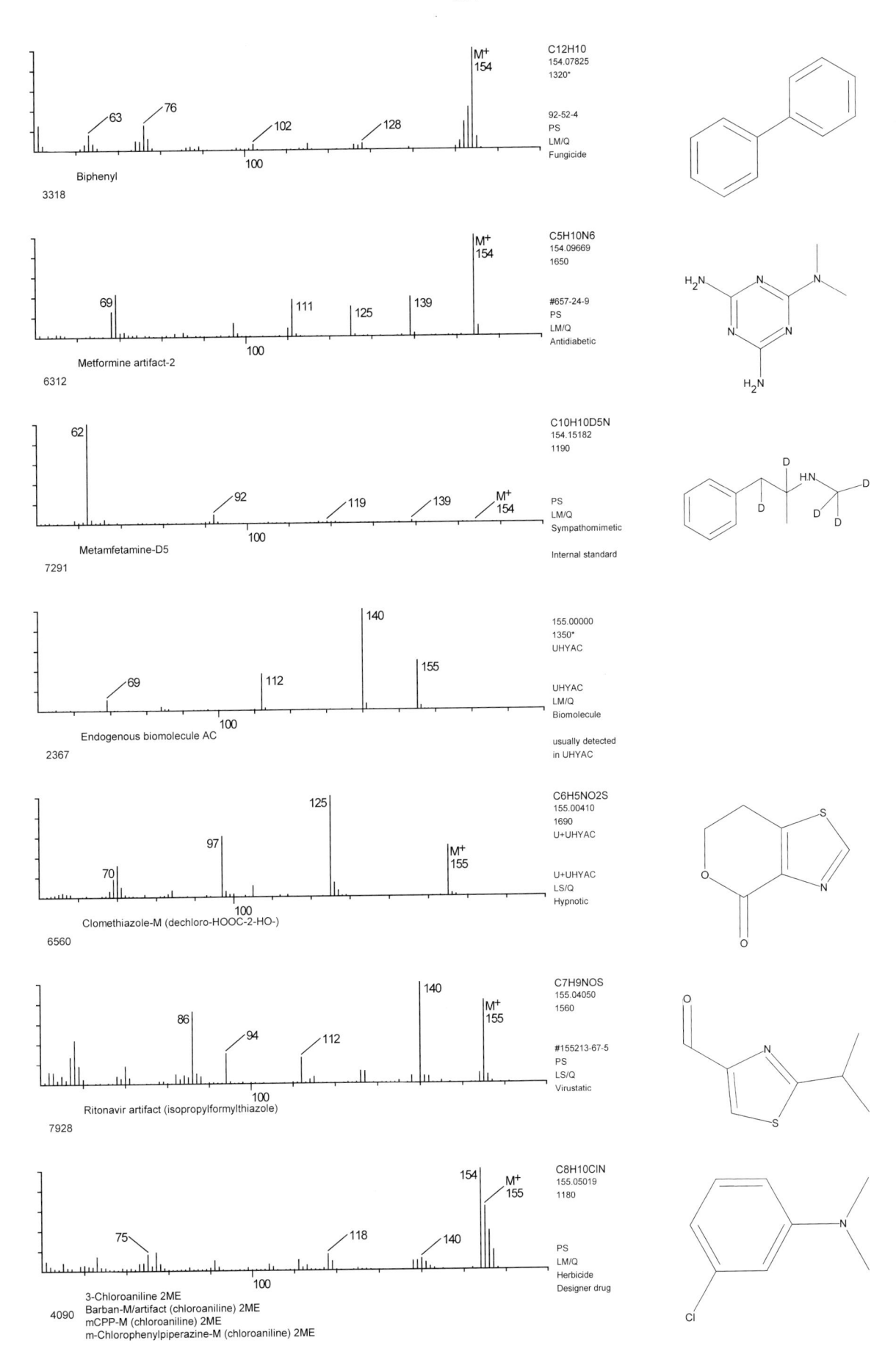

Biphenyl — 3318
C12H10
154.07825
1320*
92-52-4
PS
LM/Q
Fungicide
Peaks: 63, 76, 102, 128, M+ 154

Metformine artifact-2 — 6312
C5H10N6
154.09669
1650
#657-24-9
PS
LM/Q
Antidiabetic
Peaks: 69, 111, 125, 139, M+ 154

Metamfetamine-D5 — 7291
C10H10D5N
154.15182
1190
PS
LM/Q
Sympathomimetic
Internal standard
Peaks: 62, 92, 119, 139, M+ 154

Endogenous biomolecule AC — 2367
155.00000
1350*
UHYAC
UHYAC
LM/Q
Biomolecule
usually detected in UHYAC
Peaks: 69, 112, 140, 155

Clomethiazole-M (dechloro-HOOC-2-HO-) — 6560
C6H5NO2S
155.00410
1690
U+UHYAC
U+UHYAC
LS/Q
Hypnotic
Peaks: 70, 97, 125, M+ 155

Ritonavir artifact (isopropylformylthiazole) — 7928
C7H9NOS
155.04050
1560
#155213-67-5
PS
LS/Q
Virustatic
Peaks: 86, 94, 112, 140, M+ 155

3-Chloroaniline 2ME
Barban-M/artifact (chloroaniline) 2ME
mCPP-M (chloroaniline) 2ME
m-Chlorophenylpiperazine-M (chloroaniline) 2ME — 4090
C8H10ClN
155.05019
1180
PS
LM/Q
Herbicide
Designer drug
Peaks: 75, 118, 140, 154, M+ 155

155

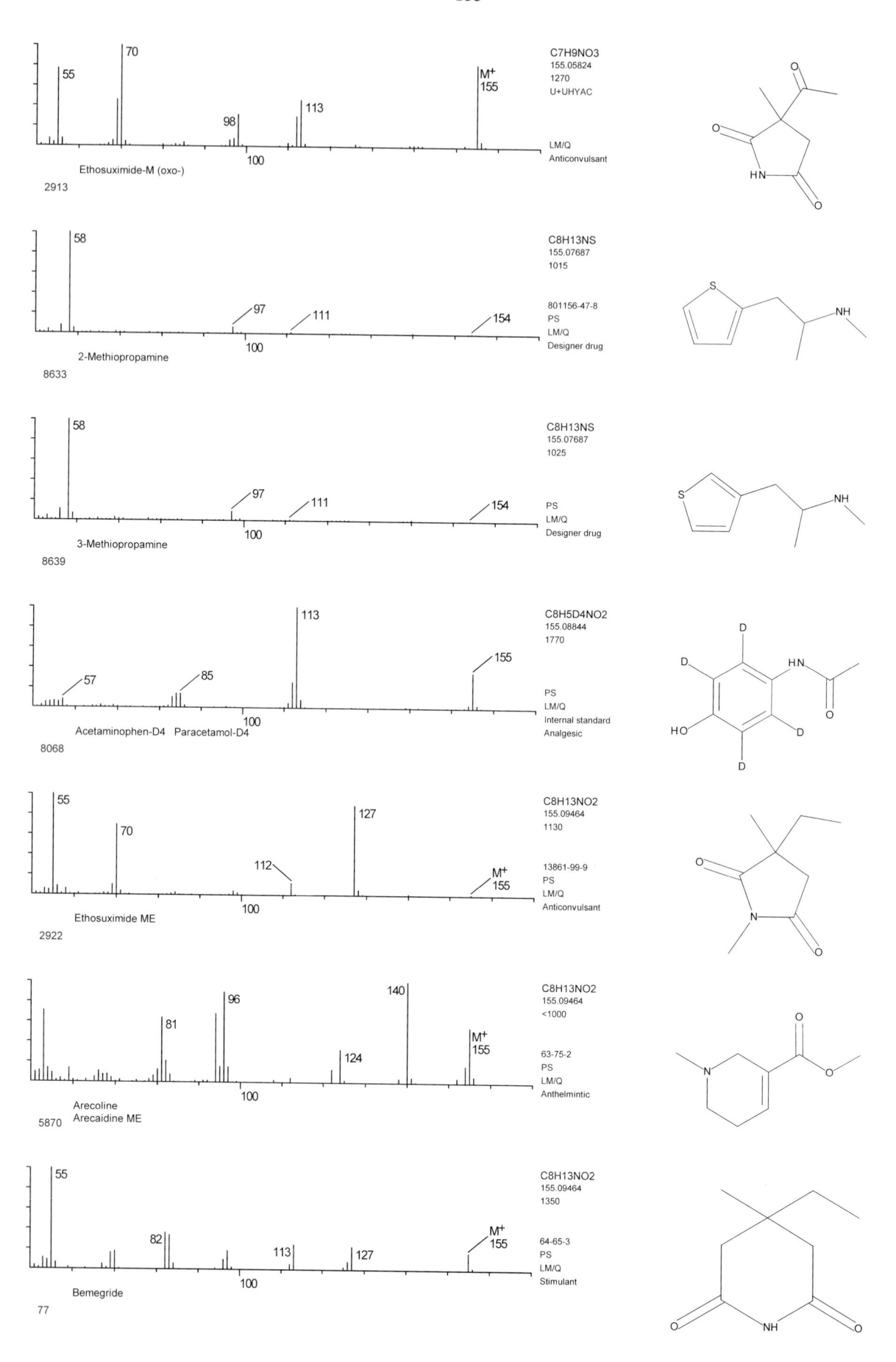

Ethosuximide-M (oxo-)
2913
Peaks: 55, 70, 98, 113, M+ 155
C7H9NO3
155.05824
1270
U+UHYAC
LM/Q
Anticonvulsant

2-Methiopropamine
8633
Peaks: 58, 97, 111, 154
C8H13NS
155.07687
1015
801156-47-8
PS
LM/Q
Designer drug

3-Methiopropamine
8639
Peaks: 58, 97, 111, 154
C8H13NS
155.07687
1025
PS
LM/Q
Designer drug

Acetaminophen-D4 Paracetamol-D4
8068
Peaks: 57, 85, 113, 155
C8H5D4NO2
155.08844
1770
PS
LM/Q
Internal standard
Analgesic

Ethosuximide ME
2922
Peaks: 55, 70, 112, 127, M+ 155
C8H13NO2
155.09464
1130
13861-99-9
PS
LM/Q
Anticonvulsant

Arecoline Arecaidine ME
5870
Peaks: 81, 96, 124, 140, M+ 155
C8H13NO2
155.09464
<1000
63-75-2
PS
LM/Q
Anthelmintic

Bemegride
77
Peaks: 55, 82, 113, 127, M+ 155
C8H13NO2
155.09464
1350
64-65-3
PS
LM/Q
Stimulant

298

155

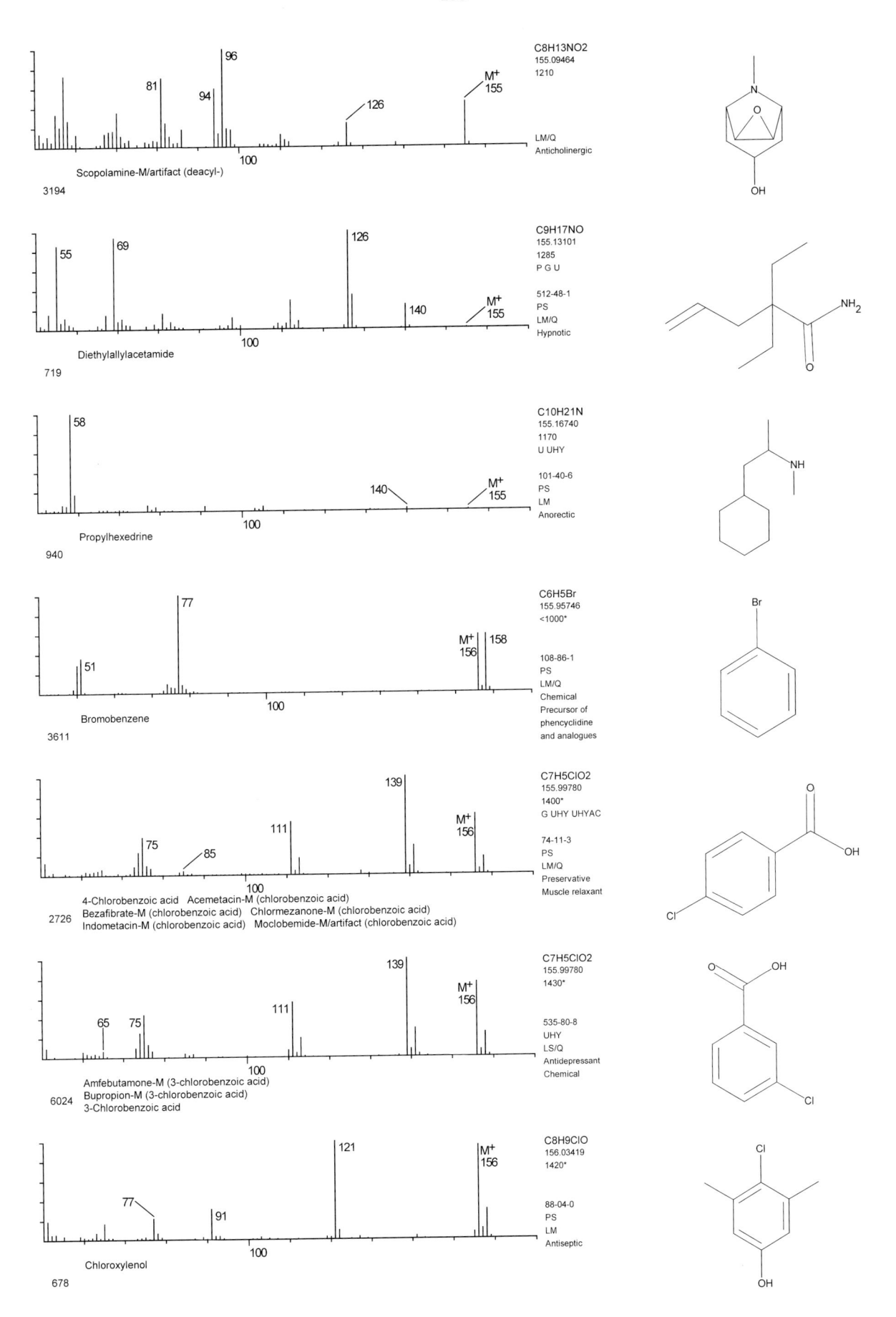

Scopolamine-M/artifact (deacyl-)
3194
C8H13NO2
155.09464
1210
LM/Q
Anticholinergic

Diethylallylacetamide
719
C9H17NO
155.13101
1285
P G U
512-48-1
PS
LM/Q
Hypnotic

Propylhexedrine
940
C10H21N
155.16740
1170
U UHY
101-40-6
PS
LM
Anorectic

Bromobenzene
3611
C6H5Br
155.95746
<1000*
108-86-1
PS
LM/Q
Chemical
Precursor of phencyclidine and analogues

4-Chlorobenzoic acid   Acemetacin-M (chlorobenzoic acid)
Bezafibrate-M (chlorobenzoic acid)   Chlormezanone-M (chlorobenzoic acid)
Indometacin-M (chlorobenzoic acid)   Moclobemide-M/artifact (chlorobenzoic acid)
2726
C7H5ClO2
155.99780
1400*
G UHY UHYAC
74-11-3
PS
LM/Q
Preservative
Muscle relaxant

Amfebutamone-M (3-chlorobenzoic acid)
Bupropion-M (3-chlorobenzoic acid)
3-Chlorobenzoic acid
6024
C7H5ClO2
155.99780
1430*
535-80-8
UHY
LS/Q
Antidepressant
Chemical

Chloroxylenol
678
C8H9ClO
156.03419
1420*
88-04-0
PS
LM
Antiseptic

156

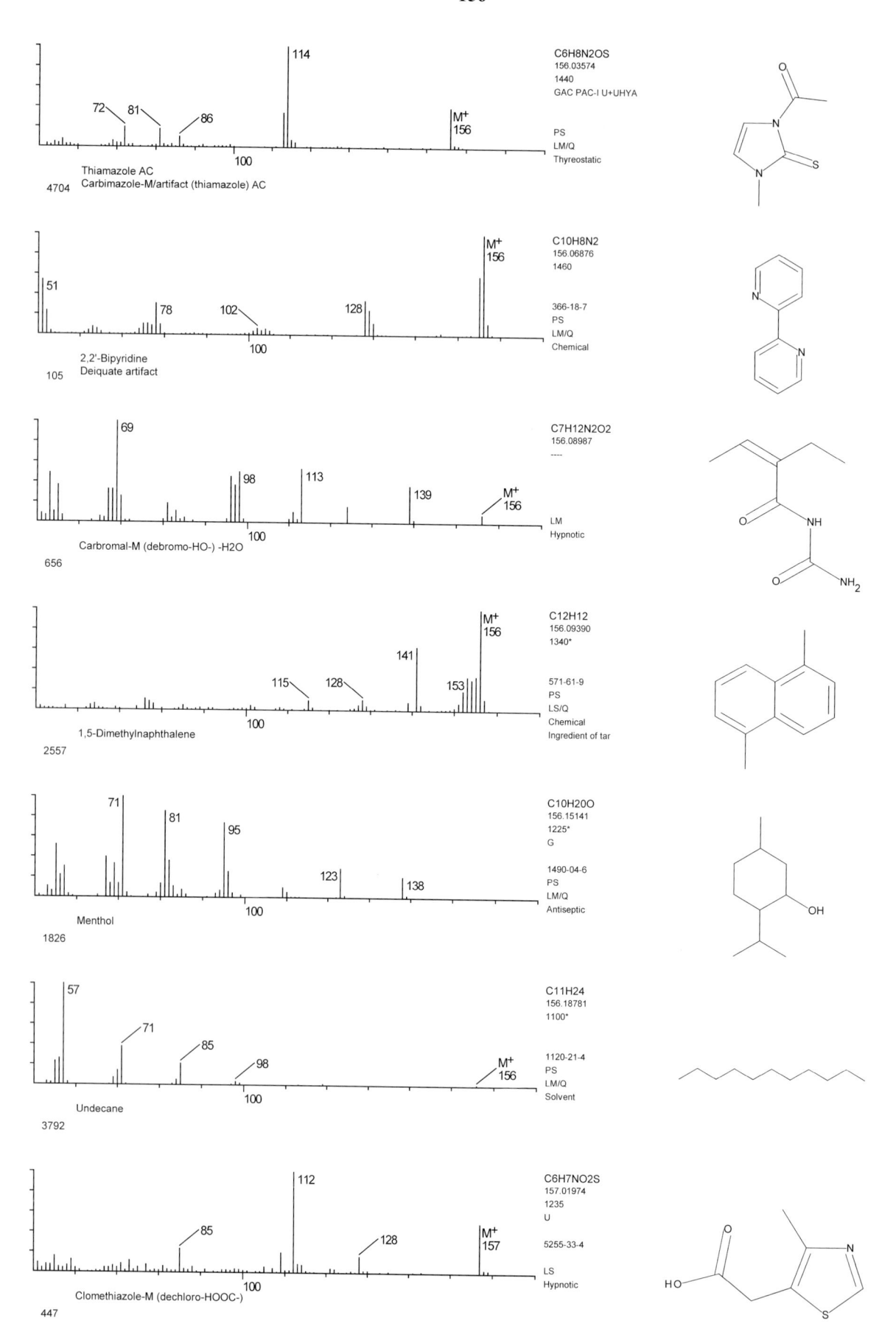

Thiamazole AC — Carbimazole-M/artifact (thiamazole) AC 4704	C6H8N2OS 156.03574 1440 GAC PAC-I U+UHYA PS LM/Q Thyreostatic	
2,2'-Bipyridine — Deiquate artifact 105	C10H8N2 156.06876 1460 366-18-7 PS LM/Q Chemical	
Carbromal-M (debromo-HO-) -H2O 656	C7H12N2O2 156.08987 ---- LM Hypnotic	
1,5-Dimethylnaphthalene 2557	C12H12 156.09390 1340* 571-61-9 PS LS/Q Chemical Ingredient of tar	
Menthol 1826	C10H20O 156.15141 1225* G 1490-04-6 PS LM/Q Antiseptic	
Undecane 3792	C11H24 156.18781 1100* 1120-21-4 PS LM/Q Solvent	
Clomethiazole-M (dechloro-HOOC-) 447	C6H7NO2S 157.01974 1235 U 5255-33-4 LS Hypnotic	

157

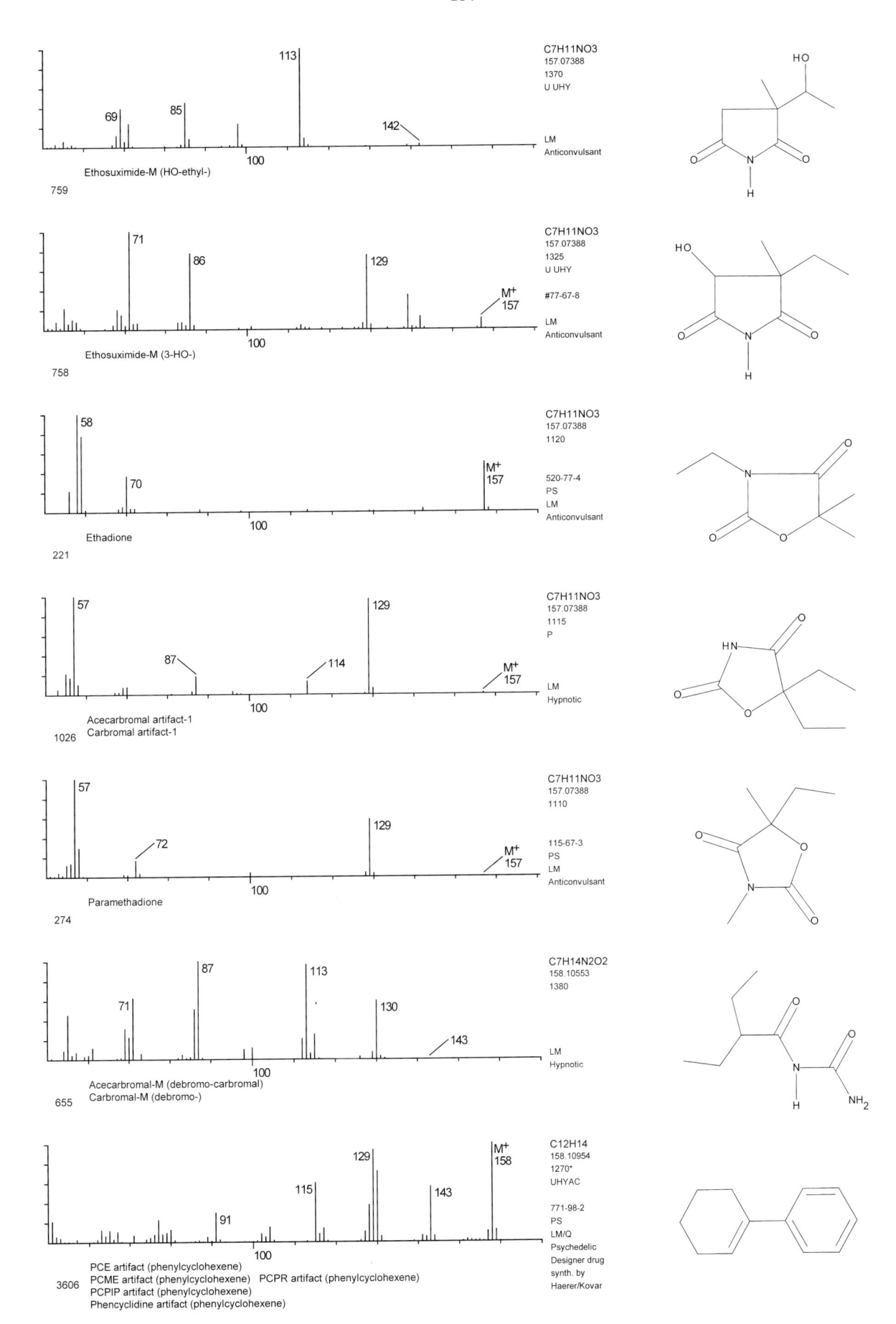

Spectrum	Formula / Info	Compound
69, 85, 113, 142	C7H11NO3 / 157.07388 / 1370 / U UHY / LM / Anticonvulsant	Ethosuximide-M (HO-ethyl-) — 759
71, 86, 129, M+ 157	C7H11NO3 / 157.07388 / 1325 / U UHY / #77-67-8 / LM / Anticonvulsant	Ethosuximide-M (3-HO-) — 758
58, 70, M+ 157	C7H11NO3 / 157.07388 / 1120 / 520-77-4 / PS / LM / Anticonvulsant	Ethadione — 221
57, 87, 114, 129, M+ 157	C7H11NO3 / 157.07388 / 1115 / P / LM / Hypnotic	Acecarbromal artifact-1 / Carbromal artifact-1 — 1026
57, 72, 129, M+ 157	C7H11NO3 / 157.07388 / 1110 / 115-67-3 / PS / LM / Anticonvulsant	Paramethadione — 274
71, 87, 113, 130, 143	C7H14N2O2 / 158.10553 / 1380 / LM / Hypnotic	Acecarbromal-M (debromo-carbromal) / Carbromal-M (debromo-) — 655
91, 115, 129, 143, M+ 158	C12H14 / 158.10954 / 1270* / UHYAC / 771-98-2 / PS / LM/Q / Psychedelic / Designer drug synth. by Haerer/Kovar	PCE artifact (phenylcyclohexene) / PCME artifact (phenylcyclohexene) / PCPR artifact (phenylcyclohexene) / PCPIP artifact (phenylcyclohexene) / Phencyclidine artifact (phenylcyclohexene) — 3606

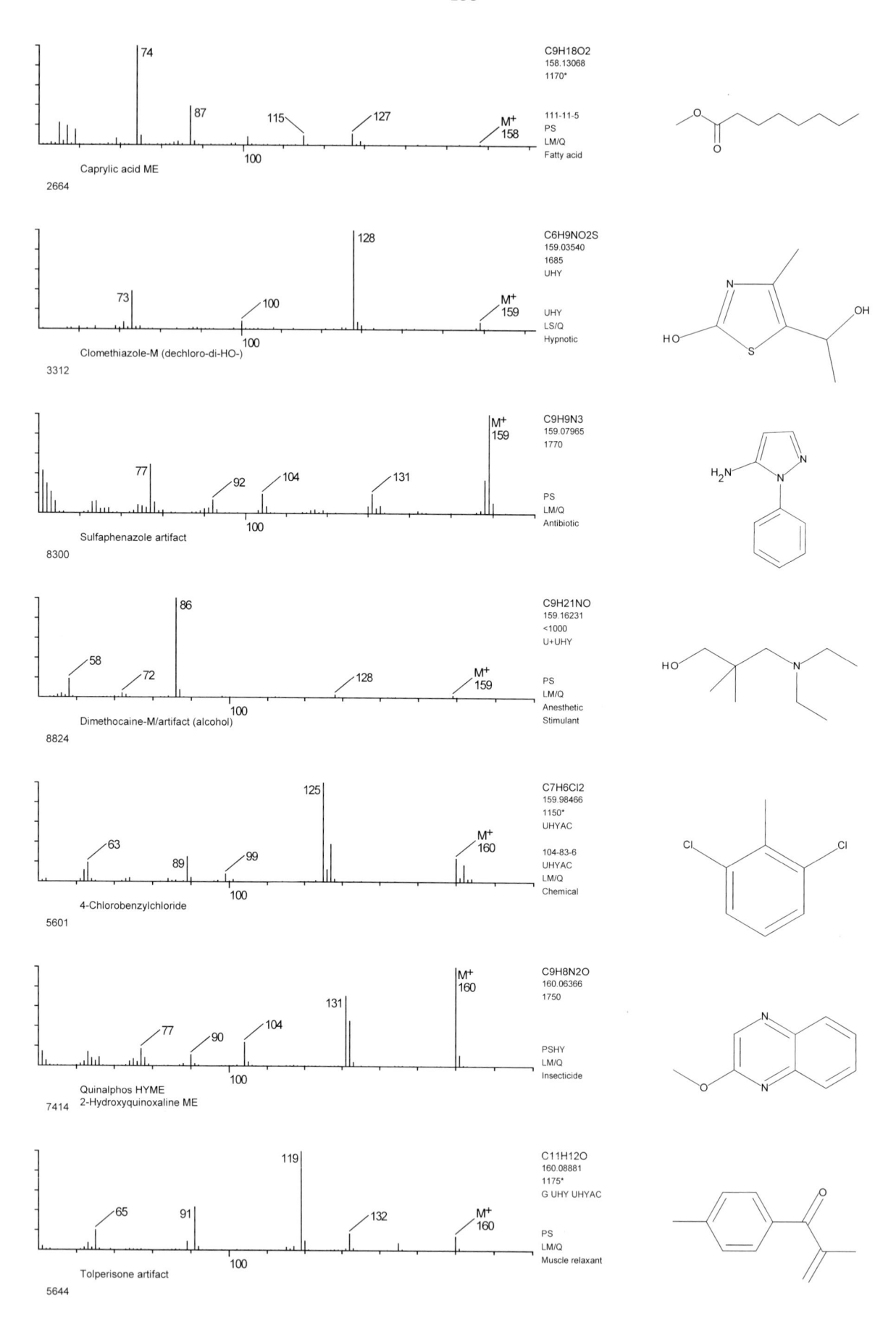

Spectrum	Compound	Formula / Data
74, 87, 115, 127, M+ 158	Caprylic acid ME — 2664	C9H18O2, 158.13068, 1170*, 111-11-5, PS, LM/Q, Fatty acid
73, 100, 128, M+ 159	Clomethiazole-M (dechloro-di-HO-) — 3312	C6H9NO2S, 159.03540, 1685, UHY, UHY, LS/Q, Hypnotic
77, 92, 104, 131, M+ 159	Sulfaphenazole artifact — 8300	C9H9N3, 159.07965, 1770, PS, LM/Q, Antibiotic
58, 72, 86, 128, M+ 159	Dimethocaine-M/artifact (alcohol) — 8824	C9H21NO, 159.16231, <1000, U+UHY, PS, LM/Q, Anesthetic, Stimulant
63, 89, 99, 125, M+ 160	4-Chlorobenzylchloride — 5601	C7H6Cl2, 159.98466, 1150*, UHYAC, 104-83-6, UHYAC, LM/Q, Chemical
77, 90, 104, 131, M+ 160	Quinalphos HYME / 2-Hydroxyquinoxaline ME — 7414	C9H8N2O, 160.06366, 1750, PSHY, LM/Q, Insecticide
65, 91, 119, 132, M+ 160	Tolperisone artifact — 5644	C11H12O, 160.08881, 1175*, G UHY UHYAC, PS, LM/Q, Muscle relaxant

160

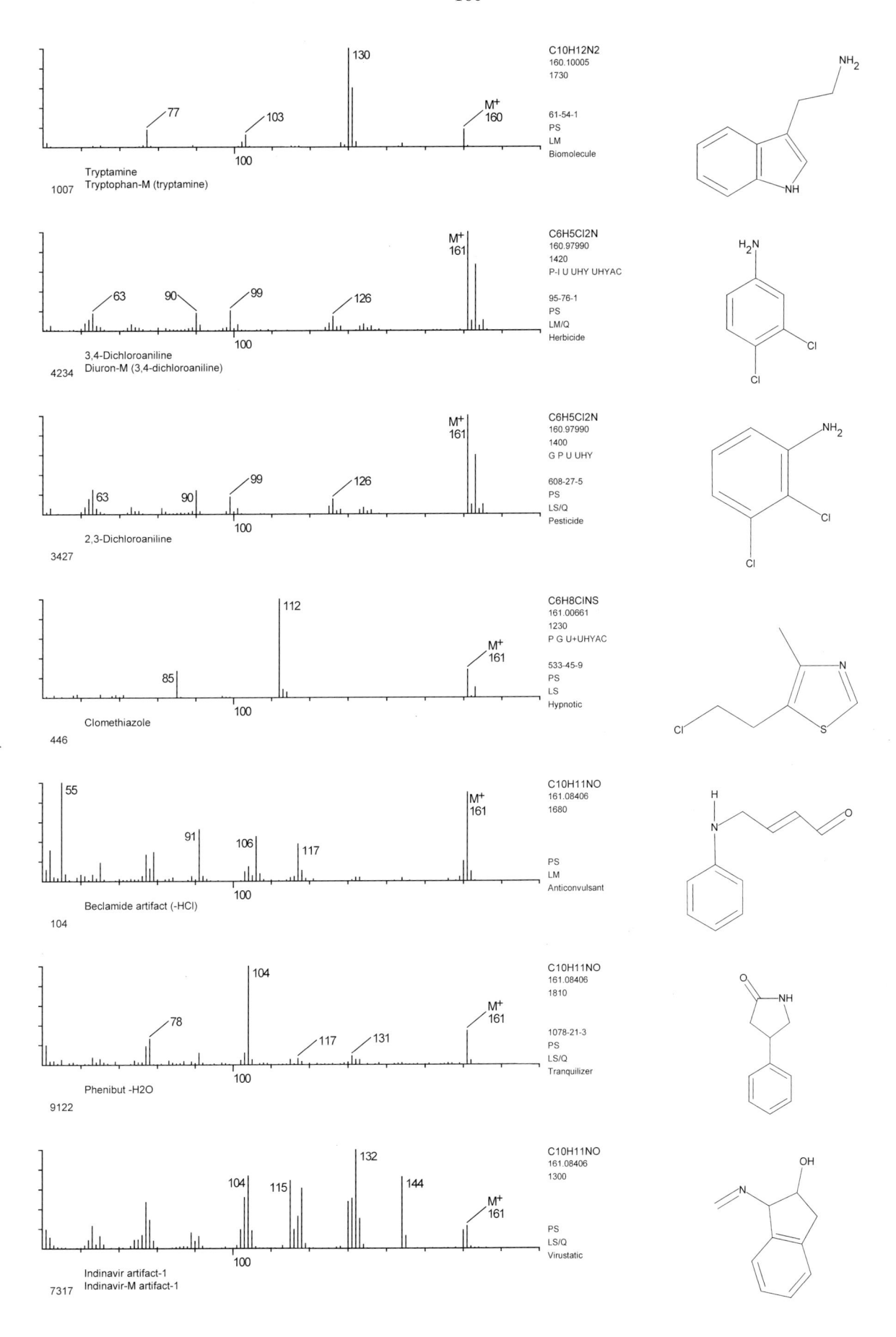

Tryptamine — 1007 — Tryptophan-M (tryptamine)	Peaks: 77, 103, 130, M+ 160	C10H12N2, 160.10005, 1730, 61-54-1, PS, LM, Biomolecule
3,4-Dichloroaniline — 4234 — Diuron-M (3,4-dichloroaniline)	Peaks: 63, 90, 99, 126, M+ 161	C6H5Cl2N, 160.97990, 1420, P-I U UHY UHYAC, 95-76-1, PS, LM/Q, Herbicide
2,3-Dichloroaniline — 3427	Peaks: 63, 90, 99, 126, M+ 161	C6H5Cl2N, 160.97990, 1400, G P U UHY, 608-27-5, PS, LS/Q, Pesticide
Clomethiazole — 446	Peaks: 85, 112, M+ 161	C6H8ClNS, 161.00661, 1230, P G U+UHYAC, 533-45-9, PS, LS, Hypnotic
Beclamide artifact (-HCl) — 104	Peaks: 55, 91, 106, 117, M+ 161	C10H11NO, 161.08406, 1680, PS, LM, Anticonvulsant
Phenibut -H2O — 9122	Peaks: 78, 104, 117, 131, M+ 161	C10H11NO, 161.08406, 1810, 1078-21-3, PS, LS/Q, Tranquilizer
Indinavir artifact-1 — 7317 — Indinavir-M artifact-1	Peaks: 104, 115, 132, 144, M+ 161	C10H11NO, 161.08406, 1300, PS, LS/Q, Virustatic

303

161

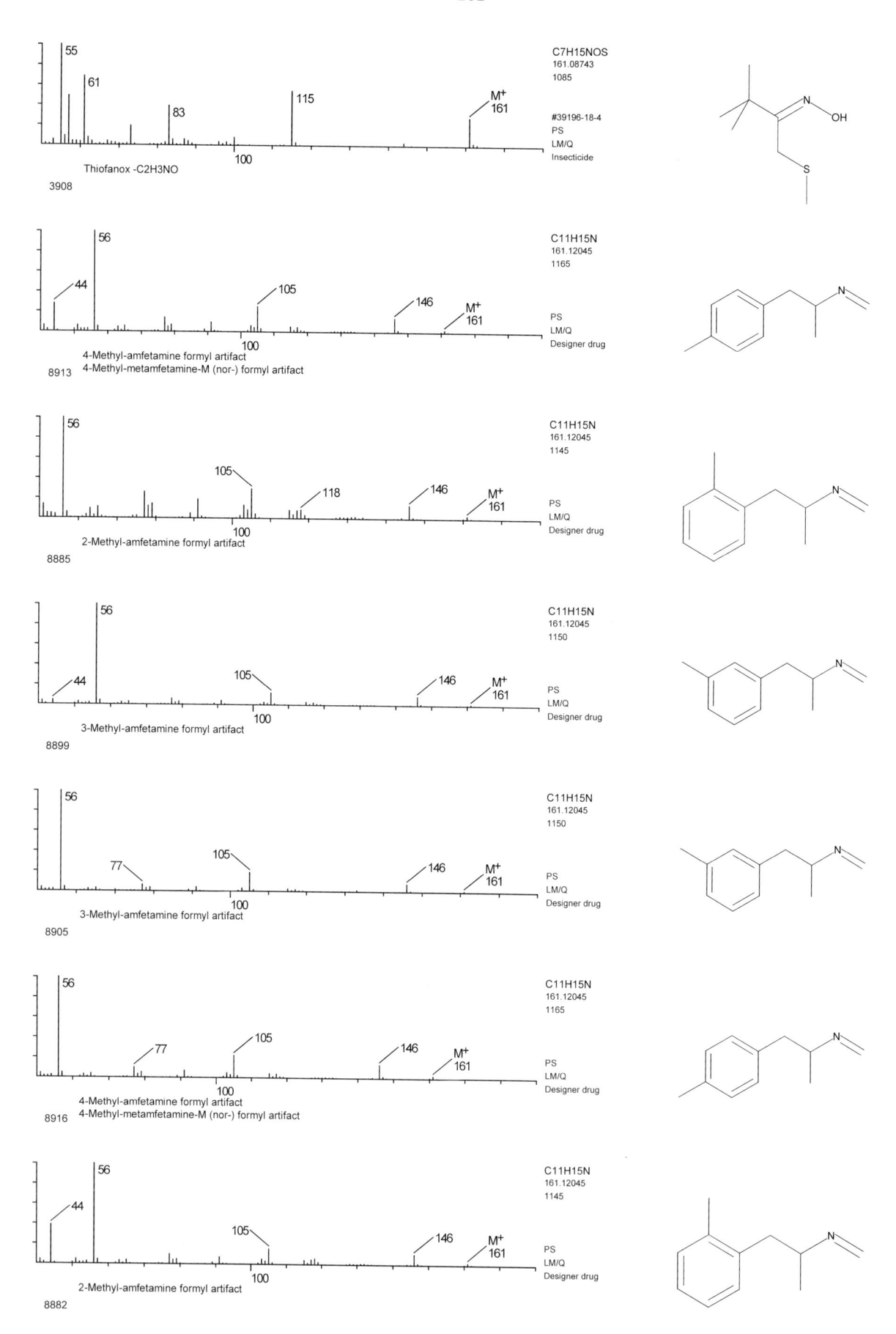

C7H15NOS
161.08743
1085
#39196-18-4
PS
LM/Q
Insecticide

Thiofanox -C2H3NO
3908

C11H15N
161.12045
1165
PS
LM/Q
Designer drug

4-Methyl-amfetamine formyl artifact
4-Methyl-metamfetamine-M (nor-) formyl artifact
8913

C11H15N
161.12045
1145
PS
LM/Q
Designer drug

2-Methyl-amfetamine formyl artifact
8885

C11H15N
161.12045
1150
PS
LM/Q
Designer drug

3-Methyl-amfetamine formyl artifact
8899

C11H15N
161.12045
1150
PS
LM/Q
Designer drug

3-Methyl-amfetamine formyl artifact
8905

C11H15N
161.12045
1165
PS
LM/Q
Designer drug

4-Methyl-amfetamine formyl artifact
4-Methyl-metamfetamine-M (nor-) formyl artifact
8916

C11H15N
161.12045
1145
PS
LM/Q
Designer drug

2-Methyl-amfetamine formyl artifact
8882

162

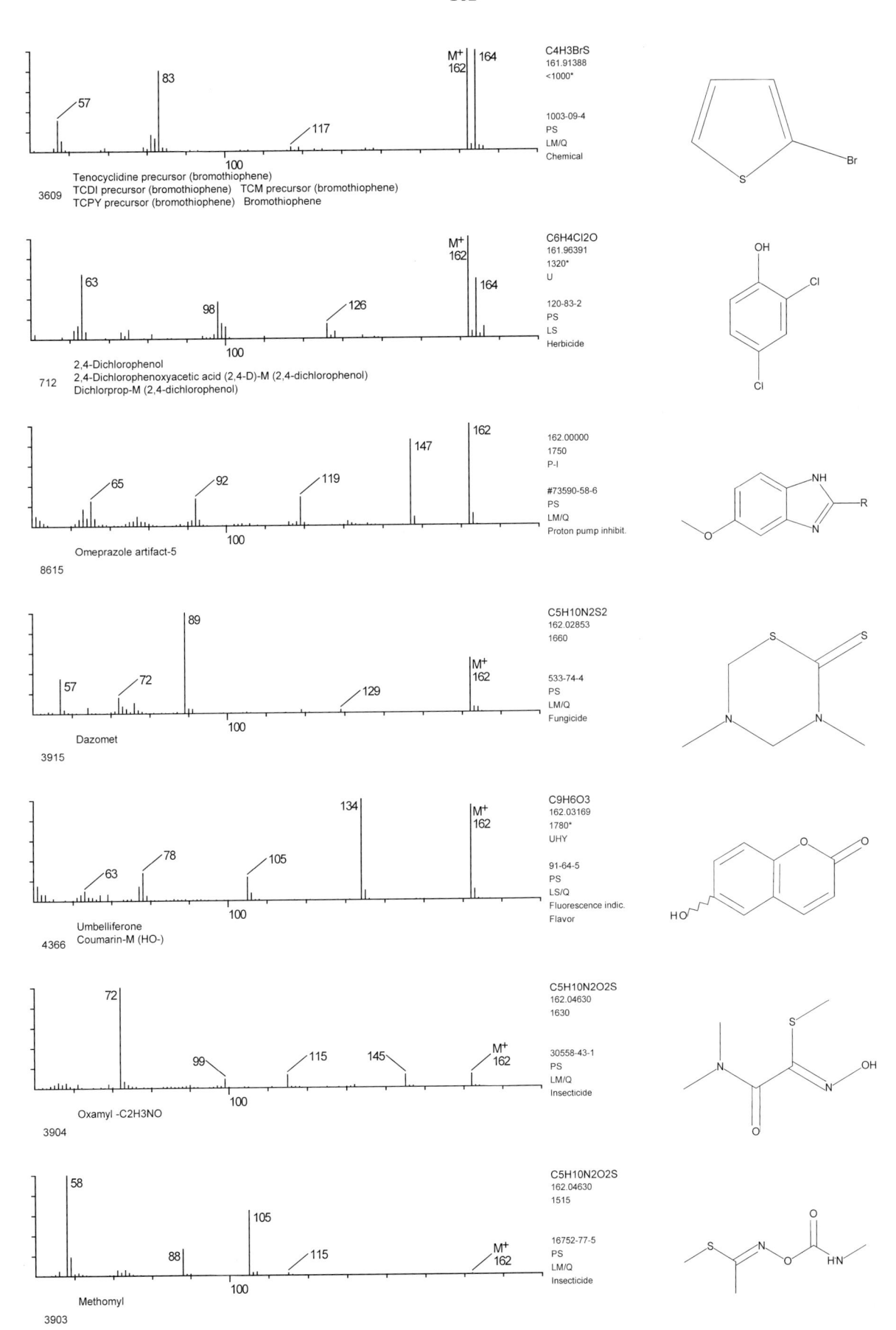

C4H3BrS
161.91388
<1000*

1003-09-4
PS
LM/Q
Chemical

Tenocyclidine precursor (bromothiophene)
TCDI precursor (bromothiophene)   TCM precursor (bromothiophene)
TCPY precursor (bromothiophene)   Bromothiophene
3609

C6H4Cl2O
161.96391
1320*
U

120-83-2
PS
LS
Herbicide

2,4-Dichlorophenol
2,4-Dichlorophenoxyacetic acid (2,4-D)-M (2,4-dichlorophenol)
Dichlorprop-M (2,4-dichlorophenol)
712

162.00000
1750
P-I

#73590-58-6
PS
LM/Q
Proton pump inhibit.

Omeprazole artifact-5
8615

C5H10N2S2
162.02853
1660

533-74-4
PS
LM/Q
Fungicide

Dazomet
3915

C9H6O3
162.03169
1780*
UHY

91-64-5
PS
LS/Q
Fluorescence indic.
Flavor

Umbelliferone
Coumarin-M (HO-)
4366

C5H10N2O2S
162.04630
1630

30558-43-1
PS
LM/Q
Insecticide

Oxamyl -C2H3NO
3904

C5H10N2O2S
162.04630
1515

16752-77-5
PS
LM/Q
Insecticide

Methomyl
3903

162

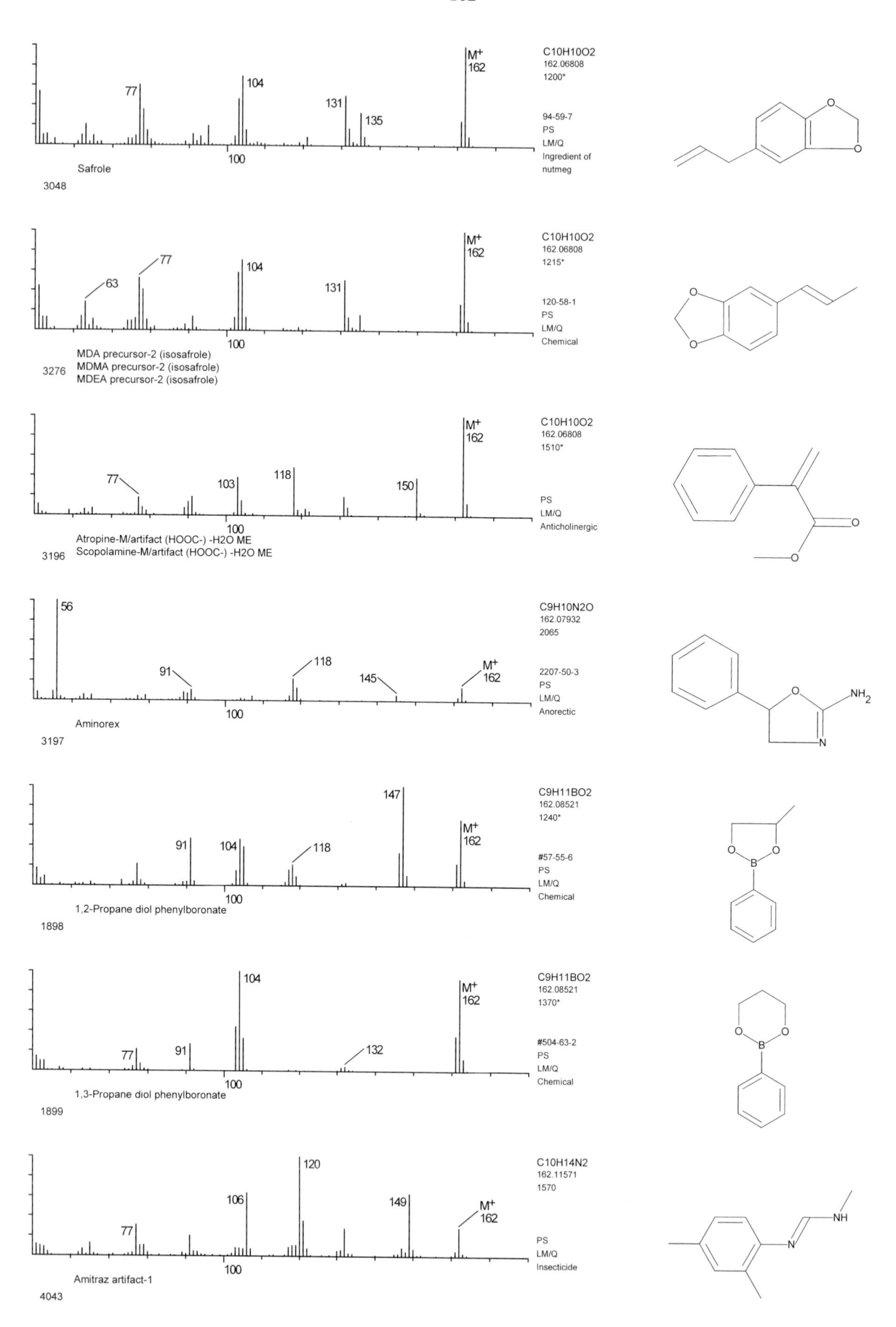

Safrole — 3048	C10H10O2 162.06808 1200* / 94-59-7 PS LM/Q Ingredient of nutmeg	m/z 77, 104, 131, 135, 162 (M+)
MDA precursor-2 (isosafrole) MDMA precursor-2 (isosafrole) MDEA precursor-2 (isosafrole) — 3276	C10H10O2 162.06808 1215* / 120-58-1 PS LM/Q Chemical	m/z 63, 77, 104, 131, 162 (M+)
Atropine-M/artifact (HOOC-) -H2O ME Scopolamine-M/artifact (HOOC-) -H2O ME — 3196	C10H10O2 162.06808 1510* PS LM/Q Anticholinergic	m/z 77, 103, 118, 150, 162 (M+)
Aminorex — 3197	C9H10N2O 162.07932 2065 / 2207-50-3 PS LM/Q Anorectic	m/z 56, 91, 118, 145, 162 (M+)
1,2-Propane diol phenylboronate — 1898	C9H11BO2 162.08521 1240* / #57-55-6 PS LM/Q Chemical	m/z 91, 104, 118, 147, 162 (M+)
1,3-Propane diol phenylboronate — 1899	C9H11BO2 162.08521 1370* / #504-63-2 PS LM/Q Chemical	m/z 77, 91, 104, 132, 162 (M+)
Amitraz artifact-1 — 4043	C10H14N2 162.11571 1570 PS LM/Q Insecticide	m/z 77, 106, 120, 149, 162 (M+)

162

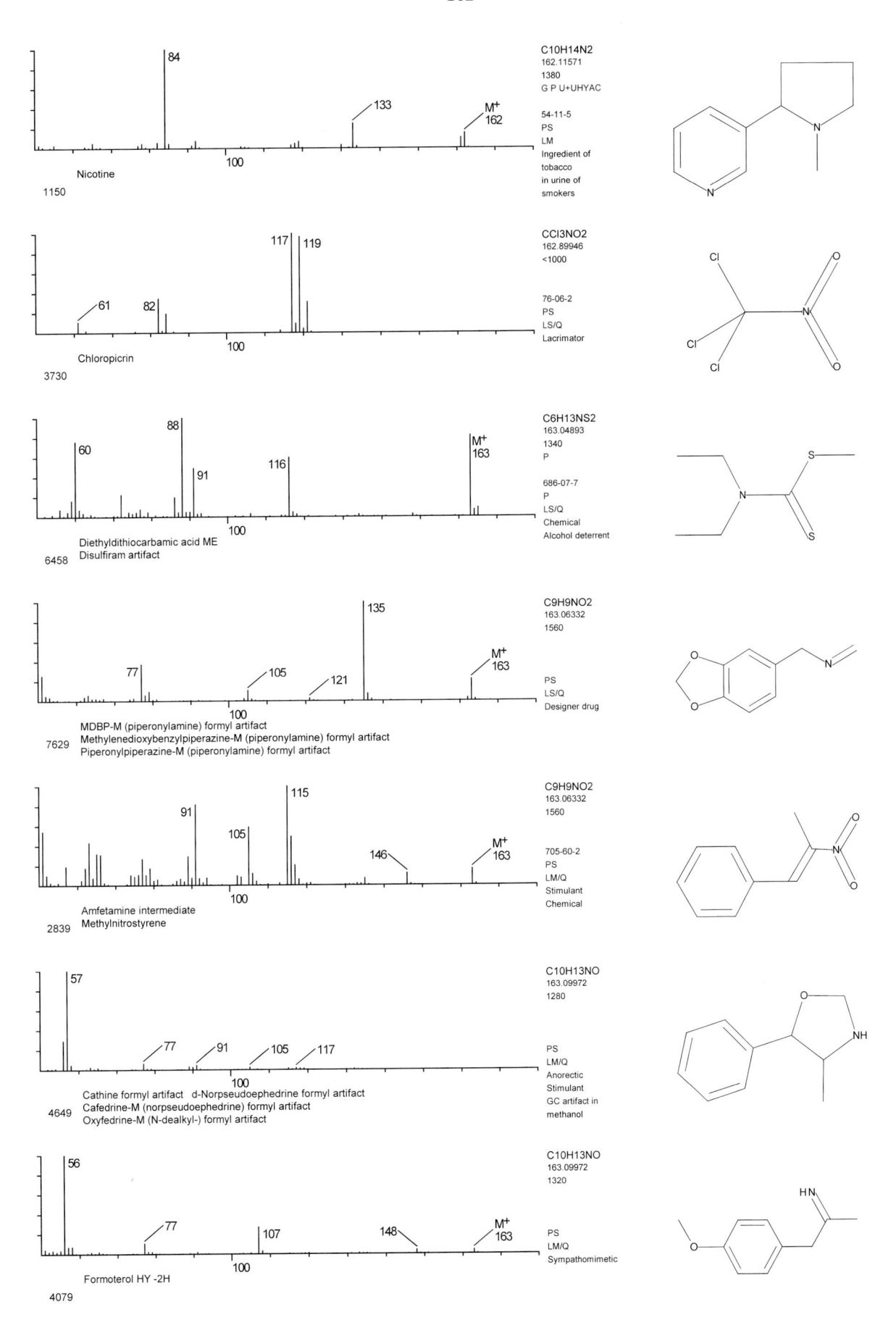

Nicotine 1150	84, 133, M+ 162	C10H14N2 162.11571 1380 G P U+UHYAC 54-11-5 PS LM Ingredient of tobacco in urine of smokers
Chloropicrin 3730	61, 82, 117, 119	CCl3NO2 162.89946 <1000 76-06-2 PS LS/Q Lacrimator
Diethyldithiocarbamic acid ME Disulfiram artifact 6458	60, 88, 91, 116, M+ 163	C6H13NS2 163.04893 1340 P 686-07-7 P LS/Q Chemical Alcohol deterrent
MDBP-M (piperonylamine) formyl artifact Methylenedioxybenzylpiperazine-M (piperonylamine) formyl artifact Piperonylpiperazine-M (piperonylamine) formyl artifact 7629	77, 105, 121, 135, M+ 163	C9H9NO2 163.06332 1560 PS LS/Q Designer drug
Amfetamine intermediate Methylnitrostyrene 2839	91, 105, 115, 146, M+ 163	C9H9NO2 163.06332 1560 705-60-2 PS LM/Q Stimulant Chemical
Cathine formyl artifact d-Norpseudoephedrine formyl artifact Cafedrine-M (norpseudoephedrine) formyl artifact Oxyfedrine-M (N-dealkyl-) formyl artifact 4649	57, 77, 91, 105, 117	C10H13NO 163.09972 1280 PS LM/Q Anorectic Stimulant GC artifact in methanol
Formoterol HY -2H 4079	56, 77, 107, 148, M+ 163	C10H13NO 163.09972 1320 PS LM/Q Sympathomimetic

307

163

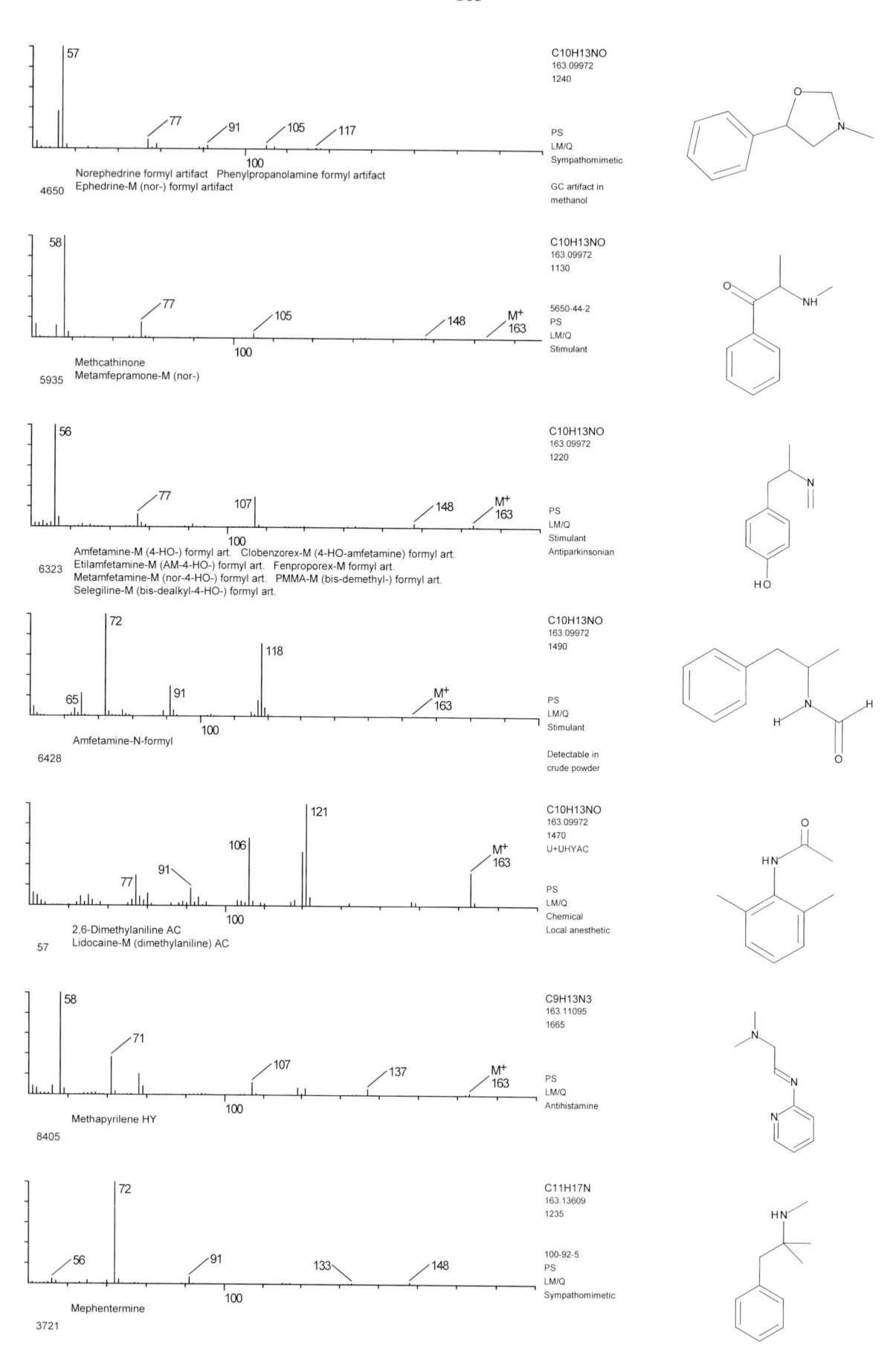

57, 77, 91, 105, 117, 100	Norephedrine formyl artifact Phenylpropanolamine formyl artifact Ephedrine-M (nor-) formyl artifact	C10H13NO 163.09972 1240 PS LM/Q Sympathomimetic GC artifact in methanol
4650		

58, 77, 105, 148, M+ 163, 100

Methcathinone
Metamfepramone-M (nor-)
5935

C10H13NO
163.09972
1130
5650-44-2
PS
LM/Q
Stimulant

56, 77, 107, 148, M+ 163, 100

Amfetamine-M (4-HO-) formyl art.   Clobenzorex-M (4-HO-amfetamine) formyl art.
Etilamfetamine-M (AM-4-HO-) formyl art.   Fenproporex-M formyl art.
Metamfetamine-M (nor-4-HO-) formyl art.   PMMA-M (bis-demethyl-) formyl art.
Selegiline-M (bis-dealkyl-4-HO-) formyl art.
6323

C10H13NO
163.09972
1220
PS
LM/Q
Stimulant
Antiparkinsonian

72, 65, 91, 118, M+ 163, 100

Amfetamine-N-formyl
6428

C10H13NO
163.09972
1490
PS
LM/Q
Stimulant
Detectable in crude powder

121, 106, 77, 91, M+ 163, 100

2,6-Dimethylaniline AC
Lidocaine-M (dimethylaniline) AC
57

C10H13NO
163.09972
1470
U+UHYAC
PS
LM/Q
Chemical
Local anesthetic

58, 71, 107, 137, M+ 163, 100

Methapyrilene HY
8405

C9H13N3
163.11095
1665
PS
LM/Q
Antihistamine

72, 56, 91, 133, 148, 100

Mephentermine
3721

C11H17N
163.13609
1235
100-92-5
PS
LM/Q
Sympathomimetic

308

163

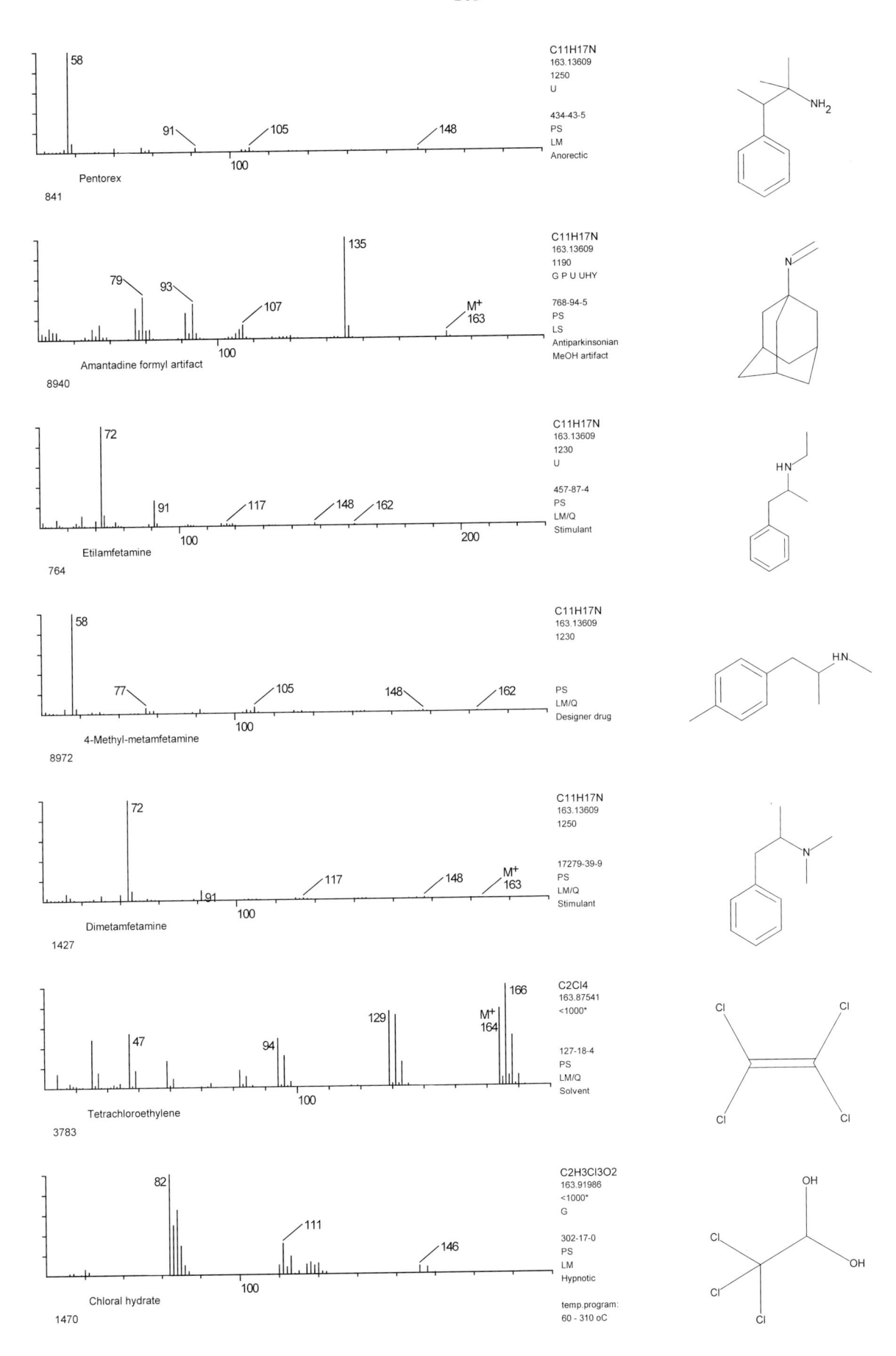

58, 91, 100, 105, 148	C11H17N 163.13609 1250 U 434-43-5 PS LM Anorectic	
Pentorex 841		
79, 93, 100, 107, 135, M+ 163	C11H17N 163.13609 1190 G P U UHY 768-94-5 PS LS Antiparkinsonian MeOH artifact	
Amantadine formyl artifact 8940		
72, 91, 100, 117, 148, 162	C11H17N 163.13609 1230 U 457-87-4 PS LM/Q Stimulant	
Etilamfetamine 764		
58, 77, 100, 105, 148, 162	C11H17N 163.13609 1230 PS LM/Q Designer drug	
4-Methyl-metamfetamine 8972		
72, 91, 100, 117, 148, M+ 163	C11H17N 163.13609 1250 17279-39-9 PS LM/Q Stimulant	
Dimetamfetamine 1427		
47, 94, 100, 129, M+ 164, 166	C2Cl4 163.87541 <1000* 127-18-4 PS LM/Q Solvent	
Tetrachloroethylene 3783		
82, 100, 111, 146	C2H3Cl3O2 163.91986 <1000* G 302-17-0 PS LM Hypnotic temp.program: 60 - 310 oC	
Chloral hydrate 1470		

309

164

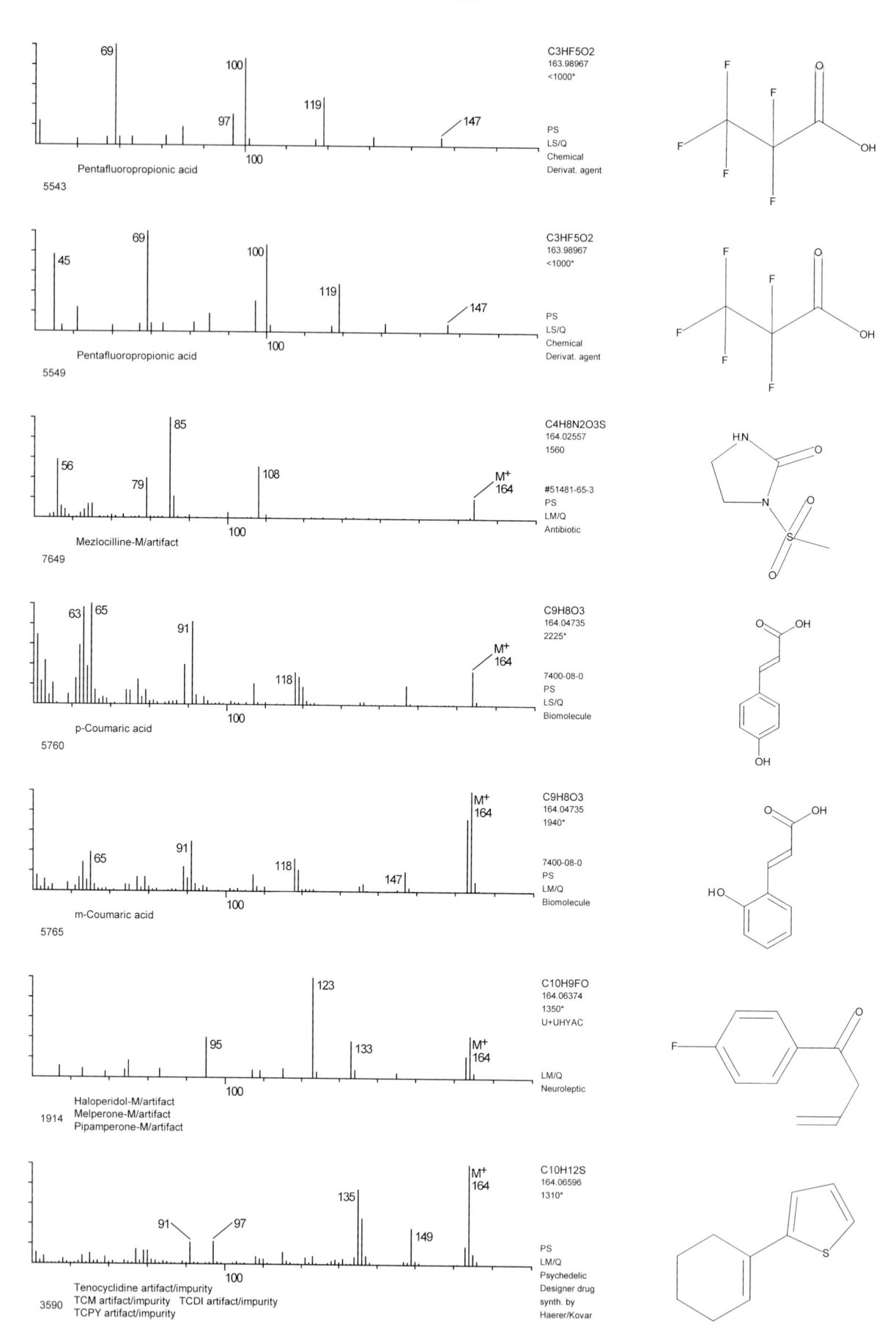

Pentafluoropropionic acid — 5543	C3HF5O2 163.98967 <1000* PS LS/Q Chemical Derivat. agent	peaks: 69, 97, 100, 119, 147
Pentafluoropropionic acid — 5549	C3HF5O2 163.98967 <1000* PS LS/Q Chemical Derivat. agent	peaks: 45, 69, 100, 119, 147
Mezlocilline-M/artifact — 7649	C4H8N2O3S 164.02557 1560 #51481-65-3 PS LM/Q Antibiotic	peaks: 56, 79, 85, 108, M+ 164
p-Coumaric acid — 5760	C9H8O3 164.04735 2225* 7400-08-0 PS LS/Q Biomolecule	peaks: 63, 65, 91, 118, M+ 164
m-Coumaric acid — 5765	C9H8O3 164.04735 1940* 7400-08-0 PS LM/Q Biomolecule	peaks: 65, 91, 118, 147, M+ 164
Haloperidol-M/artifact, Melperone-M/artifact, Pipamperone-M/artifact — 1914	C10H9FO 164.06374 1350* U+UHYAC LM/Q Neuroleptic	peaks: 95, 123, 133, M+ 164
Tenocyclidine artifact/impurity, TCM artifact/impurity, TCDI artifact/impurity, TCPY artifact/impurity — 3590	C10H12S 164.06596 1310* PS LM/Q Psychedelic Designer drug synth. by Haerer/Kovar	peaks: 91, 97, 135, 149, M+ 164

164

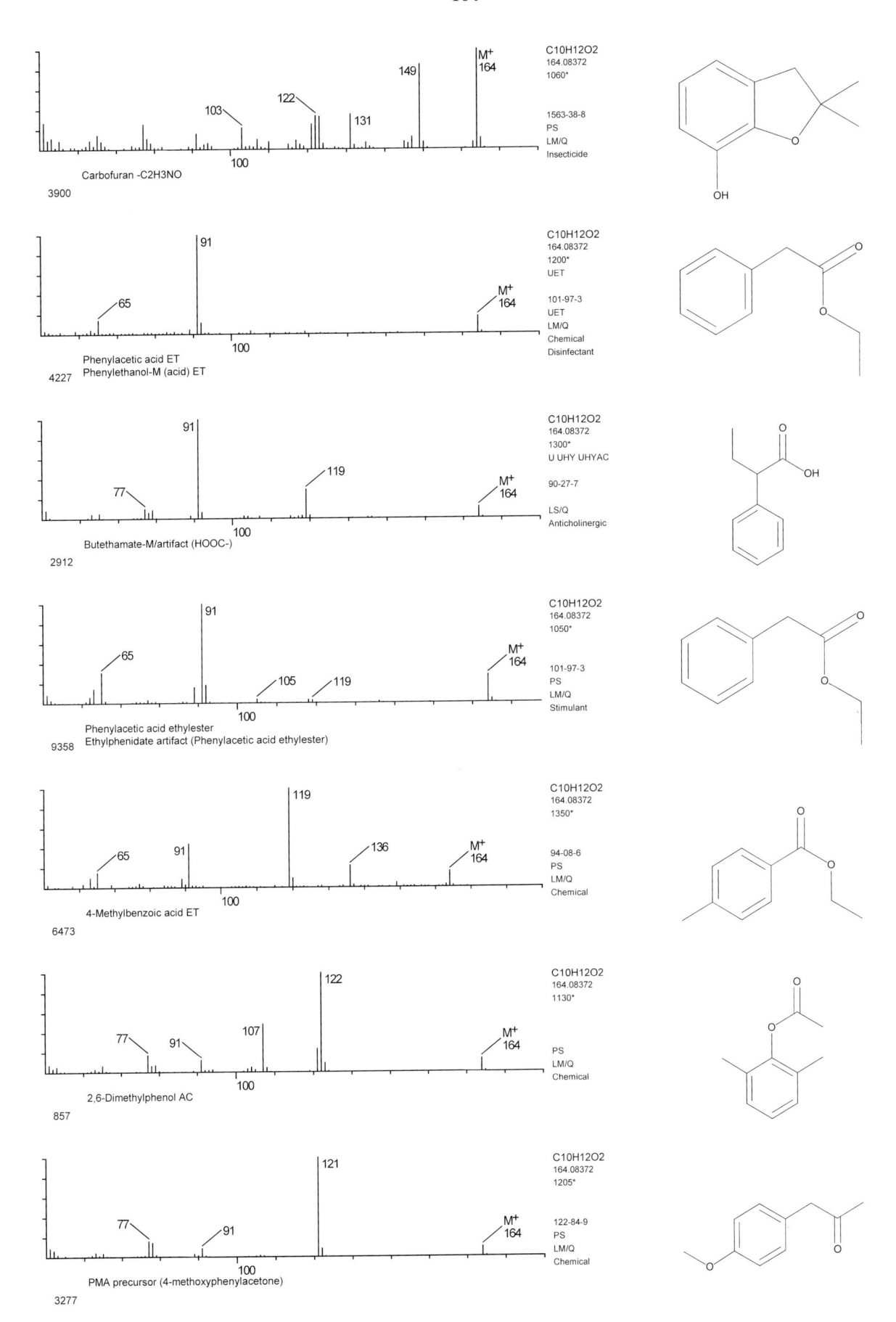

Spectrum	Compound	Formula / Info
3900	Carbofuran -C2H3NO	C10H12O2, 164.08372, 1060*, 1563-38-8, PS, LM/Q, Insecticide; peaks: 103, 122, 131, 149, M+ 164
4227	Phenylacetic acid ET / Phenylethanol-M (acid) ET	C10H12O2, 164.08372, 1200*, UET, 101-97-3, UET, LM/Q, Chemical Disinfectant; peaks: 65, 91, M+ 164
2912	Butethamate-M/artifact (HOOC-)	C10H12O2, 164.08372, 1300*, U UHY UHYAC, 90-27-7, LS/Q, Anticholinergic; peaks: 77, 91, 119, M+ 164
9358	Phenylacetic acid ethylester / Ethylphenidate artifact (Phenylacetic acid ethylester)	C10H12O2, 164.08372, 1050*, 101-97-3, PS, LM/Q, Stimulant; peaks: 65, 91, 105, 119, M+ 164
6473	4-Methylbenzoic acid ET	C10H12O2, 164.08372, 1350*, 94-08-6, PS, LM/Q, Chemical; peaks: 65, 91, 119, 136, M+ 164
857	2,6-Dimethylphenol AC	C10H12O2, 164.08372, 1130*, PS, LM/Q, Chemical; peaks: 77, 91, 107, 122, M+ 164
3277	PMA precursor (4-methoxyphenylacetone)	C10H12O2, 164.08372, 1205*, 122-84-9, PS, LM/Q, Chemical; peaks: 77, 91, 121, M+ 164

164

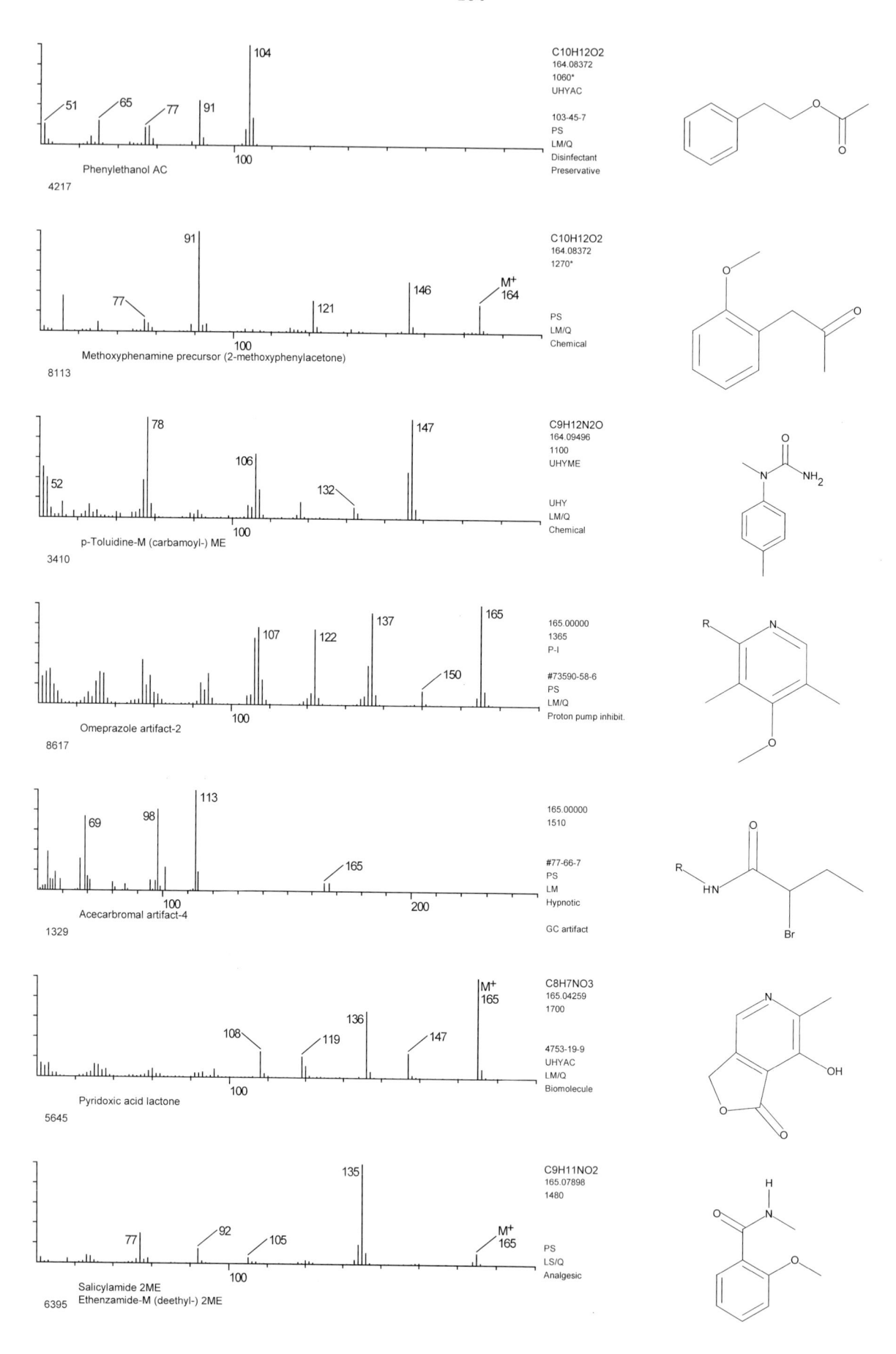

Spectrum	Compound	Formula / Info
4217	Phenylethanol AC — peaks 51, 65, 77, 91, 104	C10H12O2, 164.08372, 1060*, UHYAC, 103-45-7, PS, LM/Q, Disinfectant, Preservative
8113	Methoxyphenamine precursor (2-methoxyphenylacetone) — peaks 77, 91, 121, 146, M+ 164	C10H12O2, 164.08372, 1270*, PS, LM/Q, Chemical
3410	p-Toluidine-M (carbamoyl-) ME — peaks 52, 78, 106, 132, 147	C9H12N2O, 164.09496, 1100, UHYME, UHY, LM/Q, Chemical
8617	Omeprazole artifact-2 — peaks 107, 122, 137, 150, 165	165.00000, 1365, P-I, #73590-58-6, PS, LM/Q, Proton pump inhibit.
1329	Acecarbromal artifact-4 — peaks 69, 98, 113, 165	165.00000, 1510, #77-66-7, PS, LM, Hypnotic, GC artifact
5645	Pyridoxic acid lactone — peaks 108, 119, 136, 147, M+ 165	C8H7NO3, 165.04259, 1700, 4753-19-9, UHYAC, LM/Q, Biomolecule
6395	Salicylamide 2ME / Ethenzamide-M (deethyl-) 2ME — peaks 77, 92, 105, 135, M+ 165	C9H11NO2, 165.07898, 1480, PS, LS/Q, Analgesic

165

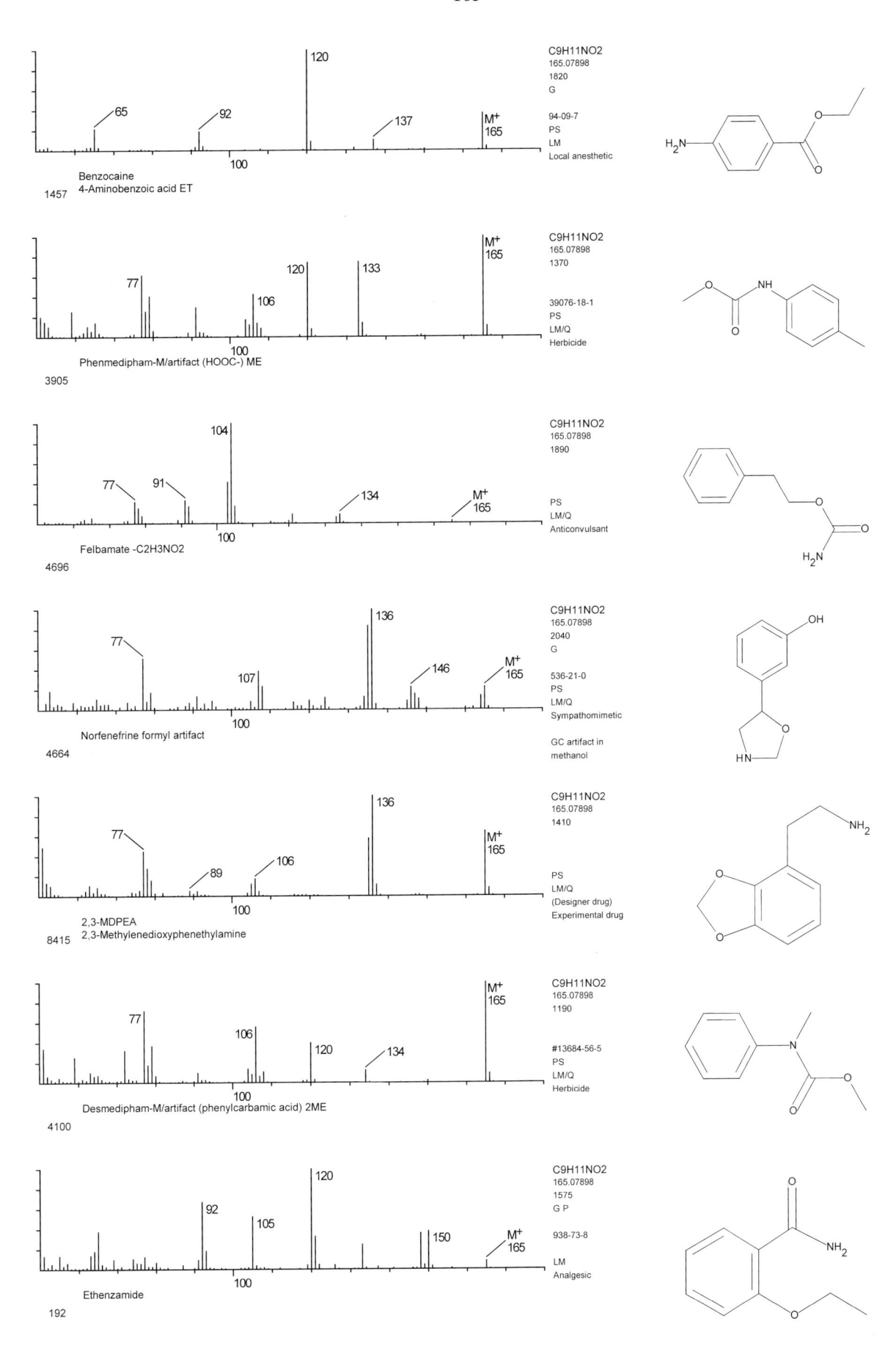

Benzocaine
4-Aminobenzoic acid ET
1457

C9H11NO2
165.07898
1820
G
94-09-7
PS
LM
Local anesthetic

Phenmedipham-M/artifact (HOOC-) ME
3905

C9H11NO2
165.07898
1370
39076-18-1
PS
LM/Q
Herbicide

Felbamate -C2H3NO2
4696

C9H11NO2
165.07898
1890
PS
LM/Q
Anticonvulsant

Norfenefrine formyl artifact
4664

C9H11NO2
165.07898
2040
G
536-21-0
PS
LM/Q
Sympathomimetic
GC artifact in methanol

2,3-MDPEA
2,3-Methylenedioxyphenethylamine
8415

C9H11NO2
165.07898
1410
PS
LM/Q
(Designer drug)
Experimental drug

Desmedipham-M/artifact (phenylcarbamic acid) 2ME
4100

C9H11NO2
165.07898
1190
#13684-56-5
PS
LM/Q
Herbicide

Ethenzamide
192

C9H11NO2
165.07898
1575
G P
938-73-8
LM
Analgesic

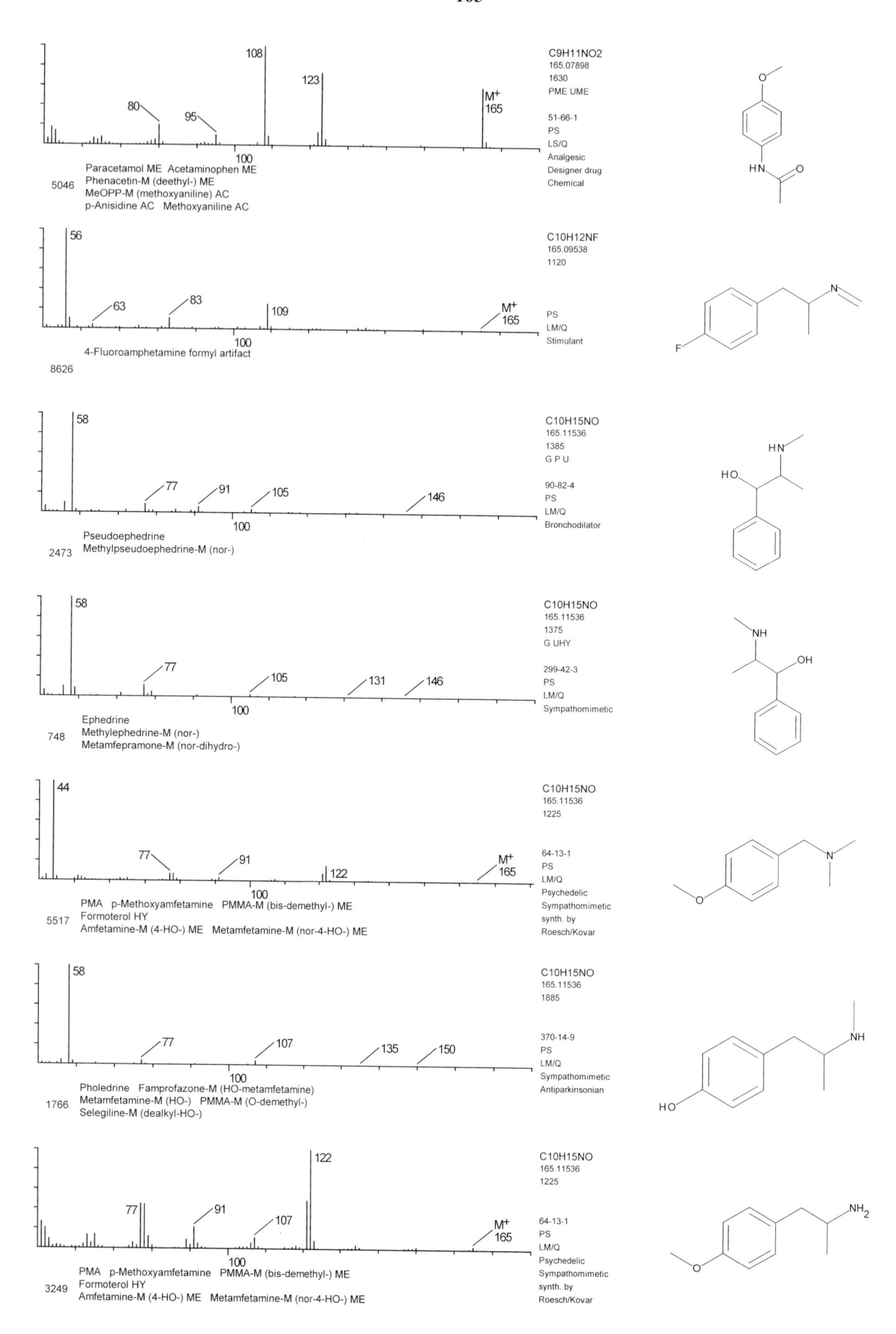

Spectrum #	Compound names	Formula / Info
5046	Paracetamol ME Acetaminophen ME Phenacetin-M (deethyl-) ME MeOPP-M (methoxyaniline) AC p-Anisidine AC Methoxyaniline AC	C9H11NO2 165.07898 1630 PME UME 51-66-1 PS LS/Q Analgesic Designer drug Chemical
8626	4-Fluoroamphetamine formyl artifact	C10H12NF 165.09538 1120 PS LM/Q Stimulant
2473	Pseudoephedrine Methylpseudoephedrine-M (nor-)	C10H15NO 165.11536 1385 G P U 90-82-4 PS LM/Q Bronchodilator
748	Ephedrine Methylephedrine-M (nor-) Metamfepramone-M (nor-dihydro-)	C10H15NO 165.11536 1375 G UHY 299-42-3 PS LM/Q Sympathomimetic
5517	PMA p-Methoxyamfetamine PMMA-M (bis-demethyl-) ME Formoterol HY Amfetamine-M (4-HO-) ME Metamfetamine-M (nor-4-HO-) ME	C10H15NO 165.11536 1225 64-13-1 PS LM/Q Psychedelic Sympathomimetic synth. by Roesch/Kovar
1766	Pholedrine Famprofazone-M (HO-metamfetamine) Metamfetamine-M (HO-) PMMA-M (O-demethyl-) Selegiline-M (dealkyl-HO-)	C10H15NO 165.11536 1885 370-14-9 PS LM/Q Sympathomimetic Antiparkinsonian
3249	PMA p-Methoxyamfetamine PMMA-M (bis-demethyl-) ME Formoterol HY Amfetamine-M (4-HO-) ME Metamfetamine-M (nor-4-HO-) ME	C10H15NO 165.11536 1225 64-13-1 PS LM/Q Psychedelic Sympathomimetic synth. by Roesch/Kovar

165

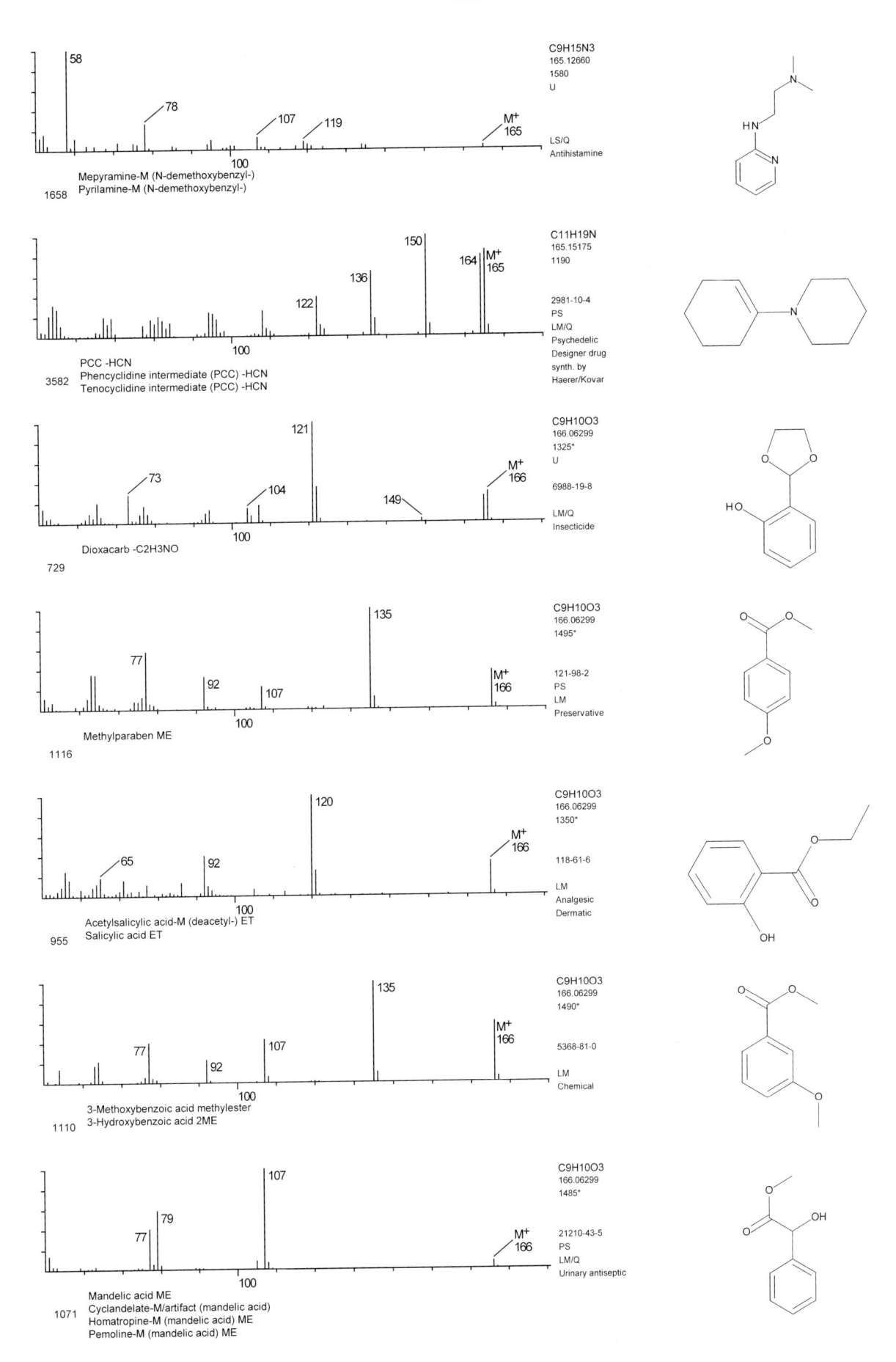

Mepyramine-M (N-demethoxybenzyl-)
Pyrilamine-M (N-demethoxybenzyl-)
1658

C9H15N3
165.12660
1580
U

LS/Q
Antihistamine

Peaks: 58, 78, 107, 119, M+ 165

PCC -HCN
Phencyclidine intermediate (PCC) -HCN
Tenocyclidine intermediate (PCC) -HCN
3582

C11H19N
165.15175
1190

2981-10-4
PS
LM/Q
Psychedelic
Designer drug
synth. by
Haerer/Kovar

Peaks: 122, 136, 150, 164, M+ 165

Dioxacarb -C2H3NO
729

C9H10O3
166.06299
1325*
U

6988-19-8

LM/Q
Insecticide

Peaks: 73, 104, 121, 149, M+ 166

Methylparaben ME
1116

C9H10O3
166.06299
1495*

121-98-2
PS
LM
Preservative

Peaks: 77, 92, 107, 135, M+ 166

Acetylsalicylic acid-M (deacetyl-) ET
Salicylic acid ET
955

C9H10O3
166.06299
1350*

118-61-6

LM
Analgesic
Dermatic

Peaks: 65, 92, 120, M+ 166

3-Methoxybenzoic acid methylester
3-Hydroxybenzoic acid 2ME
1110

C9H10O3
166.06299
1490*

5368-81-0

LM
Chemical

Peaks: 77, 92, 107, 135, M+ 166

Mandelic acid ME
Cyclandelate-M/artifact (mandelic acid)
Homatropine-M (mandelic acid) ME
Pemoline-M (mandelic acid) ME
1071

C9H10O3
166.06299
1485*

21210-43-5
PS
LM/Q
Urinary antiseptic

Peaks: 77, 79, 107, M+ 166

166

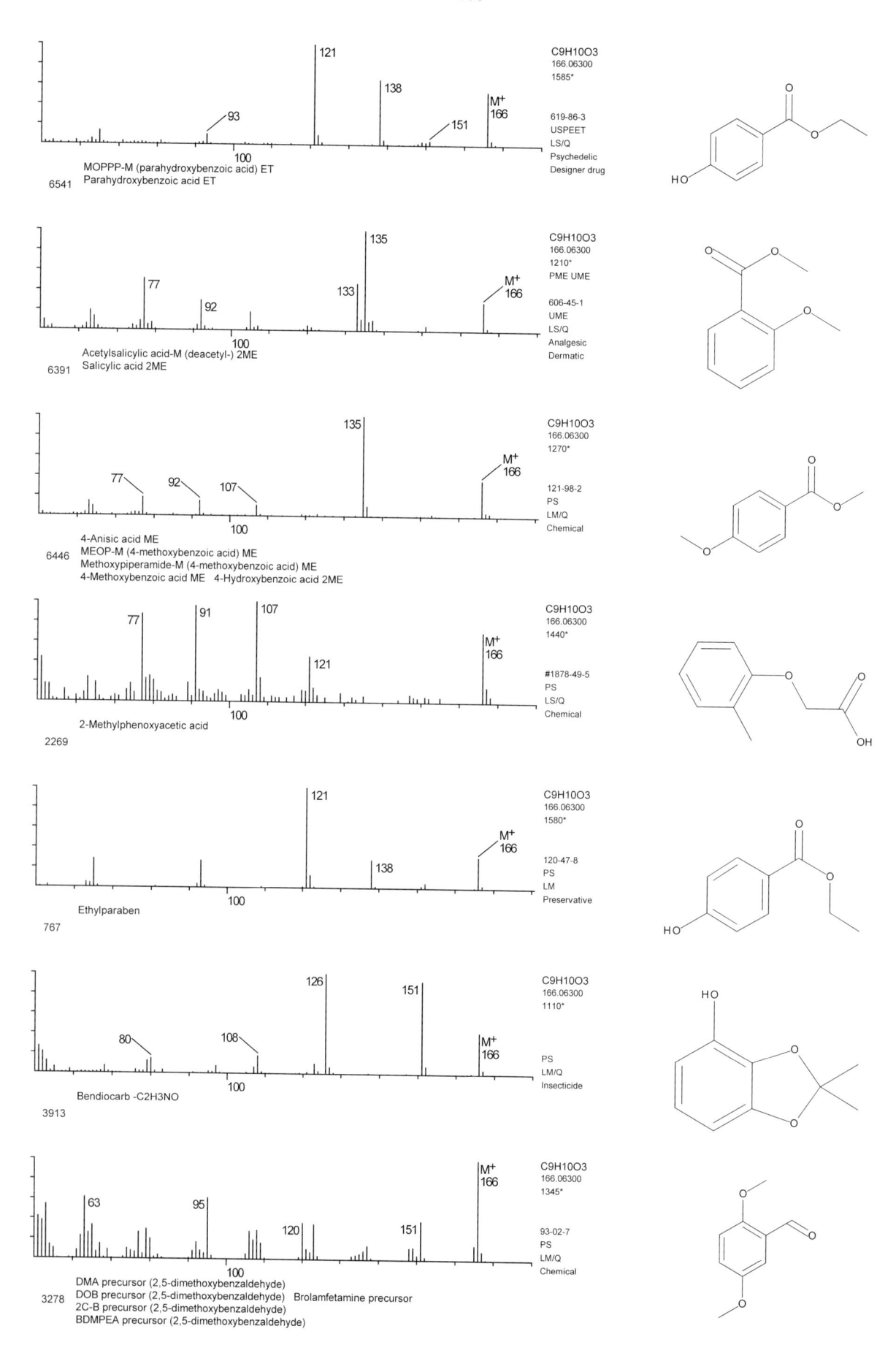

6541 MOPPP-M (parahydroxybenzoic acid) ET / Parahydroxybenzoic acid ET
Peaks: 93, 121, 138, 151, M+ 166
C9H10O3, 166.06300, 1585*
619-86-3, USPEET, LS/Q, Psychedelic Designer drug

6391 Acetylsalicylic acid-M (deacetyl-) 2ME / Salicylic acid 2ME
Peaks: 77, 92, 133, 135, M+ 166
C9H10O3, 166.06300, 1210*, PME UME
606-45-1, UME, LS/Q, Analgesic Dermatic

6446 4-Anisic acid ME / MEOP-M (4-methoxybenzoic acid) ME / Methoxypiperamide-M (4-methoxybenzoic acid) ME / 4-Methoxybenzoic acid ME / 4-Hydroxybenzoic acid 2ME
Peaks: 77, 92, 107, 135, M+ 166
C9H10O3, 166.06300, 1270*
121-98-2, PS, LM/Q, Chemical

2269 2-Methylphenoxyacetic acid
Peaks: 77, 91, 107, 121, M+ 166
C9H10O3, 166.06300, 1440*
#1878-49-5, PS, LS/Q, Chemical

767 Ethylparaben
Peaks: 121, 138, M+ 166
C9H10O3, 166.06300, 1580*
120-47-8, PS, LM, Preservative

3913 Bendiocarb -C2H3NO
Peaks: 80, 108, 126, 151, M+ 166
C9H10O3, 166.06300, 1110*
PS, LM/Q, Insecticide

3278 DMA precursor (2,5-dimethoxybenzaldehyde) / DOB precursor (2,5-dimethoxybenzaldehyde) / Brolamfetamine precursor / 2C-B precursor (2,5-dimethoxybenzaldehyde) / BDMPEA precursor (2,5-dimethoxybenzaldehyde)
Peaks: 63, 95, 120, 151, M+ 166
C9H10O3, 166.06300, 1345*
93-02-7, PS, LM/Q, Chemical

166

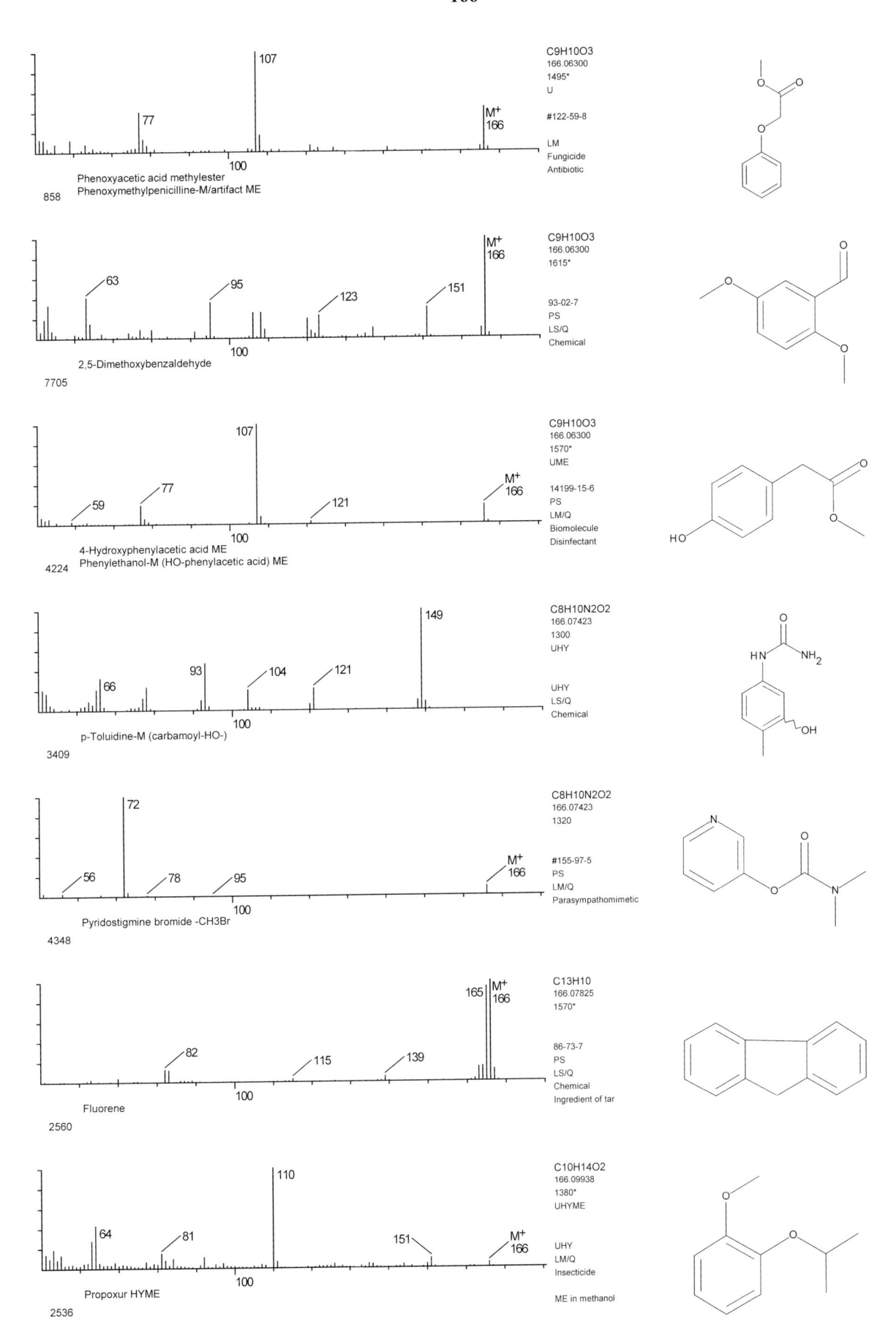

858 — Phenoxyacetic acid methylester / Phenoxymethylpenicilline-M/artifact ME
C9H10O3
166.06300
1495*
U
#122-59-8
LM
Fungicide
Antibiotic

7705 — 2,5-Dimethoxybenzaldehyde
C9H10O3
166.06300
1615*
93-02-7
PS
LS/Q
Chemical

4224 — 4-Hydroxyphenylacetic acid ME / Phenylethanol-M (HO-phenylacetic acid) ME
C9H10O3
166.06300
1570*
UME
14199-15-6
PS
LM/Q
Biomolecule
Disinfectant

3409 — p-Toluidine-M (carbamoyl-HO-)
C8H10N2O2
166.07423
1300
UHY
UHY
LS/Q
Chemical

4348 — Pyridostigmine bromide -CH3Br
C8H10N2O2
166.07423
1320
#155-97-5
PS
LM/Q
Parasympathomimetic

2560 — Fluorene
C13H10
166.07825
1570*
86-73-7
PS
LS/Q
Chemical
Ingredient of tar

2536 — Propoxur HYME
C10H14O2
166.09938
1380*
UHYME
UHY
LM/Q
Insecticide
ME in methanol

317

166

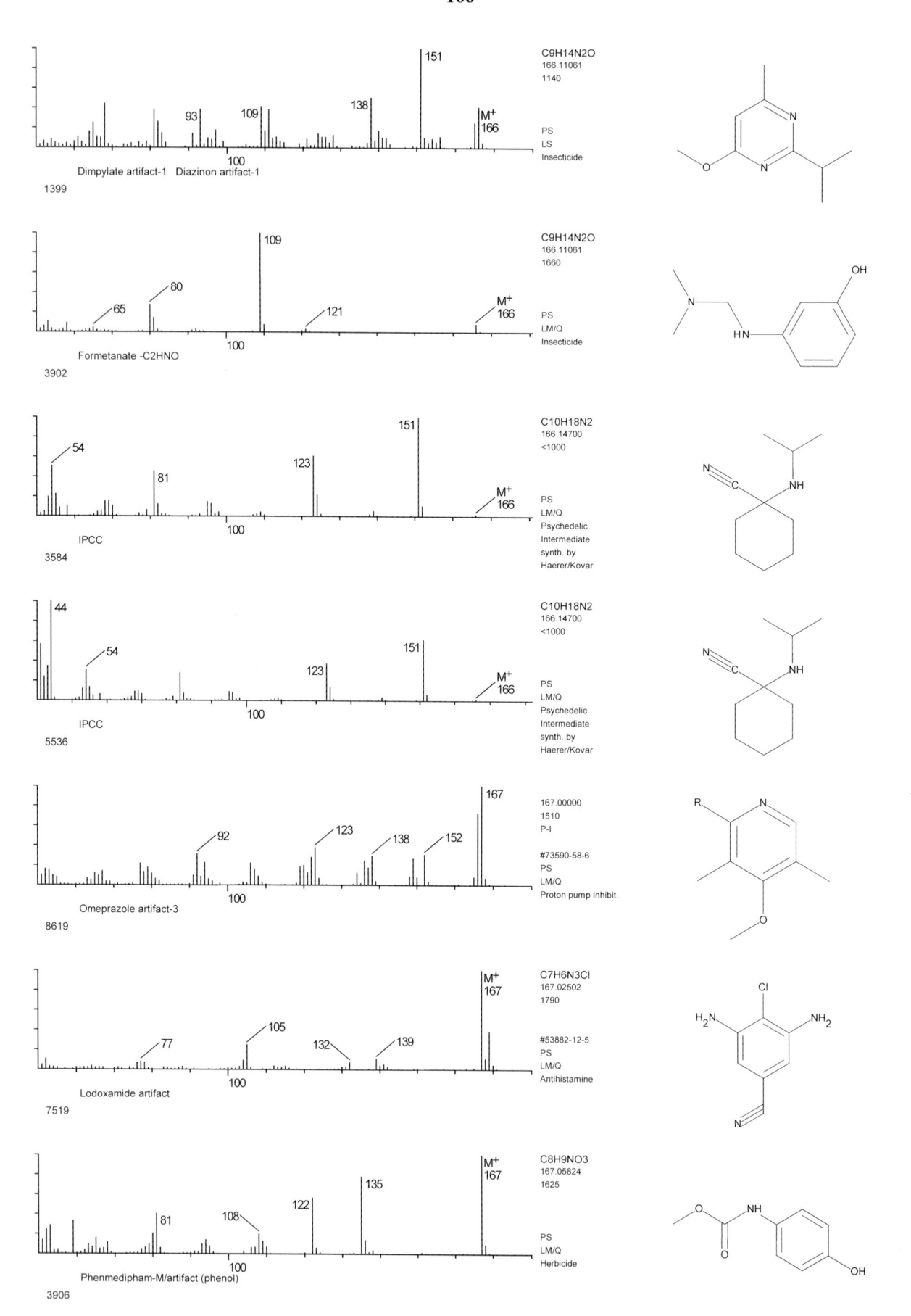

Dimpylate artifact-1 Diazinon artifact-1 1399	C9H14N2O 166.11061 1140 PS LS Insecticide
Formetanate -C2HNO 3902	C9H14N2O 166.11061 1660 PS LM/Q Insecticide
IPCC 3584	C10H18N2 166.14700 <1000 PS LM/Q Psychedelic Intermediate synth. by Haerer/Kovar
IPCC 5536	C10H18N2 166.14700 <1000 PS LM/Q Psychedelic Intermediate synth. by Haerer/Kovar
Omeprazole artifact-3 8619	167.00000 1510 P-I #73590-58-6 PS LM/Q Proton pump inhibit.
Lodoxamide artifact 7519	C7H6N3Cl 167.02502 1790 #53882-12-5 PS LM/Q Antihistamine
Phenmedipham-M/artifact (phenol) 3906	C8H9NO3 167.05824 1625 PS LM/Q Herbicide

167

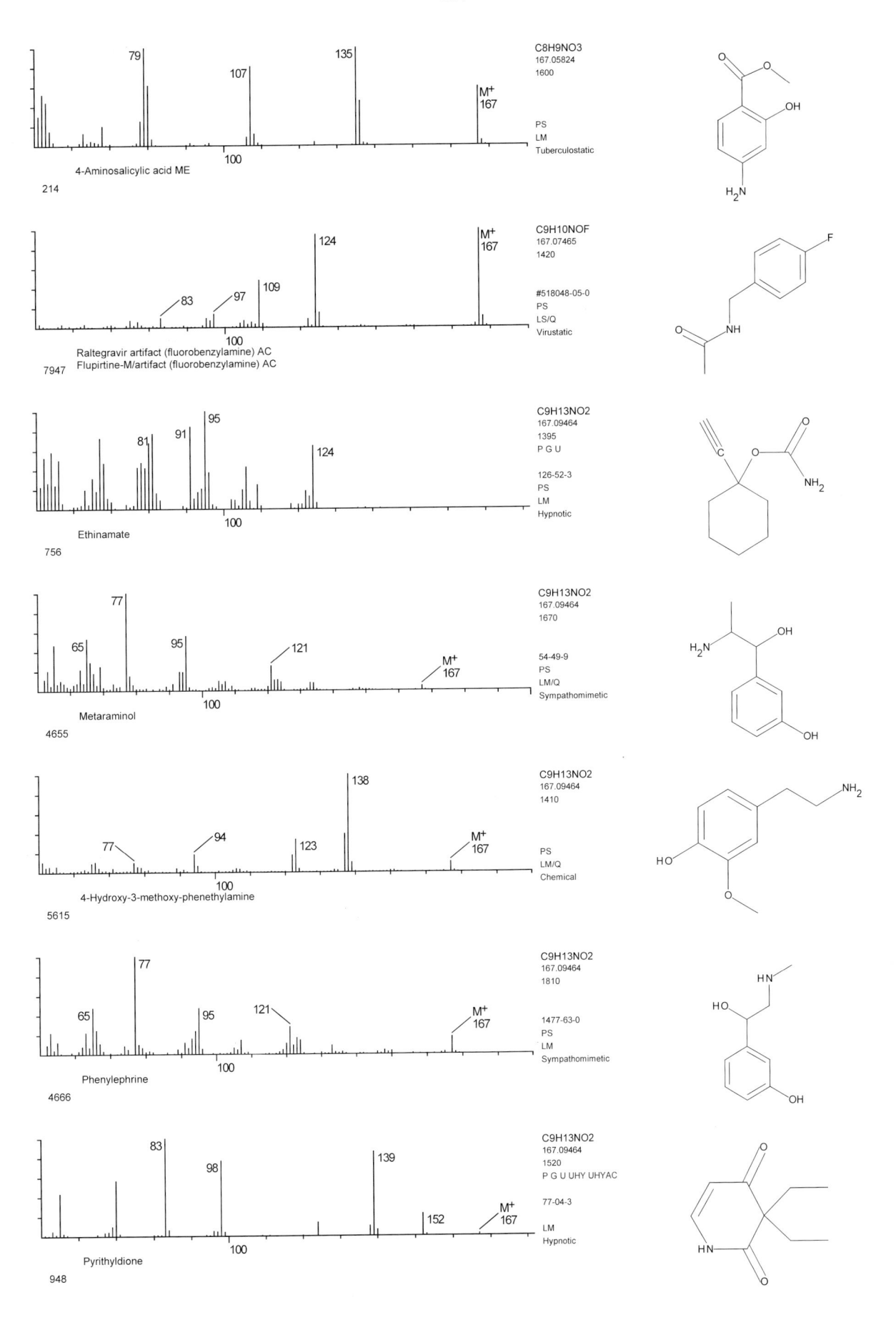

4-Aminosalicylic acid ME — peaks: 79, 107, 135, M+ 167	C8H9NO3 167.05824 1600 PS LM Tuberculostatic
Raltegravir artifact (fluorobenzylamine) AC Flupirtine-M/artifact (fluorobenzylamine) AC — peaks: 83, 97, 109, 124, M+ 167	C9H10NOF 167.07465 1420 #518048-05-0 PS LS/Q Virustatic
Ethinamate — peaks: 81, 91, 95, 124	C9H13NO2 167.09464 1395 P G U 126-52-3 PS LM Hypnotic
Metaraminol — peaks: 65, 77, 95, 121, M+ 167	C9H13NO2 167.09464 1670 54-49-9 PS LM/Q Sympathomimetic
4-Hydroxy-3-methoxy-phenethylamine — peaks: 77, 94, 123, 138, M+ 167	C9H13NO2 167.09464 1410 PS LM/Q Chemical
Phenylephrine — peaks: 65, 77, 95, 121, M+ 167	C9H13NO2 167.09464 1810 1477-63-0 PS LM Sympathomimetic
Pyrithyldione — peaks: 83, 98, 139, 152, M+ 167	C9H13NO2 167.09464 1520 P G U UHY UHYAC 77-04-3 LM Hypnotic

319

167

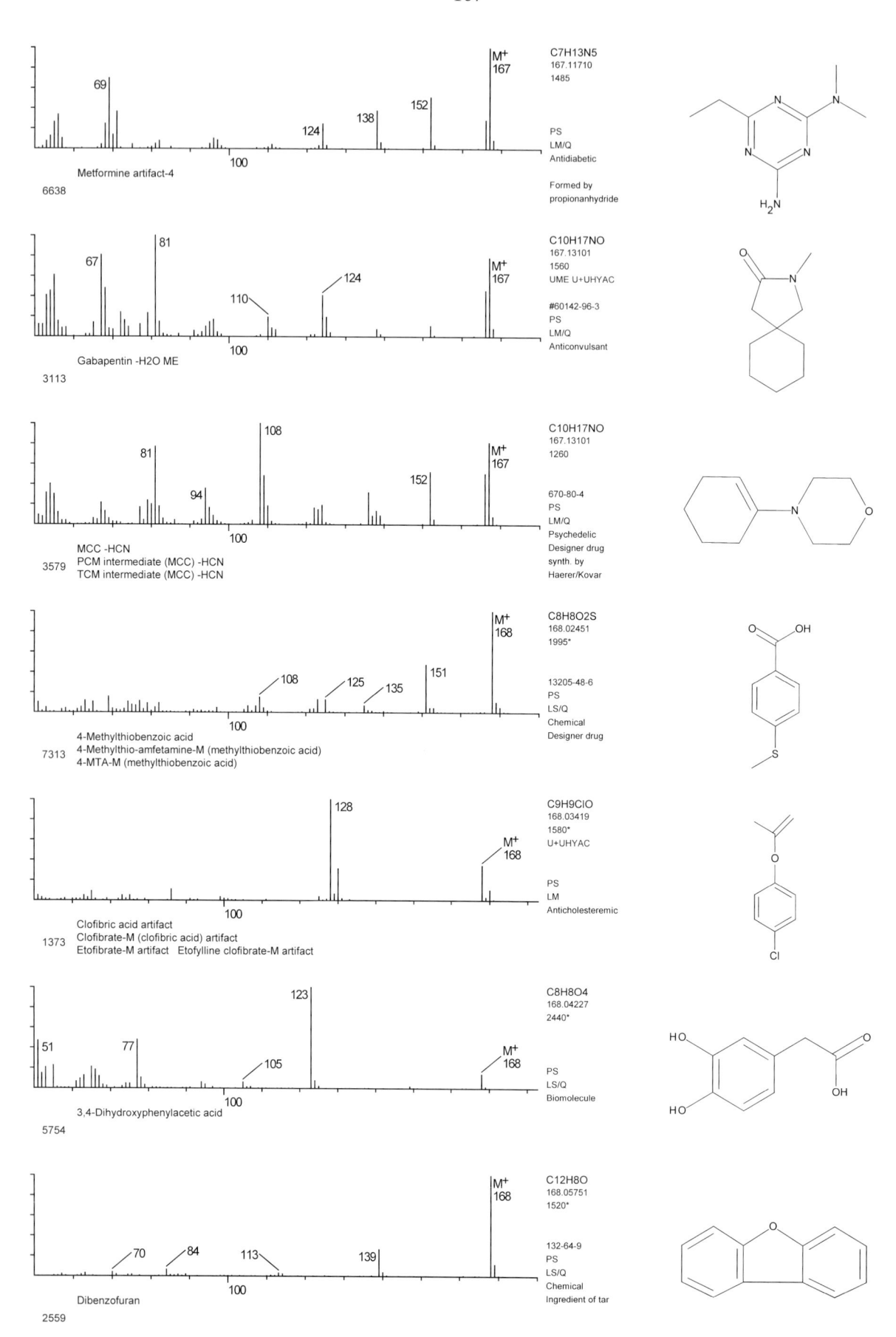

Spectrum label	Formula / Info
Metformine artifact-4 (6638)	C7H13N5, 167.11710, 1485, PS, LM/Q, Antidiabetic, Formed by propionanhydride
Gabapentin -H2O ME (3113)	C10H17NO, 167.13101, 1560, UME U+UHYAC, #60142-96-3, PS, LM/Q, Anticonvulsant
MCC -HCN / PCM intermediate (MCC) -HCN / TCM intermediate (MCC) -HCN (3579)	C10H17NO, 167.13101, 1260, 670-80-4, PS, LM/Q, Psychedelic Designer drug synth. by Haerer/Kovar
4-Methylthiobenzoic acid / 4-Methylthio-amfetamine-M (methylthiobenzoic acid) / 4-MTA-M (methylthiobenzoic acid) (7313)	C8H8O2S, 168.02451, 1995*, 13205-48-6, PS, LS/Q, Chemical Designer drug
Clofibric acid artifact / Clofibrate-M (clofibric acid) artifact / Etofibrate-M artifact Etofylline clofibrate-M artifact (1373)	C9H9ClO, 168.03419, 1580*, U+UHYAC, PS, LM, Anticholesteremic
3,4-Dihydroxyphenylacetic acid (5754)	C8H8O4, 168.04227, 2440*, PS, LS/Q, Biomolecule
Dibenzofuran (2559)	C12H8O, 168.05751, 1520*, 132-64-9, PS, LS/Q, Chemical Ingredient of tar

320

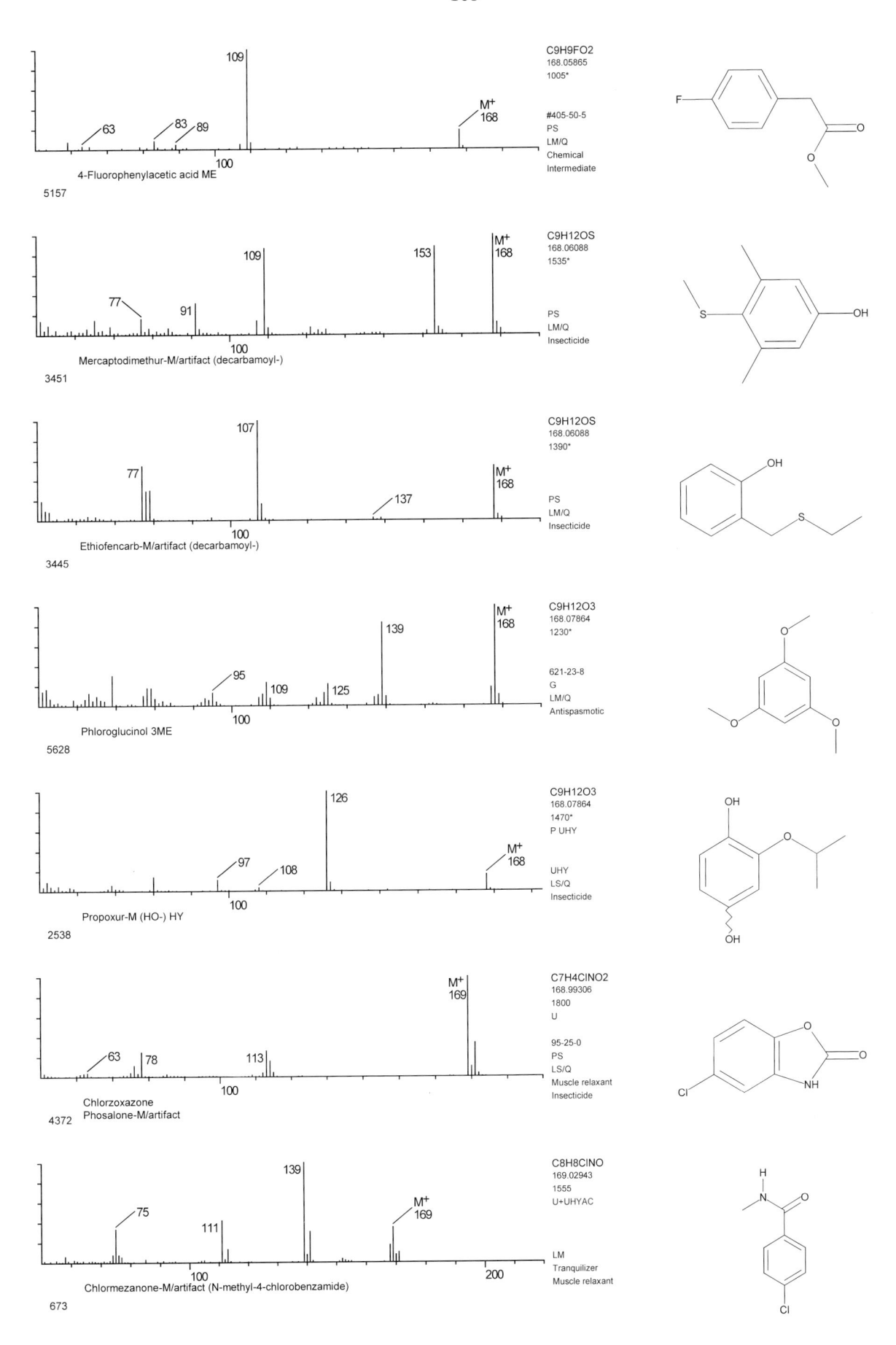

4-Fluorophenylacetic acid ME (5157)	C9H9FO2, 168.05865, 1005*, #405-50-5, PS, LM/Q, Chemical Intermediate. Peaks: 63, 83, 89, 109, M+ 168
Mercaptodimethur-M/artifact (decarbamoyl-) (3451)	C9H12OS, 168.06088, 1535*, PS, LM/Q, Insecticide. Peaks: 77, 91, 109, 153, M+ 168
Ethiofencarb-M/artifact (decarbamoyl-) (3445)	C9H12OS, 168.06088, 1390*, PS, LM/Q, Insecticide. Peaks: 77, 107, 137, M+ 168
Phloroglucinol 3ME (5628)	C9H12O3, 168.07864, 1230*, 621-23-8, G, LM/Q, Antispasmotic. Peaks: 95, 109, 125, 139, M+ 168
Propoxur-M (HO-) HY (2538)	C9H12O3, 168.07864, 1470*, P UHY, UHY, LS/Q, Insecticide. Peaks: 97, 108, 126, M+ 168
Chlorzoxazone / Phosalone-M/artifact (4372)	C7H4ClNO2, 168.99306, 1800, U, 95-25-0, PS, LS/Q, Muscle relaxant, Insecticide. Peaks: 63, 78, 113, M+ 169
Chlormezanone-M/artifact (N-methyl-4-chlorobenzamide) (673)	C8H8ClNO, 169.02943, 1555, U+UHYAC, LM, Tranquilizer, Muscle relaxant. Peaks: 75, 111, 139, M+ 169

168

321

169

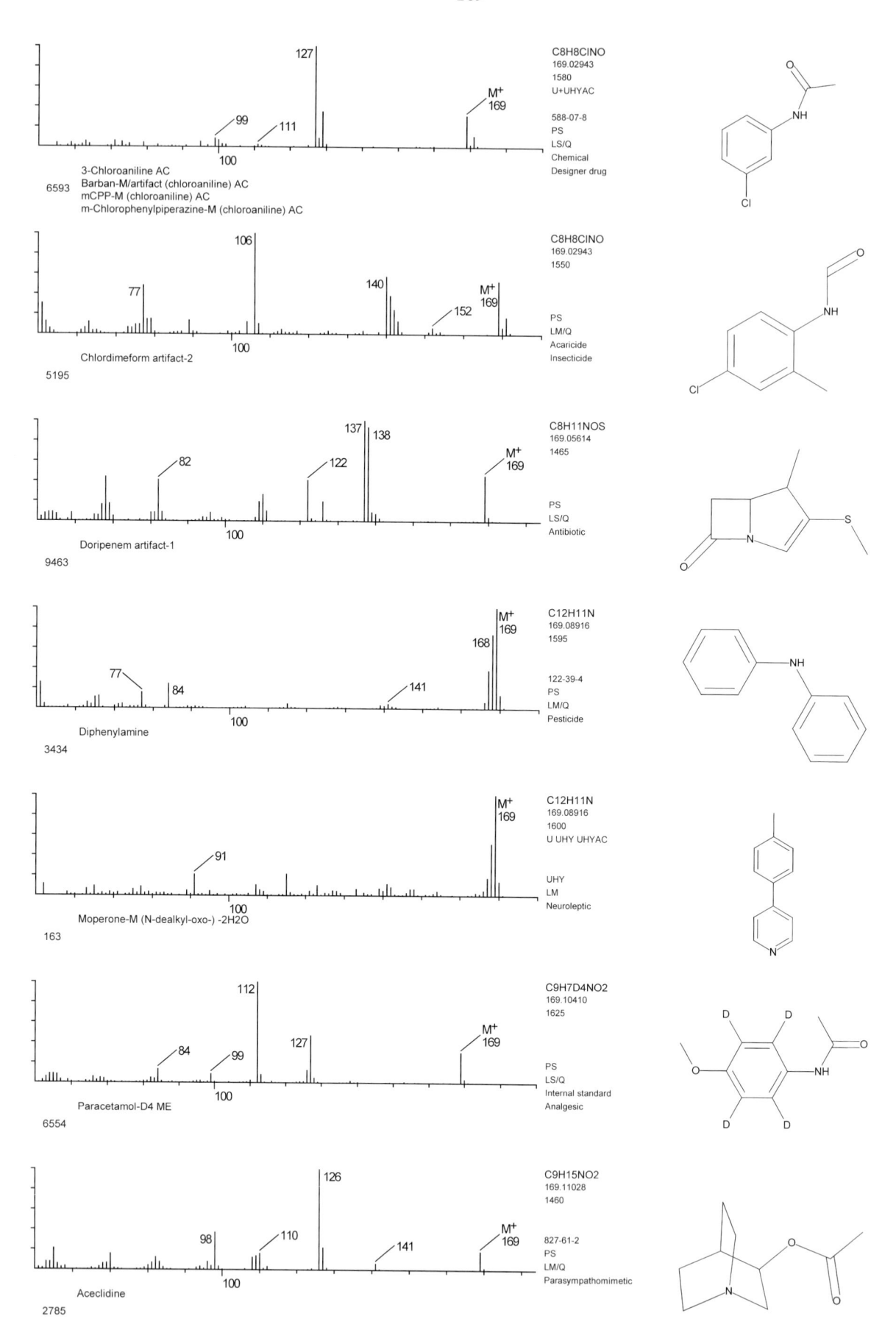

3-Chloroaniline AC
Barban-M/artifact (chloroaniline) AC
mCPP-M (chloroaniline) AC
m-Chlorophenylpiperazine-M (chloroaniline) AC
6593

C8H8ClNO
169.02943
1580
U+UHYAC

588-07-8
PS
LS/Q
Chemical
Designer drug

Chlordimeform artifact-2
5195

C8H8ClNO
169.02943
1550

PS
LM/Q
Acaricide
Insecticide

Doripenem artifact-1
9463

C8H11NOS
169.05614
1465

PS
LS/Q
Antibiotic

Diphenylamine
3434

C12H11N
169.08916
1595

122-39-4
PS
LM/Q
Pesticide

Moperone-M (N-dealkyl-oxo-) -2H2O
163

C12H11N
169.08916
1600
U UHY UHYAC

UHY
LM
Neuroleptic

Paracetamol-D4 ME
6554

C9H7D4NO2
169.10410
1625

PS
LS/Q
Internal standard
Analgesic

Aceclidine
2785

C9H15NO2
169.11028
1460

827-61-2
PS
LM/Q
Parasympathomimetic

322

169

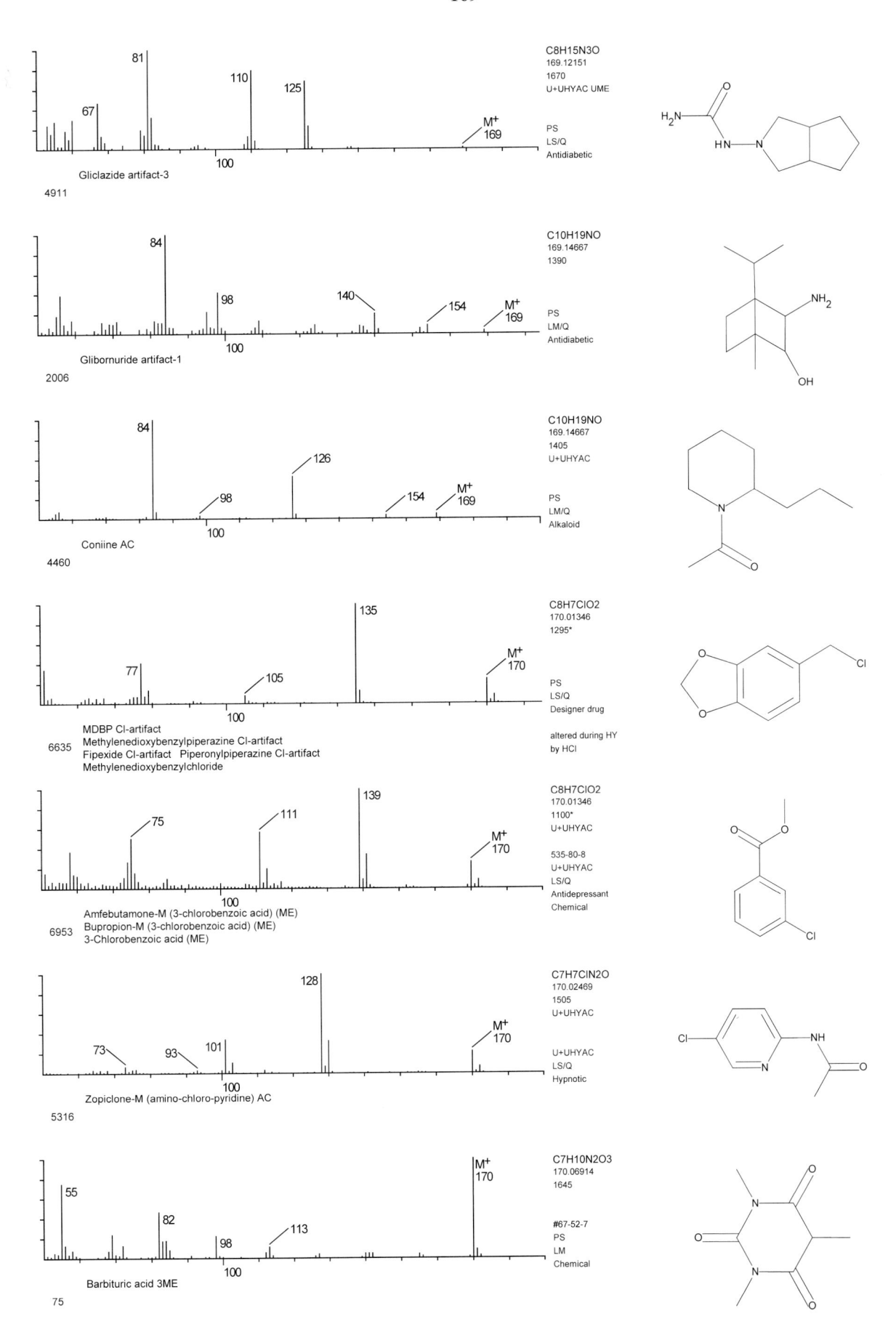

Gliclazide artifact-3
4911
81, 67, 110, 125, M+ 169
C8H15N3O
169.12151
1670
U+UHYAC UME
PS
LS/Q
Antidiabetic

Glibornuride artifact-1
2006
84, 98, 140, 154, M+ 169
C10H19NO
169.14667
1390
PS
LM/Q
Antidiabetic

Coniine AC
4460
84, 98, 126, 154, M+ 169
C10H19NO
169.14667
1405
U+UHYAC
PS
LM/Q
Alkaloid

MDBP Cl-artifact
Methylenedioxybenzylpiperazine Cl-artifact
Fipexide Cl-artifact   Piperonylpiperazine Cl-artifact
Methylenedioxybenzylchloride
6635
77, 105, 135, M+ 170
C8H7ClO2
170.01346
1295*
PS
LS/Q
Designer drug
altered during HY by HCl

Amfebutamone-M (3-chlorobenzoic acid) (ME)
Bupropion-M (3-chlorobenzoic acid) (ME)
3-Chlorobenzoic acid (ME)
6953
75, 111, 139, M+ 170
C8H7ClO2
170.01346
1100*
U+UHYAC
535-80-8
U+UHYAC
LS/Q
Antidepressant
Chemical

Zopiclone-M (amino-chloro-pyridine) AC
5316
73, 93, 101, 128, M+ 170
C7H7ClN2O
170.02469
1505
U+UHYAC
U+UHYAC
LS/Q
Hypnotic

Barbituric acid 3ME
75
55, 82, 98, 113, M+ 170
C7H10N2O3
170.06914
1645
#67-52-7
PS
LM
Chemical

323

170

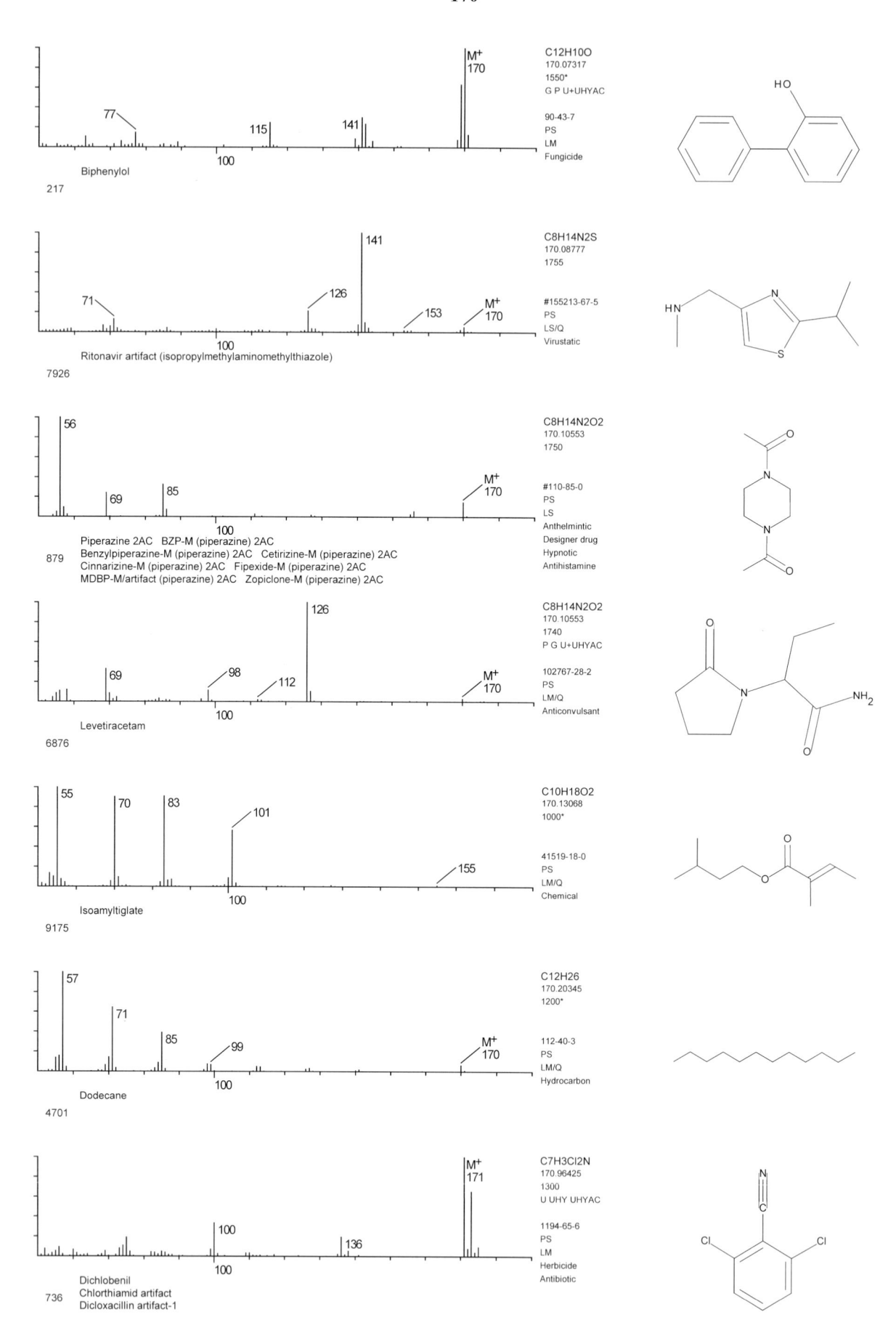

Biphenylol
C12H10O
170.07317
1550*
G P U+UHYAC

90-43-7
PS
LM
Fungicide

Peaks: 77, 115, 141, M+ 170

217

Ritonavir artifact (isopropylmethylaminomethylthiazole)
C8H14N2S
170.08777
1755

#155213-67-5
PS
LS/Q
Virustatic

Peaks: 71, 126, 141, 153, M+ 170

7926

Piperazine 2AC   BZP-M (piperazine) 2AC
Benzylpiperazine-M (piperazine) 2AC   Cetirizine-M (piperazine) 2AC
Cinnarizine-M (piperazine) 2AC   Fipexide-M (piperazine) 2AC
MDBP-M/artifact (piperazine) 2AC   Zopiclone-M (piperazine) 2AC

C8H14N2O2
170.10553
1750

#110-85-0
PS
LS
Anthelmintic
Designer drug
Hypnotic
Antihistamine

Peaks: 56, 69, 85, M+ 170

879

Levetiracetam
C8H14N2O2
170.10553
1740
P G U+UHYAC

102767-28-2
PS
LM/Q
Anticonvulsant

Peaks: 69, 98, 112, 126, M+ 170

6876

Isoamyltiglate
C10H18O2
170.13068
1000*

41519-18-0
PS
LM/Q
Chemical

Peaks: 55, 70, 83, 101, 155

9175

Dodecane
C12H26
170.20345
1200*

112-40-3
PS
LM/Q
Hydrocarbon

Peaks: 57, 71, 85, 99, M+ 170

4701

Dichlobenil
Chlorthiamid artifact
Dicloxacillin artifact-1

C7H3Cl2N
170.96425
1300
U UHY UHYAC

1194-65-6
PS
LM
Herbicide
Antibiotic

Peaks: 100, 136, M+ 171

736

171

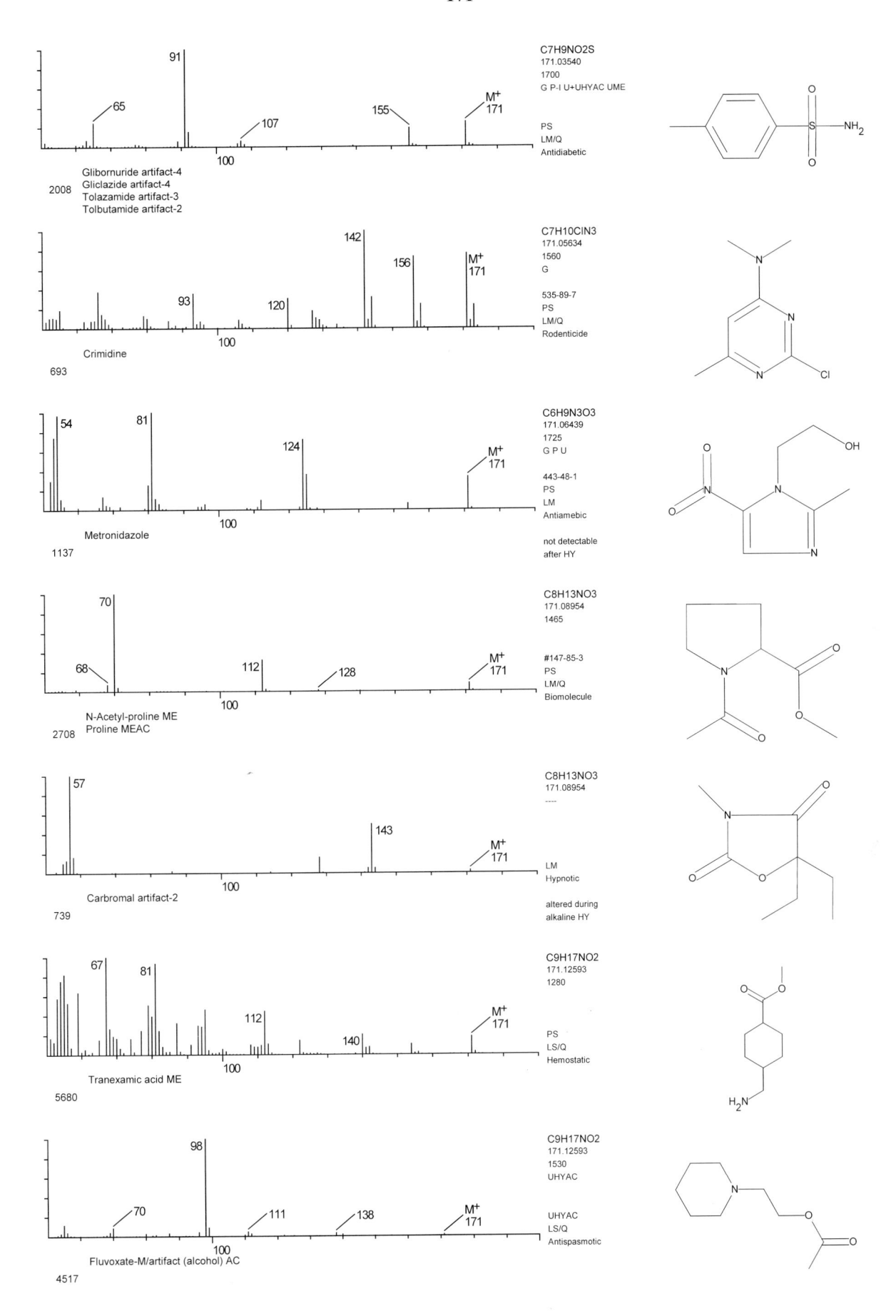

Glibornuride artifact-4
Gliclazide artifact-4
Tolazamide artifact-3
Tolbutamide artifact-2
2008

C7H9NO2S
171.03540
1700
G P-I U+UHYAC UME

PS
LM/Q
Antidiabetic

Crimidine
693

C7H10ClN3
171.05634
1560
G

535-89-7
PS
LM/Q
Rodenticide

Metronidazole
1137

C6H9N3O3
171.06439
1725
G P U

443-48-1
PS
LM
Antiamebic

not detectable after HY

N-Acetyl-proline ME
Proline MEAC
2708

C8H13NO3
171.08954
1465

#147-85-3
PS
LM/Q
Biomolecule

Carbromal artifact-2
739

C8H13NO3
171.08954

LM
Hypnotic

altered during alkaline HY

Tranexamic acid ME
5680

C9H17NO2
171.12593
1280

PS
LS/Q
Hemostatic

Fluvoxate-M/artifact (alcohol) AC
4517

C9H17NO2
171.12593
1530
UHYAC

UHYAC
LS/Q
Antispasmotic

171

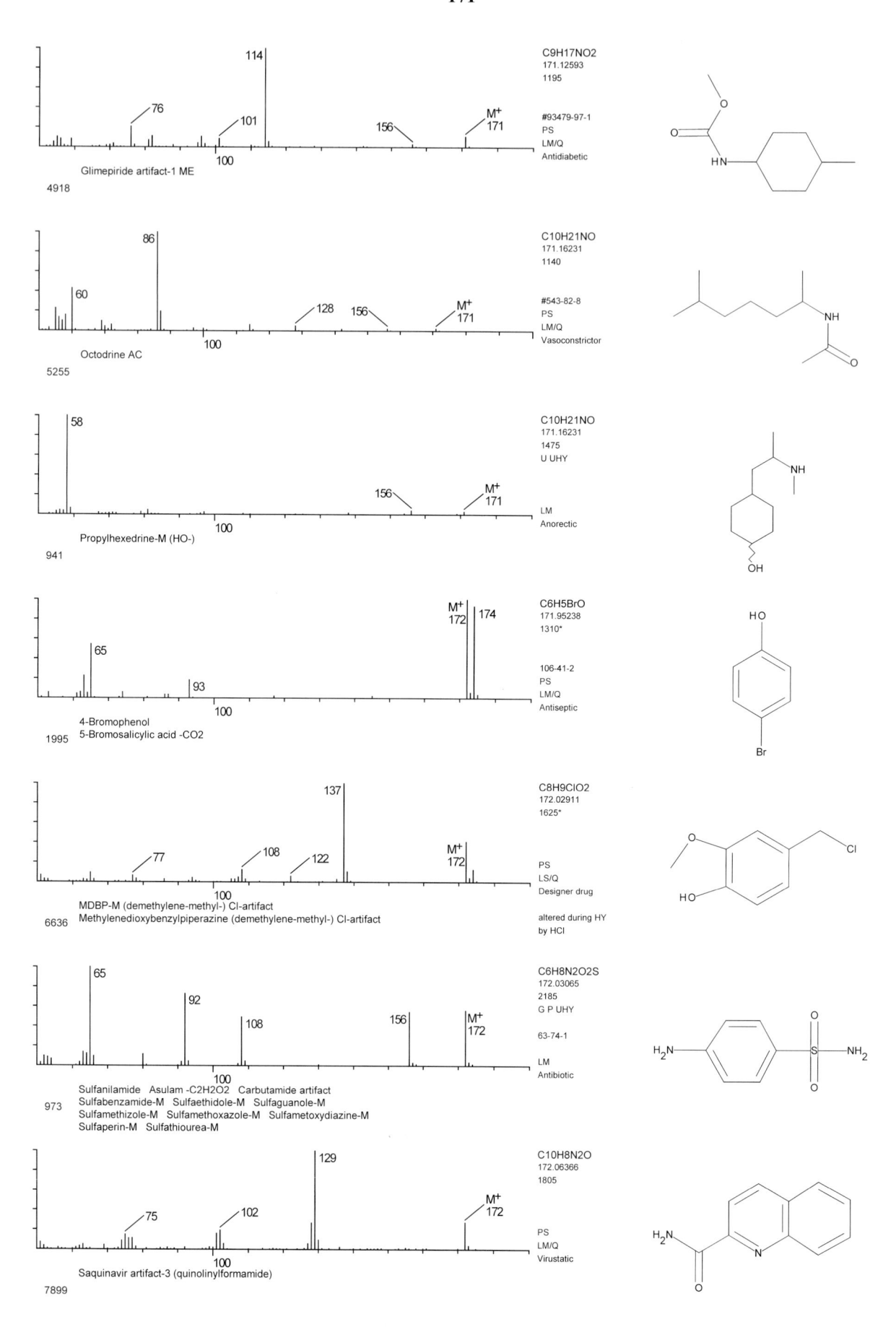

Glimepiride artifact-1 ME
4918
Peaks: 76, 101, 114, 156, M+ 171

C9H17NO2
171.12593
1195
#93479-97-1
PS
LM/Q
Antidiabetic

Octodrine AC
5255
Peaks: 60, 86, 128, 156, M+ 171

C10H21NO
171.16231
1140
#543-82-8
PS
LM/Q
Vasoconstrictor

Propylhexedrine-M (HO-)
941
Peaks: 58, 156, M+ 171

C10H21NO
171.16231
1475
U UHY

LM
Anorectic

4-Bromophenol
5-Bromosalicylic acid -CO2
1995
Peaks: 65, 93, M+ 172, 174

C6H5BrO
171.95238
1310*

106-41-2
PS
LM/Q
Antiseptic

MDBP-M (demethylene-methyl-) Cl-artifact
Methylenedioxybenzylpiperazine (demethylene-methyl-) Cl-artifact
6636
Peaks: 77, 108, 122, 137, M+ 172

C8H9ClO2
172.02911
1625*

PS
LS/Q
Designer drug
altered during HY
by HCl

Sulfanilamide   Asulam -C2H2O2   Carbutamide artifact
Sulfabenzamide-M   Sulfaethidole-M   Sulfaguanole-M
Sulfamethizole-M   Sulfamethoxazole-M   Sulfametoxydiazine-M
Sulfaperin-M   Sulfathiourea-M
973
Peaks: 65, 92, 108, 156, M+ 172

C6H8N2O2S
172.03065
2185
G P UHY
63-74-1
LM
Antibiotic

Saquinavir artifact-3 (quinolinylformamide)
7899
Peaks: 75, 102, 129, M+ 172

C10H8N2O
172.06366
1805

PS
LM/Q
Virustatic

326

172

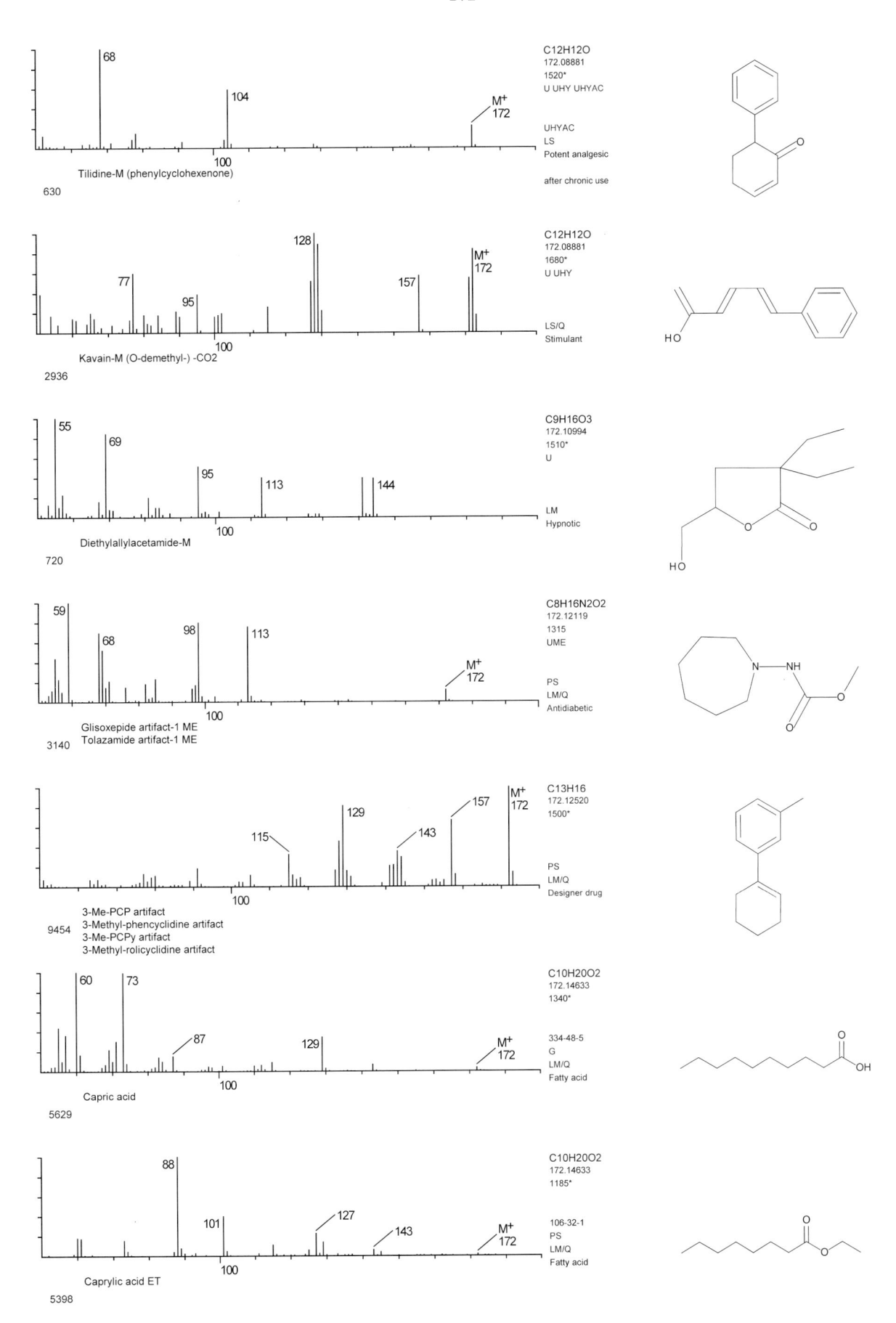

Tilidine-M (phenylcyclohexenone)
C12H12O
172.08881
1520*
U UHY UHYAC

UHYAC
LS
Potent analgesic
after chronic use

630

Kavain-M (O-demethyl-) -CO2
C12H12O
172.08881
1680*
U UHY

LS/Q
Stimulant

2936

Diethylallylacetamide-M
C9H16O3
172.10994
1510*
U

LM
Hypnotic

720

Glisoxepide artifact-1 ME
Tolazamide artifact-1 ME
C8H16N2O2
172.12119
1315
UME

PS
LM/Q
Antidiabetic

3140

3-Me-PCP artifact
3-Methyl-phencyclidine artifact
3-Me-PCPy artifact
3-Methyl-rolicyclidine artifact
C13H16
172.12520
1500*

PS
LM/Q
Designer drug

9454

Capric acid
C10H20O2
172.14633
1340*

334-48-5
G
LM/Q
Fatty acid

5629

Caprylic acid ET
C10H20O2
172.14633
1185*

106-32-1
PS
LM/Q
Fatty acid

5398

327

173

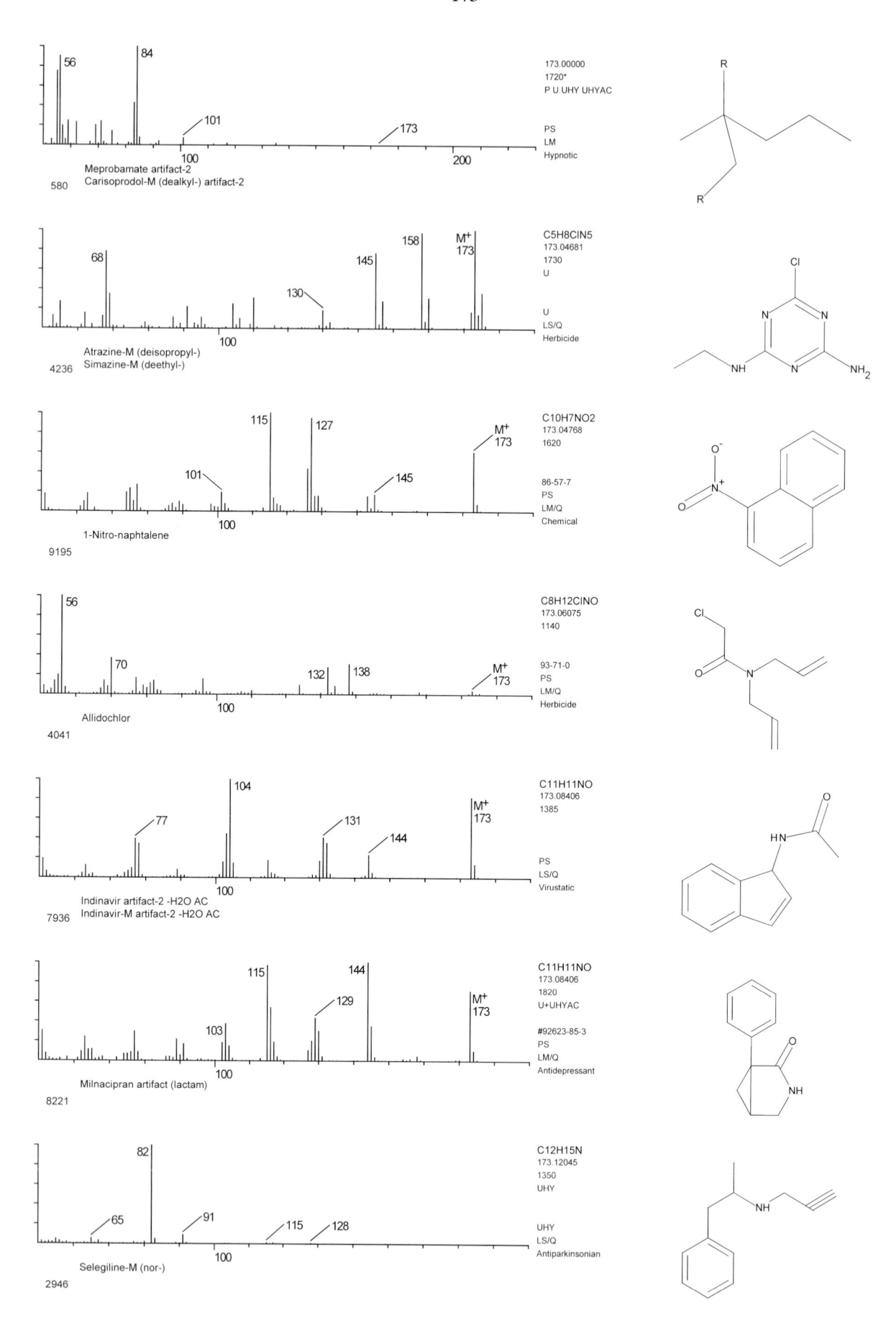

580	Meprobamate artifact-2 / Carisoprodol-M (dealkyl-) artifact-2 — 173.00000, 1720*, P U UHY UHYAC, PS, LM, Hypnotic. Peaks: 56, 84, 101, 173.
4236	Atrazine-M (deisopropyl-) / Simazine-M (deethyl-) — C5H8ClN5, 173.04681, 1730, U, LS/Q, Herbicide. Peaks: 68, 130, 145, 158, M⁺ 173.
9195	1-Nitro-naphtalene — C10H7NO2, 173.04768, 1620, 86-57-7, PS, LM/Q, Chemical. Peaks: 101, 115, 127, 145, M⁺ 173.
4041	Allidochlor — C8H12ClNO, 173.06075, 1140, 93-71-0, PS, LM/Q, Herbicide. Peaks: 56, 70, 132, 138, M⁺ 173.
7936	Indinavir artifact-2 -H2O AC / Indinavir-M artifact-2 -H2O AC — C11H11NO, 173.08406, 1385, PS, LS/Q, Virustatic. Peaks: 77, 104, 131, 144, M⁺ 173.
8221	Milnacipran artifact (lactam) — C11H11NO, 173.08406, 1820, U+UHYAC, #92623-85-3, PS, LM/Q, Antidepressant. Peaks: 103, 115, 129, 144, M⁺ 173.
2946	Selegiline-M (nor-) — C12H15N, 173.12045, 1350, UHY, UHY, LS/Q, Antiparkinsonian. Peaks: 65, 82, 91, 115, 128.

328

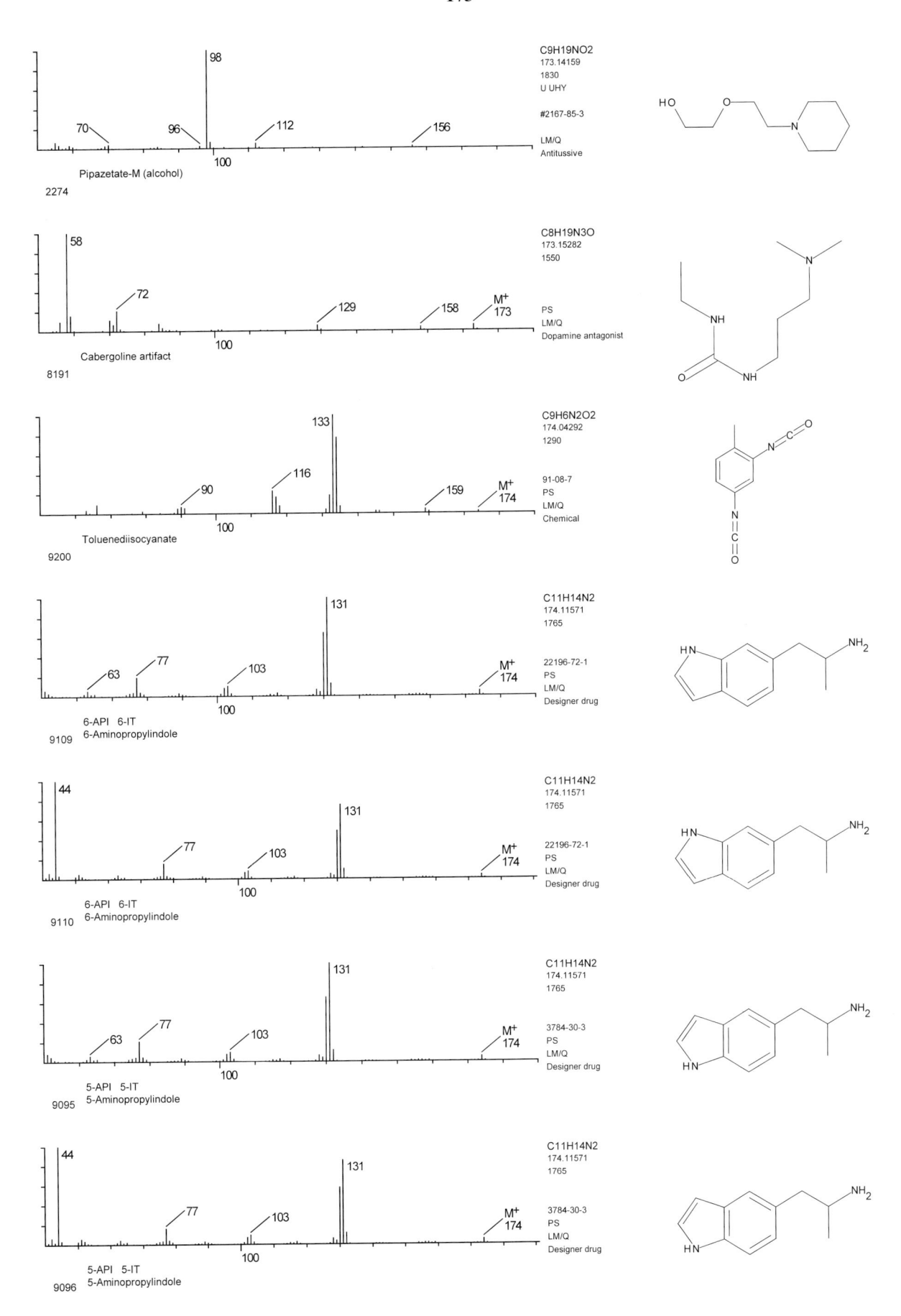

Spectrum	Formula / Data	Name
98, 70, 96, 112, 156 / 100	C9H19NO2, 173.14159, 1830, U UHY, #2167-85-3, LM/Q, Antitussive	Pipazetate-M (alcohol) 2274
58, 72, 129, 158, M+ 173 / 100	C8H19N3O, 173.15282, 1550, PS, LM/Q, Dopamine antagonist	Cabergoline artifact 8191
133, 90, 116, 159, M+ 174 / 100	C9H6N2O2, 174.04292, 1290, 91-08-7, PS, LM/Q, Chemical	Toluenediisocyanate 9200
131, 63, 77, 103, M+ 174 / 100	C11H14N2, 174.11571, 1765, 22196-72-1, PS, LM/Q, Designer drug	6-API 6-IT 6-Aminopropylindole 9109
44, 77, 103, 131, M+ 174 / 100	C11H14N2, 174.11571, 1765, 22196-72-1, PS, LM/Q, Designer drug	6-API 6-IT 6-Aminopropylindole 9110
131, 63, 77, 103, M+ 174 / 100	C11H14N2, 174.11571, 1765, 3784-30-3, PS, LM/Q, Designer drug	5-API 5-IT 5-Aminopropylindole 9095
44, 77, 103, 131, M+ 174 / 100	C11H14N2, 174.11571, 1765, 3784-30-3, PS, LM/Q, Designer drug	5-API 5-IT 5-Aminopropylindole 9096

175

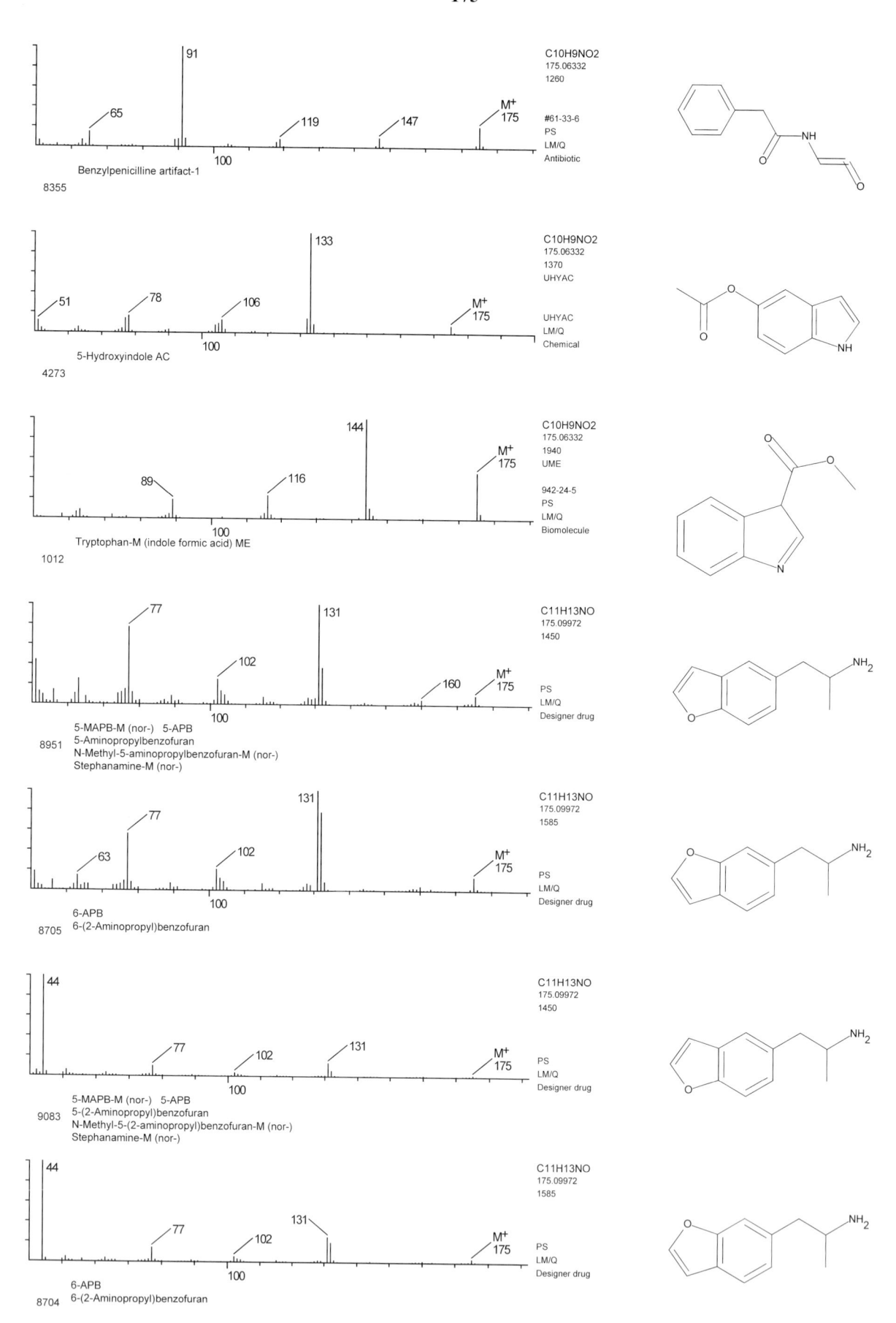

Benzylpenicilline artifact-1
8355
C10H9NO2
175.06332
1260
#61-33-6
PS
LM/Q
Antibiotic

5-Hydroxyindole AC
4273
C10H9NO2
175.06332
1370
UHYAC
UHYAC
LM/Q
Chemical

Tryptophan-M (indole formic acid) ME
1012
C10H9NO2
175.06332
1940
UME
942-24-5
PS
LM/Q
Biomolecule

5-MAPB-M (nor-)   5-APB
5-Aminopropylbenzofuran
N-Methyl-5-aminopropylbenzofuran-M (nor-)
Stephanamine-M (nor-)
8951
C11H13NO
175.09972
1450
PS
LM/Q
Designer drug

6-APB
6-(2-Aminopropyl)benzofuran
8705
C11H13NO
175.09972
1585
PS
LM/Q
Designer drug

5-MAPB-M (nor-)   5-APB
5-(2-Aminopropyl)benzofuran
N-Methyl-5-(2-aminopropyl)benzofuran-M (nor-)
Stephanamine-M (nor-)
9083
C11H13NO
175.09972
1450
PS
LM/Q
Designer drug

6-APB
6-(2-Aminopropyl)benzofuran
8704
C11H13NO
175.09972
1585
PS
LM/Q
Designer drug

330

175

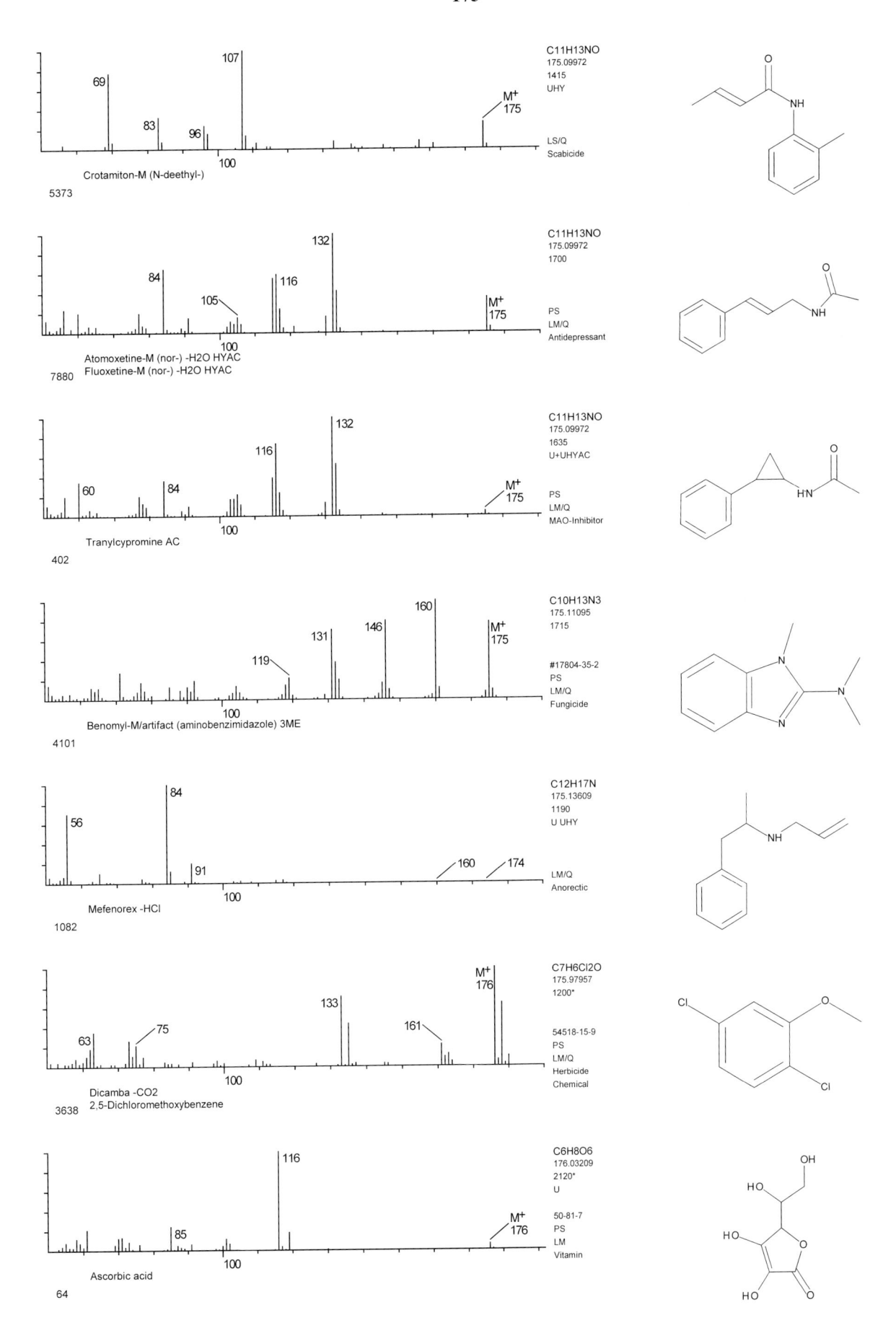

Crotamiton-M (N-deethyl-)	C11H13NO 175.09972 1415 UHY / LS/Q Scabicide
Atomoxetine-M (nor-) -H2O HYAC / Fluoxetine-M (nor-) -H2O HYAC	C11H13NO 175.09972 1700 / PS LM/Q Antidepressant
Tranylcypromine AC	C11H13NO 175.09972 1635 U+UHYAC / PS LM/Q MAO-Inhibitor
Benomyl-M/artifact (aminobenzimidazole) 3ME	C10H13N3 175.11095 1715 / #17804-35-2 PS LM/Q Fungicide
Mefenorex -HCl	C12H17N 175.13609 1190 U UHY / LM/Q Anorectic
Dicamba -CO2 / 2,5-Dichloromethoxybenzene	C7H6Cl2O 175.97957 1200* / 54518-15-9 PS LM/Q Herbicide Chemical
Ascorbic acid	C6H8O6 176.03209 2120* U / 50-81-7 PS LM Vitamin

176

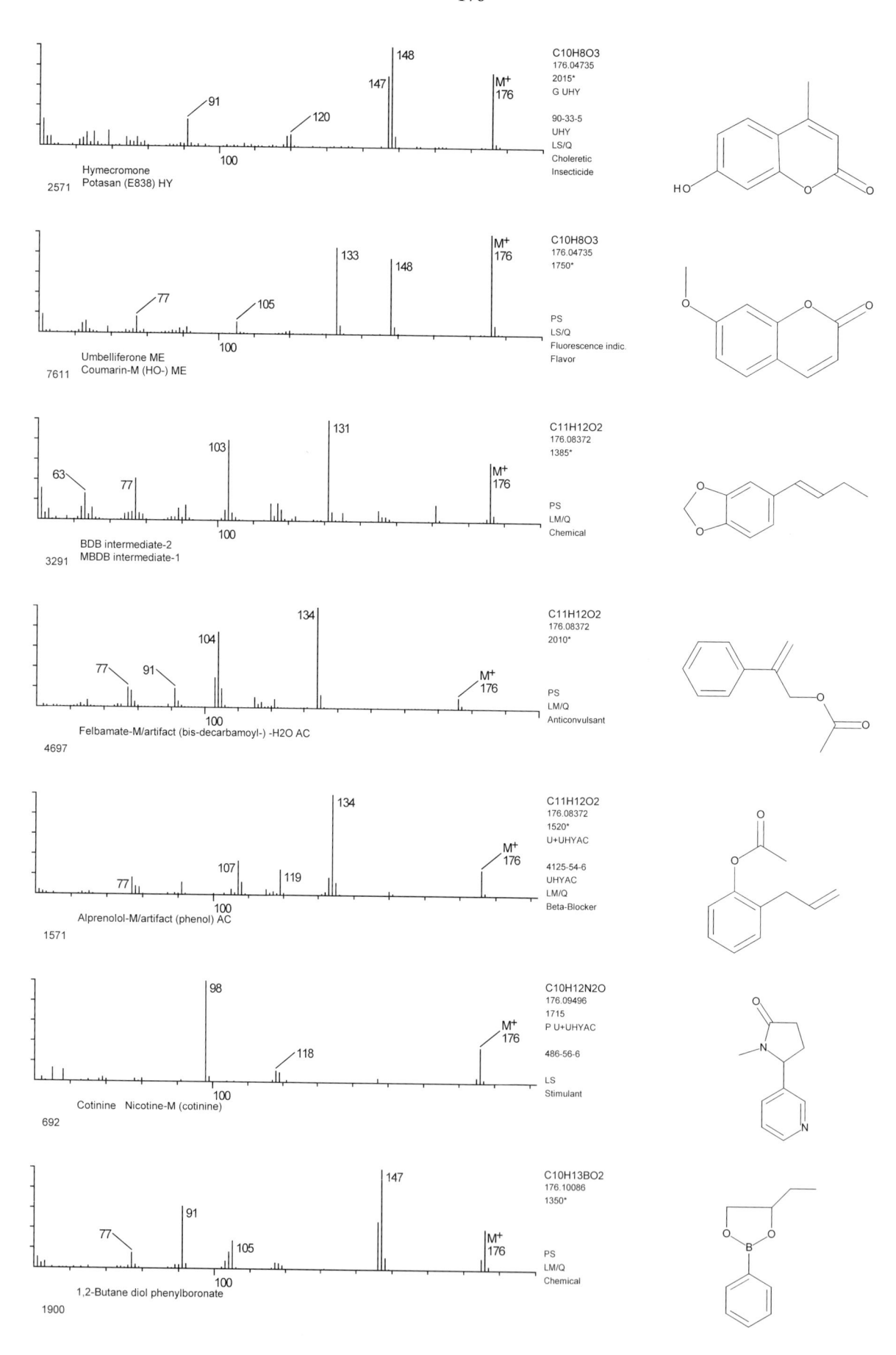

Hymecromone Potasan (E838) HY 2571	C10H8O3 176.04735 2015* G UHY 90-33-5 UHY LS/Q Choleretic Insecticide
Umbelliferone ME Coumarin-M (HO-) ME 7611	C10H8O3 176.04735 1750* PS LS/Q Fluorescence indic. Flavor
BDB intermediate-2 MBDB intermediate-1 3291	C11H12O2 176.08372 1385* PS LM/Q Chemical
Felbamate-M/artifact (bis-decarbamoyl-) -H2O AC 4697	C11H12O2 176.08372 2010* PS LM/Q Anticonvulsant
Alprenolol-M/artifact (phenol) AC 1571	C11H12O2 176.08372 1520* U+UHYAC 4125-54-6 UHYAC LM/Q Beta-Blocker
Cotinine Nicotine-M (cotinine) 692	C10H12N2O 176.09496 1715 P U+UHYAC 486-56-6 LS Stimulant
1,2-Butane diol phenylboronate 1900	C10H13BO2 176.10086 1350* PS LM/Q Chemical

176

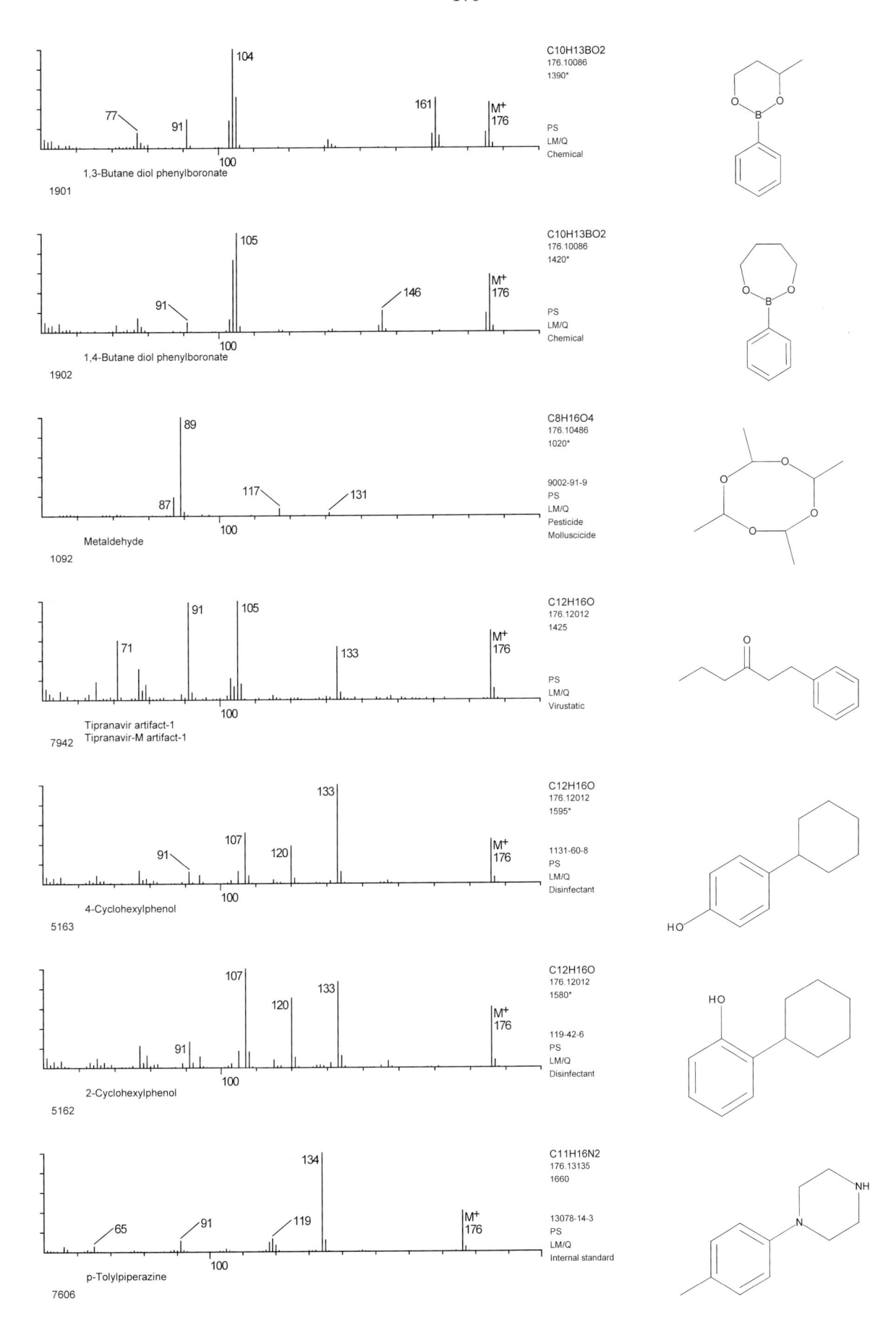

Spectrum	Compound	Formula / Data
1901	1,3-Butane diol phenylboronate	C10H13BO2, 176.10086, 1390*, PS, LM/Q, Chemical; peaks: 77, 91, 104, 161, M+ 176
1902	1,4-Butane diol phenylboronate	C10H13BO2, 176.10086, 1420*, PS, LM/Q, Chemical; peaks: 91, 105, 146, M+ 176
1092	Metaldehyde	C8H16O4, 176.10486, 1020*, 9002-91-9, PS, LM/Q, Pesticide, Molluscicide; peaks: 87, 89, 117, 131
7942	Tipranavir artifact-1 / Tipranavir-M artifact-1	C12H16O, 176.12012, 1425, PS, LM/Q, Virustatic; peaks: 71, 91, 105, 133, M+ 176
5163	4-Cyclohexylphenol	C12H16O, 176.12012, 1595*, 1131-60-8, PS, LM/Q, Disinfectant; peaks: 91, 107, 120, 133, M+ 176
5162	2-Cyclohexylphenol	C12H16O, 176.12012, 1580*, 119-42-6, PS, LM/Q, Disinfectant; peaks: 91, 107, 120, 133, M+ 176
7606	p-Tolylpiperazine	C11H16N2, 176.13135, 1660, 13078-14-3, PS, LM/Q, Internal standard; peaks: 65, 91, 119, 134, M+ 176

333

176

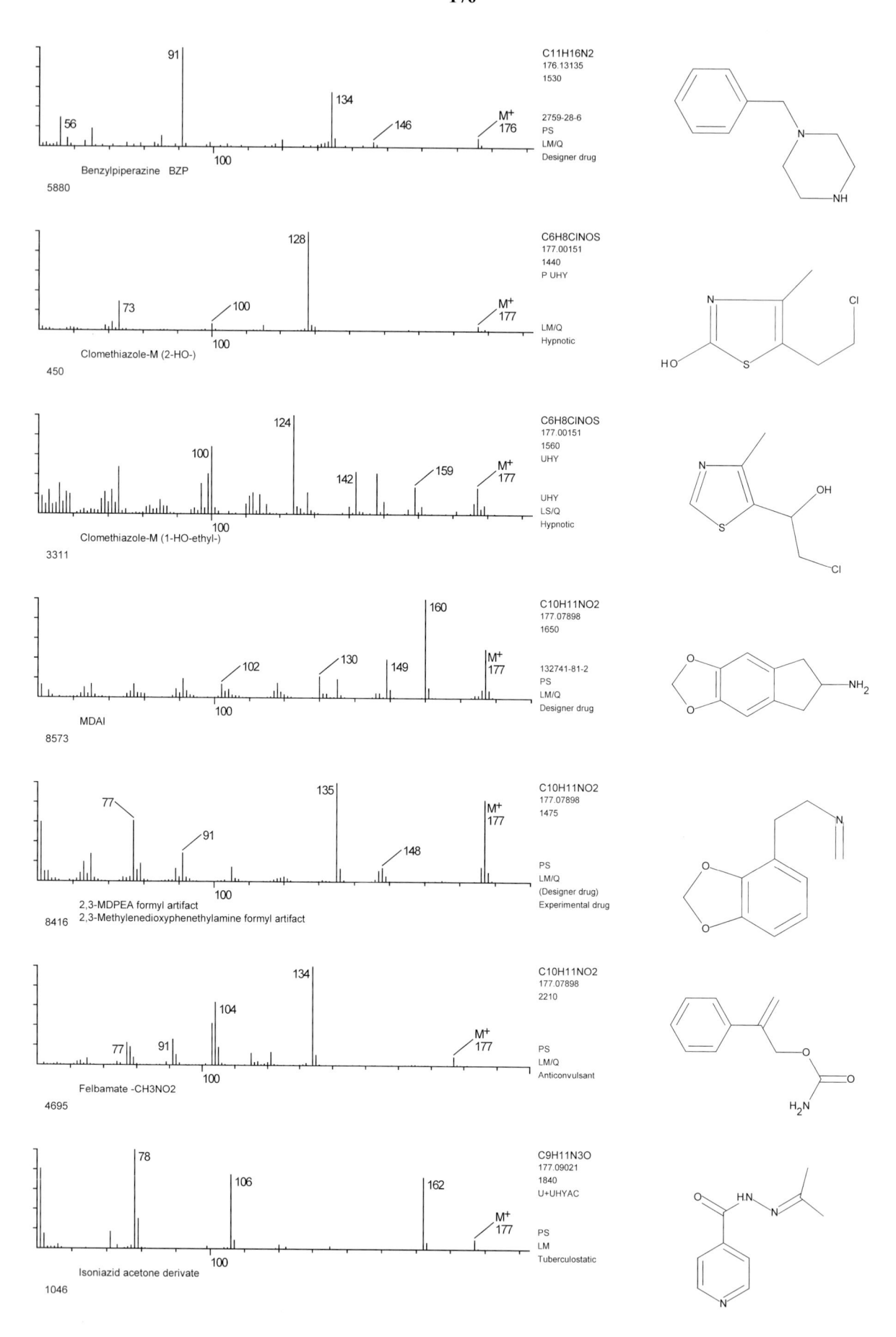

Benzylpiperazine BZP — C11H16N2, 176.13135, 1530, 2759-28-6, PS, LM/Q, Designer drug
Peaks: 56, 91, 134, 146, M+ 176
5880

Clomethiazole-M (2-HO-) — C6H8ClNOS, 177.00151, 1440, P UHY, LM/Q, Hypnotic
Peaks: 73, 100, 128, M+ 177
450

Clomethiazole-M (1-HO-ethyl-) — C6H8ClNOS, 177.00151, 1560, UHY, UHY, LS/Q, Hypnotic
Peaks: 100, 124, 142, 159, M+ 177
3311

MDAI — C10H11NO2, 177.07898, 1650, 132741-81-2, PS, LM/Q, Designer drug
Peaks: 102, 130, 149, 160, M+ 177
8573

2,3-MDPEA formyl artifact / 2,3-Methylenedioxyphenethylamine formyl artifact — C10H11NO2, 177.07898, 1475, PS, LM/Q, (Designer drug), Experimental drug
Peaks: 77, 91, 135, 148, M+ 177
8416

Felbamate -CH3NO2 — C10H11NO2, 177.07898, 2210, PS, LM/Q, Anticonvulsant
Peaks: 77, 91, 104, 134, M+ 177
4695

Isoniazid acetone derivate — C9H11N3O, 177.09021, 1840, U+UHYAC, PS, LM, Tuberculostatic
Peaks: 78, 106, 162, M+ 177
1046

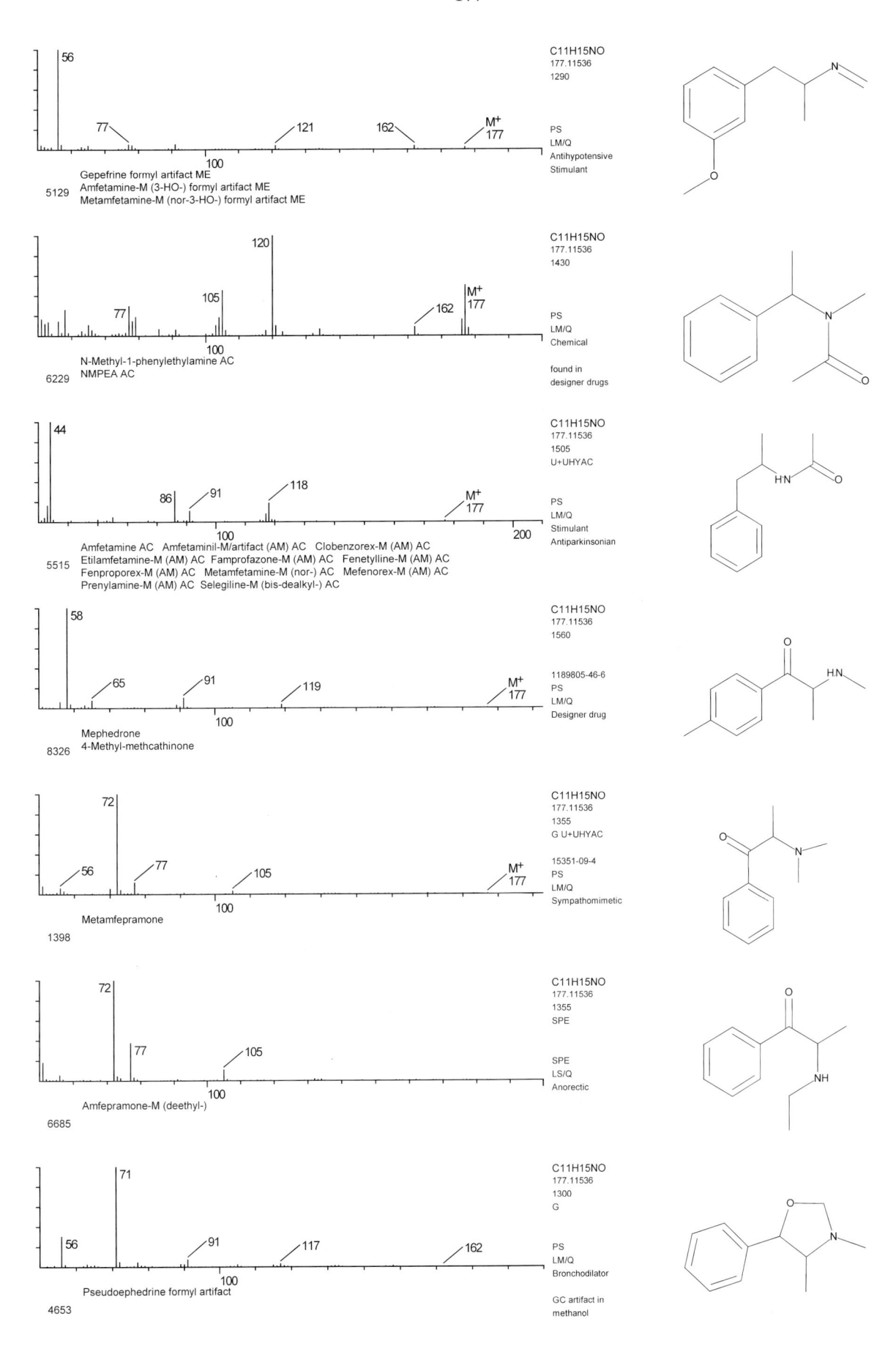

5129	Gepefrine formyl artifact ME Amfetamine-M (3-HO-) formyl artifact ME Metamfetamine-M (nor-3-HO-) formyl artifact ME	C11H15NO 177.11536 1290 PS LM/Q Antihypotensive Stimulant

Peaks: 56, 77, 121, 162, M+ 177

6229	N-Methyl-1-phenylethylamine AC NMPEA AC	C11H15NO 177.11536 1430 PS LM/Q Chemical found in designer drugs

Peaks: 77, 105, 120, 162, M+ 177

5515	Amfetamine AC Amfetaminil-M/artifact (AM) AC Clobenzorex-M (AM) AC Etilamfetamine-M (AM) AC Famprofazone-M (AM) AC Fenetylline-M (AM) AC Fenproporex-M (AM) AC Metamfetamine-M (nor-) AC Mefenorex-M (AM) AC Prenylamine-M (AM) AC Selegiline-M (bis-dealkyl-) AC	C11H15NO 177.11536 1505 U+UHYAC PS LM/Q Stimulant Antiparkinsonian

Peaks: 44, 86, 91, 118, M+ 177

8326	Mephedrone 4-Methyl-methcathinone	C11H15NO 177.11536 1560 1189805-46-6 PS LM/Q Designer drug

Peaks: 58, 65, 91, 119, M+ 177

1398	Metamfepramone	C11H15NO 177.11536 1355 G U+UHYAC 15351-09-4 PS LM/Q Sympathomimetic

Peaks: 56, 72, 77, 105, M+ 177

6685	Amfepramone-M (deethyl-)	C11H15NO 177.11536 1355 SPE SPE LS/Q Anorectic

Peaks: 72, 77, 105

4653	Pseudoephedrine formyl artifact	C11H15NO 177.11536 1300 G PS LM/Q Bronchodilator GC artifact in methanol

Peaks: 56, 71, 91, 117, 162

177

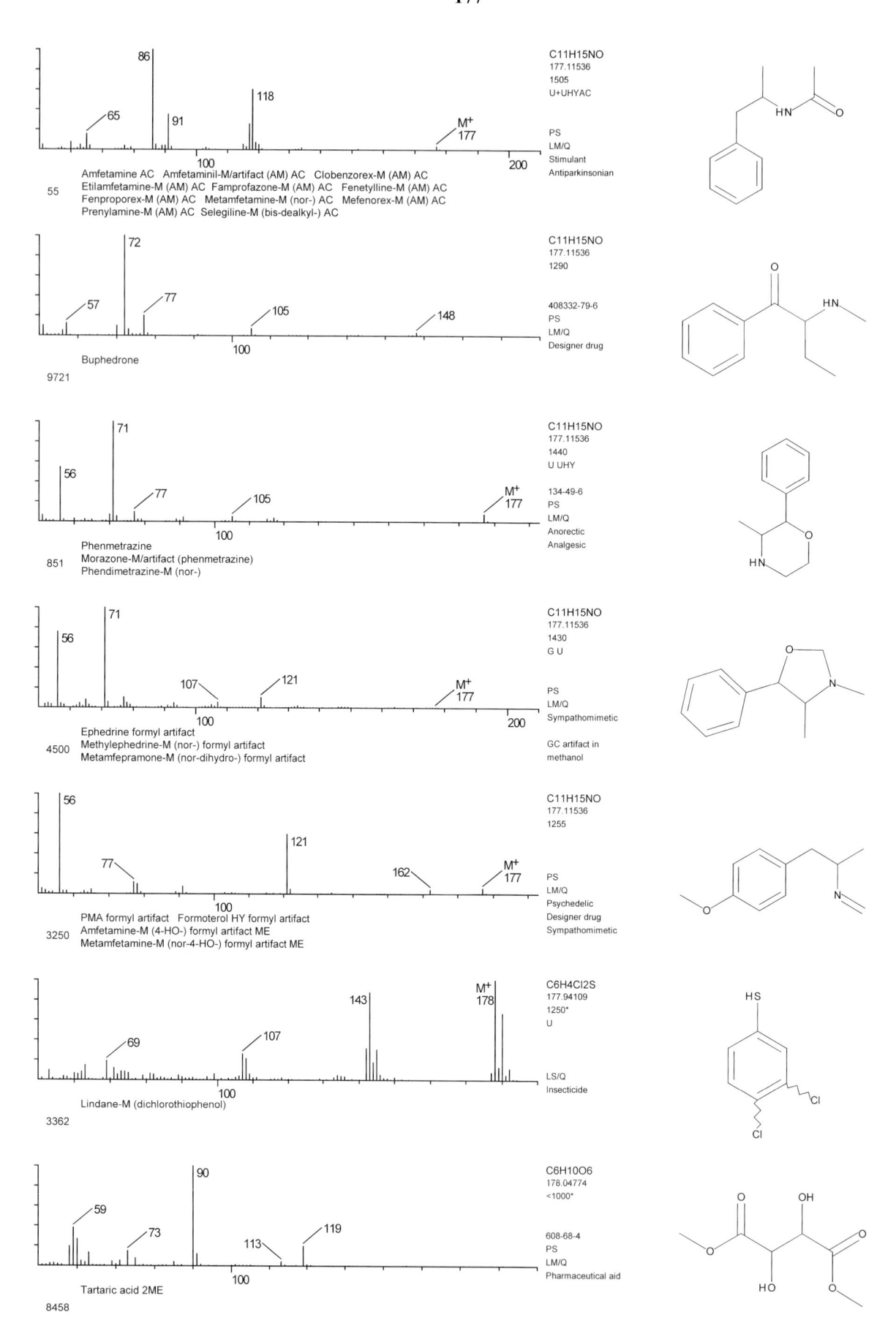

C11H15NO	
177.11536	
1505	
U+UHYAC	
PS	
LM/Q	
Stimulant	
Antiparkinsonian	

Amfetamine AC   Amfetaminil-M/artifact (AM) AC   Clobenzorex-M (AM) AC
Etilamfetamine-M (AM) AC   Famprofazone-M (AM) AC   Fenetylline-M (AM) AC
Fenproporex-M (AM) AC   Metamfetamine-M (nor-) AC   Mefenorex-M (AM) AC
Prenylamine-M (AM) AC   Selegiline-M (bis-dealkyl-) AC

55

C11H15NO
177.11536
1290

408332-79-6
PS
LM/Q
Designer drug

Buphedrone
9721

C11H15NO
177.11536
1440
U UHY

134-49-6
PS
LM/Q
Anorectic
Analgesic

Phenmetrazine
Morazone-M/artifact (phenmetrazine)
Phendimetrazine-M (nor-)
851

C11H15NO
177.11536
1430
G U

PS
LM/Q
Sympathomimetic

Ephedrine formyl artifact
Methylephedrine-M (nor-) formyl artifact
Metamfepramone-M (nor-dihydro-) formyl artifact
4500

GC artifact in methanol

C11H15NO
177.11536
1255

PS
LM/Q
Psychedelic
Designer drug
Sympathomimetic

PMA formyl artifact   Formoterol HY formyl artifact
Amfetamine-M (4-HO-) formyl artifact ME
Metamfetamine-M (nor-4-HO-) formyl artifact ME
3250

C6H4Cl2S
177.94109
1250*
U

LS/Q
Insecticide

Lindane-M (dichlorothiophenol)
3362

C6H10O6
178.04774
<1000*

608-68-4
PS
LM/Q
Pharmaceutical aid

Tartaric acid 2ME
8458

178

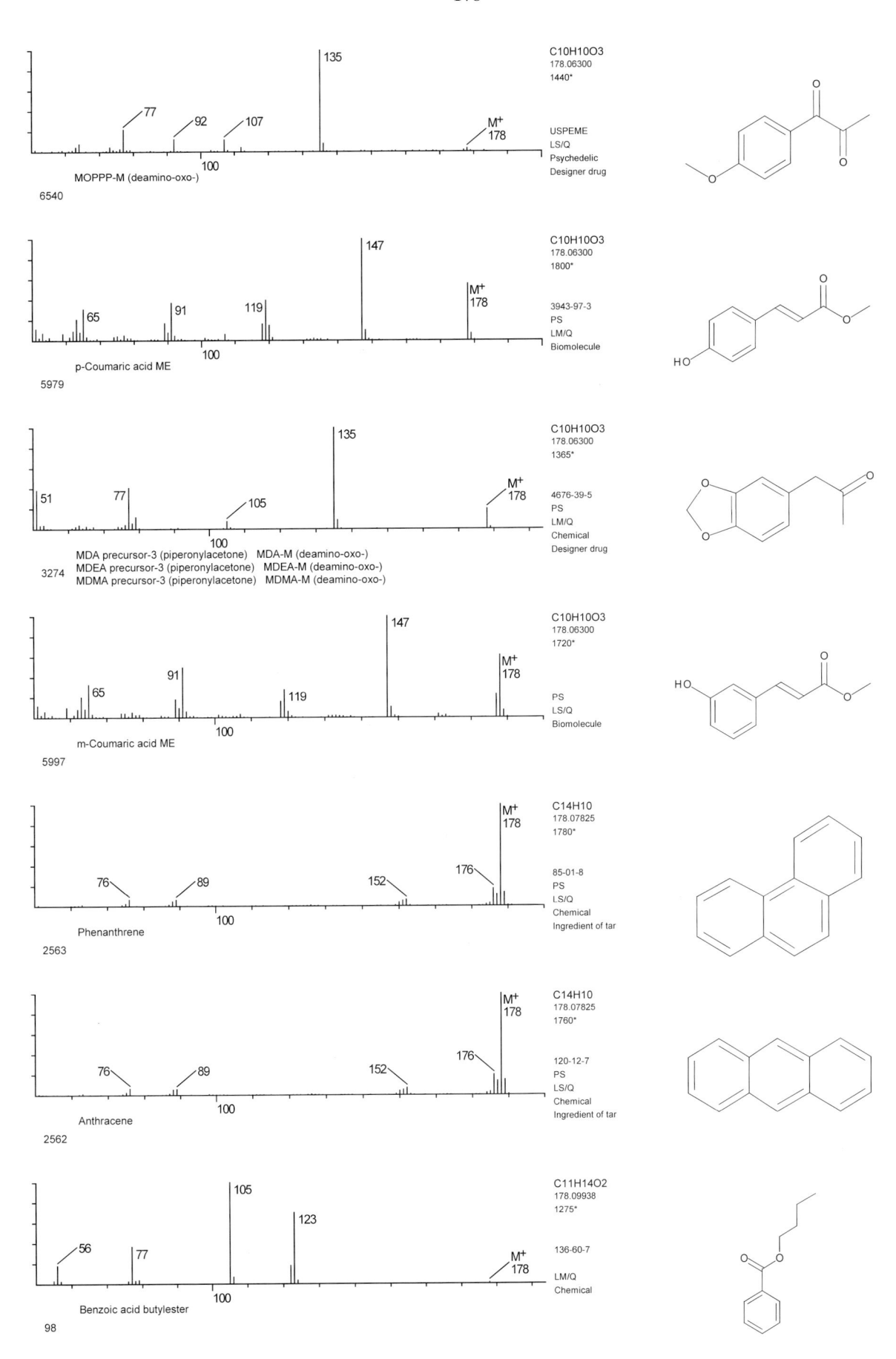

Spectrum with peaks: 77, 92, 107, 135, M+ 178	MOPPP-M (deamino-oxo-) 6540	C10H10O3 178.06300 1440* USPEME LS/Q Psychedelic Designer drug

Spectrum with peaks: 65, 91, 119, 147, M+ 178	p-Coumaric acid ME 5979	C10H10O3 178.06300 1800* 3943-97-3 PS LM/Q Biomolecule

Spectrum with peaks: 51, 77, 105, 135, M+ 178	MDA precursor-3 (piperonylacetone) MDA-M (deamino-oxo-) MDEA precursor-3 (piperonylacetone) MDEA-M (deamino-oxo-) MDMA precursor-3 (piperonylacetone) MDMA-M (deamino-oxo-) 3274	C10H10O3 178.06300 1365* 4676-39-5 PS LM/Q Chemical Designer drug

Spectrum with peaks: 65, 91, 119, 147, M+ 178	m-Coumaric acid ME 5997	C10H10O3 178.06300 1720* PS LS/Q Biomolecule

Spectrum with peaks: 76, 89, 152, 176, M+ 178	Phenanthrene 2563	C14H10 178.07825 1780* 85-01-8 PS LS/Q Chemical Ingredient of tar

Spectrum with peaks: 76, 89, 152, 176, M+ 178	Anthracene 2562	C14H10 178.07825 1760* 120-12-7 PS LS/Q Chemical Ingredient of tar

Spectrum with peaks: 56, 77, 105, 123, M+ 178	Benzoic acid butylester 98	C11H14O2 178.09938 1275* 136-60-7 LM/Q Chemical

337

178

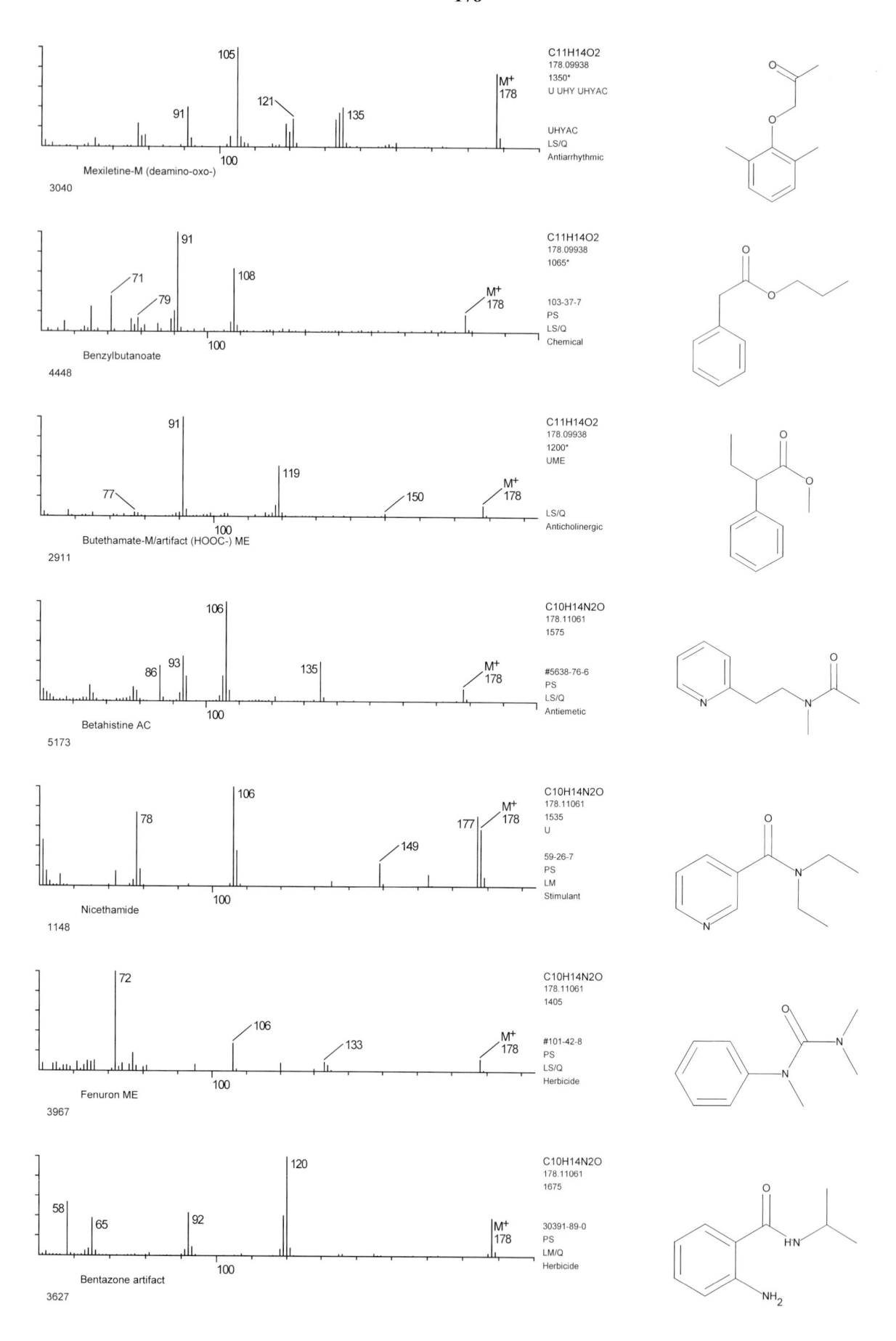

Mexiletine-M (deamino-oxo-) 3040	Peaks: 91, 105, 121, 135, M+ 178	C11H14O2 178.09938 1350* U UHY UHYAC UHYAC LS/Q Antiarrhythmic
Benzylbutanoate 4448	Peaks: 71, 79, 91, 108, M+ 178	C11H14O2 178.09938 1065* 103-37-7 PS LS/Q Chemical
Butethamate-M/artifact (HOOC-) ME 2911	Peaks: 77, 91, 119, 150, M+ 178	C11H14O2 178.09938 1200* UME LS/Q Anticholinergic
Betahistine AC 5173	Peaks: 86, 93, 106, 135, M+ 178	C10H14N2O 178.11061 1575 #5638-76-6 PS LS/Q Antiemetic
Nicethamide 1148	Peaks: 78, 106, 149, 177, M+ 178	C10H14N2O 178.11061 1535 U 59-26-7 PS LM Stimulant
Fenuron ME 3967	Peaks: 72, 106, 133, M+ 178	C10H14N2O 178.11061 1405 #101-42-8 PS LS/Q Herbicide
Bentazone artifact 3627	Peaks: 58, 65, 92, 120, M+ 178	C10H14N2O 178.11061 1675 30391-89-0 PS LM/Q Herbicide

178

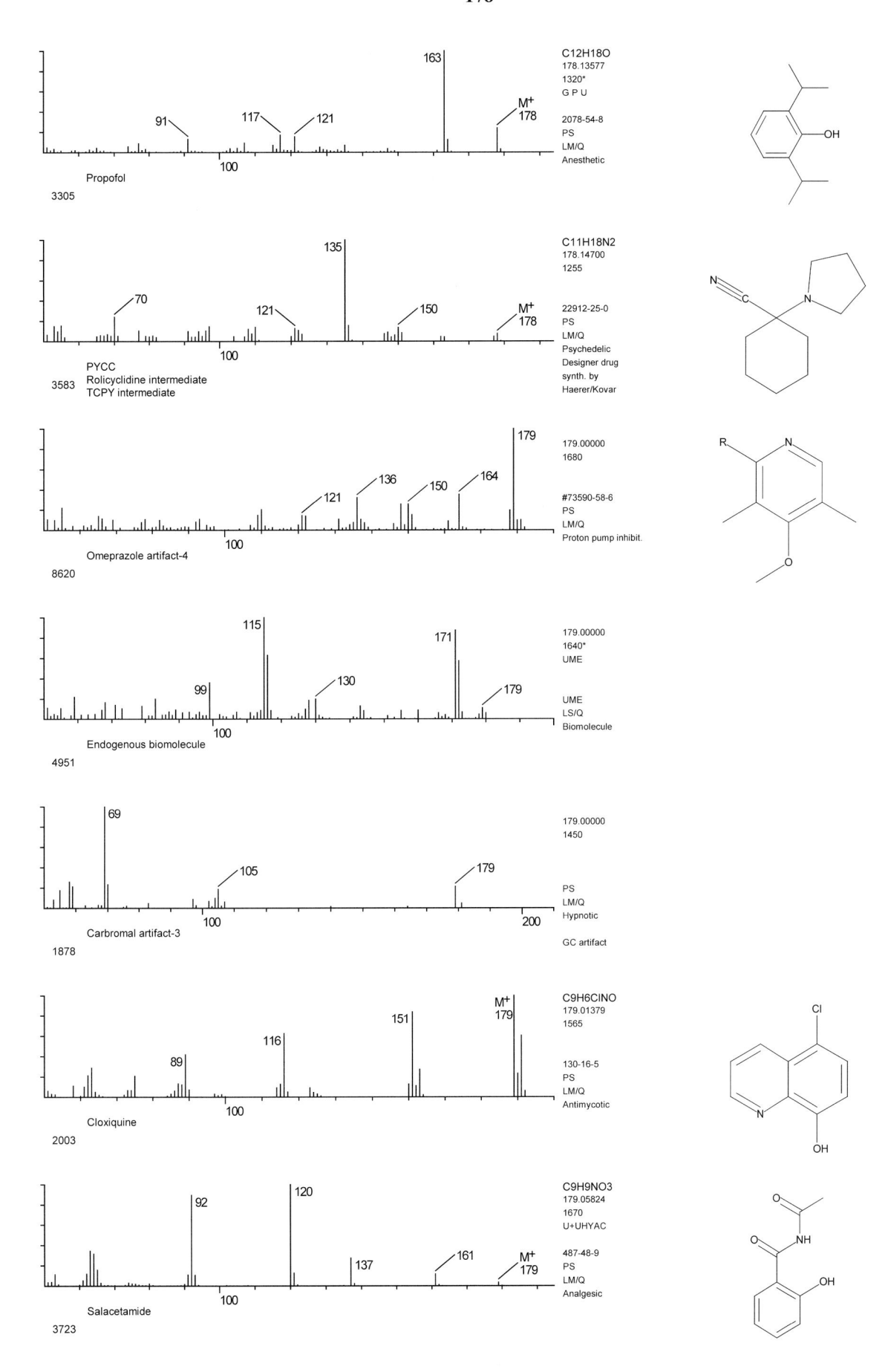

Propofol
3305
C12H18O
178.13577
1320*
G P U
2078-54-8
PS
LM/Q
Anesthetic

PYCC
Rolicyclidine intermediate
TCPY intermediate
3583
C11H18N2
178.14700
1255
22912-25-0
PS
LM/Q
Psychedelic
Designer drug synth. by Haerer/Kovar

Omeprazole artifact-4
8620
179.00000
1680
#73590-58-6
PS
LM/Q
Proton pump inhibit.

Endogenous biomolecule
4951
179.00000
1640*
UME
UME
LS/Q
Biomolecule

Carbromal artifact-3
1878
179.00000
1450
PS
LM/Q
Hypnotic
GC artifact

Cloxiquine
2003
C9H6ClNO
179.01379
1565
130-16-5
PS
LM/Q
Antimycotic

Salacetamide
3723
C9H9NO3
179.05824
1670
U+UHYAC
487-48-9
PS
LM/Q
Analgesic

339

179

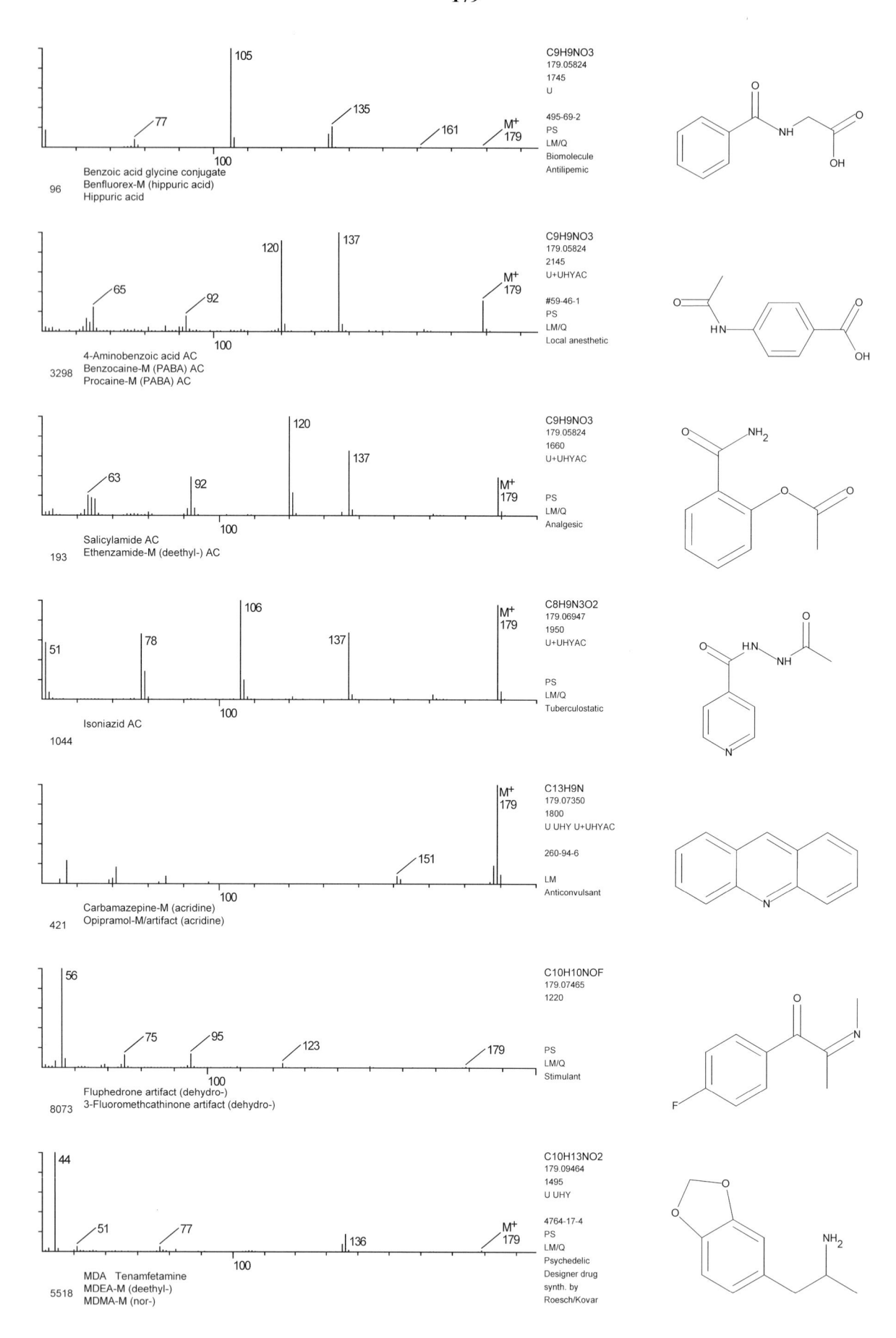

Benzoic acid glycine conjugate
Benfluorex-M (hippuric acid)
Hippuric acid
96

C9H9NO3
179.05824
1745
U

495-69-2
PS
LM/Q
Biomolecule
Antilipemic

4-Aminobenzoic acid AC
Benzocaine-M (PABA) AC
Procaine-M (PABA) AC
3298

C9H9NO3
179.05824
2145
U+UHYAC

#59-46-1
PS
LM/Q
Local anesthetic

Salicylamide AC
Ethenzamide-M (deethyl-) AC
193

C9H9NO3
179.05824
1660
U+UHYAC

PS
LM/Q
Analgesic

Isoniazid AC
1044

C8H9N3O2
179.06947
1950
U+UHYAC

PS
LM/Q
Tuberculostatic

Carbamazepine-M (acridine)
Opipramol-M/artifact (acridine)
421

C13H9N
179.07350
1800
U UHY U+UHYAC

260-94-6

LM
Anticonvulsant

Fluphedrone artifact (dehydro-)
3-Fluoromethcathinone artifact (dehydro-)
8073

C10H10NOF
179.07465
1220

PS
LM/Q
Stimulant

MDA Tenamfetamine
MDEA-M (deethyl-)
MDMA-M (nor-)
5518

C10H13NO2
179.09464
1495
U UHY

4764-17-4
PS
LM/Q
Psychedelic
Designer drug
synth. by
Roesch/Kovar

340

179

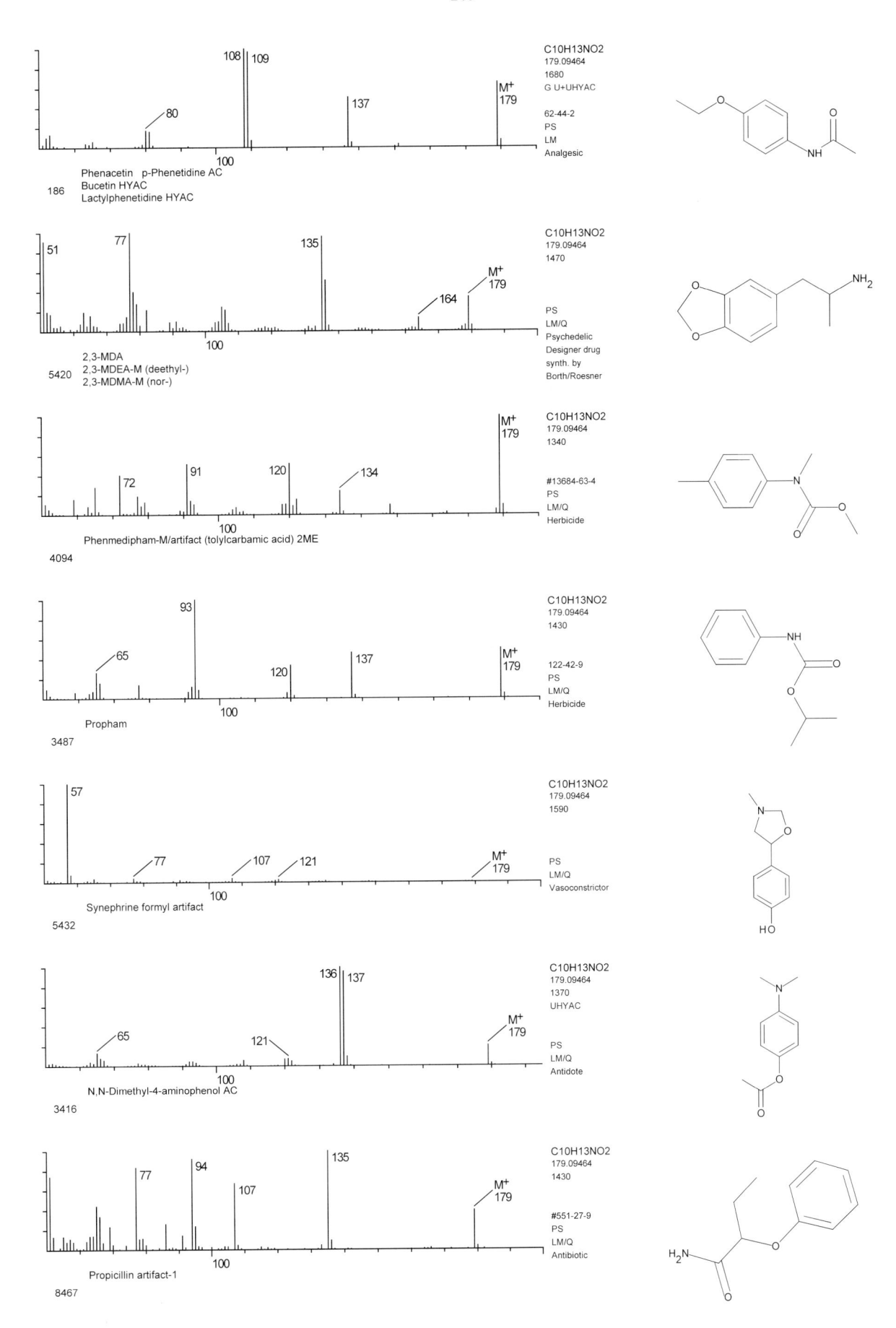

C10H13NO2
179.09464
1680
G U+UHYAC

62-44-2
PS
LM
Analgesic

Phenacetin   p-Phenetidine AC
Bucetin HYAC
Lactylphenetidine HYAC
186

C10H13NO2
179.09464
1470

PS
LM/Q
Psychedelic
Designer drug
synth. by
Borth/Roesner

2,3-MDA
2,3-MDEA-M (deethyl-)
2,3-MDMA-M (nor-)
5420

C10H13NO2
179.09464
1340

#13684-63-4
PS
LM/Q
Herbicide

Phenmedipham-M/artifact (tolylcarbamic acid) 2ME
4094

C10H13NO2
179.09464
1430

122-42-9
PS
LM/Q
Herbicide

Propham
3487

C10H13NO2
179.09464
1590

PS
LM/Q
Vasoconstrictor

Synephrine formyl artifact
5432

C10H13NO2
179.09464
1370
UHYAC

PS
LM/Q
Antidote

N,N-Dimethyl-4-aminophenol AC
3416

C10H13NO2
179.09464
1430

#551-27-9
PS
LM/Q
Antibiotic

Propicillin artifact-1
8467

341

180

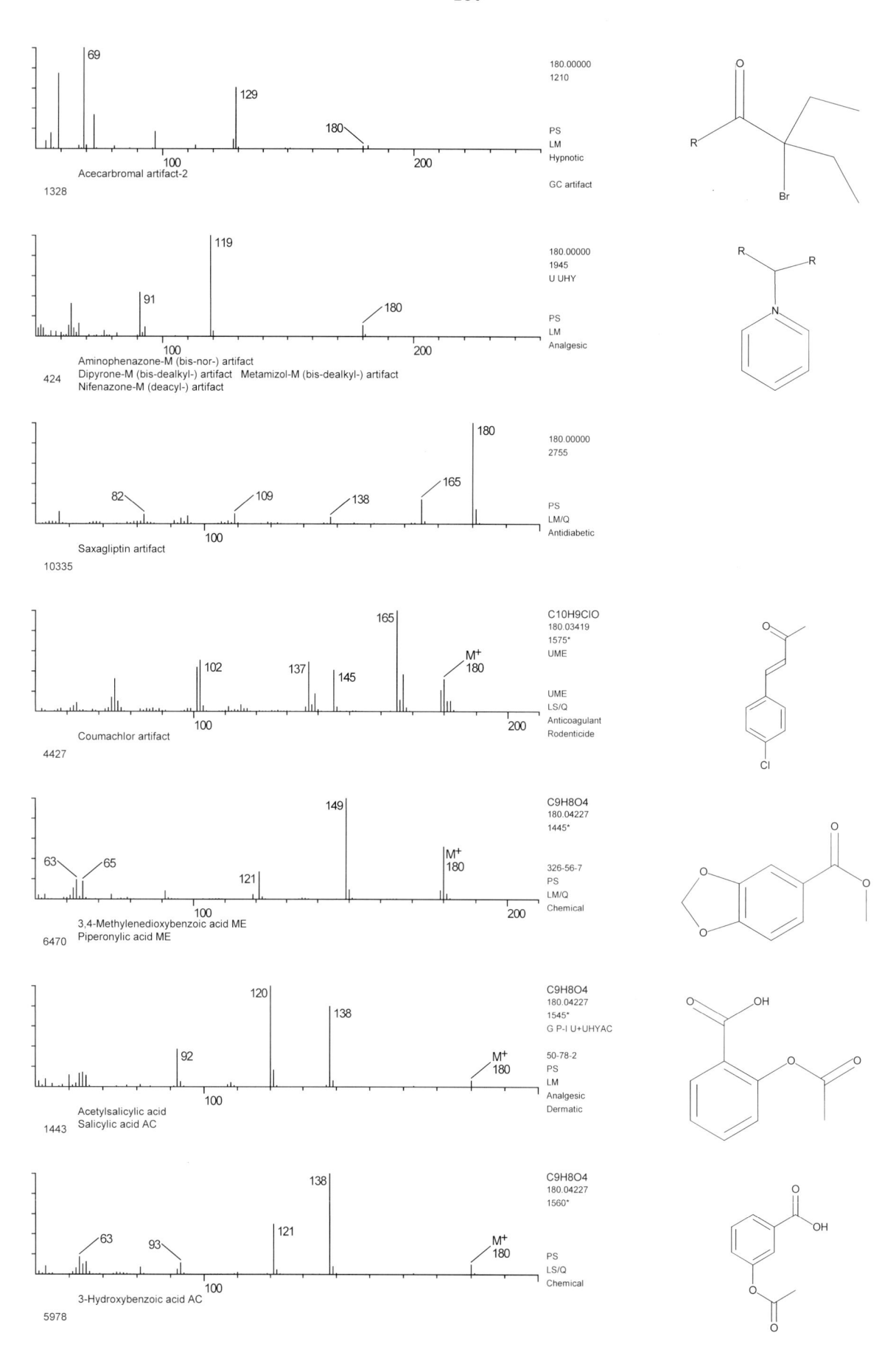

69, 129, 180 — Acecarbromal artifact-2	180.00000 1210 PS LM Hypnotic GC artifact	

1328

| 91, 119, 180 — Aminophenazone-M (bis-nor-) artifact; Dipyrone-M (bis-dealkyl-) artifact; Metamizol-M (bis-dealkyl-) artifact; Nifenazone-M (deacyl-) artifact | 180.00000 1945 U UHY PS LM Analgesic |

424

| 82, 109, 138, 165, 180 — Saxagliptin artifact | 180.00000 2755 PS LM/Q Antidiabetic |

10335

| 102, 137, 145, 165, M+ 180 — Coumachlor artifact | C10H9ClO 180.03419 1575* UME UME LS/Q Anticoagulant Rodenticide |

4427

| 63, 65, 121, 149, M+ 180 — 3,4-Methylenedioxybenzoic acid ME; Piperonylic acid ME | C9H8O4 180.04227 1445* 326-56-7 PS LM/Q Chemical |

6470

| 92, 120, 138, M+ 180 — Acetylsalicylic acid; Salicylic acid AC | C9H8O4 180.04227 1545* G P-I U+UHYAC 50-78-2 PS LM Analgesic Dermatic |

1443

| 63, 93, 121, 138, M+ 180 — 3-Hydroxybenzoic acid AC | C9H8O4 180.04227 1560* PS LS/Q Chemical |

5978

344

180

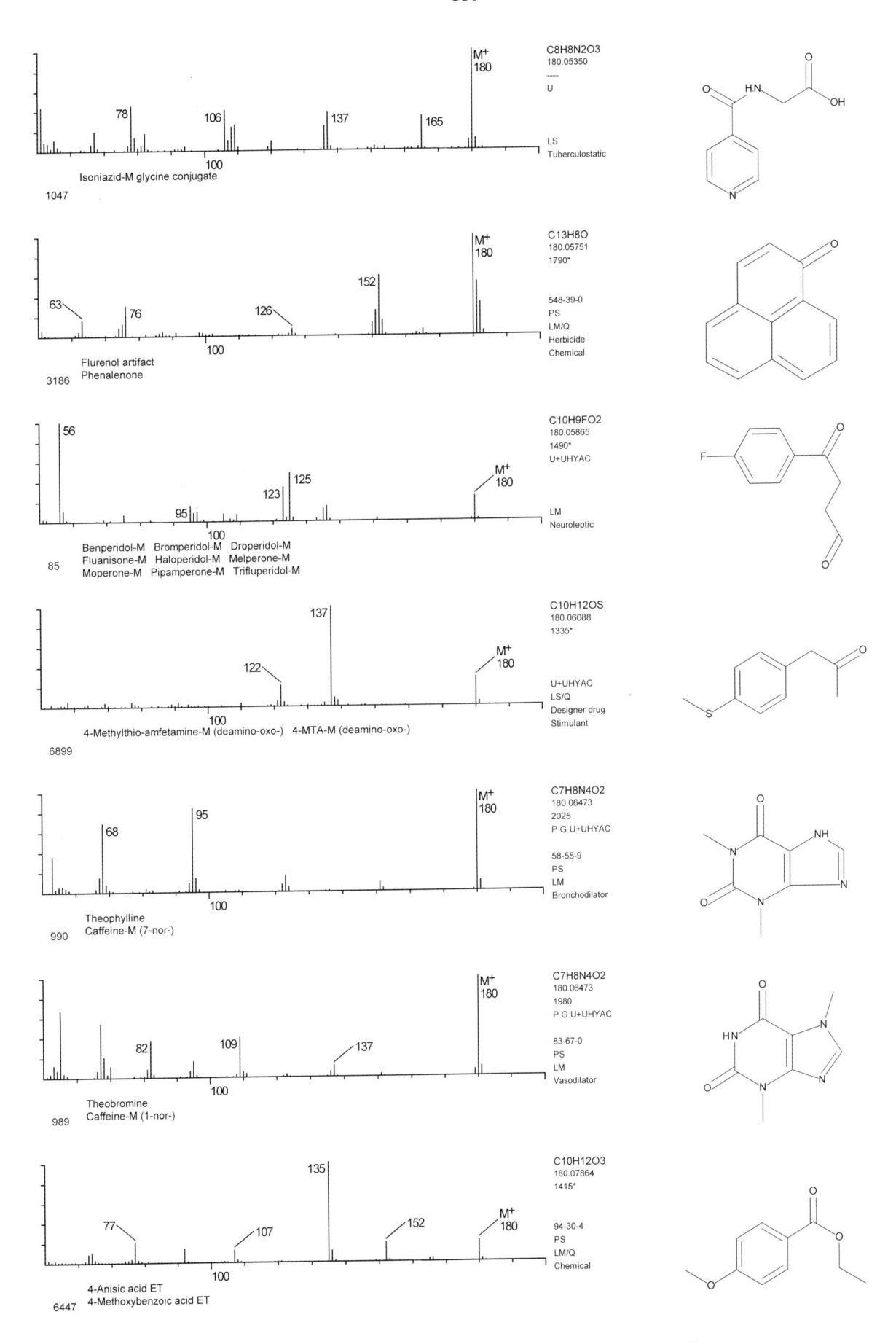

Spectrum label	Formula / data
Isoniazid-M glycine conjugate (1047) — peaks 78, 106, 137, 165, M+ 180	C8H8N2O3, 180.05350, U, LS, Tuberculostatic
Flurenol artifact / Phenalenone (3186) — peaks 63, 76, 126, 152, M+ 180	C13H8O, 180.05751, 1790*, 548-39-0, PS, LM/Q, Herbicide, Chemical
Benperidol-M, Bromperidol-M, Droperidol-M, Fluanisone-M, Haloperidol-M, Melperone-M, Moperone-M, Pipamperone-M, Trifluperidol-M (85) — peaks 56, 95, 123, 125, M+ 180	C10H9FO2, 180.05865, 1490*, U+UHYAC, LM, Neuroleptic
4-Methylthio-amfetamine-M (deamino-oxo-), 4-MTA-M (deamino-oxo-) (6899) — peaks 122, 137, M+ 180	C10H12OS, 180.06088, 1335*, U+UHYAC, LS/Q, Designer drug, Stimulant
Theophylline, Caffeine-M (7-nor-) (990) — peaks 68, 95, M+ 180	C7H8N4O2, 180.06473, 2025, P G U+UHYAC, 58-55-9, PS, LM, Bronchodilator
Theobromine, Caffeine-M (1-nor-) (989) — peaks 82, 109, 137, M+ 180	C7H8N4O2, 180.06473, 1980, P G U+UHYAC, 83-67-0, PS, LM, Vasodilator
4-Anisic acid ET, 4-Methoxybenzoic acid ET (6447) — peaks 77, 107, 135, 152, M+ 180	C10H12O3, 180.07864, 1415*, 94-30-4, PS, LM/Q, Chemical

345

180

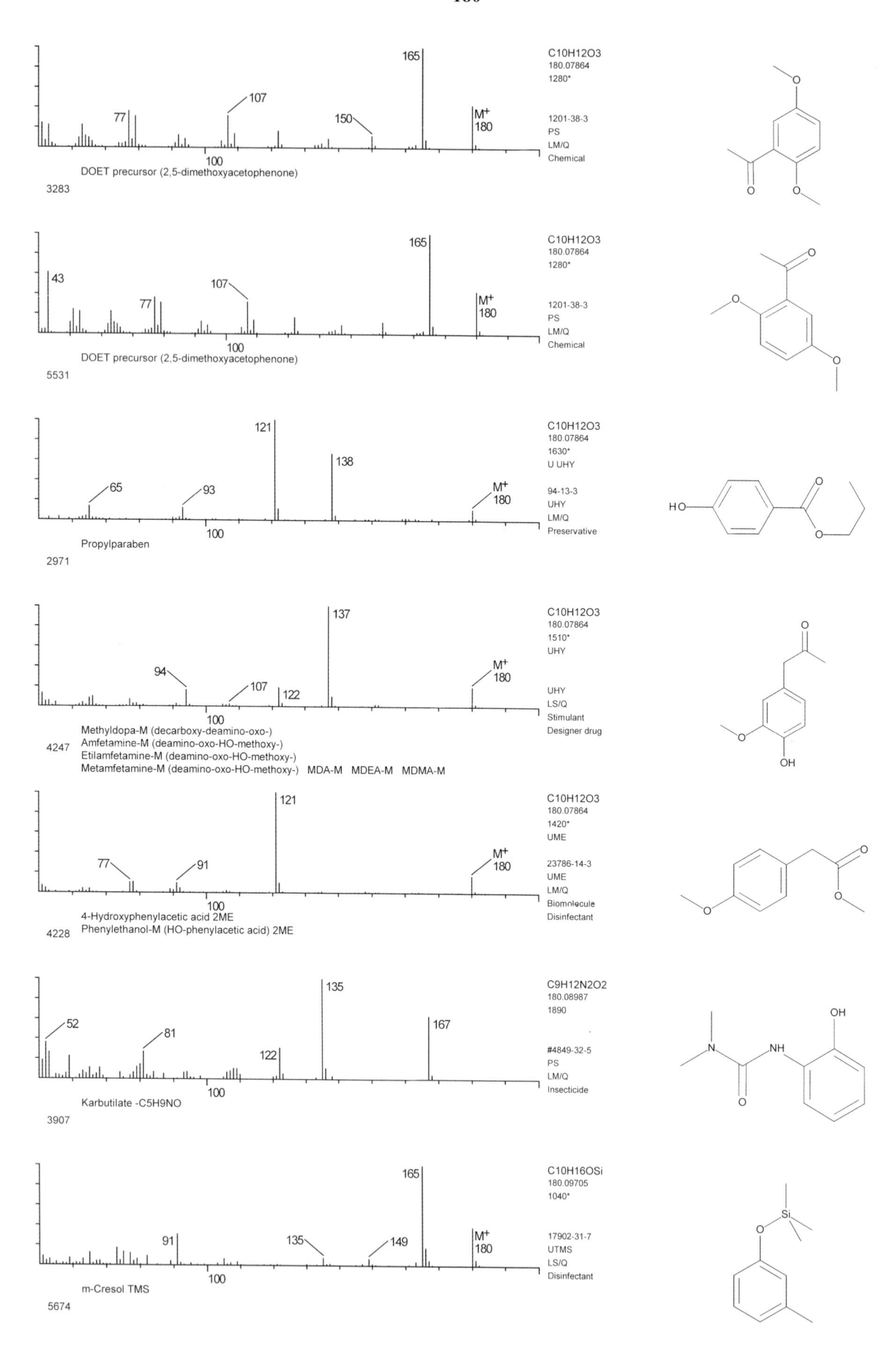

3283 — DOET precursor (2,5-dimethoxyacetophenone)
C10H12O3
180.07864
1280*

1201-38-3
PS
LM/Q
Chemical

Peaks: 77, 107, 150, 165, M+ 180

5531 — DOET precursor (2,5-dimethoxyacetophenone)
C10H12O3
180.07864
1280*

1201-38-3
PS
LM/Q
Chemical

Peaks: 43, 77, 107, 165, M+ 180

2971 — Propylparaben
C10H12O3
180.07864
1630*
U UHY

94-13-3
UHY
LM/Q
Preservative

Peaks: 65, 93, 121, 138, M+ 180

4247 — Methyldopa-M (decarboxy-deamino-oxo-)
Amfetamine-M (deamino-oxo-HO-methoxy-)
Etilamfetamine-M (deamino-oxo-HO-methoxy-)
Metamfetamine-M (deamino-oxo-HO-methoxy-) MDA-M MDEA-M MDMA-M

C10H12O3
180.07864
1510*
UHY

UHY
LS/Q
Stimulant
Designer drug

Peaks: 94, 107, 122, 137, M+ 180

4228 — 4-Hydroxyphenylacetic acid 2ME
Phenylethanol-M (HO-phenylacetic acid) 2ME

C10H12O3
180.07864
1420*
UME

23786-14-3
UME
LM/Q
Biomolecule
Disinfectant

Peaks: 77, 91, 121, M+ 180

3907 — Karbutilate -C5H9NO
C9H12N2O2
180.08987
1890

#4849-32-5
PS
LM/Q
Insecticide

Peaks: 52, 81, 122, 135, 167

5674 — m-Cresol TMS
C10H16OSi
180.09705
1040*

17902-31-7
UTMS
LS/Q
Disinfectant

Peaks: 91, 135, 149, 165, M+ 180

180

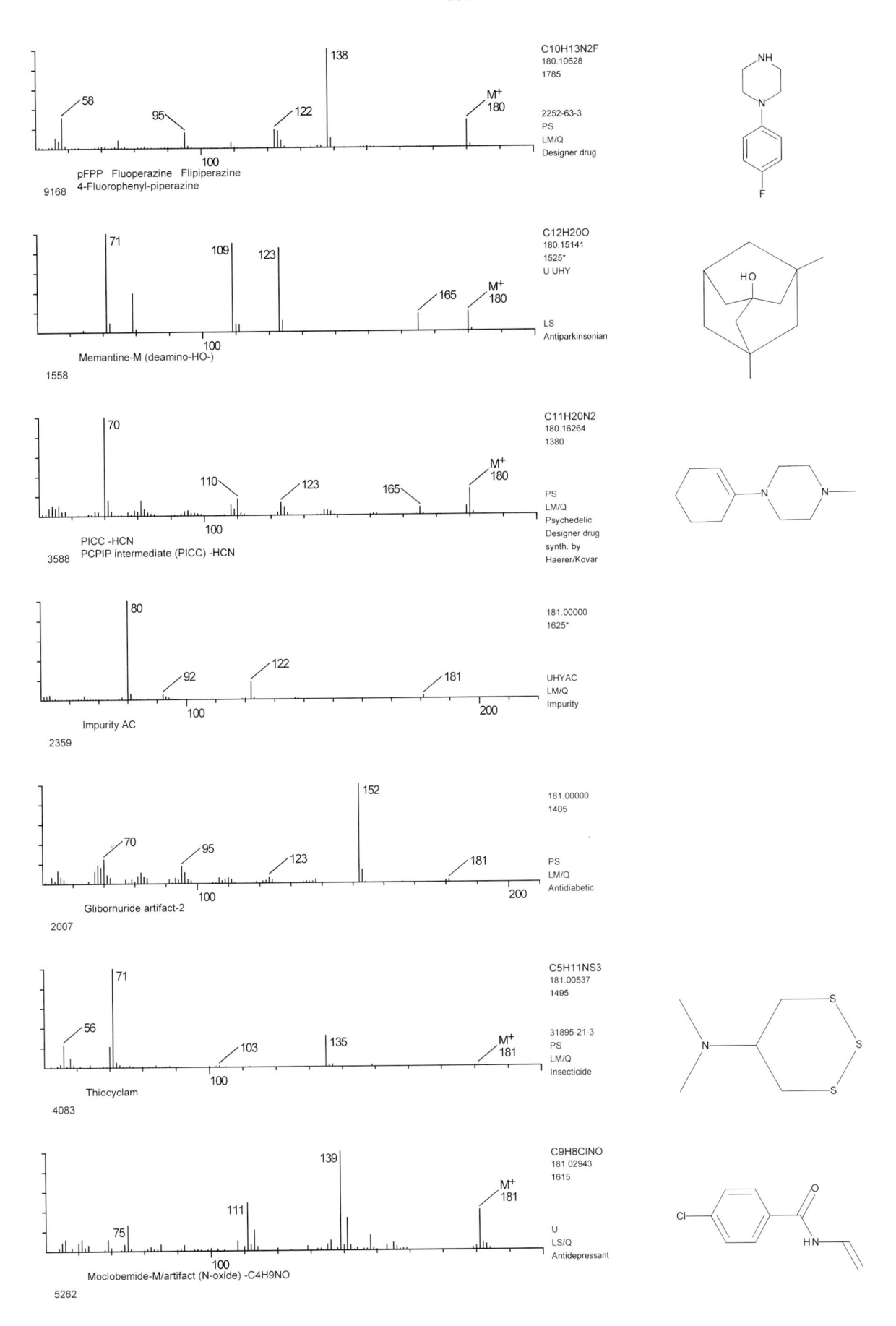

pFPP Fluoperazine Flipiperazine 4-Fluorophenyl-piperazine	C10H13N2F 180.10628 1785 / 2252-63-3 PS LM/Q Designer drug	
Memantine-M (deamino-HO-)	C12H20O 180.15141 1525* U UHY LS Antiparkinsonian	
PICC -HCN PCPIP intermediate (PICC) -HCN	C11H20N2 180.16264 1380 PS LM/Q Psychedelic Designer drug synth. by Haerer/Kovar	
Impurity AC	181.00000 1625* UHYAC LM/Q Impurity	
Glibornuride artifact-2	181.00000 1405 PS LM/Q Antidiabetic	
Thiocyclam	C5H11NS3 181.00537 1495 / 31895-21-3 PS LM/Q Insecticide	
Moclobemide-M/artifact (N-oxide) -C4H9NO	C9H8ClNO 181.02943 1615 U LS/Q Antidepressant	

181

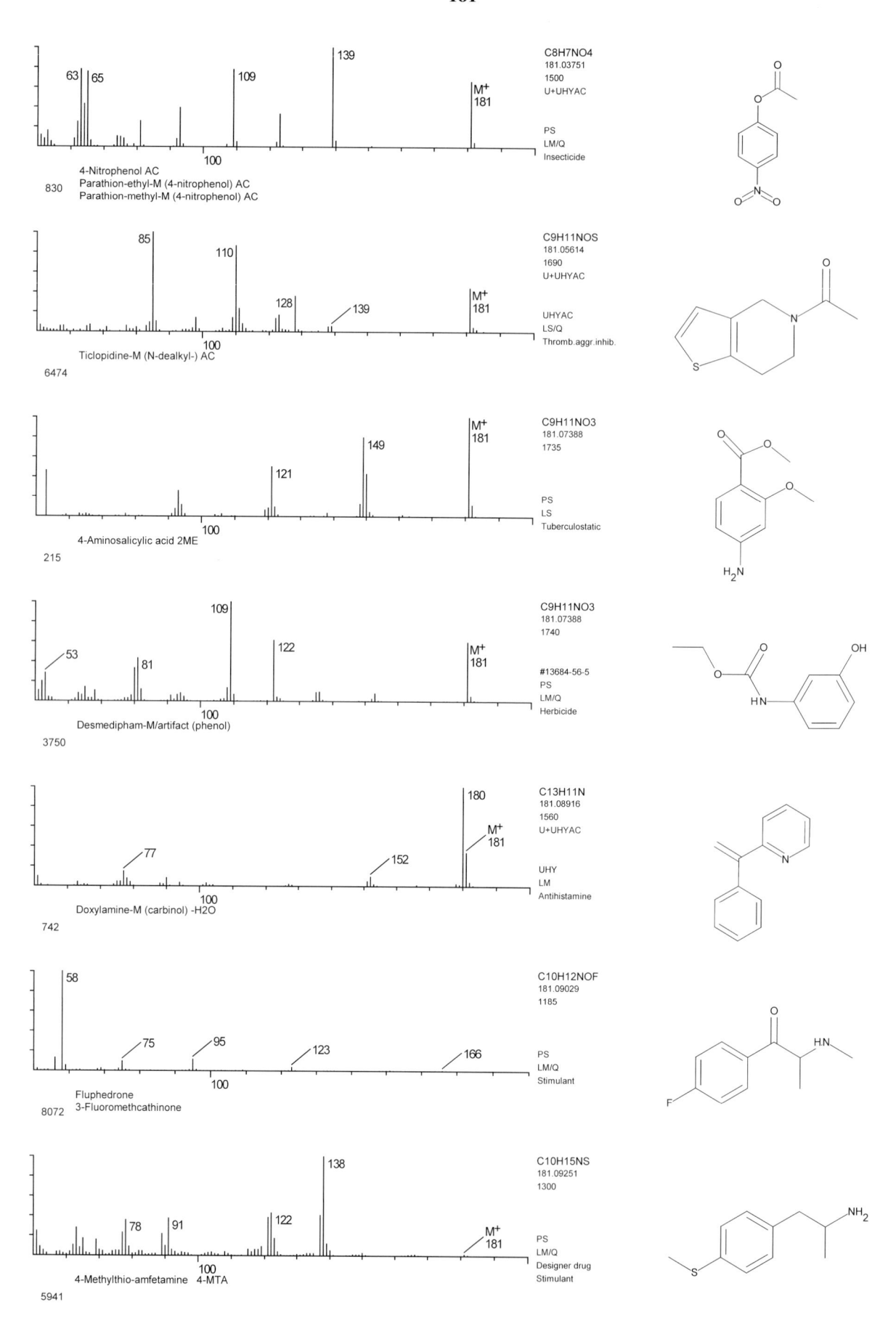

4-Nitrophenol AC Parathion-ethyl-M (4-nitrophenol) AC Parathion-methyl-M (4-nitrophenol) AC 830	C8H7NO4 181.03751 1500 U+UHYAC PS LM/Q Insecticide
Ticlopidine-M (N-dealkyl-) AC 6474	C9H11NOS 181.05614 1690 U+UHYAC UHYAC LS/Q Thromb.aggr.inhib.
4-Aminosalicylic acid 2ME 215	C9H11NO3 181.07388 1735 PS LS Tuberculostatic
Desmedipham-M/artifact (phenol) 3750	C9H11NO3 181.07388 1740 #13684-56-5 PS LM/Q Herbicide
Doxylamine-M (carbinol) -H2O 742	C13H11N 181.08916 1560 U+UHYAC UHY LM Antihistamine
Fluphedrone 3-Fluoromethcathinone 8072	C10H12NOF 181.09029 1185 PS LM/Q Stimulant
4-Methylthio-amfetamine 4-MTA 5941	C10H15NS 181.09251 1300 PS LM/Q Designer drug Stimulant

181

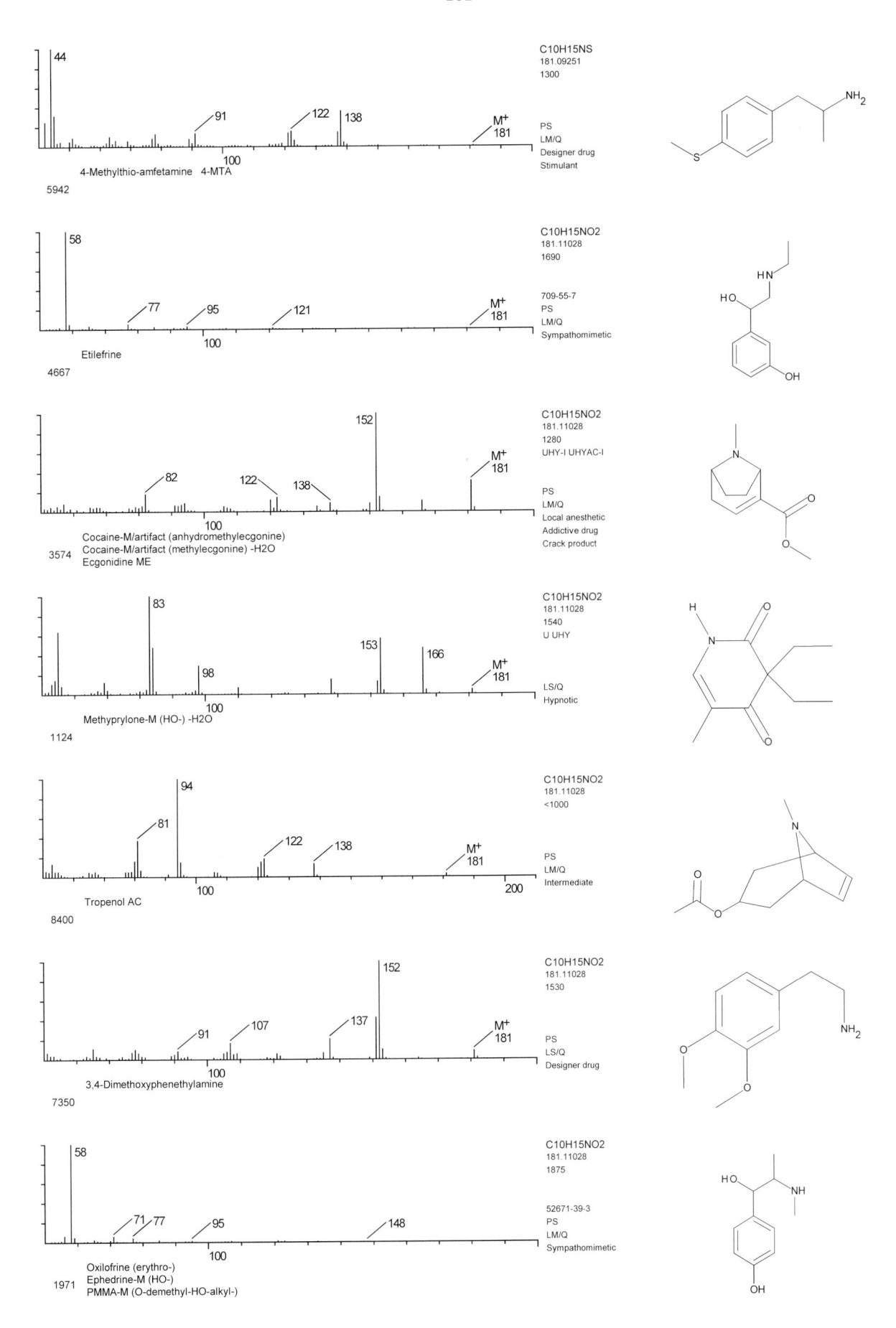

4-Methylthio-amfetamine 4-MTA	44, 91, 122, 138, M+ 181	C10H15NS 181.09251 1300 PS LM/Q Designer drug Stimulant
5942		
Etilefrine	58, 77, 95, 121, M+ 181	C10H15NO2 181.11028 1690 709-55-7 PS LM/Q Sympathomimetic
4667		
Cocaine-M/artifact (anhydromethylecgonine) Cocaine-M/artifact (methylecgonine) -H2O Ecgonidine ME	82, 122, 138, 152, M+ 181	C10H15NO2 181.11028 1280 UHY-I UHYAC-I PS LM/Q Local anesthetic Addictive drug Crack product
3574		
Methyprylone-M (HO-) -H2O	83, 98, 153, 166, M+ 181	C10H15NO2 181.11028 1540 U UHY LS/Q Hypnotic
1124		
Tropenol AC	81, 94, 122, 138, M+ 181	C10H15NO2 181.11028 <1000 PS LM/Q Intermediate
8400		
3,4-Dimethoxyphenethylamine	91, 107, 137, 152, M+ 181	C10H15NO2 181.11028 1530 PS LS/Q Designer drug
7350		
Oxilofrine (erythro-) Ephedrine-M (HO-) PMMA-M (O-demethyl-HO-alkyl-)	58, 71, 77, 95, 148	C10H15NO2 181.11028 1875 52671-39-3 PS LM/Q Sympathomimetic
1971		

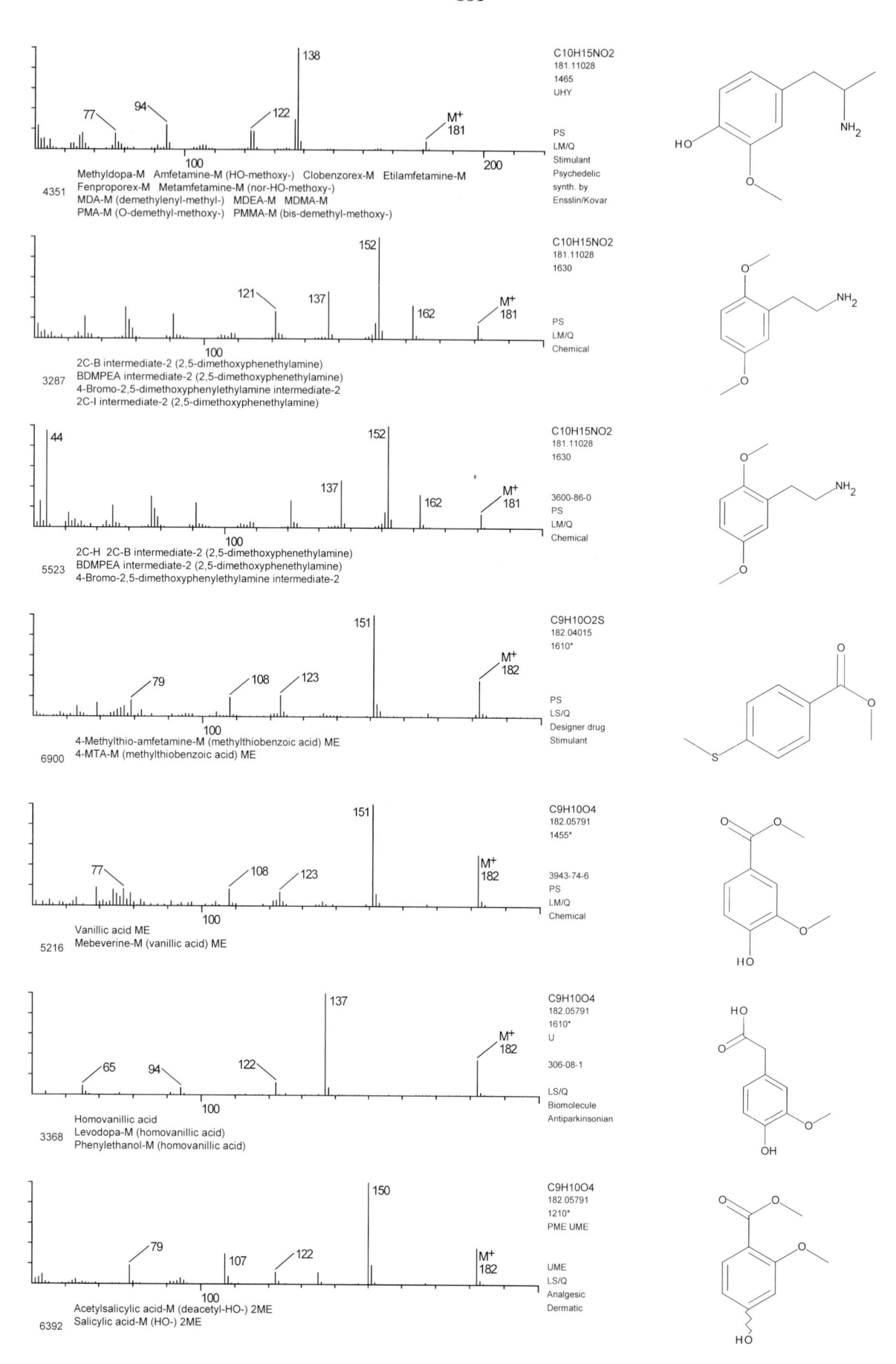

4351 Methyldopa-M Amfetamine-M (HO-methoxy-) Clobenzorex-M Etilamfetamine-M Fenproporex-M Metamfetamine-M (nor-HO-methoxy-) MDA-M (demethylenyl-methyl-) MDEA-M MDMA-M PMA-M (O-demethyl-methoxy-) PMMA-M (bis-demethyl-methoxy-) Peaks: 77, 94, 122, 138, M+ 181	C10H15NO2 181.11028 1465 UHY PS LM/Q Stimulant Psychedelic synth. by Ensslin/Kovar
3287 2C-B intermediate-2 (2,5-dimethoxyphenethylamine) BDMPEA intermediate-2 (2,5-dimethoxyphenethylamine) 4-Bromo-2,5-dimethoxyphenylethylamine intermediate-2 2C-I intermediate-2 (2,5-dimethoxyphenethylamine) Peaks: 121, 137, 152, 162, M+ 181	C10H15NO2 181.11028 1630 PS LM/Q Chemical
5523 2C-H 2C-B intermediate-2 (2,5-dimethoxyphenethylamine) BDMPEA intermediate-2 (2,5-dimethoxyphenethylamine) 4-Bromo-2,5-dimethoxyphenylethylamine intermediate-2 Peaks: 44, 137, 152, 162, M+ 181	C10H15NO2 181.11028 1630 3600-86-0 PS LM/Q Chemical
6900 4-Methylthio-amfetamine-M (methylthiobenzoic acid) ME 4-MTA-M (methylthiobenzoic acid) ME Peaks: 79, 108, 123, 151, M+ 182	C9H10O2S 182.04015 1610* PS LS/Q Designer drug Stimulant
5216 Vanillic acid ME Mebeverine-M (vanillic acid) ME Peaks: 77, 108, 123, 151, M+ 182	C9H10O4 182.05791 1455* 3943-74-6 PS LM/Q Chemical
3368 Homovanillic acid Levodopa-M (homovanillic acid) Phenylethanol-M (homovanillic acid) Peaks: 65, 94, 122, 137, M+ 182	C9H10O4 182.05791 1610* U 306-08-1 LS/Q Biomolecule Antiparkinsonian
6392 Acetylsalicylic acid-M (deacetyl-HO-) 2ME Salicylic acid-M (HO-) 2ME Peaks: 79, 107, 122, 150, M+ 182	C9H10O4 182.05791 1210* PME UME UME LS/Q Analgesic Dermatic

182

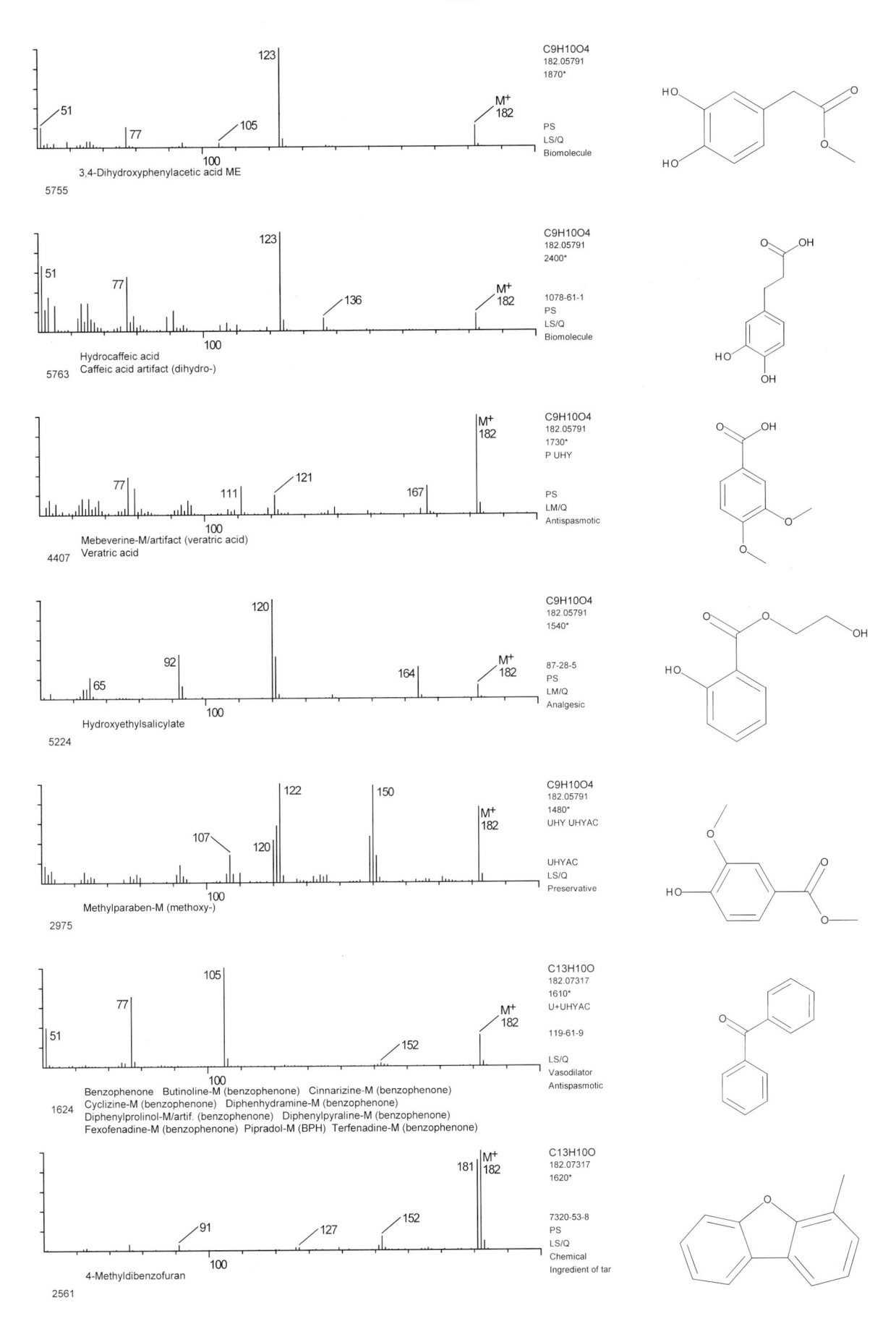

3,4-Dihydroxyphenylacetic acid ME
5755

C9H10O4
182.05791
1870*

PS
LS/Q
Biomolecule

Hydrocaffeic acid
Caffeic acid artifact (dihydro-)
5763

C9H10O4
182.05791
2400*

1078-61-1
PS
LS/Q
Biomolecule

Mebeverine-M/artifact (veratric acid)
Veratric acid
4407

C9H10O4
182.05791
1730*
P UHY

PS
LM/Q
Antispasmotic

Hydroxyethylsalicylate
5224

C9H10O4
182.05791
1540*

87-28-5
PS
LM/Q
Analgesic

Methylparaben-M (methoxy-)
2975

C9H10O4
182.05791
1480*
UHY UHYAC

UHYAC
LS/Q
Preservative

Benzophenone   Butinoline-M (benzophenone)   Cinnarizine-M (benzophenone)
Cyclizine-M (benzophenone)   Diphenhydramine-M (benzophenone)
Diphenylprolinol-M/artif. (benzophenone)   Diphenylpyraline-M (benzophenone)
Fexofenadine-M (benzophenone)   Pipradol-M (BPH)   Terfenadine-M (benzophenone)
1624

C13H10O
182.07317
1610*
U+UHYAC

119-61-9
LS/Q
Vasodilator
Antispasmotic

4-Methyldibenzofuran
2561

C13H10O
182.07317
1620*

7320-53-8
PS
LS/Q
Chemical
Ingredient of tar

182

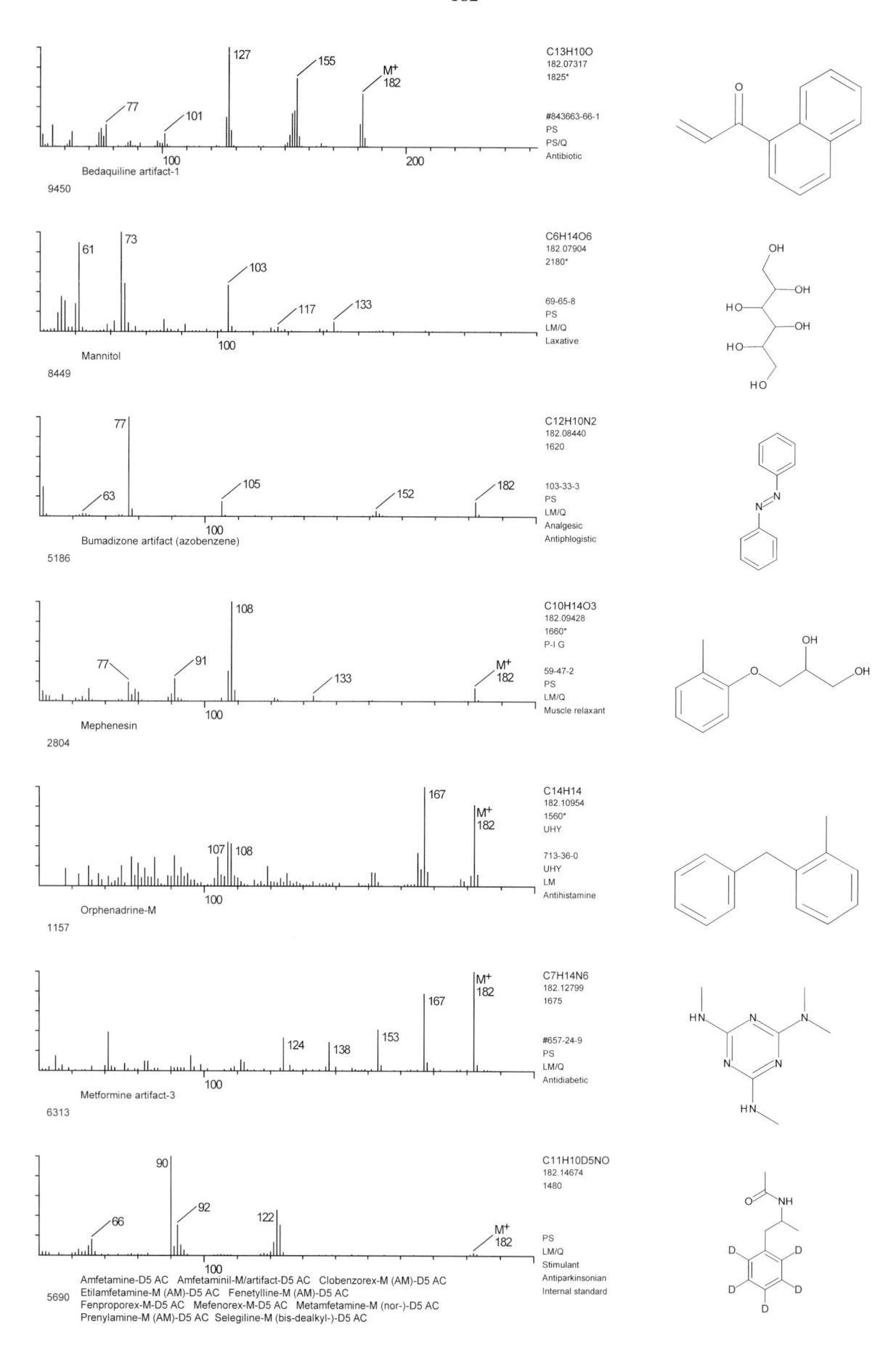

Bedaquiline artifact-1 — 9450	C13H10O / 182.07317 / 1825* / #843663-66-1 / PS / PS/Q / Antibiotic
Peaks: 77, 101, 127, 155, M+ 182	

Mannitol — 8449	C6H14O6 / 182.07904 / 2180* / 69-65-8 / PS / LM/Q / Laxative
Peaks: 61, 73, 103, 117, 133	

Bumadizone artifact (azobenzene) — 5186	C12H10N2 / 182.08440 / 1620 / 103-33-3 / PS / LM/Q / Analgesic Antiphlogistic
Peaks: 63, 77, 105, 152, 182	

Mephenesin — 2804	C10H14O3 / 182.09428 / 1660* / P-I G / 59-47-2 / PS / LM/Q / Muscle relaxant
Peaks: 77, 91, 108, 133, M+ 182	

Orphenadrine-M — 1157	C14H14 / 182.10954 / 1560* / UHY / 713-36-0 / UHY / LM / Antihistamine
Peaks: 107, 108, 167, M+ 182	

Metformine artifact-3 — 6313	C7H14N6 / 182.12799 / 1675 / #657-24-9 / PS / LM/Q / Antidiabetic
Peaks: 124, 138, 153, 167, M+ 182	

Amfetamine-D5 AC Amfetaminil-M/artifact-D5 AC Clobenzorex-M (AM)-D5 AC Etilamfetamine-M (AM)-D5 AC Fenetylline-M (AM)-D5 AC Fenproporex-M-D5 AC Mefenorex-M-D5 AC Metamfetamine-M (nor-)-D5 AC Prenylamine-M (AM)-D5 AC Selegiline-M (bis-dealkyl-)-D5 AC — 5690	C11H10D5NO / 182.14674 / 1480 / PS / LM/Q / Stimulant Antiparkinsonian Internal standard
Peaks: 66, 90, 92, 122, M+ 182	

352

183

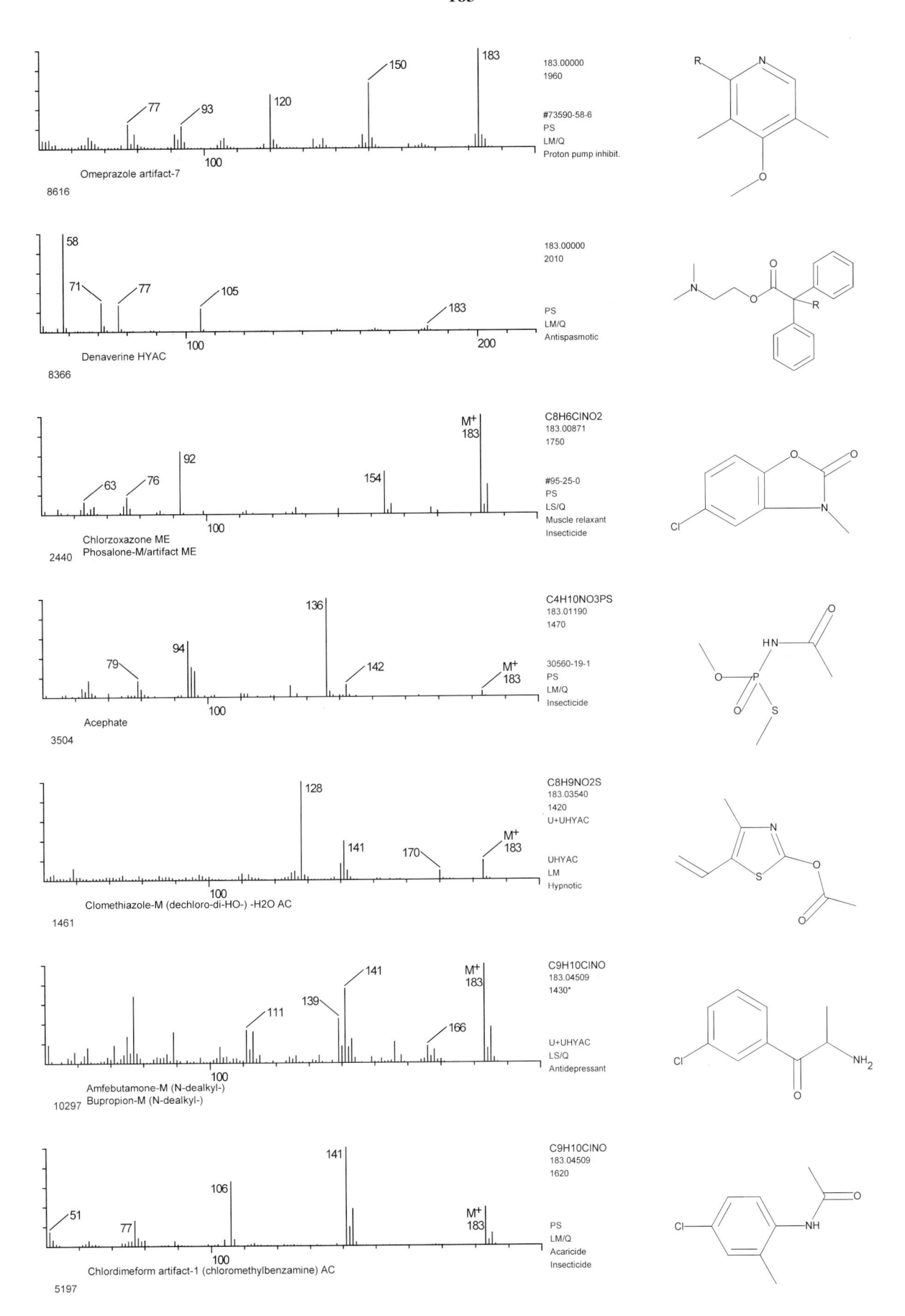

Spectrum label	m/z peaks	Formula / data
Omeprazole artifact-7 — 8616	77, 93, 120, 150, 183	183.00000 / 1960 / #73590-58-6 / PS / LM/Q / Proton pump inhibit.
Denaverine HYAC — 8366	58, 71, 77, 105, 183	183.00000 / 2010 / PS / LM/Q / Antispasmotic
Chlorzoxazone ME / Phosalone-M/artifact ME — 2440	63, 76, 92, 154, M+ 183	C8H6ClNO2 / 183.00871 / 1750 / #95-25-0 / PS / LS/Q / Muscle relaxant / Insecticide
Acephate — 3504	79, 94, 136, 142, M+ 183	C4H10NO3PS / 183.01190 / 1470 / 30560-19-1 / PS / LM/Q / Insecticide
Clomethiazole-M (dechloro-di-HO-) -H2O AC — 1461	128, 141, 170, M+ 183	C8H9NO2S / 183.03540 / 1420 / U+UHYAC / UHYAC / LM / Hypnotic
Amfebutamone-M (N-dealkyl-) / Bupropion-M (N-dealkyl-) — 10297	111, 139, 141, 166, M+ 183	C9H10ClNO / 183.04509 / 1430* / U+UHYAC / LS/Q / Antidepressant
Chlordimeform artifact-1 (chloromethylbenzamine) AC — 5197	51, 77, 106, 141, M+ 183	C9H10ClNO / 183.04509 / 1620 / PS / LM/Q / Acaricide / Insecticide

183

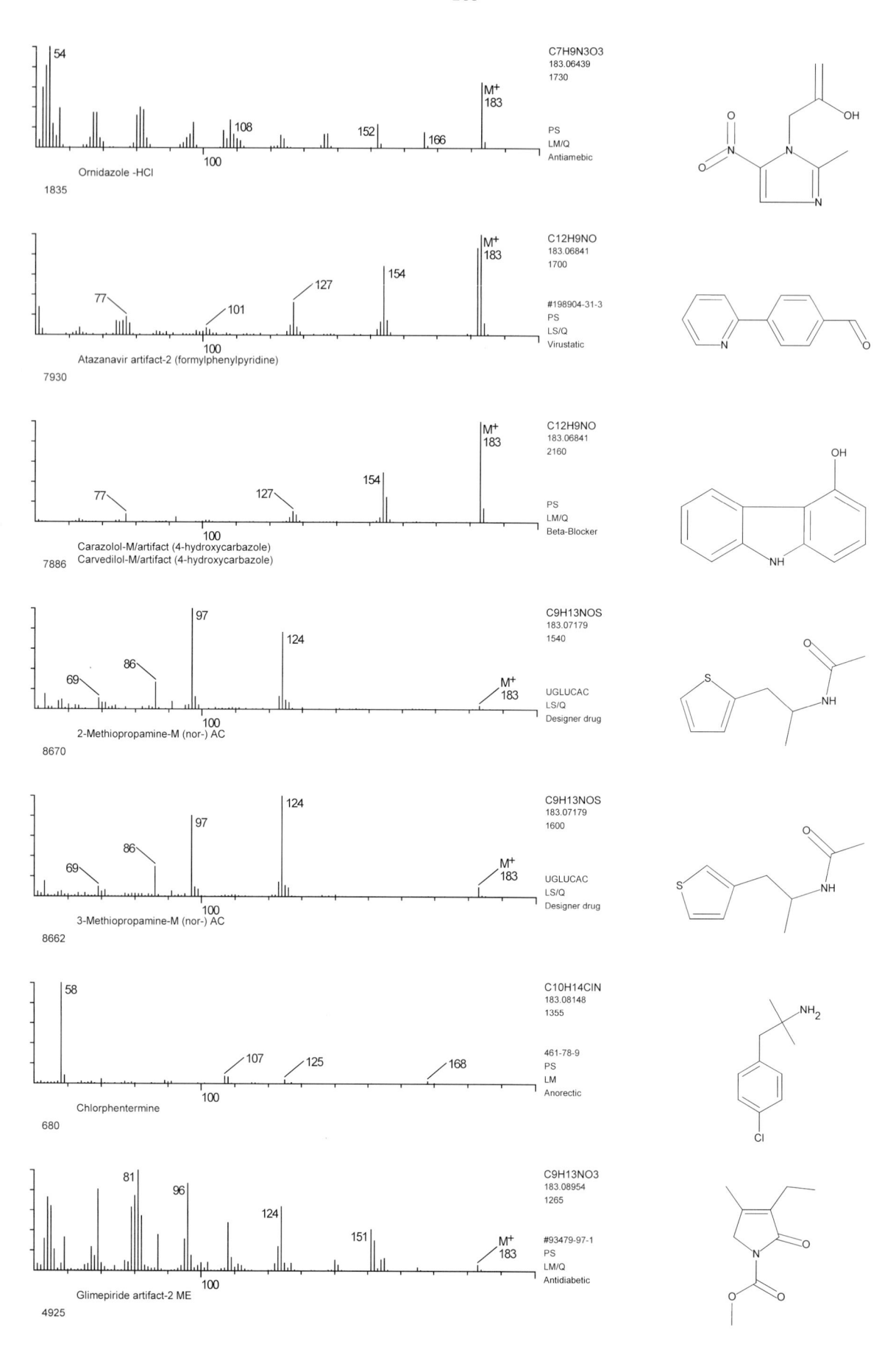

Spectrum	Compound	Formula / Data
1835	Ornidazole -HCl — peaks: 54, 100, 108, 152, 166, M+ 183	C7H9N3O3; 183.06439; 1730; PS; LM/Q; Antiamebic
7930	Atazanavir artifact-2 (formylphenylpyridine) — peaks: 77, 100, 101, 127, 154, M+ 183	C12H9NO; 183.06841; 1700; #198904-31-3; PS; LS/Q; Virustatic
7886	Carazolol-M/artifact (4-hydroxycarbazole); Carvedilol-M/artifact (4-hydroxycarbazole) — peaks: 77, 100, 127, 154, M+ 183	C12H9NO; 183.06841; 2160; PS; LM/Q; Beta-Blocker
8670	2-Methiopropamine-M (nor-) AC — peaks: 69, 86, 97, 100, 124, M+ 183	C9H13NOS; 183.07179; 1540; UGLUCAC; LS/Q; Designer drug
8662	3-Methiopropamine-M (nor-) AC — peaks: 69, 86, 97, 100, 124, M+ 183	C9H13NOS; 183.07179; 1600; UGLUCAC; LS/Q; Designer drug
680	Chlorphentermine — peaks: 58, 100, 107, 125, 168, 183	C10H14ClN; 183.08148; 1355; 461-78-9; PS; LM; Anorectic
4925	Glimepiride artifact-2 ME — peaks: 81, 96, 100, 124, 151, M+ 183	C9H13NO3; 183.08954; 1265; #93479-97-1; PS; LM/Q; Antidiabetic

183

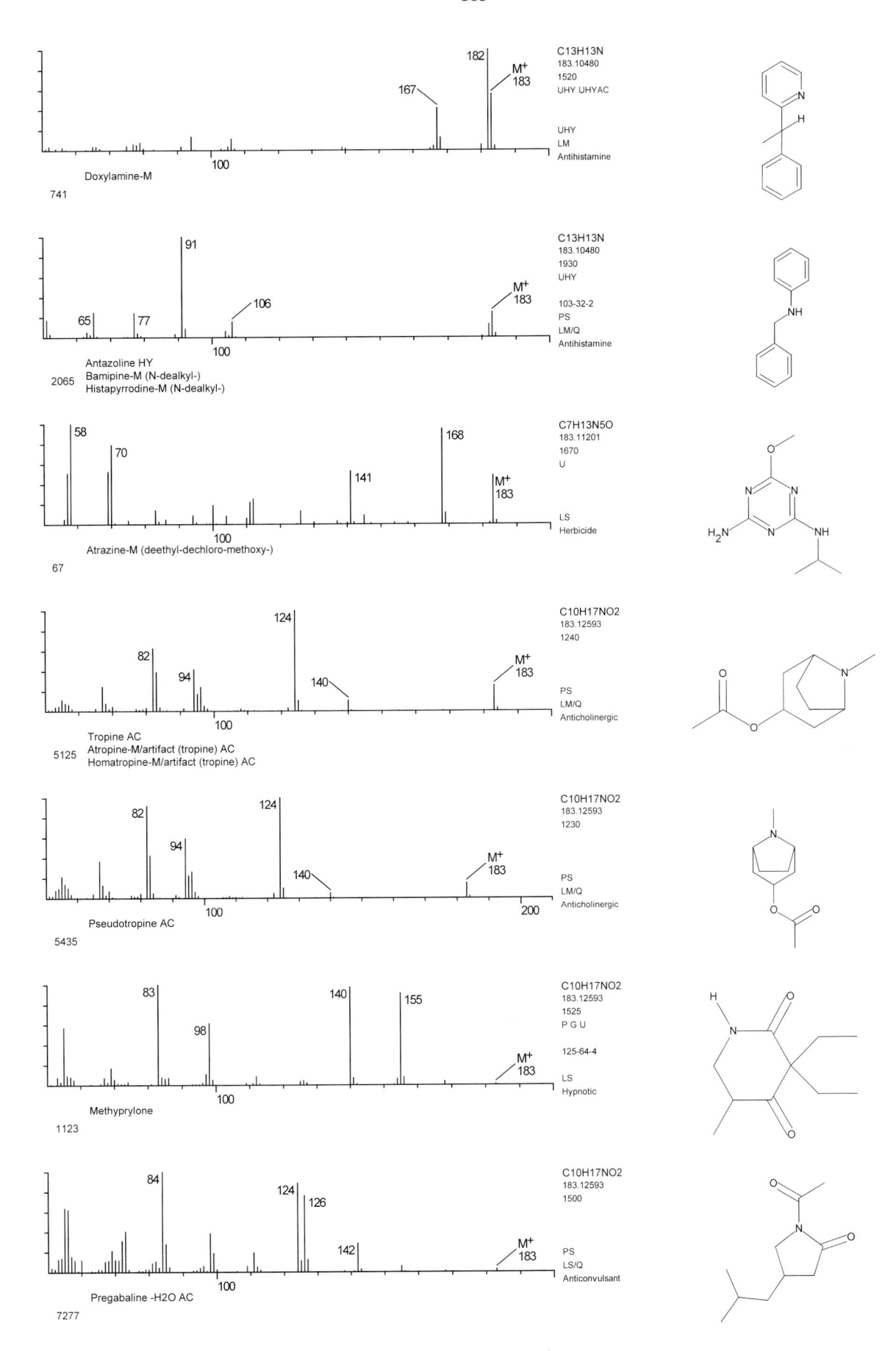

Doxylamine-M
741
C13H13N
183.10480
1520
UHY UHYAC

UHY
LM
Antihistamine

Antazoline HY
Bamipine-M (N-dealkyl-)
Histapyrrodine-M (N-dealkyl-)
2065
C13H13N
183.10480
1930
UHY

103-32-2
PS
LM/Q
Antihistamine

Atrazine-M (deethyl-dechloro-methoxy-)
67
C7H13N5O
183.11201
1670
U

LS
Herbicide

Tropine AC
Atropine-M/artifact (tropine) AC
Homatropine-M/artifact (tropine) AC
5125
C10H17NO2
183.12593
1240

PS
LM/Q
Anticholinergic

Pseudotropine AC
5435
C10H17NO2
183.12593
1230

PS
LM/Q
Anticholinergic

Methyprylone
1123
C10H17NO2
183.12593
1525
P G U

125-64-4
LS
Hypnotic

Pregabaline -H2O AC
7277
C10H17NO2
183.12593
1500

PS
LS/Q
Anticonvulsant

183

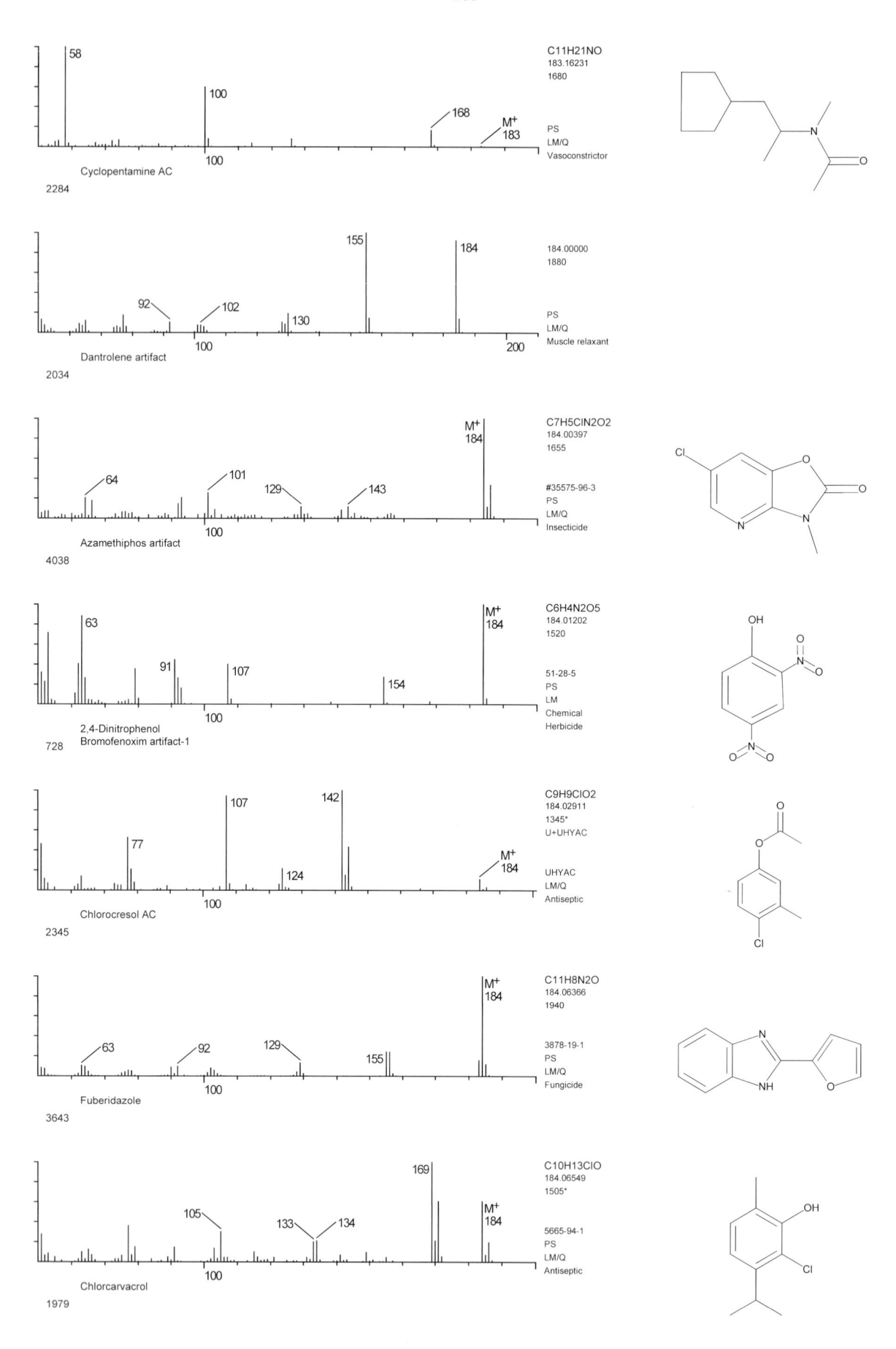

Cyclopentamine AC
2284
C11H21NO
183.16231
1680
PS
LM/Q
Vasoconstrictor
Peaks: 58, 100, 168, M+ 183

Dantrolene artifact
2034
184.00000
1880
PS
LM/Q
Muscle relaxant
Peaks: 92, 102, 130, 155, 184

Azamethiphos artifact
4038
C7H5ClN2O2
184.00397
1655
#35575-96-3
PS
LM/Q
Insecticide
Peaks: 64, 101, 129, 143, M+ 184

2,4-Dinitrophenol
Bromofenoxim artifact-1
728
C6H4N2O5
184.01202
1520
51-28-5
PS
LM
Chemical
Herbicide
Peaks: 63, 91, 107, 154, M+ 184

Chlorocresol AC
2345
C9H9ClO2
184.02911
1345*
U+UHYAC
UHYAC
LM/Q
Antiseptic
Peaks: 77, 107, 124, 142, M+ 184

Fuberidazole
3643
C11H8N2O
184.06366
1940
3878-19-1
PS
LM/Q
Fungicide
Peaks: 63, 92, 129, 155, M+ 184

Chlorcarvacrol
1979
C10H13ClO
184.06549
1505*
5665-94-1
PS
LM/Q
Antiseptic
Peaks: 105, 133, 134, 169, M+ 184

356

184

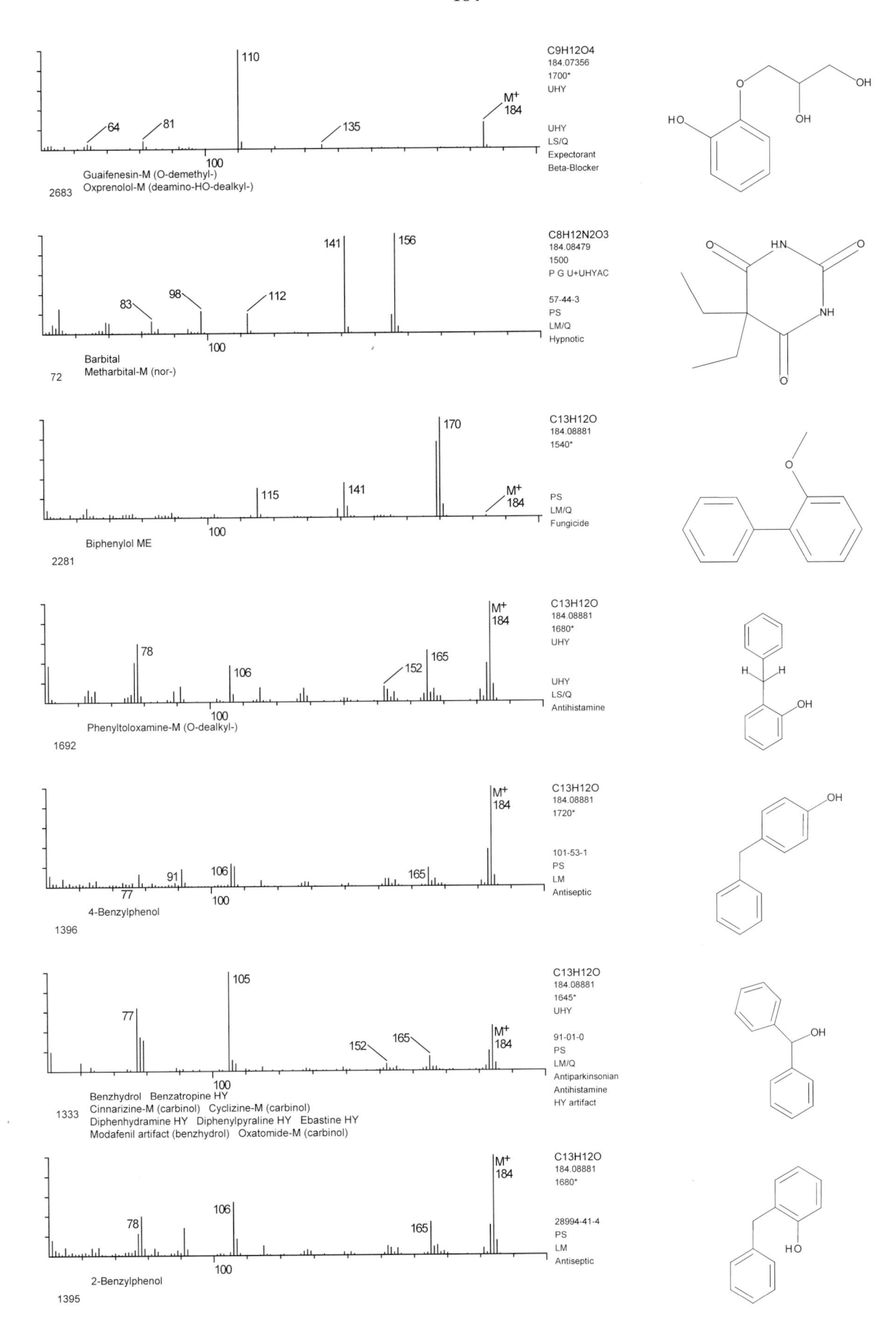

	C9H12O4 184.07356 1700* UHY UHY LS/Q Expectorant Beta-Blocker	

2683 Guaifenesin-M (O-demethyl-)
Oxprenolol-M (deamino-HO-dealkyl-)

| | C8H12N2O3
184.08479
1500
P G U+UHYAC

57-44-3
PS
LM/Q
Hypnotic | |

72 Barbital
Metharbital-M (nor-)

| | C13H12O
184.08881
1540*

PS
LM/Q
Fungicide | |

2281 Biphenylol ME

| | C13H12O
184.08881
1680*
UHY

UHY
LS/Q
Antihistamine | |

1692 Phenyltoloxamine-M (O-dealkyl-)

| | C13H12O
184.08881
1720*

101-53-1
PS
LM
Antiseptic | |

1396 4-Benzylphenol

| | C13H12O
184.08881
1645*
UHY

91-01-0
PS
LM/Q
Antiparkinsonian
Antihistamine
HY artifact | |

1333 Benzhydrol Benzatropine HY
Cinnarizine-M (carbinol) Cyclizine-M (carbinol)
Diphenhydramine HY Diphenylpyraline HY Ebastine HY
Modafenil artifact (benzhydrol) Oxatomide-M (carbinol)

| | C13H12O
184.08881
1680*

28994-41-4
PS
LM
Antiseptic | |

1395 2-Benzylphenol

184

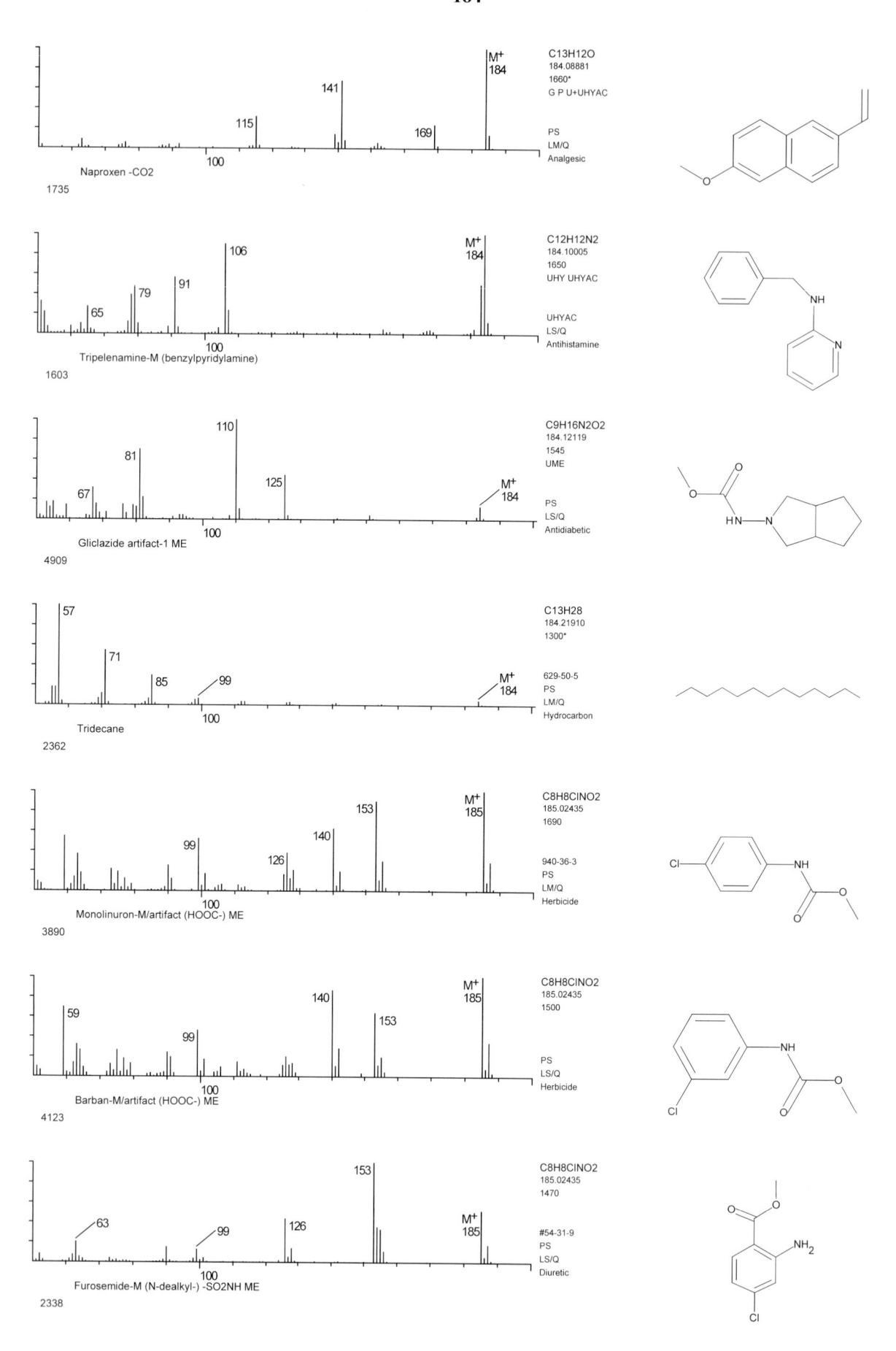

Naproxen -CO2 — 1735	C13H12O 184.08881 1660* G P U+UHYAC PS LM/Q Analgesic
Tripelenamine-M (benzylpyridylamine) — 1603	C12H12N2 184.10005 1650 UHY UHYAC UHYAC LS/Q Antihistamine
Gliclazide artifact-1 ME — 4909	C9H16N2O2 184.12119 1545 UME PS LS/Q Antidiabetic
Tridecane — 2362	C13H28 184.21910 1300* 629-50-5 PS LM/Q Hydrocarbon
Monolinuron-M/artifact (HOOC-) ME — 3890	C8H8ClNO2 185.02435 1690 940-36-3 PS LM/Q Herbicide
Barban-M/artifact (HOOC-) ME — 4123	C8H8ClNO2 185.02435 1500 PS LS/Q Herbicide
Furosemide-M (N-dealkyl-) -SO2NH ME — 2338	C8H8ClNO2 185.02435 1470 #54-31-9 PS LS/Q Diuretic

185

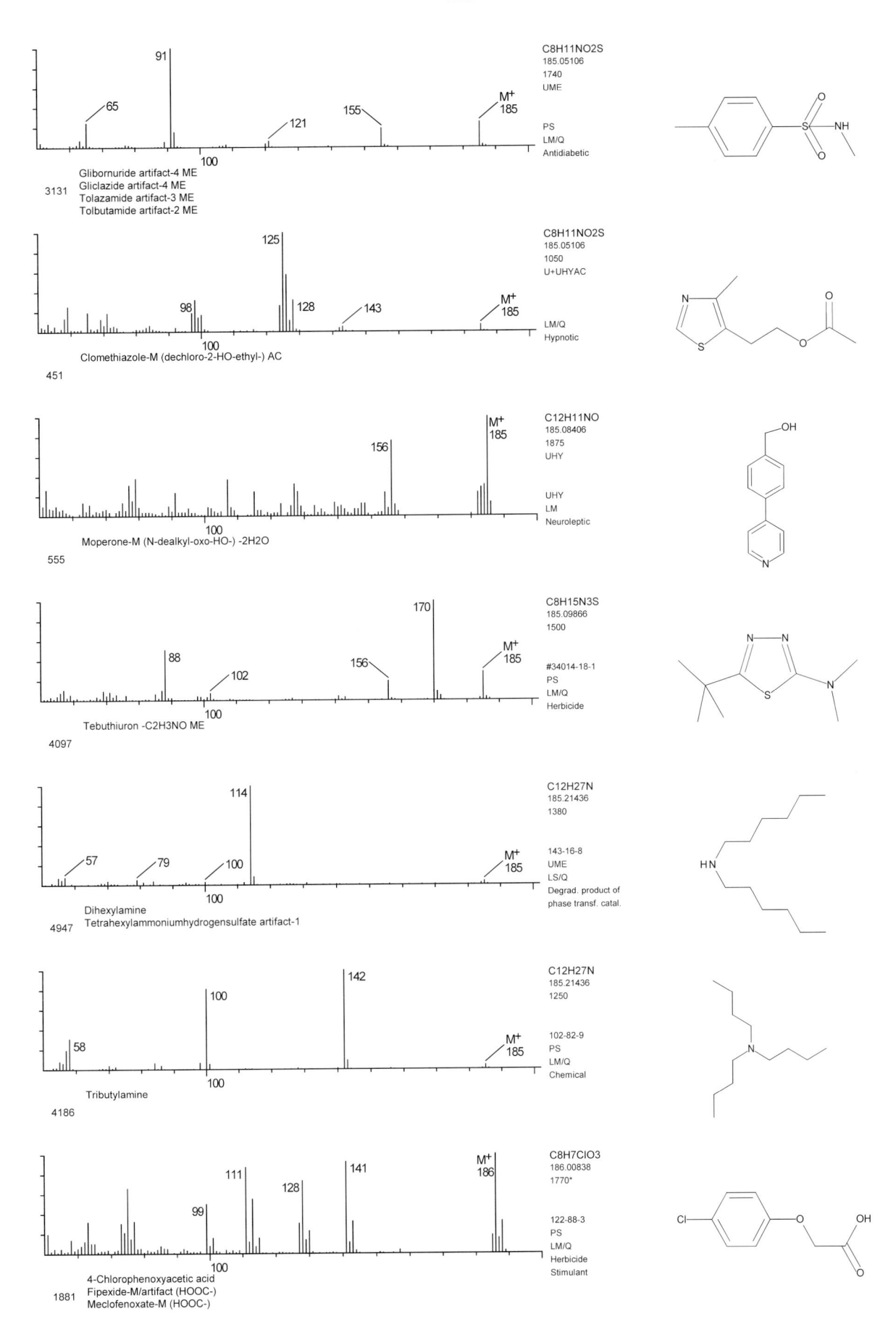

3131	Glibornuride artifact-4 ME Gliclazide artifact-4 ME Tolazamide artifact-3 ME Tolbutamide artifact-2 ME	C8H11NO2S 185.05106 1740 UME PS LM/Q Antidiabetic
451	Clomethiazole-M (dechloro-2-HO-ethyl-) AC	C8H11NO2S 185.05106 1050 U+UHYAC LM/Q Hypnotic
555	Moperone-M (N-dealkyl-oxo-HO-) -2H2O	C12H11NO 185.08406 1875 UHY UHY LM Neuroleptic
4097	Tebuthiuron -C2H3NO ME	C8H15N3S 185.09866 1500 #34014-18-1 PS LM/Q Herbicide
4947	Dihexylamine Tetrahexylammoniumhydrogensulfate artifact-1	C12H27N 185.21436 1380 143-16-8 UME LS/Q Degrad. product of phase transf. catal.
4186	Tributylamine	C12H27N 185.21436 1250 102-82-9 PS LM/Q Chemical
1881	4-Chlorophenoxyacetic acid Fipexide-M/artifact (HOOC-) Meclofenoxate-M (HOOC-)	C8H7ClO3 186.00838 1770* 122-88-3 PS LM/Q Herbicide Stimulant

186

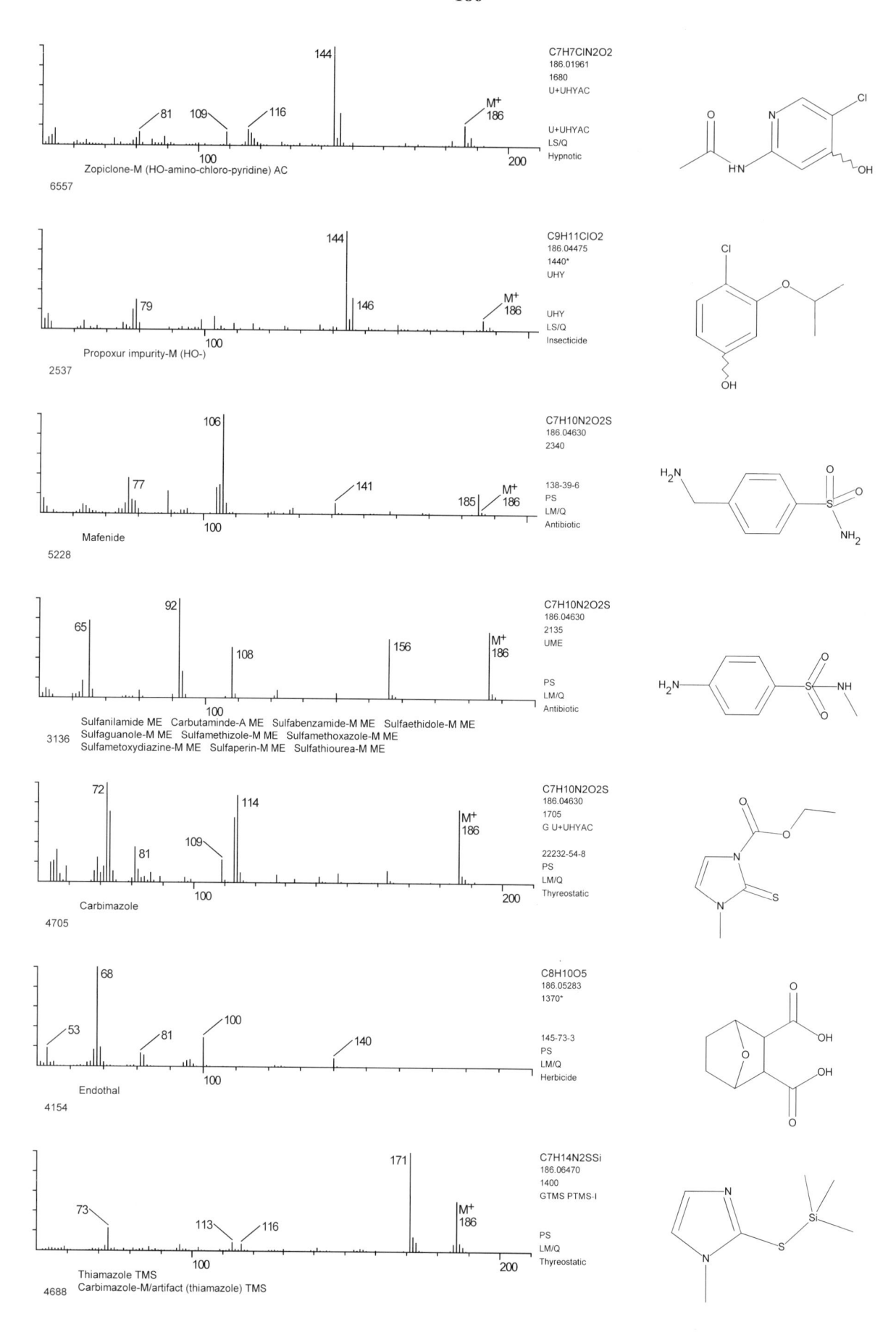

Zopiclone-M (HO-amino-chloro-pyridine) AC
6557
C7H7ClN2O2
186.01961
1680
U+UHYAC
U+UHYAC
LS/Q
Hypnotic

Propoxur impurity-M (HO-)
2537
C9H11ClO2
186.04475
1440*
UHY
UHY
LS/Q
Insecticide

Mafenide
5228
C7H10N2O2S
186.04630
2340
138-39-6
PS
LM/Q
Antibiotic

Sulfanilamide ME Carbutaminde-A ME Sulfabenzamide-M ME Sulfaethidole-M ME
Sulfaguanole-M ME Sulfamethizole-M ME Sulfamethoxazole-M ME
Sulfametoxydiazine-M ME Sulfaperin-M ME Sulfathiourea-M ME
3136
C7H10N2O2S
186.04630
2135
UME
PS
LM/Q
Antibiotic

Carbimazole
4705
C7H10N2O2S
186.04630
1705
G U+UHYAC
22232-54-8
PS
LM/Q
Thyreostatic

Endothal
4154
C8H10O5
186.05283
1370*
145-73-3
PS
LM/Q
Herbicide

Thiamazole TMS
Carbimazole-M/artifact (thiamazole) TMS
4688
C7H14N2SSi
186.06470
1400
GTMS PTMS-I
PS
LM/Q
Thyreostatic

186

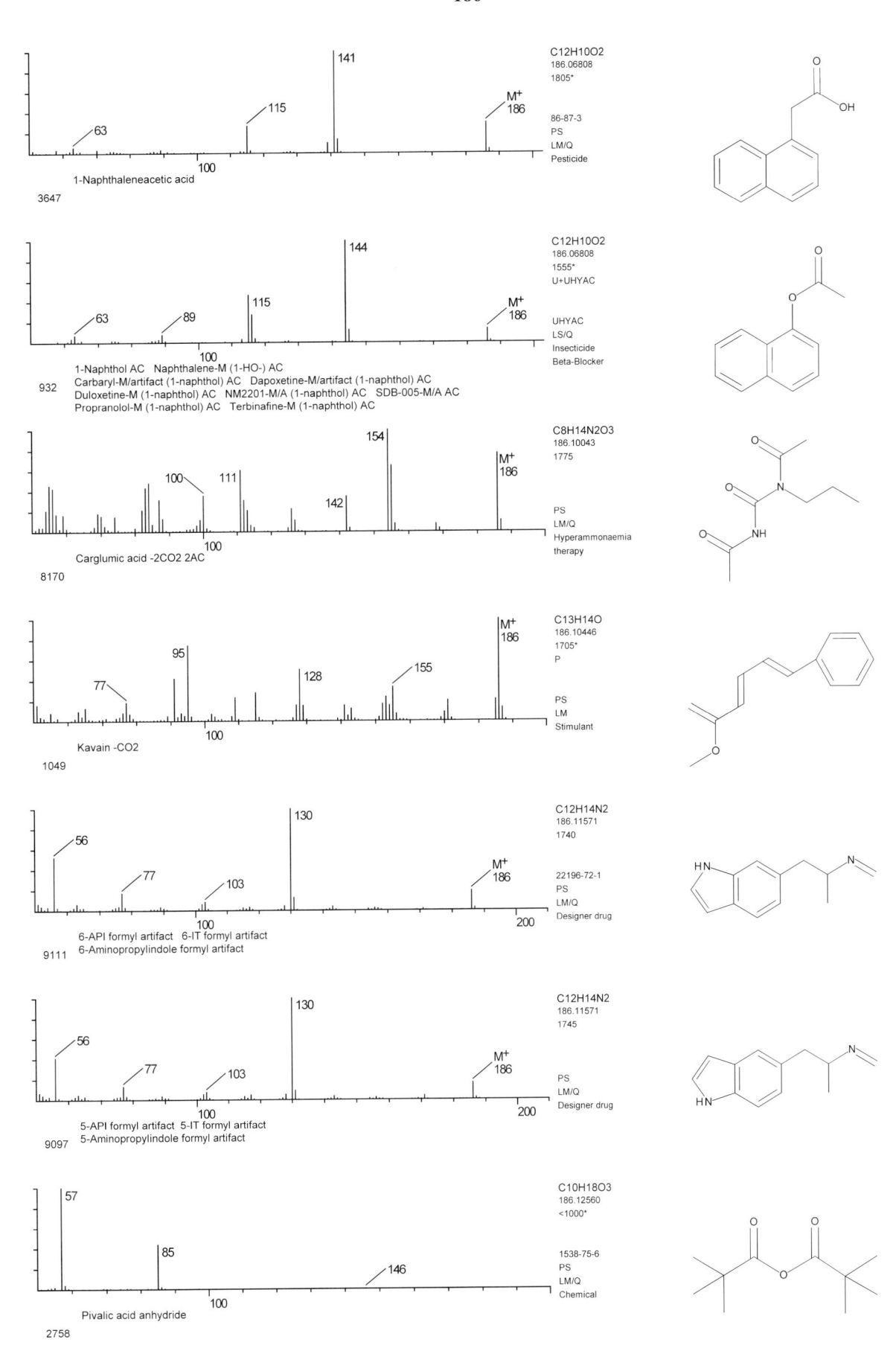

1-Naphthaleneacetic acid 3647	Peaks: 63, 115, 141, M+ 186 (100 at 141)	C12H10O2 186.06808 1805* 86-87-3 PS LM/Q Pesticide

1-Naphthol AC   Naphthalene-M (1-HO-) AC
Carbaryl-M/artifact (1-naphthol) AC   Dapoxetine-M/artifact (1-naphthol) AC
Duloxetine-M (1-naphthol) AC   NM2201-M/A (1-naphthol) AC   SDB-005-M/A AC
Propranolol-M (1-naphthol) AC   Terbinafine-M (1-naphthol) AC
932

Peaks: 63, 89, 115, 144, M+ 186
C12H10O2
186.06808
1555*
U+UHYAC
UHYAC
LS/Q
Insecticide
Beta-Blocker

Carglumic acid -2CO2 2AC
8170
Peaks: 100, 111, 142, 154, M+ 186
C8H14N2O3
186.10043
1775
PS
LM/Q
Hyperammonaemia therapy

Kavain -CO2
1049
Peaks: 77, 95, 128, 155, M+ 186
C13H14O
186.10446
1705*
P
PS
LM
Stimulant

6-API formyl artifact   6-IT formyl artifact
6-Aminopropylindole formyl artifact
9111
Peaks: 56, 77, 103, 130, M+ 186
C12H14N2
186.11571
1740
22196-72-1
PS
LM/Q
Designer drug

5-API formyl artifact   5-IT formyl artifact
5-Aminopropylindole formyl artifact
9097
Peaks: 56, 77, 103, 130, M+ 186
C12H14N2
186.11571
1745
PS
LM/Q
Designer drug

Pivalic acid anhydride
2758
Peaks: 57, 85, 146
C10H18O3
186.12560
<1000*
1538-75-6
PS
LM/Q
Chemical

361

187

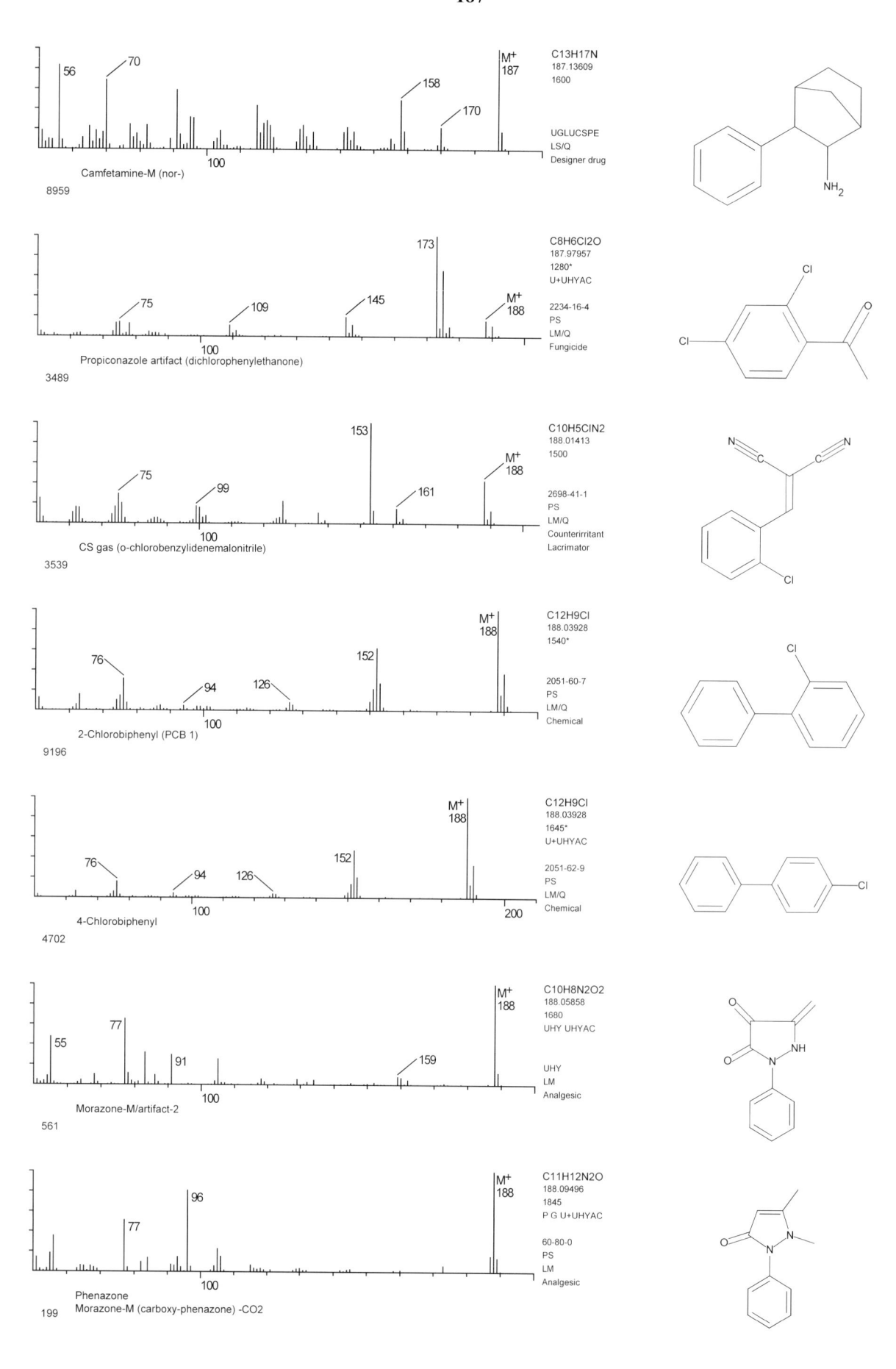

Camfetamine-M (nor-) — C13H17N, 187.13609, 1600, UGLUCSPE, LS/Q, Designer drug
Peaks: 56, 70, 100, 158, 170, M+ 187
8959

Propiconazole artifact (dichlorophenylethanone) — C8H6Cl2O, 187.97957, 1280*, U+UHYAC, 2234-16-4, PS, LM/Q, Fungicide
Peaks: 75, 109, 145, 173, M+ 188
3489

CS gas (o-chlorobenzylidenemalonitrile) — C10H5ClN2, 188.01413, 1500, 2698-41-1, PS, LM/Q, Counterirritant, Lacrimator
Peaks: 75, 99, 153, 161, M+ 188
3539

2-Chlorobiphenyl (PCB 1) — C12H9Cl, 188.03928, 1540*, 2051-60-7, PS, LM/Q, Chemical
Peaks: 76, 94, 126, 152, M+ 188
9196

4-Chlorobiphenyl — C12H9Cl, 188.03928, 1645*, U+UHYAC, 2051-62-9, PS, LM/Q, Chemical
Peaks: 76, 94, 126, 152, M+ 188
4702

Morazone-M/artifact-2 — C10H8N2O2, 188.05858, 1680, UHY UHYAC, UHY, LM, Analgesic
Peaks: 55, 77, 91, 159, M+ 188
561

Phenazone / Morazone-M (carboxy-phenazone) -CO2 — C11H12N2O, 188.09496, 1845, P G U+UHYAC, 60-80-0, PS, LM, Analgesic
Peaks: 77, 96, M+ 188
199

364

188

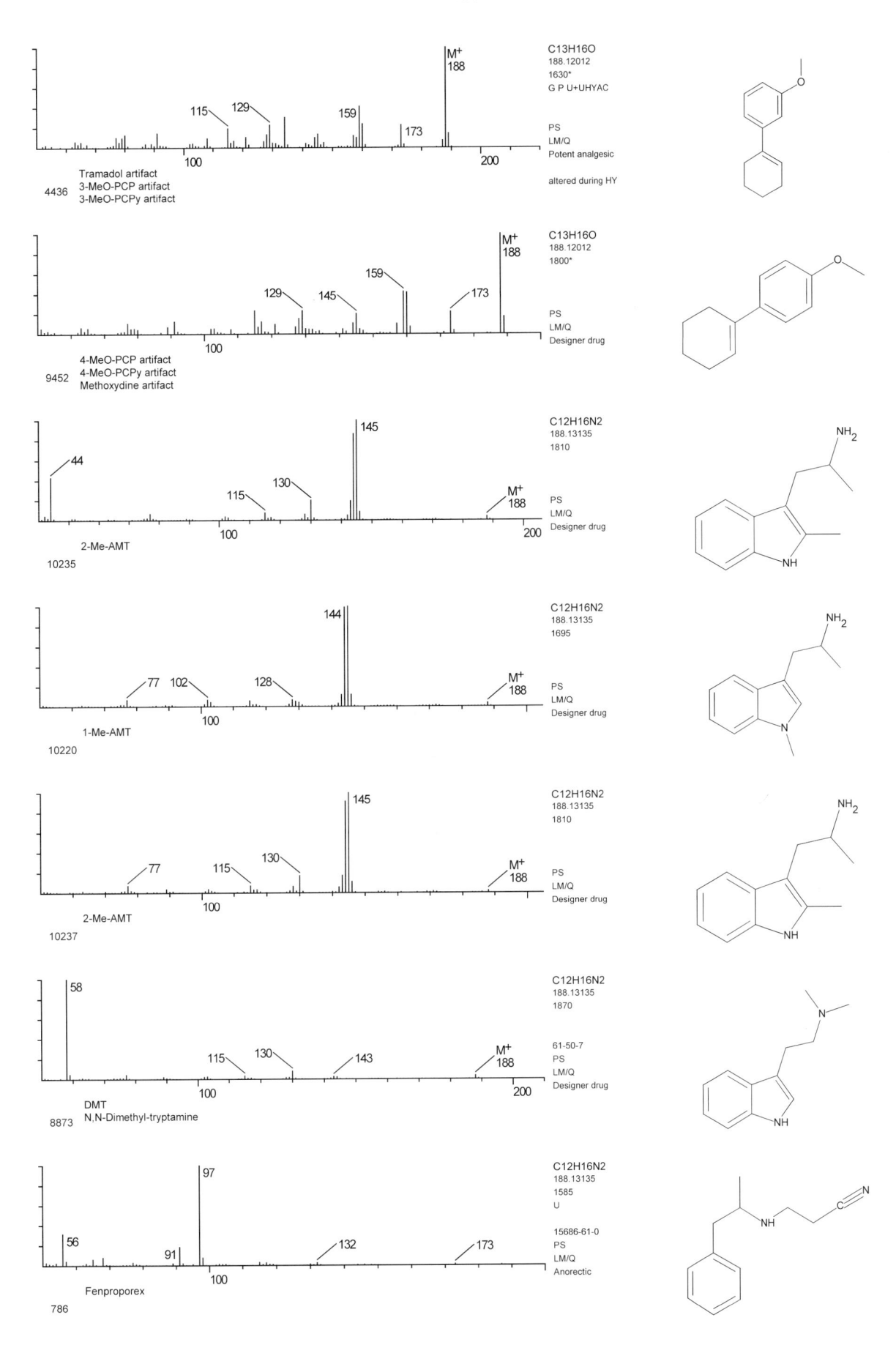

4436	Tramadol artifact 3-MeO-PCP artifact 3-MeO-PCPy artifact	C13H16O 188.12012 1630* G P U+UHYAC PS LM/Q Potent analgesic altered during HY
9452	4-MeO-PCP artifact 4-MeO-PCPy artifact Methoxydine artifact	C13H16O 188.12012 1800* PS LM/Q Designer drug
10235	2-Me-AMT	C12H16N2 188.13135 1810 PS LM/Q Designer drug
10220	1-Me-AMT	C12H16N2 188.13135 1695 PS LM/Q Designer drug
10237	2-Me-AMT	C12H16N2 188.13135 1810 PS LM/Q Designer drug
8873	DMT N,N-Dimethyl-tryptamine	C12H16N2 188.13135 1870 61-50-7 PS LM/Q Designer drug
786	Fenproporex	C12H16N2 188.13135 1585 U 15686-61-0 PS LM/Q Anorectic

365

188

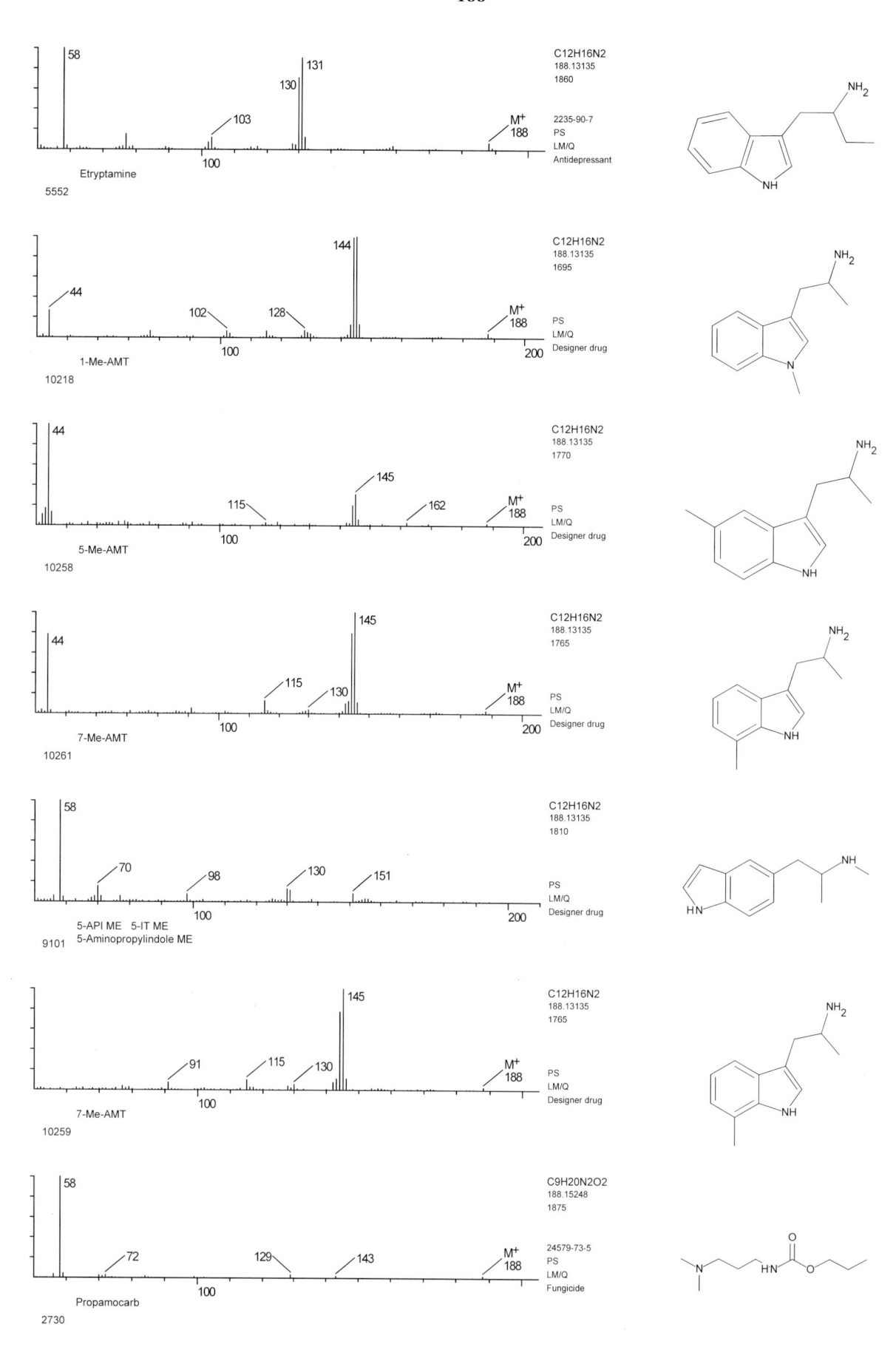

Spectrum	Compound	Formula
	Etryptamine (5552) — peaks 58, 103, 130, 131, 188 M+	C12H16N2, 188.13135, 1860, 2235-90-7, PS, LM/Q, Antidepressant
	1-Me-AMT (10218) — peaks 44, 102, 128, 144, 188 M+	C12H16N2, 188.13135, 1695, PS, LM/Q, Designer drug
	5-Me-AMT (10258) — peaks 44, 115, 145, 162, 188 M+	C12H16N2, 188.13135, 1770, PS, LM/Q, Designer drug
	7-Me-AMT (10261) — peaks 44, 115, 130, 145, 188 M+	C12H16N2, 188.13135, 1765, PS, LM/Q, Designer drug
	5-API ME / 5-IT ME / 5-Aminopropylindole ME (9101) — peaks 58, 70, 98, 130, 151	C12H16N2, 188.13135, 1810, PS, LM/Q, Designer drug
	7-Me-AMT (10259) — peaks 91, 115, 130, 145, 188 M+	C12H16N2, 188.13135, 1765, PS, LM/Q, Designer drug
	Propamocarb (2730) — peaks 58, 72, 129, 143, 188 M+	C9H20N2O2, 188.15248, 1875, 24579-73-5, PS, LM/Q, Fungicide

366

189

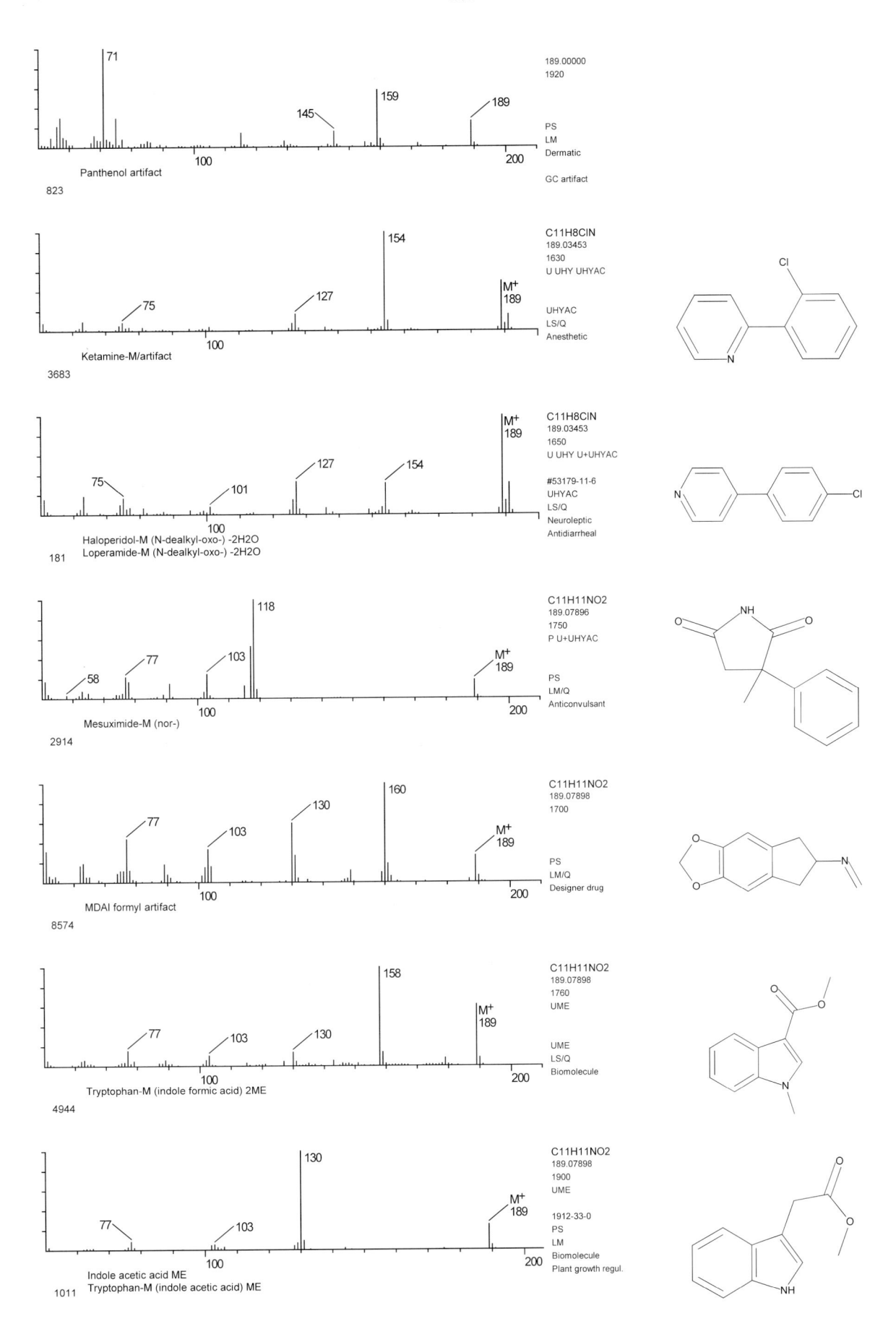

Panthenol artifact	189.00000 1920 PS LM Dermatic GC artifact
Ketamine-M/artifact	C11H8ClN 189.03453 1630 U UHY UHYAC UHYAC LS/Q Anesthetic
Haloperidol-M (N-dealkyl-oxo-) -2H2O Loperamide-M (N-dealkyl-oxo-) -2H2O	C11H8ClN 189.03453 1650 U UHY U+UHYAC #53179-11-6 UHYAC LS/Q Neuroleptic Antidiarrheal
Mesuximide-M (nor-)	C11H11NO2 189.07896 1750 P U+UHYAC PS LM/Q Anticonvulsant
MDAI formyl artifact	C11H11NO2 189.07898 1700 PS LM/Q Designer drug
Tryptophan-M (indole formic acid) 2ME	C11H11NO2 189.07898 1760 UME UME LS/Q Biomolecule
Indole acetic acid ME Tryptophan-M (indole acetic acid) ME	C11H11NO2 189.07898 1900 UME 1912-33-0 PS LM Biomolecule Plant growth regul.

189

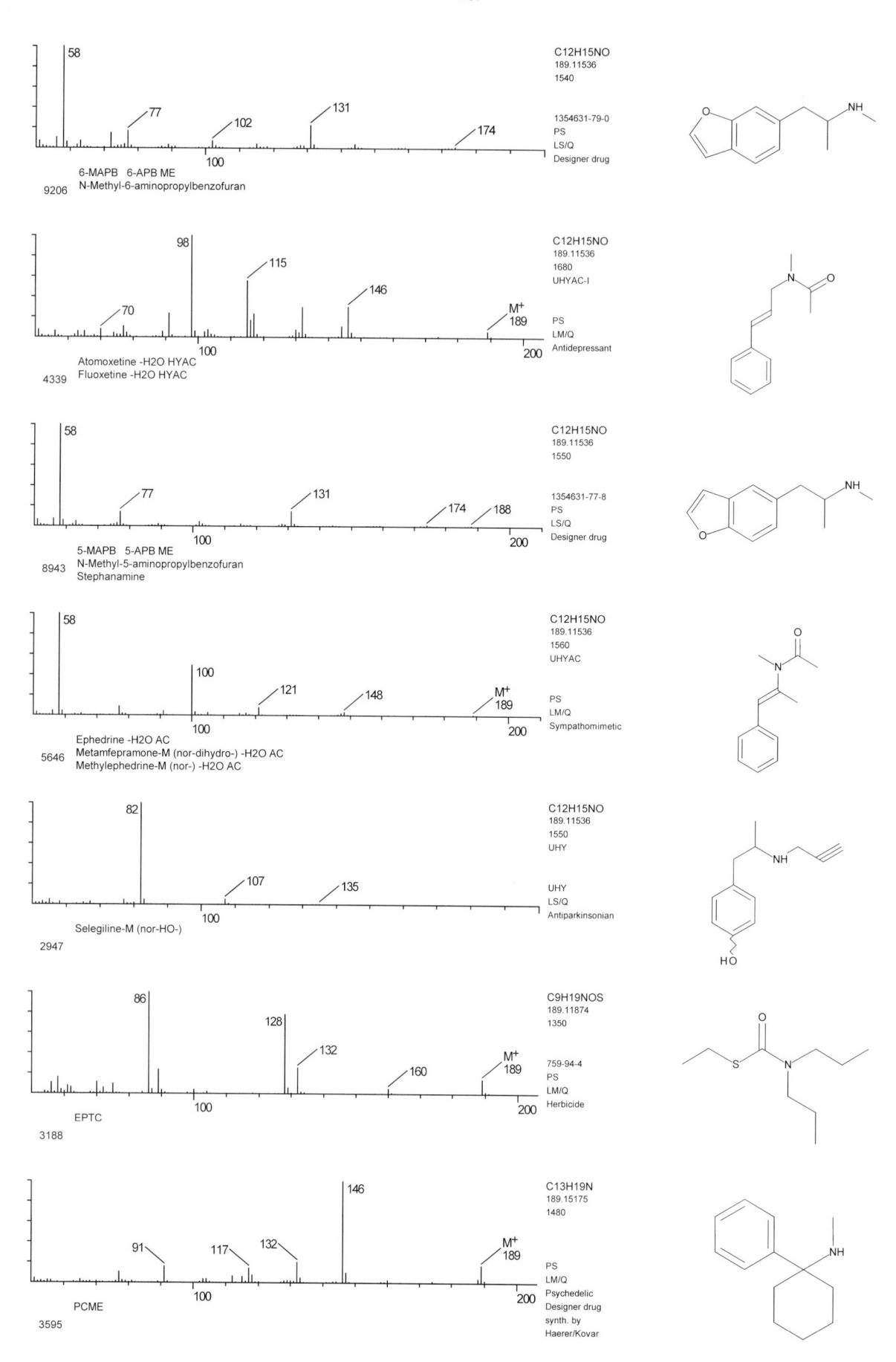

6-MAPB   6-APB ME
N-Methyl-6-aminopropylbenzofuran
9206

C12H15NO
189.11536
1540

1354631-79-0
PS
LS/Q
Designer drug

Atomoxetine -H2O HYAC
Fluoxetine -H2O HYAC
4339

C12H15NO
189.11536
1680
UHYAC-I

PS
LM/Q
Antidepressant

5-MAPB   5-APB ME
N-Methyl-5-aminopropylbenzofuran
Stephanamine
8943

C12H15NO
189.11536
1550

1354631-77-8
PS
LS/Q
Designer drug

Ephedrine -H2O AC
Metamfepramone-M (nor-dihydro-) -H2O AC
Methylephedrine-M (nor-) -H2O AC
5646

C12H15NO
189.11536
1560
UHYAC

PS
LM/Q
Sympathomimetic

Selegiline-M (nor-HO-)
2947

C12H15NO
189.11536
1550
UHY

UHY
LS/Q
Antiparkinsonian

EPTC
3188

C9H19NOS
189.11874
1350

759-94-4
PS
LM/Q
Herbicide

PCME
3595

C13H19N
189.15175
1480

PS
LM/Q
Psychedelic
Designer drug
synth. by
Haerer/Kovar

190

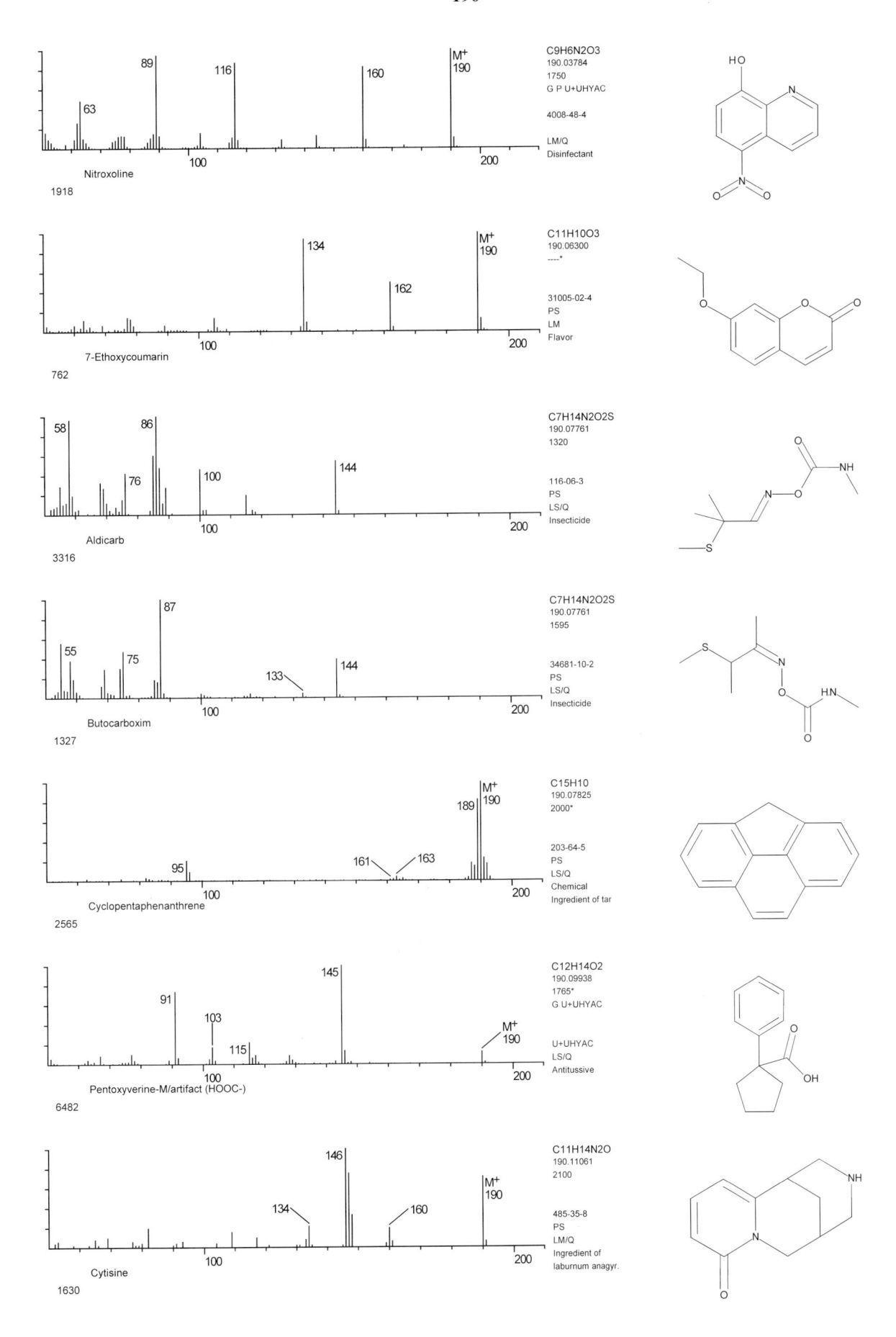

Nitroxoline 1918	peaks: 63, 89, 116, 160, M+ 190	C9H6N2O3 190.03784 1750 G P U+UHYAC 4008-48-4 LM/Q Disinfectant
7-Ethoxycoumarin 762	peaks: 134, 162, M+ 190	C11H10O3 190.06300 ----* 31005-02-4 PS LM Flavor
Aldicarb 3316	peaks: 58, 76, 86, 100, 144	C7H14N2O2S 190.07761 1320 116-06-3 PS LS/Q Insecticide
Butocarboxim 1327	peaks: 55, 75, 87, 133, 144	C7H14N2O2S 190.07761 1595 34681-10-2 PS LS/Q Insecticide
Cyclopentaphenanthrene 2565	peaks: 95, 161, 163, 189, M+ 190	C15H10 190.07825 2000* 203-64-5 PS LS/Q Chemical Ingredient of tar
Pentoxyverine-M/artifact (HOOC-) 6482	peaks: 91, 103, 115, 145, M+ 190	C12H14O2 190.09938 1765* G U+UHYAC U+UHYAC LS/Q Antitussive
Cytisine 1630	peaks: 134, 146, 160, M+ 190	C11H14N2O 190.11061 2100 485-35-8 PS LM/Q Ingredient of laburnum anagyr.

369

190

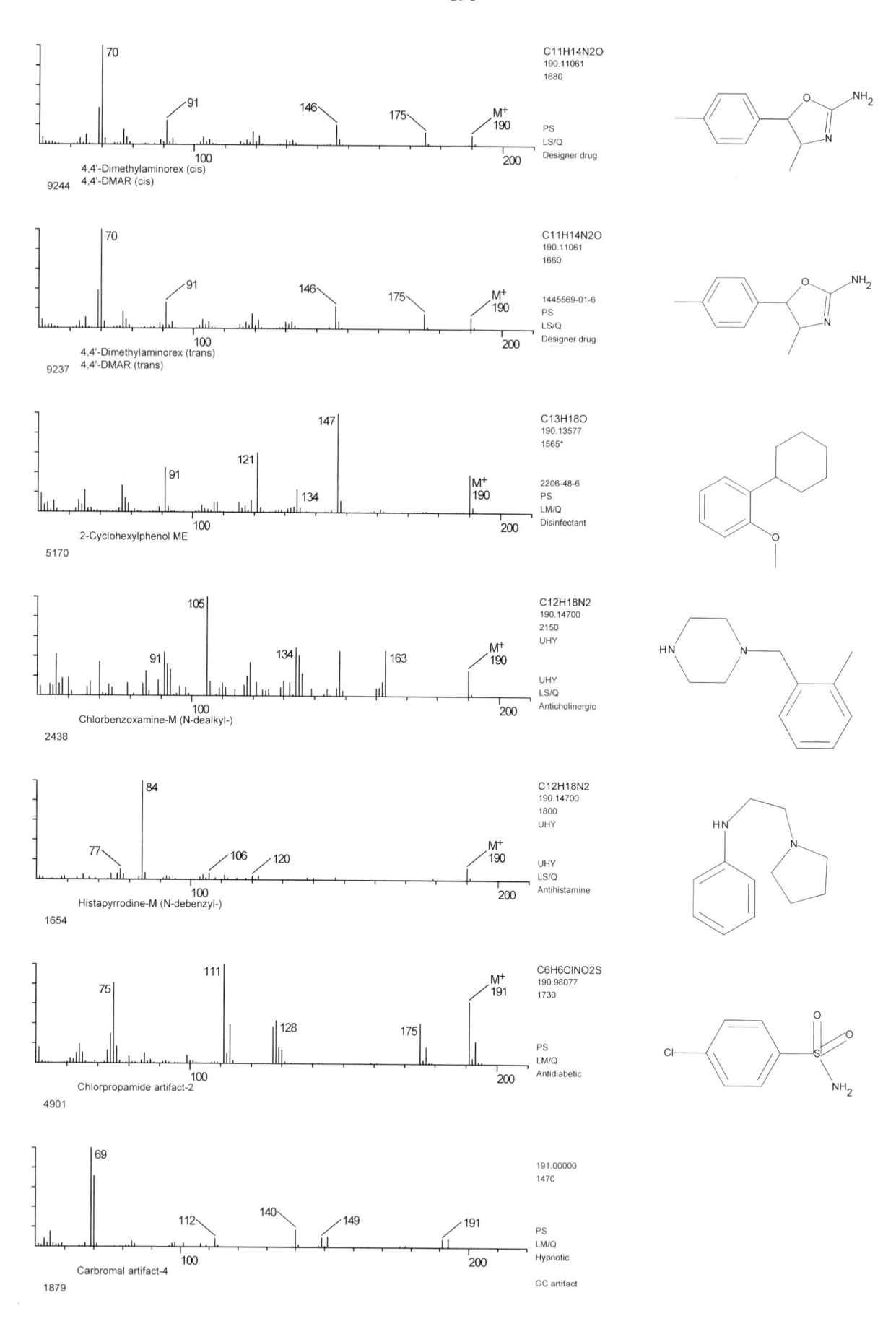

4,4'-Dimethylaminorex (cis) 4,4'-DMAR (cis) 9244	C11H14N2O 190.11061 1680 PS LS/Q Designer drug
4,4'-Dimethylaminorex (trans) 4,4'-DMAR (trans) 9237	C11H14N2O 190.11061 1660 1445569-01-6 PS LS/Q Designer drug
2-Cyclohexylphenol ME 5170	C13H18O 190.13577 1565* 2206-48-6 PS LM/Q Disinfectant
Chlorbenzoxamine-M (N-dealkyl-) 2438	C12H18N2 190.14700 2150 UHY UHY LS/Q Anticholinergic
Histapyrrodine-M (N-debenzyl-) 1654	C12H18N2 190.14700 1800 UHY UHY LS/Q Antihistamine
Chlorpropamide artifact-2 4901	C6H6ClNO2S 190.98077 1730 PS LM/Q Antidiabetic
Carbromal artifact-4 1879	191.00000 1470 PS LM/Q Hypnotic GC artifact

191

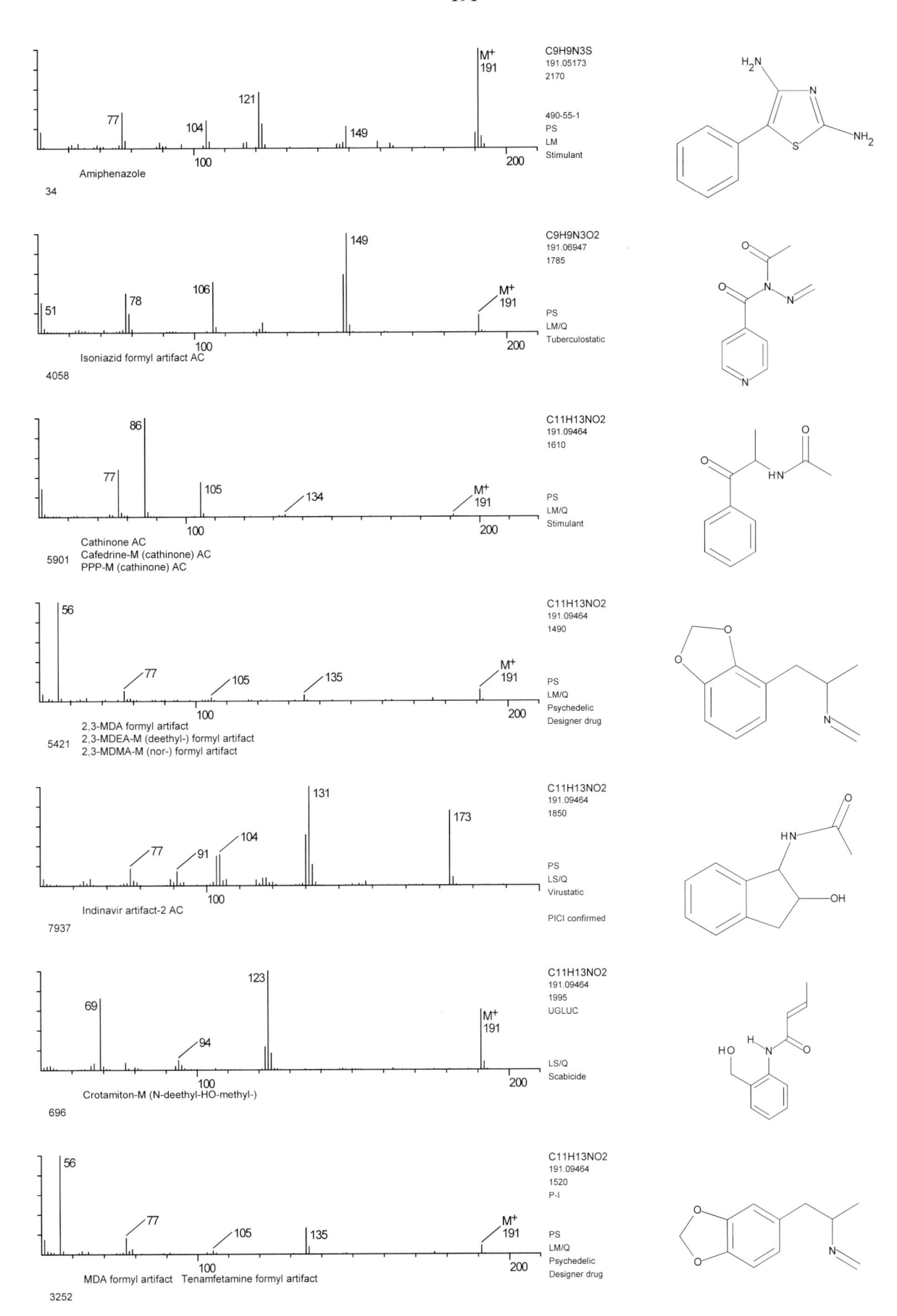

Amiphenazole	C9H9N3S 191.05173 2170 490-55-1 PS LM Stimulant	
34		
Isoniazid formyl artifact AC	C9H9N3O2 191.06947 1785 PS LM/Q Tuberculostatic	
4058		
Cathinone AC Cafedrine-M (cathinone) AC PPP-M (cathinone) AC	C11H13NO2 191.09464 1610 PS LM/Q Stimulant	
5901		
2,3-MDA formyl artifact 2,3-MDEA-M (deethyl-) formyl artifact 2,3-MDMA-M (nor-) formyl artifact	C11H13NO2 191.09464 1490 PS LM/Q Psychedelic Designer drug	
5421		
Indinavir artifact-2 AC	C11H13NO2 191.09464 1850 PS LS/Q Virustatic PICI confirmed	
7937		
Crotamiton-M (N-deethyl-HO-methyl-)	C11H13NO2 191.09464 1995 UGLUC LS/Q Scabicide	
696		
MDA formyl artifact Tenamfetamine formyl artifact	C11H13NO2 191.09464 1520 P-I PS LM/Q Psychedelic Designer drug	
3252		

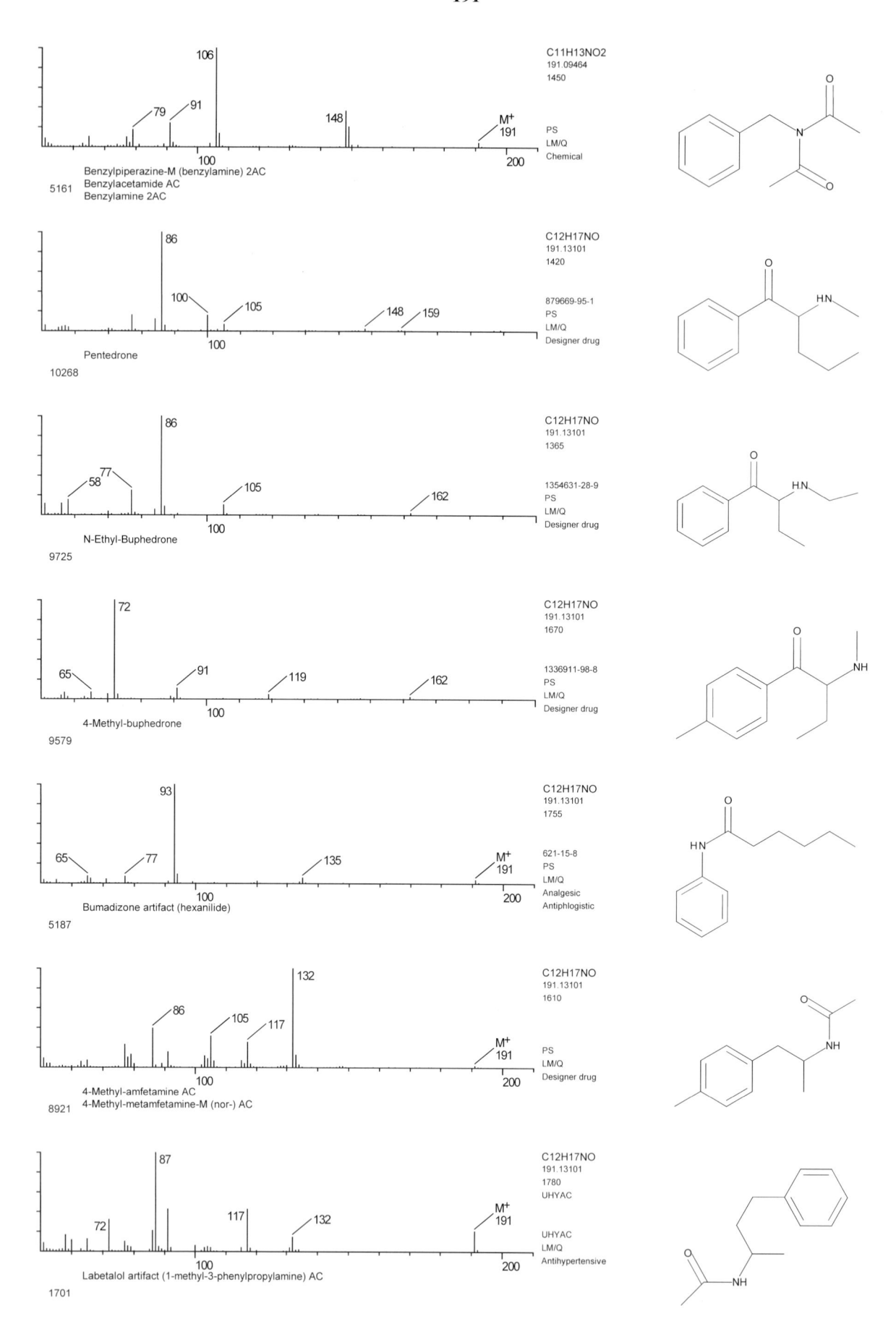

Benzylpiperazine-M (benzylamine) 2AC Benzylacetamide AC Benzylamine 2AC 5161	Peaks: 79, 91, 106, 148, M+ 191	C11H13NO2 191.09464 1450 PS LM/Q Chemical
Pentedrone 10268	Peaks: 86, 100, 105, 148, 159	C12H17NO 191.13101 1420 879669-95-1 PS LM/Q Designer drug
N-Ethyl-Buphedrone 9725	Peaks: 58, 77, 86, 105, 162	C12H17NO 191.13101 1365 1354631-28-9 PS LM/Q Designer drug
4-Methyl-buphedrone 9579	Peaks: 65, 72, 91, 119, 162	C12H17NO 191.13101 1670 1336911-98-8 PS LM/Q Designer drug
Bumadizone artifact (hexanilide) 5187	Peaks: 65, 77, 93, 135, M+ 191	C12H17NO 191.13101 1755 621-15-8 PS LM/Q Analgesic Antiphlogistic
4-Methyl-amfetamine AC 4-Methyl-metamfetamine-M (nor-) AC 8921	Peaks: 86, 105, 117, 132, M+ 191	C12H17NO 191.13101 1610 PS LM/Q Designer drug
Labetalol artifact (1-methyl-3-phenylpropylamine) AC 1701	Peaks: 72, 87, 117, 132, M+ 191	C12H17NO 191.13101 1780 UHYAC UHYAC LM/Q Antihypertensive

191

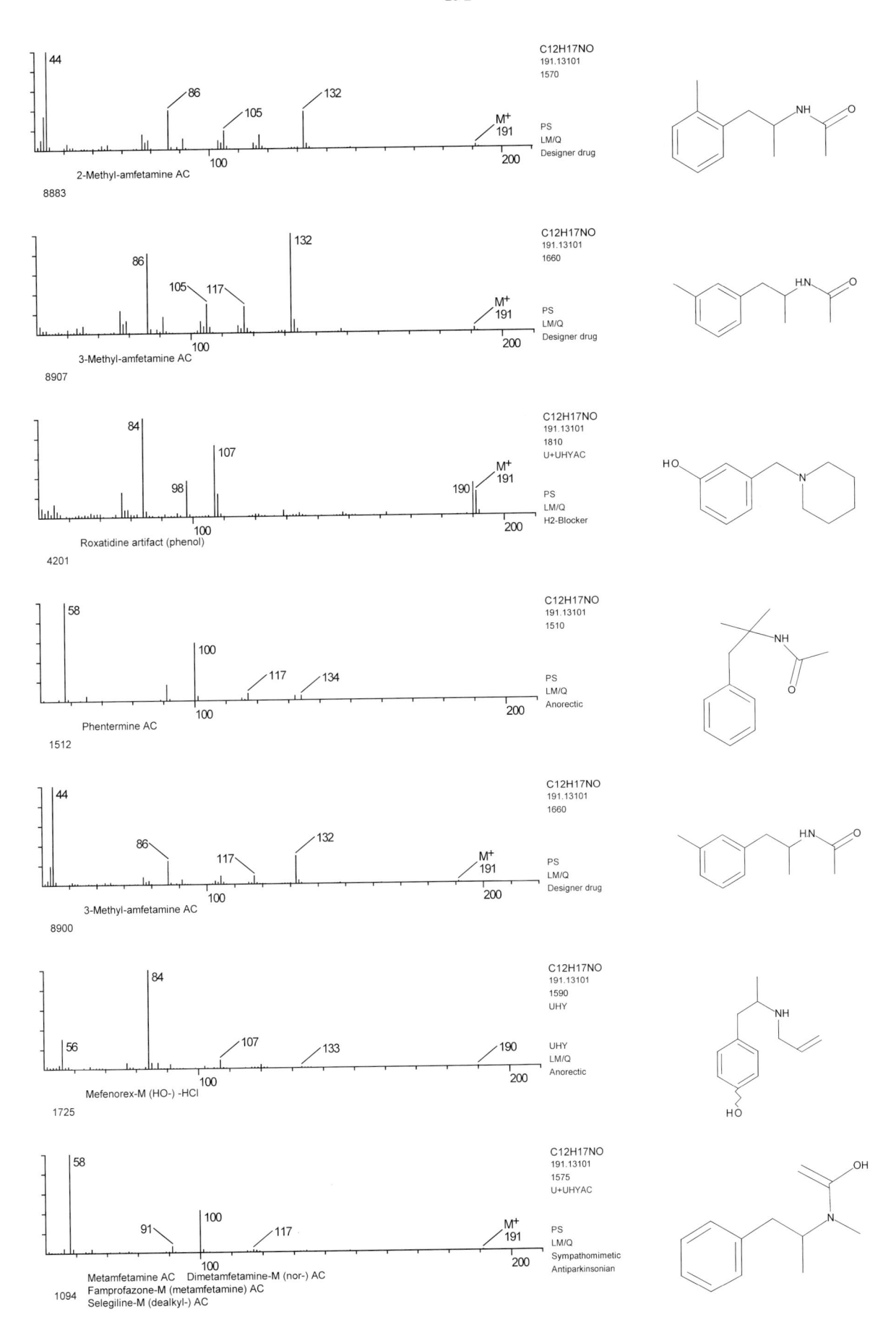

C12H17NO
191.13101
1570
PS
LM/Q
Designer drug

2-Methyl-amfetamine AC
8883

C12H17NO
191.13101
1660
PS
LM/Q
Designer drug

3-Methyl-amfetamine AC
8907

C12H17NO
191.13101
1810
U+UHYAC
PS
LM/Q
H2-Blocker

Roxatidine artifact (phenol)
4201

C12H17NO
191.13101
1510
PS
LM/Q
Anorectic

Phentermine AC
1512

C12H17NO
191.13101
1660
PS
LM/Q
Designer drug

3-Methyl-amfetamine AC
8900

C12H17NO
191.13101
1590
UHY
UHY
LM/Q
Anorectic

Mefenorex-M (HO-) -HCl
1725

C12H17NO
191.13101
1575
U+UHYAC
PS
LM/Q
Sympathomimetic
Antiparkinsonian

Metamfetamine AC   Dimetamfetamine-M (nor-) AC
Famprofazone-M (metamfetamine) AC
Selegiline-M (dealkyl-) AC
1094

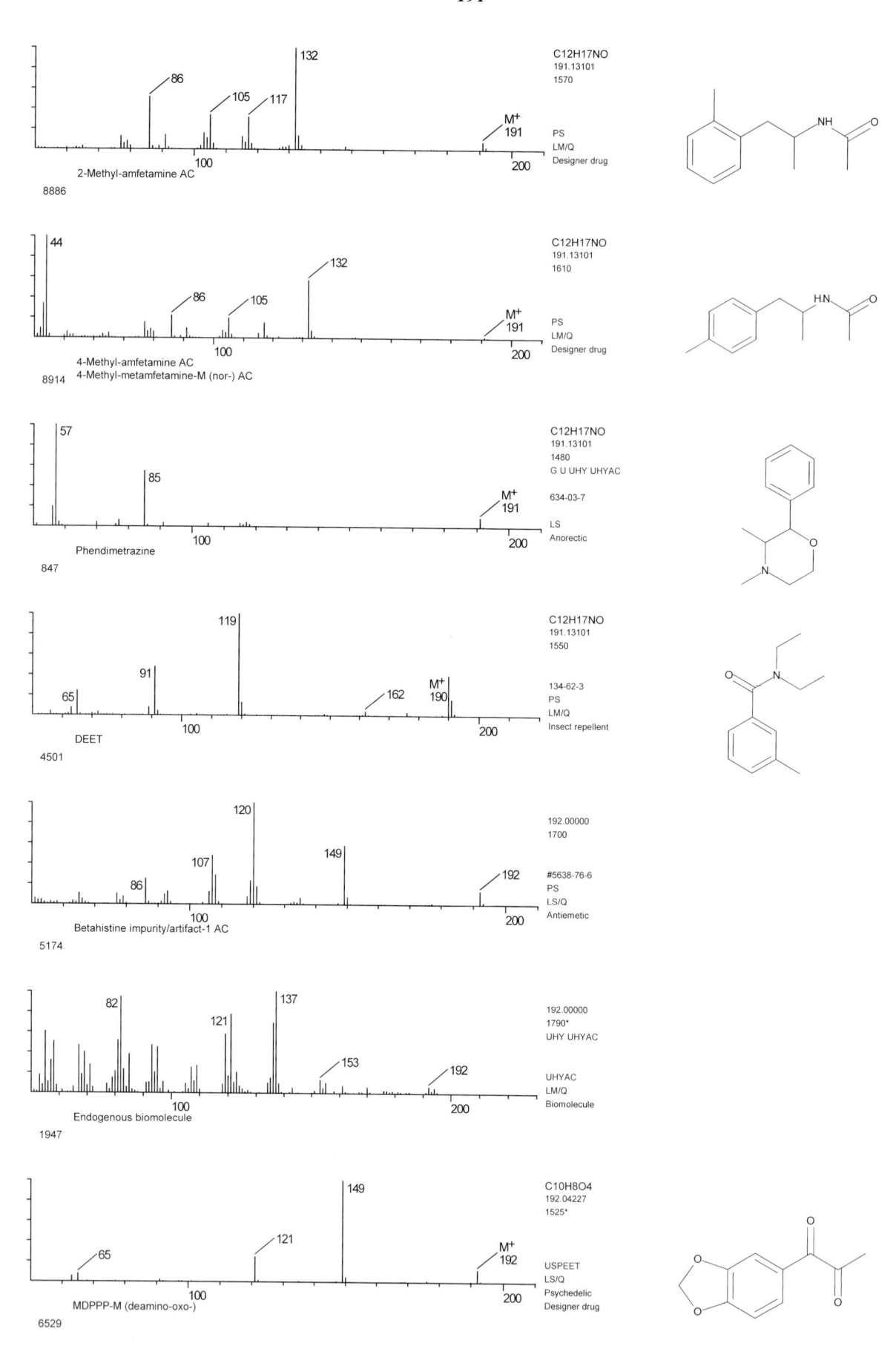

2-Methyl-amfetamine AC 8886	C12H17NO 191.13101 1570 PS LM/Q Designer drug
4-Methyl-amfetamine AC 4-Methyl-metamfetamine-M (nor-) AC 8914	C12H17NO 191.13101 1610 PS LM/Q Designer drug
Phendimetrazine 847	C12H17NO 191.13101 1480 G U UHY UHYAC 634-03-7 LS Anorectic
DEET 4501	C12H17NO 191.13101 1550 134-62-3 PS LM/Q Insect repellent
Betahistine impurity/artifact-1 AC 5174	192.00000 1700 #5638-76-6 PS LS/Q Antiemetic
Endogenous biomolecule 1947	192.00000 1790* UHY UHYAC UHYAC LM/Q Biomolecule
MDPPP-M (deamino-oxo-) 6529	C10H8O4 192.04227 1525* USPEET LS/Q Psychedelic Designer drug

192

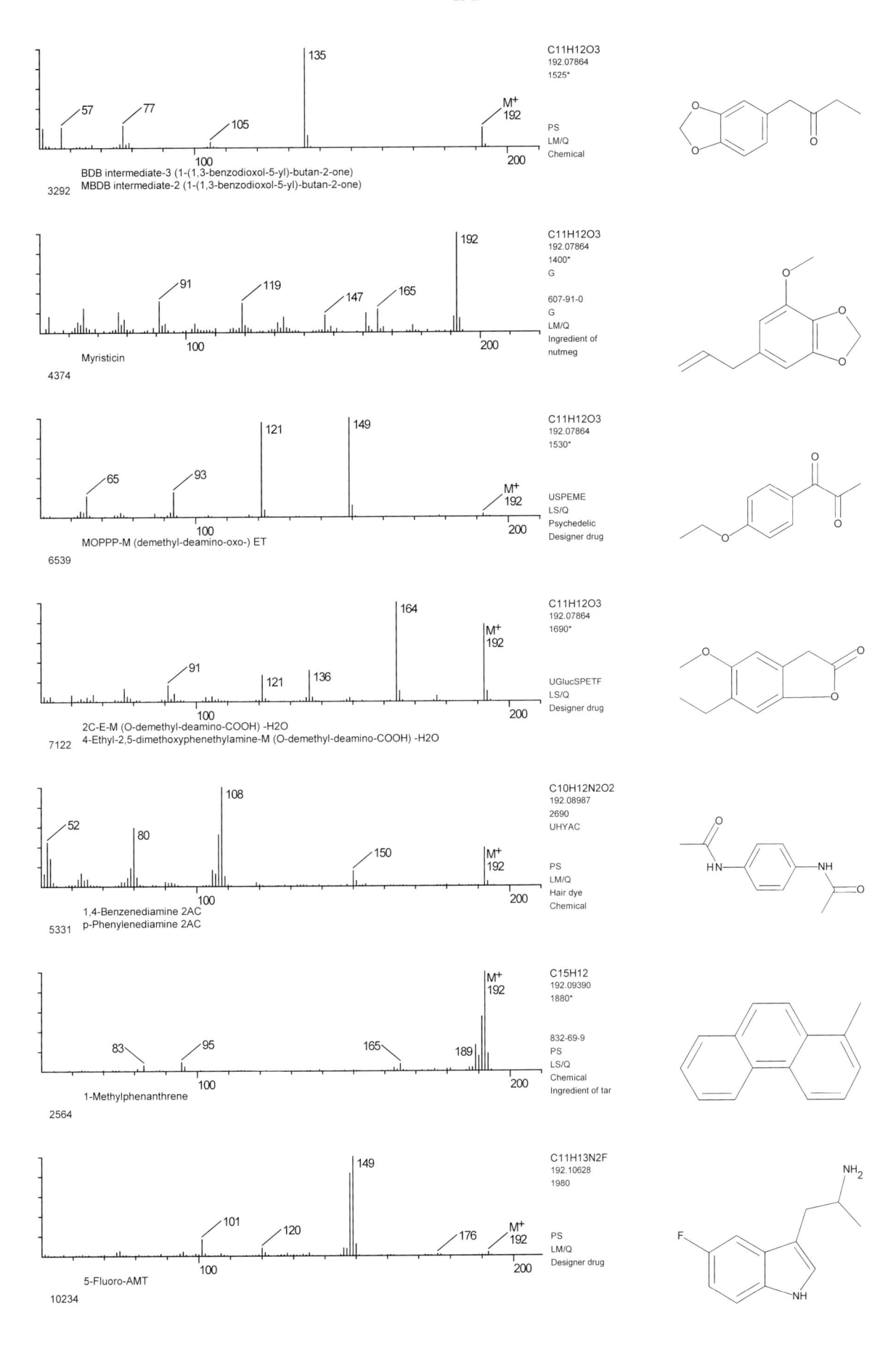

Peaks: 57, 77, 105, 135, M+ 192	C11H12O3 192.07864 1525* PS LM/Q Chemical	
BDB intermediate-3 (1-(1,3-benzodioxol-5-yl)-butan-2-one) MBDB intermediate-2 (1-(1,3-benzodioxol-5-yl)-butan-2-one) 3292		
Peaks: 91, 119, 147, 165, 192	C11H12O3 192.07864 1400* G 607-91-0 G LM/Q Ingredient of nutmeg	
Myristicin 4374		
Peaks: 65, 93, 121, 149, M+ 192	C11H12O3 192.07864 1530* USPEME LS/Q Psychedelic Designer drug	
MOPPP-M (demethyl-deamino-oxo-) ET 6539		
Peaks: 91, 121, 136, 164, M+ 192	C11H12O3 192.07864 1690* UGlucSPETF LS/Q Designer drug	
2C-E-M (O-demethyl-deamino-COOH) -H2O 4-Ethyl-2,5-dimethoxyphenethylamine-M (O-demethyl-deamino-COOH) -H2O 7122		
Peaks: 52, 80, 108, 150, M+ 192	C10H12N2O2 192.08987 2690 UHYAC PS LM/Q Hair dye Chemical	
1,4-Benzenediamine 2AC p-Phenylenediamine 2AC 5331		
Peaks: 83, 95, 165, 189, M+ 192	C15H12 192.09390 1880* 832-69-9 PS LS/Q Chemical Ingredient of tar	
1-Methylphenanthrene 2564		
Peaks: 101, 120, 149, 176, M+ 192	C11H13N2F 192.10628 1980 PS LM/Q Designer drug	
5-Fluoro-AMT 10234		

192

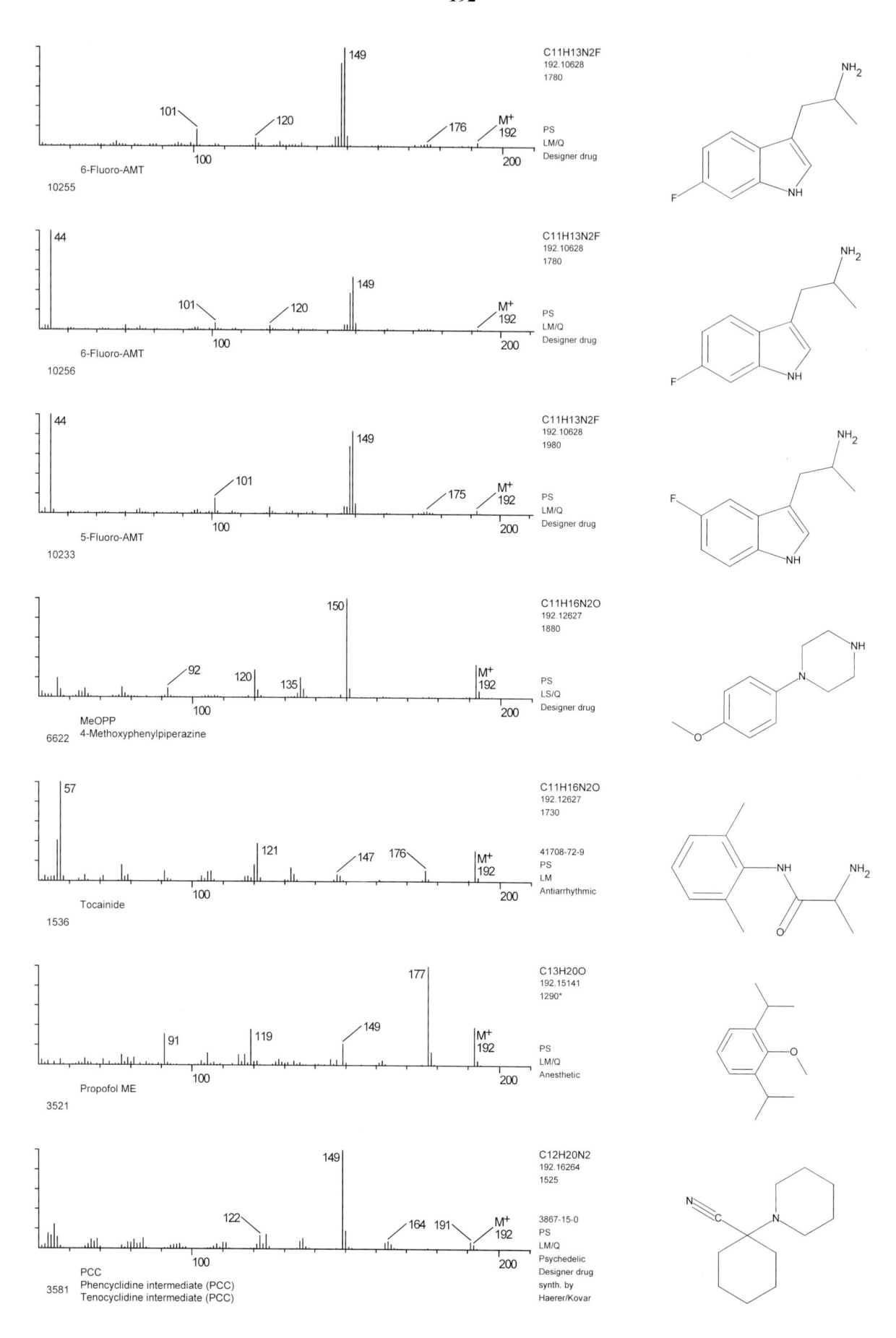

6-Fluoro-AMT 10255	149, 101, 120, 176, M+ 192	C11H13N2F 192.10628 1780 PS LM/Q Designer drug
6-Fluoro-AMT 10256	44, 101, 120, 149, M+ 192	C11H13N2F 192.10628 1780 PS LM/Q Designer drug
5-Fluoro-AMT 10233	44, 101, 149, 175, M+ 192	C11H13N2F 192.10628 1980 PS LM/Q Designer drug
MeOPP 4-Methoxyphenylpiperazine 6622	92, 120, 135, 150, M+ 192	C11H16N2O 192.12627 1880 PS LS/Q Designer drug
Tocainide 1536	57, 121, 147, 176, M+ 192	C11H16N2O 192.12627 1730 41708-72-9 PS LM Antiarrhythmic
Propofol ME 3521	91, 119, 149, 177, M+ 192	C13H20O 192.15141 1290* PS LM/Q Anesthetic
PCC Phencyclidine intermediate (PCC) Tenocyclidine intermediate (PCC) 3581	122, 149, 164, 191, M+ 192	C12H20N2 192.16264 1525 3867-15-0 PS LM/Q Psychedelic Designer drug synth. by Haerer/Kovar

193

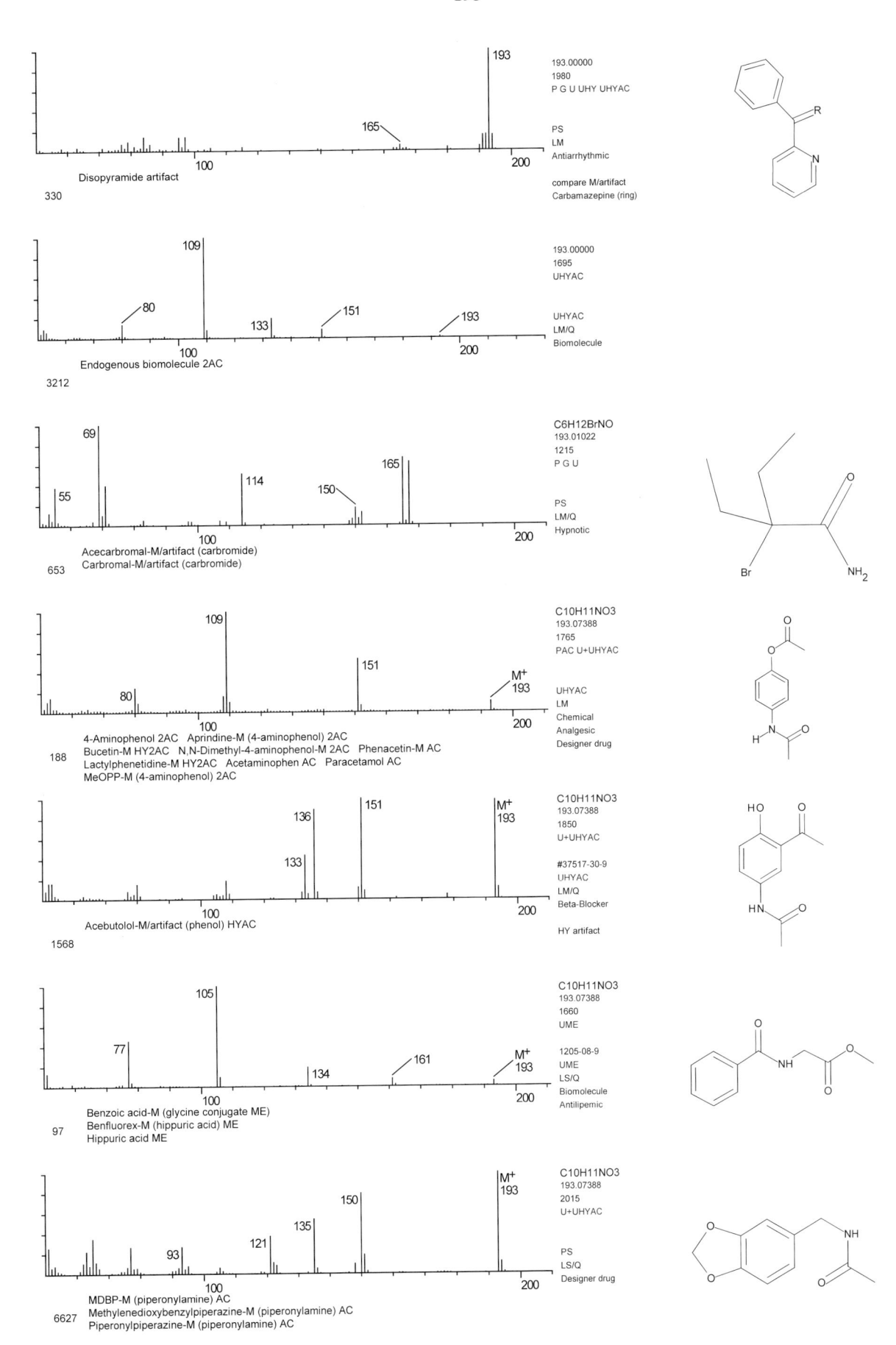

Disopyramide artifact	193.00000 1980 P G U UHY UHYAC PS LM Antiarrhythmic compare M/artifact Carbamazepine (ring)
Endogenous biomolecule 2AC	193.00000 1695 UHYAC UHYAC LM/Q Biomolecule
Acecarbromal-M/artifact (carbromide) Carbromal-M/artifact (carbromide)	C6H12BrNO 193.01022 1215 P G U PS LM/Q Hypnotic
4-Aminophenol 2AC Aprindine-M (4-aminophenol) 2AC Bucetin-M HY2AC N,N-Dimethyl-4-aminophenol-M 2AC Phenacetin-M AC Lactylphenetidine-M HY2AC Acetaminophen AC Paracetamol AC MeOPP-M (4-aminophenol) 2AC	C10H11NO3 193.07388 1765 PAC U+UHYAC UHYAC LM Chemical Analgesic Designer drug
Acebutolol-M/artifact (phenol) HYAC	C10H11NO3 193.07388 1850 U+UHYAC #37517-30-9 UHYAC LM/Q Beta-Blocker HY artifact
Benzoic acid-M (glycine conjugate ME) Benfluorex-M (hippuric acid) ME Hippuric acid ME	C10H11NO3 193.07388 1660 UME 1205-08-9 UME LS/Q Biomolecule Antilipemic
MDBP-M (piperonylamine) AC Methylenedioxybenzylpiperazine-M (piperonylamine) AC Piperonylpiperazine-M (piperonylamine) AC	C10H11NO3 193.07388 2015 U+UHYAC PS LS/Q Designer drug

193

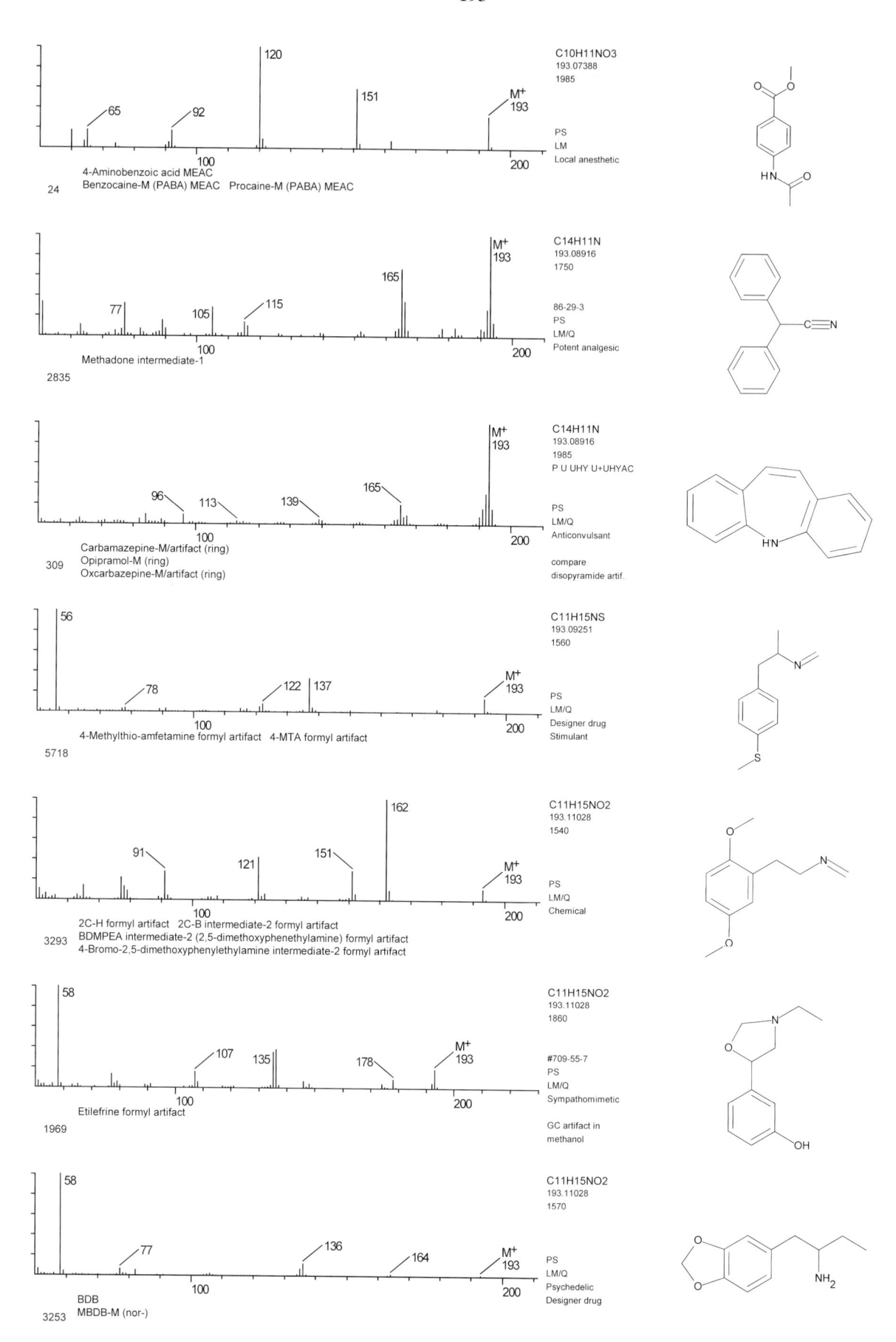

65, 92, 120, 151, M+ 193	C10H11NO3 193.07388 1985 PS LM Local anesthetic	
4-Aminobenzoic acid MEAC Benzocaine-M (PABA) MEAC Procaine-M (PABA) MEAC 24		
77, 105, 115, 165, M+ 193	C14H11N 193.08916 1750 86-29-3 PS LM/Q Potent analgesic	
Methadone intermediate-1 2835		
96, 113, 139, 165, M+ 193	C14H11N 193.08916 1985 P U UHY U+UHYAC PS LM/Q Anticonvulsant compare disopyramide artif.	
Carbamazepine-M/artifact (ring) Opipramol-M (ring) Oxcarbazepine-M/artifact (ring) 309		
56, 78, 122, 137, M+ 193	C11H15NS 193.09251 1560 PS LM/Q Designer drug Stimulant	
4-Methylthio-amfetamine formyl artifact 4-MTA formyl artifact 5718		
91, 121, 151, 162, M+ 193	C11H15NO2 193.11028 1540 PS LM/Q Chemical	
2C-H formyl artifact 2C-B intermediate-2 formyl artifact BDMPEA intermediate-2 (2,5-dimethoxyphenethylamine) formyl artifact 4-Bromo-2,5-dimethoxyphenylethylamine intermediate-2 formyl artifact 3293		
58, 107, 135, 178, M+ 193	C11H15NO2 193.11028 1860 #709-55-7 PS LM/Q Sympathomimetic GC artifact in methanol	
Etilefrine formyl artifact 1969		
58, 77, 136, 164, M+ 193	C11H15NO2 193.11028 1570 PS LM/Q Psychedelic Designer drug	
BDB MBDB-M (nor-) 3253		

193

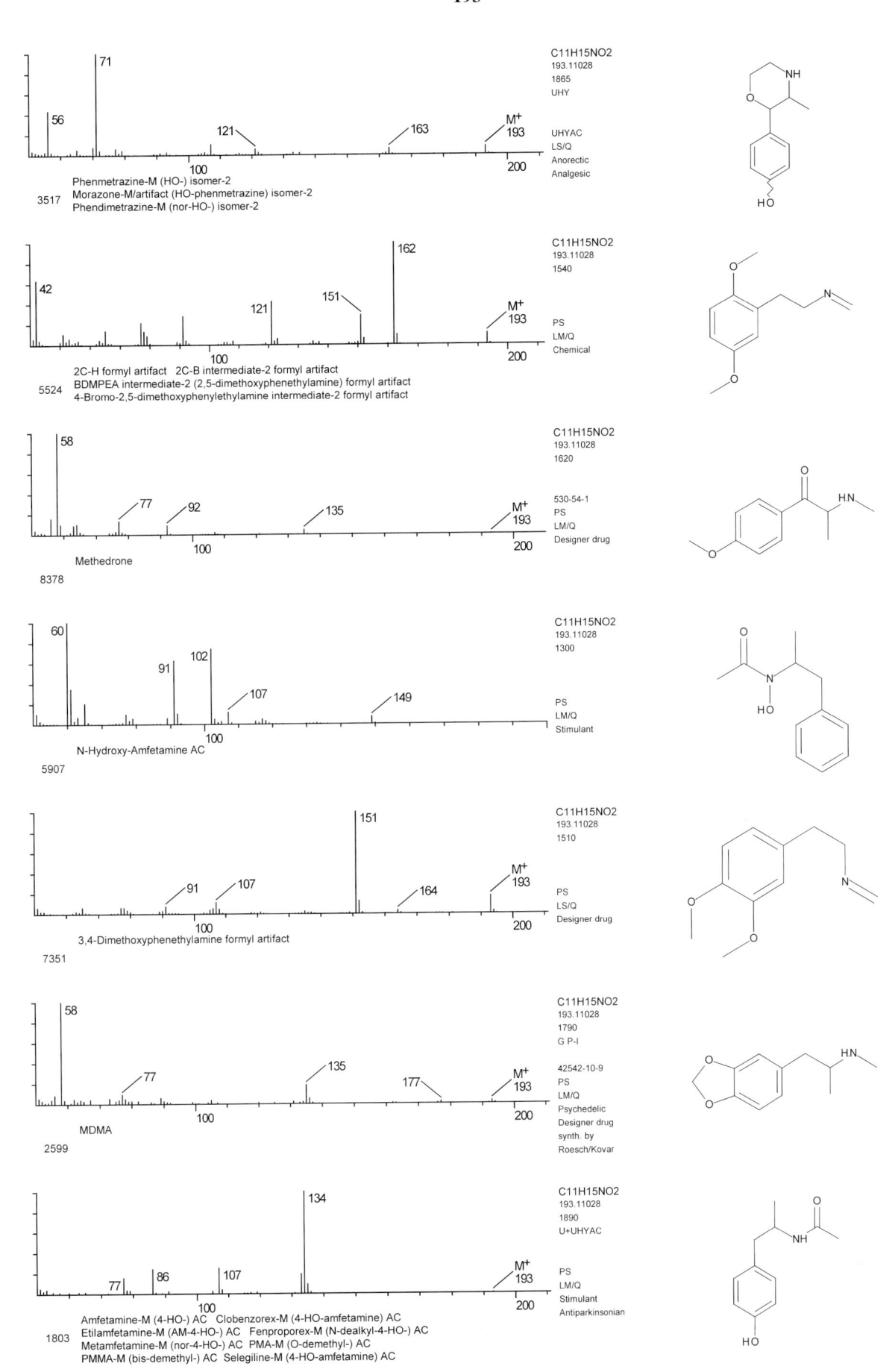

C11H15NO2
193.11028
1865
UHY
UHYAC
LS/Q
Anorectic
Analgesic

Phenmetrazine-M (HO-) isomer-2
Morazone-M/artifact (HO-phenmetrazine) isomer-2
Phendimetrazine-M (nor-HO-) isomer-2
3517

C11H15NO2
193.11028
1540
PS
LM/Q
Chemical

2C-H formyl artifact   2C-B intermediate-2 formyl artifact
BDMPEA intermediate-2 (2,5-dimethoxyphenethylamine) formyl artifact
4-Bromo-2,5-dimethoxyphenylethylamine intermediate-2 formyl artifact
5524

C11H15NO2
193.11028
1620
530-54-1
PS
LM/Q
Designer drug

Methedrone
8378

C11H15NO2
193.11028
1300
PS
LM/Q
Stimulant

N-Hydroxy-Amfetamine AC
5907

C11H15NO2
193.11028
1510
PS
LS/Q
Designer drug

3,4-Dimethoxyphenethylamine formyl artifact
7351

C11H15NO2
193.11028
1790
G P-I
42542-10-9
PS
LM/Q
Psychedelic
Designer drug
synth. by
Roesch/Kovar

MDMA
2599

C11H15NO2
193.11028
1890
U+UHYAC
PS
LM/Q
Stimulant
Antiparkinsonian

Amfetamine-M (4-HO-) AC   Clobenzorex-M (4-HO-amfetamine) AC
Etilamfetamine-M (AM-4-HO-) AC   Fenproporex-M (N-dealkyl-4-HO-) AC
Metamfetamine-M (nor-4-HO-) AC   PMA-M (O-demethyl-) AC
PMMA-M (bis-demethyl-) AC   Selegiline-M (4-HO-amfetamine) AC
1803

193

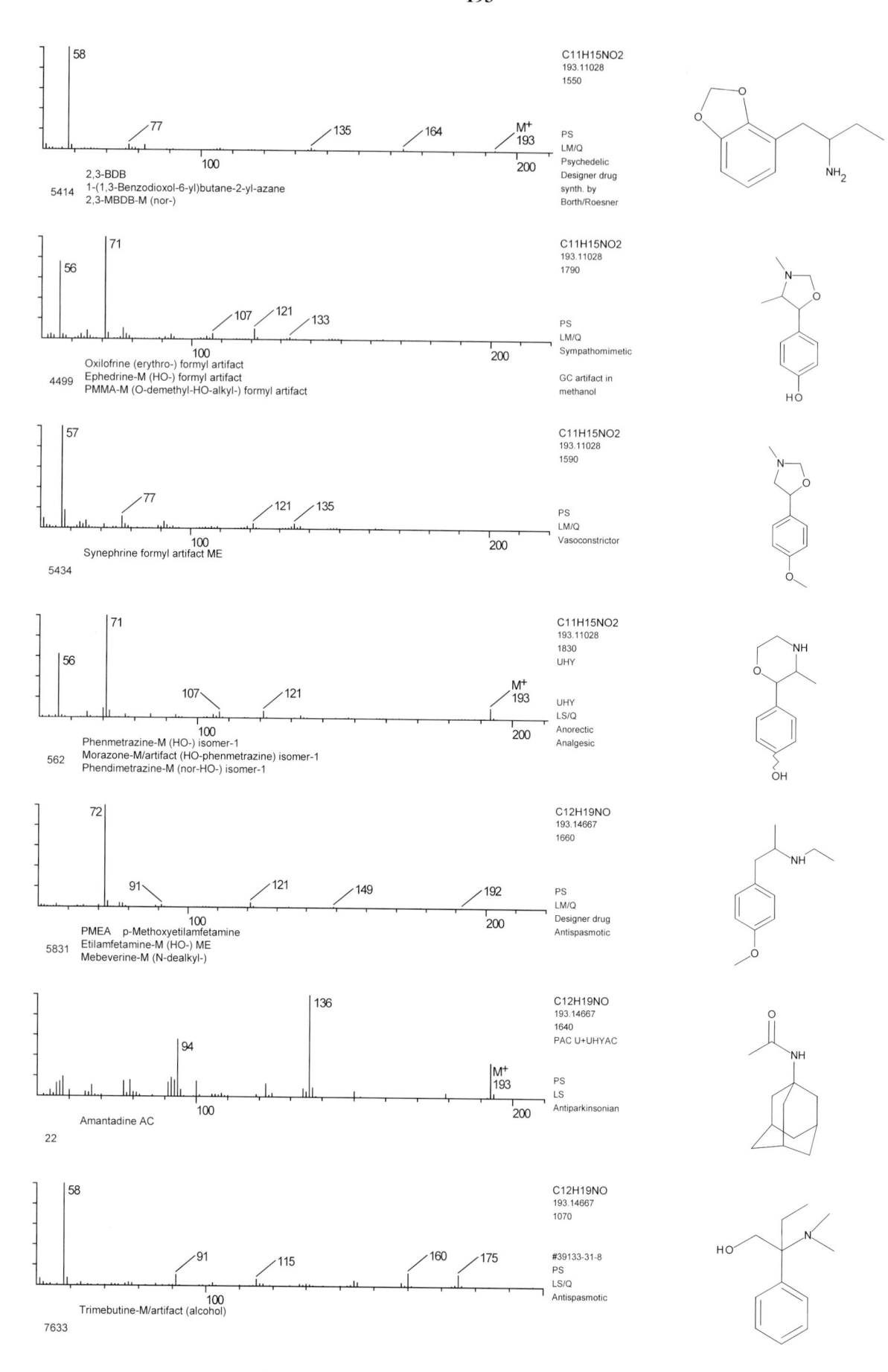

5414	2,3-BDB 1-(1,3-Benzodioxol-6-yl)butane-2-yl-azane 2,3-MBDB-M (nor-)	C11H15NO2 193.11028 1550 PS LM/Q Psychedelic Designer drug synth. by Borth/Roesner
	Peaks: 58, 77, 135, 164, M+ 193	

4499	Oxilofrine (erythro-) formyl artifact Ephedrine-M (HO-) formyl artifact PMMA-M (O-demethyl-HO-alkyl-) formyl artifact	C11H15NO2 193.11028 1790 PS LM/Q Sympathomimetic GC artifact in methanol
	Peaks: 56, 71, 107, 121, 133	

5434	Synephrine formyl artifact ME	C11H15NO2 193.11028 1590 PS LM/Q Vasoconstrictor
	Peaks: 57, 77, 121, 135	

562	Phenmetrazine-M (HO-) isomer-1 Morazone-M/artifact (HO-phenmetrazine) isomer-1 Phendimetrazine-M (nor-HO-) isomer-1	C11H15NO2 193.11028 1830 UHY UHY LS/Q Anorectic Analgesic
	Peaks: 56, 71, 107, 121, M+ 193	

5831	PMEA p-Methoxyetilamfetamine Etilamfetamine-M (HO-) ME Mebeverine-M (N-dealkyl-)	C12H19NO 193.14667 1660 PS LM/Q Designer drug Antispasmotic
	Peaks: 72, 91, 121, 149, 192	

22	Amantadine AC	C12H19NO 193.14667 1640 PAC U+UHYAC PS LS Antiparkinsonian
	Peaks: 94, 136, M+ 193	

7633	Trimebutine-M/artifact (alcohol)	C12H19NO 193.14667 1070 #39133-31-8 PS LS/Q Antispasmotic
	Peaks: 58, 91, 115, 160, 175	

193

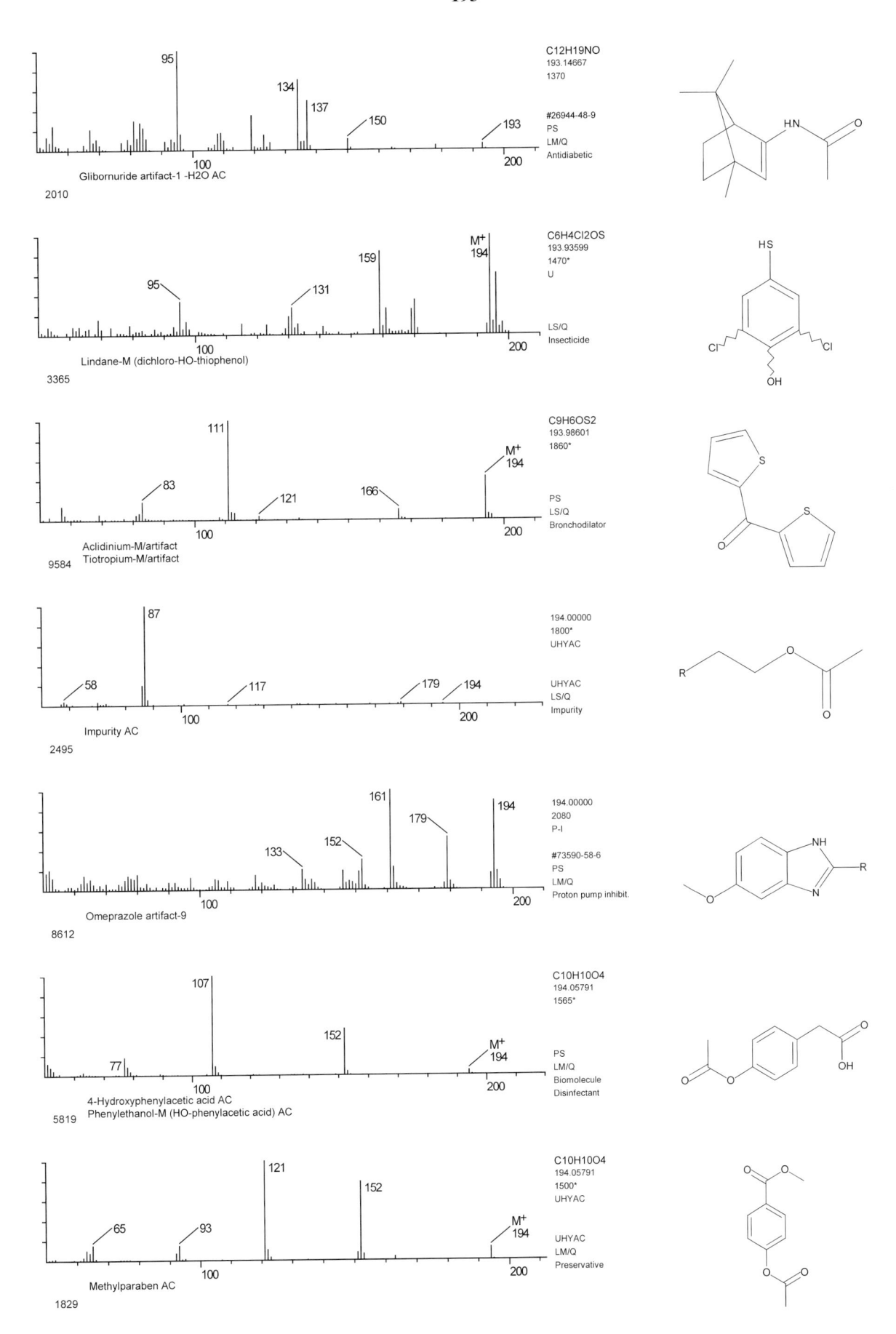

C12H19NO
193.14667
1370

#26944-48-9
PS
LM/Q
Antidiabetic

Glibornuride artifact-1 -H2O AC
2010

C6H4Cl2OS
193.93599
1470*
U

LS/Q
Insecticide

Lindane-M (dichloro-HO-thiophenol)
3365

C9H6OS2
193.98601
1860*

PS
LS/Q
Bronchodilator

Aclidinium-M/artifact
Tiotropium-M/artifact
9584

194.00000
1800*
UHYAC

UHYAC
LS/Q
Impurity

Impurity AC
2495

194.00000
2080
P-I

#73590-58-6
PS
LM/Q
Proton pump inhibit.

Omeprazole artifact-9
8612

C10H10O4
194.05791
1565*

PS
LM/Q
Biomolecule
Disinfectant

4-Hydroxyphenylacetic acid AC
Phenylethanol-M (HO-phenylacetic acid) AC
5819

C10H10O4
194.05791
1500*
UHYAC

UHYAC
LM/Q
Preservative

Methylparaben AC
1829

194

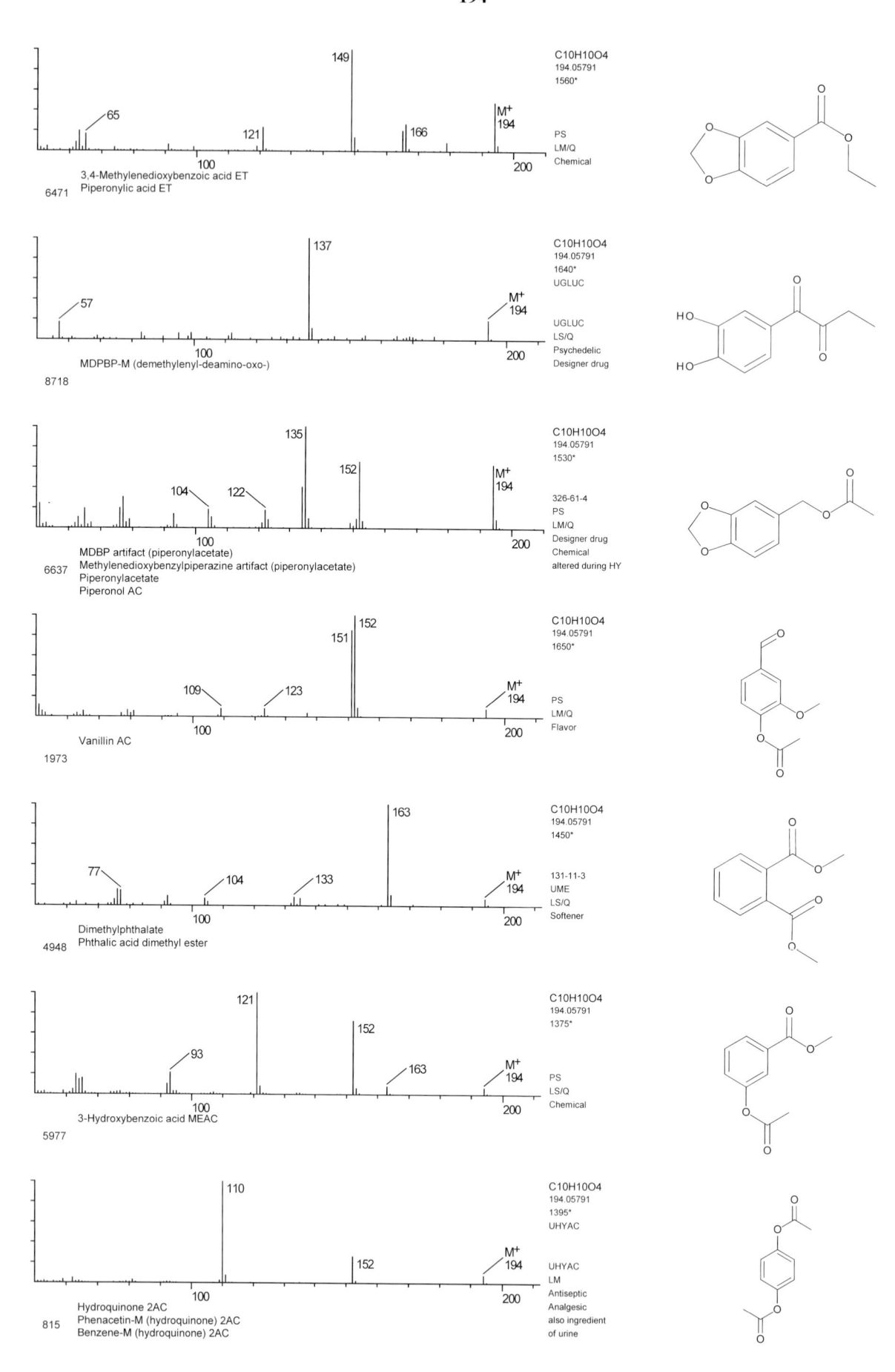

Spectrum	Name	Formula/Info
6471	3,4-Methylenedioxybenzoic acid ET / Piperonylic acid ET	C10H10O4, 194.05791, 1560*, PS, LM/Q, Chemical. Peaks: 65, 121, 149, 166, 194 M+
8718	MDPBP-M (demethylenyl-deamino-oxo-)	C10H10O4, 194.05791, 1640*, UGLUC, UGLUC, LS/Q, Psychedelic Designer drug. Peaks: 57, 137, 194 M+
6637	MDBP artifact (piperonylacetate) / Methylenedioxybenzylpiperazine artifact (piperonylacetate) / Piperonylacetate / Piperonol AC	C10H10O4, 194.05791, 1530*, 326-61-4, PS, LM/Q, Designer drug Chemical altered during HY. Peaks: 104, 122, 135, 152, 194 M+
1973	Vanillin AC	C10H10O4, 194.05791, 1650*, PS, LM/Q, Flavor. Peaks: 109, 123, 151, 152, 194 M+
4948	Dimethylphthalate / Phthalic acid dimethyl ester	C10H10O4, 194.05791, 1450*, 131-11-3, UME, LS/Q, Softener. Peaks: 77, 104, 133, 163, 194 M+
5977	3-Hydroxybenzoic acid MEAC	C10H10O4, 194.05791, 1375*, PS, LS/Q, Chemical. Peaks: 93, 121, 152, 163, 194 M+
815	Hydroquinone 2AC / Phenacetin-M (hydroquinone) 2AC / Benzene-M (hydroquinone) 2AC	C10H10O4, 194.05791, 1395*, UHYAC, UHYAC, LM, Antiseptic Analgesic also ingredient of urine. Peaks: 110, 152, 194 M+

194

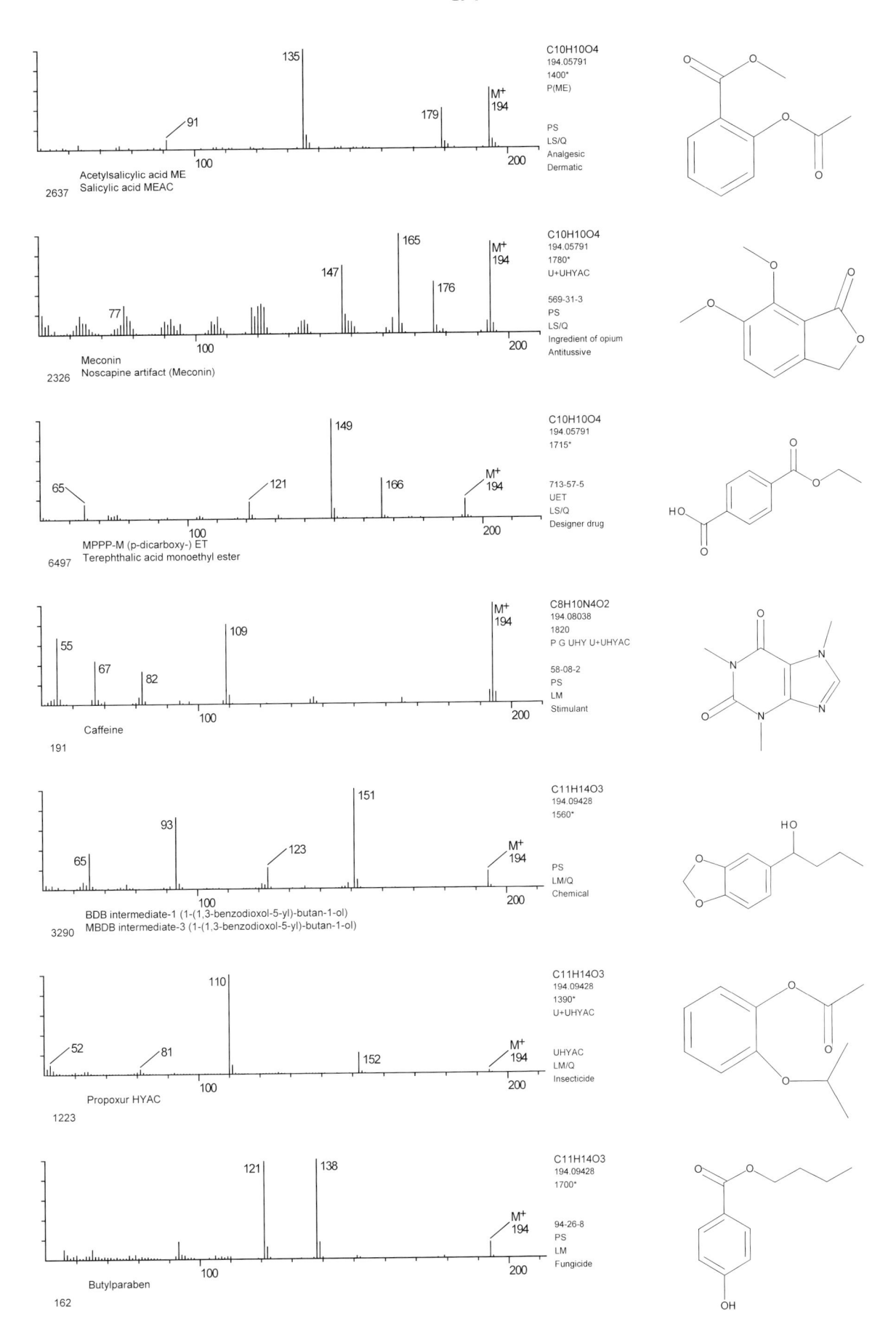

2637 Acetylsalicylic acid ME / Salicylic acid MEAC
C10H10O4
194.05791
1400*
P(ME)
M+ 194
Peaks: 91, 135, 179
PS
LS/Q
Analgesic
Dermatic

2326 Meconin / Noscapine artifact (Meconin)
C10H10O4
194.05791
1780*
U+UHYAC
569-31-3
PS
LS/Q
Ingredient of opium
Antitussive
Peaks: 77, 147, 165, 176, M+ 194

6497 MPPP-M (p-dicarboxy-) ET / Terephthalic acid monoethyl ester
C10H10O4
194.05791
1715*
713-57-5
UET
LS/Q
Designer drug
Peaks: 65, 121, 149, 166, M+ 194

191 Caffeine
C8H10N4O2
194.08038
1820
P G UHY U+UHYAC
58-08-2
PS
LM
Stimulant
Peaks: 55, 67, 82, 109, M+ 194

3290 BDB intermediate-1 (1-(1,3-benzodioxol-5-yl)-butan-1-ol) / MBDB intermediate-3 (1-(1,3-benzodioxol-5-yl)-butan-1-ol)
C11H14O3
194.09428
1560*
PS
LM/Q
Chemical
Peaks: 65, 93, 123, 151, M+ 194

1223 Propoxur HYAC
C11H14O3
194.09428
1390*
U+UHYAC
UHYAC
LM/Q
Insecticide
Peaks: 52, 81, 110, 152, M+ 194

162 Butylparaben
C11H14O3
194.09428
1700*
94-26-8
PS
LM
Fungicide
Peaks: 121, 138, M+ 194

195

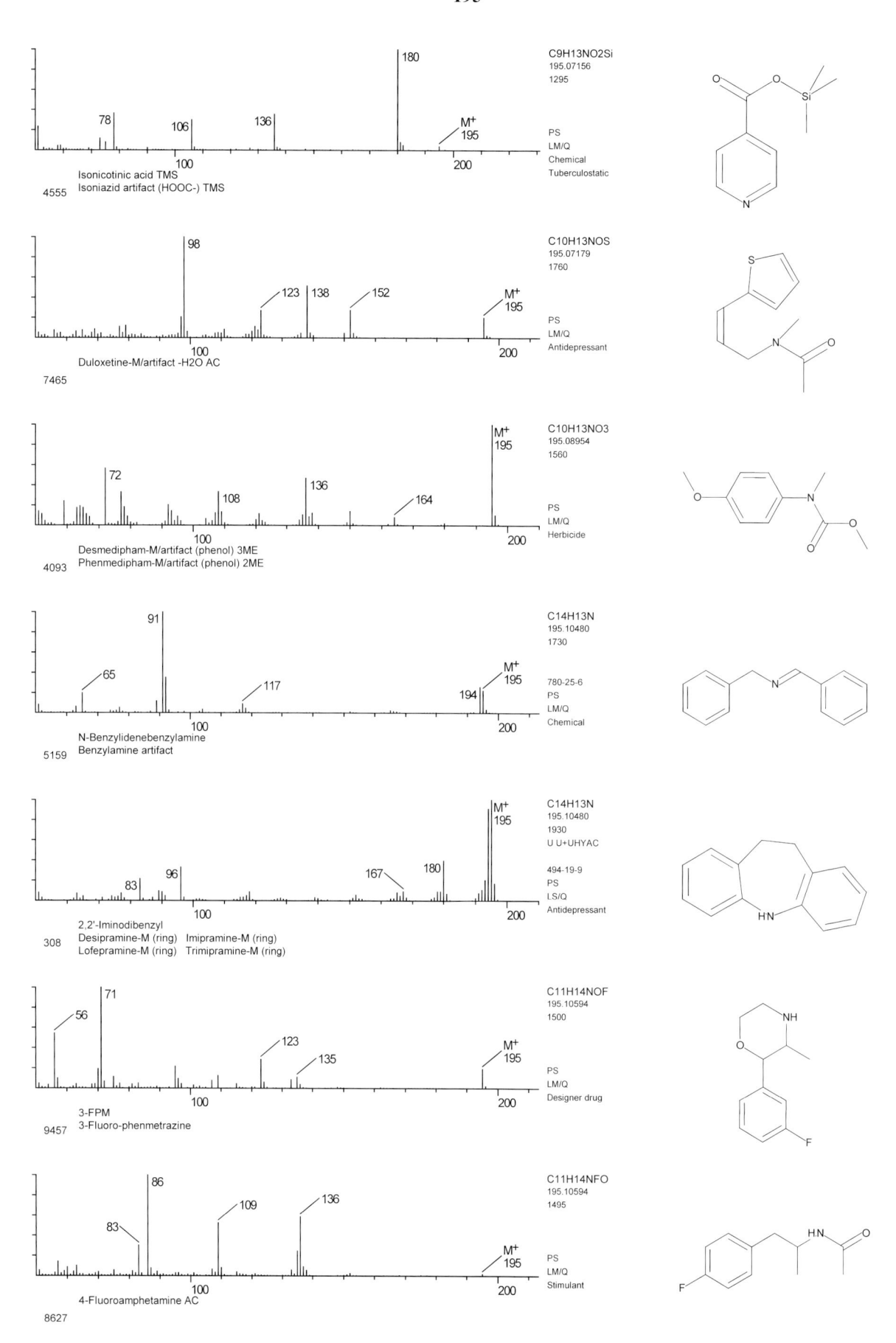

Isonicotinic acid TMS Isoniazid artifact (HOOC-) TMS 4555	C9H13NO2Si 195.07156 1295 PS LM/Q Chemical Tuberculostatic
Duloxetine-M/artifact -H2O AC 7465	C10H13NOS 195.07179 1760 PS LM/Q Antidepressant
Desmedipham-M/artifact (phenol) 3ME Phenmedipham-M/artifact (phenol) 2ME 4093	C10H13NO3 195.08954 1560 PS LM/Q Herbicide
N-Benzylidenebenzylamine Benzylamine artifact 5159	C14H13N 195.10480 1730 780-25-6 PS LM/Q Chemical
2,2'-Iminodibenzyl Desipramine-M (ring) Imipramine-M (ring) Lofepramine-M (ring) Trimipramine-M (ring) 308	C14H13N 195.10480 1930 U U+UHYAC 494-19-9 PS LS/Q Antidepressant
3-FPM 3-Fluoro-phenmetrazine 9457	C11H14NOF 195.10594 1500 PS LM/Q Designer drug
4-Fluoroamphetamine AC 8627	C11H14NFO 195.10594 1495 PS LM/Q Stimulant

195

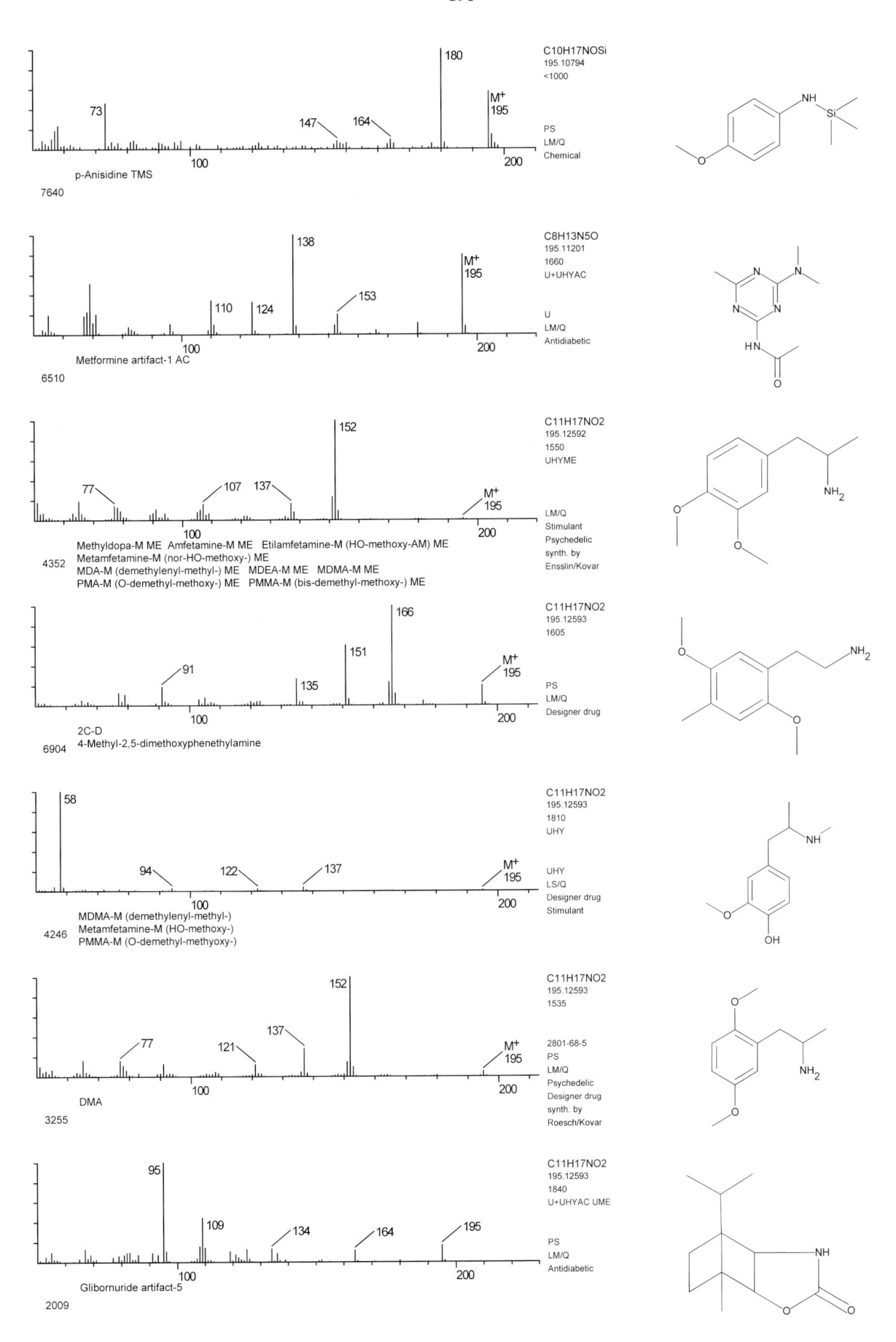

C10H17NOSi
195.10794
<1000

PS
LM/Q
Chemical

p-Anisidine TMS
7640

C8H13N5O
195.11201
1660
U+UHYAC

U
LM/Q
Antidiabetic

Metformine artifact-1 AC
6510

C11H17NO2
195.12592
1550
UHYME

LM/Q
Stimulant
Psychedelic
synth. by
Ensslin/Kovar

Methyldopa-M ME   Amfetamine-M ME   Etilamfetamine-M (HO-methoxy-AM) ME
Metamfetamine-M (nor-HO-methoxy-) ME
MDA-M (demethylenyl-methyl-) ME   MDEA-M ME   MDMA-M ME
PMA-M (O-demethyl-methoxy-) ME   PMMA-M (bis-demethyl-methoxy-) ME
4352

C11H17NO2
195.12593
1605

PS
LM/Q
Designer drug

2C-D
6904 4-Methyl-2,5-dimethoxyphenethylamine

C11H17NO2
195.12593
1810
UHY

UHY
LS/Q
Designer drug
Stimulant

MDMA-M (demethylenyl-methyl-)
Metamfetamine-M (HO-methoxy-)
4246 PMMA-M (O-demethyl-methyoxy-)

C11H17NO2
195.12593
1535

2801-68-5
PS
LM/Q
Psychedelic
Designer drug
synth. by
Roesch/Kovar

DMA
3255

C11H17NO2
195.12593
1840
U+UHYAC UME

PS
LM/Q
Antidiabetic

Glibornuride artifact-5
2009

387

195

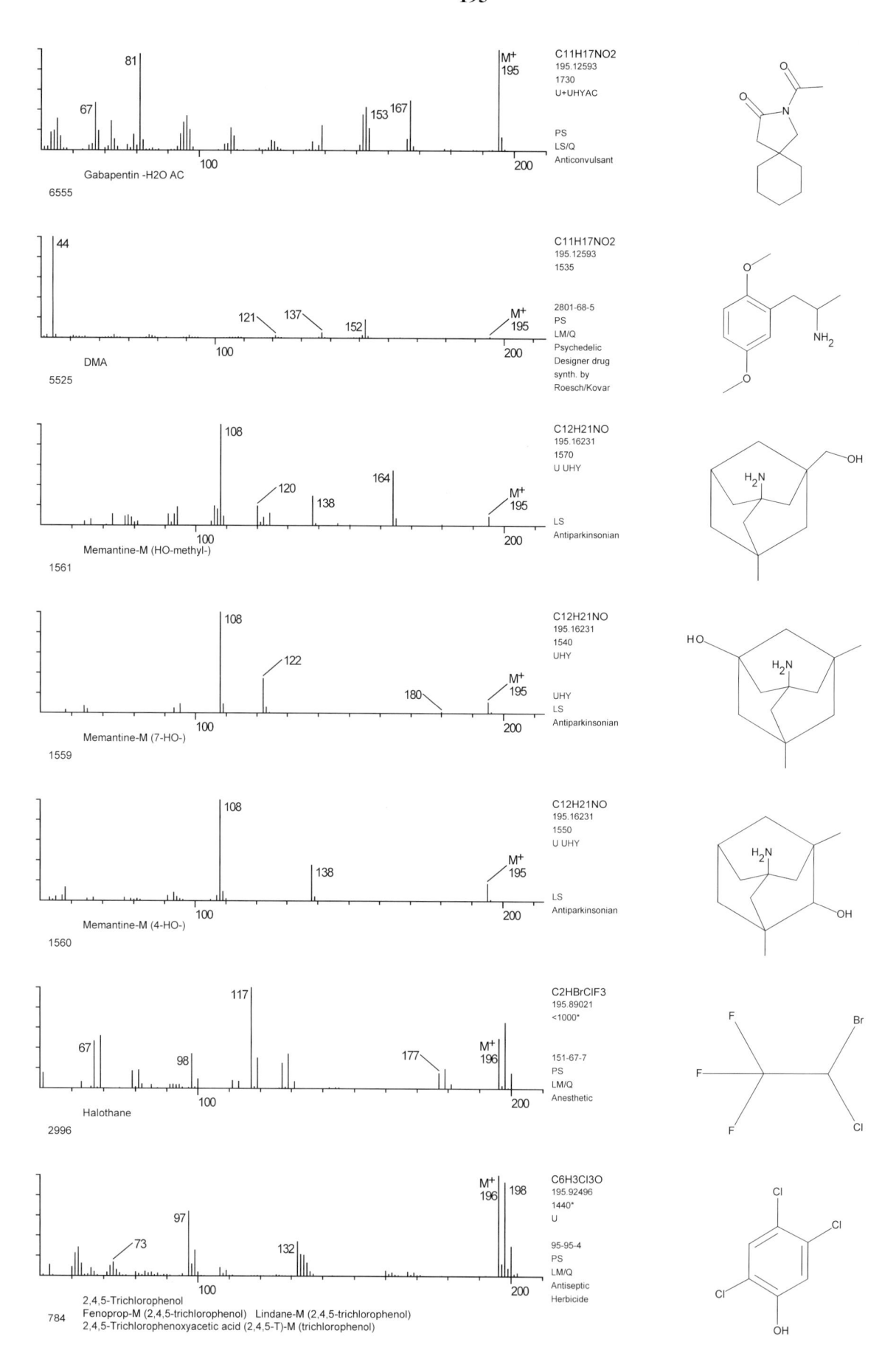

Gabapentin -H2O AC — C11H17NO2, 195.12593, 1730, U+UHYAC, PS, LS/Q, Anticonvulsant
Peaks: 67, 81, 153, 167, M+ 195
6555

DMA — C11H17NO2, 195.12593, 1535, 2801-68-5, PS, LM/Q, Psychedelic Designer drug synth. by Roesch/Kovar
Peaks: 44, 121, 137, 152, M+ 195
5525

Memantine-M (HO-methyl-) — C12H21NO, 195.16231, 1570, U UHY, LS, Antiparkinsonian
Peaks: 108, 120, 138, 164, M+ 195
1561

Memantine-M (7-HO-) — C12H21NO, 195.16231, 1540, UHY, UHY, LS, Antiparkinsonian
Peaks: 108, 122, 180, M+ 195
1559

Memantine-M (4-HO-) — C12H21NO, 195.16231, 1550, U UHY, LS, Antiparkinsonian
Peaks: 108, 138, M+ 195
1560

Halothane — C2HBrClF3, 195.89021, <1000*, 151-67-7, PS, LM/Q, Anesthetic
Peaks: 67, 98, 117, 177, M+ 196
2996

2,4,5-Trichlorophenol
Fenoprop-M (2,4,5-trichlorophenol)   Lindane-M (2,4,5-trichlorophenol)
2,4,5-Trichlorophenoxyacetic acid (2,4,5-T)-M (trichlorophenol)
C6H3Cl3O, 195.92496, 1440*, U, 95-95-4, PS, LM/Q, Antiseptic, Herbicide
Peaks: 73, 97, 132, M+ 196, 198
784

196

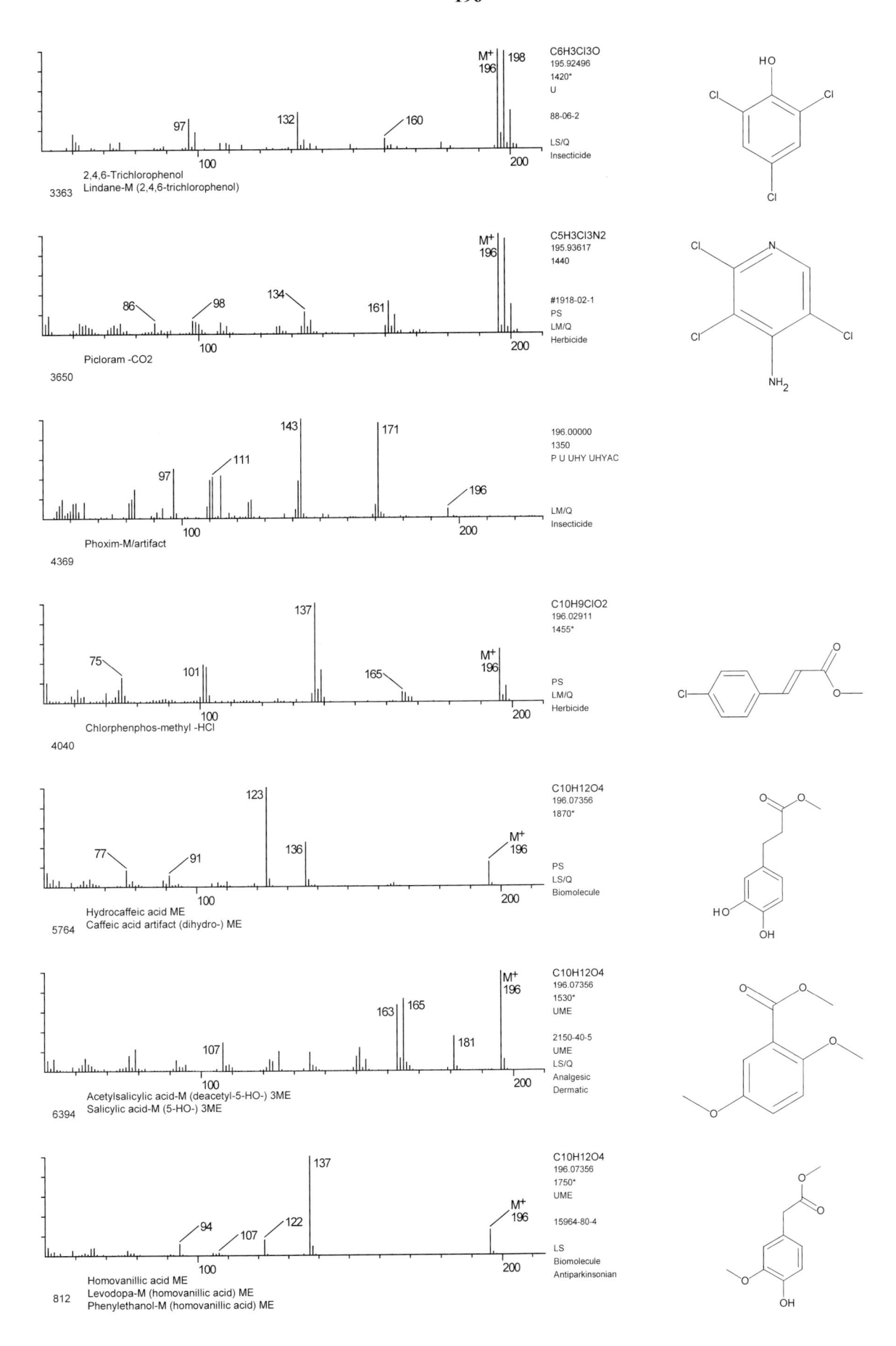

3363 — 2,4,6-Trichlorophenol / Lindane-M (2,4,6-trichlorophenol)
C6H3Cl3O, 195.92496, 1420*, U, 88-06-2, LS/Q, Insecticide
Peaks: 97, 132, 160, M+ 196, 198

3650 — Picloram -CO2
C5H3Cl3N2, 195.93617, 1440, #1918-02-1, PS, LM/Q, Herbicide
Peaks: 86, 98, 134, 161, M+ 196

4369 — Phoxim-M/artifact
196.00000, 1350, P U UHY UHYAC, LM/Q, Insecticide
Peaks: 97, 111, 143, 171, 196

4040 — Chlorphenphos-methyl -HCl
C10H9ClO2, 196.02911, 1455*, PS, LM/Q, Herbicide
Peaks: 75, 101, 137, 165, M+ 196

5764 — Hydrocaffeic acid ME / Caffeic acid artifact (dihydro-) ME
C10H12O4, 196.07356, 1870*, PS, LS/Q, Biomolecule
Peaks: 77, 91, 123, 136, M+ 196

6394 — Acetylsalicylic acid-M (deacetyl-5-HO-) 3ME / Salicylic acid-M (5-HO-) 3ME
C10H12O4, 196.07356, 1530*, UME, 2150-40-5, UME, LS/Q, Analgesic, Dermatic
Peaks: 107, 163, 165, 181, M+ 196

812 — Homovanillic acid ME / Levodopa-M (homovanillic acid) ME / Phenylethanol-M (homovanillic acid) ME
C10H12O4, 196.07356, 1750*, UME, 15964-80-4, LS, Biomolecule, Antiparkinsonian
Peaks: 94, 107, 122, 137, M+ 196

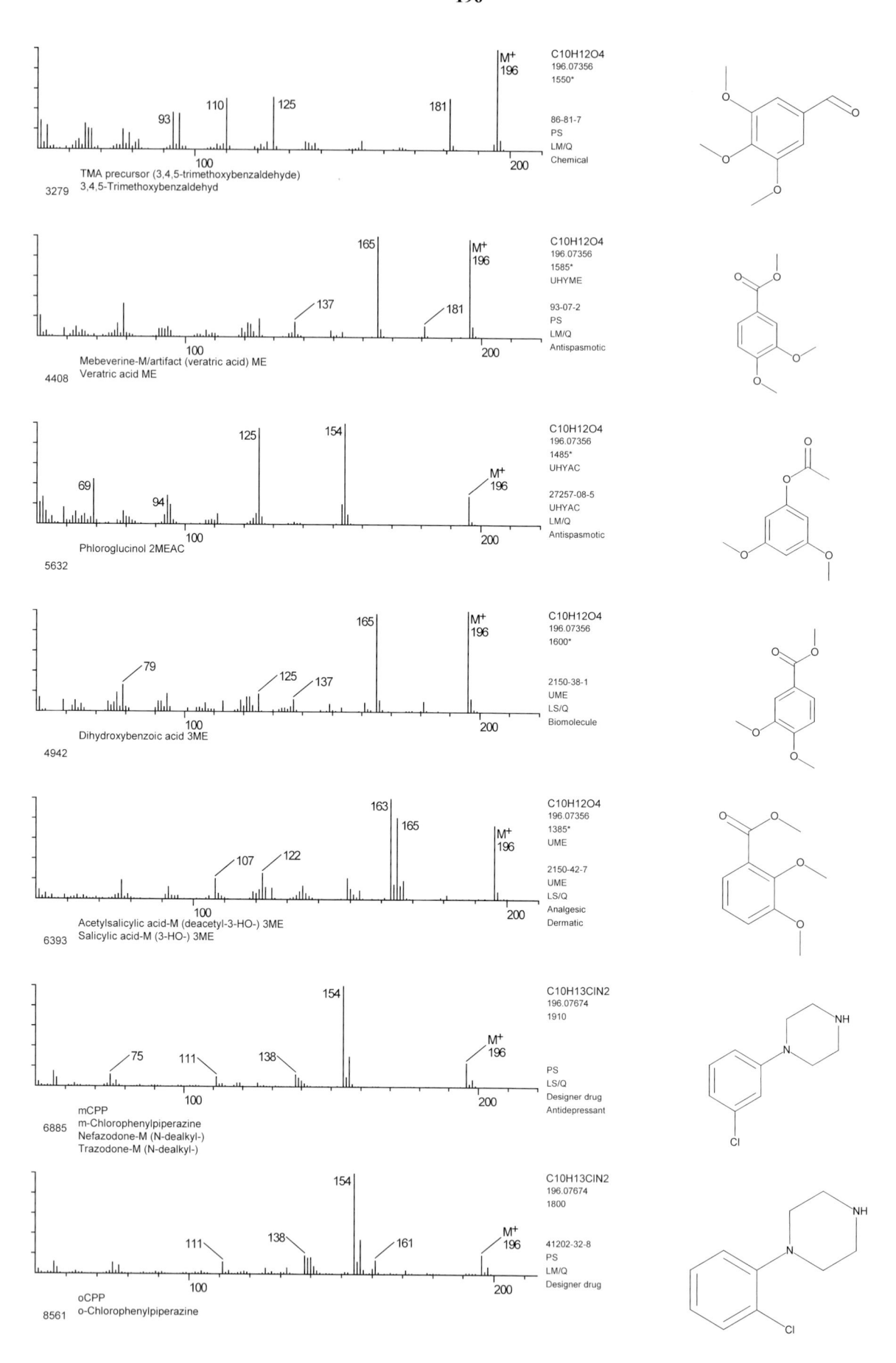

3279	TMA precursor (3,4,5-trimethoxybenzaldehyde) 3,4,5-Trimethoxybenzaldehyd	C10H12O4 196.07356 1550* 86-81-7 PS LM/Q Chemical
4408	Mebeverine-M/artifact (veratric acid) ME Veratric acid ME	C10H12O4 196.07356 1585* UHYME 93-07-2 PS LM/Q Antispasmotic
5632	Phloroglucinol 2MEAC	C10H12O4 196.07356 1485* UHYAC 27257-08-5 UHYAC LM/Q Antispasmotic
4942	Dihydroxybenzoic acid 3ME	C10H12O4 196.07356 1600* 2150-38-1 UME LS/Q Biomolecule
6393	Acetylsalicylic acid-M (deacetyl-3-HO-) 3ME Salicylic acid-M (3-HO-) 3ME	C10H12O4 196.07356 1385* UME 2150-42-7 UME LS/Q Analgesic Dermatic
6885	mCPP m-Chlorophenylpiperazine Nefazodone-M (N-dealkyl-) Trazodone-M (N-dealkyl-)	C10H13ClN2 196.07674 1910 PS LS/Q Designer drug Antidepressant
8561	oCPP o-Chlorophenylpiperazine	C10H13ClN2 196.07674 1800 41202-32-8 PS LM/Q Designer drug

196

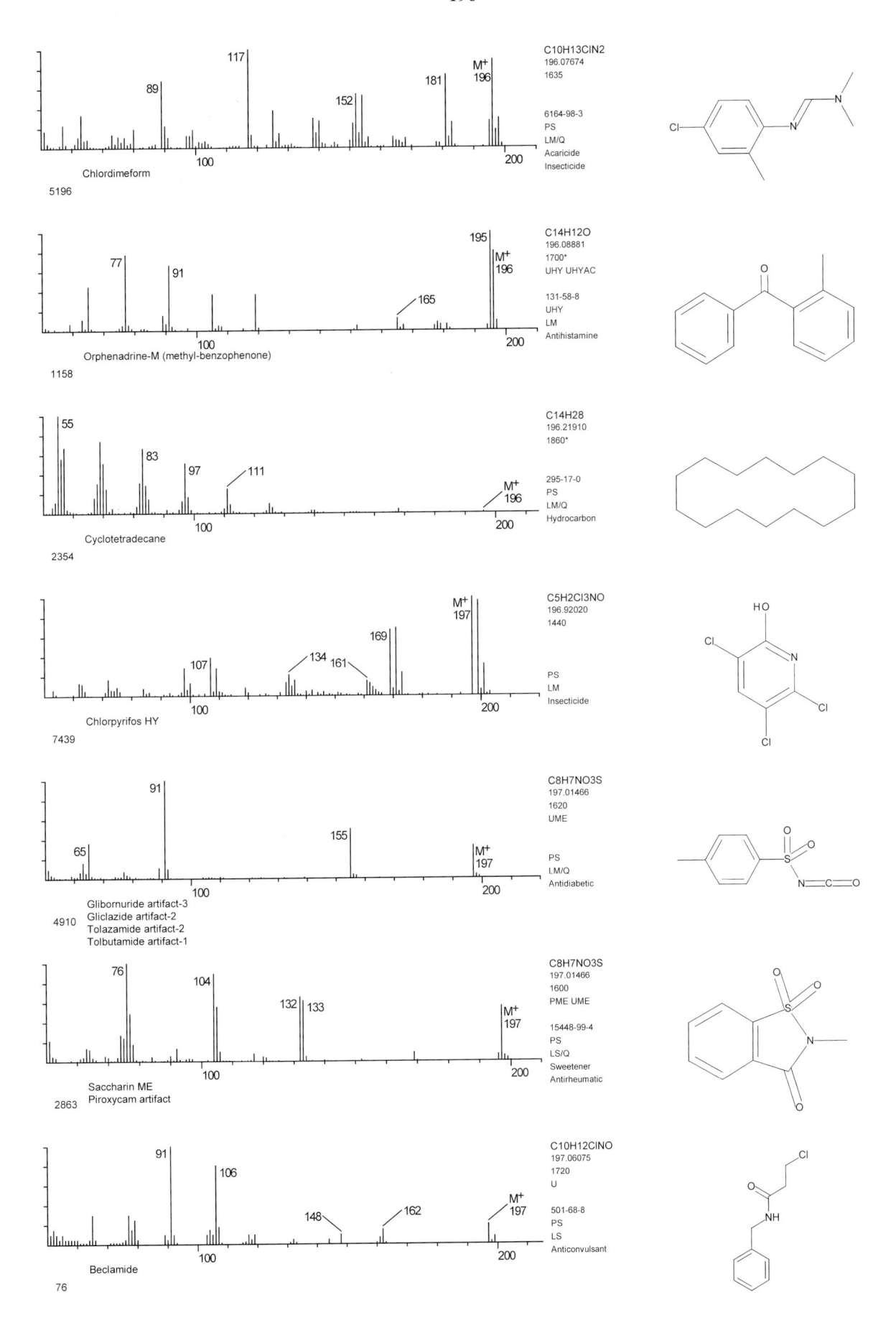

Chlordimeform
5196
C10H13ClN2
196.07674
1635
6164-98-3
PS
LM/Q
Acaricide
Insecticide

Orphenadrine-M (methyl-benzophenone)
1158
C14H12O
196.08881
1700*
UHY UHYAC
131-58-8
UHY
LM
Antihistamine

Cyclotetradecane
2354
C14H28
196.21910
1860*
295-17-0
PS
LM/Q
Hydrocarbon

Chlorpyrifos HY
7439
C5H2Cl3NO
196.92020
1440
PS
LM
Insecticide

Glibornuride artifact-3
Gliclazide artifact-2
Tolazamide artifact-2
Tolbutamide artifact-1
4910
C8H7NO3S
197.01466
1620
UME
PS
LM/Q
Antidiabetic

Saccharin ME
Piroxycam artifact
2863
C8H7NO3S
197.01466
1600
PME UME
15448-99-4
PS
LS/Q
Sweetener
Antirheumatic

Beclamide
76
C10H12ClNO
197.06075
1720
U
501-68-8
PS
LS
Anticonvulsant

197

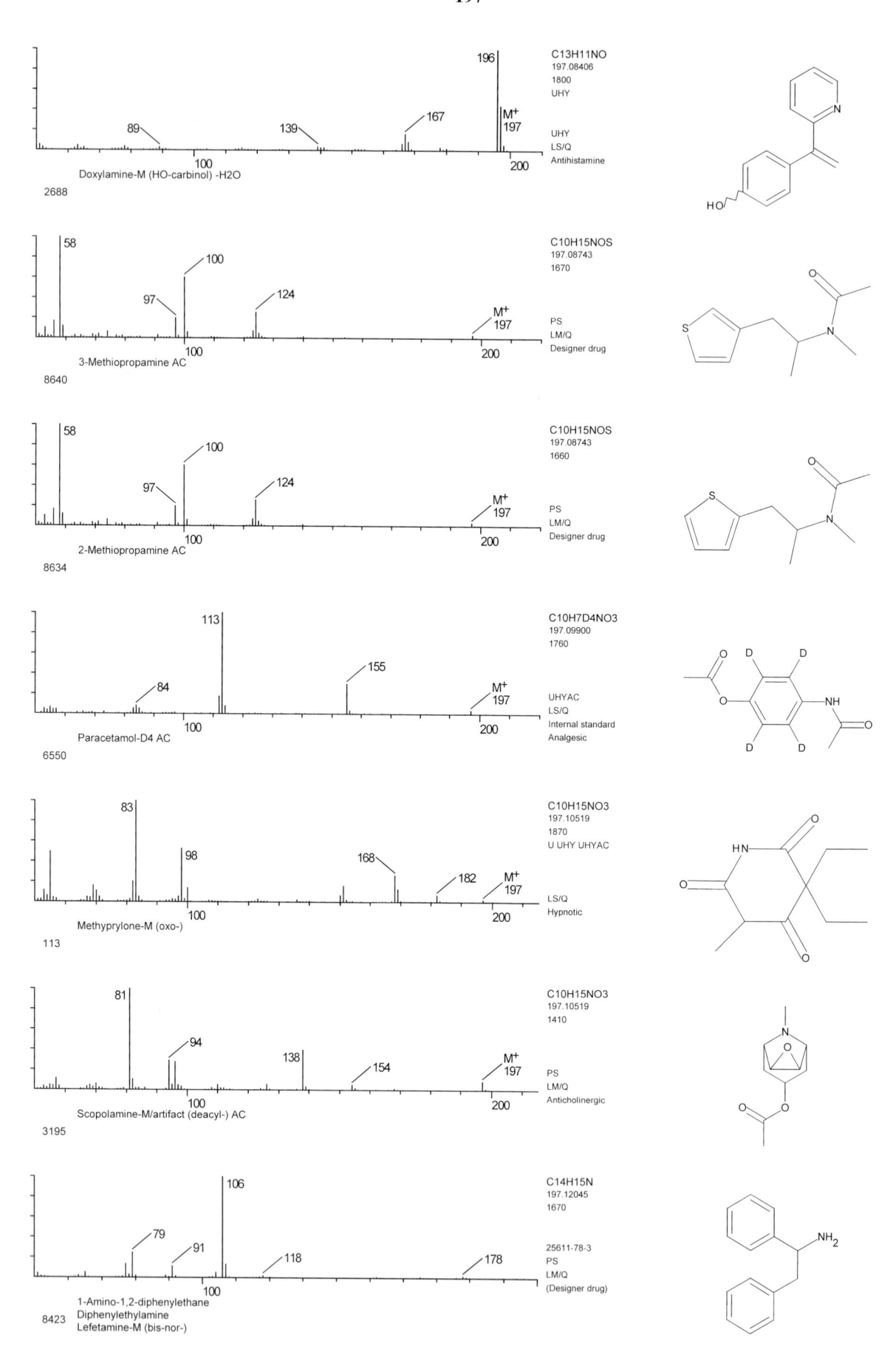

Doxylamine-M (HO-carbinol) -H2O 2688	peaks: 89, 139, 167, 196, M+ 197	C13H11NO 197.08406 1800 UHY UHY LS/Q Antihistamine
3-Methiopropamine AC 8640	peaks: 58, 97, 100, 124, M+ 197	C10H15NOS 197.08743 1670 PS LM/Q Designer drug
2-Methiopropamine AC 8634	peaks: 58, 97, 100, 124, M+ 197	C10H15NOS 197.08743 1660 PS LM/Q Designer drug
Paracetamol-D4 AC 6550	peaks: 84, 113, 155, M+ 197	C10H7D4NO3 197.09900 1760 UHYAC LS/Q Internal standard Analgesic
Methyprylone-M (oxo-) 113	peaks: 83, 98, 168, 182, M+ 197	C10H15NO3 197.10519 1870 U UHY UHYAC LS/Q Hypnotic
Scopolamine-M/artifact (deacyl-) AC 3195	peaks: 81, 94, 138, 154, M+ 197	C10H15NO3 197.10519 1410 PS LM/Q Anticholinergic
1-Amino-1,2-diphenylethane Diphenylethylamine Lefetamine-M (bis-nor-) 8423	peaks: 79, 91, 106, 118, 178	C14H15N 197.12045 1670 25611-78-3 PS LM/Q (Designer drug)

392

197

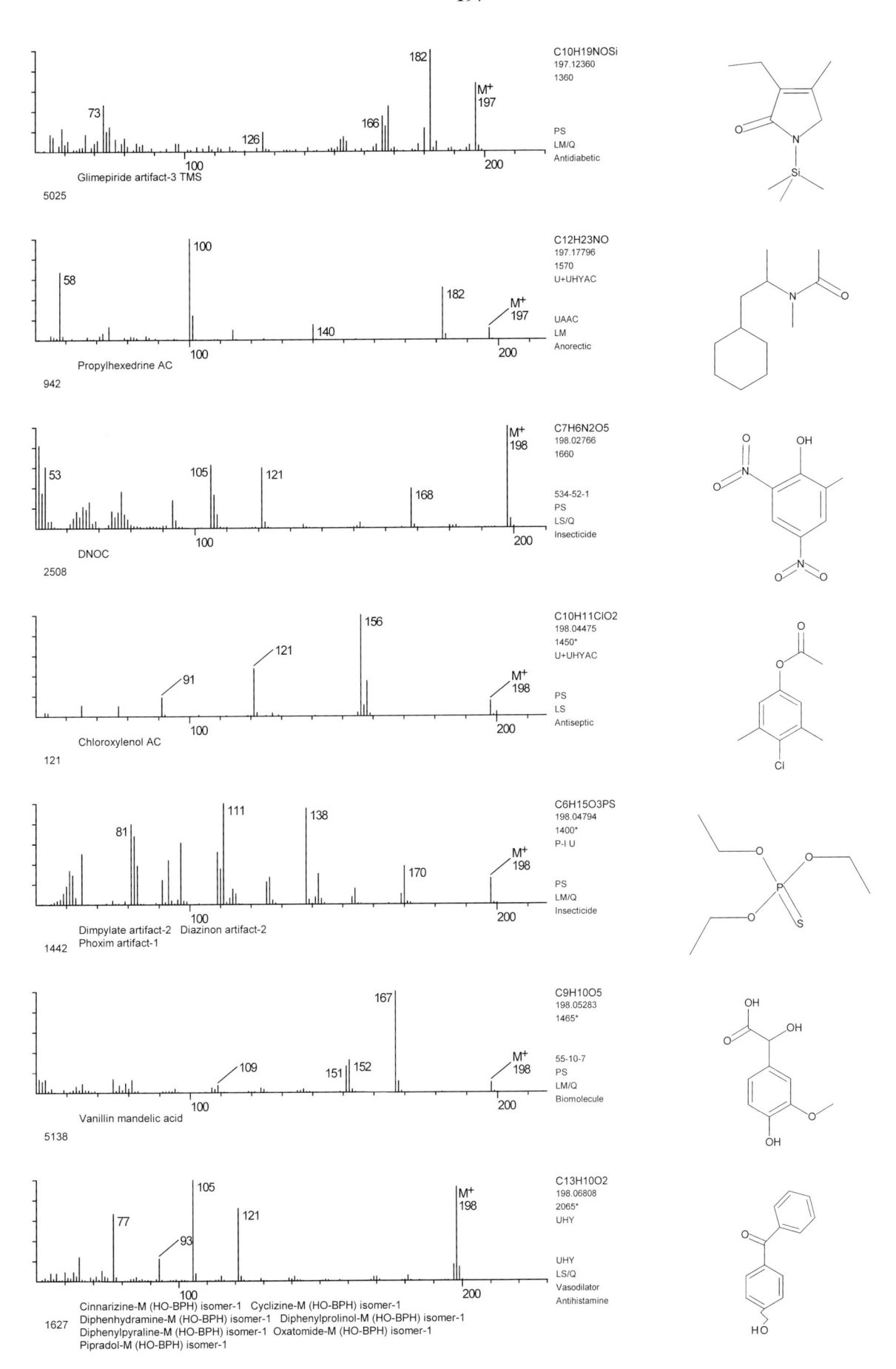

Glimepiride artifact-3 TMS 5025	73, 126, 166, 182, M+ 197	C10H19NOSi 197.12360 1360 PS LM/Q Antidiabetic
Propylhexedrine AC 942	58, 100, 140, 182, M+ 197	C12H23NO 197.17796 1570 U+UHYAC UAAC LM Anorectic
DNOC 2508	53, 105, 121, 168, M+ 198	C7H6N2O5 198.02766 1660 534-52-1 PS LS/Q Insecticide
Chloroxylenol AC 121	91, 121, 156, M+ 198	C10H11ClO2 198.04475 1450* U+UHYAC PS LS Antiseptic
Dimpylate artifact-2 Diazinon artifact-2 Phoxim artifact-1 1442	81, 111, 138, 170, M+ 198	C6H15O3PS 198.04794 1400* P-I U PS LM/Q Insecticide
Vanillin mandelic acid 5138	109, 151, 152, 167, M+ 198	C9H10O5 198.05283 1465* 55-10-7 PS LM/Q Biomolecule
Cinnarizine-M (HO-BPH) isomer-1 Cyclizine-M (HO-BPH) isomer-1 Diphenhydramine-M (HO-BPH) isomer-1 Diphenylprolinol-M (HO-BPH) isomer-1 Diphenylpyraline-M (HO-BPH) isomer-1 Oxatomide-M (HO-BPH) isomer-1 Pipradol-M (HO-BPH) isomer-1 1627	77, 93, 105, 121, M+ 198	C13H10O2 198.06808 2065* UHY UHY LS/Q Vasodilator Antihistamine

198

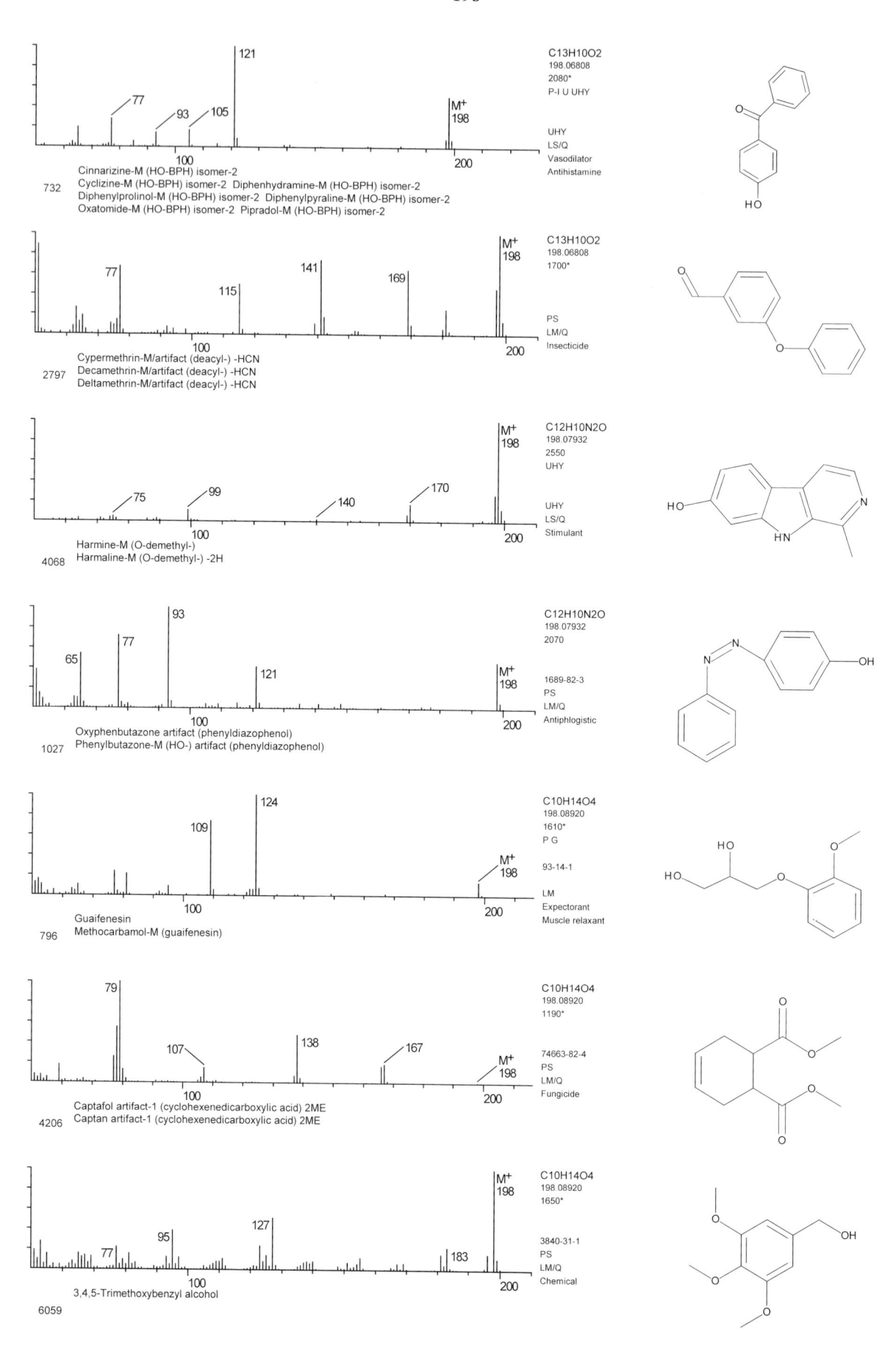

Cinnarizine-M (HO-BPH) isomer-2 Cyclizine-M (HO-BPH) isomer-2 Diphenhydramine-M (HO-BPH) isomer-2 Diphenylprolinol-M (HO-BPH) isomer-2 Diphenylpyraline-M (HO-BPH) isomer-2 Oxatomide-M (HO-BPH) isomer-2 Pipradol-M (HO-BPH) isomer-2	C13H10O2 198.06808 2080* P-I U UHY UHY LS/Q Vasodilator Antihistamine
Cypermethrin-M/artifact (deacyl-) -HCN Decamethrin-M/artifact (deacyl-) -HCN Deltamethrin-M/artifact (deacyl-) -HCN	C13H10O2 198.06808 1700* PS LM/Q Insecticide
Harmine-M (O-demethyl-) Harmaline-M (O-demethyl-) -2H	C12H10N2O 198.07932 2550 UHY UHY LS/Q Stimulant
Oxyphenbutazone artifact (phenyldiazophenol) Phenylbutazone-M (HO-) artifact (phenyldiazophenol)	C12H10N2O 198.07932 2070 1689-82-3 PS LM/Q Antiphlogistic
Guaifenesin Methocarbamol-M (guaifenesin)	C10H14O4 198.08920 1610* P G 93-14-1 LM Expectorant Muscle relaxant
Captafol artifact-1 (cyclohexenedicarboxylic acid) 2ME Captan artifact-1 (cyclohexenedicarboxylic acid) 2ME	C10H14O4 198.08920 1190* 74663-82-4 PS LM/Q Fungicide
3,4,5-Trimethoxybenzyl alcohol	C10H14O4 198.08920 1650* 3840-31-1 PS LM/Q Chemical

198

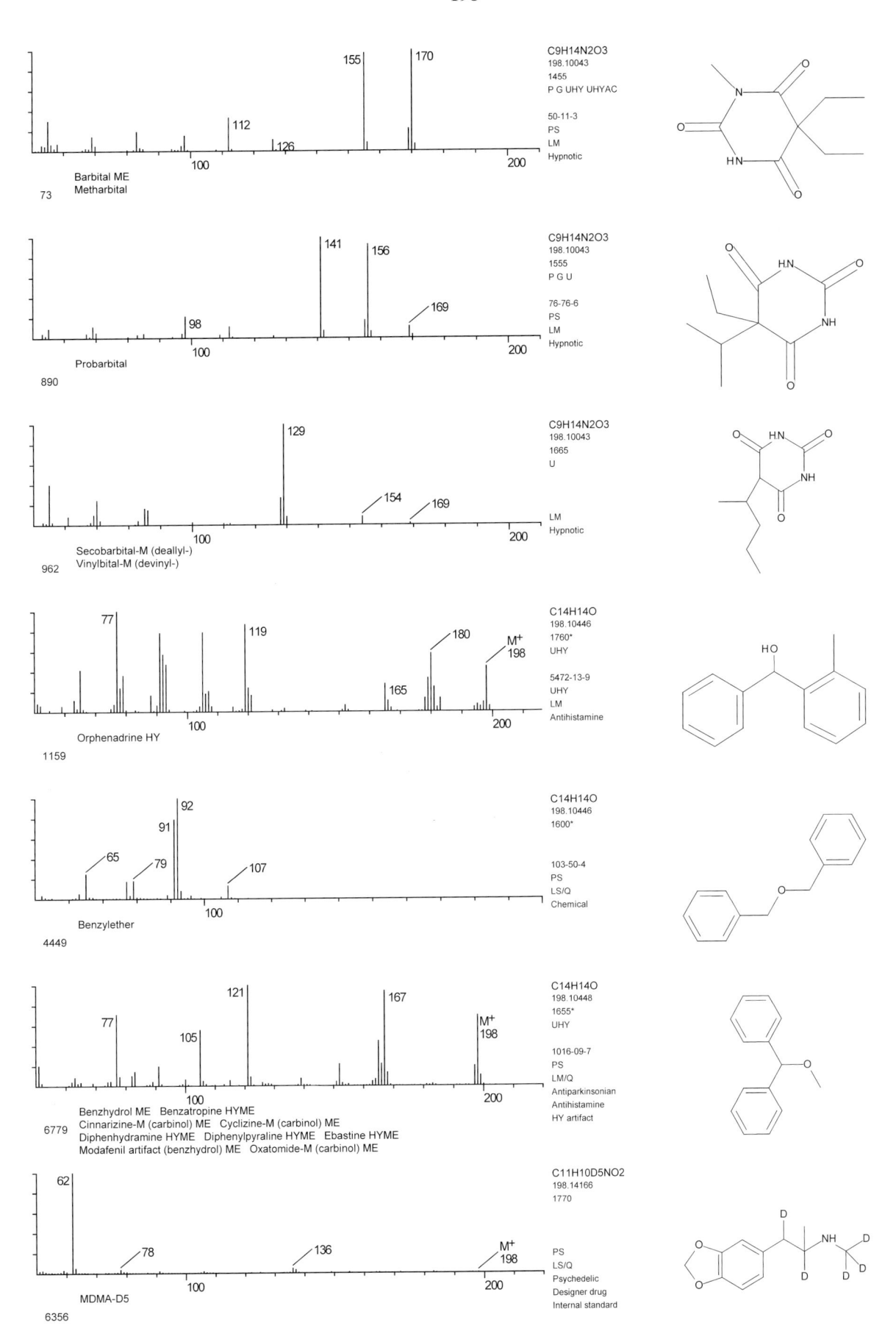

	C9H14N2O3 198.10043 1455 P G UHY UHYAC	
Peaks: 112, 126, 155, 170	50-11-3 PS LM Hypnotic	
Barbital ME Metharbital 73		

	C9H14N2O3 198.10043 1555 P G U	
Peaks: 98, 141, 156, 169	76-76-6 PS LM Hypnotic	
Probarbital 890		

	C9H14N2O3 198.10043 1665 U	
Peaks: 129, 154, 169	LM Hypnotic	
Secobarbital-M (deallyl-) Vinylbital-M (devinyl-) 962		

	C14H14O 198.10446 1760* UHY	
Peaks: 77, 119, 165, 180, M+ 198	5472-13-9 UHY LM Antihistamine	
Orphenadrine HY 1159		

	C14H14O 198.10446 1600*	
Peaks: 65, 79, 91, 92, 107	103-50-4 PS LS/Q Chemical	
Benzylether 4449		

	C14H14O 198.10448 1655* UHY	
Peaks: 77, 105, 121, 167, M+ 198	1016-09-7 PS LM/Q Antiparkinsonian Antihistamine HY artifact	
Benzhydrol ME Benzatropine HYME Cinnarizine-M (carbinol) ME Cyclizine-M (carbinol) ME Diphenhydramine HYME Diphenylpyraline HYME Ebastine HYME Modafenil artifact (benzhydrol) ME Oxatomide-M (carbinol) ME 6779		

	C11H10D5NO2 198.14166 1770	
Peaks: 62, 78, 136, M+ 198	PS LS/Q Psychedelic Designer drug Internal standard	
MDMA-D5 6356		

198

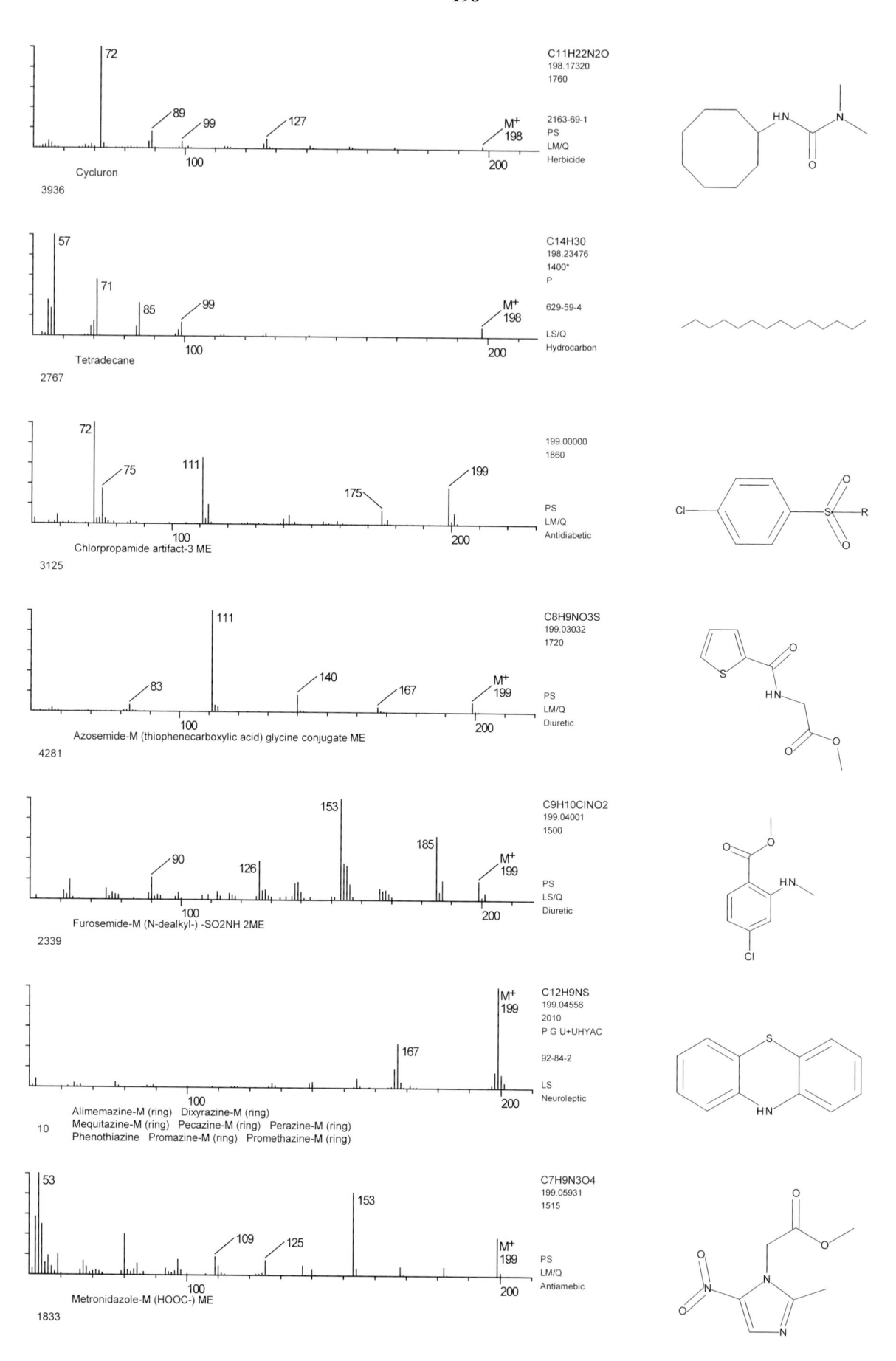

Cycluron — C11H22N2O, 198.17320, 1760, 2163-69-1, PS, LM/Q, Herbicide
Peaks: 72, 89, 99, 127, M+ 198
3936

Tetradecane — C14H30, 198.23476, 1400*, P, 629-59-4, LS/Q, Hydrocarbon
Peaks: 57, 71, 85, 99, M+ 198
2767

Chlorpropamide artifact-3 ME — 199.00000, 1860, PS, LM/Q, Antidiabetic
Peaks: 72, 75, 111, 175, 199
3125

Azosemide-M (thiophenecarboxylic acid) glycine conjugate ME — C8H9NO3S, 199.03032, 1720, PS, LM/Q, Diuretic
Peaks: 83, 111, 140, 167, M+ 199
4281

Furosemide-M (N-dealkyl-) -SO2NH 2ME — C9H10ClNO2, 199.04001, 1500, PS, LS/Q, Diuretic
Peaks: 90, 126, 153, 185, M+ 199
2339

Alimemazine-M (ring) Dixyrazine-M (ring)
Mequitazine-M (ring) Pecazine-M (ring) Perazine-M (ring)
Phenothiazine Promazine-M (ring) Promethazine-M (ring)
— C12H9NS, 199.04556, 2010, P G U+UHYAC, 92-84-2, LS, Neuroleptic
Peaks: 167, M+ 199
10

Metronidazole-M (HOOC-) ME — C7H9N3O4, 199.05931, 1515, PS, LM/Q, Antiamebic
Peaks: 53, 109, 125, 153, M+ 199
1833

396

199

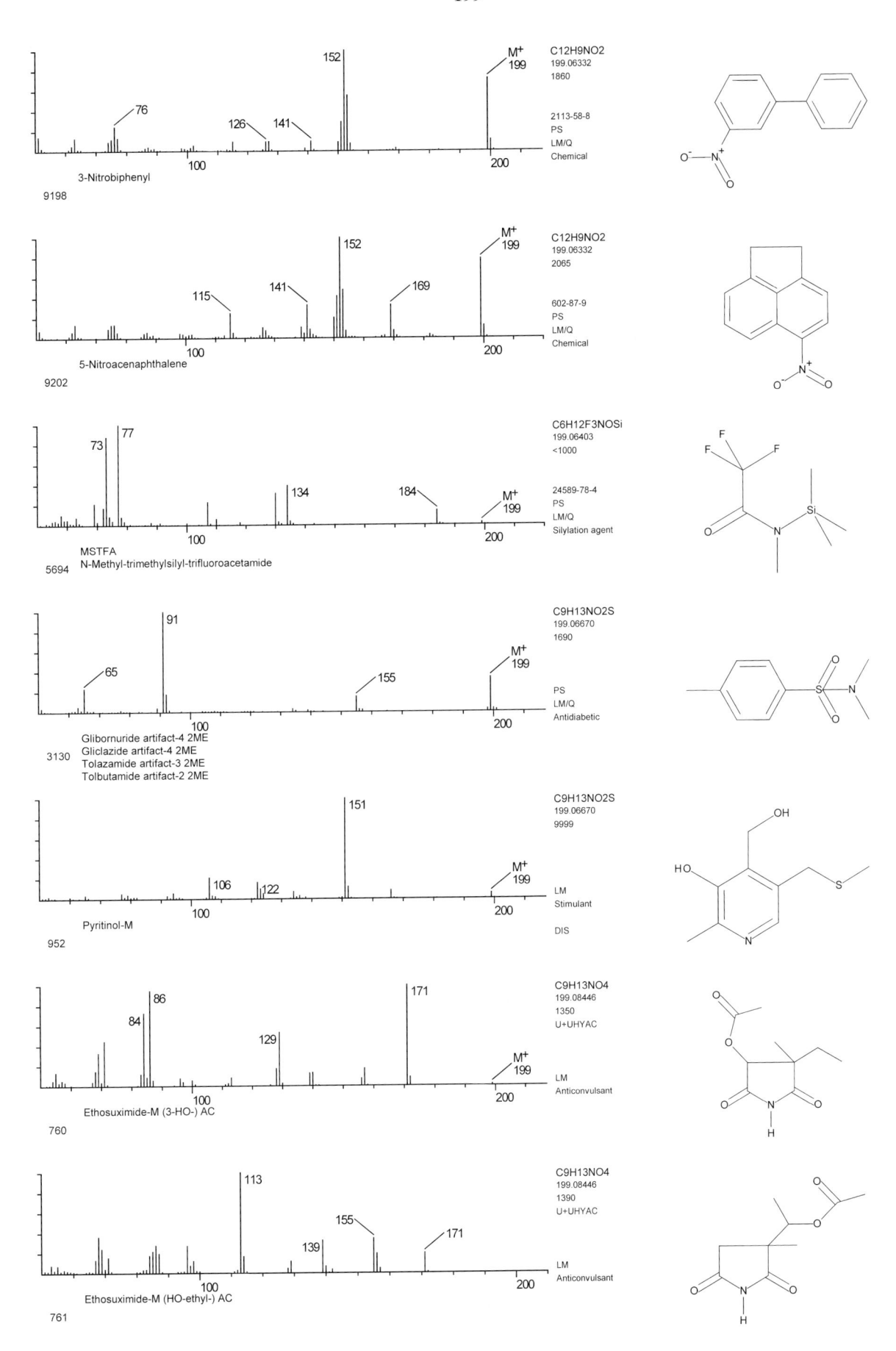

C12H9NO2
199.06332
1860

2113-58-8
PS
LM/Q
Chemical

3-Nitrobiphenyl
9198

C12H9NO2
199.06332
2065

602-87-9
PS
LM/Q
Chemical

5-Nitroacenaphthalene
9202

C6H12F3NOSi
199.06403
<1000

24589-78-4
PS
LM/Q
Silylation agent

MSTFA
N-Methyl-trimethylsilyl-trifluoroacetamide
5694

C9H13NO2S
199.06670
1690

PS
LM/Q
Antidiabetic

Glibornuride artifact-4 2ME
Gliclazide artifact-4 2ME
Tolazamide artifact-3 2ME
Tolbutamide artifact-2 2ME
3130

C9H13NO2S
199.06670
9999

LM
Stimulant
DIS

Pyritinol-M
952

C9H13NO4
199.08446
1350
U+UHYAC

LM
Anticonvulsant

Ethosuximide-M (3-HO-) AC
760

C9H13NO4
199.08446
1390
U+UHYAC

LM
Anticonvulsant

Ethosuximide-M (HO-ethyl-) AC
761

397

199

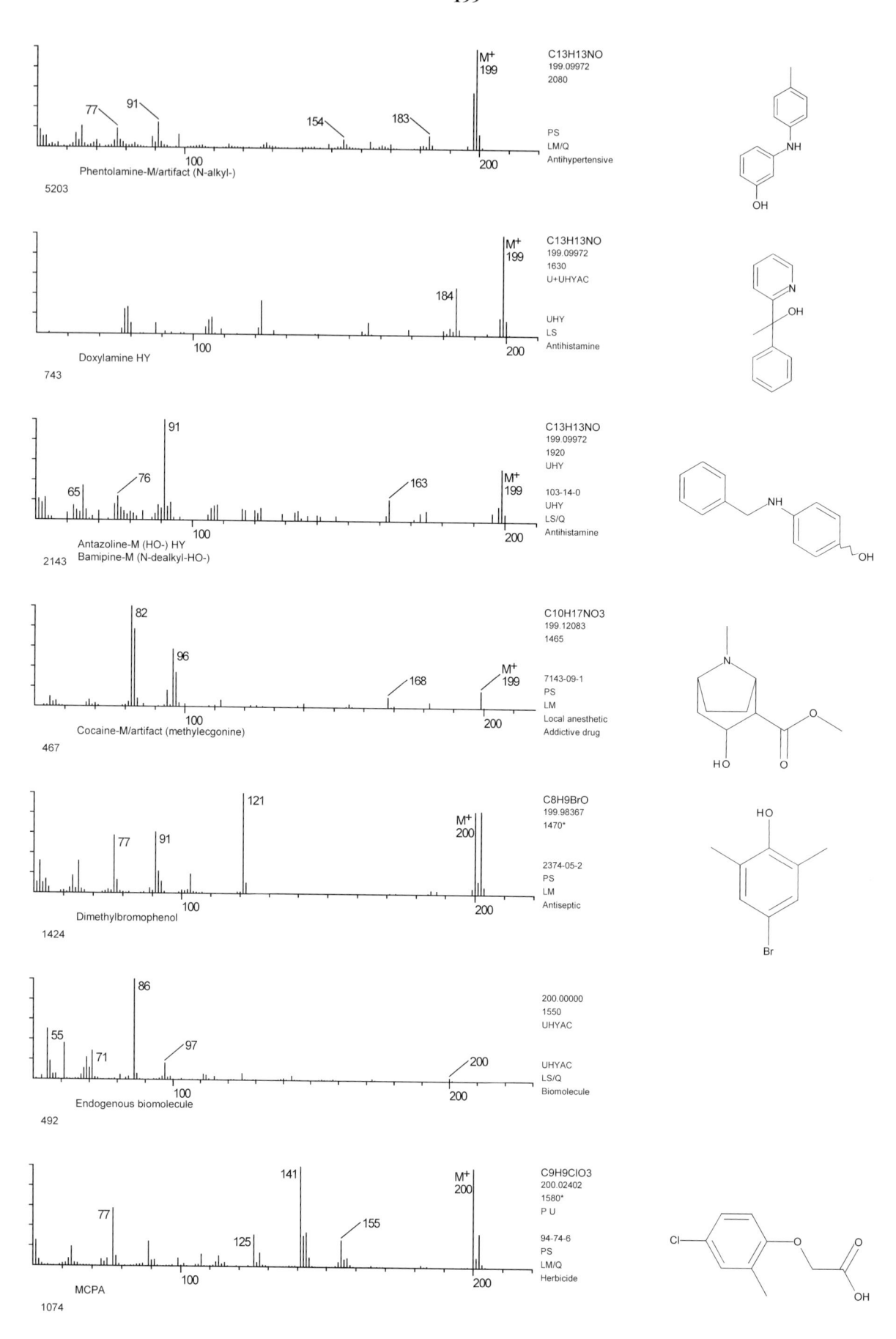

C13H13NO
199.09972
2080
PS
LM/Q
Antihypertensive

Phentolamine-M/artifact (N-alkyl-)
5203

C13H13NO
199.09972
1630
U+UHYAC
UHY
LS
Antihistamine

Doxylamine HY
743

C13H13NO
199.09972
1920
UHY
103-14-0
UHY
LS/Q
Antihistamine

Antazoline-M (HO-) HY
Bamipine-M (N-dealkyl-HO-)
2143

C10H17NO3
199.12083
1465
7143-09-1
PS
LM
Local anesthetic
Addictive drug

Cocaine-M/artifact (methylecgonine)
467

C8H9BrO
199.98367
1470*
2374-05-2
PS
LM
Antiseptic

Dimethylbromophenol
1424

200.00000
1550
UHYAC
UHYAC
LS/Q
Biomolecule

Endogenous biomolecule
492

C9H9ClO3
200.02402
1580*
P U
94-74-6
PS
LM/Q
Herbicide

MCPA
1074

200

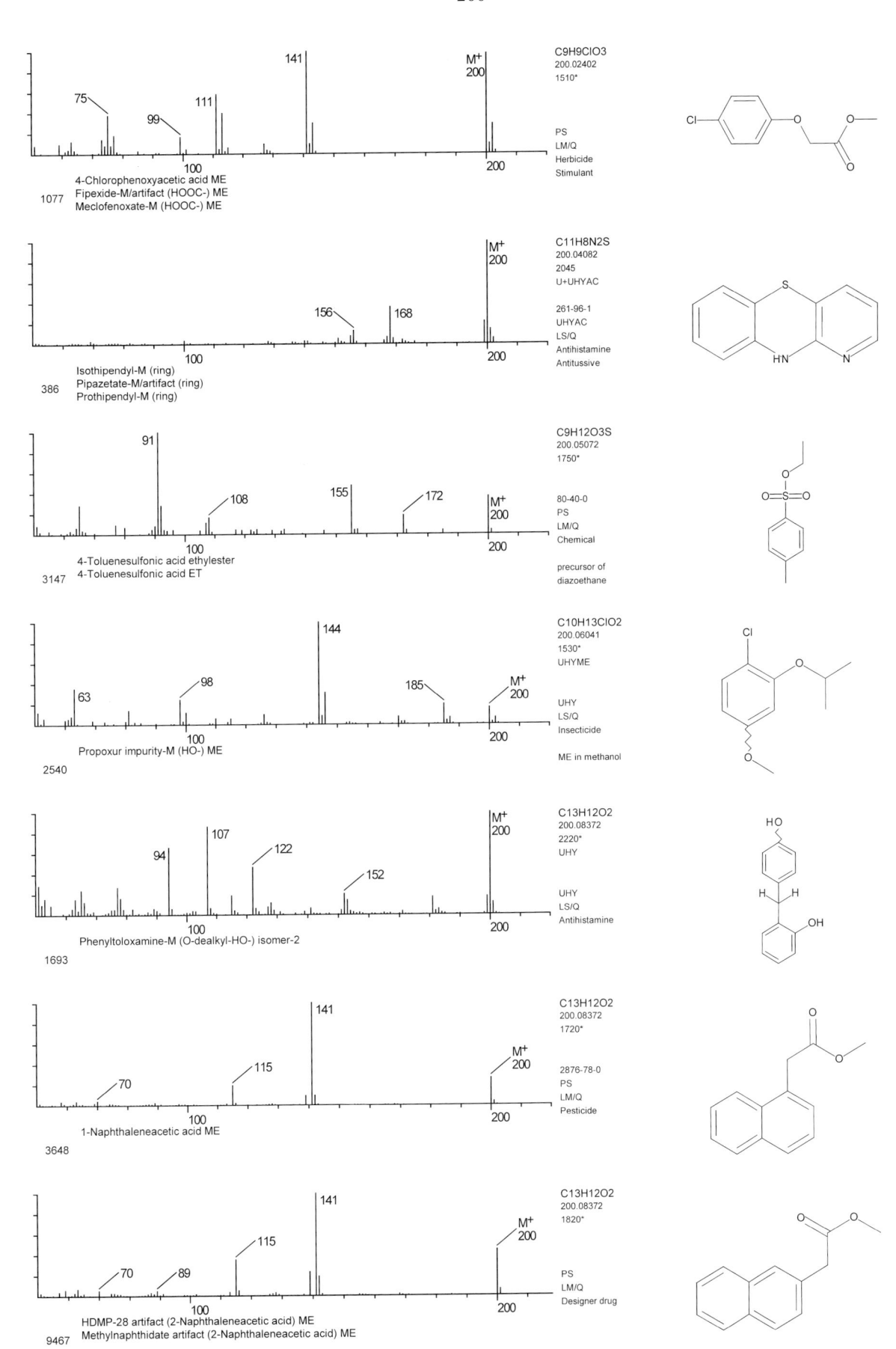

Mass spectrum with peaks at 75, 99, 111, 141, M+ 200	C9H9ClO3 200.02402 1510* PS LM/Q Herbicide Stimulant	4-chlorophenoxy methyl acetate structure

1077
4-Chlorophenoxyacetic acid ME
Fipexide-M/artifact (HOOC-) ME
Meclofenoxate-M (HOOC-) ME

Mass spectrum with peaks at 156, 168, M+ 200	C11H8N2S 200.04082 2045 U+UHYAC 261-96-1 UHYAC LS/Q Antihistamine Antitussive	Phenothiazine-pyridine structure

386
Isothipendyl-M (ring)
Pipazetate-M/artifact (ring)
Prothipendyl-M (ring)

Mass spectrum with peaks at 91, 108, 155, 172, M+ 200	C9H12O3S 200.05072 1750* 80-40-0 PS LM/Q Chemical precursor of diazoethane	Ethyl 4-toluenesulfonate structure

3147
4-Toluenesulfonic acid ethylester
4-Toluenesulfonic acid ET

Mass spectrum with peaks at 63, 98, 144, 185, M+ 200	C10H13ClO2 200.06041 1530* UHYME UHY LS/Q Insecticide ME in methanol	Chloro-isopropoxy-methoxybenzene structure

2540
Propoxur impurity-M (HO-) ME

Mass spectrum with peaks at 94, 107, 122, 152, M+ 200	C13H12O2 200.08372 2220* UHY UHY LS/Q Antihistamine	Bis(hydroxyphenyl)methane structure

1693
Phenyltoloxamine-M (O-dealkyl-HO-) isomer-2

Mass spectrum with peaks at 70, 115, 141, M+ 200	C13H12O2 200.08372 1720* 2876-78-0 PS LM/Q Pesticide	Methyl 1-naphthaleneacetate structure

3648
1-Naphthaleneacetic acid ME

Mass spectrum with peaks at 70, 89, 115, 141, M+ 200	C13H12O2 200.08372 1820* PS LM/Q Designer drug	Methyl 2-naphthaleneacetate structure

9467
HDMP-28 artifact (2-Naphthaleneacetic acid) ME
Methylnaphthidate artifact (2-Naphthaleneacetic acid) ME

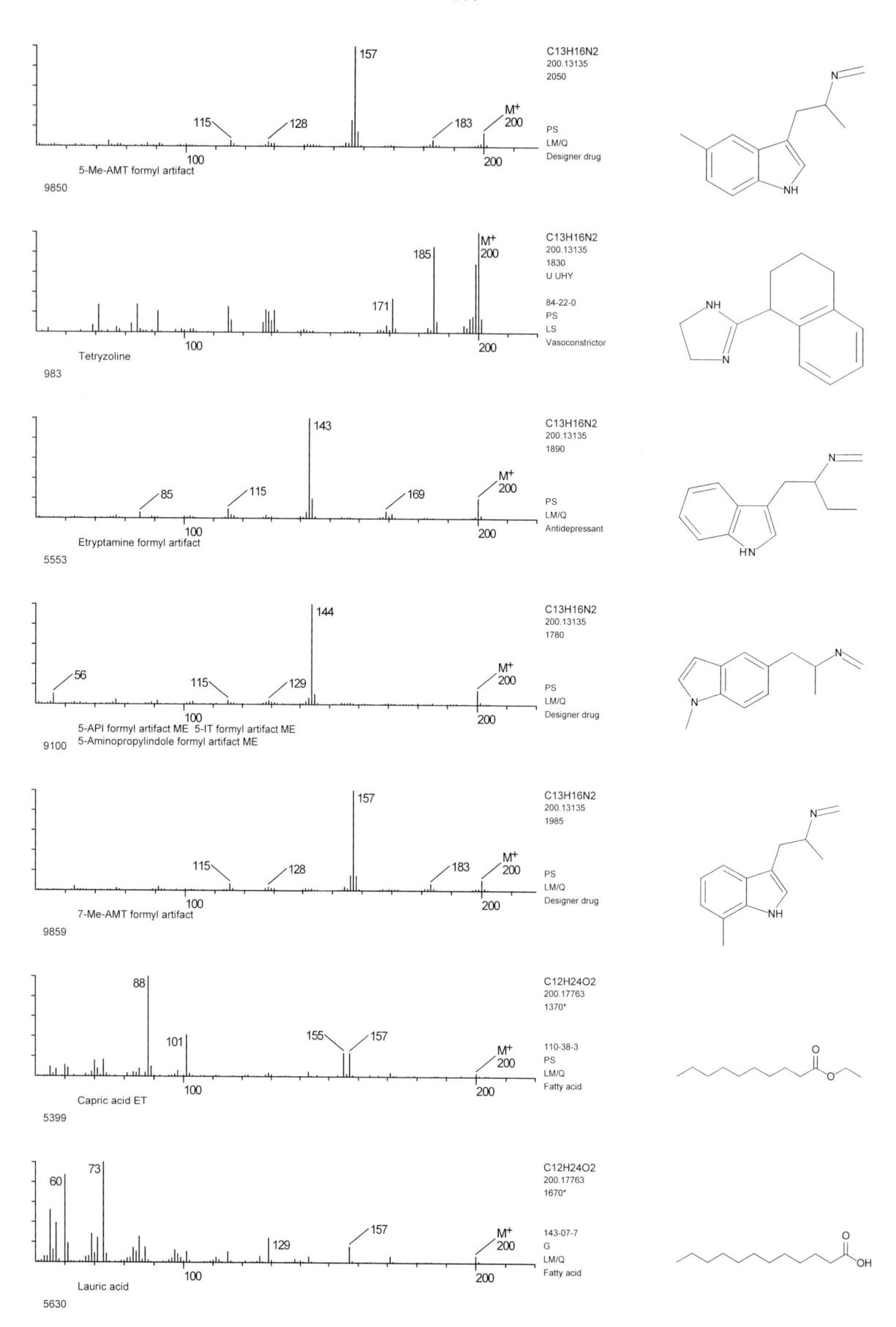

Spectrum	Compound	Formula	Mass	Info
157, 115, 128, 183, M+ 200	5-Me-AMT formyl artifact	C13H16N2	200.13135	2050, PS, LM/Q, Designer drug
M+ 200, 185, 171	Tetryzoline	C13H16N2	200.13135	1830, U UHY, 84-22-0, PS, LS, Vasoconstrictor
143, 85, 115, 169, M+ 200	Etryptamine formyl artifact	C13H16N2	200.13135	1890, PS, LM/Q, Antidepressant
144, 56, 115, 129, M+ 200	5-API formyl artifact ME / 5-IT formyl artifact ME / 5-Aminopropylindole formyl artifact ME	C13H16N2	200.13135	1780, PS, LM/Q, Designer drug
157, 115, 128, 183, M+ 200	7-Me-AMT formyl artifact	C13H16N2	200.13135	1985, PS, LM/Q, Designer drug
88, 101, 155, 157, M+ 200	Capric acid ET	C12H24O2	200.17763	1370*, 110-38-3, PS, LM/Q, Fatty acid
73, 60, 129, 157, M+ 200	Lauric acid	C12H24O2	200.17763	1670*, 143-07-7, G, LM/Q, Fatty acid

201

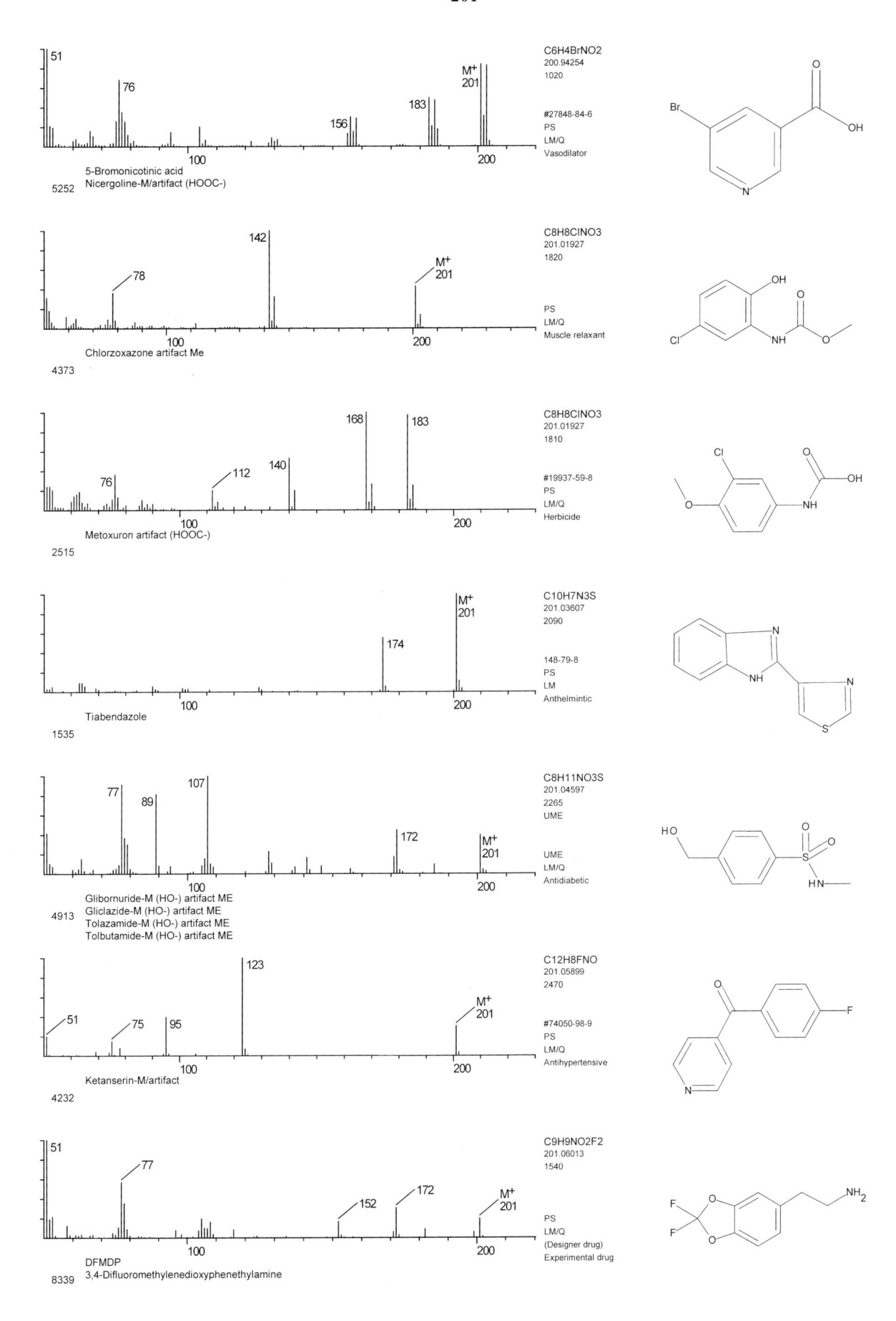

C6H4BrNO2
200.94254
1020

#27848-84-6
PS
LM/Q
Vasodilator

5-Bromonicotinic acid
Nicergoline-M/artifact (HOOC-)
5252

C8H8ClNO3
201.01927
1820

PS
LM/Q
Muscle relaxant

Chlorzoxazone artifact Me
4373

C8H8ClNO3
201.01927
1810

#19937-59-8
PS
LM/Q
Herbicide

Metoxuron artifact (HOOC-)
2515

C10H7N3S
201.03607
2090

148-79-8
PS
LM
Anthelmintic

Tiabendazole
1535

C8H11NO3S
201.04597
2265
UME

UME
LM/Q
Antidiabetic

Glibornuride-M (HO-) artifact ME
Gliclazide-M (HO-) artifact ME
Tolazamide-M (HO-) artifact ME
Tolbutamide-M (HO-) artifact ME
4913

C12H8FNO
201.05899
2470

#74050-98-9
PS
LM/Q
Antihypertensive

Ketanserin-M/artifact
4232

C9H9NO2F2
201.06013
1540

PS
LM/Q
(Designer drug)
Experimental drug

DFMDP
3,4-Difluoromethylenedioxyphenethylamine
8339

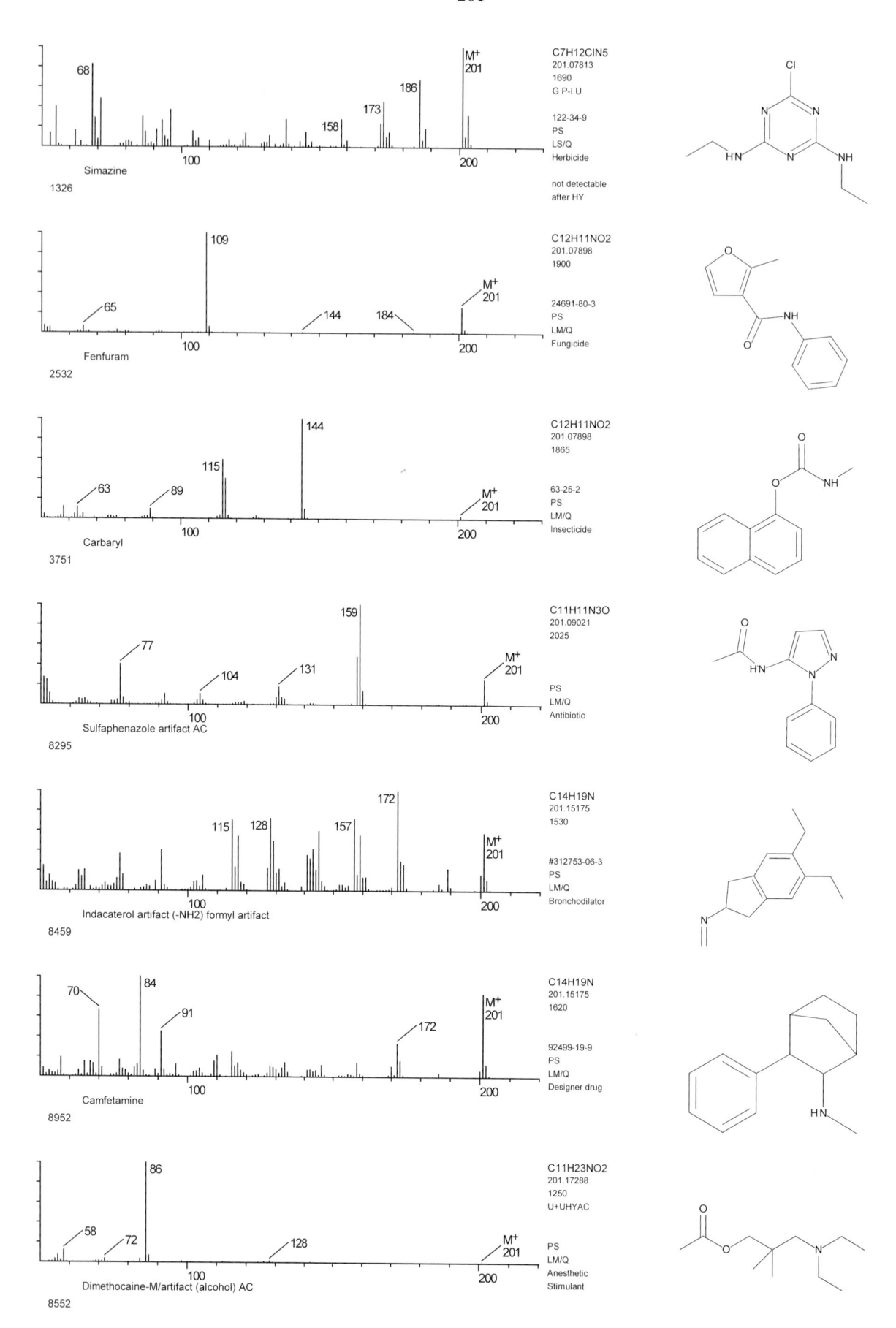

Simazine 1326
C7H12ClN5
201.07813
1690
G P-I U
122-34-9
PS
LS/Q
Herbicide
not detectable after HY

Peaks: 68, 158, 173, 186, M+ 201

Fenfuram 2532
C12H11NO2
201.07898
1900
24691-80-3
PS
LM/Q
Fungicide

Peaks: 65, 109, 144, 184, M+ 201

Carbaryl 3751
C12H11NO2
201.07898
1865
63-25-2
PS
LM/Q
Insecticide

Peaks: 63, 89, 115, 144, M+ 201

Sulfaphenazole artifact AC 8295
C11H11N3O
201.09021
2025
PS
LM/Q
Antibiotic

Peaks: 77, 104, 131, 159, M+ 201

Indacaterol artifact (-NH2) formyl artifact 8459
C14H19N
201.15175
1530
#312753-06-3
PS
LM/Q
Bronchodilator

Peaks: 115, 128, 157, 172, M+ 201

Camfetamine 8952
C14H19N
201.15175
1620
92499-19-9
PS
LM/Q
Designer drug

Peaks: 70, 84, 91, 172, M+ 201

Dimethocaine-M/artifact (alcohol) AC 8552
C11H23NO2
201.17288
1250
U+UHYAC
PS
LM/Q
Anesthetic
Stimulant

Peaks: 58, 72, 86, 128, M+ 201

202

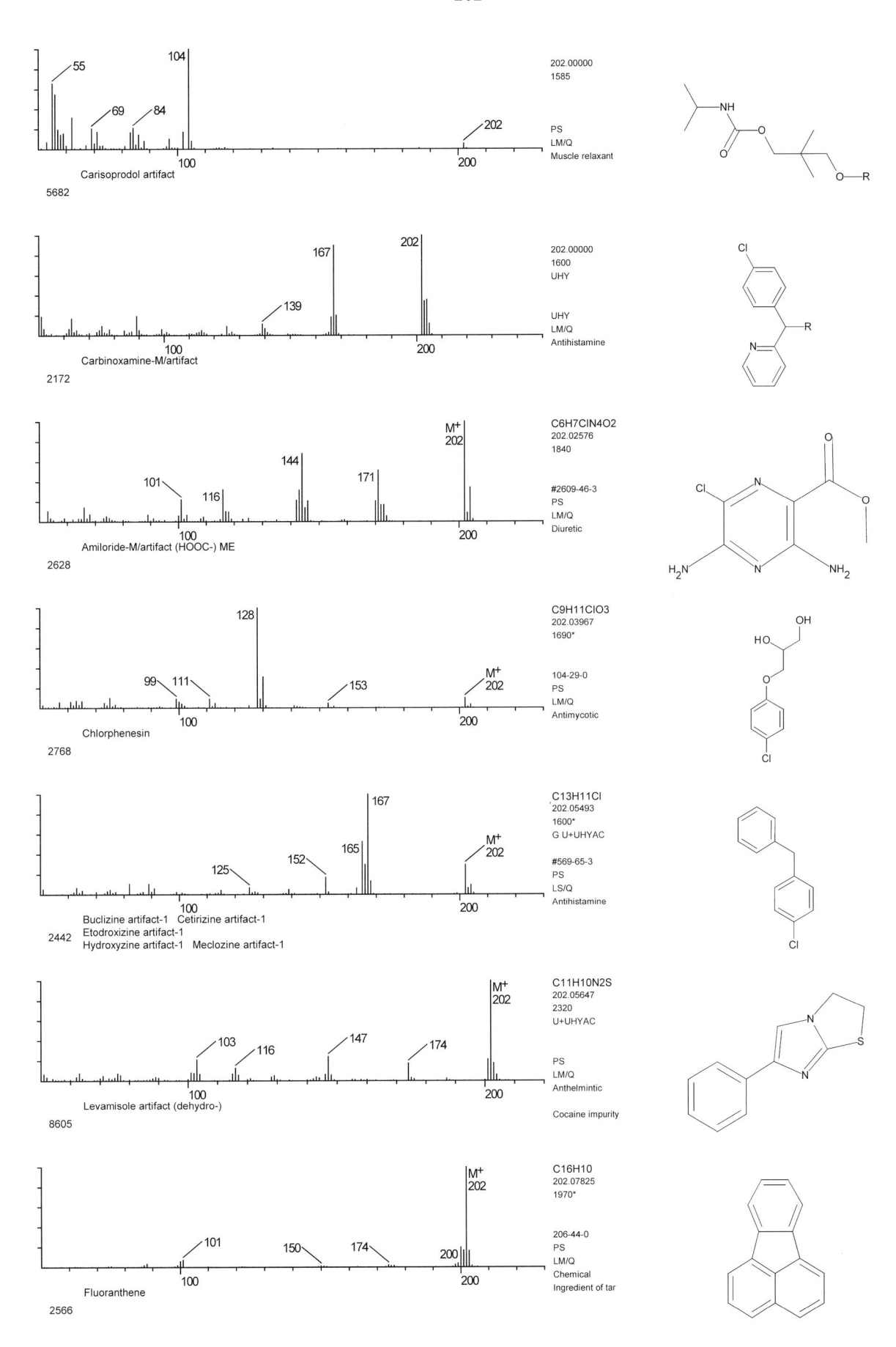

Spectrum	Peaks	Formula/Info
Carisoprodol artifact	55, 69, 84, 104, 202	202.00000, 1585, PS, LM/Q, Muscle relaxant
5682		
Carbinoxamine-M/artifact	139, 167, 202	202.00000, 1600, UHY, UHY, LM/Q, Antihistamine
2172		
Amiloride-M/artifact (HOOC-) ME	101, 116, 144, 171, M+ 202	C6H7ClN4O2, 202.02576, 1840, #2609-46-3, PS, LM/Q, Diuretic
2628		
Chlorphenesin	99, 111, 128, 153, M+ 202	C9H11ClO3, 202.03967, 1690*, 104-29-0, PS, LM/Q, Antimycotic
2768		
Buclizine artifact-1 Cetirizine artifact-1 Etodroxizine artifact-1 Hydroxyzine artifact-1 Meclozine artifact-1	125, 152, 165, 167, M+ 202	C13H11Cl, 202.05493, 1600*, G U+UHYAC, #569-65-3, PS, LS/Q, Antihistamine
2442		
Levamisole artifact (dehydro-)	103, 116, 147, 174, M+ 202	C11H10N2S, 202.05647, 2320, U+UHYAC, PS, LM/Q, Anthelmintic, Cocaine impurity
8605		
Fluoranthene	101, 150, 174, 200, M+ 202	C16H10, 202.07825, 1970*, 206-44-0, PS, LM/Q, Chemical, Ingredient of tar
2566		

202

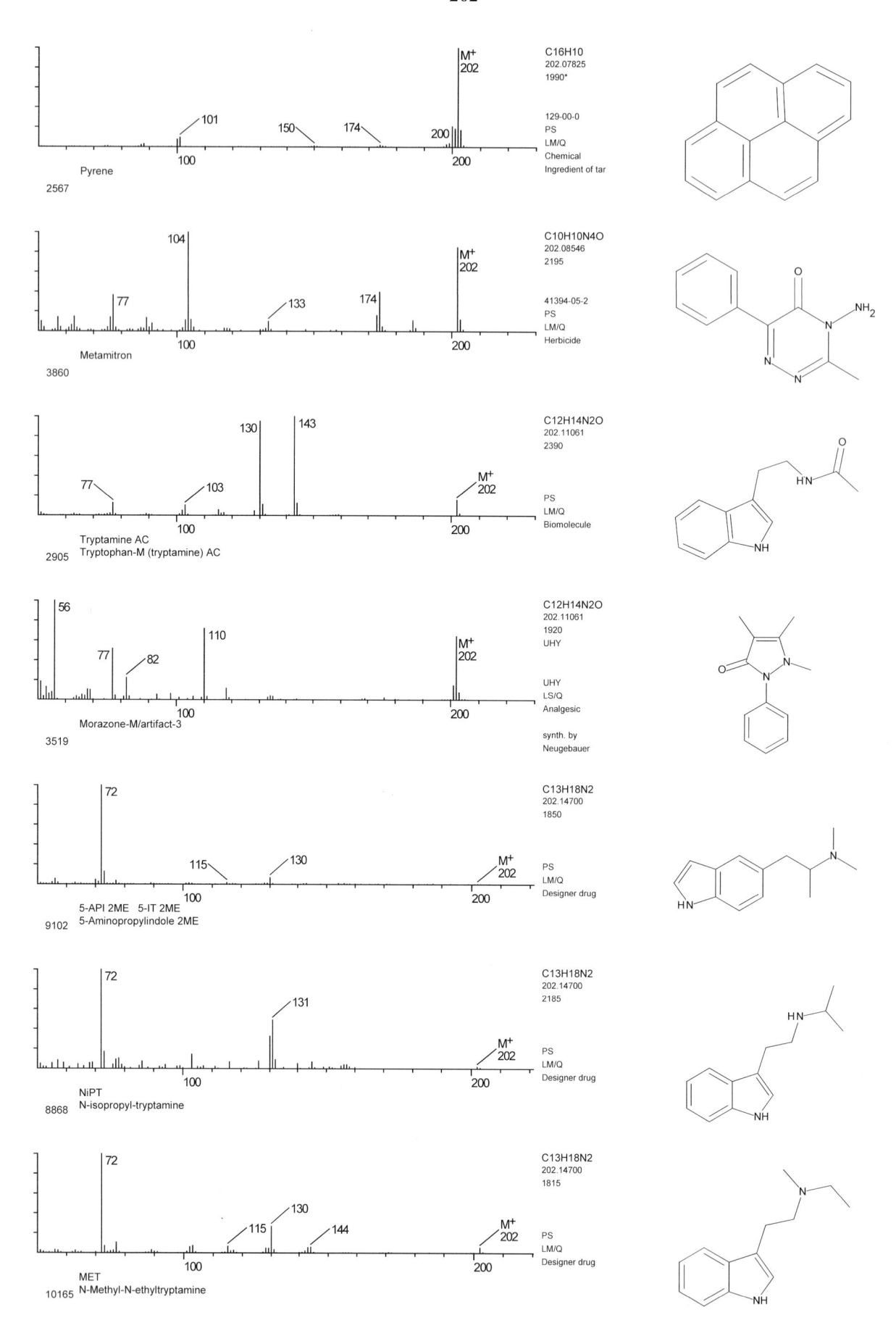

Pyrene — 2567
C16H10
202.07825
1990*
129-00-0
PS
LM/Q
Chemical
Ingredient of tar

Metamitron — 3860
C10H10N4O
202.08546
2195
41394-05-2
PS
LM/Q
Herbicide

Tryptamine AC / Tryptophan-M (tryptamine) AC — 2905
C12H14N2O
202.11061
2390
PS
LM/Q
Biomolecule

Morazone-M/artifact-3 — 3519
C12H14N2O
202.11061
1920
UHY
UHY
LS/Q
Analgesic
synth. by Neugebauer

5-API 2ME / 5-IT 2ME / 5-Aminopropylindole 2ME — 9102
C13H18N2
202.14700
1850
PS
LM/Q
Designer drug

NiPT / N-isopropyl-tryptamine — 8868
C13H18N2
202.14700
2185
PS
LM/Q
Designer drug

MET / N-Methyl-N-ethyltryptamine — 10165
C13H18N2
202.14700
1815
PS
LM/Q
Designer drug

203

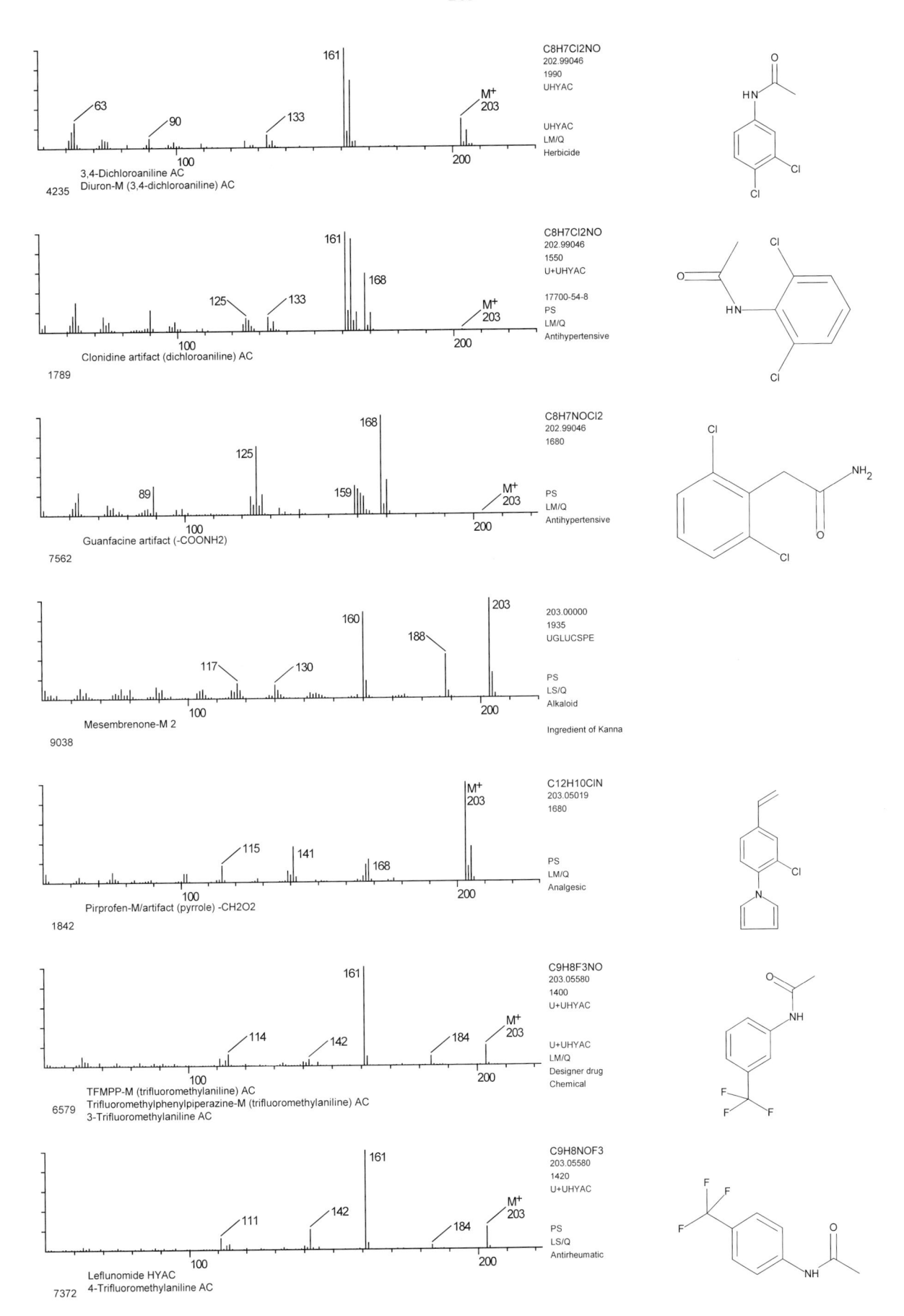

4235 — 3,4-Dichloroaniline AC / Diuron-M (3,4-dichloroaniline) AC
C8H7Cl2NO, 202.99046, 1990, UHYAC
UHYAC, LM/Q, Herbicide
Peaks: 63, 90, 133, 161, M+ 203

1789 — Clonidine artifact (dichloroaniline) AC
C8H7Cl2NO, 202.99046, 1550, U+UHYAC
17700-54-8, PS, LM/Q, Antihypertensive
Peaks: 125, 133, 161, 168, M+ 203

7562 — Guanfacine artifact (-COONH2)
C8H7NOCl2, 202.99046, 1680
PS, LM/Q, Antihypertensive
Peaks: 89, 125, 159, 168, M+ 203

9038 — Mesembrenone-M 2
203.00000, 1935, UGLUCSPE
PS, LS/Q, Alkaloid, Ingredient of Kanna
Peaks: 117, 130, 160, 188, 203

1842 — Pirprofen-M/artifact (pyrrole) -CH2O2
C12H10ClN, 203.05019, 1680
PS, LM/Q, Analgesic
Peaks: 115, 141, 168, M+ 203

6579 — TFMPP-M (trifluoromethylaniline) AC / Trifluoromethylphenylpiperazine-M (trifluoromethylaniline) AC / 3-Trifluoromethylaniline AC
C9H8F3NO, 203.05580, 1400, U+UHYAC
U+UHYAC, LM/Q, Designer drug Chemical
Peaks: 114, 142, 161, 184, M+ 203

7372 — Leflunomide HYAC / 4-Trifluoromethylaniline AC
C9H8NOF3, 203.05580, 1420, U+UHYAC
PS, LS/Q, Antirheumatic
Peaks: 111, 142, 161, 184, M+ 203

203

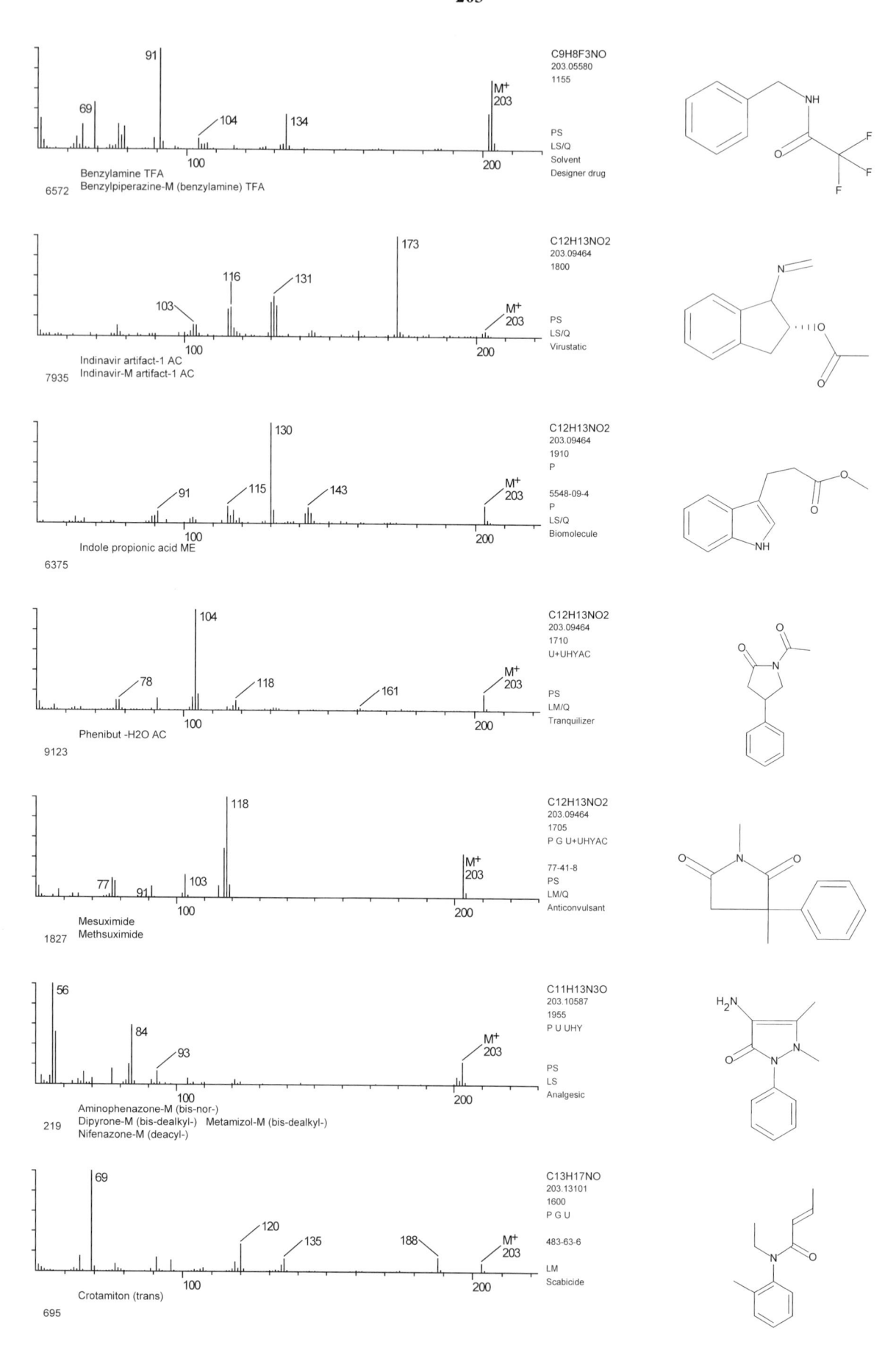

Benzylamine TFA Benzylpiperazine-M (benzylamine) TFA 6572	C9H8F3NO 203.05580 1155 PS LS/Q Solvent Designer drug
Indinavir artifact-1 AC Indinavir-M artifact-1 AC 7935	C12H13NO2 203.09464 1800 PS LS/Q Virustatic
Indole propionic acid ME 6375	C12H13NO2 203.09464 1910 P 5548-09-4 P LS/Q Biomolecule
Phenibut -H2O AC 9123	C12H13NO2 203.09464 1710 U+UHYAC PS LM/Q Tranquilizer
Mesuximide Methsuximide 1827	C12H13NO2 203.09464 1705 P G U+UHYAC 77-41-8 PS LM/Q Anticonvulsant
Aminophenazone-M (bis-nor-) Dipyrone-M (bis-dealkyl-) Metamizol-M (bis-dealkyl-) Nifenazone-M (deacyl-) 219	C11H13N3O 203.10587 1955 P U UHY PS LS Analgesic
Crotamiton (trans) 695	C13H17NO 203.13101 1600 P G U 483-63-6 LM Scabicide

203

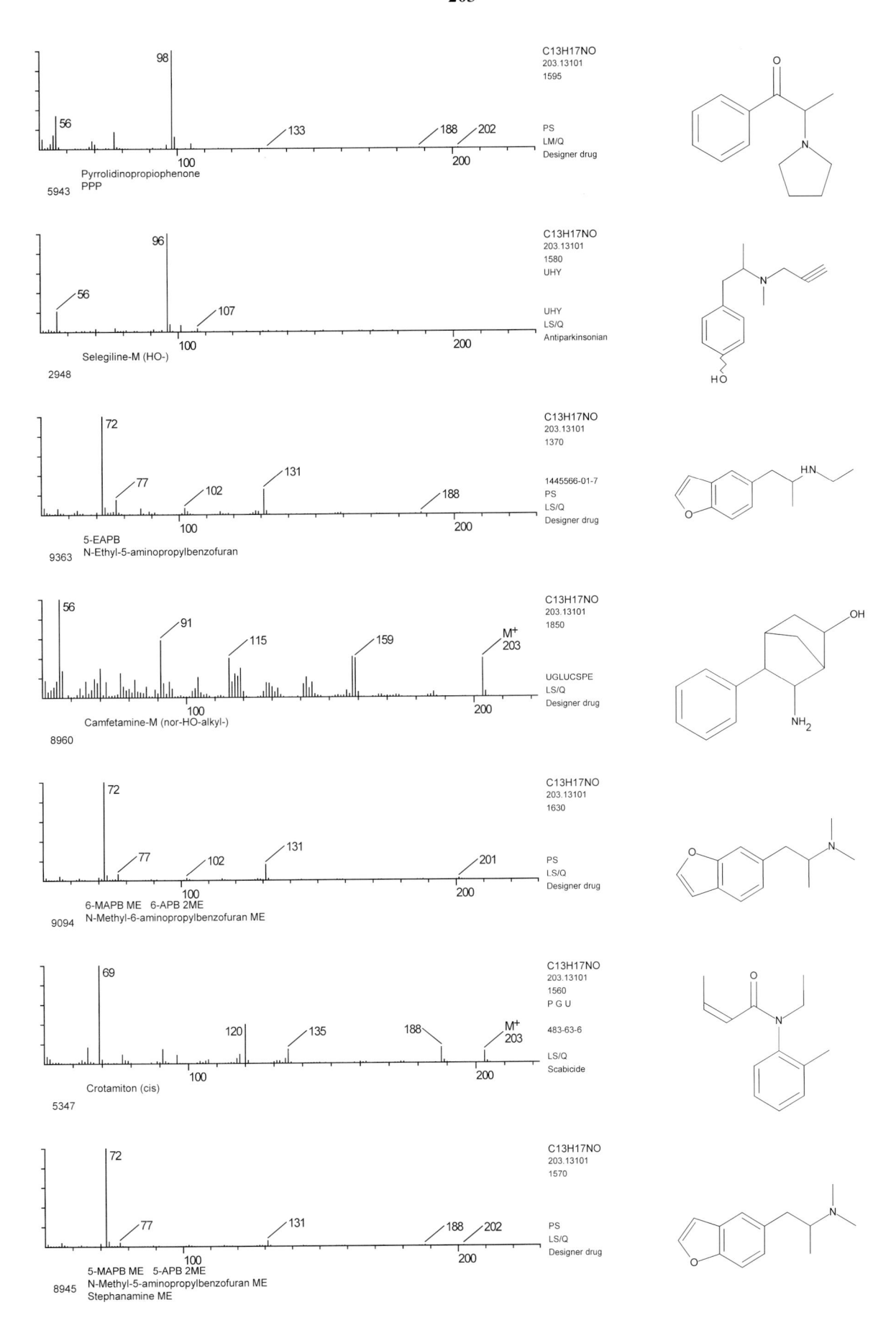

5943 Pyrrolidinopropiophenone PPP — peaks: 56, 98, 133, 188, 202	C13H17NO 203.13101 1595 PS LM/Q Designer drug
2948 Selegiline-M (HO-) — peaks: 56, 96, 107	C13H17NO 203.13101 1580 UHY UHY LS/Q Antiparkinsonian
9363 5-EAPB N-Ethyl-5-aminopropylbenzofuran — peaks: 72, 77, 102, 131, 188	C13H17NO 203.13101 1370 1445566-01-7 PS LS/Q Designer drug
8960 Camfetamine-M (nor-HO-alkyl-) — peaks: 56, 91, 115, 159, M+ 203	C13H17NO 203.13101 1850 UGLUCSPE LS/Q Designer drug
9094 6-MAPB ME 6-APB 2ME N-Methyl-6-aminopropylbenzofuran ME — peaks: 72, 77, 102, 131, 201	C13H17NO 203.13101 1630 PS LS/Q Designer drug
5347 Crotamiton (cis) — peaks: 69, 120, 135, 188, M+ 203	C13H17NO 203.13101 1560 P G U 483-63-6 LS/Q Scabicide
8945 5-MAPB ME 5-APB 2ME N-Methyl-5-aminopropylbenzofuran ME Stephanamine ME — peaks: 72, 77, 131, 188, 202	C13H17NO 203.13101 1570 PS LS/Q Designer drug

Pentoxyverine-M/artifact (alcohol) AC
6481

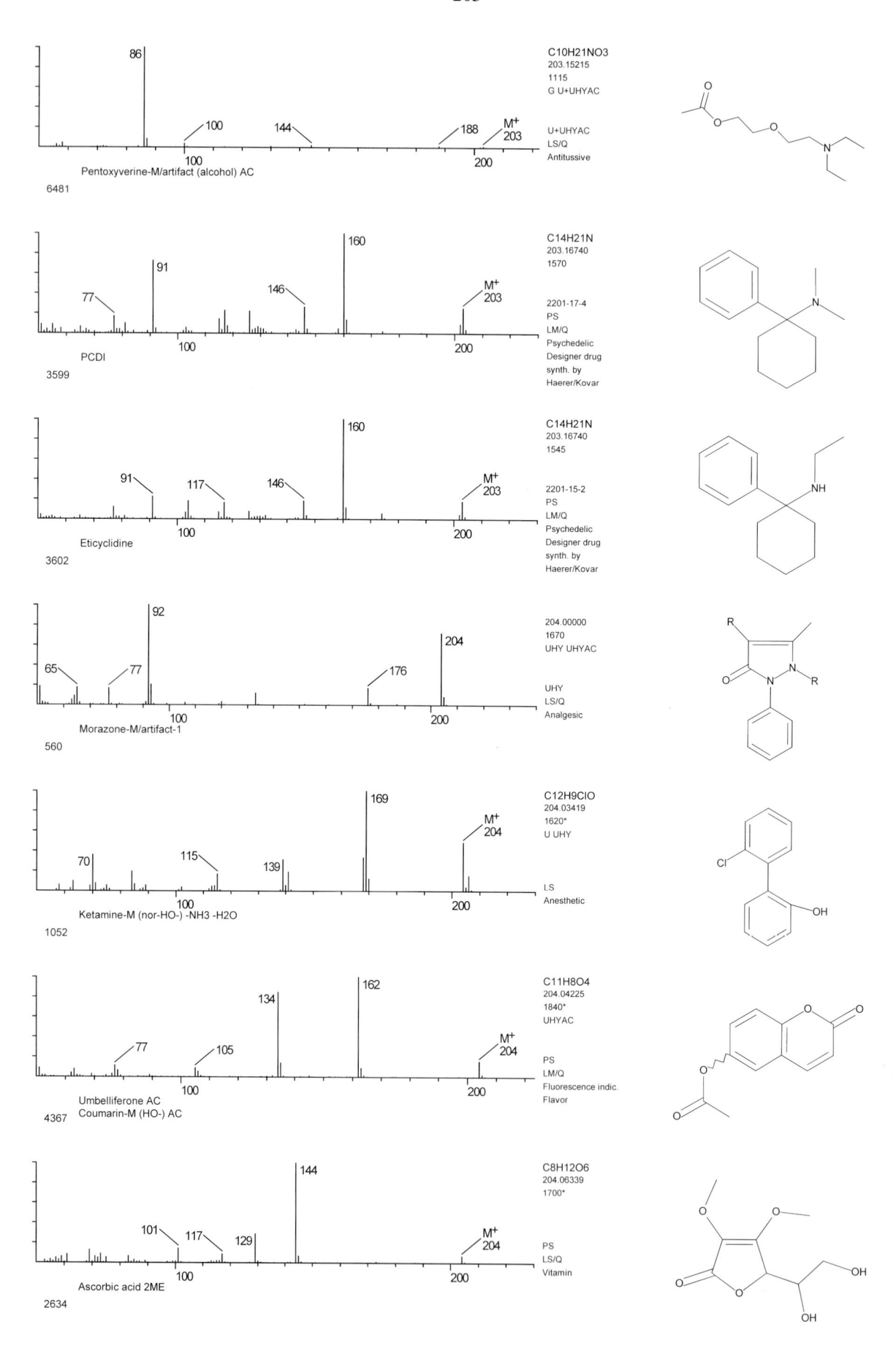

C10H21NO3
203.15215
1115
G U+UHYAC

U+UHYAC
LS/Q
Antitussive

PCDI
3599

C14H21N
203.16740
1570

2201-17-4
PS
LM/Q
Psychedelic
Designer drug
synth. by
Haerer/Kovar

Eticyclidine
3602

C14H21N
203.16740
1545

2201-15-2
PS
LM/Q
Psychedelic
Designer drug
synth. by
Haerer/Kovar

Morazone-M/artifact-1
560

204.00000
1670
UHY UHYAC

UHY
LS/Q
Analgesic

Ketamine-M (nor-HO-) -NH3 -H2O
1052

C12H9ClO
204.03419
1620*
U UHY

LS
Anesthetic

Umbelliferone AC
Coumarin-M (HO-) AC
4367

C11H8O4
204.04225
1840*
UHYAC

PS
LM/Q
Fluorescence indic.
Flavor

Ascorbic acid 2ME
2634

C8H12O6
204.06339
1700*

PS
LS/Q
Vitamin

204

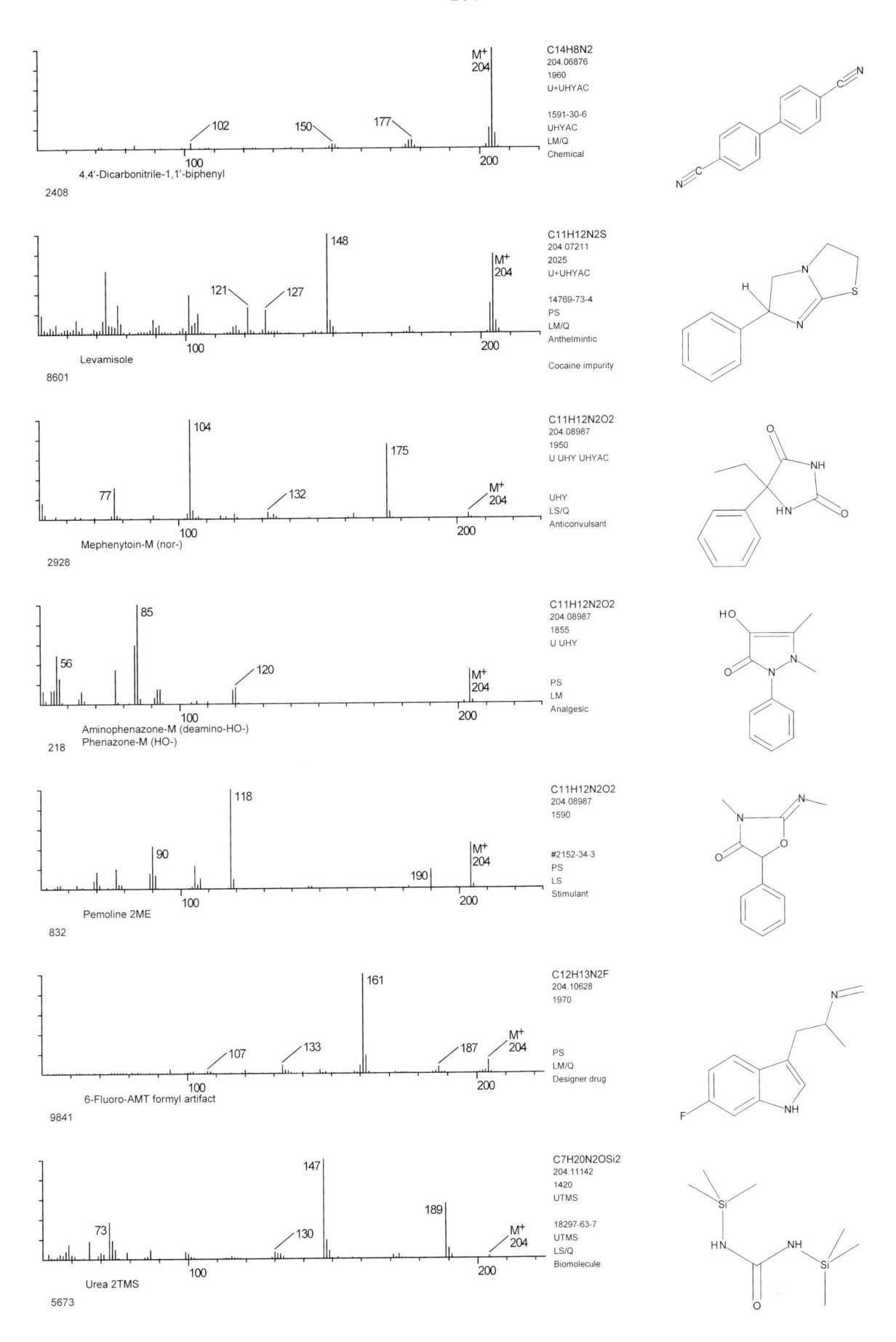

Spectrum	Formula / Info	Compound
peaks: 102, 150, 177, M+ 204	C14H8N2, 204.06876, 1960, U+UHYAC, 1591-30-6, UHYAC, LM/Q, Chemical	4,4'-Dicarbonitrile-1,1'-biphenyl (2408)
peaks: 121, 127, 148, M+ 204	C11H12N2S, 204.07211, 2025, U+UHYAC, 14769-73-4, PS, LM/Q, Anthelmintic, Cocaine impurity	Levamisole (8601)
peaks: 77, 104, 132, 175, M+ 204	C11H12N2O2, 204.08987, 1950, U UHY UHYAC, UHY, LS/Q, Anticonvulsant	Mephenytoin-M (nor-) (2928)
peaks: 56, 85, 120, M+ 204	C11H12N2O2, 204.08987, 1855, U UHY, PS, LM, Analgesic	Aminophenazone-M (deamino-HO-) / Phenazone-M (HO-) (218)
peaks: 90, 118, 190, M+ 204	C11H12N2O2, 204.08987, 1590, #2152-34-3, PS, LS, Stimulant	Pemoline 2ME (832)
peaks: 107, 133, 161, 187, M+ 204	C12H13N2F, 204.10628, 1970, PS, LM/Q, Designer drug	6-Fluoro-AMT formyl artifact (9841)
peaks: 73, 130, 147, 189, M+ 204	C7H20N2OSi2, 204.11142, 1420, UTMS, 18297-63-7, UTMS, LS/Q, Biomolecule	Urea 2TMS (5673)

205

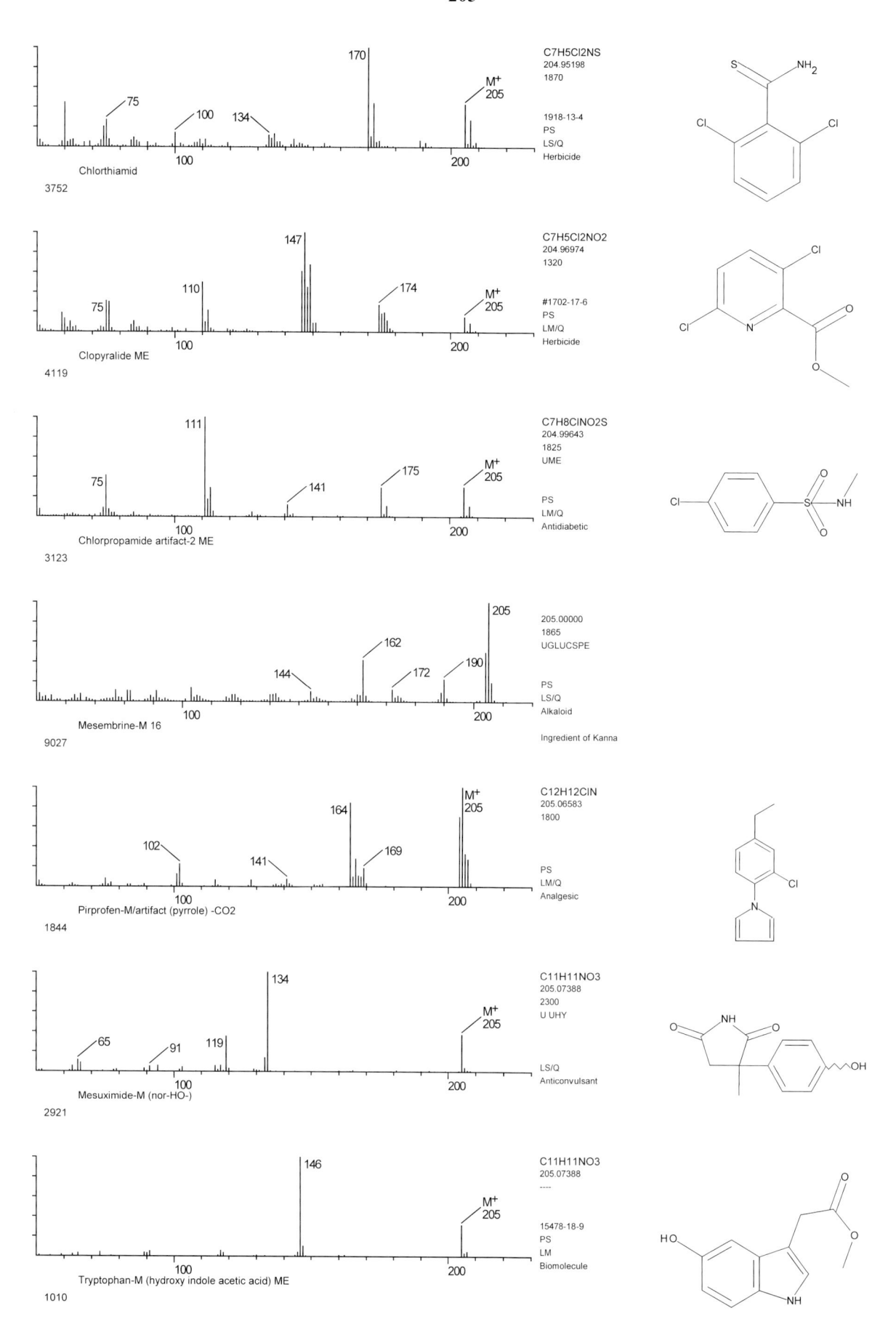

Chlorthiamid
C7H5Cl2NS
204.95198
1870
1918-13-4
PS
LS/Q
Herbicide
Peaks: 75, 100, 134, 170, M+ 205
3752

Clopyralide ME
C7H5Cl2NO2
204.96974
1320
#1702-17-6
PS
LM/Q
Herbicide
Peaks: 75, 110, 147, 174, M+ 205
4119

Chlorpropamide artifact-2 ME
C7H8ClNO2S
204.99643
1825
UME
PS
LM/Q
Antidiabetic
Peaks: 75, 111, 141, 175, M+ 205
3123

Mesembrine-M 16
205.00000
1865
UGLUCSPE
PS
LS/Q
Alkaloid
Ingredient of Kanna
Peaks: 144, 162, 172, 190, 205
9027

Pirprofen-M/artifact (pyrrole) -CO2
C12H12ClN
205.06583
1800
PS
LM/Q
Analgesic
Peaks: 102, 141, 164, 169, M+ 205
1844

Mesuximide-M (nor-HO-)
C11H11NO3
205.07388
2300
U UHY
LS/Q
Anticonvulsant
Peaks: 65, 91, 119, 134, M+ 205
2921

Tryptophan-M (hydroxy indole acetic acid) ME
C11H11NO3
205.07388

15478-18-9
PS
LM
Biomolecule
Peaks: 146, M+ 205
1010

205

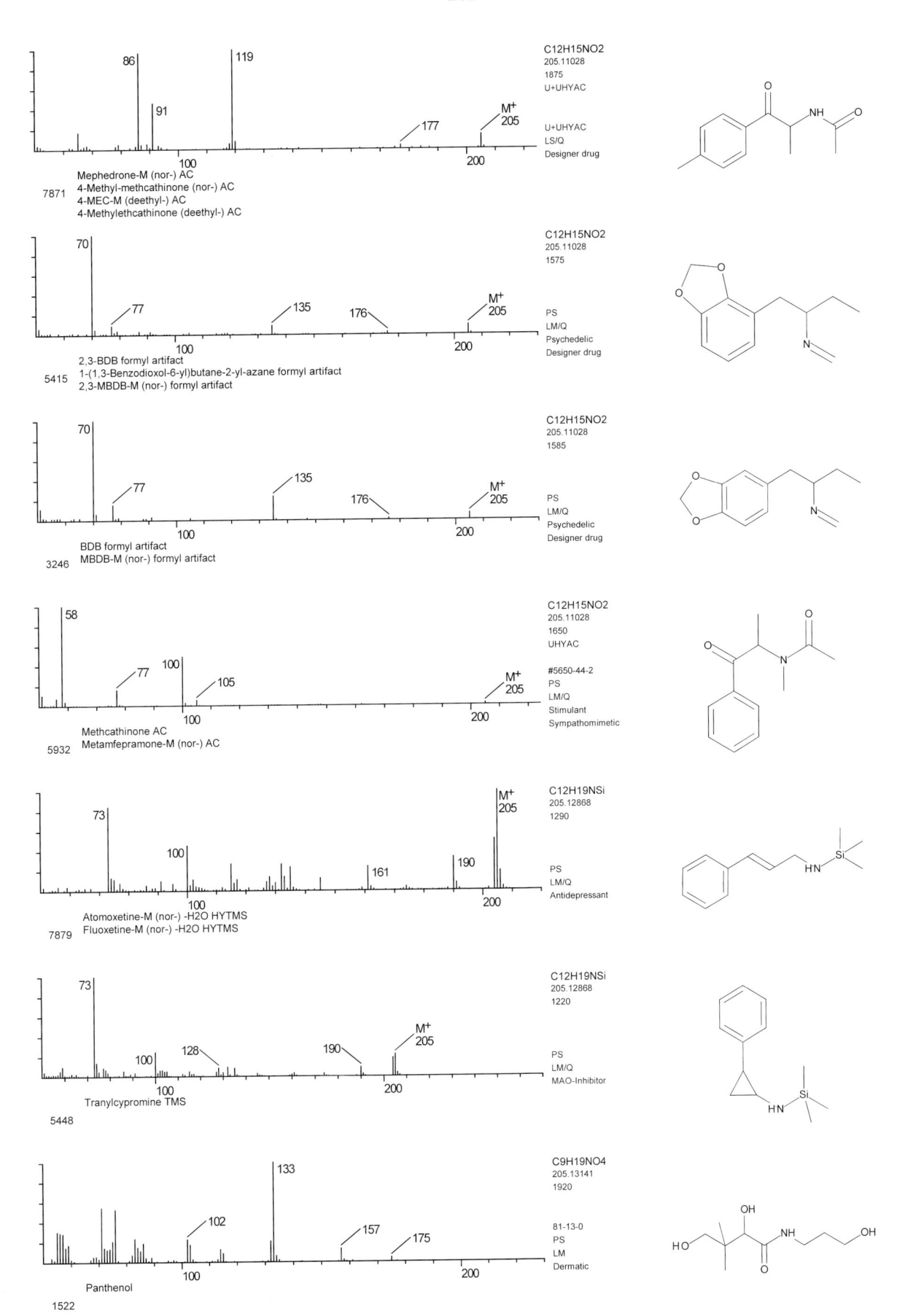

ID	Compound	Formula	Mass	RI	Category
7871	Mephedrone-M (nor-) AC / 4-Methyl-methcathinone (nor-) AC / 4-MEC-M (deethyl-) AC / 4-Methylethcathinone (deethyl-) AC	C12H15NO2	205.11028	1875	U+UHYAC / U+UHYAC / LS/Q / Designer drug
	Peaks: 86, 91, 119, 177, 205 (M+)				
5415	2,3-BDB formyl artifact / 1-(1,3-Benzodioxol-6-yl)butane-2-yl-azane formyl artifact / 2,3-MBDB-M (nor-) formyl artifact	C12H15NO2	205.11028	1575	PS / LM/Q / Psychedelic / Designer drug
	Peaks: 70, 77, 135, 176, 205 (M+)				
3246	BDB formyl artifact / MBDB-M (nor-) formyl artifact	C12H15NO2	205.11028	1585	PS / LM/Q / Psychedelic / Designer drug
	Peaks: 70, 77, 135, 176, 205 (M+)				
5932	Methcathinone AC / Metamfepramone-M (nor-) AC	C12H15NO2	205.11028	1650	UHYAC / #5650-44-2 / PS / LM/Q / Stimulant / Sympathomimetic
	Peaks: 58, 77, 100, 105, 205 (M+)				
7879	Atomoxetine-M (nor-) -H2O HYTMS / Fluoxetine-M (nor-) -H2O HYTMS	C12H19NSi	205.12868	1290	PS / LM/Q / Antidepressant
	Peaks: 73, 100, 161, 190, 205 (M+)				
5448	Tranylcypromine TMS	C12H19NSi	205.12868	1220	PS / LM/Q / MAO-Inhibitor
	Peaks: 73, 100, 128, 190, 205 (M+)				
1522	Panthenol	C9H19NO4	205.13141	1920	81-13-0 / PS / LM / Dermatic
	Peaks: 102, 133, 157, 175				

205

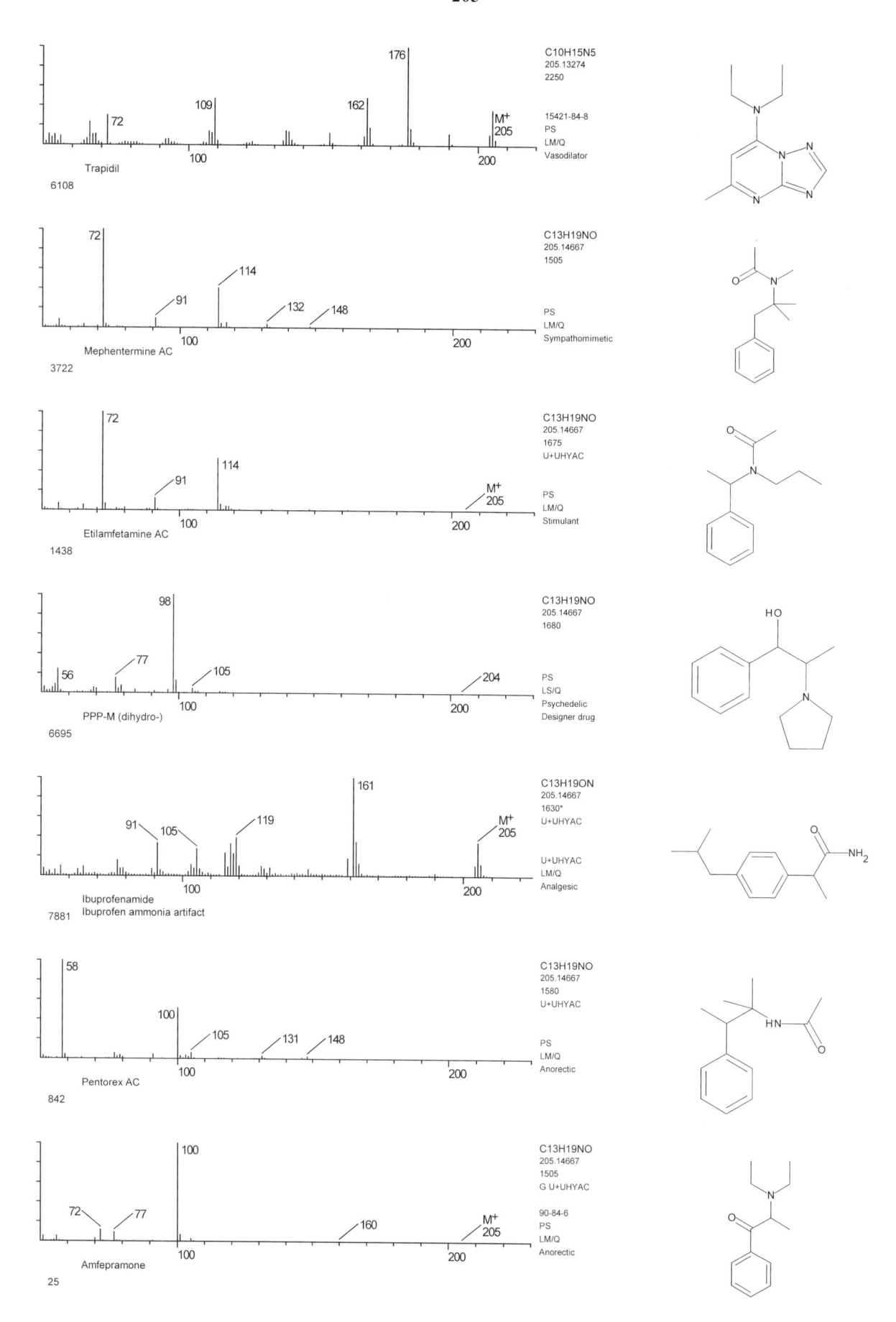

Trapidil
6108
C10H15N5
205.13274
2250
15421-84-8
PS
LM/Q
Vasodilator

Mephentermine AC
3722
C13H19NO
205.14667
1505
PS
LM/Q
Sympathomimetic

Etilamfetamine AC
1438
C13H19NO
205.14667
1675
U+UHYAC
PS
LM/Q
Stimulant

PPP-M (dihydro-)
6695
C13H19NO
205.14667
1680
PS
LS/Q
Psychedelic
Designer drug

Ibuprofenamide
Ibuprofen ammonia artifact
7881
C13H19ON
205.14667
1630*
U+UHYAC
U+UHYAC
LM/Q
Analgesic

Pentorex AC
842
C13H19NO
205.14667
1580
U+UHYAC
PS
LM/Q
Anorectic

Amfepramone
25
C13H19NO
205.14667
1505
G U+UHYAC
90-84-6
PS
LM/Q
Anorectic

205

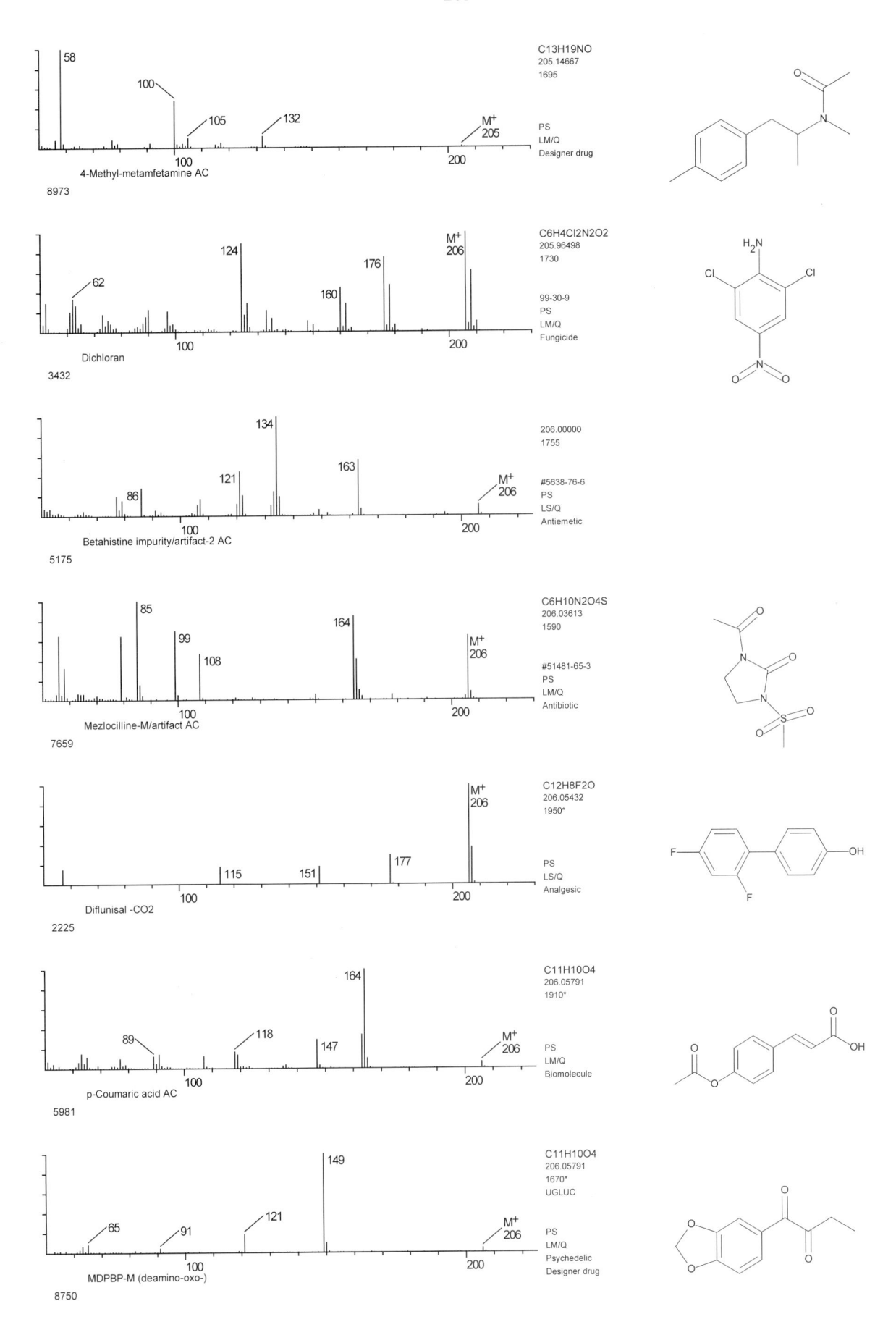

4-Methyl-metamfetamine AC
8973
C13H19NO
205.14667
1695
PS
LM/Q
Designer drug

Dichloran
3432
C6H4Cl2N2O2
205.96498
1730
99-30-9
PS
LM/Q
Fungicide

Betahistine impurity/artifact-2 AC
5175
206.00000
1755
#5638-76-6
PS
LS/Q
Antiemetic

Mezlocilline-M/artifact AC
7659
C6H10N2O4S
206.03613
1590
#51481-65-3
PS
LM/Q
Antibiotic

Diflunisal -CO2
2225
C12H8F2O
206.05432
1950*
PS
LS/Q
Analgesic

p-Coumaric acid AC
5981
C11H10O4
206.05791
1910*
PS
LM/Q
Biomolecule

MDPBP-M (deamino-oxo-)
8750
C11H10O4
206.05791
1670*
UGLUC
PS
LM/Q
Psychedelic
Designer drug

206

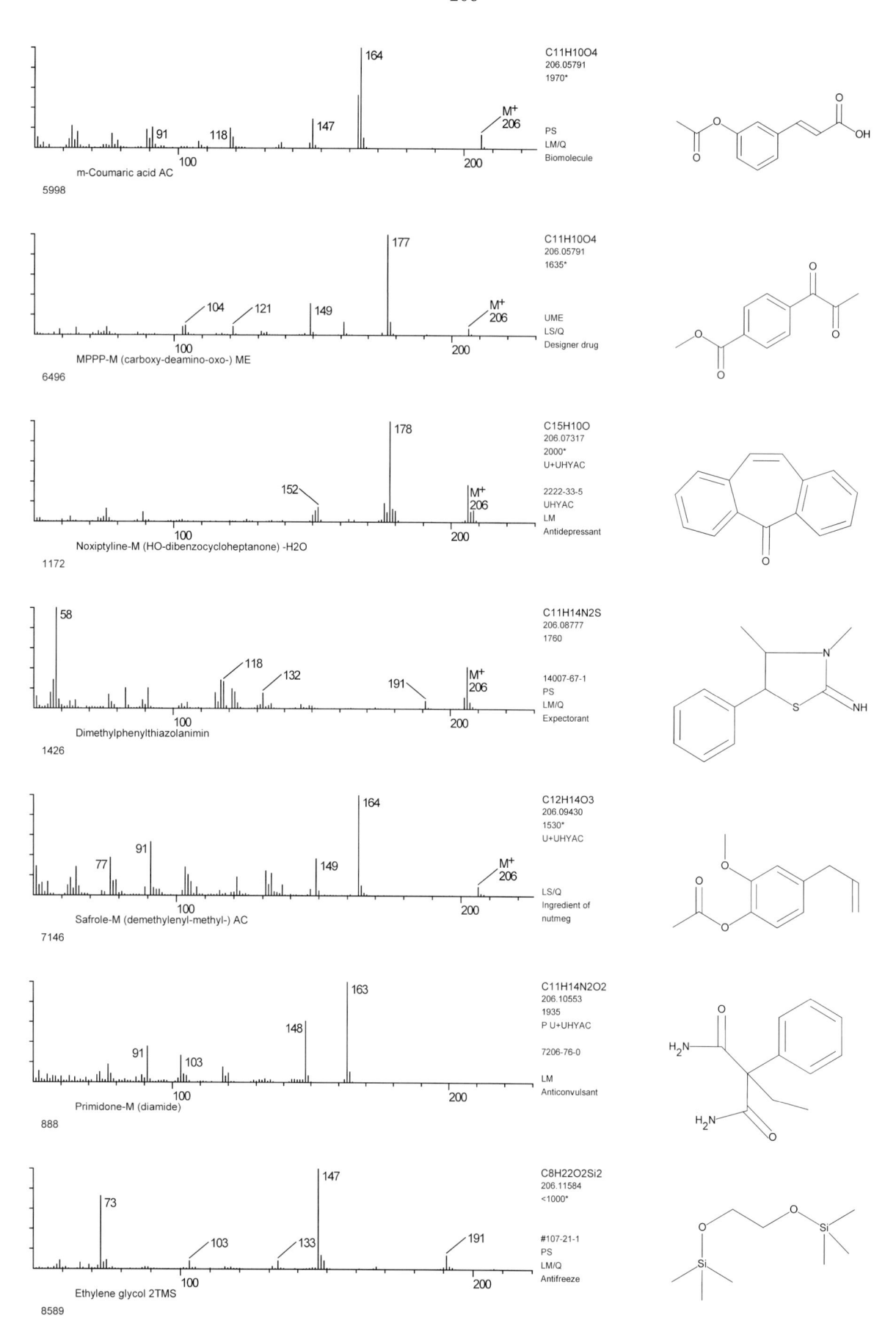

m-Coumaric acid AC
C11H10O4
206.05791
1970*
PS
LM/Q
Biomolecule
Peaks: 91, 118, 147, 164, M+ 206
5998

MPPP-M (carboxy-deamino-oxo-) ME
C11H10O4
206.05791
1635*
UME
LS/Q
Designer drug
Peaks: 104, 121, 149, 177, M+ 206
6496

Noxiptyline-M (HO-dibenzocycloheptanone) -H2O
C15H10O
206.07317
2000*
U+UHYAC
2222-33-5
UHYAC
LM
Antidepressant
Peaks: 152, 178, M+ 206
1172

Dimethylphenylthiazolanimin
C11H14N2S
206.08777
1760
14007-67-1
PS
LM/Q
Expectorant
Peaks: 58, 118, 132, 191, M+ 206
1426

Safrole-M (demethylenyl-methyl-) AC
C12H14O3
206.09430
1530*
U+UHYAC
LS/Q
Ingredient of nutmeg
Peaks: 77, 91, 149, 164, M+ 206
7146

Primidone-M (diamide)
C11H14N2O2
206.10553
1935
P U+UHYAC
7206-76-0
LM
Anticonvulsant
Peaks: 91, 103, 148, 163
888

Ethylene glycol 2TMS
C8H22O2Si2
206.11584
<1000*
#107-21-1
PS
LM/Q
Antifreeze
Peaks: 73, 103, 133, 147, 191
8589

206

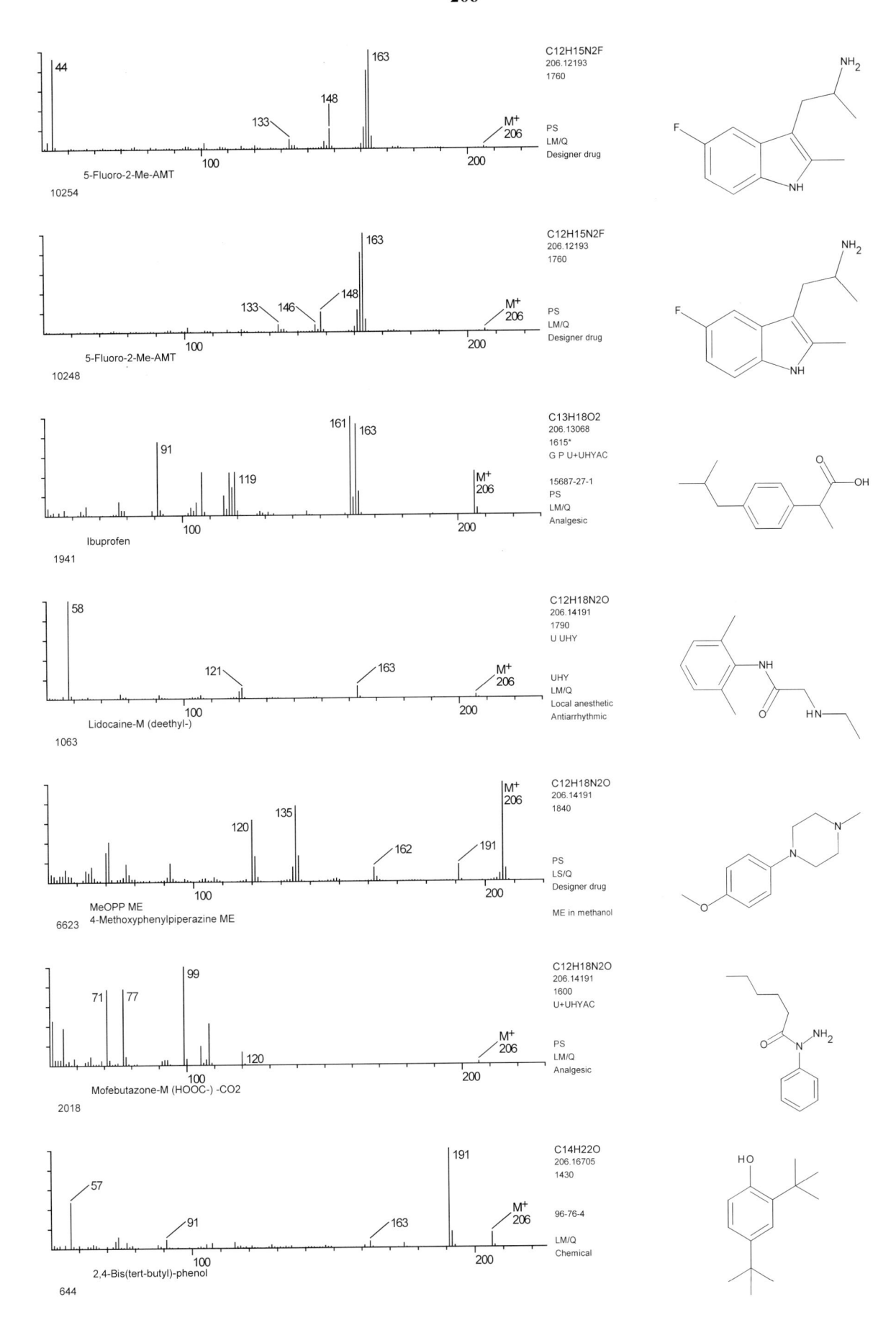

5-Fluoro-2-Me-AMT 10254	44, 133, 148, 163, M+ 206	C12H15N2F 206.12193 1760 PS LM/Q Designer drug
5-Fluoro-2-Me-AMT 10248	133, 146, 148, 163, M+ 206	C12H15N2F 206.12193 1760 PS LM/Q Designer drug
Ibuprofen 1941	91, 119, 161, 163, M+ 206	C13H18O2 206.13068 1615* G P U+UHYAC 15687-27-1 PS LM/Q Analgesic
Lidocaine-M (deethyl-) 1063	58, 121, 163, M+ 206	C12H18N2O 206.14191 1790 U UHY UHY LM/Q Local anesthetic Antiarrhythmic
MeOPP ME 4-Methoxyphenylpiperazine ME 6623	120, 135, 162, 191, M+ 206	C12H18N2O 206.14191 1840 PS LS/Q Designer drug ME in methanol
Mofebutazone-M (HOOC-) -CO2 2018	71, 77, 99, 120, M+ 206	C12H18N2O 206.14191 1600 U+UHYAC PS LM/Q Analgesic
2,4-Bis(tert-butyl)-phenol 644	57, 91, 163, 191, M+ 206	C14H22O 206.16705 1430 96-76-4 LM/Q Chemical

206

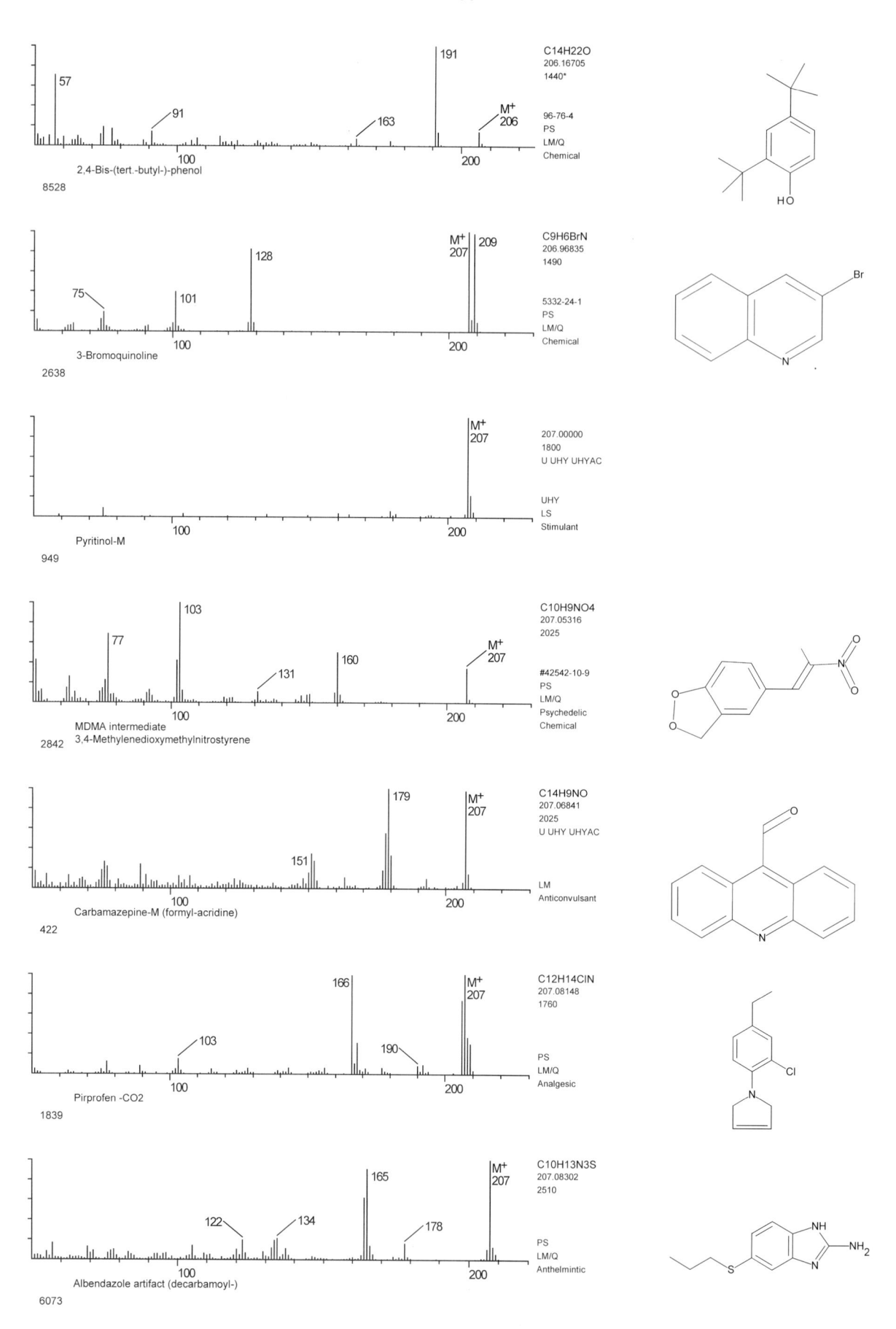

2,4-Bis-(tert.-butyl-)-phenol — peaks: 57, 91, 163, 191, M+ 206	C14H22O, 206.16705, 1440*, 96-76-4, PS, LM/Q, Chemical
3-Bromoquinoline — peaks: 75, 101, 128, M+ 207, 209	C9H6BrN, 206.96835, 1490, 5332-24-1, PS, LM/Q, Chemical
Pyritinol-M — peaks: M+ 207	207.00000, 1800, U UHY UHYAC, UHY, LS, Stimulant
MDMA intermediate 3,4-Methylenedioxymethylnitrostyrene — peaks: 77, 103, 131, 160, M+ 207	C10H9NO4, 207.05316, 2025, #42542-10-9, PS, LM/Q, Psychedelic Chemical
Carbamazepine-M (formyl-acridine) — peaks: 151, 179, M+ 207	C14H9NO, 207.06841, 2025, U UHY UHYAC, LM, Anticonvulsant
Pirprofen -CO2 — peaks: 103, 166, 190, M+ 207	C12H14ClN, 207.08148, 1760, PS, LM/Q, Analgesic
Albendazole artifact (decarbamoyl-) — peaks: 122, 134, 165, 178, M+ 207	C10H13N3S, 207.08302, 2510, PS, LM/Q, Anthelmintic

207

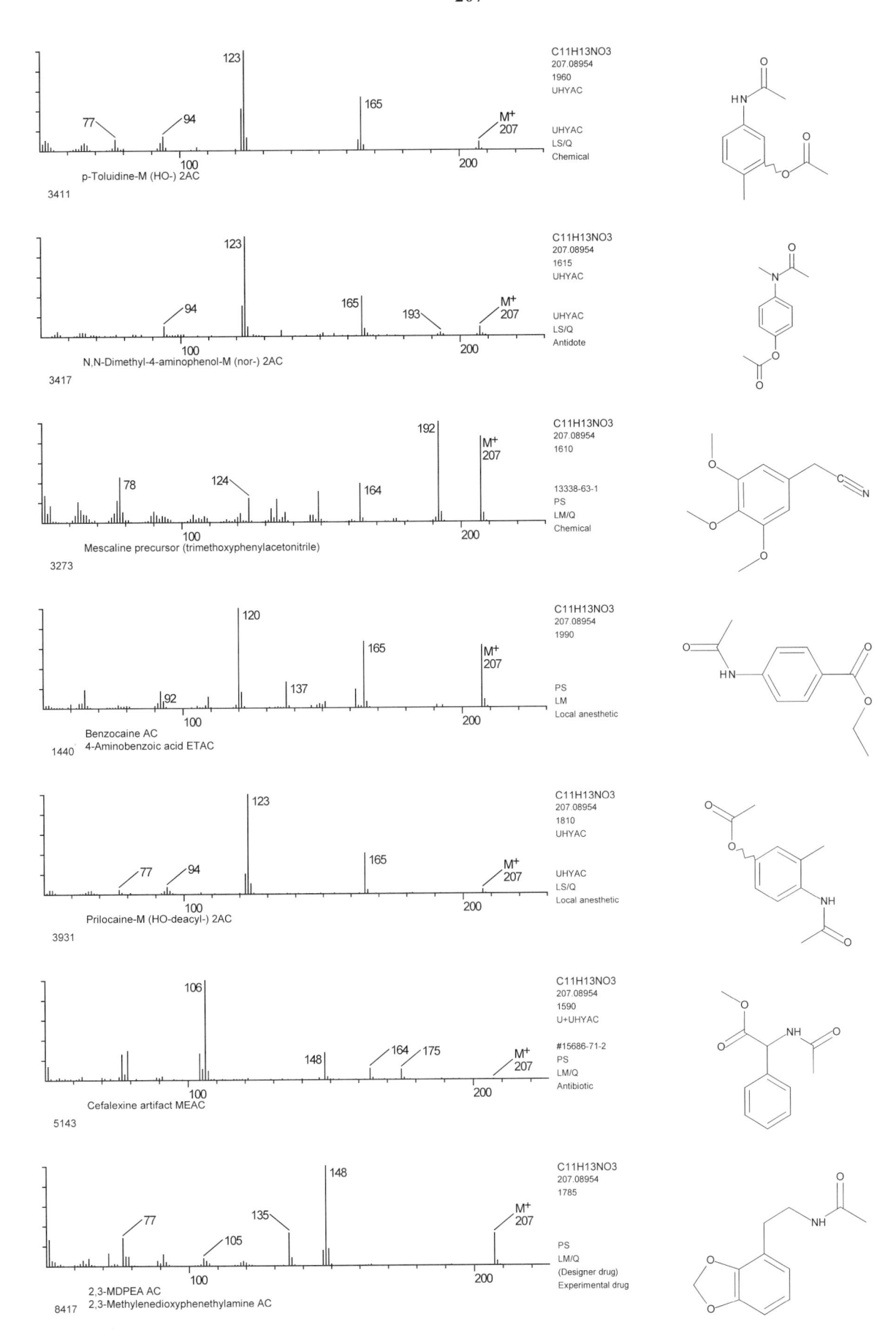

p-Toluidine-M (HO-) 2AC 3411	C11H13NO3 207.08954 1960 UHYAC UHYAC LS/Q Chemical
N,N-Dimethyl-4-aminophenol-M (nor-) 2AC 3417	C11H13NO3 207.08954 1615 UHYAC UHYAC LS/Q Antidote
Mescaline precursor (trimethoxyphenylacetonitrile) 3273	C11H13NO3 207.08954 1610 13338-63-1 PS LM/Q Chemical
Benzocaine AC 4-Aminobenzoic acid ETAC 1440	C11H13NO3 207.08954 1990 PS LM Local anesthetic
Prilocaine-M (HO-deacyl-) 2AC 3931	C11H13NO3 207.08954 1810 UHYAC UHYAC LS/Q Local anesthetic
Cefalexine artifact MEAC 5143	C11H13NO3 207.08954 1590 U+UHYAC #15686-71-2 PS LM/Q Antibiotic
2,3-MDPEA AC 2,3-Methylenedioxyphenethylamine AC 8417	C11H13NO3 207.08954 1785 PS LM/Q (Designer drug) Experimental drug

207

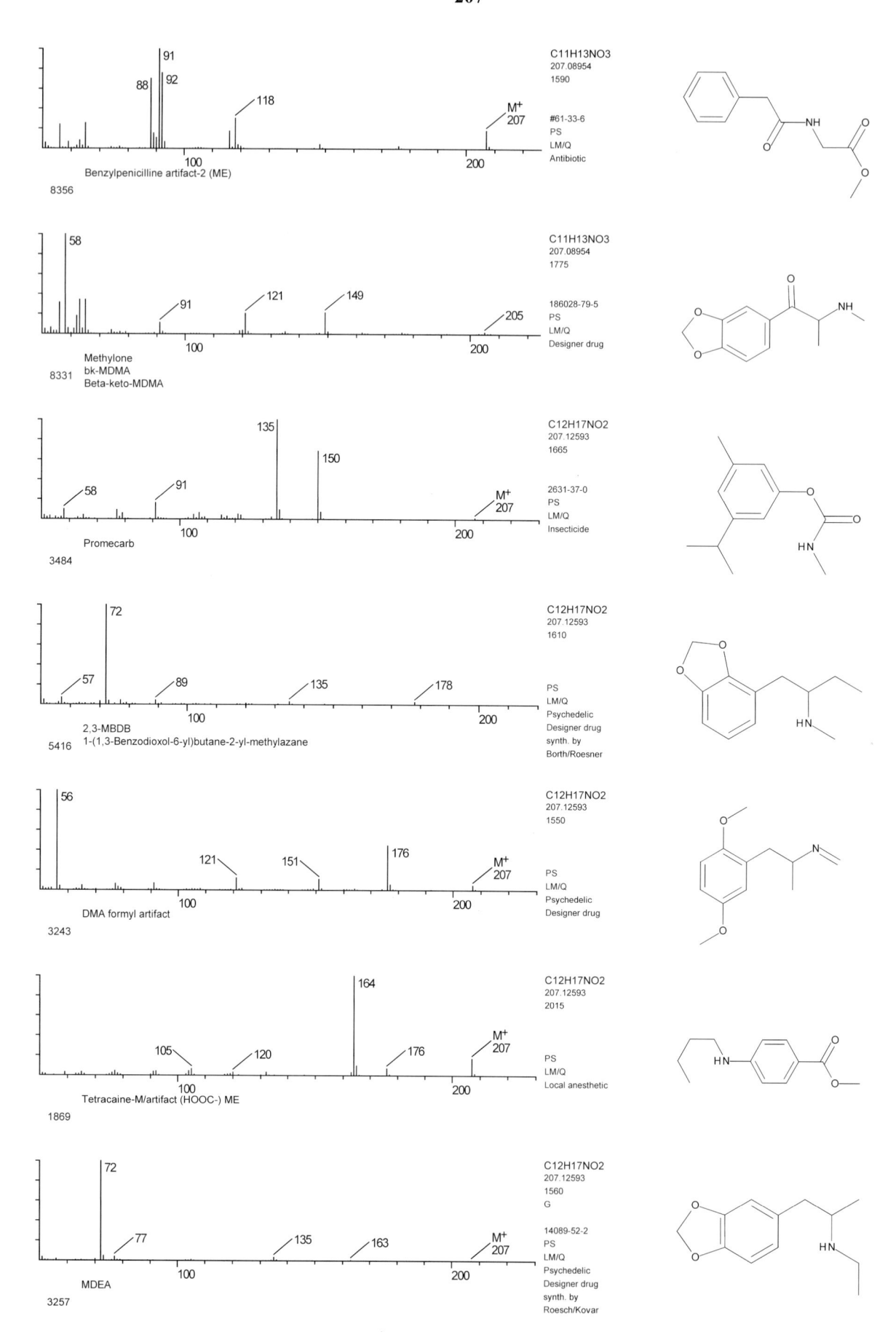

C11H13NO3
207.08954
1590
#61-33-6
PS
LM/Q
Antibiotic

Benzylpenicilline artifact-2 (ME)
8356

C11H13NO3
207.08954
1775
186028-79-5
PS
LM/Q
Designer drug

Methylone
bk-MDMA
Beta-keto-MDMA
8331

C12H17NO2
207.12593
1665
2631-37-0
PS
LM/Q
Insecticide

Promecarb
3484

C12H17NO2
207.12593
1610
PS
LM/Q
Psychedelic
Designer drug
synth. by
Borth/Roesner

2,3-MBDB
1-(1,3-Benzodioxol-6-yl)butane-2-yl-methylazane
5416

C12H17NO2
207.12593
1550
PS
LM/Q
Psychedelic
Designer drug

DMA formyl artifact
3243

C12H17NO2
207.12593
2015
PS
LM/Q
Local anesthetic

Tetracaine-M/artifact (HOOC-) ME
1869

C12H17NO2
207.12593
1560
G
14089-52-2
PS
LM/Q
Psychedelic
Designer drug
synth. by
Roesch/Kovar

MDEA
3257

420

207

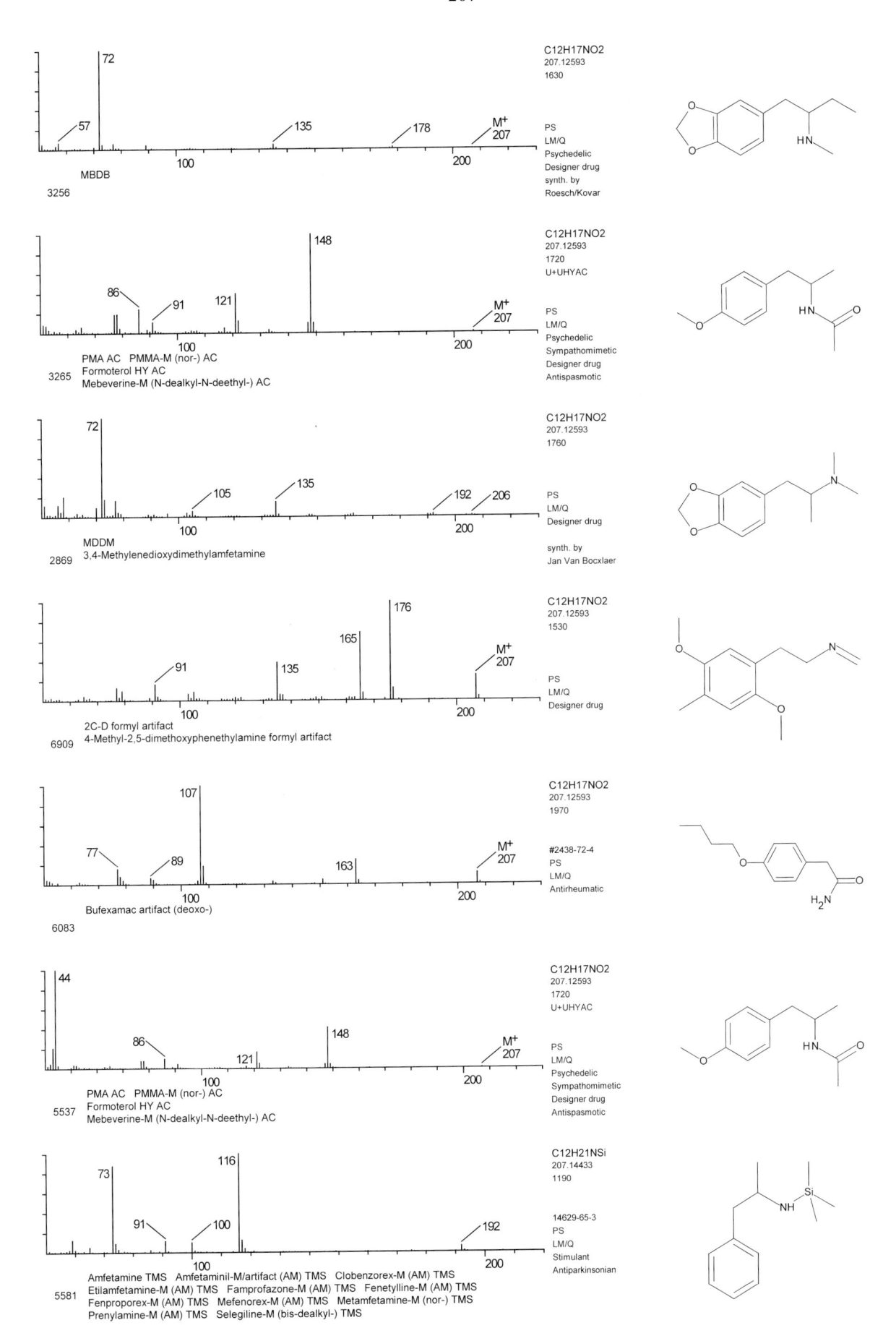

Spectrum	Peaks	Formula / Info
MBDB — 3256	57, 72, 135, 178, M⁺ 207	C12H17NO2, 207.12593, 1630, PS, LM/Q, Psychedelic, Designer drug, synth. by Roesch/Kovar
PMA AC / PMMA-M (nor-) AC / Formoterol HY AC / Mebeverine-M (N-dealkyl-N-deethyl-) AC — 3265	86, 91, 121, 148, M⁺ 207	C12H17NO2, 207.12593, 1720, U+UHYAC, PS, LM/Q, Psychedelic, Sympathomimetic, Designer drug, Antispasmotic
MDDM — 2869, 3,4-Methylenedioxydimethylamfetamine	72, 105, 135, 192, 206	C12H17NO2, 207.12593, 1760, PS, LM/Q, Designer drug, synth. by Jan Van Bocxlaer
2C-D formyl artifact — 6909, 4-Methyl-2,5-dimethoxyphenethylamine formyl artifact	91, 135, 165, 176, M⁺ 207	C12H17NO2, 207.12593, 1530, PS, LM/Q, Designer drug
Bufexamac artifact (deoxo-) — 6083	77, 89, 107, 163, M⁺ 207	C12H17NO2, 207.12593, 1970, #2438-72-4, PS, LM/Q, Antirheumatic
PMA AC / PMMA-M (nor-) AC / Formoterol HY AC / Mebeverine-M (N-dealkyl-N-deethyl-) AC — 5537	44, 86, 121, 148, M⁺ 207	C12H17NO2, 207.12593, 1720, U+UHYAC, PS, LM/Q, Psychedelic, Sympathomimetic, Designer drug, Antispasmotic
Amfetamine TMS Amfetaminil-M/artifact (AM) TMS Clobenzorex-M (AM) TMS Etilamfetamine-M (AM) TMS Famprofazone-M (AM) TMS Fenetylline-M (AM) TMS Fenproporex-M (AM) TMS Mefenorex-M (AM) TMS Metamfetamine-M (nor-) TMS Prenylamine-M (AM) TMS Selegiline-M (bis-dealkyl-) TMS — 5581	73, 91, 100, 116, 192	C12H21NSi, 207.14433, 1190, 14629-65-3, PS, LM/Q, Stimulant, Antiparkinsonian

207

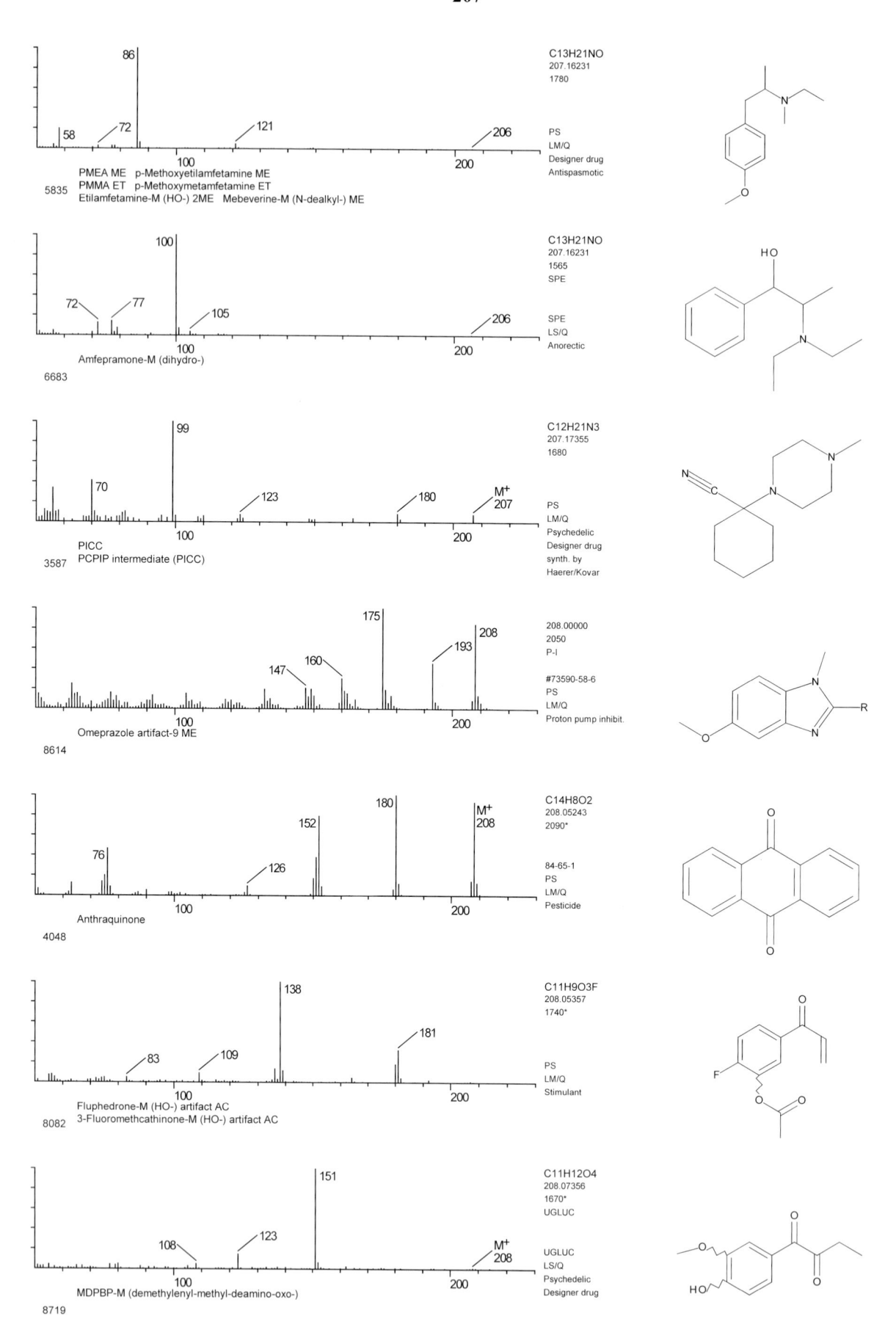

Spectrum #	Compound	Formula	Mass	Info
5835	PMEA ME p-Methoxyetilamfetamine ME PMMA ET p-Methoxymetamfetamine ET Etilamfetamine-M (HO-) 2ME Mebeverine-M (N-dealkyl-) ME	C13H21NO	207.16231	1780 PS LM/Q Designer drug Antispasmotic
6683	Amfepramone-M (dihydro-)	C13H21NO	207.16231	1565 SPE SPE LS/Q Anorectic
3587	PICC PCPIP intermediate (PICC)	C12H21N3	207.17355	1680 PS LM/Q Psychedelic Designer drug synth. by Haerer/Kovar
8614	Omeprazole artifact-9 ME		208.00000	2050 P-I #73590-58-6 PS LM/Q Proton pump inhibit.
4048	Anthraquinone	C14H8O2	208.05243	2090* 84-65-1 PS LM/Q Pesticide
8082	Fluphedrone-M (HO-) artifact AC 3-Fluoromethcathinone-M (HO-) artifact AC	C11H9O3F	208.05357	1740* PS LM/Q Stimulant
8719	MDPBP-M (demethylenyl-methyl-deamino-oxo-)	C11H12O4	208.07356	1670* UGLUC UGLUC LS/Q Psychedelic Designer drug

208

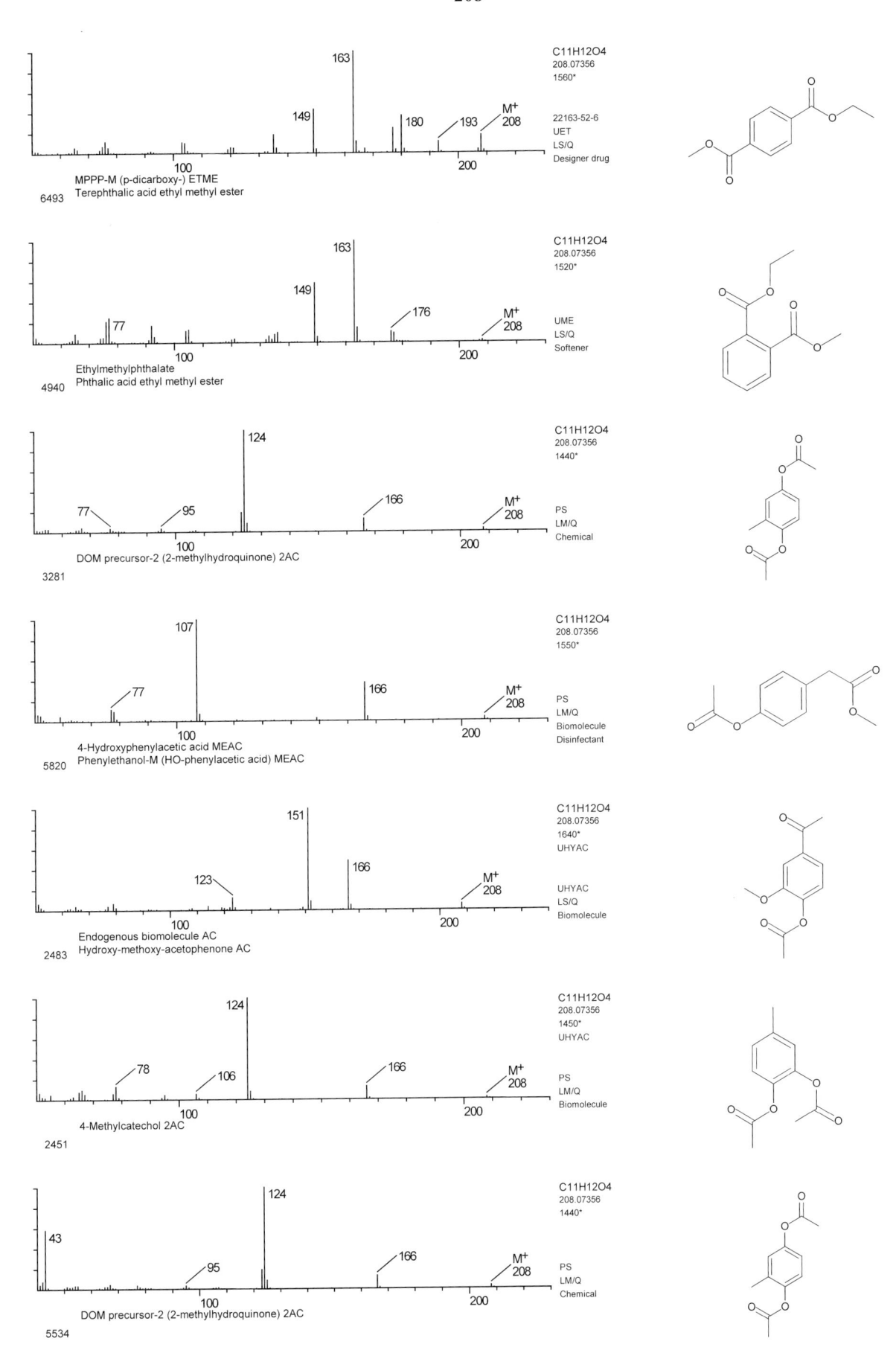

Spectrum	Compound	Formula/Info
6493	MPPP-M (p-dicarboxy-) ETME / Terephthalic acid ethyl methyl ester	C11H12O4, 208.07356, 1560*, 22163-52-6, UET, LS/Q, Designer drug. Peaks: 149, 163, 180, 193, M+ 208
4940	Ethylmethylphthalate / Phthalic acid ethyl methyl ester	C11H12O4, 208.07356, 1520*, UME, LS/Q, Softener. Peaks: 77, 149, 163, 176, M+ 208
3281	DOM precursor-2 (2-methylhydroquinone) 2AC	C11H12O4, 208.07356, 1440*, PS, LM/Q, Chemical. Peaks: 77, 95, 124, 166, M+ 208
5820	4-Hydroxyphenylacetic acid MEAC / Phenylethanol-M (HO-phenylacetic acid) MEAC	C11H12O4, 208.07356, 1550*, PS, LM/Q, Biomolecule, Disinfectant. Peaks: 77, 107, 166, M+ 208
2483	Endogenous biomolecule AC / Hydroxy-methoxy-acetophenone AC	C11H12O4, 208.07356, 1640*, UHYAC, UHYAC, LS/Q, Biomolecule. Peaks: 123, 151, 166, M+ 208
2451	4-Methylcatechol 2AC	C11H12O4, 208.07356, 1450*, UHYAC, PS, LM/Q, Biomolecule. Peaks: 78, 106, 124, 166, M+ 208
5534	DOM precursor-2 (2-methylhydroquinone) 2AC	C11H12O4, 208.07356, 1440*, PS, LM/Q, Chemical. Peaks: 43, 95, 124, 166, M+ 208

208

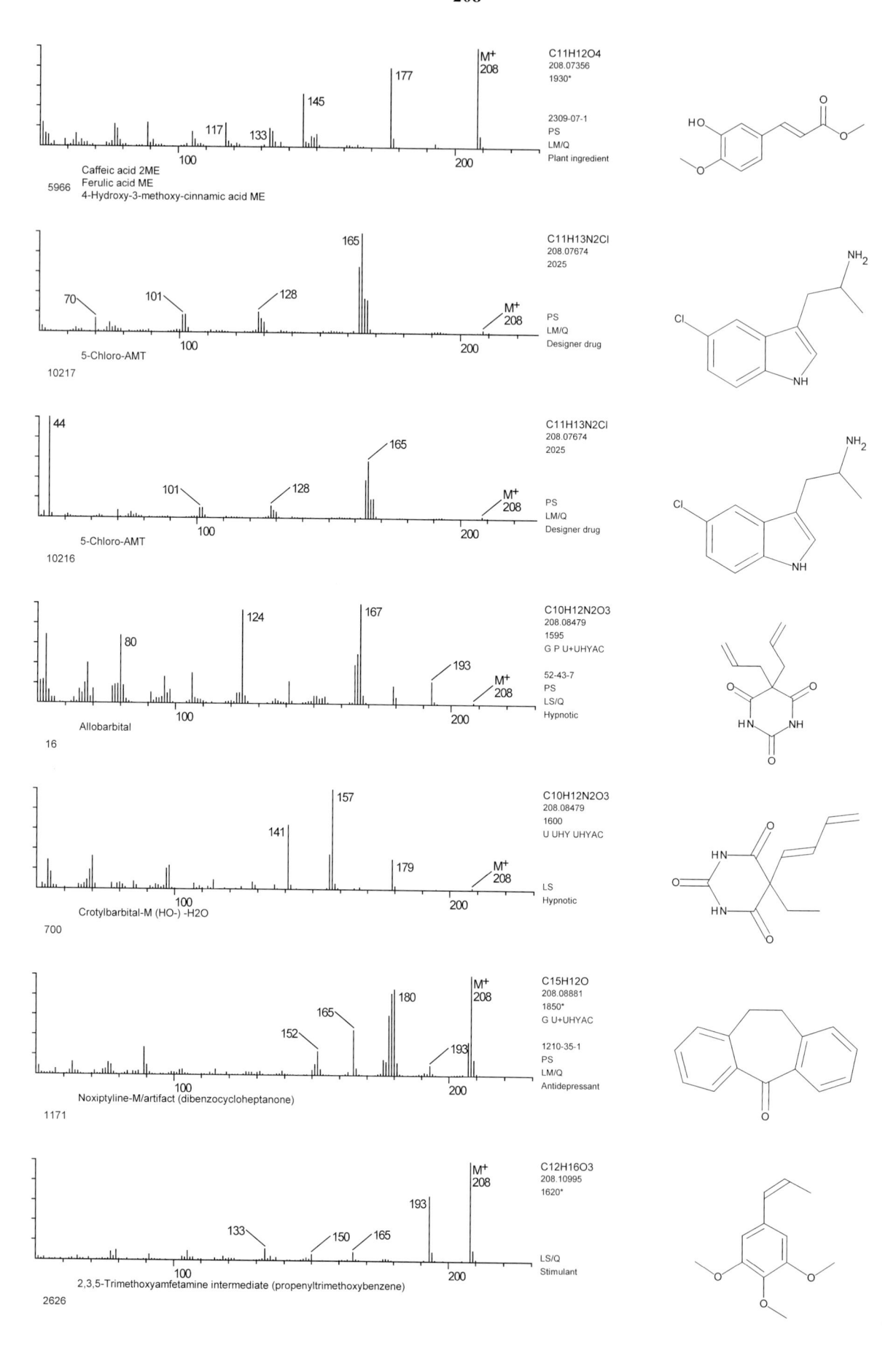

Caffeic acid 2ME
Ferulic acid ME
4-Hydroxy-3-methoxy-cinnamic acid ME

C11H12O4
208.07356
1930*

2309-07-1
PS
LM/Q
Plant ingredient

5966

5-Chloro-AMT

C11H13N2Cl
208.07674
2025

PS
LM/Q
Designer drug

10217

5-Chloro-AMT

C11H13N2Cl
208.07674
2025

PS
LM/Q
Designer drug

10216

Allobarbital

C10H12N2O3
208.08479
1595
G P U+UHYAC

52-43-7
PS
LS/Q
Hypnotic

16

Crotylbarbital-M (HO-) -H2O

C10H12N2O3
208.08479
1600
U UHY UHYAC

LS
Hypnotic

700

Noxiptyline-M/artifact (dibenzocycloheptanone)

C15H12O
208.08881
1850*
G U+UHYAC

1210-35-1
PS
LM/Q
Antidepressant

1171

2,3,5-Trimethoxyamfetamine intermediate (propenyltrimethoxybenzene)

C12H16O3
208.10995
1620*

LS/Q
Stimulant

2626

208

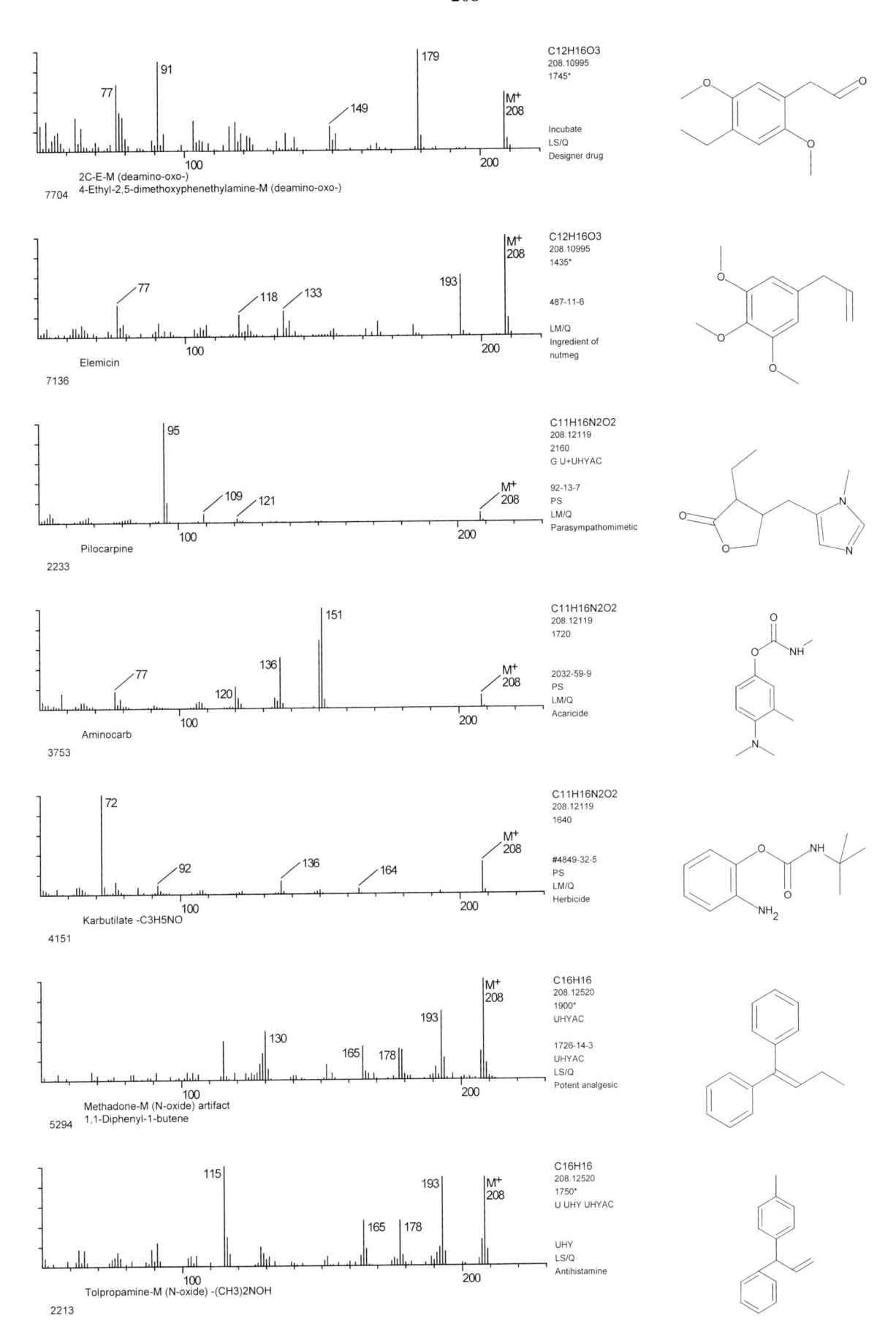

2C-E-M (deamino-oxo-) 4-Ethyl-2,5-dimethoxyphenethylamine-M (deamino-oxo-) 7704 Peaks: 77, 91, 149, 179, M+ 208	C12H16O3 208.10995 1745* Incubate LS/Q Designer drug
Elemicin 7136 Peaks: 77, 118, 133, 193, M+ 208	C12H16O3 208.10995 1435* 487-11-6 LM/Q Ingredient of nutmeg
Pilocarpine 2233 Peaks: 95, 109, 121, M+ 208	C11H16N2O2 208.12119 2160 G U+UHYAC 92-13-7 PS LM/Q Parasympathomimetic
Aminocarb 3753 Peaks: 77, 120, 136, 151, M+ 208	C11H16N2O2 208.12119 1720 2032-59-9 PS LM/Q Acaricide
Karbutilate -C3H5NO 4151 Peaks: 72, 92, 136, 164, M+ 208	C11H16N2O2 208.12119 1640 #4849-32-5 PS LM/Q Herbicide
Methadone-M (N-oxide) artifact 1,1-Diphenyl-1-butene 5294 Peaks: 130, 165, 178, 193, M+ 208	C16H16 208.12520 1900* UHYAC 1726-14-3 UHYAC LS/Q Potent analgesic
Tolpropamine-M (N-oxide) -(CH3)2NOH 2213 Peaks: 115, 165, 178, 193, M+ 208	C16H16 208.12520 1750* U UHY UHYAC UHY LS/Q Antihistamine

425

208

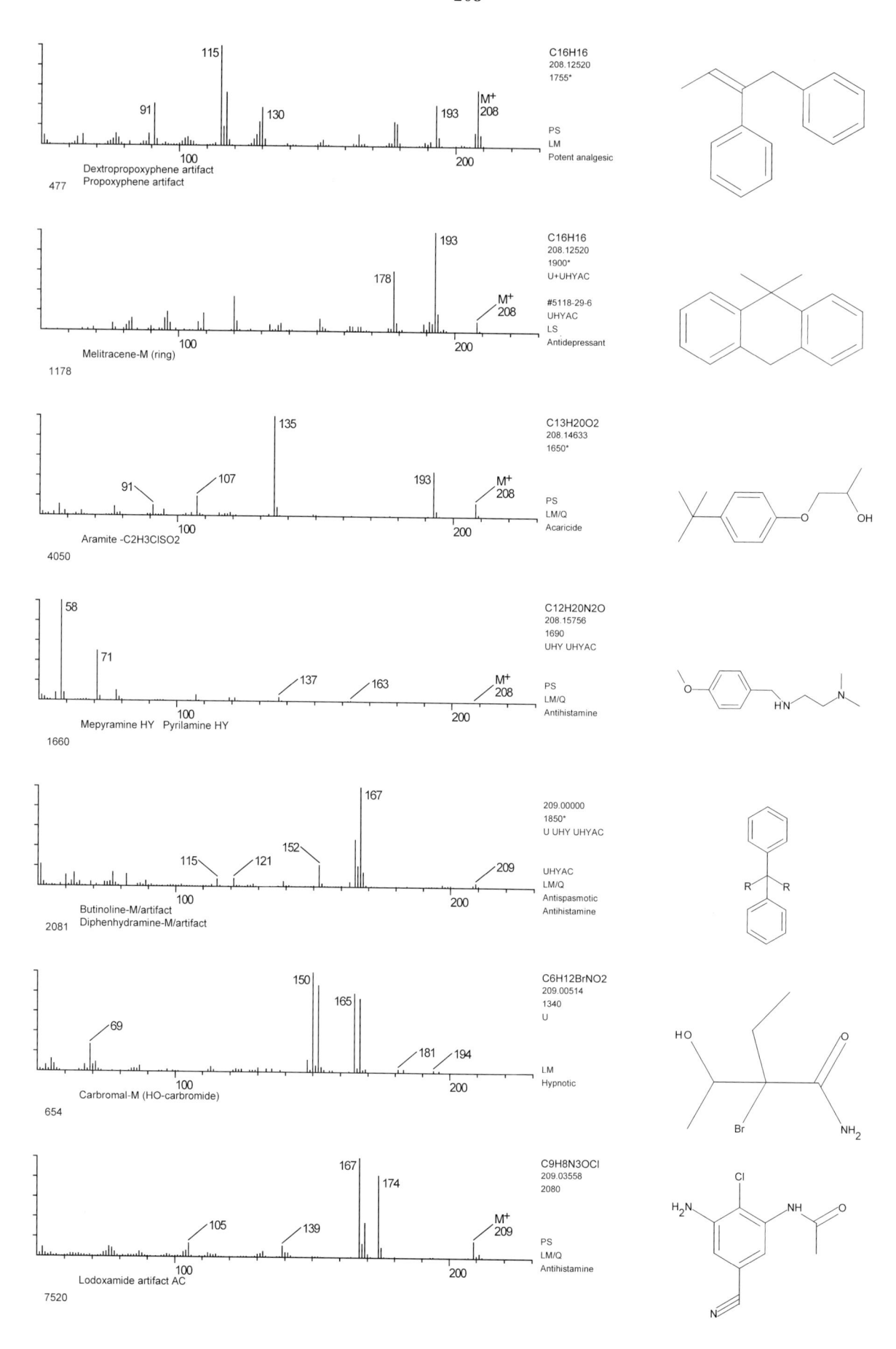

477 — Dextropropoxyphene artifact / Propoxyphene artifact
C16H16, 208.12520, 1755*, PS, LM, Potent analgesic
Peaks: 91, 115, 130, 193, M+ 208

1178 — Melitracene-M (ring)
C16H16, 208.12520, 1900*, U+UHYAC, #5118-29-6, UHYAC, LS, Antidepressant
Peaks: 178, 193, M+ 208

4050 — Aramite -C2H3ClSO2
C13H20O2, 208.14633, 1650*, PS, LM/Q, Acaricide
Peaks: 91, 107, 135, 193, M+ 208

1660 — Mepyramine HY / Pyrilamine HY
C12H20N2O, 208.15756, 1690, UHY UHYAC, PS, LM/Q, Antihistamine
Peaks: 58, 71, 137, 163, M+ 208

2081 — Butinoline-M/artifact / Diphenhydramine-M/artifact
209.00000, 1850*, U UHY UHYAC, UHYAC, LM/Q, Antispasmotic, Antihistamine
Peaks: 115, 121, 152, 167, 209

654 — Carbromal-M (HO-carbromide)
C6H12BrNO2, 209.00514, 1340, U, LM, Hypnotic
Peaks: 69, 150, 165, 181, 194

7520 — Lodoxamide artifact AC
C9H8N3OCl, 209.03558, 2080, PS, LM/Q, Antihistamine
Peaks: 105, 139, 167, 174, M+ 209

209

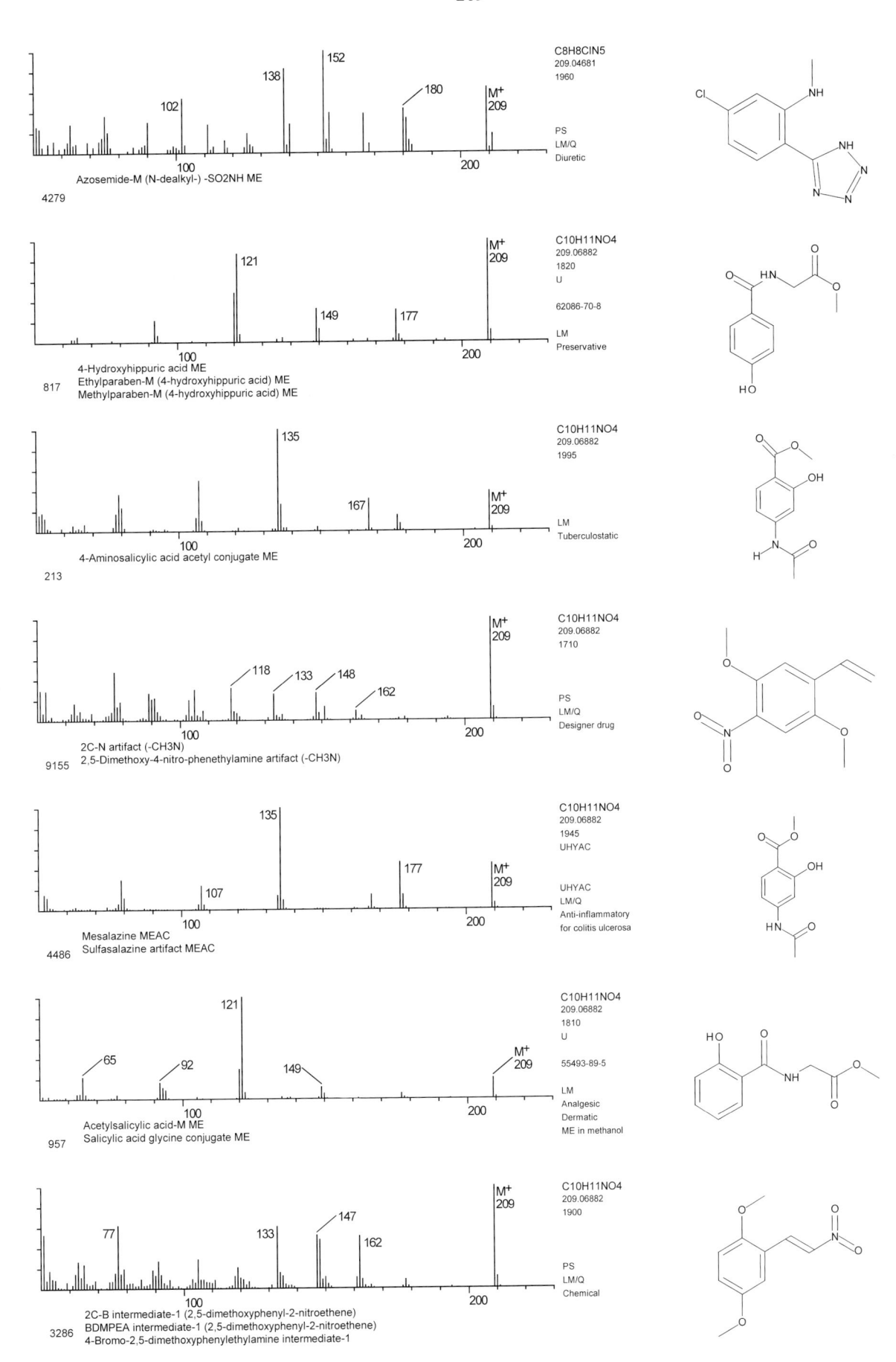

4279
Azosemide-M (N-dealkyl-) -SO2NH ME
Peaks: 102, 138, 152, 180, M+ 209
C8H8ClN5
209.04681
1960
PS
LM/Q
Diuretic

817
4-Hydroxyhippuric acid ME
Ethylparaben-M (4-hydroxyhippuric acid) ME
Methylparaben-M (4-hydroxyhippuric acid) ME
Peaks: 121, 149, 177, M+ 209
C10H11NO4
209.06882
1820
U
62086-70-8
LM
Preservative

213
4-Aminosalicylic acid acetyl conjugate ME
Peaks: 135, 167, M+ 209
C10H11NO4
209.06882
1995
LM
Tuberculostatic

9155
2C-N artifact (-CH3N)
2,5-Dimethoxy-4-nitro-phenethylamine artifact (-CH3N)
Peaks: 118, 133, 148, 162, M+ 209
C10H11NO4
209.06882
1710
PS
LM/Q
Designer drug

4486
Mesalazine MEAC
Sulfasalazine artifact MEAC
Peaks: 107, 135, 177, M+ 209
C10H11NO4
209.06882
1945
UHYAC
UHYAC
LM/Q
Anti-inflammatory
for colitis ulcerosa

957
Acetylsalicylic acid-M ME
Salicylic acid glycine conjugate ME
Peaks: 65, 92, 121, 149, M+ 209
C10H11NO4
209.06882
1810
U
55493-89-5
LM
Analgesic
Dermatic
ME in methanol

3286
2C-B intermediate-1 (2,5-dimethoxyphenyl-2-nitroethene)
BDMPEA intermediate-1 (2,5-dimethoxyphenyl-2-nitroethene)
4-Bromo-2,5-dimethoxyphenylethylamine intermediate-1
Peaks: 77, 133, 147, 162, M+ 209
C10H11NO4
209.06882
1900
PS
LM/Q
Chemical

209

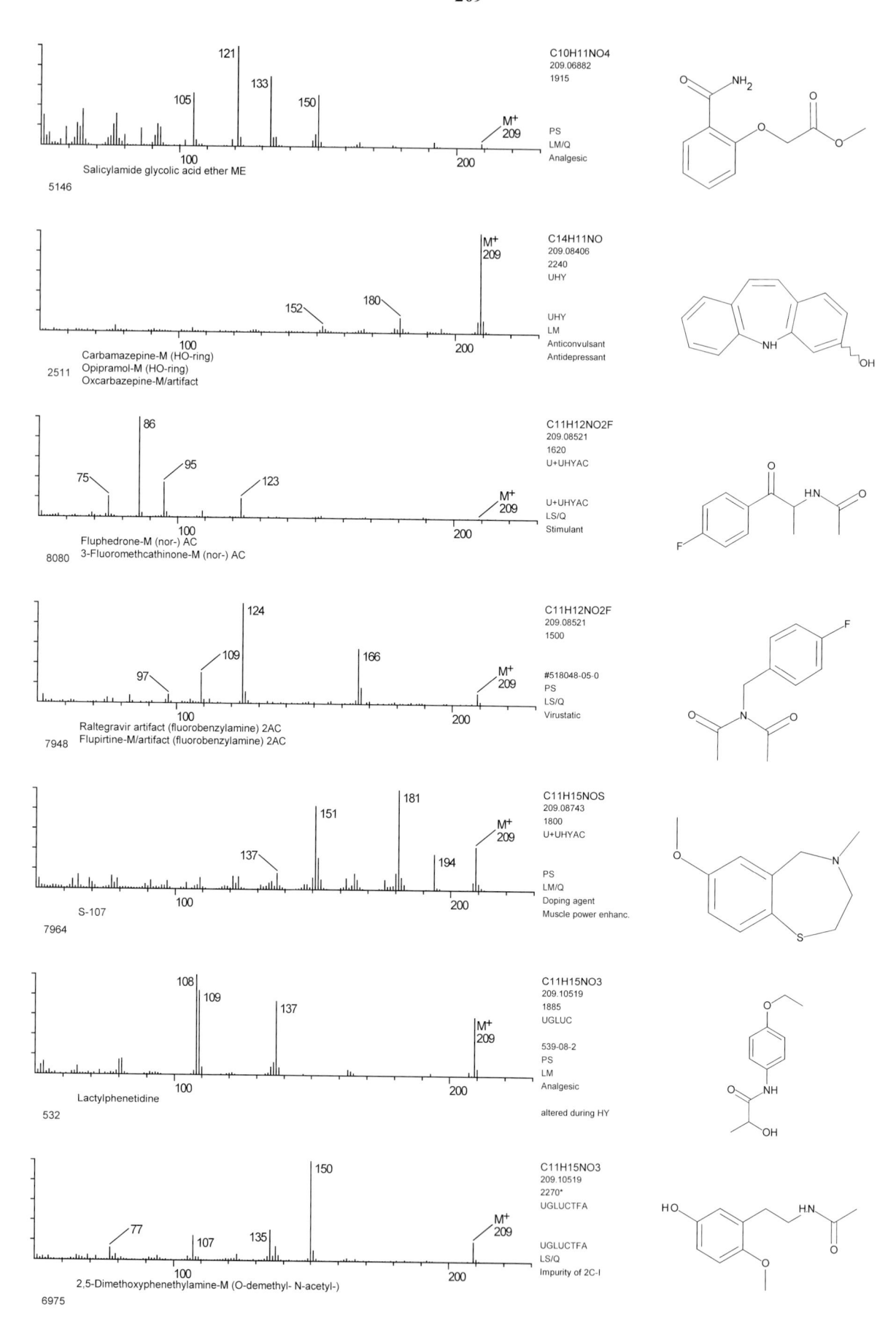

Salicylamide glycolic acid ether ME — 5146	C10H11NO4, 209.06882, 1915, PS, LM/Q, Analgesic
Carbamazepine-M (HO-ring) / Opipramol-M (HO-ring) / Oxcarbazepine-M/artifact — 2511	C14H11NO, 209.08406, 2240, UHY, UHY, LM, Anticonvulsant, Antidepressant
Fluphedrone-M (nor-) AC / 3-Fluoromethcathinone-M (nor-) AC — 8080	C11H12NO2F, 209.08521, 1620, U+UHYAC, U+UHYAC, LS/Q, Stimulant
Raltegravir artifact (fluorobenzylamine) 2AC / Flupirtine-M/artifact (fluorobenzylamine) 2AC — 7948	C11H12NO2F, 209.08521, 1500, #518048-05-0, PS, LS/Q, Virustatic
S-107 — 7964	C11H15NOS, 209.08743, 1800, U+UHYAC, PS, LM/Q, Doping agent, Muscle power enhanc.
Lactylphenetidine — 532	C11H15NO3, 209.10519, 1885, UGLUC, 539-08-2, PS, LM, Analgesic, altered during HY
2,5-Dimethoxyphenethylamine-M (O-demethyl- N-acetyl-) — 6975	C11H15NO3, 209.10519, 2270*, UGLUCTFA, UGLUCTFA, LS/Q, Impurity of 2C-I

209

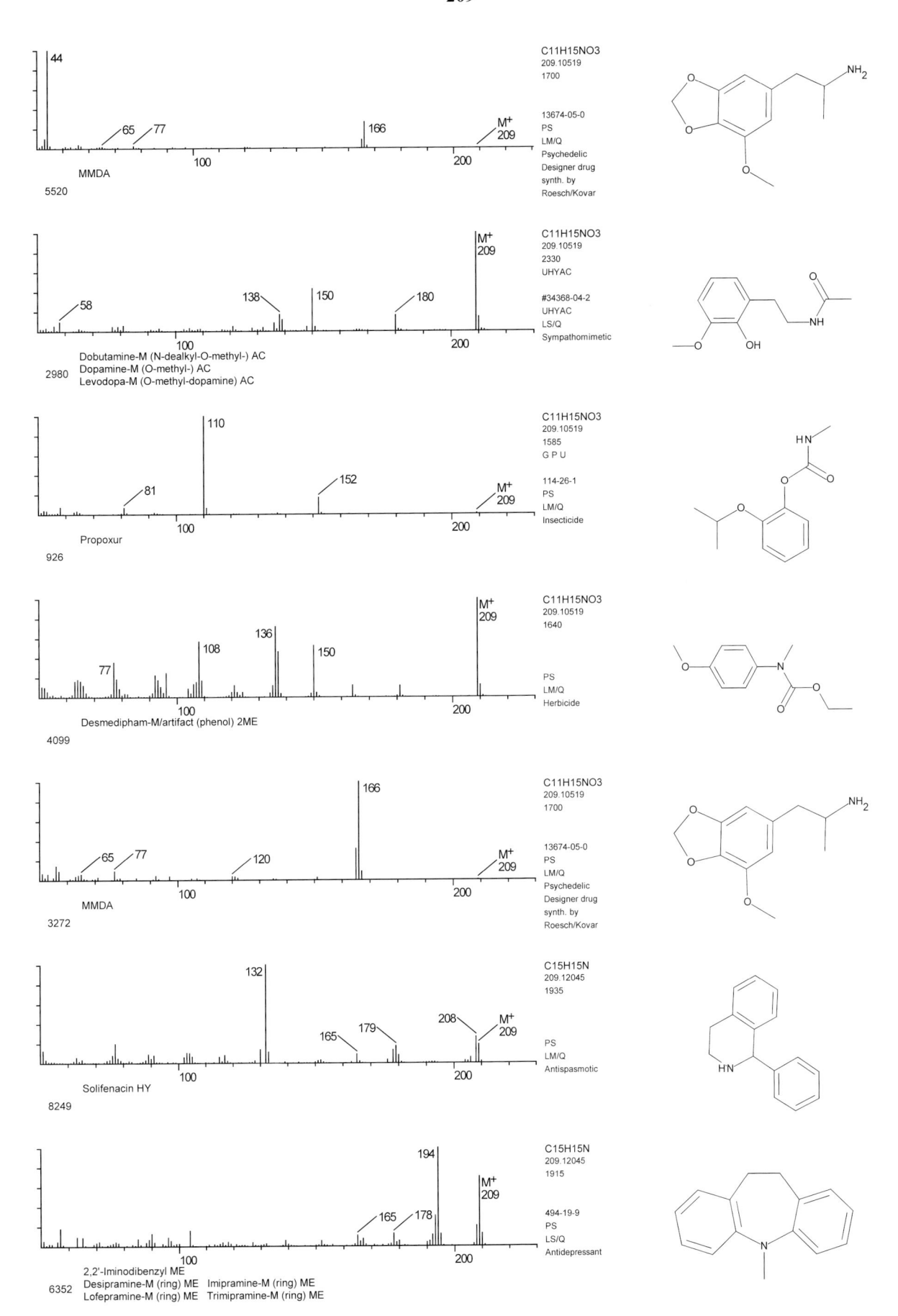

Spectrum	Compound	Formula / Info
5520	MMDA — peaks: 44, 65, 77, 166, M+ 209	C11H15NO3, 209.10519, 1700, 13674-05-0, PS, LM/Q, Psychedelic, Designer drug synth. by Roesch/Kovar
2980	Dobutamine-M (N-dealkyl-O-methyl-) AC; Dopamine-M (O-methyl-) AC; Levodopa-M (O-methyl-dopamine) AC — peaks: 58, 138, 150, 180, M+ 209	C11H15NO3, 209.10519, 2330, UHYAC, #34368-04-2, UHYAC, LS/Q, Sympathomimetic
926	Propoxur — peaks: 81, 110, 152, M+ 209	C11H15NO3, 209.10519, 1585, G P U, 114-26-1, PS, LM/Q, Insecticide
4099	Desmedipham-M/artifact (phenol) 2ME — peaks: 77, 108, 136, 150, M+ 209	C11H15NO3, 209.10519, 1640, PS, LM/Q, Herbicide
3272	MMDA — peaks: 65, 77, 120, 166, M+ 209	C11H15NO3, 209.10519, 1700, 13674-05-0, PS, LM/Q, Psychedelic, Designer drug synth. by Roesch/Kovar
8249	Solifenacin HY — peaks: 132, 165, 179, 208, M+ 209	C15H15N, 209.12045, 1935, PS, LM/Q, Antispasmotic
6352	2,2'-Iminodibenzyl ME; Desipramine-M (ring) ME; Imipramine-M (ring) ME; Lofepramine-M (ring) ME; Trimipramine-M (ring) ME — peaks: 165, 178, 194, M+ 209	C15H15N, 209.12045, 1915, 494-19-9, PS, LS/Q, Antidepressant

209

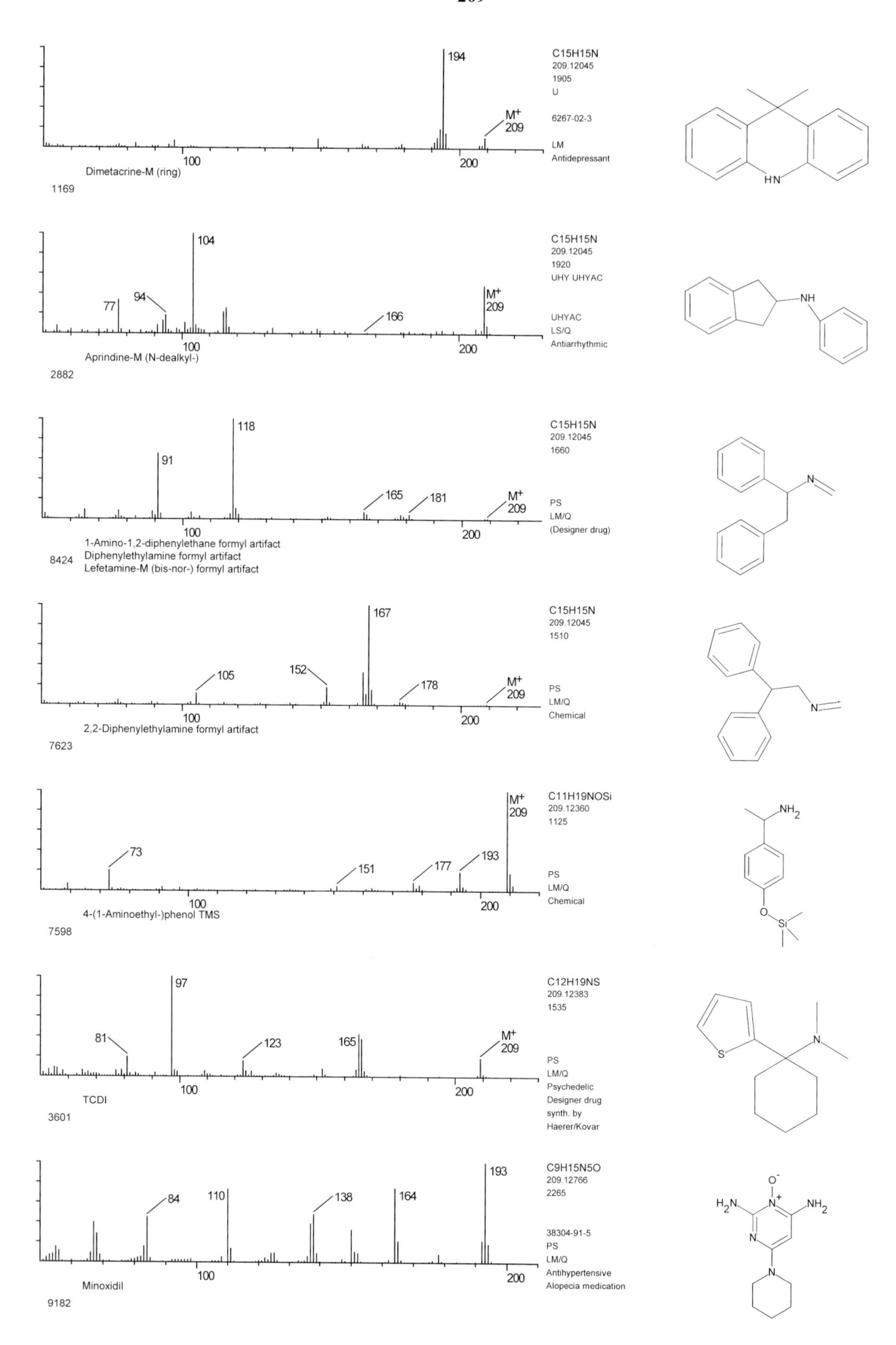

Dimetacrine-M (ring)
1169
194, M+ 209
C15H15N
209.12045
1905
U
6267-02-3
LM
Antidepressant

Aprindine-M (N-dealkyl-)
2882
77, 94, 104, 166, M+ 209
C15H15N
209.12045
1920
UHY UHYAC
UHYAC
LS/Q
Antiarrhythmic

1-Amino-1,2-diphenylethane formyl artifact
Diphenylethylamine formyl artifact
Lefetamine-M (bis-nor-) formyl artifact
8424
91, 118, 165, 181, M+ 209
C15H15N
209.12045
1660
PS
LM/Q
(Designer drug)

2,2-Diphenylethylamine formyl artifact
7623
105, 152, 167, 178, M+ 209
C15H15N
209.12045
1510
PS
LM/Q
Chemical

4-(1-Aminoethyl-)phenol TMS
7598
73, 151, 177, 193, M+ 209
C11H19NOSi
209.12360
1125
PS
LM/Q
Chemical

TCDI
3601
81, 97, 123, 165, M+ 209
C12H19NS
209.12383
1535
PS
LM/Q
Psychedelic
Designer drug
synth. by
Haerer/Kovar

Minoxidil
9182
84, 110, 138, 164, 193
C9H15N5O
209.12766
2265
38304-91-5
PS
LM/Q
Antihypertensive
Alopecia medication

209

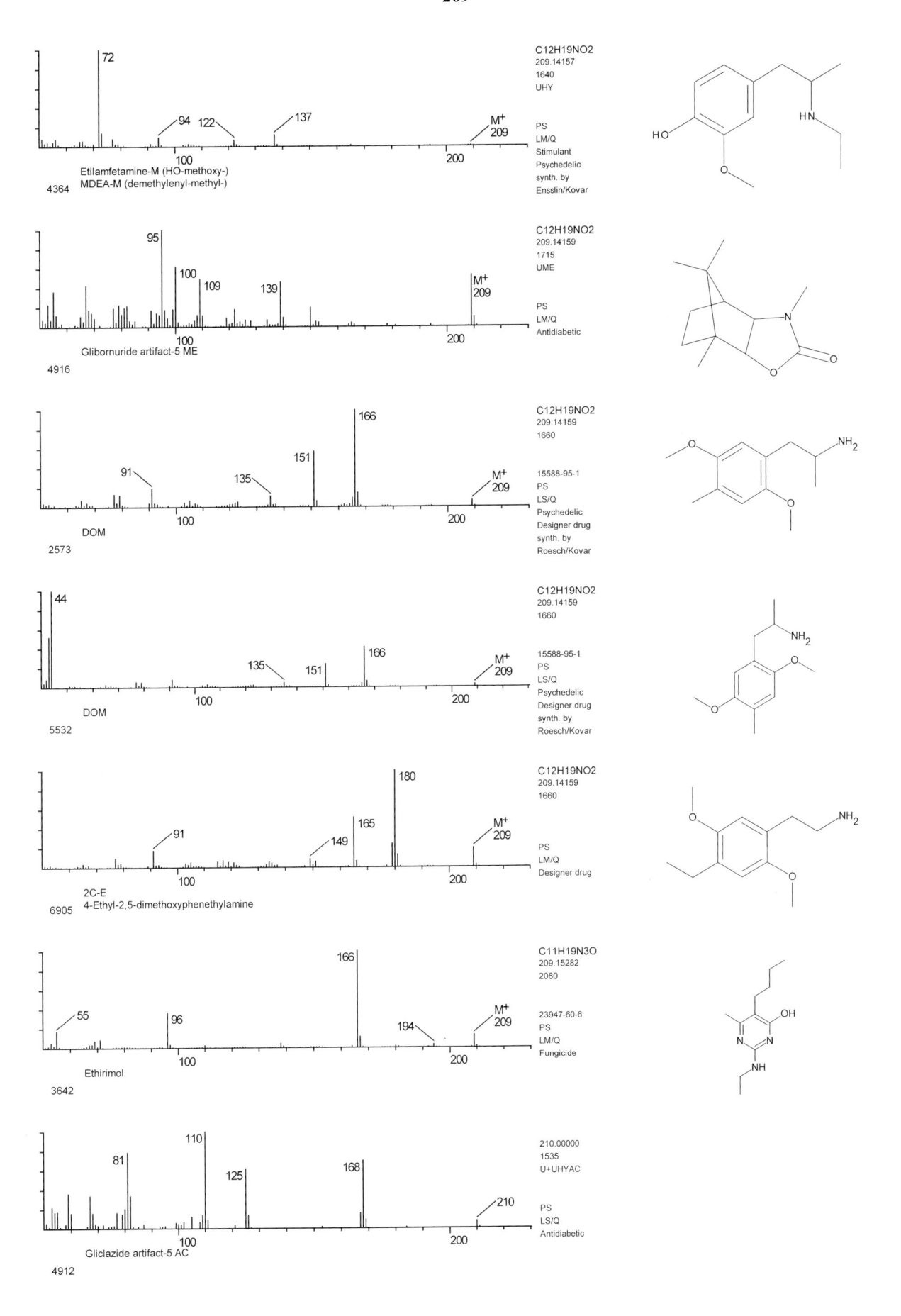

Spectrum	Compound	Formula/Info
4364	Etilamfetamine-M (HO-methoxy-) / MDEA-M (demethylenyl-methyl-) — peaks 72, 94, 122, 137, M+ 209	C12H19NO2; 209.14157; 1640; UHY; PS; LM/Q; Stimulant; Psychedelic; synth. by Ensslin/Kovar
4916	Glibornuride artifact-5 ME — peaks 95, 100, 109, 139, M+ 209	C12H19NO2; 209.14159; 1715; UME; PS; LM/Q; Antidiabetic
2573	DOM — peaks 91, 135, 151, 166, M+ 209	C12H19NO2; 209.14159; 1660; 15588-95-1; PS; LS/Q; Psychedelic; Designer drug; synth. by Roesch/Kovar
5532	DOM — peaks 44, 135, 151, 166, M+ 209	C12H19NO2; 209.14159; 1660; 15588-95-1; PS; LS/Q; Psychedelic; Designer drug; synth. by Roesch/Kovar
6905	2C-E / 4-Ethyl-2,5-dimethoxyphenethylamine — peaks 91, 149, 165, 180, M+ 209	C12H19NO2; 209.14159; 1660; PS; LM/Q; Designer drug
3642	Ethirimol — peaks 55, 96, 166, 194, M+ 209	C11H19N3O; 209.15282; 2080; 23947-60-6; PS; LM/Q; Fungicide
4912	Gliclazide artifact-5 AC — peaks 81, 110, 125, 168, 210	210.00000; 1535; U+UHYAC; PS; LS/Q; Antidiabetic

210

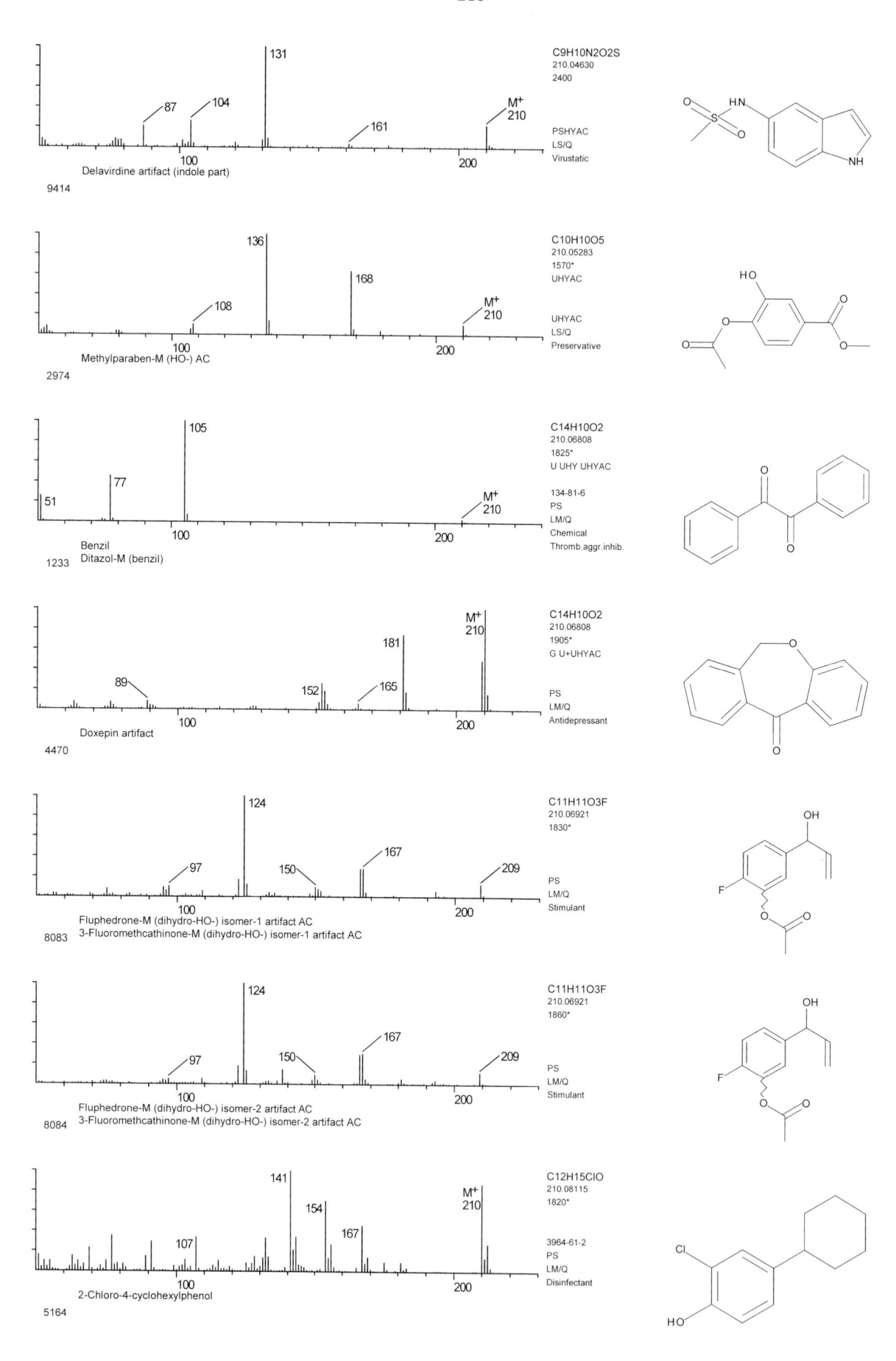

9414 — Delavirdine artifact (indole part)
C9H10N2O2S; 210.04630; 2400; PSHYAC; LS/Q; Virustatic
Peaks: 87, 104, 131, 161, M⁺ 210

2974 — Methylparaben-M (HO-) AC
C10H10O5; 210.05283; 1570*; UHYAC; UHYAC; LS/Q; Preservative
Peaks: 108, 136, 168, M⁺ 210

1233 — Benzil / Ditazol-M (benzil)
C14H10O2; 210.06808; 1825*; U UHY UHYAC; 134-81-6; PS; LM/Q; Chemical; Thromb.aggr.inhib.
Peaks: 51, 77, 105, M⁺ 210

4470 — Doxepin artifact
C14H10O2; 210.06808; 1905*; G U+UHYAC; PS; LM/Q; Antidepressant
Peaks: 89, 152, 165, 181, M⁺ 210

8083 — Fluphedrone-M (dihydro-HO-) isomer-1 artifact AC / 3-Fluoromethcathinone-M (dihydro-HO-) isomer-1 artifact AC
C11H11O3F; 210.06921; 1830*; PS; LM/Q; Stimulant
Peaks: 97, 124, 150, 167, 209

8084 — Fluphedrone-M (dihydro-HO-) isomer-2 artifact AC / 3-Fluoromethcathinone-M (dihydro-HO-) isomer-2 artifact AC
C11H11O3F; 210.06921; 1860*; PS; LM/Q; Stimulant
Peaks: 97, 124, 150, 167, 209

5164 — 2-Chloro-4-cyclohexylphenol
C12H15ClO; 210.08115; 1820*; 3964-61-2; PS; LM/Q; Disinfectant
Peaks: 107, 141, 154, 167, M⁺ 210

210

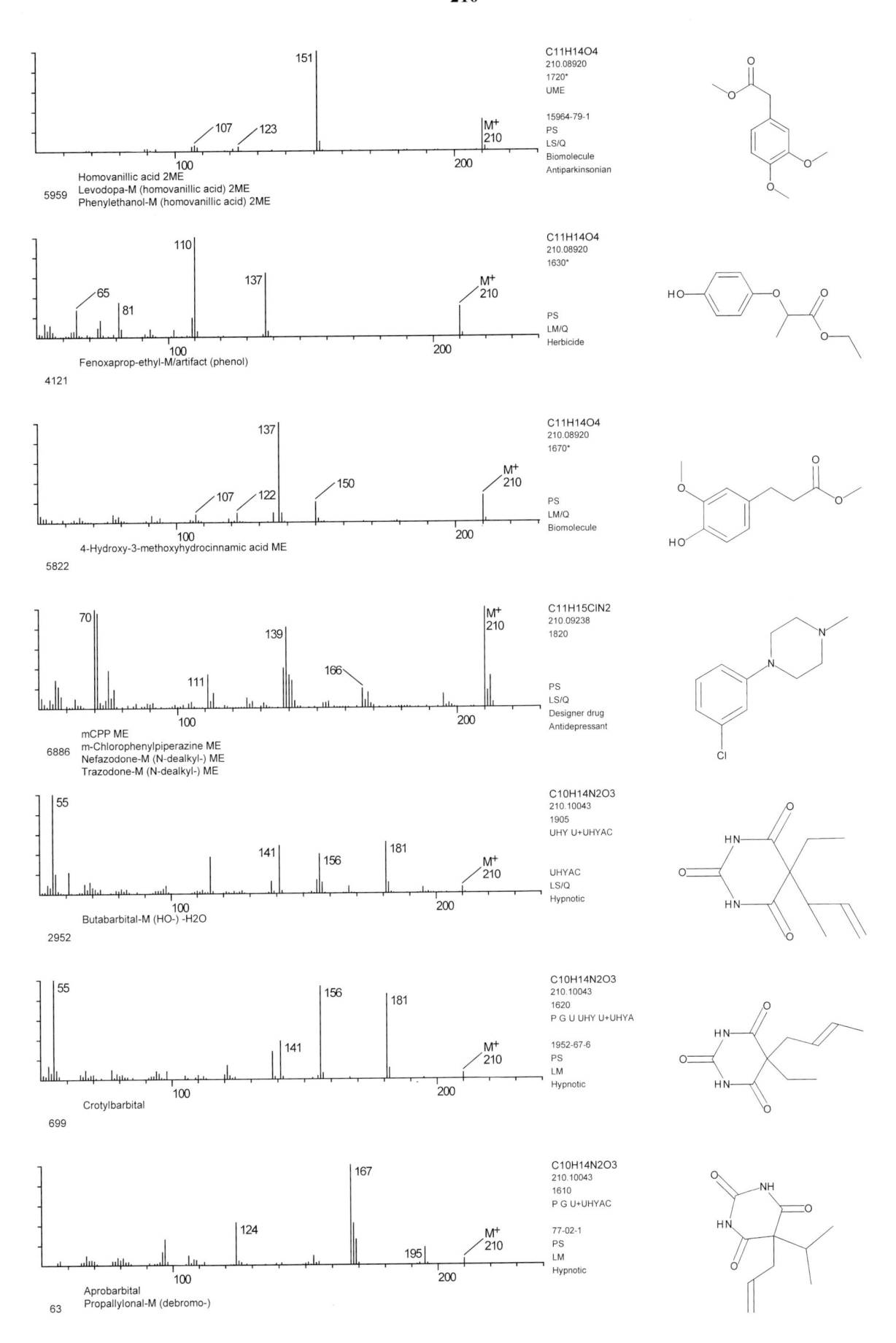

5959	Homovanillic acid 2ME Levodopa-M (homovanillic acid) 2ME Phenylethanol-M (homovanillic acid) 2ME	C11H14O4 210.08920 1720* UME 15964-79-1 PS LS/Q Biomolecule Antiparkinsonian

4121 Fenoxaprop-ethyl-M/artifact (phenol) — C11H14O4, 210.08920, 1630*, PS, LM/Q, Herbicide

5822 4-Hydroxy-3-methoxyhydrocinnamic acid ME — C11H14O4, 210.08920, 1670*, PS, LM/Q, Biomolecule

6886 mCPP ME
m-Chlorophenylpiperazine ME
Nefazodone-M (N-dealkyl-) ME
Trazodone-M (N-dealkyl-) ME — C11H15ClN2, 210.09238, 1820, PS, LS/Q, Designer drug, Antidepressant

2952 Butabarbital-M (HO-) -H2O — C10H14N2O3, 210.10043, 1905, UHY U+UHYAC, UHYAC, LS/Q, Hypnotic

699 Crotylbarbital — C10H14N2O3, 210.10043, 1620, P G U UHY U+UHYA, 1952-67-6, PS, LM, Hypnotic

63 Aprobarbital
Propallylonal-M (debromo-) — C10H14N2O3, 210.10043, 1610, P G U+UHYAC, 77-02-1, PS, LM, Hypnotic

433

210

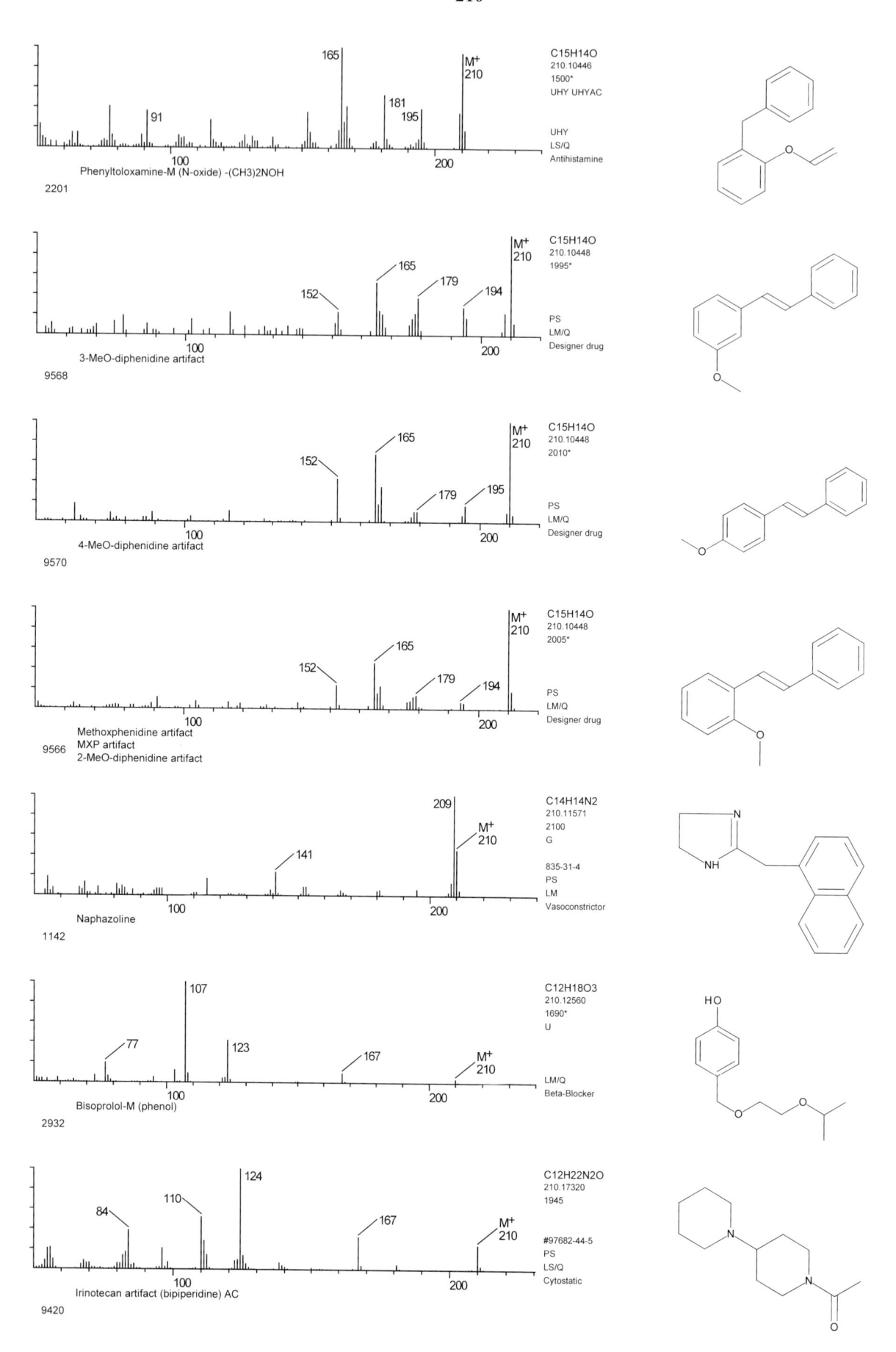

Spectrum	Compound	Formula
2201	Phenyltoloxamine-M (N-oxide) -(CH3)2NOH	C15H14O 210.10446 1500* UHY UHYAC / UHY LS/Q Antihistamine
9568	3-MeO-diphenidine artifact	C15H14O 210.10448 1995* / PS LM/Q Designer drug
9570	4-MeO-diphenidine artifact	C15H14O 210.10448 2010* / PS LM/Q Designer drug
9566	Methoxphenidine artifact / MXP artifact / 2-MeO-diphenidine artifact	C15H14O 210.10448 2005* / PS LM/Q Designer drug
1142	Naphazoline	C14H14N2 210.11571 2100 G 835-31-4 PS LM Vasoconstrictor
2932	Bisoprolol-M (phenol)	C12H18O3 210.12560 1690* U / LM/Q Beta-Blocker
9420	Irinotecan artifact (bipiperidine) AC	C12H22N2O 210.17320 1945 #97682-44-5 PS LS/Q Cytostatic

211

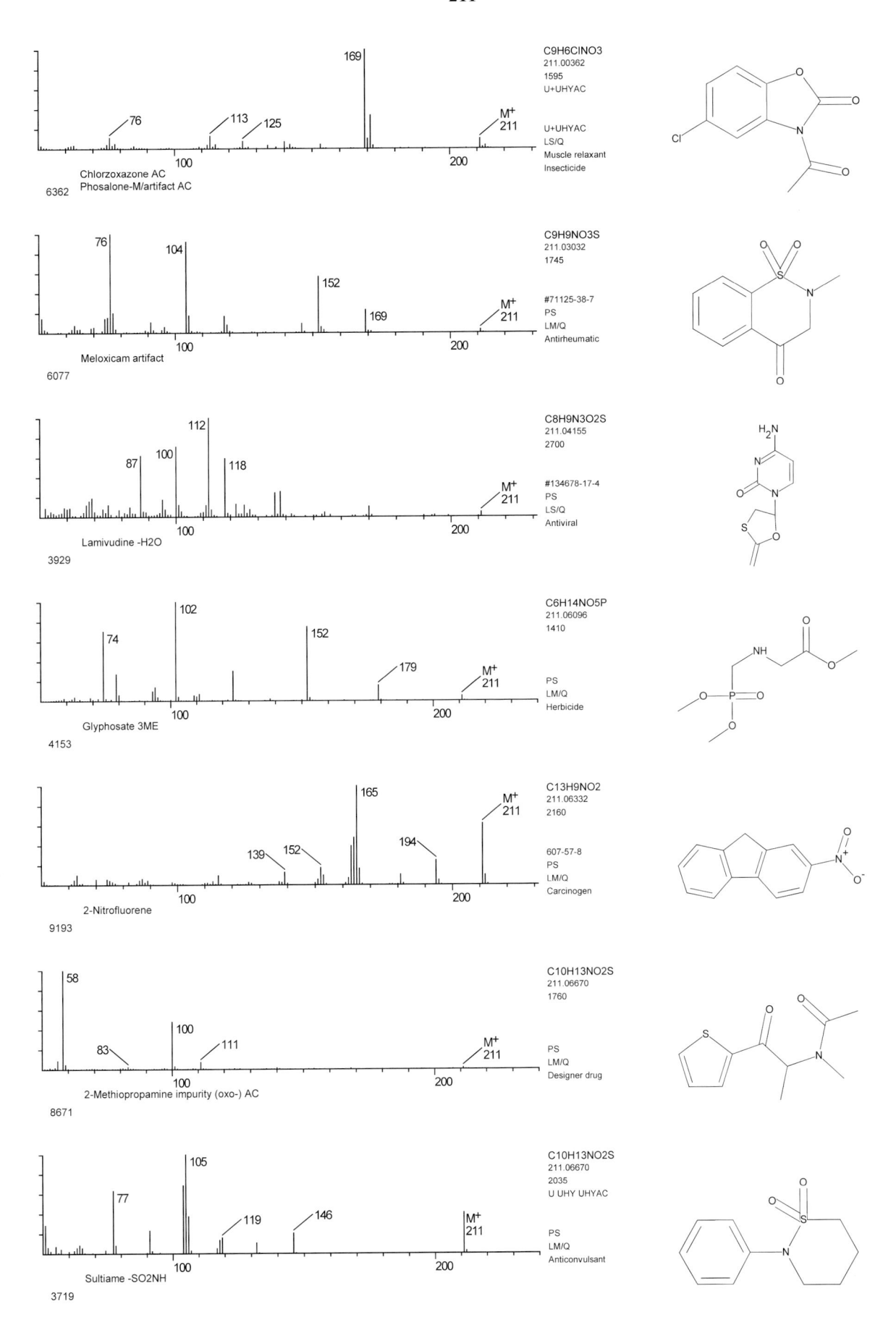

Spectrum peaks	Formula / Info	Name
76, 113, 125, 169, M+ 211	C9H6ClNO3 211.00362 1595 U+UHYAC U+UHYAC LS/Q Muscle relaxant Insecticide	Chlorzoxazone AC Phosalone-M/artifact AC 6362
76, 104, 152, 169, M+ 211	C9H9NO3S 211.03032 1745 #71125-38-7 PS LM/Q Antirheumatic	Meloxicam artifact 6077
87, 100, 112, 118, M+ 211	C8H9N3O2S 211.04155 2700 #134678-17-4 PS LS/Q Antiviral	Lamivudine -H2O 3929
74, 102, 152, 179, M+ 211	C6H14NO5P 211.06096 1410 PS LM/Q Herbicide	Glyphosate 3ME 4153
139, 152, 165, 194, M+ 211	C13H9NO2 211.06332 2160 607-57-8 PS LM/Q Carcinogen	2-Nitrofluorene 9193
58, 83, 100, 111, M+ 211	C10H13NO2S 211.06670 1760 PS LM/Q Designer drug	2-Methiopropamine impurity (oxo-) AC 8671
77, 105, 119, 146, M+ 211	C10H13NO2S 211.06670 2035 U UHY UHYAC PS LM/Q Anticonvulsant	Sultiame -SO2NH 3719

211

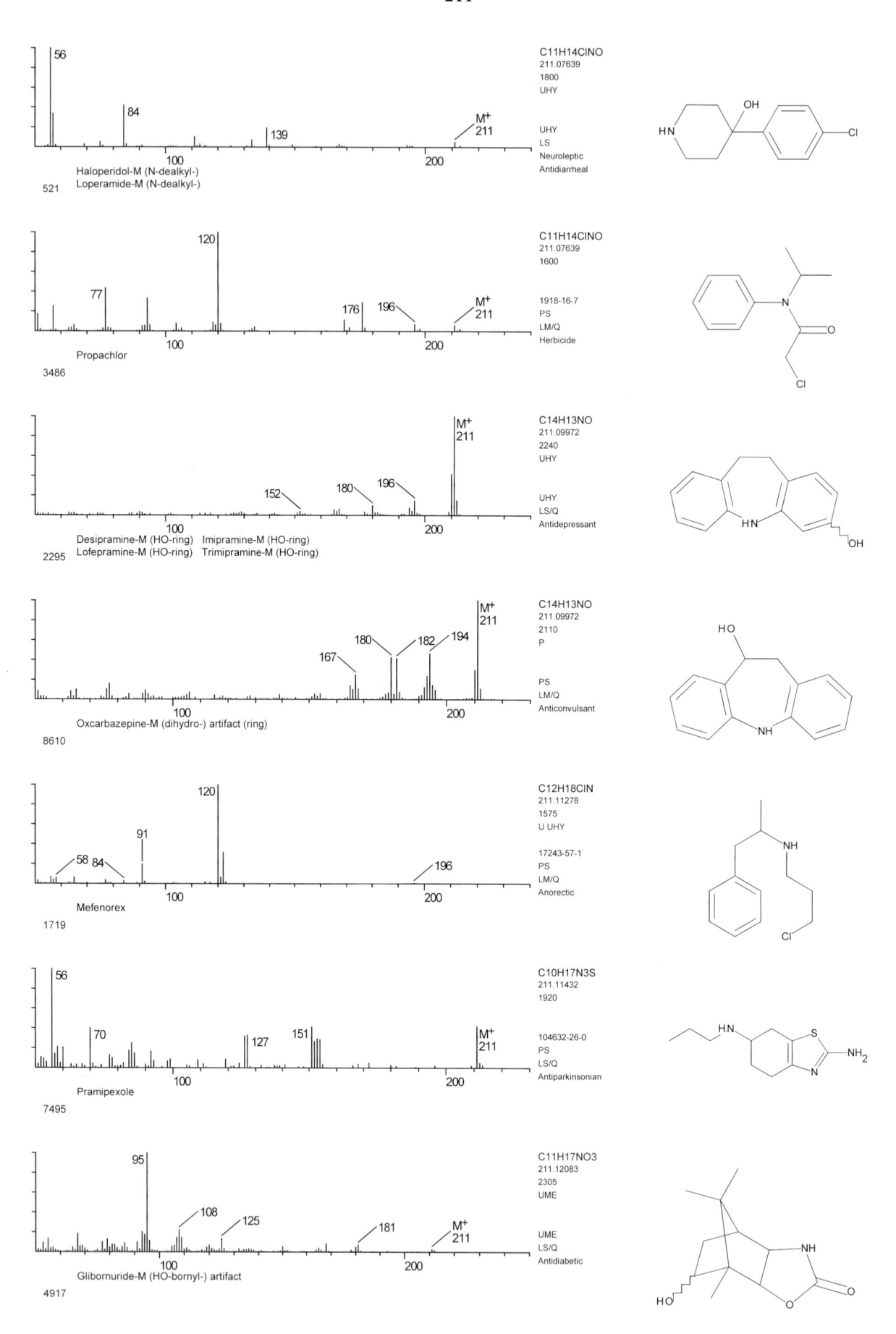

Haloperidol-M (N-dealkyl-) / Loperamide-M (N-dealkyl-)	C11H14ClNO 211.07639 1800 UHY / UHY LS Neuroleptic Antidiarrheal
Propachlor	C11H14ClNO 211.07639 1600 / 1918-16-7 PS LM/Q Herbicide
Desipramine-M (HO-ring) / Imipramine-M (HO-ring) / Lofepramine-M (HO-ring) / Trimipramine-M (HO-ring)	C14H13NO 211.09972 2240 UHY / UHY LS/Q Antidepressant
Oxcarbazepine-M (dihydro-) artifact (ring)	C14H13NO 211.09972 2110 P / PS LM/Q Anticonvulsant
Mefenorex	C12H18ClN 211.11278 1575 U UHY / 17243-57-1 PS LM/Q Anorectic
Pramipexole	C10H17N3S 211.11432 1920 / 104632-26-0 PS LS/Q Antiparkinsonian
Glibornuride-M (HO-bornyl-) artifact	C11H17NO3 211.12083 2305 UME / UME LS/Q Antidiabetic

211

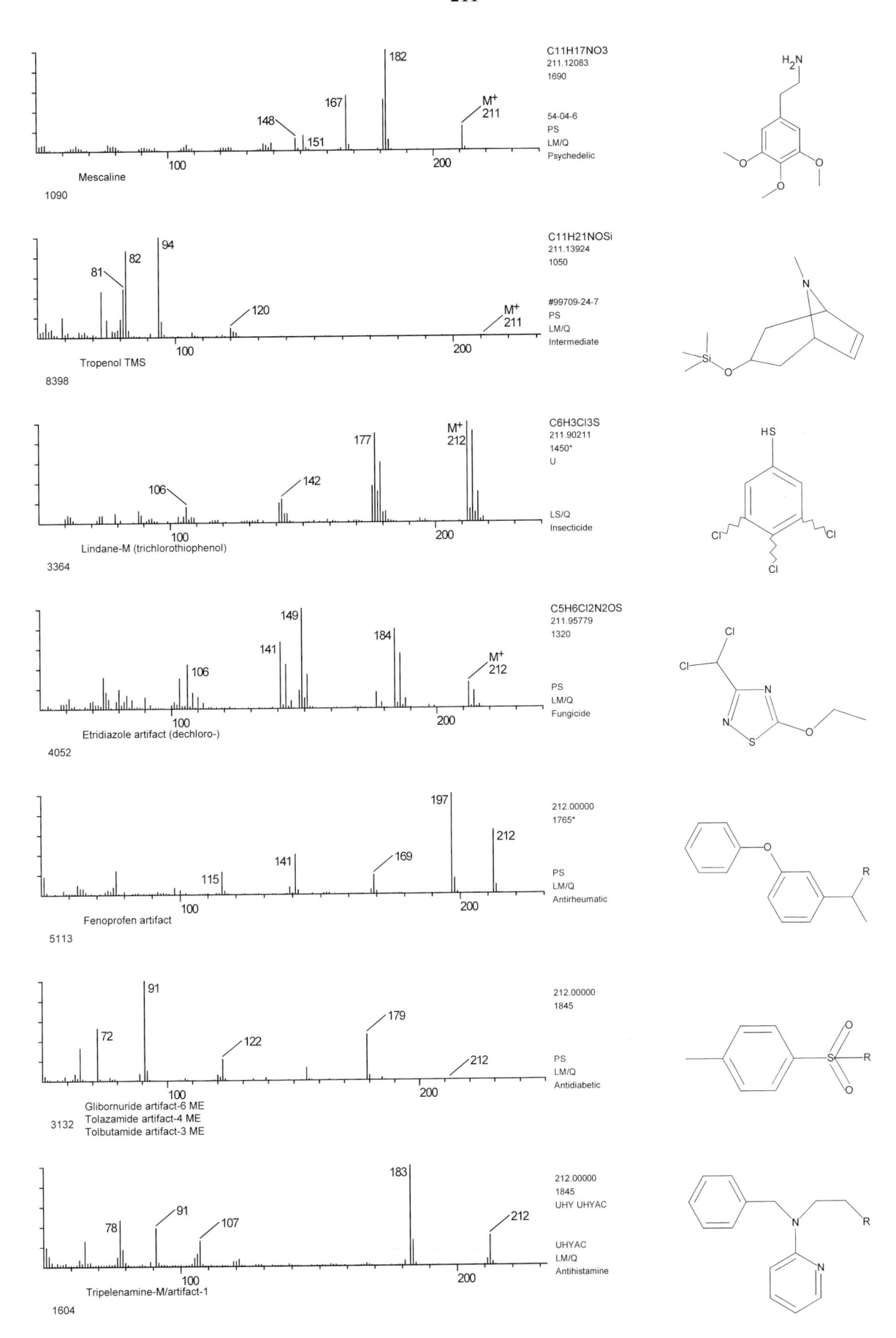

Mescaline — C11H17NO3, 211.12083, 1690, 54-04-6, PS, LM/Q, Psychedelic
Peaks: 148, 151, 167, 182, M+ 211
(1090)

Tropenol TMS — C11H21NOSi, 211.13924, 1050, #99709-24-7, PS, LM/Q, Intermediate
Peaks: 81, 82, 94, 120, M+ 211
(8398)

Lindane-M (trichlorothiophenol) — C6H3Cl3S, 211.90211, 1450*, U, LS/Q, Insecticide
Peaks: 106, 142, 177, M+ 212
(3364)

Etridiazole artifact (dechloro-) — C5H6Cl2N2OS, 211.95779, 1320, PS, LM/Q, Fungicide
Peaks: 106, 141, 149, 184, M+ 212
(4052)

Fenoprofen artifact — 212.00000, 1765*, PS, LM/Q, Antirheumatic
Peaks: 115, 141, 169, 197, 212
(5113)

Glibornuride artifact-6 ME / Tolazamide artifact-4 ME / Tolbutamide artifact-3 ME — 212.00000, 1845, PS, LM/Q, Antidiabetic
Peaks: 72, 91, 122, 179, 212
(3132)

Tripelenamine-M/artifact-1 — 212.00000, 1845, UHY UHYAC, UHYAC, LM/Q, Antihistamine
Peaks: 78, 91, 107, 183, 212
(1604)

437

212

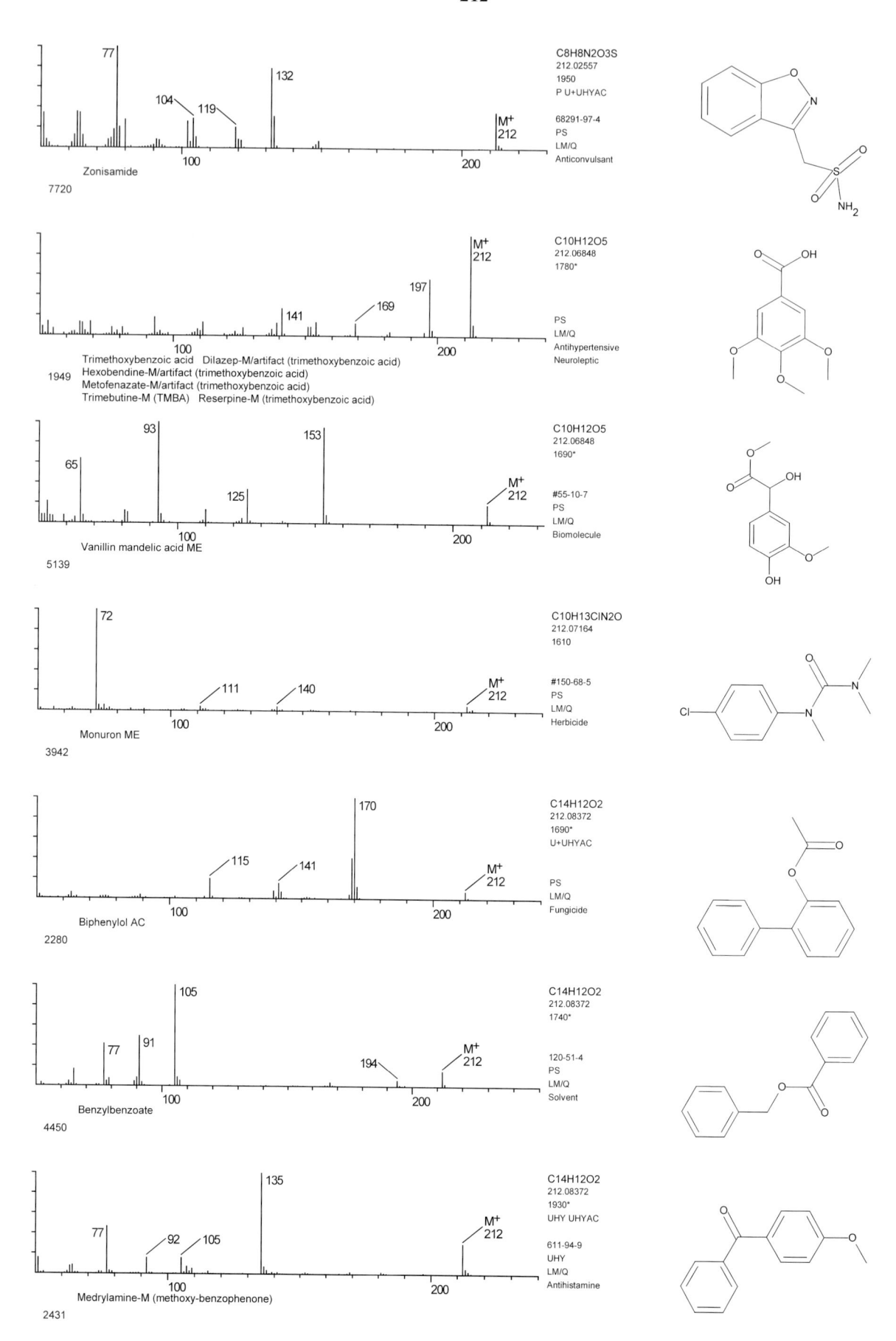

Zonisamide
7720
C8H8N2O3S
212.02557
1950
P U+UHYAC
68291-97-4
PS
LM/Q
Anticonvulsant

Trimethoxybenzoic acid   Dilazep-M/artifact (trimethoxybenzoic acid)
Hexobendine-M/artifact (trimethoxybenzoic acid)
Metofenazate-M/artifact (trimethoxybenzoic acid)
Trimebutine-M (TMBA)   Reserpine-M (trimethoxybenzoic acid)
1949
C10H12O5
212.06848
1780*
PS
LM/Q
Antihypertensive
Neuroleptic

Vanillin mandelic acid ME
5139
C10H12O5
212.06848
1690*
#55-10-7
PS
LM/Q
Biomolecule

Monuron ME
3942
C10H13ClN2O
212.07164
1610
#150-68-5
PS
LM/Q
Herbicide

Biphenylol AC
2280
C14H12O2
212.08372
1690*
U+UHYAC
PS
LM/Q
Fungicide

Benzylbenzoate
4450
C14H12O2
212.08372
1740*
120-51-4
PS
LM/Q
Solvent

Medrylamine-M (methoxy-benzophenone)
2431
C14H12O2
212.08372
1930*
UHY UHYAC
611-94-9
UHY
LM/Q
Antihistamine

212

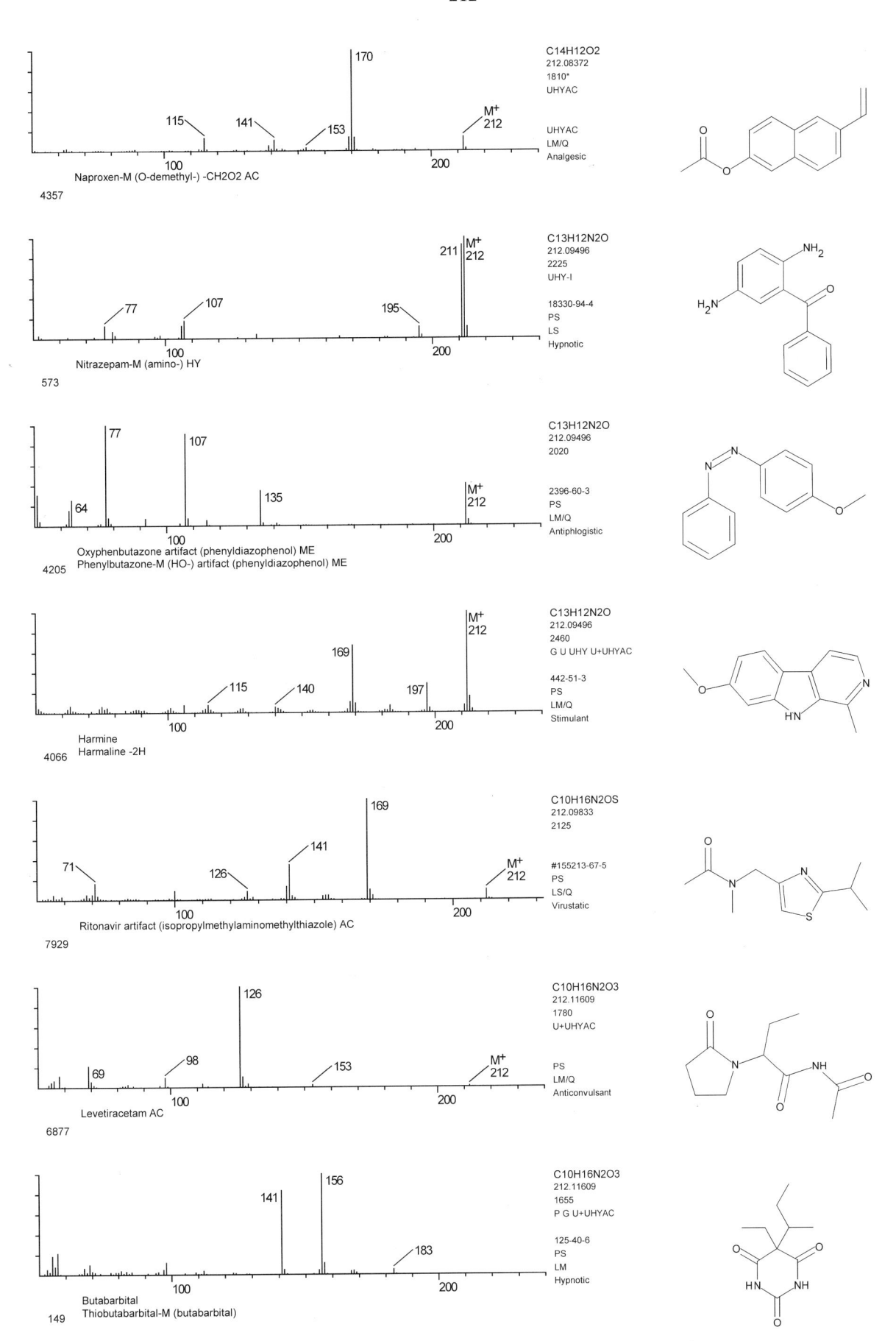

170, 115, 141, 153, M+ 212	C14H12O2 212.08372 1810* UHYAC UHYAC LM/Q Analgesic	

Naproxen-M (O-demethyl-) -CH2O2 AC
4357

77, 107, 195, 211, M+ 212	C13H12N2O 212.09496 2225 UHY-I 18330-94-4 PS LS Hypnotic	

Nitrazepam-M (amino-) HY
573

64, 77, 107, 135, M+ 212	C13H12N2O 212.09496 2020 2396-60-3 PS LM/Q Antiphlogistic	

Oxyphenbutazone artifact (phenyldiazophenol) ME
Phenylbutazone-M (HO-) artifact (phenyldiazophenol) ME
4205

115, 140, 169, 197, M+ 212	C13H12N2O 212.09496 2460 G U UHY U+UHYAC 442-51-3 PS LM/Q Stimulant	

Harmine
Harmaline -2H
4066

71, 126, 141, 169, M+ 212	C10H16N2OS 212.09833 2125 #155213-67-5 PS LS/Q Virustatic	

Ritonavir artifact (isopropylmethylaminomethylthiazole) AC
7929

69, 98, 126, 153, M+ 212	C10H16N2O3 212.11609 1780 U+UHYAC PS LM/Q Anticonvulsant	

Levetiracetam AC
6877

141, 156, 183	C10H16N2O3 212.11609 1655 P G U+UHYAC 125-40-6 PS LM Hypnotic	

Butabarbital
Thiobutabarbital-M (butabarbital)
149

212

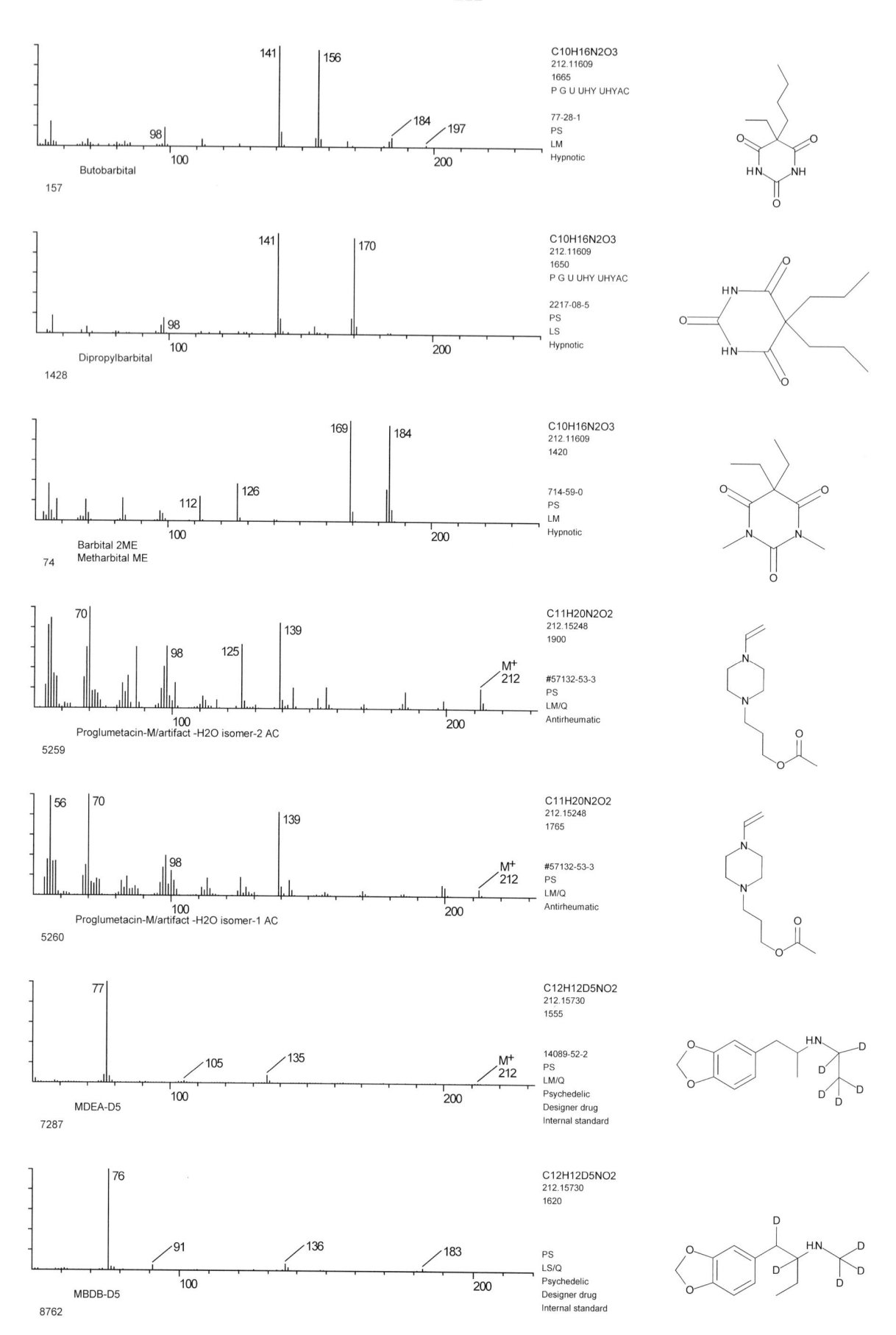

Spectrum	Compound	Formula / Data
Peaks: 98, 141, 156, 184, 197	Butobarbital	C10H16N2O3, 212.11609, 1665, P G U UHY UHYAC, 77-28-1, PS, LM, Hypnotic
Peaks: 98, 141, 170	Dipropylbarbital	C10H16N2O3, 212.11609, 1650, P G U UHY UHYAC, 2217-08-5, PS, LS, Hypnotic
Peaks: 112, 126, 169, 184	Barbital 2ME / Metharbital ME	C10H16N2O3, 212.11609, 1420, 714-59-0, PS, LM, Hypnotic
Peaks: 70, 98, 125, 139, M+ 212	Proglumetacin-M/artifact -H2O isomer-2 AC	C11H20N2O2, 212.15248, 1900, #57132-53-3, PS, LM/Q, Antirheumatic
Peaks: 56, 70, 98, 139, M+ 212	Proglumetacin-M/artifact -H2O isomer-1 AC	C11H20N2O2, 212.15248, 1765, #57132-53-3, PS, LM/Q, Antirheumatic
Peaks: 77, 105, 135, M+ 212	MDEA-D5	C12H12D5NO2, 212.15730, 1555, 14089-52-2, PS, LM/Q, Psychedelic, Designer drug, Internal standard
Peaks: 76, 91, 136, 183	MBDB-D5	C12H12D5NO2, 212.15730, 1620, PS, LS/Q, Psychedelic, Designer drug, Internal standard

212

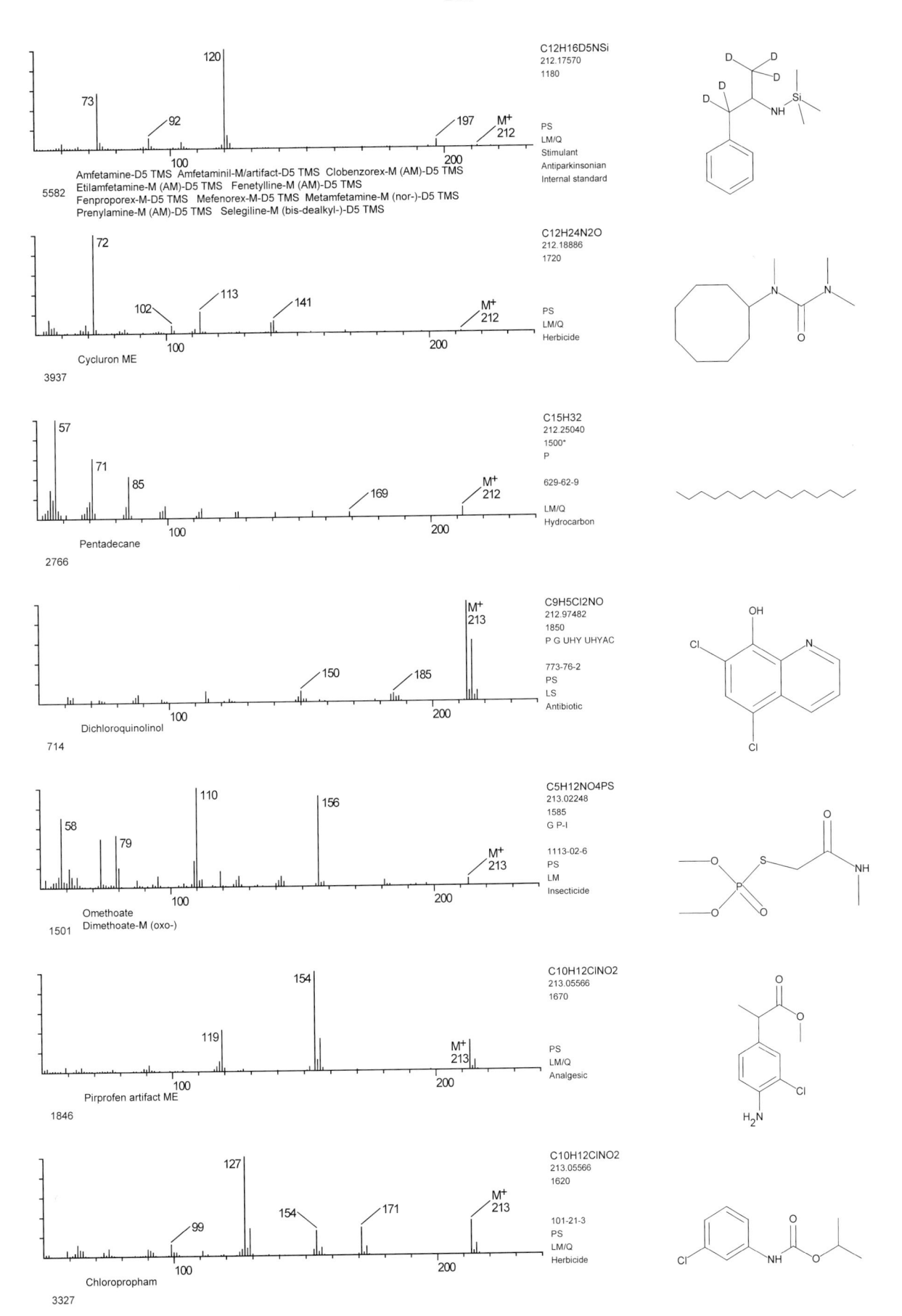

C12H16D5NSi
212.17570
1180

PS
LM/Q
Stimulant
Antiparkinsonian
Internal standard

Amfetamine-D5 TMS Amfetaminil-M/artifact-D5 TMS Clobenzorex-M (AM)-D5 TMS
Etilamfetamine-M (AM)-D5 TMS Fenetylline-M (AM)-D5 TMS
Fenproporex-M-D5 TMS Mefenorex-M-D5 TMS Metamfetamine-M (nor-)-D5 TMS
Prenylamine-M (AM)-D5 TMS Selegiline-M (bis-dealkyl-)-D5 TMS
5582

C12H24N2O
212.18886
1720

PS
LM/Q
Herbicide

Cycluron ME
3937

C15H32
212.25040
1500*

P

629-62-9

LM/Q
Hydrocarbon

Pentadecane
2766

C9H5Cl2NO
212.97482
1850

P G UHY UHYAC

773-76-2
PS
LS
Antibiotic

Dichloroquinolinol
714

C5H12NO4PS
213.02248
1585

G P-I

1113-02-6
PS
LM
Insecticide

Omethoate
Dimethoate-M (oxo-)
1501

C10H12ClNO2
213.05566
1670

PS
LM/Q
Analgesic

Pirprofen artifact ME
1846

C10H12ClNO2
213.05566
1620

101-21-3
PS
LM/Q
Herbicide

Chlorpropham
3327

441

213

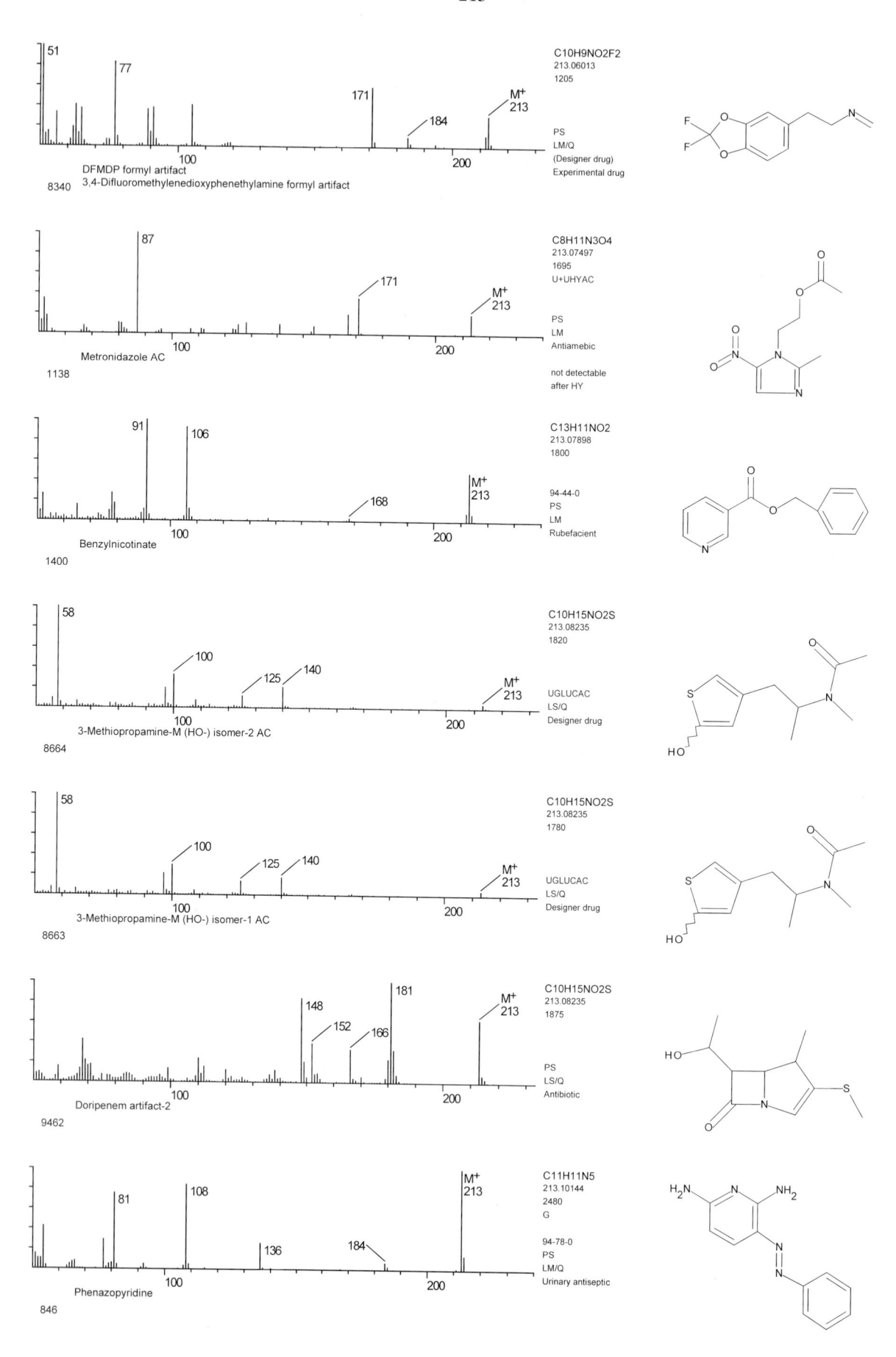

- DFMDP formyl artifact — 3,4-Difluoromethylenedioxyphenethylamine formyl artifact
  Peaks: 51, 77, 171, 184, M+ 213
  C10H9NO2F2, 213.06013, 1205, PS, LM/Q, (Designer drug) Experimental drug
  8340

- Metronidazole AC
  Peaks: 87, 171, M+ 213
  C8H11N3O4, 213.07497, 1695, U+UHYAC, PS, LM, Antiamebic, not detectable after HY
  1138

- Benzylnicotinate
  Peaks: 91, 106, 168, M+ 213
  C13H11NO2, 213.07898, 1800, 94-44-0, PS, LM, Rubefacient
  1400

- 3-Methiopropamine-M (HO-) isomer-2 AC
  Peaks: 58, 100, 125, 140, M+ 213
  C10H15NO2S, 213.08235, 1820, UGLUCAC, LS/Q, Designer drug
  8664

- 3-Methiopropamine-M (HO-) isomer-1 AC
  Peaks: 58, 100, 125, 140, M+ 213
  C10H15NO2S, 213.08235, 1780, UGLUCAC, LS/Q, Designer drug
  8663

- Doripenem artifact-2
  Peaks: 148, 152, 166, 181, M+ 213
  C10H15NO2S, 213.08235, 1875, PS, LS/Q, Antibiotic
  9462

- Phenazopyridine
  Peaks: 81, 108, 136, 184, M+ 213
  C11H11N5, 213.10144, 2480, G, 94-78-0, PS, LM/Q, Urinary antiseptic
  846

213

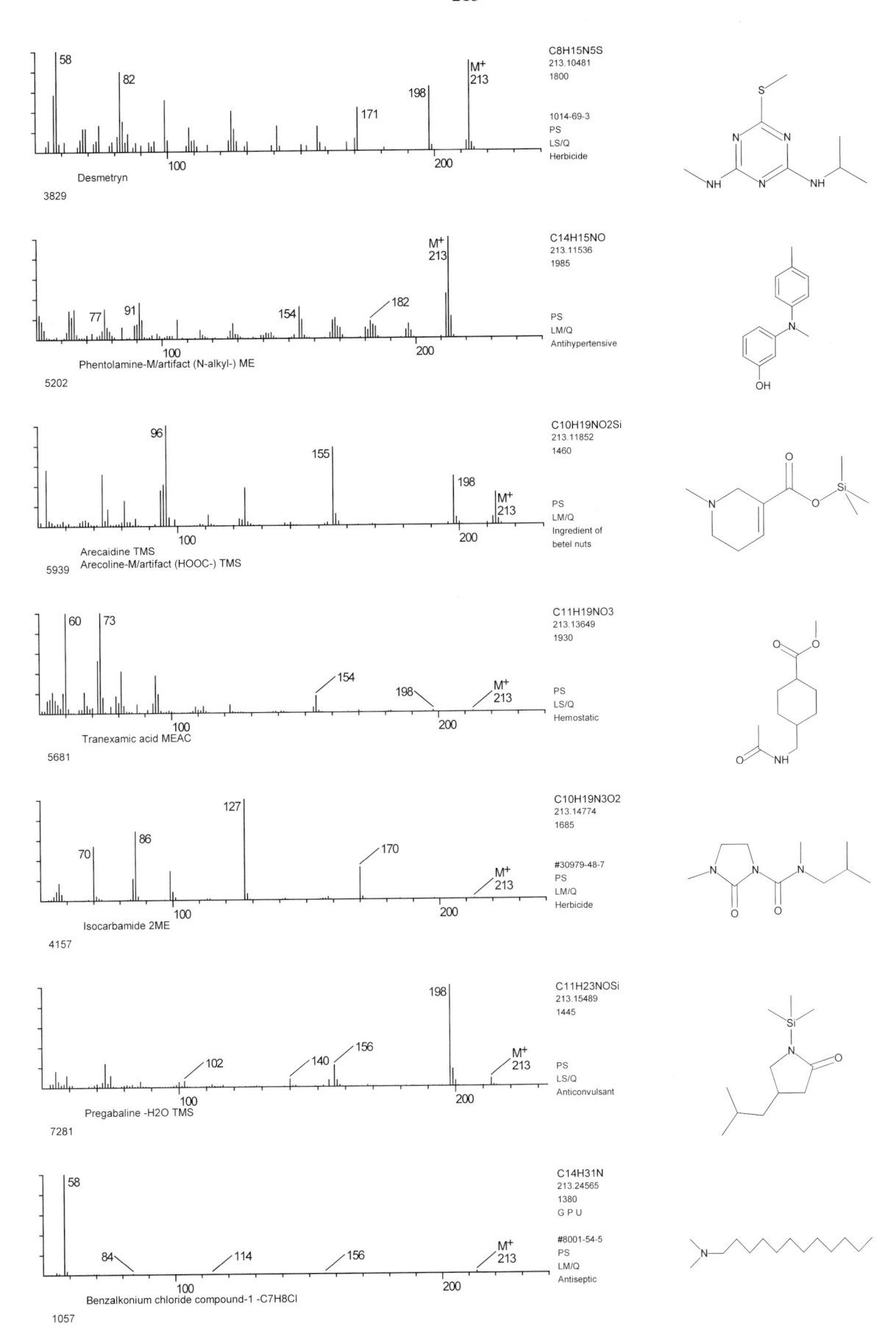

C8H15N5S
213.10481
1800

1014-69-3
PS
LS/Q
Herbicide

Desmetryn
3829

C14H15NO
213.11536
1985

PS
LM/Q
Antihypertensive

Phentolamine-M/artifact (N-alkyl-) ME
5202

C10H19NO2Si
213.11852
1460

PS
LM/Q
Ingredient of
betel nuts

Arecaidine TMS
Arecoline-M/artifact (HOOC-) TMS
5939

C11H19NO3
213.13649
1930

PS
LS/Q
Hemostatic

Tranexamic acid MEAC
5681

C10H19N3O2
213.14774
1685

#30979-48-7
PS
LM/Q
Herbicide

Isocarbamide 2ME
4157

C11H23NOSi
213.15489
1445

PS
LS/Q
Anticonvulsant

Pregabaline -H2O TMS
7281

C14H31N
213.24565
1380
G P U

#8001-54-5
PS
LM/Q
Antiseptic

Benzalkonium chloride compound-1 -C7H8Cl
1057

214

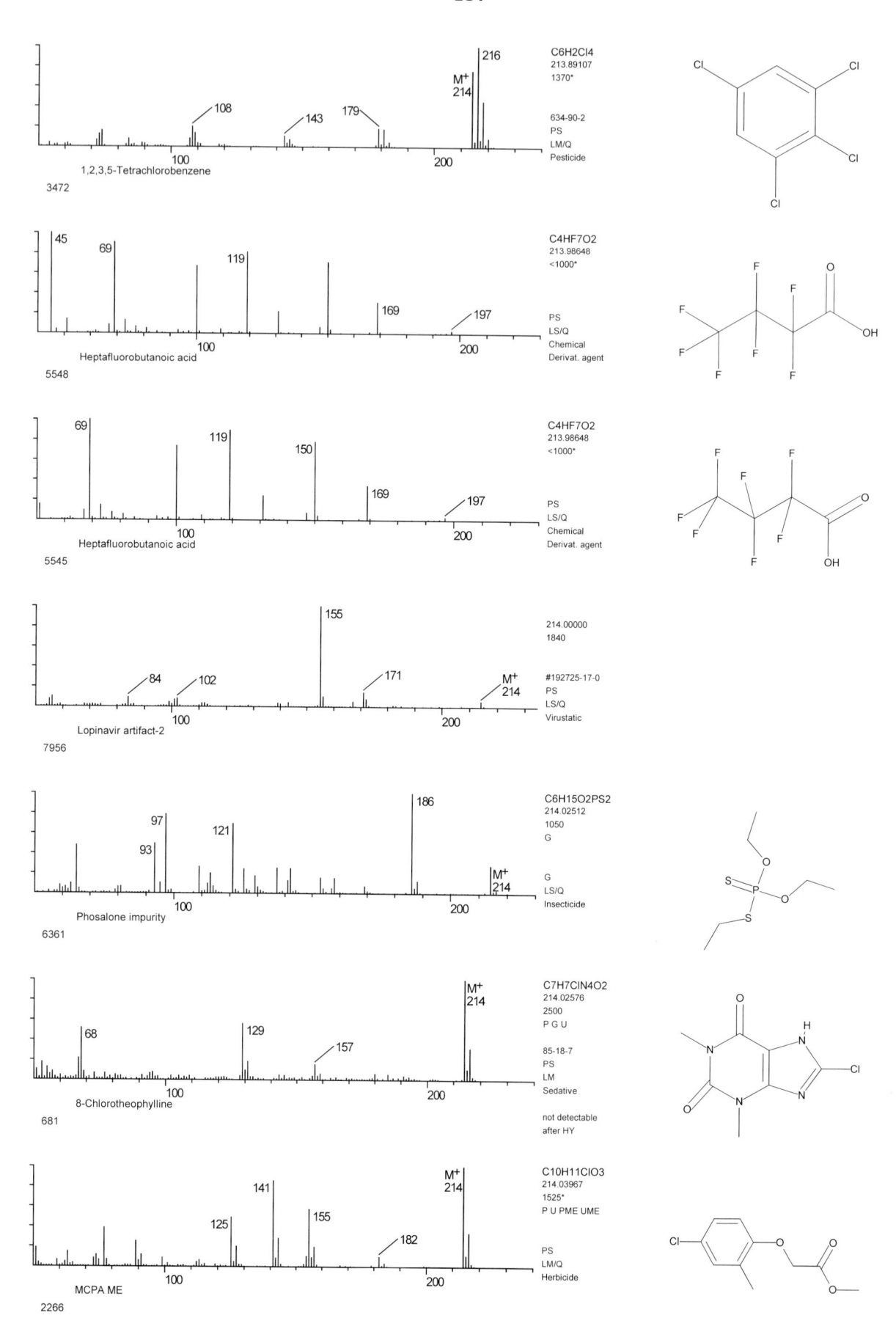

1,2,3,5-Tetrachlorobenzene — peaks: 108, 143, 179, M+ 214, 216	C6H2Cl4 213.89107 1370* 634-90-2 PS LM/Q Pesticide	
3472		
Heptafluorobutanoic acid — peaks: 45, 69, 119, 169, 197	C4HF7O2 213.98648 <1000* PS LS/Q Chemical Derivat. agent	
5548		
Heptafluorobutanoic acid — peaks: 69, 119, 150, 169, 197	C4HF7O2 213.98648 <1000* PS LS/Q Chemical Derivat. agent	
5545		
Lopinavir artifact-2 — peaks: 84, 102, 155, 171, M+ 214	214.00000 1840 #192725-17-0 PS LS/Q Virustatic	
7956		
Phosalone impurity — peaks: 93, 97, 121, 186, M+ 214	C6H15O2PS2 214.02512 1050 G G LS/Q Insecticide	
6361		
8-Chlorotheophylline — peaks: 68, 129, 157, M+ 214	C7H7ClN4O2 214.02576 2500 P G U 85-18-7 PS LM Sedative not detectable after HY	
681		
MCPA ME — peaks: 125, 141, 155, 182, M+ 214	C10H11ClO3 214.03967 1525* P U PME UME PS LM/Q Herbicide	
2266		

214

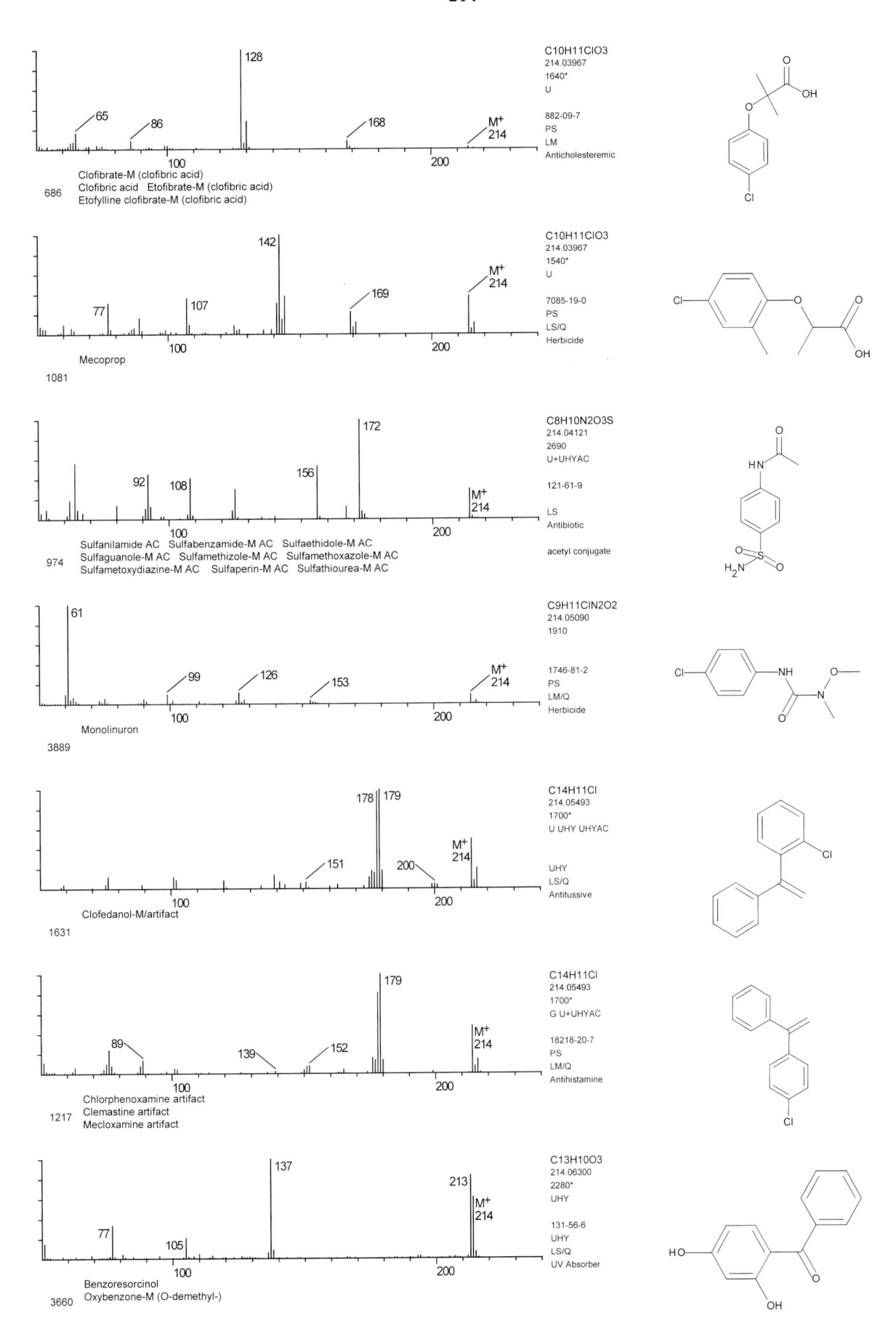

686

Clofibrate-M (clofibric acid)
Clofibric acid   Etofibrate-M (clofibric acid)
Etofylline clofibrate-M (clofibric acid)

C10H11ClO3
214.03967
1640*
U

882-09-7
PS
LM
Anticholesteremic

1081

Mecoprop

C10H11ClO3
214.03967
1540*
U

7085-19-0
PS
LS/Q
Herbicide

974

Sulfanilamide AC   Sulfabenzamide-M AC   Sulfaethidole-M AC
Sulfaguanole-M AC   Sulfamethizole-M AC   Sulfamethoxazole-M AC
Sulfametoxydiazine-M AC   Sulfaperin-M AC   Sulfathiourea-M AC

C8H10N2O3S
214.04121
2690
U+UHYAC

121-61-9
LS
Antibiotic

acetyl conjugate

3889

Monolinuron

C9H11ClN2O2
214.05090
1910

1746-81-2
PS
LM/Q
Herbicide

1631

Clofedanol-M/artifact

C14H11Cl
214.05493
1700*
U UHY UHYAC

UHY
LS/Q
Antitussive

1217

Chlorphenoxamine artifact
Clemastine artifact
Mecloxamine artifact

C14H11Cl
214.05493
1700*
G U+UHYAC

18218-20-7
PS
LM/Q
Antihistamine

3660

Benzoresorcinol
Oxybenzone-M (O-demethyl-)

C13H10O3
214.06300
2280*
UHY

131-56-6
UHY
LS/Q
UV Absorber

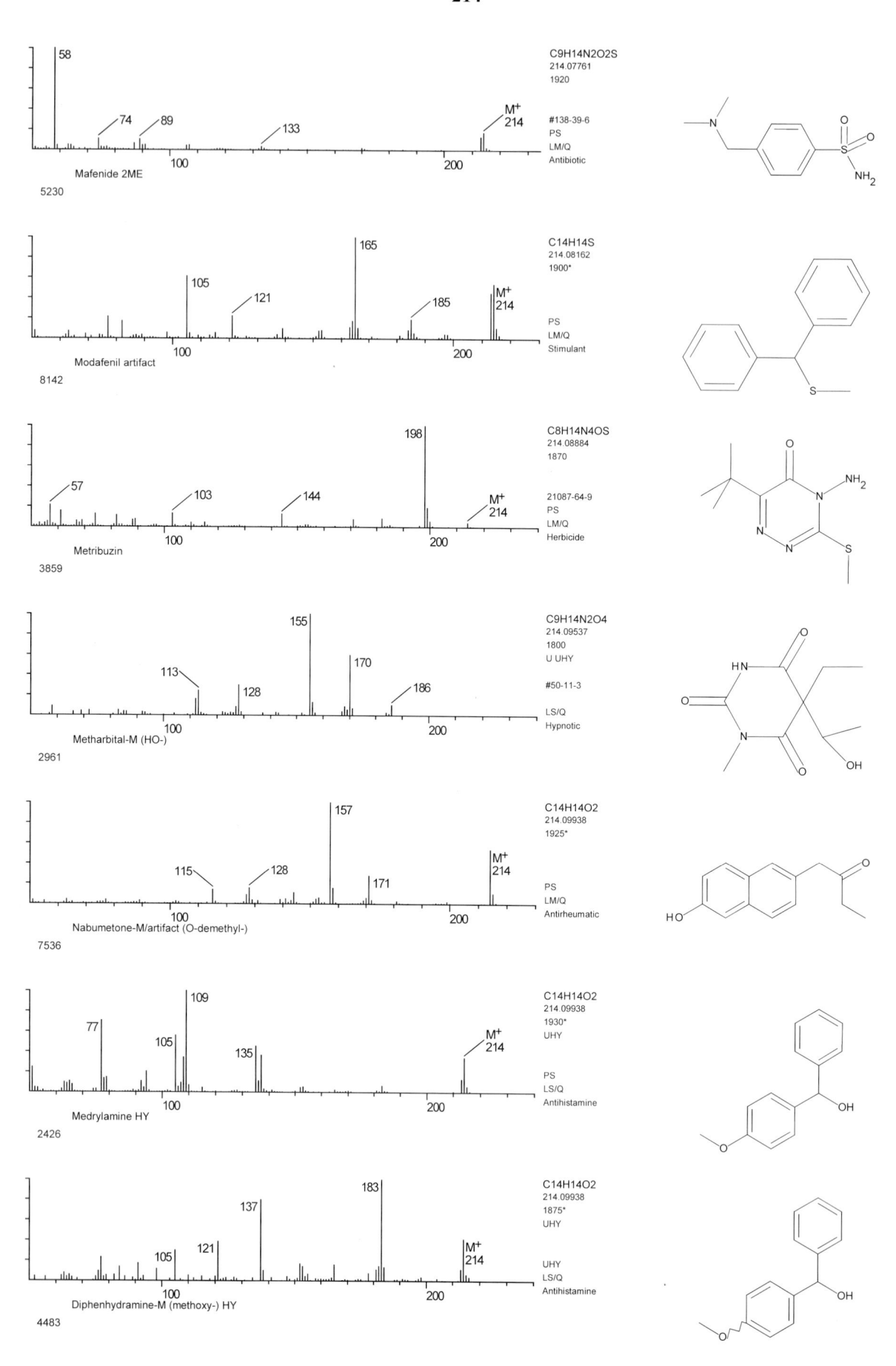

Mafenide 2ME	C9H14N2O2S 214.07761 1920 #138-39-6 PS LM/Q Antibiotic
Modafenil artifact	C14H14S 214.08162 1900* PS LM/Q Stimulant
Metribuzin	C8H14N4OS 214.08884 1870 21087-64-9 PS LM/Q Herbicide
Metharbital-M (HO-)	C9H14N2O4 214.09537 1800 U UHY #50-11-3 LS/Q Hypnotic
Nabumetone-M/artifact (O-demethyl-)	C14H14O2 214.09938 1925* PS LM/Q Antirheumatic
Medrylamine HY	C14H14O2 214.09938 1930* UHY PS LS/Q Antihistamine
Diphenhydramine-M (methoxy-) HY	C14H14O2 214.09938 1875* UHY UHY LS/Q Antihistamine

214

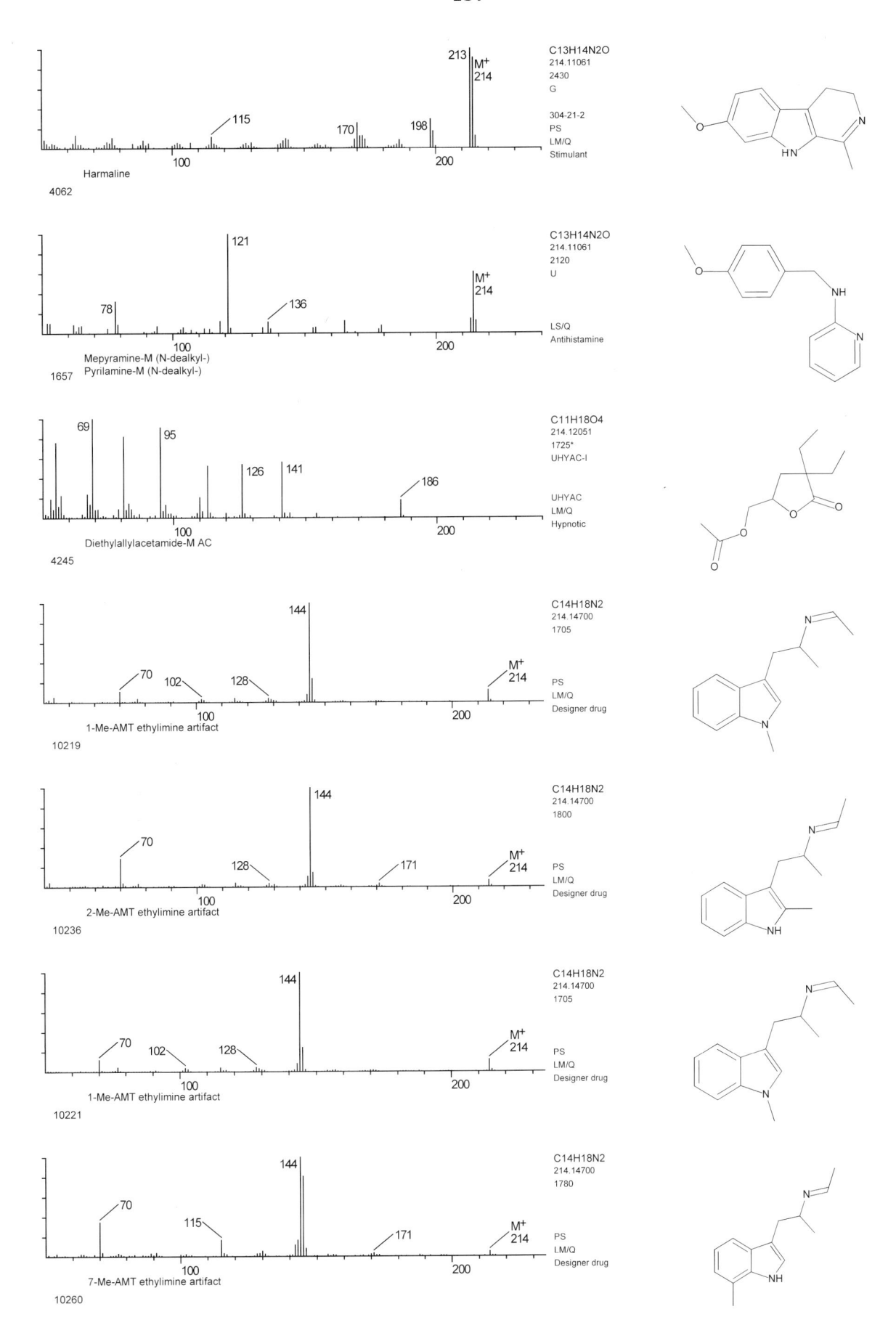

Harmaline — C13H14N2O, 214.11061, 2430, G, 304-21-2, PS, LM/Q, Stimulant
Peaks: 115, 170, 198, 213, 214 (M+)
4062

Mepyramine-M (N-dealkyl-) / Pyrilamine-M (N-dealkyl-) — C13H14N2O, 214.11061, 2120, U, LS/Q, Antihistamine
Peaks: 78, 121, 136, 214 (M+)
1657

Diethylallylacetamide-M AC — C11H18O4, 214.12051, 1725*, UHYAC-I, UHYAC, LM/Q, Hypnotic
Peaks: 69, 95, 126, 141, 186
4245

1-Me-AMT ethylimine artifact — C14H18N2, 214.14700, 1705, PS, LM/Q, Designer drug
Peaks: 70, 102, 128, 144, 214 (M+)
10219

2-Me-AMT ethylimine artifact — C14H18N2, 214.14700, 1800, PS, LM/Q, Designer drug
Peaks: 70, 128, 144, 171, 214 (M+)
10236

1-Me-AMT ethylimine artifact — C14H18N2, 214.14700, 1705, PS, LM/Q, Designer drug
Peaks: 70, 102, 128, 144, 214 (M+)
10221

7-Me-AMT ethylimine artifact — C14H18N2, 214.14700, 1780, PS, LM/Q, Designer drug
Peaks: 70, 115, 144, 171, 214 (M+)
10260

215

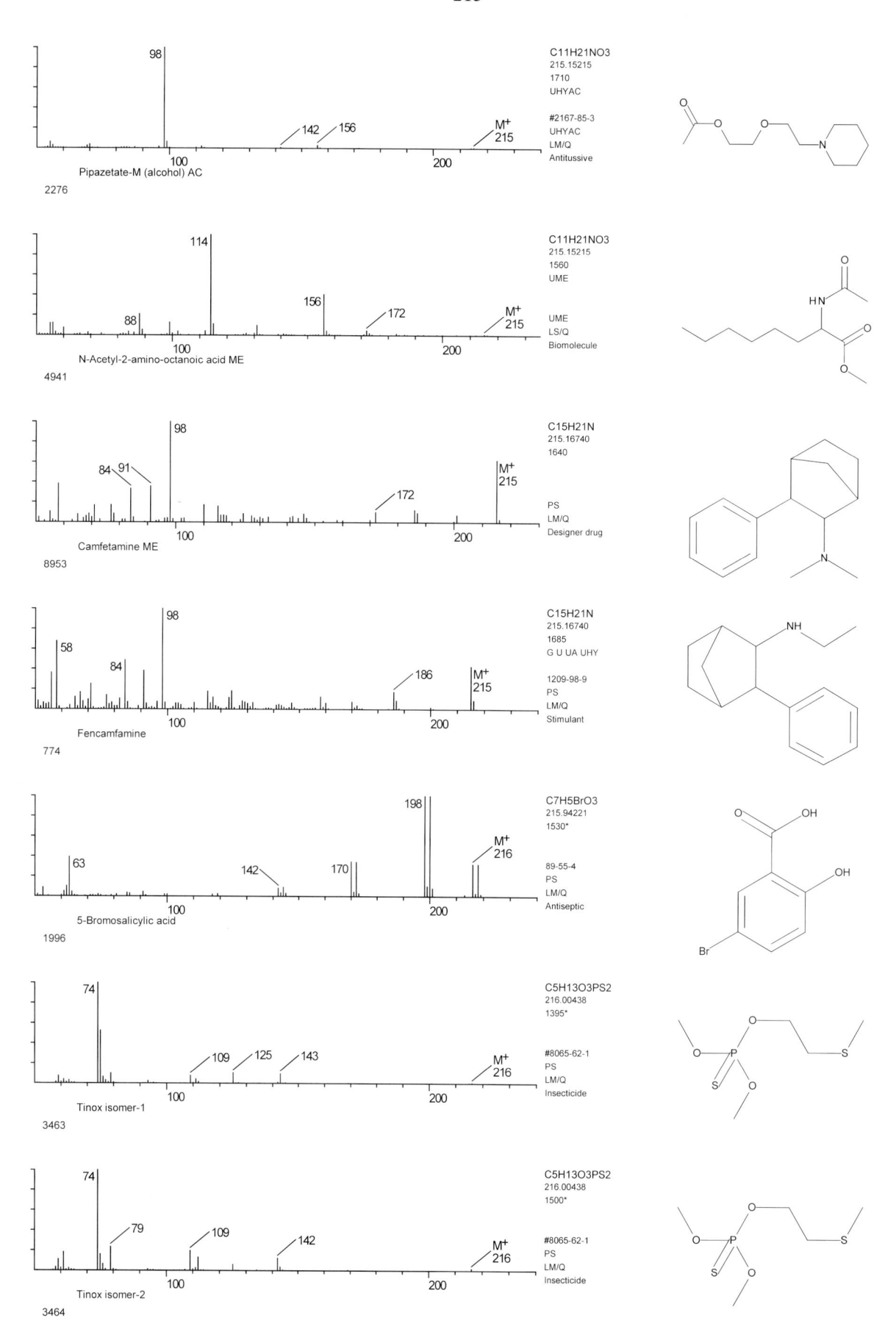

Spectrum	Compound	Formula / Data
2276	Pipazetate-M (alcohol) AC	C11H21NO3 — 215.15215 — 1710 — UHYAC — #2167-85-3 — UHYAC — LM/Q — Antitussive. Peaks: 98, 142, 156, M+ 215
4941	N-Acetyl-2-amino-octanoic acid ME	C11H21NO3 — 215.15215 — 1560 — UME — UME — LS/Q — Biomolecule. Peaks: 88, 114, 156, 172, M+ 215
8953	Camfetamine ME	C15H21N — 215.16740 — 1640 — PS — LM/Q — Designer drug. Peaks: 84, 91, 98, 172, M+ 215
774	Fencamfamine	C15H21N — 215.16740 — 1685 — G U UA UHY — 1209-98-9 — PS — LM/Q — Stimulant. Peaks: 58, 84, 98, 186, M+ 215
1996	5-Bromosalicylic acid	C7H5BrO3 — 215.94221 — 1530* — 89-55-4 — PS — LM/Q — Antiseptic. Peaks: 63, 142, 170, 198, M+ 216
3463	Tinox isomer-1	C5H13O3PS2 — 216.00438 — 1395* — #8065-62-1 — PS — LM/Q — Insecticide. Peaks: 74, 109, 125, 143, M+ 216
3464	Tinox isomer-2	C5H13O3PS2 — 216.00438 — 1500* — #8065-62-1 — PS — LM/Q — Insecticide. Peaks: 74, 79, 109, 142, M+ 216

216

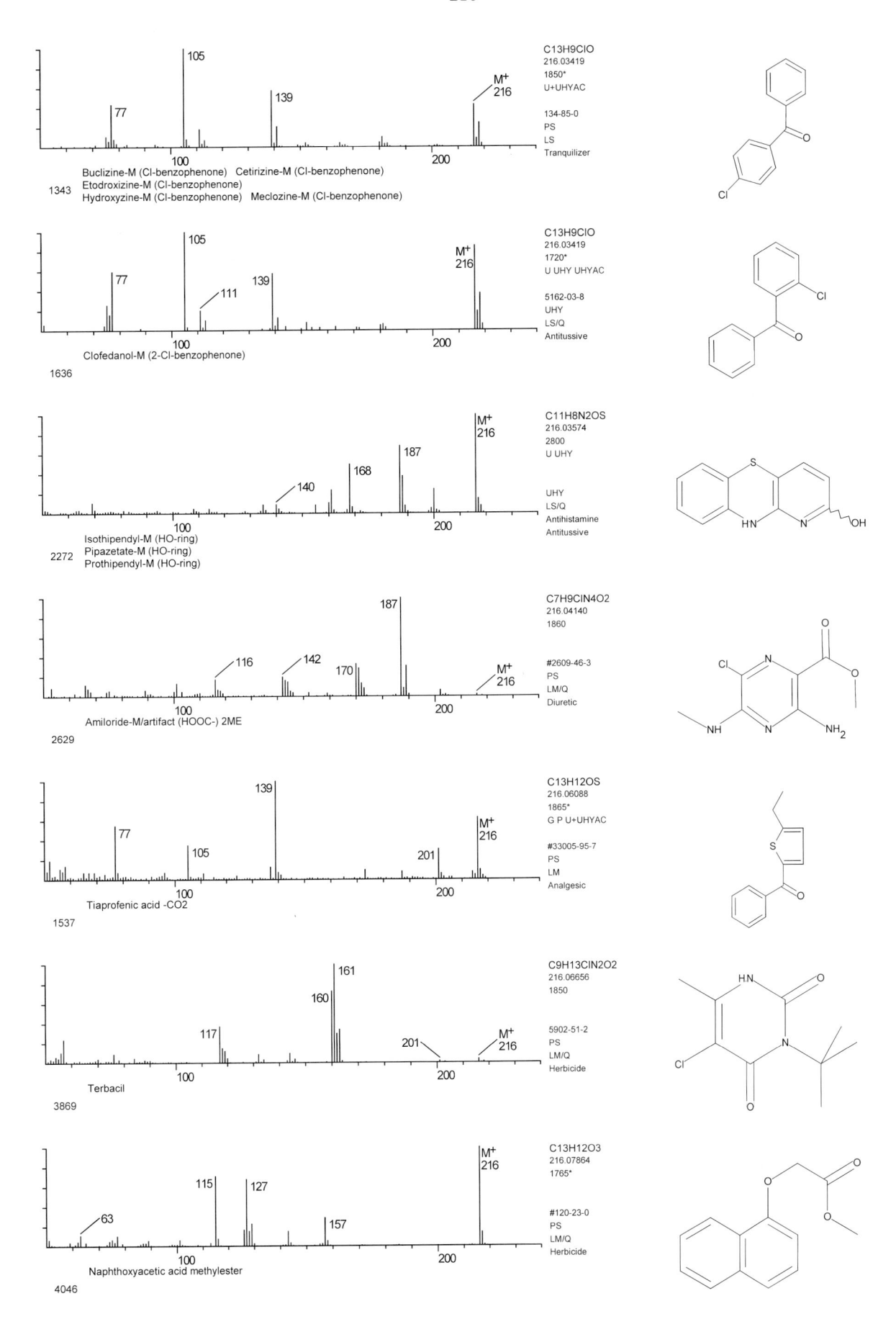

1343
C13H9ClO
216.03419
1850*
U+UHYAC

134-85-0
PS
LS
Tranquilizer

Buclizine-M (Cl-benzophenone)   Cetirizine-M (Cl-benzophenone)
Etodroxizine-M (Cl-benzophenone)
Hydroxyzine-M (Cl-benzophenone)   Meclozine-M (Cl-benzophenone)

1636
C13H9ClO
216.03419
1720*
U UHY UHYAC

5162-03-8
UHY
LS/Q
Antitussive

Clofedanol-M (2-Cl-benzophenone)

2272
C11H8N2OS
216.03574
2800
U UHY

UHY
LS/Q
Antihistamine
Antitussive

Isothipendyl-M (HO-ring)
Pipazetate-M (HO-ring)
Prothipendyl-M (HO-ring)

2629
C7H9ClN4O2
216.04140
1860

#2609-46-3
PS
LM/Q
Diuretic

Amiloride-M/artifact (HOOC-) 2ME

1537
C13H12OS
216.06088
1865*
G P U+UHYAC

#33005-95-7
PS
LM
Analgesic

Tiaprofenic acid -CO2

3869
C9H13ClN2O2
216.06656
1850

5902-51-2
PS
LM/Q
Herbicide

Terbacil

4046
C13H12O3
216.07864
1765*

#120-23-0
PS
LM/Q
Herbicide

Naphthoxyacetic acid methylester

216

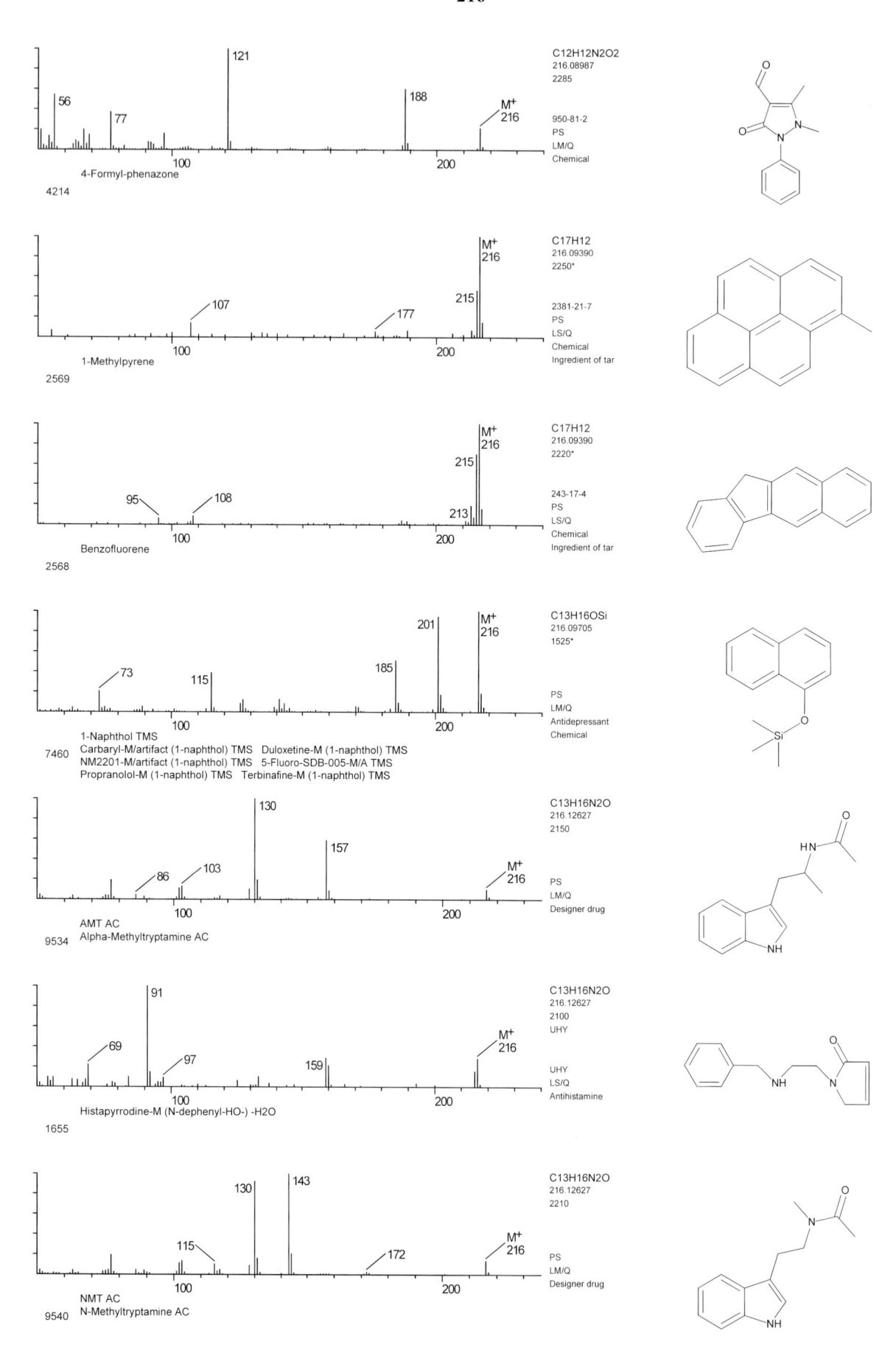

4-Formyl-phenazone
Peaks: 56, 77, 121, 188, M+ 216
C12H12N2O2
216.08987
2285
950-81-2
PS
LM/Q
Chemical

1-Methylpyrene
Peaks: 107, 177, 215, M+ 216
C17H12
216.09390
2250*
2381-21-7
PS
LS/Q
Chemical
Ingredient of tar

Benzofluorene
Peaks: 95, 108, 213, 215, M+ 216
C17H12
216.09390
2220*
243-17-4
PS
LS/Q
Chemical
Ingredient of tar

1-Naphthol TMS
Carbaryl-M/artifact (1-naphthol) TMS   Duloxetine-M (1-naphthol) TMS
NM2201-M/artifact (1-naphthol) TMS   5-Fluoro-SDB-005-M/A TMS
Propranolol-M (1-naphthol) TMS   Terbinafine-M (1-naphthol) TMS
Peaks: 73, 115, 185, 201, M+ 216
C13H16OSi
216.09705
1525*
PS
LM/Q
Antidepressant
Chemical

AMT AC / Alpha-Methyltryptamine AC
Peaks: 86, 103, 130, 157, M+ 216
C13H16N2O
216.12627
2150
PS
LM/Q
Designer drug

Histapyrrodine-M (N-dephenyl-HO-) -H2O
Peaks: 69, 91, 97, 159, M+ 216
C13H16N2O
216.12627
2100
UHY
UHY
LS/Q
Antihistamine

NMT AC / N-Methyltryptamine AC
Peaks: 115, 130, 143, 172, M+ 216
C13H16N2O
216.12627
2210
PS
LM/Q
Designer drug

216

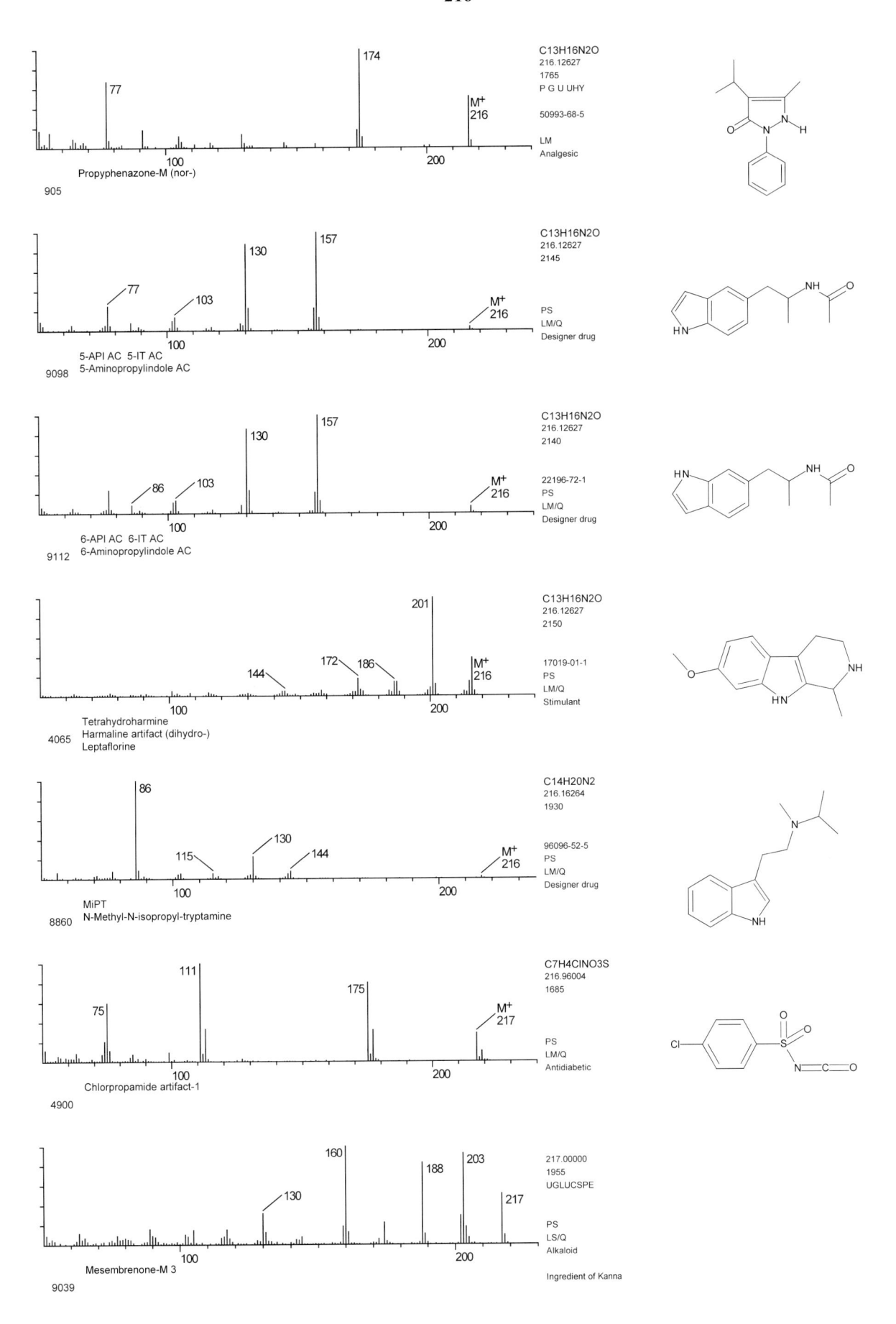

Propyphenazone-M (nor-) — 905	C13H16N2O 216.12627 1765 P G U UHY 50993-68-5 LM Analgesic Peaks: 77, 174, M+ 216
5-API AC 5-IT AC 5-Aminopropylindole AC — 9098	C13H16N2O 216.12627 2145 PS LM/Q Designer drug Peaks: 77, 103, 130, 157, M+ 216
6-API AC 6-IT AC 6-Aminopropylindole AC — 9112	C13H16N2O 216.12627 2140 22196-72-1 PS LM/Q Designer drug Peaks: 86, 103, 130, 157, M+ 216
Tetrahydroharmine Harmaline artifact (dihydro-) Leptaflorine — 4065	C13H16N2O 216.12627 2150 17019-01-1 PS LM/Q Stimulant Peaks: 144, 172, 186, 201, M+ 216
MiPT N-Methyl-N-isopropyl-tryptamine — 8860	C14H20N2 216.16264 1930 96096-52-5 PS LM/Q Designer drug Peaks: 86, 115, 130, 144, M+ 216
Chlorpropamide artifact-1 — 4900	C7H4ClNO3S 216.96004 1685 PS LM/Q Antidiabetic Peaks: 75, 111, 175, M+ 217
Mesembrenone-M 3 — 9039	217.00000 1955 UGLUCSPE PS LS/Q Alkaloid Ingredient of Kanna Peaks: 130, 160, 188, 203, 217

217

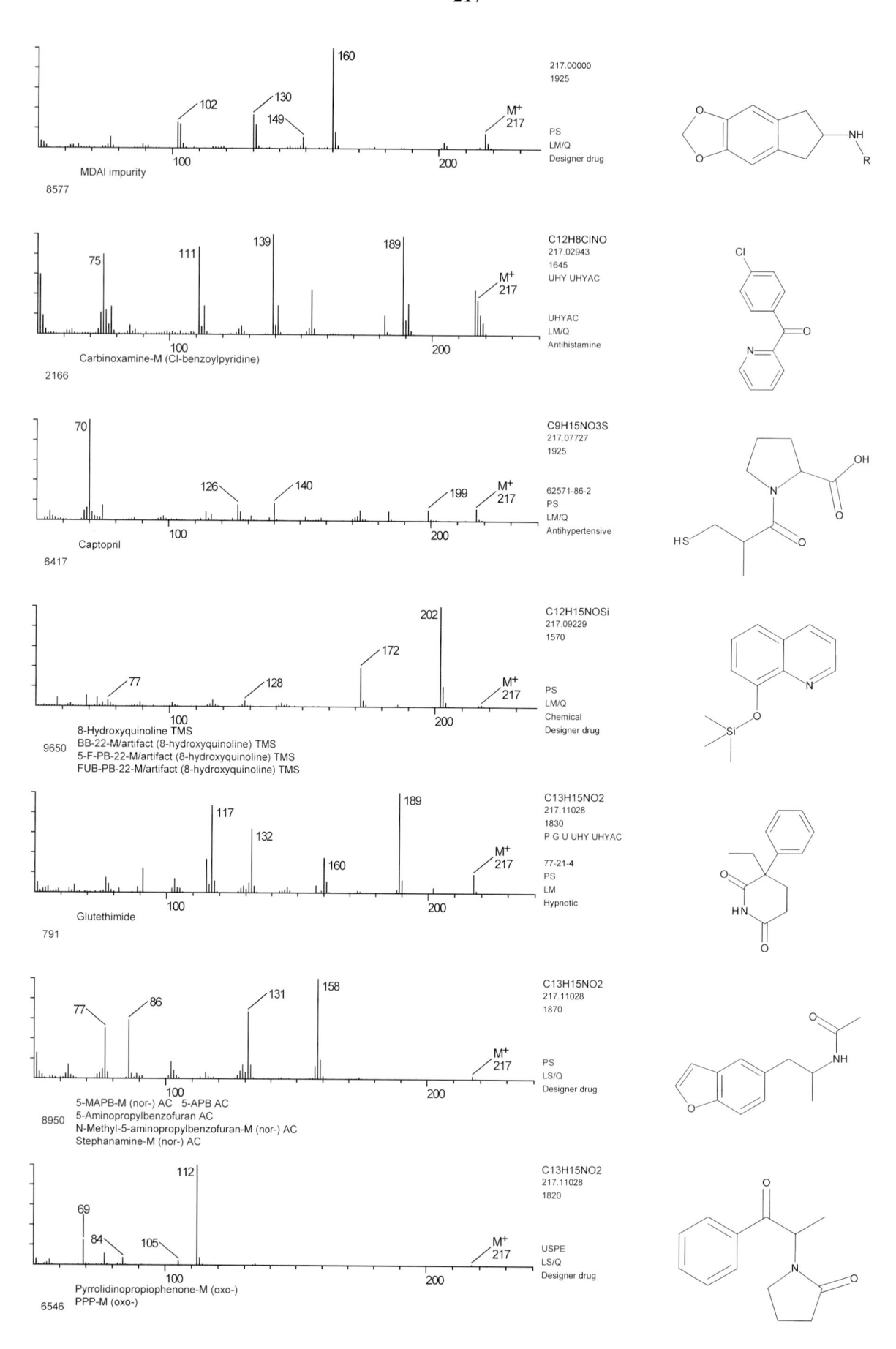

217.00000
1925

PS
LM/Q
Designer drug

MDAI impurity
8577

C12H8ClNO
217.02943
1645
UHY UHYAC

UHYAC
LM/Q
Antihistamine

Carbinoxamine-M (Cl-benzoylpyridine)
2166

C9H15NO3S
217.07727
1925
62571-86-2
PS
LM/Q
Antihypertensive

Captopril
6417

C12H15NOSi
217.09229
1570

PS
LM/Q
Chemical
Designer drug

8-Hydroxyquinoline TMS
BB-22-M/artifact (8-hydroxyquinoline) TMS
5-F-PB-22-M/artifact (8-hydroxyquinoline) TMS
FUB-PB-22-M/artifact (8-hydroxyquinoline) TMS
9650

C13H15NO2
217.11028
1830
P G U UHY UHYAC
77-21-4
PS
LM
Hypnotic

Glutethimide
791

C13H15NO2
217.11028
1870

PS
LS/Q
Designer drug

5-MAPB-M (nor-) AC   5-APB AC
5-Aminopropylbenzofuran AC
N-Methyl-5-aminopropylbenzofuran-M (nor-) AC
Stephanamine-M (nor-) AC
8950

C13H15NO2
217.11028
1820

USPE
LS/Q
Designer drug

Pyrrolidinopropiophenone-M (oxo-)
PPP-M (oxo-)
6546

454

217

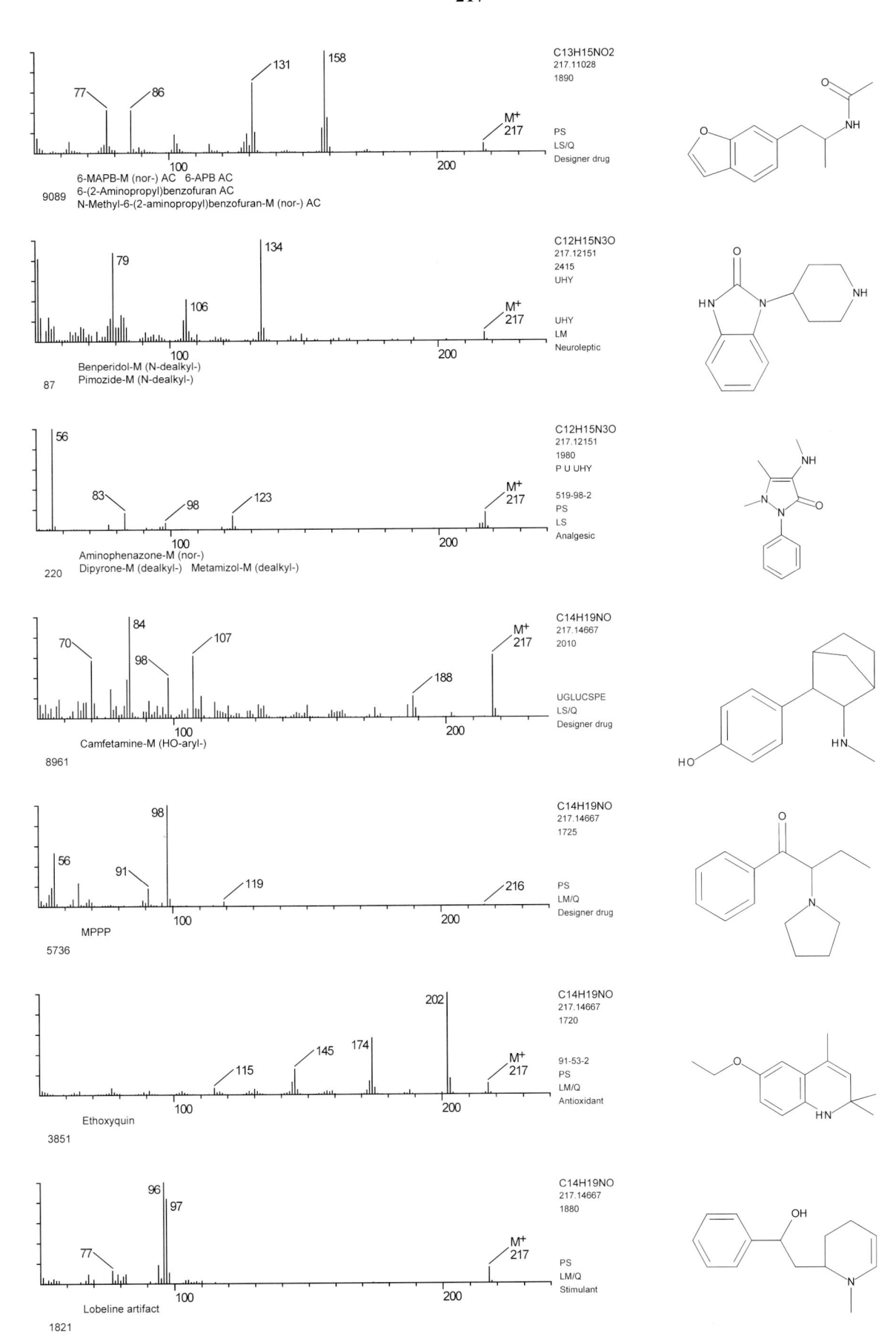

C13H15NO2
217.11028
1890
PS
LS/Q
Designer drug

6-MAPB-M (nor-) AC   6-APB AC
6-(2-Aminopropyl)benzofuran AC
N-Methyl-6-(2-aminopropyl)benzofuran-M (nor-) AC
9089

C12H15N3O
217.12151
2415
UHY
UHY
LM
Neuroleptic

Benperidol-M (N-dealkyl-)
Pimozide-M (N-dealkyl-)
87

C12H15N3O
217.12151
1980
P U UHY
519-98-2
PS
LS
Analgesic

Aminophenazone-M (nor-)
Dipyrone-M (dealkyl-)   Metamizol-M (dealkyl-)
220

C14H19NO
217.14667
2010
UGLUCSPE
LS/Q
Designer drug

Camfetamine-M (HO-aryl-)
8961

C14H19NO
217.14667
1725
PS
LM/Q
Designer drug

MPPP
5736

C14H19NO
217.14667
1720
91-53-2
PS
LM/Q
Antioxidant

Ethoxyquin
3851

C14H19NO
217.14667
1880
PS
LM/Q
Stimulant

Lobeline artifact
1821

217

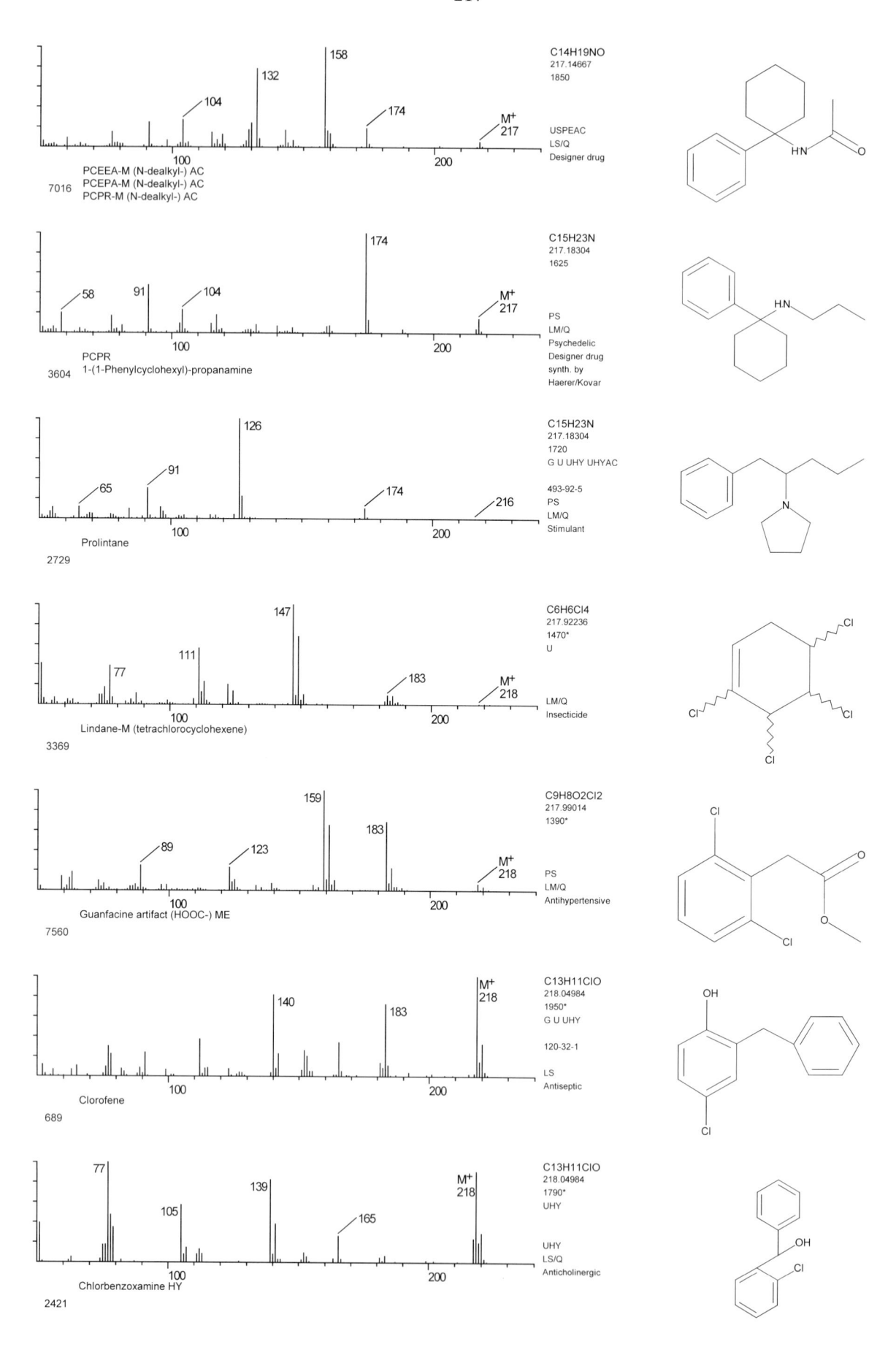

PCEEA-M (N-dealkyl-) AC
PCEPA-M (N-dealkyl-) AC
PCPR-M (N-dealkyl-) AC

C14H19NO
217.14667
1850

USPEAC
LS/Q
Designer drug

7016

PCPR 1-(1-Phenylcyclohexyl)-propanamine

C15H23N
217.18304
1625

PS
LM/Q
Psychedelic
Designer drug
synth. by
Haerer/Kovar

3604

Prolintane

C15H23N
217.18304
1720
G U UHY UHYAC

493-92-5
PS
LM/Q
Stimulant

2729

Lindane-M (tetrachlorocyclohexene)

C6H6Cl4
217.92236
1470*
U

LM/Q
Insecticide

3369

Guanfacine artifact (HOOC-) ME

C9H8O2Cl2
217.99014
1390*

PS
LM/Q
Antihypertensive

7560

Clorofene

C13H11ClO
218.04984
1950*
G U UHY

120-32-1

LS
Antiseptic

689

Chlorbenzoxamine HY

C13H11ClO
218.04984
1790*
UHY

UHY
LS/Q
Anticholinergic

2421

218

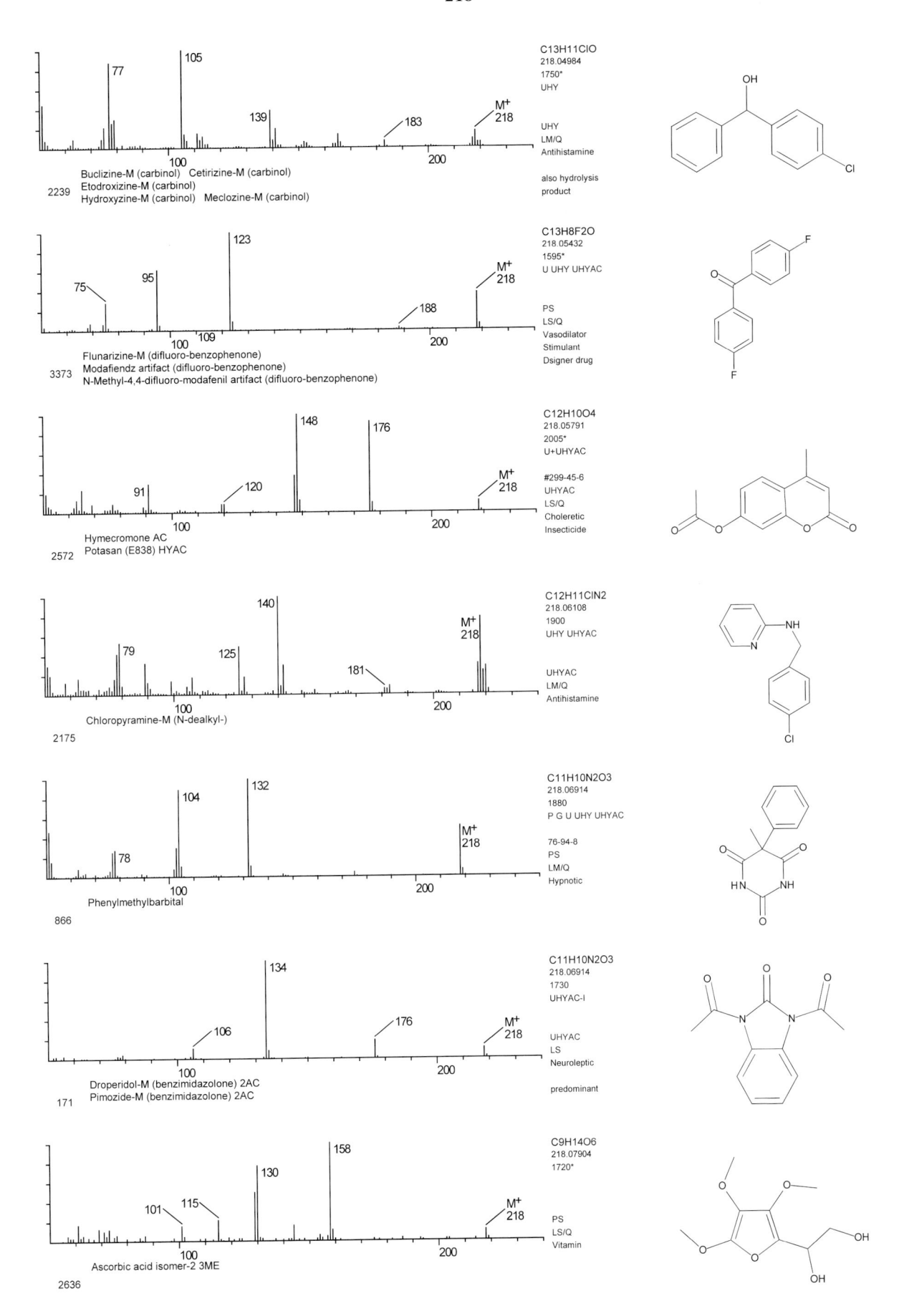

2239	77, 105, 139, 183, M+ 218 Buclizine-M (carbinol) Cetirizine-M (carbinol) Etodroxizine-M (carbinol) Hydroxyzine-M (carbinol) Meclozine-M (carbinol)	C13H11ClO 218.04984 1750* UHY UHY LM/Q Antihistamine also hydrolysis product
3373	75, 95, 123, 109, 188, M+ 218 Flunarizine-M (difluoro-benzophenone) Modafiendz artifact (difluoro-benzophenone) N-Methyl-4,4-difluoro-modafenil artifact (difluoro-benzophenone)	C13H8F2O 218.05432 1595* U UHY UHYAC PS LS/Q Vasodilator Stimulant Dsigner drug
2572	91, 120, 148, 176, M+ 218 Hymecromone AC Potasan (E838) HYAC	C12H10O4 218.05791 2005* U+UHYAC #299-45-6 UHYAC LS/Q Choleretic Insecticide
2175	79, 125, 140, 181, M+ 218 Chloropyramine-M (N-dealkyl-)	C12H11ClN2 218.06108 1900 UHY UHYAC UHYAC LM/Q Antihistamine
866	78, 104, 132, M+ 218 Phenylmethylbarbital	C11H10N2O3 218.06914 1880 P G U UHY UHYAC 76-94-8 PS LM/Q Hypnotic
171	106, 134, 176, M+ 218 Droperidol-M (benzimidazolone) 2AC Pimozide-M (benzimidazolone) 2AC	C11H10N2O3 218.06914 1730 UHYAC-I UHYAC LS Neuroleptic predominant
2636	101, 115, 130, 158, M+ 218 Ascorbic acid isomer-2 3ME	C9H14O6 218.07904 1720* PS LS/Q Vitamin

218

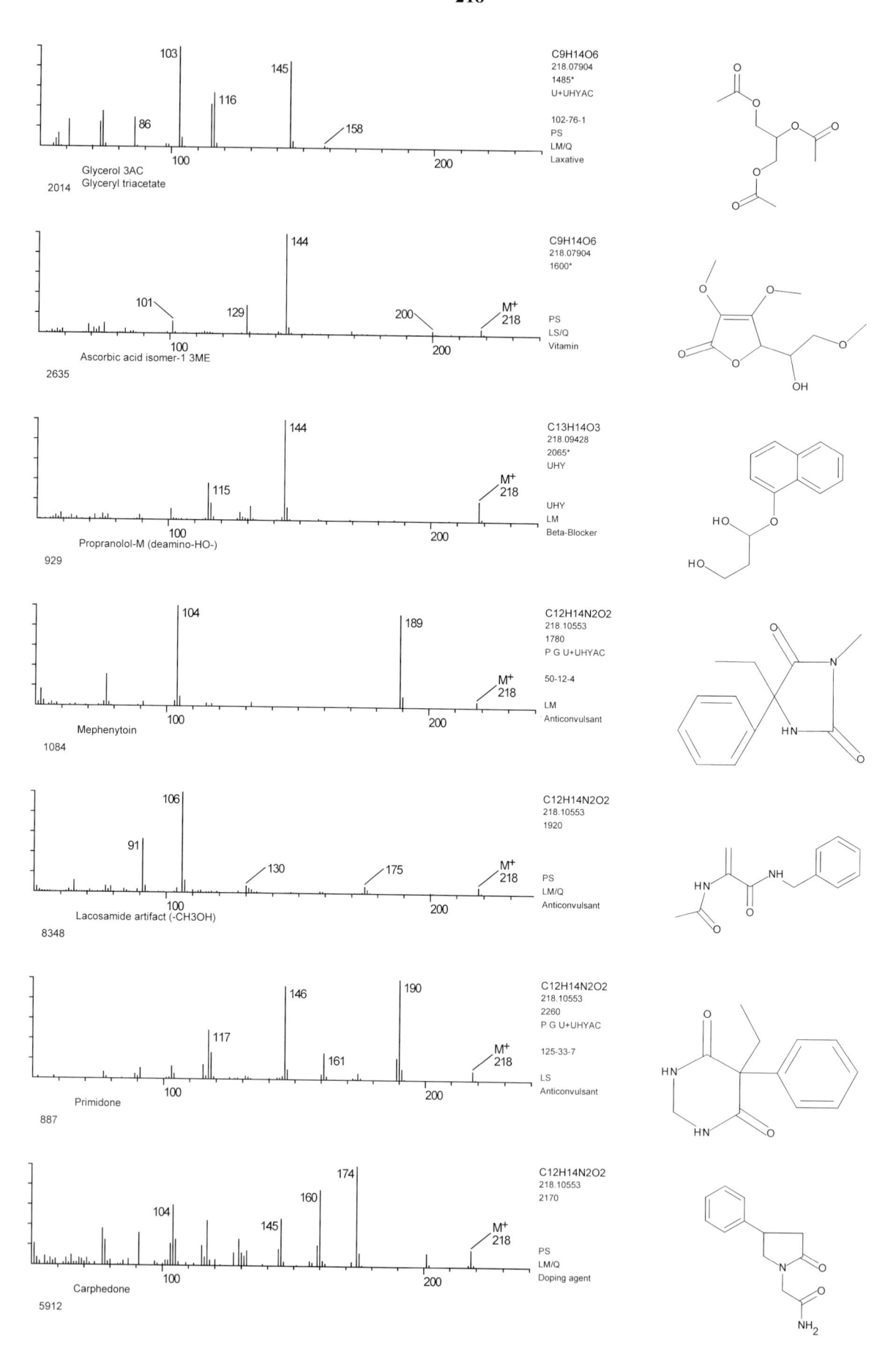

Spectrum	Compound	Formula / Info
Peaks: 86, 103, 116, 145, 158	Glycerol 3AC / Glyceryl triacetate — 2014	C9H14O6; 218.07904; 1485*; U+UHYAC; 102-76-1; PS; LM/Q; Laxative
Peaks: 101, 129, 144, 200, M+ 218	Ascorbic acid isomer-1 3ME — 2635	C9H14O6; 218.07904; 1600*; PS; LS/Q; Vitamin
Peaks: 115, 144, M+ 218	Propranolol-M (deamino-HO-) — 929	C13H14O3; 218.09428; 2065*; UHY; UHY; LM; Beta-Blocker
Peaks: 104, 189, M+ 218	Mephenytoin — 1084	C12H14N2O2; 218.10553; 1780; P G U+UHYAC; 50-12-4; LM; Anticonvulsant
Peaks: 91, 106, 130, 175, M+ 218	Lacosamide artifact (-CH3OH) — 8348	C12H14N2O2; 218.10553; 1920; PS; LM/Q; Anticonvulsant
Peaks: 117, 146, 161, 190, M+ 218	Primidone — 887	C12H14N2O2; 218.10553; 2260; P G U+UHYAC; 125-33-7; LS; Anticonvulsant
Peaks: 104, 145, 160, 174, M+ 218	Carphedone — 5912	C12H14N2O2; 218.10553; 2170; PS; LM/Q; Doping agent

218

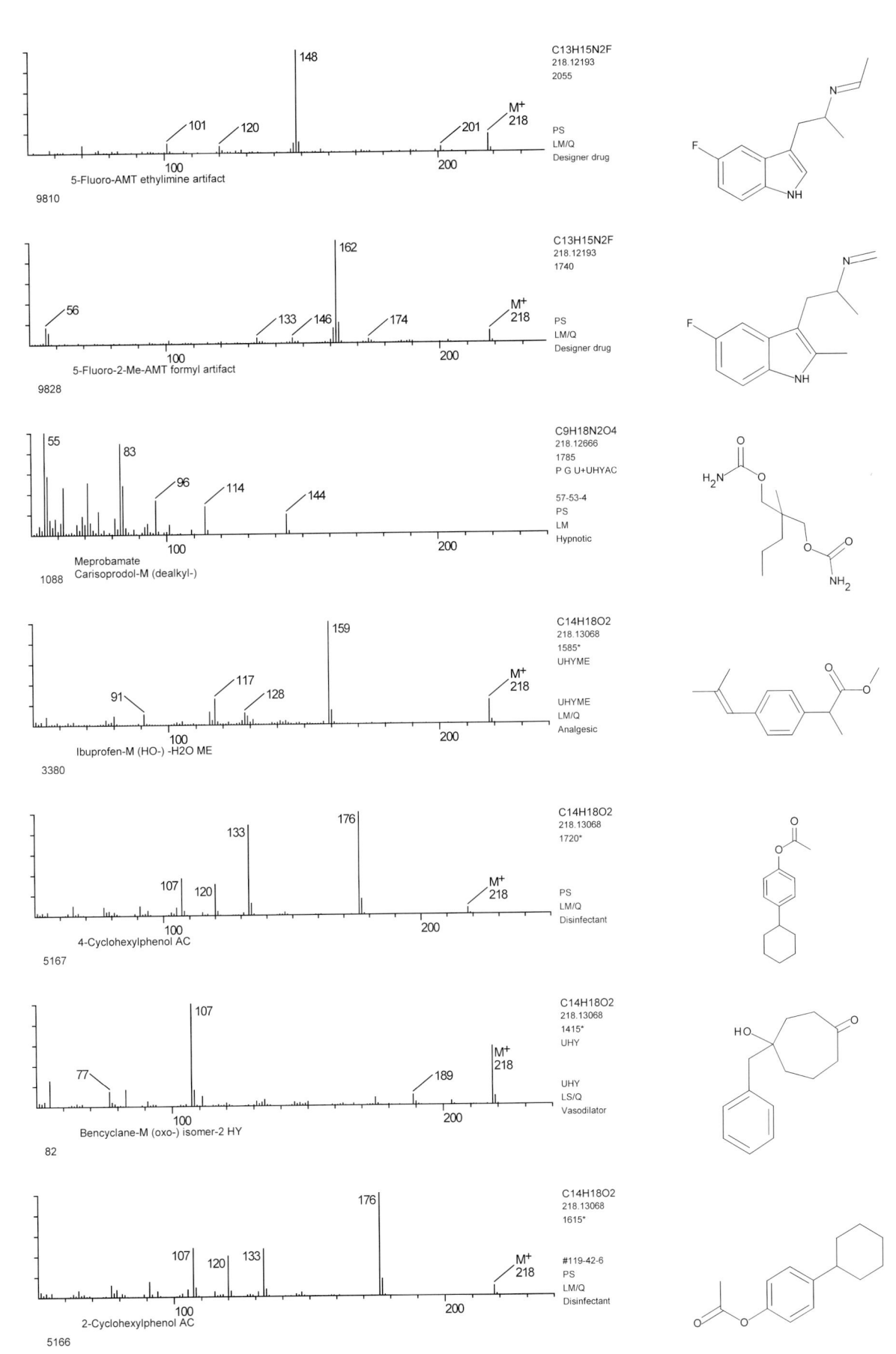

5-Fluoro-AMT ethylimine artifact	peaks: 101, 120, 148, 201, M+ 218	C13H15N2F 218.12193 2055 PS LM/Q Designer drug
9810		
5-Fluoro-2-Me-AMT formyl artifact	peaks: 56, 133, 146, 162, 174, M+ 218	C13H15N2F 218.12193 1740 PS LM/Q Designer drug
9828		
Meprobamate Carisoprodol-M (dealkyl-)	peaks: 55, 83, 96, 114, 144	C9H18N2O4 218.12666 1785 P G U+UHYAC 57-53-4 PS LM Hypnotic
1088		
Ibuprofen-M (HO-) -H2O ME	peaks: 91, 117, 128, 159, M+ 218	C14H18O2 218.13068 1585* UHYME UHYME LM/Q Analgesic
3380		
4-Cyclohexylphenol AC	peaks: 107, 120, 133, 176, M+ 218	C14H18O2 218.13068 1720* PS LM/Q Disinfectant
5167		
Bencyclane-M (oxo-) isomer-2 HY	peaks: 77, 107, 189, M+ 218	C14H18O2 218.13068 1415* UHY UHY LS/Q Vasodilator
82		
2-Cyclohexylphenol AC	peaks: 107, 120, 133, 176, M+ 218	C14H18O2 218.13068 1615* #119-42-6 PS LM/Q Disinfectant
5166		

218

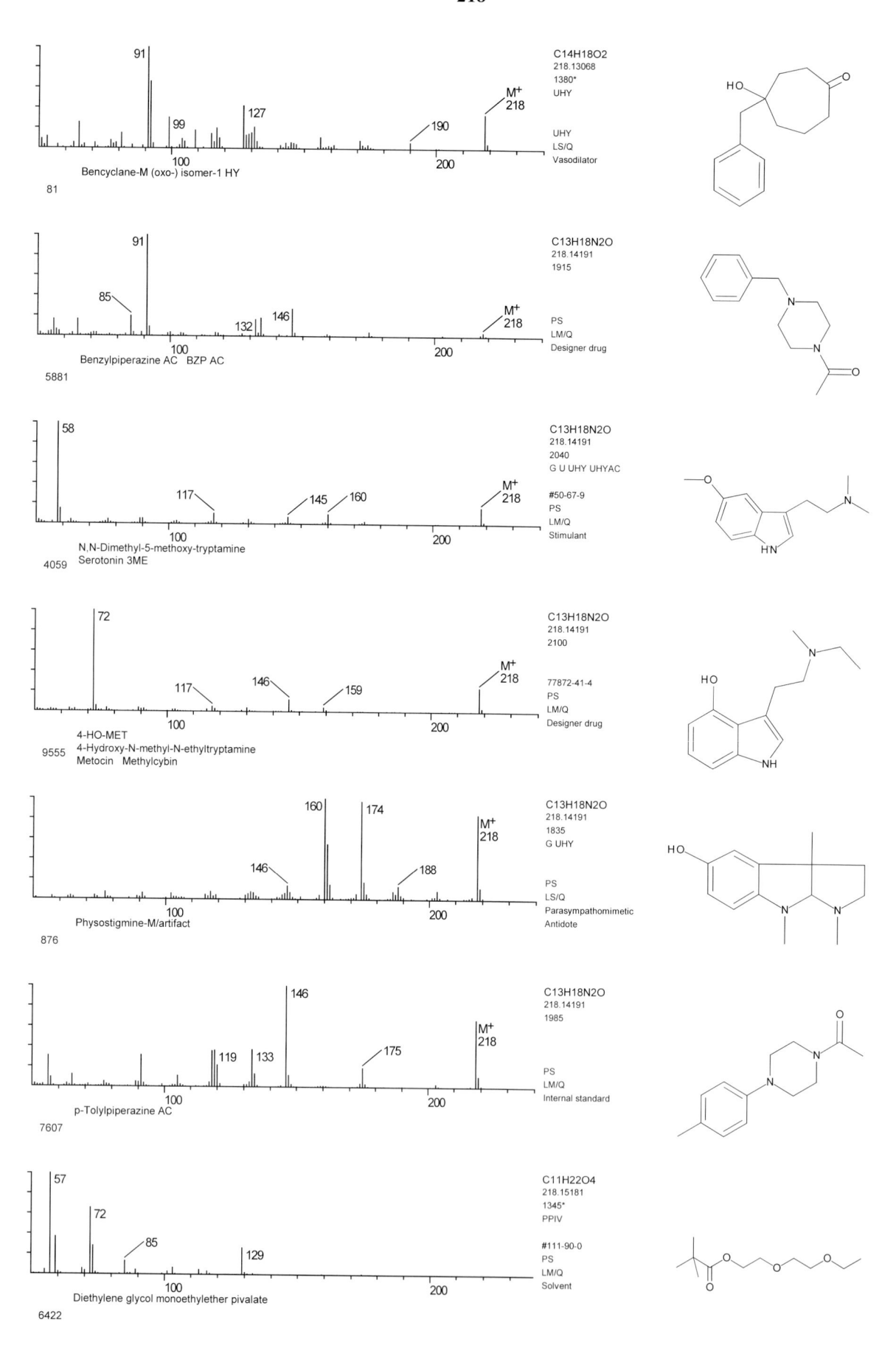

Spectrum label	Formula / Info
Bencyclane-M (oxo-) isomer-1 HY — 81	C14H18O2, 218.13068, 1380*, UHY, UHY, LS/Q, Vasodilator
Benzylpiperazine AC BZP AC — 5881	C13H18N2O, 218.14191, 1915, PS, LM/Q, Designer drug
N,N-Dimethyl-5-methoxy-tryptamine Serotonin 3ME — 4059	C13H18N2O, 218.14191, 2040, G U UHY UHYAC, #50-67-9, PS, LM/Q, Stimulant
4-HO-MET 4-Hydroxy-N-methyl-N-ethyltryptamine Metocin Methylcybin — 9555	C13H18N2O, 218.14191, 2100, 77872-41-4, PS, LM/Q, Designer drug
Physostigmine-M/artifact — 876	C13H18N2O, 218.14191, 1835, G UHY, PS, LS/Q, Parasympathomimetic Antidote
p-Tolylpiperazine AC — 7607	C13H18N2O, 218.14191, 1985, PS, LM/Q, Internal standard
Diethylene glycol monoethylether pivalate — 6422	C11H22O4, 218.15181, 1345*, PPIV, #111-90-0, PS, LM/Q, Solvent

218

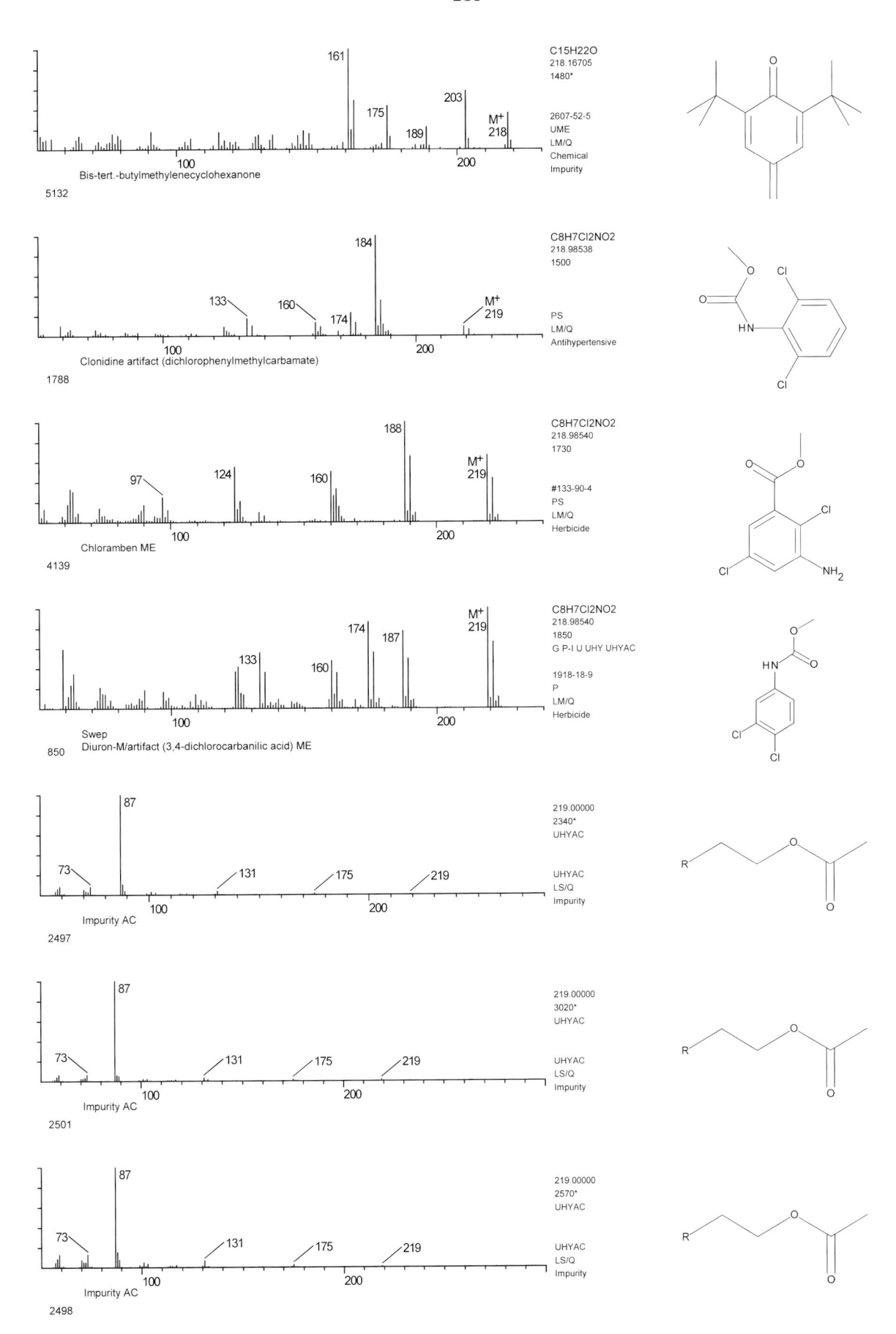

Bis-tert.-butylmethylenecyclohexanone
C15H22O
218.16705
1480*
2607-52-5
UME
LM/Q
Chemical Impurity
5132

Clonidine artifact (dichlorophenylmethylcarbamate)
C8H7Cl2NO2
218.98538
1500
PS
LM/Q
Antihypertensive
1788

Chloramben ME
C8H7Cl2NO2
218.98540
1730
#133-90-4
PS
LM/Q
Herbicide
4139

Swep / Diuron-M/artifact (3,4-dichlorocarbanilic acid) ME
C8H7Cl2NO2
218.98540
1850
G P-I U UHY UHYAC
1918-18-9
P
LM/Q
Herbicide
850

Impurity AC
219.00000
2340*
UHYAC
UHYAC
LS/Q
Impurity
2497

Impurity AC
219.00000
3020*
UHYAC
UHYAC
LS/Q
Impurity
2501

Impurity AC
219.00000
2570*
UHYAC
UHYAC
LS/Q
Impurity
2498

219

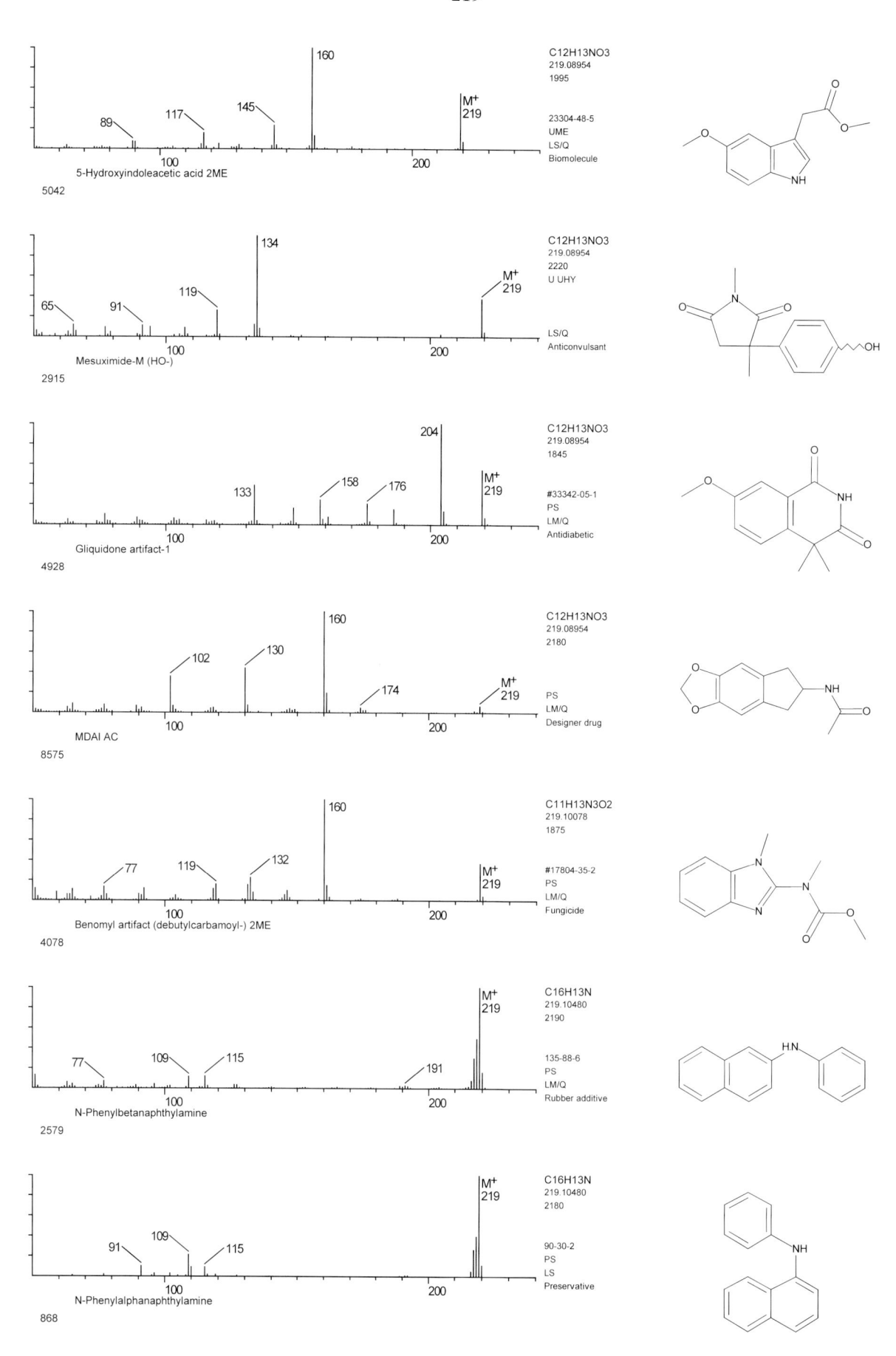

Spectrum	Compound	Formula	Details
89, 117, 145, 160, M+ 219	5-Hydroxyindoleacetic acid 2ME	C12H13NO3	219.08954 / 1995 / 23304-48-5 / UME / LS/Q / Biomolecule
65, 91, 119, 134, M+ 219	Mesuximide-M (HO-)	C12H13NO3	219.08954 / 2220 / U UHY / LS/Q / Anticonvulsant
133, 158, 176, 204, M+ 219	Gliquidone artifact-1	C12H13NO3	219.08954 / 1845 / #33342-05-1 / PS / LM/Q / Antidiabetic
102, 130, 160, 174, M+ 219	MDAI AC	C12H13NO3	219.08954 / 2180 / PS / LM/Q / Designer drug
77, 119, 132, 160, M+ 219	Benomyl artifact (debutylcarbamoyl-) 2ME	C11H13N3O2	219.10078 / 1875 / #17804-35-2 / PS / LM/Q / Fungicide
77, 109, 115, 191, M+ 219	N-Phenylbetanaphthylamine	C16H13N	219.10480 / 2190 / 135-88-6 / PS / LM/Q / Rubber additive
91, 109, 115, M+ 219	N-Phenylalphanaphthylamine	C16H13N	219.10480 / 2180 / 90-30-2 / PS / LS / Preservative

464

219

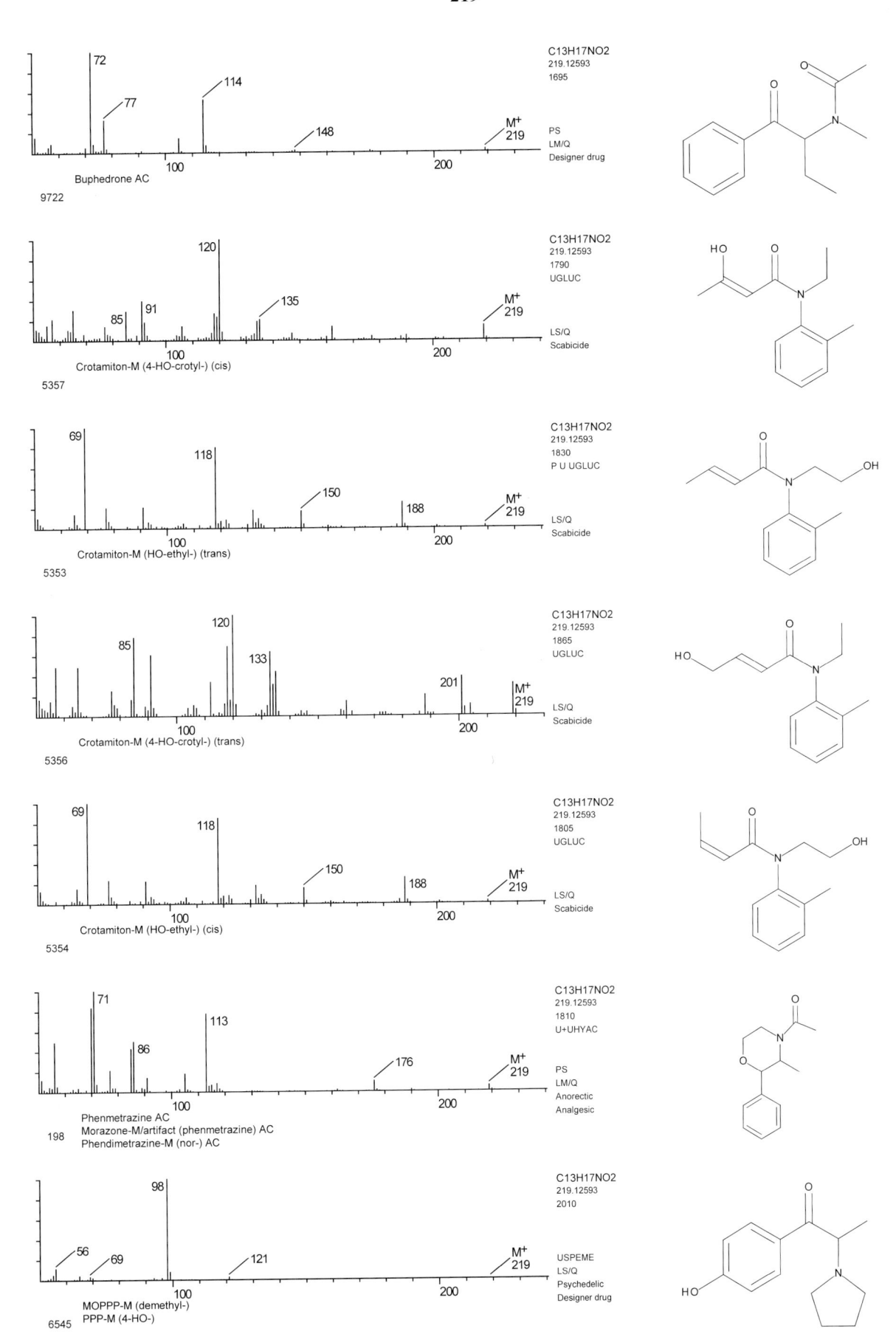

Buphedrone AC	72, 77, 114, 148, M+ 219	C13H17NO2 / 219.12593 / 1695 / PS / LM/Q / Designer drug
9722		
Crotamiton-M (4-HO-crotyl-) (cis)	85, 91, 120, 135, M+ 219	C13H17NO2 / 219.12593 / 1790 / UGLUC / LS/Q / Scabicide
5357		
Crotamiton-M (HO-ethyl-) (trans)	69, 118, 150, 188, M+ 219	C13H17NO2 / 219.12593 / 1830 / P U UGLUC / LS/Q / Scabicide
5353		
Crotamiton-M (4-HO-crotyl-) (trans)	85, 120, 133, 201, M+ 219	C13H17NO2 / 219.12593 / 1865 / UGLUC / LS/Q / Scabicide
5356		
Crotamiton-M (HO-ethyl-) (cis)	69, 118, 150, 188, M+ 219	C13H17NO2 / 219.12593 / 1805 / UGLUC / LS/Q / Scabicide
5354		
Phenmetrazine AC / Morazone-M/artifact (phenmetrazine) AC / Phendimetrazine-M (nor-) AC	71, 86, 113, 176, M+ 219	C13H17NO2 / 219.12593 / 1810 / U+UHYAC / PS / LM/Q / Anorectic / Analgesic
198		
MOPPP-M (demethyl-) / PPP-M (4-HO-)	56, 69, 98, 121, M+ 219	C13H17NO2 / 219.12593 / 2010 / USPEME / LS/Q / Psychedelic / Designer drug
6545		

219

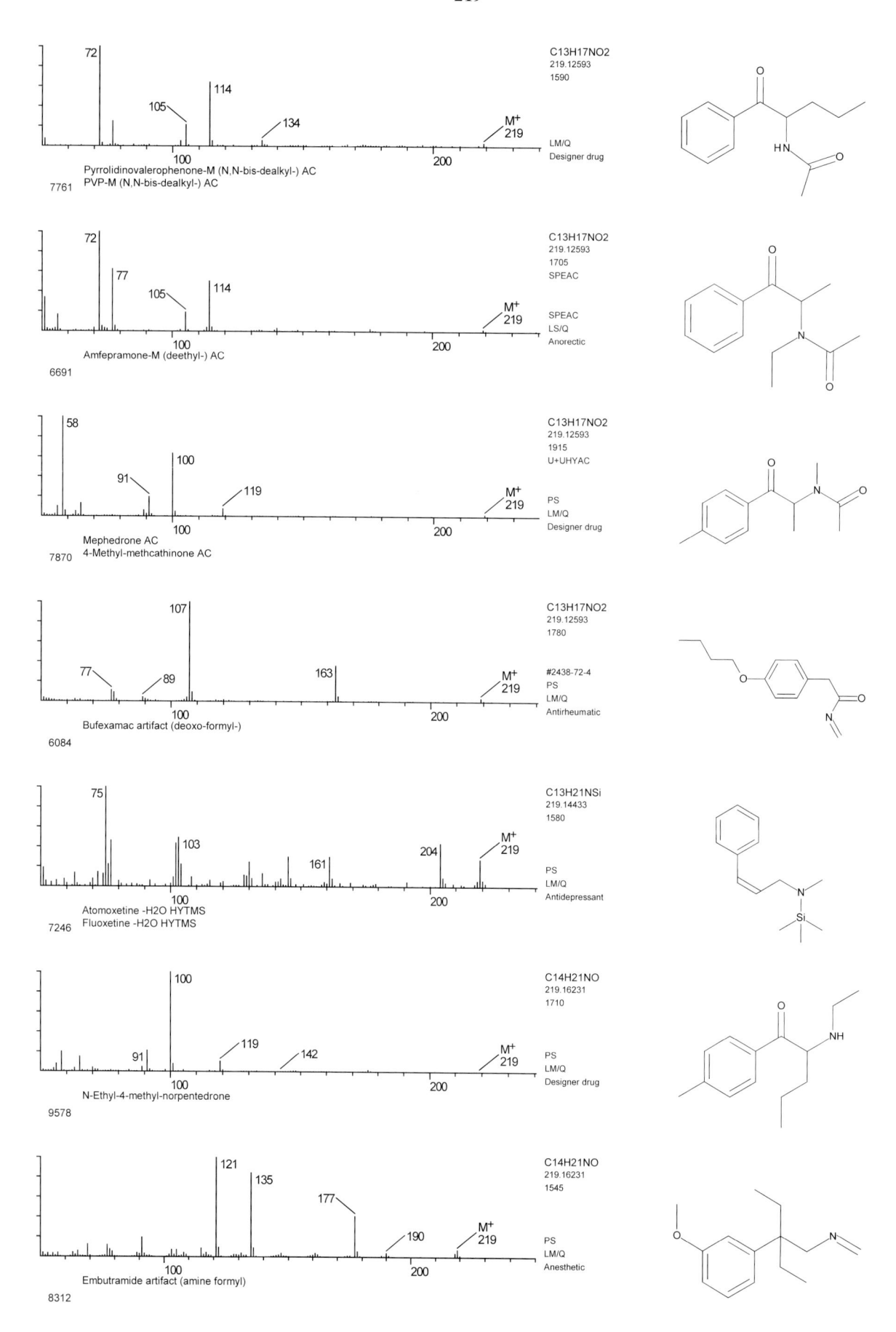

Spectrum peaks	Formula / Info
72, 105, 114, 134, M+ 219	C13H17NO2 / 219.12593 / 1590 / LM/Q / Designer drug — Pyrrolidinovalerophenone-M (N,N-bis-dealkyl-) AC / PVP-M (N,N-bis-dealkyl-) AC — 7761
72, 77, 105, 114, M+ 219	C13H17NO2 / 219.12593 / 1705 / SPEAC / SPEAC / LS/Q / Anorectic — Amfepramone-M (deethyl-) AC — 6691
58, 91, 100, 119, M+ 219	C13H17NO2 / 219.12593 / 1915 / U+UHYAC / PS / LM/Q / Designer drug — Mephedrone AC / 4-Methyl-methcathinone AC — 7870
77, 89, 107, 163, M+ 219	C13H17NO2 / 219.12593 / 1780 / #2438-72-4 / PS / LM/Q / Antirheumatic — Bufexamac artifact (deoxo-formyl-) — 6084
75, 103, 161, 204, M+ 219	C13H21NSi / 219.14433 / 1580 / PS / LM/Q / Antidepressant — Atomoxetine -H2O HYTMS / Fluoxetine -H2O HYTMS — 7246
91, 100, 119, 142, M+ 219	C14H21NO / 219.16231 / 1710 / PS / LM/Q / Designer drug — N-Ethyl-4-methyl-norpentedrone — 9578
121, 135, 177, 190, M+ 219	C14H21NO / 219.16231 / 1545 / PS / LM/Q / Anesthetic — Embutramide artifact (amine formyl) — 8312

219

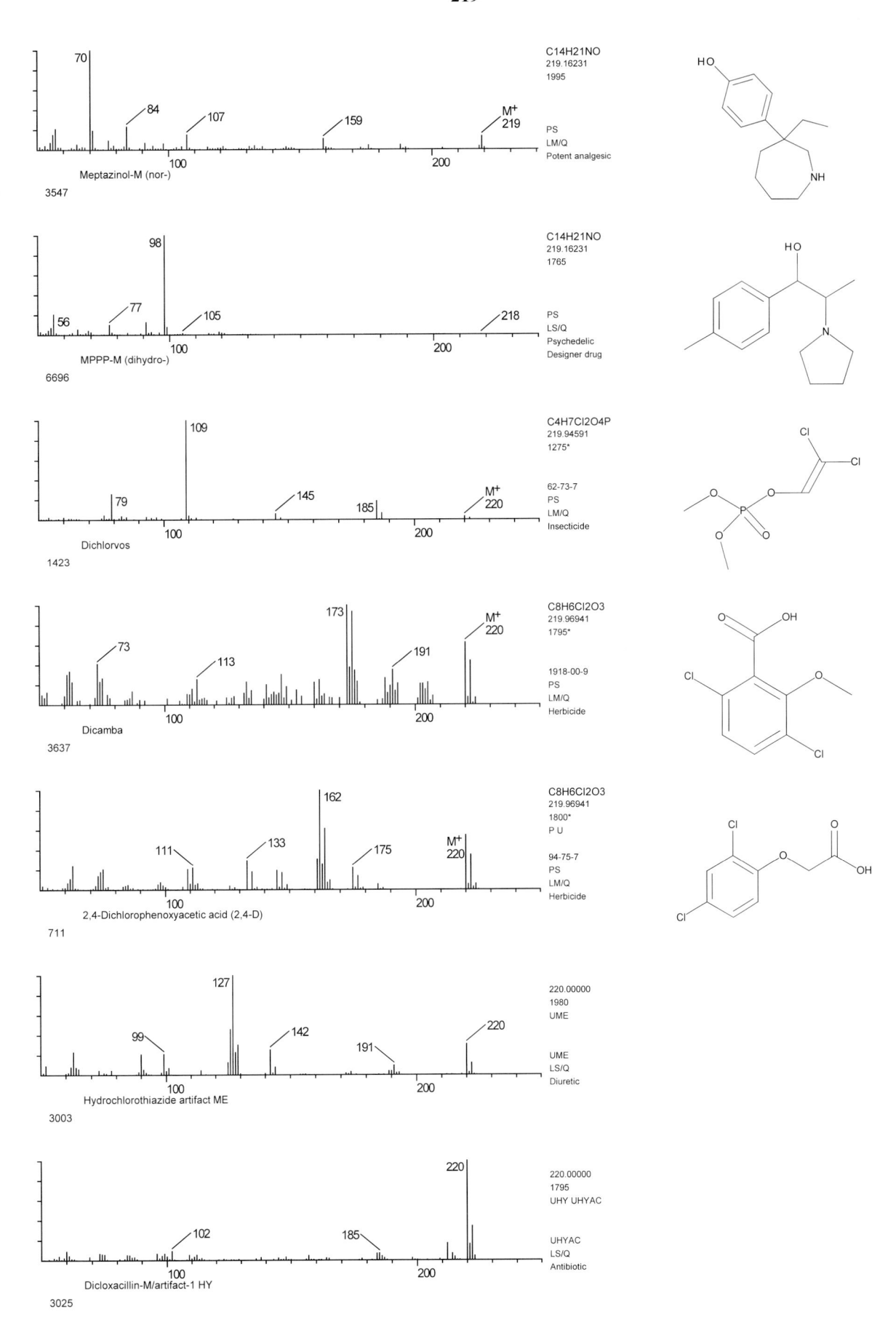

Meptazinol-M (nor-)
3547
C14H21NO
219.16231
1995
PS
LM/Q
Potent analgesic

MPPP-M (dihydro-)
6696
C14H21NO
219.16231
1765
PS
LS/Q
Psychedelic
Designer drug

Dichlorvos
1423
C4H7Cl2O4P
219.94591
1275*
62-73-7
PS
LM/Q
Insecticide

Dicamba
3637
C8H6Cl2O3
219.96941
1795*
1918-00-9
PS
LM/Q
Herbicide

2,4-Dichlorophenoxyacetic acid (2,4-D)
711
C8H6Cl2O3
219.96941
1800*
P U
94-75-7
PS
LM/Q
Herbicide

Hydrochlorothiazide artifact ME
3003
220.00000
1980
UME
UME
LS/Q
Diuretic

Dicloxacillin-M/artifact-1 HY
3025
220.00000
1795
UHY UHYAC
UHYAC
LS/Q
Antibiotic

220

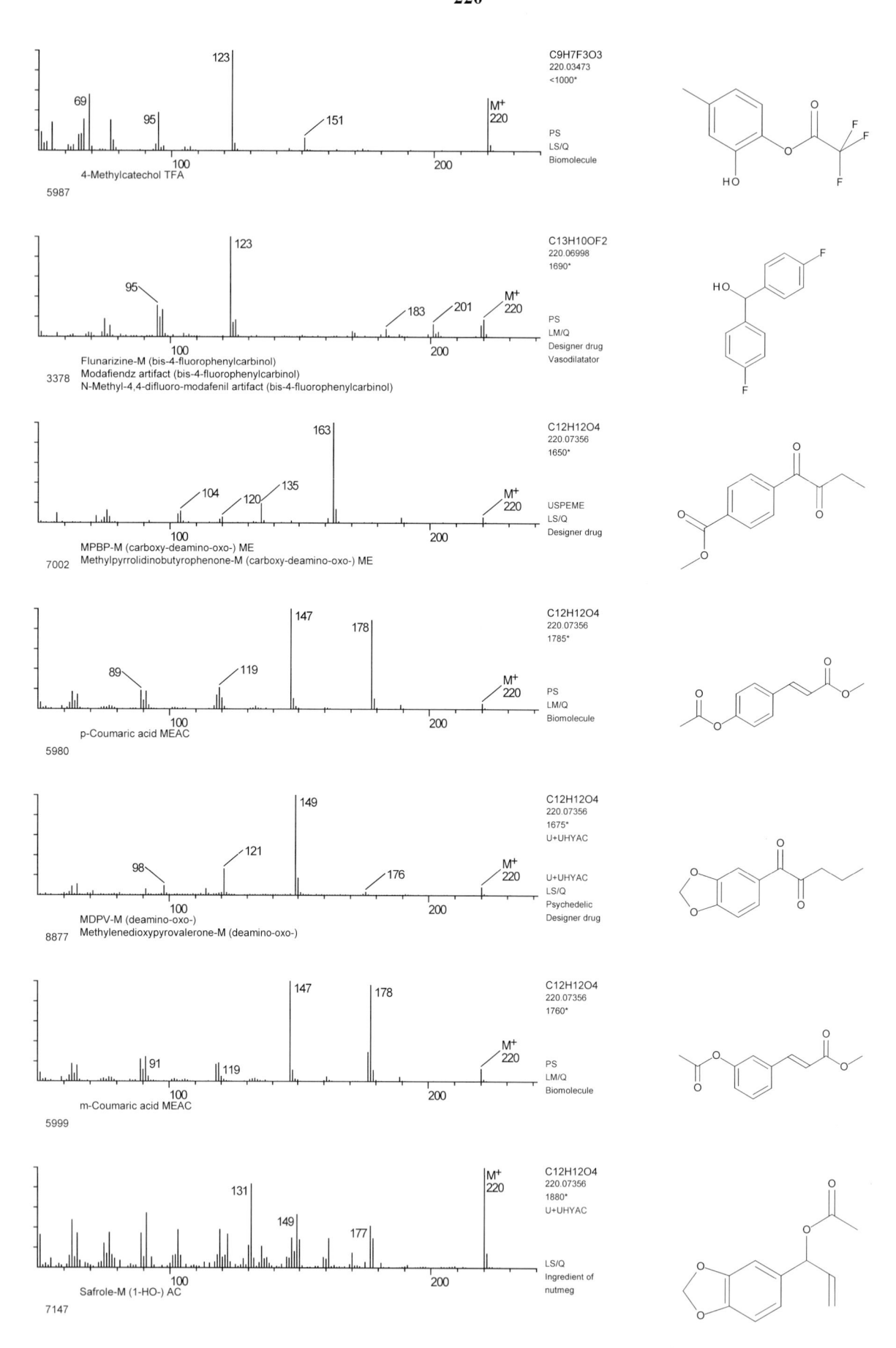

4-Methylcatechol TFA
5987
C9H7F3O3
220.03473
<1000*
PS
LS/Q
Biomolecule

Flunarizine-M (bis-4-fluorophenylcarbinol)
Modafiendz artifact (bis-4-fluorophenylcarbinol)
N-Methyl-4,4-difluoro-modafenil artifact (bis-4-fluorophenylcarbinol)
3378
C13H10OF2
220.06998
1690*
PS
LM/Q
Designer drug
Vasodilatator

MPBP-M (carboxy-deamino-oxo-) ME
Methylpyrrolidinobutyrophenone-M (carboxy-deamino-oxo-) ME
7002
C12H12O4
220.07356
1650*
USPEME
LS/Q
Designer drug

p-Coumaric acid MEAC
5980
C12H12O4
220.07356
1785*
PS
LM/Q
Biomolecule

MDPV-M (deamino-oxo-)
Methylenedioxypyrovalerone-M (deamino-oxo-)
8877
C12H12O4
220.07356
1675*
U+UHYAC
U+UHYAC
LS/Q
Psychedelic
Designer drug

m-Coumaric acid MEAC
5999
C12H12O4
220.07356
1760*
PS
LM/Q
Biomolecule

Safrole-M (1-HO-) AC
7147
C12H12O4
220.07356
1880*
U+UHYAC
LS/Q
Ingredient of nutmeg

220

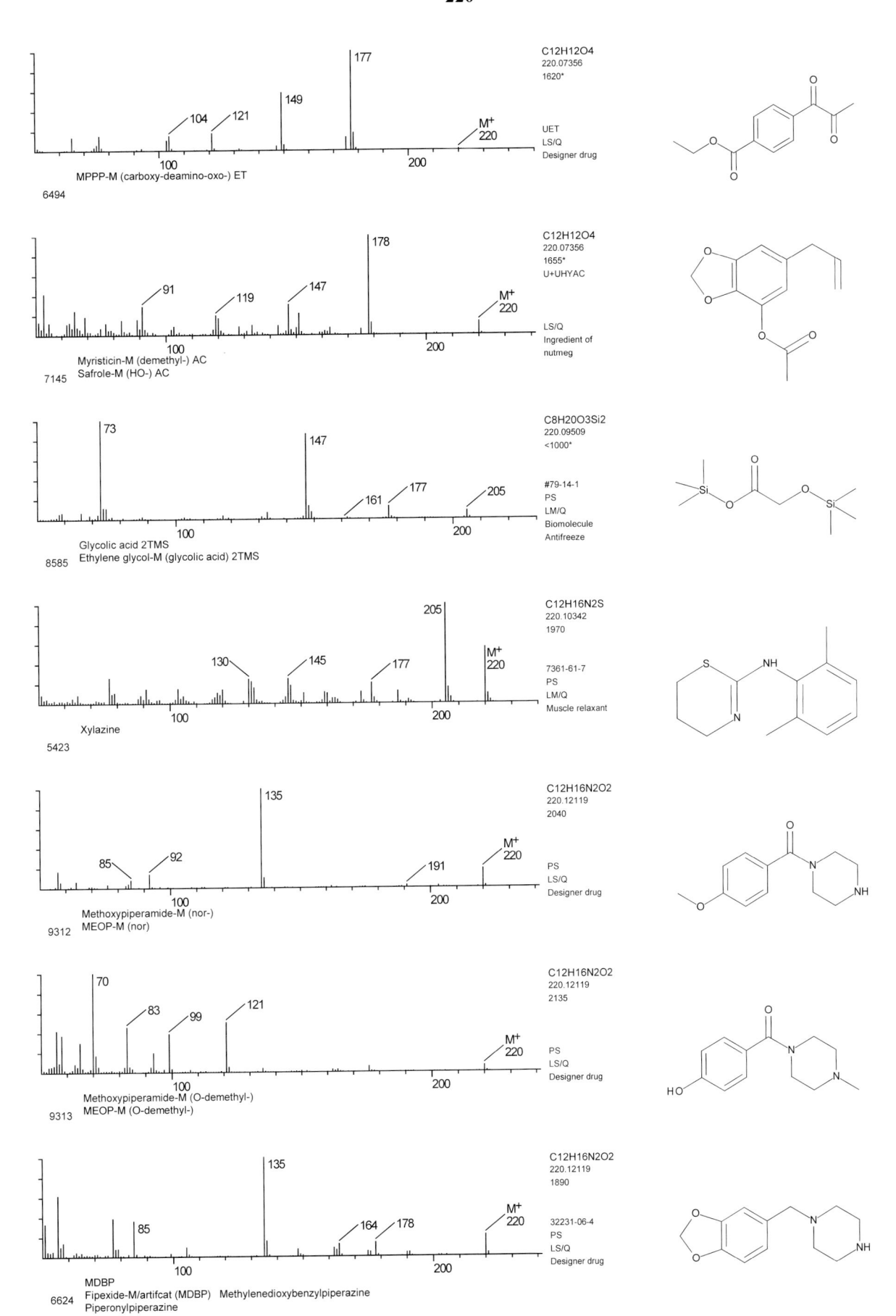

6494 — MPPP-M (carboxy-deamino-oxo-) ET
C12H12O4
220.07356
1620*
UET
LS/Q
Designer drug

7145 — Myristicin-M (demethyl-) AC / Safrole-M (HO-) AC
C12H12O4
220.07356
1655*
U+UHYAC
LS/Q
Ingredient of nutmeg

8585 — Glycolic acid 2TMS / Ethylene glycol-M (glycolic acid) 2TMS
C8H20O3Si2
220.09509
<1000*
#79-14-1
PS
LM/Q
Biomolecule
Antifreeze

5423 — Xylazine
C12H16N2S
220.10342
1970
7361-61-7
PS
LM/Q
Muscle relaxant

9312 — Methoxypiperamide-M (nor-) / MEOP-M (nor)
C12H16N2O2
220.12119
2040
PS
LS/Q
Designer drug

9313 — Methoxypiperamide-M (O-demethyl-) / MEOP-M (O-demethyl-)
C12H16N2O2
220.12119
2135
PS
LS/Q
Designer drug

6624 — MDBP / Fipexide-M/artifcat (MDBP) Methylenedioxybenzylpiperazine / Piperonylpiperazine
C12H16N2O2
220.12119
1890
32231-06-4
PS
LS/Q
Designer drug

220

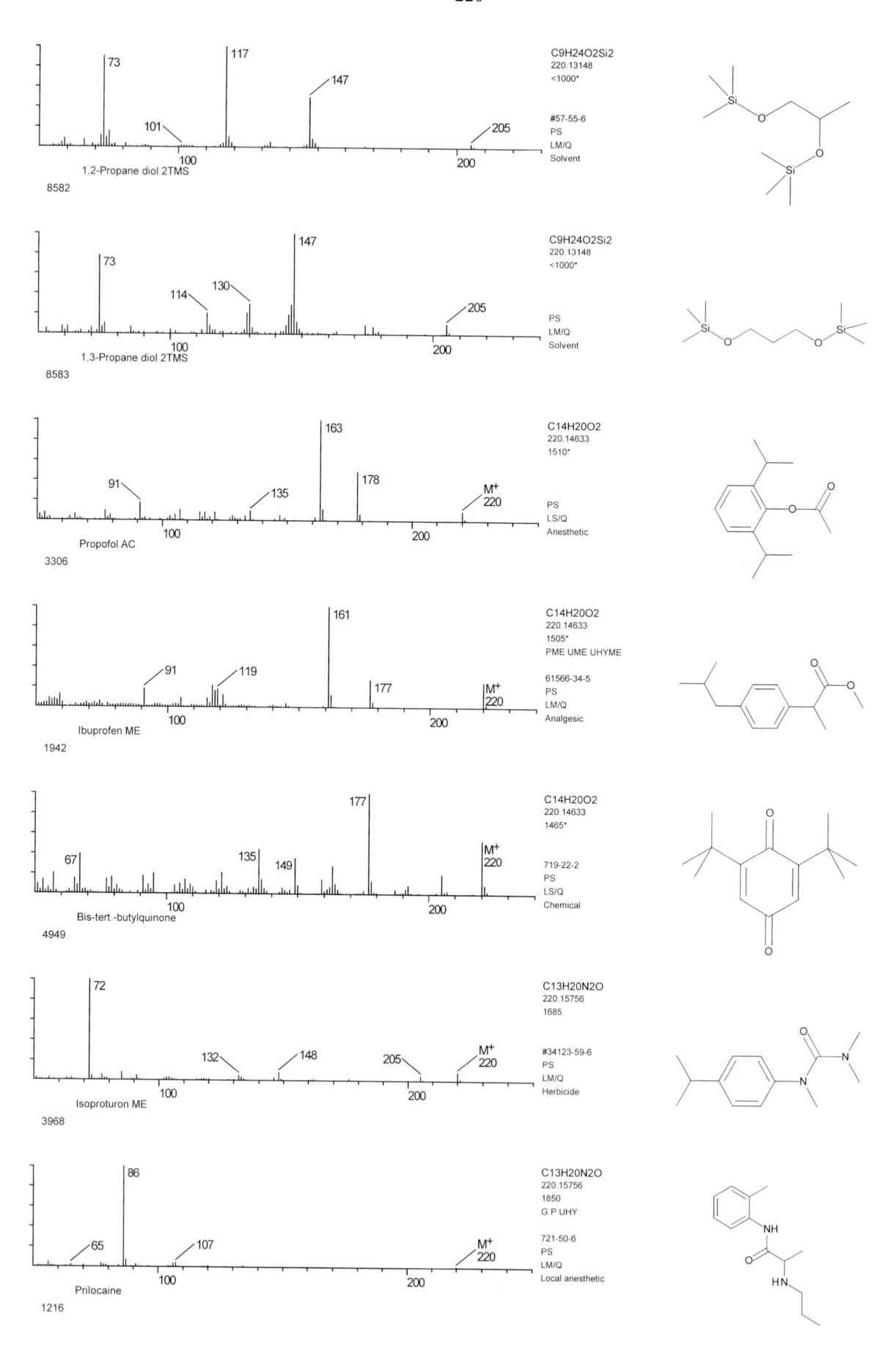

1,2-Propane diol 2TMS — 8582
C9H24O2Si2, 220.13148, <1000*
#57-55-6, PS, LM/Q, Solvent
Peaks: 73, 101, 117, 147, 205

1,3-Propane diol 2TMS — 8583
C9H24O2Si2, 220.13148, <1000*
PS, LM/Q, Solvent
Peaks: 73, 114, 130, 147, 205

Propofol AC — 3306
C14H20O2, 220.14633, 1510*
PS, LS/Q, Anesthetic
Peaks: 91, 135, 163, 178, M+ 220

Ibuprofen ME — 1942
C14H20O2, 220.14633, 1505*
PME UME UHYME
61566-34-5, PS, LM/Q, Analgesic
Peaks: 91, 119, 161, 177, M+ 220

Bis-tert.-butylquinone — 4949
C14H20O2, 220.14633, 1465*
719-22-2, PS, LS/Q, Chemical
Peaks: 67, 135, 149, 177, M+ 220

Isoproturon ME — 3968
C13H20N2O, 220.15756, 1685
#34123-59-6, PS, LM/Q, Herbicide
Peaks: 72, 132, 148, 205, M+ 220

Prilocaine — 1216
C13H20N2O, 220.15756, 1850
G P UHY
721-50-6, PS, LM/Q, Local anesthetic
Peaks: 65, 86, 107, M+ 220

220

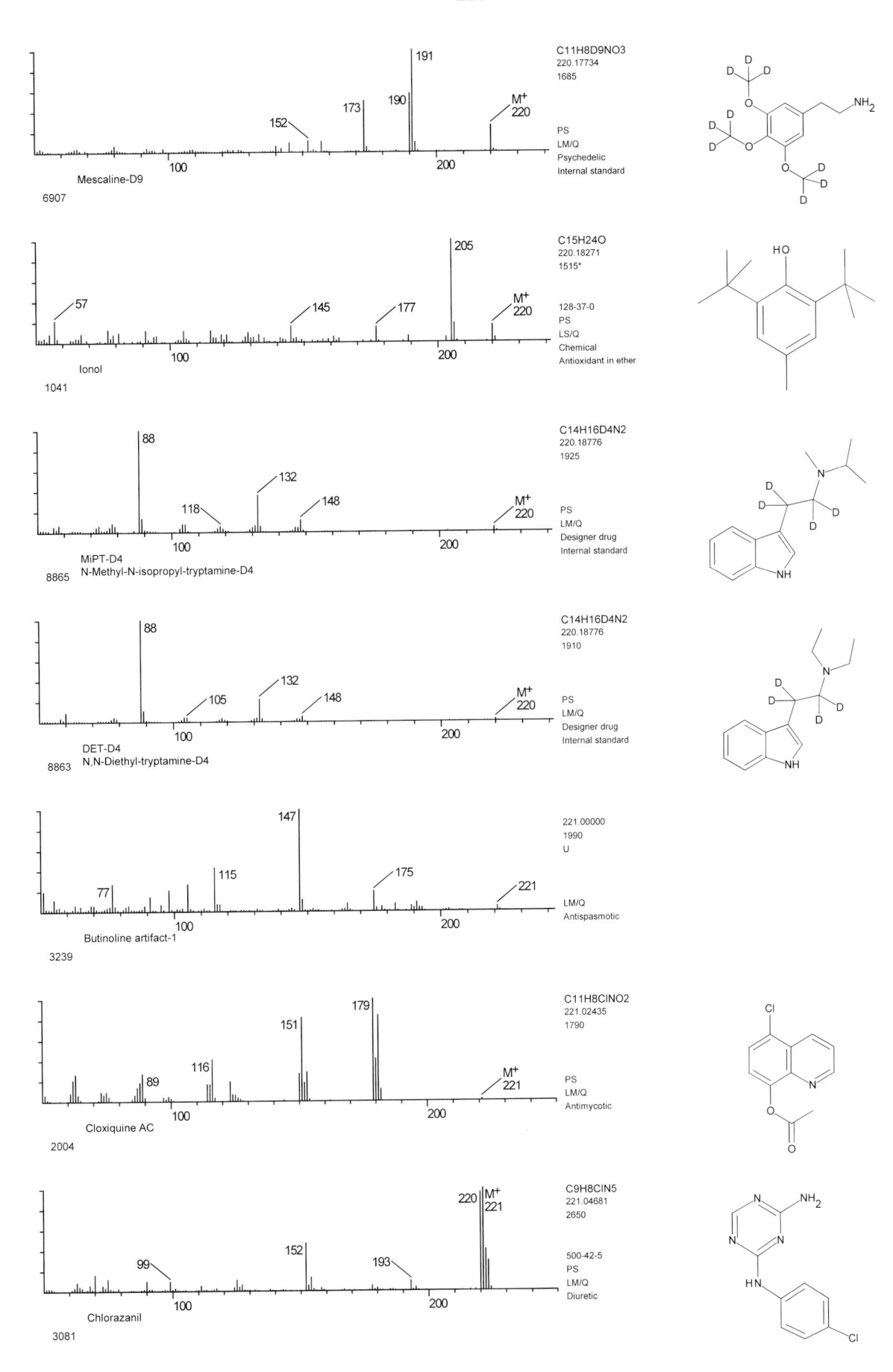

Mescaline-D9
6907
- C11H8D9NO3
- 220.17734
- 1685
- PS
- LM/Q
- Psychedelic
- Internal standard

Ionol
1041
- C15H24O
- 220.18271
- 1515*
- 128-37-0
- PS
- LS/Q
- Chemical
- Antioxidant in ether

MiPT-D4
N-Methyl-N-isopropyl-tryptamine-D4
8865
- C14H16D4N2
- 220.18776
- 1925
- PS
- LM/Q
- Designer drug
- Internal standard

DET-D4
N,N-Diethyl-tryptamine-D4
8863
- C14H16D4N2
- 220.18776
- 1910
- PS
- LM/Q
- Designer drug
- Internal standard

Butinoline artifact-1
3239
- 221.00000
- 1990
- U
- LM/Q
- Antispasmotic

Cloxiquine AC
2004
- C11H8ClNO2
- 221.02435
- 1790
- PS
- LM/Q
- Antimycotic

Chlorazanil
3081
- C9H8ClN5
- 221.04681
- 2650
- 500-42-5
- PS
- LM/Q
- Diuretic

221

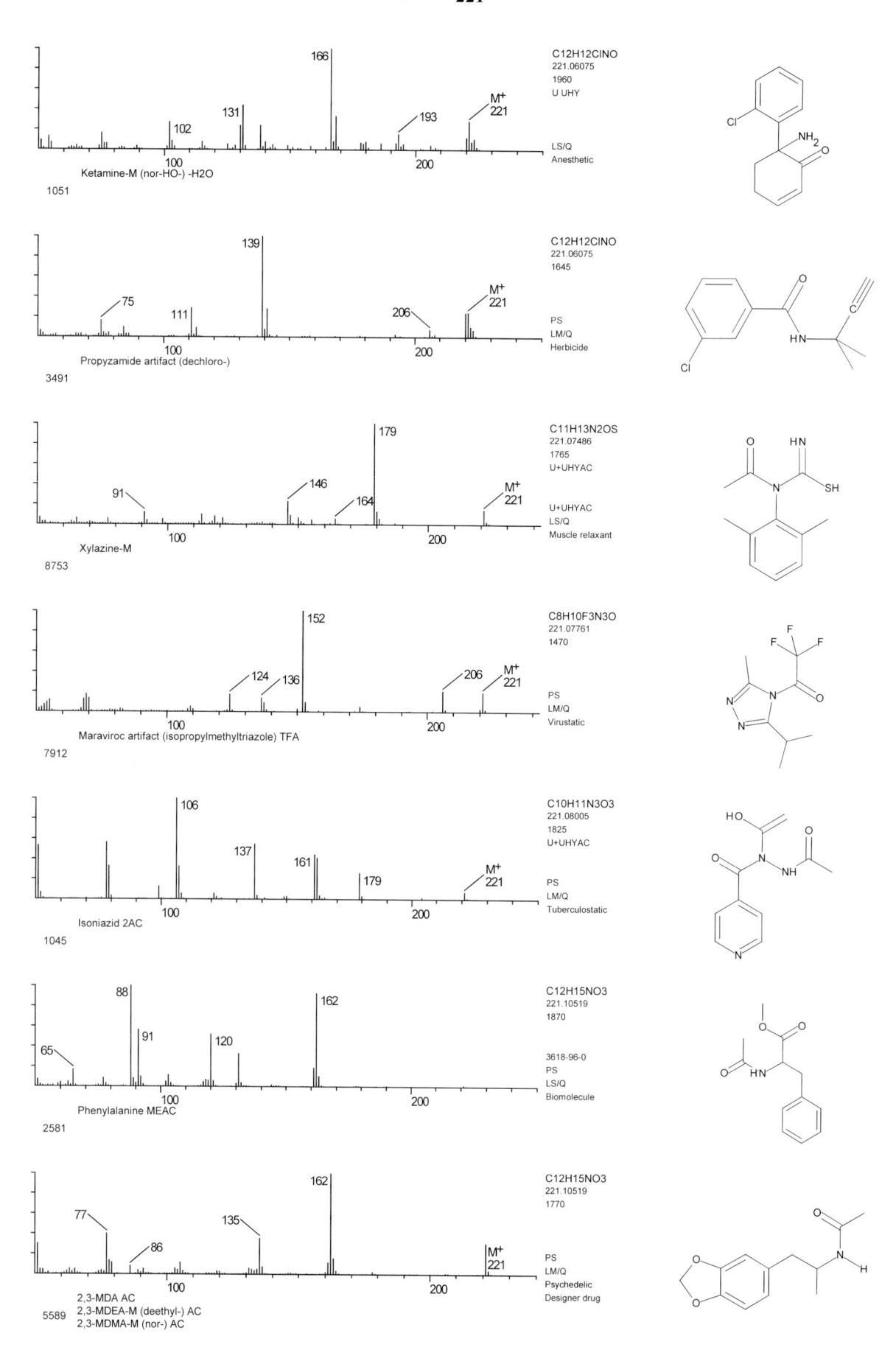

C12H12ClNO
221.06075
1960
U UHY

LS/Q
Anesthetic

Ketamine-M (nor-HO-) -H2O
1051

C12H12ClNO
221.06075
1645

PS
LM/Q
Herbicide

Propyzamide artifact (dechloro-)
3491

C11H13N2OS
221.07486
1765
U+UHYAC

U+UHYAC
LS/Q
Muscle relaxant

Xylazine-M
8753

C8H10F3N3O
221.07761
1470

PS
LM/Q
Virustatic

Maraviroc artifact (isopropylmethyltriazole) TFA
7912

C10H11N3O3
221.08005
1825
U+UHYAC

PS
LM/Q
Tuberculostatic

Isoniazid 2AC
1045

C12H15NO3
221.10519
1870

3618-96-0
PS
LS/Q
Biomolecule

Phenylalanine MEAC
2581

C12H15NO3
221.10519
1770

PS
LM/Q
Psychedelic
Designer drug

2,3-MDA AC
2,3-MDEA-M (deethyl-) AC
5589   2,3-MDMA-M (nor-) AC

221

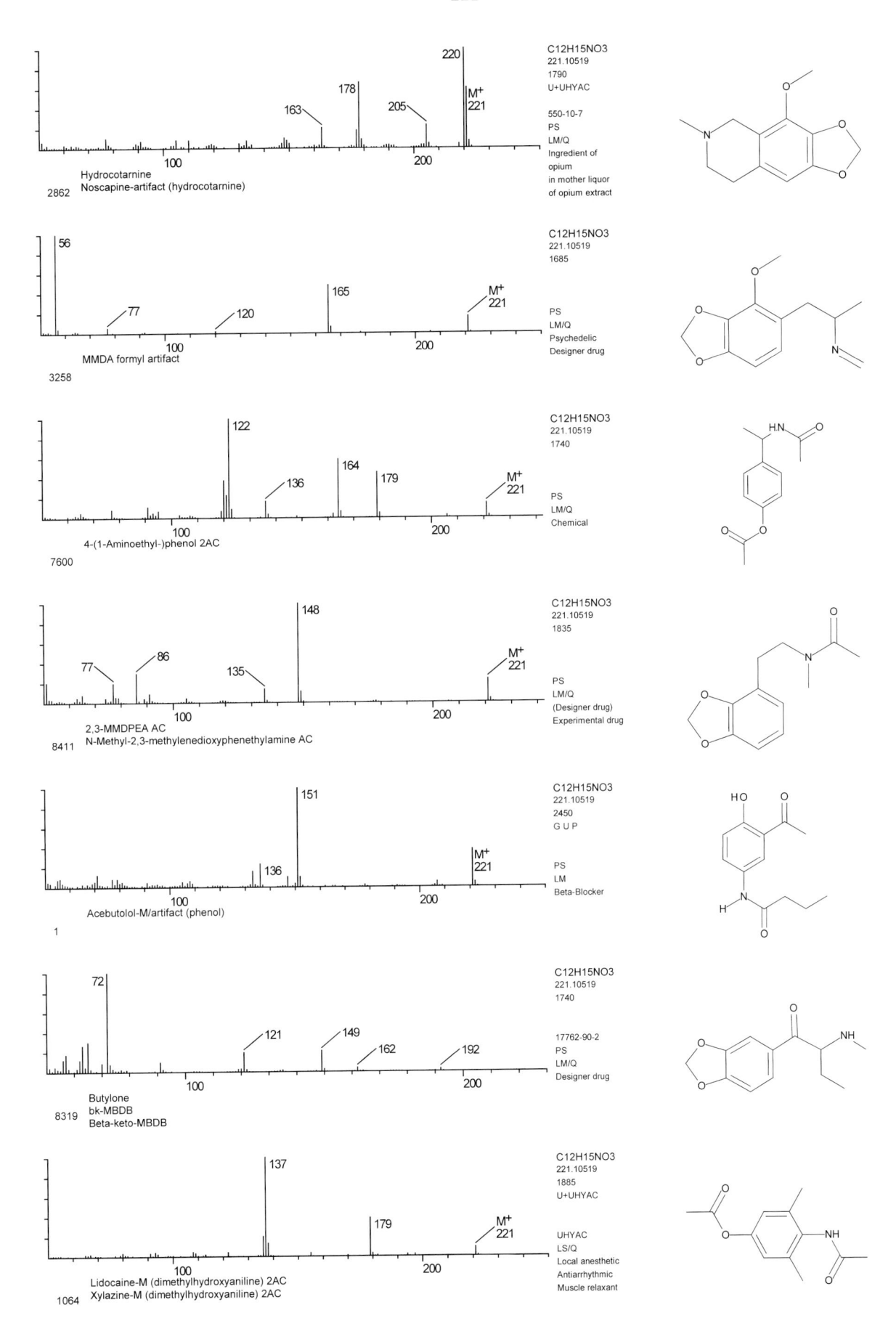

C12H15NO3
221.10519
1790
U+UHYAC

550-10-7
PS
LM/Q
Ingredient of opium
in mother liquor of opium extract

Hydrocotarnine
2862 Noscapine-artifact (hydrocotarnine)

C12H15NO3
221.10519
1685

PS
LM/Q
Psychedelic
Designer drug

MMDA formyl artifact
3258

C12H15NO3
221.10519
1740

PS
LM/Q
Chemical

4-(1-Aminoethyl-)phenol 2AC
7600

C12H15NO3
221.10519
1835

PS
LM/Q
(Designer drug)
Experimental drug

2,3-MMDPEA AC
8411 N-Methyl-2,3-methylenedioxyphenethylamine AC

C12H15NO3
221.10519
2450
G U P

PS
LM
Beta-Blocker

Acebutolol-M/artifact (phenol)
1

C12H15NO3
221.10519
1740

17762-90-2
PS
LM/Q
Designer drug

Butylone
bk-MBDB
8319 Beta-keto-MBDB

C12H15NO3
221.10519
1885
U+UHYAC

UHYAC
LS/Q
Local anesthetic
Antiarrhythmic
Muscle relaxant

Lidocaine-M (dimethylhydroxyaniline) 2AC
1064 Xylazine-M (dimethylhydroxyaniline) 2AC

221

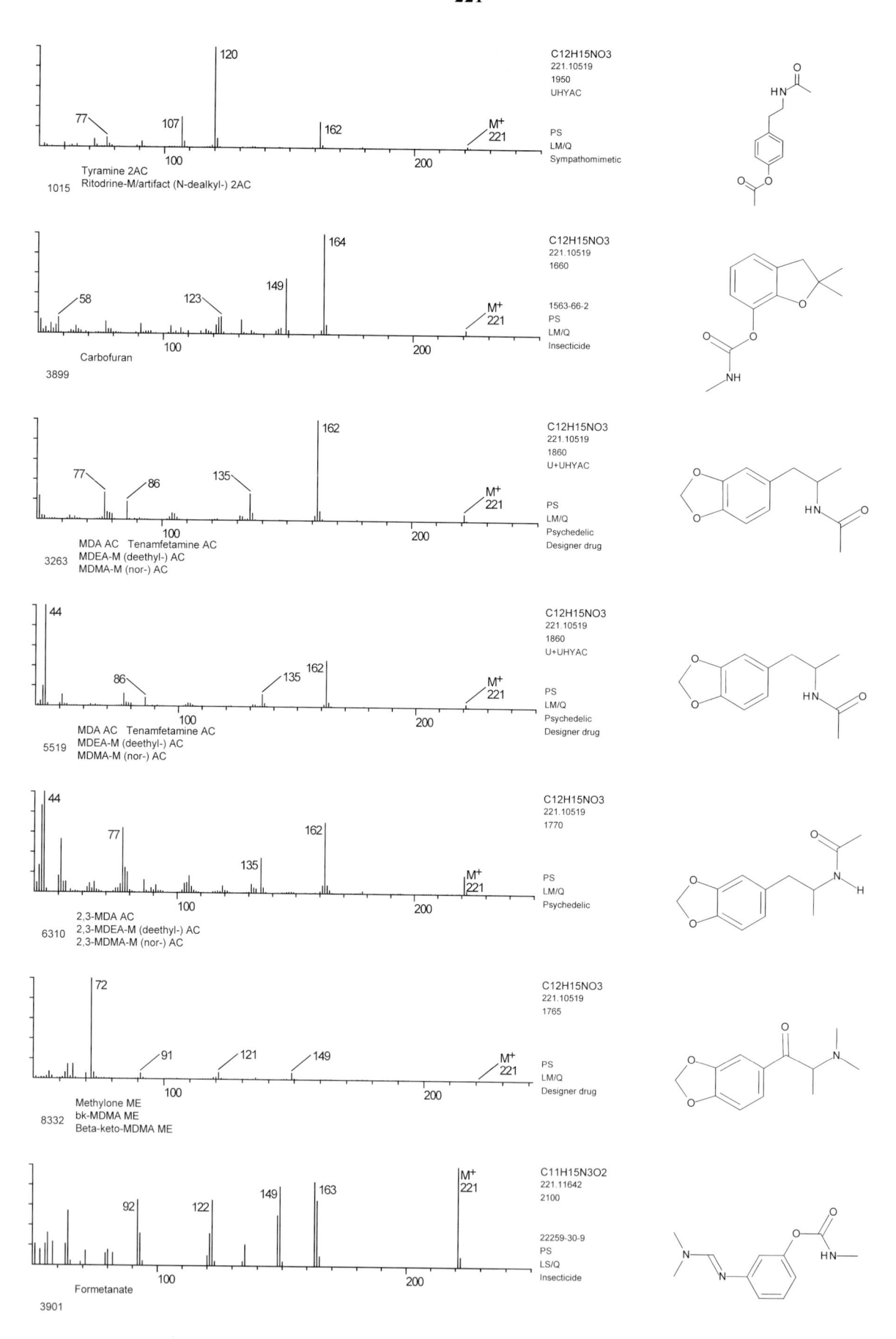

1015	Tyramine 2AC Ritodrine-M/artifact (N-dealkyl-) 2AC	C12H15NO3 221.10519 1950 UHYAC PS LM/Q Sympathomimetic
3899	Carbofuran	C12H15NO3 221.10519 1660 1563-66-2 PS LM/Q Insecticide
3263	MDA AC Tenamfetamine AC MDEA-M (deethyl-) AC MDMA-M (nor-) AC	C12H15NO3 221.10519 1860 U+UHYAC PS LM/Q Psychedelic Designer drug
5519	MDA AC Tenamfetamine AC MDEA-M (deethyl-) AC MDMA-M (nor-) AC	C12H15NO3 221.10519 1860 U+UHYAC PS LM/Q Psychedelic Designer drug
6310	2,3-MDA AC 2,3-MDEA-M (deethyl-) AC 2,3-MDMA-M (nor-) AC	C12H15NO3 221.10519 1770 PS LM/Q Psychedelic
8332	Methylone ME bk-MDMA ME Beta-keto-MDMA ME	C12H15NO3 221.10519 1765 PS LM/Q Designer drug
3901	Formetanate	C11H15N3O2 221.11642 2100 22259-30-9 PS LS/Q Insecticide

221

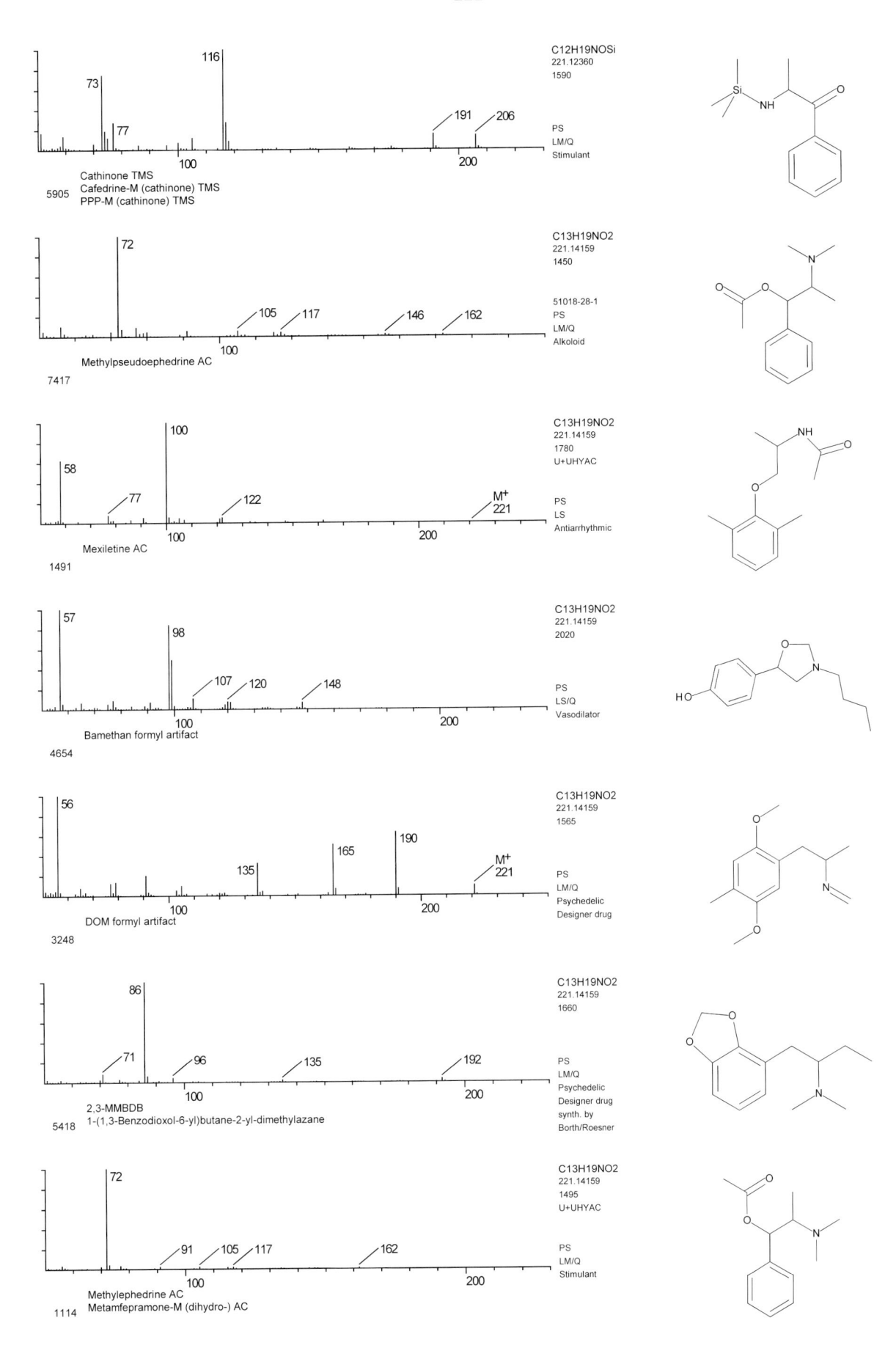

Cathinone TMS Cafedrine-M (cathinone) TMS PPP-M (cathinone) TMS 5905	Peaks: 73, 77, 116, 191, 206	C12H19NOSi 221.12360 1590 PS LM/Q Stimulant
Methylpseudoephedrine AC 7417	Peaks: 72, 105, 117, 146, 162	C13H19NO2 221.14159 1450 51018-28-1 PS LM/Q Alkoloid
Mexiletine AC 1491	Peaks: 58, 77, 100, 122, M+ 221	C13H19NO2 221.14159 1780 U+UHYAC PS LS Antiarrhythmic
Bamethan formyl artifact 4654	Peaks: 57, 98, 107, 120, 148	C13H19NO2 221.14159 2020 PS LS/Q Vasodilator
DOM formyl artifact 3248	Peaks: 56, 135, 165, 190, M+ 221	C13H19NO2 221.14159 1565 PS LM/Q Psychedelic Designer drug
2,3-MMBDB 1-(1,3-Benzodioxol-6-yl)butane-2-yl-dimethylazane 5418	Peaks: 71, 86, 96, 135, 192	C13H19NO2 221.14159 1660 PS LM/Q Psychedelic Designer drug synth. by Borth/Roesner
Methylephedrine AC Metamfepramone-M (dihydro-) AC 1114	Peaks: 72, 91, 105, 117, 162	C13H19NO2 221.14159 1495 U+UHYAC PS LM/Q Stimulant

475

221

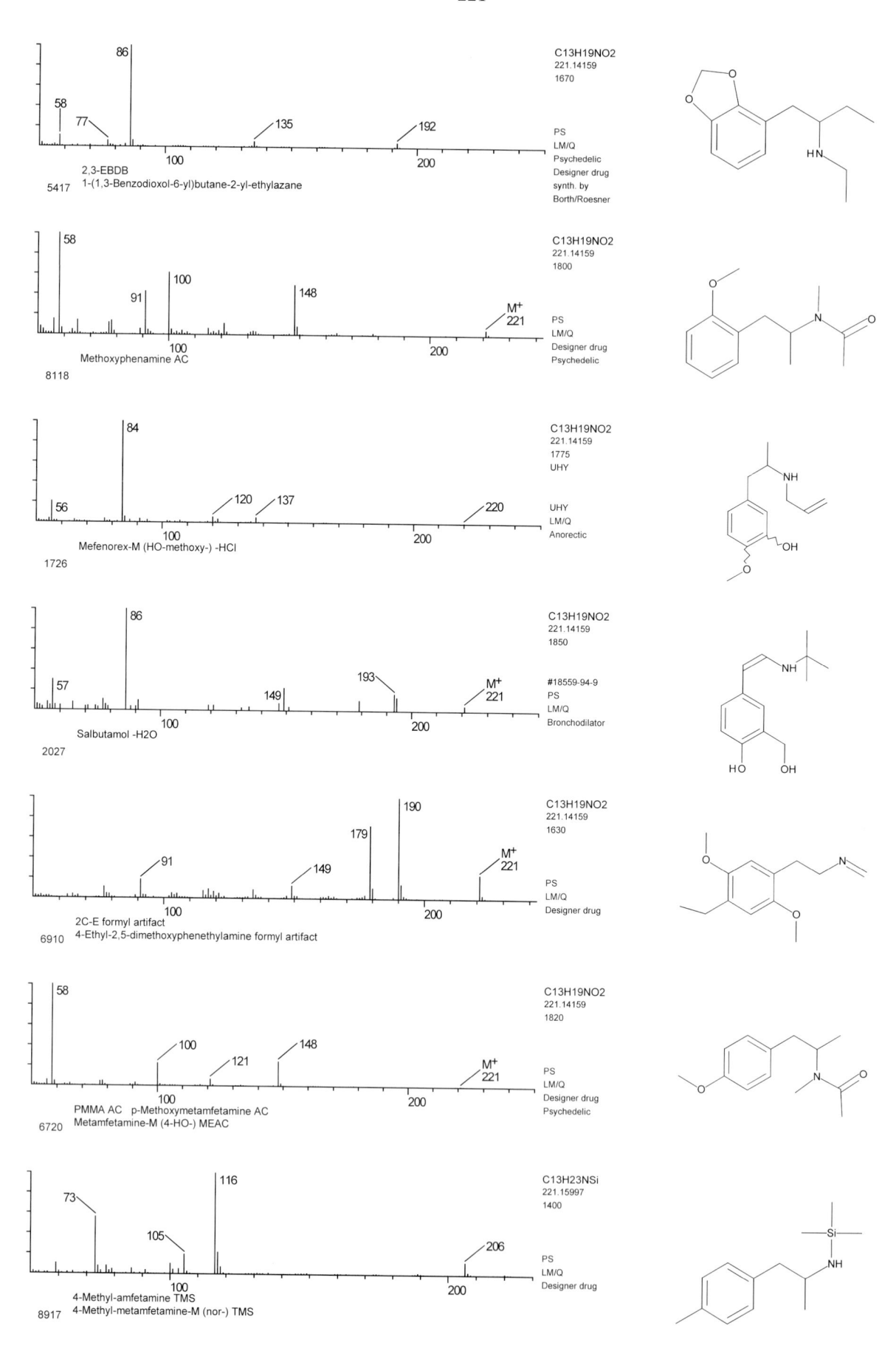

C13H19NO2
221.14159
1670
PS
LM/Q
Psychedelic
Designer drug
synth. by
Borth/Roesner

2,3-EBDB
5417 1-(1,3-Benzodioxol-6-yl)butane-2-yl-ethylazane

C13H19NO2
221.14159
1800
PS
LM/Q
Designer drug
Psychedelic

Methoxyphenamine AC
8118

C13H19NO2
221.14159
1775
UHY
UHY
LM/Q
Anorectic

Mefenorex-M (HO-methoxy-) -HCl
1726

C13H19NO2
221.14159
1850
#18559-94-9
PS
LM/Q
Bronchodilator

Salbutamol -H2O
2027

C13H19NO2
221.14159
1630
PS
LM/Q
Designer drug

2C-E formyl artifact
6910 4-Ethyl-2,5-dimethoxyphenethylamine formyl artifact

C13H19NO2
221.14159
1820
PS
LM/Q
Designer drug
Psychedelic

PMMA AC   p-Methoxymetamfetamine AC
6720 Metamfetamine-M (4-HO-) MEAC

C13H23NSi
221.15997
1400
PS
LM/Q
Designer drug

4-Methyl-amfetamine TMS
8917 4-Methyl-metamfetamine-M (nor-) TMS

476

221

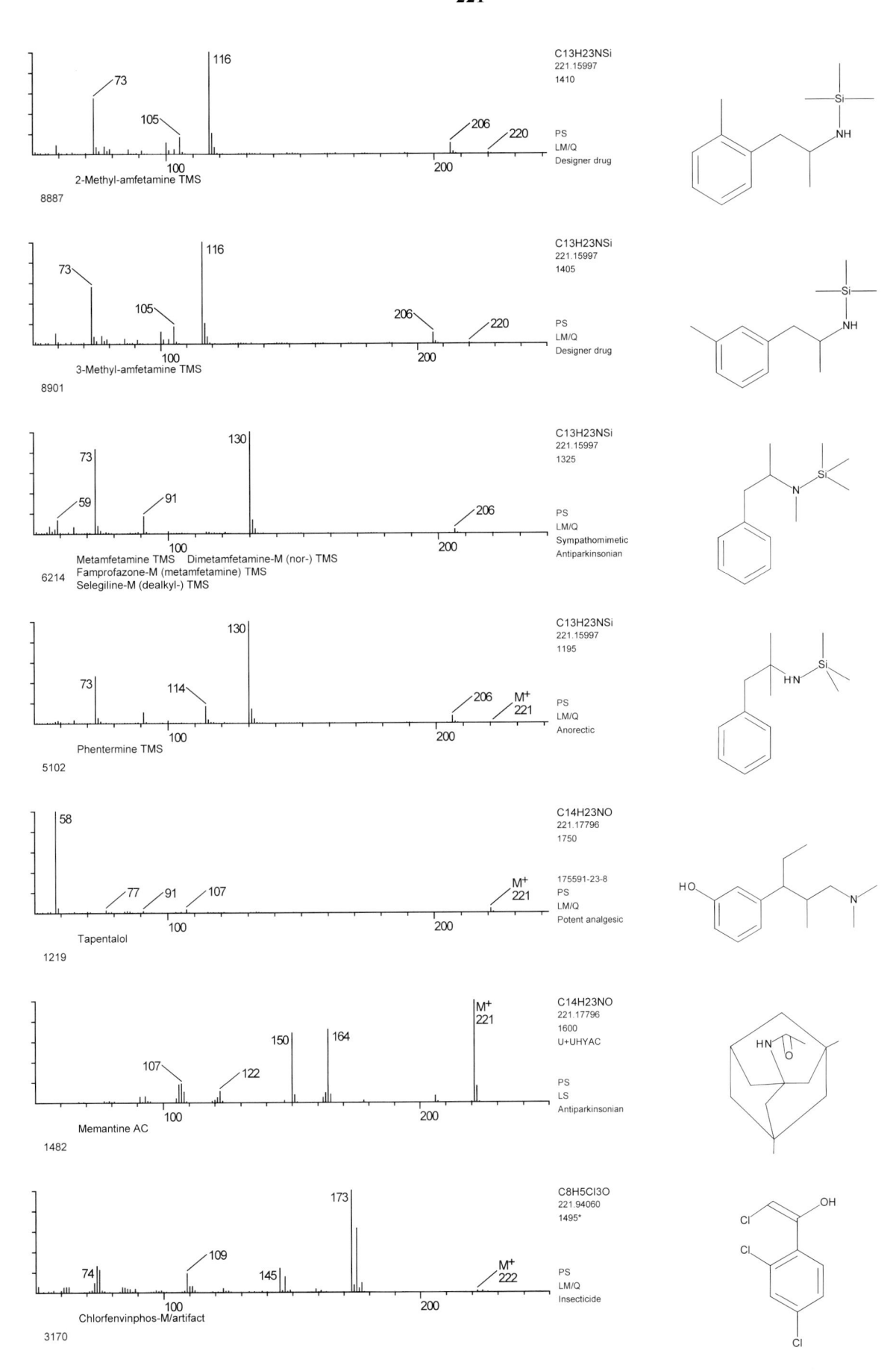

Spectrum	Formula / Info
2-Methyl-amfetamine TMS — peaks: 73, 105, 116, 206, 220 — 8887	C13H23NSi, 221.15997, 1410, PS, LM/Q, Designer drug
3-Methyl-amfetamine TMS — peaks: 73, 105, 116, 206, 220 — 8901	C13H23NSi, 221.15997, 1405, PS, LM/Q, Designer drug
Metamfetamine TMS; Dimetamfetamine-M (nor-) TMS; Famprofazone-M (metamfetamine) TMS; Selegiline-M (dealkyl-) TMS — peaks: 59, 73, 91, 130, 206 — 6214	C13H23NSi, 221.15997, 1325, PS, LM/Q, Sympathomimetic, Antiparkinsonian
Phentermine TMS — peaks: 73, 114, 130, 206, M+ 221 — 5102	C13H23NSi, 221.15997, 1195, PS, LM/Q, Anorectic
Tapentalol — peaks: 58, 77, 91, 107, M+ 221 — 1219	C14H23NO, 221.17796, 1750, 175591-23-8, PS, LM/Q, Potent analgesic
Memantine AC — peaks: 107, 122, 150, 164, M+ 221 — 1482	C14H23NO, 221.17796, 1600, U+UHYAC, PS, LS, Antiparkinsonian
Chlorfenvinphos-M/artifact — peaks: 74, 109, 145, 173, M+ 222 — 3170	C8H5Cl3O, 221.94060, 1495*, PS, LM/Q, Insecticide

222

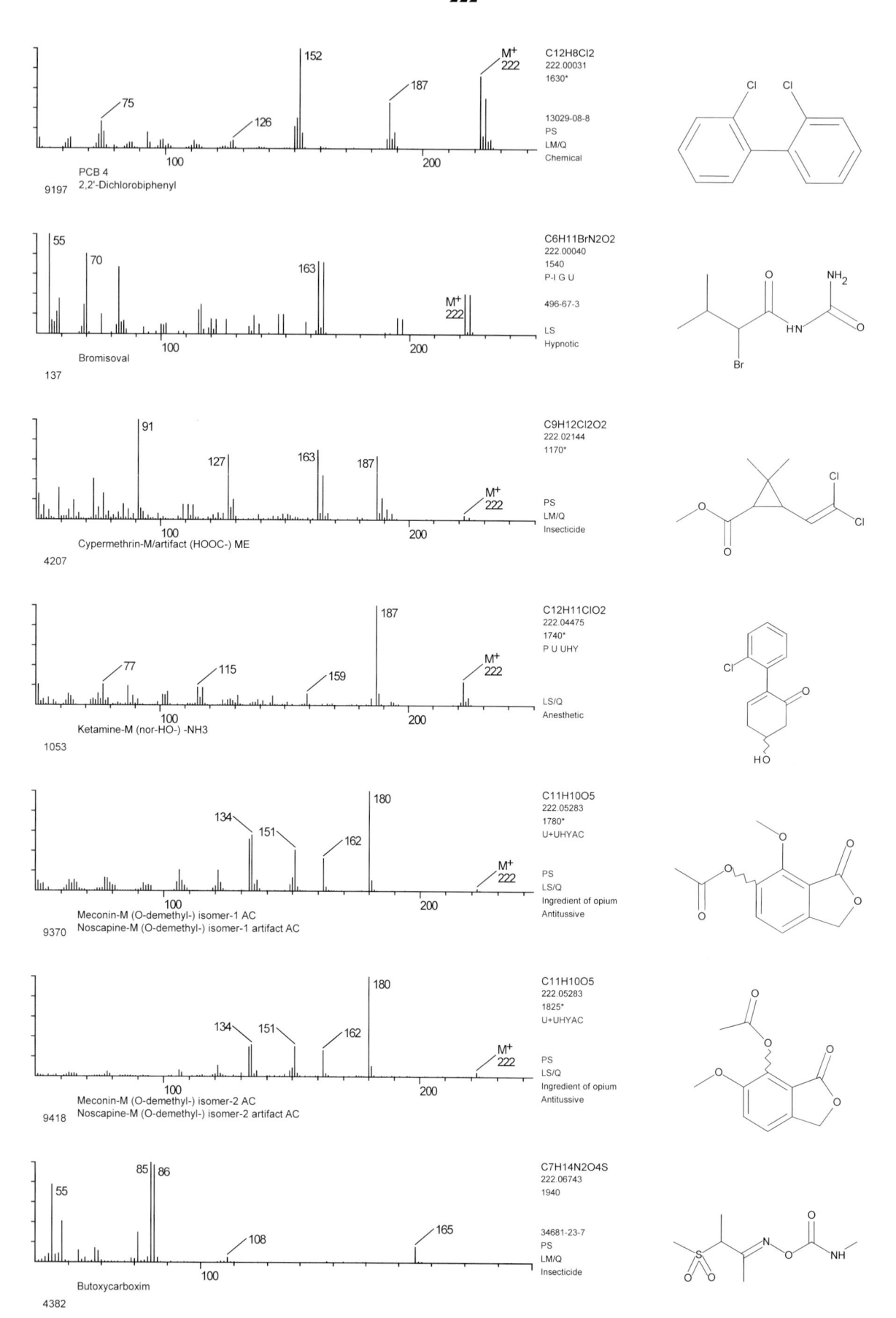

PCB 4 — 2,2'-Dichlorobiphenyl	peaks: 75, 126, 152, 187, M+ 222	C12H8Cl2, 222.00031, 1630*, 13029-08-8, PS, LM/Q, Chemical
Bromisoval	peaks: 55, 70, 163, M+ 222	C6H11BrN2O2, 222.00040, 1540, P·I G U, 496-67-3, LS, Hypnotic
Cypermethrin-M/artifact (HOOC-) ME	peaks: 91, 127, 163, 187, M+ 222	C9H12Cl2O2, 222.02144, 1170*, PS, LM/Q, Insecticide
Ketamine-M (nor-HO-) -NH3	peaks: 77, 115, 159, 187, M+ 222	C12H11ClO2, 222.04475, 1740*, P U UHY, LS/Q, Anesthetic
Meconin-M (O-demethyl-) isomer-1 AC / Noscapine-M (O-demethyl-) isomer-1 artifact AC	peaks: 134, 151, 162, 180, M+ 222	C11H10O5, 222.05283, 1780*, U+UHYAC, PS, LS/Q, Ingredient of opium, Antitussive
Meconin-M (O-demethyl-) isomer-2 AC / Noscapine-M (O-demethyl-) isomer-2 artifact AC	peaks: 134, 151, 162, 180, M+ 222	C11H10O5, 222.05283, 1825*, U+UHYAC, PS, LS/Q, Ingredient of opium, Antitussive
Butoxycarboxim	peaks: 55, 85, 86, 108, 165	C7H14N2O4S, 222.06743, 1940, 34681-23-7, PS, LM/Q, Insecticide

222

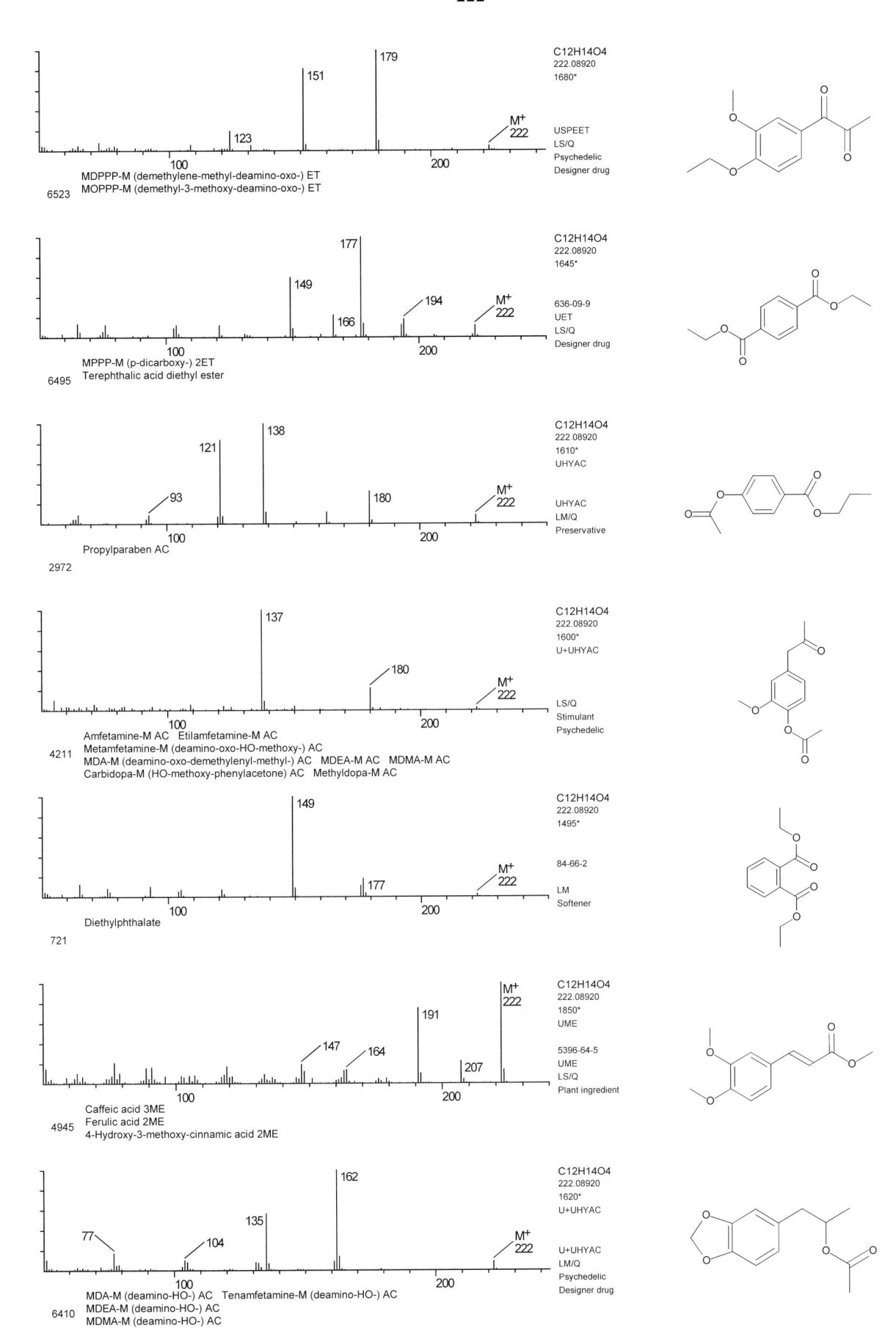

6523
MDPPP-M (demethylene-methyl-deamino-oxo-) ET
MOPPP-M (demethyl-3-methoxy-deamino-oxo-) ET
Peaks: 123, 151, 179, M+ 222
C12H14O4
222.08920
1680*
USPEET
LS/Q
Psychedelic
Designer drug

6495
MPPP-M (p-dicarboxy-) 2ET
Terephthalic acid diethyl ester
Peaks: 149, 166, 177, 194, M+ 222
C12H14O4
222.08920
1645*
636-09-9
UET
LS/Q
Designer drug

2972
Propylparaben AC
Peaks: 93, 121, 138, 180, M+ 222
C12H14O4
222.08920
1610*
UHYAC
UHYAC
LM/Q
Preservative

4211
Amfetamine-M AC   Etilamfetamine-M AC
Metamfetamine-M (deamino-oxo-HO-methoxy-) AC
MDA-M (deamino-oxo-demethylenyl-methyl-) AC   MDEA-M AC   MDMA-M AC
Carbidopa-M (HO-methoxy-phenylacetone) AC   Methyldopa-M AC
Peaks: 137, 180, M+ 222
C12H14O4
222.08920
1600*
U+UHYAC
LS/Q
Stimulant
Psychedelic

721
Diethylphthalate
Peaks: 149, 177, M+ 222
C12H14O4
222.08920
1495*
84-66-2
LM
Softener

4945
Caffeic acid 3ME
Ferulic acid 2ME
4-Hydroxy-3-methoxy-cinnamic acid 2ME
Peaks: 147, 164, 191, 207, M+ 222
C12H14O4
222.08920
1850*
UME
5396-64-5
UME
LS/Q
Plant ingredient

6410
MDA-M (deamino-HO-) AC   Tenamfetamine-M (deamino-HO-) AC
MDEA-M (deamino-HO-) AC
MDMA-M (deamino-HO-) AC
Peaks: 77, 104, 135, 162, M+ 222
C12H14O4
222.08920
1620*
U+UHYAC
U+UHYAC
LM/Q
Psychedelic
Designer drug

222

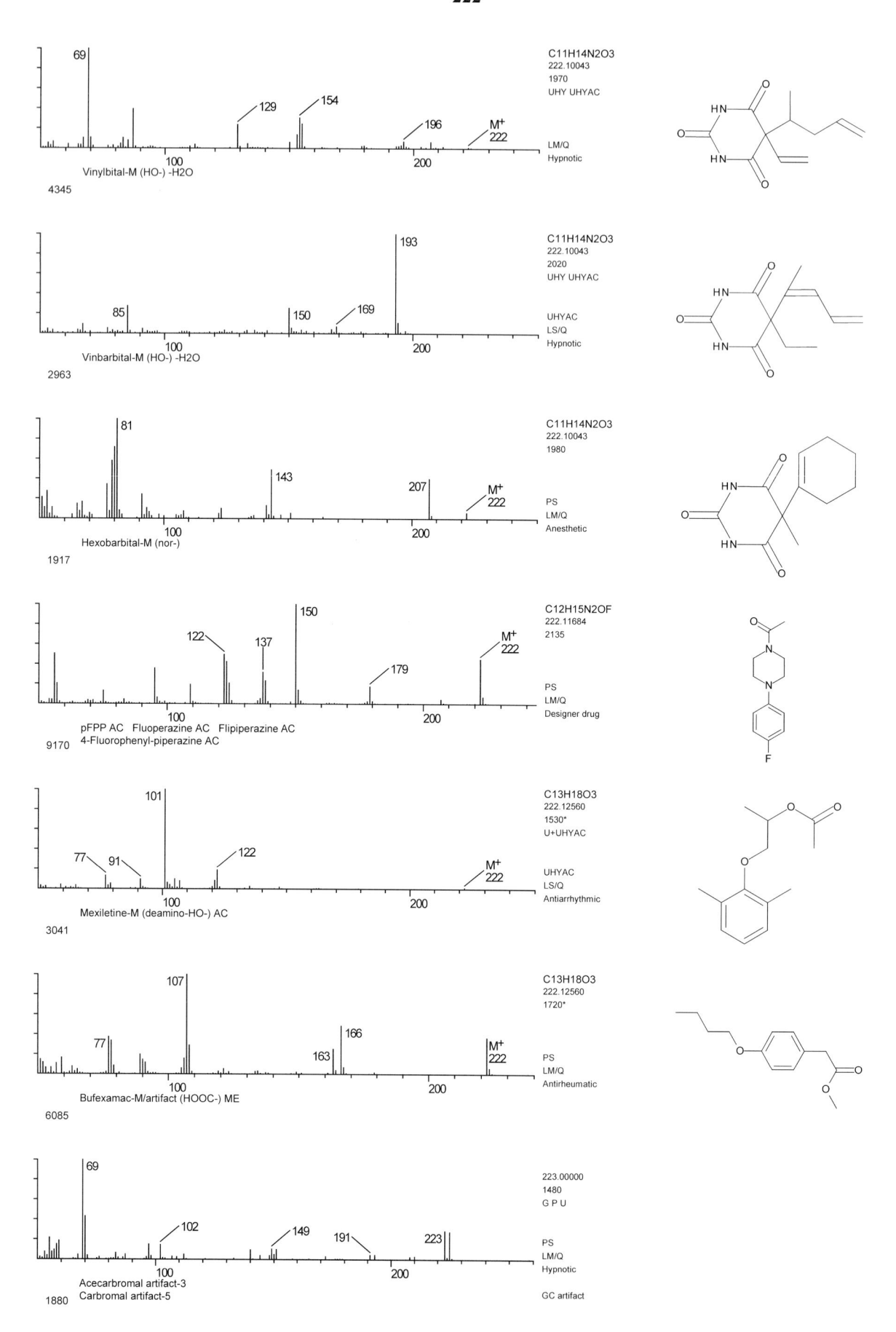

Vinylbital-M (HO-) -H2O — 4345
C11H14N2O3
222.10043
1970
UHY UHYAC
LM/Q
Hypnotic
Peaks: 69, 129, 154, 196, M+ 222

Vinbarbital-M (HO-) -H2O — 2963
C11H14N2O3
222.10043
2020
UHY UHYAC
UHYAC
LS/Q
Hypnotic
Peaks: 85, 150, 169, 193

Hexobarbital-M (nor-) — 1917
C11H14N2O3
222.10043
1980
PS
LM/Q
Anesthetic
Peaks: 81, 143, 207, M+ 222

pFPP AC   Fluoperazine AC   Flipiperazine AC
4-Fluorophenyl-piperazine AC — 9170
C12H15N2OF
222.11684
2135
PS
LM/Q
Designer drug
Peaks: 122, 137, 150, 179, M+ 222

Mexiletine-M (deamino-HO-) AC — 3041
C13H18O3
222.12560
1530*
U+UHYAC
UHYAC
LS/Q
Antiarrhythmic
Peaks: 77, 91, 101, 122, M+ 222

Bufexamac-M/artifact (HOOC-) ME — 6085
C13H18O3
222.12560
1720*
PS
LM/Q
Antirheumatic
Peaks: 77, 107, 163, 166, M+ 222

Acecarbromal artifact-3
Carbromal artifact-5 — 1880
223.00000
1480
G P U
PS
LM/Q
Hypnotic
GC artifact
Peaks: 69, 102, 149, 191, 223

223

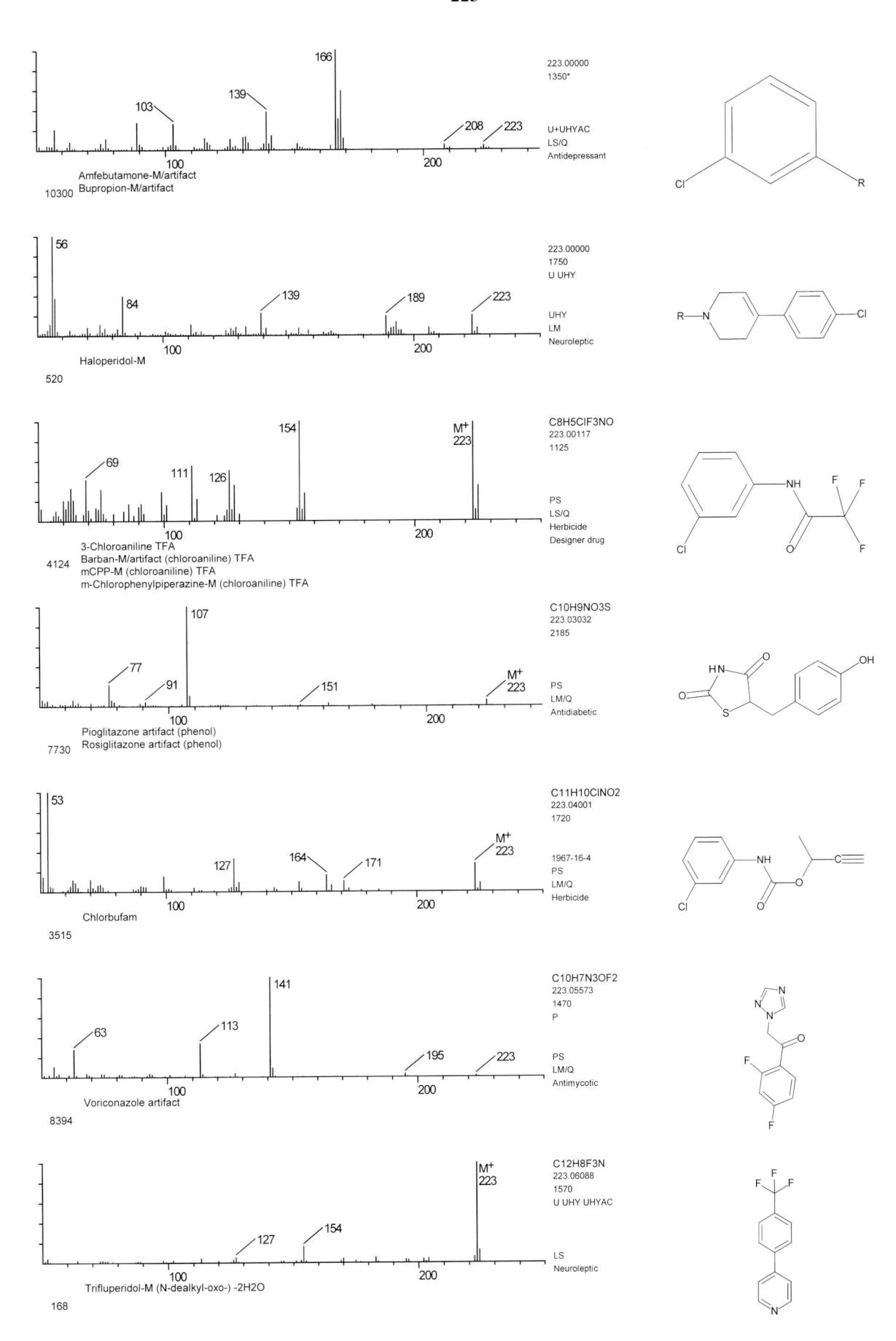

Spectrum labels	Data
Amfebutamone-M/artifact / Bupropion-M/artifact (peaks: 103, 139, 166, 208, 223)	223.00000; 1350*; U+UHYAC; LS/Q; Antidepressant
Haloperidol-M (peaks: 56, 84, 139, 189, 223)	223.00000; 1750; U UHY; UHY; LM; Neuroleptic
3-Chloroaniline TFA / Barban-M/artifact (chloroaniline) TFA / mCPP-M (chloroaniline) TFA / m-Chlorophenylpiperazine-M (chloroaniline) TFA (peaks: 69, 111, 126, 154, M+ 223)	C8H5ClF3NO; 223.00117; 1125; PS; LS/Q; Herbicide; Designer drug
Pioglitazone artifact (phenol) / Rosiglitazone artifact (phenol) (peaks: 77, 91, 107, 151, M+ 223)	C10H9NO3S; 223.03032; 2185; PS; LM/Q; Antidiabetic
Chlorbufam (peaks: 53, 127, 164, 171, M+ 223)	C11H10ClNO2; 223.04001; 1720; 1967-16-4; PS; LM/Q; Herbicide
Voriconazole artifact (peaks: 63, 113, 141, 195, 223)	C10H7N3OF2; 223.05573; 1470; P; PS; LM/Q; Antimycotic
Trifluperidol-M (N-dealkyl-oxo-) -2H2O (peaks: 127, 154, M+ 223)	C12H8F3N; 223.06088; 1570; U UHY UHYAC; LS; Neuroleptic

223

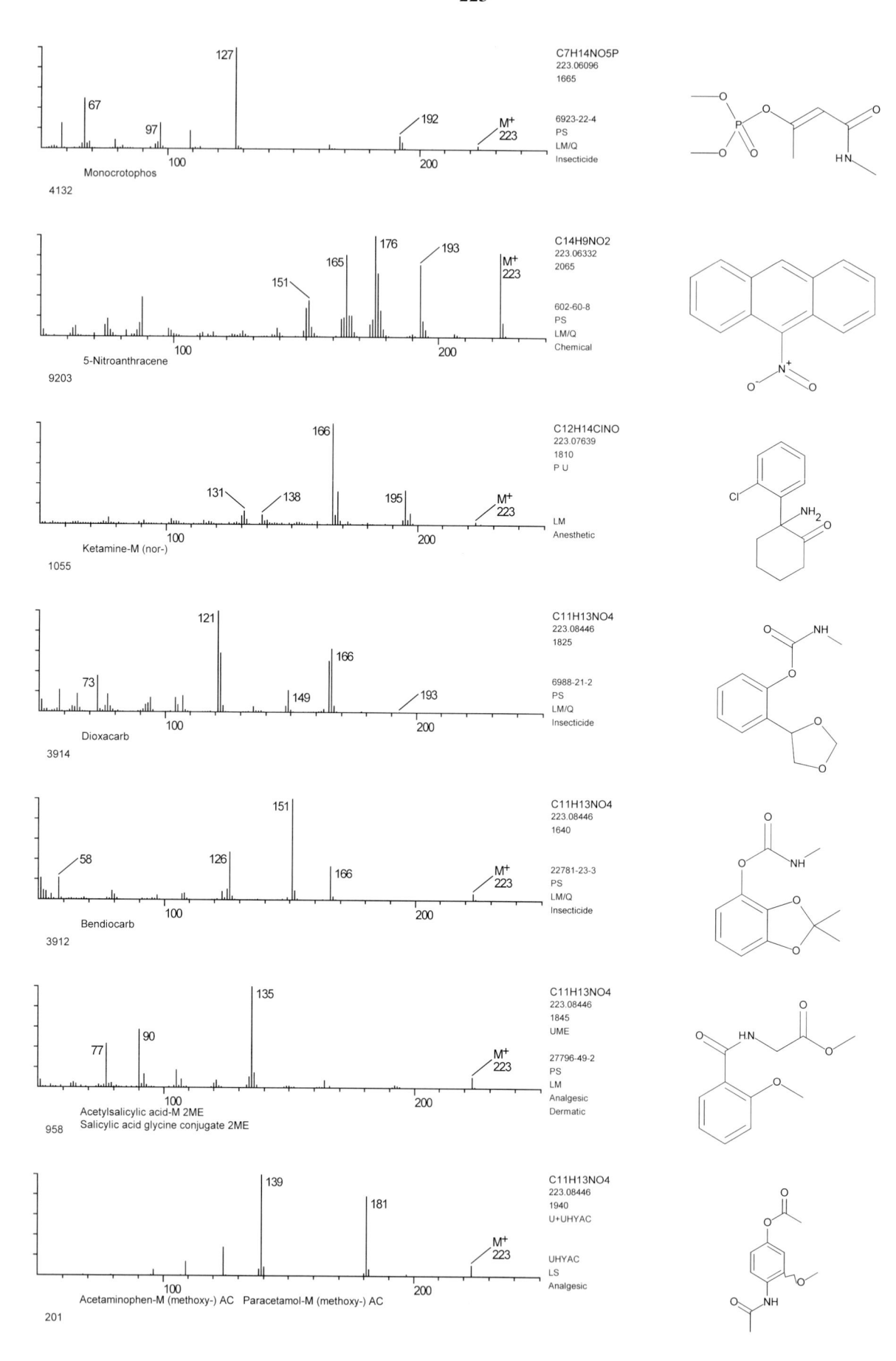

Monocrotophos — C7H14NO5P, 223.06096, 1665, 6923-22-4, PS, LM/Q, Insecticide
Peaks: 67, 97, 127, 192, M+ 223
4132

5-Nitroanthracene — C14H9NO2, 223.06332, 2065, 602-60-8, PS, LM/Q, Chemical
Peaks: 151, 165, 176, 193, M+ 223
9203

Ketamine-M (nor-) — C12H14ClNO, 223.07639, 1810, P U, LM, Anesthetic
Peaks: 131, 138, 166, 195, M+ 223
1055

Dioxacarb — C11H13NO4, 223.08446, 1825, 6988-21-2, PS, LM/Q, Insecticide
Peaks: 73, 121, 149, 166, 193
3914

Bendiocarb — C11H13NO4, 223.08446, 1640, 22781-23-3, PS, LM/Q, Insecticide
Peaks: 58, 126, 151, 166, M+ 223
3912

Acetylsalicylic acid-M 2ME / Salicylic acid glycine conjugate 2ME — C11H13NO4, 223.08446, 1845, UME, 27796-49-2, PS, LM, Analgesic, Dermatic
Peaks: 77, 90, 135, M+ 223
958

Acetaminophen-M (methoxy-) AC / Paracetamol-M (methoxy-) AC — C11H13NO4, 223.08446, 1940, U+UHYAC, UHYAC, LS, Analgesic
Peaks: 139, 181, M+ 223
201

223

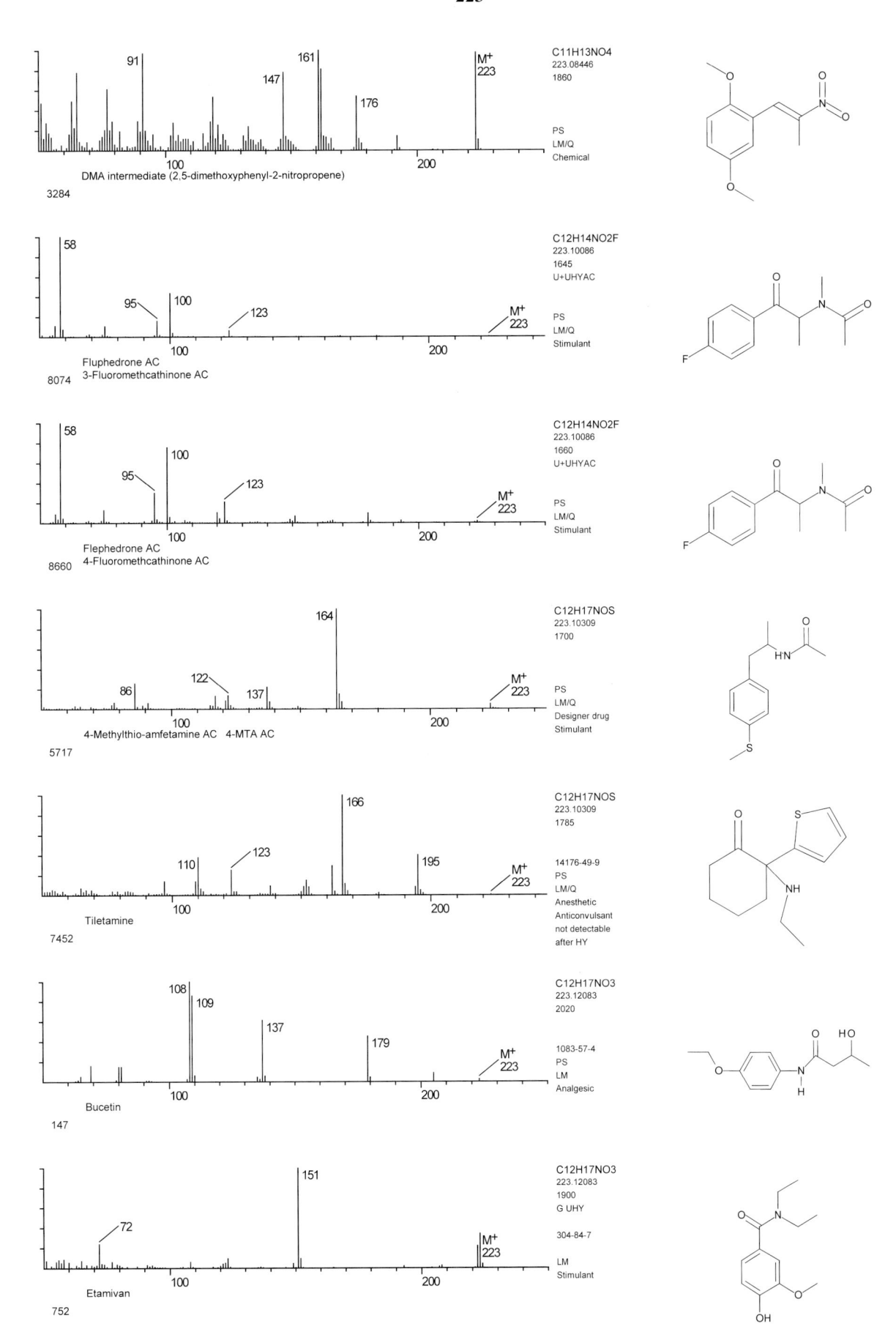

DMA intermediate (2,5-dimethoxyphenyl-2-nitropropene)
3284
Peaks: 91, 147, 161, 176, M+ 223
C11H13NO4
223.08446
1860
PS
LM/Q
Chemical

Fluphedrone AC
3-Fluoromethcathinone AC
8074
Peaks: 58, 95, 100, 123, M+ 223
C12H14NO2F
223.10086
1645
U+UHYAC
PS
LM/Q
Stimulant

Flephedrone AC
4-Fluoromethcathinone AC
8660
Peaks: 58, 95, 100, 123, M+ 223
C12H14NO2F
223.10086
1660
U+UHYAC
PS
LM/Q
Stimulant

4-Methylthio-amfetamine AC   4-MTA AC
5717
Peaks: 86, 122, 137, 164, M+ 223
C12H17NOS
223.10309
1700
PS
LM/Q
Designer drug
Stimulant

Tiletamine
7452
Peaks: 110, 123, 166, 195, M+ 223
C12H17NOS
223.10309
1785
14176-49-9
PS
LM/Q
Anesthetic
Anticonvulsant
not detectable after HY

Bucetin
147
Peaks: 108, 109, 137, 179, M+ 223
C12H17NO3
223.12083
2020
1083-57-4
PS
LM
Analgesic

Etamivan
752
Peaks: 72, 151, M+ 223
C12H17NO3
223.12083
1900
G UHY
304-84-7
LM
Stimulant

223

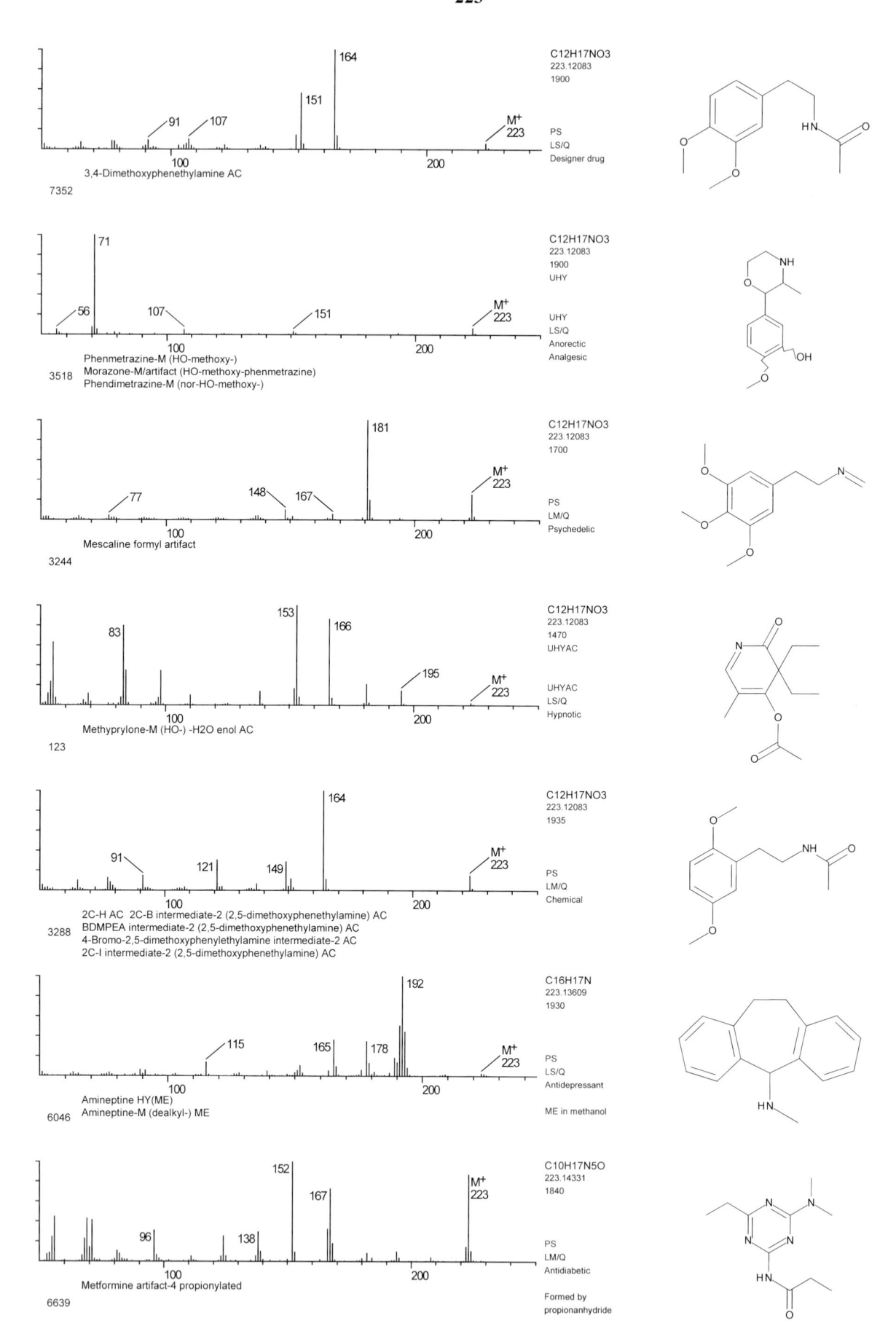

3,4-Dimethoxyphenethylamine AC 7352	Peaks: 91, 107, 151, 164, M+ 223	C12H17NO3 223.12083 1900 PS LS/Q Designer drug
Phenmetrazine-M (HO-methoxy-) Morazone-M/artifact (HO-methoxy-phenmetrazine) Phendimetrazine-M (nor-HO-methoxy-) 3518	Peaks: 56, 71, 107, 151, M+ 223	C12H17NO3 223.12083 1900 UHY UHY LS/Q Anorectic Analgesic
Mescaline formyl artifact 3244	Peaks: 77, 148, 167, 181, M+ 223	C12H17NO3 223.12083 1700 PS LM/Q Psychedelic
Methyprylone-M (HO-) -H2O enol AC 123	Peaks: 83, 153, 166, 195, M+ 223	C12H17NO3 223.12083 1470 UHYAC UHYAC LS/Q Hypnotic
2C-H AC 2C-B intermediate-2 (2,5-dimethoxyphenethylamine) AC BDMPEA intermediate-2 (2,5-dimethoxyphenethylamine) AC 4-Bromo-2,5-dimethoxyphenylethylamine intermediate-2 AC 2C-I intermediate-2 (2,5-dimethoxyphenethylamine) AC 3288	Peaks: 91, 121, 149, 164, M+ 223	C12H17NO3 223.12083 1935 PS LM/Q Chemical
Amineptine HY(ME) Amineptine-M (dealkyl-) ME 6046	Peaks: 115, 165, 178, 192, M+ 223	C16H17N 223.13609 1930 PS LS/Q Antidepressant ME in methanol
Metformine artifact-4 propionylated 6639	Peaks: 96, 138, 152, 167, M+ 223	C10H17N5O 223.14331 1840 PS LM/Q Antidiabetic Formed by propionanhydride

223

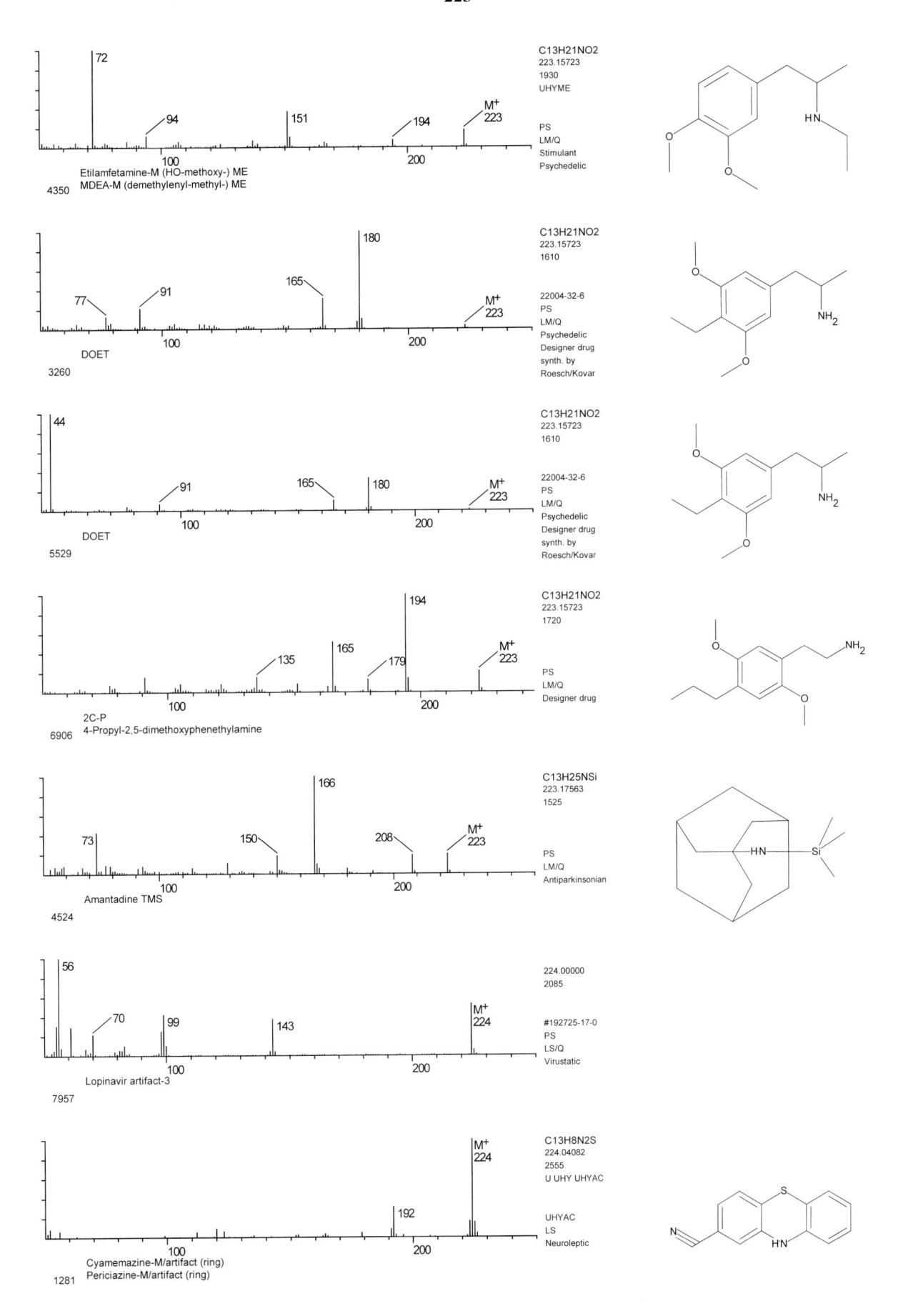

Etilamfetamine-M (HO-methoxy-) ME
MDEA-M (demethylenyl-methyl-) ME
4350
C13H21NO2
223.15723
1930
UHYME
PS
LM/Q
Stimulant
Psychedelic

Peaks: 72, 94, 151, 194, M+ 223

DOET
3260
C13H21NO2
223.15723
1610
22004-32-6
PS
LM/Q
Psychedelic
Designer drug synth. by Roesch/Kovar

Peaks: 77, 91, 165, 180, M+ 223

DOET
5529
C13H21NO2
223.15723
1610
22004-32-6
PS
LM/Q
Psychedelic
Designer drug synth. by Roesch/Kovar

Peaks: 44, 91, 165, 180, M+ 223

2C-P
4-Propyl-2,5-dimethoxyphenethylamine
6906
C13H21NO2
223.15723
1720
PS
LM/Q
Designer drug

Peaks: 135, 165, 179, 194, M+ 223

Amantadine TMS
4524
C13H25NSi
223.17563
1525
PS
LM/Q
Antiparkinsonian

Peaks: 73, 150, 166, 208, M+ 223

Lopinavir artifact-3
7957
224.00000
2085
#192725-17-0
PS
LS/Q
Virustatic

Peaks: 56, 70, 99, 143, M+ 224

Cyamemazine-M/artifact (ring)
Periciazine-M/artifact (ring)
1281
C13H8N2S
224.04082
2555
U UHY UHYAC
UHYAC
LS
Neuroleptic

Peaks: 192, M+ 224

224

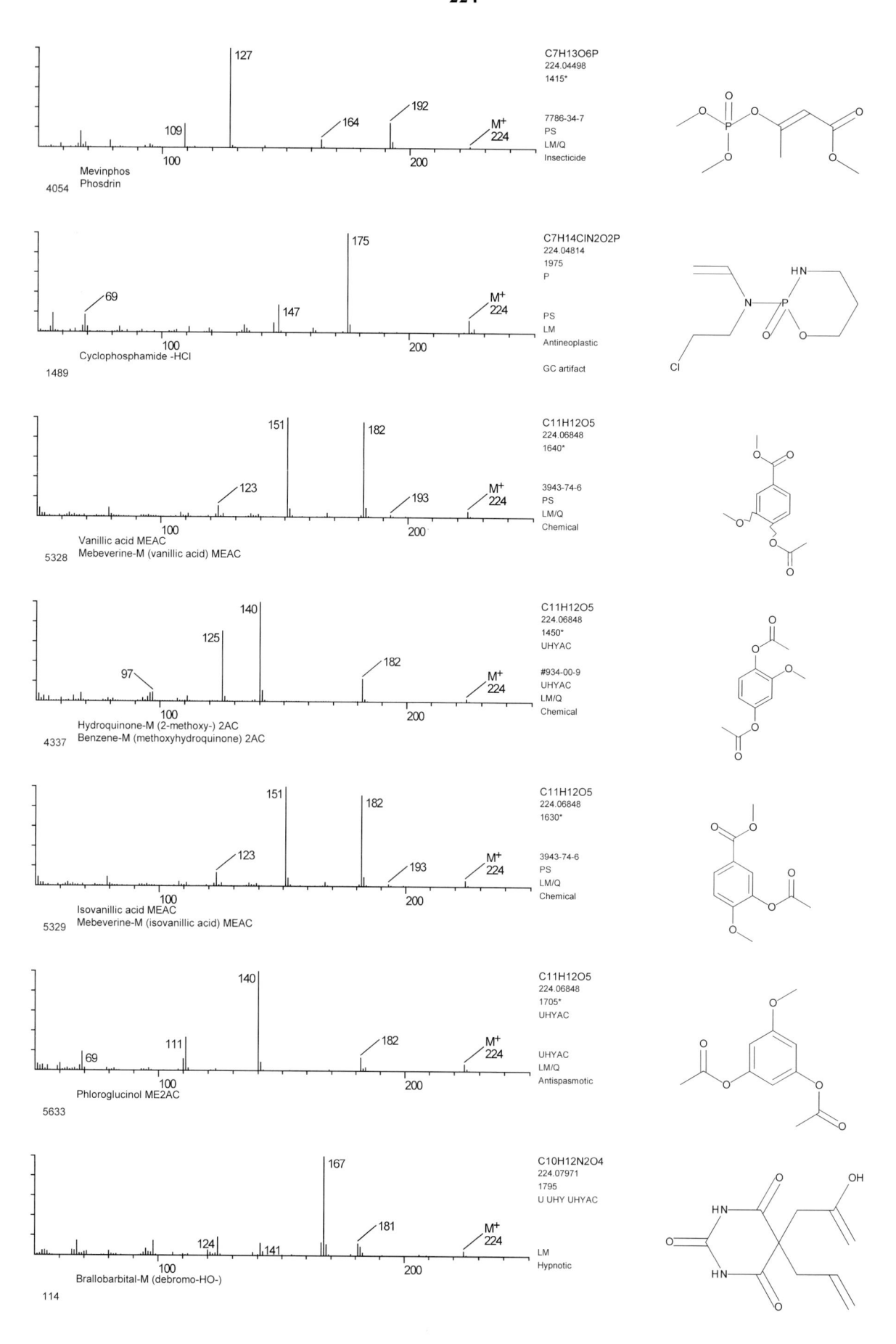

4054 Mevinphos Phosdrin
C7H13O6P
224.04498
1415*
7786-34-7
PS
LM/Q
Insecticide

1489 Cyclophosphamide -HCl
C7H14ClN2O2P
224.04814
1975
P
PS
LM
Antineoplastic
GC artifact

5328 Vanillic acid MEAC / Mebeverine-M (vanillic acid) MEAC
C11H12O5
224.06848
1640*
3943-74-6
PS
LM/Q
Chemical

4337 Hydroquinone-M (2-methoxy-) 2AC / Benzene-M (methoxyhydroquinone) 2AC
C11H12O5
224.06848
1450*
UHYAC
#934-00-9
UHYAC
LM/Q
Chemical

5329 Isovanillic acid MEAC / Mebeverine-M (isovanillic acid) MEAC
C11H12O5
224.06848
1630*
3943-74-6
PS
LM/Q
Chemical

5633 Phloroglucinol ME2AC
C11H12O5
224.06848
1705*
UHYAC
UHYAC
LM/Q
Antispasmotic

114 Brallobarbital-M (debromo-HO-)
C10H12N2O4
224.07971
1795
U UHY UHYAC
LM
Hypnotic

224

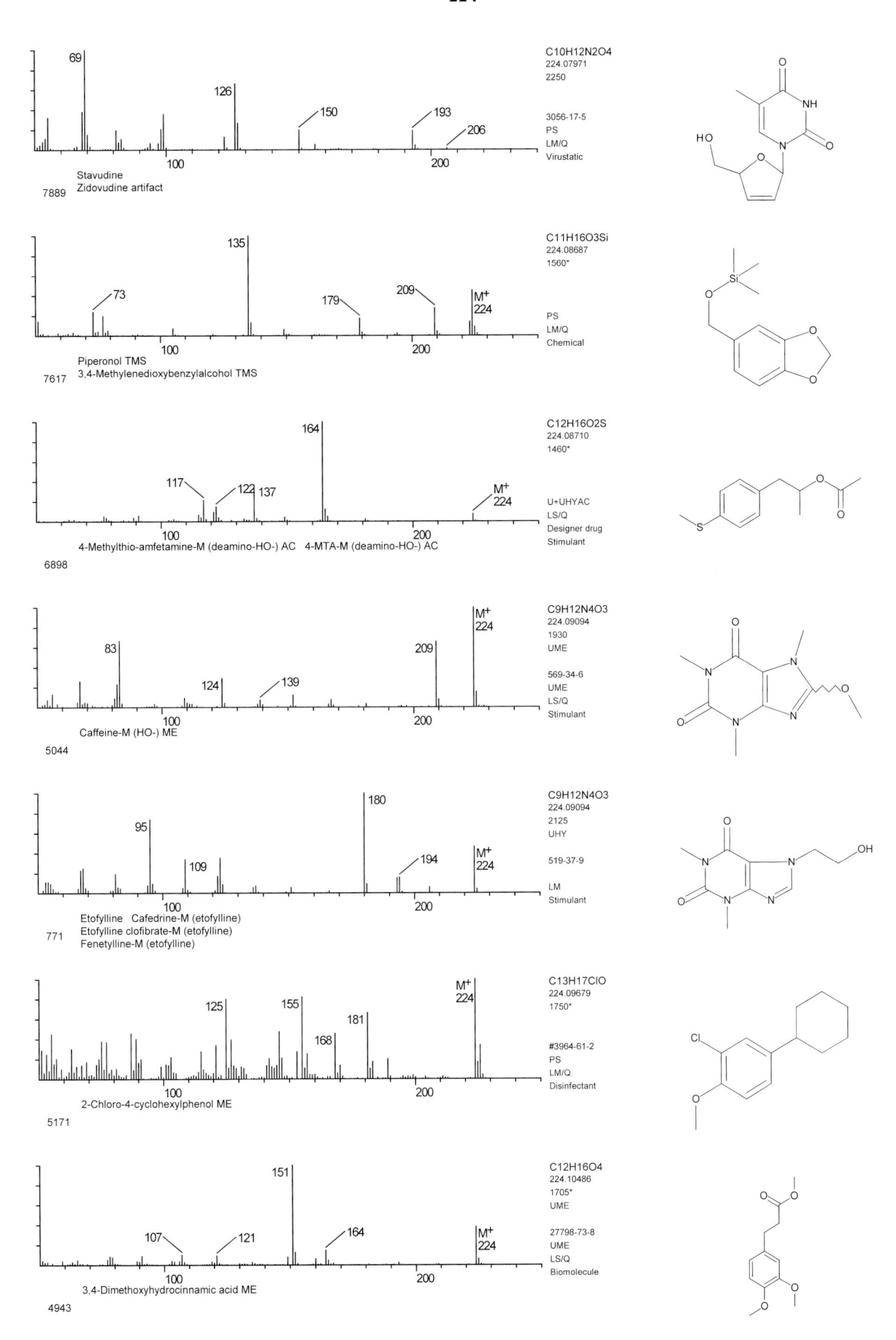

C10H12N2O4
224.07971
2250

3056-17-5
PS
LM/Q
Virustatic

Stavudine
Zidovudine artifact
7889

C11H16O3Si
224.08687
1560*

PS
LM/Q
Chemical

Piperonol TMS
3,4-Methylenedioxybenzylalcohol TMS
7617

C12H16O2S
224.08710
1460*

U+UHYAC
LS/Q
Designer drug
Stimulant

4-Methylthio-amfetamine-M (deamino-HO-) AC 4-MTA-M (deamino-HO-) AC
6898

C9H12N4O3
224.09094
1930
UME

569-34-6
UME
LS/Q
Stimulant

Caffeine-M (HO-) ME
5044

C9H12N4O3
224.09094
2125
UHY

519-37-9
LM
Stimulant

Etofylline   Cafedrine-M (etofylline)
Etofylline clofibrate-M (etofylline)
Fenetylline-M (etofylline)
771

C13H17ClO
224.09679
1750*

#3964-61-2
PS
LM/Q
Disinfectant

2-Chloro-4-cyclohexylphenol ME
5171

C12H16O4
224.10486
1705*
UME

27798-73-8
UME
LS/Q
Biomolecule

3,4-Dimethoxyhydrocinnamic acid ME
4943

487

224

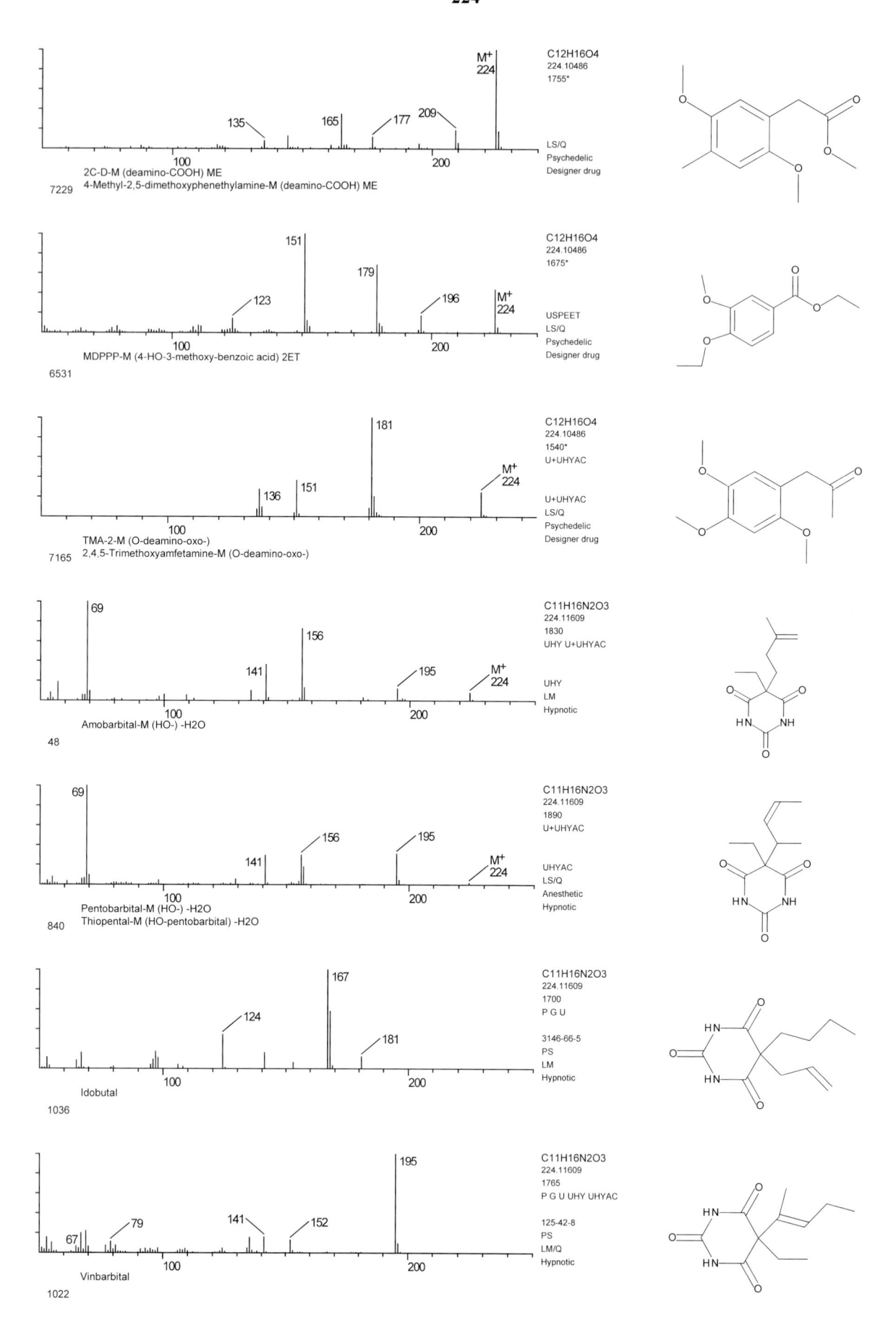

Spectrum	Compound	Formula / Data
7229	2C-D-M (deamino-COOH) ME 4-Methyl-2,5-dimethoxyphenethylamine-M (deamino-COOH) ME Peaks: 135, 165, 177, 209, M+ 224	C12H16O4 224.10486 1755* LS/Q Psychedelic Designer drug
6531	MDPPP-M (4-HO-3-methoxy-benzoic acid) 2ET Peaks: 123, 151, 179, 196, M+ 224	C12H16O4 224.10486 1675* USPEET LS/Q Psychedelic Designer drug
7165	TMA-2-M (O-deamino-oxo-) 2,4,5-Trimethoxyamfetamine-M (O-deamino-oxo-) Peaks: 136, 151, 181, M+ 224	C12H16O4 224.10486 1540* U+UHYAC U+UHYAC LS/Q Psychedelic Designer drug
48	Amobarbital-M (HO-) -H2O Peaks: 69, 141, 156, 195, M+ 224	C11H16N2O3 224.11609 1830 UHY U+UHYAC UHY LM Hypnotic
840	Pentobarbital-M (HO-) -H2O Thiopental-M (HO-pentobarbital) -H2O Peaks: 69, 141, 156, 195, M+ 224	C11H16N2O3 224.11609 1890 U+UHYAC UHYAC LS/Q Anesthetic Hypnotic
1036	Idobutal Peaks: 124, 167, 181	C11H16N2O3 224.11609 1700 P G U 3146-66-5 PS LM Hypnotic
1022	Vinbarbital Peaks: 67, 79, 141, 152, 195	C11H16N2O3 224.11609 1765 P G U UHY UHYAC 125-42-8 PS LM/Q Hypnotic

224

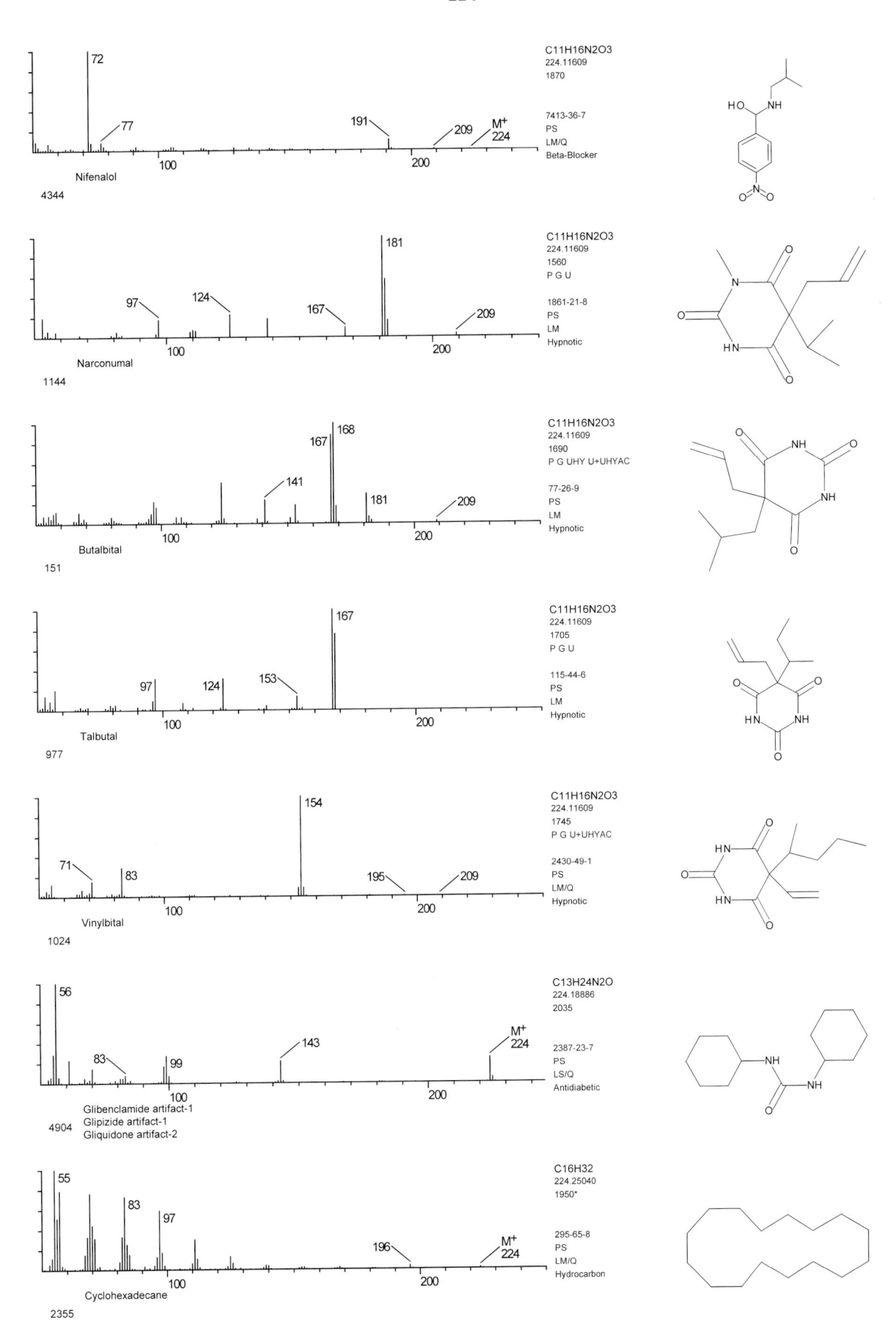

Nifenalol — 4344
C11H16N2O3
224.11609
1870
7413-36-7
PS
LM/Q
Beta-Blocker
Peaks: 72, 77, 191, 209, M+ 224

Narconumal — 1144
C11H16N2O3
224.11609
1560
P G U
1861-21-8
PS
LM
Hypnotic
Peaks: 97, 124, 167, 181, 209

Butalbital — 151
C11H16N2O3
224.11609
1690
P G UHY U+UHYAC
77-26-9
PS
LM
Hypnotic
Peaks: 141, 167, 168, 181, 209

Talbutal — 977
C11H16N2O3
224.11609
1705
P G U
115-44-6
PS
LM
Hypnotic
Peaks: 97, 124, 153, 167

Vinylbital — 1024
C11H16N2O3
224.11609
1745
P G U+UHYAC
2430-49-1
PS
LM/Q
Hypnotic
Peaks: 71, 83, 154, 195, 209

Glibenclamide artifact-1
Glipizide artifact-1
Gliquidone artifact-2 — 4904
C13H24N2O
224.18886
2035
2387-23-7
PS
LS/Q
Antidiabetic
Peaks: 56, 83, 99, 143, M+ 224

Cyclohexadecane — 2355
C16H32
224.25040
1950*
295-65-8
PS
LM/Q
Hydrocarbon
Peaks: 55, 83, 97, 196, M+ 224

225

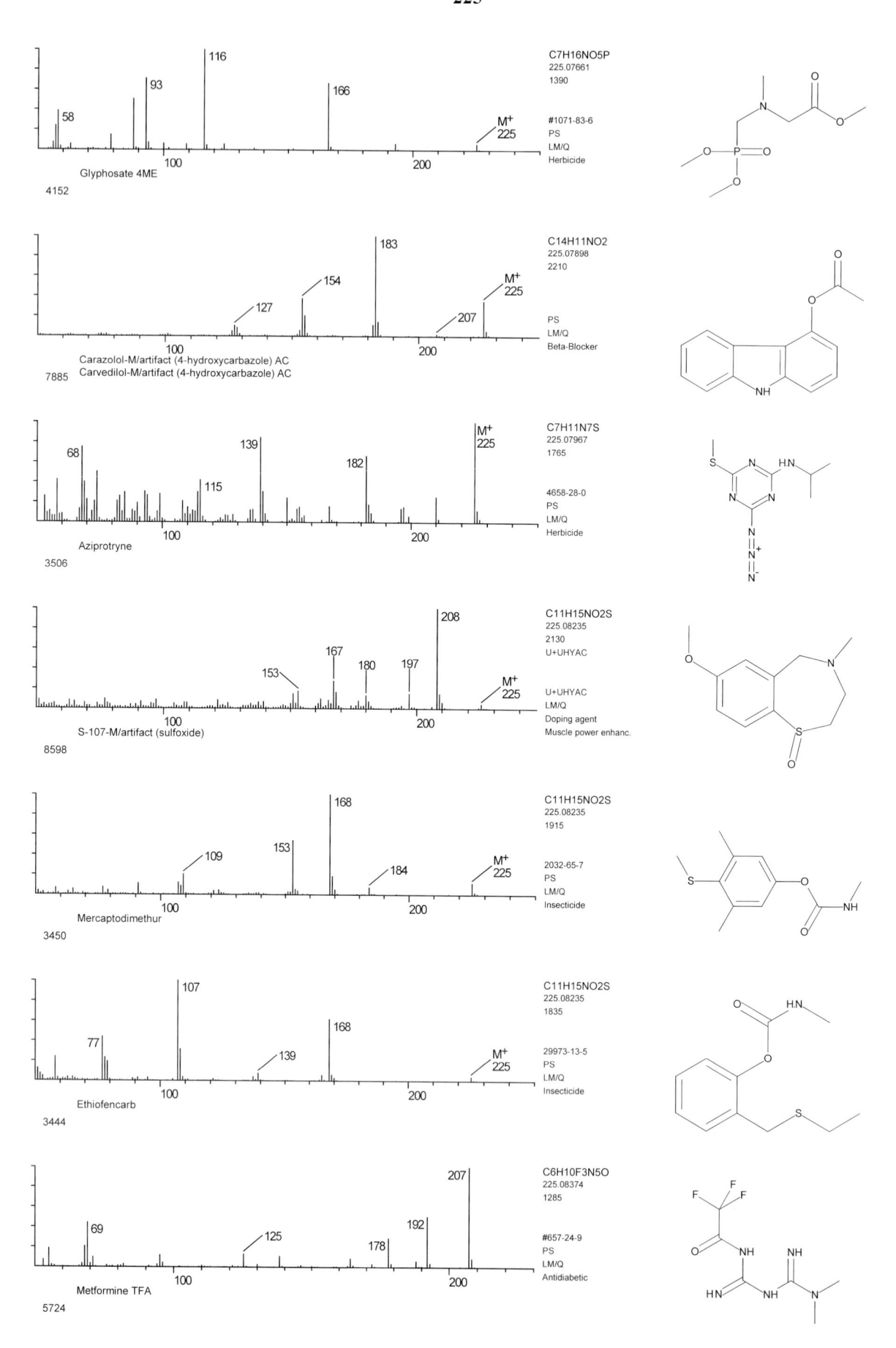

Glyphosate 4ME — 4152
C7H16NO5P
225.07661
1390
#1071-83-6
PS
LM/Q
Herbicide
Peaks: 58, 93, 116, 166, M+ 225

Carazolol-M/artifact (4-hydroxycarbazole) AC
Carvedilol-M/artifact (4-hydroxycarbazole) AC — 7885
C14H11NO2
225.07898
2210
PS
LM/Q
Beta-Blocker
Peaks: 127, 154, 183, 207, M+ 225

Aziprotryne — 3506
C7H11N7S
225.07967
1765
4658-28-0
PS
LM/Q
Herbicide
Peaks: 68, 115, 139, 182, M+ 225

S-107-M/artifact (sulfoxide) — 8598
C11H15NO2S
225.08235
2130
U+UHYAC
U+UHYAC
LM/Q
Doping agent
Muscle power enhanc.
Peaks: 153, 167, 180, 197, 208, M+ 225

Mercaptodimethur — 3450
C11H15NO2S
225.08235
1915
2032-65-7
PS
LM/Q
Insecticide
Peaks: 109, 153, 168, 184, M+ 225

Ethiofencarb — 3444
C11H15NO2S
225.08235
1835
29973-13-5
PS
LM/Q
Insecticide
Peaks: 77, 107, 139, 168, M+ 225

Metformine TFA — 5724
C6H10F3N5O
225.08374
1285
#657-24-9
PS
LM/Q
Antidiabetic
Peaks: 69, 125, 178, 192, 207

225

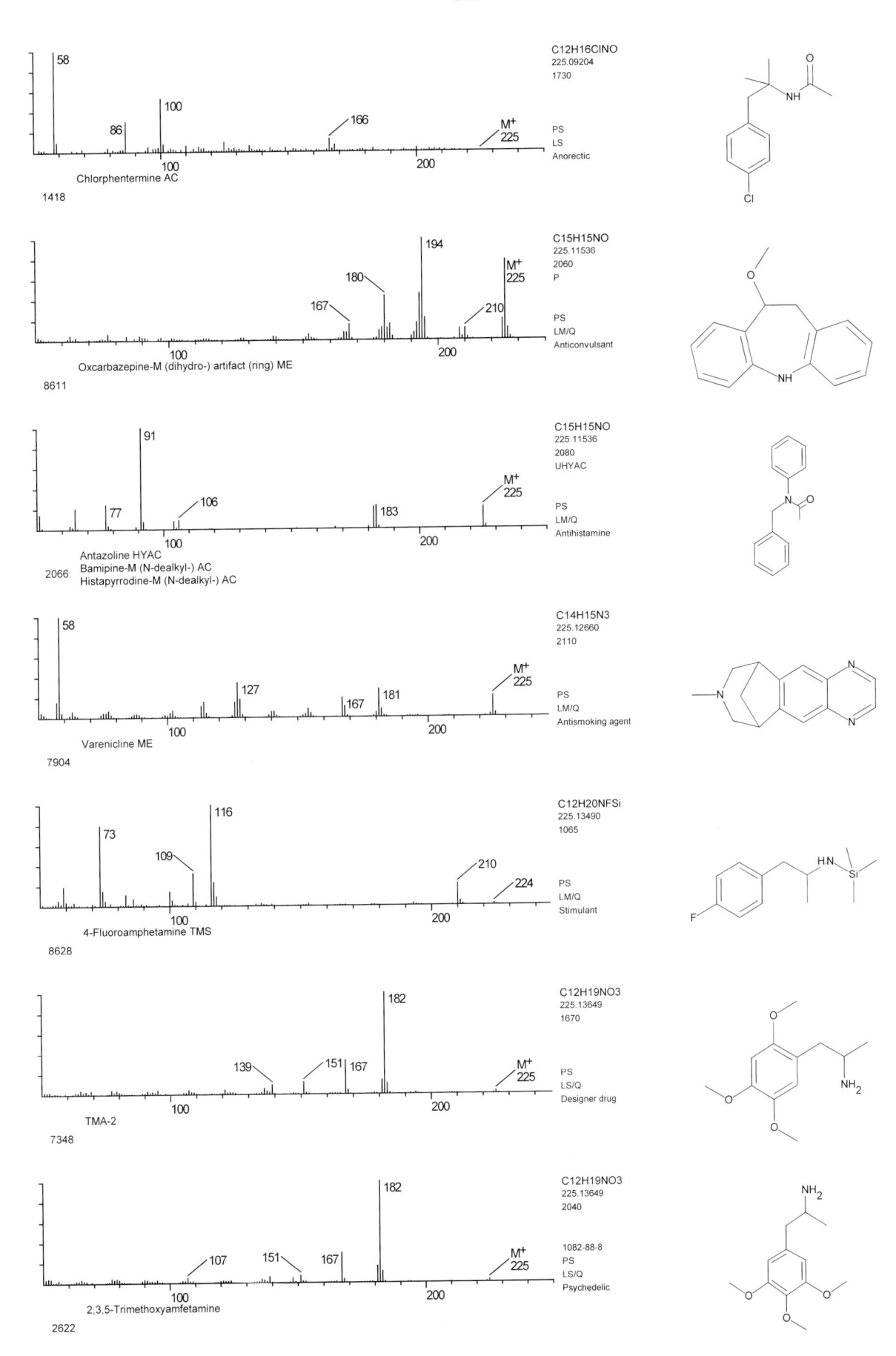

Chlorphentermine AC	C12H16ClNO 225.09204 1730	PS LS Anorectic
Oxcarbazepine-M (dihydro-) artifact (ring) ME	C15H15NO 225.11536 2060 P	PS LM/Q Anticonvulsant
Antazoline HYAC / Bamipine-M (N-dealkyl-) AC / Histapyrrodine-M (N-dealkyl-) AC	C15H15NO 225.11536 2080 UHYAC	PS LM/Q Antihistamine
Varenicline ME	C14H15N3 225.12660 2110	PS LM/Q Antismoking agent
4-Fluoroamphetamine TMS	C12H20NFSi 225.13490 1065	PS LM/Q Stimulant
TMA-2	C12H19NO3 225.13649 1670	PS LS/Q Designer drug
2,3,5-Trimethoxyamfetamine	C12H19NO3 225.13649 2040 1082-88-8	PS LS/Q Psychedelic

225

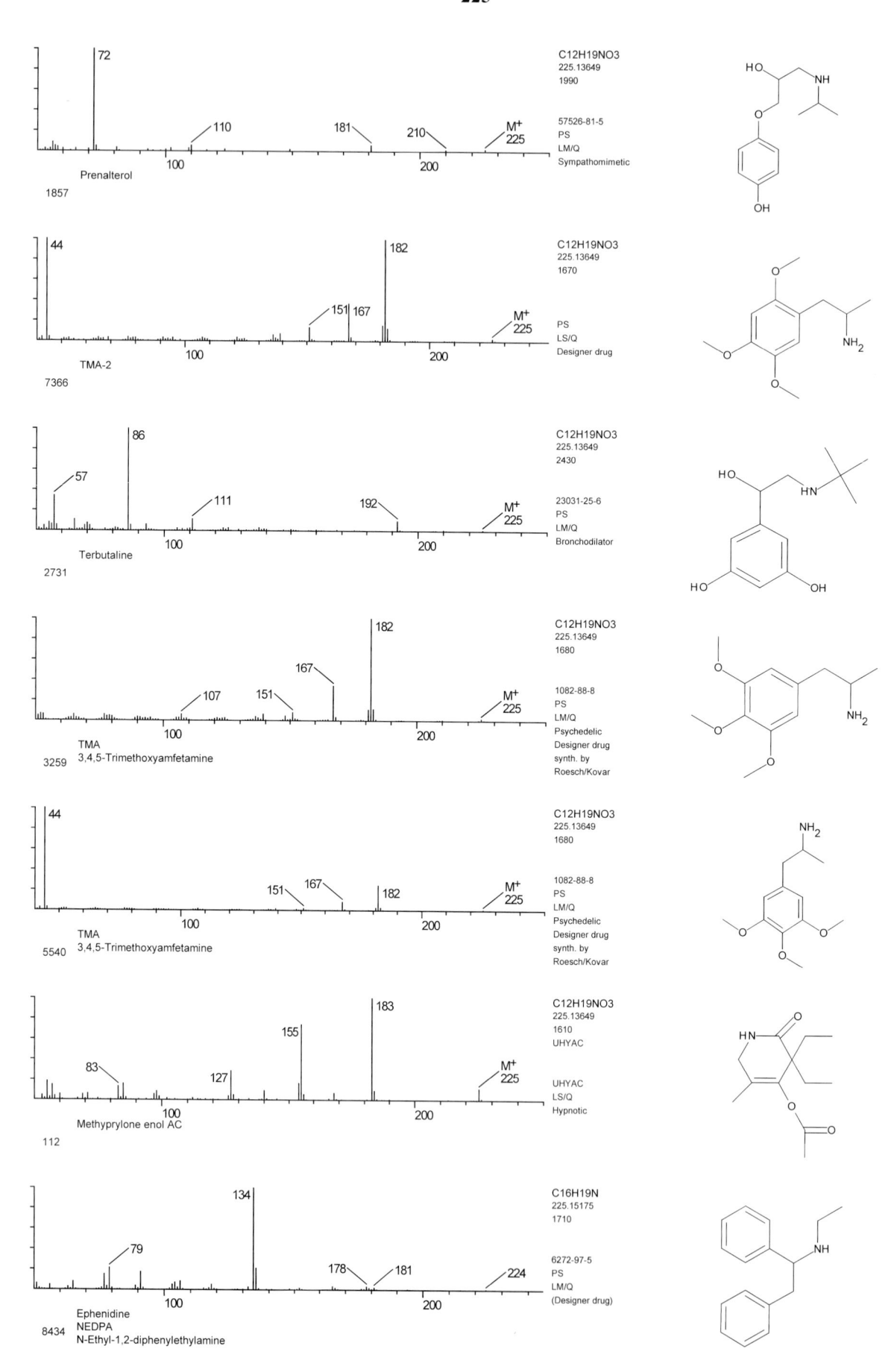

Prenalterol — 1857
C12H19NO3
225.13649
1990
57526-81-5
PS
LM/Q
Sympathomimetic
Peaks: 72, 110, 181, 210, M+ 225

TMA-2 — 7366
C12H19NO3
225.13649
1670
PS
LS/Q
Designer drug
Peaks: 44, 151, 167, 182, M+ 225

Terbutaline — 2731
C12H19NO3
225.13649
2430
23031-25-6
PS
LM/Q
Bronchodilator
Peaks: 57, 86, 111, 192, M+ 225

TMA 3,4,5-Trimethoxyamfetamine — 3259
C12H19NO3
225.13649
1680
1082-88-8
PS
LM/Q
Psychedelic
Designer drug
synth. by Roesch/Kovar
Peaks: 107, 151, 167, 182, M+ 225

TMA 3,4,5-Trimethoxyamfetamine — 5540
C12H19NO3
225.13649
1680
1082-88-8
PS
LM/Q
Psychedelic
Designer drug
synth. by Roesch/Kovar
Peaks: 44, 151, 167, 182, M+ 225

Methyprylone enol AC — 112
C12H19NO3
225.13649
1610
UHYAC
UHYAC
LS/Q
Hypnotic
Peaks: 83, 127, 155, 183, M+ 225

Ephenidine NEDPA N-Ethyl-1,2-diphenylethylamine — 8434
C16H19N
225.15175
1710
6272-97-5
PS
LM/Q
(Designer drug)
Peaks: 79, 134, 178, 181, 224

225

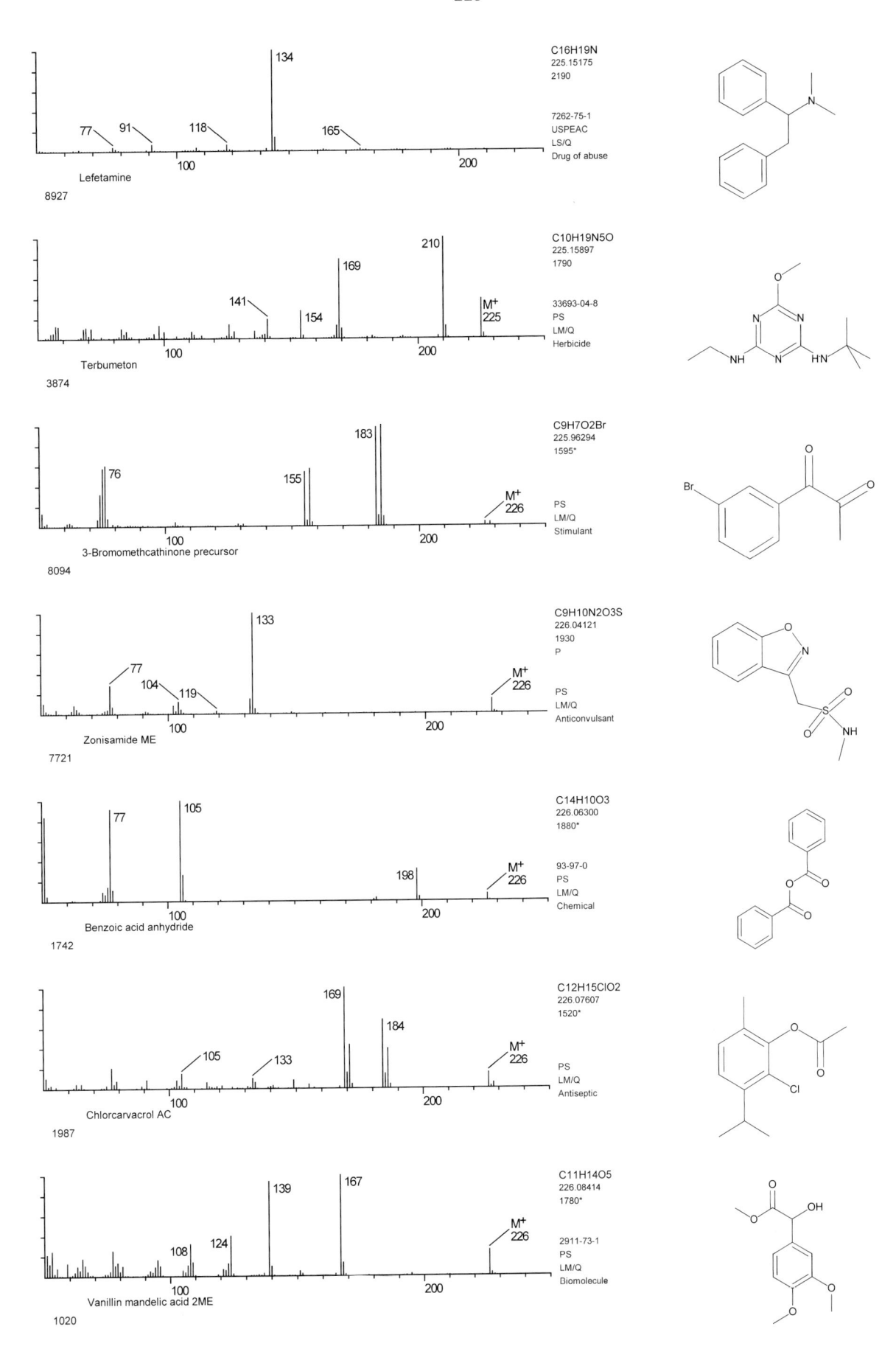

Lefetamine — 8927
C16H19N
225.15175
2190
7262-75-1
USPEAC
LS/Q
Drug of abuse
Peaks: 77, 91, 118, 134, 165

Terbumeton — 3874
C10H19N5O
225.15897
1790
33693-04-8
PS
LM/Q
Herbicide
Peaks: 141, 154, 169, 210, M+ 225

3-Bromomethcathinone precursor — 8094
C9H7O2Br
225.96294
1595*
PS
LM/Q
Stimulant
Peaks: 76, 155, 183, M+ 226

Zonisamide ME — 7721
C9H10N2O3S
226.04121
1930
P
PS
LM/Q
Anticonvulsant
Peaks: 77, 104, 119, 133, M+ 226

Benzoic acid anhydride — 1742
C14H10O3
226.06300
1880*
93-97-0
PS
LM/Q
Chemical
Peaks: 77, 105, 198, M+ 226

Chlorcarvacrol AC — 1987
C12H15ClO2
226.07607
1520*
PS
LM/Q
Antiseptic
Peaks: 105, 133, 169, 184, M+ 226

Vanillin mandelic acid 2ME — 1020
C11H14O5
226.08414
1780*
2911-73-1
PS
LM/Q
Biomolecule
Peaks: 108, 124, 139, 167, M+ 226

226

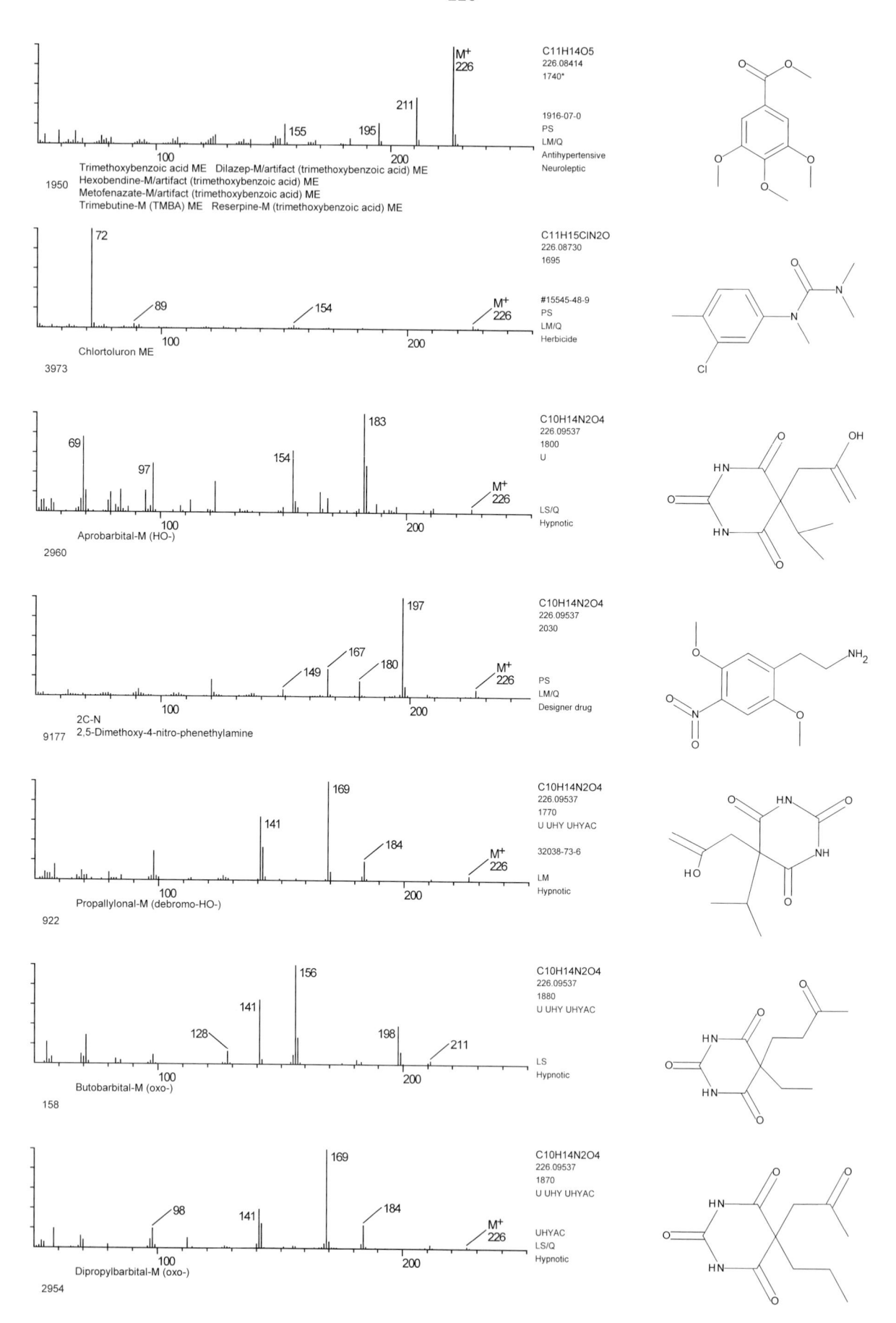

Spectrum	Formula	Details
m/z 155, 195, 211, M+ 226	C11H14O5	226.08414, 1740*, 1916-07-0, PS, LM/Q, Antihypertensive, Neuroleptic

Trimethoxybenzoic acid ME Dilazep-M/artifact (trimethoxybenzoic acid) ME
Hexobendine-M/artifact (trimethoxybenzoic acid) ME
Metofenazate-M/artifact (trimethoxybenzoic acid) ME
Trimebutine-M (TMBA) ME   Reserpine-M (trimethoxybenzoic acid) ME
1950

72, 89, 154, M+ 226 — C11H15ClN2O, 226.08730, 1695, #15545-48-9, PS, LM/Q, Herbicide
Chlortoluron ME
3973

69, 97, 154, 183, M+ 226 — C10H14N2O4, 226.09537, 1800, U, LS/Q, Hypnotic
Aprobarbital-M (HO-)
2960

149, 167, 180, 197, M+ 226 — C10H14N2O4, 226.09537, 2030, PS, LM/Q, Designer drug
2C-N 2,5-Dimethoxy-4-nitro-phenethylamine
9177

141, 169, 184, M+ 226 — C10H14N2O4, 226.09537, 1770, U UHY UHYAC, 32038-73-6, LM, Hypnotic
Propallylonal-M (debromo-HO-)
922

128, 141, 156, 198, 211 — C10H14N2O4, 226.09537, 1880, U UHY UHYAC, LS, Hypnotic
Butobarbital-M (oxo-)
158

98, 141, 169, 184, M+ 226 — C10H14N2O4, 226.09537, 1870, U UHY UHYAC, UHYAC, LS/Q, Hypnotic
Dipropylbarbital-M (oxo-)
2954

226

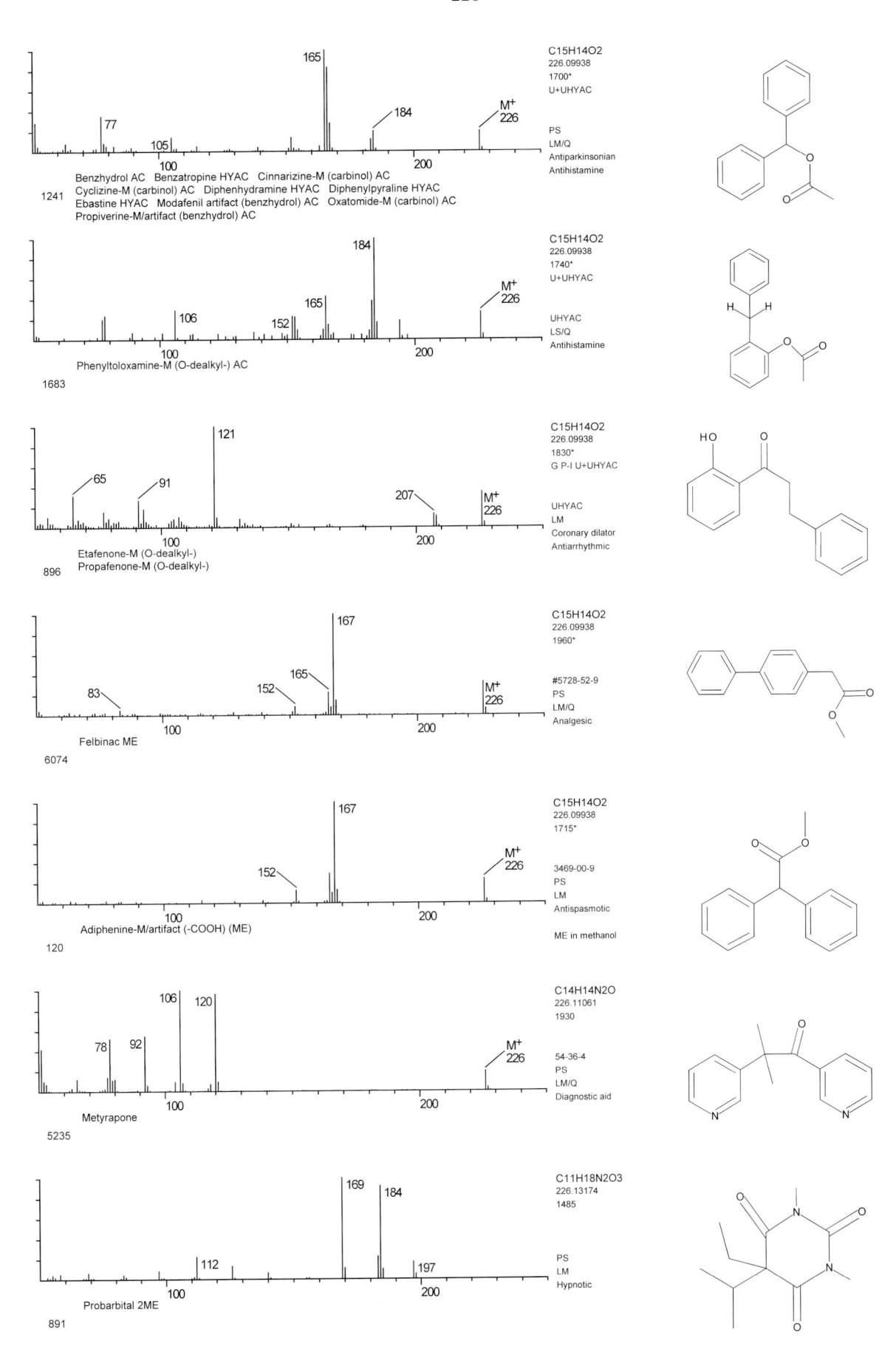

C15H14O2
226.09938
1700*
U+UHYAC

PS
LM/Q
Antiparkinsonian
Antihistamine

Benzhydrol AC   Benzatropine HYAC   Cinnarizine-M (carbinol) AC
Cyclizine-M (carbinol) AC   Diphenhydramine HYAC   Diphenylpyraline HYAC
Ebastine HYAC   Modafenil artifact (benzhydrol) AC   Oxatomide-M (carbinol) AC
Propiverine-M/artifact (benzhydrol) AC
1241

C15H14O2
226.09938
1740*
U+UHYAC

UHYAC
LS/Q
Antihistamine

Phenyltoloxamine-M (O-dealkyl-) AC
1683

C15H14O2
226.09938
1830*
G P-I U+UHYAC

UHYAC
LM
Coronary dilator
Antiarrhythmic

Etafenone-M (O-dealkyl-)
Propafenone-M (O-dealkyl-)
896

C15H14O2
226.09938
1960*

#5728-52-9
PS
LM/Q
Analgesic

Felbinac ME
6074

C15H14O2
226.09938
1715*

3469-00-9
PS
LM
Antispasmotic

ME in methanol

Adiphenine-M/artifact (-COOH) (ME)
120

C14H14N2O
226.11061
1930

54-36-4
PS
LM/Q
Diagnostic aid

Metyrapone
5235

C11H18N2O3
226.13174
1485

PS
LM
Hypnotic

Probarbital 2ME
891

226

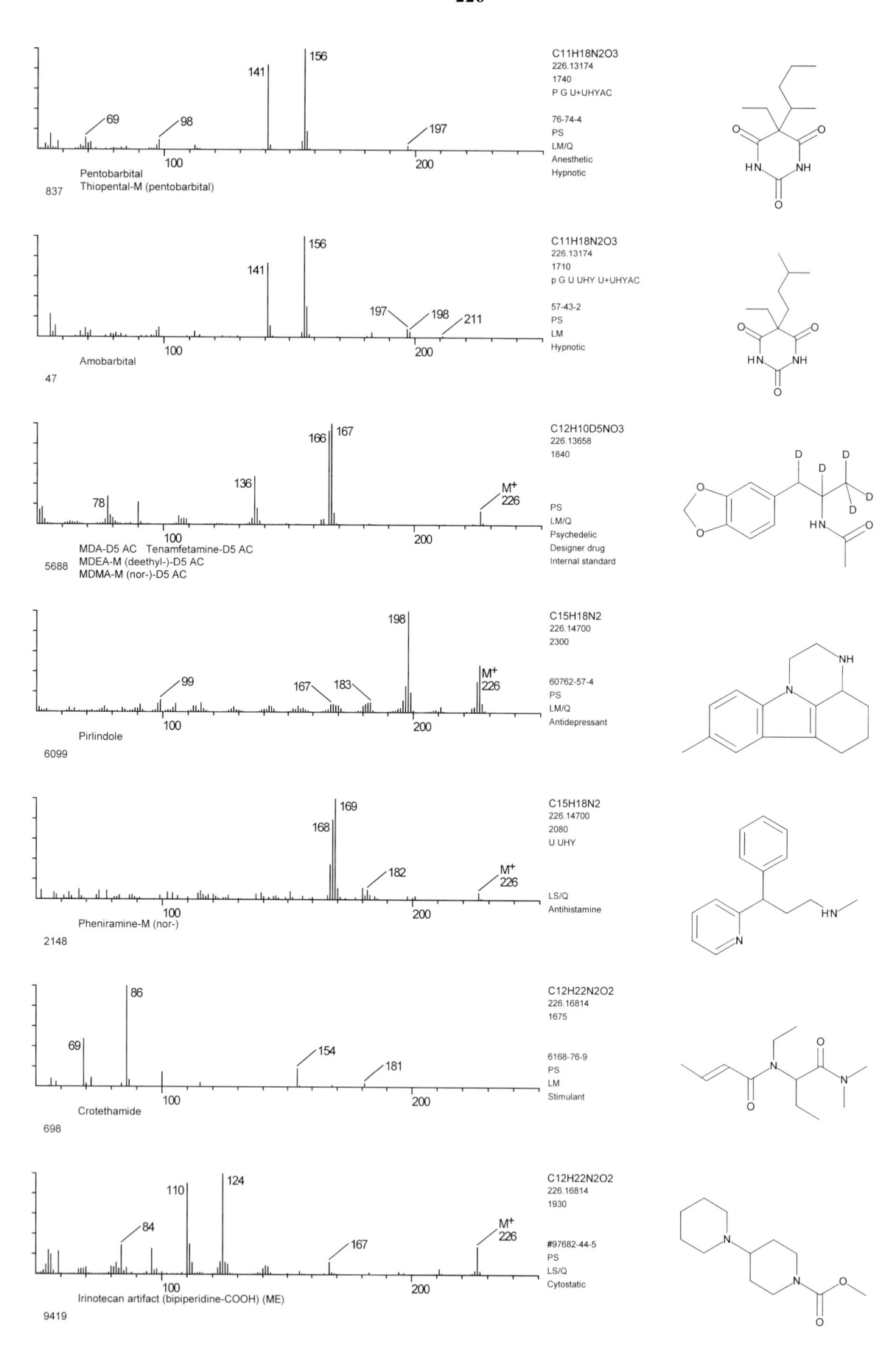

C11H18N2O3
226.13174
1740
P G U+UHYAC

76-74-4
PS
LM/Q
Anesthetic
Hypnotic

Pentobarbital
Thiopental-M (pentobarbital)

C11H18N2O3
226.13174
1710
p G U UHY U+UHYAC

57-43-2
PS
LM
Hypnotic

Amobarbital

C12H10D5NO3
226.13658
1840

PS
LM/Q
Psychedelic
Designer drug
Internal standard

MDA-D5 AC   Tenamfetamine-D5 AC
MDEA-M (deethyl-)-D5 AC
MDMA-M (nor-)-D5 AC

C15H18N2
226.14700
2300

60762-57-4
PS
LM/Q
Antidepressant

Pirlindole

C15H18N2
226.14700
2080
U UHY

LS/Q
Antihistamine

Pheniramine-M (nor-)

C12H22N2O2
226.16814
1675

6168-76-9
PS
LM
Stimulant

Crotethamide

C12H22N2O2
226.16814
1930

#97682-44-5
PS
LS/Q
Cytostatic

Irinotecan artifact (bipiperidine-COOH) (ME)

226

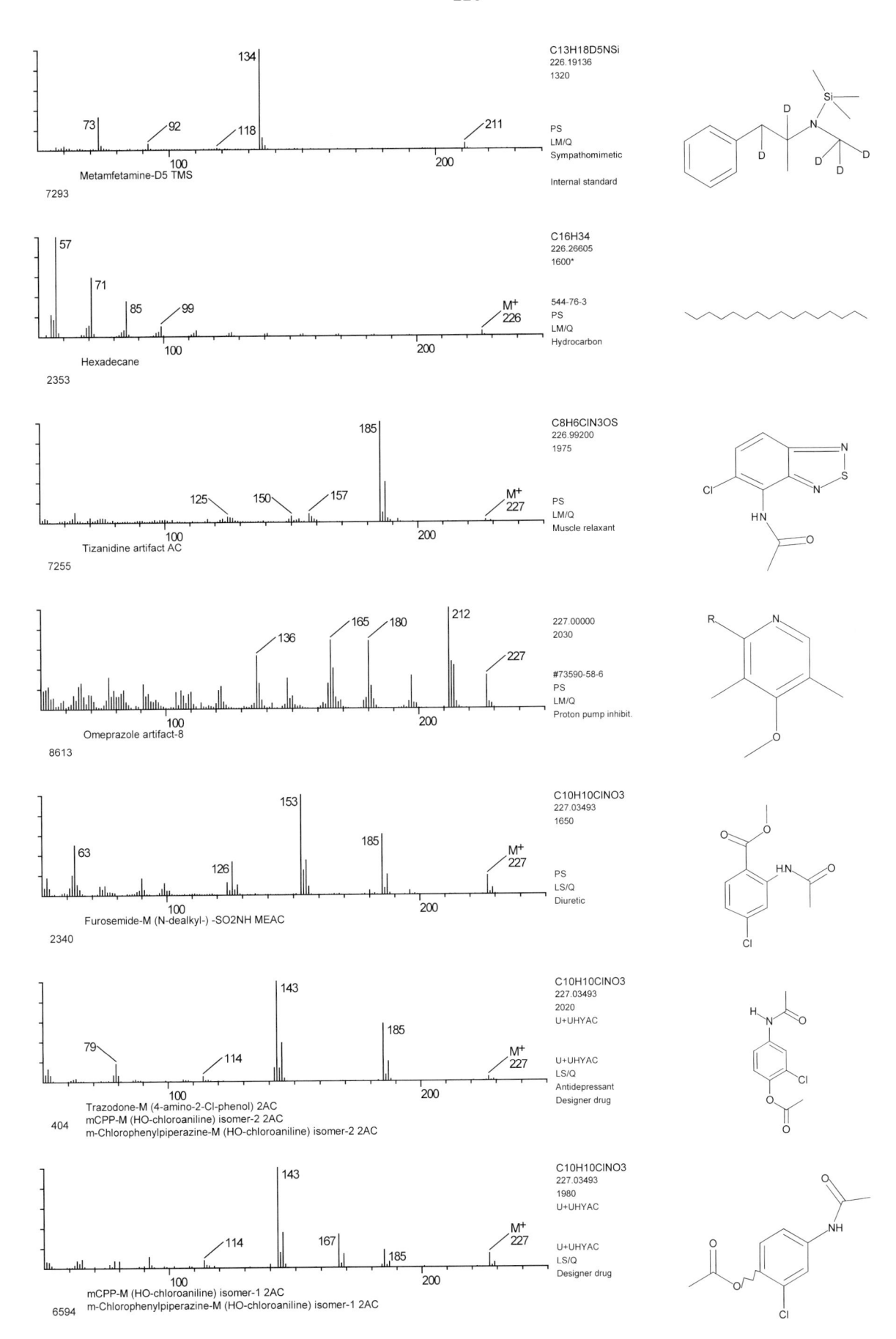

Spectrum	Formula / Data
Metamfetamine-D5 TMS — 7293	C13H18D5NSi; 226.19136; 1320; PS; LM/Q; Sympathomimetic; Internal standard
Hexadecane — 2353	C16H34; 226.26605; 1600*; 544-76-3; PS; LM/Q; Hydrocarbon
Tizanidine artifact AC — 7255	C8H6ClN3OS; 226.99200; 1975; PS; LM/Q; Muscle relaxant
Omeprazole artifact-8 — 8613	227.00000; 2030; #73590-58-6; PS; LM/Q; Proton pump inhibit.
Furosemide-M (N-dealkyl-) -SO2NH MEAC — 2340	C10H10ClNO3; 227.03493; 1650; PS; LS/Q; Diuretic
Trazodone-M (4-amino-2-Cl-phenol) 2AC / mCPP-M (HO-chloroaniline) isomer-2 2AC / m-Chlorophenylpiperazine-M (HO-chloroaniline) isomer-2 2AC — 404	C10H10ClNO3; 227.03493; 2020; U+UHYAC; U+UHYAC; LS/Q; Antidepressant; Designer drug
mCPP-M (HO-chloroaniline) isomer-1 2AC / m-Chlorophenylpiperazine-M (HO-chloroaniline) isomer-1 2AC — 6594	C10H10ClNO3; 227.03493; 1980; U+UHYAC; U+UHYAC; LS/Q; Designer drug

227

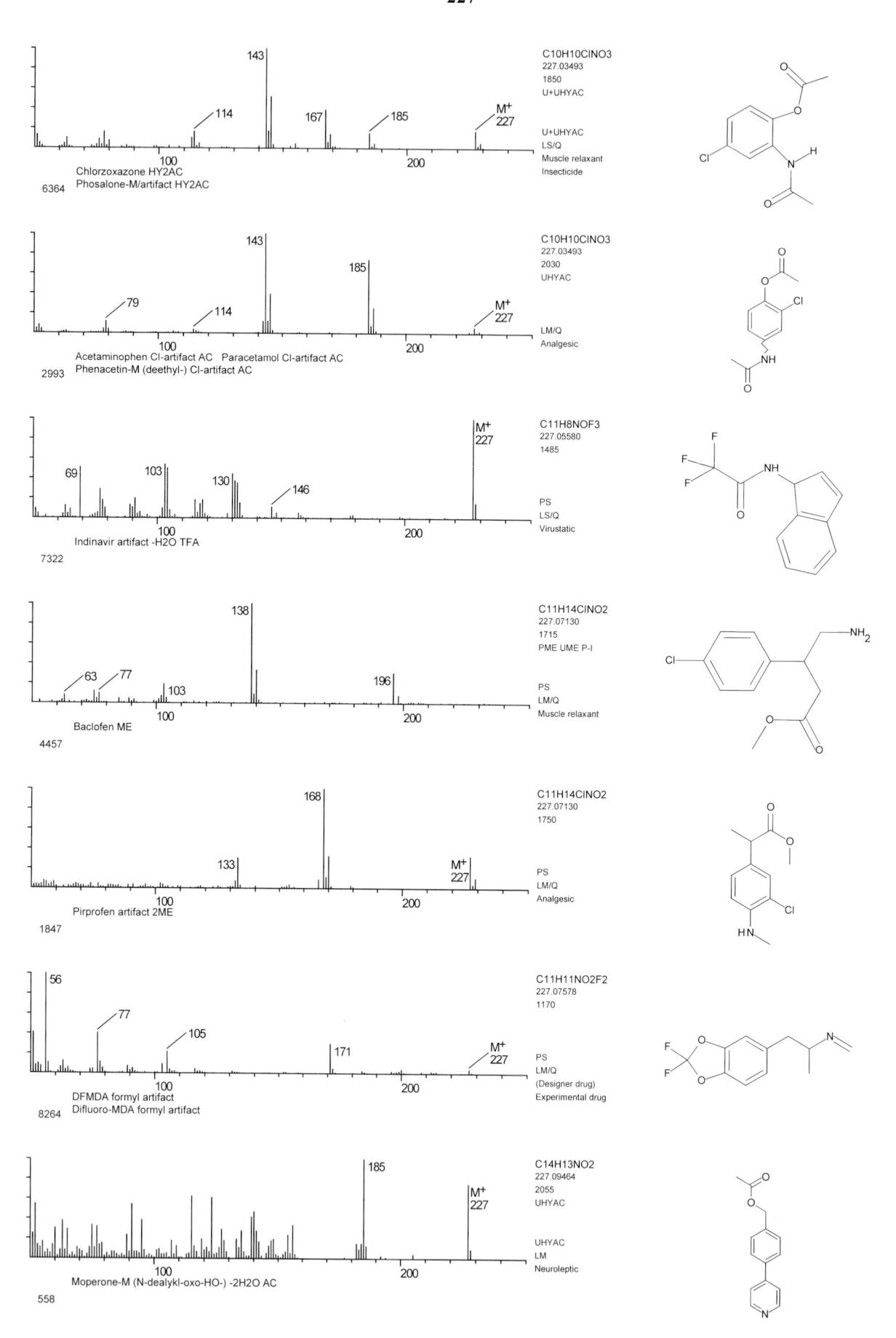

143, 114, 167, 185, M+ 227	C10H10ClNO3 227.03493 1850 U+UHYAC U+UHYAC LS/Q Muscle relaxant Insecticide	
6364 Chlorzoxazone HY2AC Phosalone-M/artifact HY2AC		
143, 79, 114, 185, M+ 227	C10H10ClNO3 227.03493 2030 UHYAC LM/Q Analgesic	
2993 Acetaminophen Cl-artifact AC Paracetamol Cl-artifact AC Phenacetin-M (deethyl-) Cl-artifact AC		
69, 103, 130, 146, M+ 227	C11H8NOF3 227.05580 1485 PS LS/Q Virustatic	
7322 Indinavir artifact -H2O TFA		
63, 77, 103, 138, 196, M+	C11H14ClNO2 227.07130 1715 PME UME P-I PS LM/Q Muscle relaxant	
4457 Baclofen ME		
133, 168, M+ 227	C11H14ClNO2 227.07130 1750 PS LM/Q Analgesic	
1847 Pirprofen artifact 2ME		
56, 77, 105, 171, M+ 227	C11H11NO2F2 227.07578 1170 PS LM/Q (Designer drug) Experimental drug	
8264 DFMDA formyl artifact Difluoro-MDA formyl artifact		
185, M+ 227	C14H13NO2 227.09464 2055 UHYAC UHYAC LM Neuroleptic	
558 Moperone-M (N-dealykl-oxo-HO-) -2H2O AC		

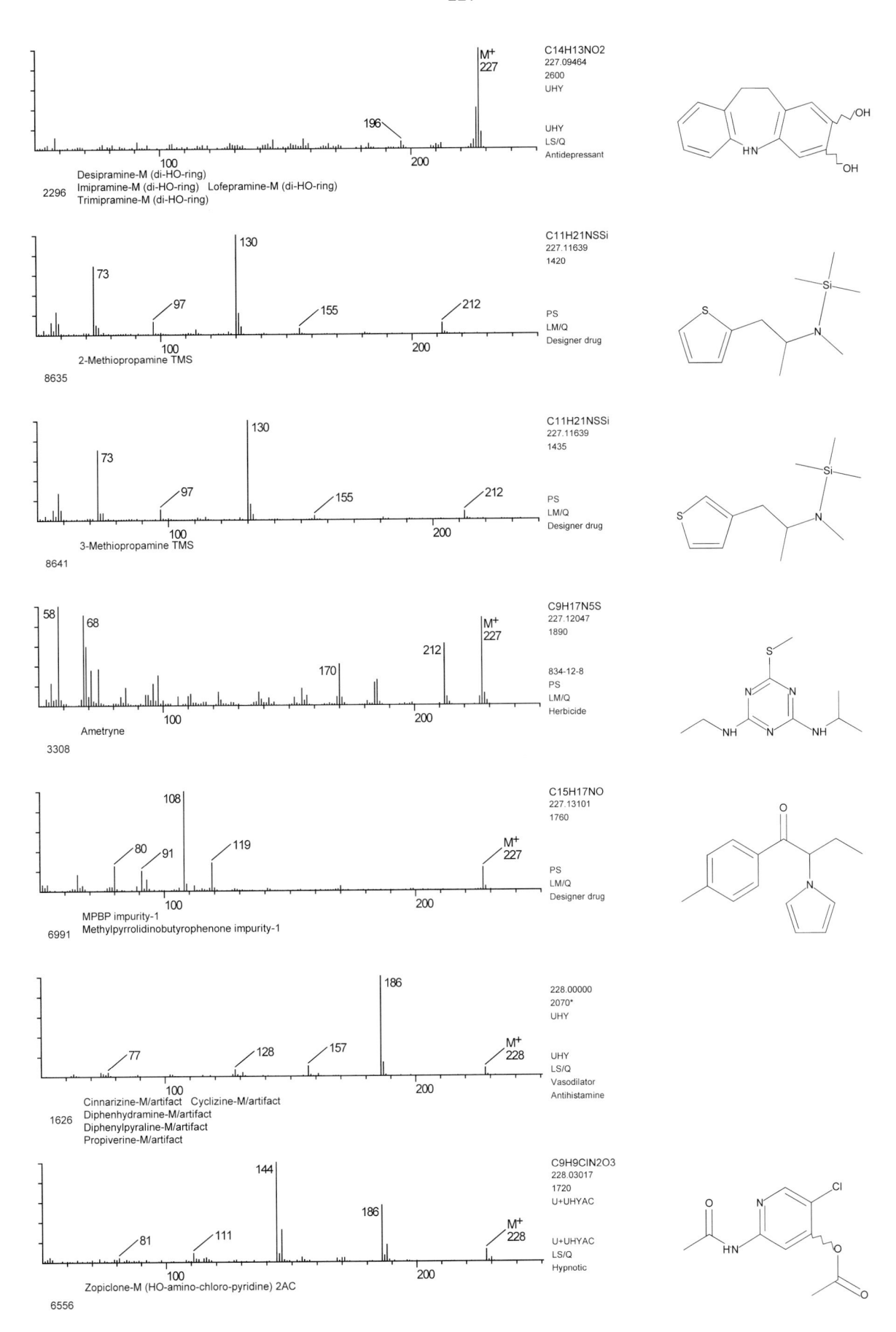

Desipramine-M (di-HO-ring) Imipramine-M (di-HO-ring) Lofepramine-M (di-HO-ring) Trimipramine-M (di-HO-ring) 2296	C14H13NO2 227.09464 2600 UHY UHY LS/Q Antidepressant
2-Methiopropamine TMS 8635	C11H21NSSi 227.11639 1420 PS LM/Q Designer drug
3-Methiopropamine TMS 8641	C11H21NSSi 227.11639 1435 PS LM/Q Designer drug
Ametryne 3308	C9H17N5S 227.12047 1890 834-12-8 PS LM/Q Herbicide
MPBP impurity-1 Methylpyrrolidinobutyrophenone impurity-1 6991	C15H17NO 227.13101 1760 PS LM/Q Designer drug
Cinnarizine-M/artifact Cyclizine-M/artifact Diphenhydramine-M/artifact Diphenylpyraline-M/artifact Propiverine-M/artifact 1626	228.00000 2070* UHY UHY LS/Q Vasodilator Antihistamine
Zopiclone-M (HO-amino-chloro-pyridine) 2AC 6556	C9H9ClN2O3 228.03017 1720 U+UHYAC U+UHYAC LS/Q Hypnotic

228

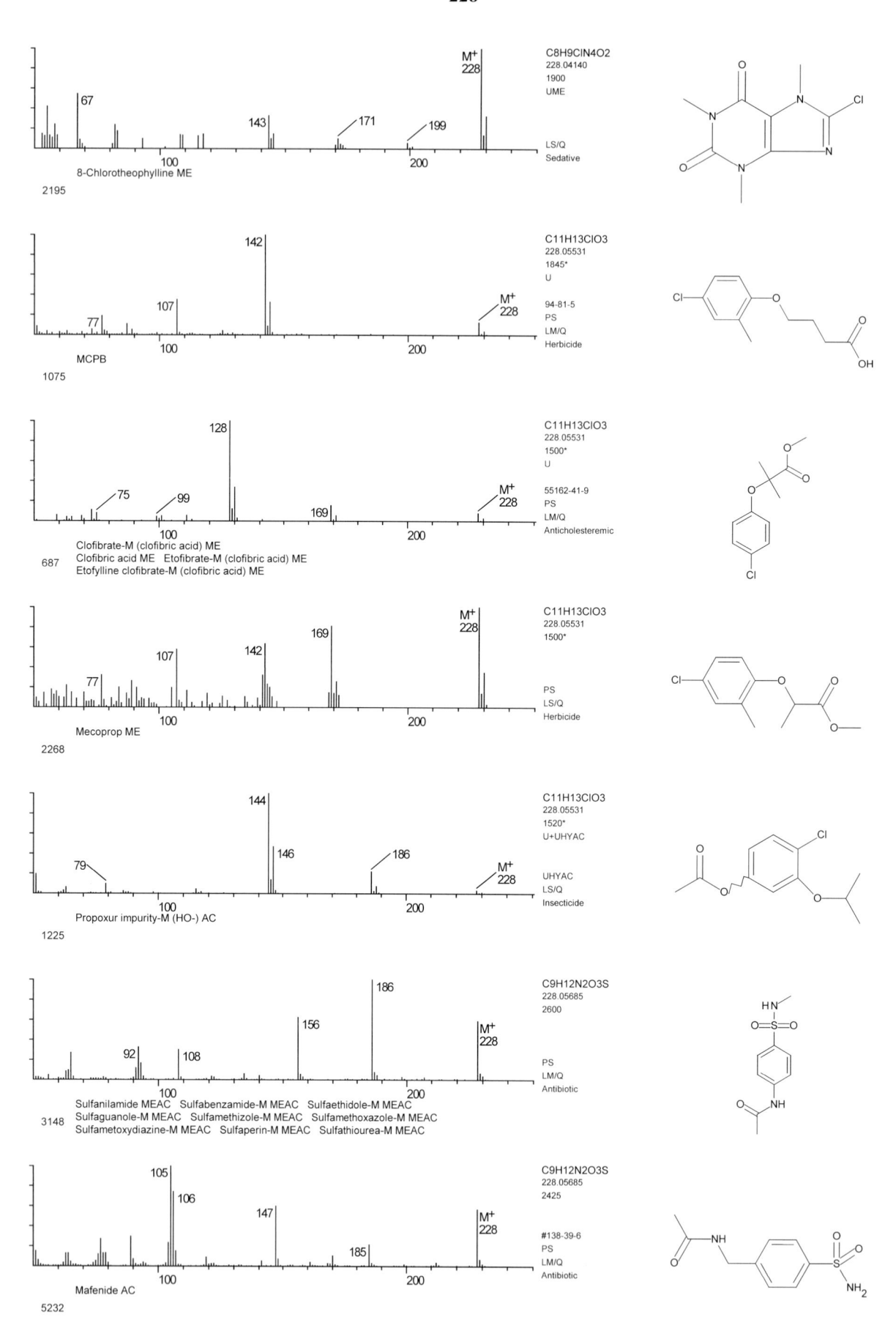

8-Chlorotheophylline ME
M+ 228, 199, 171, 143, 67
C8H9ClN4O2
228.04140
1900
UME
LS/Q
Sedative
2195

MCPB
M+ 228, 142, 107, 77
C11H13ClO3
228.05531
1845*
U
94-81-5
PS
LM/Q
Herbicide
1075

Clofibrate-M (clofibric acid) ME
Clofibric acid ME   Etofibrate-M (clofibric acid) ME
Etofylline clofibrate-M (clofibric acid) ME
M+ 228, 169, 128, 99, 75
C11H13ClO3
228.05531
1500*
U
55162-41-9
PS
LM/Q
Anticholesteremic
687

Mecoprop ME
M+ 228, 169, 142, 107, 77
C11H13ClO3
228.05531
1500*
PS
LS/Q
Herbicide
2268

Propoxur impurity-M (HO-) AC
M+ 228, 186, 146, 144, 79
C11H13ClO3
228.05531
1520*
U+UHYAC
UHYAC
LS/Q
Insecticide
1225

Sulfanilamide MEAC   Sulfabenzamide-M MEAC   Sulfaethidole-M MEAC
Sulfaguanole-M MEAC   Sulfamethizole-M MEAC   Sulfamethoxazole-M MEAC
Sulfametoxydiazine-M MEAC   Sulfaperin-M MEAC   Sulfathiourea-M MEAC
M+ 228, 186, 156, 108, 92
C9H12N2O3S
228.05685
2600
PS
LM/Q
Antibiotic
3148

Mafenide AC
M+ 228, 185, 147, 106, 105
C9H12N2O3S
228.05685
2425
#138-39-6
PS
LM/Q
Antibiotic
5232

500

228

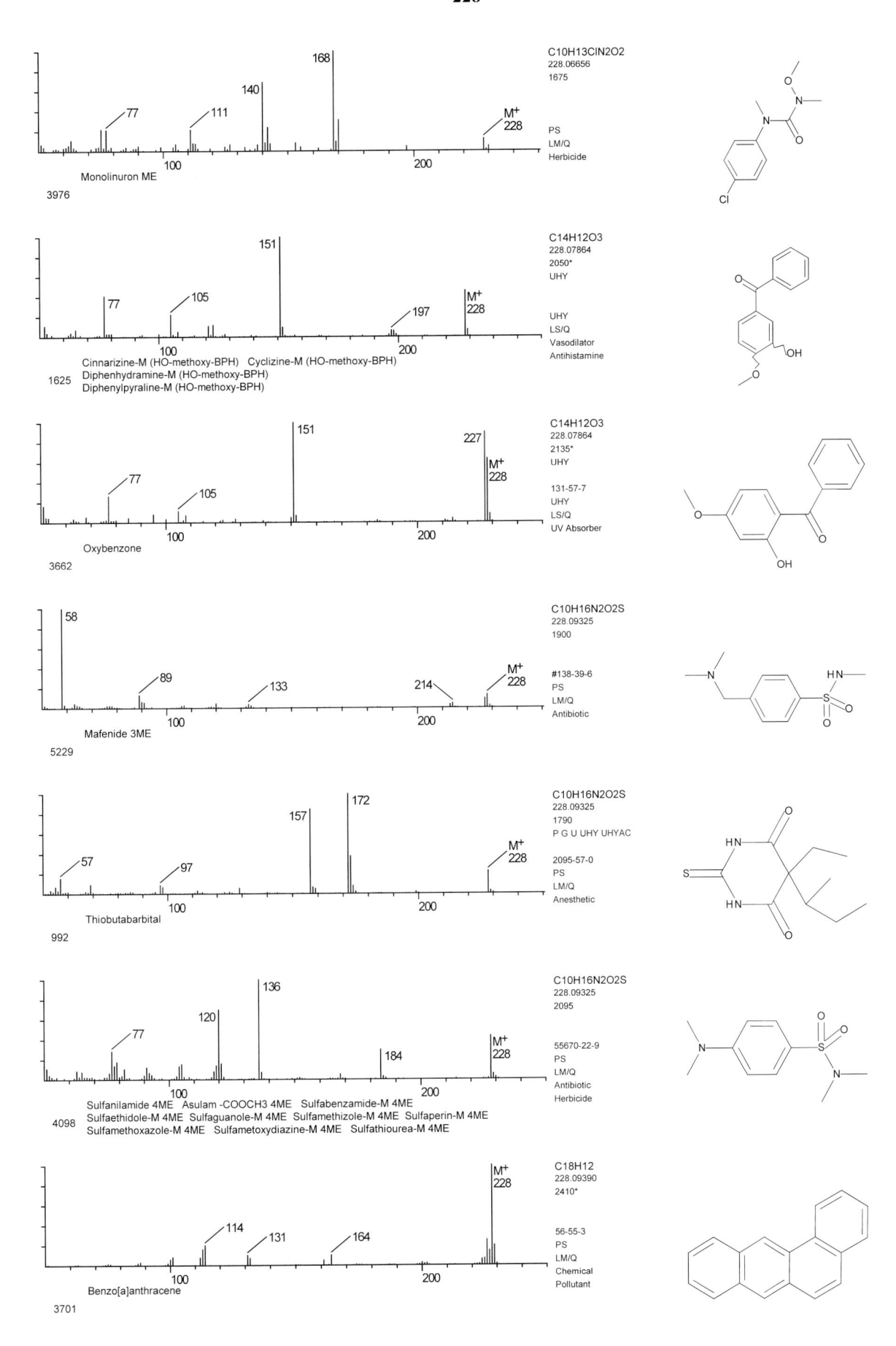

Monolinuron ME
3976
C10H13ClN2O2
228.06656
1675
PS
LM/Q
Herbicide

Cinnarizine-M (HO-methoxy-BPH)   Cyclizine-M (HO-methoxy-BPH)
Diphenhydramine-M (HO-methoxy-BPH)
Diphenylpyraline-M (HO-methoxy-BPH)
1625
C14H12O3
228.07864
2050*
UHY

UHY
LS/Q
Vasodilator
Antihistamine

Oxybenzone
3662
C14H12O3
228.07864
2135*
UHY

131-57-7
UHY
LS/Q
UV Absorber

Mafenide 3ME
5229
C10H16N2O2S
228.09325
1900

#138-39-6
PS
LM/Q
Antibiotic

Thiobutabarbital
992
C10H16N2O2S
228.09325
1790
P G U UHY UHYAC

2095-57-0
PS
LM/Q
Anesthetic

Sulfanilamide 4ME   Asulam -COOCH3 4ME   Sulfabenzamide-M 4ME
Sulfaethidole-M 4ME   Sulfaguanole-M 4ME   Sulfamethizole-M 4ME   Sulfaperin-M 4ME
4098
Sulfamethoxazole-M 4ME   Sulfametoxydiazine-M 4ME   Sulfathiourea-M 4ME
C10H16N2O2S
228.09325
2095

55670-22-9
PS
LM/Q
Antibiotic
Herbicide

Benzo[a]anthracene
3701
C18H12
228.09390
2410*

56-55-3
PS
LM/Q
Chemical
Pollutant

501

228

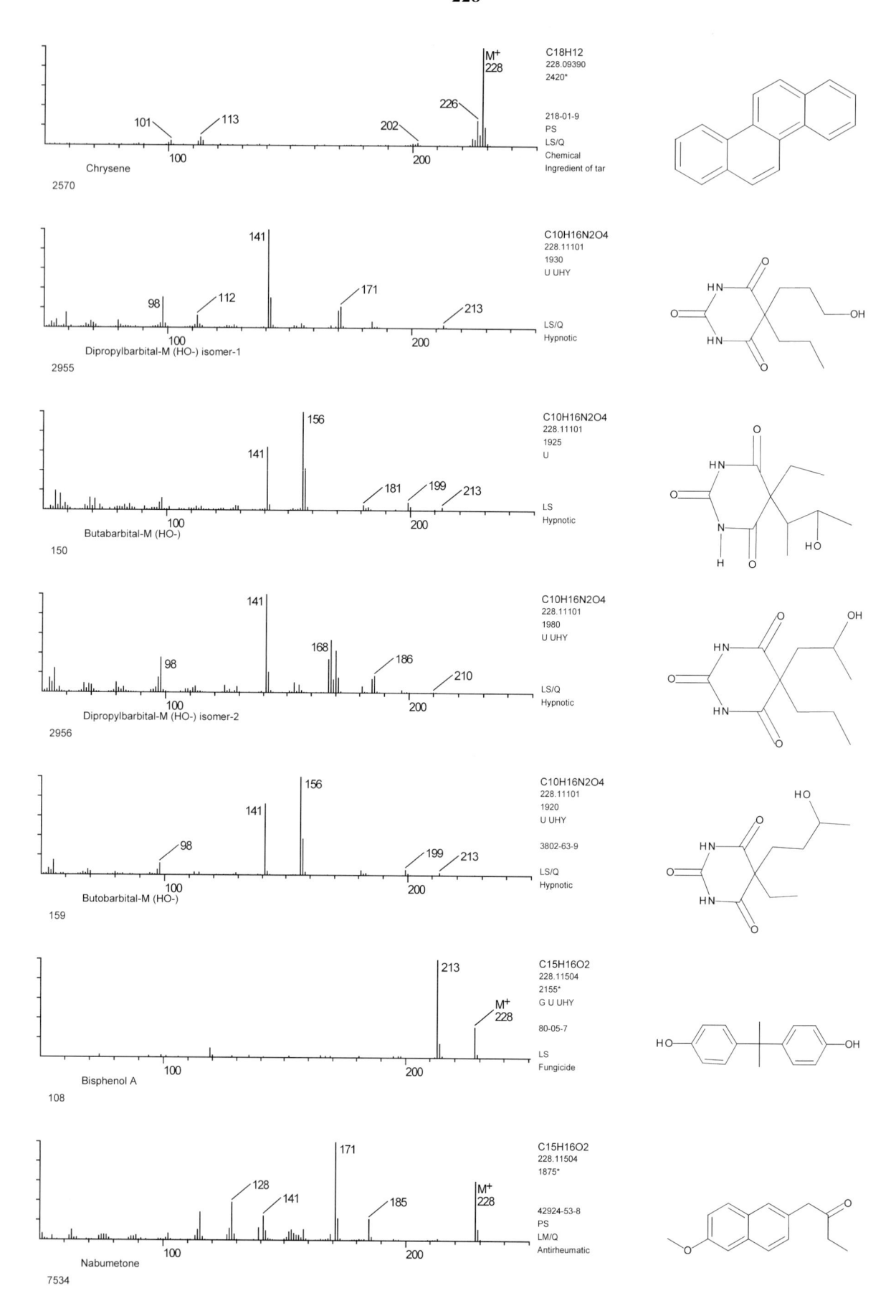

Chrysene — C18H12, 228.09390, 2420*, 218-01-9, PS, LS/Q, Chemical, Ingredient of tar
Peaks: 101, 113, 202, 226, M+ 228
2570

Dipropylbarbital-M (HO-) isomer-1 — C10H16N2O4, 228.11101, 1930, U UHY, LS/Q, Hypnotic
Peaks: 98, 112, 141, 171, 213
2955

Butabarbital-M (HO-) — C10H16N2O4, 228.11101, 1925, U, LS, Hypnotic
Peaks: 141, 156, 181, 199, 213
150

Dipropylbarbital-M (HO-) isomer-2 — C10H16N2O4, 228.11101, 1980, U UHY, LS/Q, Hypnotic
Peaks: 98, 141, 168, 186, 210
2956

Butobarbital-M (HO-) — C10H16N2O4, 228.11101, 1920, U UHY, 3802-63-9, LS/Q, Hypnotic
Peaks: 98, 141, 156, 199, 213
159

Bisphenol A — C15H16O2, 228.11504, 2155*, G U UHY, 80-05-7, LS, Fungicide
Peaks: 213, M+ 228
108

Nabumetone — C15H16O2, 228.11504, 1875*, 42924-53-8, PS, LM/Q, Antirheumatic
Peaks: 128, 141, 171, 185, M+ 228
7534

502

228

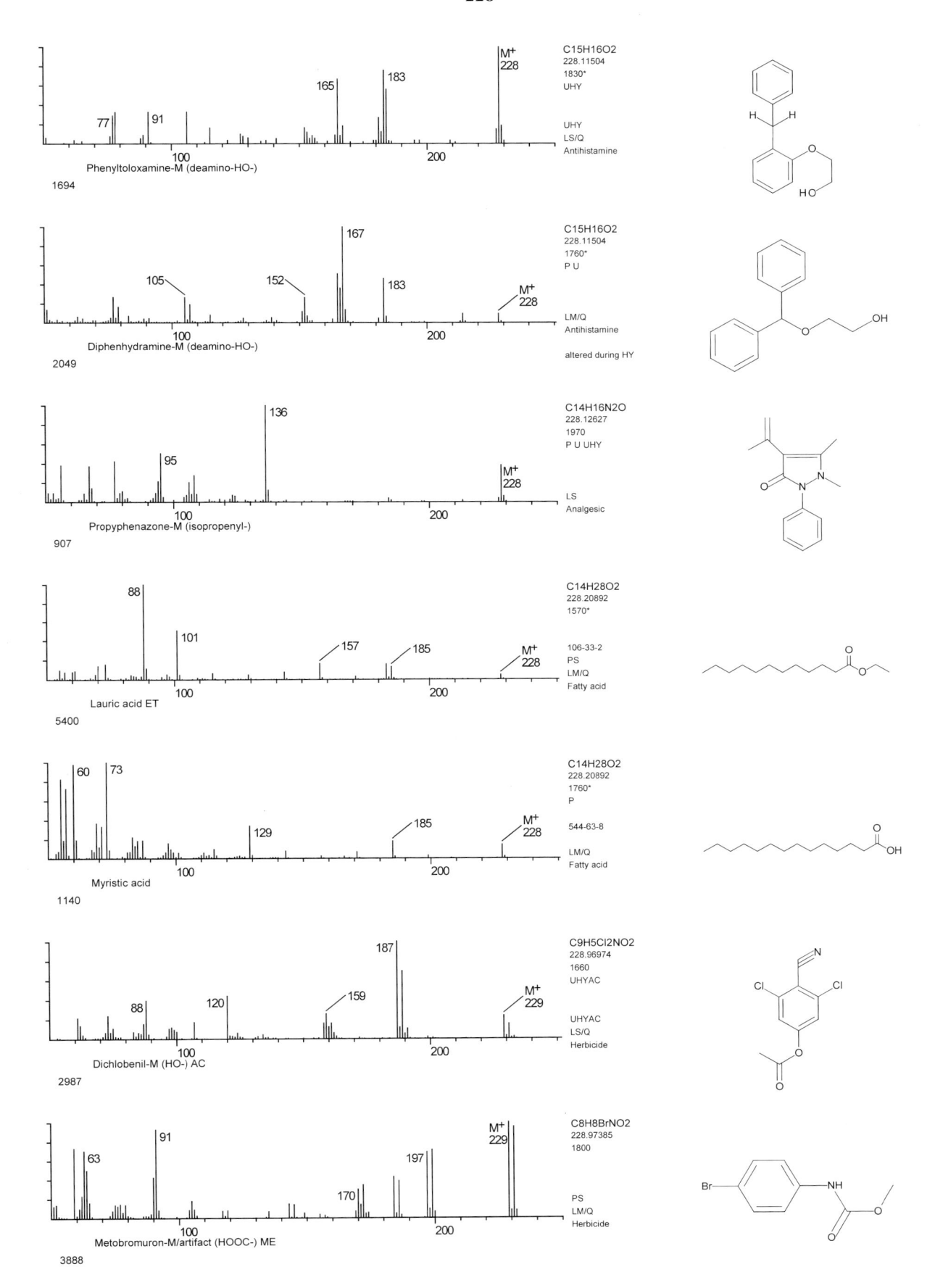

Spectrum	Formula / Info
Phenyltoloxamine-M (deamino-HO-) — 1694	C15H16O2, 228.11504, 1830*, UHY, UHY, LS/Q, Antihistamine
Diphenhydramine-M (deamino-HO-) — 2049	C15H16O2, 228.11504, 1760*, P U, LM/Q, Antihistamine, altered during HY
Propyphenazone-M (isopropenyl-) — 907	C14H16N2O, 228.12627, 1970, P U UHY, LS, Analgesic
Lauric acid ET — 5400	C14H28O2, 228.20892, 1570*, 106-33-2, PS, LM/Q, Fatty acid
Myristic acid — 1140	C14H28O2, 228.20892, 1760*, P, 544-63-8, LM/Q, Fatty acid
Dichlobenil-M (HO-) AC — 2987	C9H5Cl2NO2, 228.96974, 1660, UHYAC, UHYAC, LS/Q, Herbicide
Metobromuron-M/artifact (HOOC-) ME — 3888	C8H8BrNO2, 228.97385, 1800, PS, LM/Q, Herbicide

229

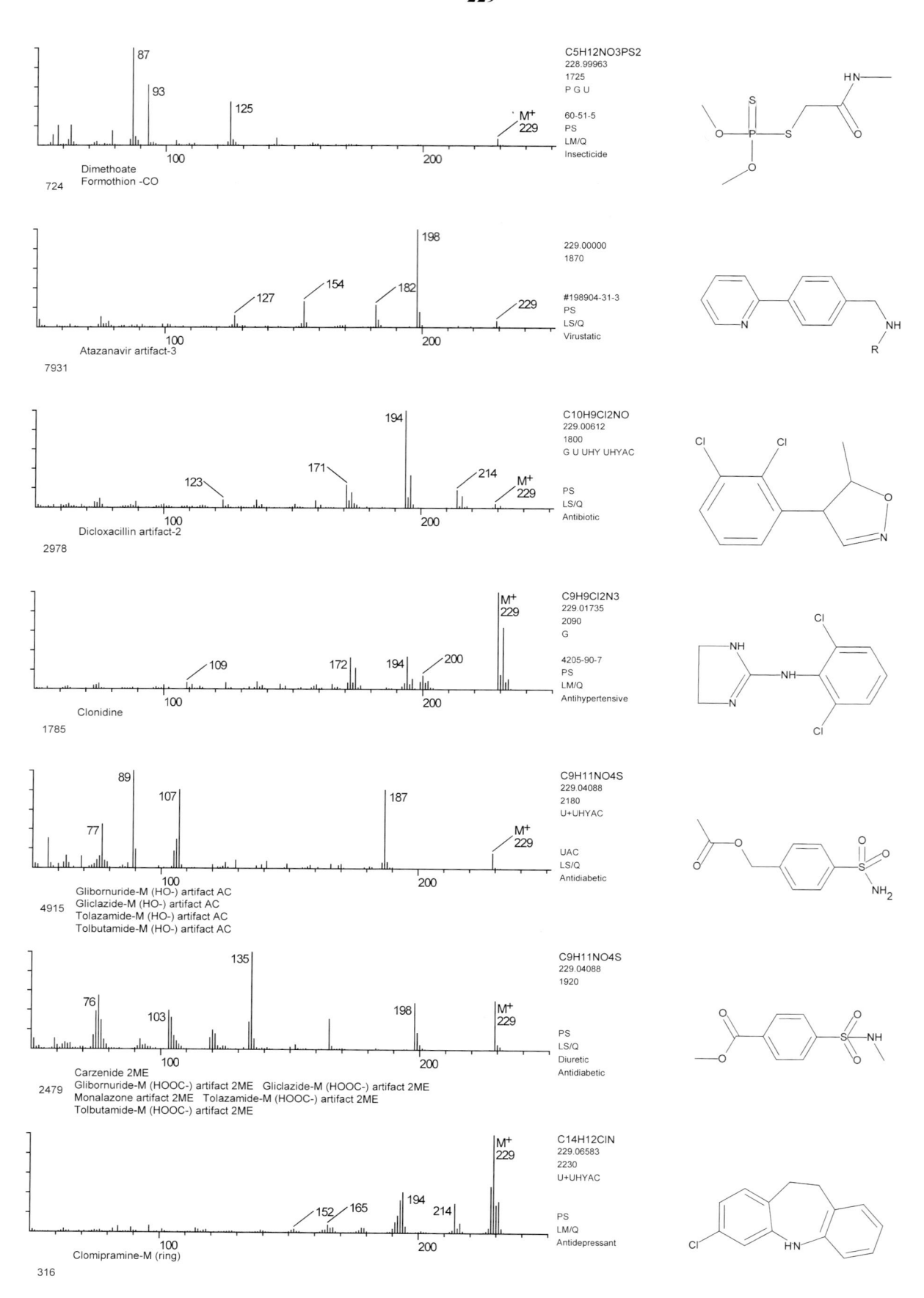

Dimethoate / Formothion -CO	C5H12NO3PS2 228.99963 1725 P G U 60-51-5 PS LM/Q Insecticide
Atazanavir artifact-3	229.00000 1870 #198904-31-3 PS LS/Q Virustatic
Dicloxacillin artifact-2	C10H9Cl2NO 229.00612 1800 G U UHY UHYAC PS LS/Q Antibiotic
Clonidine	C9H9Cl2N3 229.01735 2090 G 4205-90-7 PS LM/Q Antihypertensive
Glibornuride-M (HO-) artifact AC Gliclazide-M (HO-) artifact AC Tolazamide-M (HO-) artifact AC Tolbutamide-M (HO-) artifact AC	C9H11NO4S 229.04088 2180 U+UHYAC UAC LS/Q Antidiabetic
Carzenide 2ME Glibornuride-M (HOOC-) artifact 2ME Gliclazide-M (HOOC-) artifact 2ME Monalazone artifact 2ME Tolazamide-M (HOOC-) artifact 2ME Tolbutamide-M (HOOC-) artifact 2ME	C9H11NO4S 229.04088 1920 PS LS/Q Diuretic Antidiabetic
Clomipramine-M (ring)	C14H12ClN 229.06583 2230 U+UHYAC PS LM/Q Antidepressant

229

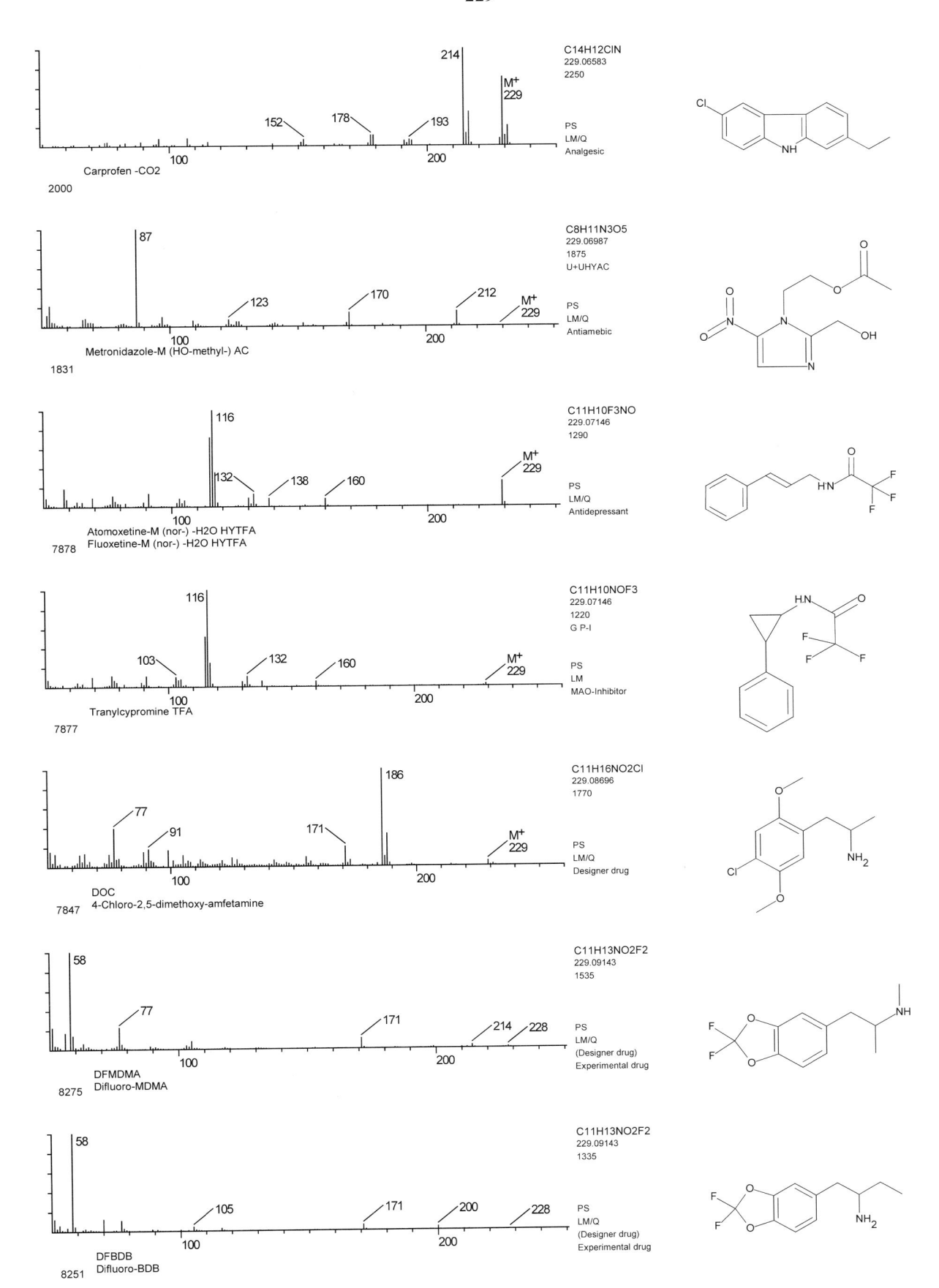

	214, 152, 178, 193, M+ 229	C14H12ClN 229.06583 2250 PS LM/Q Analgesic
Carprofen -CO2 2000		
	87, 123, 170, 212, M+ 229	C8H11N3O5 229.06987 1875 U+UHYAC PS LM/Q Antiamebic
Metronidazole-M (HO-methyl-) AC 1831		
	116, 132, 138, 160, M+ 229	C11H10F3NO 229.07146 1290 PS LM/Q Antidepressant
Atomoxetine-M (nor-) -H2O HYTFA Fluoxetine-M (nor-) -H2O HYTFA 7878		
	116, 103, 132, 160, M+ 229	C11H10NOF3 229.07146 1220 G P-I PS LM MAO-Inhibitor
Tranylcypromine TFA 7877		
	186, 77, 91, 171, M+ 229	C11H16NO2Cl 229.08696 1770 PS LM/Q Designer drug
DOC 7847 4-Chloro-2,5-dimethoxy-amfetamine		
	58, 77, 171, 214, 228	C11H13NO2F2 229.09143 1535 PS LM/Q (Designer drug) Experimental drug
DFMDMA 8275 Difluoro-MDMA		
	58, 105, 171, 200, 228	C11H13NO2F2 229.09143 1335 PS LM/Q (Designer drug) Experimental drug
DFBDB 8251 Difluoro-BDB		

229

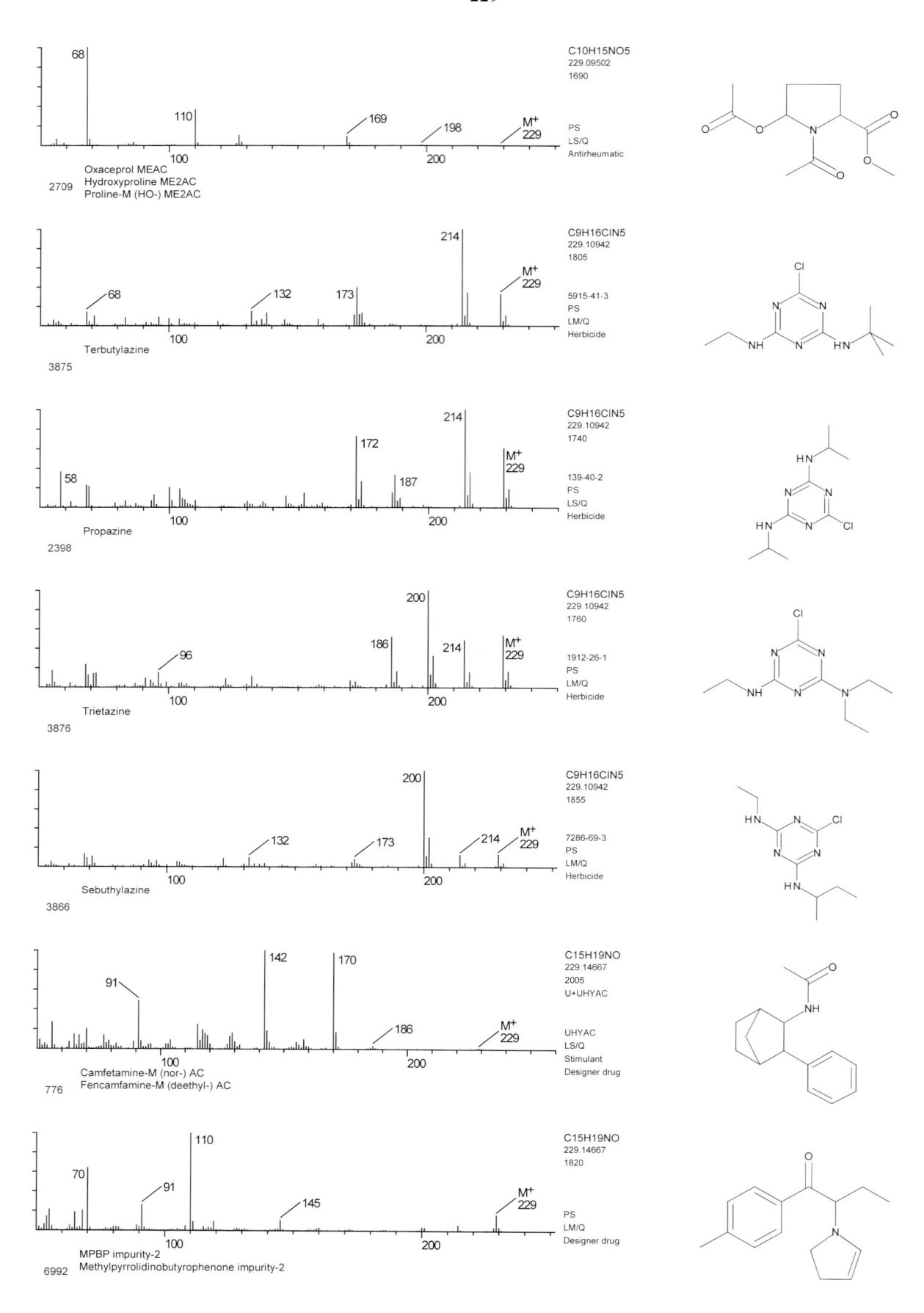

Oxaceprol MEAC / Hydroxyproline ME2AC / Proline-M (HO-) ME2AC — 2709	Peaks: 68, 110, 169, 198, M+ 229	C10H15NO5, 229.09502, 1690, PS, LS/Q, Antirheumatic
Terbutylazine — 3875	Peaks: 68, 132, 173, 214, M+ 229	C9H16ClN5, 229.10942, 1805, 5915-41-3, PS, LM/Q, Herbicide
Propazine — 2398	Peaks: 58, 172, 187, 214, M+ 229	C9H16ClN5, 229.10942, 1740, 139-40-2, PS, LS/Q, Herbicide
Trietazine — 3876	Peaks: 96, 186, 200, 214, M+ 229	C9H16ClN5, 229.10942, 1760, 1912-26-1, PS, LM/Q, Herbicide
Sebuthylazine — 3866	Peaks: 132, 173, 200, 214, M+ 229	C9H16ClN5, 229.10942, 1855, 7286-69-3, PS, LM/Q, Herbicide
Camfetamine-M (nor-) AC / Fencamfamine-M (deethyl-) AC — 776	Peaks: 91, 142, 170, 186, M+ 229	C15H19NO, 229.14667, 2005, U+UHYAC, UHYAC, LS/Q, Stimulant, Designer drug
MPBP impurity-2 / Methylpyrrolidinobutyrophenone impurity-2 — 6992	Peaks: 70, 91, 110, 145, M+ 229	C15H19NO, 229.14667, 1820, PS, LM/Q, Designer drug

229

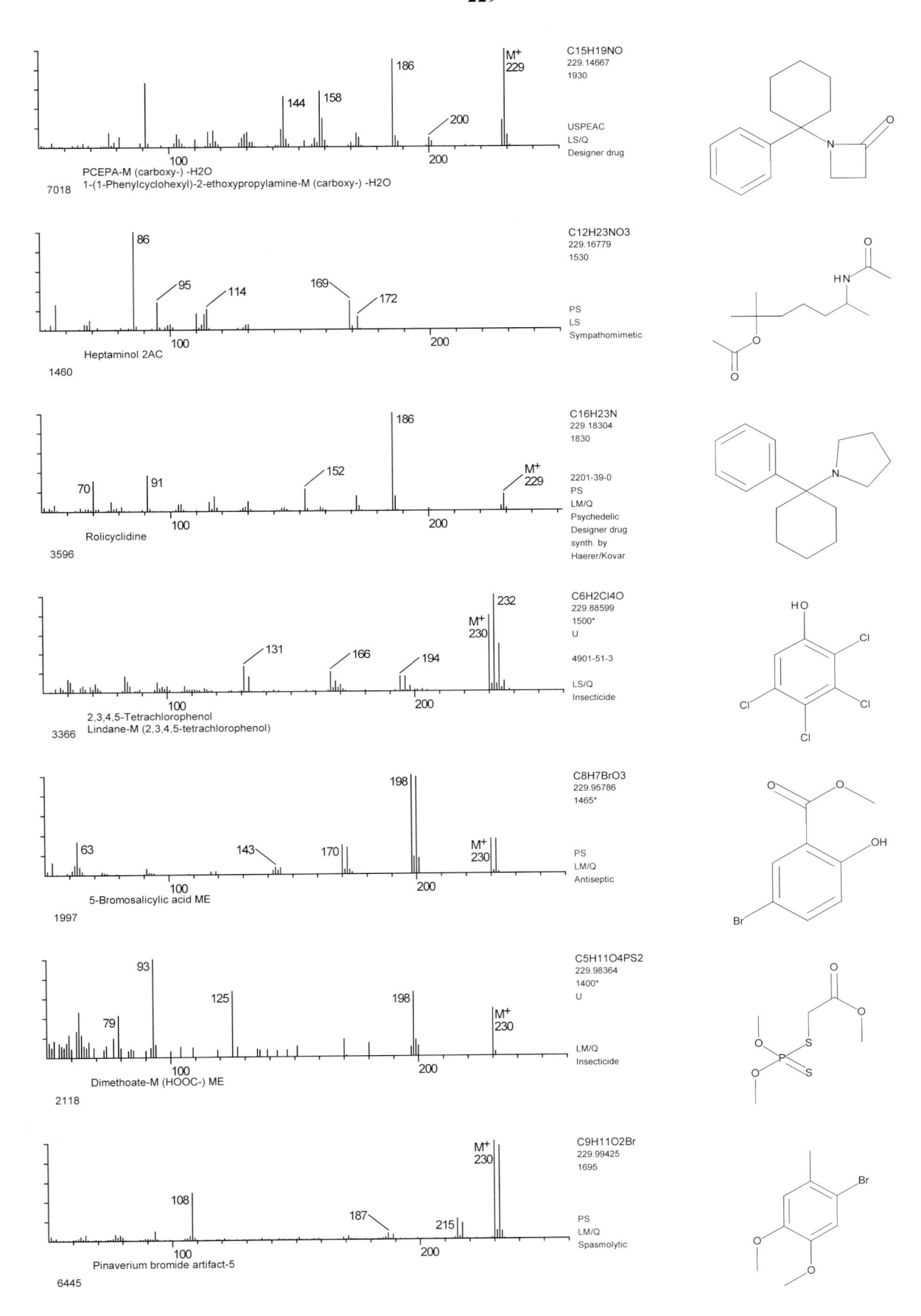

C15H19NO
229.14667
1930

USPEAC
LS/Q
Designer drug

PCEPA-M (carboxy-) -H2O
1-(1-Phenylcyclohexyl)-2-ethoxypropylamine-M (carboxy-) -H2O
7018

C12H23NO3
229.16779
1530

PS
LS
Sympathomimetic

Heptaminol 2AC
1460

C16H23N
229.18304
1830

2201-39-0
PS
LM/Q
Psychedelic
Designer drug
synth. by
Haerer/Kovar

Rolicyclidine
3596

C6H2Cl4O
229.88599
1500*
U

4901-51-3
LS/Q
Insecticide

2,3,4,5-Tetrachlorophenol
Lindane-M (2,3,4,5-tetrachlorophenol)
3366

C8H7BrO3
229.95786
1465*

PS
LM/Q
Antiseptic

5-Bromosalicylic acid ME
1997

C5H11O4PS2
229.98364
1400*
U

LM/Q
Insecticide

Dimethoate-M (HOOC-) ME
2118

C9H11O2Br
229.99425
1695

PS
LM/Q
Spasmolytic

Pinaverium bromide artifact-5
6445

230

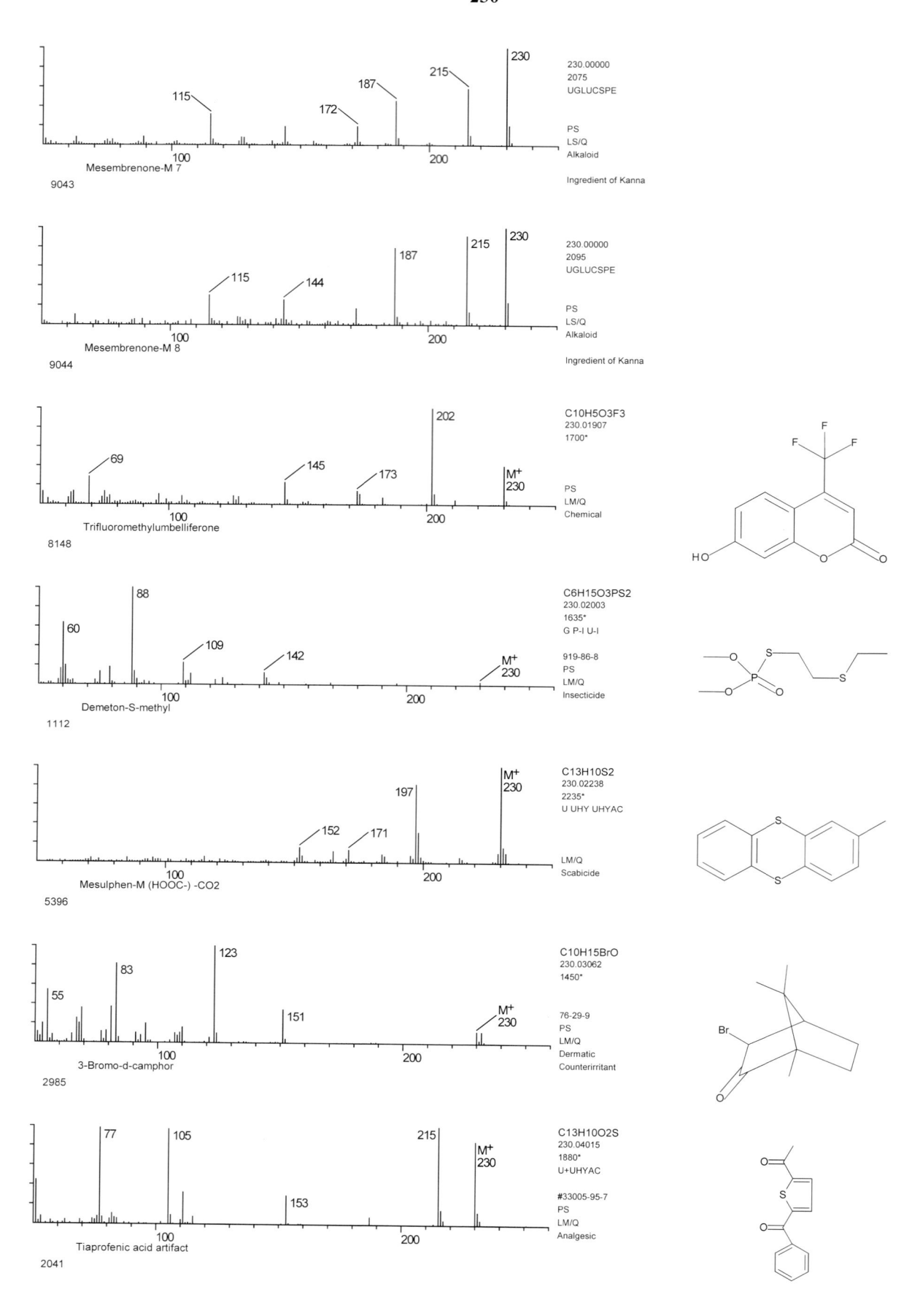

Mesembrenone-M 7	peaks: 115, 172, 187, 215, 230	230.00000 / 2075 / UGLUCSPE / PS / LS/Q / Alkaloid / Ingredient of Kanna
9043		
Mesembrenone-M 8	peaks: 115, 144, 187, 215, 230	230.00000 / 2095 / UGLUCSPE / PS / LS/Q / Alkaloid / Ingredient of Kanna
9044		
Trifluoromethylumbelliferone	peaks: 69, 145, 173, 202, M+ 230	C10H5O3F3 / 230.01907 / 1700* / PS / LM/Q / Chemical
8148		
Demeton-S-methyl	peaks: 60, 88, 109, 142, M+ 230	C6H15O3PS2 / 230.02003 / 1635* / G P-I U-I / 919-86-8 / PS / LM/Q / Insecticide
1112		
Mesulphen-M (HOOC-) -CO2	peaks: 152, 171, 197, M+ 230	C13H10S2 / 230.02238 / 2235* / U UHY UHYAC / LM/Q / Scabicide
5396		
3-Bromo-d-camphor	peaks: 55, 83, 123, 151, M+ 230	C10H15BrO / 230.03062 / 1450* / 76-29-9 / PS / LM/Q / Dermatic Counterirritant
2985		
Tiaprofenic acid artifact	peaks: 77, 105, 153, 215, M+ 230	C13H10O2S / 230.04015 / 1880* / U+UHYAC / #33005-95-7 / PS / LM/Q / Analgesic
2041		

230

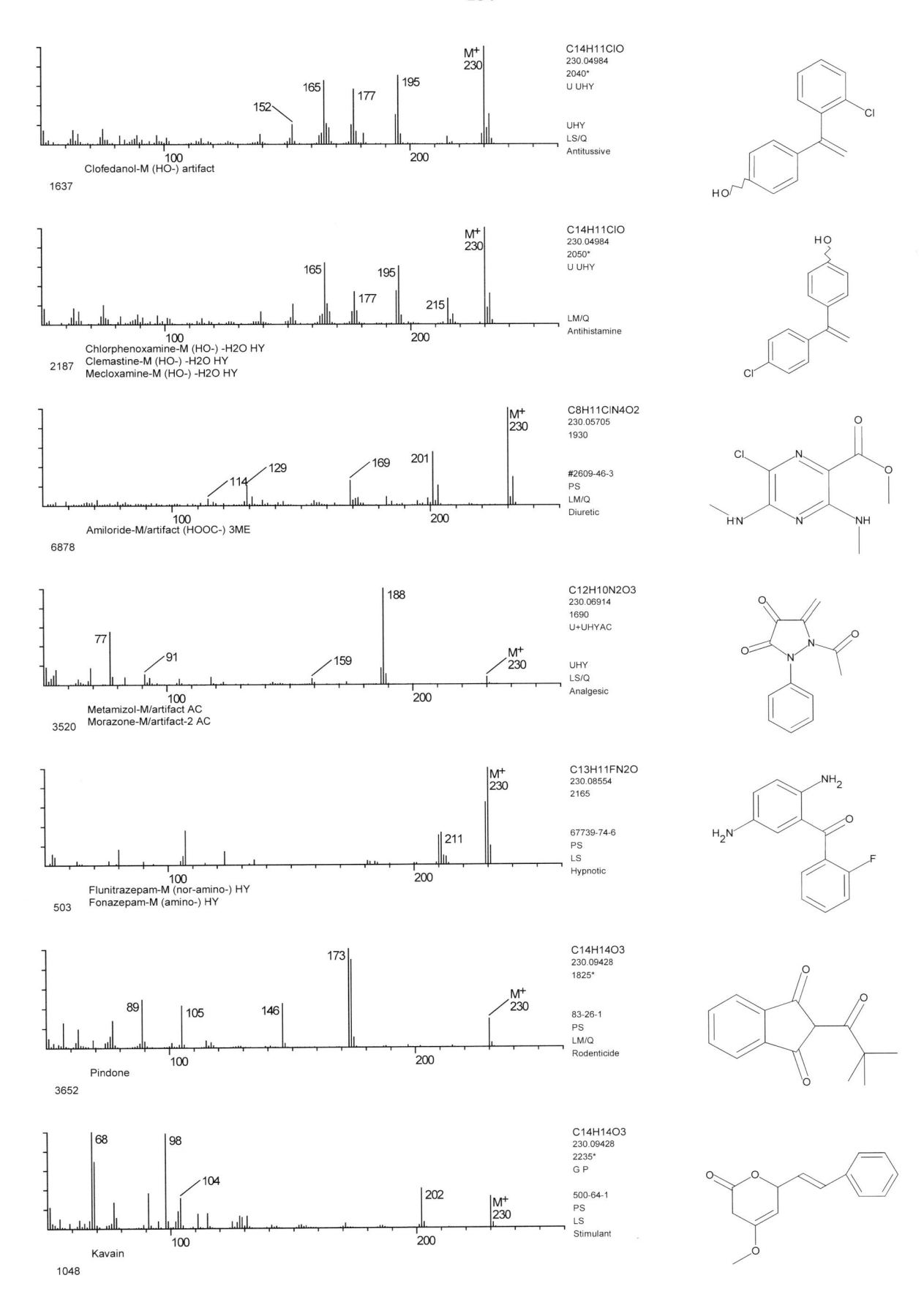

Clofedanol-M (HO-) artifact
1637

C14H11ClO
230.04984
2040*
U UHY

UHY
LS/Q
Antitussive

Chlorphenoxamine-M (HO-) -H2O HY
Clemastine-M (HO-) -H2O HY
Mecloxamine-M (HO-) -H2O HY
2187

C14H11ClO
230.04984
2050*
U UHY

LM/Q
Antihistamine

Amiloride-M/artifact (HOOC-) 3ME
6878

C8H11ClN4O2
230.05705
1930

#2609-46-3
PS
LM/Q
Diuretic

Metamizol-M/artifact AC
Morazone-M/artifact-2 AC
3520

C12H10N2O3
230.06914
1690
U+UHYAC

UHY
LS/Q
Analgesic

Flunitrazepam-M (nor-amino-) HY
Fonazepam-M (amino-) HY
503

C13H11FN2O
230.08554
2165

67739-74-6
PS
LS
Hypnotic

Pindone
3652

C14H14O3
230.09428
1825*

83-26-1
PS
LM/Q
Rodenticide

Kavain
1048

C14H14O3
230.09428
2235*
G P

500-64-1
PS
LS
Stimulant

230

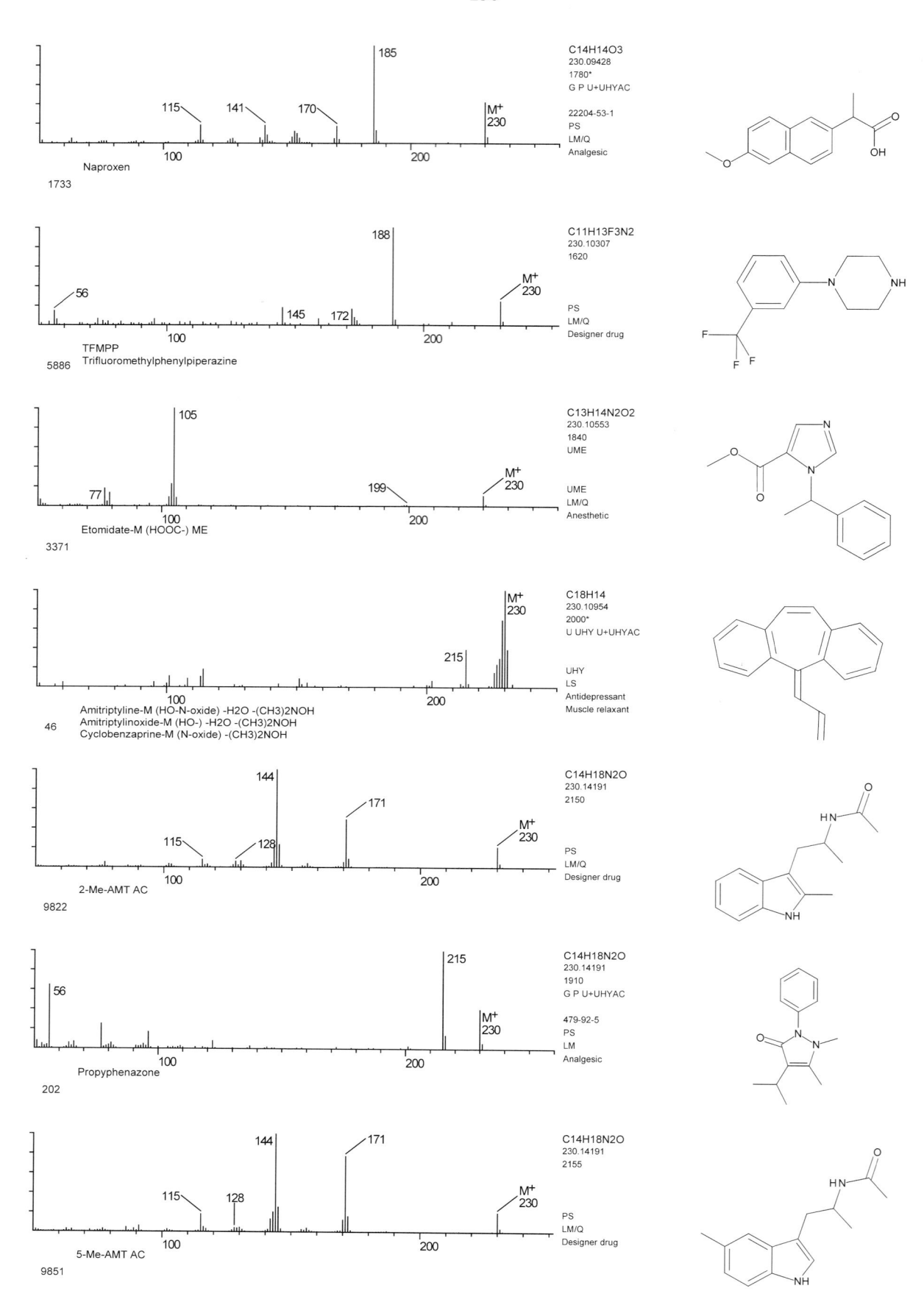

Spectrum	Compound	Formula / Info
Peaks: 115, 141, 170, 185, M+ 230	Naproxen	C14H14O3 230.09428 1780* G P U+UHYAC 22204-53-1 PS LM/Q Analgesic
Peaks: 56, 145, 172, 188, M+ 230	TFMPP Trifluoromethylphenylpiperazine	C11H13F3N2 230.10307 1620 PS LM/Q Designer drug
Peaks: 77, 105, 199, M+ 230	Etomidate-M (HOOC-) ME	C13H14N2O2 230.10553 1840 UME UME LM/Q Anesthetic
Peaks: 215, M+ 230	Amitriptyline-M (HO-N-oxide) -H2O -(CH3)2NOH Amitriptylinoxide-M (HO-) -H2O -(CH3)2NOH Cyclobenzaprine-M (N-oxide) -(CH3)2NOH	C18H14 230.10954 2000* U UHY U+UHYAC UHY LS Antidepressant Muscle relaxant
Peaks: 115, 128, 144, 171, M+ 230	2-Me-AMT AC	C14H18N2O 230.14191 2150 PS LM/Q Designer drug
Peaks: 56, 215, M+ 230	Propyphenazone	C14H18N2O 230.14191 1910 G P U+UHYAC 479-92-5 PS LM Analgesic
Peaks: 115, 128, 144, 171, M+ 230	5-Me-AMT AC	C14H18N2O 230.14191 2155 PS LM/Q Designer drug

230

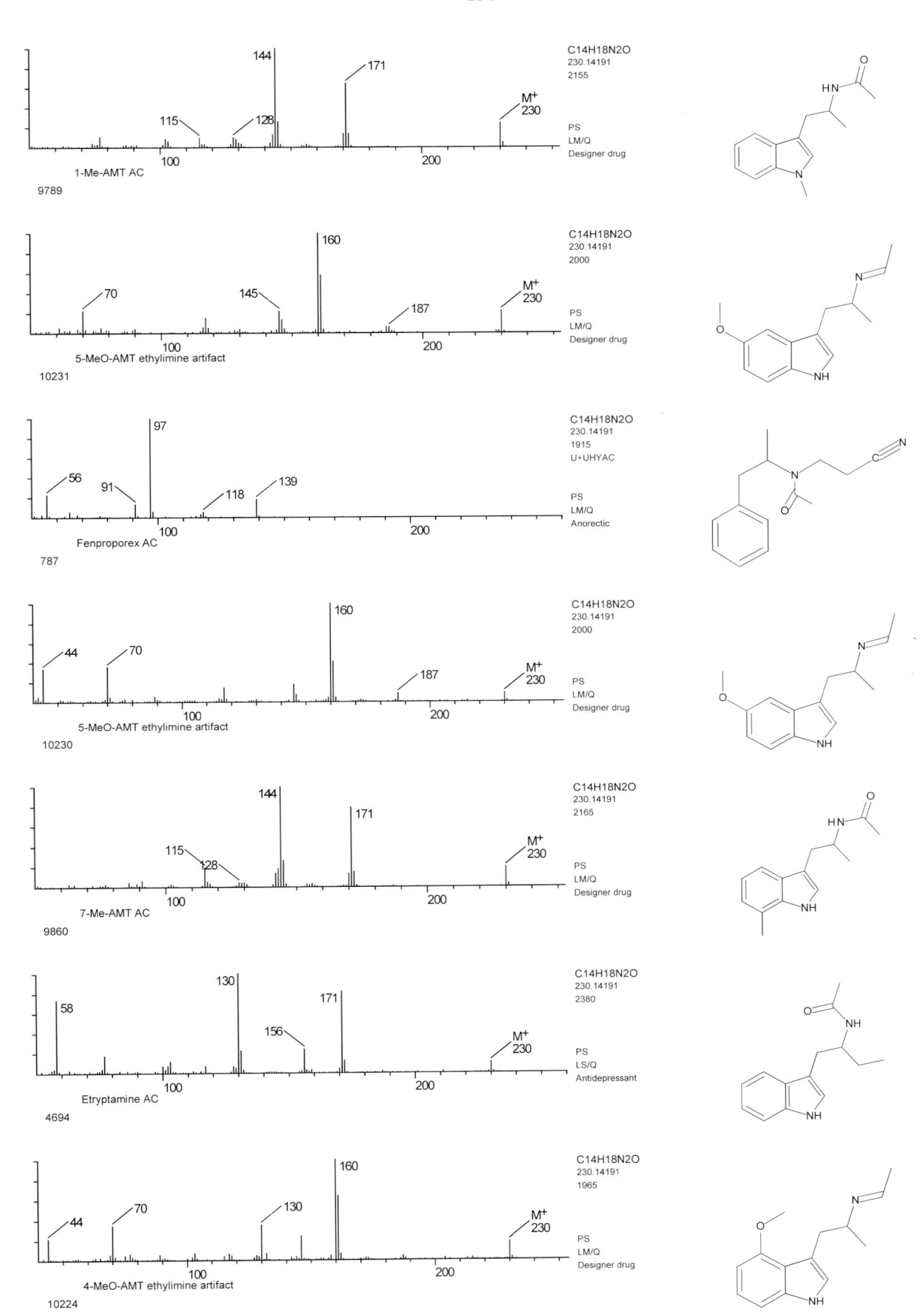

Spectrum	Formula	Compound
9789	C14H18N2O 230.14191 2155 PS LM/Q Designer drug	1-Me-AMT AC — peaks 115, 128, 144, 171, M+ 230
10231	C14H18N2O 230.14191 2000 PS LM/Q Designer drug	5-MeO-AMT ethylimine artifact — peaks 70, 145, 160, 187, M+ 230
787	C14H18N2O 230.14191 1915 U+UHYAC PS LM/Q Anorectic	Fenproporex AC — peaks 56, 91, 97, 118, 139
10230	C14H18N2O 230.14191 2000 PS LM/Q Designer drug	5-MeO-AMT ethylimine artifact — peaks 44, 70, 160, 187, M+ 230
9860	C14H18N2O 230.14191 2165 PS LM/Q Designer drug	7-Me-AMT AC — peaks 115, 128, 144, 171, M+ 230
4694	C14H18N2O 230.14191 2380 PS LS/Q Antidepressant	Etryptamine AC — peaks 58, 130, 156, 171, M+ 230
10224	C14H18N2O 230.14191 1965 PS LM/Q Designer drug	4-MeO-AMT ethylimine artifact — peaks 44, 70, 130, 160, M+ 230

230

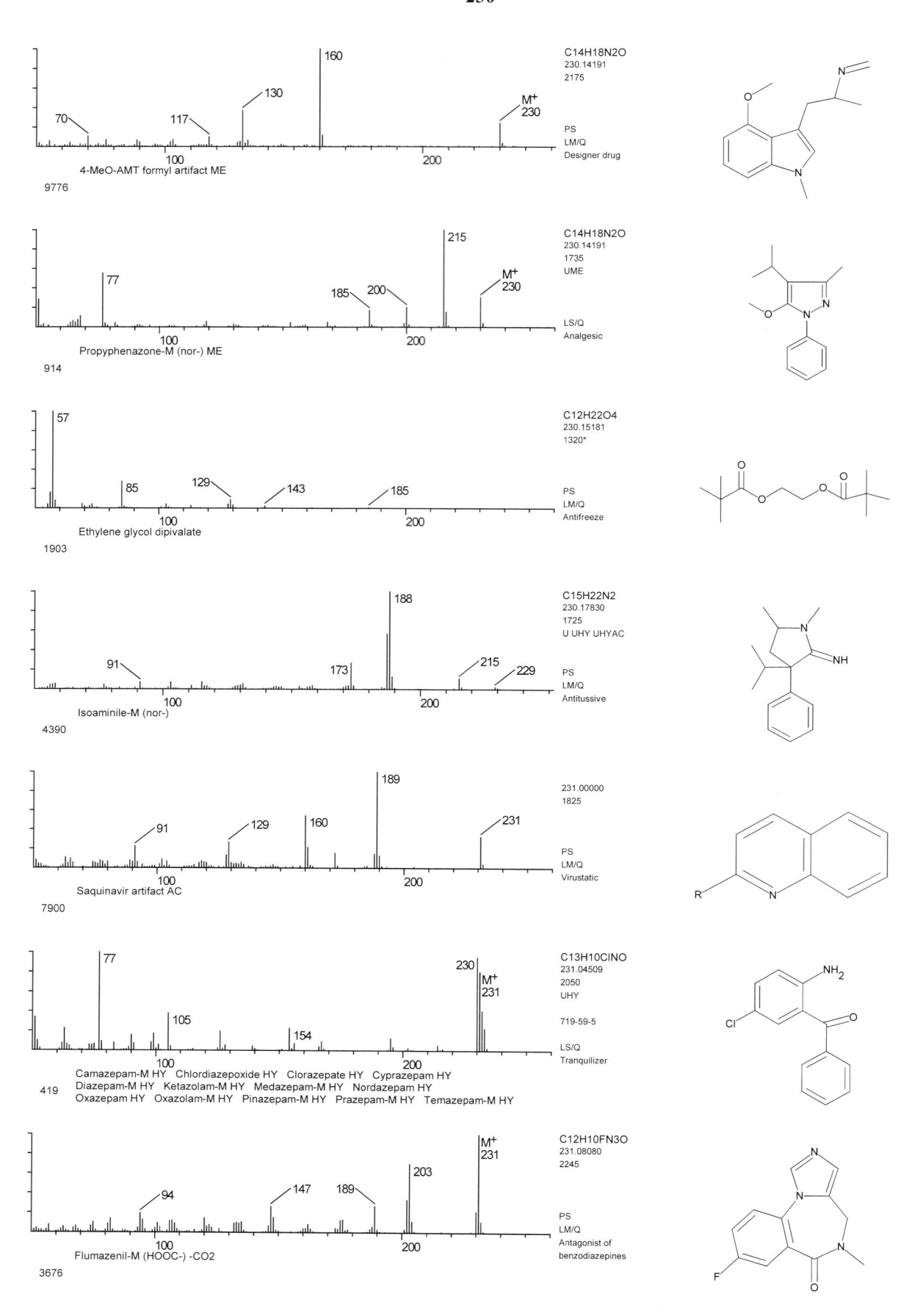

4-MeO-AMT formyl artifact ME 9776	Peaks: 70, 117, 130, 160, M+ 230	C14H18N2O 230.14191 2175 PS LM/Q Designer drug
Propyphenazone-M (nor-) ME 914	Peaks: 77, 185, 200, 215, M+ 230	C14H18N2O 230.14191 1735 UME LS/Q Analgesic
Ethylene glycol dipivalate 1903	Peaks: 57, 85, 129, 143, 185	C12H22O4 230.15181 1320* PS LM/Q Antifreeze
Isoaminile-M (nor-) 4390	Peaks: 91, 173, 188, 215, 229	C15H22N2 230.17830 1725 U UHY UHYAC PS LM/Q Antitussive
Saquinavir artifact AC 7900	Peaks: 91, 129, 160, 189, 231	231.00000 1825 PS LM/Q Virustatic
Camazepam-M HY Chlordiazepoxide HY Clorazepate HY Cyprazepam HY Diazepam-M HY Ketazolam HY Medazepam HY Nordazepam HY Oxazepam HY Oxazolam-M HY Pinazepam-M HY Prazepam-M HY Temazepam-M HY 419	Peaks: 77, 105, 154, 230, M+ 231	C13H10ClNO 231.04509 2050 UHY 719-59-5 LS/Q Tranquilizer
Flumazenil-M (HOOC-) -CO2 3676	Peaks: 94, 147, 189, 203, M+ 231	C12H10FN3O 231.08080 2245 PS LM/Q Antagonist of benzodiazepines

512

231

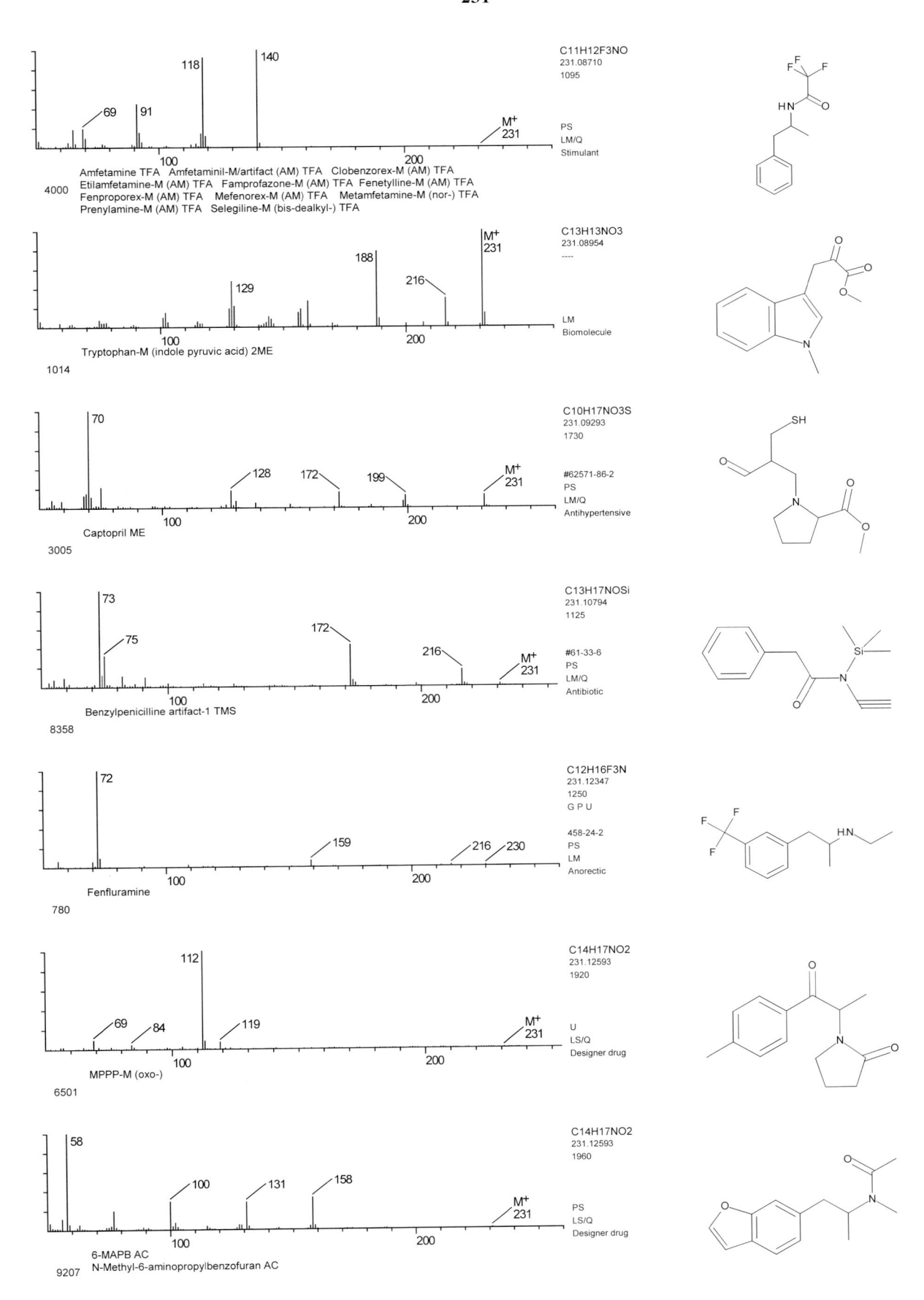

Spectrum peaks	Formula / Info	Compound
69, 91, 118, 140, M⁺ 231	C11H12F3NO — 231.08710 — 1095 — PS — LM/Q — Stimulant	Amfetamine TFA Amfetaminil-M/artifact (AM) TFA Clobenzorex-M (AM) TFA Etilamfetamine-M (AM) TFA Famprofazone-M (AM) TFA Fenetylline-M (AM) TFA Fenproporex-M (AM) TFA Mefenorex-M (AM) TFA Metamfetamine-M (nor-) TFA Prenylamine-M (AM) TFA Selegiline-M (bis-dealkyl-) TFA — 4000
129, 188, 216, M⁺ 231	C13H13NO3 — 231.08954 — ---- — LM — Biomolecule	Tryptophan-M (indole pyruvic acid) 2ME — 1014
70, 128, 172, 199, M⁺ 231	C10H17NO3S — 231.09293 — 1730 — #62571-86-2 — PS — LM/Q — Antihypertensive	Captopril ME — 3005
73, 75, 172, 216, M⁺ 231	C13H17NOSi — 231.10794 — 1125 — #61-33-6 — PS — LM/Q — Antibiotic	Benzylpenicilline artifact-1 TMS — 8358
72, 159, 216, 230	C12H16F3N — 231.12347 — 1250 — G P U — 458-24-2 — PS — LM — Anorectic	Fenfluramine — 780
69, 84, 112, 119, M⁺ 231	C14H17NO2 — 231.12593 — 1920 — U — LS/Q — Designer drug	MPPP-M (oxo-) — 6501
58, 100, 131, 158, M⁺ 231	C14H17NO2 — 231.12593 — 1960 — PS — LS/Q — Designer drug	6-MAPB AC N-Methyl-6-aminopropylbenzofuran AC — 9207

231

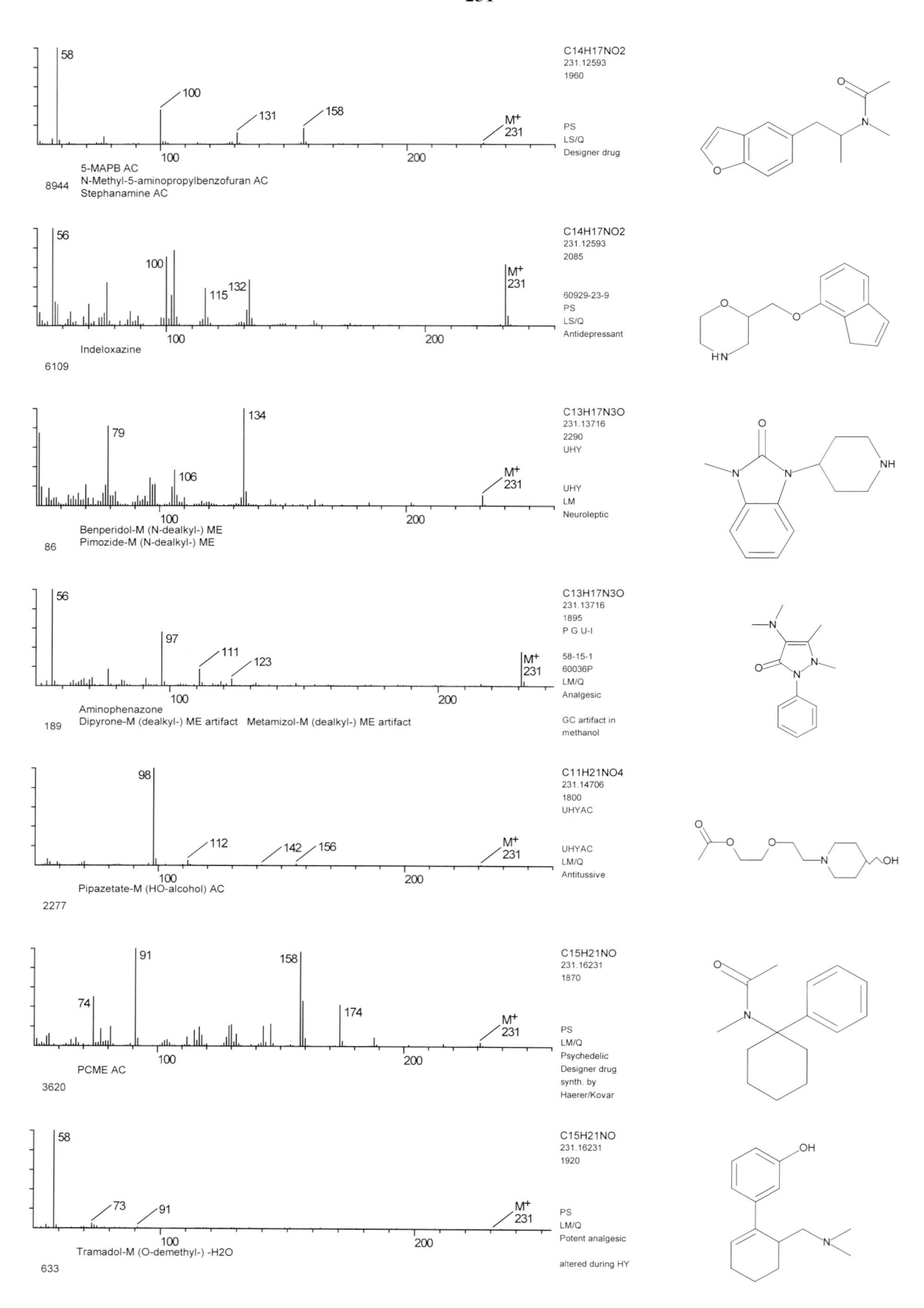

8944	58, 100, 131, 158, M+ 231 5-MAPB AC N-Methyl-5-aminopropylbenzofuran AC Stephanamine AC	C14H17NO2 231.12593 1960 PS LS/Q Designer drug
6109	56, 100, 115, 132, M+ 231 Indeloxazine	C14H17NO2 231.12593 2085 60929-23-9 PS LS/Q Antidepressant
86	79, 106, 134, M+ 231 Benperidol-M (N-dealkyl-) ME Pimozide-M (N-dealkyl-) ME	C13H17N3O 231.13716 2290 UHY UHY LM Neuroleptic
189	56, 97, 111, 123, M+ 231 Aminophenazone Dipyrone-M (dealkyl-) ME artifact Metamizol-M (dealkyl-) ME artifact	C13H17N3O 231.13716 1895 P G U-I 58-15-1 60036P LM/Q Analgesic GC artifact in methanol
2277	98, 112, 142, 156, M+ 231 Pipazetate-M (HO-alcohol) AC	C11H21NO4 231.14706 1800 UHYAC UHYAC LM/Q Antitussive
3620	74, 91, 158, 174, M+ 231 PCME AC	C15H21NO 231.16231 1870 PS LM/Q Psychedelic Designer drug synth. by Haerer/Kovar
633	58, 73, 91, M+ 231 Tramadol-M (O-demethyl-) -H2O	C15H21NO 231.16231 1920 PS LM/Q Potent analgesic altered during HY

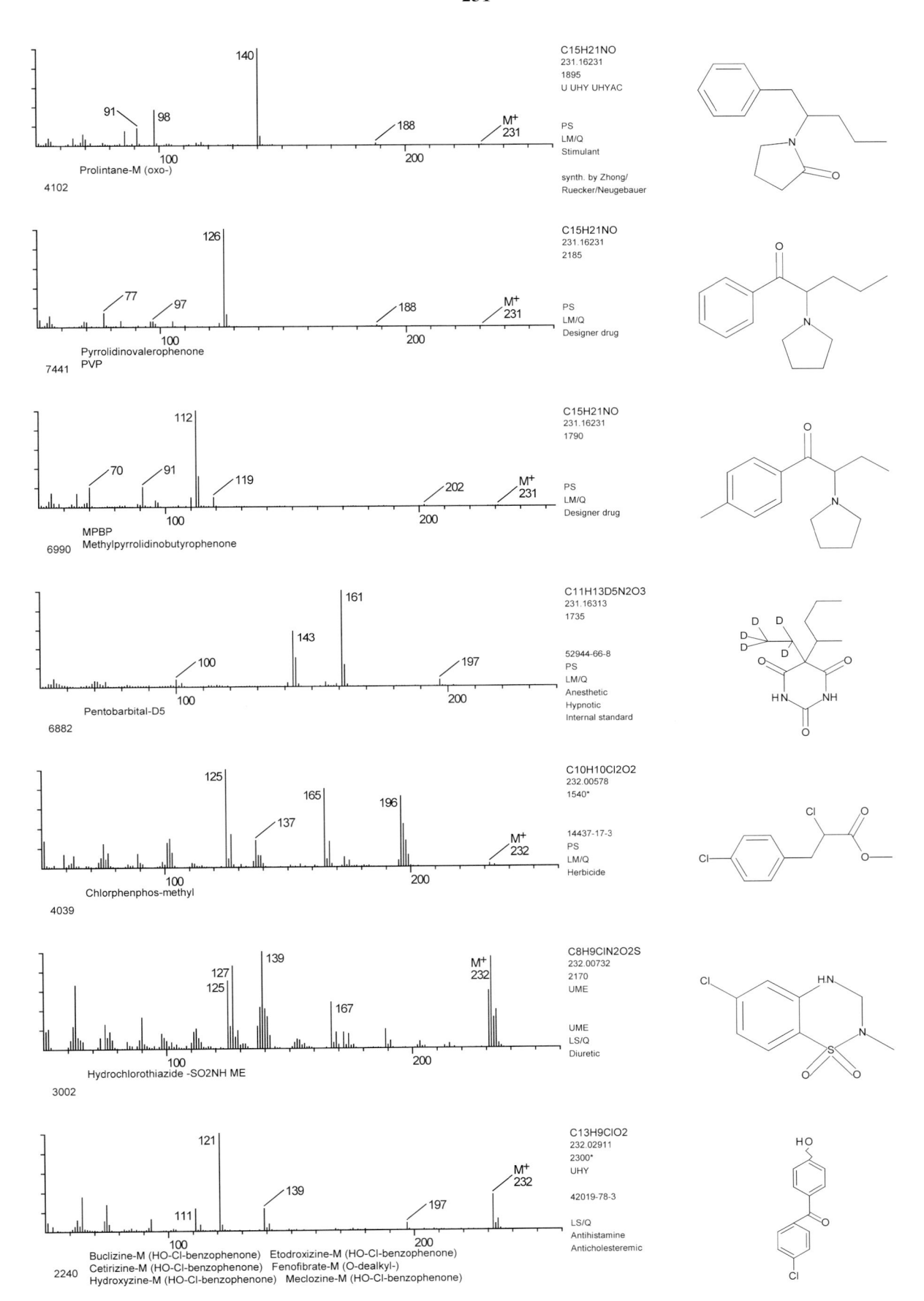

Prolintane-M (oxo-) 4102	C15H21NO 231.16231 1895 U UHY UHYAC PS LM/Q Stimulant synth. by Zhong/ Ruecker/Neugebauer
Pyrrolidinovalerophenone PVP 7441	C15H21NO 231.16231 2185 PS LM/Q Designer drug
MPBP Methylpyrrolidinobutyrophenone 6990	C15H21NO 231.16231 1790 PS LM/Q Designer drug
Pentobarbital-D5 6882	C11H13D5N2O3 231.16313 1735 52944-66-8 PS LM/Q Anesthetic Hypnotic Internal standard
Chlorphenphos-methyl 4039	C10H10Cl2O2 232.00578 1540* 14437-17-3 PS LM/Q Herbicide
Hydrochlorothiazide -SO2NH ME 3002	C8H9ClN2O2S 232.00732 2170 UME UME LS/Q Diuretic
Buclizine-M (HO-Cl-benzophenone) Etodroxizine-M (HO-Cl-benzophenone) Cetirizine-M (HO-Cl-benzophenone) Fenofibrate-M (O-dealkyl-) Hydroxyzine-M (HO-Cl-benzophenone) Meclozine-M (HO-Cl-benzophenone) 2240	C13H9ClO2 232.02911 2300* UHY 42019-78-3 LS/Q Antihistamine Anticholesteremic

232

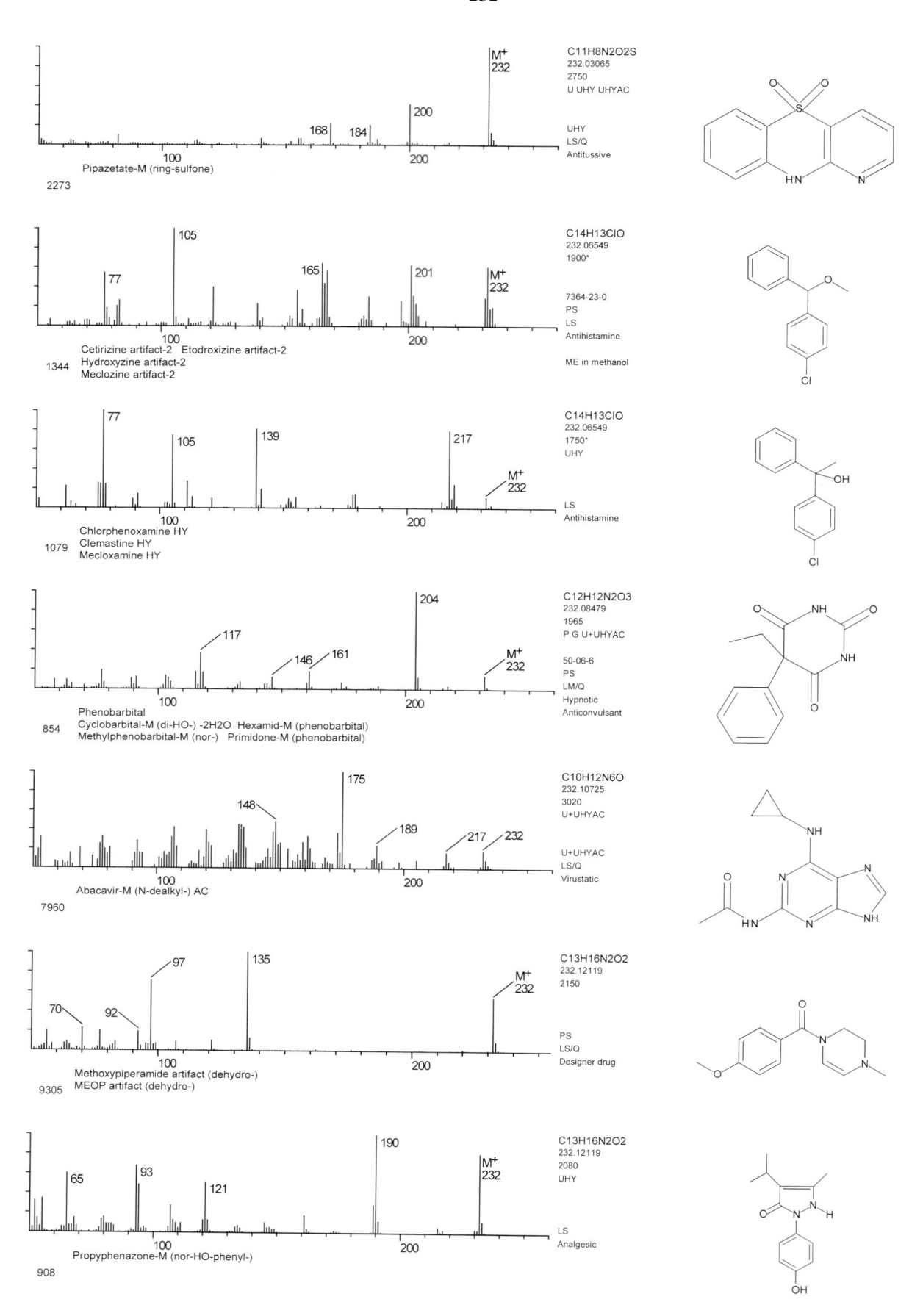

Pipazetate-M (ring-sulfone)
2273
C11H8N2O2S
232.03065
2750
U UHY UHYAC
UHY
LS/Q
Antitussive

Cetirizine artifact-2 Etodroxizine artifact-2
Hydroxyzine artifact-2
Meclozine artifact-2
1344
C14H13ClO
232.06549
1900*
7364-23-0
PS
LS
Antihistamine
ME in methanol

Chlorphenoxamine HY
Clemastine HY
Mecloxamine HY
1079
C14H13ClO
232.06549
1750*
UHY
LS
Antihistamine

Phenobarbital
Cyclobarbital-M (di-HO-) -2H2O Hexamid-M (phenobarbital)
Methylphenobarbital-M (nor-) Primidone-M (phenobarbital)
854
C12H12N2O3
232.08479
1965
P G U+UHYAC
50-06-6
PS
LM/Q
Hypnotic
Anticonvulsant

Abacavir-M (N-dealkyl-) AC
7960
C10H12N6O
232.10725
3020
U+UHYAC
U+UHYAC
LS/Q
Virustatic

Methoxypiperamide artifact (dehydro-)
MEOP artifact (dehydro-)
9305
C13H16N2O2
232.12119
2150
PS
LS/Q
Designer drug

Propyphenazone-M (nor-HO-phenyl-)
908
C13H16N2O2
232.12119
2080
UHY
LS
Analgesic

516

232

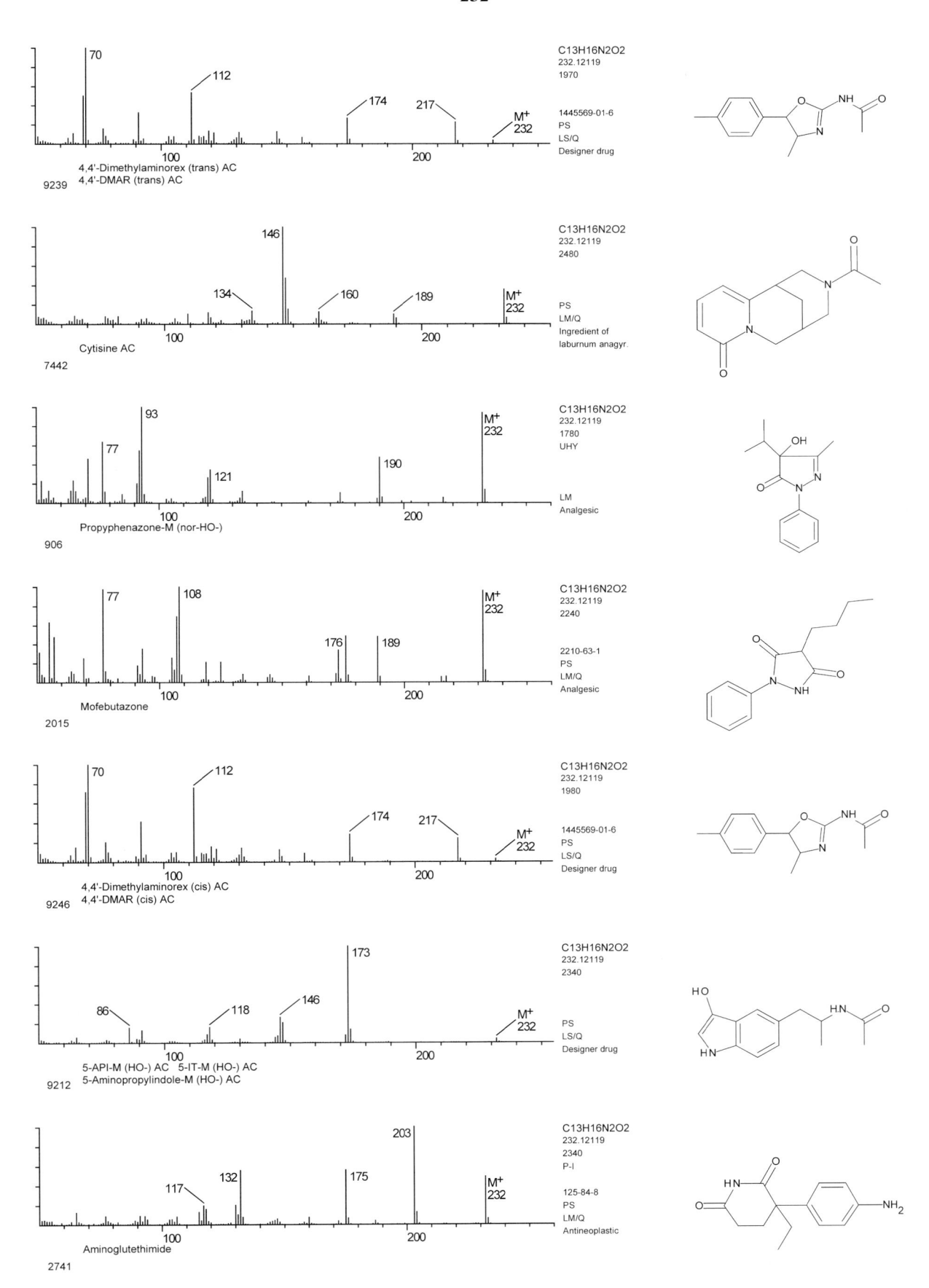

9239	4,4'-Dimethylaminorex (trans) AC 4,4'-DMAR (trans) AC Peaks: 70, 112, 174, 217, M+ 232	C13H16N2O2 232.12119 1970 1445569-01-6 PS LS/Q Designer drug
7442	Cytisine AC Peaks: 134, 146, 160, 189, M+ 232	C13H16N2O2 232.12119 2480 PS LM/Q Ingredient of laburnum anagyr.
906	Propyphenazone-M (nor-HO-) Peaks: 77, 93, 121, 190, M+ 232	C13H16N2O2 232.12119 1780 UHY LM Analgesic
2015	Mofebutazone Peaks: 77, 108, 176, 189, M+ 232	C13H16N2O2 232.12119 2240 2210-63-1 PS LM/Q Analgesic
9246	4,4'-Dimethylaminorex (cis) AC 4,4'-DMAR (cis) AC Peaks: 70, 112, 174, 217, M+ 232	C13H16N2O2 232.12119 1980 1445569-01-6 PS LS/Q Designer drug
9212	5-API-M (HO-) AC 5-IT-M (HO-) AC 5-Aminopropylindole-M (HO-) AC Peaks: 86, 118, 146, 173, M+ 232	C13H16N2O2 232.12119 2340 PS LS/Q Designer drug
2741	Aminoglutethimide Peaks: 117, 132, 175, 203, M+ 232	C13H16N2O2 232.12119 2340 P-I 125-84-8 PS LM/Q Antineoplastic

232

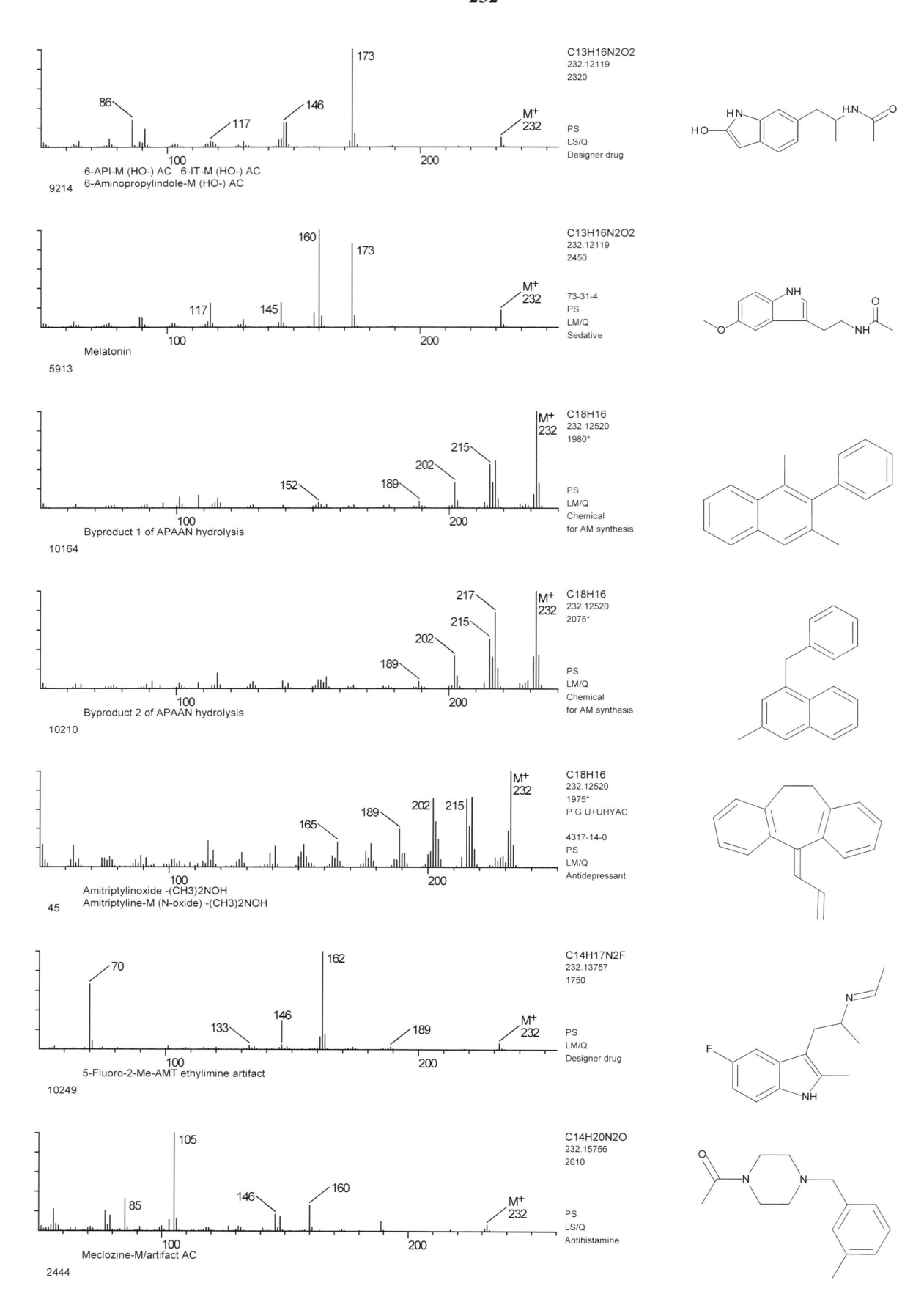

Spectrum label	Formula / info
6-API-M (HO-) AC 6-IT-M (HO-) AC 6-Aminopropylindole-M (HO-) AC 9214	C13H16N2O2 232.12119 2320 PS LS/Q Designer drug
Melatonin 5913	C13H16N2O2 232.12119 2450 73-31-4 PS LM/Q Sedative
Byproduct 1 of APAAN hydrolysis 10164	C18H16 232.12520 1980* PS LM/Q Chemical for AM synthesis
Byproduct 2 of APAAN hydrolysis 10210	C18H16 232.12520 2075* PS LM/Q Chemical for AM synthesis
Amitriptylinoxide -(CH3)2NOH Amitriptyline-M (N-oxide) -(CH3)2NOH 45	C18H16 232.12520 1975* P G U+UHYAC 4317-14-0 PS LM/Q Antidepressant
5-Fluoro-2-Me-AMT ethylimine artifact 10249	C14H17N2F 232.13757 1750 PS LM/Q Designer drug
Meclozine-M/artifact AC 2444	C14H20N2O 232.15756 2010 PS LS/Q Antihistamine

232

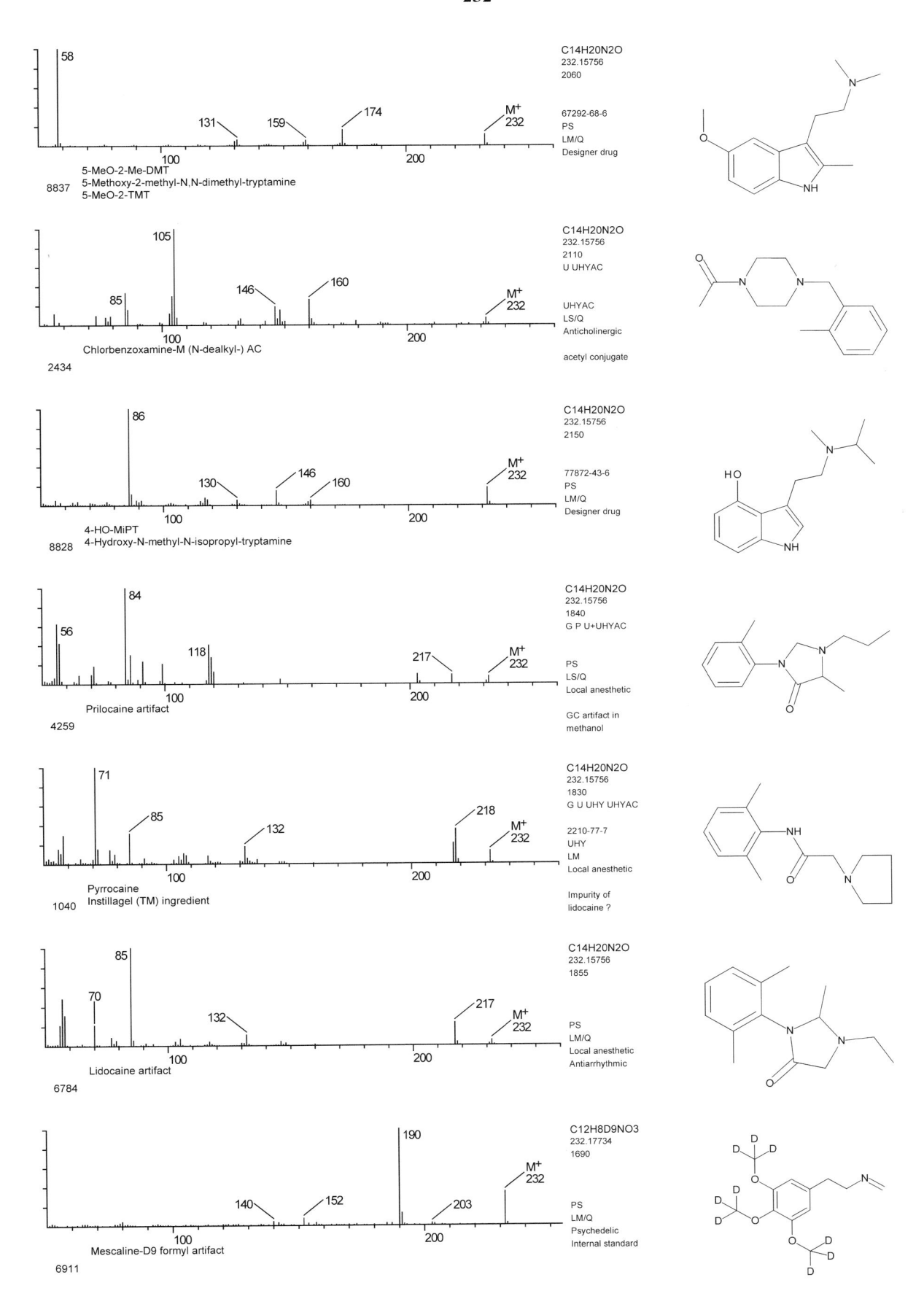

58, 131, 159, 174, M+ 232	C14H20N2O 232.15756 2060 67292-68-6 PS LM/Q Designer drug	

8837 5-MeO-2-Me-DMT
5-Methoxy-2-methyl-N,N-dimethyl-tryptamine
5-MeO-2-TMT

85, 105, 146, 160, M+ 232	C14H20N2O 232.15756 2110 U UHYAC UHYAC LS/Q Anticholinergic acetyl conjugate	

2434 Chlorbenzoxamine-M (N-dealkyl-) AC

86, 130, 146, 160, M+ 232	C14H20N2O 232.15756 2150 77872-43-6 PS LM/Q Designer drug	

8828 4-HO-MiPT
4-Hydroxy-N-methyl-N-isopropyl-tryptamine

56, 84, 118, 217, M+ 232	C14H20N2O 232.15756 1840 G P U+UHYAC PS LS/Q Local anesthetic GC artifact in methanol	

4259 Prilocaine artifact

71, 85, 132, 218, M+ 232	C14H20N2O 232.15756 1830 G U UHY UHYAC 2210-77-7 UHY LM Local anesthetic Impurity of lidocaine ?	

1040 Pyrrocaine
Instillagel (TM) ingredient

70, 85, 132, 217, M+ 232	C14H20N2O 232.15756 1855 PS LM/Q Local anesthetic Antiarrhythmic	

6784 Lidocaine artifact

140, 152, 190, 203, M+ 232	C12H8D9NO3 232.17734 1690 PS LM/Q Psychedelic Internal standard	

6911 Mescaline-D9 formyl artifact

232

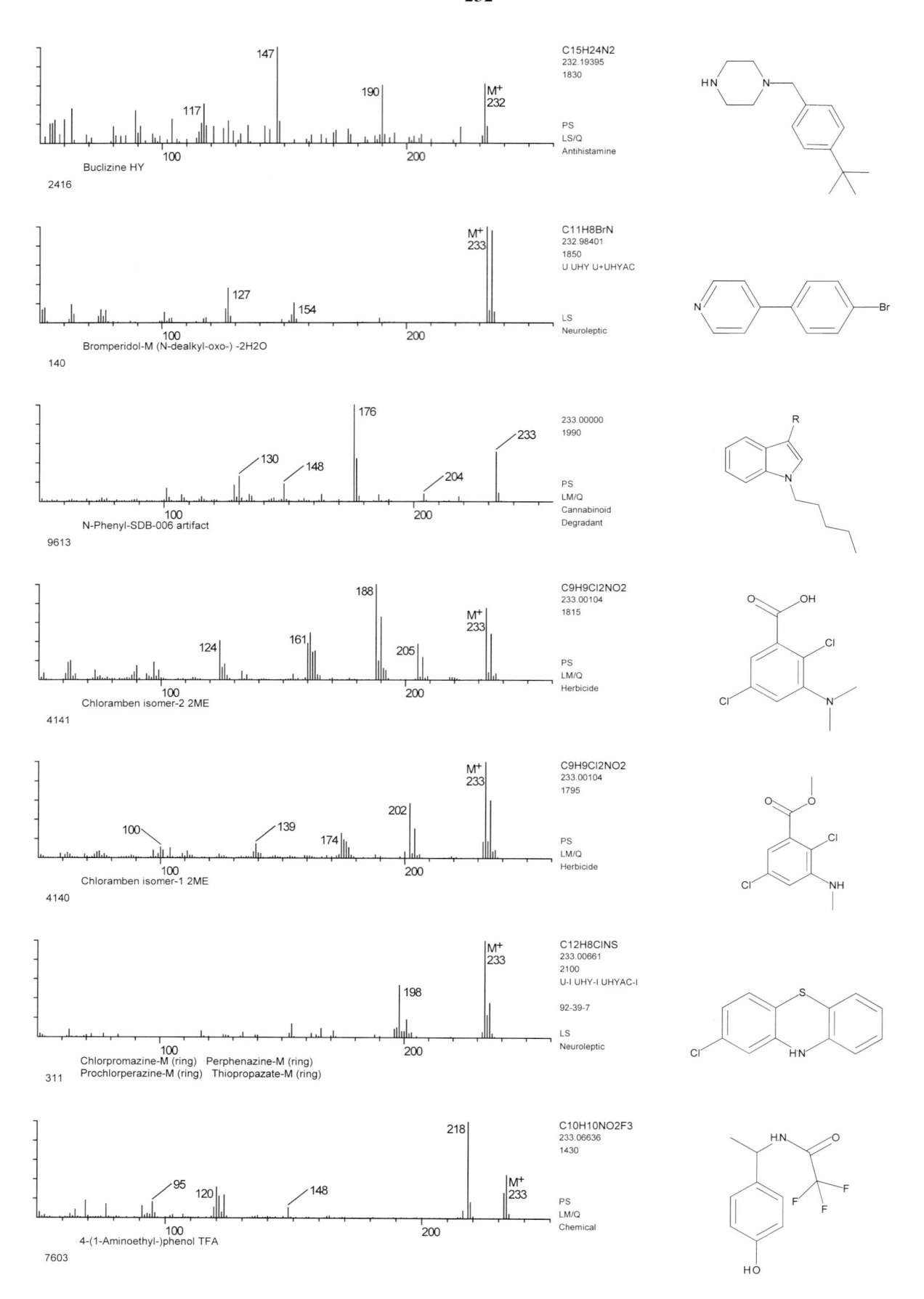

Buclizine HY — 2416
Peaks: 117, 147, 190, M+ 232
C15H24N2
232.19395
1830
PS
LS/Q
Antihistamine

Bromperidol-M (N-dealkyl-oxo-) -2H2O — 140
Peaks: 127, 154, M+ 233
C11H8BrN
232.98401
1850
U UHY U+UHYAC
LS
Neuroleptic

N-Phenyl-SDB-006 artifact — 9613
Peaks: 130, 148, 176, 204, 233
233.00000
1990
PS
LM/Q
Cannabinoid Degradant

Chloramben isomer-2 2ME — 4141
Peaks: 124, 161, 188, 205, M+ 233
C9H9Cl2NO2
233.00104
1815
PS
LM/Q
Herbicide

Chloramben isomer-1 2ME — 4140
Peaks: 100, 139, 174, 202, M+ 233
C9H9Cl2NO2
233.00104
1795
PS
LM/Q
Herbicide

**Chlorpromazine-M (ring)   Perphenazine-M (ring)
Prochlorperazine-M (ring)   Thiopropazate-M (ring)** — 311
Peaks: 198, M+ 233
C12H8ClNS
233.00661
2100
U-I UHY-I UHYAC-I
92-39-7
LS
Neuroleptic

4-(1-Aminoethyl-)phenol TFA — 7603
Peaks: 95, 120, 148, 218, M+ 233
C10H10NO2F3
233.06636
1430
PS
LM/Q
Chemical

233

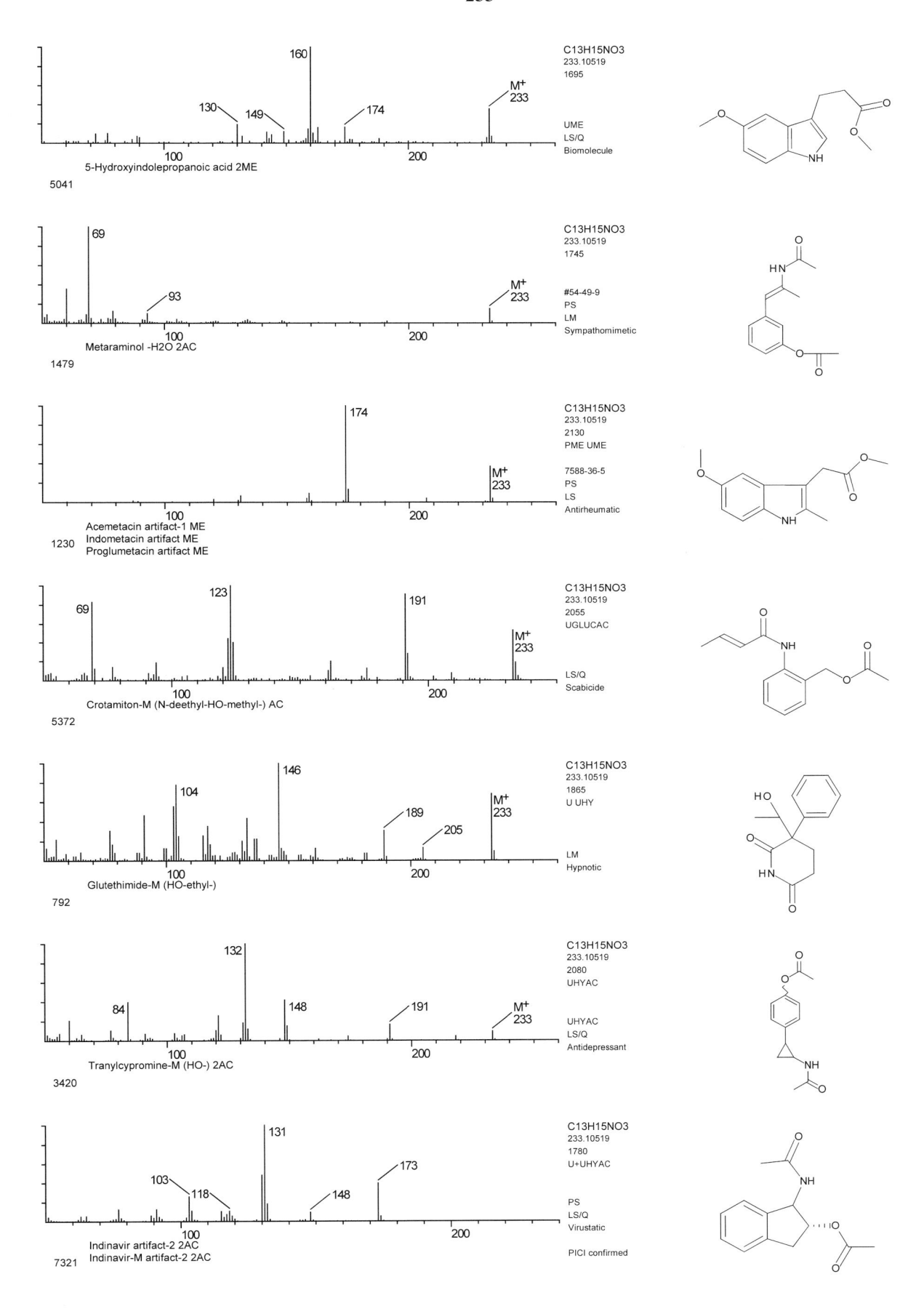

5-Hydroxyindolepropanoic acid 2ME	peaks: 130, 149, 160, 174, M+ 233	C13H15NO3 233.10519 1695 UME LS/Q Biomolecule
Metaraminol -H2O 2AC	peaks: 69, 93, M+ 233	C13H15NO3 233.10519 1745 #54-49-9 PS LM Sympathomimetic
Acemetacin artifact-1 ME Indometacin artifact ME Proglumetacin artifact ME	peaks: 174, M+ 233	C13H15NO3 233.10519 2130 PME UME 7588-36-5 PS LS Antirheumatic
Crotamiton-M (N-deethyl-HO-methyl-) AC	peaks: 69, 123, 191, M+ 233	C13H15NO3 233.10519 2055 UGLUCAC LS/Q Scabicide
Glutethimide-M (HO-ethyl-)	peaks: 104, 146, 189, 205, M+ 233	C13H15NO3 233.10519 1865 U UHY LM Hypnotic
Tranylcypromine-M (HO-) 2AC	peaks: 84, 132, 148, 191, M+ 233	C13H15NO3 233.10519 2080 UHYAC UHYAC LS/Q Antidepressant
Indinavir artifact-2 2AC Indinavir-M artifact-2 2AC	peaks: 103, 118, 131, 148, 173	C13H15NO3 233.10519 1780 U+UHYAC PS LS/Q Virustatic PICI confirmed

233

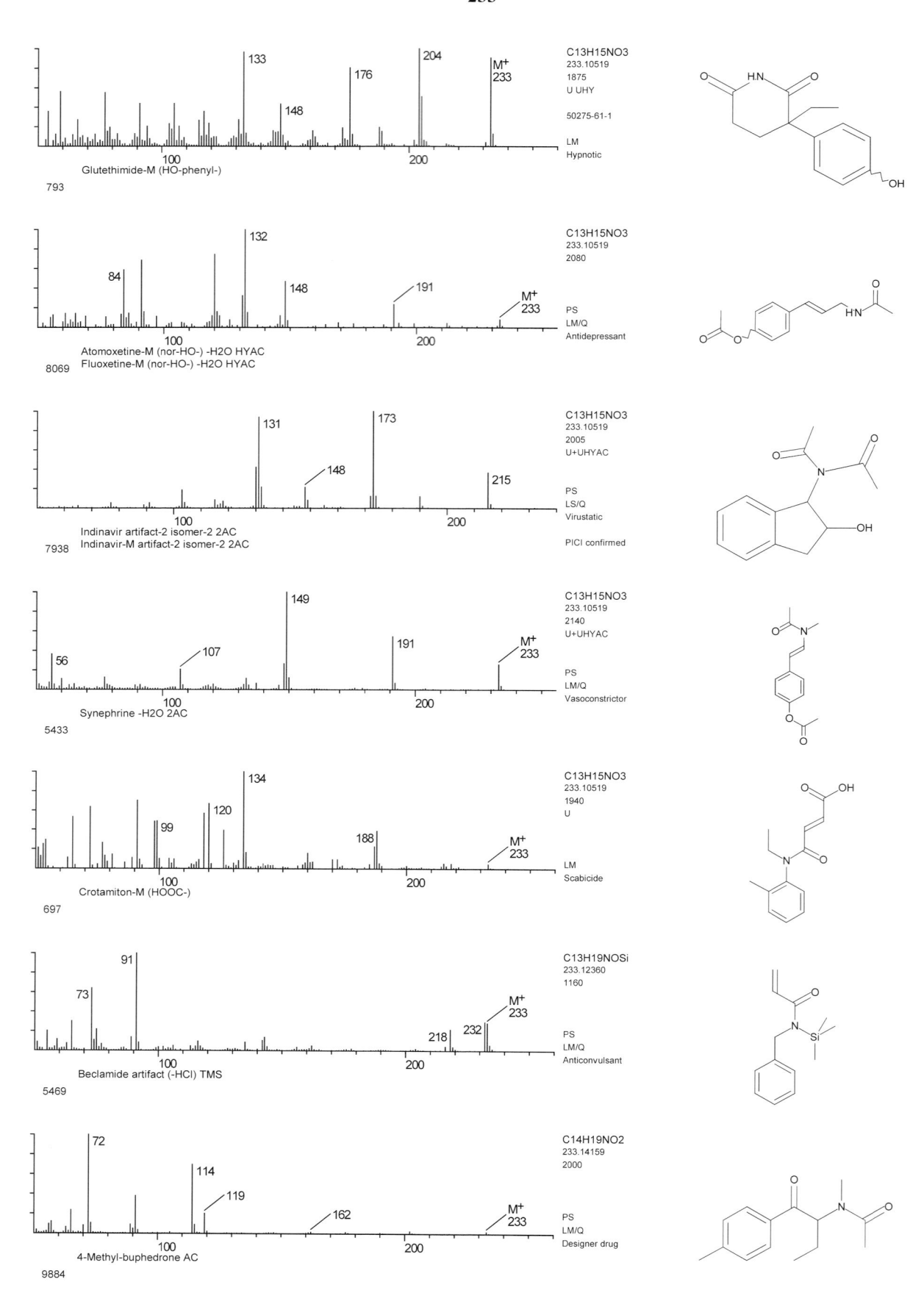

Spectrum	Compound	Formula / Info
793	Glutethimide-M (HO-phenyl-)	C13H15NO3, 233.10519, 1875, U UHY, 50275-61-1, LM, Hypnotic — peaks: 133, 148, 176, 204, M+ 233
8069	Atomoxetine-M (nor-HO-) -H2O HYAC / Fluoxetine-M (nor-HO-) -H2O HYAC	C13H15NO3, 233.10519, 2080, PS, LM/Q, Antidepressant — peaks: 84, 132, 148, 191, M+ 233
7938	Indinavir artifact-2 isomer-2 2AC / Indinavir-M artifact-2 isomer-2 2AC	C13H15NO3, 233.10519, 2005, U+UHYAC, PS, LS/Q, Virustatic, PICI confirmed — peaks: 131, 148, 173, 215
5433	Synephrine -H2O 2AC	C13H15NO3, 233.10519, 2140, U+UHYAC, PS, LM/Q, Vasoconstrictor — peaks: 56, 107, 149, 191, M+ 233
697	Crotamiton-M (HOOC-)	C13H15NO3, 233.10519, 1940, U, LM, Scabicide — peaks: 99, 120, 134, 188, M+ 233
5469	Beclamide artifact (-HCl) TMS	C13H19NOSi, 233.12360, 1160, PS, LM/Q, Anticonvulsant — peaks: 73, 91, 218, 232, M+ 233
9884	4-Methyl-buphedrone AC	C14H19NO2, 233.14159, 2000, PS, LM/Q, Designer drug — peaks: 72, 114, 119, 162, M+ 233

233

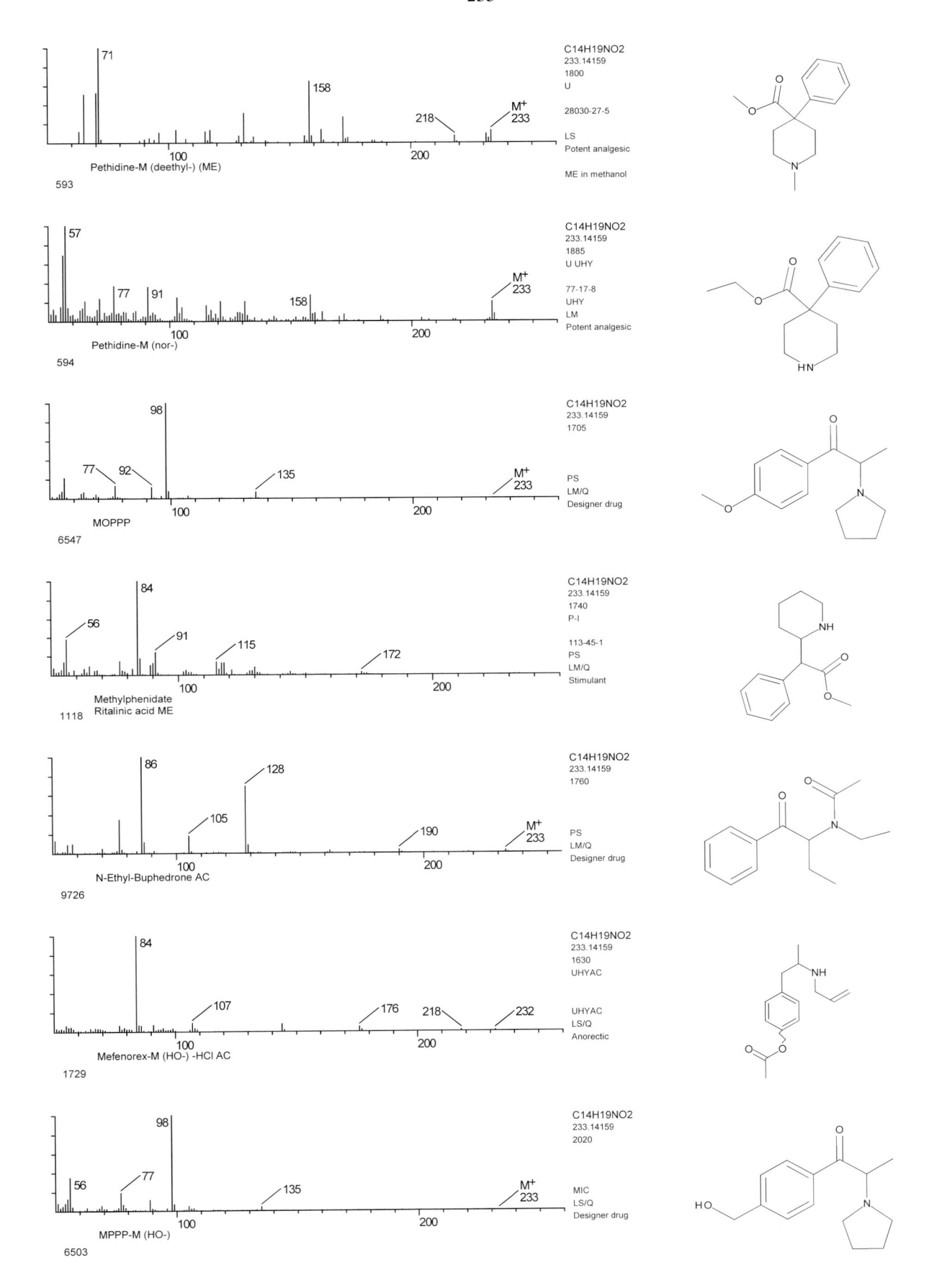

Pethidine-M (deethyl-) (ME) — 593	Peaks: 71, 158, 218, M+ 233	C14H19NO2, 233.14159, 1800, U, 28030-27-5, LS, Potent analgesic, ME in methanol
Pethidine-M (nor-) — 594	Peaks: 57, 77, 91, 158, M+ 233	C14H19NO2, 233.14159, 1885, U UHY, 77-17-8, UHY, LM, Potent analgesic
MOPPP — 6547	Peaks: 77, 92, 98, 135, M+ 233	C14H19NO2, 233.14159, 1705, PS, LM/Q, Designer drug
Methylphenidate Ritalinic acid ME — 1118	Peaks: 56, 84, 91, 115, 172	C14H19NO2, 233.14159, 1740, P-I, 113-45-1, PS, LM/Q, Stimulant
N-Ethyl-Buphedrone AC — 9726	Peaks: 86, 105, 128, 190, M+ 233	C14H19NO2, 233.14159, 1760, PS, LM/Q, Designer drug
Mefenorex-M (HO-) -HCl AC — 1729	Peaks: 84, 107, 176, 218, 232	C14H19NO2, 233.14159, 1630, UHYAC, UHYAC, LS/Q, Anorectic
MPPP-M (HO-) — 6503	Peaks: 56, 77, 98, 135, M+ 233	C14H19NO2, 233.14159, 2020, MIC, LS/Q, Designer drug

523

233

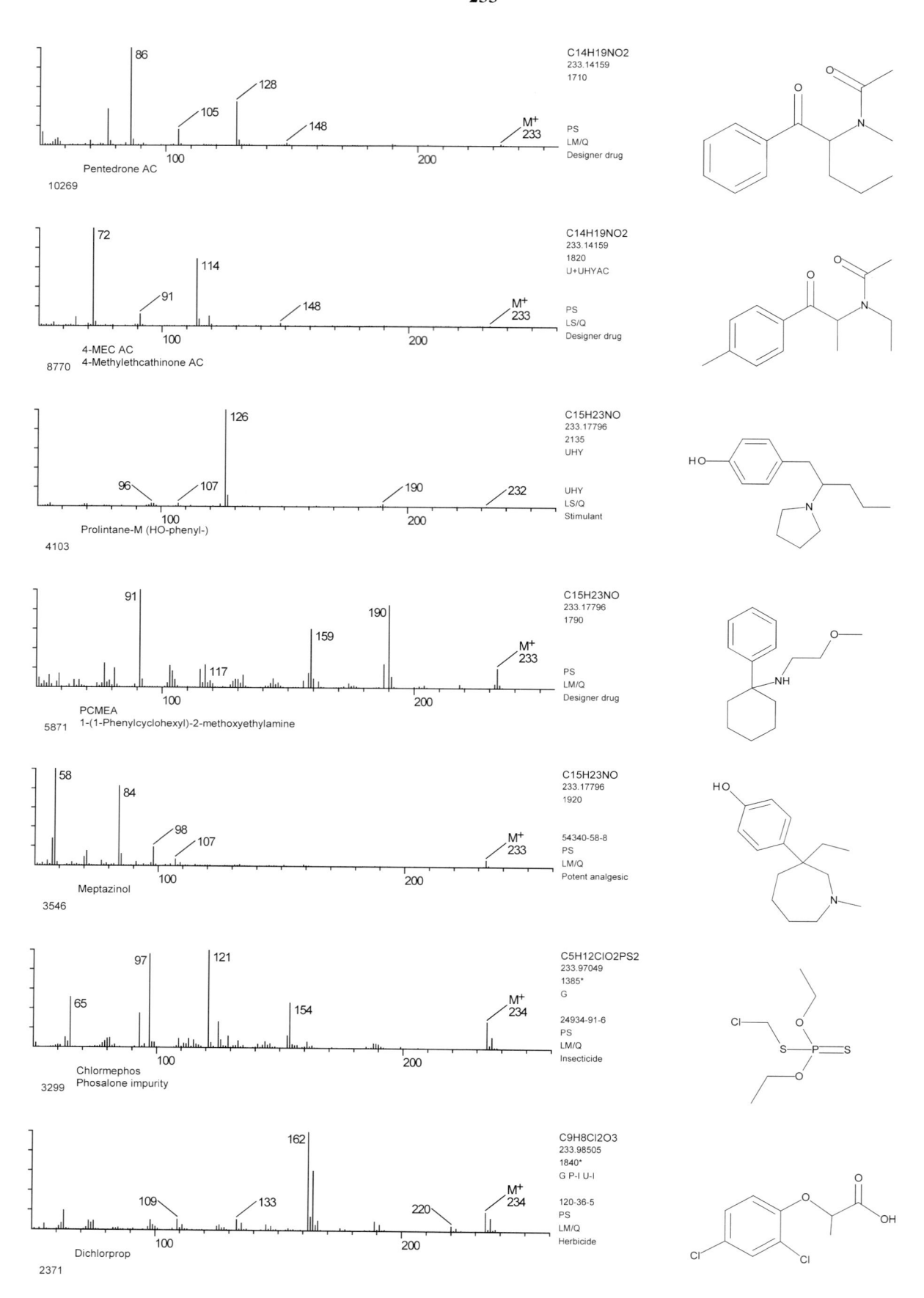

Pentedrone AC
10269
C14H19NO2
233.14159
1710
PS
LM/Q
Designer drug

4-MEC AC
4-Methylethcathinone AC
8770
C14H19NO2
233.14159
1820
U+UHYAC
PS
LS/Q
Designer drug

Prolintane-M (HO-phenyl-)
4103
C15H23NO
233.17796
2135
UHY
UHY
LS/Q
Stimulant

PCMEA
1-(1-Phenylcyclohexyl)-2-methoxyethylamine
5871
C15H23NO
233.17796
1790
PS
LM/Q
Designer drug

Meptazinol
3546
C15H23NO
233.17796
1920
54340-58-8
PS
LM/Q
Potent analgesic

Chlormephos
Phosalone impurity
3299
C5H12ClO2PS2
233.97049
1385*
G
24934-91-6
PS
LM/Q
Insecticide

Dichlorprop
2371
C9H8Cl2O3
233.98505
1840*
G P-I U-I
120-36-5
PS
LM/Q
Herbicide

234

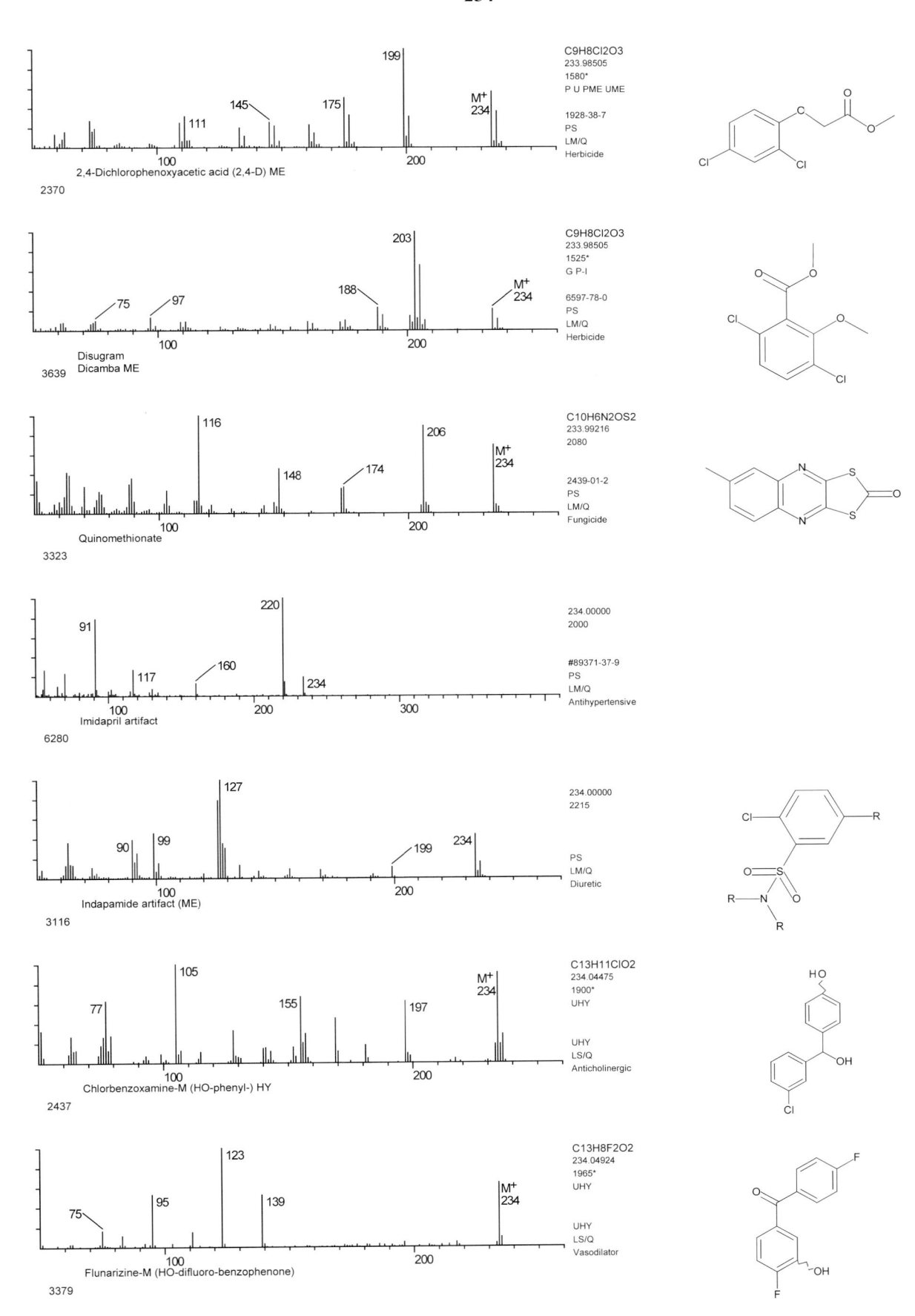

(spectrum)	C9H8Cl2O3 233.98505 1580* P U PME UME 1928-38-7 PS LM/Q Herbicide	(structure)

2,4-Dichlorophenoxyacetic acid (2,4-D) ME
2370

(spectrum)	C9H8Cl2O3 233.98505 1525* G P-I 6597-78-0 PS LM/Q Herbicide	(structure)

Disugram
Dicamba ME
3639

(spectrum)	C10H6N2OS2 233.99216 2080 2439-01-2 PS LM/Q Fungicide	(structure)

Quinomethionate
3323

(spectrum)	234.00000 2000 #89371-37-9 PS LM/Q Antihypertensive	

Imidapril artifact
6280

(spectrum)	234.00000 2215 PS LM/Q Diuretic	(structure)

Indapamide artifact (ME)
3116

(spectrum)	C13H11ClO2 234.04475 1900* UHY UHY LS/Q Anticholinergic	(structure)

Chlorbenzoxamine-M (HO-phenyl-) HY
2437

(spectrum)	C13H8F2O2 234.04924 1965* UHY UHY LS/Q Vasodilator	(structure)

Flunarizine-M (HO-difluoro-benzophenone)
3379

525

Umbelliferone TMS
Coumarin-M (HO-) TMS
7612

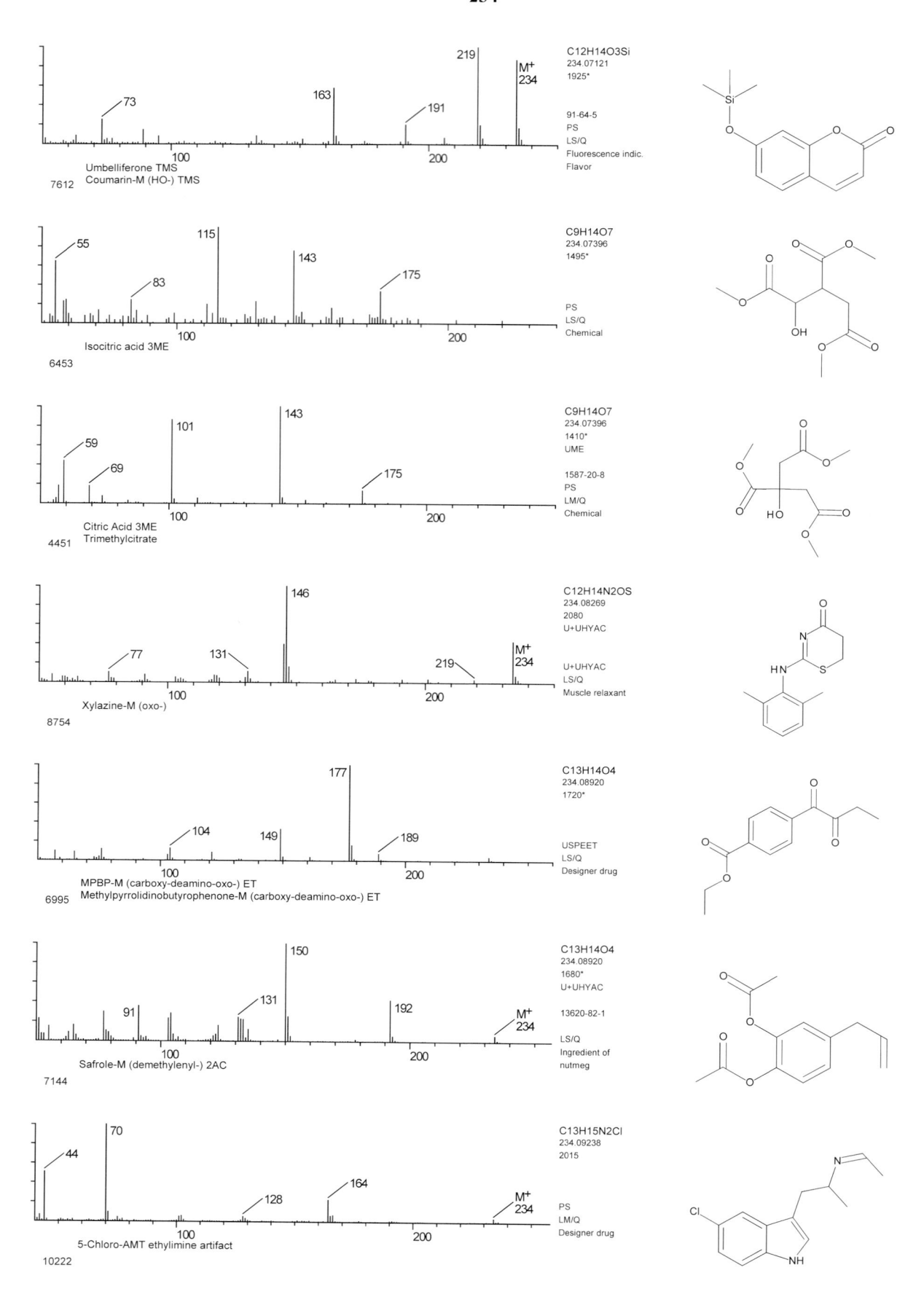

Peaks: 73, 163, 191, 219, M+ 234

C12H14O3Si
234.07121
1925*

91-64-5
PS
LS/Q
Fluorescence indic.
Flavor

Isocitric acid 3ME
6453

Peaks: 55, 83, 115, 143, 175

C9H14O7
234.07396
1495*

PS
LS/Q
Chemical

Citric Acid 3ME
Trimethylcitrate
4451

Peaks: 59, 69, 101, 143, 175

C9H14O7
234.07396
1410*
UME

1587-20-8
PS
LM/Q
Chemical

Xylazine-M (oxo-)
8754

Peaks: 77, 131, 146, 219, M+ 234

C12H14N2OS
234.08269
2080
U+UHYAC

U+UHYAC
LS/Q
Muscle relaxant

MPBP-M (carboxy-deamino-oxo-) ET
Methylpyrrolidinobutyrophenone-M (carboxy-deamino-oxo-) ET
6995

Peaks: 104, 149, 177, 189

C13H14O4
234.08920
1720*

USPEET
LS/Q
Designer drug

Safrole-M (demethylenyl-) 2AC
7144

Peaks: 91, 131, 150, 192, M+ 234

C13H14O4
234.08920
1680*
U+UHYAC

13620-82-1
LS/Q
Ingredient of nutmeg

5-Chloro-AMT ethylimine artifact
10222

Peaks: 44, 70, 128, 164, M+ 234

C13H15N2Cl
234.09238
2015

PS
LM/Q
Designer drug

234

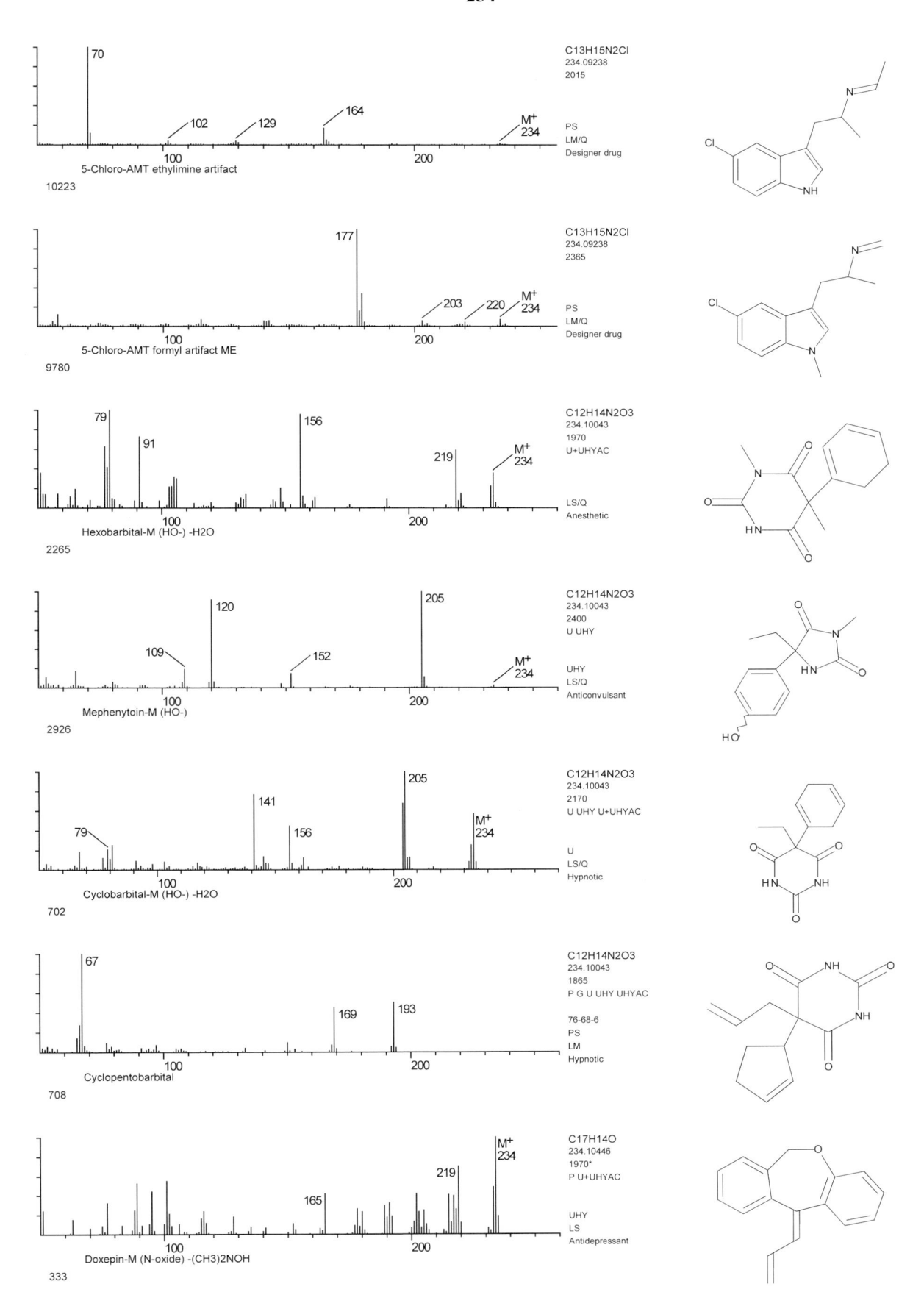

m/z peaks	Formula / Info
70, 102, 129, 164, M+ 234	C13H15N2Cl 234.09238 2015 PS LM/Q Designer drug

5-Chloro-AMT ethylimine artifact
10223

m/z peaks	Formula / Info
177, 203, 220, M+ 234	C13H15N2Cl 234.09238 2365 PS LM/Q Designer drug

5-Chloro-AMT formyl artifact ME
9780

m/z peaks	Formula / Info
79, 91, 156, 219, M+ 234	C12H14N2O3 234.10043 1970 U+UHYAC LS/Q Anesthetic

Hexobarbital-M (HO-) -H2O
2265

m/z peaks	Formula / Info
109, 120, 152, 205, M+ 234	C12H14N2O3 234.10043 2400 U UHY UHY LS/Q Anticonvulsant

Mephenytoin-M (HO-)
2926

m/z peaks	Formula / Info
79, 141, 156, 205, M+ 234	C12H14N2O3 234.10043 2170 U UHY U+UHYAC U LS/Q Hypnotic

Cyclobarbital-M (HO-) -H2O
702

m/z peaks	Formula / Info
67, 169, 193	C12H14N2O3 234.10043 1865 P G U UHY UHYAC 76-68-6 PS LM Hypnotic

Cyclopentobarbital
708

m/z peaks	Formula / Info
165, 219, M+ 234	C17H14O 234.10446 1970* P U+UHYAC UHY LS Antidepressant

Doxepin-M (N-oxide) -(CH3)2NOH
333

234

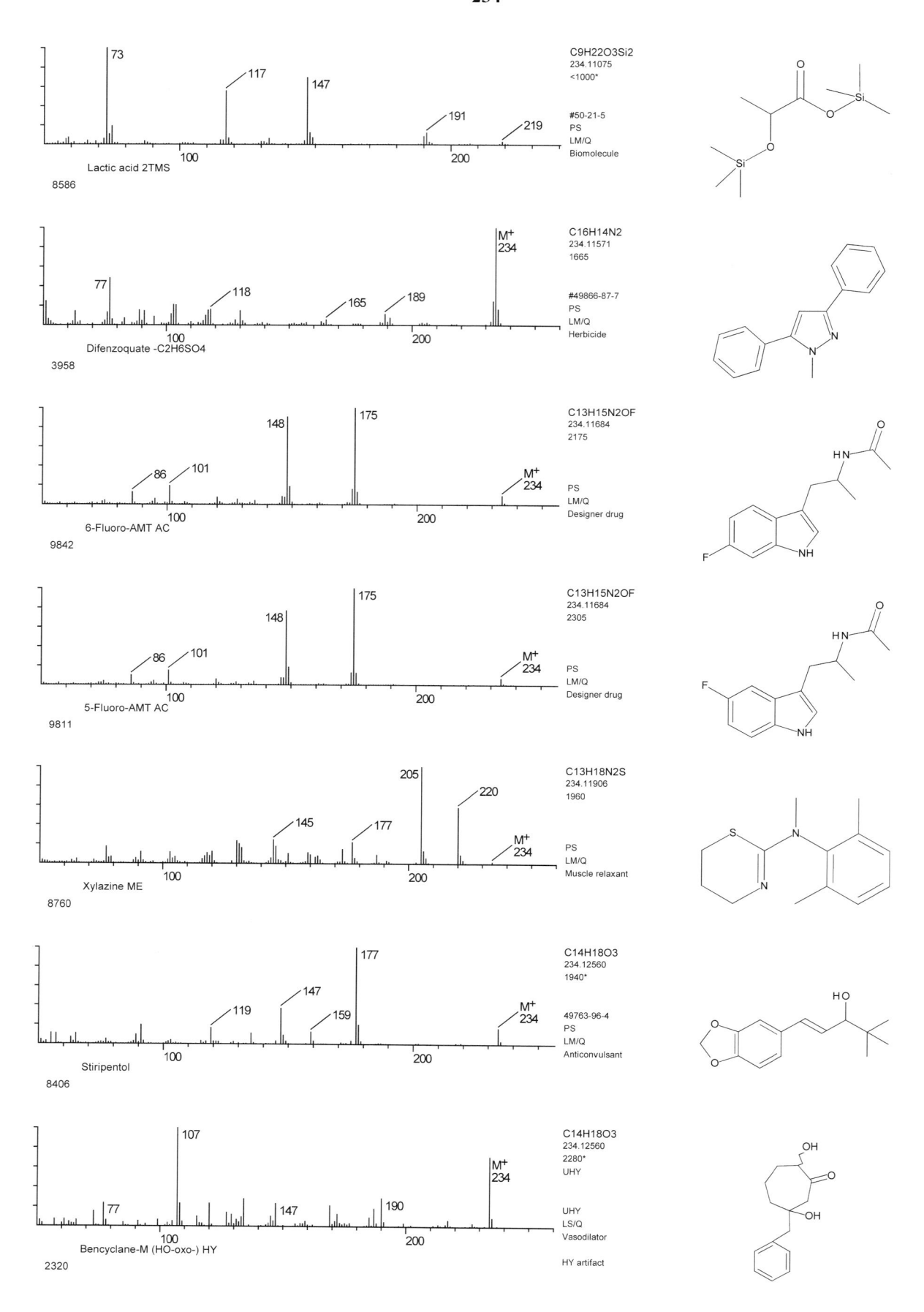

Lactic acid 2TMS — 8586
C9H22O3Si2
234.11075
<1000*
#50-21-5
PS
LM/Q
Biomolecule
Peaks: 73, 117, 147, 191, 219

Difenzoquate -C2H6SO4 — 3958
C16H14N2
234.11571
1665
#49866-87-7
PS
LM/Q
Herbicide
Peaks: 77, 118, 165, 189, M+ 234

6-Fluoro-AMT AC — 9842
C13H15N2OF
234.11684
2175
PS
LM/Q
Designer drug
Peaks: 86, 101, 148, 175, M+ 234

5-Fluoro-AMT AC — 9811
C13H15N2OF
234.11684
2305
PS
LM/Q
Designer drug
Peaks: 86, 101, 148, 175, M+ 234

Xylazine ME — 8760
C13H18N2S
234.11906
1960
PS
LM/Q
Muscle relaxant
Peaks: 145, 177, 205, 220, M+ 234

Stiripentol — 8406
C14H18O3
234.12560
1940*
49763-96-4
PS
LM/Q
Anticonvulsant
Peaks: 119, 147, 159, 177, M+ 234

Bencyclane-M (HO-oxo-) HY — 2320
C14H18O3
234.12560
2280*
UHY
UHY
LS/Q
Vasodilator
HY artifact
Peaks: 77, 107, 147, 190, M+ 234

234

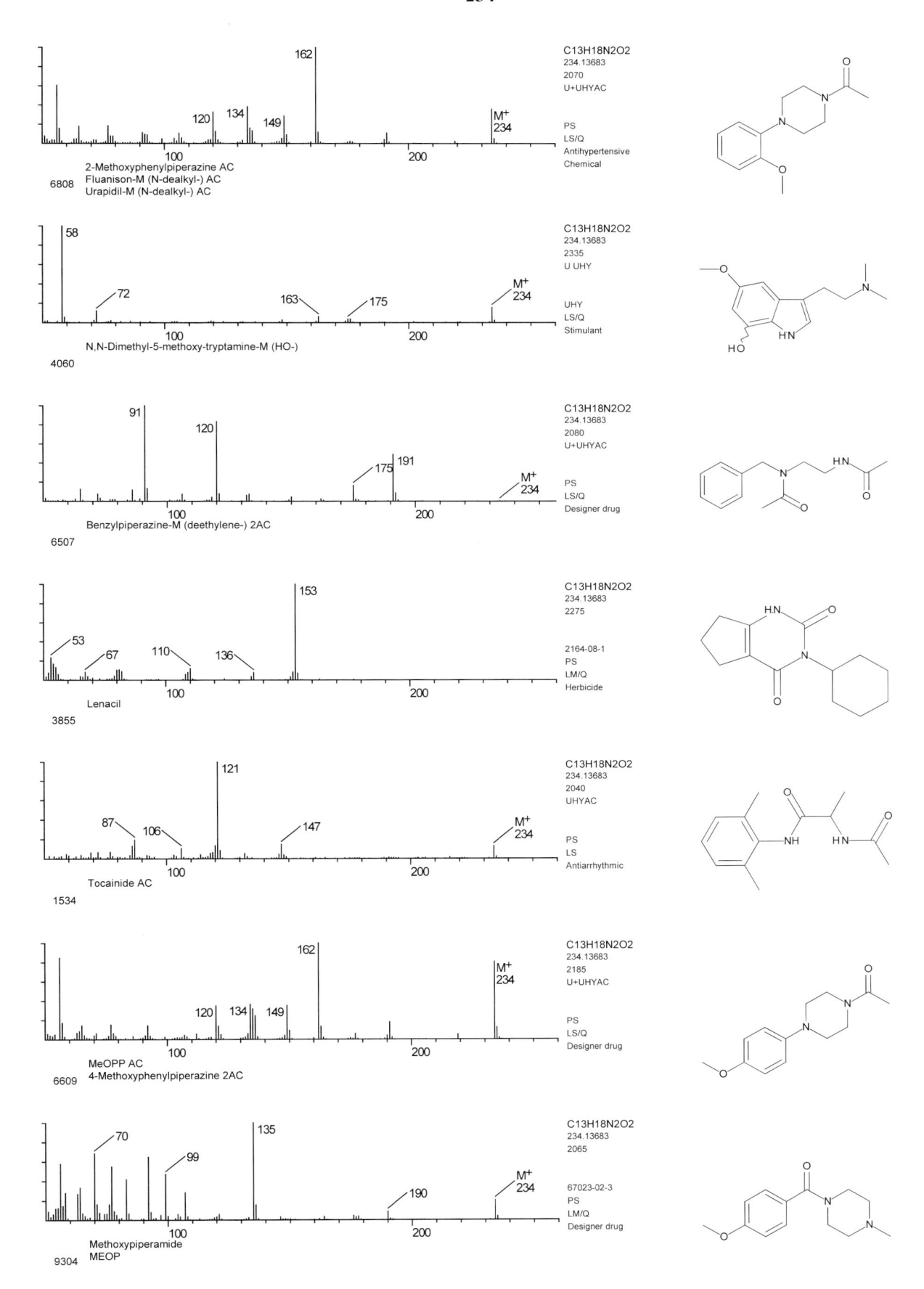

C13H18N2O2
234.13683
2070
U+UHYAC

PS
LS/Q
Antihypertensive
Chemical

2-Methoxyphenylpiperazine AC
Fluanison-M (N-dealkyl-) AC
Urapidil-M (N-dealkyl-) AC
6808

C13H18N2O2
234.13683
2335
U UHY

UHY
LS/Q
Stimulant

N,N-Dimethyl-5-methoxy-tryptamine-M (HO-)
4060

C13H18N2O2
234.13683
2080
U+UHYAC

PS
LS/Q
Designer drug

Benzylpiperazine-M (deethylene-) 2AC
6507

C13H18N2O2
234.13683
2275

2164-08-1
PS
LM/Q
Herbicide

Lenacil
3855

C13H18N2O2
234.13683
2040
UHYAC

PS
LS
Antiarrhythmic

Tocainide AC
1534

C13H18N2O2
234.13683
2185
U+UHYAC

PS
LS/Q
Designer drug

MeOPP AC
4-Methoxyphenylpiperazine 2AC
6609

C13H18N2O2
234.13683
2065

67023-02-3
PS
LM/Q
Designer drug

Methoxypiperamide MEOP
9304

234

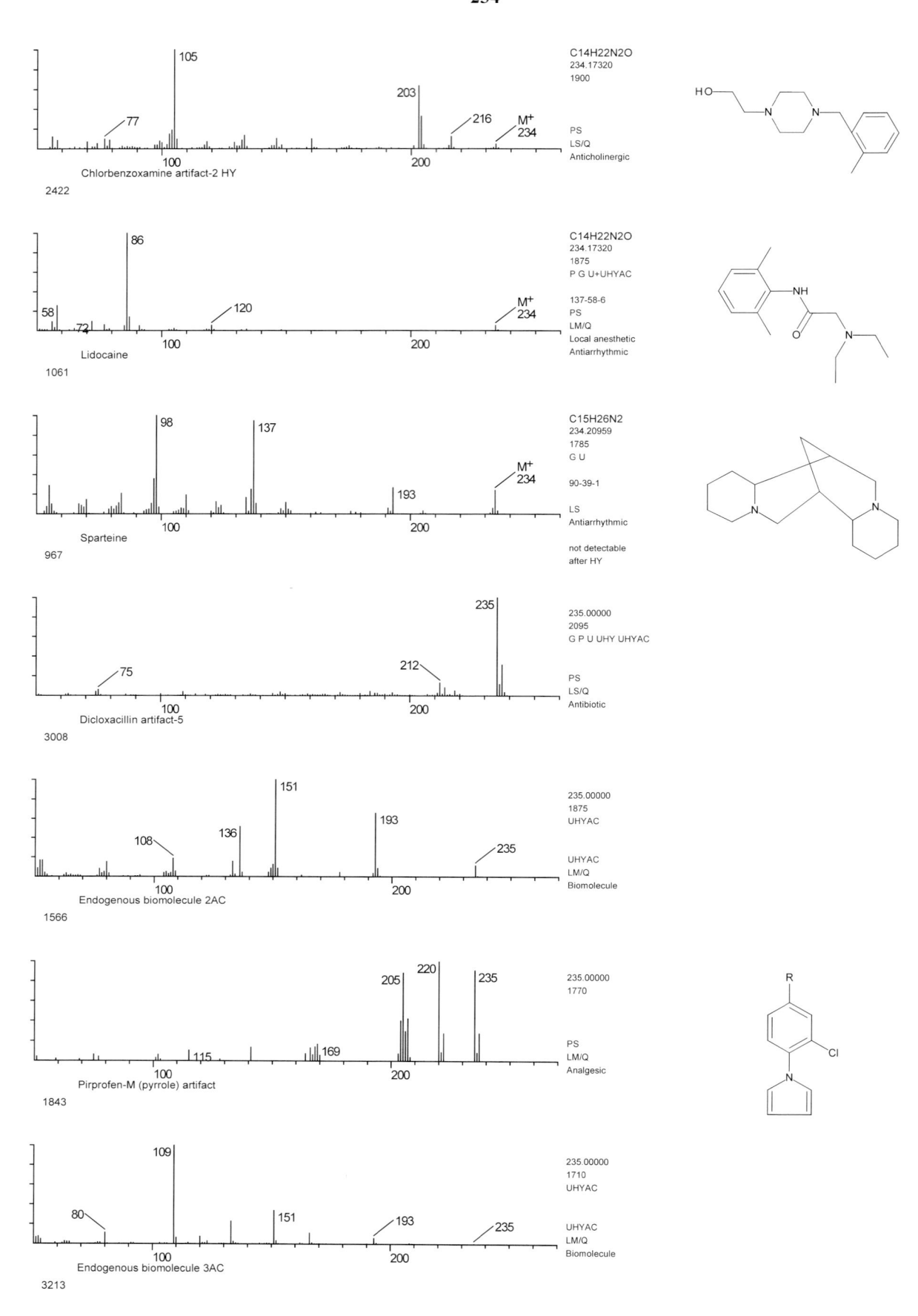

Chlorbenzoxamine artifact-2 HY — peaks 77, 105, 203, 216, M+ 234	C14H22N2O 234.17320 1900 PS LS/Q Anticholinergic	2422
Lidocaine — peaks 58, 72, 86, 120, M+ 234	C14H22N2O 234.17320 1875 P G U+UHYAC 137-58-6 PS LM/Q Local anesthetic Antiarrhythmic	1061
Sparteine — peaks 98, 137, 193, M+ 234	C15H26N2 234.20959 1785 G U 90-39-1 LS Antiarrhythmic not detectable after HY	967
Dicloxacillin artifact-5 — peaks 75, 212, 235	235.00000 2095 G P U UHY UHYAC PS LS/Q Antibiotic	3008
Endogenous biomolecule 2AC — peaks 108, 136, 151, 193, 235	235.00000 1875 UHYAC UHYAC LM/Q Biomolecule	1566
Pirprofen-M (pyrrole) artifact — peaks 115, 169, 205, 220, 235	235.00000 1770 PS LM/Q Analgesic	1843
Endogenous biomolecule 3AC — peaks 80, 109, 151, 193, 235	235.00000 1710 UHYAC UHYAC LM/Q Biomolecule	3213

530

235

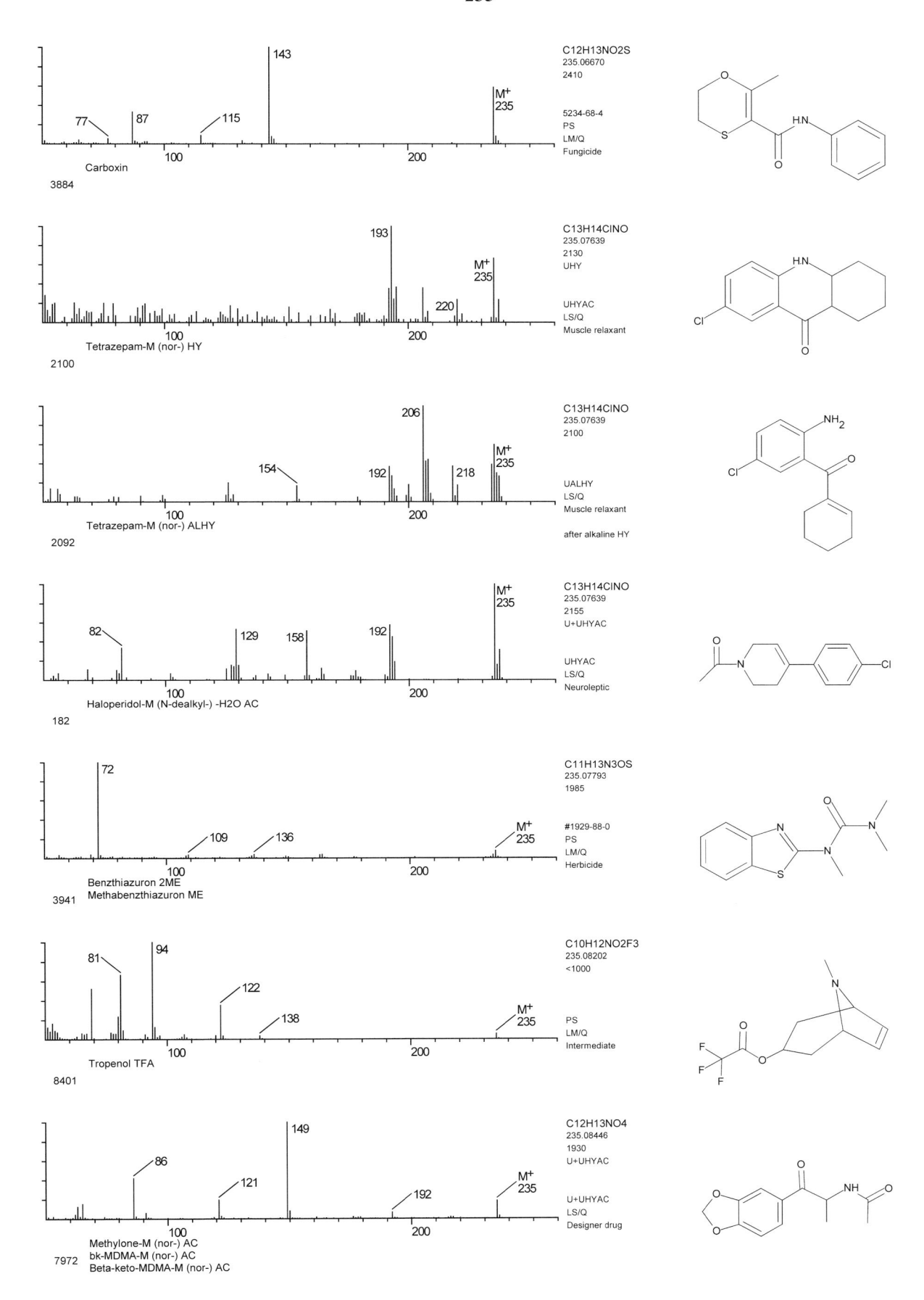

Carboxin	77, 87, 115, 143, M+ 235	C12H13NO2S 235.06670 2410 5234-68-4 PS LM/Q Fungicide
3884		

Tetrazepam-M (nor-) HY	193, 220, M+ 235	C13H14ClNO 235.07639 2130 UHY UHYAC LS/Q Muscle relaxant
2100		

Tetrazepam-M (nor-) ALHY	154, 192, 206, 218, M+ 235	C13H14ClNO 235.07639 2100 UALHY LS/Q Muscle relaxant after alkaline HY
2092		

Haloperidol-M (N-dealkyl-) -H2O AC	82, 129, 158, 192, M+ 235	C13H14ClNO 235.07639 2155 U+UHYAC UHYAC LS/Q Neuroleptic
182		

Benzthiazuron 2ME Methabenzthiazuron ME	72, 109, 136, M+ 235	C11H13N3OS 235.07793 1985 #1929-88-0 PS LM/Q Herbicide
3941		

Tropenol TFA	81, 94, 122, 138, M+ 235	C10H12NO2F3 235.08202 <1000 PS LM/Q Intermediate
8401		

Methylone-M (nor-) AC bk-MDMA-M (nor-) AC Beta-keto-MDMA-M (nor-) AC	86, 121, 149, 192, M+ 235	C12H13NO4 235.08446 1930 U+UHYAC U+UHYAC LS/Q Designer drug
7972		

235

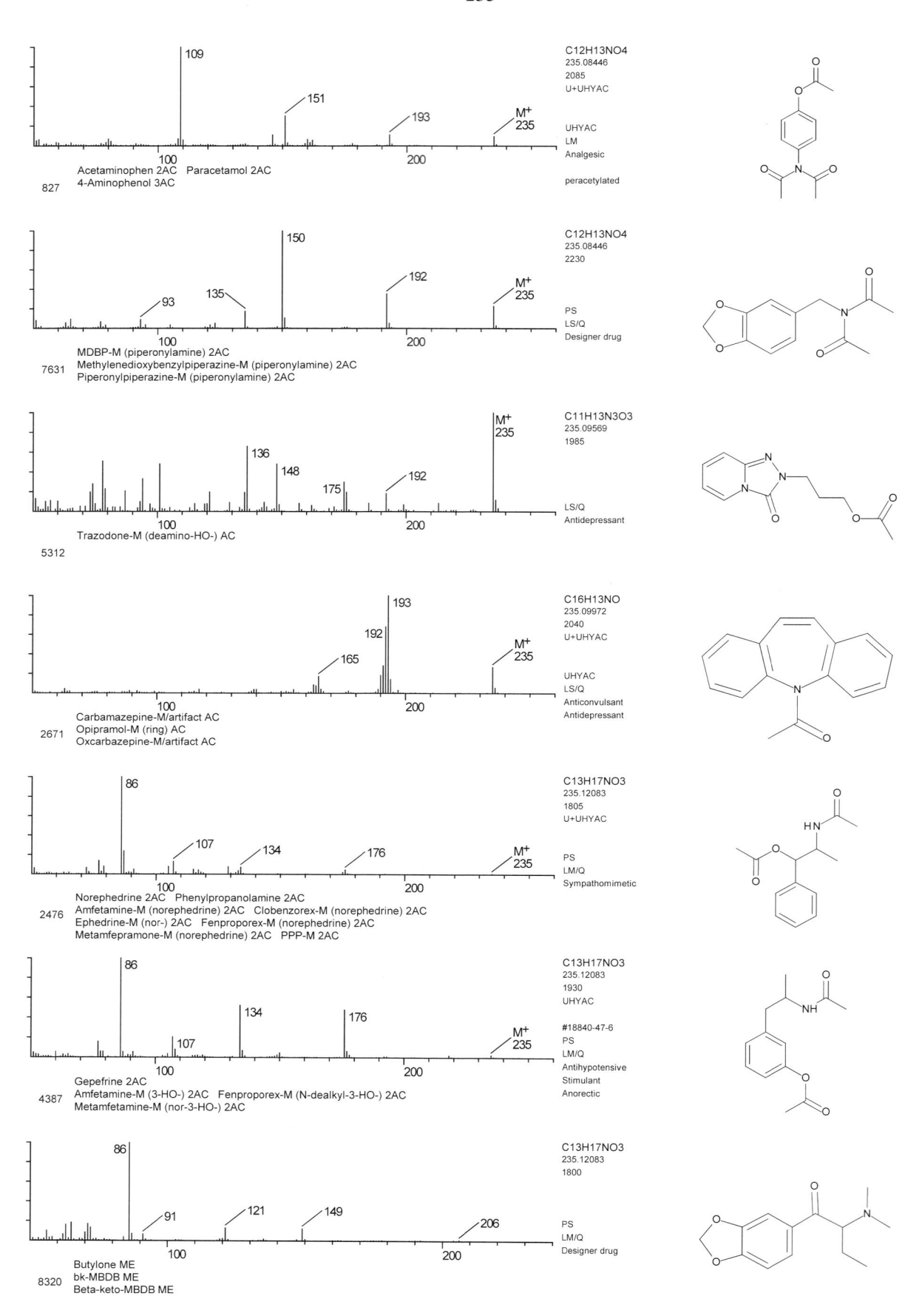

827
Acetaminophen 2AC Paracetamol 2AC
4-Aminophenol 3AC

C12H13NO4
235.08446
2085
U+UHYAC
UHYAC
LM
Analgesic
peracetylated

7631
MDBP-M (piperonylamine) 2AC
Methylenedioxybenzylpiperazine-M (piperonylamine) 2AC
Piperonylpiperazine-M (piperonylamine) 2AC

C12H13NO4
235.08446
2230
PS
LS/Q
Designer drug

5312
Trazodone-M (deamino-HO-) AC

C11H13N3O3
235.09569
1985
LS/Q
Antidepressant

2671
Carbamazepine-M/artifact AC
Opipramol-M (ring) AC
Oxcarbazepine-M/artifact AC

C16H13NO
235.09972
2040
U+UHYAC
UHYAC
LS/Q
Anticonvulsant
Antidepressant

2476
Norephedrine 2AC Phenylpropanolamine 2AC
Amfetamine-M (norephedrine) 2AC Clobenzorex-M (norephedrine) 2AC
Ephedrine-M (nor-) 2AC Fenproporex-M (norephedrine) 2AC
Metamfepramone-M (norephedrine) 2AC PPP-M 2AC

C13H17NO3
235.12083
1805
U+UHYAC
PS
LM/Q
Sympathomimetic

4387
Gepefrine 2AC
Amfetamine-M (3-HO-) 2AC Fenproporex-M (N-dealkyl-3-HO-) 2AC
Metamfetamine-M (nor-3-HO-) 2AC

C13H17NO3
235.12083
1930
UHYAC
#18840-47-6
PS
LM/Q
Antihypotensive
Stimulant
Anorectic

8320
Butylone ME
bk-MBDB ME
Beta-keto-MBDB ME

C13H17NO3
235.12083
1800
PS
LM/Q
Designer drug

235

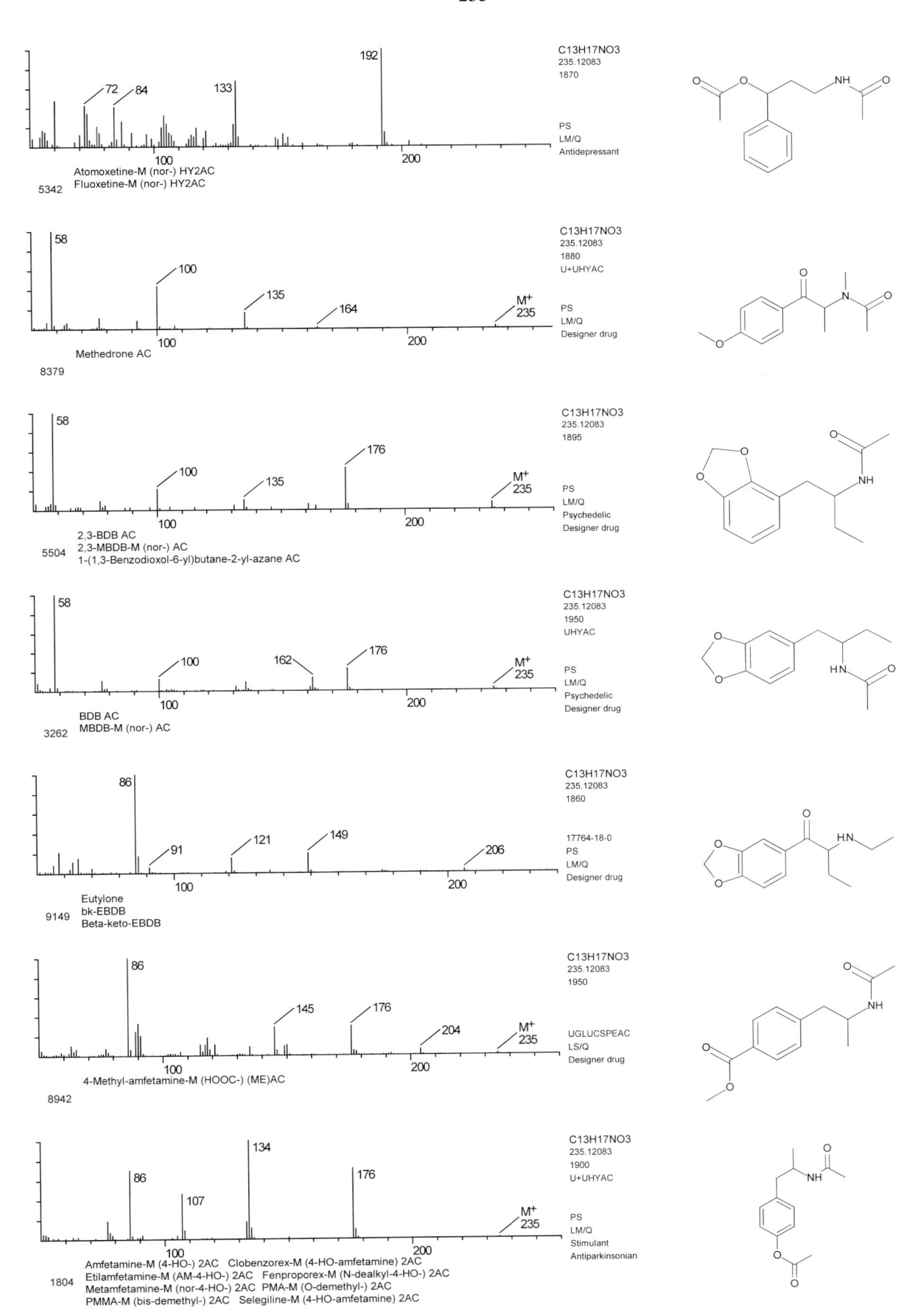

C13H17NO3
235.12083
1870

PS
LM/Q
Antidepressant

Atomoxetine-M (nor-) HY2AC
Fluoxetine-M (nor-) HY2AC
5342

C13H17NO3
235.12083
1880
U+UHYAC

PS
LM/Q
Designer drug

Methedrone AC
8379

C13H17NO3
235.12083
1895

PS
LM/Q
Psychedelic
Designer drug

2,3-BDB AC
2,3-MBDB-M (nor-) AC
1-(1,3-Benzodioxol-6-yl)butane-2-yl-azane AC
5504

C13H17NO3
235.12083
1950
UHYAC

PS
LM/Q
Psychedelic
Designer drug

BDB AC
MBDB-M (nor-) AC
3262

C13H17NO3
235.12083
1860

17764-18-0
PS
LM/Q
Designer drug

Eutylone
bk-EBDB
Beta-keto-EBDB
9149

C13H17NO3
235.12083
1950

UGLUCSPEAC
LS/Q
Designer drug

4-Methyl-amfetamine-M (HOOC-) (ME)AC
8942

C13H17NO3
235.12083
1900
U+UHYAC

PS
LM/Q
Stimulant
Antiparkinsonian

Amfetamine-M (4-HO-) 2AC   Clobenzorex-M (4-HO-amfetamine) 2AC
Etilamfetamine-M (AM-4-HO-) 2AC   Fenproporex-M (N-dealkyl-4-HO-) 2AC
Metamfetamine-M (nor-4-HO-) 2AC   PMA-M (O-demethyl-) 2AC
PMMA-M (bis-demethyl-) 2AC   Selegiline-M (4-HO-amfetamine) 2AC
1804

235

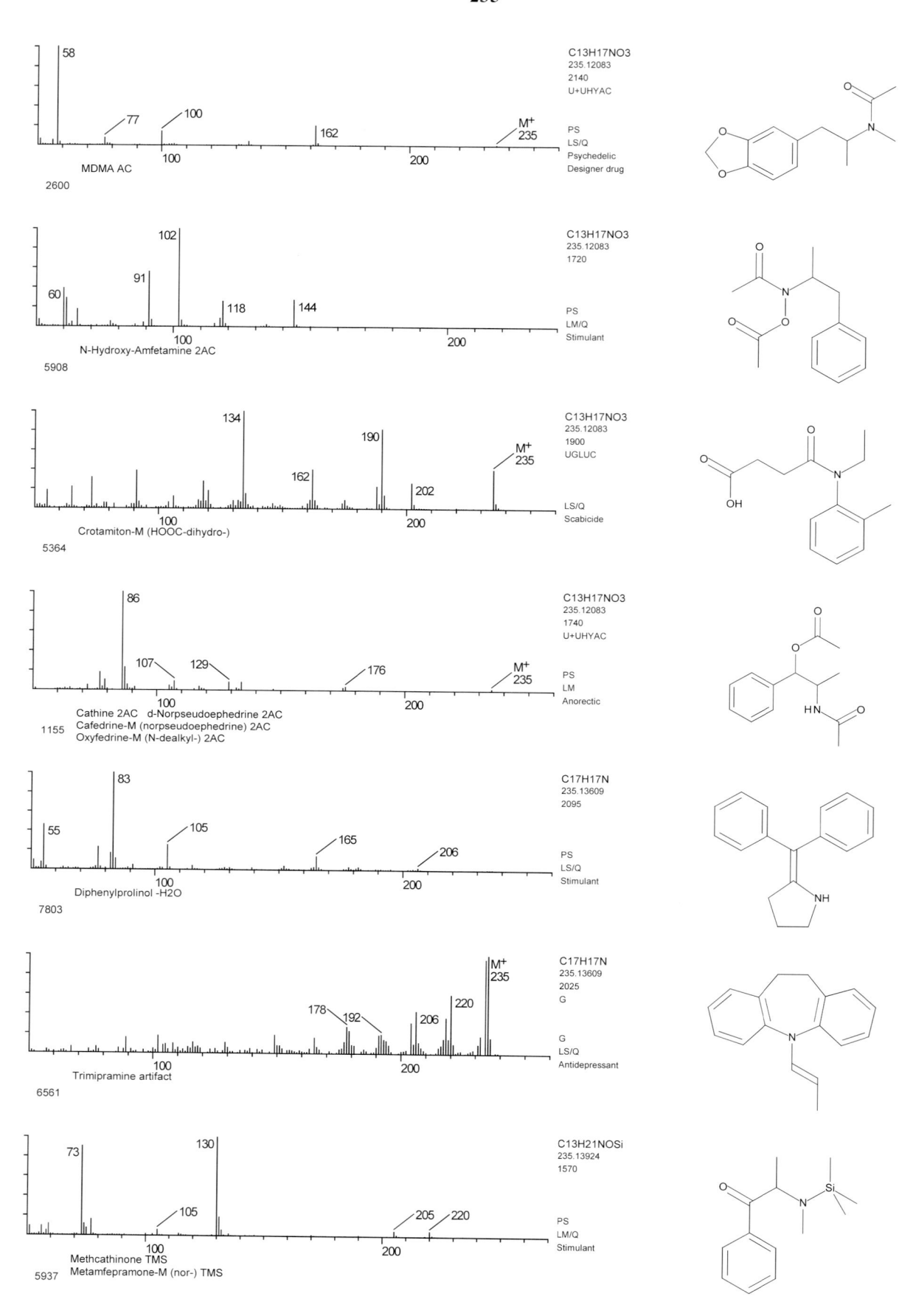

58, 77, 100, 162, M+ 235 MDMA AC 2600	C13H17NO3 235.12083 2140 U+UHYAC PS LS/Q Psychedelic Designer drug	
60, 91, 102, 118, 144 N-Hydroxy-Amfetamine 2AC 5908	C13H17NO3 235.12083 1720 PS LM/Q Stimulant	
134, 162, 190, 202, M+ 235 Crotamiton-M (HOOC-dihydro-) 5364	C13H17NO3 235.12083 1900 UGLUC LS/Q Scabicide	
86, 107, 129, 176, M+ 235 Cathine 2AC d-Norpseudoephedrine 2AC Cafedrine-M (norpseudoephedrine) 2AC Oxyfedrine-M (N-dealkyl-) 2AC 1155	C13H17NO3 235.12083 1740 U+UHYAC PS LM Anorectic	
55, 83, 105, 165, 206 Diphenylprolinol -H2O 7803	C17H17N 235.13609 2095 PS LS/Q Stimulant	
178, 192, 206, 220, M+ 235 Trimipramine artifact 6561	C17H17N 235.13609 2025 G G LS/Q Antidepressant	
73, 105, 130, 205, 220 Methcathinone TMS Metamfepramone-M (nor-) TMS 5937	C13H21NOSi 235.13924 1570 PS LM/Q Stimulant	

235

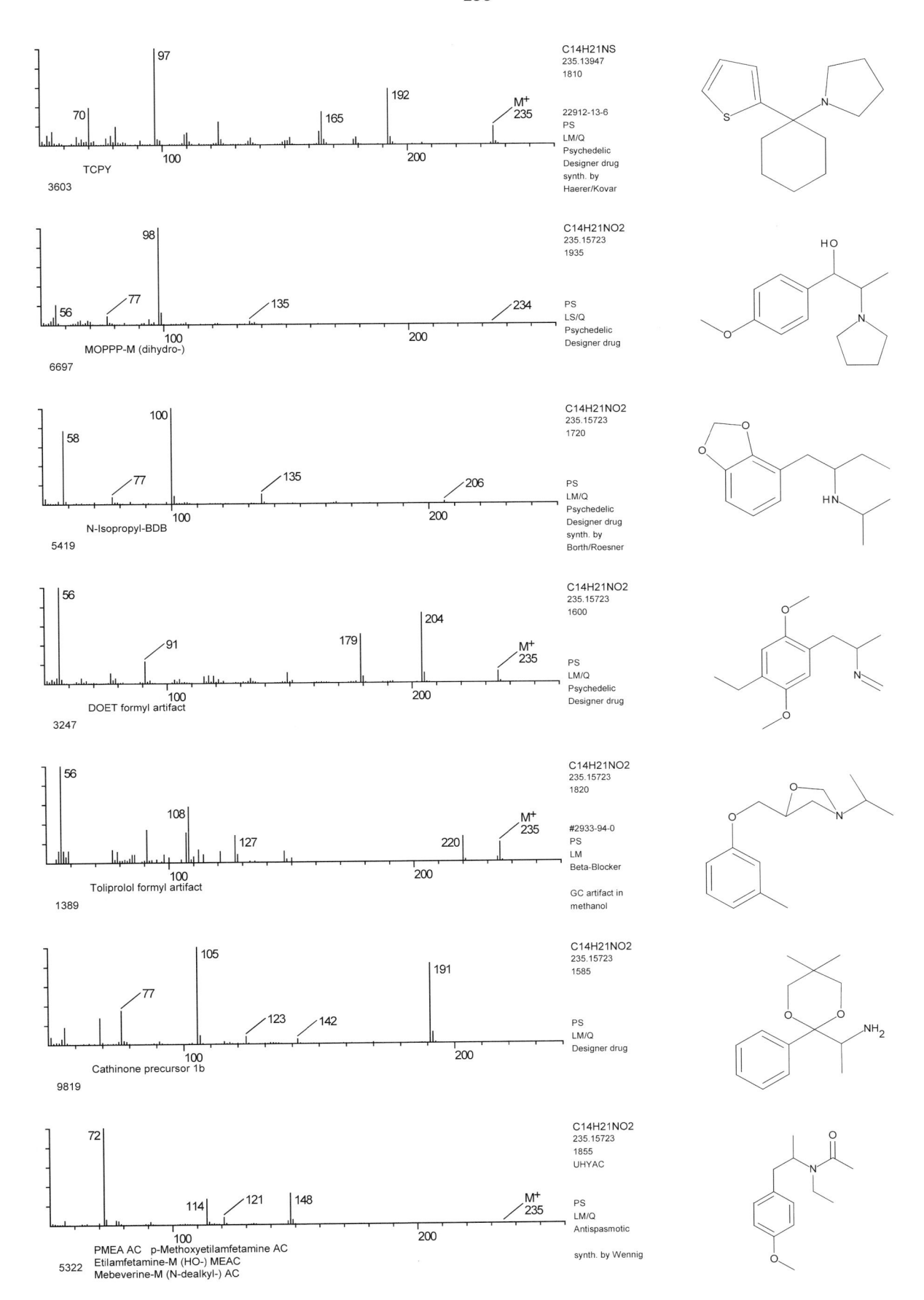

Spectrum label	Formula / info
TCPY — 3603 (peaks: 70, 97, 165, 192, 235 M+)	C14H21NS / 235.13947 / 1810 / 22912-13-6 / PS / LM/Q / Psychedelic / Designer drug / synth. by Haerer/Kovar
MOPPP-M (dihydro-) — 6697 (peaks: 56, 77, 98, 135, 234)	C14H21NO2 / 235.15723 / 1935 / PS / LS/Q / Psychedelic / Designer drug
N-Isopropyl-BDB — 5419 (peaks: 58, 77, 100, 135, 206)	C14H21NO2 / 235.15723 / 1720 / PS / LM/Q / Psychedelic / Designer drug / synth. by Borth/Roesner
DOET formyl artifact — 3247 (peaks: 56, 91, 179, 204, 235 M+)	C14H21NO2 / 235.15723 / 1600 / PS / LM/Q / Psychedelic / Designer drug
Toliprolol formyl artifact — 1389 (peaks: 56, 108, 127, 220, 235 M+)	C14H21NO2 / 235.15723 / 1820 / #2933-94-0 / PS / LM / Beta-Blocker / GC artifact in methanol
Cathinone precursor 1b — 9819 (peaks: 77, 105, 123, 142, 191)	C14H21NO2 / 235.15723 / 1585 / PS / LM/Q / Designer drug
PMEA AC p-Methoxyetilamfetamine AC / Etilamfetamine-M (HO-) MEAC / Mebeverine-M (N-dealkyl-) AC — 5322 (peaks: 72, 114, 121, 148, 235 M+)	C14H21NO2 / 235.15723 / 1855 / UHYAC / PS / LM/Q / Antispasmotic / synth. by Wennig

535

235

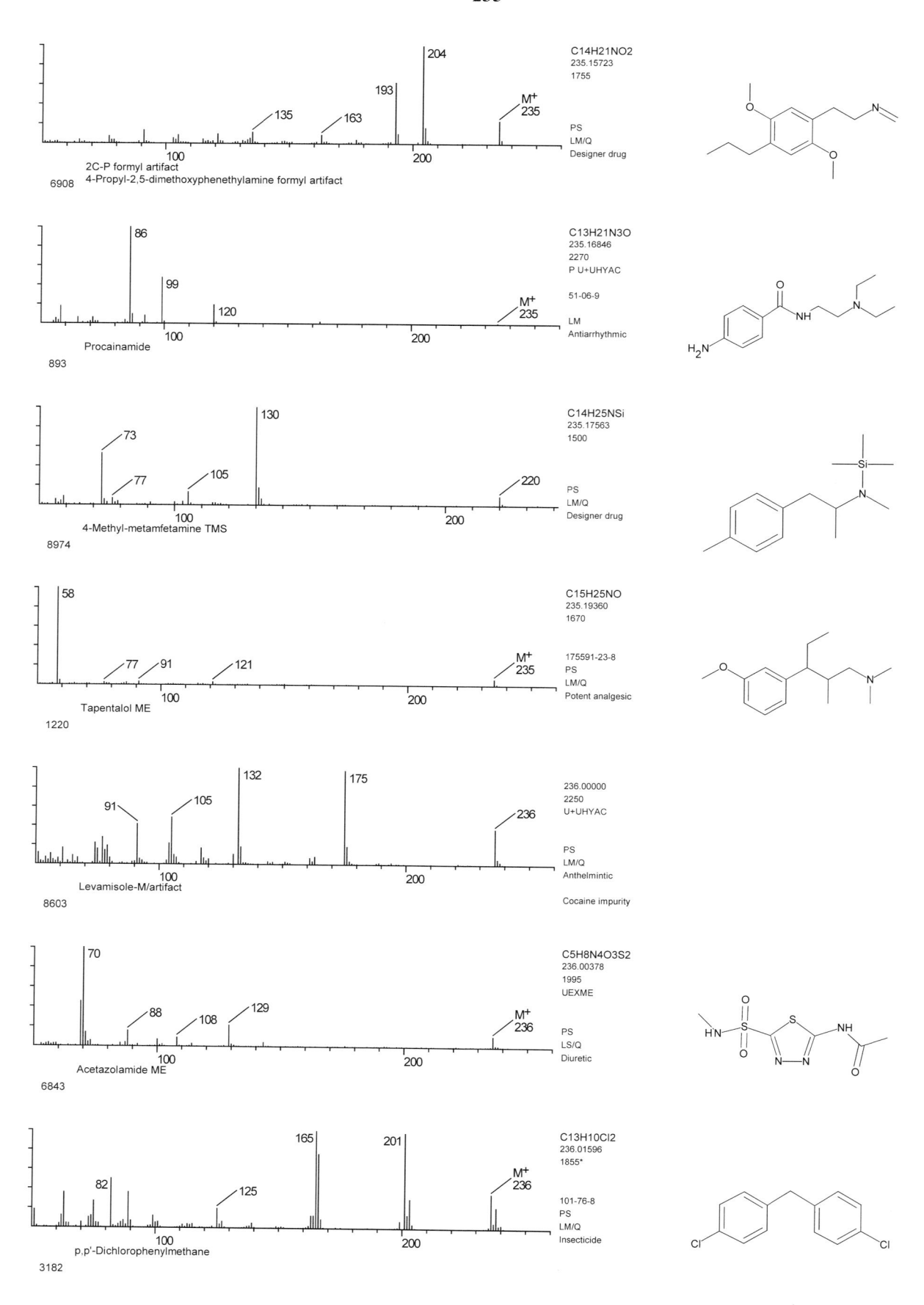

6908 — 2C-P formyl artifact / 4-Propyl-2,5-dimethoxyphenethylamine formyl artifact
Peaks: 135, 163, 193, 204, M+ 235
C14H21NO2; 235.15723; 1755; PS; LM/Q; Designer drug

893 — Procainamide
Peaks: 86, 99, 120, M+ 235
C13H21N3O; 235.16846; 2270; P U+UHYAC; 51-06-9; LM; Antiarrhythmic

8974 — 4-Methyl-metamfetamine TMS
Peaks: 73, 77, 105, 130, 220
C14H25NSi; 235.17563; 1500; PS; LM/Q; Designer drug

1220 — Tapentalol ME
Peaks: 58, 77, 91, 121, M+ 235
C15H25NO; 235.19360; 1670; 175591-23-8; PS; LM/Q; Potent analgesic

8603 — Levamisole-M/artifact
Peaks: 91, 105, 132, 175, 236
236.00000; 2250; U+UHYAC; PS; LM/Q; Anthelmintic; Cocaine impurity

6843 — Acetazolamide ME
Peaks: 70, 88, 108, 129, M+ 236
C5H8N4O3S2; 236.00378; 1995; UEXME; PS; LS/Q; Diuretic

3182 — p,p'-Dichlorophenylmethane
Peaks: 82, 125, 165, 201, M+ 236
C13H10Cl2; 236.01596; 1855*; 101-76-8; PS; LM/Q; Insecticide

236

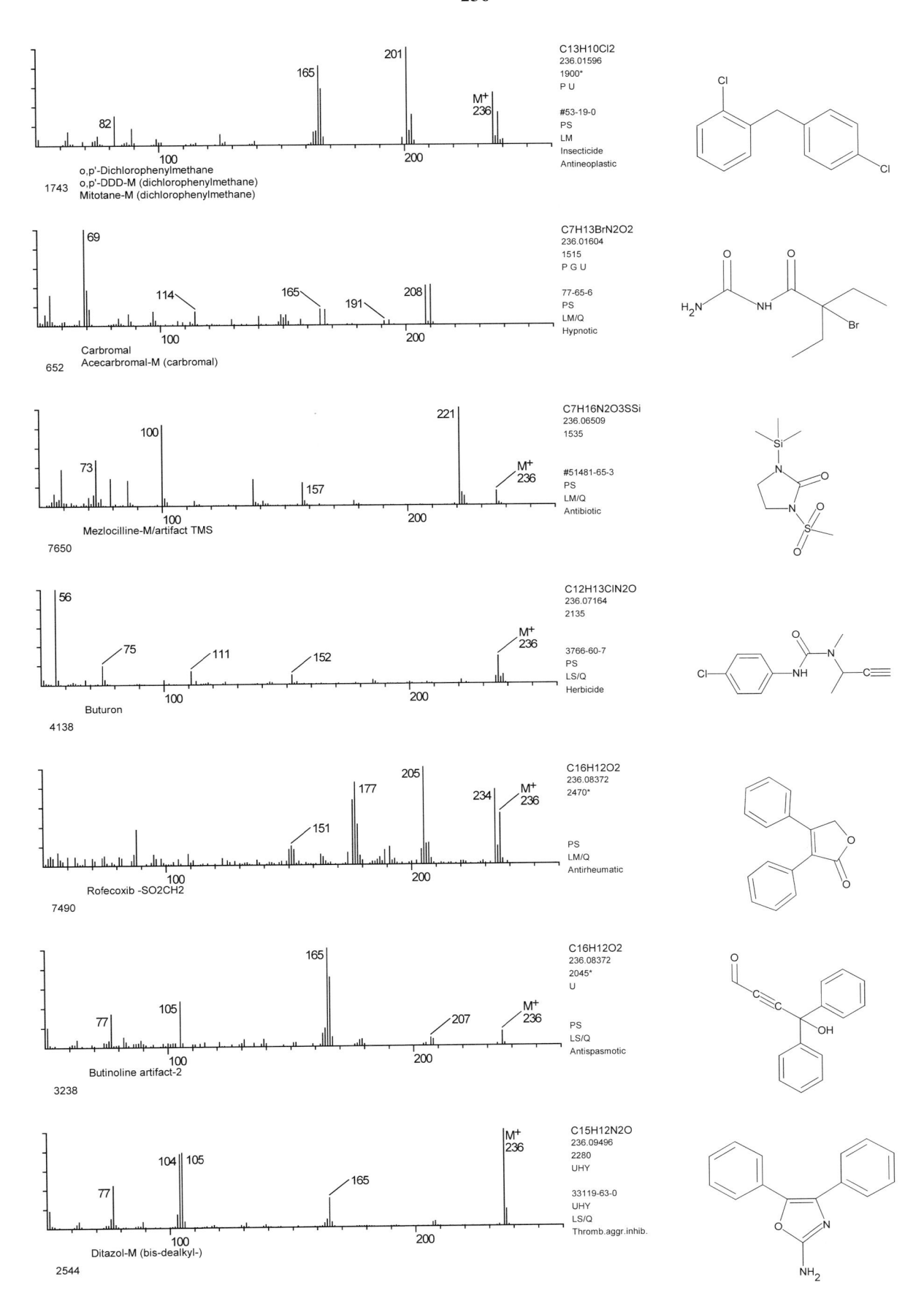

o,p'-Dichlorophenylmethane / o,p'-DDD-M (dichlorophenylmethane) / Mitotane-M (dichlorophenylmethane)
1743

C13H10Cl2
236.01596
1900*
P U
#53-19-0
PS
LM
Insecticide
Antineoplastic

Peaks: 82, 165, 201, M+ 236

Carbromal / Acecarbromal-M (carbromal)
652

C7H13BrN2O2
236.01604
1515
P G U
77-65-6
PS
LM/Q
Hypnotic

Peaks: 69, 114, 165, 191, 208

Mezlocilline-M/artifact TMS
7650

C7H16N2O3SSi
236.06509
1535
#51481-65-3
PS
LM/Q
Antibiotic

Peaks: 73, 100, 157, 221, M+ 236

Buturon
4138

C12H13ClN2O
236.07164
2135
3766-60-7
PS
LS/Q
Herbicide

Peaks: 56, 75, 111, 152, M+ 236

Rofecoxib -SO2CH2
7490

C16H12O2
236.08372
2470*
PS
LM/Q
Antirheumatic

Peaks: 151, 177, 205, 234, M+ 236

Butinoline artifact-2
3238

C16H12O2
236.08372
2045*
U
PS
LS/Q
Antispasmotic

Peaks: 77, 105, 165, 207, M+ 236

Ditazol-M (bis-dealkyl-)
2544

C15H12N2O
236.09496
2280
UHY
33119-63-0
UHY
LS/Q
Thromb.aggr.inhib.

Peaks: 77, 104, 105, 165, M+ 236

236

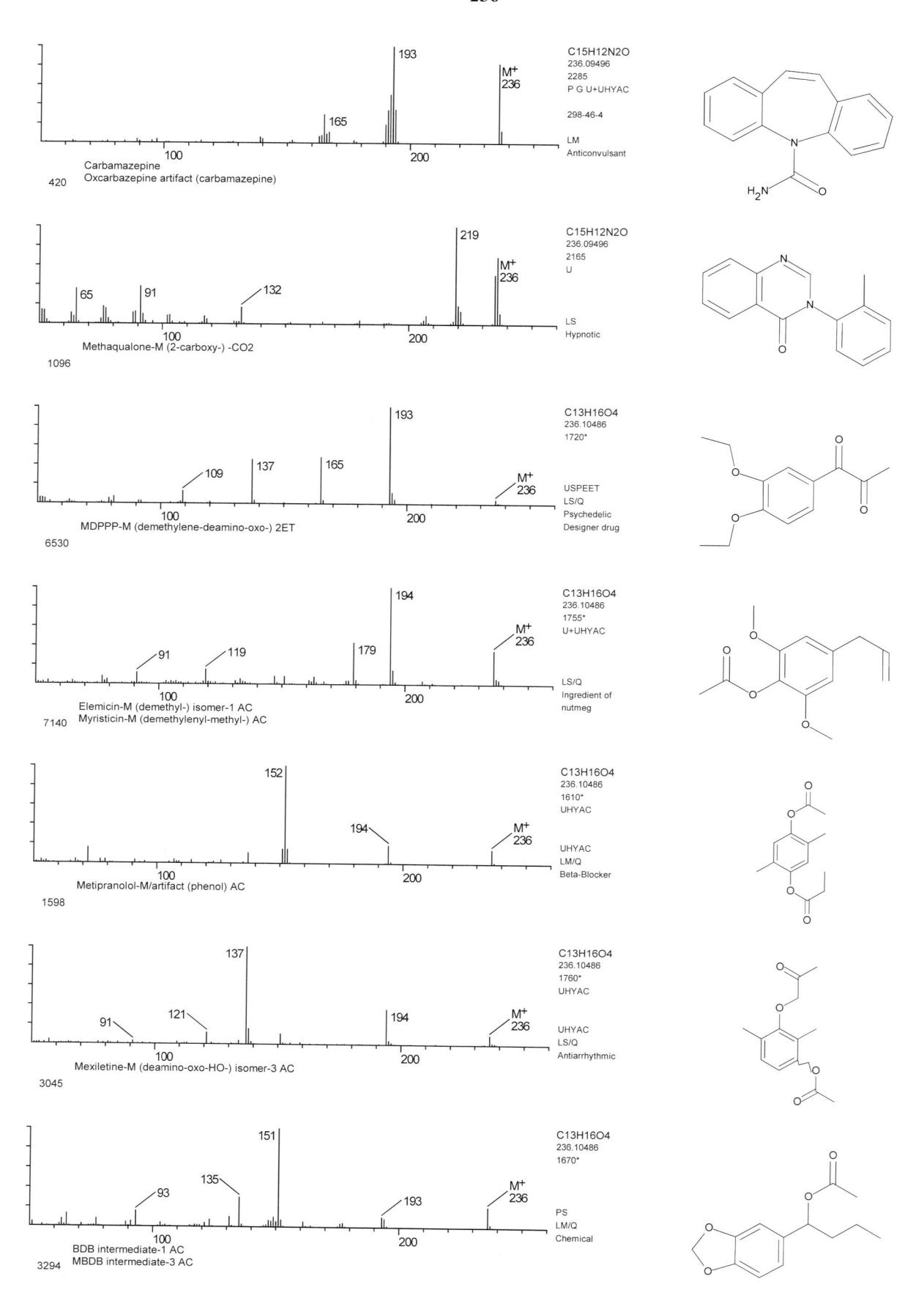

420
Carbamazepine
Oxcarbazepine artifact (carbamazepine)

C15H12N2O
236.09496
2285
P G U+UHYAC
298-46-4
LM
Anticonvulsant

1096
Methaqualone-M (2-carboxy-) -CO2

C15H12N2O
236.09496
2165
U
LS
Hypnotic

6530
MDPPP-M (demethylene-deamino-oxo-) 2ET

C13H16O4
236.10486
1720*
USPEET
LS/Q
Psychedelic
Designer drug

7140
Elemicin-M (demethyl-) isomer-1 AC
Myristicin-M (demethylenyl-methyl-) AC

C13H16O4
236.10486
1755*
U+UHYAC
LS/Q
Ingredient of nutmeg

1598
Metipranolol-M/artifact (phenol) AC

C13H16O4
236.10486
1610*
UHYAC
UHYAC
LM/Q
Beta-Blocker

3045
Mexiletine-M (deamino-oxo-HO-) isomer-3 AC

C13H16O4
236.10486
1760*
UHYAC
UHYAC
LS/Q
Antiarrhythmic

3294
BDB intermediate-1 AC
MBDB intermediate-3 AC

C13H16O4
236.10486
1670*
PS
LM/Q
Chemical

236

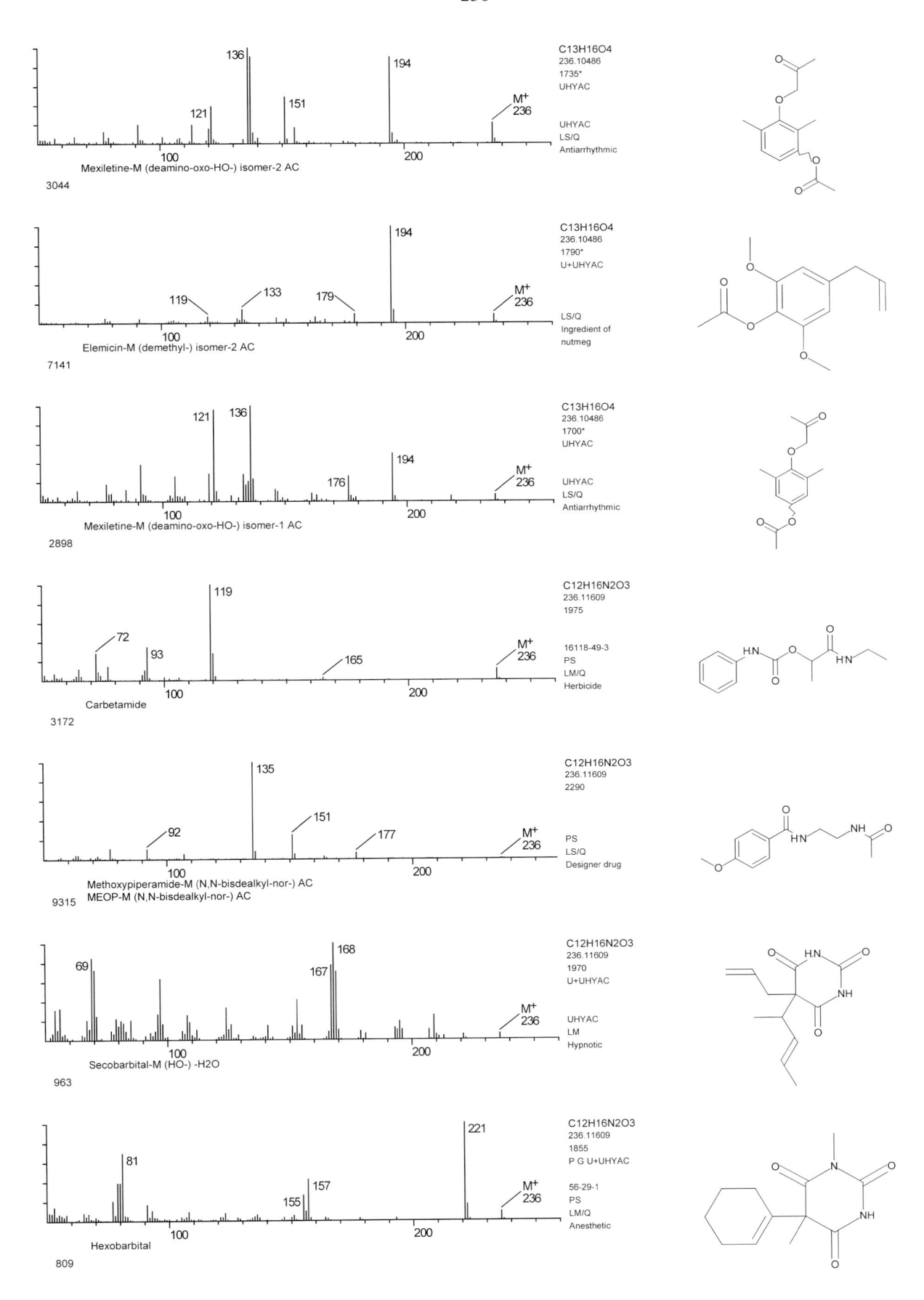

Mexiletine-M (deamino-oxo-HO-) isomer-2 AC 3044	C13H16O4 236.10486 1735* UHYAC UHYAC LS/Q Antiarrhythmic	Peaks: 121, 136, 151, 194, M+ 236
Elemicin-M (demethyl-) isomer-2 AC 7141	C13H16O4 236.10486 1790* U+UHYAC LS/Q Ingredient of nutmeg	Peaks: 119, 133, 179, 194, M+ 236
Mexiletine-M (deamino-oxo-HO-) isomer-1 AC 2898	C13H16O4 236.10486 1700* UHYAC UHYAC LS/Q Antiarrhythmic	Peaks: 121, 136, 176, 194, M+ 236
Carbetamide 3172	C12H16N2O3 236.11609 1975 16118-49-3 PS LM/Q Herbicide	Peaks: 72, 93, 119, 165, M+ 236
Methoxypiperamide-M (N,N-bisdealkyl-nor-) AC MEOP-M (N,N-bisdealkyl-nor-) AC 9315	C12H16N2O3 236.11609 2290 PS LS/Q Designer drug	Peaks: 92, 135, 151, 177, M+ 236
Secobarbital-M (HO-) -H2O 963	C12H16N2O3 236.11609 1970 U+UHYAC UHYAC LM Hypnotic	Peaks: 69, 167, 168, M+ 236
Hexobarbital 809	C12H16N2O3 236.11609 1855 P G U+UHYAC 56-29-1 PS LM/Q Anesthetic	Peaks: 81, 155, 157, 221, M+ 236

236

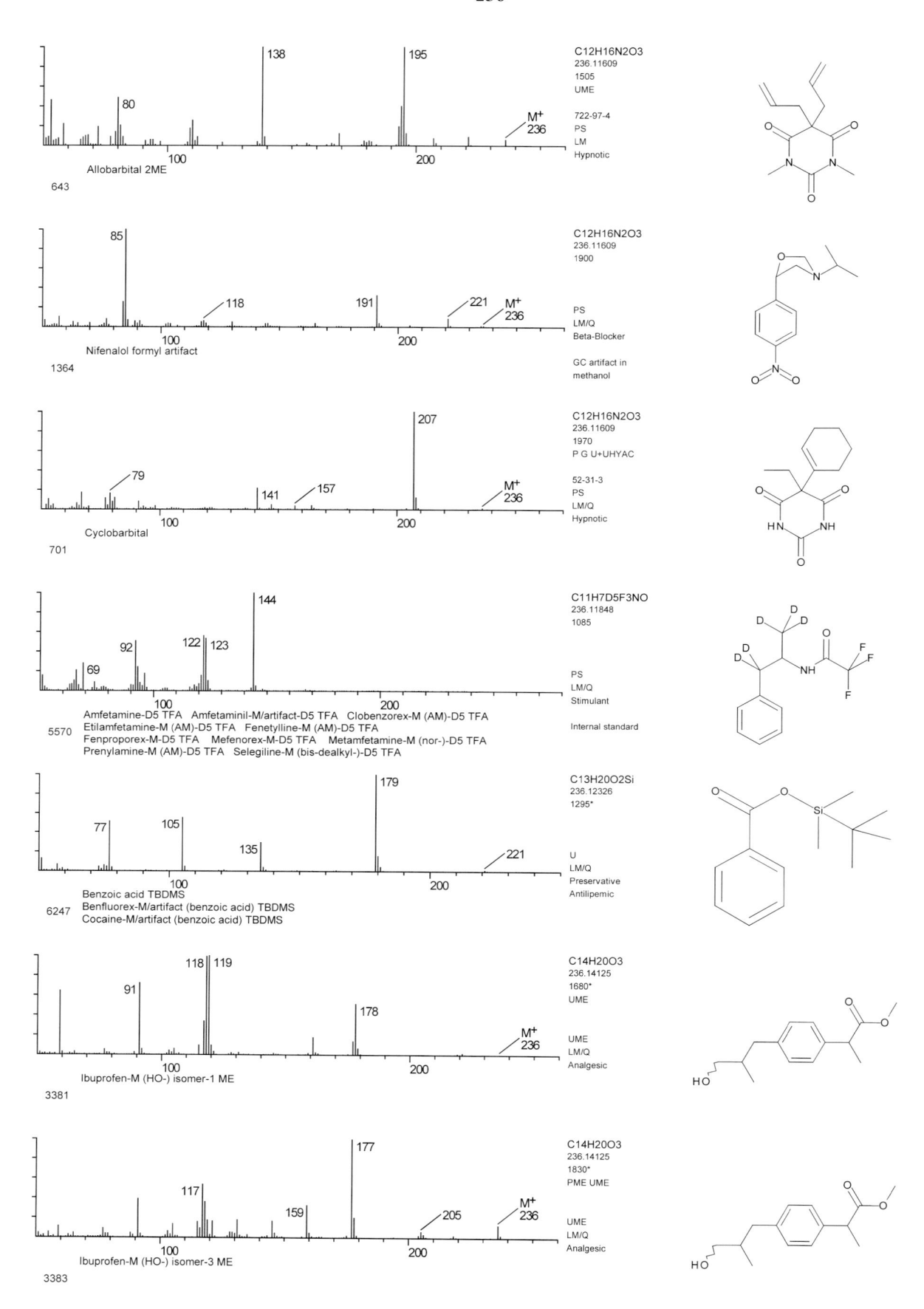

Allobarbital 2ME — peaks 80, 138, 195, M+ 236	C12H16N2O3 236.11609 1505 UME 722-97-4 PS LM Hypnotic
643	

Nifenalol formyl artifact — peaks 85, 118, 191, 221, M+ 236	C12H16N2O3 236.11609 1900 PS LM/Q Beta-Blocker GC artifact in methanol
1364	

Cyclobarbital — peaks 79, 141, 157, 207, M+ 236	C12H16N2O3 236.11609 1970 P G U+UHYAC 52-31-3 PS LM/Q Hypnotic
701	

Amfetamine-D5 TFA Amfetaminil-M/artifact-D5 TFA Clobenzorex-M (AM)-D5 TFA Etilamfetamine-M (AM)-D5 TFA Fenetylline-M (AM)-D5 TFA Fenproporex-M-D5 TFA Mefenorex-M-D5 TFA Metamfetamine-M (nor-)-D5 TFA Prenylamine-M (AM)-D5 TFA Selegiline-M (bis-dealkyl-)-D5 TFA peaks 69, 92, 122, 123, 144	C11H7D5F3NO 236.11848 1085 PS LM/Q Stimulant Internal standard
5570	

Benzoic acid TBDMS Benfluorex-M/artifact (benzoic acid) TBDMS Cocaine-M/artifact (benzoic acid) TBDMS peaks 77, 105, 135, 179, 221	C13H20O2Si 236.12326 1295* U LM/Q Preservative Antilipemic
6247	

Ibuprofen-M (HO-) isomer-1 ME — peaks 91, 118, 119, 178, M+ 236	C14H20O3 236.14125 1680* UME UME LM/Q Analgesic
3381	

Ibuprofen-M (HO-) isomer-3 ME — peaks 117, 159, 177, 205, M+ 236	C14H20O3 236.14125 1830* PME UME UME LM/Q Analgesic
3383	

236

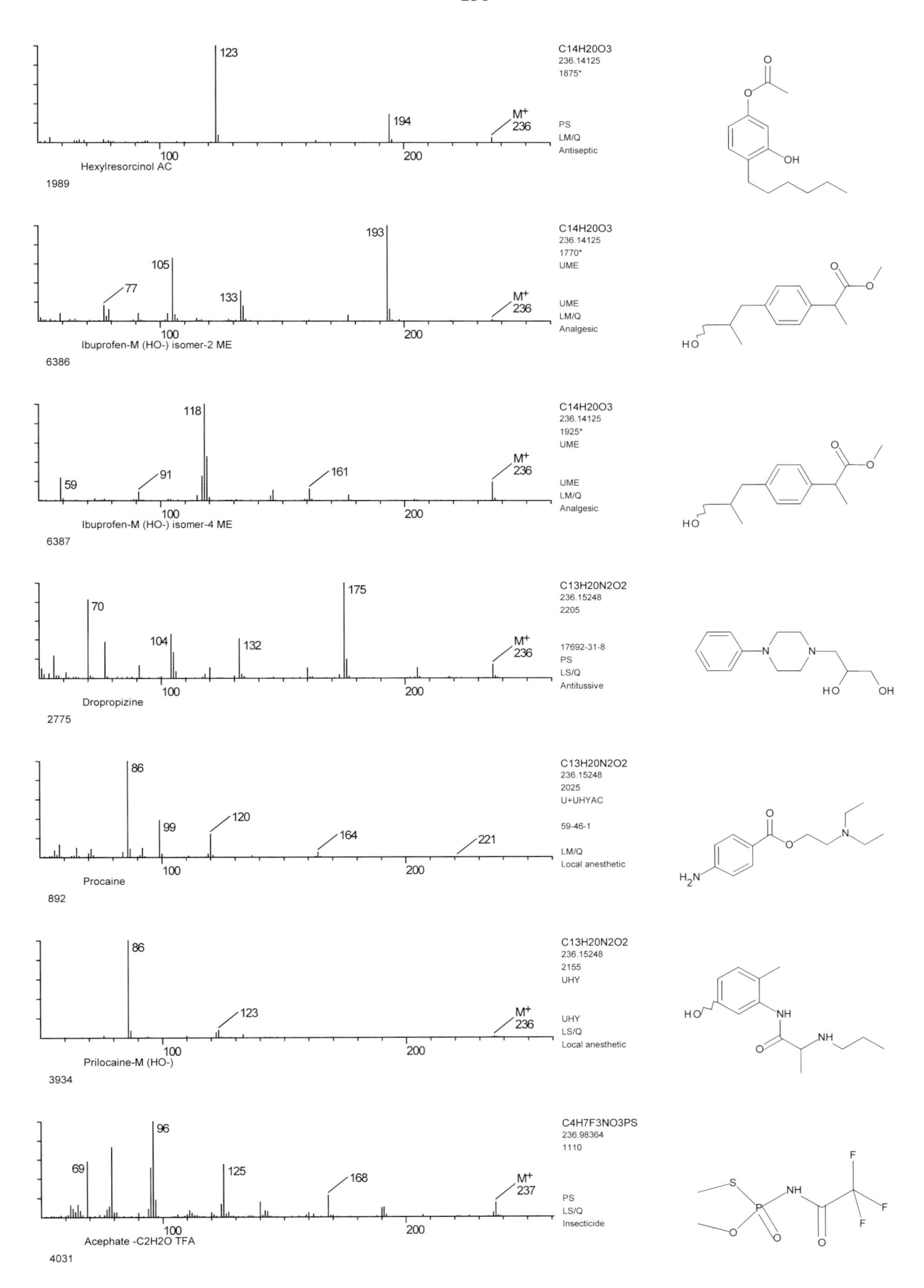

Hexylresorcinol AC
1989
C14H20O3
236.14125
1875*
PS
LM/Q
Antiseptic

Ibuprofen-M (HO-) isomer-2 ME
6386
C14H20O3
236.14125
1770*
UME
UME
LM/Q
Analgesic

Ibuprofen-M (HO-) isomer-4 ME
6387
C14H20O3
236.14125
1925*
UME
UME
LM/Q
Analgesic

Dropropizine
2775
C13H20N2O2
236.15248
2205
17692-31-8
PS
LS/Q
Antitussive

Procaine
892
C13H20N2O2
236.15248
2025
U+UHYAC
59-46-1
LM/Q
Local anesthetic

Prilocaine-M (HO-)
3934
C13H20N2O2
236.15248
2155
UHY
UHY
LS/Q
Local anesthetic

Acephate -C2H2O TFA
4031
C4H7F3NO3PS
236.98364
1110
PS
LS/Q
Insecticide

541

237

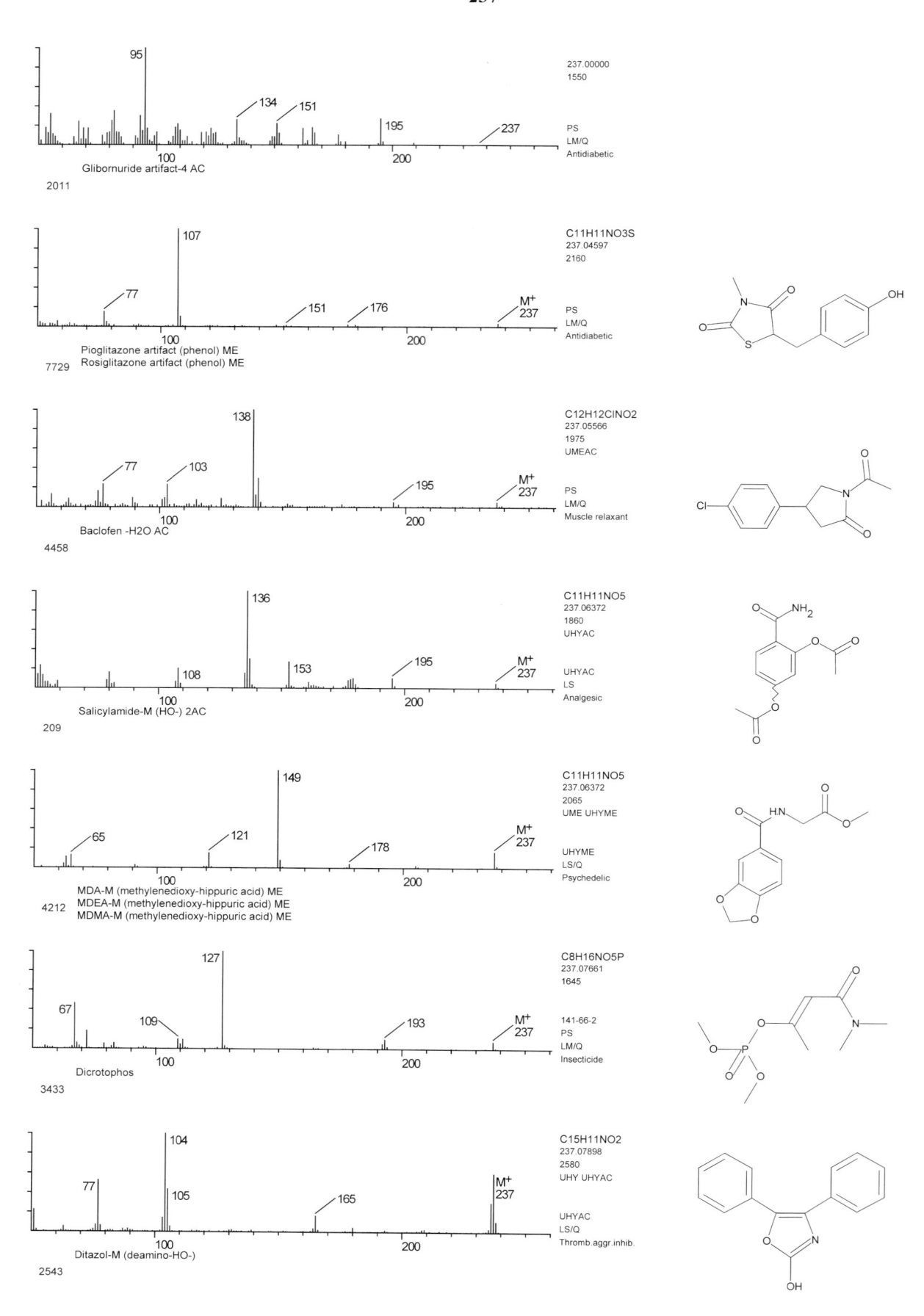

Glibornuride artifact-4 AC
237.00000
1550
PS
LM/Q
Antidiabetic
2011

Pioglitazone artifact (phenol) ME
Rosiglitazone artifact (phenol) ME
C11H11NO3S
237.04597
2160
PS
LM/Q
Antidiabetic
7729

Baclofen -H2O AC
C12H12ClNO2
237.05566
1975
UMEAC
PS
LM/Q
Muscle relaxant
4458

Salicylamide-M (HO-) 2AC
C11H11NO5
237.06372
1860
UHYAC
UHYAC
LS
Analgesic
209

MDA-M (methylenedioxy-hippuric acid) ME
MDEA-M (methylenedioxy-hippuric acid) ME
MDMA-M (methylenedioxy-hippuric acid) ME
C11H11NO5
237.06372
2065
UME UHYME
UHYME
LS/Q
Psychedelic
4212

Dicrotophos
C8H16NO5P
237.07661
1645
141-66-2
PS
LM/Q
Insecticide
3433

Ditazol-M (deamino-HO-)
C15H11NO2
237.07898
2580
UHY UHYAC
UHYAC
LS/Q
Thromb.aggr.inhib.
2543

237

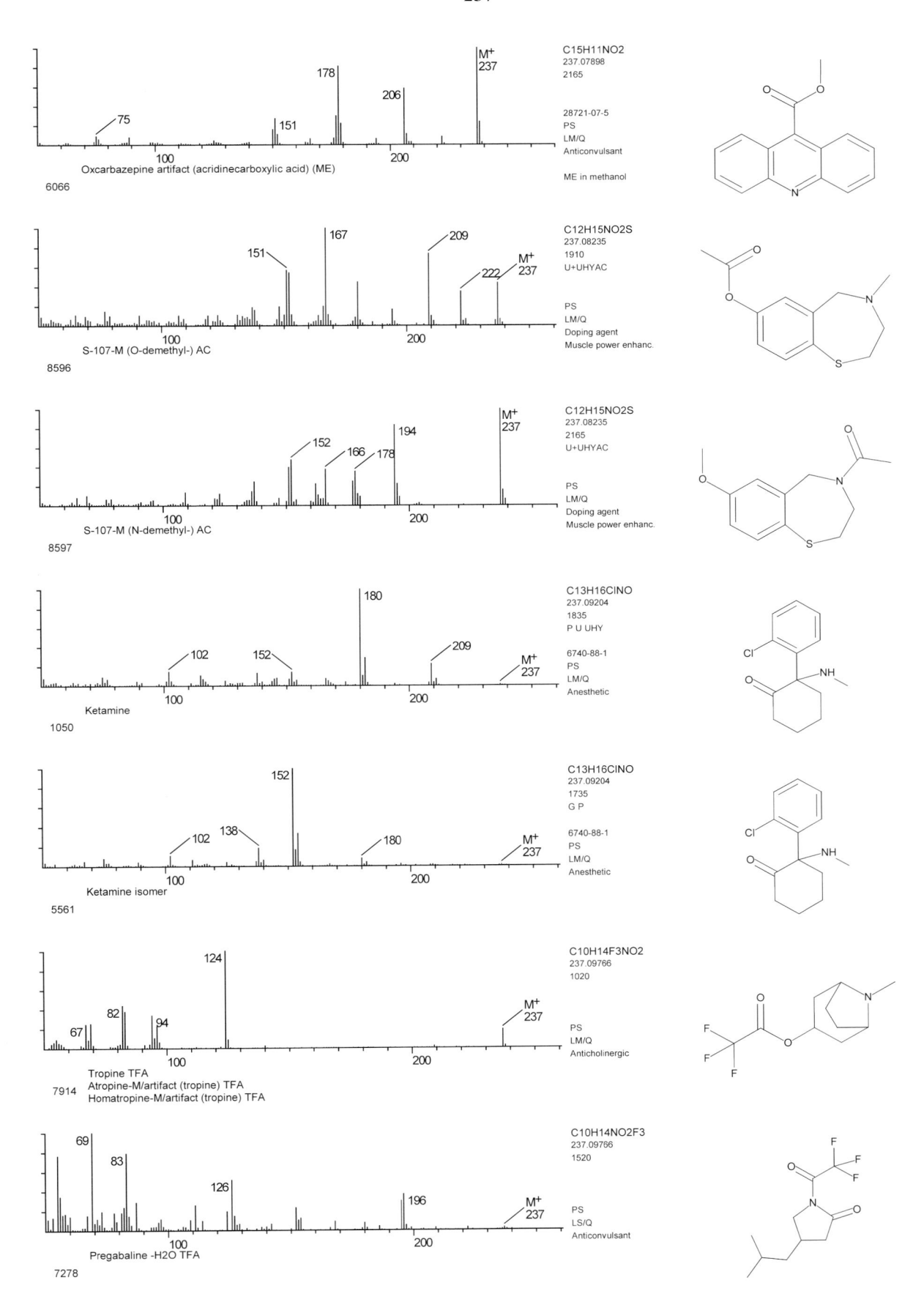

6066 — Oxcarbazepine artifact (acridinecarboxylic acid) (ME)
C15H11NO2
237.07898
2165
28721-07-5
PS
LM/Q
Anticonvulsant
ME in methanol
Peaks: 75, 151, 178, 206, M+ 237

8596 — S-107-M (O-demethyl-) AC
C12H15NO2S
237.08235
1910
U+UHYAC
PS
LM/Q
Doping agent
Muscle power enhanc.
Peaks: 151, 167, 209, 222, M+ 237

8597 — S-107-M (N-demethyl-) AC
C12H15NO2S
237.08235
2165
U+UHYAC
PS
LM/Q
Doping agent
Muscle power enhanc.
Peaks: 152, 166, 178, 194, M+ 237

1050 — Ketamine
C13H16ClNO
237.09204
1835
P U UHY
6740-88-1
PS
LM/Q
Anesthetic
Peaks: 102, 152, 180, 209, M+ 237

5561 — Ketamine isomer
C13H16ClNO
237.09204
1735
G P
6740-88-1
PS
LM/Q
Anesthetic
Peaks: 102, 138, 152, 180, M+ 237

7914 — Tropine TFA / Atropine-M/artifact (tropine) TFA / Homatropine-M/artifact (tropine) TFA
C10H14F3NO2
237.09766
1020
PS
LM/Q
Anticholinergic
Peaks: 67, 82, 94, 124, M+ 237

7278 — Pregabaline -H2O TFA
C10H14NO2F3
237.09766
1520
PS
LS/Q
Anticonvulsant
Peaks: 69, 83, 126, 196, M+ 237

237

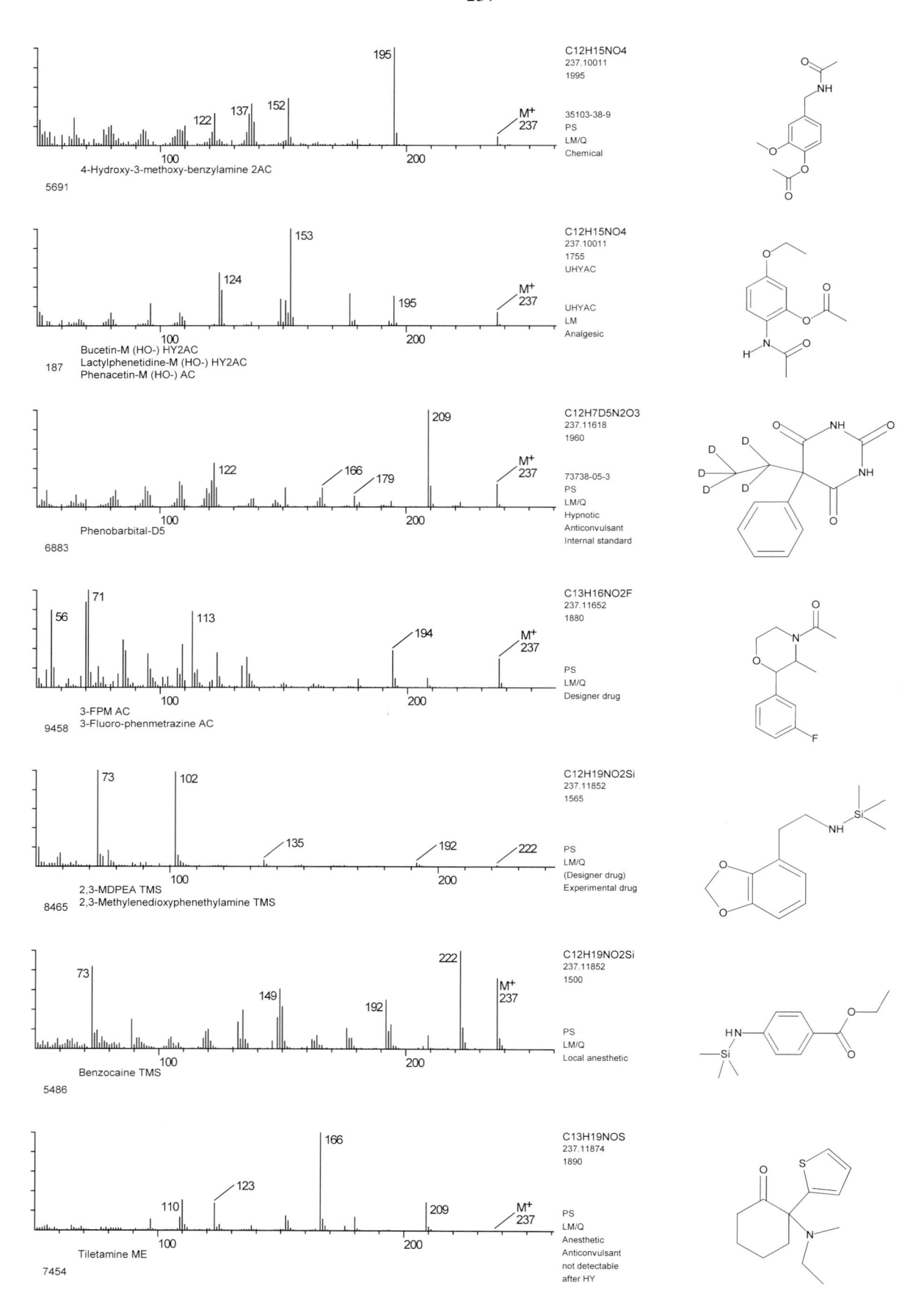

Spectrum	Formula / Info
4-Hydroxy-3-methoxy-benzylamine 2AC (5691) — peaks 122, 137, 152, 195, M+ 237	C12H15NO4; 237.10011; 1995; 35103-38-9; PS; LM/Q; Chemical
Bucetin-M (HO-) HY2AC / Lactylphenetidine-M (HO-) HY2AC / Phenacetin-M (HO-) AC (187) — peaks 124, 153, 195, M+ 237	C12H15NO4; 237.10011; 1755; UHYAC; UHYAC; LM; Analgesic
Phenobarbital-D5 (6883) — peaks 122, 166, 179, 209, M+ 237	C12H7D5N2O3; 237.11618; 1960; 73738-05-3; PS; LM/Q; Hypnotic; Anticonvulsant; Internal standard
3-FPM AC / 3-Fluoro-phenmetrazine AC (9458) — peaks 56, 71, 113, 194, M+ 237	C13H16NO2F; 237.11652; 1880; PS; LM/Q; Designer drug
2,3-MDPEA TMS / 2,3-Methylenedioxyphenethylamine TMS (8465) — peaks 73, 102, 135, 192, 222	C12H19NO2Si; 237.11852; 1565; PS; LM/Q; (Designer drug); Experimental drug
Benzocaine TMS (5486) — peaks 73, 149, 192, 222, M+ 237	C12H19NO2Si; 237.11852; 1500; PS; LM/Q; Local anesthetic
Tiletamine ME (7454) — peaks 110, 123, 166, 209, M+ 237	C13H19NOS; 237.11874; 1890; PS; LM/Q; Anesthetic; Anticonvulsant; not detectable after HY

237

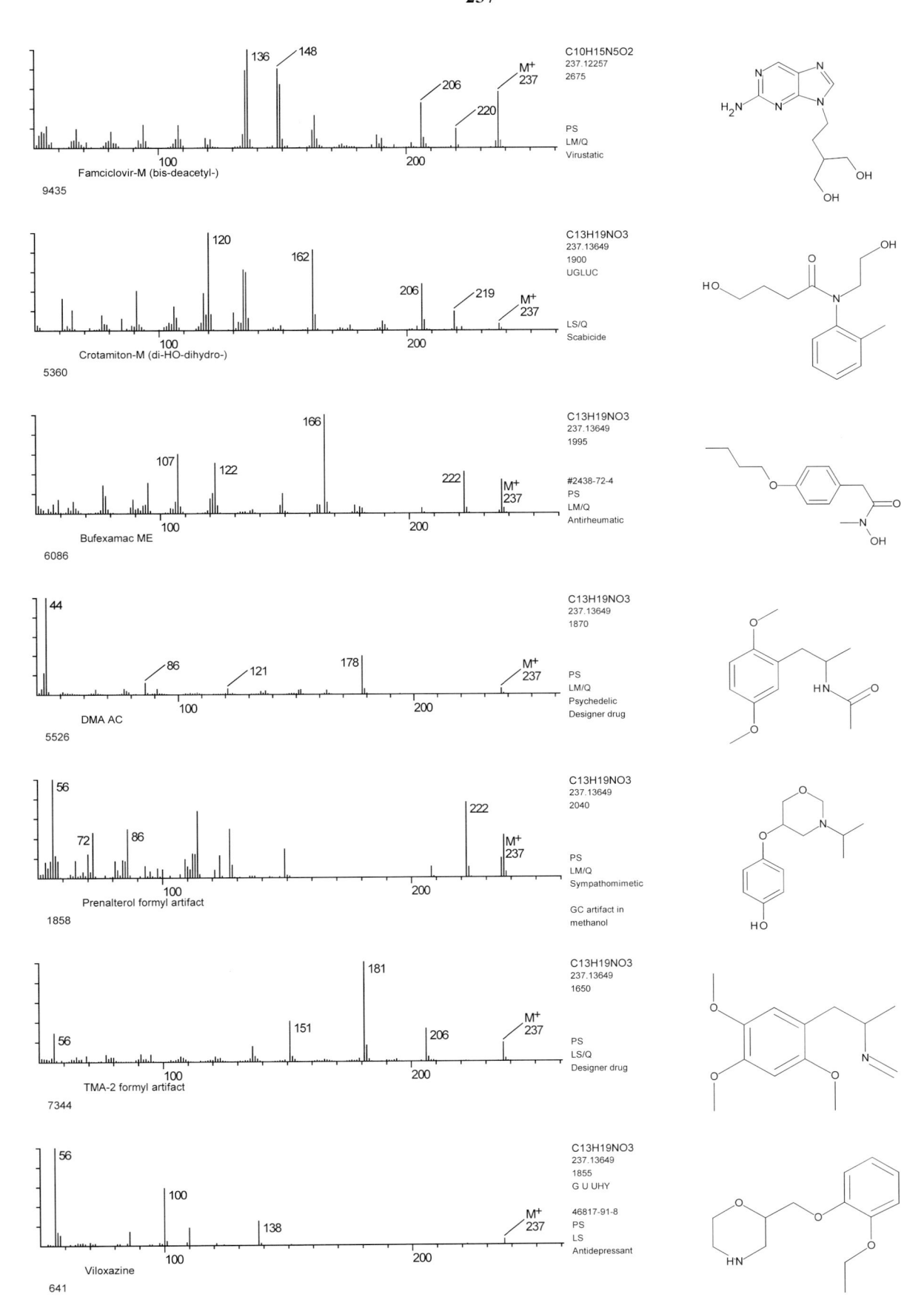

Famciclovir-M (bis-deacetyl-) 9435	Peaks: 136, 148, 206, 220, M+ 237	C10H15N5O2 237.12257 2675 PS LM/Q Virustatic
Crotamiton-M (di-HO-dihydro-) 5360	Peaks: 120, 162, 206, 219, M+ 237	C13H19NO3 237.13649 1900 UGLUC LS/Q Scabicide
Bufexamac ME 6086	Peaks: 107, 122, 166, 222, M+ 237	C13H19NO3 237.13649 1995 #2438-72-4 PS LM/Q Antirheumatic
DMA AC 5526	Peaks: 44, 86, 121, 178, M+ 237	C13H19NO3 237.13649 1870 PS LM/Q Psychedelic Designer drug
Prenalterol formyl artifact 1858	Peaks: 56, 72, 86, 222, M+ 237	C13H19NO3 237.13649 2040 PS LM/Q Sympathomimetic GC artifact in methanol
TMA-2 formyl artifact 7344	Peaks: 56, 151, 181, 206, M+ 237	C13H19NO3 237.13649 1650 PS LS/Q Designer drug
Viloxazine 641	Peaks: 56, 100, 138, M+ 237	C13H19NO3 237.13649 1855 G U UHY 46817-91-8 PS LS Antidepressant

237

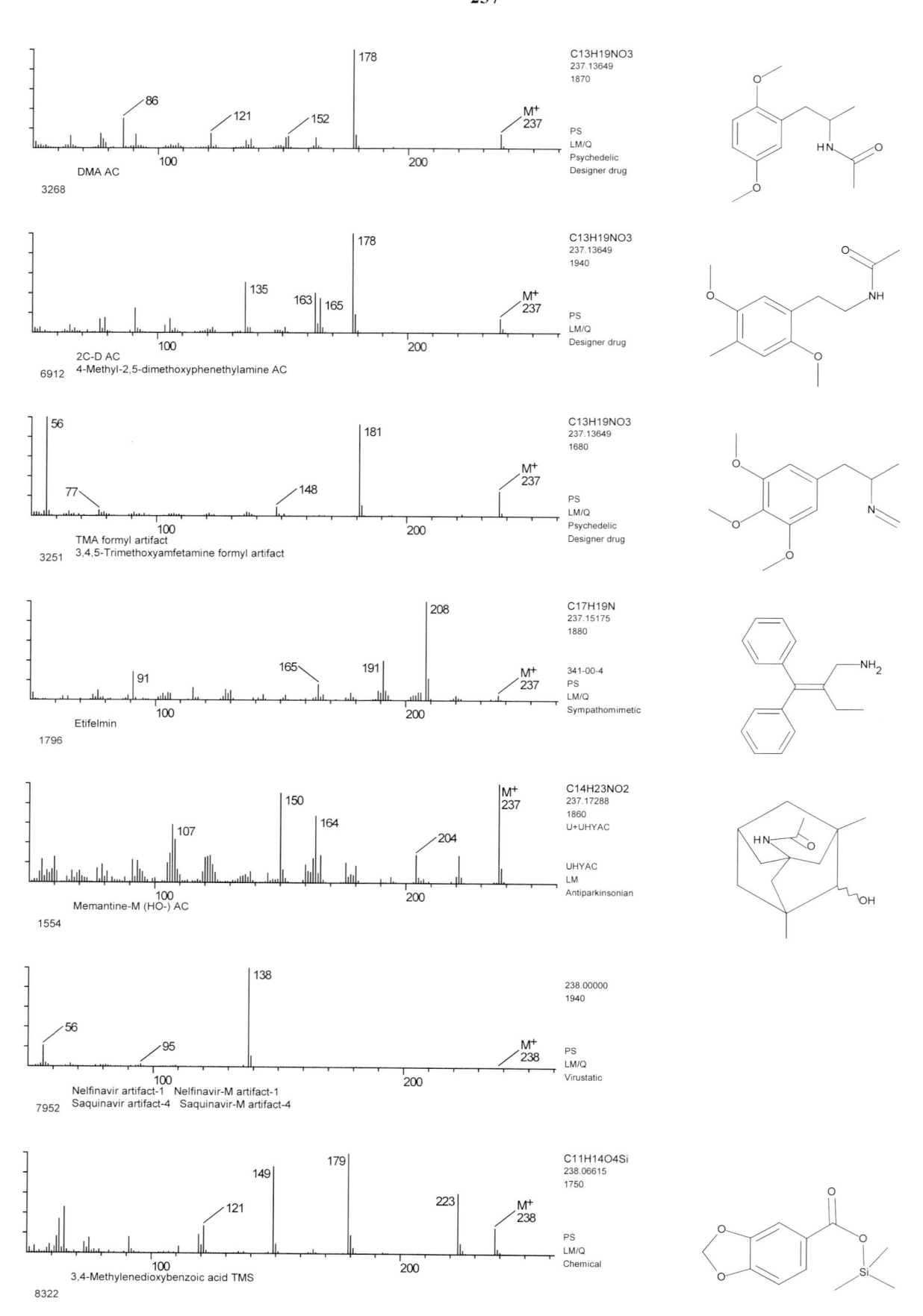

DMA AC		C13H19NO3 237.13649 1870 PS LM/Q Psychedelic Designer drug
2C-D AC 4-Methyl-2,5-dimethoxyphenethylamine AC		C13H19NO3 237.13649 1940 PS LM/Q Designer drug
TMA formyl artifact 3,4,5-Trimethoxyamfetamine formyl artifact		C13H19NO3 237.13649 1680 PS LM/Q Psychedelic Designer drug
Etifelmin		C17H19N 237.15175 1880 341-00-4 PS LM/Q Sympathomimetic
Memantine-M (HO-) AC		C14H23NO2 237.17288 1860 U+UHYAC UHYAC LM Antiparkinsonian
Nelfinavir artifact-1 Nelfinavir-M artifact-1 Saquinavir artifact-4 Saquinavir-M artifact-4		238.00000 1940 PS LM/Q Virustatic
3,4-Methylenedioxybenzoic acid TMS		C11H14O4Si 238.06615 1750 PS LM/Q Chemical

546

238

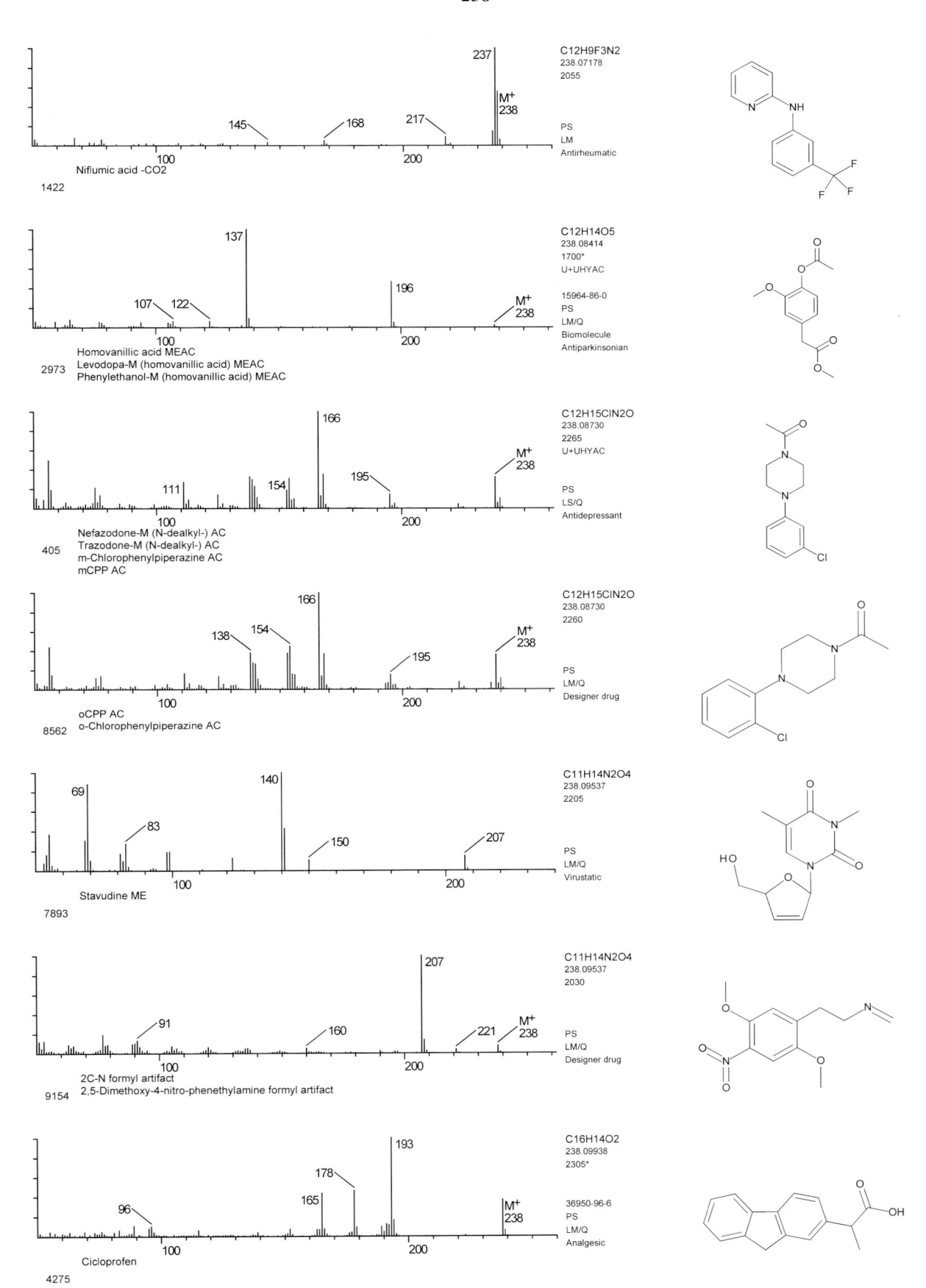

Niflumic acid -CO2 — peaks 145, 168, 217, 237, M+ 238	C12H9F3N2 238.07178 2055 PS LM Antirheumatic
1422	

Homovanillic acid MEAC Levodopa-M (homovanillic acid) MEAC Phenylethanol-M (homovanillic acid) MEAC — peaks 107, 122, 137, 196, M+ 238	C12H14O5 238.08414 1700* U+UHYAC 15964-86-0 PS LM/Q Biomolecule Antiparkinsonian
2973	

Nefazodone-M (N-dealkyl-) AC Trazodone-M (N-dealkyl-) AC m-Chlorophenylpiperazine AC mCPP AC — peaks 111, 154, 166, 195, M+ 238	C12H15ClN2O 238.08730 2265 U+UHYAC PS LS/Q Antidepressant
405	

oCPP AC o-Chlorophenylpiperazine AC — peaks 138, 154, 166, 195, M+ 238	C12H15ClN2O 238.08730 2260 PS LM/Q Designer drug
8562	

Stavudine ME — peaks 69, 83, 140, 150, 207	C11H14N2O4 238.09537 2205 PS LM/Q Virustatic
7893	

2C-N formyl artifact 2,5-Dimethoxy-4-nitro-phenethylamine formyl artifact — peaks 91, 160, 207, 221, M+ 238	C11H14N2O4 238.09537 2030 PS LM/Q Designer drug
9154	

Cicloprofen — peaks 96, 165, 178, 193, M+ 238	C16H14O2 238.09938 2305* 36950-96-6 PS LM/Q Analgesic
4275	

238

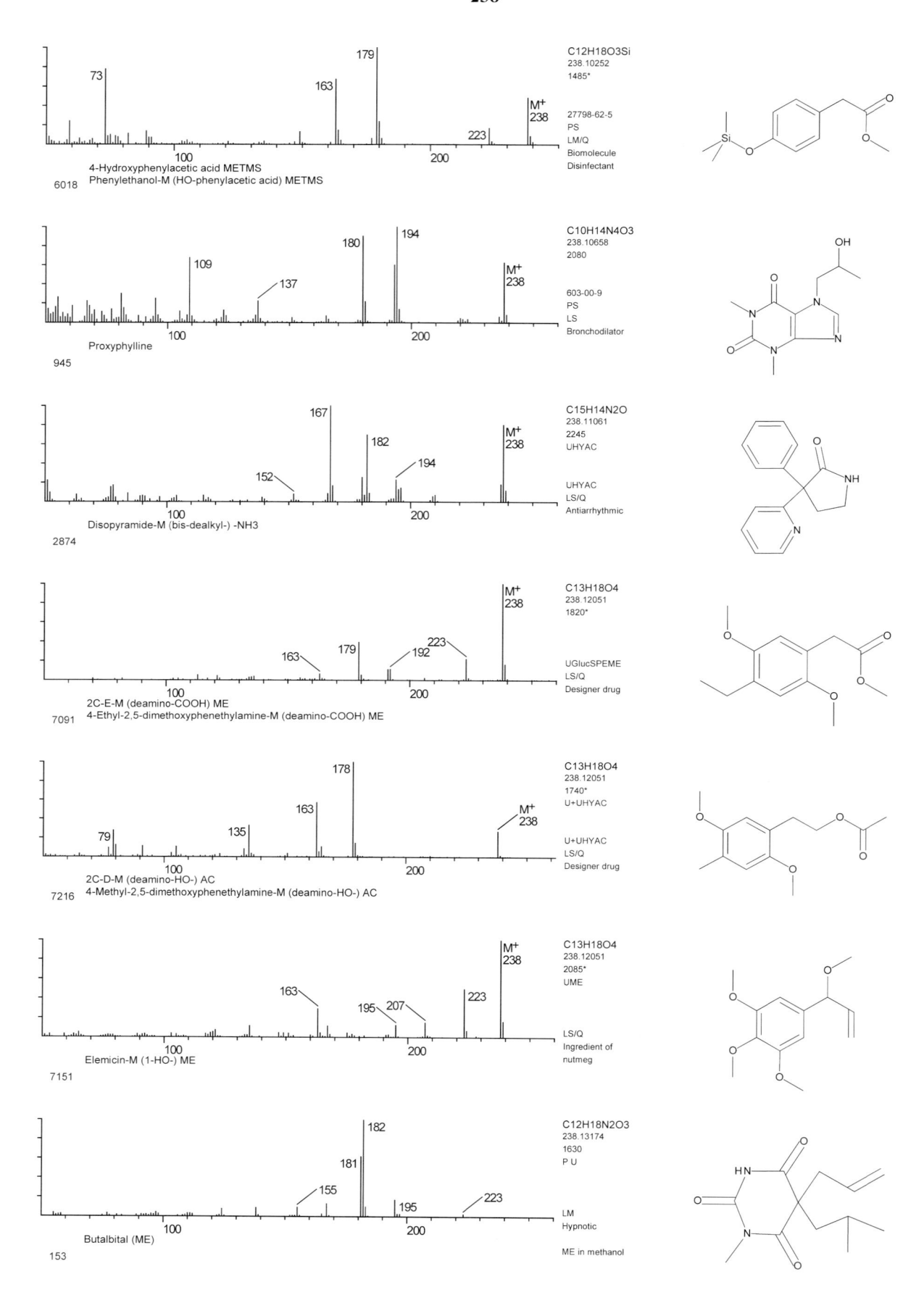

4-Hydroxyphenylacetic acid METMS Phenylethanol-M (HO-phenylacetic acid) METMS 6018	C12H18O3Si 238.10252 1485* 27798-62-5 PS LM/Q Biomolecule Disinfectant	
Proxyphylline 945	C10H14N4O3 238.10658 2080 603-00-9 PS LS Bronchodilator	
Disopyramide-M (bis-dealkyl-) -NH3 2874	C15H14N2O 238.11061 2245 UHYAC UHYAC LS/Q Antiarrhythmic	
2C-E-M (deamino-COOH) ME 4-Ethyl-2,5-dimethoxyphenethylamine-M (deamino-COOH) ME 7091	C13H18O4 238.12051 1820* UGlucSPEME LS/Q Designer drug	
2C-D-M (deamino-HO-) AC 4-Methyl-2,5-dimethoxyphenethylamine-M (deamino-HO-) AC 7216	C13H18O4 238.12051 1740* U+UHYAC U+UHYAC LS/Q Designer drug	
Elemicin-M (1-HO-) ME 7151	C13H18O4 238.12051 2085* UME LS/Q Ingredient of nutmeg	
Butalbital (ME) 153	C12H18N2O3 238.13174 1630 P U LM Hypnotic ME in methanol	

238

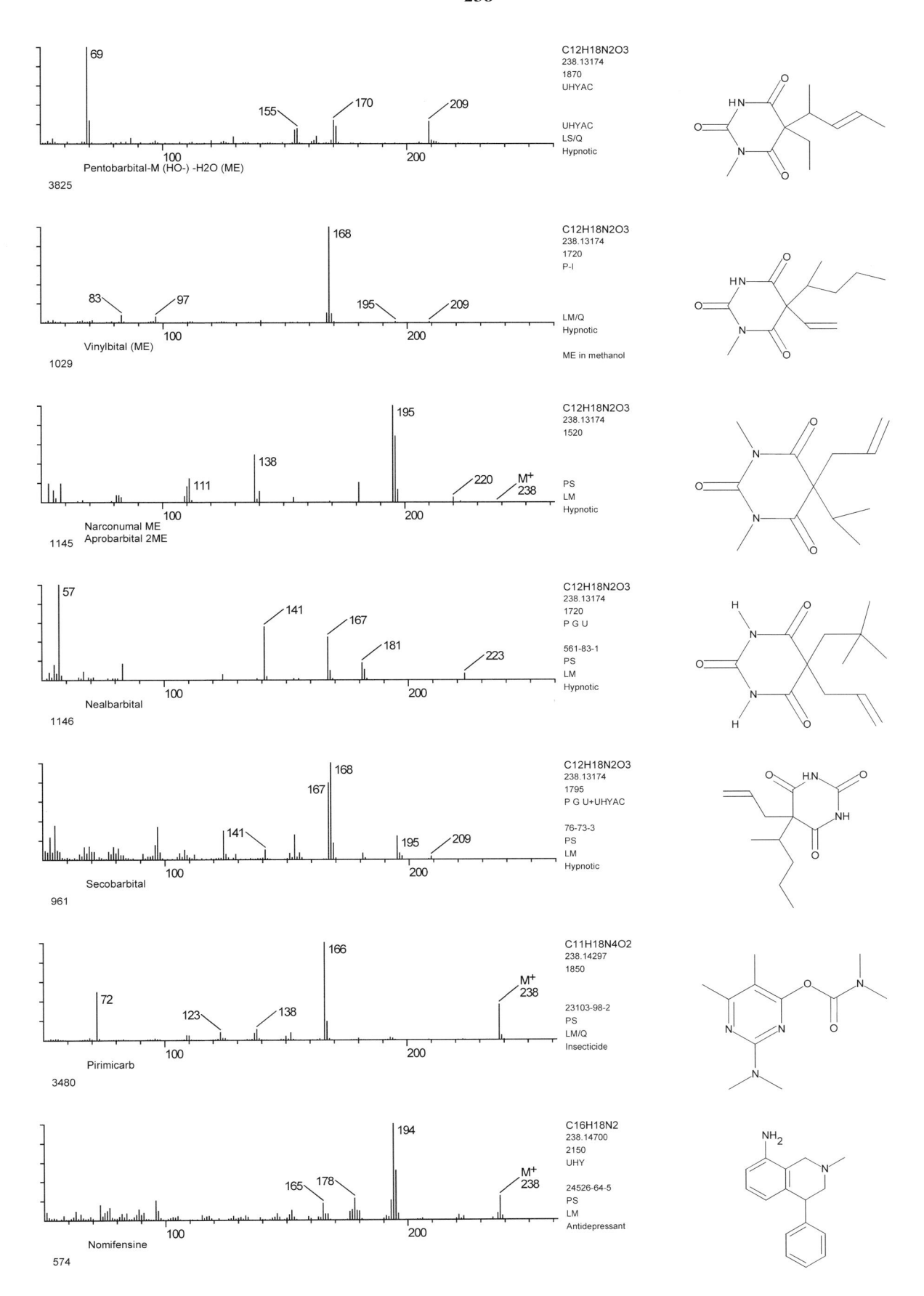

C12H18N2O3
238.13174
1870
UHYAC

UHYAC
LS/Q
Hypnotic

Pentobarbital-M (HO-) -H2O (ME)
3825

C12H18N2O3
238.13174
1720
P-I

LM/Q
Hypnotic
ME in methanol

Vinylbital (ME)
1029

C12H18N2O3
238.13174
1520

PS
LM
Hypnotic

Narconumal ME
Aprobarbital 2ME
1145

C12H18N2O3
238.13174
1720
P G U

561-83-1
PS
LM
Hypnotic

Nealbarbital
1146

C12H18N2O3
238.13174
1795
P G U+UHYAC

76-73-3
PS
LM
Hypnotic

Secobarbital
961

C11H18N4O2
238.14297
1850

23103-98-2
PS
LM/Q
Insecticide

Pirimicarb
3480

C16H18N2
238.14700
2150
UHY

24526-64-5
PS
LM
Antidepressant

Nomifensine
574

549

239

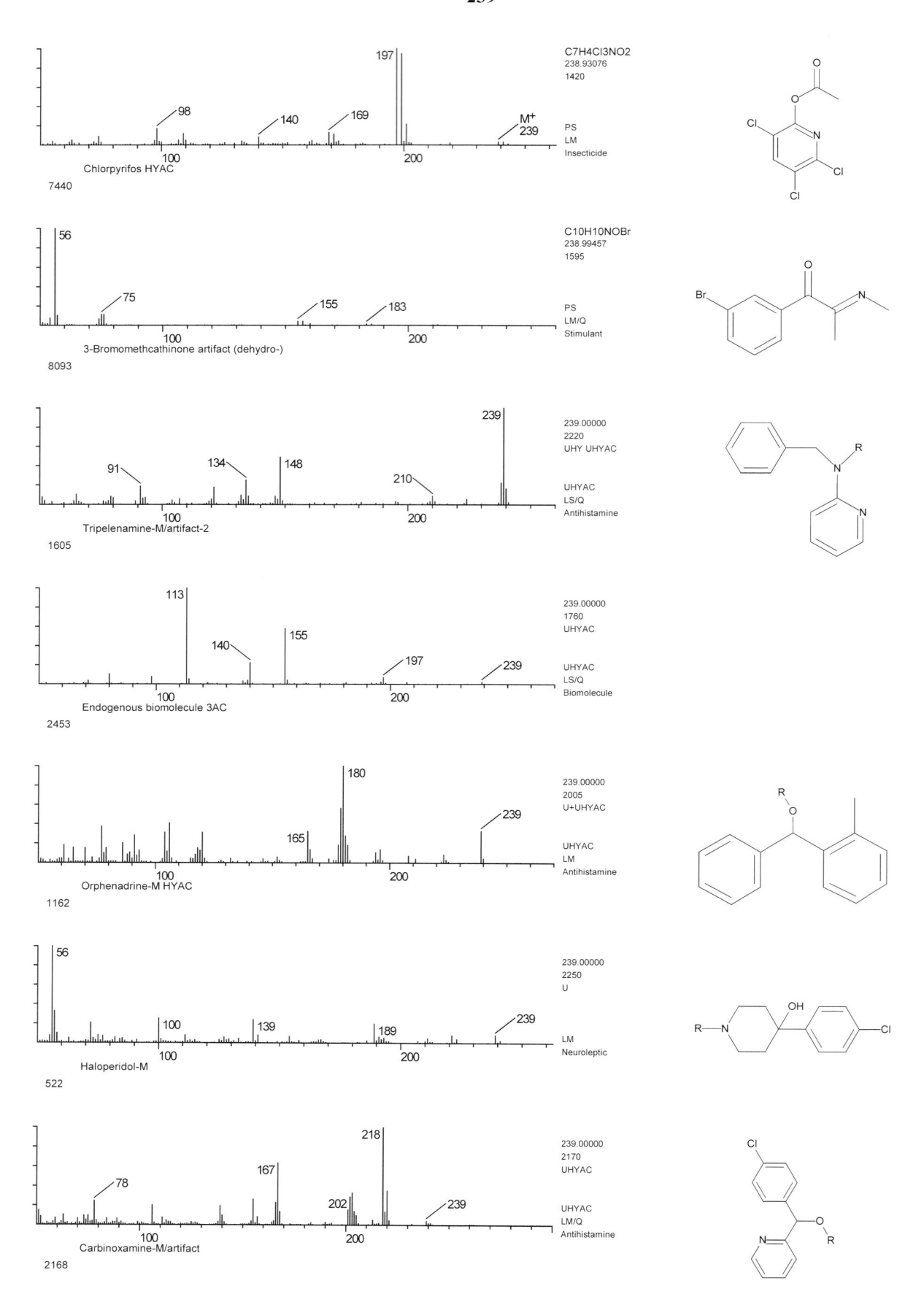

Chlorpyrifos HYAC 7440	98, 140, 169, 197, M+ 239 C7H4Cl3NO2 238.93076 1420 PS LM Insecticide
3-Bromomethcathinone artifact (dehydro-) 8093	56, 75, 155, 183 C10H10NOBr 238.99457 1595 PS LM/Q Stimulant
Tripelenamine-M/artifact-2 1605	91, 134, 148, 210, 239 239.00000 2220 UHY UHYAC UHYAC LS/Q Antihistamine
Endogenous biomolecule 3AC 2453	113, 140, 155, 197, 239 239.00000 1760 UHYAC UHYAC LS/Q Biomolecule
Orphenadrine-M HYAC 1162	165, 180, 239 239.00000 2005 U+UHYAC UHYAC LM Antihistamine
Haloperidol-M 522	56, 100, 139, 189, 239 239.00000 2250 U LM Neuroleptic
Carbinoxamine-M/artifact 2168	78, 167, 202, 218, 239 239.00000 2170 UHYAC UHYAC LM/Q Antihistamine

239

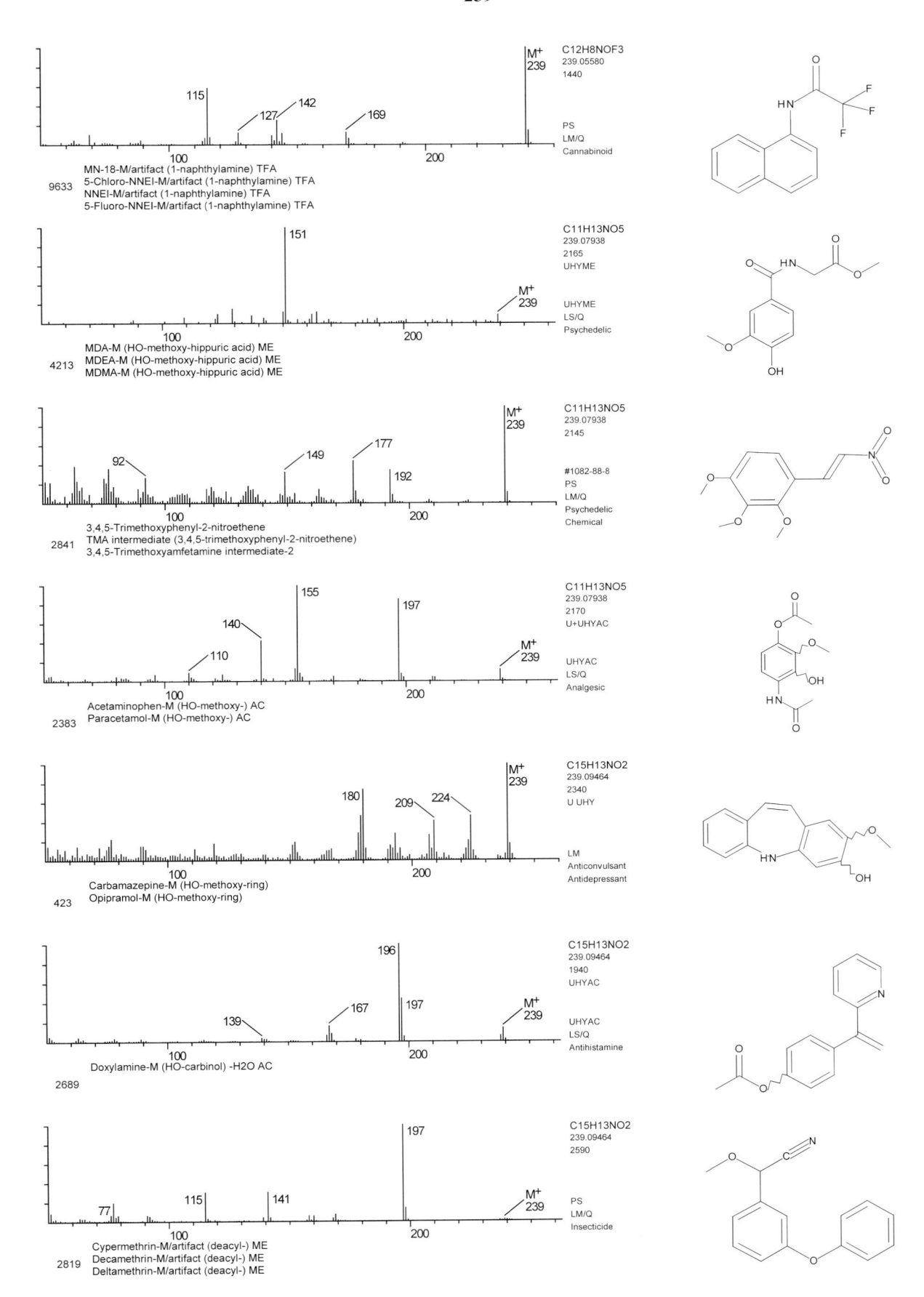

	C12H8NOF3 239.05580 1440 PS LM/Q Cannabinoid

9633 MN-18-M/artifact (1-naphthylamine) TFA
5-Chloro-NNEI-M/artifact (1-naphthylamine) TFA
NNEI-M/artifact (1-naphthylamine) TFA
5-Fluoro-NNEI-M/artifact (1-naphthylamine) TFA

	C11H13NO5 239.07938 2165 UHYME UHYME LS/Q Psychedelic

4213 MDA-M (HO-methoxy-hippuric acid) ME
MDEA-M (HO-methoxy-hippuric acid) ME
MDMA-M (HO-methoxy-hippuric acid) ME

	C11H13NO5 239.07938 2145 #1082-88-8 PS LM/Q Psychedelic Chemical

2841 3,4,5-Trimethoxyphenyl-2-nitroethene
TMA intermediate (3,4,5-trimethoxyphenyl-2-nitroethene)
3,4,5-Trimethoxyamfetamine intermediate-2

	C11H13NO5 239.07938 2170 U+UHYAC UHYAC LS/Q Analgesic

2383 Acetaminophen-M (HO-methoxy-) AC
Paracetamol-M (HO-methoxy-) AC

	C15H13NO2 239.09464 2340 U UHY LM Anticonvulsant Antidepressant

423 Carbamazepine-M (HO-methoxy-ring)
Opipramol-M (HO-methoxy-ring)

	C15H13NO2 239.09464 1940 UHYAC UHYAC LS/Q Antihistamine

2689 Doxylamine-M (HO-carbinol) -H2O AC

	C15H13NO2 239.09464 2590 PS LM/Q Insecticide

2819 Cypermethrin-M/artifact (deacyl-) ME
Decamethrin-M/artifact (deacyl-) ME
Deltamethrin-M/artifact (deacyl-) ME

239

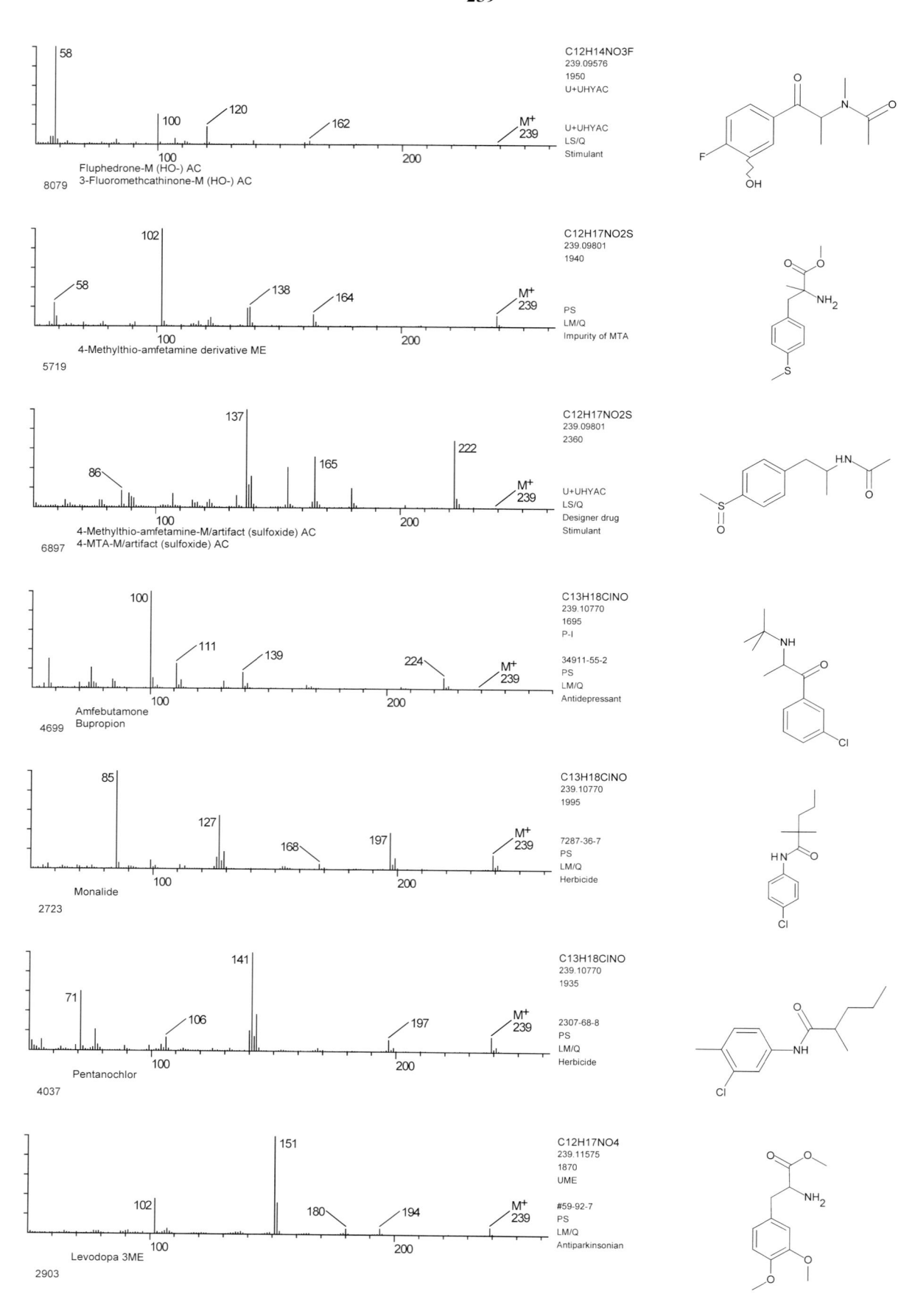

Fluphedrone-M (HO-) AC
3-Fluoromethcathinone-M (HO-) AC
8079

C12H14NO3F
239.09576
1950
U+UHYAC
U+UHYAC
LS/Q
Stimulant

4-Methylthio-amfetamine derivative ME
5719

C12H17NO2S
239.09801
1940
PS
LM/Q
Impurity of MTA

4-Methylthio-amfetamine-M/artifact (sulfoxide) AC
4-MTA-M/artifact (sulfoxide) AC
6897

C12H17NO2S
239.09801
2360
U+UHYAC
LS/Q
Designer drug
Stimulant

Amfebutamone
Bupropion
4699

C13H18ClNO
239.10770
1695
P-I
34911-55-2
PS
LM/Q
Antidepressant

Monalide
2723

C13H18ClNO
239.10770
1995
7287-36-7
PS
LM/Q
Herbicide

Pentanochlor
4037

C13H18ClNO
239.10770
1935
2307-68-8
PS
LM/Q
Herbicide

Levodopa 3ME
2903

C12H17NO4
239.11575
1870
UME
#59-92-7
PS
LM/Q
Antiparkinsonian

552

239

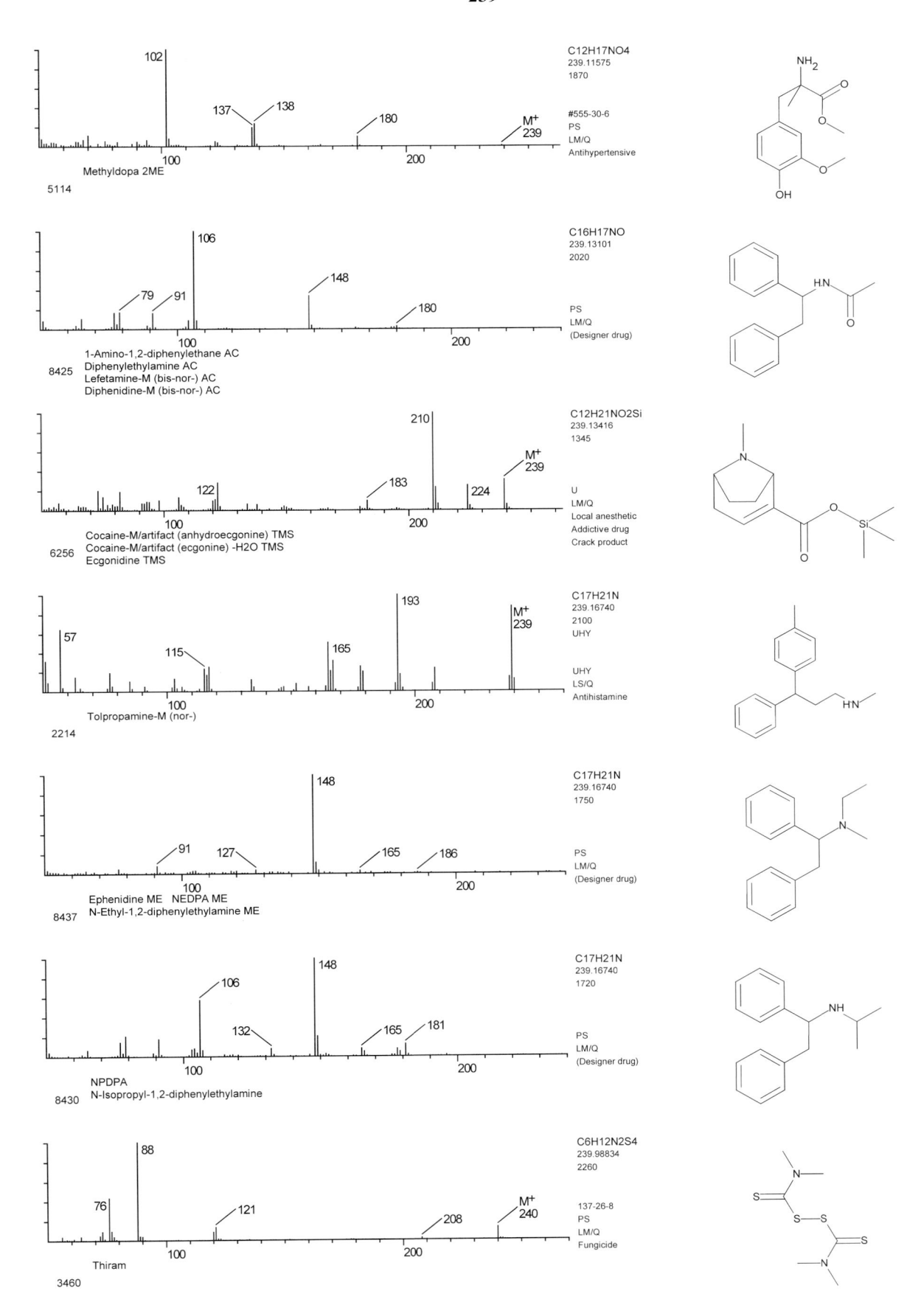

Methyldopa 2ME
5114
C12H17NO4
239.11575
1870
#555-30-6
PS
LM/Q
Antihypertensive

1-Amino-1,2-diphenylethane AC
Diphenylethylamine AC
Lefetamine-M (bis-nor-) AC
Diphenidine-M (bis-nor-) AC
8425
C16H17NO
239.13101
2020
PS
LM/Q
(Designer drug)

Cocaine-M/artifact (anhydroecgonine) TMS
Cocaine-M/artifact (ecgonine) -H2O TMS
Ecgonidine TMS
6256
C12H21NO2Si
239.13416
1345
U
LM/Q
Local anesthetic
Addictive drug
Crack product

Tolpropamine-M (nor-)
2214
C17H21N
239.16740
2100
UHY
UHY
LS/Q
Antihistamine

Ephenidine ME   NEDPA ME
N-Ethyl-1,2-diphenylethylamine ME
8437
C17H21N
239.16740
1750
PS
LM/Q
(Designer drug)

NPDPA
N-Isopropyl-1,2-diphenylethylamine
8430
C17H21N
239.16740
1720
PS
LM/Q
(Designer drug)

Thiram
3460
C6H12N2S4
239.98834
2260
137-26-8
PS
LM/Q
Fungicide

240

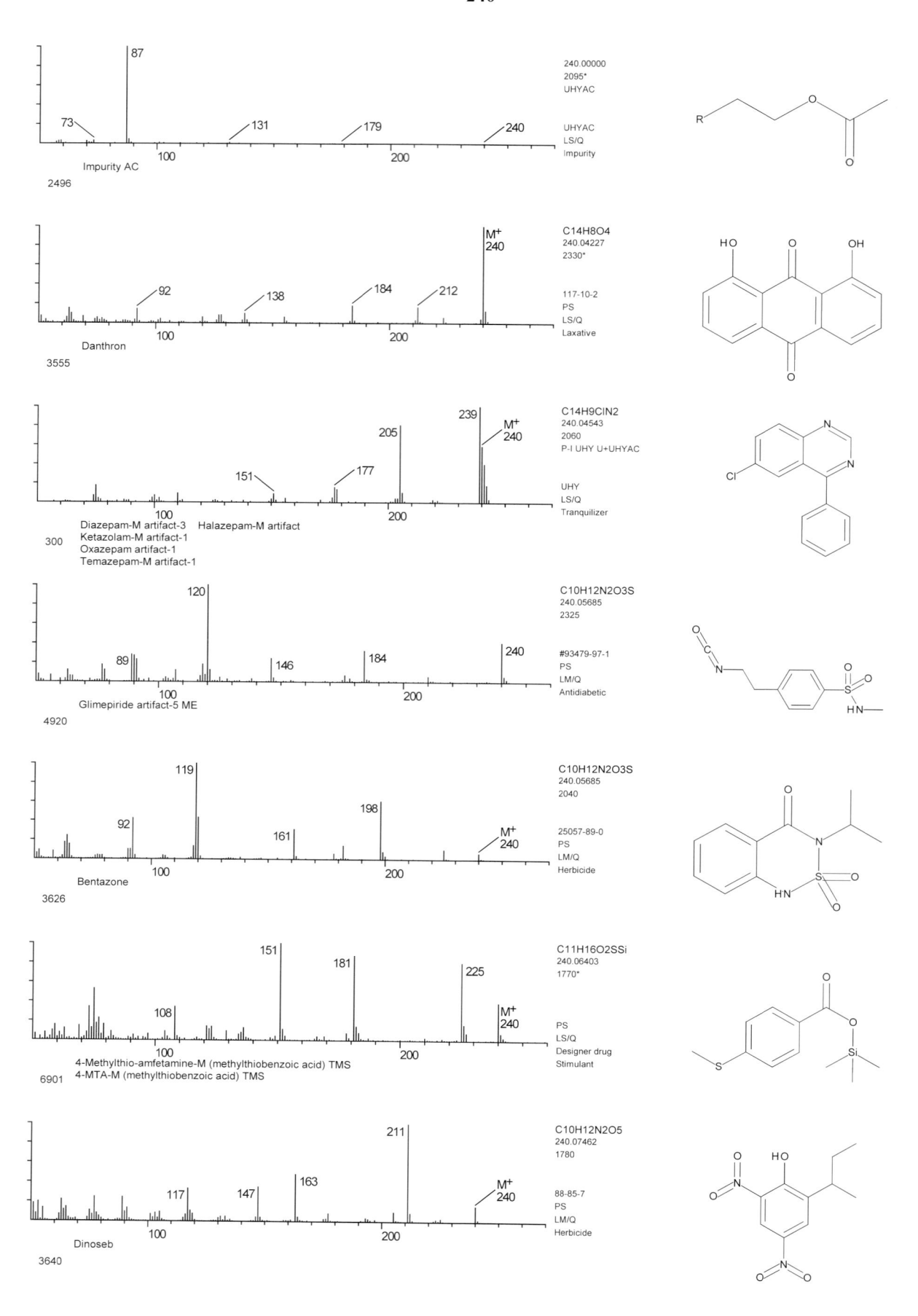

Impurity AC — peaks 73, 87, 131, 179, 240	240.00000 / 2095* / UHYAC / UHYAC / LS/Q / Impurity
2496	

Danthron — peaks 92, 138, 184, 212, 240 (M+)	C14H8O4 / 240.04227 / 2330* / 117-10-2 / PS / LS/Q / Laxative
3555	

Diazepam-M artifact-3   Halazepam-M artifact
Ketazolam-M artifact-1
Oxazepam artifact-1
Temazepam-M artifact-1
— peaks 151, 177, 205, 239, 240 (M+)
C14H9ClN2 / 240.04543 / 2060 / P-I UHY U+UHYAC / UHY / LS/Q / Tranquilizer
300

Glimepiride artifact-5 ME — peaks 89, 120, 146, 184, 240	C10H12N2O3S / 240.05685 / 2325 / #93479-97-1 / PS / LM/Q / Antidiabetic
4920	

Bentazone — peaks 92, 119, 161, 198, 240 (M+)	C10H12N2O3S / 240.05685 / 2040 / 25057-89-0 / PS / LM/Q / Herbicide
3626	

4-Methylthio-amfetamine-M (methylthiobenzoic acid) TMS / 4-MTA-M (methylthiobenzoic acid) TMS — peaks 108, 151, 181, 225, 240 (M+)	C11H16O2SSi / 240.06403 / 1770* / PS / LS/Q / Designer drug / Stimulant
6901	

Dinoseb — peaks 117, 147, 163, 211, 240 (M+)	C10H12N2O5 / 240.07462 / 1780 / 88-85-7 / PS / LM/Q / Herbicide
3640	

240

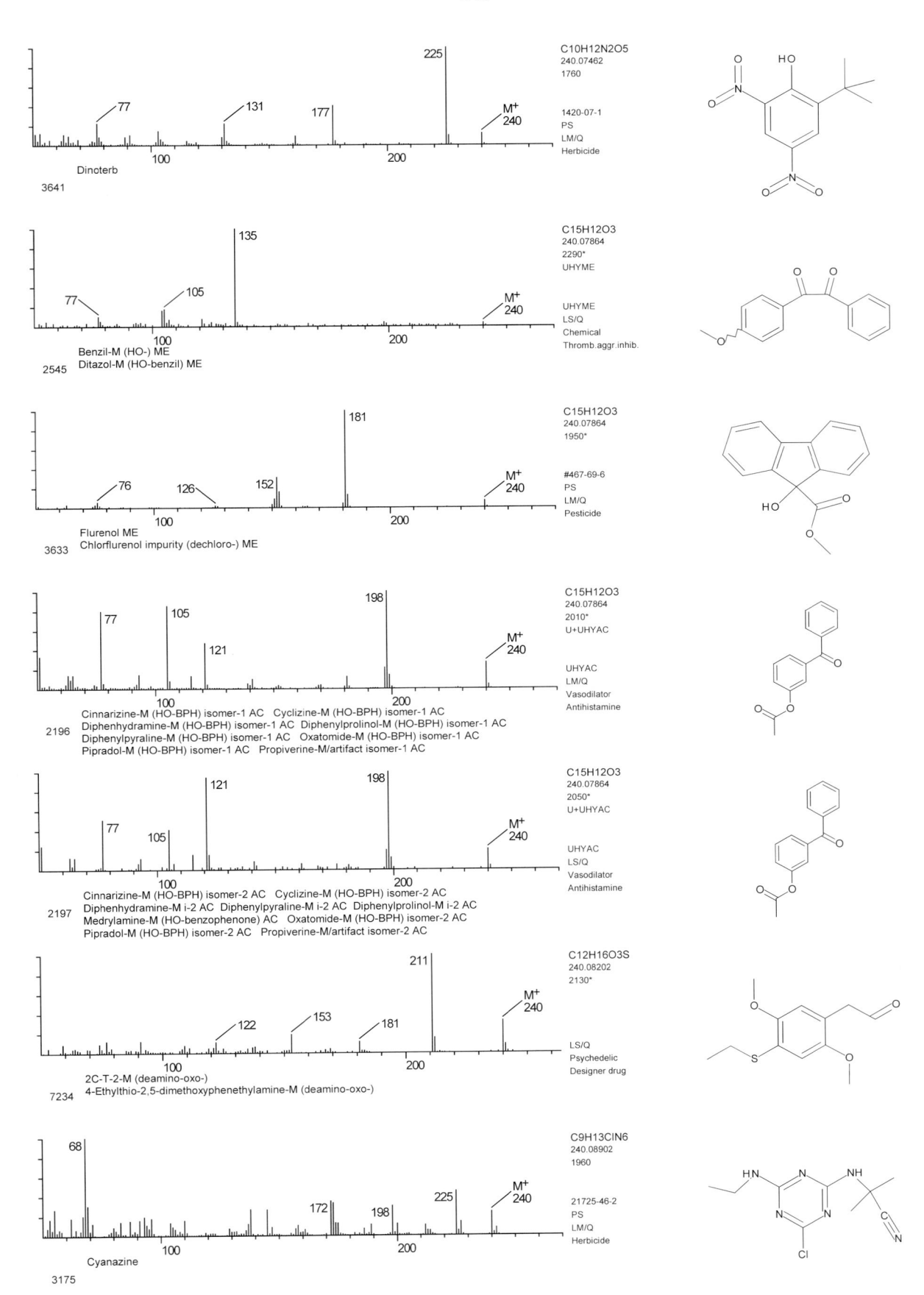

Dinoterb	77, 131, 177, 225, M+ 240	C10H12N2O5 240.07462 1760 1420-07-1 PS LM/Q Herbicide
Benzil-M (HO-) ME Ditazol-M (HO-benzil) ME	77, 105, 135, M+ 240	C15H12O3 240.07864 2290* UHYME UHYME LS/Q Chemical Thromb.aggr.inhib.
Flurenol ME Chlorflurenol impurity (dechloro-) ME	76, 126, 152, 181, M+ 240	C15H12O3 240.07864 1950* #467-69-6 PS LM/Q Pesticide
Cinnarizine-M (HO-BPH) isomer-1 AC Cyclizine-M (HO-BPH) isomer-1 AC Diphenhydramine-M (HO-BPH) isomer-1 AC Diphenylprolinol-M (HO-BPH) isomer-1 AC Diphenylpyraline-M (HO-BPH) isomer-1 AC Oxatomide-M (HO-BPH) isomer-1 AC Pipradol-M (HO-BPH) isomer-1 AC Propiverine-M/artifact isomer-1 AC	77, 105, 121, 198, M+ 240	C15H12O3 240.07864 2010* U+UHYAC UHYAC LM/Q Vasodilator Antihistamine
Cinnarizine-M (HO-BPH) isomer-2 AC Cyclizine-M (HO-BPH) isomer-2 AC Diphenhydramine-M i-2 AC Diphenylpyraline-M i-2 AC Diphenylprolinol-M i-2 AC Medrylamine-M (HO-benzophenone) AC Oxatomide-M (HO-BPH) isomer-2 AC Pipradol-M (HO-BPH) isomer-2 AC Propiverine-M/artifact isomer-2 AC	77, 105, 121, 198, M+ 240	C15H12O3 240.07864 2050* U+UHYAC UHYAC LS/Q Vasodilator Antihistamine
2C-T-2-M (deamino-oxo-) 4-Ethylthio-2,5-dimethoxyphenethylamine-M (deamino-oxo-)	122, 153, 181, 211, M+ 240	C12H16O3S 240.08202 2130* LS/Q Psychedelic Designer drug
Cyanazine	68, 172, 198, 225, M+ 240	C9H13ClN6 240.08902 1960 21725-46-2 PS LM/Q Herbicide

240

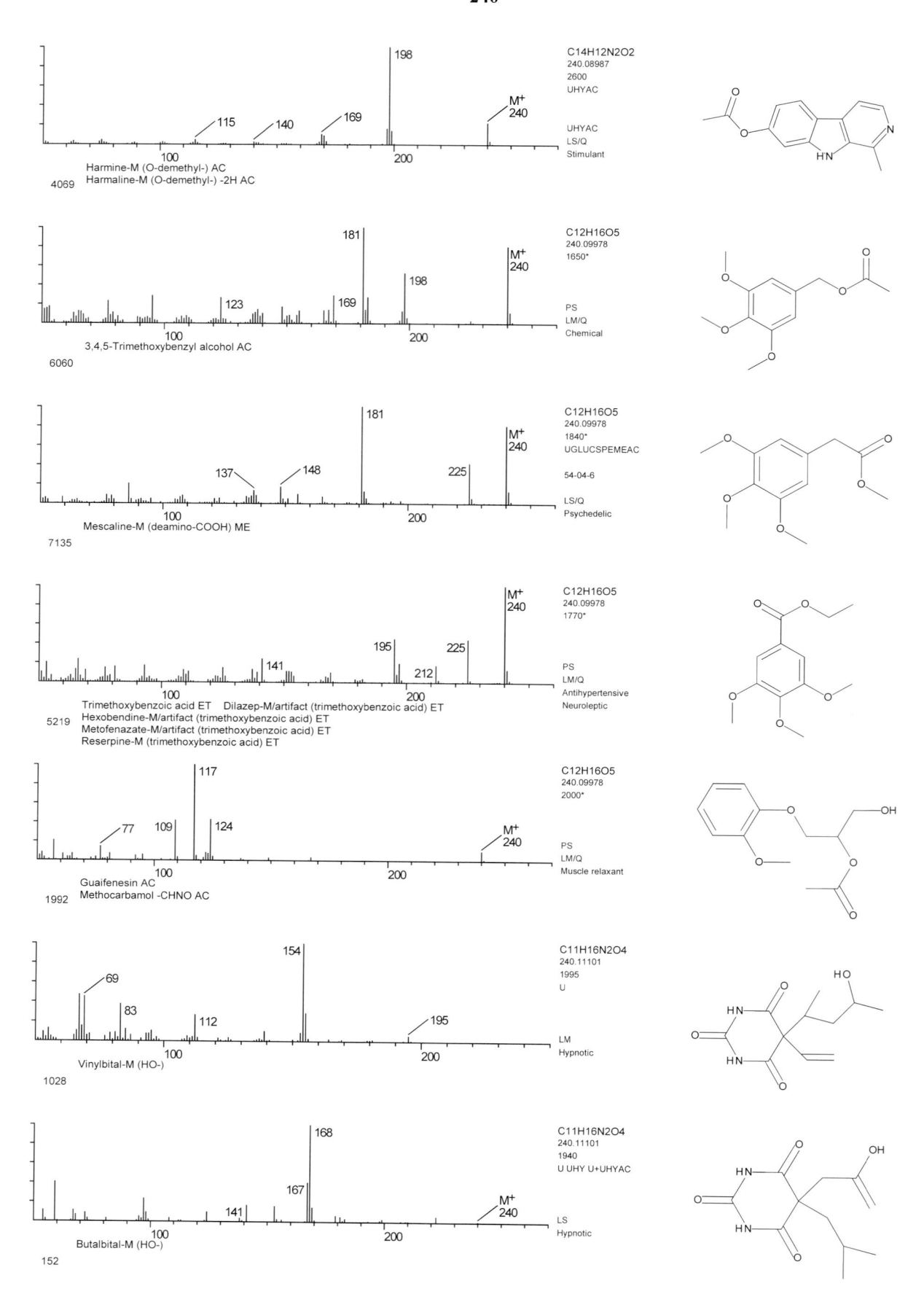

4069	Harmine-M (O-demethyl-) AC Harmaline-M (O-demethyl-) -2H AC	C14H12N2O2 240.08987 2600 UHYAC UHYAC LS/Q Stimulant
6060	3,4,5-Trimethoxybenzyl alcohol AC	C12H16O5 240.09978 1650* PS LM/Q Chemical
7135	Mescaline-M (deamino-COOH) ME	C12H16O5 240.09978 1840* UGLUCSPEMEAC 54-04-6 LS/Q Psychedelic
5219	Trimethoxybenzoic acid ET Dilazep-M/artifact (trimethoxybenzoic acid) ET Hexobendine-M/artifact (trimethoxybenzoic acid) ET Metofenazate-M/artifact (trimethoxybenzoic acid) ET Reserpine-M (trimethoxybenzoic acid) ET	C12H16O5 240.09978 1770* PS LM/Q Antihypertensive Neuroleptic
1992	Guaifenesin AC Methocarbamol -CHNO AC	C12H16O5 240.09978 2000* PS LM/Q Muscle relaxant
1028	Vinylbital-M (HO-)	C11H16N2O4 240.11101 1995 U LM Hypnotic
152	Butalbital-M (HO-)	C11H16N2O4 240.11101 1940 U UHY U+UHYAC LS Hypnotic

240

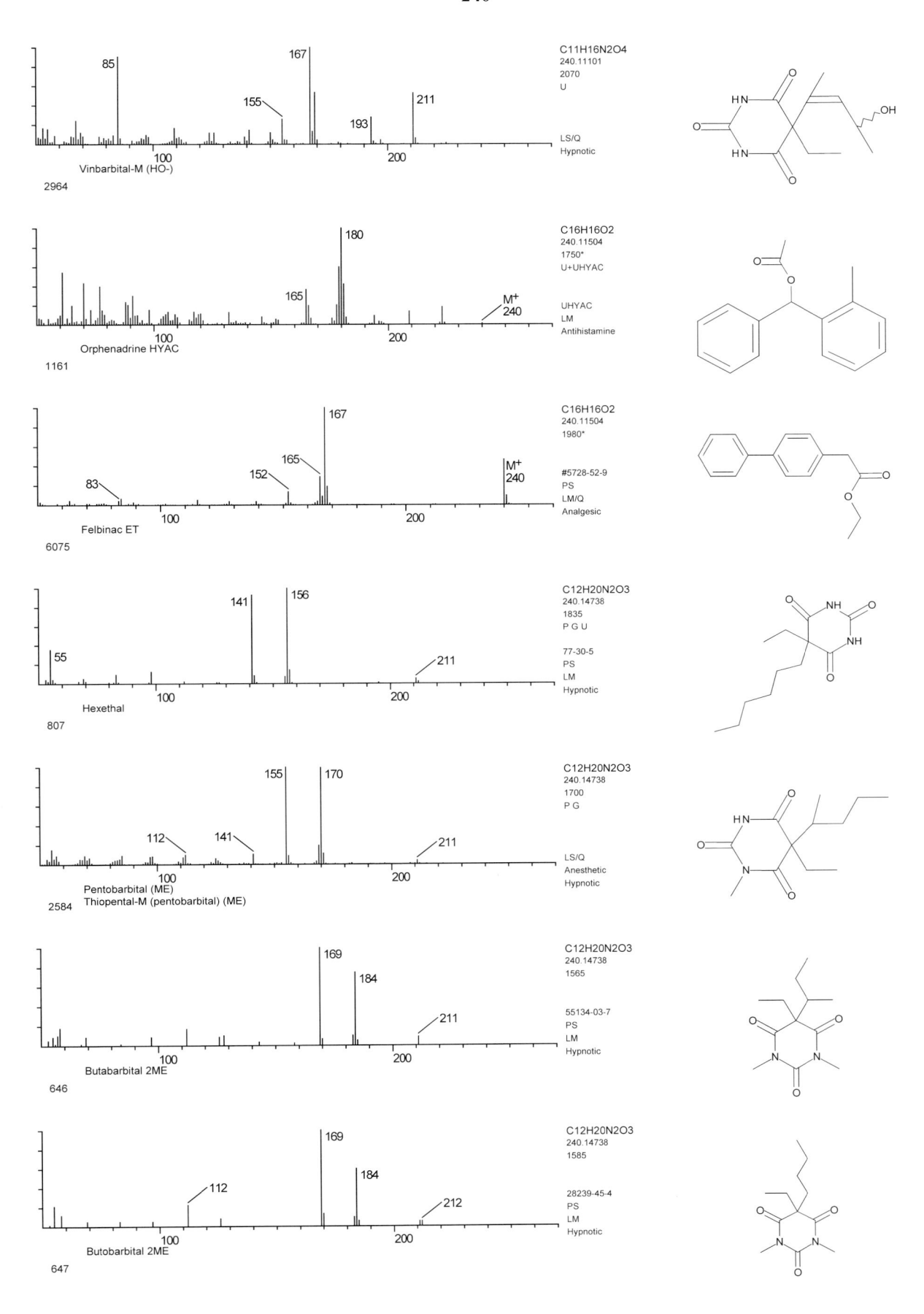

85, 155, 167, 193, 211	C11H16N2O4
Vinbarbital-M (HO-)	240.11101
2964	2070
	U
	LS/Q
	Hypnotic

165, 180, M+ 240	C16H16O2
Orphenadrine HYAC	240.11504
1161	1750*
	U+UHYAC
	UHYAC
	LM
	Antihistamine

83, 152, 165, 167, M+ 240	C16H16O2
Felbinac ET	240.11504
6075	1980*
	#5728-52-9
	PS
	LM/Q
	Analgesic

55, 141, 156, 211	C12H20N2O3
Hexethal	240.14738
807	1835
	P G U
	77-30-5
	PS
	LM
	Hypnotic

112, 141, 155, 170, 211	C12H20N2O3
Pentobarbital (ME)	240.14738
Thiopental-M (pentobarbital) (ME)	1700
2584	P G
	LS/Q
	Anesthetic
	Hypnotic

169, 184, 211	C12H20N2O3
Butabarbital 2ME	240.14738
646	1565
	55134-03-7
	PS
	LM
	Hypnotic

112, 169, 184, 212	C12H20N2O3
Butobarbital 2ME	240.14738
647	1585
	28239-45-4
	PS
	LM
	Hypnotic

240

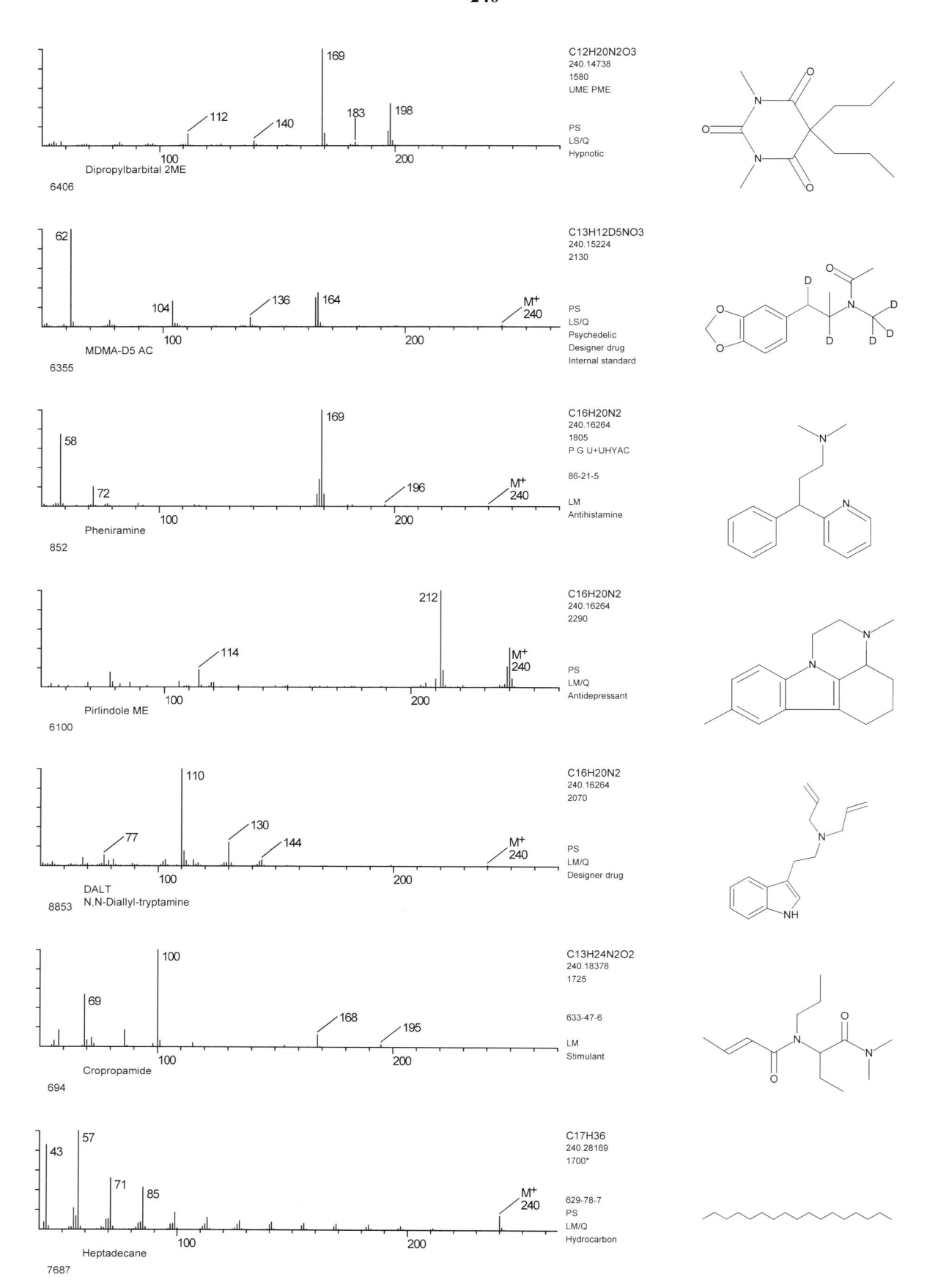

Dipropylbarbital 2ME
6406
C12H20N2O3
240.14738
1580
UME PME

PS
LS/Q
Hypnotic

MDMA-D5 AC
6355
C13H12D5NO3
240.15224
2130

PS
LS/Q
Psychedelic
Designer drug
Internal standard

Pheniramine
852
C16H20N2
240.16264
1805
P G U+UHYAC

86-21-5

LM
Antihistamine

Pirlindole ME
6100
C16H20N2
240.16264
2290

PS
LM/Q
Antidepressant

DALT
N,N-Diallyl-tryptamine
8853
C16H20N2
240.16264
2070

PS
LM/Q
Designer drug

Cropropamide
694
C13H24N2O2
240.18378
1725

633-47-6

LM
Stimulant

Heptadecane
7687
C17H36
240.28169
1700*

629-78-7
PS
LM/Q
Hydrocarbon

240

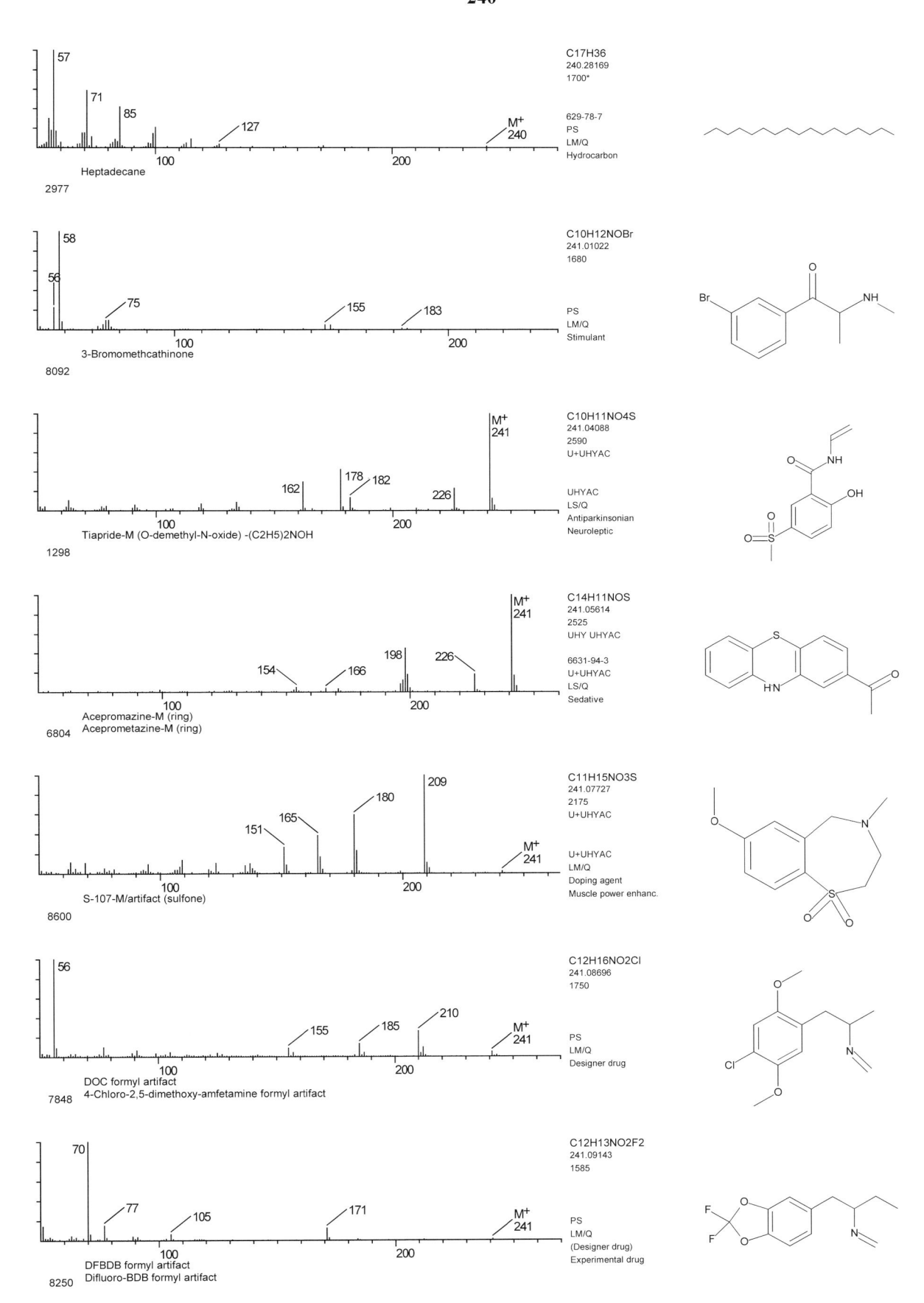

Heptadecane — 2977
C17H36
240.28169
1700*
629-78-7
PS
LM/Q
Hydrocarbon
Peaks: 57, 71, 85, 127, M+ 240

3-Bromomethcathinone — 8092
C10H12NOBr
241.01022
1680
PS
LM/Q
Stimulant
Peaks: 56, 58, 75, 155, 183

Tiapride-M (O-demethyl-N-oxide) -(C2H5)2NOH — 1298
C10H11NO4S
241.04088
2590
U+UHYAC
UHYAC
LS/Q
Antiparkinsonian
Neuroleptic
Peaks: 162, 178, 182, 226, M+ 241

Acepromazine-M (ring)
Aceprometazine-M (ring) — 6804
C14H11NOS
241.05614
2525
UHY UHYAC
6631-94-3
U+UHYAC
LS/Q
Sedative
Peaks: 154, 166, 198, 226, M+ 241

S-107-M/artifact (sulfone) — 8600
C11H15NO3S
241.07727
2175
U+UHYAC
U+UHYAC
LM/Q
Doping agent
Muscle power enhanc.
Peaks: 151, 165, 180, 209, M+ 241

DOC formyl artifact
4-Chloro-2,5-dimethoxy-amfetamine formyl artifact — 7848
C12H16NO2Cl
241.08696
1750
PS
LM/Q
Designer drug
Peaks: 56, 155, 185, 210, M+ 241

DFBDB formyl artifact
Difluoro-BDB formyl artifact — 8250
C12H13NO2F2
241.09143
1585
PS
LM/Q
(Designer drug)
Experimental drug
Peaks: 70, 77, 105, 171, M+ 241

559

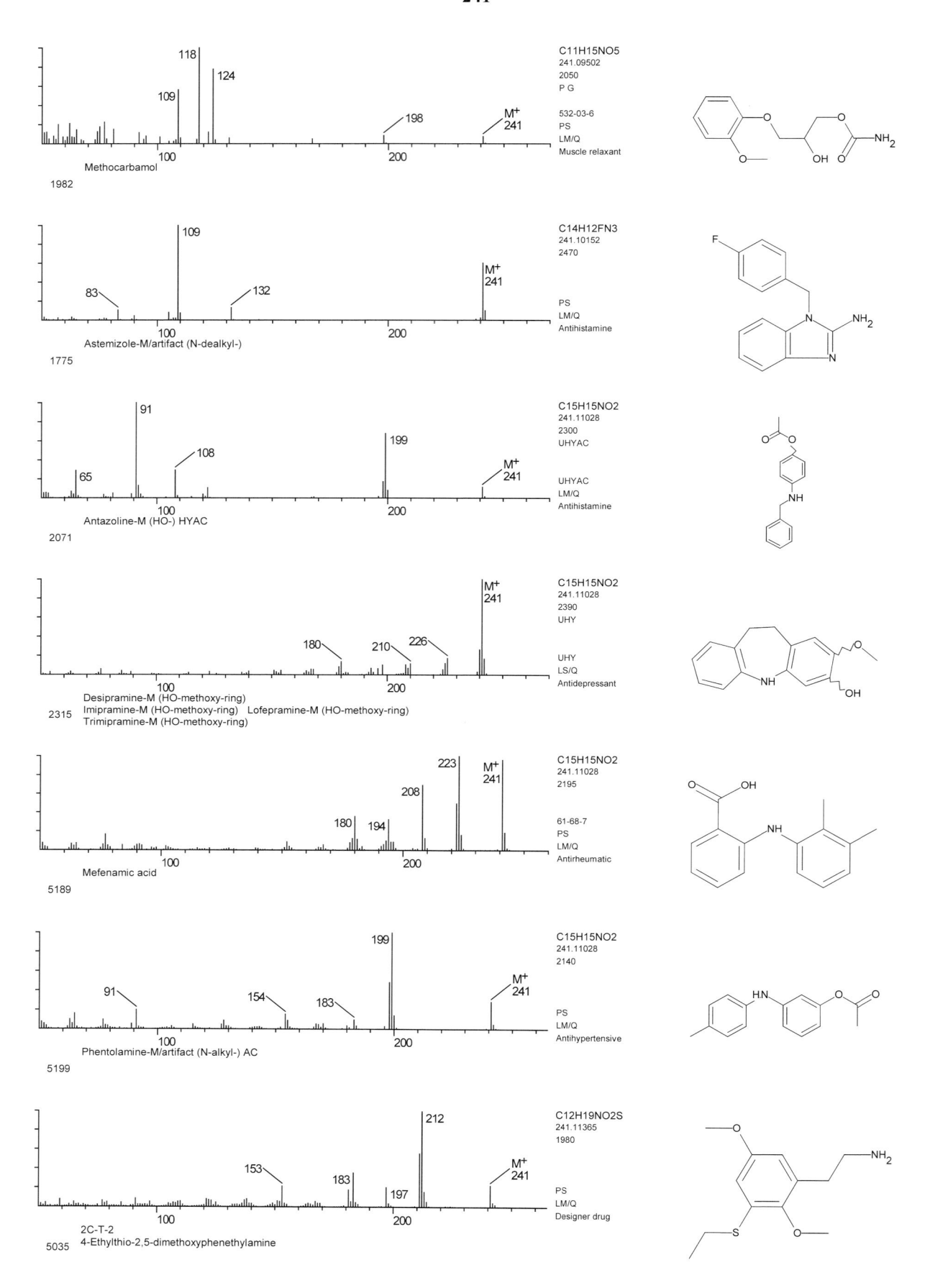

Methocarbamol	C11H15NO5 241.09502 2050 P G 532-03-6 PS LM/Q Muscle relaxant
Astemizole-M/artifact (N-dealkyl-)	C14H12FN3 241.10152 2470 PS LM/Q Antihistamine
Antazoline-M (HO-) HYAC	C15H15NO2 241.11028 2300 UHYAC UHYAC LM/Q Antihistamine
Desipramine-M (HO-methoxy-ring) Imipramine-M (HO-methoxy-ring) Lofepramine-M (HO-methoxy-ring) Trimipramine-M (HO-methoxy-ring)	C15H15NO2 241.11028 2390 UHY UHY LS/Q Antidepressant
Mefenamic acid	C15H15NO2 241.11028 2195 61-68-7 PS LM/Q Antirheumatic
Phentolamine-M/artifact (N-alkyl-) AC	C15H15NO2 241.11028 2140 PS LM/Q Antihypertensive
2C-T-2 4-Ethylthio-2,5-dimethoxyphenethylamine	C12H19NO2S 241.11365 1980 PS LM/Q Designer drug

241

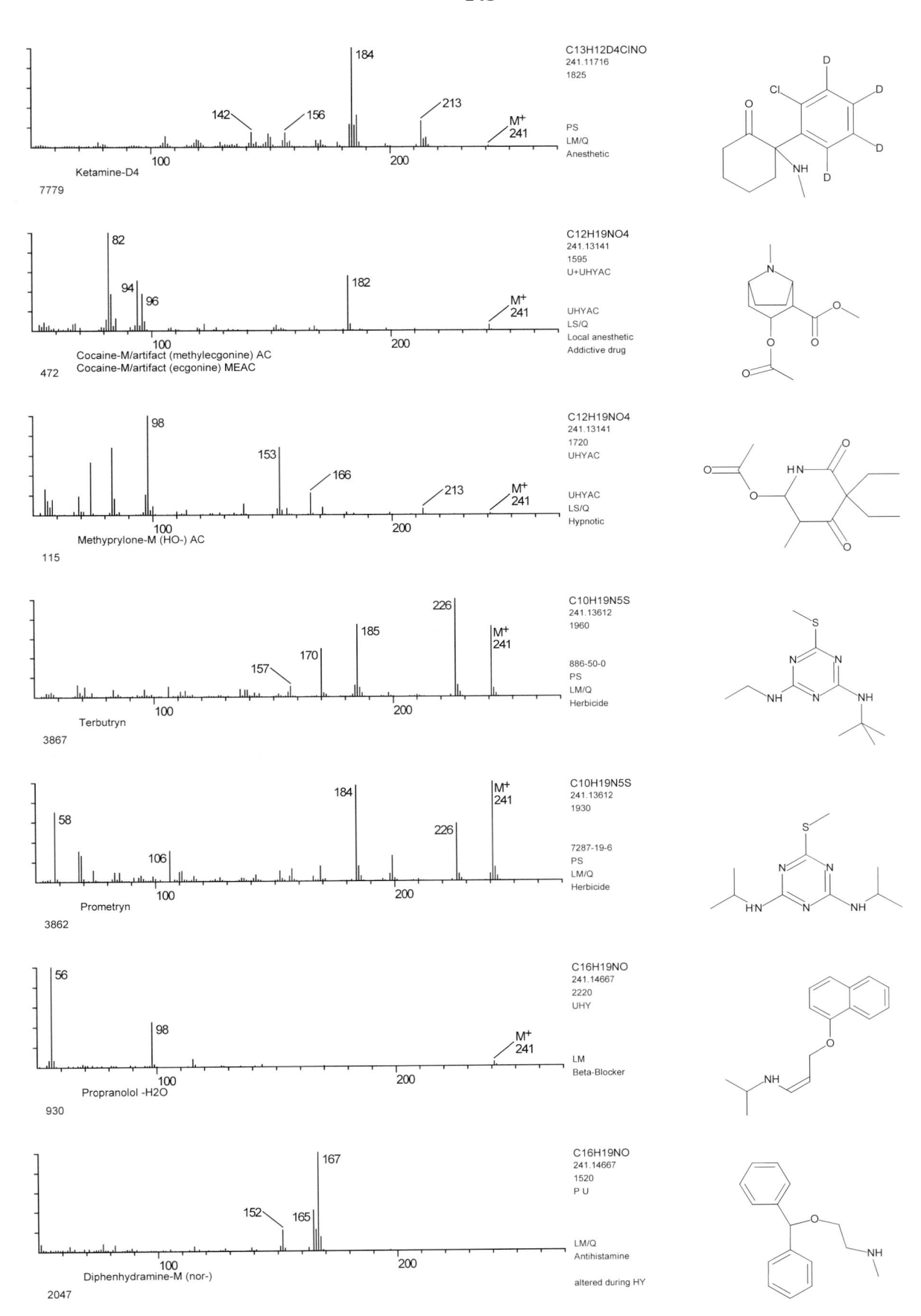

Ketamine-D4
7779
C13H12D4ClNO
241.11716
1825
PS
LM/Q
Anesthetic
Peaks: 184, 142, 156, 213, M+ 241

Cocaine-M/artifact (methylecgonine) AC
Cocaine-M/artifact (ecgonine) MEAC
472
C12H19NO4
241.13141
1595
U+UHYAC
UHYAC
LS/Q
Local anesthetic
Addictive drug
Peaks: 82, 94, 96, 182, M+ 241

Methyprylone-M (HO-) AC
115
C12H19NO4
241.13141
1720
UHYAC
UHYAC
LS/Q
Hypnotic
Peaks: 98, 153, 166, 213, M+ 241

Terbutryn
3867
C10H19N5S
241.13612
1960
886-50-0
PS
LM/Q
Herbicide
Peaks: 157, 170, 185, 226, M+ 241

Prometryn
3862
C10H19N5S
241.13612
1930
7287-19-6
PS
LM/Q
Herbicide
Peaks: 58, 106, 184, 226, M+ 241

Propranolol -H2O
930
C16H19NO
241.14667
2220
UHY
LM
Beta-Blocker
Peaks: 56, 98, M+ 241

Diphenhydramine-M (nor-)
2047
C16H19NO
241.14667
1520
P U
LM/Q
Antihistamine
altered during HY
Peaks: 152, 165, 167

241

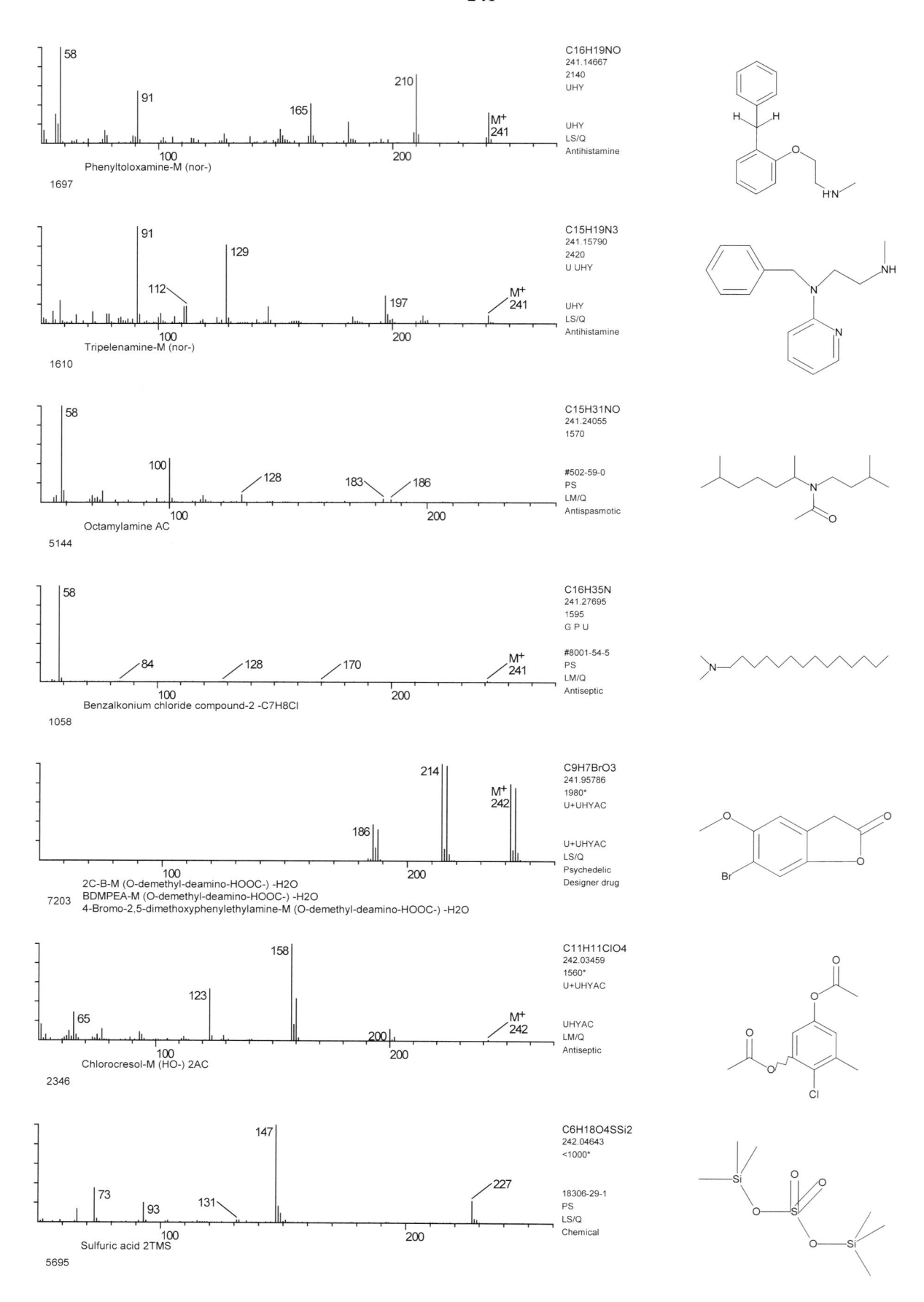

Phenyltoloxamine-M (nor-)
C16H19NO
241.14667
2140
UHY
UHY
LS/Q
Antihistamine
1697

Tripelenamine-M (nor-)
C15H19N3
241.15790
2420
U UHY
UHY
LS/Q
Antihistamine
1610

Octamylamine AC
C15H31NO
241.24055
1570
#502-59-0
PS
LM/Q
Antispasmotic
5144

Benzalkonium chloride compound-2 -C7H8Cl
C16H35N
241.27695
1595
G P U
#8001-54-5
PS
LM/Q
Antiseptic
1058

2C-B-M (O-demethyl-deamino-HOOC-) -H2O
BDMPEA-M (O-demethyl-deamino-HOOC-) -H2O
4-Bromo-2,5-dimethoxyphenylethylamine-M (O-demethyl-deamino-HOOC-) -H2O
C9H7BrO3
241.95786
1980*
U+UHYAC
U+UHYAC
LS/Q
Psychedelic
Designer drug
7203

Chlorocresol-M (HO-) 2AC
C11H11ClO4
242.03459
1560*
U+UHYAC
UHYAC
LM/Q
Antiseptic
2346

Sulfuric acid 2TMS
C6H18O4SSi2
242.04643
<1000*
18306-29-1
PS
LS/Q
Chemical
5695

562

242

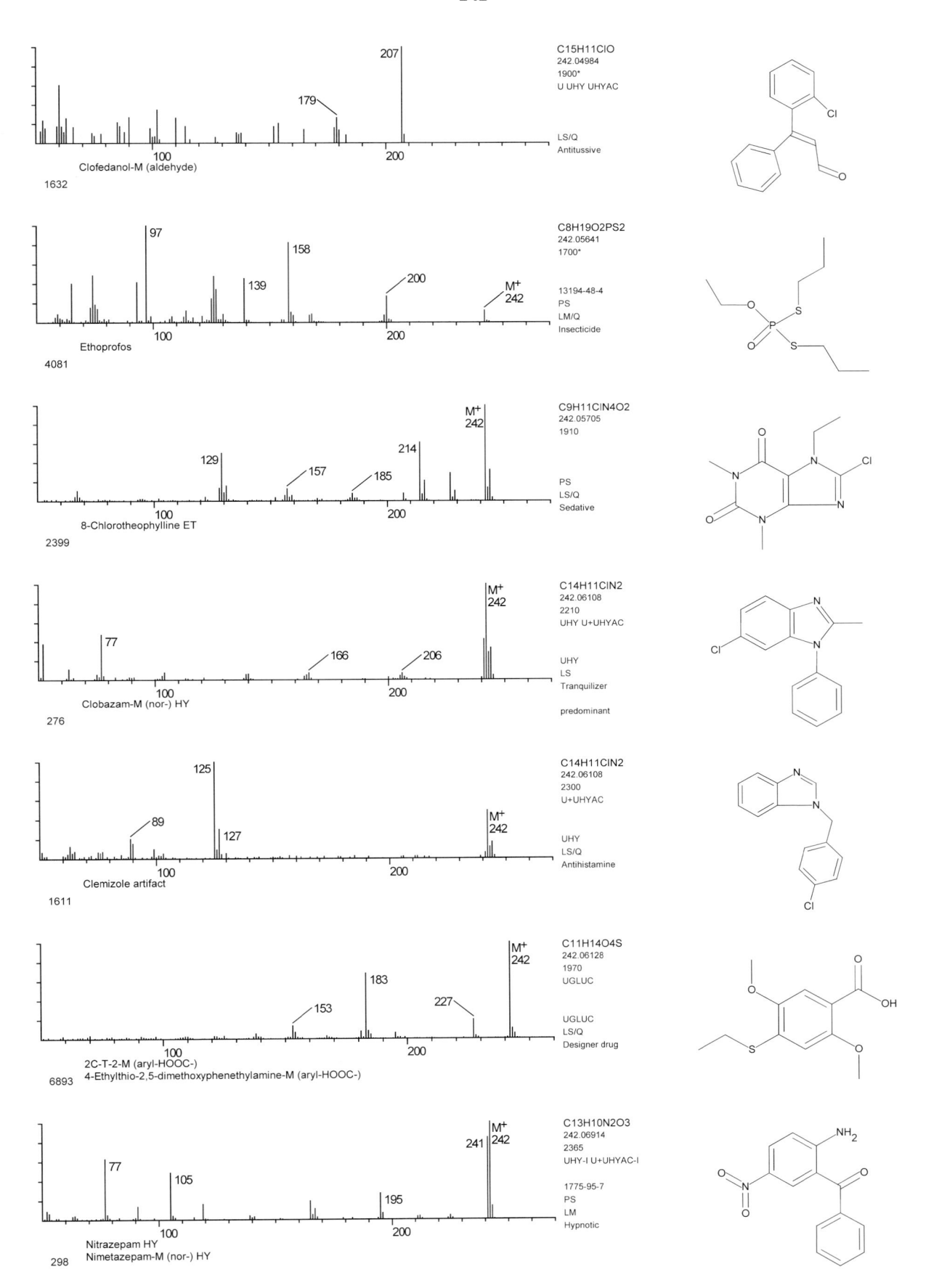

Clofedanol-M (aldehyde)
1632
C15H11ClO
242.04984
1900*
U UHY UHYAC

LS/Q
Antitussive

Peaks: 179, 207

Ethoprofos
4081
C8H19O2PS2
242.05641
1700*
13194-48-4
PS
LM/Q
Insecticide

Peaks: 97, 139, 158, 200, M+ 242

8-Chlorotheophylline ET
2399
C9H11ClN4O2
242.05705
1910
PS
LS/Q
Sedative

Peaks: 129, 157, 185, 214, M+ 242

Clobazam-M (nor-) HY
276
C14H11ClN2
242.06108
2210
UHY U+UHYAC
UHY
LS
Tranquilizer
predominant

Peaks: 77, 166, 206, M+ 242

Clemizole artifact
1611
C14H11ClN2
242.06108
2300
U+UHYAC
UHY
LS/Q
Antihistamine

Peaks: 89, 125, 127, M+ 242

2C-T-2-M (aryl-HOOC-)
4-Ethylthio-2,5-dimethoxyphenethylamine-M (aryl-HOOC-)
6893
C11H14O4S
242.06128
1970
UGLUC
UGLUC
LS/Q
Designer drug

Peaks: 153, 183, 227, M+ 242

Nitrazepam HY
Nimetazepam-M (nor-) HY
298
C13H10N2O3
242.06914
2365
UHY-I U+UHYAC-I
1775-95-7
PS
LM
Hypnotic

Peaks: 77, 105, 195, 241, M+ 242

242

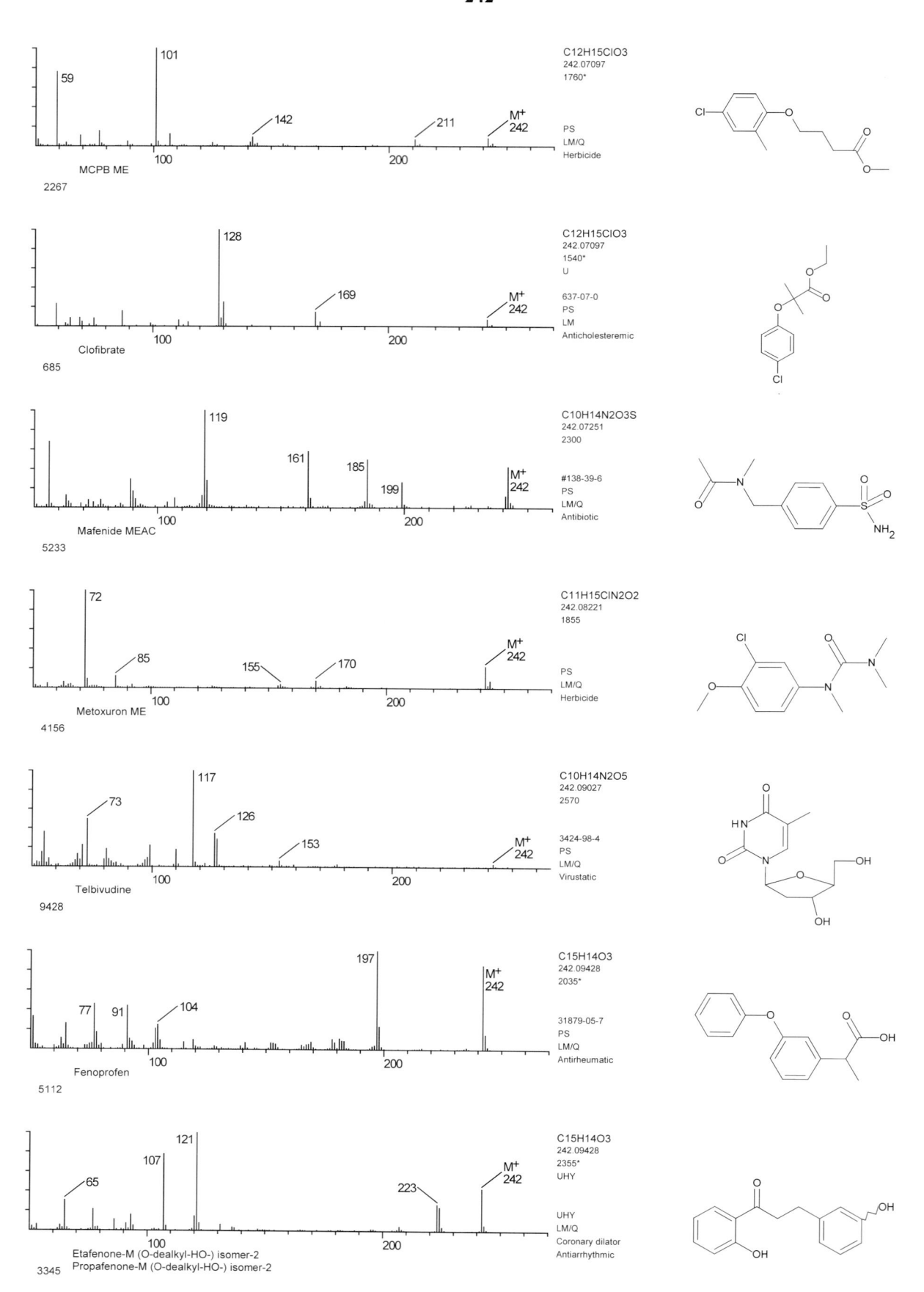

C12H15ClO3
242.07097
1760*

PS
LM/Q
Herbicide

MCPB ME
2267

C12H15ClO3
242.07097
1540*
U

637-07-0
PS
LM
Anticholesteremic

Clofibrate
685

C10H14N2O3S
242.07251
2300

#138-39-6
PS
LM/Q
Antibiotic

Mafenide MEAC
5233

C11H15ClN2O2
242.08221
1855

PS
LM/Q
Herbicide

Metoxuron ME
4156

C10H14N2O5
242.09027
2570

3424-98-4
PS
LM/Q
Virustatic

Telbivudine
9428

C15H14O3
242.09428
2035*

31879-05-7
PS
LM/Q
Antirheumatic

Fenoprofen
5112

C15H14O3
242.09428
2355*
UHY

UHY
LM/Q
Coronary dilator
Antiarrhythmic

Etafenone-M (O-dealkyl-HO-) isomer-2
Propafenone-M (O-dealkyl-HO-) isomer-2
3345

564

242

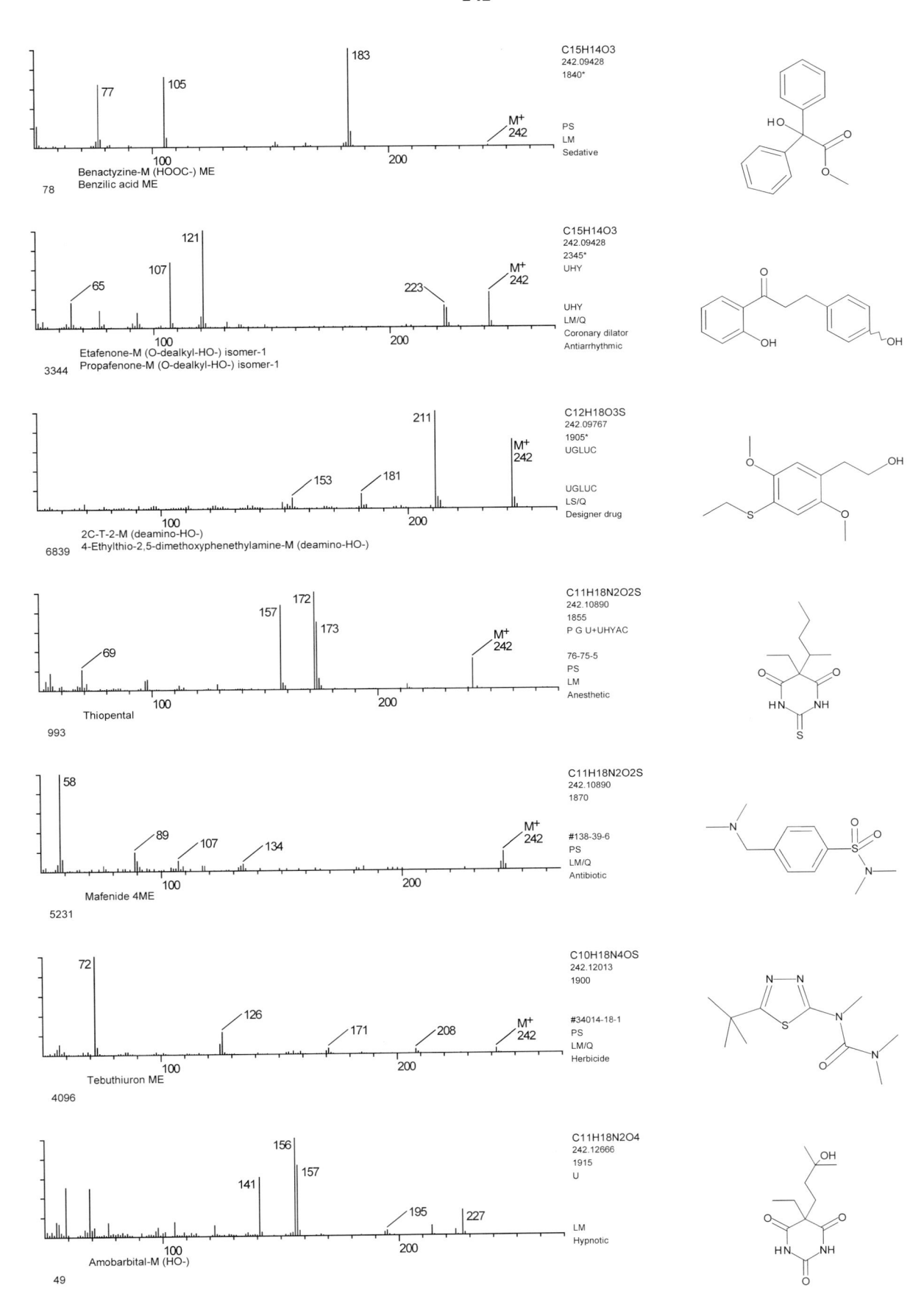

C15H14O3
242.09428
1840*

M+ 242
PS
LM
Sedative

Benactyzine-M (HOOC-) ME
Benzilic acid ME
78

C15H14O3
242.09428
2345*
UHY

M+ 242
UHY
LM/Q
Coronary dilator
Antiarrhythmic

Etafenone-M (O-dealkyl-HO-) isomer-1
Propafenone-M (O-dealkyl-HO-) isomer-1
3344

C12H18O3S
242.09767
1905*
UGLUC

M+ 242
UGLUC
LS/Q
Designer drug

2C-T-2-M (deamino-HO-)
4-Ethylthio-2,5-dimethoxyphenethylamine-M (deamino-HO-)
6839

C11H18N2O2S
242.10890
1855
P G U+UHYAC

M+ 242
76-75-5
PS
LM
Anesthetic

Thiopental
993

C11H18N2O2S
242.10890
1870

M+ 242
#138-39-6
PS
LM/Q
Antibiotic

Mafenide 4ME
5231

C10H18N4OS
242.12013
1900

M+ 242
#34014-18-1
PS
LM/Q
Herbicide

Tebuthiuron ME
4096

C11H18N2O4
242.12666
1915
U

LM
Hypnotic

Amobarbital-M (HO-)
49

242

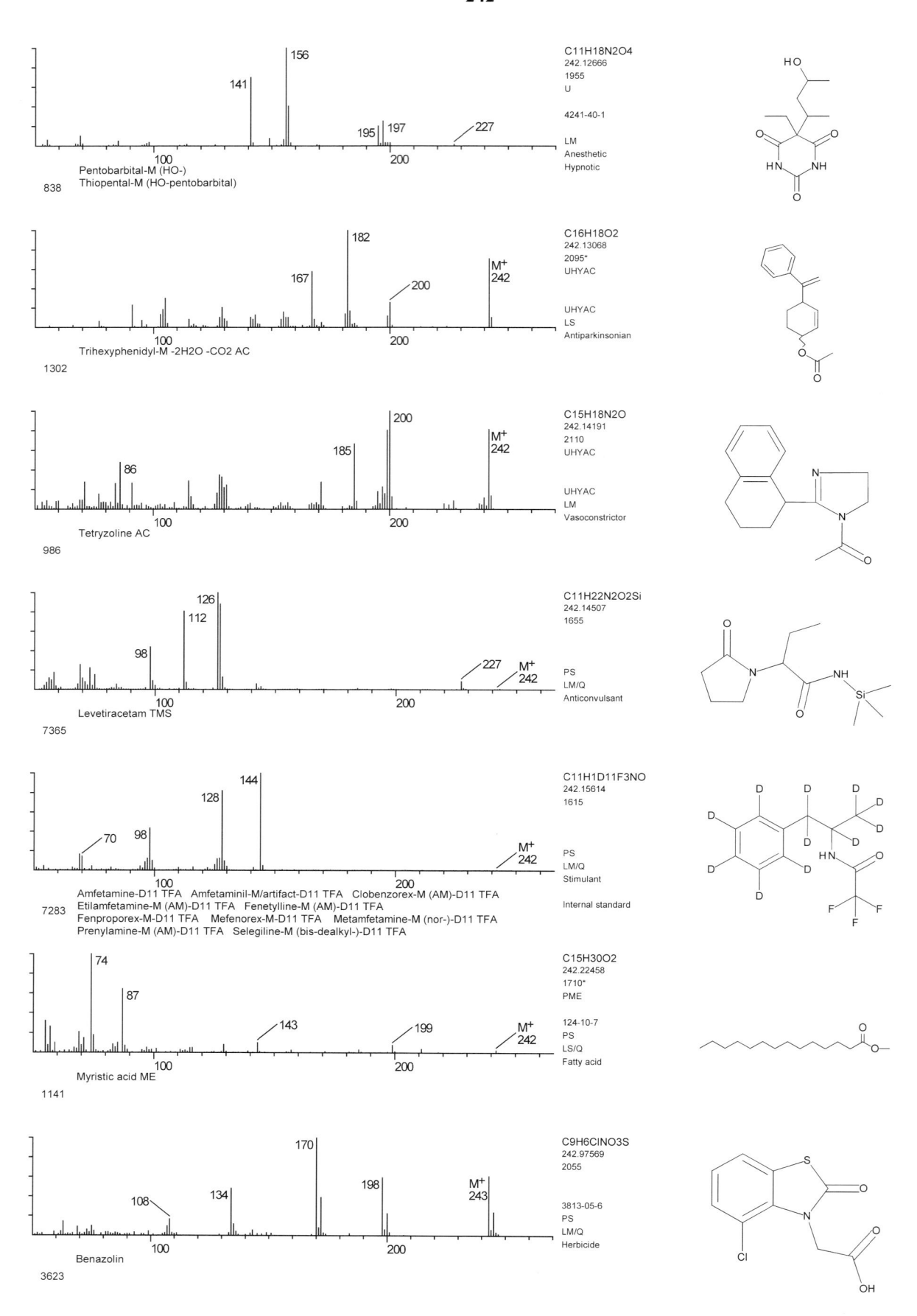

Pentobarbital-M (HO-)
Thiopental-M (HO-pentobarbital)
838

C11H18N2O4
242.12666
1955
U

4241-40-1

LM
Anesthetic
Hypnotic

Trihexyphenidyl-M -2H2O -CO2 AC
1302

C16H18O2
242.13068
2095*
UHYAC

UHYAC
LS
Antiparkinsonian

Tetryzoline AC
986

C15H18N2O
242.14191
2110
UHYAC

UHYAC
LM
Vasoconstrictor

Levetiracetam TMS
7365

C11H22N2O2Si
242.14507
1655

PS
LM/Q
Anticonvulsant

Amfetamine-D11 TFA   Amfetaminil-M/artifact-D11 TFA   Clobenzorex-M (AM)-D11 TFA
Etilamfetamine-M (AM)-D11 TFA   Fenetylline-M (AM)-D11 TFA
7283 Fenproporex-M-D11 TFA   Mefenorex-M-D11 TFA   Metamfetamine-M (nor-)-D11 TFA
Prenylamine-M (AM)-D11 TFA   Selegiline-M (bis-dealkyl-)-D11 TFA

C11H1D11F3NO
242.15614
1615

PS
LM/Q
Stimulant

Internal standard

Myristic acid ME
1141

C15H30O2
242.22458
1710*
PME

124-10-7
PS
LS/Q
Fatty acid

Benazolin
3623

C9H6ClNO3S
242.97569
2055

3813-05-6
PS
LM/Q
Herbicide

243

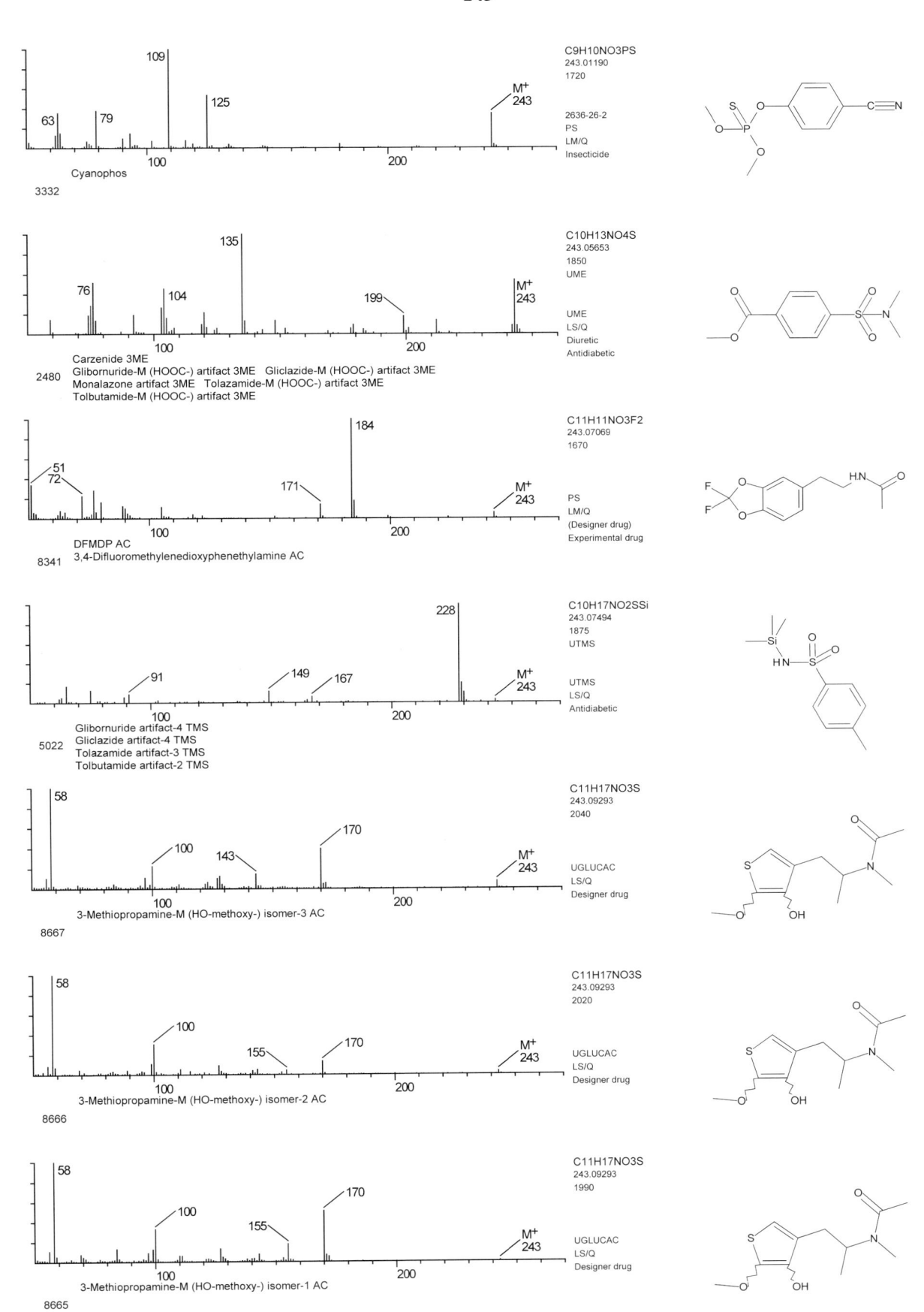

Cyanophos
3332
C9H10NO3PS
243.01190
1720
2636-26-2
PS
LM/Q
Insecticide

Carzenide 3ME
Glibornuride-M (HOOC-) artifact 3ME   Gliclazide-M (HOOC-) artifact 3ME
Monalazone artifact 3ME   Tolazamide-M (HOOC-) artifact 3ME
Tolbutamide-M (HOOC-) artifact 3ME
2480
C10H13NO4S
243.05653
1850
UME
UME
LS/Q
Diuretic
Antidiabetic

DFMDP AC
3,4-Difluoromethylenedioxyphenethylamine AC
8341
C11H11NO3F2
243.07069
1670
PS
LM/Q
(Designer drug)
Experimental drug

Glibornuride artifact-4 TMS
Gliclazide artifact-4 TMS
Tolazamide artifact-3 TMS
Tolbutamide artifact-2 TMS
5022
C10H17NO2SSi
243.07494
1875
UTMS
UTMS
LS/Q
Antidiabetic

3-Methiopropamine-M (HO-methoxy-) isomer-3 AC
8667
C11H17NO3S
243.09293
2040
UGLUCAC
LS/Q
Designer drug

3-Methiopropamine-M (HO-methoxy-) isomer-2 AC
8666
C11H17NO3S
243.09293
2020
UGLUCAC
LS/Q
Designer drug

3-Methiopropamine-M (HO-methoxy-) isomer-1 AC
8665
C11H17NO3S
243.09293
1990
UGLUCAC
LS/Q
Designer drug

243

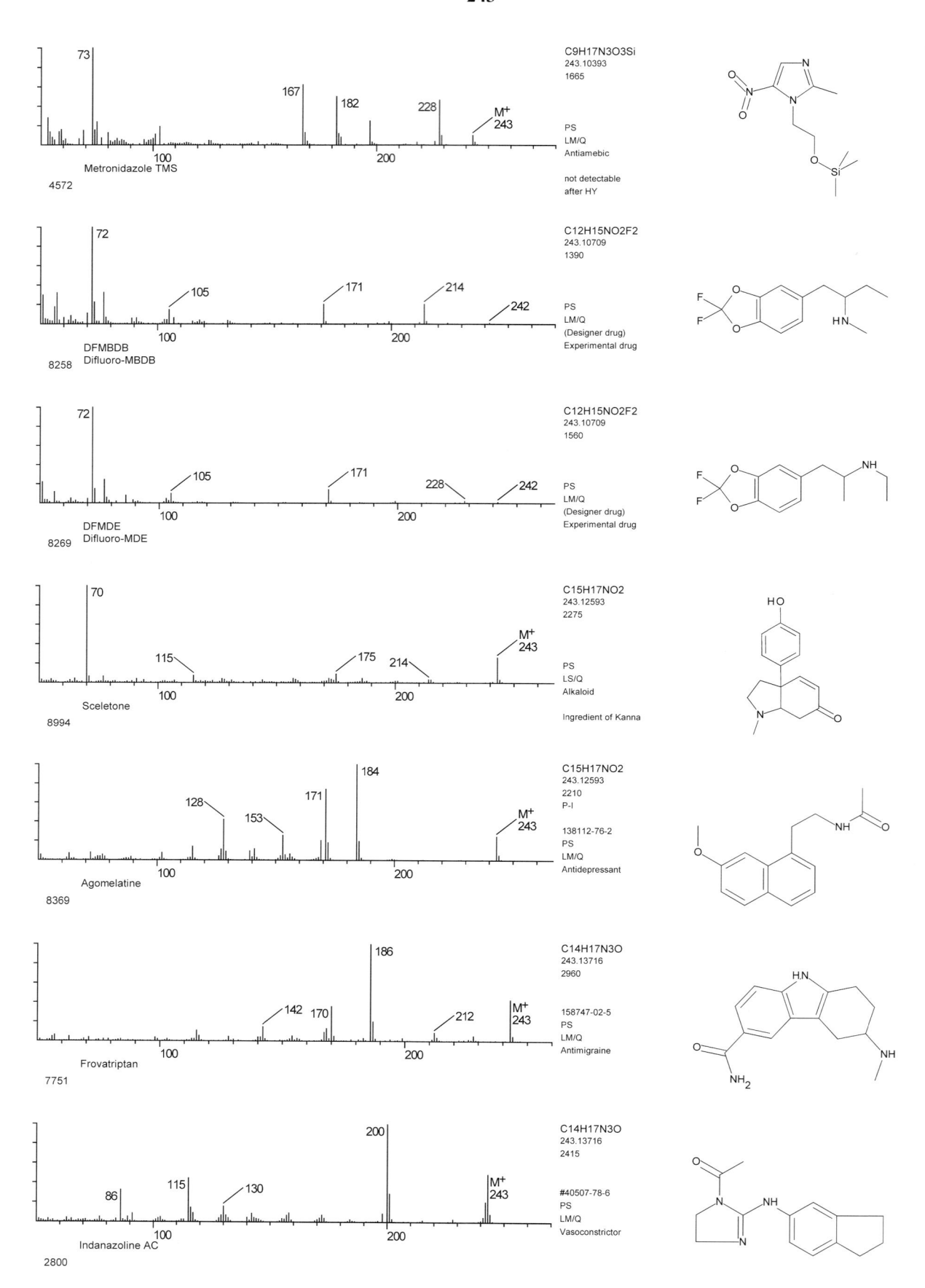

Metronidazole TMS — C9H17N3O3Si, 243.10393, 1665, PS, LM/Q, Antiamebic, not detectable after HY
Peaks: 73, 167, 182, 228, M+ 243
4572

DFMBDB Difluoro-MBDB — C12H15NO2F2, 243.10709, 1390, PS, LM/Q, (Designer drug), Experimental drug
Peaks: 72, 105, 171, 214, 242
8258

DFMDE Difluoro-MDE — C12H15NO2F2, 243.10709, 1560, PS, LM/Q, (Designer drug), Experimental drug
Peaks: 72, 105, 171, 228, 242
8269

Sceletone — C15H17NO2, 243.12593, 2275, PS, LS/Q, Alkaloid, Ingredient of Kanna
Peaks: 70, 115, 175, 214, M+ 243
8994

Agomelatine — C15H17NO2, 243.12593, 2210, P-I, 138112-76-2, PS, LM/Q, Antidepressant
Peaks: 128, 153, 171, 184, M+ 243
8369

Frovatriptan — C14H17N3O, 243.13716, 2960, 158747-02-5, PS, LM/Q, Antimigraine
Peaks: 142, 170, 186, 212, M+ 243
7751

Indanazoline AC — C14H17N3O, 243.13716, 2415, #40507-78-6, PS, LM/Q, Vasoconstrictor
Peaks: 86, 115, 130, 200, M+ 243
2800

243

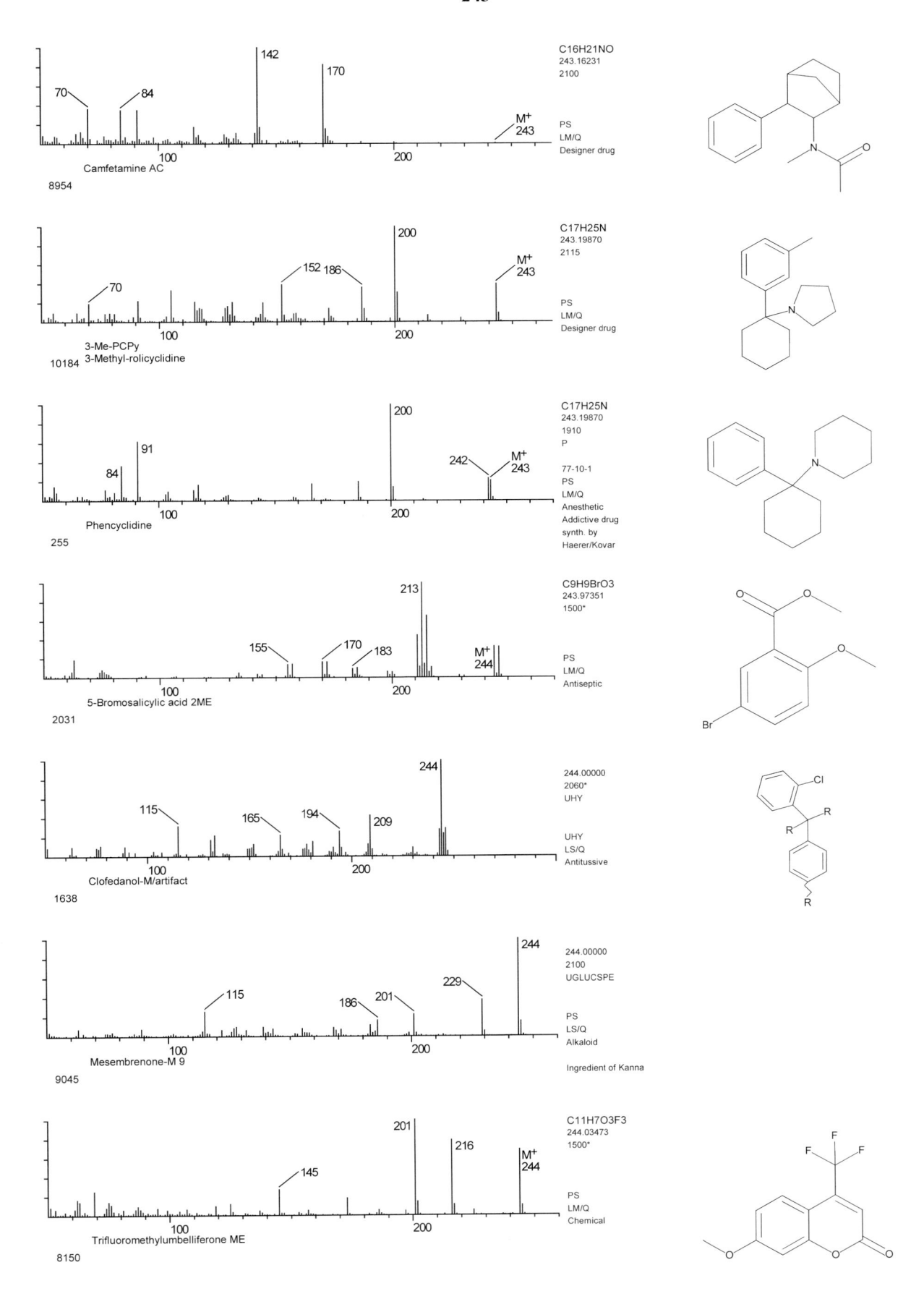

C16H21NO
243.16231
2100

PS
LM/Q
Designer drug

Camfetamine AC
8954

C17H25N
243.19870
2115

PS
LM/Q
Designer drug

3-Me-PCPy
3-Methyl-rolicyclidine
10184

C17H25N
243.19870
1910
P
77-10-1
PS
LM/Q
Anesthetic
Addictive drug
synth. by
Haerer/Kovar

Phencyclidine
255

C9H9BrO3
243.97351
1500*

PS
LM/Q
Antiseptic

5-Bromosalicylic acid 2ME
2031

244.00000
2060*
UHY

UHY
LS/Q
Antitussive

Clofedanol-M/artifact
1638

244.00000
2100
UGLUCSPE

PS
LS/Q
Alkaloid

Ingredient of Kanna

Mesembrenone-M 9
9045

C11H7O3F3
244.03473
1500*

PS
LM/Q
Chemical

Trifluoromethylumbelliferone ME
8150

244

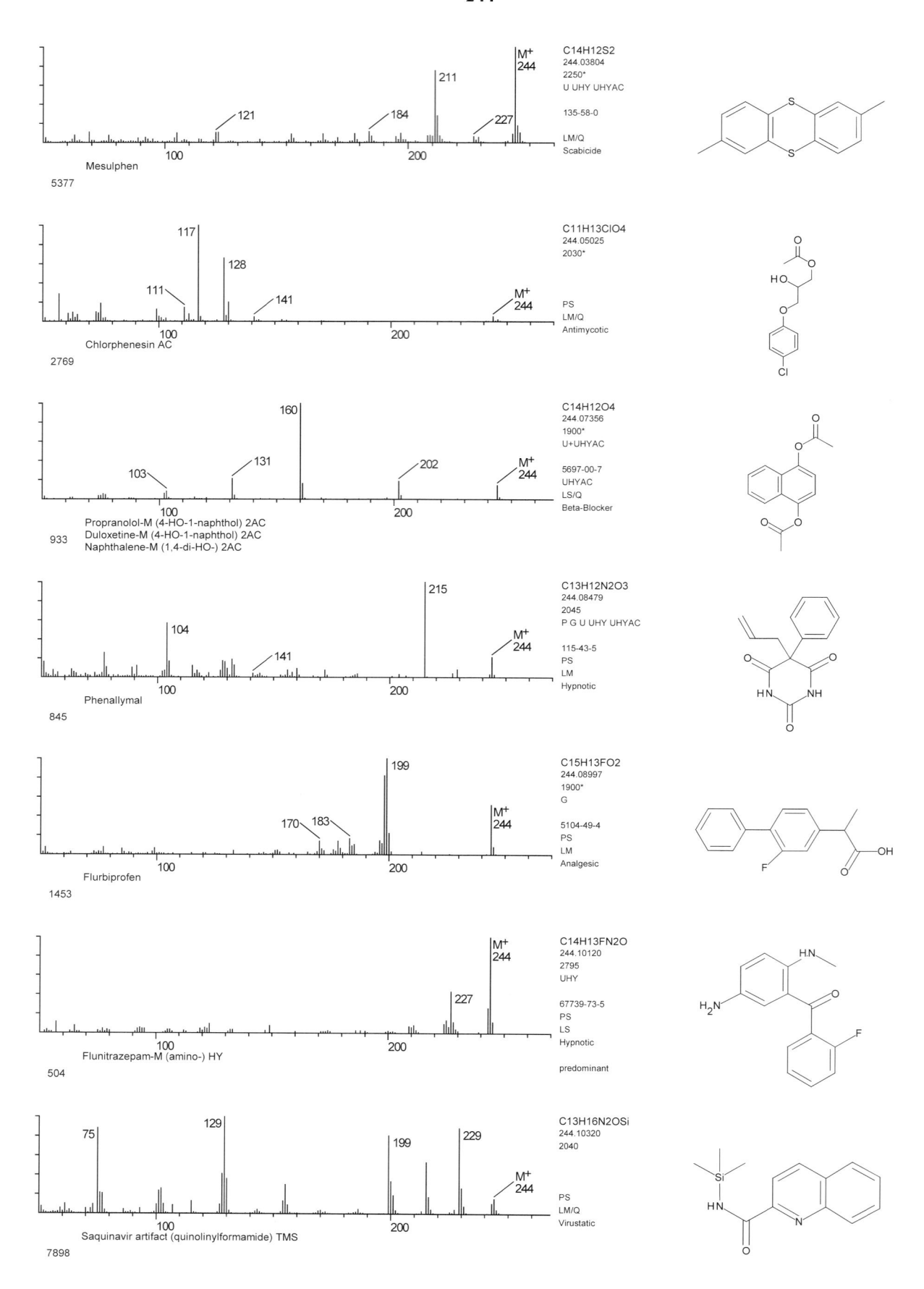

Mesulphen — 5377	m/z: 121, 184, 211, 227, M+ 244	C14H12S2 / 244.03804 / 2250* / U UHY UHYAC / 135-58-0 / LM/Q / Scabicide
Chlorphenesin AC — 2769	m/z: 111, 117, 128, 141, M+ 244	C11H13ClO4 / 244.05025 / 2030* / PS / LM/Q / Antimycotic
Propranolol-M (4-HO-1-naphthol) 2AC; Duloxetine-M (4-HO-1-naphthol) 2AC; Naphthalene-M (1,4-di-HO-) 2AC — 933	m/z: 103, 131, 160, 202, M+ 244	C14H12O4 / 244.07356 / 1900* / U+UHYAC / 5697-00-7 / UHYAC / LS/Q / Beta-Blocker
Phenallymal — 845	m/z: 104, 141, 215, M+ 244	C13H12N2O3 / 244.08479 / 2045 / P G U UHY UHYAC / 115-43-5 / PS / LM / Hypnotic
Flurbiprofen — 1453	m/z: 170, 183, 199, M+ 244	C15H13FO2 / 244.08997 / 1900* / G / 5104-49-4 / PS / LM / Analgesic
Flunitrazepam-M (amino-) HY — 504	m/z: 227, M+ 244	C14H13FN2O / 244.10120 / 2795 / UHY / 67739-73-5 / PS / LS / Hypnotic / predominant
Saquinavir artifact (quinolinylformamide) TMS — 7898	m/z: 75, 129, 199, 229, M+ 244	C13H16N2OSi / 244.10320 / 2040 / PS / LM/Q / Virustatic

570

244

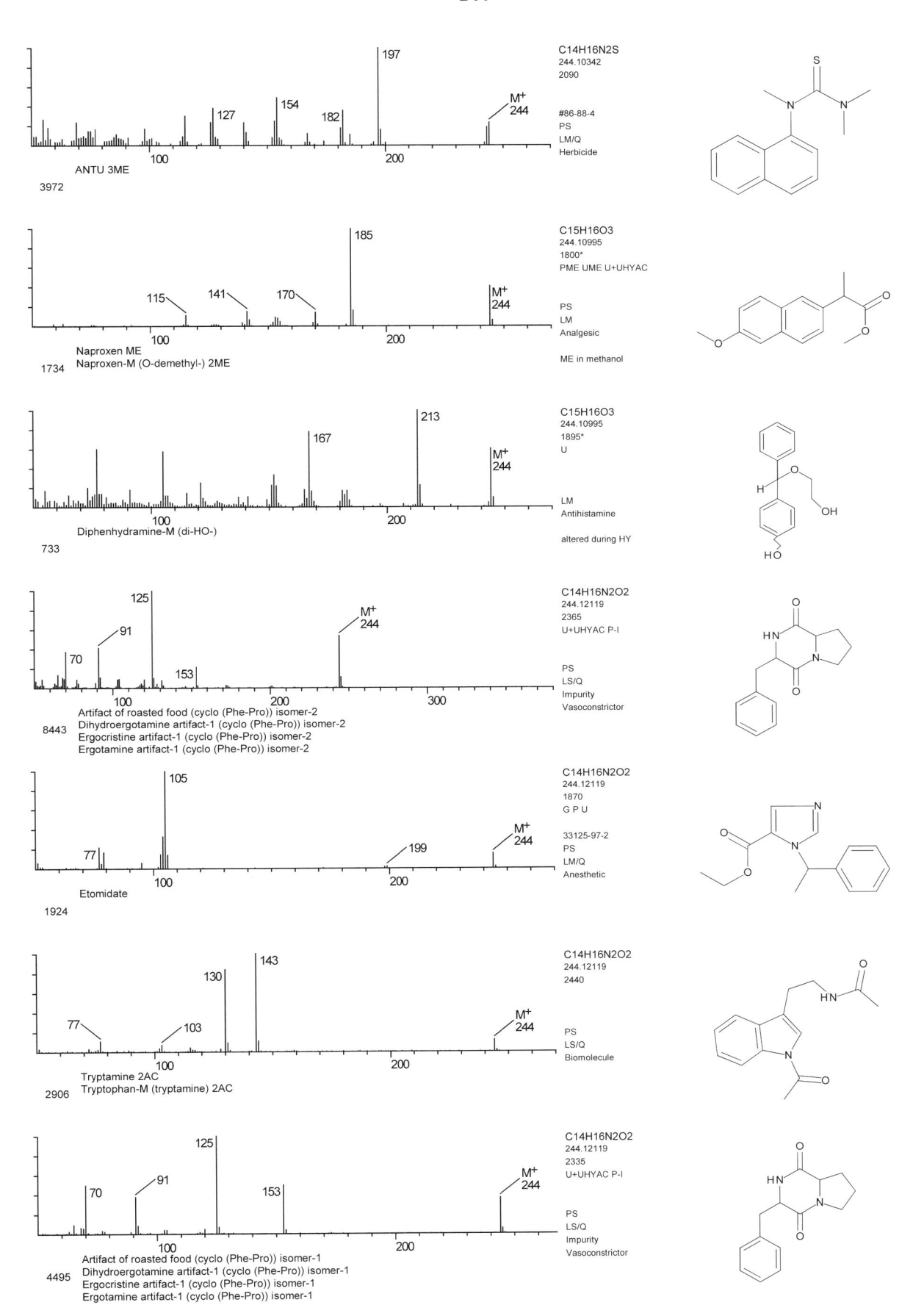

C14H16N2S
244.10342
2090

#86-88-4
PS
LM/Q
Herbicide

ANTU 3ME

C15H16O3
244.10995
1800*
PME UME U+UHYAC

PS
LM
Analgesic

Naproxen ME
Naproxen-M (O-demethyl-) 2ME   ME in methanol

C15H16O3
244.10995
1895*
U

LM
Antihistamine

Diphenhydramine-M (di-HO-)   altered during HY

C14H16N2O2
244.12119
2365
U+UHYAC P-I

PS
LS/Q
Impurity
Vasoconstrictor

Artifact of roasted food (cyclo (Phe-Pro)) isomer-2
Dihydroergotamine artifact-1 (cyclo (Phe-Pro)) isomer-2
Ergocristine artifact-1 (cyclo (Phe-Pro)) isomer-2
Ergotamine artifact-1 (cyclo (Phe-Pro)) isomer-2

C14H16N2O2
244.12119
1870
G P U

33125-97-2
PS
LM/Q
Anesthetic

Etomidate

C14H16N2O2
244.12119
2440

PS
LS/Q
Biomolecule

Tryptamine 2AC
Tryptophan-M (tryptamine) 2AC

C14H16N2O2
244.12119
2335
U+UHYAC P-I

PS
LS/Q
Impurity
Vasoconstrictor

Artifact of roasted food (cyclo (Phe-Pro)) isomer-1
Dihydroergotamine artifact-1 (cyclo (Phe-Pro)) isomer-1
Ergocristine artifact-1 (cyclo (Phe-Pro)) isomer-1
Ergotamine artifact-1 (cyclo (Phe-Pro)) isomer-1

244

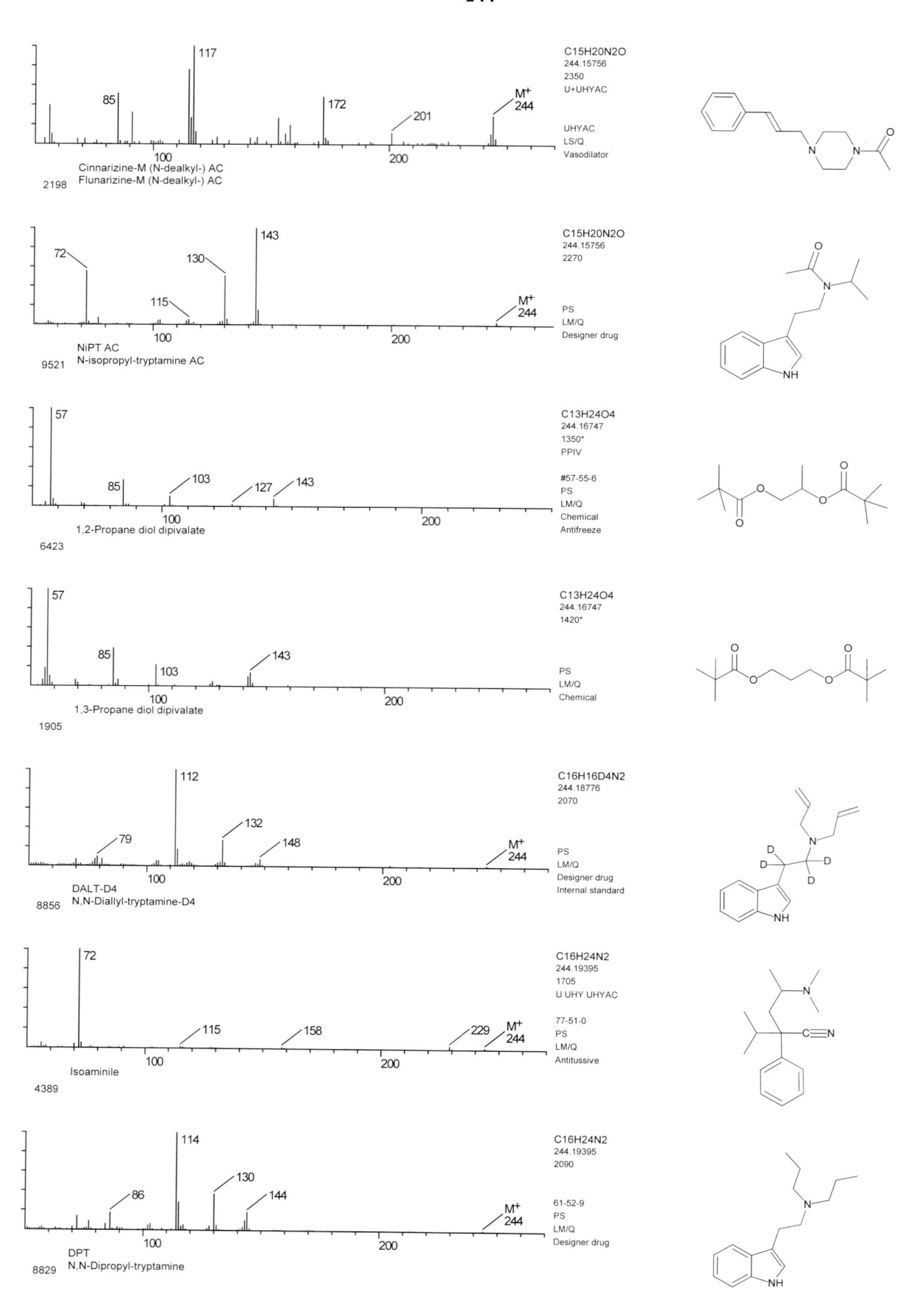

2198	Peaks: 85, 117, 172, 201, M+ 244 Cinnarizine-M (N-dealkyl-) AC Flunarizine-M (N-dealkyl-) AC	C15H20N2O 244.15756 2350 U+UHYAC UHYAC LS/Q Vasodilator
9521	Peaks: 72, 115, 130, 143, M+ 244 NiPT AC N-isopropyl-tryptamine AC	C15H20N2O 244.15756 2270 PS LM/Q Designer drug
6423	Peaks: 57, 85, 103, 127, 143 1,2-Propane diol dipivalate	C13H24O4 244.16747 1350* PPIV #57-55-6 PS LM/Q Chemical Antifreeze
1905	Peaks: 57, 85, 103, 143 1,3-Propane diol dipivalate	C13H24O4 244.16747 1420* PS LM/Q Chemical
8856	Peaks: 79, 112, 132, 148, M+ 244 DALT-D4 N,N-Diallyl-tryptamine-D4	C16H16D4N2 244.18776 2070 PS LM/Q Designer drug Internal standard
4389	Peaks: 72, 115, 158, 229, M+ 244 Isoaminile	C16H24N2 244.19395 1705 U UHY UHYAC 77-51-0 PS LM/Q Antitussive
8829	Peaks: 86, 114, 130, 144, M+ 244 DPT N,N-Dipropyl-tryptamine	C16H24N2 244.19395 2090 61-52-9 PS LM/Q Designer drug

244

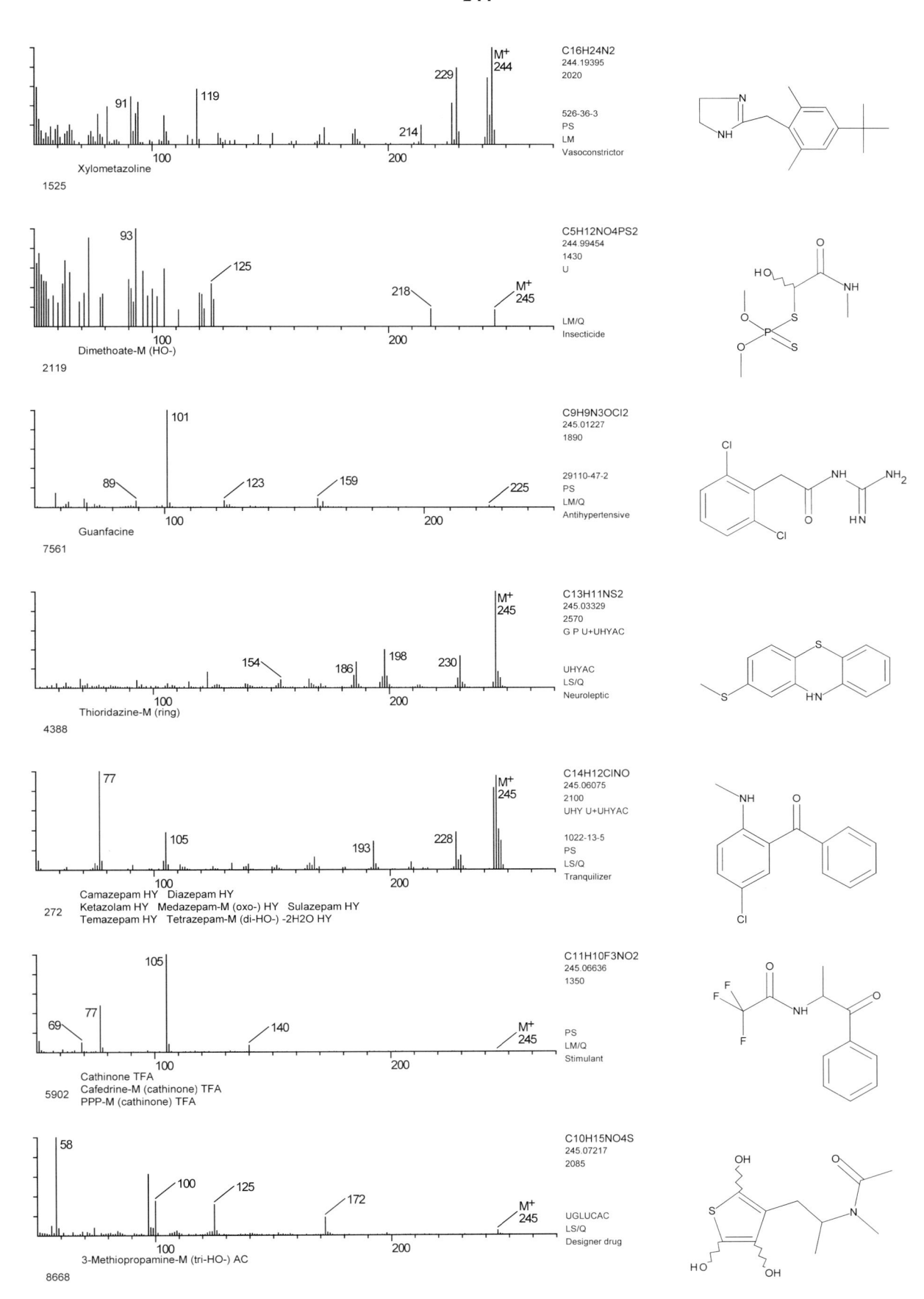

Xylometazoline — C16H24N2, 244.19395, 2020, 526-36-3, PS, LM, Vasoconstrictor
Peaks: 91, 119, 214, 229, M+ 244
1525

Dimethoate-M (HO-) — C5H12NO4PS2, 244.99454, 1430, U, LM/Q, Insecticide
Peaks: 93, 125, 218, M+ 245
2119

Guanfacine — C9H9N3OCl2, 245.01227, 1890, 29110-47-2, PS, LM/Q, Antihypertensive
Peaks: 89, 101, 123, 159, 225
7561

Thioridazine-M (ring) — C13H11NS2, 245.03329, 2570, G P U+UHYAC, UHYAC, LS/Q, Neuroleptic
Peaks: 154, 186, 198, 230, M+ 245
4388

Camazepam HY Diazepam HY
Ketazolam HY Medazepam-M (oxo-) HY Sulazepam HY
Temazepam HY Tetrazepam-M (di-HO-) -2H2O HY
— C14H12ClNO, 245.06075, 2100, UHY U+UHYAC, 1022-13-5, PS, LS/Q, Tranquilizer
Peaks: 77, 105, 193, 228, M+ 245
272

Cathinone TFA
Cafedrine-M (cathinone) TFA
PPP-M (cathinone) TFA
— C11H10F3NO2, 245.06636, 1350, PS, LM/Q, Stimulant
Peaks: 69, 77, 105, 140, M+ 245
5902

3-Methiopropamine-M (tri-HO-) AC — C10H15NO4S, 245.07217, 2085, UGLUCAC, LS/Q, Designer drug
Peaks: 58, 100, 125, 172, M+ 245
8668

245

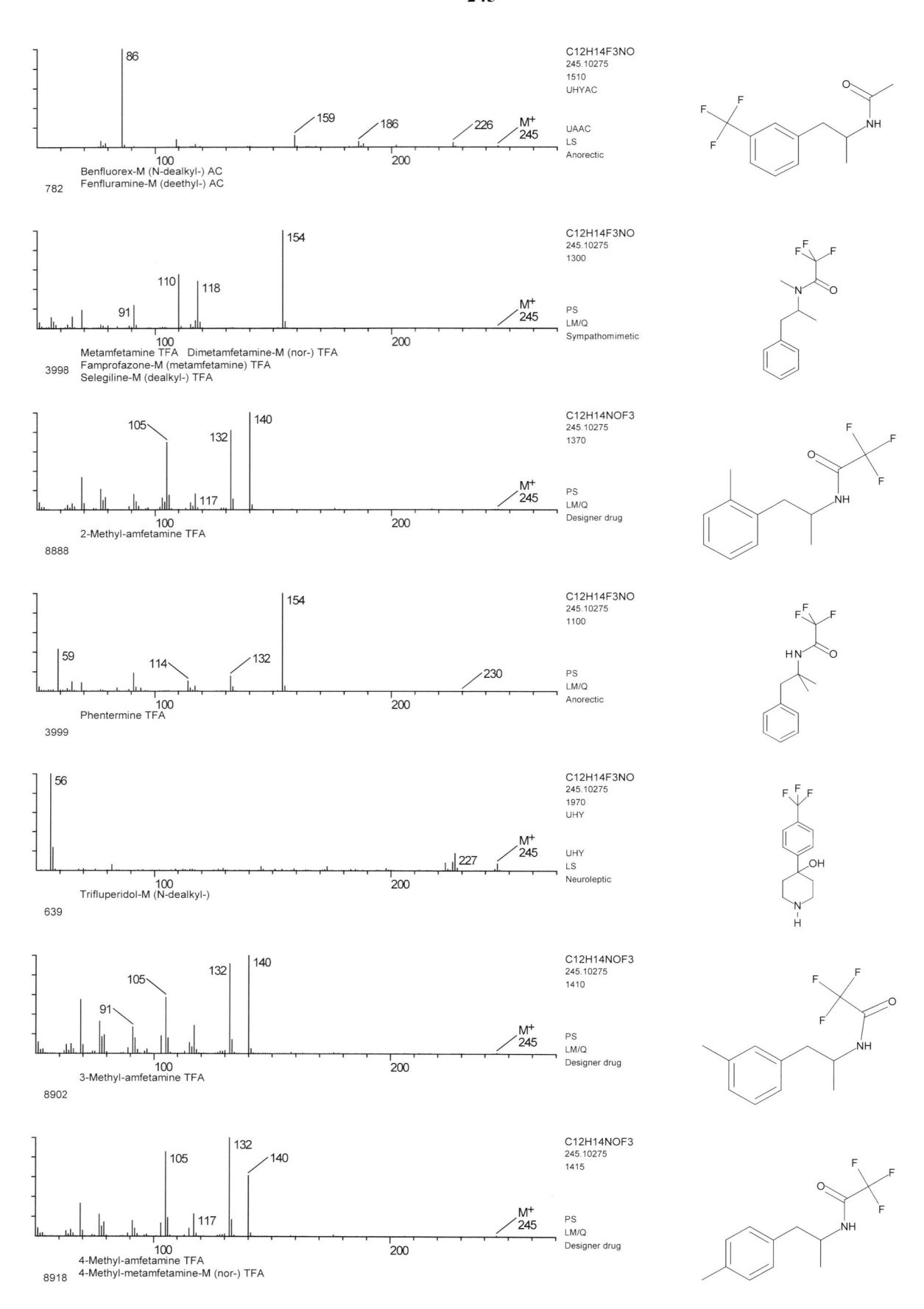

782	Benfluorex-M (N-dealkyl-) AC Fenfluramine-M (deethyl-) AC	C12H14F3NO 245.10275 1510 UHYAC UAAC LS Anorectic
3998	Metamfetamine TFA Dimetamfetamine-M (nor-) TFA Famprofazone-M (metamfetamine) TFA Selegiline-M (dealkyl-) TFA	C12H14F3NO 245.10275 1300 PS LM/Q Sympathomimetic
8888	2-Methyl-amfetamine TFA	C12H14NOF3 245.10275 1370 PS LM/Q Designer drug
3999	Phentermine TFA	C12H14F3NO 245.10275 1100 PS LM/Q Anorectic
639	Trifluperidol-M (N-dealkyl-)	C12H14F3NO 245.10275 1970 UHY UHY LS Neuroleptic
8902	3-Methyl-amfetamine TFA	C12H14NOF3 245.10275 1410 PS LM/Q Designer drug
8918	4-Methyl-amfetamine TFA 4-Methyl-metamfetamine-M (nor-) TFA	C12H14NOF3 245.10275 1415 PS LM/Q Designer drug

245

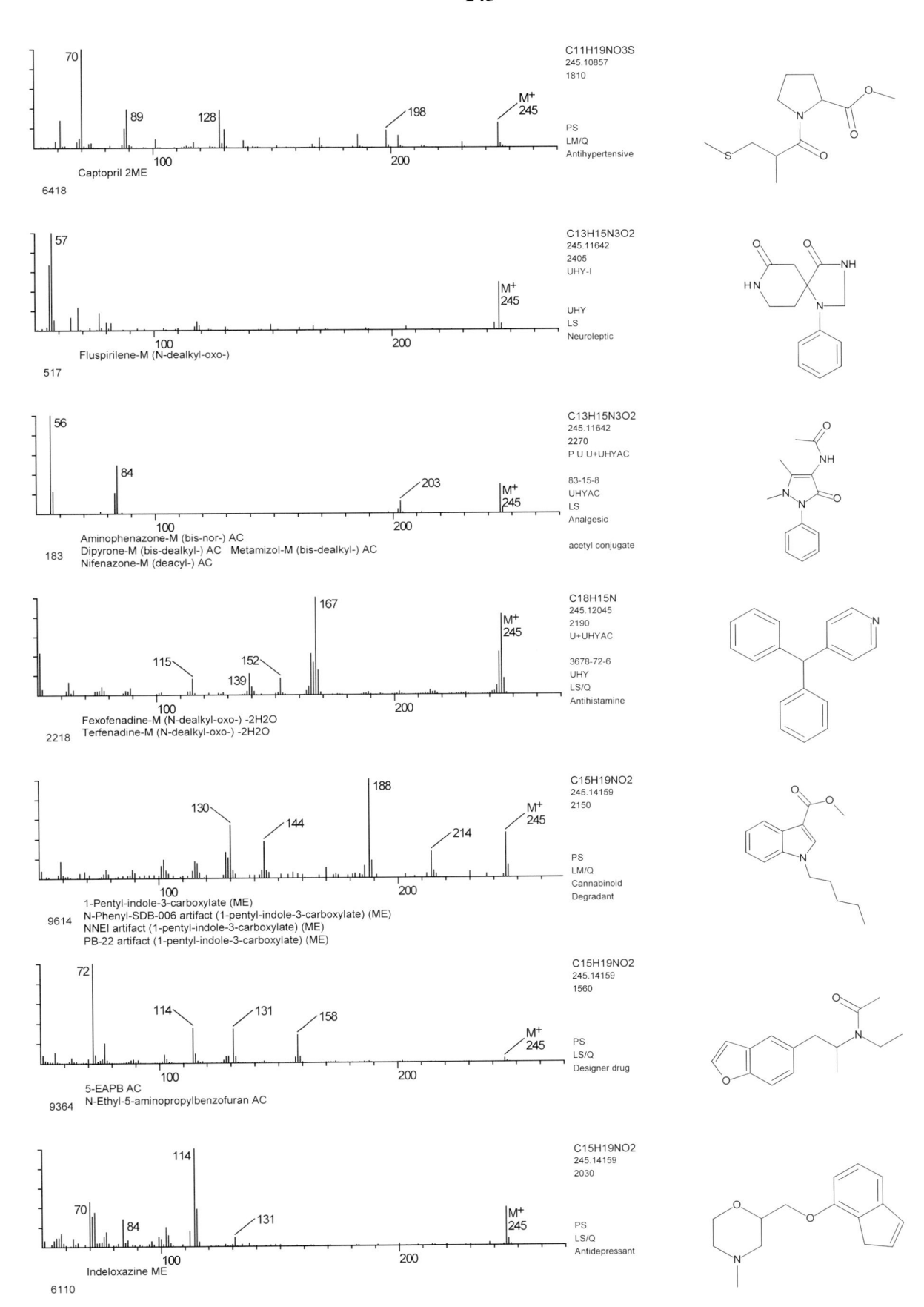

Captopril 2ME
6418

C11H19NO3S
245.10857
1810

PS
LM/Q
Antihypertensive

Fluspirilene-M (N-dealkyl-oxo-)
517

C13H15N3O2
245.11642
2405
UHY-I

UHY
LS
Neuroleptic

Aminophenazone-M (bis-nor-) AC
Dipyrone-M (bis-dealkyl-) AC   Metamizol-M (bis-dealkyl-) AC
Nifenazone-M (deacyl-) AC
183

C13H15N3O2
245.11642
2270
P U U+UHYAC

83-15-8
UHYAC
LS
Analgesic
acetyl conjugate

Fexofenadine-M (N-dealkyl-oxo-) -2H2O
Terfenadine-M (N-dealkyl-oxo-) -2H2O
2218

C18H15N
245.12045
2190
U+UHYAC

3678-72-6
UHY
LS/Q
Antihistamine

1-Pentyl-indole-3-carboxylate (ME)
N-Phenyl-SDB-006 artifact (1-pentyl-indole-3-carboxylate) (ME)
NNEI artifact (1-pentyl-indole-3-carboxylate) (ME)
PB-22 artifact (1-pentyl-indole-3-carboxylate) (ME)
9614

C15H19NO2
245.14159
2150

PS
LM/Q
Cannabinoid
Degradant

5-EAPB AC
N-Ethyl-5-aminopropylbenzofuran AC
9364

C15H19NO2
245.14159
1560

PS
LS/Q
Designer drug

Indeloxazine ME
6110

C15H19NO2
245.14159
2030

PS
LS/Q
Antidepressant

245

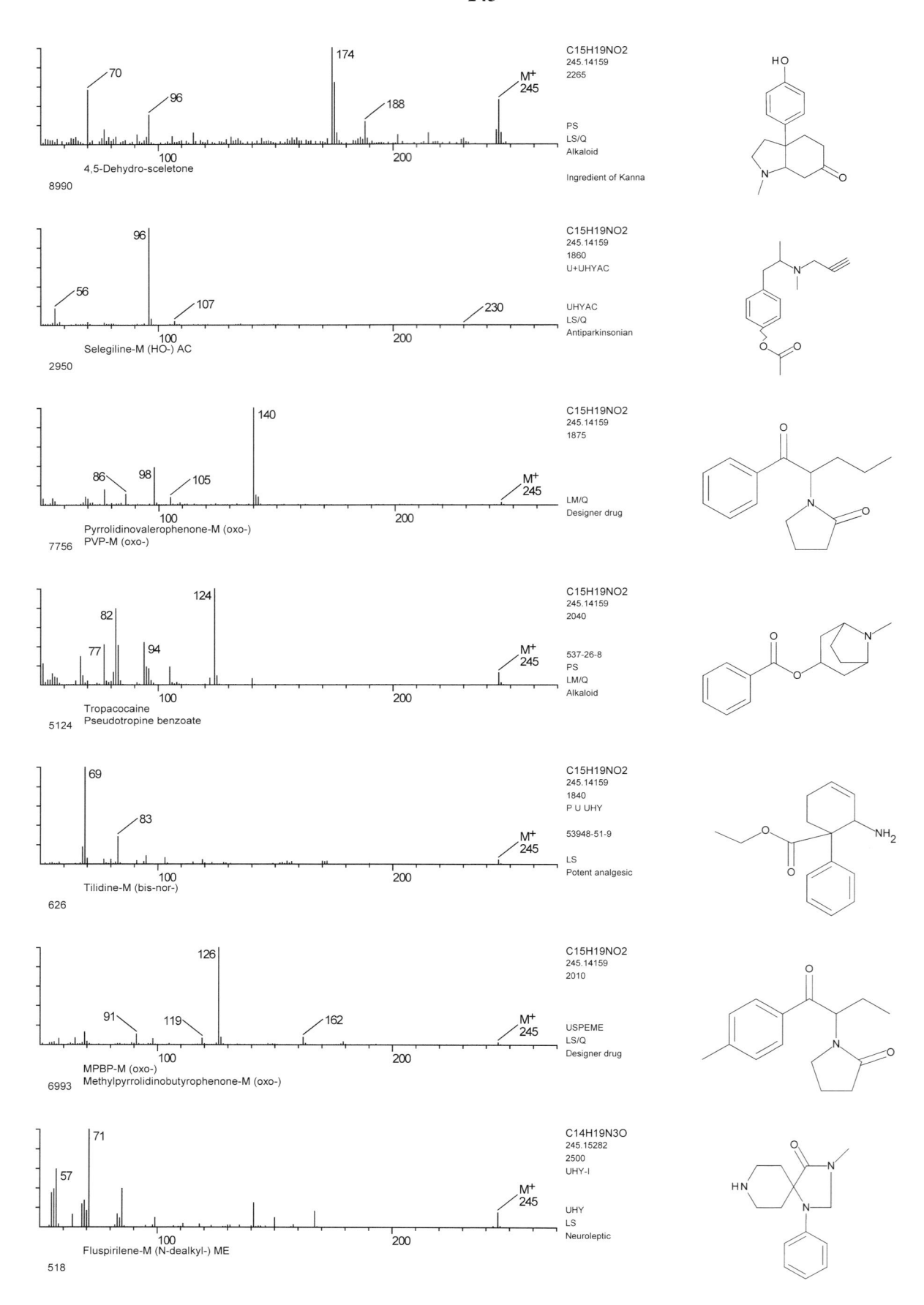

70, 96, 174, 188, M+ 245	C15H19NO2 245.14159 2265 PS LS/Q Alkaloid Ingredient of Kanna
4,5-Dehydro-sceletone 8990	

56, 96, 107, 230	C15H19NO2 245.14159 1860 U+UHYAC UHYAC LS/Q Antiparkinsonian
Selegiline-M (HO-) AC 2950	

86, 98, 105, 140, M+ 245	C15H19NO2 245.14159 1875 LM/Q Designer drug
Pyrrolidinovalerophenone-M (oxo-) 7756 PVP-M (oxo-)	

77, 82, 94, 124, M+ 245	C15H19NO2 245.14159 2040 537-26-8 PS LM/Q Alkaloid
Tropacocaine 5124 Pseudotropine benzoate	

69, 83, M+ 245	C15H19NO2 245.14159 1840 P U UHY 53948-51-9 LS Potent analgesic
Tilidine-M (bis-nor-) 626	

91, 119, 126, 162, M+ 245	C15H19NO2 245.14159 2010 USPEME LS/Q Designer drug
MPBP-M (oxo-) 6993 Methylpyrrolidinobutyrophenone-M (oxo-)	

57, 71, M+ 245	C14H19N3O 245.15282 2500 UHY-I UHY LS Neuroleptic
Fluspirilene-M (N-dealkyl-) ME 518	

576

245

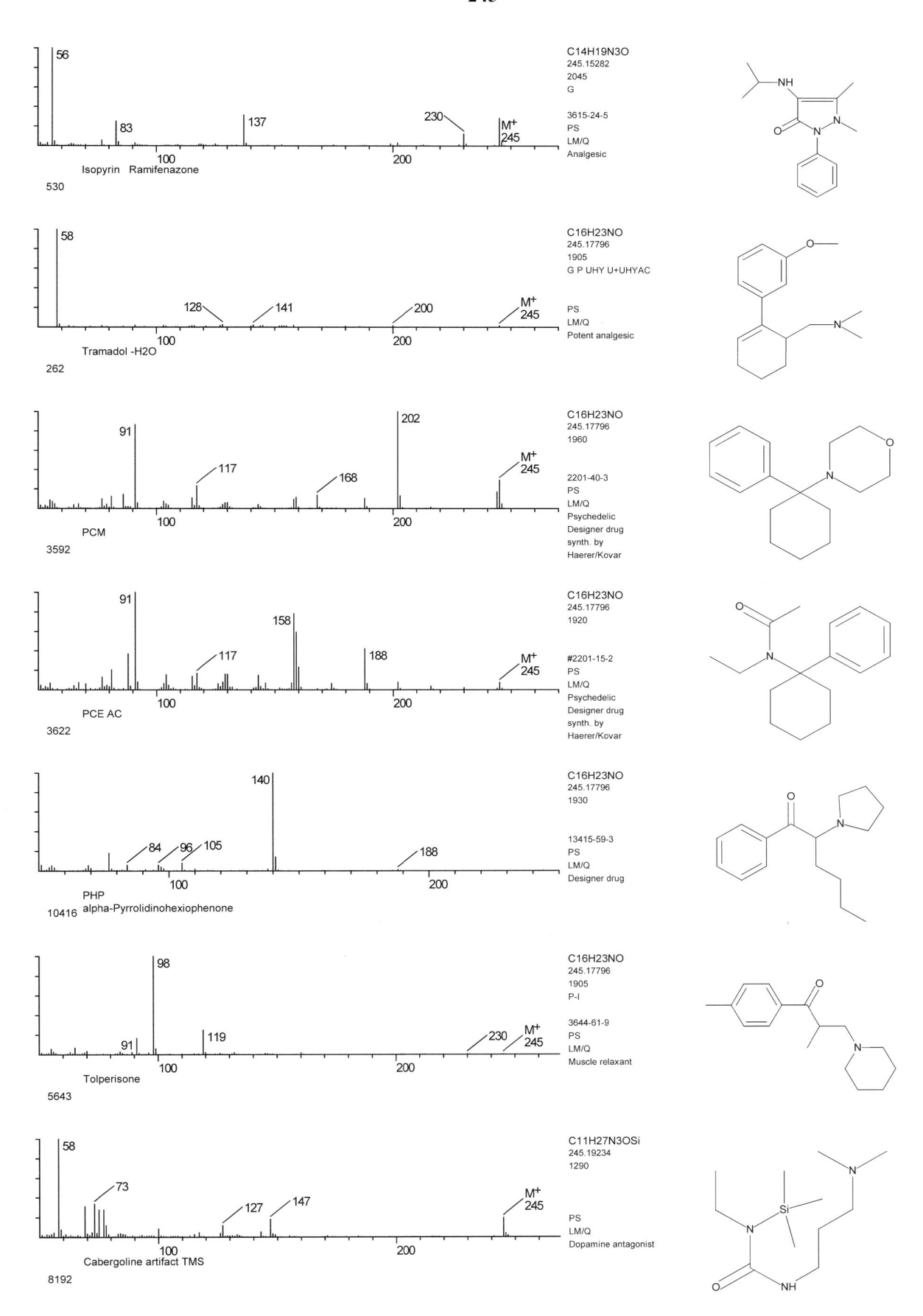

Isopyrin Ramifenazone 530	56, 83, 137, 230, M+ 245	C14H19N3O 245.15282 2045 G 3615-24-5 PS LM/Q Analgesic
Tramadol -H2O 262	58, 128, 141, 200, M+ 245	C16H23NO 245.17796 1905 G P UHY U+UHYAC PS LM/Q Potent analgesic
PCM 3592	91, 117, 168, 202, M+ 245	C16H23NO 245.17796 1960 2201-40-3 PS LM/Q Psychedelic Designer drug synth. by Haerer/Kovar
PCE AC 3622	91, 117, 158, 188, M+ 245	C16H23NO 245.17796 1920 #2201-15-2 PS LM/Q Psychedelic Designer drug synth. by Haerer/Kovar
PHP alpha-Pyrrolidinohexiophenone 10416	84, 96, 105, 140, 188	C16H23NO 245.17796 1930 13415-59-3 PS LM/Q Designer drug
Tolperisone 5643	91, 98, 119, 230, M+ 245	C16H23NO 245.17796 1905 P-I 3644-61-9 PS LM/Q Muscle relaxant
Cabergoline artifact TMS 8192	58, 73, 127, 147, M+ 245	C11H27N3OSi 245.19234 1290 PS LM/Q Dopamine antagonist

246

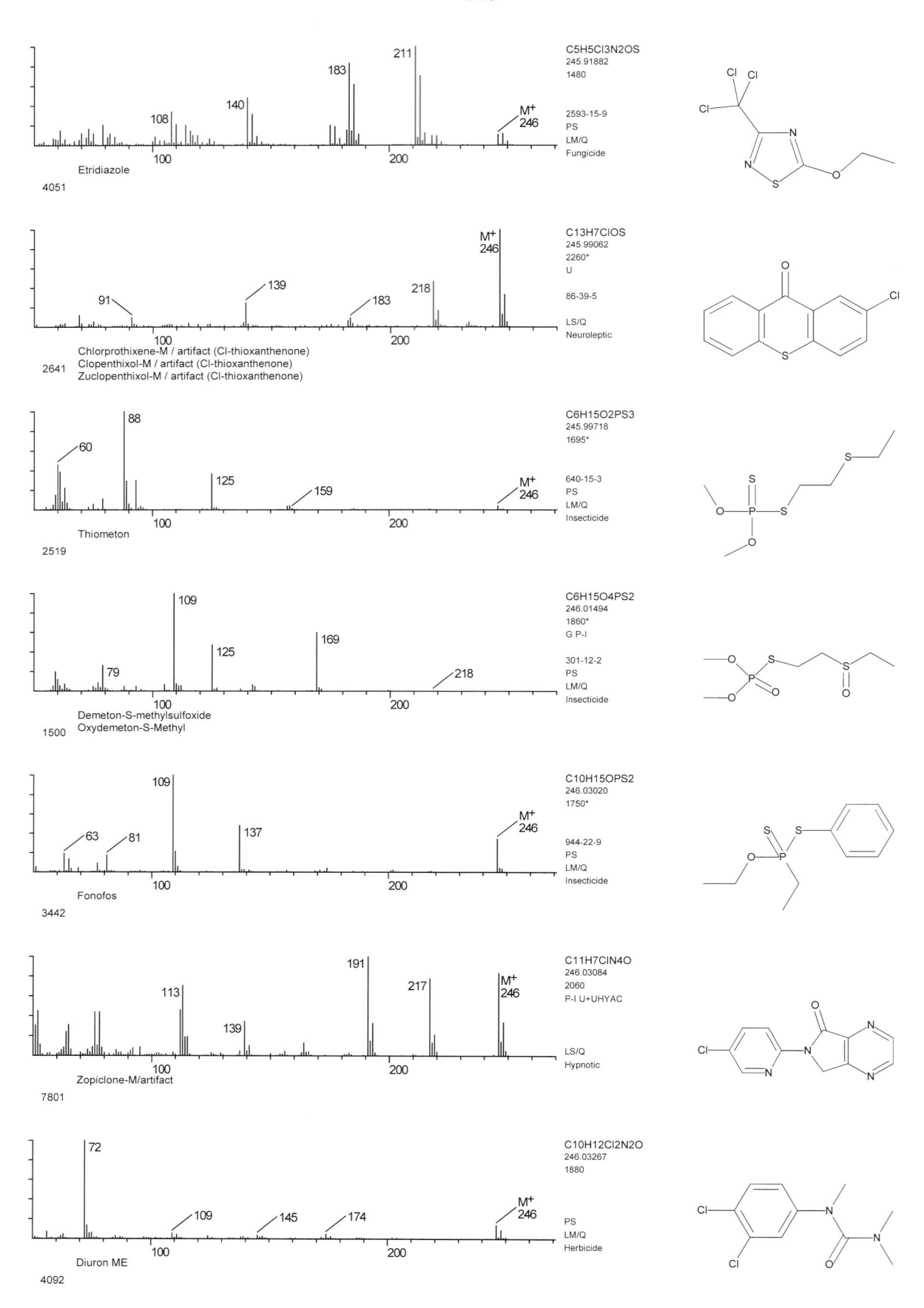

Compound	Formula	Details
Etridiazole (4051)	C5H5Cl3N2OS	245.91882, 1480, 2593-15-9, PS, LM/Q, Fungicide
Chlorprothixene-M / artifact (Cl-thioxanthenone) Clopenthixol-M / artifact (Cl-thioxanthenone) Zuclopenthixol-M / artifact (Cl-thioxanthenone) (2641)	C13H7ClOS	245.99062, 2260*, U, 86-39-5, LS/Q, Neuroleptic
Thiometon (2519)	C6H15O2PS3	245.99718, 1695*, 640-15-3, PS, LM/Q, Insecticide
Demeton-S-methylsulfoxide Oxydemeton-S-Methyl (1500)	C6H15O4PS2	246.01494, 1860*, G P-I, 301-12-2, PS, LM/Q, Insecticide
Fonofos (3442)	C10H15OPS2	246.03020, 1750*, 944-22-9, PS, LM/Q, Insecticide
Zopiclone-M/artifact (7801)	C11H7ClN4O	246.03084, 2060, P-I U+UHYAC, LS/Q, Hypnotic
Diuron ME (4092)	C10H12Cl2N2O	246.03267, 1880, PS, LM/Q, Herbicide

246

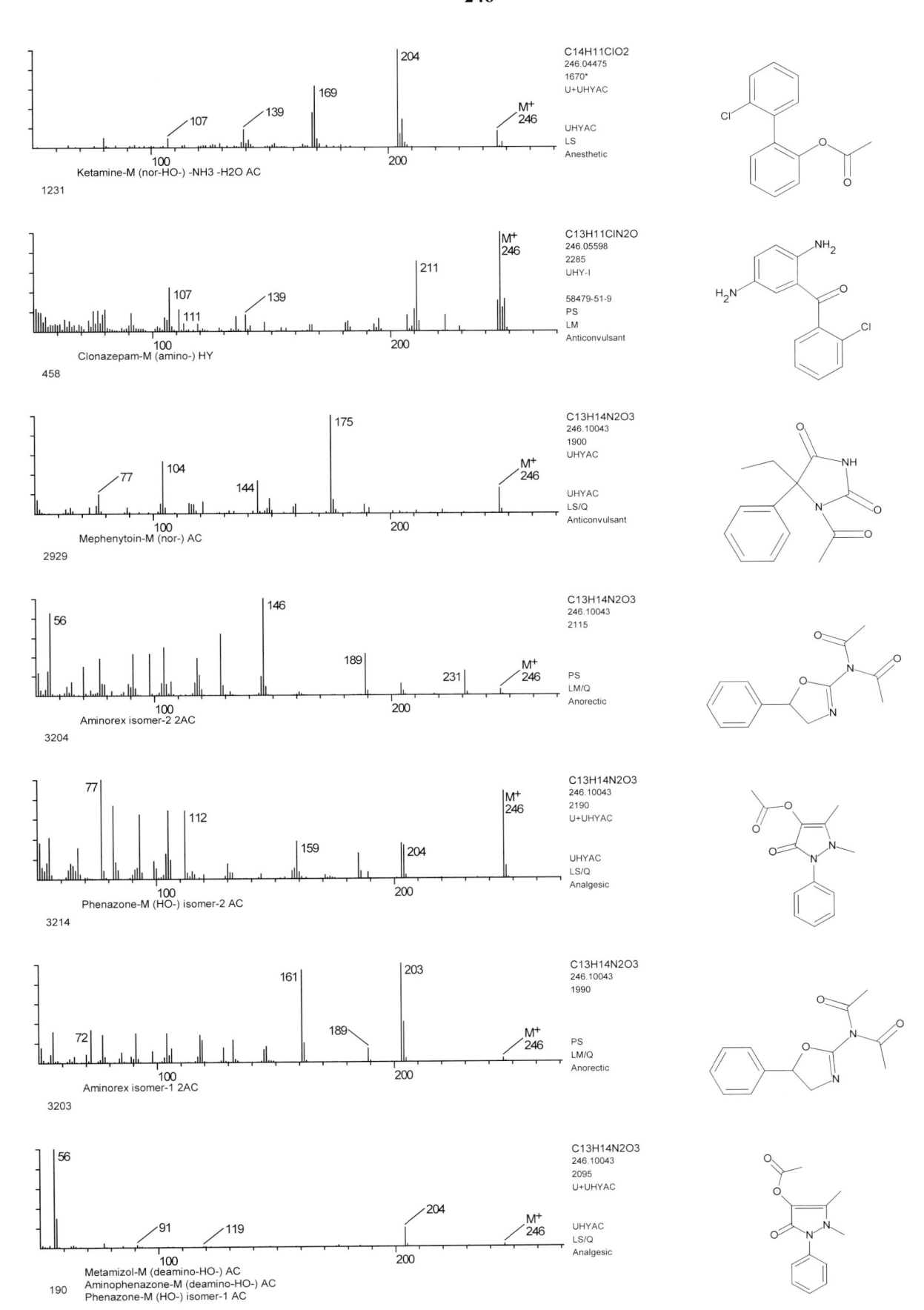

C14H11ClO2
246.04475
1670*
U+UHYAC

UHYAC
LS
Anesthetic

Ketamine-M (nor-HO-) -NH3 -H2O AC
1231

C13H11ClN2O
246.05598
2285
UHY-I

58479-51-9
PS
LM
Anticonvulsant

Clonazepam-M (amino-) HY
458

C13H14N2O3
246.10043
1900
UHYAC

UHYAC
LS/Q
Anticonvulsant

Mephenytoin-M (nor-) AC
2929

C13H14N2O3
246.10043
2115

PS
LM/Q
Anorectic

Aminorex isomer-2 2AC
3204

C13H14N2O3
246.10043
2190
U+UHYAC

UHYAC
LS/Q
Analgesic

Phenazone-M (HO-) isomer-2 AC
3214

C13H14N2O3
246.10043
1990

PS
LM/Q
Anorectic

Aminorex isomer-1 2AC
3203

C13H14N2O3
246.10043
2095
U+UHYAC

UHYAC
LS/Q
Analgesic

Metamizol-M (deamino-HO-) AC
Aminophenazone-M (deamino-HO-) AC
Phenazone-M (HO-) isomer-1 AC
190

246

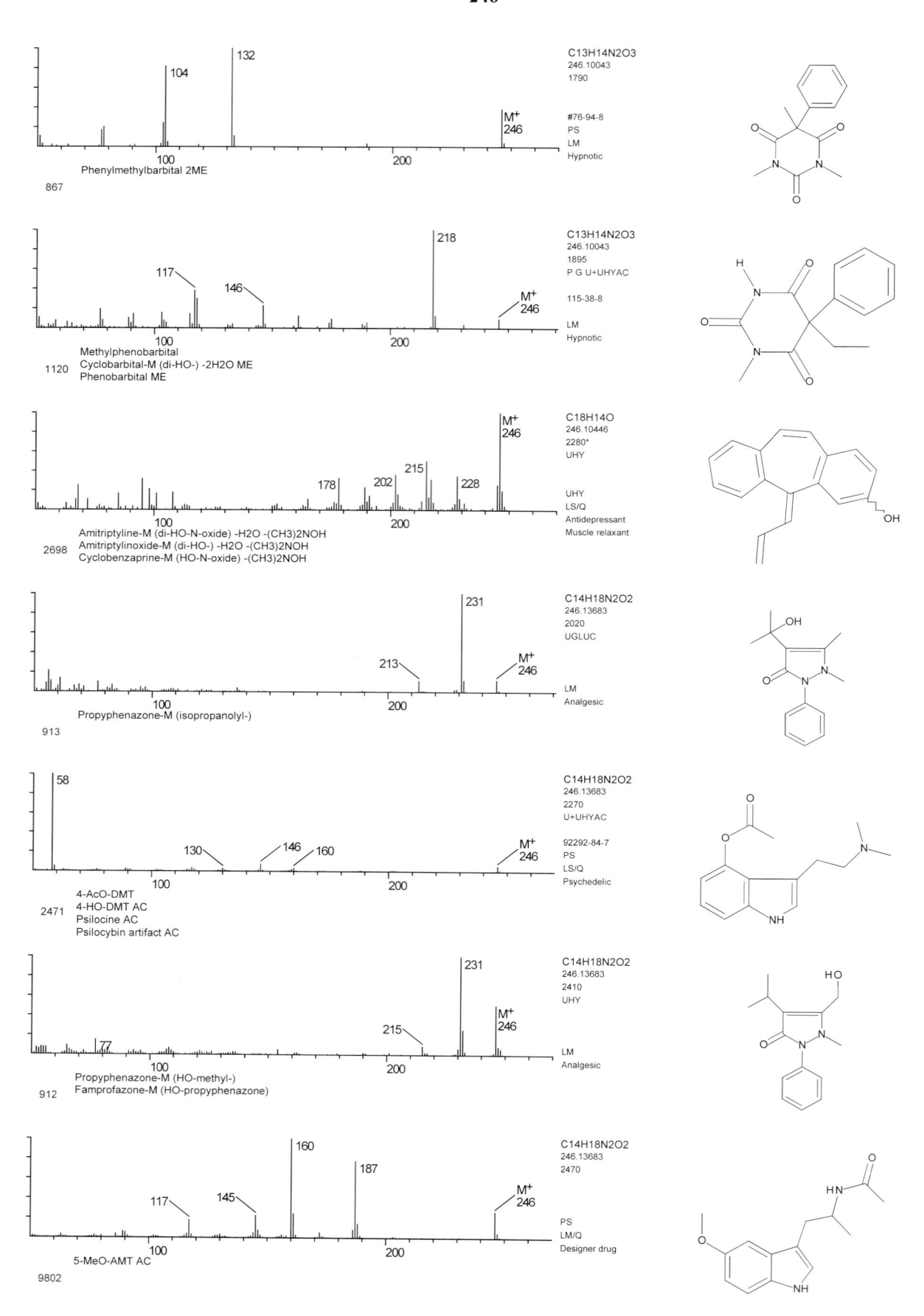

Phenylmethylbarbital 2ME	C13H14N2O3 246.10043 1790 #76-94-8 PS LM Hypnotic
867 peaks: 104, 132, M+ 246	

Methylphenobarbital Cyclobarbital-M (di-HO-) -2H2O ME Phenobarbital ME	C13H14N2O3 246.10043 1895 P G U+UHYAC 115-38-8 LM Hypnotic
1120 peaks: 117, 146, 218, M+ 246	

Amitriptyline-M (di-HO-N-oxide) -H2O -(CH3)2NOH Amitriptylinoxide-M (di-HO-) -H2O -(CH3)2NOH Cyclobenzaprine-M (HO-N-oxide) -(CH3)2NOH	C18H14O 246.10446 2280* UHY UHY LS/Q Antidepressant Muscle relaxant
2698 peaks: 178, 202, 215, 228, M+ 246	

Propyphenazone-M (isopropanolyl-)	C14H18N2O2 246.13683 2020 UGLUC LM Analgesic
913 peaks: 213, 231, M+ 246	

4-AcO-DMT 4-HO-DMT AC Psilocine AC Psilocybin artifact AC	C14H18N2O2 246.13683 2270 U+UHYAC 92292-84-7 PS LS/Q Psychedelic
2471 peaks: 58, 130, 146, 160, M+ 246	

Propyphenazone-M (HO-methyl-) Famprofazone-M (HO-propyphenazone)	C14H18N2O2 246.13683 2410 UHY LM Analgesic
912 peaks: 77, 215, 231, M+ 246	

5-MeO-AMT AC	C14H18N2O2 246.13683 2470 PS LM/Q Designer drug
9802 peaks: 117, 145, 160, 187, M+ 246	

580

246

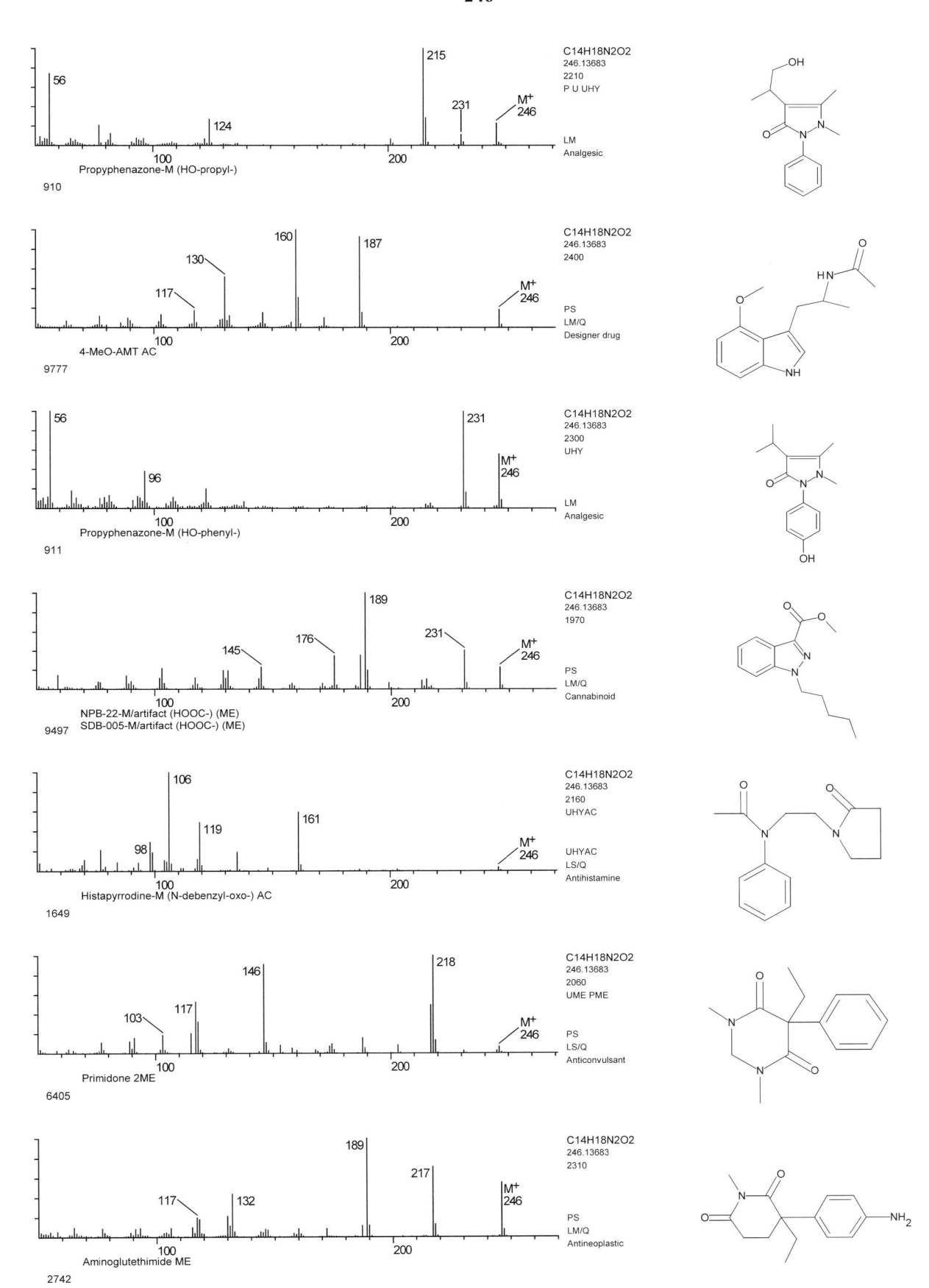

	C14H18N2O2 246.13683 2210 P U UHY LM Analgesic	
Propyphenazone-M (HO-propyl-) 910		

Peaks: 56, 124, 215, 231, M+ 246

	C14H18N2O2 246.13683 2400 PS LM/Q Designer drug	
4-MeO-AMT AC 9777		

Peaks: 117, 130, 160, 187, M+ 246

	C14H18N2O2 246.13683 2300 UHY LM Analgesic	
Propyphenazone-M (HO-phenyl-) 911		

Peaks: 56, 96, 231, M+ 246

	C14H18N2O2 246.13683 1970 PS LM/Q Cannabinoid	
NPB-22-M/artifact (HOOC-) (ME) SDB-005-M/artifact (HOOC-) (ME) 9497		

Peaks: 145, 176, 189, 231, M+ 246

	C14H18N2O2 246.13683 2160 UHYAC UHYAC LS/Q Antihistamine	
Histapyrrodine-M (N-debenzyl-oxo-) AC 1649		

Peaks: 98, 106, 119, 161, M+ 246

	C14H18N2O2 246.13683 2060 UME PME PS LS/Q Anticonvulsant	
Primidone 2ME 6405		

Peaks: 103, 117, 146, 218, M+ 246

	C14H18N2O2 246.13683 2310 PS LM/Q Antineoplastic	
Aminoglutethimide ME 2742		

Peaks: 117, 132, 189, 217, M+ 246

246

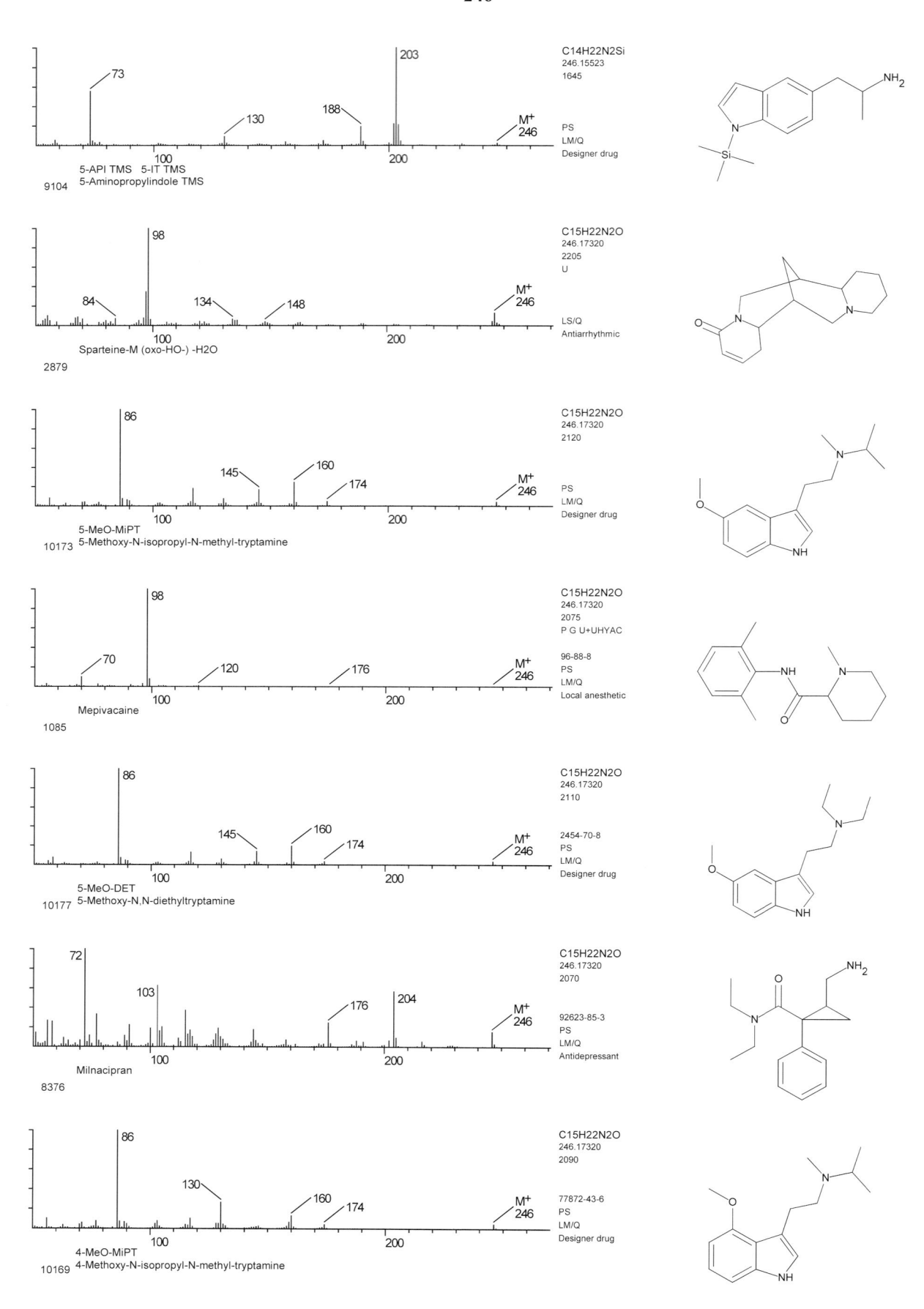

5-API TMS 5-IT TMS	C14H22N2Si
5-Aminopropylindole TMS	246.15523
(peaks: 73, 130, 188, 203, M+ 246)	1645
9104	PS / LM/Q / Designer drug

Sparteine-M (oxo-HO-) -H2O	C15H22N2O
(peaks: 84, 98, 134, 148, M+ 246)	246.17320
2879	2205
	U
	LS/Q / Antiarrhythmic

5-MeO-MiPT	C15H22N2O
5-Methoxy-N-isopropyl-N-methyl-tryptamine	246.17320
(peaks: 86, 145, 160, 174, M+ 246)	2120
10173	PS / LM/Q / Designer drug

Mepivacaine	C15H22N2O
(peaks: 70, 98, 120, 176, M+ 246)	246.17320
1085	2075
	P G U+UHYAC
	96-88-8
	PS / LM/Q / Local anesthetic

5-MeO-DET	C15H22N2O
5-Methoxy-N,N-diethyltryptamine	246.17320
(peaks: 86, 145, 160, 174, M+ 246)	2110
10177	2454-70-8
	PS / LM/Q / Designer drug

Milnacipran	C15H22N2O
(peaks: 72, 103, 176, 204, M+ 246)	246.17320
8376	2070
	92623-85-3
	PS / LM/Q / Antidepressant

4-MeO-MiPT	C15H22N2O
4-Methoxy-N-isopropyl-N-methyl-tryptamine	246.17320
(peaks: 86, 130, 160, 174, M+ 246)	2090
10169	77872-43-6
	PS / LM/Q / Designer drug

246

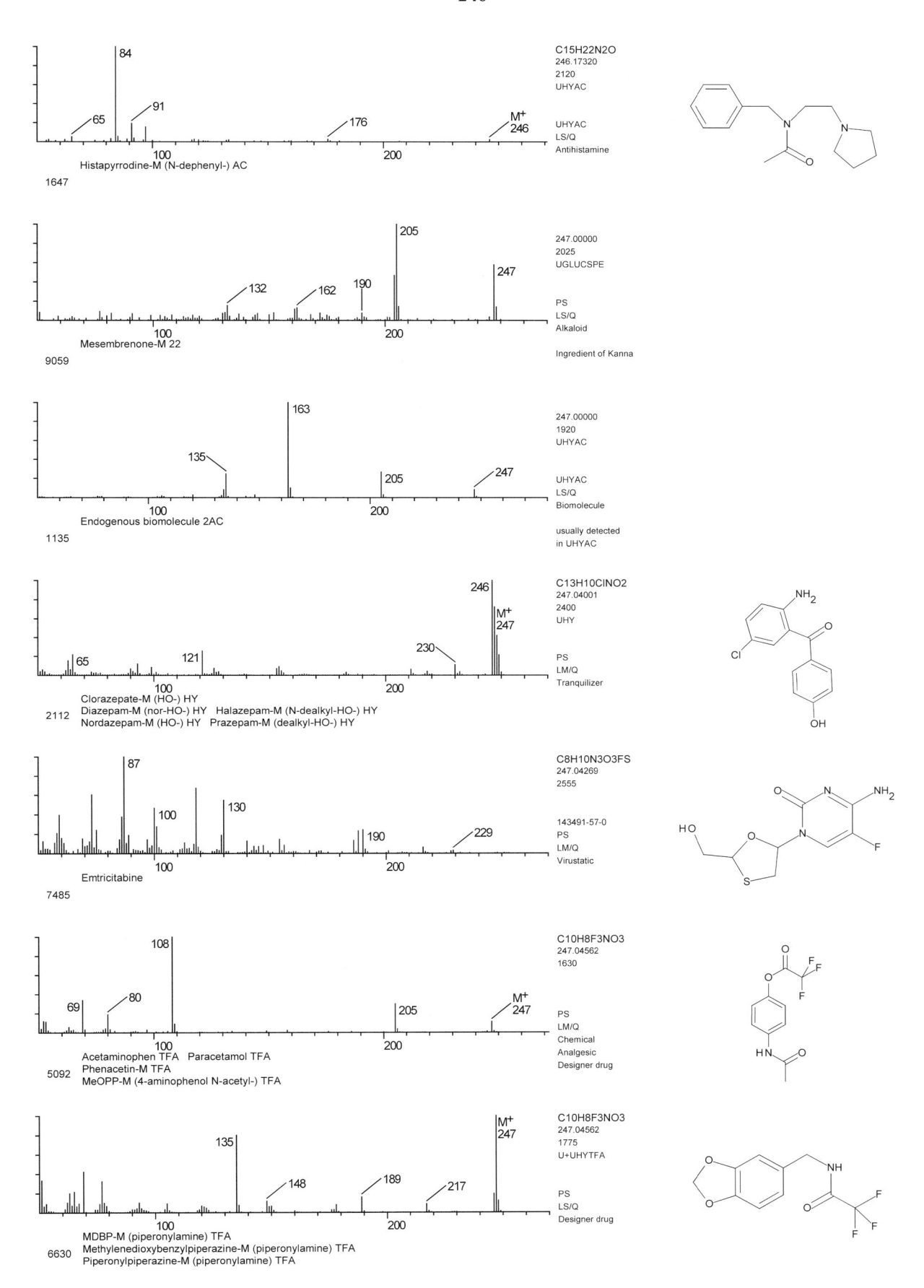

C15H22N2O
246.17320
2120
UHYAC

UHYAC
LS/Q
Antihistamine

Histapyrrodine-M (N-dephenyl-) AC
1647

247.00000
2025
UGLUCSPE

PS
LS/Q
Alkaloid

Ingredient of Kanna

Mesembrenone-M 22
9059

247.00000
1920
UHYAC

UHYAC
LS/Q
Biomolecule

usually detected in UHYAC

Endogenous biomolecule 2AC
1135

C13H10ClNO2
247.04001
2400
UHY

PS
LM/Q
Tranquilizer

Clorazepate-M (HO-) HY
Diazepam-M (nor-HO-) HY   Halazepam-M (N-dealkyl-HO-) HY
Nordazepam-M (HO-) HY   Prazepam-M (dealkyl-HO-) HY
2112

C8H10N3O3FS
247.04269
2555
143491-57-0
PS
LM/Q
Virustatic

Emtricitabine
7485

C10H8F3NO3
247.04562
1630

PS
LM/Q
Chemical
Analgesic
Designer drug

Acetaminophen TFA   Paracetamol TFA
Phenacetin-M TFA
MeOPP-M (4-aminophenol N-acetyl-) TFA
5092

C10H8F3NO3
247.04562
1775
U+UHYTFA

PS
LS/Q
Designer drug

MDBP-M (piperonylamine) TFA
Methylenedioxybenzylpiperazine-M (piperonylamine) TFA
Piperonylpiperazine-M (piperonylamine) TFA
6630

247

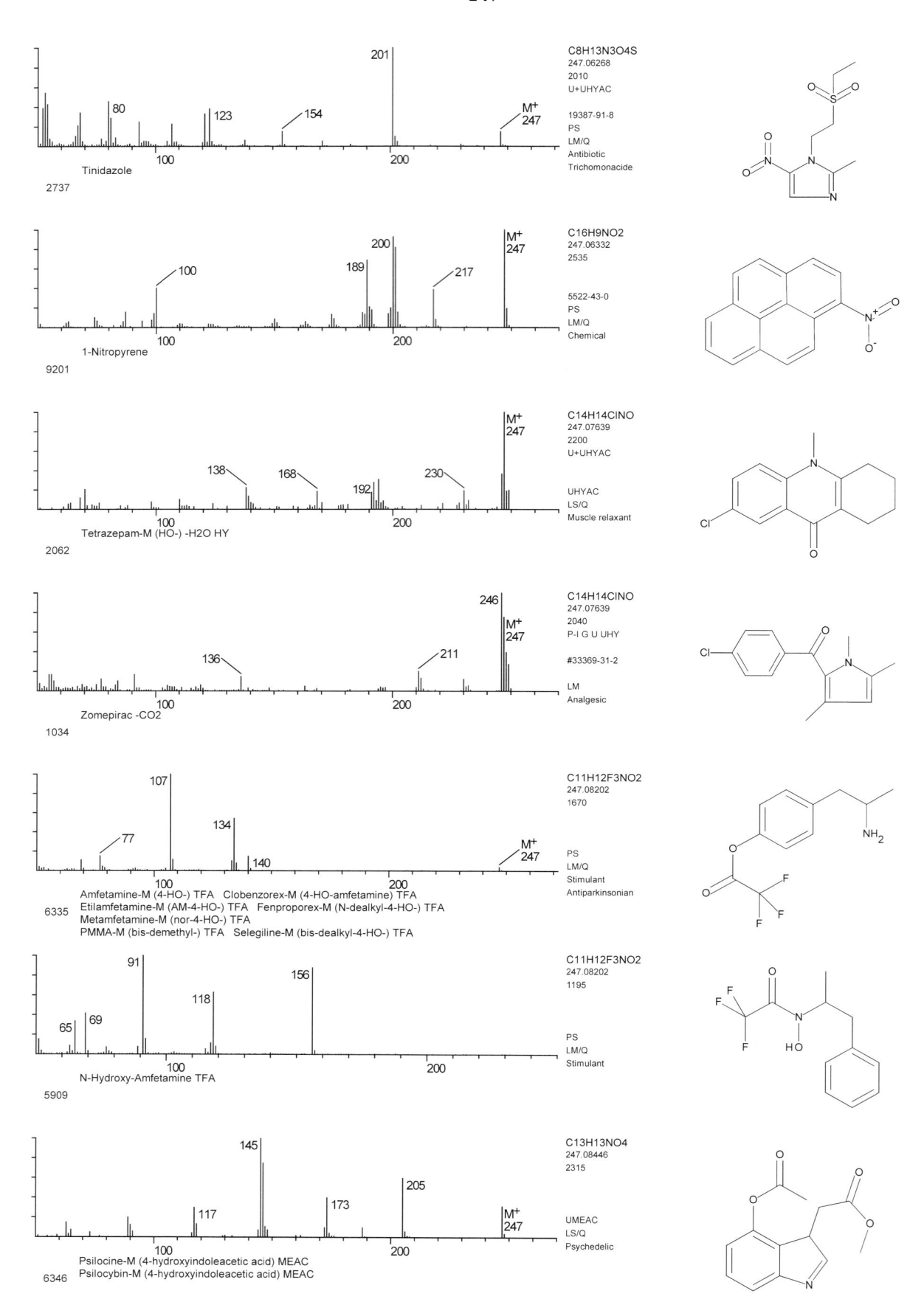

C8H13N3O4S
247.06268
2010
U+UHYAC

19387-91-8
PS
LM/Q
Antibiotic
Trichomonacide

Tinidazole
2737

C16H9NO2
247.06332
2535

5522-43-0
PS
LM/Q
Chemical

1-Nitropyrene
9201

C14H14ClNO
247.07639
2200
U+UHYAC

UHYAC
LS/Q
Muscle relaxant

Tetrazepam-M (HO-) -H2O HY
2062

C14H14ClNO
247.07639
2040
P-I G U UHY

#33369-31-2

LM
Analgesic

Zomepirac -CO2
1034

C11H12F3NO2
247.08202
1670

PS
LM/Q
Stimulant
Antiparkinsonian

Amfetamine-M (4-HO-) TFA   Clobenzorex-M (4-HO-amfetamine) TFA
Etilamfetamine-M (AM-4-HO-) TFA   Fenproporex-M (N-dealkyl-4-HO-) TFA
6335 Metamfetamine-M (nor-4-HO-) TFA
PMMA-M (bis-demethyl-) TFA   Selegiline-M (bis-dealkyl-4-HO-) TFA

C11H12F3NO2
247.08202
1195

PS
LM/Q
Stimulant

N-Hydroxy-Amfetamine TFA
5909

C13H13NO4
247.08446
2315

UMEAC
LS/Q
Psychedelic

Psilocine-M (4-hydroxyindoleacetic acid) MEAC
Psilocybin-M (4-hydroxyindoleacetic acid) MEAC
6346

584

247

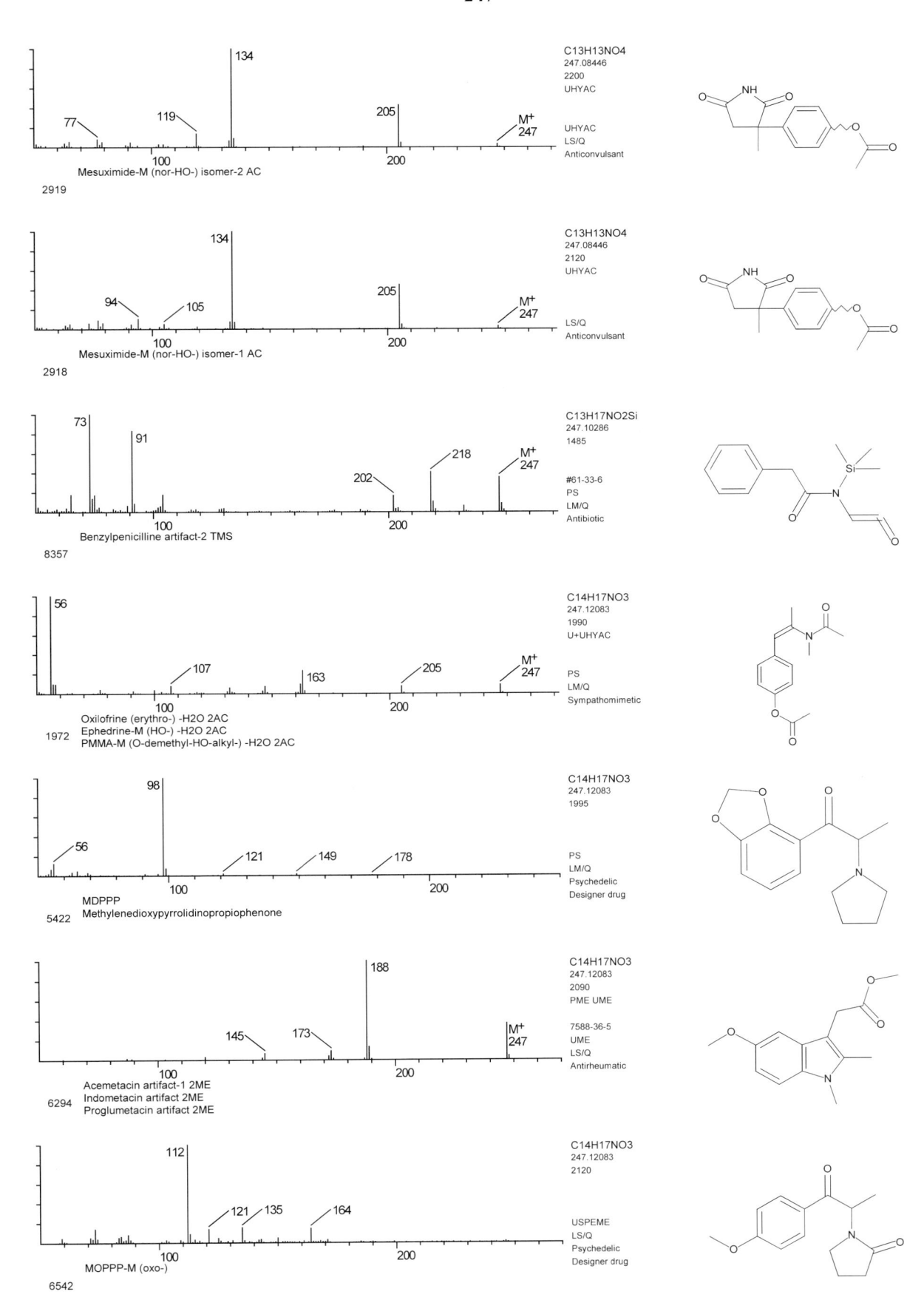

C13H13NO4
247.08446
2200
UHYAC

UHYAC
LS/Q
Anticonvulsant

Mesuximide-M (nor-HO-) isomer-2 AC
2919

C13H13NO4
247.08446
2120
UHYAC

LS/Q
Anticonvulsant

Mesuximide-M (nor-HO-) isomer-1 AC
2918

C13H17NO2Si
247.10286
1485
#61-33-6

PS
LM/Q
Antibiotic

Benzylpenicilline artifact-2 TMS
8357

C14H17NO3
247.12083
1990
U+UHYAC

PS
LM/Q
Sympathomimetic

Oxilofrine (erythro-) -H2O 2AC
Ephedrine-M (HO-) -H2O 2AC
PMMA-M (O-demethyl-HO-alkyl-) -H2O 2AC
1972

C14H17NO3
247.12083
1995

PS
LM/Q
Psychedelic
Designer drug

MDPPP
Methylenedioxypyrrolidinopropiophenone
5422

C14H17NO3
247.12083
2090
PME UME

7588-36-5
UME
LS/Q
Antirheumatic

Acemetacin artifact-1 2ME
Indometacin artifact 2ME
Proglumetacin artifact 2ME
6294

C14H17NO3
247.12083
2120

USPEME
LS/Q
Psychedelic
Designer drug

MOPPP-M (oxo-)
6542

247

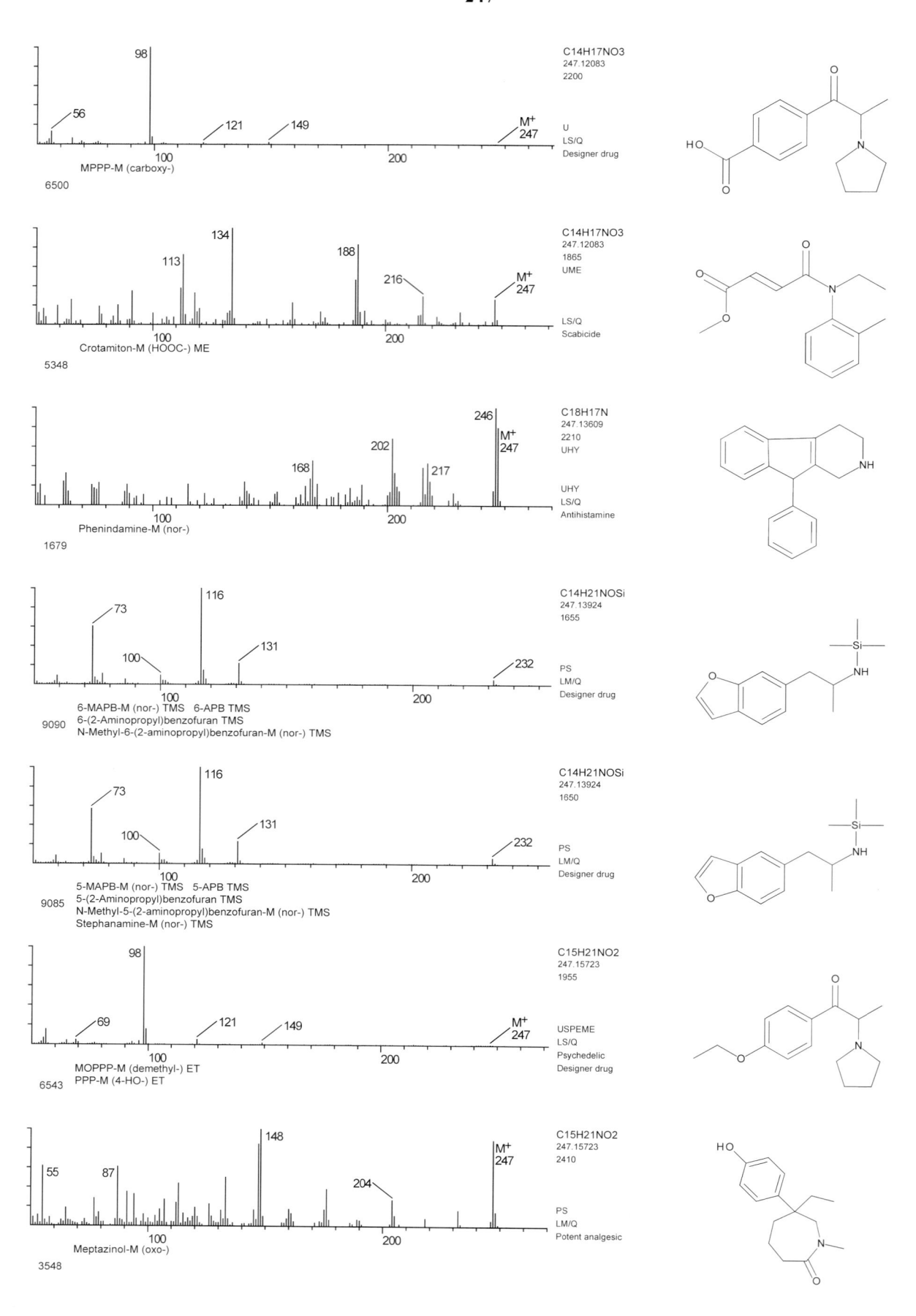

MPPP-M (carboxy-) 6500	C14H17NO3 247.12083 2200 U LS/Q Designer drug
Crotamiton-M (HOOC-) ME 5348	C14H17NO3 247.12083 1865 UME LS/Q Scabicide
Phenindamine-M (nor-) 1679	C18H17N 247.13609 2210 UHY UHY LS/Q Antihistamine
6-MAPB-M (nor-) TMS 6-(2-Aminopropyl)benzofuran TMS N-Methyl-6-(2-aminopropyl)benzofuran-M (nor-) TMS 9090	C14H21NOSi 247.13924 1655 PS LM/Q Designer drug
5-MAPB-M (nor-) TMS 5-(2-Aminopropyl)benzofuran TMS N-Methyl-5-(2-aminopropyl)benzofuran-M (nor-) TMS Stephanamine-M (nor-) TMS 9085	C14H21NOSi 247.13924 1650 PS LM/Q Designer drug
MOPPP-M (demethyl-) ET PPP-M (4-HO-) ET 6543	C15H21NO2 247.15723 1955 USPEME LS/Q Psychedelic Designer drug
Meptazinol-M (oxo-) 3548	C15H21NO2 247.15723 2410 PS LM/Q Potent analgesic

247

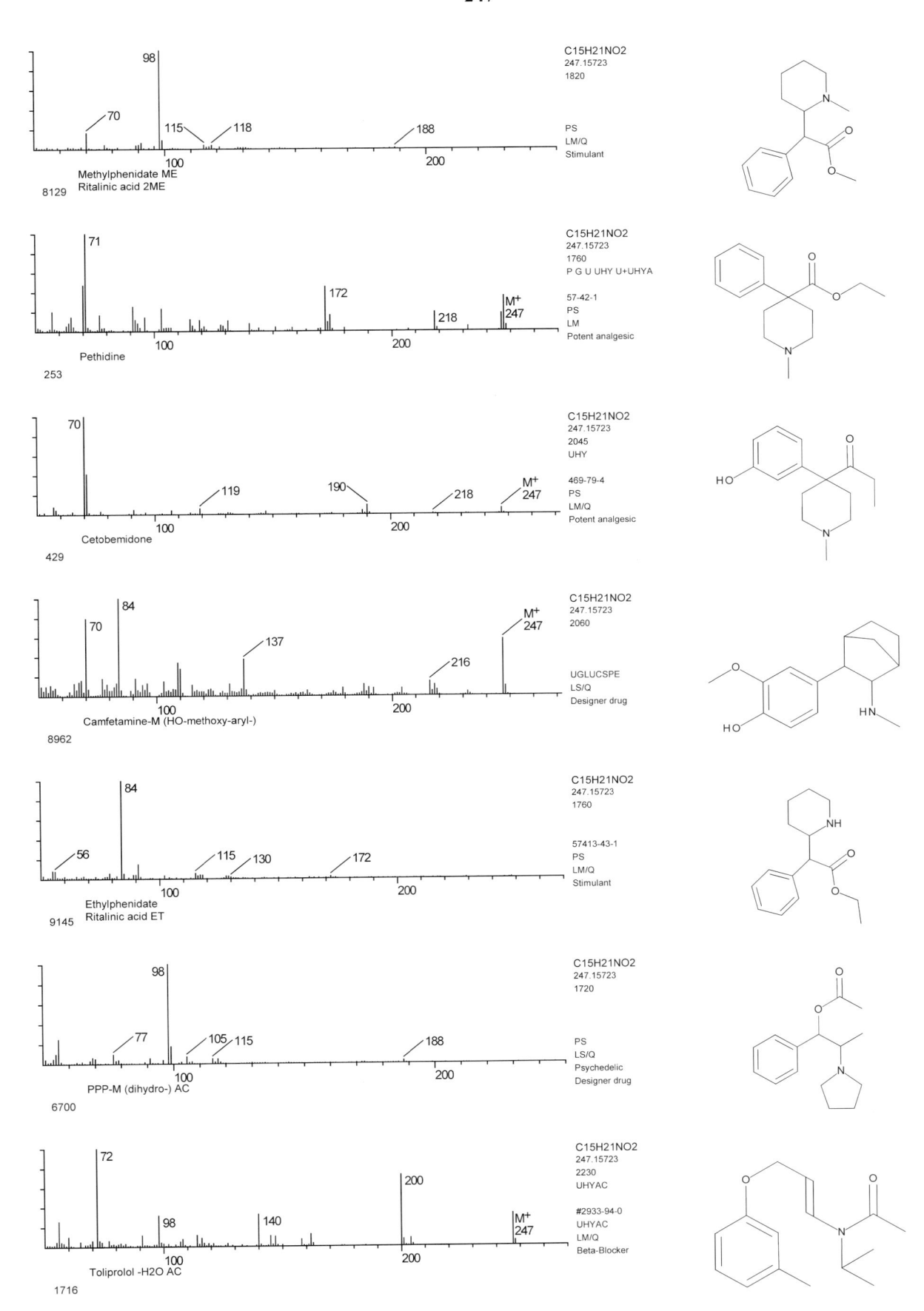

Spectrum	Compound	Formula/Info
8129	Methylphenidate ME / Ritalinic acid 2ME; peaks 70, 98, 115, 118, 188	C15H21NO2, 247.15723, 1820, PS, LM/Q, Stimulant
253	Pethidine; peaks 71, 172, 218, M+ 247	C15H21NO2, 247.15723, 1760, P G U UHY U+UHYA, 57-42-1, PS, LM, Potent analgesic
429	Cetobemidone; peaks 70, 119, 190, 218, M+ 247	C15H21NO2, 247.15723, 2045, UHY, 469-79-4, PS, LM/Q, Potent analgesic
8962	Camfetamine-M (HO-methoxy-aryl-); peaks 70, 84, 137, 216, M+ 247	C15H21NO2, 247.15723, 2060, UGLUCSPE, LS/Q, Designer drug
9145	Ethylphenidate / Ritalinic acid ET; peaks 56, 84, 115, 130, 172	C15H21NO2, 247.15723, 1760, 57413-43-1, PS, LM/Q, Stimulant
6700	PPP-M (dihydro-) AC; peaks 77, 98, 105, 115, 188	C15H21NO2, 247.15723, 1720, PS, LS/Q, Psychedelic, Designer drug
1716	Toliprolol -H2O AC; peaks 72, 98, 140, 200, M+ 247	C15H21NO2, 247.15723, 2230, UHYAC, #2933-94-0, UHYAC, LM/Q, Beta-Blocker

247

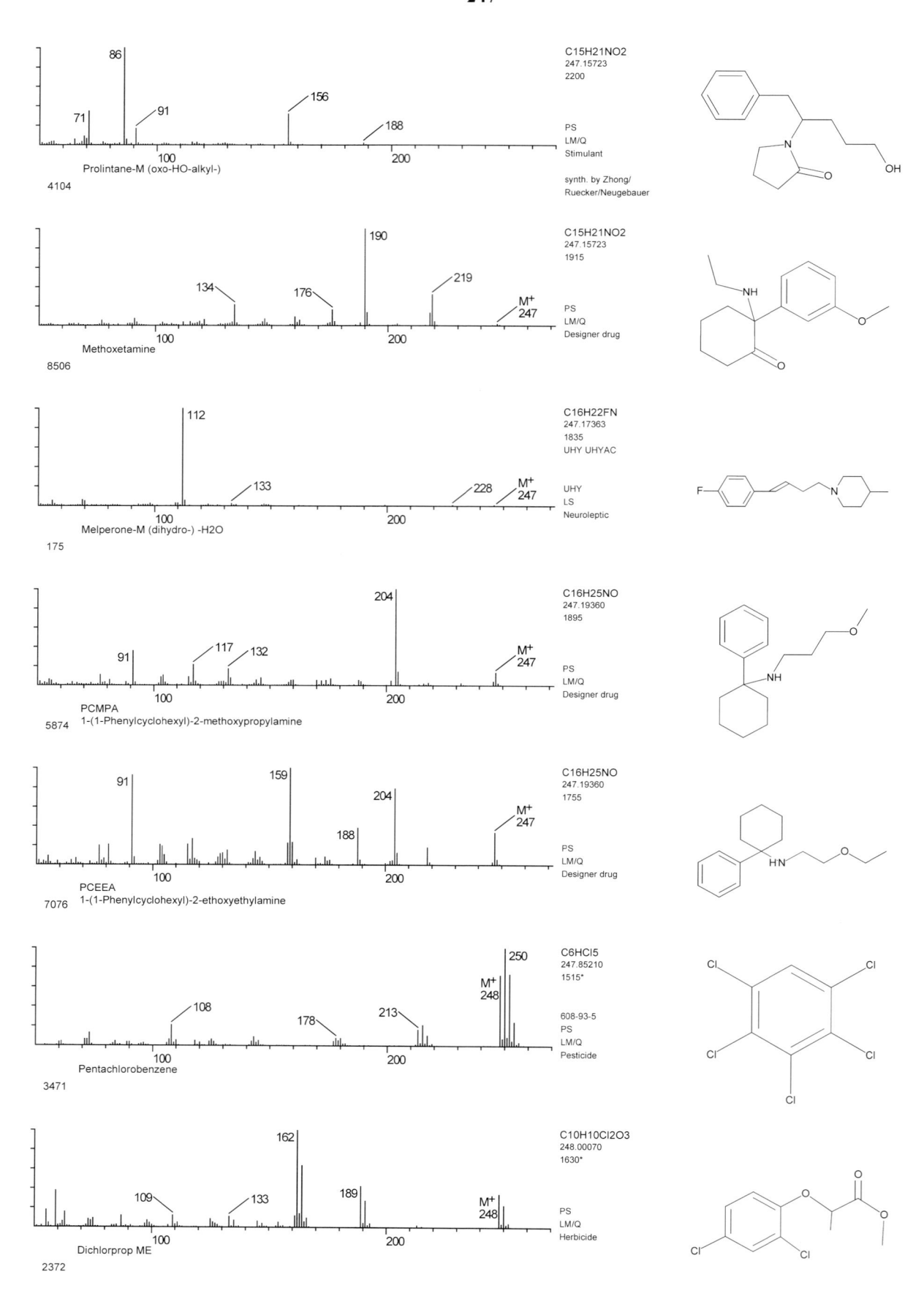

Compound	Formula / Mass	Info
Prolintane-M (oxo-HO-alkyl-) 4104	C15H21NO2 247.15723 2200	peaks: 71, 86, 91, 156, 188 PS LM/Q Stimulant synth. by Zhong/Ruecker/Neugebauer
Methoxetamine 8506	C15H21NO2 247.15723 1915	peaks: 134, 176, 190, 219, M+ 247 PS LM/Q Designer drug
Melperone-M (dihydro-) -H2O 175	C16H22FN 247.17363 1835 UHY UHYAC	peaks: 112, 133, 228, M+ 247 UHY LS Neuroleptic
PCMPA 1-(1-Phenylcyclohexyl)-2-methoxypropylamine 5874	C16H25NO 247.19360 1895	peaks: 91, 117, 132, 204, M+ 247 PS LM/Q Designer drug
PCEEA 1-(1-Phenylcyclohexyl)-2-ethoxyethylamine 7076	C16H25NO 247.19360 1755	peaks: 91, 159, 188, 204, M+ 247 PS LM/Q Designer drug
Pentachlorobenzene 3471	C6HCl5 247.85210 1515* 608-93-5	peaks: 108, 178, 213, M+ 248, 250 PS LM/Q Pesticide
Dichlorprop ME 2372	C10H10Cl2O3 248.00070 1630*	peaks: 109, 133, 162, 189, M+ 248 PS LM/Q Herbicide

588

248

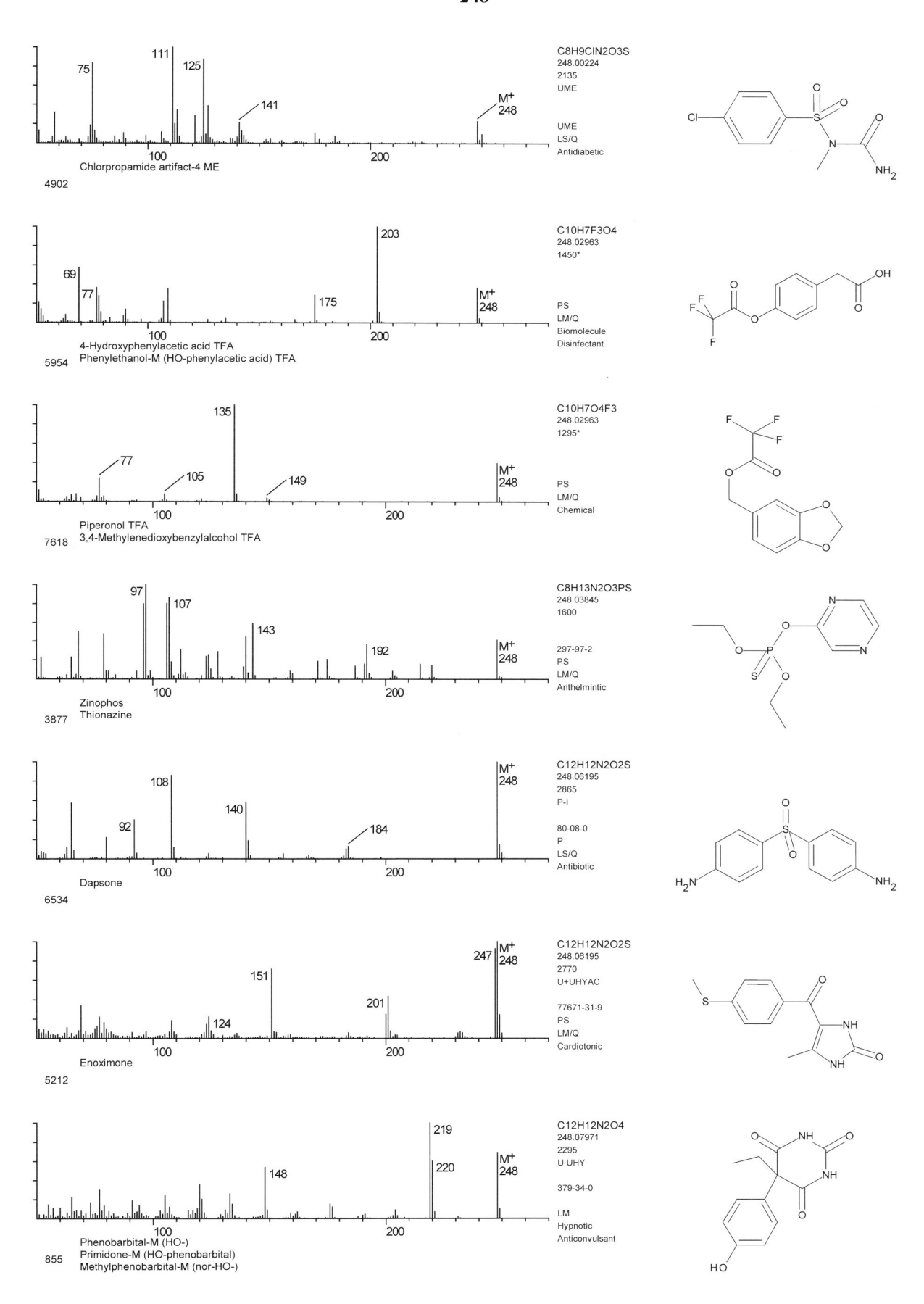

4902 Chlorpropamide artifact-4 ME
75, 111, 125, 141, M+ 248
C8H9ClN2O3S
248.00224
2135
UME

UME
LS/Q
Antidiabetic

5954 4-Hydroxyphenylacetic acid TFA
Phenylethanol-M (HO-phenylacetic acid) TFA
69, 77, 175, 203, M+ 248
C10H7F3O4
248.02963
1450*

PS
LM/Q
Biomolecule
Disinfectant

7618 Piperonol TFA
3,4-Methylenedioxybenzylalcohol TFA
77, 105, 135, 149, M+ 248
C10H7O4F3
248.02963
1295*

PS
LM/Q
Chemical

3877 Zinophos
Thionazine
97, 107, 143, 192, M+ 248
C8H13N2O3PS
248.03845
1600

297-97-2
PS
LM/Q
Anthelmintic

6534 Dapsone
92, 108, 140, 184, M+ 248
C12H12N2O2S
248.06195
2865
P-I

80-08-0
P
LS/Q
Antibiotic

5212 Enoximone
124, 151, 201, 247, M+ 248
C12H12N2O2S
248.06195
2770
U+UHYAC

77671-31-9
PS
LM/Q
Cardiotonic

855 Phenobarbital-M (HO-)
Primidone-M (HO-phenobarbital)
Methylphenobarbital-M (nor-HO-)
148, 219, 220, M+ 248
C12H12N2O4
248.07971
2295
U UHY

379-34-0
LM
Hypnotic
Anticonvulsant

248

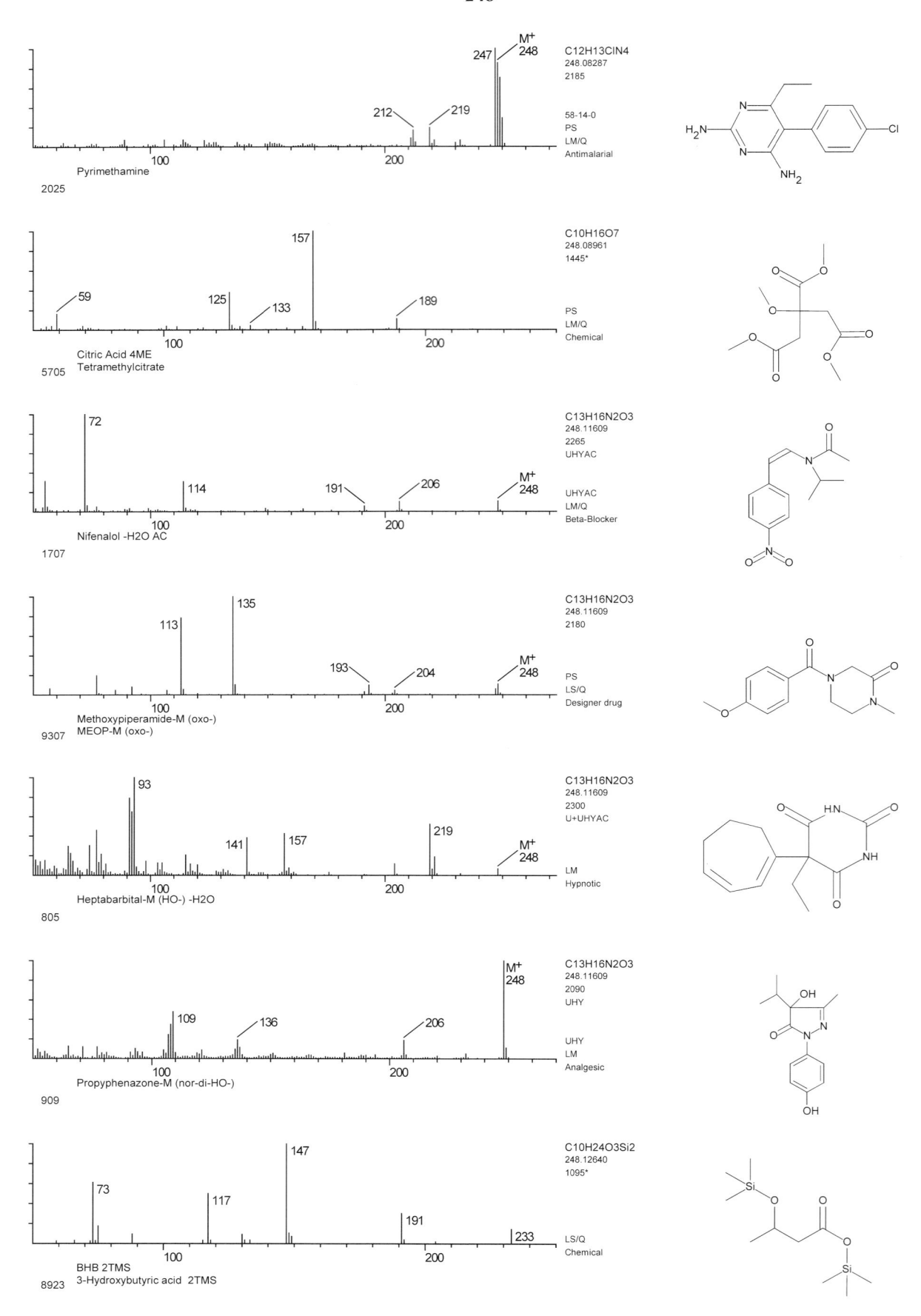

Pyrimethamine — C12H13ClN4, 248.08287, 2185, 58-14-0, PS, LM/Q, Antimalarial
Peaks: 212, 219, 247, M+ 248
2025

Citric Acid 4ME Tetramethylcitrate — C10H16O7, 248.08961, 1445*, PS, LM/Q, Chemical
Peaks: 59, 125, 133, 157, 189
5705

Nifenalol -H2O AC — C13H16N2O3, 248.11609, 2265, UHYAC, UHYAC, LM/Q, Beta-Blocker
Peaks: 72, 114, 191, 206, M+ 248
1707

Methoxypiperamide-M (oxo-) MEOP-M (oxo-) — C13H16N2O3, 248.11609, 2180, PS, LS/Q, Designer drug
Peaks: 113, 135, 193, 204, M+ 248
9307

Heptabarbital-M (HO-) -H2O — C13H16N2O3, 248.11609, 2300, U+UHYAC, LM, Hypnotic
Peaks: 93, 141, 157, 219, M+ 248
805

Propyphenazone-M (nor-di-HO-) — C13H16N2O3, 248.11609, 2090, UHY, UHY, LM, Analgesic
Peaks: 109, 136, 206, M+ 248
909

BHB 2TMS 3-Hydroxybutyric acid 2TMS — C10H24O3Si2, 248.12640, 1095*, LS/Q, Chemical
Peaks: 73, 117, 147, 191, 233
8923

590

248

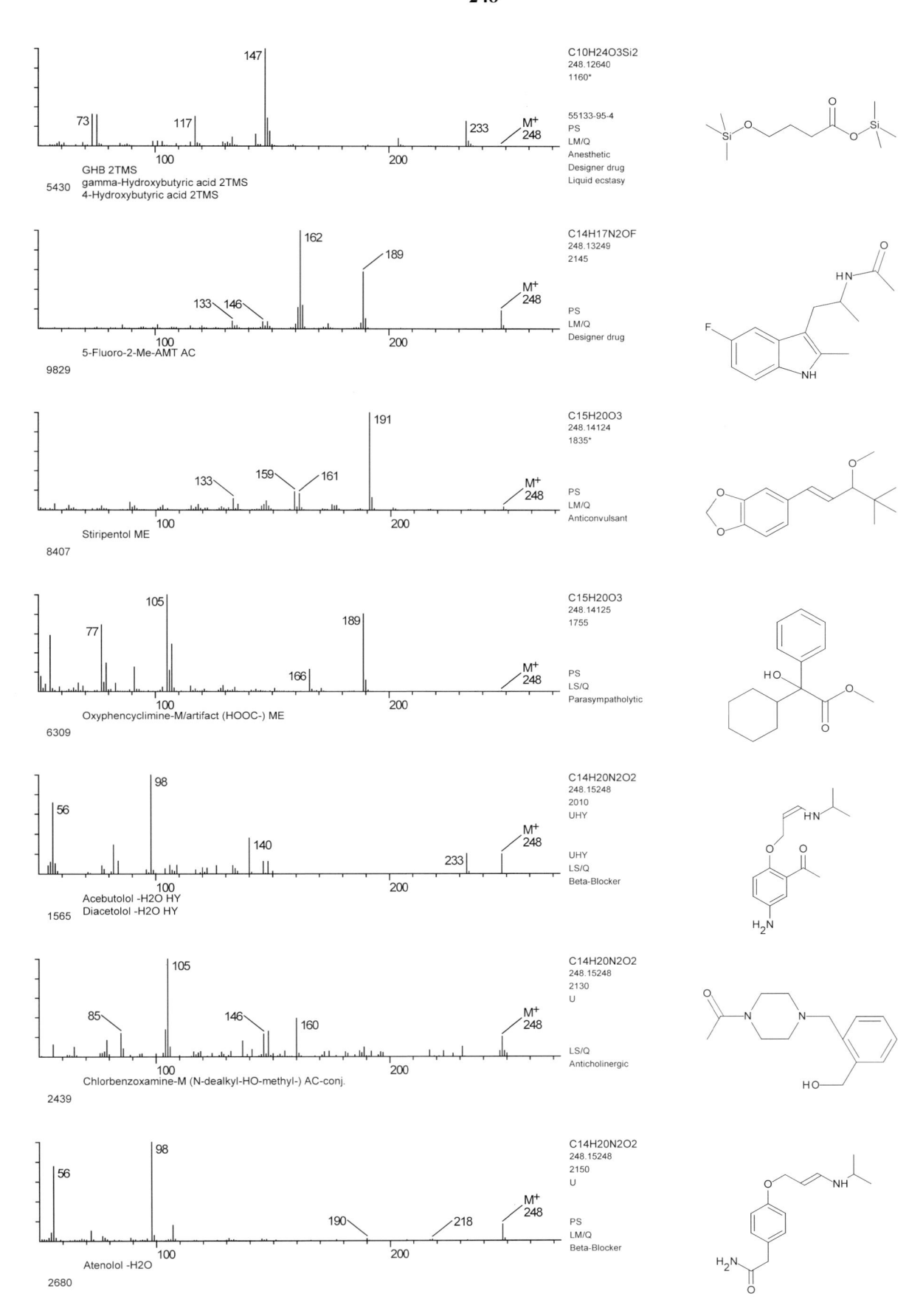

GHB 2TMS 5430 gamma-Hydroxybutyric acid 2TMS 4-Hydroxybutyric acid 2TMS Peaks: 73, 117, 147, 233, M+ 248	C10H24O3Si2 248.12640 1160* 55133-95-4 PS LM/Q Anesthetic Designer drug Liquid ecstasy
5-Fluoro-2-Me-AMT AC 9829 Peaks: 133, 146, 162, 189, M+ 248	C14H17N2OF 248.13249 2145 PS LM/Q Designer drug
Stiripentol ME 8407 Peaks: 133, 159, 161, 191, M+ 248	C15H20O3 248.14124 1835* PS LM/Q Anticonvulsant
Oxyphencyclimine-M/artifact (HOOC-) ME 6309 Peaks: 77, 105, 166, 189, M+ 248	C15H20O3 248.14125 1755 PS LS/Q Parasympatholytic
Acebutolol -H2O HY Diacetolol -H2O HY 1565 Peaks: 56, 98, 140, 233, M+ 248	C14H20N2O2 248.15248 2010 UHY UHY LS/Q Beta-Blocker
Chlorbenzoxamine-M (N-dealkyl-HO-methyl-) AC-conj. 2439 Peaks: 85, 105, 146, 160, M+ 248	C14H20N2O2 248.15248 2130 U LS/Q Anticholinergic
Atenolol -H2O 2680 Peaks: 56, 98, 190, 218, M+ 248	C14H20N2O2 248.15248 2150 U PS LM/Q Beta-Blocker

248

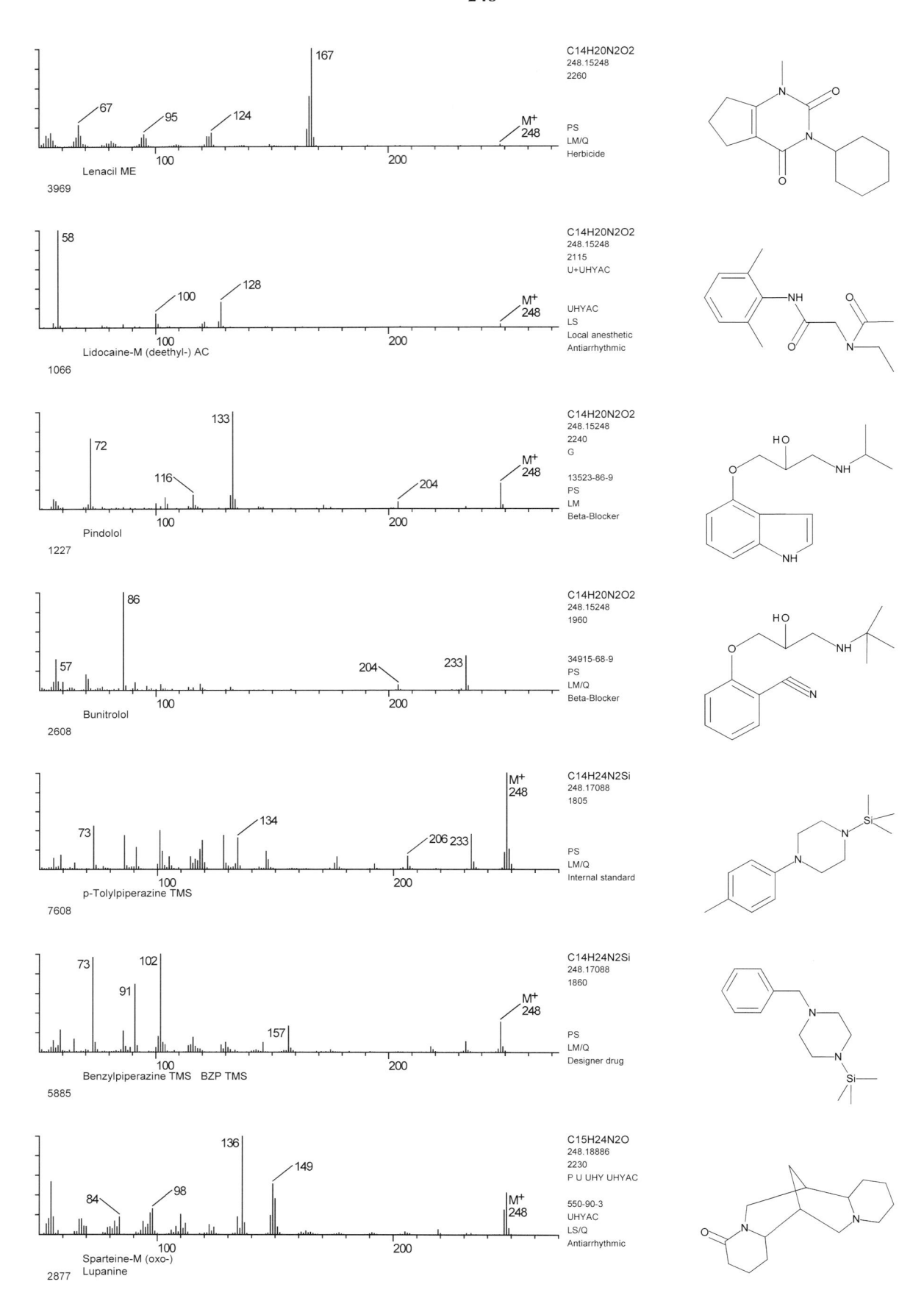

Lenacil ME — 67, 95, 124, 167, M+ 248
C14H20N2O2
248.15248
2260
PS
LM/Q
Herbicide

Lidocaine-M (deethyl-) AC — 58, 100, 128, M+ 248
C14H20N2O2
248.15248
2115
U+UHYAC
UHYAC
LS
Local anesthetic
Antiarrhythmic

Pindolol — 72, 116, 133, 204, M+ 248
C14H20N2O2
248.15248
2240
G
13523-86-9
PS
LM
Beta-Blocker

Bunitrolol — 57, 86, 204, 233
C14H20N2O2
248.15248
1960
34915-68-9
PS
LM/Q
Beta-Blocker

p-Tolylpiperazine TMS — 73, 134, 206, 233, M+ 248
C14H24N2Si
248.17088
1805
PS
LM/Q
Internal standard

Benzylpiperazine TMS BZP TMS — 73, 91, 102, 157, M+ 248
C14H24N2Si
248.17088
1860
PS
LM/Q
Designer drug

Sparteine-M (oxo-) Lupanine — 84, 98, 136, 149, M+ 248
C15H24N2O
248.18886
2230
P U UHY UHYAC
550-90-3
UHYAC
LS/Q
Antiarrhythmic

248

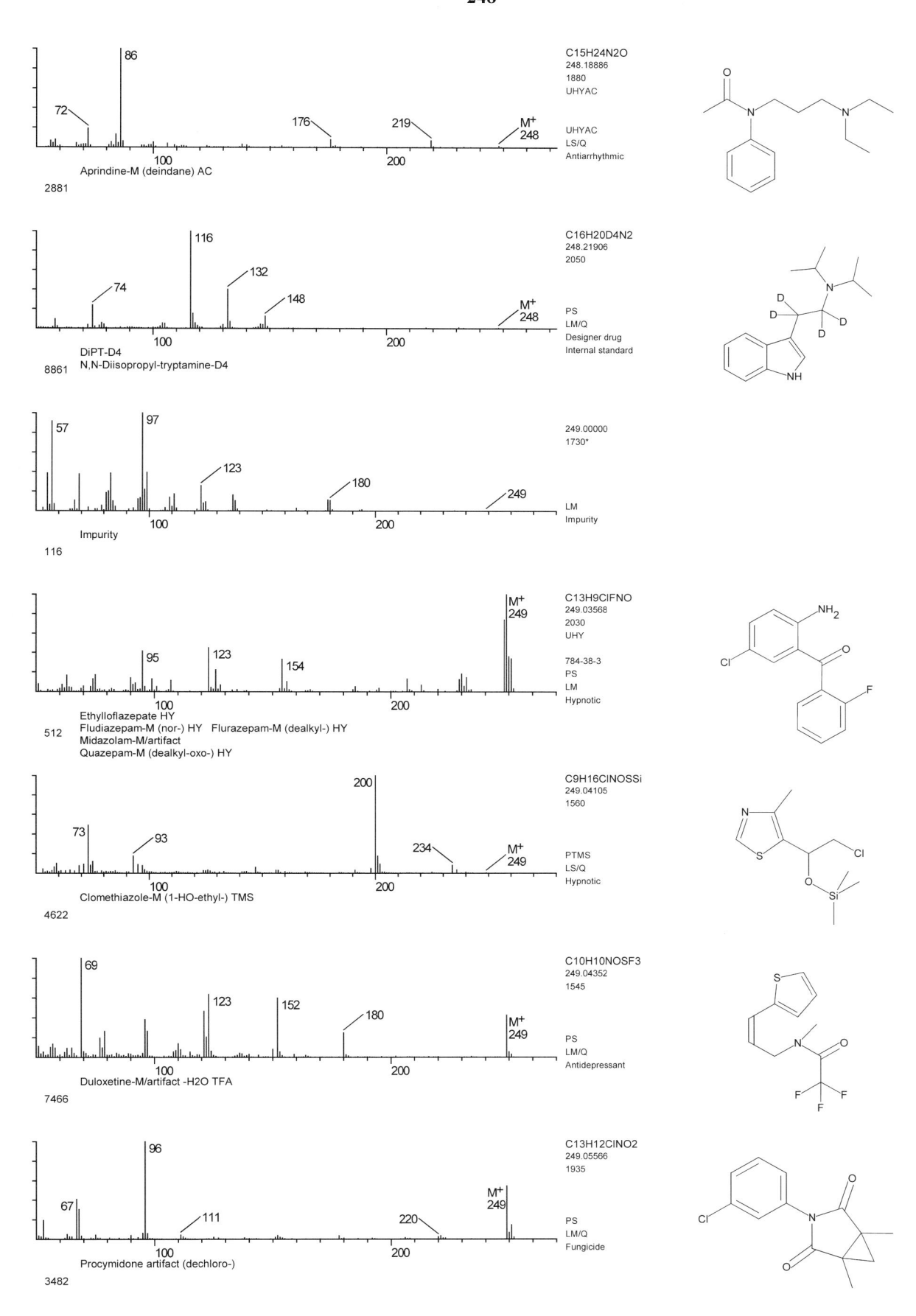

C15H24N2O
248.18886
1880
UHYAC

UHYAC
LS/Q
Antiarrhythmic

Aprindine-M (deindane) AC
2881

C16H20D4N2
248.21906
2050

PS
LM/Q
Designer drug
Internal standard

DiPT-D4
N,N-Diisopropyl-tryptamine-D4
8861

249.00000
1730*

LM
Impurity

Impurity
116

C13H9ClFNO
249.03568
2030
UHY

784-38-3
PS
LM
Hypnotic

Ethylloflazepate HY
Fludiazepam-M (nor-) HY   Flurazepam-M (dealkyl-) HY
Midazolam-M/artifact
Quazepam-M (dealkyl-oxo-) HY
512

C9H16ClNOSSi
249.04105
1560

PTMS
LS/Q
Hypnotic

Clomethiazole-M (1-HO-ethyl-) TMS
4622

C10H10NOSF3
249.04352
1545

PS
LM/Q
Antidepressant

Duloxetine-M/artifact -H2O TFA
7466

C13H12ClNO2
249.05566
1935

PS
LM/Q
Fungicide

Procymidone artifact (dechloro-)
3482

249

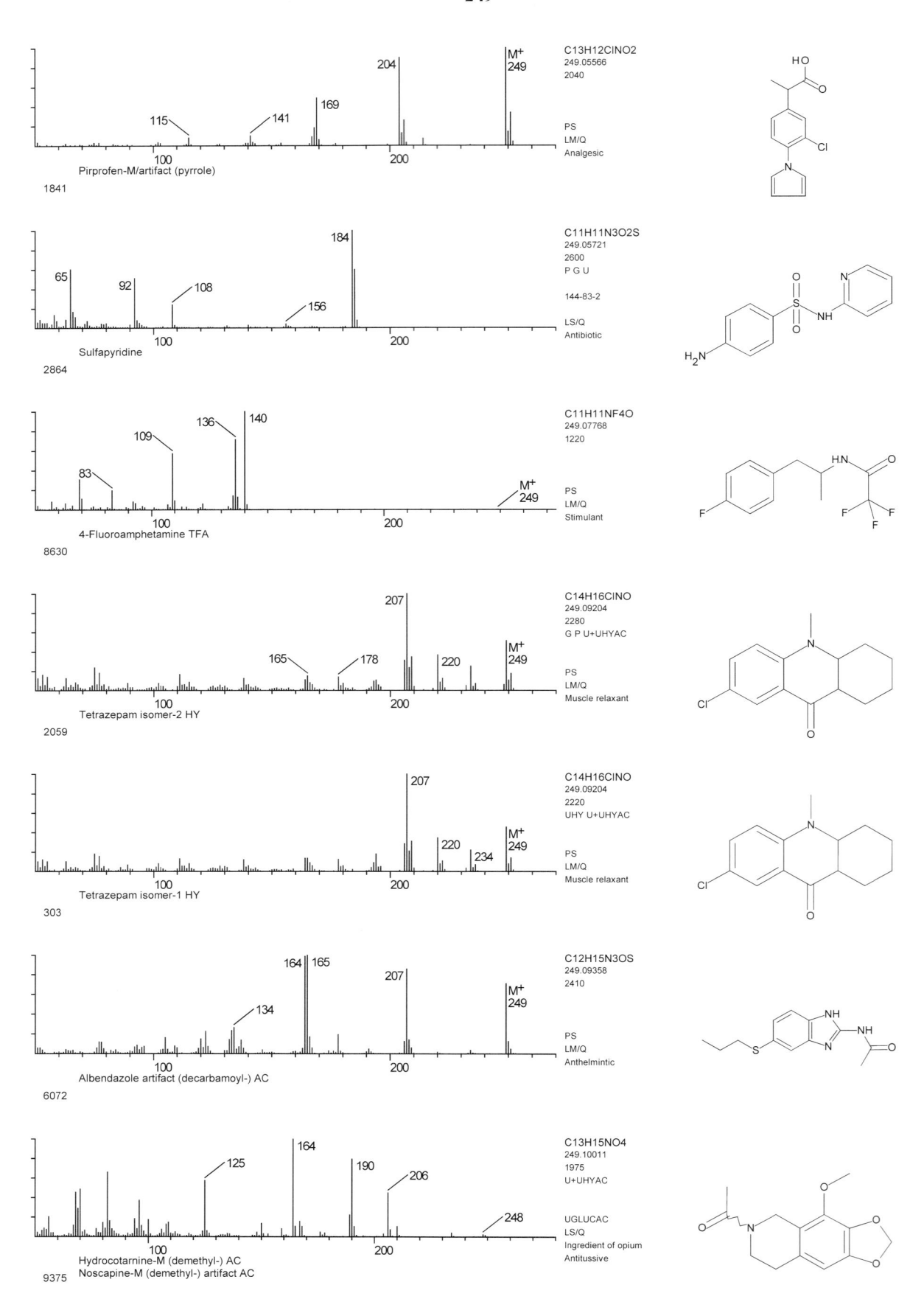

Pirprofen-M/artifact (pyrrole)
1841
115, 141, 169, 204, M+ 249
C13H12ClNO2
249.05566
2040
PS
LM/Q
Analgesic

Sulfapyridine
2864
65, 92, 108, 156, 184
C11H11N3O2S
249.05721
2600
P G U
144-83-2
LS/Q
Antibiotic

4-Fluoroamphetamine TFA
8630
83, 109, 136, 140, M+ 249
C11H11NF4O
249.07768
1220
PS
LM/Q
Stimulant

Tetrazepam isomer-2 HY
2059
165, 178, 207, 220, M+ 249
C14H16ClNO
249.09204
2280
G P U+UHYAC
PS
LM/Q
Muscle relaxant

Tetrazepam isomer-1 HY
303
207, 220, 234, M+ 249
C14H16ClNO
249.09204
2220
UHY U+UHYAC
PS
LM/Q
Muscle relaxant

Albendazole artifact (decarbamoyl-) AC
6072
134, 164, 165, 207, M+ 249
C12H15N3OS
249.09358
2410
PS
LM/Q
Anthelmintic

Hydrocotarnine-M (demethyl-) AC
Noscapine-M (demethyl-) artifact AC
9375
125, 164, 190, 206, 248
C13H15NO4
249.10011
1975
U+UHYAC
UGLUCAC
LS/Q
Ingredient of opium
Antitussive

249

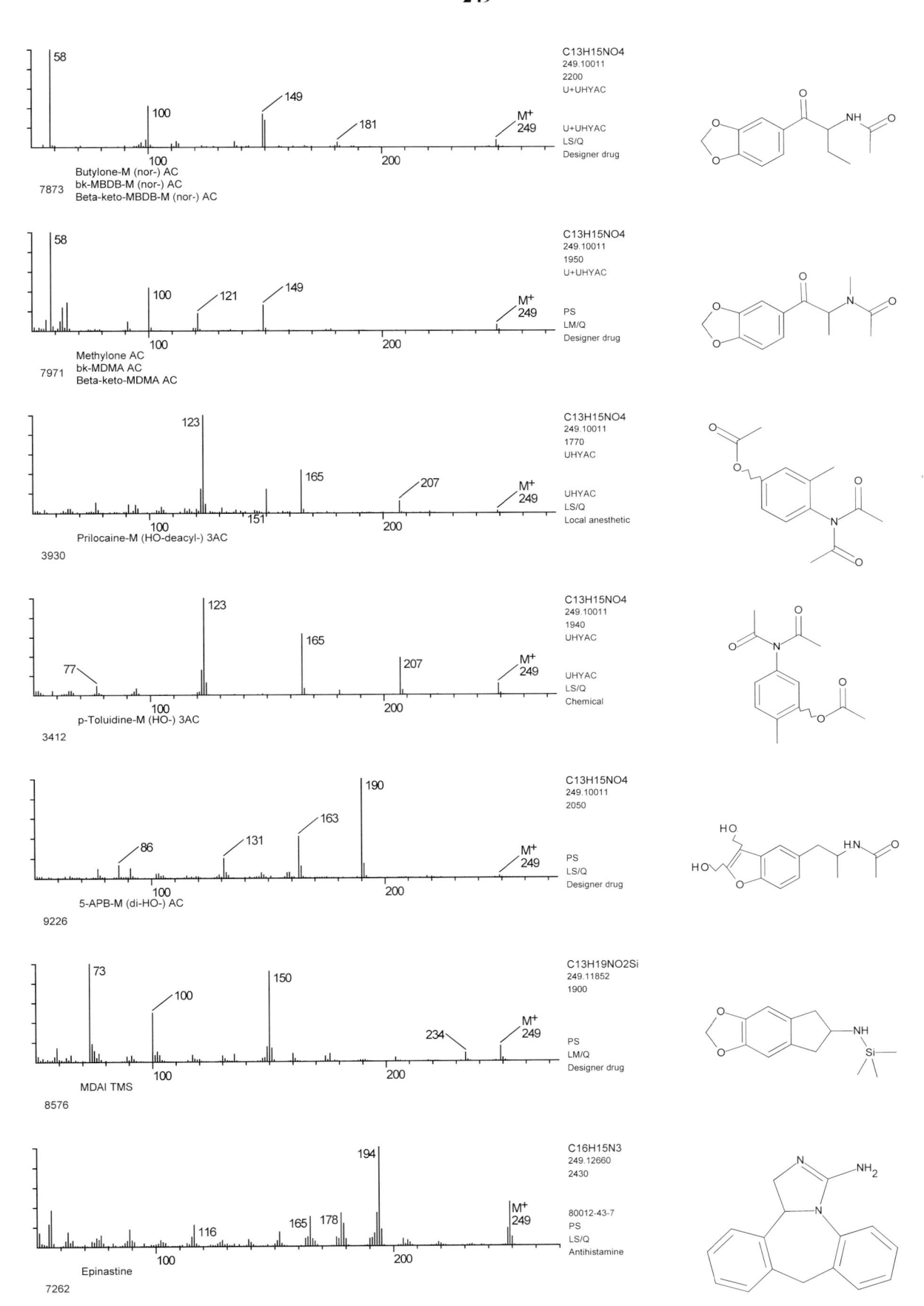

7873
Butylone-M (nor-) AC
bk-MBDB-M (nor-) AC
Beta-keto-MBDB-M (nor-) AC
C13H15NO4
249.10011
2200
U+UHYAC
U+UHYAC
LS/Q
Designer drug

7971
Methylone AC
bk-MDMA AC
Beta-keto-MDMA AC
C13H15NO4
249.10011
1950
U+UHYAC
PS
LM/Q
Designer drug

3930
Prilocaine-M (HO-deacyl-) 3AC
C13H15NO4
249.10011
1770
UHYAC
UHYAC
LS/Q
Local anesthetic

3412
p-Toluidine-M (HO-) 3AC
C13H15NO4
249.10011
1940
UHYAC
UHYAC
LS/Q
Chemical

9226
5-APB-M (di-HO-) AC
C13H15NO4
249.10011
2050
PS
LS/Q
Designer drug

8576
MDAI TMS
C13H19NO2Si
249.11852
1900
PS
LM/Q
Designer drug

7262
Epinastine
C16H15N3
249.12660
2430
80012-43-7
PS
LS/Q
Antihistamine

249

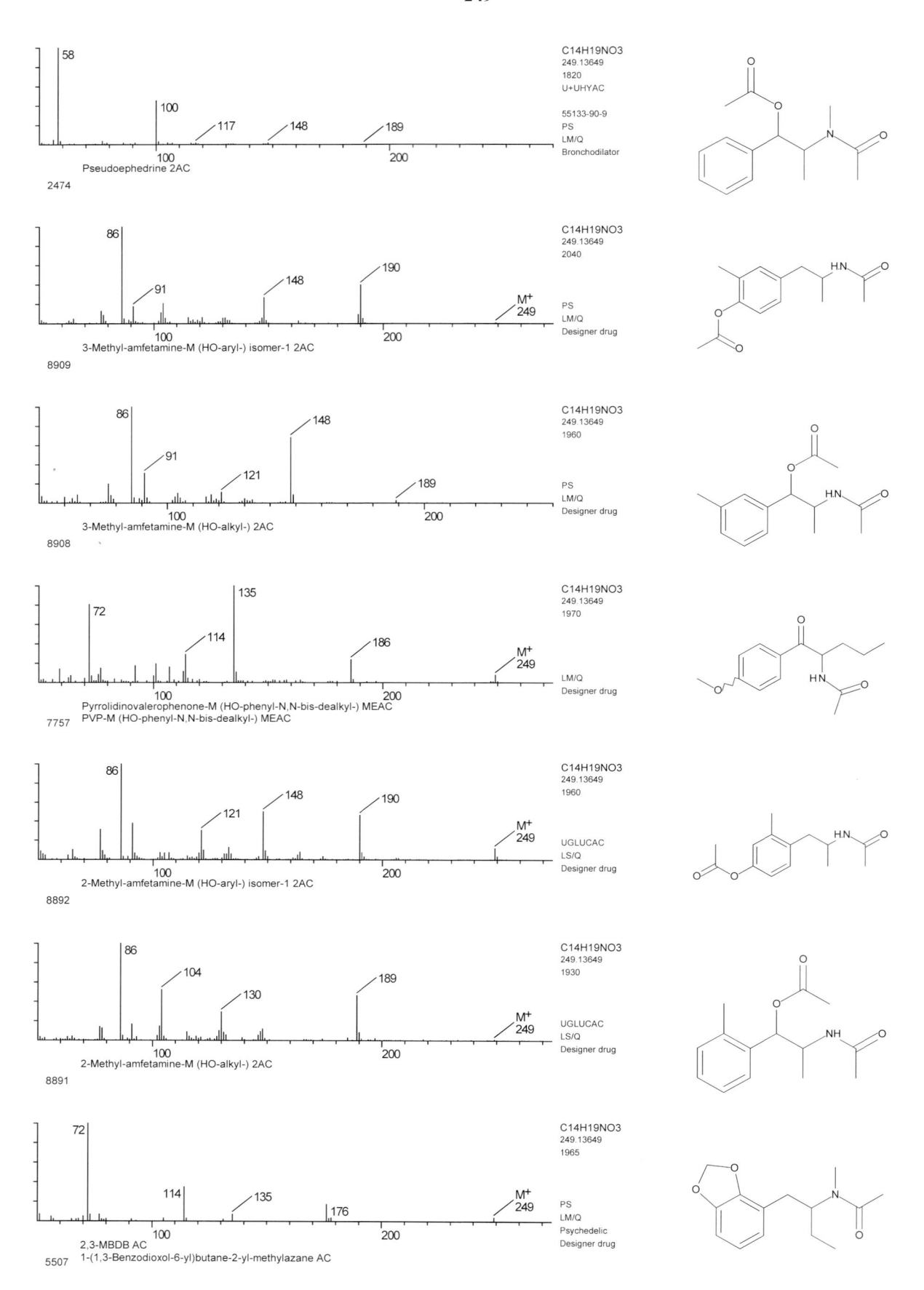

Pseudoephedrine 2AC
2474

Peaks: 58, 100, 117, 148, 189

C14H19NO3
249.13649
1820
U+UHYAC

55133-90-9
PS
LM/Q
Bronchodilator

3-Methyl-amfetamine-M (HO-aryl-) isomer-1 2AC
8909

Peaks: 86, 91, 148, 190, M+ 249

C14H19NO3
249.13649
2040

PS
LM/Q
Designer drug

3-Methyl-amfetamine-M (HO-alkyl-) 2AC
8908

Peaks: 86, 91, 121, 148, 189

C14H19NO3
249.13649
1960

PS
LM/Q
Designer drug

Pyrrolidinovalerophenone-M (HO-phenyl-N,N-bis-dealkyl-) MEAC
PVP-M (HO-phenyl-N,N-bis-dealkyl-) MEAC
7757

Peaks: 72, 114, 135, 186, M+ 249

C14H19NO3
249.13649
1970

LM/Q
Designer drug

2-Methyl-amfetamine-M (HO-aryl-) isomer-1 2AC
8892

Peaks: 86, 121, 148, 190, M+ 249

C14H19NO3
249.13649
1960

UGLUCAC
LS/Q
Designer drug

2-Methyl-amfetamine-M (HO-alkyl-) 2AC
8891

Peaks: 86, 104, 130, 189, M+ 249

C14H19NO3
249.13649
1930

UGLUCAC
LS/Q
Designer drug

2,3-MBDB AC
1-(1,3-Benzodioxol-6-yl)butane-2-yl-methylazane AC
5507

Peaks: 72, 114, 135, 176, M+ 249

C14H19NO3
249.13649
1965

PS
LM/Q
Psychedelic
Designer drug

249

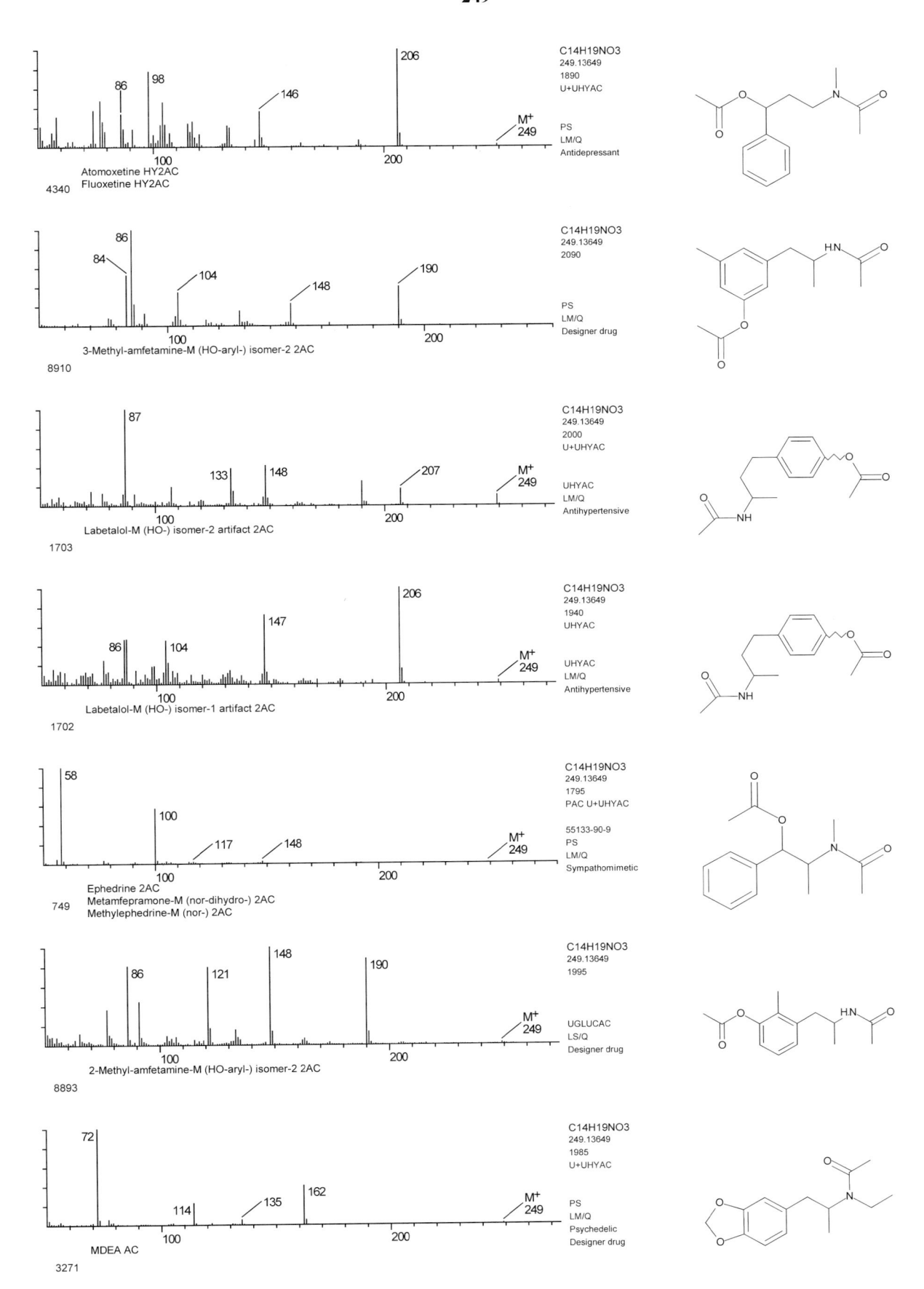

Atomoxetine HY2AC / Fluoxetine HY2AC	peaks: 86, 98, 146, 206, M+ 249	C14H19NO3 249.13649 1890 U+UHYAC PS LM/Q Antidepressant
3-Methyl-amfetamine-M (HO-aryl-) isomer-2 2AC	peaks: 84, 86, 104, 148, 190	C14H19NO3 249.13649 2090 PS LM/Q Designer drug
Labetalol-M (HO-) isomer-2 artifact 2AC	peaks: 87, 133, 148, 207, M+ 249	C14H19NO3 249.13649 2000 U+UHYAC UHYAC LM/Q Antihypertensive
Labetalol-M (HO-) isomer-1 artifact 2AC	peaks: 86, 104, 147, 206, M+ 249	C14H19NO3 249.13649 1940 UHYAC UHYAC LM/Q Antihypertensive
Ephedrine 2AC / Metamfepramone-M (nor-dihydro-) 2AC / Methylephedrine-M (nor-) 2AC	peaks: 58, 100, 117, 148, M+ 249	C14H19NO3 249.13649 1795 PAC U+UHYAC 55133-90-9 PS LM/Q Sympathomimetic
2-Methyl-amfetamine-M (HO-aryl-) isomer-2 2AC	peaks: 86, 121, 148, 190, M+ 249	C14H19NO3 249.13649 1995 UGLUCAC LS/Q Designer drug
MDEA AC	peaks: 72, 114, 135, 162, M+ 249	C14H19NO3 249.13649 1985 U+UHYAC PS LM/Q Psychedelic Designer drug

4340

8910

1703

1702

749

8893

3271

249

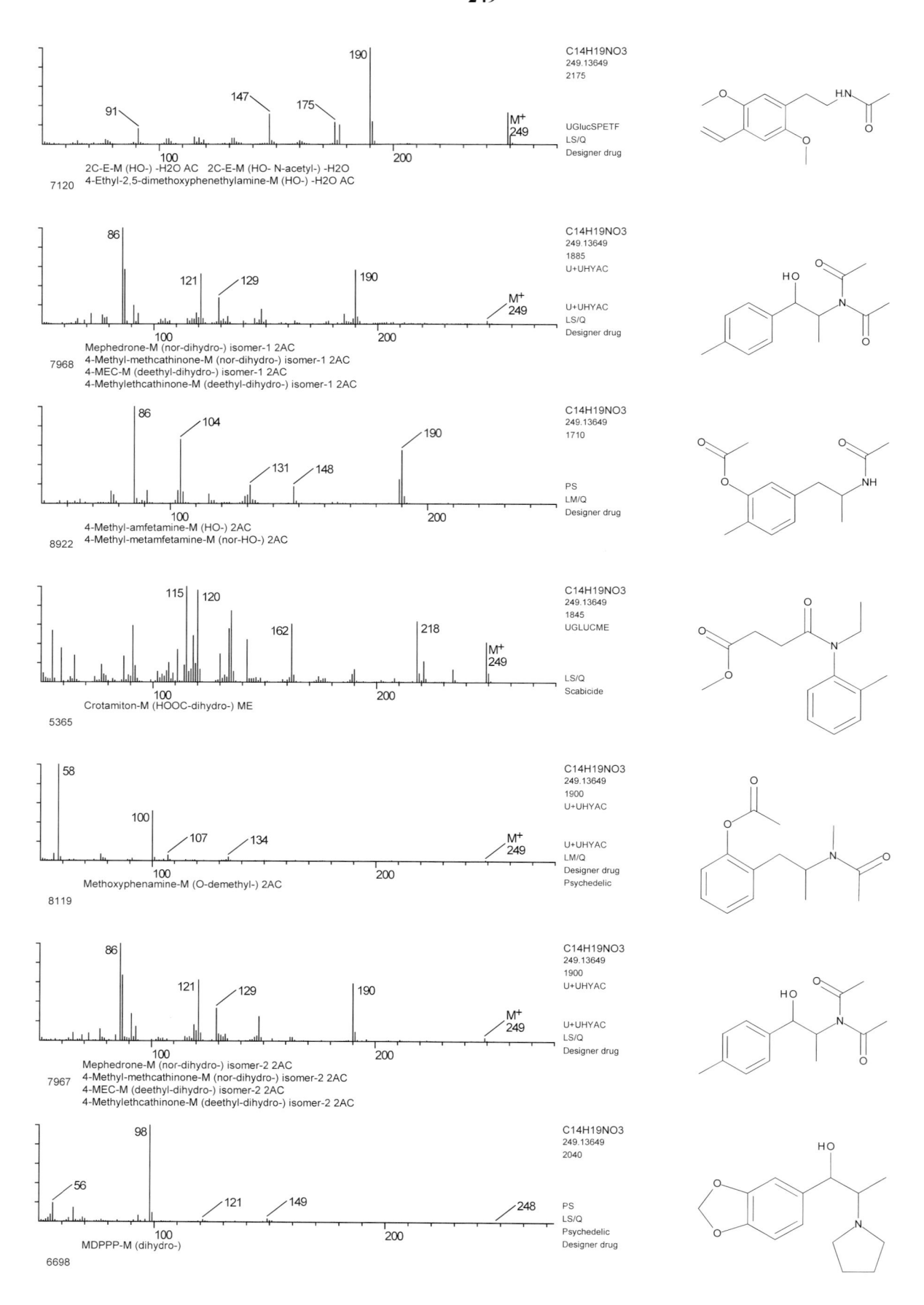

C14H19NO3
249.13649
2175
UGlucSPETF
LS/Q
Designer drug

2C-E-M (HO-) -H2O AC   2C-E-M (HO- N-acetyl-) -H2O
4-Ethyl-2,5-dimethoxyphenethylamine-M (HO-) -H2O AC
7120

C14H19NO3
249.13649
1885
U+UHYAC
U+UHYAC
LS/Q
Designer drug

Mephedrone-M (nor-dihydro-) isomer-1 2AC
4-Methyl-methcathinone-M (nor-dihydro-) isomer-1 2AC
4-MEC-M (deethyl-dihydro-) isomer-1 2AC
4-Methylethcathinone-M (deethyl-dihydro-) isomer-1 2AC
7968

C14H19NO3
249.13649
1710
PS
LM/Q
Designer drug

4-Methyl-amfetamine-M (HO-) 2AC
4-Methyl-metamfetamine-M (nor-HO-) 2AC
8922

C14H19NO3
249.13649
1845
UGLUCME
LS/Q
Scabicide

Crotamiton-M (HOOC-dihydro-) ME
5365

C14H19NO3
249.13649
1900
U+UHYAC
U+UHYAC
LM/Q
Designer drug
Psychedelic

Methoxyphenamine-M (O-demethyl-) 2AC
8119

C14H19NO3
249.13649
1900
U+UHYAC
U+UHYAC
LS/Q
Designer drug

Mephedrone-M (nor-dihydro-) isomer-2 2AC
4-Methyl-methcathinone-M (nor-dihydro-) isomer-2 2AC
4-MEC-M (deethyl-dihydro-) isomer-2 2AC
4-Methylethcathinone-M (deethyl-dihydro-) isomer-2 2AC
7967

C14H19NO3
249.13649
2040
PS
LS/Q
Psychedelic
Designer drug

MDPPP-M (dihydro-)
6698

249

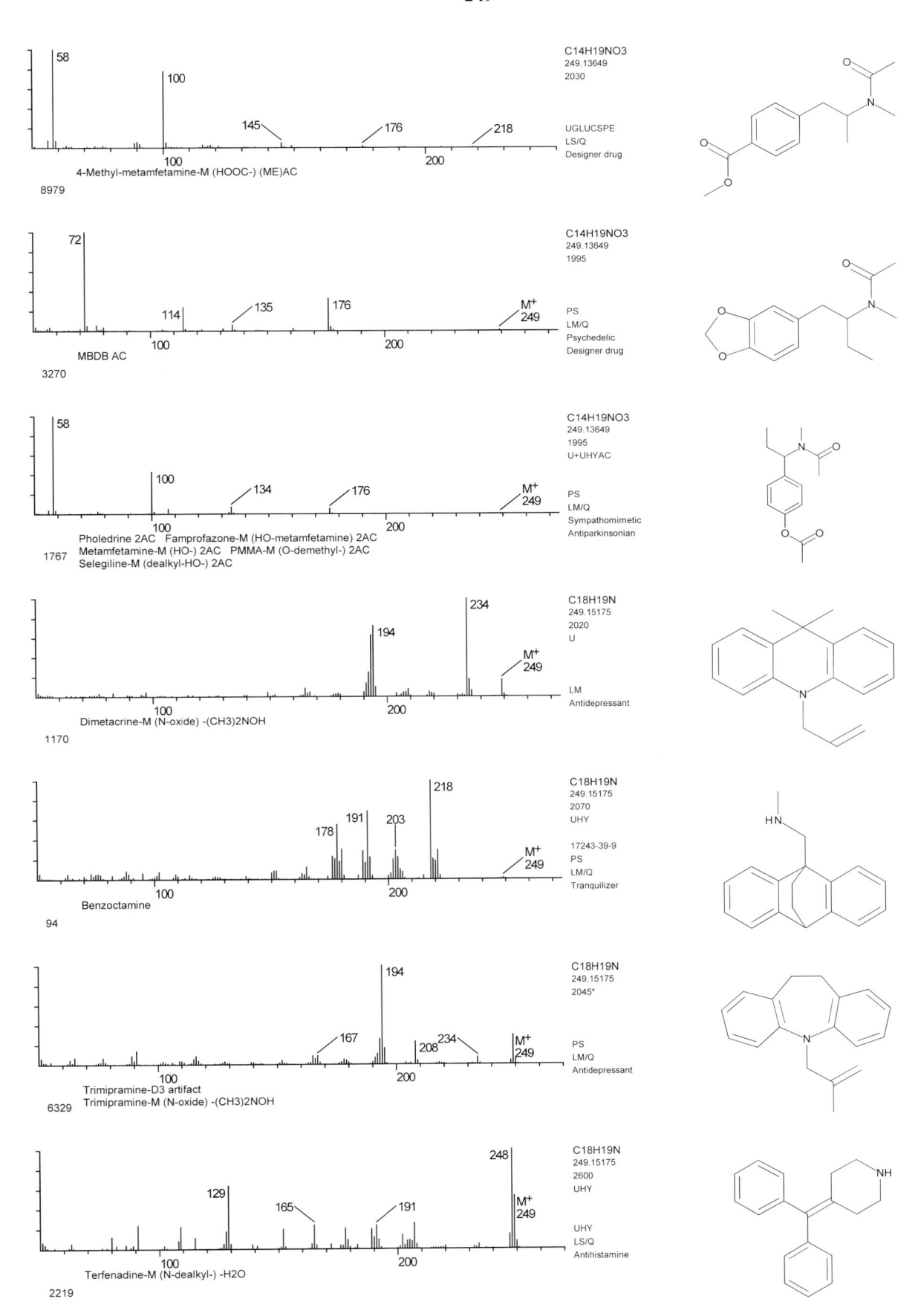

4-Methyl-metamfetamine-M (HOOC-) (ME)AC 8979	C14H19NO3 249.13649 2030 UGLUCSPE LS/Q Designer drug
MBDB AC 3270	C14H19NO3 249.13649 1995 PS LM/Q Psychedelic Designer drug
Pholedrine 2AC Famprofazone-M (HO-metamfetamine) 2AC Metamfetamine-M (HO-) 2AC PMMA-M (O-demethyl-) 2AC Selegiline-M (dealkyl-HO-) 2AC 1767	C14H19NO3 249.13649 1995 U+UHYAC PS LM/Q Sympathomimetic Antiparkinsonian
Dimetacrine-M (N-oxide) -(CH3)2NOH 1170	C18H19N 249.15175 2020 U LM Antidepressant
Benzoctamine 94	C18H19N 249.15175 2070 UHY 17243-39-9 PS LM/Q Tranquilizer
Trimipramine-D3 artifact Trimipramine-M (N-oxide) -(CH3)2NOH 6329	C18H19N 249.15175 2045* PS LM/Q Antidepressant
Terfenadine-M (N-dealkyl-) -H2O 2219	C18H19N 249.15175 2600 UHY UHY LS/Q Antihistamine

249

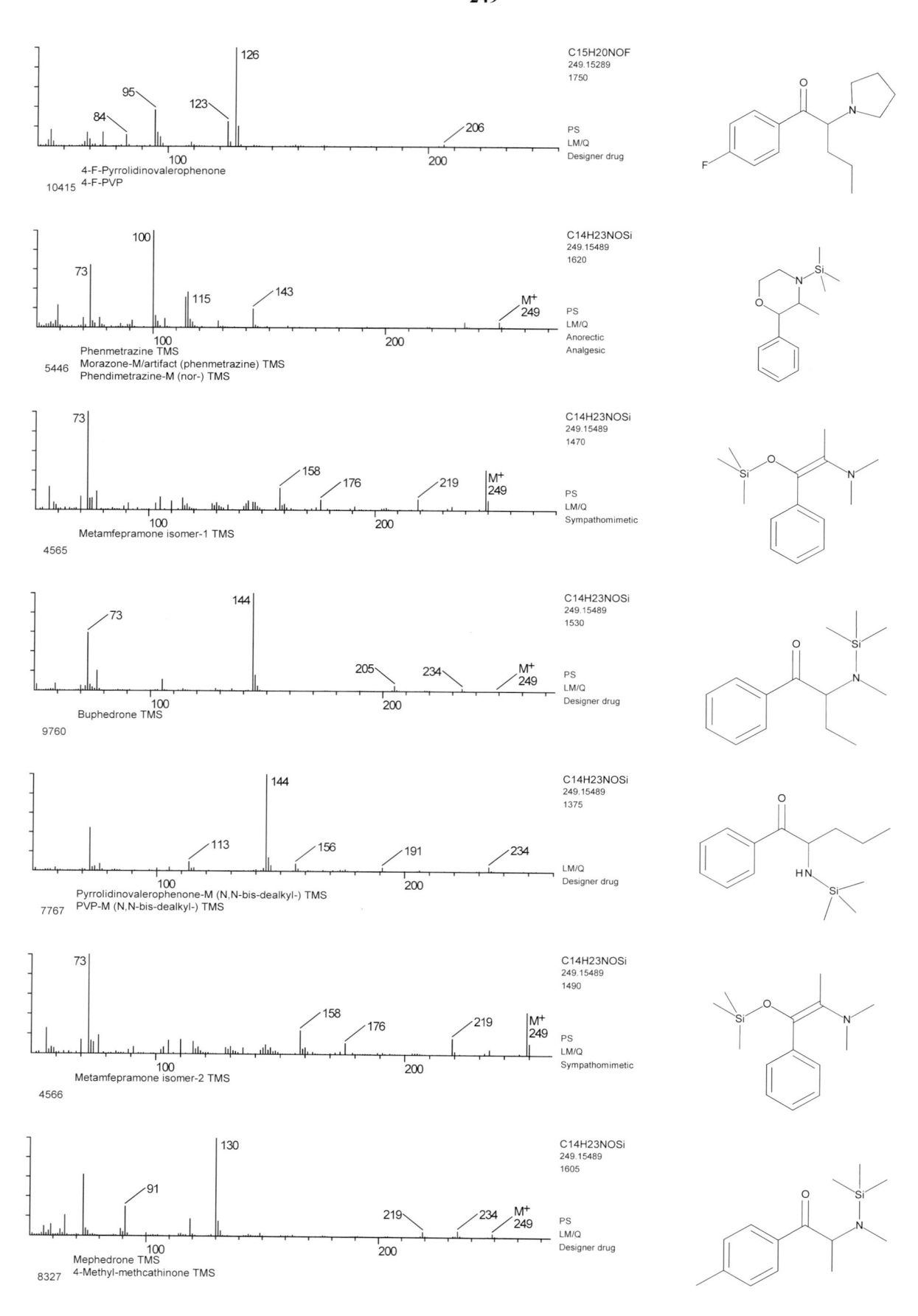

4-F-Pyrrolidinovalerophenone 10415 4-F-PVP peaks: 84, 95, 123, 126, 206	C15H20NOF 249.15289 1750 PS LM/Q Designer drug	

Phenmetrazine TMS Morazone-M/artifact (phenmetrazine) TMS 5446 Phendimetrazine-M (nor-) TMS peaks: 73, 100, 115, 143, M+ 249	C14H23NOSi 249.15489 1620 PS LM/Q Anorectic Analgesic	

Metamfepramone isomer-1 TMS 4565 peaks: 73, 158, 176, 219, M+ 249	C14H23NOSi 249.15489 1470 PS LM/Q Sympathomimetic	

Buphedrone TMS 9760 peaks: 73, 144, 205, 234, M+ 249	C14H23NOSi 249.15489 1530 PS LM/Q Designer drug	

Pyrrolidinovalerophenone-M (N,N-bis-dealkyl-) TMS 7767 PVP-M (N,N-bis-dealkyl-) TMS peaks: 113, 144, 156, 191, 234	C14H23NOSi 249.15489 1375 LM/Q Designer drug	

Metamfepramone isomer-2 TMS 4566 peaks: 73, 158, 176, 219, M+ 249	C14H23NOSi 249.15489 1490 PS LM/Q Sympathomimetic	

Mephedrone TMS 8327 4-Methyl-methcathinone TMS peaks: 91, 130, 219, 234, M+ 249	C14H23NOSi 249.15489 1605 PS LM/Q Designer drug	

249

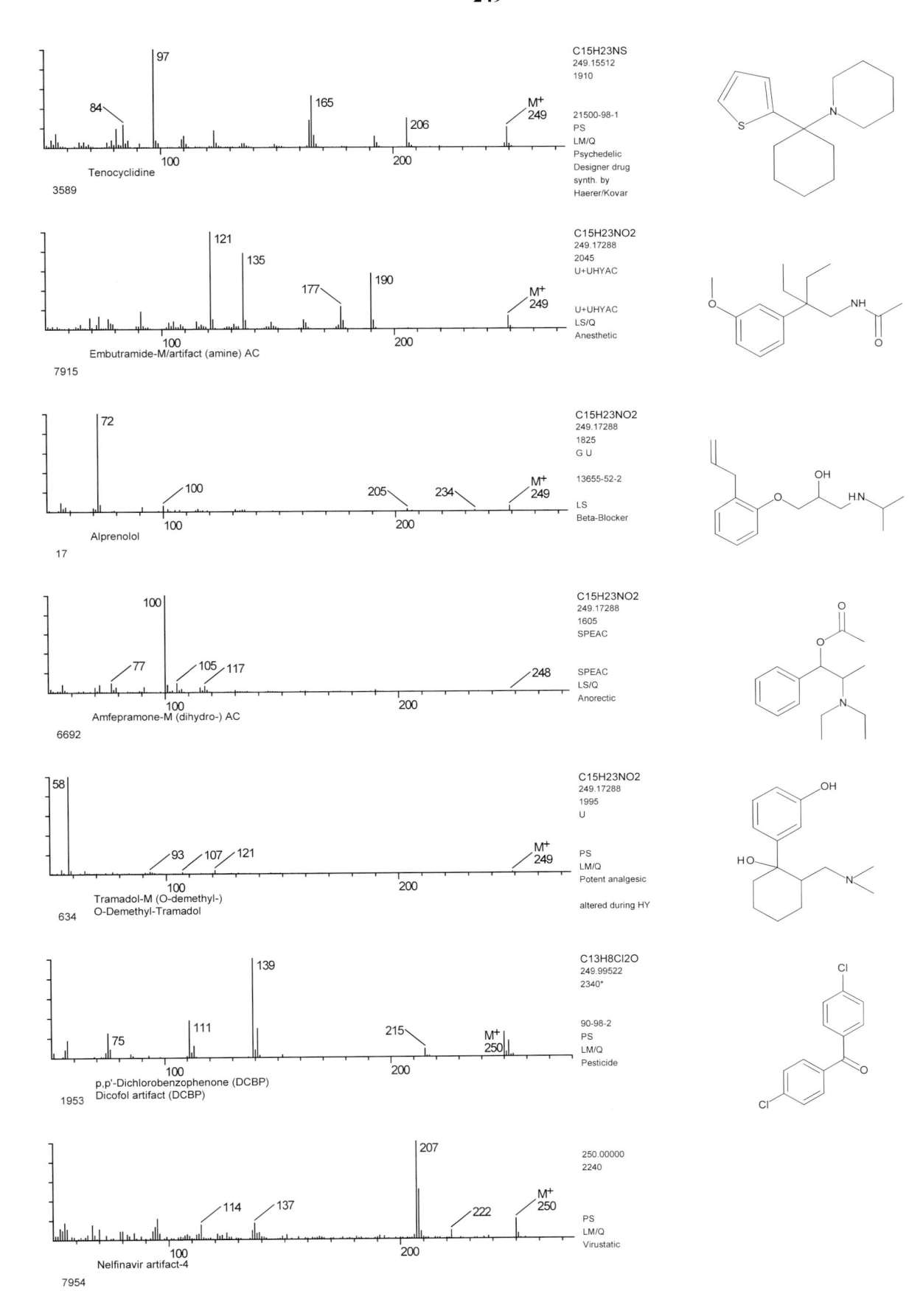

Tenocyclidine — C15H23NS, 249.15512, 1910, 21500-98-1, PS, LM/Q, Psychedelic, Designer drug synth. by Haerer/Kovar
Peaks: 84, 97, 165, 206, M+ 249
3589

Embutramide-M/artifact (amine) AC — C15H23NO2, 249.17288, 2045, U+UHYAC, U+UHYAC, LS/Q, Anesthetic
Peaks: 121, 135, 177, 190, M+ 249
7915

Alprenolol — C15H23NO2, 249.17288, 1825, G U, 13655-52-2, LS, Beta-Blocker
Peaks: 72, 100, 205, 234, M+ 249
17

Amfepramone-M (dihydro-) AC — C15H23NO2, 249.17288, 1605, SPEAC, SPEAC, LS/Q, Anorectic
Peaks: 77, 100, 105, 117, 248
6692

Tramadol-M (O-demethyl-) O-Demethyl-Tramadol — C15H23NO2, 249.17288, 1995, U, PS, LM/Q, Potent analgesic, altered during HY
Peaks: 58, 93, 107, 121, M+ 249
634

p,p'-Dichlorobenzophenone (DCBP) Dicofol artifact (DCBP) — C13H8Cl2O, 249.99522, 2340*, 90-98-2, PS, LM/Q, Pesticide
Peaks: 75, 111, 139, 215, M+ 250
1953

Nelfinavir artifact-4 — 250.00000, 2240, PS, LM/Q, Virustatic
Peaks: 114, 137, 207, 222, M+ 250
7954

250

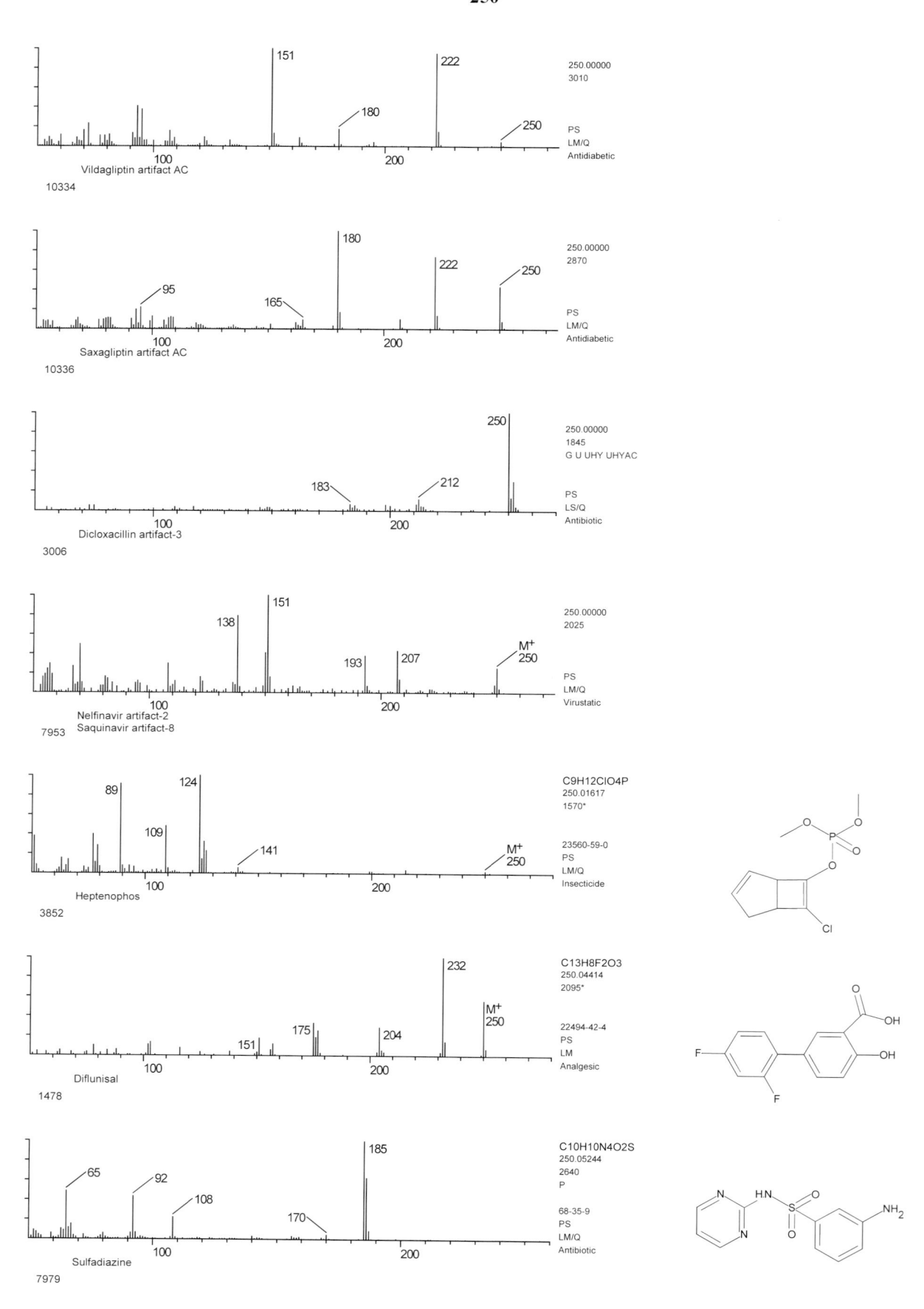

Vildagliptin artifact AC
10334
250.00000
3010
PS
LM/Q
Antidiabetic

Saxagliptin artifact AC
10336
250.00000
2870
PS
LM/Q
Antidiabetic

Dicloxacillin artifact-3
3006
250.00000
1845
G U UHY UHYAC
PS
LS/Q
Antibiotic

Nelfinavir artifact-2
Saquinavir artifact-8
7953
250.00000
2025
PS
LM/Q
Virustatic

Heptenophos
3852
C9H12ClO4P
250.01617
1570*
23560-59-0
PS
LM/Q
Insecticide

Diflunisal
1478
C13H8F2O3
250.04414
2095*
22494-42-4
PS
LM
Analgesic

Sulfadiazine
7979
C10H10N4O2S
250.05244
2640
P
68-35-9
PS
LM/Q
Antibiotic

250

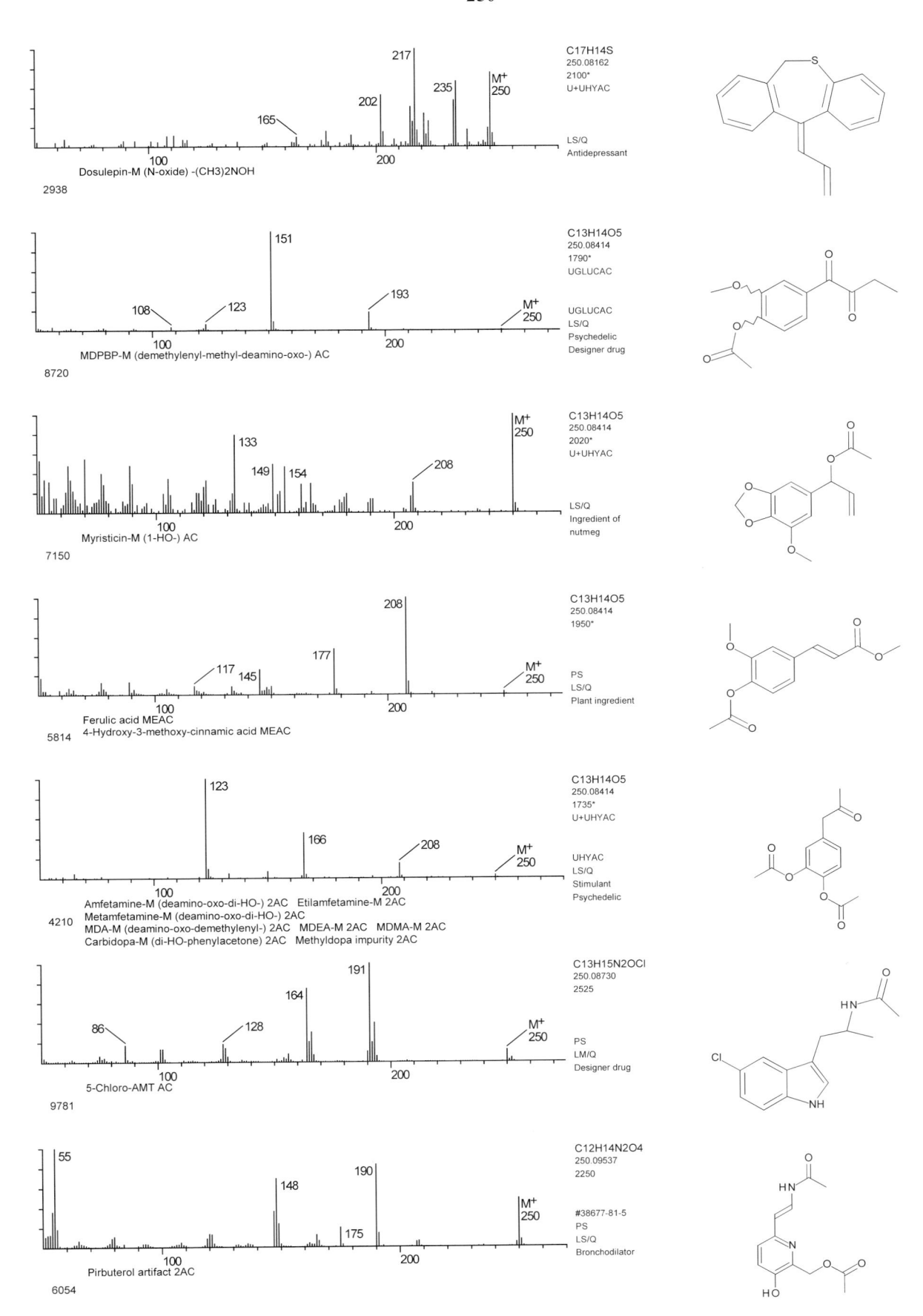

C17H14S
250.08162
2100*
U+UHYAC

LS/Q
Antidepressant

Dosulepin-M (N-oxide) -(CH3)2NOH
2938

C13H14O5
250.08414
1790*
UGLUCAC

UGLUCAC
LS/Q
Psychedelic
Designer drug

MDPBP-M (demethylenyl-methyl-deamino-oxo-) AC
8720

C13H14O5
250.08414
2020*
U+UHYAC

LS/Q
Ingredient of nutmeg

Myristicin-M (1-HO-) AC
7150

C13H14O5
250.08414
1950*

PS
LS/Q
Plant ingredient

Ferulic acid MEAC   4-Hydroxy-3-methoxy-cinnamic acid MEAC
5814

C13H14O5
250.08414
1735*
U+UHYAC

UHYAC
LS/Q
Stimulant
Psychedelic

Amfetamine-M (deamino-oxo-di-HO-) 2AC   Etilamfetamine-M 2AC
Metamfetamine-M (deamino-oxo-di-HO-) 2AC
MDA-M (deamino-oxo-demethylenyl-) 2AC   MDEA-M 2AC   MDMA-M 2AC
4210   Carbidopa-M (di-HO-phenylacetone) 2AC   Methyldopa impurity 2AC

C13H15N2OCl
250.08730
2525

PS
LM/Q
Designer drug

5-Chloro-AMT AC
9781

C12H14N2O4
250.09537
2250

#38677-81-5
PS
LS/Q
Bronchodilator

Pirbuterol artifact 2AC
6054

250

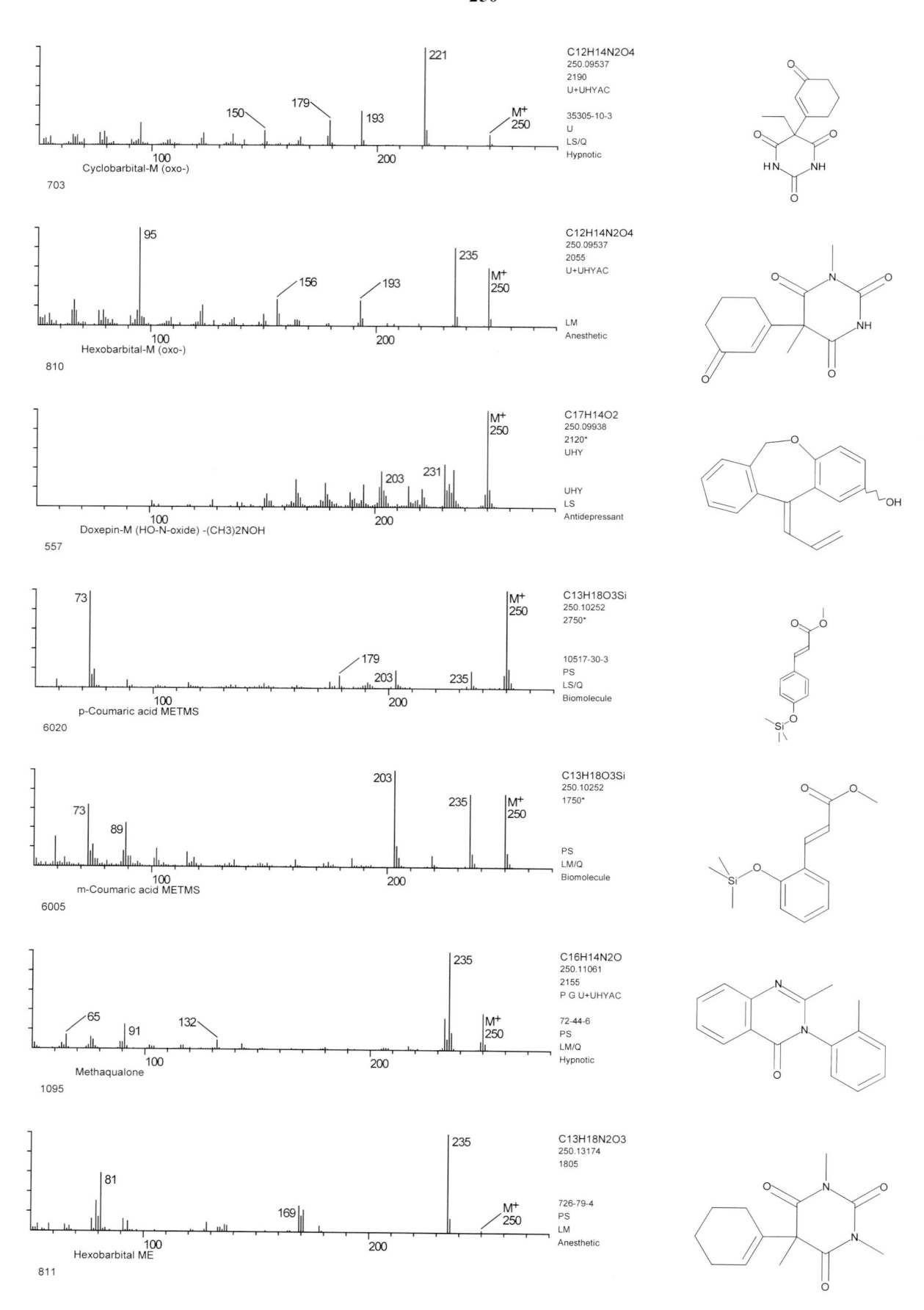

Cyclobarbital-M (oxo-)
703
Peaks: 150, 179, 193, 221, M+ 250

C12H14N2O4
250.09537
2190
U+UHYAC

35305-10-3
U
LS/Q
Hypnotic

Hexobarbital-M (oxo-)
810
Peaks: 95, 156, 193, 235, M+ 250

C12H14N2O4
250.09537
2055
U+UHYAC

LM
Anesthetic

Doxepin-M (HO-N-oxide) -(CH3)2NOH
557
Peaks: 203, 231, M+ 250

C17H14O2
250.09938
2120*
UHY

UHY
LS
Antidepressant

p-Coumaric acid METMS
6020
Peaks: 73, 179, 203, 235, M+ 250

C13H18O3Si
250.10252
2750*

10517-30-3
PS
LS/Q
Biomolecule

m-Coumaric acid METMS
6005
Peaks: 73, 89, 203, 235, M+ 250

C13H18O3Si
250.10252
1750*

PS
LM/Q
Biomolecule

Methaqualone
1095
Peaks: 65, 91, 132, 235, M+ 250

C16H14N2O
250.11061
2155
P G U+UHYAC

72-44-6
PS
LM/Q
Hypnotic

Hexobarbital ME
811
Peaks: 81, 169, 235, M+ 250

C13H18N2O3
250.13174
1805

726-79-4
PS
LM
Anesthetic

250

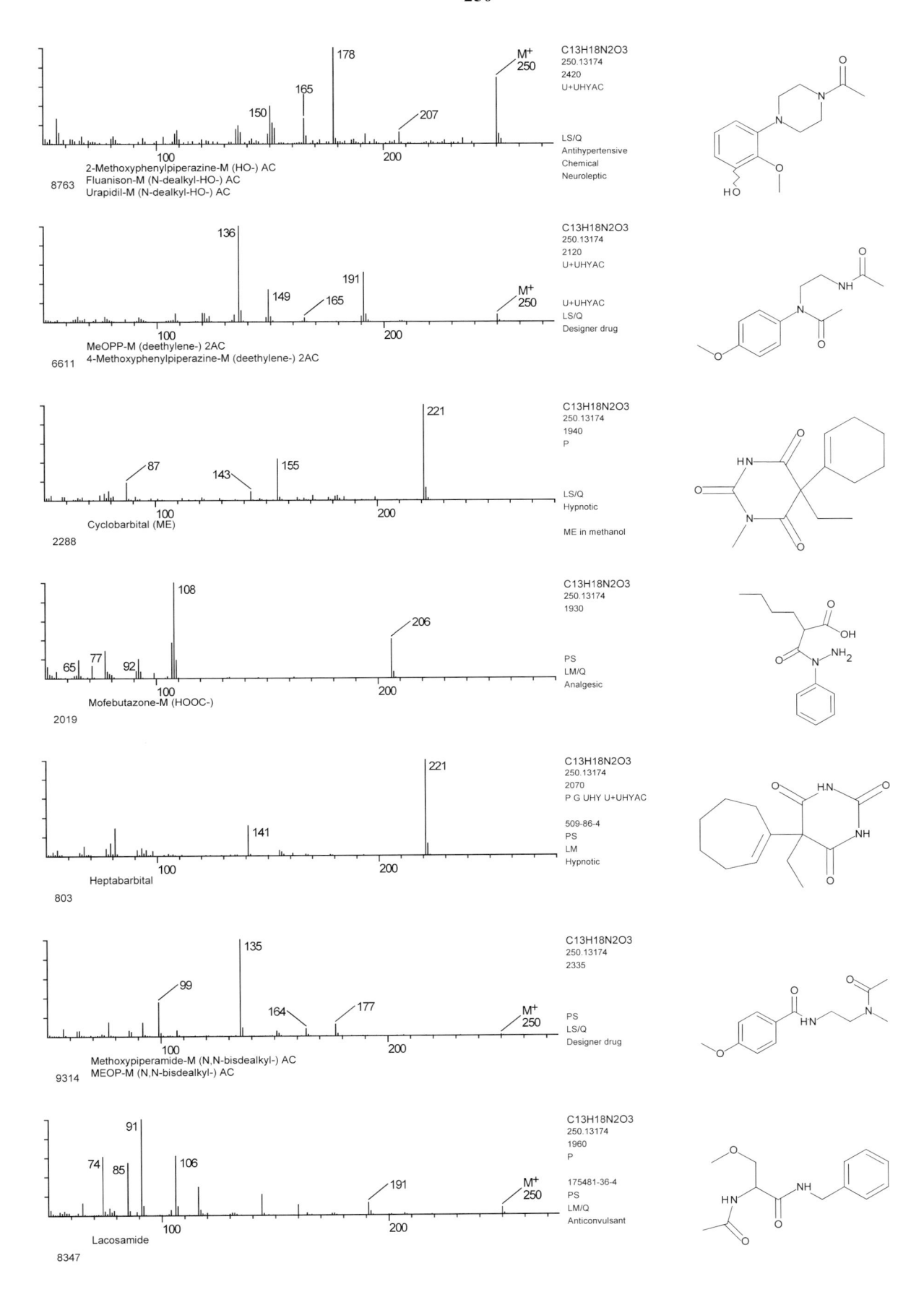

C13H18N2O3
250.13174
2420
U+UHYAC

LS/Q
Antihypertensive
Chemical
Neuroleptic

2-Methoxyphenylpiperazine-M (HO-) AC
Fluanison-M (N-dealkyl-HO-) AC
8763 Urapidil-M (N-dealkyl-HO-) AC

C13H18N2O3
250.13174
2120
U+UHYAC

U+UHYAC
LS/Q
Designer drug

MeOPP-M (deethylene-) 2AC
6611 4-Methoxyphenylpiperazine-M (deethylene-) 2AC

C13H18N2O3
250.13174
1940
P

LS/Q
Hypnotic
ME in methanol

Cyclobarbital (ME)
2288

C13H18N2O3
250.13174
1930

PS
LM/Q
Analgesic

Mofebutazone-M (HOOC-)
2019

C13H18N2O3
250.13174
2070
P G UHY U+UHYAC

509-86-4
PS
LM
Hypnotic

Heptabarbital
803

C13H18N2O3
250.13174
2335

PS
LS/Q
Designer drug

Methoxypiperamide-M (N,N-bisdealkyl-) AC
9314 MEOP-M (N,N-bisdealkyl-) AC

C13H18N2O3
250.13174
1960
P

175481-36-4
PS
LM/Q
Anticonvulsant

Lacosamide
8347

605

250

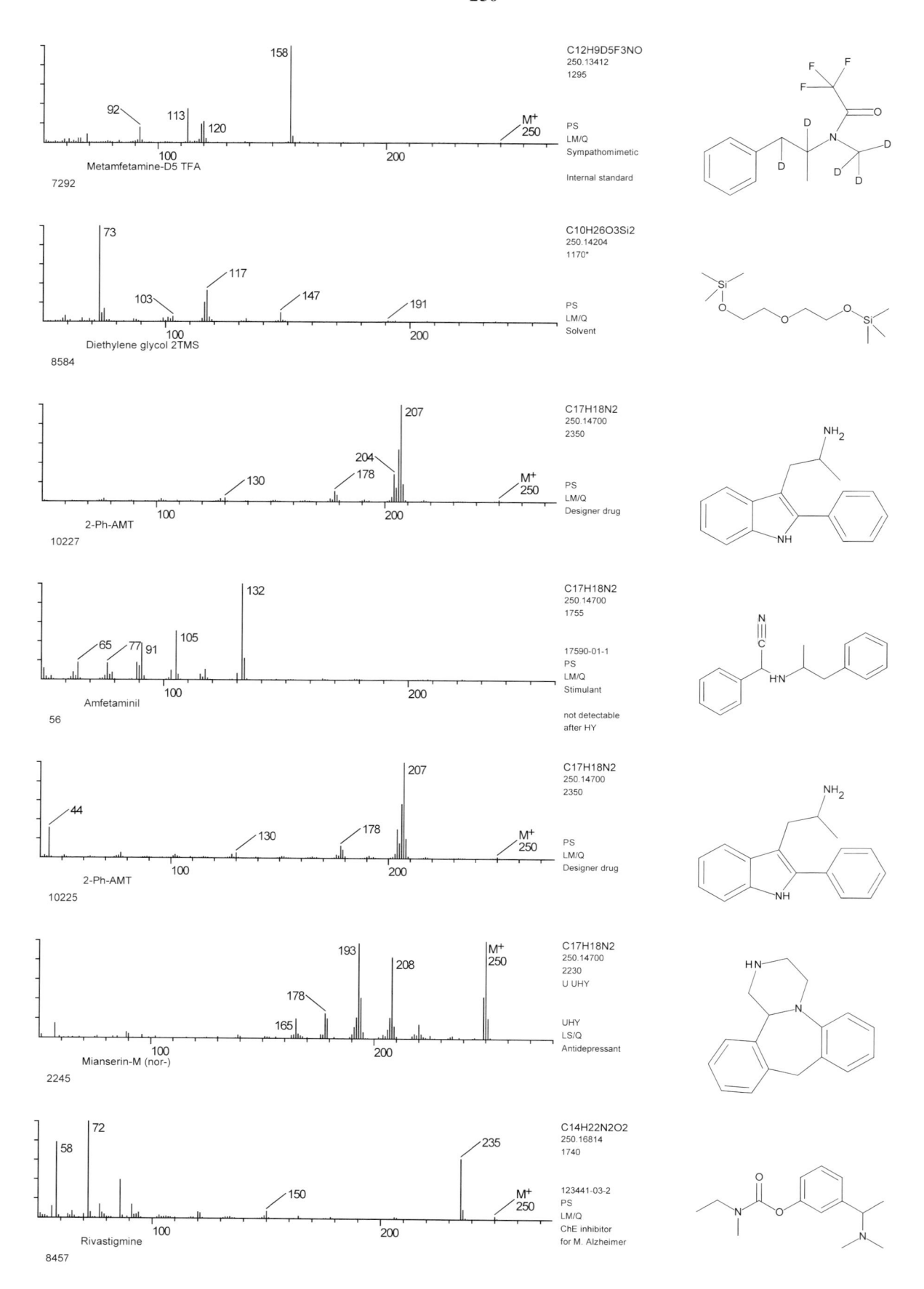

Metamfetamine-D5 TFA — peaks 92, 113, 120, 158, M+ 250 — 7292	C12H9D5F3NO 250.13412 1295 PS LM/Q Sympathomimetic Internal standard	
Diethylene glycol 2TMS — peaks 73, 103, 117, 147, 191 — 8584	C10H26O3Si2 250.14204 1170* PS LM/Q Solvent	
2-Ph-AMT — peaks 130, 178, 204, 207, M+ 250 — 10227	C17H18N2 250.14700 2350 PS LM/Q Designer drug	
Amfetaminil — peaks 65, 77, 91, 105, 132 — 56	C17H18N2 250.14700 1755 17590-01-1 PS LM/Q Stimulant not detectable after HY	
2-Ph-AMT — peaks 44, 130, 178, 207, M+ 250 — 10225	C17H18N2 250.14700 2350 PS LM/Q Designer drug	
Mianserin-M (nor-) — peaks 165, 178, 193, 208, M+ 250 — 2245	C17H18N2 250.14700 2230 U UHY UHY LS/Q Antidepressant	
Rivastigmine — peaks 58, 72, 150, 235, M+ 250 — 8457	C14H22N2O2 250.16814 1740 123441-03-2 PS LM/Q ChE inhibitor for M. Alzheimer	

250

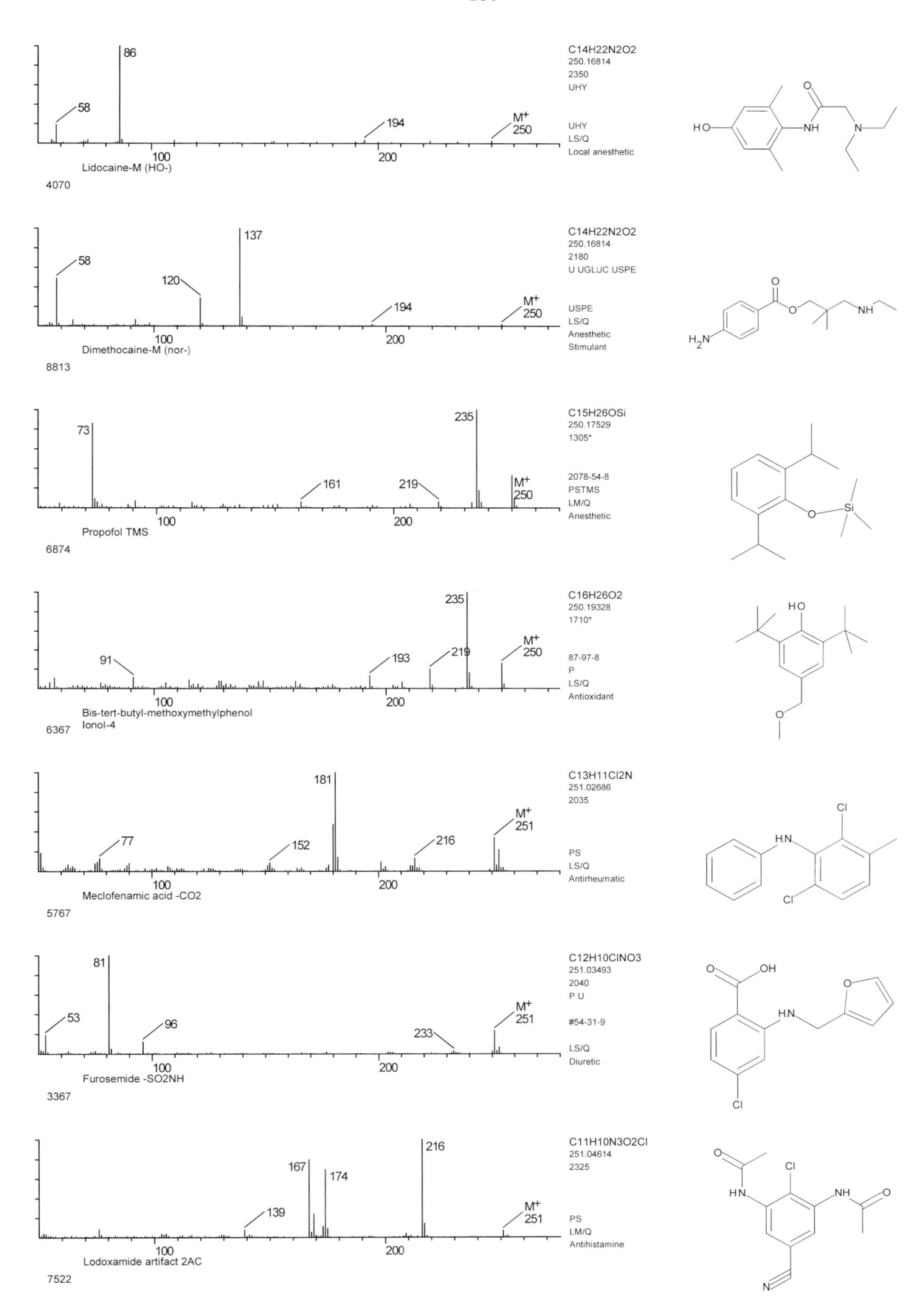

Spectrum	Compound	Formula / Info
58, 86, 194, M+ 250	Lidocaine-M (HO-) 4070	C14H22N2O2 / 250.16814 / 2350 / UHY / UHY / LS/Q / Local anesthetic
58, 120, 137, 194, M+ 250	Dimethocaine-M (nor-) 8813	C14H22N2O2 / 250.16814 / 2180 / U UGLUC USPE / USPE / LS/Q / Anesthetic Stimulant
73, 161, 219, 235, M+ 250	Propofol TMS 6874	C15H26OSi / 250.17529 / 1305* / 2078-54-8 / PSTMS / LM/Q / Anesthetic
91, 193, 219, 235, M+ 250	Bis-tert-butyl-methoxymethylphenol Ionol-4 6367	C16H26O2 / 250.19328 / 1710* / 87-97-8 / P / LS/Q / Antioxidant
77, 152, 181, 216, M+ 251	Meclofenamic acid -CO2 5767	C13H11Cl2N / 251.02686 / 2035 / PS / LS/Q / Antirheumatic
53, 81, 96, 233, M+ 251	Furosemide -SO2NH 3367	C12H10ClNO3 / 251.03493 / 2040 / P U / #54-31-9 / LS/Q / Diuretic
139, 167, 174, 216, M+ 251	Lodoxamide artifact 2AC 7522	C11H10N3O2Cl / 251.04614 / 2325 / PS / LM/Q / Antihistamine

607

251

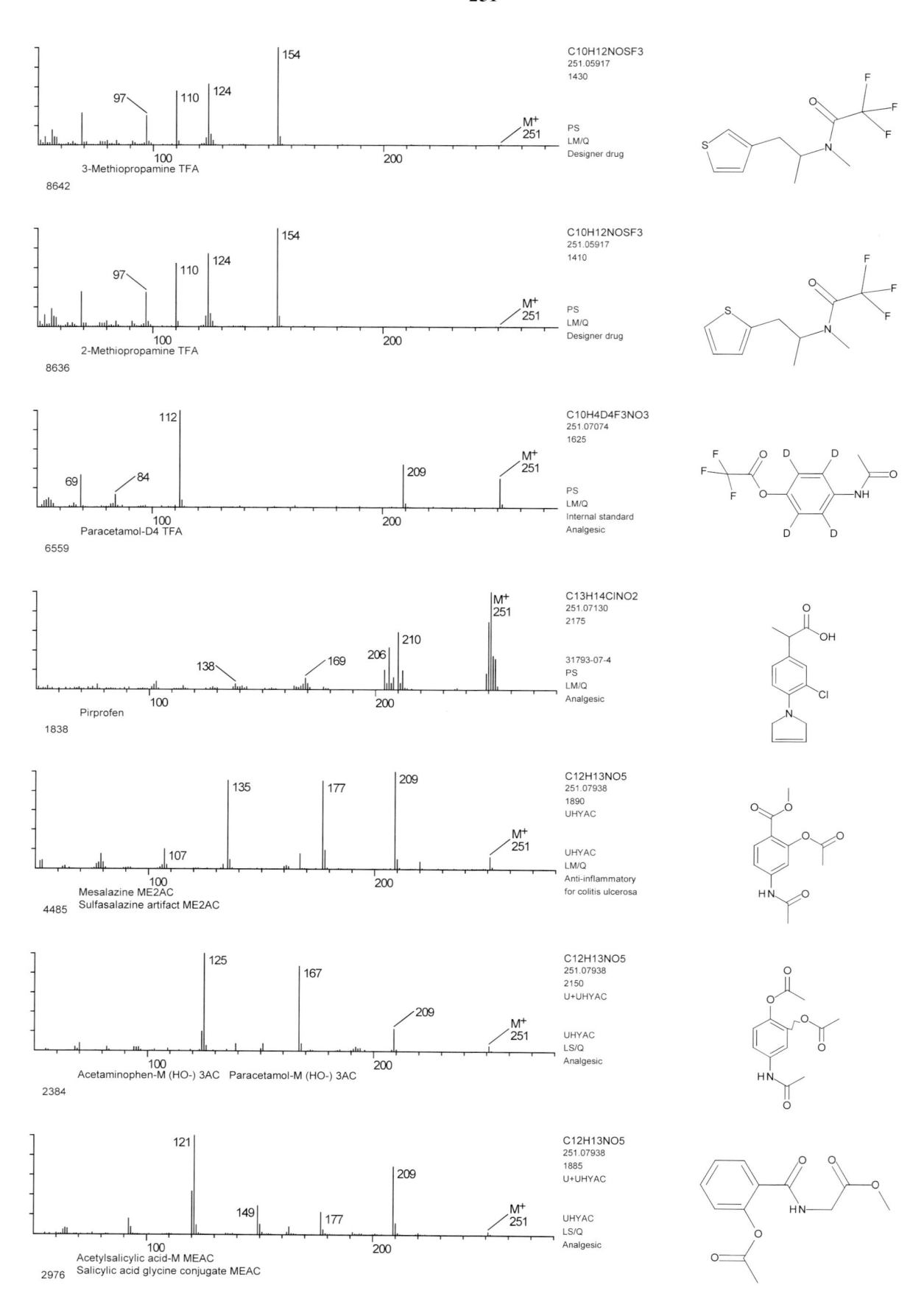

C10H12NOSF3
251.05917
1430
PS
LM/Q
Designer drug

3-Methiopropamine TFA
8642

C10H12NOSF3
251.05917
1410
PS
LM/Q
Designer drug

2-Methiopropamine TFA
8636

C10H4D4F3NO3
251.07074
1625
PS
LM/Q
Internal standard
Analgesic

Paracetamol-D4 TFA
6559

C13H14ClNO2
251.07130
2175

31793-07-4
PS
LM/Q
Analgesic

Pirprofen
1838

C12H13NO5
251.07938
1890
UHYAC

UHYAC
LM/Q
Anti-inflammatory
for colitis ulcerosa

Mesalazine ME2AC
Sulfasalazine artifact ME2AC
4485

C12H13NO5
251.07938
2150
U+UHYAC

UHYAC
LS/Q
Analgesic

Acetaminophen-M (HO-) 3AC   Paracetamol-M (HO-) 3AC
2384

C12H13NO5
251.07938
1885
U+UHYAC

UHYAC
LS/Q
Analgesic

Acetylsalicylic acid-M MEAC
Salicylic acid glycine conjugate MEAC
2976

251

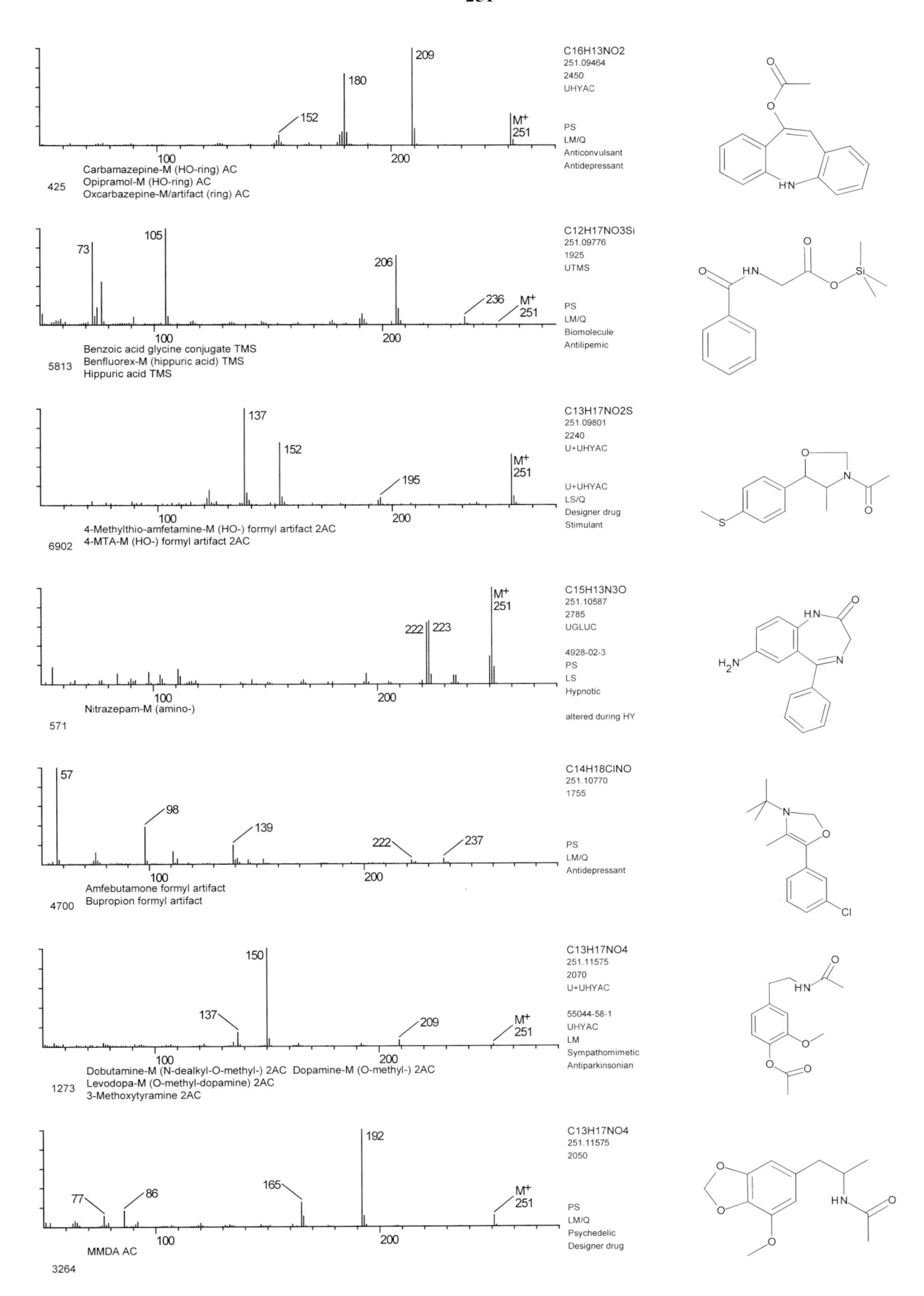

C16H13NO2
251.09464
2450
UHYAC

PS
LM/Q
Anticonvulsant
Antidepressant

Carbamazepine-M (HO-ring) AC
Opipramol-M (HO-ring) AC
Oxcarbazepine-M/artifact (ring) AC
425

C12H17NO3Si
251.09776
1925
UTMS

PS
LM/Q
Biomolecule
Antilipemic

Benzoic acid glycine conjugate TMS
Benfluorex-M (hippuric acid) TMS
Hippuric acid TMS
5813

C13H17NO2S
251.09801
2240
U+UHYAC

U+UHYAC
LS/Q
Designer drug
Stimulant

4-Methylthio-amfetamine-M (HO-) formyl artifact 2AC
4-MTA-M (HO-) formyl artifact 2AC
6902

C15H13N3O
251.10587
2785
UGLUC

4928-02-3
PS
LS
Hypnotic

altered during HY

Nitrazepam-M (amino-)
571

C14H18ClNO
251.10770
1755

PS
LM/Q
Antidepressant

Amfebutamone formyl artifact
Bupropion formyl artifact
4700

C13H17NO4
251.11575
2070
U+UHYAC

55044-58-1
UHYAC
LM
Sympathomimetic
Antiparkinsonian

Dobutamine-M (N-dealkyl-O-methyl-) 2AC Dopamine-M (O-methyl-) 2AC
Levodopa-M (O-methyl-dopamine) 2AC
3-Methoxytyramine 2AC
1273

C13H17NO4
251.11575
2050

PS
LM/Q
Psychedelic
Designer drug

MMDA AC
3264

609

251

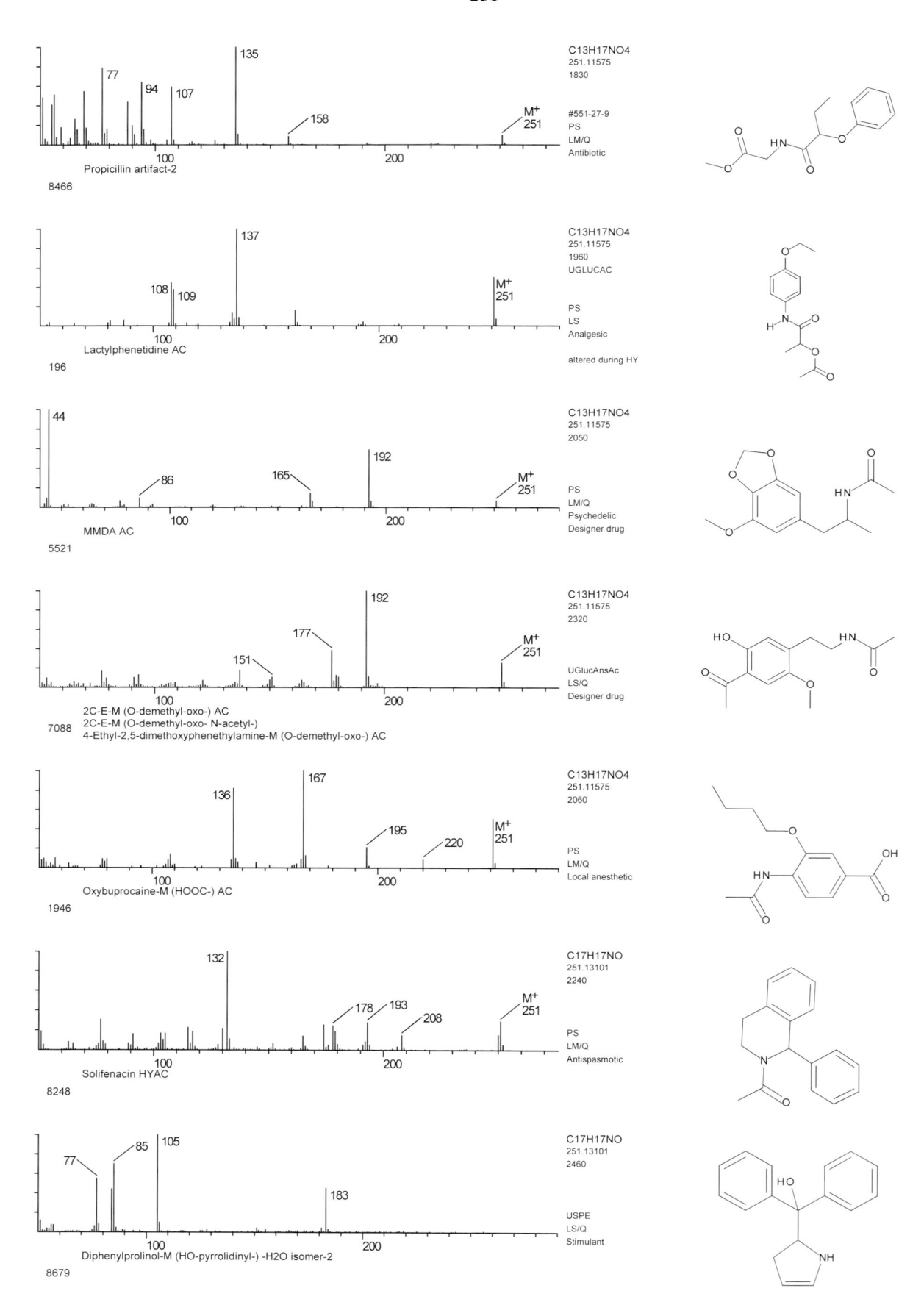

Propicillin artifact-2 — 77, 94, 107, 135, 158, M+ 251
C13H17NO4
251.11575
1830
#551-27-9
PS
LM/Q
Antibiotic
8466

Lactylphenetidine AC — 108, 109, 137, M+ 251
C13H17NO4
251.11575
1960
UGLUCAC
PS
LS
Analgesic
altered during HY
196

MMDA AC — 44, 86, 165, 192, M+ 251
C13H17NO4
251.11575
2050
PS
LM/Q
Psychedelic
Designer drug
5521

2C-E-M (O-demethyl-oxo-) AC
2C-E-M (O-demethyl-oxo- N-acetyl-)
4-Ethyl-2,5-dimethoxyphenethylamine-M (O-demethyl-oxo-) AC — 151, 177, 192, M+ 251
C13H17NO4
251.11575
2320
UGlucAnsAc
LS/Q
Designer drug
7088

Oxybuprocaine-M (HOOC-) AC — 136, 167, 195, 220, M+ 251
C13H17NO4
251.11575
2060
PS
LM/Q
Local anesthetic
1946

Solifenacin HYAC — 132, 178, 193, 208, M+ 251
C17H17NO
251.13101
2240
PS
LM/Q
Antispasmotic
8248

Diphenylprolinol-M (HO-pyrrolidinyl-) -H2O isomer-2 — 77, 85, 105, 183
C17H17NO
251.13101
2460
USPE
LS/Q
Stimulant
8679

251

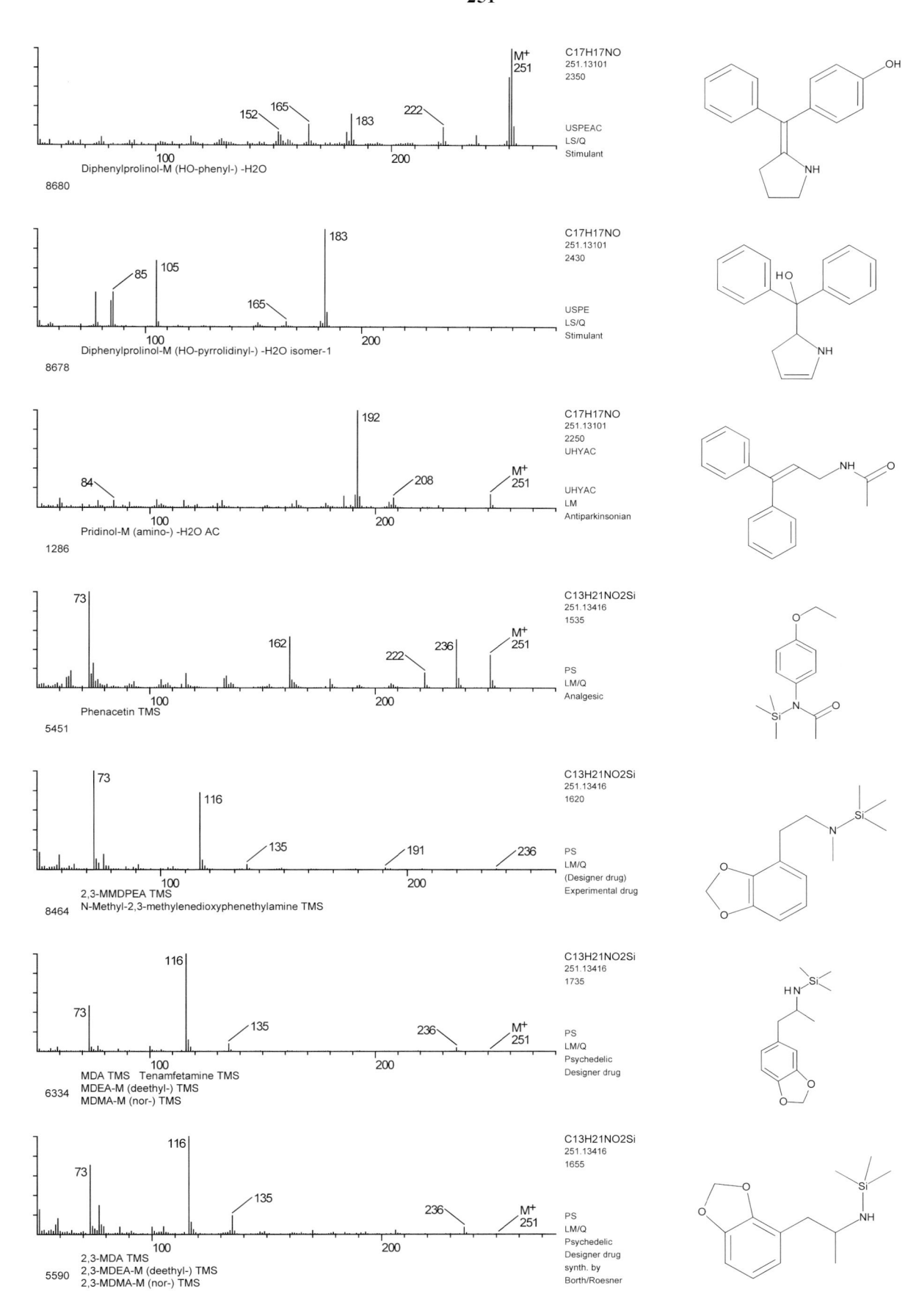

Diphenylprolinol-M (HO-phenyl-) -H2O — 8680
C17H17NO, 251.13101, 2350, USPEAC, LS/Q, Stimulant
Peaks: 152, 165, 183, 222, M+ 251

Diphenylprolinol-M (HO-pyrrolidinyl-) -H2O isomer-1 — 8678
C17H17NO, 251.13101, 2430, USPE, LS/Q, Stimulant
Peaks: 85, 105, 165, 183

Pridinol-M (amino-) -H2O AC — 1286
C17H17NO, 251.13101, 2250, UHYAC, UHYAC, LM, Antiparkinsonian
Peaks: 84, 192, 208, M+ 251

Phenacetin TMS — 5451
C13H21NO2Si, 251.13416, 1535, PS, LM/Q, Analgesic
Peaks: 73, 162, 222, 236, M+ 251

2,3-MMDPEA TMS / N-Methyl-2,3-methylenedioxyphenethylamine TMS — 8464
C13H21NO2Si, 251.13416, 1620, PS, LM/Q, (Designer drug), Experimental drug
Peaks: 73, 116, 135, 191, 236

MDA TMS / Tenamfetamine TMS / MDEA-M (deethyl-) TMS / MDMA-M (nor-) TMS — 6334
C13H21NO2Si, 251.13416, 1735, PS, LM/Q, Psychedelic, Designer drug
Peaks: 73, 116, 135, 236, M+ 251

2,3-MDA TMS / 2,3-MDEA-M (deethyl-) TMS / 2,3-MDMA-M (nor-) TMS — 5590
C13H21NO2Si, 251.13416, 1655, PS, LM/Q, Psychedelic, Designer drug synth. by Borth/Roesner
Peaks: 73, 116, 135, 236, M+ 251

251

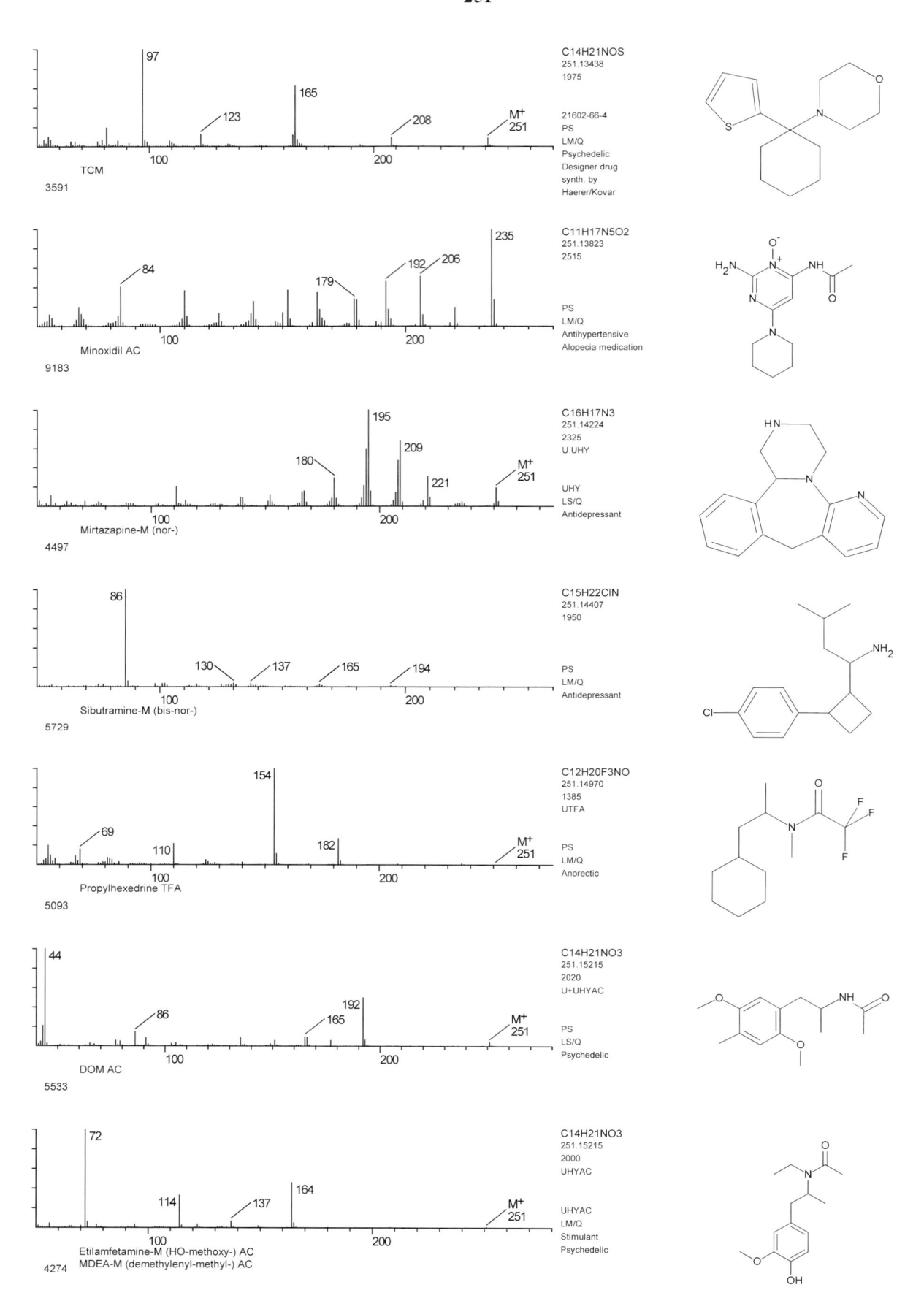

C14H21NOS
251.13438
1975

21602-66-4
PS
LM/Q
Psychedelic
Designer drug
synth. by
Haerer/Kovar

TCM
3591

Peaks: 97, 123, 165, 208, M+ 251

C11H17N5O2
251.13823
2515

PS
LM/Q
Antihypertensive
Alopecia medication

Minoxidil AC
9183

Peaks: 84, 179, 192, 206, 235

C16H17N3
251.14224
2325
U UHY

UHY
LS/Q
Antidepressant

Mirtazapine-M (nor-)
4497

Peaks: 180, 195, 209, 221, M+ 251

C15H22ClN
251.14407
1950

PS
LM/Q
Antidepressant

Sibutramine-M (bis-nor-)
5729

Peaks: 86, 130, 137, 165, 194

C12H20F3NO
251.14970
1385
UTFA

PS
LM/Q
Anorectic

Propylhexedrine TFA
5093

Peaks: 69, 110, 154, 182, M+ 251

C14H21NO3
251.15215
2020
U+UHYAC

PS
LS/Q
Psychedelic

DOM AC
5533

Peaks: 44, 86, 165, 192, M+ 251

C14H21NO3
251.15215
2000
UHYAC

UHYAC
LM/Q
Stimulant
Psychedelic

Etilamfetamine-M (HO-methoxy-) AC
MDEA-M (demethylenyl-methyl-) AC
4274

Peaks: 72, 114, 137, 164, M+ 251

251

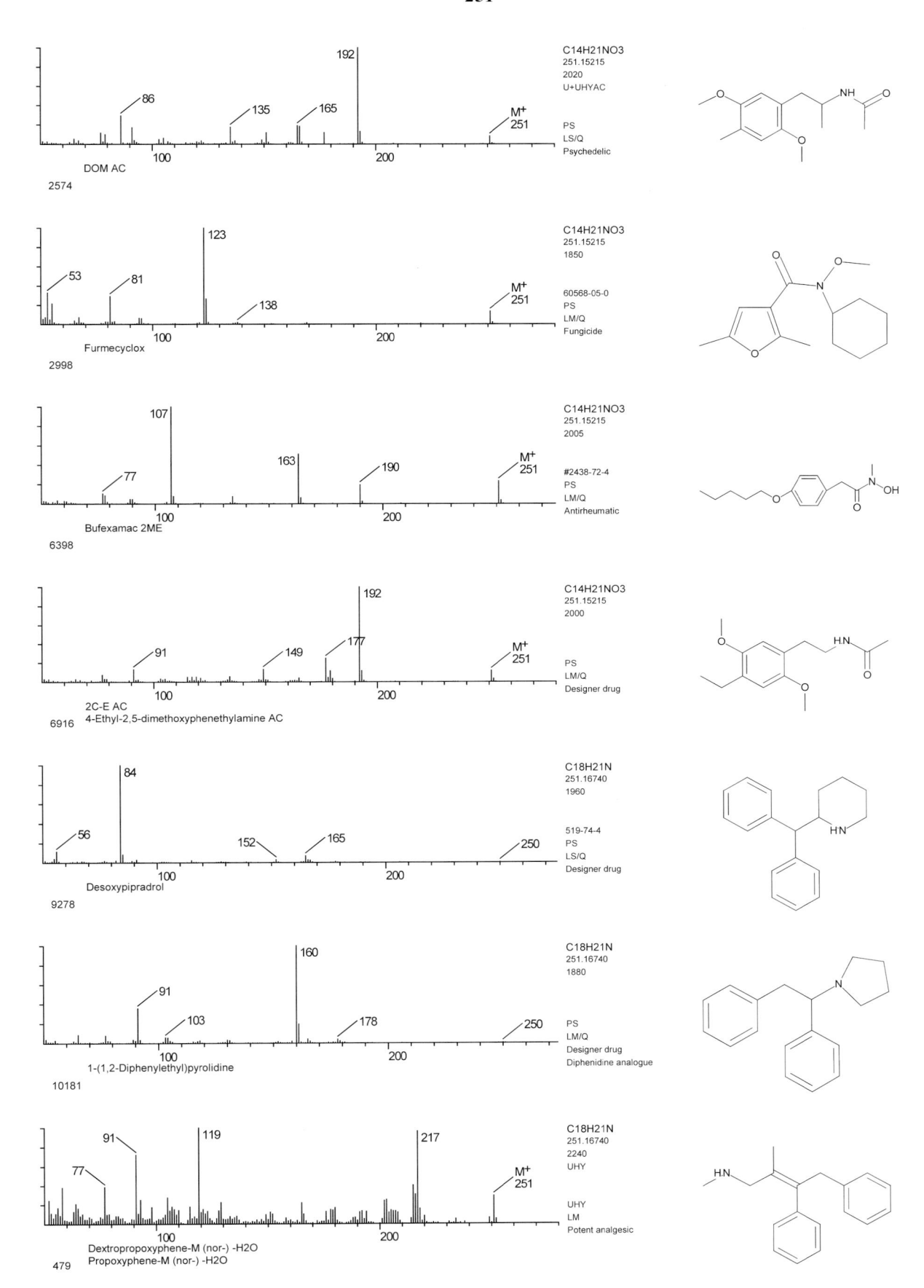

C14H21NO3
251.15215
2020
U+UHYAC
PS
LS/Q
Psychedelic

DOM AC
2574

C14H21NO3
251.15215
1850
60568-05-0
PS
LM/Q
Fungicide

Furmecyclox
2998

C14H21NO3
251.15215
2005
#2438-72-4
PS
LM/Q
Antirheumatic

Bufexamac 2ME
6398

C14H21NO3
251.15215
2000
PS
LM/Q
Designer drug

2C-E AC
4-Ethyl-2,5-dimethoxyphenethylamine AC
6916

C18H21N
251.16740
1960
519-74-4
PS
LS/Q
Designer drug

Desoxypipradrol
9278

C18H21N
251.16740
1880
PS
LM/Q
Designer drug
Diphenidine analogue

1-(1,2-Diphenylethyl)pyrolidine
10181

C18H21N
251.16740
2240
UHY
UHY
LM
Potent analgesic

Dextropropoxyphene-M (nor-) -H2O
Propoxyphene-M (nor-) -H2O
479

251

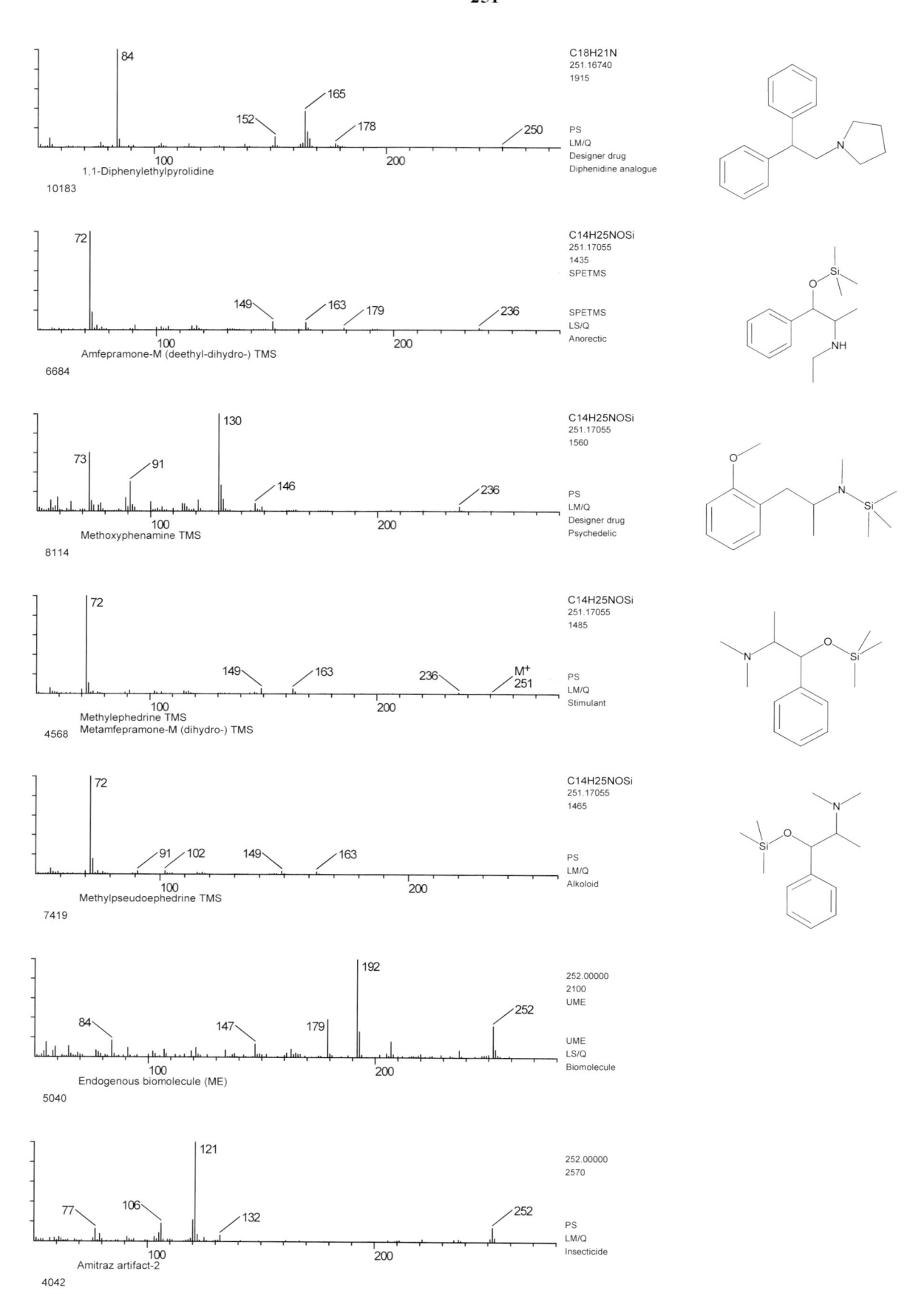

1,1-Diphenylethylpyrrolidine
Peaks: 84, 152, 165, 178, 250
C18H21N
251.16740
1915
PS
LM/Q
Designer drug
Diphenidine analogue
10183

Amfepramone-M (deethyl-dihydro-) TMS
Peaks: 72, 149, 163, 179, 236
C14H25NOSi
251.17055
1435
SPETMS
SPETMS
LS/Q
Anorectic
6684

Methoxyphenamine TMS
Peaks: 73, 91, 130, 146, 236
C14H25NOSi
251.17055
1560
PS
LM/Q
Designer drug
Psychedelic
8114

Methylephedrine TMS / Metamfepramone-M (dihydro-) TMS
Peaks: 72, 149, 163, 236, M+ 251
C14H25NOSi
251.17055
1485
PS
LM/Q
Stimulant
4568

Methylpseudoephedrine TMS
Peaks: 72, 91, 102, 149, 163
C14H25NOSi
251.17055
1465
PS
LM/Q
Alkoloid
7419

Endogenous biomolecule (ME)
Peaks: 84, 147, 179, 192, 252
252.00000
2100
UME
UME
LS/Q
Biomolecule
5040

Amitraz artifact-2
Peaks: 77, 106, 121, 132, 252
252.00000
2570
PS
LM/Q
Insecticide
4042

252

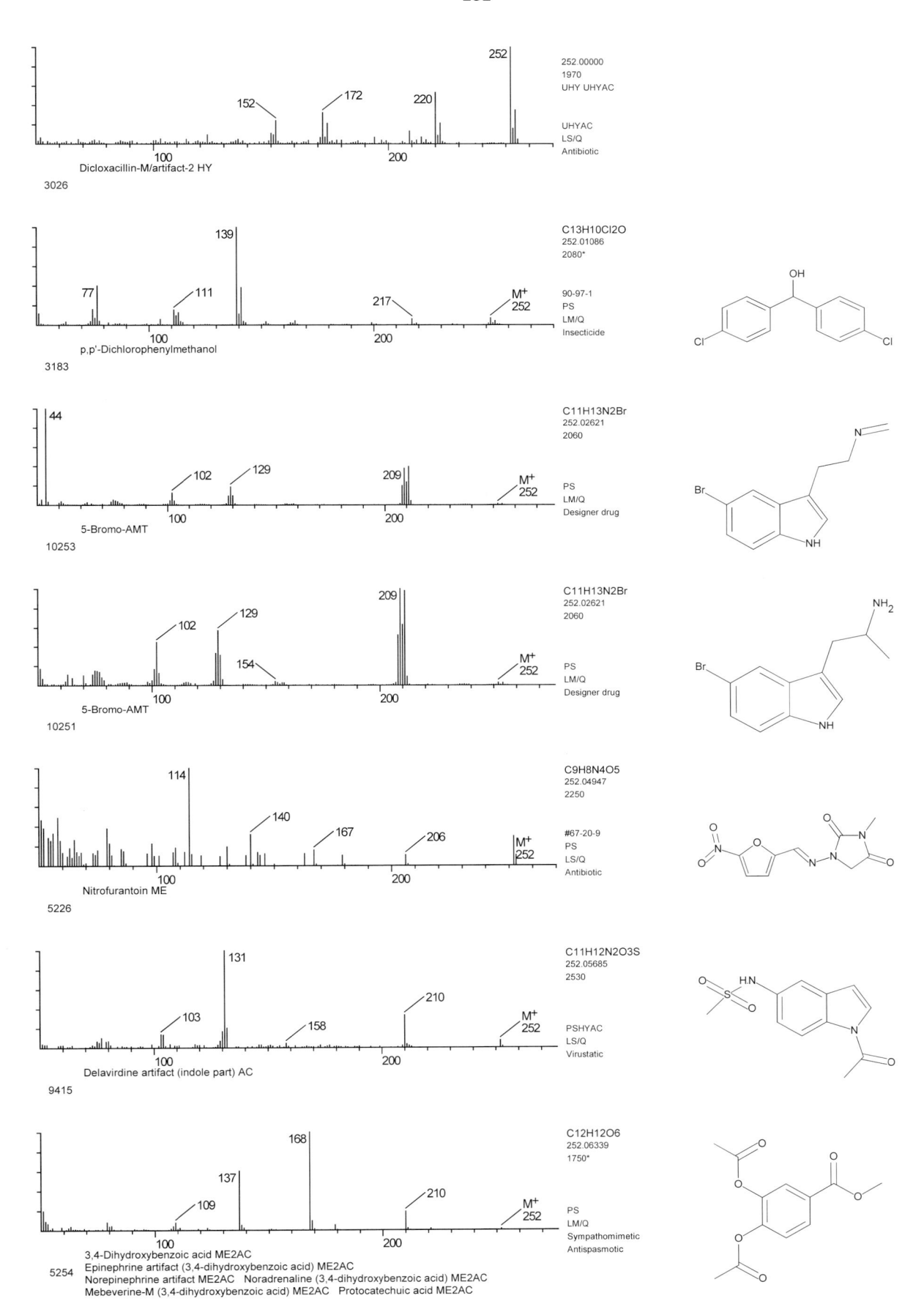

Dicloxacillin-M/artifact-2 HY
3026
252.00000
1970
UHY UHYAC
UHYAC
LS/Q
Antibiotic

p,p'-Dichlorophenylmethanol
3183
C13H10Cl2O
252.01086
2080*
90-97-1
PS
LM/Q
Insecticide

5-Bromo-AMT
10253
C11H13N2Br
252.02621
2060
PS
LM/Q
Designer drug

5-Bromo-AMT
10251
C11H13N2Br
252.02621
2060
PS
LM/Q
Designer drug

Nitrofurantoin ME
5226
C9H8N4O5
252.04947
2250
#67-20-9
PS
LS/Q
Antibiotic

Delavirdine artifact (indole part) AC
9415
C11H12N2O3S
252.05685
2530
PSHYAC
LS/Q
Virustatic

3,4-Dihydroxybenzoic acid ME2AC
Epinephrine artifact (3,4-dihydroxybenzoic acid) ME2AC
Norepinephrine artifact ME2AC   Noradrenaline (3,4-dihydroxybenzoic acid) ME2AC
Mebeverine-M (3,4-dihydroxybenzoic acid) ME2AC   Protocatechuic acid ME2AC
5254
C12H12O6
252.06339
1750*
PS
LM/Q
Sympathomimetic
Antispasmotic

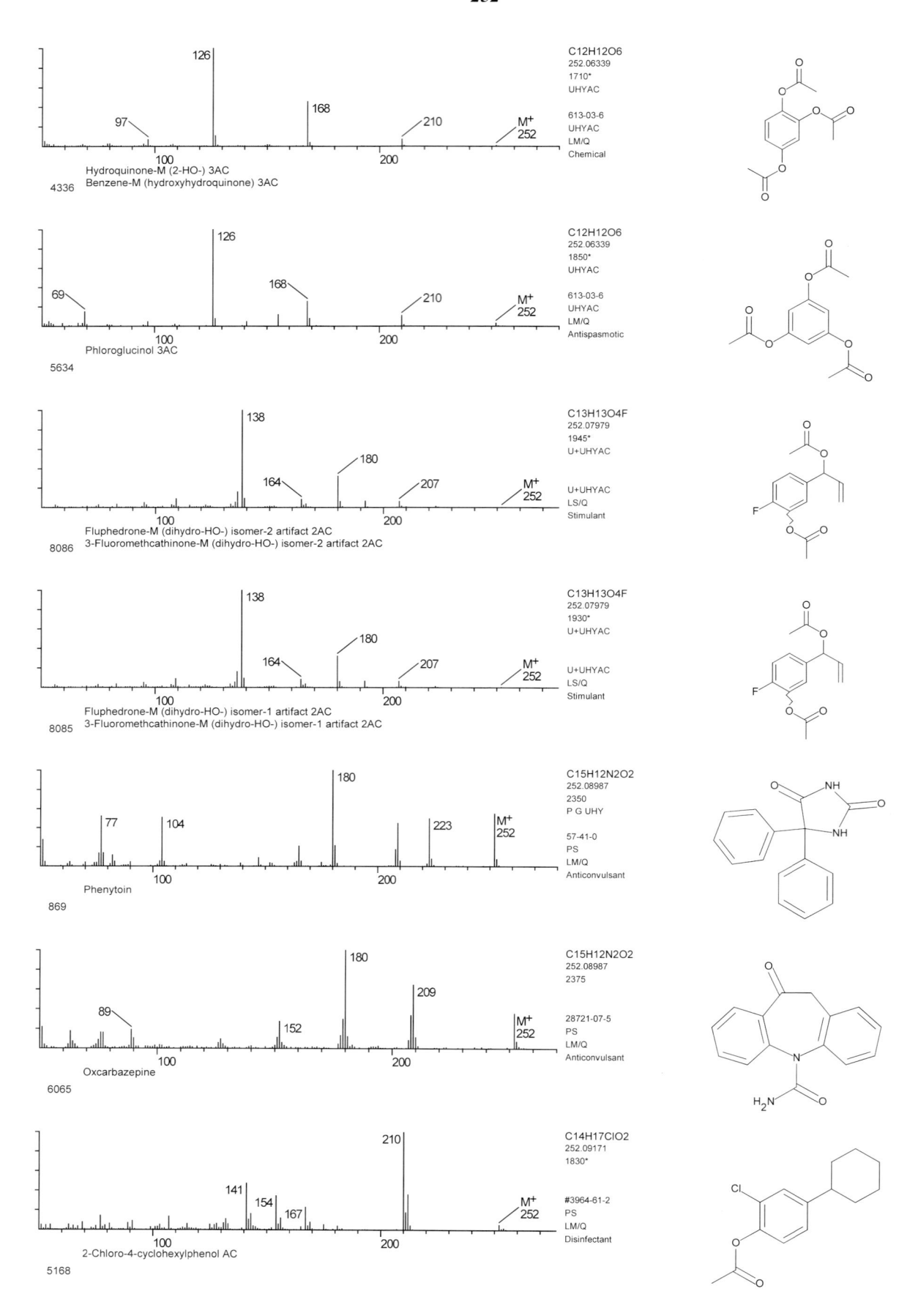

4336	Hydroquinone-M (2-HO-) 3AC Benzene-M (hydroxyhydroquinone) 3AC Peaks: 97, 126, 168, 210, 252 (M+) C12H12O6 252.06339 1710* UHYAC 613-03-6 UHYAC LM/Q Chemical
5634	Phloroglucinol 3AC Peaks: 69, 126, 168, 210, 252 (M+) C12H12O6 252.06339 1850* UHYAC 613-03-6 UHYAC LM/Q Antispasmotic
8086	Fluphedrone-M (dihydro-HO-) isomer-2 artifact 2AC 3-Fluoromethcathinone-M (dihydro-HO-) isomer-2 artifact 2AC Peaks: 138, 164, 180, 207, 252 (M+) C13H13O4F 252.07979 1945* U+UHYAC U+UHYAC LS/Q Stimulant
8085	Fluphedrone-M (dihydro-HO-) isomer-1 artifact 2AC 3-Fluoromethcathinone-M (dihydro-HO-) isomer-1 artifact 2AC Peaks: 138, 164, 180, 207, 252 (M+) C13H13O4F 252.07979 1930* U+UHYAC U+UHYAC LS/Q Stimulant
869	Phenytoin Peaks: 77, 104, 180, 223, 252 (M+) C15H12N2O2 252.08987 2350 P G UHY 57-41-0 PS LM/Q Anticonvulsant
6065	Oxcarbazepine Peaks: 89, 152, 180, 209, 252 (M+) C15H12N2O2 252.08987 2375 28721-07-5 PS LM/Q Anticonvulsant
5168	2-Chloro-4-cyclohexylphenol AC Peaks: 141, 154, 167, 210, 252 (M+) C14H17ClO2 252.09171 1830* #3964-61-2 PS LM/Q Disinfectant

252

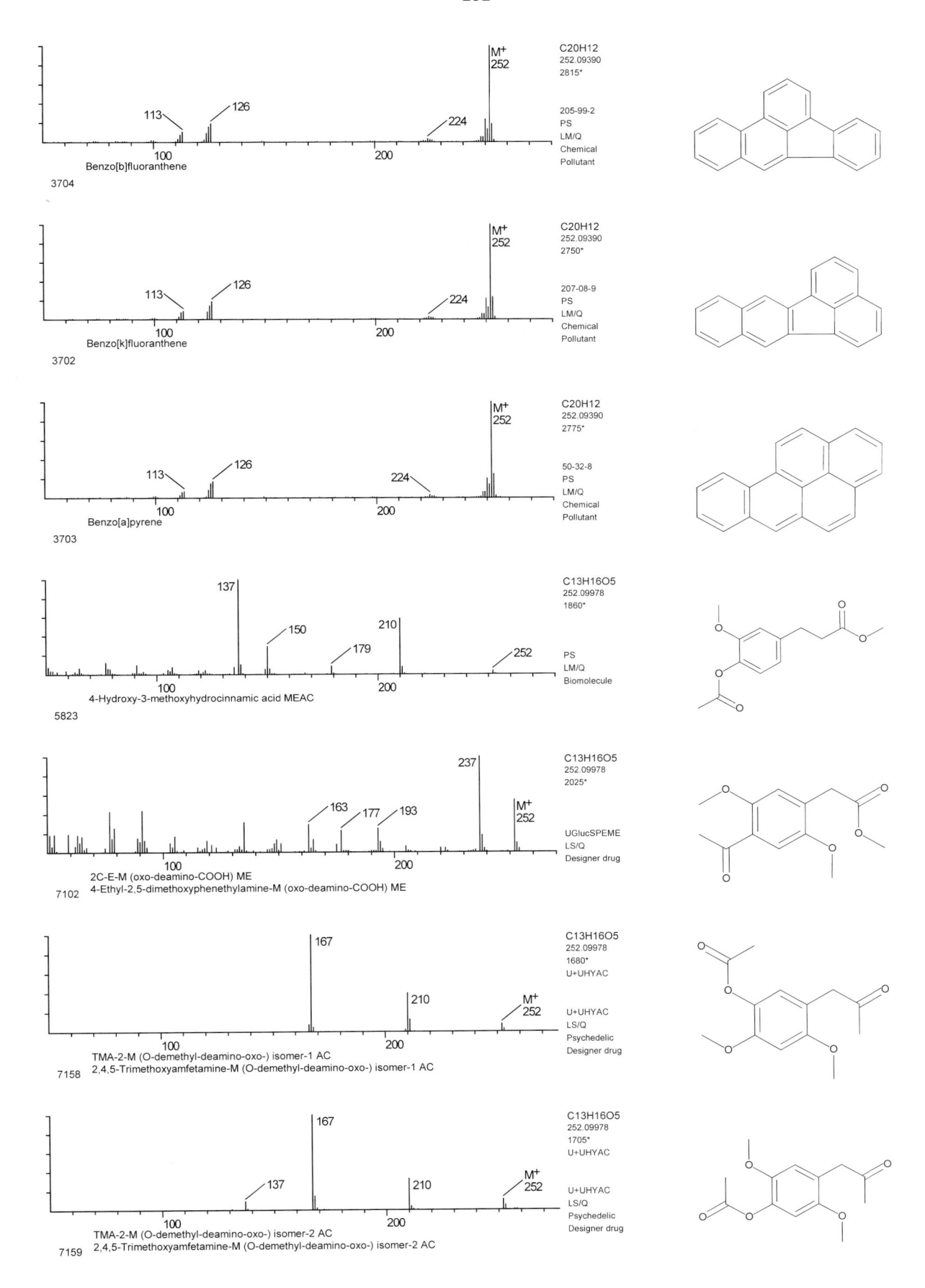

Benzo[b]fluoranthene
3704
C20H12
252.09390
2815*
205-99-2
PS
LM/Q
Chemical Pollutant

Benzo[k]fluoranthene
3702
C20H12
252.09390
2750*
207-08-9
PS
LM/Q
Chemical Pollutant

Benzo[a]pyrene
3703
C20H12
252.09390
2775*
50-32-8
PS
LM/Q
Chemical Pollutant

4-Hydroxy-3-methoxyhydrocinnamic acid MEAC
5823
C13H16O5
252.09978
1860*
PS
LM/Q
Biomolecule

2C-E-M (oxo-deamino-COOH) ME
4-Ethyl-2,5-dimethoxyphenethylamine-M (oxo-deamino-COOH) ME
7102
C13H16O5
252.09978
2025*
UGlucSPEME
LS/Q
Designer drug

TMA-2-M (O-demethyl-deamino-oxo-) isomer-1 AC
2,4,5-Trimethoxyamfetamine-M (O-demethyl-deamino-oxo-) isomer-1 AC
7158
C13H16O5
252.09978
1680*
U+UHYAC
U+UHYAC
LS/Q
Psychedelic Designer drug

TMA-2-M (O-demethyl-deamino-oxo-) isomer-2 AC
2,4,5-Trimethoxyamfetamine-M (O-demethyl-deamino-oxo-) isomer-2 AC
7159
C13H16O5
252.09978
1705*
U+UHYAC
U+UHYAC
LS/Q
Psychedelic Designer drug

252

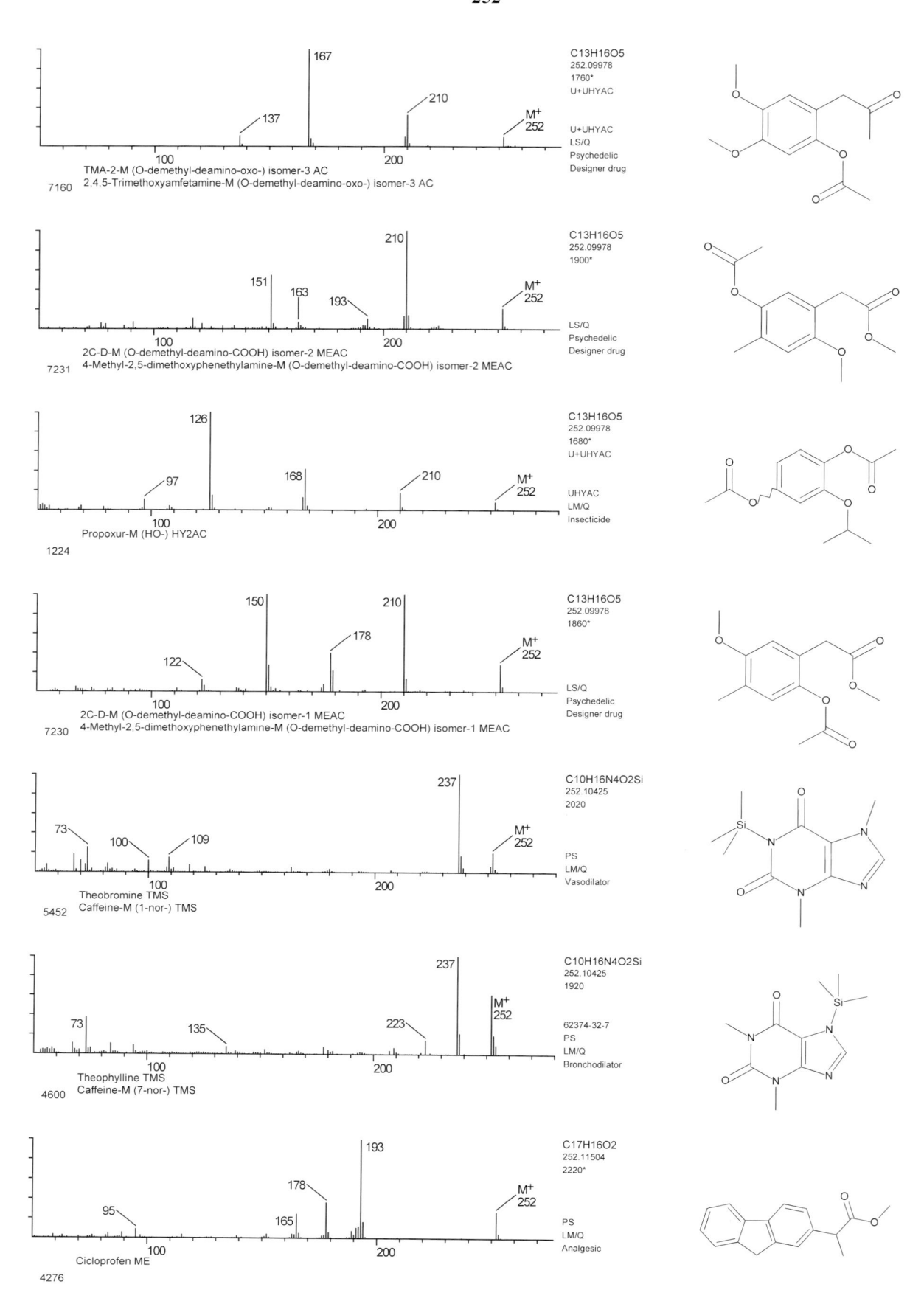

Spectrum	Formula / Info
7160 — TMA-2-M (O-demethyl-deamino-oxo-) isomer-3 AC / 2,4,5-Trimethoxyamfetamine-M (O-demethyl-deamino-oxo-) isomer-3 AC; peaks 137, 167, 210, M+ 252	C13H16O5; 252.09978; 1760*; U+UHYAC; U+UHYAC; LS/Q; Psychedelic; Designer drug
7231 — 2C-D-M (O-demethyl-deamino-COOH) isomer-2 MEAC / 4-Methyl-2,5-dimethoxyphenethylamine-M (O-demethyl-deamino-COOH) isomer-2 MEAC; peaks 151, 163, 193, 210, M+ 252	C13H16O5; 252.09978; 1900*; LS/Q; Psychedelic; Designer drug
1224 — Propoxur-M (HO-) HY2AC; peaks 97, 126, 168, 210, M+ 252	C13H16O5; 252.09978; 1680*; U+UHYAC; UHYAC; LM/Q; Insecticide
7230 — 2C-D-M (O-demethyl-deamino-COOH) isomer-1 MEAC / 4-Methyl-2,5-dimethoxyphenethylamine-M (O-demethyl-deamino-COOH) isomer-1 MEAC; peaks 122, 150, 178, 210, M+ 252	C13H16O5; 252.09978; 1860*; LS/Q; Psychedelic; Designer drug
5452 — Theobromine TMS / Caffeine-M (1-nor-) TMS; peaks 73, 100, 109, 237, M+ 252	C10H16N4O2Si; 252.10425; 2020; PS; LM/Q; Vasodilator
4600 — Theophylline TMS / Caffeine-M (7-nor-) TMS; peaks 73, 135, 223, 237, M+ 252	C10H16N4O2Si; 252.10425; 1920; 62374-32-7; PS; LM/Q; Bronchodilator
4276 — Cicloprofen ME; peaks 95, 165, 178, 193, M+ 252	C17H16O2; 252.11504; 2220*; PS; LM/Q; Analgesic

252

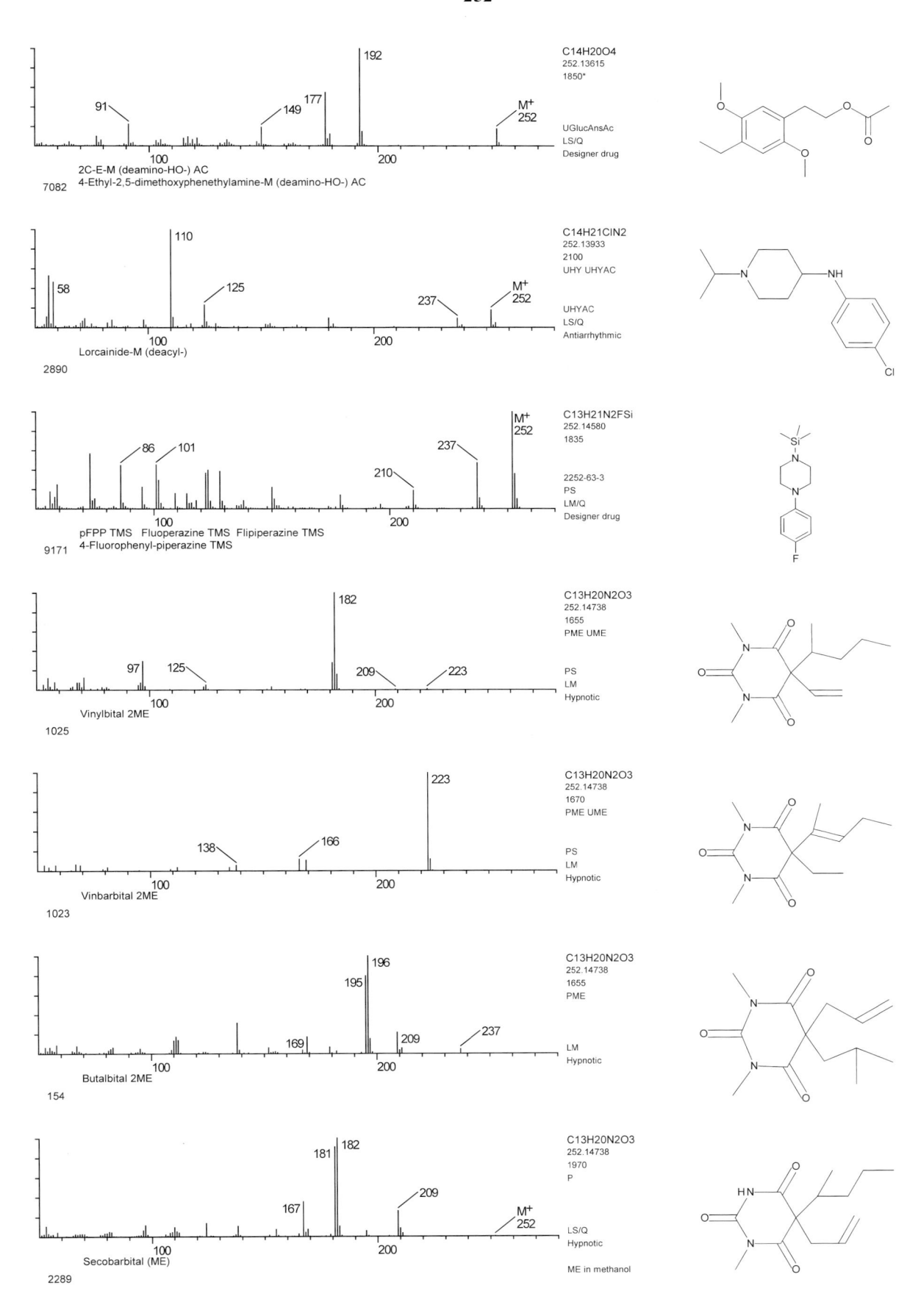

7082	2C-E-M (deamino-HO-) AC 4-Ethyl-2,5-dimethoxyphenethylamine-M (deamino-HO-) AC Peaks: 91, 149, 177, 192, M+ 252	C14H20O4 252.13615 1850* UGlucAnsAc LS/Q Designer drug

2890	Lorcainide-M (deacyl-) Peaks: 58, 110, 125, 237, M+ 252	C14H21ClN2 252.13933 2100 UHY UHYAC UHYAC LS/Q Antiarrhythmic

9171	pFPP TMS Fluoperazine TMS Flipiperazine TMS 4-Fluorophenyl-piperazine TMS Peaks: 86, 101, 210, 237, M+ 252	C13H21N2FSi 252.14580 1835 2252-63-3 PS LM/Q Designer drug

1025	Vinylbital 2ME Peaks: 97, 125, 182, 209, 223	C13H20N2O3 252.14738 1655 PME UME PS LM Hypnotic

1023	Vinbarbital 2ME Peaks: 138, 166, 223	C13H20N2O3 252.14738 1670 PME UME PS LM Hypnotic

154	Butalbital 2ME Peaks: 169, 195, 196, 209, 237	C13H20N2O3 252.14738 1655 PME LM Hypnotic

2289	Secobarbital (ME) Peaks: 167, 181, 182, 209, M+ 252	C13H20N2O3 252.14738 1970 P LS/Q Hypnotic ME in methanol

252

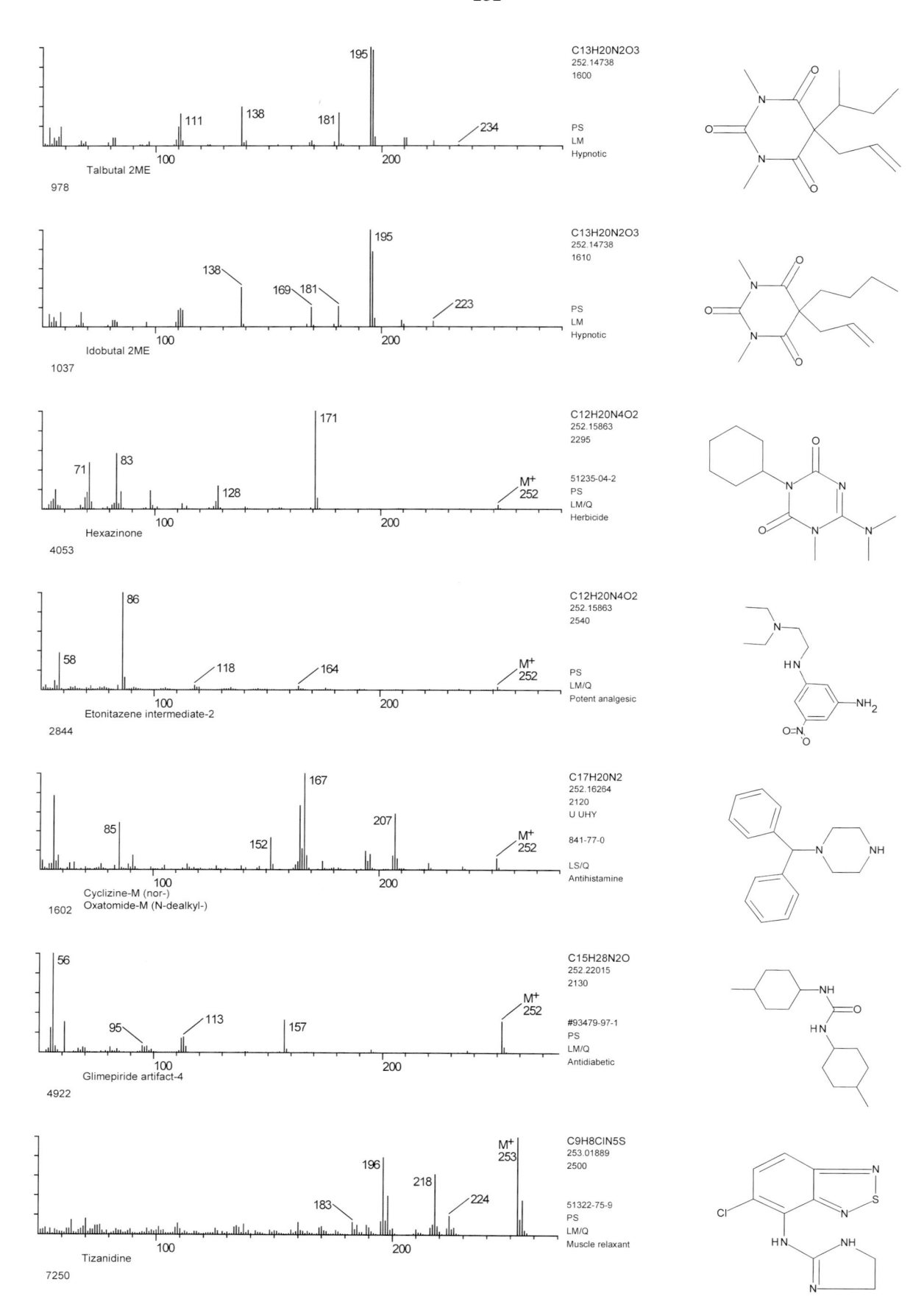

Talbutal 2ME — C13H20N2O3, 252.14738, 1600, PS, LM, Hypnotic
Peaks: 111, 138, 181, 195, 234
978

Idobutal 2ME — C13H20N2O3, 252.14738, 1610, PS, LM, Hypnotic
Peaks: 138, 169, 181, 195, 223
1037

Hexazinone — C12H20N4O2, 252.15863, 2295, 51235-04-2, PS, LM/Q, Herbicide
Peaks: 71, 83, 128, 171, M+ 252
4053

Etonitazene intermediate-2 — C12H20N4O2, 252.15863, 2540, PS, LM/Q, Potent analgesic
Peaks: 58, 86, 118, 164, M+ 252
2844

Cyclizine-M (nor-) / Oxatomide-M (N-dealkyl-) — C17H20N2, 252.16264, 2120, U UHY, 841-77-0, LS/Q, Antihistamine
Peaks: 85, 152, 167, 207, M+ 252
1602

Glimepiride artifact-4 — C15H28N2O, 252.22015, 2130, #93479-97-1, PS, LM/Q, Antidiabetic
Peaks: 56, 95, 113, 157, M+ 252
4922

Tizanidine — C9H8ClN5S, 253.01889, 2500, 51322-75-9, PS, LM/Q, Muscle relaxant
Peaks: 183, 196, 218, 224, M+ 253
7250

253

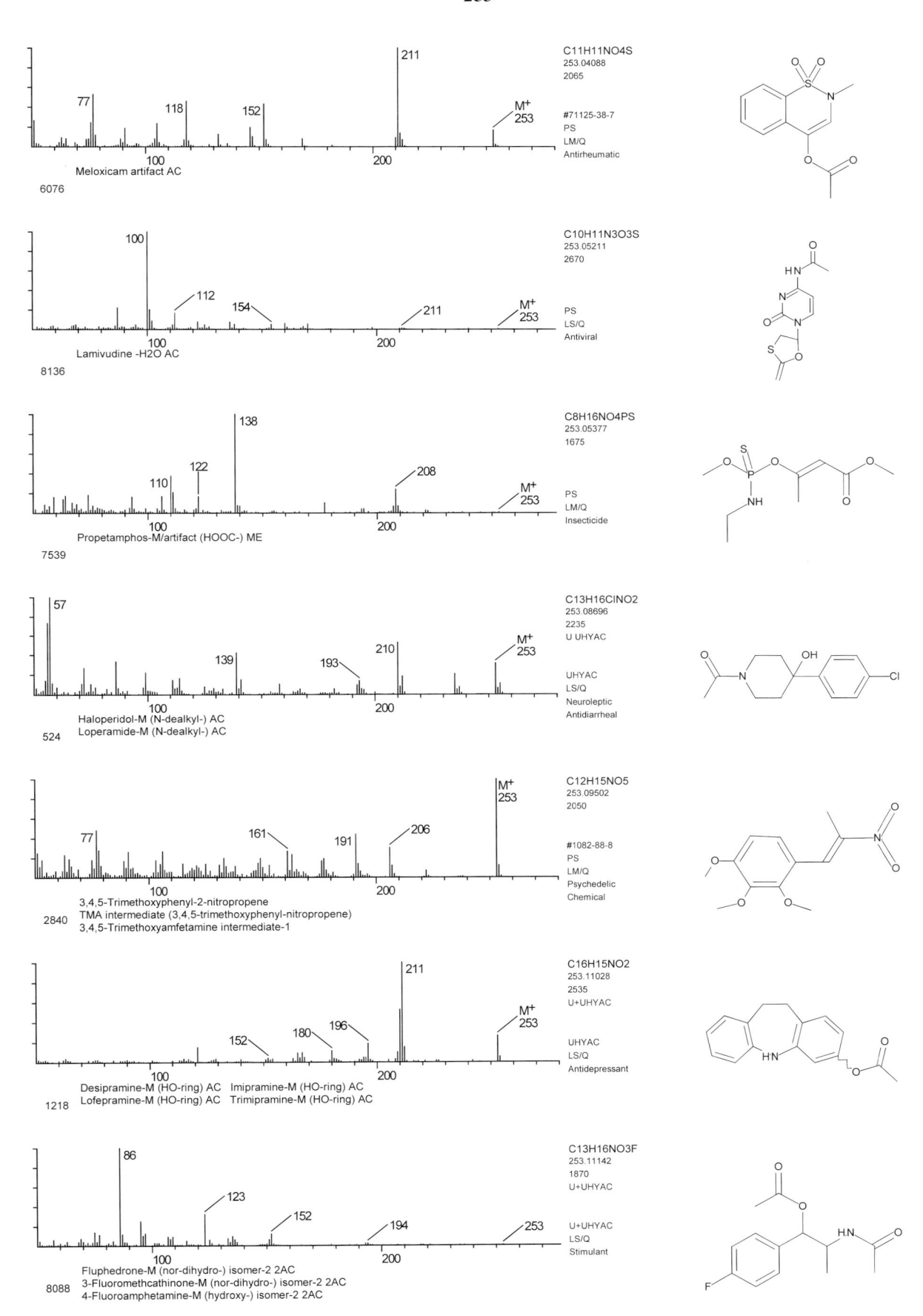

C11H11NO4S
253.04088
2065

#71125-38-7
PS
LM/Q
Antirheumatic

Meloxicam artifact AC
6076

C10H11N3O3S
253.05211
2670

PS
LS/Q
Antiviral

Lamivudine -H2O AC
8136

C8H16NO4PS
253.05377
1675

PS
LM/Q
Insecticide

Propetamphos-M/artifact (HOOC-) ME
7539

C13H16ClNO2
253.08696
2235
U UHYAC

UHYAC
LS/Q
Neuroleptic
Antidiarrheal

Haloperidol-M (N-dealkyl-) AC
Loperamide-M (N-dealkyl-) AC
524

C12H15NO5
253.09502
2050

#1082-88-8
PS
LM/Q
Psychedelic
Chemical

3,4,5-Trimethoxyphenyl-2-nitropropene
TMA intermediate (3,4,5-trimethoxyphenyl-nitropropene)
3,4,5-Trimethoxyamfetamine intermediate-1
2840

C16H15NO2
253.11028
2535
U+UHYAC

UHYAC
LS/Q
Antidepressant

Desipramine-M (HO-ring) AC   Imipramine-M (HO-ring) AC
Lofepramine-M (HO-ring) AC   Trimipramine-M (HO-ring) AC
1218

C13H16NO3F
253.11142
1870
U+UHYAC

U+UHYAC
LS/Q
Stimulant

Fluphedrone-M (nor-dihydro-) isomer-2 2AC
3-Fluoromethcathinone-M (nor-dihydro-) isomer-2 2AC
4-Fluoroamphetamine-M (hydroxy-) isomer-2 2AC
8088

253

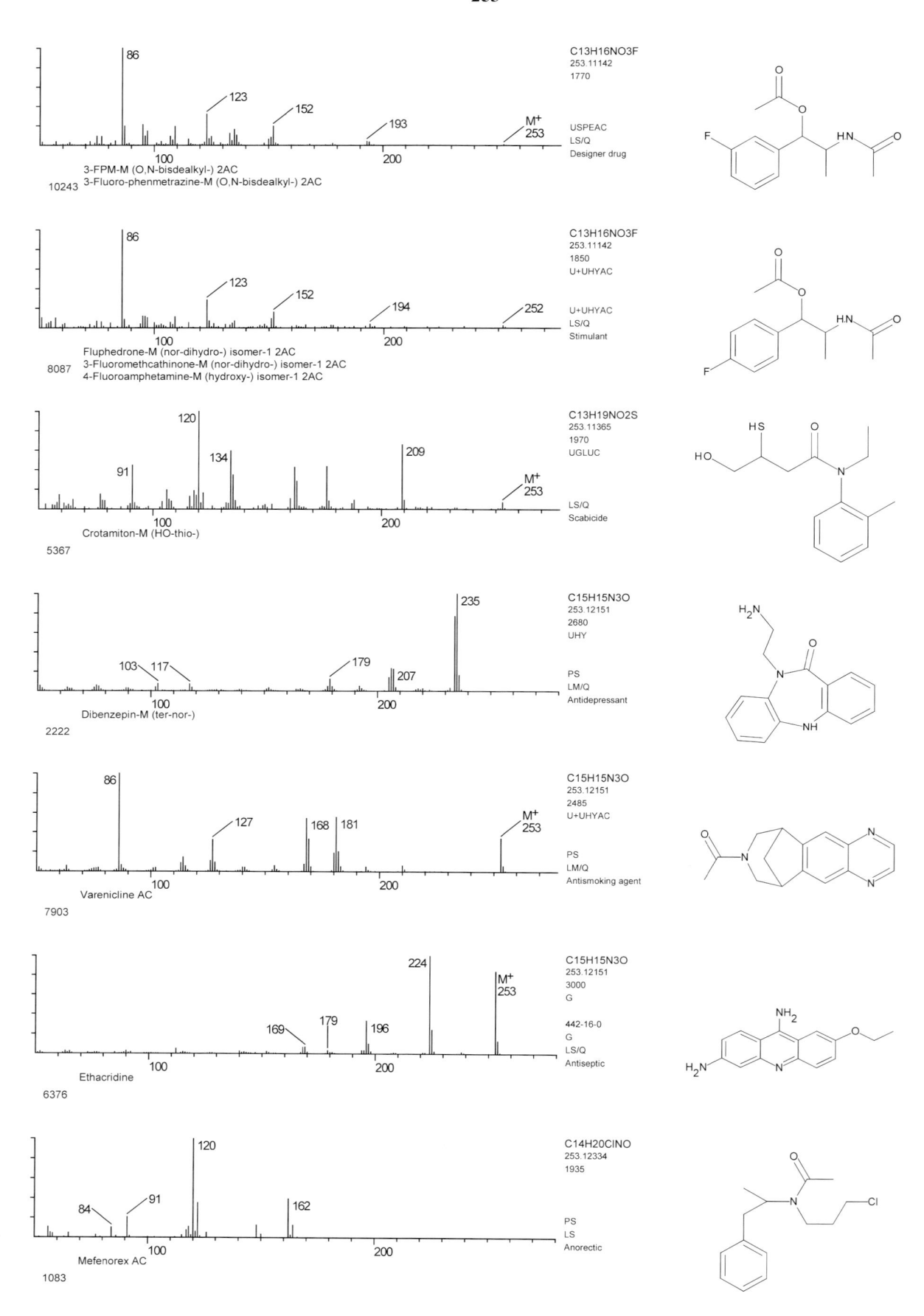

10243	3-FPM-M (O,N-bisdealkyl-) 2AC 3-Fluoro-phenmetrazine-M (O,N-bisdealkyl-) 2AC peaks: 86, 123, 152, 193, M+ 253	C13H16NO3F 253.11142 1770 USPEAC LS/Q Designer drug
8087	Fluphedrone-M (nor-dihydro-) isomer-1 2AC 3-Fluoromethcathinone-M (nor-dihydro-) isomer-1 2AC 4-Fluoroamphetamine-M (hydroxy-) isomer-1 2AC peaks: 86, 123, 152, 194, 252	C13H16NO3F 253.11142 1850 U+UHYAC U+UHYAC LS/Q Stimulant
5367	Crotamiton-M (HO-thio-) peaks: 91, 120, 134, 209, M+ 253	C13H19NO2S 253.11365 1970 UGLUC LS/Q Scabicide
2222	Dibenzepin-M (ter-nor-) peaks: 103, 117, 179, 207, 235	C15H15N3O 253.12151 2680 UHY PS LM/Q Antidepressant
7903	Varenicline AC peaks: 86, 127, 168, 181, M+ 253	C15H15N3O 253.12151 2485 U+UHYAC PS LM/Q Antismoking agent
6376	Ethacridine peaks: 169, 179, 196, 224, M+ 253	C15H15N3O 253.12151 3000 G 442-16-0 G LS/Q Antiseptic
1083	Mefenorex AC peaks: 84, 91, 120, 162	C14H20ClNO 253.12334 1935 PS LS Anorectic

253

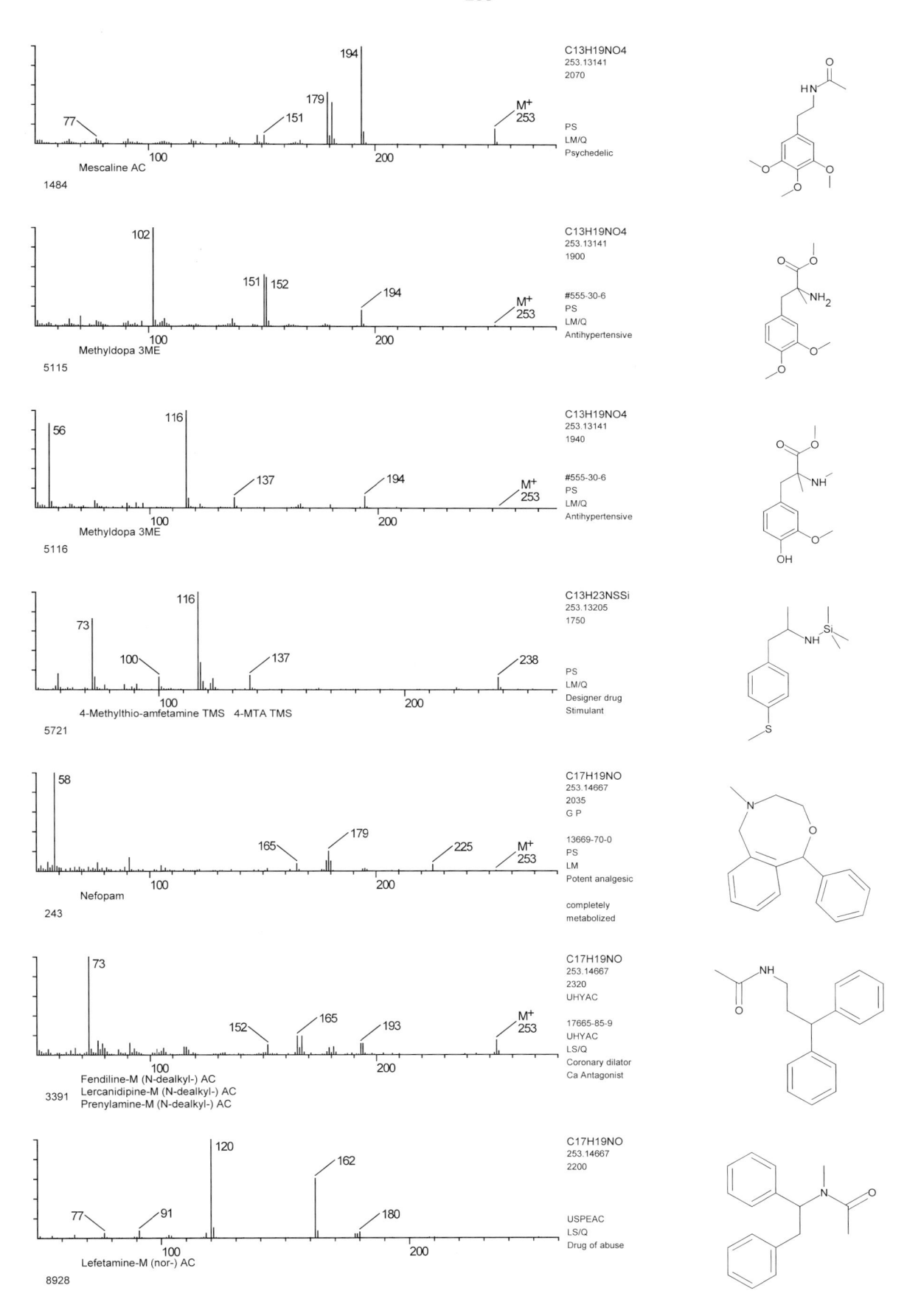

Mescaline AC	77, 151, 179, 194, M+ 253	C13H19NO4 / 253.13141 / 2070 / PS / LM/Q / Psychedelic
1484		
Methyldopa 3ME	102, 151, 152, 194, M+ 253	C13H19NO4 / 253.13141 / 1900 / #555-30-6 / PS / LM/Q / Antihypertensive
5115		
Methyldopa 3ME	56, 116, 137, 194, M+ 253	C13H19NO4 / 253.13141 / 1940 / #555-30-6 / PS / LM/Q / Antihypertensive
5116		
4-Methylthio-amfetamine TMS 4-MTA TMS	73, 100, 116, 137, 238	C13H23NSSi / 253.13205 / 1750 / PS / LM/Q / Designer drug / Stimulant
5721		
Nefopam	58, 165, 179, 225, M+ 253	C17H19NO / 253.14667 / 2035 / G P / 13669-70-0 / PS / LM / Potent analgesic / completely metabolized
243		
Fendiline-M (N-dealkyl-) AC / Lercanidipine-M (N-dealkyl-) AC / Prenylamine-M (N-dealkyl-) AC	73, 152, 165, 193, M+ 253	C17H19NO / 253.14667 / 2320 / UHYAC / 17665-85-9 / UHYAC / LS/Q / Coronary dilator / Ca Antagonist
3391		
Lefetamine-M (nor-) AC	77, 91, 120, 162, 180	C17H19NO / 253.14667 / 2200 / USPEAC / LS/Q / Drug of abuse
8928		

253

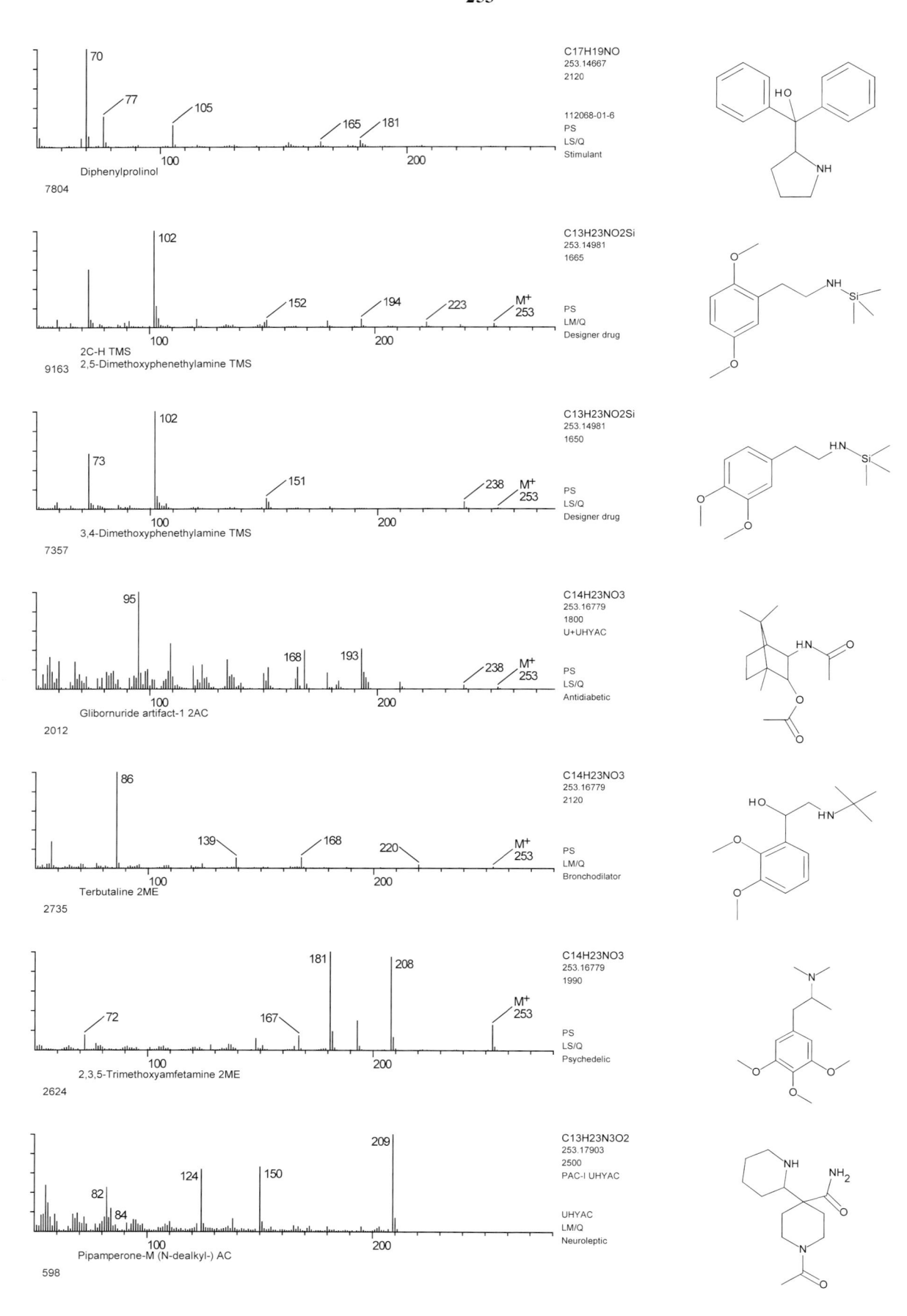

Diphenylprolinol — peaks 70, 77, 105, 165, 181	C17H19NO 253.14667 2120 112068-01-6 PS LS/Q Stimulant
7804	

2C-H TMS
2,5-Dimethoxyphenethylamine TMS — peaks 102, 152, 194, 223, 253 M+

C13H23NO2Si
253.14981
1665
PS
LM/Q
Designer drug

9163

3,4-Dimethoxyphenethylamine TMS — peaks 73, 102, 151, 238, 253 M+

C13H23NO2Si
253.14981
1650
PS
LS/Q
Designer drug

7357

Glibornuride artifact-1 2AC — peaks 95, 168, 193, 238, 253 M+

C14H23NO3
253.16779
1800
U+UHYAC
PS
LS/Q
Antidiabetic

2012

Terbutaline 2ME — peaks 86, 139, 168, 220, 253 M+

C14H23NO3
253.16779
2120
PS
LM/Q
Bronchodilator

2735

2,3,5-Trimethoxyamfetamine 2ME — peaks 72, 167, 181, 208, 253 M+

C14H23NO3
253.16779
1990
PS
LS/Q
Psychedelic

2624

Pipamperone-M (N-dealkyl-) AC — peaks 82, 84, 124, 150, 209

C13H23N3O2
253.17903
2500
PAC-I UHYAC
UHYAC
LM/Q
Neuroleptic

598

253

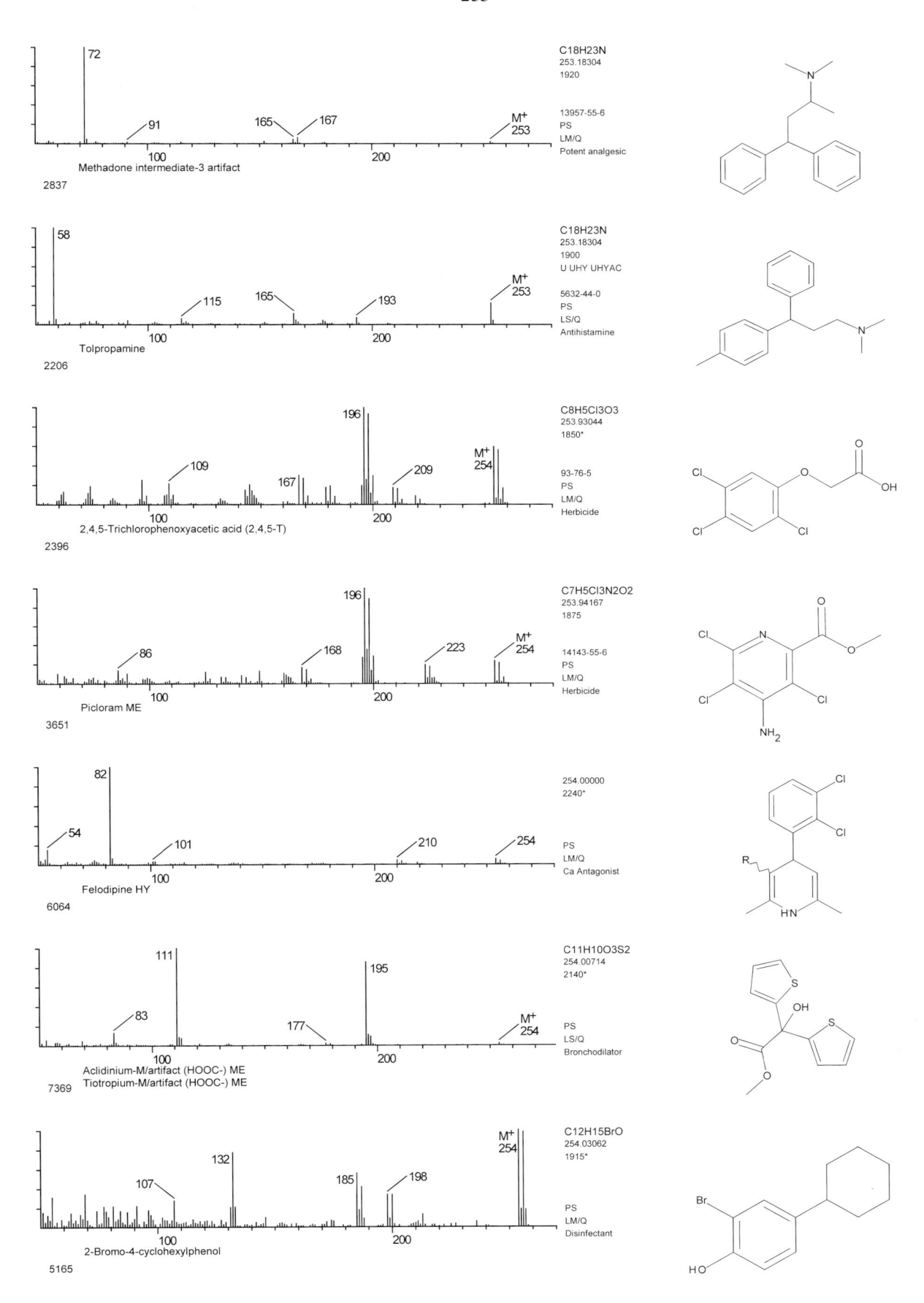

Methadone intermediate-3 artifact — 2837
C18H23N, 253.18304, 1920
13957-55-6, PS, LM/Q
Potent analgesic
Peaks: 72, 91, 165, 167, M+ 253

Tolpropamine — 2206
C18H23N, 253.18304, 1900
U UHY UHYAC
5632-44-0, PS, LS/Q
Antihistamine
Peaks: 58, 115, 165, 193, M+ 253

2,4,5-Trichlorophenoxyacetic acid (2,4,5-T) — 2396
C8H5Cl3O3, 253.93044, 1850*
93-76-5, PS, LM/Q
Herbicide
Peaks: 109, 167, 196, 209, M+ 254

Picloram ME — 3651
C7H5Cl3N2O2, 253.94167, 1875
14143-55-6, PS, LM/Q
Herbicide
Peaks: 86, 168, 196, 223, M+ 254

Felodipine HY — 6064
254.00000, 2240*
PS, LM/Q
Ca Antagonist
Peaks: 54, 82, 101, 210, 254

Aclidinium-M/artifact (HOOC-) ME / Tiotropium-M/artifact (HOOC-) ME — 7369
C11H10O3S2, 254.00714, 2140*
PS, LS/Q
Bronchodilator
Peaks: 83, 111, 177, 195, M+ 254

2-Bromo-4-cyclohexylphenol — 5165
C12H15BrO, 254.03062, 1915*
PS, LM/Q
Disinfectant
Peaks: 107, 132, 185, 198, M+ 254

254

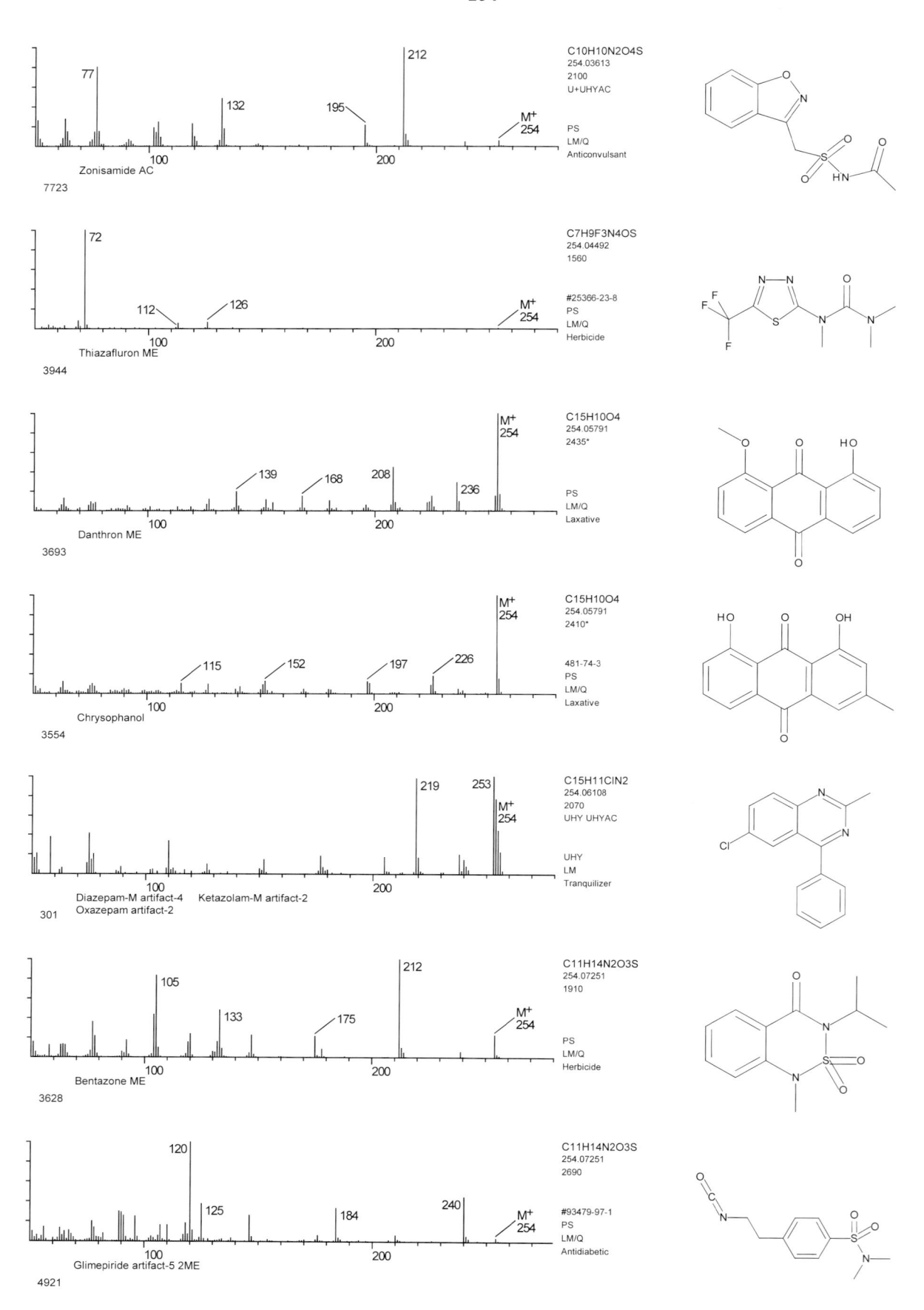

C10H10N2O4S
254.03613
2100
U+UHYAC

PS
LM/Q
Anticonvulsant

Zonisamide AC
7723

C7H9F3N4OS
254.04492
1560

#25366-23-8
PS
LM/Q
Herbicide

Thiazafluron ME
3944

C15H10O4
254.05791
2435*

PS
LM/Q
Laxative

Danthron ME
3693

C15H10O4
254.05791
2410*

481-74-3
PS
LM/Q
Laxative

Chrysophanol
3554

C15H11ClN2
254.06108
2070
UHY UHYAC

UHY
LM
Tranquilizer

Diazepam-M artifact-4 Ketazolam-M artifact-2
Oxazepam artifact-2
301

C11H14N2O3S
254.07251
1910

PS
LM/Q
Herbicide

Bentazone ME
3628

C11H14N2O3S
254.07251
2690

#93479-97-1
PS
LM/Q
Antidiabetic

Glimepiride artifact-5 2ME
4921

626

254

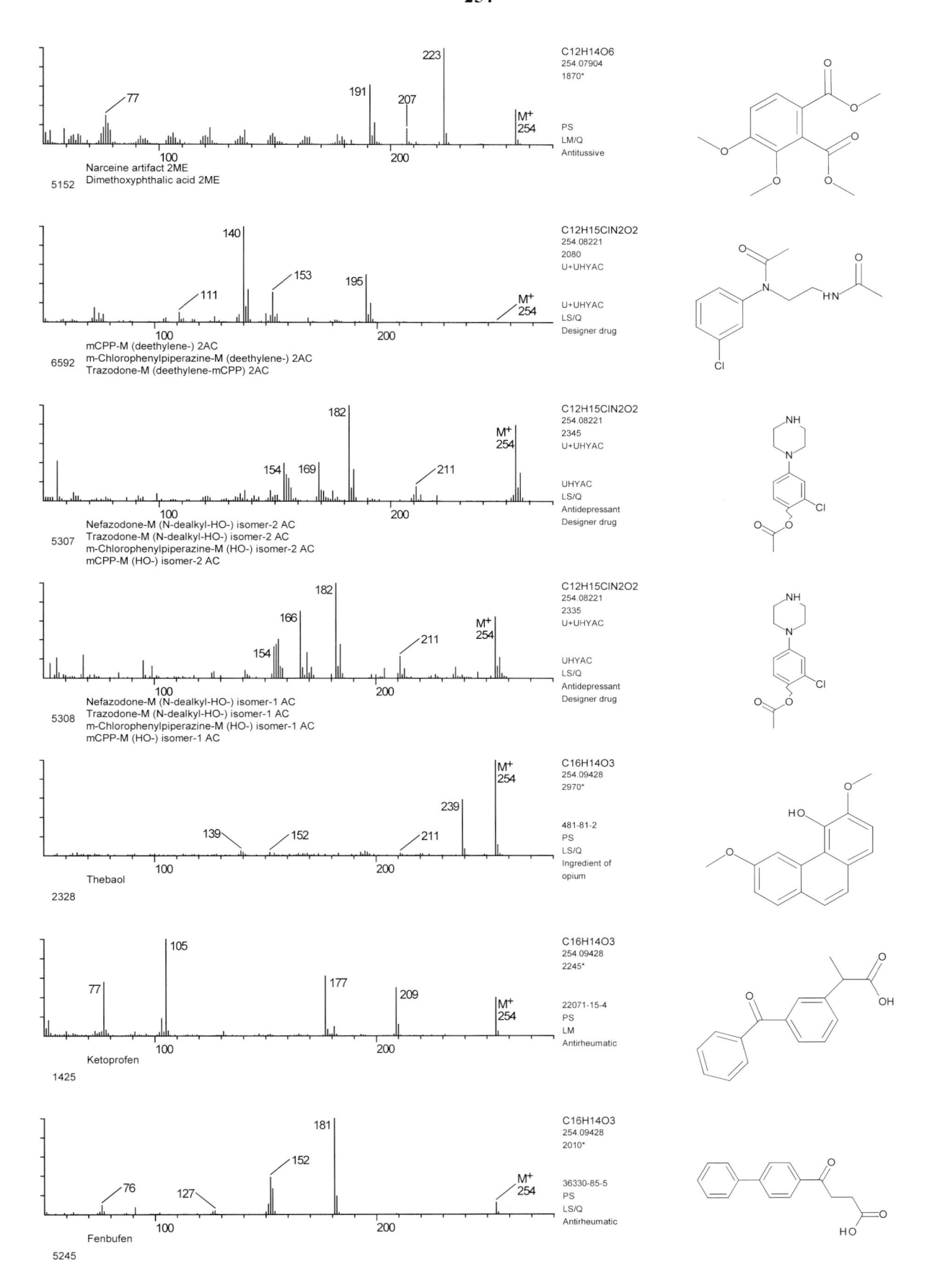

5152 Narceine artifact 2ME / Dimethoxyphthalic acid 2ME
C12H14O6
254.07904
1870*

PS
LM/Q
Antitussive

6592 mCPP-M (deethylene-) 2AC / m-Chlorophenylpiperazine-M (deethylene-) 2AC / Trazodone-M (deethylene-mCPP) 2AC
C12H15ClN2O2
254.08221
2080
U+UHYAC

U+UHYAC
LS/Q
Designer drug

5307 Nefazodone-M (N-dealkyl-HO-) isomer-2 AC / Trazodone-M (N-dealkyl-HO-) isomer-2 AC / m-Chlorophenylpiperazine-M (HO-) isomer-2 AC / mCPP-M (HO-) isomer-2 AC
C12H15ClN2O2
254.08221
2345
U+UHYAC

UHYAC
LS/Q
Antidepressant
Designer drug

5308 Nefazodone-M (N-dealkyl-HO-) isomer-1 AC / Trazodone-M (N-dealkyl-HO-) isomer-1 AC / m-Chlorophenylpiperazine-M (HO-) isomer-1 AC / mCPP-M (HO-) isomer-1 AC
C12H15ClN2O2
254.08221
2335
U+UHYAC

UHYAC
LS/Q
Antidepressant
Designer drug

2328 Thebaol
C16H14O3
254.09428
2970*

481-81-2
PS
LS/Q
Ingredient of opium

1425 Ketoprofen
C16H14O3
254.09428
2245*

22071-15-4
PS
LM
Antirheumatic

5245 Fenbufen
C16H14O3
254.09428
2010*

36330-85-5
PS
LS/Q
Antirheumatic

254

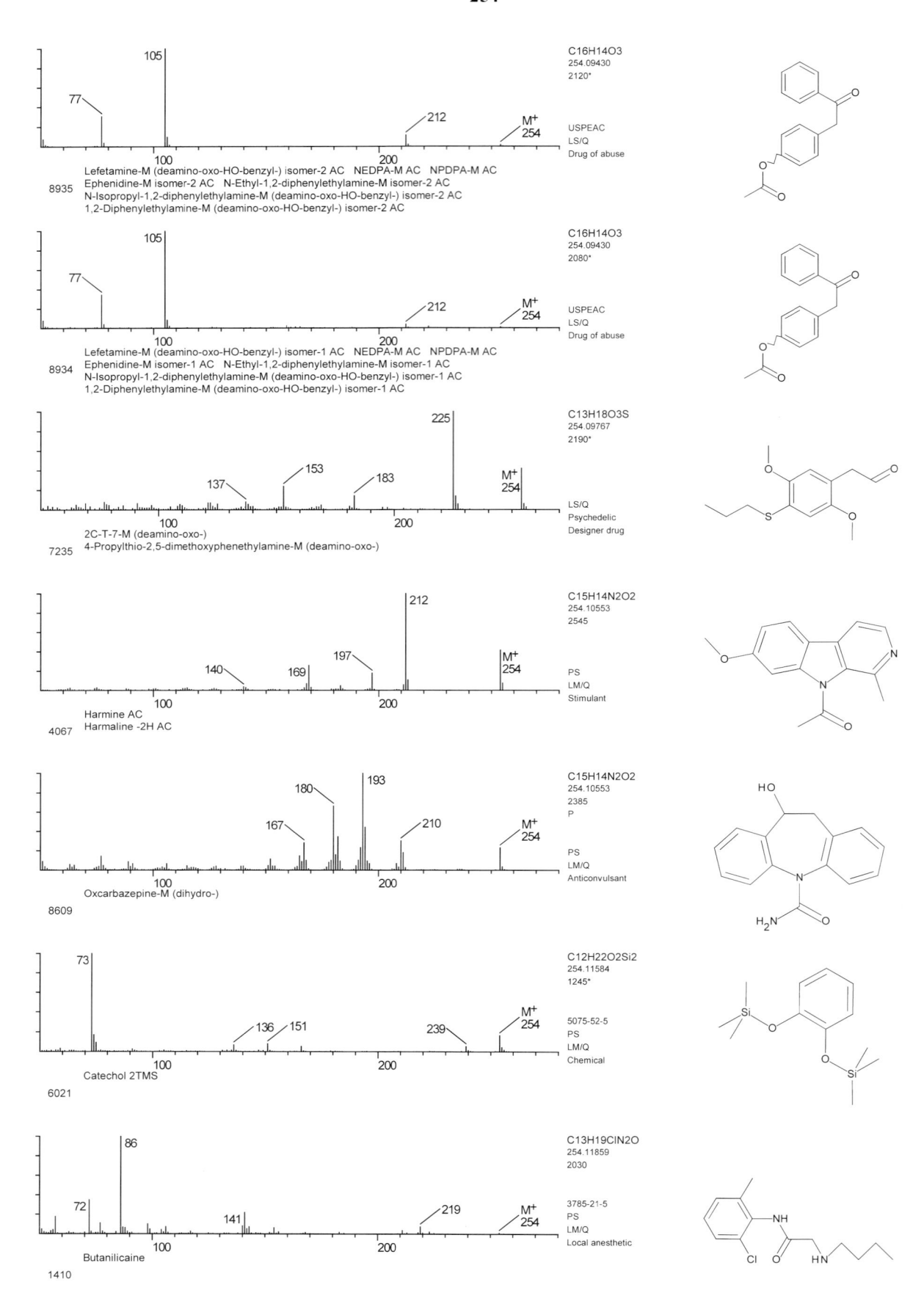

8935
Lefetamine-M (deamino-oxo-HO-benzyl-) isomer-2 AC   NEDPA-M AC   NPDPA-M AC
Ephenidine-M isomer-2 AC   N-Ethyl-1,2-diphenylethylamine-M isomer-2 AC
N-Isopropyl-1,2-diphenylethylamine-M (deamino-oxo-HO-benzyl-) isomer-2 AC
1,2-Diphenylethylamine-M (deamino-oxo-HO-benzyl-) isomer-2 AC

C16H14O3
254.09430
2120*

USPEAC
LS/Q
Drug of abuse

8934
Lefetamine-M (deamino-oxo-HO-benzyl-) isomer-1 AC   NEDPA-M AC   NPDPA-M AC
Ephenidine-M isomer-1 AC   N-Ethyl-1,2-diphenylethylamine-M isomer-1 AC
N-Isopropyl-1,2-diphenylethylamine-M (deamino-oxo-HO-benzyl-) isomer-1 AC
1,2-Diphenylethylamine-M (deamino-oxo-HO-benzyl-) isomer-1 AC

C16H14O3
254.09430
2080*

USPEAC
LS/Q
Drug of abuse

7235
2C-T-7-M (deamino-oxo-)
4-Propylthio-2,5-dimethoxyphenethylamine-M (deamino-oxo-)

C13H18O3S
254.09767
2190*

LS/Q
Psychedelic
Designer drug

4067
Harmine AC
Harmaline -2H AC

C15H14N2O2
254.10553
2545

PS
LM/Q
Stimulant

8609
Oxcarbazepine-M (dihydro-)

C15H14N2O2
254.10553
2385
P

PS
LM/Q
Anticonvulsant

6021
Catechol 2TMS

C12H22O2Si2
254.11584
1245*

5075-52-5
PS
LM/Q
Chemical

1410
Butanilicaine

C13H19ClN2O
254.11859
2030

3785-21-5
PS
LM/Q
Local anesthetic

254

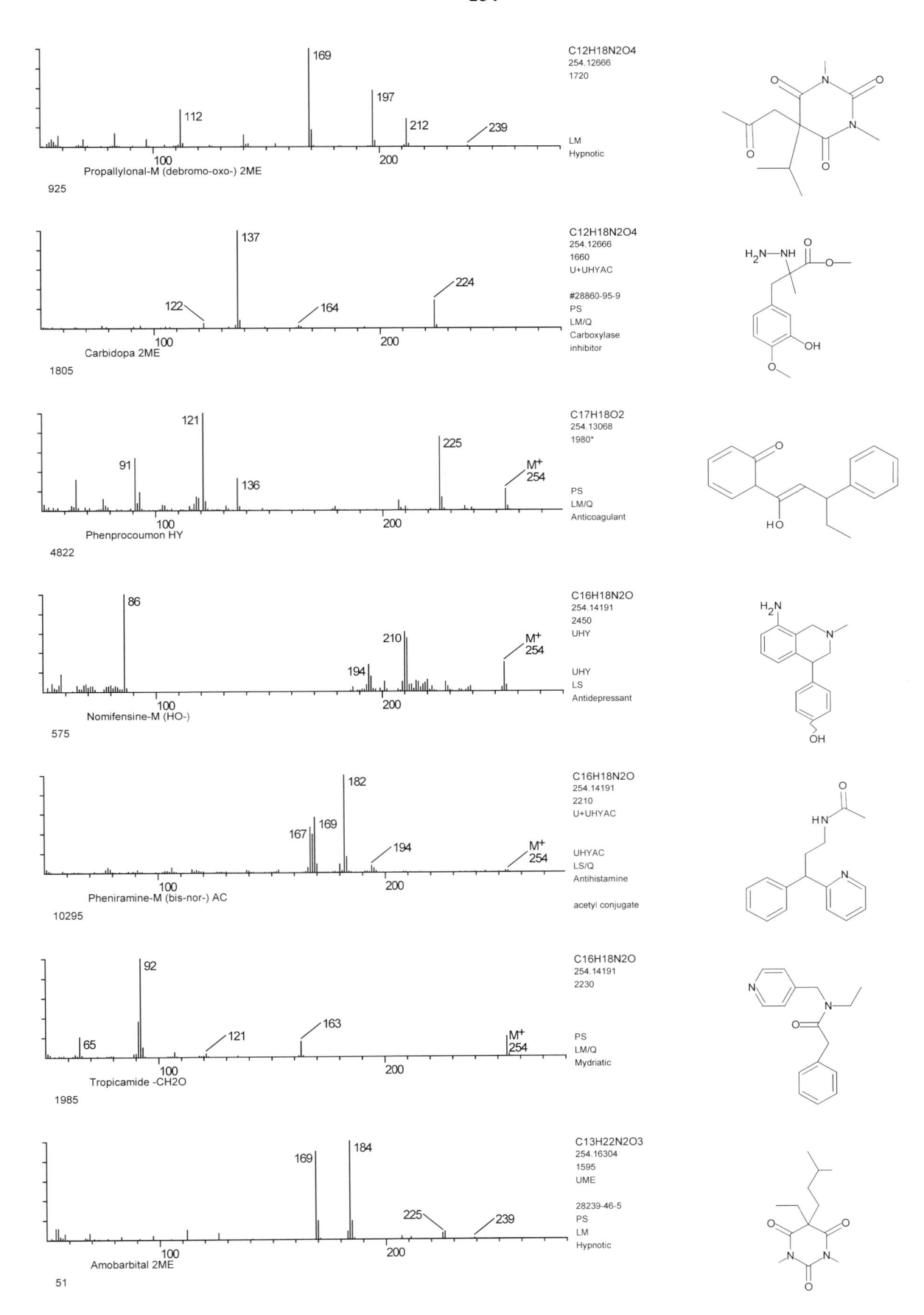

C12H18N2O4
254.12666
1720

LM
Hypnotic

Propallylonal-M (debromo-oxo-) 2ME
925

C12H18N2O4
254.12666
1660
U+UHYAC

#28860-95-9
PS
LM/Q
Carboxylase inhibitor

Carbidopa 2ME
1805

C17H18O2
254.13068
1980*

PS
LM/Q
Anticoagulant

Phenprocoumon HY
4822

C16H18N2O
254.14191
2450
UHY

UHY
LS
Antidepressant

Nomifensine-M (HO-)
575

C16H18N2O
254.14191
2210
U+UHYAC

UHYAC
LS/Q
Antihistamine

acetyl conjugate

Pheniramine-M (bis-nor-) AC
10295

C16H18N2O
254.14191
2230

PS
LM/Q
Mydriatic

Tropicamide -CH2O
1985

C13H22N2O3
254.16304
1595
UME

28239-46-5
PS
LM
Hypnotic

Amobarbital 2ME
51

629

254

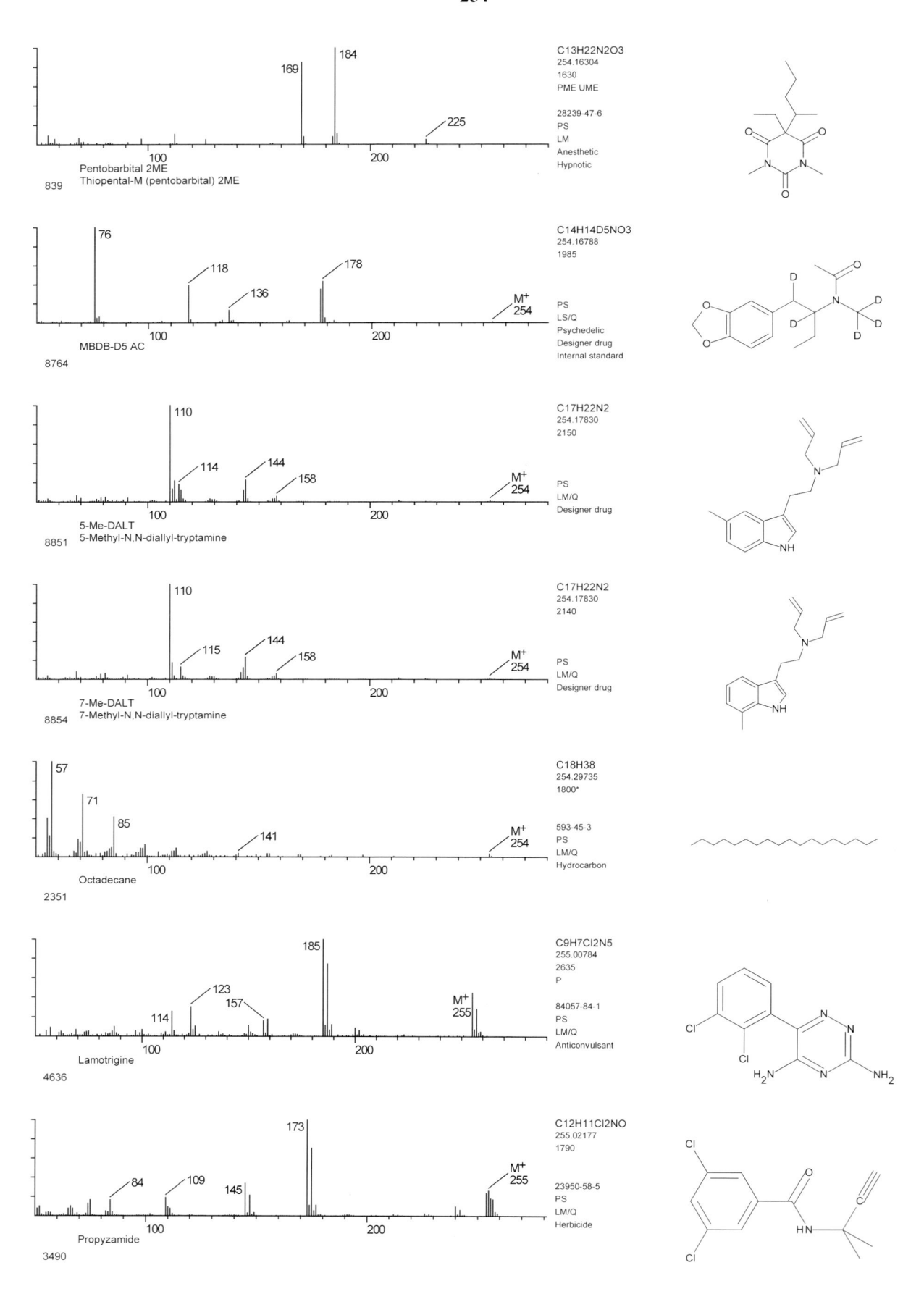

Pentobarbital 2ME
Thiopental-M (pentobarbital) 2ME
839

C13H22N2O3
254.16304
1630
PME UME

28239-47-6
PS
LM
Anesthetic
Hypnotic

MBDB-D5 AC
8764

C14H14D5NO3
254.16788
1985

PS
LS/Q
Psychedelic
Designer drug
Internal standard

5-Me-DALT
5-Methyl-N,N-diallyl-tryptamine
8851

C17H22N2
254.17830
2150

PS
LM/Q
Designer drug

7-Me-DALT
7-Methyl-N,N-diallyl-tryptamine
8854

C17H22N2
254.17830
2140

PS
LM/Q
Designer drug

Octadecane
2351

C18H38
254.29735
1800*

593-45-3
PS
LM/Q
Hydrocarbon

Lamotrigine
4636

C9H7Cl2N5
255.00784
2635
P

84057-84-1
PS
LM/Q
Anticonvulsant

Propyzamide
3490

C12H11Cl2NO
255.02177
1790

23950-58-5
PS
LM/Q
Herbicide

255

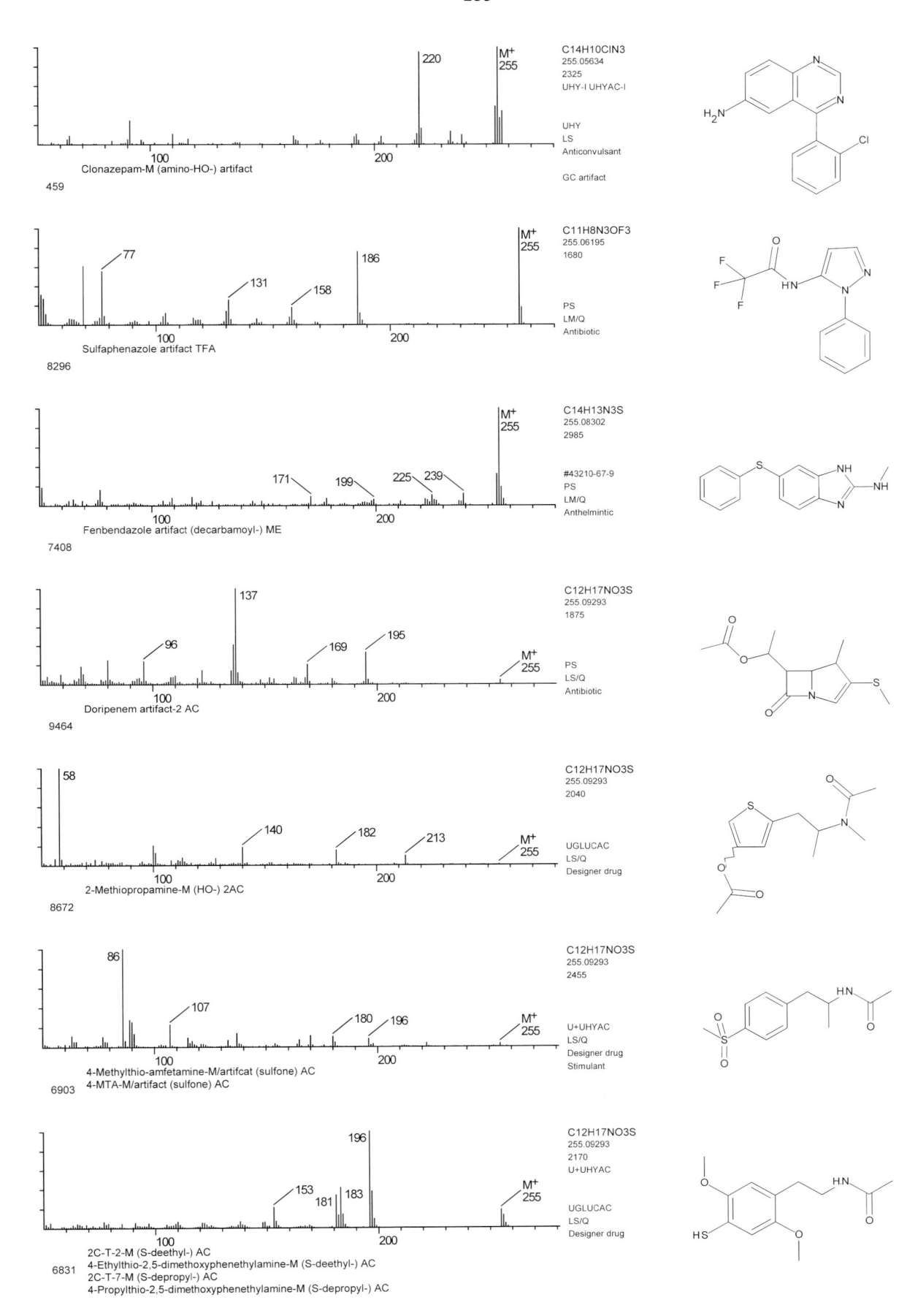

C14H10ClN3
255.05634
2325
UHY-I UHYAC-I

UHY
LS
Anticonvulsant

GC artifact

Clonazepam-M (amino-HO-) artifact
459

C11H8N3OF3
255.06195
1680

PS
LM/Q
Antibiotic

Sulfaphenazole artifact TFA
8296

C14H13N3S
255.08302
2985

#43210-67-9
PS
LM/Q
Anthelmintic

Fenbendazole artifact (decarbamoyl-) ME
7408

C12H17NO3S
255.09293
1875

PS
LS/Q
Antibiotic

Doripenem artifact-2 AC
9464

C12H17NO3S
255.09293
2040

UGLUCAC
LS/Q
Designer drug

2-Methiopropamine-M (HO-) 2AC
8672

C12H17NO3S
255.09293
2455

U+UHYAC
LS/Q
Designer drug
Stimulant

4-Methylthio-amfetamine-M/artifcat (sulfone) AC
4-MTA-M/artifact (sulfone) AC
6903

C12H17NO3S
255.09293
2170
U+UHYAC

UGLUCAC
LS/Q
Designer drug

2C-T-2-M (S-deethyl-) AC
4-Ethylthio-2,5-dimethoxyphenethylamine-M (S-deethyl-) AC
2C-T-7-M (S-depropyl-) AC
4-Propylthio-2,5-dimethoxyphenethylamine-M (S-depropyl-) AC
6831

255

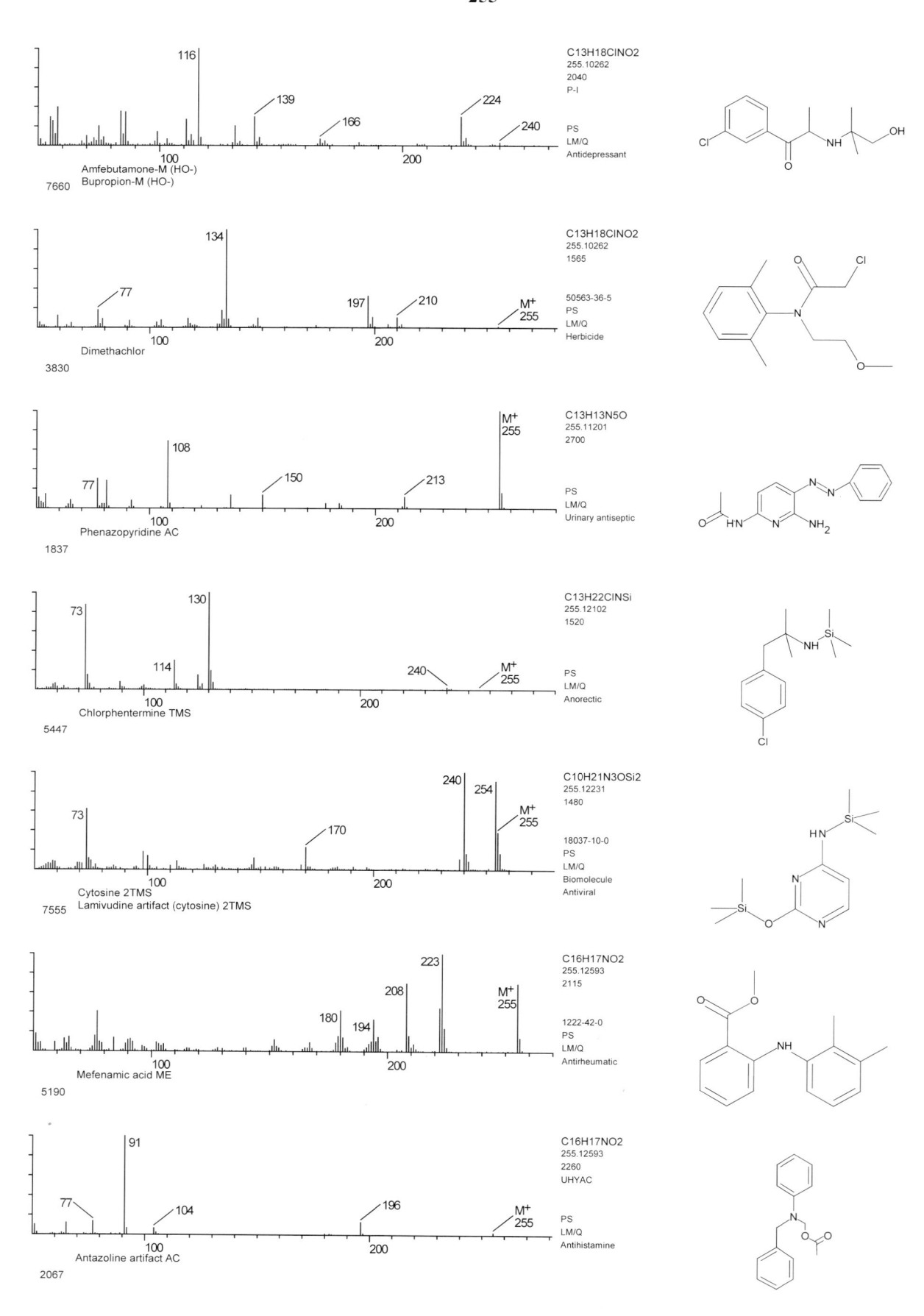

7660	Amfebutamone-M (HO-) / Bupropion-M (HO-) peaks: 116, 139, 166, 224, 240	C13H18ClNO2 255.10262 2040 P-I PS LM/Q Antidepressant
3830	Dimethachlor peaks: 77, 134, 197, 210, M+ 255	C13H18ClNO2 255.10262 1565 50563-36-5 PS LM/Q Herbicide
1837	Phenazopyridine AC peaks: 77, 108, 150, 213, M+ 255	C13H13N5O 255.11201 2700 PS LM/Q Urinary antiseptic
5447	Chlorphentermine TMS peaks: 73, 114, 130, 240, M+ 255	C13H22ClNSi 255.12102 1520 PS LM/Q Anorectic
7555	Cytosine 2TMS / Lamivudine artifact (cytosine) 2TMS peaks: 73, 170, 240, 254, M+ 255	C10H21N3OSi2 255.12231 1480 18037-10-0 PS LM/Q Biomolecule Antiviral
5190	Mefenamic acid ME peaks: 180, 194, 208, 223, M+ 255	C16H17NO2 255.12593 2115 1222-42-0 PS LM/Q Antirheumatic
2067	Antazoline artifact AC peaks: 77, 91, 104, 196, M+ 255	C16H17NO2 255.12593 2260 UHYAC PS LM/Q Antihistamine

255

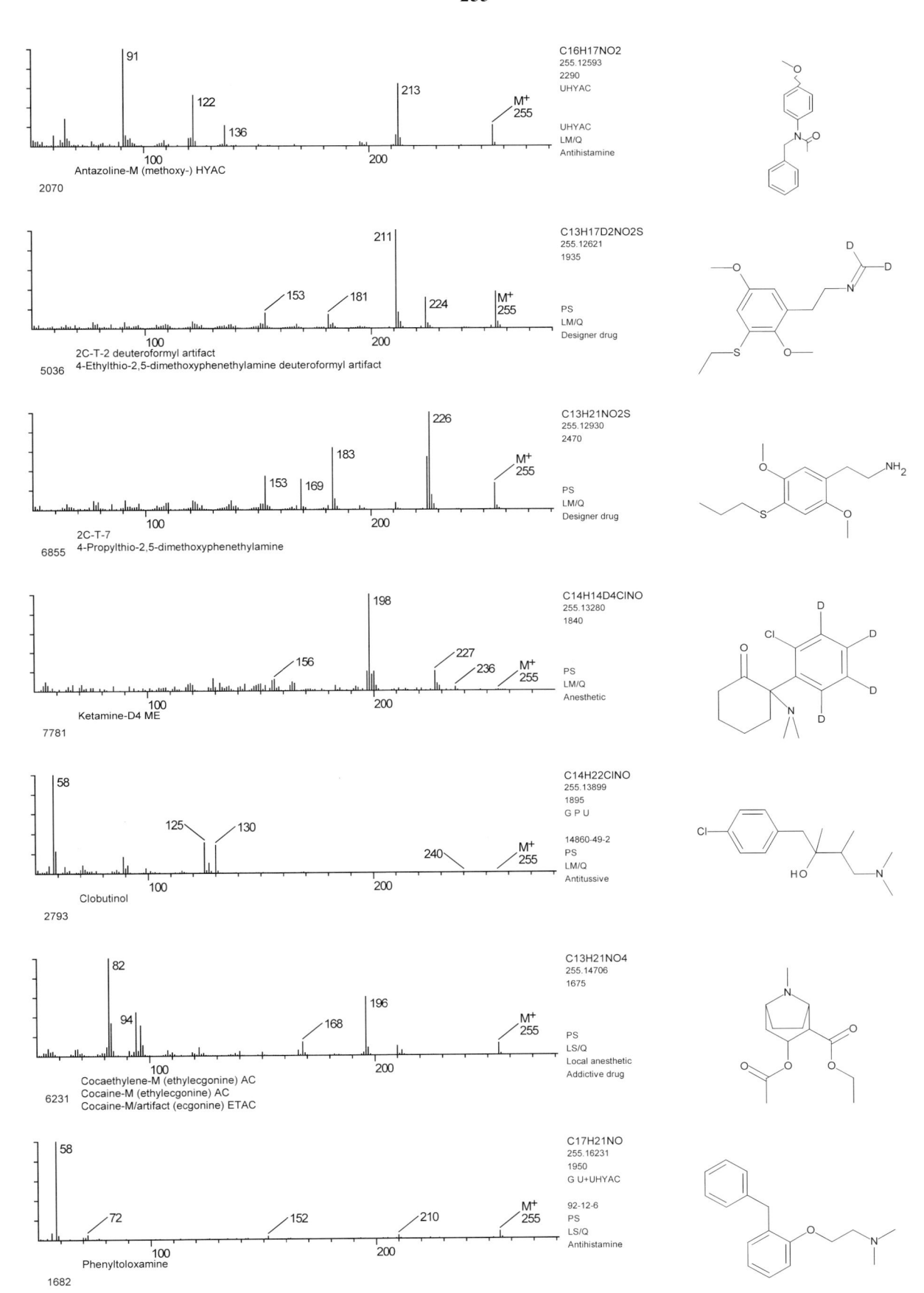

C16H17NO2
255.12593
2290
UHYAC

UHYAC
LM/Q
Antihistamine

Antazoline-M (methoxy-) HYAC
2070

C13H17D2NO2S
255.12621
1935

PS
LM/Q
Designer drug

2C-T-2 deuteroformyl artifact
5036 4-Ethylthio-2,5-dimethoxyphenethylamine deuteroformyl artifact

C13H21NO2S
255.12930
2470

PS
LM/Q
Designer drug

2C-T-7
6855 4-Propylthio-2,5-dimethoxyphenethylamine

C14H14D4ClNO
255.13280
1840

PS
LM/Q
Anesthetic

Ketamine-D4 ME
7781

C14H22ClNO
255.13899
1895
G P U

14860-49-2
PS
LM/Q
Antitussive

Clobutinol
2793

C13H21NO4
255.14706
1675

PS
LS/Q
Local anesthetic
Addictive drug

Cocaethylene-M (ethylecgonine) AC
Cocaine-M (ethylecgonine) AC
6231 Cocaine-M/artifact (ecgonine) ETAC

C17H21NO
255.16231
1950
G U+UHYAC

92-12-6
PS
LS/Q
Antihistamine

Phenyltoloxamine
1682

Orphenadrine-M (nor-)
1160

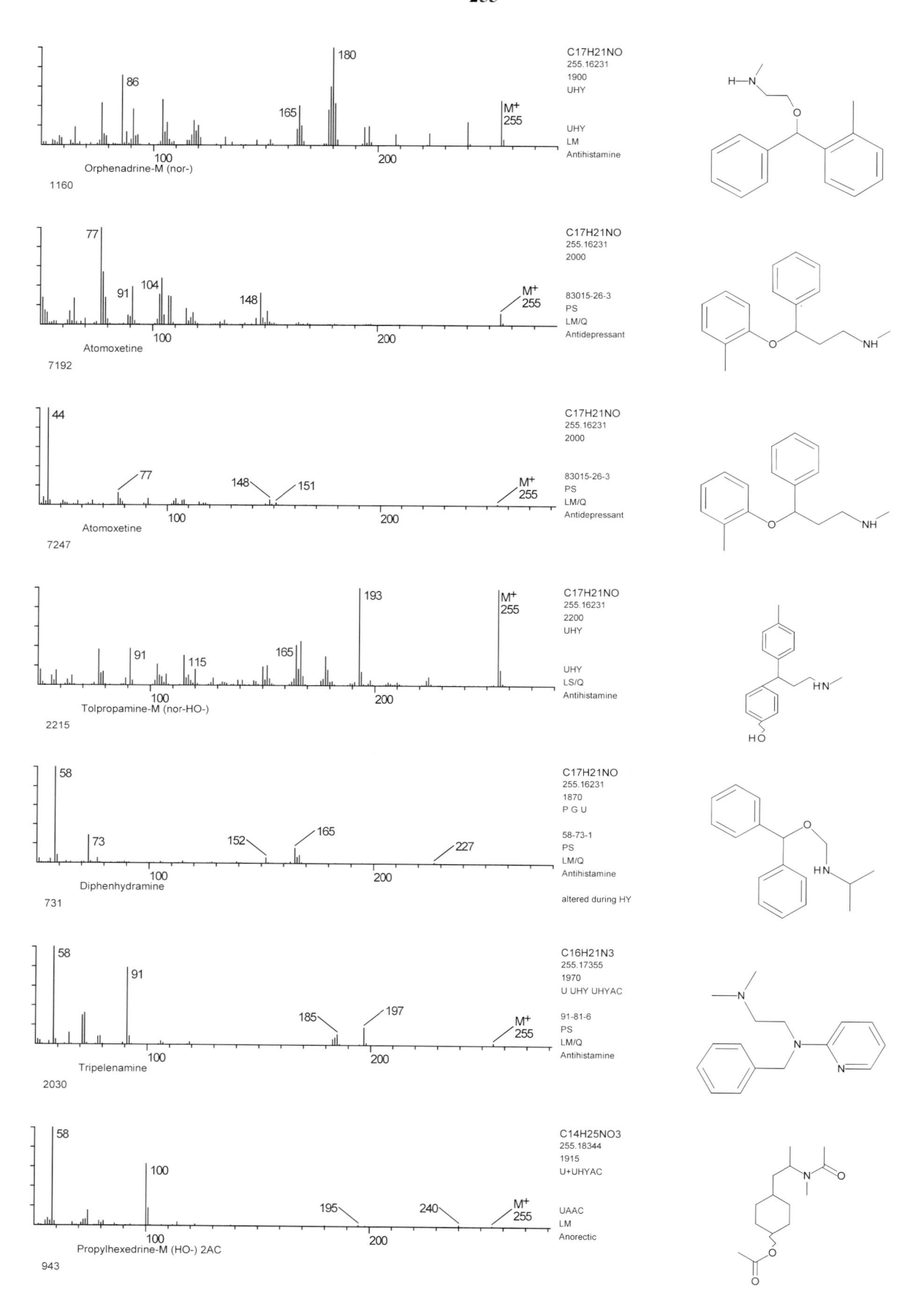

C17H21NO
255.16231
1900
UHY

UHY
LM
Antihistamine

Atomoxetine
7192

C17H21NO
255.16231
2000

83015-26-3
PS
LM/Q
Antidepressant

Atomoxetine
7247

C17H21NO
255.16231
2000

83015-26-3
PS
LM/Q
Antidepressant

Tolpropamine-M (nor-HO-)
2215

C17H21NO
255.16231
2200
UHY

UHY
LS/Q
Antihistamine

Diphenhydramine
731

C17H21NO
255.16231
1870
P G U

58-73-1
PS
LM/Q
Antihistamine

altered during HY

Tripelenamine
2030

C16H21N3
255.17355
1970
U UHY UHYAC

91-81-6
PS
LM/Q
Antihistamine

Propylhexedrine-M (HO-) 2AC
943

C14H25NO3
255.18344
1915
U+UHYAC

UAAC
LM
Anorectic

255

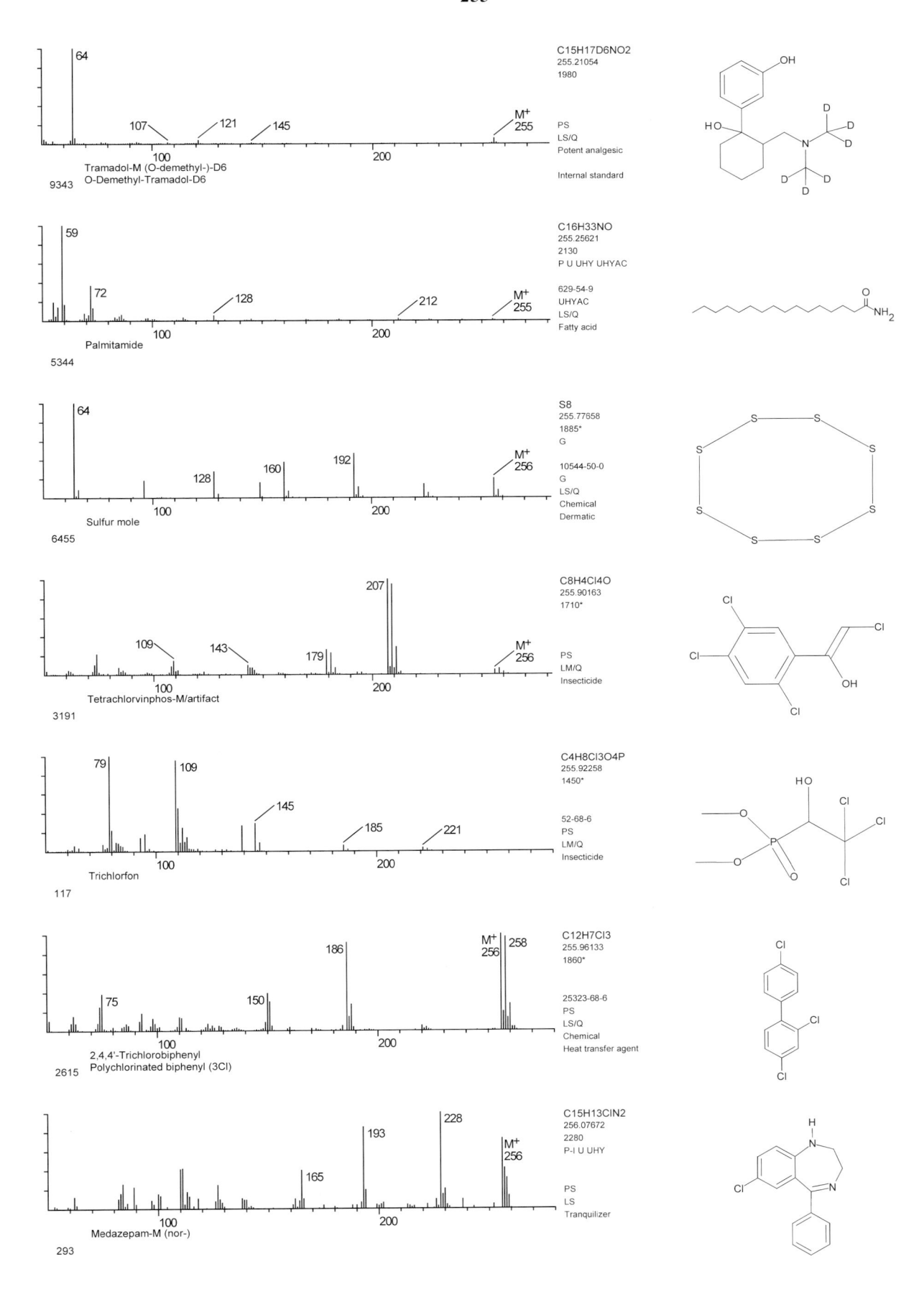

9343 Tramadol-M (O-demethyl-)-D6 / O-Demethyl-Tramadol-D6
Peaks: 64, 107, 121, 145, 255 (M+)
C15H17D6NO2
255.21054
1980
PS
LS/Q
Potent analgesic
Internal standard

5344 Palmitamide
Peaks: 59, 72, 128, 212, 255 (M+)
C16H33NO
255.25621
2130
P U UHY UHYAC
629-54-9
UHYAC
LS/Q
Fatty acid

6455 Sulfur mole
Peaks: 64, 128, 160, 192, 256 (M+)
S8
255.77658
1885*
G
10544-50-0
G
LS/Q
Chemical
Dermatic

3191 Tetrachlorvinphos-M/artifact
Peaks: 109, 143, 179, 207, 256 (M+)
C8H4Cl4O
255.90163
1710*
PS
LM/Q
Insecticide

117 Trichlorfon
Peaks: 79, 109, 145, 185, 221
C4H8Cl3O4P
255.92258
1450*
52-68-6
PS
LM/Q
Insecticide

2615 2,4,4'-Trichlorobiphenyl / Polychlorinated biphenyl (3Cl)
Peaks: 75, 150, 186, 256 (M+), 258
C12H7Cl3
255.96133
1860*
25323-68-6
PS
LS/Q
Chemical
Heat transfer agent

293 Medazepam-M (nor-)
Peaks: 165, 193, 228, 256 (M+)
C15H13ClN2
256.07672
2280
P-I U UHY
PS
LS
Tranquilizer

256

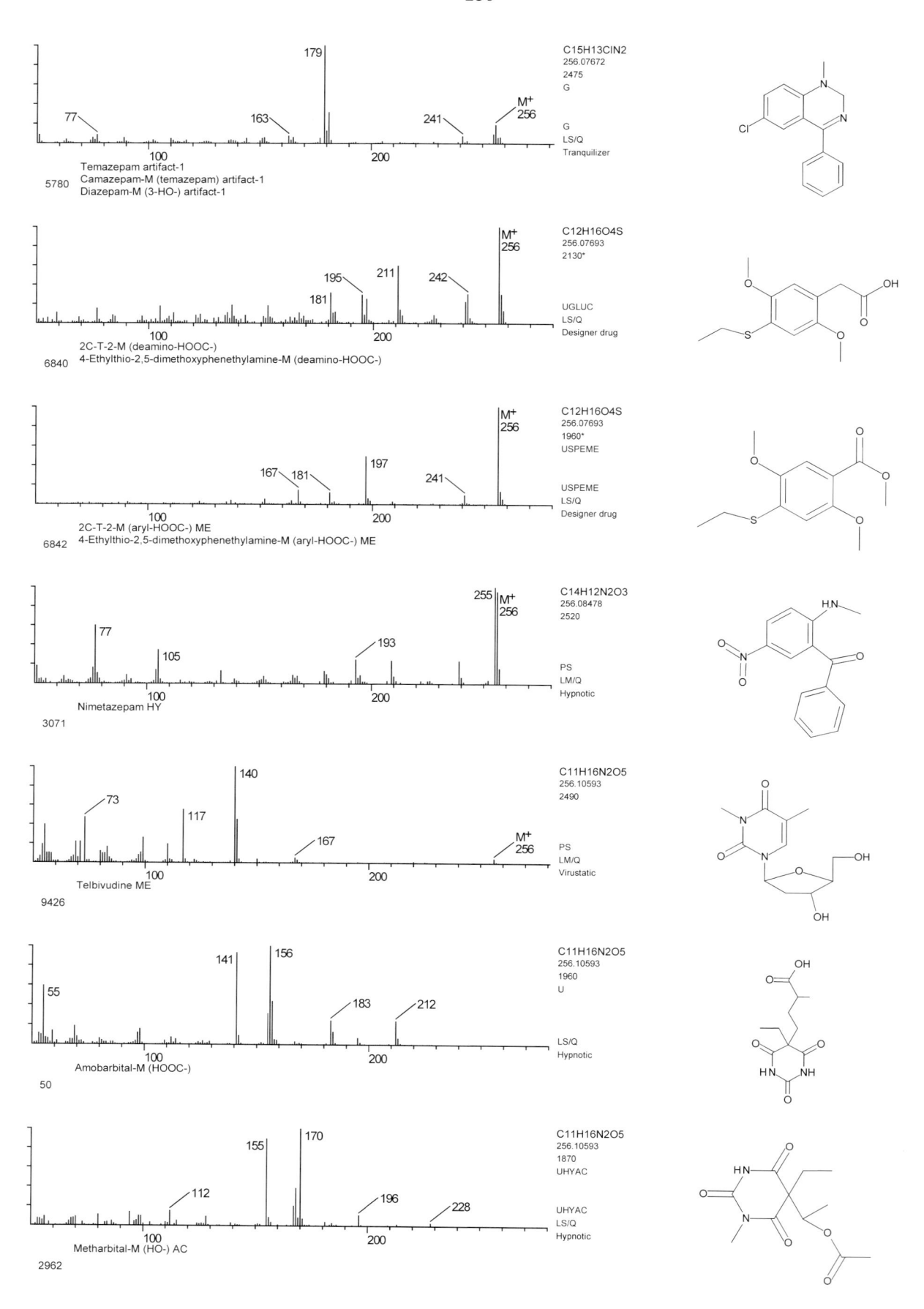

5780	Peaks: 77, 163, 179, 241, 256 (M+) Temazepam artifact-1 Camazepam-M (temazepam) artifact-1 Diazepam-M (3-HO-) artifact-1	C15H13ClN2 256.07672 2475 G G LS/Q Tranquilizer
6840	Peaks: 181, 195, 211, 242, 256 (M+) 2C-T-2-M (deamino-HOOC-) 4-Ethylthio-2,5-dimethoxyphenethylamine-M (deamino-HOOC-)	C12H16O4S 256.07693 2130* UGLUC LS/Q Designer drug
6842	Peaks: 167, 181, 197, 241, 256 (M+) 2C-T-2-M (aryl-HOOC-) ME 4-Ethylthio-2,5-dimethoxyphenethylamine-M (aryl-HOOC-) ME	C12H16O4S 256.07693 1960* USPEME USPEME LS/Q Designer drug
3071	Peaks: 77, 105, 193, 255, 256 (M+) Nimetazepam HY	C14H12N2O3 256.08478 2520 PS LM/Q Hypnotic
9426	Peaks: 73, 117, 140, 167, 256 (M+) Telbivudine ME	C11H16N2O5 256.10593 2490 PS LM/Q Virustatic
50	Peaks: 55, 141, 156, 183, 212 Amobarbital-M (HOOC-)	C11H16N2O5 256.10593 1960 U LS/Q Hypnotic
2962	Peaks: 112, 155, 170, 196, 228 Metharbital-M (HO-) AC	C11H16N2O5 256.10593 1870 UHYAC UHYAC LS/Q Hypnotic

256

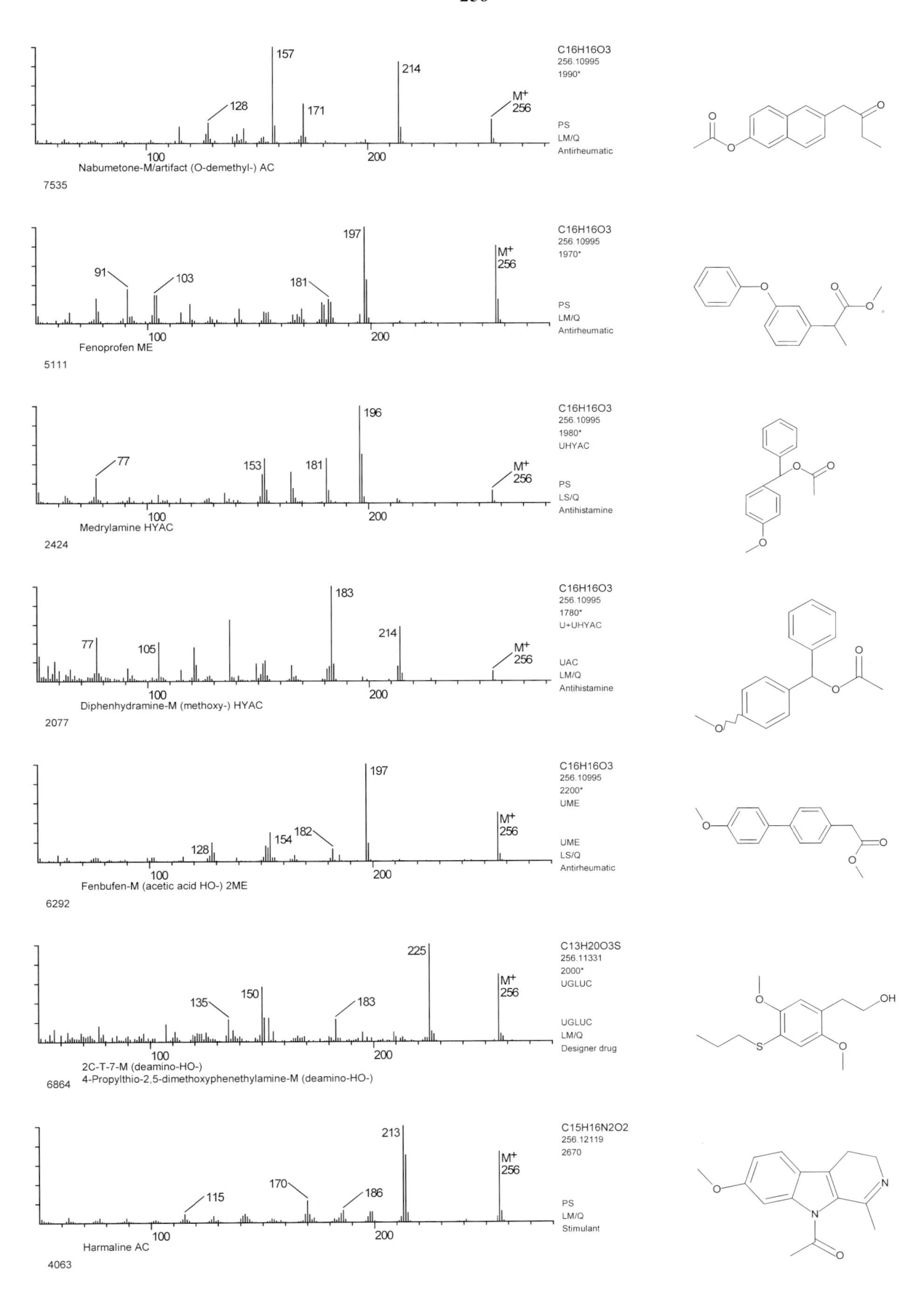

Nabumetone-M/artifact (O-demethyl-) AC
7535
C16H16O3
256.10995
1990*

PS
LM/Q
Antirheumatic

Fenoprofen ME
5111
C16H16O3
256.10995
1970*

PS
LM/Q
Antirheumatic

Medrylamine HYAC
2424
C16H16O3
256.10995
1980*
UHYAC

PS
LS/Q
Antihistamine

Diphenhydramine-M (methoxy-) HYAC
2077
C16H16O3
256.10995
1780*
U+UHYAC

UAC
LM/Q
Antihistamine

Fenbufen-M (acetic acid HO-) 2ME
6292
C16H16O3
256.10995
2200*
UME

UME
LS/Q
Antirheumatic

2C-T-7-M (deamino-HO-)
4-Propylthio-2,5-dimethoxyphenethylamine-M (deamino-HO-)
6864
C13H20O3S
256.11331
2000*
UGLUC

UGLUC
LM/Q
Designer drug

Harmaline AC
4063
C15H16N2O2
256.12119
2670

PS
LM/Q
Stimulant

256

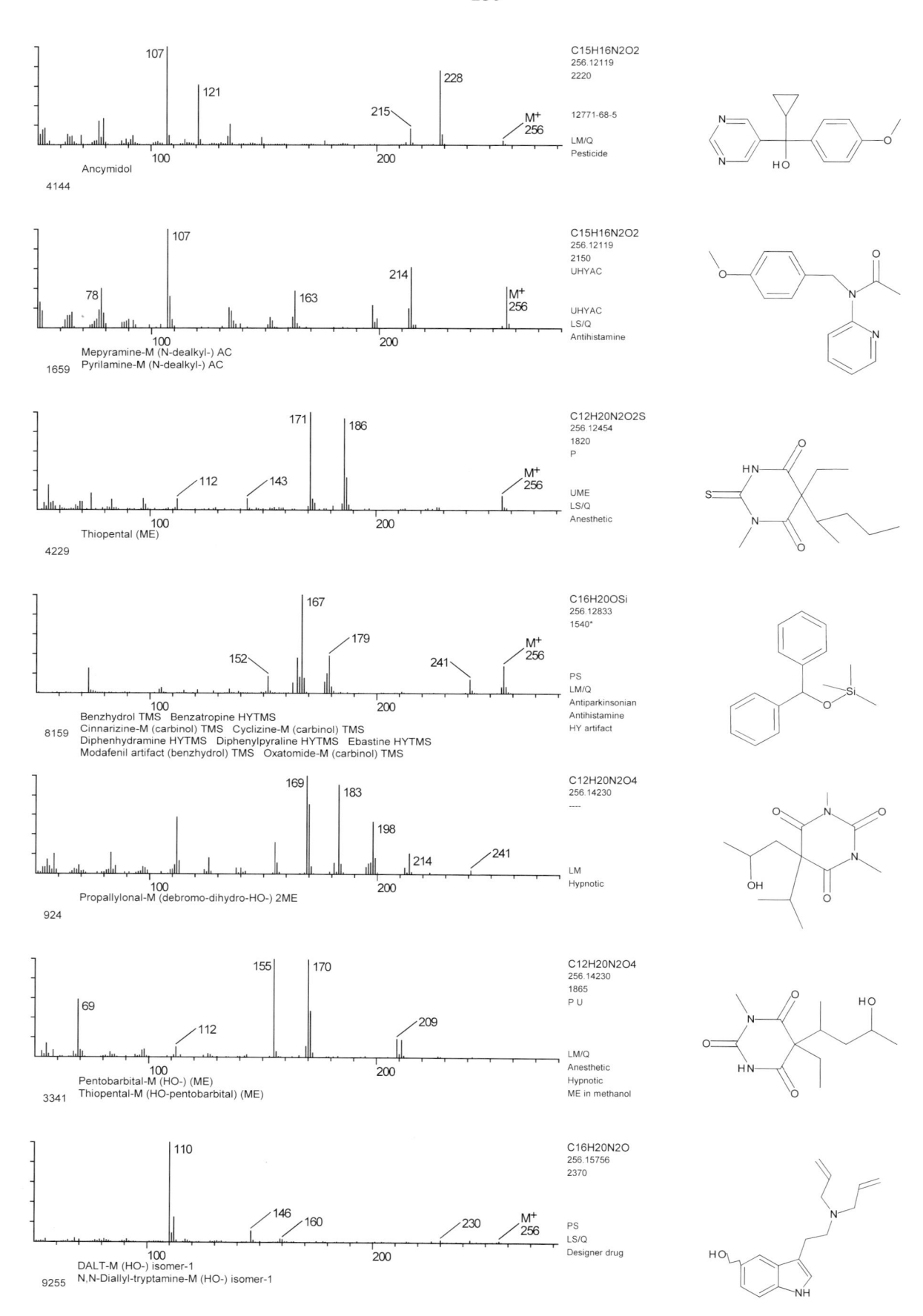

Ancymidol — peaks 107, 121, 215, 228, M+ 256	C15H16N2O2 256.12119 2220 12771-68-5 LM/Q Pesticide

4144

Mepyramine-M (N-dealkyl-) AC Pyrilamine-M (N-dealkyl-) AC — peaks 78, 107, 163, 214, M+ 256	C15H16N2O2 256.12119 2150 UHYAC UHYAC LS/Q Antihistamine

1659

Thiopental (ME) — peaks 112, 143, 171, 186, M+ 256	C12H20N2O2S 256.12454 1820 P UME LS/Q Anesthetic

4229

Benzhydrol TMS Benzatropine HYTMS Cinnarizine-M (carbinol) TMS Cyclizine-M (carbinol) TMS Diphenhydramine HYTMS Diphenylpyraline HYTMS Ebastine HYTMS Modafenil artifact (benzhydrol) TMS Oxatomide-M (carbinol) TMS — peaks 152, 167, 179, 241, M+ 256	C16H20OSi 256.12833 1540* PS LM/Q Antiparkinsonian Antihistamine HY artifact

8159

Propallylonal-M (debromo-dihydro-HO-) 2ME — peaks 169, 183, 198, 214, 241	C12H20N2O4 256.14230 ---- LM Hypnotic

924

Pentobarbital-M (HO-) (ME) Thiopental-M (HO-pentobarbital) (ME) — peaks 69, 112, 155, 170, 209	C12H20N2O4 256.14230 1865 P U LM/Q Anesthetic Hypnotic ME in methanol

3341

DALT-M (HO-) isomer-1 N,N-Diallyl-tryptamine-M (HO-) isomer-1 — peaks 110, 146, 160, 230, M+ 256	C16H20N2O 256.15756 2370 PS LS/Q Designer drug

9255

256

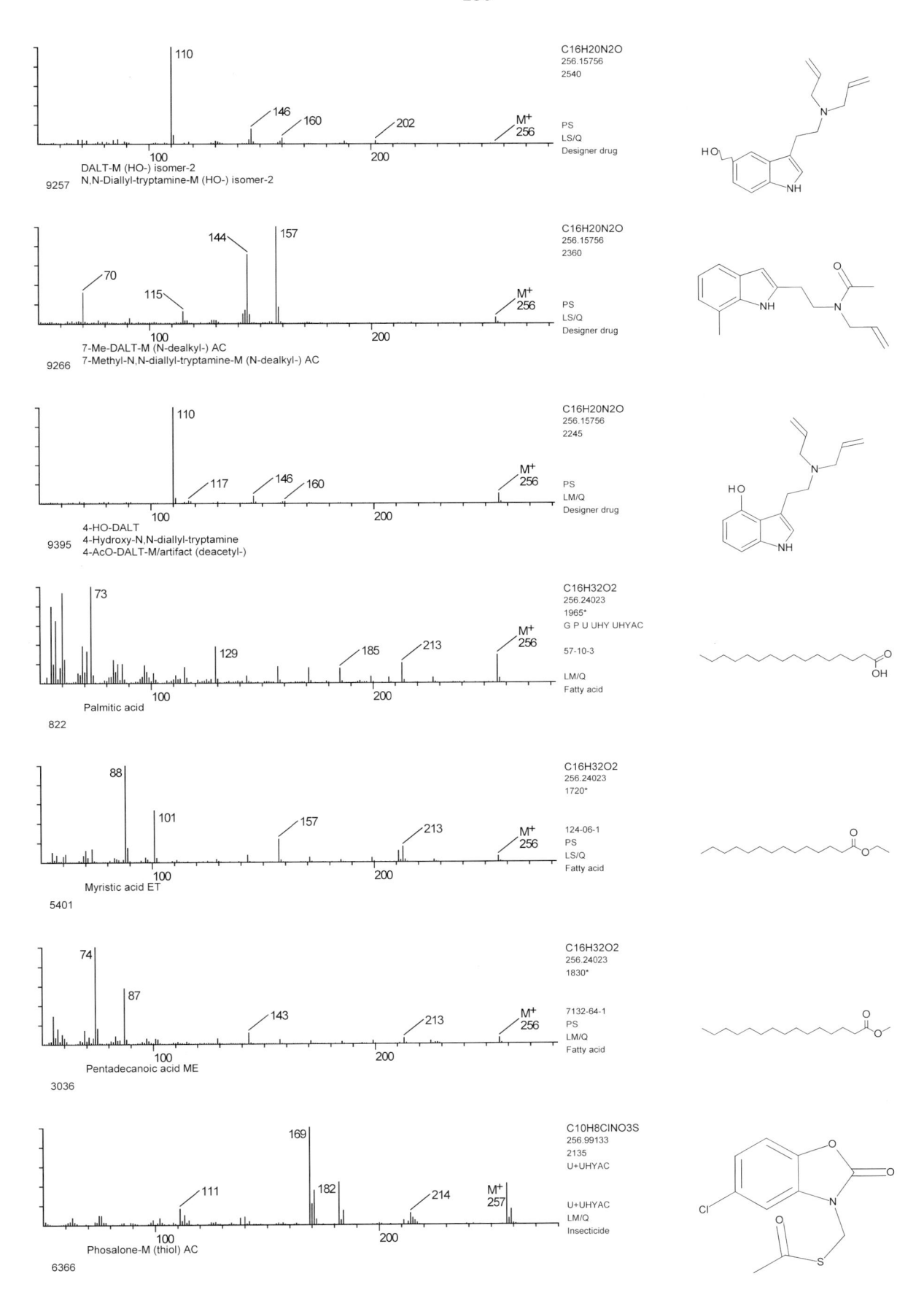

9257 DALT-M (HO-) isomer-2
N,N-Diallyl-tryptamine-M (HO-) isomer-2
C16H20N2O
256.15756
2540
PS
LS/Q
Designer drug

9266 7-Me-DALT-M (N-dealkyl-) AC
7-Methyl-N,N-diallyl-tryptamine-M (N-dealkyl-) AC
C16H20N2O
256.15756
2360
PS
LS/Q
Designer drug

9395 4-HO-DALT
4-Hydroxy-N,N-diallyl-tryptamine
4-AcO-DALT-M/artifact (deacetyl-)
C16H20N2O
256.15756
2245
PS
LM/Q
Designer drug

822 Palmitic acid
C16H32O2
256.24023
1965*
G P U UHY UHYAC
57-10-3
LM/Q
Fatty acid

5401 Myristic acid ET
C16H32O2
256.24023
1720*
124-06-1
PS
LS/Q
Fatty acid

3036 Pentadecanoic acid ME
C16H32O2
256.24023
1830*
7132-64-1
PS
LM/Q
Fatty acid

6366 Phosalone-M (thiol) AC
C10H8ClNO3S
256.99133
2135
U+UHYAC
U+UHYAC
LM/Q
Insecticide

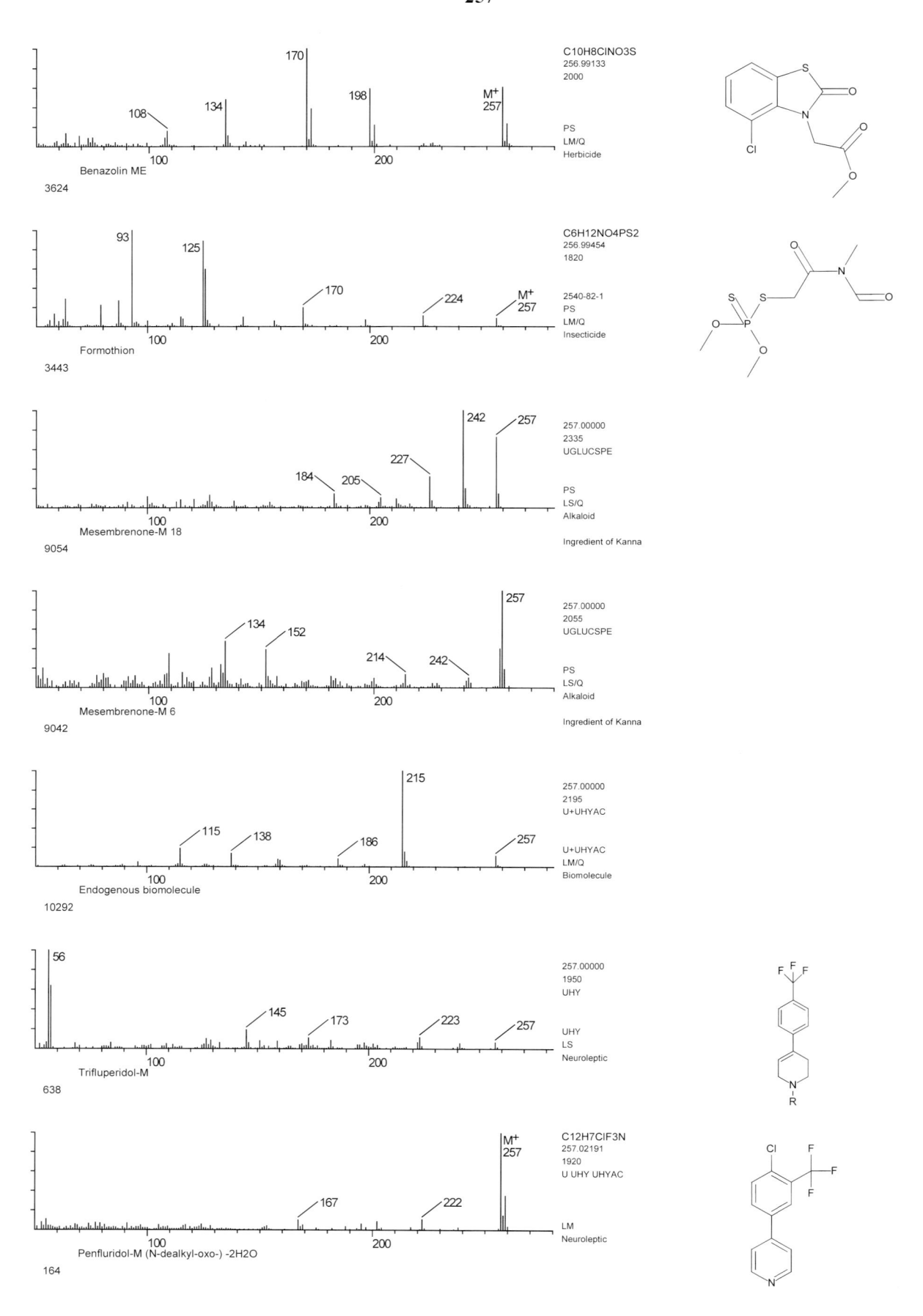

Benazolin ME	C10H8ClNO3S 256.99133 2000 PS LM/Q Herbicide
Formothion	C6H12NO4PS2 256.99454 1820 2540-82-1 PS LM/Q Insecticide
Mesembrenone-M 18	257.00000 2335 UGLUCSPE PS LS/Q Alkaloid Ingredient of Kanna
Mesembrenone-M 6	257.00000 2055 UGLUCSPE PS LS/Q Alkaloid Ingredient of Kanna
Endogenous biomolecule	257.00000 2195 U+UHYAC U+UHYAC LM/Q Biomolecule
Trifluperidol-M	257.00000 1950 UHY UHY LS Neuroleptic
Penfluridol-M (N-dealkyl-oxo-) -2H2O	C12H7ClF3N 257.02191 1920 U UHY UHYAC LM Neuroleptic

257

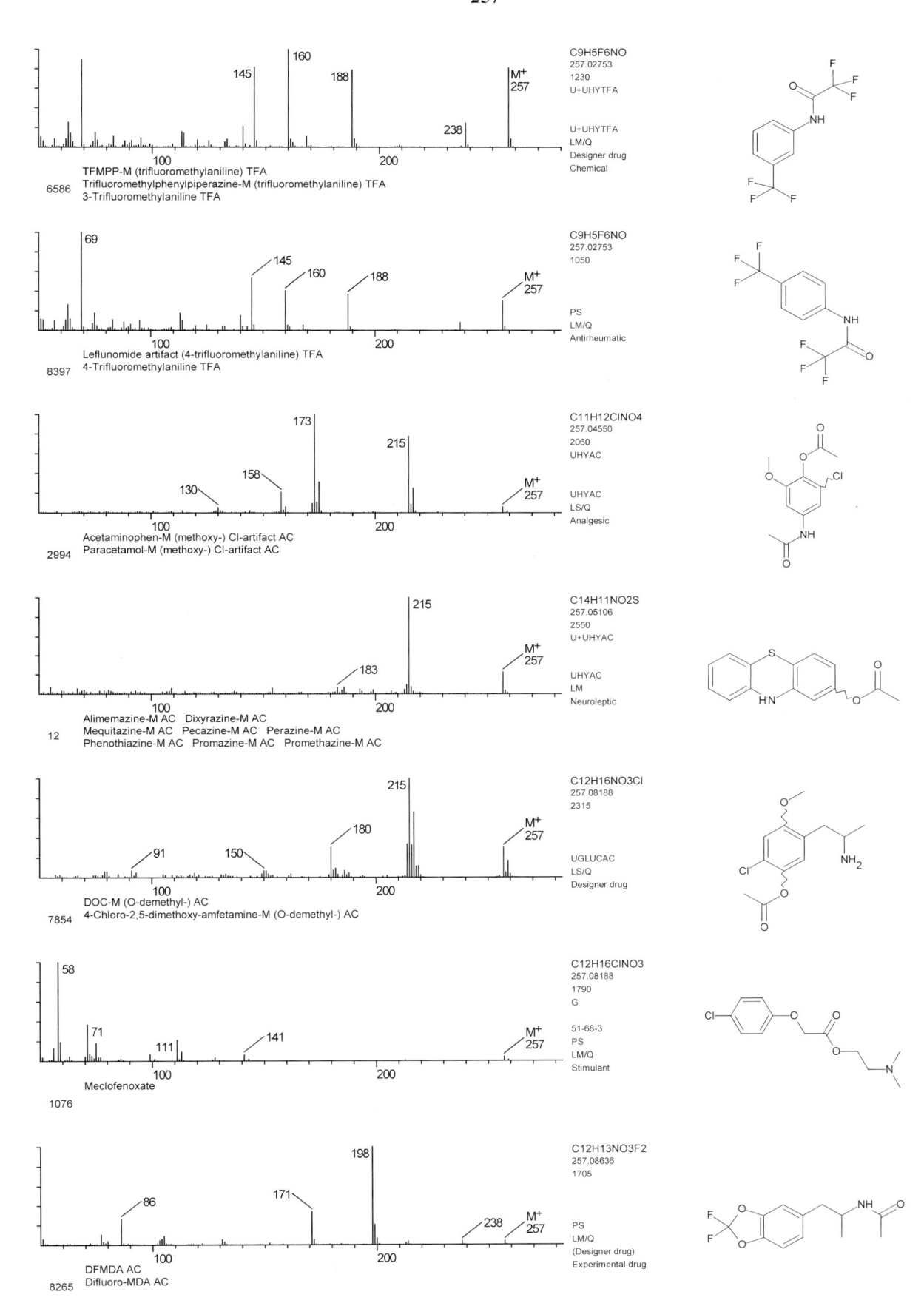

6586	TFMPP-M (trifluoromethylaniline) TFA Trifluoromethylphenylpiperazine-M (trifluoromethylaniline) TFA 3-Trifluoromethylaniline TFA	C9H5F6NO 257.02753 1230 U+UHYTFA U+UHYTFA LM/Q Designer drug Chemical

Peaks: 145, 160, 188, 238, M+ 257

8397	Leflunomide artifact (4-trifluoromethylaniline) TFA 4-Trifluoromethylaniline TFA	C9H5F6NO 257.02753 1050 PS LM/Q Antirheumatic

Peaks: 69, 145, 160, 188, M+ 257

2994	Acetaminophen-M (methoxy-) Cl-artifact AC Paracetamol-M (methoxy-) Cl-artifact AC	C11H12ClNO4 257.04550 2060 UHYAC UHYAC LS/Q Analgesic

Peaks: 130, 158, 173, 215, M+ 257

12	Alimemazine-M AC Dixyrazine-M AC Mequitazine-M AC Pecazine-M AC Perazine-M AC Phenothiazine-M AC Promazine-M AC Promethazine-M AC	C14H11NO2S 257.05106 2550 U+UHYAC UHYAC LM Neuroleptic

Peaks: 183, 215, M+ 257

7854	DOC-M (O-demethyl-) AC 4-Chloro-2,5-dimethoxy-amfetamine-M (O-demethyl-) AC	C12H16NO3Cl 257.08188 2315 UGLUCAC LS/Q Designer drug

Peaks: 91, 150, 180, 215, M+ 257

1076	Meclofenoxate	C12H16ClNO3 257.08188 1790 G 51-68-3 PS LM/Q Stimulant

Peaks: 58, 71, 111, 141, M+ 257

8265	DFMDA AC Difluoro-MDA AC	C12H13NO3F2 257.08636 1705 PS LM/Q (Designer drug) Experimental drug

Peaks: 86, 171, 198, 238, M+ 257

257

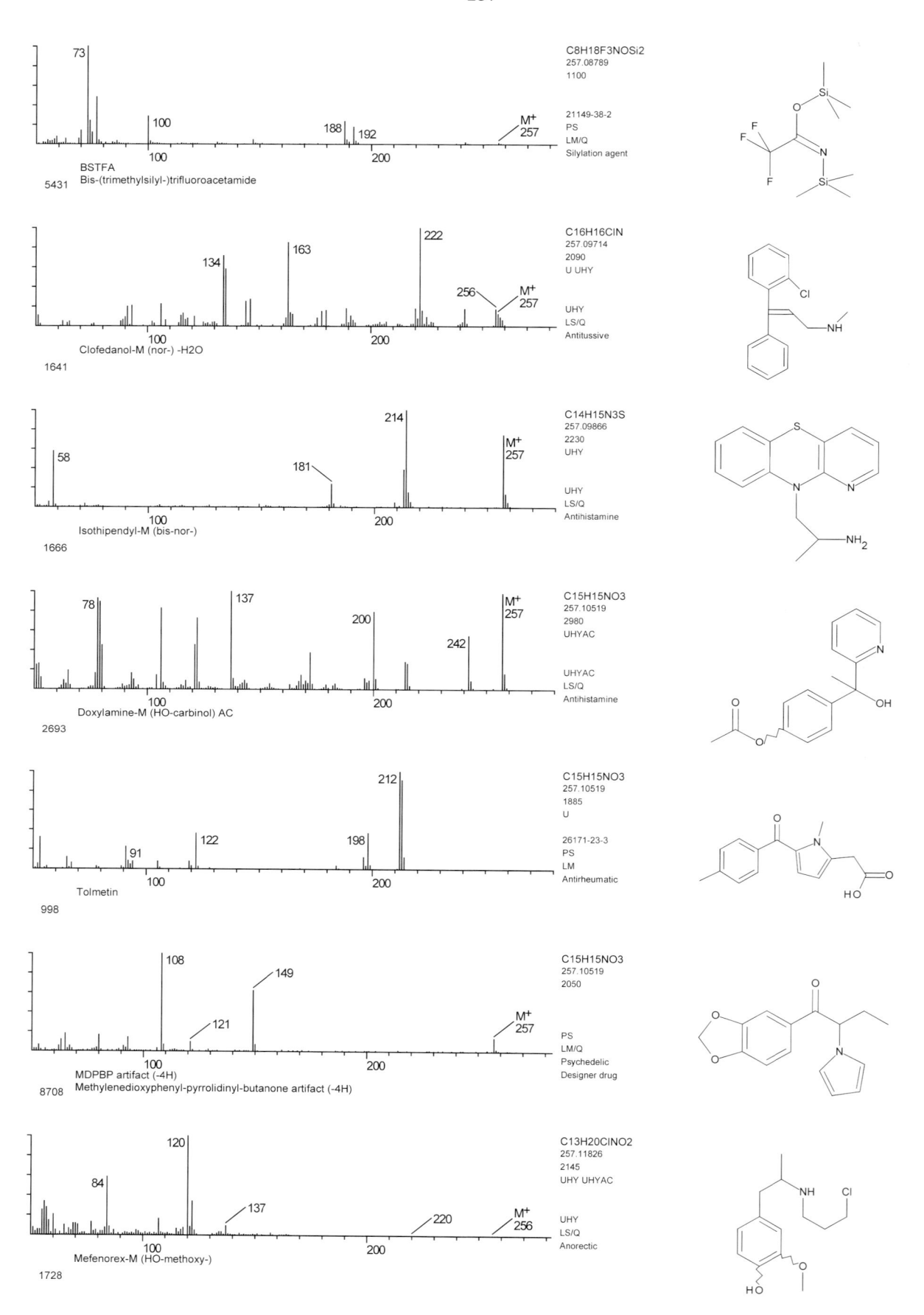

C8H18F3NOSi2
257.08789
1100

21149-38-2
PS
LM/Q
Silylation agent

BSTFA
Bis-(trimethylsilyl)-trifluoroacetamide

C16H16ClN
257.09714
2090
U UHY

UHY
LS/Q
Antitussive

Clofedanol-M (nor-) -H2O

C14H15N3S
257.09866
2230
UHY

UHY
LS/Q
Antihistamine

Isothipendyl-M (bis-nor-)

C15H15NO3
257.10519
2980
UHYAC

UHYAC
LS/Q
Antihistamine

Doxylamine-M (HO-carbinol) AC

C15H15NO3
257.10519
1885
U

26171-23-3
PS
LM
Antirheumatic

Tolmetin

C15H15NO3
257.10519
2050

PS
LM/Q
Psychedelic
Designer drug

MDPBP artifact (-4H)
Methylenedioxyphenyl-pyrrolidinyl-butanone artifact (-4H)

C13H20ClNO2
257.11826
2145
UHY UHYAC

UHY
LS/Q
Anorectic

Mefenorex-M (HO-methoxy-)

642

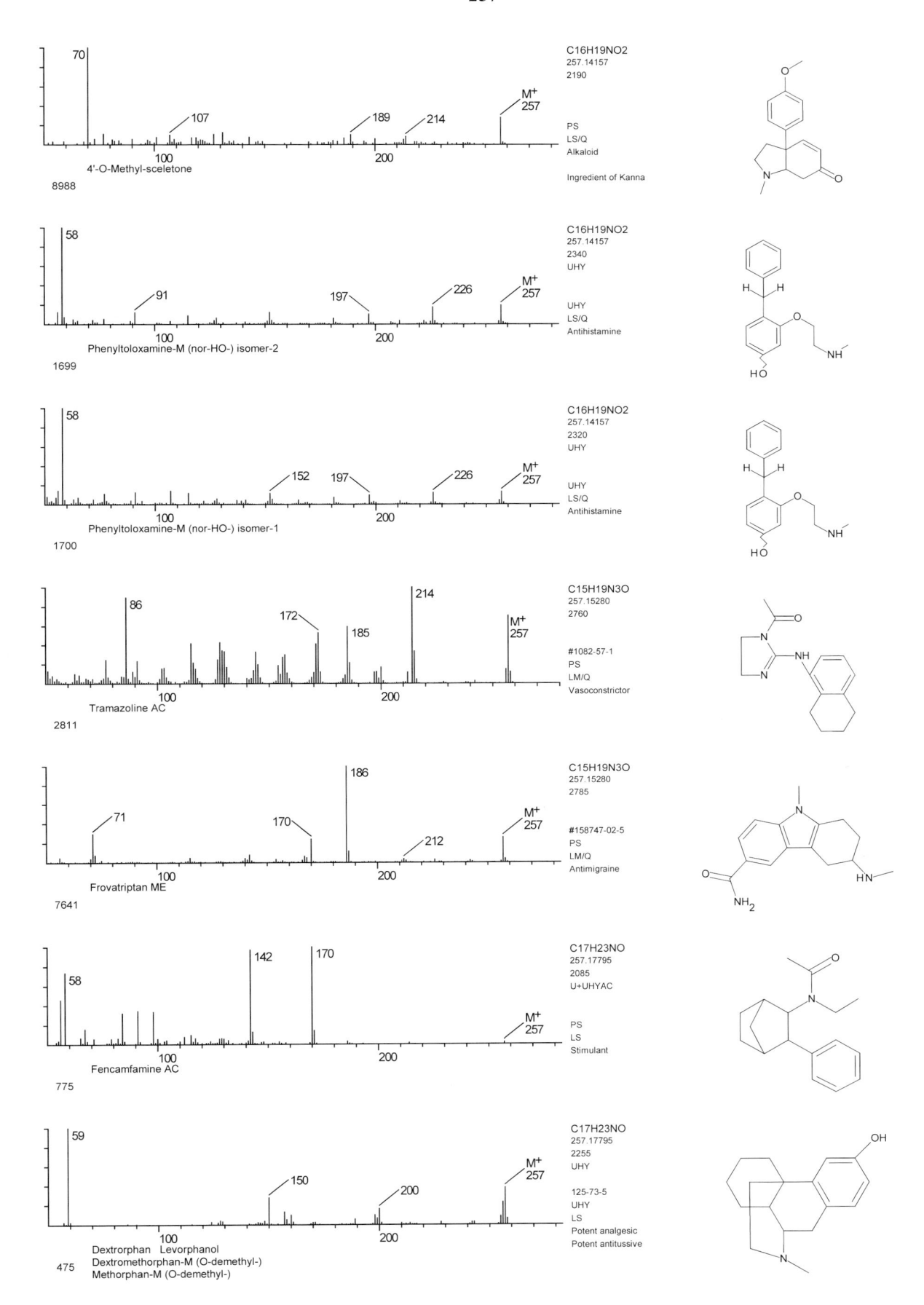

4'-O-Methyl-sceletone
8988
70, 107, 189, 214, M+ 257
C16H19NO2
257.14157
2190
PS
LS/Q
Alkaloid
Ingredient of Kanna

Phenyltoloxamine-M (nor-HO-) isomer-2
1699
58, 91, 197, 226, M+ 257
C16H19NO2
257.14157
2340
UHY
UHY
LS/Q
Antihistamine

Phenyltoloxamine-M (nor-HO-) isomer-1
1700
58, 152, 197, 226, M+ 257
C16H19NO2
257.14157
2320
UHY
UHY
LS/Q
Antihistamine

Tramazoline AC
2811
86, 172, 185, 214, M+ 257
C15H19N3O
257.15280
2760
#1082-57-1
PS
LM/Q
Vasoconstrictor

Frovatriptan ME
7641
71, 170, 186, 212, M+ 257
C15H19N3O
257.15280
2785
#158747-02-5
PS
LM/Q
Antimigraine

Fencamfamine AC
775
58, 142, 170, M+ 257
C17H23NO
257.17795
2085
U+UHYAC
PS
LS
Stimulant

Dextrorphan Levorphanol
Dextromethorphan-M (O-demethyl-)
Methorphan-M (O-demethyl-)
475
59, 150, 200, M+ 257
C17H23NO
257.17795
2255
UHY
125-73-5
UHY
LS
Potent analgesic
Potent antitussive

257

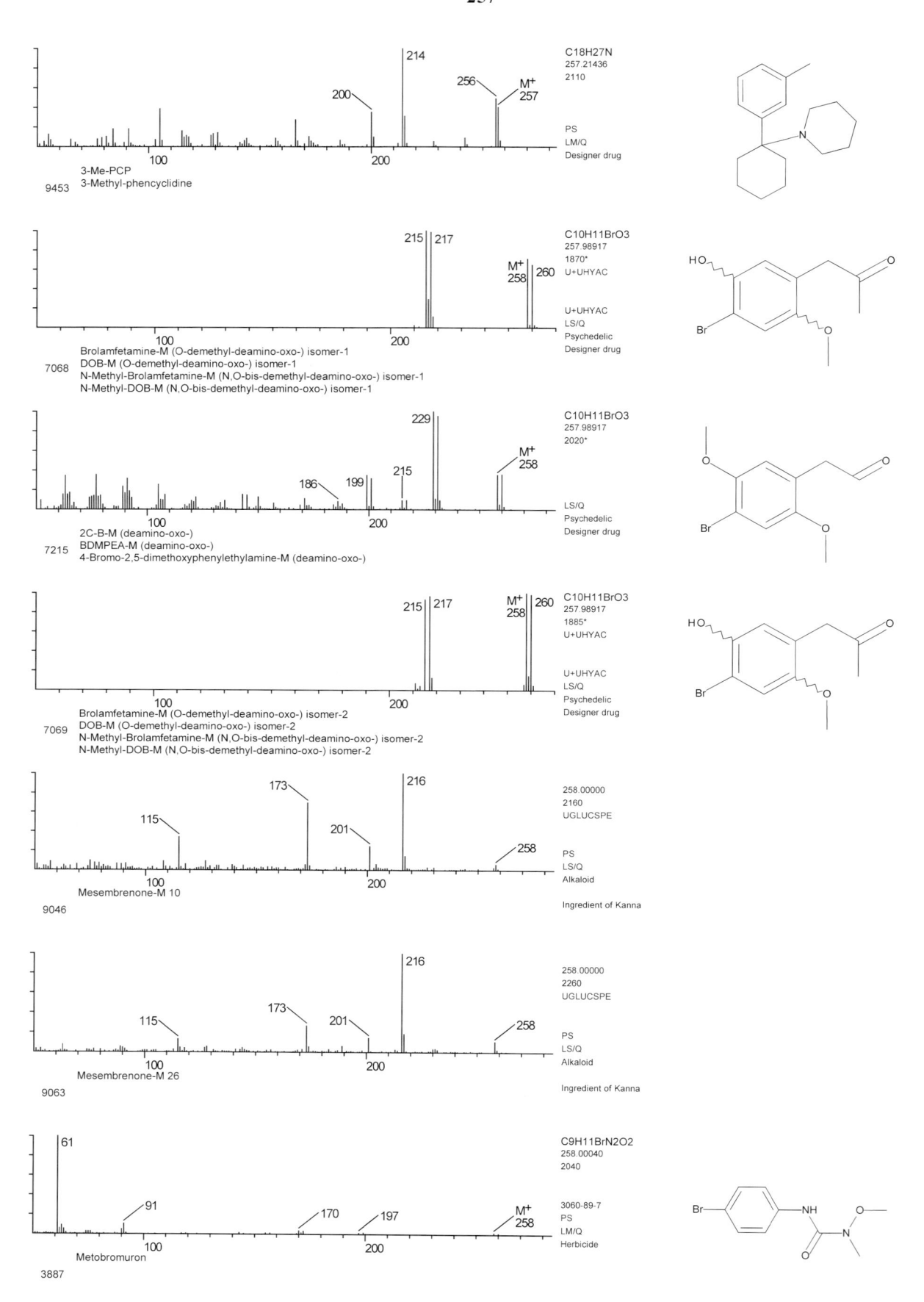

9453	3-Me-PCP 3-Methyl-phencyclidine Peaks: 214, 200, 256, M+ 257 C18H27N 257.21436 2110 PS LM/Q Designer drug
7068	Brolamfetamine-M (O-demethyl-deamino-oxo-) isomer-1 DOB-M (O-demethyl-deamino-oxo-) isomer-1 N-Methyl-Brolamfetamine-M (N,O-bis-demethyl-deamino-oxo-) isomer-1 N-Methyl-DOB-M (N,O-bis-demethyl-deamino-oxo-) isomer-1 Peaks: 215, 217, M+ 258, 260 C10H11BrO3 257.98917 1870* U+UHYAC U+UHYAC LS/Q Psychedelic Designer drug
7215	2C-B-M (deamino-oxo-) BDMPEA-M (deamino-oxo-) 4-Bromo-2,5-dimethoxyphenylethylamine-M (deamino-oxo-) Peaks: 229, 186, 199, 215, M+ 258 C10H11BrO3 257.98917 2020* LS/Q Psychedelic Designer drug
7069	Brolamfetamine-M (O-demethyl-deamino-oxo-) isomer-2 DOB-M (O-demethyl-deamino-oxo-) isomer-2 N-Methyl-Brolamfetamine-M (N,O-bis-demethyl-deamino-oxo-) isomer-2 N-Methyl-DOB-M (N,O-bis-demethyl-deamino-oxo-) isomer-2 Peaks: 215, 217, M+ 258, 260 C10H11BrO3 257.98917 1885* U+UHYAC U+UHYAC LS/Q Psychedelic Designer drug
9046	Mesembrenone-M 10 Peaks: 115, 173, 201, 216, 258 258.00000 2160 UGLUCSPE PS LS/Q Alkaloid Ingredient of Kanna
9063	Mesembrenone-M 26 Peaks: 115, 173, 201, 216, 258 258.00000 2260 UGLUCSPE PS LS/Q Alkaloid Ingredient of Kanna
3887	Metobromuron Peaks: 61, 91, 170, 197, M+ 258 C9H11BrN2O2 258.00040 2040 3060-89-7 PS LM/Q Herbicide

258

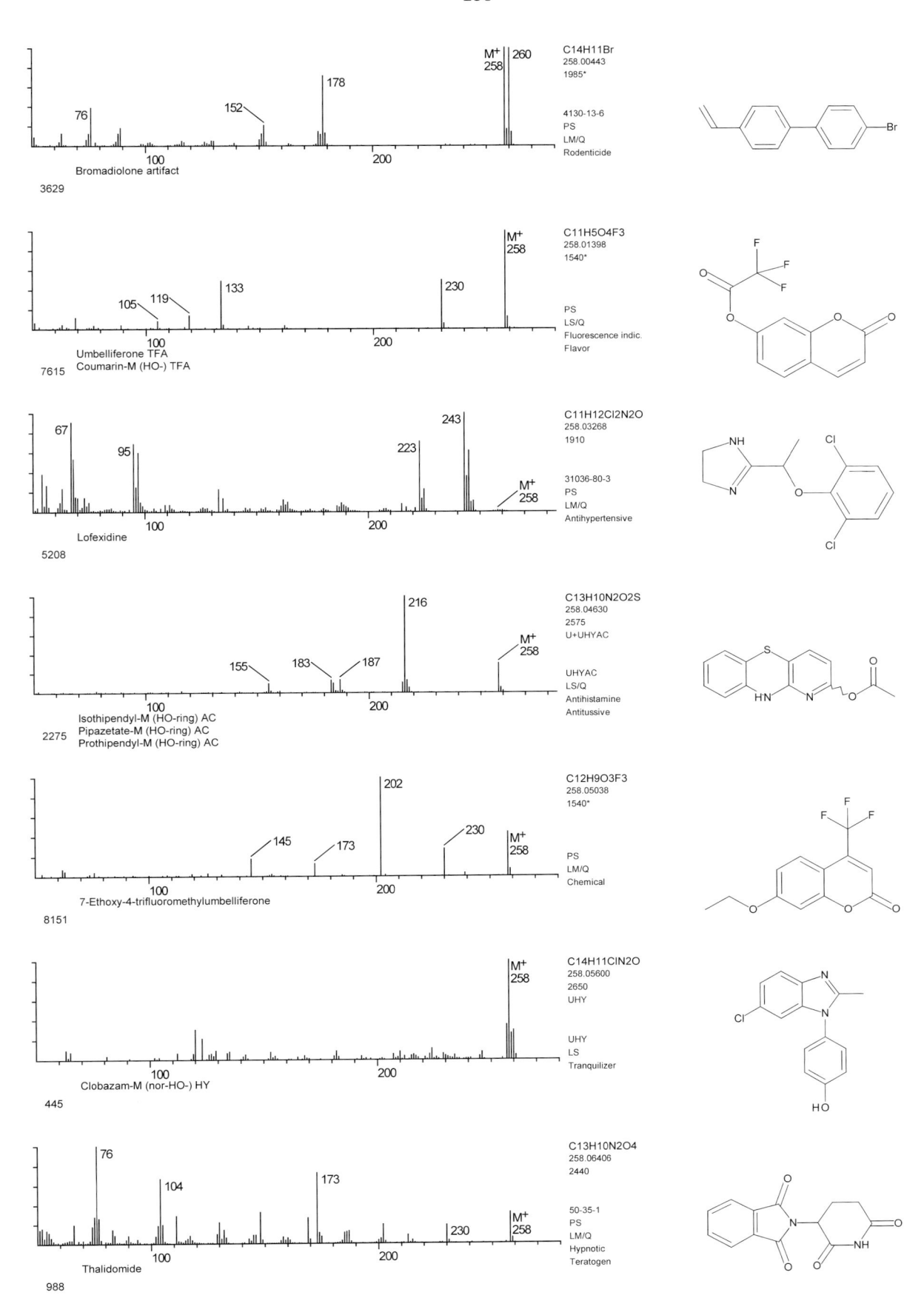

3629 — Bromadiolone artifact	C14H11Br / 258.00443 / 1985* / 4130-13-6 / PS / LM/Q / Rodenticide
7615 — Umbelliferone TFA; Coumarin-M (HO-) TFA	C11H5O4F3 / 258.01398 / 1540* / PS / LS/Q / Fluorescence indic. / Flavor
5208 — Lofexidine	C11H12Cl2N2O / 258.03268 / 1910 / 31036-80-3 / PS / LM/Q / Antihypertensive
2275 — Isothipendyl-M (HO-ring) AC; Pipazetate-M (HO-ring) AC; Prothipendyl-M (HO-ring) AC	C13H10N2O2S / 258.04630 / 2575 / U+UHYAC / UHYAC / LS/Q / Antihistamine / Antitussive
8151 — 7-Ethoxy-4-trifluoromethylumbelliferone	C12H9O3F3 / 258.05038 / 1540* / PS / LM/Q / Chemical
445 — Clobazam-M (nor-HO-) HY	C14H11ClN2O / 258.05600 / 2650 / UHY / UHY / LS / Tranquilizer
988 — Thalidomide	C13H10N2O4 / 258.06406 / 2440 / 50-35-1 / PS / LM/Q / Hypnotic / Teratogen

258

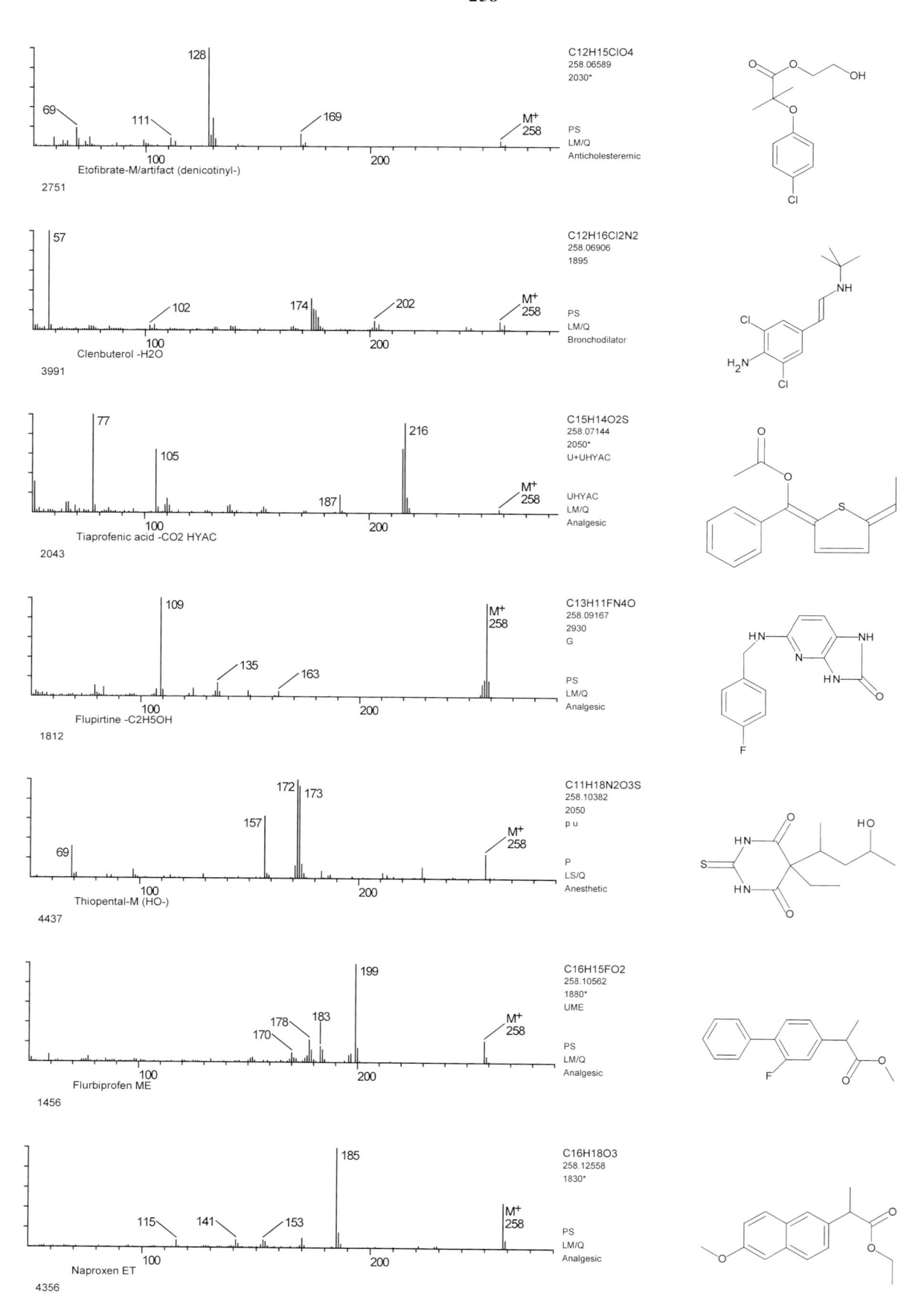

Etofibrate-M/artifact (denicotinyl-)
69, 111, 128, 169, M+ 258
C12H15ClO4
258.06589
2030*
PS
LM/Q
Anticholesteremic
2751

Clenbuterol -H2O
57, 102, 174, 202, M+ 258
C12H16Cl2N2
258.06906
1895
PS
LM/Q
Bronchodilator
3991

Tiaprofenic acid -CO2 HYAC
77, 105, 187, 216, M+ 258
C15H14O2S
258.07144
2050*
U+UHYAC
UHYAC
LM/Q
Analgesic
2043

Flupirtine -C2H5OH
109, 135, 163, M+ 258
C13H11FN4O
258.09167
2930
G
PS
LM/Q
Analgesic
1812

Thiopental-M (HO-)
69, 157, 172, 173, M+ 258
C11H18N2O3S
258.10382
2050
p u
P
LS/Q
Anesthetic
4437

Flurbiprofen ME
170, 178, 183, 199, M+ 258
C16H15FO2
258.10562
1880*
UME
PS
LM/Q
Analgesic
1456

Naproxen ET
115, 141, 153, 185, M+ 258
C16H18O3
258.12558
1830*
PS
LM/Q
Analgesic
4356

258

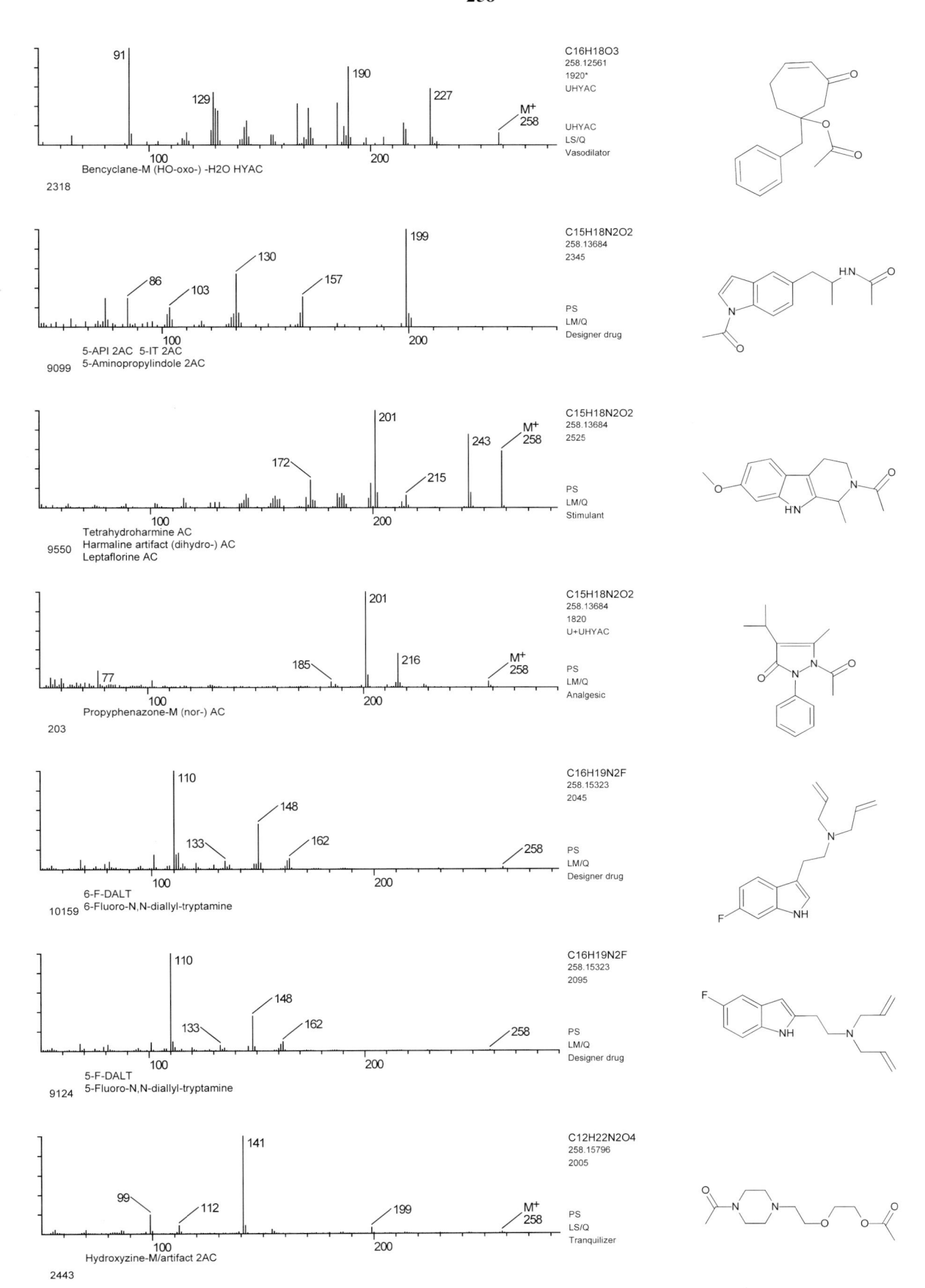

2318 — Bencyclane-M (HO-oxo-) -H2O HYAC
Peaks: 91, 129, 190, 227, 258 (M+)
C16H18O3
258.12561
1920*
UHYAC
UHYAC
LS/Q
Vasodilator

9099 — 5-API 2AC 5-IT 2AC 5-Aminopropylindole 2AC
Peaks: 86, 103, 130, 157, 199
C15H18N2O2
258.13684
2345
PS
LM/Q
Designer drug

9550 — Tetrahydroharmine AC / Harmaline artifact (dihydro-) AC / Leptaflorine AC
Peaks: 172, 201, 215, 243, 258 (M+)
C15H18N2O2
258.13684
2525
PS
LM/Q
Stimulant

203 — Propyphenazone-M (nor-) AC
Peaks: 77, 185, 201, 216, 258 (M+)
C15H18N2O2
258.13684
1820
U+UHYAC
PS
LM/Q
Analgesic

10159 — 6-F-DALT 6-Fluoro-N,N-diallyl-tryptamine
Peaks: 110, 133, 148, 162, 258
C16H19N2F
258.15323
2045
PS
LM/Q
Designer drug

9124 — 5-F-DALT 5-Fluoro-N,N-diallyl-tryptamine
Peaks: 110, 133, 148, 162, 258
C16H19N2F
258.15323
2095
PS
LM/Q
Designer drug

2443 — Hydroxyzine-M/artifact 2AC
Peaks: 99, 112, 141, 199, 258 (M+)
C12H22N2O4
258.15796
2005
PS
LS/Q
Tranquilizer

258

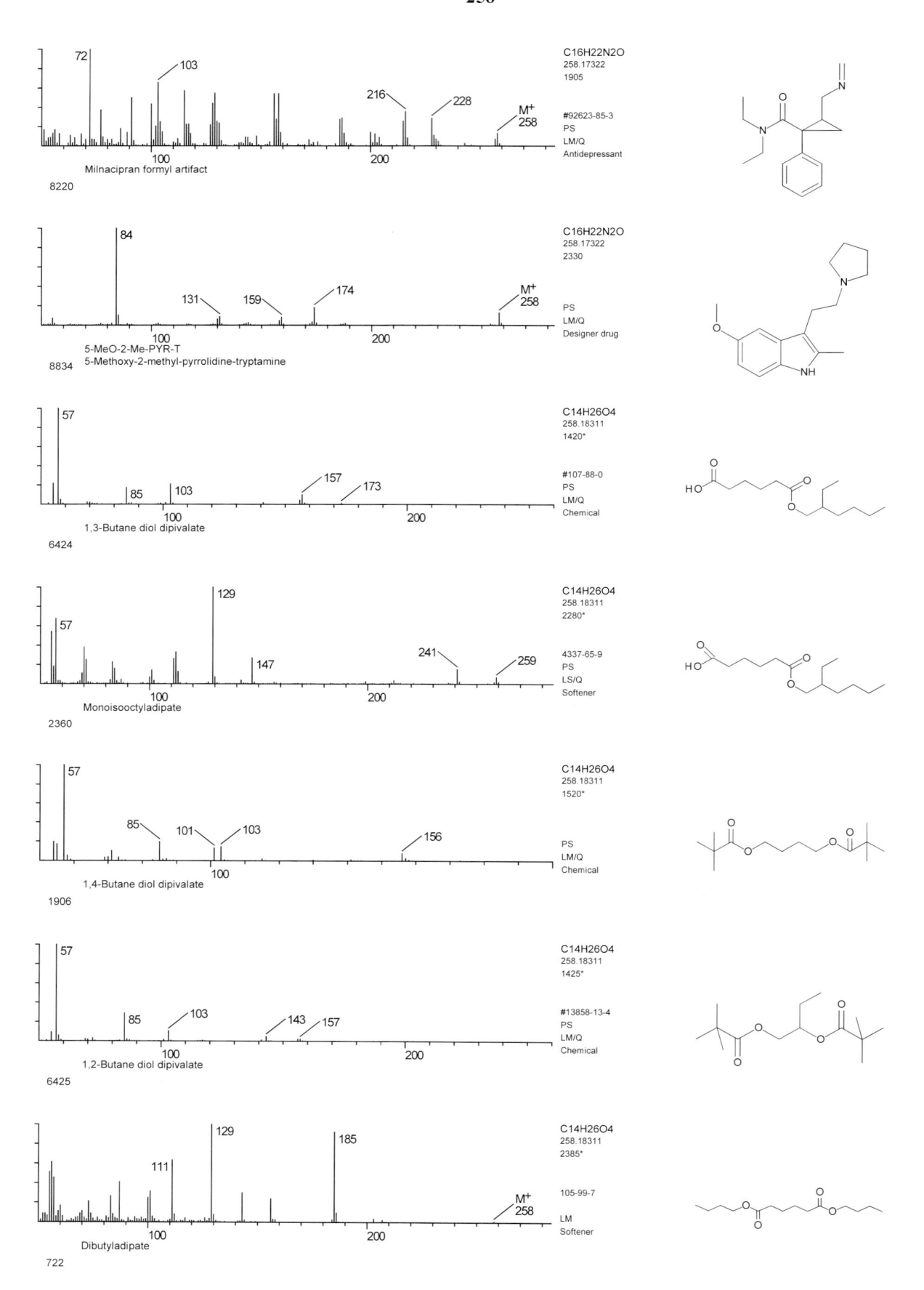

m/z peaks: 72, 103, 216, 228, M+ 258	C16H22N2O 258.17322 1905 #92623-85-3 PS LM/Q Antidepressant	
Milnacipran formyl artifact 8220		
m/z peaks: 84, 131, 159, 174, M+ 258	C16H22N2O 258.17322 2330 PS LM/Q Designer drug	
5-MeO-2-Me-PYR-T 5-Methoxy-2-methyl-pyrrolidine-tryptamine 8834		
m/z peaks: 57, 85, 103, 157, 173	C14H26O4 258.18311 1420* #107-88-0 PS LM/Q Chemical	
1,3-Butane diol dipivalate 6424		
m/z peaks: 57, 129, 147, 241, 259	C14H26O4 258.18311 2280* 4337-65-9 PS LS/Q Softener	
Monoisooctyladipate 2360		
m/z peaks: 57, 85, 101, 103, 156	C14H26O4 258.18311 1520* PS LM/Q Chemical	
1,4-Butane diol dipivalate 1906		
m/z peaks: 57, 85, 103, 143, 157	C14H26O4 258.18311 1425* #13858-13-4 PS LM/Q Chemical	
1,2-Butane diol dipivalate 6425		
m/z peaks: 111, 129, 185, M+ 258	C14H26O4 258.18311 2385* 105-99-7 LM Softener	
Dibutyladipate 722		

648

258

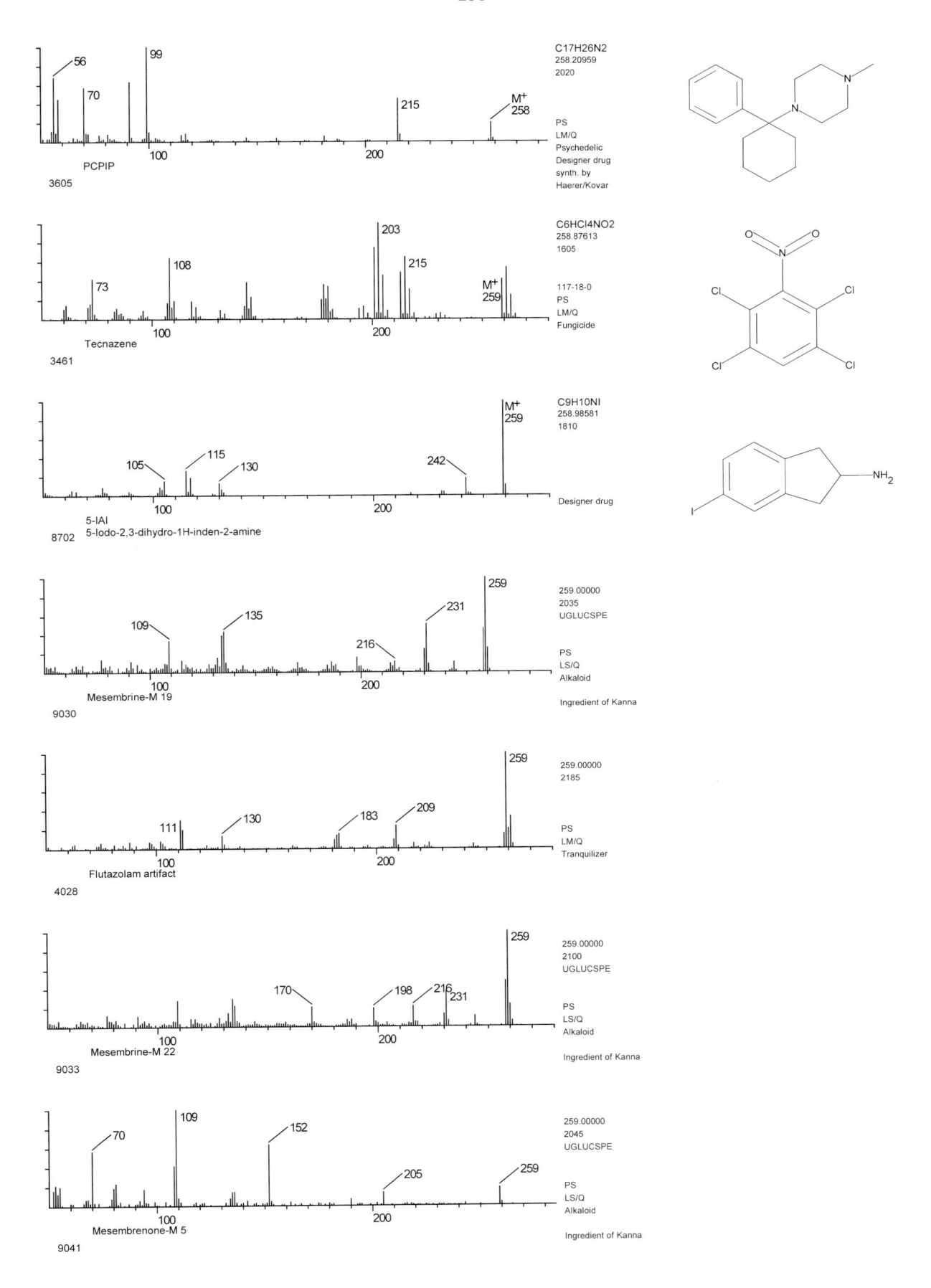

- 3605 PCPIP — C17H26N2, 258.20959, 2020, PS, LM/Q, Psychedelic, Designer drug synth. by Haerer/Kovar. Peaks: 56, 70, 99, 215, M+ 258.
- 3461 Tecnazene — C6HCl4NO2, 258.87613, 1605, 117-18-0, PS, LM/Q, Fungicide. Peaks: 73, 108, 203, 215, M+ 259.
- 8702 5-IAI, 5-Iodo-2,3-dihydro-1H-inden-2-amine — C9H10NI, 258.98581, 1810, Designer drug. Peaks: 105, 115, 130, 242, M+ 259.
- 9030 Mesembrine-M 19 — 259.00000, 2035, UGLUCSPE, PS, LS/Q, Alkaloid, Ingredient of Kanna. Peaks: 109, 135, 216, 231, 259.
- 4028 Flutazolam artifact — 259.00000, 2185, PS, LM/Q, Tranquilizer. Peaks: 111, 130, 183, 209, 259.
- 9033 Mesembrine-M 22 — 259.00000, 2100, UGLUCSPE, PS, LS/Q, Alkaloid, Ingredient of Kanna. Peaks: 170, 198, 216, 231, 259.
- 9041 Mesembrenone-M 5 — 259.00000, 2045, UGLUCSPE, PS, LS/Q, Alkaloid, Ingredient of Kanna. Peaks: 70, 109, 152, 205, 259.

259

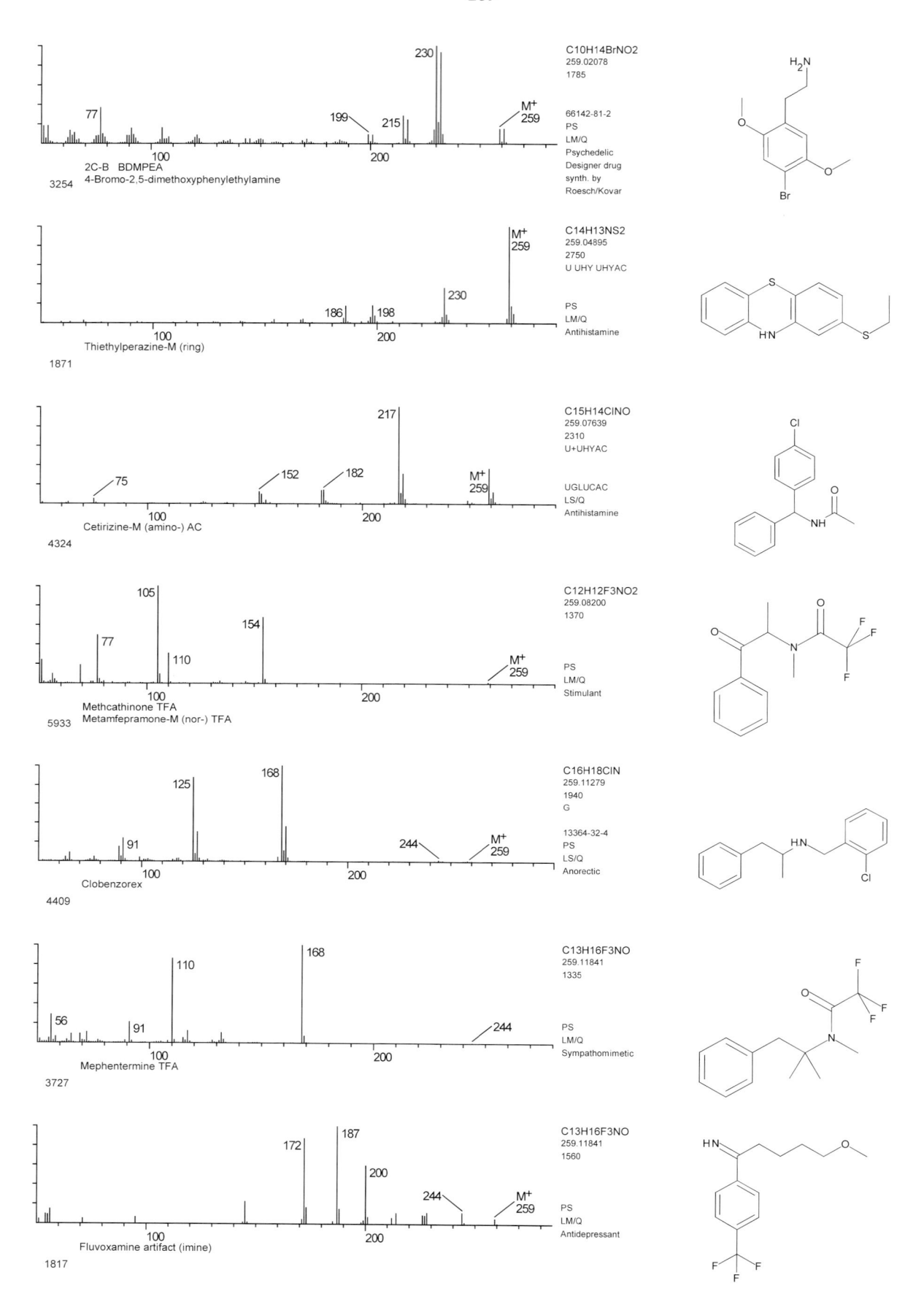

2C-B BDMPEA 4-Bromo-2,5-dimethoxyphenylethylamine 3254	C10H14BrNO2 259.02078 1785 66142-81-2 PS LM/Q Psychedelic Designer drug synth. by Roesch/Kovar	
Thiethylperazine-M (ring) 1871	C14H13NS2 259.04895 2750 U UHY UHYAC PS LM/Q Antihistamine	
Cetirizine-M (amino-) AC 4324	C15H14ClNO 259.07639 2310 U+UHYAC UGLUCAC LS/Q Antihistamine	
Methcathinone TFA Metamfepramone-M (nor-) TFA 5933	C12H12F3NO2 259.08200 1370 PS LM/Q Stimulant	
Clobenzorex 4409	C16H18ClN 259.11279 1940 G 13364-32-4 PS LS/Q Anorectic	
Mephentermine TFA 3727	C13H16F3NO 259.11841 1335 PS LM/Q Sympathomimetic	
Fluvoxamine artifact (imine) 1817	C13H16F3NO 259.11841 1560 PS LM/Q Antidepressant	

259

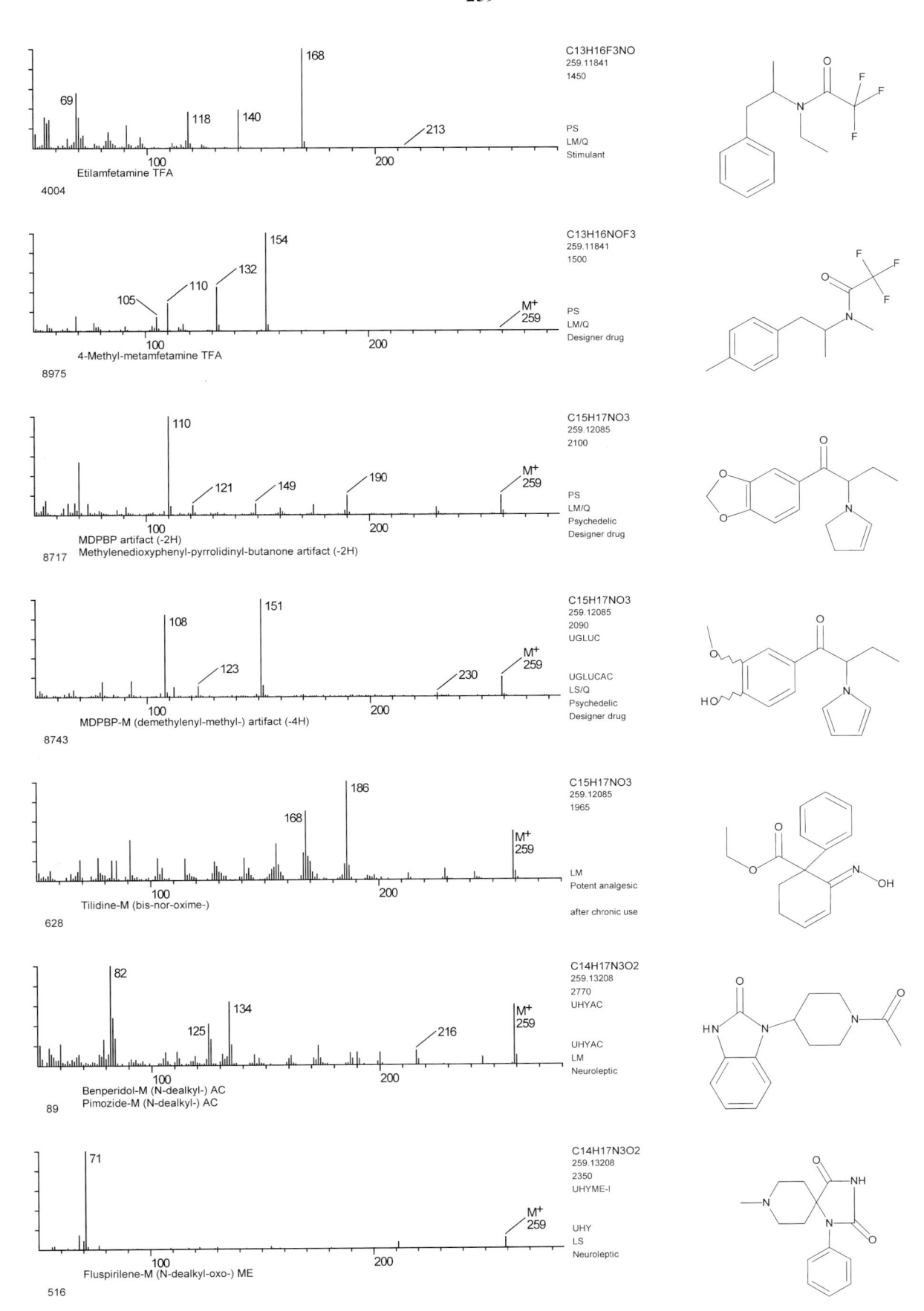

Etilamfetamine TFA
4004

C13H16F3NO
259.11841
1450

PS
LM/Q
Stimulant

4-Methyl-metamfetamine TFA
8975

C13H16NOF3
259.11841
1500

PS
LM/Q
Designer drug

MDPBP artifact (-2H)
Methylenedioxyphenyl-pyrrolidinyl-butanone artifact (-2H)
8717

C15H17NO3
259.12085
2100

PS
LM/Q
Psychedelic
Designer drug

MDPBP-M (demethylenyl-methyl-) artifact (-4H)
8743

C15H17NO3
259.12085
2090

UGLUC

UGLUCAC
LS/Q
Psychedelic
Designer drug

Tilidine-M (bis-nor-oxime-)
628

C15H17NO3
259.12085
1965

LM
Potent analgesic

after chronic use

Benperidol-M (N-dealkyl-) AC
Pimozide-M (N-dealkyl-) AC
89

C14H17N3O2
259.13208
2770
UHYAC

UHYAC
LM
Neuroleptic

Fluspirilene-M (N-dealkyl-oxo-) ME
516

C14H17N3O2
259.13208
2350
UHYME-I

UHY
LS
Neuroleptic

651

259

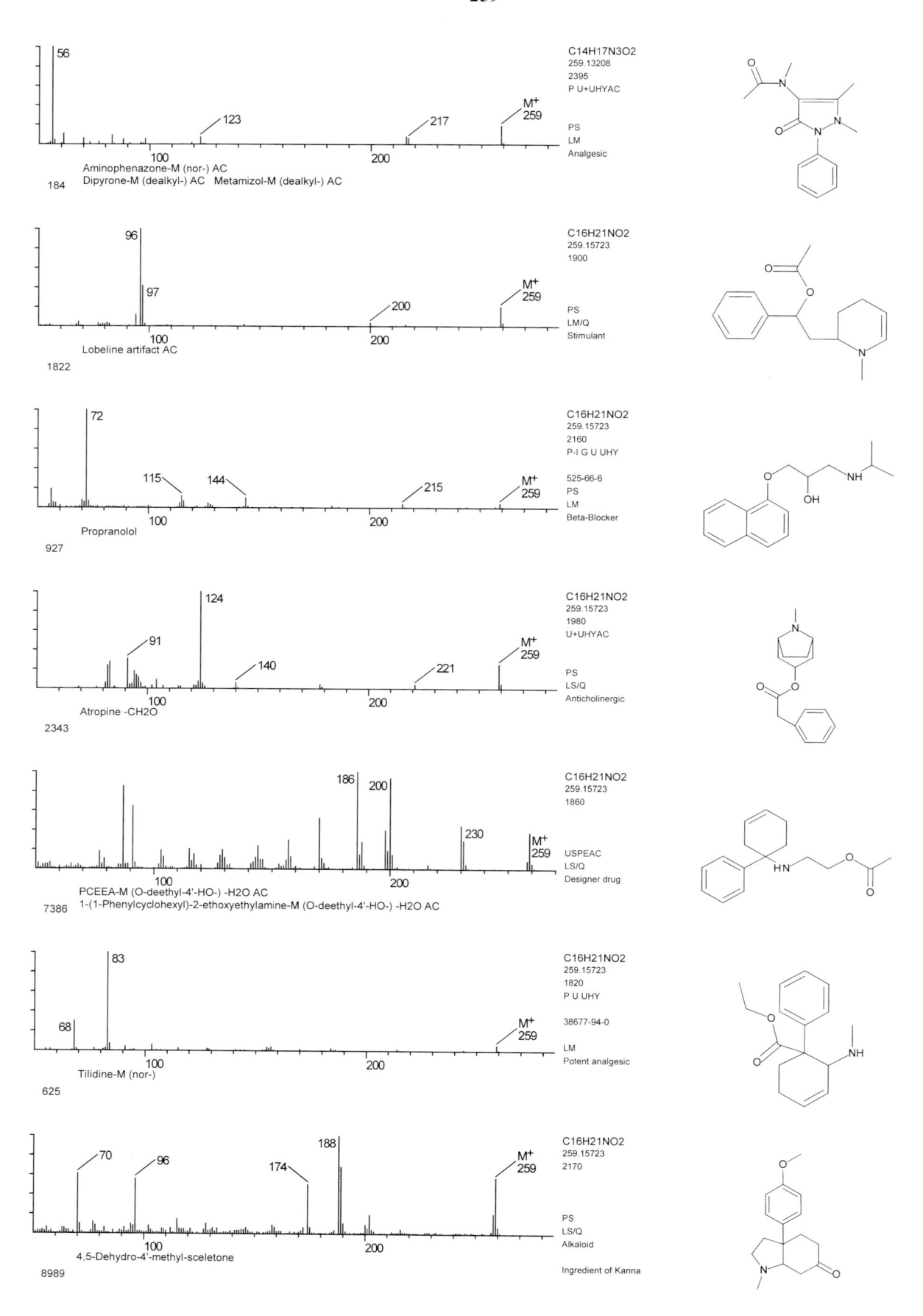

C14H17N3O2
259.13208
2395
P U+UHYAC

PS
LM
Analgesic

Aminophenazone-M (nor-) AC
Dipyrone-M (dealkyl-) AC   Metamizol-M (dealkyl-) AC
184

C16H21NO2
259.15723
1900

PS
LM/Q
Stimulant

Lobeline artifact AC
1822

C16H21NO2
259.15723
2160
P-I G U UHY

525-66-6
PS
LM
Beta-Blocker

Propranolol
927

C16H21NO2
259.15723
1980
U+UHYAC

PS
LS/Q
Anticholinergic

Atropine -CH2O
2343

C16H21NO2
259.15723
1860

USPEAC
LS/Q
Designer drug

PCEEA-M (O-deethyl-4'-HO-) -H2O AC
1-(1-Phenylcyclohexyl)-2-ethoxyethylamine-M (O-deethyl-4'-HO-) -H2O AC
7386

C16H21NO2
259.15723
1820
P U UHY

38677-94-0
LM
Potent analgesic

Tilidine-M (nor-)
625

C16H21NO2
259.15723
2170

PS
LS/Q
Alkaloid

Ingredient of Kanna

4,5-Dehydro-4'-methyl-sceletone
8989

652

259

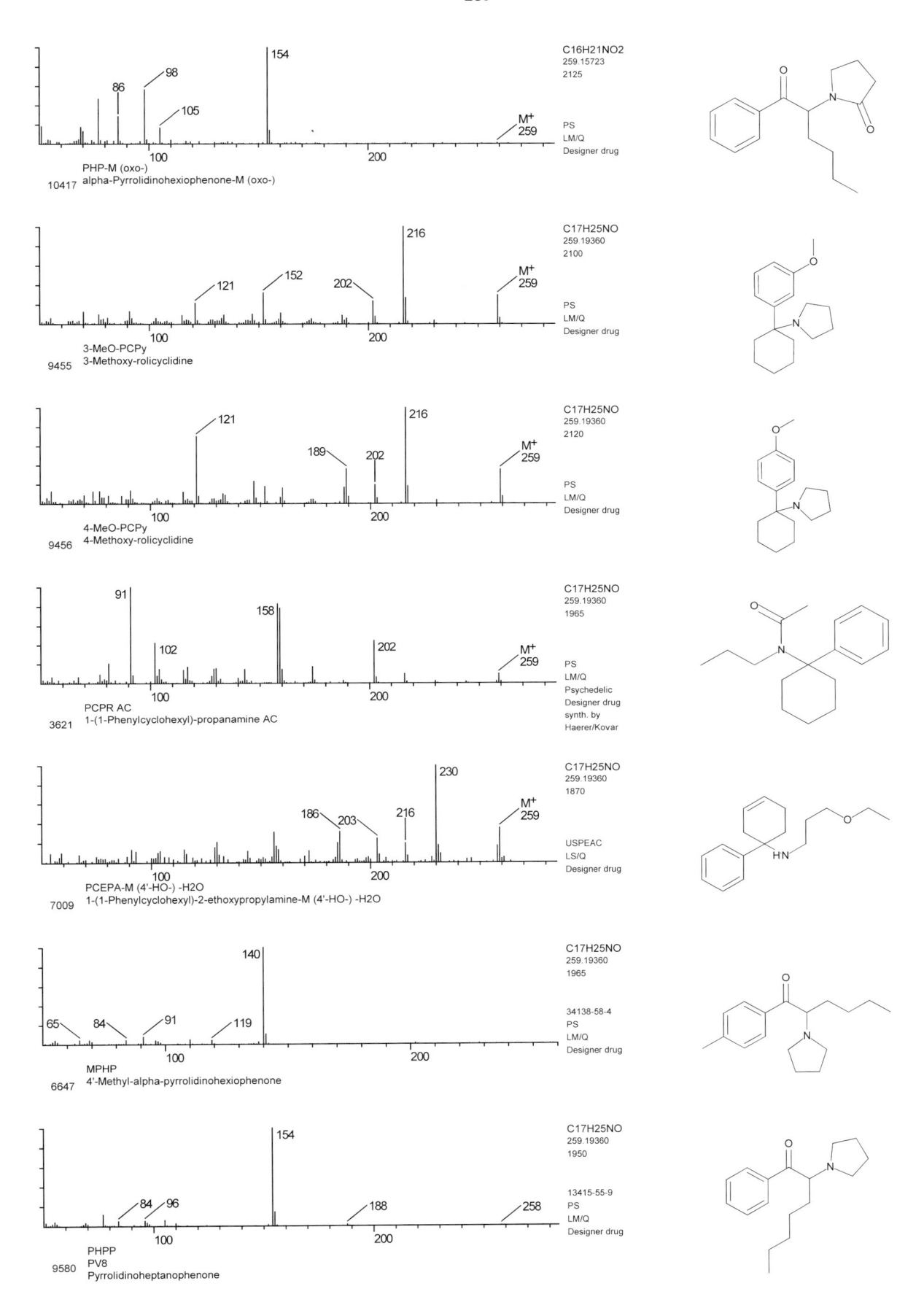

10417 PHP-M (oxo-)
alpha-Pyrrolidinohexiophenone-M (oxo-)
C16H21NO2
259.15723
2125
PS
LM/Q
Designer drug

9455 3-MeO-PCPy
3-Methoxy-rolicyclidine
C17H25NO
259.19360
2100
PS
LM/Q
Designer drug

9456 4-MeO-PCPy
4-Methoxy-rolicyclidine
C17H25NO
259.19360
2120
PS
LM/Q
Designer drug

3621 PCPR AC
1-(1-Phenylcyclohexyl)-propanamine AC
C17H25NO
259.19360
1965
PS
LM/Q
Psychedelic
Designer drug
synth. by Haerer/Kovar

7009 PCEPA-M (4'-HO-) -H2O
1-(1-Phenylcyclohexyl)-2-ethoxypropylamine-M (4'-HO-) -H2O
C17H25NO
259.19360
1870
USPEAC
LS/Q
Designer drug

6647 MPHP
4'-Methyl-alpha-pyrrolidinohexiophenone
C17H25NO
259.19360
1965
34138-58-4
PS
LM/Q
Designer drug

9580 PHPP
PV8
Pyrrolidinoheptanophenone
C17H25NO
259.19360
1950
13415-55-9
PS
LM/Q
Designer drug

259

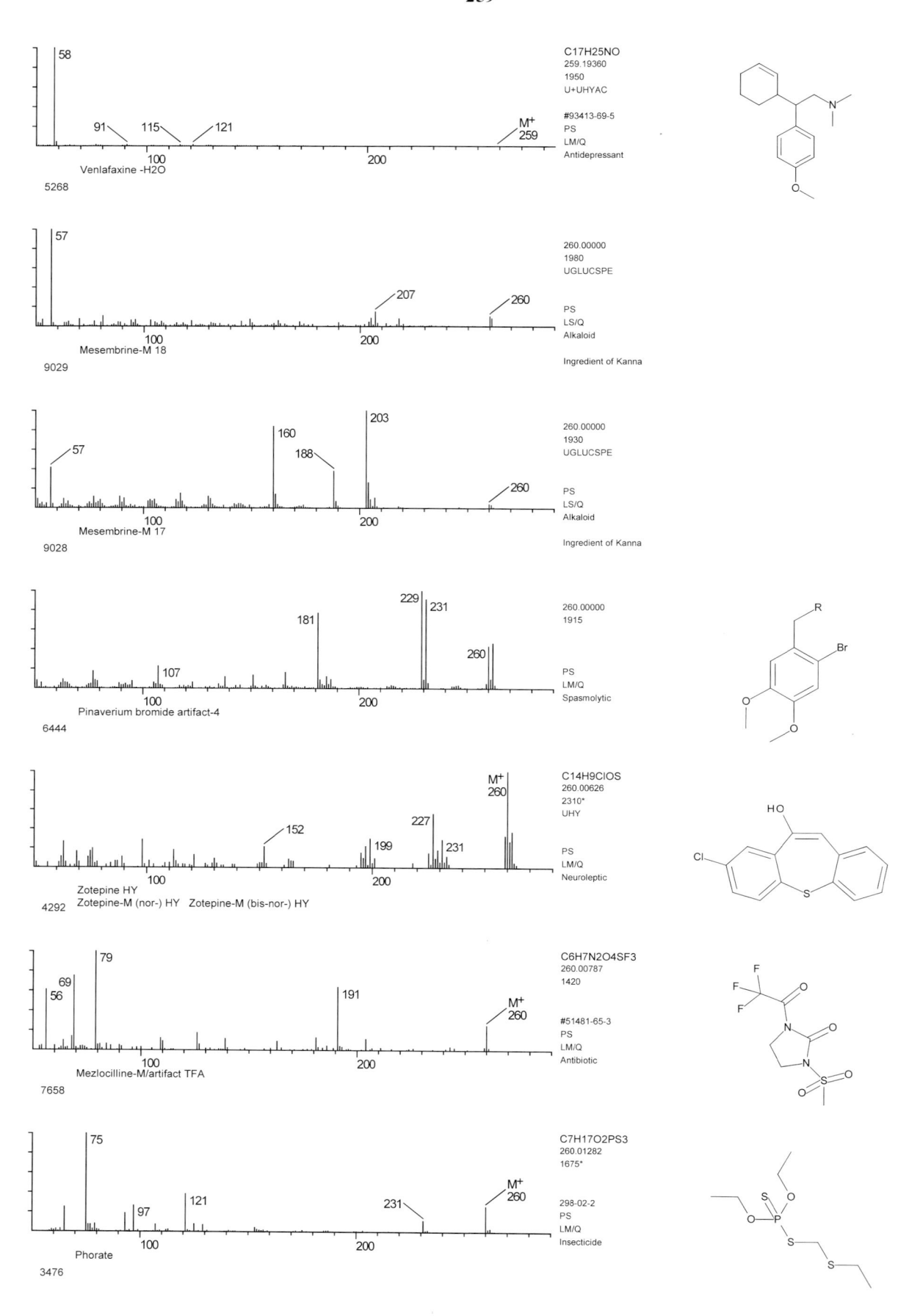

Spectrum	Formula/Info
Venlafaxine -H2O (5268) — peaks: 58, 91, 115, 121, M+ 259	C17H25NO, 259.19360, 1950, U+UHYAC, #93413-69-5, PS, LM/Q, Antidepressant
Mesembrine-M 18 (9029) — peaks: 57, 207, 260	260.00000, 1980, UGLUCSPE, PS, LS/Q, Alkaloid, Ingredient of Kanna
Mesembrine-M 17 (9028) — peaks: 57, 160, 188, 203, 260	260.00000, 1930, UGLUCSPE, PS, LS/Q, Alkaloid, Ingredient of Kanna
Pinaverium bromide artifact-4 (6444) — peaks: 107, 181, 229, 231, 260	260.00000, 1915, PS, LM/Q, Spasmolytic
Zotepine HY / Zotepine-M (nor-) HY / Zotepine-M (bis-nor-) HY (4292) — peaks: 152, 199, 227, 231, M+ 260	C14H9ClOS, 260.00626, 2310*, UHY, PS, LM/Q, Neuroleptic
Mezlocilline-M/artifact TFA (7658) — peaks: 56, 69, 79, 191, M+ 260	C6H7N2O4SF3, 260.00787, 1420, #51481-65-3, PS, LM/Q, Antibiotic
Phorate (3476) — peaks: 75, 97, 121, 231, M+ 260	C7H17O2PS3, 260.01282, 1675*, 298-02-2, PS, LM/Q, Insecticide

654

260

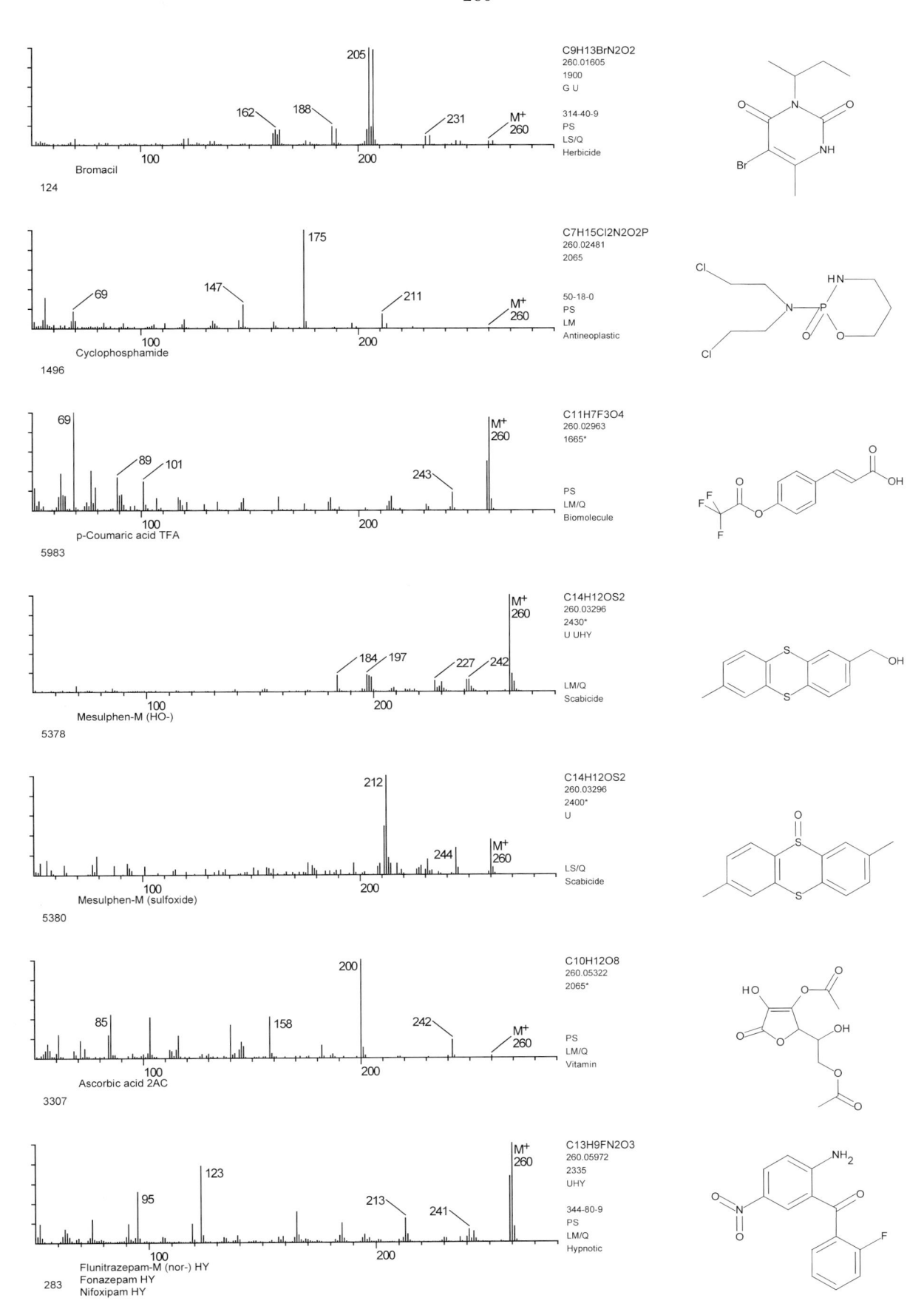

Bromacil
124
C9H13BrN2O2
260.01605
1900
G U
314-40-9
PS
LS/Q
Herbicide

Cyclophosphamide
1496
C7H15Cl2N2O2P
260.02481
2065
50-18-0
PS
LM
Antineoplastic

p-Coumaric acid TFA
5983
C11H7F3O4
260.02963
1665*
PS
LM/Q
Biomolecule

Mesulphen-M (HO-)
5378
C14H12OS2
260.03296
2430*
U UHY
LM/Q
Scabicide

Mesulphen-M (sulfoxide)
5380
C14H12OS2
260.03296
2400*
U
LS/Q
Scabicide

Ascorbic acid 2AC
3307
C10H12O8
260.05322
2065*
PS
LM/Q
Vitamin

Flunitrazepam-M (nor-) HY
Fonazepam HY
Nifoxipam HY
283
C13H9FN2O3
260.05972
2335
UHY
344-80-9
PS
LM/Q
Hypnotic

655

260

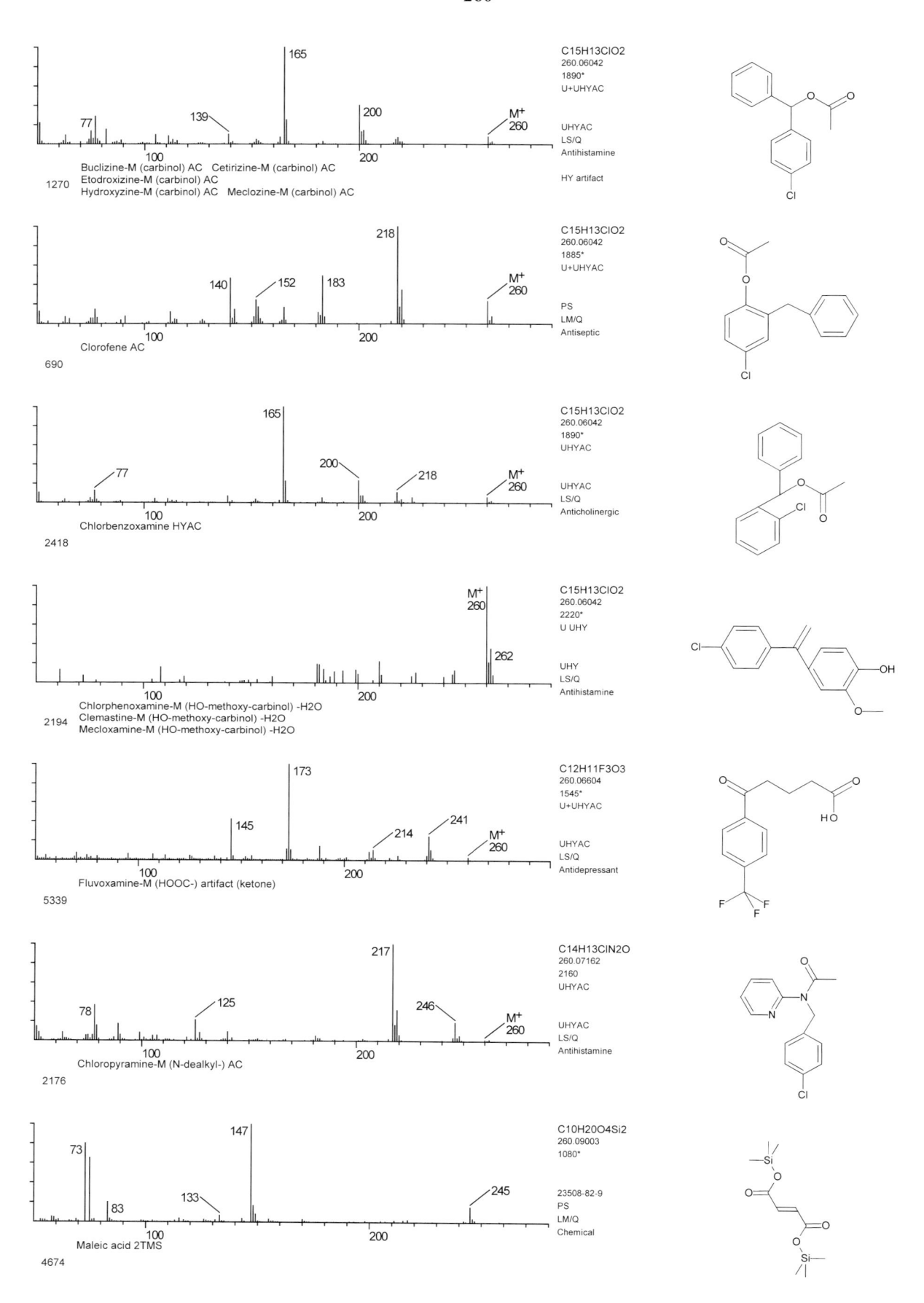

C15H13ClO2
260.06042
1890*
U+UHYAC

UHYAC
LS/Q
Antihistamine

HY artifact

Buclizine-M (carbinol) AC   Cetirizine-M (carbinol) AC
Etodroxizine-M (carbinol) AC
Hydroxyzine-M (carbinol) AC   Meclozine-M (carbinol) AC
1270

C15H13ClO2
260.06042
1885*
U+UHYAC

PS
LM/Q
Antiseptic

Clorofene AC
690

C15H13ClO2
260.06042
1890*
UHYAC

UHYAC
LS/Q
Anticholinergic

Chlorbenzoxamine HYAC
2418

C15H13ClO2
260.06042
2220*
U UHY

UHY
LS/Q
Antihistamine

Chlorphenoxamine-M (HO-methoxy-carbinol) -H2O
Clemastine-M (HO-methoxy-carbinol) -H2O
Mecloxamine-M (HO-methoxy-carbinol) -H2O
2194

C12H11F3O3
260.06604
1545*
U+UHYAC

UHYAC
LS/Q
Antidepressant

Fluvoxamine-M (HOOC-) artifact (ketone)
5339

C14H13ClN2O
260.07162
2160
UHYAC

UHYAC
LS/Q
Antihistamine

Chloropyramine-M (N-dealkyl-) AC
2176

C10H20O4Si2
260.09003
1080*

23508-82-9
PS
LM/Q
Chemical

Maleic acid 2TMS
4674

656

260

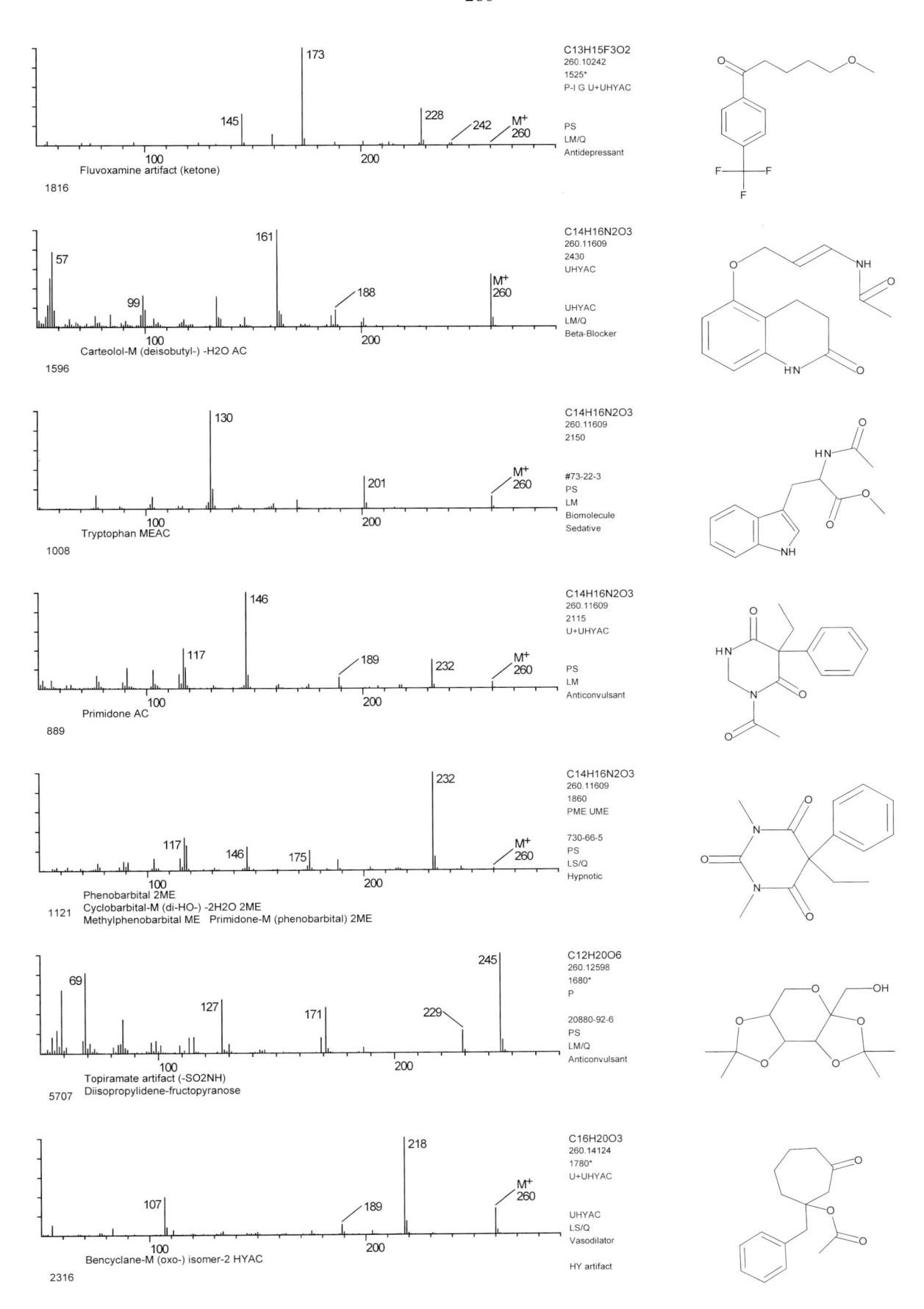

Fluvoxamine artifact (ketone)
1816
Peaks: 145, 173, 228, 242, M+ 260
C13H15F3O2
260.10242
1525*
P-I G U+UHYAC
PS
LM/Q
Antidepressant

Carteolol-M (deisobutyl-) -H2O AC
1596
Peaks: 57, 99, 161, 188, M+ 260
C14H16N2O3
260.11609
2430
UHYAC
UHYAC
LM/Q
Beta-Blocker

Tryptophan MEAC
1008
Peaks: 130, 201, M+ 260
C14H16N2O3
260.11609
2150
#73-22-3
PS
LM
Biomolecule
Sedative

Primidone AC
889
Peaks: 117, 146, 189, 232, M+ 260
C14H16N2O3
260.11609
2115
U+UHYAC
PS
LM
Anticonvulsant

Phenobarbital 2ME
Cyclobarbital-M (di-HO-) -2H2O 2ME
Methylphenobarbital ME   Primidone-M (phenobarbital) 2ME
1121
Peaks: 117, 146, 175, 232, M+ 260
C14H16N2O3
260.11609
1860
PME UME
730-66-5
PS
LS/Q
Hypnotic

Topiramate artifact (-SO2NH)
Diisopropylidene-fructopyranose
5707
Peaks: 69, 127, 171, 229, 245
C12H20O6
260.12598
1680*
P
20880-92-6
PS
LM/Q
Anticonvulsant

Bencyclane-M (oxo-) isomer-2 HYAC
2316
Peaks: 107, 189, 218, M+ 260
C16H20O3
260.14124
1780*
U+UHYAC
UHYAC
LS/Q
Vasodilator
HY artifact

260

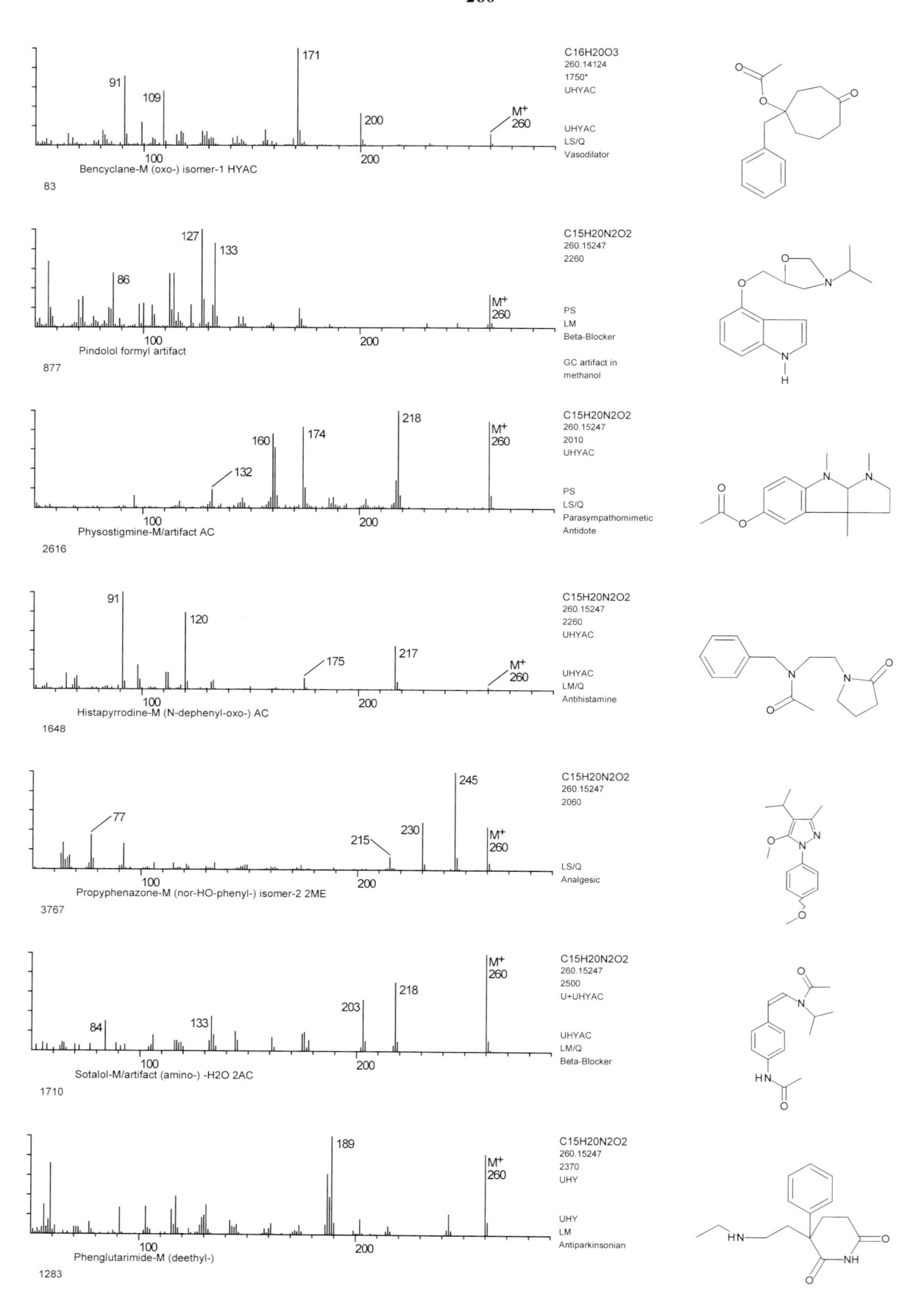

C16H20O3
260.14124
1750*
UHYAC

UHYAC
LS/Q
Vasodilator

Bencyclane-M (oxo-) isomer-1 HYAC
83

C15H20N2O2
260.15247
2260

PS
LM
Beta-Blocker

GC artifact in methanol

Pindolol formyl artifact
877

C15H20N2O2
260.15247
2010
UHYAC

PS
LS/Q
Parasympathomimetic
Antidote

Physostigmine-M/artifact AC
2616

C15H20N2O2
260.15247
2260
UHYAC

UHYAC
LM/Q
Antihistamine

Histapyrrodine-M (N-dephenyl-oxo-) AC
1648

C15H20N2O2
260.15247
2060

LS/Q
Analgesic

Propyphenazone-M (nor-HO-phenyl-) isomer-2 2ME
3767

C15H20N2O2
260.15247
2500
U+UHYAC

UHYAC
LM/Q
Beta-Blocker

Sotalol-M/artifact (amino-) -H2O 2AC
1710

C15H20N2O2
260.15247
2370
UHY

UHY
LM
Antiparkinsonian

Phenglutarimide-M (deethyl-)
1283

260

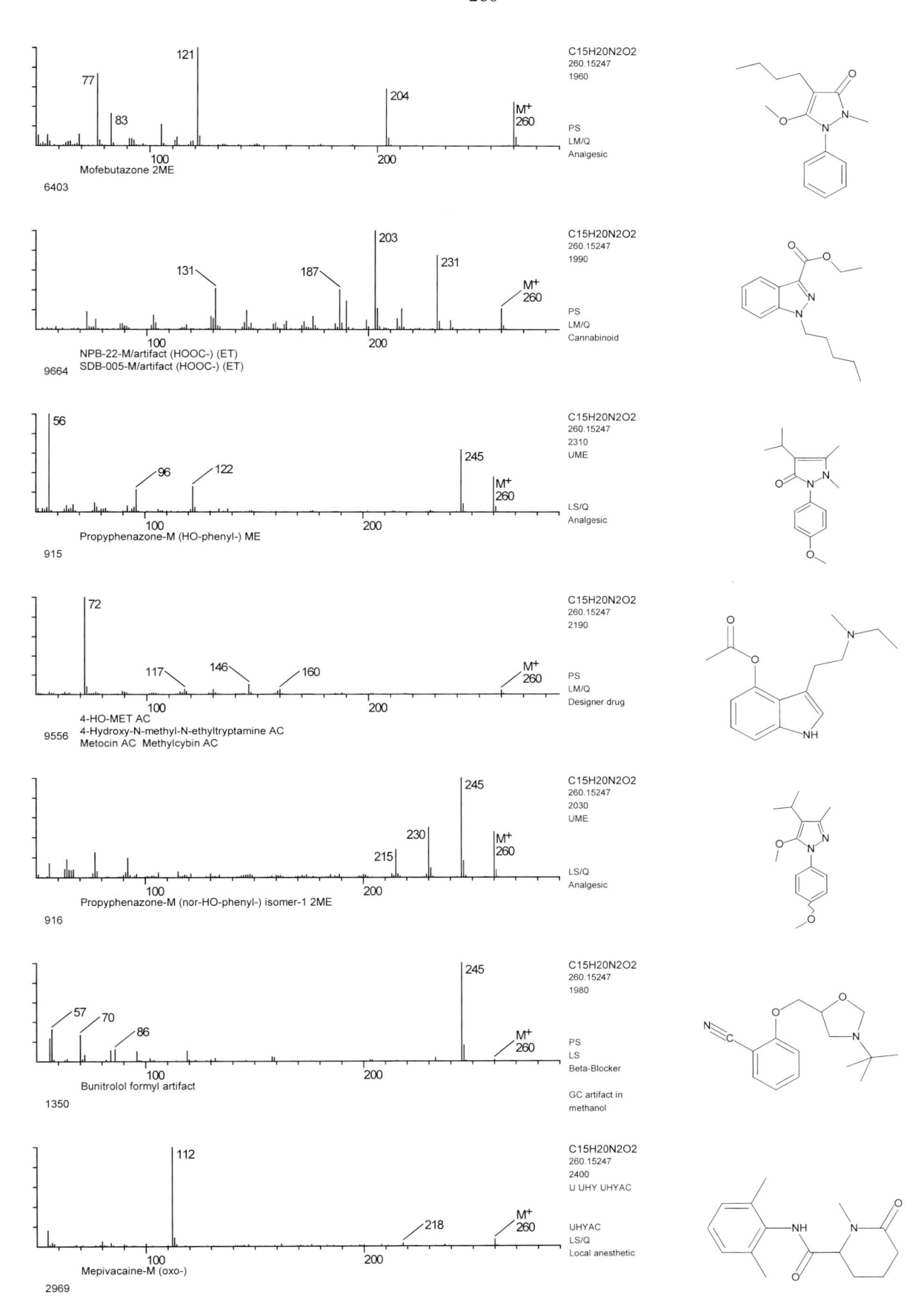

Mofebutazone 2ME
6403
77, 83, 121, 204, M+ 260
C15H20N2O2
260.15247
1960
PS
LM/Q
Analgesic

NPB-22-M/artifact (HOOC-) (ET)
SDB-005-M/artifact (HOOC-) (ET)
9664
131, 187, 203, 231, M+ 260
C15H20N2O2
260.15247
1990
PS
LM/Q
Cannabinoid

Propyphenazone-M (HO-phenyl-) ME
915
56, 96, 122, 245, M+ 260
C15H20N2O2
260.15247
2310
UME
LS/Q
Analgesic

4-HO-MET AC
4-Hydroxy-N-methyl-N-ethyltryptamine AC
Metocin AC Methylcybin AC
9556
72, 117, 146, 160, M+ 260
C15H20N2O2
260.15247
2190
PS
LM/Q
Designer drug

Propyphenazone-M (nor-HO-phenyl-) isomer-1 2ME
916
215, 230, 245, M+ 260
C15H20N2O2
260.15247
2030
UME
LS/Q
Analgesic

Bunitrolol formyl artifact
1350
57, 70, 86, 245, M+ 260
C15H20N2O2
260.15247
1980
PS
LS
Beta-Blocker
GC artifact in methanol

Mepivacaine-M (oxo-)
2969
112, 218, M+ 260
C15H20N2O2
260.15247
2400
U UHY UHYAC
UHYAC
LS/Q
Local anesthetic

260

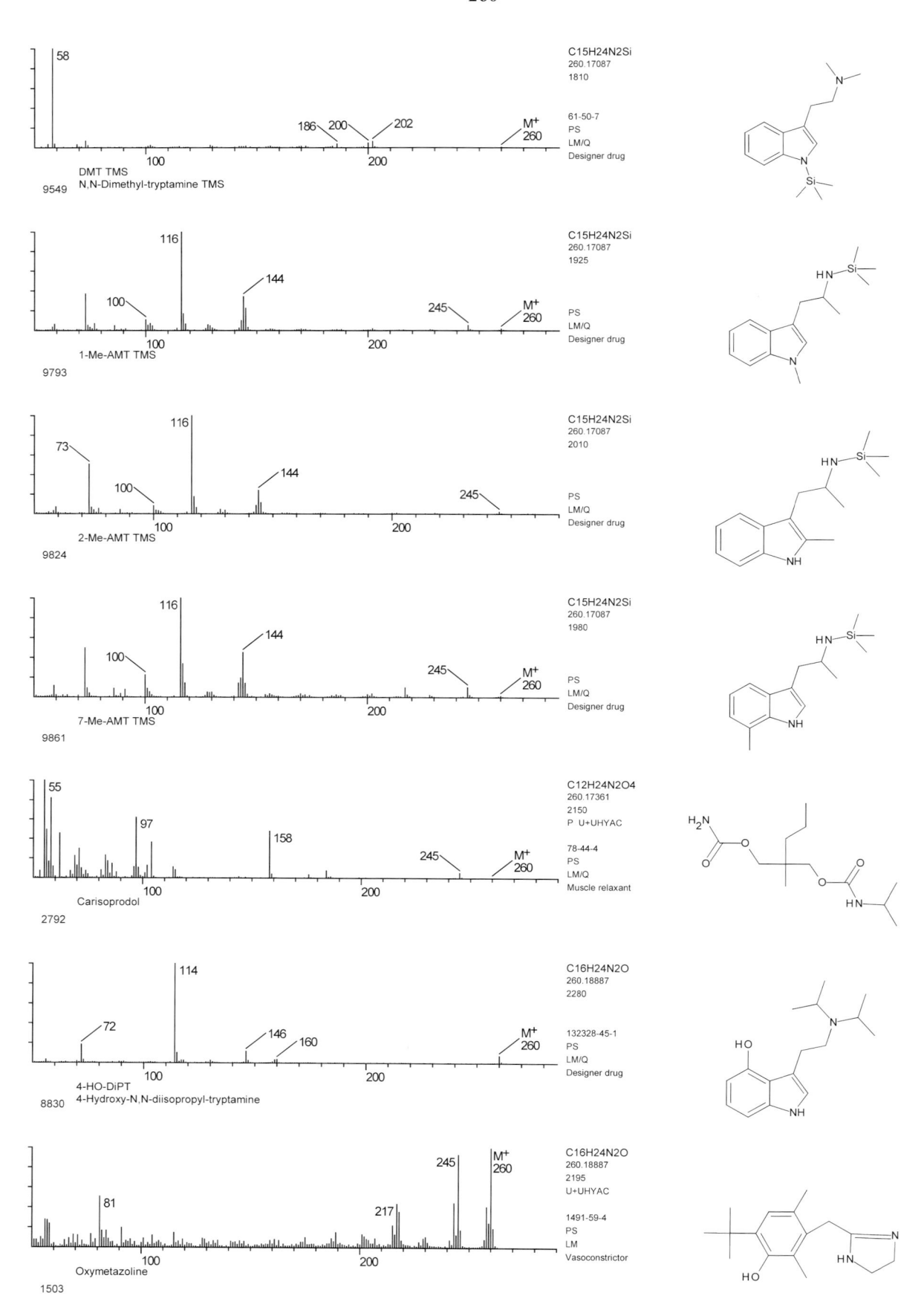

DMT TMS — N,N-Dimethyl-tryptamine TMS (9549)	C15H24N2Si, 260.17087, 1810, 61-50-7, PS, LM/Q, Designer drug. Peaks: 58, 186, 200, 202, M+ 260
1-Me-AMT TMS (9793)	C15H24N2Si, 260.17087, 1925, PS, LM/Q, Designer drug. Peaks: 100, 116, 144, 245, M+ 260
2-Me-AMT TMS (9824)	C15H24N2Si, 260.17087, 2010, PS, LM/Q, Designer drug. Peaks: 73, 100, 116, 144, 245
7-Me-AMT TMS (9861)	C15H24N2Si, 260.17087, 1980, PS, LM/Q, Designer drug. Peaks: 100, 116, 144, 245, M+ 260
Carisoprodol (2792)	C12H24N2O4, 260.17361, 2150, P U+UHYAC, 78-44-4, PS, LM/Q, Muscle relaxant. Peaks: 55, 97, 158, 245, M+ 260
4-HO-DiPT — 4-Hydroxy-N,N-diisopropyl-tryptamine (8830)	C16H24N2O, 260.18887, 2280, 132328-45-1, PS, LM/Q, Designer drug. Peaks: 72, 114, 146, 160, M+ 260
Oxymetazoline (1503)	C16H24N2O, 260.18887, 2195, U+UHYAC, 1491-59-4, PS, LM, Vasoconstrictor. Peaks: 81, 217, 245, M+ 260

660

260

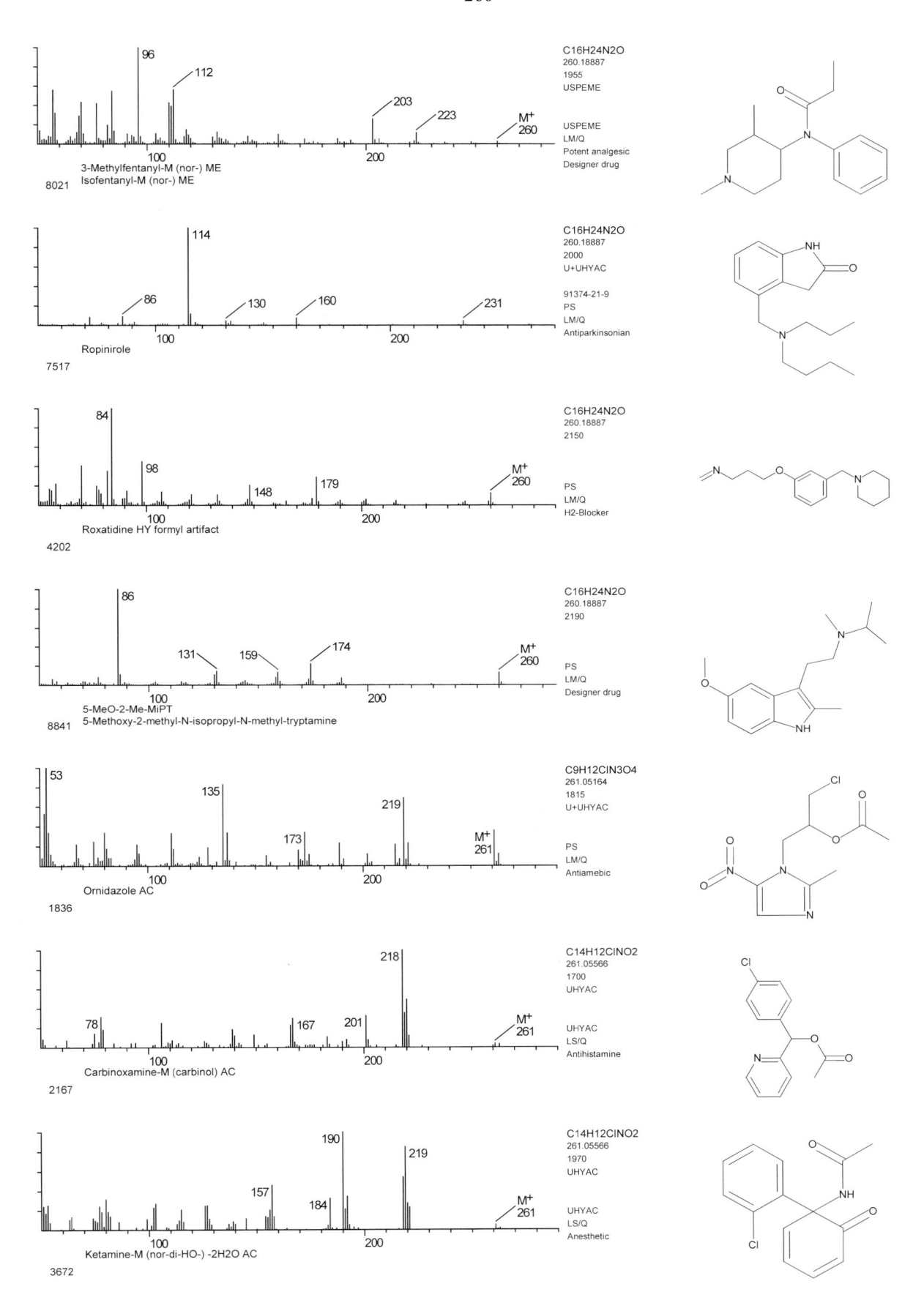

8021	3-Methylfentanyl-M (nor-) ME Isofentanyl-M (nor-) ME peaks: 96, 112, 203, 223, M+ 260	C16H24N2O 260.18887 1955 USPEME USPEME LM/Q Potent analgesic Designer drug
7517	Ropinirole peaks: 86, 114, 130, 160, 231	C16H24N2O 260.18887 2000 U+UHYAC 91374-21-9 PS LM/Q Antiparkinsonian
4202	Roxatidine HY formyl artifact peaks: 84, 98, 148, 179, M+ 260	C16H24N2O 260.18887 2150 PS LM/Q H2-Blocker
8841	5-MeO-2-Me-MiPT 5-Methoxy-2-methyl-N-isopropyl-N-methyl-tryptamine peaks: 86, 131, 159, 174, M+ 260	C16H24N2O 260.18887 2190 PS LM/Q Designer drug
1836	Ornidazole AC peaks: 53, 135, 173, 219, M+ 261	C9H12ClN3O4 261.05164 1815 U+UHYAC PS LM/Q Antiamebic
2167	Carbinoxamine-M (carbinol) AC peaks: 78, 167, 201, 218, M+ 261	C14H12ClNO2 261.05566 1700 UHYAC UHYAC LS/Q Antihistamine
3672	Ketamine-M (nor-di-HO-) -2H2O AC peaks: 157, 184, 190, 219, M+ 261	C14H12ClNO2 261.05566 1970 UHYAC UHYAC LS/Q Anesthetic

661

261

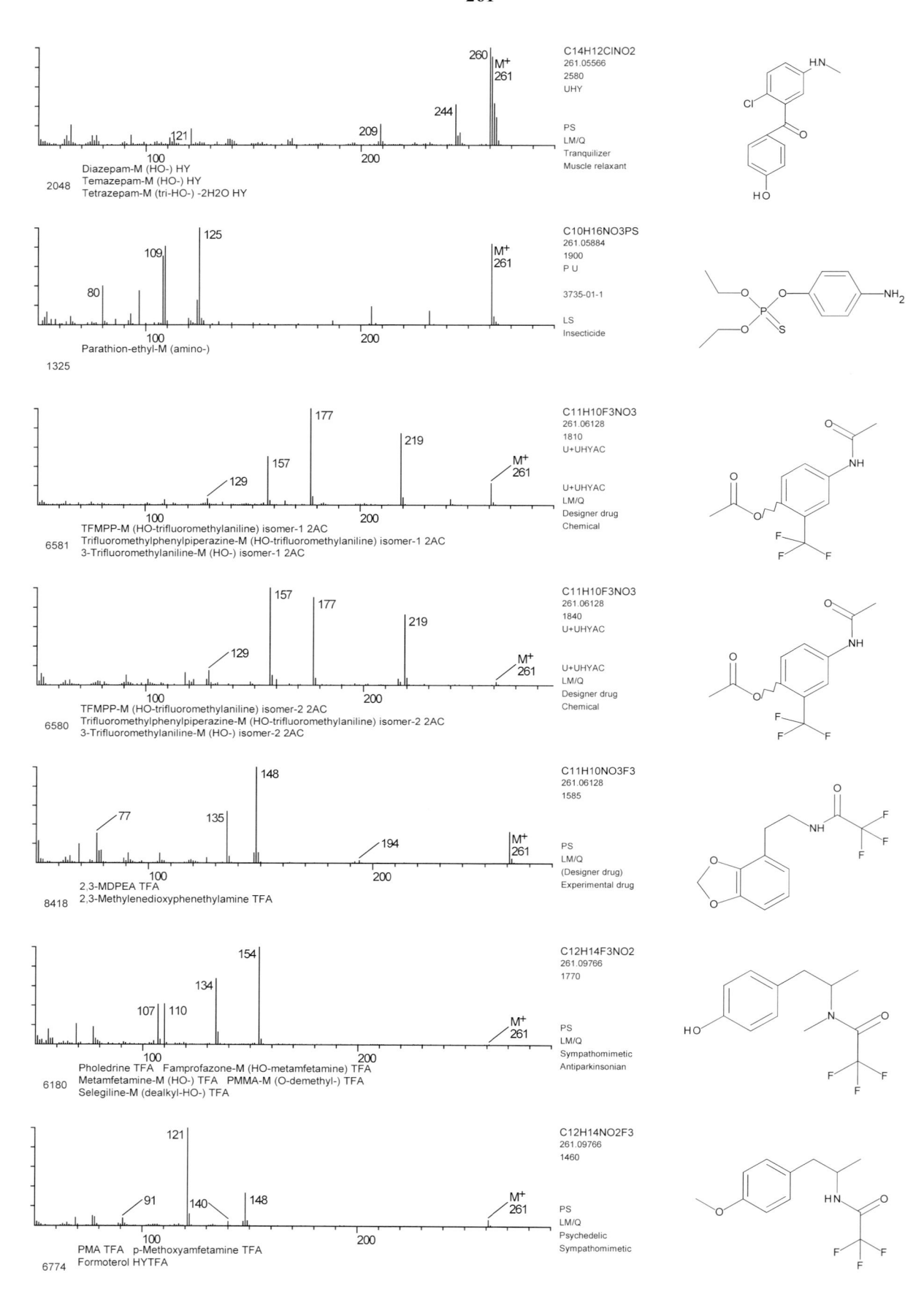

C14H12ClNO2
261.05566
2580
UHY

PS
LM/Q
Tranquilizer
Muscle relaxant

Diazepam-M (HO-) HY
Temazepam-M (HO-) HY
Tetrazepam-M (tri-HO-) -2H2O HY
2048

C10H16NO3PS
261.05884
1900
P U

3735-01-1

LS
Insecticide

Parathion-ethyl-M (amino-)
1325

C11H10F3NO3
261.06128
1810
U+UHYAC

U+UHYAC
LM/Q
Designer drug
Chemical

TFMPP-M (HO-trifluoromethylaniline) isomer-1 2AC
Trifluoromethylphenylpiperazine-M (HO-trifluoromethylaniline) isomer-1 2AC
6581   3-Trifluoromethylaniline-M (HO-) isomer-1 2AC

C11H10F3NO3
261.06128
1840
U+UHYAC

U+UHYAC
LM/Q
Designer drug
Chemical

TFMPP-M (HO-trifluoromethylaniline) isomer-2 2AC
Trifluoromethylphenylpiperazine-M (HO-trifluoromethylaniline) isomer-2 2AC
6580   3-Trifluoromethylaniline-M (HO-) isomer-2 2AC

C11H10NO3F3
261.06128
1585

PS
LM/Q
(Designer drug)
Experimental drug

2,3-MDPEA TFA
8418 2,3-Methylenedioxyphenethylamine TFA

C12H14F3NO2
261.09766
1770

PS
LM/Q
Sympathomimetic
Antiparkinsonian

Pholedrine TFA   Famprofazone-M (HO-metamfetamine) TFA
Metamfetamine-M (HO-) TFA   PMMA-M (O-demethyl-) TFA
6180   Selegiline-M (dealkyl-HO-) TFA

C12H14NO2F3
261.09766
1460

PS
LM/Q
Psychedelic
Sympathomimetic

PMA TFA   p-Methoxyamfetamine TFA
6774   Formoterol HYTFA

261

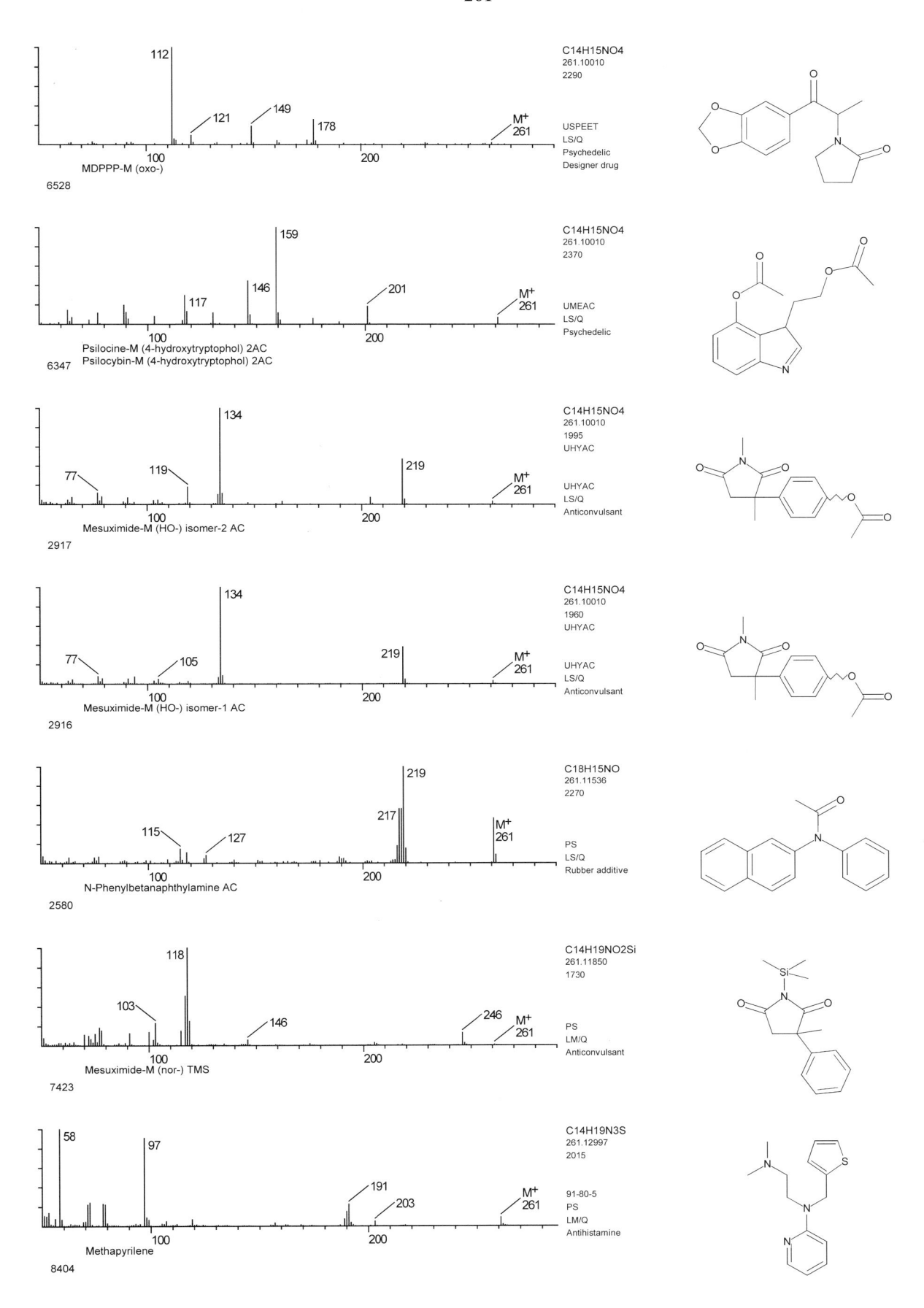

112, 121, 149, 178, M+ 261	C14H15NO4
261.10010
2290
USPEET
LS/Q
Psychedelic
Designer drug

MDPPP-M (oxo-)
6528

117, 146, 159, 201, M+ 261	C14H15NO4
261.10010
2370
UMEAC
LS/Q
Psychedelic

Psilocine-M (4-hydroxytryptophol) 2AC
Psilocybin-M (4-hydroxytryptophol) 2AC
6347

77, 119, 134, 219, M+ 261	C14H15NO4
261.10010
1995
UHYAC
UHYAC
LS/Q
Anticonvulsant

Mesuximide-M (HO-) isomer-2 AC
2917

77, 105, 134, 219, M+ 261	C14H15NO4
261.10010
1960
UHYAC
UHYAC
LS/Q
Anticonvulsant

Mesuximide-M (HO-) isomer-1 AC
2916

115, 127, 217, 219, M+ 261	C18H15NO
261.11536
2270
PS
LS/Q
Rubber additive

N-Phenylbetanaphthylamine AC
2580

103, 118, 146, 246, M+ 261	C14H19NO2Si
261.11850
1730
PS
LM/Q
Anticonvulsant

Mesuximide-M (nor-) TMS
7423

58, 97, 191, 203, M+ 261	C14H19N3S
261.12997
2015
91-80-5
PS
LM/Q
Antihistamine

Methapyrilene
8404

261

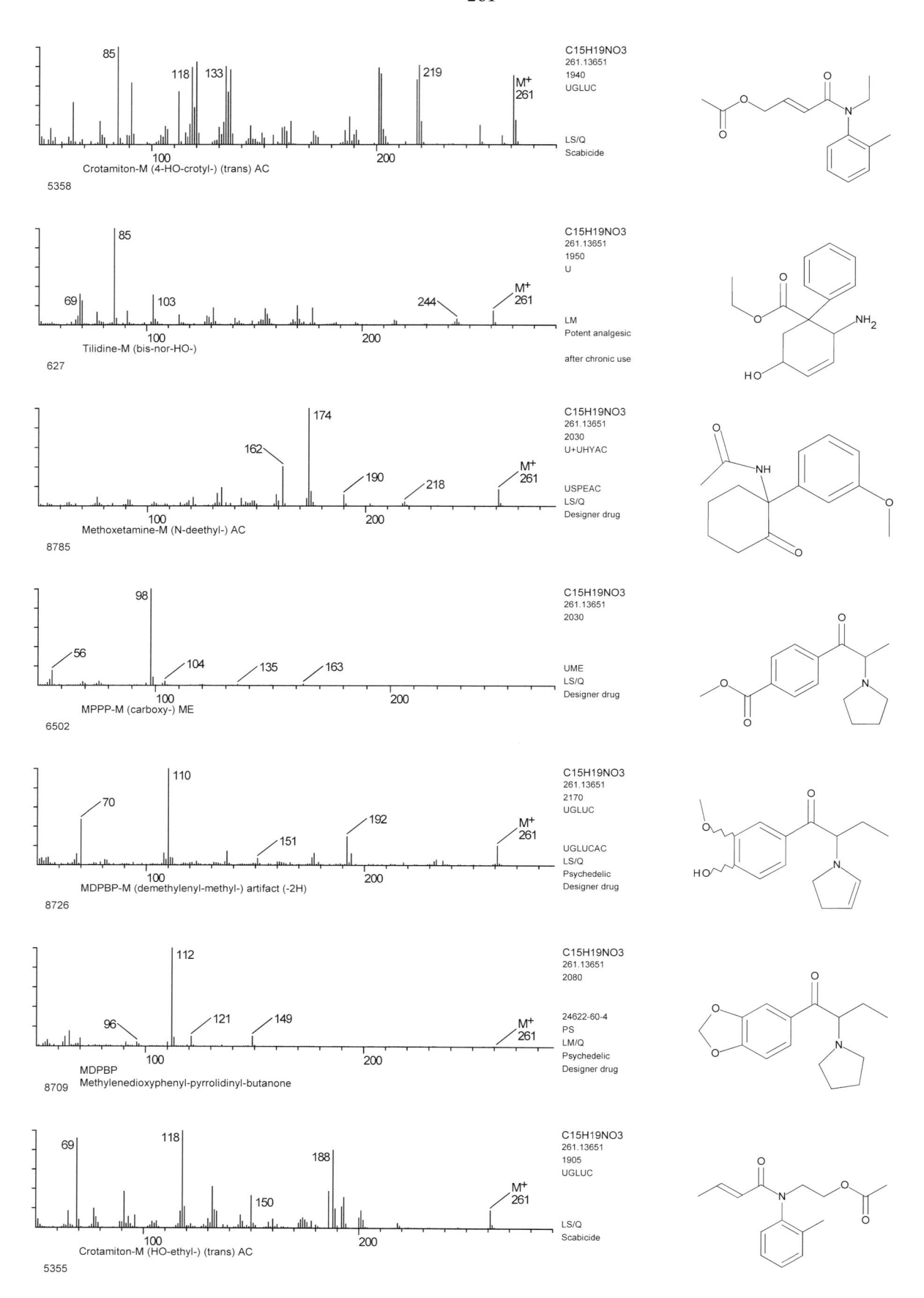

Crotamiton-M (4-HO-crotyl-) (trans) AC
5358
C15H19NO3
261.13651
1940
UGLUC
LS/Q
Scabicide

Tilidine-M (bis-nor-HO-)
627
C15H19NO3
261.13651
1950
U
LM
Potent analgesic
after chronic use

Methoxetamine-M (N-deethyl-) AC
8785
C15H19NO3
261.13651
2030
U+UHYAC
USPEAC
LS/Q
Designer drug

MPPP-M (carboxy-) ME
6502
C15H19NO3
261.13651
2030
UME
LS/Q
Designer drug

MDPBP-M (demethylenyl-methyl-) artifact (-2H)
8726
C15H19NO3
261.13651
2170
UGLUC
UGLUCAC
LS/Q
Psychedelic
Designer drug

MDPBP
Methylenedioxyphenyl-pyrrolidinyl-butanone
8709
C15H19NO3
261.13651
2080
24622-60-4
PS
LM/Q
Psychedelic
Designer drug

Crotamiton-M (HO-ethyl-) (trans) AC
5355
C15H19NO3
261.13651
1905
UGLUC
LS/Q
Scabicide

261

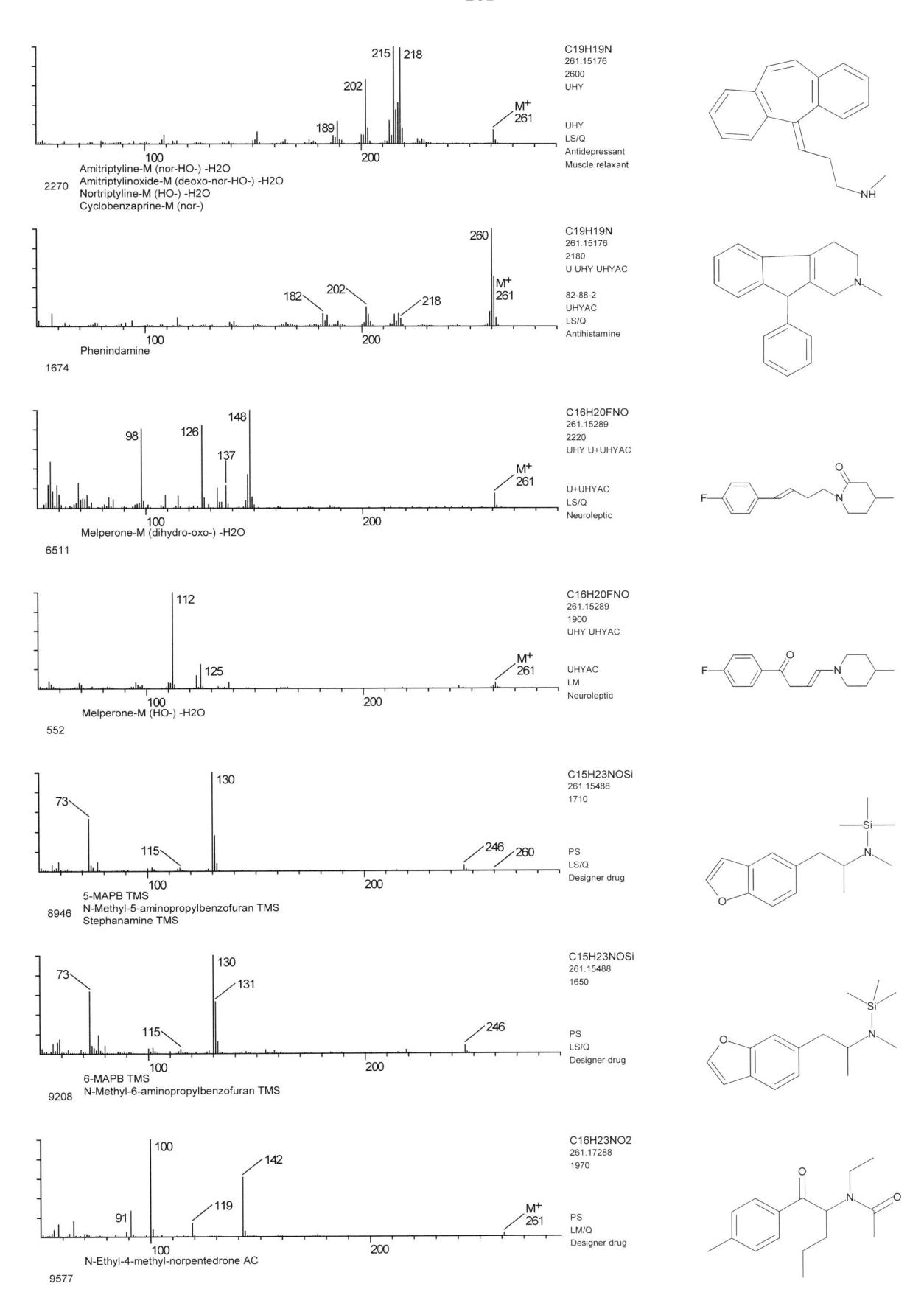

C19H19N
261.15176
2600
UHY

UHY
LS/Q
Antidepressant
Muscle relaxant

Amitriptyline-M (nor-HO-) -H2O
Amitriptylinoxide-M (deoxo-nor-HO-) -H2O
Nortriptyline-M (HO-) -H2O
Cyclobenzaprine-M (nor-)

2270

C19H19N
261.15176
2180
U UHY UHYAC

82-88-2
UHYAC
LS/Q
Antihistamine

Phenindamine
1674

C16H20FNO
261.15289
2220
UHY U+UHYAC

U+UHYAC
LS/Q
Neuroleptic

Melperone-M (dihydro-oxo-) -H2O
6511

C16H20FNO
261.15289
1900
UHY UHYAC

UHYAC
LM
Neuroleptic

Melperone-M (HO-) -H2O
552

C15H23NOSi
261.15488
1710

PS
LS/Q
Designer drug

5-MAPB TMS
N-Methyl-5-aminopropylbenzofuran TMS
Stephanamine TMS
8946

C15H23NOSi
261.15488
1650

PS
LS/Q
Designer drug

6-MAPB TMS
N-Methyl-6-aminopropylbenzofuran TMS
9208

C16H23NO2
261.17288
1970

PS
LM/Q
Designer drug

N-Ethyl-4-methyl-norpentedrone AC
9577

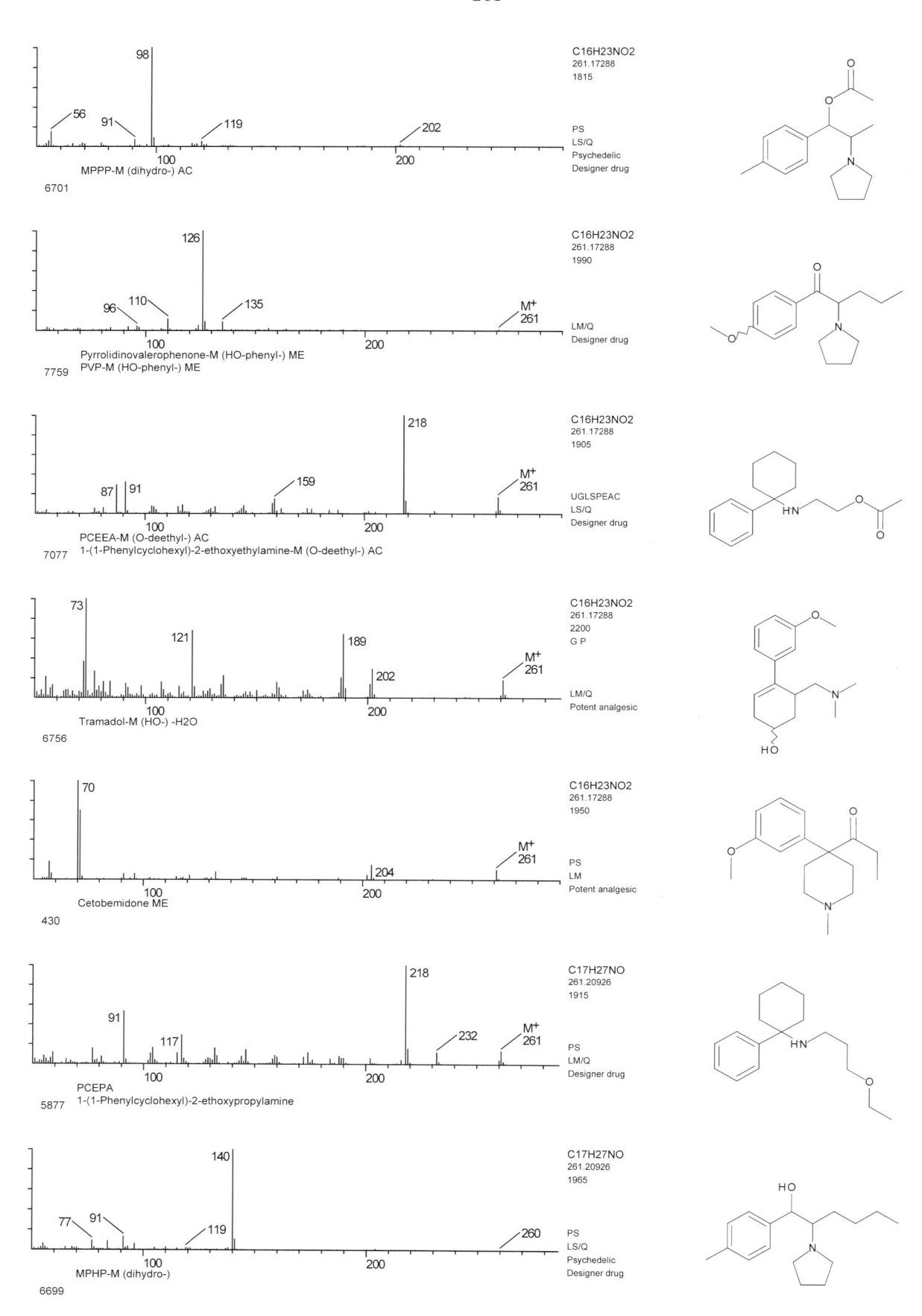

Spectrum peaks	Compound	Formula / Info
56, 91, 98, 119, 202	MPPP-M (dihydro-) AC — 6701	C16H23NO2 / 261.17288 / 1815 / PS / LS/Q / Psychedelic / Designer drug
96, 110, 126, 135, 261 M+	Pyrrolidinovalerophenone-M (HO-phenyl-) ME / PVP-M (HO-phenyl-) ME — 7759	C16H23NO2 / 261.17288 / 1990 / LM/Q / Designer drug
87, 91, 159, 218, 261 M+	PCEEA-M (O-deethyl-) AC / 1-(1-Phenylcyclohexyl)-2-ethoxyethylamine-M (O-deethyl-) AC — 7077	C16H23NO2 / 261.17288 / 1905 / UGLSPEAC / LS/Q / Designer drug
73, 121, 189, 202, 261 M+	Tramadol-M (HO-) -H2O — 6756	C16H23NO2 / 261.17288 / 2200 / G P / LM/Q / Potent analgesic
70, 204, 261 M+	Cetobemidone ME — 430	C16H23NO2 / 261.17288 / 1950 / PS / LM / Potent analgesic
91, 117, 218, 232, 261 M+	PCEPA / 1-(1-Phenylcyclohexyl)-2-ethoxypropylamine — 5877	C17H27NO / 261.20926 / 1915 / PS / LM/Q / Designer drug
77, 91, 119, 140, 260	MPHP-M (dihydro-) — 6699	C17H27NO / 261.20926 / 1965 / PS / LS/Q / Psychedelic / Designer drug

261

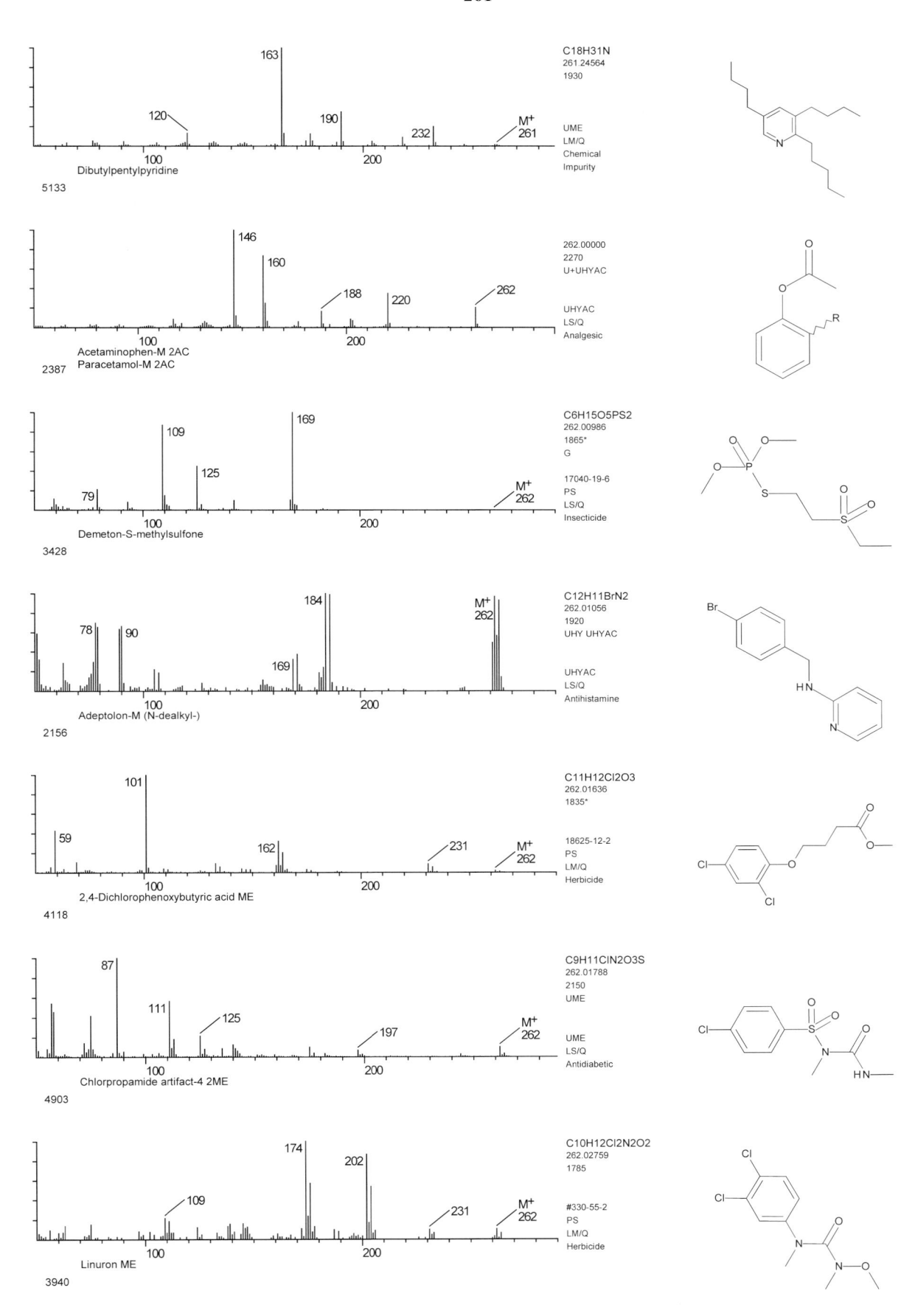

Dibutylpentylpyridine — C18H31N, 261.24564, 1930, UME, LM/Q, Chemical Impurity
Peaks: 120, 163, 190, 232, M+ 261
5133

Acetaminophen-M 2AC / Paracetamol-M 2AC — 262.00000, 2270, U+UHYAC, UHYAC, LS/Q, Analgesic
Peaks: 146, 160, 188, 220, 262
2387

Demeton-S-methylsulfone — C6H15O5PS2, 262.00986, 1865*, G, 17040-19-6, PS, LS/Q, Insecticide
Peaks: 79, 109, 125, 169, M+ 262
3428

Adeptolon-M (N-dealkyl-) — C12H11BrN2, 262.01056, 1920, UHY UHYAC, UHYAC, LS/Q, Antihistamine
Peaks: 78, 90, 169, 184, M+ 262
2156

2,4-Dichlorophenoxybutyric acid ME — C11H12Cl2O3, 262.01636, 1835*, 18625-12-2, PS, LM/Q, Herbicide
Peaks: 59, 101, 162, 231, M+ 262
4118

Chlorpropamide artifact-4 2ME — C9H11ClN2O3S, 262.01788, 2150, UME, UME, LS/Q, Antidiabetic
Peaks: 87, 111, 125, 197, M+ 262
4903

Linuron ME — C10H12Cl2N2O2, 262.02759, 1785, #330-55-2, PS, LM/Q, Herbicide
Peaks: 109, 174, 202, 231, M+ 262
3940

667

262

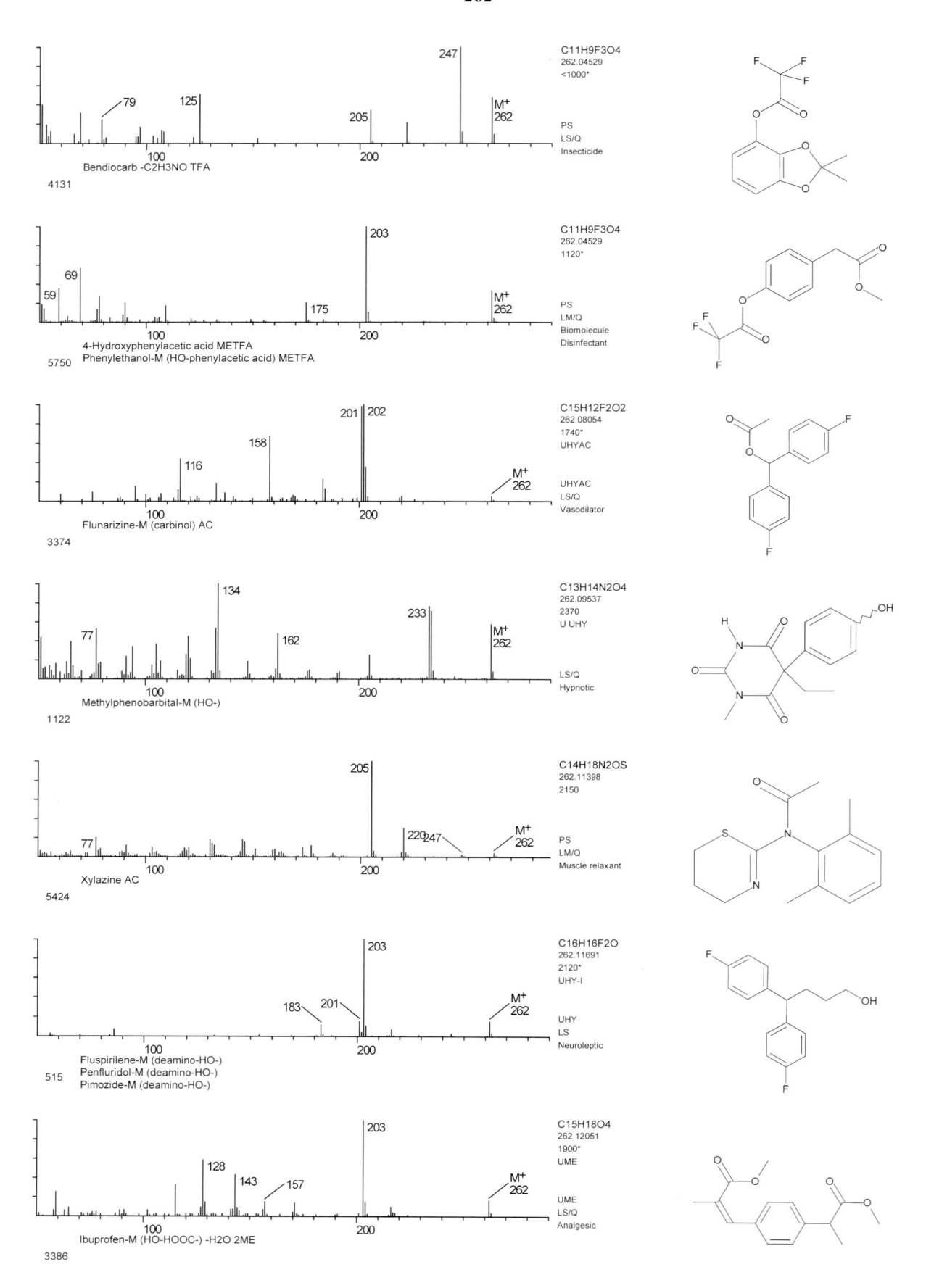

C11H9F3O4 262.04529 <1000* PS LS/Q Insecticide	Bendiocarb -C2H3NO TFA peaks: 79, 125, 205, 247, M+ 262 4131
C11H9F3O4 262.04529 1120* PS LM/Q Biomolecule Disinfectant	4-Hydroxyphenylacetic acid METFA Phenylethanol-M (HO-phenylacetic acid) METFA peaks: 59, 69, 175, 203, M+ 262 5750
C15H12F2O2 262.08054 1740* UHYAC UHYAC LS/Q Vasodilator	Flunarizine-M (carbinol) AC peaks: 116, 158, 201, 202, M+ 262 3374
C13H14N2O4 262.09537 2370 U UHY LS/Q Hypnotic	Methylphenobarbital-M (HO-) peaks: 77, 134, 162, 233, M+ 262 1122
C14H18N2OS 262.11398 2150 PS LM/Q Muscle relaxant	Xylazine AC peaks: 77, 205, 220, 247, M+ 262 5424
C16H16F2O 262.11691 2120* UHY-I UHY LS Neuroleptic	Fluspirilene-M (deamino-HO-) Penfluridol-M (deamino-HO-) Pimozide-M (deamino-HO-) peaks: 183, 201, 203, M+ 262 515
C15H18O4 262.12051 1900* UME UME LS/Q Analgesic	Ibuprofen-M (HO-HOOC-) -H2O 2ME peaks: 128, 143, 157, 203, M+ 262 3386

262

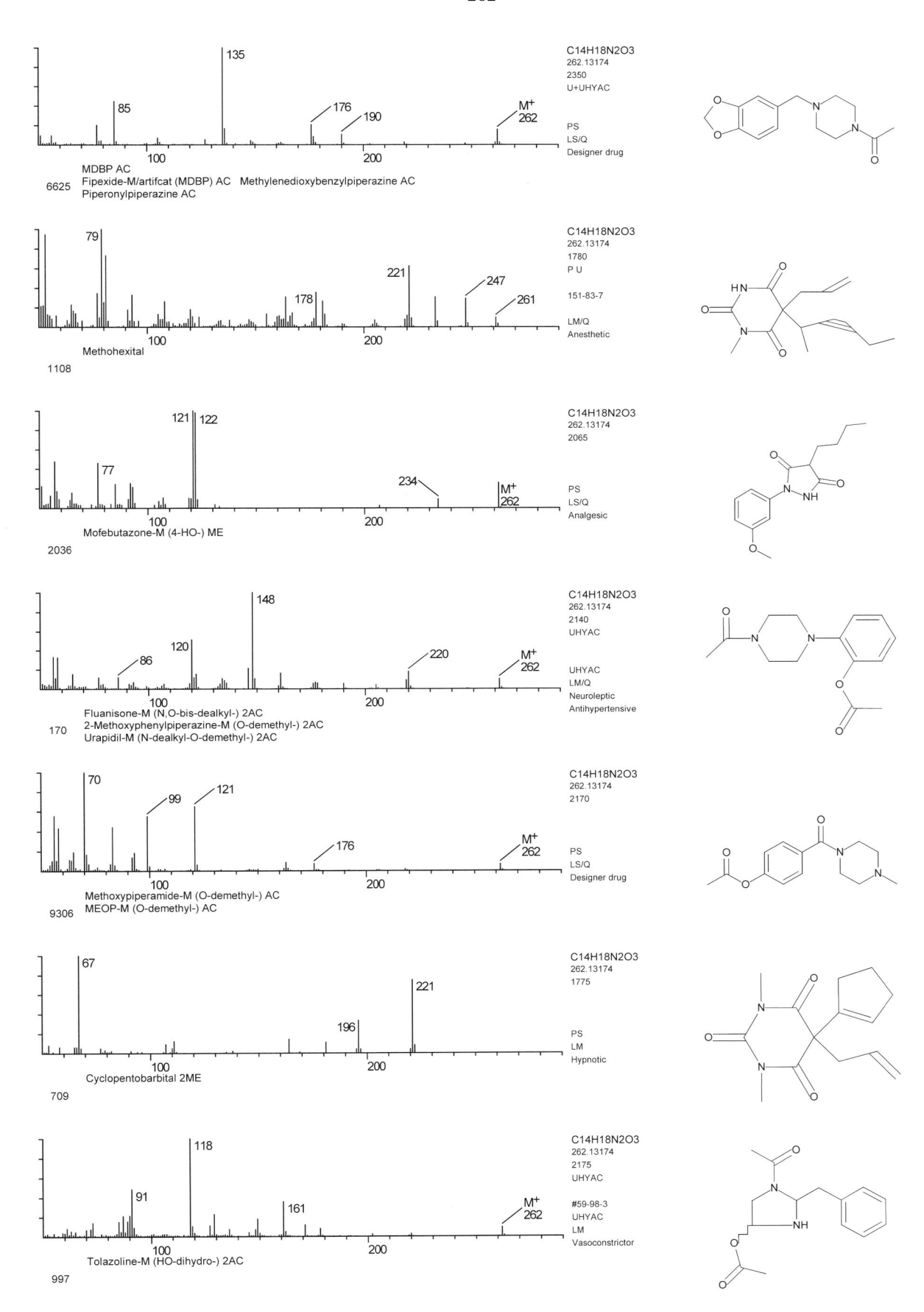

C14H18N2O3
262.13174
2350
U+UHYAC

PS
LS/Q
Designer drug

MDBP AC
Fipexide-M/artifcat (MDBP) AC   Methylenedioxybenzylpiperazine AC
Piperonylpiperazine AC
6625

C14H18N2O3
262.13174
1780
P U

151-83-7

LM/Q
Anesthetic

Methohexital
1108

C14H18N2O3
262.13174
2065

PS
LS/Q
Analgesic

Mofebutazone-M (4-HO-) ME
2036

C14H18N2O3
262.13174
2140
UHYAC

UHYAC
LM/Q
Neuroleptic
Antihypertensive

Fluanisone-M (N,O-bis-dealkyl-) 2AC
2-Methoxyphenylpiperazine-M (O-demethyl-) 2AC
Urapidil-M (N-dealkyl-O-demethyl-) 2AC
170

C14H18N2O3
262.13174
2170

PS
LS/Q
Designer drug

Methoxypiperamide-M (O-demethyl-) AC
MEOP-M (O-demethyl-) AC
9306

C14H18N2O3
262.13174
1775

PS
LM
Hypnotic

Cyclopentobarbital 2ME
709

C14H18N2O3
262.13174
2175
UHYAC

#59-98-3
UHYAC
LM
Vasoconstrictor

Tolazoline-M (HO-dihydro-) 2AC
997

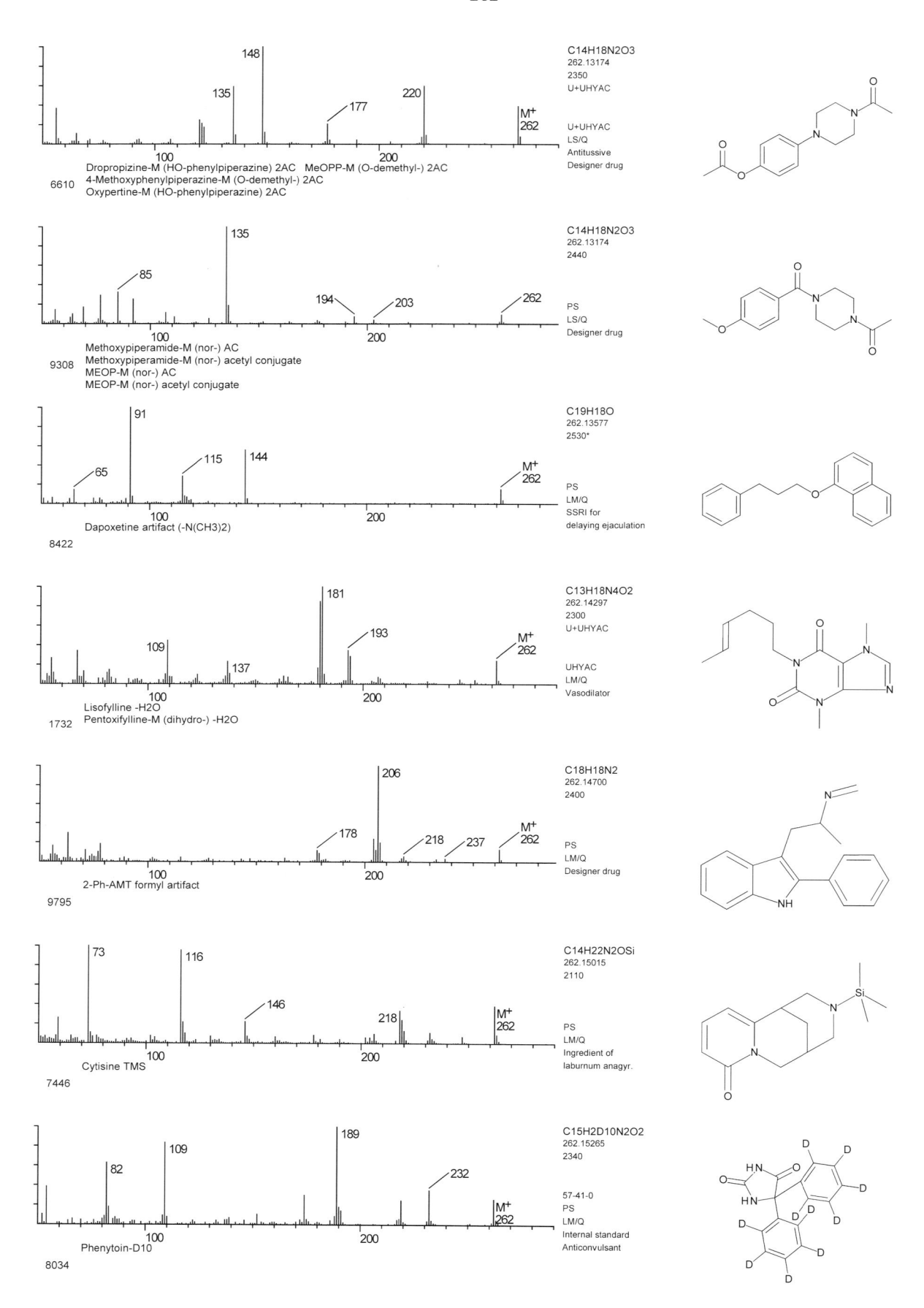

6610	C14H18N2O3 262.13174 2350 U+UHYAC U+UHYAC LS/Q Antitussive Designer drug

Dropropizine-M (HO-phenylpiperazine) 2AC   MeOPP-M (O-demethyl-) 2AC
4-Methoxyphenylpiperazine-M (O-demethyl-) 2AC
Oxypertine-M (HO-phenylpiperazine) 2AC

Peaks: 135, 148, 177, 220, M+ 262

9308	C14H18N2O3 262.13174 2440 PS LS/Q Designer drug

Methoxypiperamide-M (nor-) AC
Methoxypiperamide-M (nor-) acetyl conjugate
MEOP-M (nor-) AC
MEOP-M (nor-) acetyl conjugate

Peaks: 85, 135, 194, 203, 262

8422	C19H18O 262.13577 2530* PS LM/Q SSRI for delaying ejaculation

Dapoxetine artifact (-N(CH3)2)

Peaks: 65, 91, 115, 144, M+ 262

1732	C13H18N4O2 262.14297 2300 U+UHYAC UHYAC LM/Q Vasodilator

Lisofylline -H2O
Pentoxifylline-M (dihydro-) -H2O

Peaks: 109, 137, 181, 193, M+ 262

9795	C18H18N2 262.14700 2400 PS LM/Q Designer drug

2-Ph-AMT formyl artifact

Peaks: 178, 206, 218, 237, M+ 262

7446	C14H22N2OSi 262.15015 2110 PS LM/Q Ingredient of laburnum anagyr.

Cytisine TMS

Peaks: 73, 116, 146, 218, M+ 262

8034	C15H2D10N2O2 262.15265 2340 57-41-0 PS LM/Q Internal standard Anticonvulsant

Phenytoin-D10

Peaks: 82, 109, 189, 232, M+ 262

262

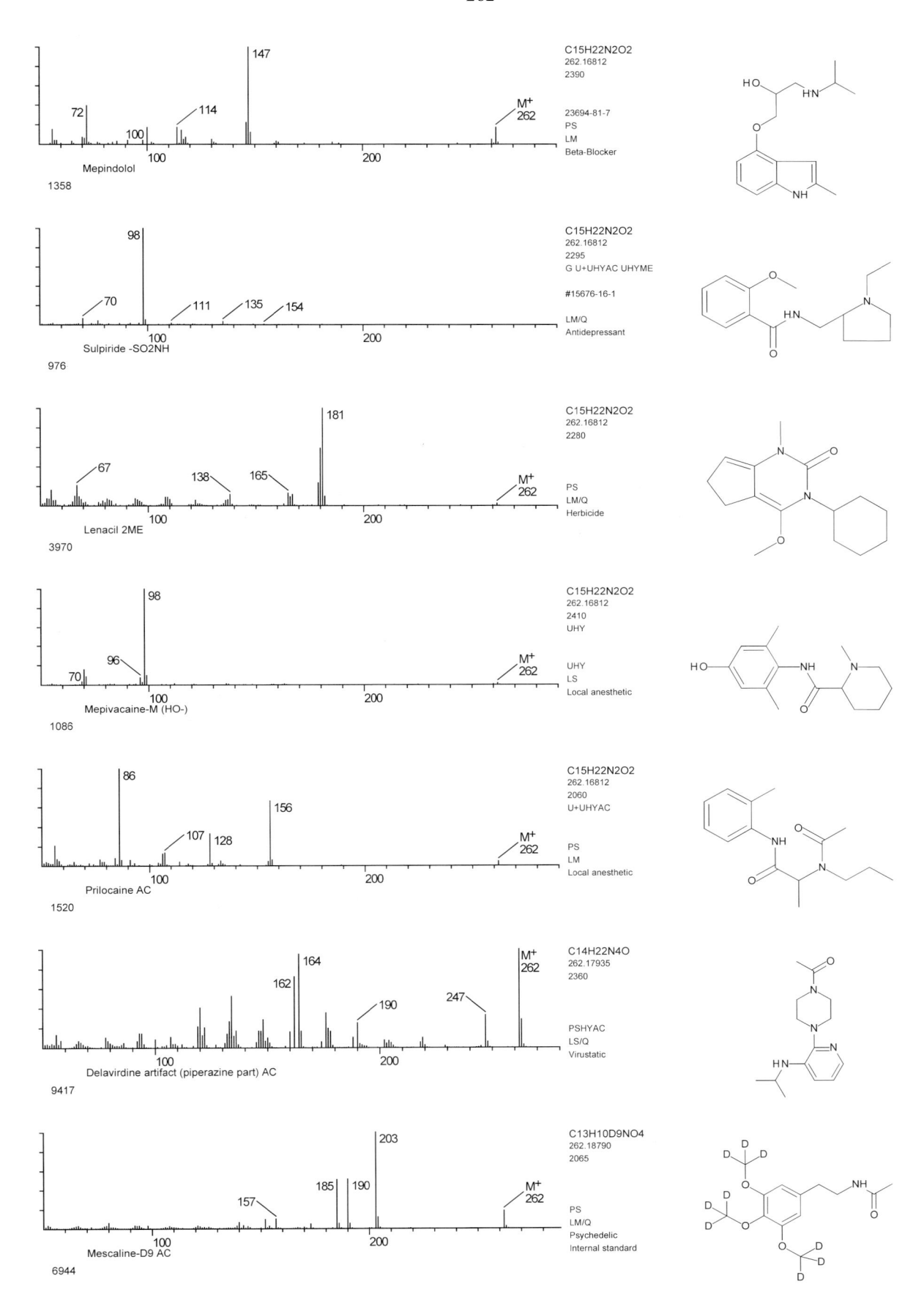

Mepindolol	C15H22N2O2 262.16812 2390	72, 100, 114, 147, M+ 262
	23694-81-7 PS LM Beta-Blocker	
1358		

Sulpiride -SO2NH	C15H22N2O2 262.16812 2295 G U+UHYAC UHYME	70, 98, 111, 135, 154
	#15676-16-1 LM/Q Antidepressant	
976		

Lenacil 2ME	C15H22N2O2 262.16812 2280	67, 138, 165, 181, M+ 262
	PS LM/Q Herbicide	
3970		

Mepivacaine-M (HO-)	C15H22N2O2 262.16812 2410 UHY	70, 96, 98, M+ 262
	UHY LS Local anesthetic	
1086		

Prilocaine AC	C15H22N2O2 262.16812 2060 U+UHYAC	86, 107, 128, 156, M+ 262
	PS LM Local anesthetic	
1520		

Delavirdine artifact (piperazine part) AC	C14H22N4O 262.17935 2360	162, 164, 190, 247, M+ 262
	PSHYAC LS/Q Virustatic	
9417		

Mescaline-D9 AC	C13H10D9NO4 262.18790 2065	157, 185, 190, 203, M+ 262
	PS LM/Q Psychedelic Internal standard	
6944		

671

263

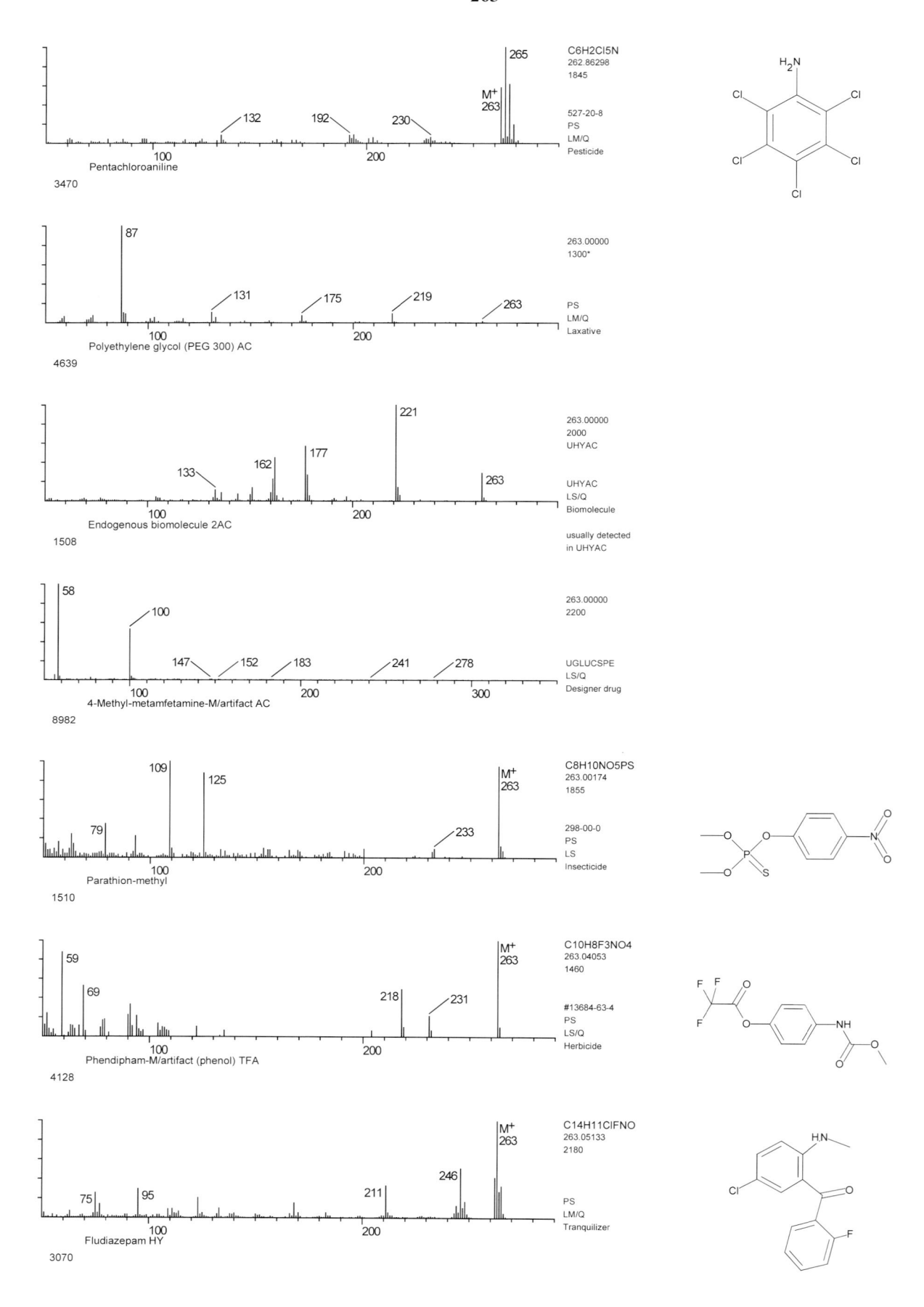

Pentachloroaniline — C6H2Cl5N, 262.86298, 1845, 527-20-8, PS, LM/Q, Pesticide
Peaks: 132, 192, 230, M+ 263, 265
3470

Polyethylene glycol (PEG 300) AC — 263.00000, 1300*, PS, LM/Q, Laxative
Peaks: 87, 131, 175, 219, 263
4639

Endogenous biomolecule 2AC — 263.00000, 2000, UHYAC, UHYAC, LS/Q, Biomolecule, usually detected in UHYAC
Peaks: 133, 162, 177, 221, 263
1508

4-Methyl-metamfetamine-M/artifact AC — 263.00000, 2200, UGLUCSPE, LS/Q, Designer drug
Peaks: 58, 100, 147, 152, 183, 241, 278
8982

Parathion-methyl — C8H10NO5PS, 263.00174, 1855, 298-00-0, PS, LS, Insecticide
Peaks: 79, 109, 125, 233, M+ 263
1510

Phendipham-M/artifact (phenol) TFA — C10H8F3NO4, 263.04053, 1460, #13684-63-4, PS, LS/Q, Herbicide
Peaks: 59, 69, 218, 231, M+ 263
4128

Fludiazepam HY — C14H11ClFNO, 263.05133, 2180, PS, LM/Q, Tranquilizer
Peaks: 75, 95, 211, 246, M+ 263
3070

263

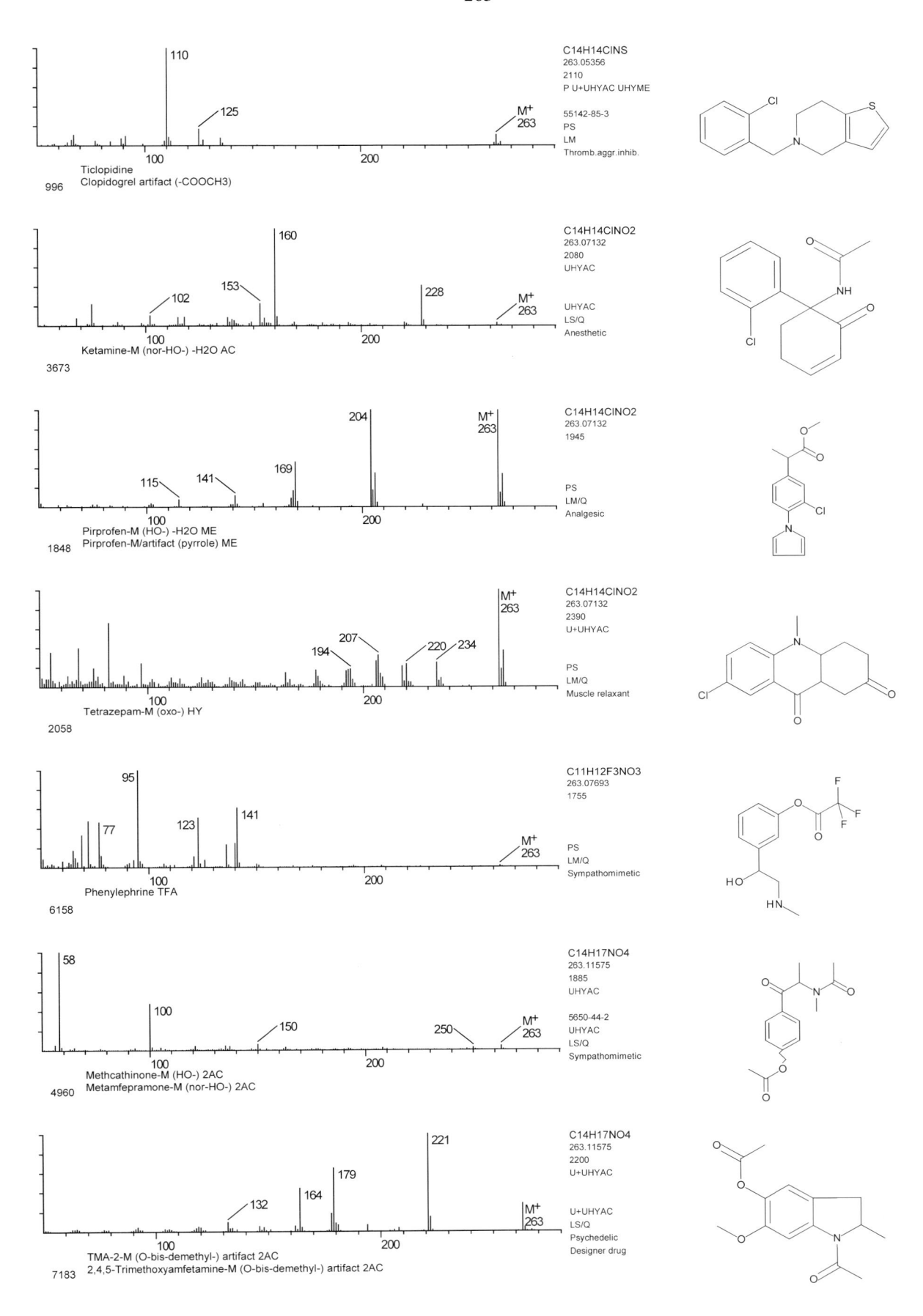

110, 125, M+ 263	Ticlopidine — Clopidogrel artifact (-COOCH3)	C14H14ClNS 263.05356 2110 P U+UHYAC UHYME 55142-85-3 PS LM Thromb.aggr.inhib.
102, 153, 160, 228, M+ 263	Ketamine-M (nor-HO-) -H2O AC	C14H14ClNO2 263.07132 2080 UHYAC UHYAC LS/Q Anesthetic
115, 141, 169, 204, M+ 263	Pirprofen-M (HO-) -H2O ME — Pirprofen-M/artifact (pyrrole) ME	C14H14ClNO2 263.07132 1945 PS LM/Q Analgesic
194, 207, 220, 234, M+ 263	Tetrazepam-M (oxo-) HY	C14H14ClNO2 263.07132 2390 U+UHYAC PS LM/Q Muscle relaxant
77, 95, 123, 141, M+ 263	Phenylephrine TFA	C11H12F3NO3 263.07693 1755 PS LM/Q Sympathomimetic
58, 100, 150, 250, M+ 263	Methcathinone-M (HO-) 2AC — Metamfepramone-M (nor-HO-) 2AC	C14H17NO4 263.11575 1885 UHYAC 5650-44-2 UHYAC LS/Q Sympathomimetic
132, 164, 179, 221, M+ 263	TMA-2-M (O-bis-demethyl-) artifact 2AC — 2,4,5-Trimethoxyamfetamine-M (O-bis-demethyl-) artifact 2AC	C14H17NO4 263.11575 2200 U+UHYAC U+UHYAC LS/Q Psychedelic Designer drug

263

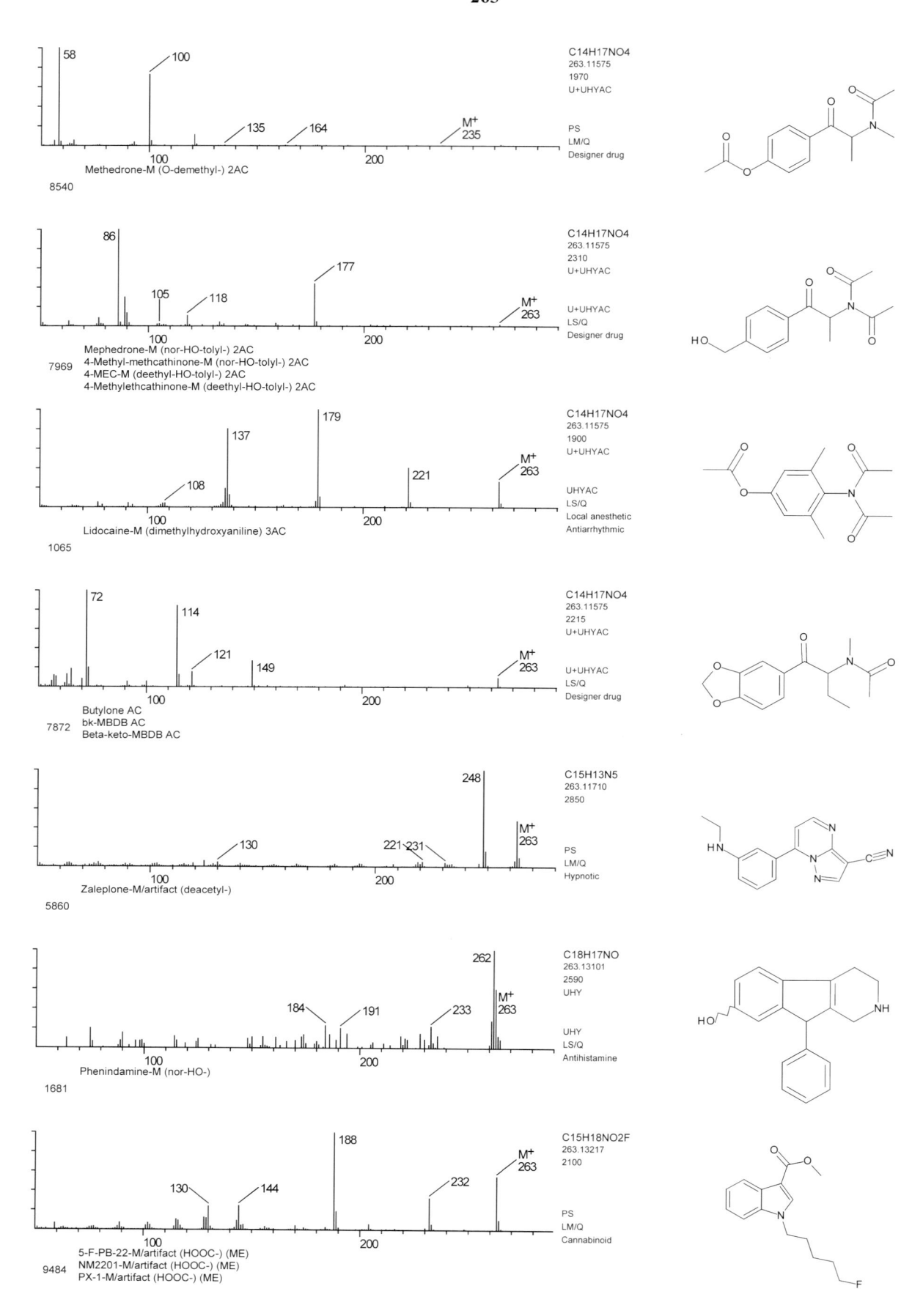

Methedrone-M (O-demethyl-) 2AC
8540
C14H17NO4
263.11575
1970
U+UHYAC
PS
LM/Q
Designer drug

Mephedrone-M (nor-HO-tolyl-) 2AC
4-Methyl-methcathinone-M (nor-HO-tolyl-) 2AC
4-MEC-M (deethyl-HO-tolyl-) 2AC
4-Methylethcathinone-M (deethyl-HO-tolyl-) 2AC
7969
C14H17NO4
263.11575
2310
U+UHYAC
U+UHYAC
LS/Q
Designer drug

Lidocaine-M (dimethylhydroxyaniline) 3AC
1065
C14H17NO4
263.11575
1900
U+UHYAC
UHYAC
LS/Q
Local anesthetic
Antiarrhythmic

Butylone AC
bk-MBDB AC
Beta-keto-MBDB AC
7872
C14H17NO4
263.11575
2215
U+UHYAC
U+UHYAC
LS/Q
Designer drug

Zaleplone-M/artifact (deacetyl-)
5860
C15H13N5
263.11710
2850
PS
LM/Q
Hypnotic

Phenindamine-M (nor-HO-)
1681
C18H17NO
263.13101
2590
UHY
UHY
LS/Q
Antihistamine

5-F-PB-22-M/artifact (HOOC-) (ME)
NM2201-M/artifact (HOOC-) (ME)
PX-1-M/artifact (HOOC-) (ME)
9484
C15H18NO2F
263.13217
2100
PS
LM/Q
Cannabinoid

674

263

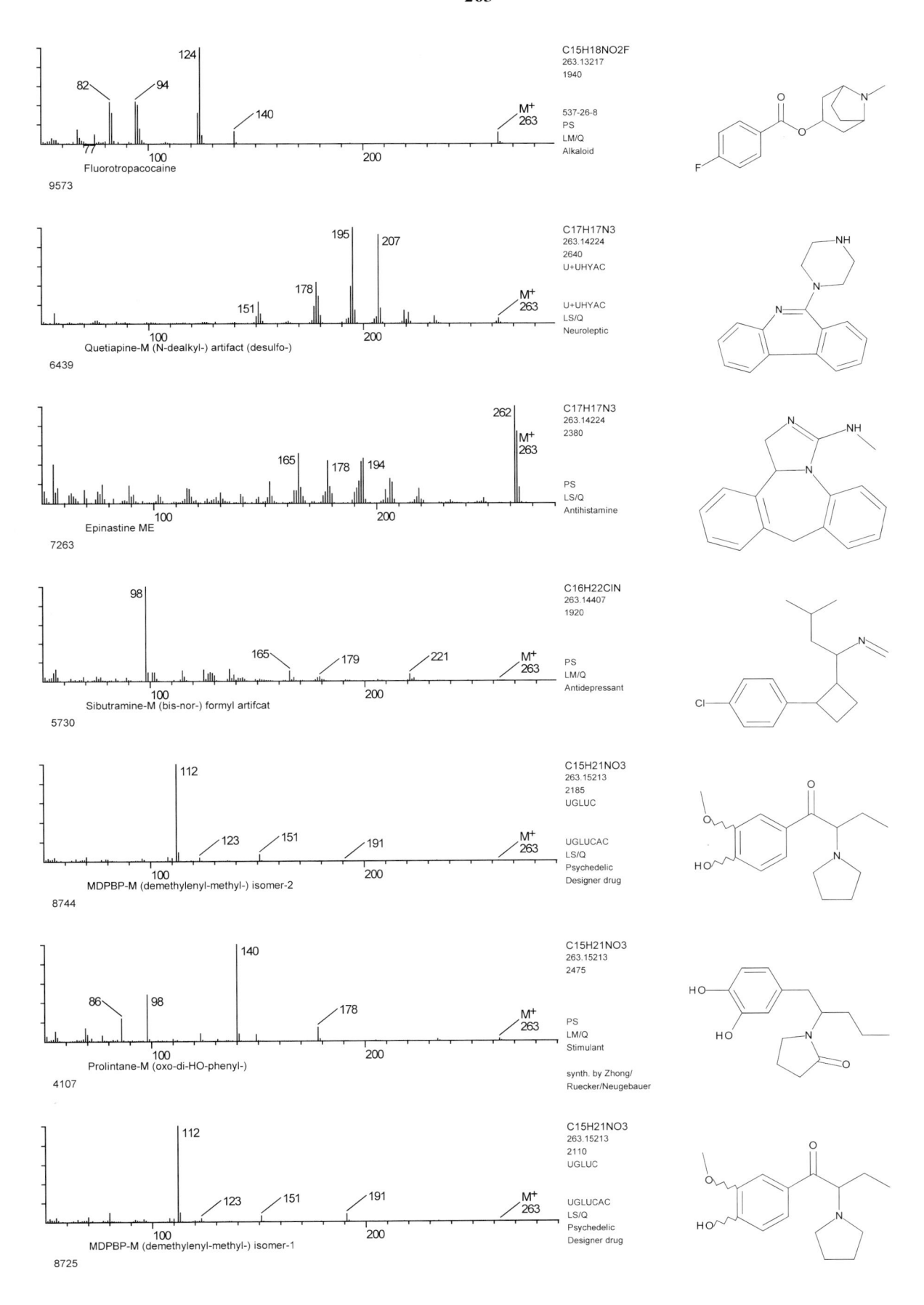

Fluorotropacocaine
9573

Peaks: 77, 82, 94, 124, 140, M+ 263

C15H18NO2F
263.13217
1940
537-26-8
PS
LM/Q
Alkaloid

Quetiapine-M (N-dealkyl-) artifact (desulfo-)
6439

Peaks: 151, 178, 195, 207, M+ 263

C17H17N3
263.14224
2640
U+UHYAC
U+UHYAC
LS/Q
Neuroleptic

Epinastine ME
7263

Peaks: 165, 178, 194, 262, M+ 263

C17H17N3
263.14224
2380
PS
LS/Q
Antihistamine

Sibutramine-M (bis-nor-) formyl artifcat
5730

Peaks: 98, 165, 179, 221, M+ 263

C16H22ClN
263.14407
1920
PS
LM/Q
Antidepressant

MDPBP-M (demethylenyl-methyl-) isomer-2
8744

Peaks: 112, 123, 151, 191, M+ 263

C15H21NO3
263.15213
2185
UGLUC
UGLUCAC
LS/Q
Psychedelic
Designer drug

Prolintane-M (oxo-di-HO-phenyl-)
4107

Peaks: 86, 98, 140, 178, M+ 263

C15H21NO3
263.15213
2475
PS
LM/Q
Stimulant

synth. by Zhong/
Ruecker/Neugebauer

MDPBP-M (demethylenyl-methyl-) isomer-1
8725

Peaks: 112, 123, 151, 191, M+ 263

C15H21NO3
263.15213
2110
UGLUC
UGLUCAC
LS/Q
Psychedelic
Designer drug

263

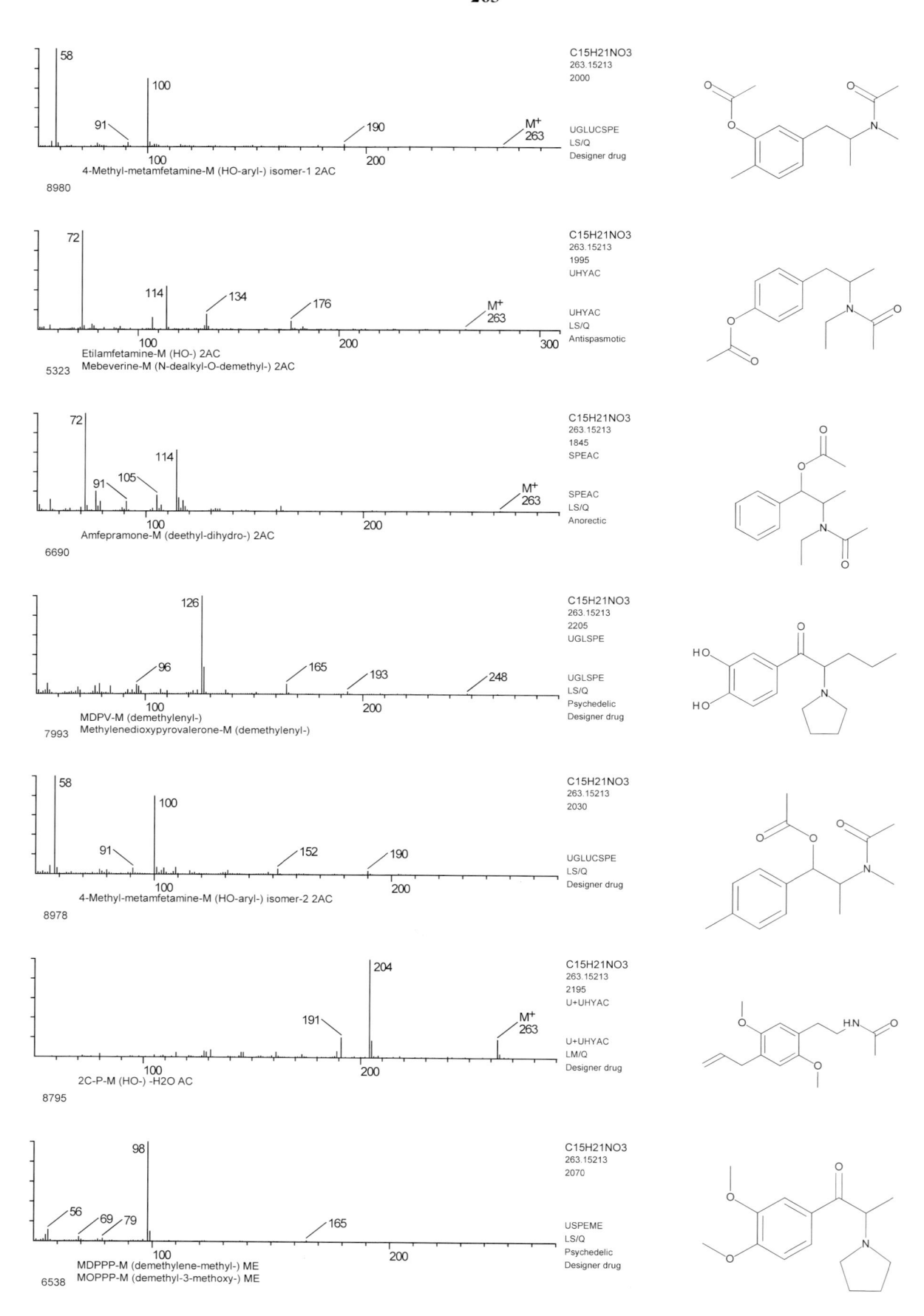

4-Methyl-metamfetamine-M (HO-aryl-) isomer-1 2AC 8980	Peaks: 58, 91, 100, 190, M+ 263	C15H21NO3 263.15213 2000 UGLUCSPE LS/Q Designer drug
Etilamfetamine-M (HO-) 2AC Mebeverine-M (N-dealkyl-O-demethyl-) 2AC 5323	Peaks: 72, 114, 134, 176, M+ 263	C15H21NO3 263.15213 1995 UHYAC UHYAC LS/Q Antispasmotic
Amfepramone-M (deethyl-dihydro-) 2AC 6690	Peaks: 72, 91, 105, 114, M+ 263	C15H21NO3 263.15213 1845 SPEAC SPEAC LS/Q Anorectic
MDPV-M (demethylenyl-) Methylenedioxypyrovalerone-M (demethylenyl-) 7993	Peaks: 96, 126, 165, 193, 248	C15H21NO3 263.15213 2205 UGLSPE UGLSPE LS/Q Psychedelic Designer drug
4-Methyl-metamfetamine-M (HO-aryl-) isomer-2 2AC 8978	Peaks: 58, 91, 100, 152, 190	C15H21NO3 263.15213 2030 UGLUCSPE LS/Q Designer drug
2C-P-M (HO-) -H2O AC 8795	Peaks: 191, 204, M+ 263	C15H21NO3 263.15213 2195 U+UHYAC U+UHYAC LM/Q Designer drug
MDPPP-M (demethylene-methyl-) ME MOPPP-M (demethyl-3-methoxy-) ME 6538	Peaks: 56, 69, 79, 98, 165	C15H21NO3 263.15213 2070 USPEME LS/Q Psychedelic Designer drug

263

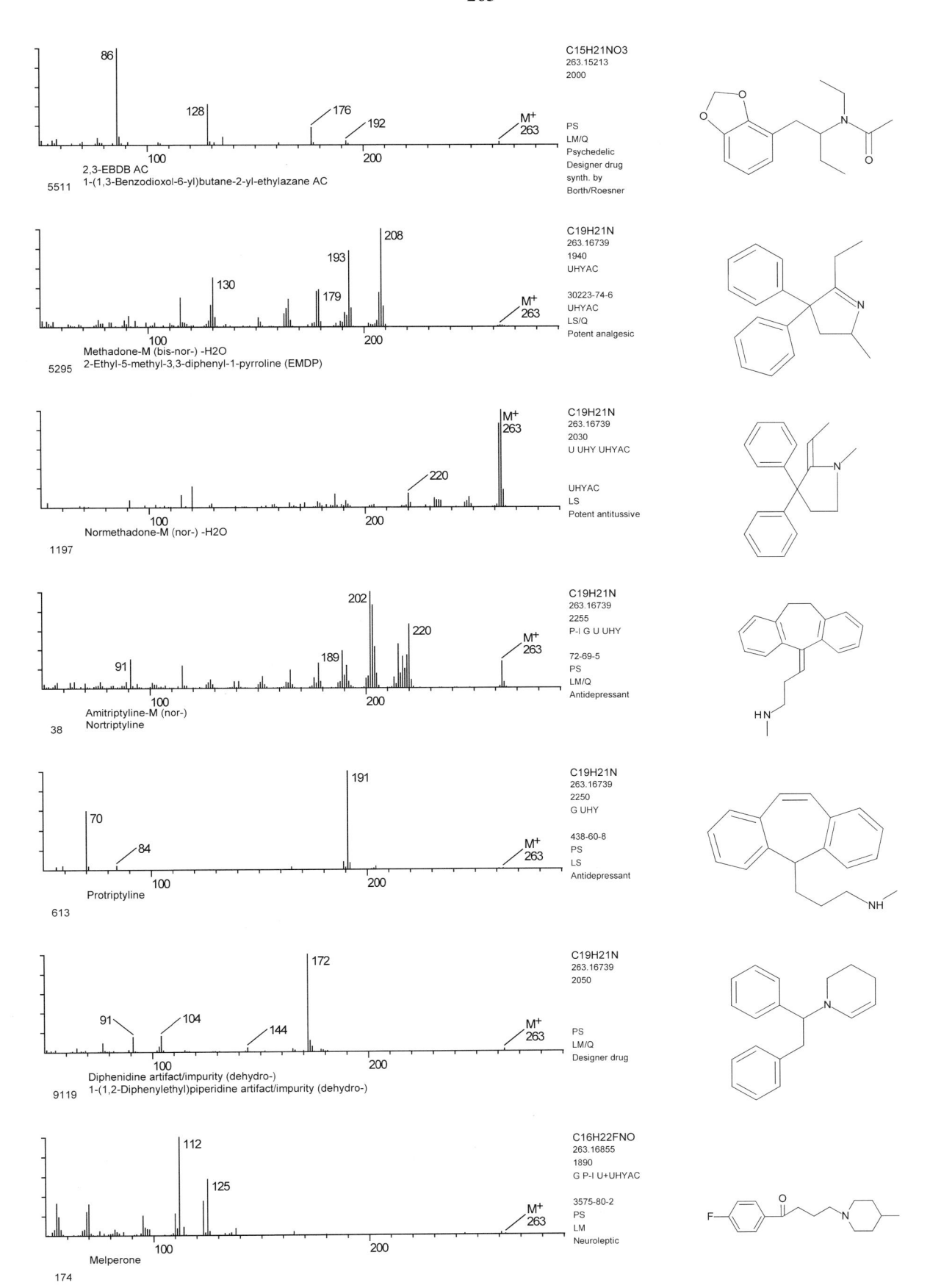

C15H21NO3
263.15213
2000

PS
LM/Q
Psychedelic
Designer drug
synth. by
Borth/Roesner

5511 2,3-EBDB AC
1-(1,3-Benzodioxol-6-yl)butane-2-yl-ethylazane AC

C19H21N
263.16739
1940
UHYAC

30223-74-6
UHYAC
LS/Q
Potent analgesic

5295 Methadone-M (bis-nor-) -H2O
2-Ethyl-5-methyl-3,3-diphenyl-1-pyrroline (EMDP)

C19H21N
263.16739
2030
U UHY UHYAC

UHYAC
LS
Potent antitussive

1197 Normethadone-M (nor-) -H2O

C19H21N
263.16739
2255
P-I G U UHY
72-69-5
PS
LM/Q
Antidepressant

38 Amitriptyline-M (nor-)
Nortriptyline

C19H21N
263.16739
2250
G UHY

438-60-8
PS
LS
Antidepressant

613 Protriptyline

C19H21N
263.16739
2050

PS
LM/Q
Designer drug

9119 Diphenidine artifact/impurity (dehydro-)
1-(1,2-Diphenylethyl)piperidine artifact/impurity (dehydro-)

C16H22FNO
263.16855
1890
G P-I U+UHYAC

3575-80-2
PS
LM
Neuroleptic

174 Melperone

677

263

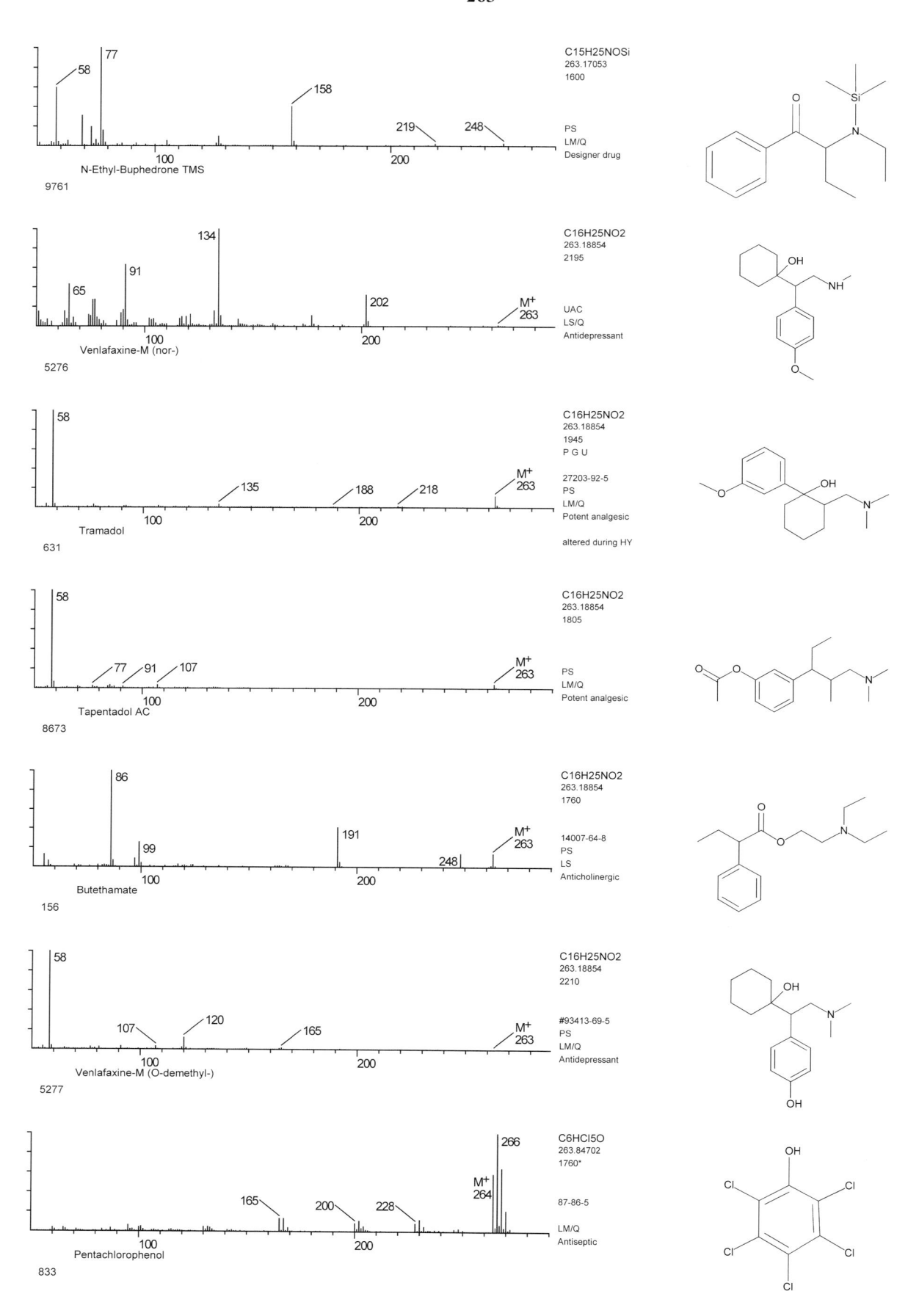

N-Ethyl-Buphedrone TMS
9761
peaks: 58, 77, 158, 219, 248
C15H25NOSi
263.17053
1600
PS
LM/Q
Designer drug

Venlafaxine-M (nor-)
5276
peaks: 65, 91, 134, 202, M+ 263
C16H25NO2
263.18854
2195
UAC
LS/Q
Antidepressant

Tramadol
631
peaks: 58, 135, 188, 218, M+ 263
C16H25NO2
263.18854
1945
P G U
27203-92-5
PS
LM/Q
Potent analgesic
altered during HY

Tapentadol AC
8673
peaks: 58, 77, 91, 107, M+ 263
C16H25NO2
263.18854
1805
PS
LM/Q
Potent analgesic

Butethamate
156
peaks: 86, 99, 191, 248, M+ 263
C16H25NO2
263.18854
1760
14007-64-8
PS
LS
Anticholinergic

Venlafaxine-M (O-demethyl-)
5277
peaks: 58, 107, 120, 165, M+ 263
C16H25NO2
263.18854
2210
#93413-69-5
PS
LM/Q
Antidepressant

Pentachlorophenol
833
peaks: 165, 200, 228, M+ 264, 266
C6HCl5O
263.84702
1760*
87-86-5
LM/Q
Antiseptic

678

264

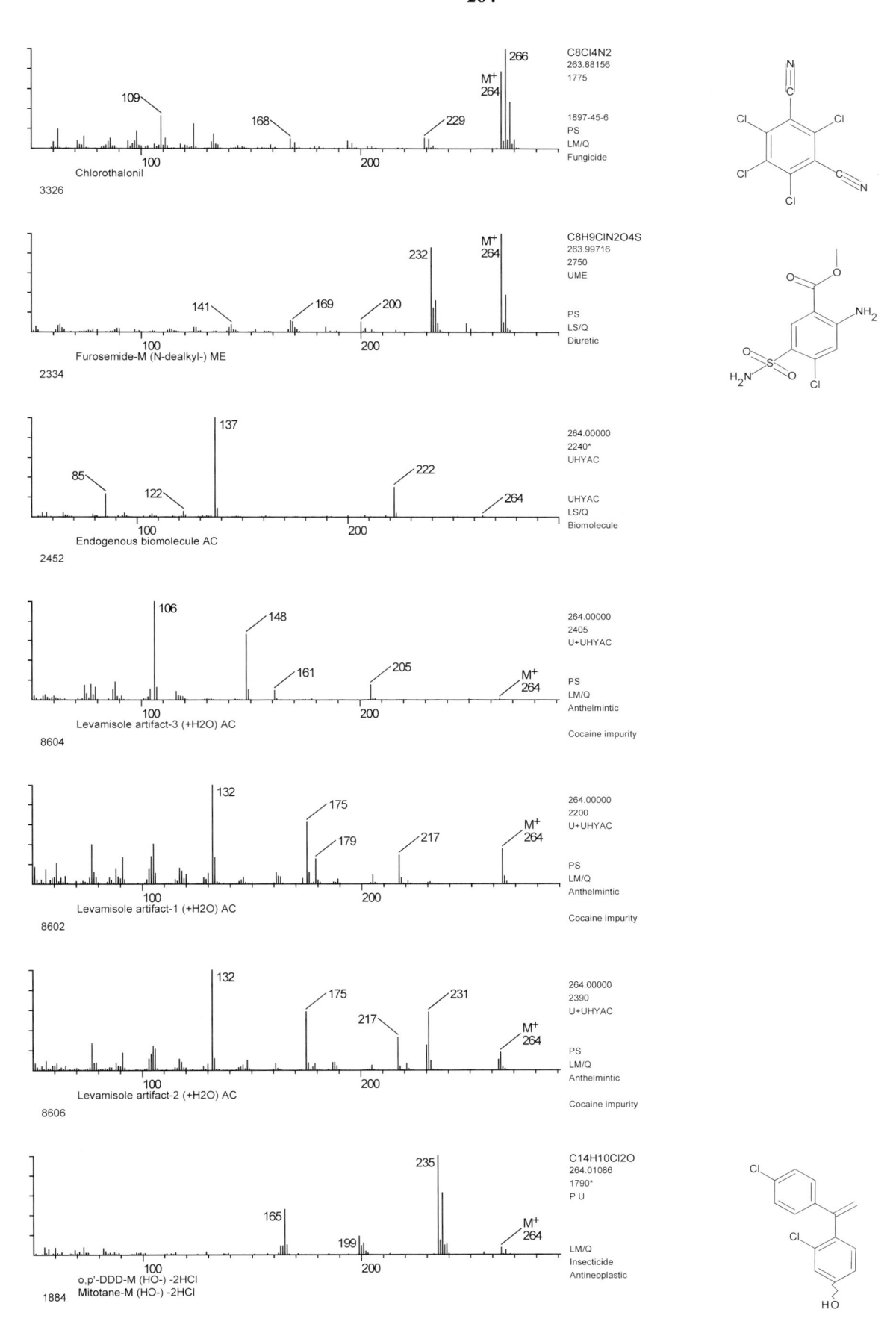

Chlorothalonil — C8Cl4N2, 263.88156, 1775, 1897-45-6, PS, LM/Q, Fungicide
Peaks: 109, 168, 229, 264 (M+), 266
3326

Furosemide-M (N-dealkyl-) ME — C8H9ClN2O4S, 263.99716, 2750, UME, PS, LS/Q, Diuretic
Peaks: 141, 169, 200, 232, 264 (M+)
2334

Endogenous biomolecule AC — 264.00000, 2240*, UHYAC, UHYAC, LS/Q, Biomolecule
Peaks: 85, 122, 137, 222, 264
2452

Levamisole artifact-3 (+H2O) AC — 264.00000, 2405, U+UHYAC, PS, LM/Q, Anthelmintic, Cocaine impurity
Peaks: 106, 148, 161, 205, 264 (M+)
8604

Levamisole artifact-1 (+H2O) AC — 264.00000, 2200, U+UHYAC, PS, LM/Q, Anthelmintic, Cocaine impurity
Peaks: 132, 175, 179, 217, 264 (M+)
8602

Levamisole artifact-2 (+H2O) AC — 264.00000, 2390, U+UHYAC, PS, LM/Q, Anthelmintic, Cocaine impurity
Peaks: 132, 175, 217, 231, 264 (M+)
8606

o,p'-DDD-M (HO-) -2HCl / Mitotane-M (HO-) -2HCl — C14H10Cl2O, 264.01086, 1790*, P U, LM/Q, Insecticide, Antineoplastic
Peaks: 165, 199, 235, 264 (M+)
1884

264

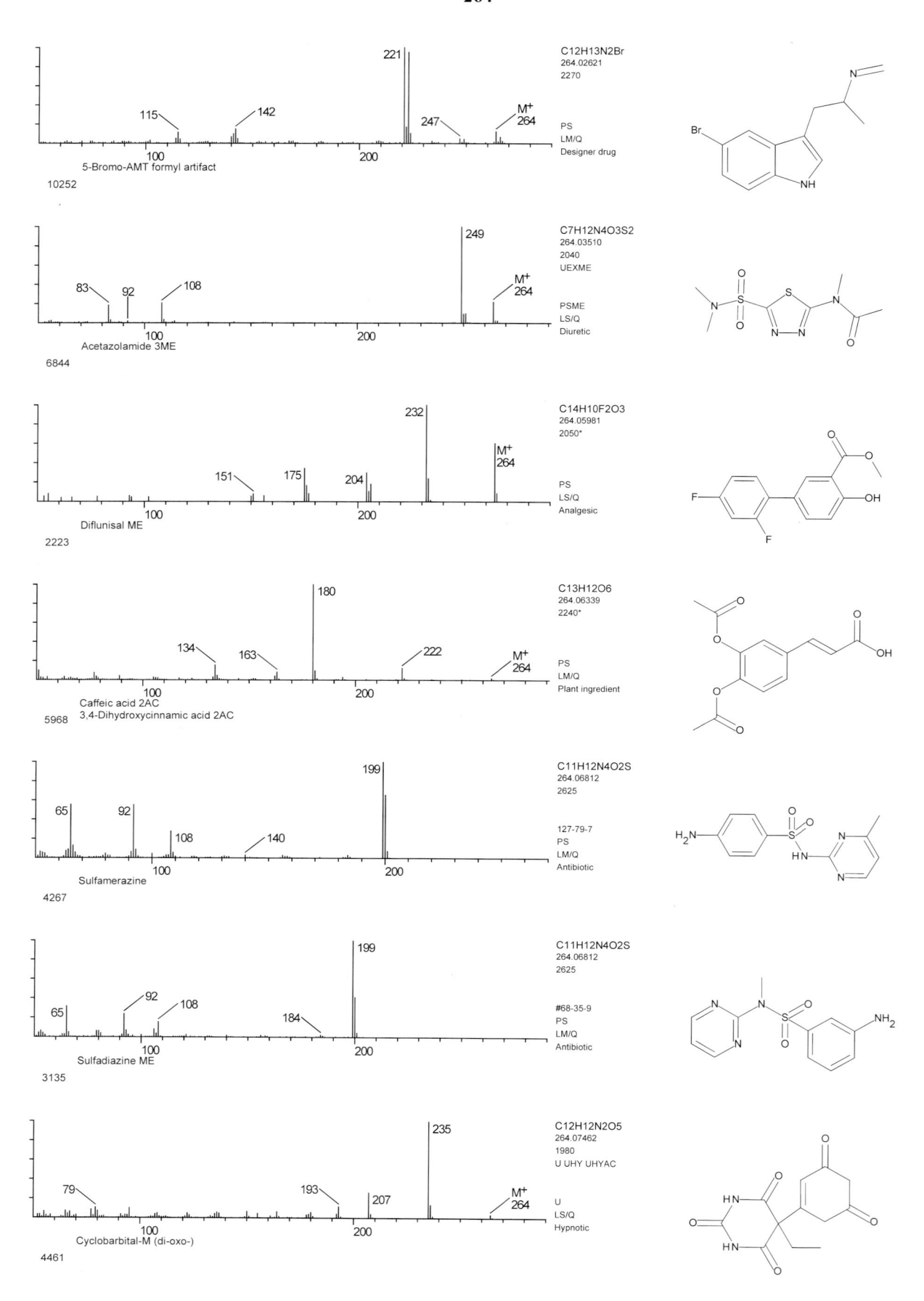

Spectrum	Compound info
Peaks: 115, 142, 221, 247, M+ 264	C12H13N2Br 264.02621 2270 PS LM/Q Designer drug 5-Bromo-AMT formyl artifact 10252
Peaks: 83, 92, 108, 249, M+ 264	C7H12N4O3S2 264.03510 2040 UEXME PSME LS/Q Diuretic Acetazolamide 3ME 6844
Peaks: 151, 175, 204, 232, M+ 264	C14H10F2O3 264.05981 2050* PS LS/Q Analgesic Diflunisal ME 2223
Peaks: 134, 163, 180, 222, M+ 264	C13H12O6 264.06339 2240* PS LM/Q Plant ingredient Caffeic acid 2AC 3,4-Dihydroxycinnamic acid 2AC 5968
Peaks: 65, 92, 108, 140, 199, M+ 264	C11H12N4O2S 264.06812 2625 127-79-7 PS LM/Q Antibiotic Sulfamerazine 4267
Peaks: 65, 92, 108, 184, 199, M+ 264	C11H12N4O2S 264.06812 2625 #68-35-9 PS LM/Q Antibiotic Sulfadiazine ME 3135
Peaks: 79, 193, 207, 235, M+ 264	C12H12N2O5 264.07462 1980 U UHY UHYAC U LS/Q Hypnotic Cyclobarbital-M (di-oxo-) 4461

264

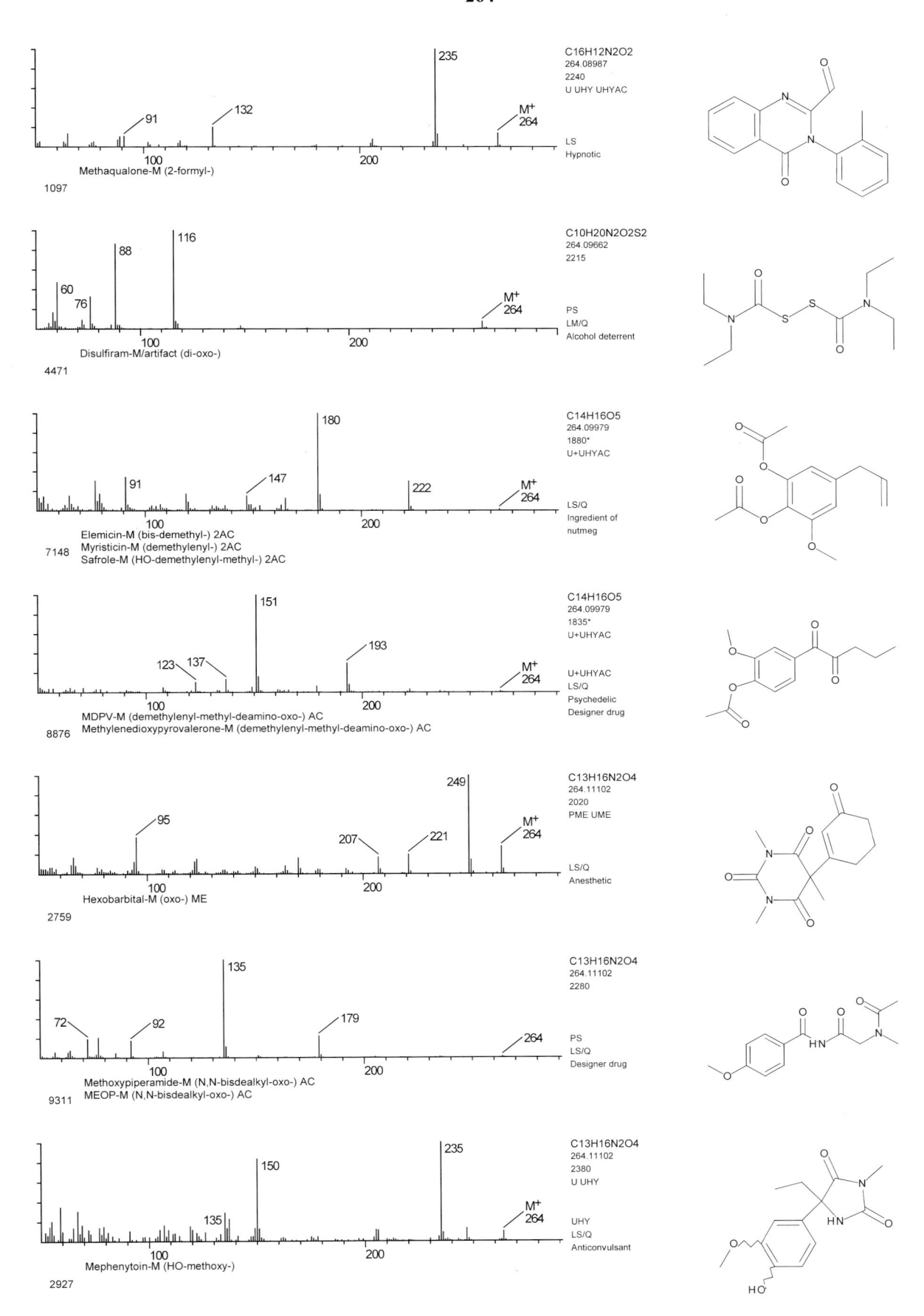

Methaqualone-M (2-formyl-)
1097
Peaks: 91, 132, 235, M+ 264
C16H12N2O2
264.08987
2240
U UHY UHYAC
LS
Hypnotic

Disulfiram-M/artifact (di-oxo-)
4471
Peaks: 60, 76, 88, 116, M+ 264
C10H20N2O2S2
264.09662
2215
PS
LM/Q
Alcohol deterrent

Elemicin-M (bis-demethyl-) 2AC
Myristicin-M (demethylenyl-) 2AC
Safrole-M (HO-demethylenyl-methyl-) 2AC
7148
Peaks: 91, 147, 180, 222, M+ 264
C14H16O5
264.09979
1880*
U+UHYAC
LS/Q
Ingredient of nutmeg

MDPV-M (demethylenyl-methyl-deamino-oxo-) AC
Methylenedioxypyrovalerone-M (demethylenyl-methyl-deamino-oxo-) AC
8876
Peaks: 123, 137, 151, 193, M+ 264
C14H16O5
264.09979
1835*
U+UHYAC
U+UHYAC
LS/Q
Psychedelic
Designer drug

Hexobarbital-M (oxo-) ME
2759
Peaks: 95, 207, 221, 249, M+ 264
C13H16N2O4
264.11102
2020
PME UME
LS/Q
Anesthetic

Methoxypiperamide-M (N,N-bisdealkyl-oxo-) AC
MEOP-M (N,N-bisdealkyl-oxo-) AC
9311
Peaks: 72, 92, 135, 179, 264
C13H16N2O4
264.11102
2280
PS
LS/Q
Designer drug

Mephenytoin-M (HO-methoxy-)
2927
Peaks: 135, 150, 235, M+ 264
C13H16N2O4
264.11102
2380
U UHY
UHY
LS/Q
Anticonvulsant

681

264

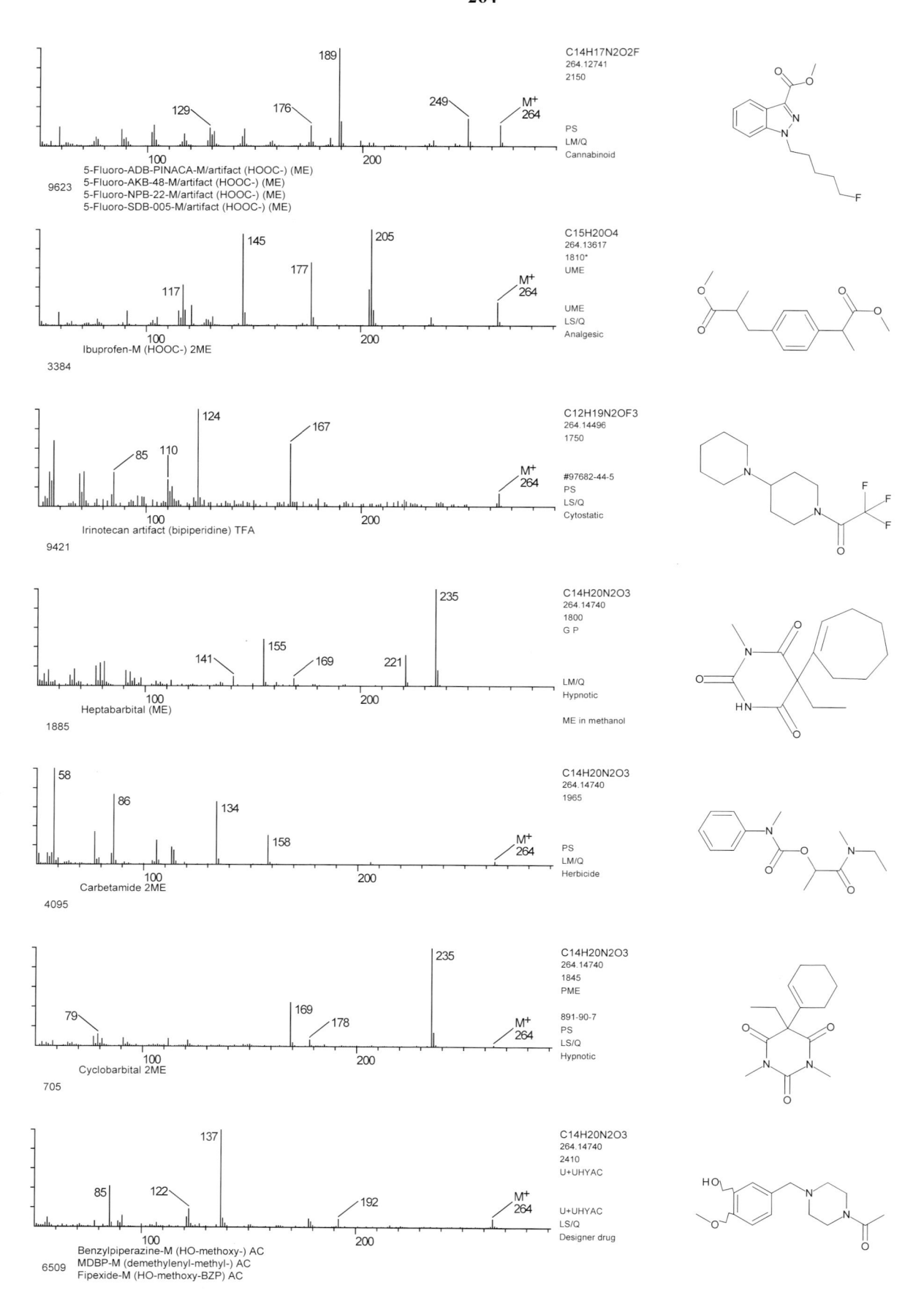

189, 129, 176, 249, M+ 264	C14H17N2O2F 264.12741 2150 PS LM/Q Cannabinoid
9623 5-Fluoro-ADB-PINACA-M/artifact (HOOC-) (ME) 5-Fluoro-AKB-48-M/artifact (HOOC-) (ME) 5-Fluoro-NPB-22-M/artifact (HOOC-) (ME) 5-Fluoro-SDB-005-M/artifact (HOOC-) (ME)	
145, 205, 117, 177, M+ 264	C15H20O4 264.13617 1810* UME UME LS/Q Analgesic
3384 Ibuprofen-M (HOOC-) 2ME	
124, 167, 85, 110, M+ 264	C12H19N2OF3 264.14496 1750 #97682-44-5 PS LS/Q Cytostatic
9421 Irinotecan artifact (bipiperidine) TFA	
235, 141, 155, 169, 221, M+ 264	C14H20N2O3 264.14740 1800 G P LM/Q Hypnotic ME in methanol
1885 Heptabarbital (ME)	
58, 86, 134, 158, M+ 264	C14H20N2O3 264.14740 1965 PS LM/Q Herbicide
4095 Carbetamide 2ME	
235, 79, 169, 178, M+ 264	C14H20N2O3 264.14740 1845 PME 891-90-7 PS LS/Q Hypnotic
705 Cyclobarbital 2ME	
137, 85, 122, 192, M+ 264	C14H20N2O3 264.14740 2410 U+UHYAC U+UHYAC LS/Q Designer drug
6509 Benzylpiperazine-M (HO-methoxy-) AC MDBP-M (demethylenyl-methyl-) AC Fipexide-M (HO-methoxy-BZP) AC	

264

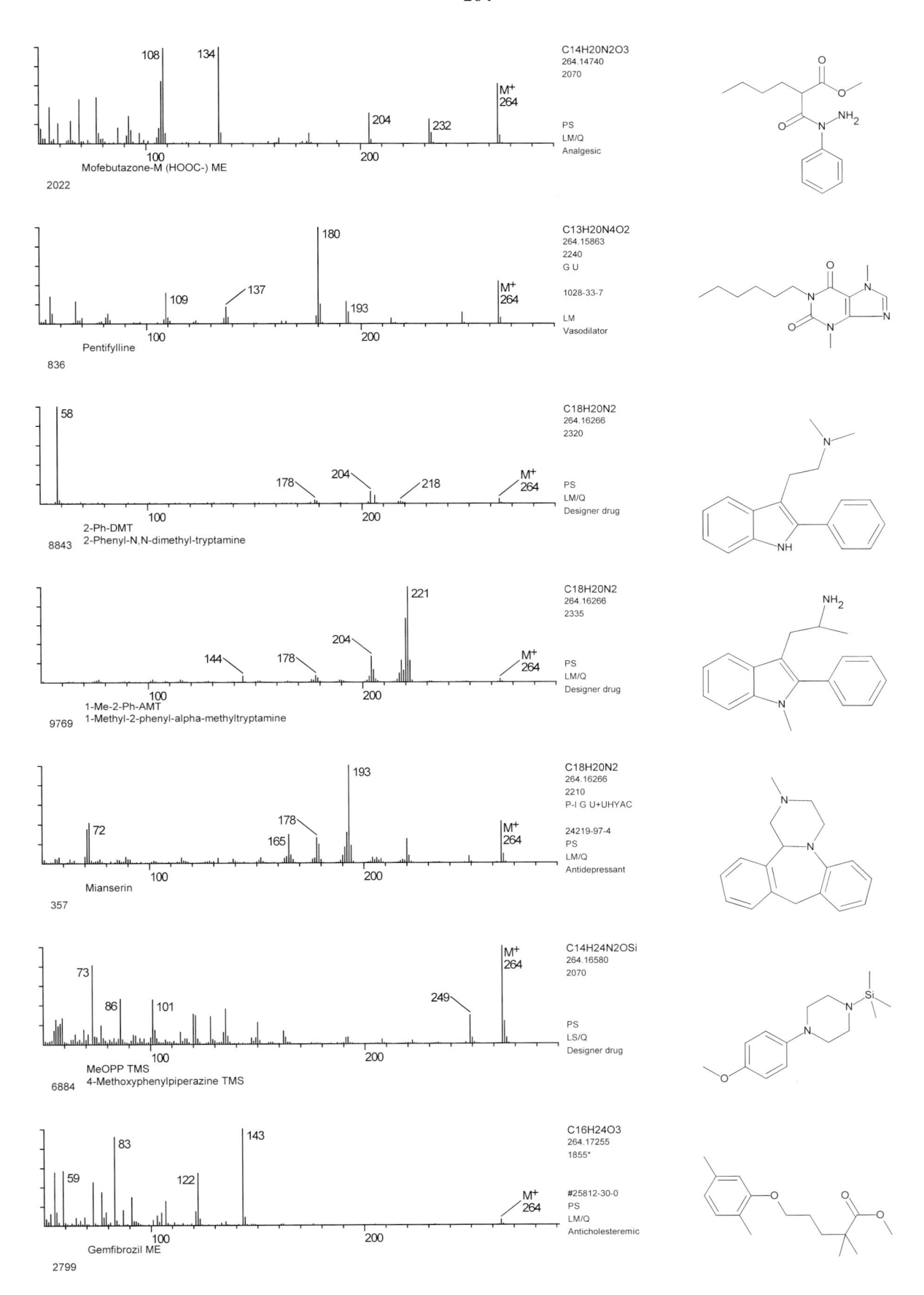

108, 134, 204, 232, M+ 264	Mofebutazone-M (HOOC-) ME 2022	C14H20N2O3 264.14740 2070 PS LM/Q Analgesic
109, 137, 180, 193, M+ 264	Pentifylline 836	C13H20N4O2 264.15863 2240 G U 1028-33-7 LM Vasodilator
58, 178, 204, 218, M+ 264	2-Ph-DMT 2-Phenyl-N,N-dimethyl-tryptamine 8843	C18H20N2 264.16266 2320 PS LM/Q Designer drug
144, 178, 204, 221, M+ 264	1-Me-2-Ph-AMT 1-Methyl-2-phenyl-alpha-methyltryptamine 9769	C18H20N2 264.16266 2335 PS LM/Q Designer drug
72, 165, 178, 193, M+ 264	Mianserin 357	C18H20N2 264.16266 2210 P-I G U+UHYAC 24219-97-4 PS LM/Q Antidepressant
73, 86, 101, 249, M+ 264	MeOPP TMS 4-Methoxyphenylpiperazine TMS 6884	C14H24N2OSi 264.16580 2070 PS LS/Q Designer drug
59, 83, 122, 143, M+ 264	Gemfibrozil ME 2799	C16H24O3 264.17255 1855* #25812-30-0 PS LM/Q Anticholesteremic

683

264

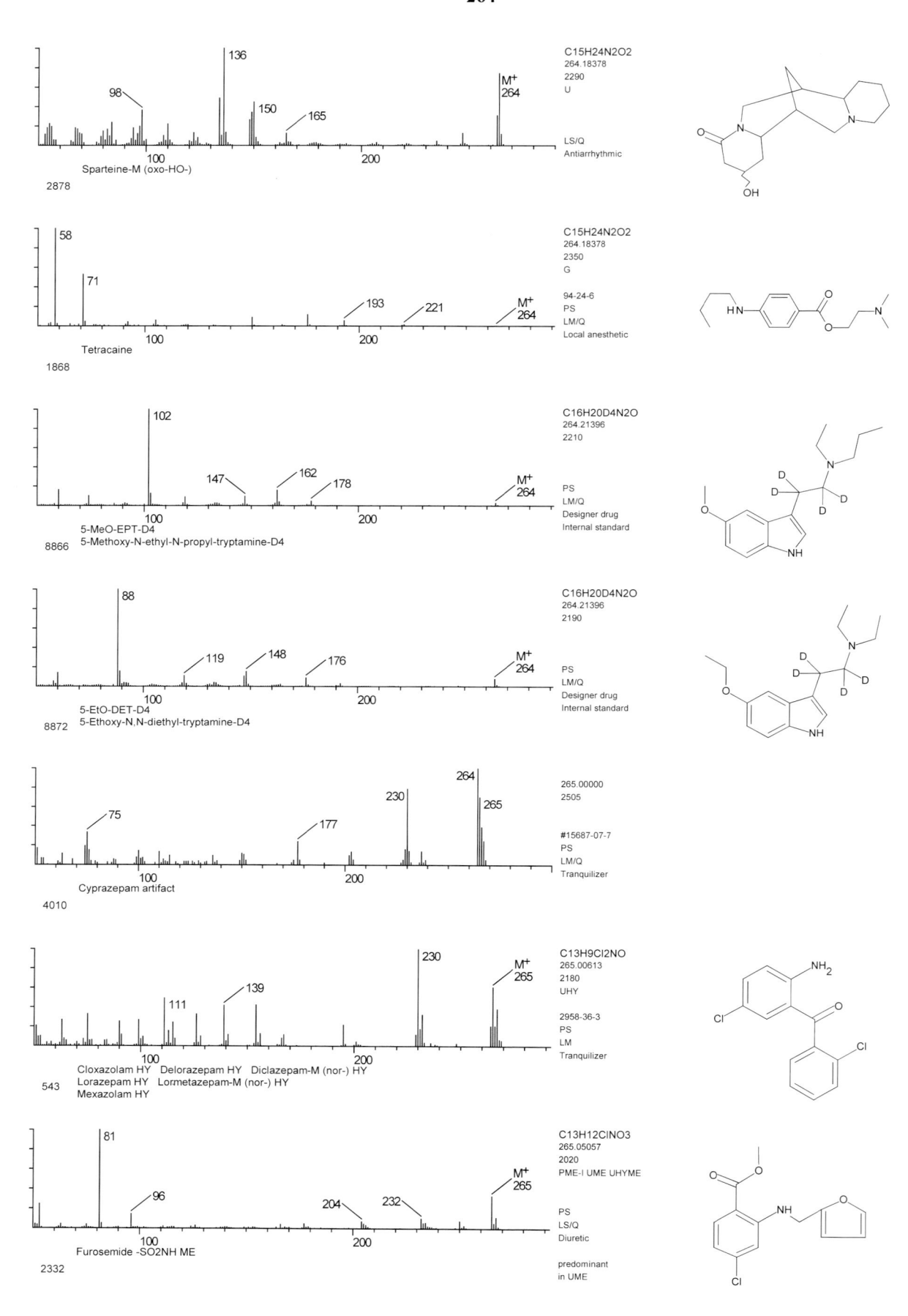

Sparteine-M (oxo-HO-)	C15H24N2O2 264.18378 2290 U LS/Q Antiarrhythmic	
2878		
Tetracaine	C15H24N2O2 264.18378 2350 G 94-24-6 PS LM/Q Local anesthetic	
1868		
5-MeO-EPT-D4 5-Methoxy-N-ethyl-N-propyl-tryptamine-D4	C16H20D4N2O 264.21396 2210 PS LM/Q Designer drug Internal standard	
8866		
5-EtO-DET-D4 5-Ethoxy-N,N-diethyl-tryptamine-D4	C16H20D4N2O 264.21396 2190 PS LM/Q Designer drug Internal standard	
8872		
Cyprazepam artifact	265.00000 2505 #15687-07-7 PS LM/Q Tranquilizer	
4010		
Cloxazolam HY Delorazepam HY Diclazepam-M (nor-) HY Lorazepam HY Lormetazepam-M (nor-) HY Mexazolam HY	C13H9Cl2NO 265.00613 2180 UHY 2958-36-3 PS LM Tranquilizer	
543		
Furosemide -SO2NH ME	C13H12ClNO3 265.05057 2020 PME-I UME UHYME PS LS/Q Diuretic predominant in UME	
2332		

684

265

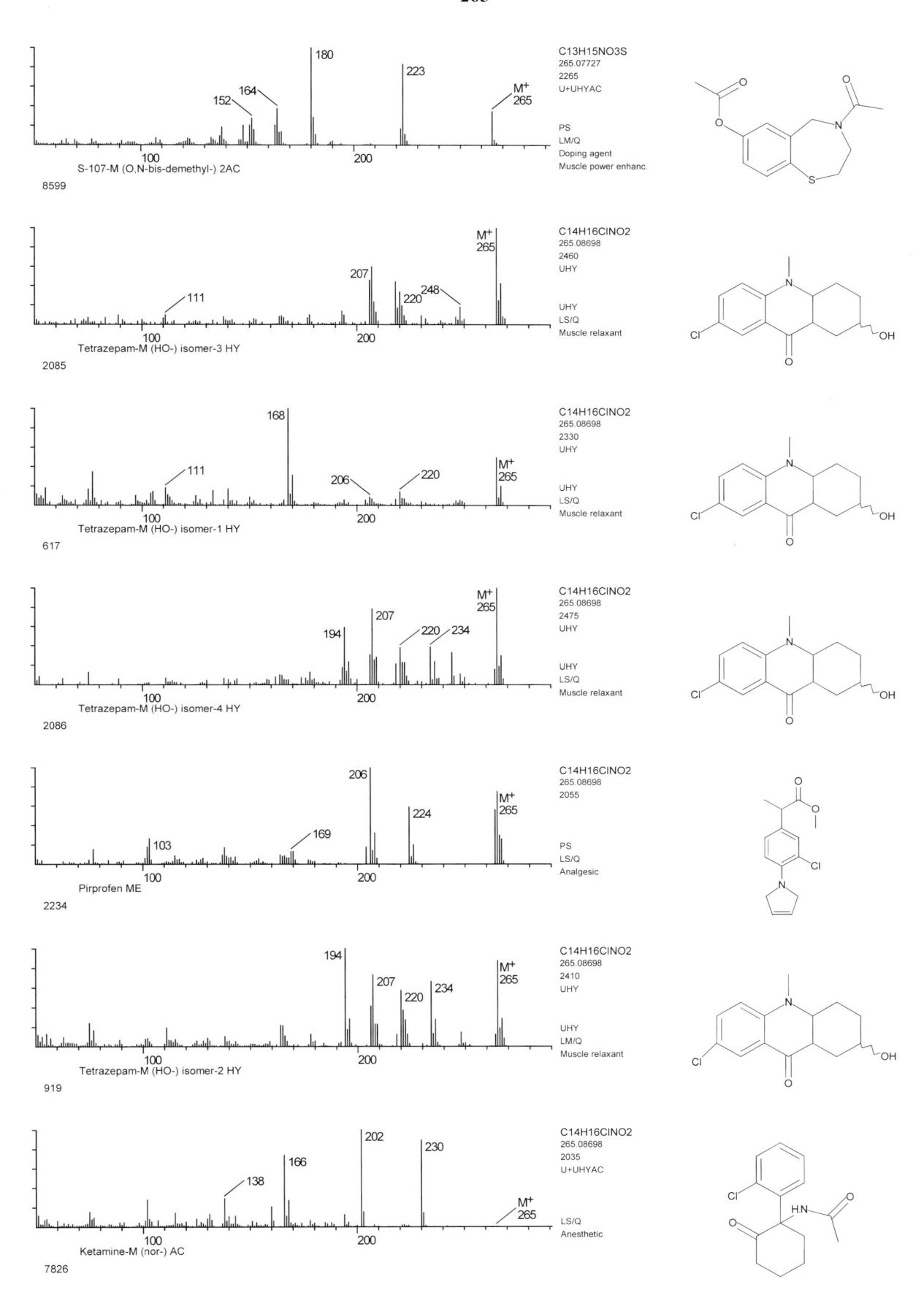

Spectrum label	Formula / data
S-107-M (O,N-bis-demethyl-) 2AC — peaks 152, 164, 180, 223, M+ 265	C13H15NO3S, 265.07727, 2265, U+UHYAC, PS, LM/Q, Doping agent, Muscle power enhanc.
Tetrazepam-M (HO-) isomer-3 HY — peaks 111, 207, 220, 248, M+ 265	C14H16ClNO2, 265.08698, 2460, UHY, UHY, LS/Q, Muscle relaxant
Tetrazepam-M (HO-) isomer-1 HY — peaks 111, 168, 206, 220, M+ 265	C14H16ClNO2, 265.08698, 2330, UHY, UHY, LS/Q, Muscle relaxant
Tetrazepam-M (HO-) isomer-4 HY — peaks 194, 207, 220, 234, M+ 265	C14H16ClNO2, 265.08698, 2475, UHY, UHY, LS/Q, Muscle relaxant
Pirprofen ME — peaks 103, 169, 206, 224, M+ 265	C14H16ClNO2, 265.08698, 2055, PS, LS/Q, Analgesic
Tetrazepam-M (HO-) isomer-2 HY — peaks 194, 207, 220, 234, M+ 265	C14H16ClNO2, 265.08698, 2410, UHY, UHY, LM/Q, Muscle relaxant
Ketamine-M (nor-) AC — peaks 138, 166, 202, 230, M+ 265	C14H16ClNO2, 265.08698, 2035, U+UHYAC, LS/Q, Anesthetic

685

265

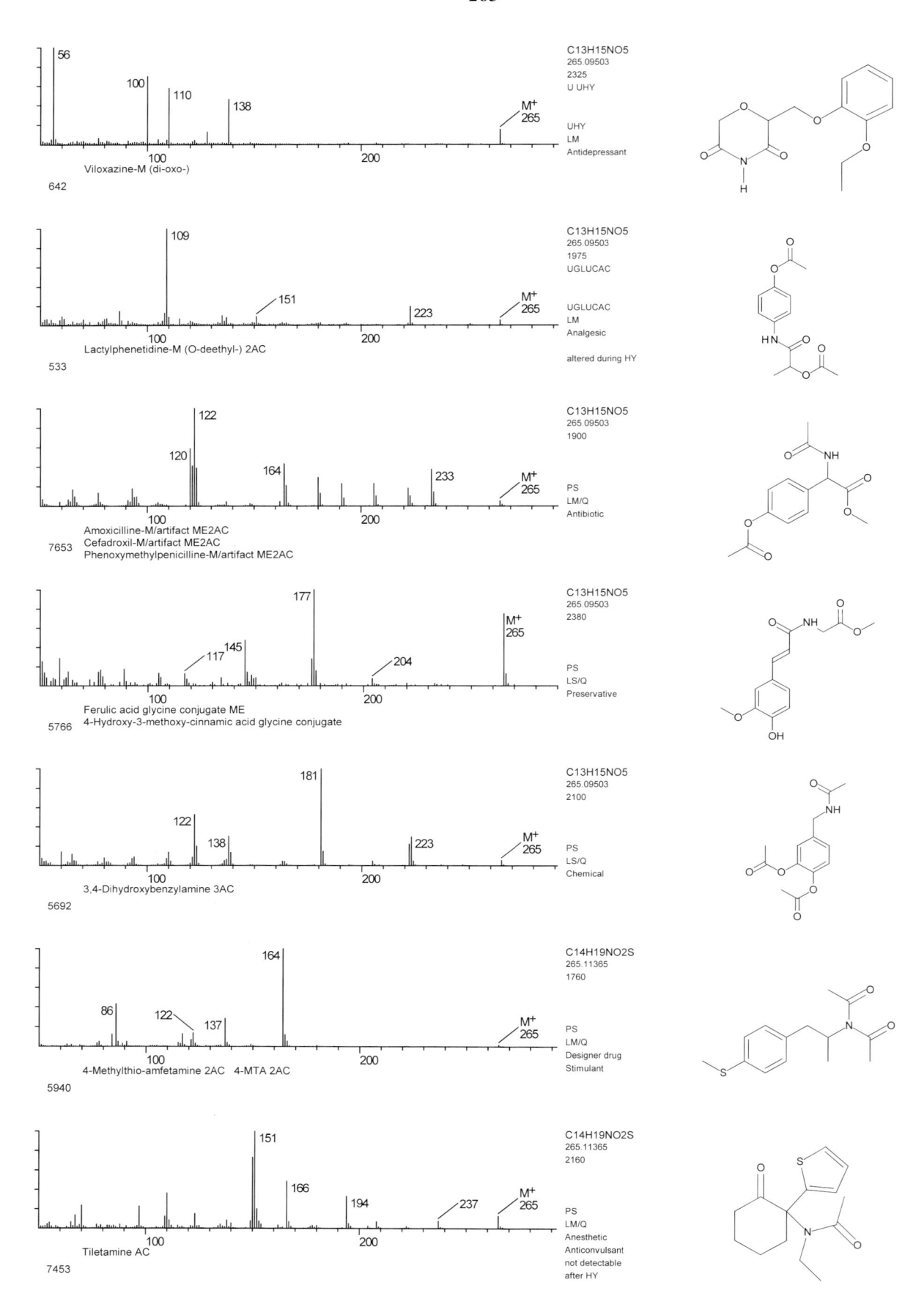

Spectrum peaks	Compound	Formula / Info
56, 100, 110, 138, M+ 265	Viloxazine-M (di-oxo-)	C13H15NO5 / 265.09503 / 2325 / U UHY / UHY / LM / Antidepressant
642		
109, 151, 223, M+ 265	Lactylphenetidine-M (O-deethyl-) 2AC	C13H15NO5 / 265.09503 / 1975 / UGLUCAC / UGLUCAC / LM / Analgesic / altered during HY
533		
120, 122, 164, 233, M+ 265	Amoxicilline-M/artifact ME2AC; Cefadroxil-M/artifact ME2AC; Phenoxymethylpenicilline-M/artifact ME2AC	C13H15NO5 / 265.09503 / 1900 / PS / LM/Q / Antibiotic
7653		
117, 145, 177, 204, M+ 265	Ferulic acid glycine conjugate ME; 4-Hydroxy-3-methoxy-cinnamic acid glycine conjugate	C13H15NO5 / 265.09503 / 2380 / PS / LS/Q / Preservative
5766		
122, 138, 181, 223, M+ 265	3,4-Dihydroxybenzylamine 3AC	C13H15NO5 / 265.09503 / 2100 / PS / LS/Q / Chemical
5692		
86, 122, 137, 164, M+ 265	4-Methylthio-amfetamine 2AC 4-MTA 2AC	C14H19NO2S / 265.11365 / 1760 / PS / LM/Q / Designer drug / Stimulant
5940		
151, 166, 194, 237, M+ 265	Tiletamine AC	C14H19NO2S / 265.11365 / 2160 / PS / LM/Q / Anesthetic / Anticonvulsant / not detectable after HY
7453		

265

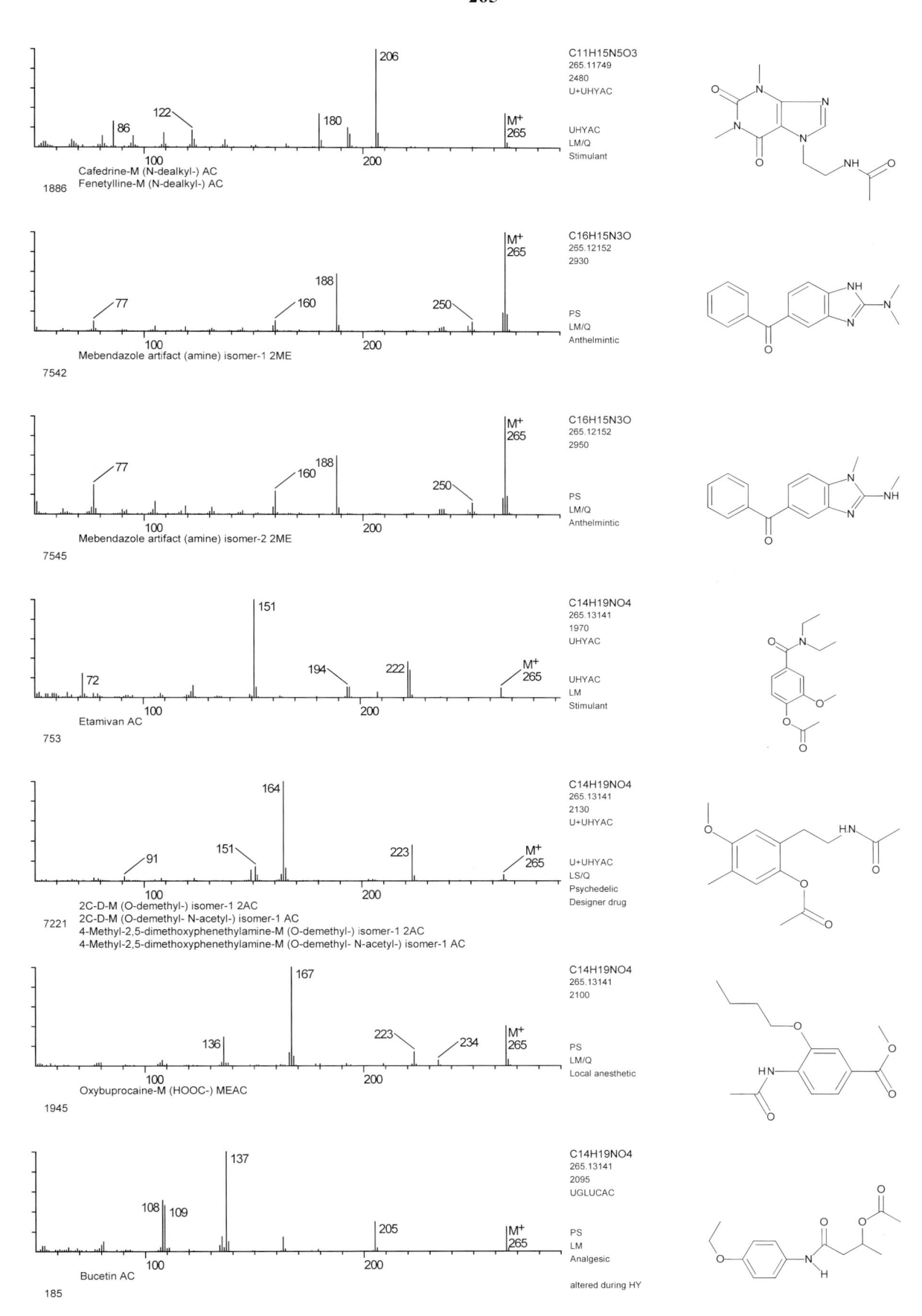

C11H15N5O3
265.11749
2480
U+UHYAC

UHYAC
LM/Q
Stimulant

Cafedrine-M (N-dealkyl-) AC
Fenetylline-M (N-dealkyl-) AC
1886

C16H15N3O
265.12152
2930

PS
LM/Q
Anthelmintic

Mebendazole artifact (amine) isomer-1 2ME
7542

C16H15N3O
265.12152
2950

PS
LM/Q
Anthelmintic

Mebendazole artifact (amine) isomer-2 2ME
7545

C14H19NO4
265.13141
1970
UHYAC

UHYAC
LM
Stimulant

Etamivan AC
753

C14H19NO4
265.13141
2130
U+UHYAC

U+UHYAC
LS/Q
Psychedelic
Designer drug

2C-D-M (O-demethyl-) isomer-1 2AC
2C-D-M (O-demethyl- N-acetyl-) isomer-1 AC
4-Methyl-2,5-dimethoxyphenethylamine-M (O-demethyl-) isomer-1 2AC
4-Methyl-2,5-dimethoxyphenethylamine-M (O-demethyl- N-acetyl-) isomer-1 AC
7221

C14H19NO4
265.13141
2100

PS
LM/Q
Local anesthetic

Oxybuprocaine-M (HOOC-) MEAC
1945

C14H19NO4
265.13141
2095
UGLUCAC

PS
LM
Analgesic

altered during HY

Bucetin AC
185

687

265

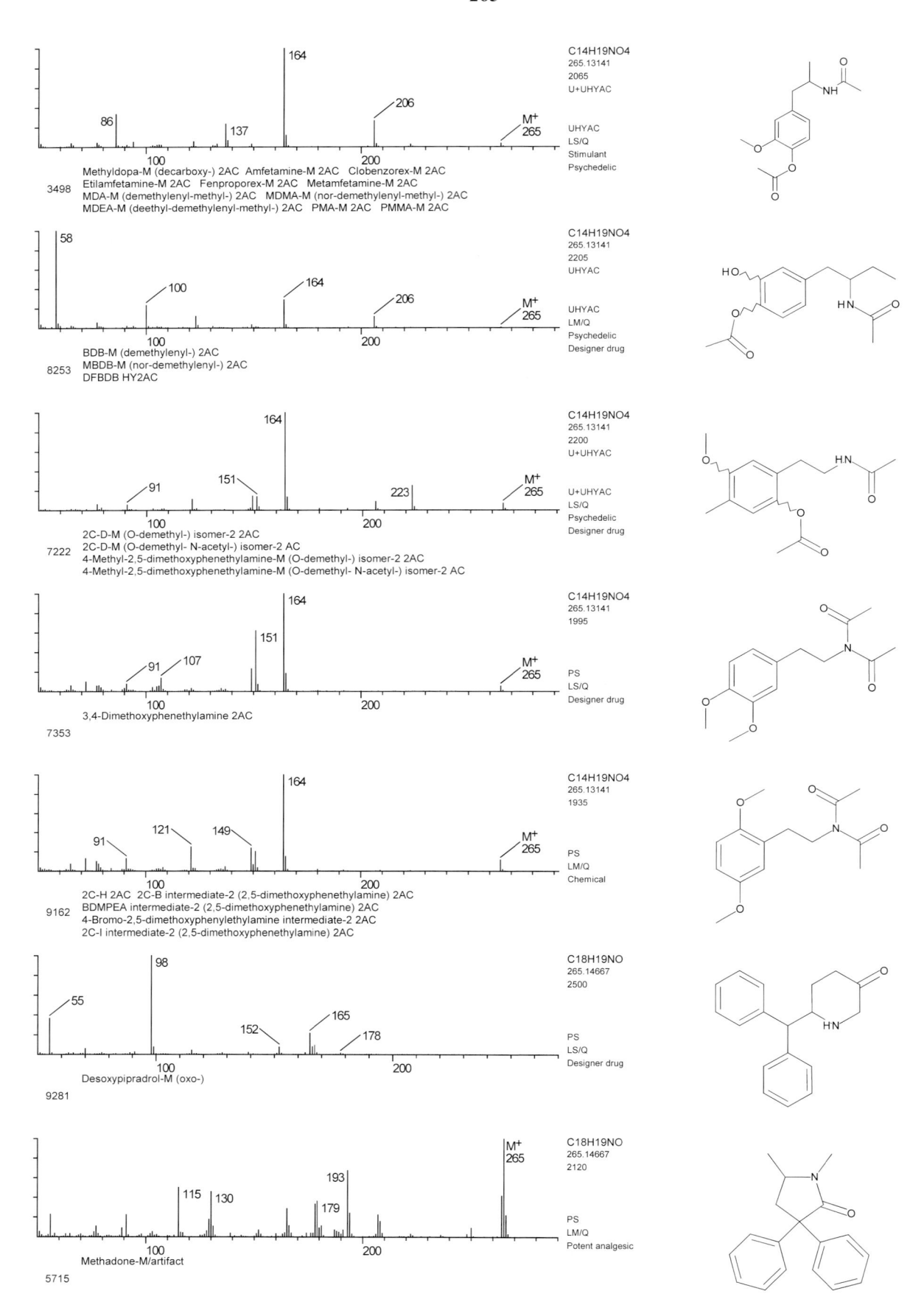

3498	86, 137, 164, 206, M+ 265 Methyldopa-M (decarboxy-) 2AC Amfetamine-M 2AC Clobenzorex-M 2AC Etilamfetamine-M 2AC Fenproporex-M 2AC Metamfetamine-M 2AC MDA-M (demethylenyl-methyl-) 2AC MDMA-M (nor-demethylenyl-methyl-) 2AC MDEA-M (deethyl-demethylenyl-methyl-) 2AC PMA-M 2AC PMMA-M 2AC	C14H19NO4 265.13141 2065 U+UHYAC UHYAC LS/Q Stimulant Psychedelic
8253	58, 100, 164, 206, M+ 265 BDB-M (demethylenyl-) 2AC MBDB-M (nor-demethylenyl-) 2AC DFBDB HY2AC	C14H19NO4 265.13141 2205 UHYAC UHYAC LM/Q Psychedelic Designer drug
7222	91, 151, 164, 223, M+ 265 2C-D-M (O-demethyl-) isomer-2 2AC 2C-D-M (O-demethyl- N-acetyl-) isomer-2 AC 4-Methyl-2,5-dimethoxyphenethylamine-M (O-demethyl-) isomer-2 2AC 4-Methyl-2,5-dimethoxyphenethylamine-M (O-demethyl- N-acetyl-) isomer-2 AC	C14H19NO4 265.13141 2200 U+UHYAC U+UHYAC LS/Q Psychedelic Designer drug
7353	91, 107, 151, 164, M+ 265 3,4-Dimethoxyphenethylamine 2AC	C14H19NO4 265.13141 1995 PS LS/Q Designer drug
9162	91, 121, 149, 164, M+ 265 2C-H 2AC 2C-B intermediate-2 (2,5-dimethoxyphenethylamine) 2AC BDMPEA intermediate-2 (2,5-dimethoxyphenethylamine) 2AC 4-Bromo-2,5-dimethoxyphenylethylamine intermediate-2 2AC 2C-I intermediate-2 (2,5-dimethoxyphenethylamine) 2AC	C14H19NO4 265.13141 1935 PS LM/Q Chemical
9281	55, 98, 152, 165, 178 Desoxypipradrol-M (oxo-)	C18H19NO 265.14667 2500 PS LS/Q Designer drug
5715	115, 130, 179, 193, M+ 265 Methadone-M/artifact	C18H19NO 265.14667 2120 PS LM/Q Potent analgesic

688

265

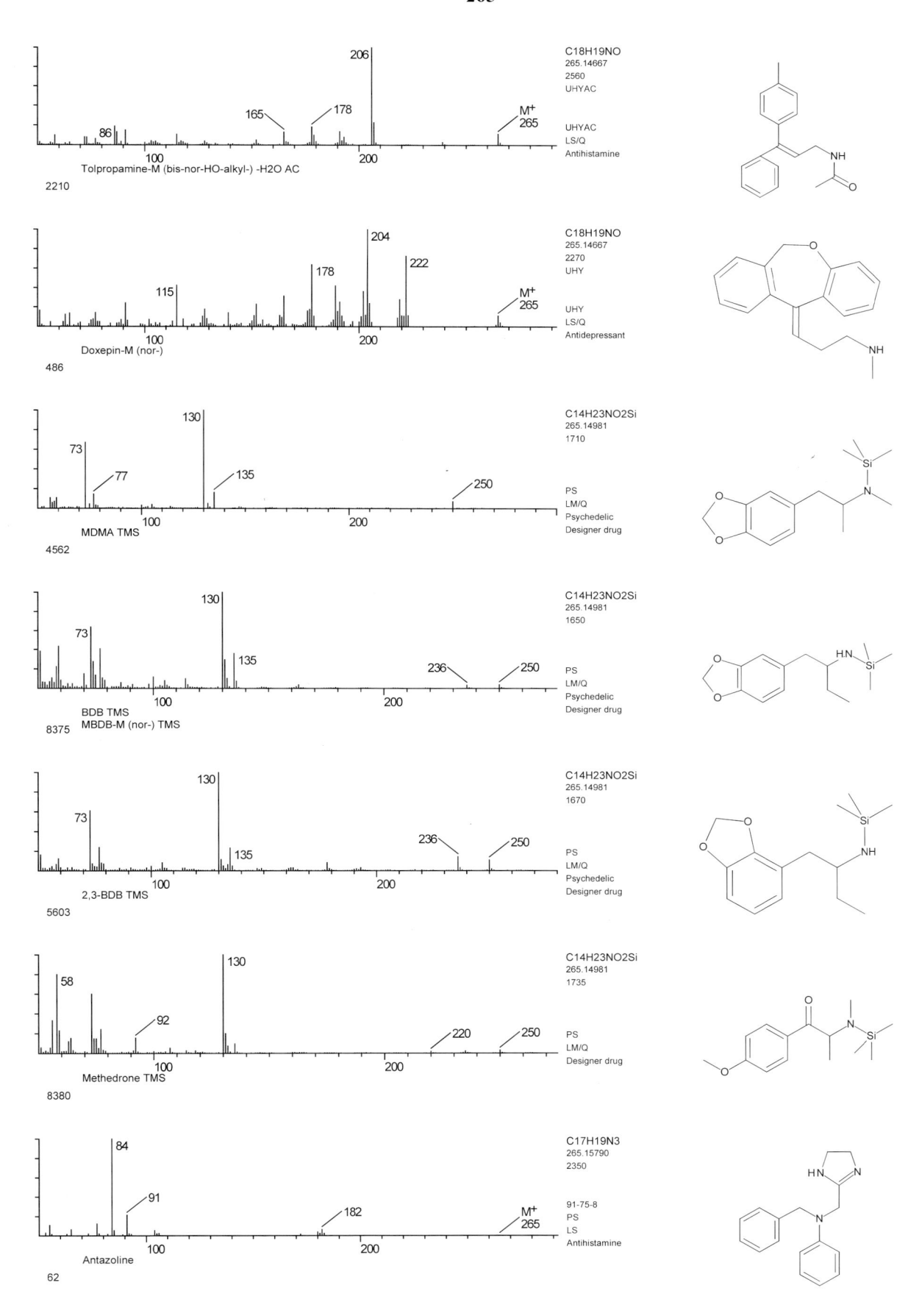

Tolpropamine-M (bis-nor-HO-alkyl-) -H2O AC
2210
Peaks: 86, 165, 178, 206, M+ 265
C18H19NO
265.14667
2560
UHYAC
UHYAC
LS/Q
Antihistamine

Doxepin-M (nor-)
486
Peaks: 115, 178, 204, 222, M+ 265
C18H19NO
265.14667
2270
UHY
UHY
LS/Q
Antidepressant

MDMA TMS
4562
Peaks: 73, 77, 130, 135, 250
C14H23NO2Si
265.14981
1710
PS
LM/Q
Psychedelic
Designer drug

BDB TMS
MBDB-M (nor-) TMS
8375
Peaks: 73, 130, 135, 236, 250
C14H23NO2Si
265.14981
1650
PS
LM/Q
Psychedelic
Designer drug

2,3-BDB TMS
5603
Peaks: 73, 130, 135, 236, 250
C14H23NO2Si
265.14981
1670
PS
LM/Q
Psychedelic
Designer drug

Methedrone TMS
8380
Peaks: 58, 92, 130, 220, 250
C14H23NO2Si
265.14981
1735
PS
LM/Q
Designer drug

Antazoline
62
Peaks: 84, 91, 182, M+ 265
C17H19N3
265.15790
2350
91-75-8
PS
LS
Antihistamine

265

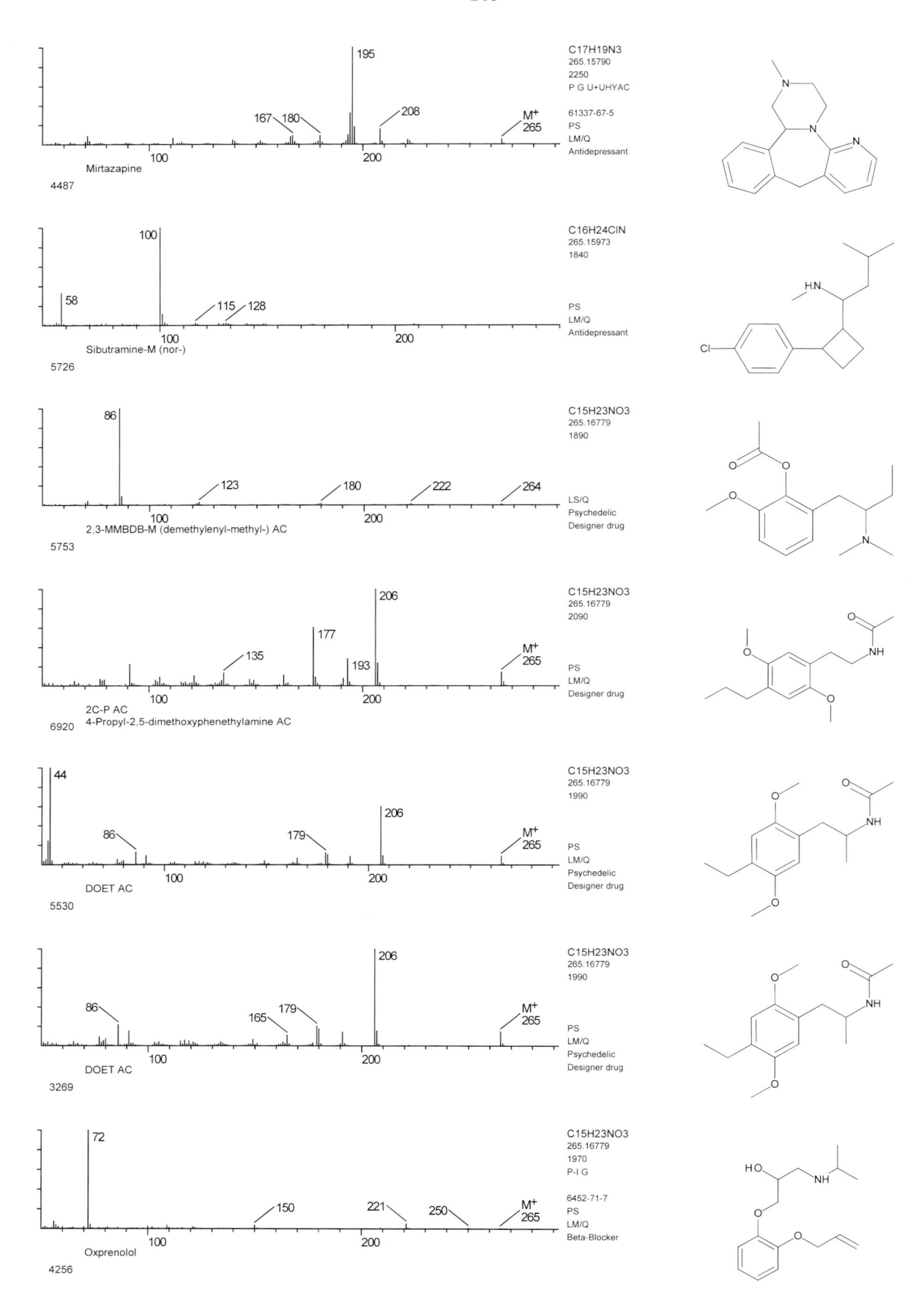

Spectrum	Compound	Formula / Info
4487	Mirtazapine — peaks 167, 180, 195, 208, M+ 265	C17H19N3, 265.15790, 2250, P G U+UHYAC, 61337-67-5, PS, LM/Q, Antidepressant
5726	Sibutramine-M (nor-) — peaks 58, 100, 115, 128	C16H24ClN, 265.15973, 1840, PS, LM/Q, Antidepressant
5753	2,3-MMBDB-M (demethylenyl-methyl-) AC — peaks 86, 123, 180, 222, 264	C15H23NO3, 265.16779, 1890, LS/Q, Psychedelic, Designer drug
6920	2C-P AC / 4-Propyl-2,5-dimethoxyphenethylamine AC — peaks 135, 177, 193, 206, M+ 265	C15H23NO3, 265.16779, 2090, PS, LM/Q, Designer drug
5530	DOET AC — peaks 44, 86, 179, 206, M+ 265	C15H23NO3, 265.16779, 1990, PS, LM/Q, Psychedelic, Designer drug
3269	DOET AC — peaks 86, 165, 179, 206, M+ 265	C15H23NO3, 265.16779, 1990, PS, LM/Q, Psychedelic, Designer drug
4256	Oxprenolol — peaks 72, 150, 221, 250, M+ 265	C15H23NO3, 265.16779, 1970, P-I G, 6452-71-7, PS, LM/Q, Beta-Blocker

265

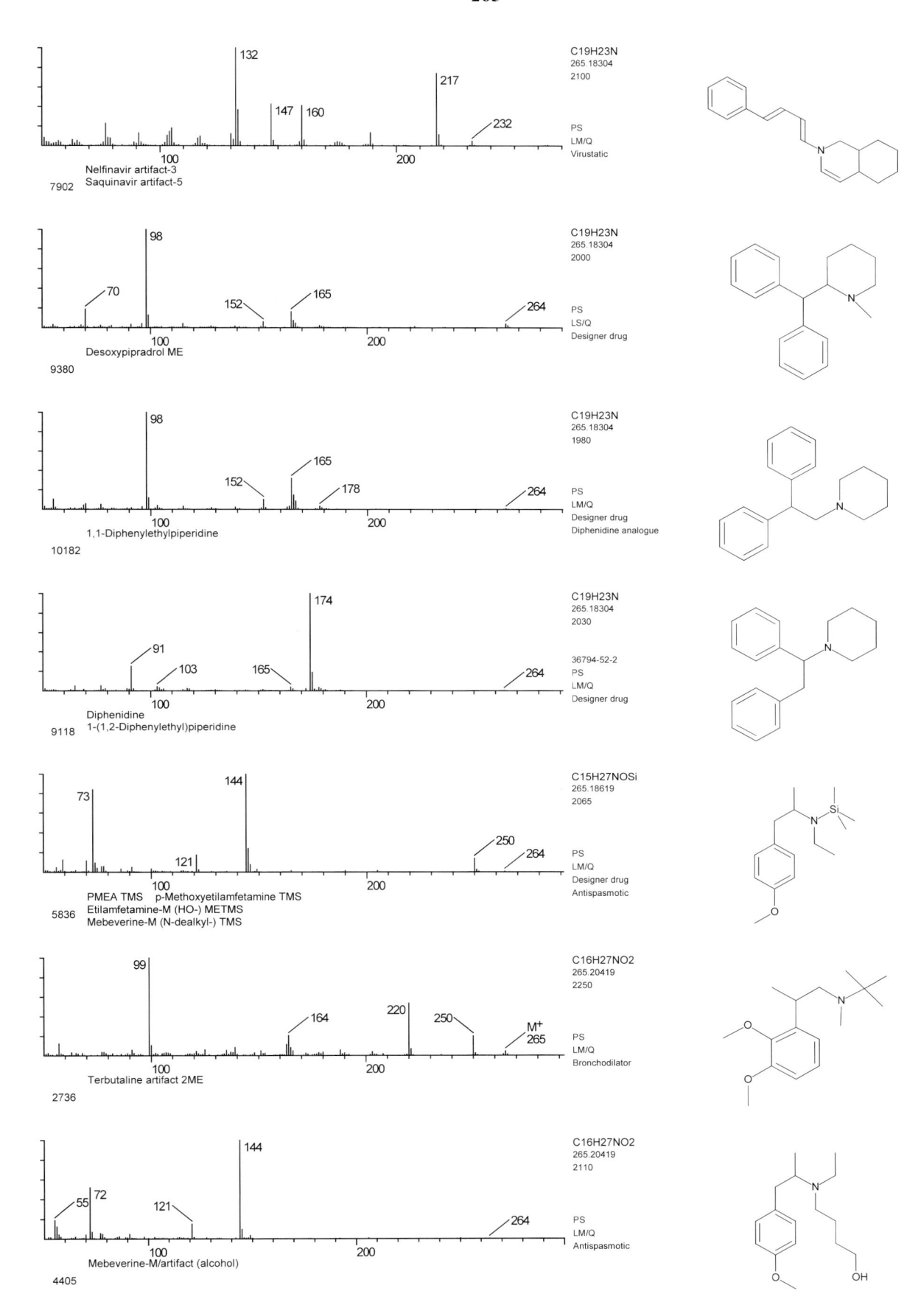

Spectrum peaks	Formula / Info
132, 147, 160, 217, 232	C19H23N 265.18304 2100 PS LM/Q Virustatic

7902 Nelfinavir artifact-3
Saquinavir artifact-5

Spectrum peaks	Formula / Info
70, 98, 152, 165, 264	C19H23N 265.18304 2000 PS LS/Q Designer drug

9380 Desoxypipradrol ME

Spectrum peaks	Formula / Info
98, 152, 165, 178, 264	C19H23N 265.18304 1980 PS LM/Q Designer drug Diphenidine analogue

10182 1,1-Diphenylethylpiperidine

Spectrum peaks	Formula / Info
91, 103, 165, 174, 264	C19H23N 265.18304 2030 36794-52-2 PS LM/Q Designer drug

9118 Diphenidine
1-(1,2-Diphenylethyl)piperidine

Spectrum peaks	Formula / Info
73, 121, 144, 250, 264	C15H27NOSi 265.18619 2065 PS LM/Q Designer drug Antispasmotic

5836 PMEA TMS p-Methoxyetilamfetamine TMS
Etilamfetamine-M (HO-) METMS
Mebeverine-M (N-dealkyl-) TMS

Spectrum peaks	Formula / Info
99, 164, 220, 250, M+ 265	C16H27NO2 265.20419 2250 PS LM/Q Bronchodilator

2736 Terbutaline artifact 2ME

Spectrum peaks	Formula / Info
55, 72, 121, 144, 264	C16H27NO2 265.20419 2110 PS LM/Q Antispasmotic

4405 Mebeverine-M/artifact (alcohol)

266

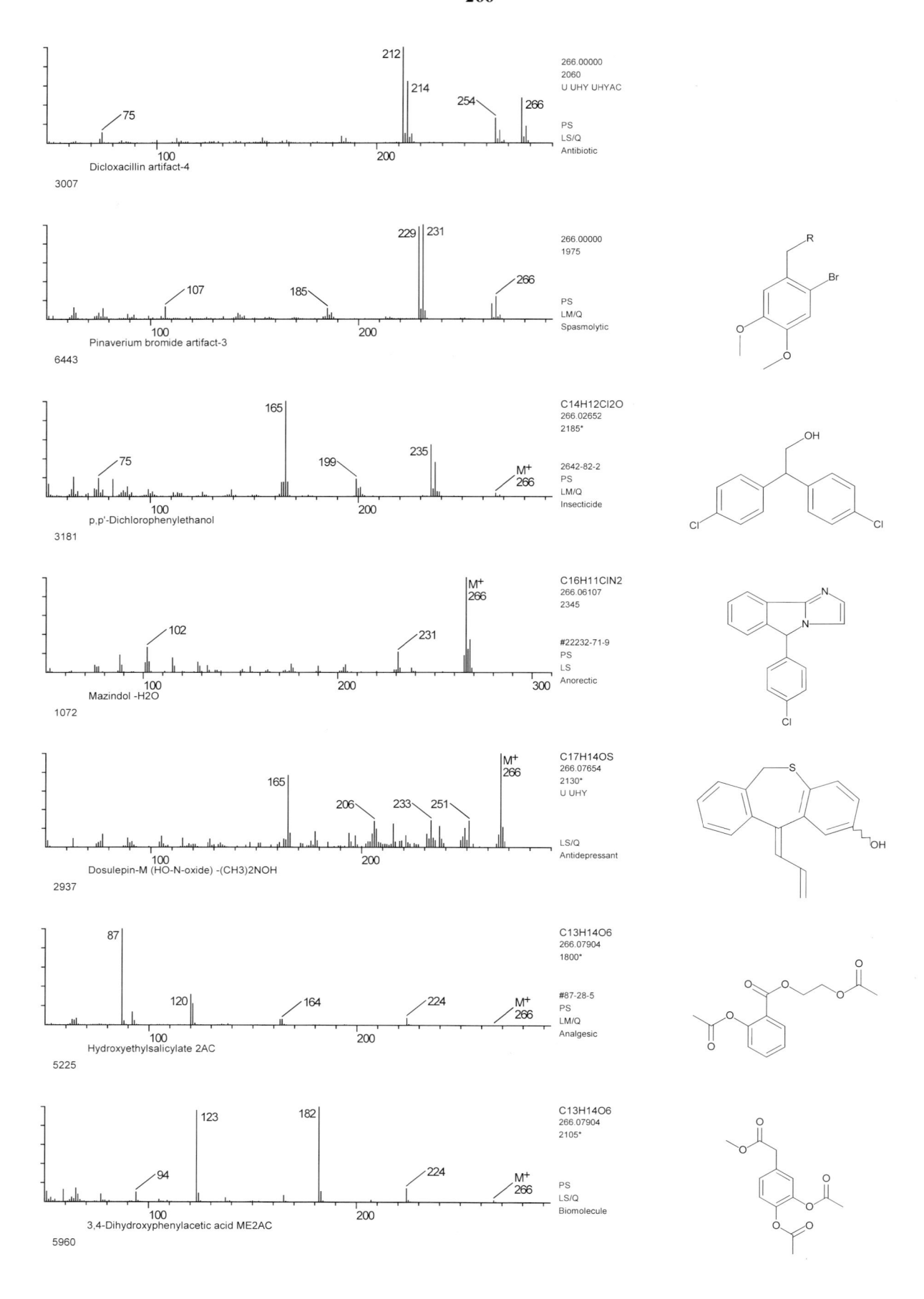

Dicloxacillin artifact-4
3007
266.00000
2060
U UHY UHYAC
PS
LS/Q
Antibiotic

Pinaverium bromide artifact-3
6443
266.00000
1975
PS
LM/Q
Spasmolytic

p,p'-Dichlorophenylethanol
3181
C14H12Cl2O
266.02652
2185*
2642-82-2
PS
LM/Q
Insecticide

Mazindol -H2O
1072
C16H11ClN2
266.06107
2345
#22232-71-9
PS
LS
Anorectic

Dosulepin-M (HO-N-oxide) -(CH3)2NOH
2937
C17H14OS
266.07654
2130*
U UHY
LS/Q
Antidepressant

Hydroxyethylsalicylate 2AC
5225
C13H14O6
266.07904
1800*
#87-28-5
PS
LM/Q
Analgesic

3,4-Dihydroxyphenylacetic acid ME2AC
5960
C13H14O6
266.07904
2105*
PS
LS/Q
Biomolecule

266

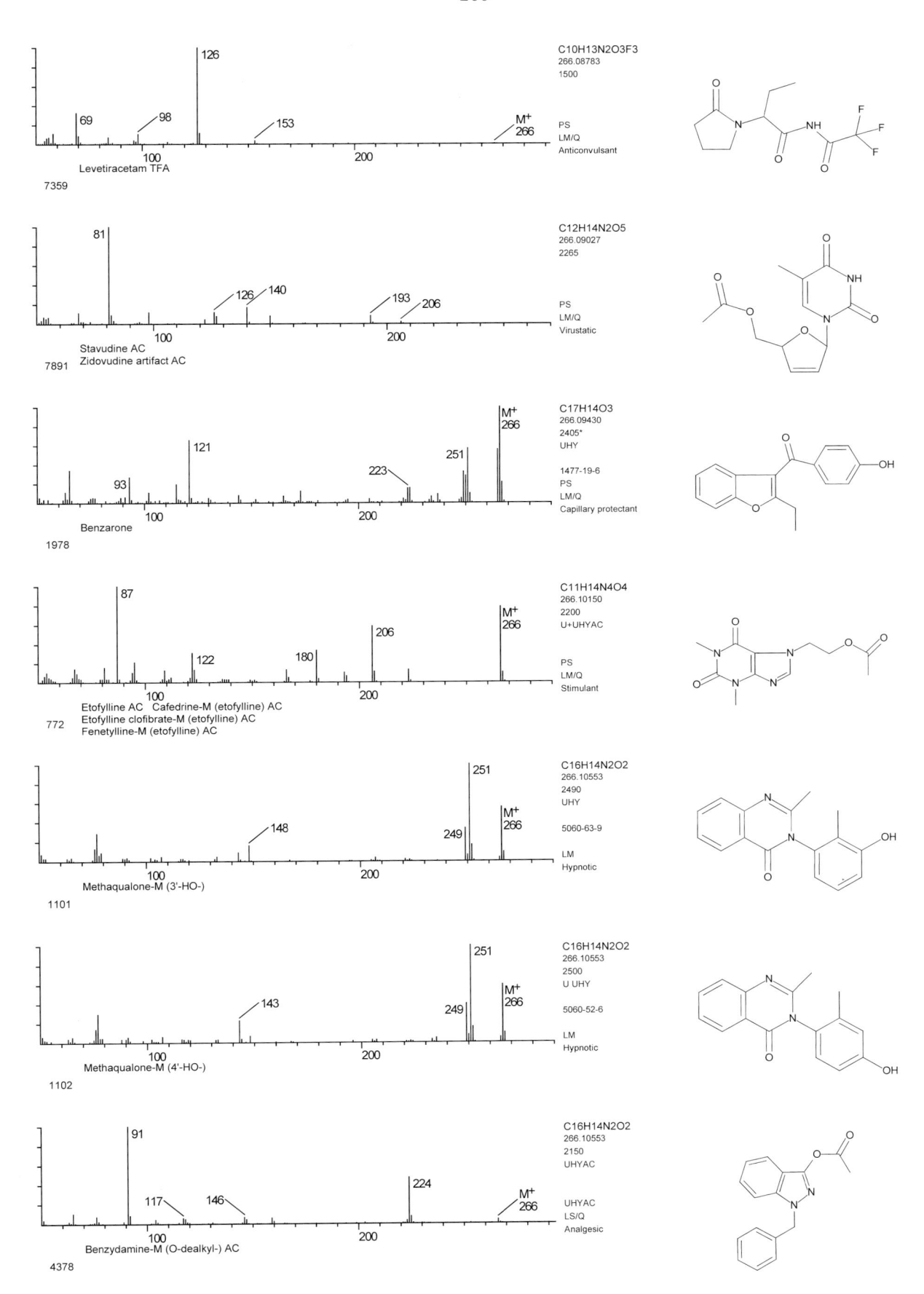

C10H13N2O3F3
266.08783
1500

PS
LM/Q
Anticonvulsant

Levetiracetam TFA
7359

C12H14N2O5
266.09027
2265

PS
LM/Q
Virustatic

Stavudine AC
Zidovudine artifact AC
7891

C17H14O3
266.09430
2405*
UHY

1477-19-6
PS
LM/Q
Capillary protectant

Benzarone
1978

C11H14N4O4
266.10150
2200
U+UHYAC

PS
LM/Q
Stimulant

Etofylline AC   Cafedrine-M (etofylline) AC
Etofylline clofibrate-M (etofylline) AC
Fenetylline-M (etofylline) AC
772

C16H14N2O2
266.10553
2490
UHY

5060-63-9

LM
Hypnotic

Methaqualone-M (3'-HO-)
1101

C16H14N2O2
266.10553
2500
U UHY

5060-52-6

LM
Hypnotic

Methaqualone-M (4'-HO-)
1102

C16H14N2O2
266.10553
2150
UHYAC

UHYAC
LS/Q
Analgesic

Benzydamine-M (O-dealkyl-) AC
4378

693

266

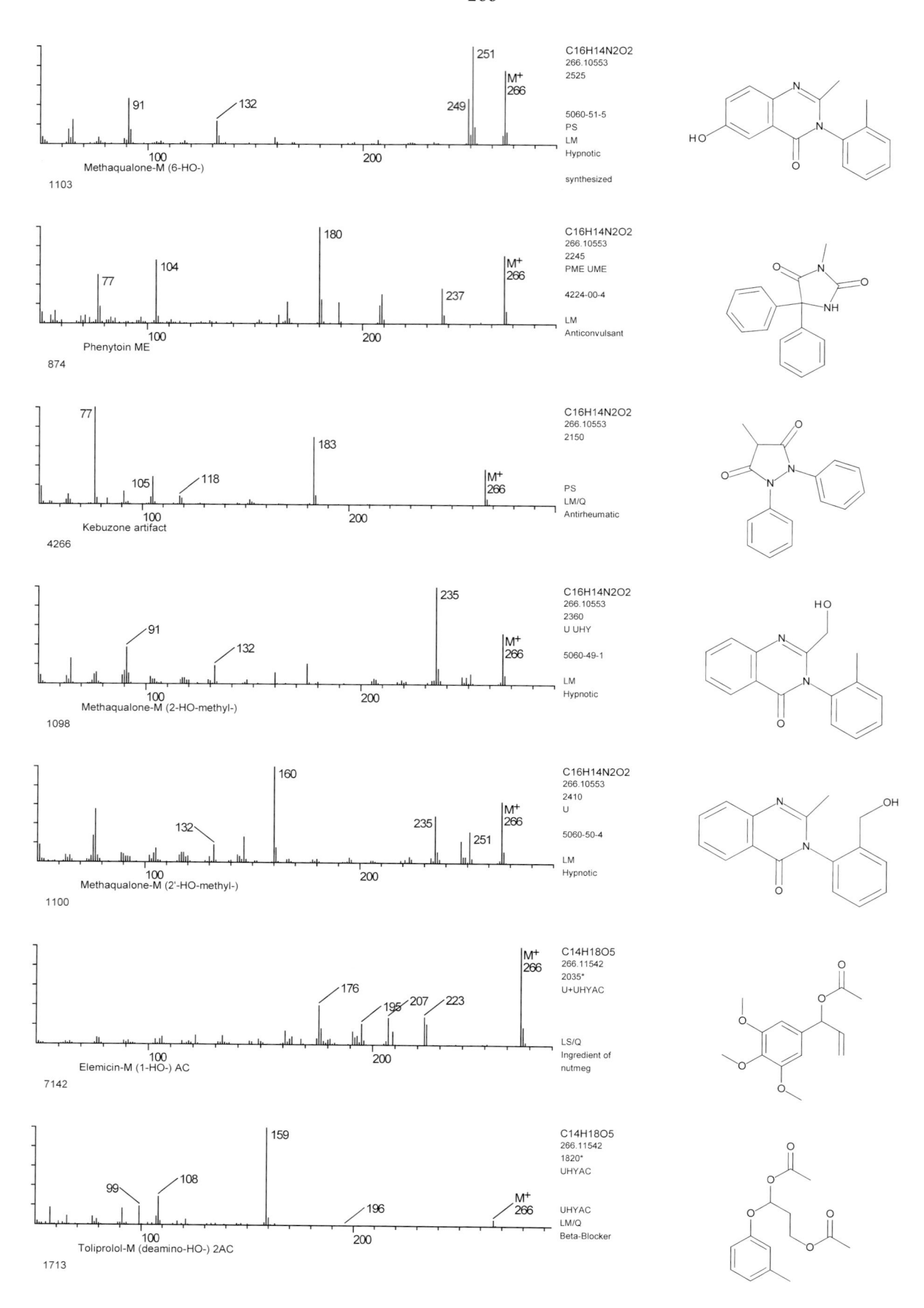

Methaqualone-M (6-HO-) 1103	91, 132, 249, 251, M+ 266	C16H14N2O2 266.10553 2525 / 5060-51-5 PS / LM / Hypnotic / synthesized
Phenytoin ME 874	77, 104, 180, 237, M+ 266	C16H14N2O2 266.10553 2245 PME UME / 4224-00-4 / LM / Anticonvulsant
Kebuzone artifact 4266	77, 105, 118, 183, M+ 266	C16H14N2O2 266.10553 2150 / PS / LM/Q / Antirheumatic
Methaqualone-M (2-HO-methyl-) 1098	91, 132, 235, M+ 266	C16H14N2O2 266.10553 2360 U UHY / 5060-49-1 / LM / Hypnotic
Methaqualone-M (2'-HO-methyl-) 1100	132, 160, 235, 251, M+ 266	C16H14N2O2 266.10553 2410 U / 5060-50-4 / LM / Hypnotic
Elemicin-M (1-HO-) AC 7142	176, 195, 207, 223, M+ 266	C14H18O5 266.11542 2035* U+UHYAC / LS/Q / Ingredient of nutmeg
Toliprolol-M (deamino-HO-) 2AC 1713	99, 108, 159, 196, M+ 266	C14H18O5 266.11542 1820* UHYAC / UHYAC / LM/Q / Beta-Blocker

694

266

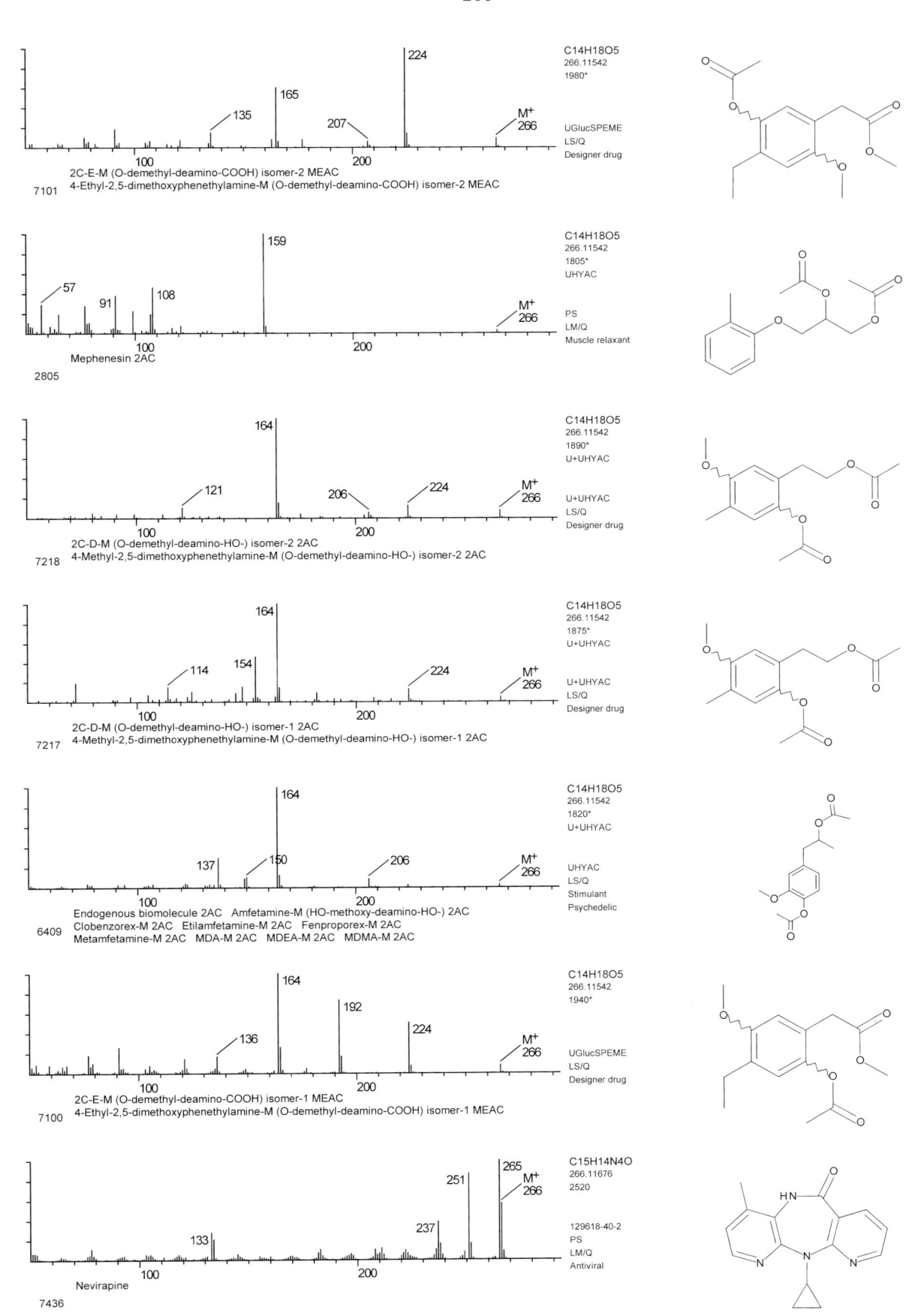

7101	Spectrum: peaks 135, 165, 207, 224, M+ 266	C14H18O5 266.11542 1980* UGlucSPEME LS/Q Designer drug
	2C-E-M (O-demethyl-deamino-COOH) isomer-2 MEAC 4-Ethyl-2,5-dimethoxyphenethylamine-M (O-demethyl-deamino-COOH) isomer-2 MEAC	
2805	Spectrum: peaks 57, 91, 108, 159, M+ 266	C14H18O5 266.11542 1805* UHYAC PS LM/Q Muscle relaxant
	Mephenesin 2AC	
7218	Spectrum: peaks 121, 164, 206, 224, M+ 266	C14H18O5 266.11542 1890* U+UHYAC U+UHYAC LS/Q Designer drug
	2C-D-M (O-demethyl-deamino-HO-) isomer-2 2AC 4-Methyl-2,5-dimethoxyphenethylamine-M (O-demethyl-deamino-HO-) isomer-2 2AC	
7217	Spectrum: peaks 114, 154, 164, 224, M+ 266	C14H18O5 266.11542 1875* U+UHYAC U+UHYAC LS/Q Designer drug
	2C-D-M (O-demethyl-deamino-HO-) isomer-1 2AC 4-Methyl-2,5-dimethoxyphenethylamine-M (O-demethyl-deamino-HO-) isomer-1 2AC	
6409	Spectrum: peaks 137, 150, 164, 206, M+ 266	C14H18O5 266.11542 1820* U+UHYAC UHYAC LS/Q Stimulant Psychedelic
	Endogenous biomolecule 2AC Amfetamine-M (HO-methoxy-deamino-HO-) 2AC Clobenzorex-M 2AC Etilamfetamine-M 2AC Fenproporex-M 2AC Metamfetamine-M 2AC MDA-M 2AC MDEA-M 2AC MDMA-M 2AC	
7100	Spectrum: peaks 136, 164, 192, 224, M+ 266	C14H18O5 266.11542 1940* UGlucSPEME LS/Q Designer drug
	2C-E-M (O-demethyl-deamino-COOH) isomer-1 MEAC 4-Ethyl-2,5-dimethoxyphenethylamine-M (O-demethyl-deamino-COOH) isomer-1 MEAC	
7436	Spectrum: peaks 133, 237, 251, 265, M+ 266	C15H14N4O 266.11676 2520 129618-40-2 PS LM/Q Antiviral
	Nevirapine	

266

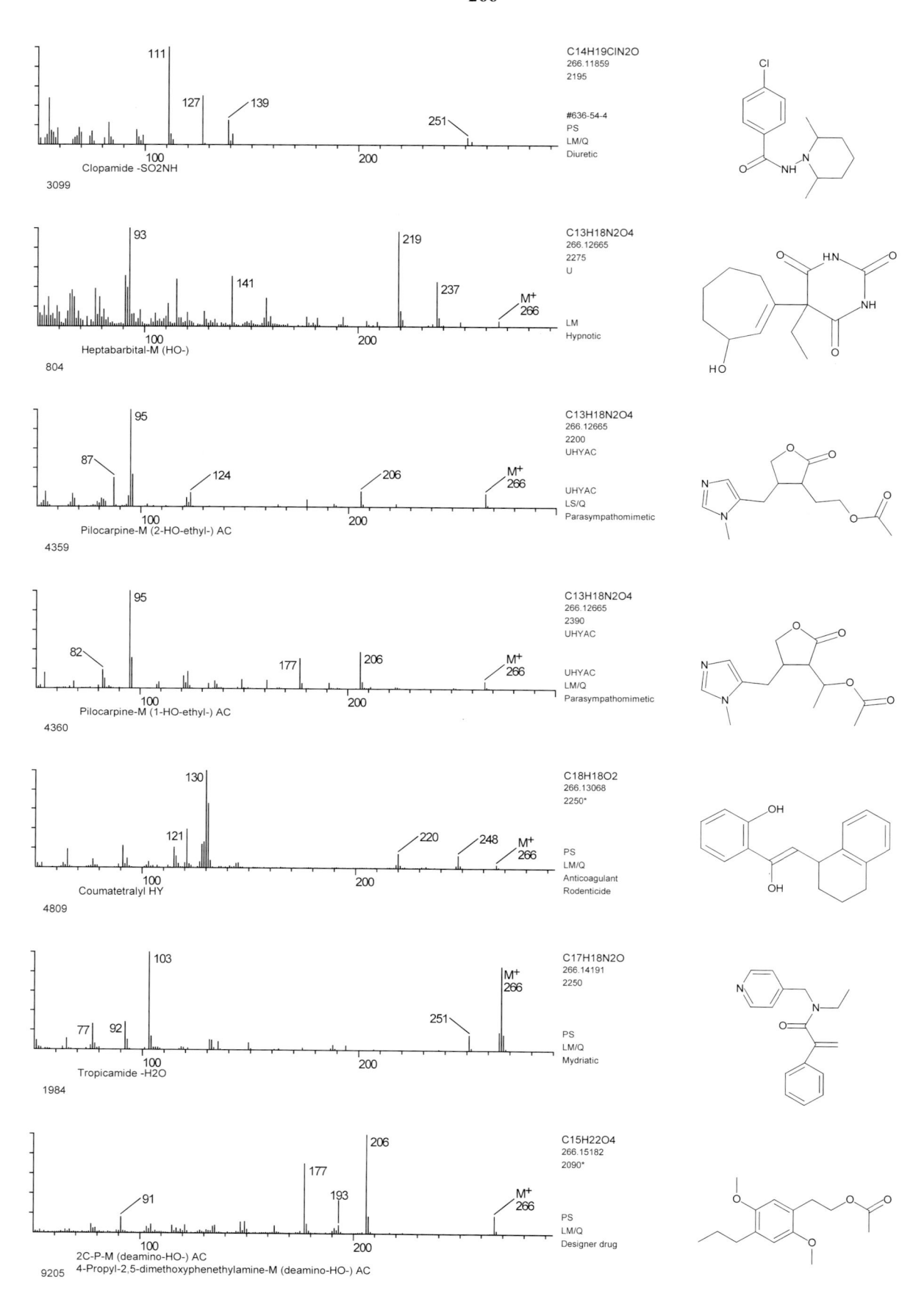

Clopamide -SO2NH
C14H19ClN2O
266.11859
2195
#636-54-4
PS
LM/Q
Diuretic

Heptabarbital-M (HO-)
C13H18N2O4
266.12665
2275
U
LM
Hypnotic

Pilocarpine-M (2-HO-ethyl-) AC
C13H18N2O4
266.12665
2200
UHYAC
UHYAC
LS/Q
Parasympathomimetic

Pilocarpine-M (1-HO-ethyl-) AC
C13H18N2O4
266.12665
2390
UHYAC
UHYAC
LM/Q
Parasympathomimetic

Coumatetralyl HY
C18H18O2
266.13068
2250*
PS
LM/Q
Anticoagulant
Rodenticide

Tropicamide -H2O
C17H18N2O
266.14191
2250
PS
LM/Q
Mydriatic

2C-P-M (deamino-HO-) AC
4-Propyl-2,5-dimethoxyphenethylamine-M (deamino-HO-) AC
C15H22O4
266.15182
2090*
PS
LM/Q
Designer drug

266

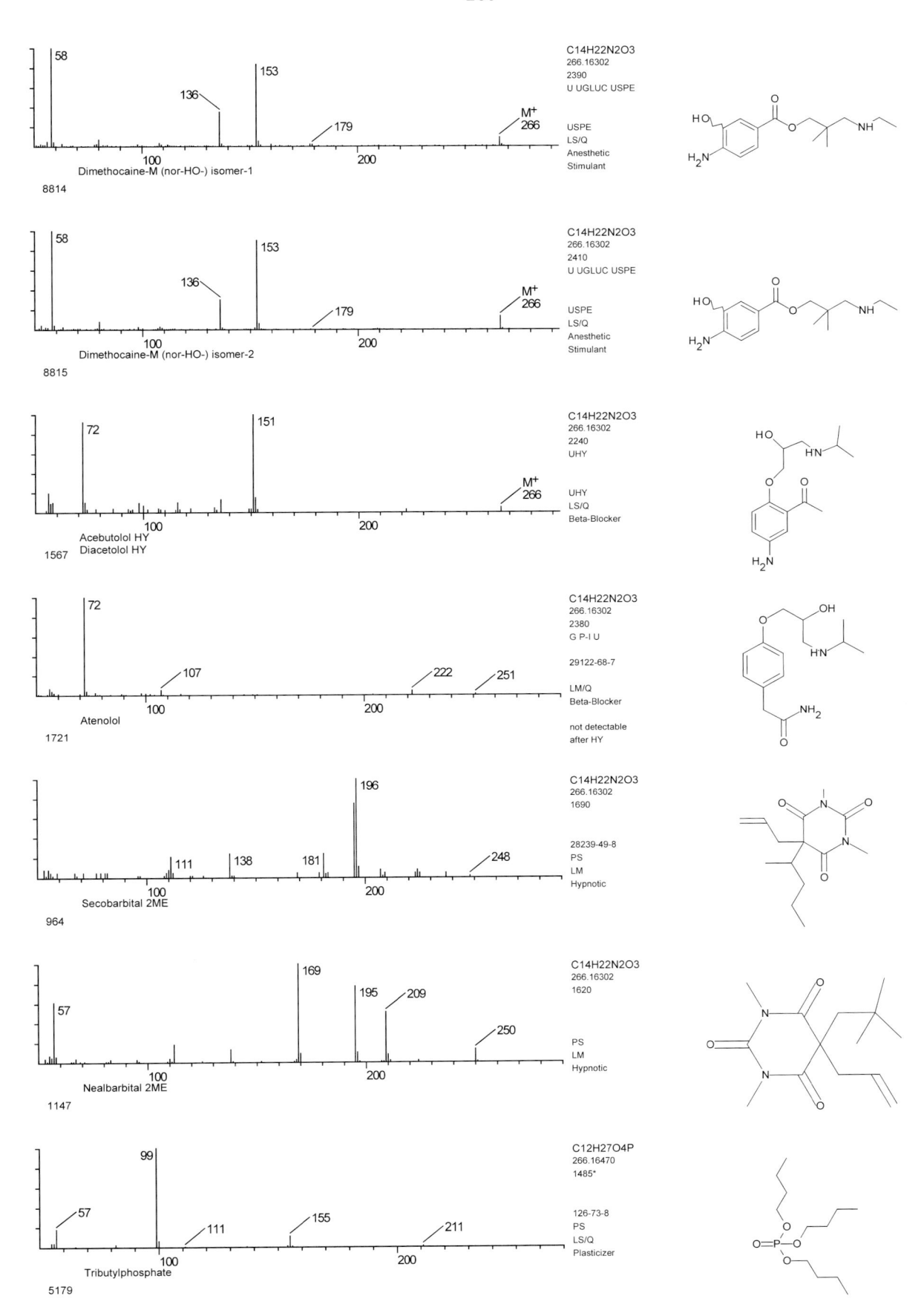

58, 136, 153, 179, M+ 266	C14H22N2O3 266.16302 2390 U UGLUC USPE USPE LS/Q Anesthetic Stimulant	Dimethocaine-M (nor-HO-) isomer-1 8814
58, 136, 153, 179, M+ 266	C14H22N2O3 266.16302 2410 U UGLUC USPE USPE LS/Q Anesthetic Stimulant	Dimethocaine-M (nor-HO-) isomer-2 8815
72, 151, M+ 266	C14H22N2O3 266.16302 2240 UHY UHY LS/Q Beta-Blocker	Acebutolol HY Diacetolol HY 1567
72, 107, 222, 251	C14H22N2O3 266.16302 2380 G P-I U 29122-68-7 LM/Q Beta-Blocker not detectable after HY	Atenolol 1721
111, 138, 181, 196, 248	C14H22N2O3 266.16302 1690 28239-49-8 PS LM Hypnotic	Secobarbital 2ME 964
57, 169, 195, 209, 250	C14H22N2O3 266.16302 1620 PS LM Hypnotic	Nealbarbital 2ME 1147
57, 99, 111, 155, 211	C12H27O4P 266.16470 1485* 126-73-8 PS LS/Q Plasticizer	Tributylphosphate 5179

266

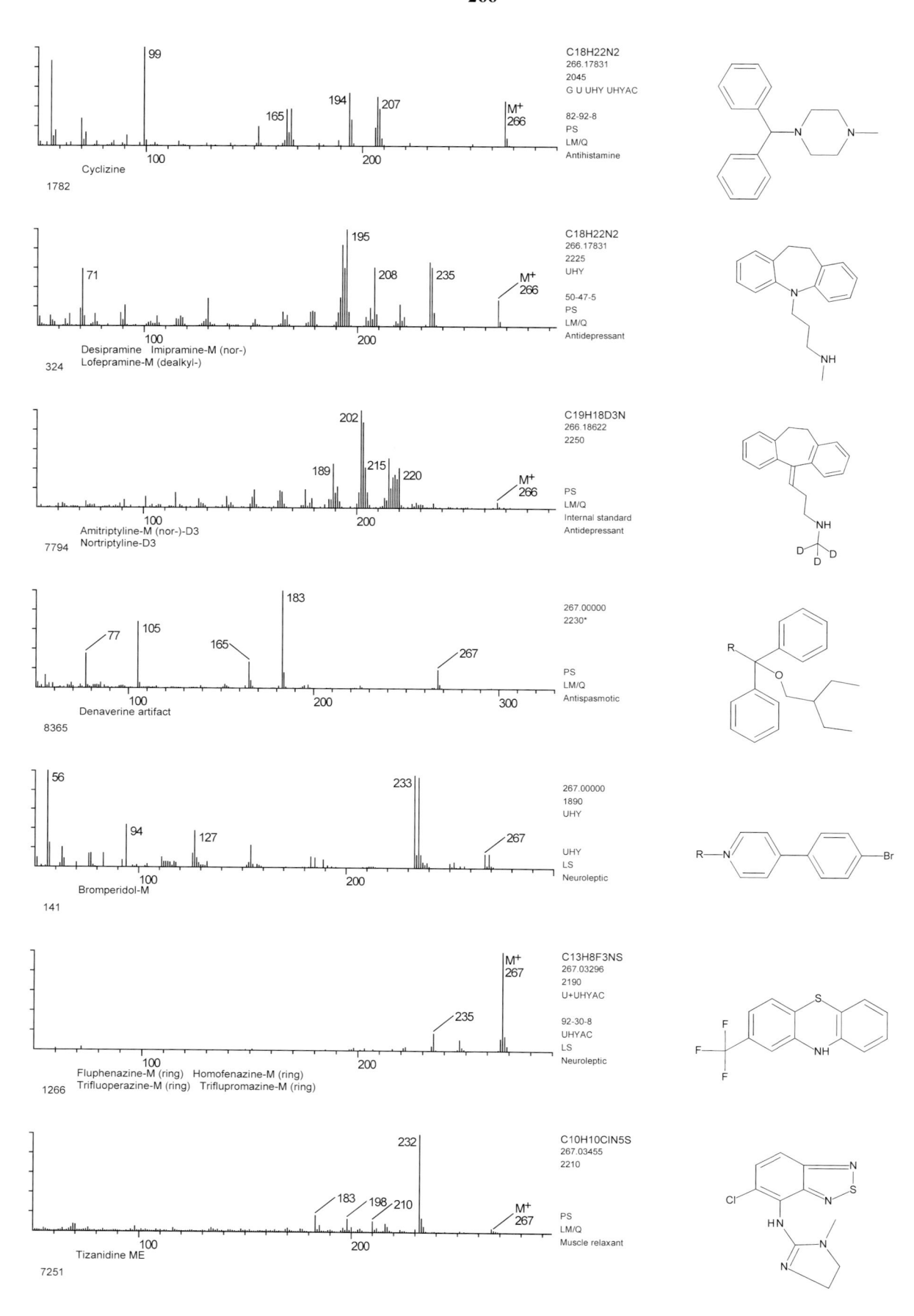

Cyclizine	99, 165, 194, 207, M+ 266	C18H22N2 266.17831 2045 G U UHY UHYAC 82-92-8 PS LM/Q Antihistamine

1782

Desipramine   Imipramine-M (nor-)
Lofepramine-M (dealkyl-) — 71, 195, 208, 235, M+ 266

C18H22N2
266.17831
2225
UHY

50-47-5
PS
LM/Q
Antidepressant

324

Amitriptyline-M (nor-)-D3
Nortriptyline-D3 — 189, 202, 215, 220, M+ 266

C19H18D3N
266.18622
2250

PS
LM/Q
Internal standard
Antidepressant

7794

Denaverine artifact — 77, 105, 165, 183, 267

267.00000
2230*

PS
LM/Q
Antispasmotic

8365

Bromperidol-M — 56, 94, 127, 233, 267

267.00000
1890
UHY

UHY
LS
Neuroleptic

141

Fluphenazine-M (ring)   Homofenazine-M (ring)
Trifluoperazine-M (ring)   Triflupromazine-M (ring) — 235, M+ 267

C13H8F3NS
267.03296
2190
U+UHYAC

92-30-8
UHYAC
LS
Neuroleptic

1266

Tizanidine ME — 183, 198, 210, 232, M+ 267

C10H10ClN5S
267.03455
2210

PS
LM/Q
Muscle relaxant

7251

698

267

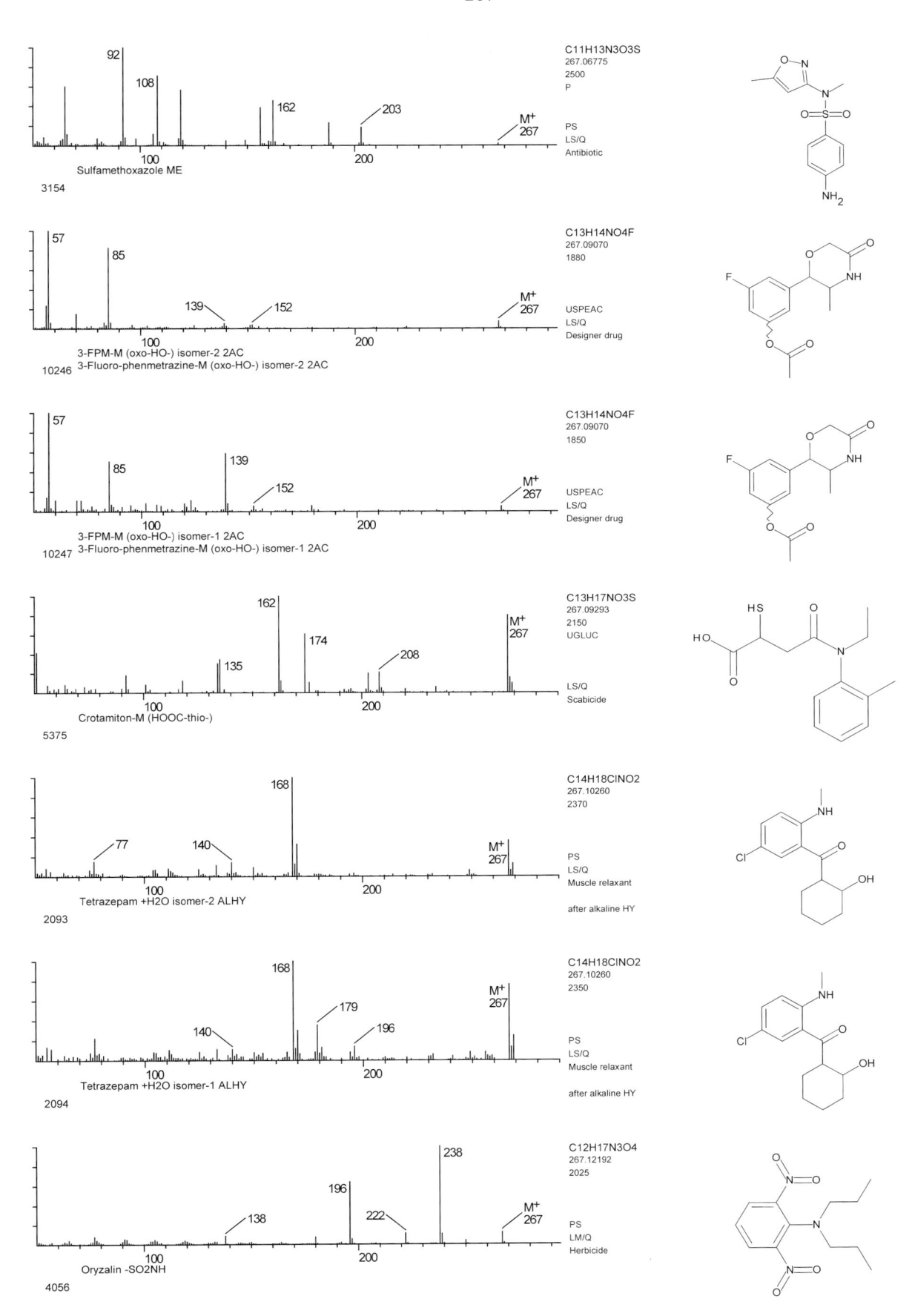

Sulfamethoxazole ME — 3154	C11H13N3O3S 267.06775 2500 P PS LS/Q Antibiotic
3-FPM-M (oxo-HO-) isomer-2 2AC — 10246 3-Fluoro-phenmetrazine-M (oxo-HO-) isomer-2 2AC	C13H14NO4F 267.09070 1880 USPEAC LS/Q Designer drug
3-FPM-M (oxo-HO-) isomer-1 2AC — 10247 3-Fluoro-phenmetrazine-M (oxo-HO-) isomer-1 2AC	C13H14NO4F 267.09070 1850 USPEAC LS/Q Designer drug
Crotamiton-M (HOOC-thio-) — 5375	C13H17NO3S 267.09293 2150 UGLUC LS/Q Scabicide
Tetrazepam +H2O isomer-2 ALHY — 2093	C14H18ClNO2 267.10260 2370 PS LS/Q Muscle relaxant after alkaline HY
Tetrazepam +H2O isomer-1 ALHY — 2094	C14H18ClNO2 267.10260 2350 PS LS/Q Muscle relaxant after alkaline HY
Oryzalin -SO2NH — 4056	C12H17N3O4 267.12192 2025 PS LM/Q Herbicide

Triamterene ME
C13H13N7
267.12323
2875
UME

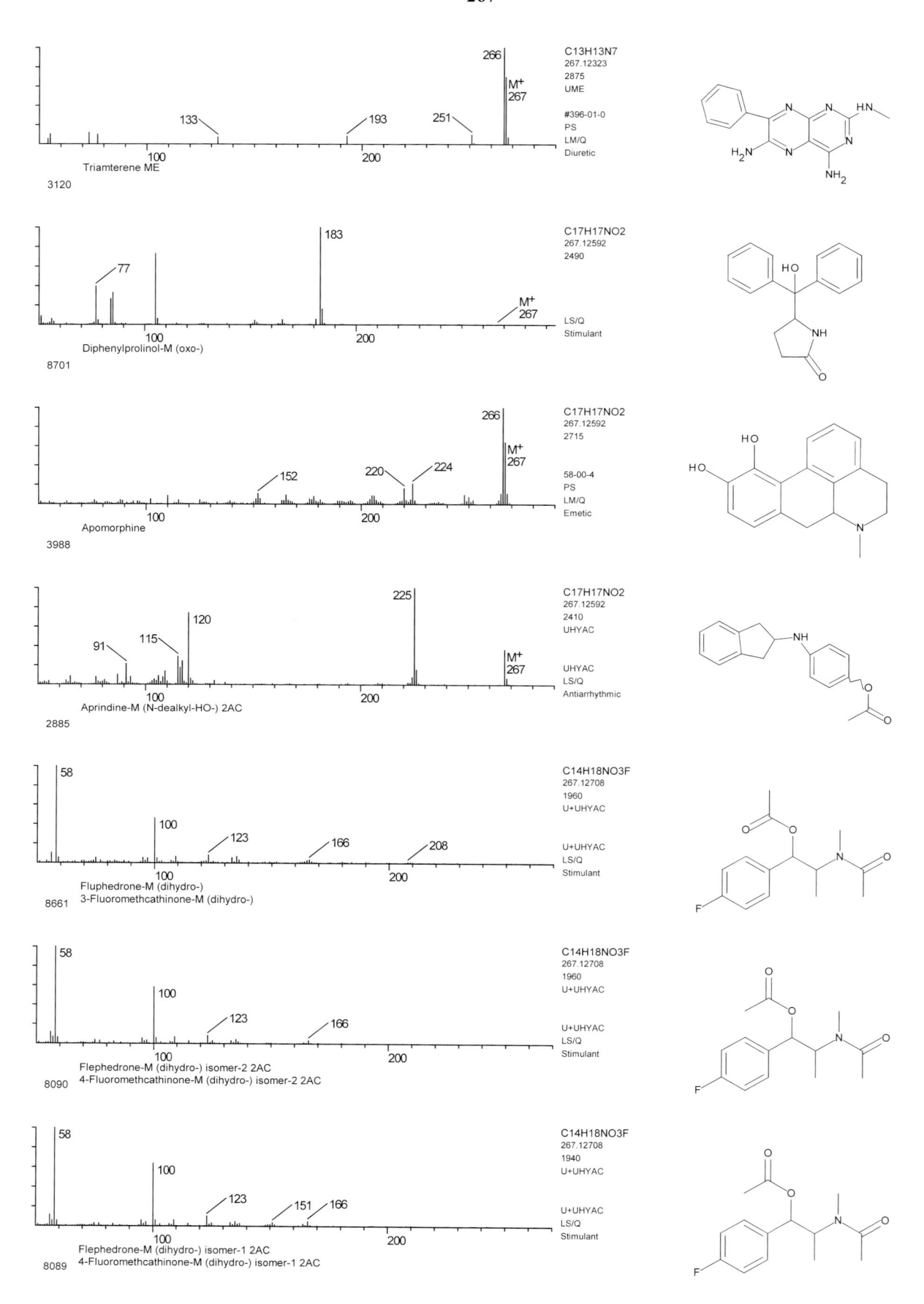

#396-01-0
PS
LM/Q
Diuretic

Peaks: 133, 193, 251, 266, M+ 267

3120

Diphenylprolinol-M (oxo-)
C17H17NO2
267.12592
2490

LS/Q
Stimulant

Peaks: 77, 183, M+ 267

8701

Apomorphine
C17H17NO2
267.12592
2715

58-00-4
PS
LM/Q
Emetic

Peaks: 152, 220, 224, 266, M+ 267

3988

Aprindine-M (N-dealkyl-HO-) 2AC
C17H17NO2
267.12592
2410
UHYAC

UHYAC
LS/Q
Antiarrhythmic

Peaks: 91, 115, 120, 225, M+ 267

2885

Fluphedrone-M (dihydro-)
3-Fluoromethcathinone-M (dihydro-)
C14H18NO3F
267.12708
1960
U+UHYAC

U+UHYAC
LS/Q
Stimulant

Peaks: 58, 100, 123, 166, 208

8661

Flephedrone-M (dihydro-) isomer-2 2AC
4-Fluoromethcathinone-M (dihydro-) isomer-2 2AC
C14H18NO3F
267.12708
1960
U+UHYAC

U+UHYAC
LS/Q
Stimulant

Peaks: 58, 100, 123, 166

8090

Flephedrone-M (dihydro-) isomer-1 2AC
4-Fluoromethcathinone-M (dihydro-) isomer-1 2AC
C14H18NO3F
267.12708
1940
U+UHYAC

U+UHYAC
LS/Q
Stimulant

Peaks: 58, 100, 123, 151, 166

8089

267

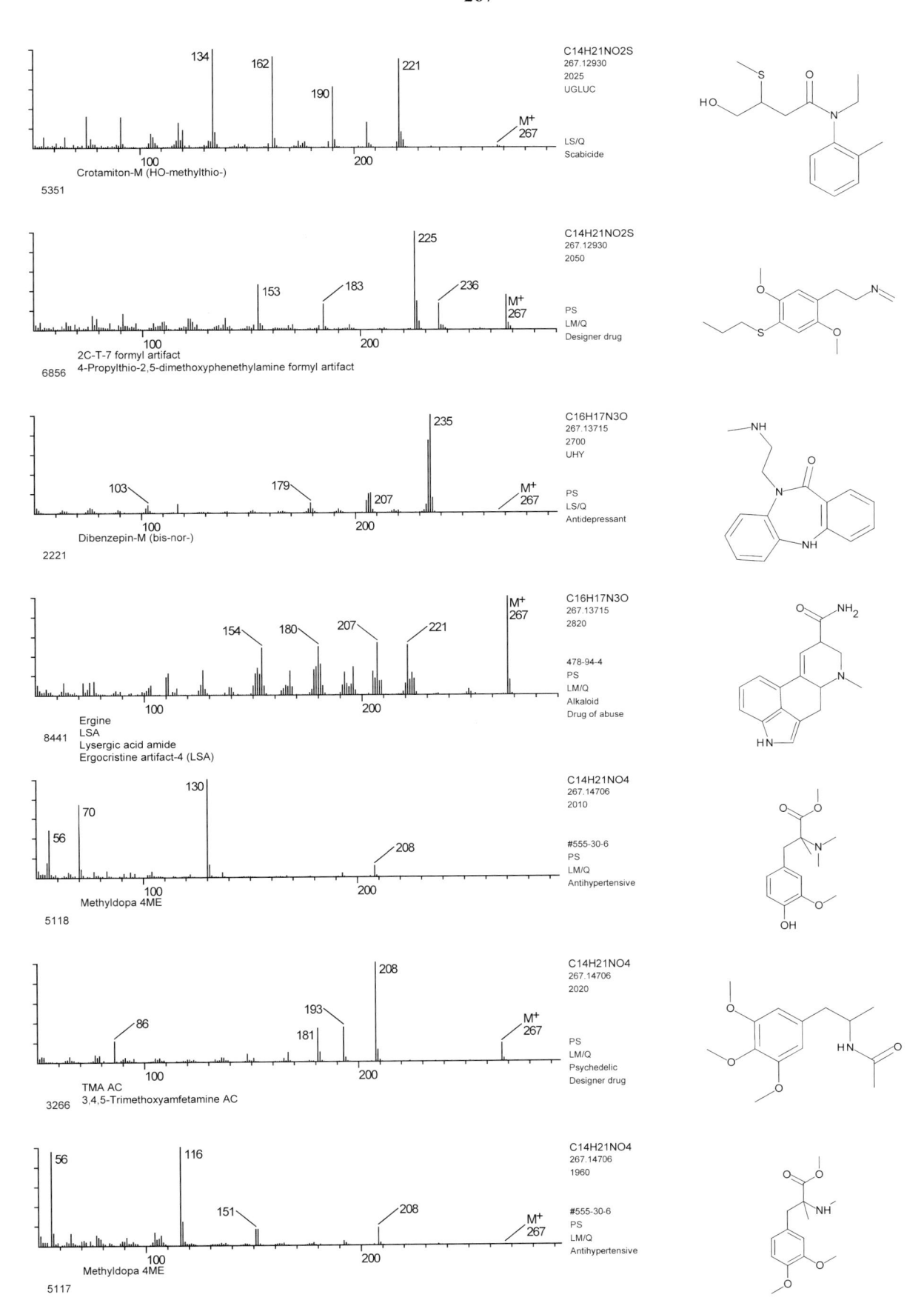

Crotamiton-M (HO-methylthio-) 5351	Peaks: 134, 162, 190, 221, M+ 267	C14H21NO2S 267.12930 2025 UGLUC LS/Q Scabicide
2C-T-7 formyl artifact 4-Propylthio-2,5-dimethoxyphenethylamine formyl artifact 6856	Peaks: 153, 183, 225, 236, M+ 267	C14H21NO2S 267.12930 2050 PS LM/Q Designer drug
Dibenzepin-M (bis-nor-) 2221	Peaks: 103, 179, 207, 235, M+ 267	C16H17N3O 267.13715 2700 UHY PS LS/Q Antidepressant
Ergine LSA Lysergic acid amide Ergocristine artifact-4 (LSA) 8441	Peaks: 154, 180, 207, 221, M+ 267	C16H17N3O 267.13715 2820 478-94-4 PS LM/Q Alkaloid Drug of abuse
Methyldopa 4ME 5118	Peaks: 56, 70, 130, 208	C14H21NO4 267.14706 2010 #555-30-6 PS LM/Q Antihypertensive
TMA AC 3,4,5-Trimethoxyamfetamine AC 3266	Peaks: 86, 181, 193, 208, M+ 267	C14H21NO4 267.14706 2020 PS LM/Q Psychedelic Designer drug
Methyldopa 4ME 5117	Peaks: 56, 116, 151, 208, M+ 267	C14H21NO4 267.14706 1960 #555-30-6 PS LM/Q Antihypertensive

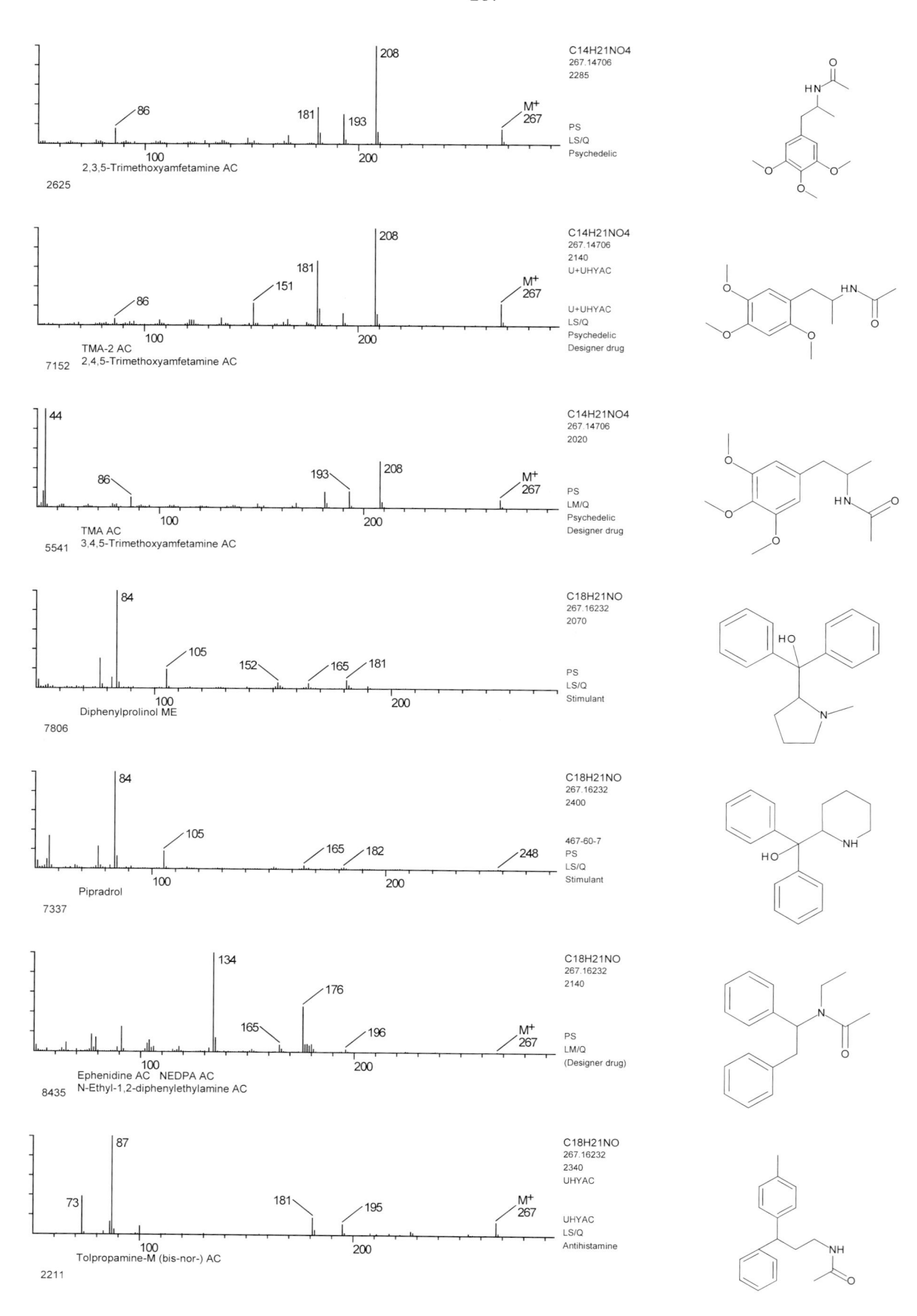

2,3,5-Trimethoxyamfetamine AC	Peaks: 86, 181, 193, 208, M+ 267	C14H21NO4 267.14706 2285 PS LS/Q Psychedelic
2625		
TMA-2 AC / 2,4,5-Trimethoxyamfetamine AC	Peaks: 86, 151, 181, 208, M+ 267	C14H21NO4 267.14706 2140 U+UHYAC U+UHYAC LS/Q Psychedelic Designer drug
7152		
TMA AC / 3,4,5-Trimethoxyamfetamine AC	Peaks: 44, 86, 193, 208, M+ 267	C14H21NO4 267.14706 2020 PS LM/Q Psychedelic Designer drug
5541		
Diphenylprolinol ME	Peaks: 84, 105, 152, 165, 181	C18H21NO 267.16232 2070 PS LS/Q Stimulant
7806		
Pipradrol	Peaks: 84, 105, 165, 182, 248	C18H21NO 267.16232 2400 467-60-7 PS LS/Q Stimulant
7337		
Ephenidine AC NEDPA AC / N-Ethyl-1,2-diphenylethylamine AC	Peaks: 134, 165, 176, 196, M+ 267	C18H21NO 267.16232 2140 PS LM/Q (Designer drug)
8435		
Tolpropamine-M (bis-nor-) AC	Peaks: 73, 87, 181, 195, M+ 267	C18H21NO 267.16232 2340 UHYAC UHYAC LS/Q Antihistamine
2211		

267

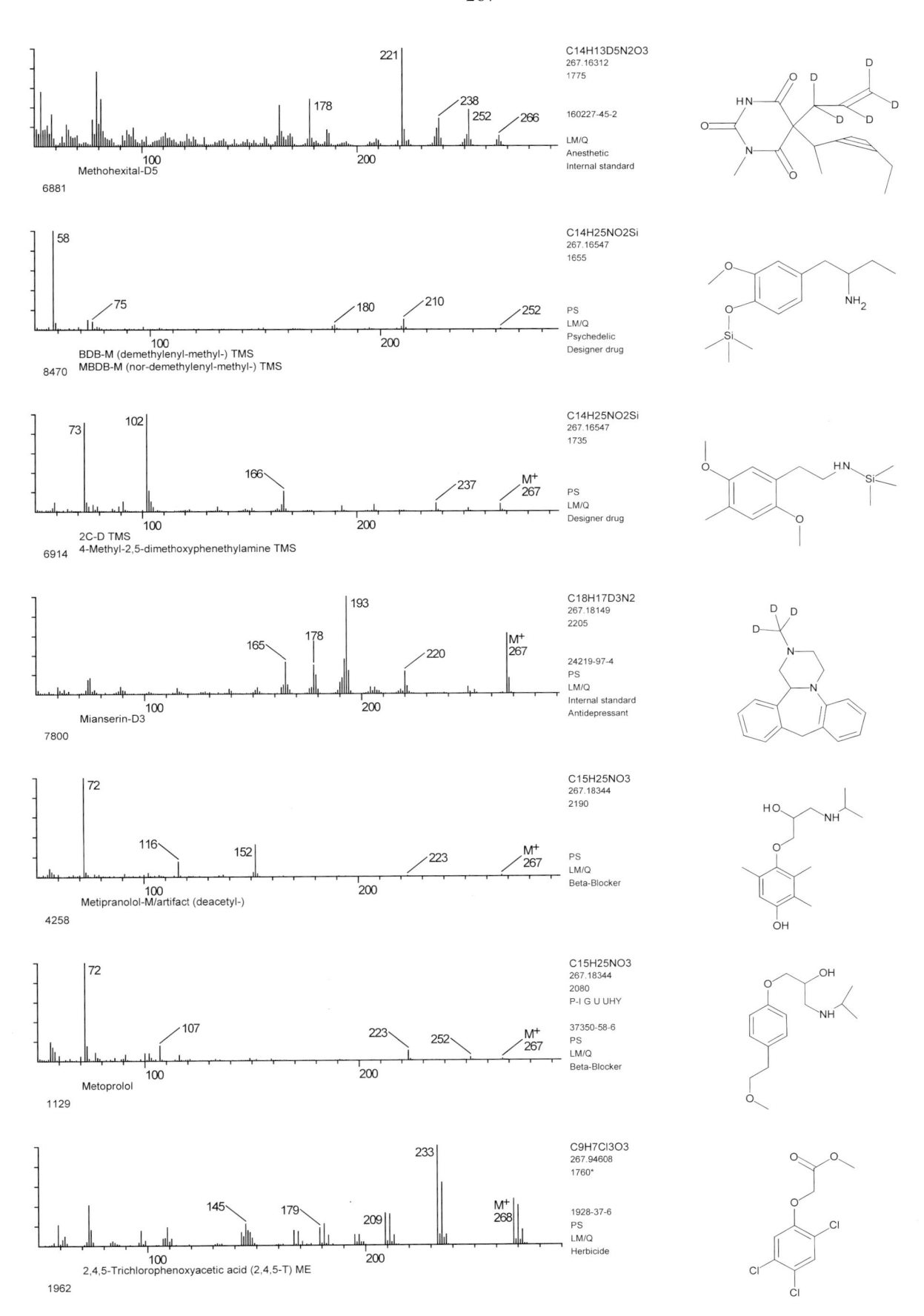

Methohexital-D5 — 6881
C14H13D5N2O3
267.16312
1775
160227-45-2
LM/Q
Anesthetic
Internal standard

BDB-M (demethylenyl-methyl-) TMS
MBDB-M (nor-demethylenyl-methyl-) TMS — 8470
C14H25NO2Si
267.16547
1655
PS
LM/Q
Psychedelic
Designer drug

2C-D TMS
4-Methyl-2,5-dimethoxyphenethylamine TMS — 6914
C14H25NO2Si
267.16547
1735
PS
LM/Q
Designer drug

Mianserin-D3 — 7800
C18H17D3N2
267.18149
2205
24219-97-4
PS
LM/Q
Internal standard
Antidepressant

Metipranolol-M/artifact (deacetyl-) — 4258
C15H25NO3
267.18344
2190
PS
LM/Q
Beta-Blocker

Metoprolol — 1129
C15H25NO3
267.18344
2080
P-I G U UHY
37350-58-6
PS
LM/Q
Beta-Blocker

2,4,5-Trichlorophenoxyacetic acid (2,4,5-T) ME — 1962
C9H7Cl3O3
267.94608
1760*
1928-37-6
PS
LM/Q
Herbicide

268

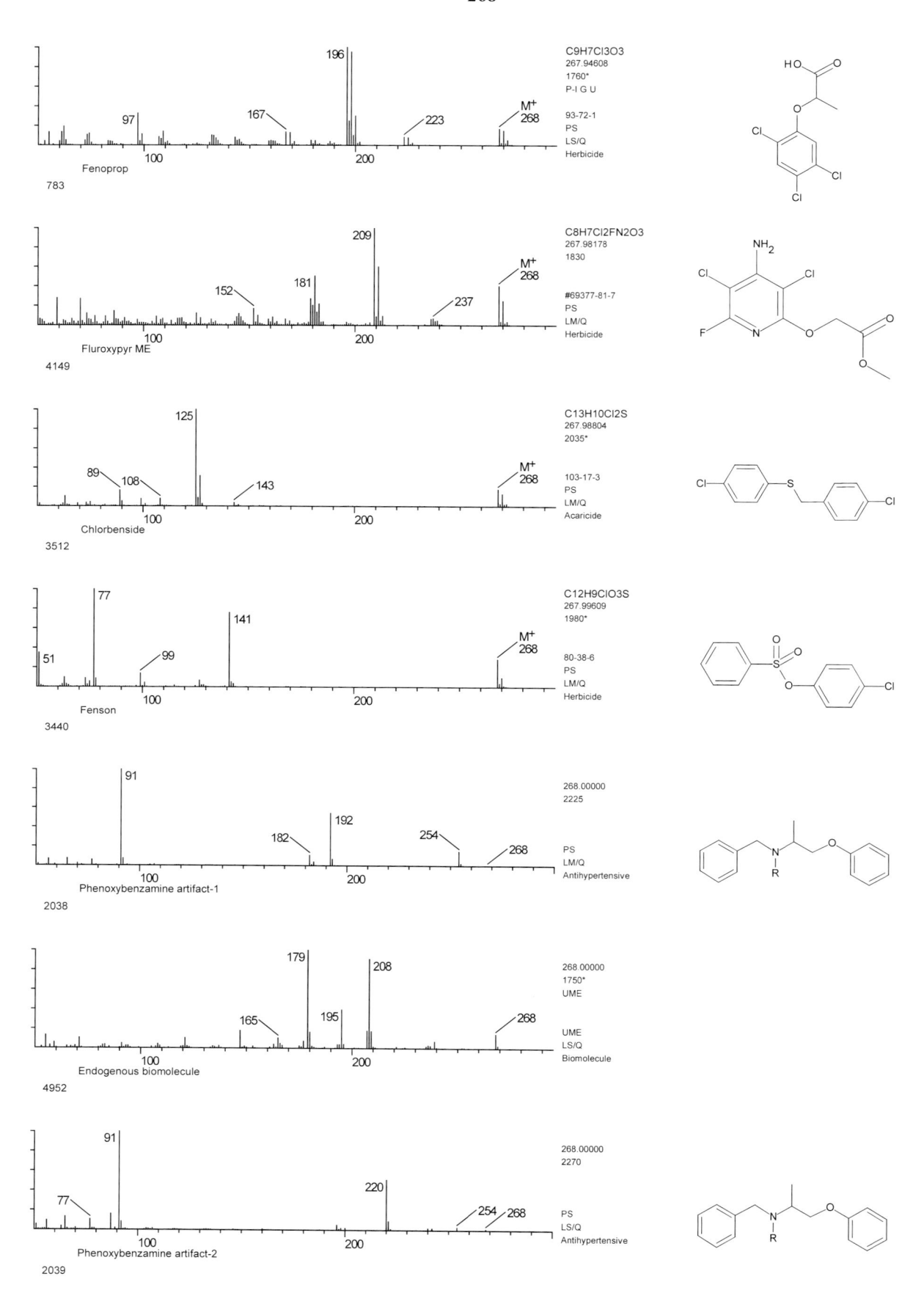

- Fenoprop — C9H7Cl3O3, 267.94608, 1760*, P-I G U, 93-72-1, PS, LS/Q, Herbicide — 783
  Peaks: 97, 167, 196, 223, M+ 268

- Fluroxypyr ME — C8H7Cl2FN2O3, 267.98178, 1830, #69377-81-7, PS, LM/Q, Herbicide — 4149
  Peaks: 152, 181, 209, 237, M+ 268

- Chlorbenside — C13H10Cl2S, 267.98804, 2035*, 103-17-3, PS, LM/Q, Acaricide — 3512
  Peaks: 89, 108, 125, 143, M+ 268

- Fenson — C12H9ClO3S, 267.99609, 1980*, 80-38-6, PS, LM/Q, Herbicide — 3440
  Peaks: 51, 77, 99, 141, M+ 268

- Phenoxybenzamine artifact-1 — 268.00000, 2225, PS, LM/Q, Antihypertensive — 2038
  Peaks: 91, 182, 192, 254, 268

- Endogenous biomolecule — 268.00000, 1750*, UME, UME, LS/Q, Biomolecule — 4952
  Peaks: 165, 179, 195, 208, 268

- Phenoxybenzamine artifact-2 — 268.00000, 2270, PS, LS/Q, Antihypertensive — 2039
  Peaks: 77, 91, 220, 254, 268

268

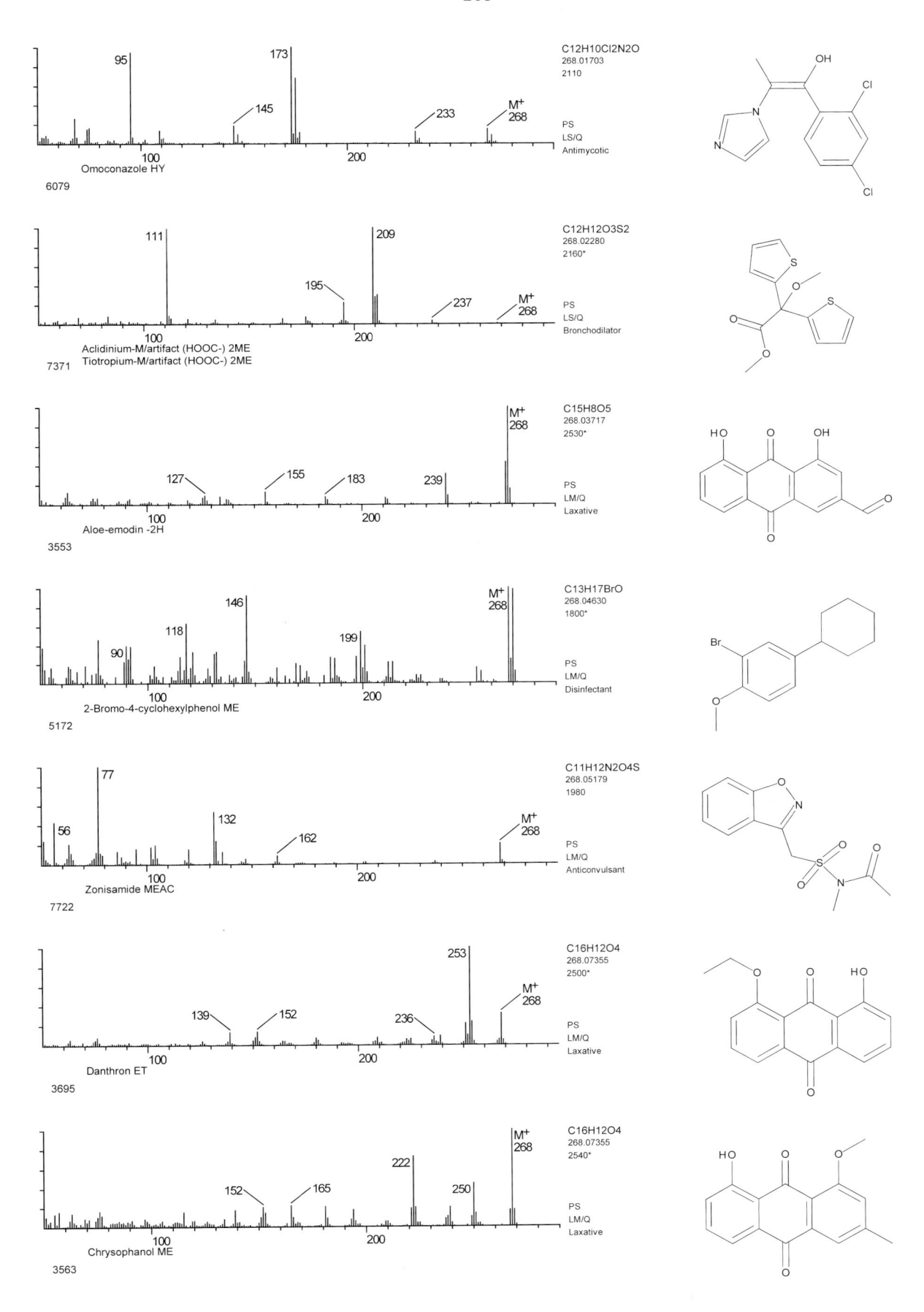

Spectrum	Formula	Mass	Info
Omoconazole HY — 6079; peaks 95, 145, 173, 233, M+ 268	C12H10Cl2N2O	268.01703	2110; PS; LS/Q; Antimycotic
Aclidinium-M/artifact (HOOC-) 2ME / Tiotropium-M/artifact (HOOC-) 2ME — 7371; peaks 111, 195, 209, 237, M+ 268	C12H12O3S2	268.02280	2160*; PS; LS/Q; Bronchodilator
Aloe-emodin -2H — 3553; peaks 127, 155, 183, 239, M+ 268	C15H8O5	268.03717	2530*; PS; LM/Q; Laxative
2-Bromo-4-cyclohexylphenol ME — 5172; peaks 90, 118, 146, 199, M+ 268	C13H17BrO	268.04630	1800*; PS; LM/Q; Disinfectant
Zonisamide MEAC — 7722; peaks 56, 77, 132, 162, M+ 268	C11H12N2O4S	268.05179	1980; PS; LM/Q; Anticonvulsant
Danthron ET — 3695; peaks 139, 152, 236, 253, M+ 268	C16H12O4	268.07355	2500*; PS; LM/Q; Laxative
Chrysophanol ME — 3563; peaks 152, 165, 222, 250, M+ 268	C16H12O4	268.07355	2540*; PS; LM/Q; Laxative

268

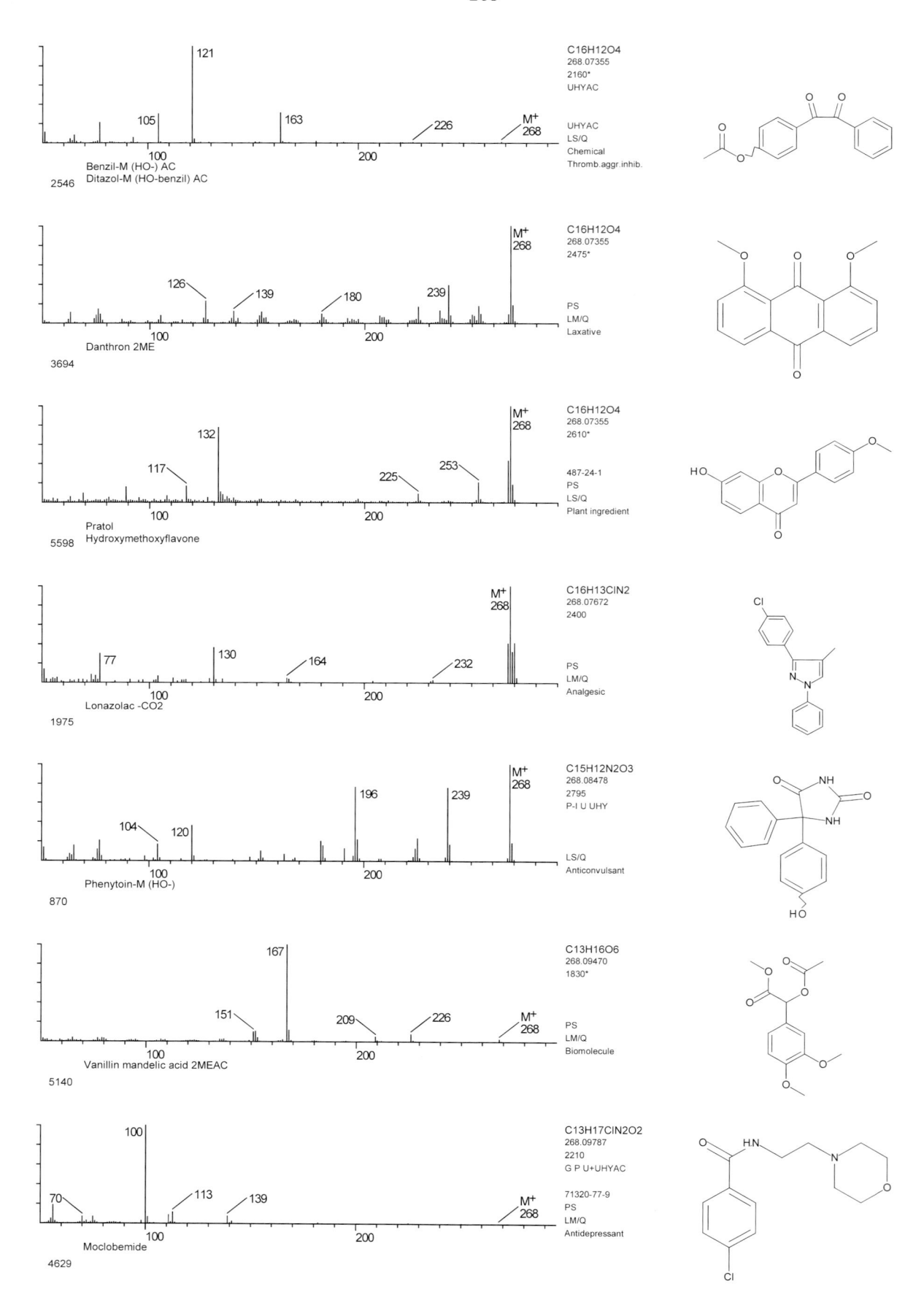

C16H12O4
268.07355
2160*
UHYAC

UHYAC
LS/Q
Chemical
Thromb.aggr.inhib.

Benzil-M (HO-) AC
Ditazol-M (HO-benzil) AC
2546

C16H12O4
268.07355
2475*

PS
LM/Q
Laxative

Danthron 2ME
3694

C16H12O4
268.07355
2610*

487-24-1
PS
LS/Q
Plant ingredient

Pratol
Hydroxymethoxyflavone
5598

C16H13ClN2
268.07672
2400

PS
LM/Q
Analgesic

Lonazolac -CO2
1975

C15H12N2O3
268.08478
2795
P-I U UHY

LS/Q
Anticonvulsant

Phenytoin-M (HO-)
870

C13H16O6
268.09470
1830*

PS
LM/Q
Biomolecule

Vanillin mandelic acid 2MEAC
5140

C13H17ClN2O2
268.09787
2210
G P U+UHYAC

71320-77-9
PS
LM/Q
Antidepressant

Moclobemide
4629

706

268

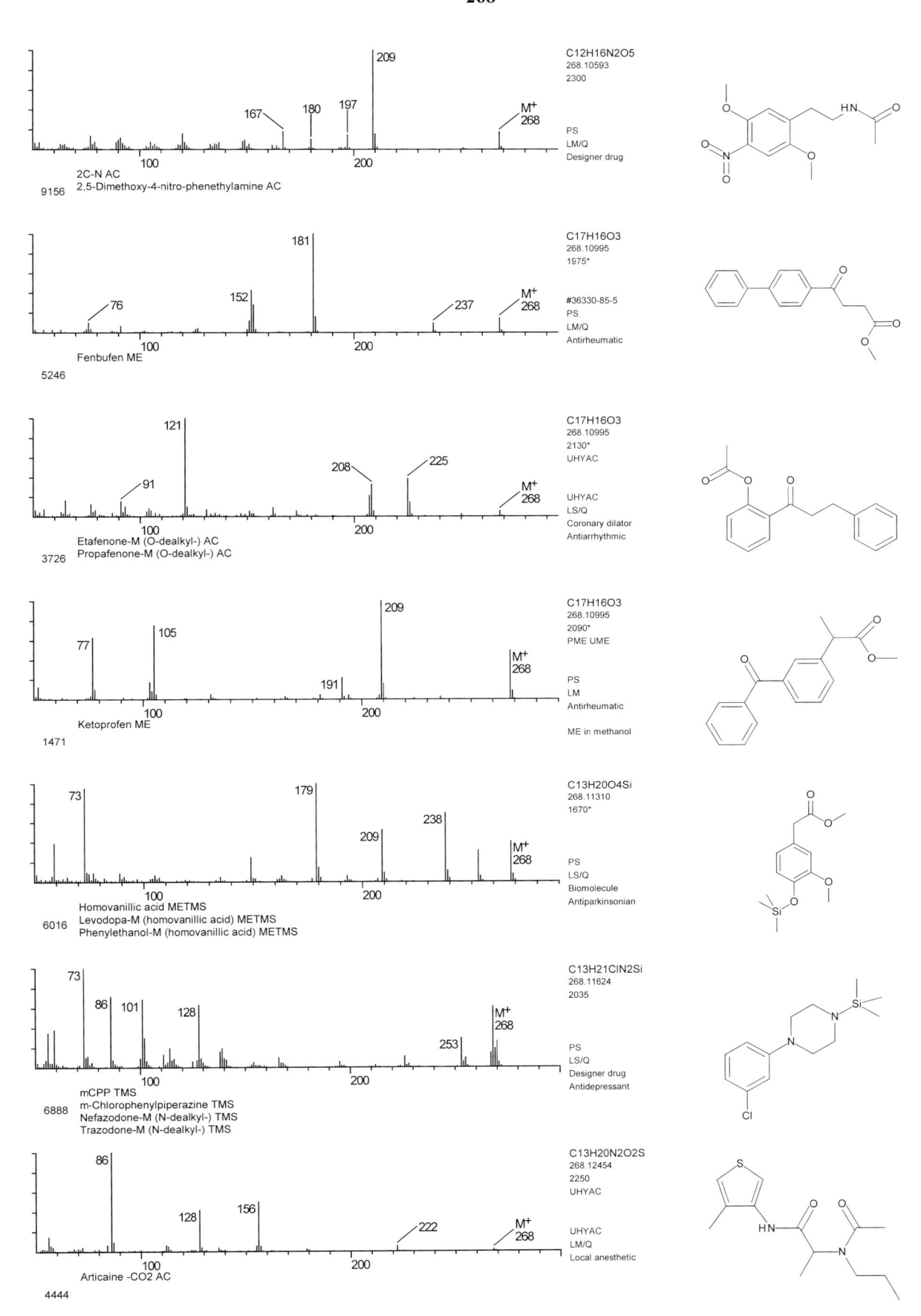

C12H16N2O5
268.10593
2300

PS
LM/Q
Designer drug

2C-N AC
2,5-Dimethoxy-4-nitro-phenethylamine AC
9156

C17H16O3
268.10995
1975*
#36330-85-5
PS
LM/Q
Antirheumatic

Fenbufen ME
5246

C17H16O3
268.10995
2130*
UHYAC

UHYAC
LS/Q
Coronary dilator
Antiarrhythmic

Etafenone-M (O-dealkyl-) AC
Propafenone-M (O-dealkyl-) AC
3726

C17H16O3
268.10995
2090*
PME UME

PS
LM
Antirheumatic

ME in methanol

Ketoprofen ME
1471

C13H20O4Si
268.11310
1670*

PS
LS/Q
Biomolecule
Antiparkinsonian

Homovanillic acid METMS
Levodopa-M (homovanillic acid) METMS
Phenylethanol-M (homovanillic acid) METMS
6016

C13H21ClN2Si
268.11624
2035

PS
LS/Q
Designer drug
Antidepressant

mCPP TMS
m-Chlorophenylpiperazine TMS
Nefazodone-M (N-dealkyl-) TMS
Trazodone-M (N-dealkyl-) TMS
6888

C13H20N2O2S
268.12454
2250
UHYAC

UHYAC
LM/Q
Local anesthetic

Articaine -CO2 AC
4444

268

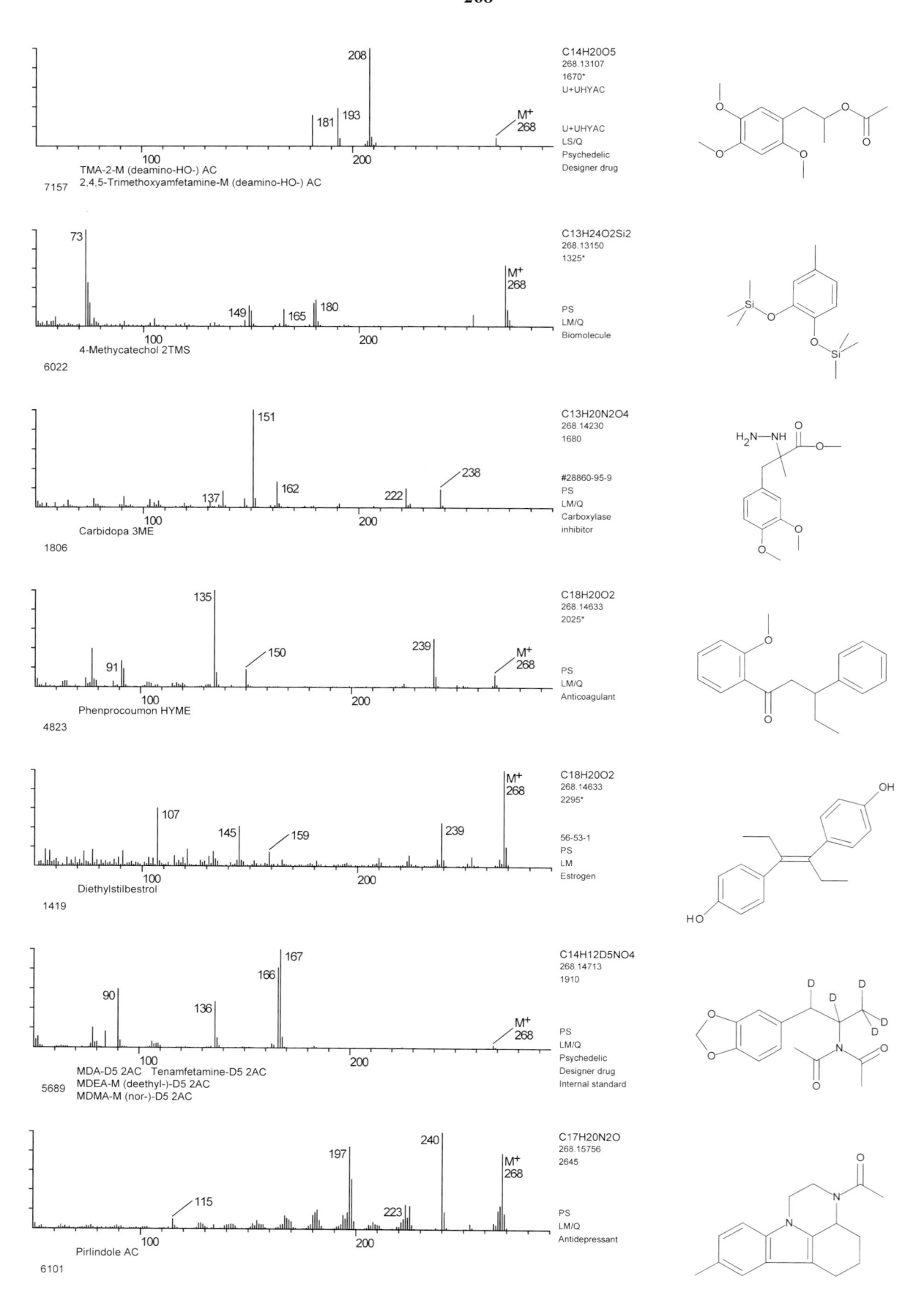

7157 TMA-2-M (deamino-HO-) AC 2,4,5-Trimethoxyamfetamine-M (deamino-HO-) AC Peaks: 181, 193, 208, M+ 268	C14H20O5 268.13107 1670* U+UHYAC U+UHYAC LS/Q Psychedelic Designer drug
6022 4-Methycatechol 2TMS Peaks: 73, 149, 165, 180, M+ 268	C13H24O2Si2 268.13150 1325* PS LM/Q Biomolecule
1806 Carbidopa 3ME Peaks: 137, 151, 162, 222, 238	C13H20N2O4 268.14230 1680 #28860-95-9 PS LM/Q Carboxylase inhibitor
4823 Phenprocoumon HYME Peaks: 91, 135, 150, 239, M+ 268	C18H20O2 268.14633 2025* PS LM/Q Anticoagulant
1419 Diethylstilbestrol Peaks: 107, 145, 159, 239, M+ 268	C18H20O2 268.14633 2295* 56-53-1 PS LM Estrogen
5689 MDA-D5 2AC Tenamfetamine-D5 2AC MDEA-M (deethyl-)-D5 2AC MDMA-M (nor-)-D5 2AC Peaks: 90, 136, 166, 167, M+ 268	C14H12D5NO4 268.14713 1910 PS LM/Q Psychedelic Designer drug Internal standard
6101 Pirlindole AC Peaks: 115, 197, 223, 240, M+ 268	C17H20N2O 268.15756 2645 PS LM/Q Antidepressant

268

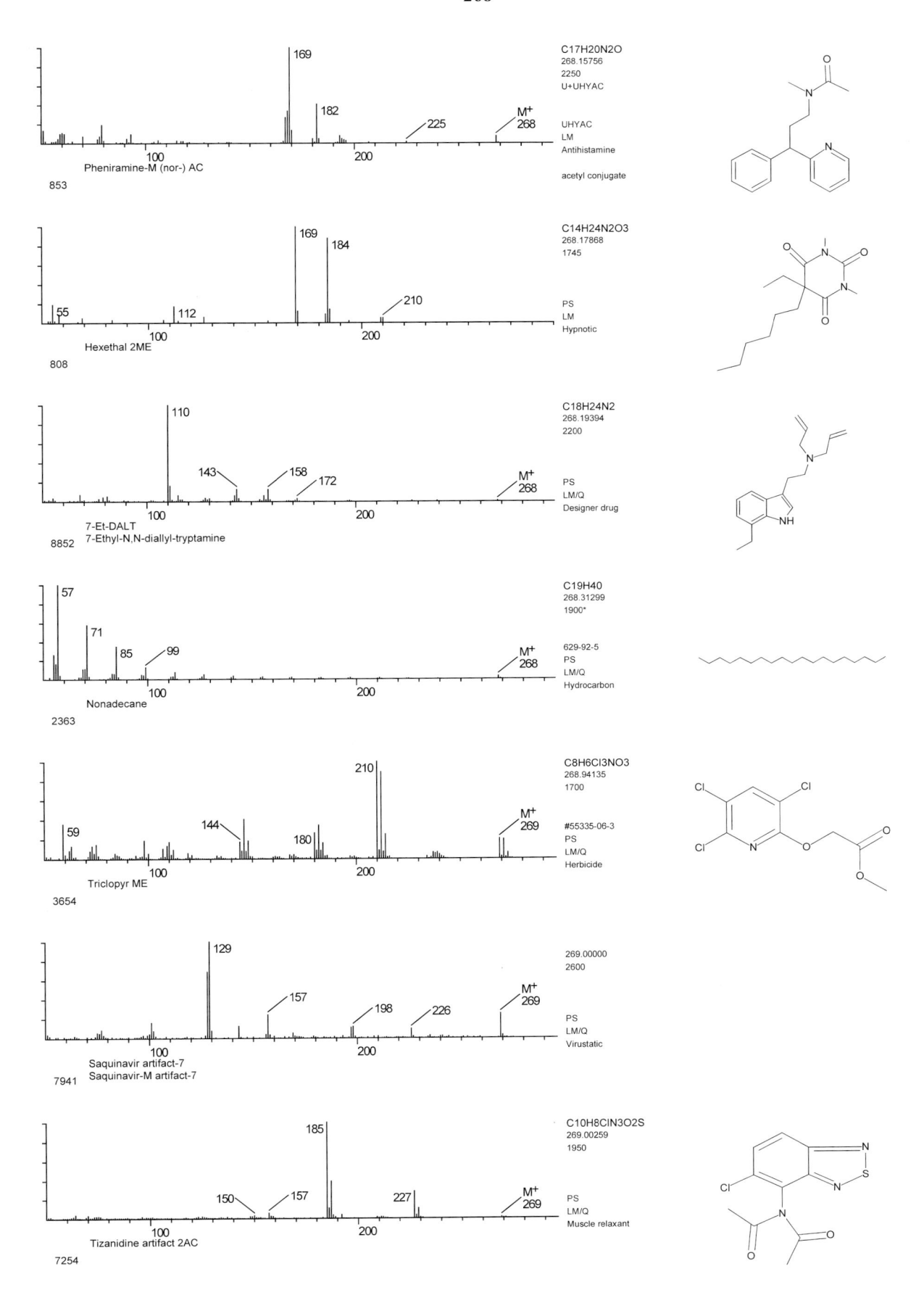

Pheniramine-M (nor-) AC
853

C17H20N2O
268.15756
2250
U+UHYAC

UHYAC
LM
Antihistamine

acetyl conjugate

Hexethal 2ME
808

C14H24N2O3
268.17868
1745

PS
LM
Hypnotic

7-Et-DALT
7-Ethyl-N,N-diallyl-tryptamine
8852

C18H24N2
268.19394
2200

PS
LM/Q
Designer drug

Nonadecane
2363

C19H40
268.31299
1900*

629-92-5
PS
LM/Q
Hydrocarbon

Triclopyr ME
3654

C8H6Cl3NO3
268.94135
1700

#55335-06-3
PS
LM/Q
Herbicide

Saquinavir artifact-7
Saquinavir-M artifact-7
7941

269.00000
2600

PS
LM/Q
Virustatic

Tizanidine artifact 2AC
7254

C10H8ClN3O2S
269.00259
1950

PS
LM/Q
Muscle relaxant

269

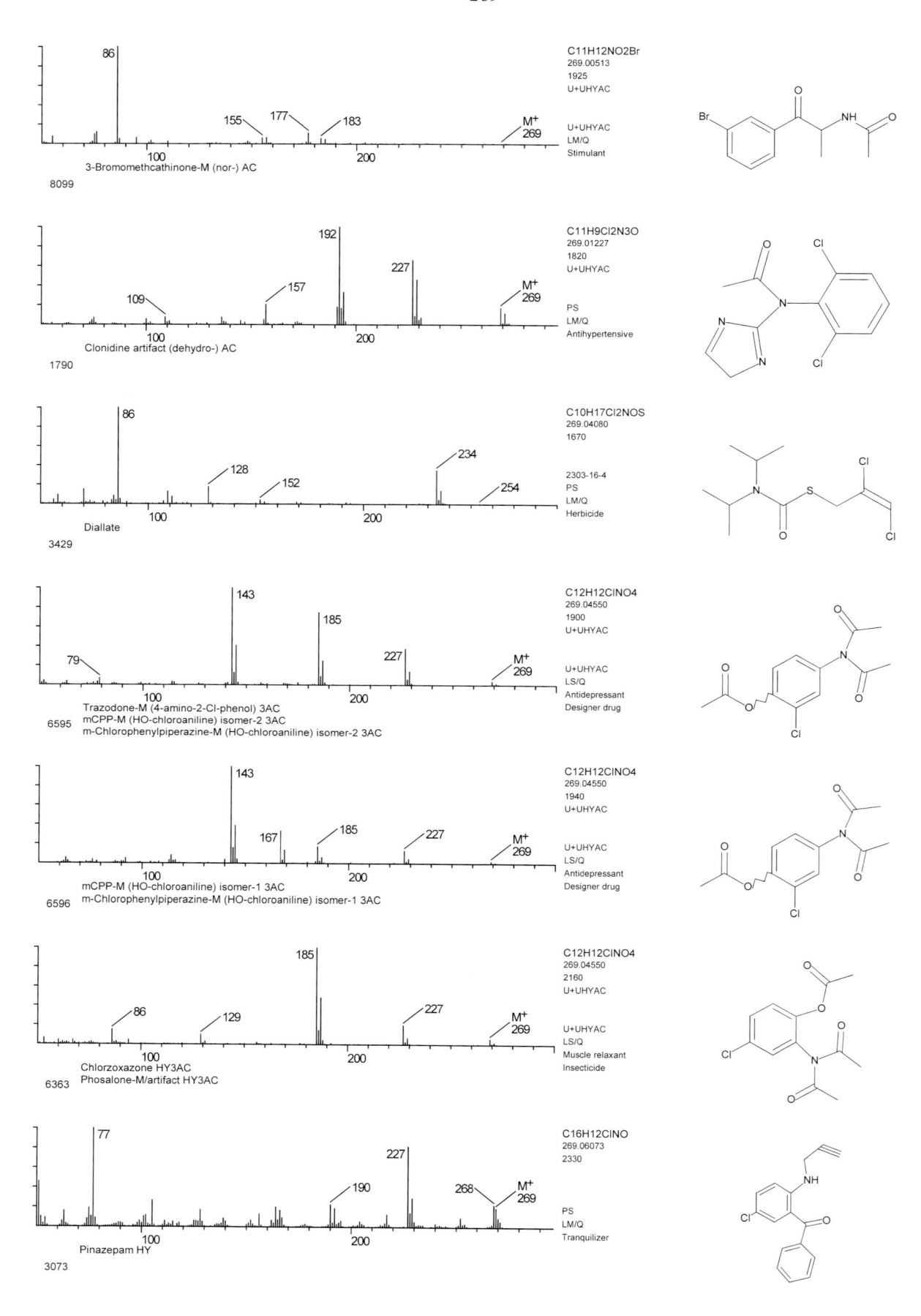

3-Bromomethcathinone-M (nor-) AC 8099	peaks: 86, 155, 177, 183, M+ 269	C11H12NO2Br 269.00513 1925 U+UHYAC U+UHYAC LM/Q Stimulant
Clonidine artifact (dehydro-) AC 1790	peaks: 109, 157, 192, 227, M+ 269	C11H9Cl2N3O 269.01227 1820 U+UHYAC PS LM/Q Antihypertensive
Diallate 3429	peaks: 86, 128, 152, 234, 254	C10H17Cl2NOS 269.04080 1670 2303-16-4 PS LM/Q Herbicide
Trazodone-M (4-amino-2-Cl-phenol) 3AC mCPP-M (HO-chloroaniline) isomer-2 3AC m-Chlorophenylpiperazine-M (HO-chloroaniline) isomer-2 3AC 6595	peaks: 79, 143, 185, 227, M+ 269	C12H12ClNO4 269.04550 1900 U+UHYAC U+UHYAC LS/Q Antidepressant Designer drug
mCPP-M (HO-chloroaniline) isomer-1 3AC m-Chlorophenylpiperazine-M (HO-chloroaniline) isomer-1 3AC 6596	peaks: 143, 167, 185, 227, M+ 269	C12H12ClNO4 269.04550 1940 U+UHYAC U+UHYAC LS/Q Antidepressant Designer drug
Chlorzoxazone HY3AC Phosalone-M/artifact HY3AC 6363	peaks: 86, 129, 185, 227, M+ 269	C12H12ClNO4 269.04550 2160 U+UHYAC U+UHYAC LS/Q Muscle relaxant Insecticide
Pinazepam HY 3073	peaks: 77, 190, 227, 268, M+ 269	C16H12ClNO 269.06073 2330 PS LM/Q Tranquilizer

269

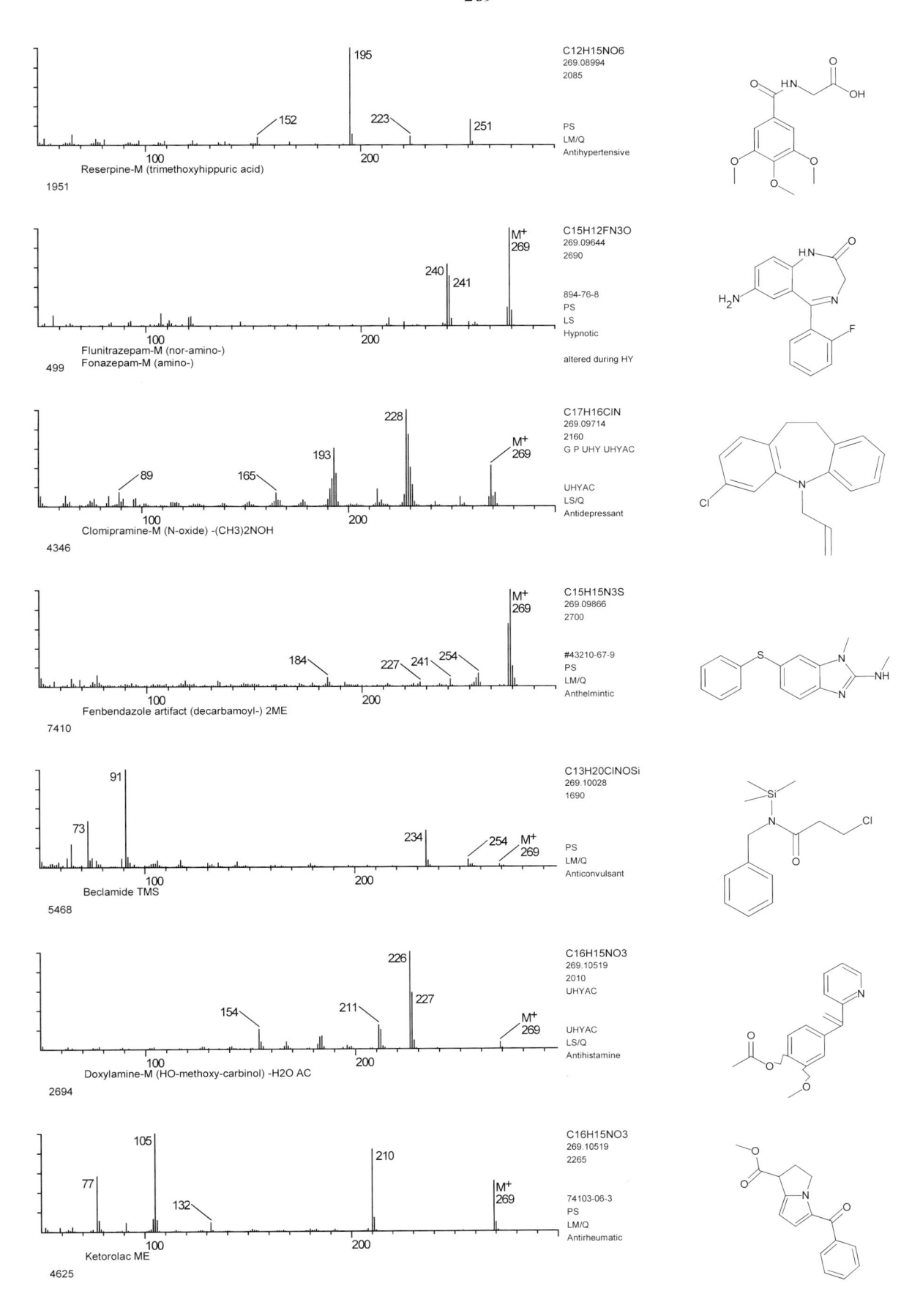

Reserpine-M (trimethoxyhippuric acid)
C12H15NO6
269.08994
2085
PS
LM/Q
Antihypertensive

Flunitrazepam-M (nor-amino-)
Fonazepam-M (amino-)
C15H12FN3O
269.09644
2690
894-76-8
PS
LS
Hypnotic
altered during HY

Clomipramine-M (N-oxide) -(CH3)2NOH
C17H16ClN
269.09714
2160
G P UHY UHYAC
UHYAC
LS/Q
Antidepressant

Fenbendazole artifact (decarbamoyl-) 2ME
C15H15N3S
269.09866
2700
#43210-67-9
PS
LM/Q
Anthelmintic

Beclamide TMS
C13H20ClNOSi
269.10028
1690
PS
LM/Q
Anticonvulsant

Doxylamine-M (HO-methoxy-carbinol) -H2O AC
C16H15NO3
269.10519
2010
UHYAC
UHYAC
LS/Q
Antihistamine

Ketorolac ME
C16H15NO3
269.10519
2265
74103-06-3
PS
LM/Q
Antirheumatic

269

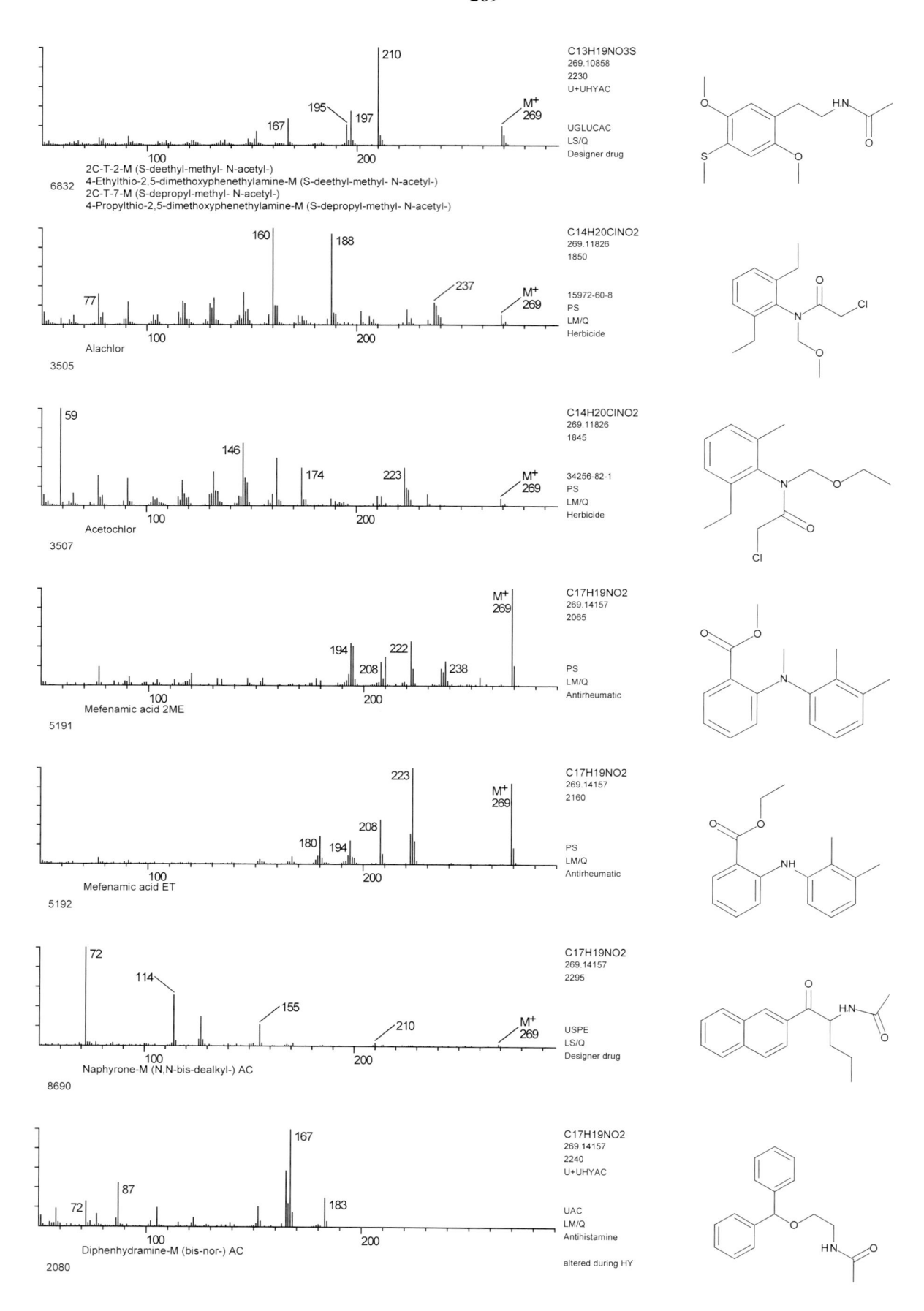

C13H19NO3S
269.10858
2230
U+UHYAC

UGLUCAC
LS/Q
Designer drug

6832
2C-T-2-M (S-deethyl-methyl- N-acetyl-)
4-Ethylthio-2,5-dimethoxyphenethylamine-M (S-deethyl-methyl- N-acetyl-)
2C-T-7-M (S-depropyl-methyl- N-acetyl-)
4-Propylthio-2,5-dimethoxyphenethylamine-M (S-depropyl-methyl- N-acetyl-)

C14H20ClNO2
269.11826
1850
15972-60-8
PS
LM/Q
Herbicide

3505 Alachlor

C14H20ClNO2
269.11826
1845
34256-82-1
PS
LM/Q
Herbicide

3507 Acetochlor

C17H19NO2
269.14157
2065
PS
LM/Q
Antirheumatic

5191 Mefenamic acid 2ME

C17H19NO2
269.14157
2160
PS
LM/Q
Antirheumatic

5192 Mefenamic acid ET

C17H19NO2
269.14157
2295
USPE
LS/Q
Designer drug

8690 Naphyrone-M (N,N-bis-dealkyl-) AC

C17H19NO2
269.14157
2240
U+UHYAC

UAC
LM/Q
Antihistamine

altered during HY

2080 Diphenhydramine-M (bis-nor-) AC

712

269

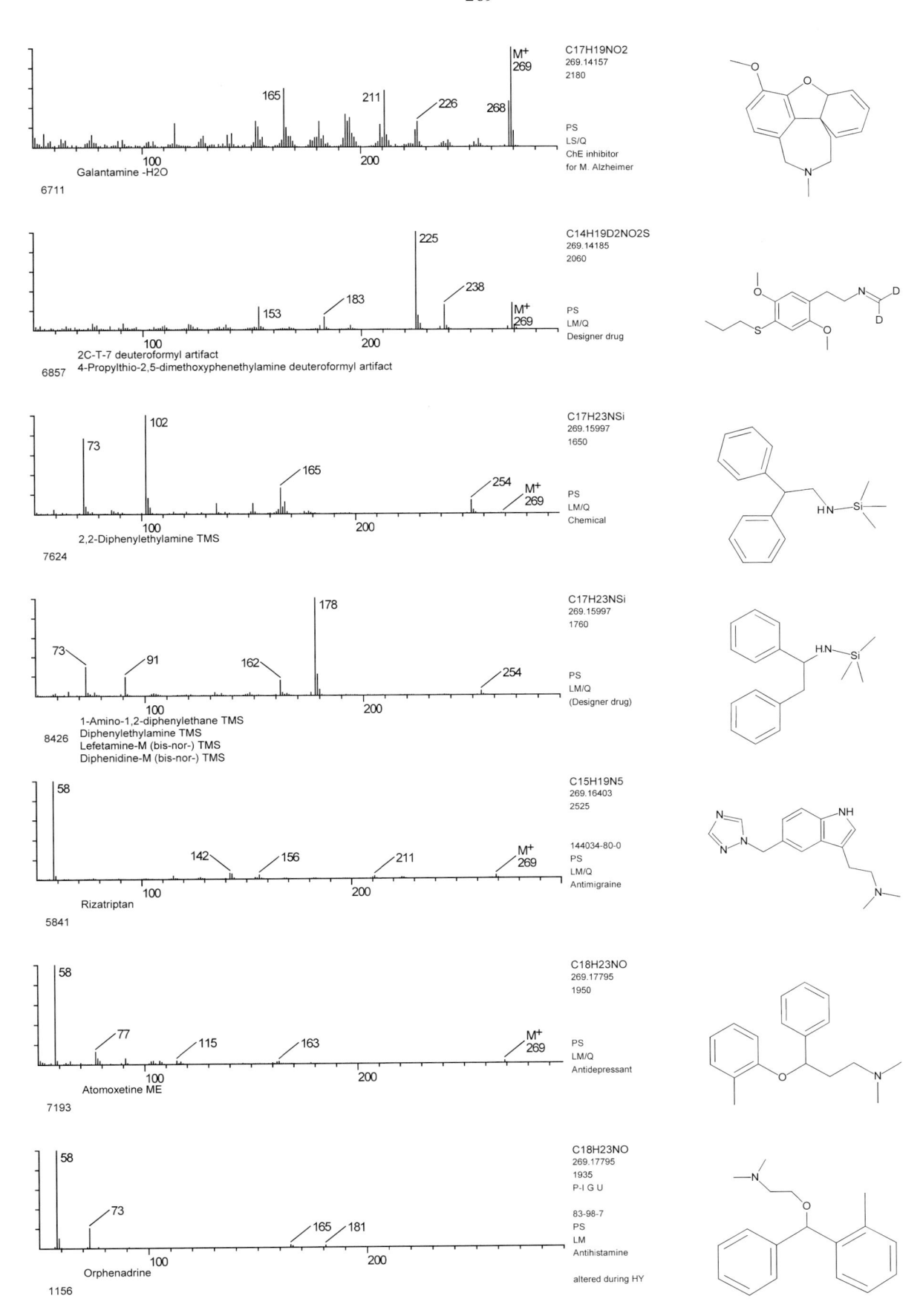

Galantamine -H2O
6711

Peaks: 165, 211, 226, 268, M+ 269

C17H19NO2
269.14157
2180

PS
LS/Q
ChE inhibitor
for M. Alzheimer

2C-T-7 deuteroformyl artifact
4-Propylthio-2,5-dimethoxyphenethylamine deuteroformyl artifact
6857

Peaks: 153, 183, 225, 238, M+ 269

C14H19D2NO2S
269.14185
2060

PS
LM/Q
Designer drug

2,2-Diphenylethylamine TMS
7624

Peaks: 73, 102, 165, 254, M+ 269

C17H23NSi
269.15997
1650

PS
LM/Q
Chemical

1-Amino-1,2-diphenylethane TMS
Diphenylethylamine TMS
Lefetamine-M (bis-nor-) TMS
Diphenidine-M (bis-nor-) TMS
8426

Peaks: 73, 91, 162, 178, 254

C17H23NSi
269.15997
1760

PS
LM/Q
(Designer drug)

Rizatriptan
5841

Peaks: 58, 142, 156, 211, M+ 269

C15H19N5
269.16403
2525

144034-80-0
PS
LM/Q
Antimigraine

Atomoxetine ME
7193

Peaks: 58, 77, 115, 163, M+ 269

C18H23NO
269.17795
1950

PS
LM/Q
Antidepressant

Orphenadrine
1156

Peaks: 58, 73, 165, 181

C18H23NO
269.17795
1935
P-I G U

83-98-7
PS
LM
Antihistamine

altered during HY

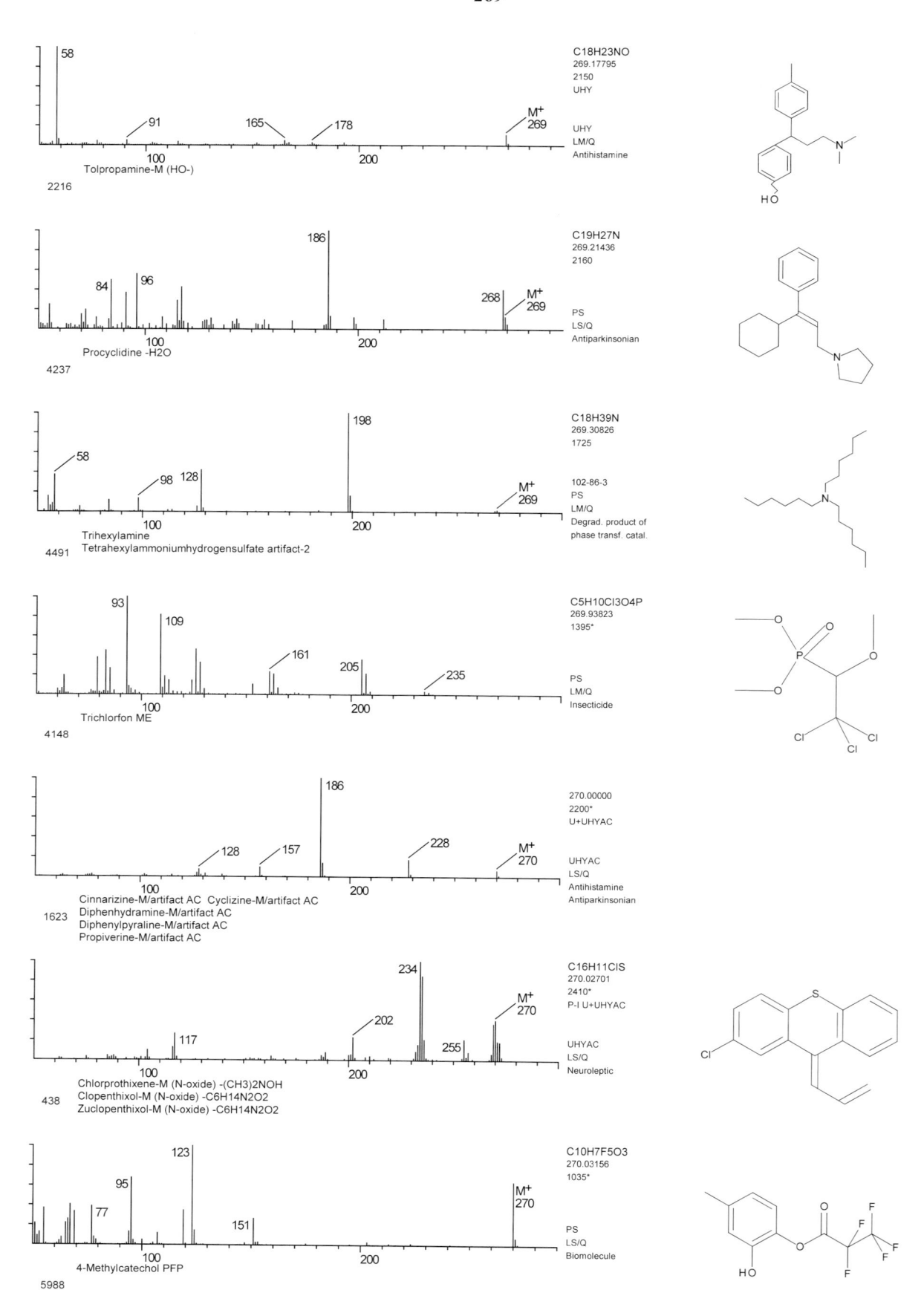

Tolpropamine-M (HO-)	peaks: 58, 91, 165, 178, M+ 269	C18H23NO 269.17795 2150 UHY / UHY LM/Q Antihistamine
2216		
Procyclidine -H2O	peaks: 84, 96, 186, 268, M+ 269	C19H27N 269.21436 2160 PS LS/Q Antiparkinsonian
4237		
Trihexylamine Tetrahexylammoniumhydrogensulfate artifact-2	peaks: 58, 98, 128, 198, M+ 269	C18H39N 269.30826 1725 102-86-3 PS LM/Q Degrad. product of phase transf. catal.
4491		
Trichlorfon ME	peaks: 93, 109, 161, 205, 235	C5H10Cl3O4P 269.93823 1395* PS LM/Q Insecticide
4148		
Cinnarizine-M/artifact AC Cyclizine-M/artifact AC Diphenhydramine-M/artifact AC Diphenylpyraline-M/artifact AC Propiverine-M/artifact AC	peaks: 128, 157, 186, 228, M+ 270	270.00000 2200* U+UHYAC UHYAC LS/Q Antihistamine Antiparkinsonian
1623		
Chlorprothixene-M (N-oxide) -(CH3)2NOH Clopenthixol-M (N-oxide) -C6H14N2O2 Zuclopenthixol-M (N-oxide) -C6H14N2O2	peaks: 117, 202, 234, 255, M+ 270	C16H11ClS 270.02701 2410* P-I U+UHYAC UHYAC LS/Q Neuroleptic
438		
4-Methylcatechol PFP	peaks: 77, 95, 123, 151, M+ 270	C10H7F5O3 270.03156 1035* PS LS/Q Biomolecule
5988		

270

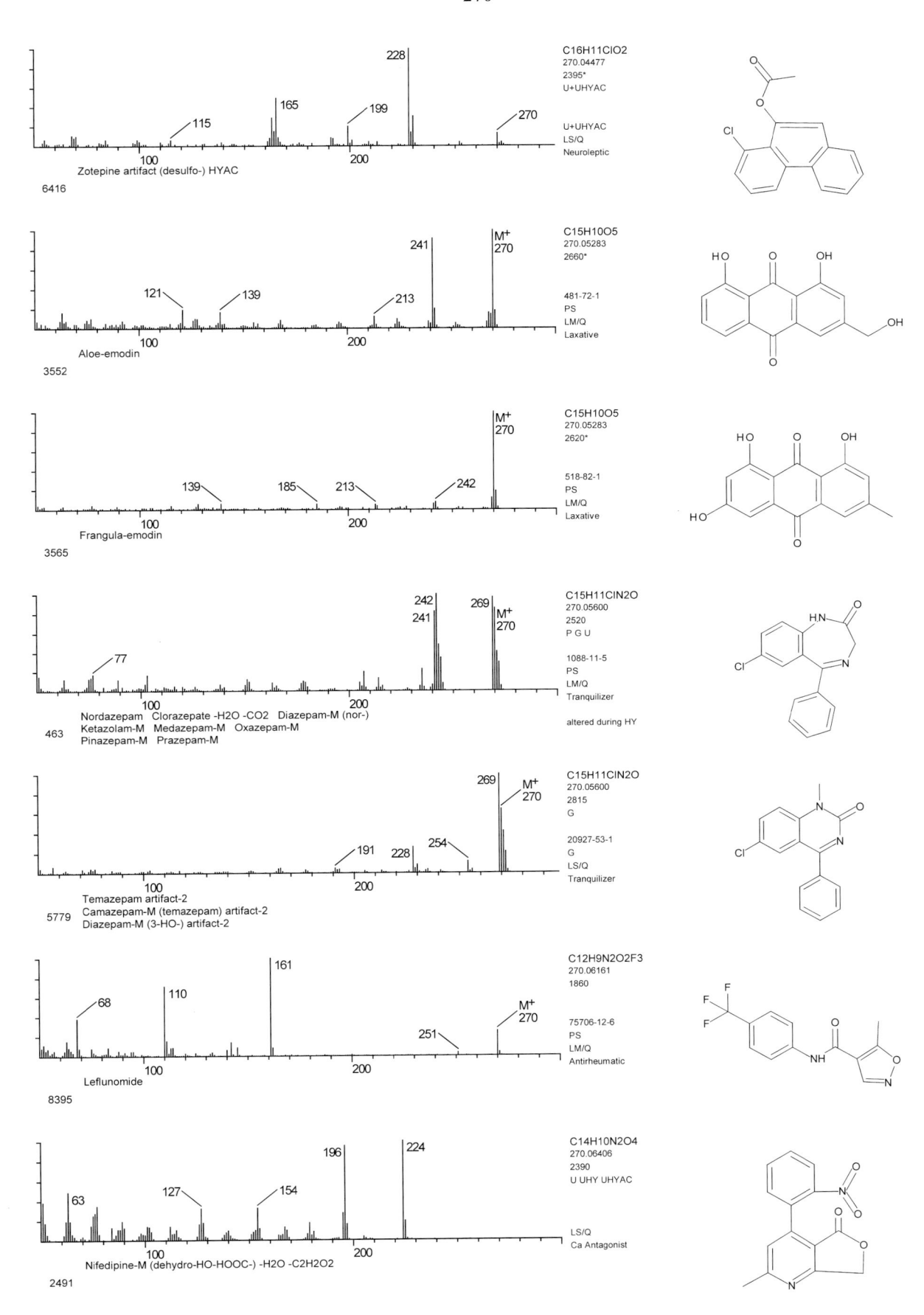

Zotepine artifact (desulfo-) HYAC — 6416	peaks: 115, 165, 199, 228, 270	C16H11ClO2 270.04477 2395* U+UHYAC U+UHYAC LS/Q Neuroleptic
Aloe-emodin — 3552	peaks: 121, 139, 213, 241, M+ 270	C15H10O5 270.05283 2660* 481-72-1 PS LM/Q Laxative
Frangula-emodin — 3565	peaks: 139, 185, 213, 242, M+ 270	C15H10O5 270.05283 2620* 518-82-1 PS LM/Q Laxative
Nordazepam Clorazepate -H2O -CO2 Diazepam-M (nor-) Ketazolam-M Medazepam-M Oxazepam-M Pinazepam-M Prazepam-M — 463	peaks: 77, 241, 242, 269, M+ 270	C15H11ClN2O 270.05600 2520 P G U 1088-11-5 PS LM/Q Tranquilizer altered during HY
Temazepam artifact-2 Camazepam-M (temazepam) artifact-2 Diazepam-M (3-HO-) artifact-2 — 5779	peaks: 191, 228, 254, 269, M+ 270	C15H11ClN2O 270.05600 2815 G 20927-53-1 G LS/Q Tranquilizer
Leflunomide — 8395	peaks: 68, 110, 161, 251, M+ 270	C12H9N2O2F3 270.06161 1860 75706-12-6 PS LM/Q Antirheumatic
Nifedipine-M (dehydro-HO-HOOC-) -H2O -C2H2O2 — 2491	peaks: 63, 127, 154, 196, 224	C14H10N2O4 270.06406 2390 U UHY UHYAC LS/Q Ca Antagonist

270

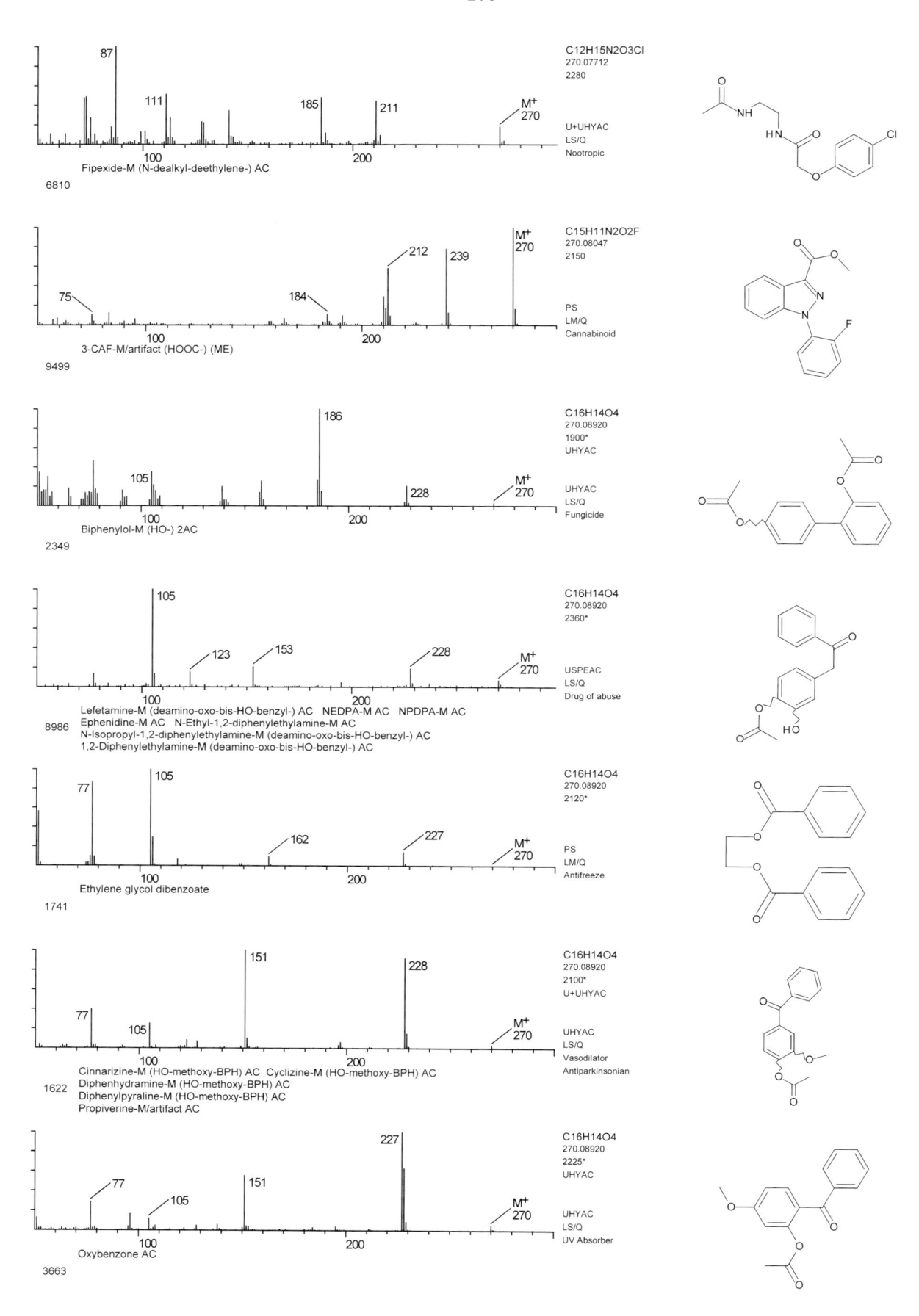

87, 111, 185, 211, M+ 270	C12H15N2O3Cl 270.07712 2280 U+UHYAC LS/Q Nootropic	
Fipexide-M (N-dealkyl-deethylene-) AC 6810		
75, 184, 212, 239, M+ 270	C15H11N2O2F 270.08047 2150 PS LM/Q Cannabinoid	
3-CAF-M/artifact (HOOC-) (ME) 9499		
105, 186, 228, M+ 270	C16H14O4 270.08920 1900* UHYAC UHYAC LS/Q Fungicide	
Biphenylol-M (HO-) 2AC 2349		
105, 123, 153, 228, M+ 270	C16H14O4 270.08920 2360* USPEAC LS/Q Drug of abuse	
Lefetamine-M (deamino-oxo-bis-HO-benzyl-) AC NEDPA-M AC NPDPA-M AC Ephenidine-M AC N-Ethyl-1,2-diphenylethylamine-M AC N-Isopropyl-1,2-diphenylethylamine-M (deamino-oxo-bis-HO-benzyl-) AC 1,2-Diphenylethylamine-M (deamino-oxo-bis-HO-benzyl-) AC 8986		
77, 105, 162, 227, M+ 270	C16H14O4 270.08920 2120* PS LM/Q Antifreeze	
Ethylene glycol dibenzoate 1741		
77, 105, 151, 228, M+ 270	C16H14O4 270.08920 2100* U+UHYAC UHYAC LS/Q Vasodilator Antiparkinsonian	
Cinnarizine-M (HO-methoxy-BPH) AC Cyclizine-M (HO-methoxy-BPH) AC Diphenhydramine-M (HO-methoxy-BPH) AC Diphenylpyraline-M (HO-methoxy-BPH) AC Propiverine-M/artifact AC 1622		
77, 105, 151, 227, M+ 270	C16H14O4 270.08920 2225* UHYAC UHYAC LS/Q UV Absorber	
Oxybenzone AC 3663		

270

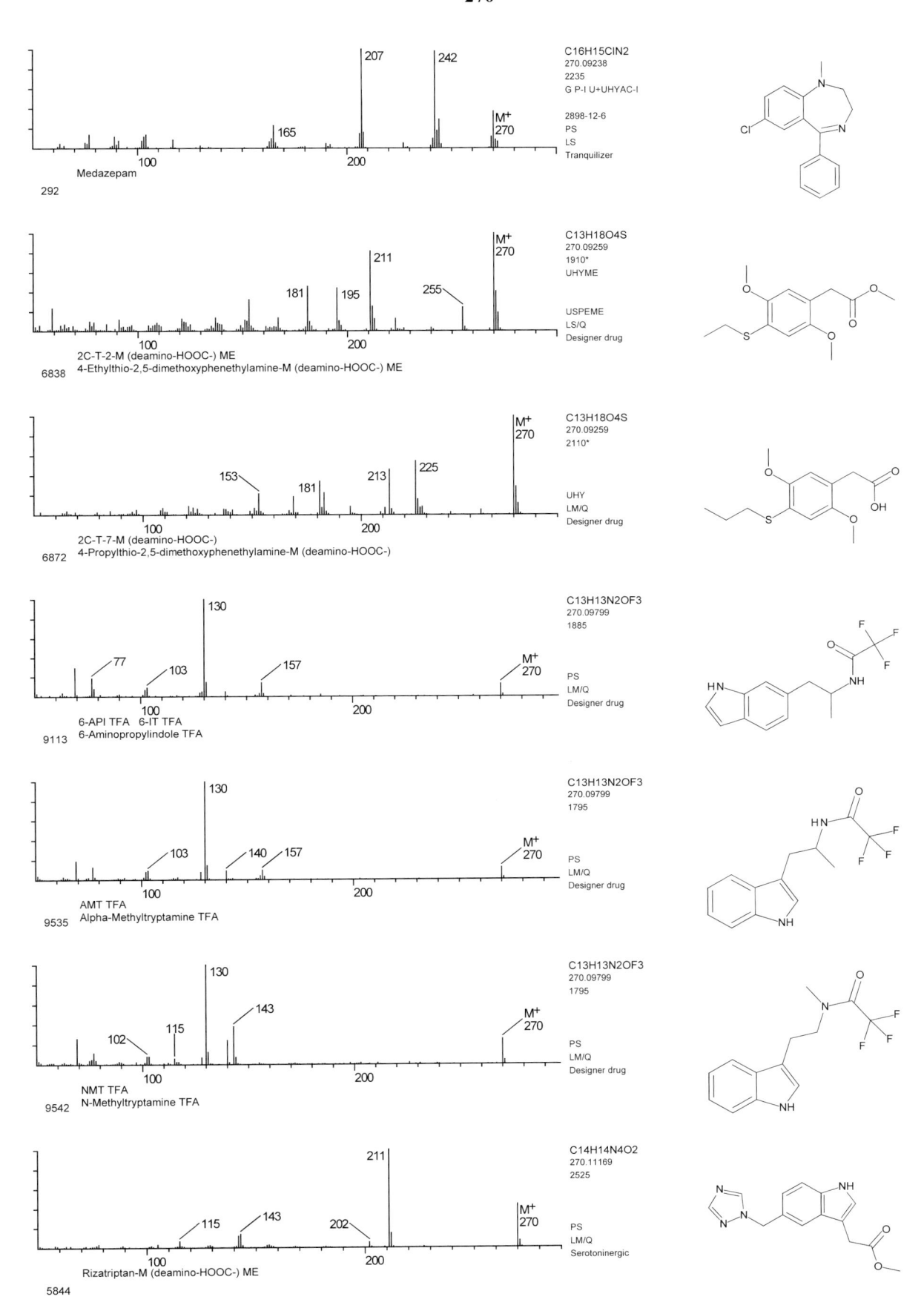

C16H15ClN2
270.09238
2235
G P-I U+UHYAC-I

2898-12-6
PS
LS
Tranquilizer

Medazepam
292

C13H18O4S
270.09259
1910*
UHYME

USPEME
LS/Q
Designer drug

2C-T-2-M (deamino-HOOC-) ME
4-Ethylthio-2,5-dimethoxyphenethylamine-M (deamino-HOOC-) ME
6838

C13H18O4S
270.09259
2110*

UHY
LM/Q
Designer drug

2C-T-7-M (deamino-HOOC-)
4-Propylthio-2,5-dimethoxyphenethylamine-M (deamino-HOOC-)
6872

C13H13N2OF3
270.09799
1885

PS
LM/Q
Designer drug

6-API TFA 6-IT TFA
6-Aminopropylindole TFA
9113

C13H13N2OF3
270.09799
1795

PS
LM/Q
Designer drug

AMT TFA
Alpha-Methyltryptamine TFA
9535

C13H13N2OF3
270.09799
1795

PS
LM/Q
Designer drug

NMT TFA
N-Methyltryptamine TFA
9542

C14H14N4O2
270.11169
2525

PS
LM/Q
Serotoninergic

Rizatriptan-M (deamino-HOOC-) ME
5844

270

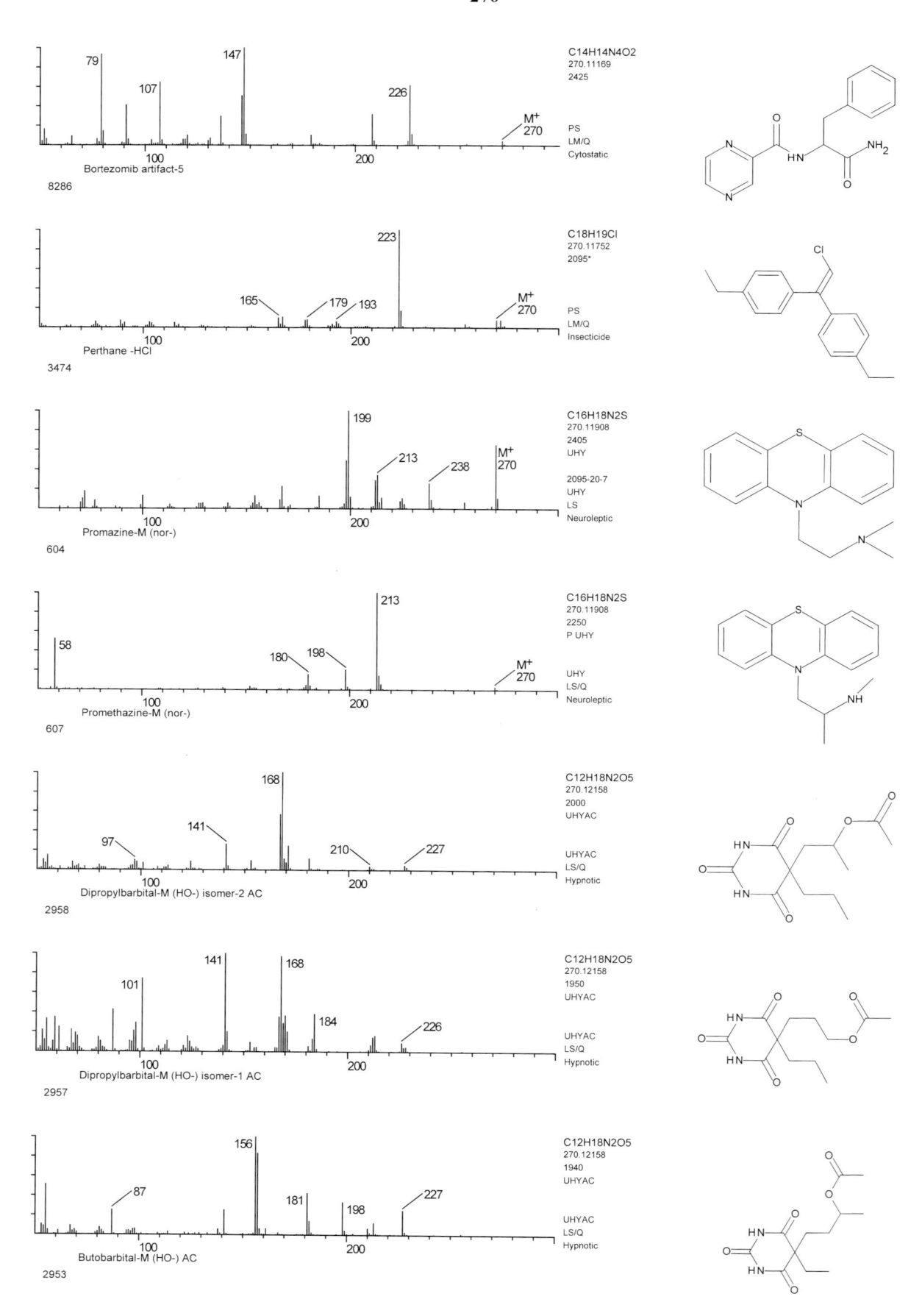

79, 107, 147, 226, M+ 270 — Bortezomib artifact-5 — 8286	C14H14N4O2 / 270.11169 / 2425 / PS / LM/Q / Cytostatic
165, 179, 193, 223, M+ 270 — Perthane -HCl — 3474	C18H19Cl / 270.11752 / 2095* / PS / LM/Q / Insecticide
199, 213, 238, M+ 270 — Promazine-M (nor-) — 604	C16H18N2S / 270.11908 / 2405 / UHY / 2095-20-7 / UHY / LS / Neuroleptic
58, 180, 198, 213, M+ 270 — Promethazine-M (nor-) — 607	C16H18N2S / 270.11908 / 2250 / P UHY / UHY / LS/Q / Neuroleptic
97, 141, 168, 210, 227 — Dipropylbarbital-M (HO-) isomer-2 AC — 2958	C12H18N2O5 / 270.12158 / 2000 / UHYAC / UHYAC / LS/Q / Hypnotic
101, 141, 168, 184, 226 — Dipropylbarbital-M (HO-) isomer-1 AC — 2957	C12H18N2O5 / 270.12158 / 1950 / UHYAC / UHYAC / LS/Q / Hypnotic
87, 156, 181, 198, 227 — Butobarbital-M (HO-) AC — 2953	C12H18N2O5 / 270.12158 / 1940 / UHYAC / UHYAC / LS/Q / Hypnotic

270

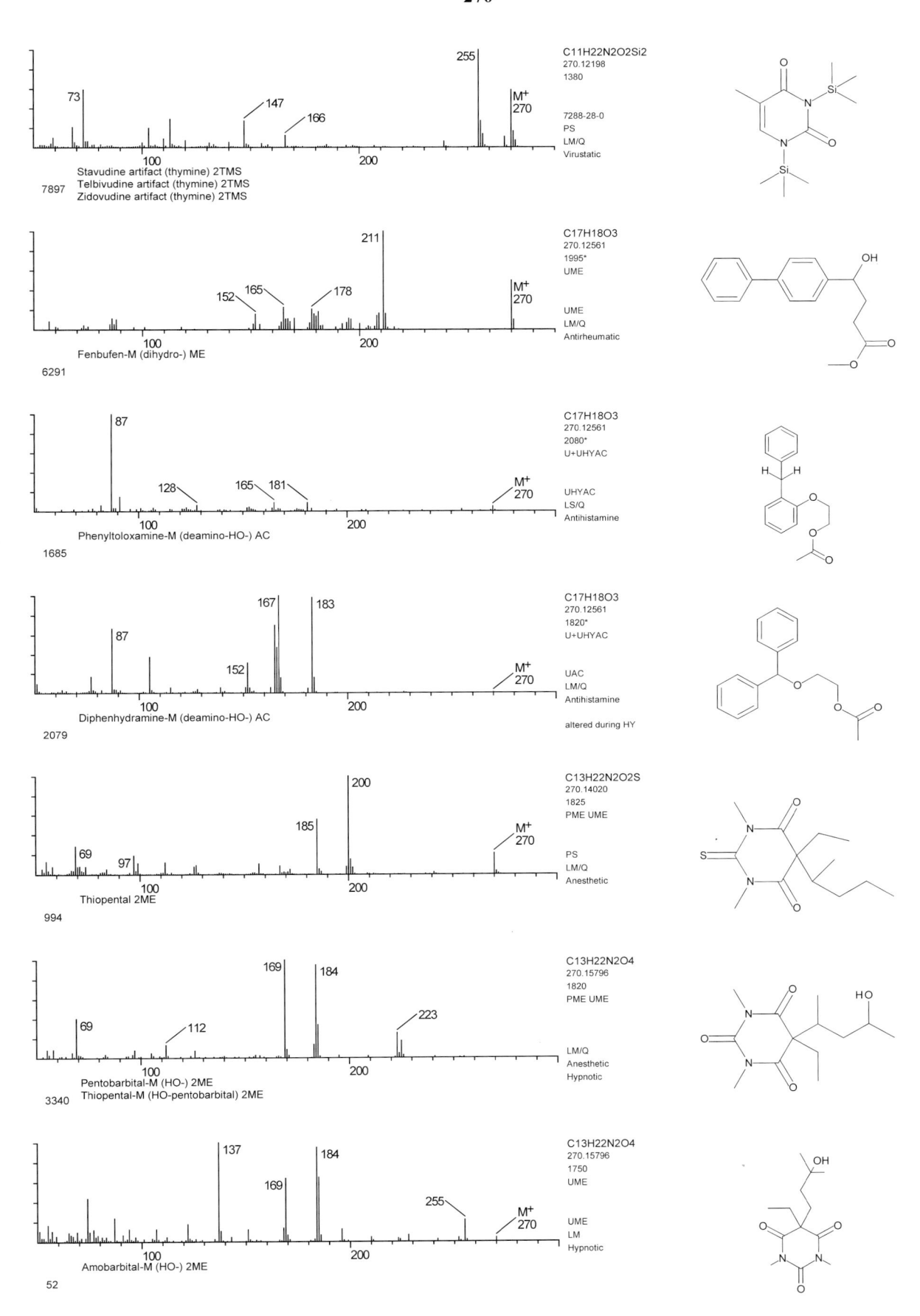

C11H22N2O2Si2
270.12198
1380

7288-28-0
PS
LM/Q
Virustatic

Stavudine artifact (thymine) 2TMS
Telbivudine artifact (thymine) 2TMS
Zidovudine artifact (thymine) 2TMS

C17H18O3
270.12561
1995*
UME

UME
LM/Q
Antirheumatic

Fenbufen-M (dihydro-) ME

C17H18O3
270.12561
2080*
U+UHYAC

UHYAC
LS/Q
Antihistamine

Phenyltoloxamine-M (deamino-HO-) AC

C17H18O3
270.12561
1820*
U+UHYAC

UAC
LM/Q
Antihistamine

altered during HY

Diphenhydramine-M (deamino-HO-) AC

C13H22N2O2S
270.14020
1825
PME UME

PS
LM/Q
Anesthetic

Thiopental 2ME

C13H22N2O4
270.15796
1820
PME UME

LM/Q
Anesthetic
Hypnotic

Pentobarbital-M (HO-) 2ME
Thiopental-M (HO-pentobarbital) 2ME

C13H22N2O4
270.15796
1750
UME

UME
LM
Hypnotic

Amobarbital-M (HO-) 2ME

270

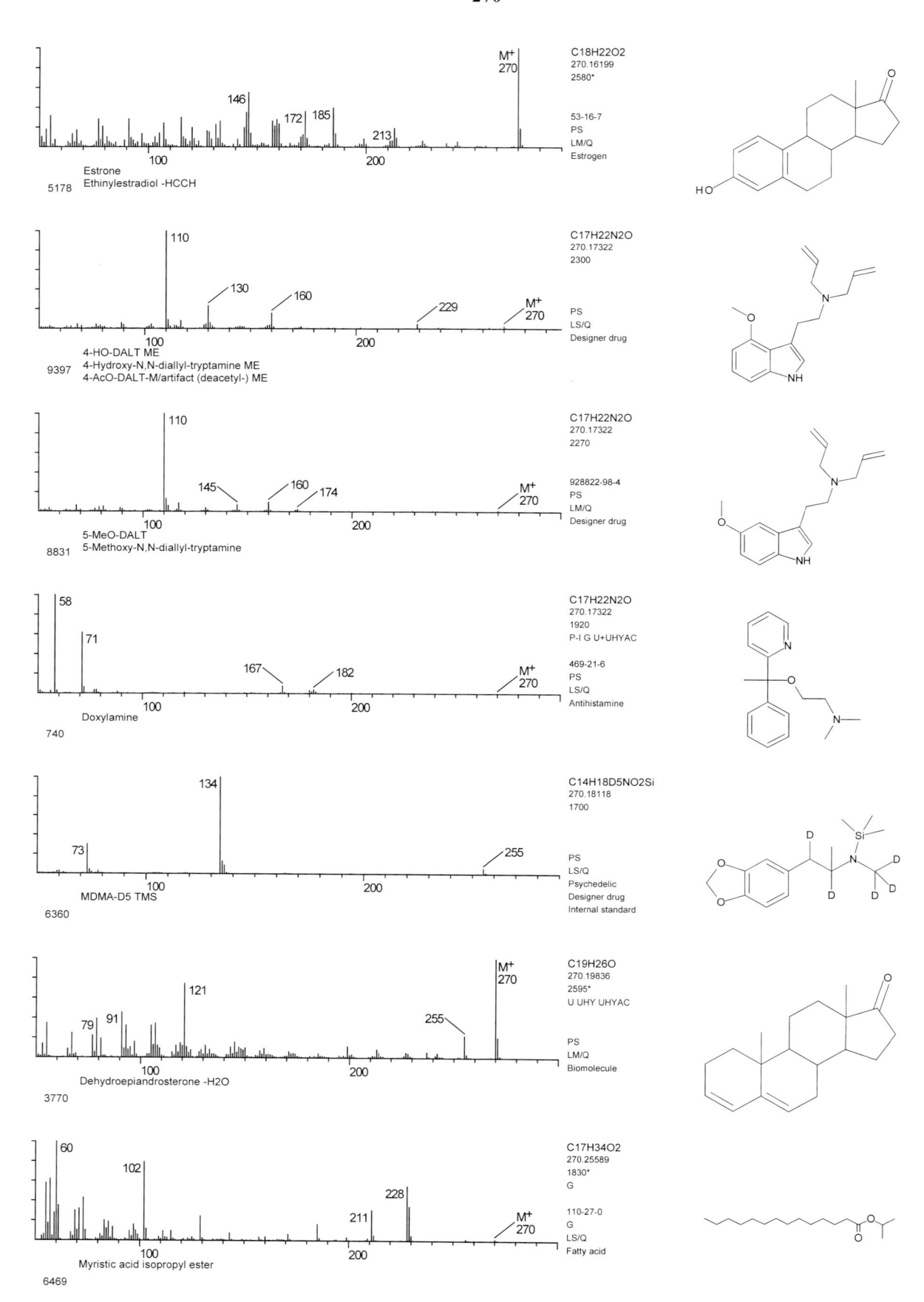

C18H22O2
270.16199
2580*

53-16-7
PS
LM/Q
Estrogen

Estrone
Ethinylestradiol -HCCH
5178

C17H22N2O
270.17322
2300

PS
LS/Q
Designer drug

4-HO-DALT ME
4-Hydroxy-N,N-diallyl-tryptamine ME
4-AcO-DALT-M/artifact (deacetyl-) ME
9397

C17H22N2O
270.17322
2270

928822-98-4
PS
LM/Q
Designer drug

5-MeO-DALT
5-Methoxy-N,N-diallyl-tryptamine
8831

C17H22N2O
270.17322
1920
P-I G U+UHYAC

469-21-6
PS
LS/Q
Antihistamine

Doxylamine
740

C14H18D5NO2Si
270.18118
1700

PS
LS/Q
Psychedelic
Designer drug
Internal standard

MDMA-D5 TMS
6360

C19H26O
270.19836
2595*
U UHY UHYAC

PS
LM/Q
Biomolecule

Dehydroepiandrosterone -H2O
3770

C17H34O2
270.25589
1830*
G

110-27-0
G
LS/Q
Fatty acid

Myristic acid isopropyl ester
6469

270

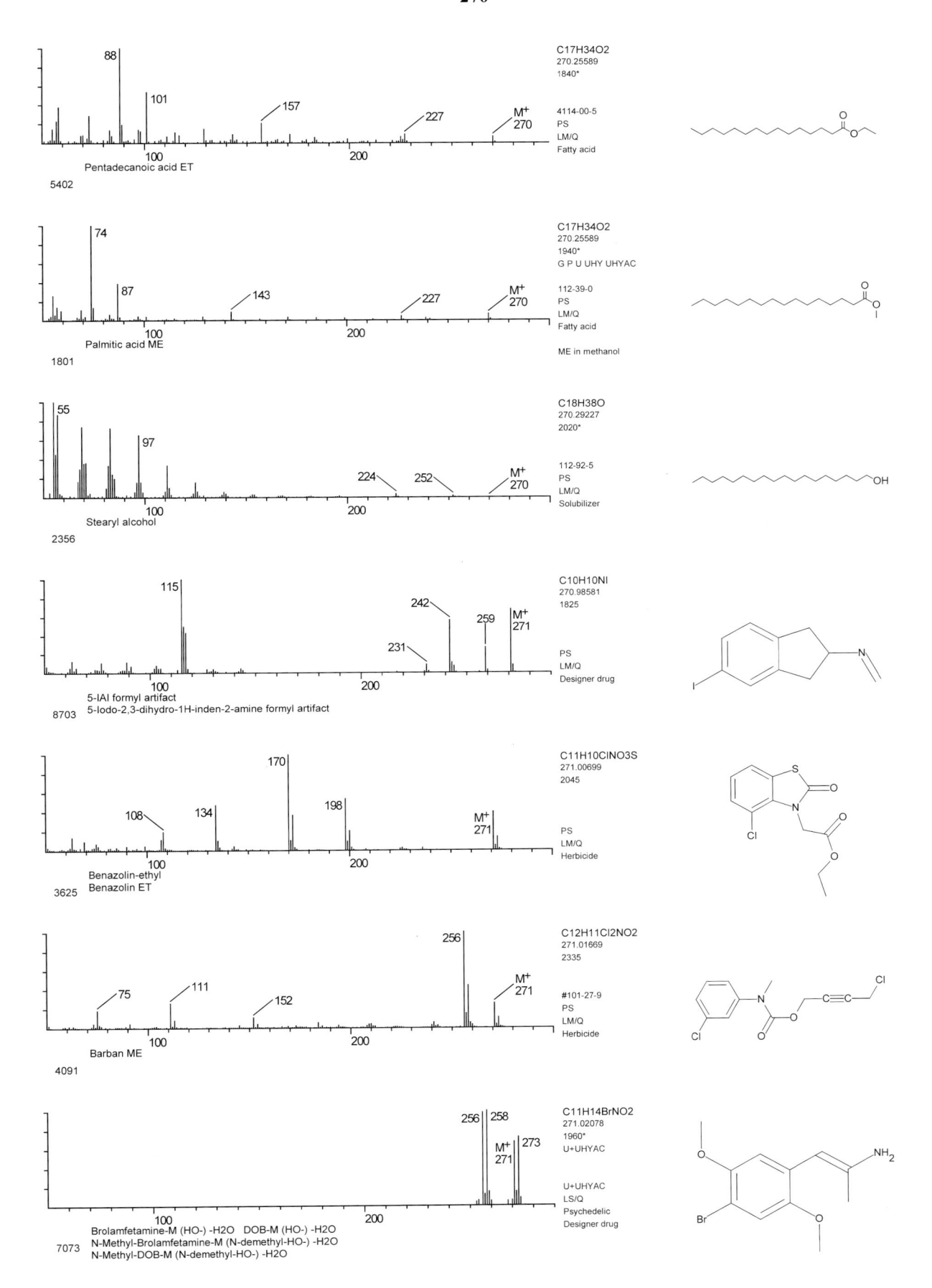

Pentadecanoic acid ET
5402
C17H34O2
270.25589
1840*
4114-00-5
PS
LM/Q
Fatty acid

Palmitic acid ME
1801
C17H34O2
270.25589
1940*
G P U UHY UHYAC
112-39-0
PS
LM/Q
Fatty acid
ME in methanol

Stearyl alcohol
2356
C18H38O
270.29227
2020*
112-92-5
PS
LM/Q
Solubilizer

5-IAI formyl artifact
5-Iodo-2,3-dihydro-1H-inden-2-amine formyl artifact
8703
C10H10NI
270.98581
1825
PS
LM/Q
Designer drug

Benazolin-ethyl
Benazolin ET
3625
C11H10ClNO3S
271.00699
2045
PS
LM/Q
Herbicide

Barban ME
4091
C12H11Cl2NO2
271.01669
2335
#101-27-9
PS
LM/Q
Herbicide

Brolamfetamine-M (HO-) -H2O   DOB-M (HO-) -H2O
N-Methyl-Brolamfetamine-M (N-demethyl-HO-) -H2O
N-Methyl-DOB-M (N-demethyl-HO-) -H2O
7073
C11H14BrNO2
271.02078
1960*
U+UHYAC
U+UHYAC
LS/Q
Psychedelic
Designer drug

721

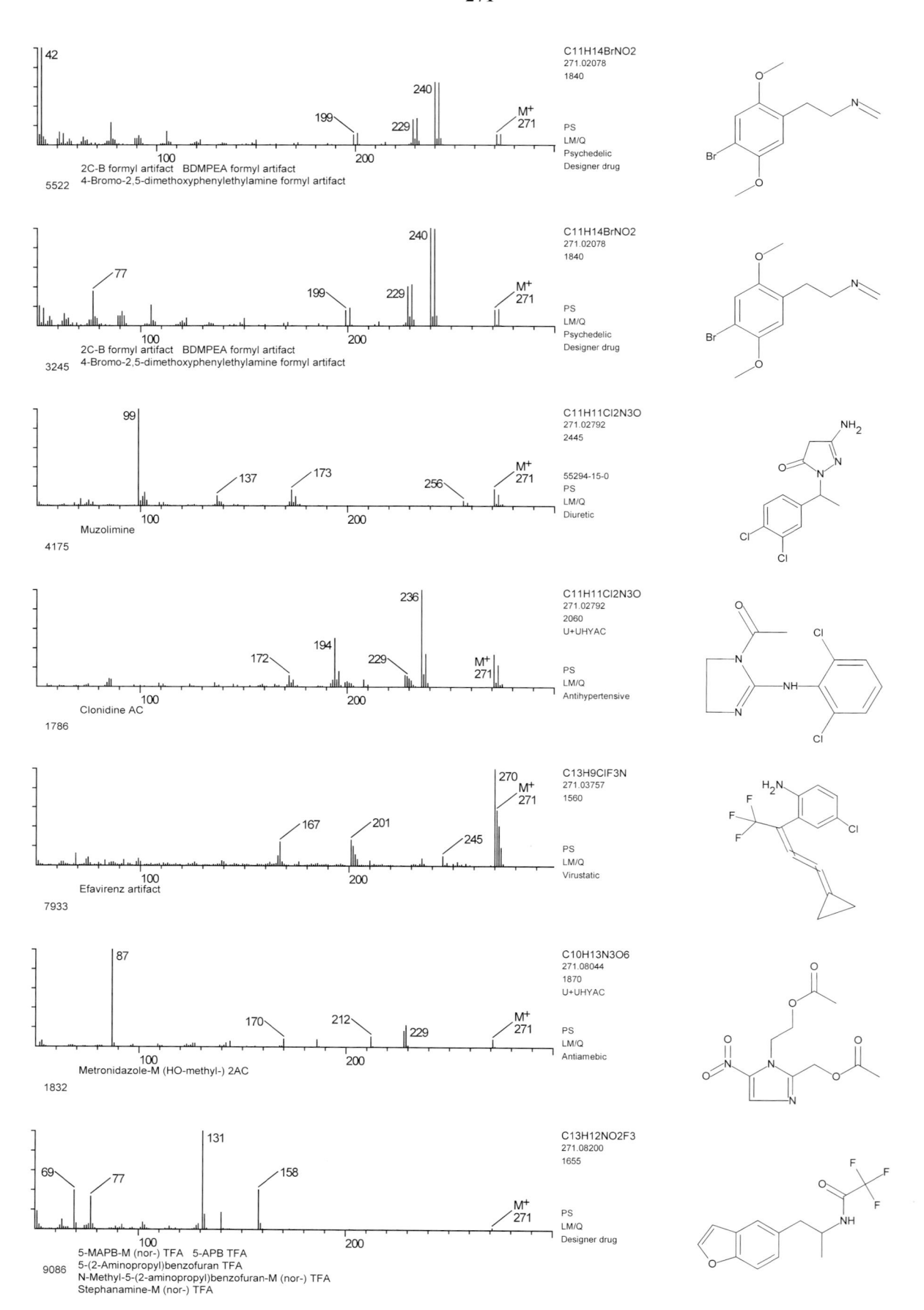

5522	2C-B formyl artifact BDMPEA formyl artifact 4-Bromo-2,5-dimethoxyphenylethylamine formyl artifact	C11H14BrNO2 271.02078 1840 PS LM/Q Psychedelic Designer drug
3245	2C-B formyl artifact BDMPEA formyl artifact 4-Bromo-2,5-dimethoxyphenylethylamine formyl artifact	C11H14BrNO2 271.02078 1840 PS LM/Q Psychedelic Designer drug
4175	Muzolimine	C11H11Cl2N3O 271.02792 2445 55294-15-0 PS LM/Q Diuretic
1786	Clonidine AC	C11H11Cl2N3O 271.02792 2060 U+UHYAC PS LM/Q Antihypertensive
7933	Efavirenz artifact	C13H9ClF3N 271.03757 1560 PS LM/Q Virustatic
1832	Metronidazole-M (HO-methyl-) 2AC	C10H13N3O6 271.08044 1870 U+UHYAC PS LM/Q Antiamebic
9086	5-MAPB-M (nor-) TFA 5-APB TFA 5-(2-Aminopropyl)benzofuran TFA N-Methyl-5-(2-aminopropyl)benzofuran-M (nor-) TFA Stephanamine-M (nor-) TFA	C13H12NO2F3 271.08200 1655 PS LM/Q Designer drug

271

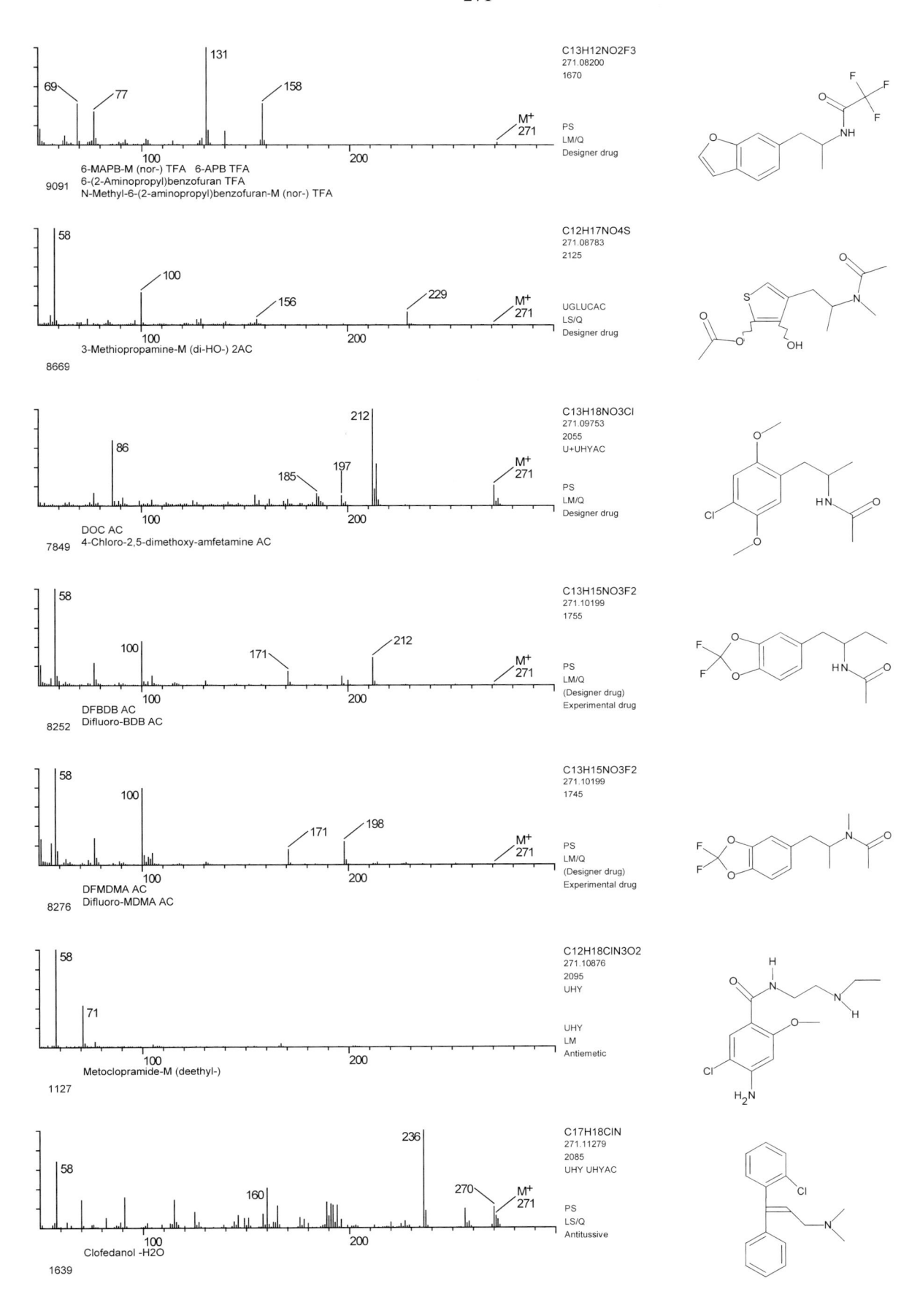

9091 — 6-MAPB-M (nor-) TFA / 6-APB TFA / 6-(2-Aminopropyl)benzofuran TFA / N-Methyl-6-(2-aminopropyl)benzofuran-M (nor-) TFA
C13H12NO2F3 — 271.08200 — 1670 — PS — LM/Q — Designer drug
Peaks: 69, 77, 131, 158, M+ 271

8669 — 3-Methiopropamine-M (di-HO-) 2AC
C12H17NO4S — 271.08783 — 2125 — UGLUCAC — LS/Q — Designer drug
Peaks: 58, 100, 156, 229, M+ 271

7849 — DOC AC / 4-Chloro-2,5-dimethoxy-amfetamine AC
C13H18NO3Cl — 271.09753 — 2055 — U+UHYAC — PS — LM/Q — Designer drug
Peaks: 86, 185, 197, 212, M+ 271

8252 — DFBDB AC / Difluoro-BDB AC
C13H15NO3F2 — 271.10199 — 1755 — PS — LM/Q — (Designer drug) — Experimental drug
Peaks: 58, 100, 171, 212, M+ 271

8276 — DFMDMA AC / Difluoro-MDMA AC
C13H15NO3F2 — 271.10199 — 1745 — PS — LM/Q — (Designer drug) — Experimental drug
Peaks: 58, 100, 171, 198, M+ 271

1127 — Metoclopramide-M (deethyl-)
C12H18ClN3O2 — 271.10876 — 2095 — UHY — UHY — LM — Antiemetic
Peaks: 58, 71

1639 — Clofedanol -H2O
C17H18ClN — 271.11279 — 2085 — UHY UHYAC — PS — LS/Q — Antitussive
Peaks: 58, 160, 236, 270, M+ 271

271

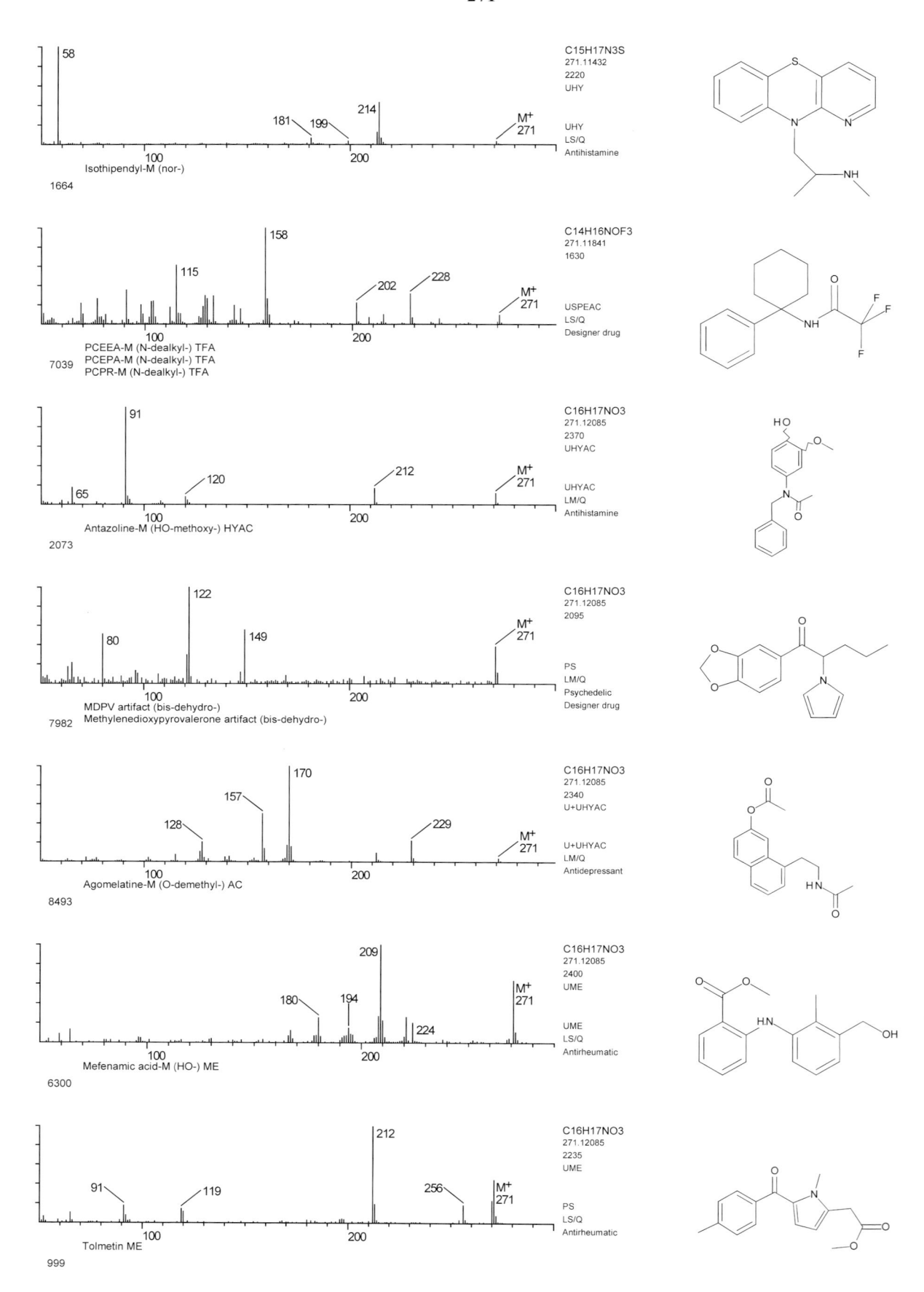

Isothipendyl-M (nor-) 1664	58, 181, 199, 214, M+ 271	C15H17N3S 271.11432 2220 UHY / UHY LS/Q Antihistamine
PCEEA-M (N-dealkyl-) TFA PCEPA-M (N-dealkyl-) TFA PCPR-M (N-dealkyl-) TFA 7039	115, 158, 202, 228, M+ 271	C14H16NOF3 271.11841 1630 USPEAC LS/Q Designer drug
Antazoline-M (HO-methoxy-) HYAC 2073	65, 91, 120, 212, M+ 271	C16H17NO3 271.12085 2370 UHYAC / UHYAC LM/Q Antihistamine
MDPV artifact (bis-dehydro-) Methylenedioxypyrovalerone artifact (bis-dehydro-) 7982	80, 122, 149, M+ 271	C16H17NO3 271.12085 2095 PS LM/Q Psychedelic Designer drug
Agomelatine-M (O-demethyl-) AC 8493	128, 157, 170, 229, M+ 271	C16H17NO3 271.12085 2340 U+UHYAC / U+UHYAC LM/Q Antidepressant
Mefenamic acid-M (HO-) ME 6300	180, 194, 209, 224, M+ 271	C16H17NO3 271.12085 2400 UME / UME LS/Q Antirheumatic
Tolmetin ME 999	91, 119, 212, 256, M+ 271	C16H17NO3 271.12085 2235 UME PS LS/Q Antirheumatic

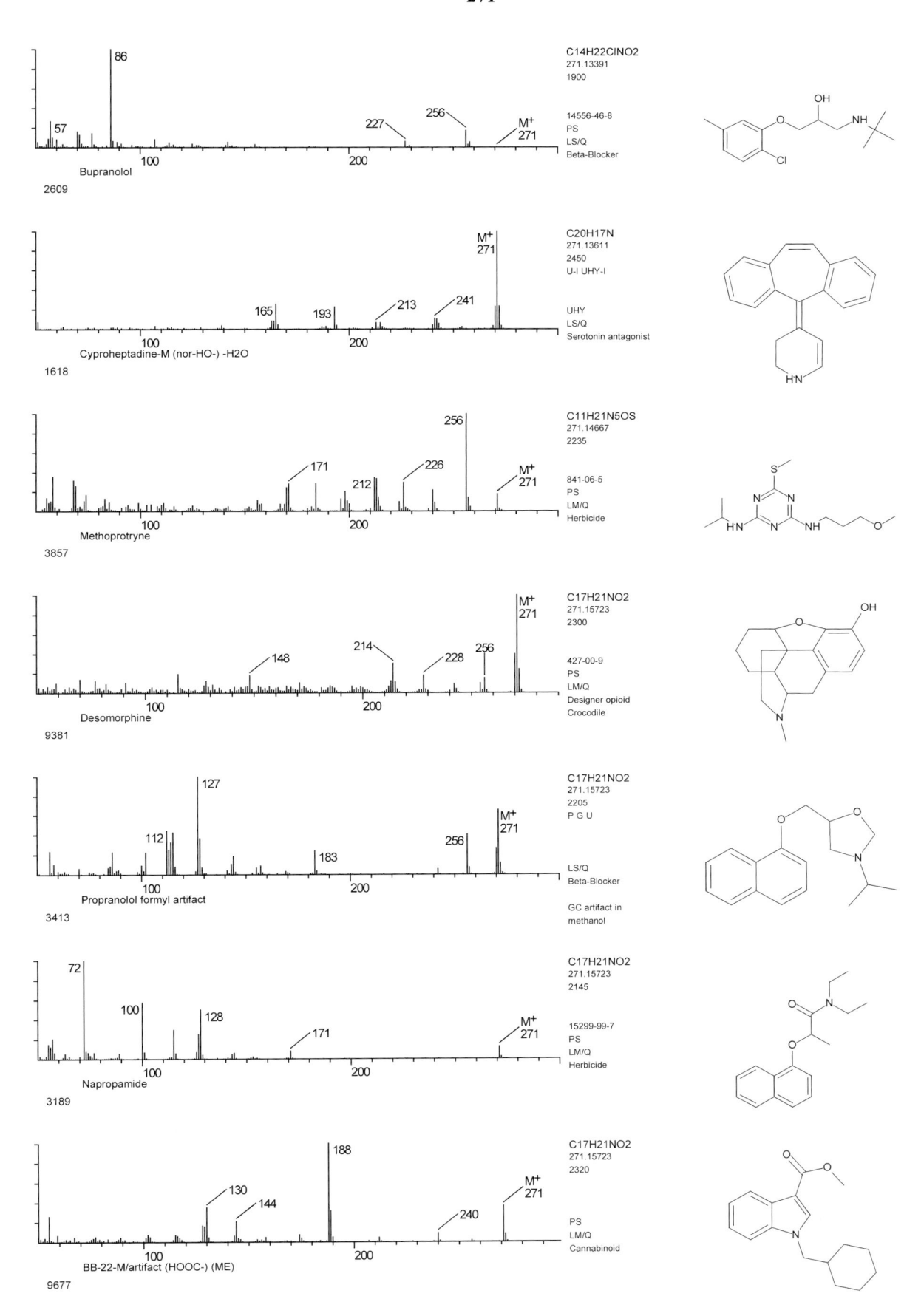

Bupranolol
2609
C14H22ClNO2
271.13391
1900
14556-46-8
PS
LS/Q
Beta-Blocker

Cyproheptadine-M (nor-HO-) -H2O
1618
C20H17N
271.13611
2450
U-I UHY-I

UHY
LS/Q
Serotonin antagonist

Methoprotryne
3857
C11H21N5OS
271.14667
2235
841-06-5
PS
LM/Q
Herbicide

Desomorphine
9381
C17H21NO2
271.15723
2300
427-00-9
PS
LM/Q
Designer opioid
Crocodile

Propranolol formyl artifact
3413
C17H21NO2
271.15723
2205
P G U

LS/Q
Beta-Blocker
GC artifact in methanol

Napropamide
3189
C17H21NO2
271.15723
2145
15299-99-7
PS
LM/Q
Herbicide

BB-22-M/artifact (HOOC-) (ME)
9677
C17H21NO2
271.15723
2320

PS
LM/Q
Cannabinoid

271

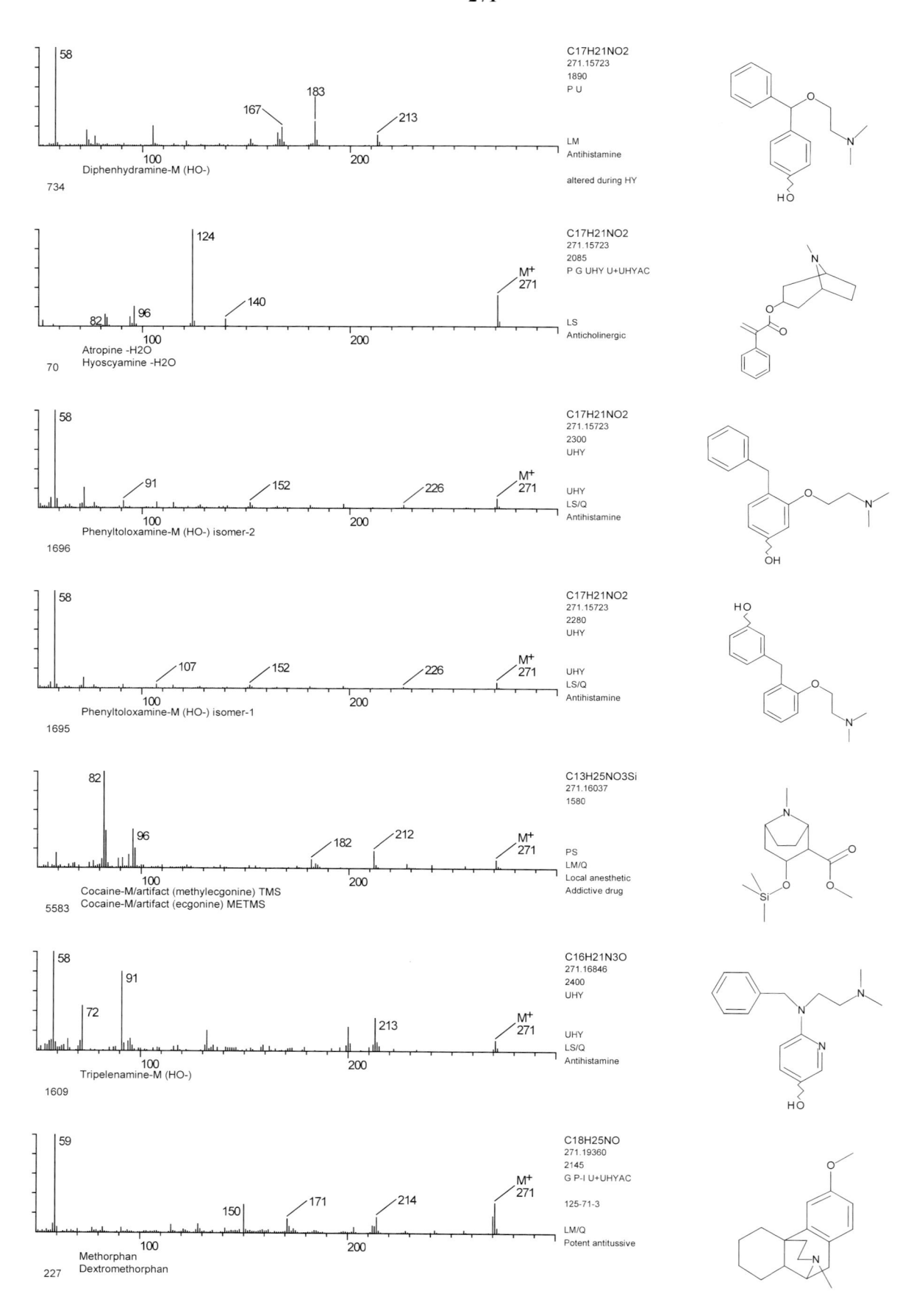

Diphenhydramine-M (HO-)
734
C17H21NO2
271.15723
1890
P U
LM
Antihistamine
altered during HY

Atropine -H2O
Hyoscyamine -H2O
70
C17H21NO2
271.15723
2085
P G UHY U+UHYAC
LS
Anticholinergic

Phenyltoloxamine-M (HO-) isomer-2
1696
C17H21NO2
271.15723
2300
UHY
UHY
LS/Q
Antihistamine

Phenyltoloxamine-M (HO-) isomer-1
1695
C17H21NO2
271.15723
2280
UHY
UHY
LS/Q
Antihistamine

Cocaine-M/artifact (methylecgonine) TMS
Cocaine-M/artifact (ecgonine) METMS
5583
C13H25NO3Si
271.16037
1580
PS
LM/Q
Local anesthetic
Addictive drug

Tripelenamine-M (HO-)
1609
C16H21N3O
271.16846
2400
UHY
UHY
LS/Q
Antihistamine

Methorphan
Dextromethorphan
227
C18H25NO
271.19360
2145
G P-I U+UHYAC
125-71-3
LM/Q
Potent antitussive

272

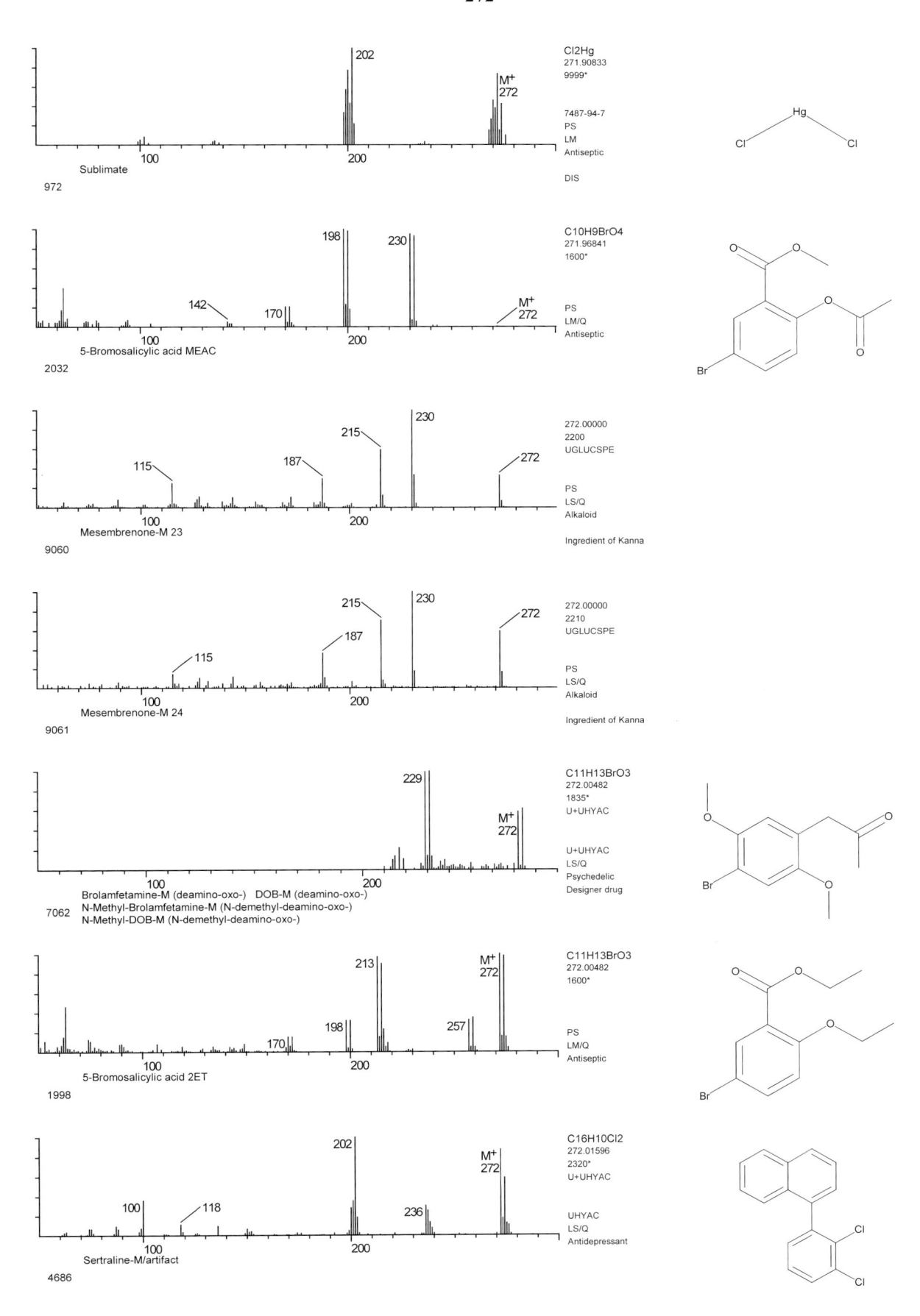

Sublimate	Cl2Hg 271.90833 9999* 7487-94-7 PS LM Antiseptic DIS
972	

5-Bromosalicylic acid MEAC	C10H9BrO4 271.96841 1600* PS LM/Q Antiseptic
2032	

Mesembrenone-M 23	272.00000 2200 UGLUCSPE PS LS/Q Alkaloid Ingredient of Kanna
9060	

Mesembrenone-M 24	272.00000 2210 UGLUCSPE PS LS/Q Alkaloid Ingredient of Kanna
9061	

Brolamfetamine-M (deamino-oxo-) DOB-M (deamino-oxo-) N-Methyl-Brolamfetamine-M (N-demethyl-deamino-oxo-) N-Methyl-DOB-M (N-demethyl-deamino-oxo-)	C11H13BrO3 272.00482 1835* U+UHYAC U+UHYAC LS/Q Psychedelic Designer drug
7062	

5-Bromosalicylic acid 2ET	C11H13BrO3 272.00482 1600* PS LM/Q Antiseptic
1998	

Sertraline-M/artifact	C16H10Cl2 272.01596 2320* U+UHYAC UHYAC LS/Q Antidepressant
4686	

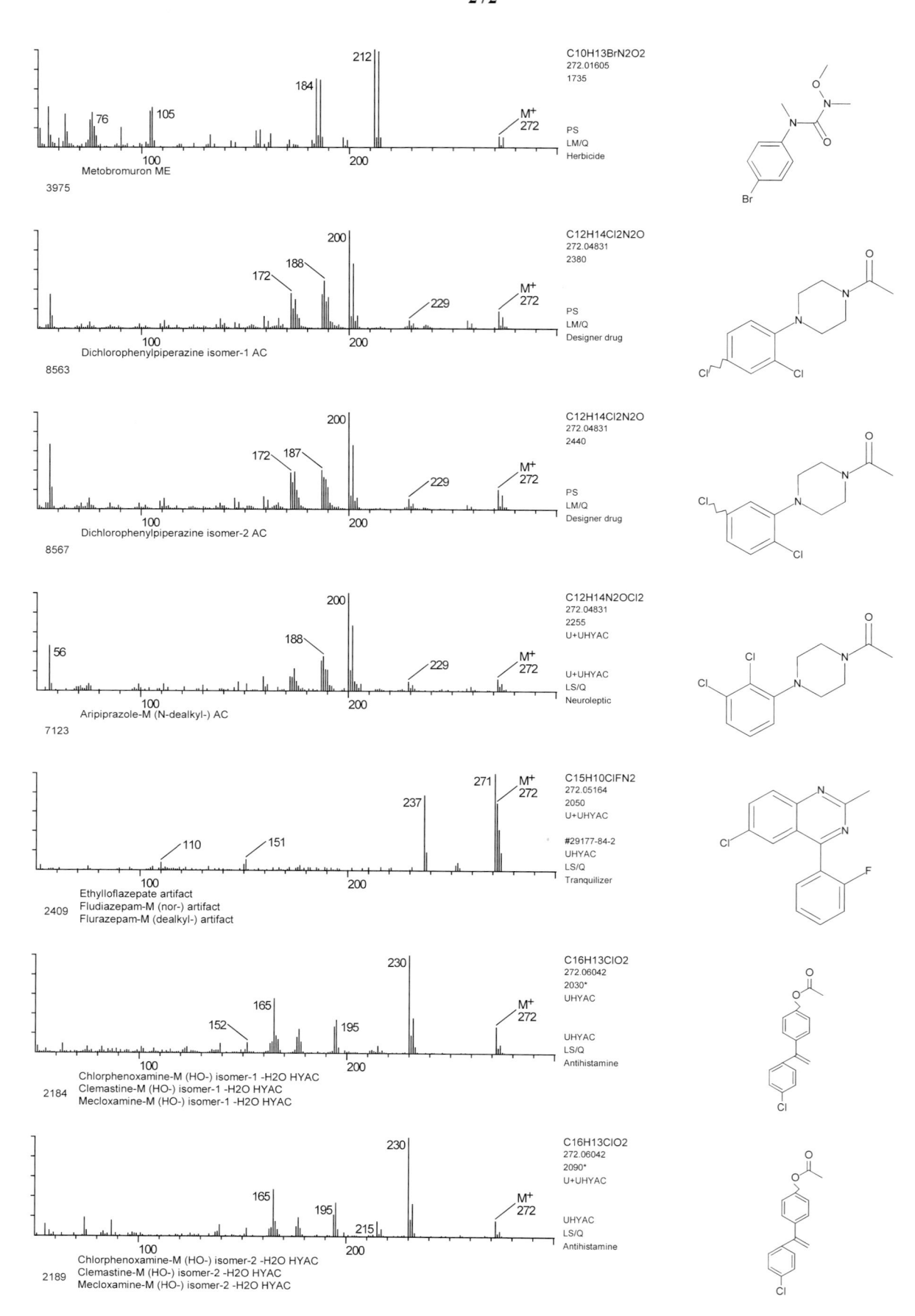

Metobromuron ME 3975 Peaks: 76, 105, 184, 212, M+ 272	C10H13BrN2O2 272.01605 1735 PS LM/Q Herbicide
Dichlorophenylpiperazine isomer-1 AC 8563 Peaks: 172, 188, 200, 229, M+ 272	C12H14Cl2N2O 272.04831 2380 PS LM/Q Designer drug
Dichlorophenylpiperazine isomer-2 AC 8567 Peaks: 172, 187, 200, 229, M+ 272	C12H14Cl2N2O 272.04831 2440 PS LM/Q Designer drug
Aripiprazole-M (N-dealkyl-) AC 7123 Peaks: 56, 188, 200, 229, M+ 272	C12H14N2OCl2 272.04831 2255 U+UHYAC U+UHYAC LS/Q Neuroleptic
Ethylloflazepate artifact Fludiazepam-M (nor-) artifact Flurazepam-M (dealkyl-) artifact 2409 Peaks: 110, 151, 237, 271, M+ 272	C15H10ClFN2 272.05164 2050 U+UHYAC #29177-84-2 UHYAC LS/Q Tranquilizer
Chlorphenoxamine-M (HO-) isomer-1 -H2O HYAC Clemastine-M (HO-) isomer-1 -H2O HYAC Mecloxamine-M (HO-) isomer-1 -H2O HYAC 2184 Peaks: 152, 165, 195, 230, M+ 272	C16H13ClO2 272.06042 2030* UHYAC UHYAC LS/Q Antihistamine
Chlorphenoxamine-M (HO-) isomer-2 -H2O HYAC Clemastine-M (HO-) isomer-2 -H2O HYAC Mecloxamine-M (HO-) isomer-2 -H2O HYAC 2189 Peaks: 165, 195, 215, 230, M+ 272	C16H13ClO2 272.06042 2090* U+UHYAC UHYAC LS/Q Antihistamine

272

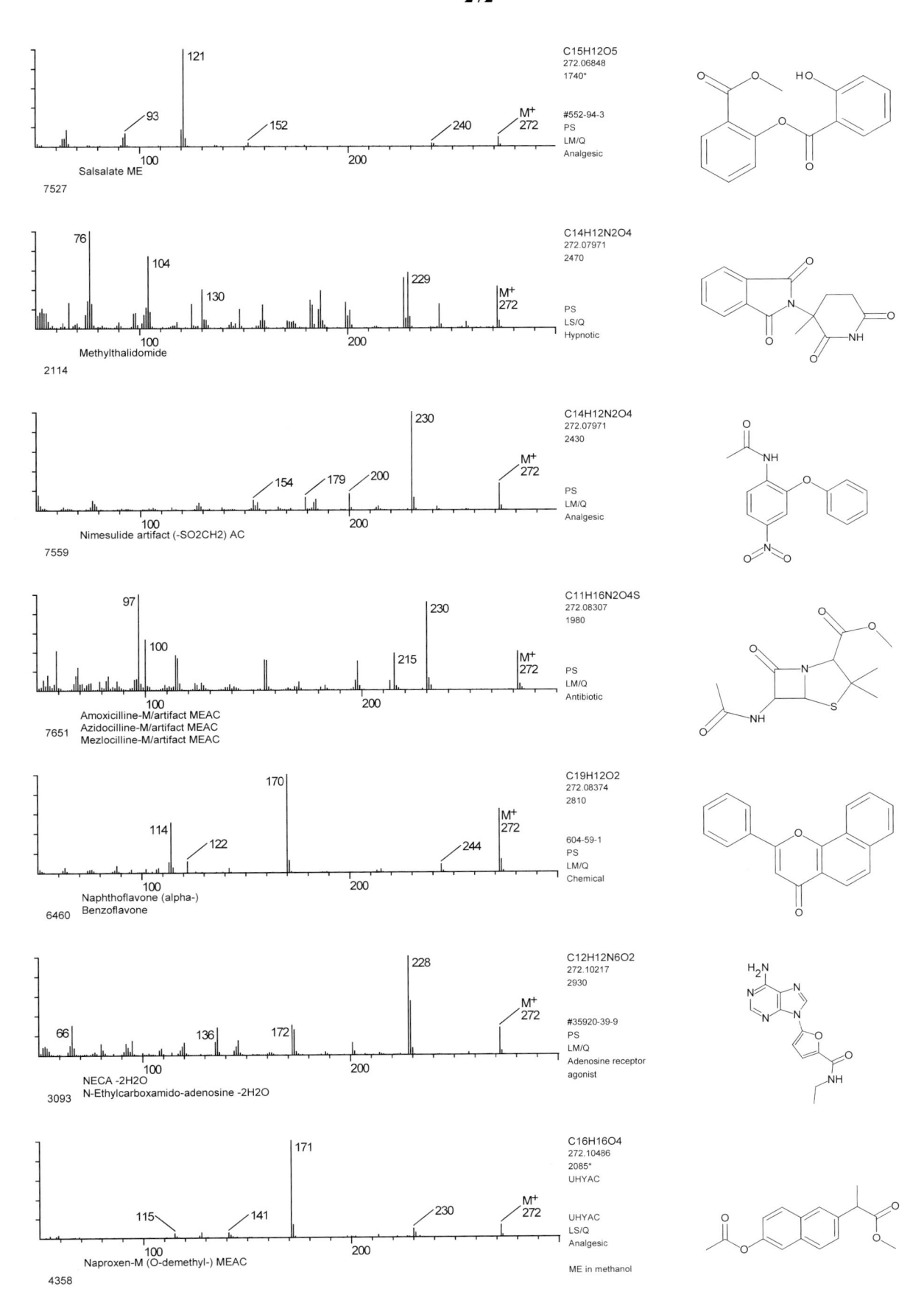

Salsalate ME — 7527
C15H12O5
272.06848
1740*
#552-94-3
PS
LM/Q
Analgesic

Peaks: 93, 121, 152, 240, M+ 272

Methylthalidomide — 2114
C14H12N2O4
272.07971
2470
PS
LS/Q
Hypnotic

Peaks: 76, 104, 130, 229, M+ 272

Nimesulide artifact (-SO2CH2) AC — 7559
C14H12N2O4
272.07971
2430
PS
LM/Q
Analgesic

Peaks: 154, 179, 200, 230, M+ 272

Amoxicilline-M/artifact MEAC
Azidocilline-M/artifact MEAC
Mezlocilline-M/artifact MEAC — 7651
C11H16N2O4S
272.08307
1980
PS
LM/Q
Antibiotic

Peaks: 97, 100, 215, 230, M+ 272

Naphthoflavone (alpha-)
Benzoflavone — 6460
C19H12O2
272.08374
2810
604-59-1
PS
LM/Q
Chemical

Peaks: 114, 122, 170, 244, M+ 272

NECA -2H2O
N-Ethylcarboxamido-adenosine -2H2O — 3093
C12H12N6O2
272.10217
2930
#35920-39-9
PS
LM/Q
Adenosine receptor agonist

Peaks: 66, 136, 172, 228, M+ 272

Naproxen-M (O-demethyl-) MEAC — 4358
C16H16O4
272.10486
2085*
UHYAC
UHYAC
LS/Q
Analgesic
ME in methanol

Peaks: 115, 141, 171, 230, M+ 272

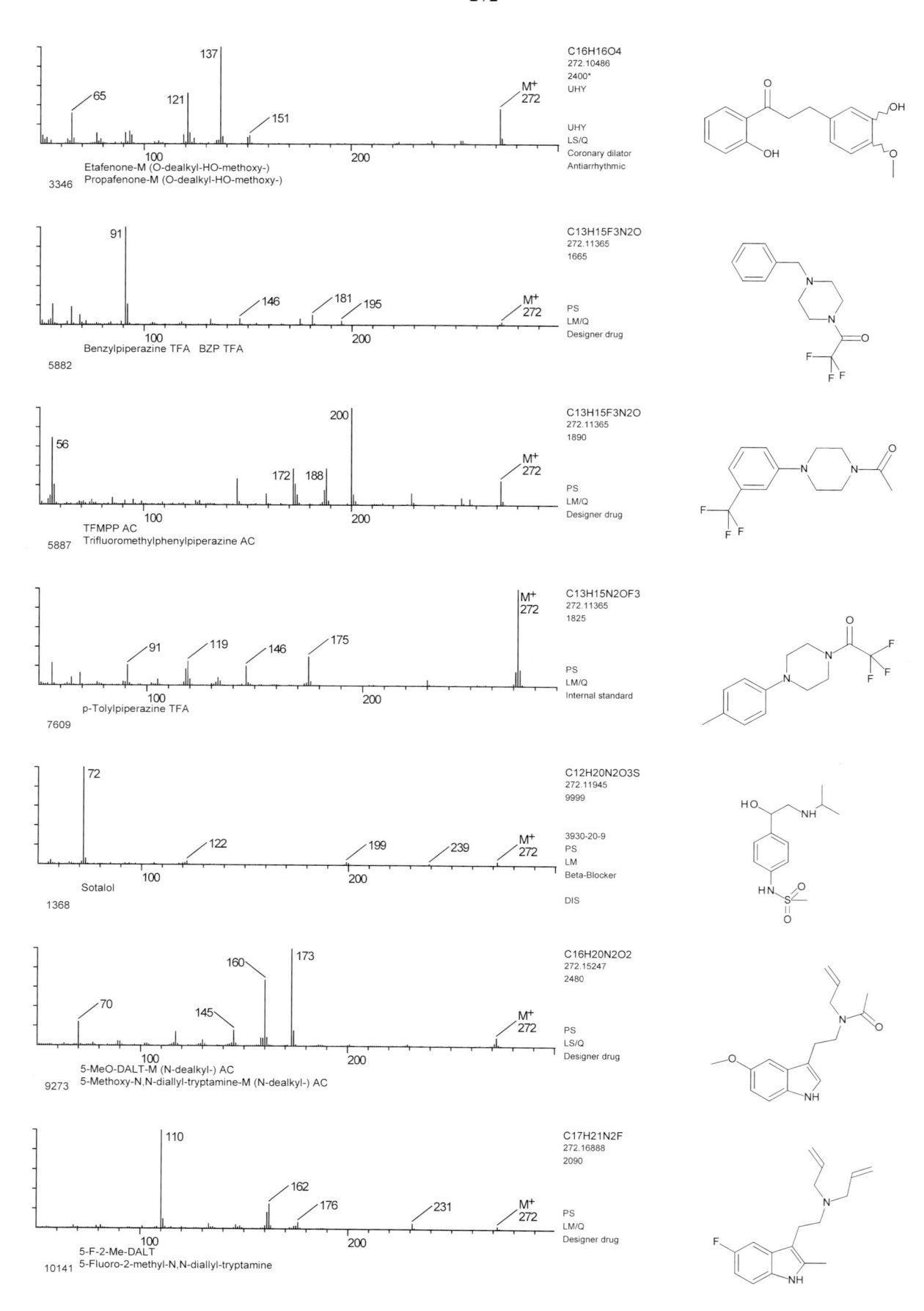

3346	Etafenone-M (O-dealkyl-HO-methoxy-) Propafenone-M (O-dealkyl-HO-methoxy-)	C16H16O4 272.10486 2400* UHY UHY LS/Q Coronary dilator Antiarrhythmic
5882	Benzylpiperazine TFA BZP TFA	C13H15F3N2O 272.11365 1665 PS LM/Q Designer drug
5887	TFMPP AC Trifluoromethylphenylpiperazine AC	C13H15F3N2O 272.11365 1890 PS LM/Q Designer drug
7609	p-Tolylpiperazine TFA	C13H15N2OF3 272.11365 1825 PS LM/Q Internal standard
1368	Sotalol	C12H20N2O3S 272.11945 9999 3930-20-9 PS LM Beta-Blocker DIS
9273	5-MeO-DALT-M (N-dealkyl-) AC 5-Methoxy-N,N-diallyl-tryptamine-M (N-dealkyl-) AC	C16H20N2O2 272.15247 2480 PS LS/Q Designer drug
10141	5-F-2-Me-DALT 5-Fluoro-2-methyl-N,N-diallyl-tryptamine	C17H21N2F 272.16888 2090 PS LM/Q Designer drug

272

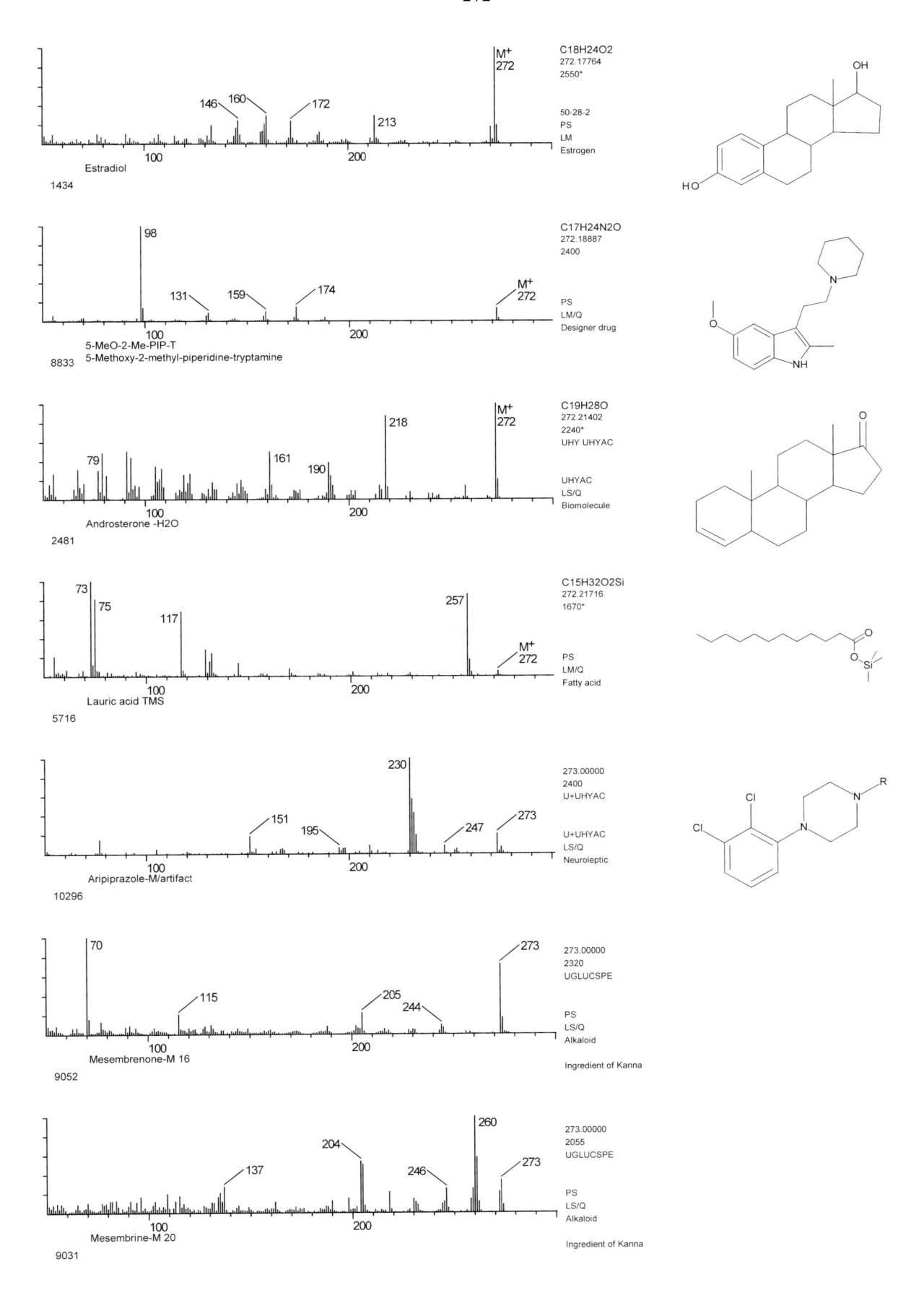

C18H24O2
272.17764
2550*

50-28-2
PS
LM
Estrogen

Estradiol
1434

C17H24N2O
272.18887
2400

PS
LM/Q
Designer drug

5-MeO-2-Me-PIP-T
5-Methoxy-2-methyl-piperidine-tryptamine
8833

C19H28O
272.21402
2240*
UHY UHYAC

UHYAC
LS/Q
Biomolecule

Androsterone -H2O
2481

C15H32O2Si
272.21716
1670*

PS
LM/Q
Fatty acid

Lauric acid TMS
5716

273.00000
2400
U+UHYAC

U+UHYAC
LS/Q
Neuroleptic

Aripiprazole-M/artifact
10296

273.00000
2320
UGLUCSPE

PS
LS/Q
Alkaloid

Ingredient of Kanna

Mesembrenone-M 16
9052

273.00000
2055
UGLUCSPE

PS
LS/Q
Alkaloid

Ingredient of Kanna

Mesembrine-M 20
9031

731

273.00000
2365
UGLUCSPE

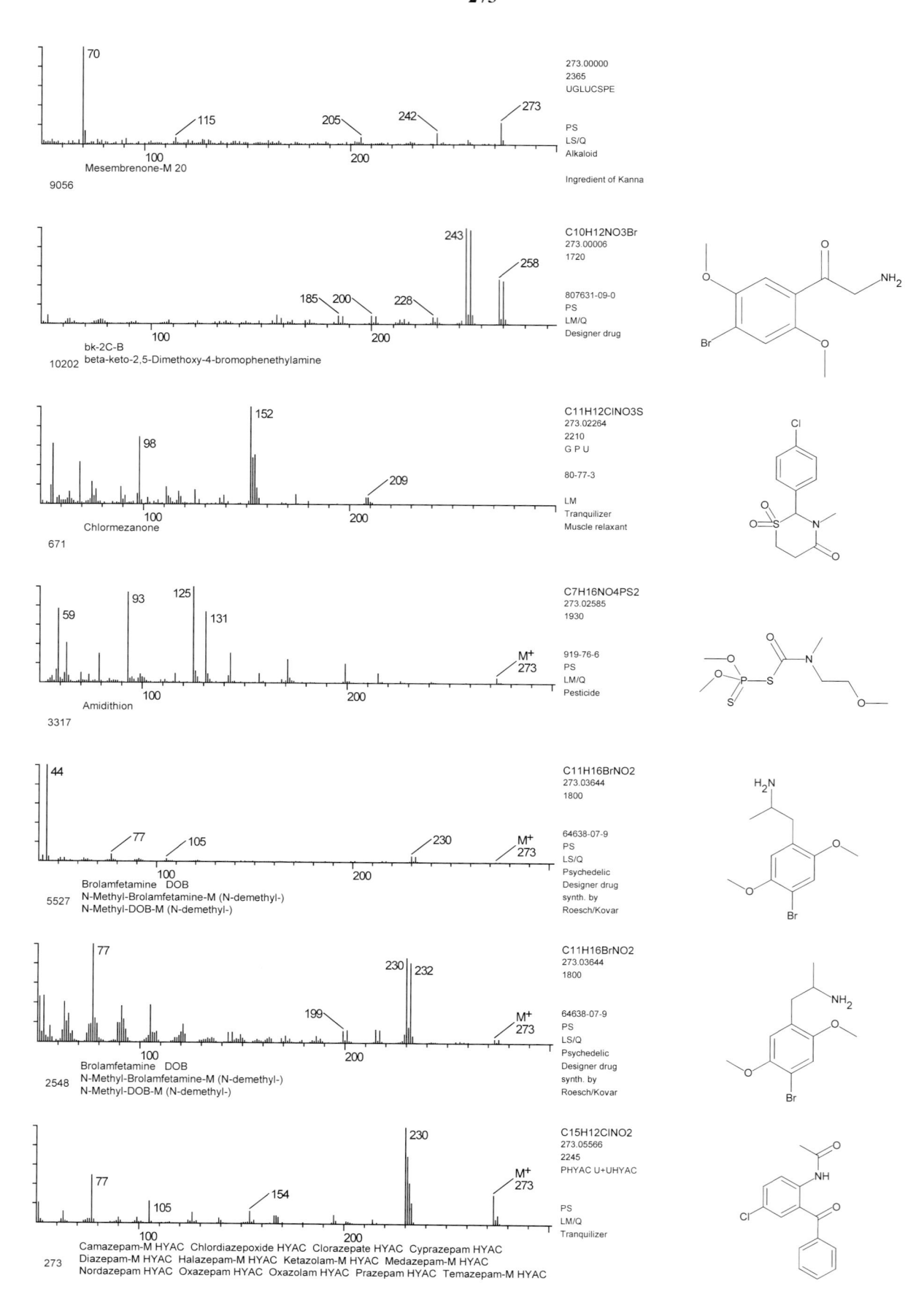

PS
LS/Q
Alkaloid

Mesembrenone-M 20
Ingredient of Kanna
9056

C10H12NO3Br
273.00006
1720

807631-09-0
PS
LM/Q
Designer drug

bk-2C-B
10202 beta-keto-2,5-Dimethoxy-4-bromophenethylamine

C11H12ClNO3S
273.02264
2210
G P U

80-77-3

LM
Tranquilizer
Muscle relaxant

Chlormezanone
671

C7H16NO4PS2
273.02585
1930

919-76-6
PS
LM/Q
Pesticide

Amidithion
3317

C11H16BrNO2
273.03644
1800

64638-07-9
PS
LS/Q
Psychedelic
Designer drug
synth. by
Roesch/Kovar

Brolamfetamine DOB
N-Methyl-Brolamfetamine-M (N-demethyl-)
5527   N-Methyl-DOB-M (N-demethyl-)

C11H16BrNO2
273.03644
1800

64638-07-9
PS
LS/Q
Psychedelic
Designer drug
synth. by
Roesch/Kovar

Brolamfetamine DOB
N-Methyl-Brolamfetamine-M (N-demethyl-)
2548   N-Methyl-DOB-M (N-demethyl-)

C15H12ClNO2
273.05566
2245
PHYAC U+UHYAC

PS
LM/Q
Tranquilizer

Camazepam-M HYAC Chlordiazepoxide HYAC Clorazepate HYAC Cyprazepam HYAC
273 Diazepam-M HYAC Halazepam-M HYAC Ketazolam HYAC Medazepam-M HYAC
Nordazepam HYAC Oxazepam HYAC Oxazolam HYAC Prazepam HYAC Temazepam-M HYAC

273

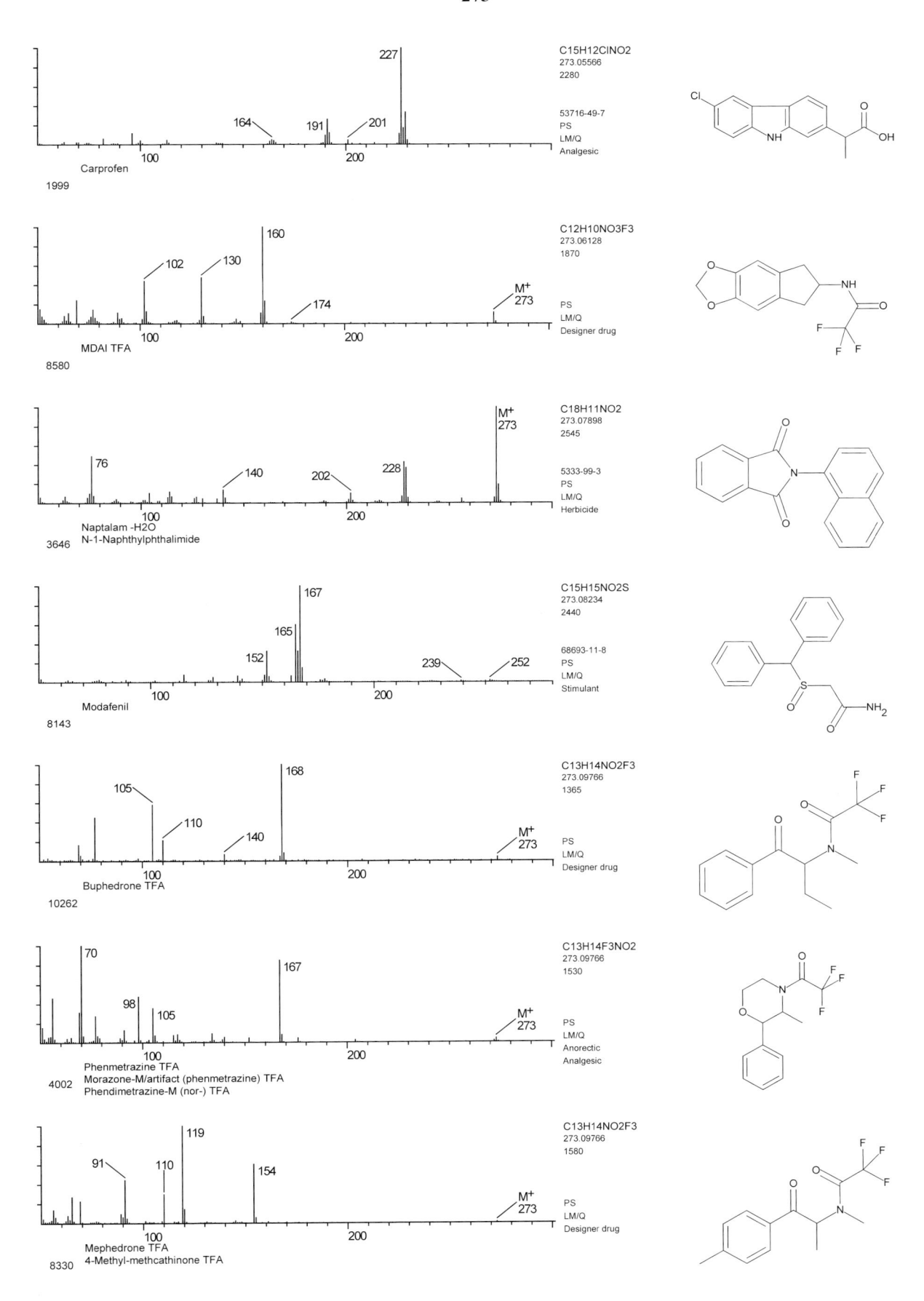

C15H12ClNO2
273.05566
2280

53716-49-7
PS
LM/Q
Analgesic

Carprofen
1999

C12H10NO3F3
273.06128
1870

PS
LM/Q
Designer drug

MDAI TFA
8580

C18H11NO2
273.07898
2545

5333-99-3
PS
LM/Q
Herbicide

Naptalam -H2O
N-1-Naphthylphthalimide
3646

C15H15NO2S
273.08234
2440

68693-11-8
PS
LM/Q
Stimulant

Modafenil
8143

C13H14NO2F3
273.09766
1365

PS
LM/Q
Designer drug

Buphedrone TFA
10262

C13H14F3NO2
273.09766
1530

PS
LM/Q
Anorectic
Analgesic

Phenmetrazine TFA
Morazone-M/artifact (phenmetrazine) TFA
Phendimetrazine-M (nor-) TFA
4002

C13H14NO2F3
273.09766
1580

PS
LM/Q
Designer drug

Mephedrone TFA
4-Methyl-methcathinone TFA
8330

273

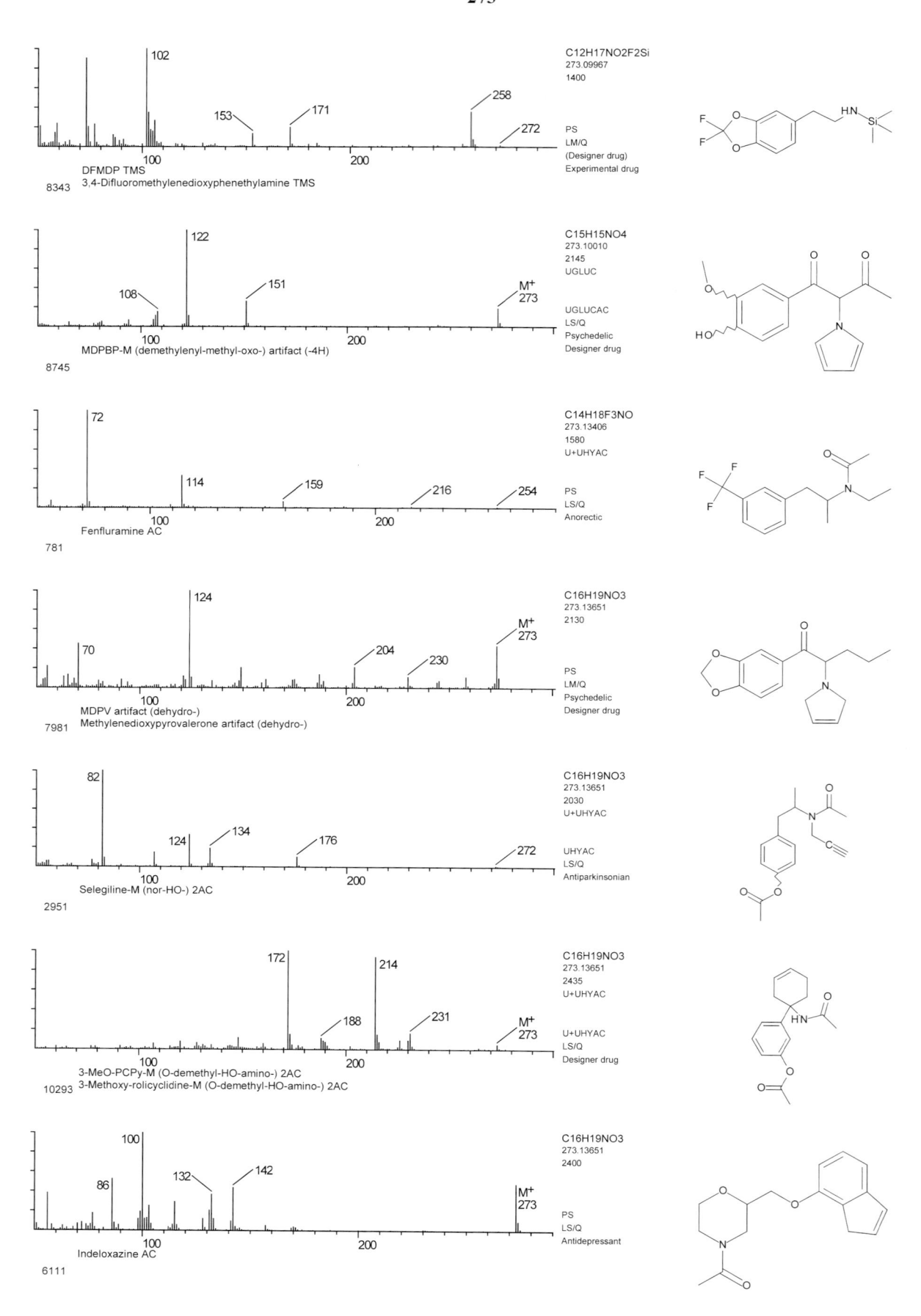

Spectrum	Compound	Formula / Info
102, 153, 171, 258, 272	DFMDP TMS — 3,4-Difluoromethylenedioxyphenethylamine TMS (8343)	C12H17NO2F2Si; 273.09967; 1400; PS; LM/Q; (Designer drug); Experimental drug
108, 122, 151, 273 (M+)	MDPBP-M (demethylenyl-methyl-oxo-) artifact (-4H) (8745)	C15H15NO4; 273.10010; 2145; UGLUC; UGLUCAC; LS/Q; Psychedelic; Designer drug
72, 114, 159, 216, 254	Fenfluramine AC (781)	C14H18F3NO; 273.13406; 1580; U+UHYAC; PS; LS/Q; Anorectic
70, 124, 204, 230, 273 (M+)	MDPV artifact (dehydro-) — Methylenedioxypyrovalerone artifact (dehydro-) (7981)	C16H19NO3; 273.13651; 2130; PS; LM/Q; Psychedelic; Designer drug
82, 124, 134, 176, 272	Selegiline-M (nor-HO-) 2AC (2951)	C16H19NO3; 273.13651; 2030; U+UHYAC; UHYAC; LS/Q; Antiparkinsonian
172, 188, 214, 231, 273 (M+)	3-MeO-PCPy-M (O-demethyl-HO-amino-) 2AC — 3-Methoxy-rolicyclidine-M (O-demethyl-HO-amino-) 2AC (10293)	C16H19NO3; 273.13651; 2435; U+UHYAC; U+UHYAC; LS/Q; Designer drug
86, 100, 132, 142, 273 (M+)	Indeloxazine AC (6111)	C16H19NO3; 273.13651; 2400; PS; LS/Q; Antidepressant

273

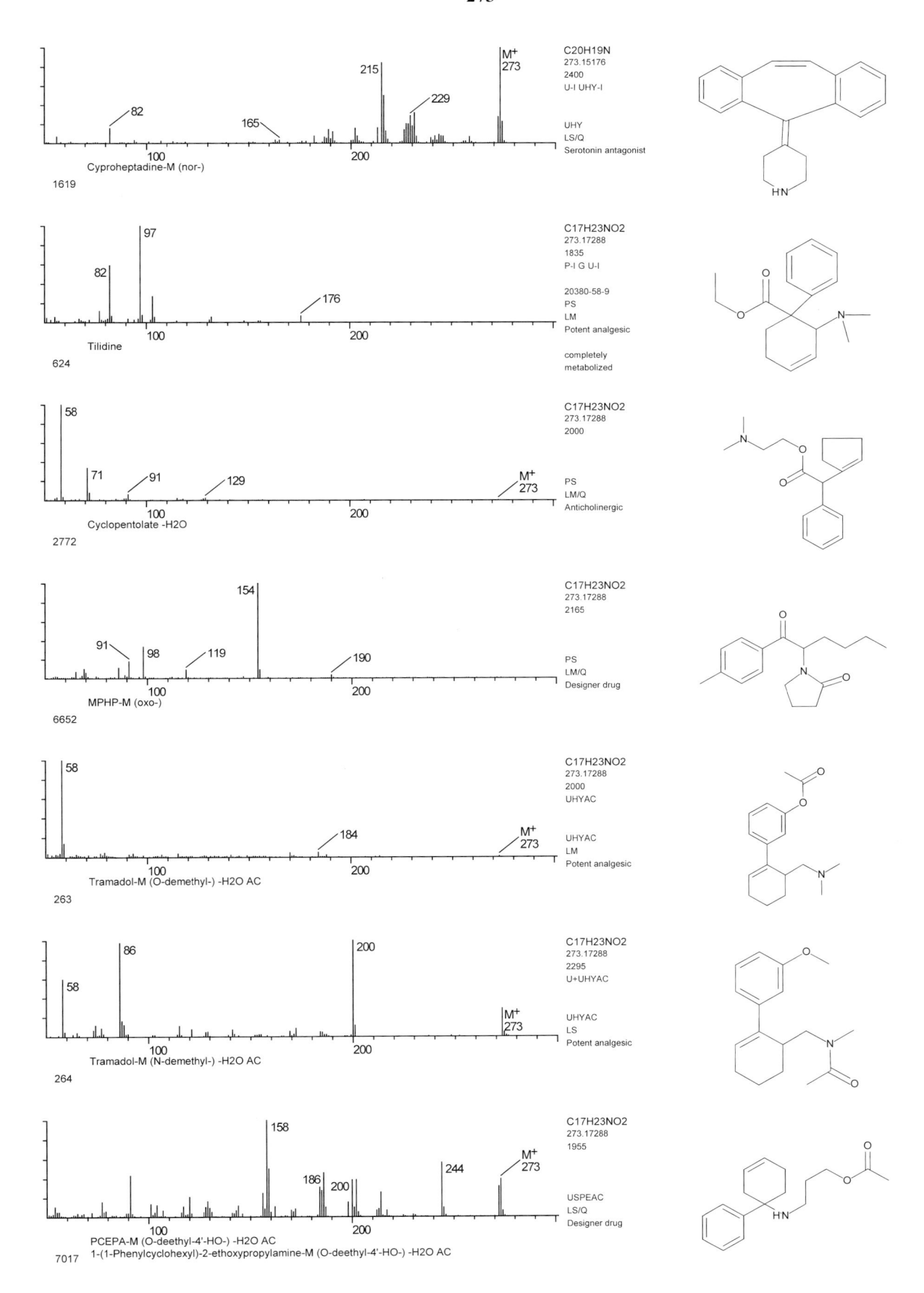

Cyproheptadine-M (nor-)
1619

C20H19N
273.15176
2400
U-I UHY-I

UHY
LS/Q
Serotonin antagonist

Tilidine
624

C17H23NO2
273.17288
1835
P-I G U-I
20380-58-9
PS
LM
Potent analgesic
completely metabolized

Cyclopentolate -H2O
2772

C17H23NO2
273.17288
2000

PS
LM/Q
Anticholinergic

MPHP-M (oxo-)
6652

C17H23NO2
273.17288
2165

PS
LM/Q
Designer drug

Tramadol-M (O-demethyl-) -H2O AC
263

C17H23NO2
273.17288
2000
UHYAC

UHYAC
LM
Potent analgesic

Tramadol-M (N-demethyl-) -H2O AC
264

C17H23NO2
273.17288
2295
U+UHYAC

UHYAC
LS
Potent analgesic

PCEPA-M (O-deethyl-4'-HO-) -H2O AC
1-(1-Phenylcyclohexyl)-2-ethoxypropylamine-M (O-deethyl-4'-HO-) -H2O AC
7017

C17H23NO2
273.17288
1955

USPEAC
LS/Q
Designer drug

273

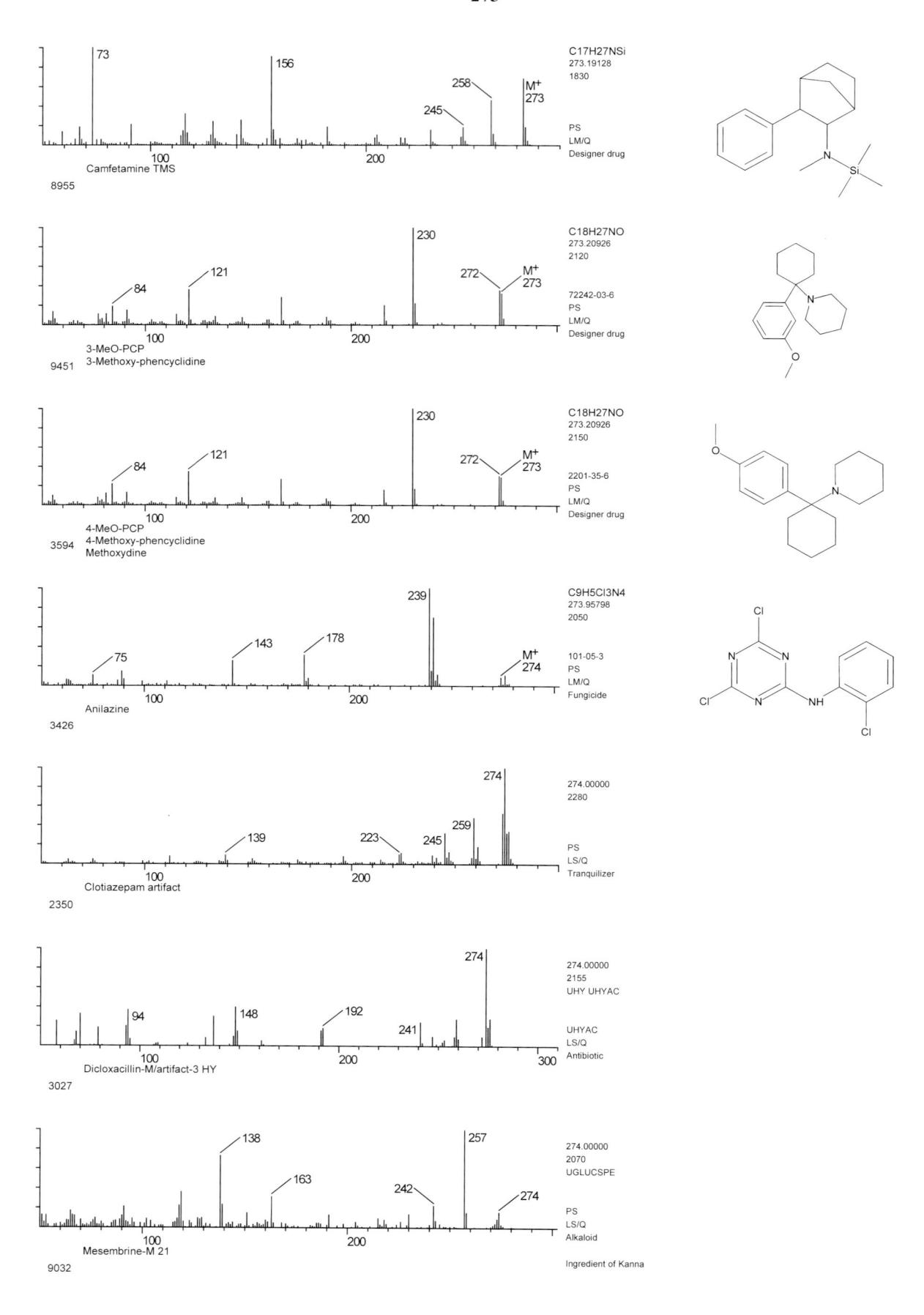

Spectrum	Formula / Info
Camfetamine TMS — peaks 73, 156, 245, 258, M+ 273	C17H27NSi; 273.19128; 1830; PS; LM/Q; Designer drug
3-MeO-PCP / 3-Methoxy-phencyclidine — peaks 84, 121, 230, 272, M+ 273	C18H27NO; 273.20926; 2120; 72242-03-6; PS; LM/Q; Designer drug
4-MeO-PCP / 4-Methoxy-phencyclidine / Methoxydine — peaks 84, 121, 230, 272, M+ 273	C18H27NO; 273.20926; 2150; 2201-35-6; PS; LM/Q; Designer drug
Anilazine — peaks 75, 143, 178, 239, M+ 274	C9H5Cl3N4; 273.95798; 2050; 101-05-3; PS; LM/Q; Fungicide
Clotiazepam artifact — peaks 139, 223, 245, 259, 274	274.00000; 2280; PS; LS/Q; Tranquilizer
Dicloxacillin-M/artifact-3 HY — peaks 94, 148, 192, 241, 274	274.00000; 2155; UHY UHYAC; UHYAC; LS/Q; Antibiotic
Mesembrine-M 21 — peaks 138, 163, 242, 257, 274	274.00000; 2070; UGLUCSPE; PS; LS/Q; Alkaloid; Ingredient of Kanna

736

274

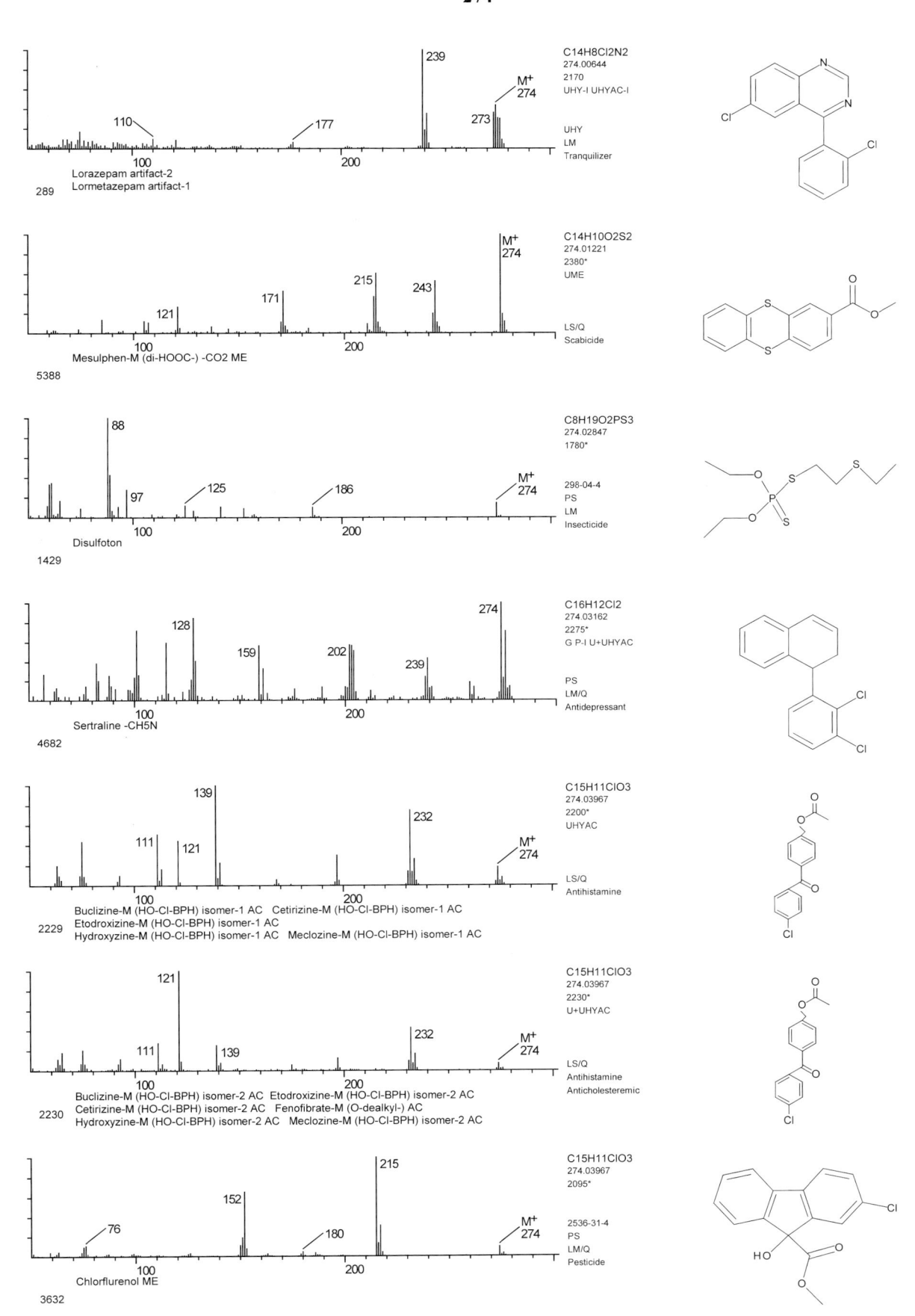

C14H8Cl2N2
274.00644
2170
UHY-I UHYAC-I

UHY
LM
Tranquilizer

Lorazepam artifact-2
Lormetazepam artifact-1
289

C14H10O2S2
274.01221
2380*
UME

LS/Q
Scabicide

Mesulphen-M (di-HOOC-) -CO2 ME
5388

C8H19O2PS3
274.02847
1780*

298-04-4
PS
LM
Insecticide

Disulfoton
1429

C16H12Cl2
274.03162
2275*
G P-I U+UHYAC

PS
LM/Q
Antidepressant

Sertraline -CH5N
4682

C15H11ClO3
274.03967
2200*
UHYAC

LS/Q
Antihistamine

Buclizine-M (HO-Cl-BPH) isomer-1 AC   Cetirizine-M (HO-Cl-BPH) isomer-1 AC
Etodroxizine-M (HO-Cl-BPH) isomer-1 AC
2229   Hydroxyzine-M (HO-Cl-BPH) isomer-1 AC   Meclozine-M (HO-Cl-BPH) isomer-1 AC

C15H11ClO3
274.03967
2230*
U+UHYAC

LS/Q
Antihistamine
Anticholesteremic

Buclizine-M (HO-Cl-BPH) isomer-2 AC   Etodroxizine-M (HO-Cl-BPH) isomer-2 AC
2230   Cetirizine-M (HO-Cl-BPH) isomer-2 AC   Fenofibrate-M (O-dealkyl-) AC
Hydroxyzine-M (HO-Cl-BPH) isomer-2 AC   Meclozine-M (HO-Cl-BPH) isomer-2 AC

C15H11ClO3
274.03967
2095*

2536-31-4
PS
LM/Q
Pesticide

Chlorflurenol ME
3632

737

274

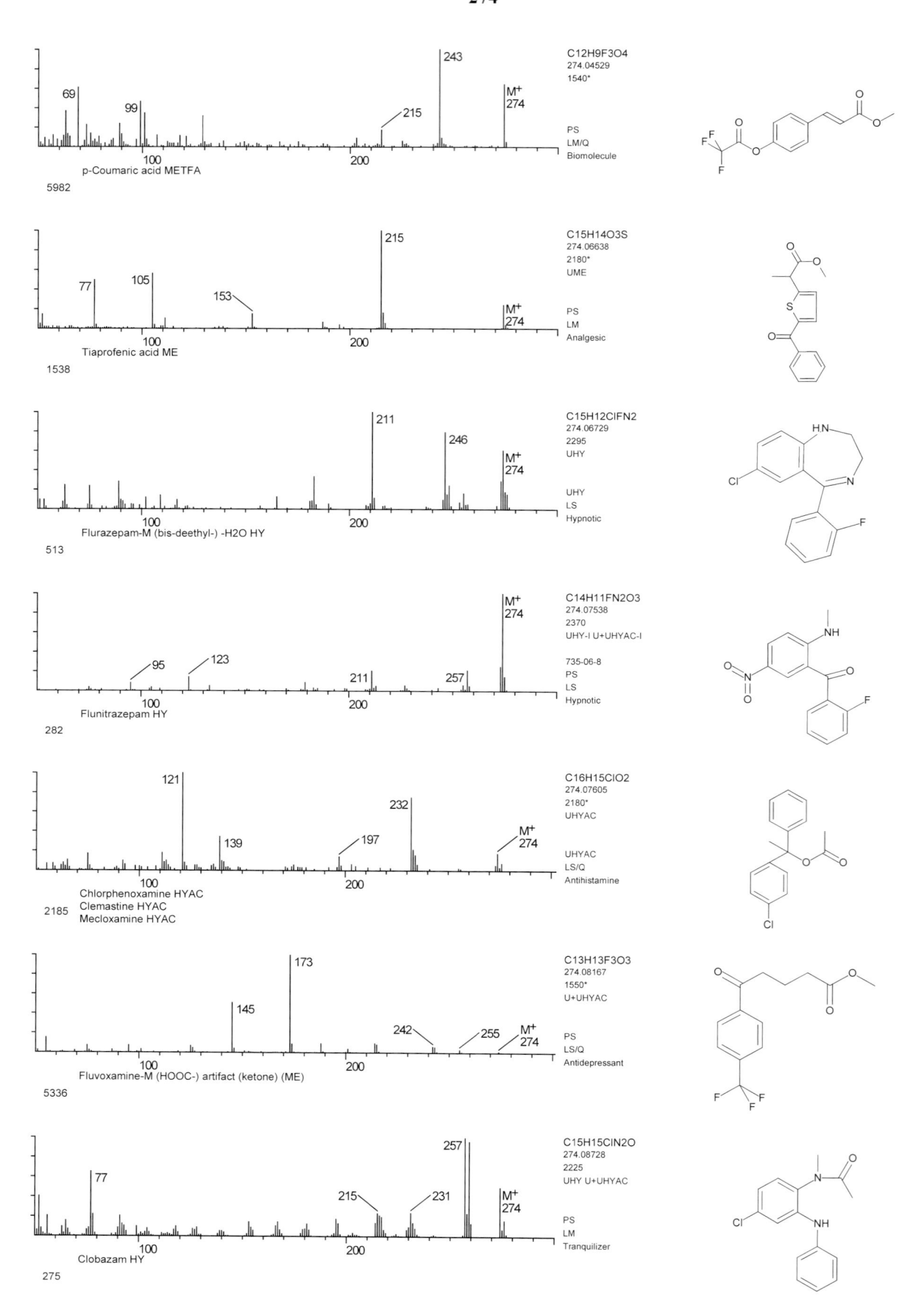

C12H9F3O4
274.04529
1540*

PS
LM/Q
Biomolecule

p-Coumaric acid METFA
5982

C15H14O3S
274.06638
2180*
UME

PS
LM
Analgesic

Tiaprofenic acid ME
1538

C15H12ClFN2
274.06729
2295
UHY

UHY
LS
Hypnotic

Flurazepam-M (bis-deethyl-) -H2O HY
513

C14H11FN2O3
274.07538
2370
UHY-I U+UHYAC-I

735-06-8
PS
LS
Hypnotic

Flunitrazepam HY
282

C16H15ClO2
274.07605
2180*
UHYAC

UHYAC
LS/Q
Antihistamine

Chlorphenoxamine HYAC
Clemastine HYAC
Mecloxamine HYAC
2185

C13H13F3O3
274.08167
1550*
U+UHYAC

PS
LS/Q
Antidepressant

Fluvoxamine-M (HOOC-) artifact (ketone) (ME)
5336

C15H15ClN2O
274.08728
2225
UHY U+UHYAC

PS
LM
Tranquilizer

Clobazam HY
275

274

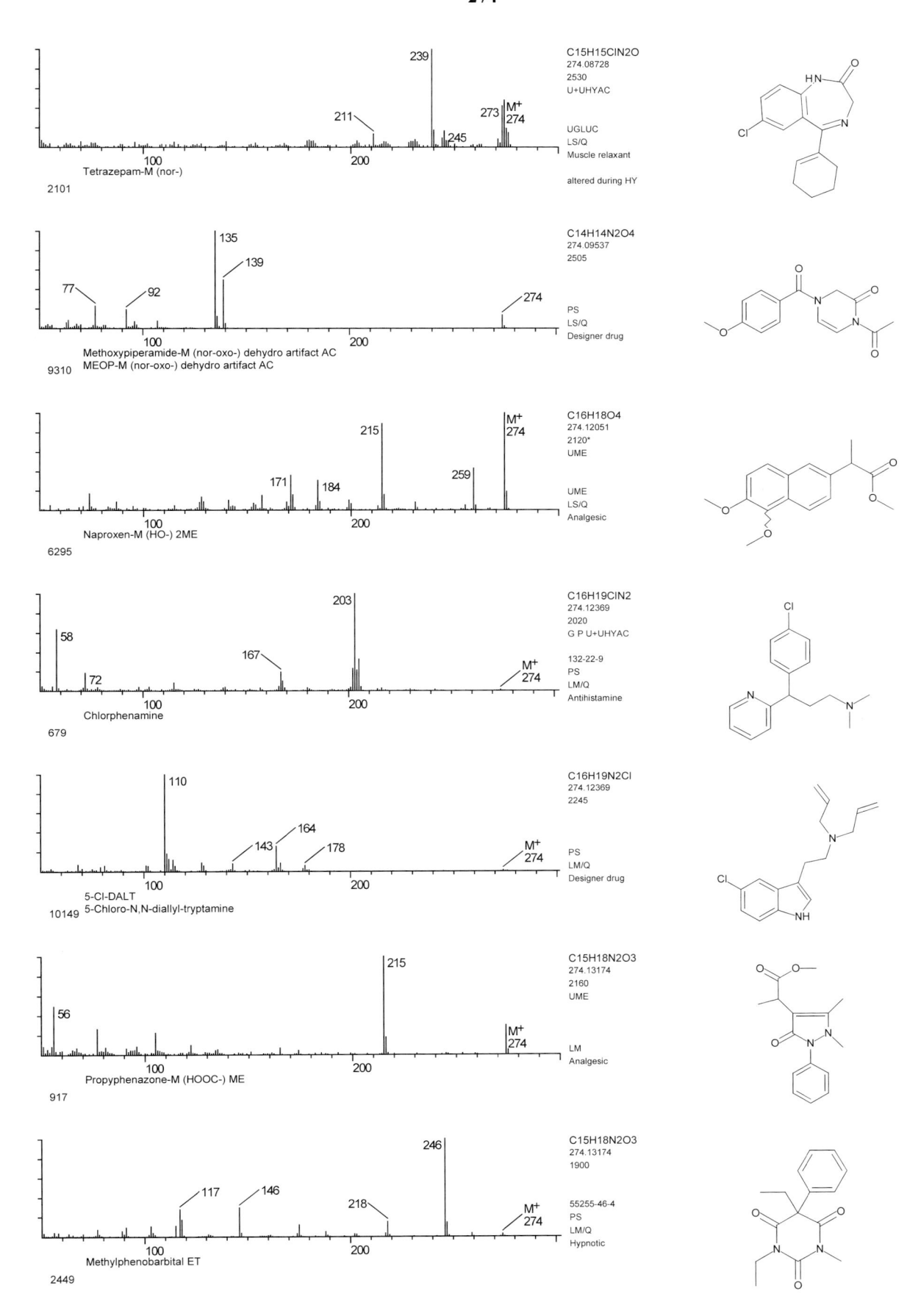

Tetrazepam-M (nor-)	C15H15ClN2O 274.08728 2530 U+UHYAC UGLUC LS/Q Muscle relaxant altered during HY
2101	
Methoxypiperamide-M (nor-oxo-) dehydro artifact AC	C14H14N2O4 274.09537 2505 PS LS/Q Designer drug
9310 MEOP-M (nor-oxo-) dehydro artifact AC	
Naproxen-M (HO-) 2ME	C16H18O4 274.12051 2120* UME UME LS/Q Analgesic
6295	
Chlorphenamine	C16H19ClN2 274.12369 2020 G P U+UHYAC 132-22-9 PS LM/Q Antihistamine
679	
5-Cl-DALT	C16H19N2Cl 274.12369 2245 PS LM/Q Designer drug
10149 5-Chloro-N,N-diallyl-tryptamine	
Propyphenazone-M (HOOC-) ME	C15H18N2O3 274.13174 2160 UME LM Analgesic
917	
Methylphenobarbital ET	C15H18N2O3 274.13174 1900 55255-46-4 PS LM/Q Hypnotic
2449	

739

274

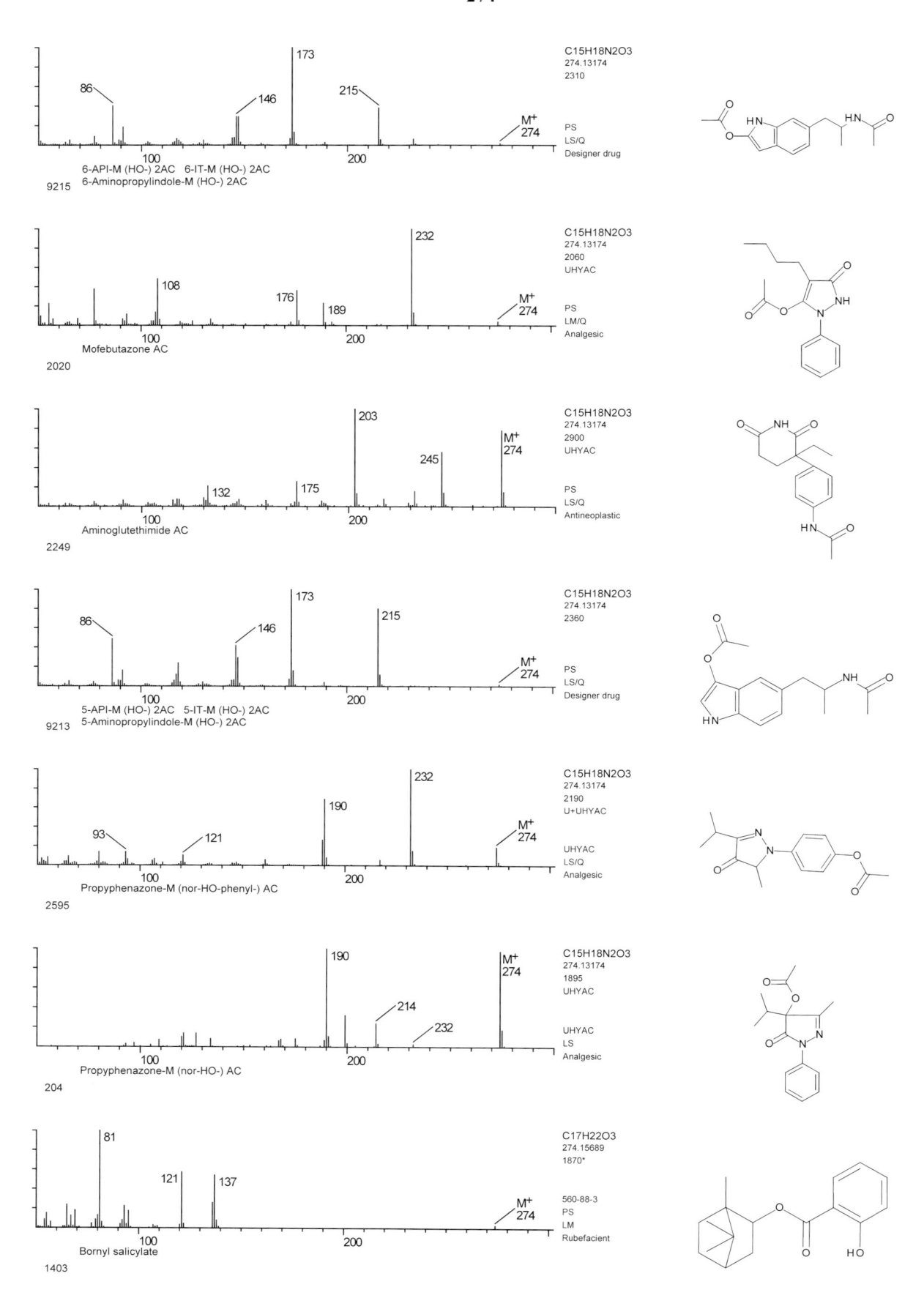

86, 146, 173, 215, M+ 274	C15H18N2O3 274.13174 2310 PS LS/Q Designer drug
6-API-M (HO-) 2AC 6-IT-M (HO-) 2AC 6-Aminopropylindole-M (HO-) 2AC 9215	

108, 176, 189, 232, M+ 274	C15H18N2O3 274.13174 2060 UHYAC PS LM/Q Analgesic
Mofebutazone AC 2020	

132, 175, 203, 245, M+ 274	C15H18N2O3 274.13174 2900 UHYAC PS LS/Q Antineoplastic
Aminoglutethimide AC 2249	

86, 146, 173, 215, M+ 274	C15H18N2O3 274.13174 2360 PS LS/Q Designer drug
5-API-M (HO-) 2AC 5-IT-M (HO-) 2AC 5-Aminopropylindole-M (HO-) 2AC 9213	

93, 121, 190, 232, M+ 274	C15H18N2O3 274.13174 2190 U+UHYAC UHYAC LS/Q Analgesic
Propyphenazone-M (nor-HO-phenyl-) AC 2595	

190, 214, 232, M+ 274	C15H18N2O3 274.13174 1895 UHYAC UHYAC LS Analgesic
Propyphenazone-M (nor-HO-) AC 204	

81, 121, 137, M+ 274	C17H22O3 274.15689 1870* 560-88-3 PS LM Rubefacient
Bornyl salicylate 1403	

274

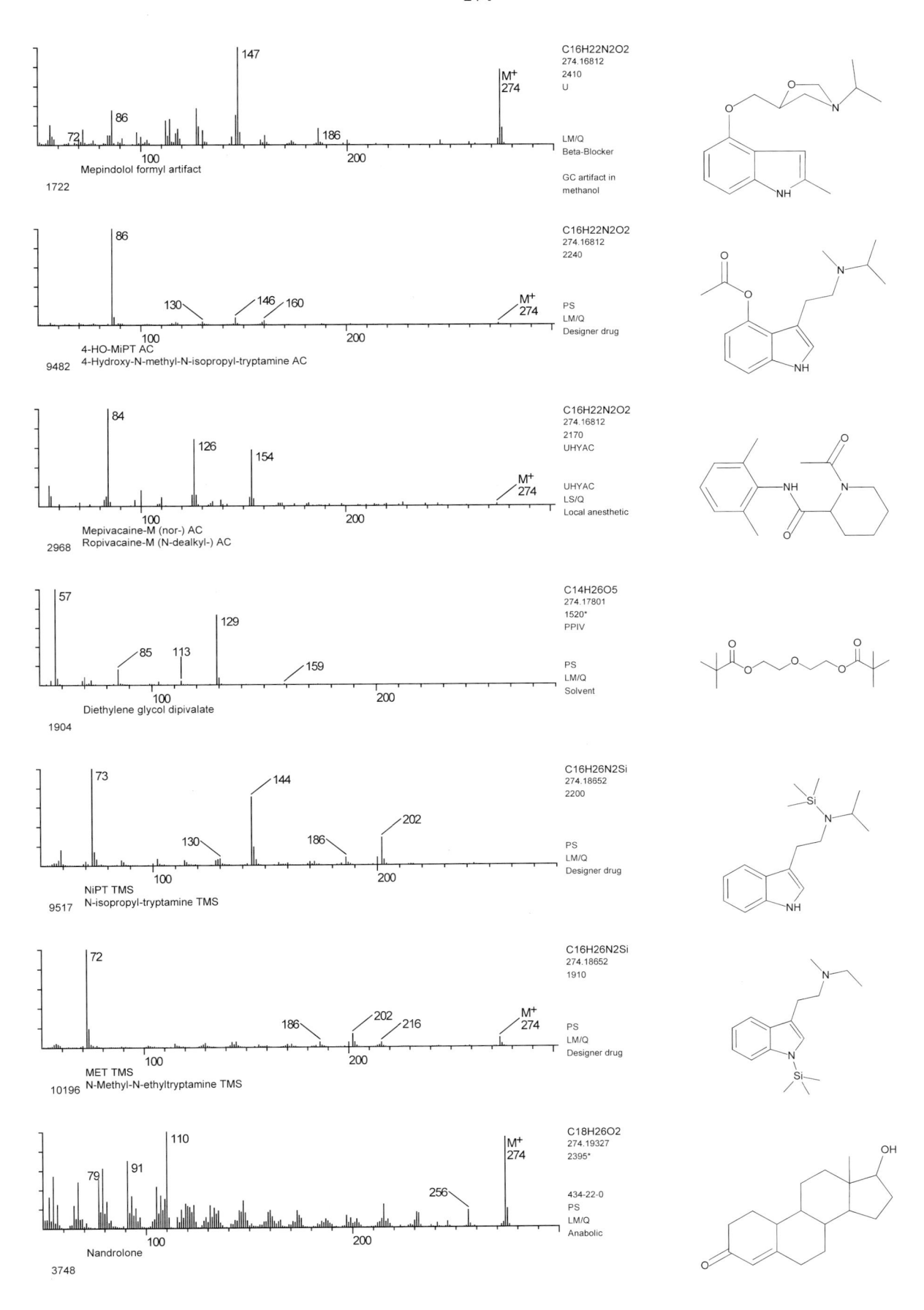

1722 — Mepindolol formyl artifact	C16H22N2O2, 274.16812, 2410, U, LM/Q, Beta-Blocker, GC artifact in methanol	
9482 — 4-HO-MiPT AC, 4-Hydroxy-N-methyl-N-isopropyl-tryptamine AC	C16H22N2O2, 274.16812, 2240, PS, LM/Q, Designer drug	
2968 — Mepivacaine-M (nor-) AC, Ropivacaine-M (N-dealkyl-) AC	C16H22N2O2, 274.16812, 2170, UHYAC, UHYAC, LS/Q, Local anesthetic	
1904 — Diethylene glycol dipivalate	C14H26O5, 274.17801, 1520*, PPIV, PS, LM/Q, Solvent	
9517 — NiPT TMS, N-isopropyl-tryptamine TMS	C16H26N2Si, 274.18652, 2200, PS, LM/Q, Designer drug	
10196 — MET TMS, N-Methyl-N-ethyltryptamine TMS	C16H26N2Si, 274.18652, 1910, PS, LM/Q, Designer drug	
3748 — Nandrolone	C18H26O2, 274.19327, 2395*, 434-22-0, PS, LM/Q, Anabolic	

274

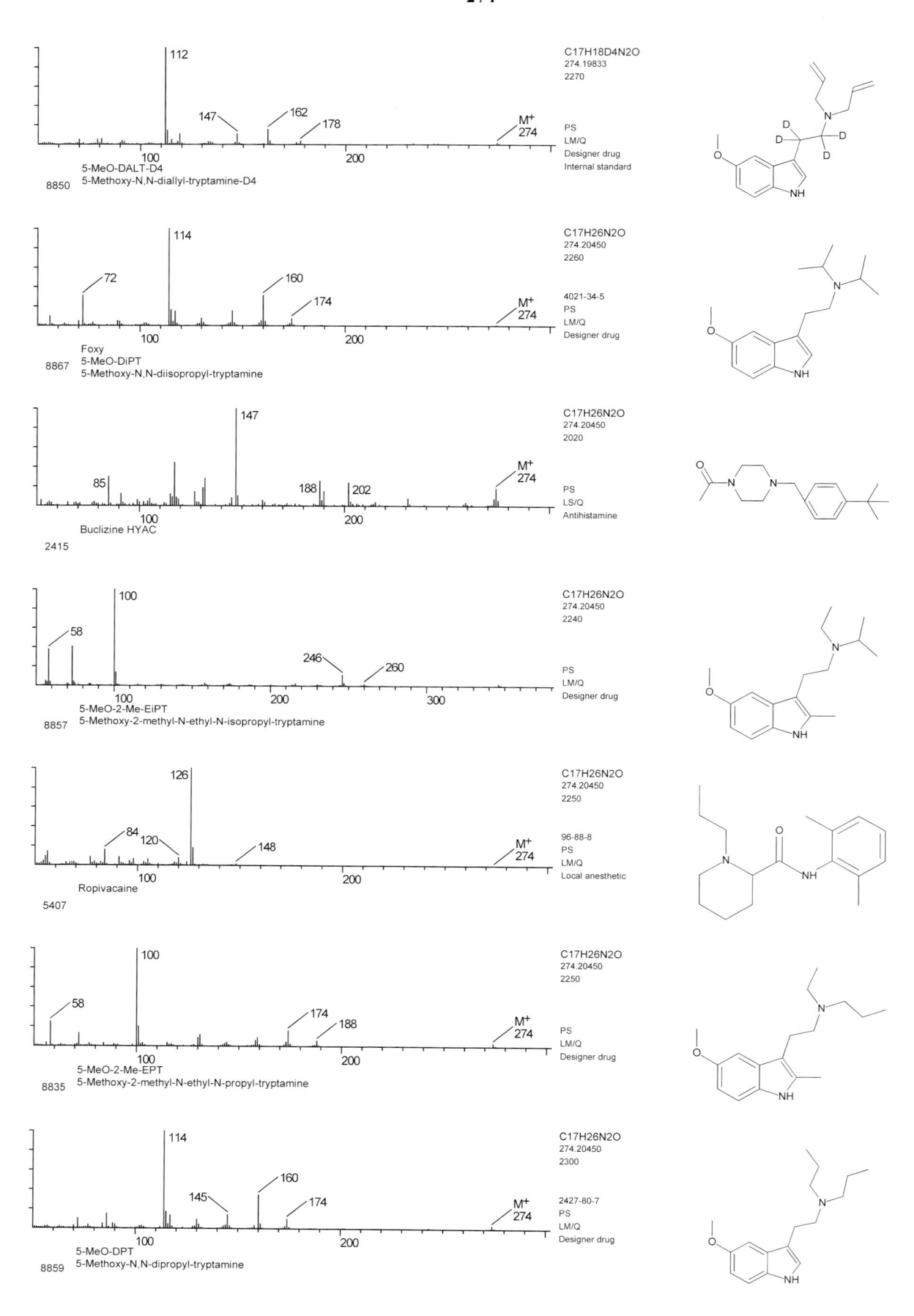

8850
5-MeO-DALT-D4
5-Methoxy-N,N-diallyl-tryptamine-D4

C17H18D4N2O
274.19833
2270

PS
LM/Q
Designer drug
Internal standard

8867
Foxy
5-MeO-DiPT
5-Methoxy-N,N-diisopropyl-tryptamine

C17H26N2O
274.20450
2260

4021-34-5
PS
LM/Q
Designer drug

2415
Buclizine HYAC

C17H26N2O
274.20450
2020

PS
LS/Q
Antihistamine

8857
5-MeO-2-Me-EiPT
5-Methoxy-2-methyl-N-ethyl-N-isopropyl-tryptamine

C17H26N2O
274.20450
2240

PS
LM/Q
Designer drug

5407
Ropivacaine

C17H26N2O
274.20450
2250

96-88-8
PS
LM/Q
Local anesthetic

8835
5-MeO-2-Me-EPT
5-Methoxy-2-methyl-N-ethyl-N-propyl-tryptamine

C17H26N2O
274.20450
2250

PS
LM/Q
Designer drug

8859
5-MeO-DPT
5-Methoxy-N,N-dipropyl-tryptamine

C17H26N2O
274.20450
2300

2427-80-7
PS
LM/Q
Designer drug

274

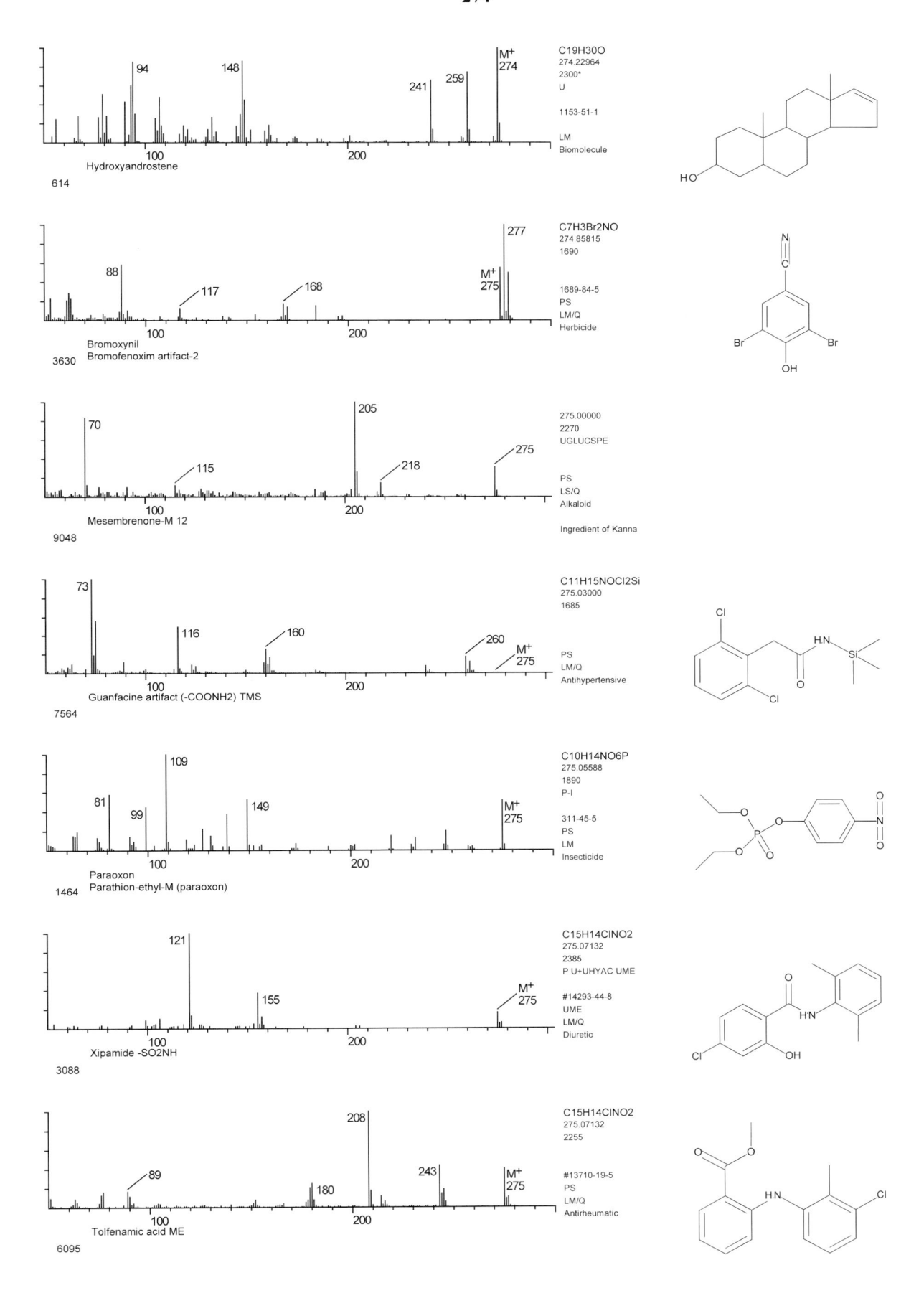

Hydroxyandrostene
614
94, 148, 241, 259, M+ 274
C19H30O
274.22964
2300*
U
1153-51-1
LM
Biomolecule

Bromoxynil
Bromofenoxim artifact-2
3630
88, 117, 168, M+ 275, 277
C7H3Br2NO
274.85815
1690
1689-84-5
PS
LM/Q
Herbicide

Mesembrenone-M 12
9048
70, 115, 205, 218, 275
275.00000
2270
UGLUCSPE
PS
LS/Q
Alkaloid
Ingredient of Kanna

Guanfacine artifact (-COONH2) TMS
7564
73, 116, 160, 260, M+ 275
C11H15NOCl2Si
275.03000
1685
PS
LM/Q
Antihypertensive

Paraoxon
Parathion-ethyl-M (paraoxon)
1464
81, 99, 109, 149, M+ 275
C10H14NO6P
275.05588
1890
P-I
311-45-5
PS
LM
Insecticide

Xipamide -SO2NH
3088
121, 155, M+ 275
C15H14ClNO2
275.07132
2385
P U+UHYAC UME
#14293-44-8
UME
LM/Q
Diuretic

Tolfenamic acid ME
6095
89, 180, 208, 243, M+ 275
C15H14ClNO2
275.07132
2255
#13710-19-5
PS
LM/Q
Antirheumatic

275

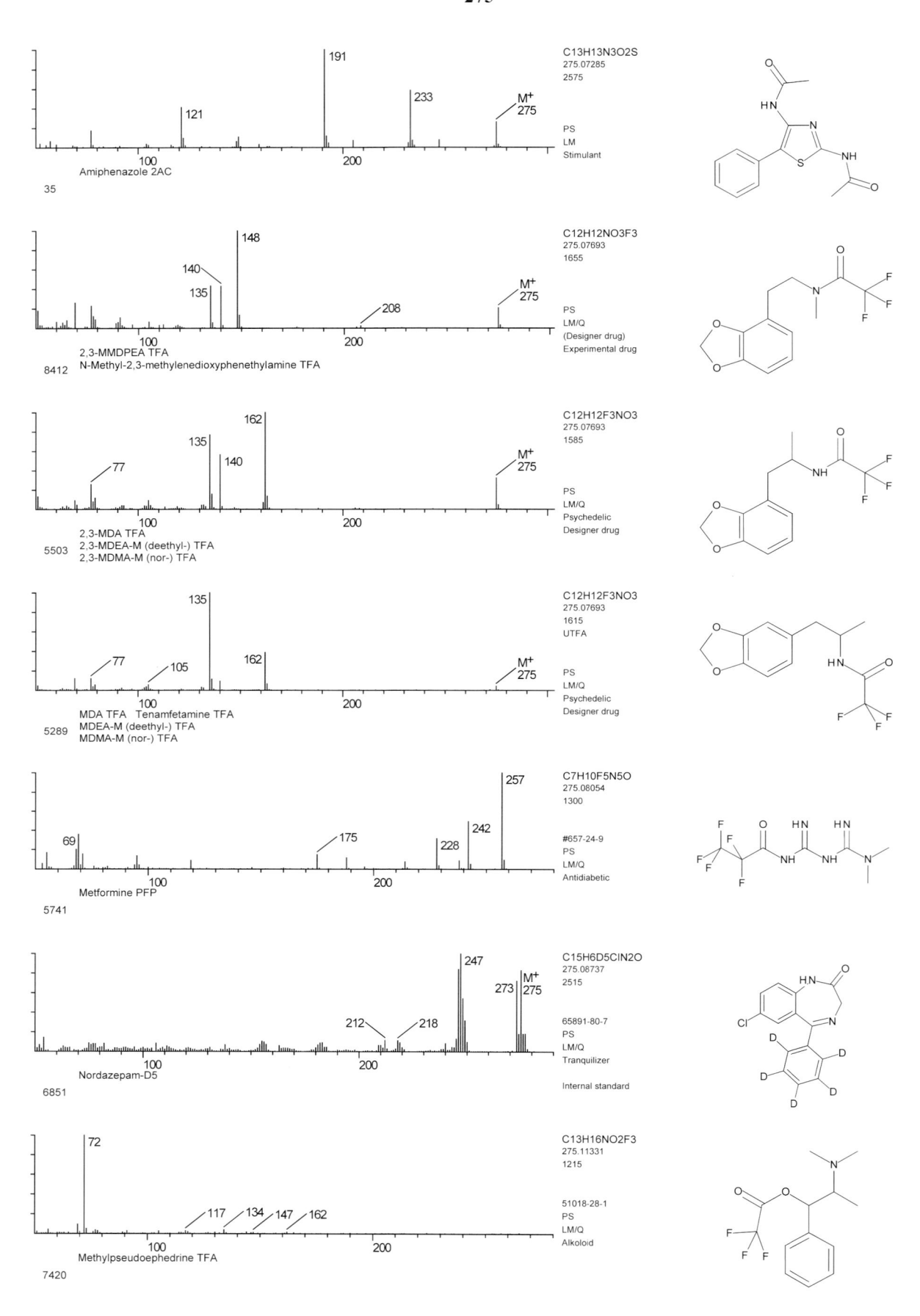

Amiphenazole 2AC — peaks 121, 191, 233, M+ 275
C13H13N3O2S
275.07285
2575
PS
LM
Stimulant
35

2,3-MMDPEA TFA
N-Methyl-2,3-methylenedioxyphenethylamine TFA — peaks 135, 140, 148, 208, M+ 275
C12H12NO3F3
275.07693
1655
PS
LM/Q
(Designer drug)
Experimental drug
8412

2,3-MDA TFA
2,3-MDEA-M (deethyl-) TFA
2,3-MDMA-M (nor-) TFA — peaks 77, 135, 140, 162, M+ 275
C12H12F3NO3
275.07693
1585
PS
LM/Q
Psychedelic
Designer drug
5503

MDA TFA Tenamfetamine TFA
MDEA-M (deethyl-) TFA
MDMA-M (nor-) TFA — peaks 77, 105, 135, 162, M+ 275
C12H12F3NO3
275.07693
1615
UTFA
PS
LM/Q
Psychedelic
Designer drug
5289

Metformine PFP — peaks 69, 175, 228, 242, 257
C7H10F5N5O
275.08054
1300
#657-24-9
PS
LM/Q
Antidiabetic
5741

Nordazepam-D5 — peaks 212, 218, 247, 273, M+ 275
C15H6D5ClN2O
275.08737
2515
65891-80-7
PS
LM/Q
Tranquilizer
Internal standard
6851

Methylpseudoephedrine TFA — peaks 72, 117, 134, 147, 162
C13H16NO2F3
275.11331
1215
51018-28-1
PS
LM/Q
Alkoloid
7420

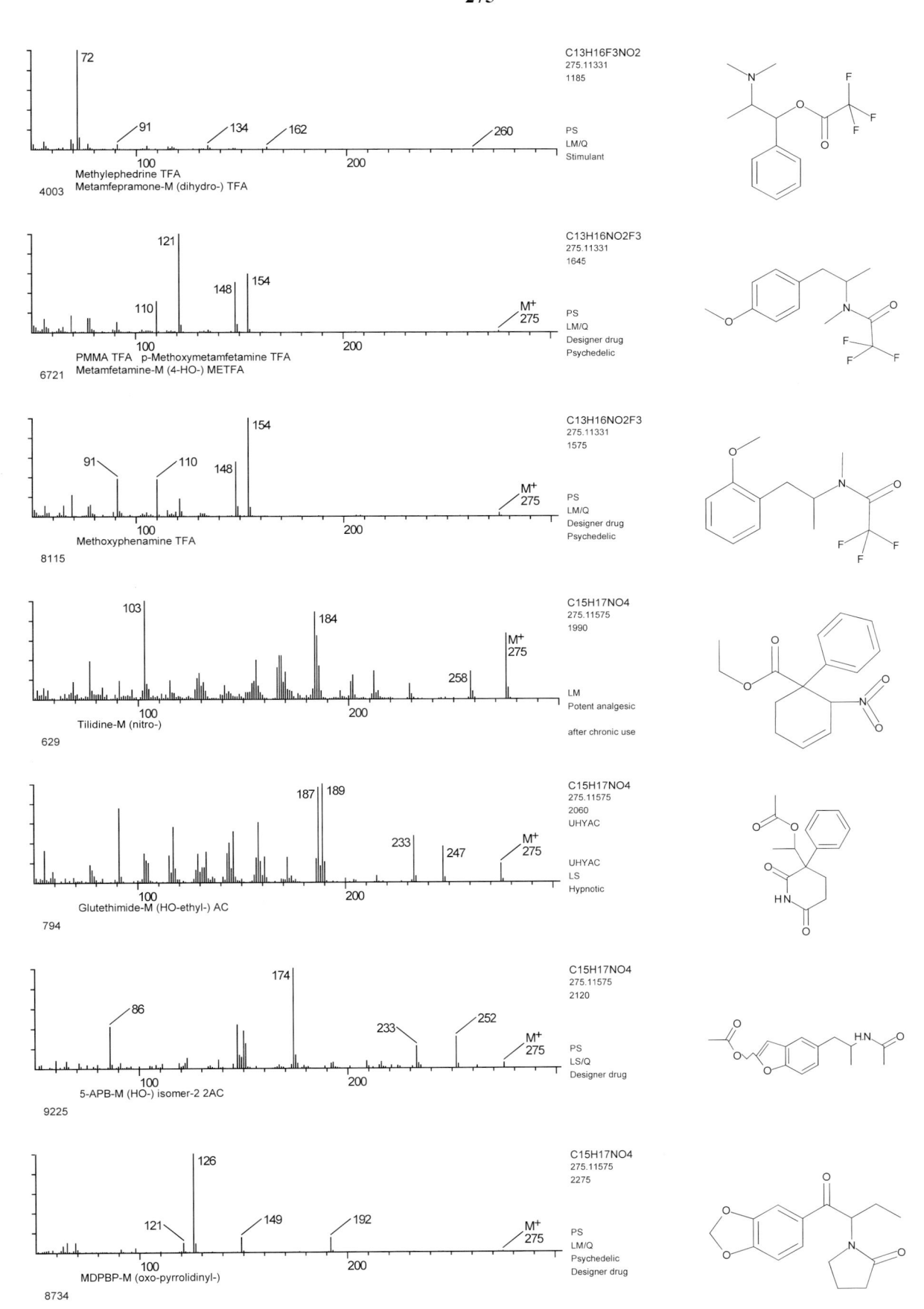

	C13H16F3NO2
	275.11331
	1185
72, 91, 134, 162, 260	
	PS
	LM/Q
	Stimulant
4003 Methylephedrine TFA / Metamfepramone-M (dihydro-) TFA	

	C13H16NO2F3
	275.11331
	1645
110, 121, 148, 154, M+ 275	
	PS
	LM/Q
	Designer drug
	Psychedelic
6721 PMMA TFA p-Methoxymetamfetamine TFA / Metamfetamine-M (4-HO-) METFA	

	C13H16NO2F3
	275.11331
	1575
91, 110, 148, 154, M+ 275	
	PS
	LM/Q
	Designer drug
	Psychedelic
8115 Methoxyphenamine TFA	

	C15H17NO4
	275.11575
	1990
103, 184, 258, M+ 275	
	LM
	Potent analgesic
	after chronic use
629 Tilidine-M (nitro-)	

	C15H17NO4
	275.11575
	2060
187, 189, 233, 247, M+ 275	UHYAC
	UHYAC
	LS
	Hypnotic
794 Glutethimide-M (HO-ethyl-) AC	

	C15H17NO4
	275.11575
	2120
86, 174, 233, 252, M+ 275	
	PS
	LS/Q
	Designer drug
9225 5-APB-M (HO-) isomer-2 2AC	

	C15H17NO4
	275.11575
	2275
121, 126, 149, 192, M+ 275	
	PS
	LM/Q
	Psychedelic
	Designer drug
8734 MDPBP-M (oxo-pyrrolidinyl-)	

275

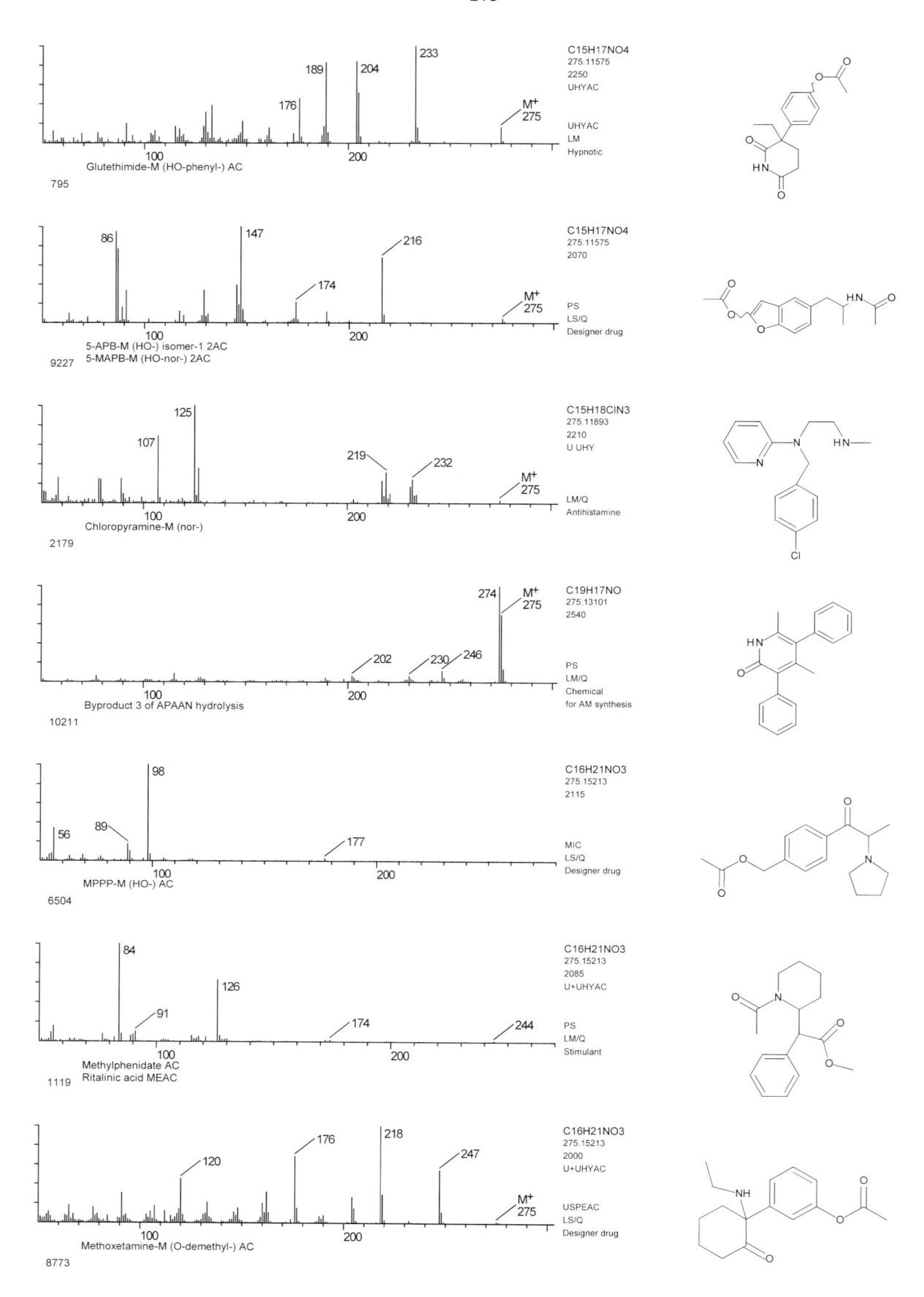

Glutethimide-M (HO-phenyl-) AC Peaks: 176, 189, 204, 233, M+ 275 795	C15H17NO4 275.11575 2250 UHYAC UHYAC LM Hypnotic

5-APB-M (HO-) isomer-1 2AC 5-MAPB-M (HO-nor-) 2AC Peaks: 86, 147, 174, 216, M+ 275 9227	C15H17NO4 275.11575 2070 PS LS/Q Designer drug

Chloropyramine-M (nor-) Peaks: 107, 125, 219, 232, M+ 275 2179	C15H18ClN3 275.11893 2210 U UHY LM/Q Antihistamine

Byproduct 3 of APAAN hydrolysis Peaks: 202, 230, 246, 274, M+ 275 10211	C19H17NO 275.13101 2540 PS LM/Q Chemical for AM synthesis

MPPP-M (HO-) AC Peaks: 56, 89, 98, 177 6504	C16H21NO3 275.15213 2115 MIC LS/Q Designer drug

Methylphenidate AC Ritalinic acid MEAC Peaks: 84, 91, 126, 174, 244 1119	C16H21NO3 275.15213 2085 U+UHYAC PS LM/Q Stimulant

Methoxetamine-M (O-demethyl-) AC Peaks: 120, 176, 218, 247, M+ 275 8773	C16H21NO3 275.15213 2000 U+UHYAC USPEAC LS/Q Designer drug

275

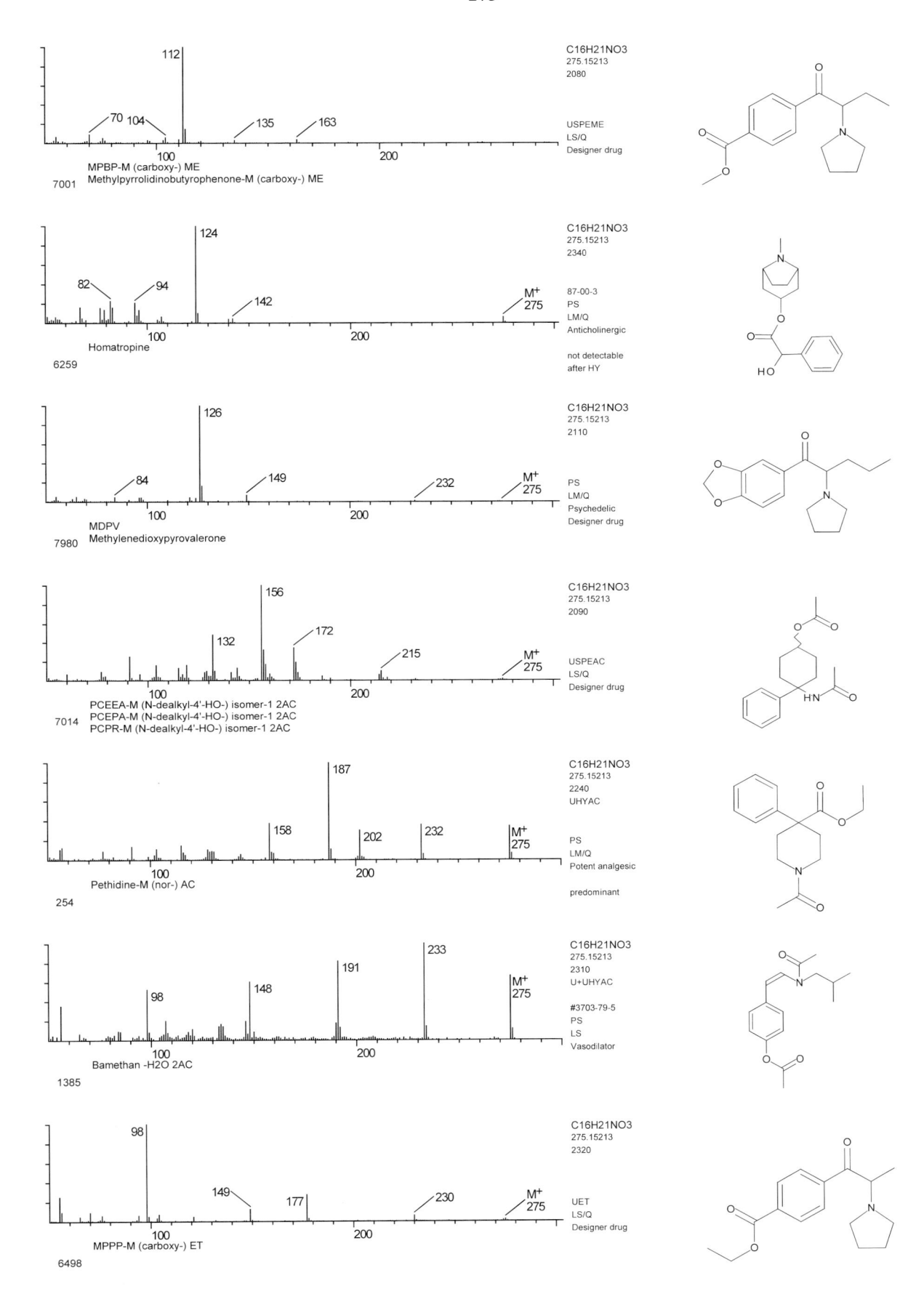

MPBP-M (carboxy-) ME 7001 Methylpyrrolidinobutyrophenone-M (carboxy-) ME peaks: 70, 104, 112, 135, 163	C16H21NO3 275.15213 2080 USPEME LS/Q Designer drug
Homatropine 6259 peaks: 82, 94, 124, 142, M+ 275	C16H21NO3 275.15213 2340 87-00-3 PS LM/Q Anticholinergic not detectable after HY
MDPV 7980 Methylenedioxypyrovalerone peaks: 84, 126, 149, 232, M+ 275	C16H21NO3 275.15213 2110 PS LM/Q Psychedelic Designer drug
PCEEA-M (N-dealkyl-4'-HO-) isomer-1 2AC PCEPA-M (N-dealkyl-4'-HO-) isomer-1 2AC 7014 PCPR-M (N-dealkyl-4'-HO-) isomer-1 2AC peaks: 132, 156, 172, 215, M+ 275	C16H21NO3 275.15213 2090 USPEAC LS/Q Designer drug
Pethidine-M (nor-) AC 254 peaks: 158, 187, 202, 232, M+ 275	C16H21NO3 275.15213 2240 UHYAC PS LM/Q Potent analgesic predominant
Bamethan -H2O 2AC 1385 peaks: 98, 148, 191, 233, M+ 275	C16H21NO3 275.15213 2310 U+UHYAC #3703-79-5 PS LS Vasodilator
MPPP-M (carboxy-) ET 6498 peaks: 98, 149, 177, 230, M+ 275	C16H21NO3 275.15213 2320 UET LS/Q Designer drug

747

275

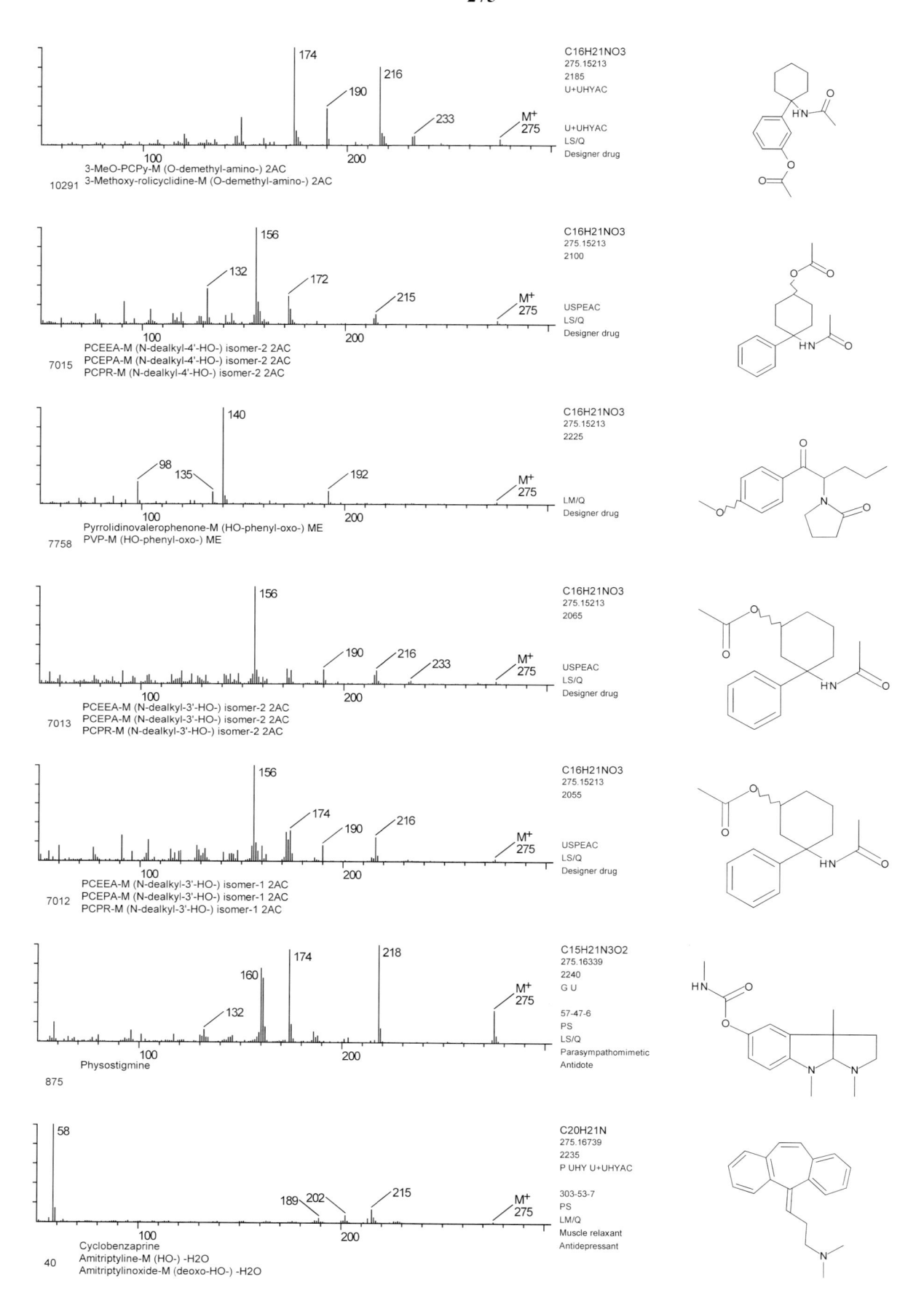

Spectrum	Compound	Formula / Data
10291	3-MeO-PCPy-M (O-demethyl-amino-) 2AC 3-Methoxy-rolicyclidine-M (O-demethyl-amino-) 2AC Peaks: 174, 190, 216, 233, M+ 275	C16H21NO3 275.15213 2185 U+UHYAC U+UHYAC LS/Q Designer drug
7015	PCEEA-M (N-dealkyl-4'-HO-) isomer-2 2AC PCEPA-M (N-dealkyl-4'-HO-) isomer-2 2AC PCPR-M (N-dealkyl-4'-HO-) isomer-2 2AC Peaks: 132, 156, 172, 215, M+ 275	C16H21NO3 275.15213 2100 USPEAC LS/Q Designer drug
7758	Pyrrolidinovalerophenone-M (HO-phenyl-oxo-) ME PVP-M (HO-phenyl-oxo-) ME Peaks: 98, 135, 140, 192, M+ 275	C16H21NO3 275.15213 2225 LM/Q Designer drug
7013	PCEEA-M (N-dealkyl-3'-HO-) isomer-2 2AC PCEPA-M (N-dealkyl-3'-HO-) isomer-2 2AC PCPR-M (N-dealkyl-3'-HO-) isomer-2 2AC Peaks: 156, 190, 216, 233, M+ 275	C16H21NO3 275.15213 2065 USPEAC LS/Q Designer drug
7012	PCEEA-M (N-dealkyl-3'-HO-) isomer-1 2AC PCEPA-M (N-dealkyl-3'-HO-) isomer-1 2AC PCPR-M (N-dealkyl-3'-HO-) isomer-1 2AC Peaks: 156, 174, 190, 216, M+ 275	C16H21NO3 275.15213 2055 USPEAC LS/Q Designer drug
875	Physostigmine Peaks: 132, 160, 174, 218, M+ 275	C15H21N3O2 275.16339 2240 G U 57-47-6 PS LS/Q Parasympathomimetic Antidote
40	Cyclobenzaprine Amitriptyline-M (HO-) -H2O Amitriptylinoxide-M (deoxo-HO-) -H2O Peaks: 58, 189, 202, 215, M+ 275	C20H21N 275.16739 2235 P UHY U+UHYAC 303-53-7 PS LM/Q Muscle relaxant Antidepressant

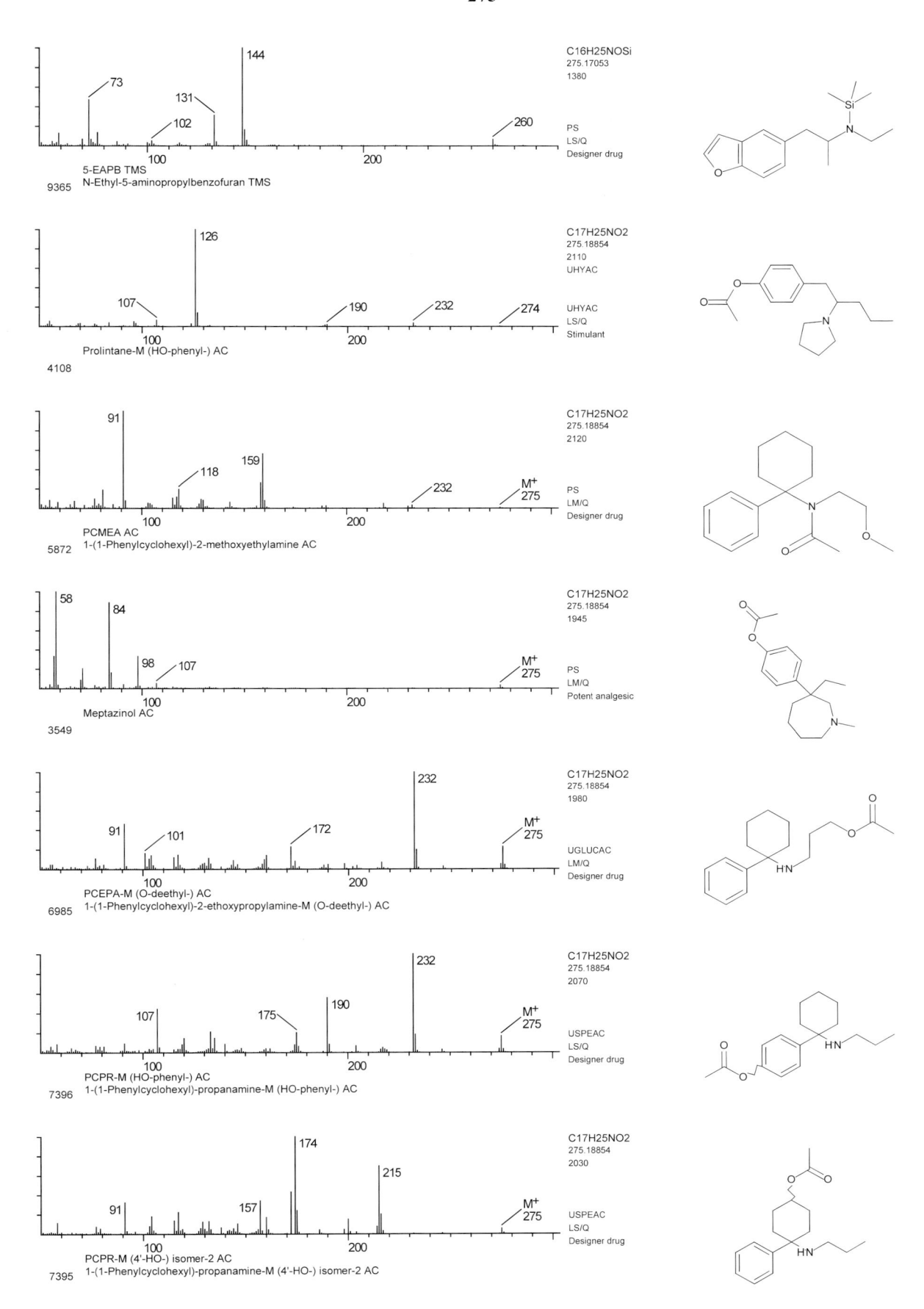

73, 102, 131, 144, 260	C16H25NOSi 275.17053 1380 PS LS/Q Designer drug	
9365	5-EAPB TMS N-Ethyl-5-aminopropylbenzofuran TMS	

107, 126, 190, 232, 274	C17H25NO2 275.18854 2110 UHYAC UHYAC LS/Q Stimulant	
4108	Prolintane-M (HO-phenyl-) AC	

91, 118, 159, 232, M+ 275	C17H25NO2 275.18854 2120 PS LM/Q Designer drug	
5872	PCMEA AC 1-(1-Phenylcyclohexyl)-2-methoxyethylamine AC	

58, 84, 98, 107, M+ 275	C17H25NO2 275.18854 1945 PS LM/Q Potent analgesic	
3549	Meptazinol AC	

91, 101, 172, 232, M+ 275	C17H25NO2 275.18854 1980 UGLUCAC LM/Q Designer drug	
6985	PCEPA-M (O-deethyl-) AC 1-(1-Phenylcyclohexyl)-2-ethoxypropylamine-M (O-deethyl-) AC	

107, 175, 190, 232, M+ 275	C17H25NO2 275.18854 2070 USPEAC LS/Q Designer drug	
7396	PCPR-M (HO-phenyl-) AC 1-(1-Phenylcyclohexyl)-propanamine-M (HO-phenyl-) AC	

91, 157, 174, 215, M+ 275	C17H25NO2 275.18854 2030 USPEAC LS/Q Designer drug	
7395	PCPR-M (4'-HO-) isomer-2 AC 1-(1-Phenylcyclohexyl)-propanamine-M (4'-HO-) isomer-2 AC	

275

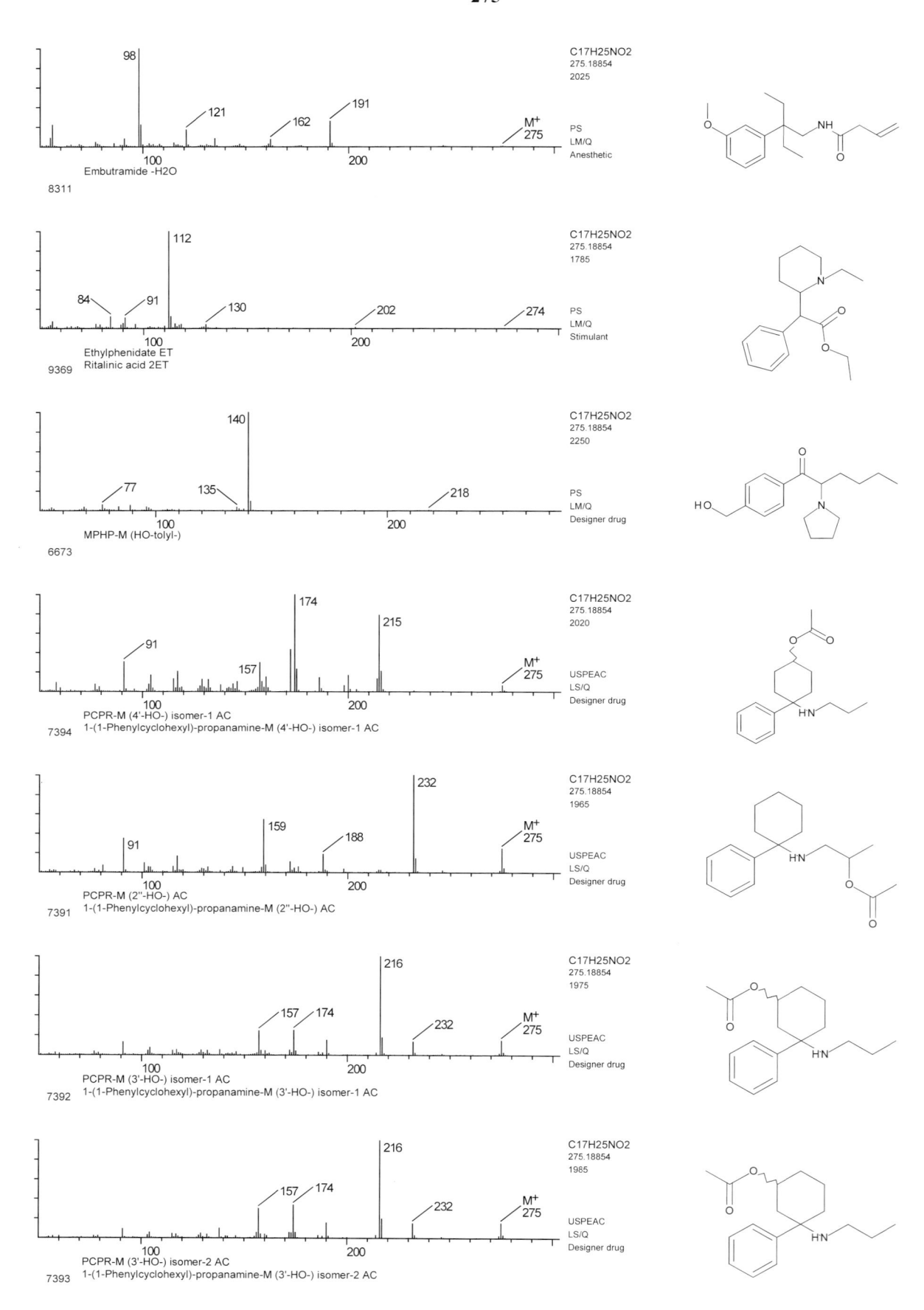

C17H25NO2
275.18854
2025

PS
LM/Q
Anesthetic

Embutramide -H2O
8311

C17H25NO2
275.18854
1785

PS
LM/Q
Stimulant

Ethylphenidate ET
Ritalinic acid 2ET
9369

C17H25NO2
275.18854
2250

PS
LM/Q
Designer drug

MPHP-M (HO-tolyl-)
6673

C17H25NO2
275.18854
2020

USPEAC
LS/Q
Designer drug

PCPR-M (4'-HO-) isomer-1 AC
1-(1-Phenylcyclohexyl)-propanamine-M (4'-HO-) isomer-1 AC
7394

C17H25NO2
275.18854
1965

USPEAC
LS/Q
Designer drug

PCPR-M (2"-HO-) AC
1-(1-Phenylcyclohexyl)-propanamine-M (2"-HO-) AC
7391

C17H25NO2
275.18854
1975

USPEAC
LS/Q
Designer drug

PCPR-M (3'-HO-) isomer-1 AC
1-(1-Phenylcyclohexyl)-propanamine-M (3'-HO-) isomer-1 AC
7392

C17H25NO2
275.18854
1965

USPEAC
LS/Q
Designer drug

PCPR-M (3'-HO-) isomer-2 AC
1-(1-Phenylcyclohexyl)-propanamine-M (3'-HO-) isomer-2 AC
7393

275

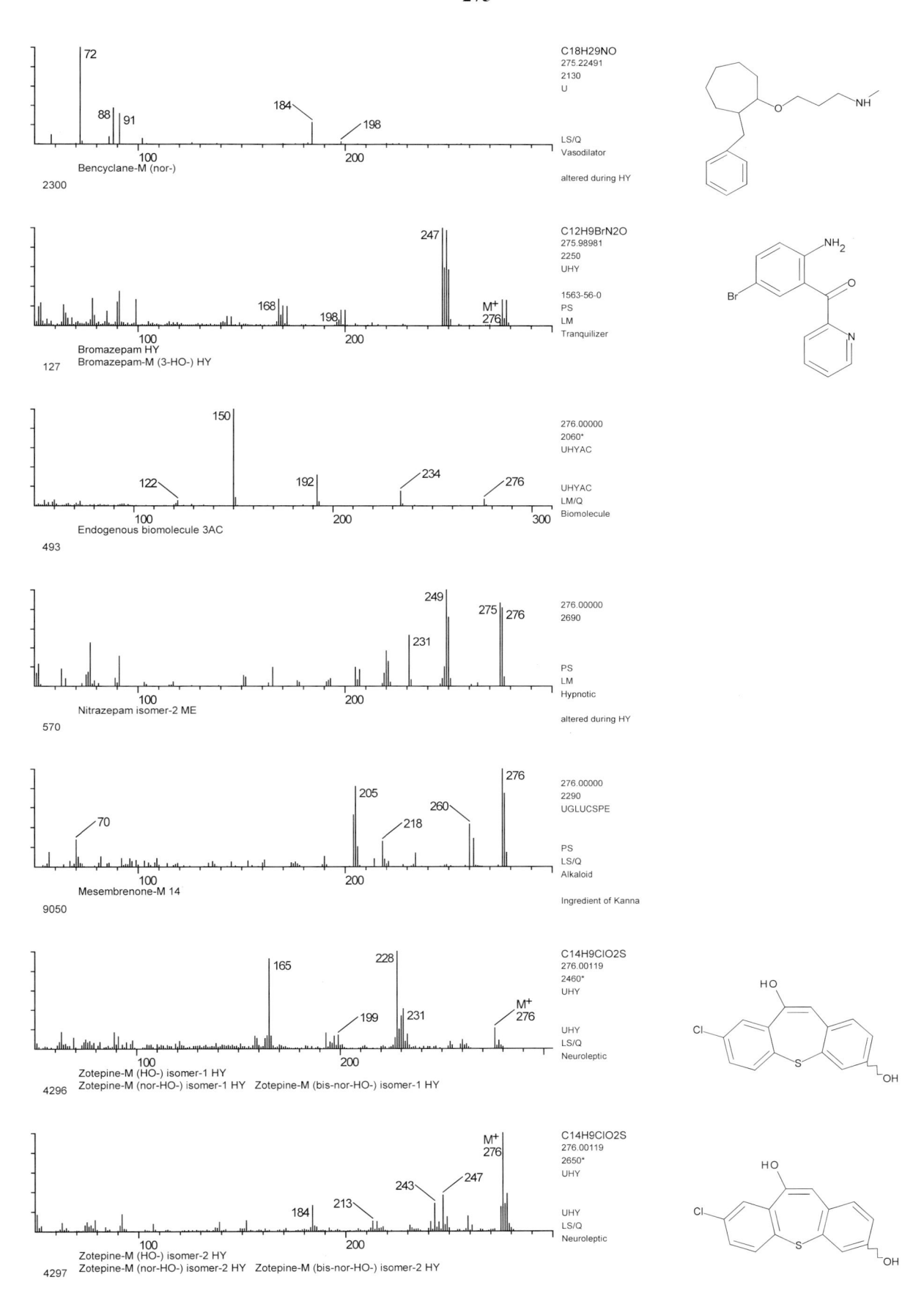

Bencyclane-M (nor-)
2300
72, 88, 91, 184, 198

C18H29NO
275.22491
2130
U

LS/Q
Vasodilator

altered during HY

Bromazepam HY
Bromazepam-M (3-HO-) HY
127
168, 198, 247, M+ 276

C12H9BrN2O
275.98981
2250
UHY

1563-56-0
PS
LM
Tranquilizer

Endogenous biomolecule 3AC
493
122, 150, 192, 234, 276

276.00000
2060*
UHYAC

UHYAC
LM/Q
Biomolecule

Nitrazepam isomer-2 ME
570
231, 249, 275, 276

276.00000
2690

PS
LM
Hypnotic

altered during HY

Mesembrenone-M 14
9050
70, 205, 218, 260, 276

276.00000
2290
UGLUCSPE

PS
LS/Q
Alkaloid

Ingredient of Kanna

Zotepine-M (HO-) isomer-1 HY
Zotepine-M (nor-HO-) isomer-1 HY Zotepine-M (bis-nor-HO-) isomer-1 HY
4296
165, 199, 228, 231, M+ 276

C14H9ClO2S
276.00119
2460*
UHY

UHY
LS/Q
Neuroleptic

Zotepine-M (HO-) isomer-2 HY
Zotepine-M (nor-HO-) isomer-2 HY Zotepine-M (bis-nor-HO-) isomer-2 HY
4297
184, 213, 243, 247, M+ 276

C14H9ClO2S
276.00119
2650*
UHY

UHY
LS/Q
Neuroleptic

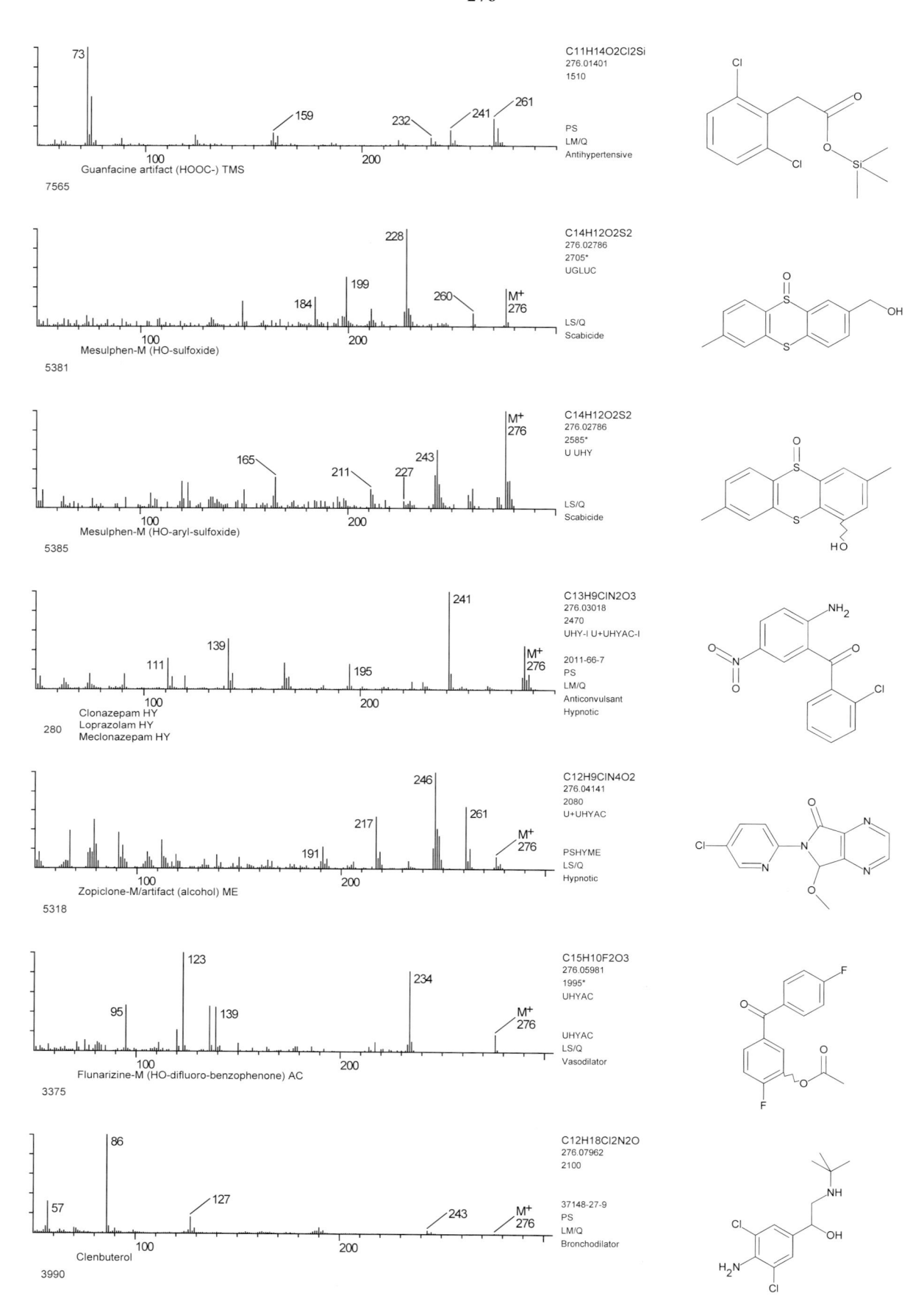

7565 Guanfacine artifact (HOOC-) TMS	C11H14O2Cl2Si 276.01401 1510 PS LM/Q Antihypertensive
5381 Mesulphen-M (HO-sulfoxide)	C14H12O2S2 276.02786 2705* UGLUC LS/Q Scabicide
5385 Mesulphen-M (HO-aryl-sulfoxide)	C14H12O2S2 276.02786 2585* U UHY LS/Q Scabicide
280 Clonazepam HY / Loprazolam HY / Meclonazepam HY	C13H9ClN2O3 276.03018 2470 UHY-I U+UHYAC-I 2011-66-7 PS LM/Q Anticonvulsant Hypnotic
5318 Zopiclone-M/artifact (alcohol) ME	C12H9ClN4O2 276.04141 2080 U+UHYAC PSHYME LS/Q Hypnotic
3375 Flunarizine-M (HO-difluoro-benzophenone) AC	C15H10F2O3 276.05981 1995* UHYAC UHYAC LS/Q Vasodilator
3990 Clenbuterol	C12H18Cl2N2O 276.07962 2100 37148-27-9 PS LM/Q Bronchodilator

276

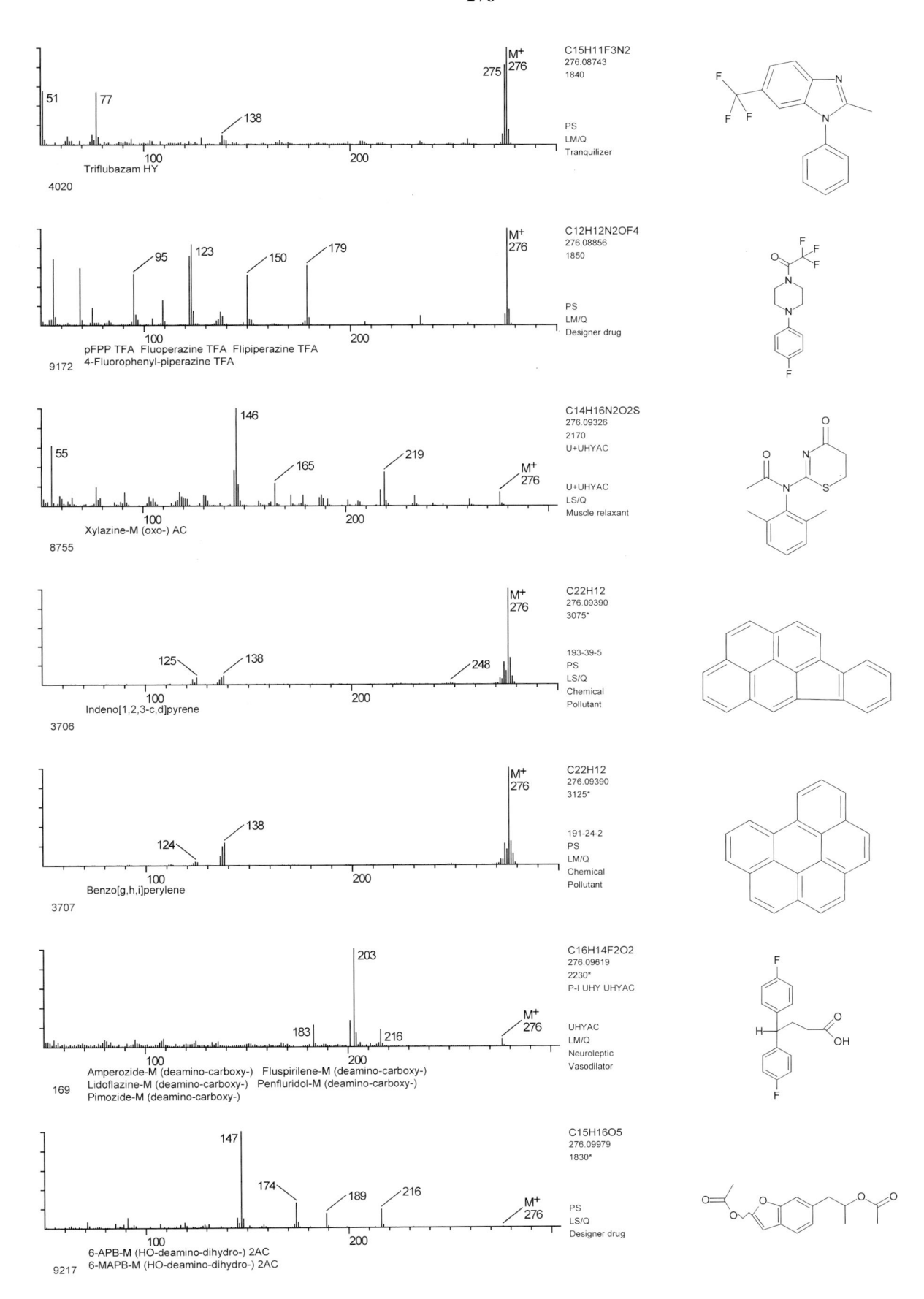

m/z peaks	Formula / Info
51, 77, 138, 275, 276 (M+)	C15H11F3N2, 276.08743, 1840, PS, LM/Q, Tranquilizer, Triflubazam HY, 4020
95, 123, 150, 179, 276 (M+)	C12H12N2OF4, 276.08856, 1850, PS, LM/Q, Designer drug, pFPP TFA Fluoperazine TFA Flipiperazine TFA, 4-Fluorophenyl-piperazine TFA, 9172
55, 146, 165, 219, 276 (M+)	C14H16N2O2S, 276.09326, 2170, U+UHYAC, U+UHYAC, LS/Q, Muscle relaxant, Xylazine-M (oxo-) AC, 8755
125, 138, 248, 276 (M+)	C22H12, 276.09390, 3075*, 193-39-5, PS, LS/Q, Chemical Pollutant, Indeno[1,2,3-c,d]pyrene, 3706
124, 138, 276 (M+)	C22H12, 276.09390, 3125*, 191-24-2, PS, LM/Q, Chemical Pollutant, Benzo[g,h,i]perylene, 3707
183, 203, 216, 276 (M+)	C16H14F2O2, 276.09619, 2230*, P-I UHY UHYAC, UHYAC, LM/Q, Neuroleptic Vasodilator, Amperozide-M (deamino-carboxy-) Fluspirilene-M (deamino-carboxy-), Lidoflazine-M (deamino-carboxy-) Penfluridol-M (deamino-carboxy-), Pimozide-M (deamino-carboxy-), 169
147, 174, 189, 216, 276 (M+)	C15H16O5, 276.09979, 1830*, PS, LS/Q, Designer drug, 6-APB-M (HO-deamino-dihydro-) 2AC, 6-MAPB-M (HO-deamino-dihydro-) 2AC, 9217

276

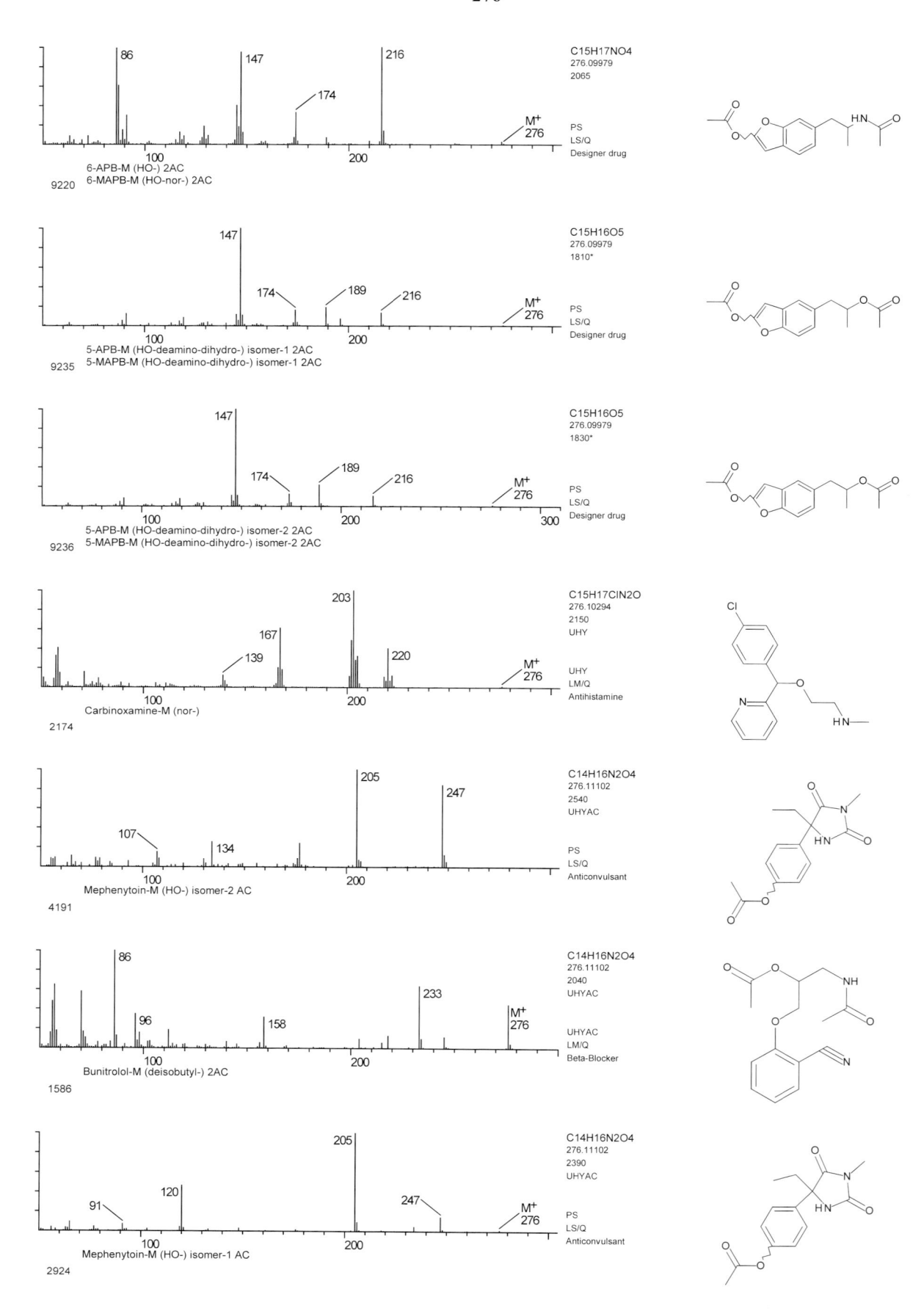

9220	C15H17NO4 / 276.09979 / 2065 / PS / LS/Q / Designer drug 6-APB-M (HO-) 2AC 6-MAPB-M (HO-nor-) 2AC Peaks: 86, 147, 174, 216, M+ 276
9235	C15H16O5 / 276.09979 / 1810* / PS / LS/Q / Designer drug 5-APB-M (HO-deamino-dihydro-) isomer-1 2AC 5-MAPB-M (HO-deamino-dihydro-) isomer-1 2AC Peaks: 147, 174, 189, 216, M+ 276
9236	C15H16O5 / 276.09979 / 1830* / PS / LS/Q / Designer drug 5-APB-M (HO-deamino-dihydro-) isomer-2 2AC 5-MAPB-M (HO-deamino-dihydro-) isomer-2 2AC Peaks: 147, 174, 189, 216, M+ 276
2174	C15H17ClN2O / 276.10294 / 2150 / UHY / UHY / LM/Q / Antihistamine Carbinoxamine-M (nor-) Peaks: 139, 167, 203, 220, M+ 276
4191	C14H16N2O4 / 276.11102 / 2540 / UHYAC / PS / LS/Q / Anticonvulsant Mephenytoin-M (HO-) isomer-2 AC Peaks: 107, 134, 205, 247
1586	C14H16N2O4 / 276.11102 / 2040 / UHYAC / UHYAC / LM/Q / Beta-Blocker Bunitrolol-M (deisobutyl-) 2AC Peaks: 86, 96, 158, 233, M+ 276
2924	C14H16N2O4 / 276.11102 / 2390 / UHYAC / PS / LS/Q / Anticonvulsant Mephenytoin-M (HO-) isomer-1 AC Peaks: 91, 120, 205, 247, M+ 276

276

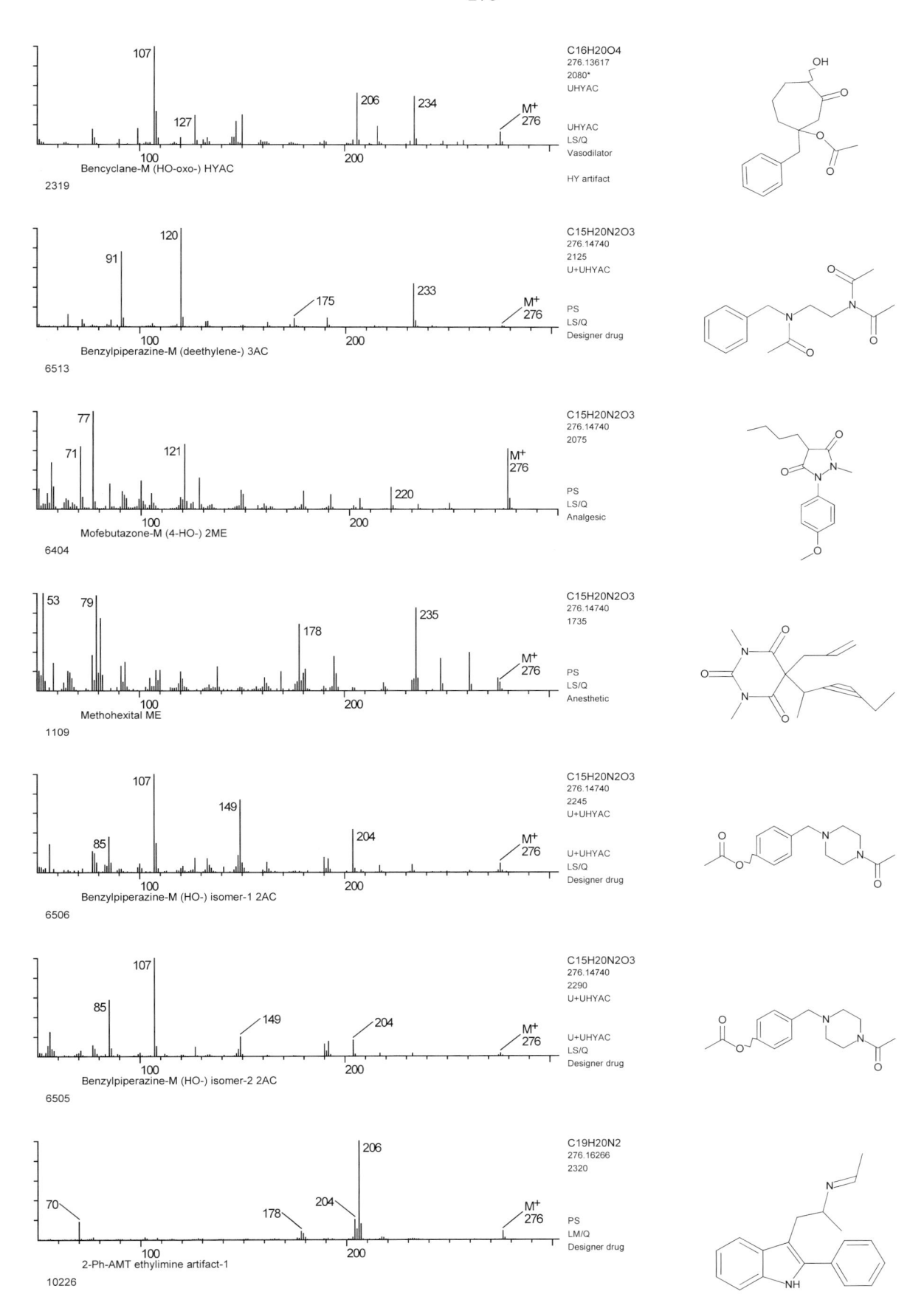

Bencyclane-M (HO-oxo-) HYAC
2319
Peaks: 107, 127, 206, 234, M+ 276
C16H20O4
276.13617
2080*
UHYAC

UHYAC
LS/Q
Vasodilator
HY artifact

Benzylpiperazine-M (deethylene-) 3AC
6513
Peaks: 91, 120, 175, 233, M+ 276
C15H20N2O3
276.14740
2125
U+UHYAC

PS
LS/Q
Designer drug

Mofebutazone-M (4-HO-) 2ME
6404
Peaks: 71, 77, 121, 220, M+ 276
C15H20N2O3
276.14740
2075

PS
LS/Q
Analgesic

Methohexital ME
1109
Peaks: 53, 79, 178, 235, M+ 276
C15H20N2O3
276.14740
1735

PS
LS/Q
Anesthetic

Benzylpiperazine-M (HO-) isomer-1 2AC
6506
Peaks: 85, 107, 149, 204, M+ 276
C15H20N2O3
276.14740
2245
U+UHYAC

U+UHYAC
LS/Q
Designer drug

Benzylpiperazine-M (HO-) isomer-2 2AC
6505
Peaks: 85, 107, 149, 204, M+ 276
C15H20N2O3
276.14740
2290
U+UHYAC

U+UHYAC
LS/Q
Designer drug

2-Ph-AMT ethylimine artifact-1
10226
Peaks: 70, 178, 204, 206, M+ 276
C19H20N2
276.16266
2320

PS
LM/Q
Designer drug

276

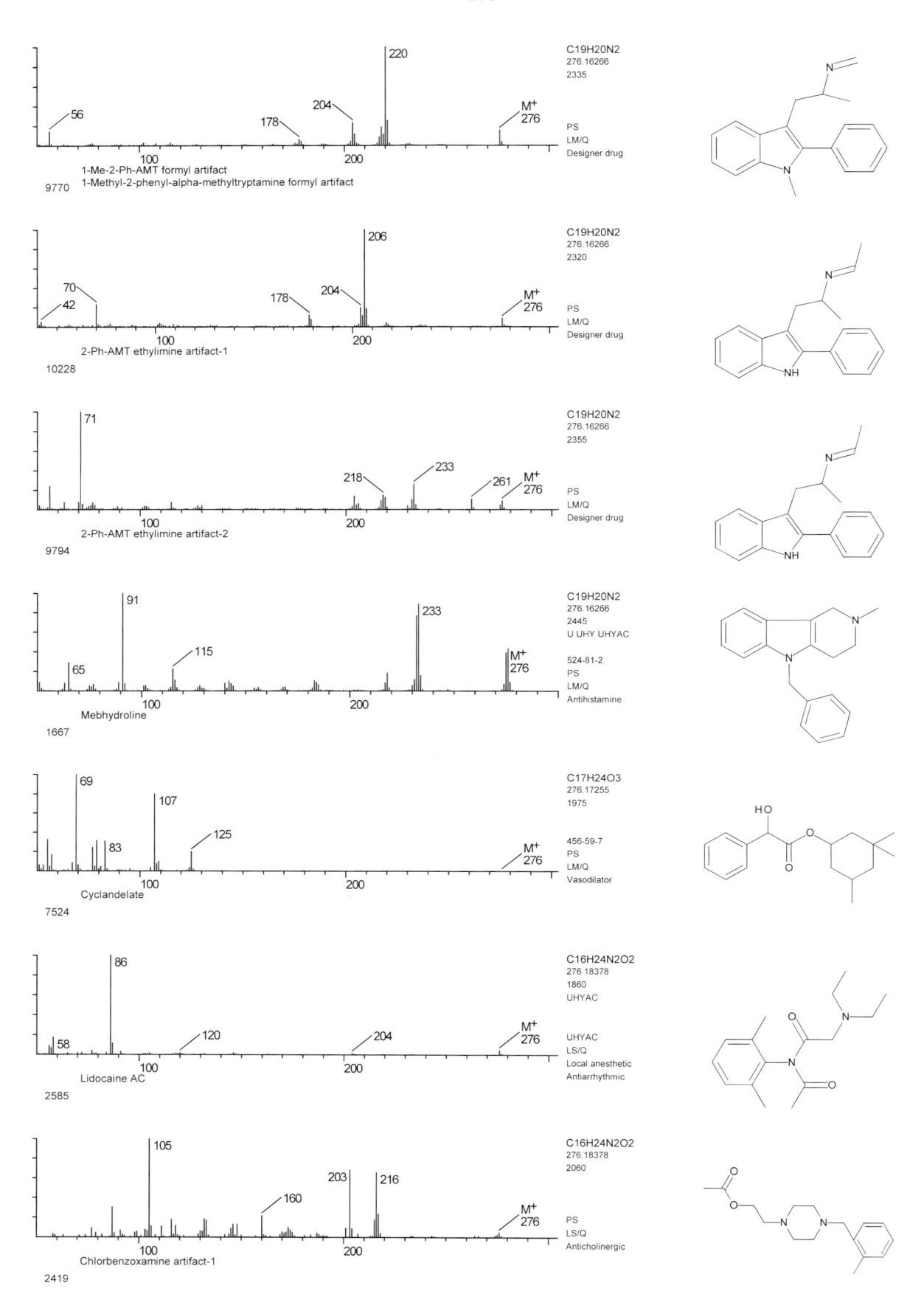

1-Me-2-Ph-AMT formyl artifact 1-Methyl-2-phenyl-alpha-methyltryptamine formyl artifact 9770	C19H20N2 276.16266 2335 PS LM/Q Designer drug	
2-Ph-AMT ethylimine artifact-1 10228	C19H20N2 276.16266 2320 PS LM/Q Designer drug	
2-Ph-AMT ethylimine artifact-2 9794	C19H20N2 276.16266 2355 PS LM/Q Designer drug	
Mebhydroline 1667	C19H20N2 276.16266 2445 U UHY UHYAC 524-81-2 PS LM/Q Antihistamine	
Cyclandelate 7524	C17H24O3 276.17255 1975 456-59-7 PS LM/Q Vasodilator	
Lidocaine AC 2585	C16H24N2O2 276.18378 1860 UHYAC UHYAC LS/Q Local anesthetic Antiarrhythmic	
Chlorbenzoxamine artifact-1 2419	C16H24N2O2 276.18378 2060 PS LS/Q Anticholinergic	

276

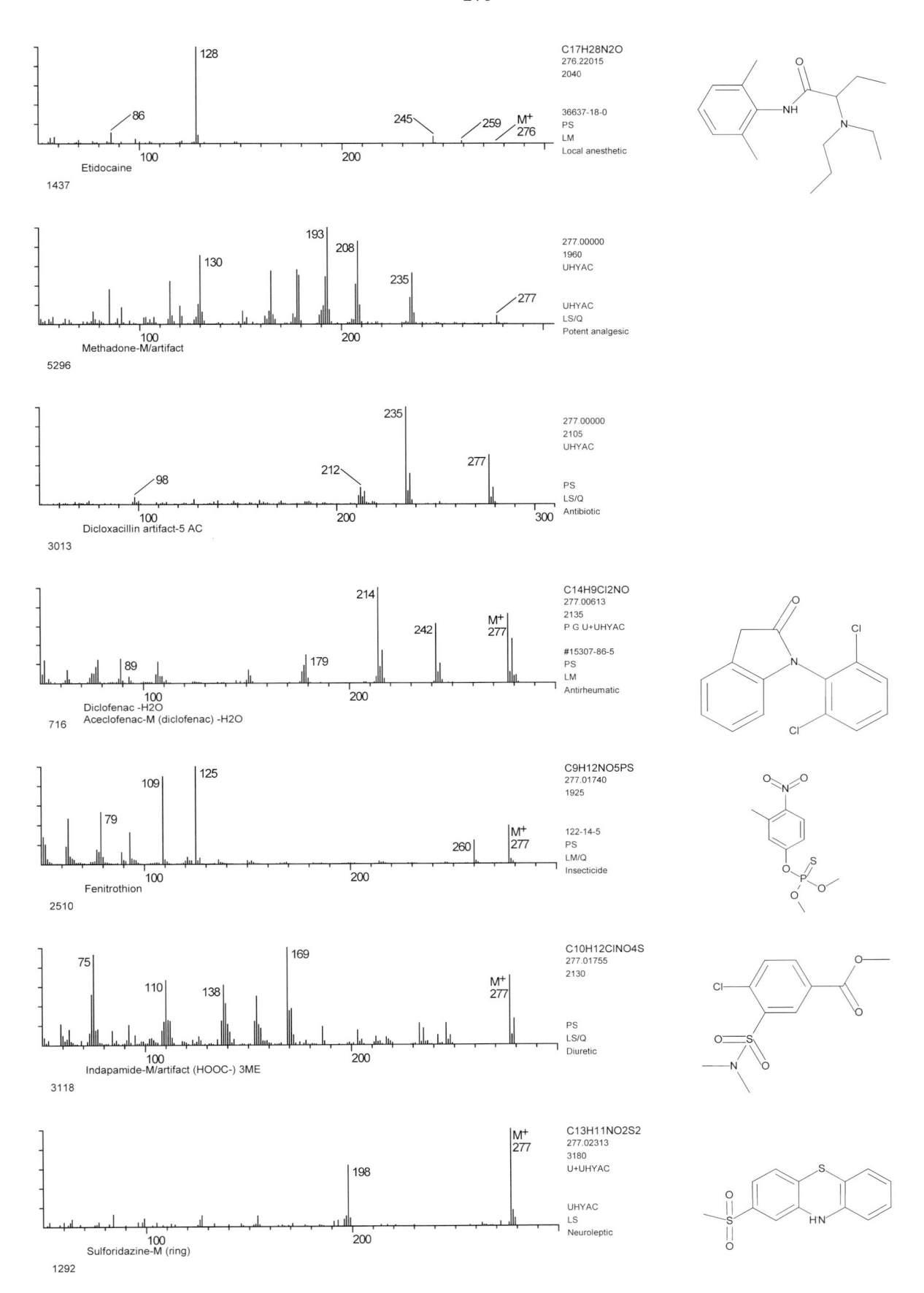

Etidocaine — peaks: 86, 128, 245, 259, M+ 276	C17H28N2O 276.22015 2040 36637-18-0 PS LM Local anesthetic
1437	

Methadone-M/artifact — peaks: 130, 193, 208, 235, 277	277.00000 1960 UHYAC UHYAC LS/Q Potent analgesic
5296	

Dicloxacillin artifact-5 AC — peaks: 98, 212, 235, 277	277.00000 2105 UHYAC PS LS/Q Antibiotic
3013	

Diclofenac -H2O / Aceclofenac-M (diclofenac) -H2O — peaks: 89, 179, 214, 242, M+ 277	C14H9Cl2NO 277.00613 2135 P G U+UHYAC #15307-86-5 PS LM Antirheumatic
716	

Fenitrothion — peaks: 79, 109, 125, 260, M+ 277	C9H12NO5PS 277.01740 1925 122-14-5 PS LM/Q Insecticide
2510	

Indapamide-M/artifact (HOOC-) 3ME — peaks: 75, 110, 138, 169, M+ 277	C10H12ClNO4S 277.01755 2130 PS LS/Q Diuretic
3118	

Sulforidazine-M (ring) — peaks: 198, M+ 277	C13H11NO2S2 277.02313 3180 U+UHYAC UHYAC LS Neuroleptic
1292	

277

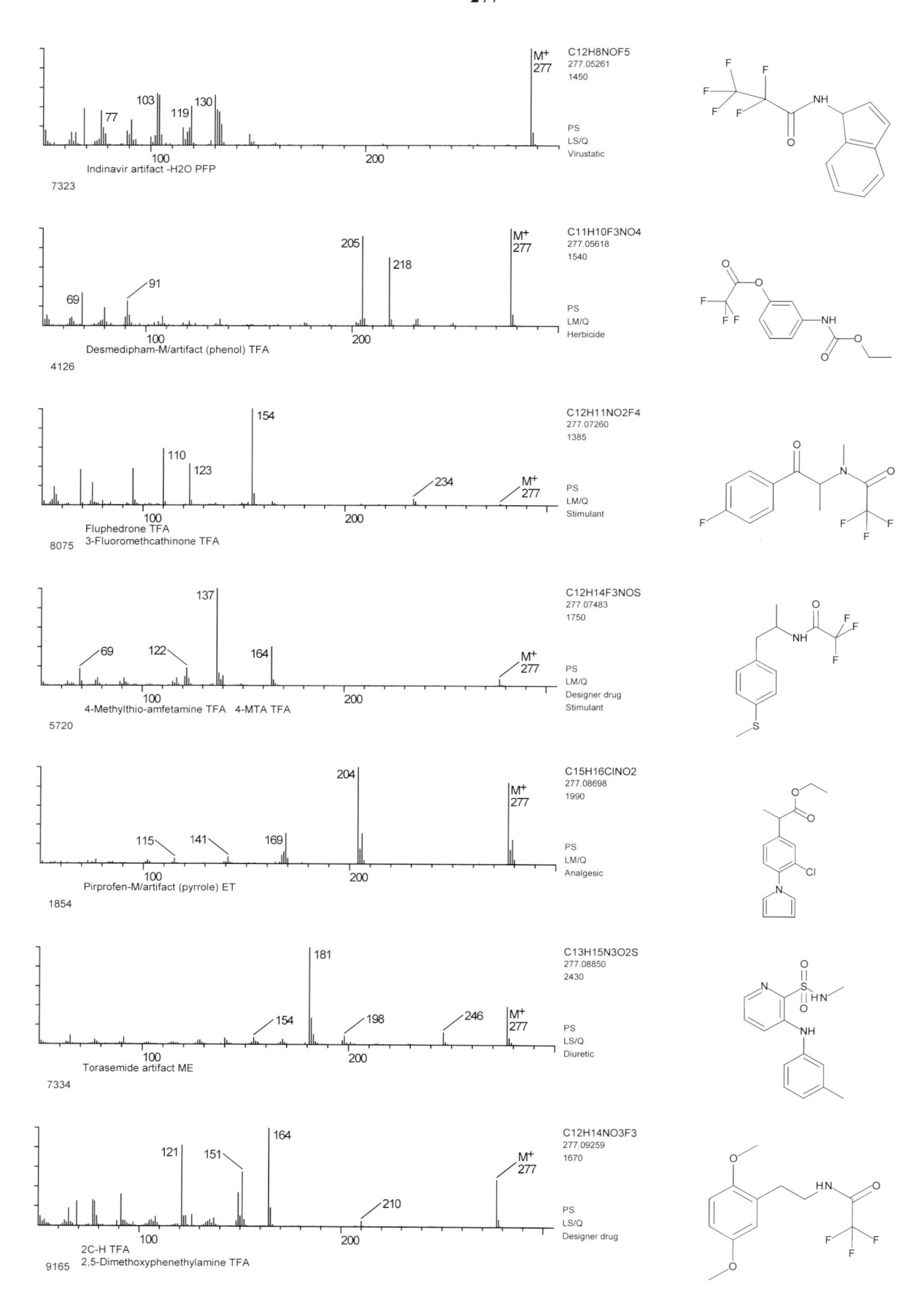

M+ 277	C12H8NOF5 277.05261 1450 PS LS/Q Virustatic	Indinavir artifact -H2O PFP (7323) peaks: 77, 103, 119, 130
M+ 277	C11H10F3NO4 277.05618 1540 PS LM/Q Herbicide	Desmedipham-M/artifact (phenol) TFA (4126) peaks: 69, 91, 205, 218
M+ 277	C12H11NO2F4 277.07260 1385 PS LM/Q Stimulant	Fluphedrone TFA / 3-Fluoromethcathinone TFA (8075) peaks: 110, 123, 154, 234
M+ 277	C12H14F3NOS 277.07483 1750 PS LM/Q Designer drug Stimulant	4-Methylthio-amfetamine TFA 4-MTA TFA (5720) peaks: 69, 122, 137, 164
M+ 277	C15H16ClNO2 277.08698 1990 PS LM/Q Analgesic	Pirprofen-M/artifact (pyrrole) ET (1854) peaks: 115, 141, 169, 204
M+ 277	C13H15N3O2S 277.08850 2430 PS LS/Q Diuretic	Torasemide artifact ME (7334) peaks: 154, 181, 198, 246
M+ 277	C12H14NO3F3 277.09259 1670 PS LS/Q Designer drug	2C-H TFA / 2,5-Dimethoxyphenethylamine TFA (9165) peaks: 121, 151, 164, 210

277

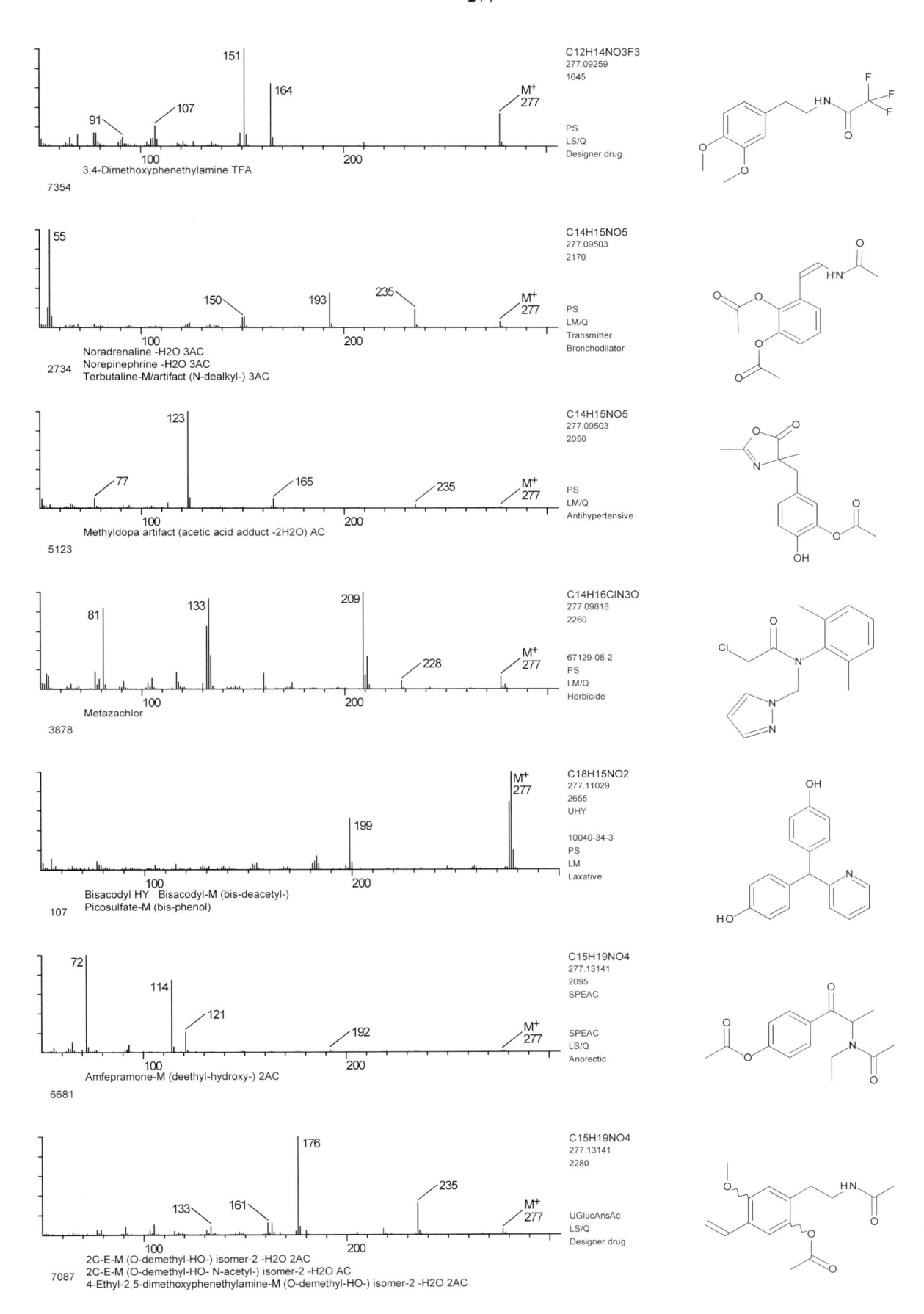

7354
3,4-Dimethoxyphenethylamine TFA
Peaks: 91, 107, 151, 164, M+ 277

C12H14NO3F3
277.09259
1645
PS
LS/Q
Designer drug

2734
Noradrenaline -H2O 3AC
Norepinephrine -H2O 3AC
Terbutaline-M/artifact (N-dealkyl-) 3AC
Peaks: 55, 150, 193, 235, M+ 277

C14H15NO5
277.09503
2170
PS
LM/Q
Transmitter
Bronchodilator

5123
Methyldopa artifact (acetic acid adduct -2H2O) AC
Peaks: 77, 123, 165, 235, M+ 277

C14H15NO5
277.09503
2050
PS
LM/Q
Antihypertensive

3878
Metazachlor
Peaks: 81, 133, 209, 228, M+ 277

C14H16ClN3O
277.09818
2260
67129-08-2
PS
LM/Q
Herbicide

107
Bisacodyl HY   Bisacodyl-M (bis-deacetyl-)
Picosulfate-M (bis-phenol)
Peaks: 199, M+ 277

C18H15NO2
277.11029
2655
UHY
10040-34-3
PS
LM
Laxative

6681
Amfepramone-M (deethyl-hydroxy-) 2AC
Peaks: 72, 114, 121, 192, M+ 277

C15H19NO4
277.13141
2095
SPEAC
SPEAC
LS/Q
Anorectic

7087
2C-E-M (O-demethyl-HO-) isomer-2 -H2O 2AC
2C-E-M (O-demethyl-HO- N-acetyl-) isomer-2 -H2O AC
4-Ethyl-2,5-dimethoxyphenethylamine-M (O-demethyl-HO-) isomer-2 -H2O 2AC
Peaks: 133, 161, 176, 235, M+ 277

C15H19NO4
277.13141
2280
UGlucAnsAc
LS/Q
Designer drug

759

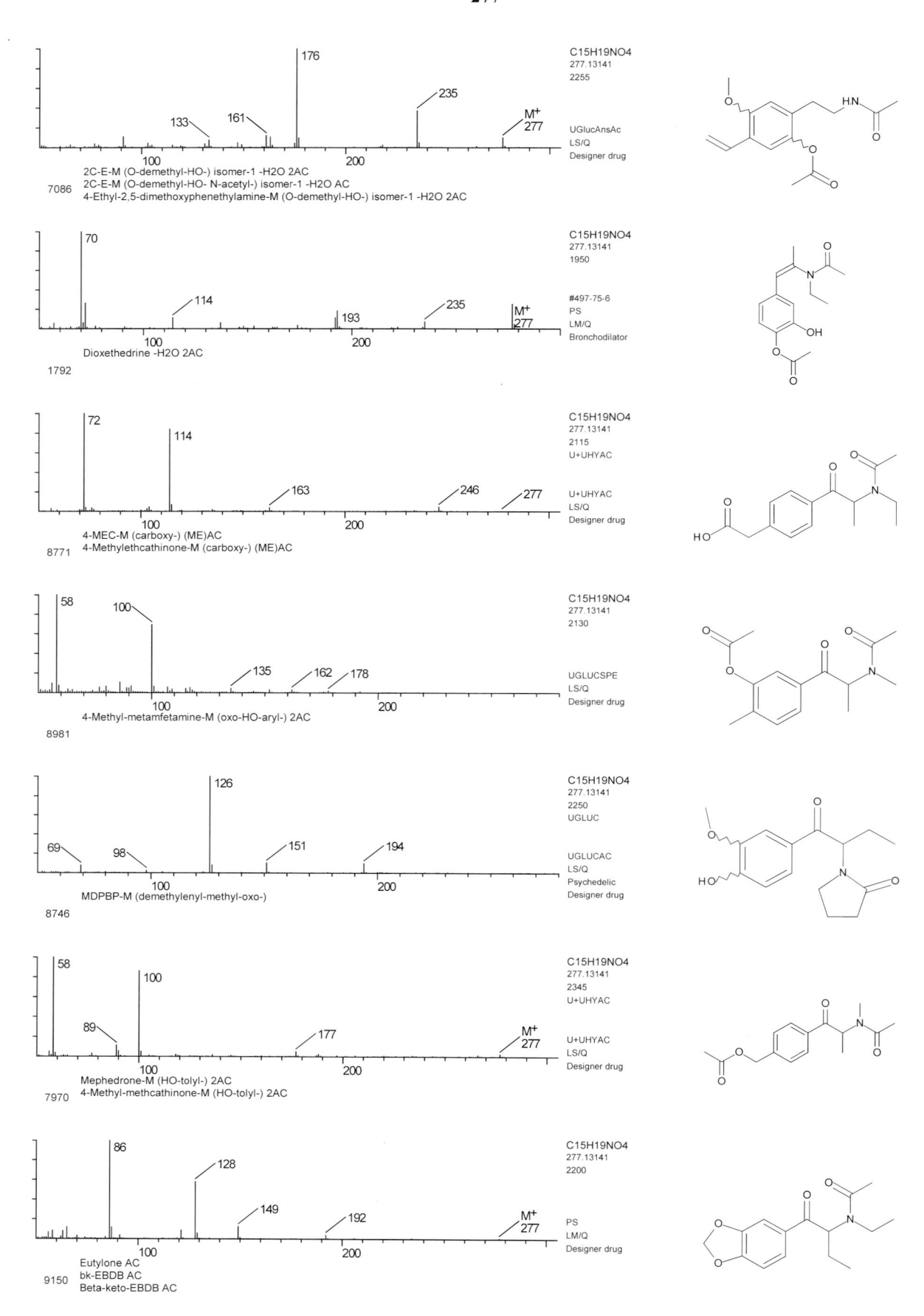

	C15H19NO4
	277.13141
	2255
	UGlucAnsAc
	LS/Q
	Designer drug

2C-E-M (O-demethyl-HO-) isomer-1 -H2O 2AC
2C-E-M (O-demethyl-HO- N-acetyl-) isomer-1 -H2O AC
4-Ethyl-2,5-dimethoxyphenethylamine-M (O-demethyl-HO-) isomer-1 -H2O 2AC

7086

	C15H19NO4
	277.13141
	1950
	#497-75-6
	PS
	LM/Q
	Bronchodilator

Dioxethedrine -H2O 2AC

1792

	C15H19NO4
	277.13141
	2115
	U+UHYAC
	U+UHYAC
	LS/Q
	Designer drug

4-MEC-M (carboxy-) (ME)AC
4-Methylethcathinone-M (carboxy-) (ME)AC

8771

	C15H19NO4
	277.13141
	2130
	UGLUCSPE
	LS/Q
	Designer drug

4-Methyl-metamfetamine-M (oxo-HO-aryl-) 2AC

8981

	C15H19NO4
	277.13141
	2250
	UGLUC
	UGLUCAC
	LS/Q
	Psychedelic
	Designer drug

MDPBP-M (demethylenyl-methyl-oxo-)

8746

	C15H19NO4
	277.13141
	2345
	U+UHYAC
	U+UHYAC
	LS/Q
	Designer drug

Mephedrone-M (HO-tolyl-) 2AC
4-Methyl-methcathinone-M (HO-tolyl-) 2AC

7970

	C15H19NO4
	277.13141
	2200
	PS
	LM/Q
	Designer drug

Eutylone AC
bk-EBDB AC
Beta-keto-EBDB AC

9150

277

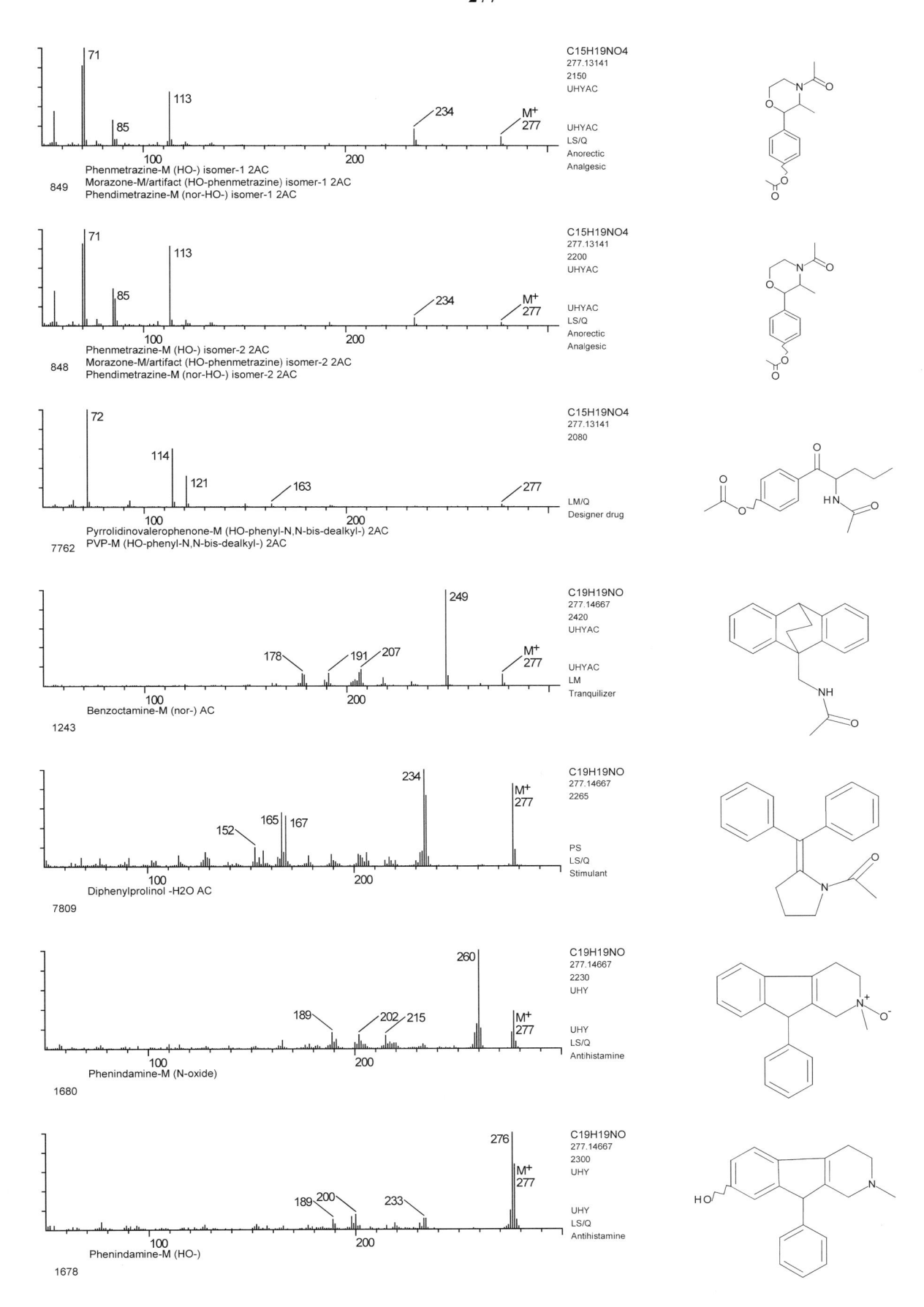

849	Phenmetrazine-M (HO-) isomer-1 2AC Morazone-M/artifact (HO-phenmetrazine) isomer-1 2AC Phendimetrazine-M (nor-HO-) isomer-1 2AC	C15H19NO4 277.13141 2150 UHYAC UHYAC LS/Q Anorectic Analgesic
848	Phenmetrazine-M (HO-) isomer-2 2AC Morazone-M/artifact (HO-phenmetrazine) isomer-2 2AC Phendimetrazine-M (nor-HO-) isomer-2 2AC	C15H19NO4 277.13141 2200 UHYAC UHYAC LS/Q Anorectic Analgesic
7762	Pyrrolidinovalerophenone-M (HO-phenyl-N,N-bis-dealkyl-) 2AC PVP-M (HO-phenyl-N,N-bis-dealkyl-) 2AC	C15H19NO4 277.13141 2080 LM/Q Designer drug
1243	Benzoctamine-M (nor-) AC	C19H19NO 277.14667 2420 UHYAC UHYAC LM Tranquilizer
7809	Diphenylprolinol -H2O AC	C19H19NO 277.14667 2265 PS LS/Q Stimulant
1680	Phenindamine-M (N-oxide)	C19H19NO 277.14667 2230 UHY UHY LS/Q Antihistamine
1678	Phenindamine-M (HO-)	C19H19NO 277.14667 2300 UHY UHY LS/Q Antihistamine

277

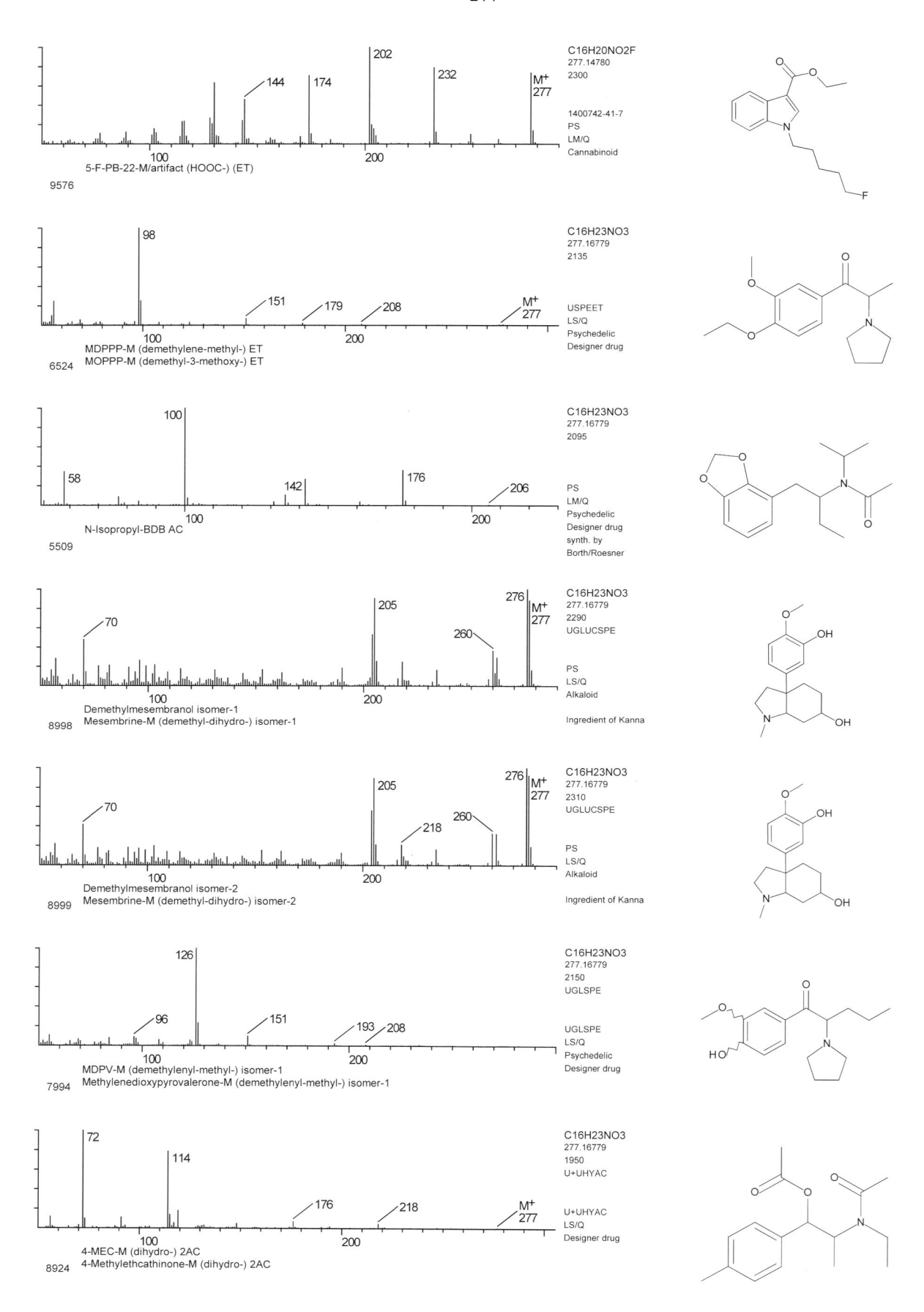

Spectrum	Formula / Info
9576 — 5-F-PB-22-M/artifact (HOOC-) (ET); peaks 144, 174, 202, 232, M+ 277	C16H20NO2F; 277.14780; 2300; 1400742-41-7; PS; LM/Q; Cannabinoid
6524 — MDPPP-M (demethylene-methyl-) ET / MOPPP-M (demethyl-3-methoxy-) ET; peaks 98, 151, 179, 208, M+ 277	C16H23NO3; 277.16779; 2135; USPEET; LS/Q; Psychedelic; Designer drug
5509 — N-Isopropyl-BDB AC; peaks 58, 100, 142, 176, 206	C16H23NO3; 277.16779; 2095; PS; LM/Q; Psychedelic; Designer drug; synth. by Borth/Roesner
8998 — Demethylmesembranol isomer-1 / Mesembrine-M (demethyl-dihydro-) isomer-1; peaks 70, 205, 260, 276, M+ 277	C16H23NO3; 277.16779; 2290; UGLUCSPE; PS; LS/Q; Alkaloid; Ingredient of Kanna
8999 — Demethylmesembranol isomer-2 / Mesembrine-M (demethyl-dihydro-) isomer-2; peaks 70, 205, 218, 260, 276, M+ 277	C16H23NO3; 277.16779; 2310; UGLUCSPE; PS; LS/Q; Alkaloid; Ingredient of Kanna
7994 — MDPV-M (demethylenyl-methyl-) isomer-1 / Methylenedioxypyrovalerone-M (demethylenyl-methyl-) isomer-1; peaks 96, 126, 151, 193, 208	C16H23NO3; 277.16779; 2150; UGLSPE; UGLSPE; LS/Q; Psychedelic; Designer drug
8924 — 4-MEC-M (dihydro-) 2AC / 4-Methylethcathinone-M (dihydro-) 2AC; peaks 72, 114, 176, 218, M+ 277	C16H23NO3; 277.16779; 1950; U+UHYAC; U+UHYAC; LS/Q; Designer drug

762

277

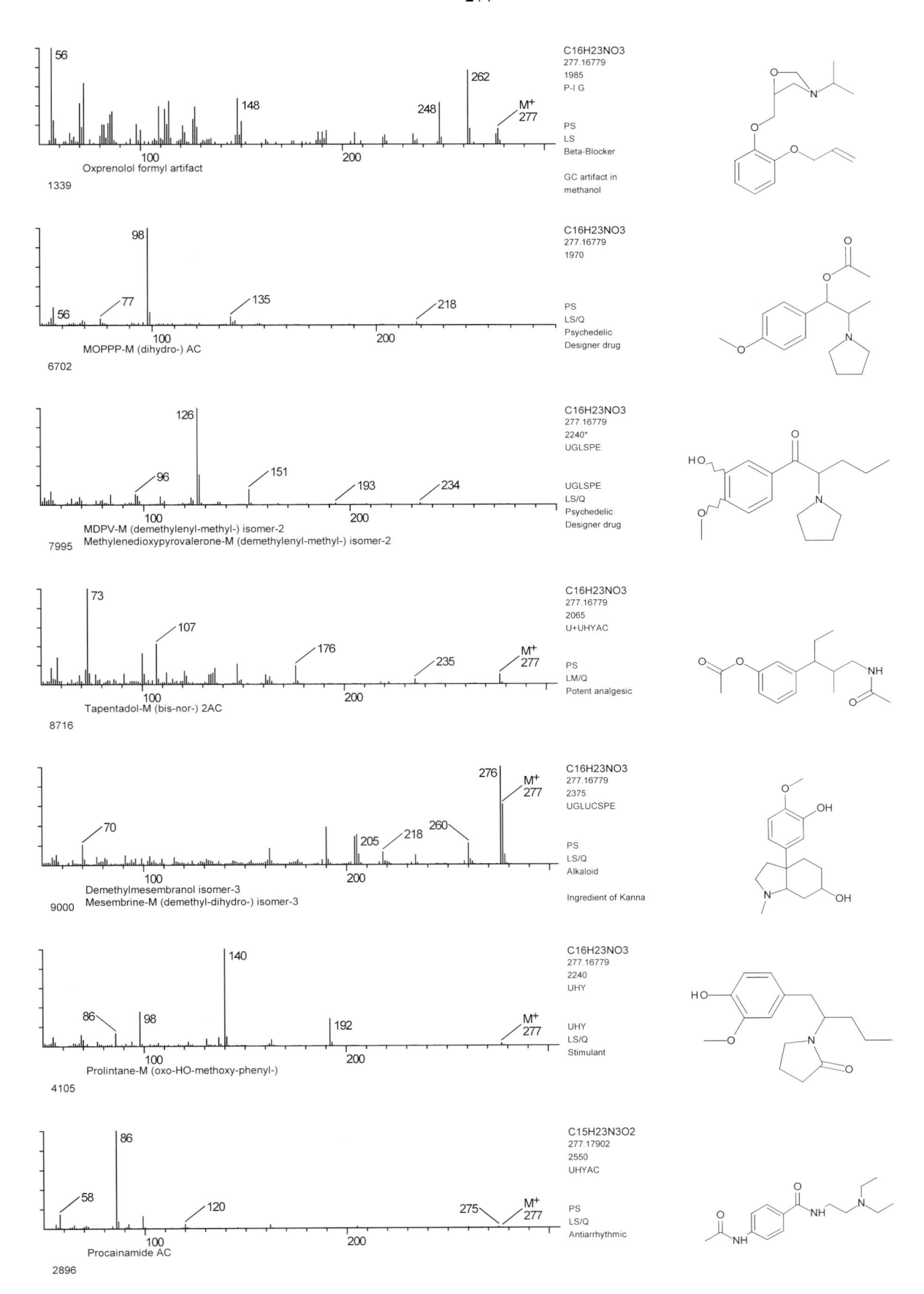

	C16H23NO3
	277.16779
	1985
	P-I G
Oxprenolol formyl artifact	PS
1339	LS
	Beta-Blocker
	GC artifact in methanol

	C16H23NO3
	277.16779
	1970
MOPPP-M (dihydro-) AC	PS
6702	LS/Q
	Psychedelic
	Designer drug

	C16H23NO3
	277.16779
	2240*
	UGLSPE
MDPV-M (demethylenyl-methyl-) isomer-2	UGLSPE
Methylenedioxypyrovalerone-M (demethylenyl-methyl-) isomer-2	LS/Q
7995	Psychedelic
	Designer drug

	C16H23NO3
	277.16779
	2065
	U+UHYAC
Tapentadol-M (bis-nor-) 2AC	PS
8716	LM/Q
	Potent analgesic

	C16H23NO3
	277.16779
	2375
	UGLUCSPE
Demethylmesembranol isomer-3	PS
Mesembrine-M (demethyl-dihydro-) isomer-3	LS/Q
9000	Alkaloid
	Ingredient of Kanna

	C16H23NO3
	277.16779
	2240
	UHY
Prolintane-M (oxo-HO-methoxy-phenyl-)	UHY
4105	LS/Q
	Stimulant

	C15H23N3O2
	277.17902
	2550
	UHYAC
Procainamide AC	PS
2896	LS/Q
	Antiarrhythmic

277

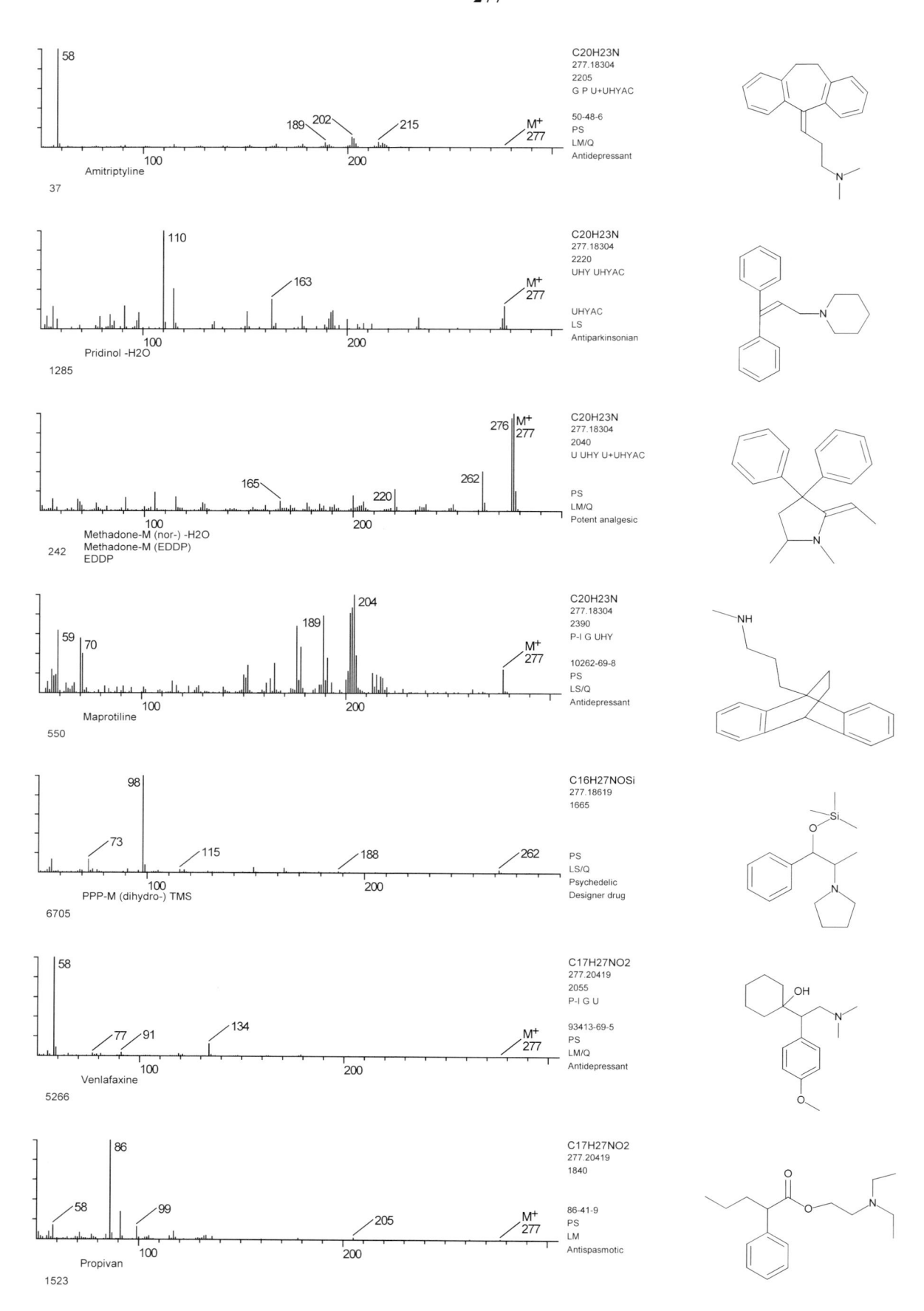

Amitriptyline — C20H23N, 277.18304, 2205, G P U+UHYAC, 50-48-6, PS, LM/Q, Antidepressant
Peaks: 58, 189, 202, 215, M+ 277

Pridinol -H2O — C20H23N, 277.18304, 2220, UHY UHYAC, UHYAC, LS, Antiparkinsonian
Peaks: 110, 163, M+ 277

Methadone-M (nor-) -H2O / Methadone-M (EDDP) / EDDP — C20H23N, 277.18304, 2040, U UHY U+UHYAC, PS, LM/Q, Potent analgesic
Peaks: 165, 220, 262, 276, M+ 277

Maprotiline — C20H23N, 277.18304, 2390, P-I G UHY, 10262-69-8, PS, LS/Q, Antidepressant
Peaks: 59, 70, 189, 204, M+ 277

PPP-M (dihydro-) TMS — C16H27NOSi, 277.18619, 1665, PS, LS/Q, Psychedelic Designer drug
Peaks: 73, 98, 115, 188, 262

Venlafaxine — C17H27NO2, 277.20419, 2055, P-I G U, 93413-69-5, PS, LM/Q, Antidepressant
Peaks: 58, 77, 91, 134, M+ 277

Propivan — C17H27NO2, 277.20419, 1840, 86-41-9, PS, LM, Antispasmotic
Peaks: 58, 86, 99, 205, M+ 277

277

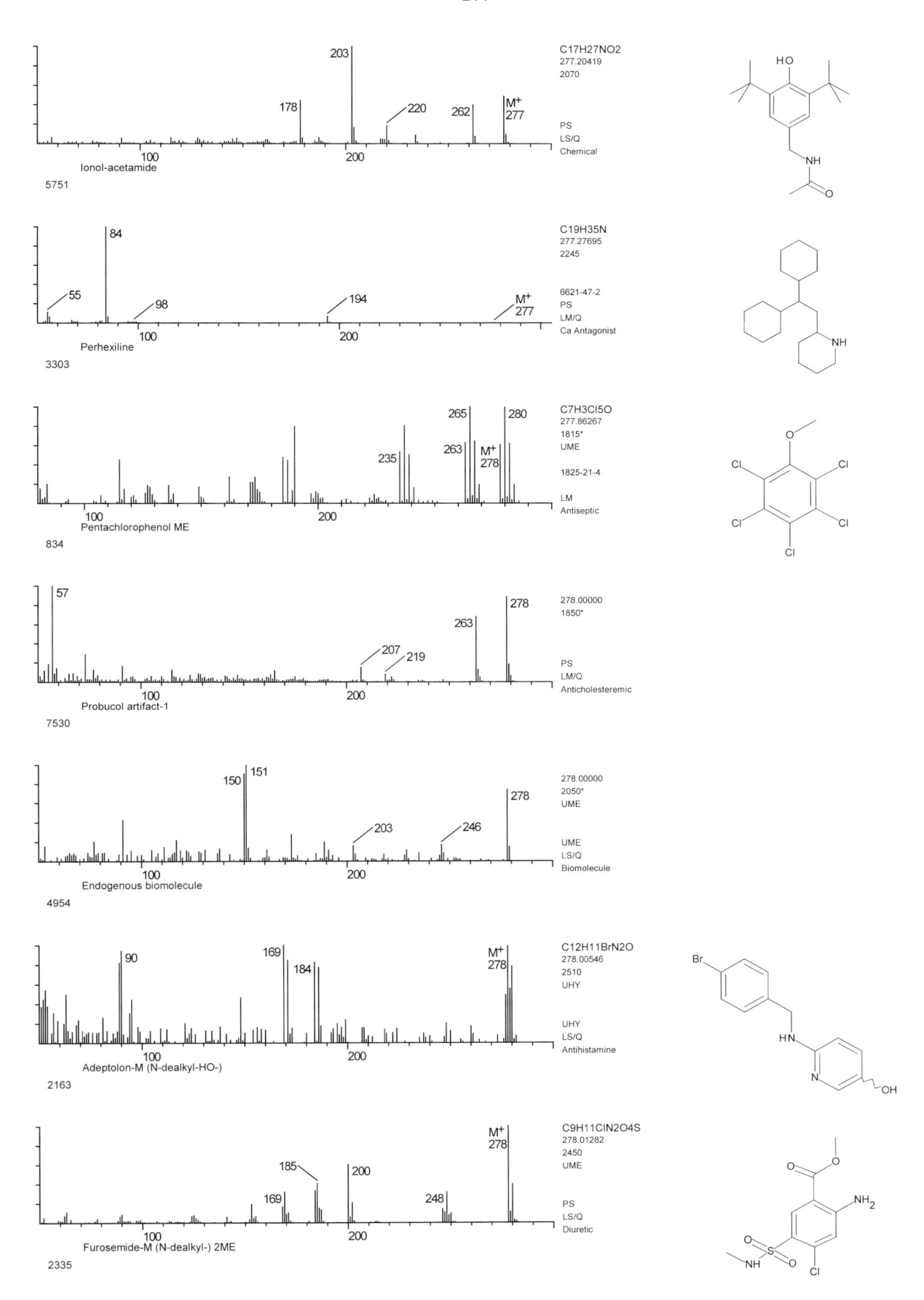

Ionol-acetamide — peaks: 178, 203, 220, 262, M+ 277	C17H27NO2 277.20419 2070 PS LS/Q Chemical 5751
Perhexiline — peaks: 55, 84, 98, 194, M+ 277	C19H35N 277.27695 2245 6621-47-2 PS LM/Q Ca Antagonist 3303
Pentachlorophenol ME — peaks: 235, 263, 265, M+ 278, 280	C7H3Cl5O 277.86267 1815* UME 1825-21-4 LM Antiseptic 834
Probucol artifact-1 — peaks: 57, 207, 219, 263, 278	278.00000 1850* PS LM/Q Anticholesteremic 7530
Endogenous biomolecule — peaks: 150, 151, 203, 246, 278	278.00000 2050* UME UME LS/Q Biomolecule 4954
Adeptolon-M (N-dealkyl-HO-) — peaks: 90, 169, 184, M+ 278	C12H11BrN2O 278.00546 2510 UHY UHY LS/Q Antihistamine 2163
Furosemide-M (N-dealkyl-) 2ME — peaks: 169, 185, 200, 248, M+ 278	C9H11ClN2O4S 278.01282 2450 UME PS LS/Q Diuretic 2335

Fenthion
3838

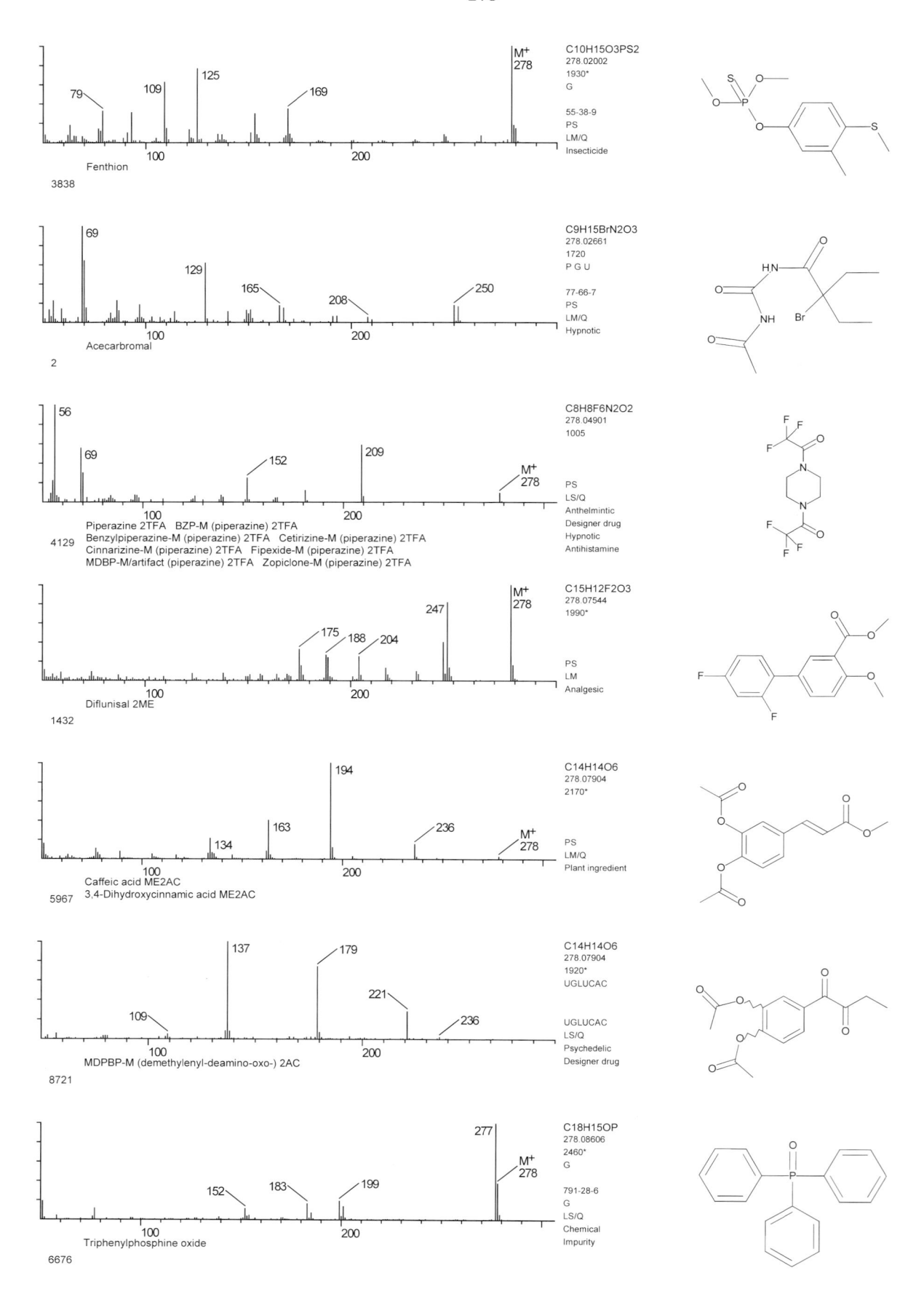

M+ 278
169
125
109
79

C10H15O3PS2
278.02002
1930*
G

55-38-9
PS
LM/Q
Insecticide

Acecarbromal
2

69
129
165
208
250

C9H15BrN2O3
278.02661
1720
P G U

77-66-7
PS
LM/Q
Hypnotic

Piperazine 2TFA   BZP-M (piperazine) 2TFA
Benzylpiperazine-M (piperazine) 2TFA   Cetirizine-M (piperazine) 2TFA
Cinnarizine-M (piperazine) 2TFA   Fipexide-M (piperazine) 2TFA
MDBP-M/artifact (piperazine) 2TFA   Zopiclone-M (piperazine) 2TFA
4129

56
69
152
209
M+ 278

C8H8F6N2O2
278.04901
1005

PS
LS/Q
Anthelmintic
Designer drug
Hypnotic
Antihistamine

Diflunisal 2ME
1432

M+ 278
247
175 188 204

C15H12F2O3
278.07544
1990*

PS
LM
Analgesic

Caffeic acid ME2AC
3,4-Dihydroxycinnamic acid ME2AC
5967

194
134 163
236
M+ 278

C14H14O6
278.07904
2170*

PS
LM/Q
Plant ingredient

MDPBP-M (demethylenyl-deamino-oxo-) 2AC
8721

137 179
109
221
236

C14H14O6
278.07904
1920*
UGLUCAC

UGLUCAC
LS/Q
Psychedelic
Designer drug

Triphenylphosphine oxide
6676

277
M+ 278
152 183 199

C18H15OP
278.08606
2460*
G

791-28-6
G
LS/Q
Chemical
Impurity

278

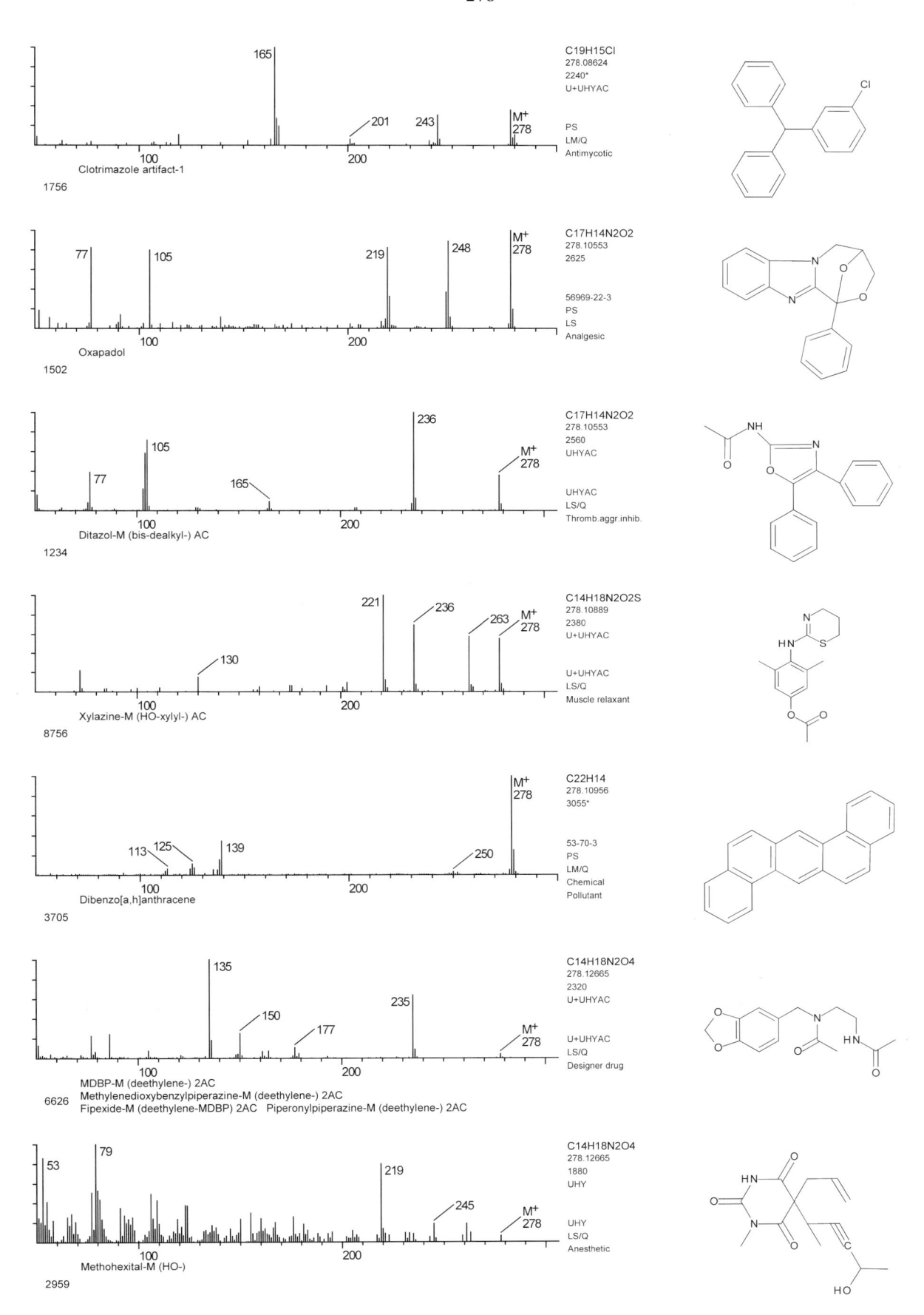

165 / 201 / 243 / M+ 278 — Clotrimazole artifact-1 — 1756	C19H15Cl 278.08624 2240* U+UHYAC PS LM/Q Antimycotic
77 / 105 / 219 / 248 / M+ 278 — Oxapadol — 1502	C17H14N2O2 278.10553 2625 56969-22-3 PS LS Analgesic
77 / 105 / 165 / 236 / M+ 278 — Ditazol-M (bis-dealkyl-) AC — 1234	C17H14N2O2 278.10553 2560 UHYAC UHYAC LS/Q Thromb.aggr.inhib.
130 / 221 / 236 / 263 / M+ 278 — Xylazine-M (HO-xylyl-) AC — 8756	C14H18N2O2S 278.10889 2380 U+UHYAC U+UHYAC LS/Q Muscle relaxant
113 / 125 / 139 / 250 / M+ 278 — Dibenzo[a,h]anthracene — 3705	C22H14 278.10956 3055* 53-70-3 PS LM/Q Chemical Pollutant
135 / 150 / 177 / 235 / M+ 278 — MDBP-M (deethylene-) 2AC / Methylenedioxybenzylpiperazine-M (deethylene-) 2AC / Fipexide-M (deethylene-MDBP) 2AC Piperonylpiperazine-M (deethylene-) 2AC — 6626	C14H18N2O4 278.12665 2320 U+UHYAC U+UHYAC LS/Q Designer drug
53 / 79 / 219 / 245 / M+ 278 — Methohexital-M (HO-) — 2959	C14H18N2O4 278.12665 1880 UHY UHY LS/Q Anesthetic

278

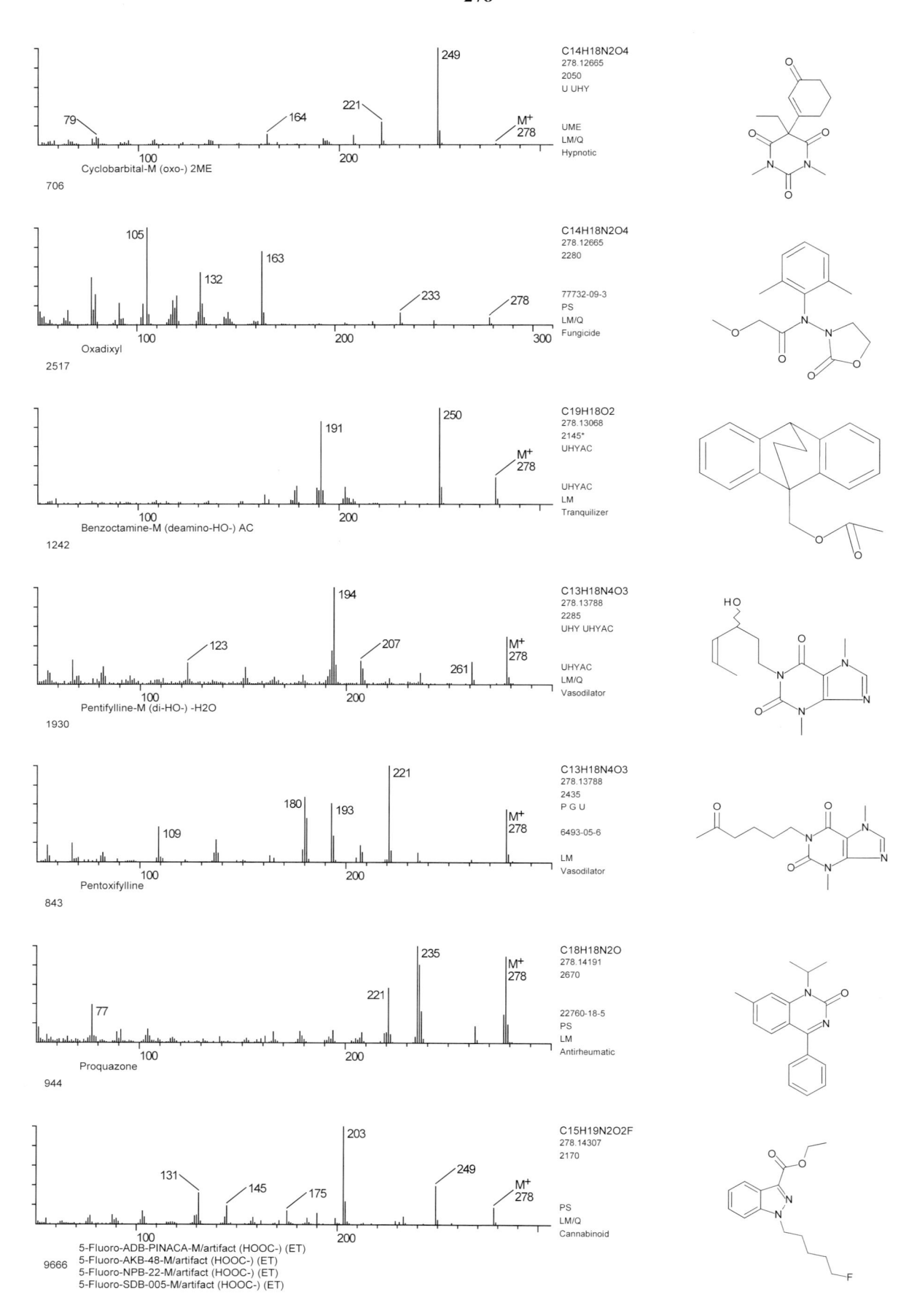

C14H18N2O4
278.12665
2050
U UHY

UME
LM/Q
Hypnotic

Cyclobarbital-M (oxo-) 2ME
706

C14H18N2O4
278.12665
2280

77732-09-3
PS
LM/Q
Fungicide

Oxadixyl
2517

C19H18O2
278.13068
2145*
UHYAC

UHYAC
LM
Tranquilizer

Benzoctamine-M (deamino-HO-) AC
1242

C13H18N4O3
278.13788
2285
UHY UHYAC

UHYAC
LM/Q
Vasodilator

Pentifylline-M (di-HO-) -H2O
1930

C13H18N4O3
278.13788
2435
P G U

6493-05-6

LM
Vasodilator

Pentoxifylline
843

C18H18N2O
278.14191
2670

22760-18-5
PS
LM
Antirheumatic

Proquazone
944

C15H19N2O2F
278.14307
2170

PS
LM/Q
Cannabinoid

5-Fluoro-ADB-PINACA-M/artifact (HOOC-) (ET)
5-Fluoro-AKB-48-M/artifact (HOOC-) (ET)
5-Fluoro-NPB-22-M/artifact (HOOC-) (ET)
5-Fluoro-SDB-005-M/artifact (HOOC-) (ET)
9666

768

278

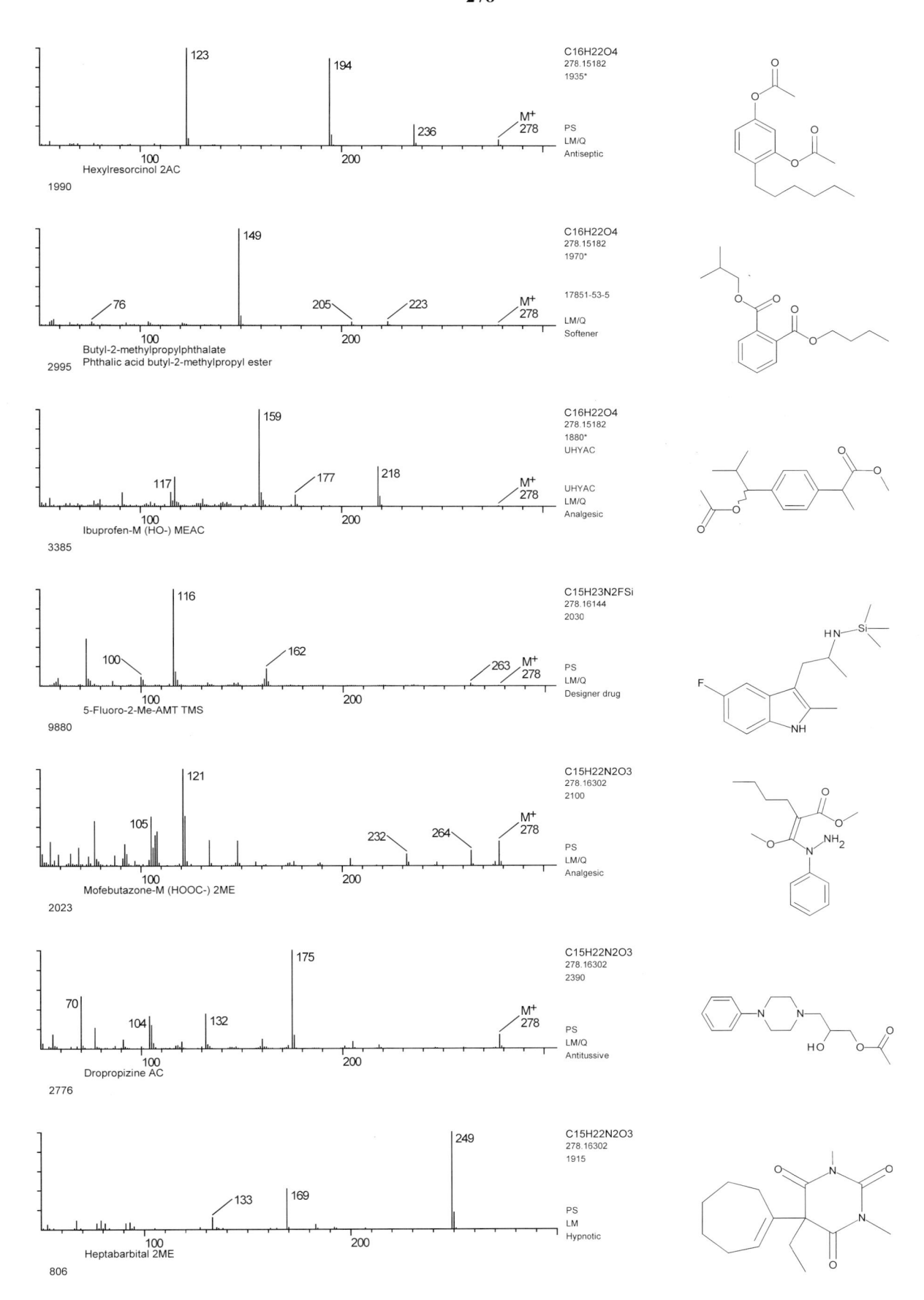

Hexylresorcinol 2AC — 123, 194, 236, M+ 278
C16H22O4
278.15182
1935*
PS
LM/Q
Antiseptic

Butyl-2-methylpropylphthalate
Phthalic acid butyl-2-methylpropyl ester — 76, 149, 205, 223, M+ 278
C16H22O4
278.15182
1970*
17851-53-5
LM/Q
Softener

Ibuprofen-M (HO-) MEAC — 117, 159, 177, 218, M+ 278
C16H22O4
278.15182
1880*
UHYAC
UHYAC
LM/Q
Analgesic

5-Fluoro-2-Me-AMT TMS — 100, 116, 162, 263, M+ 278
C15H23N2FSi
278.16144
2030
PS
LM/Q
Designer drug

Mofebutazone-M (HOOC-) 2ME — 105, 121, 232, 264, M+ 278
C15H22N2O3
278.16302
2100
PS
LM/Q
Analgesic

Dropropizine AC — 70, 104, 132, 175, M+ 278
C15H22N2O3
278.16302
2390
PS
LM/Q
Antitussive

Heptabarbital 2ME — 133, 169, 249
C15H22N2O3
278.16302
1915
PS
LM
Hypnotic

278

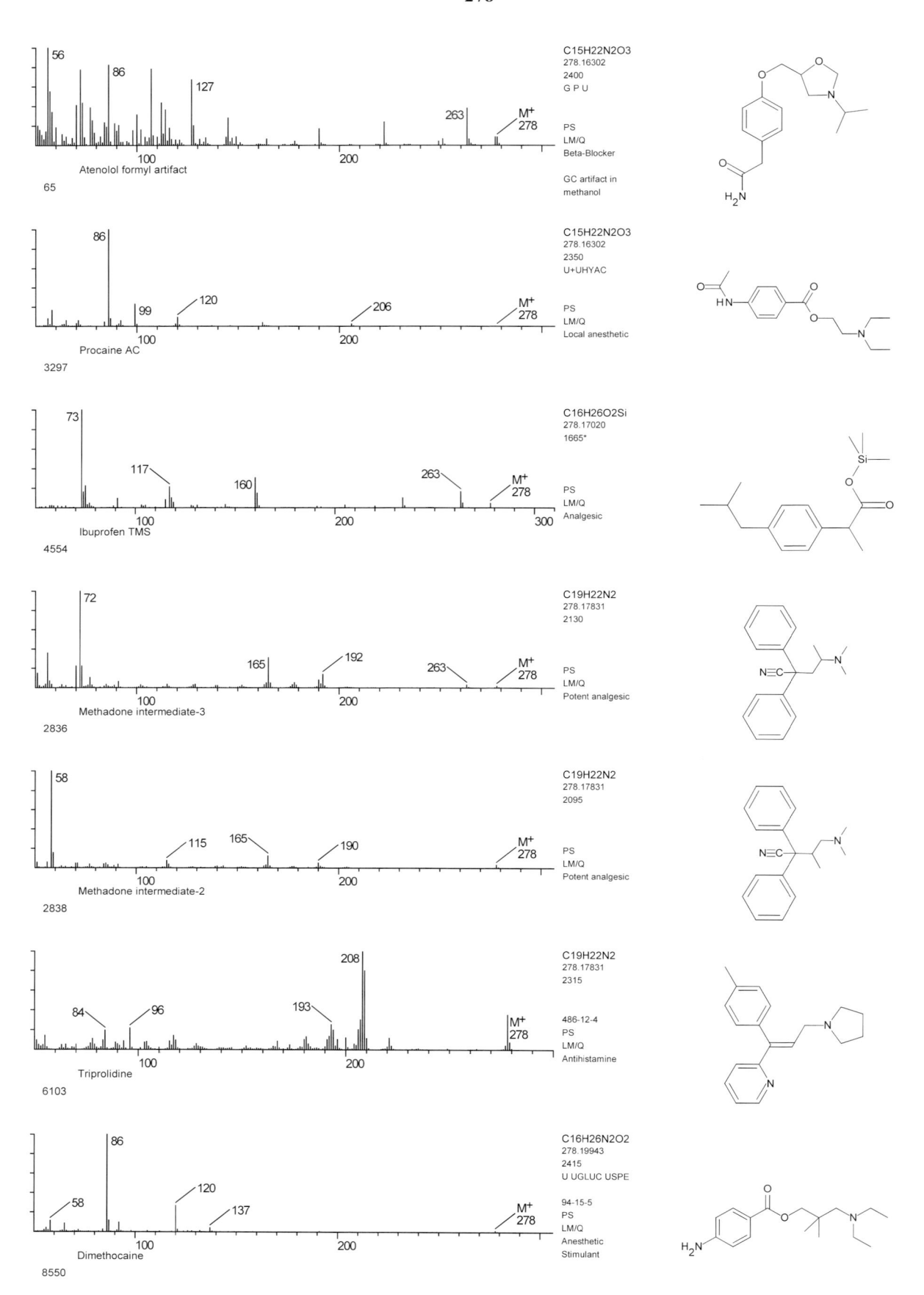

Atenolol formyl artifact
Peaks: 56, 86, 127, 263, M+ 278
C15H22N2O3
278.16302
2400
G P U
PS
LM/Q
Beta-Blocker

GC artifact in methanol

65

Procaine AC
Peaks: 86, 99, 120, 206, M+ 278
C15H22N2O3
278.16302
2350
U+UHYAC
PS
LM/Q
Local anesthetic

3297

Ibuprofen TMS
Peaks: 73, 117, 160, 263, M+ 278
C16H26O2Si
278.17020
1665*
PS
LM/Q
Analgesic

4554

Methadone intermediate-3
Peaks: 72, 165, 192, 263, M+ 278
C19H22N2
278.17831
2130
PS
LM/Q
Potent analgesic

2836

Methadone intermediate-2
Peaks: 58, 115, 165, 190, M+ 278
C19H22N2
278.17831
2095
PS
LM/Q
Potent analgesic

2838

Triprolidine
Peaks: 84, 96, 193, 208, M+ 278
C19H22N2
278.17831
2315
486-12-4
PS
LM/Q
Antihistamine

6103

Dimethocaine
Peaks: 58, 86, 120, 137, M+ 278
C16H26N2O2
278.19943
2415
U UGLUC USPE
94-15-5
PS
LM/Q
Anesthetic
Stimulant

8550

278

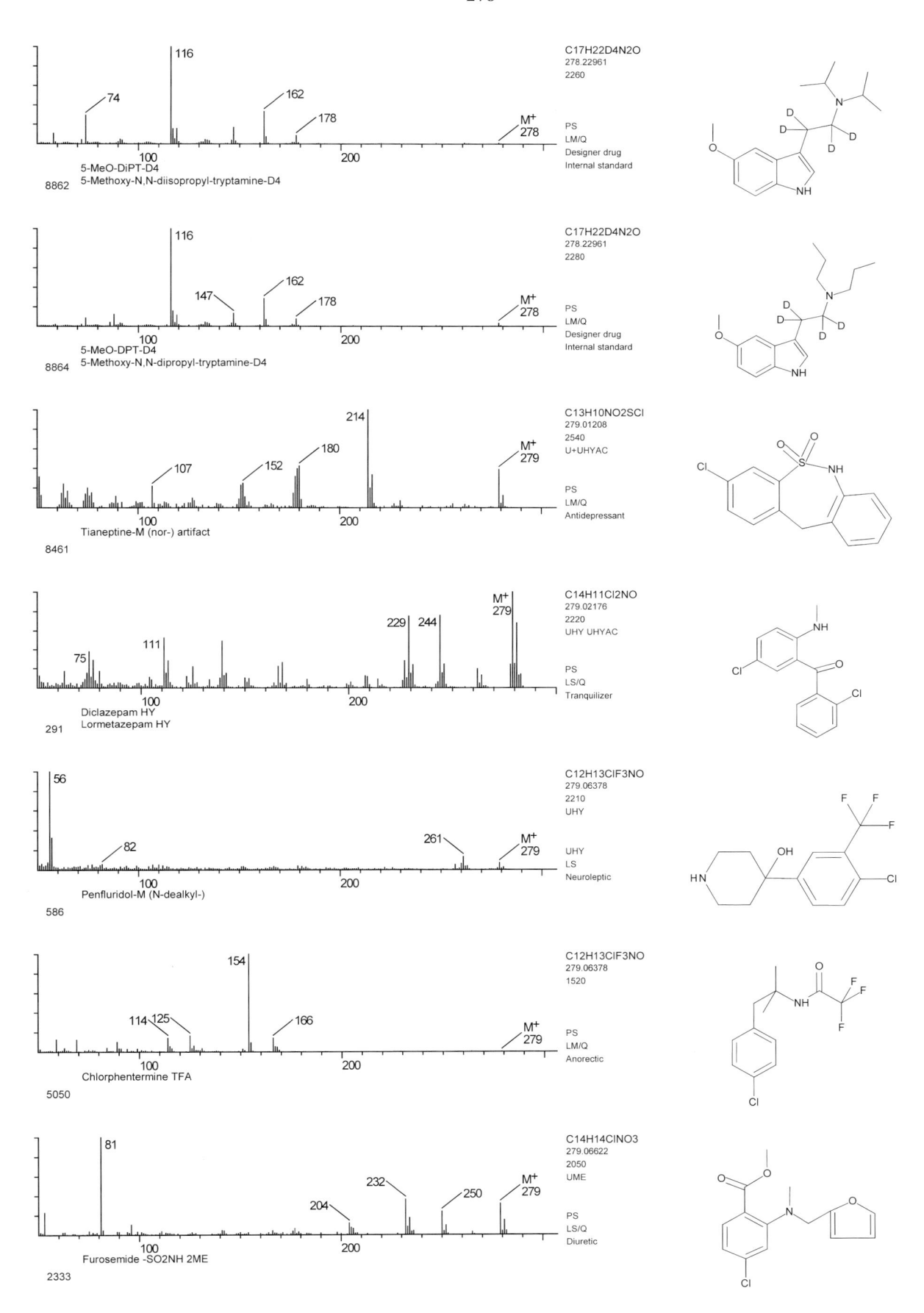

Spectrum	Formula / Info
8862 — 5-MeO-DiPT-D4, 5-Methoxy-N,N-diisopropyl-tryptamine-D4; peaks 74, 116, 162, 178, M+ 278	C17H22D4N2O; 278.22961; 2260; PS; LM/Q; Designer drug; Internal standard
8864 — 5-MeO-DPT-D4, 5-Methoxy-N,N-dipropyl-tryptamine-D4; peaks 116, 147, 162, 178, M+ 278	C17H22D4N2O; 278.22961; 2280; PS; LM/Q; Designer drug; Internal standard
8461 — Tianeptine-M (nor-) artifact; peaks 107, 152, 180, 214, M+ 279	C13H10NO2SCl; 279.01208; 2540; U+UHYAC; PS; LM/Q; Antidepressant
291 — Diclazepam HY, Lormetazepam HY; peaks 75, 111, 229, 244, M+ 279	C14H11Cl2NO; 279.02176; 2220; UHY UHYAC; PS; LS/Q; Tranquilizer
586 — Penfluridol-M (N-dealkyl-); peaks 56, 82, 261, M+ 279	C12H13ClF3NO; 279.06378; 2210; UHY; UHY; LS; Neuroleptic
5050 — Chlorphentermine TFA; peaks 114, 125, 154, 166, M+ 279	C12H13ClF3NO; 279.06378; 1520; PS; LM/Q; Anorectic
2333 — Furosemide -SO2NH 2ME; peaks 81, 204, 232, 250, M+ 279	C14H14ClNO3; 279.06622; 2050; UME; PS; LS/Q; Diuretic

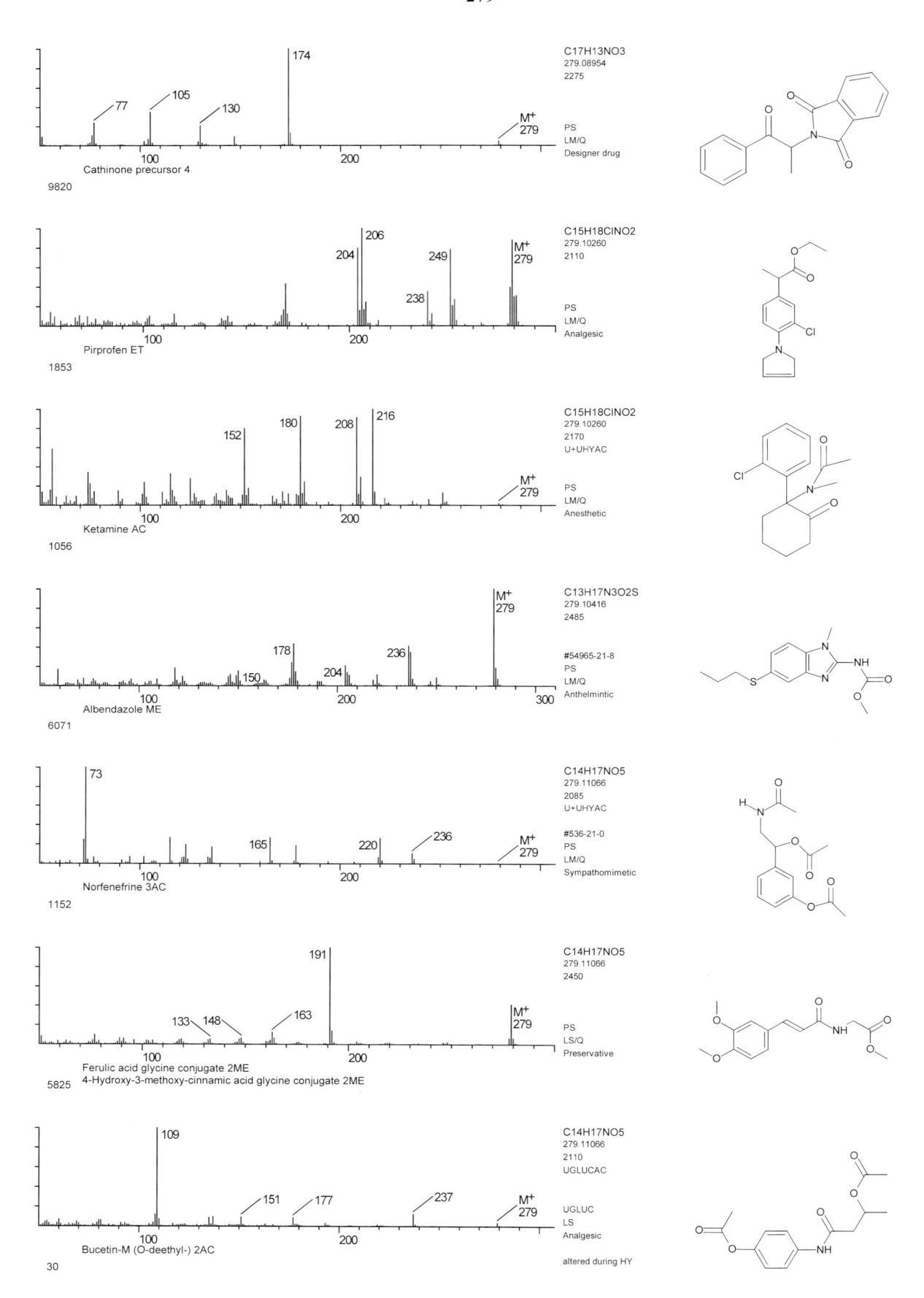

Cathinone precursor 4	C17H13NO3 279.08954 2275 PS LM/Q Designer drug
Pirprofen ET	C15H18ClNO2 279.10260 2110 PS LM/Q Analgesic
Ketamine AC	C15H18ClNO2 279.10260 2170 U+UHYAC PS LM/Q Anesthetic
Albendazole ME	C13H17N3O2S 279.10416 2485 #54965-21-8 PS LM/Q Anthelmintic
Norfenefrine 3AC	C14H17NO5 279.11066 2085 U+UHYAC #536-21-0 PS LM/Q Sympathomimetic
Ferulic acid glycine conjugate 2ME 4-Hydroxy-3-methoxy-cinnamic acid glycine conjugate 2ME	C14H17NO5 279.11066 2450 PS LS/Q Preservative
Bucetin-M (O-deethyl-) 2AC	C14H17NO5 279.11066 2110 UGLUCAC UGLUC LS Analgesic altered during HY

279

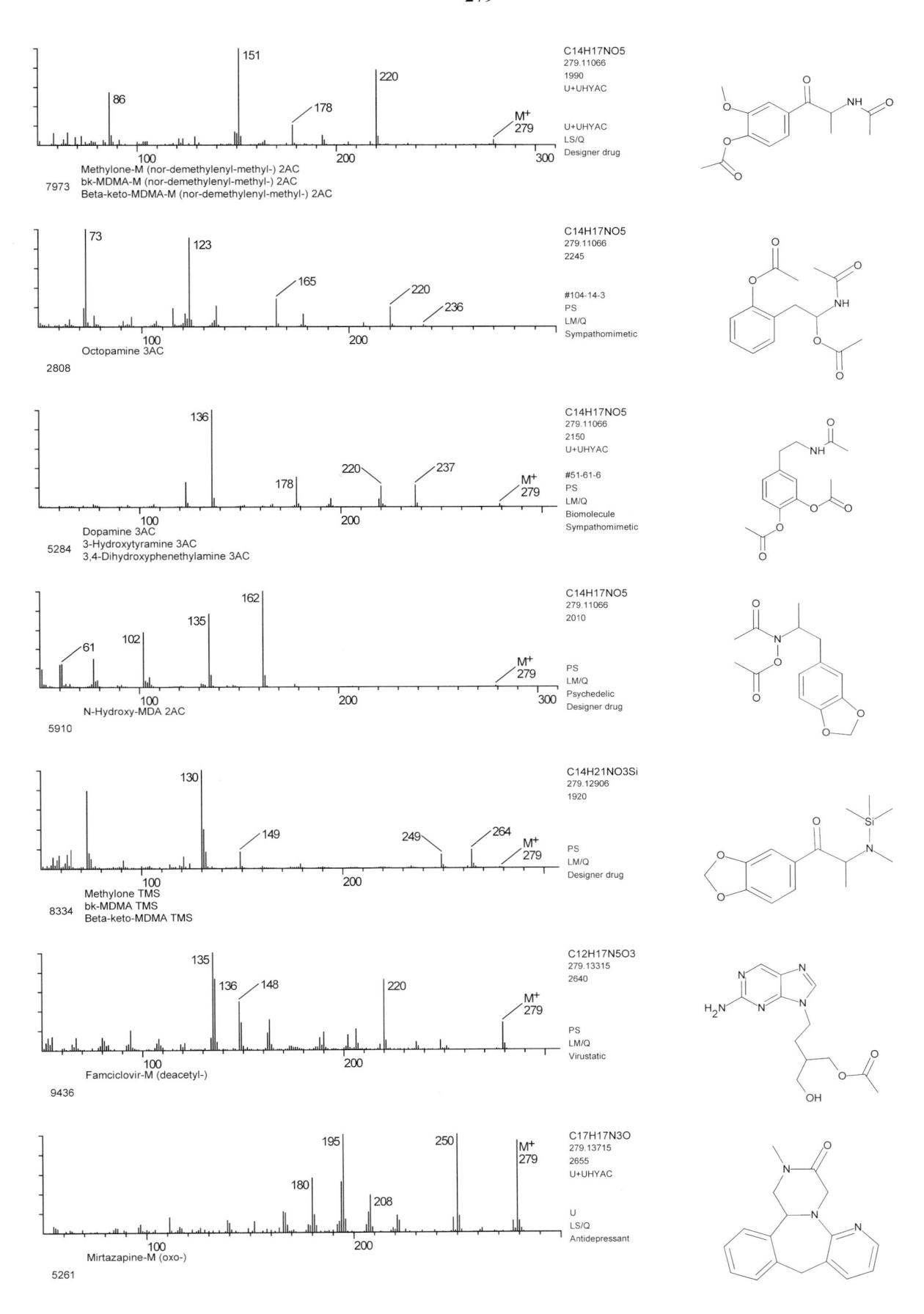

C14H17NO5
279.11066
1990
U+UHYAC

U+UHYAC
LS/Q
Designer drug

7973 Methylone-M (nor-demethylenyl-methyl-) 2AC
bk-MDMA-M (nor-demethylenyl-methyl-) 2AC
Beta-keto-MDMA-M (nor-demethylenyl-methyl-) 2AC

C14H17NO5
279.11066
2245

#104-14-3
PS
LM/Q
Sympathomimetic

2808 Octopamine 3AC

C14H17NO5
279.11066
2150
U+UHYAC

#51-61-6
PS
LM/Q
Biomolecule
Sympathomimetic

5284 Dopamine 3AC
3-Hydroxytyramine 3AC
3,4-Dihydroxyphenethylamine 3AC

C14H17NO5
279.11066
2010

PS
LM/Q
Psychedelic
Designer drug

5910 N-Hydroxy-MDA 2AC

C14H21NO3Si
279.12906
1920

PS
LM/Q
Designer drug

8334 Methylone TMS
bk-MDMA TMS
Beta-keto-MDMA TMS

C12H17N5O3
279.13315
2640

PS
LM/Q
Virustatic

9436 Famciclovir-M (deacetyl-)

C17H17N3O
279.13715
2655
U+UHYAC

U
LS/Q
Antidepressant

5261 Mirtazapine-M (oxo-)

279

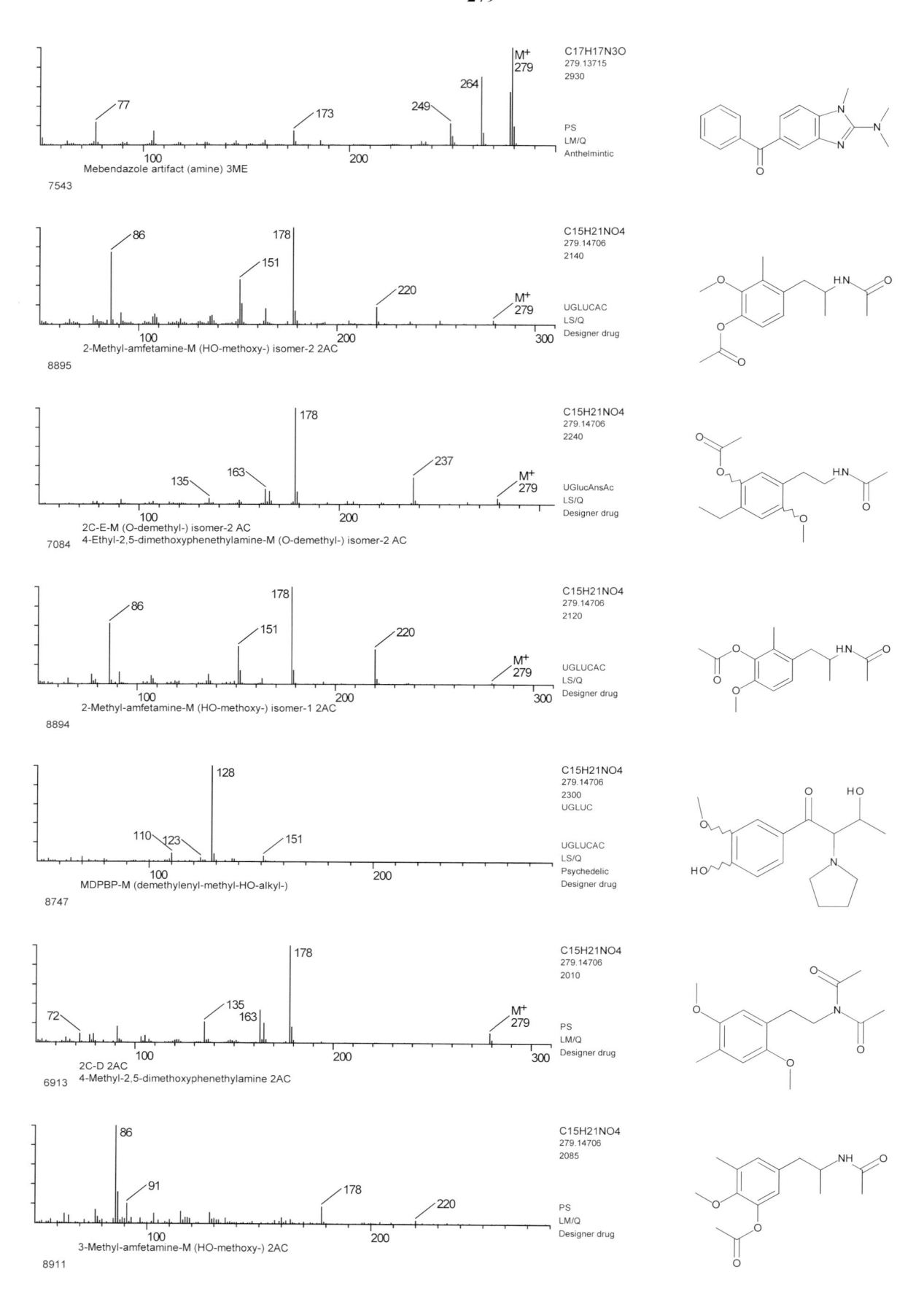

Mebendazole artifact (amine) 3ME 7543 Peaks: 77, 173, 249, 264, M+ 279	C17H17N3O 279.13715 2930 PS LM/Q Anthelmintic
2-Methyl-amfetamine-M (HO-methoxy-) isomer-2 2AC 8895 Peaks: 86, 151, 178, 220, M+ 279	C15H21NO4 279.14706 2140 UGLUCAC LS/Q Designer drug
2C-E-M (O-demethyl-) isomer-2 AC 4-Ethyl-2,5-dimethoxyphenethylamine-M (O-demethyl-) isomer-2 AC 7084 Peaks: 135, 163, 178, 237, M+ 279	C15H21NO4 279.14706 2240 UGlucAnsAc LS/Q Designer drug
2-Methyl-amfetamine-M (HO-methoxy-) isomer-1 2AC 8894 Peaks: 86, 151, 178, 220, M+ 279	C15H21NO4 279.14706 2120 UGLUCAC LS/Q Designer drug
MDPBP-M (demethylenyl-methyl-HO-alkyl-) 8747 Peaks: 110, 123, 128, 151	C15H21NO4 279.14706 2300 UGLUC UGLUCAC LS/Q Psychedelic Designer drug
2C-D 2AC 4-Methyl-2,5-dimethoxyphenethylamine 2AC 6913 Peaks: 72, 135, 163, 178, M+ 279	C15H21NO4 279.14706 2010 PS LM/Q Designer drug
3-Methyl-amfetamine-M (HO-methoxy-) 2AC 8911 Peaks: 86, 91, 178, 220	C15H21NO4 279.14706 2085 PS LM/Q Designer drug

279

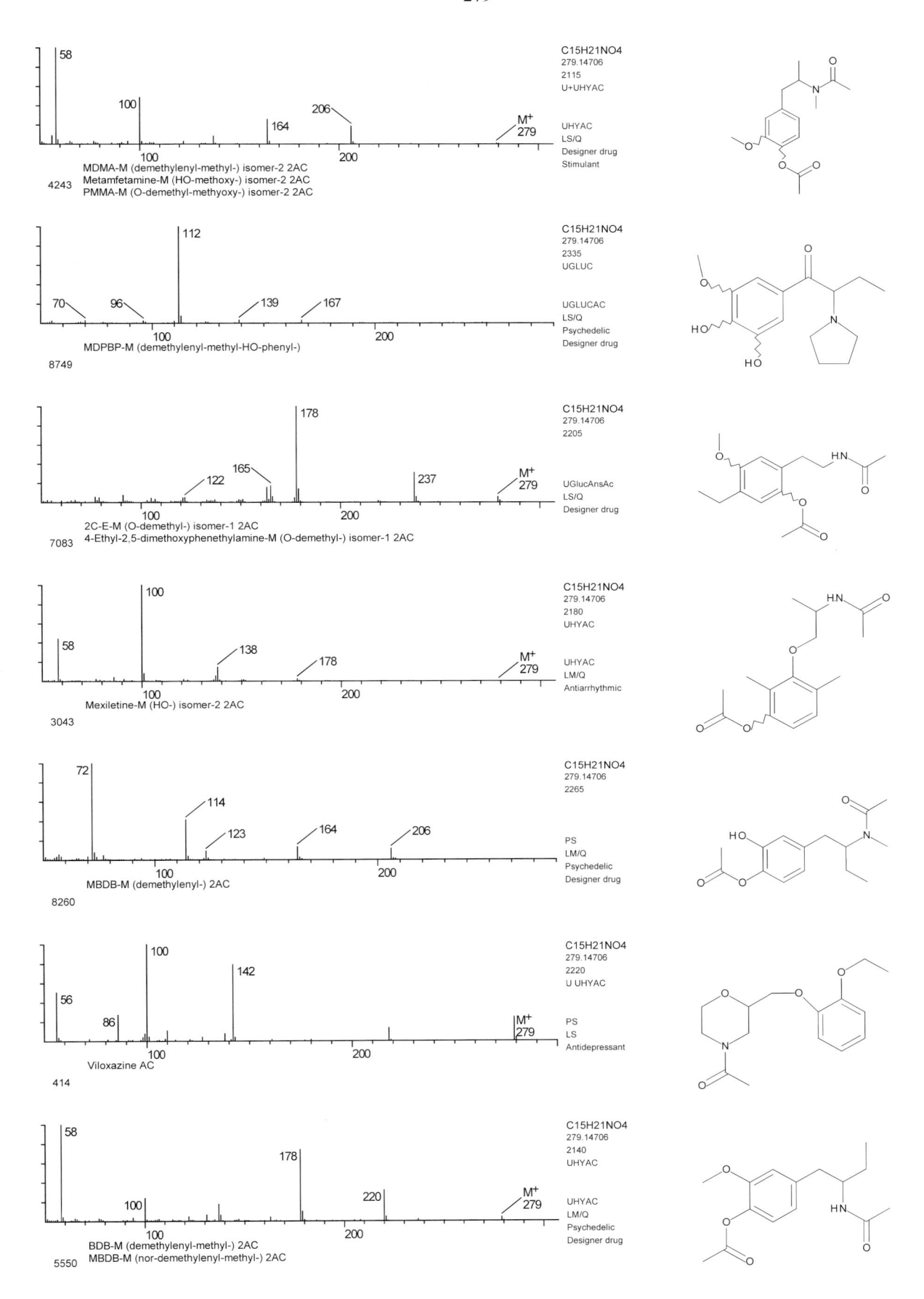

4243
MDMA-M (demethylenyl-methyl-) isomer-2 2AC
Metamfetamine-M (HO-methoxy-) isomer-2 2AC
PMMA-M (O-demethyl-methyoxy-) isomer-2 2AC

C15H21NO4
279.14706
2115
U+UHYAC
UHYAC
LS/Q
Designer drug
Stimulant

8749
MDPBP-M (demethylenyl-methyl-HO-phenyl-)

C15H21NO4
279.14706
2335
UGLUC
UGLUCAC
LS/Q
Psychedelic
Designer drug

7083
2C-E-M (O-demethyl-) isomer-1 2AC
4-Ethyl-2,5-dimethoxyphenethylamine-M (O-demethyl-) isomer-1 2AC

C15H21NO4
279.14706
2205
UGlucAnsAc
LS/Q
Designer drug

3043
Mexiletine-M (HO-) isomer-2 2AC

C15H21NO4
279.14706
2180
UHYAC
UHYAC
LM/Q
Antiarrhythmic

8260
MBDB-M (demethylenyl-) 2AC

C15H21NO4
279.14706
2265
PS
LM/Q
Psychedelic
Designer drug

414
Viloxazine AC

C15H21NO4
279.14706
2220
U UHYAC
PS
LS
Antidepressant

5550
BDB-M (demethylenyl-methyl) 2AC
MBDB-M (nor-demethylenyl-methyl-) 2AC

C15H21NO4
279.14706
2140
UHYAC
UHYAC
LM/Q
Psychedelic
Designer drug

279

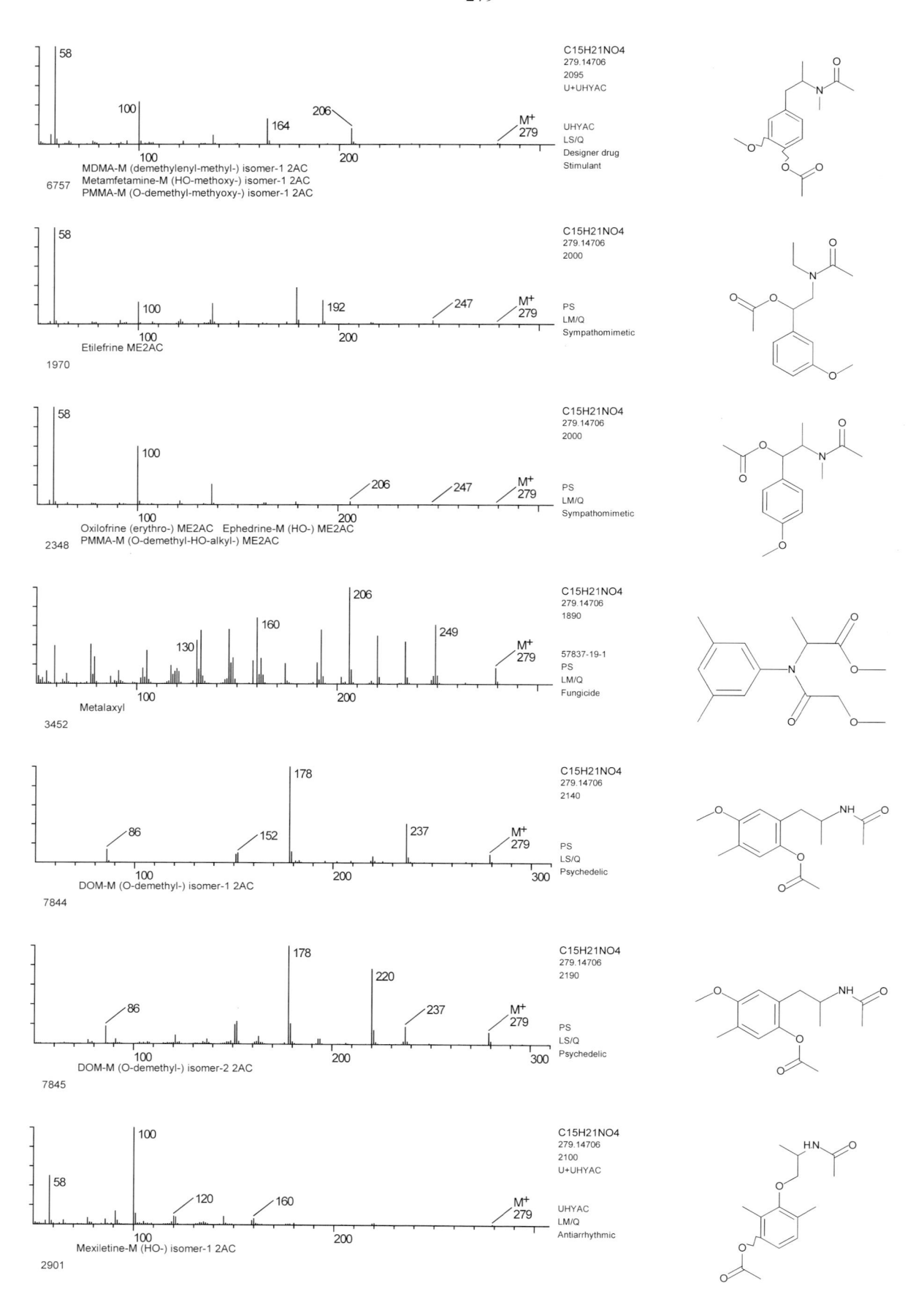

6757	MDMA-M (demethylenyl-methyl-) isomer-1 2AC Metamfetamine-M (HO-methoxy-) isomer-1 2AC PMMA-M (O-demethyl-methyoxy-) isomer-1 2AC Peaks: 58, 100, 164, 206, M+ 279 C15H21NO4 279.14706 2095 U+UHYAC UHYAC LS/Q Designer drug Stimulant
1970	Etilefrine ME2AC Peaks: 58, 100, 192, 247, M+ 279 C15H21NO4 279.14706 2000 PS LM/Q Sympathomimetic
2348	Oxilofrine (erythro-) ME2AC Ephedrine-M (HO-) ME2AC PMMA-M (O-demethyl-HO-alkyl-) ME2AC Peaks: 58, 100, 206, 247, M+ 279 C15H21NO4 279.14706 2000 PS LM/Q Sympathomimetic
3452	Metalaxyl Peaks: 130, 160, 206, 249, M+ 279 C15H21NO4 279.14706 1890 57837-19-1 PS LM/Q Fungicide
7844	DOM-M (O-demethyl-) isomer-1 2AC Peaks: 86, 152, 178, 237, M+ 279 C15H21NO4 279.14706 2140 PS LS/Q Psychedelic
7845	DOM-M (O-demethyl-) isomer-2 2AC Peaks: 86, 178, 220, 237, M+ 279 C15H21NO4 279.14706 2190 PS LS/Q Psychedelic
2901	Mexiletine-M (HO-) isomer-1 2AC Peaks: 58, 100, 120, 160, M+ 279 C15H21NO4 279.14706 2100 U+UHYAC UHYAC LM/Q Antiarrhythmic

279

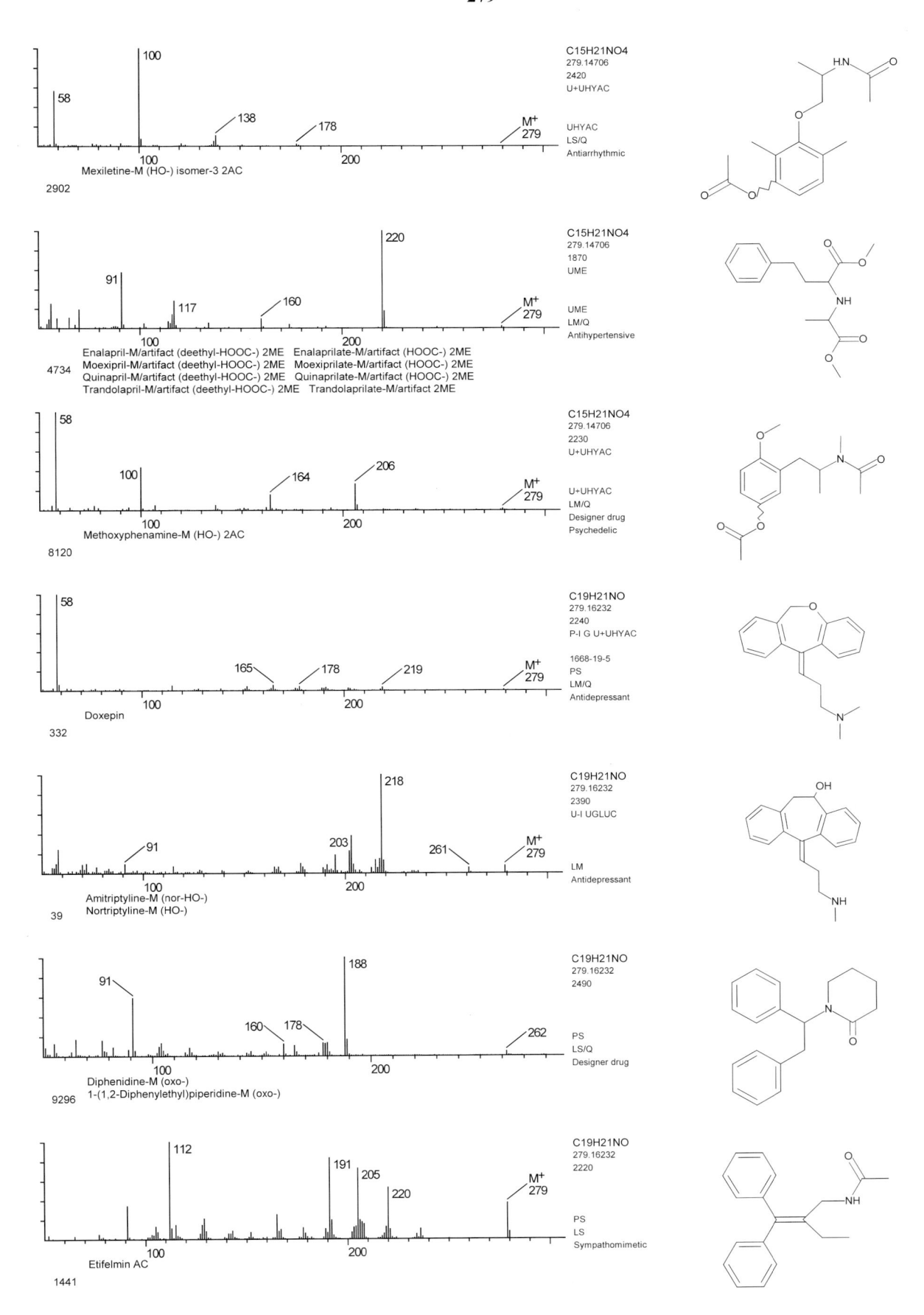

Mexiletine-M (HO-) isomer-3 2AC
2902
C15H21NO4
279.14706
2420
U+UHYAC

UHYAC
LS/Q
Antiarrhythmic

Enalapril-M/artifact (deethyl-HOOC-) 2ME   Enalaprilate-M/artifact (HOOC-) 2ME
Moexipril-M/artifact (deethyl-HOOC-) 2ME   Moexiprilate-M/artifact (HOOC-) 2ME
Quinapril-M/artifact (deethyl-HOOC-) 2ME   Quinaprilate-M/artifact (HOOC-) 2ME
Trandolapril-M/artifact (deethyl-HOOC-) 2ME Trandolaprilate-M/artifact 2ME
4734
C15H21NO4
279.14706
1870
UME

UME
LM/Q
Antihypertensive

Methoxyphenamine-M (HO-) 2AC
8120
C15H21NO4
279.14706
2230
U+UHYAC

U+UHYAC
LM/Q
Designer drug
Psychedelic

Doxepin
332
C19H21NO
279.16232
2240
P-I G U+UHYAC
1668-19-5
PS
LM/Q
Antidepressant

Amitriptyline-M (nor-HO-)
Nortriptyline-M (HO-)
39
C19H21NO
279.16232
2390
U-I UGLUC

LM
Antidepressant

Diphenidine-M (oxo-)
1-(1,2-Diphenylethyl)piperidine-M (oxo-)
9296
C19H21NO
279.16232
2490

PS
LS/Q
Designer drug

Etifelmin AC
1441
C19H21NO
279.16232
2220

PS
LS
Sympathomimetic

279

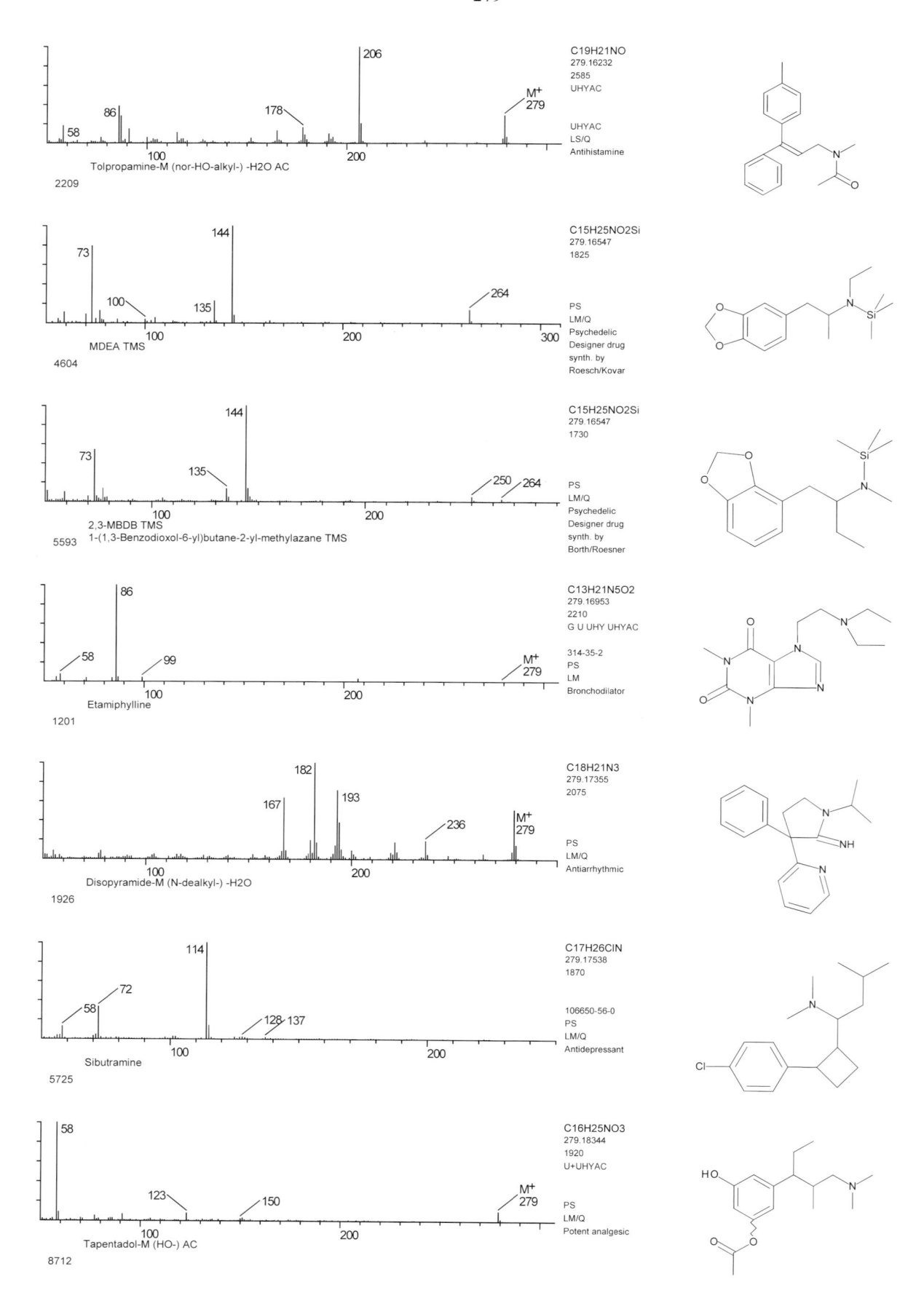

C19H21NO
279.16232
2585
UHYAC

UHYAC
LS/Q
Antihistamine

Tolpropamine-M (nor-HO-alkyl-) -H2O AC
2209

C15H25NO2Si
279.16547
1825

PS
LM/Q
Psychedelic
Designer drug
synth. by
Roesch/Kovar

MDEA TMS
4604

C15H25NO2Si
279.16547
1730

PS
LM/Q
Psychedelic
Designer drug
synth. by
Borth/Roesner

2,3-MBDB TMS
1-(1,3-Benzodioxol-6-yl)butane-2-yl-methylazane TMS
5593

C13H21N5O2
279.16953
2210
G U UHY UHYAC

314-35-2
PS
LM
Bronchodilator

Etamiphylline
1201

C18H21N3
279.17355
2075

PS
LM/Q
Antiarrhythmic

Disopyramide-M (N-dealkyl-) -H2O
1926

C17H26ClN
279.17538
1870

106650-56-0
PS
LM/Q
Antidepressant

Sibutramine
5725

C16H25NO3
279.18344
1920
U+UHYAC

PS
LM/Q
Potent analgesic

Tapentadol-M (HO-) AC
8712

279

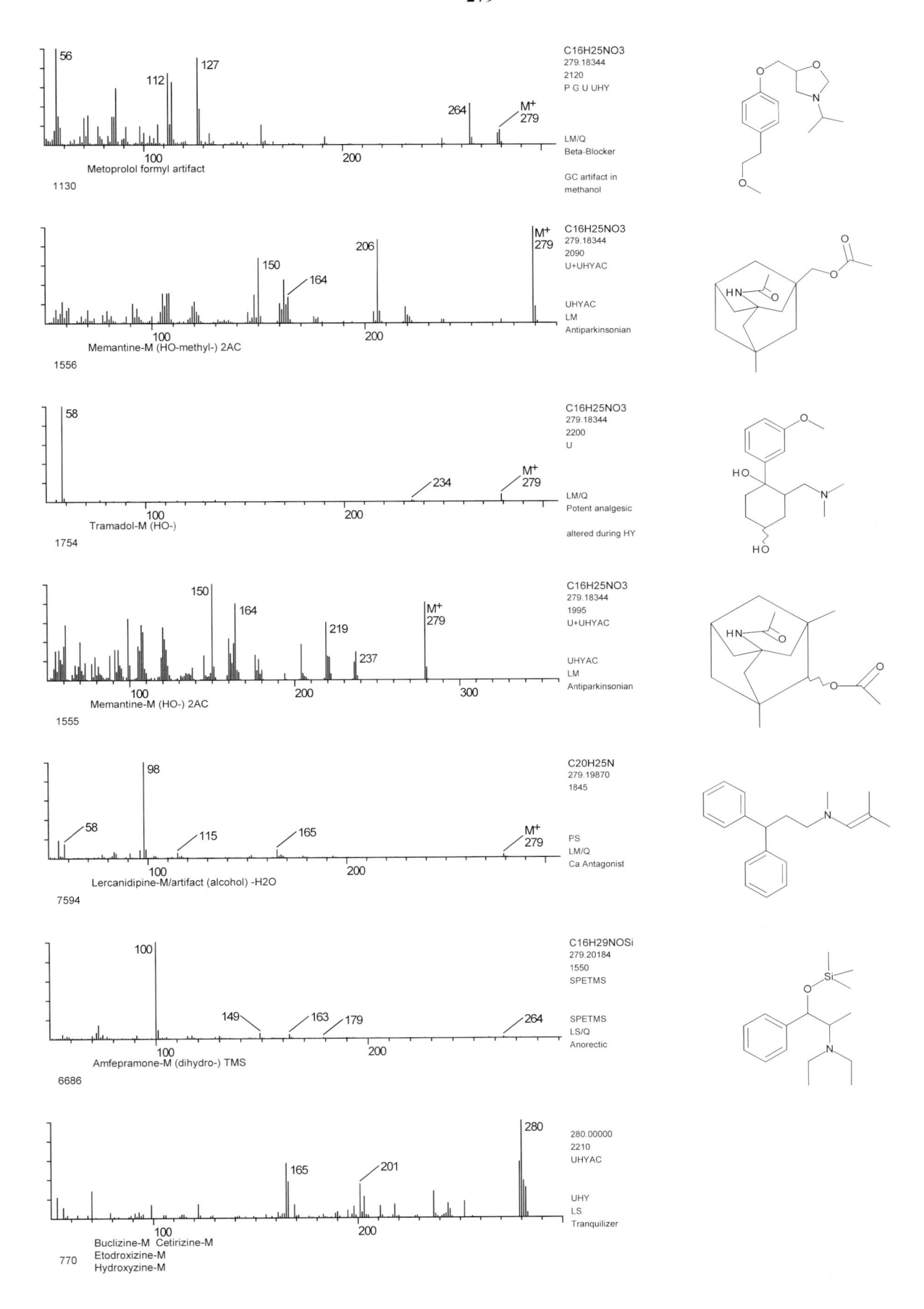

C16H25NO3
279.18344
2120
P G U UHY

LM/Q
Beta-Blocker

GC artifact in methanol

Metoprolol formyl artifact
1130

C16H25NO3
279.18344
2090
U+UHYAC

UHYAC
LM
Antiparkinsonian

Memantine-M (HO-methyl-) 2AC
1556

C16H25NO3
279.18344
2200
U

LM/Q
Potent analgesic

altered during HY

Tramadol-M (HO-)
1754

C16H25NO3
279.18344
1995
U+UHYAC

UHYAC
LM
Antiparkinsonian

Memantine-M (HO-) 2AC
1555

C20H25N
279.19870
1845

PS
LM/Q
Ca Antagonist

Lercanidipine-M/artifact (alcohol) -H2O
7594

C16H29NOSi
279.20184
1550
SPETMS

SPETMS
LS/Q
Anorectic

Amfepramone-M (dihydro-) TMS
6686

280.00000
2210
UHYAC

UHY
LS
Tranquilizer

Buclizine-M Cetirizine-M
Etodroxizine-M
Hydroxyzine-M
770

779

280

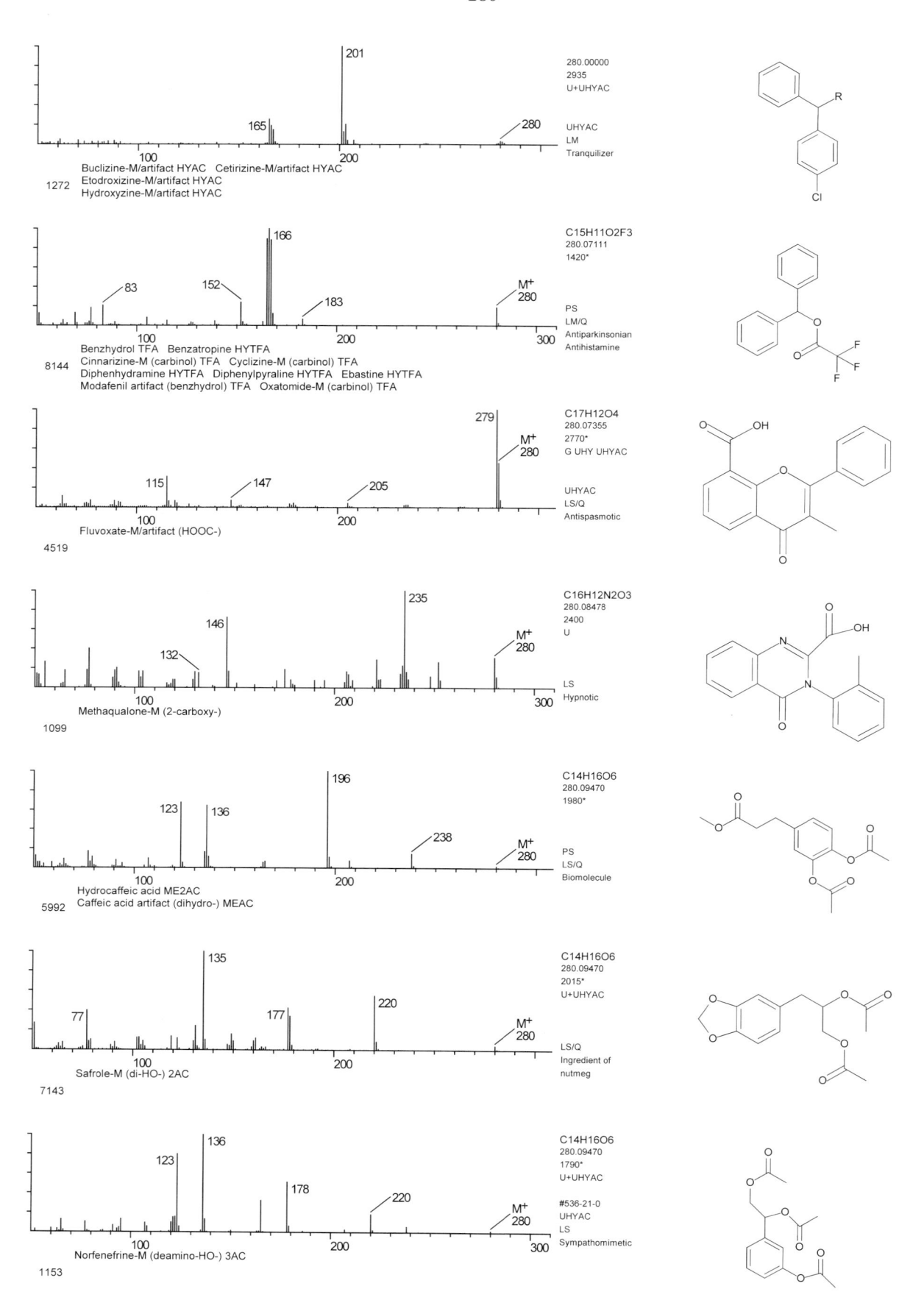

1272
Buclizine-M/artifact HYAC   Cetirizine-M/artifact HYAC
Etodroxizine-M/artifact HYAC
Hydroxyzine-M/artifact HYAC

201, 165, 280
280.00000
2935
U+UHYAC
UHYAC
LM
Tranquilizer

8144
Benzhydrol TFA   Benzatropine HYTFA
Cinnarizine-M (carbinol) TFA   Cyclizine-M (carbinol) TFA
Diphenhydramine HYTFA   Diphenylpyraline HYTFA   Ebastine HYTFA
Modafenil artifact (benzhydrol) TFA   Oxatomide-M (carbinol) TFA

166, 83, 152, 183, M+ 280
C15H11O2F3
280.07111
1420*
PS
LM/Q
Antiparkinsonian
Antihistamine

4519
Fluvoxate-M/artifact (HOOC-)

279, M+ 280, 115, 147, 205
C17H12O4
280.07355
2770*
G UHY UHYAC
UHYAC
LS/Q
Antispasmotic

1099
Methaqualone-M (2-carboxy-)

235, 146, 132, M+ 280
C16H12N2O3
280.08478
2400
U
LS
Hypnotic

5992
Hydrocaffeic acid ME2AC
Caffeic acid artifact (dihydro-) MEAC

196, 123, 136, 238, M+ 280
C14H16O6
280.09470
1980*
PS
LS/Q
Biomolecule

7143
Safrole-M (di-HO-) 2AC

135, 77, 177, 220, M+ 280
C14H16O6
280.09470
2015*
U+UHYAC
LS/Q
Ingredient of nutmeg

1153
Norfenefrine-M (deamino-HO-) 3AC

136, 123, 178, 220, M+ 280
C14H16O6
280.09470
1790*
U+UHYAC
#536-21-0
UHYAC
LS
Sympathomimetic

280

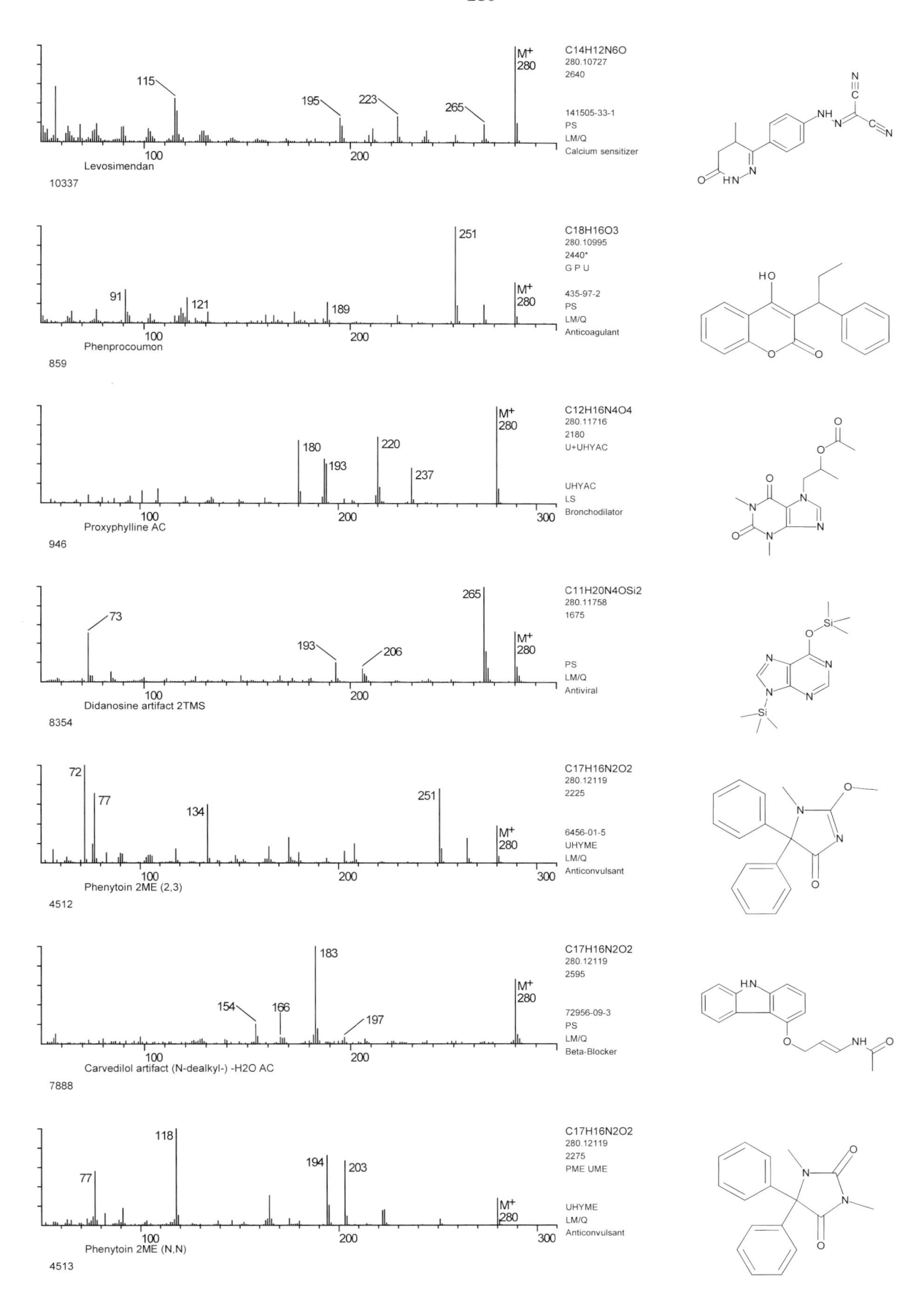

Levosimendan — 10337
C14H12N6O
280.10727
2640
141505-33-1
PS
LM/Q
Calcium sensitizer
Peaks: 115, 195, 223, 265, M+ 280

Phenprocoumon — 859
C18H16O3
280.10995
2440*
G P U
435-97-2
PS
LM/Q
Anticoagulant
Peaks: 91, 121, 189, 251, M+ 280

Proxyphylline AC — 946
C12H16N4O4
280.11716
2180
U+UHYAC
UHYAC
LS
Bronchodilator
Peaks: 180, 193, 220, 237, M+ 280

Didanosine artifact 2TMS — 8354
C11H20N4OSi2
280.11758
1675
PS
LM/Q
Antiviral
Peaks: 73, 193, 206, 265, M+ 280

Phenytoin 2ME (2,3) — 4512
C17H16N2O2
280.12119
2225
6456-01-5
UHYME
LM/Q
Anticonvulsant
Peaks: 72, 77, 134, 251, M+ 280

Carvedilol artifact (N-dealkyl-) -H2O AC — 7888
C17H16N2O2
280.12119
2595
72956-09-3
PS
LM/Q
Beta-Blocker
Peaks: 154, 166, 183, 197, M+ 280

Phenytoin 2ME (N,N) — 4513
C17H16N2O2
280.12119
2275
PME UME
UHYME
LM/Q
Anticonvulsant
Peaks: 77, 118, 194, 203, M+ 280

280

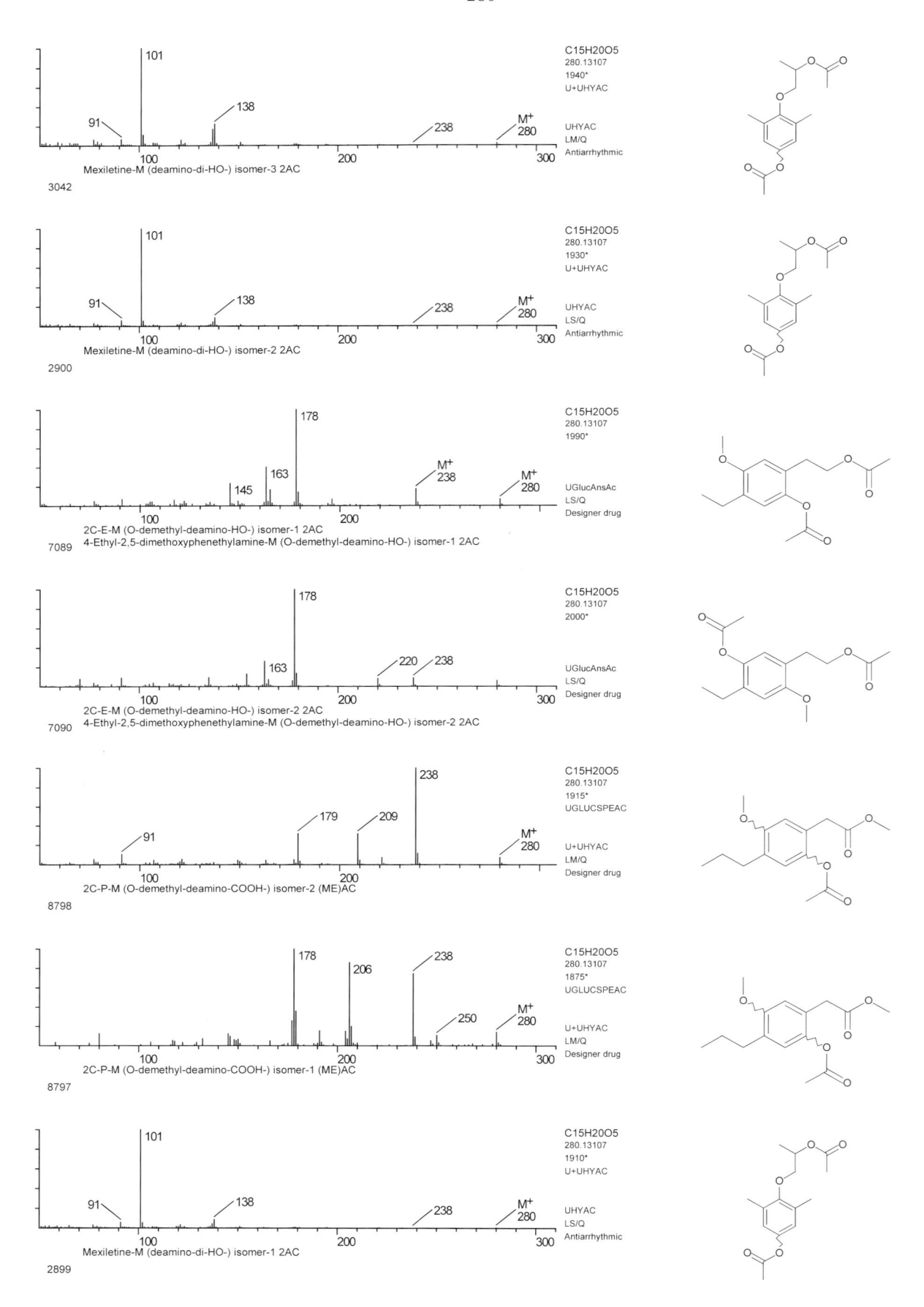

Spectrum label	Formula info
3042 — Mexiletine-M (deamino-di-HO-) isomer-3 2AC (peaks: 91, 101, 138, 238, M+ 280)	C15H20O5, 280.13107, 1940*, U+UHYAC, UHYAC, LM/Q, Antiarrhythmic
2900 — Mexiletine-M (deamino-di-HO-) isomer-2 2AC (peaks: 91, 101, 138, 238, M+ 280)	C15H20O5, 280.13107, 1930*, U+UHYAC, UHYAC, LS/Q, Antiarrhythmic
7089 — 2C-E-M (O-demethyl-deamino-HO-) isomer-1 2AC / 4-Ethyl-2,5-dimethoxyphenethylamine-M (O-demethyl-deamino-HO-) isomer-1 2AC (peaks: 145, 163, 178, M+ 238, M+ 280)	C15H20O5, 280.13107, 1990*, UGlucAnsAc, LS/Q, Designer drug
7090 — 2C-E-M (O-demethyl-deamino-HO-) isomer-2 2AC / 4-Ethyl-2,5-dimethoxyphenethylamine-M (O-demethyl-deamino-HO-) isomer-2 2AC (peaks: 163, 178, 220, 238)	C15H20O5, 280.13107, 2000*, UGlucAnsAc, LS/Q, Designer drug
8798 — 2C-P-M (O-demethyl-deamino-COOH-) isomer-2 (ME)AC (peaks: 91, 179, 209, 238, M+ 280)	C15H20O5, 280.13107, 1915*, UGLUCSPEAC, U+UHYAC, LM/Q, Designer drug
8797 — 2C-P-M (O-demethyl-deamino-COOH-) isomer-1 (ME)AC (peaks: 178, 206, 238, 250, M+ 280)	C15H20O5, 280.13107, 1875*, UGLUCSPEAC, U+UHYAC, LM/Q, Designer drug
2899 — Mexiletine-M (deamino-di-HO-) isomer-1 2AC (peaks: 91, 101, 138, 238, M+ 280)	C15H20O5, 280.13107, 1910*, U+UHYAC, UHYAC, LS/Q, Antiarrhythmic

280

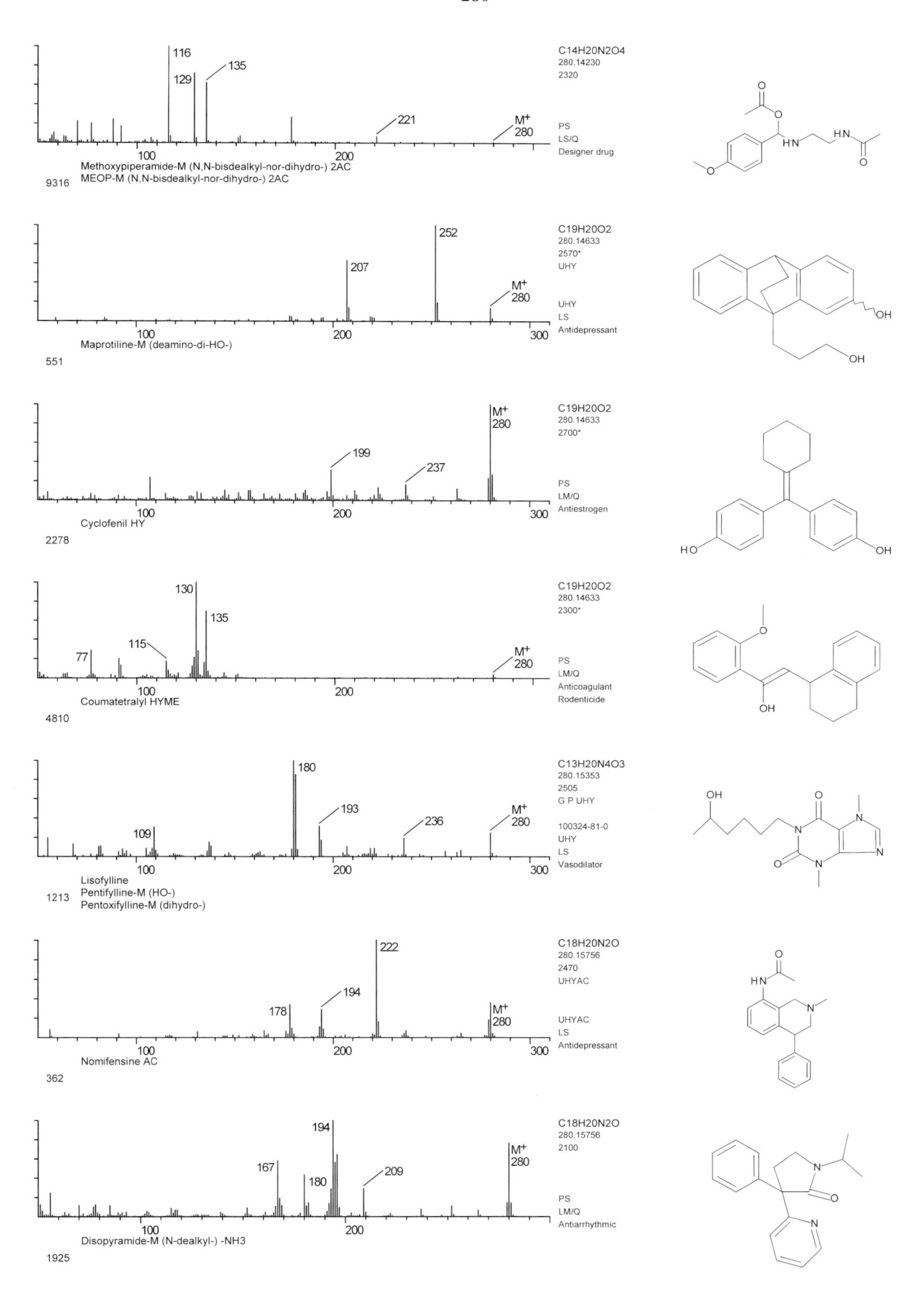

C14H20N2O4
280.14230
2320

PS
LS/Q
Designer drug

Methoxypiperamide-M (N,N-bisdealkyl-nor-dihydro-) 2AC
9316 MEOP-M (N,N-bisdealkyl-nor-dihydro-) 2AC

C19H20O2
280.14633
2570*
UHY

UHY
LS
Antidepressant

Maprotiline-M (deamino-di-HO-)
551

C19H20O2
280.14633
2700*

PS
LM/Q
Antiestrogen

Cyclofenil HY
2278

C19H20O2
280.14633
2300*

PS
LM/Q
Anticoagulant
Rodenticide

Coumatetralyl HYME
4810

C13H20N4O3
280.15353
2505
G P UHY

100324-81-0
UHY
LS
Vasodilator

Lisofylline
1213 Pentifylline-M (HO-)
Pentoxifylline-M (dihydro-)

C18H20N2O
280.15756
2470
UHYAC

UHYAC
LS
Antidepressant

Nomifensine AC
362

C18H20N2O
280.15756
2100

PS
LM/Q
Antiarrhythmic

Disopyramide-M (N-dealkyl-) -NH3
1925

280

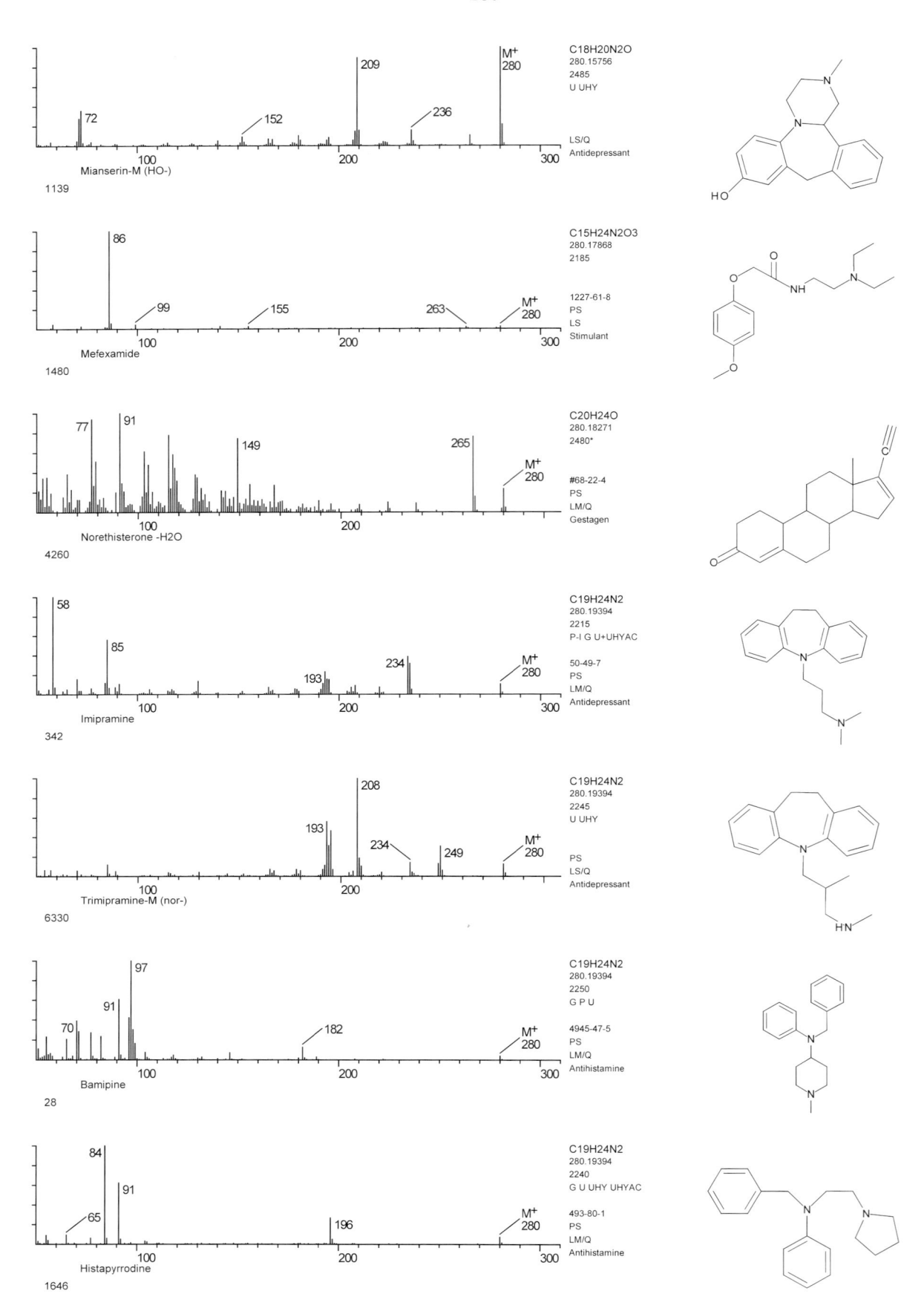

Compound	m/z peaks	Formula / Info
Mianserin-M (HO-)	72, 152, 209, 236, M+ 280	C18H20N2O, 280.15756, 2485, U UHY, LS/Q, Antidepressant, 1139
Mefexamide	86, 99, 155, 263, M+ 280	C15H24N2O3, 280.17868, 2185, 1227-61-8, PS, LS, Stimulant, 1480
Norethisterone -H2O	77, 91, 149, 265, M+ 280	C20H24O, 280.18271, 2480*, #68-22-4, PS, LM/Q, Gestagen, 4260
Imipramine	58, 85, 193, 234, M+ 280	C19H24N2, 280.19394, 2215, P-I G U+UHYAC, 50-49-7, PS, LM/Q, Antidepressant, 342
Trimipramine-M (nor-)	193, 208, 234, 249, M+ 280	C19H24N2, 280.19394, 2245, U UHY, PS, LS/Q, Antidepressant, 6330
Bamipine	70, 91, 97, 182, M+ 280	C19H24N2, 280.19394, 2250, G P U, 4945-47-5, PS, LM/Q, Antihistamine, 28
Histapyrrodine	65, 84, 91, 196, M+ 280	C19H24N2, 280.19394, 2240, G U UHY UHYAC, 493-80-1, PS, LM/Q, Antihistamine, 1646

280

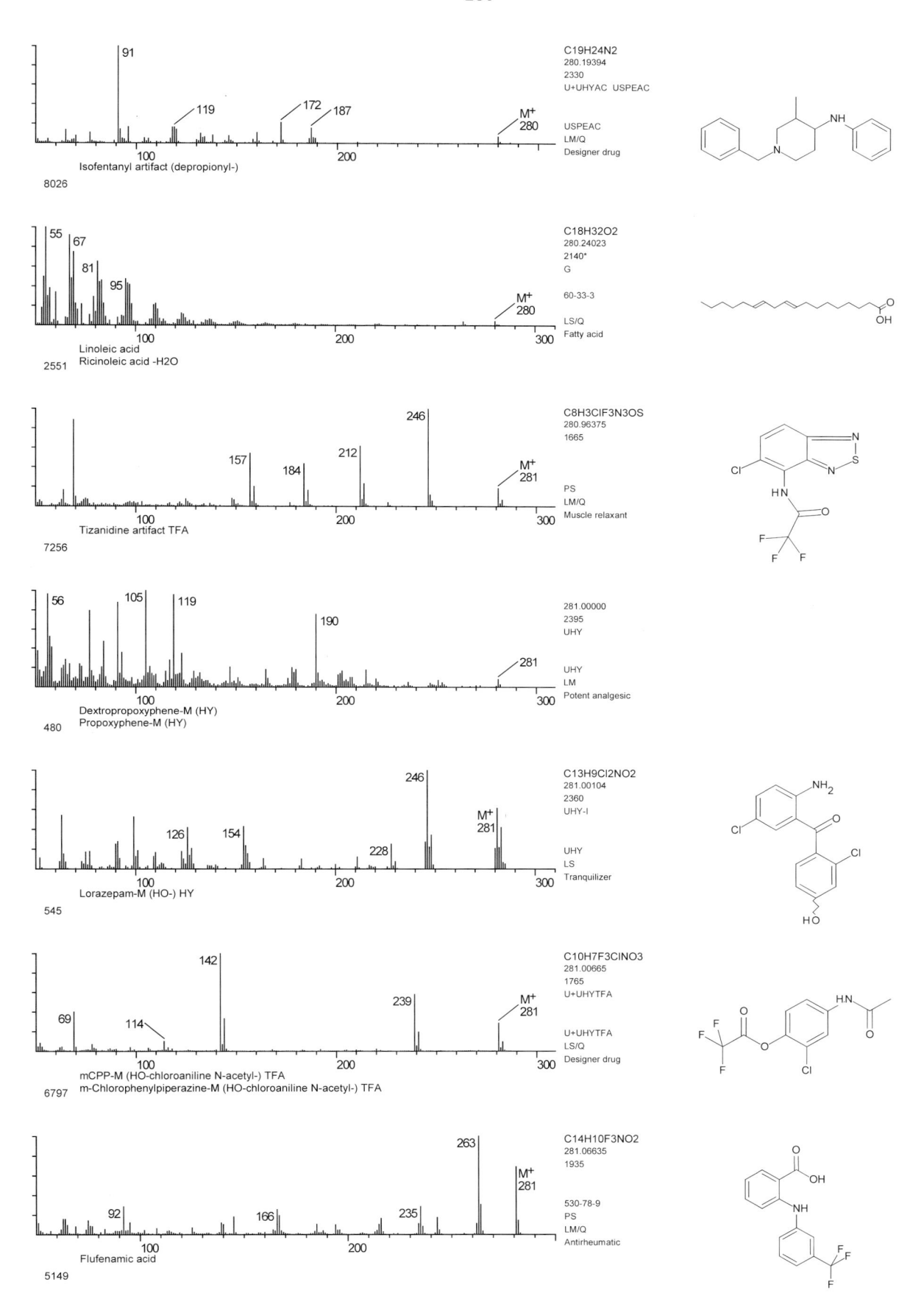

C19H24N2
280.19394
2330
U+UHYAC USPEAC

USPEAC
LM/Q
Designer drug

Isofentanyl artifact (depropionyl-)
8026

C18H32O2
280.24023
2140*
G

60-33-3
LS/Q
Fatty acid

Linoleic acid
Ricinoleic acid -H2O
2551

C8H3ClF3N3OS
280.96375
1665

PS
LM/Q
Muscle relaxant

Tizanidine artifact TFA
7256

281.00000
2395
UHY

UHY
LM
Potent analgesic

Dextropropoxyphene-M (HY)
Propoxyphene-M (HY)
480

C13H9Cl2NO2
281.00104
2360
UHY-I

UHY
LS
Tranquilizer

Lorazepam-M (HO-) HY
545

C10H7F3ClNO3
281.00665
1765
U+UHYTFA

U+UHYTFA
LS/Q
Designer drug

mCPP-M (HO-chloroaniline N-acetyl-) TFA
m-Chlorophenylpiperazine-M (HO-chloroaniline N-acetyl-) TFA
6797

C14H10F3NO2
281.06635
1935

530-78-9
PS
LM/Q
Antirheumatic

Flufenamic acid
5149

281

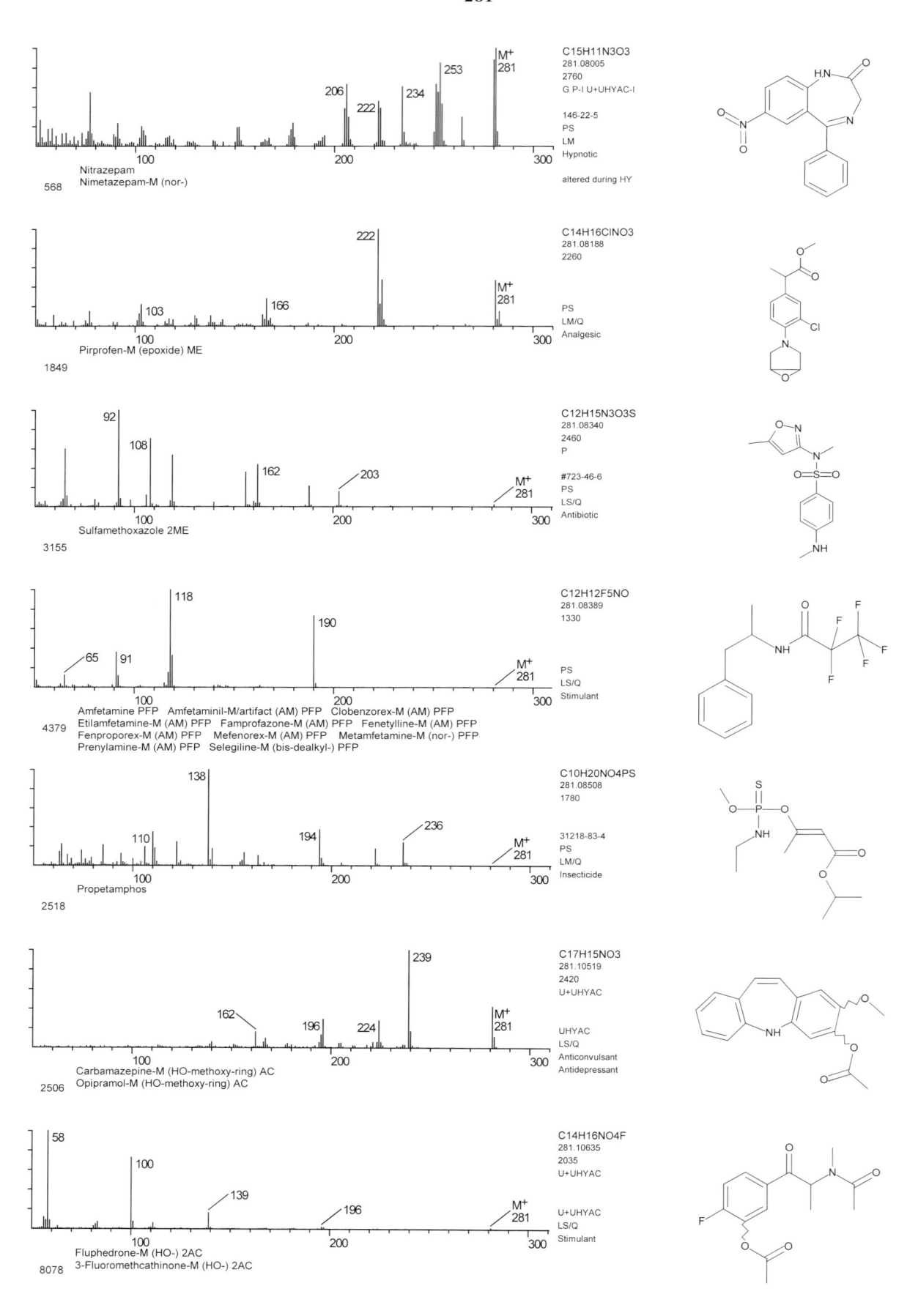

Nitrazepam
Nimetazepam-M (nor-)
568

C15H11N3O3
281.08005
2760
G P-I U+UHYAC-I

146-22-5
PS
LM
Hypnotic

altered during HY

Peaks: 206, 222, 234, 253, M+ 281

Pirprofen-M (epoxide) ME
1849

C14H16ClNO3
281.08188
2260

PS
LM/Q
Analgesic

Peaks: 103, 166, 222, M+ 281

Sulfamethoxazole 2ME
3155

C12H15N3O3S
281.08340
2460
P

#723-46-6
PS
LS/Q
Antibiotic

Peaks: 92, 108, 162, 203, M+ 281

Amfetamine PFP   Amfetaminil-M/artifact (AM) PFP   Clobenzorex-M (AM) PFP
Etilamfetamine-M (AM) PFP   Famprofazone-M (AM) PFP   Fenetylline-M (AM) PFP
Fenproporex-M (AM) PFP   Mefenorex-M (AM) PFP   Metamfetamine-M (nor-) PFP
Prenylamine-M (AM) PFP   Selegiline-M (bis-dealkyl-) PFP
4379

C12H12F5NO
281.08389
1330

PS
LS/Q
Stimulant

Peaks: 65, 91, 118, 190, M+ 281

Propetamphos
2518

C10H20NO4PS
281.08508
1780

31218-83-4
PS
LM/Q
Insecticide

Peaks: 110, 138, 194, 236, M+ 281

Carbamazepine-M (HO-methoxy-ring) AC
Opipramol-M (HO-methoxy-ring) AC
2506

C17H15NO3
281.10519
2420
U+UHYAC

UHYAC
LS/Q
Anticonvulsant
Antidepressant

Peaks: 162, 196, 224, 239, M+ 281

Fluphedrone-M (HO-) 2AC
3-Fluoromethcathinone-M (HO-) 2AC
8078

C14H16NO4F
281.10635
2035
U+UHYAC

U+UHYAC
LS/Q
Stimulant

Peaks: 58, 100, 139, 196, M+ 281

281

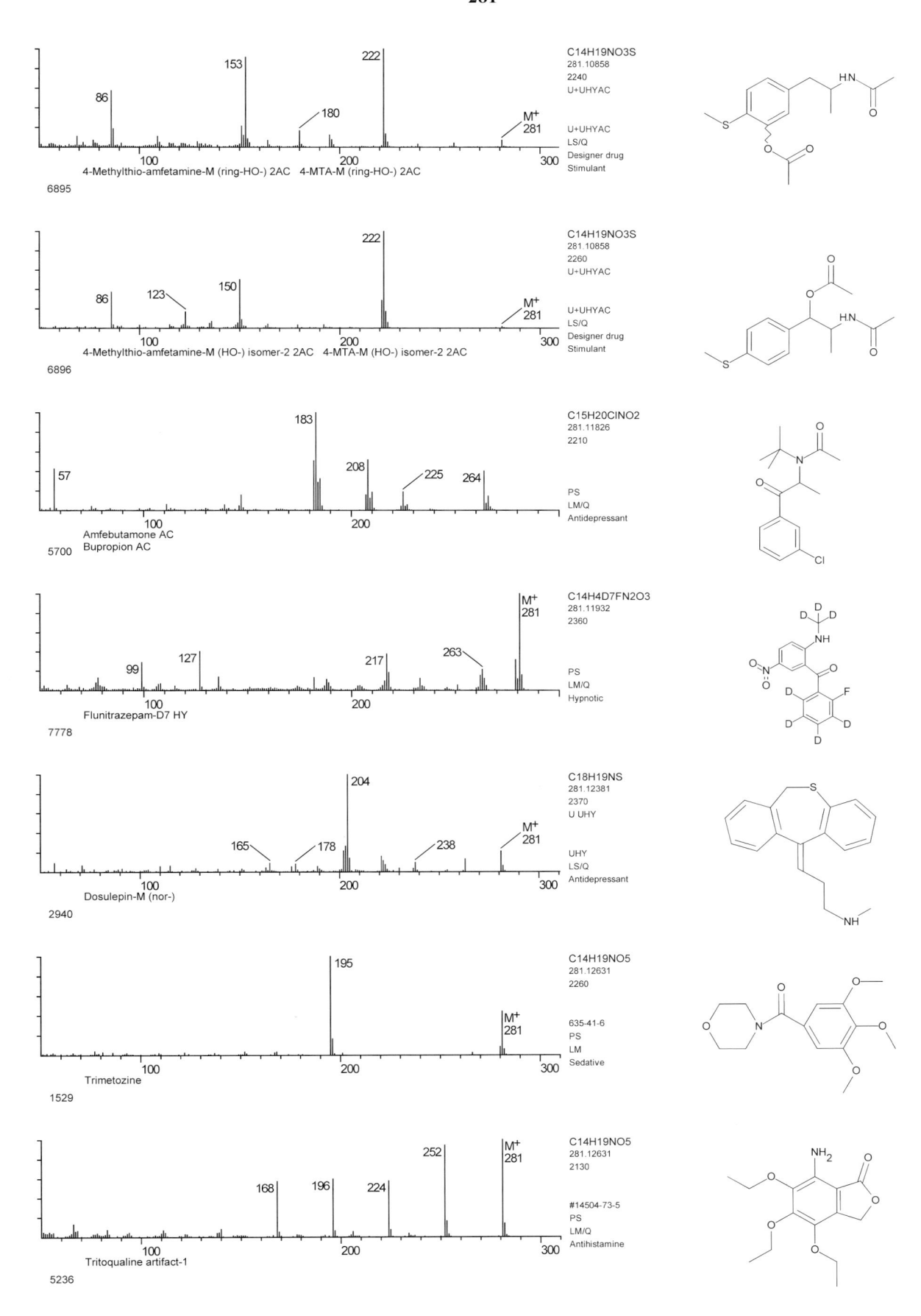

86 153 180 222 M+ 281	C14H19NO3S 281.10858 2240 U+UHYAC U+UHYAC LS/Q Designer drug Stimulant	
4-Methylthio-amfetamine-M (ring-HO-) 2AC   4-MTA-M (ring-HO-) 2AC
6895

86 123 150 222 M+ 281	C14H19NO3S 281.10858 2260 U+UHYAC U+UHYAC LS/Q Designer drug Stimulant	
4-Methylthio-amfetamine-M (HO-) isomer-2 2AC   4-MTA-M (HO-) isomer-2 2AC
6896

57 183 208 225 264	C15H20ClNO2 281.11826 2210 PS LM/Q Antidepressant	
Amfebutamone AC Bupropion AC
5700

99 127 217 263 M+ 281	C14H4D7FN2O3 281.11932 2360 PS LM/Q Hypnotic	
Flunitrazepam-D7 HY
7778

165 178 204 238 M+ 281	C18H19NS 281.12381 2370 U UHY UHY LS/Q Antidepressant	
Dosulepin-M (nor-)
2940

195 M+ 281	C14H19NO5 281.12631 2260 635-41-6 PS LM Sedative	
Trimetozine
1529

168 196 224 252 M+ 281	C14H19NO5 281.12631 2130 #14504-73-5 PS LM/Q Antihistamine	
Tritoqualine artifact-1
5236

281

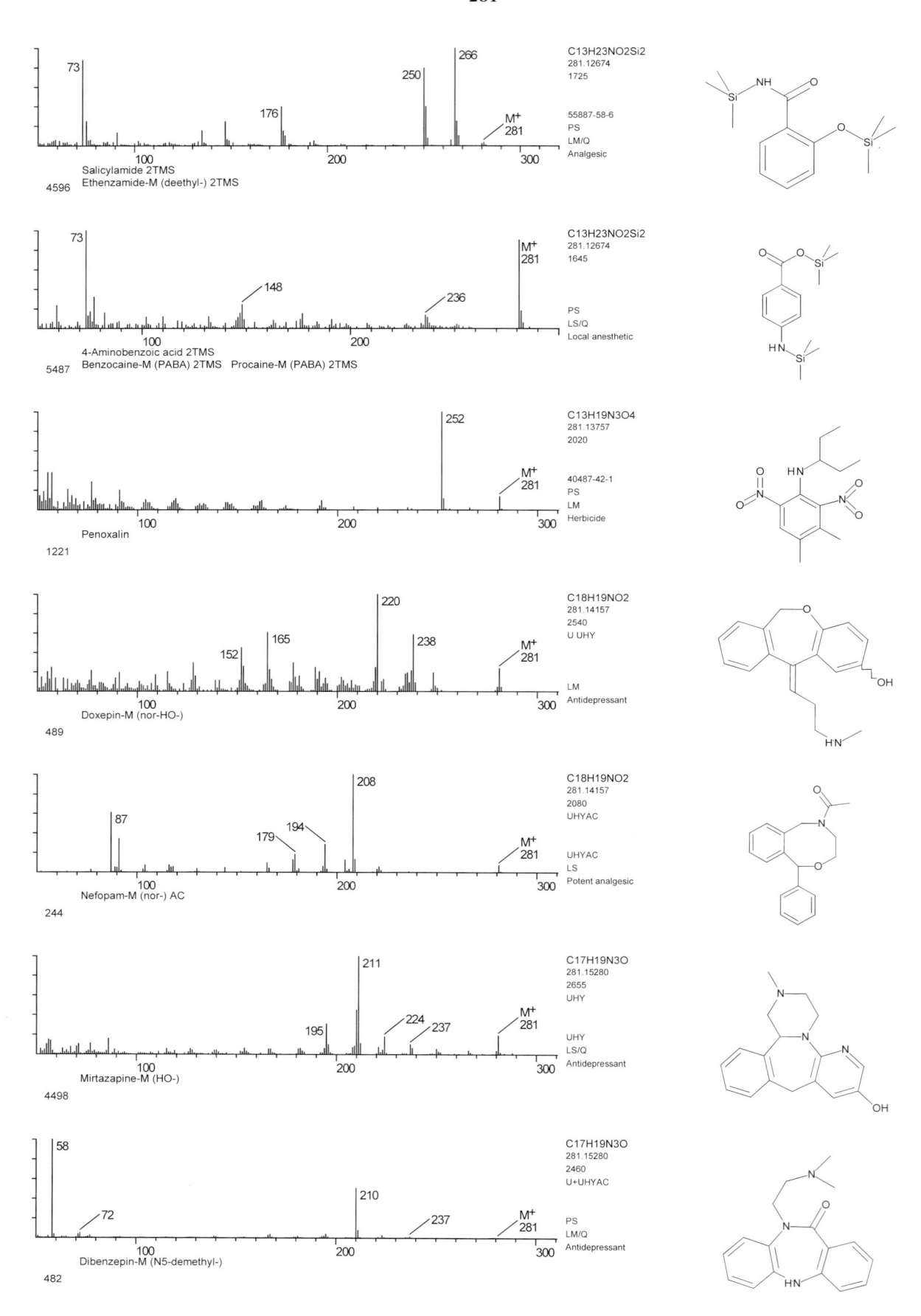

	C13H23NO2Si2 281.12674 1725 55887-58-6 PS LM/Q Analgesic	

4596 Salicylamide 2TMS
      Ethenzamide-M (deethyl-) 2TMS

| | C13H23NO2Si2
281.12674
1645
PS
LS/Q
Local anesthetic | |

5487 4-Aminobenzoic acid 2TMS
      Benzocaine-M (PABA) 2TMS   Procaine-M (PABA) 2TMS

| | C13H19N3O4
281.13757
2020
40487-42-1
PS
LM
Herbicide | |

1221 Penoxalin

| | C18H19NO2
281.14157
2540
U UHY
LM
Antidepressant | |

489   Doxepin-M (nor-HO-)

| | C18H19NO2
281.14157
2080
UHYAC
UHYAC
LS
Potent analgesic | |

244   Nefopam-M (nor-) AC

| | C17H19N3O
281.15280
2655
UHY
UHY
LS/Q
Antidepressant | |

4498 Mirtazapine-M (HO-)

| | C17H19N3O
281.15280
2460
U+UHYAC
PS
LM/Q
Antidepressant | |

482   Dibenzepin-M (N5-demethyl-)

281

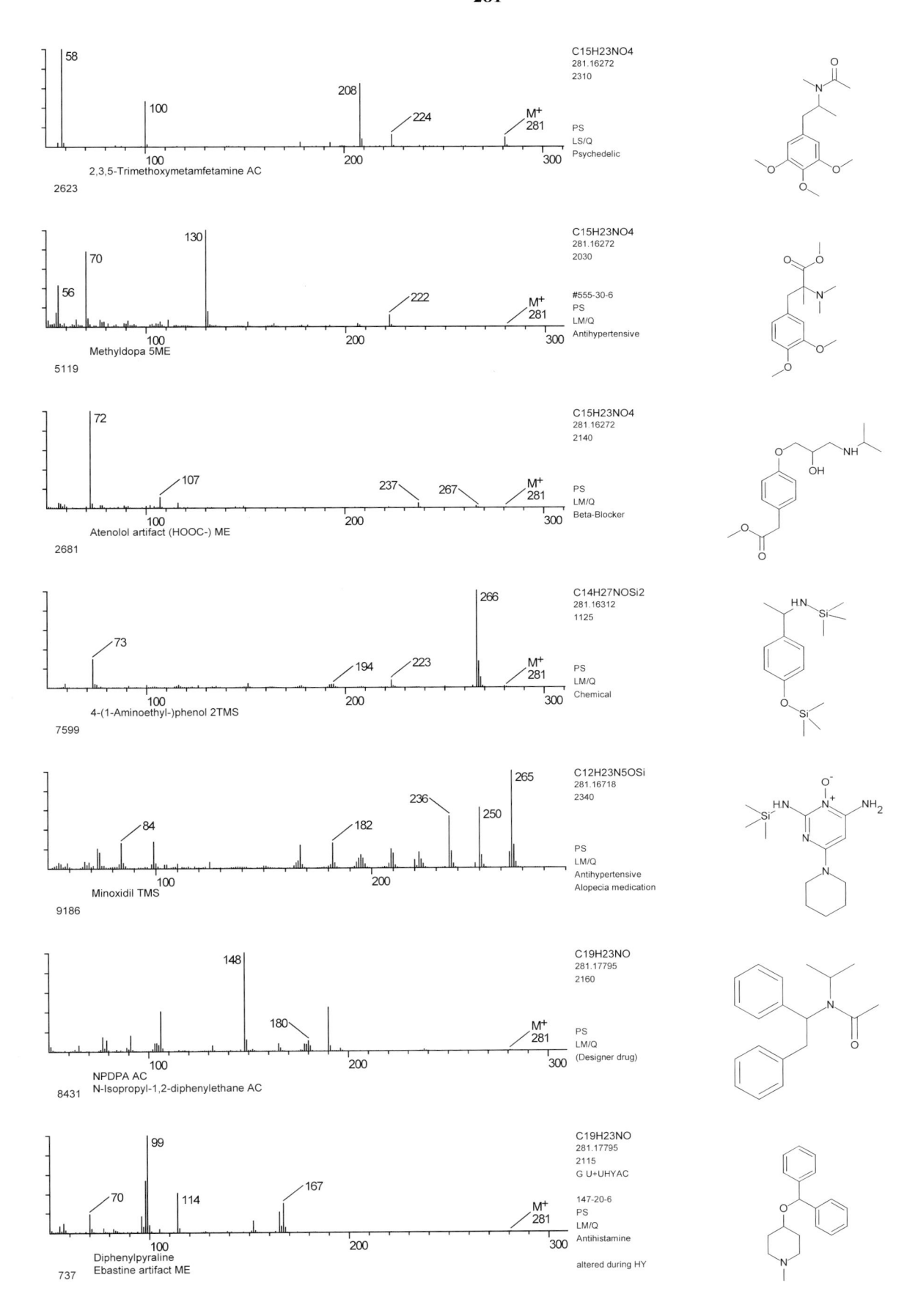

Spectrum	Compound	Formula / Info
2623	2,3,5-Trimethoxymetamfetamine AC (58, 100, 208, 224, M+ 281)	C15H23NO4, 281.16272, 2310, PS, LS/Q, Psychedelic
5119	Methyldopa 5ME (56, 70, 130, 222, M+ 281)	C15H23NO4, 281.16272, 2030, #555-30-6, PS, LM/Q, Antihypertensive
2681	Atenolol artifact (HOOC-) ME (72, 107, 237, 267, M+ 281)	C15H23NO4, 281.16272, 2140, PS, LM/Q, Beta-Blocker
7599	4-(1-Aminoethyl-)phenol 2TMS (73, 194, 223, 266, M+ 281)	C14H27NOSi2, 281.16312, 1125, PS, LM/Q, Chemical
9186	Minoxidil TMS (84, 182, 236, 250, 265, M+ 281)	C12H23N5OSi, 281.16718, 2340, PS, LM/Q, Antihypertensive, Alopecia medication
8431	NPDPA AC, N-Isopropyl-1,2-diphenylethane AC (148, 180, M+ 281)	C19H23NO, 281.17795, 2160, PS, LM/Q, (Designer drug)
737	Diphenylpyraline, Ebastine artifact ME (70, 99, 114, 167, M+ 281)	C19H23NO, 281.17795, 2115, G U+UHYAC, 147-20-6, PS, LM/Q, Antihistamine, altered during HY

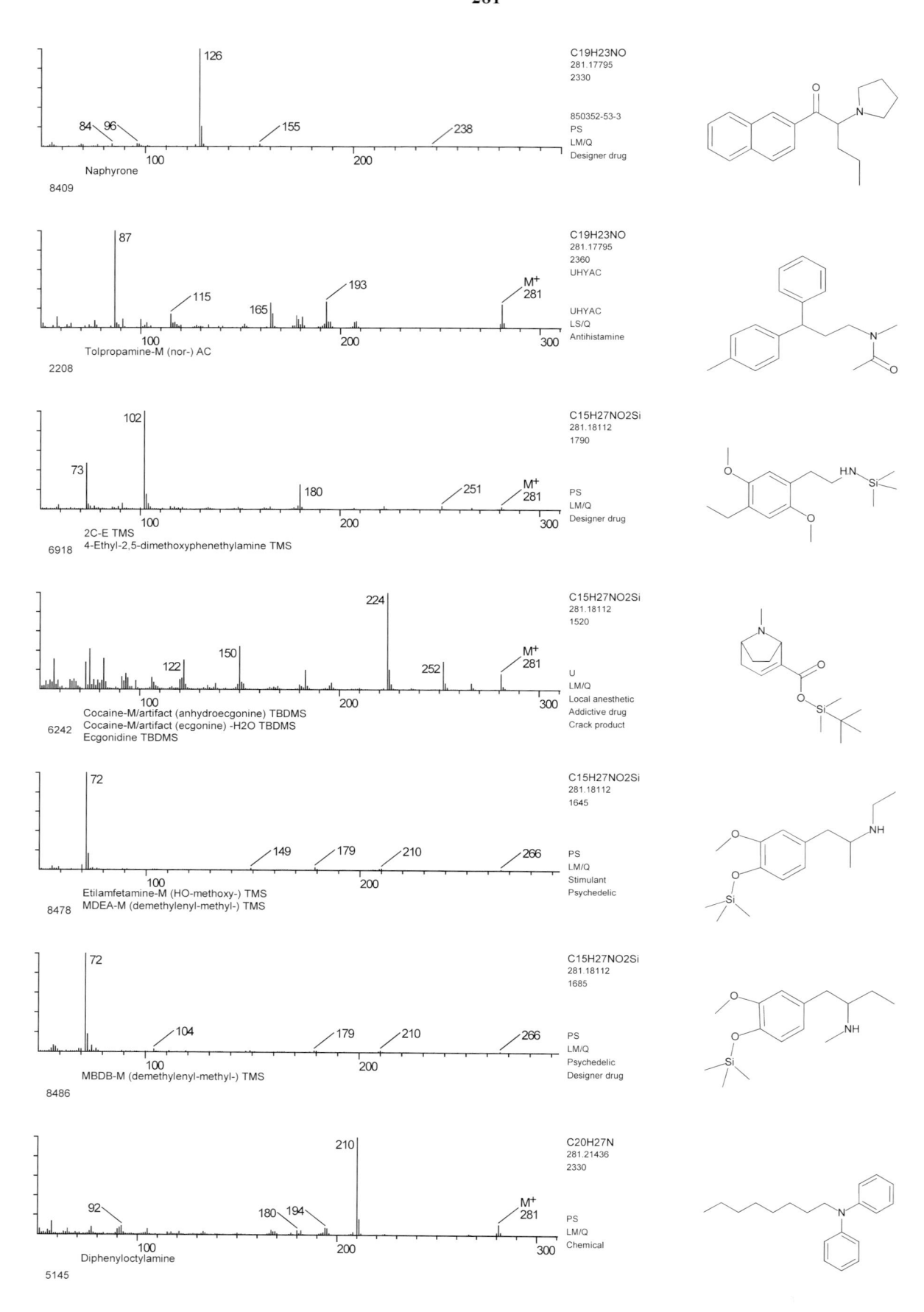

Naphyrone
8409
C19H23NO
281.17795
2330
850352-53-3
PS
LM/Q
Designer drug

Tolpropamine-M (nor-) AC
2208
C19H23NO
281.17795
2360
UHYAC
UHYAC
LS/Q
Antihistamine

2C-E TMS
4-Ethyl-2,5-dimethoxyphenethylamine TMS
6918
C15H27NO2Si
281.18112
1790
PS
LM/Q
Designer drug

Cocaine-M/artifact (anhydroecgonine) TBDMS
Cocaine-M/artifact (ecgonine) -H2O TBDMS
Ecgonidine TBDMS
6242
C15H27NO2Si
281.18112
1520
U
LM/Q
Local anesthetic
Addictive drug
Crack product

Etilamfetamine-M (HO-methoxy-) TMS
MDEA-M (demethylenyl-methyl-) TMS
8478
C15H27NO2Si
281.18112
1645
PS
LM/Q
Stimulant
Psychedelic

MBDB-M (demethylenyl-methyl-) TMS
8486
C15H27NO2Si
281.18112
1685
PS
LM/Q
Psychedelic
Designer drug

Diphenyloctylamine
5145
C20H27N
281.21436
2330
PS
LM/Q
Chemical

281

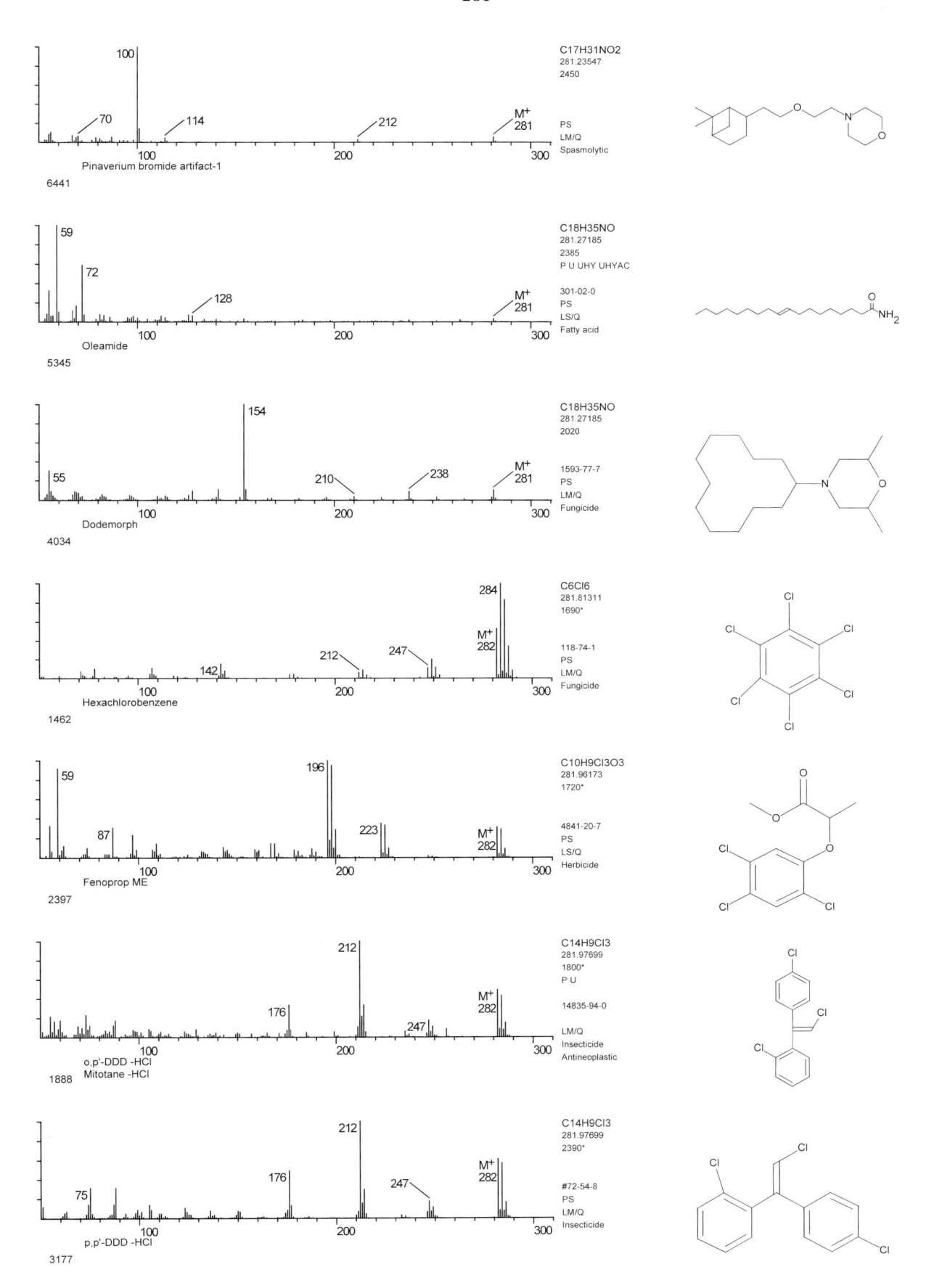

Pinaverium bromide artifact-1 6441	C17H31NO2 281.23547 2450 PS LM/Q Spasmolytic
Oleamide 5345	C18H35NO 281.27185 2385 P U UHY UHYAC 301-02-0 PS LS/Q Fatty acid
Dodemorph 4034	C18H35NO 281.27185 2020 1593-77-7 PS LM/Q Fungicide
Hexachlorobenzene 1462	C6Cl6 281.81311 1690* 118-74-1 PS LM/Q Fungicide
Fenoprop ME 2397	C10H9Cl3O3 281.96173 1720* 4841-20-7 PS LS/Q Herbicide
o,p'-DDD -HCl Mitotane -HCl 1888	C14H9Cl3 281.97699 1800* P U 14835-94-0 LM/Q Insecticide Antineoplastic
p,p'-DDD -HCl 3177	C14H9Cl3 281.97699 2390* #72-54-8 PS LM/Q Insecticide

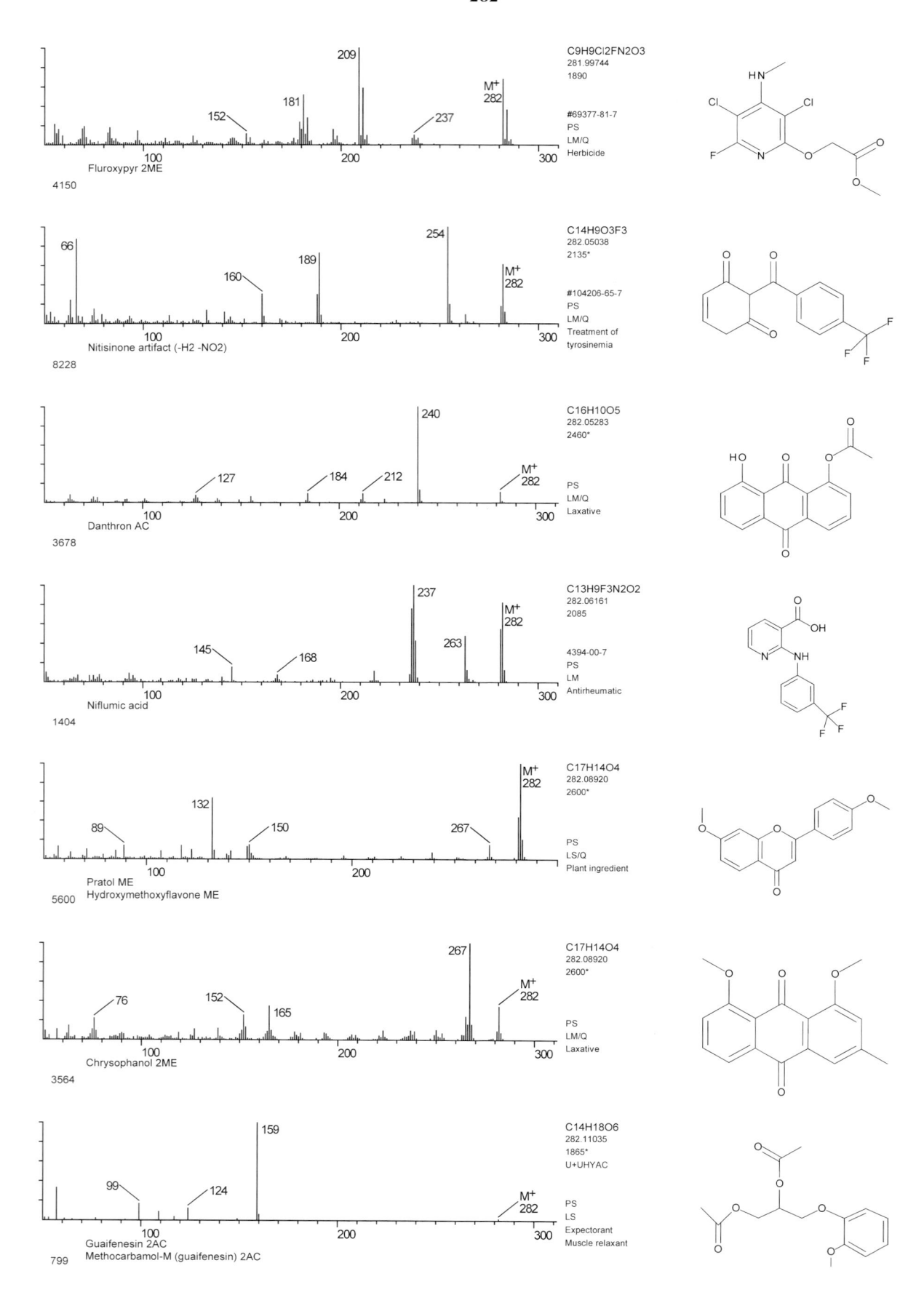

Fluroxypyr 2ME	C9H9Cl2FN2O3 281.99744 1890 #69377-81-7 PS LM/Q Herbicide
Nitisinone artifact (-H2 -NO2)	C14H9O3F3 282.05038 2135* #104206-65-7 PS LM/Q Treatment of tyrosinemia
Danthron AC	C16H10O5 282.05283 2460* PS LM/Q Laxative
Niflumic acid	C13H9F3N2O2 282.06161 2085 4394-00-7 PS LM Antirheumatic
Pratol ME Hydroxymethoxyflavone ME	C17H14O4 282.08920 2600* PS LS/Q Plant ingredient
Chrysophanol 2ME	C17H14O4 282.08920 2600* PS LM/Q Laxative
Guaifenesin 2AC Methocarbamol-M (guaifenesin) 2AC	C14H18O6 282.11035 1865* U+UHYAC PS LS Expectorant Muscle relaxant

282

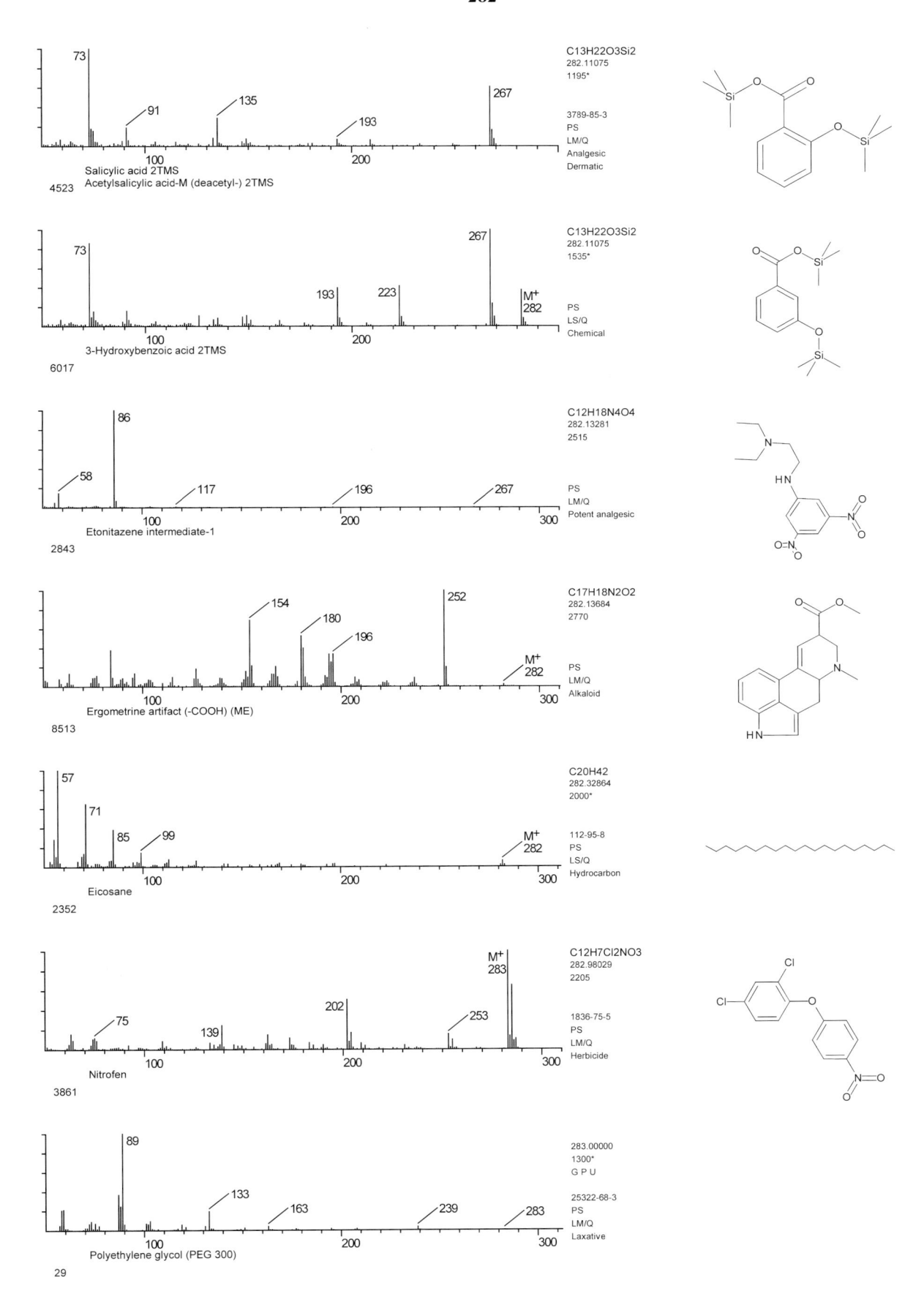

C13H22O3Si2
282.11075
1195*

3789-85-3
PS
LM/Q
Analgesic
Dermatic

Salicylic acid 2TMS
Acetylsalicylic acid-M (deacetyl-) 2TMS
4523

C13H22O3Si2
282.11075
1535*

PS
LS/Q
Chemical

3-Hydroxybenzoic acid 2TMS
6017

C12H18N4O4
282.13281
2515

PS
LM/Q
Potent analgesic

Etonitazene intermediate-1
2843

C17H18N2O2
282.13684
2770

PS
LM/Q
Alkaloid

Ergometrine artifact (-COOH) (ME)
8513

C20H42
282.32864
2000*

112-95-8
PS
LS/Q
Hydrocarbon

Eicosane
2352

C12H7Cl2NO3
282.98029
2205

1836-75-5
PS
LM/Q
Herbicide

Nitrofen
3861

283.00000
1300*
G P U

25322-68-3
PS
LM/Q
Laxative

Polyethylene glycol (PEG 300)
29

283

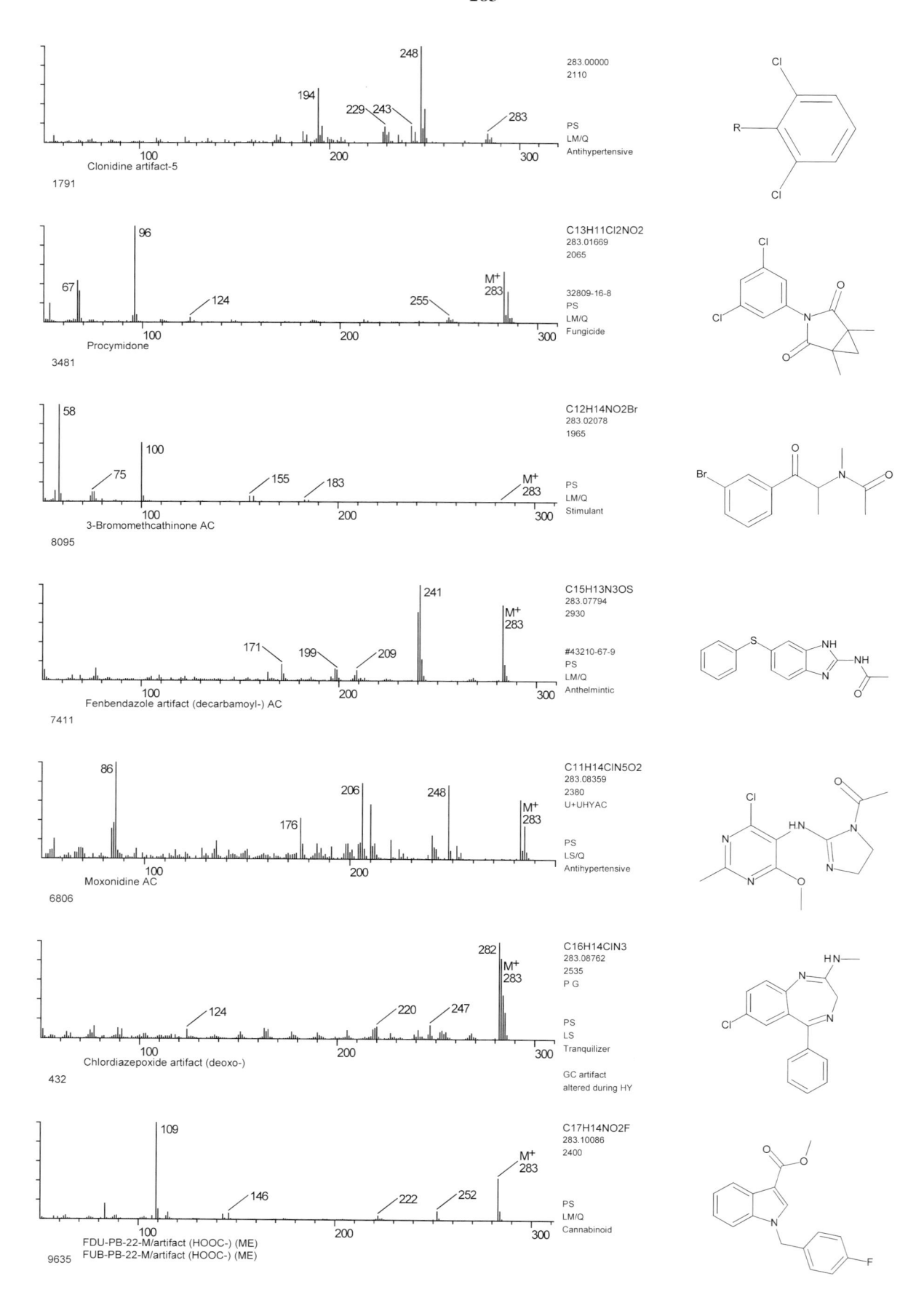

Clonidine artifact-5 — 283.00000 / 2110 / PS / LM/Q / Antihypertensive
Peaks: 194, 229, 243, 248, 283

Procymidone — C13H11Cl2NO2 / 283.01669 / 32809-16-8 / PS / LM/Q / Fungicide
Peaks: 67, 96, 124, 255, M+ 283

3-Bromomethcathinone AC — C12H14NO2Br / 283.02078 / 1965 / PS / LM/Q / Stimulant
Peaks: 58, 75, 100, 155, 183, M+ 283

Fenbendazole artifact (decarbamoyl-) AC — C15H13N3OS / 283.07794 / 2930 / #43210-67-9 / PS / LM/Q / Anthelmintic
Peaks: 171, 199, 209, 241, M+ 283

Moxonidine AC — C11H14ClN5O2 / 283.08359 / 2380 / U+UHYAC / PS / LS/Q / Antihypertensive
Peaks: 86, 176, 206, 248, M+ 283

Chlordiazepoxide artifact (deoxo-) — C16H14ClN3 / 283.08762 / 2535 / P G / PS / LS / Tranquilizer / GC artifact altered during HY
Peaks: 124, 220, 247, 282, M+ 283

FDU-PB-22-M/artifact (HOOC-) (ME)
FUB-PB-22-M/artifact (HOOC-) (ME) — C17H14NO2F / 283.10086 / 2400 / PS / LM/Q / Cannabinoid
Peaks: 109, 146, 222, 252, M+ 283

283

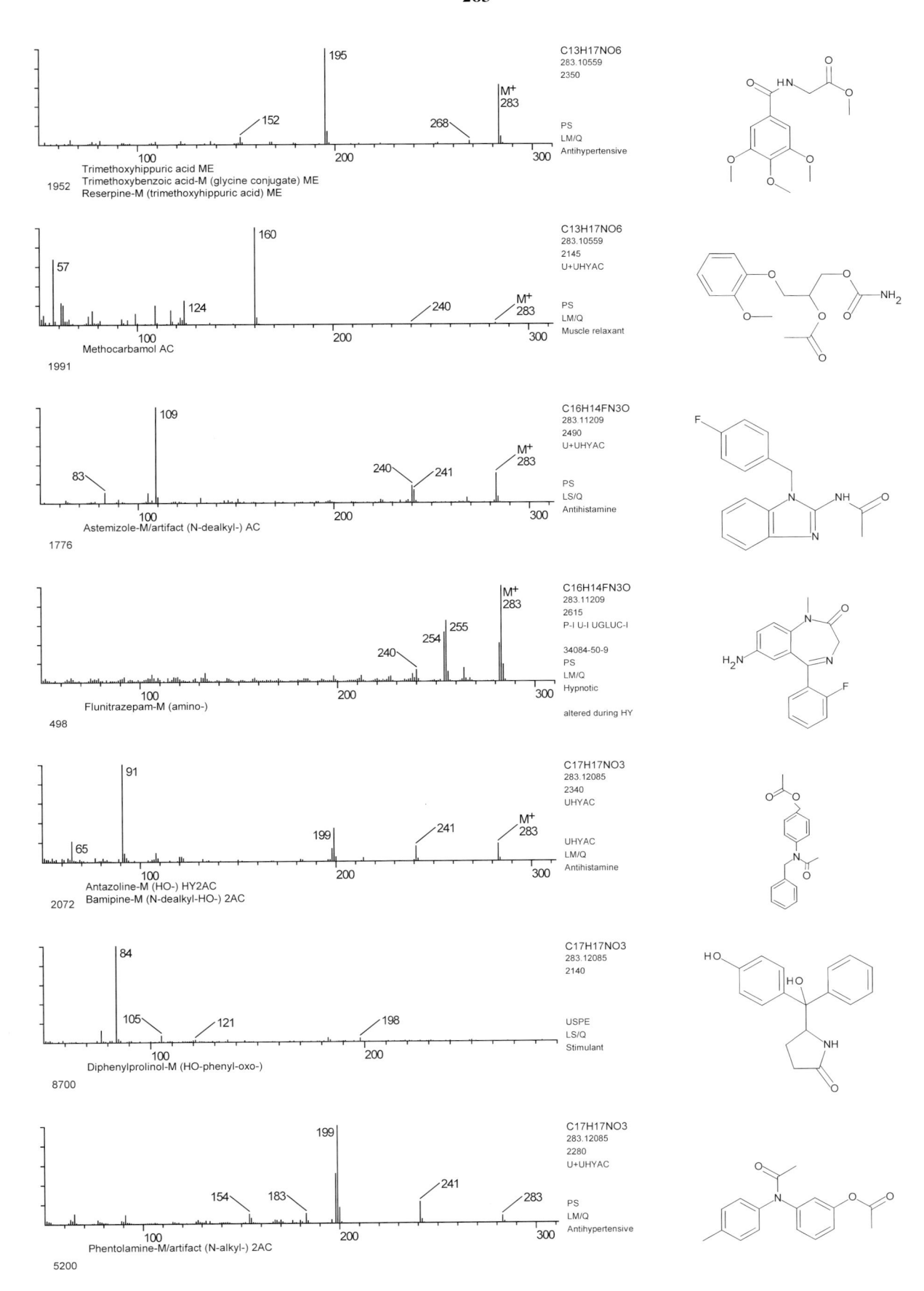

C13H17NO6
283.10559
2350

PS
LM/Q
Antihypertensive

Trimethoxyhippuric acid ME
Trimethoxybenzoic acid-M (glycine conjugate) ME
1952 Reserpine-M (trimethoxyhippuric acid) ME

C13H17NO6
283.10559
2145
U+UHYAC

PS
LM/Q
Muscle relaxant

Methocarbamol AC
1991

C16H14FN3O
283.11209
2490
U+UHYAC

PS
LS/Q
Antihistamine

Astemizole-M/artifact (N-dealkyl-) AC
1776

C16H14FN3O
283.11209
2615
P-I U-I UGLUC-I

34084-50-9
PS
LM/Q
Hypnotic

altered during HY

Flunitrazepam-M (amino-)
498

C17H17NO3
283.12085
2340
UHYAC

UHYAC
LM/Q
Antihistamine

Antazoline-M (HO-) HY2AC
2072 Bamipine-M (N-dealkyl-HO-) 2AC

C17H17NO3
283.12085
2140

USPE
LS/Q
Stimulant

Diphenylprolinol-M (HO-phenyl-oxo-)
8700

C17H17NO3
283.12085
2280
U+UHYAC

PS
LM/Q
Antihypertensive

Phentolamine-M/artifact (N-alkyl-) 2AC
5200

283

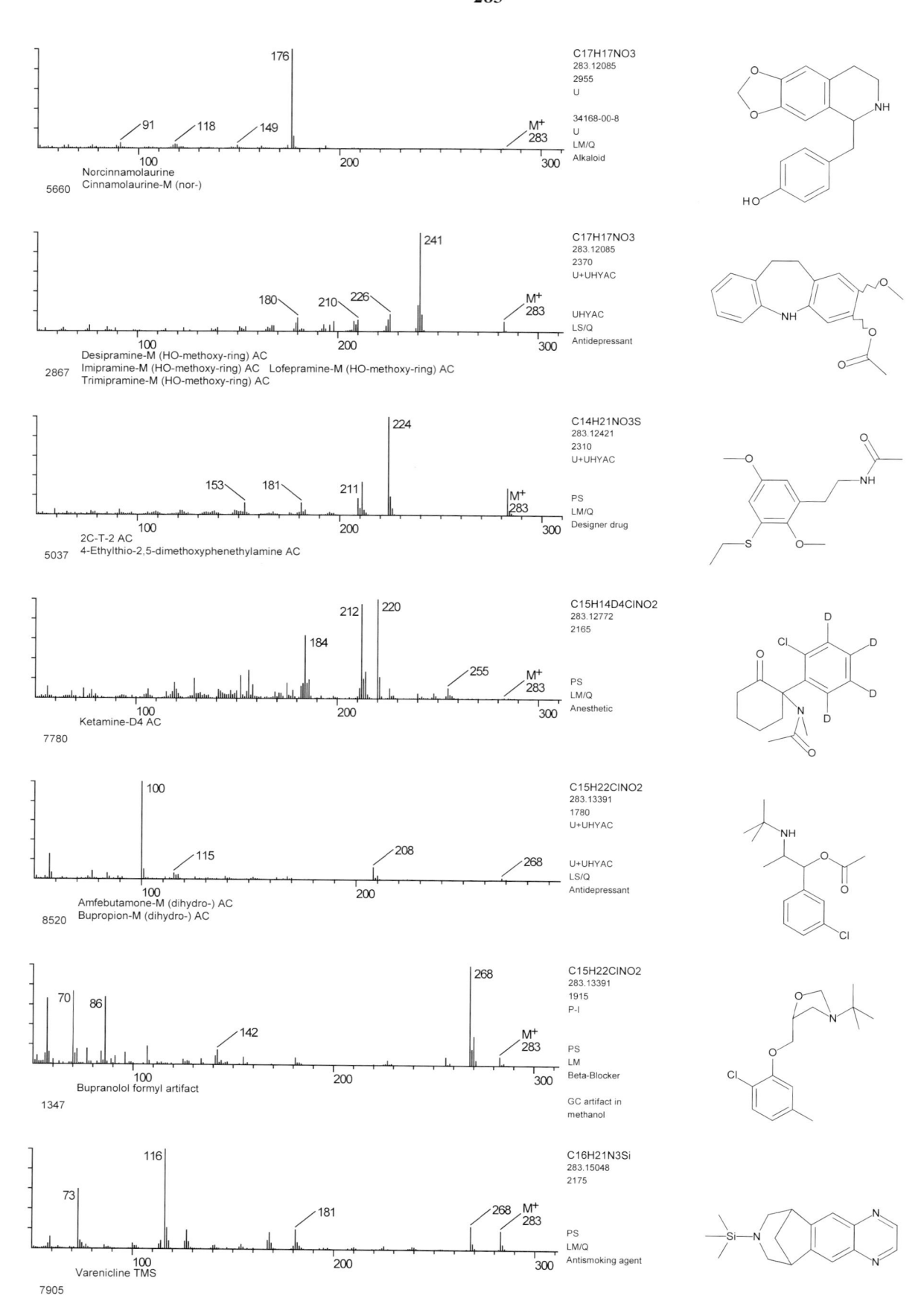

Spectrum	Formula / Info
5660 — Norcinnamolaurine / Cinnamolaurine-M (nor-); peaks 91, 118, 149, 176, M+ 283	C17H17NO3; 283.12085; 2955; U; 34168-00-8; U; LM/Q; Alkaloid
2867 — Desipramine-M (HO-methoxy-ring) AC; Imipramine-M (HO-methoxy-ring) AC; Lofepramine-M (HO-methoxy-ring) AC; Trimipramine-M (HO-methoxy-ring) AC; peaks 180, 210, 226, 241, M+ 283	C17H17NO3; 283.12085; 2370; U+UHYAC; UHYAC; LS/Q; Antidepressant
5037 — 2C-T-2 AC / 4-Ethylthio-2,5-dimethoxyphenethylamine AC; peaks 153, 181, 211, 224, M+ 283	C14H21NO3S; 283.12421; 2310; U+UHYAC; PS; LM/Q; Designer drug
7780 — Ketamine-D4 AC; peaks 184, 212, 220, 255, M+ 283	C15H14D4ClNO2; 283.12772; 2165; PS; LM/Q; Anesthetic
8520 — Amfebutamone-M (dihydro-) AC / Bupropion-M (dihydro-) AC; peaks 100, 115, 208, 268	C15H22ClNO2; 283.13391; 1780; U+UHYAC; U+UHYAC; LS/Q; Antidepressant
1347 — Bupranolol formyl artifact; peaks 70, 86, 142, 268, M+ 283	C15H22ClNO2; 283.13391; 1915; P-I; PS; LM; Beta-Blocker; GC artifact in methanol
7905 — Varenicline TMS; peaks 73, 116, 181, 268, M+ 283	C16H21N3Si; 283.15048; 2175; PS; LM/Q; Antismoking agent

796

283

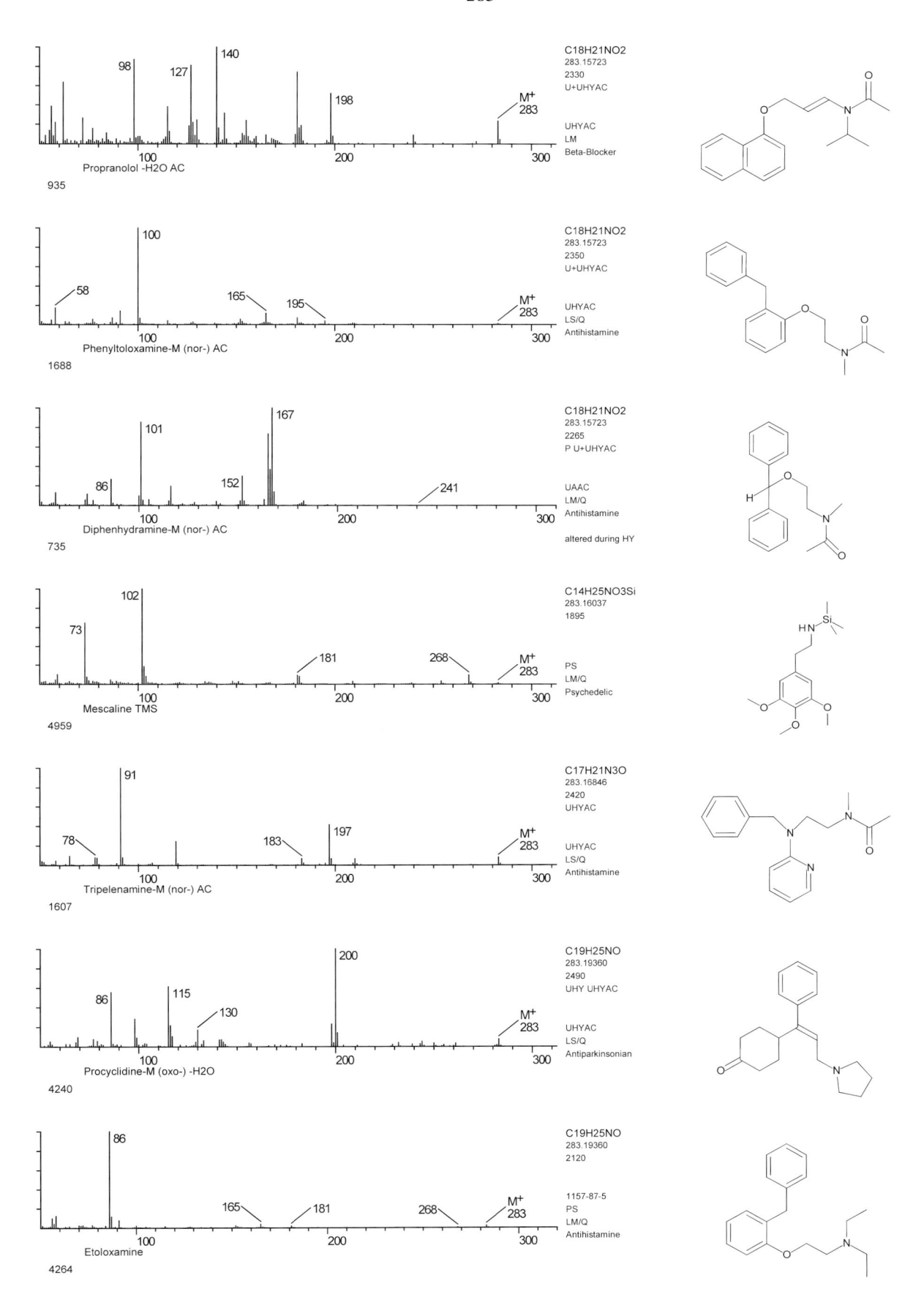

C18H21NO2
283.15723
2330
U+UHYAC

UHYAC
LM
Beta-Blocker

Propranolol -H2O AC

C18H21NO2
283.15723
2350
U+UHYAC

UHYAC
LS/Q
Antihistamine

Phenyltoloxamine-M (nor-) AC

C18H21NO2
283.15723
2265
P U+UHYAC

UAAC
LM/Q
Antihistamine

altered during HY

Diphenhydramine-M (nor-) AC

C14H25NO3Si
283.16037
1895

PS
LM/Q
Psychedelic

Mescaline TMS

C17H21N3O
283.16846
2420
UHYAC

UHYAC
LS/Q
Antihistamine

Tripelenamine-M (nor-) AC

C19H25NO
283.19360
2490
UHY UHYAC

UHYAC
LS/Q
Antiparkinsonian

Procyclidine-M (oxo-) -H2O

C19H25NO
283.19360
2120

1157-87-5
PS
LM/Q
Antihistamine

Etoloxamine

283

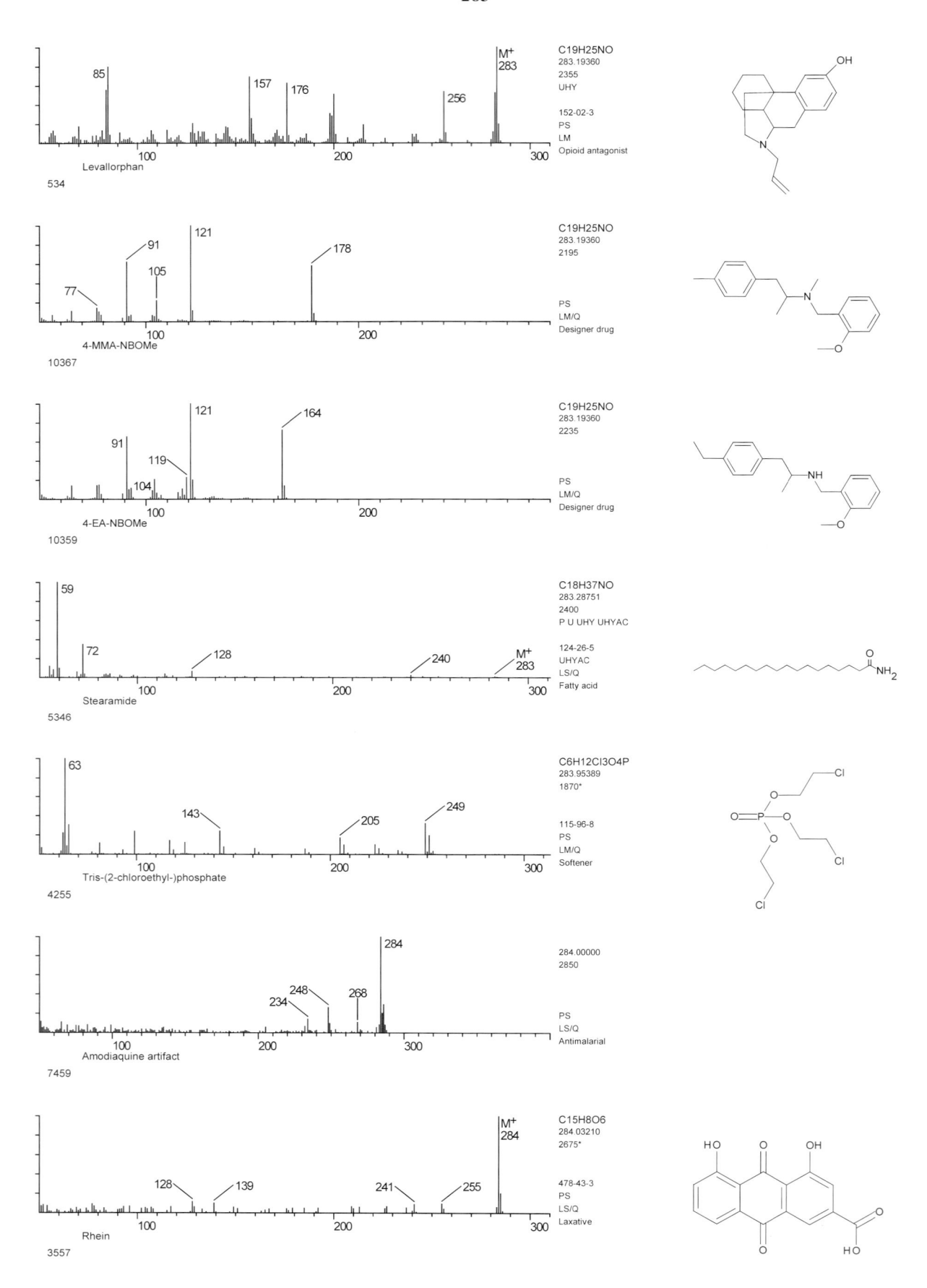

Spectrum	Formula / Data
Levallorphan — 85, 157, 176, 256, M+ 283	C19H25NO, 283.19360, 2355, UHY, 152-02-3, PS, LM, Opioid antagonist
4-MMA-NBOMe — 77, 91, 105, 121, 178	C19H25NO, 283.19360, 2195, PS, LM/Q, Designer drug
4-EA-NBOMe — 91, 104, 119, 121, 164	C19H25NO, 283.19360, 2235, PS, LM/Q, Designer drug
Stearamide — 59, 72, 128, 240, M+ 283	C18H37NO, 283.28751, 2400, P U UHY UHYAC, 124-26-5, UHYAC, LS/Q, Fatty acid
Tris-(2-chloroethyl-)phosphate — 63, 143, 205, 249	C6H12Cl3O4P, 283.95389, 1870*, 115-96-8, PS, LM/Q, Softener
Amodiaquine artifact — 234, 248, 268, 284	284.00000, 2850, PS, LS/Q, Antimalarial
Rhein — 128, 139, 241, 255, M+ 284	C15H8O6, 284.03210, 2675*, 478-43-3, PS, LS/Q, Laxative

284

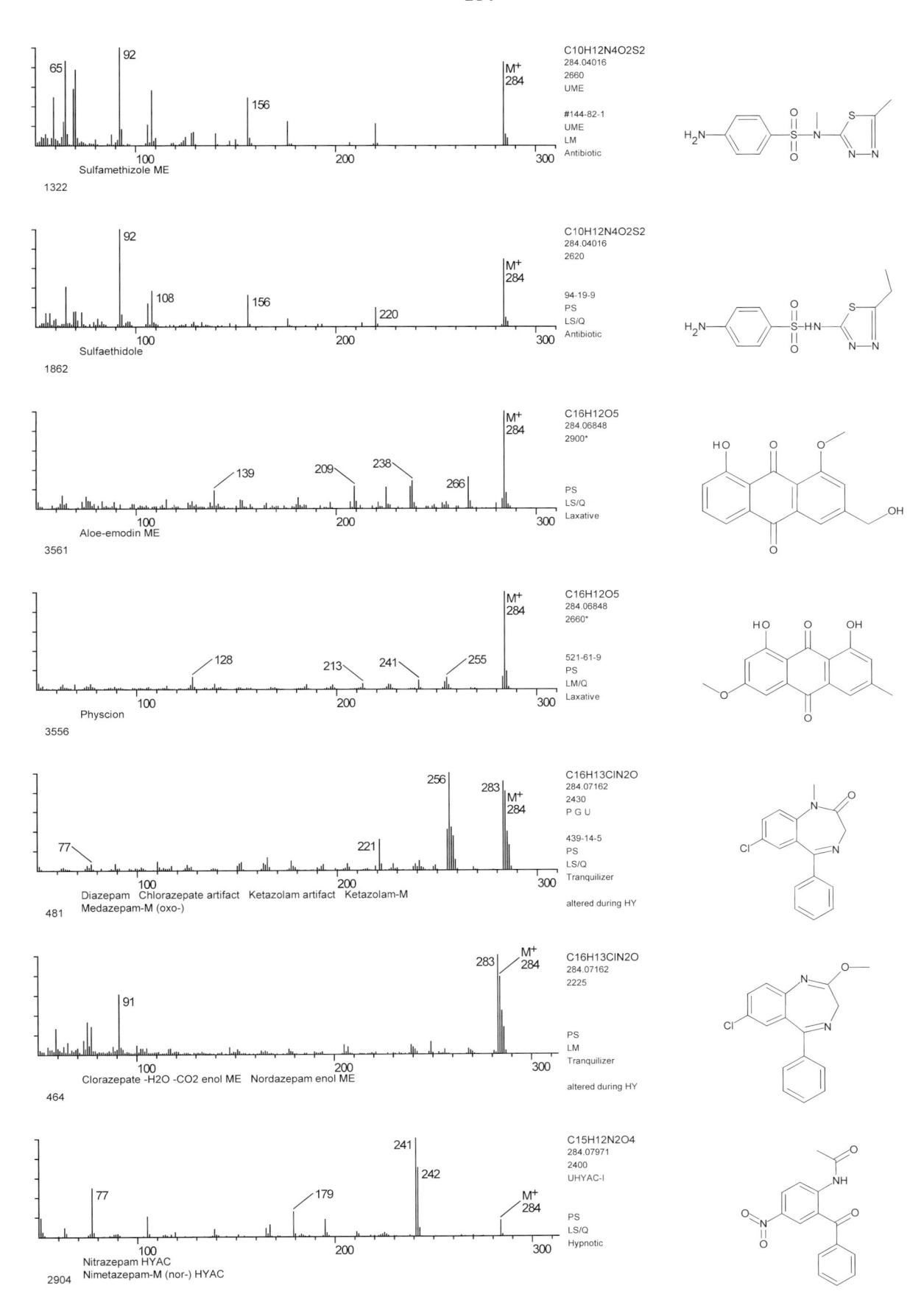

Spectrum	Formula / Info
Sulfamethizole ME (65, 92, 156, M+ 284) — 1322	C10H12N4O2S2, 284.04016, 2660, UME, #144-82-1, UME, LM, Antibiotic
Sulfaethidole (92, 108, 156, 220, M+ 284) — 1862	C10H12N4O2S2, 284.04016, 2620, 94-19-9, PS, LS/Q, Antibiotic
Aloe-emodin ME (139, 209, 238, 266, M+ 284) — 3561	C16H12O5, 284.06848, 2900*, PS, LS/Q, Laxative
Physcion (128, 213, 241, 255, M+ 284) — 3556	C16H12O5, 284.06848, 2660*, 521-61-9, PS, LM/Q, Laxative
Diazepam / Chlorazepate artifact / Ketazolam artifact / Ketazolam-M / Medazepam-M (oxo-) (77, 221, 256, 283, M+ 284) — 481	C16H13ClN2O, 284.07162, 2430, P G U, 439-14-5, PS, LS/Q, Tranquilizer, altered during HY
Clorazepate -H2O -CO2 enol ME / Nordazepam enol ME (91, 283, M+ 284) — 464	C16H13ClN2O, 284.07162, 2225, PS, LM, Tranquilizer, altered during HY
Nitrazepam HYAC / Nimetazepam-M (nor-) HYAC (77, 179, 241, 242, M+ 284) — 2904	C15H12N2O4, 284.07971, 2400, UHYAC-I, PS, LS/Q, Hypnotic

284

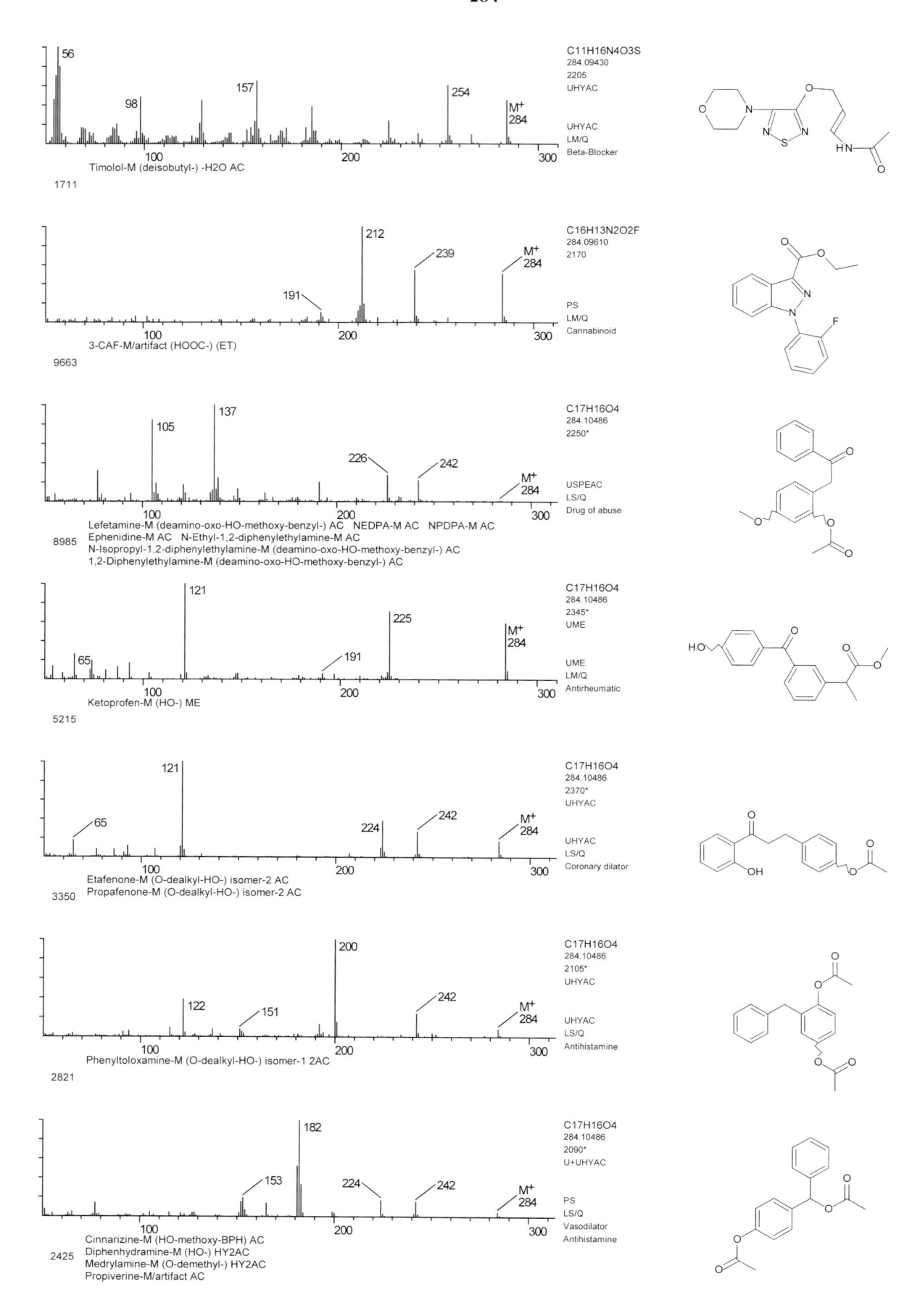

C11H16N4O3S
284.09430
2205
UHYAC

UHYAC
LM/Q
Beta-Blocker

Timolol-M (deisobutyl-) -H2O AC
1711

C16H13N2O2F
284.09610
2170

PS
LM/Q
Cannabinoid

3-CAF-M/artifact (HOOC-) (ET)
9663

C17H16O4
284.10486
2250*

USPEAC
LS/Q
Drug of abuse

Lefetamine-M (deamino-oxo-HO-methoxy-benzyl-) AC   NEDPA-M AC   NPDPA-M AC
Ephenidine-M AC   N-Ethyl-1,2-diphenylethylamine-M AC
N-Isopropyl-1,2-diphenylethylamine-M (deamino-oxo-HO-methoxy-benzyl-) AC
1,2-Diphenylethylamine-M (deamino-oxo-HO-methoxy-benzyl-) AC
8985

C17H16O4
284.10486
2345*
UME

UME
LM/Q
Antirheumatic

Ketoprofen-M (HO-) ME
5215

C17H16O4
284.10486
2370*
UHYAC

UHYAC
LS/Q
Coronary dilator

Etafenone-M (O-dealkyl-HO-) isomer-2 AC
Propafenone-M (O-dealkyl-HO-) isomer-2 AC
3350

C17H16O4
284.10486
2105*
UHYAC

UHYAC
LS/Q
Antihistamine

Phenyltoloxamine-M (O-dealkyl-HO-) isomer-1 2AC
2821

C17H16O4
284.10486
2090*
U+UHYAC

PS
LS/Q
Vasodilator
Antihistamine

Cinnarizine-M (HO-methoxy-BPH) AC
Diphenhydramine-M (HO-) HY2AC
Medrylamine-M (O-demethyl-) HY2AC
Propiverine-M/artifact AC
2425

284

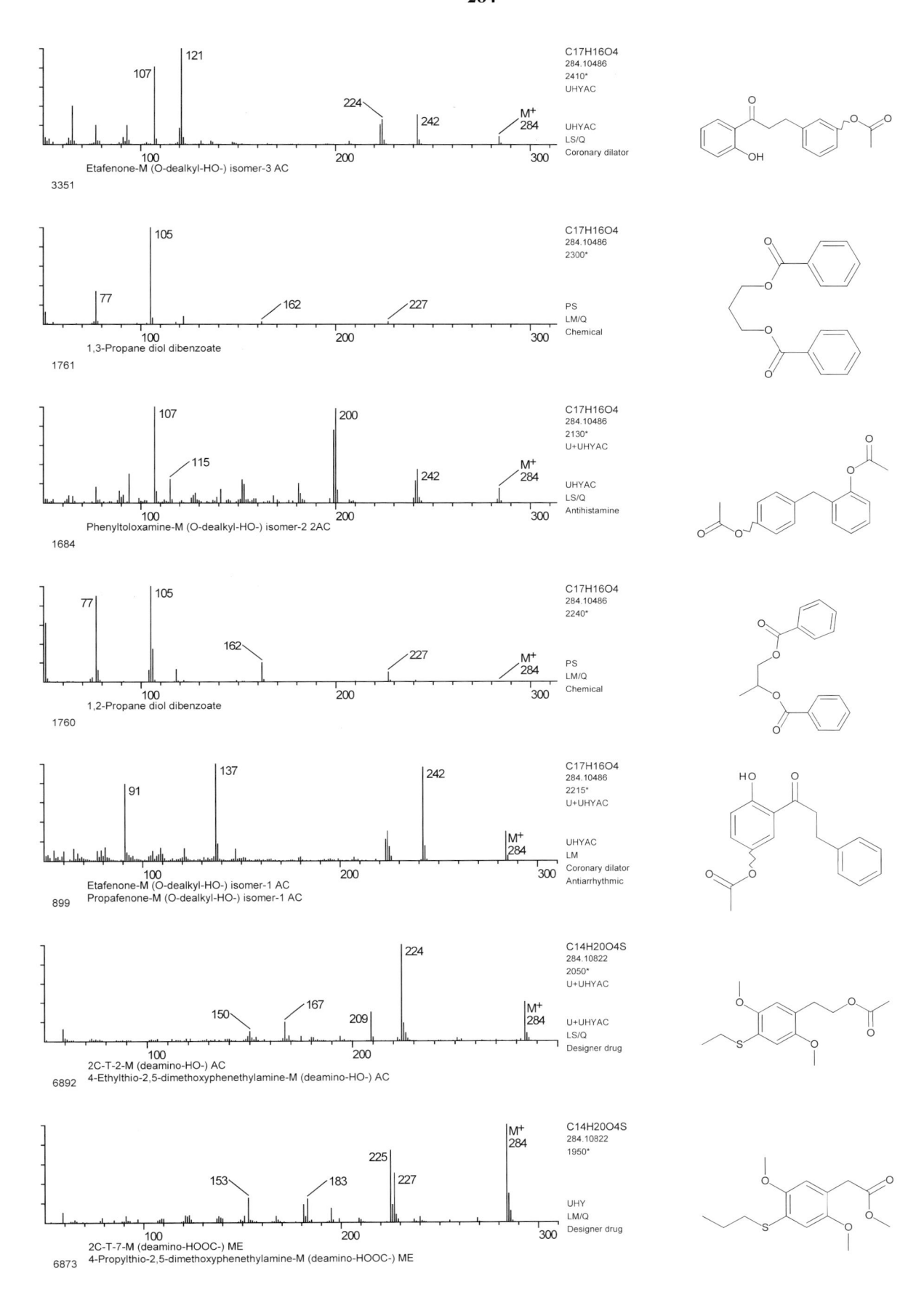

Spectrum	Compound	Formula / Info
3351	Etafenone-M (O-dealkyl-HO-) isomer-3 AC — peaks 107, 121, 224, 242, M+ 284	C17H16O4, 284.10486, 2410*, UHYAC / UHYAC, LS/Q, Coronary dilator
1761	1,3-Propane diol dibenzoate — peaks 77, 105, 162, 227	C17H16O4, 284.10486, 2300*, PS, LM/Q, Chemical
1684	Phenyltoloxamine-M (O-dealkyl-HO-) isomer-2 2AC — peaks 107, 115, 200, 242, M+ 284	C17H16O4, 284.10486, 2130*, U+UHYAC / UHYAC, LS/Q, Antihistamine
1760	1,2-Propane diol dibenzoate — peaks 77, 105, 162, 227, M+ 284	C17H16O4, 284.10486, 2240*, PS, LM/Q, Chemical
899	Etafenone-M (O-dealkyl-HO-) isomer-1 AC / Propafenone-M (O-dealkyl-HO-) isomer-1 AC — peaks 91, 137, 242, M+ 284	C17H16O4, 284.10486, 2215*, U+UHYAC / UHYAC, LM, Coronary dilator, Antiarrhythmic
6892	2C-T-2-M (deamino-HO-) AC / 4-Ethylthio-2,5-dimethoxyphenethylamine-M (deamino-HO-) AC — peaks 150, 167, 209, 224, M+ 284	C14H20O4S, 284.10822, 2050*, U+UHYAC / U+UHYAC, LS/Q, Designer drug
6873	2C-T-7-M (deamino-HOOC-) ME / 4-Propylthio-2,5-dimethoxyphenethylamine-M (deamino-HOOC-) ME — peaks 153, 183, 225, 227, M+ 284	C14H20O4S, 284.10822, 1950*, UHY, LM/Q, Designer drug

284

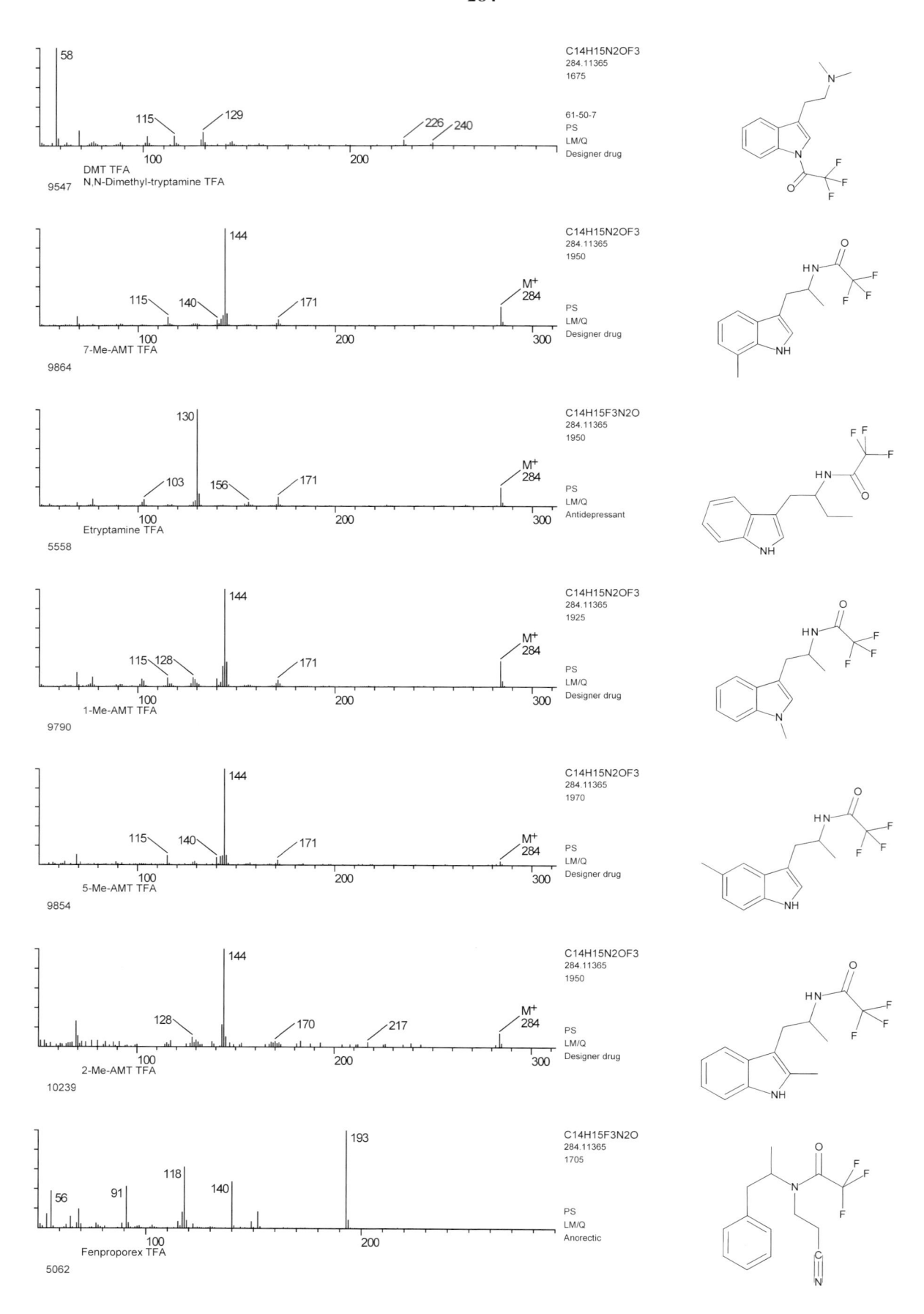

	C14H15N2OF3 284.11365 1675	
	61-50-7 PS LM/Q Designer drug	
DMT TFA N,N-Dimethyl-tryptamine TFA 9547		

	C14H15N2OF3 284.11365 1950 PS LM/Q Designer drug
7-Me-AMT TFA 9864	

	C14H15F3N2O 284.11365 1950 PS LM/Q Antidepressant
Etryptamine TFA 5558	

	C14H15N2OF3 284.11365 1925 PS LM/Q Designer drug
1-Me-AMT TFA 9790	

	C14H15N2OF3 284.11365 1970 PS LM/Q Designer drug
5-Me-AMT TFA 9854	

	C14H15N2OF3 284.11365 1950 PS LM/Q Designer drug
2-Me-AMT TFA 10239	

	C14H15F3N2O 284.11365 1705 PS LM/Q Anorectic
Fenproporex TFA 5062	

802

284

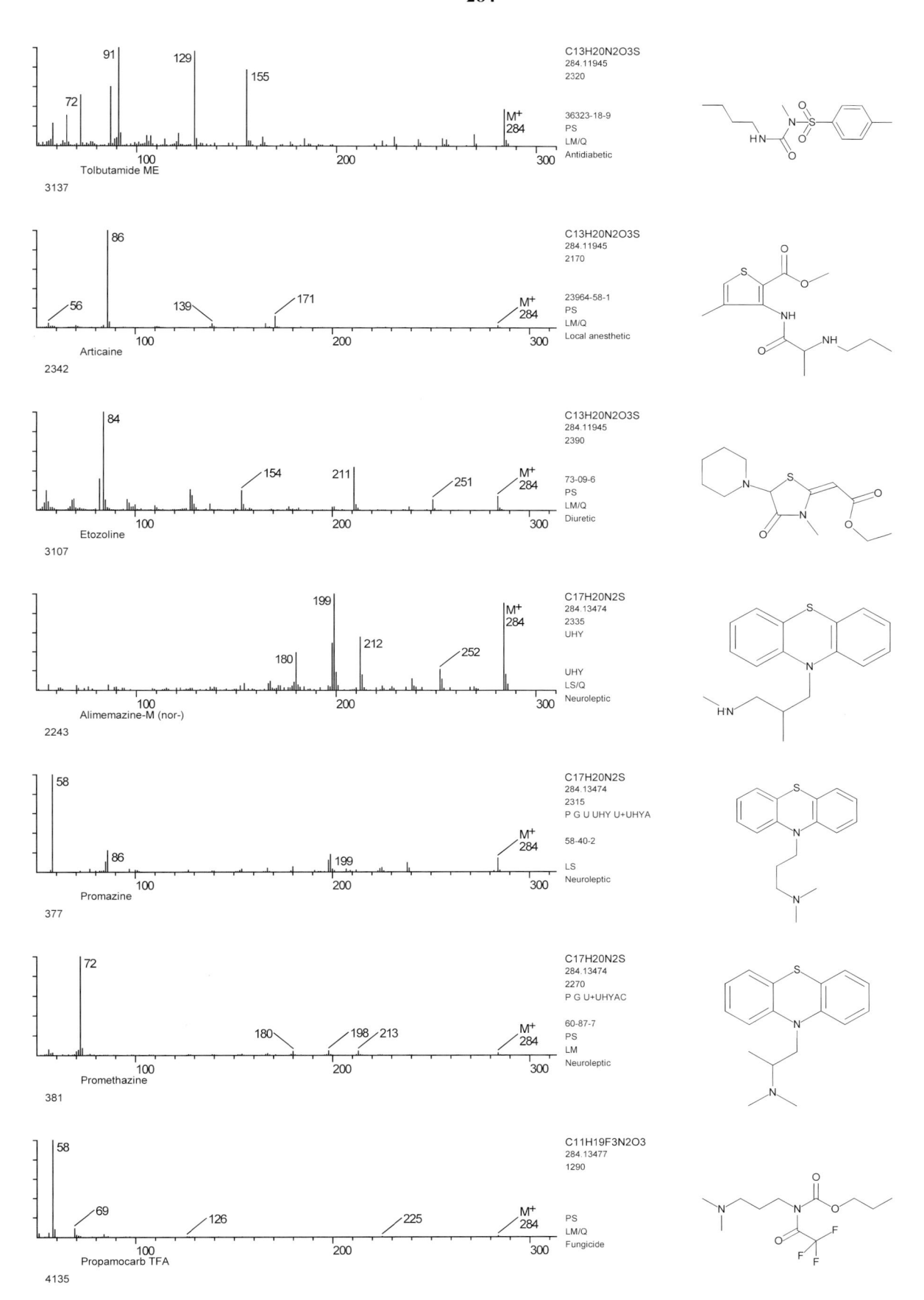

Tolbutamide ME — 3137	C13H20N2O3S 284.11945 2320 / 36323-18-9 PS LM/Q Antidiabetic	peaks: 72, 91, 129, 155, M+ 284
Articaine — 2342	C13H20N2O3S 284.11945 2170 / 23964-58-1 PS LM/Q Local anesthetic	peaks: 56, 86, 139, 171, M+ 284
Etozoline — 3107	C13H20N2O3S 284.11945 2390 / 73-09-6 PS LM/Q Diuretic	peaks: 84, 154, 211, 251, M+ 284
Alimemazine-M (nor-) — 2243	C17H20N2S 284.13474 2335 UHY / UHY LS/Q Neuroleptic	peaks: 180, 199, 212, 252, M+ 284
Promazine — 377	C17H20N2S 284.13474 2315 P G U UHY U+UHYA / 58-40-2 LS Neuroleptic	peaks: 58, 86, 199, M+ 284
Promethazine — 381	C17H20N2S 284.13474 2270 P G U+UHYAC / 60-87-7 PS LM Neuroleptic	peaks: 72, 180, 198, 213, M+ 284
Propamocarb TFA — 4135	C11H19F3N2O3 284.13477 1290 / PS LM/Q Fungicide	peaks: 58, 69, 126, 225, M+ 284

284

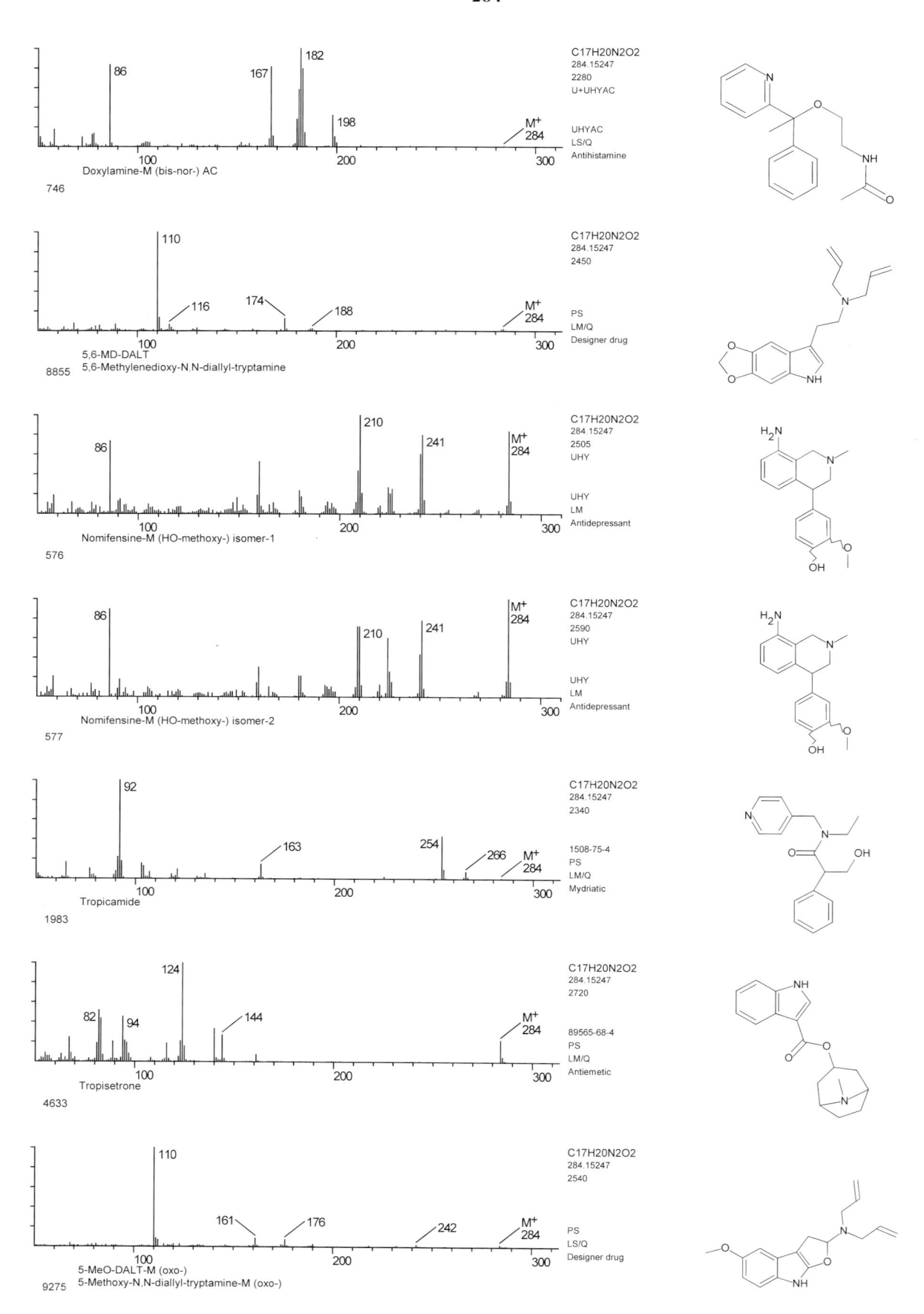

Doxylamine-M (bis-nor-) AC
746
Peaks: 86, 167, 182, 198, M+ 284
C17H20N2O2
284.15247
2280
U+UHYAC
UHYAC
LS/Q
Antihistamine

5,6-MD-DALT
5,6-Methylenedioxy-N,N-diallyl-tryptamine
8855
Peaks: 110, 116, 174, 188, M+ 284
C17H20N2O2
284.15247
2450
PS
LM/Q
Designer drug

Nomifensine-M (HO-methoxy-) isomer-1
576
Peaks: 86, 210, 241, M+ 284
C17H20N2O2
284.15247
2505
UHY
UHY
LM
Antidepressant

Nomifensine-M (HO-methoxy-) isomer-2
577
Peaks: 86, 210, 241, M+ 284
C17H20N2O2
284.15247
2590
UHY
UHY
LM
Antidepressant

Tropicamide
1983
Peaks: 92, 163, 254, 266, M+ 284
C17H20N2O2
284.15247
2340
1508-75-4
PS
LM/Q
Mydriatic

Tropisetrone
4633
Peaks: 82, 94, 124, 144, M+ 284
C17H20N2O2
284.15247
2720
89565-68-4
PS
LM/Q
Antiemetic

5-MeO-DALT-M (oxo-)
5-Methoxy-N,N-diallyl-tryptamine-M (oxo-)
9275
Peaks: 110, 161, 176, 242, M+ 284
C17H20N2O2
284.15247
2540
PS
LS/Q
Designer drug

284

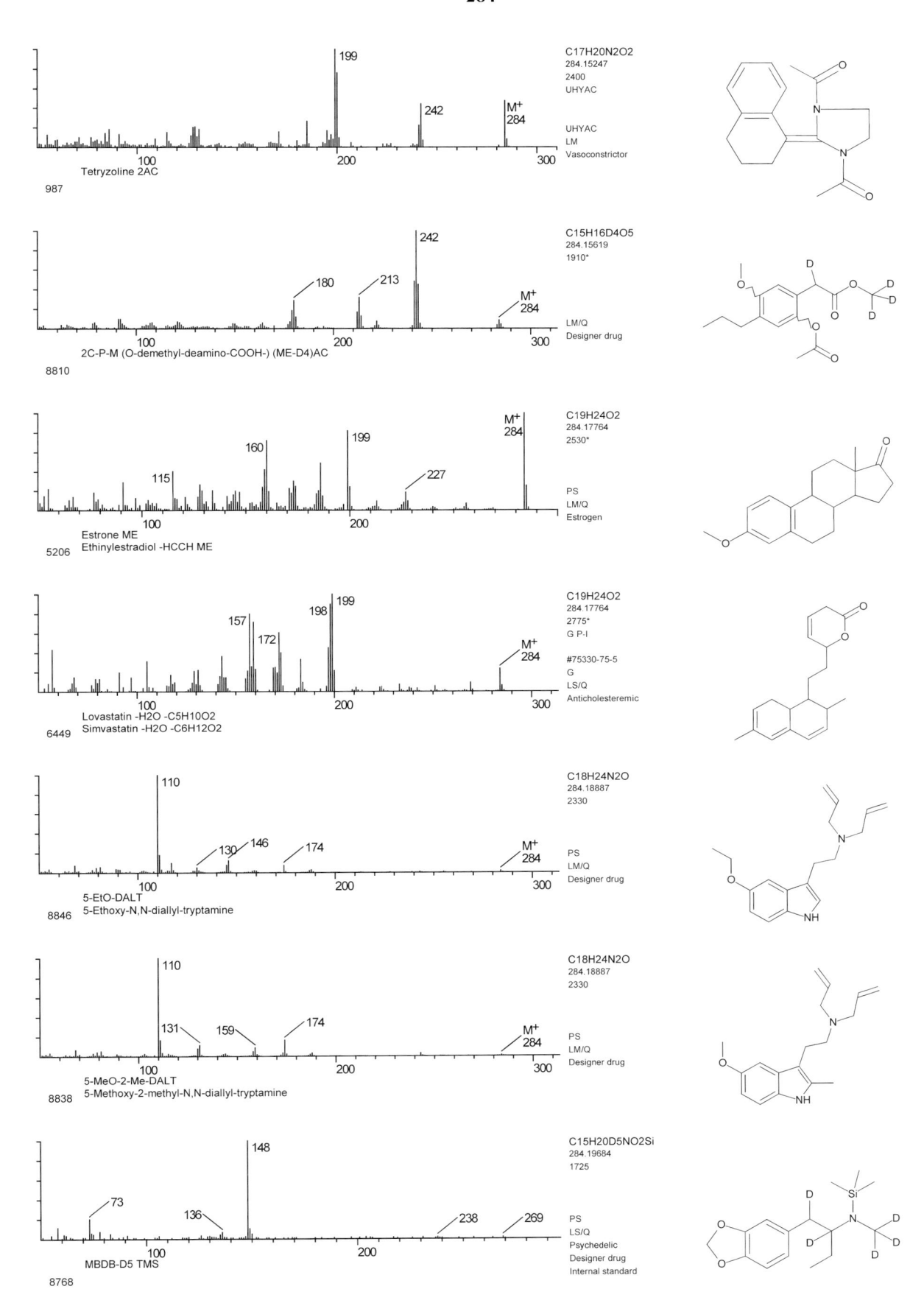

C17H20N2O2
284.15247
2400
UHYAC

UHYAC
LM
Vasoconstrictor

Tetryzoline 2AC
987

C15H16D4O5
284.15619
1910*

LM/Q
Designer drug

2C-P-M (O-demethyl-deamino-COOH-) (ME-D4)AC
8810

C19H24O2
284.17764
2530*

PS
LM/Q
Estrogen

Estrone ME
Ethinylestradiol -HCCH ME
5206

C19H24O2
284.17764
2775*
G P-I

#75330-75-5
G
LS/Q
Anticholesteremic

Lovastatin -H2O -C5H10O2
Simvastatin -H2O -C6H12O2
6449

C18H24N2O
284.18887
2330

PS
LM/Q
Designer drug

5-EtO-DALT
5-Ethoxy-N,N-diallyl-tryptamine
8846

C18H24N2O
284.18887
2330

PS
LM/Q
Designer drug

5-MeO-2-Me-DALT
5-Methoxy-2-methyl-N,N-diallyl-tryptamine
8838

C15H20D5NO2Si
284.19684
1725

PS
LS/Q
Psychedelic
Designer drug
Internal standard

MBDB-D5 TMS
8768

284

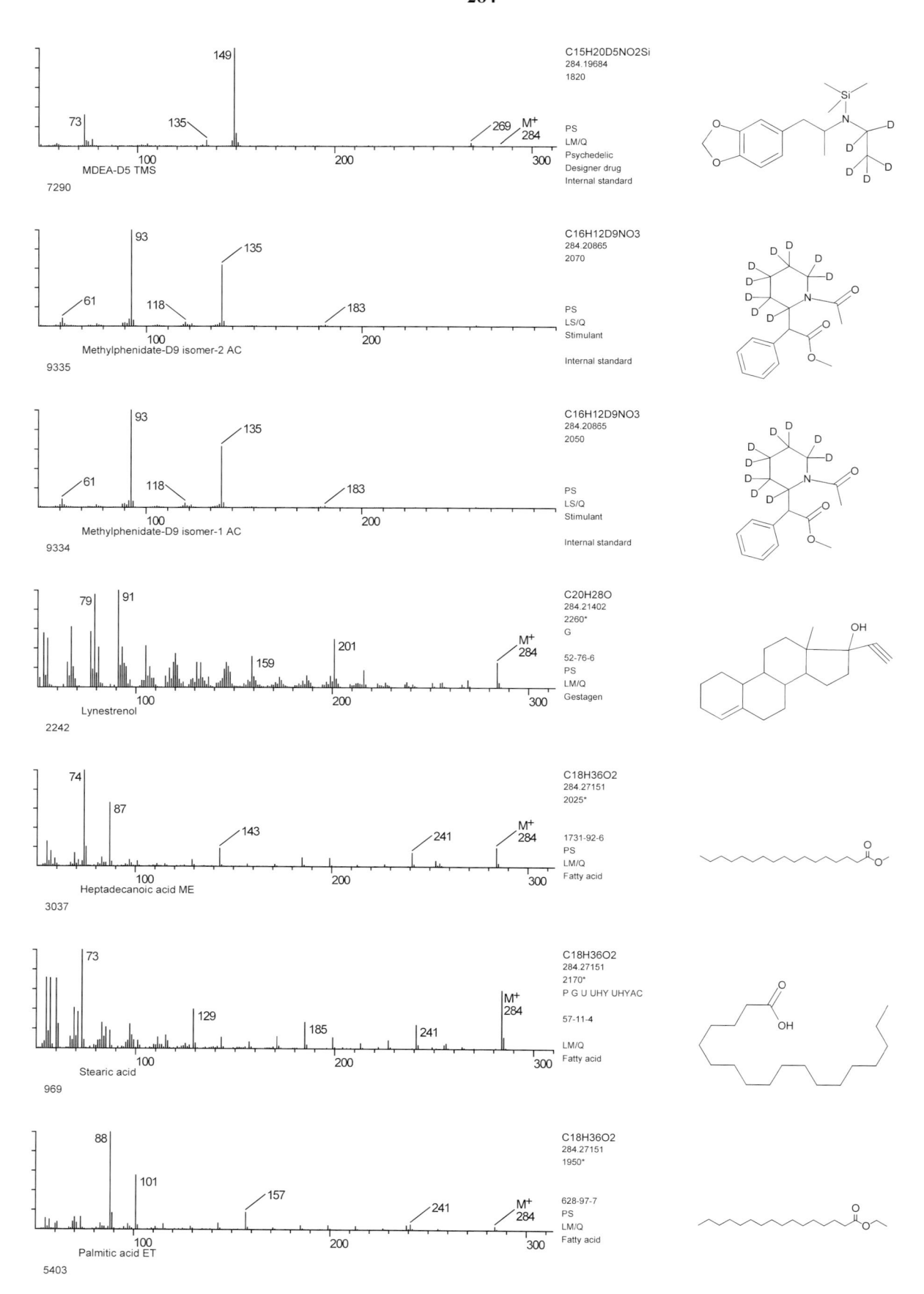

MDEA-D5 TMS
7290
C15H20D5NO2Si
284.19684
1820
PS
LM/Q
Psychedelic
Designer drug
Internal standard

Methylphenidate-D9 isomer-2 AC
9335
C16H12D9NO3
284.20865
2070
PS
LS/Q
Stimulant
Internal standard

Methylphenidate-D9 isomer-1 AC
9334
C16H12D9NO3
284.20865
2050
PS
LS/Q
Stimulant
Internal standard

Lynestrenol
2242
C20H28O
284.21402
2260*
G
52-76-6
PS
LM/Q
Gestagen

Heptadecanoic acid ME
3037
C18H36O2
284.27151
2025*
1731-92-6
PS
LM/Q
Fatty acid

Stearic acid
969
C18H36O2
284.27151
2170*
P G U UHY UHYAC
57-11-4
LM/Q
Fatty acid

Palmitic acid ET
5403
C18H36O2
284.27151
1950*
628-97-7
PS
LM/Q
Fatty acid

285

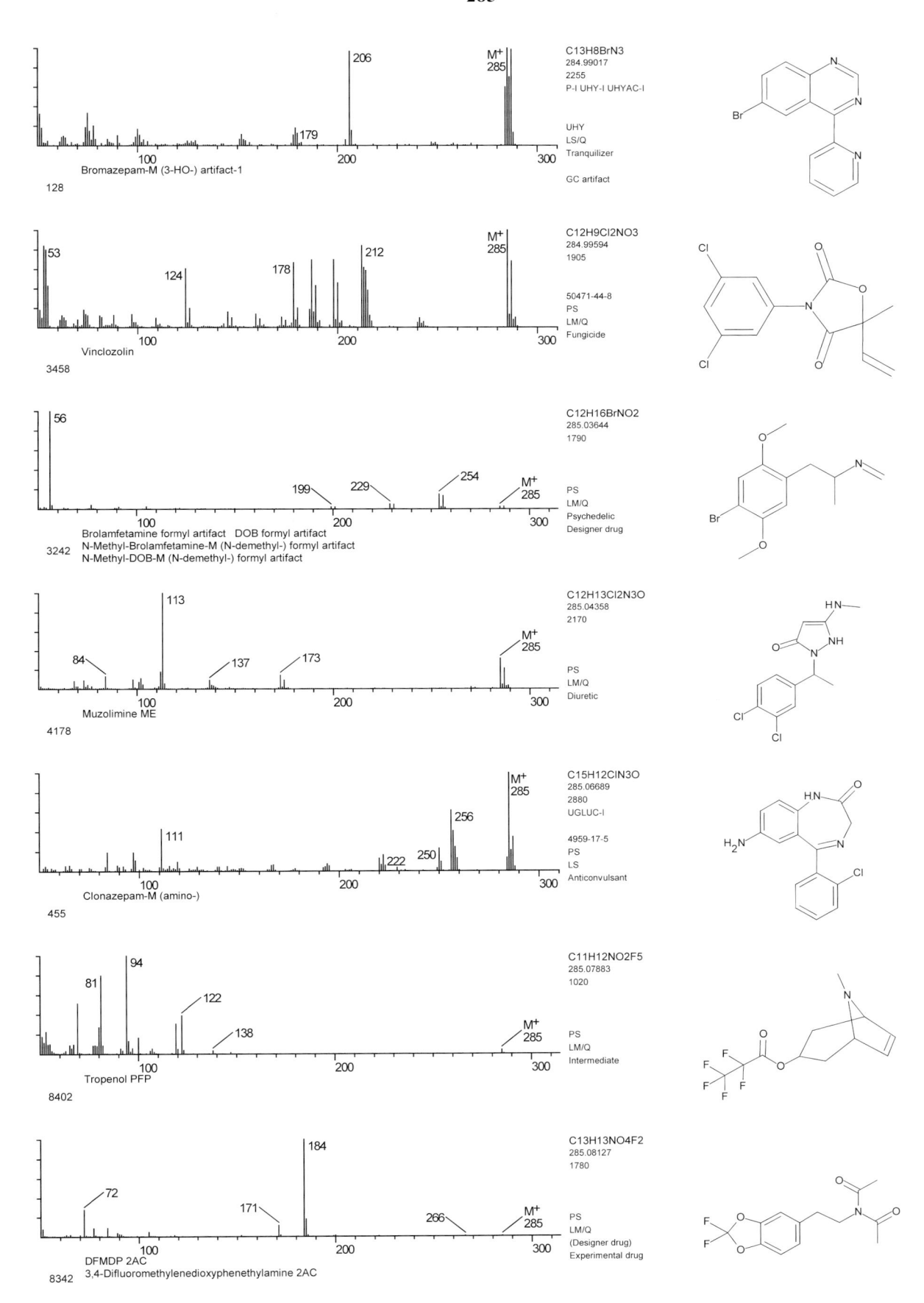

Spectrum label	Formula / data
Bromazepam-M (3-HO-) artifact-1 (128)	C13H8BrN3, 284.99017, 2255, P-I UHY-I UHYAC-I, UHY, LS/Q, Tranquilizer, GC artifact — peaks: 179, 206, M+ 285
Vinclozolin (3458)	C12H9Cl2NO3, 284.99594, 1905, 50471-44-8, PS, LM/Q, Fungicide — peaks: 53, 124, 178, 212, M+ 285
Brolamfetamine formyl artifact / DOB formyl artifact / N-Methyl-Brolamfetamine-M (N-demethyl-) formyl artifact / N-Methyl-DOB-M (N-demethyl-) formyl artifact (3242)	C12H16BrNO2, 285.03644, 1790, PS, LM/Q, Psychedelic, Designer drug — peaks: 56, 199, 229, 254, M+ 285
Muzolimine ME (4178)	C12H13Cl2N3O, 285.04358, 2170, PS, LM/Q, Diuretic — peaks: 84, 113, 137, 173, M+ 285
Clonazepam-M (amino-) (455)	C15H12ClN3O, 285.06689, 2880, UGLUC-I, 4959-17-5, PS, LS, Anticonvulsant — peaks: 111, 222, 250, 256, M+ 285
Tropenol PFP (8402)	C11H12NO2F5, 285.07883, 1020, PS, LM/Q, Intermediate — peaks: 81, 94, 122, 138, M+ 285
DFMDP 2AC / 3,4-Difluoromethylenedioxyphenethylamine 2AC (8342)	C13H13NO4F2, 285.08127, 1780, PS, LM/Q, (Designer drug), Experimental drug — peaks: 72, 171, 184, 266, M+ 285

285

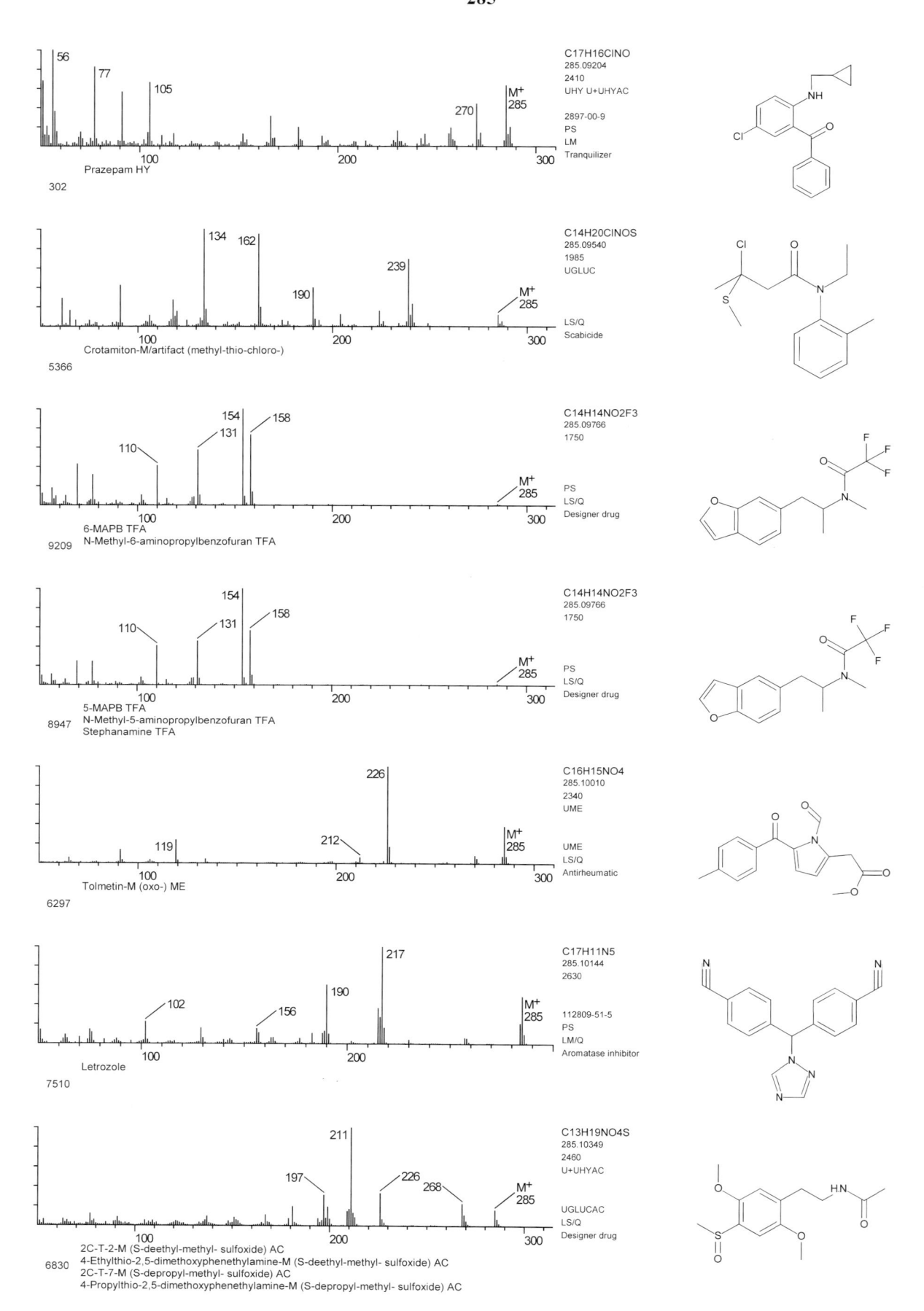

Prazepam HY
C17H16ClNO
285.09204
2410
UHY U+UHYAC
2897-00-9
PS
LM
Tranquilizer

Peaks: 56, 77, 105, 270, M+ 285
302

Crotamiton-M/artifact (methyl-thio-chloro-)
C14H20ClNOS
285.09540
1985
UGLUC
LS/Q
Scabicide

Peaks: 134, 162, 190, 239, M+ 285
5366

6-MAPB TFA
N-Methyl-6-aminopropylbenzofuran TFA
C14H14NO2F3
285.09766
1750
PS
LS/Q
Designer drug

Peaks: 110, 131, 154, 158, M+ 285
9209

5-MAPB TFA
N-Methyl-5-aminopropylbenzofuran TFA
Stephanamine TFA
C14H14NO2F3
285.09766
1750
PS
LS/Q
Designer drug

Peaks: 110, 131, 154, 158, M+ 285
8947

Tolmetin-M (oxo-) ME
C16H15NO4
285.10010
2340
UME
UME
LS/Q
Antirheumatic

Peaks: 119, 212, 226, M+ 285
6297

Letrozole
C17H11N5
285.10144
2630
112809-51-5
PS
LM/Q
Aromatase inhibitor

Peaks: 102, 156, 190, 217, M+ 285
7510

2C-T-2-M (S-deethyl-methyl- sulfoxide) AC
4-Ethylthio-2,5-dimethoxyphenethylamine-M (S-deethyl-methyl- sulfoxide) AC
2C-T-7-M (S-depropyl-methyl- sulfoxide) AC
4-Propylthio-2,5-dimethoxyphenethylamine-M (S-depropyl-methyl- sulfoxide) AC
C13H19NO4S
285.10349
2460
U+UHYAC
UGLUCAC
LS/Q
Designer drug

Peaks: 197, 211, 226, 268, M+ 285
6830

285

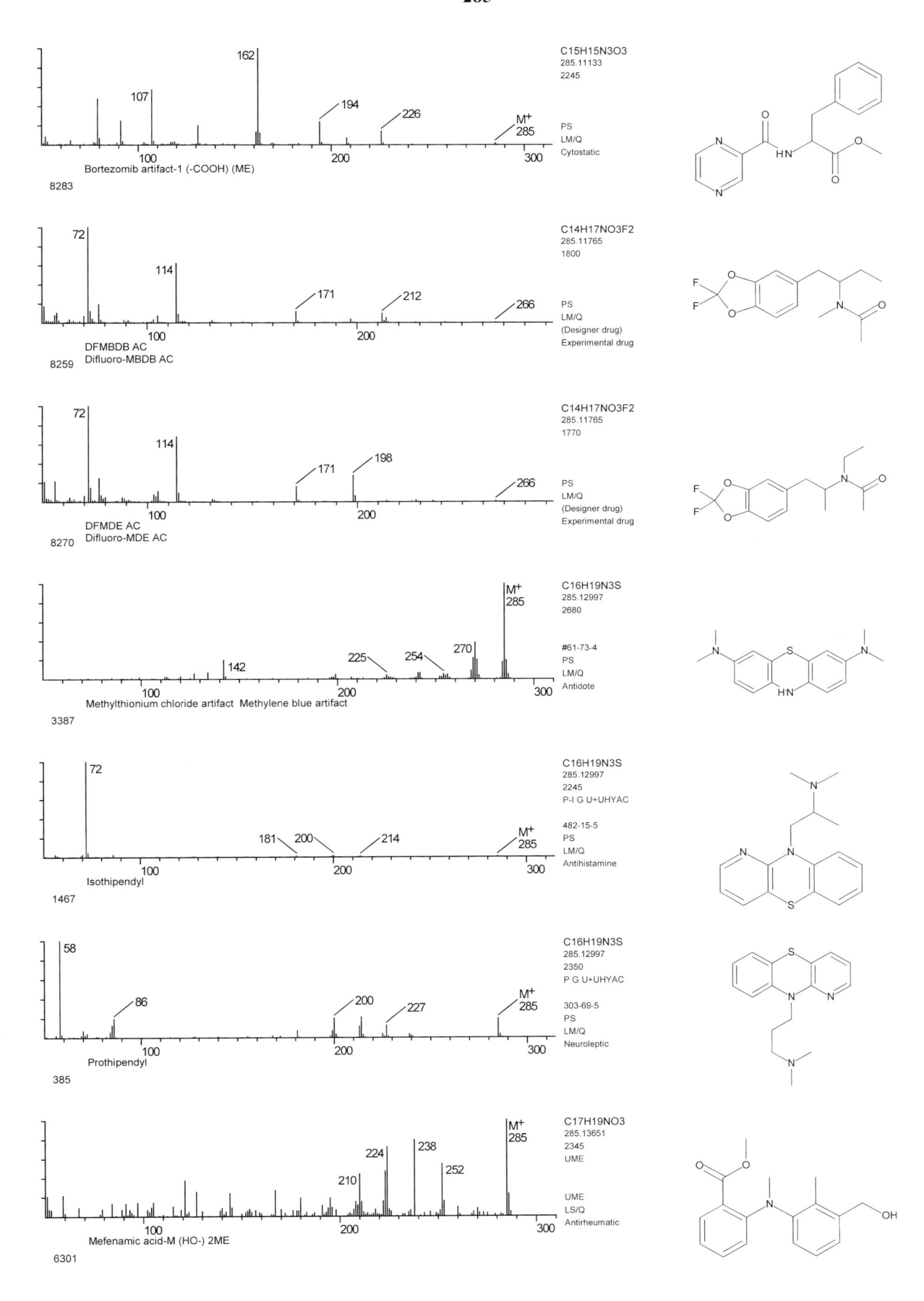

C15H15N3O3
285.11133
2245

PS
LM/Q
Cytostatic

Bortezomib artifact-1 (-COOH) (ME)
8283

C14H17NO3F2
285.11765
1800

PS
LM/Q
(Designer drug)
Experimental drug

DFMBDB AC
Difluoro-MBDB AC
8259

C14H17NO3F2
285.11765
1770

PS
LM/Q
(Designer drug)
Experimental drug

DFMDE AC
Difluoro-MDE AC
8270

C16H19N3S
285.12997
2680

#61-73-4
PS
LM/Q
Antidote

Methylthionium chloride artifact   Methylene blue artifact
3387

C16H19N3S
285.12997
2245
P-I G U+UHYAC

482-15-5
PS
LM/Q
Antihistamine

Isothipendyl
1467

C16H19N3S
285.12997
2350
P G U+UHYAC

303-69-5
PS
LM/Q
Neuroleptic

Prothipendyl
385

C17H19NO3
285.13651
2345
UME

UME
LS/Q
Antirheumatic

Mefenamic acid-M (HO-) 2ME
6301

285

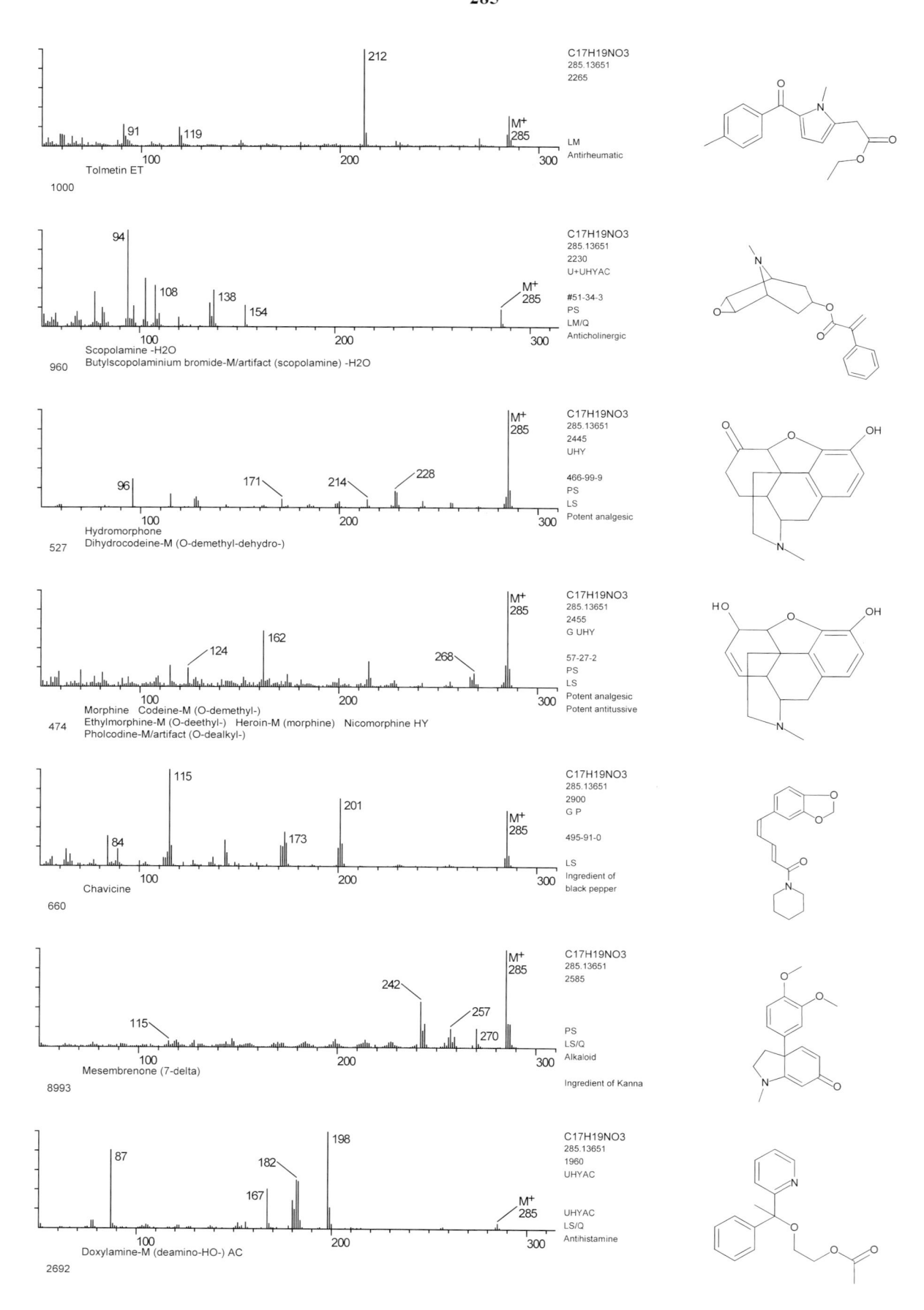

Tolmetin ET	91, 119, 212, M+ 285	C17H19NO3 285.13651 2265 LM Antirheumatic
Scopolamine -H2O Butylscopolaminium bromide-M/artifact (scopolamine) -H2O	94, 108, 138, 154, M+ 285	C17H19NO3 285.13651 2230 U+UHYAC #51-34-3 PS LM/Q Anticholinergic
Hydromorphone Dihydrocodeine-M (O-demethyl-dehydro-)	96, 171, 214, 228, M+ 285	C17H19NO3 285.13651 2445 UHY 466-99-9 PS LS Potent analgesic
Morphine Codeine-M (O-demethyl-) Ethylmorphine-M (O-deethyl-) Heroin-M (morphine) Nicomorphine HY Pholcodine-M/artifact (O-dealkyl-)	124, 162, 268, M+ 285	C17H19NO3 285.13651 2455 G UHY 57-27-2 PS LS Potent analgesic Potent antitussive
Chavicine	84, 115, 173, 201, M+ 285	C17H19NO3 285.13651 2900 G P 495-91-0 LS Ingredient of black pepper
Mesembrenone (7-delta)	115, 242, 257, 270, M+ 285	C17H19NO3 285.13651 2585 PS LS/Q Alkaloid Ingredient of Kanna
Doxylamine-M (deamino-HO-) AC	87, 167, 182, 198, M+ 285	C17H19NO3 285.13651 1960 UHYAC UHYAC LS/Q Antihistamine

285

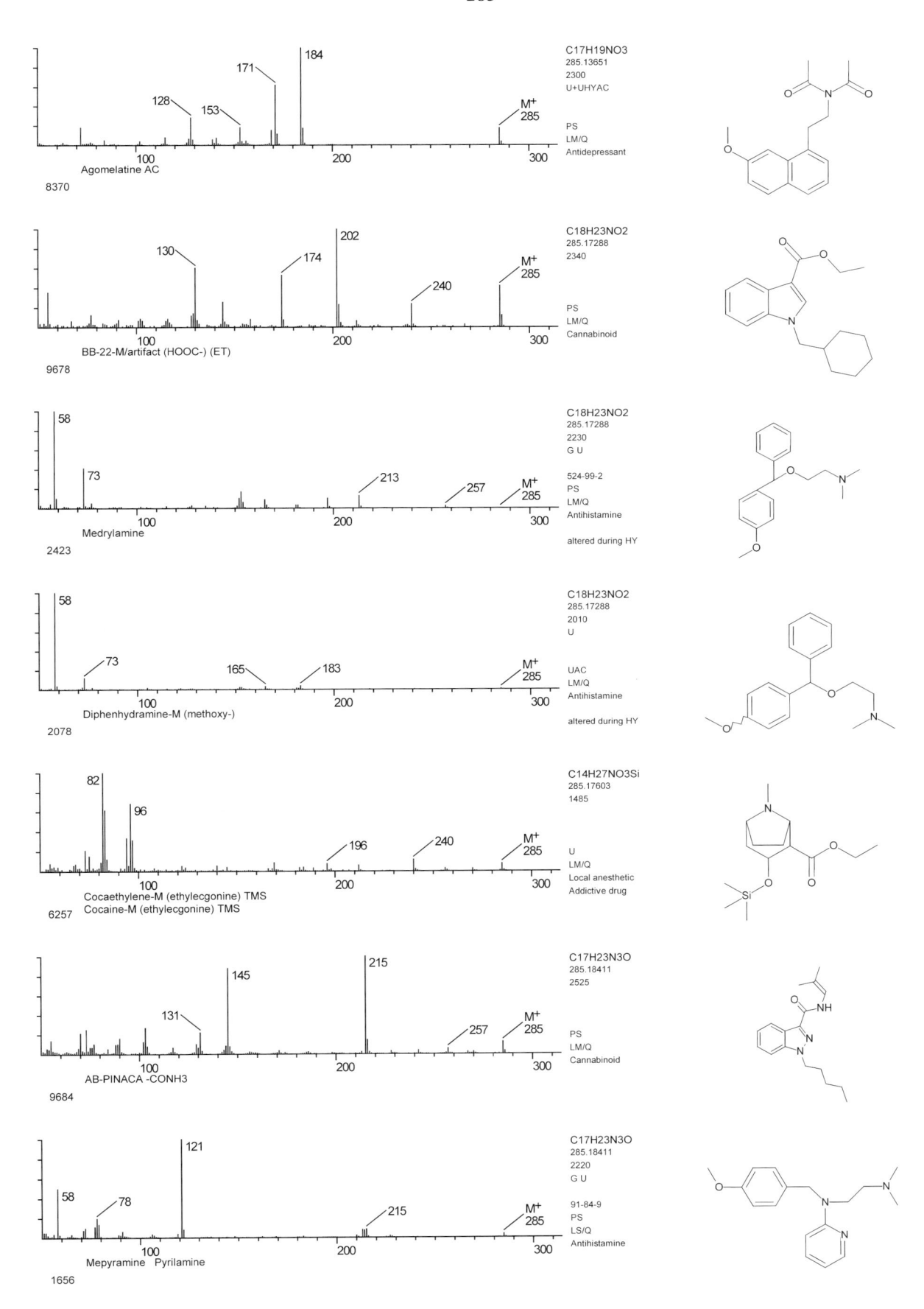

C17H19NO3 285.13651 2300 U+UHYAC PS LM/Q Antidepressant	Agomelatine AC — 8370 — peaks: 128, 153, 171, 184, M+ 285
C18H23NO2 285.17288 2340 PS LM/Q Cannabinoid	BB-22-M/artifact (HOOC-) (ET) — 9678 — peaks: 130, 174, 202, 240, M+ 285
C18H23NO2 285.17288 2230 G U 524-99-2 PS LM/Q Antihistamine altered during HY	Medrylamine — 2423 — peaks: 58, 73, 213, 257, M+ 285
C18H23NO2 285.17288 2010 U UAC LM/Q Antihistamine altered during HY	Diphenhydramine-M (methoxy-) — 2078 — peaks: 58, 73, 165, 183, M+ 285
C14H27NO3Si 285.17603 1485 U LM/Q Local anesthetic Addictive drug	Cocaethylene-M (ethylecgonine) TMS / Cocaine-M (ethylecgonine) TMS — 6257 — peaks: 82, 96, 196, 240, M+ 285
C17H23N3O 285.18411 2525 PS LM/Q Cannabinoid	AB-PINACA -CONH3 — 9684 — peaks: 131, 145, 215, 257, M+ 285
C17H23N3O 285.18411 2220 G U 91-84-9 PS LS/Q Antihistamine	Mepyramine Pyriamine — 1656 — peaks: 58, 78, 121, 215, M+ 285

285

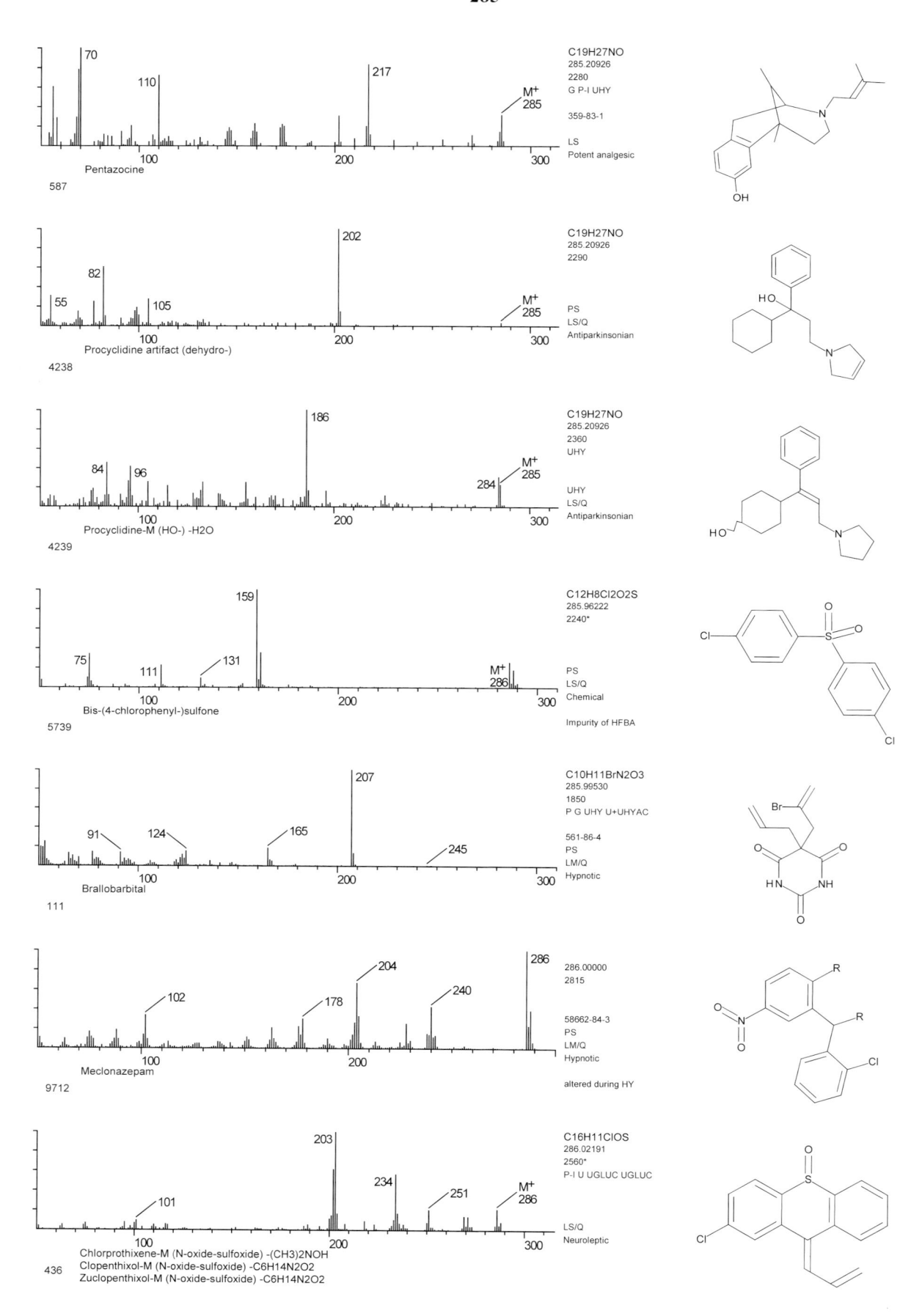

Spectrum	Compound	Formula / Info
587	Pentazocine	C19H27NO, 285.20926, 2280, G P-I UHY, 359-83-1, LS, Potent analgesic
4238	Procyclidine artifact (dehydro-)	C19H27NO, 285.20926, 2290, PS, LS/Q, Antiparkinsonian
4239	Procyclidine-M (HO-) -H2O	C19H27NO, 285.20926, 2360, UHY, UHY, LS/Q, Antiparkinsonian
5739	Bis-(4-chlorophenyl-)sulfone	C12H8Cl2O2S, 285.96222, 2240*, PS, LS/Q, Chemical, Impurity of HFBA
111	Brallobarbital	C10H11BrN2O3, 285.99530, 1850, P G UHY U+UHYAC, 561-86-4, PS, LM/Q, Hypnotic
9712	Meclonazepam	286.00000, 2815, 58662-84-3, PS, LM/Q, Hypnotic, altered during HY
436	Chlorprothixene-M (N-oxide-sulfoxide) -(CH3)2NOH / Clopenthixol-M (N-oxide-sulfoxide) -C6H14N2O2 / Zuclopenthixol-M (N-oxide-sulfoxide) -C6H14N2O2	C16H11ClOS, 286.02191, 2560*, P-I U UGLUC UGLUC, LS/Q, Neuroleptic

Key peaks:
- 587 Pentazocine: 70, 110, 217, M+ 285
- 4238 Procyclidine artifact: 55, 82, 105, 202, M+ 285
- 4239 Procyclidine-M: 84, 96, 186, 284, M+ 285
- 5739 Bis-(4-chlorophenyl-)sulfone: 75, 111, 131, 159, M+ 286
- 111 Brallobarbital: 91, 124, 165, 207, 245
- 9712 Meclonazepam: 102, 178, 204, 240, 286
- 436 Chlorprothixene-M: 101, 203, 234, 251, M+ 286

286

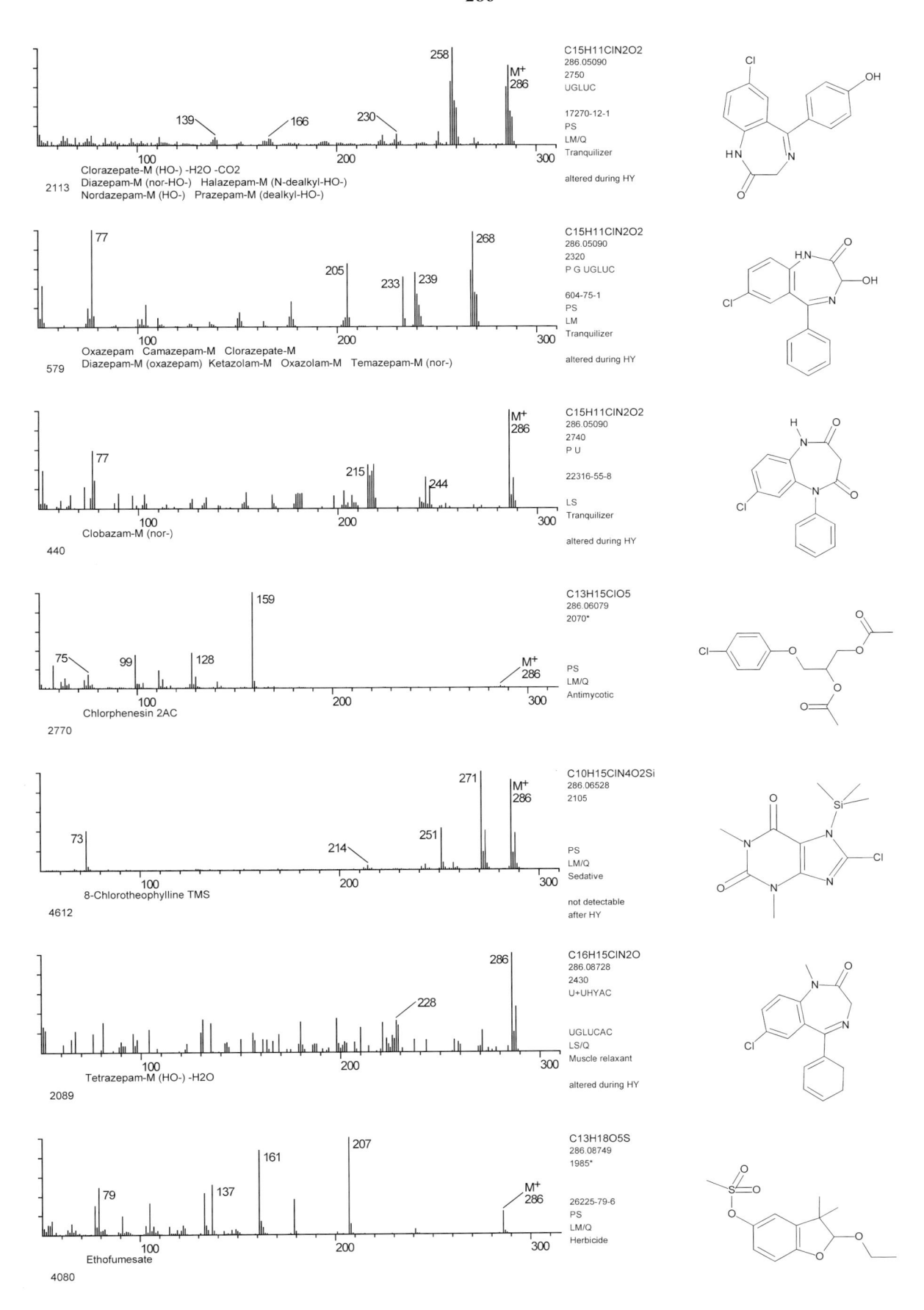

2113	Clorazepate-M (HO-) -H2O -CO2 Diazepam-M (nor-HO-) Halazepam-M (N-dealkyl-HO-) Nordazepam-M (HO-) Prazepam-M (dealkyl-HO-)	C15H11ClN2O2 286.05090 2750 UGLUC 17270-12-1 PS LM/Q Tranquilizer altered during HY
579	Oxazepam Camazepam-M Clorazepate-M Diazepam-M (oxazepam) Ketazolam-M Oxazolam-M Temazepam-M (nor-)	C15H11ClN2O2 286.05090 2320 P G UGLUC 604-75-1 PS LM Tranquilizer altered during HY
440	Clobazam-M (nor-)	C15H11ClN2O2 286.05090 2740 P U 22316-55-8 LS Tranquilizer altered during HY
2770	Chlorphenesin 2AC	C13H15ClO5 286.06079 2070* PS LM/Q Antimycotic
4612	8-Chlorotheophylline TMS	C10H15ClN4O2Si 286.06528 2105 PS LM/Q Sedative not detectable after HY
2089	Tetrazepam-M (HO-) -H2O	C16H15ClN2O 286.08728 2430 U+UHYAC UGLUCAC LS/Q Muscle relaxant altered during HY
4080	Ethofumesate	C13H18O5S 286.08749 1985* 26225-79-6 PS LM/Q Herbicide

286

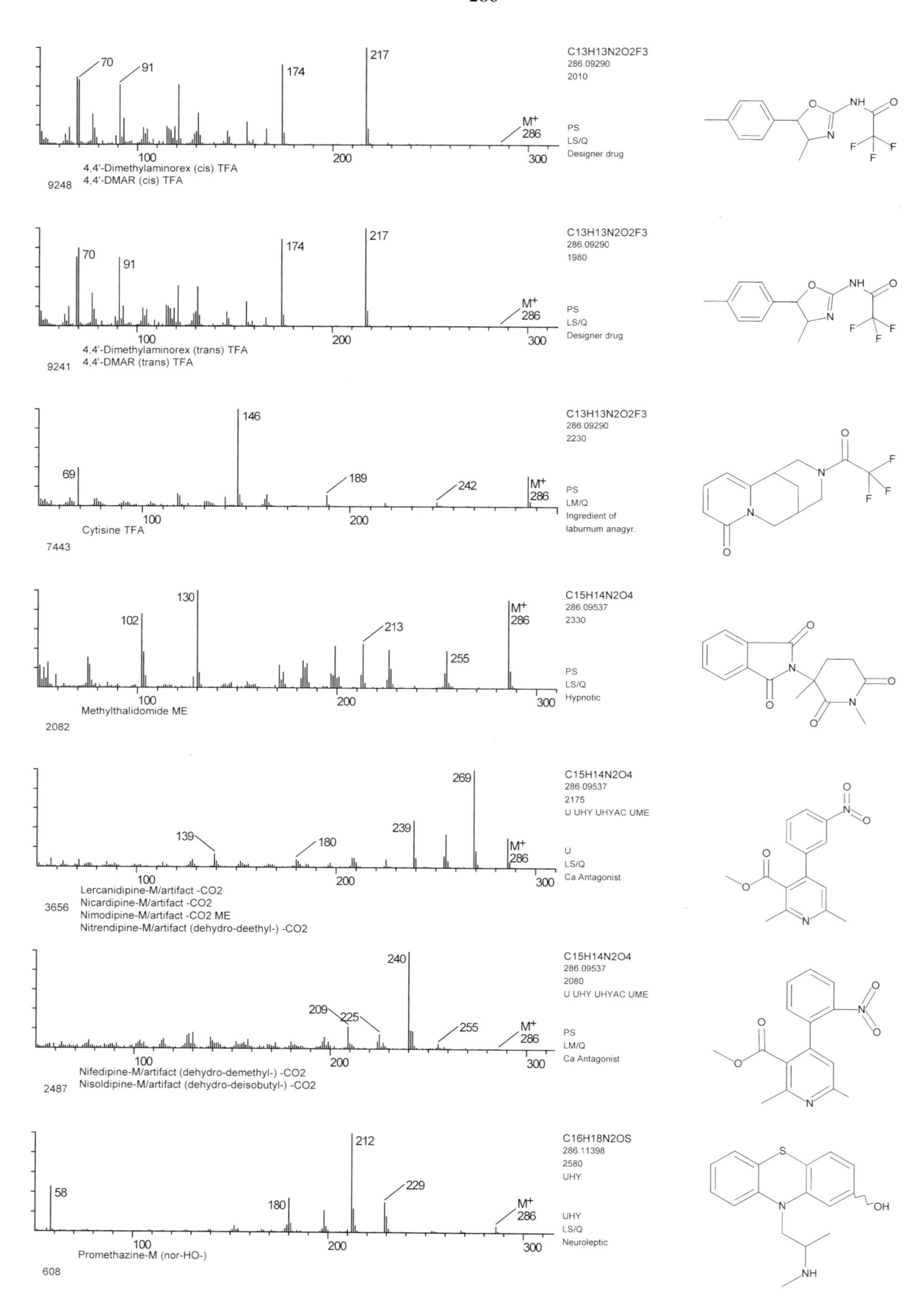

70, 91, 174, 217, M+ 286	C13H13N2O2F3 286.09290 2010 PS LS/Q Designer drug	
4,4'-Dimethylaminorex (cis) TFA 4,4'-DMAR (cis) TFA 9248		

70, 91, 174, 217, M+ 286	C13H13N2O2F3 286.09290 1980 PS LS/Q Designer drug	
4,4'-Dimethylaminorex (trans) TFA 4,4'-DMAR (trans) TFA 9241		

69, 146, 189, 242, M+ 286	C13H13N2O2F3 286.09290 2230 PS LM/Q Ingredient of laburnum anagyr.	
Cytisine TFA 7443		

102, 130, 213, 255, M+ 286	C15H14N2O4 286.09537 2330 PS LS/Q Hypnotic	
Methylthalidomide ME 2082		

139, 180, 239, 269, M+ 286	C15H14N2O4 286.09537 2175 U UHY UHYAC UME U LS/Q Ca Antagonist	
Lercanidipine-M/artifact -CO2 Nicardipine-M/artifact -CO2 Nimodipine-M/artifact -CO2 ME Nitrendipine-M/artifact (dehydro-deethyl-) -CO2 3656		

209, 225, 240, 255, M+ 286	C15H14N2O4 286.09537 2080 U UHY UHYAC UME PS LM/Q Ca Antagonist	
Nifedipine-M/artifact (dehydro-demethyl-) -CO2 Nisoldipine-M/artifact (dehydro-deisobutyl-) -CO2 2487		

58, 180, 212, 229, M+ 286	C16H18N2OS 286.11398 2580 UHY UHY LS/Q Neuroleptic	
Promethazine-M (nor-HO-) 608		

286

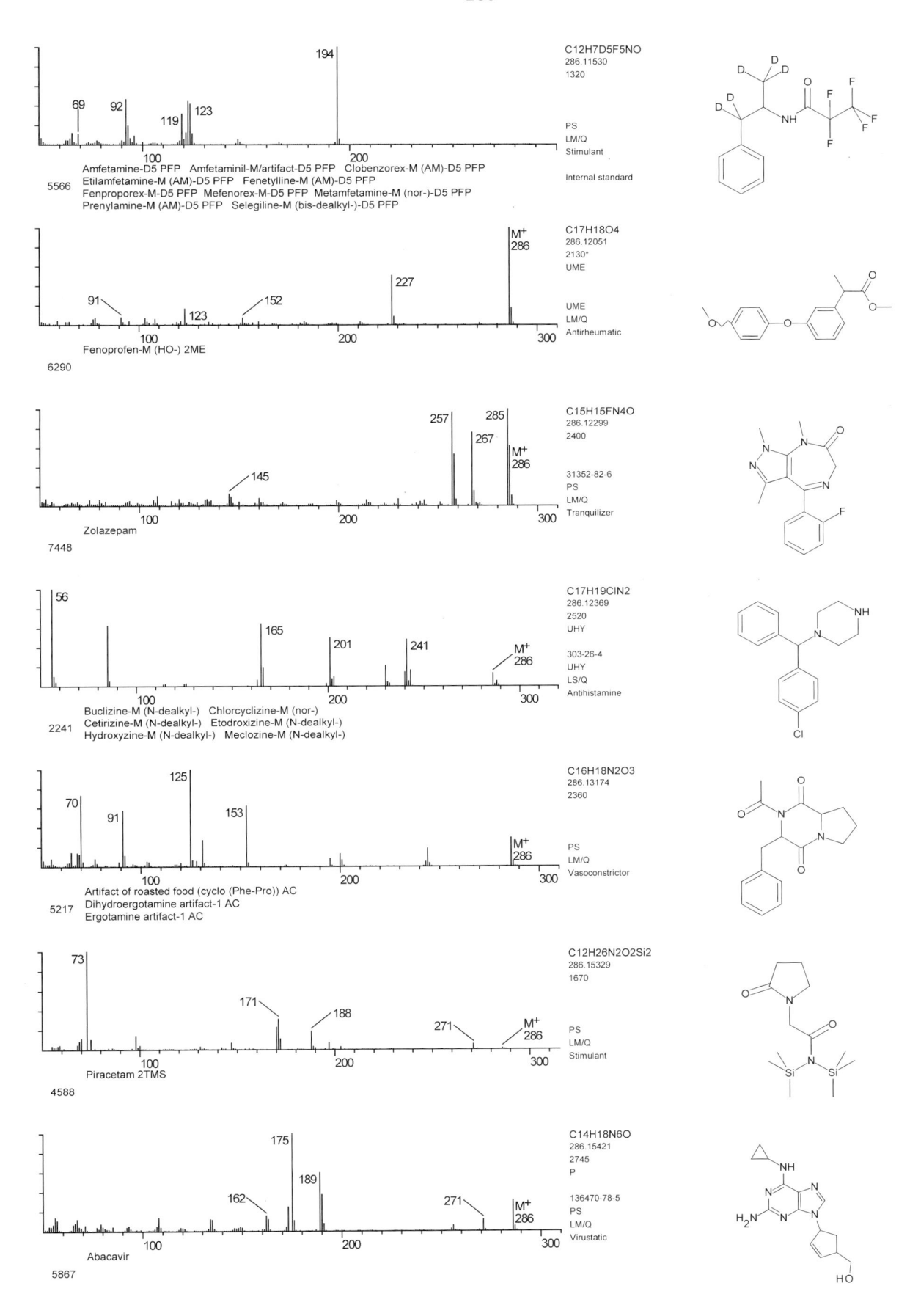

C12H7D5F5NO
286.11530
1320

PS
LM/Q
Stimulant

Internal standard

Amfetamine-D5 PFP   Amfetaminil-M/artifact-D5 PFP   Clobenzorex-M (AM)-D5 PFP
Etilamfetamine-M (AM)-D5 PFP   Fenetylline-M (AM)-D5 PFP
Fenproporex-M-D5 PFP   Mefenorex-M-D5 PFP   Metamfetamine-M (nor-)-D5 PFP
Prenylamine-M (AM)-D5 PFP   Selegiline-M (bis-dealkyl-)-D5 PFP

5566

C17H18O4
286.12051
2130*
UME

UME
LM/Q
Antirheumatic

Fenoprofen-M (HO-) 2ME

6290

C15H15FN4O
286.12299
2400

31352-82-6
PS
LM/Q
Tranquilizer

Zolazepam

7448

C17H19ClN2
286.12369
2520
UHY

303-26-4
UHY
LS/Q
Antihistamine

Buclizine-M (N-dealkyl-)   Chlorcyclizine-M (nor-)
Cetirizine-M (N-dealkyl-)   Etodroxizine-M (N-dealkyl-)
Hydroxyzine-M (N-dealkyl-)   Meclozine-M (N-dealkyl-)

2241

C16H18N2O3
286.13174
2360

PS
LM/Q
Vasoconstrictor

Artifact of roasted food (cyclo (Phe-Pro)) AC
Dihydroergotamine artifact-1 AC
Ergotamine artifact-1 AC

5217

C12H26N2O2Si2
286.15329
1670

PS
LM/Q
Stimulant

Piracetam 2TMS

4588

C14H18N6O
286.15421
2745
P

136470-78-5
PS
LM/Q
Virustatic

Abacavir

5867

815

286

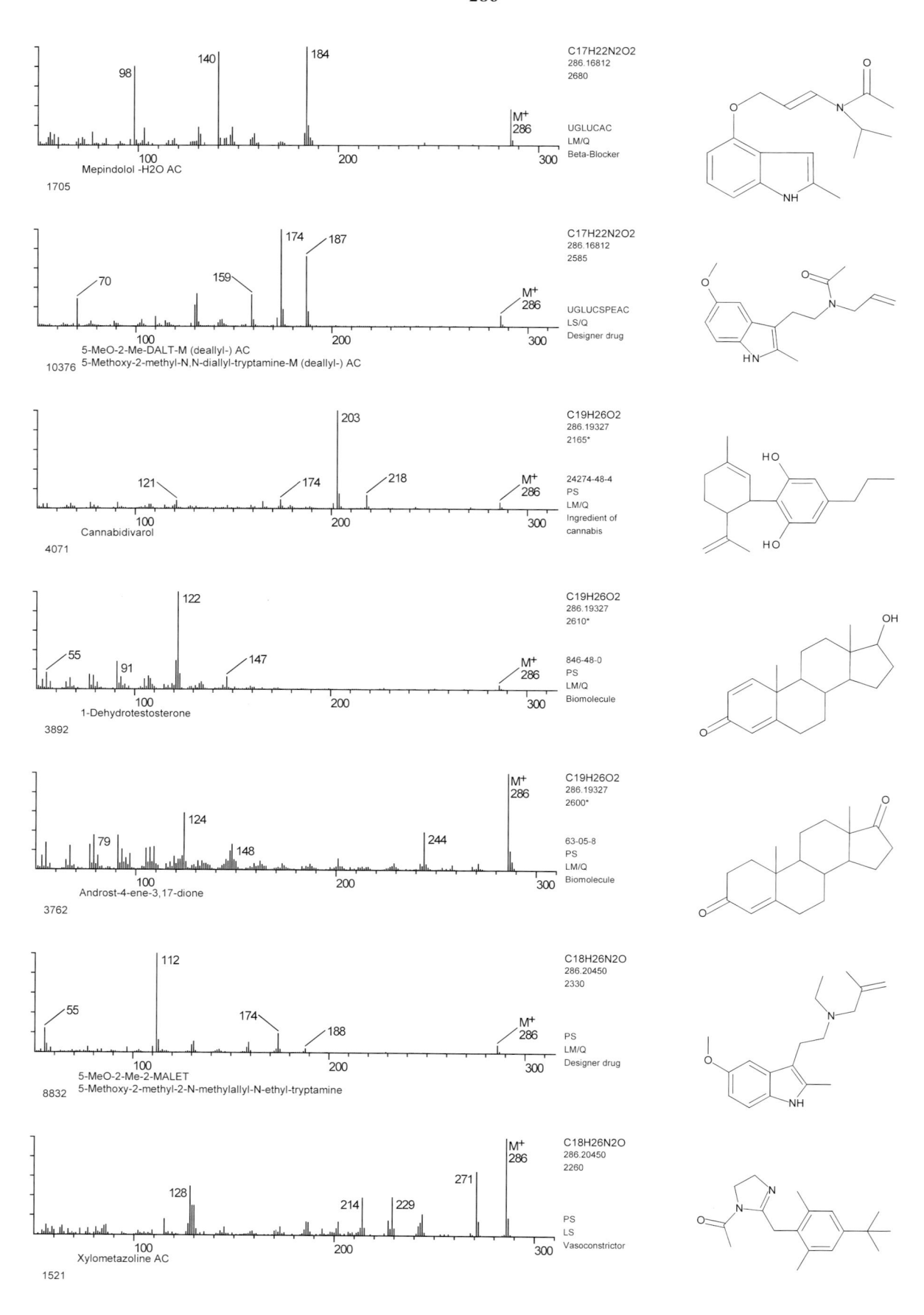

Mepindolol -H2O AC (1705)	C17H22N2O2 / 286.16812 / 2680 / UGLUCAC / LM/Q / Beta-Blocker

Peaks: 98, 140, 184, M+ 286

5-MeO-2-Me-DALT-M (deallyl-) AC / 5-Methoxy-2-methyl-N,N-diallyl-tryptamine-M (deallyl-) AC (10376)	C17H22N2O2 / 286.16812 / 2585 / UGLUCSPEAC / LS/Q / Designer drug

Peaks: 70, 159, 174, 187, M+ 286

Cannabidivarol (4071)	C19H26O2 / 286.19327 / 2165* / 24274-48-4 / PS / LM/Q / Ingredient of cannabis

Peaks: 121, 174, 203, 218, M+ 286

1-Dehydrotestosterone (3892)	C19H26O2 / 286.19327 / 2610* / 846-48-0 / PS / LM/Q / Biomolecule

Peaks: 55, 91, 122, 147, M+ 286

Androst-4-ene-3,17-dione (3762)	C19H26O2 / 286.19327 / 2600* / 63-05-8 / PS / LM/Q / Biomolecule

Peaks: 79, 124, 148, 244, M+ 286

5-MeO-2-Me-2-MALET / 5-Methoxy-2-methyl-2-N-methylallyl-N-ethyl-tryptamine (8832)	C18H26N2O / 286.20450 / 2330 / PS / LM/Q / Designer drug

Peaks: 55, 112, 174, 188, M+ 286

Xylometazoline AC (1521)	C18H26N2O / 286.20450 / 2260 / PS / LS / Vasoconstrictor

Peaks: 128, 214, 229, 271, M+ 286

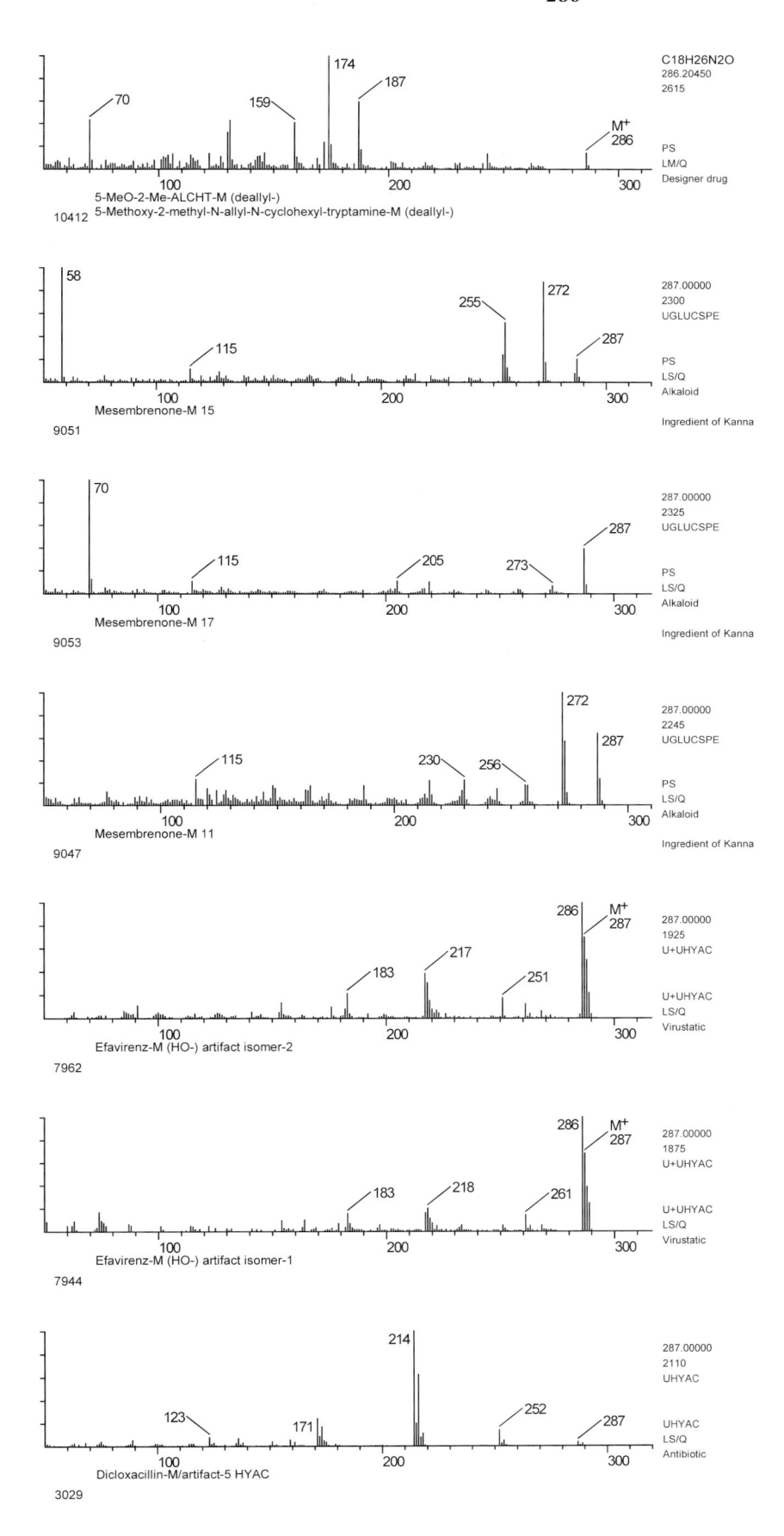

286

C18H26N2O
286.20450
2615
PS
LM/Q
Designer drug

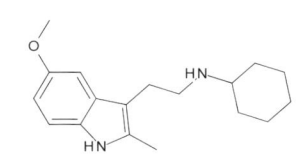

5-MeO-2-Me-ALCHT-M (deallyl-)
5-Methoxy-2-methyl-N-allyl-N-cyclohexyl-tryptamine-M (deallyl-)
10412

287.00000
2300
UGLUCSPE
PS
LS/Q
Alkaloid
Ingredient of Kanna

Mesembrenone-M 15
9051

287.00000
2325
UGLUCSPE
PS
LS/Q
Alkaloid
Ingredient of Kanna

Mesembrenone-M 17
9053

287.00000
2245
UGLUCSPE
PS
LS/Q
Alkaloid
Ingredient of Kanna

Mesembrenone-M 11
9047

287.00000
1925
U+UHYAC
U+UHYAC
LS/Q
Virustatic

Efavirenz-M (HO-) artifact isomer-2
7962

287.00000
1875
U+UHYAC
U+UHYAC
LS/Q
Virustatic

Efavirenz-M (HO-) artifact isomer-1
7944

287.00000
2110
UHYAC
UHYAC
LS/Q
Antibiotic

Dicloxacillin-M/artifact-5 HYAC
3029

287

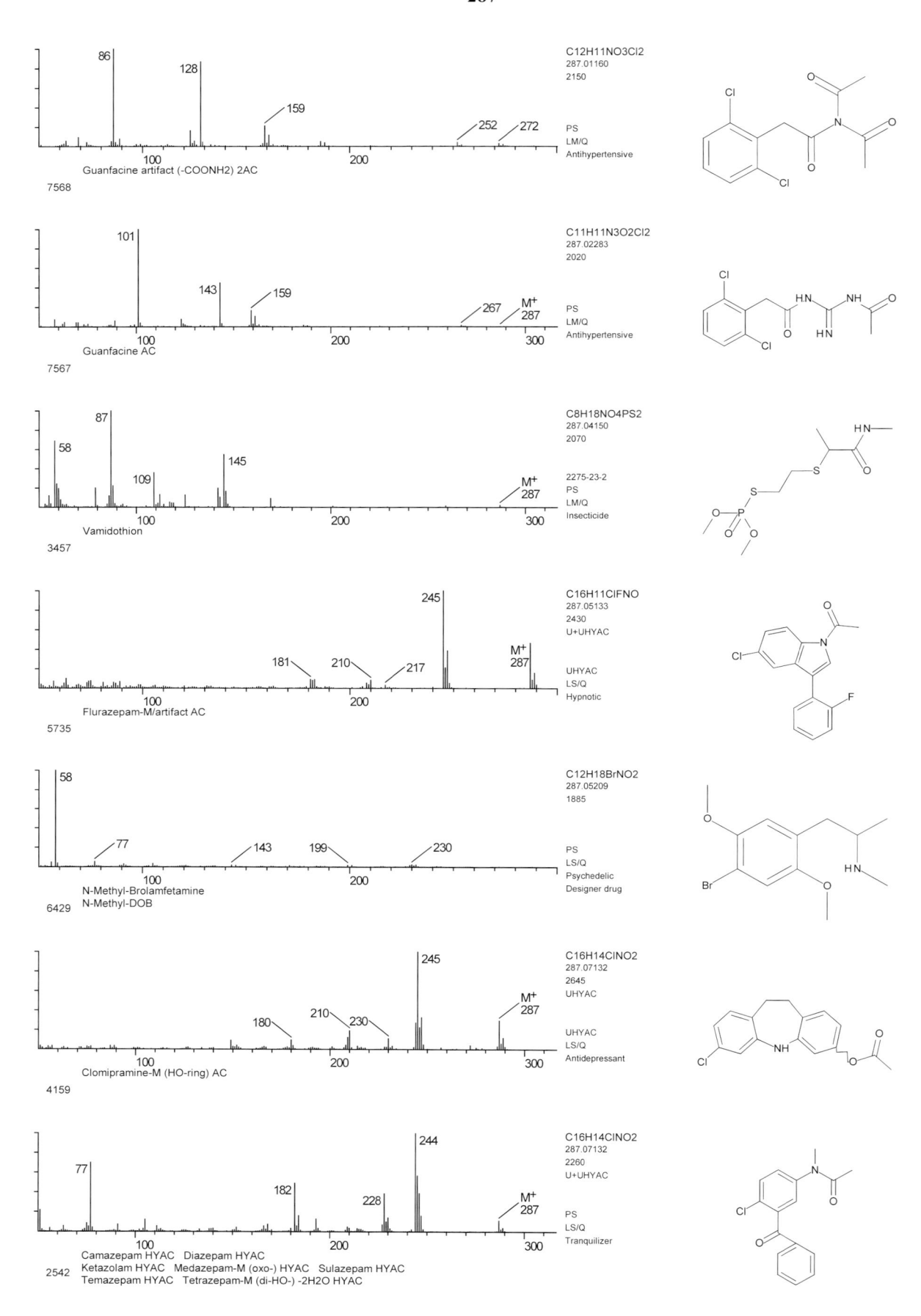

86, 128, 159, 252, 272 Guanfacine artifact (-COONH2) 2AC 7568	C12H11NO3Cl2 287.01160 2150 PS LM/Q Antihypertensive
101, 143, 159, 267, M+ 287 Guanfacine AC 7567	C11H11N3O2Cl2 287.02283 2020 PS LM/Q Antihypertensive
58, 87, 109, 145, M+ 287 Vamidothion 3457	C8H18NO4PS2 287.04150 2070 2275-23-2 PS LM/Q Insecticide
181, 210, 217, 245, M+ 287 Flurazepam-M/artifact AC 5735	C16H11ClFNO 287.05133 2430 U+UHYAC UHYAC LS/Q Hypnotic
58, 77, 143, 199, 230 N-Methyl-Brolamfetamine N-Methyl-DOB 6429	C12H18BrNO2 287.05209 1885 PS LS/Q Psychedelic Designer drug
180, 210, 230, 245, M+ 287 Clomipramine-M (HO-ring) AC 4159	C16H14ClNO2 287.07132 2645 UHYAC UHYAC LS/Q Antidepressant
77, 182, 228, 244, M+ 287 Camazepam HYAC Diazepam HYAC Ketazolam HYAC Medazepam-M (oxo-) HYAC Sulazepam HYAC Temazepam HYAC Tetrazepam-M (di-HO-) -2H2O HYAC 2542	C16H14ClNO2 287.07132 2260 U+UHYAC PS LS/Q Tranquilizer

287

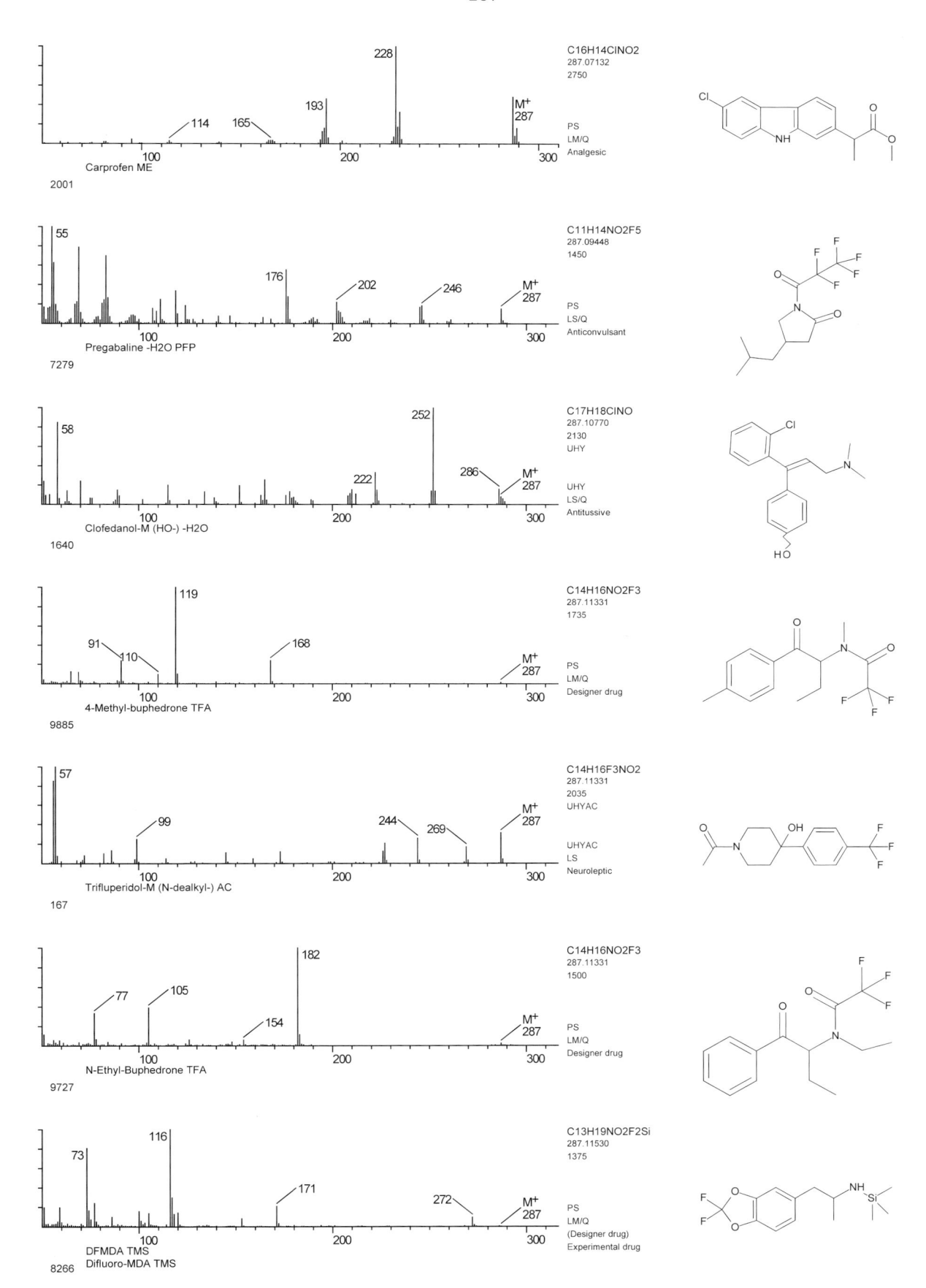

Carprofen ME
2001
114, 165, 193, 228, M+ 287
C16H14ClNO2
287.07132
2750
PS
LM/Q
Analgesic

Pregabaline -H2O PFP
7279
55, 176, 202, 246, M+ 287
C11H14NO2F5
287.09448
1450
PS
LS/Q
Anticonvulsant

Clofedanol-M (HO-) -H2O
1640
58, 222, 252, 286, M+ 287
C17H18ClNO
287.10770
2130
UHY
UHY
LS/Q
Antitussive

4-Methyl-buphedrone TFA
9885
91, 110, 119, 168, M+ 287
C14H16NO2F3
287.11331
1735
PS
LM/Q
Designer drug

Trifluperidol-M (N-dealkyl-) AC
167
57, 99, 244, 269, M+ 287
C14H16F3NO2
287.11331
2035
UHYAC
UHYAC
LS
Neuroleptic

N-Ethyl-Buphedrone TFA
9727
77, 105, 154, 182, M+ 287
C14H16NO2F3
287.11331
1500
PS
LM/Q
Designer drug

DFMDA TMS
Difluoro-MDA TMS
8266
73, 116, 171, 272, M+ 287
C13H19NO2F2Si
287.11530
1375
PS
LM/Q
(Designer drug)
Experimental drug

287

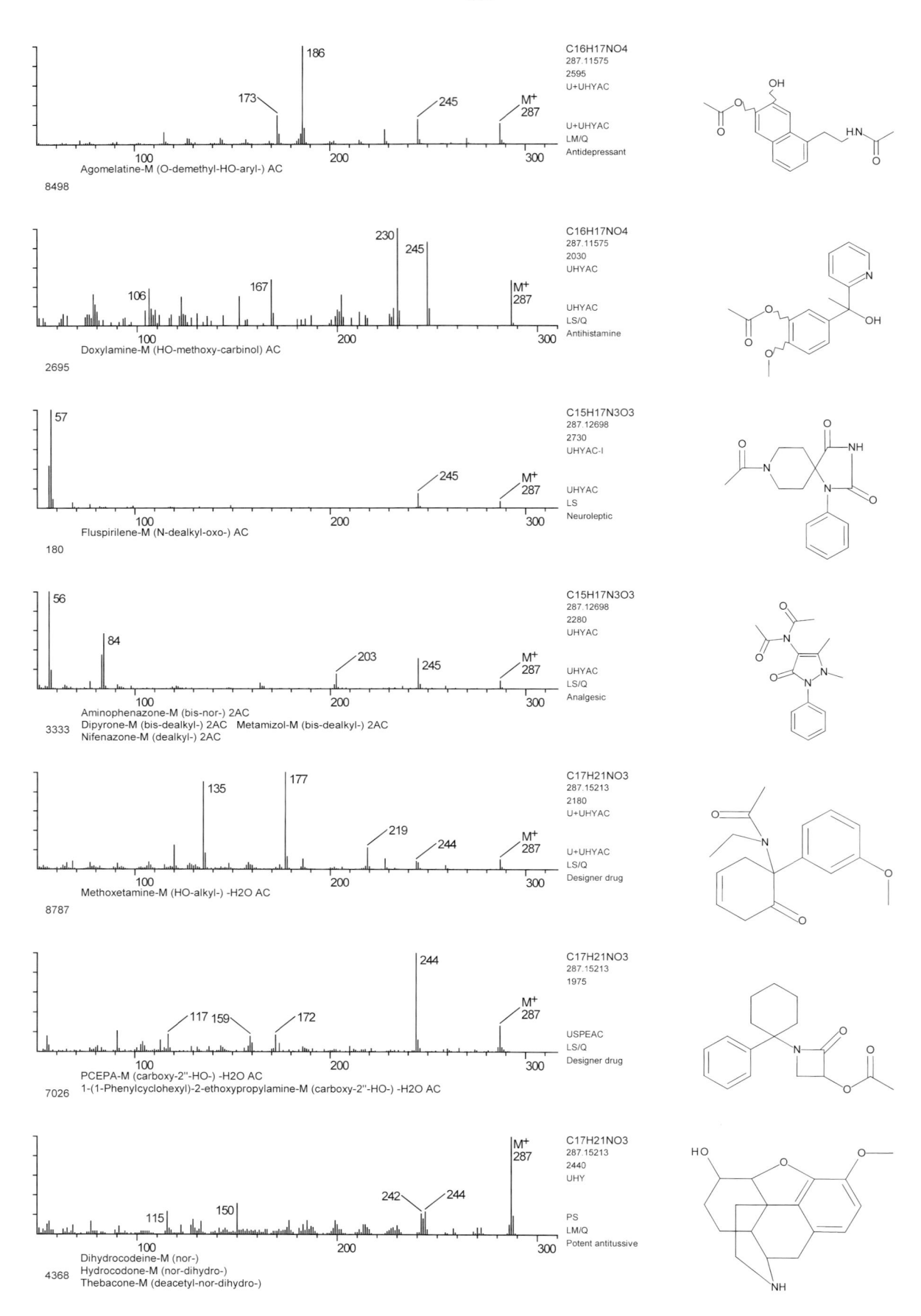

Agomelatine-M (O-demethyl-HO-aryl-) AC
8498
C16H17NO4
287.11575
2595
U+UHYAC
U+UHYAC
LM/Q
Antidepressant

Doxylamine-M (HO-methoxy-carbinol) AC
2695
C16H17NO4
287.11575
2030
UHYAC
UHYAC
LS/Q
Antihistamine

Fluspirilene-M (N-dealkyl-oxo-) AC
180
C15H17N3O3
287.12698
2730
UHYAC-I
UHYAC
LS
Neuroleptic

Aminophenazone-M (bis-nor-) 2AC
Dipyrone-M (bis-dealkyl-) 2AC   Metamizol-M (bis-dealkyl-) 2AC
Nifenazone-M (dealkyl-) 2AC
3333
C15H17N3O3
287.12698
2280
UHYAC
UHYAC
LS/Q
Analgesic

Methoxetamine-M (HO-alkyl-) -H2O AC
8787
C17H21NO3
287.15213
2180
U+UHYAC
U+UHYAC
LS/Q
Designer drug

PCEPA-M (carboxy-2''-HO-) -H2O AC
1-(1-Phenylcyclohexyl)-2-ethoxypropylamine-M (carboxy-2''-HO-) -H2O AC
7026
C17H21NO3
287.15213
1975
USPEAC
LS/Q
Designer drug

Dihydrocodeine-M (nor-)
Hydrocodone-M (nor-dihydro-)
Thebacone-M (deacetyl-nor-dihydro-)
4368
C17H21NO3
287.15213
2440
UHY
PS
LM/Q
Potent antitussive

287

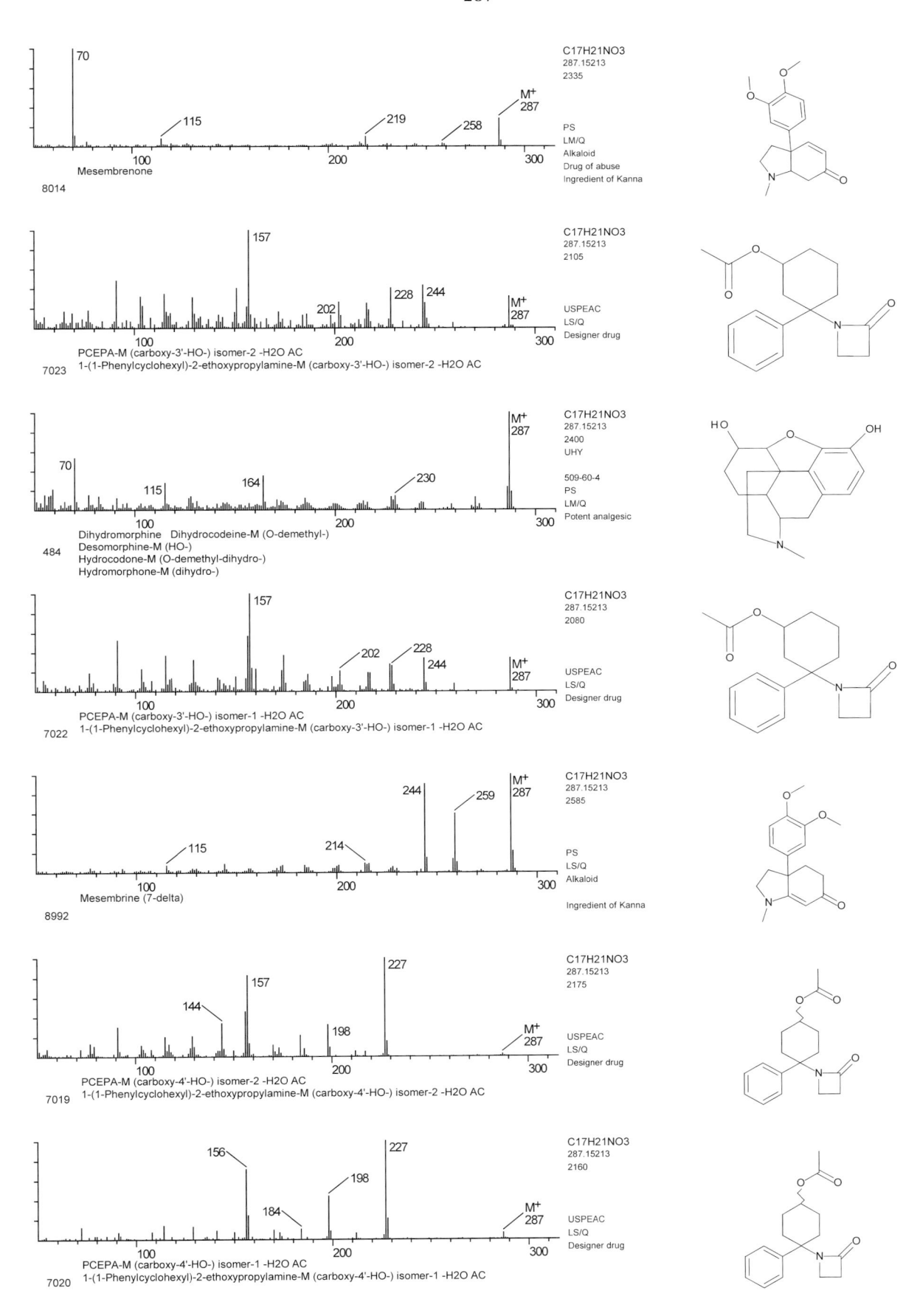

Mesembrenone 8014	Peaks: 70, 115, 219, 258, M+ 287	C17H21NO3 287.15213 2335 PS LM/Q Alkaloid Drug of abuse Ingredient of Kanna
PCEPA-M (carboxy-3'-HO-) isomer-2 -H2O AC 1-(1-Phenylcyclohexyl)-2-ethoxypropylamine-M (carboxy-3'-HO-) isomer-2 -H2O AC 7023	Peaks: 157, 202, 228, 244, M+ 287	C17H21NO3 287.15213 2105 USPEAC LS/Q Designer drug
Dihydromorphine Dihydrocodeine-M (O-demethyl-) Desomorphine-M (HO-) Hydrocodone-M (O-demethyl-dihydro-) Hydromorphone-M (dihydro-) 484	Peaks: 70, 115, 164, 230, M+ 287	C17H21NO3 287.15213 2400 UHY 509-60-4 PS LM/Q Potent analgesic
PCEPA-M (carboxy-3'-HO-) isomer-1 -H2O AC 1-(1-Phenylcyclohexyl)-2-ethoxypropylamine-M (carboxy-3'-HO-) isomer-1 -H2O AC 7022	Peaks: 157, 202, 228, 244, M+ 287	C17H21NO3 287.15213 2080 USPEAC LS/Q Designer drug
Mesembrine (7-delta) 8992	Peaks: 115, 214, 244, 259, M+ 287	C17H21NO3 287.15213 2585 PS LS/Q Alkaloid Ingredient of Kanna
PCEPA-M (carboxy-4'-HO-) isomer-2 -H2O AC 1-(1-Phenylcyclohexyl)-2-ethoxypropylamine-M (carboxy-4'-HO-) isomer-2 -H2O AC 7019	Peaks: 144, 157, 198, 227, M+ 287	C17H21NO3 287.15213 2175 USPEAC LS/Q Designer drug
PCEPA-M (carboxy-4'-HO-) isomer-1 -H2O AC 1-(1-Phenylcyclohexyl)-2-ethoxypropylamine-M (carboxy-4'-HO-) isomer-1 -H2O AC 7020	Peaks: 156, 184, 198, 227, M+ 287	C17H21NO3 287.15213 2160 USPEAC LS/Q Designer drug

287

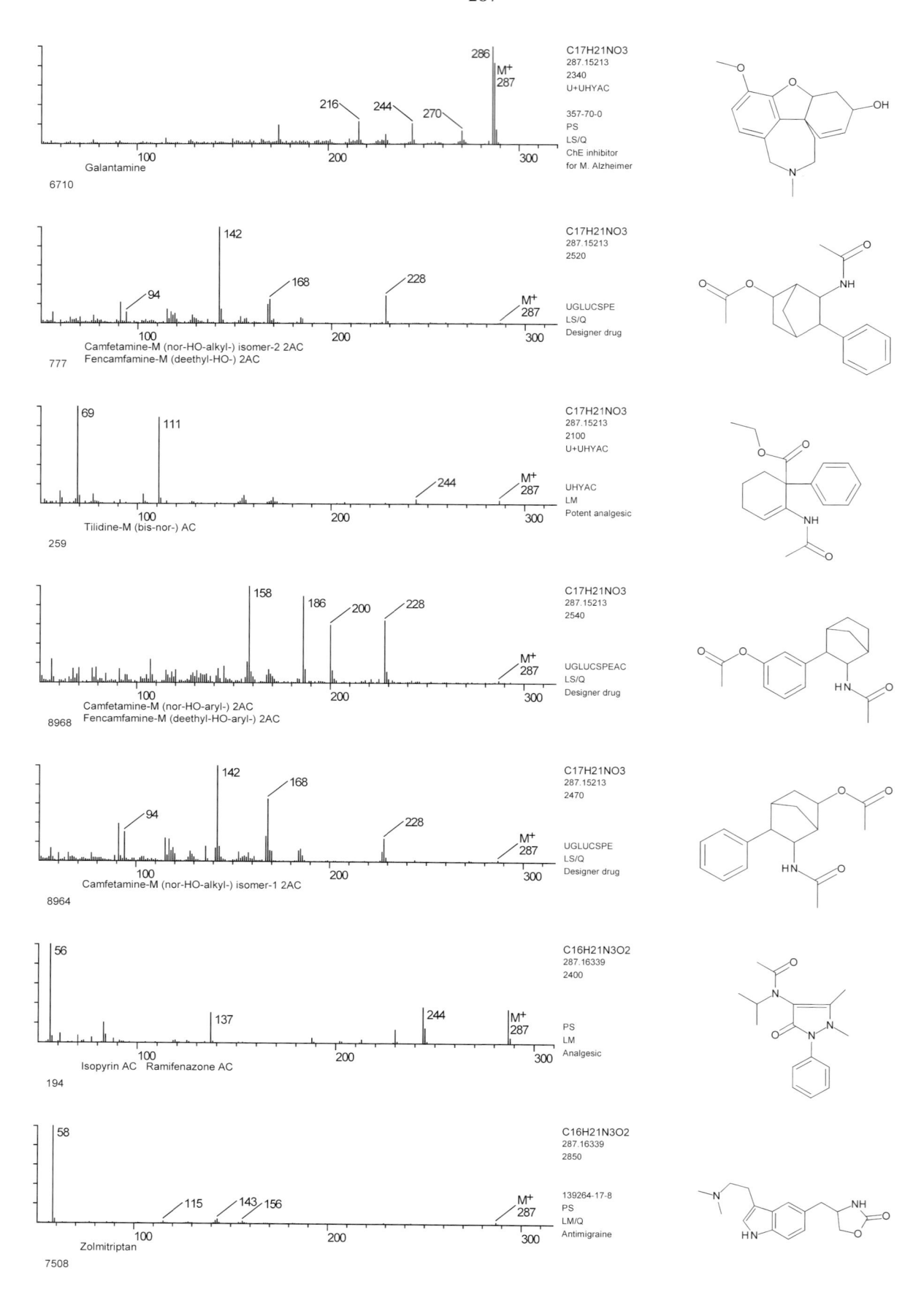

Galantamine — C17H21NO3, 287.15213, 2340, U+UHYAC, 357-70-0, PS, LS/Q, ChE inhibitor for M. Alzheimer
6710

Camfetamine-M (nor-HO-alkyl-) isomer-2 2AC
Fencamfamine-M (deethyl-HO-) 2AC — C17H21NO3, 287.15213, 2520, UGLUCSPE, LS/Q, Designer drug
777

Tilidine-M (bis-nor-) AC — C17H21NO3, 287.15213, 2100, U+UHYAC, UHYAC, LM, Potent analgesic
259

Camfetamine-M (nor-HO-aryl-) 2AC
Fencamfamine-M (deethyl-HO-aryl-) 2AC — C17H21NO3, 287.15213, 2540, UGLUCSPEAC, LS/Q, Designer drug
8968

Camfetamine-M (nor-HO-alkyl-) isomer-1 2AC — C17H21NO3, 287.15213, 2470, UGLUCSPE, LS/Q, Designer drug
8964

Isopyrin AC   Ramifenazone AC — C16H21N3O2, 287.16339, 2400, PS, LM, Analgesic
194

Zolmitriptan — C16H21N3O2, 287.16339, 2850, 139264-17-8, PS, LM/Q, Antimigraine
7508

287

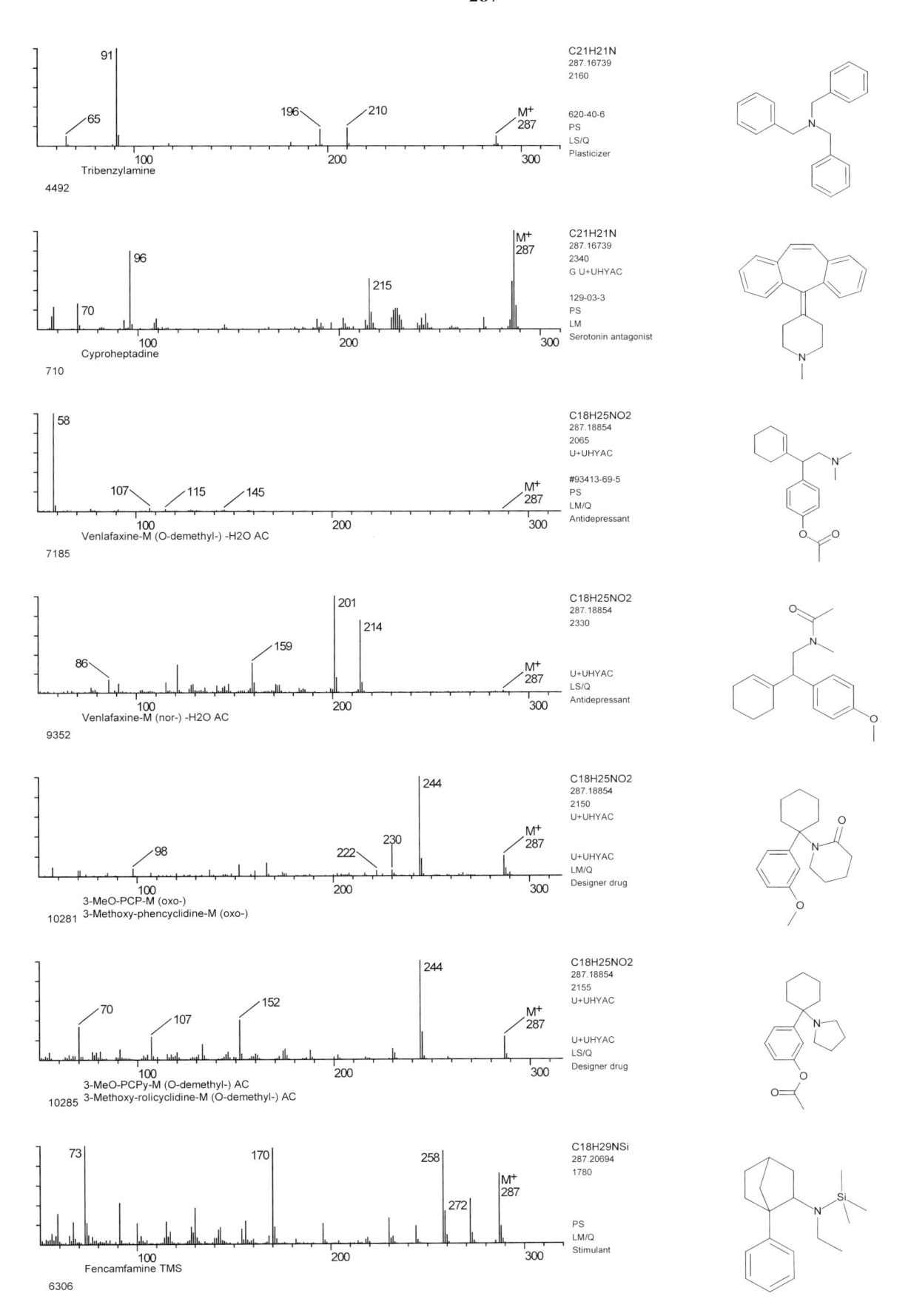

Tribenzylamine
4492
C21H21N
287.16739
2160
620-40-6
PS
LS/Q
Plasticizer

Peaks: 91, 65, 196, 210, M+ 287

Cyproheptadine
710
C21H21N
287.16739
2340
G U+UHYAC
129-03-3
PS
LM
Serotonin antagonist

Peaks: 70, 96, 215, M+ 287

Venlafaxine-M (O-demethyl-) -H2O AC
7185
C18H25NO2
287.18854
2065
U+UHYAC
#93413-69-5
PS
LM/Q
Antidepressant

Peaks: 58, 107, 115, 145, M+ 287

Venlafaxine-M (nor-) -H2O AC
9352
C18H25NO2
287.18854
2330
U+UHYAC
LS/Q
Antidepressant

Peaks: 86, 159, 201, 214, M+ 287

3-MeO-PCP-M (oxo-)
3-Methoxy-phencyclidine-M (oxo-)
10281
C18H25NO2
287.18854
2150
U+UHYAC
U+UHYAC
LM/Q
Designer drug

Peaks: 98, 222, 230, 244, M+ 287

3-MeO-PCPy-M (O-demethyl-) AC
3-Methoxy-rolicyclidine-M (O-demethyl-) AC
10285
C18H25NO2
287.18854
2155
U+UHYAC
U+UHYAC
LS/Q
Designer drug

Peaks: 70, 107, 152, 244, M+ 287

Fencamfamine TMS
6306
C18H29NSi
287.20694
1780
PS
LM/Q
Stimulant

Peaks: 73, 170, 258, 272, M+ 287

287

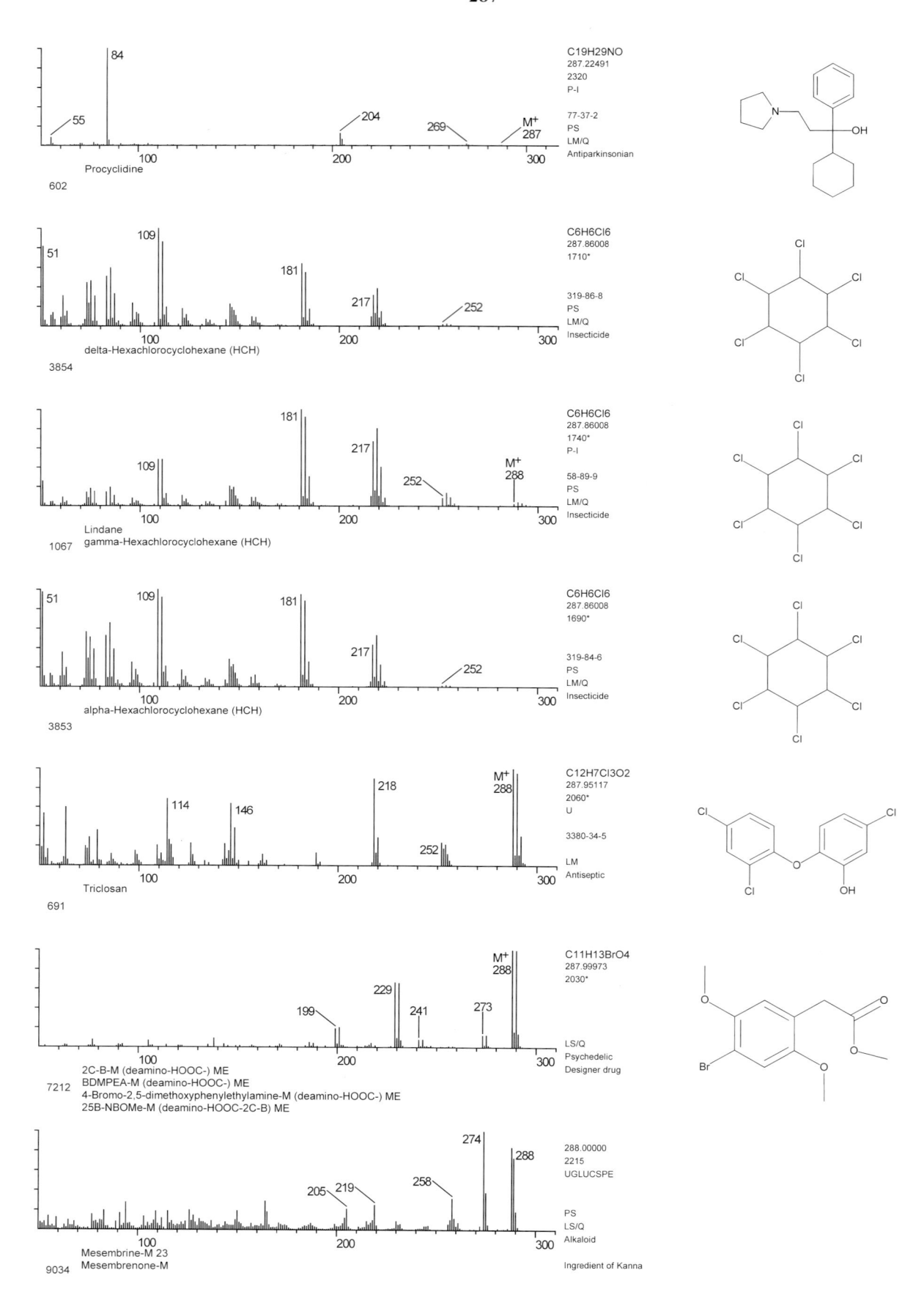

Procyclidine	C19H29NO 287.22491 2320 P-I 77-37-2 PS LM/Q Antiparkinsonian

Peaks: 55, 84, 204, 269, 287 (M+)
602

delta-Hexachlorocyclohexane (HCH)	C6H6Cl6 287.86008 1710* 319-86-8 PS LM/Q Insecticide

Peaks: 51, 109, 181, 217, 252
3854

Lindane gamma-Hexachlorocyclohexane (HCH)	C6H6Cl6 287.86008 1740* P-I 58-89-9 PS LM/Q Insecticide

Peaks: 109, 181, 217, 252, 288 (M+)
1067

alpha-Hexachlorocyclohexane (HCH)	C6H6Cl6 287.86008 1690* 319-84-6 PS LM/Q Insecticide

Peaks: 51, 109, 181, 217, 252
3853

Triclosan	C12H7Cl3O2 287.95117 2060* U 3380-34-5 LM Antiseptic

Peaks: 114, 146, 218, 252, 288 (M+)
691

2C-B-M (deamino-HOOC-) ME BDMPEA-M (deamino-HOOC-) ME 4-Bromo-2,5-dimethoxyphenylethylamine-M (deamino-HOOC-) ME 25B-NBOMe-M (deamino-HOOC-2C-B) ME	C11H13BrO4 287.99973 2030* LS/Q Psychedelic Designer drug

Peaks: 199, 229, 241, 273, 288 (M+)
7212

Mesembrine-M 23 Mesembrenone-M	288.00000 2215 UGLUCSPE PS LS/Q Alkaloid Ingredient of Kanna

Peaks: 205, 219, 258, 274, 288
9034

288

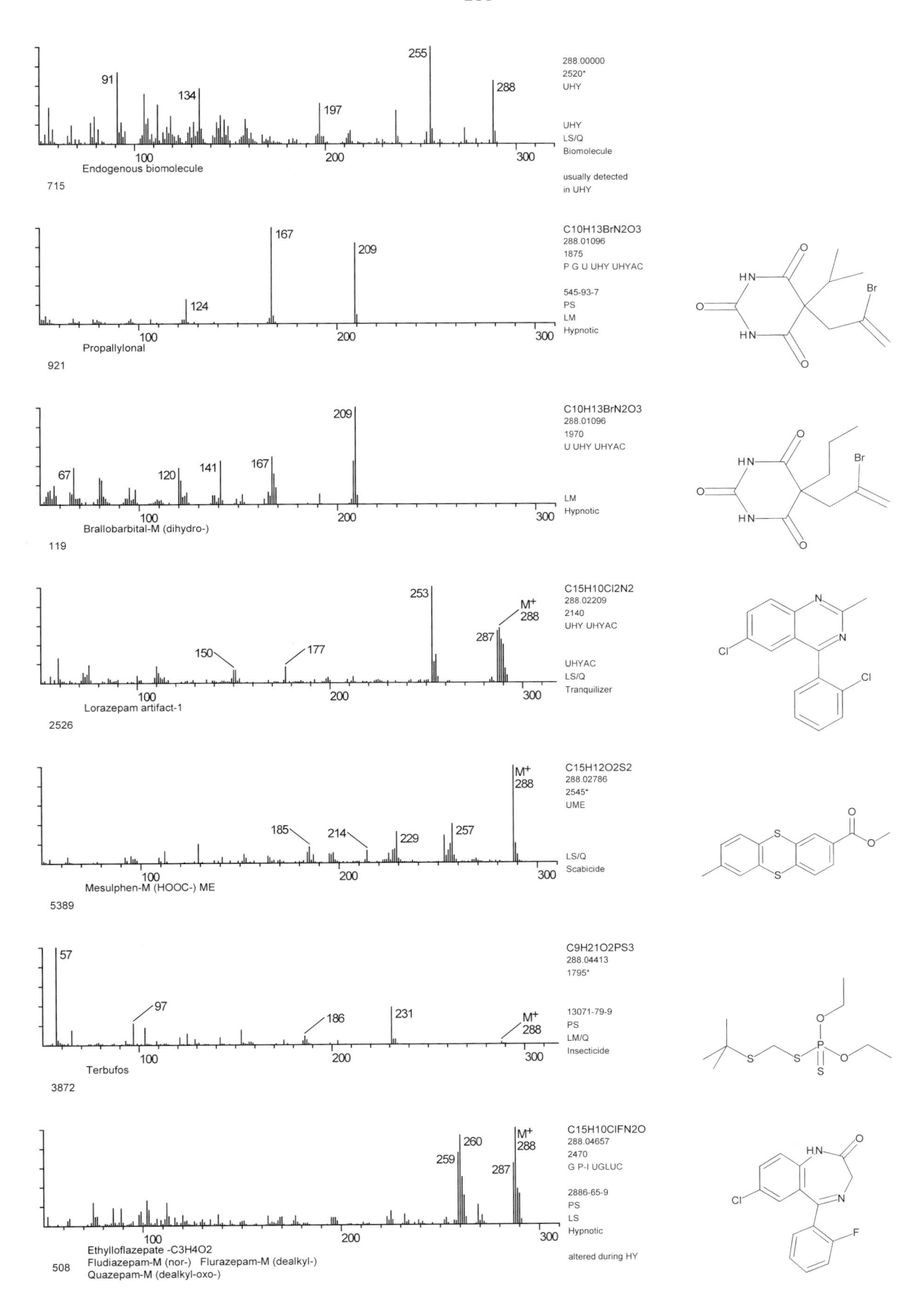

Endogenous biomolecule — 715
Peaks: 91, 134, 197, 255, 288
288.00000
2520*
UHY

UHY
LS/Q
Biomolecule

usually detected in UHY

Propallylonal — 921
Peaks: 124, 167, 209
C10H13BrN2O3
288.01096
1875
P G U UHY UHYAC

545-93-7
PS
LM
Hypnotic

Brallobarbital-M (dihydro-) — 119
Peaks: 67, 120, 141, 167, 209
C10H13BrN2O3
288.01096
1970
U UHY UHYAC

LM
Hypnotic

Lorazepam artifact-1 — 2526
Peaks: 150, 177, 253, 287, M+ 288
C15H10Cl2N2
288.02209
2140
UHY UHYAC

UHYAC
LS/Q
Tranquilizer

Mesulphen-M (HOOC-) ME — 5389
Peaks: 185, 214, 229, 257, M+ 288
C15H12O2S2
288.02786
2545*
UME

LS/Q
Scabicide

Terbufos — 3872
Peaks: 57, 97, 186, 231, M+ 288
C9H21O2PS3
288.04413
1795*

13071-79-9
PS
LM/Q
Insecticide

Ethylloflazepate -C3H4O2
Fludiazepam-M (nor-)   Flurazepam-M (dealkyl-)
Quazepam-M (dealkyl-oxo-) — 508
Peaks: 259, 260, 287, M+ 288
C15H10ClFN2O
288.04657
2470
G P-I UGLUC

2886-65-9
PS
LS
Hypnotic

altered during HY

825

288

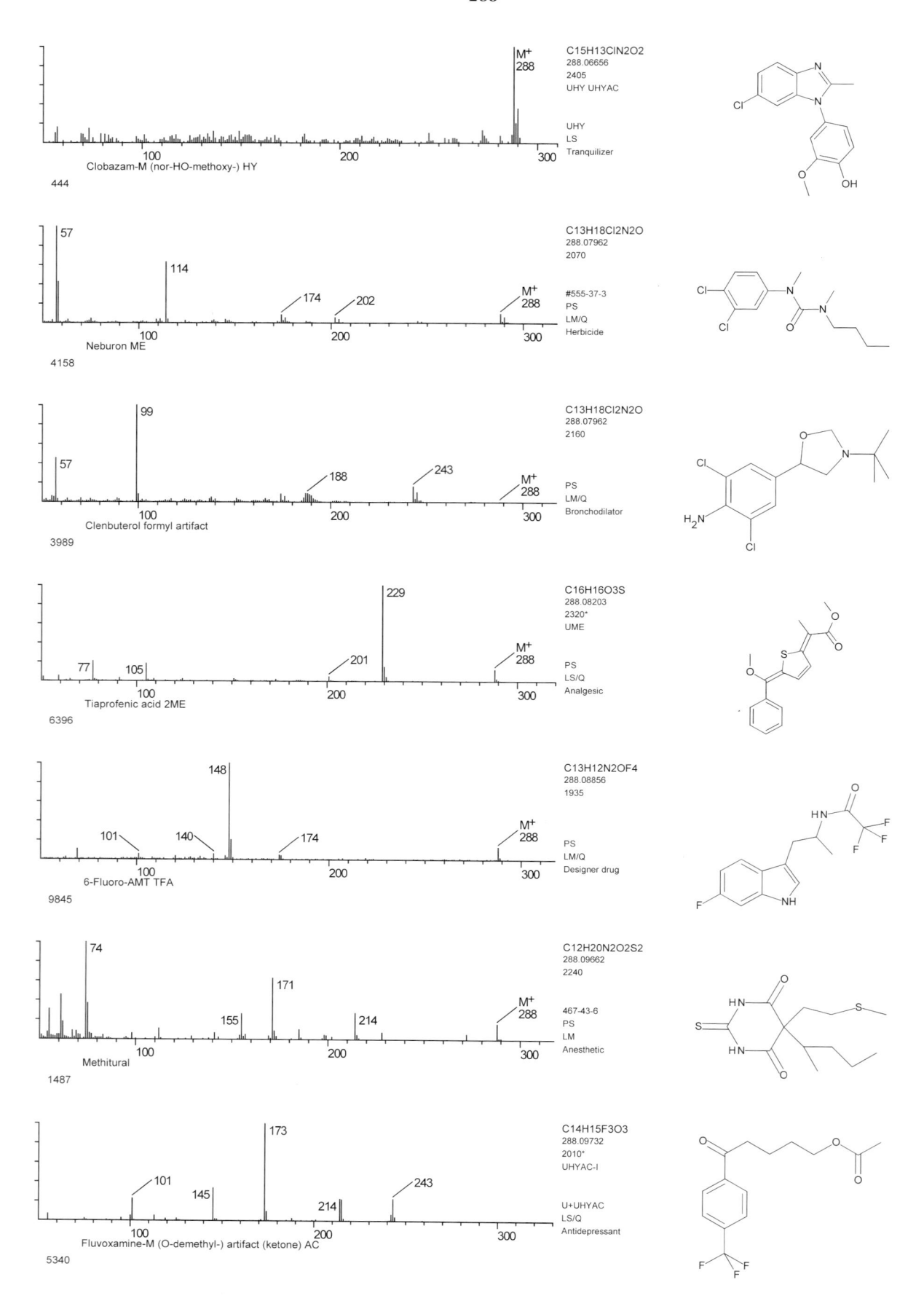

Clobazam-M (nor-HO-methoxy-) HY
444
C15H13ClN2O2
288.06656
2405
UHY UHYAC

UHY
LS
Tranquilizer

Neburon ME
4158
C13H18Cl2N2O
288.07962
2070

#555-37-3
PS
LM/Q
Herbicide

Clenbuterol formyl artifact
3989
C13H18Cl2N2O
288.07962
2160

PS
LM/Q
Bronchodilator

Tiaprofenic acid 2ME
6396
C16H16O3S
288.08203
2320*
UME

PS
LS/Q
Analgesic

6-Fluoro-AMT TFA
9845
C13H12N2OF4
288.08856
1935

PS
LM/Q
Designer drug

Methitural
1487
C12H20N2O2S2
288.09662
2240

467-43-6
PS
LM
Anesthetic

Fluvoxamine-M (O-demethyl-) artifact (ketone) AC
5340
C14H15F3O3
288.09732
2010*
UHYAC-I

U+UHYAC
LS/Q
Antidepressant

826

288

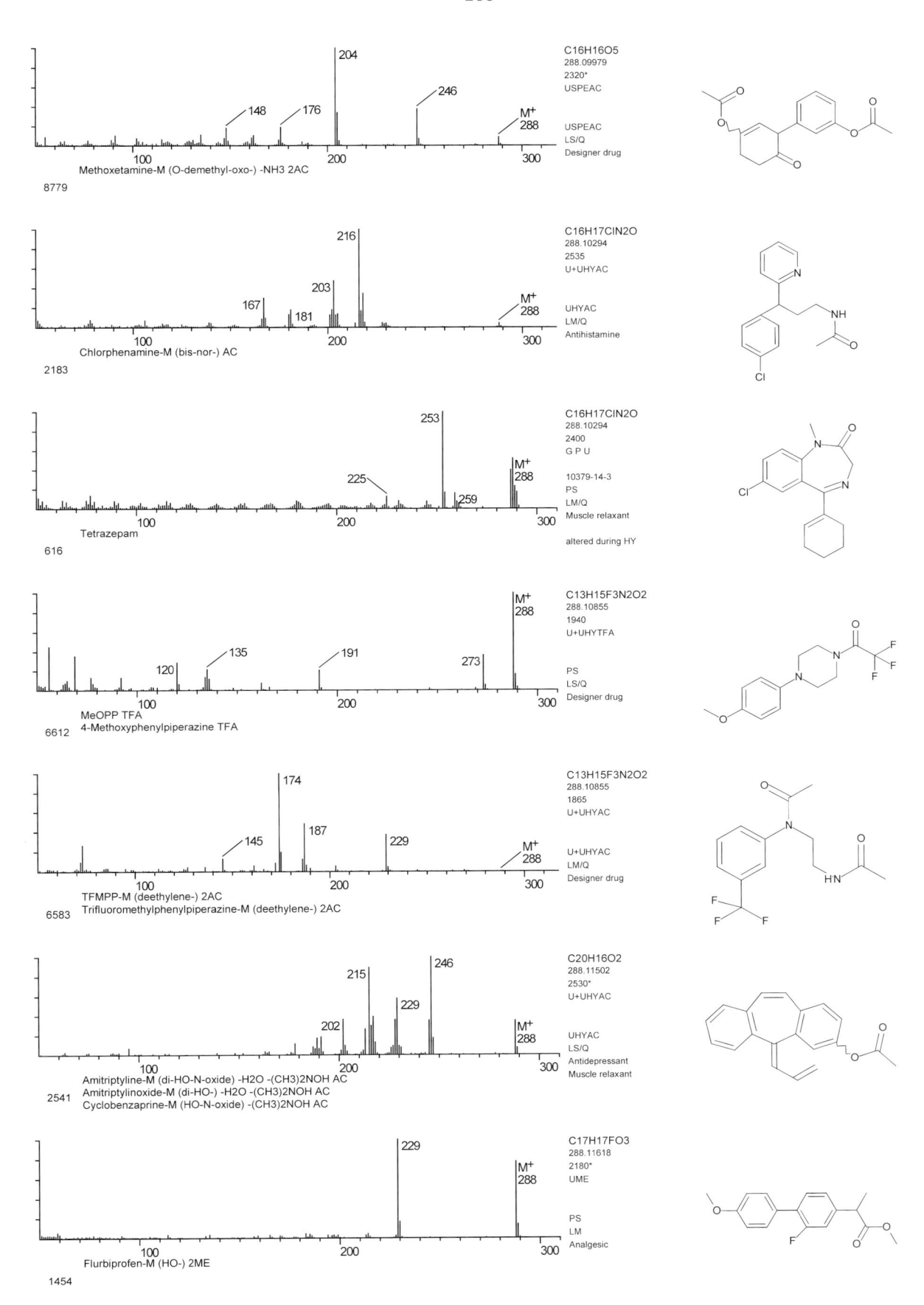

Methoxetamine-M (O-demethyl-oxo-) -NH3 2AC
8779
C16H16O5
288.09979
2320*
USPEAC
USPEAC
LS/Q
Designer drug

Chlorphenamine-M (bis-nor-) AC
2183
C16H17ClN2O
288.10294
2535
U+UHYAC
UHYAC
LM/Q
Antihistamine

Tetrazepam
616
C16H17ClN2O
288.10294
2400
G P U
10379-14-3
PS
LM/Q
Muscle relaxant
altered during HY

MeOPP TFA
4-Methoxyphenylpiperazine TFA
6612
C13H15F3N2O2
288.10855
1940
U+UHYTFA
PS
LS/Q
Designer drug

TFMPP-M (deethylene-) 2AC
Trifluoromethylphenylpiperazine-M (deethylene-) 2AC
6583
C13H15F3N2O2
288.10855
1865
U+UHYAC
U+UHYAC
LM/Q
Designer drug

Amitriptyline-M (di-HO-N-oxide) -H2O -(CH3)2NOH AC
Amitriptylinoxide-M (di-HO-) -H2O -(CH3)2NOH AC
Cyclobenzaprine-M (HO-N-oxide) -(CH3)2NOH AC
2541
C20H16O2
288.11502
2530*
U+UHYAC
UHYAC
LS/Q
Antidepressant
Muscle relaxant

Flurbiprofen-M (HO-) 2ME
1454
C17H17FO3
288.11618
2180*
UME
PS
LM
Analgesic

288

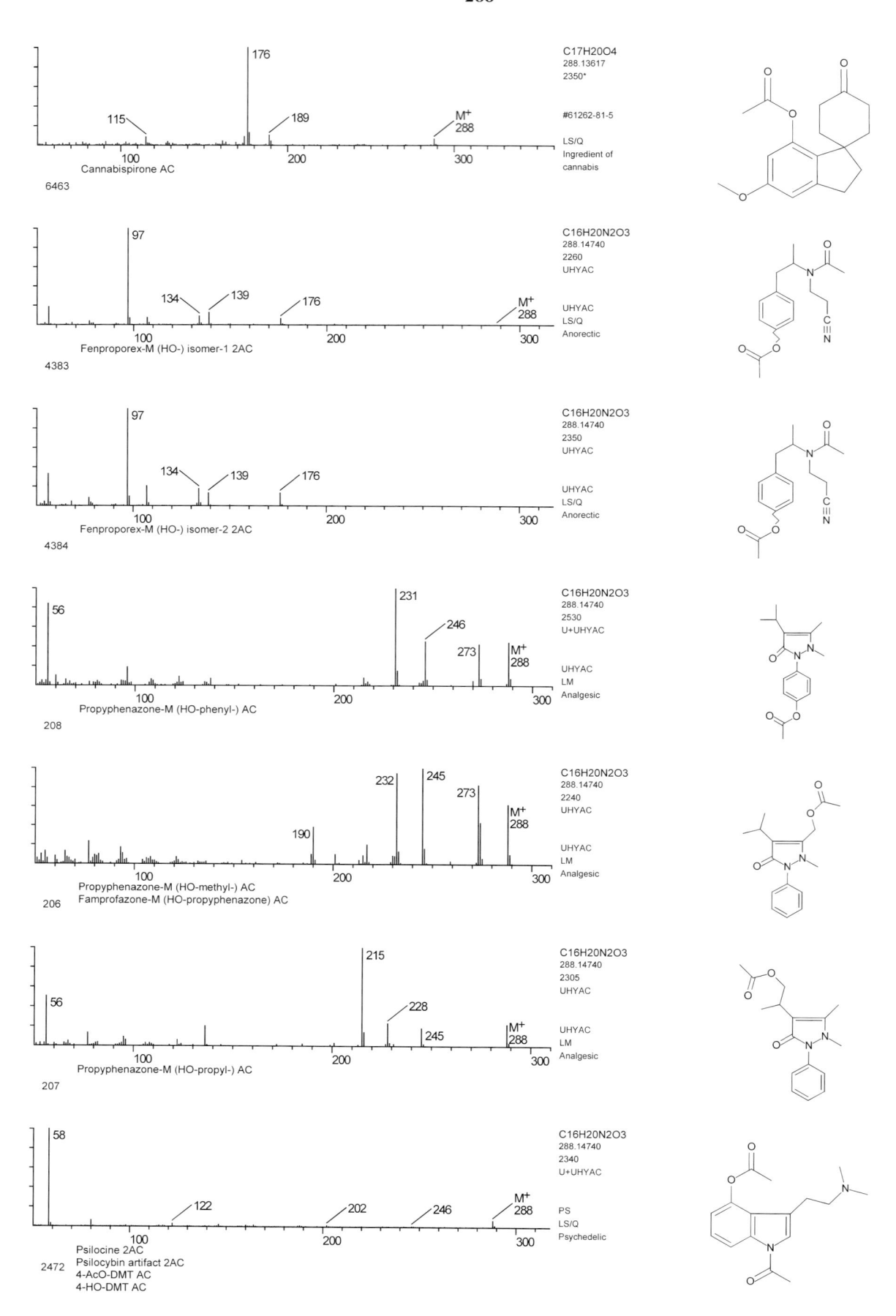

Spectrum	Peaks	Formula / Info
Cannabispirone AC — 6463	115, 176, 189, M+ 288	C17H20O4, 288.13617, 2350*, #61262-81-5, LS/Q, Ingredient of cannabis
Fenproporex-M (HO-) isomer-1 2AC — 4383	97, 134, 139, 176, M+ 288	C16H20N2O3, 288.14740, 2260, UHYAC / UHYAC, LS/Q, Anorectic
Fenproporex-M (HO-) isomer-2 2AC — 4384	97, 134, 139, 176	C16H20N2O3, 288.14740, 2350, UHYAC / UHYAC, LS/Q, Anorectic
Propyphenazone-M (HO-phenyl-) AC — 208	56, 231, 246, 273, M+ 288	C16H20N2O3, 288.14740, 2530, U+UHYAC / UHYAC, LM, Analgesic
Propyphenazone-M (HO-methyl-) AC / Famprofazone-M (HO-propyphenazone) AC — 206	190, 232, 245, 273, M+ 288	C16H20N2O3, 288.14740, 2240, UHYAC / UHYAC, LM, Analgesic
Propyphenazone-M (HO-propyl-) AC — 207	56, 215, 228, 245, M+ 288	C16H20N2O3, 288.14740, 2305, UHYAC / UHYAC, LM, Analgesic
Psilocine 2AC / Psilocybin artifact 2AC / 4-AcO-DMT AC / 4-HO-DMT AC — 2472	58, 122, 202, 246, M+ 288	C16H20N2O3, 288.14740, 2340, U+UHYAC, PS, LS/Q, Psychedelic

288

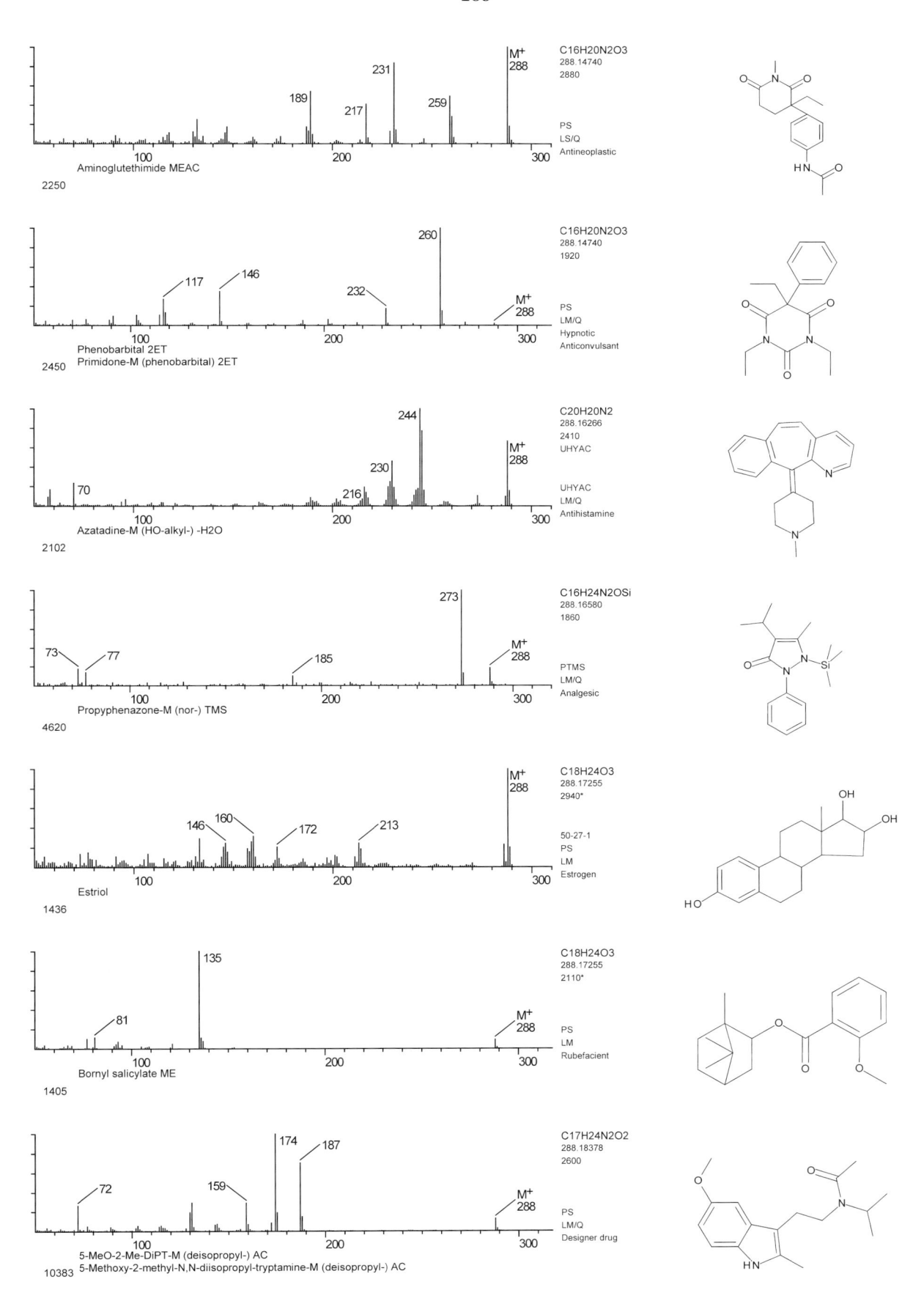

C16H20N2O3
288.14740
2880

PS
LS/Q
Antineoplastic

Aminoglutethimide MEAC
2250

C16H20N2O3
288.14740
1920

PS
LM/Q
Hypnotic
Anticonvulsant

Phenobarbital 2ET
Primidone-M (phenobarbital) 2ET
2450

C20H20N2
288.16266
2410
UHYAC

UHYAC
LM/Q
Antihistamine

Azatadine-M (HO-alkyl-) -H2O
2102

C16H24N2OSi
288.16580
1860

PTMS
LM/Q
Analgesic

Propyphenazone-M (nor-) TMS
4620

C18H24O3
288.17255
2940*

50-27-1
PS
LM
Estrogen

Estriol
1436

C18H24O3
288.17255
2110*

PS
LM
Rubefacient

Bornyl salicylate ME
1405

C17H24N2O2
288.18378
2600

PS
LM/Q
Designer drug

5-MeO-2-Me-DiPT-M (deisopropyl-) AC
5-Methoxy-2-methyl-N,N-diisopropyl-tryptamine-M (deisopropyl-) AC
10383

288

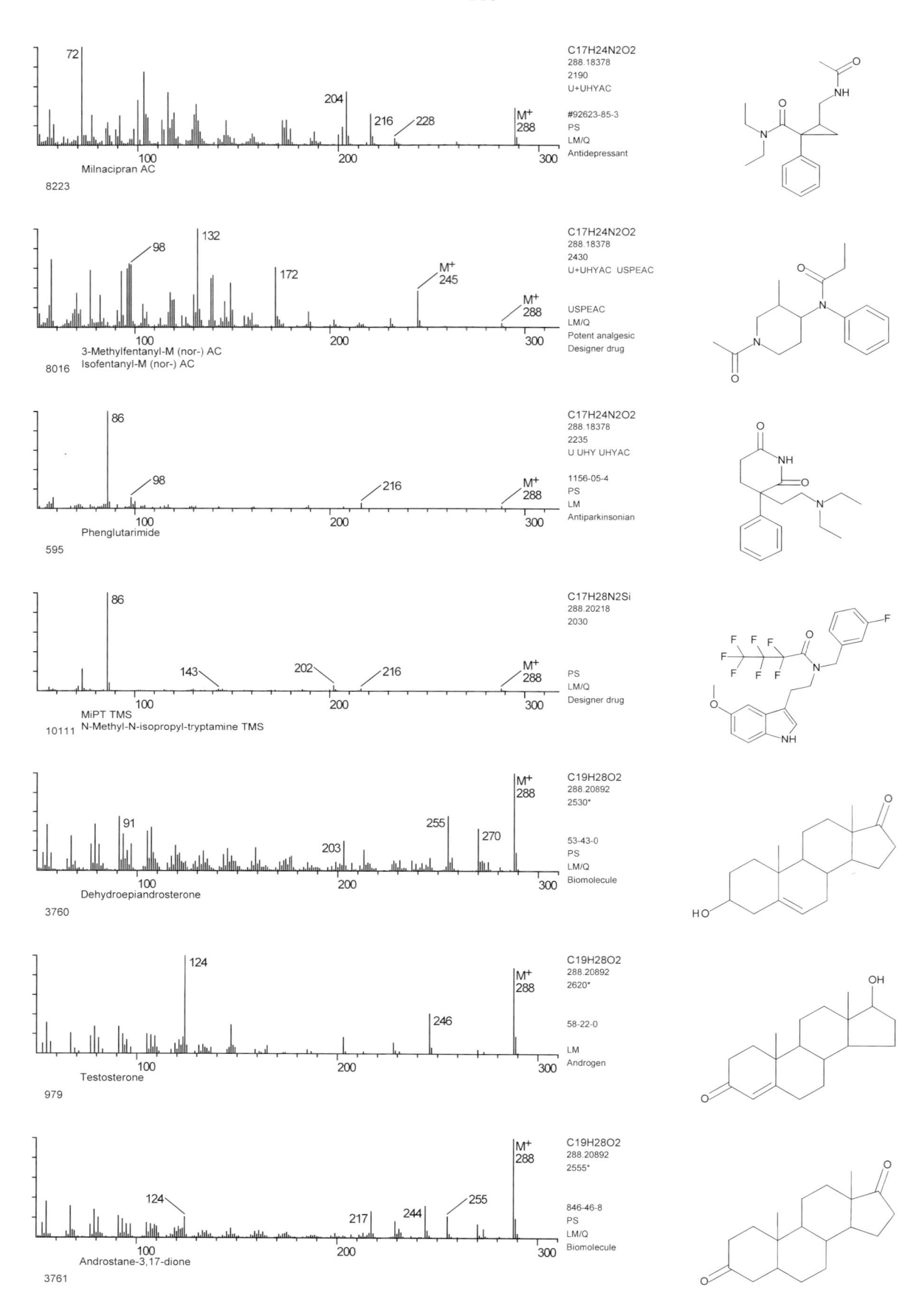

Milnacipran AC — C17H24N2O2, 288.18378, 2190, U+UHYAC, #92623-85-3, PS, LM/Q, Antidepressant
Peaks: 72, 204, 216, 228, M+ 288
8223

3-Methylfentanyl-M (nor-) AC / Isofentanyl-M (nor-) AC — C17H24N2O2, 288.18378, 2430, U+UHYAC USPEAC, USPEAC, LM/Q, Potent analgesic, Designer drug
Peaks: 98, 132, 172, M+ 245, M+ 288
8016

Phenglutarimide — C17H24N2O2, 288.18378, 2235, U UHY UHYAC, 1156-05-4, PS, LM, Antiparkinsonian
Peaks: 86, 98, 216, M+ 288
595

MiPT TMS / N-Methyl-N-isopropyl-tryptamine TMS — C17H28N2Si, 288.20218, 2030, PS, LM/Q, Designer drug
Peaks: 86, 143, 202, 216, M+ 288
10111

Dehydroepiandrosterone — C19H28O2, 288.20892, 2530*, 53-43-0, PS, LM/Q, Biomolecule
Peaks: 91, 203, 255, 270, M+ 288
3760

Testosterone — C19H28O2, 288.20892, 2620*, 58-22-0, LM, Androgen
Peaks: 124, 246, M+ 288
979

Androstane-3,17-dione — C19H28O2, 288.20892, 2555*, 846-46-8, PS, LM/Q, Biomolecule
Peaks: 124, 217, 244, 255, M+ 288
3761

288

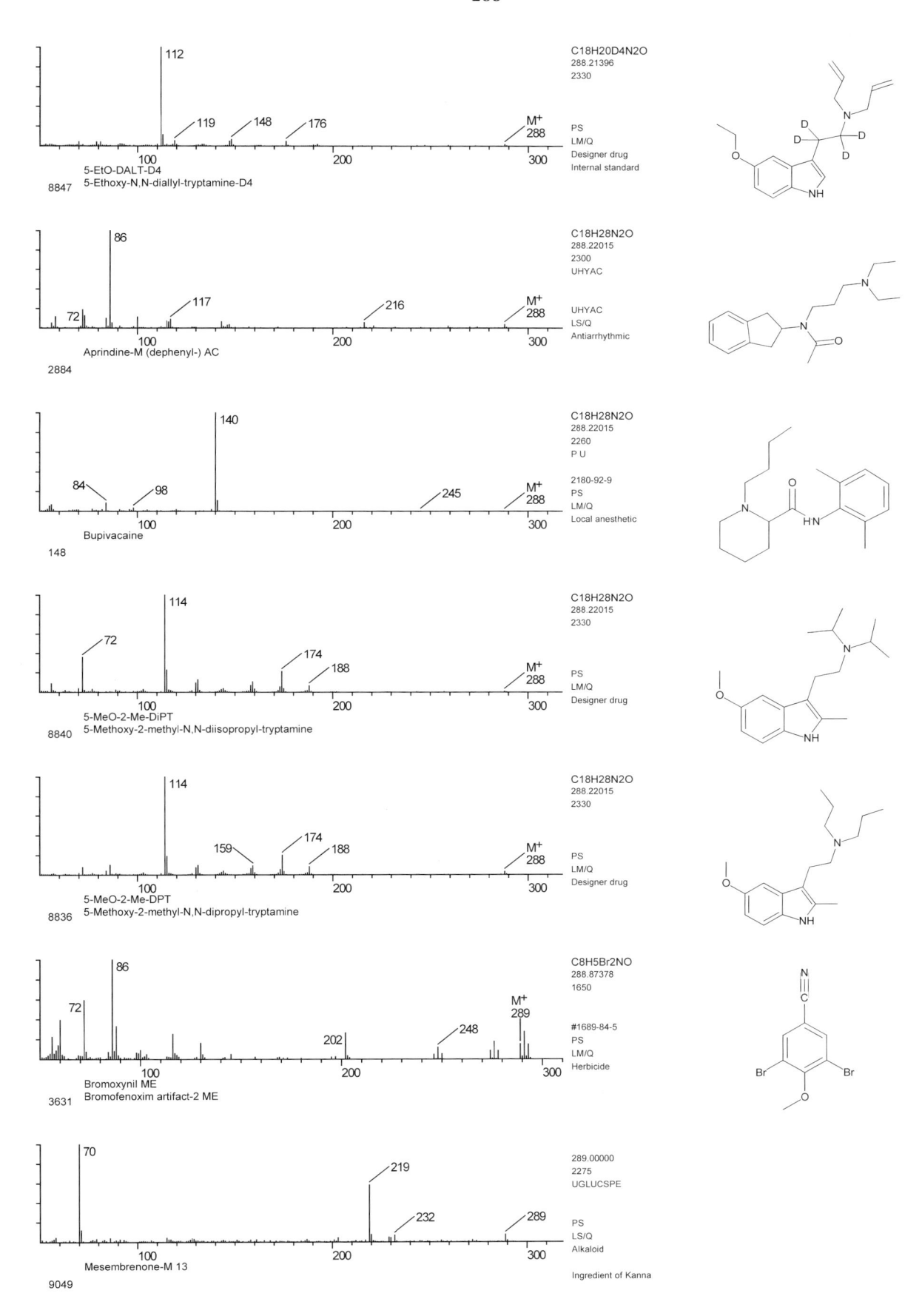

8847 5-EtO-DALT-D4
5-Ethoxy-N,N-diallyl-tryptamine-D4
Peaks: 112, 119, 148, 176, 288 (M+)
C18H20D4N2O
288.21396
2330
PS
LM/Q
Designer drug
Internal standard

2884 Aprindine-M (dephenyl-) AC
Peaks: 72, 86, 117, 216, 288 (M+)
C18H28N2O
288.22015
2300
UHYAC
UHYAC
LS/Q
Antiarrhythmic

148 Bupivacaine
Peaks: 84, 98, 140, 245, 288 (M+)
C18H28N2O
288.22015
2260
P U
2180-92-9
PS
LM/Q
Local anesthetic

8840 5-MeO-2-Me-DiPT
5-Methoxy-2-methyl-N,N-diisopropyl-tryptamine
Peaks: 72, 114, 174, 188, 288 (M+)
C18H28N2O
288.22015
2330
PS
LM/Q
Designer drug

8836 5-MeO-2-Me-DPT
5-Methoxy-2-methyl-N,N-dipropyl-tryptamine
Peaks: 114, 159, 174, 188, 288 (M+)
C18H28N2O
288.22015
2330
PS
LM/Q
Designer drug

3631 Bromoxynil ME
Bromofenoxim artifact-2 ME
Peaks: 72, 86, 202, 248, 289 (M+)
C8H5Br2NO
288.87378
1650
#1689-84-5
PS
LM/Q
Herbicide

9049 Mesembrenone-M 13
Peaks: 70, 219, 232, 289
289.00000
2275
UGLUCSPE
PS
LS/Q
Alkaloid
Ingredient of Kanna

289

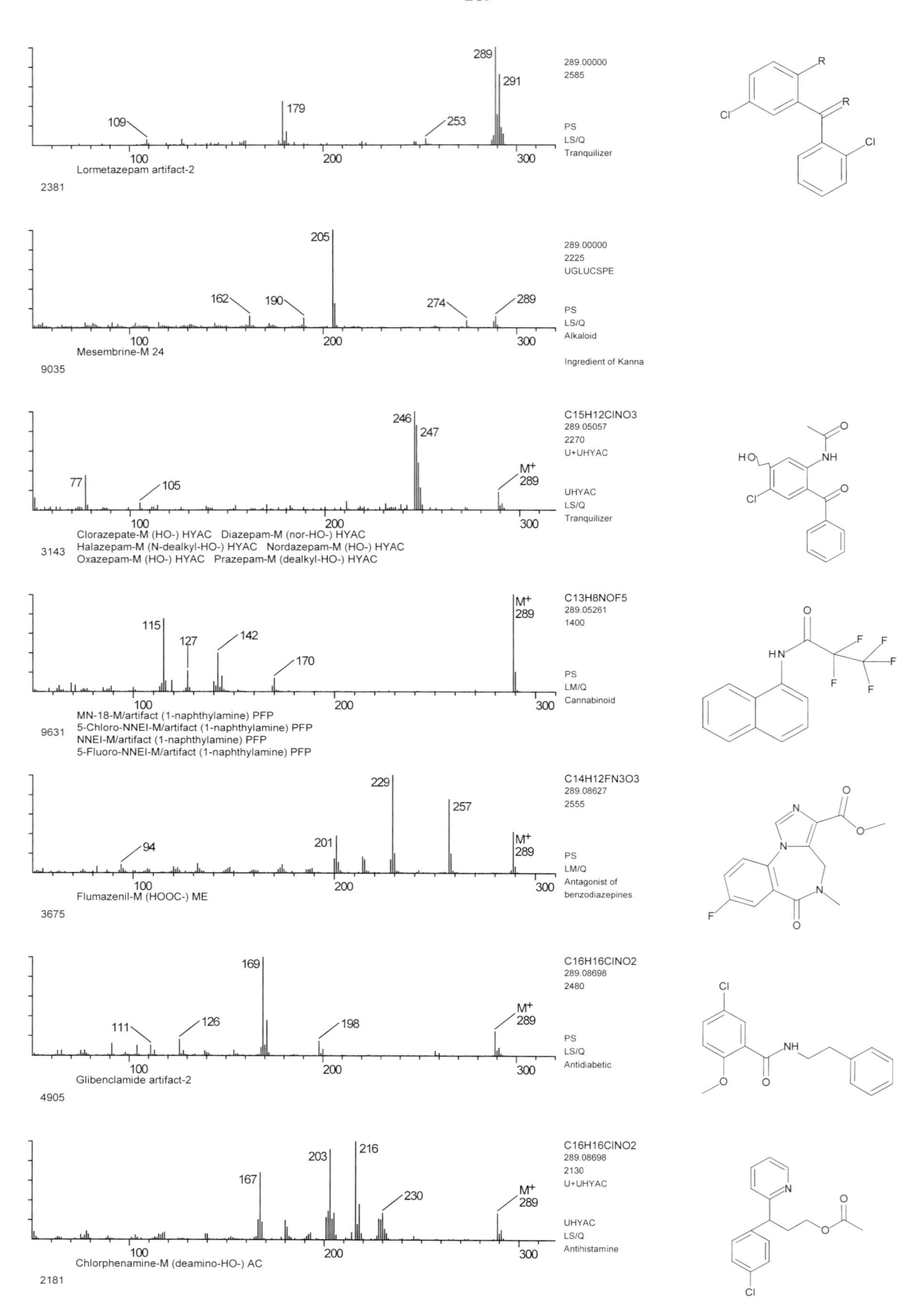

Lormetazepam artifact-2 — Tranquilizer
289.00000 / 2585 / PS / LS/Q
Peaks: 109, 179, 253, 289, 291

Mesembrine-M 24 — Alkaloid, Ingredient of Kanna
289.00000 / 2225 / UGLUCSPE / PS / LS/Q
Peaks: 162, 190, 205, 274, 289

Clorazepate-M (HO-) HYAC   Diazepam-M (nor-HO-) HYAC
Halazepam-M (N-dealkyl-HO-) HYAC   Nordazepam-M (HO-) HYAC
Oxazepam-M (HO-) HYAC   Prazepam-M (dealkyl-HO-) HYAC
C15H12ClNO3 / 289.05057 / 2270 / U+UHYAC / UHYAC / LS/Q / Tranquilizer
Peaks: 77, 105, 246, 247, M+ 289

MN-18-M/artifact (1-naphthylamine) PFP
5-Chloro-NNEI-M/artifact (1-naphthylamine) PFP
NNEI-M/artifact (1-naphthylamine) PFP
5-Fluoro-NNEI-M/artifact (1-naphthylamine) PFP
C13H8NOF5 / 289.05261 / 1400 / PS / LM/Q / Cannabinoid
Peaks: 115, 127, 142, 170, M+ 289

Flumazenil-M (HOOC-) ME — Antagonist of benzodiazepines
C14H12FN3O3 / 289.08627 / 2555 / PS / LM/Q
Peaks: 94, 201, 229, 257, M+ 289

Glibenclamide artifact-2 — Antidiabetic
C16H16ClNO2 / 289.08698 / 2480 / PS / LS/Q
Peaks: 111, 126, 169, 198, M+ 289

Chlorphenamine-M (deamino-HO-) AC — Antihistamine
C16H16ClNO2 / 289.08698 / 2130 / U+UHYAC / UHYAC / LS/Q
Peaks: 167, 203, 216, 230, M+ 289

289

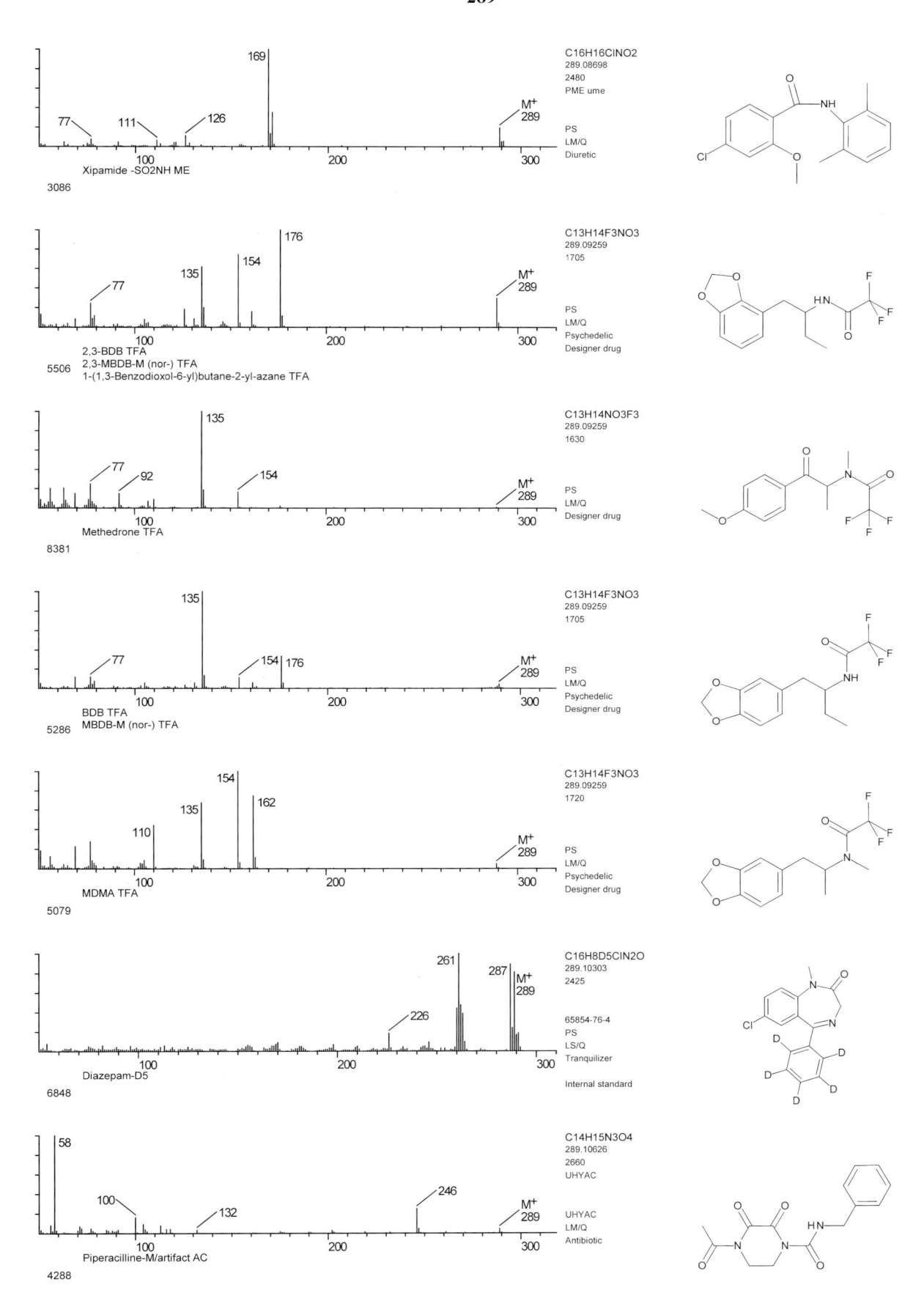

C16H16ClNO2
289.08698
2480
PME ume

M+ 289
PS
LM/Q
Diuretic

Xipamide -SO2NH ME
3086

C13H14F3NO3
289.09259
1705

M+ 289
PS
LM/Q
Psychedelic
Designer drug

2,3-BDB TFA
2,3-MBDB-M (nor-) TFA
1-(1,3-Benzodioxol-6-yl)butane-2-yl-azane TFA
5506

C13H14NO3F3
289.09259
1630

M+ 289
PS
LM/Q
Designer drug

Methedrone TFA
8381

C13H14F3NO3
289.09259
1705

M+ 289
PS
LM/Q
Psychedelic
Designer drug

BDB TFA
MBDB-M (nor-) TFA
5286

C13H14F3NO3
289.09259
1720

M+ 289
PS
LM/Q
Psychedelic
Designer drug

MDMA TFA
5079

C16H8D5ClN2O
289.10303
2425

65854-76-4
PS
LS/Q
Tranquilizer

Internal standard

Diazepam-D5
6848

C14H15N3O4
289.10626
2660
UHYAC

M+ 289
UHYAC
LM/Q
Antibiotic

Piperacilline-M/artifact AC
4288

289

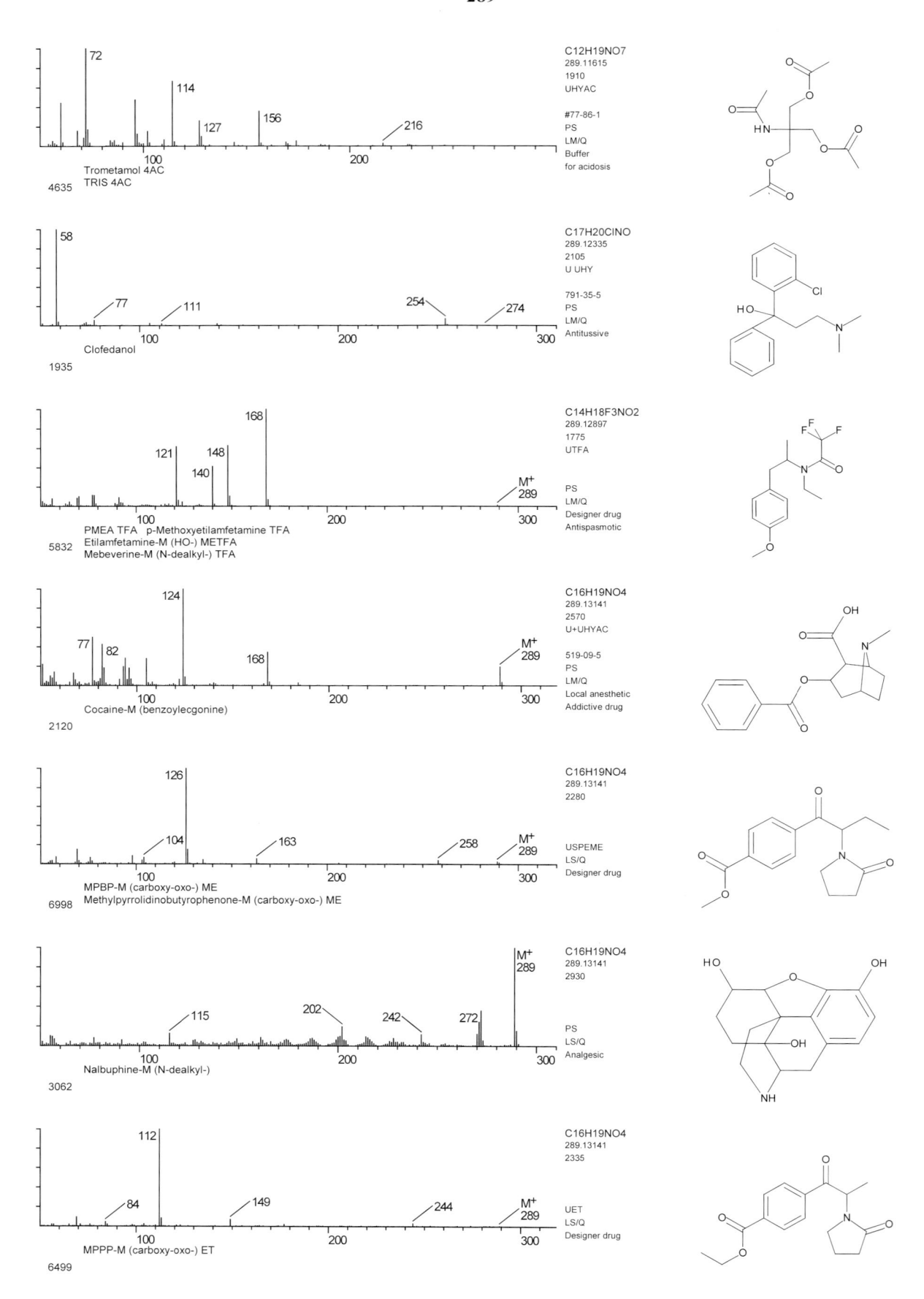

Spectrum	Compound	Formula / Info
4635	Trometamol 4AC / TRIS 4AC; peaks 72, 114, 127, 156, 216	C12H19NO7, 289.11615, 1910, UHYAC, #77-86-1, PS, LM/Q, Buffer for acidosis
1935	Clofedanol; peaks 58, 77, 111, 254, 274	C17H20ClNO, 289.12335, 2105, U UHY, 791-35-5, PS, LM/Q, Antitussive
5832	PMEA TFA p-Methoxyetilamfetamine TFA / Etilamfetamine-M (HO-) METFA / Mebeverine-M (N-dealkyl-) TFA; peaks 121, 140, 148, 168, M+ 289	C14H18F3NO2, 289.12897, 1775, UTFA, PS, LM/Q, Designer drug, Antispasmotic
2120	Cocaine-M (benzoylecgonine); peaks 77, 82, 124, 168, M+ 289	C16H19NO4, 289.13141, 2570, U+UHYAC, 519-09-5, PS, LM/Q, Local anesthetic, Addictive drug
6998	MPBP-M (carboxy-oxo-) ME / Methylpyrrolidinobutyrophenone-M (carboxy-oxo-) ME; peaks 104, 126, 163, 258, M+ 289	C16H19NO4, 289.13141, 2280, USPEME, LS/Q, Designer drug
3062	Nalbuphine-M (N-dealkyl-); peaks 115, 202, 242, 272, M+ 289	C16H19NO4, 289.13141, 2930, PS, LS/Q, Analgesic
6499	MPPP-M (carboxy-oxo-) ET; peaks 84, 112, 149, 244, M+ 289	C16H19NO4, 289.13141, 2335, UET, LS/Q, Designer drug

289

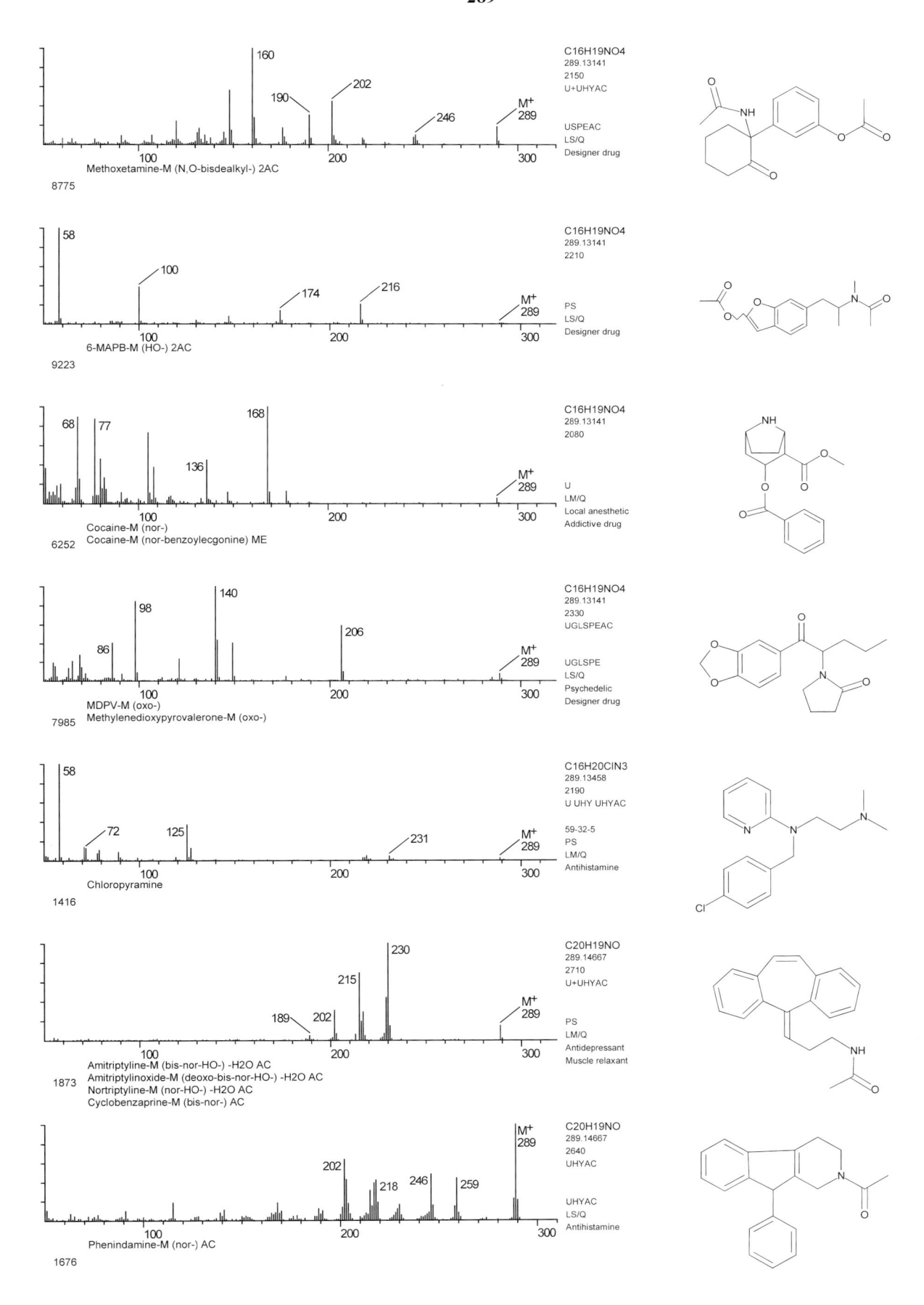

Methoxetamine-M (N,O-bisdealkyl-) 2AC — peaks: 160, 190, 202, 246, M+ 289	C16H19NO4 289.13141 2150 U+UHYAC USPEAC LS/Q Designer drug	
8775		
6-MAPB-M (HO-) 2AC — peaks: 58, 100, 174, 216, M+ 289	C16H19NO4 289.13141 2210 PS LS/Q Designer drug	
9223		
Cocaine-M (nor-) Cocaine-M (nor-benzoylecgonine) ME — peaks: 68, 77, 136, 168, M+ 289	C16H19NO4 289.13141 2080 U LM/Q Local anesthetic Addictive drug	
6252		
MDPV-M (oxo-) Methylenedioxypyrovalerone-M (oxo-) — peaks: 86, 98, 140, 206, M+ 289	C16H19NO4 289.13141 2330 UGLSPEAC UGLSPE LS/Q Psychedelic Designer drug	
7985		
Chloropyramine — peaks: 58, 72, 125, 231, M+ 289	C16H20ClN3 289.13458 2190 U UHY UHYAC 59-32-5 PS LM/Q Antihistamine	
1416		
Amitriptyline-M (bis-nor-HO-) -H2O AC Amitriptylinoxide-M (deoxo-bis-nor-HO-) -H2O AC Nortriptyline-M (nor-HO-) -H2O AC Cyclobenzaprine-M (bis-nor-) AC — peaks: 189, 202, 215, 230, M+ 289	C20H19NO 289.14667 2710 U+UHYAC PS LM/Q Antidepressant Muscle relaxant	
1873		
Phenindamine-M (nor-) AC — peaks: 202, 218, 246, 259, M+ 289	C20H19NO 289.14667 2640 UHYAC UHYAC LS/Q Antihistamine	
1676		

289

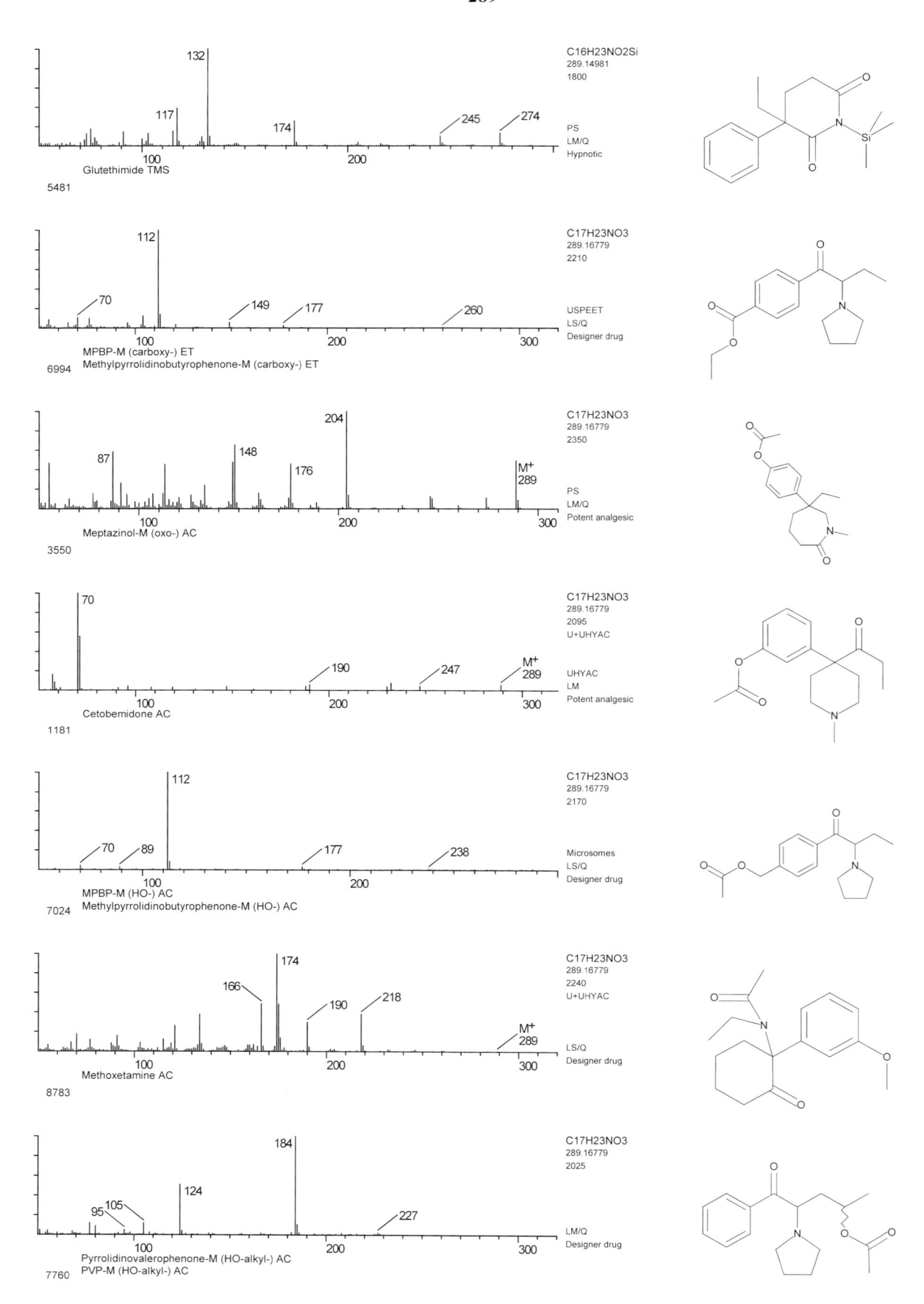

Glutethimide TMS	132, 117, 174, 245, 274	C16H23NO2Si 289.14981 1800 PS LM/Q Hypnotic
5481		
MPBP-M (carboxy-) ET Methylpyrrolidinobutyrophenone-M (carboxy-) ET	112, 70, 149, 177, 260	C17H23NO3 289.16779 2210 USPEET LS/Q Designer drug
6994		
Meptazinol-M (oxo-) AC	204, 87, 148, 176, M+ 289	C17H23NO3 289.16779 2350 PS LM/Q Potent analgesic
3550		
Cetobemidone AC	70, 190, 247, M+ 289	C17H23NO3 289.16779 2095 U+UHYAC UHYAC LM Potent analgesic
1181		
MPBP-M (HO-) AC Methylpyrrolidinobutyrophenone-M (HO-) AC	112, 70, 89, 177, 238	C17H23NO3 289.16779 2170 Microsomes LS/Q Designer drug
7024		
Methoxetamine AC	174, 166, 190, 218, M+ 289	C17H23NO3 289.16779 2240 U+UHYAC LS/Q Designer drug
8783		
Pyrrolidinovalerophenone-M (HO-alkyl-) AC PVP-M (HO-alkyl-) AC	184, 124, 95, 105, 227	C17H23NO3 289.16779 2025 LM/Q Designer drug
7760		

289

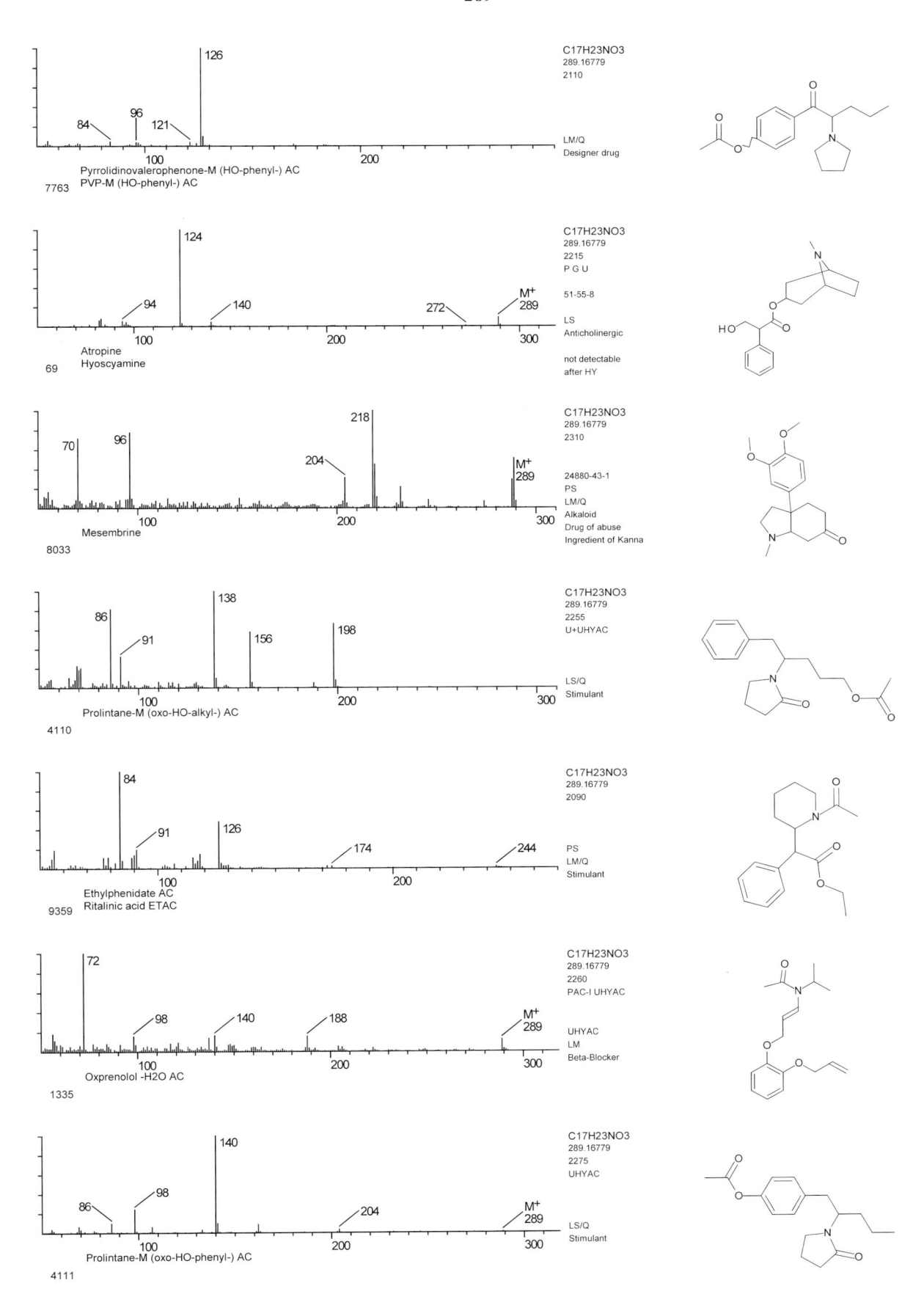

Pyrrolidinovalerophenone-M (HO-phenyl-) AC PVP-M (HO-phenyl-) AC 7763	Peaks: 84, 96, 121, 126	C17H23NO3 289.16779 2110 LM/Q Designer drug
Atropine Hyoscyamine 69	Peaks: 94, 124, 140, 272, M+ 289	C17H23NO3 289.16779 2215 P G U 51-55-8 LS Anticholinergic not detectable after HY
Mesembrine 8033	Peaks: 70, 96, 204, 218, M+ 289	C17H23NO3 289.16779 2310 24880-43-1 PS LM/Q Alkaloid Drug of abuse Ingredient of Kanna
Prolintane-M (oxo-HO-alkyl-) AC 4110	Peaks: 86, 91, 138, 156, 198	C17H23NO3 289.16779 2255 U+UHYAC LS/Q Stimulant
Ethylphenidate AC Ritalinic acid ETAC 9359	Peaks: 84, 91, 126, 174, 244	C17H23NO3 289.16779 2090 PS LM/Q Stimulant
Oxprenolol -H2O AC 1335	Peaks: 72, 98, 140, 188, M+ 289	C17H23NO3 289.16779 2260 PAC-I UHYAC UHYAC LM Beta-Blocker
Prolintane-M (oxo-HO-phenyl-) AC 4111	Peaks: 86, 98, 140, 204, M+ 289	C17H23NO3 289.16779 2275 UHYAC LS/Q Stimulant

289

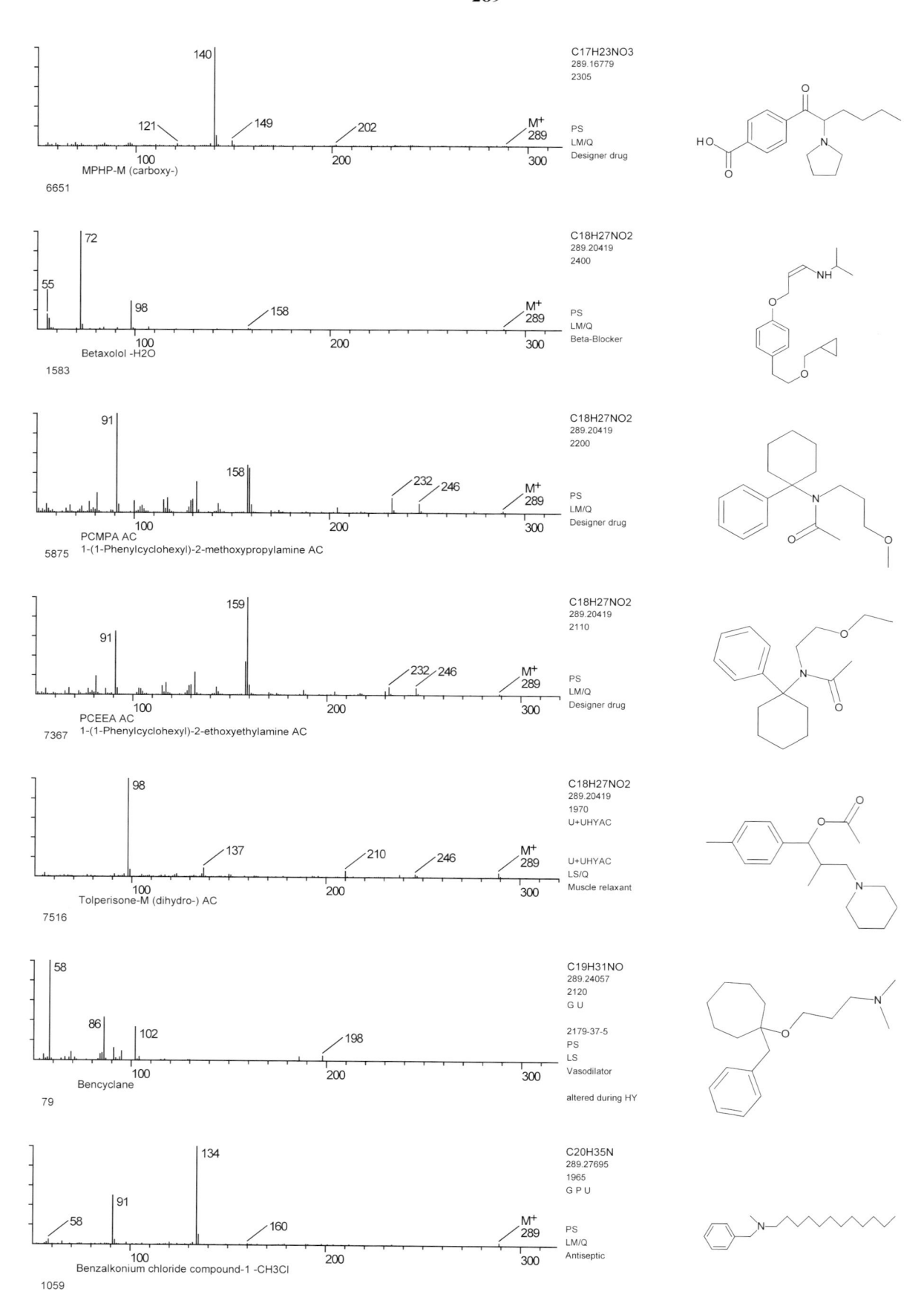

Spectrum peaks	Compound	Formula / Info
140, 121, 149, 202, M+ 289	MPHP-M (carboxy-) 6651	C17H23NO3, 289.16779, 2305, PS, LM/Q, Designer drug
72, 55, 98, 158, M+ 289	Betaxolol -H2O 1583	C18H27NO2, 289.20419, 2400, PS, LM/Q, Beta-Blocker
91, 158, 232, 246, M+ 289	PCMPA AC 1-(1-Phenylcyclohexyl)-2-methoxypropylamine AC 5875	C18H27NO2, 289.20419, 2200, PS, LM/Q, Designer drug
91, 159, 232, 246, M+ 289	PCEEA AC 1-(1-Phenylcyclohexyl)-2-ethoxyethylamine AC 7367	C18H27NO2, 289.20419, 2110, PS, LM/Q, Designer drug
98, 137, 210, 246, M+ 289	Tolperisone-M (dihydro-) AC 7516	C18H27NO2, 289.20419, 1970, U+UHYAC, U+UHYAC, LS/Q, Muscle relaxant
58, 86, 102, 198	Bencyclane 79	C19H31NO, 289.24057, 2120, G U, 2179-37-5, PS, LS, Vasodilator, altered during HY
134, 91, 58, 160, M+ 289	Benzalkonium chloride compound-1 -CH3Cl 1059	C20H35N, 289.27695, 1965, G P U, PS, LM/Q, Antiseptic

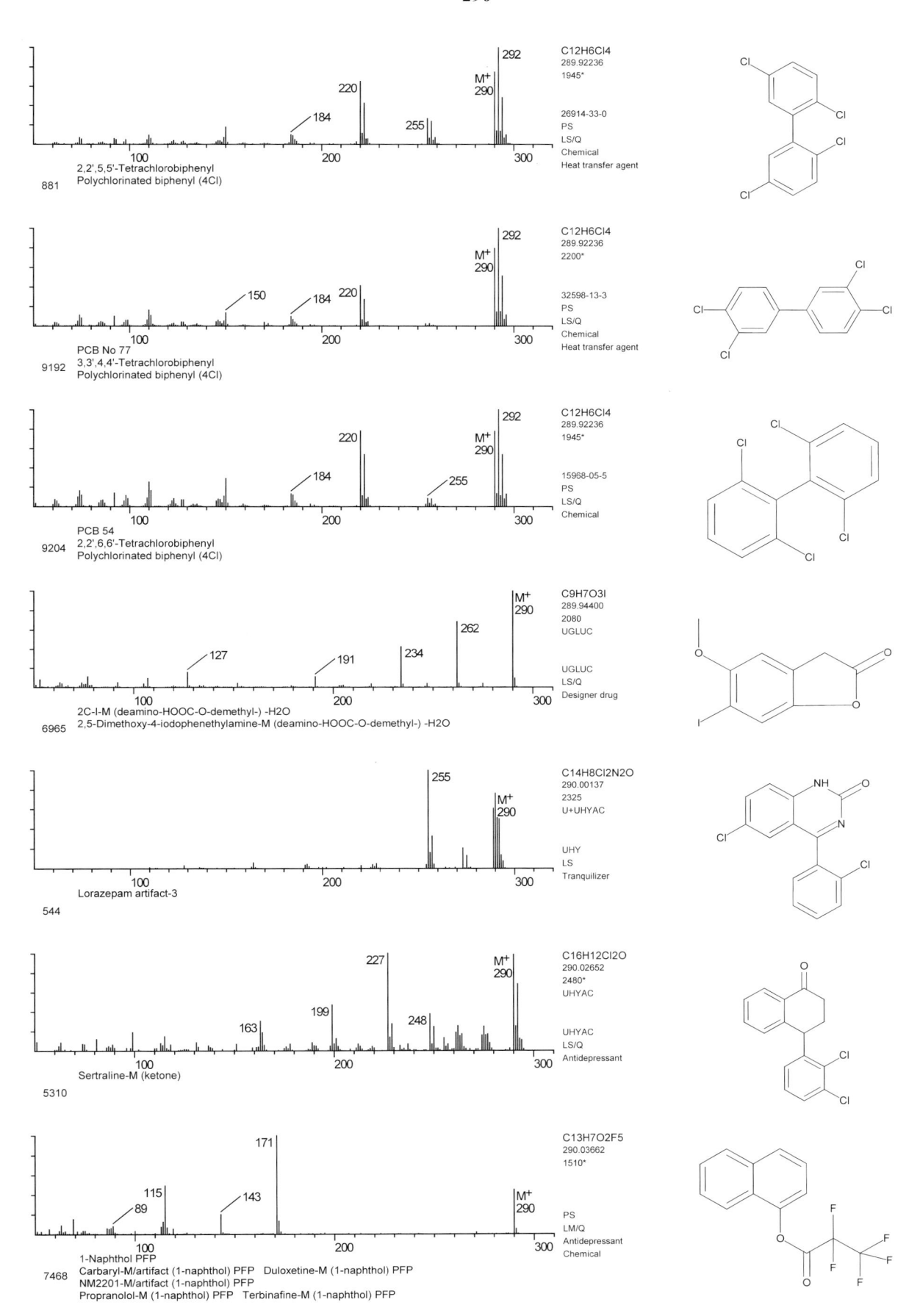

881	2,2',5,5'-Tetrachlorobiphenyl Polychlorinated biphenyl (4Cl)	C12H6Cl4 289.92236 1945* 26914-33-0 PS LS/Q Chemical Heat transfer agent
9192	PCB No 77 3,3',4,4'-Tetrachlorobiphenyl Polychlorinated biphenyl (4Cl)	C12H6Cl4 289.92236 2200* 32598-13-3 PS LS/Q Chemical Heat transfer agent
9204	PCB 54 2,2',6,6'-Tetrachlorobiphenyl Polychlorinated biphenyl (4Cl)	C12H6Cl4 289.92236 1945* 15968-05-5 PS LS/Q Chemical
6965	2C-I-M (deamino-HOOC-O-demethyl-) -H2O 2,5-Dimethoxy-4-iodophenethylamine-M (deamino-HOOC-O-demethyl-) -H2O	C9H7O3I 289.94400 2080 UGLUC UGLUC LS/Q Designer drug
544	Lorazepam artifact-3	C14H8Cl2N2O 290.00137 2325 U+UHYAC UHY LS Tranquilizer
5310	Sertraline-M (ketone)	C16H12Cl2O 290.02652 2480* UHYAC UHYAC LS/Q Antidepressant
7468	1-Naphthol PFP Carbaryl-M/artifact (1-naphthol) PFP Duloxetine-M (1-naphthol) PFP NM2201-M/artifact (1-naphthol) PFP Propranolol-M (1-naphthol) PFP Terbinafine-M (1-naphthol) PFP	C13H7O2F5 290.03662 1510* PS LM/Q Antidepressant Chemical

290

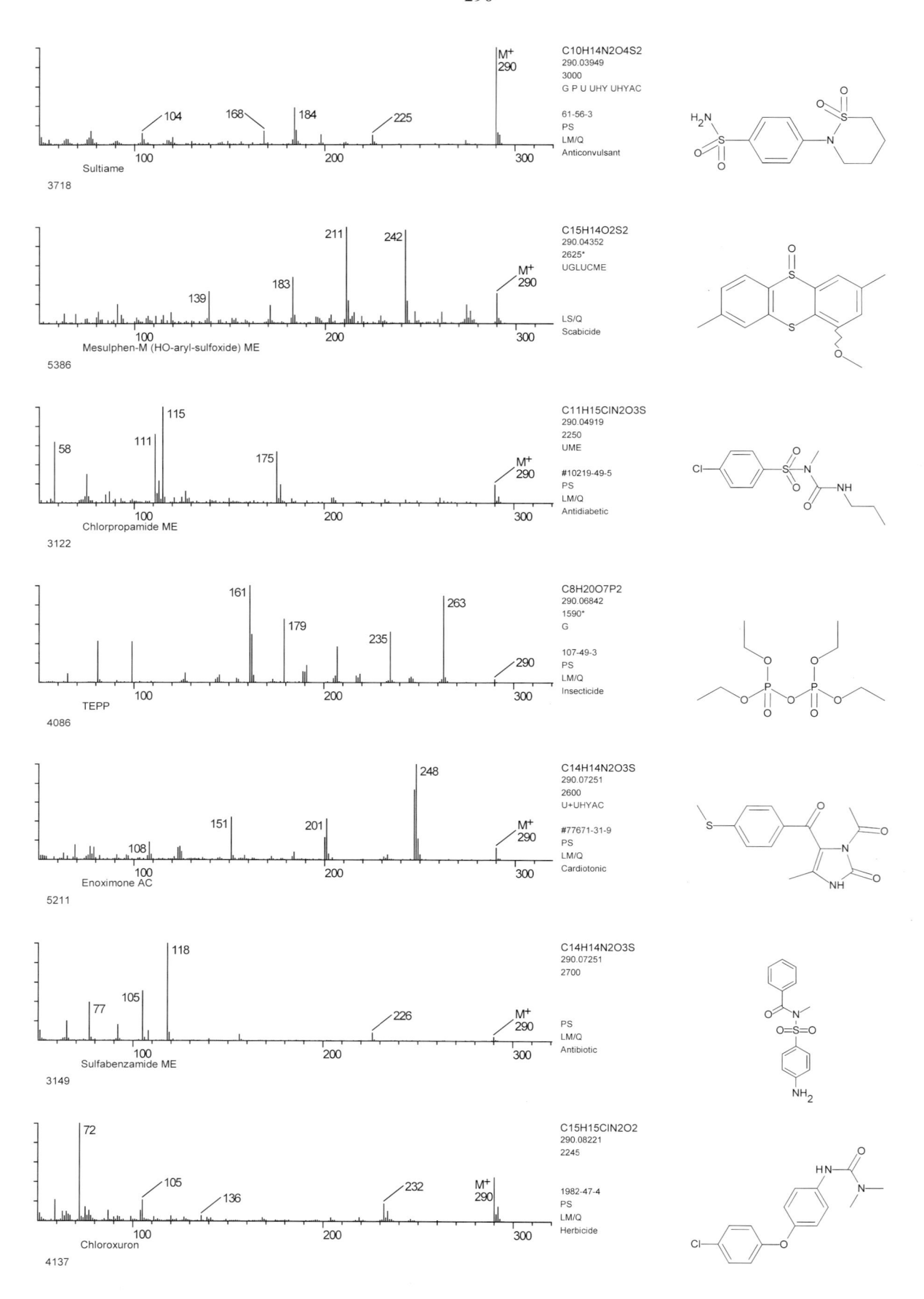

Sultiame — 3718	C10H14N2O4S2 — 290.03949 — 3000 — G P U UHY UHYAC — 61-56-3 — PS — LM/Q — Anticonvulsant
Peaks: 104, 168, 184, 225, M+ 290	

Mesulphen-M (HO-aryl-sulfoxide) ME — 5386	C15H14O2S2 — 290.04352 — 2625* — UGLUCME — LS/Q — Scabicide
Peaks: 139, 183, 211, 242, M+ 290	

Chlorpropamide ME — 3122	C11H15ClN2O3S — 290.04919 — 2250 — UME — #10219-49-5 — PS — LM/Q — Antidiabetic
Peaks: 58, 111, 115, 175, M+ 290	

TEPP — 4086	C8H20O7P2 — 290.06842 — 1590* — G — 107-49-3 — PS — LM/Q — Insecticide
Peaks: 161, 179, 235, 263, 290	

Enoximone AC — 5211	C14H14N2O3S — 290.07251 — 2600 — U+UHYAC — #77671-31-9 — PS — LM/Q — Cardiotonic
Peaks: 108, 151, 201, 248, M+ 290	

Sulfabenzamide ME — 3149	C14H14N2O3S — 290.07251 — 2700 — PS — LM/Q — Antibiotic
Peaks: 77, 105, 118, 226, M+ 290	

Chloroxuron — 4137	C15H15ClN2O2 — 290.08221 — 2245 — 1982-47-4 — PS — LM/Q — Herbicide
Peaks: 72, 105, 136, 232, M+ 290	

840

290

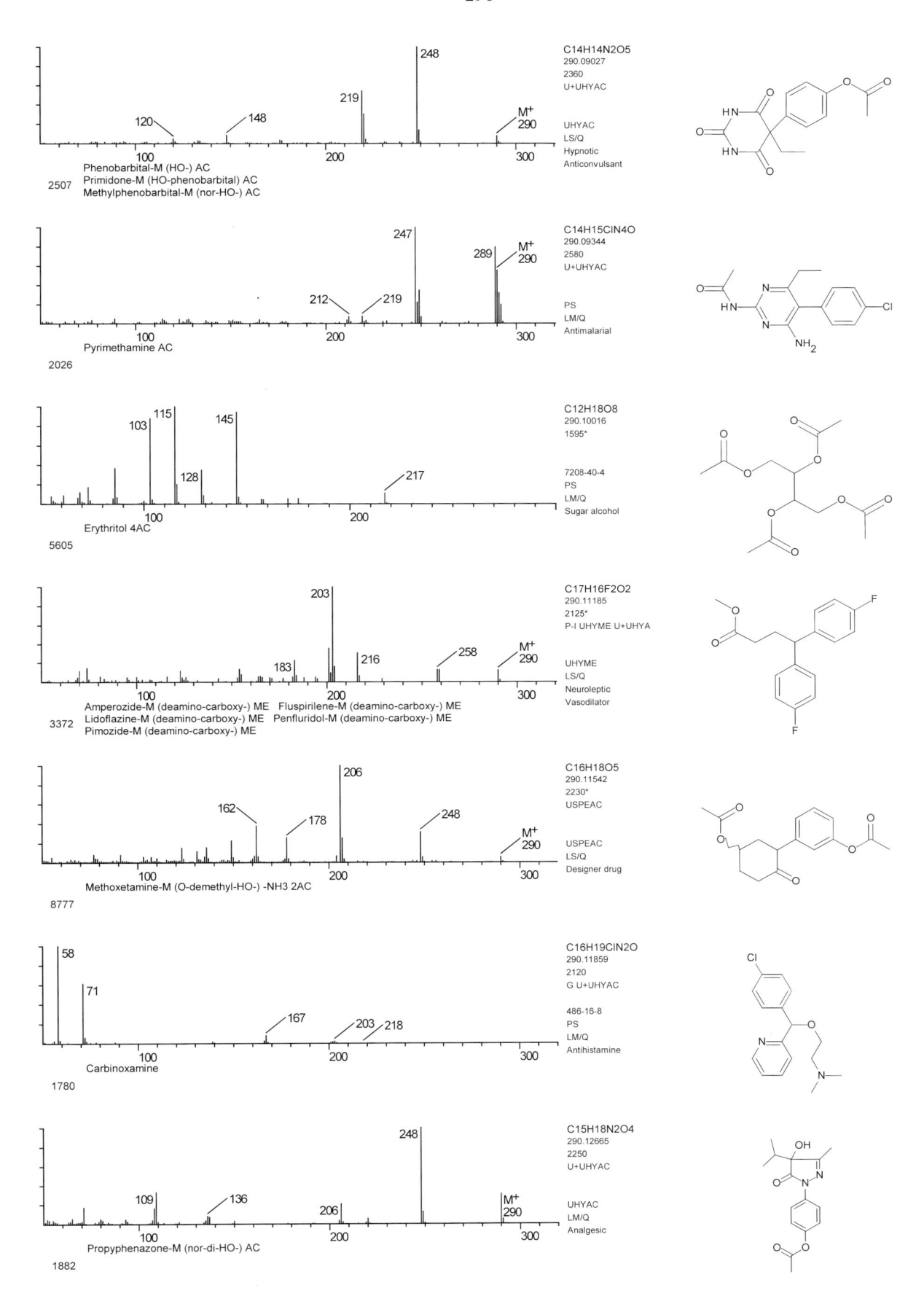

C14H14N2O5
290.09027
2360
U+UHYAC

UHYAC
LS/Q
Hypnotic
Anticonvulsant

2507
Phenobarbital-M (HO-) AC
Primidone-M (HO-phenobarbital) AC
Methylphenobarbital-M (nor-HO-) AC

C14H15ClN4O
290.09344
2580
U+UHYAC

PS
LM/Q
Antimalarial

Pyrimethamine AC
2026

C12H18O8
290.10016
1595*

7208-40-4
PS
LM/Q
Sugar alcohol

Erythritol 4AC
5605

C17H16F2O2
290.11185
2125*
P-I UHYME U+UHYA

UHYME
LS/Q
Neuroleptic
Vasodilator

3372
Amperozide-M (deamino-carboxy-) ME   Fluspirilene-M (deamino-carboxy-) ME
Lidoflazine-M (deamino-carboxy-) ME   Penfluridol-M (deamino-carboxy-) ME
Pimozide-M (deamino-carboxy-) ME

C16H18O5
290.11542
2230*
USPEAC

USPEAC
LS/Q
Designer drug

Methoxetamine-M (O-demethyl-HO-) -NH3 2AC
8777

C16H19ClN2O
290.11859
2120
G U+UHYAC

486-16-8
PS
LM/Q
Antihistamine

Carbinoxamine
1780

C15H18N2O4
290.12665
2250
U+UHYAC

UHYAC
LM/Q
Analgesic

Propyphenazone-M (nor-di-HO-) AC
1882

290

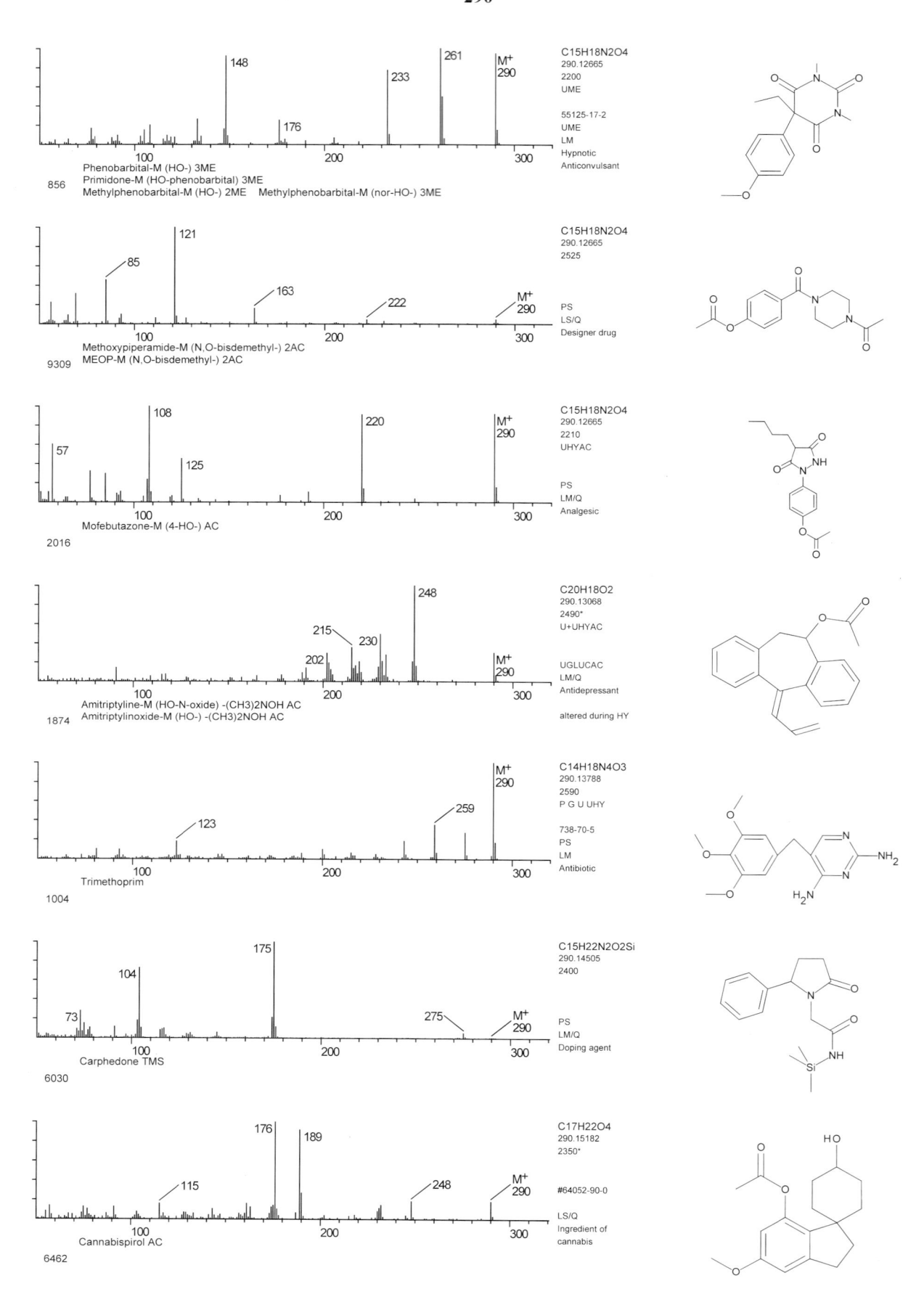

C15H18N2O4
290.12665
2200
UME

55125-17-2
UME
LM
Hypnotic
Anticonvulsant

Phenobarbital-M (HO-) 3ME
Primidone-M (HO-phenobarbital) 3ME
Methylphenobarbital-M (HO-) 2ME   Methylphenobarbital-M (nor-HO-) 3ME
856

C15H18N2O4
290.12665
2525

PS
LS/Q
Designer drug

Methoxypiperamide-M (N,O-bisdemethyl-) 2AC
9309   MEOP-M (N,O-bisdemethyl-) 2AC

C15H18N2O4
290.12665
2210
UHYAC

PS
LM/Q
Analgesic

Mofebutazone-M (4-HO-) AC
2016

C20H18O2
290.13068
2490*
U+UHYAC

UGLUCAC
LM/Q
Antidepressant

altered during HY

Amitriptyline-M (HO-N-oxide) -(CH3)2NOH AC
1874   Amitriptylinoxide-M (HO-) -(CH3)2NOH AC

C14H18N4O3
290.13788
2590
P G U UHY

738-70-5
PS
LM
Antibiotic

Trimethoprim
1004

C15H22N2O2Si
290.14505
2400

PS
LM/Q
Doping agent

Carphedone TMS
6030

C17H22O4
290.15182
2350*

#64052-90-0

LS/Q
Ingredient of cannabis

Cannabispirol AC
6462

290

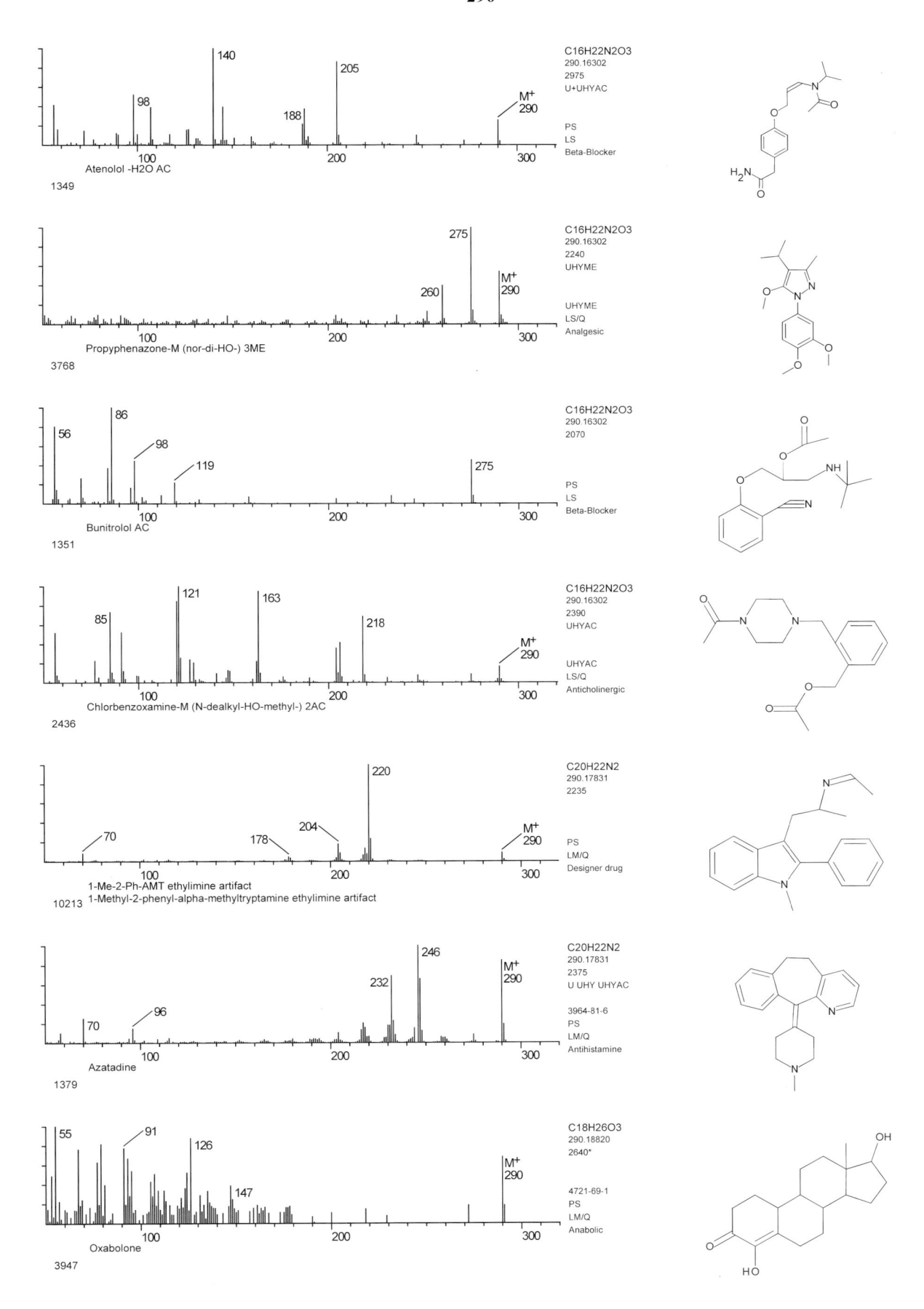

Atenolol -H2O AC	C16H22N2O3 290.16302 2975 U+UHYAC PS LS Beta-Blocker	
Propyphenazone-M (nor-di-HO-) 3ME	C16H22N2O3 290.16302 2240 UHYME UHYME LS/Q Analgesic	
Bunitrolol AC	C16H22N2O3 290.16302 2070 PS LS Beta-Blocker	
Chlorbenzoxamine-M (N-dealkyl-HO-methyl-) 2AC	C16H22N2O3 290.16302 2390 UHYAC UHYAC LS/Q Anticholinergic	
1-Me-2-Ph-AMT ethylimine artifact 1-Methyl-2-phenyl-alpha-methyltryptamine ethylimine artifact	C20H22N2 290.17831 2235 PS LM/Q Designer drug	
Azatadine	C20H22N2 290.17831 2375 U UHY UHYAC 3964-81-6 PS LM/Q Antihistamine	
Oxabolone	C18H26O3 290.18820 2640* 4721-69-1 PS LM/Q Anabolic	

290

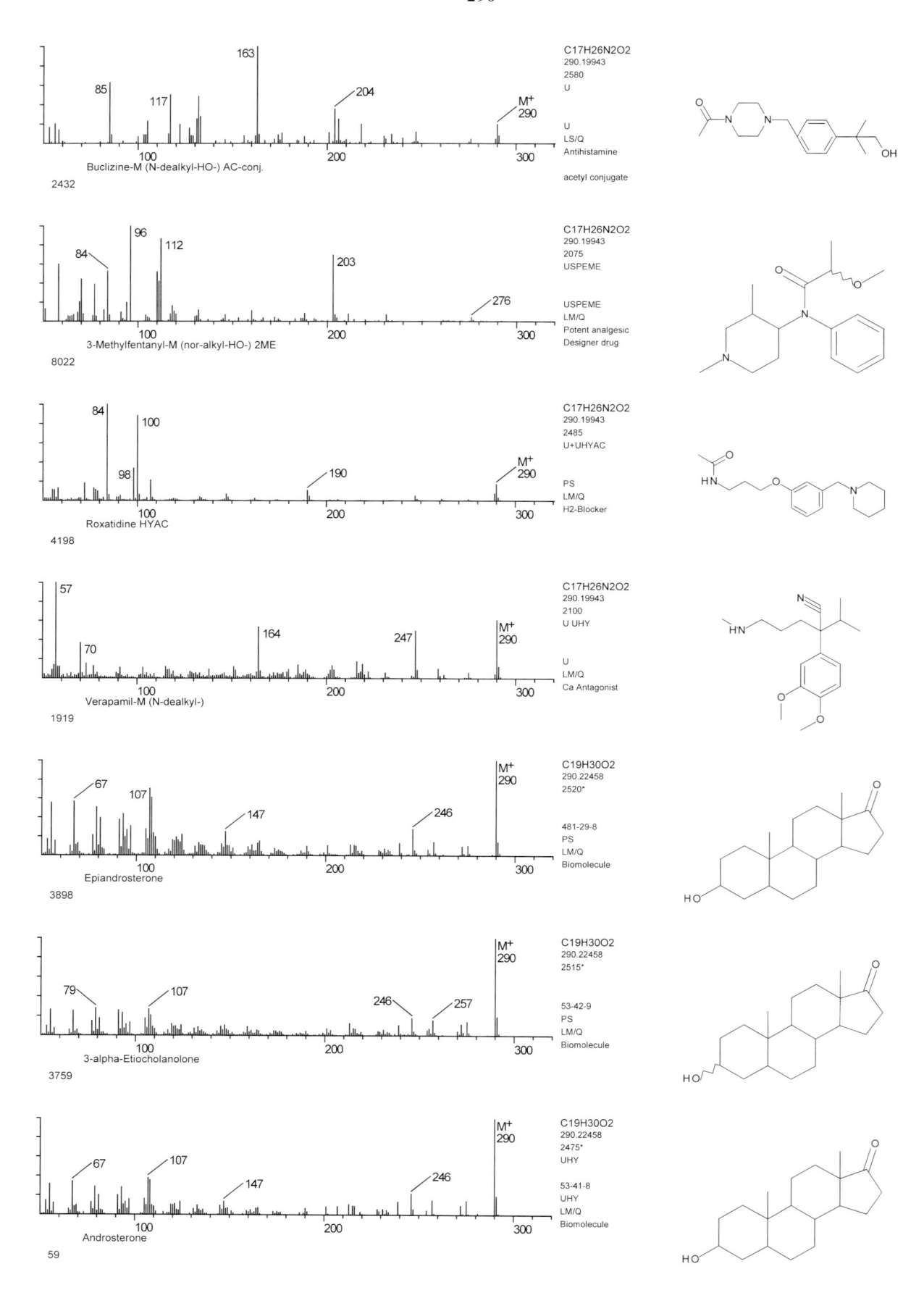

Buclizine-M (N-dealkyl-HO-) AC-conj. 2432	C17H26N2O2 290.19943 2580 U U LS/Q Antihistamine acetyl conjugate
3-Methylfentanyl-M (nor-alkyl-HO-) 2ME 8022	C17H26N2O2 290.19943 2075 USPEME USPEME LM/Q Potent analgesic Designer drug
Roxatidine HYAC 4198	C17H26N2O2 290.19943 2485 U+UHYAC PS LM/Q H2-Blocker
Verapamil-M (N-dealkyl-) 1919	C17H26N2O2 290.19943 2100 U UHY U LM/Q Ca Antagonist
Epiandrosterone 3898	C19H30O2 290.22458 2520* 481-29-8 PS LM/Q Biomolecule
3-alpha-Etiocholanolone 3759	C19H30O2 290.22458 2515* 53-42-9 PS LM/Q Biomolecule
Androsterone 59	C19H30O2 290.22458 2475* UHY 53-41-8 UHY LM/Q Biomolecule

844

290

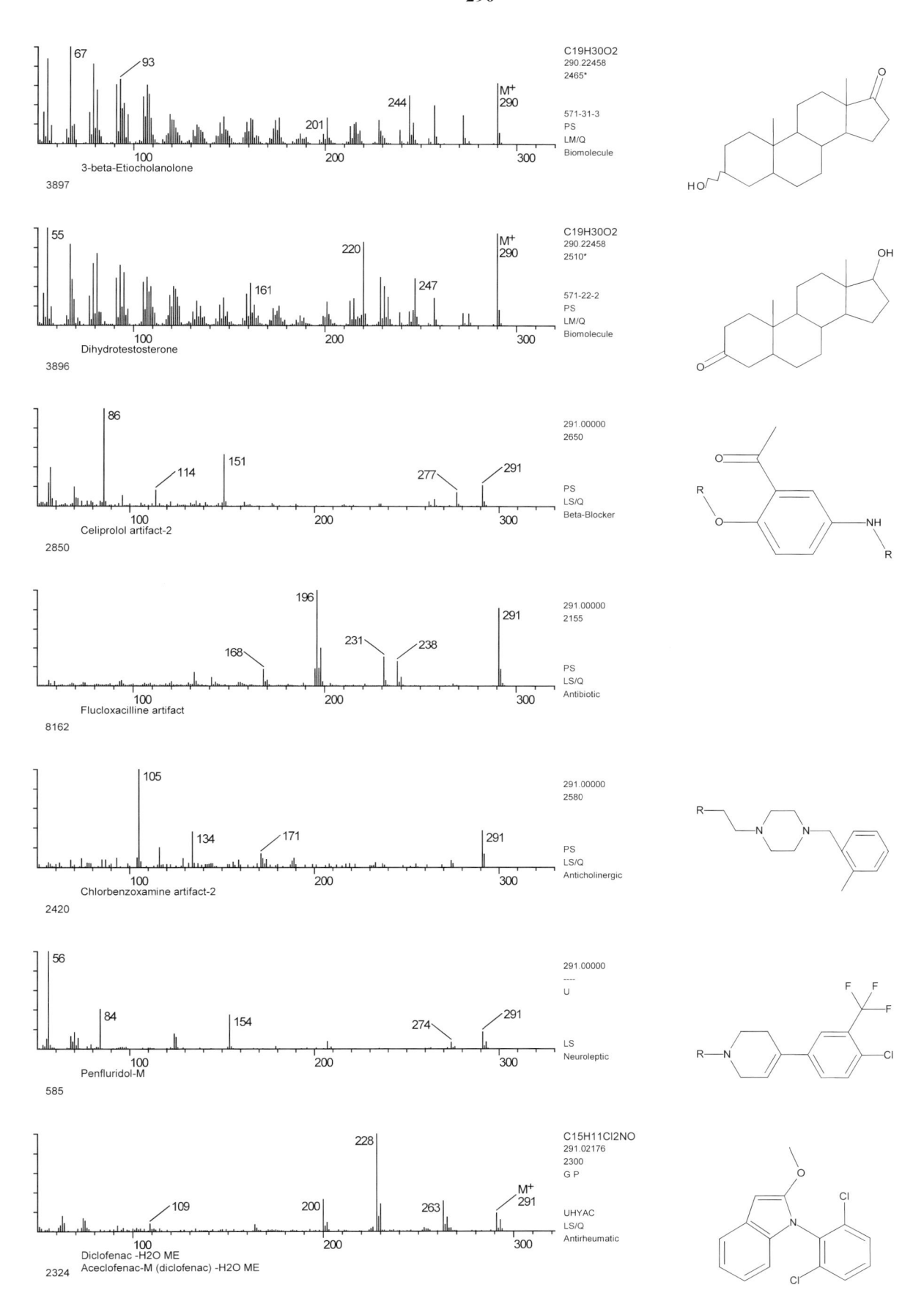

3-beta-Etiocholanolone
3897

C19H30O2
290.22458
2465*

571-31-3
PS
LM/Q
Biomolecule

Dihydrotestosterone
3896

C19H30O2
290.22458
2510*

571-22-2
PS
LM/Q
Biomolecule

Celiprolol artifact-2
2850

291.00000
2650

PS
LS/Q
Beta-Blocker

Flucloxacilline artifact
8162

291.00000
2155

PS
LS/Q
Antibiotic

Chlorbenzoxamine artifact-2
2420

291.00000
2580

PS
LS/Q
Anticholinergic

Penfluridol-M
585

291.00000

U

LS
Neuroleptic

Diclofenac -H2O ME
Aceclofenac-M (diclofenac) -H2O ME
2324

C15H11Cl2NO
291.02176
2300
G P

UHYAC
LS/Q
Antirheumatic

845

291

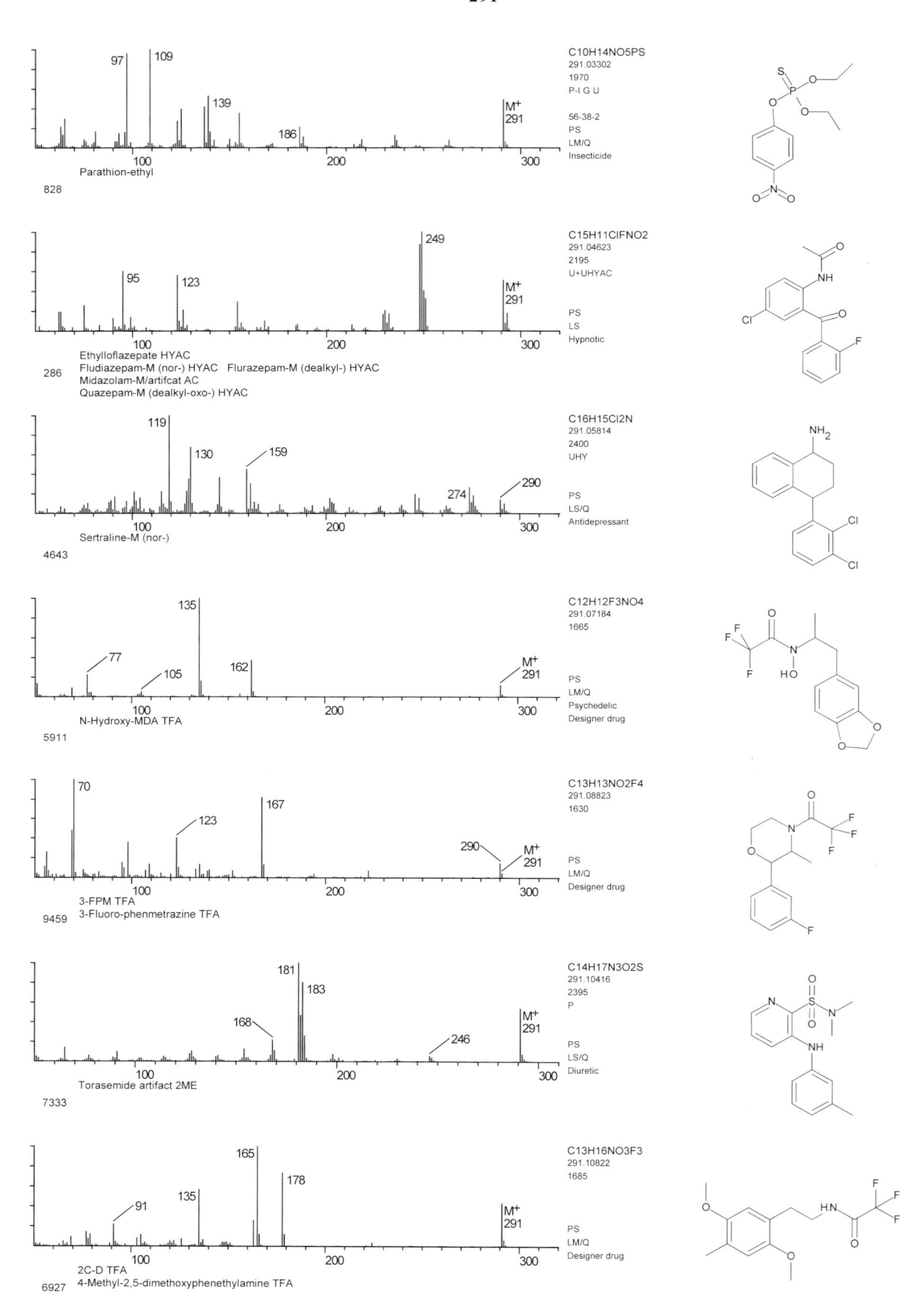

Parathion-ethyl — peaks: 97, 109, 139, 186, M+ 291	C10H14NO5PS 291.03302 1970 P-I G U 56-38-2 PS LM/Q Insecticide
828	

Ethylloflazepate HYAC Fludiazepam-M (nor-) HYAC Flurazepam-M (dealkyl-) HYAC Midazolam-M/artifcat AC Quazepam-M (dealkyl-oxo-) HYAC — peaks: 95, 123, 249, M+ 291	C15H11ClFNO2 291.04623 2195 U+UHYAC PS LS Hypnotic
286	

Sertraline-M (nor-) — peaks: 119, 130, 159, 274, 290	C16H15Cl2N 291.05814 2400 UHY PS LS/Q Antidepressant
4643	

N-Hydroxy-MDA TFA — peaks: 77, 105, 135, 162, M+ 291	C12H12F3NO4 291.07184 1665 PS LM/Q Psychedelic Designer drug
5911	

3-FPM TFA 3-Fluoro-phenmetrazine TFA — peaks: 70, 123, 167, 290, M+ 291	C13H13NO2F4 291.08823 1630 PS LM/Q Designer drug
9459	

Torasemide artifact 2ME — peaks: 168, 181, 183, 246, M+ 291	C14H17N3O2S 291.10416 2395 P PS LS/Q Diuretic
7333	

2C-D TFA 4-Methyl-2,5-dimethoxyphenethylamine TFA — peaks: 91, 135, 165, 178, M+ 291	C13H16NO3F3 291.10822 1685 PS LM/Q Designer drug
6927	

291

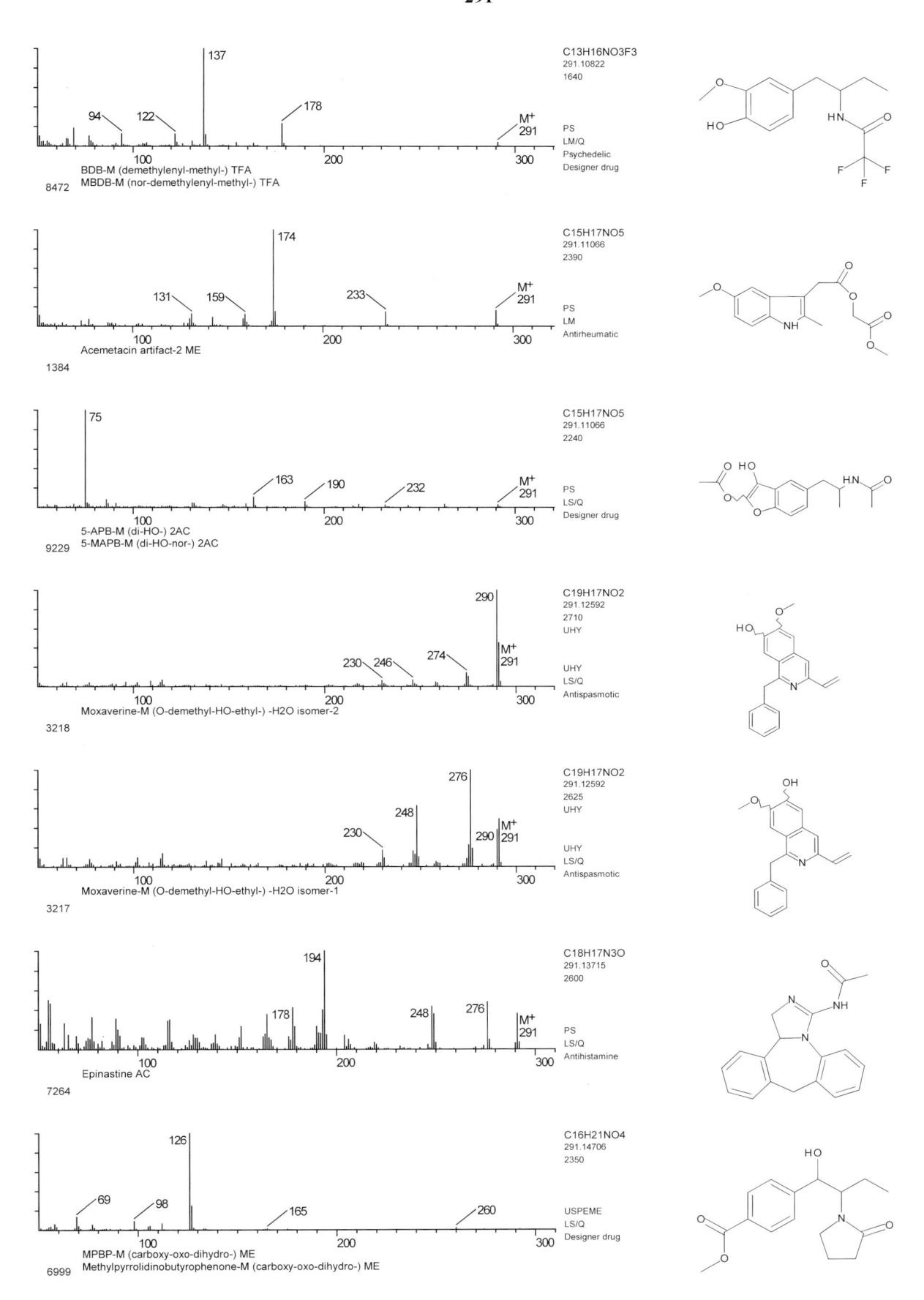

8472	BDB-M (demethylenyl-methyl-) TFA MBDB-M (nor-demethylenyl-methyl-) TFA	C13H16NO3F3 291.10822 1640 PS LM/Q Psychedelic Designer drug
1384	Acemetacin artifact-2 ME	C15H17NO5 291.11066 2390 PS LM Antirheumatic
9229	5-APB-M (di-HO-) 2AC 5-MAPB-M (di-HO-nor-) 2AC	C15H17NO5 291.11066 2240 PS LS/Q Designer drug
3218	Moxaverine-M (O-demethyl-HO-ethyl-) -H2O isomer-2	C19H17NO2 291.12592 2710 UHY UHY LS/Q Antispasmotic
3217	Moxaverine-M (O-demethyl-HO-ethyl-) -H2O isomer-1	C19H17NO2 291.12592 2625 UHY UHY LS/Q Antispasmotic
7264	Epinastine AC	C18H17N3O 291.13715 2600 PS LS/Q Antihistamine
6999	MPBP-M (carboxy-oxo-dihydro-) ME Methylpyrrolidinobutyrophenone-M (carboxy-oxo-dihydro-) ME	C16H21NO4 291.14706 2350 USPEME LS/Q Designer drug

291

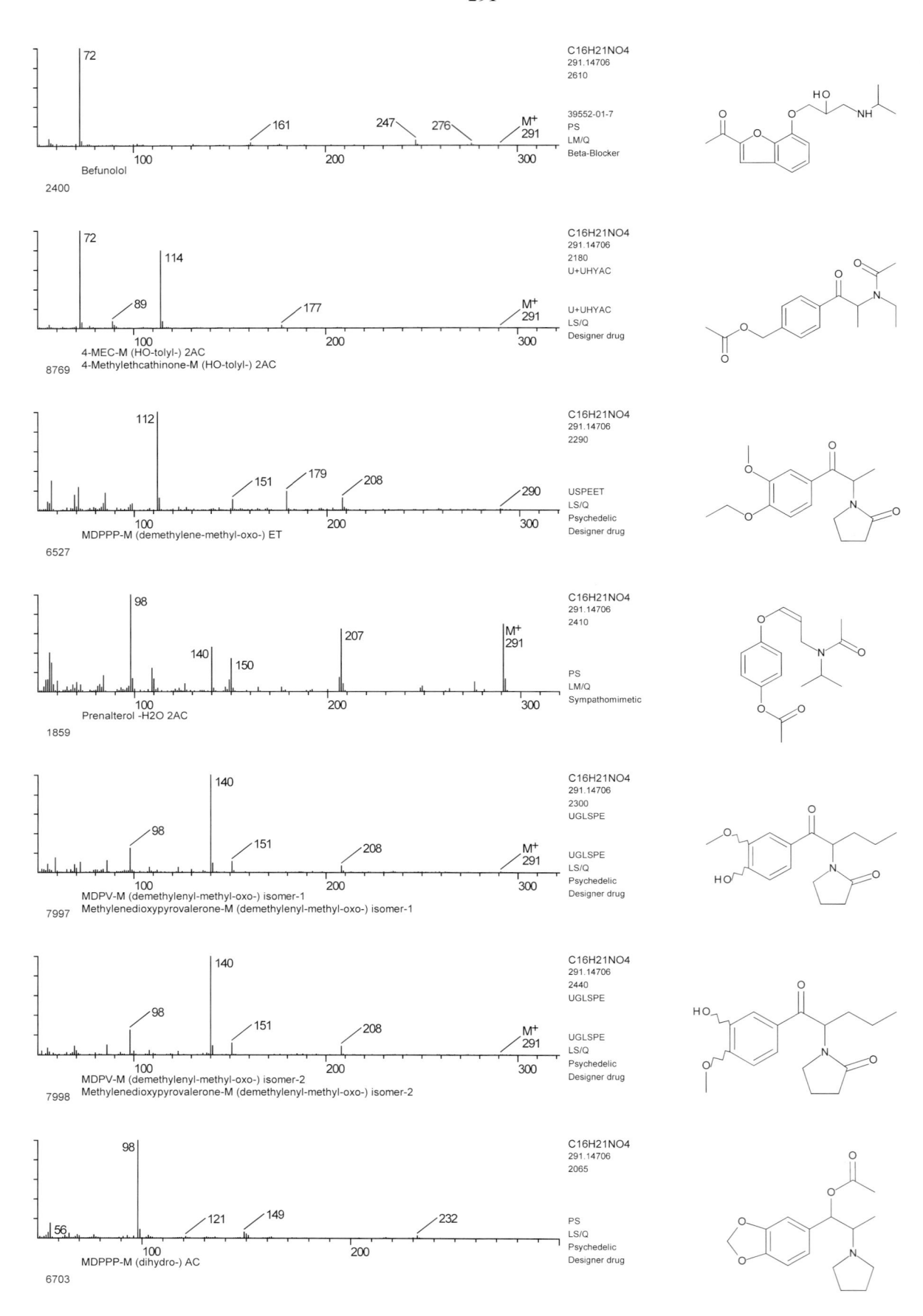

Befunolol — 2400
Peaks: 72, 161, 247, 276, M+ 291
C16H21NO4
291.14706
2610
39552-01-7
PS
LM/Q
Beta-Blocker

4-MEC-M (HO-tolyl-) 2AC — 8769
4-Methylethcathinone-M (HO-tolyl-) 2AC
Peaks: 72, 89, 114, 177, M+ 291
C16H21NO4
291.14706
2180
U+UHYAC
U+UHYAC
LS/Q
Designer drug

MDPPP-M (demethylene-methyl-oxo-) ET — 6527
Peaks: 112, 151, 179, 208, 290
C16H21NO4
291.14706
2290
USPEET
LS/Q
Psychedelic
Designer drug

Prenalterol -H2O 2AC — 1859
Peaks: 98, 140, 150, 207, M+ 291
C16H21NO4
291.14706
2410
PS
LM/Q
Sympathomimetic

MDPV-M (demethylenyl-methyl-oxo-) isomer-1 — 7997
Methylenedioxypyrovalerone-M (demethylenyl-methyl-oxo-) isomer-1
Peaks: 98, 140, 151, 208, M+ 291
C16H21NO4
291.14706
2300
UGLSPE
UGLSPE
LS/Q
Psychedelic
Designer drug

MDPV-M (demethylenyl-methyl-oxo-) isomer-2 — 7998
Methylenedioxypyrovalerone-M (demethylenyl-methyl-oxo-) isomer-2
Peaks: 98, 140, 151, 208, M+ 291
C16H21NO4
291.14706
2440
UGLSPE
UGLSPE
LS/Q
Psychedelic
Designer drug

MDPPP-M (dihydro-) AC — 6703
Peaks: 56, 98, 121, 149, 232
C16H21NO4
291.14706
2065
PS
LS/Q
Psychedelic
Designer drug

291

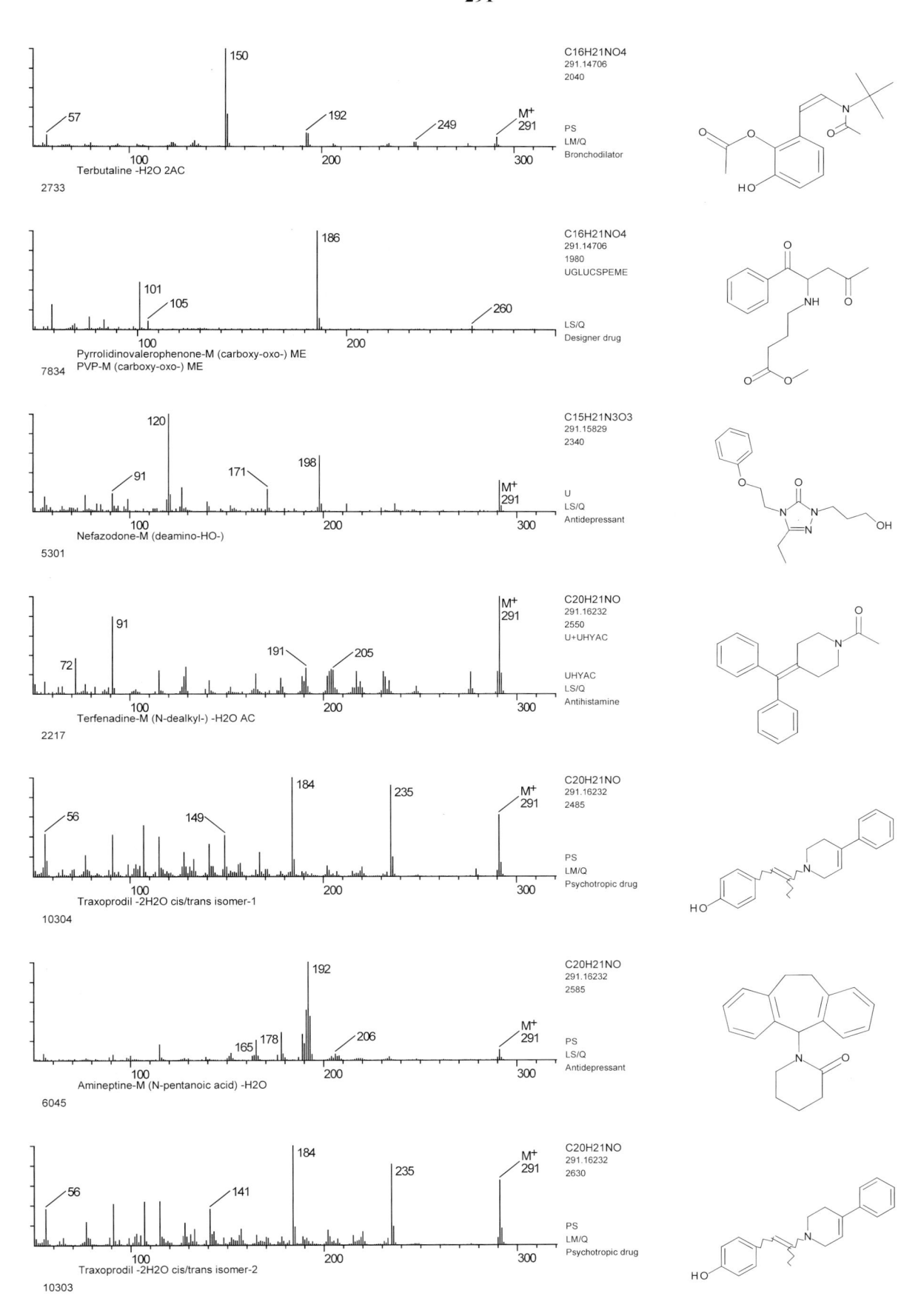

Terbutaline -H2O 2AC — peaks 57, 150, 192, 249, M+ 291	C16H21NO4 291.14706 2040 PS LM/Q Bronchodilator	2733
Pyrrolidinovalerophenone-M (carboxy-oxo-) ME / PVP-M (carboxy-oxo-) ME — peaks 101, 105, 186, 260	C16H21NO4 291.14706 1980 UGLUCSPEME LS/Q Designer drug	7834
Nefazodone-M (deamino-HO-) — peaks 91, 120, 171, 198, M+ 291	C15H21N3O3 291.15829 2340 U LS/Q Antidepressant	5301
Terfenadine-M (N-dealkyl-) -H2O AC — peaks 72, 91, 191, 205, M+ 291	C20H21NO 291.16232 2550 U+UHYAC UHYAC LS/Q Antihistamine	2217
Traxoprodil -2H2O cis/trans isomer-1 — peaks 56, 149, 184, 235, M+ 291	C20H21NO 291.16232 2485 PS LM/Q Psychotropic drug	10304
Amineptine-M (N-pentanoic acid) -H2O — peaks 165, 178, 192, 206, M+ 291	C20H21NO 291.16232 2585 PS LS/Q Antidepressant	6045
Traxoprodil -2H2O cis/trans isomer-2 — peaks 56, 141, 184, 235, M+ 291	C20H21NO 291.16232 2630 PS LM/Q Psychotropic drug	10303

291

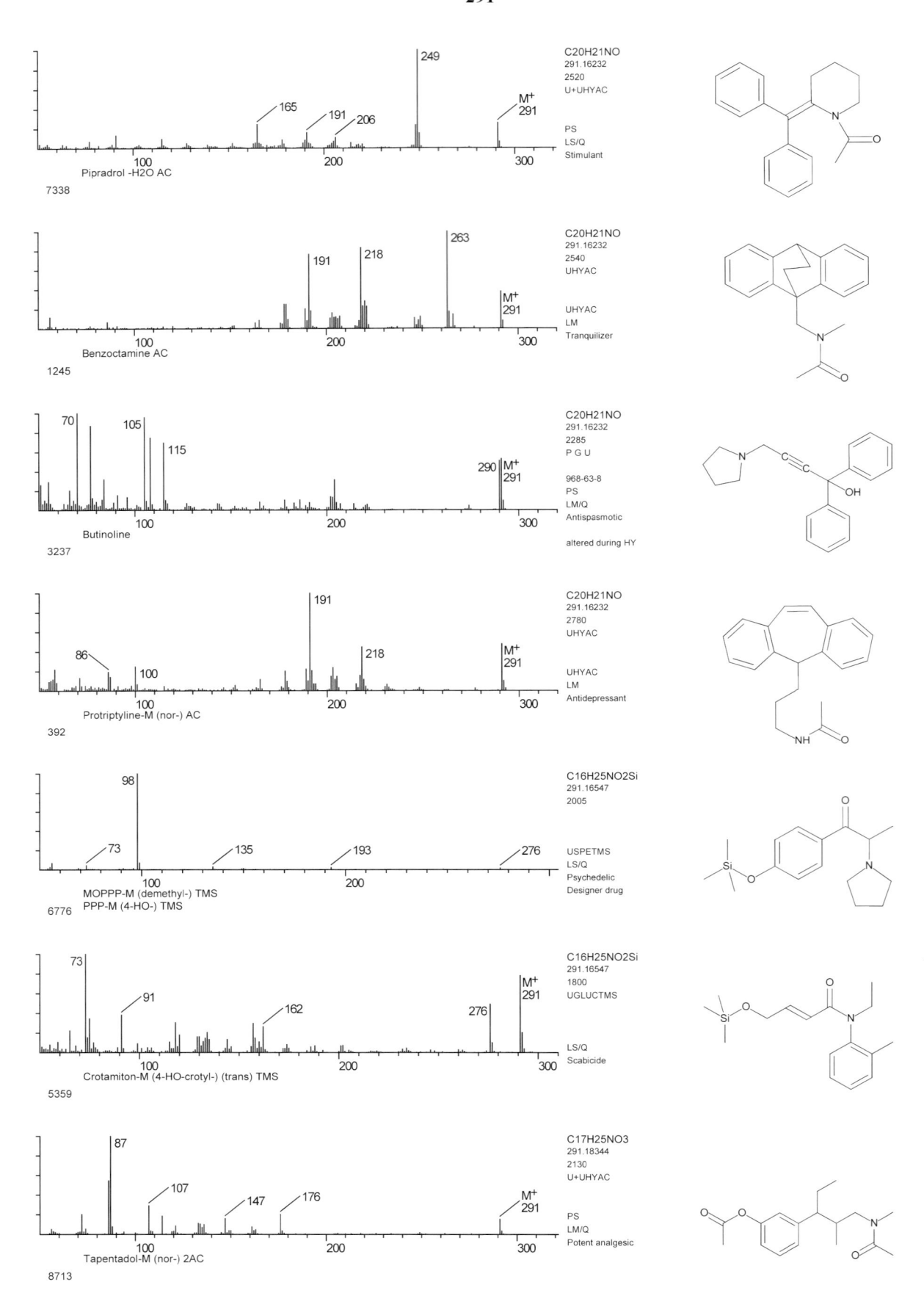

Pipradrol -H2O AC	C20H21NO 291.16232 2520 U+UHYAC PS LS/Q Stimulant
7338	
Benzoctamine AC	C20H21NO 291.16232 2540 UHYAC UHYAC LM Tranquilizer
1245	
Butinoline	C20H21NO 291.16232 2285 P G U 968-63-8 PS LM/Q Antispasmotic altered during HY
3237	
Protriptyline-M (nor-) AC	C20H21NO 291.16232 2780 UHYAC UHYAC LM Antidepressant
392	
MOPPP-M (demethyl-) TMS PPP-M (4-HO-) TMS	C16H25NO2Si 291.16547 2005 USPETMS LS/Q Psychedelic Designer drug
6776	
Crotamiton-M (4-HO-crotyl-) (trans) TMS	C16H25NO2Si 291.16547 1800 UGLUCTMS LS/Q Scabicide
5359	
Tapentadol-M (nor-) 2AC	C17H25NO3 291.18344 2130 U+UHYAC PS LM/Q Potent analgesic
8713	

291

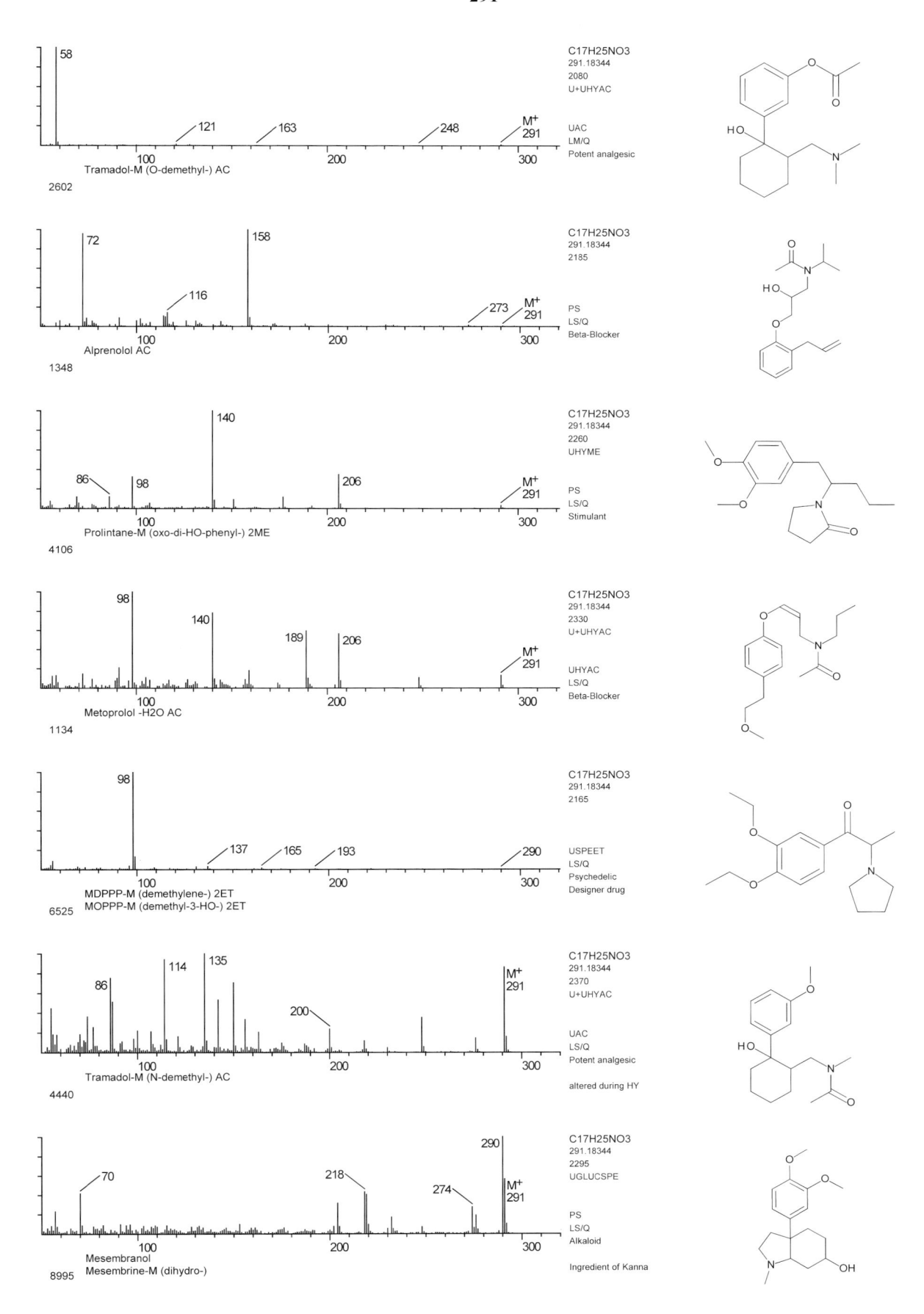

Spectrum	Compound	Formula/Info
58, 121, 163, 248, 291 M+	Tramadol-M (O-demethyl-) AC 2602	C17H25NO3 291.18344 2080 U+UHYAC UAC LM/Q Potent analgesic
72, 116, 158, 273, 291 M+	Alprenolol AC 1348	C17H25NO3 291.18344 2185 PS LS/Q Beta-Blocker
86, 98, 140, 206, 291 M+	Prolintane-M (oxo-di-HO-phenyl-) 2ME 4106	C17H25NO3 291.18344 2260 UHYME PS LS/Q Stimulant
98, 140, 189, 206, 291 M+	Metoprolol -H2O AC 1134	C17H25NO3 291.18344 2330 U+UHYAC UHYAC LS/Q Beta-Blocker
98, 137, 165, 193, 290	MDPPP-M (demethylene-) 2ET MOPPP-M (demethyl-3-HO-) 2ET 6525	C17H25NO3 291.18344 2165 USPEET LS/Q Psychedelic Designer drug
86, 114, 135, 200, 291 M+	Tramadol-M (N-demethyl-) AC 4440	C17H25NO3 291.18344 2370 U+UHYAC UAC LS/Q Potent analgesic altered during HY
70, 218, 274, 290, 291 M+	Mesembranol Mesembrine-M (dihydro-) 8995	C17H25NO3 291.18344 2295 UGLUCSPE PS LS/Q Alkaloid Ingredient of Kanna

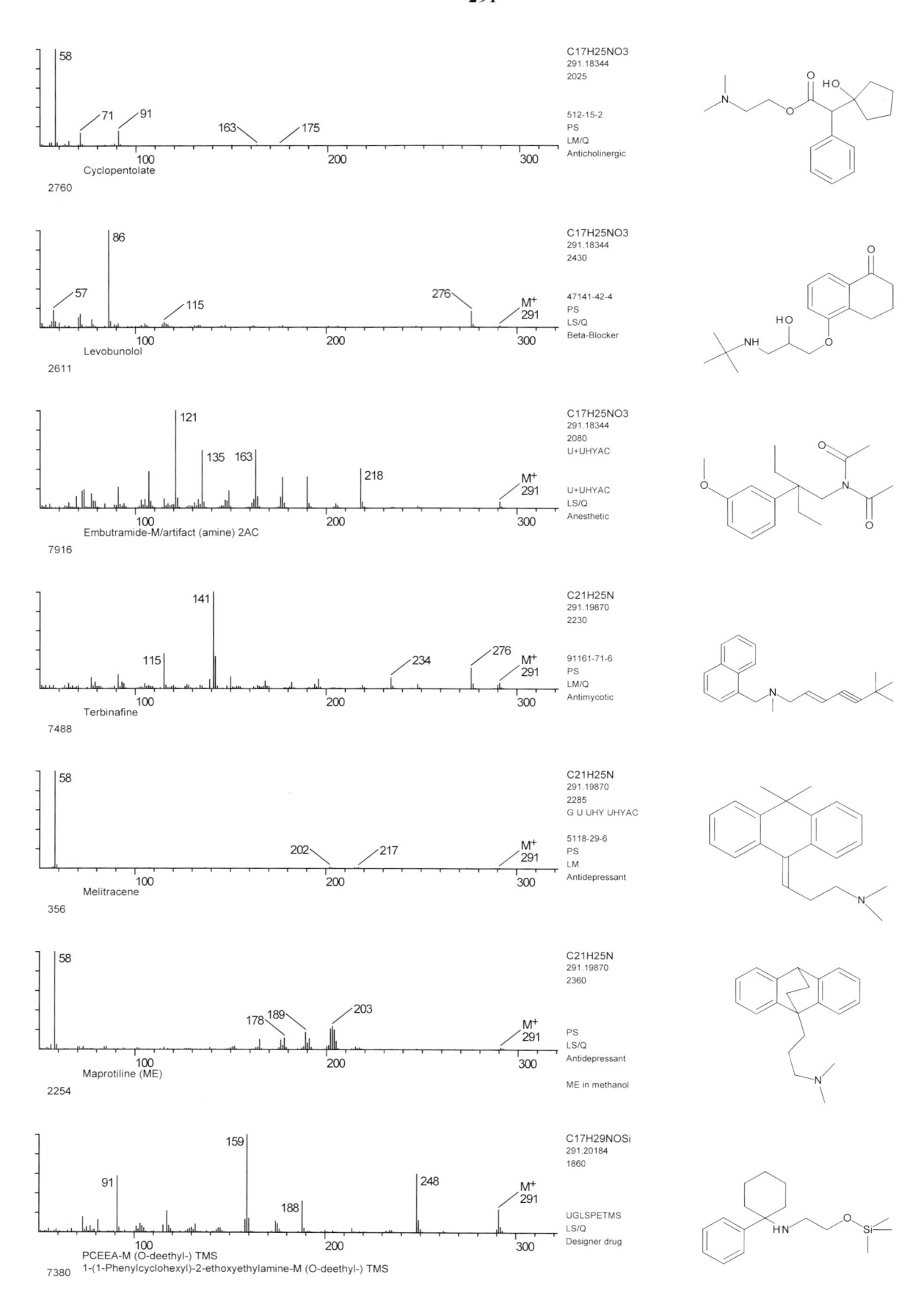

Cyclopentolate
C17H25NO3
291.18344
2025
512-15-2
PS
LM/Q
Anticholinergic
2760

Levobunolol
C17H25NO3
291.18344
2430
47141-42-4
PS
LS/Q
Beta-Blocker
2611

Embutramide-M/artifact (amine) 2AC
C17H25NO3
291.18344
2080
U+UHYAC
U+UHYAC
LS/Q
Anesthetic
7916

Terbinafine
C21H25N
291.19870
2230
91161-71-6
PS
LM/Q
Antimycotic
7488

Melitracene
C21H25N
291.19870
2285
G U UHY UHYAC
5118-29-6
PS
LM
Antidepressant
356

Maprotiline (ME)
C21H25N
291.19870
2360
PS
LS/Q
Antidepressant
ME in methanol
2254

PCEEA-M (O-deethyl-) TMS
1-(1-Phenylcyclohexyl)-2-ethoxyethylamine-M (O-deethyl-) TMS
C17H29NOSi
291.20184
1860
UGLSPETMS
LS/Q
Designer drug
7380

291

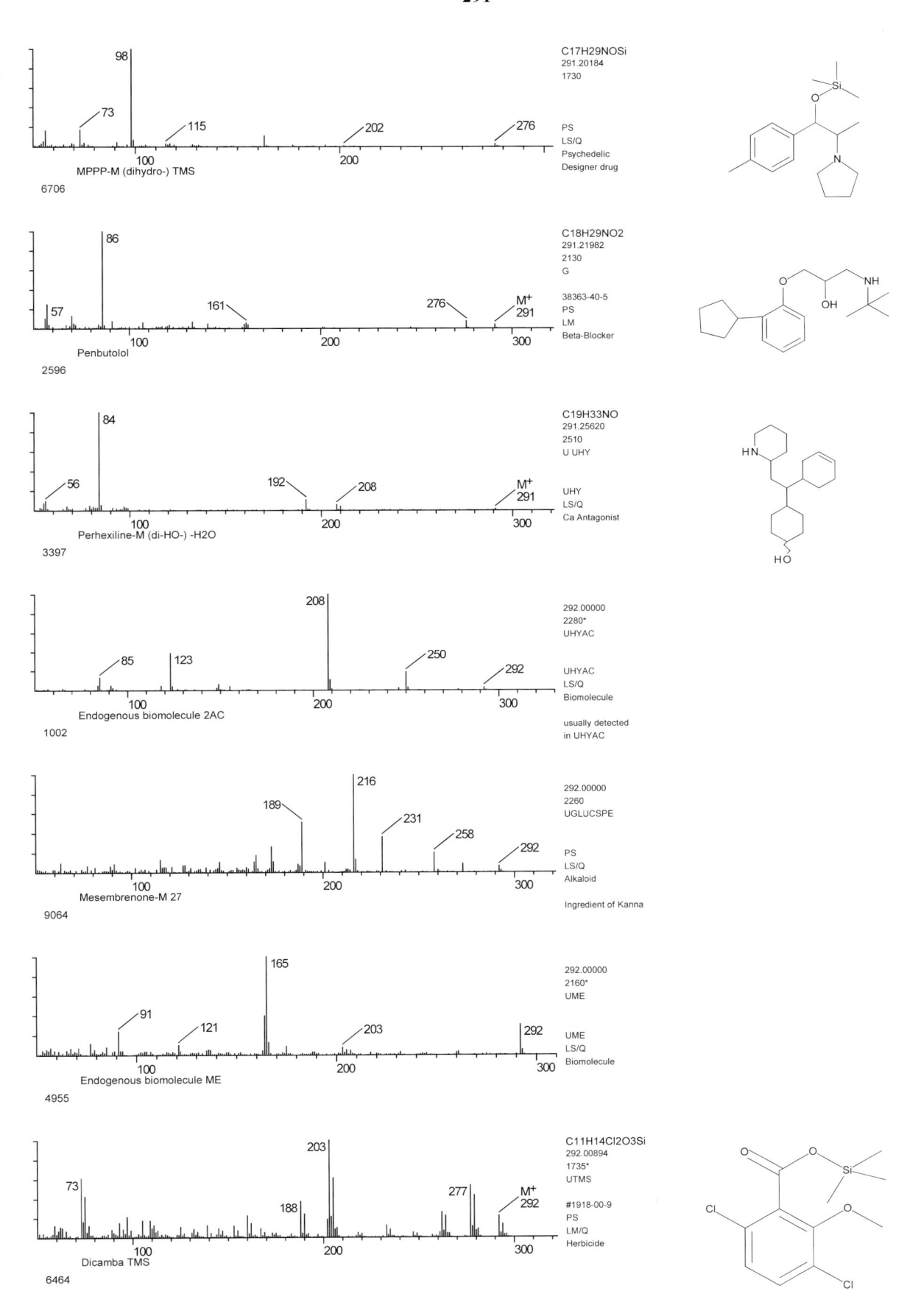

MPPP-M (dihydro-) TMS
6706
Peaks: 73, 98, 115, 202, 276
C17H29NOSi
291.20184
1730
PS
LS/Q
Psychedelic
Designer drug

Penbutolol
2596
Peaks: 57, 86, 161, 276, M+ 291
C18H29NO2
291.21982
2130
G
38363-40-5
PS
LM
Beta-Blocker

Perhexiline-M (di-HO-) -H2O
3397
Peaks: 56, 84, 192, 208, M+ 291
C19H33NO
291.25620
2510
U UHY
UHY
LS/Q
Ca Antagonist

Endogenous biomolecule 2AC
1002
Peaks: 85, 123, 208, 250, 292
292.00000
2280*
UHYAC
UHYAC
LS/Q
Biomolecule
usually detected in UHYAC

Mesembrenone-M 27
9064
Peaks: 189, 216, 231, 258, 292
292.00000
2260
UGLUCSPE
PS
LS/Q
Alkaloid
Ingredient of Kanna

Endogenous biomolecule ME
4955
Peaks: 91, 121, 165, 203, 292
292.00000
2160*
UME
UME
LS/Q
Biomolecule

Dicamba TMS
6464
Peaks: 73, 188, 203, 277, M+ 292
C11H14Cl2O3Si
292.00894
1735*
UTMS
#1918-00-9
PS
LM/Q
Herbicide

292

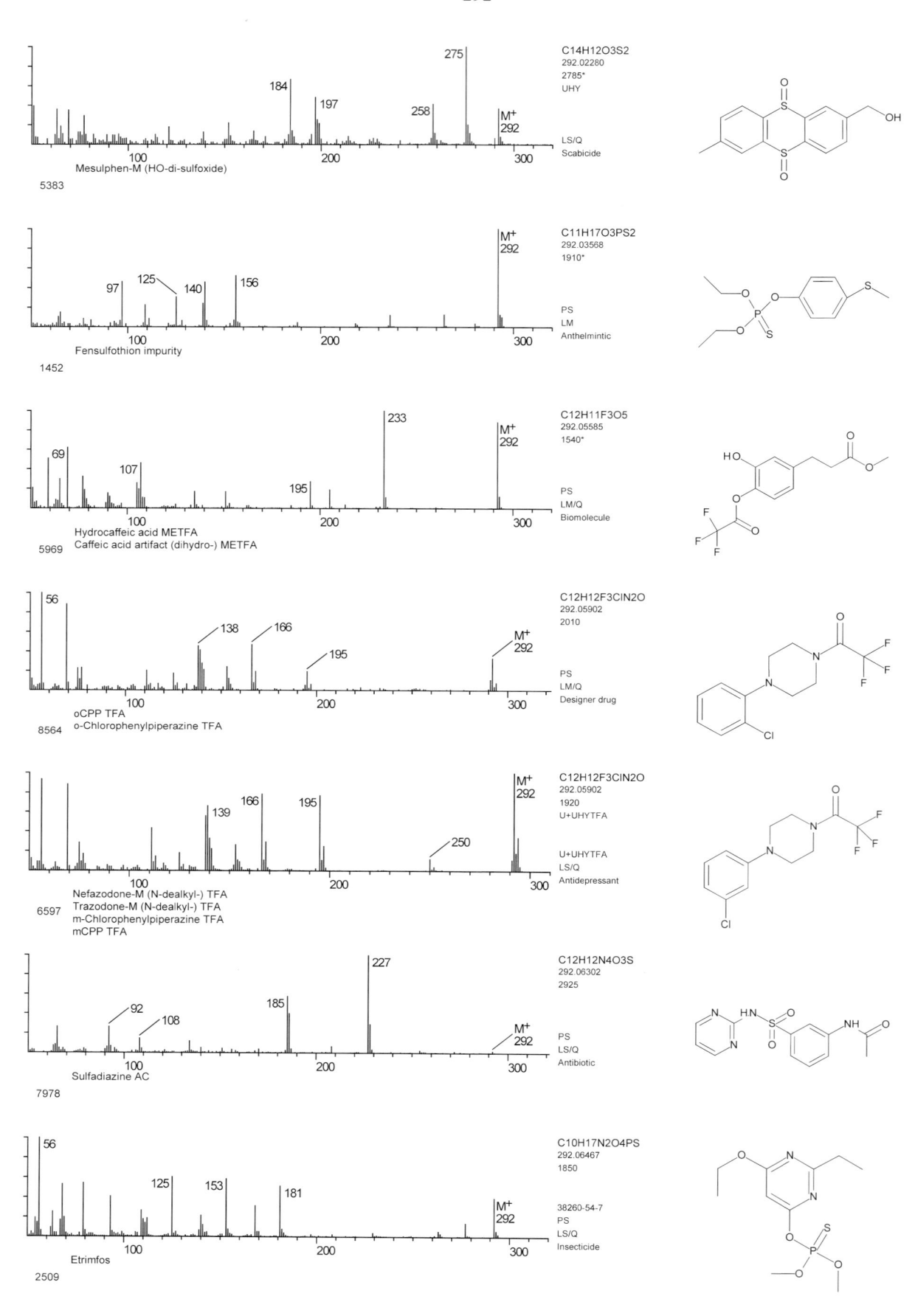

C14H12O3S2
292.02280
2785*
UHY

LS/Q
Scabicide

Mesulphen-M (HO-di-sulfoxide)
5383

C11H17O3PS2
292.03568
1910*

PS
LM
Anthelmintic

Fensulfothion impurity
1452

C12H11F3O5
292.05585
1540*

PS
LM/Q
Biomolecule

Hydrocaffeic acid METFA
Caffeic acid artifact (dihydro-) METFA
5969

C12H12F3ClN2O
292.05902
2010

PS
LM/Q
Designer drug

oCPP TFA
o-Chlorophenylpiperazine TFA
8564

C12H12F3ClN2O
292.05902
1920
U+UHYTFA

U+UHYTFA
LS/Q
Antidepressant

Nefazodone-M (N-dealkyl-) TFA
Trazodone-M (N-dealkyl-) TFA
m-Chlorophenylpiperazine TFA
mCPP TFA
6597

C12H12N4O3S
292.06302
2925

PS
LS/Q
Antibiotic

Sulfadiazine AC
7978

C10H17N2O4PS
292.06467
1850

38260-54-7
PS
LS/Q
Insecticide

Etrimfos
2509

292

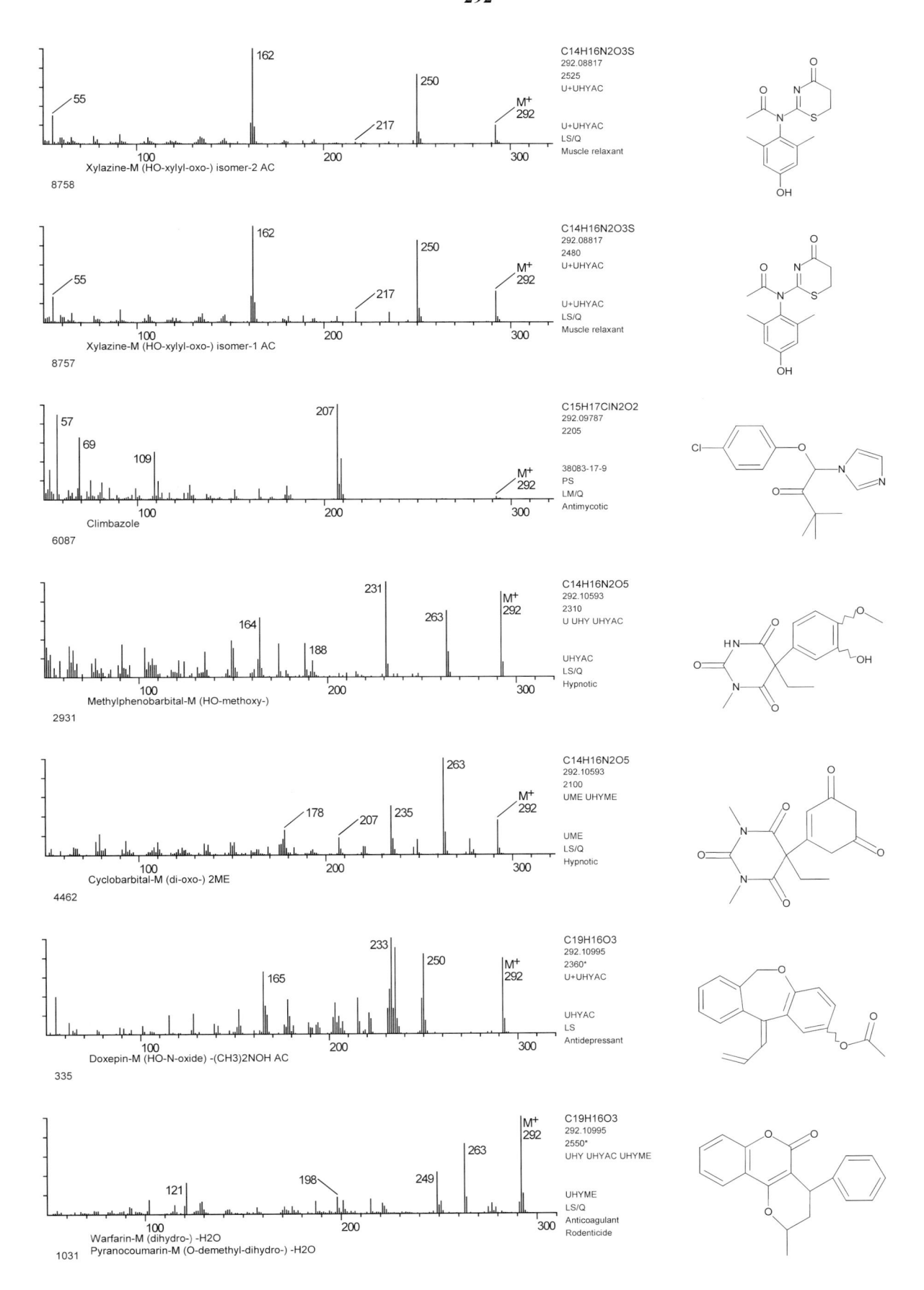

Xylazine-M (HO-xylyl-oxo-) isomer-2 AC 8758	peaks: 55, 162, 217, 250, M+ 292	C14H16N2O3S 292.08817 2525 U+UHYAC U+UHYAC LS/Q Muscle relaxant
Xylazine-M (HO-xylyl-oxo-) isomer-1 AC 8757	peaks: 55, 162, 217, 250, M+ 292	C14H16N2O3S 292.08817 2480 U+UHYAC U+UHYAC LS/Q Muscle relaxant
Climbazole 6087	peaks: 57, 69, 109, 207, M+ 292	C15H17ClN2O2 292.09787 2205 38083-17-9 PS LM/Q Antimycotic
Methylphenobarbital-M (HO-methoxy-) 2931	peaks: 164, 188, 231, 263, M+ 292	C14H16N2O5 292.10593 2310 U UHY UHYAC UHYAC LS/Q Hypnotic
Cyclobarbital-M (di-oxo-) 2ME 4462	peaks: 178, 207, 235, 263, M+ 292	C14H16N2O5 292.10593 2100 UME UHYME UME LS/Q Hypnotic
Doxepin-M (HO-N-oxide) -(CH3)2NOH AC 335	peaks: 165, 233, 250, M+ 292	C19H16O3 292.10995 2360* U+UHYAC UHYAC LS Antidepressant
Warfarin-M (dihydro-) -H2O Pyranocoumarin-M (O-demethyl-dihydro-) -H2O 1031	peaks: 121, 198, 249, 263, M+ 292	C19H16O3 292.10995 2550* UHY UHYAC UHYME UHYME LS/Q Anticoagulant Rodenticide

292

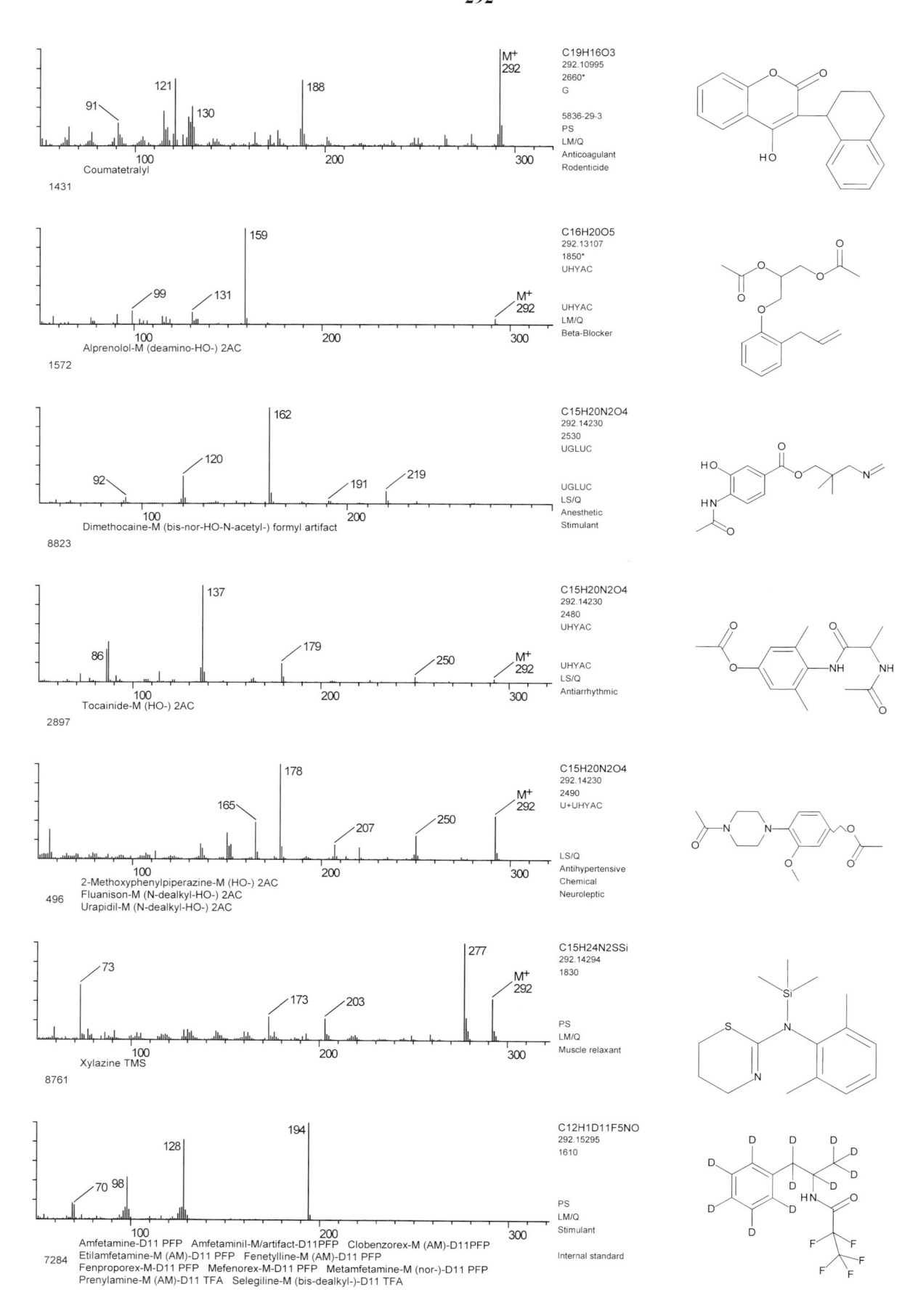

Coumatetralyl
1431
M+ 292, 188, 121, 130, 91
C19H16O3
292.10995
2660*
G
5836-29-3
PS
LM/Q
Anticoagulant
Rodenticide

Alprenolol-M (deamino-HO-) 2AC
1572
159, 99, 131, M+ 292
C16H20O5
292.13107
1850*
UHYAC
UHYAC
LM/Q
Beta-Blocker

Dimethocaine-M (bis-nor-HO-N-acetyl-) formyl artifact
8823
162, 120, 92, 191, 219
C15H20N2O4
292.14230
2530
UGLUC
UGLUC
LS/Q
Anesthetic
Stimulant

Tocainide-M (HO-) 2AC
2897
137, 86, 179, 250, M+ 292
C15H20N2O4
292.14230
2480
UHYAC
UHYAC
LS/Q
Antiarrhythmic

2-Methoxyphenylpiperazine-M (HO-) 2AC
Fluanison-M (N-dealkyl-HO-) 2AC
Urapidil-M (N-dealkyl-HO-) 2AC
496
178, 165, 207, 250, M+ 292
C15H20N2O4
292.14230
2490
U+UHYAC
LS/Q
Antihypertensive
Chemical
Neuroleptic

Xylazine TMS
8761
277, 73, 173, 203, M+ 292
C15H24N2SSi
292.14294
1830
PS
LM/Q
Muscle relaxant

Amfetamine-D11 PFP   Amfetaminil-M/artifact-D11PFP   Clobenzorex-M (AM)-D11PFP
Etilamfetamine-M (AM)-D11 PFP   Fenetylline-M (AM)-D11 PFP
Fenproporex-M-D11 PFP   Mefenorex-M-D11 PFP   Metamfetamine-M (nor-)-D11 PFP
Prenylamine-M (AM)-D11 TFA   Selegiline-M (bis-dealkyl-)-D11 TFA
7284
194, 128, 70, 98
C12H1D11F5NO
292.15295
1610
PS
LM/Q
Stimulant
Internal standard

856

292

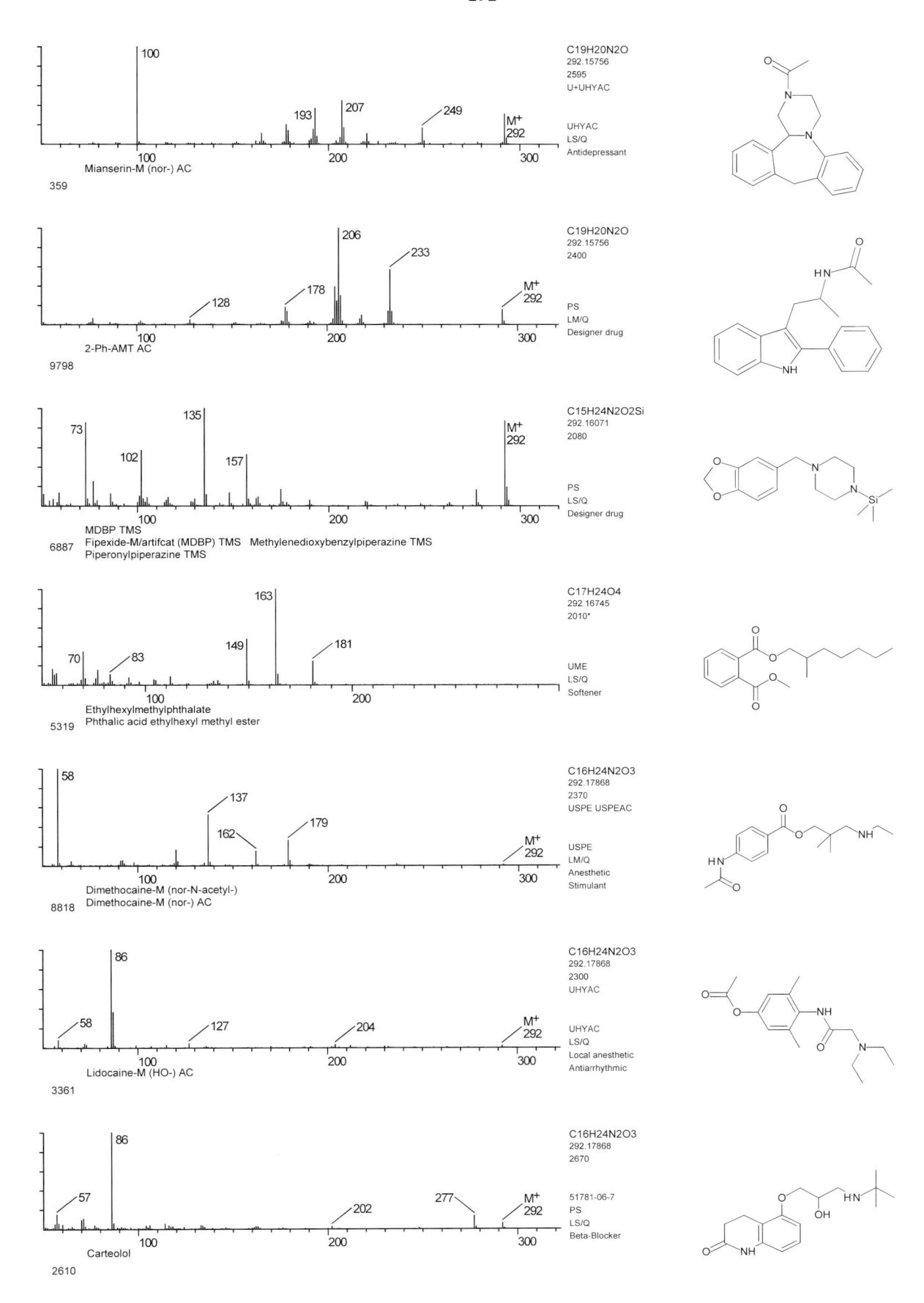

Mianserin-M (nor-) AC 359	C19H20N2O 292.15756 2595 U+UHYAC UHYAC LS/Q Antidepressant
2-Ph-AMT AC 9798	C19H20N2O 292.15756 2400 PS LM/Q Designer drug
MDBP TMS Fipexide-M/artifcat (MDBP) TMS Methylenedioxybenzylpiperazine TMS 6887 Piperonylpiperazine TMS	C15H24N2O2Si 292.16071 2080 PS LS/Q Designer drug
Ethylhexylmethylphthalate 5319 Phthalic acid ethylhexyl methyl ester	C17H24O4 292.16745 2010* UME LS/Q Softener
Dimethocaine-M (nor-N-acetyl-) 8818 Dimethocaine-M (nor-) AC	C16H24N2O3 292.17868 2370 USPE USPEAC USPE LM/Q Anesthetic Stimulant
Lidocaine-M (HO-) AC 3361	C16H24N2O3 292.17868 2300 UHYAC UHYAC LS/Q Local anesthetic Antiarrhythmic
Carteolol 2610	C16H24N2O3 292.17868 2670 51781-06-7 PS LS/Q Beta-Blocker

292

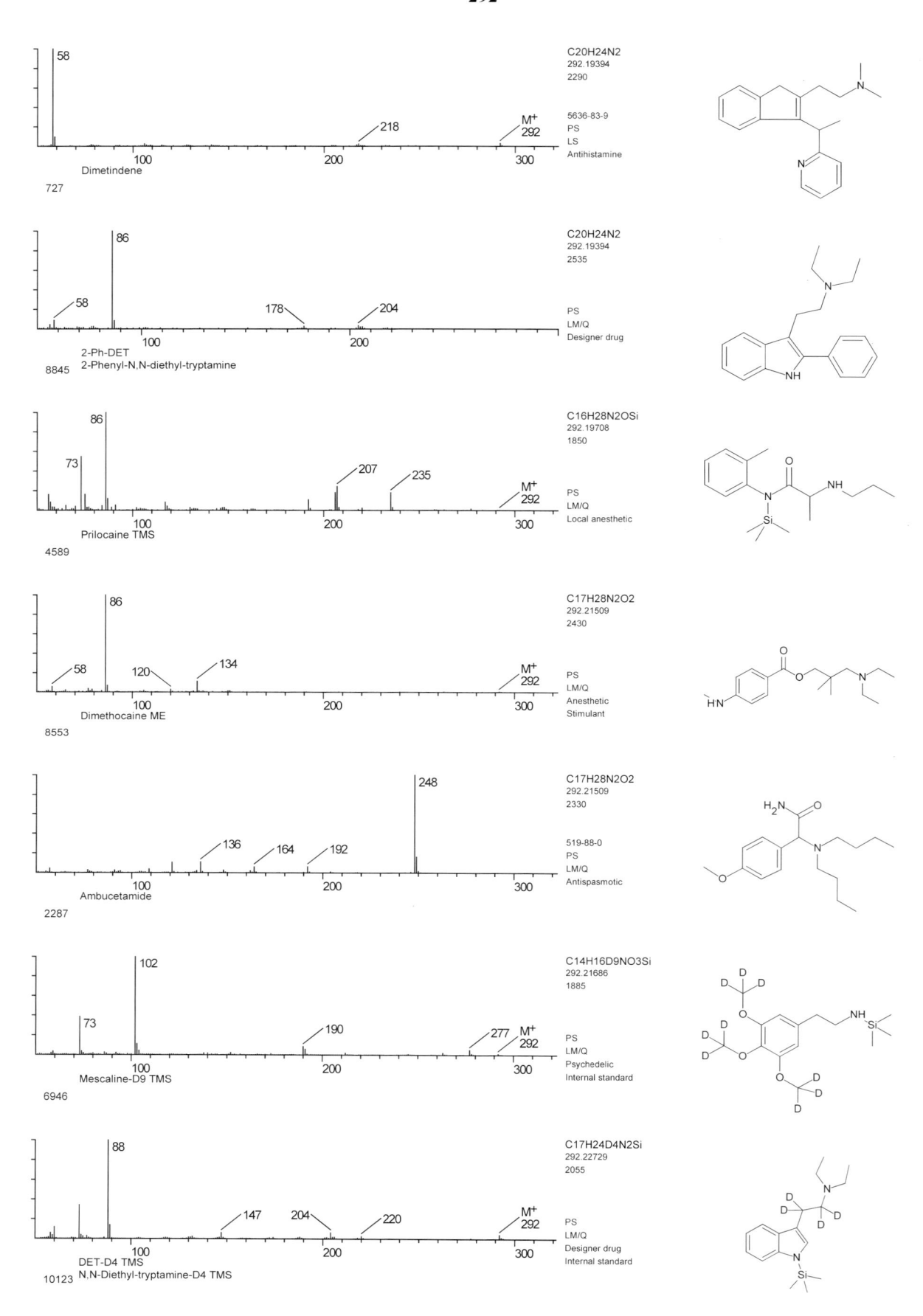

Dimetindene — 727
- C20H24N2
- 292.19394
- 2290
- 5636-83-9
- PS
- LS
- Antihistamine
- Peaks: 58, 218, M+ 292

2-Ph-DET — 8845
2-Phenyl-N,N-diethyl-tryptamine
- C20H24N2
- 292.19394
- 2535
- PS
- LM/Q
- Designer drug
- Peaks: 58, 86, 178, 204

Prilocaine TMS — 4589
- C16H28N2OSi
- 292.19708
- 1850
- PS
- LM/Q
- Local anesthetic
- Peaks: 73, 86, 207, 235, M+ 292

Dimethocaine ME — 8553
- C17H28N2O2
- 292.21509
- 2430
- PS
- LM/Q
- Anesthetic
- Stimulant
- Peaks: 58, 86, 120, 134, M+ 292

Ambucetamide — 2287
- C17H28N2O2
- 292.21509
- 2330
- 519-88-0
- PS
- LM/Q
- Antispasmotic
- Peaks: 136, 164, 192, 248

Mescaline-D9 TMS — 6946
- C14H16D9NO3Si
- 292.21686
- 1885
- PS
- LM/Q
- Psychedelic
- Internal standard
- Peaks: 73, 102, 190, 277, M+ 292

DET-D4 TMS — 10123
N,N-Diethyl-tryptamine-D4 TMS
- C17H24D4N2Si
- 292.22729
- 2055
- PS
- LM/Q
- Designer drug
- Internal standard
- Peaks: 88, 147, 204, 220, M+ 292

292

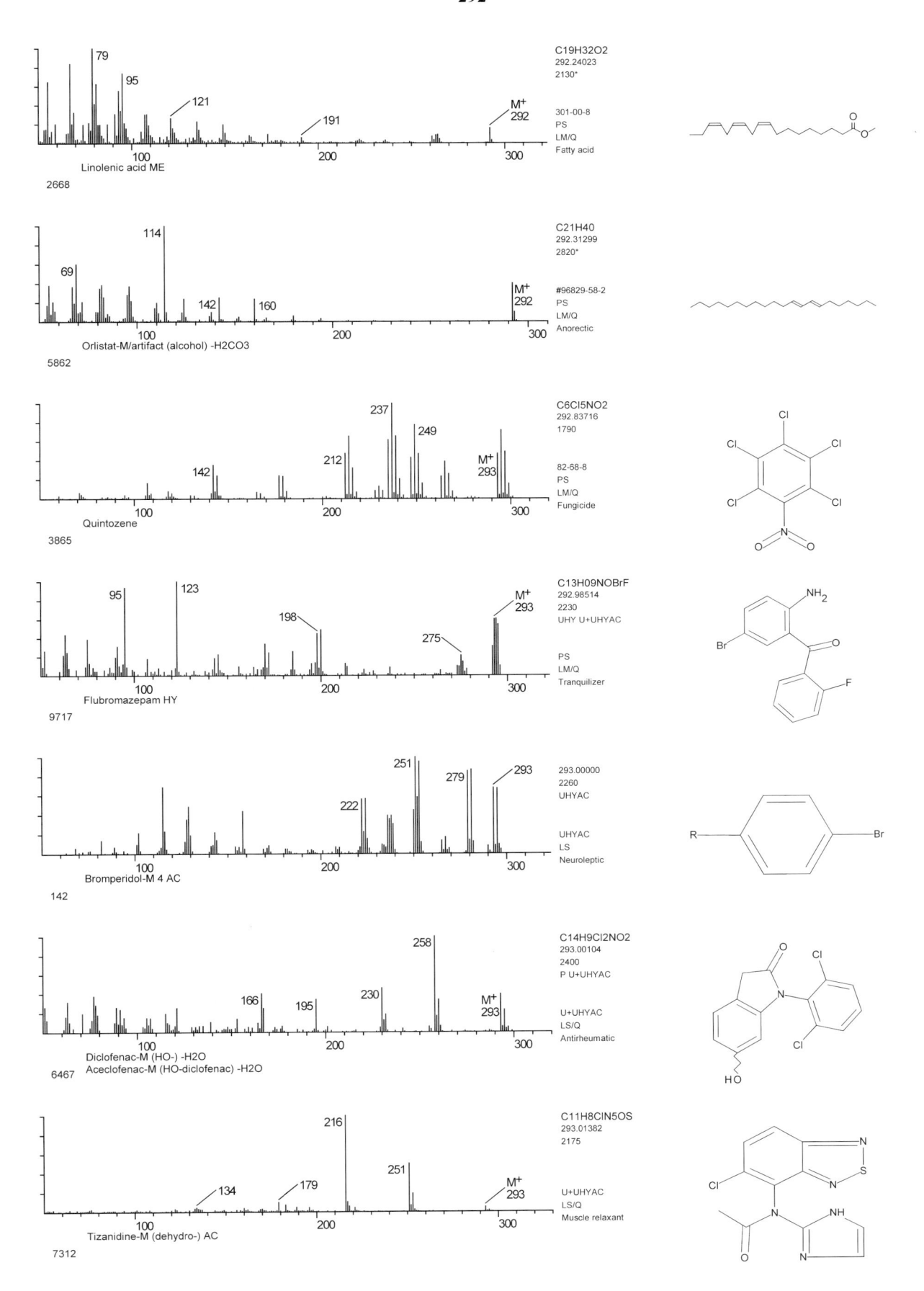

Linolenic acid ME — C19H32O2, 292.24023, 2130*, 301-00-8, PS, LM/Q, Fatty acid
Peaks: 79, 95, 121, 191, M+ 292
2668

Orlistat-M/artifact (alcohol) -H2CO3 — C21H40, 292.31299, 2820*, #96829-58-2, PS, LM/Q, Anorectic
Peaks: 69, 114, 142, 160, M+ 292
5862

Quintozene — C6Cl5NO2, 292.83716, 1790, 82-68-8, PS, LM/Q, Fungicide
Peaks: 142, 212, 237, 249, M+ 293
3865

Flubromazepam HY — C13H09NOBrF, 292.98514, 2230, UHY U+UHYAC, PS, LM/Q, Tranquilizer
Peaks: 95, 123, 198, 275, M+ 293
9717

Bromperidol-M 4 AC — 293.00000, 2260, UHYAC, UHYAC, LS, Neuroleptic
Peaks: 222, 251, 279, 293
142

Diclofenac-M (HO-) -H2O
Aceclofenac-M (HO-diclofenac) -H2O — C14H9Cl2NO2, 293.00104, 2400, P U+UHYAC, U+UHYAC, LS/Q, Antirheumatic
Peaks: 166, 195, 230, 258, M+ 293
6467

Tizanidine-M (dehydro-) AC — C11H8ClN5OS, 293.01382, 2175, U+UHYAC, LS/Q, Muscle relaxant
Peaks: 134, 179, 216, 251, M+ 293
7312

293

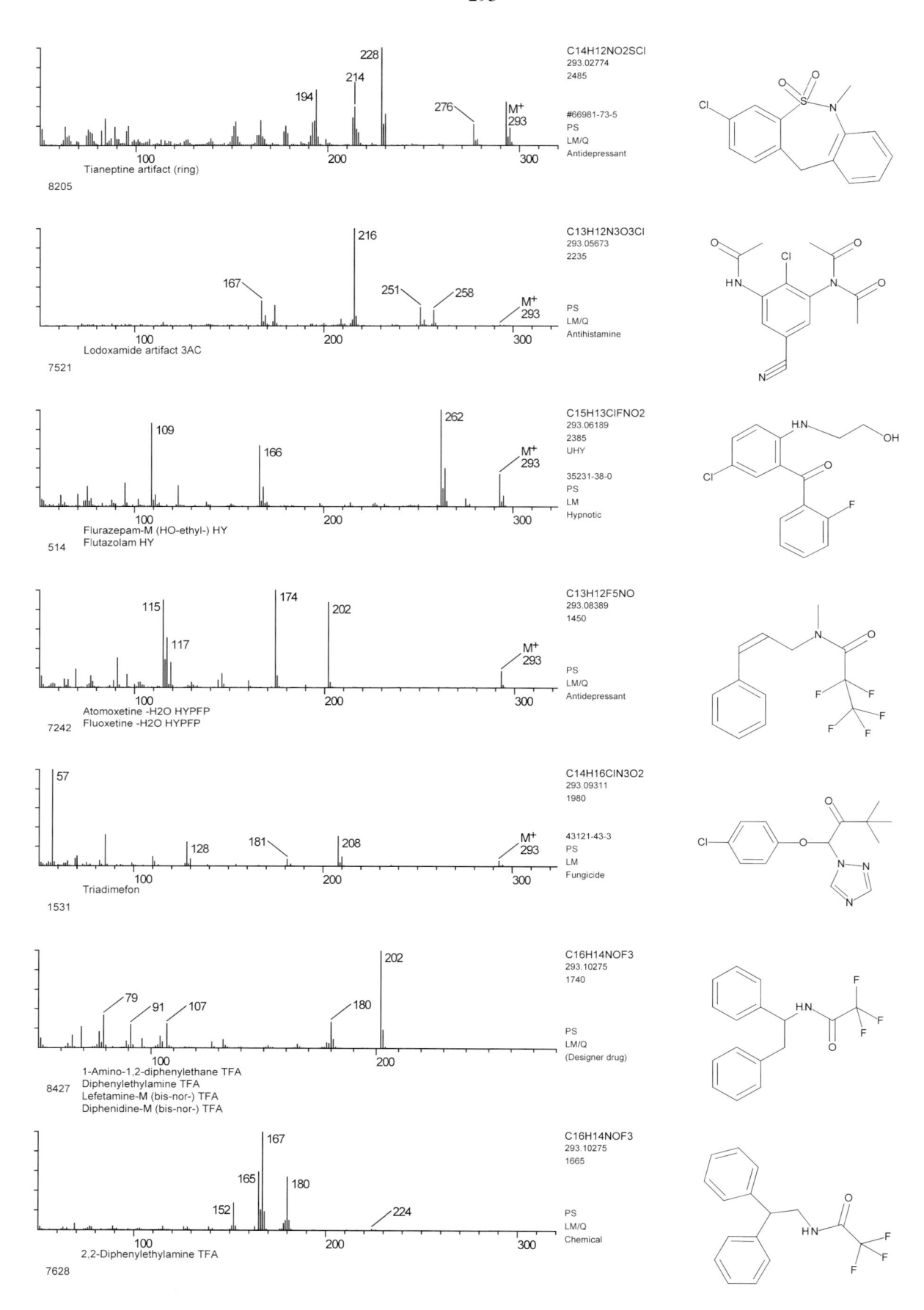

Spectrum	Formula / Info
Tianeptine artifact (ring) — peaks 194, 214, 228, 276, M+ 293	C14H12NO2SCl, 293.02774, 2485, #66981-73-5, PS, LM/Q, Antidepressant
Lodoxamide artifact 3AC — peaks 167, 216, 251, 258, M+ 293	C13H12N3O3Cl, 293.05673, 2235, PS, LM/Q, Antihistamine
Flurazepam-M (HO-ethyl-) HY / Flutazolam HY — peaks 109, 166, 262, M+ 293	C15H13ClFNO2, 293.06189, 2385, UHY, 35231-38-0, PS, LM, Hypnotic
Atomoxetine -H2O HYPFP / Fluoxetine -H2O HYPFP — peaks 115, 117, 174, 202, M+ 293	C13H12F5NO, 293.08389, 1450, PS, LM/Q, Antidepressant
Triadimefon — peaks 57, 128, 181, 208, M+ 293	C14H16ClN3O2, 293.09311, 1980, 43121-43-3, PS, LM, Fungicide
1-Amino-1,2-diphenylethane TFA / Diphenylethylamine TFA / Lefetamine-M (bis-nor-) TFA / Diphenidine-M (bis-nor-) TFA — peaks 79, 91, 107, 180, 202	C16H14NOF3, 293.10275, 1740, PS, LM/Q, (Designer drug)
2,2-Diphenylethylamine TFA — peaks 152, 165, 167, 180, 224	C16H14NOF3, 293.10275, 1665, PS, LM/Q, Chemical

293

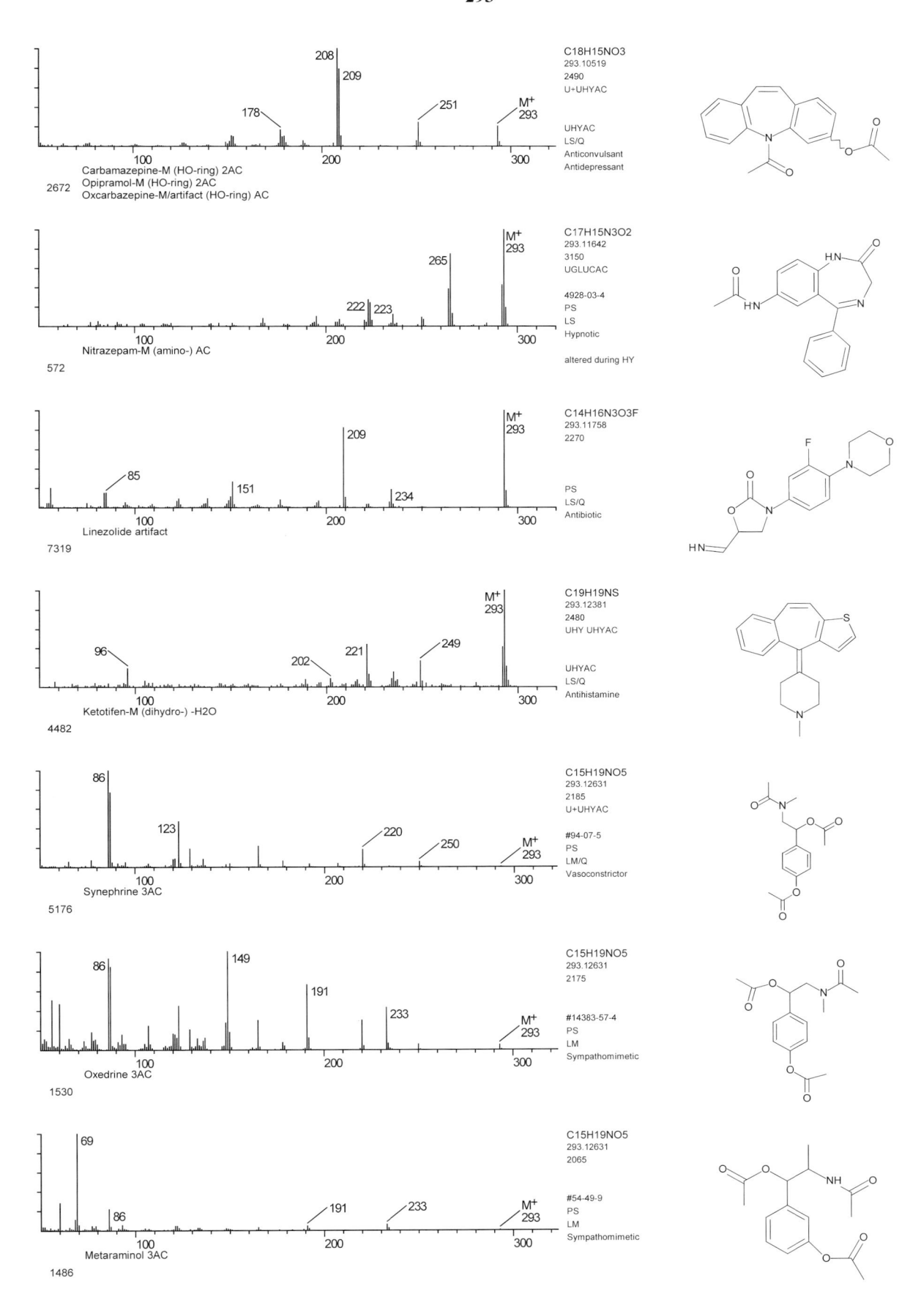

C18H15NO3
293.10519
2490
U+UHYAC

UHYAC
LS/Q
Anticonvulsant
Antidepressant

Carbamazepine-M (HO-ring) 2AC
Opipramol-M (HO-ring) 2AC
Oxcarbazepine-M/artifact (HO-ring) AC
2672

C17H15N3O2
293.11642
3150
UGLUCAC

4928-03-4
PS
LS
Hypnotic

altered during HY

Nitrazepam-M (amino-) AC
572

C14H16N3O3F
293.11758
2270

PS
LS/Q
Antibiotic

Linezolide artifact
7319

C19H19NS
293.12381
2480
UHY UHYAC

UHYAC
LS/Q
Antihistamine

Ketotifen-M (dihydro-) -H2O
4482

C15H19NO5
293.12631
2185
U+UHYAC

#94-07-5
PS
LM/Q
Vasoconstrictor

Synephrine 3AC
5176

C15H19NO5
293.12631
2175

#14383-57-4
PS
LM
Sympathomimetic

Oxedrine 3AC
1530

C15H19NO5
293.12631
2065

#54-49-9
PS
LM
Sympathomimetic

Metaraminol 3AC
1486

861

293

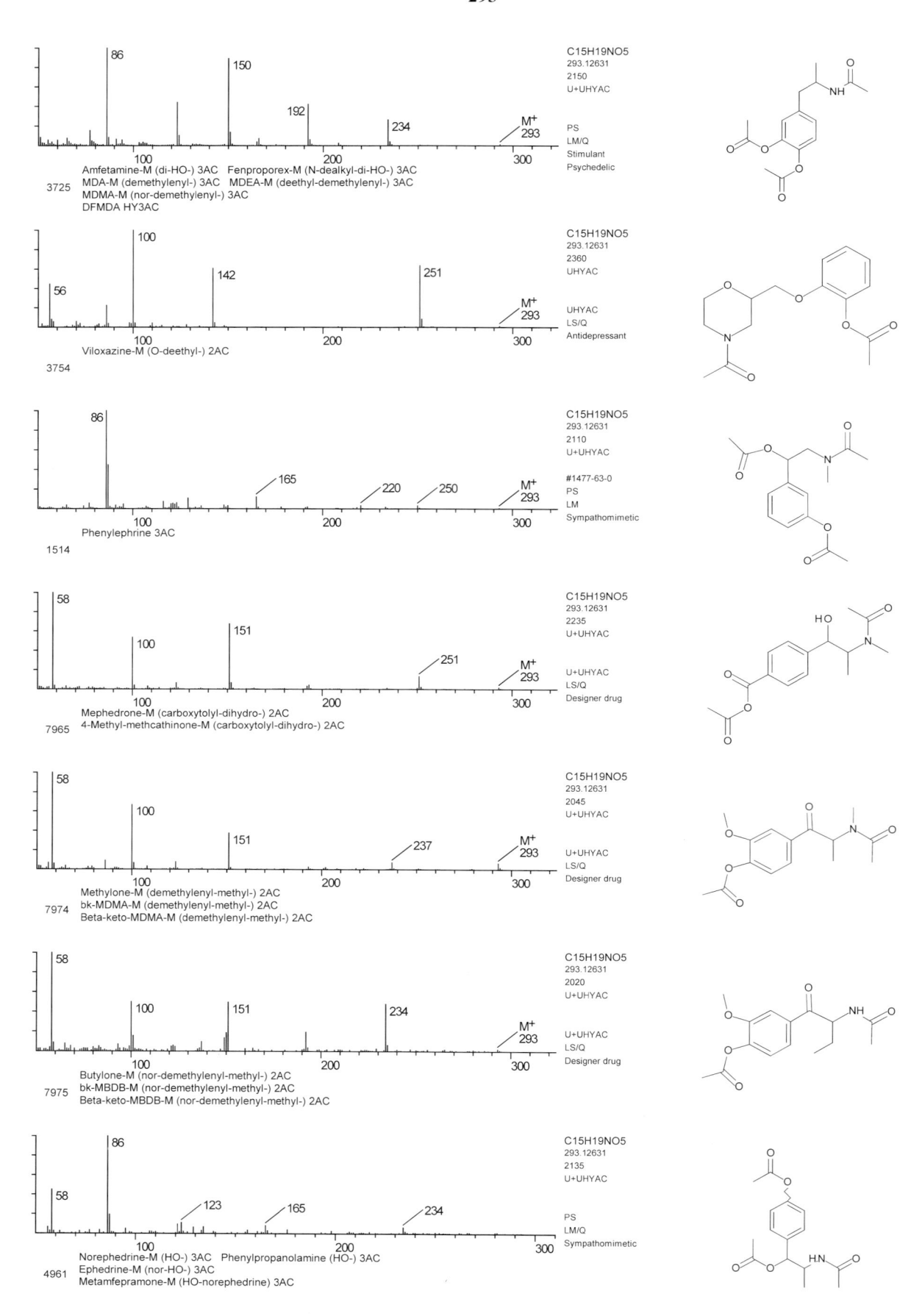

3725	Amfetamine-M (di-HO-) 3AC Fenproporex-M (N-dealkyl-di-HO-) 3AC MDA-M (demethylenyl-) 3AC MDEA-M (deethyl-demethylenyl-) 3AC MDMA-M (nor-demethylenyl-) 3AC DFMDA HY3AC

Peaks: 86, 150, 192, 234, M+ 293
C15H19NO5
293.12631
2150
U+UHYAC
PS
LM/Q
Stimulant
Psychedelic

3754 Viloxazine-M (O-deethyl-) 2AC

Peaks: 56, 100, 142, 251, M+ 293
C15H19NO5
293.12631
2360
UHYAC
UHYAC
LS/Q
Antidepressant

1514 Phenylephrine 3AC

Peaks: 86, 165, 220, 250, M+ 293
C15H19NO5
293.12631
2110
U+UHYAC
#1477-63-0
PS
LM
Sympathomimetic

7965 Mephedrone-M (carboxytolyl-dihydro-) 2AC
4-Methyl-methcathinone-M (carboxytolyl-dihydro-) 2AC

Peaks: 58, 100, 151, 251, M+ 293
C15H19NO5
293.12631
2235
U+UHYAC
U+UHYAC
LS/Q
Designer drug

7974 Methylone-M (demethylenyl-methyl-) 2AC
bk-MDMA-M (demethylenyl-methyl-) 2AC
Beta-keto-MDMA-M (demethylenyl-methyl-) 2AC

Peaks: 58, 100, 151, 237, M+ 293
C15H19NO5
293.12631
2045
U+UHYAC
U+UHYAC
LS/Q
Designer drug

7975 Butylone-M (nor-demethylenyl-methyl-) 2AC
bk-MBDB-M (nor-demethylenyl-methyl-) 2AC
Beta-keto-MBDB-M (nor-demethylenyl-methyl-) 2AC

Peaks: 58, 100, 151, 234, M+ 293
C15H19NO5
293.12631
2020
U+UHYAC
U+UHYAC
LS/Q
Designer drug

4961 Norephedrine-M (HO-) 3AC Phenylpropanolamine (HO-) 3AC
Ephedrine-M (nor-HO-) 3AC
Metamfepramone-M (HO-norephedrine) 3AC

Peaks: 58, 86, 123, 165, 234
C15H19NO5
293.12631
2135
U+UHYAC
PS
LM/Q
Sympathomimetic

862

293

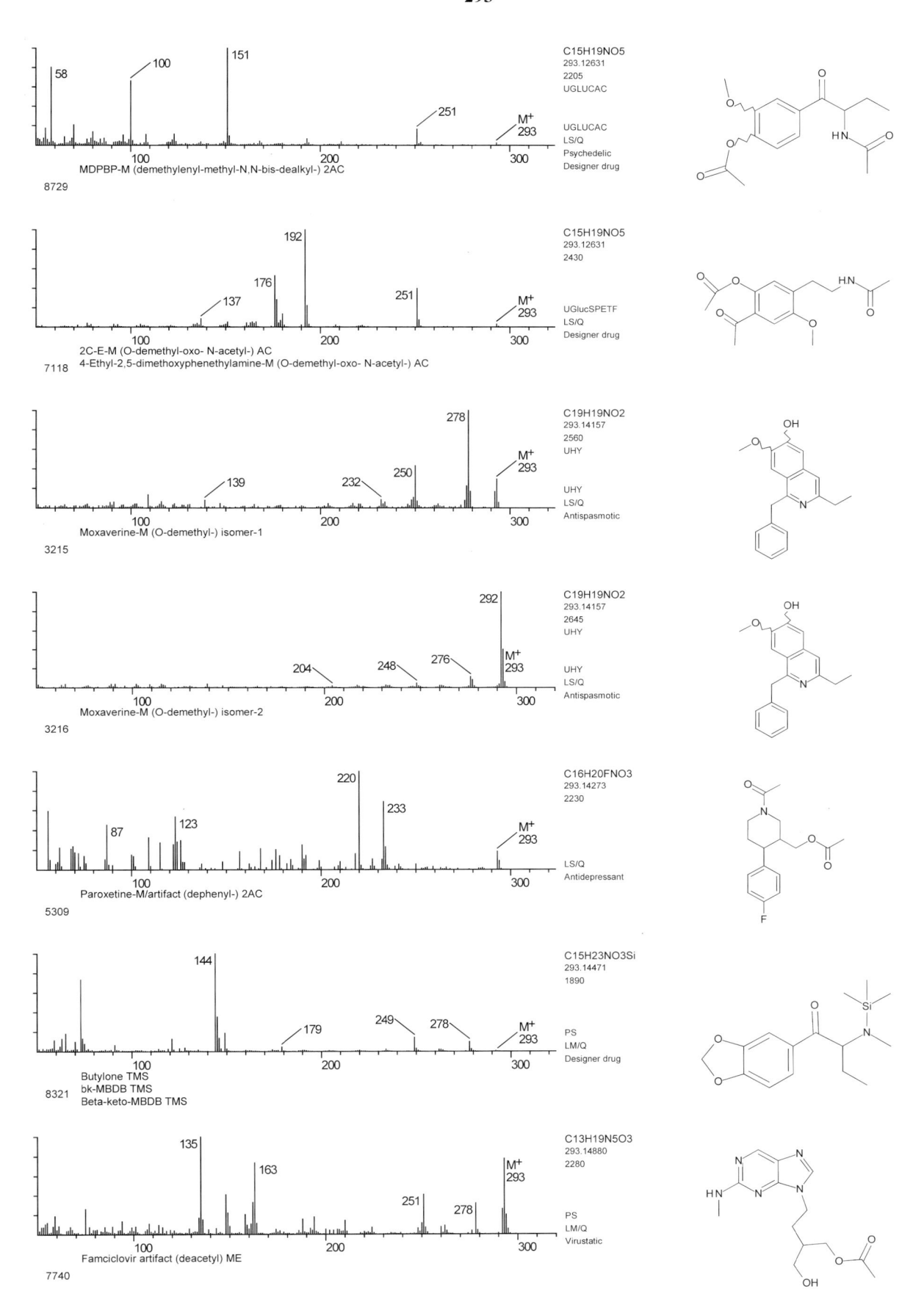

Spectrum	Formula / Info
MDPBP-M (demethylenyl-methyl-N,N-bis-dealkyl-) 2AC — peaks 58, 100, 151, 251, M+ 293 — 8729	C15H19NO5; 293.12631; 2205; UGLUCAC; UGLUCAC; LS/Q; Psychedelic; Designer drug
2C-E-M (O-demethyl-oxo- N-acetyl-) AC / 4-Ethyl-2,5-dimethoxyphenethylamine-M (O-demethyl-oxo- N-acetyl-) AC — peaks 137, 176, 192, 251, M+ 293 — 7118	C15H19NO5; 293.12631; 2430; UGlucSPETF; LS/Q; Designer drug
Moxaverine-M (O-demethyl-) isomer-1 — peaks 139, 232, 250, 278, M+ 293 — 3215	C19H19NO2; 293.14157; 2560; UHY; UHY; LS/Q; Antispasmotic
Moxaverine-M (O-demethyl-) isomer-2 — peaks 204, 248, 276, 292, M+ 293 — 3216	C19H19NO2; 293.14157; 2645; UHY; UHY; LS/Q; Antispasmotic
Paroxetine-M/artifact (dephenyl-) 2AC — peaks 87, 123, 220, 233, M+ 293 — 5309	C16H20FNO3; 293.14273; 2230; LS/Q; Antidepressant
Butylone TMS / bk-MBDB TMS / Beta-keto-MBDB TMS — peaks 144, 179, 249, 278, M+ 293 — 8321	C15H23NO3Si; 293.14471; 1890; PS; LM/Q; Designer drug
Famciclovir artifact (deacetyl) ME — peaks 135, 163, 251, 278, M+ 293 — 7740	C13H19N5O3; 293.14880; 2280; PS; LM/Q; Virustatic

293

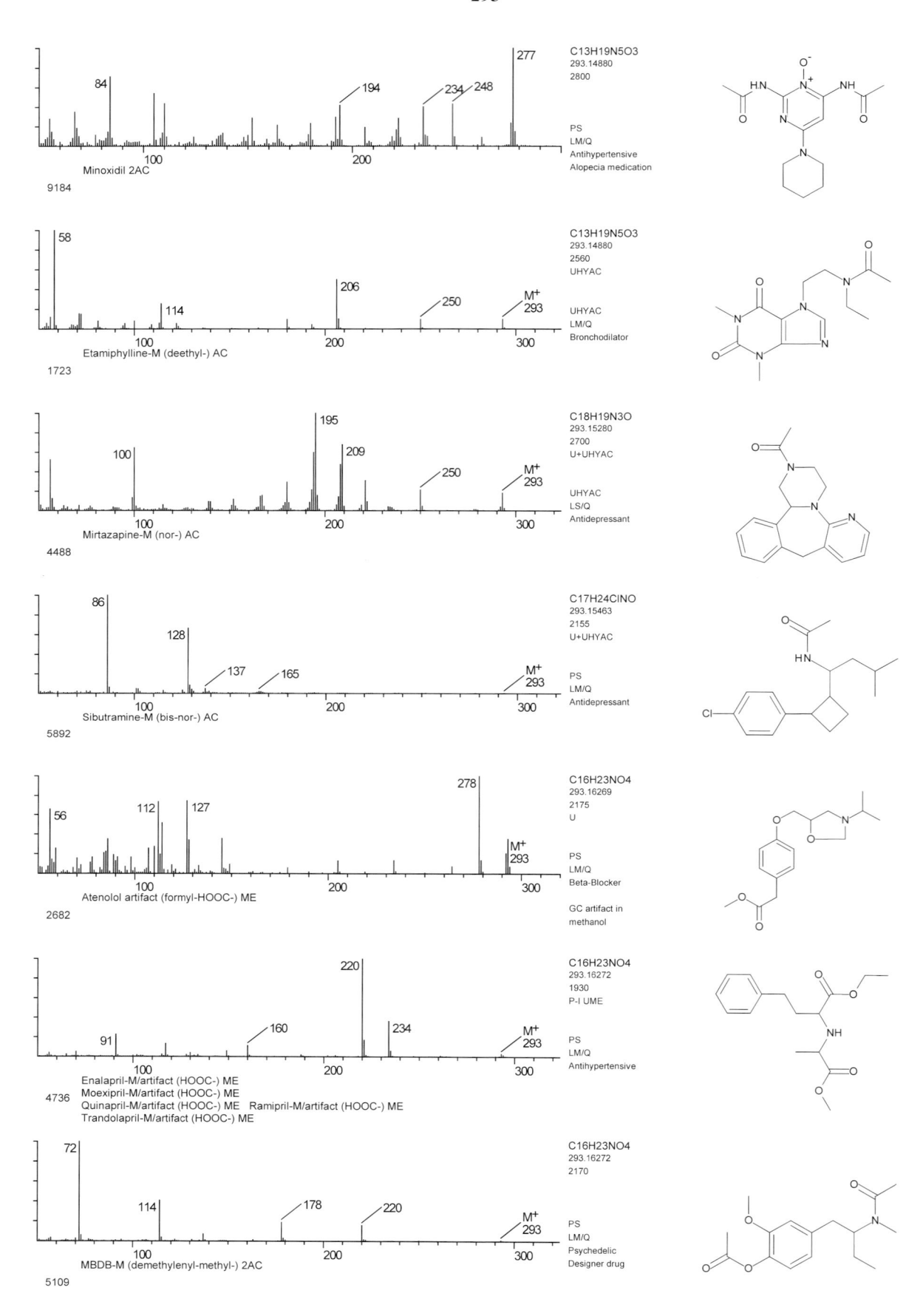

Minoxidil 2AC — 9184
Peaks: 84, 194, 234, 248, 277
C13H19N5O3
293.14880
2800
PS
LM/Q
Antihypertensive
Alopecia medication

Etamiphylline-M (deethyl-) AC — 1723
Peaks: 58, 114, 206, 250, M+ 293
C13H19N5O3
293.14880
2560
UHYAC
UHYAC
LM/Q
Bronchodilator

Mirtazapine-M (nor-) AC — 4488
Peaks: 100, 195, 209, 250, M+ 293
C18H19N3O
293.15280
2700
U+UHYAC
UHYAC
LS/Q
Antidepressant

Sibutramine-M (bis-nor-) AC — 5892
Peaks: 86, 128, 137, 165, M+ 293
C17H24ClNO
293.15463
2155
U+UHYAC
PS
LM/Q
Antidepressant

Atenolol artifact (formyl-HOOC-) ME — 2682
Peaks: 56, 112, 127, 278, M+ 293
C16H23NO4
293.16269
2175
U
PS
LM/Q
Beta-Blocker
GC artifact in methanol

Enalapril-M/artifact (HOOC-) ME
Moexipril-M/artifact (HOOC-) ME
Quinapril-M/artifact (HOOC-) ME **Ramipril-M/artifact (HOOC-) ME**
Trandolapril-M/artifact (HOOC-) ME — 4736
Peaks: 91, 160, 220, 234, M+ 293
C16H23NO4
293.16272
1930
P-I UME
PS
LM/Q
Antihypertensive

MBDB-M (demethylenyl-methyl-) 2AC — 5109
Peaks: 72, 114, 178, 220, M+ 293
C16H23NO4
293.16272
2170
PS
LM/Q
Psychedelic Designer drug

293

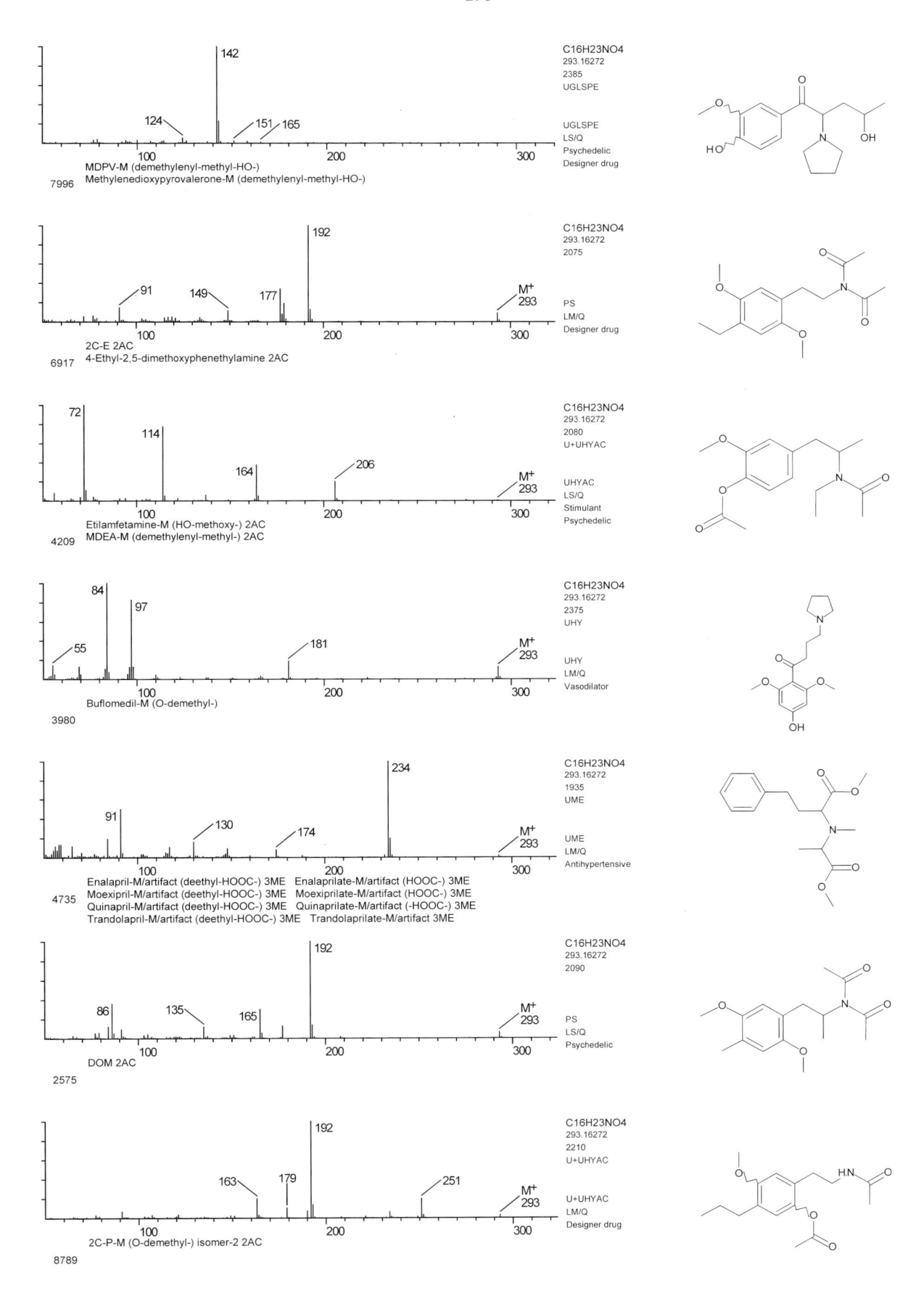

142 124 151 165 100 200 300 MDPV-M (demethylenyl-methyl-HO-) 7996 Methylenedioxypyrovalerone-M (demethylenyl-methyl-HO-)	C16H23NO4 293.16272 2385 UGLSPE UGLSPE LS/Q Psychedelic Designer drug	

192 91 149 177 M+ 293 100 200 300 2C-E 2AC 6917 4-Ethyl-2,5-dimethoxyphenethylamine 2AC	C16H23NO4 293.16272 2075 PS LM/Q Designer drug	

72 114 164 206 M+ 293 100 200 300 Etilamfetamine-M (HO-methoxy-) 2AC 4209 MDEA-M (demethylenyl-methyl-) 2AC	C16H23NO4 293.16272 2080 U+UHYAC UHYAC LS/Q Stimulant Psychedelic	

84 97 55 181 M+ 293 100 200 300 Buflomedil-M (O-demethyl-) 3980	C16H23NO4 293.16272 2375 UHY UHY LM/Q Vasodilator	

234 91 130 174 M+ 293 100 200 300 Enalapril-M/artifact (deethyl-HOOC-) 3ME Enalapril-M/artifact (HOOC-) 3ME Moexipril-M/artifact (deethyl-HOOC-) 3ME Moexiprilate-M/artifact (HOOC-) 3ME 4735 Quinapril-M/artifact (deethyl-HOOC-) 3ME Quinaprilate-M/artifact (-HOOC-) 3ME Trandolapril-M/artifact (deethyl-HOOC-) 3ME Trandolaprilate-M/artifact 3ME	C16H23NO4 293.16272 1935 UME UME LM/Q Antihypertensive	

192 86 135 165 M+ 293 100 200 300 DOM 2AC 2575	C16H23NO4 293.16272 2090 PS LS/Q Psychedelic	

192 163 179 251 M+ 293 100 200 300 2C-P-M (O-demethyl-) isomer-2 2AC 8789	C16H23NO4 293.16272 2210 U+UHYAC U+UHYAC LM/Q Designer drug	

293

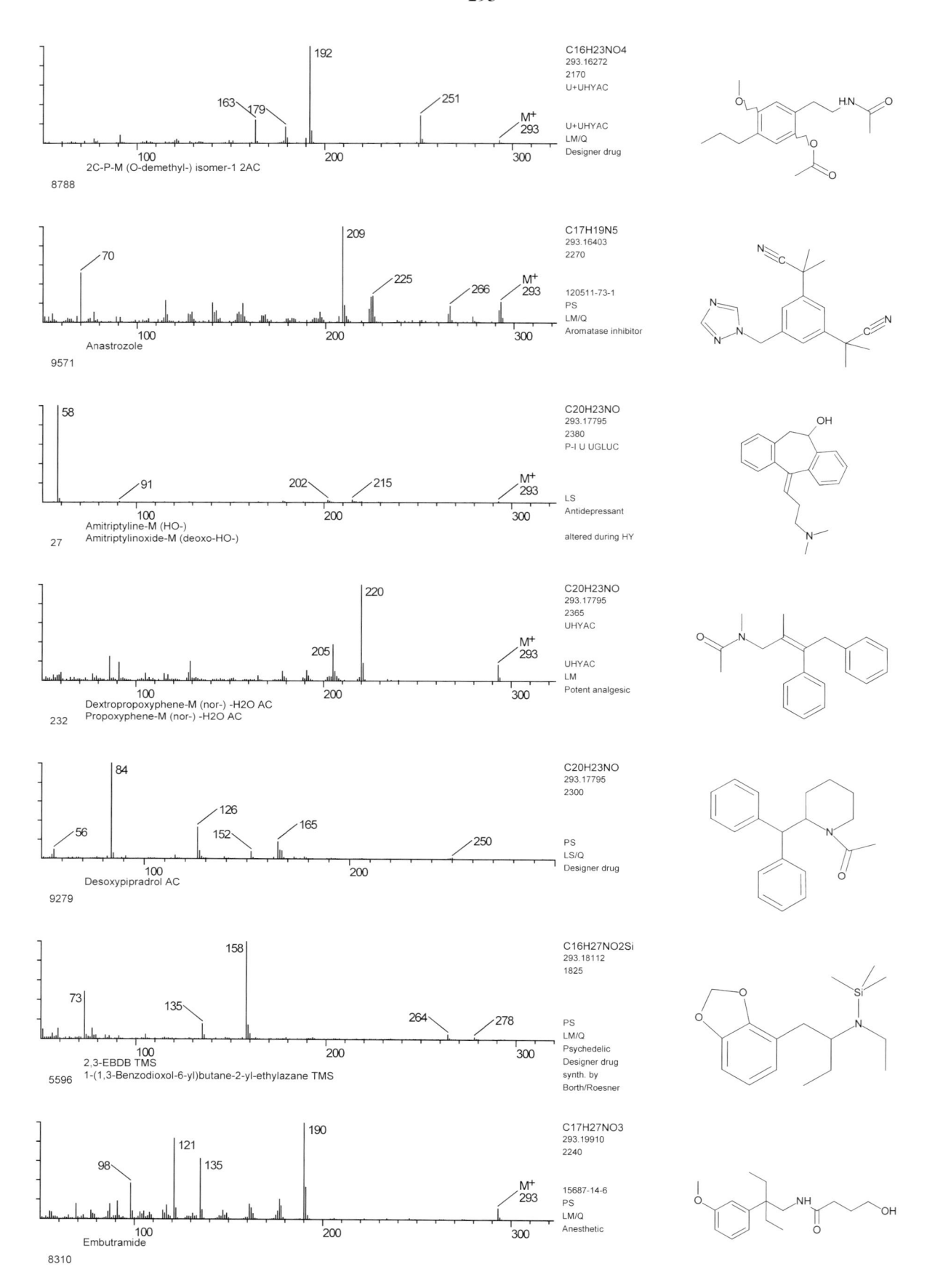

2C-P-M (O-demethyl-) isomer-1 2AC — peaks: 163, 179, 192, 251, M+ 293	C16H23NO4 293.16272 2170 U+UHYAC U+UHYAC LM/Q Designer drug
8788	
Anastrozole — peaks: 70, 209, 225, 266, M+ 293	C17H19N5 293.16403 2270 120511-73-1 PS LM/Q Aromatase inhibitor
9571	
Amitriptyline-M (HO-) Amitriptylinoxide-M (deoxo-HO-) — peaks: 58, 91, 202, 215, M+ 293	C20H23NO 293.17795 2380 P-I U UGLUC LS Antidepressant altered during HY
27	
Dextropropoxyphene-M (nor-) -H2O AC Propoxyphene-M (nor-) -H2O AC — peaks: 205, 220, M+ 293	C20H23NO 293.17795 2365 UHYAC UHYAC LM Potent analgesic
232	
Desoxypipradrol AC — peaks: 56, 84, 126, 152, 165, 250	C20H23NO 293.17795 2300 PS LS/Q Designer drug
9279	
2,3-EBDB TMS 1-(1,3-Benzodioxol-6-yl)butane-2-yl-ethylazane TMS — peaks: 73, 135, 158, 264, 278	C16H27NO2Si 293.18112 1825 PS LM/Q Psychedelic Designer drug synth. by Borth/Roesner
5596	
Embutramide — peaks: 98, 121, 135, 190, M+ 293	C17H27NO3 293.19910 2240 15687-14-6 PS LM/Q Anesthetic
8310	

293

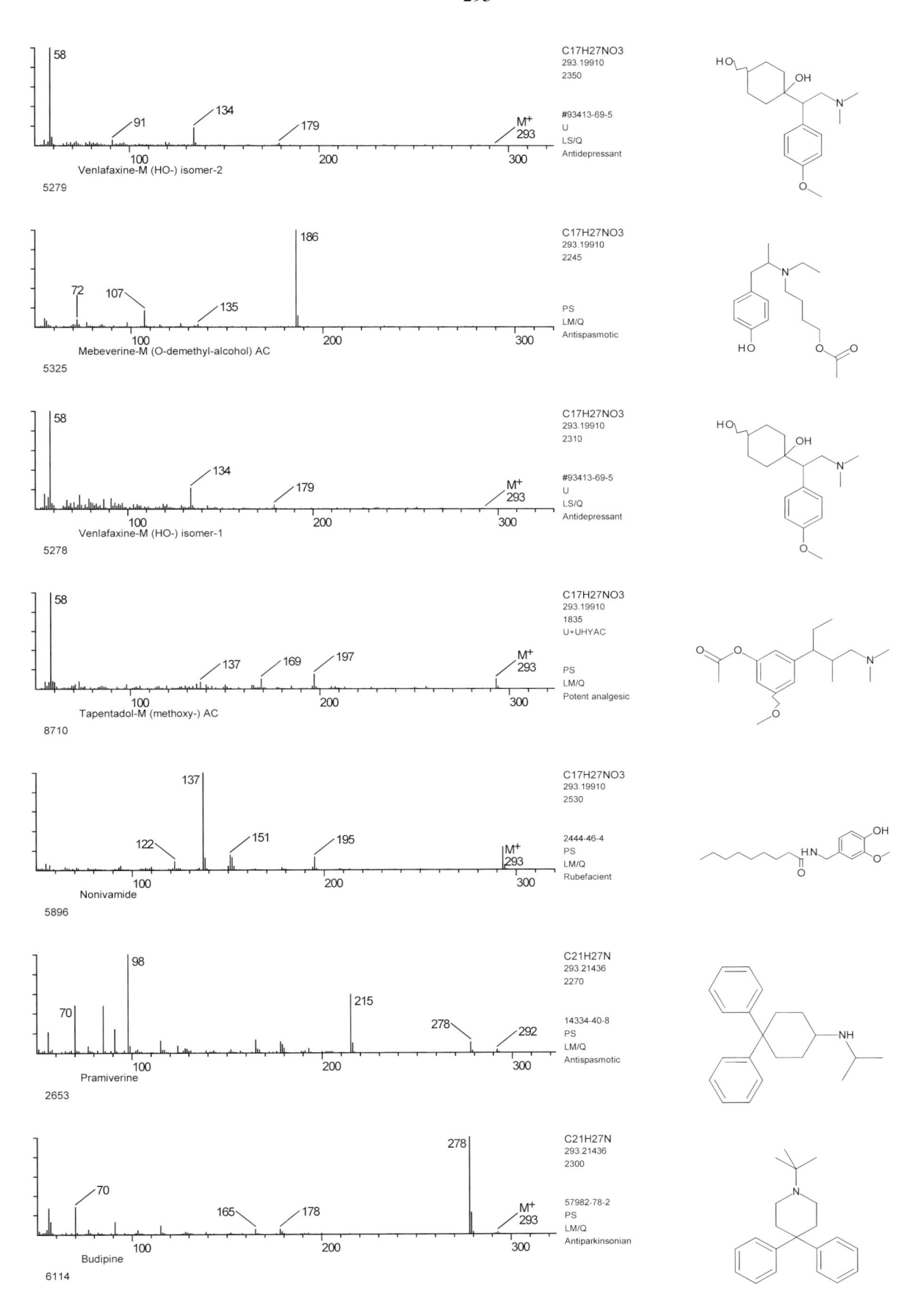

58, 91, 134, 179, M+ 293	C17H27NO3 293.19910 2350 #93413-69-5 U LS/Q Antidepressant	
Venlafaxine-M (HO-) isomer-2 5279		
72, 107, 135, 186	C17H27NO3 293.19910 2245 PS LM/Q Antispasmotic	
Mebeverine-M (O-demethyl-alcohol) AC 5325		
58, 134, 179, M+ 293	C17H27NO3 293.19910 2310 #93413-69-5 U LS/Q Antidepressant	
Venlafaxine-M (HO-) isomer-1 5278		
58, 137, 169, 197, M+ 293	C17H27NO3 293.19910 1835 U+UHYAC PS LM/Q Potent analgesic	
Tapentadol-M (methoxy-) AC 8710		
122, 137, 151, 195, M+ 293	C17H27NO3 293.19910 2530 2444-46-4 PS LM/Q Rubefacient	
Nonivamide 5896		
70, 98, 215, 278, 292	C21H27N 293.21436 2270 14334-40-8 PS LM/Q Antispasmotic	
Pramiverine 2653		
70, 165, 178, 278, M+ 293	C21H27N 293.21436 2300 57982-78-2 PS LM/Q Antiparkinsonian	
Budipine 6114		

293

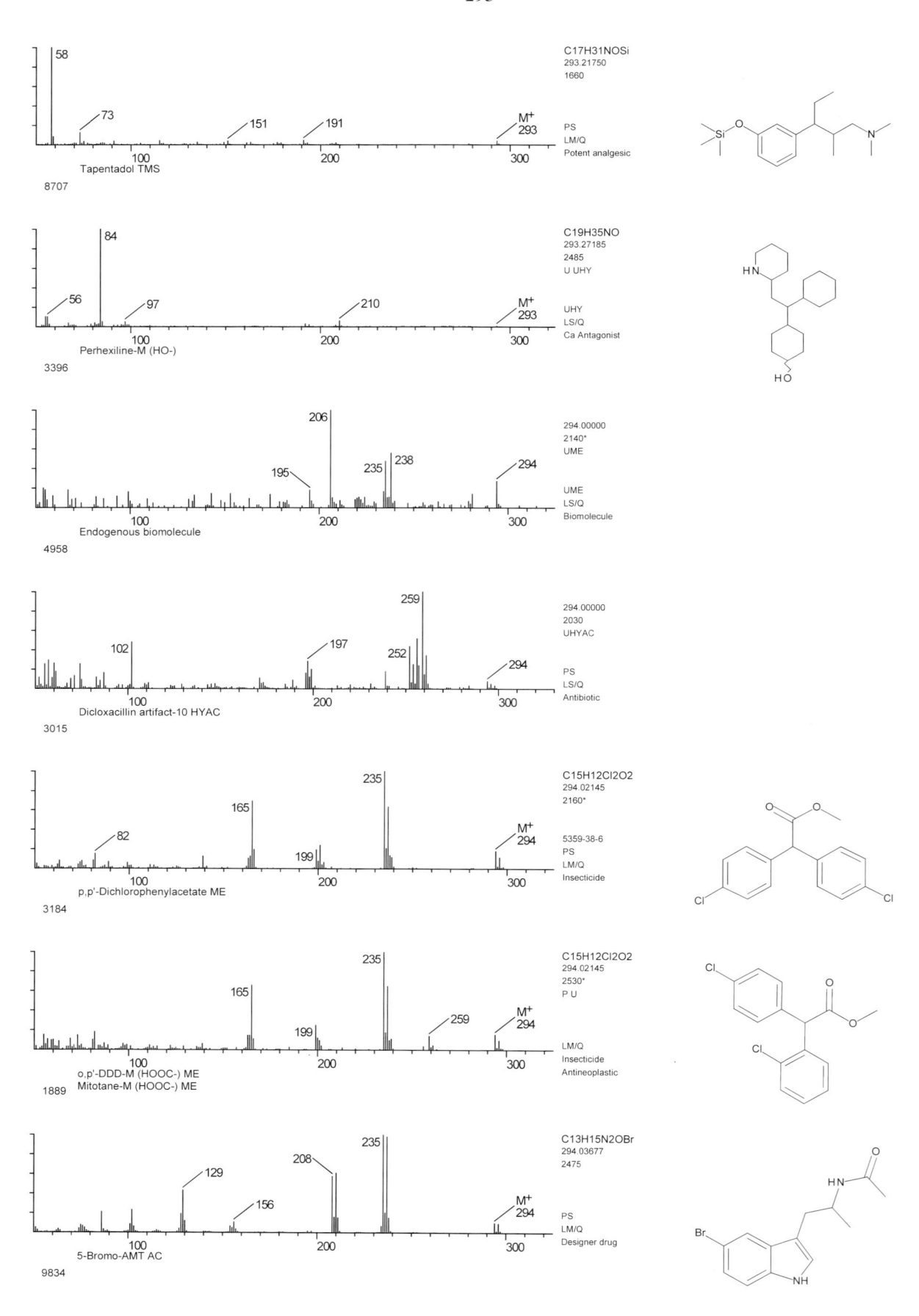

Tapentadol TMS — 8707
58, 73, 151, 191, M+ 293
C17H31NOSi
293.21750
1660
PS
LM/Q
Potent analgesic

Perhexiline-M (HO-) — 3396
56, 84, 97, 210, M+ 293
C19H35NO
293.27185
2485
U UHY
UHY
LS/Q
Ca Antagonist

Endogenous biomolecule — 4958
195, 206, 235, 238, 294
294.00000
2140*
UME
UME
LS/Q
Biomolecule

Dicloxacillin artifact-10 HYAC — 3015
102, 197, 252, 259, 294
294.00000
2030
UHYAC
PS
LS/Q
Antibiotic

p,p'-Dichlorophenylacetate ME — 3184
82, 165, 199, 235, M+ 294
C15H12Cl2O2
294.02145
2160*
5359-38-6
PS
LM/Q
Insecticide

o,p'-DDD-M (HOOC-) ME / Mitotane-M (HOOC-) ME — 1889
165, 199, 235, 259, M+ 294
C15H12Cl2O2
294.02145
2530*
P U
LM/Q
Insecticide
Antineoplastic

5-Bromo-AMT AC — 9834
129, 156, 208, 235, M+ 294
C13H15N2OBr
294.03677
2475
PS
LM/Q
Designer drug

294

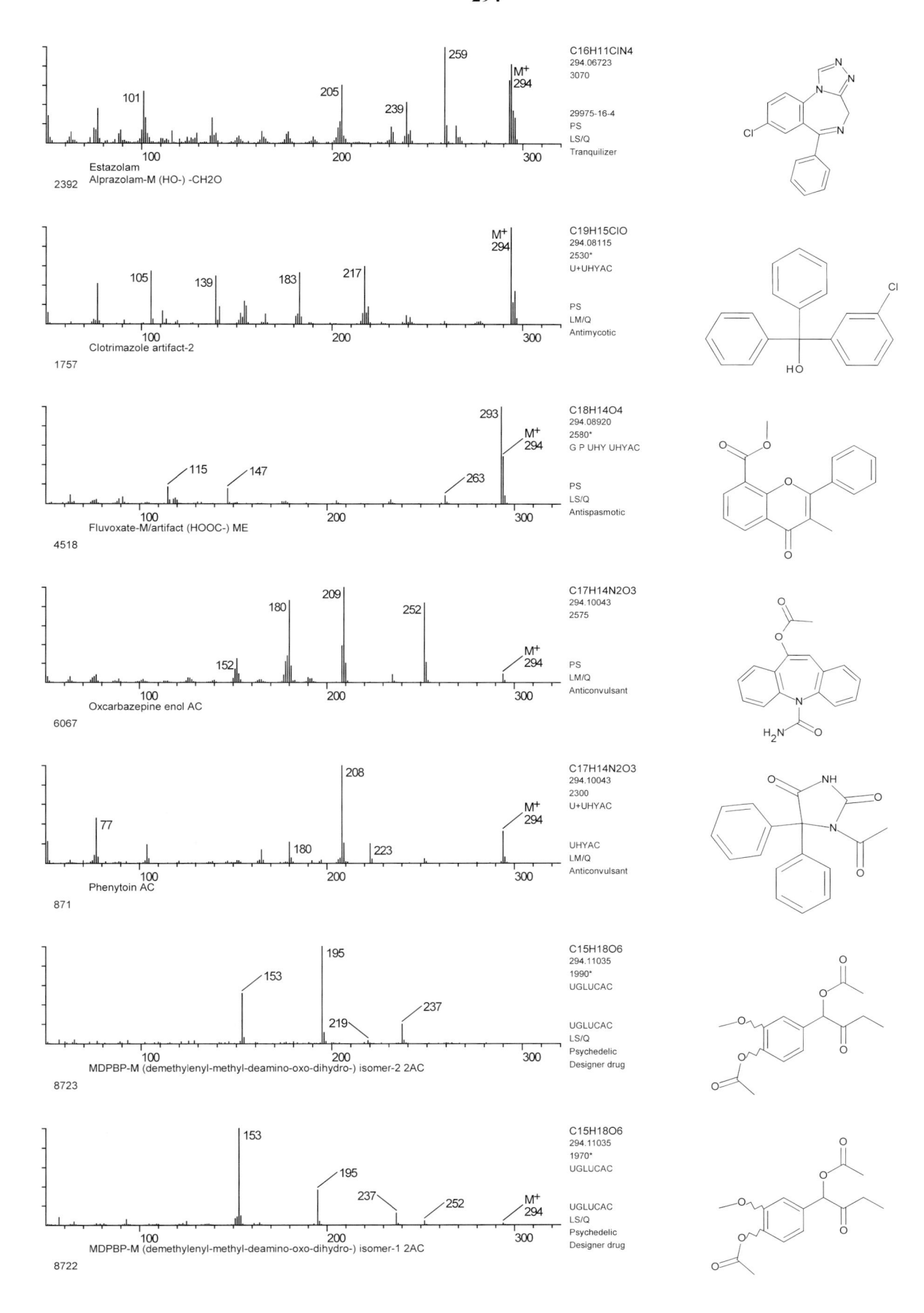

C16H11ClN4
294.06723
3070

29975-16-4
PS
LS/Q
Tranquilizer

Estazolam
Alprazolam-M (HO-) -CH2O
2392

C19H15ClO
294.08115
2530*
U+UHYAC

PS
LM/Q
Antimycotic

Clotrimazole artifact-2
1757

C18H14O4
294.08920
2580*
G P UHY UHYAC

PS
LS/Q
Antispasmotic

Fluvoxate-M/artifact (HOOC-) ME
4518

C17H14N2O3
294.10043
2575

PS
LM/Q
Anticonvulsant

Oxcarbazepine enol AC
6067

C17H14N2O3
294.10043
2300
U+UHYAC

UHYAC
LM/Q
Anticonvulsant

Phenytoin AC
871

C15H18O6
294.11035
1990*
UGLUCAC

UGLUCAC
LS/Q
Psychedelic
Designer drug

MDPBP-M (demethylenyl-methyl-deamino-oxo-dihydro-) isomer-2 2AC
8723

C15H18O6
294.11035
1970*
UGLUCAC

UGLUCAC
LS/Q
Psychedelic
Designer drug

MDPBP-M (demethylenyl-methyl-deamino-oxo-dihydro-) isomer-1 2AC
8722

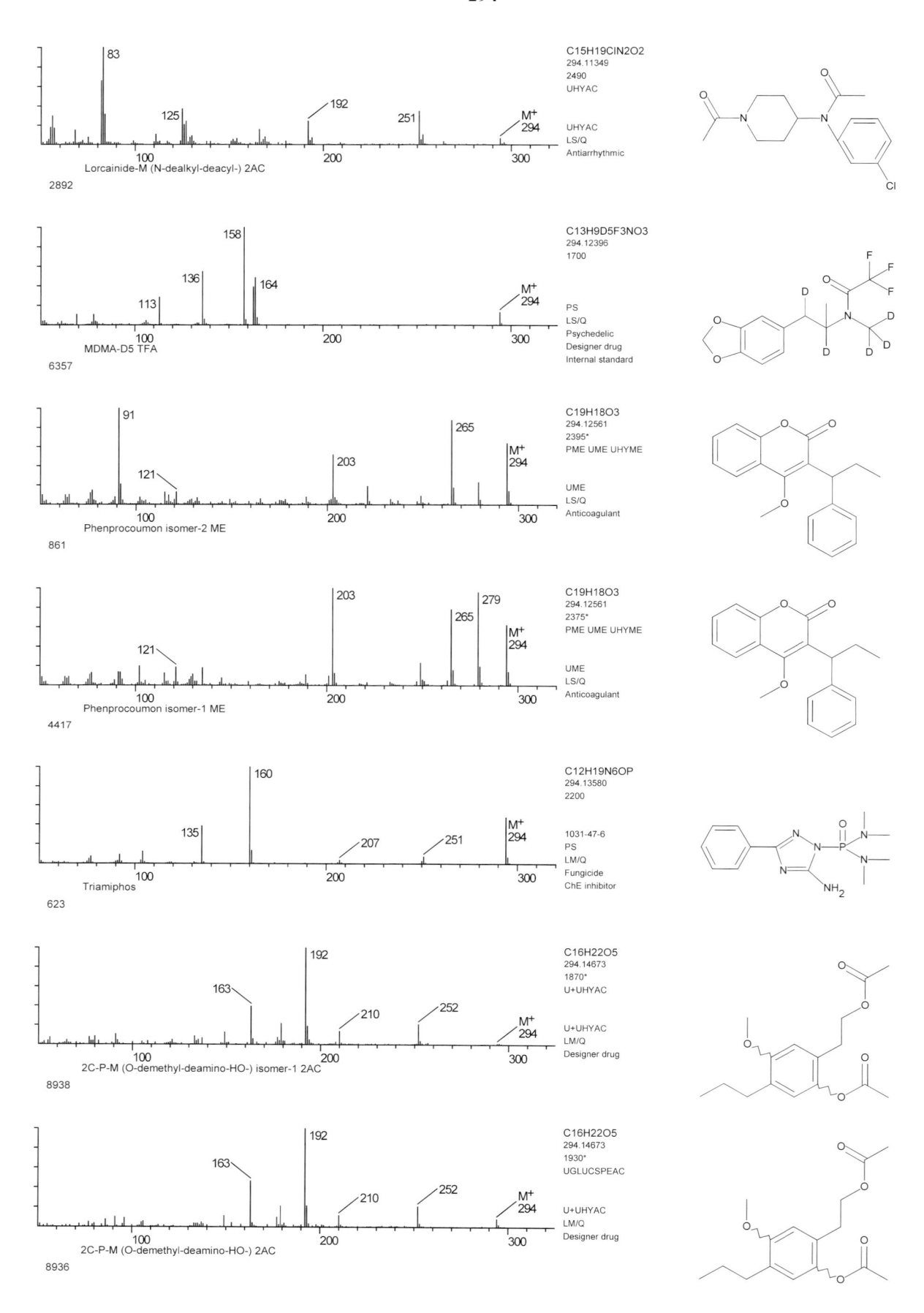

Lorcainide-M (N-dealkyl-deacyl-) 2AC 2892	C15H19ClN2O2 294.11349 2490 UHYAC UHYAC LS/Q Antiarrhythmic
MDMA-D5 TFA 6357	C13H9D5F3NO3 294.12396 1700 PS LS/Q Psychedelic Designer drug Internal standard
Phenprocoumon isomer-2 ME 861	C19H18O3 294.12561 2395* PME UME UHYME UME LS/Q Anticoagulant
Phenprocoumon isomer-1 ME 4417	C19H18O3 294.12561 2375* PME UME UHYME UME LS/Q Anticoagulant
Triamiphos 623	C12H19N6OP 294.13580 2200 1031-47-6 PS LM/Q Fungicide ChE inhibitor
2C-P-M (O-demethyl-deamino-HO-) isomer-1 2AC 8938	C16H22O5 294.14673 1870* U+UHYAC U+UHYAC LM/Q Designer drug
2C-P-M (O-demethyl-deamino-HO-) 2AC 8936	C16H22O5 294.14673 1930* UGLUCSPEAC U+UHYAC LM/Q Designer drug

294

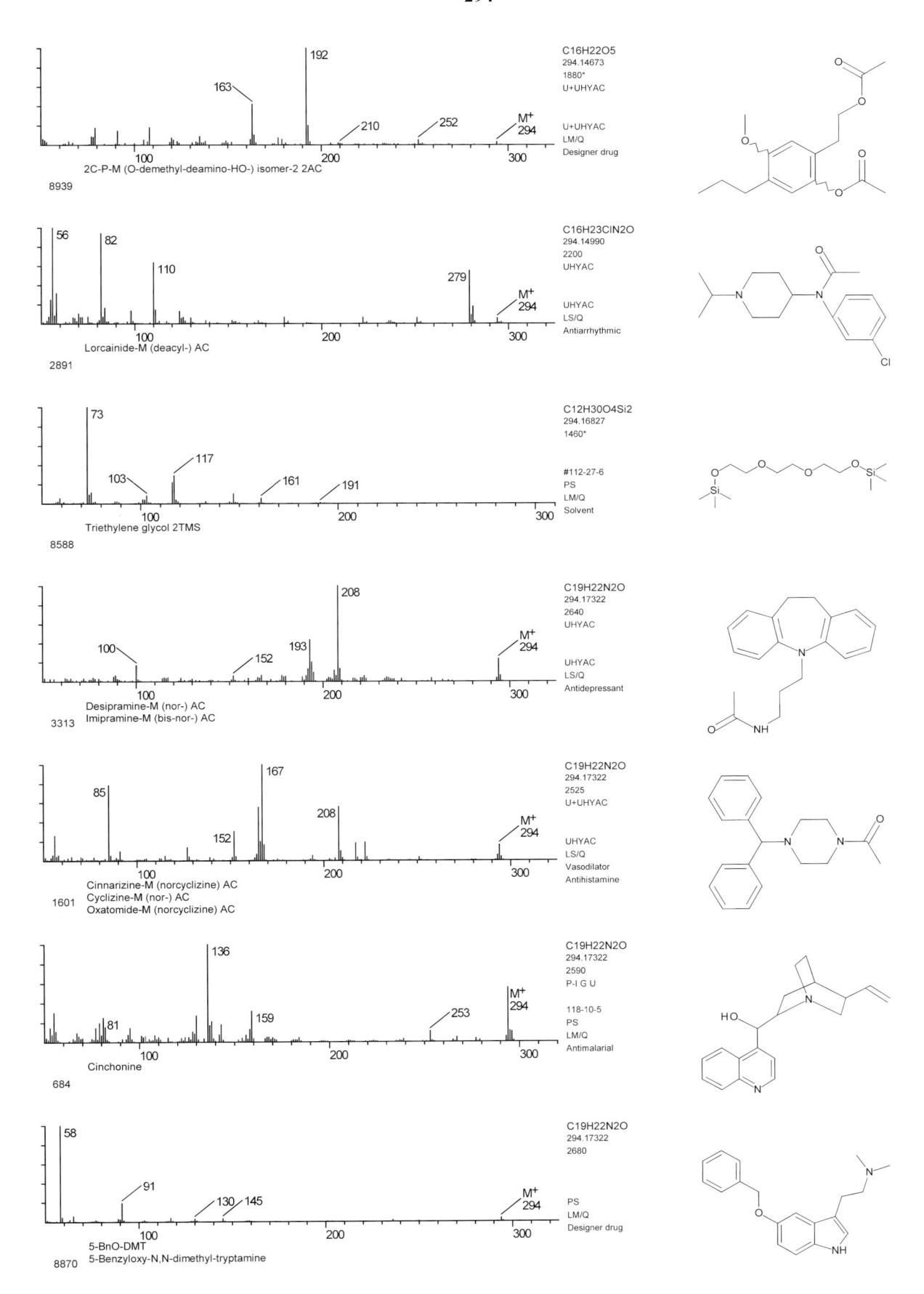

192, 163, 210, 252, M+ 294 2C-P-M (O-demethyl-deamino-HO-) isomer-2 2AC 8939	C16H22O5 294.14673 1880* U+UHYAC U+UHYAC LM/Q Designer drug
56, 82, 110, 279, M+ 294 Lorcainide-M (deacyl-) AC 2891	C16H23ClN2O 294.14990 2200 UHYAC UHYAC LS/Q Antiarrhythmic
73, 103, 117, 161, 191 Triethylene glycol 2TMS 8588	C12H30O4Si2 294.16827 1460* #112-27-6 PS LM/Q Solvent
100, 152, 193, 208, M+ 294 Desipramine-M (nor-) AC Imipramine-M (bis-nor-) AC 3313	C19H22N2O 294.17322 2640 UHYAC UHYAC LS/Q Antidepressant
85, 152, 167, 208, M+ 294 Cinnarizine-M (norcyclizine) AC Cyclizine-M (nor-) AC Oxatomide-M (norcyclizine) AC 1601	C19H22N2O 294.17322 2525 U+UHYAC UHYAC LS/Q Vasodilator Antihistamine
81, 136, 159, 253, M+ 294 Cinchonine 684	C19H22N2O 294.17322 2590 P-I G U 118-10-5 PS LM/Q Antimalarial
58, 91, 130, 145, M+ 294 5-BnO-DMT 5-Benzyloxy-N,N-dimethyl-tryptamine 8870	C19H22N2O 294.17322 2680 PS LM/Q Designer drug

294

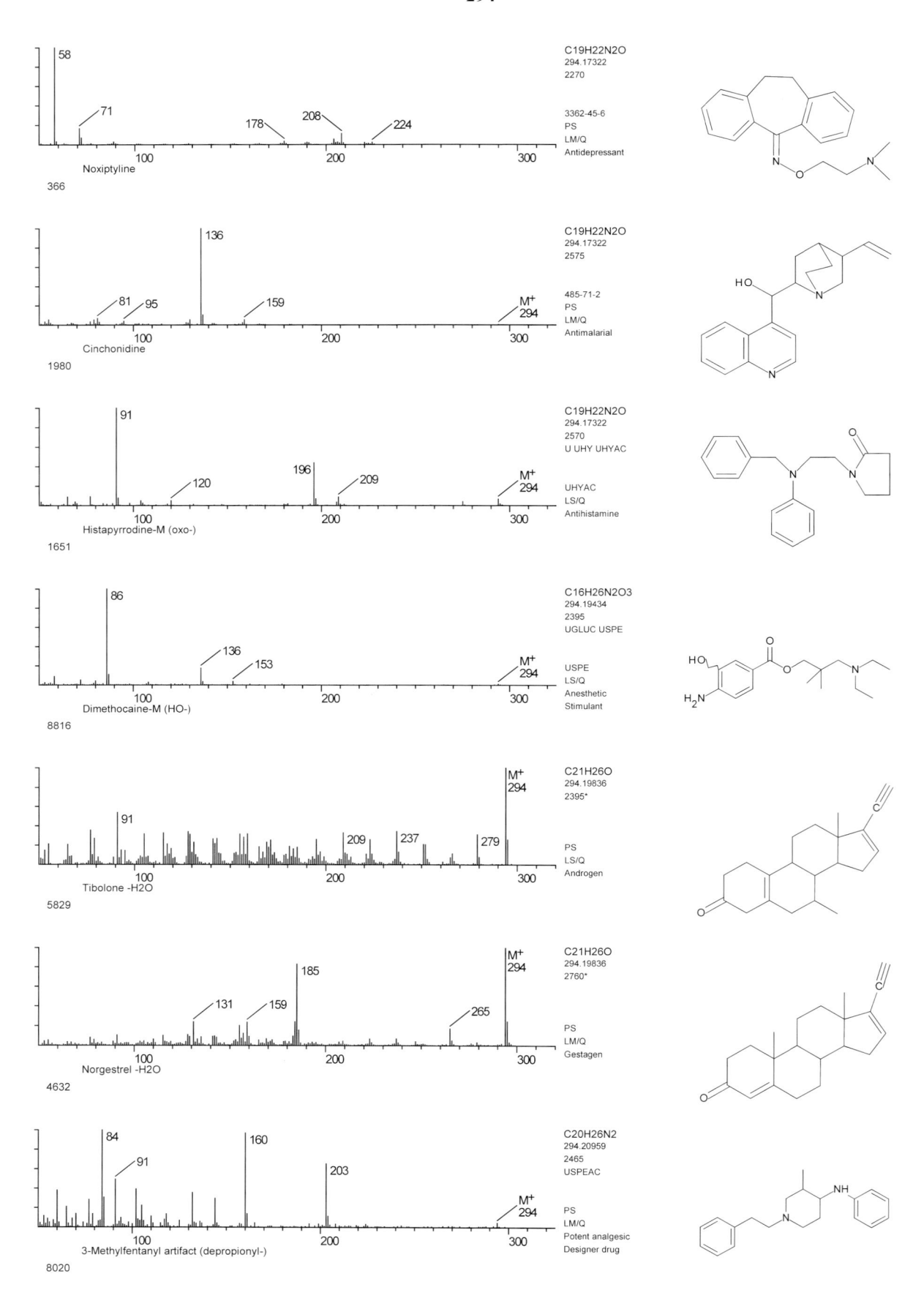

Noxiptyline — peaks: 58, 71, 178, 208, 224	C19H22N2O 294.17322 2270 3362-45-6 PS LM/Q Antidepressant
366	
Cinchonidine — peaks: 81, 95, 136, 159, M+ 294	C19H22N2O 294.17322 2575 485-71-2 PS LM/Q Antimalarial
1980	
Histapyrrodine-M (oxo-) — peaks: 91, 120, 196, 209, M+ 294	C19H22N2O 294.17322 2570 U UHY UHYAC UHYAC LS/Q Antihistamine
1651	
Dimethocaine-M (HO-) — peaks: 86, 136, 153, M+ 294	C16H26N2O3 294.19434 2395 UGLUC USPE USPE LS/Q Anesthetic Stimulant
8816	
Tibolone -H2O — peaks: 91, 209, 237, 279, M+ 294	C21H26O 294.19836 2395* PS LS/Q Androgen
5829	
Norgestrel -H2O — peaks: 131, 159, 185, 265, M+ 294	C21H26O 294.19836 2760* PS LM/Q Gestagen
4632	
3-Methylfentanyl artifact (depropionyl-) — peaks: 84, 91, 160, 203, M+ 294	C20H26N2 294.20959 2465 USPEAC PS LM/Q Potent analgesic Designer drug
8020	

294

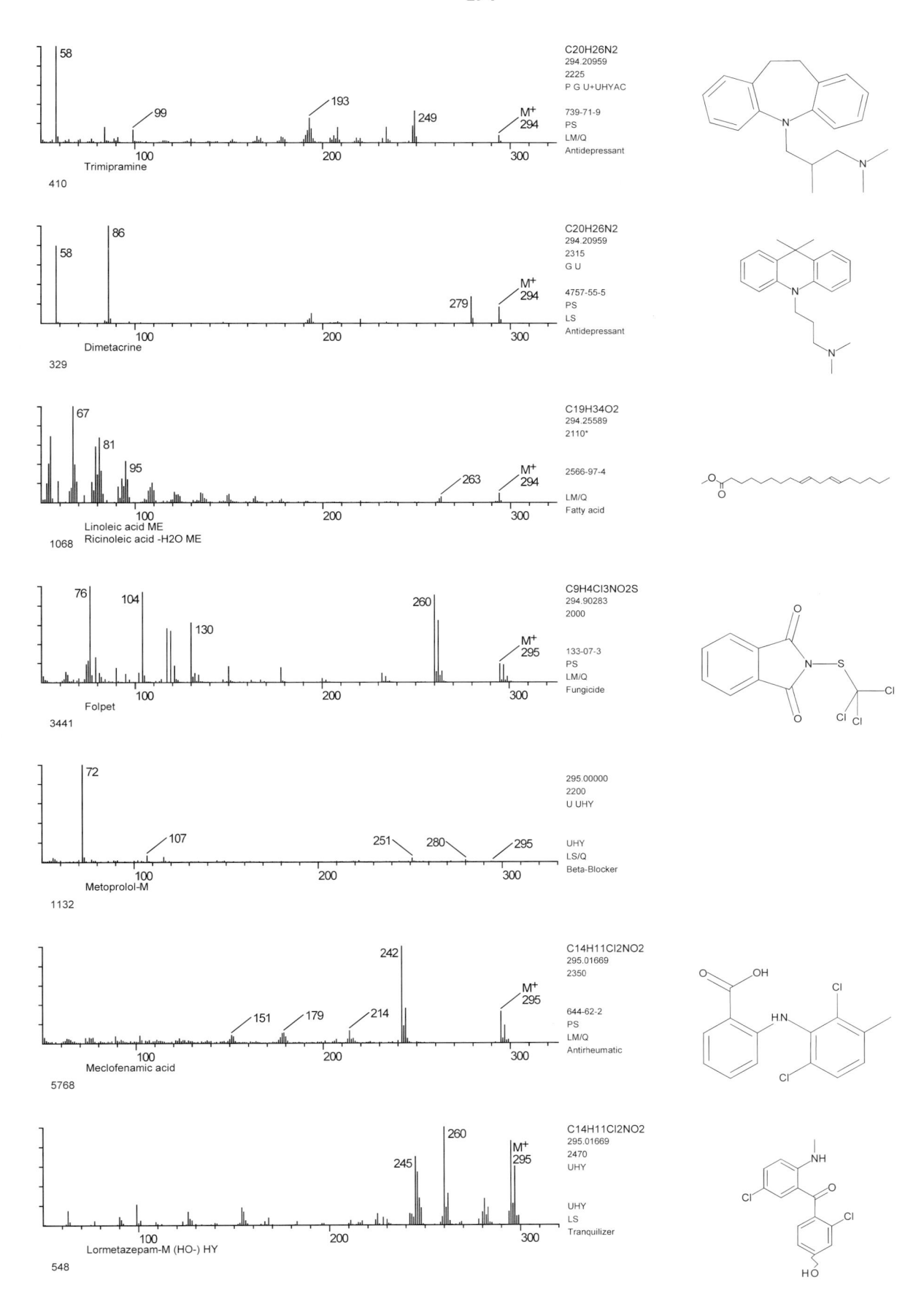

58, 99, 193, 249, M+ 294 — Trimipramine (410)	C20H26N2 294.20959 2225 P G U+UHYAC 739-71-9 PS LM/Q Antidepressant	
58, 86, 279, M+ 294 — Dimetacrine (329)	C20H26N2 294.20959 2315 G U 4757-55-5 PS LS Antidepressant	
67, 81, 95, 263, M+ 294 — Linoleic acid ME / Ricinoleic acid -H2O ME (1068)	C19H34O2 294.25589 2110* 2566-97-4 LM/Q Fatty acid	
76, 104, 130, 260, M+ 295 — Folpet (3441)	C9H4Cl3NO2S 294.90283 2000 133-07-3 PS LM/Q Fungicide	
72, 107, 251, 280, 295 — Metoprolol-M (1132)	295.00000 2200 U UHY UHY LS/Q Beta-Blocker	
151, 179, 214, 242, M+ 295 — Meclofenamic acid (5768)	C14H11Cl2NO2 295.01669 2350 644-62-2 PS LM/Q Antirheumatic	
245, 260, M+ 295 — Lormetazepam-M (HO-) HY (548)	C14H11Cl2NO2 295.01669 2470 UHY UHY LS Tranquilizer	

295

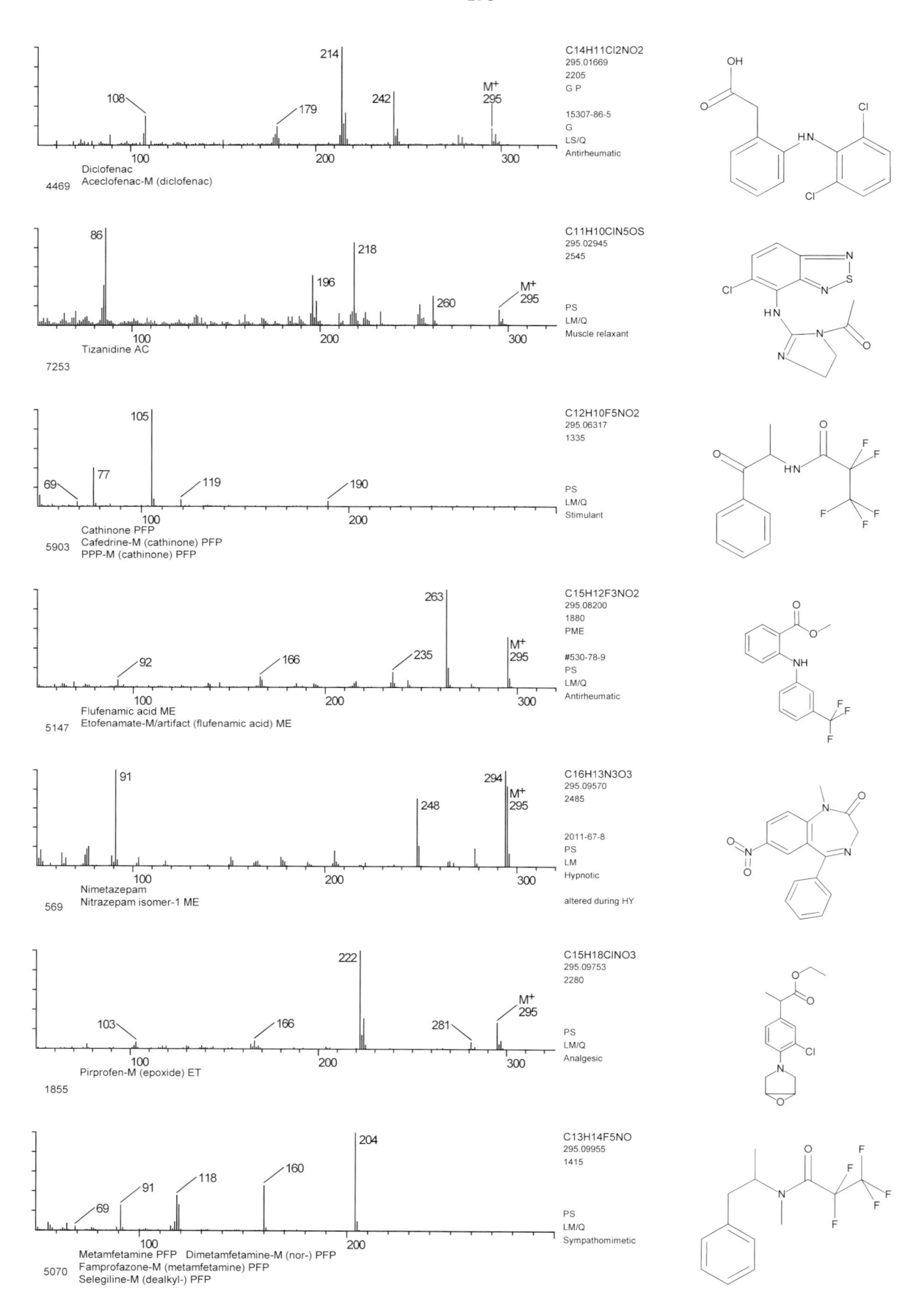

4469	Diclofenac Aceclofenac-M (diclofenac) peaks: 108, 179, 214, 242, M+ 295	C14H11Cl2NO2 295.01669 2205 G P 15307-86-5 G LS/Q Antirheumatic
7253	Tizanidine AC peaks: 86, 196, 218, 260, M+ 295	C11H10ClN5OS 295.02945 2545 PS LM/Q Muscle relaxant
5903	Cathinone PFP Cafedrine-M (cathinone) PFP PPP-M (cathinone) PFP peaks: 69, 77, 105, 119, 190	C12H10F5NO2 295.06317 1335 PS LM/Q Stimulant
5147	Flufenamic acid ME Etofenamate-M/artifact (flufenamic acid) ME peaks: 92, 166, 235, 263, M+ 295	C15H12F3NO2 295.08200 1880 PME #530-78-9 PS LM/Q Antirheumatic
569	Nimetazepam Nitrazepam isomer-1 ME peaks: 91, 248, 294, M+ 295	C16H13N3O3 295.09570 2485 2011-67-8 PS LM Hypnotic altered during HY
1855	Pirprofen-M (epoxide) ET peaks: 103, 166, 222, 281, M+ 295	C15H18ClNO3 295.09753 2280 PS LM/Q Analgesic
5070	Metamfetamine PFP Dimetamfetamine-M (nor-) PFP Famprofazone-M (metamfetamine) PFP Selegiline-M (dealkyl-) PFP peaks: 69, 91, 118, 160, 204	C13H14F5NO 295.09955 1415 PS LM/Q Sympathomimetic

295

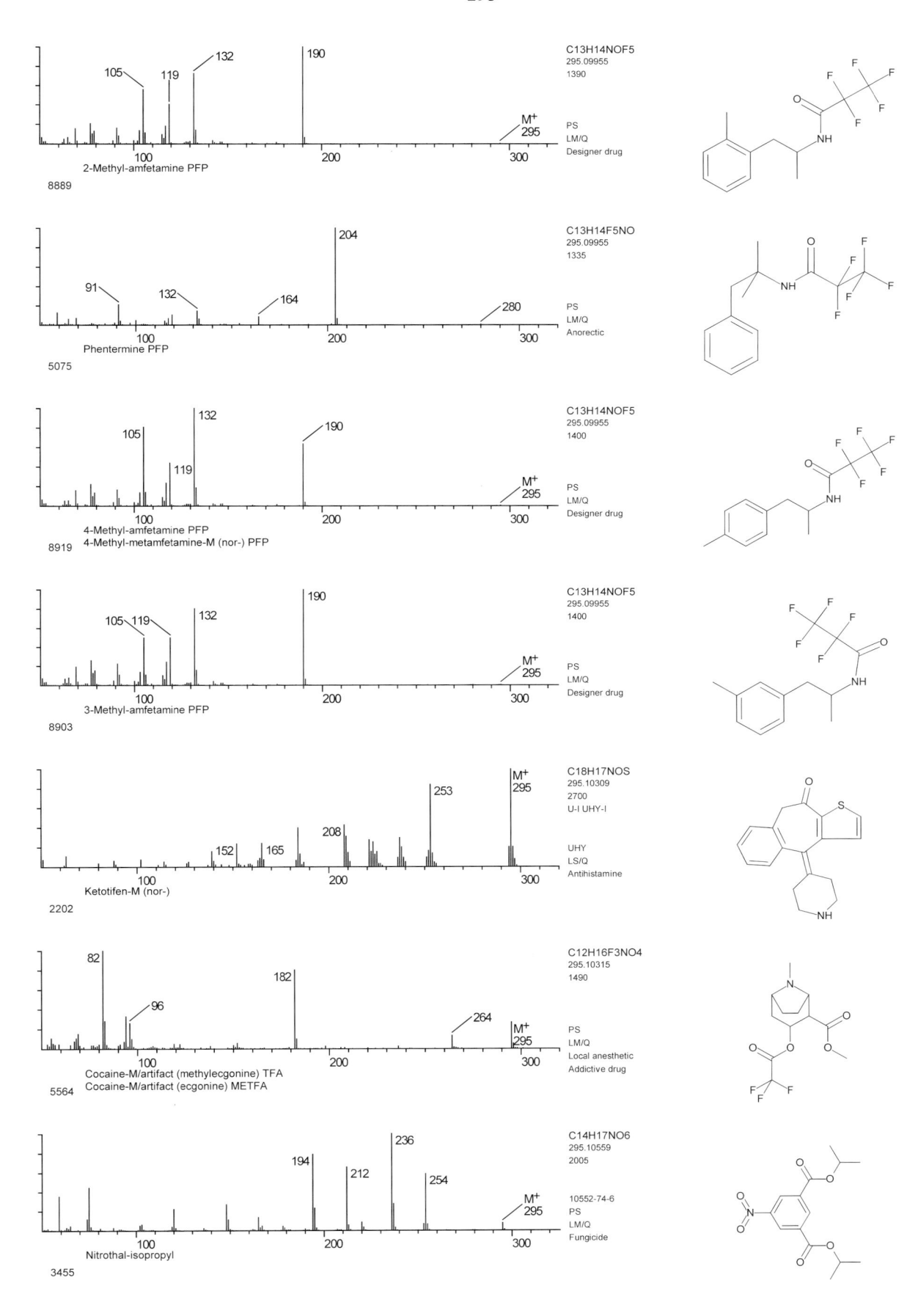

Spectrum: peaks 105, 119, 132, 190, M+ 295	C13H14NOF5 295.09955 1390 PS LM/Q Designer drug	2-Methyl-amfetamine PFP
8889		
Spectrum: peaks 91, 132, 164, 204, 280	C13H14F5NO 295.09955 1335 PS LM/Q Anorectic	Phentermine PFP
5075		
Spectrum: peaks 105, 119, 132, 190, M+ 295	C13H14NOF5 295.09955 1400 PS LM/Q Designer drug	4-Methyl-amfetamine PFP / 4-Methyl-metamfetamine-M (nor-) PFP
8919		
Spectrum: peaks 105, 119, 132, 190, M+ 295	C13H14NOF5 295.09955 1400 PS LM/Q Designer drug	3-Methyl-amfetamine PFP
8903		
Spectrum: peaks 152, 165, 208, 253, M+ 295	C18H17NOS 295.10309 2700 U-I UHY-I UHY LS/Q Antihistamine	Ketotifen-M (nor-)
2202		
Spectrum: peaks 82, 96, 182, 264, M+ 295	C12H16F3NO4 295.10315 1490 PS LM/Q Local anesthetic Addictive drug	Cocaine-M/artifact (methylecgonine) TFA / Cocaine-M/artifact (ecgonine) METFA
5564		
Spectrum: peaks 194, 212, 236, 254, M+ 295	C14H17NO6 295.10559 2005 10552-74-6 PS LM/Q Fungicide	Nitrothal-isopropyl
3455		

295

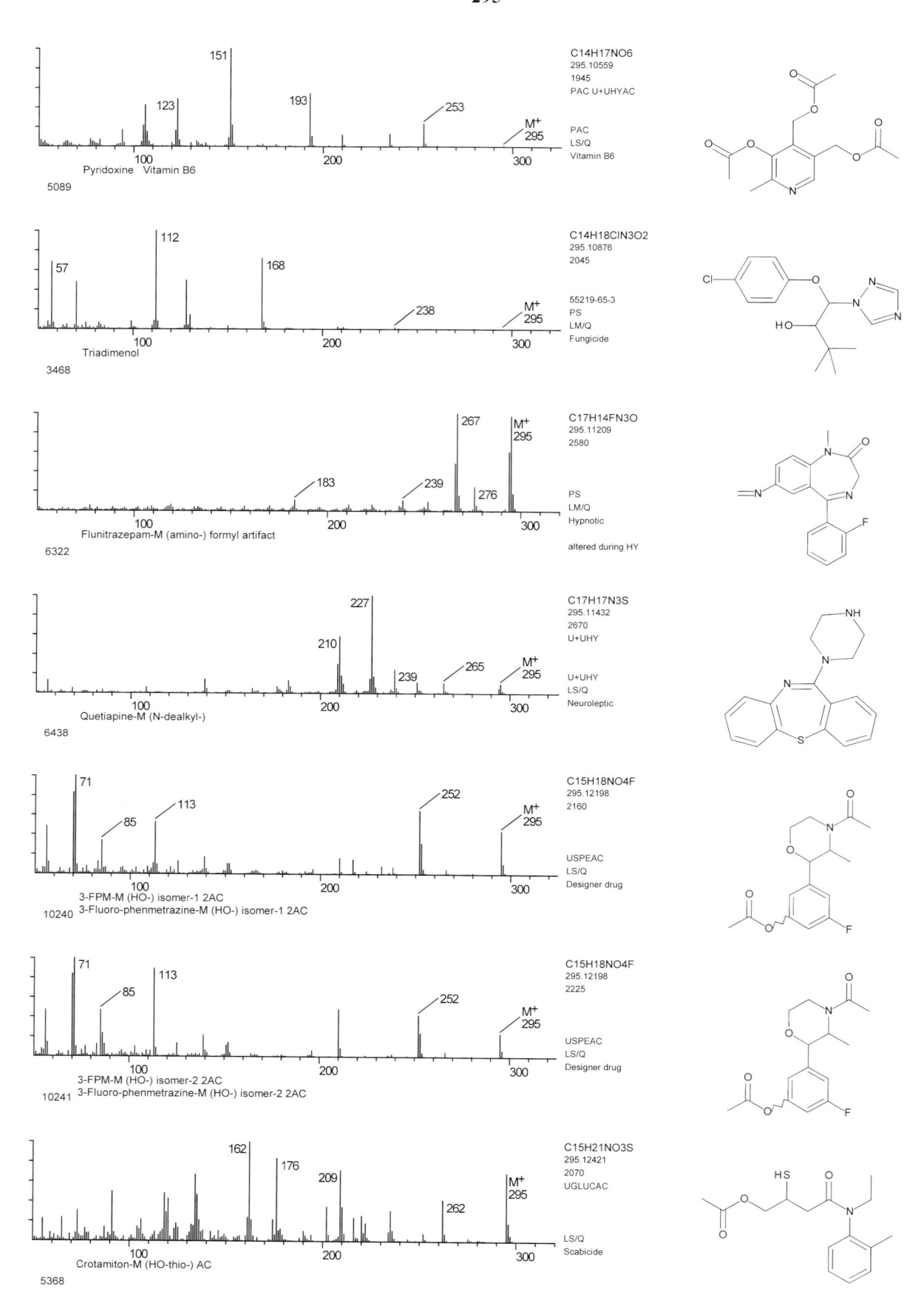

C14H17NO6
295.10559
1945
PAC U+UHYAC

PAC
LS/Q
Vitamin B6

Pyridoxine Vitamin B6
5089

C14H18ClN3O2
295.10876
2045

55219-65-3
PS
LM/Q
Fungicide

Triadimenol
3468

C17H14FN3O
295.11209
2580

PS
LM/Q
Hypnotic

altered during HY

Flunitrazepam-M (amino-) formyl artifact
6322

C17H17N3S
295.11432
2670
U+UHY

U+UHY
LS/Q
Neuroleptic

Quetiapine-M (N-dealkyl-)
6438

C15H18NO4F
295.12198
2160

USPEAC
LS/Q
Designer drug

3-FPM-M (HO-) isomer-1 2AC
3-Fluoro-phenmetrazine-M (HO-) isomer-1 2AC
10240

C15H18NO4F
295.12198
2225

USPEAC
LS/Q
Designer drug

3-FPM-M (HO-) isomer-2 2AC
3-Fluoro-phenmetrazine-M (HO-) isomer-2 2AC
10241

C15H21NO3S
295.12421
2070
UGLUCAC

LS/Q
Scabicide

Crotamiton-M (HO-thio-) AC
5368

876

295

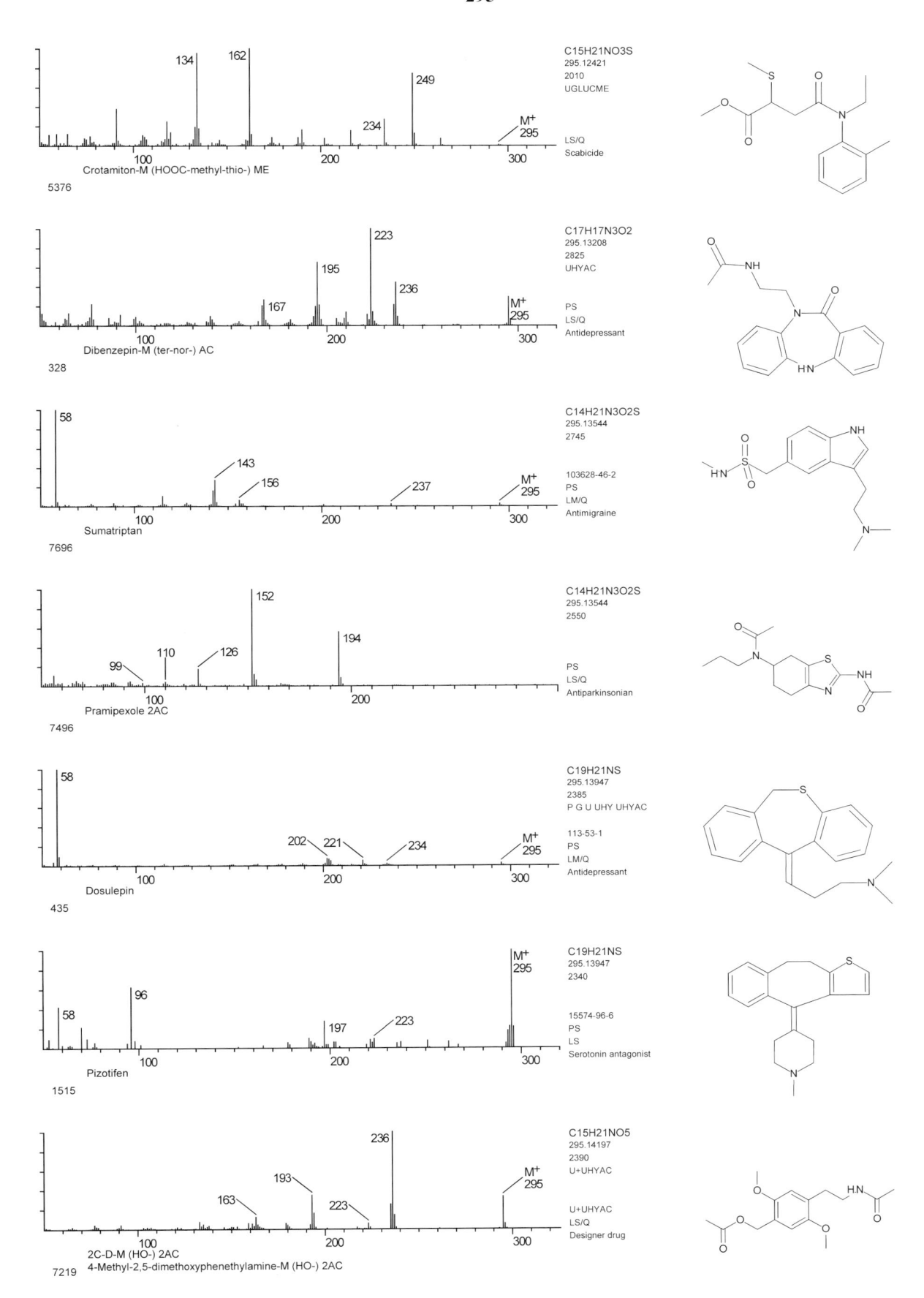

C15H21NO3S
295.12421
2010
UGLUCME

LS/Q
Scabicide

Crotamiton-M (HOOC-methyl-thio-) ME
5376

C17H17N3O2
295.13208
2825
UHYAC

PS
LS/Q
Antidepressant

Dibenzepin-M (ter-nor-) AC
328

C14H21N3O2S
295.13544
2745

103628-46-2
PS
LM/Q
Antimigraine

Sumatriptan
7696

C14H21N3O2S
295.13544
2550

PS
LS/Q
Antiparkinsonian

Pramipexole 2AC
7496

C19H21NS
295.13947
2385
P G U UHY UHYAC

113-53-1
PS
LM/Q
Antidepressant

Dosulepin
435

C19H21NS
295.13947
2340

15574-96-6
PS
LS
Serotonin antagonist

Pizotifen
1515

C15H21NO5
295.14197
2390
U+UHYAC

U+UHYAC
LS/Q
Designer drug

2C-D-M (HO-) 2AC
4-Methyl-2,5-dimethoxyphenethylamine-M (HO-) 2AC
7219

295

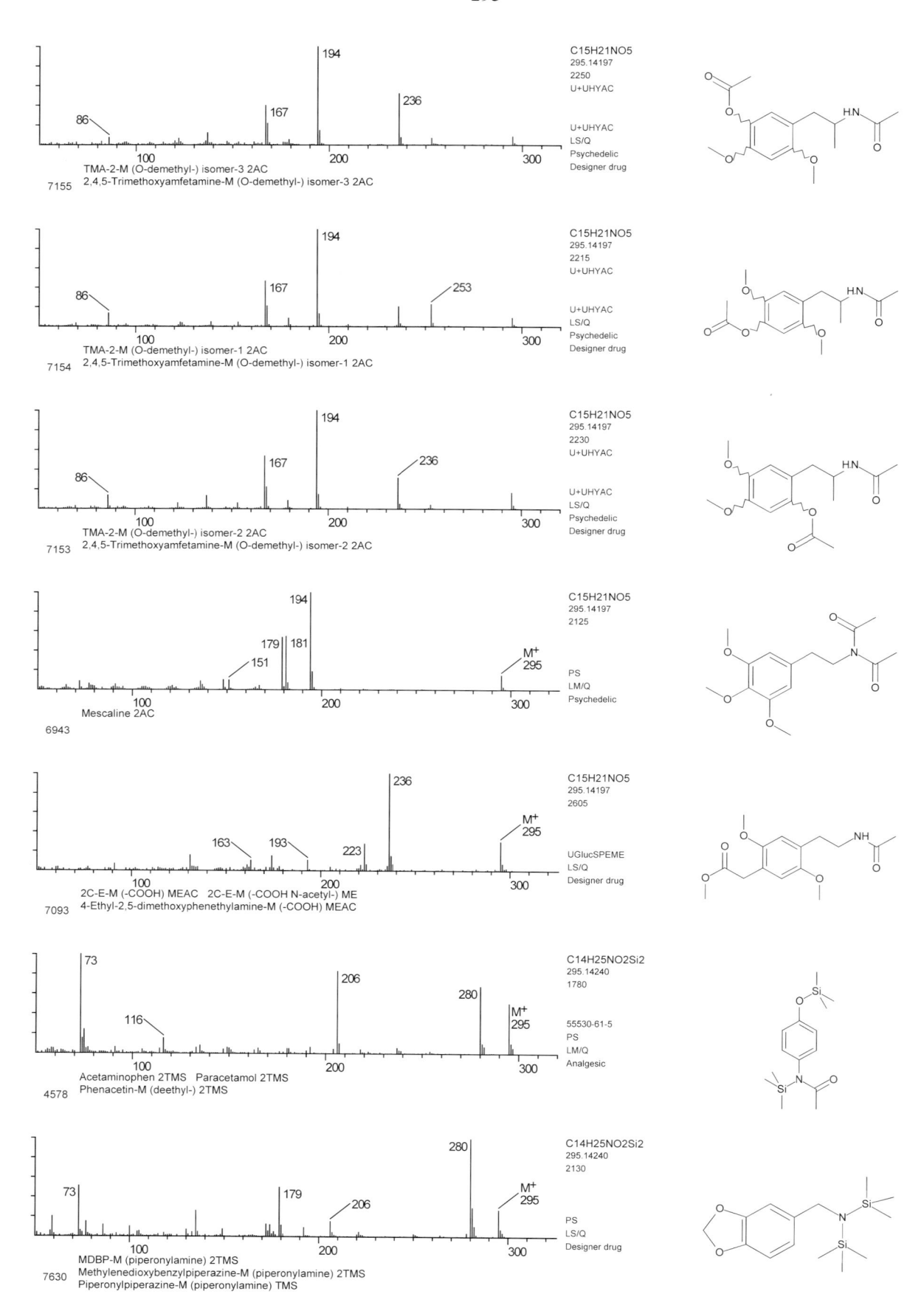

7155 TMA-2-M (O-demethyl-) isomer-3 2AC
2,4,5-Trimethoxyamfetamine-M (O-demethyl-) isomer-3 2AC
Peaks: 86, 167, 194, 236
C15H21NO5
295.14197
2250
U+UHYAC
U+UHYAC
LS/Q
Psychedelic
Designer drug

7154 TMA-2-M (O-demethyl-) isomer-1 2AC
2,4,5-Trimethoxyamfetamine-M (O-demethyl-) isomer-1 2AC
Peaks: 86, 167, 194, 253
C15H21NO5
295.14197
2215
U+UHYAC
U+UHYAC
LS/Q
Psychedelic
Designer drug

7153 TMA-2-M (O-demethyl-) isomer-2 2AC
2,4,5-Trimethoxyamfetamine-M (O-demethyl-) isomer-2 2AC
Peaks: 86, 167, 194, 236
C15H21NO5
295.14197
2230
U+UHYAC
U+UHYAC
LS/Q
Psychedelic
Designer drug

6943 Mescaline 2AC
Peaks: 151, 179, 181, 194, M+ 295
C15H21NO5
295.14197
2125
PS
LM/Q
Psychedelic

7093 2C-E-M (-COOH) MEAC   2C-E-M (-COOH N-acetyl-) ME
4-Ethyl-2,5-dimethoxyphenethylamine-M (-COOH) MEAC
Peaks: 163, 193, 223, 236, M+ 295
C15H21NO5
295.14197
2605
UGlucSPEME
LS/Q
Designer drug

4578 Acetaminophen 2TMS   Paracetamol 2TMS
Phenacetin-M (deethyl-) 2TMS
Peaks: 73, 116, 206, 280, M+ 295
C14H25NO2Si2
295.14240
1780
55530-61-5
PS
LM/Q
Analgesic

7630 MDBP-M (piperonylamine) 2TMS
Methylenedioxybenzylpiperazine-M (piperonylamine) 2TMS
Piperonylpiperazine-M (piperonylamine) TMS
Peaks: 73, 179, 206, 280, M+ 295
C14H25NO2Si2
295.14240
2130
PS
LS/Q
Designer drug

295

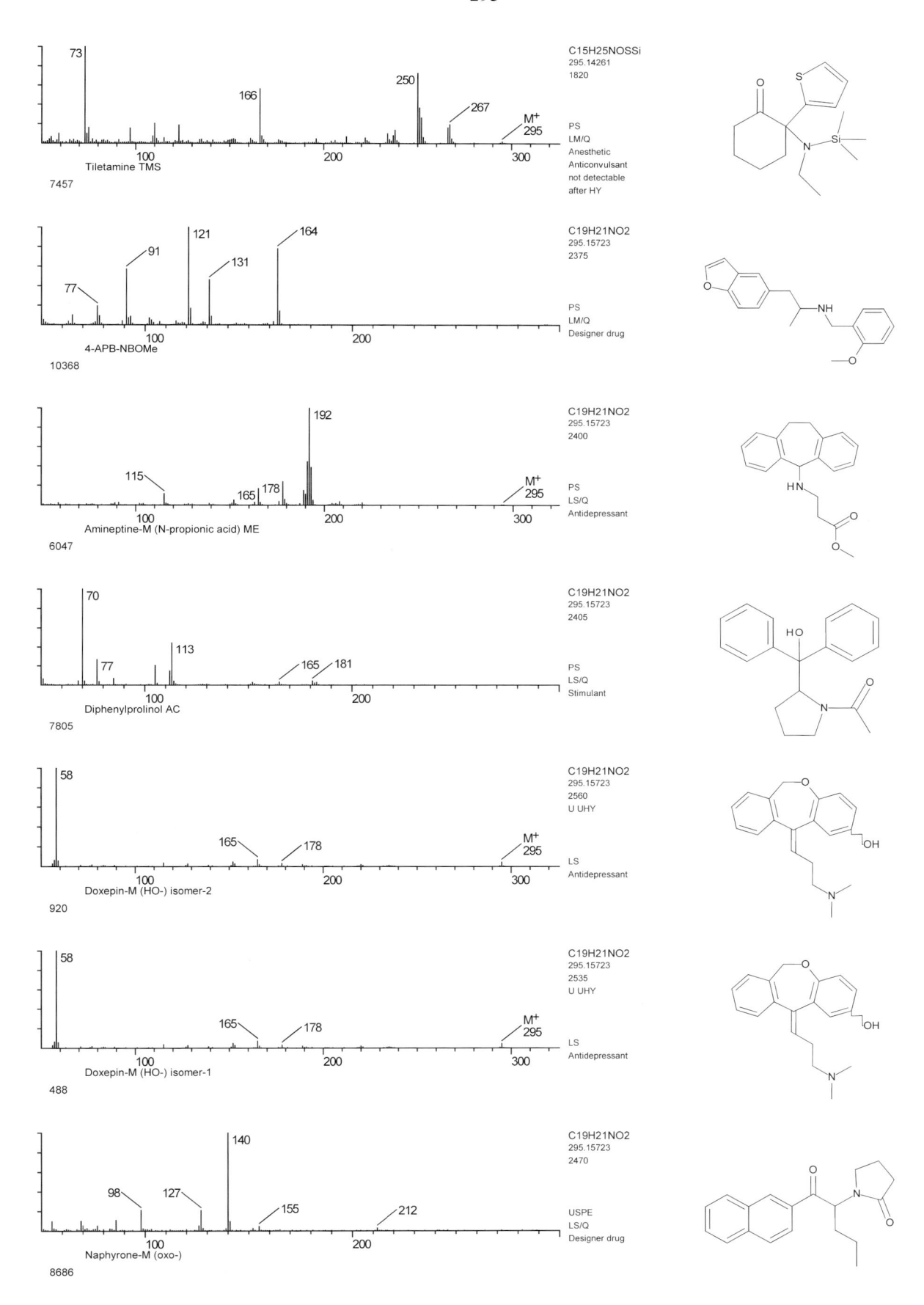

Tiletamine TMS
7457
Peaks: 73, 166, 250, 267, M+ 295
C15H25NOSSi
295.14261
1820
PS
LM/Q
Anesthetic
Anticonvulsant
not detectable
after HY

4-APB-NBOMe
10368
Peaks: 77, 91, 121, 131, 164
C19H21NO2
295.15723
2375
PS
LM/Q
Designer drug

Amineptine-M (N-propionic acid) ME
6047
Peaks: 115, 165, 178, 192, M+ 295
C19H21NO2
295.15723
2400
PS
LS/Q
Antidepressant

Diphenylprolinol AC
7805
Peaks: 70, 77, 113, 165, 181
C19H21NO2
295.15723
2405
PS
LS/Q
Stimulant

Doxepin-M (HO-) isomer-2
920
Peaks: 58, 165, 178, M+ 295
C19H21NO2
295.15723
2560
U UHY
LS
Antidepressant

Doxepin-M (HO-) isomer-1
488
Peaks: 58, 165, 178, M+ 295
C19H21NO2
295.15723
2535
U UHY
LS
Antidepressant

Naphyrone-M (oxo-)
8686
Peaks: 98, 127, 140, 155, 212
C19H21NO2
295.15723
2470
USPE
LS/Q
Designer drug

295

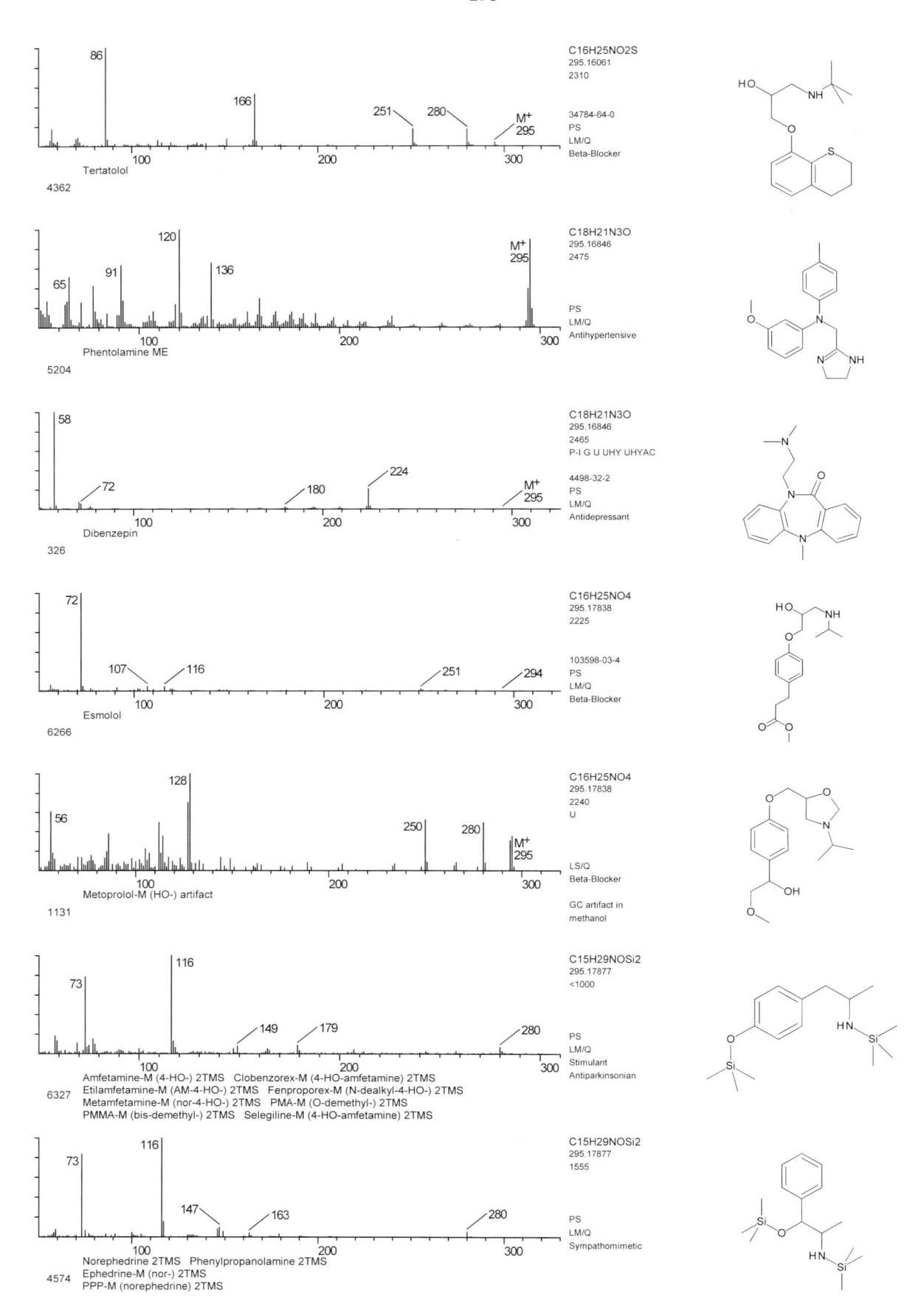

m/z peaks	Compound	Formula	Info
86, 166, 251, 280, 295 (M+)	Tertatolol	C16H25NO2S	295.16061; 2310; 34784-64-0; PS; LM/Q; Beta-Blocker; 4362
65, 91, 120, 136, 295 (M+)	Phentolamine ME	C18H21N3O	295.16846; 2475; PS; LM/Q; Antihypertensive; 5204
58, 72, 180, 224, 295 (M+)	Dibenzepin	C18H21N3O	295.16846; 2465; P-I G U UHY UHYAC; 4498-32-2; PS; LM/Q; Antidepressant; 326
72, 107, 116, 251, 294	Esmolol	C16H25NO4	295.17838; 2225; 103598-03-4; PS; LM/Q; Beta-Blocker; 6266
56, 128, 250, 280, 295 (M+)	Metoprolol-M (HO-) artifact	C16H25NO4	295.17838; 2240; U; LS/Q; Beta-Blocker; GC artifact in methanol; 1131
73, 116, 149, 179, 280	Amfetamine-M (4-HO-) 2TMS Clobenzorex-M (4-HO-amfetamine) 2TMS Etilamfetamine-M (AM-4-HO-) 2TMS Fenproporex-M (N-dealkyl-4-HO-) 2TMS Metamfetamine-M (nor-4-HO-) 2TMS PMA-M (O-demethyl-) 2TMS PMMA-M (bis-demethyl-) 2TMS Selegiline-M (4-HO-amfetamine) 2TMS	C15H29NOSi2	295.17877; <1000; PS; LM/Q; Stimulant Antiparkinsonian; 6327
73, 116, 147, 163, 280	Norephedrine 2TMS Phenylpropanolamine 2TMS Ephedrine-M (nor-) 2TMS PPP-M (norephedrine) 2TMS	C15H29NOSi2	295.17877; 1555; PS; LM/Q; Sympathomimetic; 4574

295

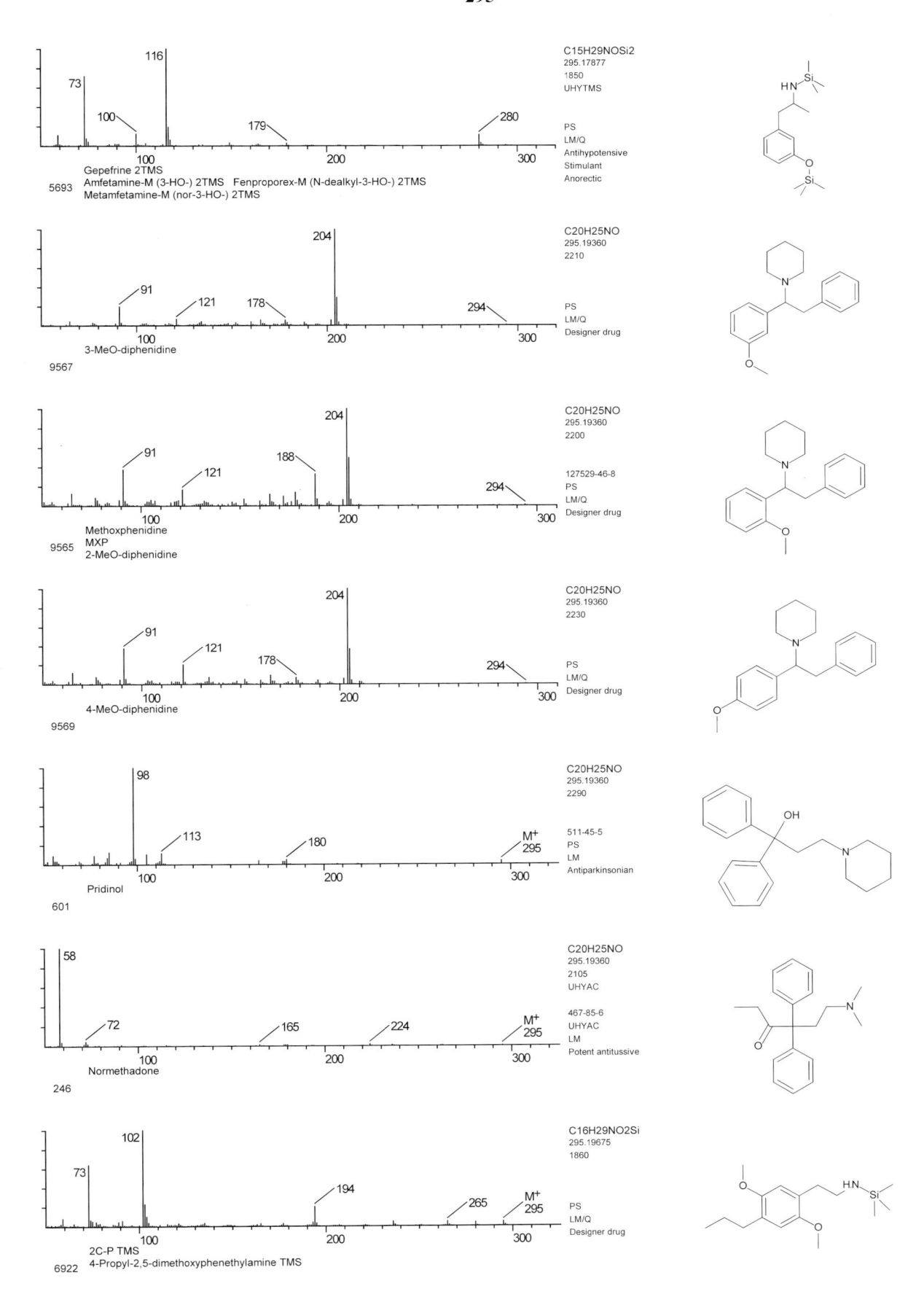

5693
Gepefrine 2TMS
Amfetamine-M (3-HO-) 2TMS   Fenproporex-M (N-dealkyl-3-HO-) 2TMS
Metamfetamine-M (nor-3-HO-) 2TMS

C15H29NOSi2
295.17877
1850
UHYTMS

PS
LM/Q
Antihypotensive
Stimulant
Anorectic

9567
3-MeO-diphenidine

C20H25NO
295.19360
2210

PS
LM/Q
Designer drug

9565
Methoxphenidine
MXP
2-MeO-diphenidine

C20H25NO
295.19360
2200

127529-46-8
PS
LM/Q
Designer drug

9569
4-MeO-diphenidine

C20H25NO
295.19360
2230

PS
LM/Q
Designer drug

601
Pridinol

C20H25NO
295.19360
2290

511-45-5
PS
LM
Antiparkinsonian

246
Normethadone

C20H25NO
295.19360
2105
UHYAC

467-85-6
UHYAC
LM
Potent antitussive

6922
2C-P TMS
4-Propyl-2,5-dimethoxyphenethylamine TMS

C16H29NO2Si
295.19675
1860

PS
LM/Q
Designer drug

296

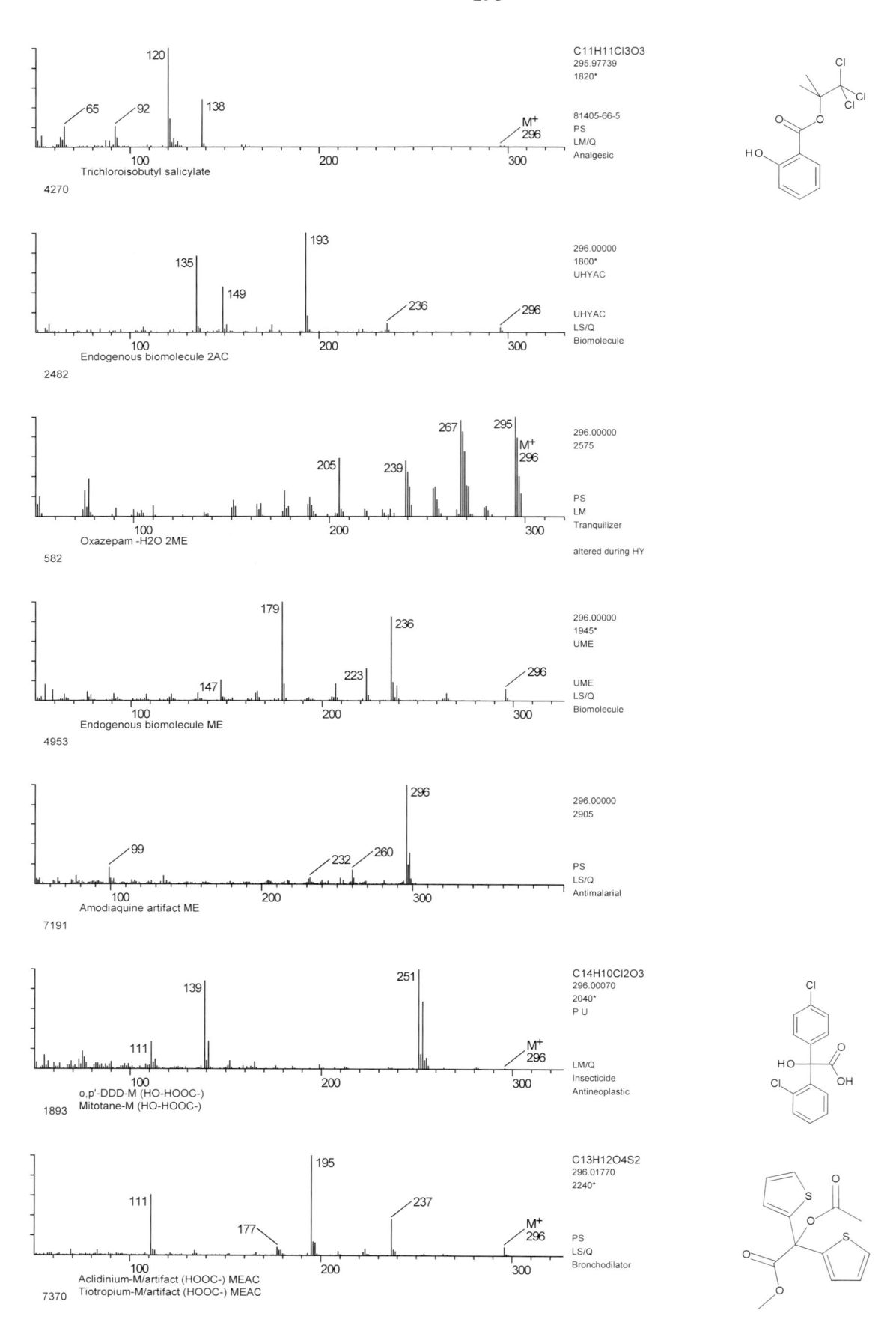

Trichloroisobutyl salicylate — 4270
Peaks: 65, 92, 120, 138, M+ 296
C11H11Cl3O3
295.97739
1820*
81405-66-5
PS
LM/Q
Analgesic

Endogenous biomolecule 2AC — 2482
Peaks: 135, 149, 193, 236, 296
296.00000
1800*
UHYAC
UHYAC
LS/Q
Biomolecule

Oxazepam -H2O 2ME — 582
Peaks: 205, 239, 267, 295, M+ 296
296.00000
2575
PS
LM
Tranquilizer
altered during HY

Endogenous biomolecule ME — 4953
Peaks: 147, 179, 223, 236, 296
296.00000
1945*
UME
UME
LS/Q
Biomolecule

Amodiaquine artifact ME — 7191
Peaks: 99, 232, 260, 296
296.00000
2905
PS
LS/Q
Antimalarial

o,p'-DDD-M (HO-HOOC-)
Mitotane-M (HO-HOOC-) — 1893
Peaks: 111, 139, 251, M+ 296
C14H10Cl2O3
296.00070
2040*
P U
LM/Q
Insecticide
Antineoplastic

Aclidinium-M/artifact (HOOC-) MEAC
Tiotropium-M/artifact (HOOC-) MEAC — 7370
Peaks: 111, 177, 195, 237, M+ 296
C13H12O4S2
296.01770
2240*
PS
LS/Q
Bronchodilator

296

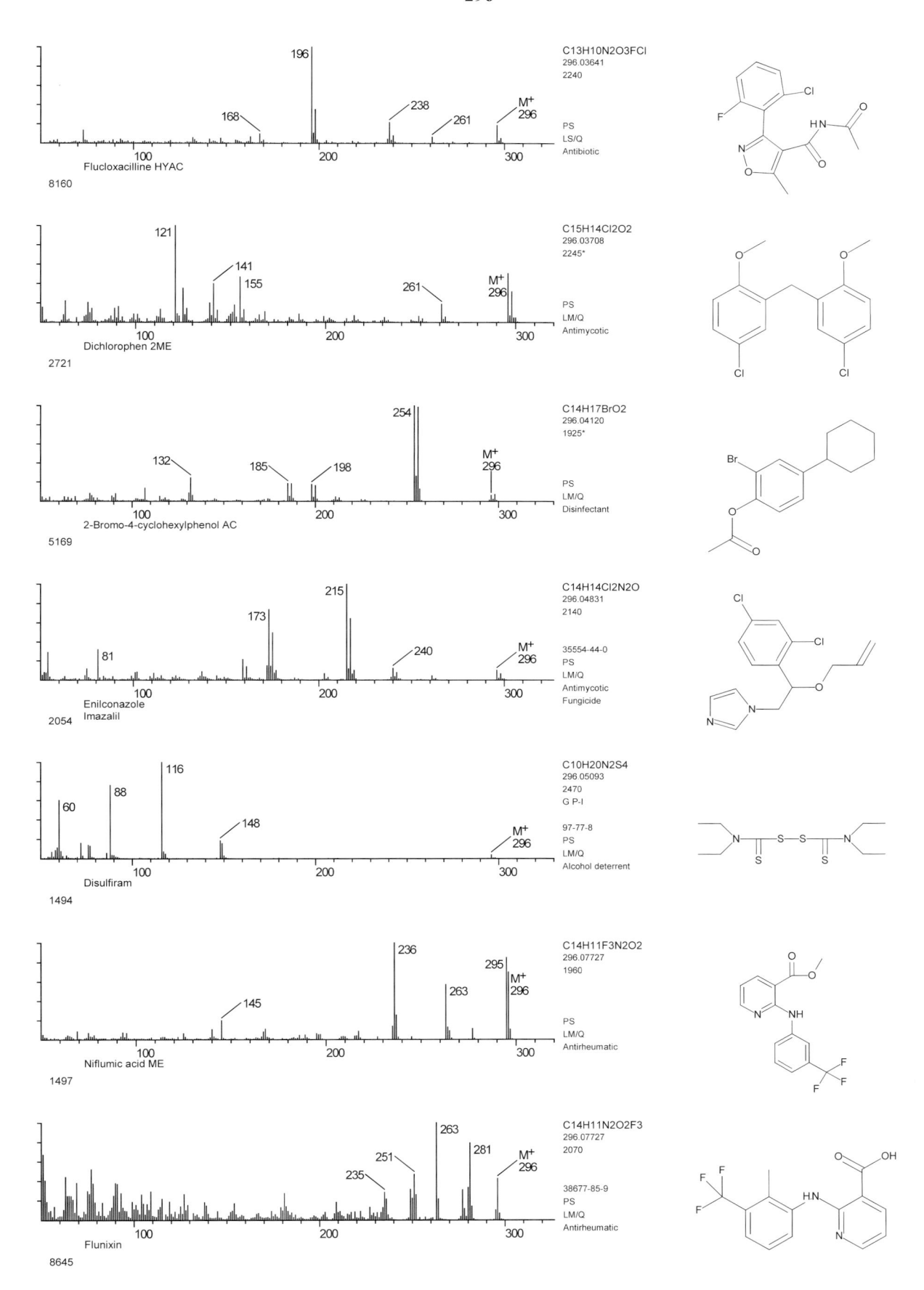

Spectrum peaks	Formula / Info
168, 196, 238, 261, M+ 296	C13H10N2O3FCl / 296.03641 / 2240 / PS / LS/Q / Antibiotic — Flucloxacilline HYAC / 8160
121, 141, 155, 261, M+ 296	C15H14Cl2O2 / 296.03708 / 2245* / PS / LM/Q / Antimycotic — Dichlorophen 2ME / 2721
132, 185, 198, 254, M+ 296	C14H17BrO2 / 296.04120 / 1925* / PS / LM/Q / Disinfectant — 2-Bromo-4-cyclohexylphenol AC / 5169
81, 173, 215, 240, M+ 296	C14H14Cl2N2O / 296.04831 / 2140 / 35554-44-0 / PS / LM/Q / Antimycotic Fungicide — Enilconazole Imazalil / 2054
60, 88, 116, 148, M+ 296	C10H20N2S4 / 296.05093 / 2470 / G P-I / 97-77-8 / PS / LM/Q / Alcohol deterrent — Disulfiram / 1494
145, 236, 263, 295, M+ 296	C14H11F3N2O2 / 296.07727 / 1960 / PS / LM/Q / Antirheumatic — Niflumic acid ME / 1497
235, 251, 263, 281, M+ 296	C14H11N2O2F3 / 296.07727 / 2070 / 38677-85-9 / PS / LM/Q / Antirheumatic — Flunixin / 8645

883

296

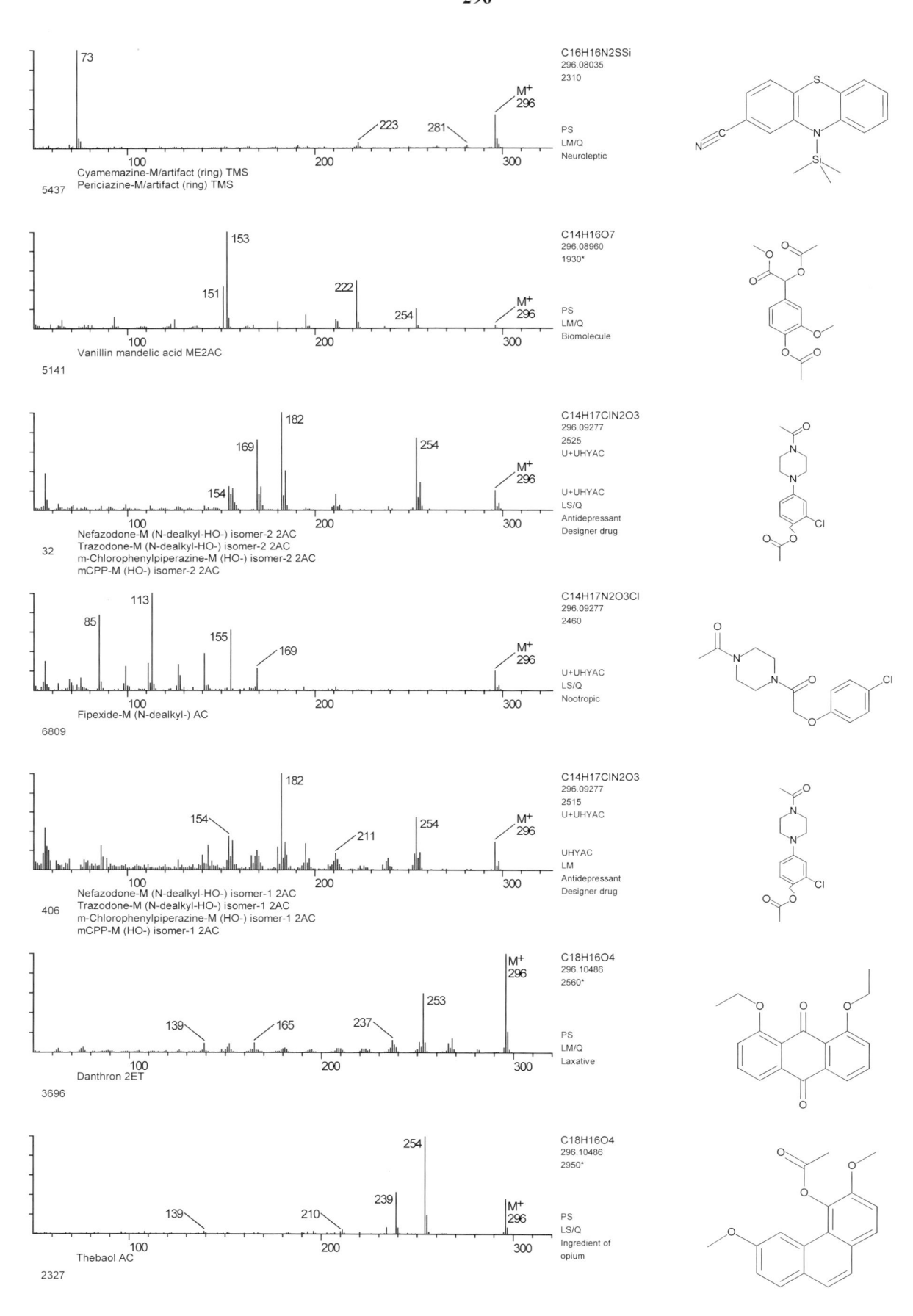

5437	Cyamemazine-M/artifact (ring) TMS Periciazine-M/artifact (ring) TMS C16H16N2SSi 296.08035 2310 PS LM/Q Neuroleptic
5141	Vanillin mandelic acid ME2AC C14H16O7 296.08960 1930* PS LM/Q Biomolecule
32	Nefazodone-M (N-dealkyl-HO-) isomer-2 2AC Trazodone-M (N-dealkyl-HO-) isomer-2 2AC m-Chlorophenylpiperazine-M (HO-) isomer-2 2AC mCPP-M (HO-) isomer-2 2AC C14H17ClN2O3 296.09277 2525 U+UHYAC LS/Q Antidepressant Designer drug
6809	Fipexide-M (N-dealkyl-) AC C14H17N2O3Cl 296.09277 2460 U+UHYAC LS/Q Nootropic
406	Nefazodone-M (N-dealkyl-HO-) isomer-1 2AC Trazodone-M (N-dealkyl-HO-) isomer-1 2AC m-Chlorophenylpiperazine-M (HO-) isomer-1 2AC mCPP-M (HO-) isomer-1 2AC C14H17ClN2O3 296.09277 2515 U+UHYAC UHYAC LM Antidepressant Designer drug
3696	Danthron 2ET C18H16O4 296.10486 2560* PS LM/Q Laxative
2327	Thebaol AC C18H16O4 296.10486 2950* PS LS/Q Ingredient of opium

296

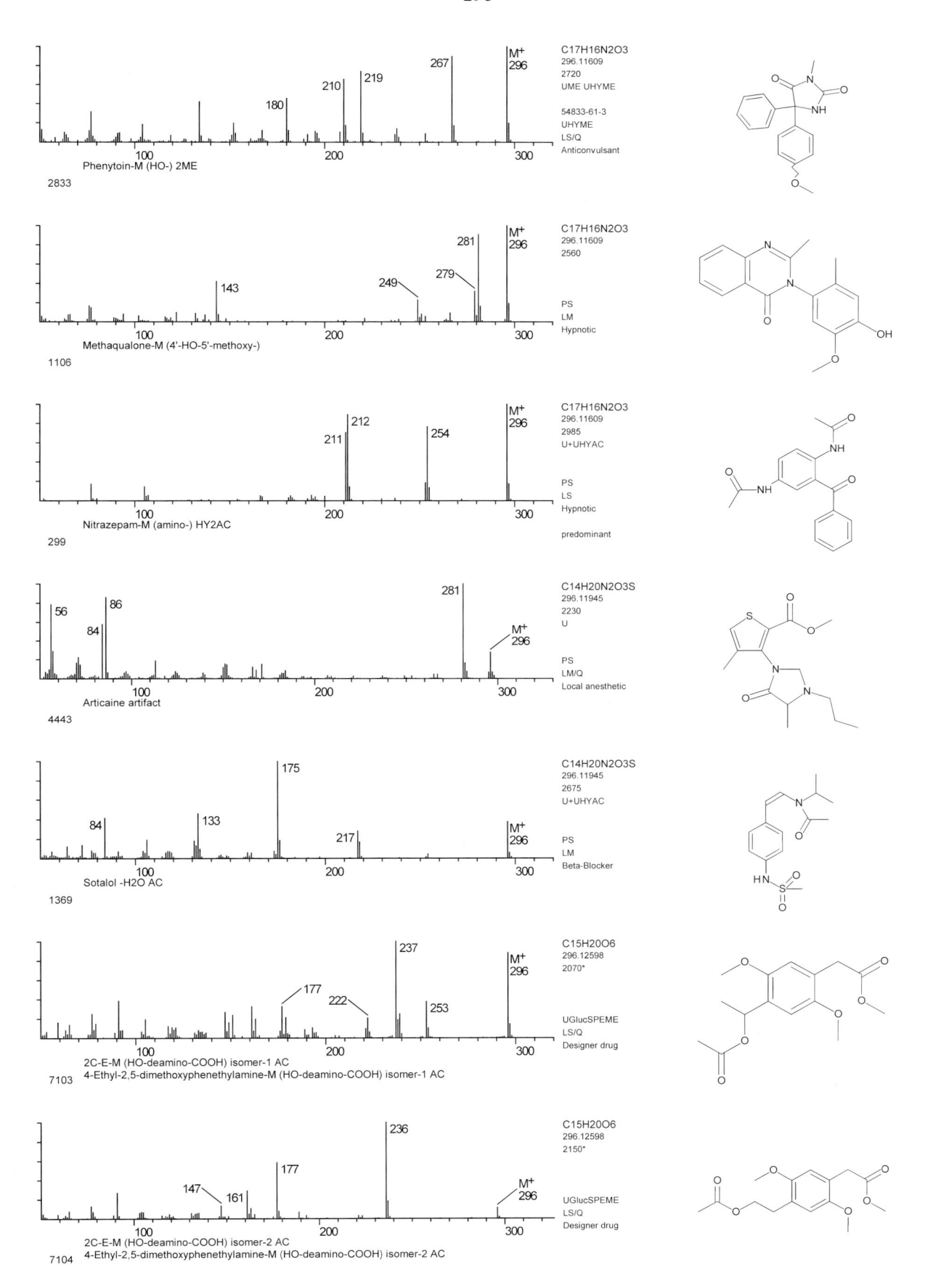

Phenytoin-M (HO-) 2ME
C17H16N2O3
296.11609
2720
UME UHYME

54833-61-3
UHYME
LS/Q
Anticonvulsant

Methaqualone-M (4'-HO-5'-methoxy-)
C17H16N2O3
296.11609
2560

PS
LM
Hypnotic

Nitrazepam-M (amino-) HY2AC
C17H16N2O3
296.11609
2985
U+UHYAC

PS
LS
Hypnotic
predominant

Articaine artifact
C14H20N2O3S
296.11945
2230
U

PS
LM/Q
Local anesthetic

Sotalol -H2O AC
C14H20N2O3S
296.11945
2675
U+UHYAC

PS
LM
Beta-Blocker

2C-E-M (HO-deamino-COOH) isomer-1 AC
4-Ethyl-2,5-dimethoxyphenethylamine-M (HO-deamino-COOH) isomer-1 AC
C15H20O6
296.12598
2070*

UGlucSPEME
LS/Q
Designer drug

2C-E-M (HO-deamino-COOH) isomer-2 AC
4-Ethyl-2,5-dimethoxyphenethylamine-M (HO-deamino-COOH) isomer-2 AC
C15H20O6
296.12598
2150*

UGlucSPEME
LS/Q
Designer drug

296

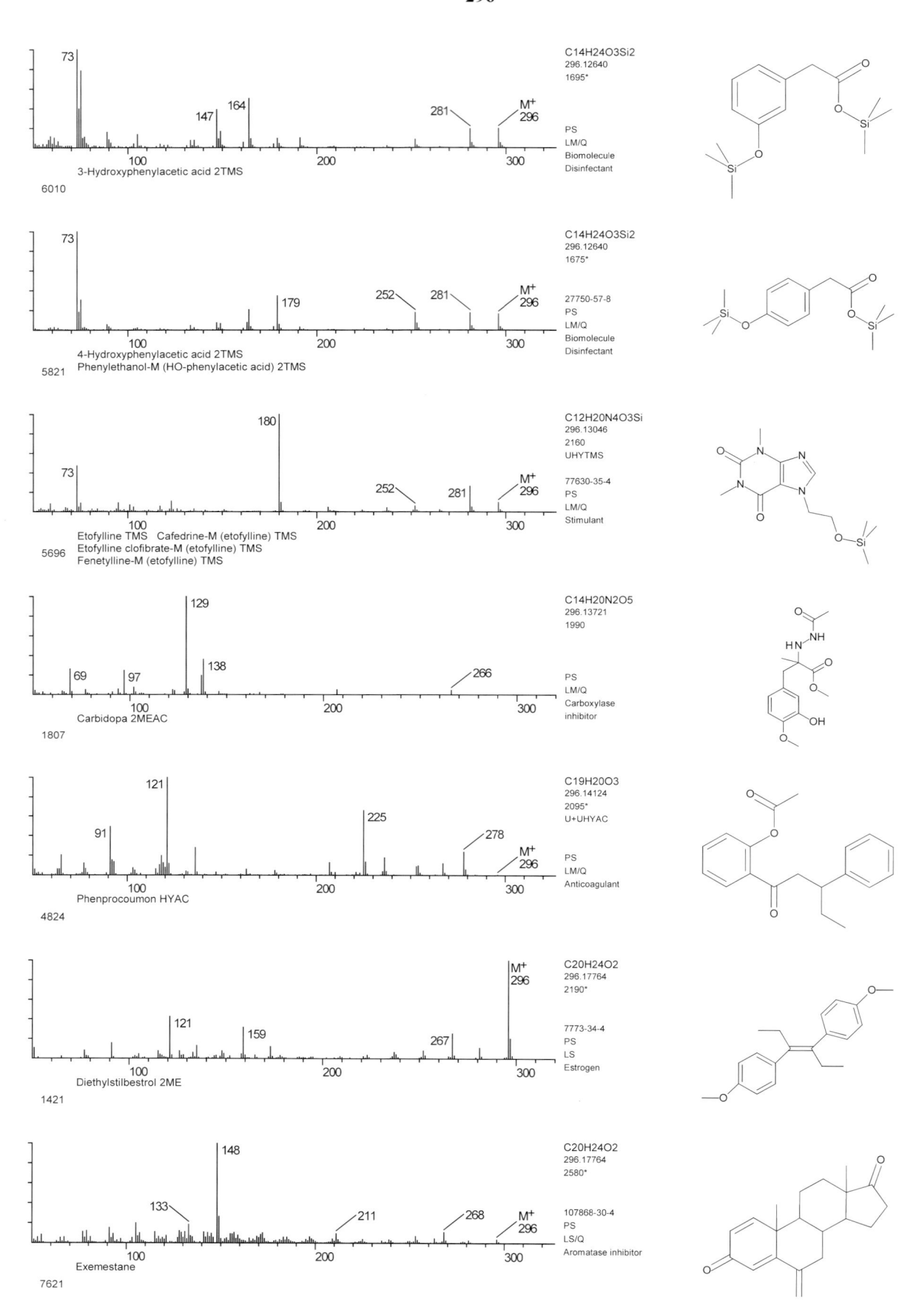

C14H24O3Si2
296.12640
1695*

PS
LM/Q
Biomolecule
Disinfectant

3-Hydroxyphenylacetic acid 2TMS
6010

C14H24O3Si2
296.12640
1675*

27750-57-8
PS
LM/Q
Biomolecule
Disinfectant

4-Hydroxyphenylacetic acid 2TMS
Phenylethanol-M (HO-phenylacetic acid) 2TMS
5821

C12H20N4O3Si
296.13046
2160
UHYTMS
77630-35-4
PS
LM/Q
Stimulant

Etofylline TMS Cafedrine-M (etofylline) TMS
Etofylline clofibrate-M (etofylline) TMS
Fenetylline-M (etofylline) TMS
5696

C14H20N2O5
296.13721
1990

PS
LM/Q
Carboxylase
inhibitor

Carbidopa 2MEAC
1807

C19H20O3
296.14124
2095*
U+UHYAC

PS
LM/Q
Anticoagulant

Phenprocoumon HYAC
4824

C20H24O2
296.17764
2190*

7773-34-4
PS
LS
Estrogen

Diethylstilbestrol 2ME
1421

C20H24O2
296.17764
2580*

107868-30-4
PS
LS/Q
Aromatase inhibitor

Exemestane
7621

886

296

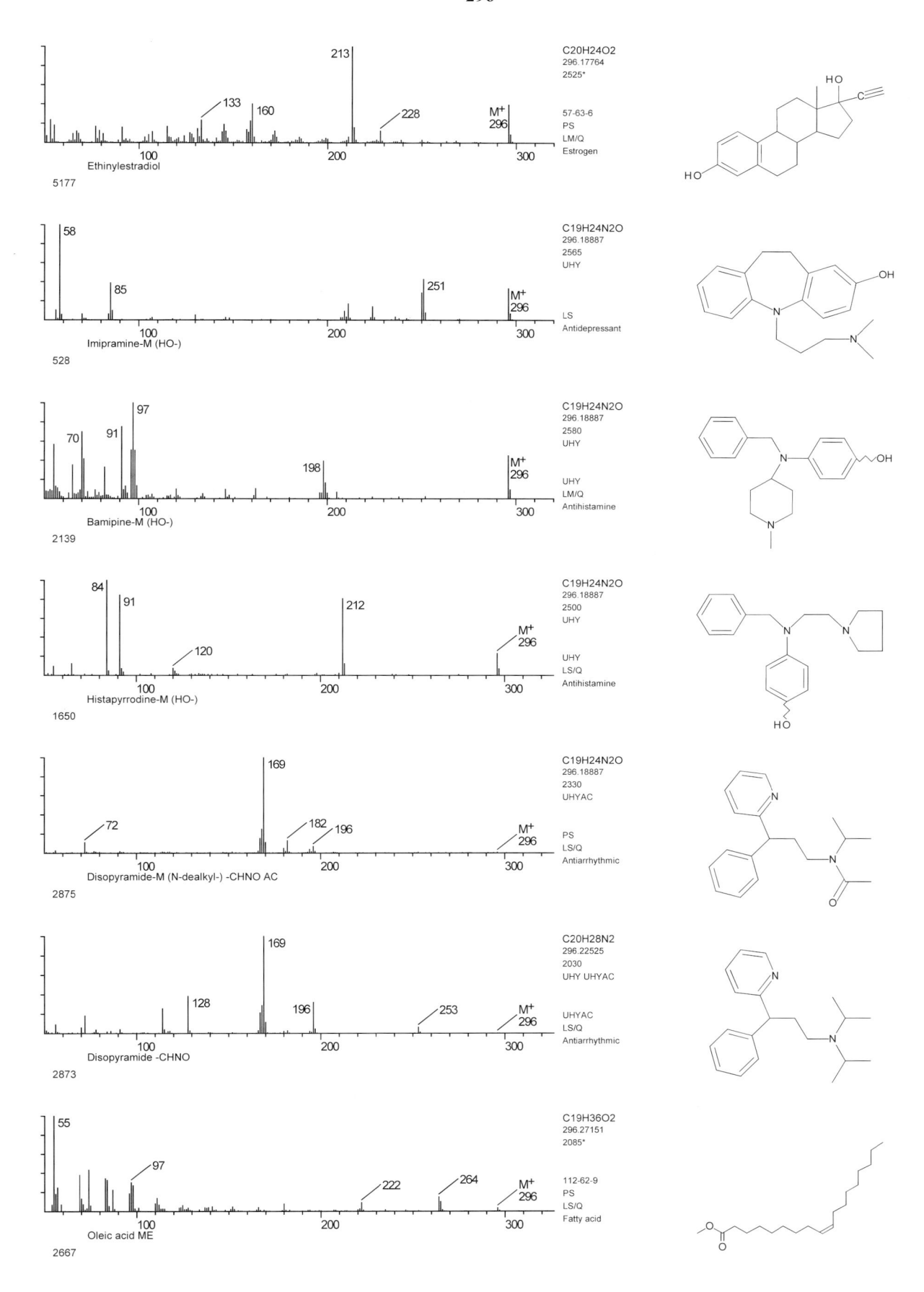

Ethinylestradiol — peaks: 133, 160, 213, 228, M+ 296	C20H24O2 296.17764 2525* 57-63-6 PS LM/Q Estrogen
5177	

Imipramine-M (HO-) — peaks: 58, 85, 251, M+ 296	C19H24N2O 296.18887 2565 UHY LS Antidepressant
528	

Bamipine-M (HO-) — peaks: 70, 91, 97, 198, M+ 296	C19H24N2O 296.18887 2580 UHY UHY LM/Q Antihistamine
2139	

Histapyrrodine-M (HO-) — peaks: 84, 91, 120, 212, M+ 296	C19H24N2O 296.18887 2500 UHY UHY LS/Q Antihistamine
1650	

Disopyramide-M (N-dealkyl-) -CHNO AC — peaks: 72, 169, 182, 196, M+ 296	C19H24N2O 296.18887 2330 UHYAC PS LS/Q Antiarrhythmic
2875	

Disopyramide -CHNO — peaks: 128, 169, 196, 253, M+ 296	C20H28N2 296.22525 2030 UHY UHYAC UHYAC LS/Q Antiarrhythmic
2873	

Oleic acid ME — peaks: 55, 97, 222, 264, M+ 296	C19H36O2 296.27151 2085* 112-62-9 PS LS/Q Fatty acid
2667	

297

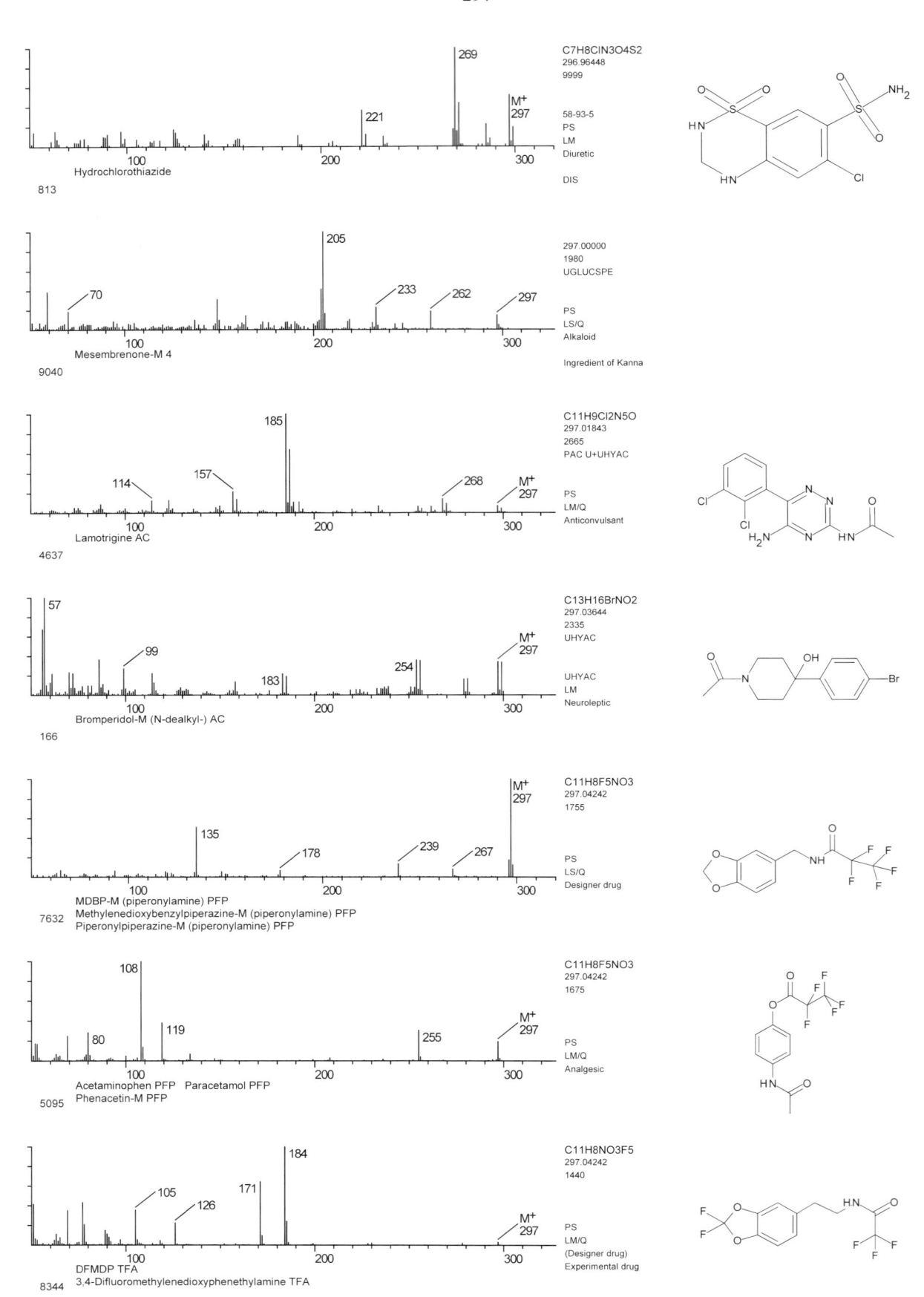

Hydrochlorothiazide
813
C7H8ClN3O4S2
296.96448
9999
58-93-5
PS
LM
Diuretic
DIS

Mesembrenone-M 4
9040
297.00000
1980
UGLUCSPE
PS
LS/Q
Alkaloid
Ingredient of Kanna

Lamotrigine AC
4637
C11H9Cl2N5O
297.01843
2665
PAC U+UHYAC
PS
LM/Q
Anticonvulsant

Bromperidol-M (N-dealkyl-) AC
166
C13H16BrNO2
297.03644
2335
UHYAC
UHYAC
LM
Neuroleptic

MDBP-M (piperonylamine) PFP
Methylenedioxybenzylpiperazine-M (piperonylamine) PFP
Piperonylpiperazine-M (piperonylamine) PFP
7632
C11H8F5NO3
297.04242
1755
PS
LS/Q
Designer drug

Acetaminophen PFP   Paracetamol PFP
Phenacetin-M PFP
5095
C11H8F5NO3
297.04242
1675
PS
LM/Q
Analgesic

DFMDP TFA
3,4-Difluoromethylenedioxyphenethylamine TFA
8344
C11H8NO3F5
297.04242
1440
PS
LM/Q
(Designer drug)
Experimental drug

297

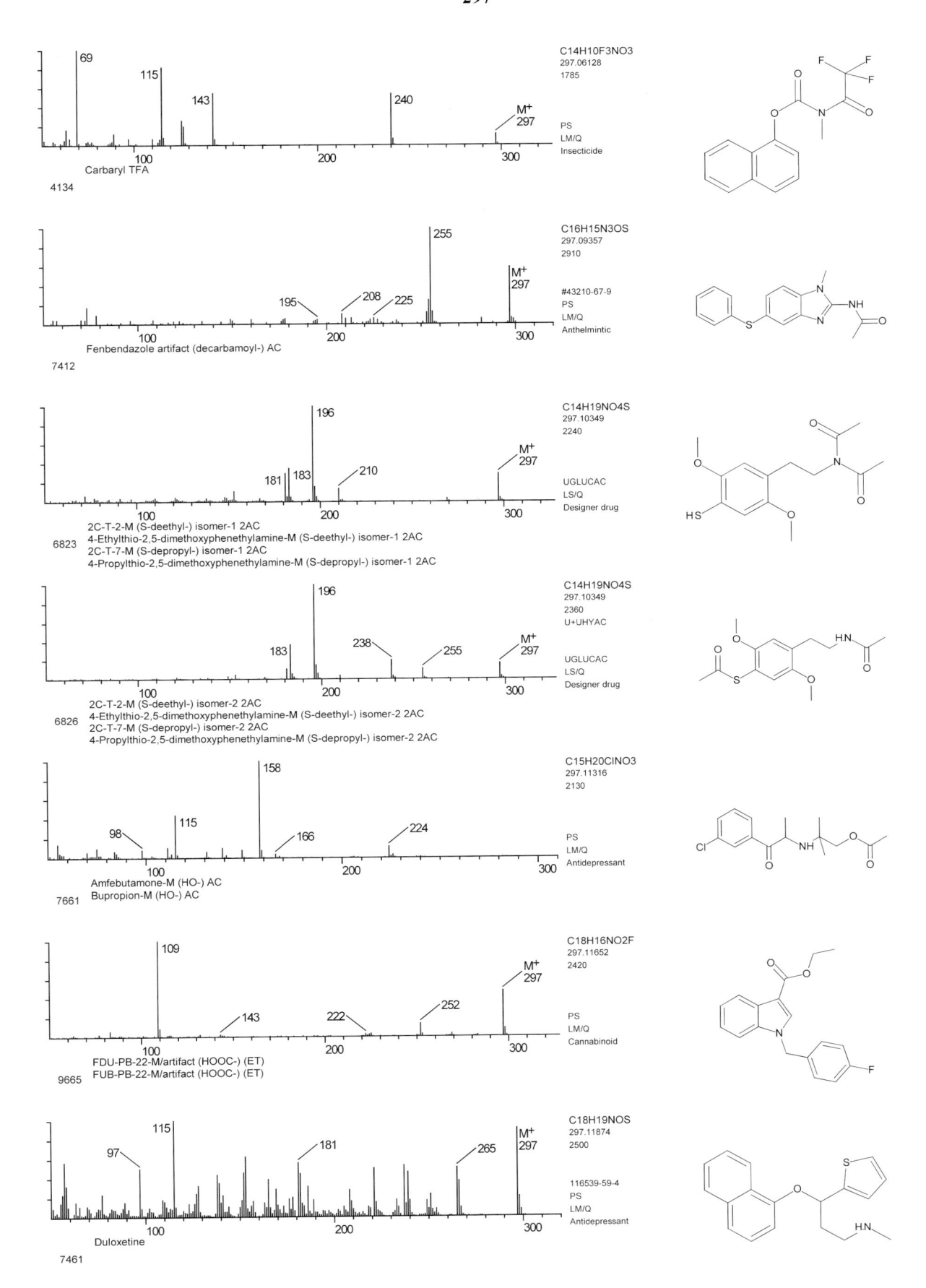

Carbaryl TFA — peaks: 69, 115, 143, 240, M+ 297	C14H10F3NO3 297.06128 1785 PS LM/Q Insecticide
4134	

Fenbendazole artifact (decarbamoyl-) AC — peaks: 195, 208, 225, 255, M+ 297	C16H15N3OS 297.09357 2910 #43210-67-9 PS LM/Q Anthelmintic
7412	

2C-T-2-M (S-deethyl-) isomer-1 2AC 4-Ethylthio-2,5-dimethoxyphenethylamine-M (S-deethyl-) isomer-1 2AC 2C-T-7-M (S-depropyl-) isomer-1 2AC 4-Propylthio-2,5-dimethoxyphenethylamine-M (S-depropyl-) isomer-1 2AC peaks: 181, 183, 196, 210, M+ 297	C14H19NO4S 297.10349 2240 UGLUCAC LS/Q Designer drug
6823	

2C-T-2-M (S-deethyl-) isomer-2 2AC 4-Ethylthio-2,5-dimethoxyphenethylamine-M (S-deethyl-) isomer-2 2AC 2C-T-7-M (S-depropyl-) isomer-2 2AC 4-Propylthio-2,5-dimethoxyphenethylamine-M (S-depropyl-) isomer-2 2AC peaks: 183, 196, 238, 255, M+ 297	C14H19NO4S 297.10349 2360 U+UHYAC UGLUCAC LS/Q Designer drug
6826	

Amfebutamone-M (HO-) AC Bupropion-M (HO-) AC peaks: 98, 115, 158, 166, 224	C15H20ClNO3 297.11316 2130 PS LM/Q Antidepressant
7661	

FDU-PB-22-M/artifact (HOOC-) (ET) FUB-PB-22-M/artifact (HOOC-) (ET) peaks: 109, 143, 222, 252, M+ 297	C18H16NO2F 297.11652 2420 PS LM/Q Cannabinoid
9665	

Duloxetine — peaks: 97, 115, 181, 265, M+ 297	C18H19NOS 297.11874 2500 116539-59-4 PS LM/Q Antidepressant
7461	

297

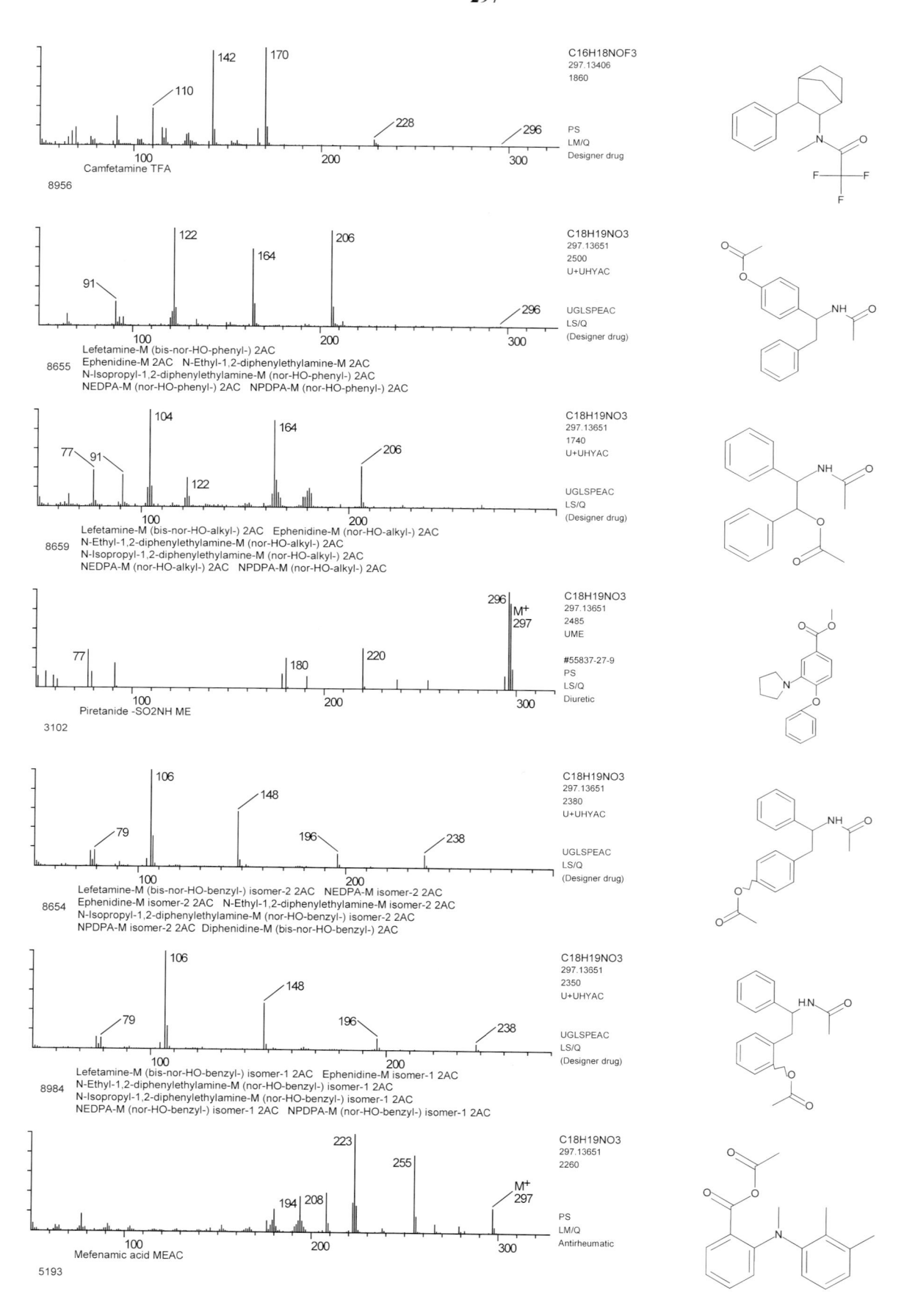

C16H18NOF3
297.13406
1860
PS
LM/Q
Designer drug

Camfetamine TFA
8956

Peaks: 110, 142, 170, 228, 296

C18H19NO3
297.13651
2500
U+UHYAC
UGLSPEAC
LS/Q
(Designer drug)

Lefetamine-M (bis-nor-HO-phenyl-) 2AC
Ephenidine-M 2AC   N-Ethyl-1,2-diphenylethylamine-M 2AC
N-Isopropyl-1,2-diphenylethylamine-M (nor-HO-phenyl-) 2AC
NEDPA-M (nor-HO-phenyl-) 2AC   NPDPA-M (nor-HO-phenyl-) 2AC
8655

Peaks: 91, 122, 164, 206, 296

C18H19NO3
297.13651
1740
U+UHYAC
UGLSPEAC
LS/Q
(Designer drug)

Lefetamine-M (bis-nor-HO-alkyl-) 2AC   Ephenidine-M (nor-HO-alkyl-) 2AC
N-Ethyl-1,2-diphenylethylamine-M (nor-HO-alkyl-) 2AC
N-Isopropyl-1,2-diphenylethylamine-M (nor-HO-alkyl-) 2AC
NEDPA-M (nor-HO-alkyl-) 2AC   NPDPA-M (nor-HO-alkyl-) 2AC
8659

Peaks: 77, 91, 104, 122, 164, 206

C18H19NO3
297.13651
2485
UME
#55837-27-9
PS
LS/Q
Diuretic

Piretanide -SO2NH ME
3102

Peaks: 77, 180, 220, 296 M+, 297

C18H19NO3
297.13651
2380
U+UHYAC
UGLSPEAC
LS/Q
(Designer drug)

Lefetamine-M (bis-nor-HO-benzyl-) isomer-2 2AC   NEDPA-M isomer-2 2AC
Ephenidine-M isomer-2 2AC   N-Ethyl-1,2-diphenylethylamine-M isomer-2 2AC
N-Isopropyl-1,2-diphenylethylamine-M (nor-HO-benzyl-) isomer-2 2AC
NPDPA-M isomer-2 2AC   Diphenidine-M (bis-nor-HO-benzyl-) 2AC
8654

Peaks: 79, 106, 148, 196, 238

C18H19NO3
297.13651
2350
U+UHYAC
UGLSPEAC
LS/Q
(Designer drug)

Lefetamine-M (bis-nor-HO-benzyl-) isomer-1 2AC   Ephenidine-M isomer-1 2AC
N-Ethyl-1,2-diphenylethylamine-M (nor-HO-benzyl-) isomer-1 2AC
N-Isopropyl-1,2-diphenylethylamine-M isomer-1 2AC
NEDPA-M (nor-HO-benzyl-) isomer-1 2AC   NPDPA-M (nor-HO-benzyl-) isomer-1 2AC
8984

Peaks: 79, 106, 148, 196, 238

C18H19NO3
297.13651
2260
PS
LM/Q
Antirheumatic

Mefenamic acid MEAC
5193

Peaks: 194, 208, 223, 255, 297 M+

297

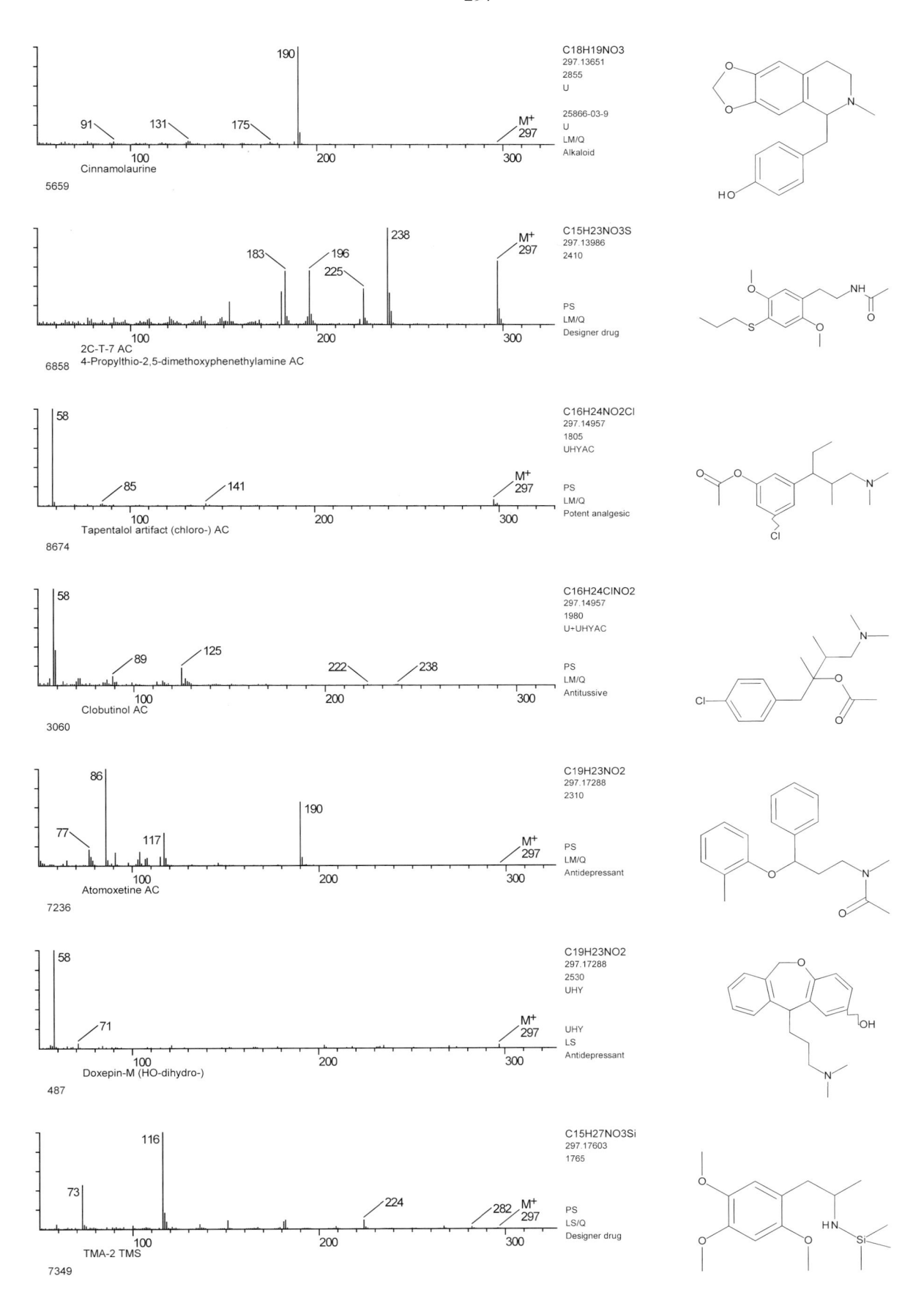

Cinnamolaurine
C18H19NO3
297.13651
2855
U
25866-03-9
U
LM/Q
Alkaloid
5659

2C-T-7 AC
4-Propylthio-2,5-dimethoxyphenethylamine AC
C15H23NO3S
297.13986
2410
PS
LM/Q
Designer drug
6858

Tapentalol artifact (chloro-) AC
C16H24NO2Cl
297.14957
1805
UHYAC
PS
LM/Q
Potent analgesic
8674

Clobutinol AC
C16H24ClNO2
297.14957
1980
U+UHYAC
PS
LM/Q
Antitussive
3060

Atomoxetine AC
C19H23NO2
297.17288
2310
PS
LM/Q
Antidepressant
7236

Doxepin-M (HO-dihydro-)
C19H23NO2
297.17288
2530
UHY
UHY
LS
Antidepressant
487

TMA-2 TMS
C15H27NO3Si
297.17603
1765
PS
LS/Q
Designer drug
7349

891

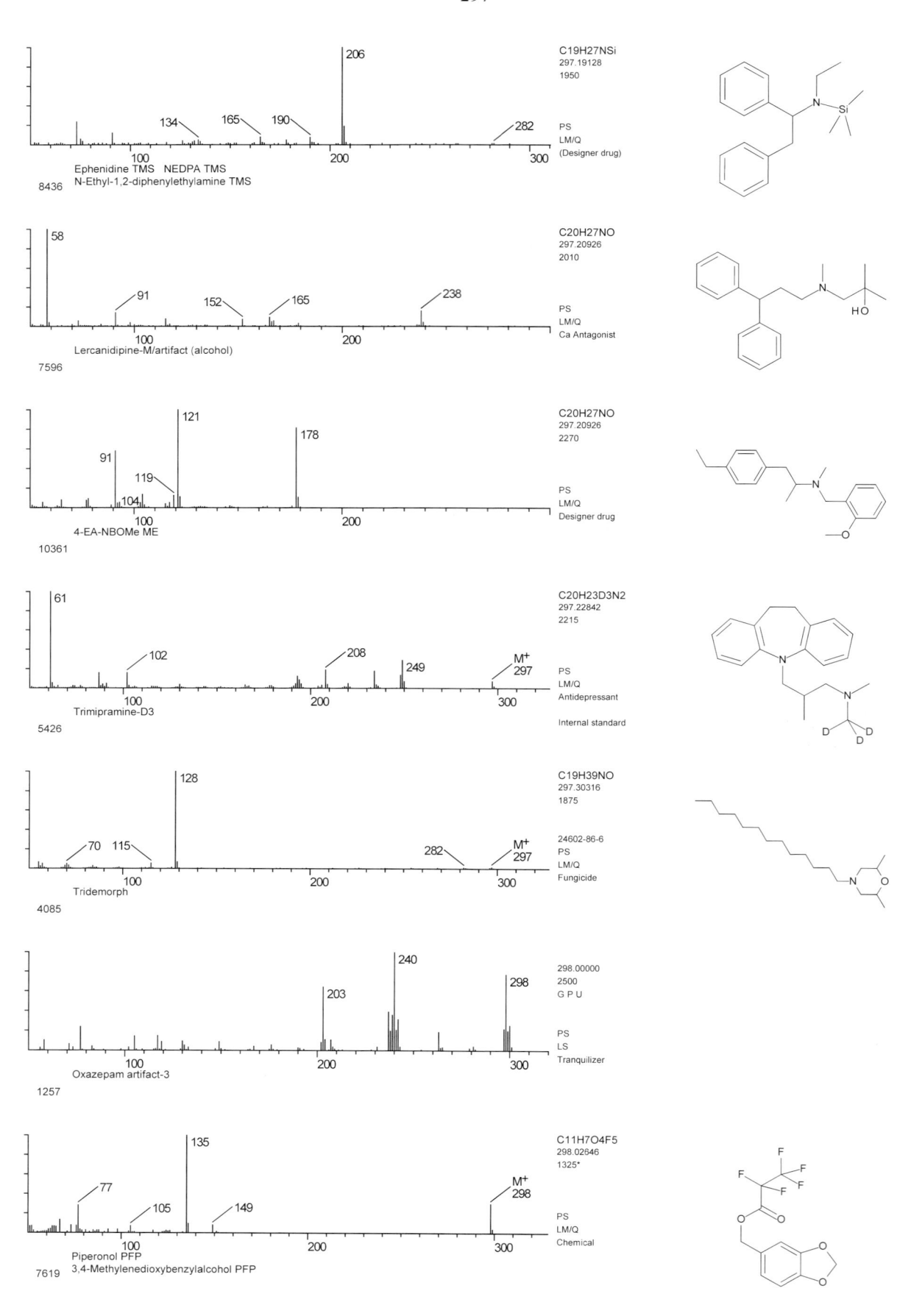

Spectrum	Formula / Data
Ephenidine TMS / NEDPA TMS / N-Ethyl-1,2-diphenylethylamine TMS — 8436	C19H27NSi; 297.19128; 1950; PS; LM/Q; (Designer drug); peaks 134, 165, 190, 206, 282
Lercanidipine-M/artifact (alcohol) — 7596	C20H27NO; 297.20926; 2010; PS; LM/Q; Ca Antagonist; peaks 58, 91, 152, 165, 238
4-EA-NBOMe ME — 10361	C20H27NO; 297.20926; 2270; PS; LM/Q; Designer drug; peaks 91, 104, 119, 121, 178
Trimipramine-D3 — 5426	C20H23D3N2; 297.22842; 2215; PS; LM/Q; Antidepressant; Internal standard; peaks 61, 102, 208, 249, M+ 297
Tridemorph — 4085	C19H39NO; 297.30316; 1875; 24602-86-6; PS; LM/Q; Fungicide; peaks 70, 115, 128, 282, M+ 297
Oxazepam artifact-3 — 1257	298.00000; 2500; G P U; PS; LS; Tranquilizer; peaks 203, 240, 298
Piperonol PFP / 3,4-Methylenedioxybenzylalcohol PFP — 7619	C11H7O4F5; 298.02646; 1325*; PS; LM/Q; Chemical; peaks 77, 105, 135, 149, M+ 298

298

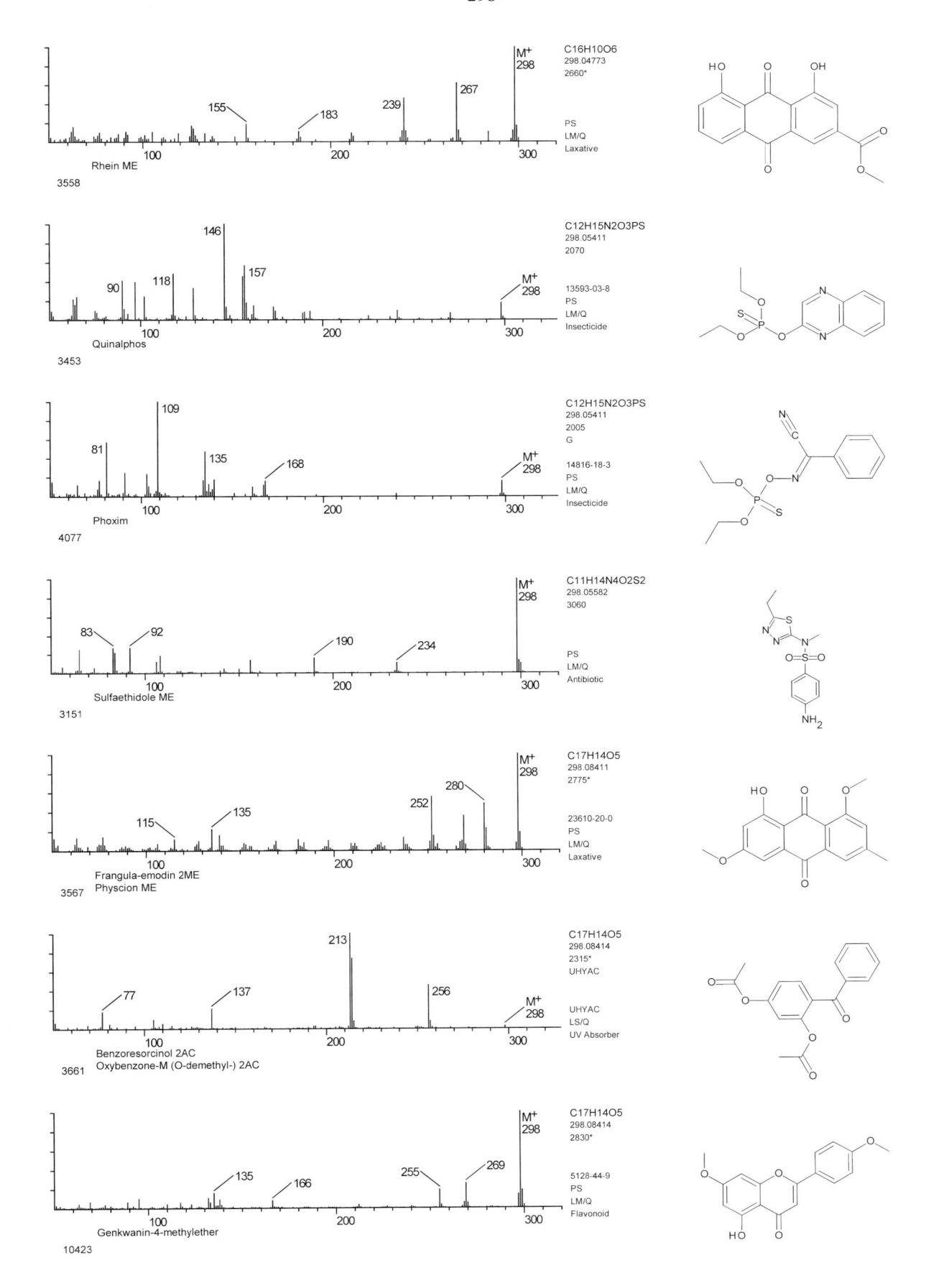

Spectrum	Formula / Data
Rhein ME — 3558	C16H10O6, 298.04773, 2660*, PS, LM/Q, Laxative; peaks 155, 183, 239, 267, M+ 298
Quinalphos — 3453	C12H15N2O3PS, 298.05411, 2070, 13593-03-8, PS, LM/Q, Insecticide; peaks 90, 118, 146, 157, M+ 298
Phoxim — 4077	C12H15N2O3PS, 298.05411, 2005, G, 14816-18-3, PS, LM/Q, Insecticide; peaks 81, 109, 135, 168, M+ 298
Sulfaethidole ME — 3151	C11H14N4O2S2, 298.05582, 3060, PS, LM/Q, Antibiotic; peaks 83, 92, 190, 234, M+ 298
Frangula-emodin 2ME / Physcion ME — 3567	C17H14O5, 298.08411, 2775*, 23610-20-0, PS, LM/Q, Laxative; peaks 115, 135, 252, 280, M+ 298
Benzoresorcinol 2AC / Oxybenzone-M (O-demethyl-) 2AC — 3661	C17H14O5, 298.08414, 2315*, UHYAC, UHYAC, LS/Q, UV Absorber; peaks 77, 137, 213, 256, M+ 298
Genkwanin-4-methylether — 10423	C17H14O5, 298.08414, 2830*, 5128-44-9, PS, LM/Q, Flavonoid; peaks 135, 166, 255, 269, M+ 298

298

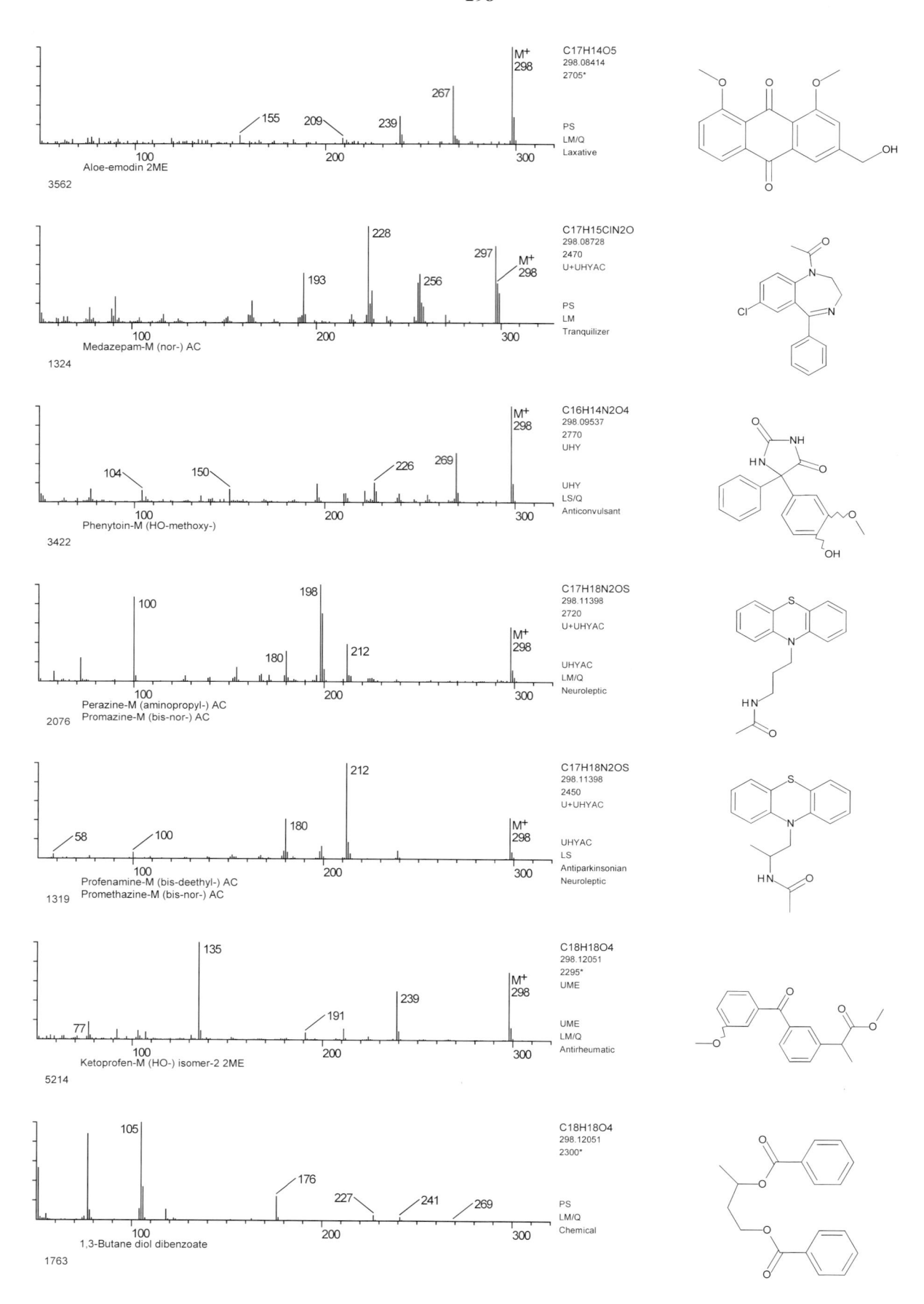

Aloe-emodin 2ME
C17H14O5
298.08414
2705*
PS
LM/Q
Laxative

Medazepam-M (nor-) AC
C17H15ClN2O
298.08728
2470
U+UHYAC
PS
LM
Tranquilizer

Phenytoin-M (HO-methoxy-)
C16H14N2O4
298.09537
2770
UHY
UHY
LS/Q
Anticonvulsant

Perazine-M (aminopropyl-) AC
Promazine-M (bis-nor-) AC
C17H18N2OS
298.11398
2720
U+UHYAC
UHYAC
LM/Q
Neuroleptic

Profenamine-M (bis-deethyl-) AC
Promethazine-M (bis-nor-) AC
C17H18N2OS
298.11398
2450
U+UHYAC
UHYAC
LS
Antiparkinsonian
Neuroleptic

Ketoprofen-M (HO-) isomer-2 2ME
C18H18O4
298.12051
2295*
UME
UME
LM/Q
Antirheumatic

1,3-Butane diol dibenzoate
C18H18O4
298.12051
2300*
PS
LM/Q
Chemical

298

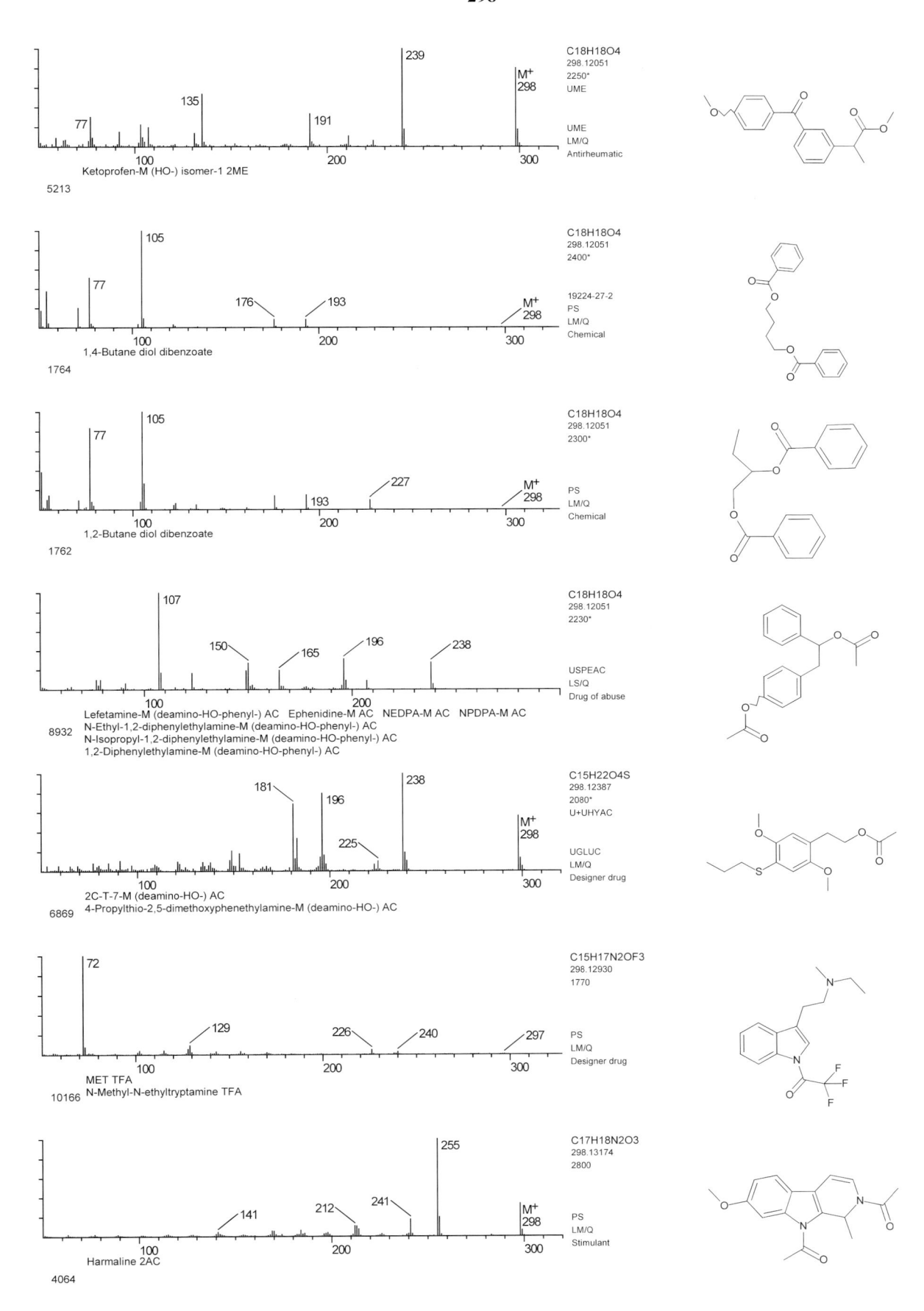

Ketoprofen-M (HO-) isomer-1 2ME — peaks 77, 135, 191, 239, M+ 298	C18H18O4 298.12051 2250* UME UME LM/Q Antirheumatic	
5213		

1,4-Butane diol dibenzoate — peaks 77, 105, 176, 193, M+ 298	C18H18O4 298.12051 2400* 19224-27-2 PS LM/Q Chemical	
1764		

1,2-Butane diol dibenzoate — peaks 77, 105, 193, 227, M+ 298	C18H18O4 298.12051 2300* PS LM/Q Chemical	
1762		

Lefetamine-M (deamino-HO-phenyl-) AC Ephenidine-M AC NEDPA-M AC NPDPA-M AC N-Ethyl-1,2-diphenylethylamine-M (deamino-HO-phenyl-) AC N-Isopropyl-1,2-diphenylethylamine-M (deamino-HO-phenyl-) AC 1,2-Diphenylethylamine-M (deamino-HO-phenyl-) AC — peaks 107, 150, 165, 196, 238	C18H18O4 298.12051 2230* USPEAC LS/Q Drug of abuse	
8932		

2C-T-7-M (deamino-HO-) AC 4-Propylthio-2,5-dimethoxyphenethylamine-M (deamino-HO-) AC — peaks 181, 196, 225, 238, M+ 298	C15H22O4S 298.12387 2080* U+UHYAC UGLUC LM/Q Designer drug	
6869		

MET TFA N-Methyl-N-ethyltryptamine TFA — peaks 72, 129, 226, 240, 297	C15H17N2OF3 298.12930 1770 PS LM/Q Designer drug	
10166		

Harmaline 2AC — peaks 141, 212, 241, 255, M+ 298	C17H18N2O3 298.13174 2800 PS LM/Q Stimulant	
4064		

298

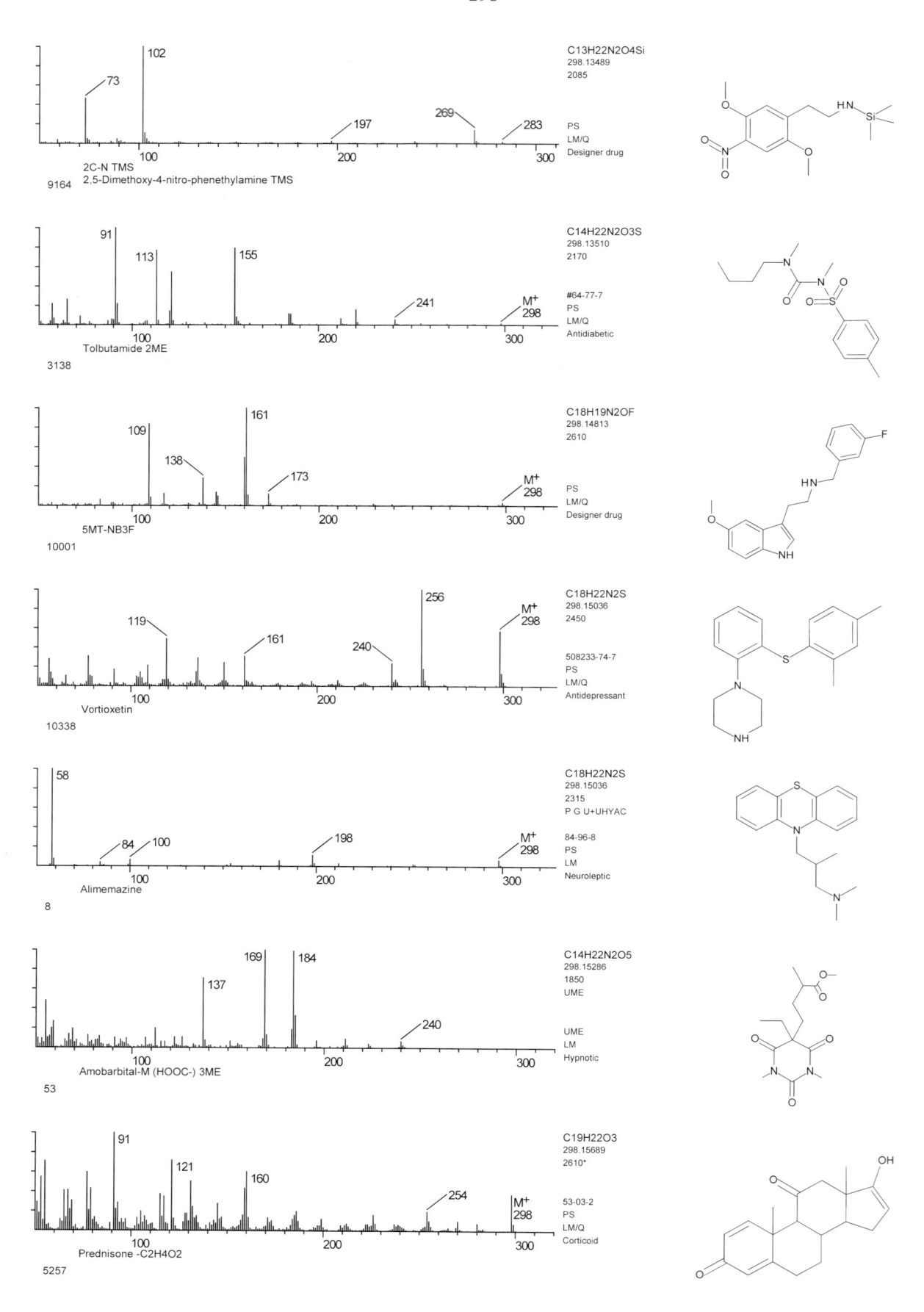

Spectrum	Formula / Info
2C-N TMS — 2,5-Dimethoxy-4-nitro-phenethylamine TMS (9164); peaks 73, 102, 197, 269, 283	C13H22N2O4Si; 298.13489; 2085; PS; LM/Q; Designer drug
Tolbutamide 2ME (3138); peaks 91, 113, 155, 241, M+ 298	C14H22N2O3S; 298.13510; 2170; #64-77-7; PS; LM/Q; Antidiabetic
5MT-NB3F (10001); peaks 109, 138, 161, 173, M+ 298	C18H19N2OF; 298.14813; 2610; PS; LM/Q; Designer drug
Vortioxetin (10338); peaks 119, 161, 240, 256, M+ 298	C18H22N2S; 298.15036; 2450; 508233-74-7; PS; LM/Q; Antidepressant
Alimemazine (8); peaks 58, 84, 100, 198, M+ 298	C18H22N2S; 298.15036; 2315; P G U+UHYAC; 84-96-8; PS; LM; Neuroleptic
Amobarbital-M (HOOC-) 3ME (53); peaks 137, 169, 184, 240	C14H22N2O5; 298.15286; 1850; UME; UME; LM; Hypnotic
Prednisone -C2H4O2 (5257); peaks 91, 121, 160, 254, M+ 298	C19H22O3; 298.15689; 2610*; 53-03-2; PS; LM/Q; Corticoid

896

298

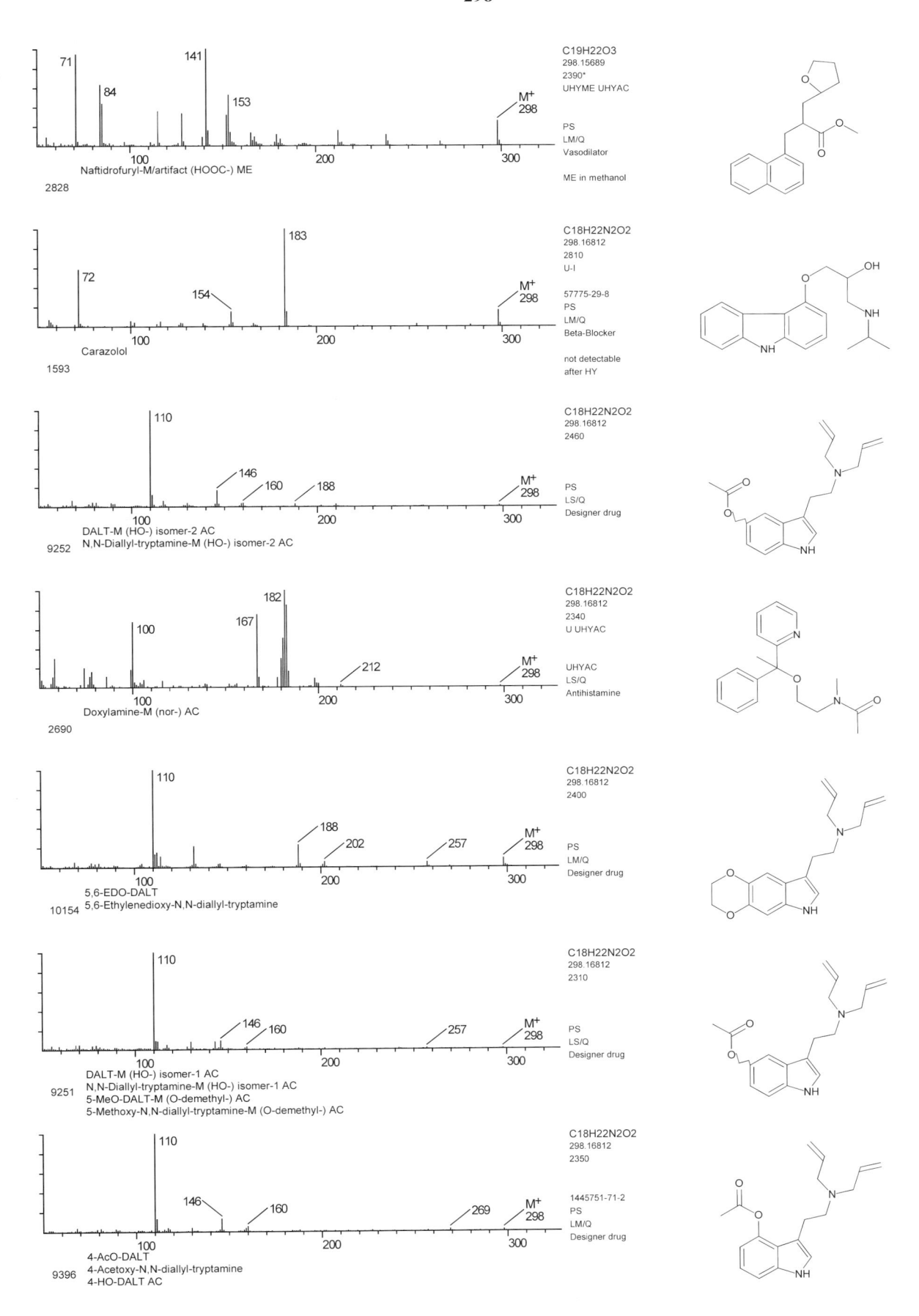

Naftidrofuryl-M/artifact (HOOC-) ME
2828
Peaks: 71, 84, 141, 153, M+ 298
C19H22O3
298.15689
2390*
UHYME UHYAC
PS
LM/Q
Vasodilator

ME in methanol

Carazolol
1593
Peaks: 72, 154, 183, M+ 298
C18H22N2O2
298.16812
2810
U-I
57775-29-8
PS
LM/Q
Beta-Blocker

not detectable after HY

DALT-M (HO-) isomer-2 AC
N,N-Diallyl-tryptamine-M (HO-) isomer-2 AC
9252
Peaks: 110, 146, 160, 188, M+ 298
C18H22N2O2
298.16812
2460
PS
LS/Q
Designer drug

Doxylamine-M (nor-) AC
2690
Peaks: 100, 167, 182, 212, M+ 298
C18H22N2O2
298.16812
2340
U UHYAC
UHYAC
LS/Q
Antihistamine

5,6-EDO-DALT
5,6-Ethylenedioxy-N,N-diallyl-tryptamine
10154
Peaks: 110, 188, 202, 257, M+ 298
C18H22N2O2
298.16812
2400
PS
LM/Q
Designer drug

DALT-M (HO-) isomer-1 AC
N,N-Diallyl-tryptamine-M (HO-) isomer-1 AC
5-MeO-DALT-M (O-demethyl-) AC
5-Methoxy-N,N-diallyl-tryptamine-M (O-demethyl-) AC
9251
Peaks: 110, 146, 160, 257, M+ 298
C18H22N2O2
298.16812
2310
PS
LS/Q
Designer drug

4-AcO-DALT
4-Acetoxy-N,N-diallyl-tryptamine
4-HO-DALT AC
9396
Peaks: 110, 146, 160, 269, M+ 298
C18H22N2O2
298.16812
2350
1445751-71-2
PS
LM/Q
Designer drug

298

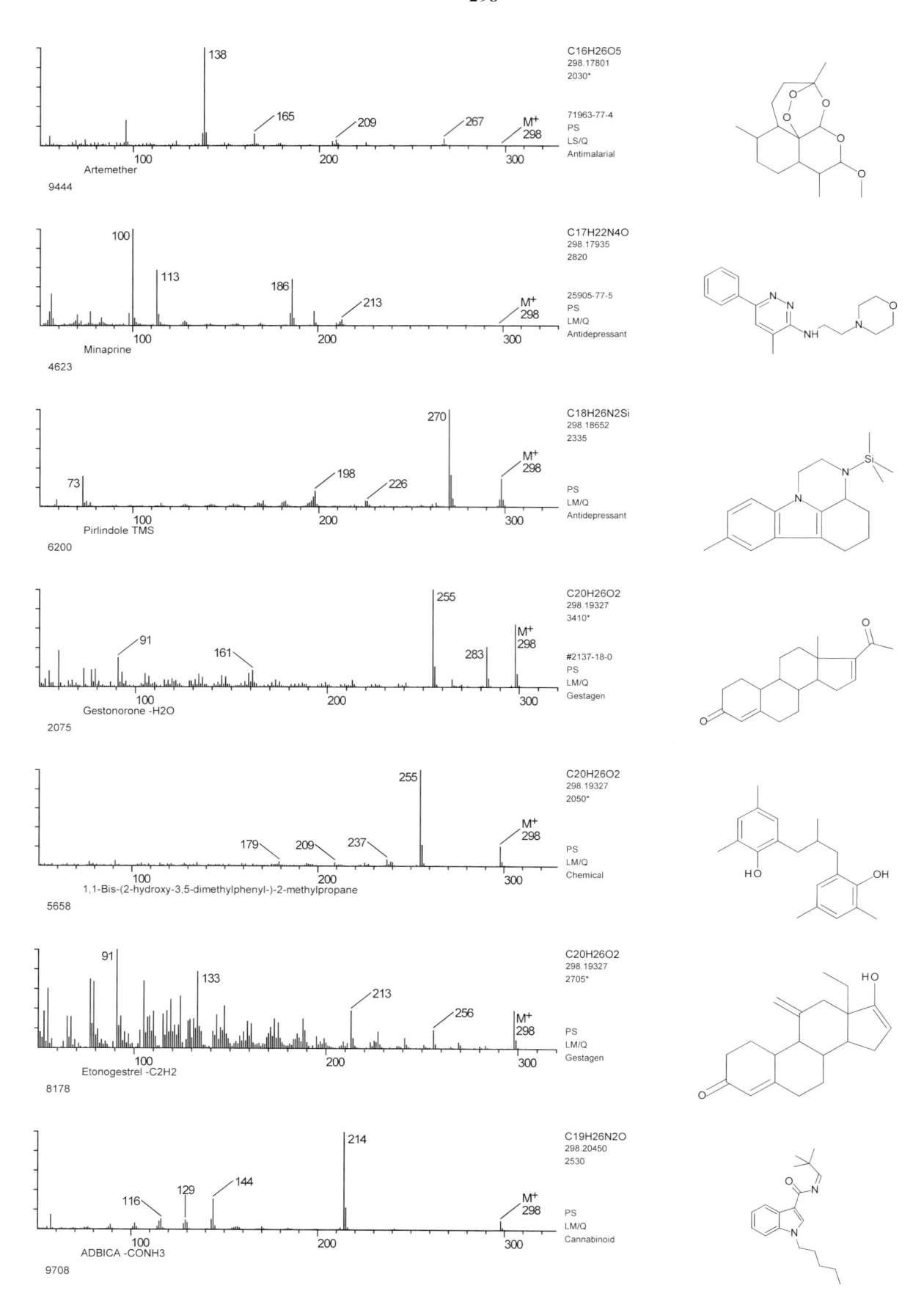

Artemether — 9444	C16H26O5 298.17801 2030* 71963-77-4 PS LS/Q Antimalarial Peaks: 138, 165, 209, 267, M+ 298
Minaprine — 4623	C17H22N4O 298.17935 2820 25905-77-5 PS LM/Q Antidepressant Peaks: 100, 113, 186, 213, M+ 298
Pirlindole TMS — 6200	C18H26N2Si 298.18652 2335 PS LM/Q Antidepressant Peaks: 73, 198, 226, 270, M+ 298
Gestonorone -H2O — 2075	C20H26O2 298.19327 3410* #2137-18-0 PS LM/Q Gestagen Peaks: 91, 161, 255, 283, M+ 298
1,1-Bis-(2-hydroxy-3,5-dimethylphenyl-)-2-methylpropane — 5658	C20H26O2 298.19327 2050* PS LM/Q Chemical Peaks: 179, 209, 237, 255, M+ 298
Etonogestrel -C2H2 — 8178	C20H26O2 298.19327 2705* PS LM/Q Gestagen Peaks: 91, 133, 213, 256, M+ 298
ADBICA -CONH3 — 9708	C19H26N2O 298.20450 2530 PS LM/Q Cannabinoid Peaks: 116, 129, 144, 214, M+ 298

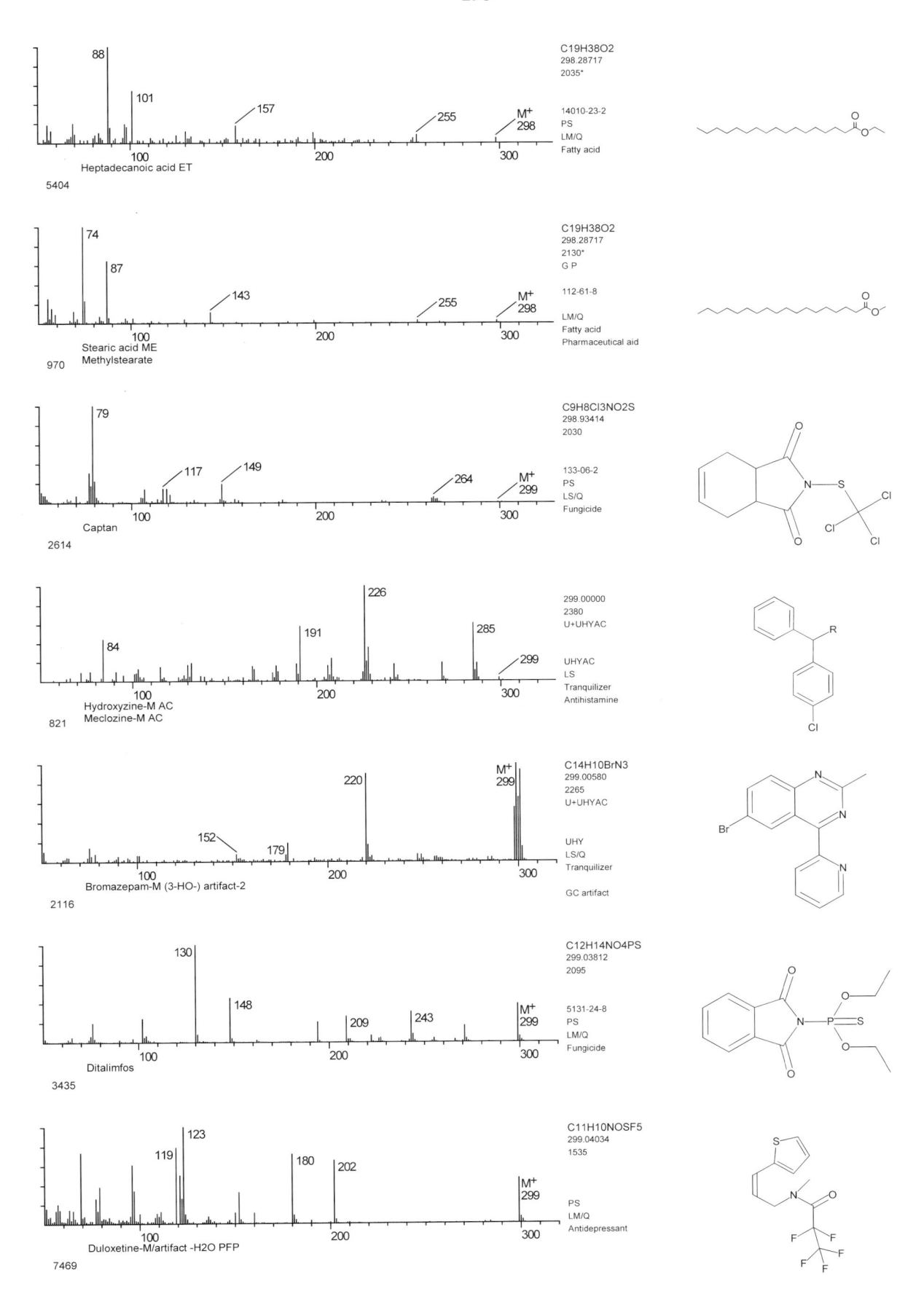

Compound	Formula	Mass	CAS
Heptadecanoic acid ET	C19H38O2	298.28717	14010-23-2
Stearic acid ME / Methylstearate	C19H38O2	298.28717	112-61-8
Captan	C9H8Cl3NO2S	298.93414	133-06-2
Hydroxyzine-M AC / Meclozine-M AC		299.00000	
Bromazepam-M (3-HO-) artifact-2	C14H10BrN3	299.00580	
Ditalimfos	C12H14NO4PS	299.03812	5131-24-8
Duloxetine-M/artifact -H2O PFP	C11H10NOSF5	299.04034	

299

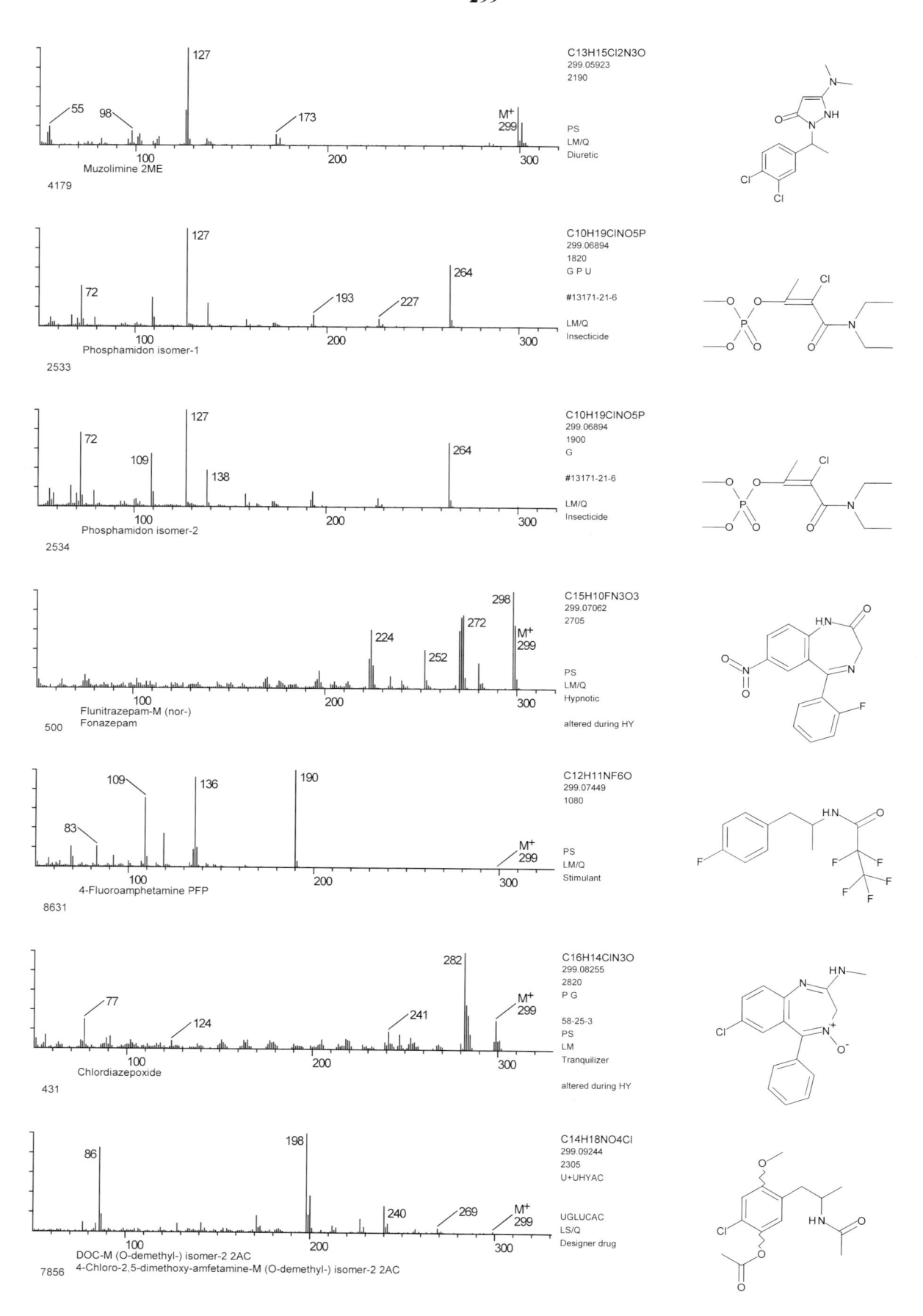

Muzolimine 2ME — peaks: 55, 98, 127, 173, M+ 299	C13H15Cl2N3O 299.05923 2190 PS LM/Q Diuretic
4179	

Phosphamidon isomer-1 — peaks: 72, 127, 193, 227, 264	C10H19ClNO5P 299.06894 1820 G P U #13171-21-6 LM/Q Insecticide
2533	

Phosphamidon isomer-2 — peaks: 72, 109, 127, 138, 264	C10H19ClNO5P 299.06894 1900 G #13171-21-6 LM/Q Insecticide
2534	

Flunitrazepam-M (nor-) Fonazepam — peaks: 224, 252, 272, 298, M+ 299	C15H10FN3O3 299.07062 2705 PS LM/Q Hypnotic altered during HY
500	

4-Fluoroamphetamine PFP — peaks: 83, 109, 136, 190, M+ 299	C12H11NF6O 299.07449 1080 PS LM/Q Stimulant
8631	

Chlordiazepoxide — peaks: 77, 124, 241, 282, M+ 299	C16H14ClN3O 299.08255 2820 P G 58-25-3 PS LM Tranquilizer altered during HY
431	

DOC-M (O-demethyl-) isomer-2 2AC 4-Chloro-2,5-dimethoxy-amfetamine-M (O-demethyl-) isomer-2 2AC — peaks: 86, 198, 240, 269, M+ 299	C14H18NO4Cl 299.09244 2305 U+UHYAC UGLUCAC LS/Q Designer drug
7856	

900

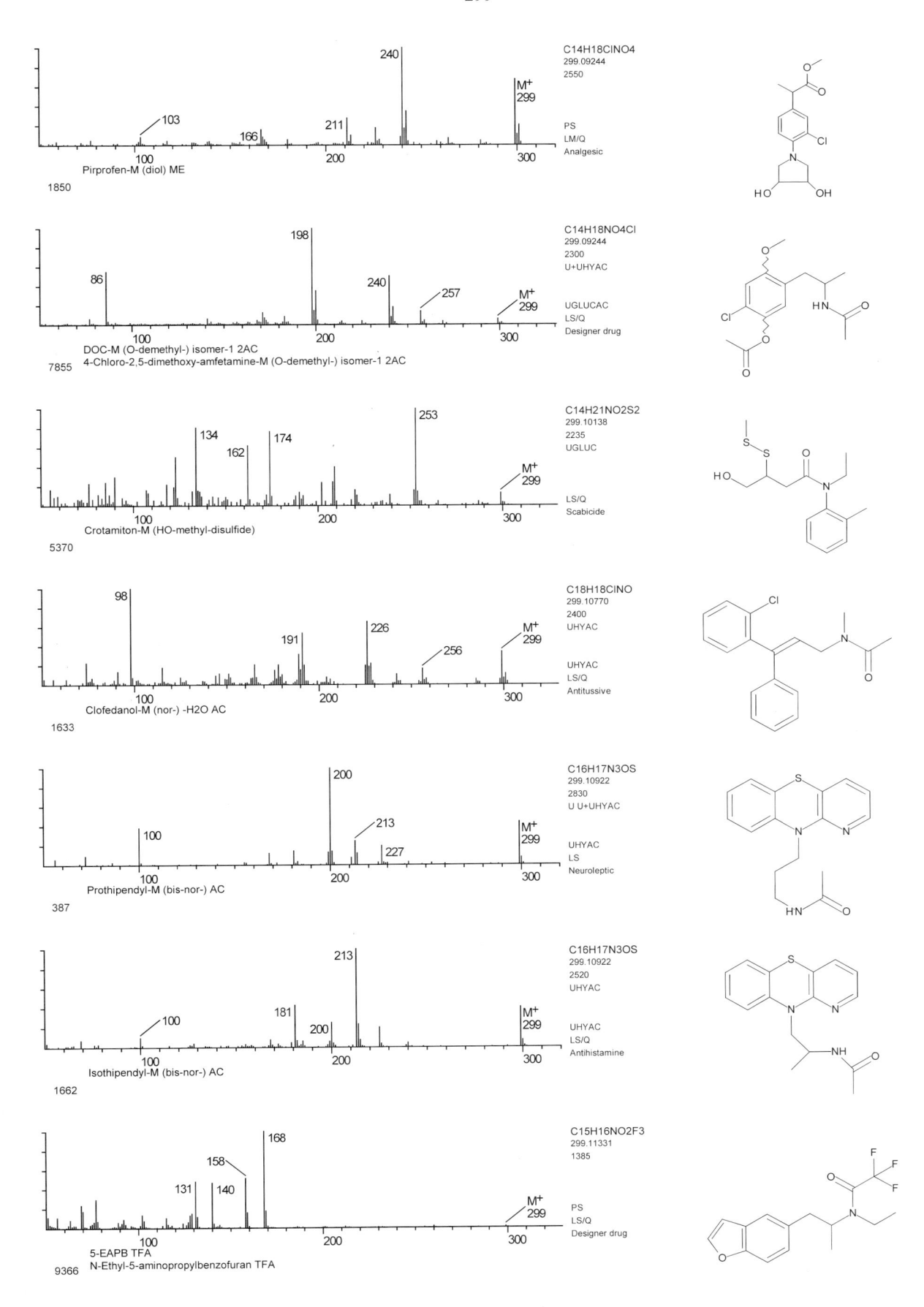

Pirprofen-M (diol) ME	C14H18ClNO4 299.09244 2550 PS LM/Q Analgesic	
DOC-M (O-demethyl-) isomer-1 2AC 4-Chloro-2,5-dimethoxy-amfetamine-M (O-demethyl-) isomer-1 2AC	C14H18NO4Cl 299.09244 2300 U+UHYAC UGLUCAC LS/Q Designer drug	
Crotamiton-M (HO-methyl-disulfide)	C14H21NO2S2 299.10138 2235 UGLUC LS/Q Scabicide	
Clofedanol-M (nor-) -H2O AC	C18H18ClNO 299.10770 2400 UHYAC UHYAC LS/Q Antitussive	
Prothipendyl-M (bis-nor-) AC	C16H17N3OS 299.10922 2830 U U+UHYAC UHYAC LS Neuroleptic	
Isothipendyl-M (bis-nor-) AC	C16H17N3OS 299.10922 2520 UHYAC UHYAC LS/Q Antihistamine	
5-EAPB TFA N-Ethyl-5-aminopropylbenzofuran TFA	C15H16NO2F3 299.11331 1385 PS LS/Q Designer drug	

299

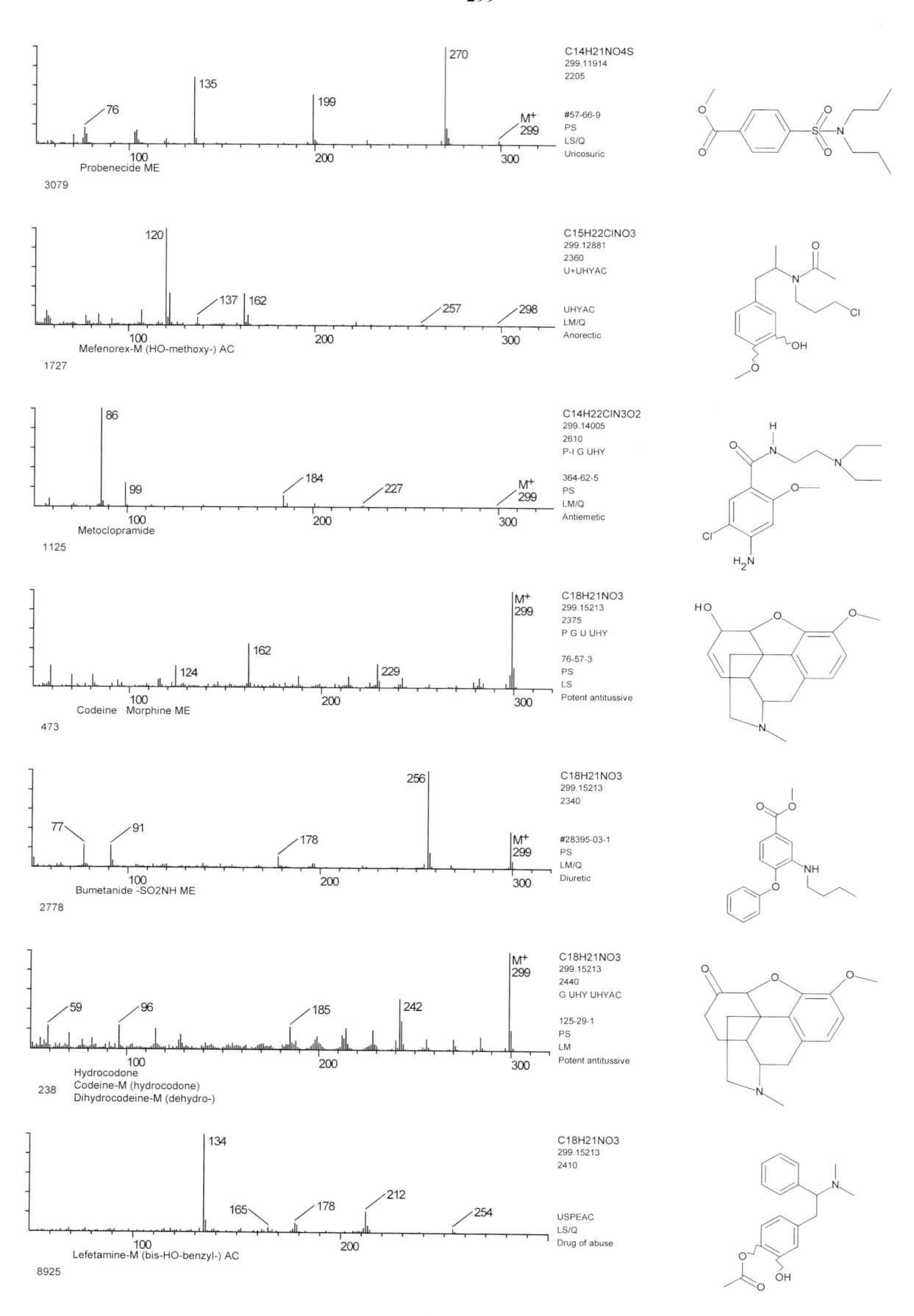

Probenecide ME	C14H21NO4S 299.11914 2205 #57-66-9 PS LS/Q Uricosuric	Peaks: 76, 135, 199, 270, M+ 299
3079		

Mefenorex-M (HO-methoxy-) AC	C15H22ClNO3 299.12881 2360 U+UHYAC UHYAC LM/Q Anorectic	Peaks: 120, 137, 162, 257, 298
1727		

Metoclopramide	C14H22ClN3O2 299.14005 2610 P-I G UHY 364-62-5 PS LM/Q Antiemetic	Peaks: 86, 99, 184, 227, M+ 299
1125		

Codeine Morphine ME	C18H21NO3 299.15213 2375 P G U UHY 76-57-3 PS LS Potent antitussive	Peaks: 124, 162, 229, M+ 299
473		

Bumetanide -SO2NH ME	C18H21NO3 299.15213 2340 #28395-03-1 PS LM/Q Diuretic	Peaks: 77, 91, 178, 256, M+ 299
2778		

Hydrocodone Codeine-M (hydrocodone) Dihydrocodeine-M (dehydro-)	C18H21NO3 299.15213 2440 G UHY UHYAC 125-29-1 PS LM Potent antitussive	Peaks: 59, 96, 185, 242, M+ 299
238		

Lefetamine-M (bis-HO-benzyl-) AC	C18H21NO3 299.15213 2410 USPEAC LS/Q Drug of abuse	Peaks: 134, 165, 178, 212, 254
8925		

299

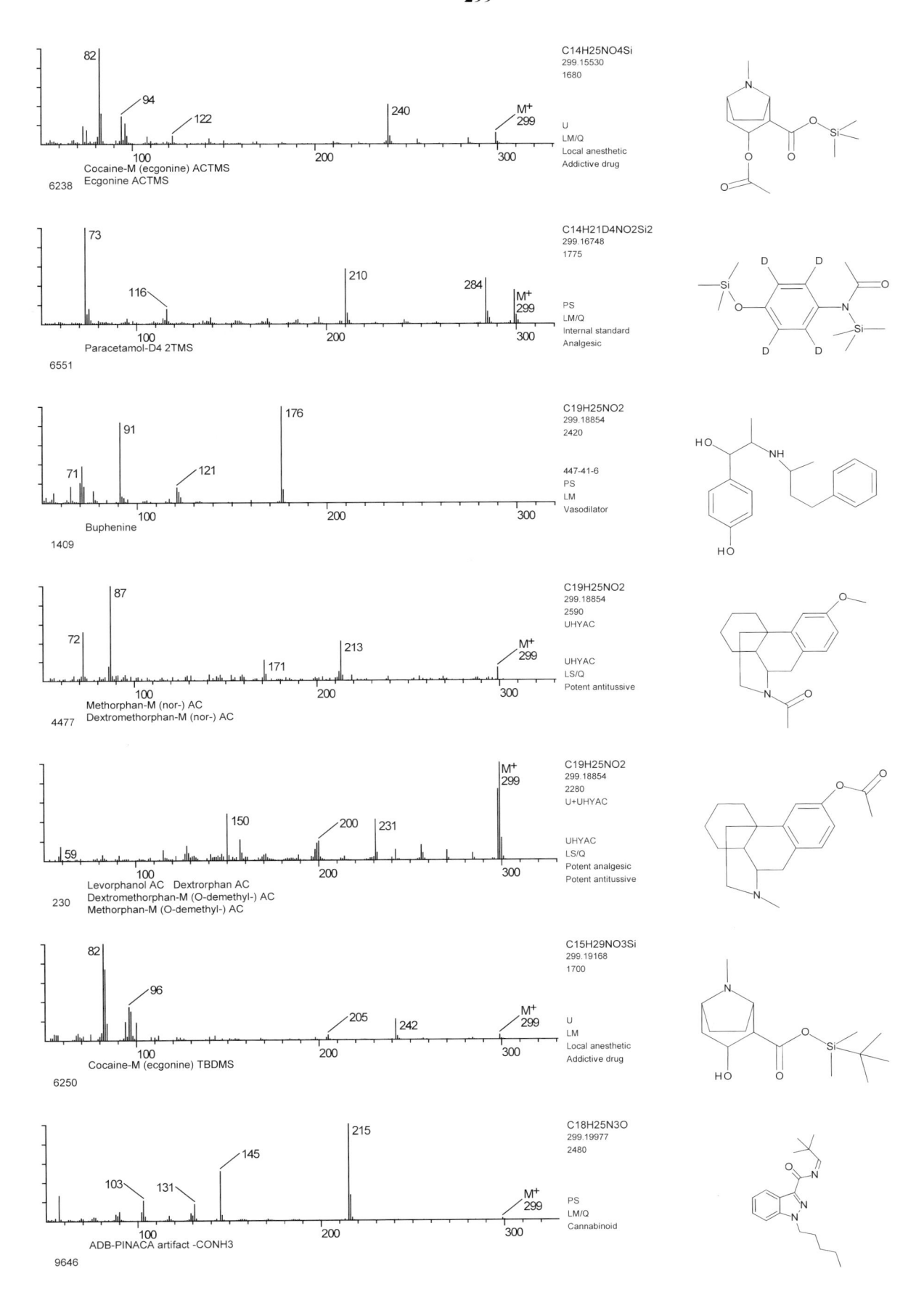

Cocaine-M (ecgonine) ACTMS
Ecgonine ACTMS
6238
Peaks: 82, 94, 122, 240, M+ 299
C14H25NO4Si
299.15530
1680
U
LM/Q
Local anesthetic
Addictive drug

Paracetamol-D4 2TMS
6551
Peaks: 73, 116, 210, 284, M+ 299
C14H21D4NO2Si2
299.16748
1775
PS
LM/Q
Internal standard
Analgesic

Buphenine
1409
Peaks: 71, 91, 121, 176
C19H25NO2
299.18854
2420
447-41-6
PS
LM
Vasodilator

Methorphan-M (nor-) AC
Dextromethorphan-M (nor-) AC
4477
Peaks: 72, 87, 171, 213, M+ 299
C19H25NO2
299.18854
2590
UHYAC
UHYAC
LS/Q
Potent antitussive

Levorphanol AC   Dextrorphan AC
Dextromethorphan-M (O-demethyl-) AC
Methorphan-M (O-demethyl-) AC
230
Peaks: 59, 150, 200, 231, M+ 299
C19H25NO2
299.18854
2280
U+UHYAC
UHYAC
LS/Q
Potent analgesic
Potent antitussive

Cocaine-M (ecgonine) TBDMS
6250
Peaks: 82, 96, 205, 242, M+ 299
C15H29NO3Si
299.19168
1700
U
LM
Local anesthetic
Addictive drug

ADB-PINACA artifact -CONH3
9646
Peaks: 103, 131, 145, 215, M+ 299
C18H25N3O
299.19977
2480
PS
LM/Q
Cannabinoid